The Theory of
LINEAR ANTENNAS

The Theory of
LINEAR ANTENNAS

with

Charts and Tables for Practical Applications

Ronold W. P. King

Gordon McKay Professor of Applied Physics
Harvard University

Harvard University Press
Cambridge, Massachusetts
1956

©
Copyright 1956 by the
President and Fellows of Harvard College
All rights reserved

Distributed in Great Britain by
Geoffrey Cumberlege
Oxford University Press
London

LIBRARY OF CONGRESS CATALOG CARD NUMBER 56-5354
PRINTED IN THE UNITED STATES OF AMERICA

Experiment escorts us last—
His pungent company
Will not allow an Axiom
An Opportunity.
     Emily Dickinson

PREFACE

This book had its inception in the spring of 1939 in a graduate course in radio communication which included a variety of topics, among them antennas. The fact that it proved to be impossible to present this subject in a scientific manner led first to a series of individual investigations and ultimately to a systematic program of research. It is a pleasure to acknowledge that an important early stimulus to the scientific analysis of antennas was a paper entitled "Theoretical investigations into the transmitting and receiving qualities of antennae," by Professor Erik Hallén. In addition, the criticism of other workers in the field, especially that of Dr. S. A. Schelkunoff, provided a continuing and helpful challenge.

Originally it was hoped to publish an advanced textbook on the broad subject of antennas as Volume II of *Electromagnetic Engineering*. Volume I, "Fundamentals," treated general electromagnetic theory. This plan was interrupted by the war, which brought with it the need for a comprehensive but elementary treatment of antennas for use in training programs in radar. Extensive research was devoted to the preparation and continual improvement of mimeographed notes for this purpose and they were ultimately published as a chapter in the book *Transmission lines, antennas, and wave guides*. Following the war the generous support of a Joint Services Contract for fundamental research under the Office of Naval Research opened the way for an extensive and systematic study of problems in electromagnetic radiation and microwave optics. A sequence of graduate courses, including one on electromagnetic radiation, provided the background in the training of students planning to write doctoral dissertations in the field.

In 1948 a set of "Notes on antennas" was prepared and used as a text in the graduate course. These notes were expanded and revised and ultimately developed into the manuscript for this book. However, instead of a textbook on the broad subject of antennas, it had become a treatise on the linear radiator. This evolution in its scope and level necessitated a change in publication plans and in title.

The preparation of this volume has been a long and in some of its details a tedious task. Primarily, however, it has been a pleasure and a privilege. In many respects it has been a coöperative enterprise. Some of the theoretical and much of the experimental work reported was done by students working for their doctorate and by other members of the research and academic staffs in the Cruft Laboratory. Although many of their individual contributions are recognized throughout the book, it is appropriate to acknowledge here the generous coöperation and valuable contributions of the following present or one-time students and members of the Cruft Laboratory research group: A. L. Aden, H. W. Andrews, D. J. Angelakos, T. Chang, P. Conley, B. C. Dunn, C. E. Faflick, J. V. N. Granger, C. W. Harrison, Jr., E. O. Hartig,

R. M. Hatch, W. Kelvin, Phyllis Kennedy, D. D. King, R. D. Kodis, D. Middleton, T. Morita, C. Moritz, W. E. Owen, C. H. Papas, D. K. Reynolds, T. E. Roberts, R. V. Row, J. Sevick, S. Stein, J. E. Storer, C. T. Tai, C. C. Tang, J. Taylor, K. Tomiyasu, D. G. Wilson, T. W. Winternitz, T. T. Wu. Credit is also due to Professors L. Brillouin and P. Le Corbeiller for helpful discussions and to Professor E. L. Chaffee, Director of Cruft Laboratory and of the Joint Services Contract. Special credit belongs to Dr. C. T. Tai for graciously preparing a first draft of Secs. 5 through 8 in Chap. VIII.

The computation of numerical results was carried out by Misses Betty Finn, Phyllis Kennedy, Julie Klimas, Suzanne Knight, and Mary Lee Richardson.

Most of the drafting was done by Mr. Elmer Rising and his assistants, Misses Polly Horan, Sylvia Fry, and especially Adela Pokorna. A number of drawings were made by Mrs. Thais Carter and Miss Phyllis Kennedy.

The manuscript was typed in part by Mrs. Virginia Haydock and Miss Mary O'Neill but principally and most expertly by Miss Phyllis Kennedy. Valuable editorial assistance for the first part of the manuscript was given by Miss M. K. Ahern and for the whole book by Mr. Joseph D. Elder, science editor of the Harvard University Press.

The proofs of the entire volume were read by Mr. Tai T. Wu. Thanks to his careful and constructive work, both the accuracy and the clarity of the presentation have been enhanced.

A major part of the illustrations was first published in Cruft Laboratory Technical Reports under Contract N5ORI-76, T.O.1, between the President and Fellows of Harvard College and the Office of Naval Research or in Doctoral Dissertations at Harvard University. Subsequently many appeared in papers published in technical and scientific journals. The writer is indebted to the editors of the *Journal of Applied Physics* and the *Proceedings of the Institute of Radio Engineers* for permission to reproduce figures appearing in papers that originated at Harvard University and to make this general acknowledgment. Actually, all such illustrations were made from originals retained at Harvard, and none was obtained by direct reproduction of illustrations in journals.

The following figures are reproduced by permission of the Controller of H.M. Stationery Office, and British Crown copyright is reserved: in Chapter VI, Figs. 4.9a, 5.3, 5.4, 5.5, 5.6, 5.7, 5.8, 5.9, 5.10a, 5.10b, 5.13, 5.14, 5.15, 5.16, 5.17, 5.18, 5.19, 7.2, 7.3, 7.4; in Chapter VII, Fig. 27.2.

Acknowledgment is made to the Institution of Electrical Engineers for permission to reproduce Figs. 6.2, 6.3, and 6.4 in Chapter VI.

Figures 10.1, 10.2, 10.3, 10.4, 10.5, 10.6 and 11.1 in Chapter VI are based on or reproduced from figures in the *Bell System Technical Journal* for January 1943, by permission of the associate editor.

Figures 4.9b and 4.13 are taken from the *Wireless Engineer* by permission of the editor.

The writer is grateful to the Navy Department (Office of Naval Research), the Signal Corps of the U.S. Army, and the U.S. Air Force for supporting his research program and to Harvard University for a sabbatical leave of absence which permitted him to complete most of the final draft of the manuscript. He is also most appreciative of the help and cheer of Justine and Christopher King who lived with the ever-growing manuscript and skilfully and happily kept it from becoming a tyrant.

RONOLD W. P. KING

CONTENTS

Note Concerning the Numbering of Equations, Figures, and References; Notation, xxi

Introduction: A Bridge from Mathematics to Engineering in Antenna Theory, 1

Isolated Antennas

1. Free Oscillations, 3
2. Forced Oscillations and the Poynting-Vector Theorem, 3
3. Cylindrical Antennas and Equivalent Transmission Lines, 4
4. The Antenna as a Boundary-Value Problem, 5
5. Retarded Potentials and Integral Equations, 7

Coupled Antennas

6. The emf Method, 8
7. The Retarded-Potential Method, 8
8. Antenna Over Conducting Earth, 9

Elements of a Consistent Theory

9. Theory and Practice, 10

I. Essentials of Electromagnetic Theory, 12

1. Density Functions, 12
2. Essential Density Functions, 12
3. The Equation of Continuity, 12
4. The Field Equations and the Boundary Conditions, 13
5. Auxiliary Field Vectors and Constants, 13
6. The Force and Torque Equations, 14
7. Potential Functions; Potential of Axial Current in Cylinder, 14
8. Polarization and Magnetization Potentials, 20
9. Integrals of the Field Equations, 21
10. Energy Functions, 21
11. Simple Media, 21
12. The Magnetic Dipole, 22

II. Linear Radiators as Circuit Elements, 24

Antenna as Transmitting System

1. Theory and Experiment: Discussion of the Problem, 24
2. Definition of an Antenna, 25
3. Antenna as a Circuit Element, 26
4. Transmitting System with Simple External Generator, 26

Antenna and Transmission Line

    5. Two-Wire Line with Symmetric Impedance as End-Load, 31
    6. Two-Wire Transmission-Line Theory; Generalized Equations, 33
    7. Approximate Solution of the Generalized Transmission-Line Equations, 44
    8. Antenna Terminating Two-Wire Line, 50
    9. Antenna as Mid-Point Load for Symmetrically Driven Line; Antenna as End Load with Stub Support, 55
  10. Antenna Center-Driven from a Two-Wire Line with Minimum Coupling; Image Methods of Driving an Antenna Using Open-Wire and Coaxial Lines, 59

Hallén's Theory of Cylindrical Antennas

  11. Cylindrical Antenna, 69
  12. Formal Solution of Differential Equation for Vector Potential of Antenna. General Symmetry Conditions, 72
  13. Vector and Scalar Potentials for the Symmetric Antenna, 74
  14. Hallén's Integral Equation, 76
  15. Series Solution of the Integral Equation, 81

Distributions of Current and Charge

  16. The Zeroth-Order Solutions for Current and Charge, 86
  17. Simplification for Small Base Separation, 87
  18. Hallén's Expansion, 89
  19. Hallén's Solution: Evaluation of First-Order Integrals, 94
  20. Expansion of King and Middleton; Parameters, 101
  21. King-Middleton Solution: General Formula for Current, 107
  22. Axial Distribution of Current, 110
  23. Axial Distribution of Charge, 119
  24. The Instantaneous Current and Charge, 123
  25. An Alternative Method of Solving Hallén's Integral Equation, 123
  26. Experimental Determination of the Distributions of Current and Charge, 127

Theoretical Impedance and Admittance of Cylindrical Antenna

  27. General Formula for Impedance and Admittance of Center-Driven Antenna, 141
  28. Impedances of Zeroth and Modified Zeroth Order, 144
  29. Admittance and Impedance Factors for Small Base Separation, 149
  30. Impedance and Admittance for Zero Base Separation; First- and Second-order Solutions, 151
  31. Impedance and Current for the Electrically Short Antenna, 184
  32. Impedance of Antenna with Small Base Separation, 193

Apparent Impedance of Cylindrical Antenna as Load on a Transmission Line

33. Antenna with Positive or Negative Capacitance in Shunt, 193
34. Apparent Load Impedance of Antenna; Summaries of Theoretical and Experimental Results, 198
35. Impedance Measurements with Open-Wire Lines; the Image-Plane Line, 215
36. Impedance Measurements with a Long, Two-Wire Line, 219
37. Impedance Measurements with a Vertical Coaxial Line; Effect of Dielectric Bead, 221
38. Effect of Transmission-Line Dimensions on Impedance of Antennas Measured with Coaxial Line, 227

Other Formulations of the Cylindrical Antenna Problem

39. Storer's Variational Modification of Hallén's Integral Equation, 237
40. The emf Method and the Poynting-Vector Method for Determining the Impedance of Cylindrical Antennas, 258
41. The Integral-Equation Formulation of L. V. King, 259
42. Storm's Method of Undetermined Coefficients, 261

III. Circuit Properties of Arrays of Linear Radiators, 263

Theory of Two Identical, Parallel, Nonstaggered, Center-Driven Antennas

1. General Discussion of the Problem, 264
2. Coupled Antennas with Small Base-Separations; Arbitrary Driving Voltages, 264
3. Symmetrically Driven Antennas, 267
4. Antisymmetrically and Arbitrarily Driven Antennas, 269
5. Closely Spaced Antennas: Cage Antenna; Two-Wire Line, 273
6. First-Order Distributions of Current in Symmetrically and Antisymmetrically Driven Antenna, 275
7. Impedances of Symmetrically and Antisymmetrically Driven Antennas; Antenna Parallel to Conducting Plane, 277
8. Impedance of Antennas with Arbitrary Voltages: Self- and Mutual Impedances, 286
9. Coupling and End Effects for Antennas with Individual Feed Lines, 312
10. H-Arrays, 318
11. Driven Antenna with Parasite, 322
12. Transmission-Line Radiators: Antenna with Parasite; Folded Dipole, 334
13. Experimental Determination of Self- and Mutual Impedance of Parallel Antennas, 346

Theory of Identical, Parallel, Nonstaggered Antennas
  14. General Theory, 351
  15. Three Antennas at the Vertices of an Equilateral Triangle, 361
  16. Four Antennas at the Corners of a Square, 366
  17. Cage Antennas, 369
  18. Current and Impedance of Antenna with Corner Reflector, 370
  19. Parallel Arrays with All Elements Driven; Broadside and End-Fire Arrays; Circuit Properties, 371
  20. Parallel Arrays with Parasitic Elements, 377

Arrays with All Units in Neutral Planes
  21. Two Mutually Perpendicular Antennas, 378
  22. Turnstile Antennas and Arrays, 381

V-Antennas
  23. Integral Equation for Symmetric, Center-Driven V-Antenna, 381
  24. Solution of the Integral Equation for the V-Antenna, 384
  25. Impedance and Current for V-Antenna, 386
  26. Crossed Antennas, 396

Asymmetrically Driven Linear Antennas
  27. Current and Impedance for Asymmetrically Driven Cylindrical Antennas, 397
  28. Functions and Parameters for Asymmetrically Driven Antennas, 402
  29. Approximate Impedance of Asymmetrically Driven Antennas, 403
  30. Theory of the Sleeve Dipole, 407
  31. Half-Dipole with Multielement Counterpoise; Ground-Plane Antenna, 418

Collinear Arrays
  32. Three-Element Collinear Array; General Analysis of Central Antenna, 422
  33. Three-Element Collinear Array; General Analysis of Outer Antennas, 430
  34. Collinear Array of Three Half-Wave Dipoles, 436
  35. The Center-Driven Collinear Array with Phase-Reversing Stubs, 442
  36. Experimental Study of the Collinear Array; Lumped Capacitive Coupling, 447

IV. The Receiving Antenna as a Circuit Element, 456
  General Theory of Two Different, Arbitrarily Oriented Antennas
    1. Integral Equations with Arbitrary Spacing, 456
    2. Specialization of the General Case to Great Separation—Equations for Vector Potential, 459

The Center-Loaded Receiving Antenna

3. General Formula for Current in the Center-Loaded Antenna, 461
4. Equivalent Circuit of a Receiving Antenna; Effective Length, 466
5. Zeroth-Order Solution for Center-Loaded Receiving Antenna, 468
6. The Expansion Parameter for the Receiving Antenna, 470
7. First-Order Current in Center-Loaded Receiving Antenna, 473
8. Experimental Determination of Distribution of Current and Impedance for a Base-Loaded Receiving Antenna on a Conducting Plane, 479
9. Effective Length, 486
10. Power in the Load; Effective Cross Section, 496

Parasitic Antennas as Scatterers and Reflectors in the Radiation Zone

11. Reradiation from Parasitic Antennas—General Theory, 501
12. Current Distribution in Reradiating Antenna, 503
13. Reflecting or Back-Scattering Cross Section of Reradiating Antennas, 506
14. Experimental Determination of Reflecting Cross Section, 512
15. Two Parallel Nonstaggered Receiving and Scattering Antennas, 517

V. The Electromagnetic Field of Center-Driven and Multiple Half-Wave Antennas, 522

Infinitely Thin Center-Driven Antennas—Cylindrical Coördinates

1. The Vector Potential, 523
2. The Magnetic Field, 524
3. The Electric Field, 526
4. The Radiation Field, 528
5. Field Patterns, 530

Theory of Thin Center-Driven and Multiple Half-Wave Antennas—Confocal Coördinates

6. The Complete Field of Multiple Half-Wave Antennas in Confocal Coördinates, 532
7. The Radiation Field of a Multiple Half-Wave Antenna, 535
8. Phase Relations in the Field of Multiple Half-Wave Antennas, 538
9. Field of Antiresonant Antenna and Antennas of Arbitrary Length, 546
10. Graphs of Instantaneous Electric and Magnetic Fields for Multiple Half-Wave Antennas, 549
11. Correlation of Theoretical Near-Zone Field of Multiple Half-Wave Antenna with Experiment, 555

Radiation Factors

12. Radiation Function, Radiation Resistance, Directivity of Center-Driven and Multiple Half-Wave Antennas, 560
13. The Approximate Representation of the Radiation Field of a Center-Driven Antenna; Effective Length of a Driven Antenna, 566

Center-Driven Antennas of Nonvanishing Cross Section

14. The Radiation Field of a Symmetric Center-Driven Antenna of Nonvanishing Radius—Application of the Reciprocal Theorem, 568
15. The Complete Field of a Center-Driven Antenna of Small Radius—Approximate Analysis, 571
16. The Radiation Field of a Center-Driven Antenna of Small Radius—Approximate Analysis, 574
17. The Electric Field Near the End of an Antenna; Spark Discharges, 575

Center-Driven Antennas with Unequal Currents in the Halves

18. Radiation Field of an Unbalanced Center-Driven Antenna, 576

VI. Electromagnetic Fields of Antenna Arrays, 579

Uniform Parallel Arrays

1. The Diffraction Formula and the General Array Factor, 579
2. The Collinear Array; The Marconi-Franklin Antenna, 587
3. Parallel Arrays—the Broadside and the End-Fire Array, 595

Parasitic Parallel Arrays

4. Field of Driven Antenna with Single Parasite—Approximate Second-Order Theory, 622
5. Field of Driven Antenna with Several Parasites—Zeroth-Order Solution; Arrays of Yagi-Uda Type, 635
6. Yagi-Uda Arrays—Experimental Investigation, 645
7. Broadside Array with Parasitic Curtain, 647

Nonuniform Parallel Arrays

8. Complex Array Polynomials, 651
9. Binomial End-Fire Arrays, 652
10. Directive End-Fire Arrays with Assigned Nulls, 654
11. Nonuniform Broadside Arrays Derived from Uniform Arrays, 660
12. Real Polynomials for Symmetric Nonuniform Broadside Arrays, 661
13. Optimum Currents for Symmetric Broadside Arrays; Tchebyscheff Polynomials, 665
14. Closely Spaced Tchebyscheff Arrays, 676

Arrays with Omnidirectional Properties

  15. Circular Array, 679
  16. Mutually Perpendicular Antennas and Turnstile Array, 684
  17. Resonant V-Antenna, 687

The Reciprocal Theorem and the Properties of Arrays

  18. Transmitting Arrays with Elements of Finite Cross Section, 689
  19. Receiving Arrays, 690

VII. Antennas Over a Conducting Region, 695

  1. Hertzian Potentials, 696

Vertical Dipoles Over Conducting Earth; General Formulation

  2. The Polarization Potential of a Vertical Electric Dipole or of an Element of a Vertical Antenna, 698
  3. The Magnetization Potential of a Vertical Magnetic Dipole or of an Element of a Horizontal Loop Antenna, 701
  4. Representation of Spherical Waves as a Bundle of Plane Waves, 703
  5. Boundary Conditions for Vertical Dipoles; Incident, Reflected, and Refracted Waves, 707

Far-Zone Fields of Vertical Dipoles Over a Conducting Earth

  6. Asymptotic Integration of the Hertzian Field of a Dipole; Field Patterns, 715
  7. Dipoles in Air Over Dielectrics and Conductors, 726
  8. The Far-Zone Field of an Antenna with Sinusoidally Distributed Current Over a Conducting Earth, 743

Quasi-Near-Zone Fields of Vertical Dipoles Over a Conducting Earth

  9. Van der Pol's Integrals for the Vertical Dipole Over a Plane Earth, 746
  10. Approximate Integration of Van der Pol's Formulas, 754
  11. Norton's Formulas for Practical Evaluation, 759
  12. The Electromagnetic Field of a Vertical Dipole; Summary, 760
  13. The Vertical Electromagnetic Field of a Vertical Dipole, 763
  14. The Radial Electromagnetic Fields, 768
  15. Complete Electromagnetic Fields of Vertical Dipoles, 770
  16. Polarization and Tilt of the Surface Waves, 773
  17. The Field of a Vertical Electric Dipole at Large Numerical Distances, 774
  18. Quasi-Near-Zone Fields of Vertical Antennas with Sinusoidal Currents Over a Conducting Earth, 775

Horizontal Dipoles Over a Conducting Earth

19. Hertzian Potentials of Horizontal Dipoles Over a Conducting Earth, 779
20. Far-Zone Hertzian Potentials of Horizontal Dipoles Over a Conducting Earth, 783
21. The Electromagnetic Field of a Horizontal Dipole Over a Conducting Earth, 785
22. Comparison of the Fields of Horizontal Electric Dipoles with Those of Vertical Dipoles, 791
23. Horizontal Antennas with Sinusoidal Currents Over a Conducting Earth, 794
24. Currents Excited on a Perfectly Conducting Plane by a Parallel Antenna, 794

Impedance and Radiation Resistance of Antennas Over Conducting Planes

25. Impedance of Antennas Over Infinite, Perfectly Conducting Planes, 799
26. Radiation Resistance of a Vertical Electric Dipole Over Plane Earth, 800
27. Radiation Resistance of a Horizontal Electric Dipole Over a Plane Earth, 804
28. Impedance of Base-Driven Antenna on a Ground Screen of Finite Size, 808
29. Impedance of Base-Driven Antenna on a Ground Screen of Finite Size on an Infinite Plane Imperfectly Conducting Earth, 815

VIII. The Antenna as a Boundary-Value Problem. 818

1. Hemispheroidal and Conical Antennas, 818
2. Equations for Spherical Waves with Rotational Symmetry, 820

Conical Antennas

3. Boundary Conditions and Equations for the Symmetric, Spherically Capped Biconical Antenna, 824
4. The Dominant Mode; Apparent Terminal Admittance, 826
5. Interior Complementary Modes; Input Impedance, 828
6. Exterior Complementary Modes; Matching of Fields, 830
7. Integrals of Products of Legendre Functions, 831
8. General Solution for Y_{la} and the Infinite Set of Linear Equations, 832
9. Solution for the Apparent Terminal Admittance of a Biconical Antenna with Small Angles, 833
10. Impedance of a Biconical Antenna with Small Angle, 836

Cylindrical Antennas

11. Thin Antennas of Arbitrary Cross Section; Schelkunoff's Theory, 838
12. The Gap Problem: Cylindrical Antenna with Biconical Transmission Line, 844
13. Zuhrt's Analysis of Cylindrical Antennas, 848

Appendix: Tables of Generalized Sine and Cosine Integrals, 857

Problems, 881

Bibliography, 901

List of Principal Symbols, 913

Index, 927

NOTE CONCERNING THE NUMBERING OF EQUATIONS, FIGURES, AND REFERENCES; NOTATION

Chapters are numbered in roman; sections are numbered in arabic beginning with 1 in each chapter; equations are numbered consecutively (1), (2), . . . in each section, with no reference to the section number. When reference is made to an equation in the same section, only the equation number is given, *e.g.*, (5). When reference is made to an equation in another section in the same chapter, the section and equation numbers are given in the form (7.14). When reference is made to an equation in another chapter, the chapter number, section number, and equation number are given, *e.g.*, (II.7.14). Figures are given both section and figure numbers; thus, Fig. 6.2 is the second figure in Sec. 6. Reference to figures in another chapter includes the chapter number, *e.g.*, Fig. II.6.6. Chapter and section numbers appear at the top of each page, and by reference to these numbers any equation or figure can be quickly found.

Bibliographic references for each chapter are arranged alphabetically and numbered consecutively. General references and textbooks are included in the bibliography for Chapter I. Reference to the bibliography for the same chapter is by arabic number only, *e.g.*, Schelkunoff[24]; a reference to the bibliography for another chapter includes the chapter number, *e.g.*, Hallén[II.15].

The notation distinguishes among vector, real scalar, and complex scalar quantities. In the text all vectors, whether real or complex, are in boldface roman type, except that a few unit vectors are represented by boldface Greek letters with circumflex; real scalars are in lightface italic or lightface Greek; complex scalars are in boldface italic or boldface Greek. A comparison of the several kinds of type used is available in the List of Symbols. In figures, vectors are indicated by arrowheads on the appropriate line segments; complex quantities are distinguished from real quantities by an underline.

The Theory of
LINEAR ANTENNAS

The Theory of
LINEAR ANTENNAS

INTRODUCTION*

A BRIDGE FROM MATHEMATICS TO ENGINEERING IN ANTENNA THEORY

The first radio transmitting system was constructed in 1887 by Heinrich Hertz for the purpose of verifying experimentally the existence of the electromagnetic waves that had been predicted mathematically by Maxwell two decades earlier. Hertz's transmitter consisted of a straight copper wire end-loaded with large spheres or cylinders of metal and driven by a spark discharge across a gap at its center; its resonant frequency was 53.5 Mhz. As a receiver Hertz used a small rectangular loop of wire with a micrometer spark gap in the middle of one side. A faint discharge across the gap as observed in a darkened room constituted "reception." With this simple equipment Hertz verified Maxwell's predictions and laid the experimental foundations for practical radio transmission, which Marconi introduced to the world in 1896. Thus, inspired and advised by Helmholtz and guided by his own theoretical insight and experimental skill, Hertz built a bridge from mathematics to engineering, from the differential equations and boundary conditions of Maxwell to the wireless transmission of Rutherford and Marconi.

Marconi's transmitter resembled that of Hertz. It consisted of a grounded vertical wire several meters long and driven by a spark discharge across a gap near its base. The receiver was a similar wire with the gap replaced by a magnetic detector. Since the only conducting path from the transmitting antenna to ground was by way of a spark across the gap, the oscillations in the antenna were highly damped. This undesirable feature was improved by F. Braun in 1898 when he patented a circuit in which the spark gap was in a separate primary circuit in series with an appropriate coil and condenser. This tank circuit was coupled inductively to a secondary consisting of the antenna in series with a coupling coil in which the driving electromotive force was induced and which provided a continuous conducting path from the antenna to ground. Except for the later insertion of a transmission line between the antenna and its coupling coil, in order to permit the generator to be located at a convenient distance from the antenna, the Braun transmitter provided the complete electrical equivalent of the modern base-driven broadcast antenna.

In view of the fact that the fundamental electrical structure of the simplest radiating circuit has been known essentially in its present form for over fifty years, it seems natural to assume that all of its properties have long ago been investigated in complete detail both theoretically and experimentally. But this is not the case. Indeed, the publication of this volume was postponed for several years in order to achieve a hitherto unknown correlation between theory and experiment and with it to obtain a more complete understanding of the wire antenna driven from a transmission line. The reasons for this long delay in the acquisition of fundamental knowledge about antennas appear to be twofold. First, the structural simplicity of a wire driven from a transmission line is misleading; an exact analysis of this circuit is extremely difficult both theoretically and experimentally. Second, there have been few serious attempts until recently to bridge in a complete, quantitative sense the gulf between the extensive and highly successful practical know-how of the engineer and the radio amateur on the one hand, and the intricate and idealized analyses of theoretical physicists and applied mathematicians (with results that are neither consistent nor operationally significant) on the other.

An illuminating example of the manner in which the interests of the mathematician may be at cross purposes with the needs of the practical engineer may be found in the problem of the antenna gap. Although the

---

* Papers referred to in the introduction are included in the bibliographies at the ends of the appropriate chapters.

original antennas of Hertz and Marconi included a series gap across which the driving voltage was impressed, this gap was eliminated by Braun, so that since 1898 practical transmitting systems have consisted of a continuous conducting path from one end of the antenna, around the transmission line (if there is one) and a coupling coil, back to the other end of the antenna or to ground. Notwithstanding this complete absence of a gap in actual antennas, some contemporary mathematicians maintain that a gap is "an essential part" of every antenna. Indeed, they go so far as to state that no antenna can radiate without a gap. Upon this hypothesis they have based extensive mathematical studies of the so-called "gap problem in antenna theory" which are interesting but somewhat removed from reality.

To be sure, the gulf between theory and practice is seldom as great as in the example of the gap. However, a common characteristic of the purely mathematical approach is to regard physical reality as much less important than considerations of mathematical convenience, rigor, or even interest. After all, the mathematician is accustomed to predetermine the boundary conditions of his problems in a manner that facilitates their solution according to his own high standards of rigor. And he prefers a rigorous solution subject to a hypothetical set of boundary conditions to an approximate solution subject to the actual boundaries of the problem requiring solution. Since practical circuits involving transmitting systems are far too intricate to permit of rigorous solutions, the method of the mathematician is, in effect, to substitute a similar, but actually different and much simpler, problem for the actual one. This he solves rigorously and then assumes or implies that his results apply to all similar practical arrangements for which he would substitute the same mathematical model. Since differences in actual circuits are not contained in his idealized boundary conditions, their effects can not appear in the final solution, its rigor notwithstanding. Indeed, he may even regard a method that takes account of such differences as inferior to his apparently more general theory. This is illustrated, for example, in the introduction to one theoretical paper in which may be found reference to "another method with the unfortunate property that the fundamental integral equation requires re-solution for each new way of driving the antenna."

Although the comment is not strictly true for the method it seeks to criticize, the implication is clear: if differences in the driving conditions enter into the formulation of an antenna problem, it must be a fault of the method of analysis. If this were true, which it is not, it would follow as a necessary consequence that general formulas and universal curves that do not depend on how an antenna is driven are theoretically meaningful and, therefore, a logical goal of mathematical investigations and a legitimate request of the practical engineer. Actually, the *measurable* characteristics of an antenna depend greatly on the detailed construction of the driving circuit, and a theory that takes account of this fact of observation is superior to one that does not. To be sure, such a theory can not provide a general formula, but this does not mean that it is more special. Surely, the most specialized result is the one that is limited in its "generality" to the boundless range of the hypothetical but has not even a single actual and physically meaningful application.

Since the ideal of complete exactness in the formulation and in the solution of practical antenna problems is unattainable, an alternative to the mathematical technique of simply solving a different problem must be found. Evidently, what is required is a method that sacrifices a measure of rigor in the solution in order to achieve greater accuracy in the formulation of the boundary conditions. It is only by requiring the assumed boundary conditions to fit the actual conditions closely that account can be taken of essential differences characteristic of particular structures. For example, antennas driven from a variety of transmission lines with different connections must *not* be replaced one and all by antennas uniformly driven by a hypothetical voltage maintained across an equally hypothetical gap. It is quite natural that this method does not appeal to the mathematician, since usually he is not prepared to introduce *physically reasonable* simplifying approximations *at the proper place*. The requisite judgment to make such approximations requires a careful study and complete understanding of the experimental problem of making measurements on each particular circuit. The plotting of experimentally determined points with a theoretical curve is largely meaningless unless a study has been made of the correlation between the hypothesis underlying the theoretical analysis and the physical structure on which measurements were made. In particular, the operational significance of theoretically defined quantities for which experimental analogs are to be measured must be assured. The admonition attributed to Hansen, that "the impedance concept is no substitute for thought," must always be borne in mind in the very general sense that no mathematically convenient definition of any concept or quantity is physically meaningful unless it is operationally significant, that is, unless the definition includes an expressed or implied method of measurement.

Both the theory and the practice of radio transmission have made great strides in the last half-century at the hands of brilliant mathematicians and skilled and inventive engineers. There have been many disciples of Maxwell and of Marconi, but few of Hertz. Consequently, relatively few bridges have been built to join the realm of mathematical models with the world of actual structures made of copper wire. In order to illustrate that the basic problem of systematic coördination of theory and practice has been neglected, and at the same time to furnish a critical and historical background for the material presented in this volume, the following discussion is provided. It must be emphasized that no complete history of the antenna in radio communication is to be outlined. What is intended is a summary of the highlights in the development of those phases only that have a direct bearing on the relatively restricted subject matter of this book.

ISOLATED ANTENNAS

1. *Free Oscillations*

Since the original radiators of Hertz and Marconi were excited in their natural modes, it was to be expected that early studies of antennas should concern themselves primarily with free oscillations, with no concern for the method of excitation. The first antenna investigated was the *Hertzian dipole*, consisting essentially of an oscillating filament of current of very short length between two metal spheres. Hertz himself determined the complete electromagnetic field of such an oscillator in 1888 under the assumption that its dimensions were vanishingly small so that it became what might be called a mathematical doublet. The field of Marconi's antenna was approximated by that of half of a Hertzian dipole over a perfectly conducting infinite plane, a so-called Abraham oscillator.

The analysis of cylindrical wires of finite length rather than of infinitesimal dipoles was begun with the determination by Abraham in 1898 of the natural modes of a thin ellipsoid. For obvious mathematical reasons in formulating simple boundary conditions this shape was preferred to the cylinder. Six years later a complete discussion and numerical tabulation of the proper vibrations of a prolate spheroid over a wide range of eccentricities from the sphere to the thin ellipsoid was prepared by Marcel Brillouin. In the same year, 1904, Hack represented the electromagnetic field of thin ellipsoids oscillating in each of several modes in the form of expanding waves corresponding to Hertz's representation of the field of the doublet. Other analyses of the free oscillations of spheres were made by Mie in 1908 and Debye in 1909. In all of these studies of proper modes the theory naturally was based on a determination of the complete field in the vicinity of the antenna with proper boundary conditions on the surface.

Owing to its less attractive shape from the mathematical point of view of formulating simple boundary conditions and determining the complete field as a set of characteristic functions, the cylinder was analyzed more advantageously using the retarded-potential method described by Pocklington in 1897, and discussed by Rayleigh in 1912 and by L. Brillouin in its application to radiation in 1922. The natural oscillations of cylinders were treated by Oseen in 1913–14 and by Hallén in 1930. A rigorous justification of the use of the retarded-potential method with cylindrical conductors of finite radius with currents distributed transversely according to normal skin effect was given by Zinke in 1941.

2. *Forced Oscillations and the Poynting-Vector Theorem*

The development of powerful generators of undamped oscillations over a wide band of frequencies brought with it the possibility of maintaining large currents in antennas at frequencies other than the natural frequency of the antenna circuit itself. The problem of forced oscillations in cylindrical conductors thus arose. Since it involves not only the boundary conditions on the metal surfaces but also the driving conditions, this problem is much more formidable than that of free oscillations. Indeed, the simplest practical circuit, consisting of a cylindrical antenna driven from an open-wire or coaxial line, is so uninviting from the point of view of formulating boundary conditions that no attempt was made to analyze it. In its place a variety of mathematically more acceptable, physically hypothetical sets of boundary and driving conditions were studied. These were all more or less closely related to the actual antenna–transmission-line problem, yet differed from it in essential details. It is instructive to consider some of them in turn.

Perhaps the simplest and most powerful theoretical tool for determining the power radiated from a transmitting system is the Poynting-vector theorem. This expresses the total power transferred from sources within a closed surface of arbitrary shape to the universe outside in terms of the integral of the normal component of the Poynting vector over the *closed* surface. It is this theorem that served as a basis for many methods used in the study of antennas in which forced oscillations were maintained. Since the current in the antenna is the actual unknown, the Poynting-vector theorem logically leads to an integral equation in the current. But this is not the traditional manner of applying the theorem to radiating circuits. The usual procedure was to replace the actual antenna with its driving circuit by a superficially similar but fundamentally different one in which the driving mechanism was ignored and the total power radiated was expressed in terms of a radiation resistance and an assumed current in the antenna at a convenient reference point.

Since the electromagnetic field at great distances from an antenna is much simpler than in its vicinity, the Poynting-vector theorem is most conveniently applied to a great sphere enclosing the transmitting system in its radiation zone. Using this method, the power radiated by a Hertzian dipole was determined by Hertz in 1888, and the power radiated from a thin ellipsoid with an essentially sinusoidal distribution of current by Abraham in 1898 and 1901. In 1908 Rüdenberg derived a formula for a top-loaded antenna that bears his name, and this was generalized in 1917 by Van der Pol. In 1920 G. W. Pierce computed the radiation resistance of an inverted L-antenna erected on a perfectly conducting plane earth. He assumed the antenna to be so driven that the distribution of current along the horizontal and vertical members was a continuing sinusoid.

When the Poynting-vector theorem is applied to the cylindrical surface of an antenna instead of to a great sphere, it is called the emf method. This was employed by Pistolkors in 1929, by Bechmann in 1931, by Carter in 1932, and by Labus in 1933 to determine the self-impedance of a linear radiator with an *assumed sinusoidal distribution* of current. In defining the self-impedance the antenna was assumed to be center-driven from a transmission line. Actually, a sinusoidal distribution of current requires a continuous distribution of sources of electromotive force along the entire antenna, with amplitudes and phases adjusted properly for each size and shape of antenna. On the other hand, a center-driven antenna can not have a sinusoidally distributed current since this distribution leads to an electromagnetic field on the surface of the antenna that violates the boundary conditions. Since it was precisely the field on the surface that was used in the evaluation of the complex power by the emf method, it is not surprising that only a rough approximation of the input impedance was achieved. This approximation was best when the input current was large, as with antennas near resonance, and poorest when the input current was small.

The complete electromagnetic field of an antenna with a sinusoidal distribution of current was given by Bechmann in 1931, and in more complete form by Riazin and Brown in 1937. A detailed study of the far-zone field of a linear radiator with sinusoidal current was made by Carter, Hansell, and Lindenblad in 1931.

Since the far-zone field of an antenna is much less sensitive to differences in the assumed distribution of current than the field at the surface of the antenna, far-zone field patterns and the radiation resistance referred to the maximum current of antennas driven from transmission lines are determined more accurately from the sinusoidal distribution than is the input impedance. Note, however, that the input resistance of a resonant, perfectly conducting antenna is equal to the radiation resistance referred to maximum current.

3. Cylindrical Antennas and Equivalent Transmission Lines

Since the circuit properties of a center-driven antenna could not be determined accurately by the emf method and an assumed sinusoidal current except when the antenna was very thin and had a length near resonance, other methods of investigation were sought.

In particular, the attempt was made to replace the center-driven antenna by an "equivalent" transmission line in determining its impedance properties. Thus, in 1934 Siegel and Labus distributed the power radiated from an antenna along an open-end section of two-wire line as if it were the power dissipated in the ohmic resistance of the wires. They were able to show that the input impedance of such a section of line with artificially increased attenuation resembled that of an antenna of equal length more closely than did the input impedance calculated by the emf method, especially near antiresonance where the latter method fails completely. In some particulars the agreement was quite good, in others, notably the resonant and antiresonant lengths, it was poor. A similar transmission-line formula was devised by Wells in 1941.

Considerably greater success in constructing a hypothetical transmission line with impedance properties that resembled those of a centre-driven antenna was achieved by Schelkunoff in 1941. Using as a point of departure the thin biconical antenna (for which a rigorous representation in transmission-line form is possible), Schelkunoff investigated the properties of a tapered two-wire line with a varying characteristic impedance and an end-load appropriately defined to dissipate the radiated power. He found that his first-order approximation of the impedance properties of this line resembled those of the cylindrical antenna quite closely, notably near antiresonance. Near resonance the correspondence was less satisfactory. It was shown by Tai in 1950 that when Schelkunoff's first-order formula for impedance was expanded by including second-order terms, it led to impedance curves that were quite different in magnitude and even in shape for moderately thick antennas from those of the first-order theory.

In 1948 Hallén made a direct analysis of the cylindrical antenna using the retarded-potential method in which he expressed the distribution of current in the well-known transmission-line form of traveling waves multiply reflected at the ends. This investigation showed that reflection coefficients which are *constants* characteristic of the ends *in a transmission line* are *functions of the number of reflections* experienced by the traveling waves *in the antenna*. This indicates that, since the cylindrical antenna evidently is not equivalent to a transmission line with constant reflection coefficients, it is likewise not equivalent to a transmission line with a lumped load at the ends.

4. *The Antenna as a Boundary-Value Problem*

The several methods of treating a driven antenna so far discussed have been concerned primarily with obtaining satisfactory engineering formulas to characterize the impedance and field properties of an antenna. In order to treat the forced oscillations in an antenna according to the rigorous methods of solving boundary-value problems, it was natural that a thin ellipsoidal antenna should be chosen, just as in the study of free oscillations, because of its mathematically attractive boundaries. However, in order to maintain forced oscillations in a prolate spheroid and preserve spheroidal symmetry for the complete electromagnetic field in the vicinity of the antenna, the method of driving had to be selected with care. Two mathematically acceptable methods of excitation were suggested. Of these the simplest was to immerse the entire spheroid in a uniform electric field parallel to its long axis. The second method assumed an impressed, rotationally symmetric electric field maintained across a gap or belt at its center.

Page and Adams in 1938 and Ryder in 1942 studied the prolate spheroid immersed in an axial electric field. It is evident that an antenna excited in this manner differs fundamentally from an antenna center-driven from a transmission line. In particular, no input impedance can be defined in the conventional sense of electric-circuit theory, since the antenna has no terminals. The analysis of such an antenna also has no bearing on the related problem of reception. Although a receiving antenna is immersed in a uniform field if it is at some distance from the transmitter, this field need not be parallel to its axis. Moreover, in order to be used to receive a signal the antenna must be connected to a load by a transmission line, and the presence of such a line and load alters completely the distribution of current in the antenna. It follows that the analysis of the spheroid in a uniform field has little application to the practical problem of transmission and reception. On the other hand, it does apply to the unloaded antenna used as a scatterer or reradiator. But for this purpose one isolated antenna is not very useful. A single parasitic antenna must be quite close to a transmitter in order to be effective, and for scattering or reradiating many antennas more or less close together must be used. In both of these cases the presence of other antennas destroys the symmetry and makes the results inapplicable.

The spheroid center-driven by a belt or gap at the equator across which a rotationally symmetric field is impressed was analyzed by Stratton and Chu in 1941. This method of driving obviously does *not* correspond to the picture of two half spheroids at the ends of a two-wire line which actually illustrated the theoretical paper. A rotationally symmetric electric field at the edge of a gap can be maintained only if the antenna is driven from within by a radial transmission line. Such a transmission line was not considered by Stratton and Chu, who defined the impedance of the antenna at the outer edge of the gap without specifying how the voltage was maintained or how the current crossed it. It was this "gap problem" that was discussed by Infeld in 1947. If it is recognized that the sides of the supposed gap are necessarily the two conductors of a radial transmission line leading to a generator in the interior of the antenna, the gap problem is replaced by a transmission-line problem with end-effect. This radial transmission-line problem is as essential a part of the analysis of an internally driven antenna as is the two-wire line of an externally driven one. If the radial transmission line is made biconical, a mathematically attractive situation is achieved in that the generator may be assumed concentrated at the apex of the two cones forming the conductors of the transmission line. Needless to say, such a generator has no practical counterpart. But for the mathematician a source that can be represented by a singularity in the electric field is ideal. By using a sphere instead of a prolate spheroid the metal surfaces of both transmission line and antenna are represented conveniently in spherical coördinates. By opening the biconical line wider a biconical horn or antenna is obtained. Such a horn was investigated by Barrow in 1939.

The analytical advantages of the apex-driven biconical structure was recognized by Schelkunoff, who studied the biconical antenna without end caps in 1941 under the assumption that the current vanished at the edges of the cones. This condition is never exactly satisfied, but it is a satisfactory approximation if the cones are thin enough. The formulation was developed further by Schelkunoff in 1946 and especially by Smith and Tai in 1948, who formulated the problem rigorously, taking into account all boundaries including spherical end-surfaces. A particularly interesting feature of the biconical antenna is the fact that its dominant mode is a true transmission-line mode with constant characteristic impedance. By taking advantage of this fact Schelkunoff was able to formulate the driving-point impedance at the apices of the cones in the form appropriate for an end-loaded transmission line. (His formula for the end-loaded, tapered transmission line as an equivalent for the cylindrical antenna was based on this representation.)

Although the biconical structure has great mathematical advantages in permitting the convenient formulation of both boundary and driving conditions, the application of the theory to a practical antenna is not obvious. Generators that are equivalent to a singularity in the electric field do not exist and are difficult to approximate. Needless to say, the pictures of two cones driven from a two-wire line that illustrated the theoretical papers have little connection with the problem analyzed. The finite separation of the cones and the presence of the two conductors of the line destroys the biconical-spheroidal symmetry and introduces end- and coupling effects that are ignored completely in the theory. An important step toward obtaining an operationally significant impedance for a transmitting system in which a cone is the radiating element was made by Papas in 1949, following a suggestion by L. Brillouin. It consisted in driving the cone from a coaxial line over a conducting plane, with the dominant mode in the coaxial line matched to the dominant mode in the biconical line-antenna by making their characteristic impedances equal. By determining the apparent load impedance as seen from the coaxial line a *measurable quantity* is defined.

The direct analysis of the cylindrical antenna as a boundary-value problem was made by Zuhrt in 1944 but not published until 1950. The structure analyzed was an infinite collinear array of cylinders, each center-driven by a discontinuity in scalar potential. The individual units were separated by a finite distance, which ultimately was allowed to become infinite to leave only the central unit. The driving voltages were equal in magnitude and alternated in sign from one unit to the next adjacent ones. By representing the current distribution in each antenna by an odd cosine series with undetermined coefficients, expressing the field inside and outside an infinite cylinder of radius equal to that of the antennas, and matching these fields across the cylinder, Zuhrt determined the coefficients and thus solved the problem. Since he did not take account of the fields inside the tubular antennas but assumed the current to vanish at their edges, his analysis is limited to thin cylinders. Moreover, since the distribution of current on cylindrical antennas can be represented by a small number of terms in an odd cosine series *only* when the antenna is short or its length is near resonance, and very many terms are required near antiresonance, the method is practical only in a restricted sense.

5. Retarded Potentials and Integral Equations

The first attempts to solve the problem of the cylindrical antenna directly rather than by substituting for it a different problem were made independently by L. V. King in 1937 and by Hallén in 1938. King analyzed a vertical cylinder separated from a perfectly conducting horizontal plane by a narrow gap. It was driven by a voltage maintained between the ground and a hypothetical ring around the antenna at a short distance above the gap. Since no mechanism was specified for maintaining the voltage and no conducting path was provided for the flow of charges from the base of the antenna to ground, the generator was a mathematical abstraction resembling that assumed by Stratton and Chu in the analysis of the spheroidal antenna. By assuming a sinusoidal current plus an added term with coefficients to be determined so that the field maintained by the total current satisfied the boundary conditions, King derived an integral equation which he solved approximately essentially in reciprocal powers of the quantity $2ln(2h/a)$, where h is the length and a the radius of the antenna. If an error in neglecting significant terms is corrected, L. V. King's solution corresponds to that derived by Hallén a year later by a different method.

Hallén derived an integral equation for the axial current in a cylinder, using essentially the retarded-potential method of Pocklington. Beginning with an antenna center-driven from a two-wire line, Hallén eliminated the line by assuming, in effect, the antenna to be an unbroken conductor with a short section at the center coupled to a parallel, conducting bridge terminating the feeding line. The induced emf was then assumed to be concentrated as a discontinuity in scalar potential instead of distributed over a short but finite length. Note that there is no gap in the antenna and that the emf is induced from the outside. Simultaneously, Hallén treated the center-loaded cylindrical antenna in an arbitrarily oriented uniform electric field. Hallén solved his integral equation by a method of iteration in reciprocal powers of the same parameter used by L. V. King. Since no account was taken of the ends of the antenna, the analysis applied to thin cylinders. A simplification of Hallén's formulation and extensive computations from his formulas were made by R. King and Blake and by R. King and Harrison as applied to both the transmitting and the receiving antenna in 1942–1945. The integral-equation formulation was investigated by Synge in 1942 and, with special consideration of the problem of the end-surfaces, by Brillouin in 1943. Other methods of carrying out the iteration in Hallén's integral equation using different expansion parameters in order to achieve more rapid convergence were introduced by Miss Gray in 1944 and by R. King and Middleton in 1946 as applied to the transmitting antenna. A similar modification for the unloaded receiving antenna was made by Van Vleck in 1947. A variational method for solving Hallén's integral equation was developed by Storer and extended by Tai in 1950. The results of this method were in close agreement with those of King and Middleton.

The failure of all existing analyses to take account of the actual driving conditions of practical antennas driven from transmission lines, and the interesting but disconcerting fact that there seemed to be little difficulty in obtaining carefully measured experimental data to substantiate the widely divergent results of different theories, led to a systematic theoretical and experimental study of antennas driven from various types of transmission lines in different connections and orientations by R. King and Winternitz in 1948 and 1950, by R. King, Tomiyasu, and Conley in 1949, and by Hartig in 1950. These investigators found that transmission-line end-effects and the coupling between antenna and line affect greatly the apparent impedance of an antenna as a load terminating a line. They achieved complete quantitative correlation between antenna theory and experiment by combining theoretical impedances determined from the expansion of R. King and Middleton or the variational solution of Storer and Tai with lumped, terminal-zone networks designed for each type of transmission line and each different connection. Widely different measured impedances for the same antenna but connected to a different line or to the same line with a different orientation were thus correlated with theory. It was shown that if the apparent impedance of an antenna terminating a transmission line *of any type* and connected to the antenna in any manner is measured repeatedly as the spacing of the conductors of the line is decreased progressively, and the values so obtained are extrapolated to zero line spacing, the values at zero spacing may be identified with those calculated from either the King-Middleton or the variational formula when the driving source is assumed to be a discontinuity in scalar potential, that is, a so-called slice generator. In this way an operational significance is given to the properties of antennas driven by slice generators without gap. Since the impedance of an antenna driven by such a hypothetical generator is independent of a transmission line, it may be represented by general formulas and universal curves which depend only on the length and diameter of the cylinder. Although they do not represent measurable quantities, since a discontinuity in scalar potential does not exist as a practical driving source, they are useful in conjunction with appropriate terminal-zone networks which transform the ideal theoretical value into a measurable apparent value.

The problem of determining the electromagnetic field of a linear radiator in terms of the actual distribution of current instead of an assumed sinusoidal current has received relatively little attention. This is owing in part to the fact that far-zone fields computed from a sinusoidal current are quite satisfactory, in part to the presence of other factors that are at least as significant in their effect on the field. An example is the earth with its finite rather than infinite conductivity. The first far-zone field pattern for a linear radiator with finite cross section was derived by L. V. King in 1937. In 1943 Harrison and R. King determined the far-zone field of a linear radiator using an approximate distribution of current that included a sinusoidal quadrature component and a component in phase with the driving voltage. An accurate representation of the far-zone field was achieved by the same investigators in 1944 by application of the reciprocal theorem to obtain the far-zone field characteristic of an antenna with finite radius from its complex effective length.

COUPLED ANTENNAS

6. *The emf Method*

The analysis of coupled antennas may be divided into two principal groups: those which assume a sinusoidal distribution of current in all antennas in an array in order to determine mutual impedances and electromagnetic fields, and those which seek to determine the actual distribution of current and the associated self- and mutual impedances and electromagnetic field.

Most of the early and many of the more recent studies of coupled antennas in a variety of configurations fall into the first group. This includes in particular the extensive work of Carter in 1932 and of Brown in 1937. More recent work using the sinusoidal assumption was carried out by Walkinshaw in 1946, by Cox in 1947, and by Barzilai and by Starnecki and Fitch in 1948. Most of these investigators made use of the emf method. Since the current is accurately sinusoidal only when each antenna is driven by a continuous distribution of electromotive forces of proper amplitude and phase along the entire length, and since it is a moderately good approximation for antennas driven from or loaded by transmission lines only when the antennas are short or near resonance and quite thin, the practical application of self- and mutual impedances determined by the emf method is limited. However, most practical multielement arrays make use of half-wave elements for which the results of the sinusoidal theory, although quantitatively approximate, are useful.

The far-zone fields of uniform arrays of linear radiators of short or half-wave elements with sinusoidally distributed currents were studied particularly by Southworth in 1930 and by Sterba in 1931. The radiation resistances of parallel arrays referred to maximum current as obtained from the far-zone fields by application of the Poynting-vector theorem were determined by Bontsch-Bruewitch in 1926 and by Papas and R. King in 1948. Nonuniform arrays were treated in a comprehensive paper by Schelkunoff in 1943, by Dolph in 1946, and by Taylor and Whinnery in 1951.

7. *The Retarded-Potential Method*

The mathematically convenient boundaries of isolated spheroidal or biconical antennas cease to be attractive when two or more arbitrarily placed antennas are involved. Indeed, the entire method of determining a complete field in terms of proper functions and satisfying appropriately chosen boundary conditions is unavailable for coupled antennas. The fundamental reason for this is that there are no systems of coördinates in which the complicated boundary conditions can be expressed simply and in which the wave equation is separable. Therefore, analyses of coupled antennas paralleling the mathematically elegant studies of isolated spheroidal and biconical structures have not been carried out.

It is particularly in the analysis of coupled antennas that the retarded-potential method of Pocklington demonstrates its power. This method was first applied to two parallel antennas, each center-driven by a discontinuity in scalar potential of arbitrary magnitude and phase, by R. King and Harrison in 1944. Using the method of symmetric components, they determined the distributions of current in the antennas and the self- and mutual impedances. By replacing one generator by a voltage drop across an arbitrary impedance, the case of a driven antenna with a single tuned parasite was analyzed. The same method was applied to the folded dipole by R. King in 1945. Essentially the same problem of two coupled antennas was solved independently by Bouwkamp in 1948. The formulation of R. King and Harrison was improved by Tai in 1948 and extended by him to three antennas at the vertices of an equilateral triangle and to the corner reflector. The analysis was generalized further to N parallel antennas by R. King in 1950. The vector-potential method was applied to two collinear antennas by Harrison in 1945; the formulation was improved and extended to include antennas coupled by sections of transmission line by R. King in 1950, by Andrews in 1953, and by Faflick and Tang in 1954. The practical problem of coupled antennas individually driven from transmission lines with finite spacing is treated in this volume for the first time.

The theory of the V-antenna and the practical problem of driving it from a two-wire line was developed by R. King in 1950. The asymmetrically driven antenna was analyzed by Synge in 1942 and by Hallén in 1948, using the method of traveling waves. An alternative formulation and its application to the sleeve dipole was given by R. King in 1950. An extensive study of the sleeve dipole was made by Taylor in the same year. He determined distributions of current and impedances experimentally and theoretically, using Storer's variational method.

Experimental studies of coupled antennas and the correlation of theory with experiment have not progressed as far as in the case of the single isolated antenna. McPetrie and Saxton in 1946 and Starkey and Fitch in 1950 measured the front-to-back ratio for an antenna with a single parasite and compared their results with the corresponding ratios determined theoretically by R. King in 1948 using the first-order self- and mutual impedances computed by Tai. The apparently poor agreement led these investigators to develop an "engineering" formula based on an arbitrarily end-loaded transmission line that has no theoretical foundation. Actually, the front-to-back ratio is so sensitive to small changes in reactance that account must be taken of the finite transmission-line spacing and of end-effects if satisfactory agreement is to be obtained. It is shown in this volume that theoretical results corrected to approximate the actual conditions of the apparatus are in good agreement with experimental values. This is another illustration of the importance of taking account of actual rather than idealized boundaries and driving conditions. The correct correlation of theory with experiment involves more than the placing in juxtaposition of the measured results for a practical circuit and the theoretical results of an abstract set of boundary and driving conditions.

Measurements of the distribution of current and the impedance of an antenna with a coupled parasite and of a folded dipole were made by Morita and Faflick in 1949. Their results are in good agreement with the corresponding theoretical quantities when account is taken of the properties of the actual circuits, including especially transmission-line end-effect and coupling effects. More extensive measurements on coupled antennas of various types driven from different transmission lines in several connections and orientations are required before a complete understanding of the significance of the many complications arising in the coupled-antenna problem is achieved. A theoretical study of a large variety of folded antennas was made by Harrison in 1953.

8. Antenna Over Conducting Earth

The practically important problem of the electromagnetic field of an antenna over the earth is complicated by the curvature of the earth and by the ionosphere. Since these factors are a part of the study of wave propagation around the earth, which is a major field in itself, and since they may be treated in a manner quite independent of the antenna problem, it is sufficient in considering this latter to deal with antennas over a plane earth. The first study of horizontal and vertical Hertzian dipoles over a plane conducting earth was made by Sommerfeld in 1909 and extended by von Hoerschelmann in 1912. In 1919 Weyl reanalyzed the same problem by a different method and obtained Sommerfeld's solution but without one term that was included in Sommerfeld's result and that was represented as a surface wave—the so-called Zenneck surface wave. Since that time the problem has been examined by many investigators, including especially Sommerfeld, Strutt, Van der Pol, and Bremmer. The question as to the existence of the surface wave that occurred in Sommerfeld's solution has been the subject of much controversy. It has finally been resolved by Ott, who showed that the solution may be expressed in a form involving a term that is a space wave at sufficiently great distances but becomes a Zenneck surface wave very near the source. This conclusion has been confirmed and supplemented by Baños and Wesley. However, the contribution to the electromagnetic field by the surface wave is negligible even close to the antenna. The reduction of Sommerfeld's and Van der Pol's theoretical results to practical use has been carried out primarily by Norton, who has prepared extensive charts dealing with ground-wave propagation. Norton's formulation includes a term designated as a surface wave, but this is not the Zenneck surface wave.

Most of the investigations of antennas over a conducting earth have been concerned with the determination of the potential or the electromagnetic field of Hertzian dipoles or of antennas with sinusoidally distributed currents. In only a few papers, notably in those by Niessen in 1935 and by Sommerfeld and Renner in 1942, has the radiation resistance of a Hertzian dipole over a plane conducting earth been investigated. Experimental measurements of the radiation resistance of antennas of finite length over a conducting earth were made by Proctor in 1950. However, there have been no fundamental theoretical studies to determine the input impedance of an antenna over or on a conducting earth. No analysis of the impedance of a base-driven antenna with a ground system buried in a conducting earth is available. Obviously, these are mathematically very intricate problems, and it can not be anticipated if or how soon they may be solved.

Closely related to the problem of the antenna erected on the earth is the problem of the vertical antenna base-driven over a highly conducting ground screen of *finite* size. Measurements of impedances of antennas erected on circular and square ground planes by Meier and Summers in 1949 showed that both resistance and reactance vary significantly as the size of the plane and its shape are changed. Leitner and Spence studied the problem theoretically in 1950, using a quarter-wave antenna over an oblate spheroid. Since their solution was in series form that converged slowly for large screens, Storer in the same year derived a simple formula for the difference between the impedance of an antenna of arbitrary length on a large but finite ground screen and on an infinite ground plane.

ELEMENTS OF A CONSISTENT THEORY

9. *Theory and Practice*

Problems in electromagnetic radiation are difficult from both the experimental and the theoretical points of view because infinity must be one of the boundaries and because essential structures must have dimensions of the same order of magnitude as the wavelength. These requirements make unavailable the analytical simplifications that are possible in the theory of wave guides and cavities, where the entire field is enclosed in highly conducting walls, or in conventional electric-circuit theory, where the dimensions of the circuit are small compared with the wavelength. As a consequence, the general problem of radiating circuits that are unrestricted in shape or size has hardly been touched. The study of antennas that are not required to be sufficiently thin to permit an essentially one-dimensional analysis so far has been limited to isolated structures of very simple shape with highly idealized driving mechanisms. On the other hand, the problem of two or more coupled radiators has been handled only for antennas that are sufficiently thin to make a quasi-one-dimensional analysis a satisfactory approximation. It is clear, therefore, that a consistent theory of antennas which is to include a complete quantitative treatment of radiating and receiving systems formed of single antennas or arrays is restricted to an essentially one-dimensional analysis of linear radiators. By a complete quantitative treatment is meant an analysis that begins with the Maxwell field equations and ends with the quantitative determination of operationally significant quantities in important practical systems. What is desired is a theory for determining the power transferred to the load of a transmission line that is driven from a receiving antenna in terms of the power supplied to another transmission line that feeds an antenna or array of antennas. Since the formulation and development of such a theory is an extensive and complicated task even for linear radiators, the scope of this volume has been appropriately restricted. This does not mean that types of antennas other than those discussed are less important or less interesting. It means simply that that type of antenna that permits a systematic and comprehensive investigation has been selected and the investigation carried out as completely as the present state of theoretical and experimental knowledge permits.

Although other methods of analysis are included, the essential formulation of the theory of linear radiators necessarily is made in terms of the retarded-potential method of Pocklington, which leads to an integral equation. No other approach permits the analysis of isolated antennas, coupled antennas of various types, receiving antennas, and the associated transmission-line problems by a single straightforward procedure. In spite of the advantages of variational methods in solving integral equations, the method of solution preferred in the systematic analysis is the iterative procedure of successive approximations. This has the attractive property of leading directly and simply to a zeroth-order solution which corresponds to the results obtained by the emf method using a sinusoidal current, and at the same time providing a procedure for obtaining higher-order solutions of greater accuracy.

The systematic formulation of antenna theory and its application presented in this volume is directed neither to the mathematician whose primary interest is rigor, nor to the practical engineer who desires only a final working formula or a set of charts. Much as Hertz created the original connecting link from Maxwell to Marconi, so this volume seeks to provide a bridge from the mathematician to the practical engineer. As such, it is addressed to the applied scientist who is concerned with physical phenomena of practical importance and their mathematical representation in a form that provides both an insight into the physical aspects of the problem and reasonable quantitative accuracy in their numerical evaluation. It is to be anticipated, nevertheless, that the obvious inadequacy of available mathematical techniques in the solution of integral equations, and in particular in the solution of simultaneous integral equations, may prove a challenge to the mathematician. Similarly, the emphasis on experimental measurement and the correlation of theory with experiment should appeal to the practical engineer, if only in terms of the utility of the extensive, mutually consistent, and well-integrated charts and numerical tables of values of important quantities as determined both theoretically and experimentally.

CHAPTER I

ESSENTIALS OF ELECTROMAGNETIC THEORY

The systematic study of antennas presented in this volume depends upon general macroscopic electrodynamics as formulated in detail in the literature.[31, 43, 49, 51] For convenience, important symbols, formulas, and equations (written in terms of complex amplitudes for a periodic dependence upon the time in the form $\rho_{inst} = \rho e^{j\omega t}$) are given in this chapter together with brief definitions or descriptions.

1. Density Functions

The average electrical properties of matter are described mathematically in terms of six characteristic functions or densities that are assumed to be slowly varying from point to point in a body or region. The six densities with a qualitative description of their significance follow.

The volume density of charge, ρ, in coulombs per cubic meter, is a scalar point function that measures the average density of charge in the neighborhood of every point in the interior of the body or region. Its complex amplitude is ρ in $\rho_{inst} = \rho e^{j\omega t}$.

The surface density of charge, η, in coulombs per square meter, is a scalar point function that measures the average density of charge in a thin surface or boundary layer of atomic thickness. Its complex amplitude is η.

The volume density of polarization, **P**, in coulombs per square meter, is a polar vector point function that measures the average density and direction of a distribution of dipoles or their equivalent in the interior of the body or region. Its complex amplitude is **P**.

The volume density of moving charge or of current, **i**, in amperes per square meter, is a polar vector point function that measures the average magnitude and direction of nonrandom flow of charges across each unit area in the interior of the body or region. Its complex amplitude is **i**.

The surface density of moving charge or of current, **l**, in amperes per meter, is a polar vector point function that measures the average magnitude and direction of nonrandom flow of charge across each unit width of a thin surface or boundary layer of atomic thickness. Its complex amplitude is **l**.

The volume density of magnetization, **M**, in amperes per meter, is an axial vector point function that measures the average magnitude and direction of microscopic current whirls in the neighborhood of each point in the interior of a body or region. Its complex amplitude is **M**.

2. Essential Density Functions

Since the volume and surface densities ρ, η, **P**; **i**, **l**, **M**, are not all independent, but actually involve in their statistical definitions the manner in which the body or region is subdivided into volume and surface cells, it is desirable to introduce essential densities or characteristics that are independent of the mode of subdivision. Their complex amplitudes are defined as follows:

$$\bar{\rho} \equiv \rho - \operatorname{div} \mathbf{P}, \tag{1}$$

$$\overline{\rho_m \mathbf{v}} \equiv \mathbf{i} + \operatorname{curl} \mathbf{M} + j\omega \mathbf{P}, \tag{2}$$

$$\bar{\eta} \equiv \eta + \hat{\mathbf{n}} \cdot \mathbf{P}, \tag{3}$$

$$\overline{\eta_m \mathbf{v}} \equiv \mathbf{l} - \hat{\mathbf{n}} \times \mathbf{M}, \tag{4}$$

where $\hat{\mathbf{n}}$ is a unit *outwardly* directed normal, $\hat{\mathbf{n}} \cdot \mathbf{P} \equiv (\hat{\mathbf{n}}, \mathbf{P})$ is the scalar or dot product, and $\hat{\mathbf{n}} \times \mathbf{M} \equiv [\hat{\mathbf{n}}, \mathbf{M}]$ is the vector or cross-product. The vector operators used in (1) and (3) are the divergence and the curl, defined by

$$\operatorname{div} \mathbf{P} \equiv \nabla \cdot \mathbf{P} \equiv \lim_{\Delta\tau \to 0} \frac{\int_\Sigma \hat{\mathbf{n}} \cdot \mathbf{P}\, d\sigma}{\Delta\tau}, \tag{5}$$

$$\operatorname{curl} \mathbf{M} \equiv \nabla \times \mathbf{M} \equiv \lim_{\Delta\tau \to 0} \frac{\int_\Sigma \hat{\mathbf{n}} \times \mathbf{M}\, d\sigma}{\Delta\tau}. \tag{6}$$

3. The Equation of Continuity

The fundamental postulate of conservation of electric charge is expressed mathematically in the following complex form for periodic phenomena at points in the interior of a body or region:

$$\text{div } \mathbf{i} + j\omega\eta = 0. \tag{1}$$

In a *surface* or boundary layer of atomic thickness the corresponding equation is

$$\text{div } \mathbf{l} + j\omega\eta - \hat{\mathbf{n}} \cdot \mathbf{i} = 0. \tag{2}$$

The shorthand notation

$$\mathbf{l} \equiv \mathbf{l}_1 + \mathbf{l}_2, \tag{3a}$$

$$\eta \equiv \eta_1 + \eta_2, \tag{3b}$$

$$\mathbf{i} \equiv \mathbf{i}_1 + \mathbf{i}_2 \tag{3c}$$

is used, the subscript 1 referring to the region 1 on one side of the boundary, the subscript 2 to the region 2 on the other side. The unit vector $\hat{\mathbf{n}}$ carries the same subscript as $\mathbf{i}$; it is *external* to the region *indicated by the subscript*. In longhand form (2) is

$$\text{div } (\mathbf{l}_1 + \mathbf{l}_2) + j\omega(\eta_1 + \eta_2) - \hat{\mathbf{n}}_1 \cdot \mathbf{i}_1 - \hat{\mathbf{n}}_2 \cdot \mathbf{i}_2 = 0. \tag{4}$$

Equations (1) and (2) may be expressed in terms of the essential densities as follows.

$$\text{div } \overline{\rho_m \mathbf{v}} + j\omega \overline{\rho} = 0, \tag{5}$$

$$\text{div } \overline{\eta_m \mathbf{v}} + j\omega \overline{\eta} - \hat{\mathbf{n}} \cdot \overline{\rho_m \mathbf{v}} = 0. \tag{6}$$

In a cylindrical conductor carrying a total axial current of complex amplitude I_z and a charge per unit length of complex amplitude q, the equations of continuity combine into

$$\frac{\partial I_z}{\partial z} + j\omega q = 0. \tag{7}$$

4. The Field Equations and the Boundary Conditions

The action of charges in one body or region on those in another is expressed in two steps, of which the first defines the *electromagnetic field* at all points in space due to the charges and currents in one body, and the second defines the *force* and *torque* on the second body in terms of this field. The electromagnetic field consists of the vector point functions $\mathbf{E}$ in volts per meter and $\mathbf{B}$ in volt seconds (or webers) per square meter. Their complex amplitudes are defined by

$$\epsilon_0 \text{ div } \mathbf{E} = \overline{\rho}, \tag{1}$$

$$\text{curl } \mathbf{E} = -j\omega \mathbf{B}, \tag{2}$$

$$\nu_0 \text{ curl } \mathbf{B} = \overline{\rho_m \mathbf{v}} + j\omega \epsilon_0 \mathbf{E}, \tag{3}$$

$$\text{div } \mathbf{B} = 0, \tag{4}$$

where ϵ_0 and ν_0 are universal constants determined experimentally. ϵ_0 is the fundamental electric constant (dielectric constant or permittivity of space); ν_0 is the fundamental magnetic constant (diamagnetic constant or reluctivity of space). The reciprocal of the magnetic constant is the permeability, denoted by μ_0. Numerical values are

$$\epsilon_0 = 8.854 \times 10^{-12} \text{ farad/meter}$$

$$\doteq \frac{1}{36\pi} \times 10^{-9} \text{ farad/meter}, \tag{5}$$

$$\nu_0 = 7.95 \times 10^5 \text{ meter/henry}$$

$$\doteq \frac{1}{4\pi} \times 10^7 \text{ meter/henry}, \tag{6}$$

$$\frac{1}{\epsilon_0} = 1.129 \times 10^{11} \text{ meter/farad}$$

$$\doteq 36\pi \times 10^9 \text{ meter/farad}, \tag{7}$$

$$\frac{1}{\nu_0} = \mu_0 = 1.257 \times 10^{-6} \text{ henry/meter}$$

$$= 4\pi \times 10^{-7} \text{ henry/meter}. \tag{8}$$

At a boundary between regions 1 and 2 where $\overline{\rho}$ and $\overline{\rho_m \mathbf{v}}$ are discontinuous the complex field equations for amplitudes reduce to

$$\epsilon_0 \hat{\mathbf{n}} \cdot \mathbf{E} = -\overline{\eta}, \tag{9}$$

$$\hat{\mathbf{n}} \times \mathbf{E} = 0, \tag{10}$$

$$\nu_0 \hat{\mathbf{n}} \times \mathbf{B} = -\overline{\eta_m \mathbf{v}}, \tag{11}$$

$$\hat{\mathbf{n}} \cdot \mathbf{B} = 0. \tag{12}$$

In longhand these are

$$\epsilon_0 \hat{\mathbf{n}}_1 \cdot \mathbf{E}_1 + \epsilon_0 \hat{\mathbf{n}}_2 \cdot \mathbf{E}_2 = -\overline{\eta}_1 - \overline{\eta}_2, \tag{13}$$

$$\hat{\mathbf{n}}_1 \times \mathbf{E}_1 + \hat{\mathbf{n}}_2 \times \mathbf{E}_2 = 0, \tag{14}$$

$$\nu_0 \hat{\mathbf{n}}_1 \times \mathbf{B}_1 + \nu_0 \hat{\mathbf{n}}_2 \times \mathbf{B}_2 = -\overline{\eta_m \mathbf{v}}_1 - \overline{\eta_m \mathbf{v}}_2, \tag{15}$$

$$\hat{\mathbf{n}}_1 \cdot \mathbf{B}_1 + \hat{\mathbf{n}}_2 \cdot \mathbf{B}_2 = 0. \tag{16}$$

The normal component of the electric vector and the tangential component of the magnetic vector are discontinuous, while the tangential component of $\mathbf{E}$ and the normal component of $\mathbf{B}$ are continuous across a boundary.

5. Auxiliary Field Vectors and Constants

The auxiliary electric vector $\mathbf{D}$ in coulombs per square meter and the auxiliary magnetic vector $\mathbf{H}$ in amperes per meter are often convenient shorthand symbols. They are defined by

$$\mathbf{D} \equiv \epsilon_0 \mathbf{E} + \mathbf{P}, \tag{1}$$

$$\mathbf{H} \equiv \nu_0 \mathbf{B} + \mathbf{M}. \tag{2}$$

At points in space where **P** and **M** vanish, **D** and **H** differ from **E** and **B** by constant factors only.

The following combinations of the universal constants ϵ_0 and ν_0 are also useful:

$$v_0 \equiv \sqrt{\frac{\nu_0}{\epsilon_0}} = \sqrt{\frac{1}{\mu_0 \epsilon_0}}$$
$$\doteq 3 \times 10^8 \text{ meter/second}, \quad (3)$$

$$\zeta_0 \equiv \sqrt{\frac{1}{\nu_0 \epsilon_0}} = \sqrt{\frac{\mu_0}{\epsilon_0}}$$
$$= 376.7 \text{ ohm} \doteq 120\pi \text{ ohm}, \quad (4)$$

where v_0 is a velocity said to be characteristic of space, and ζ_0 is a resistance, also said to be characteristic of space.

6. The Force and Torque Equations

In most circuit problems it is not necessary to calculate forces acting on charges because interest is primarily in the pointer readings of meters which, while specifying a condition of equilibrium between mechanical and electrical forces, are calibrated directly in terms of electrical quantities such as current and potential difference. That is, the final step in a calculation that determines the condition of equilibrium of electrical and mechanical forces and torques in terms of currents and the constants of springs or the expansion of heated wires is contained in the calibration of the meter. It is important to bear in mind that the calculation of the current or potential difference must take into account the effect of the meter in the circuit. If this is expressible as a simple equivalent impedance, this is readily accomplished and the calculation of current or potential difference is sufficient for comparison with experimental pointer readings. In all cases the calibration of a meter and a specification of its impedance apply to a particular frequency or a more or less limited range of frequencies. If a pointer reading is noted at frequencies outside this range, it is necessary to reëxamine the condition of equilibrium and its effect in the network in terms of the construction of the meter and the physical and geometrical properties of the network, as discussed in chap. VI of reference 31.

The equations of equilibrium are

$$\mathbf{F}_m + \mathbf{F} = 0, \quad (1)$$
$$\mathbf{T}_m + \mathbf{T} = 0. \quad (2)$$

The electrical force **F** and torque **T** on a volume τ characterized by the essential densities $\bar{\rho}$, $\bar{\eta}$, $\overline{\rho_m \mathbf{v}}$, and $\overline{\eta_m \mathbf{v}}$ are

$$\mathbf{F} = \int_\tau (\bar{\rho}\mathbf{E} + \overline{\rho_m \mathbf{v}} \times \mathbf{B}) \, d\tau$$
$$+ \int_\Sigma (\bar{\eta}\mathbf{E} + \overline{\eta_m \mathbf{v}} \times \mathbf{B}) \, d\sigma, \quad (3)$$

$$\mathbf{T} = \int_\tau \mathbf{r} \times d\mathbf{F}_\tau \, d\tau + \int_\Sigma \mathbf{r} \times d\mathbf{F}_\sigma \, d\sigma, \quad (4)$$

where **E** and **B** are the electromagnetic fields at the elements of integration due to all charges and currents except those in and on τ and **r** is the vector from an arbitrary origin to $d\tau$ or $d\sigma$. In (4) $d\mathbf{F}_\tau$ is the integrand of the first integral in (3), $d\mathbf{F}_\sigma$ the integrand of the second integral in (3). $\mathbf{F}_m$ and $\mathbf{T}_m$ are mechanical force and torque.

7. Potential Functions; Potential of Axial Current in Cylinder

The complex amplitudes of the scalar potential ϕ in volts and of the vector potential **A** in volt seconds (or webers) per meter are defined by

$$-\text{grad } \phi = \mathbf{E} + j\omega \mathbf{A}, \quad (1)$$

$$\begin{cases} \text{curl } \mathbf{A} = \mathbf{B}, & (2a) \\ \text{div } \mathbf{A} = -j\frac{\beta_0^2}{\omega}\phi, & (2b) \end{cases}$$

with

$$\beta_0^2 = \omega^2 \epsilon_0/\nu_0 = (\omega/v_0)^2. \quad (3)$$

The potential functions so defined satisfy the general wave equations,

$$\nabla^2 \phi + \beta_0^2 \phi = -\bar{\rho}/\epsilon_0, \quad (4)$$
$$\nabla^2 \mathbf{A} + \beta_0^2 \mathbf{A} = -\overline{\rho_m \mathbf{v}}/\nu_0. \quad (5)$$

Particular integrals of these equations are called Helmholtz integrals. They are

$$\phi = \frac{1}{4\pi\epsilon_0} \left(\int_\tau \bar{\rho}' \frac{e^{-j\beta_0 R}}{R} \, d\tau' \right.$$
$$\left. + \int_\Sigma \bar{\eta}' \frac{e^{-j\beta_0 R}}{R} \, d\sigma' \right), \quad (6)$$

$$\mathbf{A} = \frac{1}{4\pi\nu_0} \left(\int_\tau \overline{\rho_m \mathbf{v}}' \frac{e^{-j\beta_0 R}}{R} \, d\tau' \right.$$
$$\left. + \int_\Sigma \overline{\eta_m \mathbf{v}}' \frac{e^{-j\beta_0 R}}{R} \, d\sigma' \right), \quad (7)$$

where $R \, [= \sqrt{(x-x')^2+(y-y')^2+(z-z')^2}]$ is the distance between the element of integration ($d\tau'$ or $d\sigma'$) at point P' (primed coördinates) and the point P (unprimed coördinates) where the potential is calculated.

If (6) and (7) are applied to a cylindrical conductor of length $2h$ and radius a placed along the z-axis with a total axial current of amplitude I'_z and charge per unit length q', they reduce to

$$\Phi = \frac{1}{4\pi\epsilon_0} \int_{-h}^{h} \frac{q'}{R} e^{-j\beta_0 R} dz', \tag{8}$$

$$\mathbf{A} = \hat{\mathbf{z}} A_z; \quad A_z = \frac{1}{4\pi\nu_0} \int_{-h}^{h} \frac{I'_z}{R} e^{-j\beta_0 R} dz'. \tag{9}$$

The total axial current and charge per unit length in the conductor of cross-sectional area S and circumference s are given by

$$I'_z = \int_S \mathbf{i}'_{fz} dS' = 2\pi \int_0^a \mathbf{i}'_{fz} r \, dr$$

for rotational symmetry, (10)

$$q' = \int_s \eta'_f \, ds' = 2\pi a \eta'_f$$

for rotational symmetry. (11)

The subscript f denotes free charge. The distance R is measured to the element dz' on the axis. It is assumed that the following inequality is satisfied:

$$\beta_0 a \ll 1. \tag{12}$$

In the far zone, wave zone, or radiation zone defined by

$$\beta_0 R \gg 1, \tag{13a}$$

$$R \gg h, \tag{13b}$$

and denoted by a superscript r, (8) and (9) reduce to

$$\Phi^r = \frac{1}{4\pi\epsilon_0} \frac{e^{-j\beta_0 R_0}}{R_0} \int_{-h}^{h} q' e^{j\beta_0 \hat{\mathbf{R}}_0 \cdot \mathbf{z}'} dz', \tag{14}$$

$$A_z^r = \frac{1}{4\pi\nu_0} \frac{e^{-j\beta_0 R_0}}{R_0} \int_{-h}^{h} I'_z e^{j\beta_0 \hat{\mathbf{R}}_0 \cdot \mathbf{z}'} dz', \tag{15}$$

where R_0 is measured to the center of the conductor.

The polar components A_Θ^r, A_Φ^r, A_R^r of the vector potential in the far zone are

$$A_\Theta^r = -A_z^r \sin \Theta, \tag{16a}$$

$$A_\Phi^r = 0, \tag{16b}$$

$$A_R^r = A_z^r \cos \Theta. \tag{16c}$$

The electromagnetic vectors can be calculated from the potential functions. Appropriate general formulas are

$$\mathbf{E} = -\operatorname{grad} \Phi - j\omega \mathbf{A}$$

$$= \frac{-j\omega}{\beta_0^2} (\operatorname{grad} \operatorname{div} \mathbf{A} + \beta_0^2 \mathbf{A}), \tag{17}$$

$$\mathbf{B} = \operatorname{curl} \mathbf{A}. \tag{18}$$

It is seen from (17) and (18) that the complex amplitudes of $\mathbf{E}$ and $\mathbf{B}$ can be determined entirely from the vector potential.

In the important special case (9) when all currents that contribute significantly to the vector potential are parallel to the z-axis the vector potential has only a z-component. Thus, $\mathbf{A} \doteq \hat{\mathbf{z}} A_z$. Then

$$\mathbf{E} = \frac{-j\omega}{\beta_0^2} \left(\operatorname{grad} \frac{\partial A_z}{\partial z} + \hat{\mathbf{z}} \beta_0^2 A_z \right) \tag{19}$$

and

$$\mathbf{B} = \operatorname{curl} \hat{\mathbf{z}} A_z. \tag{20}$$

The components of the field in cylindrical coördinates are

$$E_r = \frac{-j\omega}{\beta_0^2} \frac{\partial^2 A_z}{\partial r \, \partial z}, \tag{21a}$$

$$E_\theta = \frac{-j\omega}{\beta_0^2} \frac{1}{r} \frac{\partial^2 A_z}{\partial \theta \, \partial z}, \tag{21b}$$

$$E_z = \frac{-j\omega}{\beta_0^2} \left(\frac{\partial^2 A_z}{\partial z^2} + \beta_0^2 A_z \right), \tag{21c}$$

$$B_r = \frac{1}{r} \frac{\partial A_z}{\partial \theta}, \tag{22a}$$

$$B_\theta = -\frac{\partial A_z}{\partial r}, \tag{22b}$$

$$B_z = 0. \tag{22c}$$

When rotational symmetry obtains, the terms involving differentiation with respect to θ vanish.

In deriving the one-dimensional integrals (8) and (9), which involve the charge per unit length q' and the total axial current I'_z, from the volume and surface integrals (6) and (7), which are expressed in terms of the volume and surface densities of charge and current, the condition $R \gg a$ was originally imposed in addition to (12). It was then shown (ref. 31, Sec. IV.4) that for a sufficiently long conductor (8) and (9) are good approximations even at points just outside the surface of the conductor at $r = a$ if (12) is satisfied and rotational symmetry obtains. Although the proof in (ref. 31, Sec. IV.4) is quite adequate, it does not show how the contributions to A_z at a point z depend upon the distance $|z - z'|$ when this point is just outside the surface at $r = a$ of the conductor. A detailed study of this dependence has been made by Zinke[8] for a cylindrical conductor in which the radial distribution of current density is that required by skin-effect theory (ref. 31, chap. V). The results of this intricate analysis, which involves integrals of elliptic integrals, show that (9) is indeed valid just outside the surface of the antenna if R has the form

$$R = R_e = \sqrt{(z - z')^2 + r_e^2} \quad (19)$$

instead of

$$R = \sqrt{(z - z')^2 + a^2}, \quad (20)$$

where r_e is an effective radius that depends upon $(z - z')$ and the skin depth. For $|z - z'| > 6a$ this effective radius is independent of $|z - z'|$ and is given by $r_e = \sqrt{2a}$, but since $R = R_e \doteq |z - z'|$ for values as large as this, it is immaterial whether the effective or the actual radius is used. At $|z - z'| = 0.45a$, $r_e = a$. At $|z - z'| = 0$, r_e depends upon the skin depth, d_s. Thus for $d_s/a = 10^{-1}, 10^{-2}, 10^{-3}$, respectively, $r_e/a = 0.62, 0.43, 0.33$. Curves showing a/R_e for $d_s/a = 10^{-1}$ and 10^{-2} are shown in Fig. 7.1 together with the curve a/R. Clearly, the vector potential at z is actually more completely determined by the current I_z than is indicated if R is used instead of R_e. However, subject to (12), variations in the current I_z', are essentially negligible over distances $|z - z'|$ that do not exceed small multiples of the radius a, specifically for $|z - z'| \leq 10a$. Since retardation also is negligible over such distances, the correct value of A_z as given by

$$A_z = \frac{1}{4\pi v_0} \int_{-h}^{h} \frac{I_z' e^{-j\beta_0 \sqrt{(z-z')^2 + r_e^2}}}{\sqrt{(z - z')^2 + r_e^2}} dz', \quad (21)$$

is well approximated by

$$A_z \doteq \frac{1}{4\pi v_0} \int_{-h}^{z-10a} \frac{I_z' e^{-j\beta_0 |z - z'|}}{|z - z'|} dz'$$

$$+ \frac{I_z}{4\pi v_0} \int_{z-10a}^{z+10a} \frac{dz'}{\sqrt{(z - z')^2 + r_e^2}}$$

$$+ \frac{1}{4\pi v_0} \int_{z+10a}^{h} \frac{I_z' e^{-j\beta_0 |z-z'|}}{|z - z'|} dz'. \quad (22)$$

Alternatively, if A_z is approximated by

$$A_z \doteq \frac{1}{4\pi v_0} \int_{-h}^{h} \frac{I_z' e^{-j\beta_0 \sqrt{(z-z')^2 + a^2}}}{\sqrt{(z - z')^2 + a^2}} dz', \quad (23)$$

the corresponding expansion is

$$A_z \doteq \frac{1}{4\pi v_0} \int_{-h}^{z-10a} \frac{I_z' e^{-j\beta_0 |z - z'|}}{|z - z'|} dz'$$

$$+ \frac{I_z}{4\pi v_0} \int_{z-10a}^{z+10a} \frac{dz'}{\sqrt{(z - z')^2 + a^2}}$$

$$+ \frac{1}{4\pi v_0} \int_{z+10a}^{h} \frac{I_z' e^{-j\beta_0 |z-z'|}}{|z - z'|} dz'. \quad (24)$$

The difference between (22) and (24) reduces to the difference between the second integrals in (22) and (24). This is simply the difference between the areas under the curves for a/R_e and a/R in Fig. 7.1. For two values of d_s/a shown, this difference is less than 1 percent for $|z - z'| \gtrsim 2.5$. It follows that for lengths $2h$ that are large compared with a, (24) is an excellent approximation to (22), which is equivalent to stating that (23) is an excellent approximation to (21). In the analysis of antennas and circuits involving conductors that are very long compared with the radius, (23) will be used as a good approximation of (9). Similarly, (8) will be used with R given by (20).

It has been shown that the vector potential at a point P on the surface just outside a straight cylindrical conductor of radius a is independent of the cross-sectional distribution of the axial current provided the point P is not within a distance $5a$ of the end of the wire. It will be shown now that if the cylinder is not of circular cross section but has an arbitrary shape, the total axial current is the same as that in an *equivalent circular cylinder* of radius a where the *equivalent radius a* is defined in terms of the cross-sectional dimensions and shape of the actual cylinder. Evidently, if this can be proved, the solution for the total current in a circular cylinder of radius a solves at the same time the problem of cylinders of all simple cross-sectional shapes. It must be required, however, that the maximum cross-sectional dimension $2u$ satisfies the conditions,

$$\beta_0 u \ll 1; \quad u \ll h, \quad (25)$$

where h is the half-length of the cylinder.

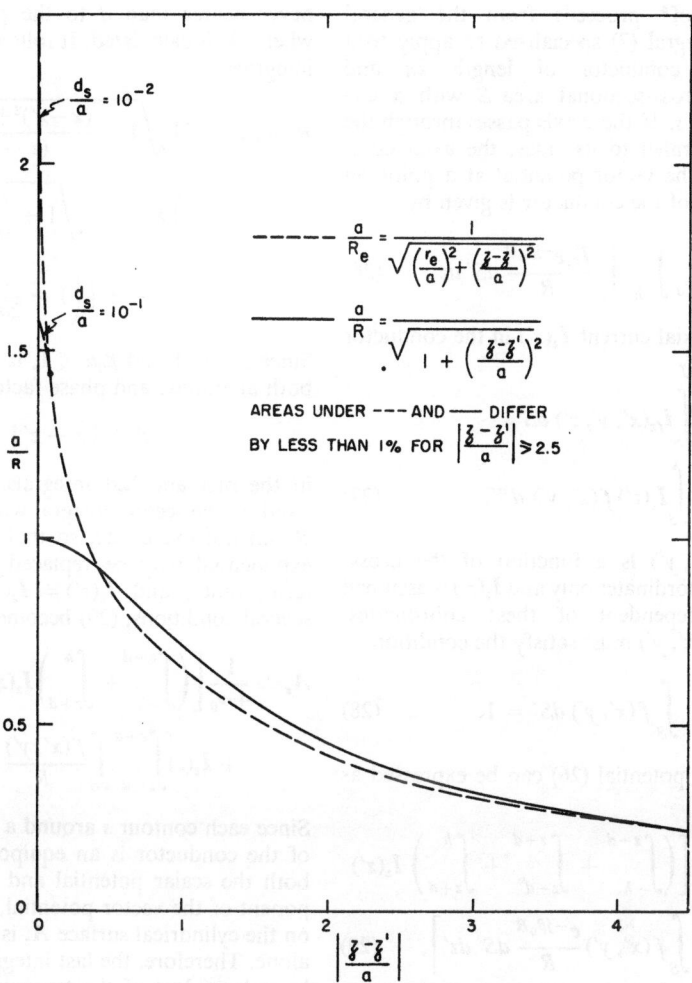

Fig. 7.1. Curves of a/R_e and a/R.

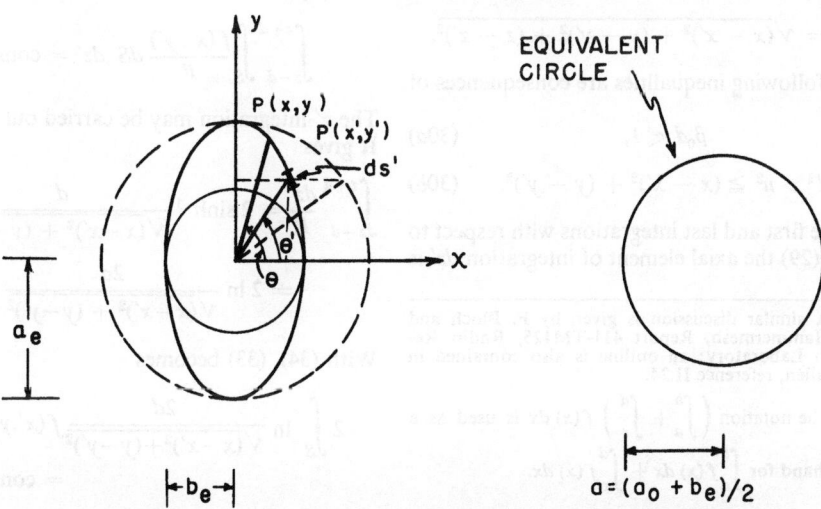

Fig. 7.2. Equivalent circle for elliptical cross section.

The proof* proceeds from the general volume integral (7) specialized to apply to a cylindrical conductor of length $2h$ and arbitrary cross-sectional area S with a circumference s. If the z-axis passes through the cylinder parallel to its sides, the axial component of the vector potential at a point on the surface of the conductor is given by

$$A_z = \frac{1}{4\pi\nu_0} \int_{-h}^{h} \int_S \frac{i'_{fz} e^{-j\beta_0 R}}{R} \, dS' \, dz'. \quad (26)$$

The total axial current $I_z(z')$ in the conductor is defined by

$$I_z(z') = \int_S i_{fz}(x', y', z') \, dS'$$

$$= \int_S I_z(z') f(x', y') \, dS', \quad (27)$$

where $f(x', y')$ is a function of the cross-sectional coördinates only and $I_z(z')$ is assumed to be independent of these coördinates. Clearly, $f(x', y')$ must satisfy the condition

$$\int_S f(x', y') \, dS' = 1. \quad (28)$$

The vector potential (26) can be expressed as follows:†

$$A_z = \frac{1}{4\pi\nu_0} \left[\left(\int_{-h}^{z-d} + \int_{z-d}^{z+d} + \int_{z+d}^{h} \right) I_z(z') \right.$$

$$\left. \times \int_S f(x', y') \frac{e^{-j\beta_0 R}}{R} \, dS' \, dz' \right], \quad (29)$$

where $d \sim 10u$ and

$$R = \sqrt{(x-x')^2 + (y-y')^2 + (z-z')^2}.$$

The following inequalities are consequences of (25):

$$\beta_0 d \ll 1, \quad (30a)$$

$$d^2 \gg u^2 \geq (x-x')^2 + (y-y')^2. \quad (30b)$$

In the first and last integrations with respect to z' in (29) the axial element of integration dz' is

---

* A similar discussion is given by F. Bloch and M. Hammermesh, Report 411-TM125, Radio Research Laboratory; an outline is also contained in E. Hallén, reference II.24.

† The notation $\left(\int_a^b + \int_c^d \right) f(x) \, dx$ is used as a shorthand for $\int_a^b f(x) \, dx + \int_c^d f(x) \, dx$.

never nearer than d to the point $P(x, y, z)$ where A_z is calculated. It follows that in these integrals

$$R = |z - z'| \sqrt{1 + \frac{(x-x')^2 + (y-y')^2}{(z-z')^2}}$$

$$\doteq |z - z'| \sqrt{1 + \frac{u^2}{d^2}}$$

$$\doteq |z - z'| \left(1 + \frac{u^2}{2d^2}\right). \quad (31a)$$

Since $u/d \ll 1$ and $\beta_0 u \ll 1$, it follows that in both amplitude and phase factors

$$R \doteq |z - z'| \quad (31b)$$

in the first and last integrals. On the other hand, in the second integral with respect to z', R can not exceed $2d$, so that with (30a) the exponential may be replaced by its leading term, unity, and $I_z(z') \doteq I_z(z)$. With these several conditions, (29) becomes

$$A_z \doteq \frac{1}{4\pi\nu_0} \left[\left(\int_{-h}^{z-d} + \int_{z+d}^{h} \right) I_z(z') \frac{e^{-j\beta_0|z-z'|}}{|z-z'|} \, dz' \right.$$

$$\left. + I_z(z) \int_{z-d}^{z+d} \int_S \frac{f(x', y')}{R} \, dS' \, dz' \right]. \quad (32)$$

Since each contour s around a cross section S of the conductor is an equipotential line for both the scalar potential and the axial component of the vector potential, it follows that on the cylindrical surface A_z is a function of z alone. Therefore, the last integral in (32) must be independent of the transverse coördinates x' and y'. That is,

$$\int_{z-d}^{z+d} \int_S \frac{f(x', y')}{R} \, dS' \, dz' = \text{const.} \quad (33)$$

The z'-integration may be carried out directly. It gives

$$\int_{z-d}^{z+d} \frac{dz'}{R} = 2 \sinh^{-1} \frac{d}{\sqrt{(x-x')^2 + (y-y')^2}}$$

$$\doteq 2 \ln \frac{2d}{\sqrt{(x-x')^2 + (y-y')^2}}. \quad (34)$$

With (34), (33) becomes

$$2 \int_S \ln \frac{2d}{\sqrt{(x-x')^2 + (y-y')^2}} f(x', y') \, dS'$$

$$= \text{const.} \quad (35)$$

Since the constant on the right is independent of the shape of the cross section, it may be evaluated for a circular cylinder of radius a. If the z-axis coincides with the axis of the cylinder, rotational symmetry obtains. Moreover, it has been shown that the vector potential on the surface is independent of the radial distribution of the axial current. In particular, therefore, it may be assumed concentrated along the axes. In this case $P(x, y)$ is on the surface, $P'(x', y')$ on the axis, so that $\sqrt{(x-x')^2 + (y-y')^2} = \sqrt{x^2 + y^2} = a$. With this value and (28), (35) reduces to $2 \ln 2d/a$. It follows that the *equivalent radius* a of the circular cylinder that may be substituted for a cylinder of arbitrary cross section S is defined by the relations

$$\ln a = \int_S \ln \sqrt{(x-x')^2+(y-y')^2}\, f(x', y')\, dS', \quad (36a)$$

$$\int_S f(x', y')\, dS' = 1. \quad (36b)$$

Once the equivalent radius a has been determined, the analysis of the axial distribution of total current I_z in a cylindrical cross section of arbitrary shape is reduced to the analysis of an equivalent cylinder using the formulas (8) through (11).

The solution of (36a) subject to (36b) is most readily carried out if it is assumed that the axial current is confined to a thin sheet along the surface. Owing to skin effect this closely approximates the situation usually encountered in practice. This means that $f(x', y')$ is zero except along the contour, where it has the value $f(s')$. The region of integration reduces to the contour s of the cross section S. With this change (36a) and (36b) become

$$2 \ln a = \oint \ln [(x-x')^2+(y-y')^2] f(s')\, ds', \quad (37a)$$

$$\oint f(s')\, ds' = 1. \quad (37b)$$

A generally useful cross-sectional shape is the ellipse of semimajor axis a_e along the y-axis, semiminor axis b_e along the x-axis, and eccentricity $e_e = b_e/a_e$. If the cylindrical coördinate θ is measured from the positive x-axis, the coördinates (x, y) of a point P on the ellipse may be expressed in terms of θ using the following transformations, which characterize an ellipse and are readily verified in Fig. 7.2:

$$x = b_e \cos \theta, \quad y = a_e \sin \theta. \quad (38a)$$

Similarly x', y' may be expressed in terms of θ':

$$x' = b_e \cos \theta', \quad y' = a_e \sin \theta', \quad (38b)$$

$$ds' = \sqrt{dx'^2 + dy'^2}$$
$$= \sqrt{b_e^2 \sin^2 \theta' + a_e^2 \cos^2 \theta'}\, d\theta'. \quad (38c)$$

With $f(s') = g(\theta')$, (37a) and (37b) become:

$$2 \ln a = \int_0^{2\pi} \ln [b_e^2 (\cos \theta - \cos \theta')^2 + a_e^2 (\sin \theta - \sin \theta')^2]$$
$$g(\theta') \sqrt{b_e^2 \sin^2 \theta' + a_e^2 \cos^2 \theta'}\, d\theta', \quad (39a)$$

$$\int_0^{2\pi} g(\theta') \sqrt{b_e^2 \sin^2 \theta' + a_e^2 \cos^2 \theta'}\, d\theta' = 1. \quad (39b)$$

It is evident that the following function satisfies (39b):

$$g(\theta') = 1/2\pi \sqrt{b_e^2 \sin^2 \theta' + a_e^2 \cos^2 \theta'}. \quad (40)$$

If (40), substituted in (39a), is independent of θ, it satisfies all conditions and is the true cross-sectional distribution function. The following formulas defining ϕ and ψ are needed:

$$\cos \theta - \cos \theta' = -2 \sin \tfrac{1}{2}(\theta+\theta') \sin \tfrac{1}{2}(\theta-\theta')$$
$$\equiv -2 \sin \phi \sin \psi, \quad (41a)$$

$$\sin \theta - \sin \theta' = 2 \sin \tfrac{1}{2}(\theta-\theta') \cos \tfrac{1}{2}(\theta+\theta')$$
$$\equiv 2 \sin \psi \cos \phi. \quad (41b)$$

With (40) and (41a, b), (39a) becomes

$$4\pi \ln a = \int_0^\pi \ln (4 \sin^2 \phi)\, d\phi$$
$$+ 2 \int_0^\pi \ln (b_e^2 \sin^2 \psi + a_e^2 \cos^2 \psi)\, d\psi. \quad (42)$$

By the standard formula

$$\int_0^{\pi/2} \ln \sin x\, dx = -(\pi/2) \ln 2,$$

the first integral in (42) vanishes. The second integral is readily evaluated as follows:

$$2\pi \ln a = \int_0^\pi \ln a_e^2[(1-e_e^2) \cos^2 \psi + e_e^2]\, d\psi,$$

so that with the standard formula

$$\int_0^\pi \ln (a + b \cos x)\, dx = \pi \ln \tfrac{1}{2}(a+\sqrt{a^2+b^2}),$$

$$\ln \frac{a}{a_e} = \frac{1}{2\pi} \int_0^{2\pi} \ln [\tfrac{1}{2}(1 + e_e^2)$$
$$+ \tfrac{1}{2}(1-e_e^2) \cos u]\, du = \ln \tfrac{1}{2}(1+e_e). \quad (43)$$

Thus, the equivalent radius of the ellipse is

$$a = \tfrac{1}{2}a_e(1+e_e) = \tfrac{1}{2}(a_e+b_e). \quad (44)$$

For a circle, this reduces to $a = a_e$ as it should. For a flat strip with rounded edges $b_e \ll a_e$,

$$a = \tfrac{1}{2}a_e = \tfrac{1}{4}w, \quad (45)$$

where w is the width of the strip.

The equivalent radii of other cross-sectional shapes may be obtained in a similar manner or by conformal transformation. For regular polygons of n sides each of length w they are given in Table 7.1.

Table 7.1

n	2	3	4	5	6
a/w	0.25	0.42	0.59	0.76	0.92

8. Polarization and Magnetization Potentials

The polarization or Hertzian potential $\mathbf{\Pi}_e$ (the symbol $\mathbf{Z}$ also is used[31]) and the magnetization* or Fitzgerald potential $\mathbf{\Pi}_m$ (the symbol $\mathbf{Y}$ also is used[31]) are defined in general to satisfy the following equations:

$$\text{curl curl } \mathbf{\Pi}_e - \beta_0^2 \mathbf{\Pi}_e + \text{grad } f = \mathbf{P}/\epsilon_0, \quad (1)$$

$$\text{curl curl } \mathbf{\Pi}_m - \beta_0^2 \mathbf{\Pi}_m + \text{grad } g = \mathbf{M}/\nu_0. \quad (2)$$

The scalar functions f and g are arbitrary. The polarization and magnetization potentials are related to the electromagnetic field vectors by the formulas

$$\mathbf{E} = -\text{grad } f + \beta_0^2 \mathbf{\Pi}_e - j\omega \text{ curl } \mathbf{\Pi}_m, \quad (3)$$

$$\mathbf{B} = -\text{grad } g + \beta_0^2 \mathbf{\Pi}_m + j\frac{\beta_0^2}{\omega} \text{ curl } \mathbf{\Pi}_e. \quad (4)$$

The part of the field computed from $\mathbf{\Pi}_e$ is said to be of *electric type*, the part computed from $\mathbf{\Pi}_m$ of *magnetic type*. Since $\mathbf{\Pi}_e$ and $\mathbf{\Pi}_m$ are mutually independent, fields of electric and magnetic types are independent.

Except for use in a few problems, the general definition of the polarization and magnetization potentials implied in (1) and (2) may be specialized as follows:

$$-\text{div } \mathbf{\Pi}_e = f = \phi, \quad (5)$$

$$-\text{div } \mathbf{\Pi}_m = g, \quad (6)$$

---

* This potential is also called a Hertzian potential by many writers.

where ϕ is the scalar potential. Subject to (5) and (6), (1) and (2) reduce to the wave equation if use is made of the vector identity

$$\text{grad div } \mathbf{\Pi} - \text{curl curl } \mathbf{\Pi} \equiv \nabla^2 \mathbf{\Pi}. \quad (7)$$

Thus

$$\nabla^2 \mathbf{\Pi}_e + \beta_0^2 \mathbf{\Pi}_e = -\mathbf{P}/\epsilon_0, \quad (8)$$

$$\nabla^2 \mathbf{\Pi}_m + \beta_0^2 \mathbf{\Pi}_m = -\mathbf{M}/\nu_0. \quad (9)$$

Particular integrals of these equations are

$$\mathbf{\Pi}_e = \frac{1}{4\pi\epsilon_0} \int_\tau \frac{\mathbf{P}'}{R} e^{-j\beta_0 R} d\tau', \quad (10)$$

$$\mathbf{\Pi}_m = \frac{1}{4\pi\nu_0} \int_\tau \frac{\mathbf{M}'}{R} e^{-j\beta_0 R} d\tau'. \quad (11)$$

It is implied in (1) to (11) that by appropriate choice of subdivision or of physical model it is possible to set

$$\bar{\rho}' = -\text{div } \mathbf{P}', \quad (12)$$

$$\overline{\rho_m \mathbf{v}'} = \text{curl } \mathbf{M}' + j\omega \mathbf{P}'. \quad (13)$$

The specialized polarization and magnetization potentials that satisfy the wave equation (8, 9) are related to the scalar and vector potentials by the following relations:

$$\phi = -\text{div } \mathbf{\Pi}_e, \quad (14)$$

$$\mathbf{A} = j\frac{\beta_0^2}{\omega} \mathbf{\Pi}_e + \text{curl } \mathbf{\Pi}_m. \quad (15)$$

The electric and magnetic vectors $\mathbf{E}$ and $\mathbf{B}$ are derived using the following formulas; the two types of field are written separately this time:

Electric type

$$\mathbf{E} = \text{grad div } \mathbf{\Pi}_e + \beta_0^2 \mathbf{\Pi}_e, \quad (16a)$$

$$\mathbf{B} = j\frac{\beta_0^2}{\omega} \text{ curl } \mathbf{\Pi}_e; \quad (16b)$$

Magnetic type

$$\mathbf{E} = -j\omega \text{ curl } \mathbf{\Pi}_m, \quad (17a)$$

$$\mathbf{B} = \text{grad div } \mathbf{\Pi}_m + \beta_0^2 \mathbf{\Pi}_m. \quad (17b)$$

The H-field at points in empty space is obtained from the B-field by the simple relation

$$\mathbf{H} = \nu_0 \mathbf{B}. \quad (18)$$

9. Integrals of the Field Equations

General integrals of the field equations (4.1)–(4.4) are obtained using (7.1) and (7.2a) with (7.6) and (7.7). They are

$$\mathbf{E} = \frac{1}{4\pi\epsilon_0} \int_\tau e^{-j\beta_0 R} \left\{ \frac{\overline{\rho}'}{R^2} \hat{\mathbf{R}} + \frac{j\beta_0}{R^2} (\overline{\rho}' \hat{\mathbf{R}} - \overline{\rho_m \mathbf{v}'}/v_0) \right\} d\tau' + \mathbf{S}_E, \quad (1)$$

$$\mathbf{B} = -\frac{1}{4\pi v_0} \int_\tau e^{-j\beta_0 R} \left(\frac{1}{R^2} + \frac{j\beta_0}{R^2} \right) \hat{\mathbf{R}} \times \overline{\rho_m \mathbf{v}'} \, d\tau' + \mathbf{S}_B. \quad (2)$$

In these equations S_E and S_B are surface integrals obtained from the respective preceding volume integral by writing σ for τ and η for ρ. Corresponding expressions for the electromagnetic field of a cylindrical conductor with total axial current I'_s and charge per unit length q' are:

$$\mathbf{E} = \frac{1}{4\pi\epsilon_0} \int_{-h}^{h} e^{-j\beta_0 R} \left\{ \frac{q'}{R^2} \hat{\mathbf{R}} + \frac{j\beta_0}{R} \left(q'\hat{\mathbf{R}} - \frac{I'}{v_0} \right) \right\} ds', \quad (3)$$

$$\mathbf{B} = -\frac{1}{4\pi v_0} \int_{-h}^{h} e^{-j\beta_0 R} \left(\frac{1}{R^2} + \frac{j\beta_0}{R} \right) \hat{\mathbf{R}} \times \mathbf{I}' \, ds'. \quad (4)$$

The near-zone or induction-zone field is given by the $1/R^2$ terms in (3) and (4). The far-zone or radiation-zone field is given by

$$\mathbf{E}^r = v_0 \mathbf{B}^r \times \hat{\mathbf{R}}, \quad (5)$$
$$\mathbf{B}^r = -j\beta_0 \hat{\mathbf{R}} \times \mathbf{A}^r, \quad \beta_0 R \gg 1 \quad (6)$$

with $\mathbf{A}^r$ defined by (7.15). In polar coördinates, with Θ measured from the axis, R from the center of the conductor,

$$\mathbf{E}^r = -j\omega(A_\Theta^r \hat{\boldsymbol{\Theta}} + A_\Phi^r \hat{\boldsymbol{\Phi}}), \quad (7)$$
$$\mathbf{B}^r = j\beta_0(A_\Phi^r \hat{\boldsymbol{\Theta}} - A_\Theta^r \hat{\boldsymbol{\Phi}}). \quad (8)$$

If rotational symmetry obtains about the axis of the conductor, A_Φ^r vanishes in (7) and (8).

10. Energy Functions

The total time-average power transferred across a *closed* surface Σ in space is given by the real part $\overline{\overline{T}}_r$ of the complex energy transfer function T:

$$T = \int_{\Sigma \text{ (closed)}} \hat{\mathbf{n}} \cdot \mathbf{S} \, d\sigma. \quad (1)$$

Here $\mathbf{S}$ is the complex Poynting vector given by

$$\mathbf{S} = \frac{v_0}{2} \mathbf{E} \times \mathbf{B}^* = \tfrac{1}{2} \mathbf{E} \times \mathbf{H}^*, \quad (2)$$

where $\mathbf{B}^*$ is the complex conjugate of $\mathbf{B}$. In terms of the far-zone field of one or more cylindrical conductors, $\mathbf{S}$ is real and is given by

$$\mathbf{S} = \frac{\omega^2}{2\zeta_0} \{A_\Theta^r A_\Theta^{r*} + A_\Phi^r A_\Phi^{r*}\}\hat{\mathbf{R}}_0. \quad (3)$$

A dimensionless radiation function referred to a reference current I_p is defined by

$$K_p^2(\Theta, \Phi) \equiv \left(\frac{4\pi\omega R_0}{\zeta_0} \right)^2 \left(\frac{1}{I_p I_p^*} \right) (A_\Theta^r A_\Theta^{r*} + A_\Phi^r A_\Phi^{r*}). \quad (4)$$

In terms of the radiation function, the total time-average transfer of power across a great sphere of radius R in the far zone is

$$\overline{\overline{T}}_r = \frac{I_p^2 \zeta_0}{32\pi^2} \int_0^{2\pi} \int_0^\pi K_p^2(\Theta, \Phi) \sin\Theta \, d\Theta \, d\Phi. \quad (5)$$

The radiation resistance R_p^e referred to I_p is

$$R_p^e = \frac{2\overline{\overline{T}}_r}{I_p^2} = \frac{\zeta_0}{16\pi^2} \int_0^{2\pi} \int_0^\pi K_p^2(\Theta, \Phi) \sin\Theta \, d\Theta \, d\Phi. \quad (6)$$

The absolute directivity in the direction Θ_m, Φ_m is

$$D = \frac{\zeta_0}{4\pi} \frac{K_p^2(\Theta_m, \Phi_m)}{R_p^e}. \quad (7)$$

11. Simple Media

A simple medium is homogeneous and isotropic and electrically so constituted that the following constitutive relations between complex amplitudes of the density functions and the field vectors are satisfied.

$$\mathbf{P} = (\epsilon_r - 1)\epsilon_0 \mathbf{E} = (\epsilon - \epsilon_0)\mathbf{E}, \quad (1)$$
$$-\mathbf{M} = (\nu_r - 1)\nu_0 \mathbf{B} = (\nu - \nu_0)\mathbf{B}, \quad (2)$$
$$\mathbf{i}_f = \sigma \mathbf{E}. \quad (3)$$

The dimensionless complex parameter

$$\epsilon_r = \epsilon'_r - j\epsilon''_r \quad (4)$$

is the relative permittivity or relative dielectric constant of the simple medium; the product $\epsilon = \epsilon_0 \epsilon_r$ is called the absolute permittivity or absolute dielectric constant.

The dimensionless complex parameter

$$\nu_r = \nu'_r - j\nu''_r \quad (5)$$

is the relative reluctivity of the simple medium; the product $\nu = \nu_0 \nu_r$ is called the absolute reluctivity. In dielectric media encountered in the study of antennas, ν_r is real and practically equal to unity. The complex parameter

$$\sigma = \sigma' - j\sigma'' \tag{6}$$

is the complex conductivity of the simple medium. It is measured in reciprocal ohm-meters or in mhos per meter.

The combination $\sigma + j\omega\epsilon$ occurs frequently. It is denoted by

$$j\omega\xi \equiv \sigma + j\omega\epsilon \tag{7a}$$

or

$$\xi = \epsilon - \frac{j\sigma}{\omega} = \left(\epsilon' - \frac{\sigma''}{\omega}\right) - \frac{j}{\omega}(\sigma' + \omega\epsilon''); \tag{7b}$$

ξ is called the complex dielectric factor of the medium. The real part of (7b) is the real effective permittivity or dielectric constant ϵ_e; the factor $\sigma' + \omega\epsilon''$ is the real effective conductivity. Thus,

$$\epsilon_e \equiv \epsilon' - \frac{\sigma''}{\omega}, \qquad \epsilon_{er} = \epsilon'_r - \frac{\sigma''}{\omega\epsilon_0}, \tag{8a}$$

$$\sigma_e \equiv \sigma' + \omega\epsilon'', \qquad \sigma_e = \sigma' + \omega\epsilon_0\epsilon''_r. \tag{8b}$$

All equations and solutions written for points in empty, unbounded space are correct for an unbounded simple medium if ν is written for ν_0, ξ for ϵ_0. The complex phase constant

$$\beta = \beta_s - j\alpha_s \equiv \omega\sqrt{\xi/\nu} \tag{9}$$

appears in the formulas for simple media wherever the real phase factor

$$\beta_0 \equiv \omega\sqrt{\epsilon_0/\nu_0} \tag{10}$$

occurs in the formulas for points in empty space. In most media it is possible to assume ν_r to be real. In this case

$$\beta = \beta_s - j\alpha_s = \omega\sqrt{\epsilon_e/\nu}\sqrt{1 - jh_e}, \tag{11}$$

where

$$h_e \equiv \frac{\sigma_e}{\omega\epsilon_e} = \frac{\sigma_e}{\omega\epsilon_0\epsilon_{er}} \tag{12}$$

is called the loss tangent. The factor $\sqrt{1-jh_e}$ may be separated into real and imaginary parts using the $f(h)$- and $g(h)$-functions tabulated in ref. 31, Appendix II. Thus

$$\beta = \omega\sqrt{\epsilon_e/\nu}\,[f(h_e) - jg(h_e)], \tag{13}$$

with

$$f(h) = [\tfrac{1}{2}(\sqrt{1+h^2}+1)]^{\frac{1}{2}} = \cosh(\tfrac{1}{2}\sinh^{-1}h), \tag{14a}$$

$$g(h) = [\tfrac{1}{2}(\sqrt{1+h^2}-1)]^{\frac{1}{2}} = \sinh(\tfrac{1}{2}\sinh^{-1}h), \tag{14b}$$

so that the real phase constant β_s and the real attenuation constant α_s are

$$\beta_s = \omega\sqrt{\epsilon_e/\nu}\,f(h_e), \tag{15}$$

$$\alpha_s = \omega\sqrt{\epsilon_e/\nu}\,g(h_e). \tag{16}$$

Good dielectrics are defined by

$$h_e^2 \ll 1, \tag{17}$$

whence

$$f(h_e) \doteq 1; \qquad g(h_e) \doteq \frac{h_e}{2}, \tag{18}$$

so that

$$\beta_s \doteq \omega\sqrt{\epsilon_e/\nu} = \omega/v_e, \tag{19}$$

$$\alpha_s \doteq \frac{\sigma_e}{2}\frac{1}{\sqrt{\nu\epsilon_e}} = \tfrac{1}{2}\sigma_e\zeta_e. \tag{20}$$

The effective phase-velocity characteristic of the simple medium is given by

$$v_e = \sqrt{\nu/\epsilon_e}\,; \tag{21}$$

the effective characteristic resistance of the simple medium is

$$\zeta_e = \frac{1}{\sqrt{\nu\epsilon_e}}. \tag{22}$$

Good conductors are defined by

$$h_e \gg 1; \qquad \epsilon_r \doteq 1, \tag{23}$$

whence

$$f(h_e) \doteq g(h_e) \doteq \sqrt{\frac{h_e}{2}}, \tag{24}$$

so that

$$\alpha_s \doteq \beta_s \doteq \sqrt{\frac{\omega\sigma_e}{2\nu}}. \tag{25}$$

In a charge-separating region (or an electrical generator) the constitutive relation (3) is conveniently replaced by

$$i_f = \sigma(E + E^e) = \sigma E + i^e, \tag{26}$$

where E^e is the impressed field maintained by the generator. Independently maintained densities of current are called *impressed* densities of current and are denoted by i^e.

12. The Magnetic Dipole

In the investigation of the electromagnetic field of linear radiators over a conducting earth in Chapter VII the magnetic analog of the Hertzian or electric dipole or doublet is a convenient tool. Physically, the magnetic dipole or doublet consists of a loop of wire that is sufficiently small compared with the wavelength so that the amplitude of the current around its contour is essentially uniform and equal to I_θ. The complex magnetic moment of such a closed ring of current is given by [ref. 31, Eq. (I.26.6)],

$$m_z = I_0 S, \qquad (1)$$

where S is the area of the loop which lies in the xy-plane. The electromagnetic field of a magnetic dipole or small loop antenna is determined in Chapter VII, beginning in Sec. 3.

The impedance of a small loop is given by (ref. 31, chap. VI),

$$Z = R + jX = R^e + R^i + j(\omega L^e + X^i), \quad(2)$$

where R^e is the radiation resistance, R^i the ohmic resistance, L^e the external inductance, and X^i the internal reactance. For a single-turn loop these quantities are given by

$$R^e = 20\beta_0^4 S^2, \qquad R^i = \oint r^i \, ds, \qquad (3)$$

$$L^e = \oint\oint \frac{ds \cdot ds'}{R}, \qquad X^i = \oint x^i \, ds, \quad (4)$$

where $\beta_0 = \omega/v_0 = 2\pi/\lambda_0$, S is the area enclosed by the loop, r^i is the ohmic resistance per unit length, x^i is the internal reactance per unit length of the conductor forming the loop, and R is the distance from the element ds' on the axis of the cylindrical conductor to the element ds on the surface of the conductor. The integrals are taken completely around the closed loop. At high frequencies $r^i \doteq x^i \doteq (1/2\pi a)\sqrt{\omega/2v_1\sigma_1}$, where a is the radius of the conductor, $v_1 = 1/\mu_1$ is the absolute reluctivity (or reciprocal permeability), and σ_1 is the conductivity of the wire.

The following conditions of the quasi-near zone are assumed to be satisfied:

$$\beta_0 a \ll 1; \qquad \beta_0^2 R_{\max}^2 \ll 1. \qquad (4)$$

CHAPTER II

LINEAR RADIATORS AS CIRCUIT ELEMENTS

A transmitting antenna is a device in which periodically varying distributions of current and charge are so disposed that significant electromagnetic forces are exerted on electric charges at great distances, as, for example, in the wires of receiving antennas. Usually an antenna consists of an arrangement of conductors of appropriate shape and adequate size. In order to maintain alternating currents in the conductors of an antenna these must be connected to a circuit including an electric charge-separating region or generator, or they may be coupled as a secondary to a current-inducing circuit excited by a generator. An antenna with its driving generator and the necessary connecting network constitutes a *complete transmitting system*.

Important properties of an antenna as a *circuit element* are the distributions of current and charge in it. The current at specified terminals is proportional to the admittance looking into those terminals. The currents throughout the conductors of the antenna determine the electromagnetic field, and it is this field at distance points which characterizes the radiating properties of the antenna. In its simplest form, the linear radiator, it is a conductor or wire of sufficiently small cross-sectional size to make a one-dimensional analysis, analogous to that familiar in conventional electric-circuit theory, a satisfactory approximation. This chapter is concerned with the behavior of a single linear radiator as an element in an electric circuit of conventional type—in particular, as the load for a transmission line.

Antennas with cross sections that are large compared with the wavelength require a three-dimensional analysis. They are not investigated in this volume.

ANTENNA AS TRANSMITTING SYSTEM

1. *Theory and Experiment: Discussion of the Problem*

The complete study of a problem in applied electromagnetism, such as the determination of the electrical properties of an antenna, is composed of three parts. These are, first, a theoretical analysis consisting of a more or less exact solution of the Maxwell-Lorentz equations subject to appropriate boundary conditions; second, an experimental investigation involving a suitably designed apparatus and tables of observed pointer readings; and third, a coördination of the postulates and predictions of the theory with the conditions and observations of experiment. Ideally, the boundary conditions in the mathematical model must correspond closely to the geometrical and physical properties of the apparatus used in experimental measurements if a precise check of the theory is desired. Unfortunately, such close correspondence is seldom achieved in the study of radiation, in which both mathematical and experimental complexities are great. This serves to emphasize that on the one hand the boundary conditions postulated in the theory, and on the other hand the design of the apparatus and the technique of measurement, are not only significant *individually* in providing, respectively, solvable mathematical and experimental problems, but of primary importance *jointly* when the attempt is made to establish a direct and simple correspondence between predicted and observed pointer readings. In most problems in applied physics a compromise must be sought which assures (1) a theoretical problem that is not so intricate that it defies solution with available mathematical methods and computational facilities; (2) experimental requirements for which adequate apparatus and accurate techniques of measurement can be developed; and (3) a correspondence between the boundary conditions of theory and the geometry and physical properties of the apparatus that does not imply excessive idealization on the one hand or too great a departure from the practically important and generally useful on the other.

Specifically, for the problem of determining the circuit properties of an antenna, a structure must be selected that is typical and generally important in practice and that may be analyzed in terms of a theoretical model which fulfills the following requirements: it must be a fair representation of physical reality; it must be susceptible to mathematical analysis that is sufficiently simple in its final form to permit (a) the computation of useful results, and (b) the interpretation of these results in terms of the essential geometrical and physical characteristics of the antenna. While the immediate purpose of a mathematical theory for an antenna is to compute theoretical pointer readings that agree with the experimental observations on a particular and carefully arranged physical model, in broader perspective such agreement is merely verification that the theoretical model does approximate the physical one and hence that the theory *has useful application*. The broader function of an experimentally checked mathematical theory is to provide insight into the problem as a whole, in both a qualitative and a quantitative sense. The purpose is to find answers for questions like these: What is the *general* significance of the several variables and parameters involved? How may they be modified to achieve special conditions appropriate to a variety of practical needs? How can the results of the simple structure *actually* studied be generalized, at least qualitatively, in order to gain some understanding of the properties of more complex arrangements that do not lend themselves so well to mathematical investigation, or, at least, that have not yet been analyzed?

The general procedure to be used as a guide in the systematic study of the circuit properties of linear radiators ideally involves the following:

1. The selection of one or more relatively simple structures with associated circuits and generators that are suitable for both theoretical and experimental investigations.

2. The formulation of the mathematical problem in terms of fundamental electromagnetic principles involving, in particular, appropriate boundary conditions which describe a more or less idealized model including a "theoretical" antenna with its associated circuit; the analytical solution of the mathematical problem; and the evaluation, display, and interpretation of significant theoretical results.

3. The design of an "experimental" antenna and of associated apparatus for obtaining useful pointer readings; a description of experimental procedures and techniques with a discussion of the accuracy of measurements; and the display and interpretation of significant experimental results.

4. A critical review of the correlation between theory and experiment for the particular structure investigated. Such a correlation is not necessarily or advisedly merely a concluding retrospect. It should be a continuous process illuminating the entire study.

5. Generalization of the theoretical and experimental methods and results obtained for a particular structure and circuit to a class of related structures and circuits.

2. Definition of an Antenna

The essential purpose of a radiating circuit is to maintain a significant electromagnetic field at distant points. Every periodically varying electromagnetic field is *defined* in terms of associated distributions of current and charge. Thus, the electromagnetic field of a circuit is due to the currents and charges strictly in *all parts* of the circuit. However, in different circuits the contributions to the electromagnetic field in the radiation zone due to currents in the several parts may vary widely. In some circuits the currents in *all* parts may contribute about equally so that the entire circuit is both an antenna and a complete transmitting system. In others, the contributions to the far-zone field by currents in most of the circuit may cancel and so become negligible compared with the contributions from currents in a readily specified part which is designated as the antenna; obviously, this part is not in itself a complete transmitting system. Thus, the *criterion defining an antenna is the requirement that it contain distributions of current that contribute directly and significantly to the electromagnetic field at distant points.*

It is important to note that the electromagnetic energy-transfer function (defined as the real part of the integral

$$T = \int_{\Sigma \text{ (closed)}} (\hat{\mathbf{n}}, \mathbf{S}) \, d\sigma \qquad (1)$$

over a completely closed surface of the normal component of the complex Poynting vector) may not be used to define or locate an antenna. For, if the surface of integration, Σ in (1), is progressively contracted and moved to keep the real part of T the same as when Σ is a great sphere, the discovery is made that it is *always the generator* and *never the antenna* that is located. The real part of T when Σ is a surface that completely encloses an antenna but does not contain a generator is always only the net energy transferred *into* the antenna to supply dissipation in ohmic resistance. Clearly the transfer function T is useful in specifying the net transfer of energy from or into a closed region. It may be used to define a complete transmitting system but not to locate an antenna, if this is only a part of such a system, as is usually the case.

3. *Antenna as a Circuit Element*

An antenna of very simple structure consists of two straight, collinear cylindrical conductors each of length $\bar{h}$ and small radius a, separated in the middle by a gap of half-length δ. At the center of the structure is located the origin O of a system of cylindrical coördinates, r, θ, z, which has its z-axis coincident with the axis of rotational symmetry of the conductors. The axial distance from the origin to each extremity is denoted by h, so that $h = \bar{h} + \delta$. The geometrical arrangement is shown in Fig. 3.1.

The antenna of Fig. 3.1 is a circuit element in the same sense as the combination of coil and condenser in Fig. 3.2. If the ends A and B in each of the two elements are maintained at different alternating potentials ϕ_A and ϕ_B by a generator connected to them in some manner, there will be currents in the conductors. In Fig. 3.2 the currents are in the conductors of the coil and are directed to charge alternately and oppositely the adjacent surfaces of the condenser; in Fig. 3.1 the currents are in the conductors of the antenna so directed that the outer surfaces of the halves are alternately and oppositely charged in a manner yet to be determined, but necessarily in accordance with the general principles deducible from the equation of continuity (ref. I.31, Sec. I.25).

The properties of the coil and condenser as a circuit element include a determination of the current in the coil, the charge on the condenser, and the impedance Z_{AB}. In the conventional literature the impedance of a circuit element between points A and B is made to depend only on the geometric configuration and the properties of the materials between these points. It may be shown (ref. I.31, chap. VI) that this is true approximately in the more common special cases, but that rigorously and in general the current in such an element, and hence the very definition of its impedance, is a function of the shape and structure of the *entire* circuit of which it forms a part, including the generator. Therefore, it is clear that the phrase, *circuit properties of an element*, actually is meaningful only if the element is connected in an appropriately designed circuit. This presents no real problem in near-zone circuits in which the amplitude of the current in a series circuit is the same at all points (ref. I.31, chap. VI). On the other hand, if a two-wire line is in the circuit and its length is appreciable in terms of the wavelength, the circuit properties of an element depend greatly upon whether it terminates the line or whether one of the conductors of the line is cut and the element is connected in series. Indeed, a meaningful impedance can be defined in general only if the element is a symmetrical load at the end of the line so that the currents at its terminals A and B are equal and opposite.

Since an antenna cannot be restricted to the near zone, and its structure must be one that distributes currents and charges in a manner to maintain relatively large forces at great distances, the question whether it can be treated as a circuit element with properties that are independent of the circuit that maintains the potential difference across its terminals requires careful study. The properties in question include the distribution of current and charge, and the impedance. To be sure, the analysis of the antenna of Fig. 3.1 may be attempted simply by postulating a potential difference $V_{BA} = \phi_B - \phi_A$ without specifying how it is maintained. Alternatively, a rotationally symmetric electric field E^e_{AB} may be assumed to be maintained between A and B, by an unspecified mechanism as shown in Fig. 3.3. The physical and practical significance of such analyses then depends upon whether it is possible and practicable to maintain the postulated potentials or the assumed electric field even approximately. If not, the analyses are at best mathematically interesting exercises. Here, therefore, is an illustration of the importance of studying the coördination of theory with experiment *before* an analysis is undertaken. Is it possible to drive the simple symmetric antenna of Fig. 3.1 in such a manner that its properties as a circuit element are independent of the connecting circuit and the generator?

4. *Transmitting System with Simple External Generator*

A practical *transmitting system* consists of an antenna, a generator, and the connecting network necessary to maintain currents in the conductors of the antenna. The *antenna* is that part of the system in which are all the currents that contribute significantly to the radiation

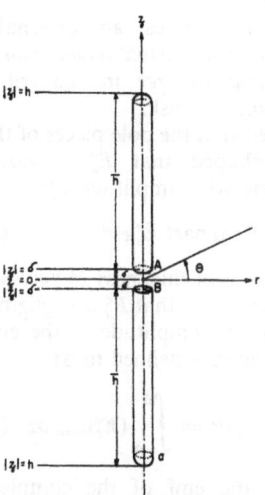

Fig. 3.1. Cylindrical antenna as a circuit element.

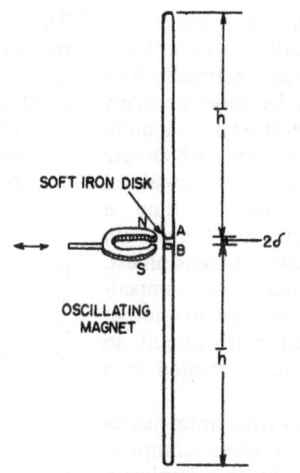

Fig. 4.1. Antenna with oscillating magnet as generator.

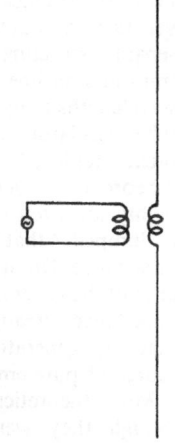

Fig. 4.2. Antenna coupled to primary circuit.

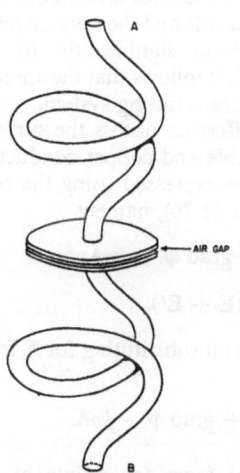

Fig. 3.2. Coil in series with condenser as a circuit element.

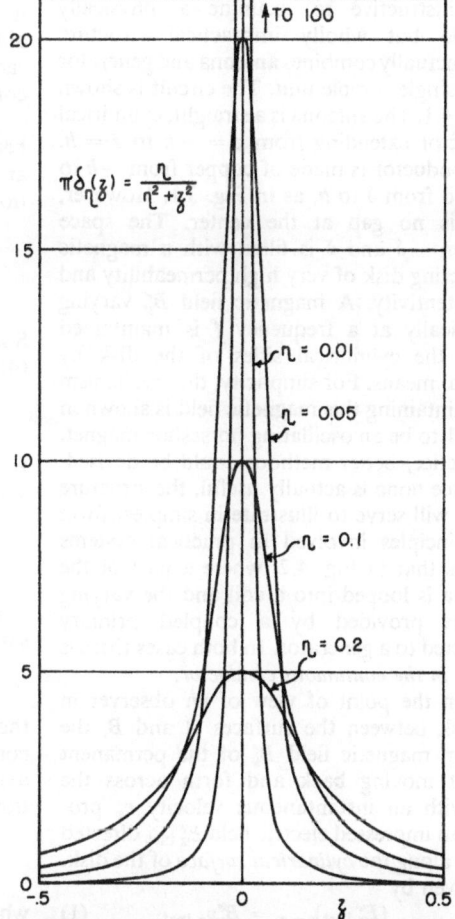

Fig. 4.3. Graph of function $\pi\delta_\eta(z) = \eta/(\eta^2+z^2)$.

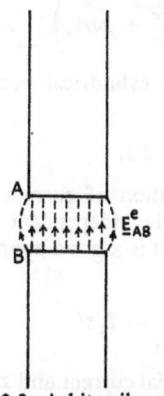

Fig. 3.3. Arbitrarily assumed field E^e_{AB} in gap at center of antenna.

field. In practical transmitting systems, the generator and antenna usually are completely separate structures connected electrically by a transmission line that may be short or many wavelengths long. The generators are vacuum-tube oscillators of various types which are often replaced analytically by idealized "theoretical" generators that combine a highly localized or "concentrated" emf with a "lumped," that is, physically dimensionless, impedance. Often the equivalent series impedance of the generator is combined with another impedance already included in the circuit, so that the generator is virtually treated as a source of pure emf.

Many theoretical analyses treat antennas as though they were in themselves complete transmitting systems. Actually this implies a self-contained generator which, in effect, is equivalent to a source of emf. Before considering the important and intricate problem of an antenna driven from a transmission line, it is instructive to examine a physically possible but wholly impractical structure which actually combines antenna and generator into a single, simple unit. The circuit is shown in Fig. 4.1. The antenna is a straight, cylindrical conductor extending from $z = -h$ to $z = h$. The conductor is made of copper from $-h$ to $-\delta$ and from δ to h, as in Fig. 3.1. However, there is no gap at the center. The space between $-\delta$ and δ is filled with a magnetic conducting disk of very high permeability and low retentivity. A magnetic field B_y^e varying periodically at a frequency f is maintained across the cylindrical sides of the disk by external means. For simplicity, the mechanism for maintaining this magnetic field is shown in Fig. 4.1 to be an oscillating horseshoe magnet. In practice, better methods could be devised, but since none is actually useful, the structure shown will serve to illustrate in simplest form the principles involved in practical systems such as that in Fig. 4.2, where a part of the antenna is looped into a coil and the varying field is provided by a coupled primary connected to a generator. In both cases there is *no gap in the continuous conductor*.

From the point of view of an observer in the disk between the surfaces A and B, the uniform magnetic field B_y^e of the permanent magnet moving back and forth across the disk with an instantaneous velocity v_x produces an impressed electric field $E_{z\,\text{inst}}^e$ directed axially along the *cylindrical surface* of the disk. It is given by

$$(E_z^e\,\text{inst})_{r=a} = B_y^e v_x\,\text{inst}. \quad (1)$$

The superscript e denotes an externally maintained or *impressed field* that is *independent of the currents and charges* in the entire conductor, including the disk.

Let it be assumed that the pole pieces of the magnet are so shaped that $E_{z\,\text{inst}}^e$ varies sinusoidally in time with amplitude E_z^e:

$$E_{z\,\text{inst}}^e = \text{Real part } E_z^e\, e^{j\omega t}. \quad (2)$$

If the magnetic disk is of thickness 2δ between A and B and it is assumed that B_y^e is negligible except in the disk, the amplitude of the emf induced in the antenna is defined to be

$$V^e = \int_{-h}^{h} (E_z^e)_{r=a}\, dz \doteq \int_{-\delta}^{\delta} (E_z^e)_{r=a}\, dz. \quad (3)$$

Note that V^e is the emf of the complete transmitting system; it is not the potential difference between the surfaces A and B at $z = \pm\delta$. However, since the system is a continuous straight conductor and the currents in both the magnetizable and the nonmagnetizable parts contribute significantly to the radiation-zone field, it follows that the antenna coincides with the transmitting system.

The potential difference across the surfaces between magnetizable and copper conductors at A and B may be expressed using the relations (I.7.17) and (I.11.26), namely,

$$\mathbf{E} = -\text{grad } \phi - j\omega \mathbf{A}, \quad (4)$$

$$\mathbf{i}_f = \sigma(\mathbf{E} + \mathbf{E}^e). \quad (5)$$

Solving (5) for $\mathbf{E}^e$ and substituting for $\mathbf{E}$ from (4) leaves

$$\mathbf{E}^e = \frac{\mathbf{i}_f}{\sigma} + \text{grad } \phi + j\omega \mathbf{A}. \quad (6)$$

In one-dimensional form for points at the surface $r = a$, (6) is

$$\left(E_z^e = \frac{i_{fz}}{\sigma} + \frac{\partial \phi}{\partial z} + j\omega A_z\right)_{r=a}. \quad (7)$$

If the radius of a cylindrical conductor satisfies the inequality

$$\beta_0 a \ll 1, \quad (8)$$

the transverse distribution of current in the conductor is essentially independent of the axial distribution and it is a good approximation to set

$$\left(\frac{i_{fz}}{\sigma}\right)_{r=a} = I_z z^i, \quad (9)$$

where I_z is the total axial current and z^i is the

internal impedance per unit length. If rotational symmetry obtains,

$$I_z = 2\pi \int_0^a i_{fz} r \, dr. \tag{10}$$

General formulas for z^i are given in ref. I.31, Sec. V.7. At the high frequencies used in antennas, the Rayleigh formula [ref. I.31, Eq. (V.7.13)] is a good approximation. It is

$$z^i = \frac{1+j}{2\pi a} \sqrt{\frac{\omega}{2\nu\sigma}}, \tag{11}$$

where ν is the absolute reluctivity and σ is the conductivity of the conductor. With (9) and (7), the general, one-dimensional equation that is the electromagnetic foundation of electric-circuit analysis is

$$E_z^e = I_z z^i + \frac{d\Phi}{dz} + j\omega A_z. \tag{12}$$

If (12) is substituted in (3) the result is

$$(V^e)_{r=a} = \int_{-\delta}^{\delta} I_z z^i \, dz + \int_{-\delta}^{\delta} (d\Phi)_{r=a}$$

$$+ j\omega \int_{-\delta}^{\delta} (A_z)_{r=a} \, dz. \tag{13}$$

The scalar potential difference between the surfaces A and B at $z = \pm\delta$ is

$$(V_\delta)_{r=a} \equiv [\Phi(\delta) - \Phi(-\delta)]_{r=a}$$

$$= (V^e)_{r=a} - \int_{-\delta}^{\delta} I_z z^i \, dz$$

$$- j\omega \int_{-\delta}^{\delta} (A_z)_{r=a} \, dz. \tag{14}$$

Let the internal and external impedances of the generator be defined by

$$Z_g^i = \frac{1}{I(\delta)} \int_{-\delta}^{\delta} I_z z^i \, dz \tag{15a}$$

and

$$Z_g^e = \frac{j\omega}{I(\delta)} \int_{-\delta}^{\delta} (A_z)_{r=a} \, dz, \tag{15b}$$

respectively. Note that Z_g^e in (15b) depends on the z-component of the vector potential on the surface at $r = a$ between $z = -\delta$ and $z = +\delta$. This component is determined primarily by the current I_z between $z = -\delta$ and $z = +\delta$, but not entirely. The currents in the conductor beyond $z = \pm\delta$ also contribute, so that the impedance

$$Z_g = Z_g^i + Z_g^e \tag{15c}$$

is not a property of the structure of the generator alone but depends upon the external circuit.

With (15a, b, c), (14) reduces to

$$V_\delta = V^e - I_\delta Z_g, \tag{16}$$

which states that the potential difference at the terminals AB of the generator is equal to the generated emf minus the voltage drop in the generator. If (16) is divided by I_δ, and the thickness 2δ of the generating region is sufficiently small to make

$$I_\delta \doteq I_0 \tag{17}$$

a good approximation, the result is

$$Z_\delta \equiv \frac{V_\delta}{I_\delta} = \frac{V^e}{I_0} - Z_g = Z_0 - Z_g, \tag{18a}$$

where Z_δ is the impedance of the *external* circuit and Z_0 is the impedance of the *complete transmitting system*, given by

$$Z_0 = Z_\delta + Z_g. \tag{18b}$$

Note that this definition of Z_0 is independent of the thickness or length 2δ of the generating region in which the externally maintained field E^e is active so long as 2δ is small enough to make (17) a good approximation. Hence, if V^e is defined by the first integral in (3), that is, by $V^e = \int_{-h}^{h} (E_z^e)_{r=a} \, dz$, δ in (15a) through (18b) serves merely to denote the boundary between the part of the circuit that is defined to be the generating region and the part designated as the external circuit. Actually, from the point of view of the radiation field this is an artificial distinction that may be eliminated simply by shifting the boundary by setting $\delta = 0$. As a consequence Z_g vanishes and Z_δ, since it now involves the entire circuit from $-h$ to h, increases to Z_0.

An alternative and mathematically convenient interpretation of Z_0 is useful. Suppose the magnetizable disk of thickness 2δ and the oscillating magnet could be made thinner and thinner while the permeability of the disk and the field B_z^e due to the magnet are increased so as to keep $V^e = \int_{-\delta}^{\delta} (E_z^e) \, dz$ constant. Although physically unattainable, the following can be defined in the limit as $\delta \to 0$:

$$V_\delta \to V_0 - V^e, \quad Z_g \to 0, \quad Z_\delta \to Z_0. \tag{19}$$

The impedance Z_0 in (19) is the same as that in (18b). On the other hand, V^e, instead of being interpreted as due to a physically realizable impressed field maintained over a small distance 2δ, is interpreted as due to an equivalent fictitious impressed field which is maintained at an infinite magnitude in a region of zero length, and which has a line integral across the region that is *finite and equal to* V^e. A disk generator of vanishingly small thickness corresponds to an ideal, completely localized emf. It is called a belt or slice generator or a delta-function generator. The name delta-function generator for an idealized emf originates with a convenient mathematical definition of a function like V_0 which has a finite magnitude and is defined in an infinitely narrow region. This definition depends on the Dirac delta function which, for the case at hand, is denoted by $\delta(z)$. The essential properties of this function may be represented by a number of mathematical expressions. A convenient one for present purposes is

$$\delta(z) = \lim_{\eta \to 0} \delta_\eta(z), \qquad (20a)$$

where

$$\delta_\eta(z) = \frac{1}{\pi}\frac{\eta}{\eta^2 + z^2}. \qquad (20b)$$

Evidently $\delta_\eta(z)$ is a function of the type shown in Fig. 4.3. As η is made smaller and smaller, $\delta_\eta(0)$ becomes greater and greater. However,

$$\int_{-h}^{h} \delta_\eta(z)\, dz = \frac{1}{\pi}\int_{-h}^{h} \frac{\eta\, dz}{\eta^2 + z^2}$$

$$= \frac{1}{\pi}\int_{-h/\eta}^{h/\eta} \frac{du}{1+u^2} = \frac{2}{\pi}\tan^{-1}\frac{h}{\eta}. \qquad (21)$$

Therefore, as η approaches zero,

$$\int_{-h}^{h} \delta(z)\, dz = \lim_{\eta \to 0}\int_{-h}^{h} \delta_\eta(z)\, dz = 1, \qquad (22a)$$

$$\delta(z) = \lim_{\eta \to 0} \delta_\eta(z) = \begin{cases} 0 & \text{if } z \neq 0, \\ \infty & \text{if } z = 0. \end{cases} \qquad (22b)$$

The function $\delta(z)$ is convenient in the definition of the driving voltage or emf of the idealized slice generator. By setting

$$(E_z^e)_{r=a} = V^e \delta_\eta(z), \qquad (23a)$$

an impressed field is defined which, for sufficiently small values of η in (20b), rises very rapidly to a high maximum at $z = 0$. In the limit of a slice generator,

$$(E_z^e)_{r=a} = V^e \delta(z), \qquad (23b)$$

which defines an impressed field along a distance of zero length that is infinite in magnitude but that is so constituted that its integral along any path including $z = 0$ is finite. Thus, with (3) and (23b),

$$V_0 = \int_{-h}^{h} (E_z^e)_{r=a}\, dz = \int_{-h}^{h} V^e \delta(z)\, dz = V^e. \qquad (24)$$

This is a correct analytical representation of the emf of a slice generator.

The impedance of the external circuit is defined by

$$Z_\delta = V_\delta/I_\delta \qquad (25a)$$

when δ is finite. For a slice generator, $\delta = 0$ so that

$$Z_0 = V_0/I_0. \qquad (25b)$$

I_δ is obtained by setting $z = \delta$, I_0 by setting $z = 0$ in the expression for current. General equations for determining this current are formulated readily using (12) and the special form

$$\frac{\partial A_z}{\partial z} + \frac{j\beta_0^2}{\omega}\Phi = 0 \qquad (26a)$$

of the general equation of continuity for the potential functions, namely,

$$\operatorname{div} \mathbf{A} + \frac{j\beta_0^2}{\omega}\Phi = 0. \qquad (26b)$$

Since the only significant currents in the simple transmitting system studied in this section are axial, $A_\theta = 0$, $A_r \doteq 0$, so that (26b) reduces to (26a). By eliminating Φ from (12) with the aid of (26a), the following equation is obtained:

$$E_z^e = I_z z^i + \frac{j\omega}{\beta_0^2}\left(\frac{\partial^2 A_z}{\partial z^2} + \beta_0^2 A_z\right). \qquad (27)$$

By solving this differential equation for A_z and equating the solution to the integral

$$A_z = \frac{1}{4\pi v_0}\int_{-h}^{h} I_z' \frac{e^{-j\beta_0 R}}{R}\, dz';$$

$$R = \sqrt{(z-z')^2 + a^2}, \qquad (28)$$

an integral equation is obtained for the current. The solution of this equation is begun in Sec. 11.

An equivalent but formally somewhat different integral equation is obtained by integrating (27) directly, to give

$$\int_{-h}^{h} E_z^e\, dz = V_0 = \int_{-h}^{h} z^i I_z\, dz$$

$$+ \frac{j\omega}{\beta_0^2}\int_{-h}^{h}\left(\frac{\partial^2}{\partial z^2} + \beta_0^2\right) A_z\, dz. \qquad (29)$$

By substituting the integral (28) for A_z in (29) an integral equation for the current is obtained. Alternatively, (27) may be multiplied on both sides by I_z and then integrated. With (22a, b) and (24) the result is:

$$\int_{-h}^{h} E_z^e I_z \, dz = V_0 I_0 = \int_{-h}^{h} z^i I_z^2 \, dz$$
$$+ \frac{j\omega}{\beta_0^2} \int_{-h}^{h} I_z \left(\frac{\partial^2}{\partial z^2} + \beta_0^2 \right) A_z \, dz. \quad (30)$$

By substituting (28) for A_z, another somewhat different integral equation is obtained for the current. Division of (29) by I_0 and (30) by I_0^2 transforms these equations into expressions for evaluating the impedance. The solution of (30), which has important advantages compared with (29), is considered in Sec. 39.

Since antennas with simple slice or delta-function generators are not available in practice, the complications introduced by transmission-line feeders must be considered in order that a practically significant problem may be formulated and solved. A study of this problem when the transmission lines are conventional open-wire or coaxial lines is carried out in the following sections. The theoretically interesting problem of an idealized complete transmitting system consisting of a cylindrical antenna *driven from within* by a short biconical transmission line excited by a point generator is considered in Sec. 12 of Chapter VIII.

ANTENNA AND TRANSMISSION LINE

5. Two-Wire Line with Symmetric Impedance as End Load

A simple and useful method for driving a symmetric load is with a two-wire line coupled to a balanced generator. Several arrangements are shown in Fig. 5.1: in (a), generator and line are coupled by coils; in (b), by single-turn loops; and in (c), directly to a symmetric u.h.f. oscillator. Application of Thévenin's theorem* at CD or use of the general coupling theorem[37] shows that all the circuits are equivalent, under suitably fixed conditions, to the symmetrical circuit of Fig. 5.1(d), in which impedanceless slice-generators each with emf $\frac{1}{2}V_0^e$ are concentrated at C and D with an appropriately defined impedance Z_0 in series with them. For the present, let the load impedance Z_s be conventional in the sense that an internal impedance per unit length, z^i, can be defined

---

* See, for example, W. L. Everitt, *Communication engineering* (McGraw-Hill, New York, ed. 2, 1937).

continuously along its contour between A and B. An antenna as load (Fig. 5.2) is not of this type and is considered at the end of the section.

The general equation fundamental to one-dimensional circuit analysis is[14] (ref. I.31, chap. VI)

$$E_s^e = I_s z^i + \frac{d\Phi}{ds} + j\omega A_s, \quad (1)$$

where E_s^e is the impressed field maintained along a specified section of the conducting surface at $r = a$, I_s is the total current in the conductor, z^i is its internal impedance per unit length, A_s is the component of vector potential tangent to the surface at $r = a$, and s is a variable along the surface parallel to the axis of the conductor. Equation (1) is the same as (4.12), with the general contour variable s replacing z.

If (1) is integrated completely around the circuit of Fig. 5.1(d), and note is taken of

$$\oint d\Phi = 0, \quad (2)$$

the result for counterclockwise integration is

$$V^e = \oint E_s^e \, ds = \oint I_s z^i \, ds$$
$$+ j\omega \oint A_s \, ds. \quad (3)$$

If the integration is only from C to D, and since E_s^e vanishes except between C and D, the following is obtained:

$$V^e = \int_C^D E_s^e \, ds = \int_C^D I_s z^i \, ds$$
$$+ \int_C^D d\Phi + j\omega \int_C^D A_s \, ds. \quad (4)$$

The impedance Z_0 between C and D is defined by

$$Z_0 = Z_0^i + Z_0^e, \quad (5)$$

with

$$Z_0^i = \frac{1}{I_{sc}} \int_C^D I_s z^i \, ds \quad (6)$$

and

$$Z_0^e = \frac{j\omega}{I_{sc}} \int_C^D (A_s)_{r=a} \, ds. \quad (7)$$

The integration is carried out along the surface of the actual contour of the conductor forming Z_0. It is assumed that the currents at C and D are equal and in the *same* direction around the contour; hence, they are equal and *opposite* transmission-line currents. For the transmission line, $I_{sc} = I_0$. The impedance Z_0 is formally independent of the current in the load $Z_{BA} = Z_s$ if the line is long enough so that the contribution to A_s in (7) by this current is negligible compared with the contribution by the current in Z_0 itself. Similarly, Z_0 is formally independent of the currents in the two-wire line if the actual geometrical structure of Z_0 is sufficiently symmetrical so that the contributions to A_s in (7) by the currents in the two wires of the line are zero or cancel in the integration.

Using (5) to (7) in (4), the result is,

$$V^e = I_{sc} Z_0 + V_{DC}, \qquad (8)$$

where V_{DC} is the scalar potential difference across CD and is defined by

$$V_{DC} \equiv \Phi_D - \Phi_C = -\int_D^C d\Phi. \qquad (9)$$

If (4) is subtracted from (3), and (9) is used,

$$V_{DC} = \oint I_s z^i \, ds - \int_C^D I_s z^i \, ds$$

$$+ j\omega \oint A_s \, ds - j\omega \int_C^D A_s \, ds \qquad (10)$$

or

$$V_{DC} = \int_D^C I_s z^i \, ds + j\omega \int_D^C A_s \, ds, \qquad (11)$$

where the integrals in (11) are evaluated along the path $DBAC$. The input impedance of the line is defined by

$$Z_{DC} = Z_{DC}^i + Z_{DC}^e \equiv \frac{V_{DC}}{I_{sc}}, \qquad (12)$$

where

$$Z_{DC}^i = \frac{1}{I_{sc}} \int_D^C I_s z^i \, ds, \qquad (13)$$

$$Z_{DC}^e = \frac{j\omega}{I_{sc}} \int_D^C A_s \, ds. \qquad (14)$$

The input impedance Z_{DC} of the line is formally independent of the impedance Z_0 under the same conditions, stated above, that make Z_0 formally independent of the currents in the outside circuit.

The terminal impedance of the line, $Z_{BA} \equiv Z_s$, is defined by repeating (4) to (14) with the integration from A to B. Thus, integrating along the path $ACDB$,

$$V^e = \int_A^B E_s^e \, ds = \int_A^B I_s z^i \, ds$$

$$+ \int_A^B d\Phi + j\omega \int_A^B A_s \, ds. \qquad (15)$$

By subtracting (15) from (3) the following is obtained:

$$V_{BA} \equiv \Phi_B - \Phi_A = -\int_B^A d\Phi, \qquad (16)$$

$$V_{BA} = \oint I_s z^i \, ds - \int_A^B I_s z^i \, ds$$

$$+ j\omega \oint A_s \, ds - j\omega \int_A^B A_s \, ds, \qquad (17)$$

or

$$V_{BA} = \int_B^A I_s z^i \, ds + j\omega \int_B^A A_s \, ds, \qquad (18)$$

where the path of integration is from B to A along the surface of the conductor in Fig. 5.1(d).

The impedance is given by

$$Z_s = Z_{BA} = Z_{BA}^i + Z_{BA}^e \equiv \frac{V_{BA}}{I_{sA}}, \qquad (19)$$

with

$$Z_s^i = Z_{BA}^i = \frac{1}{I_{sA}} \int_B^A I_s z^i \, ds, \qquad (20)$$

$$Z_s^e = Z_{BA}^e = \frac{j\omega}{I_{sA}} \int_B^A A_s \, ds. \qquad (21)$$

It is assumed that complete symmetry is maintained so that $I_{sB} = I_{sA} = I_s$. It is significant to note that the impedance $Z_{BA} = Z_s$ can be defined entirely in terms of the currents in the conductors between B and A *only if* these are sufficiently far from Z_0 at the other end of the line and, in addition, are all perpendicular to the conductors of the line, or are symmetrically placed to provide cancellation in the integration.

An internal impedance per unit length, z^i, can be defined for a continuous conductor connecting A and B, and for a conductor in series with a parallel-plate condenser, the plates of which are sufficiently closely spaced so that the electric field due to the charges on their adjacent surfaces is confined essentially to the dielectric between them. If the terminating impedor is so constructed that there exists no continuous path between A and B along which z^i may be defined, it is not correct to write (3) and (4). This means that (16) to (21) are no longer useful for impedors such as the antenna in Fig. 5.2.

An alternative procedure that is applicable to terminations in general, including those for which a continuous z^i does not exist, may be developed as follows. Instead of integrating (1) around a closed contour as in (3) for the circuit of Fig. 5.1(d), (1) may be integrated only along the surface of the conductor on the path $ACDB$ in Fig. 5.2. This gives

$$V^e = \int_A^B (I_s z^i + j\omega A_s)\, ds + \Phi_B - \Phi_A \quad (22)$$

The scalar potential difference $\Phi_B - \Phi_A$ can be expressed in terms of the distributions of charge per unit length, $q(s)$, in all conductors. Thus,

$$V_{BA} \equiv \Phi_B - \Phi_A = \frac{1}{4\pi\varepsilon_0}\left\{\int q(s')\frac{e^{-j\beta_0 R_B}}{R_B}\, ds' \right.$$
$$\left. - \int q(s')\frac{e^{-j\beta_0 R_A}}{R_A}\, ds'\right\}, \quad (23)$$

where R_A is the distance from the point A to the element ds', and R_B is the distance from B to ds'. The integration is extended over all conductors in the entire circuit, including the antenna (Fig. 5.2) or other load; ds' is an element along the axis of the conductor. The charge per unit length in (23) may be expressed in terms of the total current $I_s(s')$ using the equation of continuity in the form

$$\frac{dI_s(s')}{ds'} + j\omega q(s') = 0. \quad (24)$$

With $q(s')$ replaced by $\frac{j}{\omega}\frac{dI_s(s')}{ds'}$ in (23), and this substituted in (22), an integral equation is obtained in the total current I_s. If this is known, the impedance of the entire circuit is readily defined as the ratio of V^e to the current entering or leaving the generator.

A formal definition for the impedance of the termination is

$$Z_{BA} = V_{BA}/I_A. \quad (25)$$

In general, Z_{BA} is not a characteristic of the termination alone, but depends on the circuit of which it is a part. However, if the distance AB (which is the separation b of the conductors of the transmission line, Fig. 5.2) is made sufficiently small ($\beta_0 b = 2\pi b/\lambda_0 \ll 1$), the coupling between the charges in the antenna and those in adjacent parts of the line is reduced so that Z_{BA} for the antenna reduces approximately to Z_δ in (4.18a) with $\delta \to 0$. A general condition is formulated as follows: Z_{BA} may be treated as an impedance that is independent of the circuit to which it is connected when AB is sufficiently small so that Z_{BA} differs negligibly from $Z_{B'A'}$, where $Z_{B'A'}$ is the impedance of the termination (antenna) in series with a short section of line of length comparable with AB (Fig. 5.2). Note that this condition is always implied in the definition of so-called "lumped" circuit elements in conventional electric circuits. (Ref. I.31, Sec. VI.18). In effect, the condition is equivalent to *replacing the scalar potential difference across terminals that are separated by a finite distance by a discontinuity in scalar potential across terminals that are separated by a vanishingly small distance.* All loads are thus assumed to be equivalent to "lumped" loads.

The formal analysis of electric circuits by the contour-integral method based on the fundamental relation (1) is unique and unambiguous. It is as exact a formulation as is possible in one-dimensional form. Except when applied to circuits in which the current amplitude is the same at all points as in conventional circuit theory, the actual solution of an integral equation of the form (3), (4), (18), or (22) with (23) and (24) for the distribution of current is difficult, and each circuit configuration must be analyzed separately. For a few circuits, other, more or less approximate, methods are available. An important example is the two-wire line. In the following analysis of the antenna center-driven from a two-wire line as in Fig. 5.2, (22) is replaced by a transmission-line equation involving V_{BA} which, in turn, is expressed by an integral equation essentially equivalent to (22).

6. Two-Wire Transmission-Line Theory; Generalized Equations

Instead of attempting to evaluate the intricate integral equation (5.3) or (5.4) for the current in the transmission line of Fig. 5.1(d), or Fig. 5.2, it is possible to derive a pair of differential equations—the conventional transmission-line equations—that may be solved for the current. However, the rigorous derivation of these equations from fundamental electromagnetic theory involves assumptions and approximations that are often overlooked but that are of great importance in the definition, the theoretical

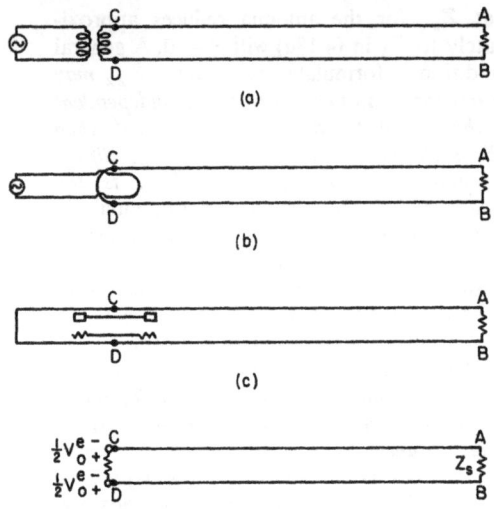

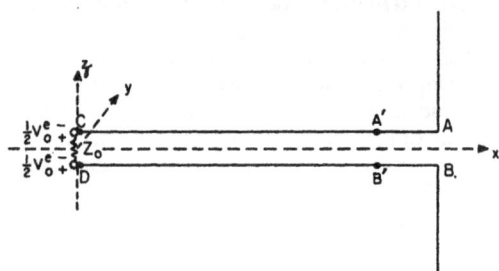

Fig. 5.2. Antenna as load for two-wire line.

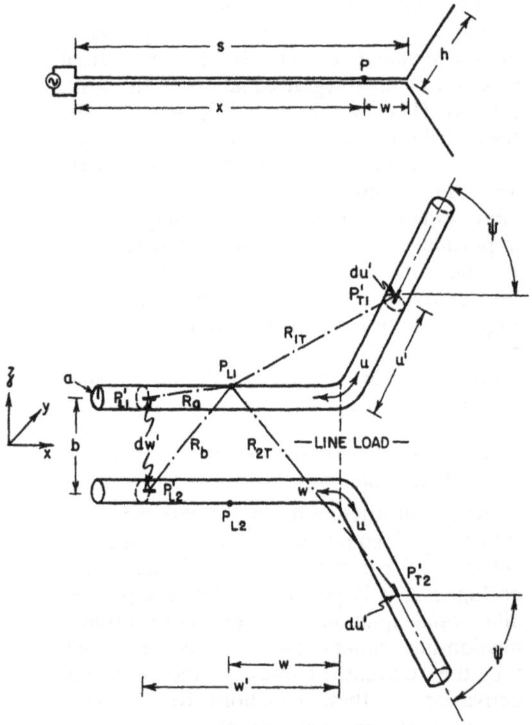

Fig. 5.1. Coupled generators for two-wire lines.

Fig. 6.1. Two-wire line with V antenna as load.

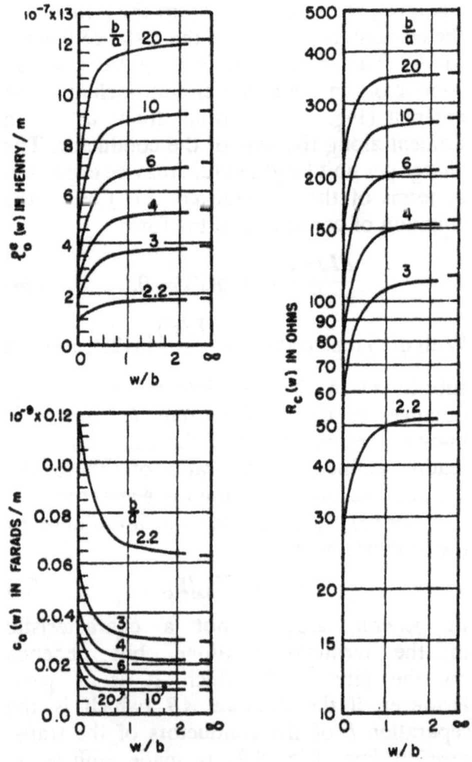

Fig. 6.2. Graphs of $l_0^e(w)$, $c_0(w)$, and $R_c(w)$ as functions of w/b with b/a as parameter.

determination, and the experimental measurement of impedances terminating the line, especially if these are antennas. One method of obtaining the transmission-line equations is given in ref. I.31, chap. VI for a two-wire line of finite conductivity immersed in an imperfect dielectric. While entirely adequate for deriving the conventional equations and the conditions limiting their validity, this derivation does not separate end effects and coupling effects between the line and a load from radiation, since all of these are neglected. In order to analyze conditions near the end of a line terminated in an impedor such as an antenna, a more detailed analysis is required in which only radiation is made negligible. Although present interest is in lines in air, the more general formulation for lines immersed in an imperfect dielectric may be carried out with negligible added complication.

The electromagnetic foundations of the transmission-line equations are the defining equation for the scalar potential ϕ,

$$\mathbf{E} = -\mathrm{grad}\,\phi - j\omega\mathbf{A}, \quad (1)$$

and the equation of continuity for potentials,

$$\mathrm{div}\,\mathbf{A} + j\frac{\beta^2}{\omega}\phi = 0, \quad (2)$$

where

$$\beta^2 = \omega^2 \xi/\nu; \quad \xi = \epsilon_e - j\sigma_e/\omega.$$

If the two-wire line is parallel to the x-axis in the xz-plane with the x-axis midway between its two conductors, as shown in Fig. 5.2, only the x-component of (1) is required, namely,

$$E_x = z^i I_x = -\frac{\partial \phi}{\partial x} - j\omega A_x. \quad (3)$$

In Cartesian coördinates (2) is

$$\frac{\partial A_x}{\partial x} + \frac{\partial A_y}{\partial y} + \frac{\partial A_z}{\partial z} + \frac{j\beta^2}{\omega}\phi = 0. \quad (4)$$

The components A_y, A_z, and their derivatives $\partial A_y/\partial y$, $\partial A_z/\partial z$ are zero everywhere on the surfaces of the two conductors of the line *if this is infinitely long or has open ends*. The components A_y, A_z, or both A_y and A_z and their derivatives *may* differ from zero near the end of a line terminated in a coil, a straight wire, an antenna, etc., in which there are currents perpendicular to the x-axis. Hence, $\partial A_y/\partial y$ and $\partial A_z/\partial z$ may not be omitted from (4) without first verifying that they are zero or negligible, if the resulting equations are to be valid near the end of the line.

Let (3) and (4) be written successively with subscripts 1 and 2 for the two conductors of the line and subtracted, to give

$$-\frac{\partial}{\partial x}(\phi_1 - \phi_2) = z_1^i I_{1x} - z_2^i I_{2x} + j\omega(A_{1x} - A_{2x}), \quad (5)$$

$$-\left[\left(\frac{\partial A_x}{\partial x}\right)_1 - \left(\frac{\partial A_x}{\partial x}\right)_2\right] - \left[\left(\frac{\partial A_y}{\partial y}\right)_1 - \left(\frac{\partial A_y}{\partial y}\right)_2\right]$$

$$-\left[\left(\frac{\partial A_z}{\partial z}\right)_1 - \left(\frac{\partial A_z}{\partial z}\right)_2\right] = j\frac{\beta^2}{\omega}\left[\phi_1 - \phi_2\right] \quad (6)$$

The potentials and their derivatives in (5) and (6) are defined on the equipotential surfaces of the two conductors 1 and 2 with their respective centers at $(x, 0, b/2)$, $(x, 0, -b/2)$. Let it be required that the two conductors of the line be *identical* and *symmetrically driven* and *loaded*. That is, the impedors terminating both ends of the line must be so symmetrical in their structures that the currents in the two conductors are equal and opposite. It follows that

$$z_1^i = z_2^i \equiv z^i/2, \quad (7)$$

$$I_{2x} = -I_{1x} \equiv -I_x, \quad (8)$$

where z^i is the internal impedance per loop unit length of the two-conductor line. The x-component of the vector potential is calculated from

$$A_x(x) \equiv A_x = \frac{1}{4\pi\nu}\int I'_x \frac{\exp(-j\beta R)}{R}dx', \quad (9a)$$

so that with (8),

$$A_{2x} = -A_{1x}. \quad (9b)$$

If there is a z-component of current in the symmetrical terminations, it must be an even function of z, and hence the z-component of the vector potential must be an even function. Thus,

$$I_z(-z) = I_z(z); \quad A_z(-z) = A_z(z). \quad (10a)$$

This means that interchanging $-z$ and z has no effect in (10a), so that

$$\frac{\partial A_z(z)}{\partial z} = \frac{\partial A_z(-z)}{-\partial z} = -\frac{\partial A_z(-z)}{\partial z}. \quad (10b)$$

It follows that

$$\left(\frac{\partial A_z}{\partial z}\right)_1 - \left(\frac{\partial A_z}{\partial z}\right)_2 = 2\left(\frac{\partial A_z}{\partial z}\right)_1. \quad (10c)$$

Note that (10b) requires that

$$\left(\frac{\partial A_z}{\partial z}\right)_{z=0} = -\left(\frac{\partial A_z}{\partial z}\right)_{z=0} = 0, \quad (10d)$$

since only zero is equal to its negative.

In addition to (8), let the symmetry of the load permit the following condition:

$$\left(\frac{\partial A_y}{\partial y}\right)_1 = \left(\frac{\partial A_y}{\partial y}\right)_2. \quad (11)$$

For the antenna in Fig. 5.2, A_y and its derivative are zero everywhere. With the notation*

$$V = \Phi_1 - \Phi_2 = 2\Phi_1, \quad (12a)$$

$$W_x = A_{1x} - A_{2x} = 2A_{1x}, \quad (12b)$$

$$W_z = A_{1z} - A_{2z} = 2A_{1z}, \quad (12c)$$

and with (10c) and (11), the equations (5) and (6) become:

$$-\frac{\partial V}{\partial x} = z^i I_x + j\omega W_x, \quad (13a)$$

$$-\frac{\partial W_x}{\partial x} = j\frac{\beta^2}{\omega} V + \frac{\partial W_z}{\partial z}. \quad (13b)$$

The general application of these equations to the terminal zone of a transmission line may be studied using the circuit of Fig. 6.1, in which the termination begins as a V, without specializing the formulation to this particular configuration. For convenience, the distance $w = s - x$ measured back along the line from the line-load junction is used as independent variable instead of x, the distance from the generator. Since $\partial/\partial w = -\partial/\partial x$, (13a) and (13b) become

$$\frac{\partial V(w)}{\partial w} = z^i I_{xL}(w) + j\omega W_x(w), \quad (13c)$$

$$\frac{\partial W_x(w)}{\partial w} = j\frac{\beta^2}{\omega} V(w) + \frac{\partial W_z(w)}{\partial z}. \quad (13d)$$

The subscript L is added to I_x, and the point w at which it is evaluated is indicated explicitly, in order to distinguish the total current in the positive x-direction on the transmission line at a distance w from the load from the x-component of the current in the load. Note that $V(w)$ and $W_x(w)$ are, in general,

---

*Conventional transmission-line theory arbitrarily assigns a positive direction to its current that is the opposite of that logically derived from the field equations. Hence, it is necessary to set $V = \Phi_1 - \Phi_2$ instead of $V = \Phi_2 - \Phi_1$ as is usual.

potential differences due to charges and currents on *both* the line and the termination, so that

$$V(w) = V_L(w) + V_T(w);$$

$$W_x(w) = W_{xL}(w) + W_{xT}(w). \quad (14)$$

Since there are no currents in the z-direction in the line, $W_z(w) = W_{zT}(w)$. The subscript L denotes the transmission line, the subscript T the termination. Referring to Fig. 6.1, the components of the vector potential differences between the points P_{L1} and P_{L2} on the equipotential surfaces of the line conductors at a distance w from the load are given by

$$W_{xL}(w) = 2A_{1xL}(w)$$

$$= \frac{1}{2\pi v}\int_0^s I_{xL}(w')P_L(w, w')\,dw', \quad (15a)$$

$$W_{xT}(w) = 2A_{1xT}(w)$$

$$= \frac{\cos\psi}{2\pi v}\int_0^h I_{uT}(u')P_T(w, u')\,du', \quad (15b)$$

$$W_{zT}(w) = 2A_{1zT}(w)$$

$$= \frac{\sin\psi}{2\pi v}\int_0^h I_{uT}(u')P_T(w, u')\,du'. \quad (15c)$$

Note that

$$W_{uT}(w) = \sqrt{W_{xT}^2(w) + W_{zT}^2(w)}$$

$$= \frac{1}{2\pi v}\int_0^h I_{uT}(u')P_T(w, u')\,du'. \quad (15d)$$

The corresponding scalar potential differences are

$$V_L(w) = 2\Phi_{1L}(w)$$

$$= \frac{1}{2\pi\xi}\int_0^s q_L(w')P_L(w, w')\,dw', \quad (15e)$$

$$V_T(w) = 2\Phi_{1T}(w)$$

$$= \frac{1}{2\pi\xi}\int_0^h q_T(u')P_T(u, u')\,du'. \quad (15f)$$

In (15a–f),

$$P_L(w, w') = \frac{\exp(-j\beta R_a)}{R_a} - \frac{\exp(-j\beta R_b)}{R_b};$$

$$P_T(w, u') = \frac{\exp(-j\beta R_{1T})}{R_{1T}} - \frac{\exp(-j\beta R_{2T})}{R_{2T}}. \quad (15g)$$

In general, the angle ψ is a function of u'

and must be written under the signs of integration in (15b, c). In the circuit in Fig. 6.1 and in applications to be studied in this volume, ψ is a constant. The distances from the respective elements of current $I_{xL}(w')\,dw'$ on each line wire and $I_{uT}(u')\,du' = \sqrt{I_{xT}^2(u') + I_{zT}^2(u')}\,du'$ on the termination to the point of calculation at w on the line are

$$R_a = \sqrt{(w' - w)^2 + a^2};$$

$$R_b = \sqrt{(w' - w)^2 + b^2}, \quad (16a)$$

$$R_{1T} = \sqrt{(w + u'\cos\psi)^2 + (u'\sin\psi)^2 + a^2};$$

$$R_{2T} = \sqrt{(w + u'\cos\psi)^2 + (u'\sin\psi + b)^2 + a^2}, \quad (16b)$$

where $z = 0$ in R_{1T} and $z = -b$ in R_{2T}. In R_{2T}, a^2 is negligible compared with b^2 and it is assumed that points P_{L1} and P_{L2} in Fig. 6.1, where the potentials are calculated, are at coördinates $(x;\ y = a;\ z = \mp b/2)$ instead of at $(x;\ y = 0;\ z = \mp b/2 \mp a)$ as shown. The formulas in (16a) for R_a and R_b are the same for all terminations, but R_{1T} and R_{2T} depend upon the configuration of the particular load. The formulas in (16b) apply specifically to the V section shown in Fig. 6.1. With $\psi = \pi/2$ they apply to the antenna in Fig. 5.2. The length s of the transmission line is assumed to be very great, so that

$$R_a(w' = s) \doteq R_b(w' = s) \text{ so that } P_L(w, w' = s) \doteq 0. \quad (17a)$$

The length of one half of the termination is h. It is assumed that *one* of the following conditions obtains at $u' = h$:

$$I_{zT}(u' = h) = 0 \quad \text{or} \quad R_{1T}(u' = h) = R_{2T}(u' = h)$$

so that $\quad P_T(w, u' = h) = 0. \quad (17b)$

At the line-load junction $(w' = 0, u' = 0)$, charges and currents are continuous, so that

$$q_\delta \equiv q_L(w' \to 0) = q_T(u' \to 0);$$

$$I_\delta \equiv I_{xL}(w' \to 0) = I_{uT}(u' \to 0). \quad (17c)$$

The scalar potential differences in (15e, f) could be evaluated from (13d) by setting

$$V_L(w) = -\frac{j\omega}{\beta^2}\frac{\partial W_{xL}(w)}{\partial w}$$

$$= \frac{j}{\omega}\frac{1}{2\pi\xi}\int_0^s I_{xL}(w')\frac{\partial}{\partial w'}P_L(w, w')\,dw' \quad (18a)$$

and

$$V_T(w) = -\frac{j\omega}{\beta^2}\left[\frac{\partial W_{xT}(w)}{\partial w} - \frac{\partial W_{zT}(w)}{\partial z}\right] \quad (18b)$$

$$= -\frac{j}{\omega}\frac{1}{2\pi\xi}\int_0^h I_{uT}(u')\left[\cos\psi\frac{\partial}{\partial w} - \sin\psi\frac{\partial}{\partial z}\right]P_T(w, u')\,du'$$

$$= -\frac{j}{\omega}\frac{1}{2\pi\xi}\int_0^h I_{uT}(u')\frac{\partial}{\partial u'}P_T(w, u')\,du'. \quad (18c)$$

Use has been made of the relation $\partial/\partial w = -\partial/\partial w'$ in (18a) and of $\cos\psi\,\partial/\partial w - \sin\psi\,\partial/\partial z = (\partial/\partial w)(\partial w/\partial u') + (\partial/\partial z)(\partial z/\partial u') = \partial/\partial u'$ in obtaining (18c). These relations follow from (16a, b). Integration of (18a) by parts, using (17a) and the equation of continuity in the form $q_L(w') = \frac{1}{j\omega}\frac{dI_{xL}(w')}{dw'}$, gives

$$V_L(w) = \frac{1}{2\pi\xi}\int_0^s q_L(w')P_L(w, w')\,dw'$$

$$+ \frac{Q(0)}{2\pi\xi}P_L(w, 0). \quad (19a)$$

Similarly, integration by parts of (18c), using (17b) and the equation of continuity in the form $q_T(u') = -\frac{1}{j\omega}\frac{dI_{uT}(u')}{du'}$, gives

$$V_T(w) = \frac{1}{2\pi\xi}\int_0^h q_T(u')P_T(w, u')\,du'$$

$$- \frac{Q(0)}{2\pi\xi}P_T(w, 0). \quad (19b)$$

In (19a) and (19b), with (17c),

$$Q(0) = \frac{I_{xL}(0)}{j\omega} = \frac{I_{uT}(0)}{j\omega}, \quad (20a)$$

$$P_L(w, 0) = P_T(w, 0). \quad (20b)$$

Comparison shows that $V_L(w)$ in (19a) differs from $V_L(w)$ in (15e) by the *added constant* term $Q(0)P_L(w, 0)/2\pi\xi$. Correspondingly, $V_T(w)$ in (19b) differs from $V_T(w)$ in (15f) by the *subtracted constant* term $Q(0)P_T(w, 0)/2\pi\xi$. Since the constant terms are equal, the sum $V(w) = V_L(w) + V_T(w)$ is the same.

The appearance of the extra constant terms in the expressions for $V_L(w)$ and $V_T(w)$ when these are calculated from (13d) arises from the fact—already explained in ref. I.31,

Sec. IV.11—that integrals like those in (15) are physically meaningful only when applied to an electrically complete system enclosed by boundaries that are not traversed by currents. This is not true of the separate integrals for $V_L(w)$ and $V_T(w)$; it is true for their sum. The contributions which appear in (19a) and (19b) are those of fictitious, infinitely thin, adjacent layers of charge of opposite sign and magnitude Q on each side of the plane through the conductors at $w = 0$, $u = 0$. With these layers the line and the termination are independently electrically complete, and no current crosses the boundary between the infinite capacitance they charge. Since these charges do not actually exist, they can contribute nothing to the solution of the problem of the line and its termination. Whether they are included or omitted in the expressions $V_T(w)$ and $V_L(w)$ is immaterial in the final results.

Partial differentiation with respect to w of the first-order equations (13c) and (13d) and use of (18) yields the following second-order equations:

$$\frac{\partial^2 V(w)}{\partial w^2} + \beta^2 V_L(w)$$
$$= \frac{\partial}{\partial w}[z^i I_{xL}(w) + j\omega W_{xT}(w)], \quad (21a)$$

$$\frac{\partial^2 W_{xL}(w)}{\partial w^2} + \beta^2 W_x(w)$$
$$= \frac{j\beta^2}{\omega}\left[z^i I_{xL}(w) - \frac{\partial V_T(w)}{\partial w}\right]. \quad (21b)$$

Equations (21a) and (21b) are integro-differential equations for determining the current $I_{xL}(w)$ in the transmission line. Since the current $I_{uT}(u)$ in the termination is involved in $W_{xT}(w)$ and $V_T(w)$, it is not possible to determine $I_{xL}(w)$ without a knowledge of $I_{uT}(u)$. Since—as is shown in later sections—$I_{uT}(u)$ must be determined by solving similar equations that involve $I_{xL}(w)$, it is clear that the determination of the currents in the line and the load involves the solution of a pair of simultaneous integro-differential equations. An *exact* solution of such a pair of equations is no less involved than a solution of the integral equation obtained from the contour integral (5.4) when applied to the transmission line and its load.

It is significant to note that for perfect conductors for which $z^i = 0$ *and* at distances from the termination at which the terms in $W_T(w)$ and $V_T(w)$ are negligible, the vector and the scalar potential differences $W_x(w)$ and $V(w)$ individually satisfy the *homogeneous* wave equation, since the right-hand members of (21a) and (21b) vanish and $V(w) = V_L(w)$, $W_x(w) = W_{xL}(w)$. Note that no restriction has been introduced to make radiation negligible.

Unlike the potential functions, the total current $I_{xL}(w)$ satisfies the homogeneous wave equation *only if radiation is made negligible* by imposing the following condition on the distance b between centers of the conductors of the transmission line:

$$|\beta b| \ll 1. \quad (22a)$$

This restriction is sufficiently severe to make radiation play a negligible part in determining both the amplitude and the distribution of voltage and current along the line even when not loaded. That is, radiated power is negligible compared with power dissipated in heating the conductors. For a line with a load impedance, and even for determining the approximate distribution of voltage or current on an unloaded line, condition (22a) is more severe than necessary. So long as the radiated power does not exceed that dissipated in heat, and the ohmic resistance per loop unit length is small compared with the inductive reactance per loop unit length, the effect of radiation is quite small. This may be expressed quantitatively by requiring the average radiation resistance per loop unit length to be small compared with the inductive reactance per loop unit length. Thus, since the radiation resistance of a resonant line does not exceed $(\zeta/4\pi)\beta^2 b^2$ for a section of length $\lambda/2$ or longer, the condition for a line in air is

$$\frac{\zeta_0}{4\pi}\beta_0^2 b^2 \cdot \frac{2}{\lambda_0} \ll \omega l_0^e = \frac{\omega}{\pi v}\ln\frac{b}{a}.$$

This reduces to

$$\beta_0^2 b^2 \ll 4\pi \ln\frac{b}{a}, \quad (22b)$$

which is less severe than (22a).

In carrying out the analysis of the transmission line it is assumed that (22a) must be satisfied, but that for qualitative purposes, and quantitatively for estimating orders of magnitude, larger values of βb may be used than are actually permitted by the restriction (22a). This restriction is equivalent to putting all parts of the transmission line *effectively* in the near zone with respect to one another regardless of the length of the line. As a result of the requirement (8) that the currents in the two conductors be equal in magnitude and opposite in direction, it follows that the electromagnetic forces on the current and charge in any element dz' at z' in one conductor due to the currents and charges in an opposite pair of elements dz_1 and dz_2 at z, are virtually equal and opposite for all distances $|z - z'|$ that exceed a few multiples of b. Hence, elements separated by distances $[(z-z')^2+b^2]^{\frac{1}{2}}$, which are comparable with b or small multiples thereof, are in the near zone as a result of (22a). Elements for which $|z - z'|$ exceeds small multiples of b contribute forces that *cancel* for opposite pairs. Thus, if (22a) is satisfied, it is a good approximation to apply conventional near-zone circuit theory to differential elements of a two-wire line. This is the basis of the usual methods of deriving the transmission-line equations by treating each differential length of the line as equivalent to a lumped-constant circuit. This may be shown to follow directly from (13a, b).

In the derivation in ref. I.31, chap. VI, an expression like (15a) for $W_x(x)$ is expanded in the form

$$W_x(x) = I_x(x) z^e(x)/j\omega, \qquad (23a)$$

with

$$z^e(x) = \frac{j\omega}{2\pi v} \int_0^s f(x, x') \left[\frac{\exp(-j\beta R_a)}{R_a} - \frac{\exp(-j\beta R_b)}{R_b} \right] dx' \qquad (23b)$$

and

$$f(x, x') = I_x(x')/I_x(x). \qquad (23c)$$

The integral is then written in two parts as follows:

$$z^e(x) = \frac{j\omega}{2\pi v} \left\{ \int_0^s \left(\frac{1}{R_a} - \frac{1}{R_b} \right) dx' \right.$$
$$+ \int_0^s \left[\frac{f(x, x') \exp(-j\beta R_a) - 1}{R_a} \right.$$
$$\left. \left. - \frac{f(x, x') \exp(-j\beta R_b) - 1}{R_b} \right] dx' \right\}. \qquad (23d)$$

In the first integral the distribution function is assumed to be unity and retardation is neglected. The second integral takes account of both nonuniformity in the distribution of current and retardation. To derive the conventional line equations the second integral is neglected and the first integral is approximated by its value at distances from the ends of the line that are large compared with the separation of its conductors.

In the present analysis a more general procedure is followed. Since the contributions to $W_x(w)$ and $V(w)$ at a distance w from the end of the line due, respectively, to currents and charges at w' in the line and at u' in the termination are significant only for distances $|w' - w|$ and $u' + w$ that are not large compared with b, it is a good approximation to expand the currents $I_{xL}(w')$ and $I_{uT}(u')$ in terms of $I_{xL}(w)$ and the charges per unit length $q_L(w')$ and $q_T(u')$ in terms of $q_L(w)$ and retain only the first two terms. Thus,

$$q_L(w') \doteq q_L(w) + (w' - w) \frac{\partial q_L(w)}{\partial w}, \qquad (24a)$$

$$I_{xL}(w') \doteq I_{xL}(w) + (w' - w) \frac{\partial I_{xL}(w)}{\partial w}, \qquad (24b)$$

$$q_T(u') \doteq q_L(w) - (u' + w) \frac{\partial q_L(w)}{\partial w}, \qquad (24c)$$

$$I_{uT}(u') \doteq I_{xL}(w) - (u' + w) \frac{\partial I_{xL}(w)}{\partial w}. \qquad (24d)$$

Note that

$$I_{xT}(u') = I_{uT}(u') \cos \psi, \qquad (24e)$$

where $I_{uT}(u')$ is the total axial current at u' in the termination and ψ is the angle between the direction of the current at u' and the z-axis. The derivatives in (24) may be replaced by introducing the equation of continuity,

$$\frac{\partial I_{xL}(w)}{\partial w} - j\omega q_L(w) = 0, \qquad (25a)$$

and its first derivative,

$$\frac{\partial^2 I_{xL}(w)}{\partial w^2} - j\omega \frac{\partial q_L(w)}{\partial w} = 0. \qquad (25b)$$

Since the first-order terms in (24) are small if βb is as small as required, it follows that only the leading terms need be considered in evaluating $\partial q_L(w)/\partial w$ in (24a) and (24c). That is, for use in a small correction term it is adequate to use the value of $I_{xL}(w)$ obtained by neglecting ohmic losses in the line, end effect, and inductive coupling to the termination. Thus, in (21b) all terms on the right may be neglected, and $W_z(w)$ may be approximated by $W_{xL}(w)$. The principal part of $W_{xL}(w)$ is obtained from (15a) by replacing $I_{xL}(w')$ by $I_{xL}(w)$ and treating the integral as a constant; this yields the conventional transmission-line equations, as is shown later. With $W_x(w) \doteq W_{xL}(w) \sim I_{xL}(w)$, it follows that (21b) reduces to

$$\frac{\partial^2 I_{xL}(w)}{\partial w^2} + \beta^2 I_{xL}(w) \doteq 0. \qquad (26a)$$

This equation is only a rough approximation, since it corresponds to lossless-line theory, but it is adequate for use in a correction term. With (26a) it follows from (25b) that

$$\partial q_L(w)/\partial w = \frac{j\beta^2}{\omega} I_{zL}(w). \quad (26b)$$

Substitution of (25a) and (26b) in (24) gives:

$$j\omega q_L(w') = j\omega q_L(w) - (w' - w)\beta^2 I_{zL}(w), \quad (27a)$$

$$I_{zL}(w') = I_{zL}(w) + (w' - w)j\omega q_L(w), \quad (27b)$$

$$j\omega q_u(u') = j\omega q_L(w) + (u' + w)\beta^2 I_{zL}(w), \quad (27c)$$

$$I_{uT}(u') = I_{zL}(w) - (u' + w)j\omega q_L(w). \quad (27d)$$

Substitution of (27) in (15a, b) and (15e, f) gives the following expressions for $W_x(w) = W_{xL}(w) + W_{xT}(w)$ and $V(w) = V_L(w) + V_T(w)$:

$$W_x(w) = \frac{1}{2\pi v} \Big\{ I_{zL}(w)[k_0(w) \mp k_{0T}(w)] + j\omega q_L(w)[k_1(w) + k_{1T}(w)]/\beta \Big\}, \quad (28a)$$

$$V(w) = \frac{1}{2\pi \xi} \Big\{ q_L(w)[k_0(w) + k'_{0T}(w)] + \frac{j\beta}{\omega} I_{zL}(w)[k_1(w) + k'_{1T}(w)] \Big\}, \quad (28b)$$

where,

$$k_0(w) = \int_0^s P_L(w, w')\, dw'$$

$$= \int_0^s \left(\frac{1}{R_a} - \frac{1}{R_b} \right) dw'$$

$$+ \int_0^s \left[\frac{\exp(-j\beta R_a) - 1}{R_a} - \frac{\exp(-j\beta R_b) - 1}{R_b} \right] dw', \quad (29a)$$

$$k_1(w) = \beta \int_0^s (w' - w) P_L(w, w')\, dw'$$

$$= \beta \int_0^s (w' - w) \left(\frac{1}{R_a} - \frac{1}{R_b} \right) dw'$$

$$+ \beta \int_0^s (w' - w) \left[\frac{\exp(-j\beta R_a) - 1}{R_a} - \frac{\exp(-j\beta R_b) - 1}{R_b} \right] dw', \quad (29b)$$

$$k'_{0T}(w) = \int_0^h P_T(w, u')\, du'$$

$$= \int_0^h \left(\frac{1}{R_{1T}} - \frac{1}{R_{2T}} \right) du'$$

$$+ \int_0^h \left[\frac{\exp(-j\beta R_{1T}) - 1}{R_{1T}} - \frac{\exp(-j\beta R_{2T}) - 1}{R_{2T}} \right] du', \quad (29c)$$

$$k'_{1T}(w) = -\beta \int_0^h (u' + w) P_T(w, u')\, du'$$

$$= -\beta \int_0^h (u' + w) \left(\frac{1}{R_{1T}} - \frac{1}{R_{2T}} \right) du'$$

$$- \beta \int_0^h (u' + w) \left[\frac{\exp(-j\beta R_{1T}) - 1}{R_{1T}} - \frac{\exp(-j\beta R_{2T}) - 1}{R_{2T}} \right] du'. \quad (29d)$$

With ψ a constant, as here assumed,

$$k_{0T}(w) = k'_{0T}(w) \cos \psi;$$
$$k_{1T}(w) = k'_{1T}(w) \cos \psi. \quad (29e)$$

The second integrals on the right in (29a–d) take account of radiation from the transmission line. They are negligibly small when $(w' - w)$ or $(u' + w)$ is large compared with b, since $R_a \doteq R_b$ or $R_{1T} \doteq R_{2T}$. On the other hand, when $(w' - w)$ or $(u' + w)$ is of the order of magnitude of small multiples of b or less, $|\beta R_a|$ and $|\beta R_b|$ or $|\beta R_{1T}|$ and $|\beta R_{2T}|$ necessarily are small compared with unity if (22a) is satisfied. If the exponentials are expanded in series, all terms below those of third power cancel. Hence, subject to (22a), these integrals and, hence, radiation are always negligible compared with the first integrals in (29a–d). This is shown by direct calculation in ref. I.31, Sec. VI.24.

The first integrals on the right in (29a) and (29b) are readily evaluated. Since the second integrals are negligible, the results are:

$$k_0(w) \doteq k_0(w) = \sinh^{-1} \frac{w}{a} - \sinh^{-1} \frac{w}{b}$$

$$+ \sinh^{-1} \frac{x}{a} - \sinh^{-1} \frac{x}{b}. \quad (30a)$$

Or, alternatively,

$$k_0(w) = 2 \ln \frac{b}{a} - \ln F(w), \quad (30b)$$

where

$$F(w) = \frac{(\sqrt{w^2 + b^2} + w)(\sqrt{x^2 + b^2} + x)}{(\sqrt{w^2 + a^2} + w)(\sqrt{x^2 + a^2} + x)}. \quad (30c)$$

Similarly,

$$k_1(w) = \beta(\sqrt{w^2 + b^2} - \sqrt{w^2 + a^2} - \sqrt{x^2 + b^2} + \sqrt{x^2 + a^2}). \quad (31)$$

Note that for

$$\left.\begin{array}{l} w^2 \gg b^2 \\ x^2 \gg b^2 \end{array}\right\} : k_0(w) \doteq k_0(\infty) \equiv 2\ln\frac{b}{a};$$

$$k_1(w) \doteq 0, \quad (32a)$$

whereas for

$$\left.\begin{array}{l} w = 0 \\ x^2 = s^2 \gg b^2 \end{array}\right\} : k_0(0) \doteq \tfrac{1}{2}k_0(\infty) = \ln\frac{b}{a};$$

$$k_1(0) \doteq \beta(b - a). \quad (32b)$$

Evidently the largest possible value of the ratio $k_1(w)/k_0(w)$ is for $w = 0$. Hence,

$$p_0(w) \equiv \frac{k_1(w)}{k_0(w)}; \quad \frac{k_1(0)}{k_0(0)} = \frac{\beta(b-a)}{\ln b/a}. \quad (32c)$$

Since $\ln b/a$ always exceeds $\ln 2$, the ratio $|k_1(w)/k_0(w)|$ is always small if the condition $|\beta b| \ll 1$ is satisfied. Even subject only to (22b), it is usually possible to neglect $|k_1^2(w)/k_0^2(w)|$ compared with unity, that is,

$$|p_0^2(w)| \equiv \left|\frac{k_1^2(w)}{k_0^2(w)}\right| \ll 1. \quad (32d)$$

Let the following symbolism be introduced:

$$l^e(w) = l_0^e(w) + l_T^e(w)$$
$$= [k_0(w) + k_{0T}(w)]/2\pi v, \quad (33a)$$

$$j\omega y^{-1}(w) = j\omega[y_0^{-1}(w) + y_T^{-1}(w)]$$
$$= [k_0(w) + k'_{0T}(w)]/2\pi\xi, \quad (33b)$$

$$y(w) = g(w) + j\omega c(w), \quad (33c)$$

$$p(w) \equiv \frac{k_1(w) + k_{1T}(w)}{k_0(w) + k_{0T}(w)}, \quad (33d)$$

$$p'(w) \equiv \frac{k_1(w) + k'_{1T}(w)}{k_0(w) + k'_{0T}(w)}. \quad (33e)$$

Note that

$$\beta^2 \equiv \omega^2\xi/v = -j\omega l_0^e(w)y_0(w). \quad (33f)$$

With this symbolism substituted in (28a) and (28b), these become:

$$V(w) = y^{-1}(w)[j\omega q_L(w) - \beta I_{xL}(w)p'(w)], \quad (34a)$$

$$W_x(w) = l^e(w)[I_{xL}(w) + j\omega q_L(w)p(w)/\beta]. \quad (34b)$$

Note that

$$V_L(w) = y_0^{-1}(w)[j\omega q_L(w) - \beta I_{xL}(w)p_0(w)], \quad (34c)$$

$$W_{xL}(w) = l_0^e(w)[I_{xL}(w) + j\omega q_L(w)p_0(w)/\beta]. \quad (34d)$$

The following functions for a line in air ($\xi = \epsilon_0$, $v = v_0$) are shown in Fig. 6.2:*
$l_0^e(w) = k_0(w)/2\pi v_0$; $c_0(w) = 2\pi\epsilon_0/k_0(w)$, $R_c(w) = [l_0^e(w)/c_0(w)]^{\frac{1}{2}} = \frac{\zeta_0}{2\pi}k_0(w)$. The functions $k_0(w)$, $k_1(w)/\beta_0 b$, and $p_0(w)/\beta_0 b$ are shown in Fig. 6.3.†

With the formulas (34a) and (34b) for $V(w)$ and $W_x(w)$ the general nonhomogeneous equations (21a) and (21b) may be reduced to approximate homogeneous forms in $V(w)$ and $W_{xL}(w)$. This is accomplished by introducing the following dimensionless ratio functions to characterize the capacitive and inductive coupling between the line and the load. Thus, let

$$\varphi_1 \equiv \varphi_1(w) \equiv V_L(w)/V(w);$$
$$a_1 \equiv a_1(w) \equiv W_x(w)/W_{xL}(w). \quad (35a)$$

Note that when $\varphi_1(w) = 1$, $V(w) = V_L(w)$, $V_T(w) = 0$, so that there is no capacitive coupling. Similarly, when $a_1(w) = 1$, $W_x(w) = W_{xL}(w)$, $W_{xT}(w) = 0$, and there is no inductive coupling.

$$\varphi_1(w) = \frac{y(w)}{y_0(w)}\left[\frac{1 + p'(w)H(w)}{1 + p_0(w)H(w)}\right];$$

$$H(w) \equiv -\frac{\beta I_{xL}(w)}{j\omega q_L(w)} \doteq \frac{\partial q_L(w)/\partial w}{\beta q_L(w)}; \quad (35b)$$

$$a_1(w) = \frac{l^e(w)}{l_0^e(w)}\left[\frac{1 + p(w)G(w)}{1 + p_0(w)G(w)}\right];$$

$$G(w) \equiv \frac{j\omega q_L(w)}{\beta I_{xL}(w)} = \frac{\partial I_{xL}(w)/\partial w}{\beta I_{xL}(w)}. \quad (35c)$$

If the rates of change of charge and current within distances $w \leq 10b$ along the line and $u \leq 10b$ along the termination are sufficiently small, satisfactory approximations of $\varphi_1(w)$ and $a_1(w)$ are obtained from their leading terms. These are equivalent to assuming charge and current sensibly constant along the line

---

* See p. 34.
† See p. 48.

and the termination near their junction.

$$\varphi_1(w) \doteq y(w)/y_0(w)$$
$$\doteq k_0(w)/[k_0(w) + k'_{0T}(w)] = \varphi_1(w)$$
$$\doteq c(w)/c_0(w) \text{ for } g \text{ small}, \quad (35d)$$

$$a_1(w) \doteq l^e(w)/l_0^e(w)$$
$$= [k_0(w) + k_{0T}(w)]/k_0(w) = a_1(w)$$
$$\doteq z(w)/z_0(w) \text{ for } z^i \text{ small}. \quad (35e)$$

If the charges and currents differ considerably in the adjacent parts of the line and the termination, (35b) and (35c) may be used with approximate values of $F(w)$ and $G(w)$. Alternatively, values of $V_L(w)$ and $V_T(w)$, or of $W_{xL}(w)$ and $W_{xT}(w)$ may be calculated using approximate *average* values of $q_L(w')$ and $q_T(u')$ or of $I_{xL}(w')$ and $I_{xT}(u')$ that are not necessarily equal. From these the ratio functions are readily evaluated.

Substitution of (35a) in (21a, b) gives:

$$\frac{\partial^2 V(w)}{\partial w^2} + \beta^2 \varphi_1(w) V(w) = \frac{\partial}{\partial w} \left\{ z^i I_{xL}(w) \right.$$
$$\left. + j\omega W_x(w)[1 - 1/a_1] \right\}, \quad (36a)$$

$$\frac{\partial^2 W_{xL}(w)}{\partial w^2} + \beta^2 a_1(w) W_{xL}(w)$$
$$= \frac{j\beta^2}{\omega} \left\{ z^i I_{xL}(w) - \frac{\partial}{\partial w}[(1 - \varphi_1) V(w)] \right\}. \quad (36b)$$

Since the terms with z^i, the internal impedance per unit length, as a factor are extremely small for good conductors, it is adequate if only the leading part of (34b) is used *in such terms* and if coupling effects are neglected. That is, let

$$z^i I_{xL}(w) \doteq z^i W_x(w)/l_0^e(w). \quad (37)$$

Substitution of (37) in (13c) and (36a) and the subsequent elimination of $W_x(w)$ from (36a) gives for the expression in braces in (36a):

$$\left\{ \frac{\partial V(w)}{\partial w} \left[\frac{z^i}{j\omega l_0^e(w)} + 1 - \frac{1}{a_1} \right] \middle/ \left[\frac{z^i}{j\omega l_0^e(w)} + 1 \right] \right\}$$
$$= \left\{ \left[1 - \frac{j\omega l_0^e(w)}{a_1 z_0(w)} \right] \frac{\partial V(w)}{\partial w} \right\}, \quad (38a)$$

where

$$z_0(w) \equiv z^i + j\omega l_0^e(w). \quad (38b)$$

In differentiating (38a) partially with respect to w, it is a satisfactory approximation to neglect variations in correction terms. That is, $a_1(w) = a_1$ may be treated as a constant, so that

$$\frac{\partial}{\partial w} \left\{ \left[1 - \frac{j\omega l_0^e(w)}{a_1 z_0(w)} \right] \frac{\partial V(w)}{\partial w} \right\}$$
$$\doteq \left[1 - \frac{j\omega l_0^e(w)}{a_1 z_0(w)} \right] \frac{\partial^2 V(w)}{\partial w^2}. \quad (38c)$$

Substitution of (38c) for the right-hand member of (36a) and rearrangement of terms gives at once

$$\frac{\partial^2 V(w)}{\partial w^2} - \gamma^2(w) V(w) = 0, \quad (39a)$$

where, using (33f),

$$\gamma^2(w) \equiv -\frac{\beta^2 a_1(w) \varphi_1(w) z_0(w)}{j\omega l_0^e(w)}$$
$$= z_0(w) y_0(w) a_1(w) \varphi_1(w). \quad (39b)$$

A similar manipulation of (36b) gives

$$\frac{\partial^2 W_{xL}(w)}{\partial w^2} - \gamma^2(w) W_{xL}(w) = 0, \quad (39c)$$

where $\gamma^2(w)$ is as in (39b).

Instead of determining $I_{xL}(w)$ from $W_{xL}(w)$ as given in (39c), it is convenient to obtain an explicit expression for $I_{xL}(w)$ in terms of $V(w)$, so that $I_{xL}(w)$ may be derived directly from solutions of (39a). The desired formula for $I_{xL}(w)$ is derived by substituting $j\omega q_L(w)$ from (34a) into (34b) to obtain $W_x(w)$:

$$W_x(w) = l_0^e(w)\{[1 + p(w)p'(w)]I_{xL}(w)$$
$$+ y(w)p(w)V(w)/\beta\}. \quad (40a)$$

Substitution of (40a) into (13c) and solution for $I_{xL}(w)$ using (33f) gives

$$I_{xL}(w) = \left[\frac{1}{z(w)[1 + p(w)p'(w)]} \right]$$
$$\left[\frac{\partial V(w)}{\partial w} + \frac{l^e(w)y(w)}{l_0^e(w)y_0(w)} \beta p(w) V(w) \right], \quad (40b)$$

where

$$z(w) = z^i + j\omega l^e(w). \quad (40c)$$

With the notation in (35d) and (35e) and subject to the condition

$$|p(w)p'(w)| \ll 1, \quad (41)$$

$I_{xL}(w)$ in (40b) reduces to

$$I_{xL}(w) = \frac{1}{z(w)}\left[\frac{\partial V(w)}{\partial w} + \beta a_1(w)\varphi_1(w)p(w)V(w)\right]. \quad (42)$$

This equation [or (40b)] and (39a) are the final formulas for the loaded transmission line. They are valid at all points, including the region near the load and the generator. A similar expression for $W_x(w)$ in terms of $V(w)$ may be obtained by substituting (42) in (13c). However, since the term in z^i is very small in good conductors, (13c) without this term is a good approximation.

At distances from the ends of the line specified by

$$w^2 \gg b^2, \quad x^2 \gg b^2, \quad (43)$$

the general equations (39a), (39c) and (42) reduce to conventional form. With (43) the following relations are good approximations:

$$k_0(w) \doteq k_0(\infty) = 2\ln\frac{b}{a}; \quad k_1(w) \doteq 0; \quad (44a)$$

$$k_{0T}(w) \doteq 0; \quad k'_{0T}(w) \doteq 0; \quad k_{1T}(w) \doteq 0; \quad (44b)$$

$$p_0(w) \doteq 0; \quad p(w) \doteq 0; \quad p'(w) \doteq 0; \quad (44c)$$

$$\varphi_1(w) \doteq 1; \quad V_L(w) \doteq V(w); \quad (45a)$$

$$a_1(w) \doteq 1; \quad W_{xL}(w) \doteq W_x(w). \quad (45b)$$

Hence,

$$\frac{\partial^2 V(w)}{\partial w^2} - \gamma^2 V(w) = 0, \quad (46a)$$

$$\frac{\partial^2 W_x(w)}{\partial w^2} - \gamma^2 W_x(w) = 0, \quad (46b)$$

$$I_{xL}(w) = \frac{1}{z}\frac{\partial V(w)}{\partial w}, \quad (47)$$

where

$$\gamma^2 = zy = (z^i + j\omega l_0^e)(g_0 + j\omega c_0), \quad (48a)$$

$$l_0^e = \frac{1}{\pi\nu}\ln\frac{b}{a}; \quad c_0 = \frac{\pi\epsilon_e}{\ln(b/a)}; \quad g_0 = \frac{\pi\sigma_e}{\ln(b/a)} \quad (48b)$$

These are the conventional equations and constants for a two-wire line. It follows directly from (34a, b) that

$$V(w) = j\omega q_L(w)/y_0 = q_L(w)/c_0 \text{ if } g_0 = 0, \quad (49)$$

$$W_x(w) = I_{xL}(w)l_0^e. \quad (50)$$

Hence, the scalar potential difference is linearly related to the charge per unit length, and the axial component of the vector potential difference is proportional to the current. Evidently, $q_L(w)$ may be substituted for $V(w)$ in (46a) and $I_{xL}(w)$ for $W_x(w)$ in (46b). The distributions of charge and scalar potential difference are the same, as are the distributions of current and axial vector potential difference *except near the ends of the line where these simple proportionalities are not true.*

Subject to the condition

$$a^2/b^2 \ll 1, \quad (51)$$

(46)–(50) are the formulas for conventional transmission-line theory. The results obtained subject to (51) are readily generalized to permit unrestricted values of a/b by substituting b_e for b as given in ref. I.31, p. 468:

$$b_e = \tfrac{1}{2}b[1 + \sqrt{1 - (2a/b)^2}].$$

Conventional line theory assumes (46)–(50) to be valid for all values of w and x including those that violate (43). That this leads to serious errors whenever βb is not vanishingly small is shown in the following sections.

Expressed in terms of the first-order equation (13b), the conditions (43) upon which conventional line theory depends are related to the following inequality:

$$\left|\frac{\partial A_z}{\partial z}\right|_1 \ll \left|\frac{\beta^2}{2\omega}V\right| = \left|\frac{\beta^2}{\omega}\Phi_1\right|. \quad (52)$$

Since A_z is to contribute nothing to the transmission-line problem, it may be assumed that the derivatives of the two sides of (52) must satisfy the same inequality. Thus,

$$\left|\frac{\partial^2 A_z}{\partial z^2}\right|_1 \ll \left|\frac{\beta^2}{\omega}\frac{\partial \Phi}{\partial z}\right|_1. \quad (53)$$

The general equation satisfied by A_z in the dielectric is [ref. I.31, Eq. (III.14.34b)],

$$\frac{\partial^2 A_z}{\partial x^2} + \frac{\partial^2 A_z}{\partial y^2} + \frac{\partial^2 A_z}{\partial z^2} + \beta^2 A_z = 0. \quad (54)$$

For most terminations, including especially antennas, the term $(\partial^2 A_z/\partial z^2)_1$ is larger than or at least as large as $(\partial^2 A_z/\partial x^2)_1$ or $(\partial^2 A_z/\partial y^2)_1$ at $z = b/2$ near the termination. Therefore, the correct order of magnitude is given by:

$$\left(\frac{\partial^2 A_z}{\partial z^2}\right)_1 \doteq -\beta^2 A_{1z}. \qquad (55)$$

With (55), (53) reduces to

$$|\omega A_{1z}| \ll \left|\frac{\partial \Phi}{\partial z}\right|_1. \qquad (56)$$

But since

$$E_{1z} = -\left(\frac{\partial \Phi}{\partial z}\right)_1 - j\omega A_{1z}, \qquad (57)$$

(56) is equivalent to

$$|\omega A_{1z}| \ll \left|\sqrt{E_{1z}^2 + \omega^2 A_{1z}^2}\right| \qquad (58)$$

or

$$|\omega A_{1z}| \ll |E_{1z}|. \qquad (59)$$

It follows from (59) that conventional transmission-line theory is valid only at distances $w = s - x$ from the termination that are sufficiently great so that the vector potential contributes negligibly to the scalar potential difference on the line, that is,

$$V = \Phi_1 - \Phi_2 = \int_{-b/2}^{b/2} (E_z + j\omega A_z)\, dz$$

$$\doteq \int_{-b/2}^{b/2} E_z\, dz. \qquad (60)$$

Evidently, this is only another way of requiring the coupling of the line to the termination to be negligible in those parts of the line where conventional line theory is assumed to apply.

7. Approximate Solution of the Generalized Transmission-Line Equations

The generalized transmission-line equations that are to be solved for the scalar potential difference $V(w)$ and the current $I_{xL}(w)$ are summarized below:

$$\frac{\partial^2 V(w)}{\partial w^2} - \gamma^2(w)V(w) = 0, \qquad (1)$$

$$I_{xL}(w) = \frac{1}{z(w)}\left[\frac{\partial V(w)}{\partial w} + \beta p(w)a_1(w)\varphi_1(w)V(w)\right]. \qquad (2)$$

The generalized propagation "constant" is

$$\gamma^2(w) = \gamma^2 a_1(w)\varphi_1(w), \qquad (3a)$$

where

$$\gamma^2 = z_0(w)y_0(w) = zy;$$

$$a_1(w) = \frac{W_x(w)}{W_{xL}(w)}; \quad \varphi_1(w) = \frac{V_z(w)}{V(w)} \qquad (3b)$$

$$z_0(w) = z^i + j\omega l_0^e(w) \doteq (z^i + j\omega l_0^e)\frac{k_0(w)}{k_0(\infty)}$$

$$\equiv z\frac{k_0(w)}{k_0(\infty)}, \qquad (4a)$$

$$y_0(w) = g_0(w) + j\omega c_0(w) = (g_0 + j\omega c_0)\frac{k_0(\infty)}{k_0(w)}$$

$$\equiv y\frac{k_0(\infty)}{k_0(w)}, \qquad (4b)$$

$$z(w) = z^i + j\omega l^e(w)$$

$$\doteq (z^i + j\omega l_0^e)\left[\frac{k_0(w) + k_{0T}(w)}{k_0(\infty)}\right]$$

$$\doteq z_0(w)\left[\frac{k_0(w) + k_{0T}(w)}{k_0(w)}\right]$$

$$= z_0(w)a_1(w), \qquad (5a)$$

$$y(w) = g(w) + j\omega c(w)$$

$$= (g_0 + j\omega c_0)\left[\frac{k_0(\infty)}{k_0(w) + k'_{0T}(w)}\right]$$

$$= y_0(w)\left[\frac{k_0(w)}{k_0(w) + k'_{0T}(w)}\right]$$

$$= y_0(w)\varphi_1(w). \qquad (5b)$$

When z^i and g_0 are very small, as with good conductors immersed in a near-perfect dielectric,

$$z(w) \doteq j\omega l^e(w) = j\omega l_0^e(w)a_1(w), \qquad (5c)$$

$$y(w) \doteq j\omega c(w) = j\omega c_0(w)\varphi_1(w). \qquad (5d)$$

Throughout, the variable $w = s - x$ is measured from the actual impedance $Z_\delta = V_\delta/I_\delta$ at $x = s$ or $w = 0$. Note that Z_δ is defined as the ratio of scalar potential difference across the terminals when separated by a distance $2\delta = b$ to the current I_δ entering and leaving these terminals.

The solution of (1) cannot be carried out directly, since $\gamma^2(w)$ is not a constant parameter independent of w in the terminal zones of length d defined by

$$0 \leq x \leq d \sim 10b,$$
$$0 \leq w \leq d \sim 10b, \qquad (6)$$

but is a complicated function of x or w, different for each termination. However, $\gamma^2(w)$ does reduce to the constant γ^2 at all points outside the terminal zones. For distances b between the conductors of the line that satisfy (6.22a), namely, $|\beta b| \ll 1$, the length d of the terminal zone does not exceed small multiples of b. Accordingly, if the solution of (1) along the principal line outside the terminal zone is expressed in the form

$$V(w) = A \cosh \gamma w + B \sinh \gamma w, \qquad (7)$$

A, B, and γ are constants outside the terminal zones. In the terminal zones a solution in the form (7) is not helpful, since A, B, and γ would have to be complicated functions of the variable w (or x). This suggests the following method for determining an *approximate* solution of (1) at all points outside the terminal zones. The physical basis of the method is straightforward. It involves reducing (1) and (2) to conventional form at *all* points along the line by replacing $\gamma^2(w)$, $z(w)$, and $p(w)$ by the constant values these functions have outside the terminal zone, namely, γ^2, z, and zero. This substitution is exact outside the terminal zone; it is increasingly in error as the terminal zone is entered and the terminated end of the line is approached. This error is distributed over a distance that is short compared with the wavelength,* so that it is a good approximation to *compensate for it by using appropriately defined lumped elements connected in parallel or series with the actual terminal impedance Z_δ*. (The discussion is confined to the load at $w = 0$ or $x = s$.) In other words, the analytical procedure is equivalent to replacing the actual terminal zone with its terminating impedance Z_δ by an equal length of *conventional line with constant parameters* and a fictitious, apparent load Z_{sa} that is a combination of lumped elements in parallel or series with Z_δ. The next step is to define these lumped elements.

The effect of the scalar potential difference that is maintained across the conductors of the line by the charges in the termination is equivalent to a change in the admittance per unit length of the line from $y_0(w)$ to $y(w) = y_0(w)\varphi_1(w)$. If y_0 is substituted for $y(w)$, no significant error is made outside the terminal zone. In the terminal zone the error per unit length is given by $y(w) - y_0$. The total error in admittance in the entire terminal zone of length d is

$$Y_T = \int_0^d [y(w) - y_0] \, dw$$

$$= \int_0^d [y_0(w)\varphi_1(w) - y_0] \, dw. \qquad (8a)$$

In most practical cases g_0 is negligible so that Y_T in (8a) is equivalent to $j\omega C_T$, where

$$C_T = \int_0^d [c(w) - c_0] \, dw$$

$$= \int_0^d [c_0(w)\varphi_1(w) - c_0] \, dw. \qquad (8b)$$

That is, a constant distributed capacitance per unit length c_0 and a lumped capacitance C_T at $w = 0$ are assumed to be approximately equivalent to a distributed capacitance per unit length $c(w)$. Since C_T takes account of the coupling between the termination and the line, it can be evaluated only if the structure of the termination is given. Its evaluation for simple antennas is in Secs. 8 and 9.

The next step is to replace $z(w)$ by z and evaluate an appropriate lumped element to be connected in series with Z_δ at $w = 0$. Thus, if z is substituted for $z(w)$, no error is made outside the terminal zone; in the terminal zone the error in impedance per unit length is $z(w) - z$, and the total error in impedance is

$$Z_T = \int_0^d [z(w) - z] \, dw$$

$$= \int_0^d [z_0(w)a_1(w) - z] \, dw, \qquad (9a)$$

Since a small error in z^i is insignificant if confined to the terminal zone, Z_T in (9a) is practically equivalent to $j\omega L_T$, where

$$L_T = \int_0^d [l_0^e(w) - l_0^e] \, dw$$

$$= \int_0^d [l_0^e(w)a_1(w) - l_0^e] \, dw. \qquad (9b)$$

Thus a lumped inductance L_T must be assumed connected in series with Z_δ; or, more symmetrically, a lumped inductance $\tfrac{1}{2}L_T$ must be

---

* In (6) the length of the terminal zone is chosen to be $d \doteq 10b$. This is to make terms of the form $(d^2 + a^2)^{1/2} - (d^2 + b^2)^{1/2}$ negligible. However, coupling between the load (for instance, an antenna) and the line may extend beyond $10b$. Coupling is best approximated when $d \doteq 0.1\lambda$ for the antenna load in Fig. 5.2, as discussed in Sec. 8.

connected in series with *each* of the two conductors of the line at $w = 0$ if the constant parameter z is substituted for the actual, variable parameter in $z(w)$.

In general, the lumped series inductance defined in (9b) depends upon the termination. Only if this is at right angles, so that $\psi = \pi/2$ (as in the case of the simple antennas considered in Secs. 8 and 9), or if all currents in the termination are distributed to provide canceling vector potentials, is L_T readily evaluated. With (5a) and (6.30a), (9b) has the following value for $a_1(w) = 1$ as with $\psi = \pi/2$; it is assumed that the dielectric is air so that $v = v_0$:

$$L_T = \frac{1}{2\pi v_0} \int_0^d \left[\sinh^{-1}\frac{w}{a} - \sinh^{-1}\frac{w}{b} \right.$$
$$+ \sinh^{-1}\frac{x}{a} - \sinh^{-1}\frac{x}{b}$$
$$\left. - 2\ln\frac{b}{a} \right] dw. \quad (10a)$$

Near the end $w = 0$ or $x = s$, x is very great compared with b and a, so that

$$L_T = \frac{1}{2\pi v_0} \left\{ \int_0^d \left[\sinh^{-1}\frac{w}{a} - \sinh^{-1}\frac{w}{b} \right] dw \right.$$
$$\left. - d\ln\frac{b}{a} \right\}. \quad (10b)$$

The integration gives

$$L_T = \frac{1}{2\pi v_0} \left\{ d\ln\frac{b(d + \sqrt{d^2 + a^2})}{a(d + \sqrt{d^2 + b^2})} - \sqrt{d^2 + a^2} \right.$$
$$\left. + \sqrt{d^2 + b^2} + a - b - d\ln\frac{b}{a} \right\}. \quad (10c)$$

For

$$d^2 \gg b^2, \quad (10d)$$

this reduces to the simple form:

$$L_T \doteq -\left(\frac{b-a}{2\pi v_0}\right). \quad (\psi = \pi/2) \quad (10e)$$

Up to this point, the approximate method of solving (1) and (2) is equivalent to replacing the actual transmission line with variable parameters by a fictitious line with fixed parameters and a modified termination. The two lines are completely equivalent outside the terminal zone and in maintaining the same ratio of scalar potential difference to current across the load.

The equations for the fictitious line are

$$\frac{\partial^2 V(w)}{\partial w^2} - \gamma^2 V(w) = 0, \quad (11)$$

$$I_x(w) = \frac{1}{z}\left[\frac{\partial V(w)}{\partial w}\right], \quad (12)$$

where

$$\gamma^2 = yz \doteq j\omega c_0 z, \quad z = z^i + j\omega l_0^e,$$
$$z^i = r^i + j\omega l^i. \quad (13)$$

Evidently, the substitution of l_0^e for $l^e(w)$ and c_0 for $c(w)$ implies the establishment of conditions for which 0 must be substituted for $p(w)$.

The scalar potential difference at w is given by (7). The current is obtained from (12) using (7). In differentiating $V(w)$, A and B are assumed to be constant, since $V(w)$ already is in the form appropriate to conventional line theory. Thus

$$I_x(w) = \frac{\gamma}{z}[A \sinh \gamma w + B \cosh \gamma w]. \quad (14)$$

In conventional line theory the ratio γ/z is, by definition, the characteristic admittance Y_c; its reciprocal is the characteristic impedance Z_c. For wires in empty space,

$$Z_c \equiv \frac{z}{Y} = \frac{z}{\sqrt{j\omega c_0 z}} = \sqrt{\frac{z}{j\omega c_0}} = R_c(1 - j\phi_c), \quad (15)$$

where

$$R_c \equiv \sqrt{l_0^e/c_0}; \quad \phi_c \doteq r^i/2\omega l_0^e$$
$$\text{when } (r^i/\omega l_0^e)^2 \ll 1. \quad (16)$$

Except for certain problems involving very short lengths of line, the distortion factor ϕ_c is negligible for good conductors and good dielectrics, as here assumed. Accordingly,

$$Z_c \doteq R_c, \quad Y_c \doteq G_c. \quad (17)$$

The input admittance, $Y_{in}(w) = G_{in}(w) + jB_{in}(w)$, of a section of line of length w, is obtained by dividing (14) by (7):

$$Y_{in}(w) = \frac{I_x(w)}{V(w)}$$
$$= G_c \left[\frac{A \sinh \gamma w + B \cosh \gamma w}{A \cosh \gamma w + B \sinh \gamma w}\right]. \quad (18)$$

The potential difference across the *apparent* terminal admittance Y_{sa} consisting of the load admittance Y_δ in parallel with C_T and in series with L_T is obtained from (7) by setting $w=0$. Thus

$$V_{sa} = A. \tag{19}$$

The apparent terminal current into the admittance Y_{sa} is obtained from (14) with (15) and (17) by setting $w=0$. Thus,

$$I_x(w=0) \equiv I_{sa} = G_c B,$$
$$B = I_{sa}R_c = V_{sa}Y_{sa}R_c. \tag{20}$$

Substitution of (19) and (20) in (7), (14), and (18) gives the following conventional formulas:

$$V(w) = V_{sa}[\cosh \gamma w + Y_{sa}R_c \sinh \gamma w], \tag{21}$$

$$I_x(w) = V_{sa}G_c[\sinh \gamma w + Y_{sa}R_c \cosh \gamma w], \tag{22}$$

$$Y_{in}(w) \equiv \frac{I_x(w)}{V(w)}$$

$$= G_c \left[\frac{\sinh \gamma w + Y_{sa}R_c \cosh \gamma w}{\cosh \gamma w + Y_{sa}R_c \sinh \gamma w}\right]. \tag{23}$$

These relations are valid for $w \geq d$ where $d \sim 10b$. The corresponding formulas for the axial vector potential difference and the charge per unit length are derivable from conventional equations. Thus,

$$W_x(w) = I_x(w)l_0^e$$

$$= \frac{V_{sa}}{v_0}[\sinh \gamma w + Y_{sa}R_c \cosh \gamma w], \tag{24}$$

$$q(w) = V(w)c_0$$
$$= V_{sa}c_0[\cosh \gamma w + Y_{sa}R_c \sinh \gamma w]. \tag{25}$$

It is to be noted that the load admittance Y_δ and the apparent terminal admittance Y_{sa} consisting of Y_δ in combination with L_T and C_T are not in themselves independent, measurable quantities. They are merely convenient symbols to represent ratios of scalar potential differences to current that are involved in determining the measurable input admittance of a section of line that is longer than the terminal zone and is terminated in Y_δ. Specifically, if the length of the section is $\lambda/2$, and its losses are negligible compared with those in Y_δ, the *input admittance is precisely* Y_{sa}.

For determining the scalar potential difference and current for the line *outside* the terminal zone, or for the actual admittance Y_δ, it is possible to treat the terminal zone as though conventional line theory with constant parameters applied, provided a lumped capacitance C_T is connected in parallel with the series combination of the actual load Z_δ and a lumped inductance L_T. The equivalent circuit of the terminal zone is shown in Fig. 7.1a. The complete *apparent* load admittance $Y_{sa} = 1/Z_{sa}$, which is the effective termination of the fictitious line with uniform parameters throughout its length, is related to the load impedance $Z_\delta = 1/Y_\delta$ by the following formulas:

$$Y_{sa} = \frac{1}{Z_\delta + j\omega L_T} + j\omega C_T, \tag{26}$$

$$Z_\delta = \frac{1}{Y_{sa} - j\omega C_T} - j\omega L_T. \tag{27}$$

Rationalization of (26) and solution for R_{sa} and X_{sa} gives

$$R_{sa} = R_\delta \{1 - 2\omega C_T(X_\delta + \omega L_T) + \omega^2 C_T^2[R_\delta^2 + (X_\delta + \omega L_T)^2]\}^{-1}, \tag{28}$$

$$X_{sa} = \frac{(X_\delta + \omega L_T) - \omega C_T[R_\delta^2 + (X_\delta + \omega L_T)^2]}{1 - 2\omega C_T(X_\delta + \omega L_T) + \omega^2 C_T^2[R_\delta^2 + (X_\delta + \omega L_T)^2]}. \tag{29}$$

For a line with sufficiently small spacing, $L_T \doteq 0$, $C_T \doteq 0$,

$$R_{sa} \doteq R_\delta, \quad X_{sa} \doteq X_\delta. \tag{30}$$

For the circuit of Fig. 7.1b the equations corresponding to (26) and (27) are obtained from (26) and (27) by setting $L_T = 0$ and adding ωL_T to the right-hand member of (29).

For use in conjunction with an "image line" as discussed in Sec. 10, the circuit of Fig. 7.2 is convenient. It is equivalent to Fig. 7.1b, but with elements so arranged that they are symmetrical with respect to a plane midway between the wires of the line.

The lumped series inductance L_T compensates for the use of a constant inductance per unit length l_0^e in the terminal zone in place of the variable $l^e(w)$. It includes the variation in inductance per unit length near the end of the line as contained in $l_0^e(w)$ and the effect of inductive coupling as represented by $a_1(w)$. When $\psi = \pi/2$, $a_1(w) = 1$ and L_T is negative. The lumped parallel capacitance C_T may be positive or negative, depending upon the structure of the termination. For Fig. 5.2 it is negative. It corrects for the use of a constant capacitance per unit length c_0 in the terminal zone instead of the variable $c(w)$. The function $c(w)$ includes the variation in

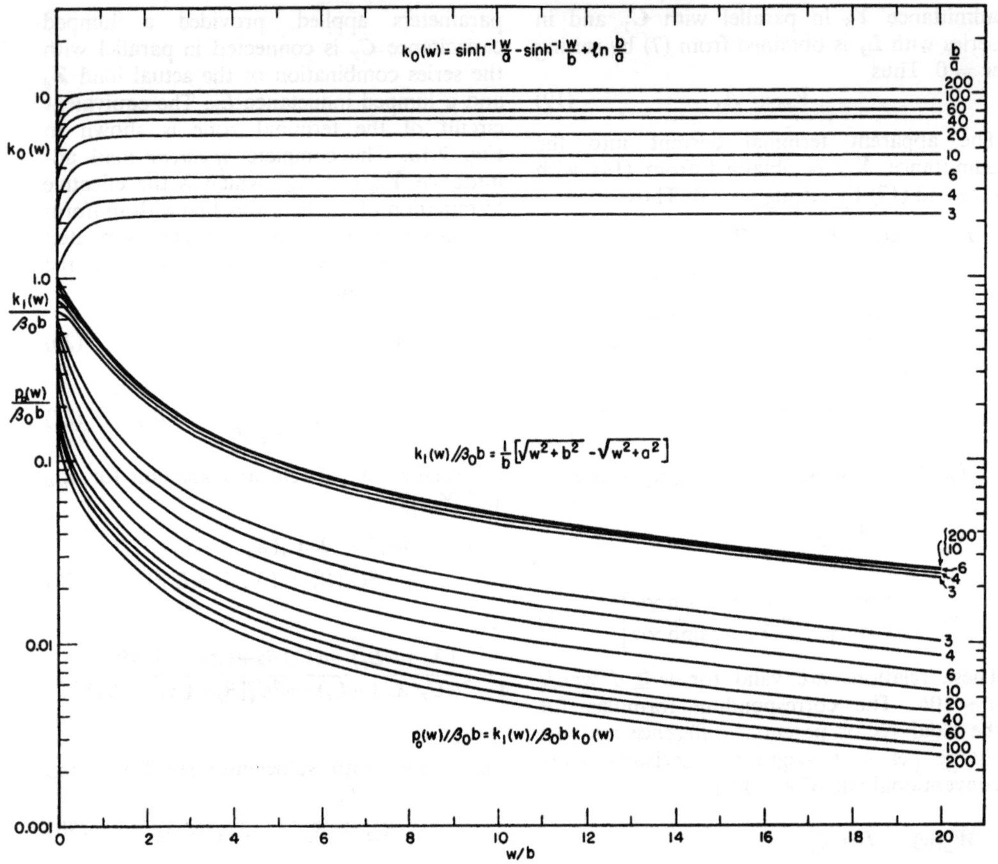

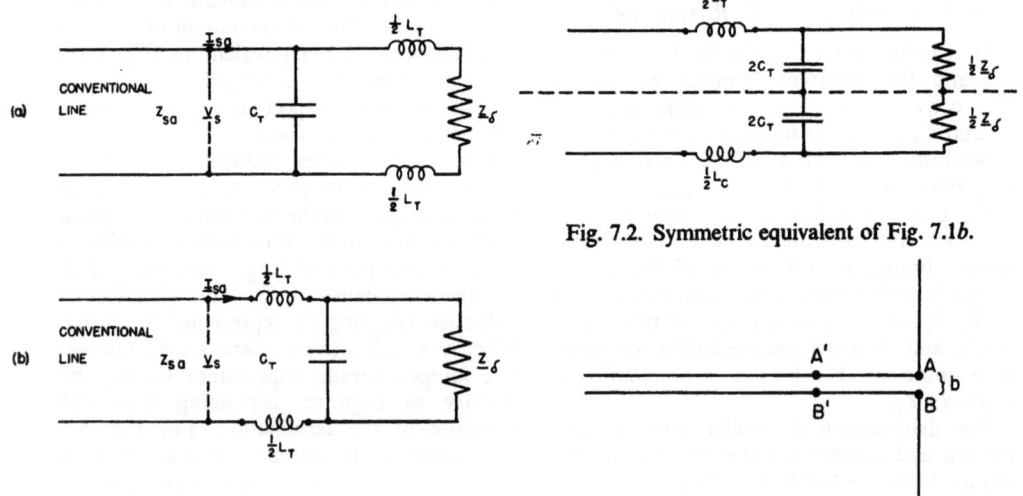

Fig. 6.3. The functions $k_0(w)$, $k_1(w)/\beta_0 b$, and $p_0(w)/\beta_0 b$.

Fig. 7.1. Equivalent circuits for terminal zone.

Fig. 7.2. Symmetric equivalent of Fig. 7.1b.

Fig. 8.1. Symmetric center-driven antenna.

capacitance per unit length near the end of the line as contained in $c_0(w)$ and, in addition, the effect of coupling between the line and the termination.

To summarize the approximate solution of the generalized transmission-line equations, the following steps may be listed:

1. Replace the terminal zone of the actual line by an approximately equivalent lumped-constant network consisting of an inductance L_T in series with the actual load impedance Z_δ at $w = 0$ and a capacitance C_T in parallel with it as shown in Fig. 7.1a. The arrangement shown in Fig. 7.1b may be used if more convenient.

2. Use the resultant impedance Z_{sa} (composed of Z_δ in series with L_T and in parallel with C_T) as the *apparent* terminal impedance for a fictitious line that has no terminal zone.

3. Apply conventional line theory with Z_{sa} as the terminal impedance at $w = 0$.

4. Use (26) or (27) to obtain Y_δ from Z_{sa} or Z_{sa} from Y_δ. Note that theoretical analyses of impedance usually determine $Z_\delta = 1/Y_\delta$, whereas all experimental methods based on line theory determine $Z_{sa} = 1/Y_{sa}$.

It is interesting to interpret the approximate solutions obtained using a lumped, terminal-zone network in terms of the general equations (1) and (2) in the special case with $\psi = \pi/2$, $a_1(w) = 1$. The most significant observation is that account has been taken of the variable propagation constant $\gamma(w)$ in the terminal zone by a shift in the entire voltage-distribution curve (corresponding to the added effect of combining L_T and C_T as lumped elements with the load) while retaining its essentially sinusoidal nature. Actually, the distribution curve is distorted from the sinusoidal in the terminal zone in a manner to provide the same voltage across the load, when C_T and L_T are not present. Since the vector potential difference is given quite accurately by (6.13c) in the form $W_x(w) \doteq (1/j\omega)[\partial V(w)/\partial w]$, the behavior of $W_x(w)$ is readily visualized from that of $V(w)$.

The behavior of the current in the terminal zone is more intricate. Its principal term varies as $\partial V(w)/\partial w$ divided by $z(w) \doteq j\omega l_0^e(w)$. Since $l_0^e(w)$ is given by l_0^e outside the terminal zone and by $\frac{1}{2} l_0^e$ at the load, it is clear that this part of the current increases to double the value it would have if linearly related to $W_x(w)$. The second term of the current in (2) has $V(w)$ as a factor, but the product $p(w)\varphi_1(w)$ that increases rapidly from zero outside the terminal zone to a rather large value at the load is primarily significant in determining the contribution to the current in the terminal zone. Usually, this term is important only when the amplitude of the current in the terminal zone is small compared with that in other, relatively near parts of the line. This is true whenever the terminal impedance is large compared with the characteristic impedance of the line. The simplest case is the open end. For this, $\varphi_1(w) = 0$ and the current actually vanishes at $w = 0$. Neglecting line losses, the approximate solution includes

$$V(w) \doteq V_{sa} \cos(\beta_0 w + \psi_T), \quad (31)$$
$$\partial V(w)/\partial w \doteq -V_{sa}\beta_0 \sin(\beta_0 w + \psi_T), \quad (32)$$

where ψ_T is determined by the *positive* value of C_T for the open end. The approximate solution for the apparent terminal impedance assumes that the current in the terminal zone is sinusoidal and proportional to $\partial V(w)/\partial w$; it does not vanish at $w = 0$. In so far as the current and voltage in the line outside the terminal zone are concerned, this solution is entirely adequate, since C_T has been chosen to make them approximately correct there. In effect, the correct sinusoidal distributions outside the terminal zones have merely been extended through the terminal zones and a termination provided to fit.

A picture of the actual distribution of current in the terminal zone is obtained when both terms in (2) are considered. Evidently, the second term contributes precisely that part required to reduce the sinusoidal current of the first term to zero at $w = 0$ as is actually the case. Vanishing current at $w = 0$ gives

$$p(0) = \tan \psi_T. \quad (33)$$

When the termination is a high but not infinite impedance, the second term in (2) reduces the current to the actual value entering Y_δ at $w = 0$ and eliminates the need for C_T.

The term with $\partial V(w)/\partial w$ in (2) is proportional to the vector potential $W_x(w)$ which, in turn, is determined largely but not entirely by $I_x(w)$. If the current $I_x(w)$ is small at a point w in the terminal zone, $W_x(w)$ and $\partial V(w)/\partial w$ at this point are larger compared with $I_x(w)$ than at points on the line where the current is great. Thus, the second term in (2) compensates for the fact that $I_x(w) \doteq W_x(w)/l_0^e(w)$ gives too large a value where $I_x(w)$ is small.

The difference between the current given by (2) and that given by $I_x(w) \doteq W_x(w)/l_0^e$ is essentially the current into C_T. This includes the effect of the second term in (2) as well as of the variable $l_0^e(w)$.

The distribution of charge is readily visualized from $I_x(w)$ since $q(w) = (-j/\omega)[\partial I_x(w)/\partial w]$. Clearly, $q(w)$ is not proportional to $V(w)$ in the terminal zone.

8. Antenna Terminating Two-Wire Line

The rigorous one-dimensional analysis of the circuit consisting of a completely balanced two-wire line terminated in a symmetric center-driven antenna (Figs. 5.2 and 8.1*) requires the solution of simultaneous integro-differential equations for the distributions of current both in the antenna and in the two-wire line. Since the antenna and the transmission line are significantly coupled over distances along the antenna and along the line comparable with small multiples of the finite separation b of the conductors of the line, the definition of an input impedance Z_{BA} (Fig. 8.1) that is characteristic of the *antenna alone* and independent of the feeding line is not possible. An impedance in the sense of conventional transmission-line theory can be defined only for a sufficiently long section of line terminated in the antenna. Thus, $Z_{B'A'}$ can be defined if the distance $A'A$ in Fig. 8.1 is large compared with the spacing b. In the language of the preceding section, a true transmission-line impedance can be defined only for points outside the terminal zone. Note, however, that the impedance $Z_{B'A'}$ of the section of terminated line to the right of $B'A'$ is not independent of the presence of the line to the left of $B'A'$. The assumption is implicit in the definition of every impedance on a transmission line that the uniform line continues in *both directions* from the points of definition $B'A'$.

In order to obtain an *approximate* expression for the impedance of an antenna terminating a two-wire line, the method described in the preceding section may be used. In effect, this considers the characteristics of the section of line to the right of $B'A'$ (Fig. 8.1) in three parts, which, although interrelated, do have a measure of independence. These parts concern themselves, respectively, with characteristics that depend primarily upon (1) the transmission line, (2) the antenna, and (3) the coupling between them.

Since all significant coupling is confined to distances from the junction of antenna and line that are of order of magnitude b, the condition $\beta_0 b \ll 1$ assures essentially near-zone coupling, which, in general, may be inductive, capacitive, or both. In the present problem of the symmetrical center-driven antenna, conditions of symmetry and perpendicularity completely eliminate *inductive* coupling between the currents in the line and in the antenna. The principal approximation in the actual procedure is to replace the equivalent of a *distributed* capacitive coupling between line and antenna by a *concentrated* coupling in the form of a lumped capacitance in parallel with the antenna at its junction with the line. In the preceding section, the effect of capacitive coupling between antenna and line is eliminated and, in addition, the capacitance per unit length of the line is corrected to take account of the fact that the line ends at the junction instead of continuing indefinitely, as implied in conventional line theory. These are both accomplished by setting $\varphi_1(w) \equiv V_L(w)/V(w)$ equal to unity along the entire line up to the junction point and introducing the lumped capacitance C_T in parallel with the antenna across the junction of antenna and line. With this substitution for the distributed coupling of an approximately equivalent effect localized at the junction in the form of a lumped capacitance, the transmission-line problem (1) and the antenna problem (2) are made effectively independent of each other except for the common potential difference across the terminals at the junction points and the requirement of continuity for the current at these terminals.

In the preceding section, the quasi-independent transmission-line problem (1) is analyzed approximately. It involves corrections to take account of the ending of the transmission line and to permit the use of conventional formulas. These corrections are contained in the lumped elements L_T and C_T. This last also includes the effect of coupling. It is the purpose of this section to evaluate C_T for the center-driven antenna as end load for a two-wire line.

Beginning in Sec. 11, the antenna problem (2) is analyzed by determining the input admittance Y_δ of the antenna. This is defined as the ratio of current I_δ entering and leaving the terminals of the antenna from and to the line at $w = 0$, $z = \pm\delta = \pm\frac{1}{2}b$, to the scalar potential difference V_δ maintained across them, $Y_\delta = I_\delta/V_\delta$ (Fig. 8.2).

The determination of the lumped capacitance C_T defined by

$$C_T = \int_a^d [c_0(w)\varphi_1(w) - c_0]\,dw \qquad (1)$$

---

* See p. 48.

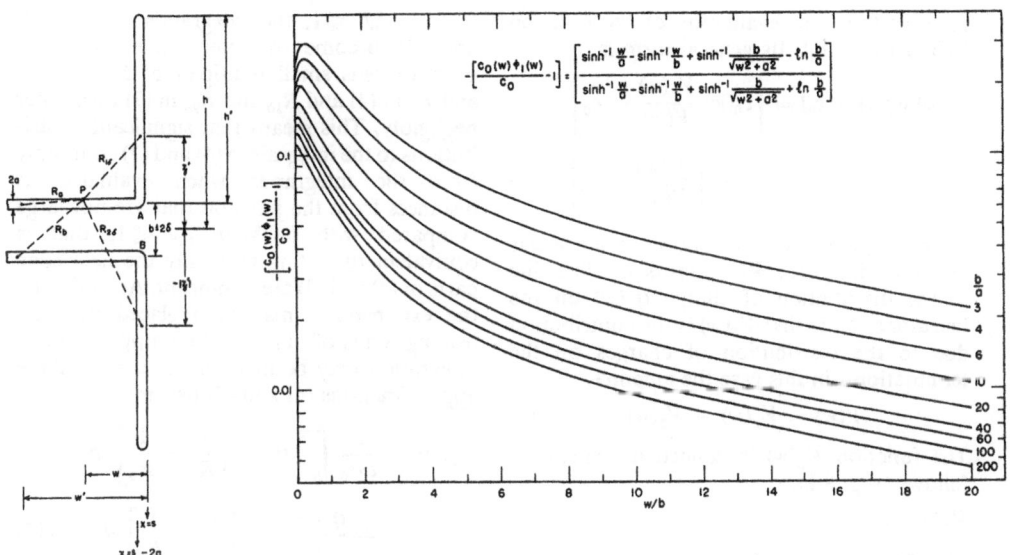

Fig. 8.2. Notation for center-driven antenna.

Fig. 8.3. Graphs of $-[c_0(w)\varphi_1(w) - c_0]/c_0$ as a function of w/b with b/a as parameter.

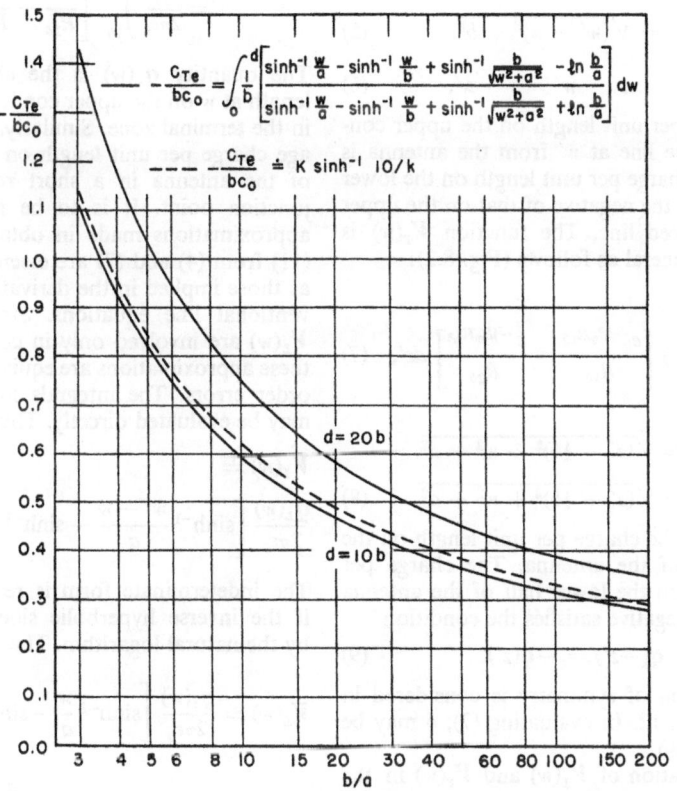

Fig. 8.4. Graphs of $-C_{Te}/bc_0$ as a function of b/a.

involves the prior evaluation of the function $[c_0(w)\varphi_1(w) - c_0]$. Its general form is

$$[c_0(w)\varphi_1(w) - c_0] = \left[c_0(w)\frac{V_L(w)}{V(w)} - c_0\right]$$

$$= \left[\frac{q_L(w)}{V(w)} - c_0\right], \quad (2)$$

where $V_L(w)$ is the contribution to the scalar potential difference $V(w)$ across the line due to the distribution of charge $q_L(w)$ on the line alone. Note that $V_T(w)$ is the contribution due to the distribution of charges on the termination—in this case the antenna:

$$V(w) = V_L(w) + V_T(w). \quad (3)$$

The function $V_L(w)$ is defined in general as follows (Fig. 8.2):

$$V_L(w) =$$

$$\frac{1}{2\pi\varepsilon_0}\int_0^s q(w')\left[\frac{e^{-j\beta_0 R_a}}{R_a} - \frac{e^{-j\beta_0 R_b}}{R_b}\right]dw', \quad (4)$$

where

$$R_a = \sqrt{(w'-w)^2 + a^2},$$

$$R_b = \sqrt{(w'-w)^2 + b^2}, \quad (5)$$

and

$$w \equiv s - x, \quad w' \equiv s - x'. \quad (6)$$

The charge per unit length on the upper conductor of the line at w' from the antenna is $q(w')$. The charge per unit length on the lower conductor is the negative of that on the upper for a balanced line. The function $V_T(w)$ is defined in general as follows (Fig. 8.2):

$$V_T(w) =$$

$$\frac{1}{2\pi\varepsilon_0}\int_\delta^h q(z')\left[\frac{e^{-j\beta_0 R_{1\delta}}}{R_{1\delta}} - \frac{e^{-j\beta_0 R_{2\delta}}}{R_{2\delta}}\right]dz', \quad (7)$$

where

$$R_{1\delta} \doteq \sqrt{(z' - \tfrac{1}{2}b)^2 + w^2 + a^2},$$

$$R_{2\delta} \doteq \sqrt{(z' + \tfrac{1}{2}b)^2 + w^2 + a^2}, \quad (8)$$

and $q(z')$ is the charge per unit length on the upper half of the antenna. The charge per unit length on the lower half of the antenna where z' is negative satisfies the condition

$$q(-z') = -q(z'). \quad (9)$$

The condition of symmetry is considered in detail in Sec. 12. In evaluating (7), δ may be replaced by $\tfrac{1}{2}b$.

The evaluation of $V_L(w)$ and $V_T(w)$ in the general formulas given in (4) and (7) is intricate. Fortunately, subject to the condition (6.22a), $\beta_0 b \ll 1$, the integrands in both (4) and (7) become extremely small when w' or z' exceeds small multiples of b, since R_a and R_b in (4) and $R_{1\delta}$ and $R_{2\delta}$ in (7) then differ negligibly. This means that significant contributions to the integrals in (4) and (7) come only from the integrands when evaluated at distances from the junction that are not large compared with b. With $\beta_0 b \ll 1$, that is equivalent to stating that only the near-zone parts of (4) and (7) are significant. Accordingly, the exponentials may be replaced by their leading terms of unity, and the upper limit of integration may be made infinite. With these approximations (4) and (7) become

$$V_L(w) \doteq \frac{1}{2\pi\varepsilon_0}\int_0^\infty q(w')\left[\frac{1}{R_a} - \frac{1}{R_b}\right]dw'$$

$$\doteq \frac{q_L(w)}{2\pi\varepsilon_0}\int_0^\infty \left[\frac{1}{R_a} - \frac{1}{R_b}\right]dw', \quad (10)$$

$$V_T(w) \doteq \frac{1}{2\pi\varepsilon_0}\int_{\tfrac{1}{2}b}^\infty q(z')\left[\frac{1}{R_{1\delta}} - \frac{1}{R_{2\delta}}\right]dz'$$

$$\doteq \frac{\bar{q}_T}{2\pi\varepsilon_0}\int_{\tfrac{1}{2}b}^\infty \left[\frac{1}{R_{1\delta}} - \frac{1}{R_{2\delta}}\right]dz'. \quad (11)$$

The quantity $q_L(w)$ is the charge per unit length at w on the upper conductor of the line in the terminal zone. Similarly, $\bar{q}_T$ is the average charge per unit length on the upper half of the antenna in a short region near the junction point. It is to be noted that the approximations made in obtaining (10) and (11) from (4) and (7) are essentially the same as those implied in the derivation of the conventional line equations. Since $V_T(w)$ and $V_L(w)$ are involved only in correction terms, these approximations are equivalent to higher-order errors. The integrals in (10) and (11) may be evaluated directly. Thus,

$$V_L(w) \doteq$$

$$\frac{q_L(w)}{2\pi\varepsilon_0}\left[\sinh^{-1}\frac{w'-w}{a} - \sinh^{-1}\frac{w'-w}{b}\right]_0^\infty. \quad (12)$$

The indeterminate form is readily evaluated if the inverse hyperbolic sines are replaced by the natural logarithm. The result is

$$V_L(w) \doteq \frac{q_L(w)}{2\pi\varepsilon_0}\left[\sinh^{-1}\frac{w}{a} - \sinh^{-1}\frac{w}{b} + \ln\frac{b}{a}\right]. \quad (13)$$

$$V_L(w) \doteq \frac{\bar{q}_L}{c_0(w)}, \quad (14)$$

where $\bar{q}_L$ is an average value of $q_L(w)$ in the terminal zone. Similarly,

$$V_T(w) \doteq \frac{\bar{q}_T}{2\pi\varepsilon_0} \left[\sinh^{-1}\frac{z' - \frac{1}{2}b}{\sqrt{w^2 + a^2}} - \sinh^{-1}\frac{z' + \frac{1}{2}b}{\sqrt{w^2 + a^2}} \right]_{\frac{1}{2}b}^{\infty}, \quad (15)$$

or

$$V_T(w) \doteq \frac{\bar{q}_T}{2\pi\varepsilon_0} \sinh^{-1}\frac{b}{\sqrt{w^2 + a^2}}. \quad (16)$$

In order to evaluate $[c_0(w)\varphi_1(w) - c_0]$ it is necessary to substitute (14) and (16) in (2). For convenience, let the charge-ratio factor k_q be defined as follows:

$$k_q \equiv \bar{q}_T/\bar{q}_L. \quad (17)$$

Thus,

$$[c_0(w)\varphi_1(w) - c_0]/c_0 =$$

$$-\left[\frac{\sinh^{-1}\frac{w}{a} - \sinh^{-1}\frac{w}{b} + k_q \sinh^{-1}\frac{b}{\sqrt{w^2+a^2}} - \ln\frac{b}{a}}{\sinh^{-1}\frac{w}{a} - \sinh^{-1}\frac{w}{b} + k_q \sinh^{-1}\frac{b}{\sqrt{w^2+a^2}} + \ln\frac{b}{a}}\right]$$
(18)

This may be written in logarithmic form using the relation between the inverse hyperbolic sine and the natural logarithm, namely,

$$\sinh^{-1}\frac{w}{a} = \ln\left[\frac{w + \sqrt{w^2 + a^2}}{a}\right]. \quad (19)$$

The evaluation of (18) is possible only if k_q is known. In the following it is assumed to be real, $k_q \doteq k_q$; then $\varphi_1(w) \doteq \varphi_1(w)$.

Although the order of magnitude of k_q is readily established, the determination of an approximate numerical value involves certain difficulties, since it really presupposes at least a fair estimate of the distributions of charge on the conductors of the antenna and the line near their junctions, where the simple equations do not apply. The simple and obvious approximation, which is certainly satisfactory in all cases subject to $\beta_0 b \ll 1$, is to assume that the average charge per unit length on the line is approximately the same in line and antenna over the relatively short distances from the junction that yield significant contributions to $V_L(w)$ and $V_T(w)$. This is equivalent to the approximation involved in conventional line theory and gives directly

$$k_q \doteq 1. \quad (20)$$

A better approximation is to use the distribution of charge obtained in a later section for the isolated cylindrical antenna together with the conventional transmission-line distribution on the line. With these values it is found that when the concentration of charge near the junction is a maximum and the effect of capacitance in parallel is large, the approximation $\bar{q}_L \doteq \bar{q}_T$ is a good one. When the maximum charge is not at or near the junction, values of k_q differing somewhat from unity are obtained. In order to permit the use of other values of k_q than unity, the symbol k_q is carried along explicitly in the formulas. In the following discussion numerical calculations are made only for $k_q = 1$.

The function $-[c_0(w)\varphi_1(w) - c_0]/c_0$ as defined in (18) is shown plotted in Fig. 8.3 with b/a as parameter and w/b as independent variable, in the range from $w/b = 0$ to $w/b = 20$, with $k_q = 1$. The area under each of these curves from $w = 0$ to $w = d$ multiplied by $-c_0 b = -b\pi\varepsilon_0/(\ln b/a)$ is the desired lumped capacitance C_{Te} for each value of b/a and an appropriate choice of d when plotted against a linear scale. Note that $\varphi_1(w) \doteq \varphi_1(w)$.

$$C_{Te} = \int_0^d [c_0(w)\varphi_1(w) - c_0]\, dw. \quad (21)$$

The choice of the length d of the terminal zone is complicated by the coexistence of both a transmission-line end effect and of coupling between the antenna and the line. Whereas the former becomes negligible for $d \geq 5b$, as can be seen from the manner in which $c_0(w)$ approaches c_0 in Fig. 6.2, this is not true of the effect of coupling. Note that in (18) the bracket on the right is well approximated by $b/[2w \ln(b/a)]$ when $w \geq 5b$. If this quantity is substituted in (21), C_{Te} is found to be given by a logarithm that increases indefinitely with d. As suggested by T. T. Wu, this is avoided by retaining the exponentials in (7) in the form, $\exp(-j\beta_0 R_{1b}) \doteq \exp(-j\beta_0 R_{2b}) \doteq \exp(-j\beta_0 w)$ when $w \geq 5b$. The real part, $\cos \beta_0 w$, may be retained as a factor in (18) when w equals or exceeds $5b$, so that (18) becomes

$$\left[\frac{\beta_0 w}{2 \ln(b/a)}\right]\left[\frac{\cos \beta_0 w}{\beta_0 w}\right].$$

(The imaginary part, $\sin \beta_0 w$, contributes to a mutual resistance term, not to the capacitance.) If this expression is substituted in (21), the contribution to the integral by the range between $\beta_0 w = 0.6165$ (or $w = 0.095\lambda_0$) and infinity is zero. Therefore, it is necessary merely to integrate over a distance $d = 0.095\lambda_0 \doteq 0.1\lambda_0$. Thus, when $b = 0.01\lambda_0$, $d \doteq 10b$, when $b = 0.005\lambda_0$, $d \doteq 20b$. Values of C_{Te} obtained from (21) by numerical methods are in Fig. 8.4 as a function of b/a for $d = 10b$ and $d = 20b$.

For the rapid evaluation of C_T, a closed formula is desirable. Since for many purposes only the correct order of magnitude of C_T is required, it is proposed to replace $[c_0(w)\varphi_1(w) - c_0]/c_0$ with $\varphi_1(w)$ real by a simple approximate formula that is directly integrable. This can be done very readily. The shapes of the curves obtained in Fig. 8.3 suggest a function of the form

$$[c_0(w)\varphi_1(w) - c_0]/c_0 \doteq -\frac{K}{\sqrt{(w/b)^2 + k^2}} \quad (22)$$

where K and k are arbitrary constants to be assigned values that will make (22) a good approximation.

If (22) is substituted in (1), a simple expression for C_{Te} is obtained at once. Thus,

$$C_{Te} \doteq -c_0 \int_0^d \frac{Kb}{\sqrt{w^2 + k^2 b^2}}\, dw$$

$$= -c_0 bK \sinh^{-1} \frac{d}{Kb}. \quad (23)$$

With $d = 10b$,

$$C_{Te} \doteq -c_0 bK \sinh^{-1}(10/k). \quad (24)$$

In order to determine K and k, the function $-kb/\sqrt{w^2 + k^2 b^2}$ must be fitted to $[c_0(w)\varphi_1(w) - c_0]/c_0$ in the range of w from zero to $d = 10b$. This may be done by requiring (24) to be satisfied exactly at two suitably chosen points in the range. Reference to Fig. 8.3 suggests that since the principal contributions to the integral are for values of w less than $5b$, reasonable points for exact matching are $w = a$ and $w = 3b$.

Let the value of the bracket in (18) be denoted by H_1 when $w = a$ and by H_2 when $w = 3b$. With $k_q \doteq k_q$ real,

$$H_1 = \left[\frac{A - \ln(b/a)}{A + \ln(b/a)}\right], \quad H_2 = \left[\frac{B - \ln(b/a)}{B + \ln(b/a)}\right], \quad (25a)$$

where

$$A \equiv \sinh^{-1} 1 - \sinh^{-1}\frac{a}{b} + k_q \sinh^{-1}\frac{b}{a\sqrt{2}}, \quad (25b)$$

$$B \equiv \sinh^{-1}\frac{3b}{a} - \sinh^{-1} 3$$

$$+ k_q \sinh^{-1}\frac{b}{\sqrt{9b^2 + a^2}}. \quad (25c)$$

For $k_q = 1$ these reduce to

$$H_1 = \frac{1.228 - a/b}{2\ln b/a + 1.228 - a/b}, \quad (26a)$$

$$H_2 = \frac{0.327}{2\ln b/a + 0.327}. \quad (26b)$$

Equating $-H_1$ and $-H_2$ to the right-hand member of (22) with $w = a$ and $w = 3b$, respectively, gives

$$K = H_1 \sqrt{(a/b)^2 + k^2}, \quad (27)$$

$$K = H_2 \sqrt{9 + k^2}. \quad (28)$$

These two equations are readily solved for k. The result is

$$k = \sqrt{\frac{(3H_2/H_1)^2 - (a/b)^2}{1 - (H_2/H_1)^2}}. \quad (29a)$$

Substitution in (27) and neglect of $(a/b)^2$ compared with 9 gives

$$K = 3H_2/\sqrt{1 - (H_2/H_1)^2}. \quad (29b)$$

For given values of b/a, H_1 and H_2 and from these k and K are readily evaluated for use in (23). Curves showing the approximate function $-C_{Te}/bc_0$ as evaluated from (24) are given in Fig. 8.4 together with the same function determined from (18) by numerical integration. The agreement is seen to be fair, with the approximate curve falling between the curves for terminal zones of length $d = 10b$ and $d = 20b$. Since the proposed representation of a distributed effect by a lumped capacitance is in any case far from exact, the values of C_{Te} taken from the simple integrated formula (24) are satisfactory.

If $\ln(b/a)$ is sufficiently great so that it is the leading term in the denominator of (18), (21) may be integrated into

$$C_{Te} \doteq -3c_0 b/[2\ln(b/a)] \quad (30)$$

As a specific numerical example, consider a line for which $b/a = 7\cdot 4$ or $\ln b/a = 2$. Then, with $k_q = 1$,

$$H_1 = 0\cdot 215, \qquad H_2 = 0\cdot 0755, \qquad (31a)$$

$$k = 1\cdot 11, \qquad K = 0\cdot 24. \qquad (31b)$$

Substitution of these values in (24) gives

$$C_{Te} \doteq -0\cdot 70 c_0 b. \qquad (32)$$

Evidently the order of magnitude of C_{Te} is that of a section of uniform line of length equal to the line spacing b. Even though $\ln (b/a)$ is only 2, (30) gives $C_{Te} = -0\cdot 75 c_0 b$, which is a good approximation of (32).

Throughout the above discussion the condition $a^2 \ll b^2$ was assumed. This condition may be removed and the final results generalized approximately to more closely spaced lines by substituting

$$b_e = \tfrac{1}{2} b [1 + \sqrt{1 - (2a/b)^2}] \qquad (33)$$

for b.

Summarizing the analysis of the two-wire line terminated in a center-driven antenna, the following may be said. The apparent impedance Z_{sa} terminating the two-wire line—which is the actual impedance looking toward the antenna from a point on the line (assumed lossless) one-half-wavelength from the antenna—may be determined using the circuit of Fig. 7.1a and formulas (7.28, 29) if the ratio of scalar potential difference to current at the terminals of the antenna, namely, $Z_\delta = V_\delta/I_\delta$, is known. In Fig. 7.1a or Eqs. (7.28, 29),

$$L_{Te} = -\frac{(b-a)}{2\pi v_0}, \qquad (34)$$

$$C_{Te} = \int_0^d [c_0(w)\varphi_1(w) - c_0]\,dw$$

$$\doteq -Kc_0 b \sinh^{-1}(d/kb), \qquad (35)$$

where k and K are given in (29a) and (29b). The subscript e on L_T and C_T identifies the values for an end-loaded line with antennas in the plane of the line; L_{Te} in (34) is from (7.10e).

The remaining problem is to evaluate $Z_\delta = V_\delta/I_\delta$ by treating the antenna as if isolated with a scalar potential difference V_δ maintained across rings around the adjacent ends and currents I_δ entering and leaving the symmetrical halves. This analysis is begun in Sec. 11. In Sec. 9 the theory developed in this section is modified to apply to arrangements that are experimentally more convenient but that do not greatly change the relative positions of, and hence the coupling between, the transmission line and the antenna. In Sec. 10 other methods of center-driving an antenna from a two-wire line are considered, including arrangements in which the line is not coupled to the antenna directly.

9. Antenna as Mid-Point Load for Symmetrically Driven Line; Antenna as End Load with Stub Support

The circuit of Fig. 8.1, in which a center-driven antenna terminates a balanced two-wire line, has several practical shortcomings if interest lies in the accurate measurement of the impedance of the antenna. One of the difficulties in this apparently simple circuit is to support the antenna and line at or near their junction. A common method is to use a piece of dielectric material as shown in Fig. 9.1. However, a piece of dielectric at this critical location necessarily introduces added complications* which, in the simplest case of a sufficiently thin piece, involve the determination of an equivalent added lumped capacitance in parallel with the antenna. The theoretical or experimental determination of the *actual* added capacitance due to a dielectric support and not merely the *apparent* added capacitance is sufficiently complex to make the accurate evaluation of the contribution to the load by the dielectric support difficult. In practice, to be sure, it is usually only the *apparent combined* impedance terminating the line that is of interest, and this is readily measured directly in each case. On the other hand, a complete, even though perhaps quantitatively only approximate, understanding of the several factors that contribute to the apparent impedance of a termination requires a separation of the effect of a dielectric support from those effects characteristic of the configuration of conductors.

An interesting and very flexible structure which provides both a means of support for the line at its junction with the antenna and complete symmetry is the circuit shown in Fig. 9.2. Instead of being connected to the line as an end load, as in Fig. 8.1, the antenna is attached to the center of a symmetrical line

---

* In the laboratory these complications may be avoided if materials like Styrofoam or Polyfoam are used, but these materials lack the strength required in most practical installations.

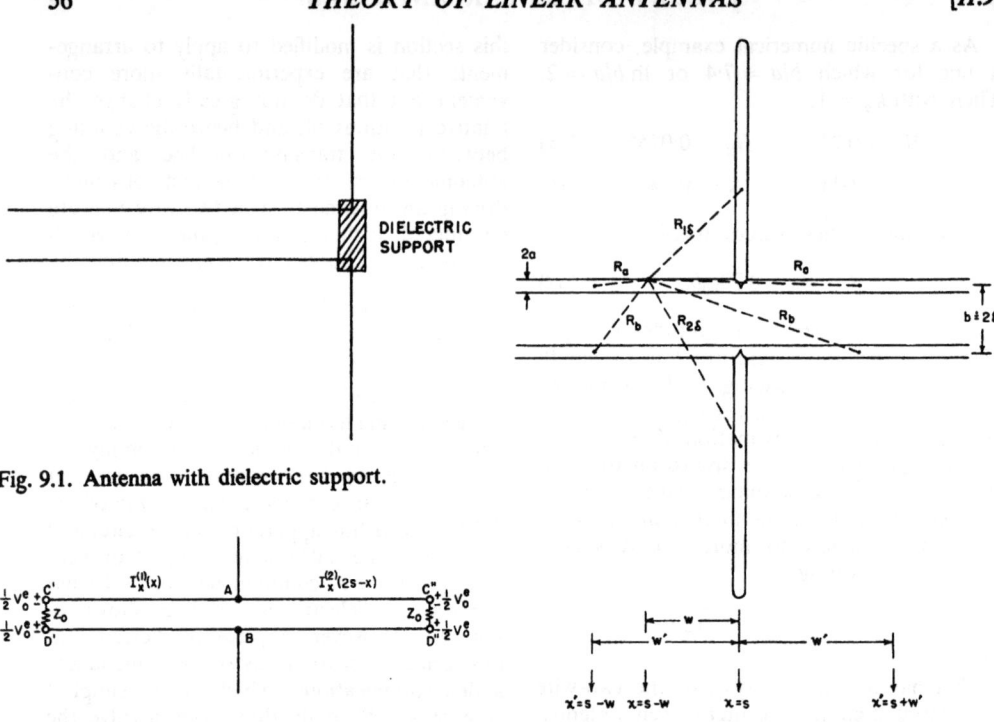

Fig. 9.1. Antenna with dielectric support.

Fig. 9.2. Antenna as center load for symmetric line.

Fig. 9.3. Notation for antenna as center load.

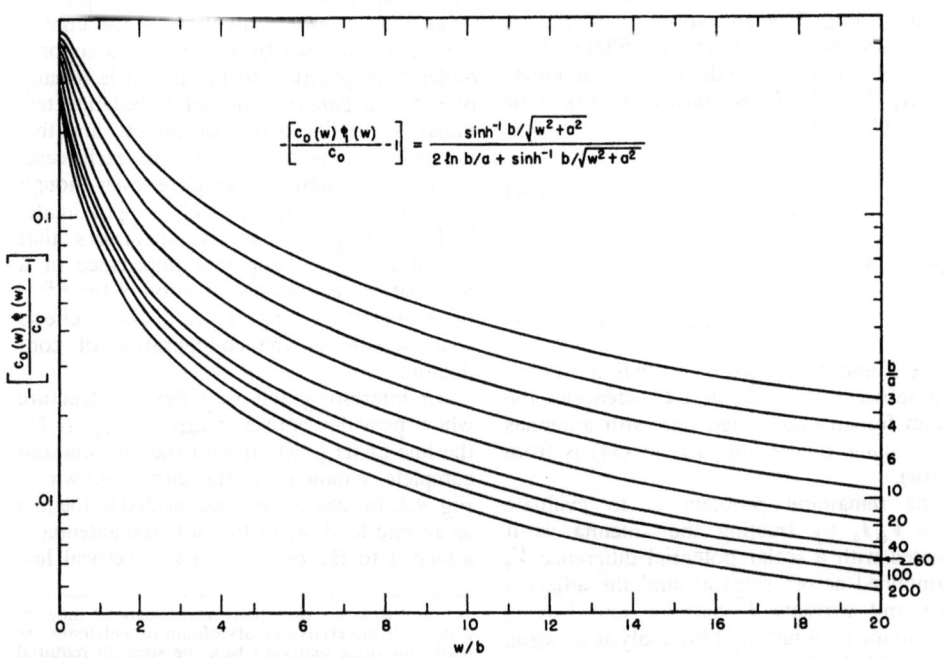

Fig. 9.4. Graphs of $-[c_0(w)\varphi_1(w) - c_0]/c_0$ as a function of w/b with b/a as parameter.

that is driven from *both ends by identical generators*. In this circuit the currents in the two lines are equal in magnitude and opposite in direction at corresponding points measured from the antenna. On the other hand, the charges per unit length at these points are the same in magnitude and sign. Specifically,

$$I_x^{(1)}(x) = -I_x^{(2)}(2s - x), \quad (1a)$$

$$q^{(1)}(x) = q^{(2)}(2s - x), \quad (1b)$$

where $I_x^{(1)}(x)$ is the current in the positive x-direction in the left-hand line, numbered (1), at distance x from the origin, and $I_x^{(2)}(2s - x)$ is the current in the x-direction in the right-hand line, numbered (2), at distance $2s - x$ from the origin. The corresponding charges per unit length are $q^{(1)}(x)$ and $q^{(2)}(2s - x)$. With distances measured from the junction point $x = s$, expressed in terms of $w = s - x$, (1a, b) are equivalent to

$$I_x^{(1)}(w) = -I_x^{(2)}(-w), \quad (2a)$$

$$q^{(1)}(w) = q^{(2)}(-w). \quad (2b)$$

The currents at the junction point $w = 0$ combine into the antenna current $I_z(\delta)$. Thus,

$$I_z(\delta) = I_x^{(1)}(w = 0) - I_x^{(2)}(w = 0)$$
$$= 2I_x^{(1)}(w = 0). \quad (3a)$$

On the other hand, the charge per unit length on the antenna very near the junction point is equal to the charges per unit length on each of the two line conductors very near the junction:

$$q(z \doteq \delta) \doteq q^{(1)}(w \doteq 0) = q^{(2)}(w \doteq 0). \quad (3b)$$

If the impedance Z_δ of the antenna is defined formally as the ratio of the scalar potential difference V_δ at the junction point to the current $I_z(\delta)$,

$$Z_\delta = \frac{V_\delta}{I_z(\delta)} = \frac{\Phi(\delta) - \Phi(-\delta)}{I_z(\delta)}, \quad (4)$$

it follows with (3a) that

$$Z_\delta = \frac{V_\delta}{2I_x^{(1)}(w = 0)}. \quad (5)$$

Hence, if complete symmetry is maintained, each line may be treated as though terminated in an impedance

$$2Z_\delta = \frac{V_\delta}{I_x^{(1)}(w = 0)} = \frac{V_\delta}{-I_x^{(2)}(w = 0)}. \quad (6)$$

Therefore, the entire analysis of Secs. 6–8 applies, except that the lumped elements L_T and C_T must be reevaluated since the two parts of the line are coupled not only to the antenna but also to each other.

In order to determine L_T for the symmetrically driven line it is first necessary to evaluate $l_0^e(w)$, which is given by

$$l_0^e(w) \doteq \frac{W_x(w)}{I_x(w)} \doteq \frac{1}{2\pi v_0} \left[\int_0^s \left(\frac{1}{R_a} - \frac{1}{R_b} \right) dw' - \int_{-s}^0 \left(\frac{1}{R_a} - \frac{1}{R_b} \right) dw' \right], \quad (7)$$

where

$$R_a = \sqrt{(w' - w)^2 + a^2},$$
$$R_b = \sqrt{(w' - w)^2 + b^2},$$

as shown in Fig. 9.3. The negative sign in front of the second integral in (7) is a consequence of the fact that the current in the right-hand line is opposite to that in the left-hand line. Since the current actually leaves the conductors of the transmission line to enter the antenna between $w = a$ and $w = -a$, the limit 0 in the integrals in (7) is an approximation. Probably it would be more accurate to use average distances between $w = 0$ and $w = \pm a$, but the error introduced by using 0 instead of a distance less than a is negligible. Since (7) is readily shown to reduce, subject to the condition $s^2 \gg b^2$, to the form

$$l_0^e(w) \doteq \frac{1}{\pi v_0} \left(\sinh^{-1} \frac{w}{a} - \sinh^{-1} \frac{w}{b} \right), \quad (8)$$

it follows from (7.10a) that L_T for the symmetrically driven, center-loaded line is just double the value for the end-loaded line. Subject to the same approximation (7.10d) on the size of the terminal zone, the result is

$$L_{TC} = -\left(\frac{b - a}{\pi v_0} \right). \quad (9)$$

Note that L_{TC} in (9) is the lumped inductance to be connected in series with *each* half of the symmetrical line with uniform constants. The subscript C identifies the center-loaded line.

It remains to evaluate C_{TC}. This is done as in Sec. 8, but with the following modified expressions for $V_L(w)$ and $V_T(w)$, which must now include the effect of the additional feeding line:

$$V_L(w) \doteq \frac{q_L}{2\pi\epsilon_0} \int_{-\infty}^{\infty} \left(\frac{1}{R_a} - \frac{1}{R_b} \right) dw', \quad (10)$$

$$V_T(w) \doteq \frac{q_T}{2\pi\epsilon_0} \int_0^{\infty} \left(\frac{1}{R_{1\delta}} - \frac{1}{R_{2\delta}} \right) dz'. \quad (11)$$

These integrals are

$$V_L(w) \doteq \frac{\bar{q}_L}{\pi\varepsilon_0} \ln\frac{b}{a}, \qquad (12)$$

$$V_T(w) \doteq \frac{\bar{q}_T}{2\pi\varepsilon_0} \sinh^{-1}\frac{b}{\sqrt{w^2+a^2}}. \qquad (13)$$

As before, let

$$k_q \equiv \bar{q}_T/\bar{q}_L, \qquad (14)$$

where $\bar{q}_T$ is the average charge per unit length on the antenna near the junction with the line, and $\bar{q}_L$ is the average charge per unit length on one conductor of each of the two lines near the junction. As for the end-loaded antenna, numerical evaluation is limited to

$$k_q \doteq 1. \qquad (15)$$

Substitution of (12) and (13) in the fundamental formula (8.2),

$$c_0(w)\varphi_1(w) - c_0 \doteq \frac{q_L(w)}{V(w)} - c_0, \qquad (16)$$

gives

$$\frac{c_0(w)\varphi_1(w) - c_0}{c_0} = -\left[\frac{k_q \sinh^{-1} b/\sqrt{w^2+a^2}}{2\ln(b/a) + k_q \sinh^{-1} b/\sqrt{w^2+a^2}}\right]. \qquad (17)$$

Formula (17) for the symmetrically driven, center-loaded line corresponds to (8.18) for the end-loaded line. Curves of $-[c_0(w)\varphi_1(w) - c_0]/c_0$ for $k_q = 1$ are shown in Fig. 9.4. Comparison with Fig. 8.3 shows that for the center-loaded line $[c_0(w)\varphi_1(w) - c_0]/c_0$ is approximately equal to its value for the end-loaded line except at very small values of w/b. For k_q and $\varphi_1(w)$ real, C_{TC} is given by

$$C_{TC} = \int_0^d [c_0(w)\varphi_1(w) - c_0]\, dw. \qquad (18)$$

Curves giving C_{TC} as a function of b/a are shown in Fig. 9.5 as evaluated from (18) with (17) by numerical methods with $d = 10b$ and $d = 20b$, corresponding to $b \doteq 0.1\lambda_0$ and $b \doteq 0.05\lambda_0$ since, in the absence of end effects, $d \doteq 0.1\lambda$ as shown in Sec. 8. Fig. 8.4 shows that C_{Te} and C_{TC} are roughly equal.

Since the general shapes of the curves of $[c_0(w)\varphi_1(w) - c_0]/c_0$ for the center-loaded line are the same as for the end-loaded line except near $w/b = 0$, they may be evaluated as before by setting

$$[c_0(w)\varphi_1(w) - c_0]/c_0 \doteq -K/\sqrt{(w/b)^2 + k^2}. \qquad (19)$$

Here K and k are obtained by evaluating (17) in (19) at $w = 0$ and $w = 3b$.

If the bracket in (17) is denoted by H_1 and H_2 when $w = 0$ and $w = 3b$, the following formulas may be derived:

$$H_1 = \frac{k_q \ln 2b/a}{(k_q + 2) \ln 2b/a - 2\ln 2}; \qquad (20)$$

for $k_q = 1$ (20) reduces to

$$H_1 = \frac{\ln b/a + 0.693}{3 \ln b/a + 0.693}; \qquad (21)$$

and

$$H_2 = \frac{0.329 k_q}{2 \ln b/a + 0.329 k_q}. \qquad (22)$$

The formulas for K and k are like those of (8.29) with $a = 0$:

$$k = (3H_2/H_1)/\sqrt{1 - (H_2/H_1)^2}, \qquad (23)$$

$$K = kH_1. \qquad (24)$$

The integral of (19) is

$$C_{TC} \doteq -c_0 \int_0^d K/\sqrt{(w/b)^2 + k^2}\, dw$$

$$\doteq -c_0 bK \sinh^{-1}(d/kb). \qquad (25a)$$

With $d = 10b$,

$$C_{TC} \doteq -c_0 bK \sinh^{-1}(10/k). \qquad (25b)$$

Curves of $-C_{TC}/bc_0$ as calculated from (25b) are shown in Fig. 9.5 with the same function calculated from (18) with (17) and $k_q = 1$.

When $2\ln(b/a)$ is sufficiently great so that it is the leading term in the denominator of (17), (18) may be integrated to give the following approximate value:

$$C_{TC} \doteq -2c_0 b/[\ln(b/a)]. \qquad (25c)$$

As a numerical example, consider a center-loaded, symmetrically driven line for which $\ln b/a = 2$ or $b/a = 7.4$. With $k_q = 1$, (20) to (24) give

$$H_1 = 0.402, \qquad H_2 = 0.076, \qquad (26a)$$

$$k = 0.585, \qquad K = 0.236. \qquad (26b)$$

Substitution in (25b) gives

$$C_{TC} \doteq -0.83 c_0 b. \qquad (27)$$

This is numerically only slightly greater than the value $-0.70 c_0 b$ given in (8.32) for C_{Te} for the corresponding end-loaded line. The approximate value given by (25c) is $C_{TC} \doteq -c_0 b$.

To conclude the discussion of the symmetrically driven line, attention is called to the circuit of Fig. 9.6a, which must replace the antenna as center load if conventional transmission-line theory is to be a satisfactory approximation. In actual use, the circuit of Fig. 9.6b, which is equivalent to Fig. 9.6a, is convenient. Since no current crosses the central axis of symmetry, the two connections between the two loads $2Z_\delta$ may be removed and each side treated separately. Alternatively, the circuit may be arranged as in Fig. 9.6c, which is convenient when an "image line" is used, as discussed in Sec. 10.

Since L_{Tc} for the center-loaded line is exactly twice L_{Te} for the end-loaded line (Sec. 8), but is connected in series with $2Z_\delta$ in Fig. 9.6b, the inductive corrections are the same for the center-loaded and end-loaded lines. Since C_{Tc} is approximately the same for the center-loaded as C_{Te} for the end-loaded line, C_{Tc} in parallel with $2Z_\delta$ for the center-loaded line is *effectively double* C_{Te} in parallel with Z_δ for the end-loaded line. The apparent terminal impedance Z_{sa} of a *particular* antenna must, therefore, be different when determined for the antenna as end load from its value for the same antenna as center load, unless C_T is negligible.

With all elements in the equivalent circuit of both end-loaded and center-loaded lines determined, it remains to evaluate the impedance $Z_\delta = V_\delta/I_\delta$ of an antenna with gap 2δ across which a scalar potential difference V_δ is maintained and in the terminals of which is a current I_δ. Note that account has been taken of the coupling between antenna and line in C_T, so that the effect of the charges in the line on the charges in the antenna (which is simply the other half of the effect of the charges in the antenna on the charges in the line), may be ignored in evaluating Z_δ.

Instead of the dielectric support mentioned at the beginning of this section, a so-called "stub" support consisting of a high-impedance, antiresonant section of transmission line may be used (Fig. 9.7). In order to achieve a sufficiently high impedance, losses in the stub must be very small and hence the adjustment to antiresonance is very critical. A practical procedure is to determine accurately with a sensitive probe the location of the voltage maximum with the antenna absent, and then to attach the antenna exactly at this point. If the stub is so constructed that the impedance of the stub is very large compared with that of the antenna, the current in the stub near the junction with antenna and line is very small, so that the inductance per unit length of the transmission line differs only negligibly from that determined in Sec. 7 for the unsupported line and antenna. Hence, the same value of $l_0^e(w)$ and $L_T = L_{Te}$ may be used. However, the concentrations of charge on the sections of the stub near the junction point are by no means negligible compared with those on adjacent sections of the antenna and the line proper, so that a considerable modification in $c(w)$ and, hence, of C_T is required. Owing to lack of symmetry, the charge-ratio factors involved in their calculation are more difficult to determine. Since the capacitive correction is most effective and important when there is maximum voltage across the terminals of the antenna, the estimation of an approximate value is best carried out for this condition. When there is maximum voltage at the terminals of the antenna, the distribution of voltage and charge along the transmission line including the stub is essentially the same as the distribution for the symmetrical, center-loaded line discussed earlier in this section. Accordingly, the value of C_T must be the same and equal to C_{Tc} for the coupling between antenna and line on *each* side of the antenna. However, in this instance, the impedance terminating the line is not $2Z_\delta$ but simply Z_δ. This means that the equivalent terminal-zone network is that shown in Fig. 9.7b. No inductances are required on the right, since the current into the high-impedance stub is always small. Since L_{Te} is small, the two capacitances may be combined as in Fig. 9.7c without significant error. It is well to bear in mind that the capacitance C_{Tc} on the right in Fig. 9.7b is due entirely to coupling between the antenna and the stub, so that it can *not* be compensated by adjusting the stub with the antenna removed. It is for this reason that it must be included in the circuit even if the stub is adjusted experimentally to put a voltage maximum at the load terminals before the antenna is connected.

10. *Antenna Center-Driven from a Two-Wire Line with Minimum Coupling; Image Methods of Driving an Antenna Using Open-Wire and Coaxial Lines*

The analysis of an antenna that is center-driven from a two-wire line may involve all or several of the following complications in the terminal zone.

1. The inductance per unit length is not

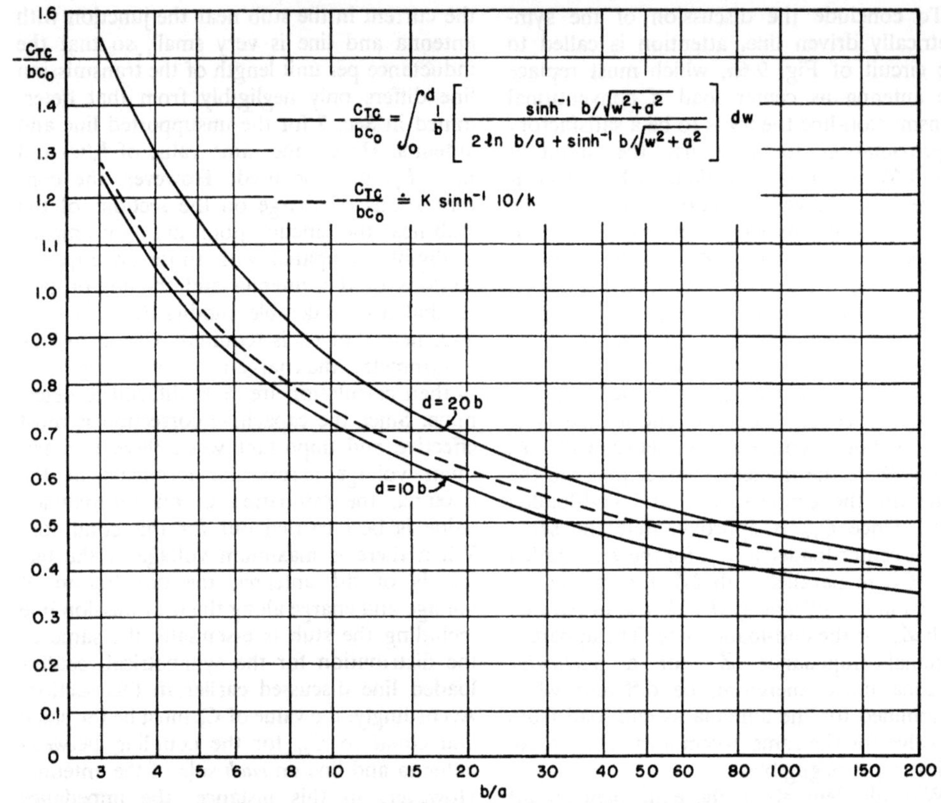

Fig. 9.5. Graphs of $-C_{Tc}/bc_0$ as a function of b/a.

$$-\frac{C_{TC}}{bc_0} = \int_0^d \frac{1}{b}\left[\frac{\sinh^{-1} b/\sqrt{w^2+a^2}}{2\ln b/a + \sinh^{-1} b/\sqrt{w^2+a^2}}\right] dw$$

$$-\frac{C_{TC}}{bc_0} \doteq K \sinh^{-1} 10/k$$

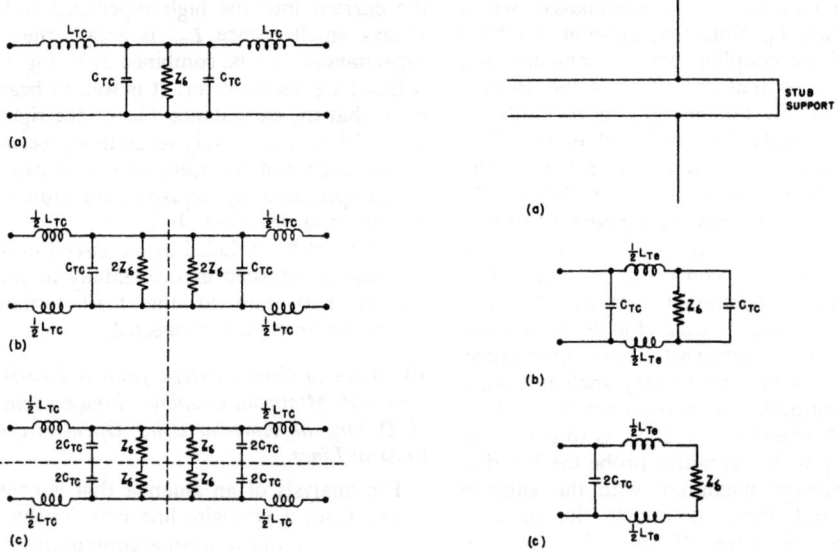

Fig. 9.6. Equivalent circuits for terminal zone of line with antenna as center load.

Fig. 9.7. Equivalent circuits for terminal zone of line with antenna and stub support as end load.

constant and equal to l_0^e as assumed in line theory. This is due to the difference between the forces acting on the currents in the line in the terminal zone and at a distance $n\lambda_0/2$ back along the line where conventional line theory applies. Such differences may be ascribed (a) to the line alone if the currents in the antenna maintain no component of vector potential axially along the conductors of the line, as when antenna and line are mutually perpendicular; (b) to the line and inductive coupling between currents in the line and antenna if the conditions in (a) are not satisfied. Correction for a variable inductance per unit length involves a lumped series inductance L_T, which may be negative.

2. The capacitance per unit length is not constant and equal to c_0 as assumed in line theory. This is due to differences between the forces acting on the charges in the line in the terminal zone and at a distance $n\lambda_0/2$ back along the line. Such differences may be ascribed (a) to the transmission line alone if $c_0(w) \neq c_0$ in the terminal zone and the charges in the antenna do *not* contribute to the scalar potential difference across the conductors of the line as in arrangements described later in this section; (b) entirely to capacitive coupling between the antenna and the line if $c_0(w) = c_0$ in the terminal zone and the charges in the antenna contribute significantly to the scalar potential difference across the conductors of the line as in the symmetrically driven, center-loaded line described in Sec. 9; (c) to both the line and capacitive coupling between antenna and line if $c_0(w) \neq c_0$ and the charges in the antenna contribute to the potential difference across the line as in the end-loaded line discussed in Sec. 8. Correction for a variable capacitance per unit length involves a lumped shunt capacitance C_T that may be negative.

3. The antenna and line are supported at the junction points by a piece of dielectric, a high-impedance stub, or both. Dielectric supports introduce the equivalent of a lumped capacitance in parallel with the load. Such a capacitance is difficult to calculate and usually must be determined experimentally. A stub support involves changes in capacitance and inductance per unit length of line so that its effect is normally included in (1) and (2). Since currents in the stub near the junction points with the antenna and the line proper are very small, their effect usually is insignificant so that the contribution to L_T is negligible. On the other hand, the charge per unit length on the stub near the junction is large so that the contribution to C_T is important. If the impedance of the stub is sufficiently high, its presence may just compensate for the capacitive end effect of the line alone, so that C_T reduces essentially to the capacitive coupling between antenna and line. This condition may be achieved with both a dielectric support and a stub so adjusted that the impedance of the parallel combination is very high.

The conventional method of center-driving a symmetrical antenna as an end load on a two-wire line (Sec. 8) involves all the complications as listed except that inductive coupling is eliminated by having the antenna perpendicular to the line. The symmetrically driven, center-loaded two-wire line (Sec. 9) avoids complications (3) and (4), reduces (2c) to (2b), and leaves (1) unchanged. Thus, capacitive coupling between antenna and line is highly significant with the antenna both as end load and as center load. In both, inductive coupling is absent if the antenna is perpendicular to the line.

Antennas with no coupling to the line. In order to eliminate all coupling between antenna and transmission line it is necessary to orient the antenna so that the vector and scalar potentials *due to currents and charges on the transmission line* vanish at *all* points along the antenna. In the end-loaded and center-loaded lines the vector potential due to currents in the line is zero on the surface of the antenna, since antenna and line are mutually perpendicular; however, the scalar potential on the surface of the antenna due to charges on the line does not vanish. In order to make this zero on the antenna, opposite points on the two conductors of the line must be made equidistant from each point on the axis of the antenna. If this is done, the contribution to the scalar potential at any cross section of the antenna by the charges in one conductor of the line is canceled by the scalar potential due to an identical distribution of charge of *opposite sign* on the second conductor of the line. One method of accomplishing this cancellation is shown in Fig. 10.1 for an antenna driven as an end load. In the circuit of Fig. 10.1 the antenna is perpendicular to all other conductors, so that only currents in the antenna itself contribute to the component of the vector potential tangent to the surface of the antenna. Accordingly, there is *no inductive coupling* between the antenna and the line or the antenna and the short bridges

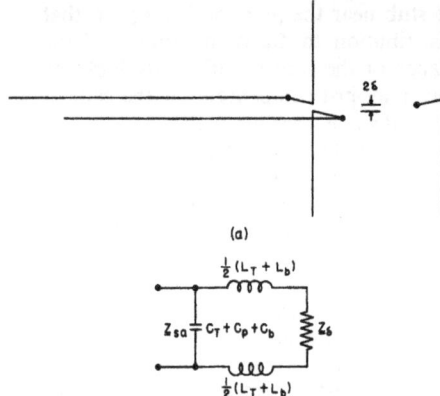

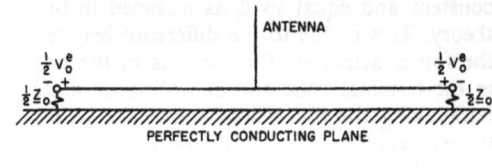

Fig. 10.4. Antenna and single line over an infinite conducting plane.

Fig. 10.1. Antenna as end load in plane perpendicular to two-wire line.

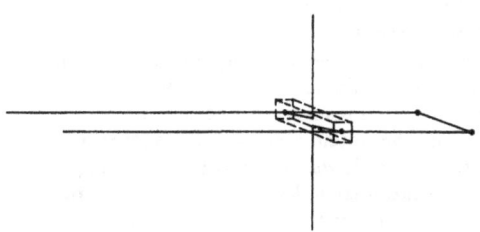

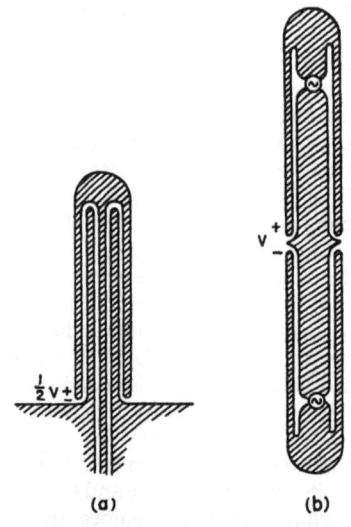

Fig. 10.2. Antenna with dielectric and inductive stub support.

Fig. 10.5. Antenna base-driven from coaxial line over conducting plane with symmetric equivalent.

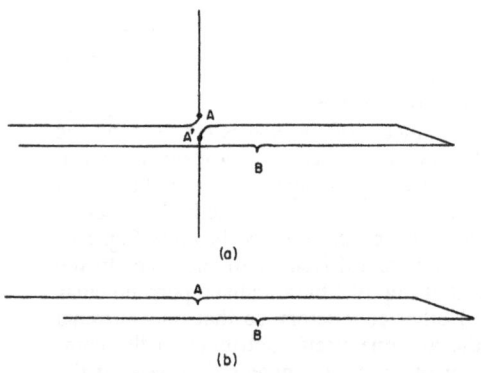

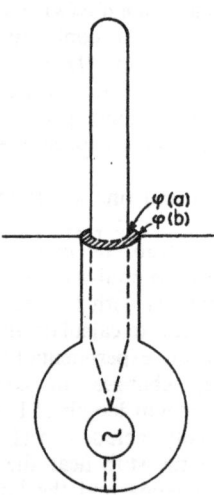

Fig. 10.3. Antenna in series with one conductor of line.

Fig. 10.6. Conventional method of base-driving antenna from coaxial line.

connecting the antenna to the line. This means that the inductive end effect is due exclusively to the absence of an infinite line just as in the more conventional circuit analyzed in Sec. 8. Therefore, the same value of L_T may be used.

In the conventional end-loaded line the correcting capacitance C_T combines the effect of nonuniformity in the capacitance per unit length of line with coupling between line and load. In the arrangement of Fig. 10.1 this latter effect vanishes so that C_T is simplified to the extent that direct coupling between the two parallel conductors of the line and the antenna is eliminated. If the two short conductors connecting the ends of the line to the terminals of the antenna were exactly perpendicular to the antenna the scalar potentials due to the equal charges of opposite sign on them would cancel exactly. Evidently this is possible only in the idealized case of zero gap. For a finite gap the potentials due to the charges on the short connecting sections do not quite cancel, so that a small capacitive coupling between antenna and connecting conductors exists. Since each of the conductors is essentially of length $b/2$, the magnitude of the coupling is very small and approximately equivalent to a small lumped capacitance connected across the terminals of the antenna.

The equivalent lumped constants for the compensating terminal-zone network for the circuit of Fig. 10.1a include L_T just as given by (7.10e) and C_T, which differs from (8.1) in that $\varphi_1(w) = 1$. This is equivalent to $V_T = 0$ or $k_q = 0$. Hence, (8.18) becomes

$$C_T = \int_0^d [c_0(w) - c_0]\, dw = c_0 \int_0^d f\, dw, \quad (1)$$

where

$$f = \left[\frac{\ln(b/a) - \sinh^{-1}(w/a) + \sinh^{-1}(w/b)}{\ln(b/a) + \sinh^{-1}(w/a) - \sinh^{-1}(w/b)}\right].$$

Note that the integrand in (1) is a *positive* quantity, so that C_T is also positive. The function $[c_0(w) - c_0]$ in (1) is the difference between $c_0(w)$ and c_0 in Fig. 6.2. The area between the curve $[c_0(w) - c_0]$ and the straight line c_0 is C_T when a linear scale is used. It may be obtained by numerical or other approximate methods. The conductors connecting the antenna to the line are not included in Z_δ or in the transmission line. The inductive contribution by these conductors with co-directional currents is given by [Eq. (VI.11.22) of ref. I.31]

$$L_b = \frac{b}{2\pi v_0}\left(\ln \frac{2b}{a_b} - 1\right), \quad (2)$$

where a_b is the radius of the conductor forming the connecting bridges. This inductance may be combined with L_T. The capacitive contribution may be approximated by calculating the static capacitance between the two connectors and adding this as C_b to C_T. Since this usually involves a dielectric support or spacer, C_b is not evaluated readily. On the other hand, since none of the three capacitances involves significant coupling to the antenna, their combined capacitance may be determined experimentally if the antenna is removed. Application of conventional transmission-line techniques and equations gives $C_T + C_b$ directly. With $C_T + C_b$ and $L_T + L_b$ known quantities, the apparent terminating impedance Z_{sa} may be determined from Z_δ using the equivalent network in Fig. 10.1b.

Instead of measuring $C_T + C_b$, their shunting effect may be canceled effectively by connecting an inductive stub across the line as in Fig. 10.2. By adjustment of the stub with the antenna removed, the shunting impedance of the capacitance $C_T + C_b$ and the inductive stub may be made very high compared with the maximum impedance of the antenna by locating a voltage maximum across the line at the connectors. It is then not necessary to determine $C_T + C_b$ explicitly. Furthermore, since L_T is a small negative inductance and L_b is positive, a proper choice of the cross-sectional size of the connectors between antenna and line makes $L_T + L_b$ negligibly small. This is true when

$$b[\ln(2b/a_b) - 1] = b - a, \quad (3a)$$

or

$$a_b = 2be^{-(2-a/b)} \quad (\doteq 0.27b \text{ for } a \ll b). \quad (3b)$$

These adjustments serve to compensate for all end effects and the effect of the connectors. Hence,

$$Z_{sa} \doteq Z_\delta. \quad (4)$$

Another method of driving the antenna with very much reduced coupling to the line is shown in Fig. 10.3. This circuit is first connected as in Fig. 10.3b, with the antenna removed and the points AA' in Fig. 10.3a connected together at A in Fig. 10.3b. The stub length is then so adjusted that a current maximum occurs at A and B, so that the impedance looking to the right at AB is extremely small compared with the impedance of the antenna for all except very short lengths. With this adjustment completed the antenna is connected as in Fig. 10.3a where, in effect, the antenna and the stub are in series and not in parallel as in Fig. 10.2. The impedance measured from the main line is again essentially Z_δ, except for the small coupling between the antenna and the short curved sections which connect it to the line. Unfortunately, the series arrangement of Fig. 10.3b does not satisfy the fundamental requirement of symmetry, so that the currents in the line are unbalanced and the line contributes to the radiation.

The experimental study of antennas center-driven from a balanced two-wire line may be facilitated in some instances if use is made of the theorem of images (see ref. I.31, Sec. IV.20) to provide equivalent structures which provide a shield for the observer. For the conventional end-loaded line (Sec. 8) or the center-loaded line (Sec. 9) (but not for the circuits of Figs. 10.1 and 10.2), the plane $z = 0$ perpendicular to and bisecting the rectangle formed by the line is a plane of symmetry for the entire circuit (Fig. 10.4). On this plane the tangential component of the electric field is zero everywhere, so that nothing is changed electromagnetically if a perfectly conducting sheet is placed to coincide with the plane $z = 0$. Such a sheet completely isolates the halves of the circuit and the fields associated with their charges and currents, so that one half may be removed without disturbing the other. Each then consists of one conductor of the original two-wire line parallel to a perfectly conducting sheet at a distance $b/2$ and one half of the original antenna perpendicular to the sheet as shown in Fig. 10.4 for the center-loaded line. According to the theorem of images, the electromagnetic field and the distributions of current and charges in the upper half-space, $z = 0$, are unchanged. However, the theorem of images does not apply to the scalar and vector potentials, as is readily verified. For the two-wire line, ϕ and the components A_x and A_y vanish in the plane $z = 0$ as a result of cancellation of contributions from equal and opposite charges and currents on the symmetrical halves of the line. For the single line over the infinite conducting plane (Fig. 10.4) ϕ, A_x, and A_y may differ from zero along the perfectly conducting plane owing to contributions from surface layers of charges and current on the plane. Since

$$E_x = -\frac{\partial \phi}{\partial x} - j\omega A_x, \qquad (5)$$

$$E_y = -\frac{\partial \phi}{\partial y} - j\omega A_y, \qquad (6)$$

both E_x and E_y can be zero everywhere at $z = 0$, while ϕ, A_x, and A_y are not.

The scalar potential difference at the cross section x between the two conductors of the line is given by

$$V_b = \phi(x, 0, b/2) - \phi(x, 0, -b/2)$$
$$= \int_{-b/2}^{b/2} E_z \, dz + j\omega \int_{-b/2}^{b/2} A_z \, dz. \qquad (7)$$

The scalar potential difference between the plane $z = 0$ and the conductor at $z = b/2$ is one-half this value. It is

$$V_{b/2} = \phi(x, 0, b/2) - \phi(x, 0, 0)$$
$$= \int_0^{b/2} E_z \, dz + j\omega \int_0^{b/2} A_z \, dz. \qquad (8)$$

This formula also gives the potential difference between the single conductor and the perfectly conducting image plane. Since the electric field for $z \geqq 0$ is the same for the two-conductor line as for the single conductor over the conducting plane, the first integral in (8) is the same in the two cases. But this is not true of the integral of A_z. At $z = 0$, the value of A_z for the single line over the conducting plane is exactly half that for the two-conductor line. For $0 < z \leqq b/2$, there is no simple relation between the values of A_z in the two cases. The difference is merely the contribution to A_z by the current in the lower half of the antenna. Although this is the same as the current in the upper half, the distances from corresponding elements to points in the range $0 < z \leqq b/2$ are not the same, so that the resultant A_z differs in phase and magnitude. It follows that in general $V_{b/2}$ for the two-wire line is not the same as $V_{b/2}$ for the single line over the conducting plane. The two are approximately the same at distances $w = s - x$ from the antenna (at $x = s$, $w = 0$) that are sufficiently great so that the following inequality is true:

$$\left| \int_0^{b/2} E_z \, dz \right| \gg \left| \int_0^{b/2} j\omega A_z \, dz \right|. \qquad (9)$$

Accordingly, the impedance of a section of two-wire line of length w terminated in the symmetrical antenna is double the impedance measured on the line with half the antenna over the conducting plane *only if w is sufficiently large* to make the contribution of $j\omega \int_0^{b/2} A_z \, dz$ to the scalar potential difference negligible. But this restriction is the same as that obtained in (6.60) for the two-conductor line. It is nothing more than the statement already made in Secs. 8 and 9 that an independent impedance cannot be defined for an antenna terminating a two-wire line, since the antenna is coupled to the adjacent parts of the line. An impedance in the sense of conventional transmission-line theory can be defined for a section of line, only with respect to points that are sufficiently far from the antenna to make the direct coupling between the main line and the antenna negligible. The condition (9) is essentially equivalent to this, so that the impedance on the line with image plane is one-half the impedance on the two-wire line for the entire range over which an independent impedance can be defined for the two-wire line itself. The use of the single line over a highly conducting plane in making impedance measurements is discussed in Sec. 35.

Base-driven antenna over conducting plane. Instead of using half of a two-wire line over a conducting image plane to drive an antenna perpendicular to this plane, it is often convenient to use a coaxial line of which the inner conductor and the dielectric pierce the plane while the outer conductor is connected to it. A possible arrangement is shown in Fig. 10.5a, and a more usual one in Fig. 10.6. According to the theorem of images, the electromagnetic field and the distributions of current and charge in the upper half-space $z \geq 0$ are the same for the circuit of Fig. 10.5a as for the physically impractical, symmetrical, center-driven circuit of Fig. 10.5b. In order that the scalar potential difference driving the antenna in the circuit of Fig. 10.5b be twice that in Fig. 10.5a, the terminal surfaces must be sufficiently close together to make a condition corresponding to (9) a good approximation. Such a condition is

$$V_{\delta/2} = \Phi(\delta) - \Phi(0) \doteq \int_0^{\delta} E_z \, dz \quad (10)$$

Under these conditions it is immaterial whether the terminal surfaces actually have the form of a hole in the conducting plane as in Fig. 10.6 instead of a slit around the surface of the antenna as in Fig. 10.5a. The corresponding requirement in this case is

$$V_{\delta/2} = \Phi(b) - \Phi(a) \doteq \int_a^b E_r \, dr \quad (11)$$

Since the circuit of Fig. 10.6 is of great practical importance and the equation (11) often is not satisfied, an investigation of the contribution to the apparent impedance of the half-dipole antenna by a finite line spacing, $b - a$, is desirable. Although a rigorous mathematical analysis of this problem is theoretically much more attractive than the corresponding problem for the dipole center-driven from an open-wire line, it is nevertheless so intricate that it has not been achieved. For present purposes an approximate analysis paralleling that carried out for a dipole as end load on a two-wire line is adequate and may be carried out by a simple extension of the method of Secs. 7, 8, and 9. This involves an approximate determination of the inductance $l^e(w)$ and capacitance $c(w)$ per unit length of the coaxial line as a function of the distance w from the end where it joins the antenna and the conducting plane. In the terminal zone $w \leq d \doteq 10(b - a)$ transmission-line end effects and coupling between the antenna and the line may be significant. Outside the terminal zone, $l^e(w)$ and $c(w)$ are well approximated by the conventional constant values l_0^e and c_0 for a uniform line obtained by allowing w to approach infinity. The integrated differences

$$L_T = \int_0^d [l^e(w) - l_0^e] \, dw, \quad C_T = \int_0^d [c(w) - c_0] \, dw \quad (12)$$

are the lumped series inductance and shunt capacitance of the terminal-zone network which, when combined with the theoretical impedance of the half-dipole when isolated, constitutes the *apparent* impedance of the antenna as measured on the line at a distance $w = \lambda_0/2$. In the circuit of Fig. 10.6 the theoretical impedance is simply $Z_0/2$, where Z_0 is the impedance of the isolated antenna with its image when center-driven by a discontinuity in scalar potential, that is, by a slice generator. In order to determine L_T, it is necessary to evaluate $l^e(w)$ as given by

$$l^e(w) \doteq W_z(w)/I_{1zL}(w), \quad (13)$$

where $W_z(w) = A_{1z}(w) - A_{2z}(w)$ is the vector

potential difference between the two conductors of the coaxial line at $r = a$ and $r = b$, and $I_{1zL}(w)$ is the total current in the inner conductor of the line. The vector potential in the coaxial line of length s due to the current $I_{1zL}(w')$ in the inner conductor of the line ($0 < w' < s$) and due to the current $I_{1zT}(w')$ in the cylindrical antenna of length h ($0 > w' > -h$) is given by

$$A_z(w) = \frac{1}{4\pi v_0} \int_{-h}^{s} I_{1z}(w') \frac{e^{-j\beta_0 R_r}}{R_r} dw', \quad (14)$$

where

$$R_r = \sqrt{(w' - w)^2 + r^2}. \quad (0 \leq w \leq s; \quad a \leq r \leq b) \quad (15)$$

Since the inner radius b of the outer conductor of the coaxial line is assumed to be small compared with the wavelength, so that the inequality

$$\beta_0^2 b^2 \ll 1 \quad (16)$$

is a good approximation, the principal contributions to the vector potential difference between the inner and outer conductors of the coaxial line come from elements of current for which $(w' - w)$ is small. Since R_a and R_b approach each other as $w' - w$ is increased to greater and greater values, the contributions of distant currents are negligible, no significant error is made if retardation is neglected, if the current amplitude at all points w' on the antenna and line is assumed to be equal to that in the line at w, and if the limits of integration are replaced by infinity. Thus, the vector potential difference at w due to the *currents in the inner conductor* and on the antenna is

$$W_z(w) = A_{1z}(w) - A_{2z}(w)$$
$$\doteq \frac{I_{1zL}(w)}{4\pi v_0} \int_{-\infty}^{\infty} \left(\frac{1}{R_a} - \frac{1}{R_b}\right) dw'$$
$$= \frac{I_{1zL}(w)}{2\pi v_0} \ln \frac{b}{a}. \quad (17)$$

The vector potential difference (18) is the same in form as that between an infinitely long conductor of radius a carrying a *uniform* current and a coaxial infinite cylinder of radius b. It is immaterial whether this is a conducting cylinder, or simply a mathematical cylinder in space, since the vector potential due to currents on the inner surface of a metal cylinder is constant with respect to r for $a \leq r \leq b$, so that they contribute nothing to the vector potential difference (or to the magnetic field). It may be concluded that currents on the sheath of the coaxial line contribute nothing to $W_z(w)$. Since the currents on the conducting image plane are radial they contribute nothing to $W_z(w)$. Hence (18) gives the entire z-component of the vector potential difference. Comparison with (13) shows that

$$l^e(w) \doteq l_0^e = \frac{1}{2\pi v_0} \ln \frac{b}{a}. \quad (18)$$

Hence,

$$L_T \doteq 0. \quad (19)$$

The evaluation of C_T may be carried out in an analogous manner by evaluating $c(w)$ as defined by

$$c(w) \equiv q_L(w)/V(w), \quad (20)$$

where

$$\Phi_1(w) - \Phi_2(w) \equiv V(w) = V_L(w) + V_T(w) \quad (21)$$

is the total potential difference at the cross section w due to charges $q_L(w)$ on the line and charges $q_T(w)$ on the termination, which includes the cylindrical antenna and the metal image plane.

Corresponding to (17), and subject to the same approximations, the scalar potential difference between the two conductors due to charges on the inner conductor of the line and on the antenna is

$$V_1(w) = \Phi_{1a}(w) - \Phi_{1b}(w)$$
$$\doteq \frac{q_{1L}(w)}{4\pi \epsilon_0} \int_{-\infty}^{\infty} \left(\frac{1}{R_a} - \frac{1}{R_b}\right) dw'$$
$$= \frac{q_{1L}(w)}{2\pi \epsilon_0} \ln \frac{b}{a} = \frac{q_{1L}(w)}{c_0}. \quad (22)$$

The scalar potential difference in (22) is the same in form as that between an infinitely long cylindrical conductor of radius a with a static charge and a coaxial infinite cylinder of radius b. It is immaterial whether this is a conducting cylinder or merely a mathematical envelope in space, since the scalar potential due to a static charge on the inner surface of a metal cylinder is constant with respect to r for $a \leq r \leq b$, so that it contributes nothing to the scalar potential difference (or to the radial electric field). It may be concluded that the charges on the sheath of the coaxial line contribute nothing to $V_1(w)$.

Unlike the currents on the image plane,

which contributed nothing to the axial vector potential difference, the charges on the image plane do contribute to the scalar potential difference. The scalar potential at a radius r at a distance w from the end of the coaxial line due to the total charge $q_{2T}\,dr'$ in all rings of radius r' and width dr' on the plane is

$$\Phi_{2T}(w) = \frac{1}{4\pi\epsilon_0}\int_0^{2\pi}\int_b^\infty \frac{q_{2T}(r')}{2\pi r'}\frac{e^{-j\beta_0 R_T}}{R_T}$$
$$\times r'\,dr'\,d\theta', \quad (23)$$

where R_T is drawn from dr' in Fig. 10.7 to a point at radius r between P_1 and P_2. If it is assumed that the total charge on a ring of radius r' and of unit width on the image plane is equal to the total charge per unit length on the inner surface of the coaxial sheath at w, it follows that

$$q_{2T}(r') \doteq q_{2L}(w) = -q_{1L}(w), \quad (24)$$

so that

$$\Phi_{2T}(w) \doteq -\frac{q_{1L}(w)}{8\pi^2\epsilon_0}\int_0^{2\pi}\int_b^\infty \frac{e^{-j\beta_0 R_T}}{R_T}\,dr'\,d\theta'. \quad (25)$$

The quasi-stationary potential difference at w in the line due to the charges in the image plane is

$$V_2(w) = \Phi_{2a}(w) - \Phi_{2b}(w)$$
$$\doteq -\frac{q_{1L}(w)}{8\pi^2\epsilon_0}\int_0^{2\pi}\int_b^\infty \left(\frac{1}{R_{1T}} - \frac{1}{R_{2T}}\right)dr'\,d\theta', \quad (26)$$

where

$$R_{2T} = \sqrt{r'^2 + b^2 - 2r'b\cos\theta' + w^2}, \quad (27)$$
$$R_{1T} = \sqrt{r'^2 + a^2 - 2r'a\cos\theta' + w^2}. \quad (28)$$

It follows that the total potential difference between the two conductors of the line at a distance w from the load end is the sum of (22) and (26). Accordingly,

$$\frac{V(w)}{q_{1L}(w)} = \frac{1}{c(w)} = \left(\frac{1}{c_0} - \frac{\psi}{4\pi\epsilon_0}\right)$$
$$= \frac{1}{4\pi\epsilon_0}\left(2\ln\frac{b}{a} + \psi\right), \quad (29)$$

where

$$\psi \equiv \frac{1}{2\pi}\int_0^{2\pi}\int_b^\infty \left(\frac{1}{R_{2T}} - \frac{1}{R_{1T}}\right)dr'\,d\theta'. \quad (30)$$

When (29) is solved for $c(w)$ it follows directly that

$$\frac{c(w) - c_0}{c_0} = -\left[\frac{\psi}{\psi + 2\ln(b/a)}\right], \quad (31)$$

so that the lumped shunt capacitance C_T is a negative quantity given by

$$C_T = \int_0^d [c(w) - c_0]\,dw$$
$$= -c_0\int_0^d \left[\frac{\psi}{\psi + 2\ln(b/a)}\right]dw. \quad (32)$$

The numerical determination of C_T requires first the evaluation of the integral for ψ and then the integration of (32).

The exact evaluation of ψ as defined in (30) leads to complicated integrals of elliptic integrals. Fortunately, an approximate evaluation of ψ is possible. This is accomplished by dividing the range of integration with respect to θ into four regions in each of which the function is assumed to be constant at a middle value. Thus, choosing the middle values in the full range of 2π at $\theta = 0$, $\pi/2$, π, $3\pi/2$, the integral in (30) may be approximated as follows:

$$\psi \doteq \frac{1}{\pi}\int_b^\infty \left\{\left[\frac{1}{R_{2T}(0)} - \frac{1}{R_{1T}(0)}\right]\int_0^{\pi/4}d\theta'\right.$$
$$+ \left[\frac{1}{R_{2T}(\frac{1}{2}\pi)} - \frac{1}{R_{1T}(\frac{1}{2}\pi)}\right]\int_{\pi/4}^{3\pi/4}d\theta'$$
$$\left. + \left[\frac{1}{R_{2T}(\pi)} - \frac{1}{R_{1T}(\pi)}\right]\int_{3\pi/4}^\pi d\theta'\right\}dr'$$

$$= \frac{1}{4}\int_b^\infty \left[\frac{1}{\sqrt{(r'-b)^2 + w^2}}\right.$$
$$\left. - \frac{1}{\sqrt{(r'-a)^2 + w^2}}\right]dr'$$
$$+ \frac{1}{2}\int_b^\infty \left[\frac{1}{\sqrt{r'^2 + b^2 + w^2}}\right.$$
$$\left. - \frac{1}{\sqrt{r'^2 + a^2 + w^2}}\right]dr'$$
$$+ \frac{1}{4}\int_b^\infty \left[\frac{1}{\sqrt{(r'+b)^2 + w^2}}\right.$$
$$\left. - \frac{1}{\sqrt{(r'+a)^2 + w^2}}\right]dr'. \quad (33)$$

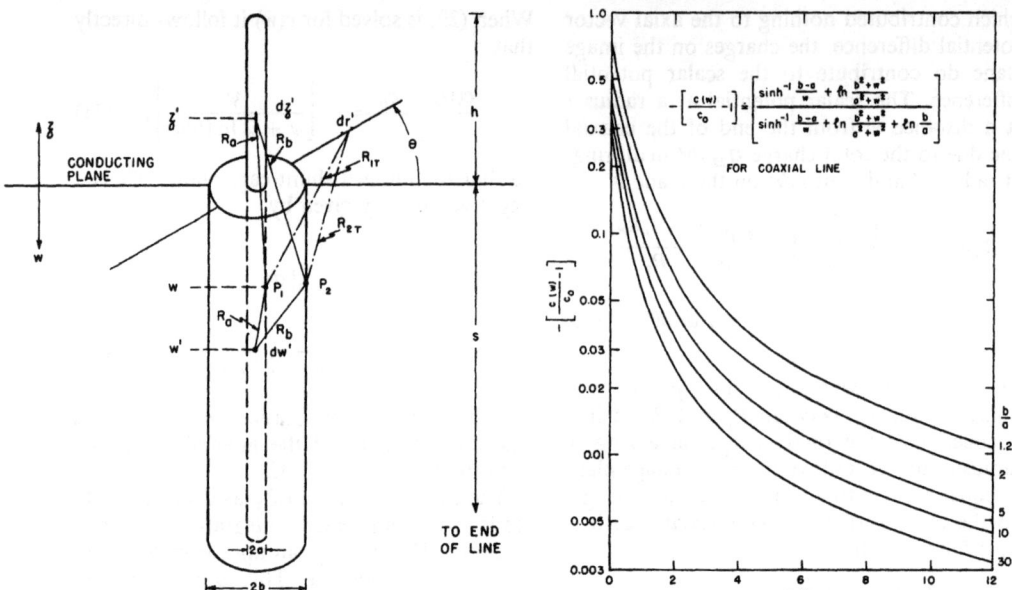

Fig. 10.7. Notation for antenna with coaxial drive over conducting plane.

Fig. 10.8. Graphs of $-[c(w) - c_0]/c_0$ as a function of w/b with b/a as parameter for the coaxial line.

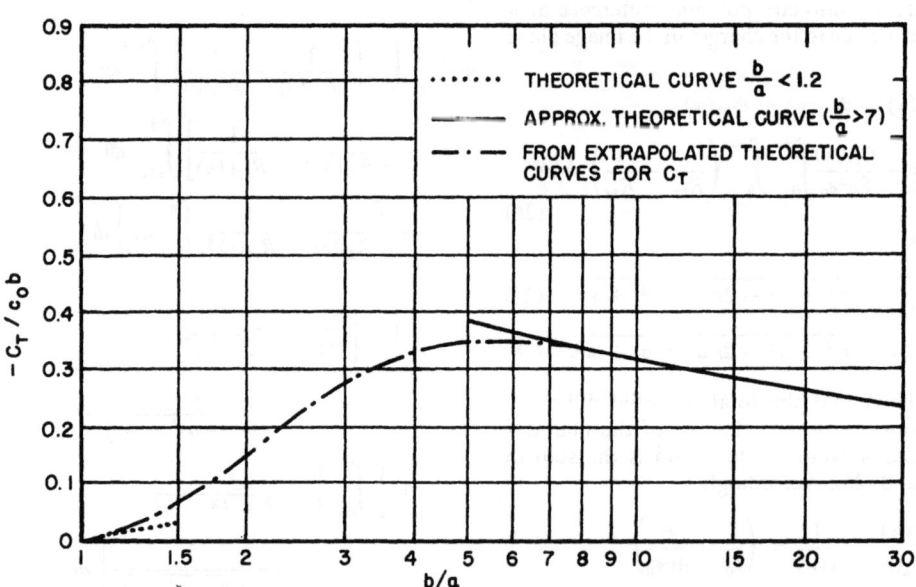

Fig. 10.9. $-C_T/bc_0$ as a function of b/a for the coaxial line.

These integrals may be evaluated directly. Thus,

$$\psi \doteq \frac{1}{4}\sinh^{-1}\frac{b-a}{w} - \frac{1}{2}\sinh^{-1}\frac{b}{\sqrt{b^2+w^2}}$$
$$+ \frac{1}{2}\sinh^{-1}\frac{b}{\sqrt{a^2+w^2}} - \frac{1}{4}\sinh^{-1}\frac{2b}{w}$$
$$+ \frac{1}{4}\sinh^{-1}\frac{b+a}{w}. \quad (34a)$$

By transforming to the natural logarithm, using $\sinh^{-1}(x/a) = \ln(x + \sqrt{x^2 + a^2})$, the following alternative expression is obtained:

$$\psi \doteq \frac{1}{4}\sinh^{-1}\frac{b-a}{w} + \frac{1}{2}\ln\sqrt{\frac{b^2+w^2}{a^2+w^2}}$$
$$- \frac{1}{2}\ln\left[\frac{b + \sqrt{2b^2+w^2}}{b + \sqrt{b^2+a^2+w^2}}\right]$$
$$- \frac{1}{4}\ln\left[\frac{2b + \sqrt{4b^2+w^2}}{b+a+\sqrt{(b+a)^2+w^2}}\right].$$
$$(34b)$$

Since the arguments of the last two logarithms differ only slightly from unity even when $w = 0$, these terms may be neglected. The resulting final expression is:

$$\psi \doteq \frac{1}{4}\left[\sinh^{-1}\frac{b-a}{w} + \ln\frac{b^2+w^2}{a^2+w^2}\right]$$
$$= \frac{1}{4}\left[\sinh^{-1}\frac{1-(a/b)}{w/b} + \ln\left(1 + \frac{w^2}{b^2}\right)\right.$$
$$\left. - \ln\left(\frac{a^2}{b^2} + \frac{w^2}{b^2}\right)\right]. \quad (35)$$

The function $-\left[\frac{c(w)}{c_0} - 1\right]$ obtained by substituting (35) in (31) is shown in Fig. 10.8 with w/b as variable and b/a as parameter. Although curves are shown for values of $b/a < 7$, these are not accurate for determining C_T, since (22) is adequate only when the integration is carried over distances that are at least five times the radius a, as is shown in Chapter I. Since $c(w)$ differs significantly from c_0 over distances of the order of magnitude of $b - a$, this distance must be greater than $5a$ or the principal contributions to C_T will be derived from a range of integration for (22) that is too short to yield accurate results.

By determining the areas under the curves in Fig. 10.8 in the range $w/b = 0$ to $w/b = 10$ by numerical methods, C_T as defined in (32) may be determined. It is represented in Fig. 10.9 in the form $-C_T/bc_0$ for $b/a > 7$. For values of b/a near unity, the coaxial line with its circular inner conductor was approximated by two parallel planes separated by a distance $(b - a)$ to the beginning of the antenna, where one of the planes is bent down at right angles to form the conducting ground screen. The charge on the straight plane, corresponding to the inner conductor, may be determined electrostatically using the Schwarz-Christoffel mapping. The ratio of this charge as a function of the distance w from the end of the parallel-plate region to the charge at an infinite distance from the end is proportional to $c(w)/c_0$. Using this ratio, the formula $-C_T/bc_0 \doteq 0.068(1 - a/b)$ was determined for $b/a < 1.2$. The corresponding curve is shown in Fig. 10.9. In order to join the approximate curve for $b/a > 7$ with that for $b/a < 1.2$, $-C_T$ itself was plotted (using Fig. 10.8 for all values of b/a) for two values of a. The curves so obtained decreased smoothly as b/a was reduced but reached a minimum between $b/a = 4$ and $b/a = 5$, and then increased sharply. By continuing the smoothly decreasing parts of the curves from above $b/a = 5$ toward $b/a = 1$ where $C_T \doteq -0.068(2\pi a \epsilon_0)$, using these extended portions of the curves to evaluate $-C_T/bc_0$, and joining the single curve so obtained smoothly to the curve for $b/a < 1.2$, the complete curve in Fig. 10.9 was obtained.[41]

HALLÉN'S THEORY OF CYLINDRICAL ANTENNAS*

11. Cylindrical Antenna

The analytical problem to be studied is the determination of the axial distribution of the total current I_z and the charge per unit length q for a conducting cylinder of *arbitrary cross section* that extends along the z-axis from $z = -h$ to $z = -\delta$ and from $z = \delta$ to $z = h$. Since it is proved in Sec. I.7 that the distribution of axial current is independent of the shape of the cross section if the maximum dimension is small compared with the wavelength, it is necessary only to analyze the circular cylinder of radius a. The total axial current in a cylinder of arbitrary cross-sectional shape is the same as the axial current in a *circular* cylinder of *equivalent* radius a as defined in Sec. I.7. An actual or

---
* The analysis does not follow Hallén's original formulation.

equivalent circular cylinder is shown in Fig. 11.1. It is postulated that a scalar potential difference $V_\delta = \Phi(\delta) - \Phi(-\delta)$ is maintained across the edges of the cylindrical envelope at $z = \pm\delta$. For the idealized slice or delta-function generator described in Sec. 4, the cylindrical conducting surface continues unbroken at $z = 0$ and $V_0 = \lim_{\delta \to 0} [\Phi(\delta) - \Phi(-\delta)]$ $= V_0 \delta(z)$, where $\delta(z)$ is the Dirac delta function. In the circuits of Secs. 8, 9, and 10 the antenna joins the conductors of the transmission line between $z = \pm\frac{1}{2}b$ and $z = \pm\delta$, so that there are small junction regions of dimensions comparable with the radius a of the conductor. In a one-dimensional analysis it is not possible to take accurate account of such regions. For small values of the radius a their effect may be approximated by assuming the halves of the antenna to extend to the axes of the conductors of the transmission line, so that $b \doteq 2\delta$, and by measuring the length of the line to the axis of the antenna. For larger values of a, additional lumped capacitances in parallel with the antenna may be required to represent adequately the effect of such junction regions. In the theoretical analysis that follows it is assumed that all such effects are either negligible or taken into account separately, so that the only chargeable surface of the antenna near $z = \pm\delta$ is the *cylindrical envelope*. The analysis is restricted to antennas for which

$$\beta_0 a \ll 1. \qquad (1)$$

The antenna in Fig. 11.1 is shown with hemispherical ends rather than with flat or hollow ones. Since a one-dimensional analysis such as the one contemplated cannot take accurate account of the hemispherical end surfaces, a brief discussion of these is required in conjunction with the condition $I_z = 0$ at $z = h$ that is imposed in the analysis. With hemispherical ends $I_z(h)$ is strictly zero. With the flat metal ends of a solid cylinder or the open ends of a tube $I_z(h)$ can not be zero. With a solid cylinder there is a flow of electric charges onto the flat ends since the axial currents I_z at $z = \pm h$ become radial currents I_r at $r = a$. Although these currents are small if the chargeable area πa^2 of the end is small, they are not negligible. If the cylinder is hollow, I_z at $z = h$ is a flow of electric charges around the sharp edge onto the inner surface of the cylinder. This surface is charged with a high density at the sharp edge and with an appreciable but diminishing density inward to a distance from the end comparable with the radius of the tube. Beyond this distance the density of charge is extremely minute. Evidently, with either a solid cylinder or a tube the effective half-length of the antenna exceeds h by an amount that is difficult to determine accurately, but that is of the order of magnitude of a.

With hemispherical ends $I_z(h) = 0$ and, in addition, the total chargeable surface of the antenna remains the same. Thus, referring to Fig. 11.2, the area of a hemisphere of radius a is $2\pi a^2$ in (a); the area of the cylindrical surface of length a is also $2\pi a^2$ in (b). Since a continuous change in the radius of the antenna for a short distance from each end has a negligible effect in the analysis, it may be assumed that the boundary condition for the tangential component of the electric field, namely,

$$\hat{n}_1 \times \mathbf{E}_1 + \hat{n}_2 \times \mathbf{E}_2 = 0 \qquad (2a)$$

(where the subscript 1 refers to the conductor, subscript 2 to surrounding space), is well approximated by

$$E_{z1} = E_{z2} \text{ at } r = a. \qquad (-h \leq z \leq -\delta,$$
$$\delta \leq z \leq h) \quad (2b)$$

This is actually the condition for a physically unavailable "one-dimensional" cylinder of radius a and half-length h that has no chargeable surfaces beyond the edges at $z = \pm h$. Its approximate physical equivalent is a cylinder of the same length along the axis, but with hemispherical ends.

The electric field in the conductor satisfies the constitutive relation

$$i_f = \sigma_c \mathbf{E}_1, \qquad (-h \leq z \leq -\delta,$$
$$\delta \leq z \leq h), \quad (3)$$

where σ_c is the real conductivity of the conductors.

In the analysis it is assumed that the cross-sectional distribution of current in the conductor is independent of the axial distribution. If the diameter of the conductor is small compared with its length, as is here supposed, this is an excellent approximation. Since rotational symmetry obtains, the total axial current I_z in the conductor is related to the axial component of the electric field at the surface of the cylinder by the formula defining the internal impedance $z^i = r^i + jx^i$ per unit length of the conductor (ref. I.31, Sec. V.8). It is

$$z^i = \frac{(E_{z1})_{r=a}}{I_z} \doteq \frac{1+j}{2\pi a}\sqrt{\frac{\omega}{2\sigma_c \nu}}. \qquad (4)$$

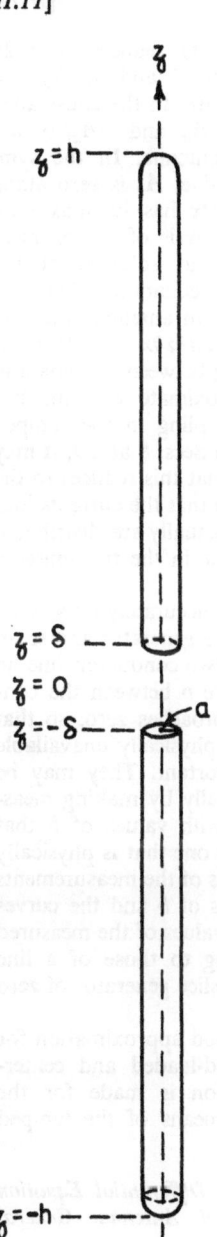

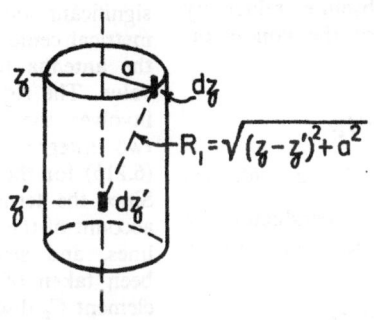

Fig. 11.1. Actual or equivalent circular cylinder.

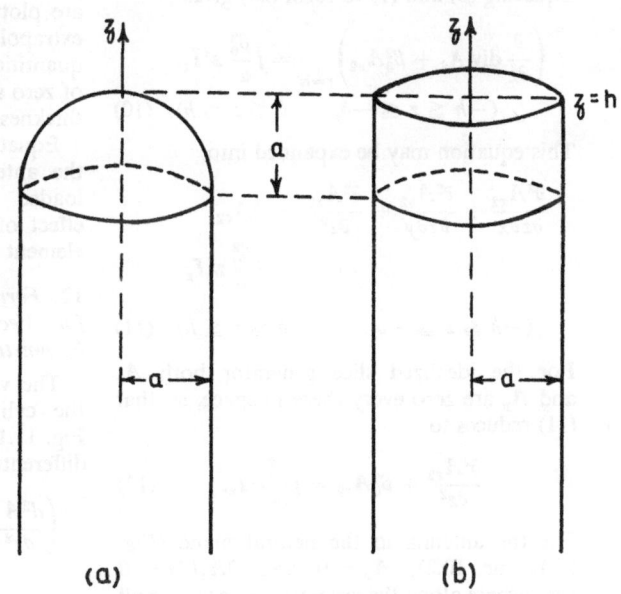

Fig. 11.2. Antennas with hemispherical and plane ends.

The formula on the right is valid at frequencies for which

$$a\sqrt{\omega\sigma_c/\nu} \geq 10. \quad (5)$$

Here $\nu = 1/\mu$ is the absolute reluctivity (reciprocal permeability) of the conductor. Accordingly,

$$(E_{z1})_{r=a} = \left(\frac{i_z}{\sigma_c}\right)_{r=a} = z^i I_z.$$

$$(-h \leq z \leq -\delta, \quad \delta \leq z \leq h). \quad (6)$$

At all points outside the conductor the electric field may be expressed in terms of the scalar and vector potentials:

$$\mathbf{E}_2 = -\text{grad } \Phi_2 - j\omega \mathbf{A}_2$$

$$= -\frac{j\omega}{\beta_0^2}(\text{grad div } \mathbf{A}_2 + \beta_0^2 \mathbf{A}_2), \quad (7)$$

where $\beta_0 = \omega/v_0 = 2\pi/\lambda_0$, and where use has been made of the definition of div $\mathbf{A}_2$,

$$\text{div } \mathbf{A}_2 + j\frac{\beta_0^2}{\omega}\Phi_2 \equiv 0. \quad (8)$$

The z-component of (7) on the surface of the antenna is

$$(E_{z2})_{r=a} = -\frac{j\omega}{\beta_0^2}\left(\frac{\partial}{\partial z}\text{div } \mathbf{A}_2 + \beta_0^2 A_{z2}\right)_{r=a}.$$

$$(-h \leq z \leq -\delta, \quad \delta \leq z \leq h) \quad (9)$$

Equating (6) and (9) to form (2b) gives

$$\left(\frac{\partial}{\partial z}\text{div } \mathbf{A}_2 + \beta_0^2 A_{z2}\right)_{r=a} = j\frac{\beta_0^2}{\omega} z^i I_z.$$

$$(-h \leq z \leq -\delta, \quad \delta \leq z \leq h) \quad (10)$$

This equation may be expanded into

$$\frac{\partial^2 A_{x2}}{\partial z \partial x} + \frac{\partial^2 A_{y2}}{\partial z \partial y} + \frac{\partial^2 A_{z2}}{\partial z^2} + \beta_0^2 A_{z2}$$

$$= j\frac{\beta_0^2}{\omega} z^i I_z.$$

$$(-h \leq z \leq -\delta, \quad \delta \leq z \leq h) \quad (11)$$

For the idealized slice generator both A_x and A_y are zero everywhere in space, so that (11) reduces to

$$\frac{\partial^2 A_{z2}}{\partial z^2} + \beta_0^2 A_{z2} = j\frac{\beta_0^2}{\omega}z^i I_z. \quad (12)$$

For the antenna in the neutral plane (Fig. 10.1a or 10.2), $A_x = 0$ and $\partial A_x/\partial x = 0$ everywhere along the antenna; A_y has a small but maximum value with respect to changes in y over short distances from the line, so that $\partial A_y/\partial y = 0$. Hence, (11) reduces to (12). In the circuits of Secs. 8 and 9, A_y and $\partial A_y/\partial y$ vanish everywhere. In the end-loaded line of Sec. 8, both A_x and $\partial A_x/\partial x$ are significant along the antenna. In the symmetrical center-loaded line, A_x is zero along the antenna but $\partial A_x/\partial x$ has its maximum value. The rigorous analysis of the problem involves the simultaneous solution of the two integrodifferential equations (11) and (6.21b) for the currents in antenna and line. Since the term in $\partial^2 A_{x2}/\partial z \partial x$ in (11) takes account of the coupling between antenna and lines, and since approximate account has been taken of this coupling in the lumped element C_T discussed in Secs. 8 and 9, it may be ignored in (11), so that this reduces to the simpler form (12). Note that the currents into and the charges on C_T actually are distributed in a short terminal zone in the transmission line *and* in the antenna.

Equation (12) applies accurately only to an antenna driven by a slice generator and to an antenna driven by a two-conductor line in the limit as the distance b between the conductors of the line approaches zero, so that $\delta \to 0$. These cases are physically unavailable but none the less important. They may be approached experimentally by making measurements successively with values of b that decrease to the smallest one that is physically practicable. If the results of the measurements are plotted as functions of b and the curves extrapolated to $b \to 0$, values of the measured quantities corresponding to those of a line of zero spacing or of a slice generator of zero thickness are obtained.

Equation (12) is a good approximation for the antenna in the end-loaded and center-loaded line if correction is made for the effect of coupling by means of the lumped element C_T.

12. Formal Solution of Differential Equation for Vector Potential of Antenna. General Symmetry Conditions

The vector potential in space just outside the cylindrical surface of the antenna in Fig. 11.1 satisfies the nonhomogeneous linear differential equation

$$\left(\frac{d^2 A_z}{dz^2} + \beta_0^2 A_z\right)_{r=a} = j\frac{\beta_0^2}{\omega} z^i I_z.$$

$$(-h \leq z \leq -\delta, \quad \delta \leq z \leq h) \quad (1)$$

Since no confusion can arise, the subscript 2 is omitted from the potentials.

The general solution of (1) is the sum of a complementary function A_{zc} and a particular integral A_{zp}. The complementary function is a solution of the homogeneous equation, which is also the complete equation if the conductor is perfect and $z^i = 0$. The homogeneous equation is

$$\frac{d^2 A_z}{dz^2} + \beta_0^2 A_z = 0. \qquad (2)$$

It is readily verified that

$$A_{zc} = \frac{-j}{v_0}(C_1 \cos \beta_0 z + C_2 \sin \beta_0 z)$$

$$(r = a) \qquad (3)$$

is a solution of (2); C_1 and C_2 are arbitrary constants of integration. The factor $-j/v_0$ is included for later convenience in order to make the C's have the dimensions of scalar potentials measured in volts.

A particular integral in (1) for the upper half of the antenna is

$$A_{zp} = \frac{jz^i}{v_0} \int_\delta^z I(s) \sin \beta_0(z-s)\, ds, \qquad (4)$$

as may be verified by direct substitution using the following rule for differentiating a definite integral:

$$\psi = \int_a^b f(s, z)\, ds, \qquad (5)$$

$$\frac{d\psi}{dz} = \int_a^b \frac{df}{dz}\, ds + f(b, z) \frac{db}{dz} - f(a, z) \frac{da}{dz}. \qquad (6)$$

The general solution of (1) is the sum of (3) and (4):

$$(A_z)_{r=a} = \frac{-j}{v_0}\bigg[C_1 \cos \beta_0 z + C_2 \sin \beta_0 z$$
$$- z^i \int_\delta^z I(s) \sin \beta_0(z-s)\, ds\bigg].$$

$$(\delta \leq z \leq h) \qquad (7)$$

A particular integral for the lower half of the antenna is like (4) but with $-\delta$ instead of δ written in the lower limit. The resulting expression for the vector potential on the lower half of the antenna is given in (8). In general, the arbitrary constants C_3 and C_4 in (8) are not the same as C_1 and C_2 in (7).

The relations between C_3, C_4 and C_1, C_2 must be determined from symmetry conditions.

$$(A_z)_{r=a} = \frac{-j}{v_0}\bigg[C_3 \cos \beta_0 z + C_4 \sin \beta_0 z$$
$$- z^i \int_{-\delta}^z I(s) \sin \beta_0(z-s)\, ds\bigg].$$

$$(-\delta \leq z \leq -h) \qquad (8)$$

The determination of the axial distribution of current in the cylindrical antenna from (7) is more convenient if the current $I_z \equiv I(z)$ is expressed as the sum of even (symmetric) and odd (antisymmetric) parts, respectively, $I^s(z)$ and $I^a(z)$. Thus, it is always possible to consider the current at a point z in two parts, as follows:

$$I(z) = \tfrac{1}{2}[I(z) + I(-z)] + \tfrac{1}{2}[I(z) - I(-z)], \qquad (9)$$

or,

$$I_z \equiv I(z) = I^s(z) + I^a(z), \qquad (10)$$

where the quantity in the first bracket is by definition even, so that

$$I^s(-z) = I^s(z), \qquad (11)$$

while the quantity in the second bracket is odd, so that

$$I^a(-z) = -I^a(z). \qquad (12)$$

Symmetry conditions for the distribution of charge are easily obtained using the equation of continuity in the form (I.3.7),

$$\frac{dI_z}{dz} + j\omega q = 0. \qquad (13)$$

Thus

$$q(z) = \frac{-j}{\omega} \frac{dI_z(z)}{dz}, \qquad (14)$$

$$q(-z) = \frac{-j}{\omega} \frac{dI_z(-z)}{(-dz)}. \qquad (15)$$

In the symmetric case, currents are even functions of z, charges odd functions:

$$I(-z) = I(z), \qquad q(-z) = -q(z). \qquad (16a)$$

In the antisymmetric case, currents are odd functions, charges even functions:

$$I(-z) = -I(z), \qquad q(-z) = q(z). \qquad (16b)$$

The total charge is

$$q(z) = \tfrac{1}{2}[q(z) - q(-z)] + \tfrac{1}{2}[q(z) + q(-z)]$$
$$= q^s(z) + q^a(z). \qquad (17)$$

The scalar and vector potentials in space (region 2) due to the current and charge in a cylindrical conductor (region 1) are given by (I.7.8,9),

$$\Phi = \frac{1}{4\pi\varepsilon_0}\left(\int_{-h}^{-\delta}+\int_{\delta}^{h}\right)\frac{q(z')}{R_1}e^{-j\beta_0 R_1}\,dz', \quad (18a)$$

$$A_z = \frac{1}{4\pi v_0}\left(\int_{-h}^{-\delta}+\int_{\delta}^{h}\right)\frac{I(z')}{R_1}e^{-j\beta_0 R_1}\,dz', \quad (18b)$$

where

$$R_1 = \sqrt{(z-z')^2 + a^2}. \quad (19)$$

It is proved in I.7 and reference I.31, Sec. IV.4 that (18a, b) are good approximations at points just outside the surface of the conductor, $r = a$.

The scalar and vector potentials given by (18a) and (18b) may be expressed in a form like (10) and (17), since with (11.8) in the form

$$\frac{dA_z}{dz} + \frac{j\beta_0^2}{\omega}\Phi = 0 \quad (20)$$

they satisfy a relation corresponding exactly to (13). Hence,

$$A(z) = \tfrac{1}{2}[A(z) + A(-z)] + \tfrac{1}{2}[A(z) - A(-z)]$$
$$= A^s(z) + A^a(z), \quad (21a)$$

$$\Phi(z) = \tfrac{1}{2}[\Phi(z) - \Phi(-z)] + \tfrac{1}{2}[\Phi(z) + \Phi(-z)]$$
$$= \Phi^s(z) + \Phi^a(z), \quad (21b)$$

where the symmetric conditions

$$A^s(-z) = A^s(z), \quad \Phi^s(-z) = -\Phi^s(-z) \quad (22a)$$

apply in the first brackets in (21a, b), and the antisymmetric conditions

$$A^a(-z) = -A^a(z), \quad \Phi^a(-z) = \Phi^a(z) \quad (22b)$$

apply in the second brackets. It is now verified readily that (18a) and (18b) satisfy (22a) for the symmetric currents and charges defined by (16a), and that they satisfy (22b) for the antisymmetric currents and charges defined by (16b).

Summarizing, the conditions of symmetry are as follows:

Symmetric case: current and vector potential even; charge and scalar potential odd.

$$I(-z) = I(z), \quad q(-z) = -q(z),$$
$$A(-z) = A(z), \quad \Phi(-z) = -\Phi(z). \quad (23a)$$

Antisymmetric case: current and vector potential odd; charge and scalar potential even.

$$I(-z) = -I(z), \quad q(-z) = q(z),$$
$$A(-z) = -A(z), \quad \Phi(-z) = \Phi(z). \quad (23b)$$

The general case: superposition of odd and even functions.

$$I(z) = I^s(z) + I^a(z),$$
$$q(z) = q^s(z) + q^a(z),$$
$$A(z) = A^s(z) + A^a(z), \quad (24)$$
$$\Phi(z) = \Phi^s(z) + \Phi^a(z).$$

Symmetric currents and vector potentials are even, equal in magnitude, and *codirectional* at corresponding points defined by $\pm z$; antisymmetric currents and vector potentials are odd, equal in magnitude, and *oppositely* directed at corresponding points. Symmetric charges per unit length and scalar potentials are odd, equal in magnitude, and *opposite* in sign; antisymmetric charges per unit length are even, equal in magnitude, and *alike* in sign. Distributions of both symmetric and antisymmetric types are excited readily in practice. Note that in geometrically unsymmetric structures the *even* currents may contribute to the *odd* part of the vector potential. This is true, for example, in the collinear array analyzed in Chapter III.

13. Vector and Scalar Potentials for the Symmetric Antenna

The analysis in this chapter is concerned exclusively with the center-driven symmetric antenna. Odd currents do not exist because they are not excited by the symmetrically placed driving potential difference, nor required by the equally symmetric boundary conditions. Accordingly, in the remainder of this chapter it may be assumed that

$$I(z) = I^s(z), \quad I^a(z) = 0, \quad (1)$$

so that the symmetric case alone obtains:

$$I(-z) = I(z), \quad q(-z) = -q(z), \quad (2a)$$
$$A(-z) = A(z), \quad \Phi(-z) = -\Phi(z). \quad (2b)$$

The solutions (12.7) and (12.8) for the vector potential must be made to satisfy the conditions (2). This requires that

$$C_1 \cos\beta_0 z + C_2 \sin\beta_0 z$$
$$-z^i \int_\delta^z I(s)\sin\beta_0(z-s)\,ds$$
$$= C_3 \cos(-\beta_0 z) + C_4 \sin(-\beta_0 z)$$
$$-z^i \int_{-\delta}^{-z} I(s)\sin\beta_0(-z-s)\,ds. \quad (3)$$

This is satisfied for all values of z in so far as the trigonometric factors are concerned if

$$C_3 = C_1, \quad C_4 = -C_2. \quad (4)$$

It is readily verified that the two integrals are the same in any case. Thus, writing $u = -s$, $du = -ds$, and changing the limits appropriately — when $s = -z$, $u = z$; when $s = -\delta$, $u = \delta$ — the integral on the right becomes

$$\int_\delta^z I(-u) \sin \beta_0(-z+u)(-du). \quad (5)$$

With $I(-u) = I(u)$ from (2), (5) becomes

$$\int_\delta^z I(u) \sin \beta_0(z-u) \, du. \quad (6)$$

Clearly (6) is identical with the integral on the left in (3), so that (4) is sufficient to make (12.7) and (12.8) satisfy the symmetry condition (1). Accordingly,

$$[A_z(z)]_{r=a} = \frac{-j}{v_0} \left[C_1 \cos \beta_0 z + C_2 \sin \beta_0 z \right.$$
$$\left. - z^i \int_\delta^z I(s) \sin \beta_0(z-s) \, ds \right],$$
$$(\delta \leq z \leq h) \quad (7)$$

$$[A_z(z)]_{r=a} = \frac{-j}{v_0} \left[C_1 \cos \beta_0 z - C_2 \sin \beta_0 z \right.$$
$$\left. - z^i \int_{-\delta}^z I(s) \sin \beta_0(z-s) \, ds \right].$$
$$(-h \leq z \leq -\delta) \quad (8)$$

The scalar potential for the symmetric antenna is obtained from (7) and (8) using the equation of continuity for potential functions,

$$\text{div } \mathbf{A} + j \frac{\beta_0^2}{\omega} \Phi = 0, \quad (9)$$

in the form

$$\frac{\partial A_z}{\partial z} + j \frac{\beta_0^2}{\omega} \Phi = 0, \quad (10)$$

to which (9) reduces when $\mathbf{A} = \hat{z} A_z$. Thus,

$$\Phi(z) = j \frac{\omega}{\beta_0^2} \frac{\partial A_z(z)}{\partial z}, \quad (11)$$

$$\Phi(-z) = j \frac{\omega}{\beta_0^2} \frac{\partial A_z(-z)}{-\partial z} = -j \frac{\omega}{\beta_0^2} \frac{\partial A_z(z)}{\partial z}. \quad (12)$$

Upon differentiating (12.1) with respect to z and using (11) or (12), Φ_2 is seen to satisfy the scalar wave equation,

$$\left(\frac{d^2\Phi}{dz^2} + \beta_0^2 \Phi \right)_{r=a} = -z^i \frac{dI_z(z)}{dz} = j\omega z^i q(z). \quad (13)$$

Use has been made of the equation of continuity $dI_z(z)/dz + j\omega q(z) = 0$. In (13), $q(z)$ is the charge per unit length.

Using (11) and (7), the solution of (13) corresponding to (7) is

$$[\Phi(z)]_{r=a} = \left[-C_1 \sin \beta_0 z + C_2 \cos \beta_0 z \right.$$
$$\left. - z^i \int_\delta^z I(s) \cos \beta_0(z-s) \, ds \right].$$
$$(\delta \leq z \leq h) \quad (14)$$

Using (12) and (8), the solution of (13) corresponding to (8) is

$$[\Phi(z)]_{r=a} = \left[-C_1 \sin \beta_0 z - C_2 \cos \beta_0 z \right.$$
$$\left. - z^i \int_{-\delta}^z I(s) \cos \beta_0(z-s) \, ds \right].$$
$$(-h \leq z \leq -\delta) \quad (15)$$

It is readily shown that (14) and (15) satisfy the relation $\Phi(-z) = -\Phi(z)$ by writing $(-z)$ for z in (15), with z understood to be the positive value written in (14). Thus,

$$[\Phi(-z)]_{r=a} = \left[-C_1 \sin \beta_0(-z) \right.$$
$$- C_2 \cos \beta_0(-z)$$
$$\left. - z^i \int_{-\delta}^{-z} I(s) \cos \beta_0(-z-s) \, ds \right].$$
$$(16)$$

Noting that, with the same change in variable used in (5) and the condition $I(-z) = I(z)$,

$$\int_{-\delta}^{-z} I(s) \cos \beta_0(-z-s) \, ds$$
$$= -\int_\delta^z I(u) \cos \beta_0(z-u) \, du, \quad (17)$$

it follows that

$$[\Phi(-z)]_{r=a} = \left[C_1 \sin \beta_0 z - C_2 \cos \beta_0 z \right.$$
$$\left. + z^i \int_\delta^z I(s) \cos \beta_0(z-s) \, ds \right].$$
$$(18)$$

Comparison of (14) and (18) shows that $\Phi(-z) = -\Phi(z)$ as required. It follows that

$$[\Phi(+z) - \Phi(-z) = 2\Phi(z)]_{r=a}. \quad (19)$$

The arbitrary constant of integration C_2 in (7) may be expressed in terms of the driving potential difference V_δ maintained across the cylindrical ends of the antenna at $z = \pm\delta$. From (19) it follows that, with $z = \delta$,

$$[V_\delta = \Phi(\delta) - \Phi(-\delta) = 2\Phi(\delta)]_{r=a}. \quad (20)$$

At $z = \delta$, (14) becomes:

$$[\Phi(\delta)]_{r=a} = -C_1 \sin \beta_0 \delta + C_2 \cos \beta_0 \delta. \quad (21)$$

Hence, substituting (21) in (20),

$$V_\delta = 2[\Phi(\delta)]_{r=a} = 2(-C_1 \sin \beta_0 \delta + C_2 \cos \beta_0 \delta), \quad (22)$$

so that

$$C_2 = (\tfrac{1}{2}V_\delta + C_1 \sin \beta_0 \delta)/\cos \beta_0 \delta. \quad (23)$$

This value of C_2 can be substituted in (7) to give

$$[A_z(z)]_{r=a} = \frac{-j}{v_0 \cos \beta_0 \delta}[C_1 \cos \beta_0(z-\delta) + \tfrac{1}{2}V_\delta \sin \beta_0 z]$$

$$+ \frac{jz^i}{v_0}\int_\delta^z I(s) \sin \beta_0(z-s)\, ds.$$

$$(\delta \leq z \leq h). \quad (24)$$

Similarly,

$$[A_z(z)]_{r=a} = \frac{-j}{v_0 \cos \beta_0 \delta}[C_1 \cos \beta_0(z+\delta) - \tfrac{1}{2}V_\delta \sin \beta_0 z]$$

$$+ \frac{jz^i}{v_0}\int_{-\delta}^z I(s) \sin \beta_0(z-s)\, ds.$$

$$(-h \leq z \leq -\delta) \quad (25)$$

The substitution of (23) in (14) and (15) gives

$$[\Phi(z)]_{r=a} = \frac{1}{\cos \beta_0 \delta}[-C_1 \sin \beta_0(z-\delta) + \tfrac{1}{2}V_\delta \cos \beta_0 z]$$

$$- z^i \int_\delta^z I(s) \cos \beta_0(z-s)\, ds.$$

$$(\delta \leq z \leq h) \quad (26)$$

$$[\Phi(z)]_{r=a} = \frac{1}{\cos \beta_0 \delta}[-C_1 \sin \beta_0(z+\delta) - \tfrac{1}{2}V_\delta \cos \beta_0 z]$$

$$- z^i \int_{-\delta}^z I(s) \cos \beta_0(z-s)\, ds.$$

$$(-h \leq z \leq -\delta) \quad (27)$$

Formulas (24) to (27) are the general expressions for the vector and scalar potentials just outside the surface of the cylindrical antenna.

14. Hallén's Integral Equation

The vector potential at a point (a, z) just outside the surface of the cylindrical conductor is given approximately by

$$(A_z)_{r=a} = \frac{1}{4\pi v_0}\left(\int_{-h}^{-\delta} \frac{I_z'}{R_1} e^{-j\beta_0 R_1}\, dz' + \int_\delta^h \frac{I_z'}{R_1} e^{-j\beta_0 R_1}\, dz'\right), \quad (1)$$

where, as shown in Fig. 11.1,

$$R_1 = \sqrt{(z-z')^2 + a^2}. \quad (2)$$

Substitution of (1) in (13.7) gives the *general integral equation for the current in the antenna*. Noting that $v_0/\nu_0 = 1/\sqrt{\varepsilon_0 v_0} = \zeta_0 \doteq 376.7\,\text{ohm}$, it is

$$\left(\int_{-h}^{-\delta} + \int_\delta^h\right) I_z' K_1(z, z')\, dz'$$

$$= \frac{-j4\pi}{\zeta_0 \cos \beta_0 \delta}\left[C_1 \cos \beta_0(z-\delta) + \tfrac{1}{2}V_\delta \sin \beta_0 z\right]$$

$$+ \frac{j4\pi z^i}{\zeta_0}\int_\delta^z I(s) \sin \beta_0(z-s)\, ds,$$

$$(\delta \leq z \leq h) \quad (3)$$

where

$$K_1(z, z') \equiv \frac{e^{-j\beta_0 R_1}}{R_1} \quad (4)$$

is the kernel of the integral equation.

A similar equation may be written for the lower half of the antenna. Since $I(-z) = I(z)$, it is unnecessary to carry through the detailed analysis for $I(-z)$. A solution of the integral (3) in closed form is not known. A solution in the form of a series may be obtained by expressing I_z' in terms of a convenient reference current such as I_δ and a distribution function $f(z)$ and then improving the approximation by iteration.*

---

* The question whether the integral equation (3) is in the best form from the point of view of having the reference current I_δ insensitive to small changes in the distribution function is considered in Sec. 39. It is shown there that by application of the calculus of variations an improved form is available. It is shown also that (3) leads to results in excellent agreement with those obtained by variational methods if the iteration is carried out twice, *and if a good choice is made for $f(z)$.*

Let
$$I_z = I_\delta f(z). \quad (5a)$$
Then
$$I'_z = I_\delta f(z'), \quad (5b)$$
so that
$$I'_z = I_z \frac{f(z')}{f(z)} \equiv I_z g(z, z'). \quad (6)$$

Now let a complex function $\Psi_\delta(z)$ be defined as follows:

$$\Psi_\delta(z) \equiv \left(\int_{-h}^{-\delta} + \int_{\delta}^{h}\right) g(z, z') K_1(z, z') \, dz'. \quad (7)$$

If the distribution function $g(z, z') \equiv f(z')/f(z)$ is the *actual* one, it is correct to set

$$\left(\int_{-h}^{-\delta} + \int_{\delta}^{h}\right) I'_z K_1(z, z') \, dz' = I_z \Psi_\delta(z). \quad (8)$$

If $g(z, z')$ is known only approximately, it is possible to write without approximation

$$\left(\int_{-h}^{-\delta} + \int_{\delta}^{h}\right) I'_z K_1(z, z') \, dz' = I_z \Psi_\delta(z)$$
$$+ \left(\int_{-h}^{-\delta} + \int_{\delta}^{h}\right) [I'_z - I_z g(z, z')]$$
$$\times K_1(z, z') \, dz'. \quad (9)$$

The more nearly $g(z, z')$ approximates the correct distribution, the smaller are the difference integrals on the right in (9). If $g(z, z')$ can be chosen accurately enough so that the sum of the integrals on the right in (9) is considerably smaller than the term $I_z \Psi_\delta(z)$, this term may be treated as the principal part and the sum of the difference integrals as a small correction.

If $g(z, z')$ is the actual distribution function, so that (8) is true, it follows that $\Psi_\delta(z)$ is given by

$$\Psi_\delta(z) = 4\pi\nu_0 \frac{(A_z)_{r=a}}{I_z}. \quad (10)$$

That is, $\Psi_\delta(z)$ is proportional to the ratio of the vector potential at any point (a, z) just outside the surface of the antenna to the actual, total axial current in the conductor in the cross section at z. It is clear from (1) that $(A_z)_{r=a}$ is determined largely by the current at and near the cross section at z. Owing to the factor R_1 in the denominator, the contribution to the integral by the current I'_z in an element dz' at z' that is at a considerable axial distance $|z' - z|$ from the point (a, z) where $(A_z)_{r=a}$ is calculated, is necessarily small unless z happens to be a point where I_z is much lesst han I'_z at z'. In general, it is reasonable to expect the ratio (10) to be moderately constant and predominantly real over all parts of the antenna except where the current is small or zero. Since I_z is zero at the ends, $\Psi_\delta(z)$ is infinite at $z = \pm h$. On the other hand, the product $I_z \Psi_\delta(z)$ remains small.

Although there is no current between $z = \pm \delta$, the current at $z = \pm \delta$ is always finite, so that $\Psi_\delta(\delta)$ is not infinite, but has the same order of magnitude as if the current I_z continued instead of turning and becoming I_x on the transmission line. $\Psi_\delta(z)$ is not defined between $z = -\delta$ and $z = \delta$.

Assuming that $\Psi_\delta(z)$ is predominantly real and sensibly constant over the greater part of the surface of the antenna, this constant value must be well approximated by $\Psi_\delta(z)$ at some reference point z_r.

Let the magnitude of $\Psi_\delta(z_r)$ be denoted by

$$|\Psi_\delta(z_r)| \equiv \Psi. \quad (11a)$$

Also let

$$\Psi_\delta(z) = \Psi + \gamma(z), \quad (11b)$$

where $\gamma(z)$ is a complex correction term defined so that (11b) is true. For most of the range of z between δ and h, or $-\delta$ and $-h$, $\gamma(z)$ should be very small. At the ends $z = \pm h$, where $I_z = 0$, $\gamma_\delta(z)$ is infinite, but $[I_z \gamma_\delta(z)]_{z \to \pm h}$ remains finite and small, since it is proportional to the vector potential at the ends of the antenna near which the current and hence the vector potential are always small.

Clearly, if $g(z, z')$ is only a good approximation of the distribution of current instead of the actual relative distribution function, (11a) and (11b) may be written nevertheless.

If (11b) is substituted in (9), the result is

$$\left(\int_{-h}^{-\delta} + \int_{\delta}^{h}\right) I'_z K_1(z, z') \, dz' = I_z \Psi + I_z \gamma(z)$$
$$+ \left(\int_{-h}^{-\delta} + \int_{\delta}^{h}\right) [I'_z - I_z g(z, z')]$$
$$\times K_1(z, z') \, dz'. \quad (12)$$

With $g(z, z')$ properly chosen and Ψ correctly defined, the only large term on the right in (12) is $I_z \Psi$. Near the ends, where $\gamma(z)$ becomes large, all terms become small because I_z is extremely small. Accordingly, if (12) is substituted in (3) and Ψ is finite, it is possible to solve formally for I_z in $I_z\Psi$. Thus,

$$I_z = \frac{-j4\pi}{\zeta_0 \Psi \cos \beta_0 \delta} \left[C_1 \cos \beta_0(z-\delta) \right.$$
$$\left. + \frac{1}{2} V_\delta \sin \beta_0 z \right]$$
$$+ \frac{j4\pi z^i}{\zeta_0 \Psi} \int_\delta^z I(s) \sin \beta_0(z-s)\, ds$$
$$- \frac{1}{\Psi} \left\{ I_z \gamma(z) + \left(\int_{-h}^\delta + \int_\delta^h \right) \right.$$
$$\left. [I_z' - I_z g(z, z')] K_1(z, z')\, dz' \right\}. \quad (13)$$

This equation is exact. Like (3) it is an integral equation for the current. However, it differs from (3) in that the current itself appears on the left and is contained on the right only in a difference integral that is small if $g(z, z')$ is properly chosen, and in a term that is negligible except very near the ends at $z = \pm h$. Since it is postulated that the current vanishes at $z = \pm h$, it is usually of no importance to know exactly what the distribution of current is very near the ends, if it is known everywhere else. It is in any event readily determined by interpolation.

The equation (13) is put in a preferred form for subsequent evaluation of the constant C_1 if (3) is written with $z = h$ and divided by Ψ, namely,

$$0 = \frac{-j4\pi}{\zeta_0 \Psi \cos \beta_0 \delta} \left[C_1 \cos \beta_0(h-\delta) \right.$$
$$\left. + \frac{1}{2} V_\delta \sin \beta_0 h \right]$$
$$+ \frac{j4\pi z^i}{\zeta_0 \Psi} \int_\delta^h I(s) \sin \beta_0(h-s)\, ds$$
$$- \frac{1}{\Psi} \left(\int_{-h}^{-\delta} + \int_\delta^h \right) I_z' K_1(h, z')\, dz', \quad (14)$$

with

$$K_1(h, z') \equiv \frac{e^{-j\beta_0 R_{1h}}}{R_{1h}} \quad (15a)$$

and

$$R_{1h} \equiv \sqrt{(h-z')^2 + a^2}, \quad (15b)$$

and (14) is subtracted from (13). This gives

$$I_z = \frac{-j4\pi}{\zeta_0 \Psi \cos \beta_0 \delta}$$
$$\times \left\{ C_1[\cos \beta_0(z-\delta) - \cos \beta_0(h-\delta)] \right.$$
$$\left. + \frac{1}{2} V_\delta [\sin \beta_0 z - \sin \beta_0 h] \right\}$$
$$+ \frac{j4\pi z^i}{\zeta_0 \Psi} \left\{ \int_\delta^z I(s) \sin \beta_0 (z-s)\, ds \right.$$
$$\left. - \int_\delta^h I(s) \sin \beta_0(h-s)\, ds \right\}$$
$$- \frac{I_z \gamma(z)}{\Psi} - \frac{1}{\Psi} \left\{ \left(\int_{-h}^{-\delta} + \int_\delta^h \right) \right.$$
$$[I_z' - I_z g(z, z')] K_1(z, z')\, dz'$$
$$\left. - \left(\int_{-h}^{-\delta} + \int_\delta^h \right) I_z' K_1(h, z')\, dz' \right\},$$
$$(\delta \leq z \leq h) \quad (16)$$

This expression is preferred because the right-hand member vanishes for *all values* of $\beta_0 h$ when $z = \pm h$ as is required, whereas the right-hand member of (13) cannot be made to vanish at $z = \pm h$ when $\cos \beta_0(h-\delta) = 0$, since the arbitrary constant C_1 disappears completely. Equation (16) is the final, exact form of the original integral equation (3). Its advantage over the apparently simpler form (3) lies in the fact that all terms on the right involving the current are small compared with I_z itself (except near the ends where $I_h = 0$) if the conductor is good (z^i small) and the distribution function $g(z, z')$ is well chosen to make the difference terms small.

It is to be noted that the form of (16) and the definition of Ψ in (11) are essentially arbitrary. An alternative procedure in defining Ψ is to use the real part instead of the magnitude of $\Psi_\delta'(z)$. Another possibility is described in Sec. 31, where it is shown that for an electrically short antenna (16) is not as satisfactory as a similar equation obtained by subtracting from each member of (3) the same expression with $z = h$, and introducing a difference kernel of the form

$$L_1(z, z') = K_1(z, z') - K_1(h, z') \quad (17)$$

in place of $K_1(z, z')$. A yet more refined procedure is to introduce a complex distribution function $g(z, z')$, separate the integral on the left in (3) into its real and imaginary parts, and so obtain two integral equations from the real and imaginary parts of the original, complex equation. Use of this procedure is made in Secs. 25 and 31. In general, the relatively simple procedure described in this section is satisfactory for antennas that are neither extremely short nor excessively long. Exceptions are considered in due course.

Before considering the solution of (16) it is well to recall that the original integral equation (3), of which (16) is a special form, is a quasi-one-dimensional *approximation* of a three-dimensional problem. On the whole the approximation is a good one except very near the ends of the antenna or its junction with the driving transmission line, where the integral (1) is not an accurate representation of the vector potential. As shown in Sec. 7 of Chapter I, the vector potential on the surface of a cylindrical conductor is well represented by (1) only at distances from the ends of the antenna or from sharp bends that are not less than about five times the radius of the conductor. Errors that are roughly equivalent to a change in the axial coördinate by an amount equal to the radius a must be expected very near ends or sharp bends. Such errors can not be eliminated so long as an essentially one-dimensional equation is used to solve a three-dimensional problem. In most practical applications involving thin conductors driven from actual transmission lines they are unimportant.

Actually, two kinds of approximation are involved in the integral equation (3). On the one hand, the quasi-one-dimensional mathematical formulation can represent the physical problem only approximately; on the other hand, the two expressions for the vector potential that are equated to form (3) are not rigorously equal very near $z = \pm h$ and $z = \pm \delta$. The integral on the left in (3) when multiplied by $1/4\pi\nu_0$ is the vector potential A_z at $r = a$ of a *line source* of current I'_z along the z-axis, and only approximately the vector potential of a rotationally symmetric distribution of current in the volume $r \leq a$. Accordingly, $\phi = (j\omega/\beta_0^2)(\partial A_z/\partial z)$ is the scalar potential at $r = a$ of a line source of charge per unit length $q(z')$, which satisfies the equation of continuity, $(dI(z')/dz') + j\omega q(z') = 0$, and only approximately the scalar potential of a surface density of charge on the cylindrical conductor of radius a. Moreover, the contributions to the scalar potential near $z = \pm \delta$ by charges on surfaces other than the cylindrical envelope of the antenna itself are ignored, and some such surfaces *must exist* in every physically possible radiating system. Evidently, the driving voltage $V_\delta = \phi(\delta) - \phi(-\delta)$, as defined in Sec. 13 in terms of $(\partial A_z/\partial z)$, is in error for two reasons. The first is that rigorously V_δ is defined in terms of a line distribution of charge along the axis instead of a surface distribution at $r = a$, the second is that in practice there are always charged surfaces other than the cylindrical envelope of the antenna that contribute to $\phi(\pm \delta)$. If the transmission-line spacing is greater than the radius a, as is usual in practice, the effect on the potential difference V_δ of locating the charges on the axis instead of at $r = a$ is not great. However, account must be taken of the contributions by charges on all surfaces near $z = \pm \delta$. This may be done with an appropriate terminal-zone network, as described in Secs. 8, 9, and 10.

The integral equation (3) reduces to the simple form originally derived by Hallén when $\delta = 0$, and it is this form that is usually discussed in the literature. The simplified equation for $\delta = 0$ is

$$4\pi\nu_0 A_z = \int_{-h}^{h} I'_z K_1(z, z') \, dz'$$
$$= \frac{-j4\pi}{\zeta_0}\left[C_1 \cos \beta_0 z + \frac{1}{2} V_0 \sin \beta_0 |z|\right], \quad (18)$$

where

$$V_0 = \lim_{\delta \to 0} [\phi(\delta) - \phi(-\delta)] \quad (19)$$

is a discontinuity in scalar potential. From the physical point of view, (18) is a simpler, but necessarily poorer, approximation than (3) with terminal-zone network. Indeed, the current obtained from (18) is at best the first term in a Maclaurin expansion in powers of $\beta_0 \delta$ of the current in (3). This conclusion follows from the fact that a discontinuity in scalar potential along a conductor is physically unavailable as a driving mechanism.

The integral equation (18) is also unsatisfactory from the mathematical point of view, since at $z = 0$ the left-hand member has a continuous derivative with respect to z, whereas the right-hand member has a discontinuous derivative. This is a consequence of the fact that the derivative of the right-hand member involves by definition a discontinuity in scalar potential at $r = a$, $z = 0$, whereas the derivative of the left-hand member rigorously defines the scalar potential at $r = a$, $z = 0$ due to a discontinuous line source of charge at $r = 0$, $-h \leq z \leq h$. This potential is continuous at $r = a$, $z = 0$. It is to be expected that when a physically possible driving mechanism—as with a transmission line of some kind—is selected and potentials appropriate to this are calculated, the entire difficulty will disappear. This is illustrated in Sec. 12 of Chapter VIII, using a biconical transmission line within the radius a. It can be demonstrated here using a very short radial transmission line, also within the radius a.

Consider a physically complete transmitting system consisting of the cylindrical halves of the antenna of radius a and half-length h with flat conducting ends at $z = \pm \delta$; a wire of radius $a_1 (\ll a)$ connecting the centers of these flat ends; and a point-generator at the center of the thin wire that maintains a potential difference V_δ across the flat ends (see Fig. 12.6 of Chapter VIII). For simplicity, let the conductors all be perfect, so that only surface densities of current exist, I_z on the cylindrical surfaces of radii a and a_1, and I_r on the flat ends. The problem is to derive the integral equation for the total current, $I_z = 2\pi a I_z$, in the cylindrical antenna in terms of the scalar potential difference $V_\delta = \Phi(\delta) - \Phi(-\delta)$ maintained across the surfaces at $z = \pm \delta$. Since the radius a of these surfaces is very small compared with the wavelength, the potential difference on the radial transmission line from $r = a_1$ to $r = a$ may be assumed to be approximately uniform. (The more general case is analyzed in Sec. 12, Chapter VIII with a biconical line.)

The general equation satisfied by the vector potential at all points outside the conductors is, from (I.7,5),

$$\nabla^2 A_z + \beta_0^2 A_z = 0. \quad (20)$$

The solution of this equation may be obtained by applying Green's symmetric theorem,[I.31]

$$\int_\tau [u \nabla^2 v - v \nabla^2 u] \, d\tau'$$
$$= \int_\Sigma \left[u \frac{\partial v}{\partial n} - v \frac{\partial u}{\partial n} \right] d\sigma', \quad (21)$$

with $u = A_z$ and $v = G$, where G is the free-space Green's function that satisfies the equation

$$\nabla^2 G + \beta_0^2 G = -\delta(R). \quad (22)$$

In (22), R is the distance between the point where G is calculated and a point source. The delta function satisfies the relations:

$$\delta(R) = 0, \quad R \neq 0, \quad (23a)$$

$$\int_\tau \delta(R) \, d\tau = 1 \text{ if } \tau \text{ includes } R = 0$$
$$= 0 \text{ if } \tau \text{ does not include } R = 0. \quad (23b)$$

The solution of (22) is

$$G = \frac{e^{-j\beta_0 R}}{4\pi R} = \frac{K(z, z')}{4\pi}. \quad (24)$$

Note that G is continuous with its normal derivative across the boundaries of all conductors.

After adding and subtracting $\beta_0^2 uv$ on the left in (21), the use of (20), (22), and (23b) leads to

$$A_z = \left(\int_{-h}^{-\delta} + \int_\delta^h \right) G \left(\frac{\partial A_z}{\partial r} \right)_{r=a} 2\pi a \, dz'$$
$$+ \int_{-\delta}^\delta G \left(\frac{\partial A_z}{\partial r} \right)_{r=a_1} 2\pi a_1 \, dz'$$
$$- \int_0^{2\pi} \int_{a_1}^a G \left[\left(\frac{\partial A_z}{\partial z} \right)_{z=\delta} \right.$$
$$\left. - \left(\frac{\partial A_z}{\partial z} \right)_{z=-\delta} \right] r' dr' d\theta'. \quad (25)$$

The boundary conditions on the derivatives of A_z are

$$\left(\frac{\partial A_z}{\partial r} \right)_{r=a} = I_z/v_0; \quad \left(\frac{\partial A_z}{\partial z} \right)_{z=\delta} = \frac{\beta_0^2}{j\omega} \Phi(\delta). \quad (26)$$

Of these the first is in ref. I.31, p. 171; the second is like (13.11).

With (24) and (26) and the total axial current,

$$I'_z = 2\pi a I'_z = 2\pi a_1 I'_{1z}, \quad (27)$$

the general formula (25) reduces to the following:

$$A_z = \frac{1}{4\pi v_0} \int_{-h}^h I'_z K_1(z, z') \, dz'$$
$$+ \frac{j\beta_0^2}{4\pi \omega} \int_0^{2\pi} \int_{a_1}^a \left[\Phi(\delta) \frac{e^{-j\beta_0 R_{1\delta}}}{R_{1\delta}} \right.$$
$$\left. - \Phi(-\delta) \frac{e^{-j\beta_0 R_{2\delta}}}{R_{2\delta}} \right] r' dr' d\theta', \quad (28)$$

where, with A_z evaluated at $r = a$, z, on the surface of the antenna, $K_1(z, z')$ is as in (4) and R_1 is as in (2); $R_{1\delta}$ is the distance from the element $r'dr'd\theta'$ on the surface at $z = \delta$, $a_1 \leq r \leq a$, to the point of calculation at $r = a$, z. Similarly, $R_{2\delta}$ is the distance from the element $r'dr'd\theta'$ on the conducting surface at $z = -\delta$, $a_1 \leq r \leq a$, to the same point of calculation.

As a consequence of symmetry, $\phi(-\delta) = -\phi(\delta) = -V_\delta/2$; also $\phi(\delta)$ may be assumed approximately independent of r' and θ' as long as $\beta_0 a$ is sufficiently small compared with unity. With these considerations the general formula (28) is approximated as follows:

$$A_z \doteq \frac{1}{4\pi v_0} \int_{-h}^{h} I'_z K(z, z') \, dz'$$
$$+ \frac{j\beta_0^2 V_\delta}{8\pi\omega} \int_0^{2\pi} \int_{a_1}^{a} \left(\frac{e^{-j\beta_0 R_{1\delta}}}{R_{1\delta}} + \frac{e^{-j\beta_0 R_{2\delta}}}{R_{2\delta}} \right) r' dr' d\theta'. \quad (29)$$

An approximate evaluation of the last integral in (29) may be carried out by noting that each of the two parts is equivalent to determining the potential at $r = a$, z due to a disk of radius a with a uniformly distributed charge density. This potential is only slightly smaller than the potential on the axis at $r = 0$, z, which is readily evaluated in closed form. Thus, let

$$J = \int_0^{2\pi} \int_{a_1}^{a} \left(\frac{e^{-j\beta_0 R_{1\delta}}}{R_{1\delta}} + \frac{e^{-j\beta_0 R_{2\delta}}}{R_{2\delta}} \right) r' dr' d\theta'$$
$$\doteq 2\pi \int_{a_1}^{a} \left(\frac{e^{-j\beta_0 R_{10}}}{R_{10}} + \frac{e^{-j\beta_0 R_{20}}}{R_{20}} \right) r' dr', \quad (30)$$

where
$$R_{10} = \sqrt{r'^2 + (z-\delta)^2},$$
$$R_{20} = \sqrt{r'^2 + (z+\delta)^2}.$$

For simplicity, let $a_1 \doteq 0$ and $\delta \doteq 0$ (but so that the two adjacent end surfaces of the antenna halves are *not* actually in contact). Since $R_{10} \doteq R_{20} = R$, and $dR = r'dr'/R$, it follows that

$$J \doteq \frac{j4\pi}{\beta_0} (e^{-j\beta_0 \sqrt{a^2+z^2}} - e^{-j\beta_0 |z|}). \quad (31)$$

With (31) it follows from (29) that

$$A_z \doteq \frac{1}{4\pi v_0} \int_{-h}^{h} I'_z K(z, z') \, dz'$$
$$- \frac{V_\delta}{2v_0} (e^{-j\beta_0 \sqrt{a^2+z^2}} - e^{-j\beta_0 |z|}). \quad (32)$$

If this quantity is equated to (13.7) with $\delta = 0$ and $z^i = 0$, the result is

$$\int_{-h}^{h} I'_z K(z, z') \, dz' = \frac{-j4\pi}{\zeta_0} \Big[C_1 \cos \beta_0 z$$
$$+ \frac{jV_0}{2} (\cos \beta_0 \sqrt{z^2 + a^2} - \cos \beta_0 z)$$
$$+ \frac{V_0}{2} \sin \beta_0 \sqrt{z^2 + a^2} \Big]. \quad (33)$$

This equation was first obtained by Gans and Bemporad.[19] Note that the derivatives with respect to z of both members of (33) are continuous at $z = 0$. Note also that (33) differs from (18) only very near $z = 0$. When z^2 is large compared with a^2, for example, when $z \geq 5a$, (33) reduces to (18). At $z = 0$,

$$\int_{-h}^{h} I'_z K(0, z') \, dz'$$
$$= \frac{-j4\pi}{\zeta_0} \Big[C_1 + \frac{1}{2} V_0 \beta_0 a (1 - j\beta_0 a/2) \Big]. \quad (34)$$

Since C_1 also has the factor $\frac{1}{2} V_0$ (Sec. 15) and is necessarily large compared with $\frac{1}{2} V_0 \beta_0 a$ when the assumed condition, $\beta_0 a \ll 1$, is satisfied, there is actually little difference between (18) and (33) except that the latter is mathematically more attractive near and at $z = 0$. It is verified readily that when δ is greater than a the correction near $z = \pm \delta$ becomes even less significant.

For use with actual transmission lines, the general integral equation (3) and its modification (16) are entirely adequate for the determination of Z_δ, provided account is taken of the transmission line and any other chargeable surfaces near $z = \pm \delta$ by means of an appropriate terminal-zone network as discussed in Sec. 8, 9, and 10.

15. Series Solution of the Integral Equation

It has been assumed that the antenna is highly conducting so that the terms in z^i contribute in a negligible degree to the general formula (14.16). Let it be assumed in addition that the relative distribution function $g(z, z')$ will be determined so that it is a sufficiently good approximation of the actual distribution of current that the difference terms on the right in (14.16) are small compared with the trigonometric terms, at least for all values of z for which I_z does not itself become small.

If this is true, a good approximate solution for the current is given by the trigonometric terms alone. Let them be called a *zeroth-order solution* and denoted by $(I_z)_0$. Thus,

$$I_z \doteq (I_z)_0 = \frac{-j4\pi}{\zeta_0 \Psi \cos\beta_0 \delta}$$
$$\times \{C_1[\cos\beta_0(z-\delta) - \cos\beta_0(h-\delta)]$$
$$+ \tfrac{1}{2}V_\delta[\sin\beta_0 z - \sin\beta_0 h]\}.$$
$$(\delta \leq z \leq h) \quad (1a)$$

Since

$$\cos\beta_0(z-\delta) = \cos\beta_0 z \cos\beta_0 \delta + \sin\beta_0 z \sin\beta_0 \delta$$

it is correct to write (1a) as follows:

$$(I_z)_0 = \frac{-j4\pi}{\zeta_0 \Psi \cos\beta_0 \delta}$$
$$\times [C_1 \cos\beta_0 \delta (\cos\beta_0 z - \cos\beta_0 h)$$
$$+ (C_1 \sin\beta_0 \delta + \tfrac{1}{2}V_\delta)(\sin\beta_0 |z| - \sin\beta_0 h)]. \quad (1b)$$

Insertion of the absolute-value sign in $\sin\beta_0|z|$ makes (1b) correct for $-h \leq z \leq -\delta$ as well as for $\delta \leq z \leq h$. Clearly, $I(-z) = I(z)$. This form is required when (1b) is used in integrals evaluated over the entire antenna.

For convenience, let the following notation be introduced:

$$F_0(z) \equiv \cos\beta_0 z; \qquad F_0(h) \equiv \cos\beta_0 h;$$
$$F_{0z} \equiv F_0(z) - F_0(h); \quad (2a)$$
$$G_0(z) \equiv \sin\beta_0 |z|; \qquad G_0(h) \equiv \sin\beta_0 h;$$
$$G_{0z} \equiv G_0(z) - G_0(h). \quad (2b)$$

With (2a, b), (1b) becomes

$$(I_z)_0 = \frac{-j4\pi}{\zeta_0 \Psi \cos\beta_0 \delta} [C_1 \cos\beta_0 \delta \, F_{0z}$$
$$+ (C_1 \sin\beta_0 \delta + \tfrac{1}{2}V_\delta)G_{0z}]. \quad (3)$$

The actual current is

$$I_z = (I_z)_0 + (I_z)_c, \quad (4)$$

where $(I_z)_c$ is a correction term consisting of all of (14.16) except the terms corresponding to (1). This correction term can be evaluated approximately by substituting $(I_z)_0$ given by (3) wherever I_z (or I'_z) appears. Let this approximate correction term be denoted by $(I_z)_{c1}$. It is

$$(I_z)_{c1} = \frac{-j4\pi}{\zeta_0 \Psi^2 \cos\beta_0 \delta} [C_1 \cos\beta_0 \delta \, F_{1z}$$
$$+ (C_1 \sin\beta_0 \delta + \tfrac{1}{2}V_\delta)G_{1z}], \quad (5)$$

where

$$F_{1z} = F_1(z) - F_1(h), \quad (6a)$$
$$G_{1z} = G_1(z) - G_1(h), \quad (6b)$$

and where,

$$F_1(z) = \frac{j4\pi z^i}{\zeta_0} \int_\delta^z F_{0s} \sin\beta_0(z-s)\, ds$$
$$- F_{0z}\gamma(z) - \left(\int_{-h}^h - \int_{-\delta}^\delta\right)$$
$$[F_{0z'} - F_{0z} g(z, z')] K_1(z, z')\, dz', \quad (7a)$$

$$F_1(h) = \frac{j4\pi z^i}{\zeta_0} \int_\delta^h F_{0s} \sin\beta_0(h-s)\, ds$$
$$- \left(\int_{-h}^h - \int_{-\delta}^\delta\right) F_{0z'} K_1(h, z')\, dz', \quad (7b)$$

$$G_1(z) = \frac{j4\pi z^i}{\zeta_0} \int_\delta^z G_{0s} \sin\beta_0(z-s)\, ds$$
$$- G_{0z}\gamma(z) - \left(\int_{-h}^h - \int_{-\delta}^\delta\right)$$
$$[G_{0z'} - G_{0z} g(z, z')] K_1(z, z')\, dz', \quad (7c)$$

$$G_1(h) = \frac{j4\pi z^i}{\zeta_0} \int_\delta^h G_{0s} \sin\beta_0(h-s)\, ds$$
$$- \left(\int_{-h}^h - \int_{-\delta}^\delta\right) G_{0z'} K_1(h, z')\, dz'. \quad (7d)$$

It is to be noted that the G-functions differ from the F functions only in having G_{0z} appear throughout in place of F_{0z}.

Pairs of integrals in (6) in the form

$$\int_{-h}^{-\delta} \{\}\, dz' + \int_\delta^h \{\}\, dz' \equiv \left(\int_{-h}^{-\delta} + \int_\delta^h\right)\{\}\, dz' \quad (8a)$$

have been replaced in (7a)–(7d) by

$$\int_{-h}^h \{\}\, dz' - \int_{-\delta}^\delta \{\}\, dz' \equiv \left(\int_{-h}^h - \int_{-\delta}^\delta\right)\{\}\, dz'. \quad (8b)$$

Since the integrands in all integrals in the form (8a) involve the current as a factor and since this is actually zero from $-\delta$ to δ, the first integral in (8b) reduces to the sum in (8a), while the second integral in (8b) vanishes. However, if desired, the current may be assumed to continue from $-\delta$ to δ with any convenient distribution, since whatever is added in the first integral in (8b) is subtracted in the second integral. For purposes of evaluating the integrals, it is more convenient to assume the distribution function $I_z = I_\delta f(z)$ in (14.5) to be defined continuously from $-h$ to h including $-\delta$ to δ, so that the form in (8b) is used in (7a)–(7d).

The expressions (7a) and (7c) may be written in the following more compact form, in which the difference integral is not shown explicitly:

$$F_1(z) = \frac{j4\pi z^i}{\zeta_0} \int_\delta^z F_{0s} \sin \beta_0(z-s) \, ds$$

$$- \left(\int_{-h}^h - \int_{-\delta}^\delta \right) F_{0z'} K_1(z, z') \, dz + F_{0z} \Psi, \tag{9a}$$

$$G_1(z) = \frac{j4\pi z^i}{\zeta_0} \int_\delta^z G_{0s} \sin \beta_0(z-s) \, ds$$

$$- \left(\int_{-h}^h - \int_{-\delta}^\delta \right) G_{0z'} K_1(z, z') \, dz' + G_{0z} \Psi. \tag{9b}$$

While formally simpler, the actual evaluation of (9a) and (9b) reduces to the evaluation of (7a) and (7c). With the proper choice of $g(z, z')$, the complex correction function $\gamma(z)$ in (7a) and (7c) is negligible except near the ends, where the current is known to vanish. It follows that $\gamma(z)$ need not be evaluated.

A better approximation of the actual current than that given by the zeroth-order solution (3) is obtained by combining the zeroth-order solution with the first-order correction term to give the following first-order solution:

$$I_z \doteq (I_z)_1 = (I_z)_0 + (I_z)_{c1}. \tag{10a}$$

With (3) and (5) substituted in (10a), the first-order current is

$$(I_z)_1 = \frac{-j4\pi}{\zeta_0 \Psi \cos \beta_0 \delta} \left[C_1 \cos \beta_0 \delta \left(F_{0z} + \frac{F_{1z}}{\Psi} \right) \right.$$

$$\left. + (C_1 \sin \beta_0 \delta + \tfrac{1}{2} V_\delta) \left(G_{0z} + \frac{G_{1z}}{\Psi} \right) \right]. \tag{10b}$$

A second-order correction factor may be evaluated by substituting (10b) instead of (5) in I_c given by all terms in (14.16) except $(I_z)_0$. The correction current so obtained is I_{c1} plus a new set of terms denoted by I_{c2}. Thus,

$$(I_z)_c \doteq (I_z)_{c1} + (I_z)_{c2}, \tag{11a}$$

where

$$(I_z)_{c2} = \frac{j4\pi}{\zeta_0 \Psi^3} [C_1 \cos \beta_0 \delta \, F_{2z} + (C_1 \sin \beta_0 \delta$$

$$+ \tfrac{1}{2} V_\delta) G_{2z}]. \tag{11b}$$

In (11b) the following notation is used:

$$F_{2z} \equiv F_2(z) - F_2(h), \tag{12a}$$

$$G_{2z} \equiv G_2(z) - G_2(h). \tag{12b}$$

The functions $F_2(z)$, $F_2(h)$, $G_2(z)$, and $G_2(h)$ are given by (7a, b, c, d) with $F_1(z)$ written for $F_0(z)$, $F_1(h)$ for $F_0(h)$, $G_1(z)$ for $G_0(z)$, and $G_1(h)$ for $G_0(h)$.

A second-order solution is given by

$$I_z \doteq (I_z)_2 = (I_z)_0 + (I_z)_{c1} + (I_z)_{c2}. \tag{13}$$

The process of successive approximation in determining the correction factor $(I_z)_c$ can be continued indefinitely in the form of a series. The nth-order correction current is

$$(I_z)_c \doteq (I_z)_{c1} + (I_z)_{c2} + (I_z)_{c3} + \cdots$$

$$+ (I_z)_{cm} + \cdots + (I_z)_{cn}. \tag{14}$$

The nth-order solution for the current is

$$I_z \doteq (I_z)_n = (I_z)_0 + \sum_{m=1}^n (I_z)_{cm}, \tag{15}$$

with

$$(I_z)_{cm} = \frac{-j4\pi}{\zeta_0 \Psi^{m+1}} [C_1 \cos \beta_0 \delta \, F_{mz}$$

$$+ (C_1 \sin \beta_0 \delta + \tfrac{1}{2} V_\delta) G_{mz}], \tag{16}$$

where

$$F_{mz} = F_m(z) - F_m(h), \tag{17a}$$

$$G_{mz} = G_m(z) - G_m(h), \tag{17b}$$

and where

$$F_m(z) = \frac{j4\pi z^i}{\zeta_0} \int_\delta^z F_{m-1,s} \sin \beta_0(z-s) \, ds$$

$$- F_{m-1,z} \gamma(z) - \left(\int_{-h}^h - \int_{-\delta}^\delta \right) [F_{m-1,z'}$$

$$- F_{m-1,z} \, g(z, z')] K_1(z, z') \, dz', \tag{18a}$$

$$F_m(h) = \frac{j4\pi z^i}{\zeta_0} \int_\delta^h F_{m-1,s} \sin \beta_0(h-s) \, ds$$

$$- \left(\int_{-h}^h - \int_{-\delta}^\delta \right) F_{m-1,z'} K_1(h, z') \, dz', \tag{18b}$$

$$G_m(z) = \frac{j4\pi z^i}{\zeta_0} \int_\delta^z G_{m-1,s} \sin \beta_0(z-s) \, ds$$

$$- G_{m-1,z} \gamma(z) - \left(\int_{-h}^h - \int_{-\delta}^\delta \right) [G_{m-1,z'}$$

$$- G_{m-1,z} \, g(z, z')] K_1(z, z') \, dz', \tag{18c}$$

$$G_m(h) = \frac{j4\pi z^i}{\zeta_0} \int_\delta^h G_{m-1,s} \sin \beta_0(h-s) \, ds$$

$$- \left(\int_{-h}^h - \int_{-\delta}^\delta \right) G_{m-1,z'} K_1(h, z') \, dz'. \tag{18d}$$

Alternative forms of (18a) and (18c) are

$$F_m(z) = \frac{j4\pi z^i}{\zeta_0} \int_\delta^z F_{m-1,s} \sin \beta_0(z-s)\, ds$$
$$+ F_{m-1,z}\Psi$$
$$- \left(\int_{-h}^h - \int_{-\delta}^\delta\right) F_{m-1,z'} K_1(z, z')\, dz', \tag{18e}$$

$$G_m(z) = \frac{j4\pi z^i}{\zeta_0} \int_\delta^z G_{m-1,s} \sin \beta_0(z-s)\, ds$$
$$+ G_{m-1,z}\Psi$$
$$- \left(\int_{-h}^h - \int_{-\delta}^\delta\right) G_{m-1,z'} K_1(z, z')\, dz'. \tag{18f}$$

Using (3) and (16) in (15), the result is

$$I_z \doteq (I_z)_n$$
$$= \frac{-j4\pi}{\zeta_0 \Psi \cos\beta_0\delta} \left[C_1 \cos \beta_0 \delta \sum_{m=0}^n \frac{F_{mz}}{\Psi^m} \right.$$
$$\left. + (C_1 \sin \beta_0\delta + \tfrac{1}{2}V_\delta) \sum_{m=0}^n \frac{G_{mz}}{\Psi^m} \right]. \tag{19}$$

The arbitrary constant C_1 can be evaluated in series by substituting (19) in (14.14). Using (2a, b) and expanding $\cos\beta_0(h-\delta)$, (14.14) can be put in the following form:

$$\frac{1}{\cos\beta_0\delta}[C_1 \cos\beta_0\delta F_0(h) + (C_1 \sin\beta_0\delta$$
$$+ \tfrac{1}{2}V_\delta) G_0(h)] - z^i \int_\delta^h I(s) \sin \beta_0(z-s)\, ds$$
$$- \frac{j\zeta_0}{4\pi}\left(\int_{-h}^h - \int_{-\delta}^\delta\right) I_z' K_1(h, z')\, dz' = 0. \tag{20}$$

If (19) is now substituted for the current wherever this appears in (20), integrals are obtained that are like (18b) and (18d) with m running from 1 to n. Hence (20) becomes

$$C_1 \cos\beta_0\delta \sum_{m=0}^n \frac{F_m(h)}{\Psi^m} + (C_1 \sin\beta_0\delta$$
$$+ \tfrac{1}{2}V_\delta) \sum_{m=0}^n \frac{G_m(h)}{\Psi^m} = 0. \tag{21}$$

Solving for C_1 gives

$$C_1 = -\tfrac{1}{2}V_\delta$$
$$\times \left\{ \frac{\sum_{m=0}^n G_m(h)/\Psi^m}{\cos\beta_0\delta \sum_{m=0}^n F_m(h)/\Psi^m + \sin\beta_0\delta \sum_{m=0}^n G_m(h)/\Psi^m} \right\}. \tag{22}$$

If (22) is substituted in (19), terms in the numerator with $\sin\beta_0\delta$ as a factor cancel and the current is:

$$I_z \doteq (I_z)_n = \frac{j2\pi V_\delta}{\zeta_0 \Psi}$$
$$\times \left\{ \frac{\sum_{m=0}^n G_m(h)/\Psi^m \cdot \sum_{m=0}^n F_{mz}/\Psi^m - \sum_{m=0}^n F_m(h)/\Psi^m \cdot \sum_{m=0}^n G_{mz}/\Psi^m}{\cos\beta_0\delta \sum_{m=0}^n F_m(h)/\Psi^m + \sin\beta_0\delta \sum_{m=0}^n G_m(h)/\Psi^m} \right\} \tag{23a}$$

If (17a) and (17b) are introduced in (23a), the terms independent of z in the numerator cancel. The simpler result is

$$I_z \doteq (I_z)_n = \frac{j2\pi V_\delta}{\zeta_0 \Psi}$$
$$\times \left\{ \frac{\sum_{m=0}^n G_m(h)/\Psi^m \cdot \sum_{m=0}^n F_m(z)/\Psi^m - \sum_{m=0}^n F_m(h)/\Psi^m \cdot \sum_{m=0}^n G_m(z)/\Psi^m}{\cos\beta_0\delta \sum_{m=0}^n F_m(h)/\Psi^m + \sin\beta_0\delta \sum_{m=0}^n G_m(h)/\Psi^m} \right\} \tag{23b}$$

This is the complete solution for the current. Depending upon the accuracy required and the rapidity of convergence, a larger or smaller number of terms must be used in order that $(I_z)_n$ be a satisfactory approximation of I_z. As will become apparent, the zeroth-order term is sufficient for some purposes, whereas for others a very much higher degree of accuracy is required. Clearly, the larger Ψ, the better is the approximation for a fixed number of terms. The solution (23b) may be written in the following shorthand form:

$$I_z = \frac{j2\pi V_\delta}{\zeta_0 \Psi} \times \left\{ \frac{\sum_{j=1}^{n}\sum_{i=1}^{n}[G_j(h)F_i(z) - F_j(h)G_i(z)]/\Psi^j\Psi^i}{\cos\beta_0\delta \sum_{j=1}^{n} F_j(h)/\Psi^j + \sin\beta_0\delta \sum_{j=1}^{n} G_j(h)/\Psi^j} \right\}. \quad (24)$$

It may be expanded into

$$I_z = \frac{j2\pi V_\delta}{\zeta_0 \Psi D} \Big\{[G_0(h)F_0(z) - F_0(h)G_0(z)]$$
$$+ \frac{1}{\Psi}[G_0(h)F_1(z) - F_0(h)G_1(z)]$$
$$+ G_1(h)F_0(z) - F_1(h)G_0(z)]$$
$$+ \frac{1}{\Psi^2}[G_0(h)F_2(z) - F_0(h)G_2(z)]$$
$$+ G_1(h)F_1(z) - F_1(h)G_1(z)$$
$$+ G_2(h)F_0(z) - F_2(h)G_0(z)] + \cdots \Big\}, \quad (25a)$$

where

$$D \equiv \cos\beta_0\delta \left[F_0(h) + \frac{F_1(h)}{\Psi} + \frac{F_2(h)}{\Psi^2} + \cdots \right]$$
$$+ \sin\beta_0\delta \left[G_0(h) + \frac{G_1(h)}{\Psi} + \frac{G_2(h)}{\Psi^2} + \cdots \right]. \quad (25b)$$

Using (2a, b), (25) may be rearranged into:

$$I_z = \frac{j2\pi V_\delta}{\zeta_0 \Psi} \times$$
$$\left[\frac{\sin\beta_0(h-z) + M_1(z)/\Psi + M_2(z)/\Psi^2 + \cdots}{\cos\beta_0(h-\delta) + A_1/\Psi + A_2/\Psi^2 + \cdots} \right]$$
$$(\delta \leq z \leq h). \quad (26)$$

The following symbolism is used in the numerator of (26):

$$M_1(z) \equiv M_1^I(z) + jM_1^{II}(z)$$
$$= F_1(z) \sin\beta_0 h - G_1(z) \cos\beta_0 h$$
$$+ G_1(h) \cos\beta_0 z - F_1(h) \sin\beta_0 z, \quad (27)$$

$$M_2(z) \equiv M_2^I(z) + jM_2^{II}(z)$$
$$= F_{2z}G_0(h) - G_{2z}F_0(h) + F_{1z}G_1(h)$$
$$- G_{1z}F_1(h) + F_{0z}G_2(h) - G_{0z}F_2(h). \quad (28a)$$

Using (2a, b) and (6a, b) in (28a), this becomes:

$$M_2(z) \equiv M_2^I(z) + jM_2^{II}(z)$$
$$= F_2(z) \sin\beta_0 h - G_2(z) \cos\beta_0 h$$
$$+ F_1(z)G_1(h) - G_1(z)F_1(h)$$
$$+ \cos\beta_0 z G_2(h) - \sin\beta_0 z F_2(h). \quad (28b)$$

In the denominator of (26),

$$A_1 \equiv A_1^I + jA_1^{II}$$
$$= F_1(h) \cos\beta_0\delta + G_1(h) \sin\beta_0\delta, \quad (29a)$$

$$A_2 \equiv A_2^I + jA_2^{II}$$
$$= F_2(h) \cos\beta_0\delta + G_2(h) \sin\beta_0\delta. \quad (29b)$$

The complex amplitude I_z in (26) can be expressed in its real and imaginary parts as follows. Let

$$I_z = I_z'' + jI_z', \quad (30)$$

$$N^I = \sin\beta_0(h-z) + M_1^I(z)/\Psi$$
$$+ M_2^I(z)/\Psi^2 + \cdots, \quad (31a)$$

$$N^{II} = M_1^{II}(z)/\Psi + M_2^{II}(z)/\Psi^2 + \cdots, \quad (31b)$$

$$D^I = \cos\beta_0 h + A_1^I/\Psi + A_2^I/\Psi^2 + \cdots, \quad (32a)$$

$$D^{II} = A_1^{II}/\Psi + A_2^{II}/\Psi^2 + \cdots. \quad (32b)$$

Also let

$$N = Ne^{j\phi_N} = N^I + jN^{II},$$
$$N = \sqrt{(N^I)^2 + (N^{II})^2},$$
$$\phi_N = \tan^{-1} N^{II}/N^I, \quad (33a)$$

$$D = De^{j\phi_D} = D^I + jD^{II},$$
$$D = \sqrt{(D^I)^2 + (D^{II})^2},$$
$$\phi_D = \tan^{-1} D^{II}/D^I. \quad (33b)$$

With this notation the complex amplitude of the current in the antenna is

$$I_z = I_z'' + jI_z' = \frac{j2\pi V_\delta}{\Psi\zeta_0} \left(\frac{N^I + jN^{II}}{D^I + jD^{II}} \right)$$
$$= \frac{j2\pi V_\delta}{\Psi\zeta_0} \left(\frac{Ne^{-j(\phi_D - \phi_N)}}{D} \right). \quad (34)$$

Accordingly, referring phase to the driving potential difference by requiring V_δ to be real,

$V_\delta = V_\delta$, it follows that the component of the current in phase with V_δ is

$$I''_z = V_\delta \left(\frac{2\pi N}{\Psi \zeta_0 D}\right) \sin(\phi_D - \phi_N). \quad (35a)$$

The component of current in phase quadrature with V_δ is

$$I'_z = V_\delta \left(\frac{2\pi N}{\Psi \zeta_0 D}\right) \cos(\phi_D - \phi_N). \quad (35b)$$

In polar form the complex current is

$$I_z = I_z e^{j\theta_I}, \quad (36)$$

where

$$I_z = \sqrt{(I'_z)^2 + (I''_z)^2} = V_\delta \left(\frac{2\pi N}{\Psi \zeta_0 D}\right), \quad (37)$$

$$\theta_I = \tan^{-1}(\phi_D - \phi_N). \quad (38)$$

The instantaneous current referred to a driving potential difference

$$V_{\delta\,\text{inst}} = V_\delta \sin \omega t = \text{imaginary part of } V_\delta e^{j\omega t} \quad (39)$$

is obtained from (36) by first multiplying by $e^{j\omega t}$ and then selecting the imaginary part. It is

$$I_{z\,\text{inst}} = V_\delta \left(\frac{2\pi N}{\Psi \zeta_0 D}\right) \sin(\omega t + \theta_I). \quad (40)$$

The imaginary part is here chosen in (39) and (40) for later convenience in following the conventional procedure of referring phases to the current I'_z as the *real* part of $I'_z e^{j\omega t}$ in the calculation of electromagnetic fields.

DISTRIBUTIONS OF CURRENT AND CHARGE

16. The Zeroth-Order Solutions for Current and Charge

The solution for the current in the form (15.26) is useful only if the parameter Ψ is sufficiently large so that

$$\left|\sum_{m=1}^{n} \frac{M_m(z)}{\Psi^m}\right|^2 < 1, \quad (1a)$$

$$\left|\sum_{m=1}^{n} \frac{A_m}{\Psi^m}\right|^2 < 1. \quad (1b)$$

If these conditions are satisfied, the leading term in the current is the *zeroth-order current* given by

$$(I_z)_0 = \frac{j2\pi V_\delta}{\zeta_0 \Psi} \frac{\sin \beta_0(h-z)}{\cos \beta_0(h-\delta)}, \quad \delta \le z \le h. \quad (2)$$

The actual current is approximated by (2) in the ranges

$$\left|\sin \beta_0(h-z)\right|^2 \gg \left|\sum_{m=1}^{n} \frac{M_m(z)}{\Psi^m}\right|^2, \quad (3a)$$

$$\left|\cos \beta_0(h-\delta)\right|^2 \gg \left|\sum_{m=1}^{n} \frac{A_m}{\Psi^m}\right|^2. \quad (3b)$$

In the ranges that satisfy *both* (3a) and (3b) the approximate current is

$$I_z \doteq (I_z)_0 = \frac{j2\pi V_\delta}{\zeta_0 \Psi} \frac{\sin \beta_0(h-z)}{\cos \beta_0(h-\delta)}. \quad (4)$$

The input current is at $z = \delta$. In the ranges satisfying (3a, b) it is

$$(I_\delta)_0 \doteq \frac{j2\pi V_\delta}{\zeta_0 \Psi} \tan \beta_0(h-\delta). \quad (5)$$

Clearly (5) has *no* application when $\beta_0(h-\delta)$ is near or at $n\pi/2$ with $n = 1, 2, \cdots$. The maximum current is at $\beta_0(h-z) = [(2n+1)/2]\pi$ or at $h - z = [(2n+1)/4]\lambda_0$. Subject to (3a, b) and with (5),

$$(I_m)_0 = \frac{j2\pi V_\delta}{\zeta_0 \Psi} \frac{1}{\cos \beta_0(h-\delta)} = \frac{(I_\delta)_0}{\sin \beta_0(h-\delta)}. \quad (6)$$

If $\beta_0(h-\delta) < \pi/2$, $(h-\delta) < \lambda_0/4$, I_m is a fictitious reference current.

Upon combining (2) and (3),

$$(I_z)_0 = (I_0)_0 \frac{\sin \beta_0(h-z)}{\sin \beta_0(h-\delta)}$$

$$= (I_m)_0 \sin \beta_0(h-z). \quad (\delta \le z \le h) \quad (7a)$$

For the symmetric antenna under consideration, $I(z)$ is an even function, so that

$$(I_z)_0 = (I_m)_0 \sin \beta_0(h+z). \quad (-h \le z \le -\delta) \quad (7b)$$

A convenient shorthand for (7a) and (7b) is

$$(I_z)_0 = (I_m)_0 \sin \beta_0(h - |z|). \quad (7c)$$

The distribution (7c) is the well-known sinusoidal current usually assumed for symmetric, center-driven antennas. Clearly it is an approximation subject to the restrictions (3a, b). Since (15.26) reduces to (7a) strictly only when

$$\Psi \to \infty, \quad (8)$$

it is clear from (2) that a nonvanishing current is possible only when V: is infinite.

It is shown later that (8) is true for an infinitely thin and, hence, physically unavailable antenna. It is also shown that Ψ increases as the ratio h/a increases.

Although (7a) is the limiting form of (15.26) as the radius of an antenna of finite length is allowed to approach zero, it is not a correct solution of the original integral equation (14.3) for an antenna of zero radius. Expansion in powers of $1/\Psi$ is permissible only when Ψ is finite. It is shown in Chapter V that the axial component of the electric field, E_z, due to a sinusoidally distributed current is not zero in the range $-h \leq z \leq -\delta$, $\delta \leq z \leq h$, but has a finite value. Actually, *a sinusoidal distribution of current is not possible even for an infinitely thin antenna if there is only a single generator at the center. A continuous distribution of generators along the entire antenna is required with emf's of proper amplitudes and phases.* However, for large values of Ψ corresponding to very large ratios of h/a, the sinusoidal distribution given in (7a) is a fair approximation over most of the antenna if this is not too long.

Curves for $(I_z)_0/(I_m)_0$ computed from (7a) are shown in Fig. 16.1. It is clear from (3) that, for a given applied voltage and a fixed value of Ψ, I_m is smallest for $\beta_0(h-\delta) = 0$, π, 2π, etc.; largest (infinite) for $\beta_0(h-\delta) = \pi/2$, $3\pi/2$, $5\pi/2$.

The zeroth-order distribution of charge on the antenna is readily obtained using the equation of continuity

$$\frac{dI_z}{dz} + j\omega q_z = 0 \qquad (9)$$

with (7a). Thus,

$$(q_z)_0 = \frac{j}{\omega}\frac{d}{dz}[(I_m)_0 \sin \beta_0(h-z)]$$

$$= \frac{-j(I_m)_0}{v_0} \cos \beta_0(h-z), \quad (z \geq 0) \qquad (10a)$$

$$(q_z)_0 = \frac{j}{\omega}\frac{d}{dz}[(I_m)_0 \sin \beta_0(h+z)]$$

$$= \frac{j(I_m)_0}{v_0} \cos \beta_0(h+z). \quad (z \leq 0) \qquad (10b)$$

For convenience, let

$$(q_z)_0 = \frac{-j(I_m)_0}{v_0} = \frac{2\pi\varepsilon_0 V_\delta}{\Psi \cos \beta_0(h-\delta)}. \qquad (11)$$

Then

$$(q_z)_0 = \pm(q_h)_0 \cos \beta_0(h - |z|);$$

upper sign for $z \geq \delta$,

lower sign for $z \leq -\delta$. (12)

The distribution (12) is shown in Fig. 16.2.

17. Simplification for Small Base Separation

The general solution obtained in Sec. 15 for the distribution of current in a symmetric center-driven antenna involves no restriction on the base separation 2δ. However, if the antenna is driven from a two-wire line, and it is desired to analyze the line using conventional theory which neglects radiation from the line, the distance b between centers of the conductors is necessarily limited by the inequality (6.22a), which for air as dielectric is

$$\beta_0 b \ll 1, \qquad (1a)$$

or at least by (6.22b),

$$\beta_0^2 b^2 \ll 4\pi \ln(b/a). \qquad (1b)$$

If the transmission line satisfies (1a) and if 2δ is also very small compared with the length of the antenna, the following inequalities apply:

$$\beta_0 \delta \ll 1, \qquad (2)$$

$$\delta \ll h. \qquad (3)$$

In practice, many antenna systems satisfy (2) and (3).

It is to be expected that if 2δ is as short as required in (2) and (3), considerable simplification should be possible in the intricate general formulas for the current in Sec. 15. The nature of the simplification is suggested in the zeroth-order solution in Sec. 16. From this it is clear that the zeroth-order current, including especially the input current which defines the impedance, is the same for an antenna with a base separation 2δ as for an antenna of the *same length of conductor* with $\delta = 0$.

Referring to (16.2), the current at a distance $\bar{z} = z - \delta$ from the driving terminal $z = \delta$, $\bar{z} = 0$, in an antenna of half-length of conductor $\bar{h} = h - \delta$ and separation half-length δ, is

$$(I_z)_0 = \frac{j2\pi V_\delta}{\zeta_0 \Psi}\frac{\sin \beta_0[(h-\delta)-(z-\delta)]}{\cos \beta_0(h-\delta)}$$

$$= \frac{j2\pi V_\delta}{\zeta_0 \Psi}\frac{\sin \beta_0(\bar{h}-\bar{z})}{\cos \beta_0 \bar{h}}. \qquad (4)$$

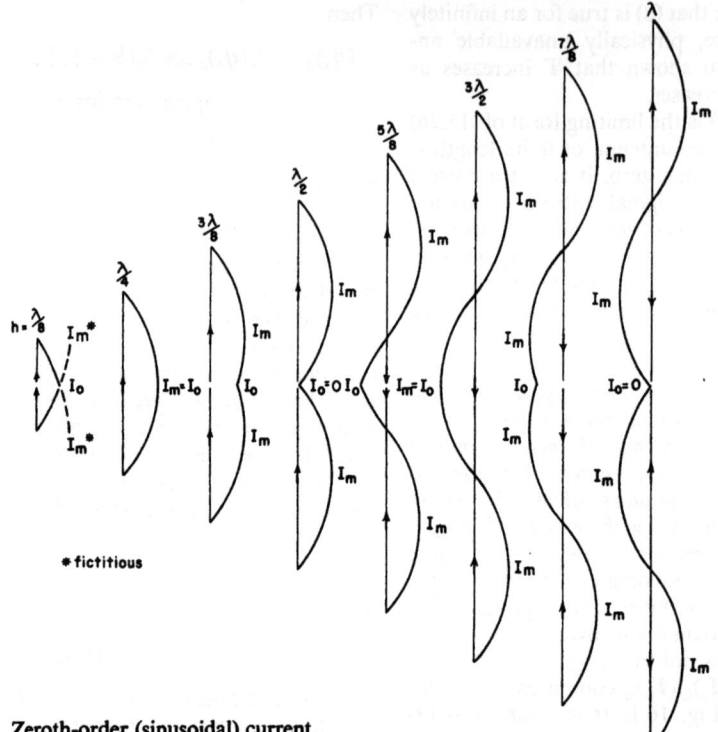

Fig. 16.1. Zeroth-order (sinusoidal) current.

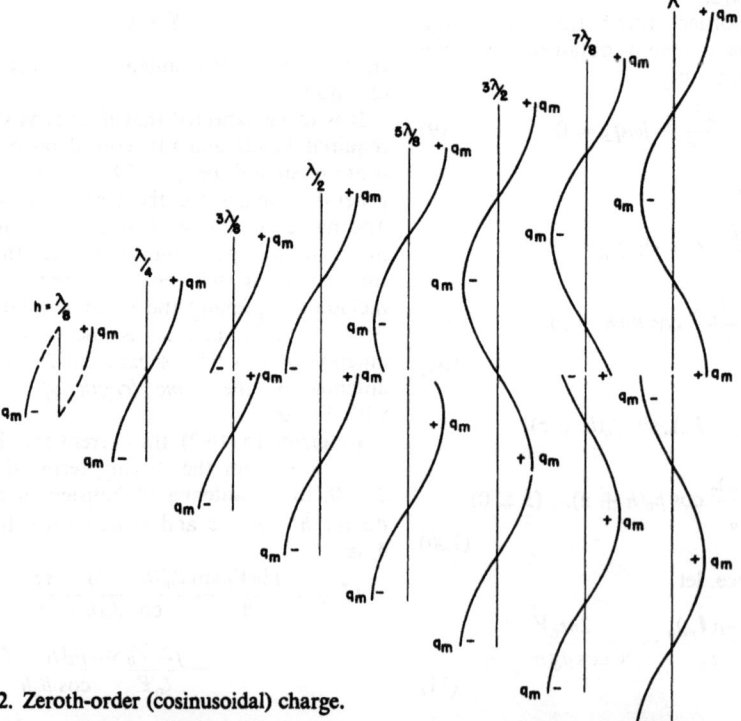

Fig. 16.2. Zeroth-order (cosinusoidal) charge.

The current at a distance $z = \bar{z}$ from the driving terminal $z = \bar{z} = 0$ in an antenna of half-length of conductor $h = \bar{h}$ and with $\delta = 0$ is

$$(I_z)_0 = \frac{j2\pi V_{\bar{0}}}{\zeta_0 \Psi} \frac{\sin \beta_0(\bar{h} - \bar{z})}{\cos \beta_0 \bar{h}}$$

$$= \frac{j2\pi V_0}{\zeta_0 \Psi} \frac{\sin \beta_0(h - z)}{\cos \beta_0 h}. \quad (5)$$

Clearly, the distribution along the conductor as measured from the ends is the same for small values of δ as for $\delta = 0$ if $V_\delta = V_{\bar{0}} = V_0$.

This identity of distribution of zeroth-order current along the conductor, even for quite large values of δ, shows that the effect of the base separation in determining the distribution and magnitude of the current is less important than the length of the conductor. This is reasonable and could have been anticipated directly from physical arguments. Thus, the length 2δ in no way affects the interaction of elements of current that are *both* on the same half of the antenna. It plays a part only in the interaction of two elements of current of which the first is on one half, the second on the other half of the antenna. The over-all effect of the currents in the lower half of the antenna acting upon the current in the upper half is diminished a very little in amplitude and retarded slightly in phase as 2δ is increased but kept small. However, if 2δ is no longer than permitted by (2) and (3), the over-all change in amplitude and shift in phase are both essentially negligible. Thus, over most of the length of the antenna,

$$\frac{1}{R + 2\delta} \doteq \frac{1}{R}, \quad e^{-j\beta_0(R+2\delta)} \doteq e^{-j\beta_0 R} \quad (6)$$

if (2) and (3) are valid.

To neglect the effect of the base separation in the interaction of currents in the halves of the antenna is analytically equivalent to setting $\delta = 0$ in the general formulas and then writing $\bar{h} = h - \delta$ for h, $\bar{z} = z - \delta$ for z, and $\bar{z}' = z' - \delta$ for z'. Specifically,

$$\int_{-h}^{-\delta} \frac{I_z'}{R_1} e^{-j\beta_0 R_1} dz' + \int_{\delta}^{h} \frac{I_z'}{R_1} e^{-j\beta_0 R_1} dz'$$

$$\doteq \int_{-\bar{h}}^{\bar{h}} \frac{I_z(\bar{z}')}{R_1} e^{-j\beta_0 R_1} d\bar{z}'. \quad (7)$$

Thus the general formula (15.26) reduces exactly to the same form as for $\delta = 0$ except that $\bar{h}$, $\bar{z}$, I_z, and $V_{\bar{0}} = V_\delta$ occur in place of h, z, I_z, and V_δ. The solution for a small base separation thus reduces to the solution for zero separation if distances are measured along the conductor and not from the center.

In the following sections the analysis is carried out in general without restricting δ to be sufficiently small to permit the approximation (7). It is understood, however, that the solution for very small separations may be determined approximately where convenient by setting $\delta = 0$ and substituting $\bar{h} = h - \delta$, $\bar{z} = z - \delta$, $\bar{z}' = z' - \delta$, and $V_{\bar{0}} = V_\delta$ for h, z, z', and V_δ. It is shown later that this is a better approximation for antennas for which h is near $n\lambda_0/4$, n odd, than for antennas with h near $n\lambda_0/4$, n even.

18. Hallén's Expansion

The evaluation of (15.26) presupposes the determination of a suitable expansion parameter Ψ, and this, in turn, depends upon the selection of an appropriate distribution function $g(z, z')$ in the integrals (14.7), namely,

$$\Psi_\delta(z) = 4\pi v_0 \frac{(A_z)_{r=a}}{I_z}$$

$$= \left(\int_{-h}^{-\delta} + \int_{\delta}^{h} \right) g(z, z') K_1(z, z') \, dz', \quad (1)$$

where

$$K_1(z, z') \equiv \frac{e^{-j\beta_0 R_1}}{R_1}, \quad R_1 = \sqrt{(z-z')^2 + a^2}, \quad (2)$$

and

$$g(z, z') = I(z')/I(z) \equiv I_z'/I_z. \quad (3)$$

The selection of a function $g(z, z')$ is equivalent physically to the choice of an approximate current as a function of z' in the integral for the vector potential, that is, in

$$(A_z)_{r=a} = \frac{1}{4\pi v_0} \left(\int_{-h}^{-\delta} + \int_{\delta}^{h} \right) I_z' K_1(z, z') \, dz'. \quad (4)$$

It is well known that a process of iteration, such as that used in obtaining the series (15.26) for the current, is improved in accuracy the better the arbitrarily selected approximate distribution represents the true but unknown distribution. Since the evaluation of many terms in the series (15.26) is laborious, it is important to select a distribution that approximates I_z' closely. On the other hand, since the function $\Psi_\delta(z)$ must be evaluated, the approximate current chosen must be both a good representation of the actual current and at the same time not so complex as to make impossible the determination of $\Psi_\delta(z)$ in a reasonably simple form. Roughly speaking, the more closely the current in (4) is approximated, the more intricate will be the evaluation of and the final formula for the expansion parameter Ψ, and the more will the series (15.26) be improved in accuracy.

Of the several approximate distributions of current that have been used, two are described: that used by Hallén,[25] which is both the original one and the simplest, and that introduced by King and Middleton,[48] which seeks to approximate closely the actual current. In view of the fact that the representation due to King and Middleton may be expressed conveniently in terms of the integrals used in the solution in the Hallén form, the latter is carried out first. This achieves the double purpose of preparing the way for the more intricate solution, and simultaneously and appropriately of emphasizing the pioneer work done by Hallén.

The distribution function implicit in the analysis of Hallén is the same as that used in the derivation of the transmission-line equations in ref. I.31, chap. VI. It is the function that would be used most naturally in order to solve directly the integral in (14.3), whereas the distribution function introduced by King and Middleton follows naturally from the expanded form (15.26). The function involved in the solution by Hallén may be justified physically by the following simple argument.

The vector potential at a point z on the surface $r = a$ of the antenna is expressed in (4) as the sum of the retarded effects of the elements of current at all points z' on the axis divided by the distances between them and z. The principal contribution to $(A_z)_{r=a}$ is from currents in the adjacent parts of the conductors for which R_1 is of the order of magnitude of small multiples of the small quantity a, and the exponential is practically unity. This is true except at and near points where I_z is itself very small compared with I_z' elsewhere along the antenna. Since the significant contributions are due to the current in adjacent elements of the conductor, a rough approximation is to assume that the current is everywhere the same as at the point where $(A_z)_{r=a}$ is evaluated, so that for all values of z', $I_z' \doteq I_z$. If the currents at points not near z are relatively so unimportant that this representation is at all possible, the effect of retardation must be negligible as well. The suggested representation leads to the following very simple equivalent of (14.9):

$$\left(\int_{-h}^{-\delta} + \int_{-\delta}^{h}\right) I_z' K_1(z, z') \, dz'$$

$$= I_z \left(\int_{-h}^{h} - \int_{-\delta}^{\delta}\right) \frac{dz'}{R_1}$$

$$+ \left(\int_{-h}^{h} - \int_{-\delta}^{\delta}\right) [I_z' e^{-j\beta_0 R_1} - I_z] \frac{dz'}{R_1}. \tag{5}$$

The relative distribution function implicit in (5) is

$$g_H(z, z') = e^{j\beta_0 R_1}. \tag{6}$$

This function obviously does not represent or in any measure approximate the actual distribution of current. With (6),

$$\Psi_{\delta H}(z) = \left(\int_{-h}^{h} - \int_{-\delta}^{\delta}\right) \frac{dz'}{R_1}$$

$$\equiv \Omega(h, z) - \Omega(\delta, z), \tag{7}$$

where

$$\Omega(h, z) \equiv \sinh^{-1}\left(\frac{h+z}{a}\right) + \sinh^{-1}\left(\frac{h-z}{a}\right)$$

$$= \ln \frac{z + h + \sqrt{(z+h)^2 + a^2}}{z - h + \sqrt{(z-h)^2 + a^2}}, \tag{8}$$

$$\Omega(\delta, z) \equiv \sinh^{-1}\left(\frac{z+\delta}{a}\right) - \sinh^{-1}\left(\frac{z-\delta}{a}\right). \tag{9}$$

Alternatively, in terms of the natural logarithm,

$$\Omega(h, z) = \Omega + \ln\left(1 - \frac{z^2}{h^2}\right) + \Delta(z), \tag{10}$$

where

$$\Omega \equiv 2 \ln \frac{2h}{a}, \tag{11}$$

$$\Delta(z) = \ln \tfrac{1}{4}\left\{\left[\sqrt{1 + \left(\frac{a}{h-z}\right)^2} + 1\right] \right.$$

$$\left. \times \left[\sqrt{1 + \left(\frac{a}{h+z}\right)^2} + 1\right]\right\}. \tag{12}$$

Figure 18.1 shows Ω plotted as a function of h/a. The value of $\Omega(h, z)$ at $z = \pm h$ is

$$\Omega(h, \pm h) = \sinh^{-1} \frac{2h}{a} \doteq \ln \frac{4h}{a} = \frac{1}{2}\Omega + \ln 2. \tag{13}$$

Terms in a^2 are neglected compared with h^2 in the logarithmic form of (13).

The function $\Omega(\delta, z)$ is shown in Fig. 18.2 for the values $\delta/a = 0.5, 1, 2, 5, 10$. The part

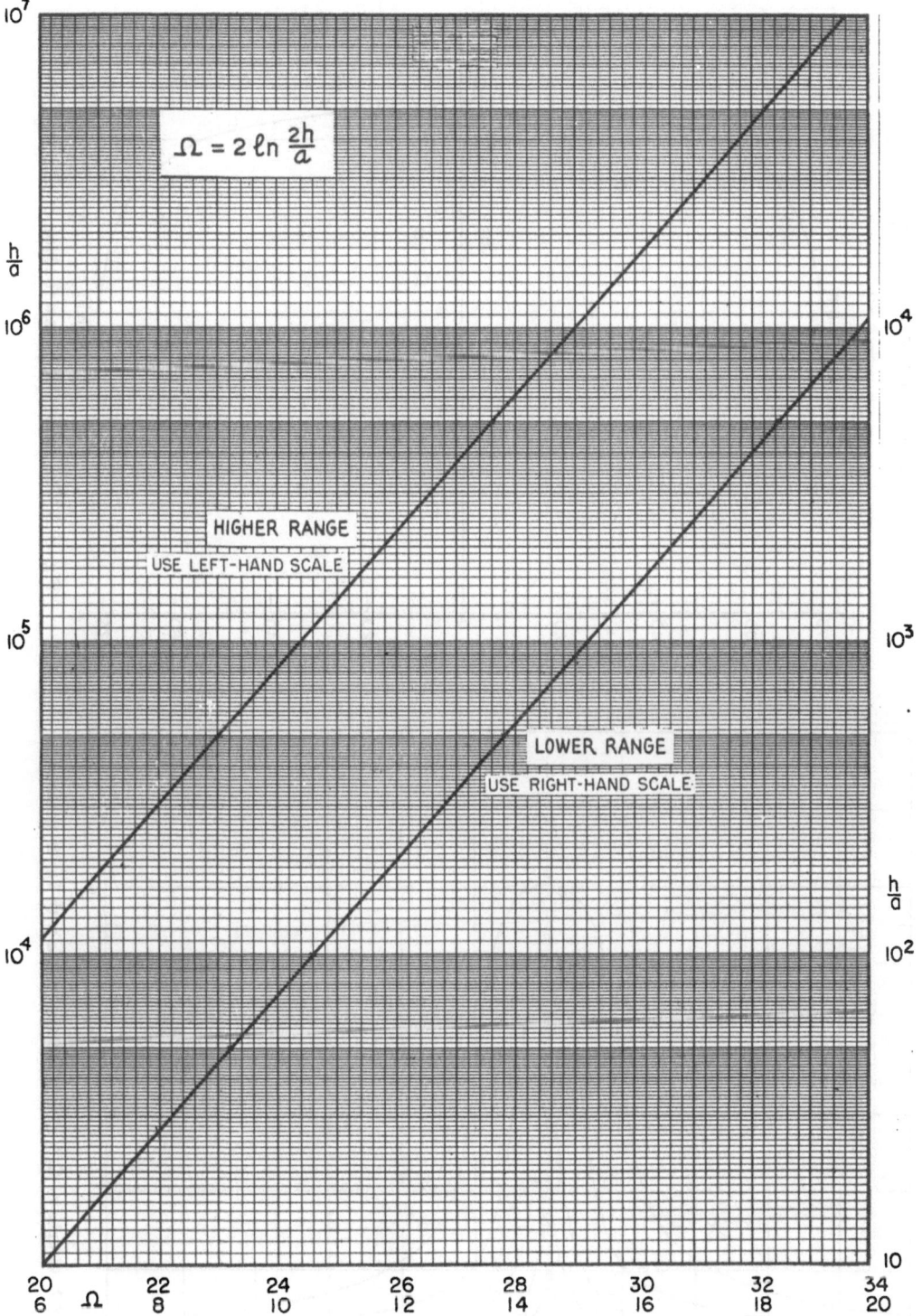

Fig. 18.1. Graphs of Ω as a function of h/a.

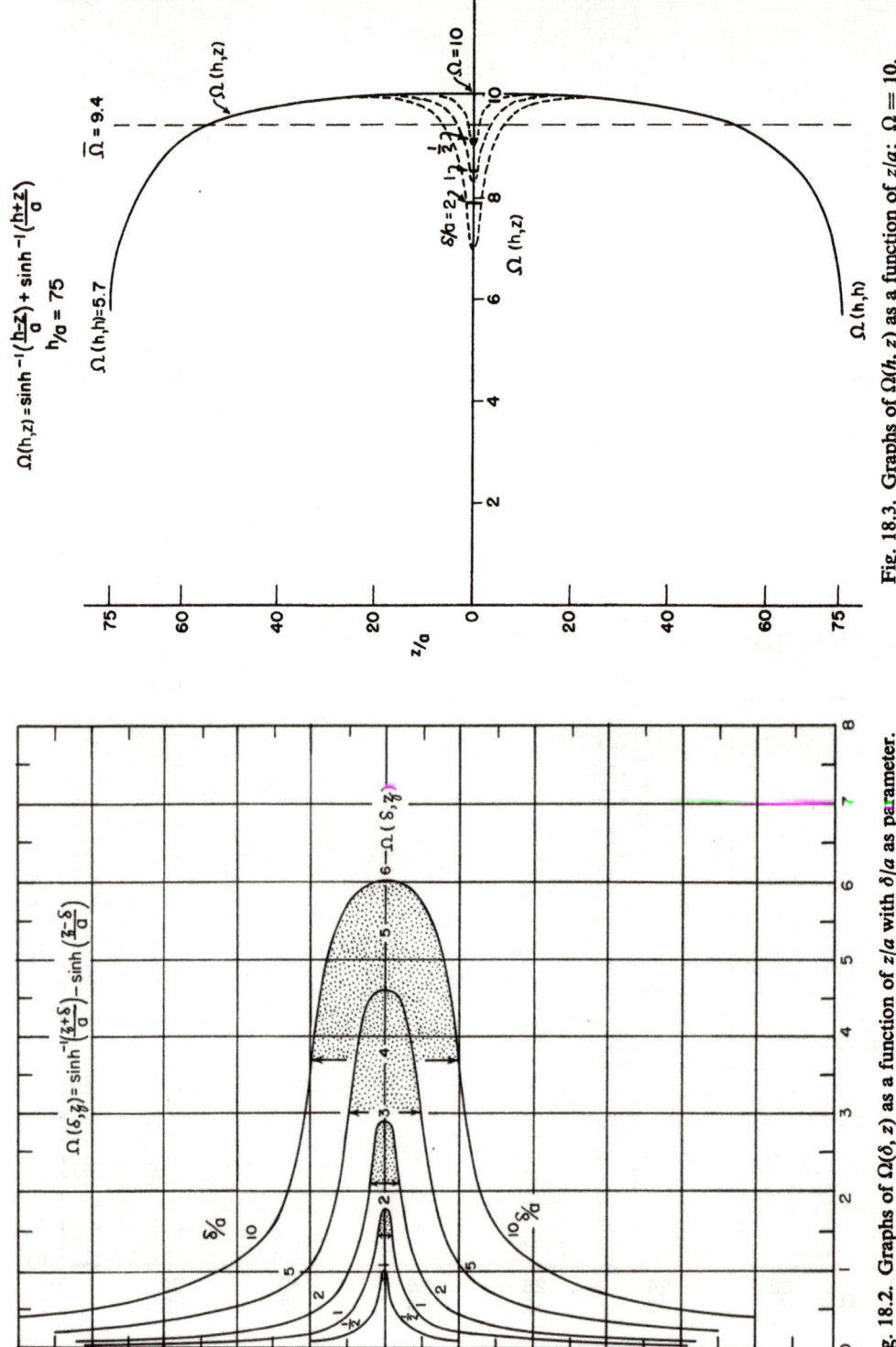

Fig. 18.3. Graphs of $\Omega(h, z)$ as a function of z/a; $\Omega = 10$.

Fig. 18.2. Graphs of $\Omega(\delta, z)$ as a function of z/a with δ/a as parameter.

of the function between $z = \delta$ and $z = -\delta$ is between the conducting halves of the antenna.

The function $\Omega(h, z)$ is shown in Fig. 18.3 for $\Omega = 10$, or $h/a \doteq 75$. It is seen that $\Omega(h, z)$ is reasonably constant over the greater part of the length of the antenna. At $z = \pm h$ it is not infinite, as it would have to be if the assumed distribution function were accurate. In fact, $\Omega(h, h)$ is the smallest value of $\Omega(h, z)$. The decrease in $\Omega(h, z)$ due to $\Omega(\delta, z)$ is also shown in Fig. 18.3 for $\delta/a = 0.5, 1, 2$ corresponding to values of $h/\delta \doteq 150, 75, 37.5$. Even for the smallest of these values, which does not actually satisfy $\delta \ll h$, the curve is modified significantly only quite near the gap in the conductors. For values of h/δ that satisfy $\delta \ll h$, the effect of $\Omega(\delta, z)$ in determining a value of $\Psi_{\delta H}(z)$ which is approximately constant along the antenna is not significant.

For sufficiently large values of h/a and h/δ the function $\Psi_{\delta H}(z)$ may be approximated by $\Omega(h, z)$ and this may be represented by a constant value which might perhaps be chosen as the average value of $\Omega(h, z)$, given by $\bar{\Omega} = \Omega - 0.614$. For large values of h/a the difference between the average and the maximum value Ω is not large. Following the original choice by Hallén, Ω rather than $\bar{\Omega}$ is used as the expansion parameter. Thus,

$$\Psi_{\delta H}(z) = \Psi_H + \gamma_H(z) \qquad (14)$$

is represented by

$$\Omega(h, z) = \Omega + \ln\left(1 - \frac{z^2}{h^2}\right) + \Delta(z)$$
$$- \Omega(\delta, z), \qquad (15a)$$

so that

$$\Psi = \Psi_H = \Omega, \qquad (15b)$$

$$\gamma(z) = \gamma_H(z) = \ln\left(1 - \frac{z^2}{h^2}\right) + \Delta(z)$$
$$- \Omega(\delta, z). \qquad (15c)$$

The parameter $\Omega = 2 \ln (2h/a)$ is a convenient measure of the thickness of an antenna.

It is to be noted that if the substitution suggested in Sec. 17 for very small values of δ is made, so that $\bar{h} = h - \delta$ replaces h, the difference between the new and the original is negligible if $\delta \ll h$. Thus

$$\Omega = 2\ln\frac{2h}{a} \doteq 2\ln\frac{2(\bar{h} + \delta)}{a}$$
$$= 2\ln\frac{2\bar{h}}{a}\left(1 + \frac{\delta}{\bar{h}}\right) = 2\left(\ln\frac{2\bar{h}}{a} + \frac{\delta}{\bar{h}}\right). \qquad (16a)$$

Since Ω must be of magnitude 10 or greater, and δ/h should not exceed 0.01 for the approximation made, it follows that

$$\Omega = 2\ln\frac{2h}{a} \doteq 2\ln\frac{2\bar{h}}{a}. \qquad (16b)$$

With (15b) in (15.26), the solution due to Hallén[*] is

$$I_z = \frac{j2\pi V_\delta}{\Omega \zeta_0}$$

$$\times \begin{bmatrix} \sin\beta_0(h-z) + (1/\Omega)(M^{\mathrm{I}}_{1H} + jM^{\mathrm{II}}_{1H}) \\ + (1/\Omega^2)(M^{\mathrm{I}}_{2H} + jM^{\mathrm{II}}_{2H}) + \cdots \\ \hline \cos\beta_0(h-\delta) + (1/\Omega)(A^{\mathrm{I}}_{1H} + jA^{\mathrm{II}}_{1H}) \\ + (1/\Omega^2)(A^{\mathrm{I}}_{2H} + jA^{\mathrm{II}}_{2H}) + \cdots \end{bmatrix}.$$
$$(17)$$

The functions of M and A are defined in terms of the F- and G-functions as in (15.27). A subscript H is added to indicate the Hallén form. The functions $F_m(z)$ and $F_m(h)$, as defined in (15.18), become:

$$F_{mH}(z) = \frac{j4\pi z^i}{\zeta_0}\int_\delta^z F_{m-1,s}\sin\beta_0(z-s)\,ds$$
$$- F_{m-1,z}\left[\ln\left(1 - \frac{z^2}{h^2}\right) + \Delta(z) - \Omega(\delta, z)\right]$$
$$- \left(\int_{-h}^h - \int_{-\delta}^\delta\right)[F_{m-1,z'}e^{-j\beta_0 R_1} - F_{m-1,z}]\frac{dz'}{R_1},$$
$$(18a)$$

$$F_{mH}(h) = \frac{j4\pi z^i}{\zeta_0}\int_\delta^h F_{m-1,s}\sin\beta_0(h-s)\,ds$$
$$- \left(\int_{-h}^h - \int_{-\delta}^\delta\right)F_{m-1,z'}K_1(h, z')\,dz',$$
$$(18b)$$

$$G_{mH}(z) = \frac{j4\pi z^i}{\zeta_0}\int_\delta^z G_{m-1,s}\sin\beta_0(z-s)\,ds$$
$$- G_{m-1,z}\left[\ln\left(1 - \frac{z^2}{h^2}\right) + \Delta(z) - \Omega(\delta, z)\right]$$
$$- \left(\int_{-h}^h - \int_{-\delta}^\delta\right)[G_{m-1,z'}e^{-j\beta_0 R_1} - G_{m-1,z}]\frac{dz'}{R_1},$$
$$(18c)$$

$$G_{mH}(h) = \frac{j4\pi z^i}{\zeta_0}\int_\delta^h G_{m-1,s}\sin\beta_0(h-s)\,ds$$
$$- \left(\int_{-h}^h - \int_{-\delta}^\delta\right)G_{m-1,z'}K_1(h, z')\,dz'.$$
$$(18d)$$

---

[*] Hallén's solution is for $\delta = 0$; see ref. 25.

Alternatively,

$$F_{mH}(z) = \frac{j4\pi z^i}{\zeta_0} \int_\delta^z F_{m-1,s} \sin\beta_0(z-s)\, ds$$
$$+ F_{m-1,z}\Omega$$
$$- \left(\int_{-h}^h - \int_{-\delta}^\delta\right) F_{m-1,z'} K_1(z, z')\, dz', \quad (18e)$$

$$G_{mH}(z) = \frac{j4\pi z^i}{\zeta_0} \int_\delta^z G_{m-1,s} \sin\beta_0(z-s)\, ds$$
$$+ G_{m-1,z}\Omega$$
$$- \left(\int_{-h}^h - \int_{-\delta}^\delta\right) G_{m-1,z'} K_1(z, z')\, dz'. \quad (18f)$$

19. Hallén's Solution: Evaluation of First-Order Integrals

The first-order functions appearing in the Hallén form of the solution with $\delta = 0$ are given below.

$$F_{1H}(z) = \frac{j4\pi z^i}{\zeta_0} \int_0^z F_{0s} \sin\beta_0(z-s)\, ds$$
$$- F_{0z}\left[\ln\left(1 - \frac{z^2}{h^2}\right) + \Delta(z)\right]$$
$$- \int_{-h}^h \left[F_{0z'} K_1(z, z') - \frac{F_{0z}}{R_1}\right] dz', \quad (1a)$$

$$F_{1H}(h) = \frac{j4\pi z^i}{\zeta_0} \int_0^h F_{0s} \sin\beta_0(h-s)\, ds$$
$$- \int_{-h}^h F_{0z'} K_1(h, z')\, dz', \quad (1b)$$

$$G_{1H}(z) = \frac{j4\pi z^i}{\zeta_0} \int_0^z G_{0s} \sin\beta_0(z-s)\, ds$$
$$- G_{0z}\left[\ln\left(1 - \frac{z^2}{h^2}\right) + \Delta(z)\right]$$
$$- \int_{-h}^h \left[G_{0z'} K_1(z, z') - \frac{G_{0z}}{R_1}\right] dz', \quad (1c)$$

$$G_{1H}(h) = \frac{j4\pi z^i}{\zeta_0} \int_0^h G_{0s} \sin\beta_0(z-s)\, ds$$
$$- \int_{-h}^h G_{0z'} K_1(h, z')\, dz'. \quad (1d)$$

As before,
$$F_{0z} = F_0(z) - F_0(h) = \cos\beta_0 z - \cos\beta_0 h, \quad (2a)$$
$$G_{0z} = G_0(z) - G_0(h) = \sin\beta_0|z| - \sin\beta_0 h. \quad (2b)$$

For convenience in the evaluation, let the following functions be introduced:

$$C_a(h, z) = \int_{-h}^h \cos\beta_0 z' \frac{e^{-j\beta_0 R_1}}{R_1}\, dz'$$
$$= \int_0^h \cos\beta_0 z' \left(\frac{e^{-j\beta_0 R_1}}{R_1} + \frac{e^{-j\beta_0 R_2}}{R_2}\right) dz', \quad (3)$$

$$S_a(h, z) = \int_{-h}^h \sin\beta_0|z'| \frac{e^{-j\beta_0 R_1}}{R_1}\, dz'$$
$$= \int_0^h \sin\beta_0 z' \left(\frac{e^{-j\beta_0 R_1}}{R_1} + \frac{e^{-j\beta_0 R_2}}{R_2}\right) dz', \quad (4)$$

$$E_a(h, z) = \int_{-h}^h \frac{e^{-j\beta_0 R_1}}{R_1}\, dz'$$
$$= \int_0^h \left(\frac{e^{-j\beta_0 R_1}}{R_1} + \frac{e^{-j\beta_0 R_2}}{R_2}\right) dz', \quad (5)$$

with
$$R_1 = \sqrt{(z-z')^2 + a^2}, \quad (6)$$
$$R_2 = \sqrt{(z+z')^2 + a^2}. \quad (7)$$

Also let
$$J_c(z) \equiv \int_0^z \cos\beta_0 s \sin\beta_0(z-s)\, ds$$
$$- \cos\beta_0 h \int_0^z \sin\beta_0(z-s)\, ds, \quad (8)$$

$$J_s(z) \equiv \int_0^z \sin\beta_0|s| \sin\beta_0(z-s)\, ds$$
$$- \sin\beta_0 h \int_0^z \sin\beta_0(z-s)\, ds. \quad (9)$$

From (18.11)
$$\Omega = 2\ln(2h/a). \quad (10)$$

In terms of the integrals (3)–(5)

$$F_{1H}(z) = \frac{j4\pi z^i}{\zeta_0} J_c(z) + \Omega F_{0z} - C_a(h, z)$$
$$+ E_a(h, z)\cos\beta_0 h, \quad (11a)$$

$$F_{1H}(h) = \frac{j4\pi z^i}{\zeta_0} J_c(h) - C_a(h, h)$$
$$+ E_a(h, h)\cos\beta_0 h, \quad (11b)$$

$$G_{1H}(z) = \frac{j4\pi z^i}{\zeta_0} J_s(z) + \Omega G_{0z} - S_a(h, z)$$
$$+ E_a(h, z)\sin\beta_0 h, \quad (11c)$$

$$G_{1H}(h) = \frac{j4\pi z^i}{\zeta_0} J_s(h) - S_a(h, h)$$
$$+ E_a(h, h)\sin\beta_0 h. \quad (11d)$$

Integrals in (8) and (9) may be evaluated directly. There is no difficulty with (8) since no magnitude signs appear, and it is valid for positive and negative values of z and s. The result is

$$J_c(z) = \tfrac{1}{2} z \sin \beta_0 z + \frac{\cos \beta_0 h}{\beta_0} (1 - \cos \beta_0 z). \tag{12}$$

The magnitude sign in (9) may be removed as long as integration is exclusively over positive values of z. If this is done, the integral is

$$J_s(z) = -\tfrac{1}{2} z \cos \beta_0 z + \frac{\sin \beta_0 z}{2\beta_0}$$
$$- \frac{\sin \beta_0 h}{\beta_0}(1 - \cos \beta_0 z), \quad z \geq 0. \tag{13}$$

In order to evaluate (9) for negative values of z and hence of s, it is necessary to preserve the positive sign of $\sin \beta_0|s|$ by writing it $(-\sin \beta_0 s)$ with $s \leq 0$. This need not be done for other terms involving s. Hence, with $-z$ written for z,

$$J_s(-z) = -\int_0^{-z} \sin \beta_0 s \sin \beta_0(-z - s)\, ds$$
$$- \sin \beta_0 h \int_0^{-z} \sin \beta_0(-z - s)\, ds.$$
$$(z \geq 0) \quad (14)$$

After change of the variable according to $u = -s$, $du = -ds$; $u = z$ when $s = -z$; $u = 0$ when $s = 0$, (14) becomes,

$$J_s(-z) = -\int_0^z \sin \beta_0 u \sin \beta_0(z - u)\, du$$
$$- \sin \beta_0 h \int_0^z \sin \beta_0(z - u)\, du. \tag{15}$$

The integral is

$$J_s(-z) = \tfrac{1}{2} z \cos \beta_0 z - \frac{\sin \beta_0 z}{2\beta_0}$$
$$- \frac{\sin \beta_0 h}{\beta_0}(1 - \cos \beta_0 z). \quad (z \geq 0) \tag{16}$$

A shorthand for (13) and (16) is

$$J_s(z) = -\tfrac{1}{2}|z| \cos \beta_0 z + \frac{\sin \beta_0 |z|}{2\beta_0}$$
$$- \frac{\sin \beta_0 h}{\beta_0}(1 - \cos \beta_0 z). \tag{17}$$

The integrals (8)–(10) may be expressed in terms of the following generalized integral sines and cosines, for which short tables are given in the Appendix and extensive tables are to be found in the literature (see ref. I.25). In the integrals

$$W = \sqrt{U^2 + A^2}. \tag{18}$$

The integrals are

$$S(A, U) = \int_0^U \frac{\sin W}{W}\, dU, \tag{19a}$$

$$C(A, U) = \int_0^U \frac{1 - \cos W}{W}\, dU, \tag{19b}$$

$$Ss(A, U) = \int_0^U \frac{\sin W}{W} \sin U\, dU, \tag{20a}$$

$$Cs(A, U) = \int_0^U \frac{\cos W}{W} \sin U\, dU, \tag{20b}$$

$$Sc(A, U) = \int_0^U \frac{\sin W}{W} \cos U\, dU, \tag{20c}$$

$$Cc(A, U) = \int_0^U \frac{\cos W}{W} (1 - \cos U)\, dU. \tag{20d}$$

Integrals that are closely related and that are expressible in terms of the above integrals and the inverse hyperbolic sine (see ref. I.24) or the natural logarithm are

$$C_i(A, U) \equiv \int_0^U \frac{\cos W}{W}\, dU$$
$$= \int_0^U \frac{dU}{W} - \int_0^U \frac{1 - \cos W}{W}\, dU$$
$$= \ln\left(\frac{U + W}{A}\right) - C(A, U), \tag{21}$$

$$Cc_i(A, U) \equiv \int_0^U \frac{\cos W}{W} \cos U\, dU$$
$$= \int_0^U \frac{\cos W}{W}\, dU$$
$$- \int_0^U \frac{\cos W}{W}(1 - \cos U)\, dU$$
$$= C_i(A, U) - Cc(A, U)$$
$$= \ln\left(\frac{U + W}{A}\right) - C(A, U)$$
$$- Cc(A, U). \tag{22}$$

If desired, the logarithm may be expressed in the following equivalent form:

$$\int_0^U \frac{dU}{W} = \ln\left(\frac{U+W}{A}\right) = \sinh^{-1}\frac{U}{A}. \quad (23)$$

The conventionally defined integral sines and cosines are given by

$$Si(U) = \int_0^U \frac{\sin U}{U} dU = S(0, U), \quad (24)$$

$$Ci(U) = \int_\infty^U \frac{\cos U}{U} d = C + \ln U - Cin U, \quad (25)$$

where

$$Cin(U) = \int_0^U \frac{1-\cos U}{U} dU = C(0, U) \quad (26a)$$

and

$$C = \lim_{m \to \infty}\left(1 + \frac{1}{2} + \frac{1}{3} + \cdots + \frac{1}{m} - \ln m\right)$$
$$= 0.5772 \cdots \quad (26b)$$

is Euler's constant.

The following relations exist among the above functions:

$$Ss(A, U) = \tfrac{1}{2}[Cin(\sqrt{A^2+U^2}+U)$$
$$+ Cin(\sqrt{A^2+U^2}-U)] - Cin A, \quad (27a)$$

$$= -\tfrac{1}{2}[Ci(\sqrt{A^2+U^2}+U)$$
$$+ Ci(\sqrt{A^2+U^2}-U)] + Ci A, \quad (27b)$$

$$Cs(A, U) = \tfrac{1}{2}[Si(\sqrt{A^2+U^2}+U)$$
$$+ Si(\sqrt{A^2+U^2}-U)] - Si A, \quad (27c)$$

$$Sc(A, U) = \tfrac{1}{2}[Si(\sqrt{A^2+U^2}+U)$$
$$- Si(\sqrt{A^2+U^2}-U)], \quad (27d)$$

$$Cc(A, U) = \tfrac{1}{2}[Cin(\sqrt{A^2+U^2}+U)$$
$$- Cin(\sqrt{A^2+U^2}-U)] - Cin U, \quad (27e)$$

$$Cc_i(A, U) = \tfrac{1}{2}[Ci(\sqrt{A^2+U^2}+U)$$
$$- Ci(\sqrt{A^2+U^2}-U)]. \quad (27f)$$

When $A = 0$ these reduce to
$$Sc(0, U) = Cs(0, U)$$
$$= \tfrac{1}{2}S(0, 2U) = \tfrac{1}{2}Si\, 2U, \quad (27g)$$
$$Ss(0, U) = \tfrac{1}{2}C(0, 2U) = \tfrac{1}{2}Cin\, 2U, \quad (27h)$$
$$Cc(0, U) = \tfrac{1}{2}C(0, 2U) - C(0, U)$$
$$= \tfrac{1}{2}Cin\, 2U - Cin\, U. \quad (27i)$$

With (22) and (27e),

$$Cc_i(0, U) = \sinh^{-1}\frac{U}{A} - \tfrac{1}{2}Cin\, 2U. \quad (27j)$$

Important symmetry relations for the generalized integral sines and cosines are:

$$S(A, -U) = -S(A, U), \quad (28a)$$
$$C(A, -U) = -C(A, U). \quad (28b)$$
$$Ss(A, -U) = Ss(A, U), \quad (28c)$$
$$Sc(A, -U) = -Sc(A, U), \quad (28d)$$
$$Cs(A, -U) = Cs(A, U), \quad (28e)$$
$$Cc(A, -U) = -Cc(A, U). \quad (28f)$$
$$C_i(A, -U) = -C_i(A, U), \quad (28g)$$
$$Cc_i(A, -U) = -Cc_i(A, U). \quad (28h)$$

Evidently, the only even functions are those involving $\sin U$ in the integrand.

The functions $C_a(h, z)$ and $S_a(h, z)$ in (3) and (4) may be expressed in the form

$$C_a(h, z) = \tfrac{1}{2}[I_1 + I_2 + I_3 + I_4],$$
$$S_a(h, z) = \frac{-j}{2}[I_1 - I_2 + I_3 - I_4], \quad (29)$$

where

$$I_1 = e^{j\beta_0 z}\int_0^h e^{j\beta_0(z'-z)}\frac{e^{-j\beta_0 R_1}}{R_1}dz'$$
$$= e^{jU_0}\int_{-U_0}^{U_1} e^{jU}\frac{e^{-jW}}{W}dU \quad (30a)$$

$$I_2 = e^{-j\beta_0 z}\int_0^h e^{-j\beta_0(z'-z)}\frac{e^{-j\beta_0 R_1}}{R_1}dz'$$
$$= e^{-jU_0}\int_{-U_0}^{U_1} e^{-jU}\frac{e^{-jW}}{W}dU, \quad (30b)$$

$$I_3 = e^{-j\beta_0 z}\int_0^h e^{j\beta_0(z'+z)}\frac{e^{-j\beta_0 R_2}}{R_2}dz'$$
$$= e^{-jU_0}\int_{U_0}^{U_2} e^{jU}\frac{e^{-jW}}{W}dU, \quad (30c)$$

$$I_4 = e^{j\beta_0 z}\int_0^h e^{-j\beta_0(z'+z)}\frac{e^{-j\beta_0 R_2}}{R_2}dz'$$
$$= e^{jU_0}\int_{U_0}^{U_2} e^{-jU}\frac{e^{-jW}}{W}dU, \quad (30d)$$

where
$$W = \sqrt{U^2 + A^2}, \quad A = \beta_0 a, \quad (31a)$$
$$U = \beta_0(z' - z) \text{ in } I_1 \text{ and } I_2,$$
$$U = \beta_0(z' + z) \text{ in } I_3 \text{ and } I_4, \quad (31b)$$
$$U_0 = \beta_0 z, \quad U_1 = \beta_0(h - z), \quad U_2 = \beta_0(h + z). \quad (31c)$$

The integrals (30a) to (30d) are readily reduced to the forms (19) to (22). Thus,
$$I_1 = e^{jU_0}[Cc_i(A, U) + Ss(A, U)$$
$$+ jCs(A, U) - jSc(A, U)]\Big|_{-U_0}^{U_1}, \quad (32a)$$
$$I_2 = e^{-jU_0}[Cc_i(A, U) - Ss(A, U)$$
$$- jCs(A, U) - jSc(A, U)]\Big|_{-U_0}^{U_1}, \quad (32b)$$
$$I_3 = e^{-jU_0}[Cc_i(A, U) + Ss(A, U)$$
$$+ jCs(A, U) - jSc(A, U)]\Big|_{U_0}^{U_2}, \quad (32c)$$
$$I_4 = e^{jU_0}[Cc_i(A, U) - Ss(A, U)$$
$$- jCs(A, U) - jSc(A, U)]\Big|_{U_0}^{U_2}. \quad (32d)$$

Substitution of (32a–d) in (29), using (28a–h), gives
$$C_a(h, z) = \cos U_0[Cc_i(A, U_2) + Cc_i(A, U_1)$$
$$- jSc(A, U_2) - jSc(A, U_1)]$$
$$+ \sin U_0[Cs(A, U_2) - Cs(A, U_1)$$
$$- jSs(A, U_2) + jSs(A, U_1)], \quad (33)$$
$$S_a(h, z) = \cos U_0[Cs(A, U_2) + Cs(A, U_1)$$
$$- 2Cs(A, U_0) - jSs(A, U_2)$$
$$- jSs(A, U_1) + j2Ss(A, U_0)]$$
$$+ \sin U_0[-Cc_i(A, U_2)$$
$$+ Cc_i(A, U_1) + 2Cc_i(A, U_0)$$
$$+ jSc(A, U_2) - jSc(A, U_1)$$
$$- j2Sc(A, U_0)]. \quad (34)$$

With (22) in (33) and (34), $C_a(h, z)$ and $S_a(h, z)$ are expressed entirely in terms of tabulated functions. Note that these expressions are exact.

The integral $E_a(h, z)$ in (5) is expressed in terms of tabulated functions as follows:
$$E_a(h, z) = \int_{-U_0}^{U_1} \frac{e^{-jW}}{W} dU + \int_{U_0}^{U_2} \frac{e^{-jW}}{W} dU$$
$$= \int_{-U_0}^{U_1} \frac{\cos W}{W} dU - j \int_{-U_0}^{U_1} \frac{\sin W}{W} dU$$
$$+ \int_{U_0}^{U_2} \frac{\cos W}{W} dU - j \int_{U_0}^{U_2} \frac{\sin W}{W} dU$$
$$= C_i(A, U_2) + C_i(A, U_1)$$
$$- jS(A, U_2) - jS(A, U_1). \quad (35)$$

With (21) and (23),
$$E_a(h, z) = \sinh^{-1}\frac{U_2}{A} + \sinh^{-1}\frac{U_1}{A}$$
$$- C(A, U_2) - C(A, U_1)$$
$$- jS(A, U_2) - jS(A, U_1). \quad (36)$$

In (33), (34), and (36),
$$U_0 = \beta_0 z, \quad U_1 = \beta_0(h - z),$$
$$U_2 = \beta_0(h + z), \quad A = \beta_0 a. \quad (37)$$

The functions $C_a(h, h)$, $S_a(h, h)$, $E_a(h, h)$ are obtained from (33), (34), (36), with $z = h$ or with
$$U_0 = \beta_0 h, \quad U_1 = 0, \quad U_2 = 2\beta_0 h. \quad (38)$$

The functions $C_a(h, z)$, $S_a(h, z)$, $E_a(h, z)$ are shown in Figs. 19.1–19.6 for $\beta_0 h = \pi/2$ and π and for $\Omega = 10$ and 20.

When $\beta_0^2 a^2 = A^2$ is very small compared with $\beta_0^2 h^2$, the use of (27g)–(27j) in (33), (34), and (36) leads to the following approximate formulas:
$$C_a(h, z) \doteq -\tfrac{1}{2}\cos U_0(\operatorname{Cin} 2U_2 + \operatorname{Cin} 2U_1$$
$$+ j\operatorname{Si} 2U_2 + j\operatorname{Si} 2U_1)$$
$$+ \tfrac{1}{2}\sin U_0(\operatorname{Si} 2U_2 - \operatorname{Si} 2U_1$$
$$- j\operatorname{Cin} 2U_2 + j\operatorname{Cin} 2U_1)$$
$$+ \cos U_0(\sinh^{-1} U_2/A + \sinh^{-1} U_1/A), \quad (39)$$
$$S_a(h, z) \doteq \tfrac{1}{2}\cos U_0(\operatorname{Si} 2U_2 + \operatorname{Si} 2U_1$$
$$- 2\operatorname{Si} 2U_0 - j\operatorname{Cin} 2U_2$$
$$- j\operatorname{Cin} 2U_1 + j2\operatorname{Cin} 2U_0)$$
$$+ \tfrac{1}{2}\sin U_0(\operatorname{Cin} 2U_2 - \operatorname{Cin} 2U_1$$
$$- 2\operatorname{Cin} 2U_0 + j\operatorname{Si} 2U_2$$
$$- j\operatorname{Si} 2U_1 - j2\operatorname{Si} 2U_0)$$
$$- \sin U_0(\sinh^{-1} U_2/A - \sinh^{-1} U_1/A$$
$$- 2\sinh^{-1} U_0/A), \quad (40)$$
$$E_a(h, z) \doteq \sinh^{-1} U_2/A + \sinh^{-1} U_1/A$$
$$- \operatorname{Cin} U_2 - \operatorname{Cin} U_1 - j\operatorname{Si} U_2 - j\operatorname{Si} U_1, \quad (41)$$

where U_0, U_1, and U_2 are given in (37).

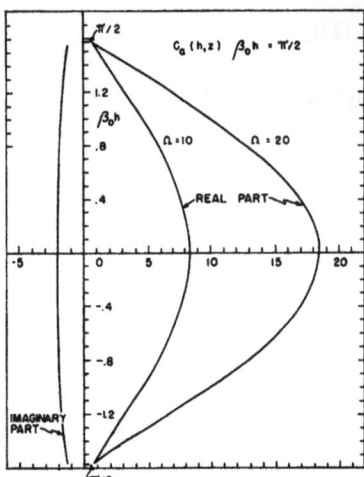

Fig. 19.1. Real and imaginary parts of $C_a(h, z)$ for $\beta_0 h = \pi/2$, $\Omega = 10, 20$.

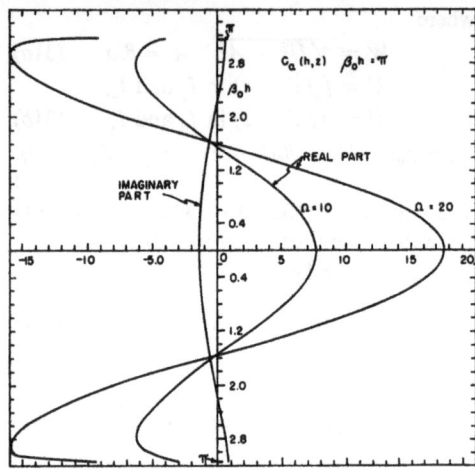

Fig. 19.2. Real and imaginary parts of $C_a(h, z)$ for $\beta_0 h = \pi$, $\Omega = 10, 20$.

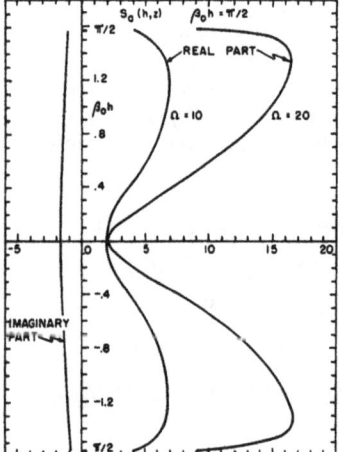

Fig. 19.3. Real and imaginary parts of $S_a(h, z)$ for $\beta_0 h = \pi/2$, $\Omega = 10, 20$.

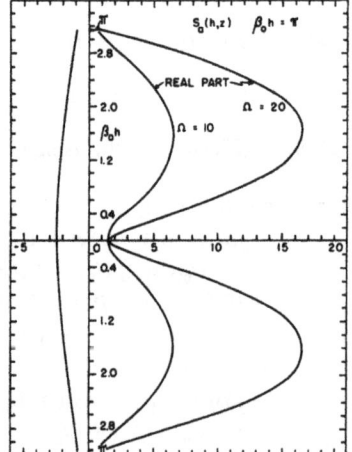

Fig. 19.4. Real and imaginary parts of $S_a(h, z)$ for $\beta_0 h = \pi$, $\Omega = 10, 20$.

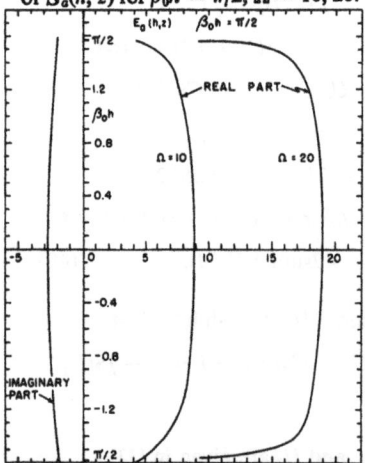

Fig. 19.5. Real and imaginary parts of $E_a(h, z)$ for $\beta_0 h = \pi/2$, $\Omega = 10, 20$.

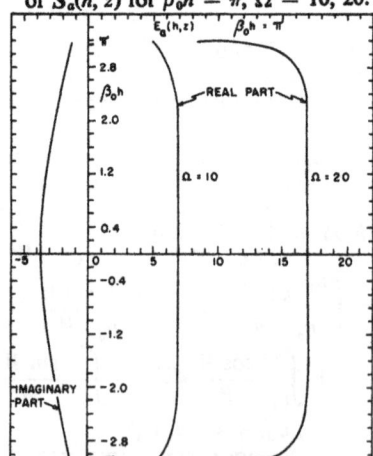

Fig. 19.6. Real and imaginary parts of $E_a(h, z)$ for $\beta_0 h = \pi$, $\Omega = 10, 20$.

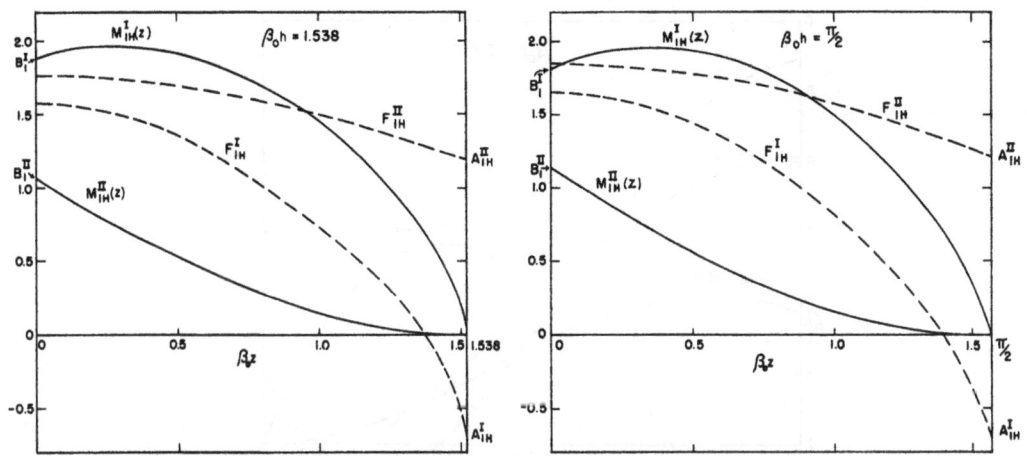

Fig. 19.7. Real and imaginary parts of $M_{1H}(z)$ and $F_{1H}(z)$ for $\beta_0 h = 1.538$.

Fig. 19.8. Real and imaginary parts of $M_{1H}(z)$ and $F_{1H}(z)$ for $\beta_0 h = \pi/2$.

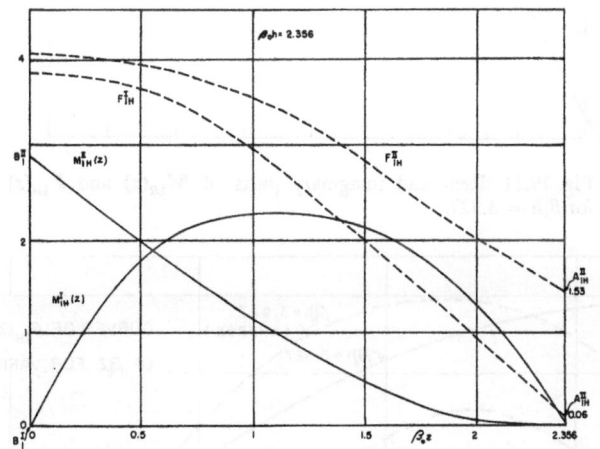

Fig. 19.9. Real and imaginary parts of $M_{1H}(z)$ and $F_{1H}(z)$ for $\beta_0 h = 2.356$.

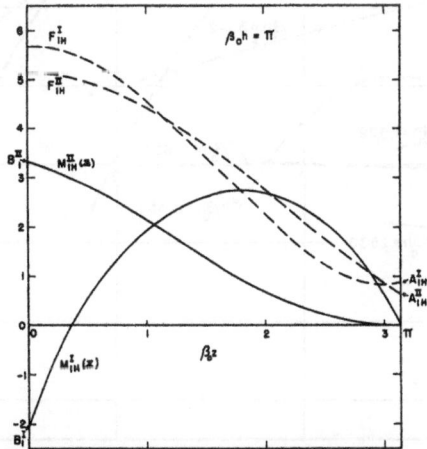

Fig. 19.10. Real and imaginary parts of $M_{1H}(z)$ and $F_{1H}(z)$ for $\beta_0 h = \pi$.

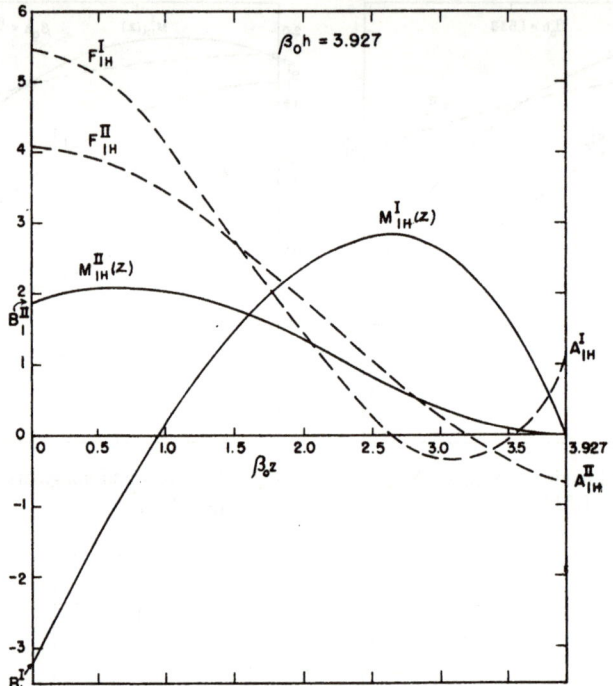

Fig. 19.11. Real and imaginary parts of $M_{1H}(z)$ and $F_{1H}(z)$ for $\beta_0 h = 3.927$.

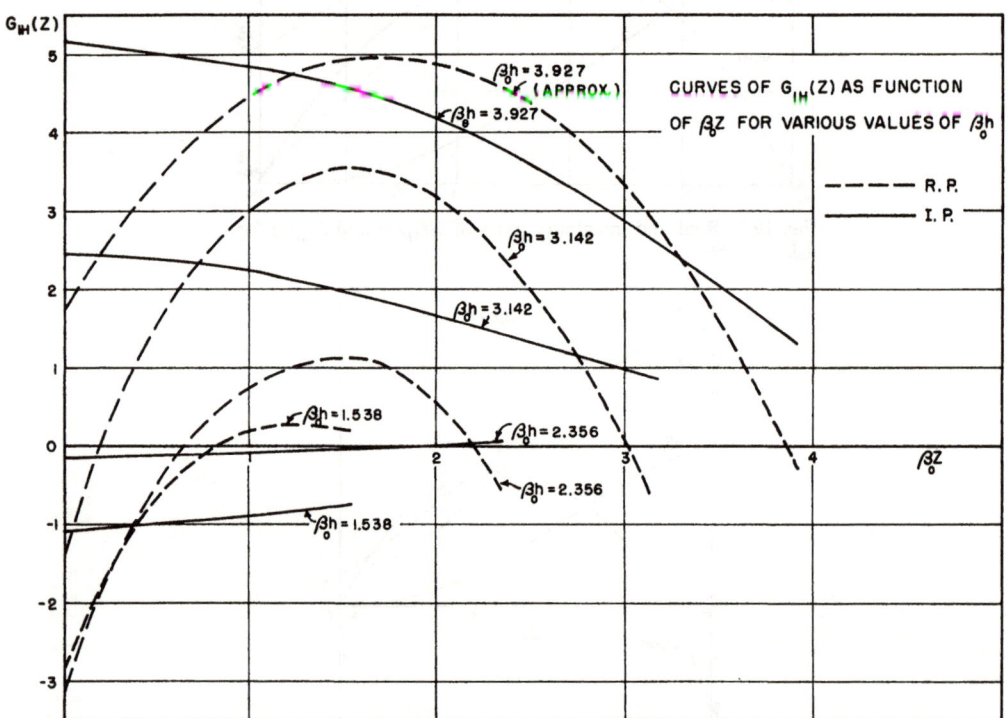

Fig. 19.12. Curves of $G_{1H}(z)$ as a function of $\beta_0 z$ for various values of $\beta_0 h$.

With (10), (12), (17), (33), (34), and (36), the functions $F_{1H}(z)$, $F_{1H}(h)$, $G_{1H}(z)$, $G_{1H}(h)$ as defined in (11a)–(11d) are expressed entirely in terms of tabulated functions, so that their numerical computation is possible. The function $M_{1H}(z) = M_{1H}^{I}(z) + jM_{1H}^{II}(z)$ as defined in (15.27) with $\delta = 0$ has been evaluated using (11a)–(11d). It is shown graphically in Figs. 19.7–19.11 together with $F_{1H}(z) = F_{1H}^{I}(z) + jF_{1H}^{II}(z)$ for several values of $\beta_0 h$. The corresponding values of $F_{1H}(h) = A_{1H} = A_{1H}^{I} + jA_{1H}^{II}$ as defined in (15.29a) with $\delta = 0$ are indicated. The terms in z^i were neglected in these calculations, since their contributions are negligible if the conductor is copper. This is shown explicitly in calculating the impedance in a later section. The function $G_{1H}(z)$ is represented in Fig. 19.12.

20. Expansion of King and Middleton; Parameters

It is pointed out in Sec. 18 that the relative distribution function

$$g_H(z, z') = e^{j\beta_0 R_1} \qquad (1)$$

leads to the simplest and the most obvious expansion of the original integral equation (14.3). On the other hand, if the formal solution is carried through to obtain (15.26) without previously selecting $g(z, z')$, it is clear that the leading term in the distribution of current for any value of Ψ must be of the form

$$I_z = K f_0(z), \qquad (2a)$$

with

$$f_0(z) \equiv \sin \beta_0(h - |z|); \qquad (2b)$$

K is an amplitude factor independent of z. Accordingly, an approximate relative distribution function is

$$g_{K0}(z, z') \equiv \frac{\sin \beta_0(h - |z'|)}{\sin \beta_0(h - |z|)} = \frac{f_0(z')}{f_0(z)}. \qquad (3)$$

This function is known to approximate the true current. Use of (3) in the general formula (14.7) defining the function $\Psi(z)$ gives

$$\Psi_{K1}(z) \equiv \left(\int_{-h}^{h} - \int_{-\delta}^{\delta} \right) g_{K0}(z, z') K_1(z, z') \, dz'. \qquad (4)$$

This function involves the factor $f_0(z) = \sin \beta_0(h - |z|)$ in the denominator of the integrand. Since this is not a function of z', it is convenient to introduce the function

$$\psi_1(z) = \left(\int_{-h}^{h} - \int_{-\delta}^{\delta} \right) f_0(z') K_1(z, z') \, dz', \qquad (5)$$

so that

$$\Psi_{K1}(z) = \frac{\psi_1(z)}{f_0(z)}. \qquad (6)$$

The function $\psi_1(z)$ has the form

$$\psi_1(z) = [C_a(h, z) - C_a(\delta, z)] \sin \beta_0 h \\ - [S_a(h, z) - S_a(\delta, z)] \cos \beta_0 h, \qquad (7)$$

where $C_a(h, z)$ and $S_a(h, z)$ are defined in (19.3) and (19.4), and where $C_a(\delta, z)$ and $S_a(\delta, z)$ are obtained from (19.3) and (19.4) by substituting δ for h in the upper limits of the integrals.

Approximate formulas for $C_a(\delta, z)$ and $S_a(\delta, z)$ may be evaluated for two important ranges if δ is small. If the value of z is large compared with δ, the integrals may be expanded in a Maclaurin series in powers of δ and only the leading term retained. Thus,

$$C_a(\delta, z) = [C_a(\delta, z)]_{\delta=0} \\ + \delta \left[\frac{\partial C_a(\delta, z)}{\partial \delta} \right]_{\delta=0} + \cdots, \qquad (8)$$

$$S_a(\delta, z) = [S_a(\delta, z)]_{\delta=0} \\ + \delta \left[\frac{\partial S_a(\delta, z)}{\partial \delta} \right]_{\delta=0} + \cdots. \qquad (9)$$

Using the general formula (12.6) for differentiating a definite integral, one obtains:

$$C_a(\delta, z) \doteq \delta \left\{ \cos \beta_0 \delta \left[\frac{e^{-j\beta_0 \sqrt{(z-\delta)^2 + a^2}}}{\sqrt{(z-\delta)^2 + a^2}} \right. \right. \\ \left. \left. + \frac{e^{-j\beta_0 \sqrt{(z+\delta)^2 + a^2}}}{\sqrt{(z+\delta)^2 + a^2}} \right] \right\}_{\delta=0} \\ \doteq \frac{2\delta}{R_0} e^{-j\beta_0 R_0}, \qquad (10)$$

$$S_a(\delta, z) \doteq \delta \left\{ \sin \beta_0 \delta \left[\frac{e^{-j\beta_0 \sqrt{(z-\delta)^2 + a^2}}}{\sqrt{(z-\delta)^2 + a^2}} \right. \right. \\ \left. \left. + \frac{e^{-j\beta_0 \sqrt{(z-\delta)^2 + a^2}}}{\sqrt{z-\delta)^2 + a^2}} \right] \right\}_{\delta=0} = 0, \qquad (11)$$

where

$$R_0 = \sqrt{z^2 + a^2}, \qquad (12)$$

and where it is assumed that the following inequality is satisfied:

$$\delta/R_0 \ll 1. \qquad (13)$$

If both δ and the value of z are sufficiently small that

$$(\beta_0 \delta)^2 \ll 1 \tag{14}$$

and

$$(\beta_0 R_0)^2 \ll 1, \quad (\beta_0 R_1)^2 \ll 1, \quad (\beta_0 R_2)^2 \ll 1, \tag{15}$$

the exponentials in (19.3) and (19.4) may be expanded and only the leading terms retained. Thus,

$$e^{-j\beta_0 R_1} \doteq 1 - j\beta_0 R_1; \quad e^{-j\beta_0 R_2} \doteq 1 - j\beta_0 R_2. \tag{16}$$

Since $z' \leqq \delta$, it follows with (14) that

$$\cos \beta_0 z' \doteq 1, \tag{17}$$

$$\sin \beta_0 z' \doteq \beta_0 z'. \tag{18}$$

Accordingly, from (19.3),

$$C_a(\delta, z) \doteq \int_0^\delta \left(\frac{1}{R_1} + \frac{1}{R_2}\right) dz' - 2j\beta_0 \int_0^\delta dz', \tag{19}$$

$$= \sinh^{-1} \frac{z+\delta}{a} - \sinh^{-1} \frac{z-\delta}{a} - 2j\beta_0 \delta$$

$$= \Omega(\delta, z) - 2j\beta_0 \delta. \tag{20}$$

Similarly, from (19.4),

$$S_a(\delta, z) \doteq \beta_0 \int_0^\delta \frac{z' \, dz'}{\sqrt{(z-z')^2 + a^2}}$$

$$+ \beta_0 \int_0^\delta \frac{z' \, dz'}{\sqrt{(z+z')^2 + a^2}}$$

$$- 2j\beta_0^2 \int_0^\delta z' \, dz', \tag{21}$$

$$\doteq \beta_0 \int_{z-\delta}^{z+\delta} \frac{u \, du}{\sqrt{u^2 + a^2}},$$

$$= \beta_0[\sqrt{(z+\delta)^2 + a^2} - \sqrt{(z-\delta)^2 + a^2}]. \tag{22}$$

Small terms in $\beta_0^2 \delta^2$ have been neglected. At $z = \delta$,

$$C_a(\delta, \delta) \doteq \sinh^{-1}(2\delta/a) - 2j\beta_0 \delta, \tag{23}$$

$$S_a(\delta, \delta) \doteq \beta_0(\sqrt{4\delta^2 + a^2} - a). \tag{24}$$

The only term in $C_a(\delta, z)$ in (20) that is large is $\Omega(\delta, z)$. As discussed in Sec. 18, the magnitude of $\Omega(\delta, z)$ diminishes very rapidly to small values within relatively short distances along the antenna from the end of the antenna at $z = \delta$ if the conditions

$$h \gg \delta, \quad h \gg a \tag{25}$$

are satisfied. This is verified for $\psi_1(z)$ and $\Psi_{K1}(z)$ later in the section, Figs. 20.8 and 20.9. Accordingly, since $C_a(\delta, z)$ does not significantly modify $C_a(h, z)$ except very near $z = \delta$, and $S_a(\delta, z)$ is very small even at $z = \delta$, it is permissible to omit the terms in δ in determining $|\Psi(z_r)|$ but to include them in $\gamma(z)$. Omitting them from (7), it follows that

$$\psi_1(z) \doteq C_a(h, z) \sin \beta_0 h - S_a(h, z) \cos \beta_0 h. \tag{26}$$

In particular,

$$\psi_1(z) \doteq C_a(\lambda_0/4, z), \quad \beta_0 h = \pi/2, \tag{27}$$

$$\psi_1(z) \doteq S_a(\lambda_0/2, z), \quad \beta_0 h = \pi. \tag{28}$$

Evidently, the graphs of $C_a(h, z)$ with $\beta_0 h = \pi/2$ are also graphs of $\psi_1(z)$; these are given in Figs. 20.1 and 20.2 for $\Omega = 10$ and 20. Similarly, graphs of $S_a(h, z)$ with $\beta_0 h = \pi$ are also graphs of $\psi_1(z)$; these are given in Figs. 20.3 and 20.4 for $\Omega = 10$ and 20. The function $\psi_1(z)$ is seen to have a very small imaginary part, so that it and $\Psi_{K1}(z) = \psi_1(z)/f_0(z)$ are predominantly real, in confirmation of the assumption made in introducing Ψ in Sec. 14. Accordingly, the parameter $\Psi = |\Psi(z_r)|$ defined in (14.11a) may be chosen to be the real quantities*

$$\Psi_{K1} = |\Psi_{K1}(0)| = |\psi_1(0)|/\sin \beta_0 h,$$
$$(\beta_0 h \leqq \pi/2) \quad (29a)$$

$$\Psi_{K1} = |\Psi_{K1}(h - \lambda_0/4)| = |\psi_1(h - \lambda_0/4)|.$$
$$(\beta_0 h \geqq \pi/2) \quad (29b)$$

The real functions

$$|\Psi_{K1}(z)| = \frac{|\psi_1(z)|}{f_0(z)} = \frac{|\psi_1(z)|}{\sin \beta_0(h - |z|)} \tag{30}$$

and

$$|\Psi_{K1}(0)| \sin \beta_0(h - |z|) \quad \text{or}$$
$$|\Psi_{K1}(h - \lambda_0/4)| \sin \beta_0(h - |z|) \tag{31}$$

also are plotted in Figs. 20.1 to 20.4. For $\beta_0 h = \pi/2$, and with both $\Omega = 10$ and 20, $|\Psi_{K1}(z)|$ is seen to be quite constant over the entire length of the antenna except near the ends, where it becomes infinite, as it should. The excellence of the representation is well shown by comparing the curves for the actual function $|\psi_1(z)|$ with the approximate equivalent $\psi_1(0) \sin \beta_0(h - |z|)$. If this function is allowed to replace $|\psi_1(z)|$, then

---
* A more accurate treatment of electrically short antennas with $\beta_0 h \leqq 0.5$ is given in Sec. 31.

[II.20] THEORY OF LINEAR ANTENNAS 103

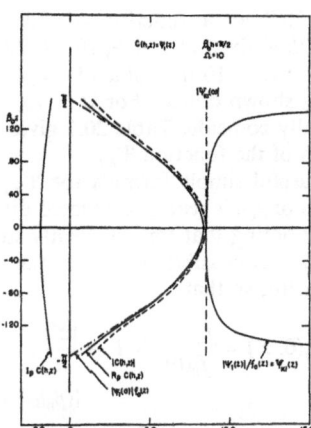

Fig. 20.1. The functions $\psi_1(z)$ and $|\Psi_{K1}(z)|$, also $|\Psi_{K1}(0)| = \Psi_{K1}$, for $\beta_0 h = \pi/2$, $\Omega = 10$.

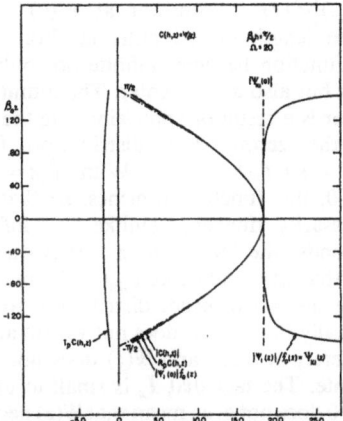

Fig. 20.2. The functions $\psi_1(z)$ and $|\Psi_{K1}(z)|$, also $|\Psi_{K1}(0)| = \Psi_{K1}$, for $\beta_0 h = \pi/2$, $\Omega = 20$.

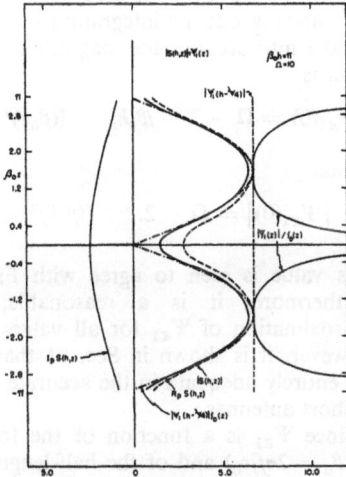

Fig. 20.3. The functions $\psi_1(z)$ and $|\Psi_{K1}(z)|$, also $|\Psi_{K1}(h - \tfrac{1}{4}\lambda_0)| = \Psi_{K1}$, for $\beta_0 h = \pi$, $\Omega = 10$.

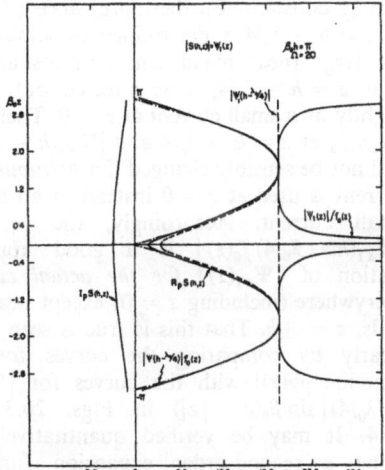

Fig. 20.4. The functions $\psi_1(z)$ and $|\Psi_{K1}(z)|$, also $|\Psi_{K1}(h - \tfrac{1}{4}\lambda_0)| = \Psi_{K1}$, for $\beta_0 h = \pi$, $\Omega = 20$.

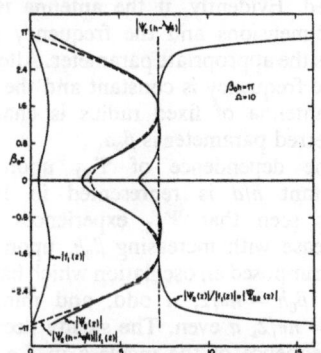

Fig. 20.5. The functions $\psi_2(z)$ and $|\Psi_{K2}(z)|$, also $|\Psi_2(h - \tfrac{1}{4}\lambda_0)| = \Psi_{K2}$, for $\beta_0 h = \pi$, $\Omega = 10$.

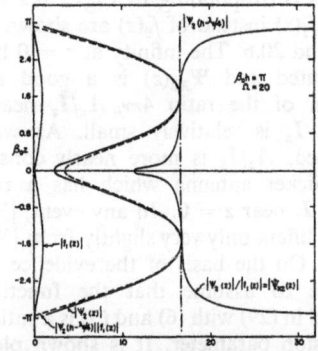

Fig. 20.6. The functions $\psi_2(z)$ and $|\Psi_{K2}(z)|$, also $|\Psi_2(h - \tfrac{1}{4}\lambda_0)| = \Psi_{K2}$, for $\beta_0 h = \pi$, $\Omega = 20$.

$|\psi_1(z)|/f_0(z)$ is constant at $\psi_1(0)$ over the entire length of the antenna. For $\beta_0 h = \pi$, the function becomes infinite not only at the ends but also at the center. The infinity at the center is a result of approximating the current by the zeroth-order distribution function $f_0(z) = \sin\beta_0(h - |z|)$. With $\beta_0 h = \pi$ and $z = 0$, this function vanishes, so that $\Psi_{K1}(z)$ necessarily diverges. Unlike the infinity at the ends, the infinity at $z = 0$ is due to the fact that $f_0(z)$ and hence $g_{K0}(z, z')$ are approximate instead of exact distribution functions. Actually the current does not vanish at $z = 0$; it merely is small and $\Psi(z)$ does not become infinite. The fact that I_z is small at and near $z = 0$ does not even mean that $\Psi(z)$ necessarily becomes very large; $\Psi(z)$ is by definition proportional to the ratio $(A_z)_{r=a}/I_z$, and $(A_z)_{r=a}$ is determined largely by the current at and near z. Hence $(A_0)_{r=a}$ tends to be small if I_0 is, and their ratio should remain moderately constant. Furthermore, since $(A_z)_{r=a}$ at $z = h - \lambda_0/4$ is determined principally by the large (near maximum) currents at and near $z = h - \lambda_0/4$, it is affected only very slightly by a small current at $z = 0$. Therefore $(A_z)_{r=a}$ at $z = h - \lambda_0/4$ and $|\Psi_{K1}(h - \lambda_0/4)|$ will not be sensibly changed if a *fictitious* zero current is used at $z = 0$ instead of an actual small current. Accordingly, the function $|\Psi_{K1}(h - \lambda_0/4) f_0(z)|$ is a good approximation of $|\Psi_{K1}(z)|$ for the actual current everywhere (including $z = 0$) except near the ends, $z = \pm h$. That this is true is seen more clearly by comparing the curves for the function $|\psi(z)|$ with the curves for $|\Psi_{K1}(h - \lambda_0/4)|\sin\beta_0(h - |z|)$ in Figs. 20.3 and 20.4. It may be verified quantitatively by using a second-order expansion function $\Psi_{K2}(z)$ evaluated using an approximation of the first-order function $f_1(z)$ as calculated in Sec. 25 instead of the sinusoidal, zeroth-order distribution function $f_0(z) = \sin\beta_0(h - |z|)$. Curves corresponding to Figs. 20.3 and 20.4 using $f_1(z)$ instead of $f_0(z)$ are shown in Figs. 20.5 and 20.6. The infinity at $z = 0$ has been eliminated and $\Psi_{K2}(z)$ is a good approximation of the ratio $4\pi v_0 A_z/I_z$ near $z = 0$ where I_z is relatively small. As would be expected, A_z/I_z is more nearly constant for the thicker antenna which has a relatively larger I_z near $z = 0$. In any event, $|\Psi_{K2}(h - \lambda_0/4)|$ differs only very slightly from $|\Psi_{K1}(h - \lambda_0/4)|$. On the basis of the evidence given it is safe to assume that the function Ψ_{K1} defined in (29) with (6) and (5) is a satisfactory expansion parameter. It is shown plotted in Fig. 20.7 as a function of $\beta_0 h$ with $\Omega = 2\ln(2h/a)$ fixed at 7, 8, 9, 10, 11, 12.5, 15, and 20. For $\Omega = 10$ the value of Ψ_{K1}' is given and Ψ_{K2}' is shown dotted. For $\Omega = 15, 20$ the two virtually coincide. Table 20.1 gives computed values of the function Ψ_{K1}'.

A useful simple formula for Ψ_{K1} for small values of $\beta_0 h$ is readily evaluated directly from (4) by noting that for sufficiently small values of $\beta_0 h$ it is possible to replace the sine by its argument, so that

$$g_{K0}(0, z') = \frac{f_0(z')}{f_0(0)} \doteq 1 - \frac{|z'|}{h}.$$
$$[(\beta_0 h)^2 \ll 1] \quad (32)$$

Furthermore, at $z = 0$,

$$e^{-j\beta_0 R_1} \doteq 1 - j\beta_0 R_1, \quad (33a)$$

where

$$R_1 = \sqrt{z'^2 + a^2}. \quad (33b)$$

Upon substituting (32) and (33a, b) in (4) and carrying out the integration in two steps to take into account the magnitude sign, the result is

$$\Psi_{K1}(0) \doteq \Omega - 2 - j\beta_0 h, \quad [(\beta_0 h)^2 \ll 1] \quad (34)$$

so that

$$|\Psi_{K1}(0)| \doteq \Omega - 2. \quad [(\beta_0 h)^2 \ll 1] \quad (35)$$

This value is seen to agree with Fig. 20.5. Furthermore, it is a reasonable, rough approximation of Ψ_{K1} for all values of $\beta_0 h$. However, it is shown in Sec. 31 that (35) is not entirely adequate in the accurate analysis of short antennas.

Since Ψ_{K1} is a function of the frequency (in $\beta_0 = 2\pi f/v_0$) and of the half-length h and radius a of the antenna in the dimensionless combinations h/a and $\beta_0 h$ or $\beta_0 a$ and $\beta_0 h$, it is possible to select either h/a [in $\Omega = 2\ln(2h/a)$] or $\beta_0 a = 2\pi a/\lambda_0$ as a parameter while $\beta_0 h$ is varied. Evidently, if the antenna is fixed in its dimensions and the frequency is varied, h/a is the appropriate parameter. Alternatively, if the frequency is constant and the length of an antenna of fixed radius is changed, the preferred parameter is $\beta_0 a$.

The dependence of Ψ_{K1} upon $\beta_0 h$ for constant h/a is represented in Fig. 20.7. It is seen that Ψ_{K1} experiences a steady decrease with increasing $\beta_0 h$, upon which is superimposed an oscillation which has maxima near $\beta_0 h = n\pi/2$, n odd, and minima near $\beta_0 h = n\pi/2$, n even. The steady decrease is a consequence of the increase in $\beta_0 a$ as $\beta_0 h$ is

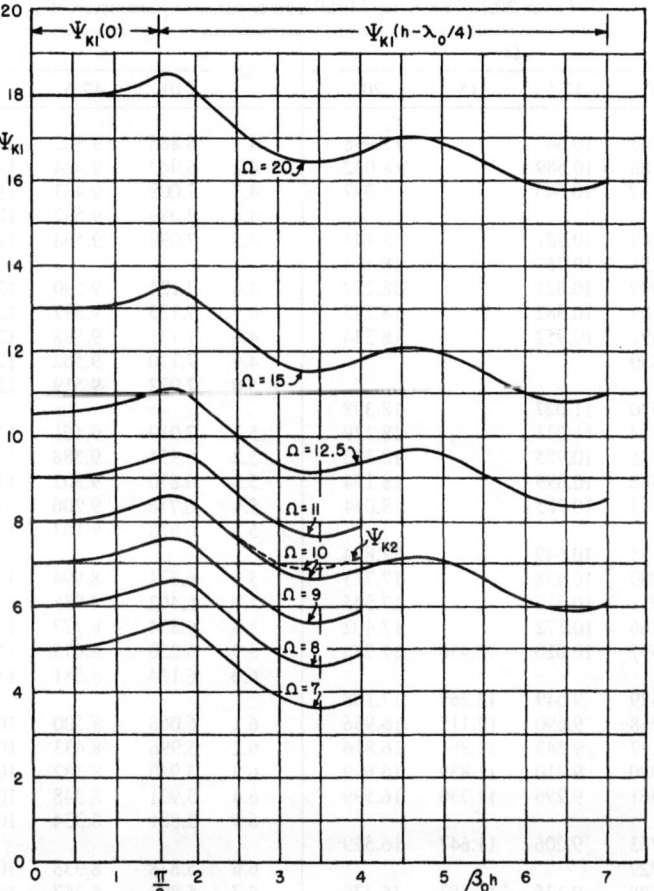

Fig. 20.7. Expansion parameters Ψ'_{K1} and Ψ'_{K2} as functions of $\beta_0 h$ with Ω as parameter.

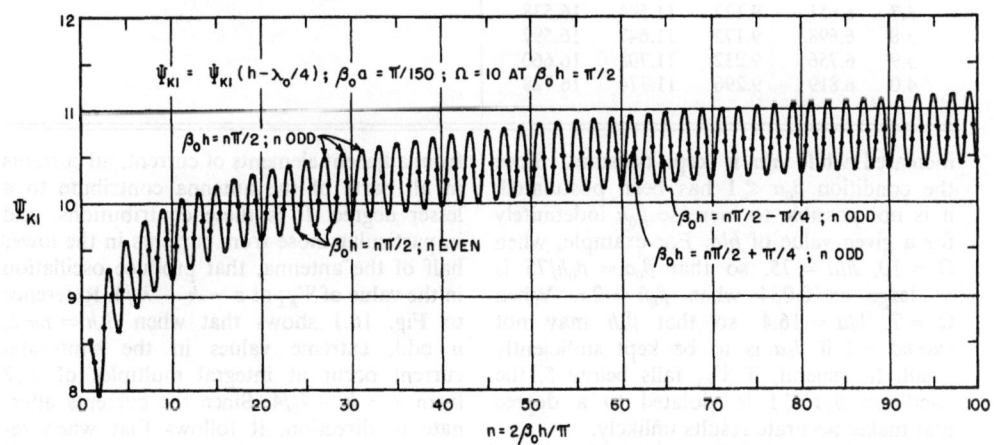

Fig. 20.8. Ψ'_{K1} as a function of $n = 2\beta_0 h/\pi$ with $\beta_0 a$ constant.

TABLE 20.1. The King-Middleton expansion parameter, Ψ_{K1}.

$\beta_0 h$	Ω				$\beta_0 h$	Ω			
	10	12.5	15	20		10	12.5	15	20
0.5	8.063	10.547		17.998	4.1	6.885	9.362	11.839	16.794
0.7	8.108	10.589		18.032	4.2	6.949	9.424	11.899	16.855
0.9	8.167	10.645		18.077	4.3	7.008	9.483	11.958	16.907
					4.4	7.059	9.532	12.005	16.951
1.1	8.247	10.721		18.143	4.5	7.098	9.568	12.038	16.978
1.2	8.295	10.767		18.183					
1.3	8.349	10.821		18.227	4.6	7.124	9.590	12.056	16.988
1.4	8.409	10.882		18.278	4.7	7.135	9.597	12.059	16.983
1.5	8.491	10.952		18.334	4.8	7.131	9.588	12.046	16.964
$\pi/2$	8.550				4.9	7.110	9.562	12.014	16.919
					5.0	7.072	9.519	11.965	16.859
1.6	8.570	11.027		18.398					
1.7	8.584	11.033		18.379	5.1	7.019	9.461	11.902	16.786
1.8	8.545	10.985		18.305	5.2	6.951	9.388	11.824	16.697
1.9	8.468	10.899		18.194	5.3	6.870	9.302	11.734	16.597
2.0	8.363	10.785		18.054	5.4	6.778	9.206	11.634	16.490
					5.5	6.678	9.102	11.527	16.377
2.1	8.235	10.649		17.894					
2.2	8.089	10.498		17.723	5.6	6.571	8.994	11.417	16.264
2.3	7.931	10.335		17.545	5.7	6.462	8.885	11.308	16.153
2.4	7.766	10.172		17.402	5.8	6.354	8.777	11.201	16.047
2.5	7.597	10.010	12.436	17.275	5.9	6.250	8.675	11.100	15.950
					6.0	6.154	8.581	11.009	15.864
2.6	7.429	9.849	12.269	17.108					
2.7	7.268	9.690	12.112	16.956	6.1	6.068	8.500	10.931	15.794
2.8	7.117	9.543	11.967	16.816	6.2	5.996	8.433	10.870	15.743
2.9	6.980	9.410	11.839	16.699	6.3	5.940	8.382	10.823	15.710
3.0	6.861	9.296	11.730	16.599	6.4	5.901	8.348	10.795	15.689
					6.5	5.881	8.334	10.786	15.695
3.1	6.763	9.206	11.647	16.529					
π	6.729				6.6	5.878	8.338	10.799	15.720
3.2	6.688	9.135	11.583	16.478	6.7	5.893	8.357	10.821	15.750
3.3	6.636	9.089	11.542	16.448	6.8	5.922	8.390	10.857	15.791
3.4	6.608	9.108	11.525	16.443	6.9	5.965	8.435	10.906	15.846
3.5	6.602	9.102	11.530	16.457	7.0	6.019	8.491	10.963	15.908
3.6	6.618	9.086	11.554	16.491					
3.7	6.651	9.122	11.594	16.538					
3.8	6.698	9.173	11.647	16.597					
3.9	6.756	9.232	11.708	16.660					
4.0	6.819	9.296	11.774	16.728					

increased while h/a is kept constant. Since the condition $\beta_0 a \ll 1$ has been postulated, it is not possible to increase $\beta_0 h$ indefinitely for a given value of h/a. For example, when $\Omega = 10$, $h/a = 75$, so that $\beta_0 a = \beta_0 h/75$ is as large as 0.084 when $\beta_0 h = 2\pi$. When $\Omega = 7$, $h/a = 16.4$, so that $\beta_0 h$ may not exceed $\pi/2$ if $\beta_0 a$ is to be kept sufficiently small. In general, if Ψ_{K1} falls below 5, the condition $\beta_0 a \ll 1$ is violated to a degree that makes accurate results unlikely.

Although the contributions to the vector potential at a point z are primarily derived from adjacent elements of current, all currents in all parts of the antenna contribute to a lesser degree. It is these contributions, and in particular those from currents in the lower half of the antenna, that produce oscillation in the value of Ψ_{K1} at $z = h - \lambda_0/4$. Reference to Fig. 16.1 shows that when $\beta_0 h = n\pi/2$, n odd, extreme values in the sinusoidal current occur at integral multiples of $\lambda_0/2$ from $z = h - \lambda_0/4$. Since the currents alternate in direction, it follows that when retardation is taken into account they all contribute *in phase* to the vector potential at $z = h - \lambda_0/4$. Alternatively, when $\beta_0 h = n\pi/2$, n even, the current maxima in the lower half of the antenna are at such distances and the currents are in such directions that their contributions to the vector potential at $z = h - \lambda_0/4$ are essentially in phase opposition to the corresponding contributions from the upper half of the antenna.

In the analysis of very long antennas it is necessary to select $\beta_0 a$ as the constant parameter. Let it be assumed that $\beta_0 a$ is so small compared with $\beta_0 h$ that (19.39) and (19.40) may be used in (26) and (29b). Also let $\beta_0 h$ be sufficiently large that with $z = h - \lambda_0/4$ in (19.39, 40) the following approximations are justified:

$$Si\, 2U_2 \doteq Si\, 2U_0 \doteq \pi/2, \quad (36a)$$

$$Cin\, 2U_2 \doteq C + \ln(4\beta_0 h - \pi), \quad (36b)$$

$$Cin\, 2U_0 \doteq C + \ln(2\beta_0 h - \pi), \quad (36c)$$

$$\sinh^{-1}(U_2/A) \doteq \ln(4\beta_0 h - \pi) - \ln \beta_0 a, \quad (36d)$$

$$\sinh^{-1}(U_1/A) \doteq \ln(\pi/\beta_0 a), \quad (36e)$$

$$\sinh^{-1}(U_0/A) \doteq \ln(2\beta_0 h - \pi) - \ln \beta_0 a. \quad (36f)$$

With these relations the complex expansion parameter becomes:

$$\Psi_{K1}(h - \lambda_0/4) = -\tfrac{1}{2}[\ln(4\beta_0 h - \pi) \cos 2\beta_0 h$$
$$+ 4 \ln \beta_0 a + Cin\, \pi - 2 \ln \pi + C]$$
$$+ \ln(2\beta_0 h - \pi) \cos^2 \beta_0 h$$
$$+ \tfrac{1}{2}j[\ln(4\beta_0 h - \pi) - \ln(2\beta_0 h - \pi)]$$
$$- \tfrac{1}{2}j[Si\, \pi + \pi/2]. \quad (37)$$

Using this formula and setting $\beta_0 a = \pi/150$ so that $\Omega = 10$ at $\beta_0 h = \pi/2$, the curve of $|\Psi_{K1}(h - \lambda_0/4)|$ in Fig. 20.8 is determined. It is seen that with constant $\beta_0 a$ the expansion parameter *increases* steadily with $\beta_0 h$. The oscillations superimposed upon the steady increase have the same explanation given above for Fig. 20.7.

The effect of a finite gap in determining $\psi_1(z)$ from (7) is shown in Fig. 20.9 for $\beta_0 h = \pi/2$. Evidently, the effect of small gaps on Ψ_{K1} is unimportant. For $\beta_0 h > \pi/2$ there is no effect, since the expansion parameter is not defined near $z = 0$.

21. King-Middleton Solution: General Formula for Current

The distribution of current in the King-Middleton form is obtained from (15.26) with appropriate subscripts. Thus,

$$I_z = \frac{j2\pi V_\delta}{\zeta_0 \Psi_{K1}} \left[\frac{\sin \beta_0(h-z) + M_{1K}(z)/\Psi_{K1}}{\cos \beta_0(h-\delta) + A_{1K}/\Psi_{K1}} \right.$$
$$\left. + \frac{M_{2K}(z)/\Psi_{K1}^2 + \cdots}{+ A_{2K}/\Psi_{K1}^2 + \cdots} \right], \quad (1)$$

where $M_{1K}(z)$ is given by (15.27), $M_{2K}(z)$ by (15.28b), A_{1K} by (15.29a), and A_{2K} by (15.29b) with subscripts K added to the functions M, A, F, and G.

The zeroth-order functions $F_0(z)$, $F_0(h)$, $G_0(z)$, and $G_0(h)$ are the same in the King-Middleton analysis as in the Hallén form defined in (15.2a, b).

The first-order functions are obtained from (15.9a) and (15.7b). Thus,

$$F_{1K}(z) = \frac{j4\pi z^i}{\zeta_0} \int_\delta^z F_{0s} \sin \beta_0(z-s)\, ds$$
$$+ F_{0z}\Psi_{K1}$$
$$- \left(\int_{-h}^h - \int_{-\delta}^\delta \right) F_{0z'} K_1(z, z')\, dz'$$
$$= \frac{j4\pi z^i}{\zeta_0} J_c(z) + F_{0z}\Psi_{K1}$$
$$- [C_a(h, z) - C_a(\delta, z)]$$
$$+ \cos \beta_0 h[E_a(h, z) - E_a(\delta, z)], \quad (1a)$$

$$F_{1K}(h) = \frac{j4\pi z^i}{\zeta_0} \int_\delta^h F_{0s} \sin \beta_0(h-s)\, ds$$
$$- \left(\int_{-h}^h - \int_{-\delta}^\delta \right) F_{0z'} K_1(h, z')\, dz'$$
$$= \frac{j4\pi z^i}{\zeta_0} J_c(h) - [C_a(h, h) - C_a(\delta, h)]$$
$$+ \cos \beta_0 h[E_a(h, h) - E_a(\delta, h)]. \quad (1b)$$

The functions $G_{1K}(z)$ and $G_{1K}(h)$ are obtained from (1a) and (1b) by writing G for F. The explicit formulas are:

$$G_{1K}(z) = \frac{j4\pi z^i}{\zeta_0} J_s(z) + G_{0z}\Psi_{K1}$$
$$- [S_a(h, z) - S_a(\delta, z)]$$
$$+ \sin \beta_0 h[E_a(h, z) - E_a(\delta, z)], \quad (1c)$$

$$G_{1K}(h) = \frac{j4\pi z^i}{\zeta_0} J_s(h) - [S_a(h, h) - S_a(\delta, h)]$$
$$+ \sin \beta_0 h[E_a(h, h) - E_a(\delta, h)]. \quad (1d)$$

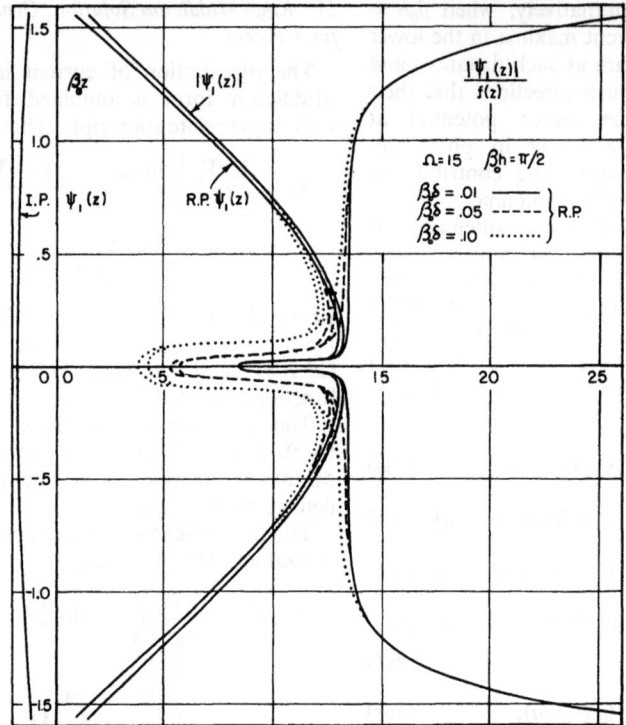

Fig. 20.9. The function $\Psi'_1(z)$ with $\beta_0\delta$ as parameter, $\Omega = 15$, $\beta_0 h = \pi/2$ (Winternitz).

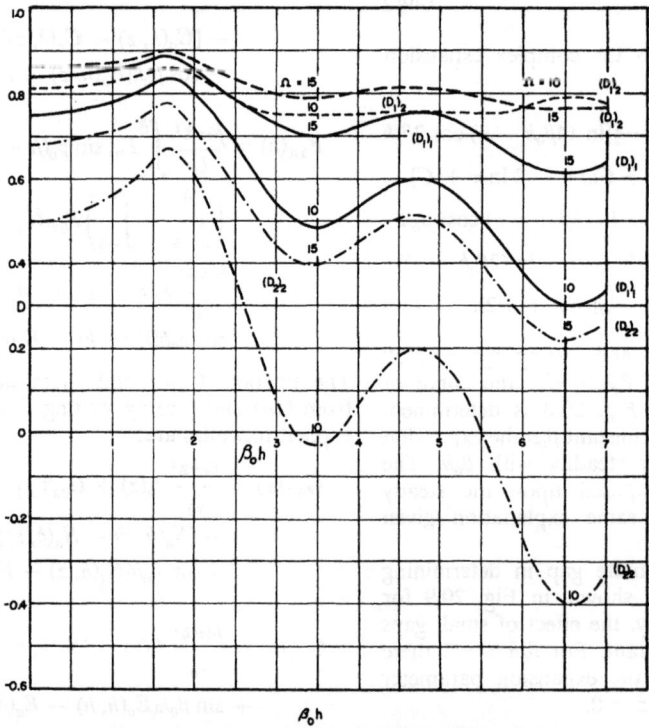

Fig. 21.1. D-factors for King-Middleton solution.

The second-order functions are obtained from the integral forms of the first-order functions by replacing $F_0(z)$ by $F_{1K}(z)$, $F_0(h)$ by $F_{1K}(h)$, $G_0(z)$ by $G_{1K}(z)$, and $G_0(h)$ by $G_{1K}(h)$, as given in (1a)–(1d).

An alternative form of the King-Middleton solution (1) in which use is made of the functions $F_{1H}(z)$, $G_{1H}(z)$, etc., which occur in the Hallén solution, is readily evaluated as follows. If (1a) is compared with (18.18e) written with $m = 1$, it is seen that

$$F_{1K}(z) = F_{1H}(z) + (\Psi_{K1} - \Omega) F_{0z}. \quad (2a)$$

Similarly, if (1b) is compared with (18.18b) with $m = 1$, it is clear that

$$F_{1K}(h) = F_{1H}(h). \quad (2b)$$

Also,

$$(F_{1z})_K = F_{1K}(z) - F_{1K}(h)$$
$$= (F_{1z})_H + (\Psi_{K1} - \Omega) F_{0z}. \quad (3)$$

The second-order functions are given by (18.18e) and (18.18b) with $m = 2$:

$$F_{2K}(z) = \frac{j4\pi z^i}{\zeta_0} \int_0^z (F_{1s})_K \sin \beta_0(z - s)\, ds$$
$$+ (F_{1z})_K \Psi_{K1}$$
$$- \left(\int_{-h}^h - \int_{-\delta}^\delta \right) (F_{1z'})_K K_1(z, z')\, dz', \quad (4a)$$

$$F_{2K}(h) = \frac{j4\pi z^i}{\zeta_0} \int_0^h (F_{1s})_K \sin \beta_0(h - s)\, ds$$
$$- \left(\int_{-h}^h - \int_{-\delta}^\delta \right) (F_{1z'})_K K_1(h, z')\, dz' \quad (4b)$$

$G_{2K}(z)$ is the same as (4a) with G written for F, (4c)

$G_{2K}(h)$ is the same as (4b) with G written for F. (4d)

After using (3) in (4a) and introducing (18.18e) with $m = 2$, it follows that

$$F_{2K}(z) = F_{2H}(z) + (\Psi_{K1} - \Omega)(F_{1z})_H$$
$$+ (\Psi_{K1} - \Omega) \left\{ \frac{j4\pi z^i}{\zeta_0} \int_0^z F_{0s} \sin \beta_0(z - s)\, ds \right.$$
$$\left. + F_{0z} \Psi_{K1} - \left(\int_{-h}^h - \int_{-\delta}^\delta \right) F_{0z'} K_1(z, z')\, dz' \right\}. \quad (5)$$

If (18.18e) is again used, but this time with $m = 1$,

$$F_{2K}(z) = F_{2H}(h) + (\Psi_{K1} - \Omega)(F_{1z})_H$$
$$+ (\Psi_{K1} - \Omega) F_{1H}(z)$$
$$+ (\Psi_{K1} - \Omega)^2 F_{0z}. \quad (6)$$

Substitution of (3) in (4b) gives

$$F_{2K}(h) = F_{2H}(h) + (\Psi_{K1} - \Omega) F_{1H}(h). \quad (7)$$

Subtraction of (7) from (6) leads to the final result

$$(F_{2z})_K \equiv F_{2K}(z) - F_{2K}(h)$$
$$= (F_{2z})_H + 2(\Psi_{K1} - \Omega)(F_{1z})_H$$
$$+ (\Psi_{K1} - \Omega)^2 F_{0z}. \quad (8)$$

The above procedure may be repeated in order to evaluate $(F_{nz})_K$. The result is

$$(F_{nz})_K = \sum_{i=0}^n \frac{n!}{(n-i)!\, i!} (\Psi_{K1} - \Omega)^i (F_{n-i,z})_H. \quad (9)$$

The functions $G_{nK}(z)$, $G_{nK}(h)$, $(G_{nz})_K$ may be obtained from the above expression by writing G for F.

If (2a) and (2b) and the corresponding expressions with G written for F are substituted in (15.27), the result is

$$M_{1K}(z) = M_{1H}(z)$$
$$+ (\Psi_{K1} - \Omega) \sin \beta_0(h - |z|). \quad (10)$$

Similarly, using (4a, b) and expressions like them, but with G written for F in (15.27), the following expression is obtained:

$$M_{2K}(z) = M_{2H}(z) + 2(\Psi_{K1} - \Omega) M_{1H}(z)$$
$$+ (\Psi_{K1} - \Omega)^2 \sin \beta_0(h - |z|). \quad (11)$$

In general,

$$M_{nK}(z) = \sum_{i=0}^n \frac{n!}{(n-i)!\, i!}$$
$$\times (\Psi_{K1} - \Omega)^i M_{n-i,H}(z), \quad (n \geq 1) \quad (12)$$

where it is understood that

$$M_{0H}(z) = \sin \beta_0(h - |z|). \quad (13)$$

Similarly,

$$F_{nK}(h) = \sum_{i=0}^{n-1} \frac{(n-1)!}{(n-i-1)!\, i!}$$
$$\times (\Psi_{K1} - \Omega)^i F_{n-i,H}(h), \quad (n \geq 1) \quad (14)$$

where it is understood that

$$F_{0H}(h) = F_0(h) = \cos \beta_0 h. \quad (15)$$

Upon substituting (12) and (14) in the general formula for current (15.26), this becomes

$$(I_z)_m = \frac{j2\pi V_\delta}{\zeta_0 \Psi_{K1}}$$

$$\times \left\{ \begin{array}{l} (D_1)_m \sin \beta_0(h - |z|) + \dfrac{(D_2)_m}{\Psi_{K1}} [M_{1H}^I \\ + jM_{1H}^{II}] + \dfrac{(D_3)_m}{\Psi_{K1}^2} [M_{2H}^I + jM_{2H}^{II}] + \cdots \\ \cos \beta_0(h - \delta) + \dfrac{(D_1)_{m-1}}{\Psi_{K1}} [A_{1H}^I + jA_{1H}^{II}] \\ + \dfrac{(D_2)_{m-1}}{\Psi_{K1}^2} [A_{2H}^I + jA_{2H}^{II}] + \cdots \end{array} \right\}.$$

(16)

With the shorthand

$$x = 1 - \frac{\Omega}{\Psi_{K1}} \quad (17)$$

the real D factors in (16) are obtained from the following table:

Order $m =$	0	1	2	3	4	
$(D_1)_m =$	1	$+x$	$+x^2$	$+x^3$	$+x^4$	$+\cdots$
$(D_2)_m =$		1	$+2x$	$+3x^2$	$+4x^3$	$+\cdots$
$(D_3)_m =$			1	$+3x$	$+6x^2$	$+\cdots$ (18)
$(D_4)_m =$				1	$+4x$	$+\cdots$
$(D_5)_m =$					1	$+\cdots$

The structure of the D factors is readily seen from (18). For example, $(D_2)_3 = 1 + 2x + 3x^2$. Each column has the same coefficients as $(1 + x)^n$.

It is interesting and significant that when the series in (18) are summed for an infinite number of terms, that is, $m \to \infty$, the result is

$$D_n = \frac{1}{(n-1)!} \frac{d^{n-1}}{dx^{n-1}} \frac{1}{1-x} = \frac{1}{(1-x)^n}$$

$$= \left(\frac{\Psi_{K1}}{\Omega}\right)^n. \quad (x < 1, \quad n \geq 1) \quad (19)$$

With (19), (16) is identically (18.17). Furthermore, if $a \to 0$, $\Psi_{K1}/\Omega \to 1$ for all values of $\beta_0 h$, so that (16) approaches (18.17) as the radius a approaches zero.

It is important to note that if a finite number of terms is used in (16), all terms belonging to a given order m of the solution must be retained and no others. That is, if an mth-order solution is evaluated, only terms contributed by $M_{nK}(z)$ and $F_{nK}(h)$ with $n = 0, 1, 2, \cdots, m$ are used. It is readily verified that this is equivalent to writing

$$(D_1)_0 = 1,$$

$$(D_1)_1 = 1 + \left(1 - \frac{\Omega}{\Psi}\right), \quad (20)$$

$$(D_2)_1 = 1.$$

Similarly,

$$(D_1)_2 = 1 + \left(1 - \frac{\Omega}{\Psi}\right) + \left(1 - \frac{\Omega}{\Psi}\right)^2,$$

$$(D_2)_2 = 1 + 2\left(1 - \frac{\Omega}{\Psi}\right), \quad (21)$$

$$(D_3)_2 = 1.$$

The factors $(D_1)_1$, $(D_1)_2$, and $(D_2)_2$ are shown in Fig. 21.1 for $\Omega = 10$ and 15, as functions of $\beta_0 h$.

22. Axial Distribution of Current

The numerical evaluation of the axial distribution of current in a cylindrical, center-driven antenna from the general formula (21.16) is laborious if higher-order terms are to be included. But the fact that a zeroth-order current was used in deriving the expansion parameter, which is itself a reasonably good approximation of the current, as is known from experimental measurements, suggests that a first-order current is a satisfactory approximation. This should be true certainly for the distribution along the antenna, where precise quantitative information is relatively unimportant. The input current, on the other hand, determines the impedance, for which greater quantitative accuracy is required. It would appear, therefore, that a determination of first-order current at all points along the antenna and second-order input current and impedance should be adequate. The evaluation of first- and second-order impedance and input current is considered in detail in Sec. 27. It is shown that the curves for the first-order impedance as a function of $\beta_0 h$ with $\Omega = 2 \ln (2h/a)$ as parameter closely approximate the curves for second-order impedance in general shape and in relative magnitude. In fact, the only significant difference between the two sets of curves is a greater shift of the entire second-order impedance curve to smaller values of $\beta_0 h$. This suggests that the second-order distribution of current may be estimated from the first-order distribution of an antenna that has the same input current as required by the second-order impedance, but a slightly shorter length. Thus, a knowledge of the first-order distribution of current and the second-order input current permits a very satisfactory approximate representation of the second-order distribution.

The distribution of current given by the general formula (21.16) depends upon the base separation. Comparison of this formula with the same expression for $\delta = 0$ shows that the effect of the base separation is determined primarily by a number of integrals of the general form

$$\int_{-\delta}^{\delta} \frac{I_z'}{R} e^{-j\beta_0 R} \, dz', \qquad (1)$$

where $R = R_1 = \sqrt{(z - z')^2 + a^2}$ or $R = R_{1h} = \sqrt{(h - z')^2 + a^2}$. It was shown in Sec. 20 in the discussion of (20.10) and (20.11) that integrals of this type are significant only for values of z near δ; that is, the base separation has a negligible effect on the current along most of the antenna, so that the general distribution for $\delta = 0$ is a good approximation for antennas with δ small. Only in determining the input current must account be taken of the base separation.

In summary, it may be concluded that a first-order solution for current with $\delta = 0$ should be adequate in determining magnitude and phase at all points along the antenna except near the driving terminals. The input current and the impedance require a second-order solution that takes account of the base-separation.

The *first-order* distribution of current for $\delta = 0$, as calculated from (21.16), is shown in Figs. 22.1 to 22.6. The curves show the components* I_z''/V_0, I_z'/V_0, the magnitude $|I_z|/V_0$, and the phase angle θ_I referred to V_0 as functions of $\beta_0 z$, with $\beta_0 h = 1.538$, $\pi/2$, $3\pi/4$, 2.75, π, and $5\pi/4$ for $\Omega = 10$ and 20.

Examination of Figs. 22.1 to 22.6 shows that both components of current are significant. Although in the *first-order* solution (in which terms of order higher than $1/\Psi_{K1}$ are neglected) the curves for $\Omega = 20$ are better approximations than the curves for $\Omega = 10$, the effect of decreasing the ratio h/a on the distribution of current is correctly indicated and the order of magnitude is right. The value of $\beta_0 h$ for the curve $\Omega = 20$ in Fig. 22.1 is chosen so that the component of current I_z', which is in phase-quadrature with V_0, is nearly zero. If Ω is made a little greater or $\beta_0 h$ slightly smaller, the current at all cross sections of the antenna is nearly in phase with V_0 for a thin antenna. If Ω is made considerably greater or $\beta_0 h$ much smaller, the component I_z' leads V_0 instead of lagging, as in Fig. 22.1. An antenna for which $\beta_0 h = \pi/2$ always has a lagging quadrature component of current.

If an antenna is adjusted so that $\beta_0 h$ is less than π by just the right amount for the particular value of h/a, it is possible to have the *input* current in phase with V_0. However, the component of current in phase quadrature with V_0 is then not near zero anywhere on the antenna except at $z = 0$ and, of course, at $z = \pm h$. In fact, for thin antennas this quadrature component is the principal current everywhere except near $z = 0$. Its distribution is sketched for $\beta_0 h = 2.75$ in Fig. 22.4, in which the input current is computed from first-order theory and the rest of the curves are estimated using Fig. 22.5.

In Figs. 22.3 to 22.6 it is important to note the large increase in the magnitude of the component of current I_z'' in phase with the driving voltage as the ratio h/a decreases and Ω is thereby reduced. Since I_z'' is larger near the center of the antenna, it might be supposed that the magnitude $I_z = \sqrt{I_z'^2 + I_z''^2}$ near the center $z = 0$ must be relatively very much greater for antennas of large radius than for antennas of small radius. However, the increase near the center in I_z'' is accompanied by an increase in the component I_z' that is relatively much greater in the *outer* parts of the antenna as the radius is increased. Thus, whereas both I_z'' and I_z' increase greatly as Ω is decreased, the distribution of I_z' is characterized in addition by a change suggesting an outward pushing of current along the antenna. This is readily seen in Figs. 22.3 to 22.6 in which $I_{z\max}'$ occurs at larger values of $\beta_0 z$ for $\Omega = 10$ than for $\Omega = 20$, and in which the ratio $I_{\max}'/I_0'$ is much larger for $\Omega = 10$ than for $\Omega = 20$. The net effect on the magnitude I_z may be seen in Figs. 22.3 to 22.6. As Ω is decreased, I_z increases everywhere in the antenna. In addition, the entire distribution is moved outward along the antenna and there is an additional marked increase near the center for antennas of half-length near $\lambda_0/2$.

---

* The use of a prime to denote a component of current as in (15.30) and a current at a primed coordinate in an integral should lead to no confusion.

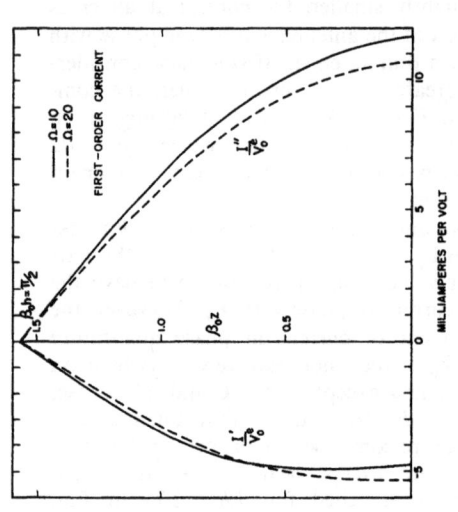

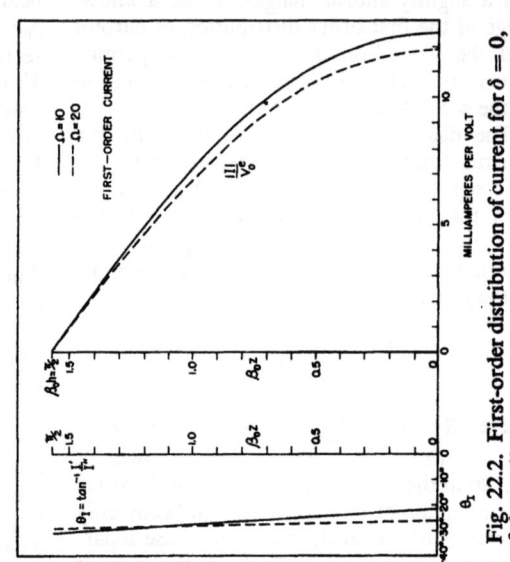

Fig. 22.2. First-order distribution of current for $\delta = 0$, $\beta_0 h = \pi/2$.

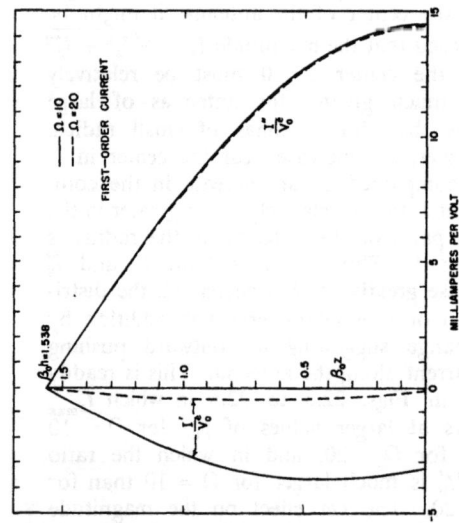

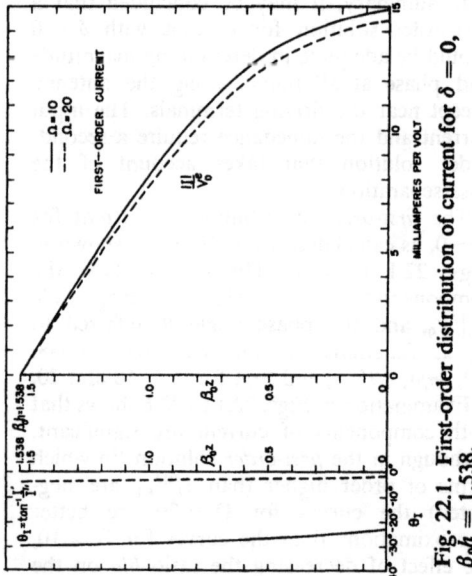

Fig. 22.1. First-order distribution of current for $\delta = 0$, $\beta_0 h = 1.538$.

[II.22] THEORY OF LINEAR ANTENNAS 113

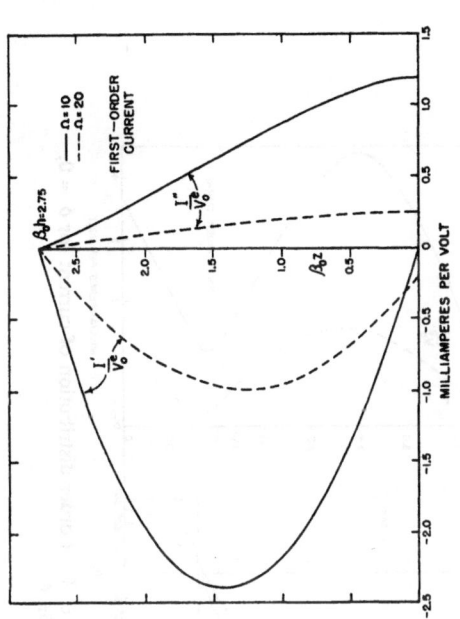

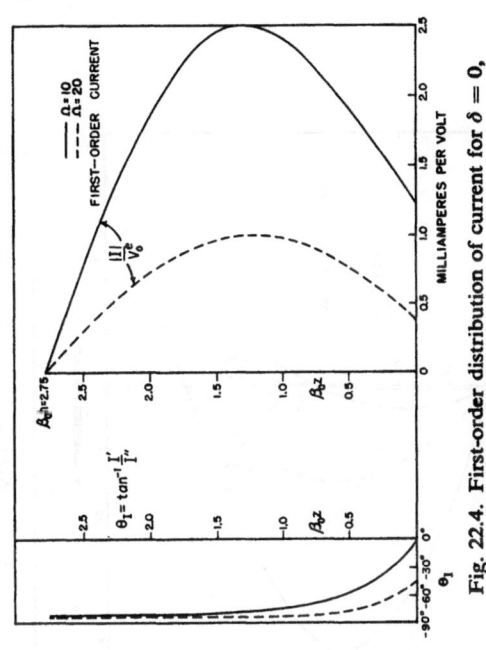

Fig. 22.4. First-order distribution of current for $\delta = 0$, $\beta_0 h = 2.75$.

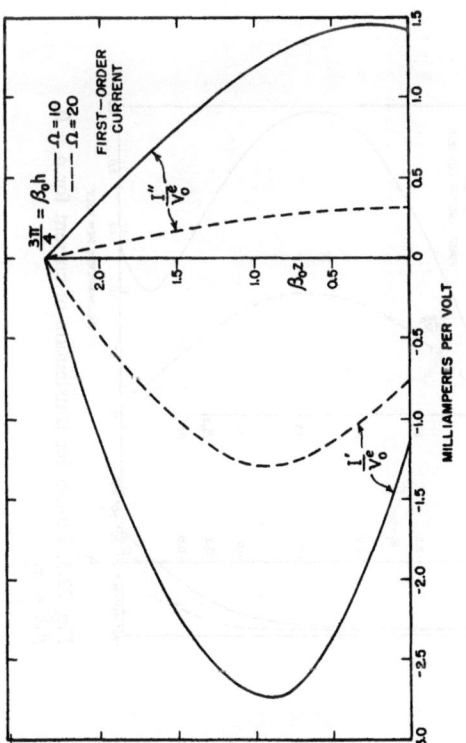

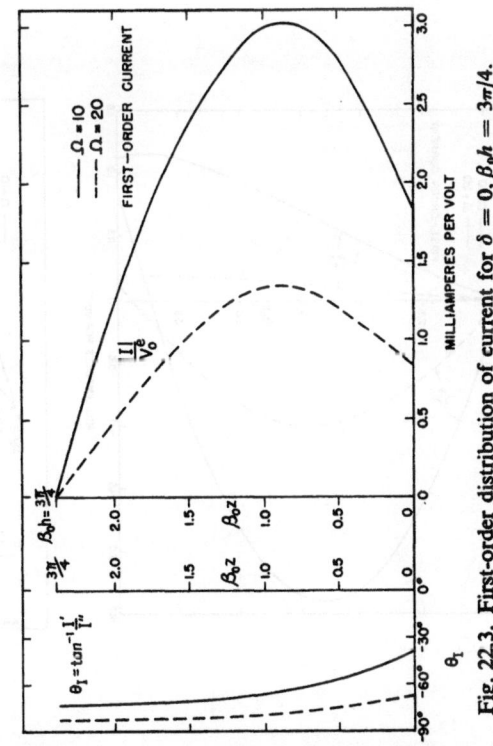

Fig. 22.3. First-order distribution of current for $\delta = 0$, $\beta_0 h = 3\pi/4$.

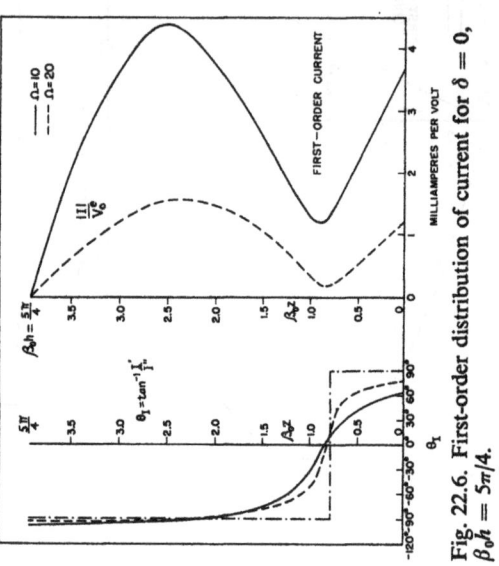

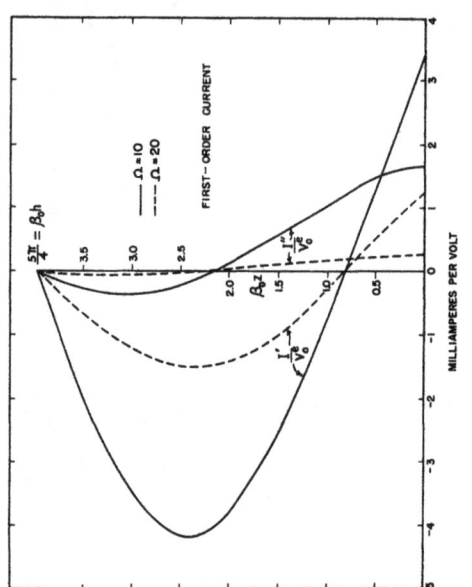

Fig. 22.6. First-order distribution of current for $\delta = 0$, $\beta_0 h = 5\pi/4$.

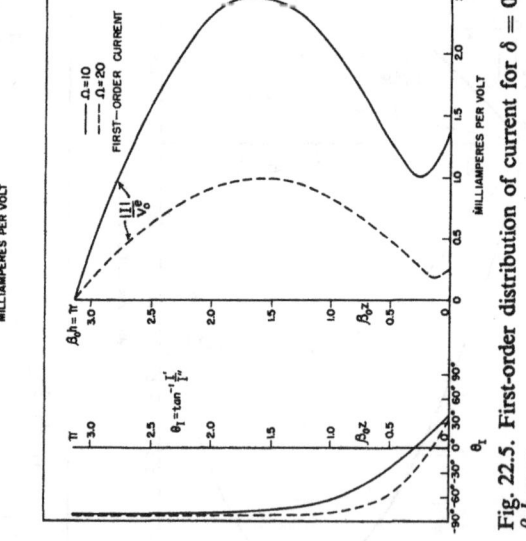

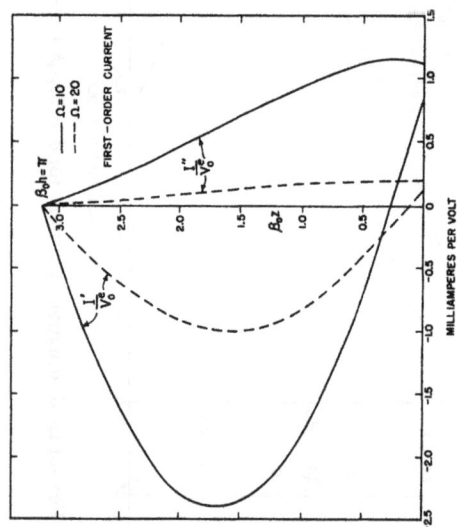

Fig. 22.5. First-order distribution of current for $\delta = 0$, $\beta_0 h = \pi$.

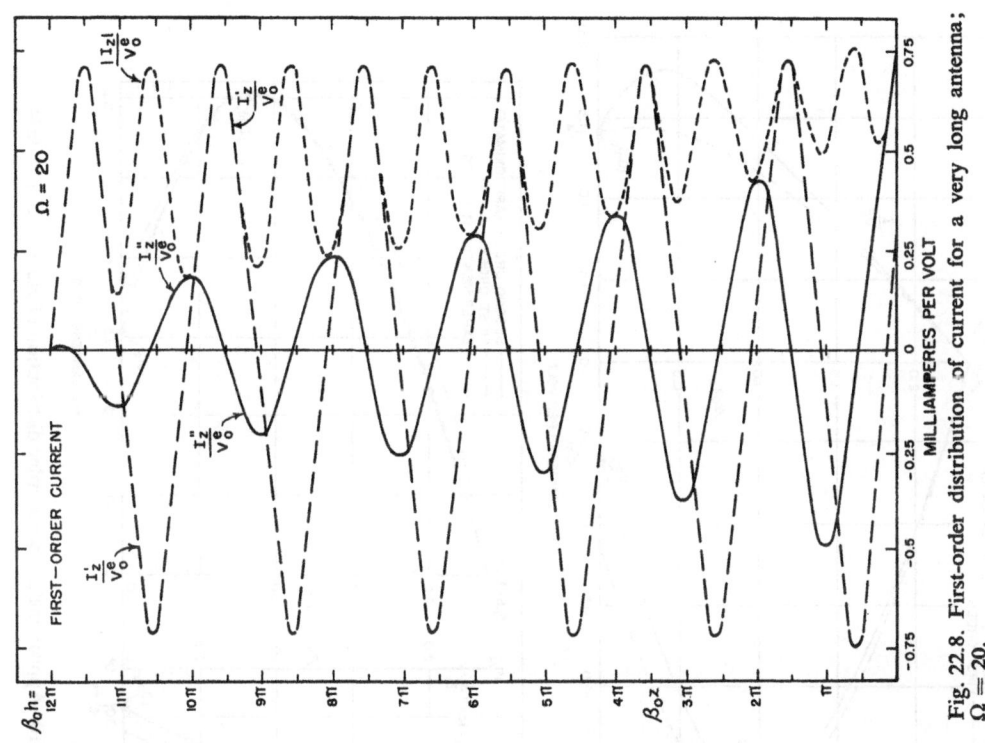

Fig. 22.8. First-order distribution of current for a very long antenna; $\Omega = 20$.

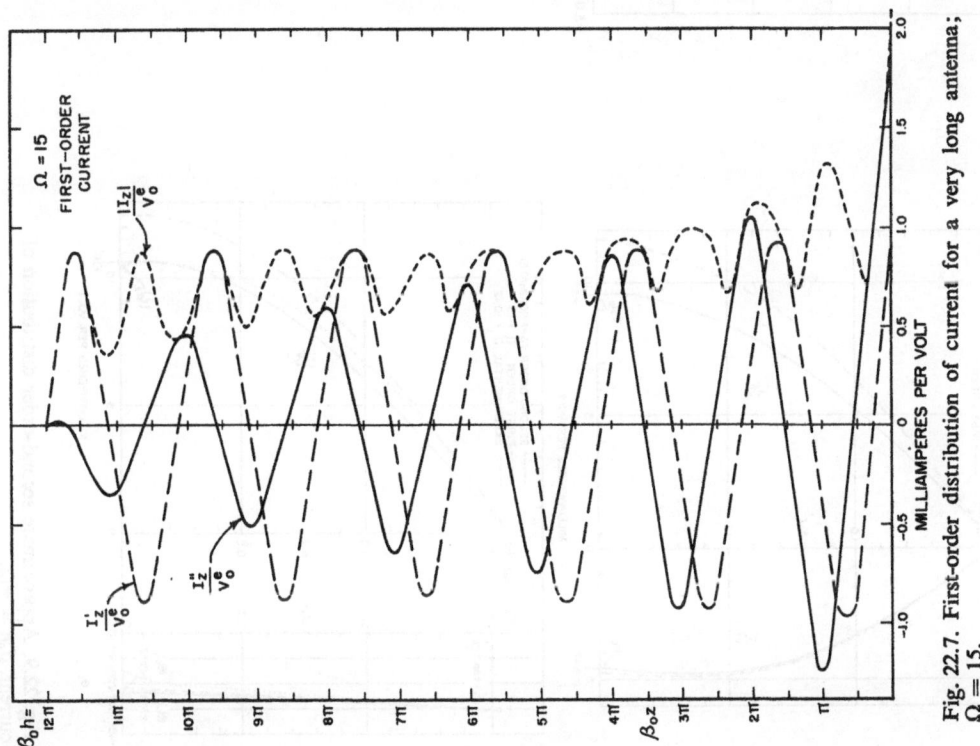

Fig. 22.7. First-order distribution of current for a very long antenna; $\Omega = 15$.

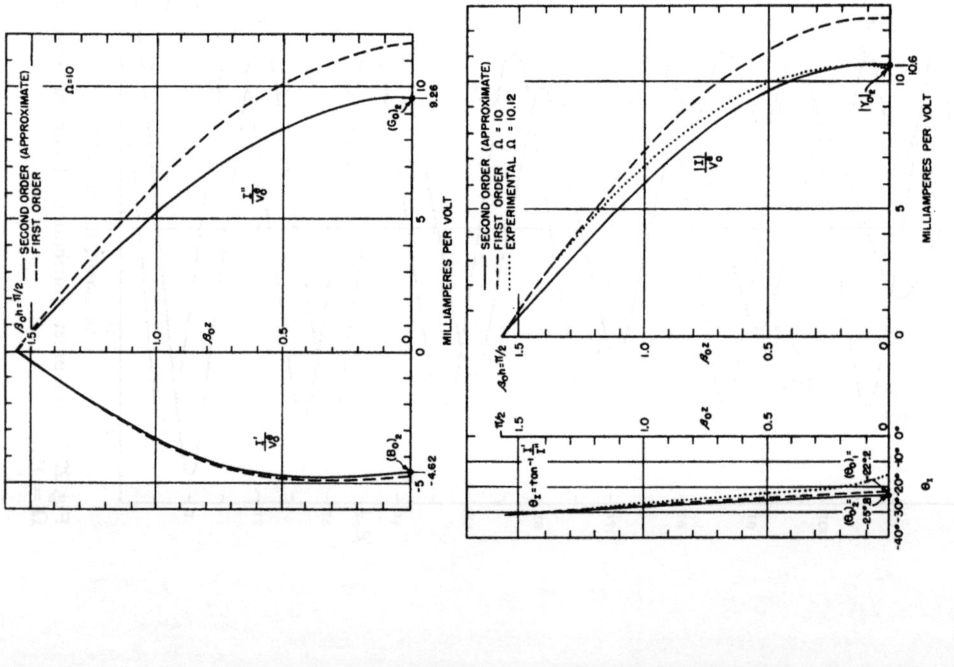

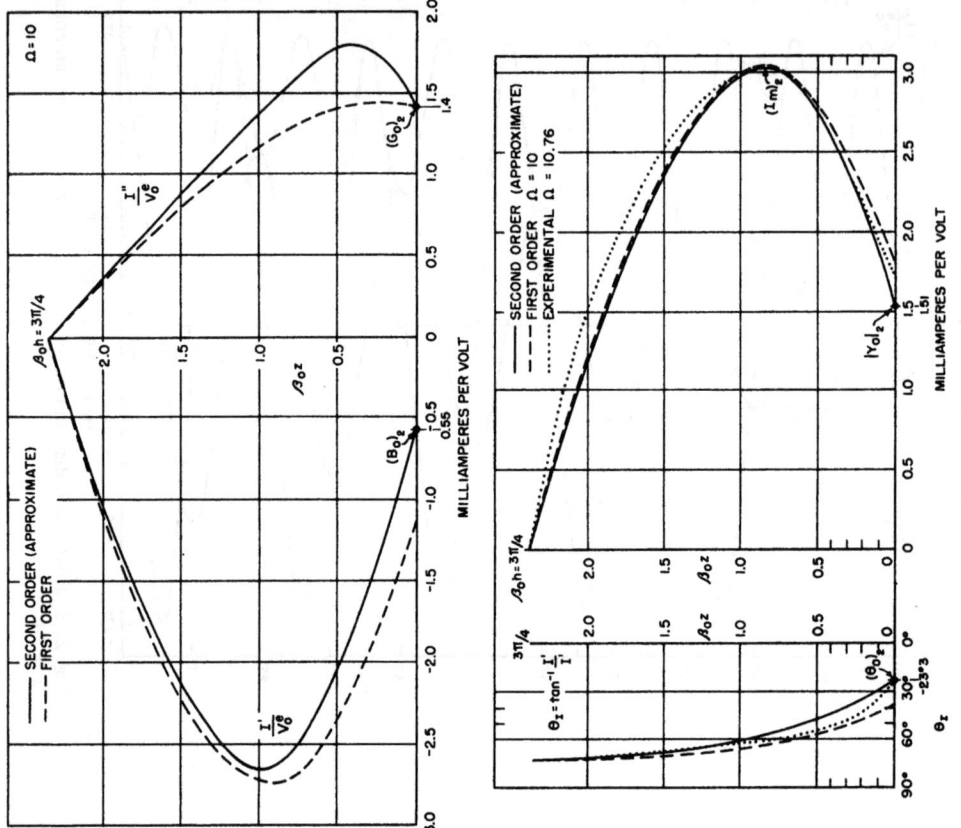

Fig. 22.10. Approximate second-order distribution of current, $\beta_0 h = 3\pi/4$.

Fig. 22.9. Approximate second-order distribution of current, $\beta_0 h = \pi/2$.

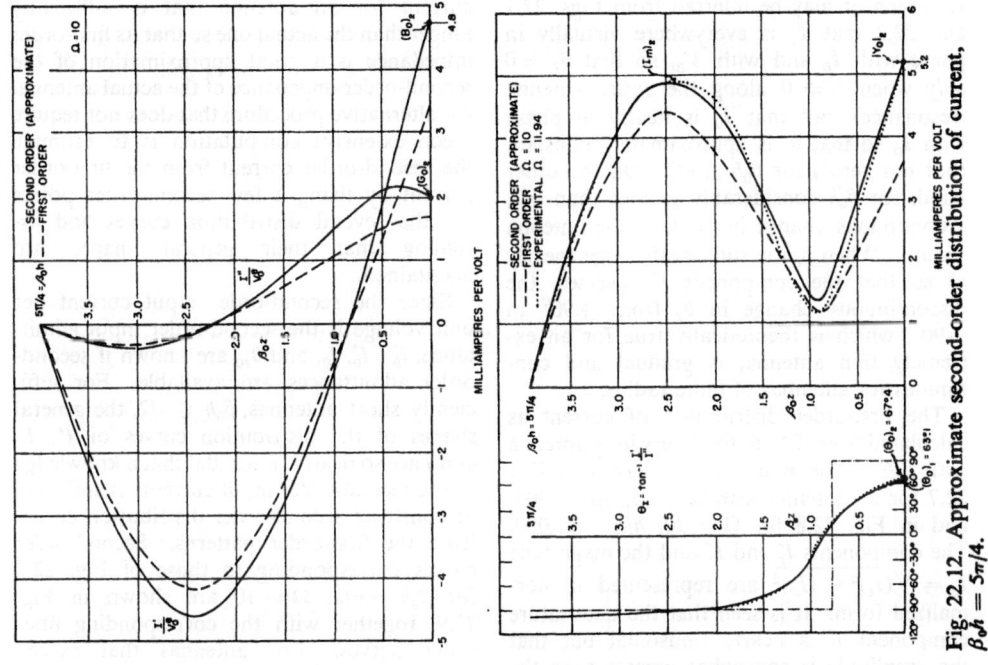

Fig. 22.12. Approximate second-order distribution of current, $\beta_0 h = 5\pi/4$.

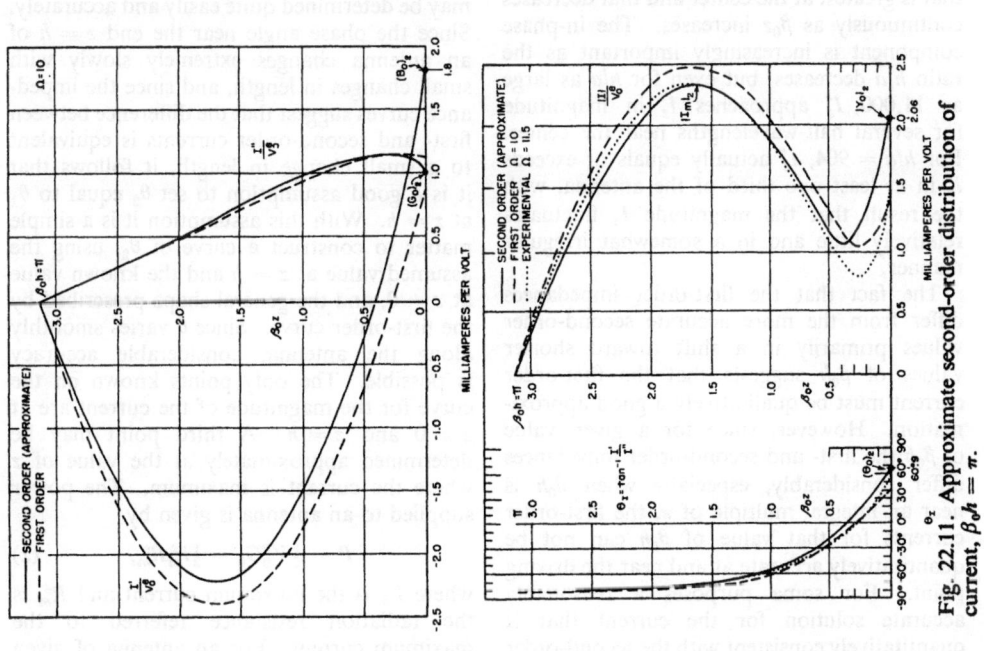

Fig. 22.11. Approximate second-order distribution of current, $\beta_0 h = \pi$.

The behavior of the phase angle θ_I of the complex current I_z referred to V_0 is also shown. It is seen or may be inferred from Figs. 22.1 and 22.2 that I_z is everywhere virtually in phase with I_0 and with V_0, so that $\theta_I \doteq 0$ only when $I_z' \doteq 0$ along the entire antenna (resonance), but that I_z is nearly in phase with I_0 so that θ_I is approximately constant (but not zero!) for $\beta_0 h \leq \pi/2$. On the other hand, for $\beta_0 h$ considerably greater than $\pi/2$, a continuous change in θ_I along the antenna occurs. When $\beta_0 h$ is sufficiently large (nearly π) so that the component I_z' reverses, the discontinuous change in θ_I from $+90°$ to $-90°$, which is theoretically true for an extremely thin antenna, is gradual and continuous for antennas of finite radius.

The first-order distribution of current as calculated from (21.16) for a very long antenna with $\beta_0 h = 12\pi$ or $h = 6\lambda_0$ is shown in Fig. 22.7 for an antenna with $\Omega = 15$, $h/a = 904$, and in Fig. 22.8 for $\Omega = 20$, $h/a = 11{,}000$. The components I_z' and I_z'' and the magnitude $I_z = \sqrt{(I_z')^2 + (I_z'')^2}$ are represented in normalized form. It is seen that the quadrature component I_z' is nearly sinusoidal but that the amplitude is somewhat greater near the center and the ends of the antenna than elsewhere. The in-phase component I_z'', on the other hand, is characterized by an amplitude that is greatest at the center and that decreases continuously as $\beta_0 z$ increases. The in-phase component is increasingly important as the ratio h/a decreases, but even for h/a as large as 11,000, I_z'' approaches I_z' in magnitude for several half-wavelengths near the center. For $h/a = 904$, I_z'' actually equals or exceeds I_z' in at least one third of the antenna, with the result that the magnitude I_z fluctuates relatively little and in a somewhat irregular manner.

The fact that the first-order impedances differ from the more accurate second-order values primarily in a shift toward shorter values of $\beta_0 h$ suggests that the first-order current must be qualitatively a good approximation. However, since for a given value of $\beta_0 h$ the first- and second-order impedances differ considerably, especially when $\beta_0 h$ is near an integral multiple of π, the first-order currents for that value of $\beta_0 h$ can not be quantitatively accurate at and near the driving point. For some purposes a reasonably accurate solution for the current that is quantitatively consistent with the second-order impedances is required. One method of obtaining such a distribution has been suggested, namely, to determine the first-order current for an antenna that is sufficiently longer than the actual one so that its first-order impedance is a good approximation of the second-order impedance of the actual antenna. An alternative procedure that does not require much extensive computation is to estimate the second-order current from the first-order current by fixing a few second-order points for the several distribution curves and assuming that their general shapes are maintained.

Since the second-order input current per unit voltage is the second-order input admittance, I_0'', I_0', I_0, and θ_{I0} are known if second-order admittances are available. For sufficiently short antennas, $\beta_0 h \leq \pi/2$, the general shapes of the distribution curves of I'', I', and I are so nearly sinusoidal that a knowledge of the second-order input currents is sufficient to construct second-order distribution curves from the first-order patterns. Second-order curves corresponding to those of Fig. 22.2 for $\beta_0 h = \pi/2$, $\Omega = 10$ are shown in Fig. 22.9, together with the corresponding first-order curves. For antennas that exceed $\pi/2$ in their electrical half-length, the input current is not sufficient to obtain even a rough estimate of the distribution curves. Fortunately, the second-order phase angle may be determined quite easily and accurately. Since the phase angle near the end $z = h$ of an antenna changes extremely slowly with small changes in length, and since the impedance curves suggest that the difference between first- and second-order currents is equivalent to a small change in length, it follows that it is a good assumption to set θ_2 equal to θ_1 at $z = h$. With this assumption it is a simple matter to construct a curve of θ_2, using the assumed value at $z = h$ and the known value at $z = 0$, and the general shape prescribed by the first-order curve. Since θ varies smoothly along the antenna, considerable accuracy is possible. The only points known on the curve for the magnitude of the current are at $z = 0$ and $z = h$. A third point may be determined approximately at the value of z where the current is maximum. The power supplied to an antenna is given by

$$P = \tfrac{1}{2} V_0^2 G_0 = \tfrac{1}{2} I_m^2 R_m^e, \qquad (2)$$

where I_m is the maximum current and R_m^e is the radiation resistance referred to the maximum current. For an antenna of given length and radius, R_m^e is an essentially constant quantity that is insensitive to small changes in current distribution which do not affect the far-zone field significantly. It follows that for an antenna with constant applied voltage, I_m is proportional to $\sqrt{G_0}$. Accordingly,

$$\frac{(I_m)_2}{(I_m)_1} \doteq \sqrt{\frac{(G_0)_2}{(G_0)_1}}. \qquad (3)$$

Thus, with the maximum magnitude of the first-order current available, together with the first- and second-order input conductances, the maximum magnitude of the second-order current may be determined. Since the location of maximum current is very near $z = h - \lambda_0/4$ on all antennas, a third point is available on the second-order current-magnitude curve. For an antenna with $\beta_0 h$ sufficiently short that $|I|$ has no minimum, the entire distribution of $|I|$ is readily drawn as in Fig. 22.10 for $\beta_0 h = 3\pi/4$. With the curve for θ_2 drawn smoothly between $\theta_2 = \theta_1$ at $z = h$ and the known value of θ_2 at $z = 0$, the components I' and I'' may be computed. The resulting curves are in Fig. 22.10.

When the current-magnitude curve has a minimum, the construction of the approximate second-order curve from I_0, $I_h = 0$, and $I_{\max}$ is less accurate. However, the minimum must occur very near the point where $\theta_2 = 0$, and it may be assumed that it does not differ greatly in depth from the first-order value. After an estimate has been made, as in Figs. 22.11 and 22.12, a check is obtained when I' and I'' are computed. There may be no irregularities in these curves near the minimum. By adjusting this minimum until both I' and I'' curves are smooth, a satisfactory result is obtained. This was done in Figs. 22.11 and 22.12, where the component curves are seen to be smooth and reasonable in the region of minimum I.

The approximate second-order distributions of current shown in Figs. 22.9 to 22.12 should be quantitatively reasonably correct. They differ from the first-order distributions primarily near the driving point and in a small difference in the relative magnitudes of the in-phase component I'' compared with the quadrature component I'. In the second-order distributions I'' is relatively somewhat greater than in the first-order distribution, since in all cases the phase angle θ_2 is shifted more or less toward greater positive values as compared with θ_1. The experimental distributions also shown in these figures are discussed in Sec. 26, where the effect of the base separation also is considered.

23. Axial Distribution of Charge

The distribution of charge on an antenna is readily derived from the current by the method used in Sec. 16 for the zeroth-order charge. Since the charge is known to be an odd function, the distribution in the lower half is obtained directly from that in the upper half. The equation of continuity in the form

$$\frac{dI_z}{dz} + j\omega q_z = 0, \qquad (1)$$

combined with

$$I_z = I_z'' + jI_z', \qquad (2)$$

gives the following distribution of charge:

$$q_z = j(q_z'' + jq_z') = \frac{j}{\omega}\left(\frac{dI_z''}{dz} + j\frac{dI_z'}{dz}\right), \qquad (3)$$

so that

$$q_z'' = \frac{1}{\omega}\frac{dI_z''}{dz}, \qquad (4)$$

$$q_z' = \frac{1}{\omega}\frac{dI_z'}{dz}. \qquad (5)$$

Since the distribution of current is conveniently plotted in amperes per input volt against electrical distance from the origin at $z = 0$ in radians, (4) and (5) may be rewritten in the form

$$(q_z''/V_0) = \frac{1}{v_0}\frac{d(I_z''/V_0)}{d(\beta_0 z)}, \qquad (6)$$

$$(q_z'/V_0) = \frac{1}{v_0}\frac{d(I_z'/V_0)}{d(\beta_0 z)}. \qquad (7)$$

The derivatives in (6) and (7) are the slopes of the curves for distribution of current given in Figs. 22.1 to 22.6. Hence, if the numerical values obtained for the slopes are divided by $v_0 = 3 \times 10^8$ m/sec, the charge in coulombs per meter per volt input is obtained. Curves obtained graphically from Figs. 22.1 to 22.6 for q_z'', q_z', $q_z = \sqrt{(q_z'')^2 + (q_z')^2}$, and $\theta_q = \tan^{-1} q_z'/q_z''$ are given in Figs. 23.1 to 23.5 for $\beta_0 h = 1.538$, $3\pi/4$, 2.75, π, and $5\pi/4$, with $\Omega = 10$ and 20. The distributions q_z resemble the magnitudes of the corresponding zeroth-order distributions in Fig. 16.2 in general outline, but differ from them in the absence of sharp zeros and in a relatively very much greater concentration of charge near the ends of the antenna, especially for

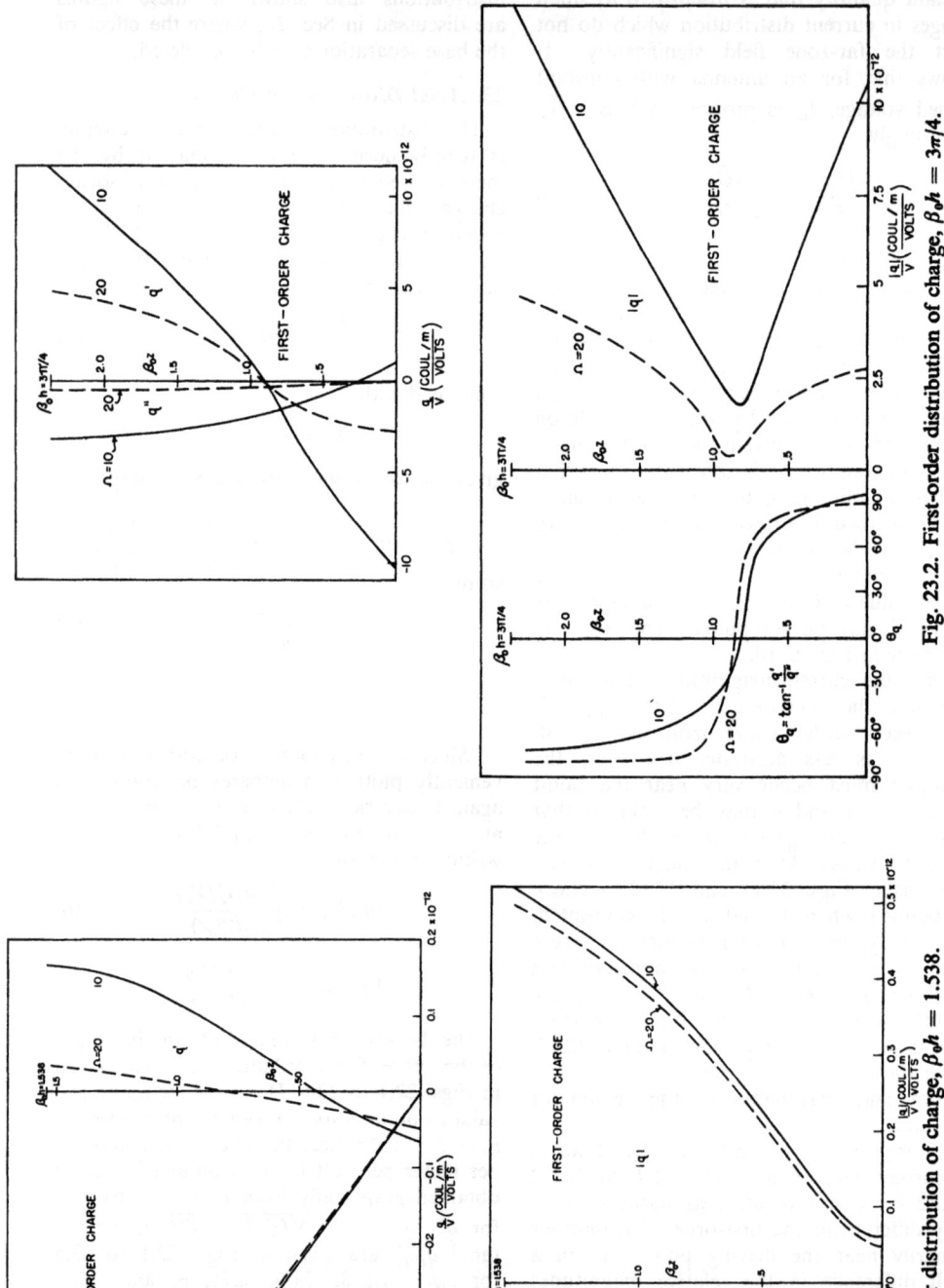

Fig. 23.2. First-order distribution of charge, $\beta_0 h = 3\pi/4$.

Fig. 23.1. First-order distribution of charge, $\beta_0 h = 1.538$.

[II.23]   THEORY OF LINEAR ANTENNAS   121

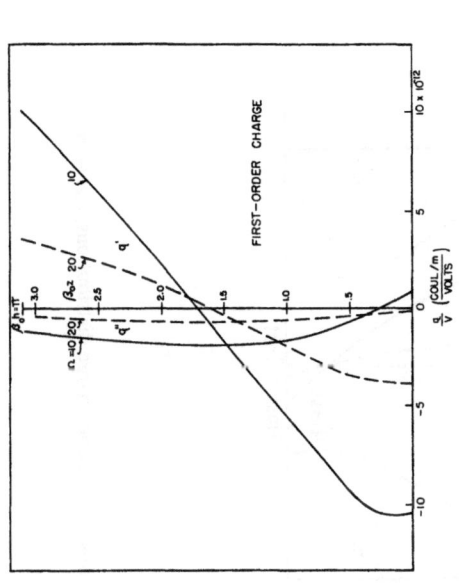

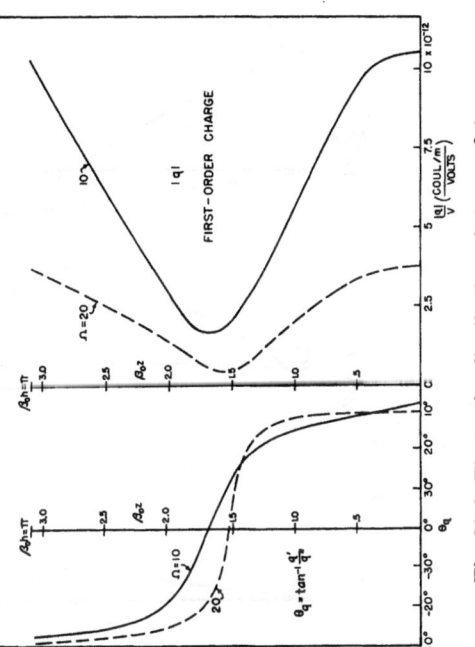

Fig. 23.4. First-order distribution of charge, $\beta_0 h = \pi$.

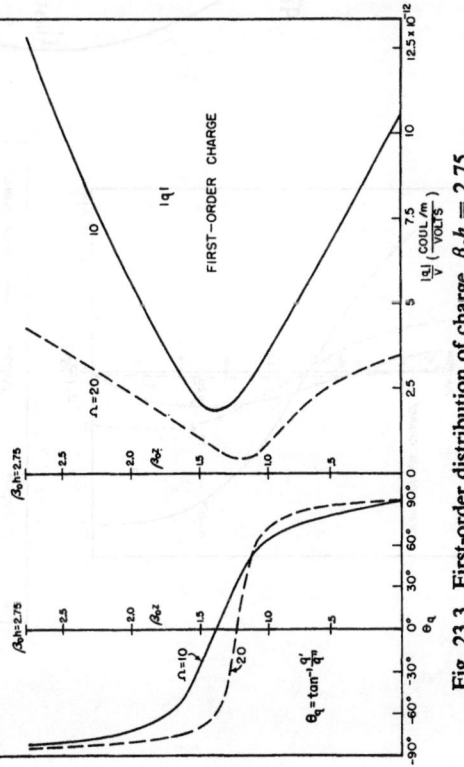

Fig. 23.3. First-order distribution of charge, $\beta_0 h = 2.75$.

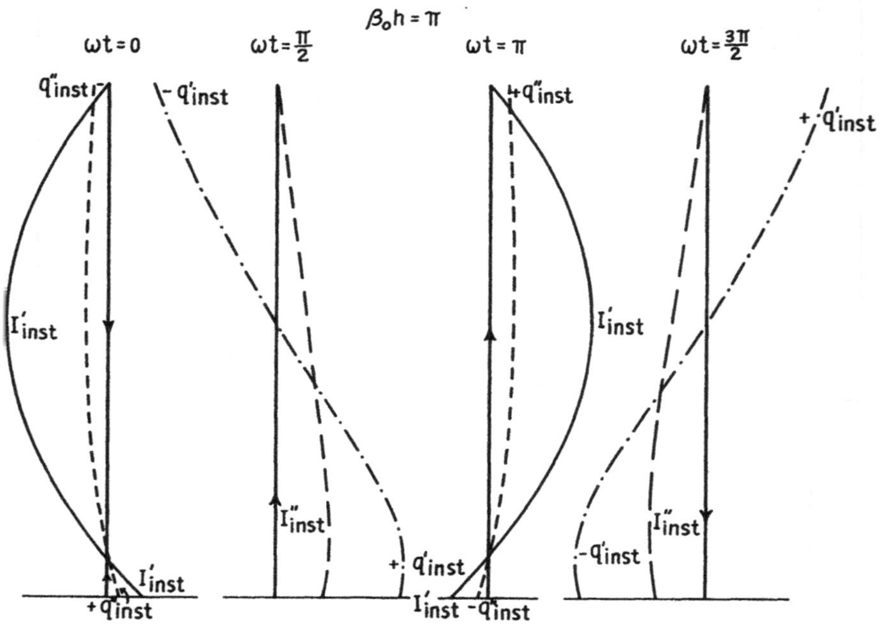

Fig. 24.1. Instantaneous current and charge distributions, $\beta_0 h = \pi$.

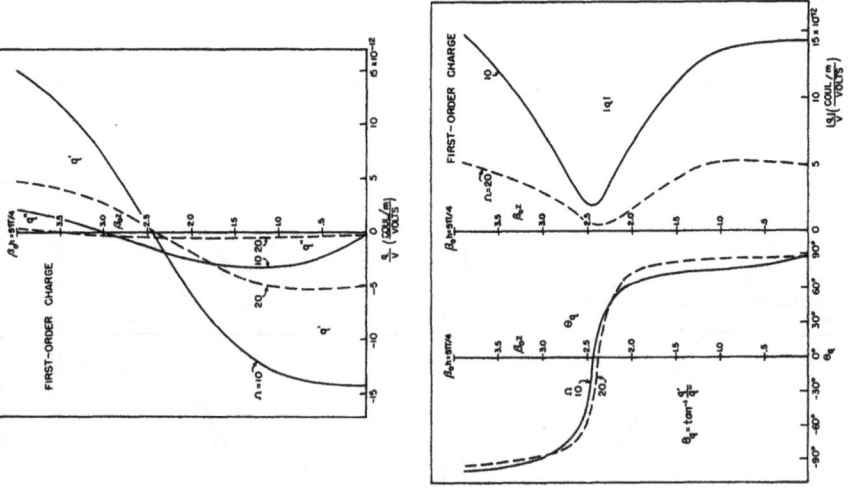

Fig. 23.5. First-order distribution of charge, $\beta_0 h = 5\pi/4$.

the thicker antenna with $\Omega = 10$. The rectified cosine curves of the zeroth-order distribution are considerably distorted for $\Omega = 20$ and even more for $\Omega = 10$. In Fig. 23.2, for example, q'_z approximates a cosine curve with maximum at the top for $\Omega = 20$, whereas for $\Omega = 10$ the concentration of charge at the top is greatly increased so that the rounded cosine curve is replaced by a line that is almost straight. This effect is also evident in the curves for $|q_z|$.

24. The Instantaneous Current and Charge

The instantaneous current is in general most conveniently studied in terms of its quadrature components I'_z and I''_z. Thus, with the complex amplitude given by

$$I_z = I''_z + jI'_z, \quad (1)$$

the complex instantaneous current $I_{z_{\text{inst}}}$ is obtained by multiplying both sides of (1) by $e^{j\omega t}$. The real instantaneous current $I_{z_{\text{inst}}}$ is obtained by taking the imaginary part of each side corresponding to

$$V^e_{0_{\text{inst}}} = V^e_0 \sin \omega t. \quad (2)$$

It is

$$I_{z_{\text{inst}}} = \text{Imaginary part of } (I''_z + jI'_z) e^{j\omega t}, \quad (3a)$$

so that

$$I_{z_{\text{inst}}} = I'_z \cos \omega t + I''_z \sin \omega t. \quad (3b)$$

Typical distributions of the real magnitudes I''_z and I'_z are given in Figs. 22.1 and 22.2.

The instantaneous charge per unit length is readily obtained from (23.3) or directly from (3b) using the equation of continuity for total current:

$$\frac{dI_{z_{\text{inst}}}}{dz} + \frac{dq_{z_{\text{inst}}}}{dt} = 0. \quad (4)$$

Thus,

$$\frac{dq_{z_{\text{inst}}}}{dt} = -\left(\frac{dI'_z}{dz} \cos \omega t + \frac{dI''_z}{dz} \sin \omega t\right), \quad (5)$$

so that integration with respect to the time gives

$$q_{z_{\text{inst}}} = -(q'_z \sin \omega t - q''_z \cos \omega t), \quad (6)$$

where q'_z and q''_z are given by (23.4) and (23.5) and where the constant of integration has been set equal to zero since the system as a whole is electrically neutral.

The instantaneous currents and charges therefore may be described in terms of the distributions I'_z and q'_z, I''_z and q''_z. These distributions are shown in Fig. 24.1 for $\beta_0 h = \pi$ at the successive instants $\omega t = 0, \pi/2, \pi, 3\pi/2$.

For $\beta_0 h = \pi/2$ the distributions I'_z and I''_z are nearly alike, so that (3b) is conveniently replaced by

$$I_{z_{\text{inst}}} = \sqrt{(I'_z)^2 + (I''_z)^2} \sin (\omega t + \theta_I)$$
$$= I_z \sin (\omega t + \theta_I), \quad (7)$$

where

$$\theta_I = \tan^{-1} (I'_z/I''_z) \doteq \text{const.} \quad (8)$$

It is then unnecessary to plot $I'_{z_{\text{inst}}}, I''_{z_{\text{inst}}}, q'_{z_{\text{inst}}}$, and $q''_{z_{\text{inst}}}$ separately, since a graph of $I_{z_{\text{inst}}}$ and $q_{z_{\text{inst}}}$ can be made so long as θ_I is constant or nearly so.

25. An Alternative Method of Solving Hallén's Integral Equation

Although in the integral equation (14.3) for the current in a cylindrical antenna the current and the kernel are complex, the solution by iteration described in the preceding sections makes use of a real expansion parameter and a zeroth-order current that consists of a single component in phase quadrature with the driving voltage V^e_δ. It follows that the component of current *in phase* with V^e_δ always is one step lower in the iteration than the quadrature current. This is not unreasonable since, in general, the quadrature current predominates except when the electrical half-length $\beta_0 h$ is at or near an odd multiple of $\pi/2$. Nevertheless, all quantities that depend primarily on the in-phase component are determined less accurately in a solution of given order than quantities that depend upon the quadrature current.* In particular, the entire solution is less accurate when $\beta_0 h = n\pi/2$, n odd, than for other lengths.

This difficulty can be eliminated by introducing both components of current directly in the complex integral equation (14.3) and separating this into two simultaneous real integral equations.

For simplicity let $\delta = 0$, $z^i = 0$, and let phase be referred to the driving voltage, so that this is real. In order to avoid ambiguity with primed coördinates, let a new notation be introduced temporarily for the complex current as follows:

$$I_z = I''_z + jI'_z = I_r(z) + jI_i(z), \quad (1)$$

---

* An important example of this fact is developed in Sec. 31.

where $I_r(z) = I_z''$ is the component in phase with V_0 and $I_t(z) = I_z'$ is the component in phase-quadrature. Also, let the following real kernels be introduced:

$$K_c(z, z') \equiv \frac{\cos \beta_0 R_1}{R_1} + \frac{\cos \beta_0 R_2}{R_2}, \quad (2a)$$

$$K_s(z, z') \equiv -\left(\frac{\sin \beta_0 R_1}{R_1} + \frac{\sin \beta_0 R_2}{R_2}\right). \quad (2b)$$

With (1) and (2), (14.3) may be separated into two real integral equations. They are

$$\int_0^h [I_r(z')K_c(z, z') - I_t(z')K_s(z, z')] \, dz'$$

$$= \frac{4\pi}{\zeta_0} C'F_0(z), \quad (3a)$$

$$\int_0^h [I_t(z')K_c(z, z') + I_r(z')K_s(z, z')] \, dz'$$

$$= -\frac{4\pi}{\zeta_0} [C''F_0(z) + \tfrac{1}{2}V_0 G_0(z)]. \quad (3b)$$

As defined in (15.2a, b), $F_0(z) = \cos \beta_0 z$, $G_0(z) = \sin \beta_0 z$; C'' and C' are the real and imaginary parts of the complex constant C_1. In order to illustrate the use of difference kernels of the type defined in (14.17), equations like (3a, b) with $z = h$ may be subtracted from (3a, b) to give

$$\int_0^h [I_r(z')L_c(z, z') - I_t(z')L_s(z, z')] \, dz'$$

$$= \frac{4\pi}{\zeta_0} C'F_{0z}, \quad (4a)$$

$$\int_0^h [I_r(z')L_c(z, z') + I_t(z')L_s(z, z')] \, dz'$$

$$= -\frac{4\pi}{\zeta_0} [C''F_{0z} + \tfrac{1}{2}V_0 G_{0z}], \quad (4b)$$

where $L(z, z') = K(z, z') - K(h, z')$ as in (14.17), and F_{0z} and G_{0z} are as in (15.2a, b).

Actually, an important advantage for present purposes is gained by the use of the difference kernels. Since $K_s(z, z')$ reduces to extremely small values at $z' = z$, whereas $K_c(z', z)$ has its maximum values at $z' = z$, it follows that $\int_0^h I(z')K_s(z, z') \, dz'$ is relatively small and slowly varying as a function of z, whereas $\int_0^h I(z')K_c(z, z') \, dz'$ varies approximately with the current and thus assumes values at points along the antenna that greatly exceed the small value at its end, $z = h$. Accordingly, an integral involving the difference kernel, $L_s(z, z')$, is relatively much smaller compared with an integral involving $L_c(z, z')$ than when $K_s(z, z')$ and $K_c(z, z')$ are used in their places. An approximate solution of (4a, b) for the components of current, therefore, is to be preferred to a corresponding solution of (3a, b). In particular, the solution of (4a) as an equation in $I_r(z)$ with a less essential contribution from $I_t(z)$, and of (4b) as an equation in $I_t(z)$ with only a small contribution from $I_r(z)$, is suggested.

After the introduction of a distribution function for each component of current in the manner of Sec. 14,

$$g_j(z, z') = I_j(z')/I_j(z), \quad (j = r \text{ or } t) \quad (5)$$

and the definition of

$$W_j(z) = W_j + \sigma_j(z) \equiv \int_0^h g_j(z, z')L_c(z, z') \, dz',$$
$$(j = r \text{ or } t) \quad (6)$$

where W_j is the sensibly constant value of $W_j(z)$ at an appropriate reference point, $z = z_r$, and $\sigma_j(z)$ is a correction term that is very small for most values of z, the following equations are obtained:

$$I_r(z) = \frac{4\pi}{\zeta_0 W_r} C'F_{0z} + \frac{1}{W_r}[D_r(z) - E_t(z)], \quad (7a)$$

$$I_t(z) = -\frac{4\pi}{\zeta_0 W_t}(C''F_{0z} + \tfrac{1}{2}V_0 G_{0z})$$
$$+ \frac{1}{W_t}[D_t(z) + E_r(z)], \quad (7b)$$

where the difference integrals are

$$D_j(z) \equiv -I_j(z)\,\sigma_j(z)$$
$$- \int_0^h [I_j(z') - I_j(z)g_j(z, z')] L_c(z, z') \, dz',$$
$$(j = r \text{ or } t) \quad (8a)$$

$$E_j(z) \equiv -\int_0^h I_j(z')L_s(z, z') \, dz'.$$
$$(j = r \text{ or } t) \quad (8b)$$

Solutions of (7a) and (7b) may be derived using the method of iteration described in Sec. 15, with zeroth-order solutions given by

$$[I_r(z)]_0 = \frac{4\pi}{\zeta_0 W_r} C'F_{0z}, \quad (9a)$$

$$[I_t(z)]_0 = -\frac{4\pi}{\zeta_0 W_t}[C''F_{0z} + \tfrac{1}{2}V_0 G_{0z}]. \quad (9b)$$

The corresponding expressions after a first iteration are:

$$[I_r(z)]_1 = \frac{4\pi}{\zeta_0 W_r}[C'(F_{0z} + Fc_{1zr}W_r^{-1})$$
$$+ W_t^{-1}(C''Fs_{1z} + \tfrac{1}{2}V_0 Gs_{1z})] \quad (10a)$$

$$[I_t(z)]_1 = -\frac{4\pi}{\zeta_0 W_t}[C''(F_{0z} + Fc_{1zt}W_t^{-1})$$
$$+ \tfrac{1}{2}V_0(G_{0z} + Gc_{1zt}W_t^{-1})$$
$$- C'Fs_{1z}W_r^{-1}], \quad (10b)$$

where Fc_{1zj} and Gc_{1zj}, $j = r$ or t, are given, respectively, by (8a) with F_{0z} and G_{0z} substituted for $I_j(z)$; and Fs_{1z} and Gs_{1z} are given by (8b) with F_{0z} and G_{0z} substituted for $I_j(z)$. Higher-order terms may be obtained by continuing the iteration. Note that Fc_{1z} and Fs_{1z} may be separated so that $Fc_{1z} = Fc_1(z) - Fc_1(h)$, $Fs_{1z} = Fs_1(z) - Fs_1(h)$, where $Fc_1(z)$ and $Fs_1(z)$ are given by (8a) and (8b), respectively, with $K_c(z, z')$ and $K_s(z, z')$ substituted for $L_c(z, z')$ and $L_s(z, z')$. When $z = h$ these reduce to

$$Fc_1(h) \equiv Fc_1 = -\int_0^h F_{0z'}K_c(h, z')\,dz', \quad (11a)$$

$$Fs_1(h) \equiv Fs_1 = -\int_0^h F_{0z'}K_s(h, z')\,dz'. \quad (11b)$$

Similar expressions are obtained by replacing F by G.

The constants C' and C'' may be evaluated from (3a, b) by setting* $z = h$ and using the currents (9a, b) for first-order values, (10a, b) for second-order values, and so on. The first-order values are

$$C' = \tfrac{1}{2}V_0 N'/W_t \Delta, \quad C'' = -\tfrac{1}{2}V_0 N''/\Delta, \quad (12a)$$

where

$$N' = Fs_1(G_0 + Gc_1 W_t^{-1})$$
$$- Gs_1(F_0 + Fc_1 W_t^{-1}), \quad (12b)$$

$$N'' = (F_0 + Fc_1 W_r^{-1})(G_0 + Gc_1 W_t^{-1})$$
$$+ Fs_1 W_r^{-1} Gs_1 W_r^{-1}, \quad (12c)$$

$$\Delta = (F_0 + Fc_1 W_r^{-1})(F_0 + Fc_1 W_t^{-1})$$
$$+ Fs_1^2 W_r^{-1} W_t^{-1}. \quad (12d)$$

---

* Any other convenient value of z may be used. In particular, for very short antennas, $z = 0$ is appropriate, as shown in Sec. 31. For long antennas, $z = h - \lambda_0/4$ may be desirable.

With (12a)–(12d), (10a) and (10b) may be expanded into the following for $0 \leq z \leq h$:

$$I_r(z) = \frac{2\pi V_0}{\zeta_0 W_r W_t D_r}[\{F_0(z)[G_0 Fs_1 - F_0 Gs_1]$$
$$- F_0[Fs_1(z)G_0 - Gs_1(z)F_0]\}$$
$$+ W_r^{-1}\{Fc_{1r}(z)[G_0 Fs_1 - F_0 Gs_1]$$
$$- Fc_1[Fs_1(z)G_0 - Gs_1(z)F_0]\}$$
$$+ W_t^{-1}\{F_0(z)[Fs_1 Gc_1 - Gs_1 Fc_1]$$
$$- F_0[Fs_1(z)Gc_1 - Gs_1(z)Fc_1]\}$$
$$+ W_r^{-1}W_t^{-1}\{\cdots\} + \cdots], \quad (13a)$$

$$I_t(z) = \frac{2\pi V_0}{\zeta_0 W_t} \times$$
$$\left\{\frac{\sin\beta_0(h-z) + W_t^{-1}M_{1t}^I(z) + W_t^{-2}\{\cdots\} + \cdots}{[F_0 + W_t^{-1}Fc_1 + W_r^{-1}W_t^{-1}Fs_1^2[F_0 + Fc_1 W_t^{-1}]^{-1}}\right.$$
$$\left. + \frac{W_r^{-1}W_t^{-1}\{\cdots\} + \cdots}{D_r}\right\}, \quad (13b)$$

where

$$D_r \equiv F_0^2 + (W_r^{-1} + W_t^{-1})F_0 Fc_1$$
$$+ W_r^{-1}W_t^{-1}(Fc_1^2 + Fs_1^2 + \cdots) + \cdots \quad (13c)$$

and

$$M_{1t}^I(z) \equiv F_0(z)Gc_1 - G_0(z)Fc_1$$
$$+ Fc_{1t}(z)G_0 - Gc_{1t}(z)F_0. \quad (13d)$$

A comparison of (13b) with the imaginary part of (15.26) shows that the zeroth- and first-order terms are, in fact, identical except for the appearance of W_t as expansion parameter instead of Ψ, since difference kernels were used. It may be concluded that the solution for the quadrature current $I_t(z) \equiv I_z'$ obtained previously is entirely adequate.

The component of current in phase with the driving voltage is seen to have a leading term that is essentially of first order, so that the simpler method of iteration in Sec. 15 (in which the zeroth-order current in the complex integral equation is assumed to be a quadrature current) is justified. Moreover, the leading term in the numerator of (13a) differs relatively little from the leading real term in (15.26), which is $M_1^{II}(z)$ in (15.27). However, when $\beta_0 h$ is sufficiently near an odd multiple of $\pi/2$, the zeroth- and first-order terms in the denominator of (13a) become small or vanish so that the remaining *leading terms actually are of second order*. Specifically, when $\beta_0 h = n\pi/2$, n odd, the zeroth-order distributions of current for $I_r(z)$ and $I_t(z)$ are alike, so that $W_r = W_t = W$, and $g(z, z')$

in (5) is equal to $\cos \beta_0 z'/\cos \beta_0 z$. It follows that (13a) reduces to

$$I_r(z) = \frac{2\pi V_0}{\zeta_0}$$

$$\times \left\{ \frac{[Fs_1 + W^{-1}(Fs_1 Gc_1 - Gs_1 Fc_1)] \cos \beta_0 z}{Fc_1^2 + Fs_1^2 + 2W^{-1}[Fc_1(Fc_2 - Fss_2)]} \right\}, \quad (14)$$

where $(Fc_1^2 + Fs_1^2)$ in (14) is the complete second-order term and the rest of the denominator is the complete third-order term when $\beta_0 h = n\pi/2$, n odd. This latter is obtained by substituting (10a, b) in (3a, b) in order to evaluate Δ in (12d) to a higher order. Since any contributions to the third-order value of Δ, obtained by substituting third-order currents in (3a, b), must have $F_0 = \cos \beta_0 h$ as a factor, they contribute nothing to Δ when $\beta_0 h = n\pi/2$, n odd.

The following formulas are valid when $\beta_0 h = n\pi/2$, n odd:

$$F_{0z} = F_0(z) - F_0(h) = \cos \beta_0 z,$$

$$Fc_{1z} = Fc_1(z) - Fc_1(h), \quad (15)$$

$$Fc_1(z) = W \cos \beta_0 z - \int_0^h K_c(z, z') \cos \beta_0 z' \, dz'$$

$$= W \cos \beta_0 z - \text{R.P.} \, C_a(h, z), \quad (16a)$$

$$Fs_1(z) = -\int_0^h K_s(z, z') \cos \beta_0 z' \, dz'$$

$$= -\text{I.P.} \, C_a(h, z), \quad (16b)$$

$$Gc_1(z) = W(\sin \beta_0 z - 1)$$

$$- \int_0^h K_c(z, z')(\sin \beta_0 z' - 1) \, dz'$$

$$= W[\sin \beta_0 z - 1]$$

$$- \text{R.P.}[S_a(h, z) - E_a(h, z)], \quad (16c)$$

$$Gs_1(z) = -\int_0^h K_s(z, z') \sin \beta_0 z' \, dz'$$

$$= -\text{I.P.}[S_a(h, z) - E_a(h, z)]. \quad (16d)$$

By setting $z = h$ in (16a)–(16d) the following are obtained:

$$Fc_1 \equiv Fc_1(h) = -\text{R.P.} \, C_a(h, h)$$
$$\doteq -\tfrac{1}{2} Si \, 4\beta_0 h, \quad (17a)$$

$$Fs_1 \equiv Fs_1(h) = -\text{I.P.} \, C_a(h, h)$$
$$\doteq +\tfrac{1}{2} Cin \, 4\beta_0 h, \quad (17b)$$

$$Gc_1 \equiv Gc_1(h) = -\text{R.P.}[S_a(h, h) - E_a(h, h)]$$
$$\doteq -\tfrac{1}{2} Cin \, 4\beta_0 h + 2 \ln 2, \quad (17c)$$

$$Gs_1 \equiv Gs_1(h) = -\text{I.P.}[S_a(h, h) - E_a(h, h)]$$
$$\doteq -\tfrac{1}{2} Si \, 4\beta_0 h. \quad (17d)$$

The second-order functions that occur in (14) are defined as follows:

$$Fc_2 \equiv Fc_2(h) = -\int_0^h K_c(h, z') Fc_{1z'} \, dz', \quad (18a)$$

$$Fs_2 \equiv Fs_2(h) = -\int_0^h K_s(h, z') Fc_{1z'} \, dz', \quad (18b)$$

$$Fcs_2 \equiv Fcs_2(h) = -\int_0^h K_c(h, z') Fs_{1z'} \, dz', \quad (18c)$$

$$Fss_2 \equiv Fss_2(h) = -\int_0^h K_s(h, z') Fs_{1z'} \, dz'. \quad (18d)$$

The expansion parameter is defined at $z_r = 0$ for $\beta_0 h = \pi/2$. It is

$$W = W(0) = \int_0^h L_c(0, z') \cos \beta_0 z' \, dz'$$

$$= \text{R.P.}[C_a(h, 0) - C_a(h, h)]. \quad (19)$$

The real and imaginary parts of the functions $C_a(h, z)$, $S_a(h, z)$, and $E_a(h, z)$ are represented in Figs. 19.1, 19.3, and 19.5. It is evident from (14) that an accurate determination of the magnitude of $I_r(z)$ when $\beta_0 h = n\pi/2$, n odd, requires at least third-order terms in the denominator. Since the relative distribution of current is not affected by the denominator, consideration of this problem is appropriately reserved for later study in conjunction with the input current and admittance.

For some purposes, for example, the determination of the expansion parameter or electromagnetic fields, an analytically simple approximate representation of the current in a cylindrical antenna is needed. Evidently the leading terms in the two components of current are given by the following for $0 \leq z \leq h$:

$$\frac{I_r(z)}{I_r(0)} \equiv \frac{I_z''}{I_0''} =$$

$$\left\{ \frac{\begin{array}{l} [Fs_1 \sin \beta_0 h - Gs_1 \cos \beta_0 h] \cos \beta_0 z \\ - [Fs_1(z) \sin \beta_0 h - Gs_1(z) \cos \beta_0 h] \cos \beta_0 h \end{array}}{\begin{array}{l}[Fs_1 \sin \beta_0 h - Gs_1 \cos \beta_0 h] \\ - [Fs_1(0) \sin \beta_0 h - Gs_1(0) \cos \beta_0 h] \cos \beta_0 h \end{array}} \right\},$$

(20)

$$\frac{I_t(z)}{I_t(h - \lambda_0/4)} \equiv \frac{I_z'}{I_m'} = \sin \beta_0(h - z). \quad (21)$$

Except when the antenna involved is long, (20) is unnecessarily complicated and the simpler shifted cosine used as a zeroth-order current in (9a) is adequate. That is, for $\beta_0 h < 2\pi$,

$$\frac{I_r(z)}{I_r(0)} \equiv \frac{I_z''}{I_0''} = \frac{\cos \beta_0 z - \cos \beta_0 h}{1 - \cos \beta_0 h}. \quad (22)$$

The ratios (21) and (22) are represented in Figs. 25.1 and 25.2, together with the corresponding first-order components as obtained from Figs. 22.1–22.6 for $\Omega = 10$. The agreement is moderately good. In special cases when $h \doteq \lambda_0/2$ and the zeroth-order value of I_0'' is zero, it may be preferable to use (22) with h increased sufficiently above $\lambda_0/2$ that the correct value of I_0'' is obtained as is shown in Fig. 25.1. The small error resulting from the extra length usually is outweighed by the advantage of having the correct input current.

It may be concluded that a relatively simple approximation of the distribution of current in a cylindrical antenna of small radius with $\beta_0 h < 2\pi$ is given by $I_z = I_z'' + jI_z'$, using (21) and (22). If desired, the ratio of I_0'' to I_m' may be obtained from first-order data. When $\beta_0 h \leq \pi/2$, the distributions (21) and (22) are sufficiently alike that the single formula

$$I_z = I_0 \frac{\sin \beta_0(h - z)}{\sin \beta_0 h} \quad (23)$$

is satisfactory for the *complex* current.

Throughout this section, currents in the range $0 \leq z \leq h$ have been determined. Currents in the range $-h \leq z \leq 0$ are obtained directly, using the symmetry condition $I(-z) = I(z)$.

Distributions of charge corresponding to the simple currents (21) and (22) may be obtained by using the equation of continuity (16.9).

It is shown in this section, using an alternative and more accurate solution of Hallén's integral equation, that the previous analysis is satisfactory in general, but that the accuracy of a given order of solution may be less when $\beta_0 h$ is near odd multiples of $\pi/2$ than for other values. Convenient zeroth-order formulas are obtained for the two components of current.

26. Experimental Determination of the Distributions of Current and Charge

The experimental measurement of the relative magnitude and phase of the axial distributions of current and charge in a cylindrical antenna requires a sensitive detecting circuit with a coupling element that may be moved parallel to the axis of the antenna. The coupling element must satisfy three requirements: (a) its dimensions must be small compared with the length of the antenna and with the wavelength; (b) it must be close to the antenna compared with the wavelength; (c) it must be sufficiently sensitive that it may be coupled very loosely. The first two requirements are necessary in order that the voltage or current induced in the coupling element may be due primarily to the charges or current in a section of the antenna that is short compared with the wavelength. The third condition insures that the presence of the coupling element disturbs the distribution of current or charge as little as possible. Both conditions (b) and (c) may be achieved by making the size of the coupling element small.

Useful coupling elements are a small loop for measuring current and a short probe or antenna for measuring charge. In conventional form, the loop or probe is supported near the antenna and connected to the detector-load by a section of transmission line that is outside the antenna, as shown in Fig. 26.1. For rough measurements, arrangements of this sort are quite satisfactory. For more accurate work, they have the great disadvantage that the entire detector circuit and the observer are also coupled to the antenna more or less closely. The disturbing effect of this coupling is difficult to determine. A better arrangement is shown in Fig. 26.2, where the antenna consists of a tube with a narrow slot parallel to the axis extending from one end to the other. Since all currents are axial, this slot produces

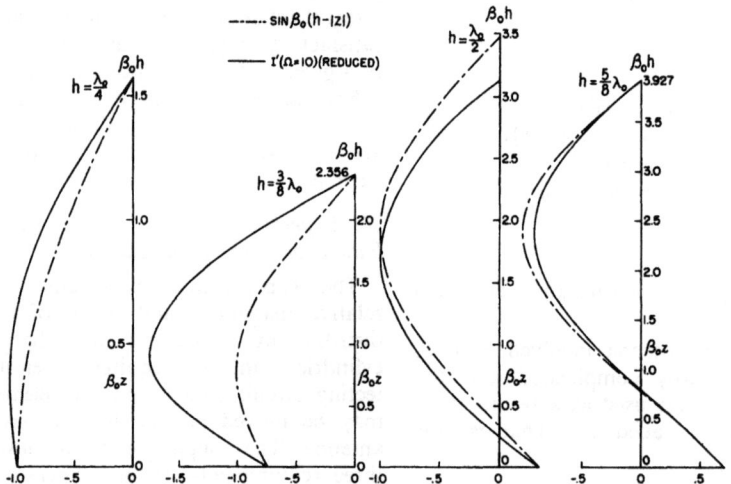

Fig. 25.1. Curves of I'_z/I'_m for different values of h.

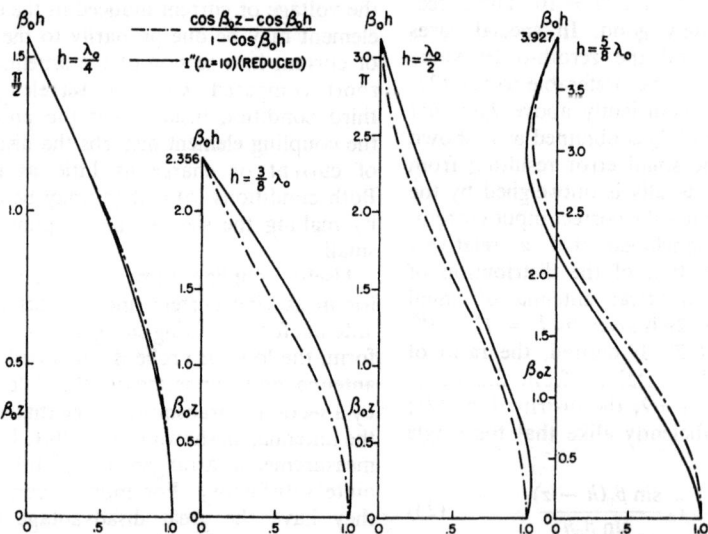

Fig. 25.2. Curves of I''_z/I''_0 for different values of h.

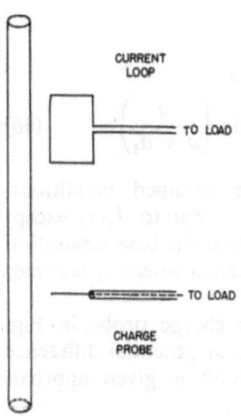

Fig. 26.1. External current and charge probes.

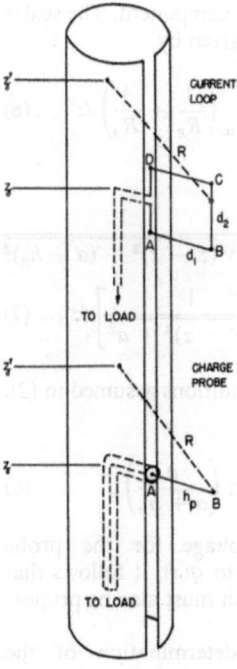

Fig. 26.2. Internally connected current and charge probes.

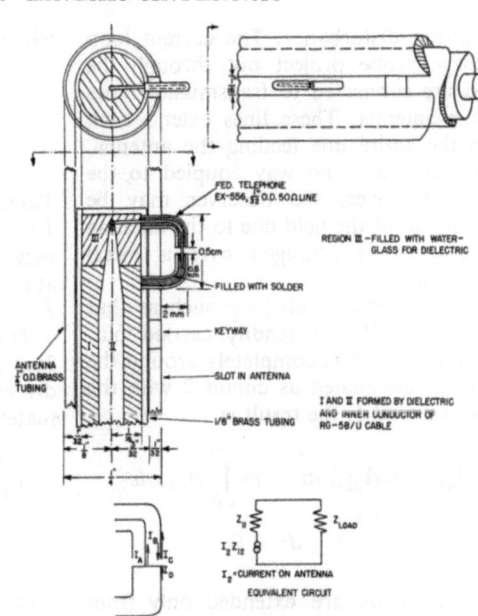

Fig. 26.3. Shielded-loop probe.

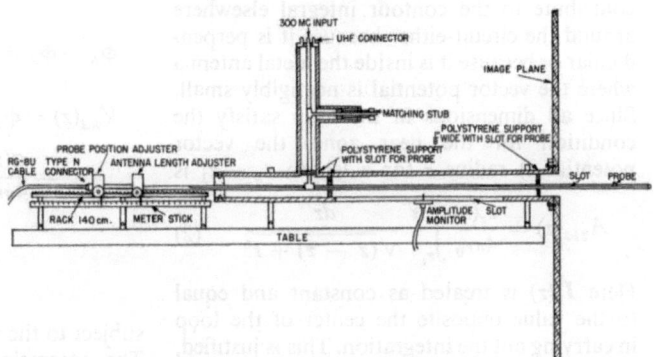

Fig. 26.4. Schematic diagram of measuring line.

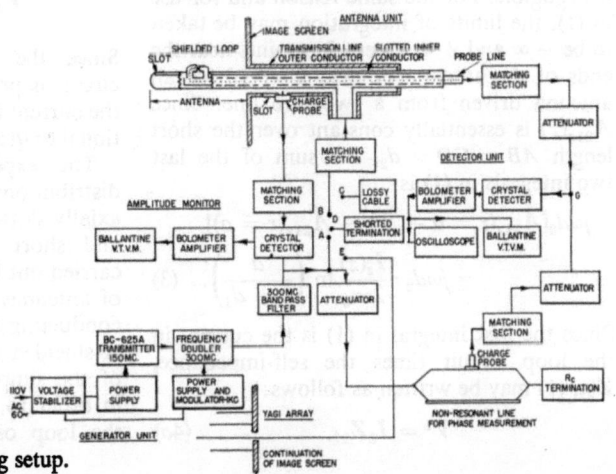

Fig. 26.5. Block diagram of measuring setup.

no significant disturbance. The current loop and charge probe project out through the slot and are connected to transmission lines inside the antenna. These lines extend back through the entire line feeding the antenna, so that they are in no way coupled to the antenna. Moreover, the observer may be completely out of the field due to the antenna if the antenna is arranged over a metal image plane.

The analysis of a small loop such as that shown in Fig. 26.2 is readily carried out. Thus, integrating (5.1) completely around the loop circuit, designated as circuit 2 with the antenna as circuit 1, the result is,

$$\oint (z^i I_{2s} + j\omega A_{22s}) \, ds + j\omega \int_B^C A_{21z} \, dz$$
$$+ j\omega \int_D^A A_{21z} \, dz = 0. \quad (1)$$

The last integrals are extended only from B to C and D to A, since the vector potential A_{21z} due to I_z in the antenna does not contribute to the contour integral elsewhere around the circuit either because it is perpendicular or because it is inside the metal antenna where the vector potential is negligibly small. Since all dimensions in the loop satisfy the condition for the near zone, the vector potential at radius r for a length $z_2 - z_1$ is

$$A_{21z}(z) \doteq \frac{I_z(z)}{4\pi v_0} \int_{z_1}^{z_2} \frac{dz'}{\sqrt{(z'-z)+r^2}} \quad (2)$$

Here $I_z(z)$ is treated as constant and equal to the value opposite the center of the loop in carrying out the integration. This is justified, since the contribution to $A_{21z}(r = a + d)$ $- A_{21z}(r = a)$ by currents far from this point is negligible. For the same reason and for use in (1), the limits of integration may be taken to be $-\infty$ and $+\infty$ except for points near the ends of the antenna or near $z = \pm \delta$ for an antenna driven from a two-wire line. Since $A_{21z}(z)$ is essentially constant over the short length $AB = CD = d_2$, the sum of the last two integrals in (1) is

$$j\omega d_2 [A_{21z}(r = a + d_1) - A_{21z}(r = a)]$$
$$= j\omega d_2 \cdot \frac{I_z(z)}{2\pi v_0} \ln\left(\frac{a}{a + d_1}\right). \quad (3)$$

Since the first integral in (1) is the current in the loop circuit times the self-impedance, Z_{22}, (1) may be written as follows:

$$V^i = I_2 Z_{22}, \quad (4a)$$

where

$$V^i = -I_z(z) \cdot Z_{12}$$
$$= I_z(z) \cdot \frac{j\omega d_2}{2\pi v_0} \ln\left(\frac{a}{a + d_1}\right). \quad (4b)$$

Thus, subject to the assumed conditions, V^i and I_2 are proportional to $I_z(z)$ except very near the ends or near the base separation at the center of the antenna where I_z becomes I_x.

The analysis of the charge probe in Fig. 26.2 is similar. The scalar potential difference driving the probe circuit is given approximately by

$$V_{BA} = \Phi_B - \Phi_A = \int_a^{a+h_p} d\Phi = \int_a^{a+h_p} E_r \, dr, \quad (5)$$

where h_p is the length of the probe. There is no contribution to V from the vector potential, since this has no radial component. The scalar potential difference is given by

$$\Phi_B - \Phi_A \doteq \frac{q(z)}{4\pi\varepsilon_0} \int_{-\infty}^{\infty} \left(\frac{1}{R_B} - \frac{1}{R_A}\right) dz', \quad (6)$$

$$V_{BA}(z) = \Phi_B - \Phi_A$$
$$\doteq \frac{q(z)}{4\pi\varepsilon_0} \int_{-\infty}^{\infty} \left[\frac{1}{\sqrt{(z'-z)^2 + (a+h_p)^2}}\right.$$
$$\left. - \frac{1}{\sqrt{(z'-z)^2 + a^2}}\right] dz', \quad (7)$$

subject to the same conditions assumed in (2). The integration gives

$$V_{BA} = \frac{q(z)}{2\pi\varepsilon_0} \ln\left(\frac{a}{a + h_p}\right). \quad (8)$$

Since the driving voltage for the probe circuit is proportional to $q(z)$, it follows that the current in this circuit must also be proportional to $q(z)$.

The experimental determination of the distributions of current and charge along an axially slotted cylindrical antenna using loops and short antennas as probes has been carried out by Morita[60] for a range of lengths of antennas driven from a coaxial line over a conducting plane at a frequency of 300 Mhz. A shielded loop (Fig. 26.3) was used instead of the simpler structure shown in Fig. 26.2 in order to eliminate unbalanced currents in the loop oscillating in the so-called dipole mode. In so far as the *desired* currents are concerned, a shielded loop[1.32] behaves essentially like the simpler one in Fig. 26.2. A schematic diagram of the measuring line used by Morita is given in Fig. 26.4; a block diagram of the entire measuring setup is shown in Fig. 26.5, including parts used for receiving antennas to which reference is made in Chapter IV. The image plane involved was $15 \times 4\frac{1}{2}$ m. The length of the antenna was varied by moving the entire slotted inner conductor of the coaxial feed-line. The slot in the outer conductor of the line permits the use of a loosely coupled probe connected to an amplitude monitor, which is required in order to maintain a constant power level.

Since an amplitude measurement is no more accurate than the calibration of the detecting system, great care was exercised in obtaining a calibration curve for the crystal with its amplifier. In addition, it was verified that the shielded-loop probe actually had no significant error due to unbalance, and that the effect of the axial slot in the antenna was negligible. In Figs. 26.6 to 26.11 are shown experimentally determined curves of the magnitude of the current (marked $|I|/V_0^e$) and in Figs. 26.12 to 26.14, curves of the relative magnitude of the charge for antennas of different lengths from resonance to $\beta_0 h = 3\pi/2$. With the experimental curves of current are also shown the zeroth-order (sinusoidal) current and the first-order King-Middleton current. With the measured curves for charge are the first-order King-Middleton curves, and curves obtained from the slopes of the experimental curves (Figs. 26.6–26.11) for current.

In order to measure the relative phases of current and charge per unit length at a given cross section of the antenna, an auxiliary *nonresonant* line was used in which current and voltage distributions ideally have the form $A = A_0 e^{-j\beta x}$, neglecting losses. The signal from the probe in the antenna was mixed with the signal from a sensitive, very loosely coupled probe on the auxiliary slotted line which did not introduce a significant departure in the standing-wave ratio of unity. By moving this probe along the nonresonant line to a point where the two signals were 180° out of phase, a minimum indication was obtained. This established a reference point for comparing phases, since variations in the phase of the signal from the probe in the antenna as this probe is moved may be compensated by a measurable displacement Δx in the position of the probe in the auxiliary line. The difference in phase between the first and second minimum settings is $\Delta \phi = \beta \Delta x$. Graphs of the relative phases of current and charge are in Figs. 26.6 to 26.15.

In Figs. 26.7, 26.9, 26.11, 26.12, 26.13, 26.14, the distributions are also given in the junction region and in the coaxial line; note that the scale is smaller over most of this range. The behavior at the junction between antenna and line is well shown, as is the behavior at the dielectric support placed in the line near, but not at, the junction with the antenna. As could have been predicted, the current is continuous, the charge discontinuous. In Figs. 26.6 to 26.11 the two components I'' and I' of the current also are plotted.

Comparison of the several experimental curves of current and charge with the first-order King-Middleton distributions shows good agreement except near $\beta_0 z = 0$, where at least a second-order theory is required for a satisfactory approximation. The assumption that a knowledge of the first-order current and the second-order current at the driving point is adequate is thus confirmed. If base-separation and terminal-zone effects are small, the normalized second-order current at the driving point is given by the input admittance of the isolated antenna. If these effects are large, currents in the terminal zones cannot, in general, be determined accurately.

The experimental results of Morita are compared with the approximate *second-order* curves for current in Figs. 22.9b, 22.10b, 22.11b, and 22.12b. The agreement between theory and experiment is seen to be entirely satisfactory in view of the relatively large base separation at the driving point of the antenna. Note that significant differences between measured and calculated *input* currents occur principally in the case of $\beta_0 h = \pi$, where terminal-zone effects are greatest. This phase of the problem is considered in Sec. 34 when the impedance is investigated.

The distribution of current along an antenna driven from an *open-wire* line was determined experimentally by Angelakos[4] using techniques similar to those of Morita[60]. In order to provide a completely balanced system and eliminate the disturbing effect of apparatus and the observer in the field of the antenna, an image method was used with a single conductor of radius $a = 0.317$ cm driving an antenna parallel to a large, highly conducting

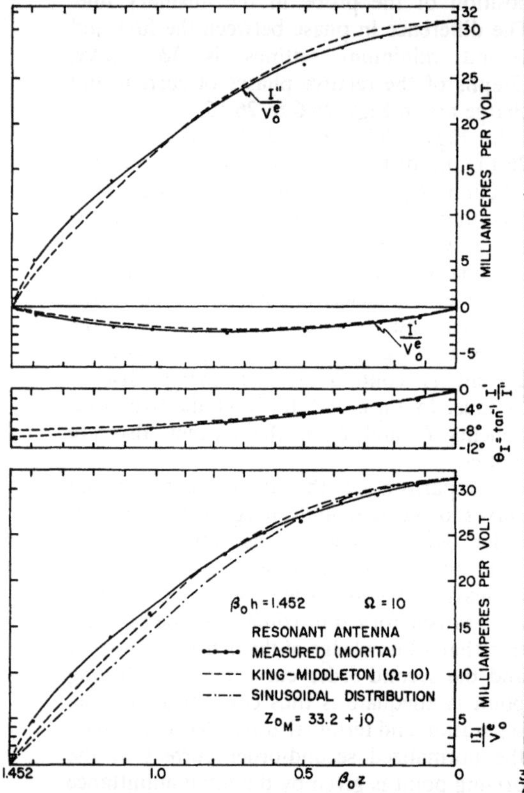

Fig. 26.6. Experimentally determined curves of magnitude of current; $\beta_0 h = 1.452$, $\Omega = 10$ (Morita).

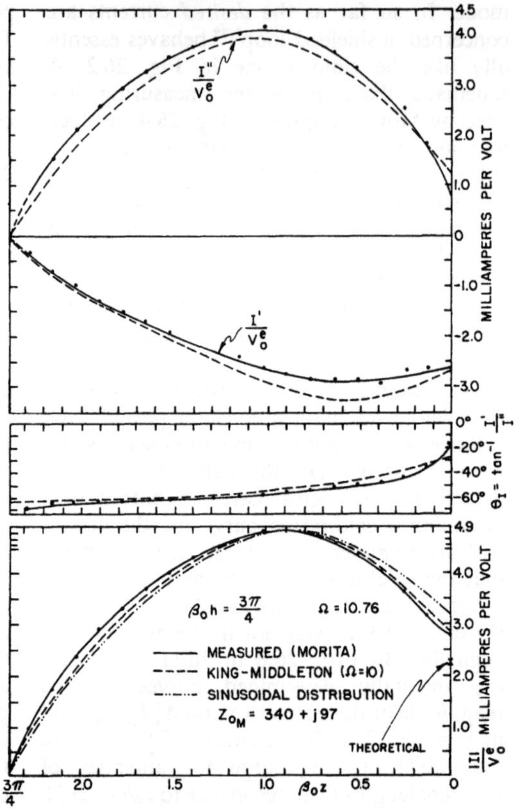

Fig. 26.7. Experimentally determined curves of magnitude of current; $\beta_0 h = 3\pi/4$, $\Omega = 10.76$ (Morita).

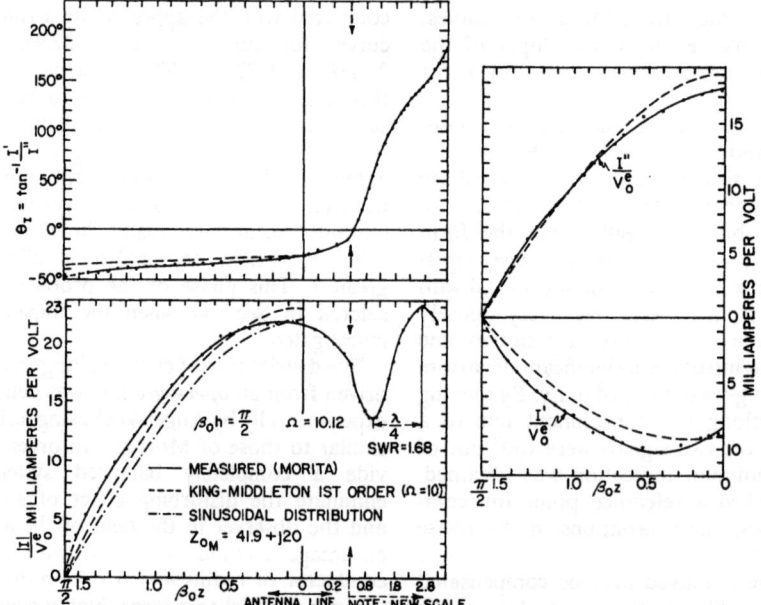

Fig. 26.8. Experimentally determined curves of magnitude of current; $\beta_0 h = \pi/2$, $\Omega = 10.12$ (Morita).

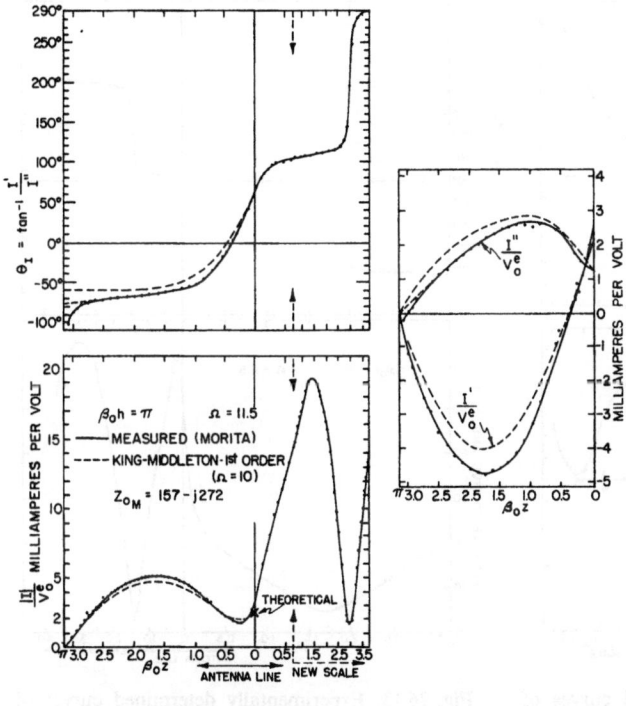

Fig. 26.9. Experimentally determined curves of magnitude of current; $\beta_0 h = \pi$, $\Omega = 11.5$ (Morita).

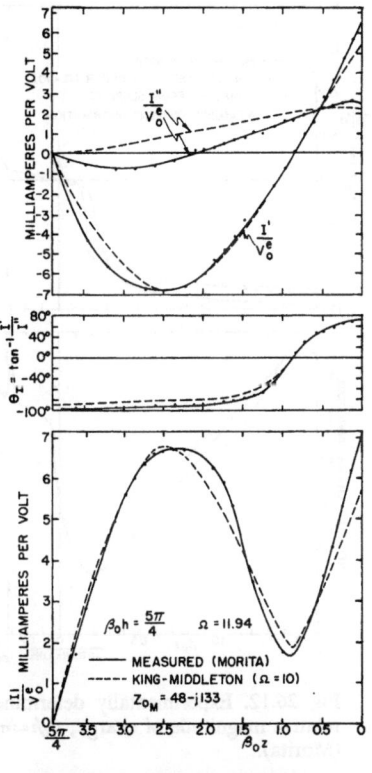

Fig. 26.10. Experimentally determined curves of magnitude of current; $\beta_0 h = 5\pi/4$, $\Omega = 11.94$ (Morita).

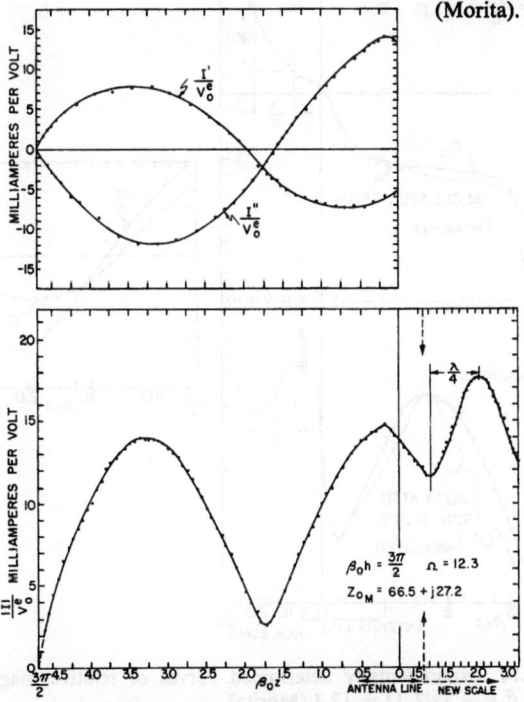

Fig. 26.11. Experimentally determined curves of current; $\beta_0 h = 3\pi/2$, $\Omega = 12.3$ (Morita).

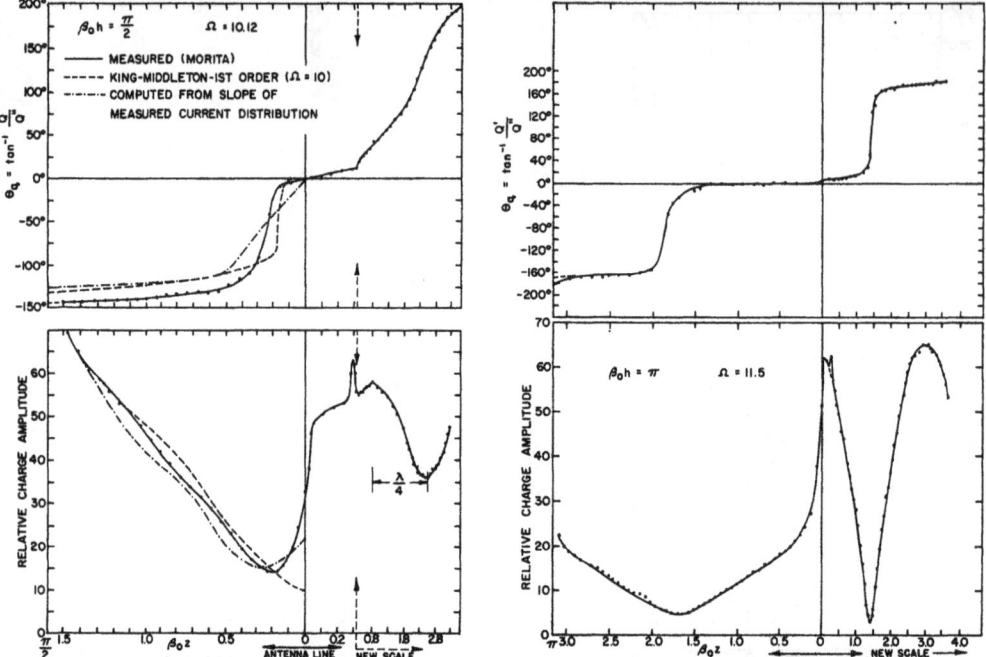

Fig. 26.12. Experimentally determined curves of relative magnitude of charge; $\beta_0 h = \pi/2$, $\Omega = 10.12$ (Morita).

Fig. 26.13. Experimentally determined curves of relative magnitude of charge; $\beta_0 h = \pi$, $\Omega = 11.5$ (Morita).

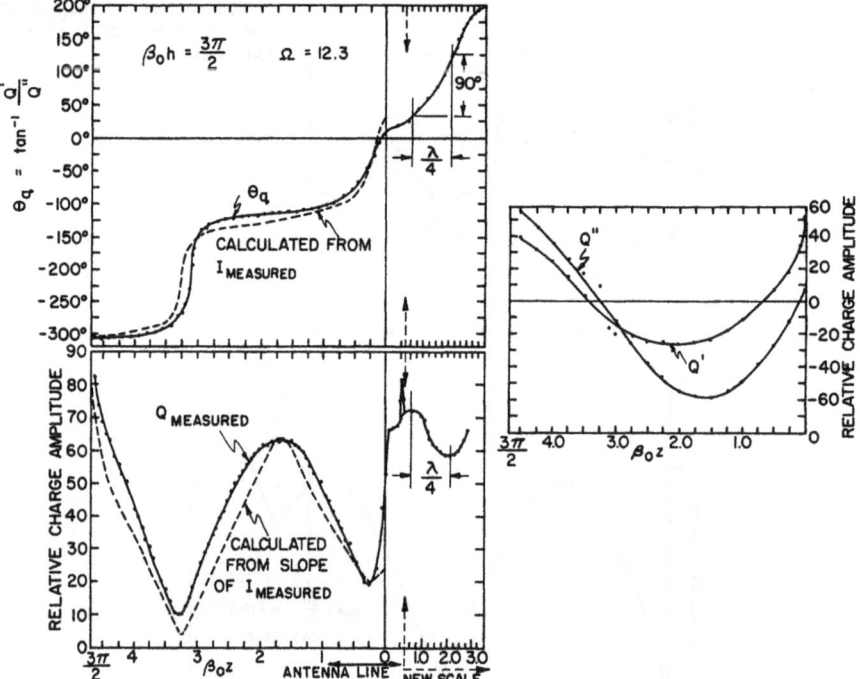

Fig. 26.14. Experimentally determined curves of relative magnitude of charge; $\beta_0 h = 3\pi/2$, $\Omega = 12.3$ (Morita).

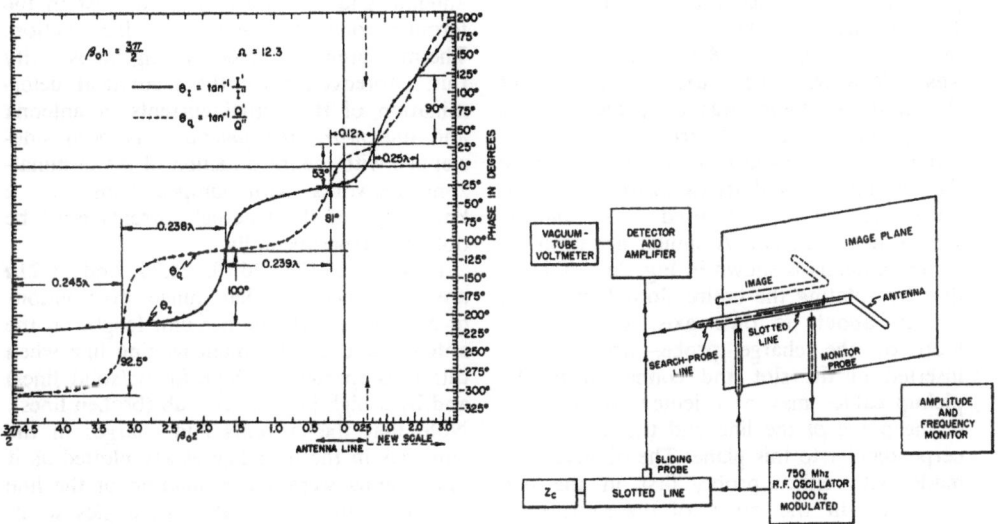

Fig. 26.15. Phase of current and charge distributions (Morita).

Fig. 26.16. Schematic diagram of open-wire measuring unit.

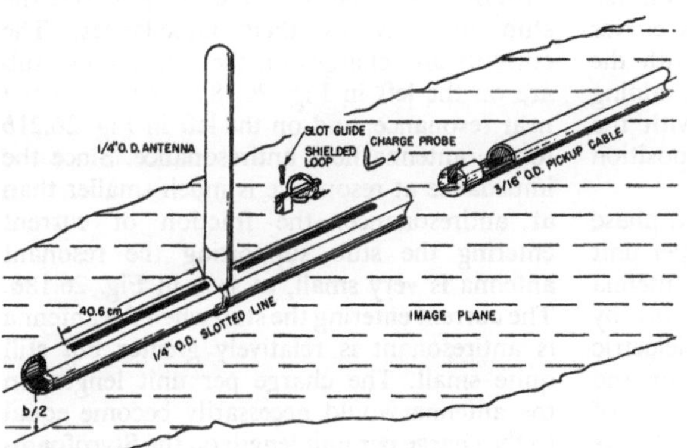

Fig. 26.17a. Current and charge probes for line.

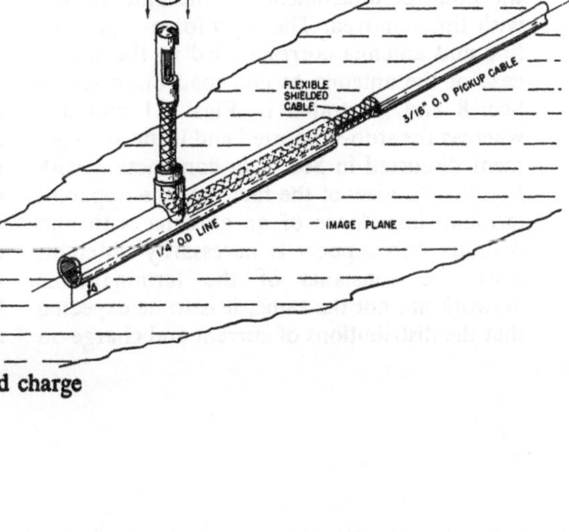

Fig. 26.17b. Structure for mounting current and charge probes in antenna.

plane as shown schematically in Fig. 26.16. The distance from the plane to the axis of the line was $b/2 = 0.98$ cm. The frequency was 750 Mhz. The auxiliary equipment shown in the figure was all placed behind the ground screen. Current and charge probes were arranged to move in slots along the driving line and its extension in a supporting stub if this was used in the manner shown in Fig. 26.17a. A similar arrangement for the antenna is shown in Fig. 26.17b. Note that by rotating the entire slotted line (Fig. 26.17a) about its own axis, the shielded loop or the charge probe—whichever is inserted in the slot and connected to the pickup cable—may be oriented so that it is in the plane of the line and the antenna or perpendicular to this plane. The observations made with either probe, first in the one and then in the other of these positions, differ negligibly for locations along the line that are not too near the antenna. However, very close to the junction of antenna and line large differences are to be expected, especially with the loop as probe, since when this is in the plane of antenna and line it is coupled to the currents in the antenna as well as in the line. For this reason, all distribution curves in Figs. 26.18 to 26.21 were taken with the probes perpendicular to the plane containing the antenna and the line, that is, with the slot rotated through 90° from the position shown in Fig. 26.17b.

Measurements of the magnitude and phase of the current ($I = I e^{j\theta_I}$) and charge per unit length ($q = q e^{j\theta_q}$) were made with the antenna and the end of the line supported (a) by Styrofoam blocks which have a dielectric constant that differs negligibly from air and (b) by a high-impedance stub consisting of an extension of the driving line beyond its junction with the antenna and a connection to the ground screen at a point such that a voltage maximum was maintained exactly at the point of attachment for the antenna but with this removed. The Styrofoam-supported line and antenna corresponded to the simple case of the antenna as end load discussed in Sec. 8 and illustrated in Figs. 8.1 and 8.2, whereas the stub-supported end is the arrangement discussed in Sec. 9 in conjunction with Fig. 9.7. In view of the fact that the apparent terminal impedance of an antenna with and without stub support is necessarily different, since the constants of the terminal-zone network are not the same, it is to be expected that the distributions of current and charge on antenna and line must differ at least in the terminal zone near the antenna–line junction. The measurements made by Angelakos verify this. Moreover, since the theoretical determination of the actual currents in antenna and line *near their junction* has been sidestepped with the introduction of the terminal-zone networks with lumped constants, a knowledge of these actual currents must be obtained experimentally.

In Figs. 26.18a, 26.19, 26.20, and 26.21a are experimentally determined distributions of current and charge per unit length on the antenna and on the main feeding line when this is supported by Styrofoam (solid lines) and by a high-impedance stub (broken lines). Note that the currents and charges in the antennas in the four figures are plotted as if the antenna were a continuation of the line conductor instead of at right angles to it. This presentation permits convenient comparison of the magnitudes and the continuity of current and charge as the junction is approached from antenna and line. With the Styrofoam supports, the current is continuous at the junction; with the stub support, the current divides between the antenna and the stub inversely as their impedances. The currents and charges on the rather long stub are on the left in Fig. 26.18b for an antenna near resonance, and on the left in Fig. 26.21b for an antenna near antiresonance. Since the impedance at resonance is much smaller than at antiresonance, the fraction of current entering the stub supporting the resonant antenna is very small, as seen in Fig. 26.18b. The current entering the stub when the antenna is antiresonant is relatively greater but still quite small. The charge per unit length on the antenna would necessarily become equal to the charge per unit length on the Styrofoam-supported line as the junction is approached if they actually joined without providing additional chargeable surfaces not included in the cylindrical envelopes of the antenna and the line conductor. Since with a right-angled bend in a conductor of finite cross section the corner includes metal surfaces that in part belong to *both* the theoretical antenna and the theoretical line and in part to neither, and since a charge probe actually measures the average field due to all charges near it, it is not possible in a quasi-one-dimensional analysis to determine theoretically the distribution of charge per unit axial length on the surfaces forming the corner, nor is it possible to measure such distributions.

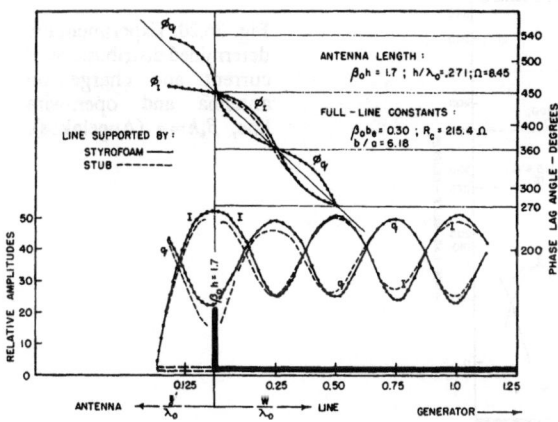

Fig. 26.18a. Experimentally determined distributions of current and charge on antenna and open-wire line, $\beta_0 h = 1.7$ (Angelakos).

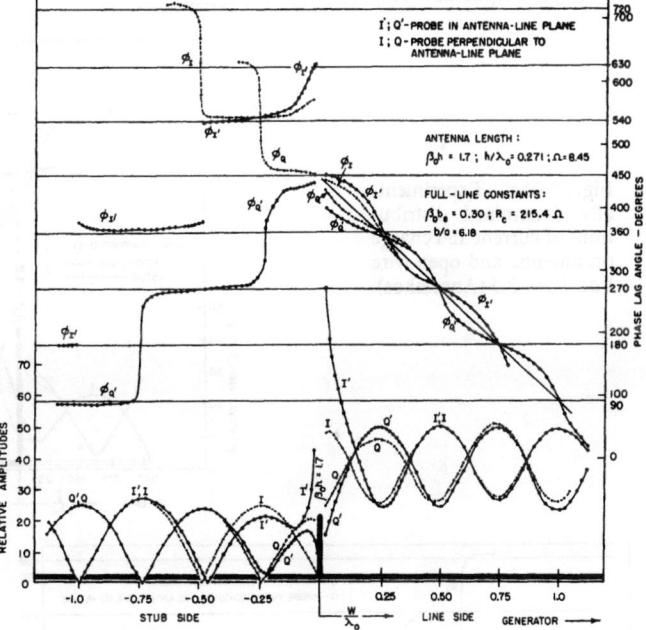

Fig. 26.18b. Experimentally determined distributions of current and charge on line and supporting stub when electrical length of antenna is $\beta_0 h = 1.7$ (Angelakos).

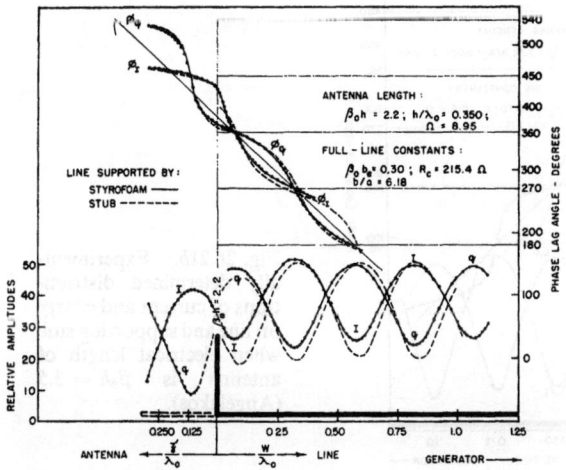

Fig. 26.19. Experimentally determined distributions of current and charge on antenna and open-wire line, $\beta_0 h = 2.2$ (Angelakos).

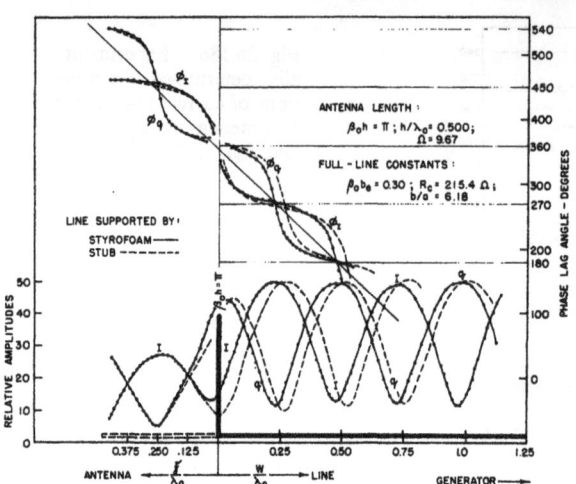

Fig. 26.20. Experimentally determined distributions of current and charge on antenna and open-wire line, $\beta_0 h = \pi$ (Angelakos).

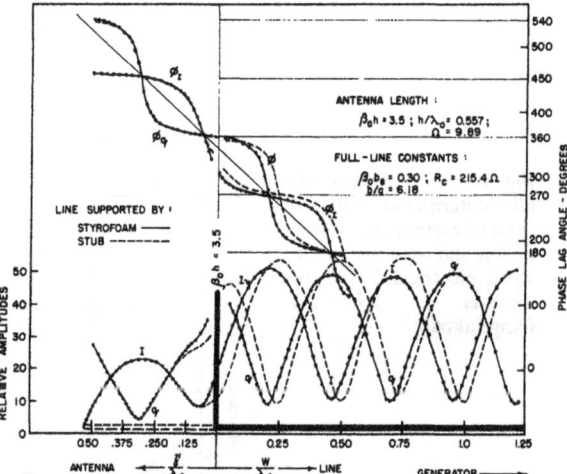

Fig. 26.21a. Experimentally determined distributions of current and charge on antenna and open-wire line, $\beta_0 h = 3.5$ (Angelakos).

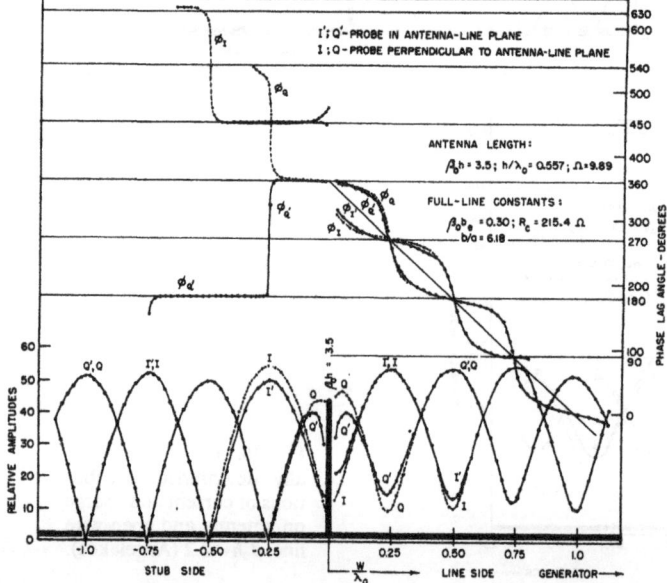

Fig. 26.21b. Experimentally determined distributions of current and charge on line and supporting stub when electrical length of antenna is $\beta_0 h = 3.5$ (Angelakos).

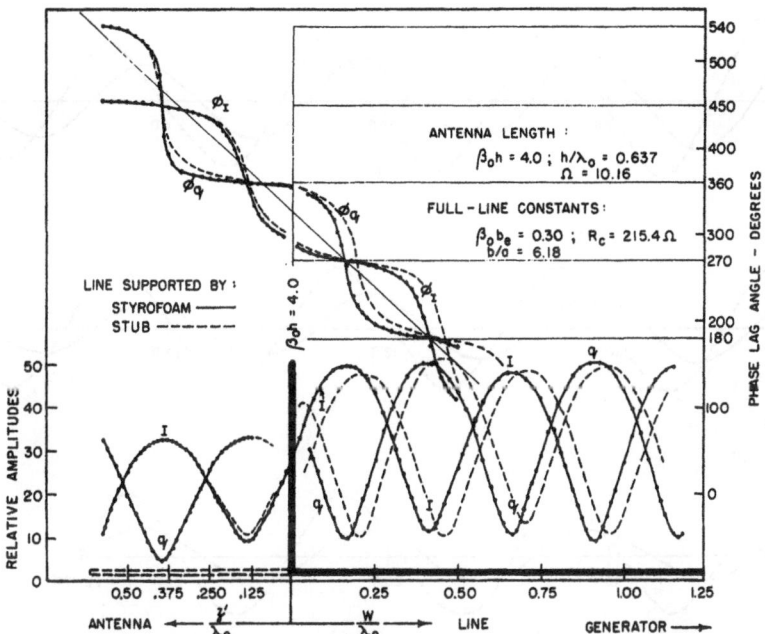

Fig. 26.22. Experimentally determined distributions of current and charge on antenna and open-wire line, $\beta_0 h = 4.0$ (Angelakos).

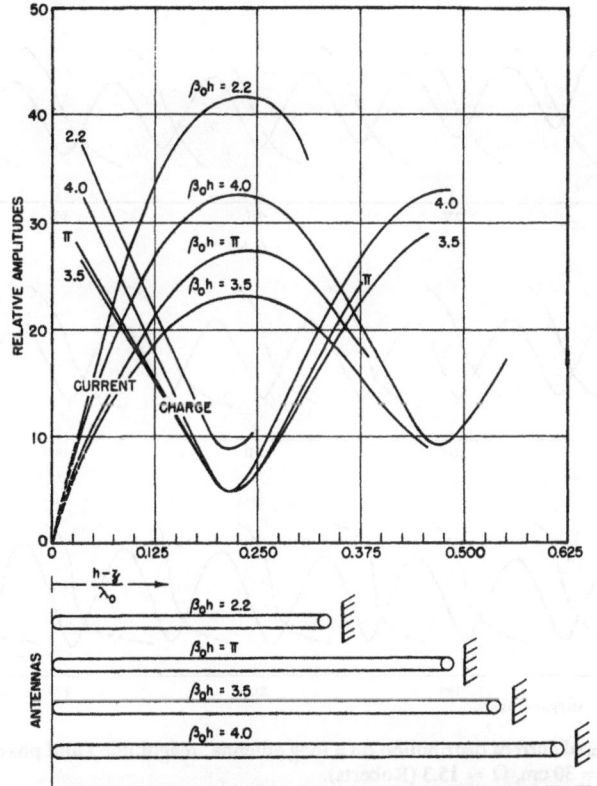

Fig. 26.23. Experimentally determined relative magnitudes of currents and charges on antennas of different lengths (Angelakos).

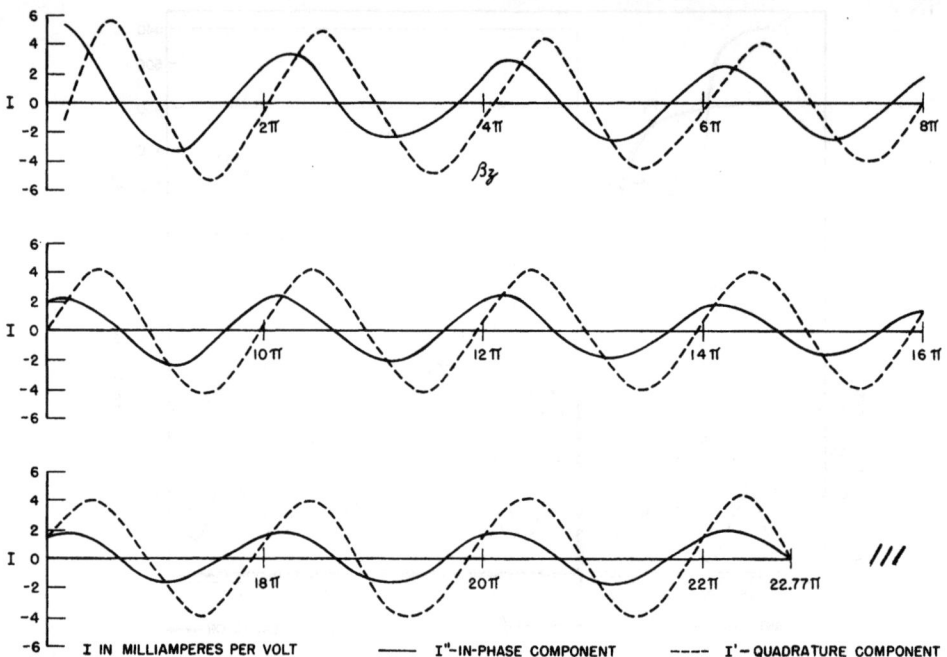

Fig. 26.24. Measured current distribution on a long antenna: the components I'' and I'; $\beta_0 a = 0.0665$, $\beta_0 h = 22.77\pi$, $\Omega = 15.3$ (Roberts).

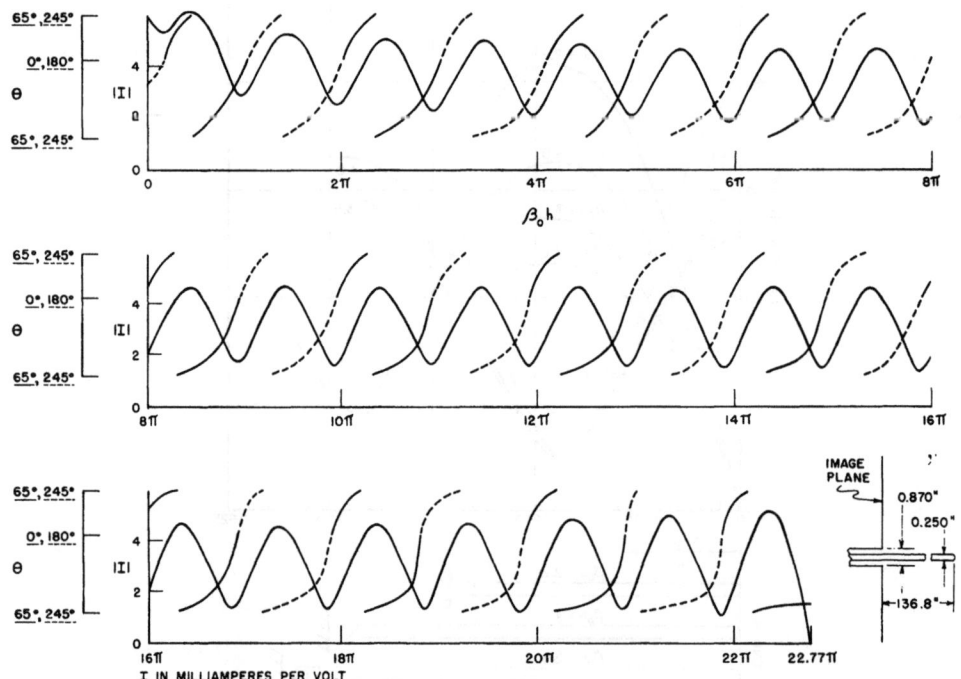

Fig. 26.25. Measured current distribution on a long antenna: magnitude I and phase θ_I; $\beta_0 a = 0.0665$, $\beta_0 h = 22.77\pi$, $\lambda_0 = 30$ cm, $\Omega = 15.3$ (Roberts).

What is required is a three-dimensional analysis and an experiment with sufficiently small and mobile probes to determine the charge density. This is no less true of the stub-supported antenna, where the part of the line conductor of length $2a$ at the base of the antenna cannot be assigned to line, stub, or antenna unambiguously. Hence a charge per unit length for antenna, line, or stub can neither be defined nor measured. Thus, the curves for q in Figs. 26.18 to 26.22 can not in general be extrapolated across the junction nor are their values quite close to the junction particularly meaningful. Unlike the loop for current measurement, the charge probe can not be oriented to respond only to the charge on antenna, line, or stub. It responds to all charges that are sufficiently near. Nevertheless, some indication of the nature of the distributions of charge is given in Figs. 26.18 to 26.22.

A comparison of Fig. 26.18 for the resonant antenna with Fig. 26.21 for the antiresonant antenna is interesting. Note that near resonance the distributions in Fig. 26.18a with the stub support differ negligibly from the corresponding distributions with the Styrofoam supports. Since the inductive end correction L_T is about equal and small in the two cases, and voltages are small so that the effect of the shunt capacitance C_T is negligible, it follows that $Z_{sa} \doteq Z_\delta$ when $\beta_0 h$ is adjusted near resonance, that is, when $\beta_0 h = 1.7$ or $\beta_0(h - \frac{1}{2}b) = \beta_0(h - \delta) = 1.52$. Hence, differences in distribution with the two types of support must be confined to the terminal zone and there can be no difference in the location of charge maxima and current maxima on the line. This situation is altered greatly when $\beta_0 h = 3.5$, a value near antiresonance, as shown in Fig. 26.21a. The distribution curves on the line for the two types of support, while similar in shape, are shifted significantly in their relative positions along the line, indicating an appreciable change in the apparent reactance. Since the general distributions of current on the antenna are essentially the same in the two cases, it follows that these distributions must differ greatly *very near* the junction of the line with the antenna.

A comparison of the magnitudes of current and charge on the outer parts of antennas of different lengths as determined experimentally by Angelakos is shown in Fig. 26.23. The distributions are represented only as near to the line as the curves for Styrofoam and stub supports coincide. Corresponding theoretical values are not plotted in the same figure, since accurate values are not available with $\Omega < 10$, but the agreement in general shape is seen to be very satisfactory. Note particularly the *linear* rather than cosinusoidal distribution of charge near the end of the antenna.

The distribution of current in a very long antenna ($\beta_0 h = 22.77\pi$; $\Omega = 15.3$) is shown in Figs. 26.24 and 26.25 as measured by Roberts, using the method of Morita described earlier in this section. It is seen that the two components of current behave essentially in the manner predicted theoretically in Fig. 22.8. However, as pointed out near the end of Sec. 22, the large in-phase component of current I_z'' makes the current in very long antennas depart sufficiently from the zeroth-order distribution to make advisable the use of two real integral equations in the two components of current, instead of one complex equation as in Sec. 25.

THEORETICAL IMPEDANCE AND ADMITTANCE OF CYLINDRICAL ANTENNA

27. General Formula for Impedance and Admittance of Center-Driven Antenna

An independent impedance characteristic of a symmetrical antenna center-driven from a two-wire line can be defined in an operationally significant sense only for the antenna *and a section of the transmission line* that is longer than the terminal zone $d \sim 10b$. That is, only the input impedance of a section of the line of length $w > d$ terminated in the antenna is measurable. Even this impedance is not independent in the sense that it has a meaning when the particular section of line is detached from the long driving line or when it is arranged in any other way than as a smooth continuation of the driving line. However, it does have the important physical significance that it satisfies the well-known power relations of the transmission line, so that a knowledge of the impedance and the current at or the potential difference across its terminal points is sufficient to determine the transfer of power to it and the load connected to it.

The input impedance of a section of line terminated in an antenna or other load may be measured by conventional transmission-line techniques, using observations made on the line outside the terminal zones. The impedance so determined then may be used in conventional transmission-line formulas to calculate the *apparent* terminal impedance Z_{sa}. This impedance, in turn, may be considered to be that of a network consisting of an impedance Z_δ characteristic of the load in series with an inductance L_T and in parallel with a capacitance C_T.

Although the elements Z_δ, L_T, C_T are individually fictitious in a physical sense, their combination in the prescribed network terminating a *fictitious* line with a terminal zone of zero length permits the calculation of a measurable transmission-line impedance. Since L_T and C_T have been determined for important types of lines, it remains to determine Z_δ, which is defined as the ratio of scalar potential to current at the edges of the cylindrical antenna with base separation 2δ, under the assumption that the only chargeable surfaces are those of the cylindrical envelope of length h for each half. As has been discussed, the hemispherical ends at top and bottom approximate fictitious cylindrical ends of an equal axial length that have no end surfaces or inside surfaces. Approximate account is taken of the junction region of antenna and line by measuring the length of the antenna from the axis of the line and the length of the line to the axis of the antenna. If additional correction is required, it must be made by experimental determination of effective lumped capacitances. The problem of determining Z_δ reduces to setting $z = \delta$ in the expression for the current I_z in (21.16) and solving for the ratio $V_\delta/I_z(\delta)$. In spite of the serious physical limitations of the significance of this ratio, it is called the impedance of the antenna, with the understanding that it is a fictitious impedance for use in a specified equivalent network only. It approaches the *independent* impedance of the idealized "cylindrical-surface" antenna as 2δ is reduced. A true, independent impedance of the antenna can be *defined* only in the limit as δ approaches zero and the structure of the antenna is such that only its cylindrical surface is available for charges and currents. This is theoretically true when the antenna is a complete transmitting system driven by an idealized slice-generator as discussed in Sec. 4. In this case, Z_0, although physically unavailable, is nevertheless useful for studying the properties characteristic of an independent antenna. The impedance Z_0 may be defined operationally as the impedance obtained by measuring the apparent terminating impedance Z_{sa} on a transmission line for successively smaller line spacings and then extrapolating the results to zero line spacing. The impedance Z_δ is approximated by the impedance of an antenna perpendicular to the plane of a two-wire line, as discussed in Sec. 10 and represented in Fig. 10.2.

The mth-order impedance $(Z_\delta)_m$ is obtained from (21.1) by setting $z = \delta$. Thus,

$$(Z_\delta)_m = V_\delta/(I_\delta)_m, \qquad (1a)$$

where phase is referred to the voltage with

$$(V_\delta)_{\text{inst}} = V_\delta \sin \omega t = \text{I.P. } V_\delta e^{j\omega t}, \qquad (1b)$$

$$(Z_\delta)_m = \frac{-j\zeta_0 \Psi_{K1}}{2\pi}$$

$$\times \left[\frac{\cos \beta_0(h - \delta) + \sum_{n=1}^{m} A_{nK}/\Psi_{K1}^n}{\sin \beta_0(h - \delta) + \sum_{n=1}^{m} B_{nK}/\Psi_{K1}^n} \right]. \qquad (2a)$$

Alternatively, in the form of (21.16),

$$(Z_\delta)_m = \frac{-j\zeta_0 \Psi_{K1}}{2\pi}$$

$$\times \left[\frac{\cos \beta_0(h - \delta) + \sum_{n=1}^{m}(D_n)_{m-1} A_{nH}/\Psi_{K1}^n}{(D_1)_m \sin \beta_0(h - \delta) + \sum_{n=1}^{m}(D_{n+1})_m B_{nH}/\Psi_{K1}^n} \right]$$

$$(2b)$$

The functions A_1 and A_2 are given in (15.29) with an identifying subscript K or H affixed to each of the A's, F's, and G's. The functions

$$B_n \equiv M_n(\delta) \qquad (3)$$

are given by $M_n(z)$ in (15.27) and (15.28) for $n = 1, 2, \cdots$ by setting $z = \delta$ and adding subscripts K or H to each of the M's, F's, and G's. The factors $(D_n)_m$ in (2b) are defined in (21.17, 18) and shown graphically in Fig. 21.1. The expansion parameter Ψ_{K1} is given graphically in Fig. 20.7. The original Hallén solution is obtainable from (2b) by substituting Ω for Ψ_{K1} and setting all D factors equal to unity.

In practice, only a small number of terms can be evaluated. While there is no difficulty in breaking off the series in (2a), some care is required in (2b). The rule is that all terms belonging to a given order of solution must be retained and no others. The order of the solution is determined by the number of substitutions made in the method of successive approximations used in deriving the general formula for current. The mth-order solution is given by (2) if terms are included in both numerator and denominator which have $1/\Psi^n$, with $n = 0, 1, 2, \cdots, m$, as an explicitly written factor, and all higher-powered terms are omitted. The number of terms to be retained in the several D factors in (2b) is obtained from (21.18). All terms up to and including those in column m are retained in an mth-order solution; all others are omitted. For example, with $\zeta_0/2\pi \doteq 60$ ohm, zeroth-, first-, and second-order solutions in the forms (2a) and (2b) are, respectively,

$$(\dot{Z}_\delta)_0 = -j60\Psi_{K1} \cot \beta_0(h - \delta), \quad (4)$$

$$(Z_\delta)_1 = -j60\Psi_{K1} \frac{\cos \beta_0(h-\delta) + A_{1H}/\Psi_{K1}}{(D_1)_1 \sin \beta_0(h-\delta) + B_{1H}/\Psi_{K1}},$$
(5a)

$$(Z_\delta)_2 = -j60\Psi_{K1} \times$$
$$\frac{\cos \beta_0(h - \delta) + (D_1)_1 A_{1H}/\Psi_{K1} + A_{2H}/\Psi_{K1}^2}{(D_1)_2 \sin \beta_0(h - \delta) + (D_2)_2 B_{1H}/\Psi_{K1} + B_{2H}/\Psi_{K1}^2}.$$
(5b)

Note that Z_δ in (4) and (5) is in ohms.

The functions A and B in (2a, b) are defined as follows:

$$A_1 = A_1^I + jA_1^{II} \equiv F_1(h) \cos \beta_0 \delta$$
$$+ G_1(h) \sin \beta_0 \delta, \quad (6a)$$

$$B_1 = B_1^I + jB_1^{II} \equiv F_1(0) \sin \beta_0 h$$
$$+ G_1(h) - G_1(0) \cos \beta_0 h, \quad (6b)$$

$$A_2 = A_2^I + jA_2^{II} \equiv F_2(h) \cos \beta_0 \delta$$
$$+ G_2(h) \sin \beta_0 \delta, \quad (7a)$$

$$B_2 = B_2^I + jB_2^{II} \equiv F_2(0) \sin \beta_0 h$$
$$+ G_1(h)F_1(0) - G_1(0)F_1(h)$$
$$+ G_2(h) - G_2(0) \cos \beta_0 h. \quad (7b)$$

With appropriate subscripts K and H on A, B, F, and G, these formulas apply to either (2a) or (2b). Specific formulas for $F_{1K}(0)$, $F_{1K}(h)$, $G_{1K}(0)$, and $G_{1K}(h)$ may be obtained from (21.1a–d). Similar formulas for $F_{1H}(0)$, $F_{1H}(h)$, $G_{1H}(0)$, and $G_{1H}(h)$ with $\delta = 0$ are in (19.11a–d).

Numerical determination of $(Z_\delta)_1$ and $(Z_\delta)_2$ has been carried out by evaluating $(Z_0)_1$ and $(Z_0)_2$ and providing correction factors to obtain $(Z_\delta)_1$ and $(Z_\delta)_2$. This is done in Sec. 29. The specific forms in which $(Z_0)_1$ and $(Z_0)_2$ have been computed are (5a) and (5b) with

$$A_{1H} \equiv F_{1H}(h) = \alpha_1^I + j\alpha_1^{II}$$
$$+ jhr_{\alpha 1}(r^i + jx^i), \quad (8a)$$

$$B_{1H} = F_{1H}(0) \sin \beta_0 h + G_{1H}(h)$$
$$- G_{1H}(0) \cos \beta_0 h$$
$$= \beta_1^I + j\beta_1^{II} + jhr_{\beta_1}(r^i + jx^i), \quad (8b)$$

where

$$\alpha_1^I = \tfrac{1}{2}[\cos \beta_0 h(Cin\, 4\beta_0 h - 2Cin\, 2\beta_0 h)$$
$$- \sin \beta_0 h\, Si\, 4\beta_0 h], \quad (9a)$$

$$\alpha_1^{II} = \tfrac{1}{2}[\cos \beta_0 h(Si\, 4\beta_0 h - 2Si\, 2\beta_0 h)$$
$$+ \sin \beta_0 h\, Cin\, 4\beta_0 h], \quad (9b)$$

$$\beta_1^I = \tfrac{1}{2}[\cos \beta_0 h(4Si\, 2\beta_0 h - Si\, 4\beta_0 h)$$
$$+ \sin \beta_0 h(2Cin\, 2\beta_0 h$$
$$- Cin\, 4\beta_0 h + 4 \ln 2)], \quad (9c)$$

$$\beta_1^{II} = \tfrac{1}{2}[\cos \beta_0 h(Cin\, 4\beta_0 h - 4Cin\, 2\beta_0 h)$$
$$+ \sin \beta_0 h(2Si\, 2\beta_0 h - Si\, 4\beta_0 h)], \quad (9d)$$

$$r_{\alpha 1} = \tfrac{1}{2} \sin \beta_0 h - \cos \beta_0 h(1 - \cos \beta_0 h)/\beta_0 h, \quad (9e)$$

$$r_{\beta 1} = (\sin 2\beta_0 h - \sin \beta_0 h)/2\beta_0 h - \tfrac{1}{2} \cos \beta_0 h, \quad (9f)$$

$$r^i \doteq x^i \doteq \frac{1}{2\pi a}\sqrt{\frac{\omega}{2v\sigma}} \doteq \frac{1}{2\pi a}\sqrt{\frac{\omega\mu}{2\sigma}}. \quad (10)$$

Numerical values of the first-order functions, α_1^I, α_1^{II}, β_1^I, β_1^{II}, $r_{\alpha 1}$, $r_{\beta 1}$ are given in Table 27.1 and in Fig. 27.1 and Fig. 27.3 for $\beta_0 h \leq 7$.*

---

* For $\beta_0 h > 7$ the formulas $Cin\, \chi \doteq 0.5772 + \ln \chi - \frac{\sin \chi}{\chi}$, $Si\, \chi \doteq \frac{\pi}{2} - \frac{\cos \chi}{\chi}$, $|\chi| \gg 1$, may be used to simplify (9a)–(9d).

The functions $A_{2H} = A_{2H}^{I} + jA_{2H}^{II}$, $B_{2H} = B_{2H}^{I} + jB_{2H}^{II}$ may be expressed as follows:

$$A_{2H}^{I} + jA_{2H}^{II} \equiv F_{2H}(h) = \alpha_2^{I} + j\alpha_2^{II}$$
$$+ jhr_{\alpha 2}(r^i + jx^i), \quad (11a)$$

$$B_{2H}^{I} + jB_{2H}^{II} = F_{2H}(0)\sin\beta_0 h + G_{1H}(h)F_{1H}(0)$$
$$- G_{1H}(0)F_{1H}(h) + G_{2H}(h)$$
$$- G_{2H}(0)\cos\beta_0 h = \beta_2^{I} + j\beta_2^{II}$$
$$+ jhr_{\beta 2}(r^i + jx^i). \quad (11b)$$

Numerical values of the second-order functions, α_2^{I}, α_2^{II}, β_2^{I}, β_2^{II}, are given in Table 27.1 and in Fig. 27.2. The functions $r_{\alpha 2}$ and $r_{\beta 2}$ have not been computed, but since the contributions of the terms with z^i as a factor are usually negligible, this is of no importance.

In order to evaluate the impedance from (2) or from (5) it is convenient to write numerator and denominator separately in polar form. Thus, omitting many subscripts,

$$Z_\delta = \frac{\zeta_0 \Psi}{2\pi}\left[\frac{n^{I} + jn^{II}}{d^{I} + jd^{II}}\right] = \frac{\zeta_0 \Psi}{2\pi}\frac{n}{d}e^{j(\theta_n - \theta_d)}, \quad (12)$$

where

$$n^2 = (n^{I})^2 + (n^{II})^2, \quad (13a)$$
$$d^2 = (d^{I})^2 + (d^{II})^2, \quad (13b)$$
$$\theta_n = \tan^{-1}(n^{I}/n^{II}), \quad (13c)$$
$$\theta_d = \tan^{-1}(d^{I}/d^{II}), \quad (13d)$$

and

$$n^{I} = \frac{D_1 A_1^{II}}{\Psi} + \frac{D_2 A_2^{II}}{\Psi^2} + \cdots, \quad (14a)$$

$$n^{II} = -\left[\cos\beta_0(h - \delta) + \frac{D_1 A_1^{I}}{\Psi}\right.$$
$$\left. + \frac{D_2 A_2^{I}}{\Psi^2} + \cdots\right], \quad (14b)$$

$$d^{I} = D_1 \sin\beta_0(h - \delta) + \frac{D_2 B_1^{I}}{\Psi}$$
$$+ \frac{D_3 B_2^{I}}{\Psi^2} + \cdots, \quad (14c)$$

$$d^{II} = \frac{D_2 B_1^{II}}{\Psi} + \frac{D_3 B_2^{II}}{\Psi^2} + \cdots. \quad (14d)$$

The number of terms in the series and the values of the D-factors depend upon the order of the solution, as already explained.

The resistance R_δ and reactance X_δ are the real and imaginary parts of (12). Thus,

$$Z_\delta = R_\delta + jX_\delta, \quad (15)$$

where

$$R_\delta = \frac{\zeta_0 \Psi}{2\pi}\frac{n}{d}\cos(\theta_n - \theta_d), \quad (16)$$

$$X_\delta = \frac{\zeta_0 \Psi}{2\pi}\frac{n}{d}\sin(\theta_n - \theta_d). \quad (17)$$

A simple and general representation of the resistance and reactance given in (16) and (17) as functions of $\beta_0 h$ and h/a is possible only if the terms involving the internal impedance per unit length of the conductor, that is, z^i, are negligible. This is due to the fact that the expression for z^i does not involve the frequency in the combination of $\beta_0 h$, but directly and under a radical. Fortunately, z^i is so small for all good conductors, in particular for copper which is the usual material for antennas, that the contributions to the impedance from terms in z^i are no greater than at most 3 percent in a well-designed antenna. Accordingly, these terms may be omitted in the calculations. This is accomplished by writing α_n^{I}, α_n^{II}, β_n^{I}, β_n^{II} respectively for A_n^{I}, A_n^{II}, B_n^{I}, B_n^{II} in (14a, b, c, d).

The general formula for the admittance is the reciprocal of (2). Thus,

$$Y_\delta = G_\delta + jB_\delta = \frac{1}{Z_\delta}$$
$$= \frac{R_\delta}{R_\delta^2 + X_\delta^2} - j\frac{X_\delta}{R_\delta^2 + X_\delta^2}. \quad (18)$$

28. Impedances of Zeroth and Modified Zeroth Order

The simple zeroth-order solution for the impedance of the center-driven cylindrical antenna of half length h is given by (27.4). It is a pure reactance. For simplicity, let the subscripts be omitted from the expansion parameter. Thus,

$$(R_\delta)_0 = 0, \quad (1)$$
$$(Z_\delta)_0 = j(X_\delta)_0 = \frac{-j\zeta_0 \Psi}{2\pi}\cot\beta_0(h - \delta).$$

Evidently this is only a very rough approximation of the impedance except for values of $\beta_0(h - \delta)$ for which the higher-order terms are negligible. This is true subject to the following conditions:

$$\left|\cos\beta_0(h - \delta)\right| \gg \left|\sum_{n=1}^{m}(D_n)_{m-1}A_n/\Psi^n\right|, \quad (2a)$$

$$\left|\sin\beta_0(h - \delta)\right| \gg \frac{1}{(D_1)_m}\left|\sum_{n=1}^{m}(D_{n+1})_n B_n/\Psi^n\right|. \quad (2b)$$

TABLE 27.1. The functions α_1, α_2, β_1, β_2.*

$\beta_0 h$	$\cos \beta_0 h$	$\alpha_1 = \alpha_1^I + j\alpha_1^{II}$	$\alpha_2 = \alpha_2^I + j\alpha_2^{II}$	$\sin \beta_0 h$	$\beta_1 = \beta_1^I + j\beta_1^{II}$	$\beta_2 = \beta_2^I + j\beta_2^{II}$
0.0	+1.000000	−0.0000 +0.0000j	−0.0000 +0.0000j	+0.000000	+0.0000 +0.0000j	+0.0000 +0.0000j
0.1	+0.995004	−0.0100 +0.0006j	−0.0360 +0.0022j	+0.099833	+0.3374 +0.0000j	+1.3462 +0.0002j
0.2	+0.980067	−0.0393 +0.0053j	−0.1415 +0.0171j	+0.198669	+0.6667 +0.0006j	+2.6426 +0.0039j
0.3	+0.955336	−0.0864 +0.0175j	−0.3100 +0.0563j	+0.295520	+0.9802 +0.0028j	+3.8420 +0.0192j
0.4	+0.921061	−0.1490 +0.0407j	−0.5311 +0.1285j	+0.389418	+1.2710 +0.0083j	+4.9014 +0.0591j
0.5	+0.877583	−0.2234 +0.0773j	−0.7920 +0.2392j	+0.479426	+1.5326 +0.0197j	+5.7838 +0.1399j
0.6	+0.825336	−0.3061 +0.1293j	−1.0784 +0.3901j	+0.564642	+1.7594 +0.0397j	+6.4599 +0.2792j
0.7	+0.764842	−0.3925 +0.1974j	−1.3762 +0.5790j	+0.644218	+1.9475 +0.0713j	+6.9079 +0.4946j
0.8	+0.696707	−0.4781 +0.2816j	−1.6723 +0.8004j	+0.717356	+2.0938 +0.1176j	+7.1141 +0.8018j
0.9	+0.621610	−0.5583 +0.3812j	−1.9559 +1.0462j	+0.783327	+2.1960 +0.1811j	+7.0723 +1.2134j
1.0	+0.540302	−0.6291 +0.4935j	−2.2191 +1.3069j	+0.841471	+2.2540 +0.2641j	+6.7825 +1.7374j
1.1	+0.453596	−0.6866 +0.6157j	−2.4564 +1.5728j	+0.891207	+2.2682 +0.3681j	+6.2504 +2.3770j
1.2	+0.362358	−0.7278 +0.7450j	−2.6654 +1.8344j	+0.932039	+2.2403 +0.4941j	+5.4859 +3.1306j
1.3	+0.267499	−0.7504 +0.8778j	−2.8455 +2.0844j	+0.963558	+2.1725 +0.6423j	+4.5019 +3.9915j
1.4	+0.169967	−0.7527 +1.0090j	−2.9977 +2.3171j	+0.985450	+2.0681 +0.8112j	+3.3135 +4.9495j
1.5	+0.070737	−0.7345 +1.1351j	−3.1232 +2.5291j	+0.997495	+1.9308 +0.9996j	+1.9376 +5.9908j
1.6	−0.029200	−0.6957 +1.2517j	−3.2229 +2.7192j	+0.999574	+1.7644 +1.2042j	+0.3921 +7.0994j
1.7	−0.128844	−0.6377 +1.3550j	−3.2969 +2.8875j	+0.991665	+1.5731 +1.4216j	−1.3036 +8.2580j
1.8	−0.227202	−0.5619 +1.4419j	−3.3436 +3.0354j	+0.973848	+1.3601 +1.6477j	−3.1291 +9.4484j
1.9	−0.323290	−0.4708 +1.5097j	−3.3604 +3.1643j	+0.946300	+1.1301 +1.8781j	−5.0620 +10.6524j
2.0	−0.416147	−0.3673 +1.5562j	−3.3431 +3.2752j	+0.909297	+0.8864 +2.1071j	−7.0786 +11.8517j
2.1	−0.504846	−0.2541 +1.5805j	−3.2867 +3.3680j	+0.863209	+0.6317 +2.3304j	−9.1530 +13.0282j
2.2	−0.588501	−0.1343 +1.5819j	−3.1862 +3.4413j	+0.808496	+0.3691 +2.5431j	−11.2576 +14.1641j
2.3	−0.666276	−0.0108 +1.5605j	−3.0369 +3.4922j	+0.745705	+0.1011 +2.7393j	−13.3631 +15.2414j
2.4	−0.737394	+0.1134 +1.5170j	−2.8350 +3.5166j	+0.675463	−0.1701 +2.9154j	−15.4390 +16.2421j
2.5	−0.801144	+0.2360 +1.4528j	−2.5787 +3.5091j	+0.598472	−0.4425 +3.0676j	−17.4539 +17.1481j
2.6	−0.856889	+0.3552 +1.3695j	−2.2678 +3.4640j	+0.515501	−0.7142 +3.1931j	−19.3764 +17.9411j
2.7	−0.904072	+0.4687 +1.2691j	−1.9045 +3.3755j	+0.427380	−0.9833 +3.2883j	−21.1756 +18.6030j
2.8	−0.942222	+0.5756 +1.1534j	−1.4931 +3.2385j	+0.334988	−1.2479 +3.3520j	−22.8220 +19.1159j
2.9	−0.970958	+0.6750 +1.0247j	−1.0397 +3.0487j	+0.239249	−1.5065 +3.3828j	−24.2878 +19.4628j
3.0	−0.989992	+0.7662 +0.8851j	−0.5524 +2.8034j	+0.141120	−1.7560 +3.3799j	−25.5479 +19.6277j
3.1	−0.999135	+0.8487 +0.7362j	−0.0402 +2.5018j	+0.041581	−1.9946 +3.3422j	−26.5798 +19.5964j
3.2	−0.998295	+0.9225 +0.5800j	+0.4869 +2.1448j	−0.058374	−2.2200 +3.2732j	−27.3644 +19.3568j
3.3	−0.987480	+0.9876 +0.4181j	+1.0187 +1.7349j	−0.157746	−2.4292 +3.1705j	−27.8859 +18.8991j
3.4	−0.966798	+1.0435 +0.2515j	+1.5450 +1.2770j	−0.255541	−2.6196 +3.0361j	−28.1322 +18.2165j
3.5	−0.936457	+1.0905 +0.0816j	+2.0560 +0.7771j	−0.350783	−2.7884 +2.8713j	−28.0945 +17.3054j
3.6	−0.896758	+1.1280 −0.0903j	+2.5430 +0.2427j	−0.442520	−2.9325 +2.6772j	−27.7679 +16.1655j
3.7	−0.848100	+1.1556 −0.2634j	+2.9979 −0.3177j	−0.529836	−3.0495 +2.4552j	−27.1510 +14.7998j
3.8	−0.790968	+1.1731 −0.4364j	+3.4141 −0.8948j	−0.611858	−3.1375 +2.2070j	−26.2459 +13.2150j
3.9	−0.725932	+1.1795 −0.6082j	+3.7862 −1.4792j	−0.687766	−3.1945 +1.9344j	−25.0578 +11.4213j
4.0	−0.653644	+1.1743 −0.7772j	+4.1100 −2.0610j	−0.756802	−3.2188 +1.6393j	−23.5957 +9.4320j
4.1	−0.574824	+1.1569 −0.9422j	+4.3828 −2.6312j	−0.818277	−3.2096 +1.3235j	−21.8709 +7.2642j
4.2	−0.490261	+1.1265 −1.1017j	+4.6030 −3.1808j	−0.871576	−3.1662 +0.9889j	−19.8980 +4.9376j
4.3	−0.400799	+1.0829 −1.2535j	+4.7700 −3.7021j	−0.916166	−3.0891 +0.6383j	−17.6939 +2.4748j
4.4	−0.307333	+1.0253 −1.3960j	+4.8839 −4.1880j	−0.951602	−2.9790 +0.2743j	−15.2780 −0.0091j
4.5	−0.210796	+0.9541 −1.5271j	+4.9459 −4.6330j	−0.977530	−2.8369 −0.1003j	−12.6714 −2.7570j
4.6	−0.112153	+0.8695 −1.6448j	+4.9568 −5.0323j	−0.993691	−2.6647 −0.4823j	−9.8976 −5.4705j
4.7	−0.012389	+0.7719 −1.7467j	+4.9180 −5.3825j	−0.999923	−2.4645 −0.8681j	−6.9814 −8.2099j
4.8	+0.087499	+0.6625 −1.8316j	+4.8308 −5.6811j	−0.996165	−2.2390 −1.2535j	−3.9490 −10.9449j
4.9	+0.186512	+0.5424 −1.8974j	+4.6959 −5.9263j	−0.982453	−1.9903 −1.6345j	−0.8285 −13.6451j
5.0	+0.283662	+0.4129 −1.9426j	+4.5142 −6.1171j	−0.958924	−1.7214 −2.0072j	+2.3511 −16.2805j
5.1	+0.377978	+0.2763 −1.9665j	+4.2860 −6.2526j	−0.925815	−1.4356 −2.3666j	+5.5596 −18.8212j
5.2	+0.468517	+0.1345 −1.9683j	+4.0117 −6.3321j	−0.883455	−1.1361 −2.7086j	+8.7656 −21.2388j
5.3	+0.554374	−0.0108 −1.9475j	+3.6915 −6.3547j	−0.832267	−0.8252 −3.0290j	+11.9367 −23.5055j
5.4	+0.634693	−0.1572 −1.9044j	+3.3261 −6.3196j	−0.772764	−0.5062 −3.3234j	+15.0407 −25.5949j
5.5	+0.708670	−0.3026 −1.8394j	+2.9167 −6.2254j	−0.705540	−0.1824 −3.5880j	+18.0441 −27.4821j
5.6	+0.775566	−0.4448 −1.7536j	+2.4651 −6.0711j	−0.631267	+0.1437 −3.8194j	+20.9140 −29.1438j
5.7	+0.834713	−0.5822 −1.6480j	+1.9742 −5.8555j	−0.550686	+0.4696 −4.0142j	+23.6183 −30.5577j
5.8	+0.885520	−0.7131 −1.5241j	+1.4480 −5.5774j	−0.464602	+0.7922 −4.1701j	+26.1252 −31.7044j
5.9	+0.927478	−0.8359 −1.3873j	+0.8916 −5.2364j	−0.373877	+1.1087 −4.2854j	+28.4052 −32.5655j
6.0	+0.960170	−0.9496 −1.2287j	+0.3111 −4.8327j	−0.279415	+1.4166 −4.3583j	+30.4303 −33.1255j
6.1	+0.983268	−1.0528 −1.0608j	−0.2862 −4.3674j	−0.182163	+1.7134 −4.3874j	+32.1751 −33.3710j
6.2	+0.996542	−1.1450 −0.8821j	−0.8922 −3.8426j	−0.083089	+1.9963 −4.3725j	+33.6173 −33.2915j
6.3	+0.999859	−1.2256 −0.6946j	−1.4983 −3.2620j	+0.016814	+2.2623 −4.3145j	+34.7378 −32.8797j
6.4	+0.993185	−1.2940 −0.5002j	−2.0957 −2.6302j	+0.116549	+2.5087 −4.2133j	+35.5210 −32.1313j
6.5	+0.976588	−1.3497 −0.3005j	−2.6753 −1.9533j	+0.215120	+2.7331 −4.0697j	+35.9550 −31.0456j
6.6	+0.950233	−1.3926 −0.0976j	−3.2286 −1.2385j	+0.311541	+2.9325 −3.8852j	+36.0321 −29.6257j
6.7	+0.914383	−1.4224 +0.1067j	−3.7474 −0.4941j	+0.404850	+3.1054 −3.6621j	+35.7484 −27.8787j
6.8	+0.869397	−1.4387 +0.3111j	−4.2243 +0.2707j	+0.494113	+3.2485 −3.4022j	+35.1043 −25.8153j
6.9	+0.815725	−1.4412 +0.5135j	−4.6530 +1.0462j	+0.578440	+3.3602 −3.1072j	+34.1037 −23.4505j
7.0	+0.753902	−1.4301 +0.7123j	−5.0281 +1.8222j	+0.656987	+3.4392 −2.7801j	+32.7545 −20.8028j

* From King and Blake, *Proc. I.R.E.* **30**, 337 (1942) and E. Hallén, *Trans. Roy. Inst. Technol.*, Stockholm, No. 13, p. 10 (1947).

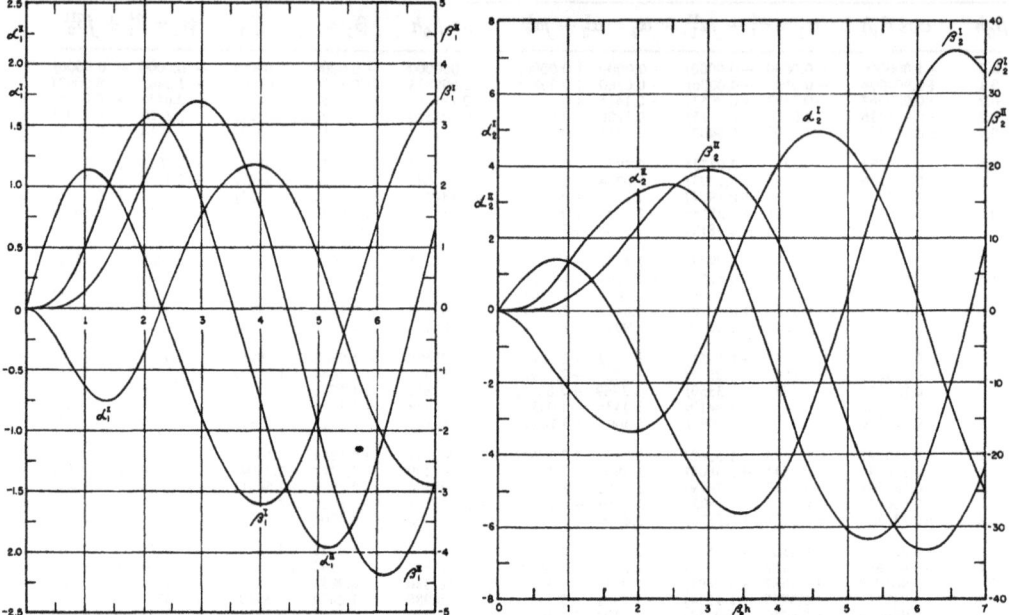

Fig. 27.1. First-order functions $\alpha_1^I, {}_I^{II}\alpha, \beta_1^I, \beta_1^{II}$.

Fig. 27.2. Second-order functions $\alpha_2^I, \alpha_2^{II}, \beta_2^I, \beta_2^{II}$.

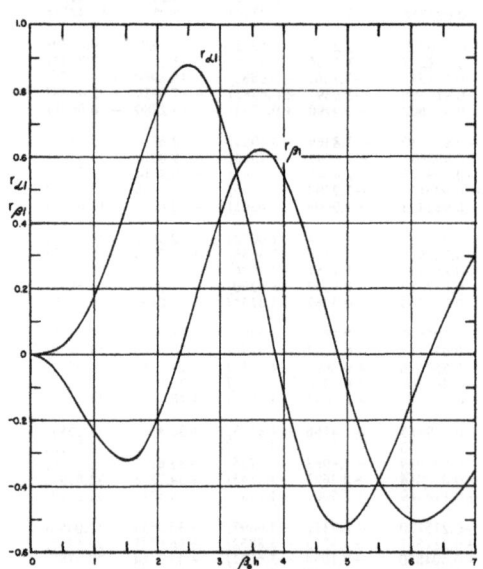

Fig. 27.3. First-order functions $r_{\alpha 1}, r_{\beta 1}$.

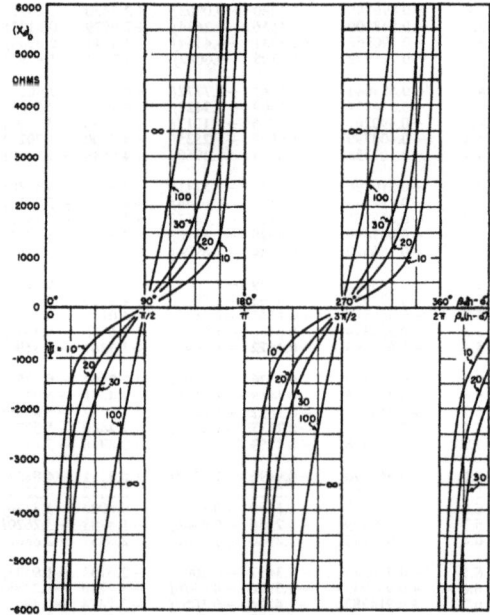

Fig. 28.1. The function $(X_\delta)_0 = 60\,\Psi \cot \beta_0(h-\delta)$.

For very large values of Ψ these conditions should be satisfied for $\beta_0(h-\delta)$ not too near $n\pi/2$, with n integral. As the radius a of the antenna is made smaller and smaller, the parameter Ψ becomes larger and larger, so that the ranges of $\beta_0(h-\delta)$ for which (2a) and (2b) are satisfied increase until only narrow bands at $\beta_0(h-\delta) = n\pi/2$ are excluded. The impedance given by (1) is then accurate for virtually all values of $\beta_0(h-\delta)$ except $\beta_0(h-\delta) = n\pi/2$. It consists of practically vertical lines from enormous negative values to enormous positive values at $\beta_0(h-\delta) = n\pi/2$, $n = 0, 1, 2, \cdots$. The reactance is very large and negative between $\beta_0(h-\delta) = 0$ and $\beta_0(h-\delta) = \pi/2$; it is positive between $\beta_0(h-\delta) = \pi/2$ and $\beta_0(h-\delta) = \pi$; and so on. Curves of the function $(\zeta_0\Psi/2\pi)\cot\beta_0(h-\delta)$ with $\zeta_0/2\pi \doteq 60$ ohm are shown in Fig. 28.1 for $\Psi = 10, 20, 30, 100$, and ∞.

The zeroth-order formula (1) is not a good approximation near $\beta_0(h-\delta) = n\pi/2$, even for Ψ very large. Furthermore, since $(Z_\delta)_0$ as given in (1) is a pure reactance, the leading real term is not available. This means that over the ranges specified by (2a, b) the resistance is small compared with the reactance. Since the reactance can be tuned out, and the resistance is involved in the radiated power, it is important to determine the leading real term in the impedance. This is readily done by suitable rearrangement of the first-order solution.

$$(Z_\delta)_1 = \frac{-j\zeta_0\Psi}{2\pi}$$
$$\times \left\{\frac{[\cos\beta_0(h-\delta) + A_1^I/\Psi] + j(A_1^{II}/\Psi)}{[\sin\beta_0(h-\delta) + B_1^I/\Psi] + j[B_1^{II}/\Psi]}\right\}. \quad (3)$$

Rationalization of this expression to separate real and imaginary parts leads to

$$(R_\delta)_1 = \frac{\zeta_0}{2\pi}$$
$$\times \left[\frac{A_1^{II}[\sin\beta_0(h-\delta) + B_1^I/\Psi] - B_1^{II}[\cos\beta_0(h-\delta) + A_1^I/\Psi]}{[\sin\beta_0(h-\delta) + B_1^I/\Psi]^2 + (B_1^{II}/\Psi)^2}\right], \quad (4a)$$

$$(X_\delta)_1 = -\frac{\zeta_0\Psi}{2\pi}$$
$$\times \left[\frac{[\cos\beta_0(h-\delta) + A_1^I/\Psi] \times [\sin\beta_0(h-\delta) + B_1^I/\Psi] - A_1^{II}B_1^{II}/\Psi^2}{[\sin\beta_0(h-\delta) + B_1^I/\Psi]^2 + [B_1^{II}/\Psi]^2}\right]. \quad (4b)$$

The leading terms in (4a, b) constitute an improved zeroth-order solution. They are obtained by neglecting all terms in the square brackets which have $1/\Psi$ or $1/\Psi^2$ as a factor. Thus,

$$(R_\delta)_{01} = \frac{\zeta_0}{2\pi}\left[\frac{A_1^{II} - B_1^{II}\cot\beta_0(h-\delta)}{\sin\beta_0(h-\delta)}\right], \quad (5a)$$

$$(X_\delta)_0 = -\frac{\zeta_0\Psi}{2\pi}\cot\beta_0(h-\delta). \quad (5b)$$

A modified form of (5b) is often given. It is obtained from (4b) by retaining the next-order terms in powers of $1/\Psi$ in the numerator, but not in the denominator. Hence, it is a better approximation than (5b) near and at $\beta_0(h-\delta) = \pi/2$, but no better at or near $\beta_0(h-\delta) = \pi$. It is

$$(X_\delta)_{01} = -\frac{\zeta_0}{2\pi}[\Psi\cot\beta_0(h-\delta)$$
$$+ A_1^I/\sin\beta_0(h-\delta)$$
$$+ B_1^I\cot\beta_0(h-\delta)/\sin\beta_0(h-\delta)]. \quad (5c)$$

Formulas (5a) and (5c) are frequently given in the literature as the resistance and reactance of a cylindrical antenna using Hallén's parameter $\Psi = \Omega$ and with $\delta = 0$. If these substitutions are made and the explicit formulas for $A_1^I, A_1^{II}, B_1^I, B_1^{II}$ are used with

$$\Omega = 2\ln\frac{2h}{a} = 2\ln 2 + 2\ln\frac{h}{a}, \quad (6)$$

it is possible to express (5a) and (5c) as follows:

$$(R_0)_{01} = R_0^e + R_0^i, \quad (7)$$
$$(X_0)_{01} = X_0^e + X_0^i, \quad (8)$$

with

$$R_0^e = \frac{\zeta_0}{4\pi}[(1 - \cot^2\beta_0 h)\operatorname{Cin} 4\beta_0 h$$
$$+ 4\cot^2\beta_0 h\operatorname{Cin} 2\beta_0 h$$
$$+ 2\cot\beta_0 h(\operatorname{Si} 4\beta_0 h - 2\operatorname{Si} 2\beta_0 h)], \quad (9)$$

$$X_0^e = \frac{\zeta_0}{4\pi}\Big\{(1 - \cot^2\beta_0 h)\operatorname{Si} 4\beta_0 h$$
$$+ 4\cot^2\beta_0 h\operatorname{Si} 2\beta_0 h$$
$$+ 2\cot\beta_0 h\Big[2\operatorname{Cin} 2\beta_0 h - \operatorname{Cin} 4\beta_0 h$$
$$- 2\ln\frac{h}{a}\Big]\Big\}, \quad (10)$$

$$R_0^i = (r^i/\beta_0)(\beta_0 h \csc^2\beta_0 h - \cot\beta_0 h), \quad (11)$$
$$X_0^i = (x^i/\beta_0)(\beta_0 h \csc^2\beta_0 h - \cot\beta_0 h). \quad (12)$$

Note that $\zeta_0/4\pi = 30$ ohms. It is significant that (9) is independent of the radius a.

It is important to note that in the modified zeroth-order forms (5a) and (5c), when written as in (7) to (12), it is possible to separate the resistance $(R_0)_{01}$ into an external resistance R_0^e due to radiation and an internal resistance R_0^i due to dissipation in the conductor. These resistances are additive and independent. In the general form (2) or in approximations of higher order than the modified zeroth, such a separation is not possible. This means that the distribution of current is determined by radiation and thermal dissipation in a complex manner. This is not of great significance in practice when good conductors are used, since the effect of ohmic dissipation usually is negligible. The form (7) with (9) and (11) is useful in comparing orders or magnitude of R^i and R^e.

Since the zeroth-order distribution of current is sinusoidal, the external resistance and reactance given in (9) and (10) must correspond to such a distribution. When given in the literature these formulas often are not derived from a general formula such as (2). They are obtained by integrating the normal component of the Poynting vector in the form (I.10.1) over a cylindrical envelope of radius a and length $2h$ that completely encloses the antenna. Since the antenna is driven by a discontinuity in scalar potential, it constitutes a complete transmitting system. This is the so-called emf method; it is discussed in Sec. 40. In carrying out the integration, a sinusoidally distributed current is assumed to exist. Strictly, such a current can be maintained only by a distribution of generators along the entire length of the antenna, and not by a single generator at its center. An equivalent integration for R^e (but not for X^e) is carried out in Chapter V, where the electromagnetic field and the Poynting vector due to a sinusoidally distributed current are derived. In evaluating R^e by either of these methods, it is possible to refer the distribution of current to any desired reference current other than the input current $I_z(0)$ implied in (5a). In particular, the maximum current in the sinusoidal distribution may be chosen. If this is done, the resistance obtained is denoted by R_m^e; it is related to R_0^e by the formula

$$P = I_z^2(0) R_0^e = I_{z\max}^2 R_m^e, \quad (13)$$

where P is the total power radiated. Since

$$I_{z\max} \equiv I_m = I_z(0)/\sin \beta_0 h \quad (14)$$

if the distribution is sinusoidal of zeroth order, it follows that with (9) and (13)

$$R_m^e = \frac{\zeta_0}{4\pi}[-\cos 2\beta_0 h \; \text{Cin } 4\beta_0 h$$
$$+ 4\cos^2 \beta_0 h \; \text{Cin } 2\beta_0 h$$
$$+ \sin 2\beta_0 h(\text{Si } 4\beta_0 h - 2\text{Si } 2\beta_0 h)]. \quad (15)$$

An important special case is for $\beta_0 h = n\pi/2$ with n odd. Then

$$R_0^e = R_m^e = \frac{\zeta_0}{4\pi} \text{Cin } 2n\pi. \quad (n \text{ odd}) \quad (16)$$

Note that $\zeta_0 \doteq 120\pi$ ohm, so that for

$$n = 1, \quad R_m^e = 73.13 \text{ ohm}, \quad (17a)$$
$$n = 3, \quad R_m^e = 105.5 \text{ ohm}. \quad (17b)$$

A curve of (15) for a range of values of $h/\lambda_0 = \beta_0 h/2\pi$ is shown in Fig. V.12.1.

Corresponding to (16), (10) reduces to

$$X_0^e = \frac{\zeta_0}{4\pi} \text{Si } 2n\pi. \quad (n \text{ odd}) \quad (18)$$

Specifically,

$$n = 1, \quad X_0^e = 42.5 \text{ ohm}, \quad (19a)$$
$$n = 3, \quad X_0^e = 45.8 \text{ ohm}. \quad (19b)$$

The significance of these values is considered after the second-order impedance has been discussed.

The formulas (16)–(19) may be derived directly from the first-order form (3) without introducing the modified zeroth-order solution. By setting $\delta = 0$, $\beta_0 h = n\pi/2$, n odd, in (3) this becomes

$$(Z_0)_1 = \frac{-j\zeta_0}{2\pi}\left(\frac{A_1}{1 + B_1/\Psi}\right). \quad (20)$$

For sufficiently thin antennas, it is a good approximation to set

$$|B_1/\Psi| \ll 1, \quad (21)$$

so that with (27.6a) and (21.1b)

$$(Z_0)_{01} \doteq \frac{-j\zeta_0}{2\pi} A_1 = \frac{-j\zeta_0}{2\pi} C_a(h, h). \quad (22)$$

This is a convenient simple form. It is reduced further by using (19.33) with $\beta_0 h = n\pi/2$, n odd. Thus,

$$(Z_0)_{01} \doteq \frac{-j\zeta_0}{2\pi}[Cs(\beta_0 a, 2\beta_0 h)$$
$$- jSs(\beta_0 a, 2\beta_0 h)]. \quad (23)$$

Finally, with $\beta_0 a = 0$, and (19.27a, c),

$$(Z_0)_{01} \doteq \frac{\zeta_0}{4\pi} [\operatorname{Cin} 4\beta_0 h - j \operatorname{Si} 4\beta_0 h],$$

$$\beta_0 h = n\pi/2, \quad n = 1, 3, 5, \cdots. \quad (24)$$

This coincides with (16) and (18).

In dealing with antennas and arrays with $\beta_0 h = n\pi/2$, n odd, it is often convenient to make use of (22) or its equivalent. Correspondingly simple formulas are obtained in Chapter III for coupled antennas.

29. Admittance and Impedance Factors for Small Base Separation

In the general formula (27.2) for Z_δ there is no restriction on the magnitude of the base separation 2δ. Actually, since Z_δ is useful only in conjunction with an appropriate network involving L_T and C_T as a termination for a two-wire line which must satisfy a condition of the form

$$\beta_0 b \ll 1, \quad (1)$$

the separation 2δ is limited. For the conventional method of driving the antenna as an end load or as a center load, 2δ is taken to be equal to the line spacing b. For other arrangements, including those described in Sec. 10 for minimum coupling, 2δ may be considerably less than b. Accordingly, little, if anything, is gained by permitting 2δ in the expression for Z_δ to exceed b for a two-wire line. Hence, conditions like (1) may be written with b replaced by 2δ or, since only the order of magnitude is involved, by δ. If $\beta_0 \delta$ is required to be small compared with unity, it is possible to express Y_δ (or Z_δ) in a Maclaurin series in powers of the small quantity $\beta_0 \delta$ and retain only the first correction term. Thus,

$$Y_\delta \doteq Y_0 + \beta_0 \delta \left[\frac{\partial Y_\delta}{\partial \beta_0 \delta}\right]_{\delta=0} = Y_0(1 - \epsilon), \quad (2)$$

where*

$$\epsilon = \epsilon'' - j\epsilon' \equiv -\frac{\beta_0 \delta}{Y_0}\left[\frac{\partial Y_\delta}{\partial \beta_0 \delta}\right]_{\delta=0}. \quad (3)$$

With $Y = G + jB$, it follows that

$$G_\delta = G_0(1 - \epsilon'') - B_0 \epsilon', \quad (4)$$

$$B_\delta = B_0(1 - \epsilon'') + G_0 \epsilon'. \quad (5)$$

---

* $\epsilon = \epsilon'' - j\epsilon'$ in this section is not to be confused with the dielectric constant.

The impedance is given by

$$Z_\delta = Z_0/(1 - \epsilon), \quad (6)$$

so that

$$R_\delta = \frac{R_0(1 - \epsilon'') + X_0 \epsilon'}{(1 - \epsilon'')^2 + (\epsilon')^2}, \quad (7)$$

$$X_\delta = \frac{X_0(1 - \epsilon'') - R_0 \epsilon'}{(1 - \epsilon'')^2 + (\epsilon')^2}. \quad (8)$$

In order to evaluate ϵ'' and ϵ' it is necessary to differentiate the following expression for $(Y_\delta)_m$ with respect to $\beta_0 \delta$:

$$(Y_\delta)_m = \frac{j2\pi}{\zeta_0 \Psi_{K1}}$$

$$\times \left[\frac{(D_1)_m \sin \beta_0(h - \delta) + \sum_{n=1}^{m} (D_{n+1})_m B_n^\delta/\Psi_{K1}^n}{\cos \beta_0(h - \delta) + \sum_{n=1}^{m} (D_n)_{m-1} A_n^\delta/\Psi_{K1}^n}\right], \quad (9)$$

and use this derivative in (3) in conjunction with $(Y_0)_m$ obtained from (9) by setting $\delta = 0$. With subscripts on Ψ omitted, the expression for ϵ is

$$\epsilon = \epsilon'' - j\epsilon' = \beta_0 \delta$$

$$\times \left\{ \left[\frac{\sin \beta_0 h + G_1(h)/\Psi + G_2(h)/\Psi^2 + \cdots}{\cos \beta_0 h + F_1(h)/\Psi + F_2(h)/\Psi^2 + \cdots}\right] \right.$$

$$+ \frac{2}{\Psi} \frac{e^{-j\beta_0 h}}{\beta_0 h}$$

$$\times \left[\frac{(1 - \cos \beta_0 h) + [F_1(0) - F_1(h)]/\Psi + \cdots}{\cos \beta_0 h + F_1(h)/\Psi + F_2(h)/\Psi^2 + \cdots}\right]$$

$$+ \frac{2}{\Psi} \frac{e^{-j\beta_0 h}}{\beta_0 h}$$

$$\left. \times \left[\frac{\sin \beta_0 h + G_1(h)/\Psi + \cdots}{\sin \beta_0 h + M_1(0)/\Psi + M_2(0)/\Psi^2 + \cdots}\right] \right\}. \quad (10)$$

Curves of the second-order factors $(h/\delta)\epsilon''$ and $(h/\delta)\epsilon'$ as computed by Winternitz from (10) are shown in Figs. 29.1 and 29.2 for $\Omega = 10, 15,$ and 20 in the range of $\beta_0 h$ from 1.1 to 4.0. These factors are required in obtaining the second-order admittance $(Y_\delta)_2$ or impedance $(Z_\delta)_2$ from $(Z_0)_2$. The functions $(h/\delta)\epsilon''$ and $(h/\delta)\epsilon'$ are shown instead of ϵ'' and ϵ', since they are more conveniently represented graphically.

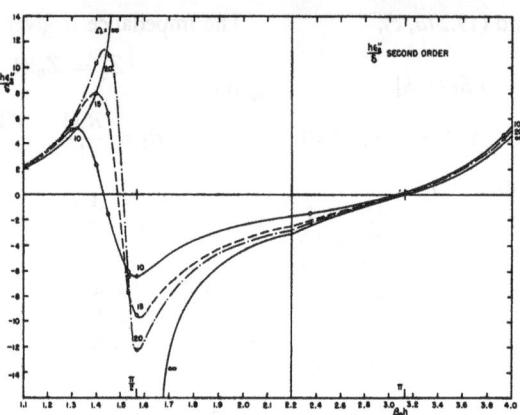

Fig. 29.1. Second-order curves for $h\varepsilon_2''/\delta$.

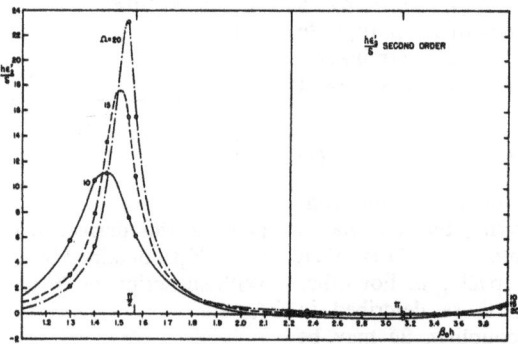

Fig. 29.2. Second-order curves for $h\varepsilon_2'/\delta$.

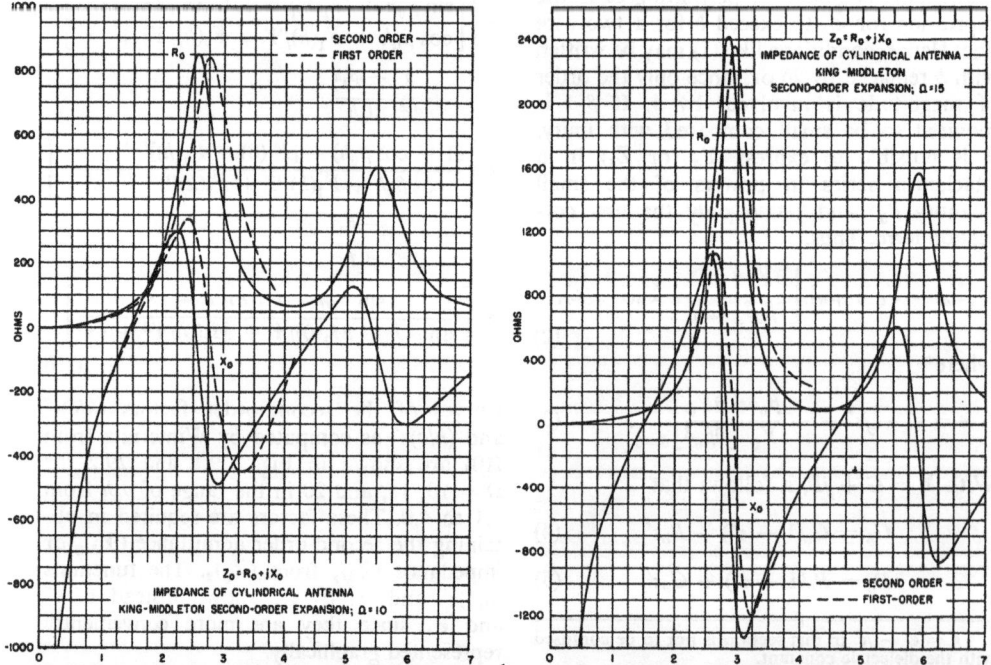

Fig. 30.1a. Impedance of cylindrical antenna, King-Middleton second-order expansion; $\Omega = 10$.

Fig. 30.1b. Impedance of cylindrical antenna, King-Middleton second-order expansion; $\Omega = 15$.

The problem of determining Y_δ or Z_δ for antennas with a small base separation 2δ that satisfies the condition

$$\beta_0\delta \ll 1 \quad \text{or} \quad \beta_0^2\delta^2 \ll \left(\ln\frac{2\delta}{a}\right)^2 \quad (11)$$

has been reduced to evaluating Y_0 or Z_0.

30. Impedance and Admittance for Zero Base Separation; First- and Second-Order Solutions

The impedance Z_δ of an antenna with a small base separation 2δ may be expressed in terms of the impedance Z_0 of an antenna with zero base separation (Sec. 29). This makes Z_0 a convenient reference quantity for expressing the general circuit properties of a cylindrical antenna and studying the significance of the several variables and parameters. Whenever reference is made to the impedance of a cylindrical antenna without specific reference to a finite base separation 2δ, the value Z_0 is meant.

General formulas for impedance and admittance with $\delta = 0$ are obtained from (27.2) and (29.9). In order to evaluate the impedance and the admittance numerically and so obtain a graphical representation, the functions A_{nH} and B_{nH} as well as the D-factors and the expansion parameter must be known to the desired order of approximation. The functions A_{nH} and B_{nH} are given by

$$A_{nH} = \alpha_n^I + j\alpha_n^{II} + jhr_{n\alpha}(r^i + jx^i)/30, \quad (1)$$

$$B_{nH} = \beta_n^I + j\beta_n^{II} + jhr_{n\beta}(r^i + jx^i)/30. \quad (2)$$

Typical D-factors are shown in Fig. 21.1; Ψ_{K1} is shown in Fig. 20.7. Numerical tables have been computed for $\alpha_n^I + j\alpha_n^{II}$, $\beta_n^I + j\beta_n^{II}$ for $n = 1$ and $n = 2$. The functions $r_{n\alpha}$ and $r_{n\beta}$ are available only for $n = 1$. These tables are given in Sec. 27, together with graphical representations. Since the correction factors ε'' and ε' involved in the solution for small values of 2δ are available in second-order form, the apparent impedance of an antenna terminating a two-wire line can be computed if terms with the internal impedance per unit length z^i are neglected. These terms *must* be neglected if a general representation of the impedance as a function of $\beta_0 h$ is to be possible. That they are negligible for good conductors may be verified in terms of the first-order solution for appropriately chosen special cases.

Formulas for the impedance and admittance of first and second order for $z^i = 0$ and $\delta = 0$ are as follows. The numerical value, $\zeta_0/2\pi \doteq 60$ ohm, has been inserted, so that Z_0 is in ohms.

$$(Z_0)_1 = 1/(Y_0)_1$$
$$= -j60\Psi\left[\frac{\cos\beta_0 h + (\alpha_1^I + j\alpha_1^{II})/\Psi}{(D_1)_1\sin\beta_0 h + (\beta_1^I + j\beta_1^{II})/\Psi}\right], \quad (3)$$

$$(Z_0)_2 = 1/(Y_0)_2$$
$$= -j60\Psi\left[\frac{\cos\beta_0 h + (D_1)_1(\alpha_1^I + j\alpha_1^{II})/\Psi}{+ (\alpha_2^I + j\alpha_2^{II})/\Psi^2}\right], \quad (4)$$

where

$$(D_1)_1 = \left(2 - \frac{\Omega}{\Psi}\right), \quad (5a)$$

$$(D_1)_2 = \left(3 - 3\frac{\Omega}{\Psi} + \frac{\Omega^2}{\Psi^2}\right), \quad (5b)$$

$$(D_2)_2 = \left(3 - 2\frac{\Omega}{\Psi}\right). \quad (5c)$$

For the King-Middleton solution, numerical values of the D-factors are given in Fig. 21.1, and those of $\Psi = \Psi_{K1}$ in Fig. 20.7. For the Hallén solution, $\Psi = \Omega$, so that all D-factors are equal to unity. For purposes of computation, (3) and (4) are more convenient in polar form:

$$(Z_0)_m = 1/(Y_0)_m$$
$$= 60\Psi\left|\frac{a_m + jb_m}{c_m + jd_m}\right|e^{j[\tan^{-1}(b_m/a_m) - \tan^{-1}(d_m/c_m)]} \quad (6)$$

where, for the first-order King-Middleton solution, $m = 1$,

$$a_1 = \alpha_1^{II}, \quad (7a)$$

$$b_1 = -(\Psi\cos\beta_0 h + \alpha_1^I), \quad (7b)$$

$$c_1 = \Psi(D_1)_1\sin\beta_0 h + \beta_1^I, \quad (7c)$$

$$d_1 = \beta_1^{II}, \quad (7d)$$

and for the second-order King-Middleton solution, $m = 2$,

$$a_2 = \alpha_1^{II}(D_1)_1 + \alpha_2^{II}/\Psi, \quad (8a)$$

$$b_2 = -[\Psi\cos\beta_0 h + \alpha_1^I(D_1)_1 + \alpha_2^I/\Psi], \quad (8b)$$

$$c_2 = \Psi(D_1)_2\sin\beta_0 h + \beta_1^I(D_2)_2 + \beta_2^I/\Psi, \quad (8c)$$

$$d_2 = \beta_1^{II}(D_2)_2 + \beta_2^{II}/\Psi. \quad (8d)$$

For the simpler Hallén solution, the corresponding factors for the first-order solution are

$$a_1 = \alpha_1^{II}, \quad (9a)$$

$$b_1 = -(\Omega \cos \beta_0 h + \alpha_1^{I}), \quad (9b)$$

$$c_1 = \Omega \sin \beta_0 h + \beta_1^{I}, \quad (9c)$$

$$d_1 = \beta_1^{II}. \quad (9d)$$

For the second-order Hallén solution,

$$a_2 = \alpha_1^{II} + \alpha_2^{II}/\Omega, \quad (10a)$$

$$b_2 = -(\Omega \cos \beta_0 h + \alpha_1^{I} + \alpha_2^{I}/\Omega), \quad (10b)$$

$$c_2 = \Omega \sin \beta_0 h + \beta_1^{I} + \beta_2^{I}/\Omega, \quad (10c)$$

$$d_2 = \beta_1^{II} + \beta_2^{II}/\Omega. \quad (10d)$$

Graphical representations of the King-Middleton impedances and admittances computed from (6) are shown in Figs. 30.1–30.10. Numerical values are given in Tables 30.1–30.11.* The general shapes of the curves are best represented using a linear scale as in Figs. 30.1a and 30.2a, b, where R_0 and X_0 in $Z_0 = R_0 + jX_0$ and G_0 and B_0 in $Y_0 = G_0 + jB_0$ are plotted as functions of $\beta_0 h$ for several values of $\Omega \equiv 2 \ln 2h/a$. Since $\beta_0 = \omega/v_0$, $\beta_0 h$ may be varied either by changing the length of the antenna at a fixed frequency or by changing the driving frequency for an antenna of fixed length. In both cases, the ratio h/a is kept constant for each curve. A very compact representation is given in Fig. 30.3a, b, c, where Z_0 is plotted in the complex plane with Ω as parameter. Similar graphs of Y_0 are shown in Figs. 30.4a, b. In Figs. 30.5a, b, R_0 and $|X_0|$ are shown on a logarithmic scale; the values of X_0 near its zeros are shown in Fig. 30.5c. Corresponding logarithmic curves for G_0 and $|B_0|$ and linear curves of B_0 near its zeros are in Figs. 30.6a, b, c.

The general discussion of the characteristics of the curves for the impedance $Z_0 = R_0 + jX_0$ of a perfectly conducting cylindrical antenna driven from a line with almost zero spacing or by a slice generator is centered conveniently about certain critical points. Of these, the most important are the values of $\beta_0 h$ for which the reactance vanishes. These may be divided into *two distinct* groups designated as *input resonance* and *input antiresonance*.

---

* As Ω is reduced below 10, the accuracy decreases rapidly. With $\Omega = 7$, $h/a = 16.4$, little more than the correct order of magnitude is assured. This is considered in Sec. 38.

Input resonance is characterized by

$$X_0 = 0, \quad \beta_0 h = (\beta_0 h)_{\text{res}} \text{ is near } n\pi/2 \text{ with}$$

$$n = 1, 3, 5, 7, \cdots, \quad (11a)$$

$R_0 = (R_0)_{\text{res}}$ is near minimum except for

$$n = 1. \quad (11b)$$

As $\beta_0 h$ is increased from zero, the condition of input resonance occurs first when $\beta_0 h$ is somewhat smaller than $\pi/2$ and is repeated at intervals that are slightly less than π. The quantities $\pi/2 - (\beta_0 h)_{\text{res}}$ and $3\pi/2 - (\beta_0 h)_{\text{res}}$ are shown in Fig. 30.7a as functions of $\Omega = 2 \ln 2h/a$. It is seen that the resonant length of an antenna approaches $n\pi/2$, n odd as the ratio h/a and hence as Ω is increased. For h/a sufficiently large, $(\beta_0 h)_{\text{res}}$ differs negligibly from $n\pi/2$, n odd. For all finite values of Ω, $(\beta_0 h)_{\text{res}}$ is less than $n\pi/2$, n odd, by an amount that is appreciable in the range of practical antennas for which Ω usually is less than 15. The resonant resistances $(R_0)_{\text{res}}$ are near minimum values except for $n = 1$. Their values for each n are virtually independent of the ratio h/a in the range $h/a \geq 75$ ($\Omega \geq 10$) for which the present quasi-one-dimensional theory is a good approximation as seen in Fig. 30.7b. For $n = 1$ and 3, $(R_0)_{\text{res}}$ (Fig. 30.7b) is very close to the values 73.13 ohm and 105.5 ohm obtained by the Poynting-vector method (see Chapter V) for antennas with an assumed sinusoidal distribution of current. This is reasonable, since at and near resonance the distribution of current even in a moderately thick antenna does not differ significantly from the sinusoidal, as is pointed put in Sec. 16. Actually, a truly sinusoidal distribution is possible only with a distribution of generators along the antenna. It cannot be maintained by a single generator, regardless of the radius of the antenna.

The behavior of X_0 and R_0 near resonance and particularly between $(\beta_0 h)_{\text{res}}$ and $n\pi/2$, n odd, is interesting. The variation of X_0 with $\beta_0 h$ and its dependence upon the ratio h/a in Ω are well shown in Fig. 30.5c. It is seen that the slopes of the reactance curves for different values of Ω increase with Ω. This steepening of slope is accompanied by a shift of the point of zero reactance from smaller values of $\beta_0 h$ towards $n\pi/2$. The points of intersection of the reactance curves with one another and with the abscissas $\beta_0 h = \pi/2$, $3\pi/2$ occur at values of X_0 between 40 and 50 ohm. Curves of X_0 as a function of Ω

for $\beta_0 h = \pi/2, 3\pi/2$ are shown in Fig. 30.7b. For antennas in the practical range of Ω, these do not differ greatly from the values 42.5 and 45 ohm determined by the Poynting-vector method for a physically unavailable antenna of radius a with a sinusoidally distributed current.

The behavior of R_0 near resonance may be determined from Fig. 30.5a. It is seen that in a range extending above and below $\beta_0 h = n\pi/2$, n odd, the resistances of thicker antennas, with smaller values of Ω are *larger* than for thinner antennas. Curves of R_0 as a function of Ω for $\beta_0 h = \pi/2, 3\pi/2$ are shown in Fig. 30.7b. It is seen that R_0 decreases with increasing Ω. For sufficiently large values of Ω, $R_0(\beta_0 h = \pi/2)$ must be near $(R_0)_{\text{res}}$, which remains sensibly constant near 72 ohm. Thus the curves for R_0 at $\beta_0 h = \pi/2, 3\pi/2$ should approach $(R_0)_{\text{res}}$, $(n = 1, 3)$ as Ω increases. The constancy of $(R_0)_{\text{res}}$ as h/a or Ω is varied is the result of the simultaneous increase in R_0 and the shift of $(R_0)_{\text{res}}$ to smaller values of $\beta_0 h$ as Ω is decreased. The two changes effectively cancel.

Input antiresonance is characterized by

$$X_0 = 0, \quad \beta_0 h = (\beta_0 h)_{\text{antires}} \text{ is near } n\pi/2,$$
$$\text{with } n = 2, 4, 6, 8, \cdots, \quad (12a)$$

$$R_0 = (R_0)_{\text{antires}} \text{ is very near a maximum.}$$
$$(12b)$$

It occurs at values of $\beta_0 h$ that are less than $\pi, 2\pi, 3\pi, \cdots$ by an amount that increases with decreasing Ω in a manner resembling the occurrence of resonance near $\pi/2, 3\pi/2, \cdots$. However, the magnitudes of the shifts from the values $n\pi/2$, n even, at antiresonance very much exceed corresponding shifts from $n\pi/2$, n odd, at resonance. This is shown in Fig. 30.7a and also in Fig. 30.5c. From the latter figure it is clear that the slope of the X_0-curves near antiresonance is much steeper than, and of course opposite to, that near resonance. The steepness of the curves increases with Ω and the location of its zero value shifts nearer $\beta_0 h = n\pi/2$, n even.

The reactance X_0 has a maximum for $\beta_0 h$ slightly less than $(\beta_0 h)_{\text{antires}}$ and a minimum at $\beta_0 h$ slightly greater than $(\beta_0 h)_{\text{antires}}$. The resistance R_0 has its maximum very nearly at antiresonance; its magnitude at antiresonance together with $X_{0_{\text{min}}}$ and $X_{0_{\text{max}}}$ are shown in Fig. 30.7b as a function of Ω. The ratio $|X_{\text{min}}|/|X_{\text{max}}|_n$ is an important quantity characterizing the behavior of the reactance near the nth antiresonance. It is always greater than unity; it decreases with increasing Ω; it increases as n is increased. In effect this means that the capacitive lobe of the reactance curve is always larger than the associated inductive lobe, and that the reactance becomes increasingly capacitive as the length of the antenna is increased. This is shown well in Fig. 30.5b, and also in Fig. 30.3 by the tilt in the diameters of the circles. The quantity $|X_{\text{min}}|/|X_{\text{max}}|_n$ for $n = 1, 2$ is shown in Fig. 30.8.

It is significant that for each n

$$|X_{\text{min}}| + |X_{\text{max}}| \doteq R_{\text{max}}. \quad (13)$$

For convenient reference, numerical values of important critical quantities are given in Table 30.1. Curves of R_0 and X_0 as functions of Ω are given in Figs. 30.9, 30.10a, b.

Note that the input impedance at antiresonance as determined by the King-Middleton theory cannot be compared with a quantity at $\beta_0 h = \pi$ corresponding to $Z_0 = 73.13 + j42.5$ at $\beta_0 h = \pi/2$ determined by the Poynting-vector method with an assumed sinusoidal distribution of current. At $\beta_0 h = \pi$ the input current is zero for a sinusoidally distributed current, so that Z_0 is necessarily infinite. This failure of a method based on an assumed sinusoidal current is readily understood in the light of the discussion of Sec. 16, from which it is clear that a simple sinusoidal distribution is a good approximation only for antenna half-lengths for which $\beta_0 h$ is near $n\pi/2$, n odd.

A discussion of critical points in the admittance curves paralleling this discussion of the impedance is readily carried out using the appropriate Figs. 30.2a, b, 30.4, 30.6, and Table 30.2. Perhaps the most significant point, aside from the generally reciprocal nature of the curves, is the relative constancy of the maximum values of conductance as Ω is varied. Since these maxima occur very near resonance, so that $G_0 \doteq 1/R_0$, the explanation is the same as that given for the constancy of R_0 at resonance. Since G_0 is unaffected by capacitance in parallel with the antenna at its terminals, it follows that G_0 should be relatively insensitive to transmission-line spacing (so that G_δ should not differ greatly from G_0) and be quite unaffected by the capacitance C_T in the lumped-constant network required to correct for line theory and antenna-line coupling. Since the effect of the series inductance L_T is relatively very small, it follows that the conductance G_0 should be

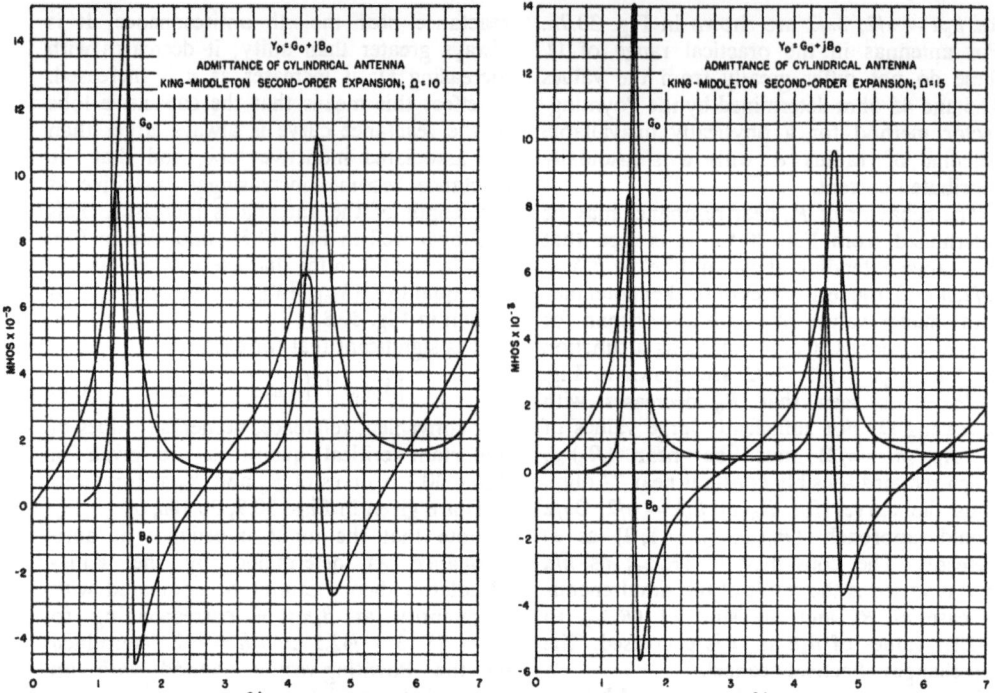

Fig. 30.2a. Admittance of cylindrical antenna, King-Middleton second-order expansion; $\Omega = 10$.

Fig. 30.2b. Admittance of cylindrical antenna, King-Middleton second-order expansion; $\Omega = 15$.

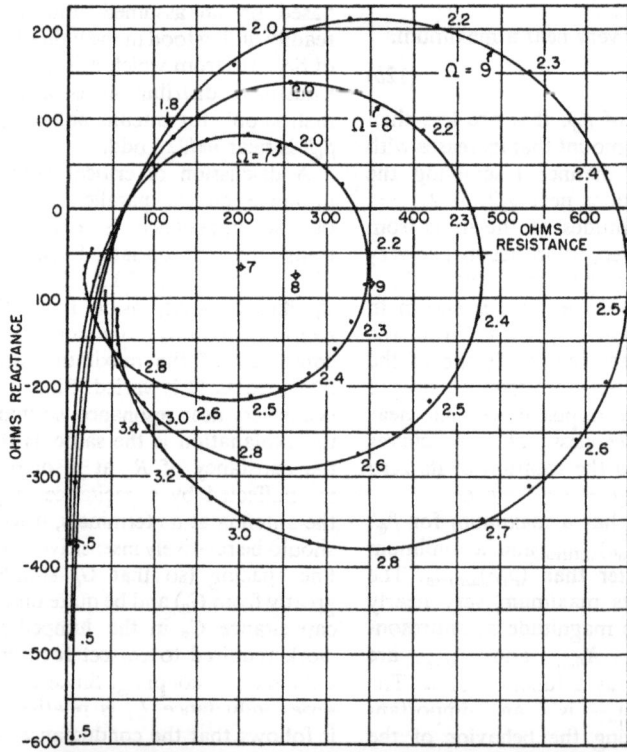

Fig. 30.3a. Circular graphs of King-Middleton second-order impedance; $\Omega = 7, 8, 9$.

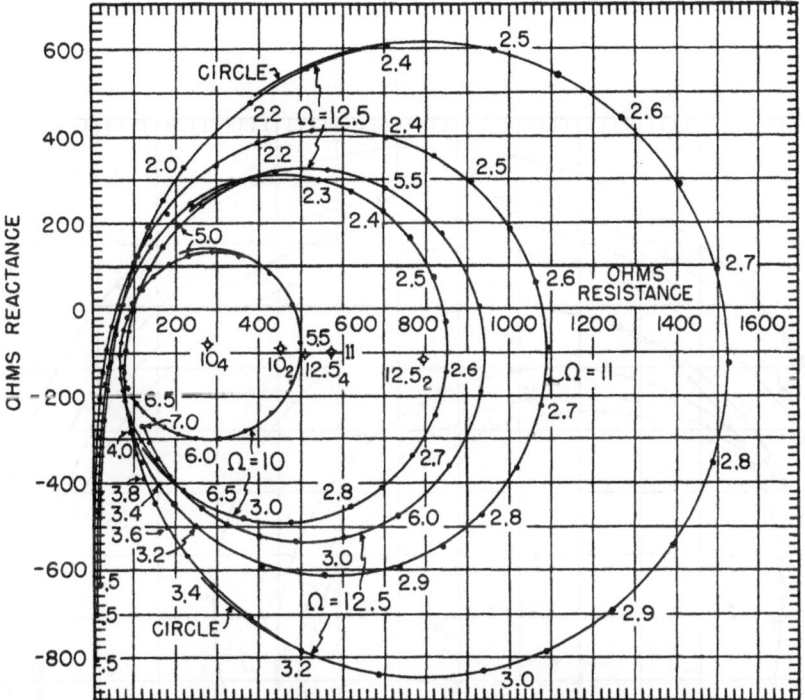

Fig. 30.3b. Circular graphs of King-Middleton second-order impedance; $\Omega = 10, 11, 12.5$.

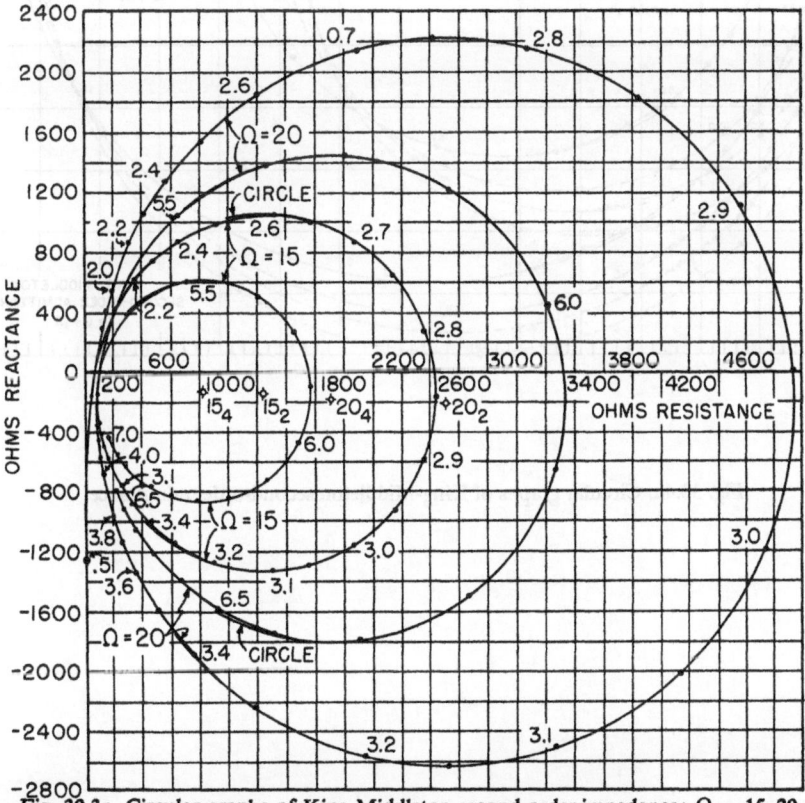

Fig. 30.3c. Circular graphs of King-Middleton second-order impedance; $\Omega = 15, 20$.

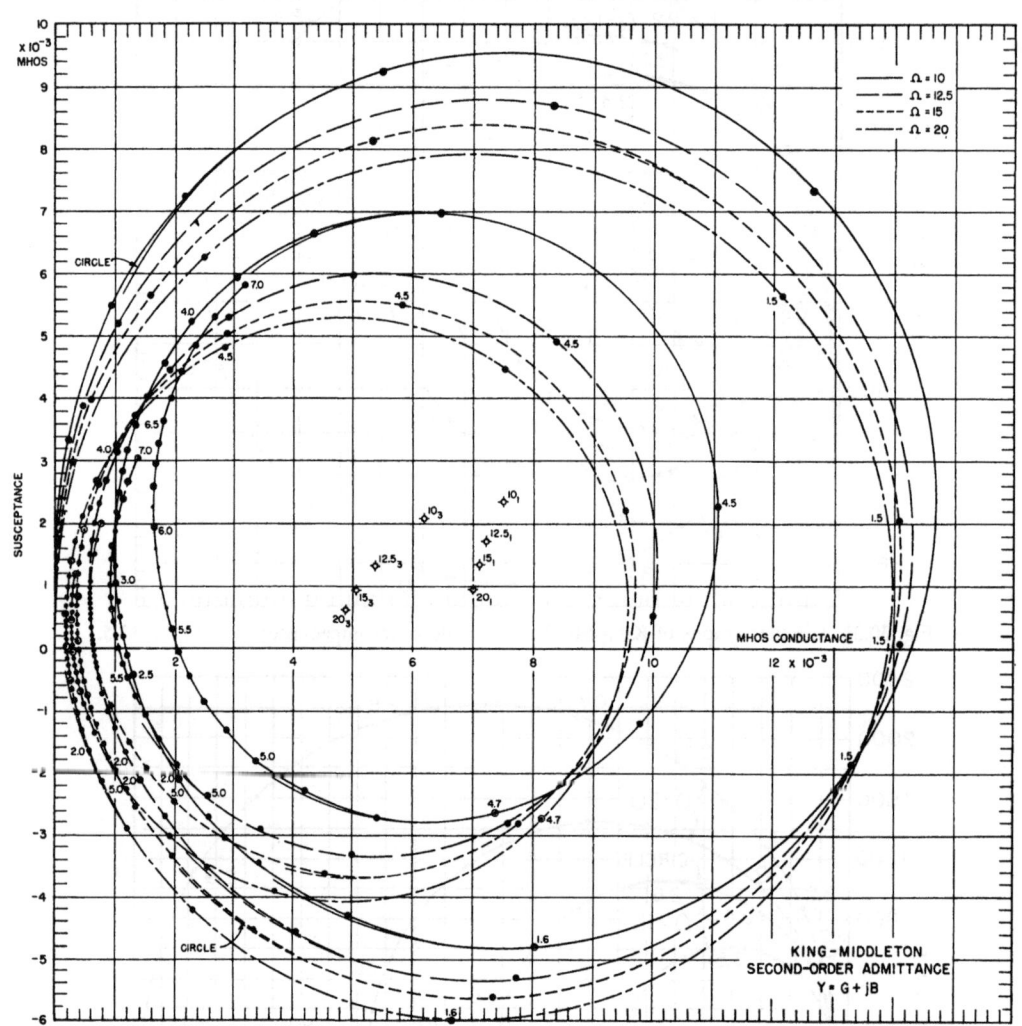

Fig. 30.4a. Circular graphs of King-Middleton second-order admittance.

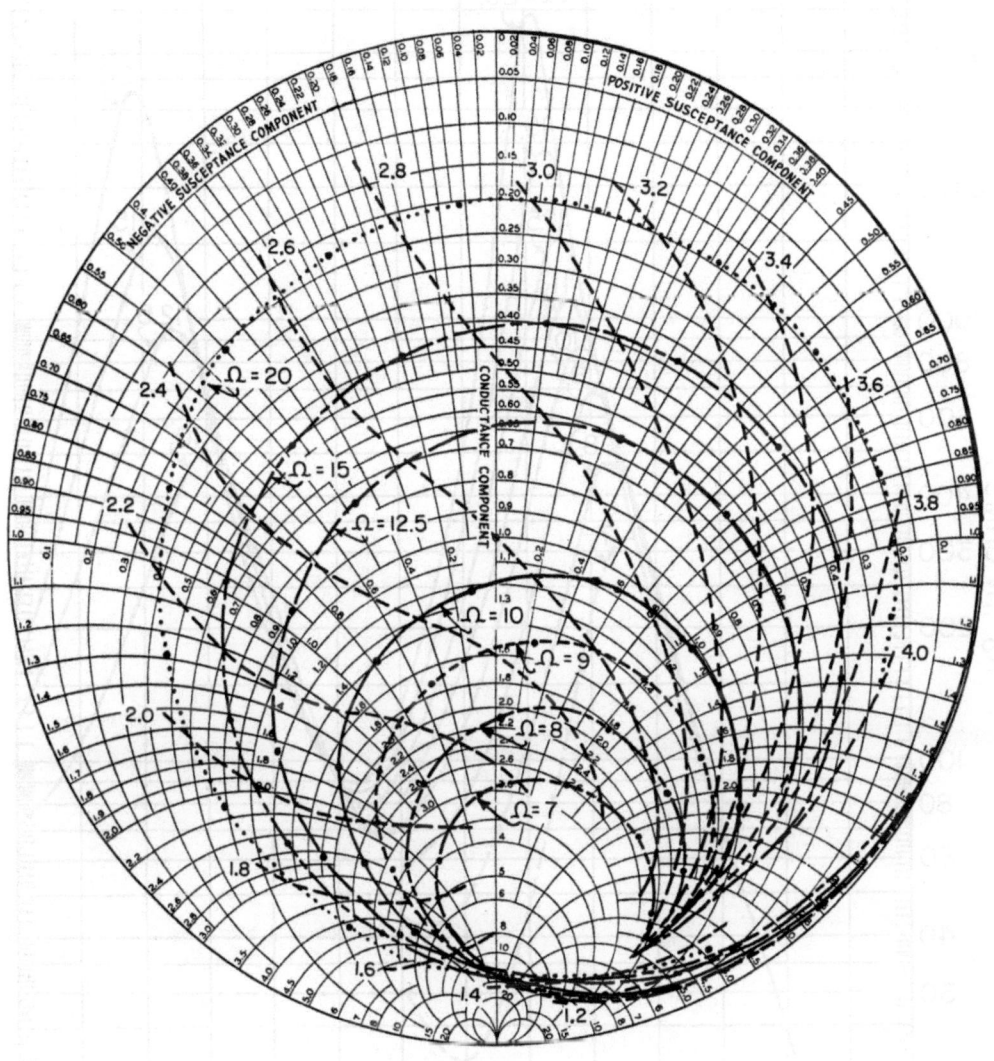

Fig. 30.4b. Second-order input admittance of center-driven dipole; Mhos × 10⁻³ vs. $\beta_0 h$; after King-Middleton.

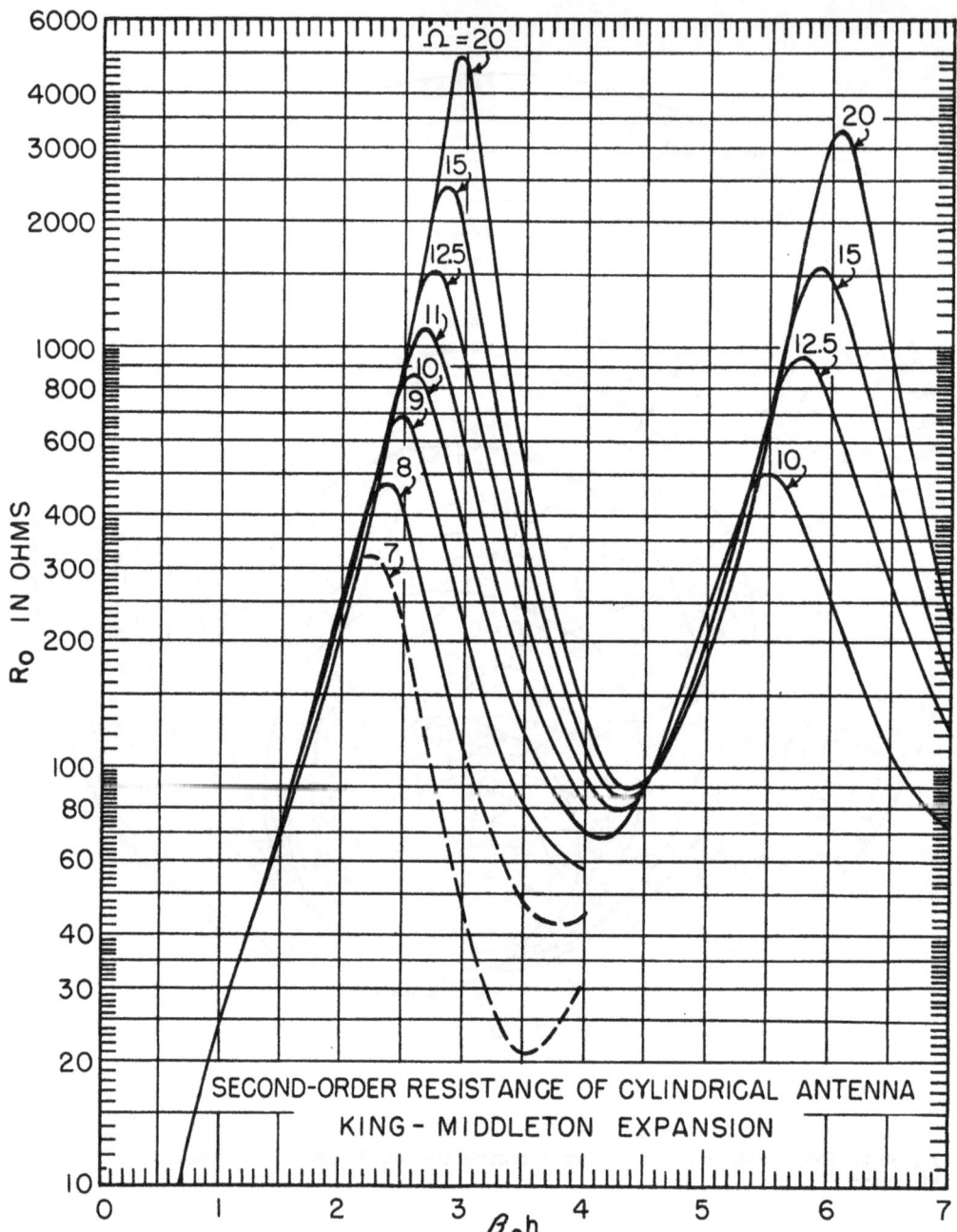

Fig. 30.5a. Resistance R_0 of cylindrical antenna, King-Middleton second-order expansion.

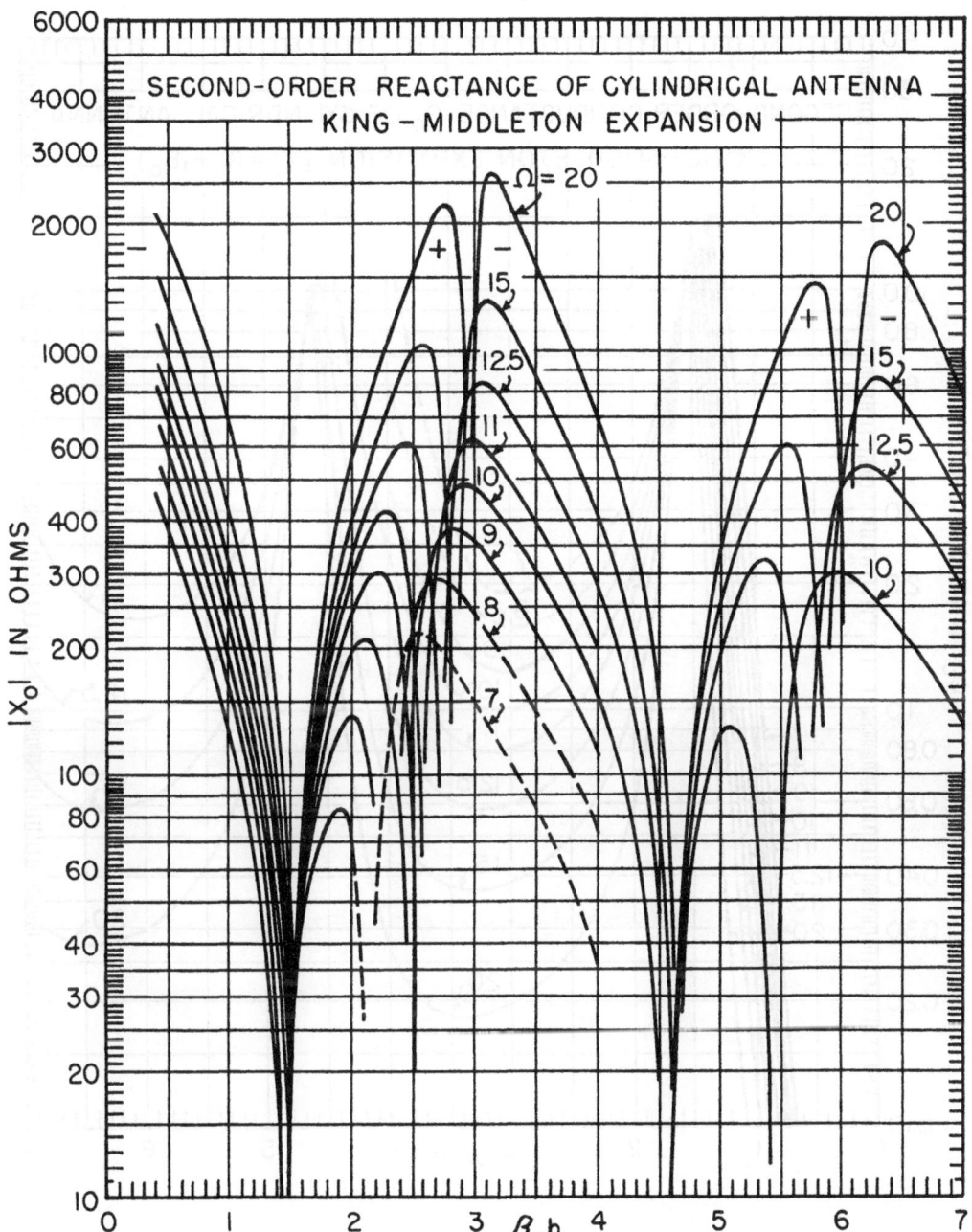

Fig. 30.5b. Reactance X_0 of cylindrical antenna, King-Middleton second-order expansion.

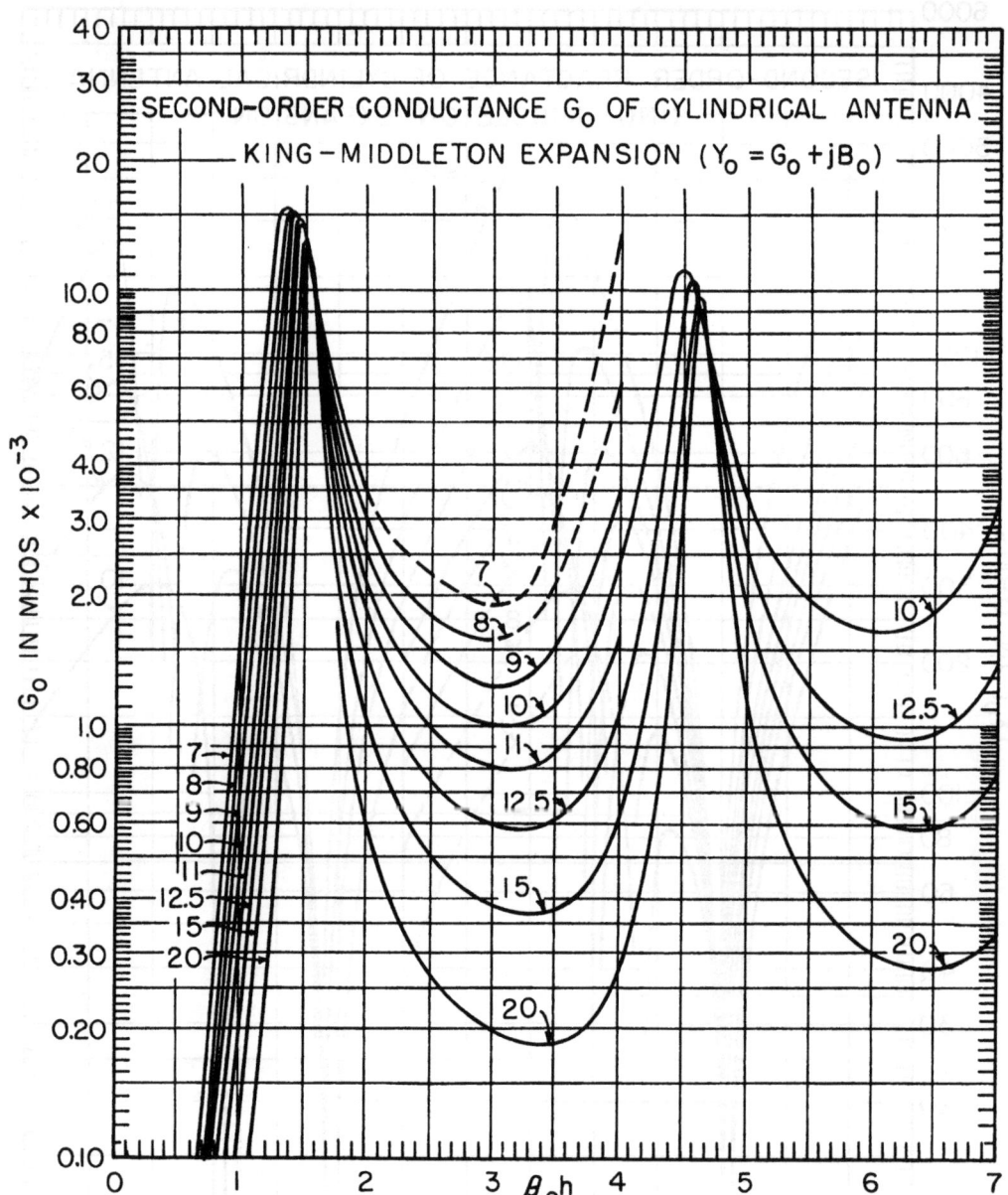

Fig. 30.5c. Conductance G_0 of cylindrical antenna, King-Middleton second-order expansion.

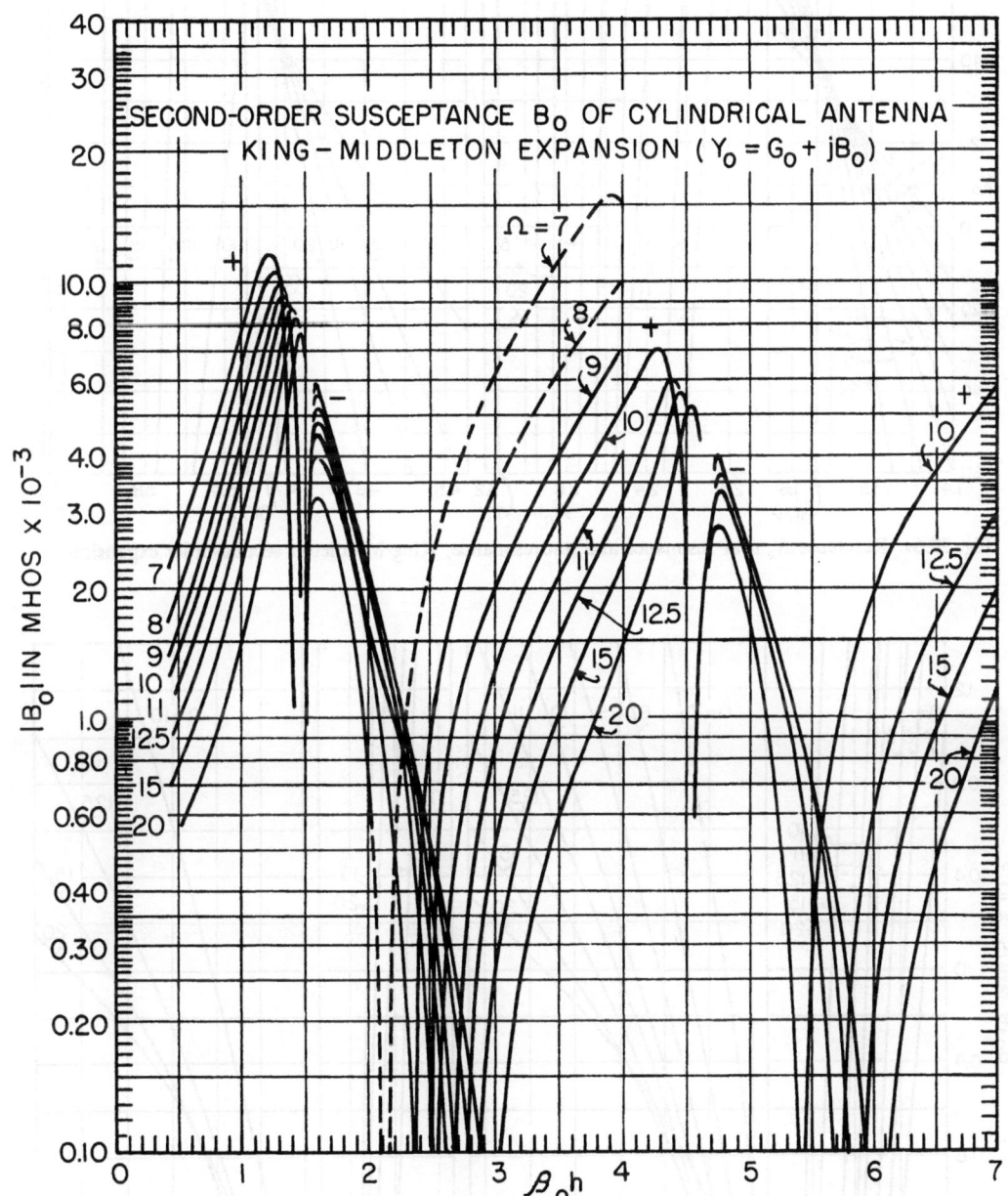

Fig. 30.5d. Susceptance B_0 of cylindrical antenna, King-Middleton second-order expansion.

KING-MIDDLETON SECOND-ORDER THEORY

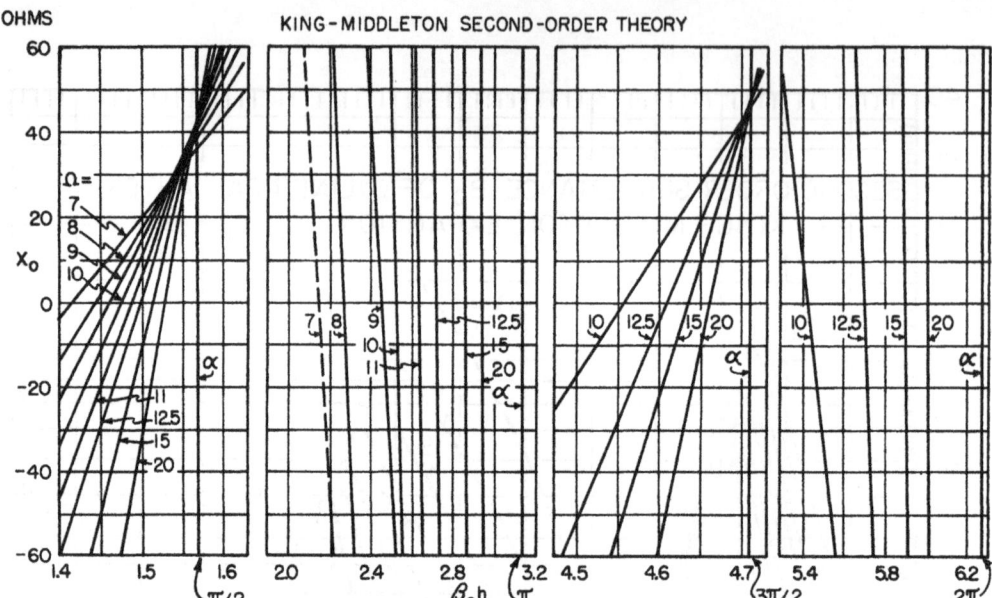

Fig. 30.6a. Reactance X_0 near resonance and antiresonance, King-Middleton second-order expansion.

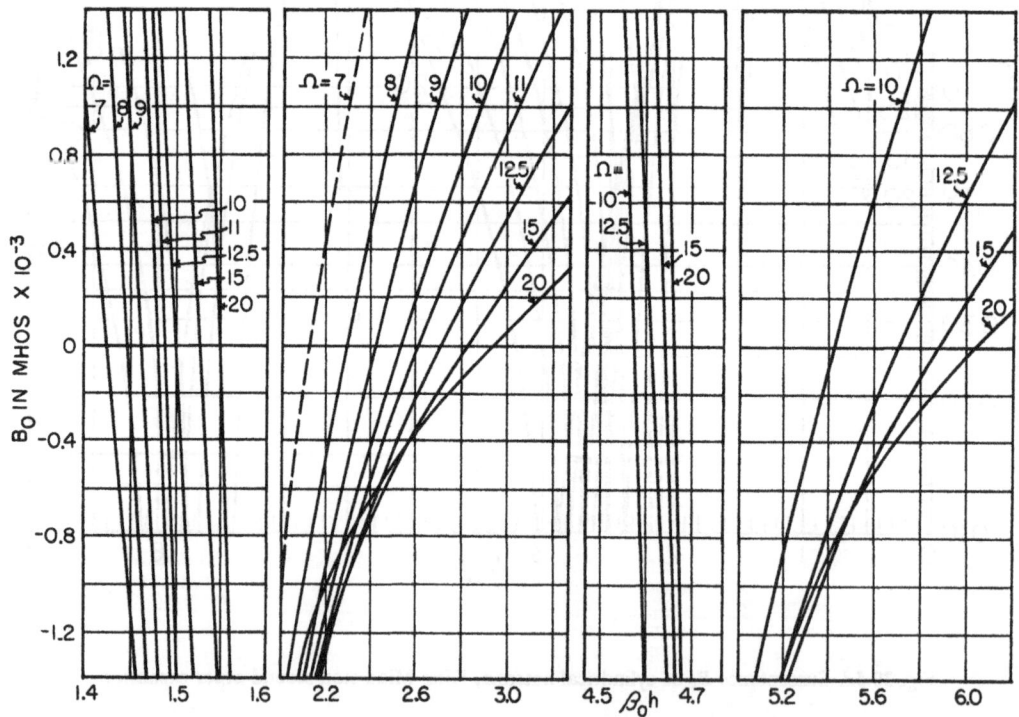

Fig. 30.6b. Susceptance B_0 of cylindrical antenna near resonance and antiresonance, King-Middleton second-order expansion.

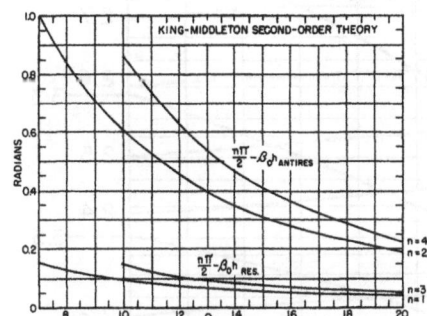

Fig. 30.7a. Resonant and antiresonant lengths from King-Middleton second-order expansion.

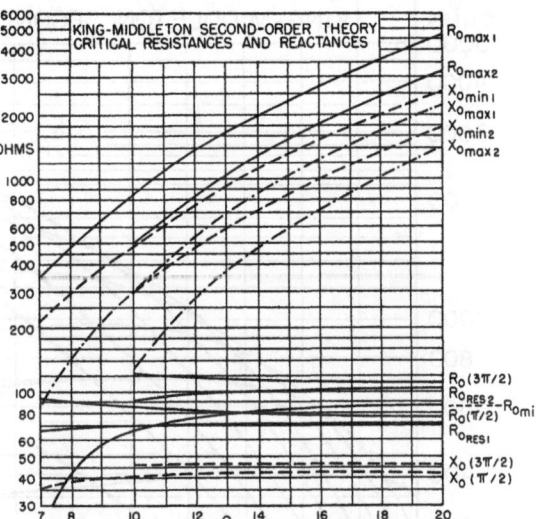

Fig. 30.7b. Critical resistances and reactances from King-Middleton second-order expansion.

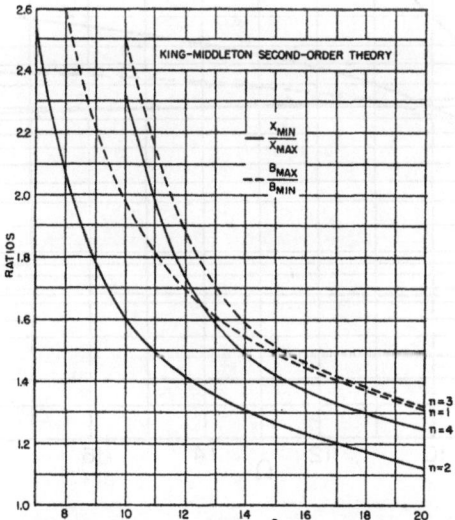

Fig. 30.8. Curves of X_{min}/X_{max}, King-Middleton second-order expansion.

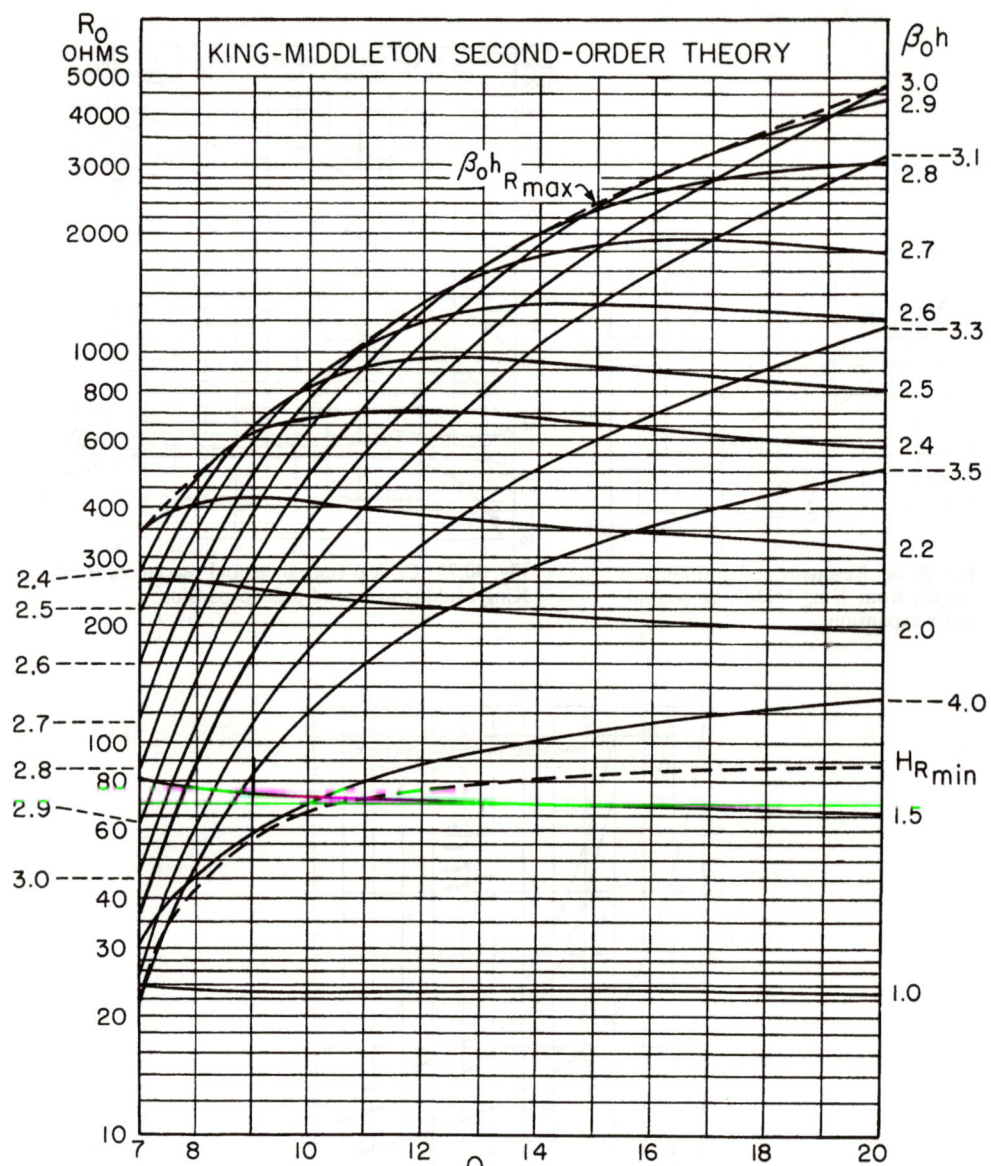

Fig. 30.9. Resistance as a function of Ω with $\beta_0 h$ as parameter, King-Middleton second-order expansion.

[II.30] THEORY OF LINEAR ANTENNAS 165

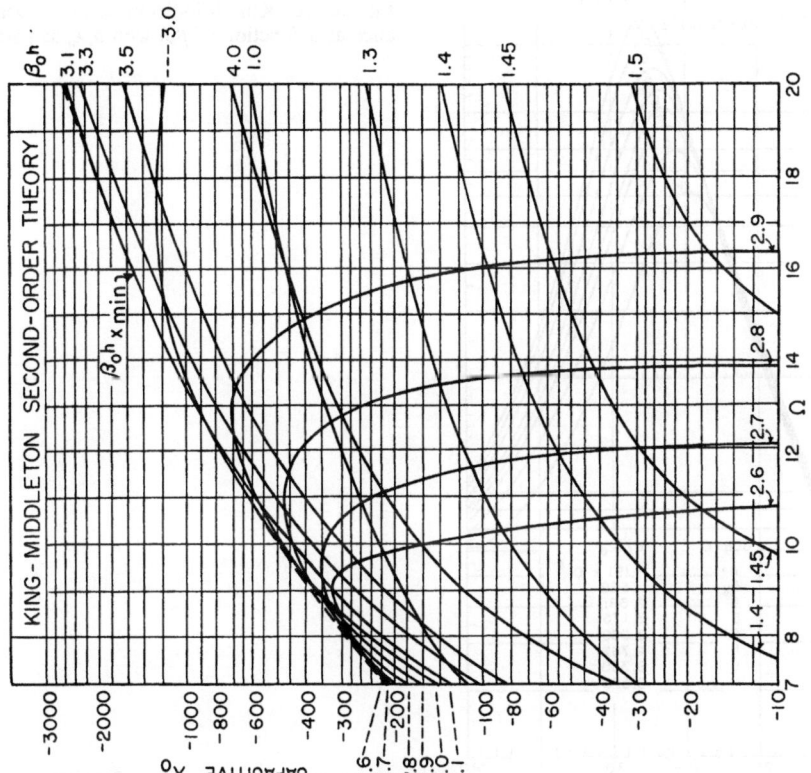

Fig. 30.10b. Capacitive reactance as a function of Ω with $\beta_0 h$ as parameter, King-Middleton second-order expansion.

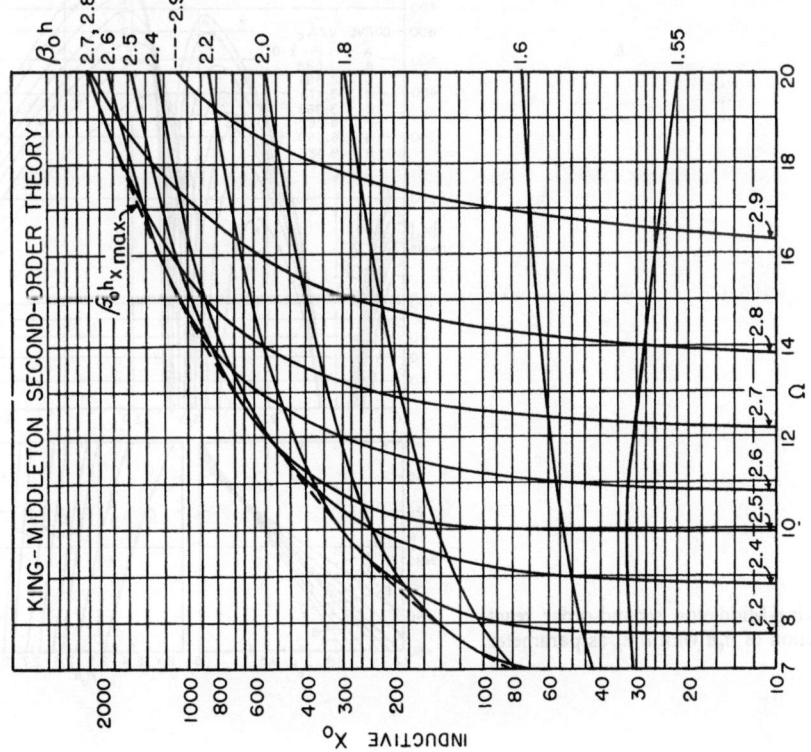

Fig. 30.10a. Inductive reactance as a function of Ω with $\beta_0 h$ as parameter, King-Middleton second-order expansion.

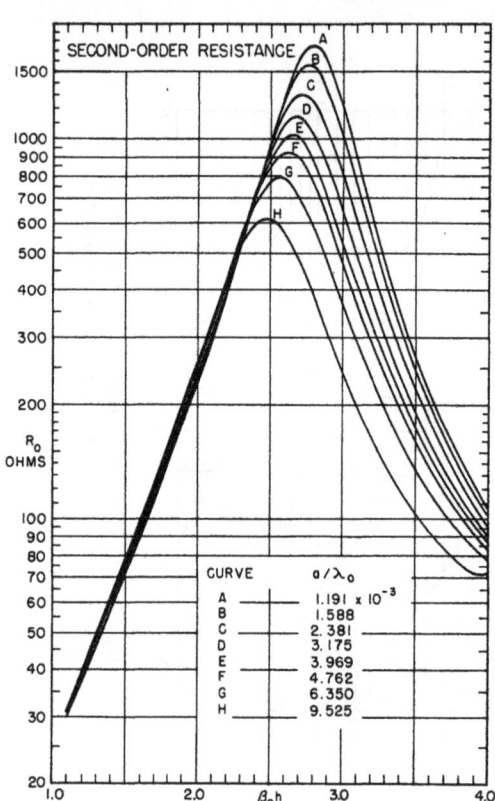

Fig. 30.11a. King-Middleton second-order resistance as a function of $\beta_0 h$ with a/λ_0 as parameter.

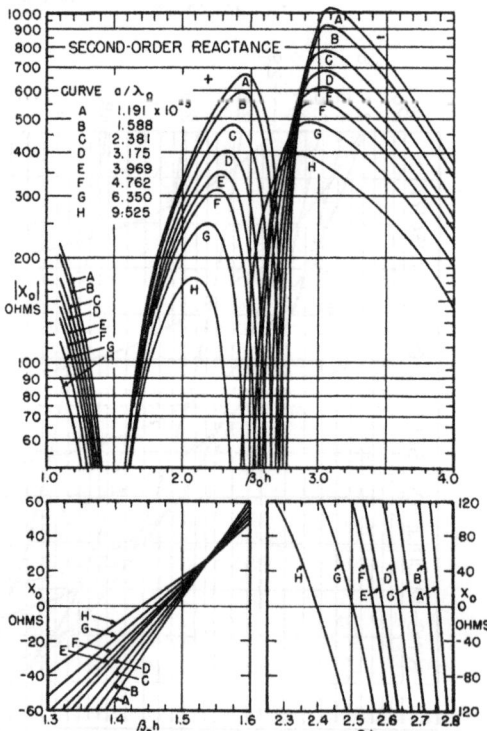

Fig. 30.11b. King-Middleton second-order reactance as a function of $\beta_0 h$ with a/λ_0 as parameter.

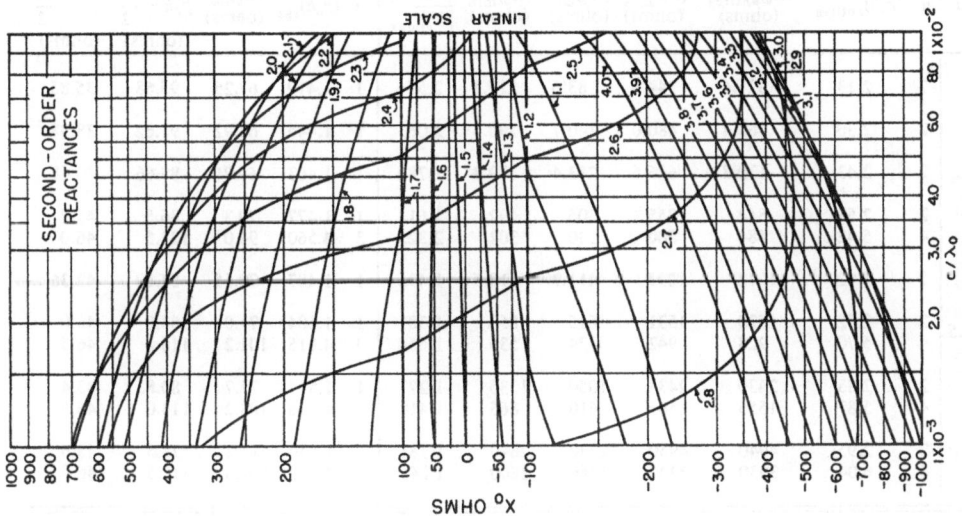

Fig. 30.12b. King-Middleton second-order reactance as a function of a/λ_0 with $\beta_0 h$ as parameter.

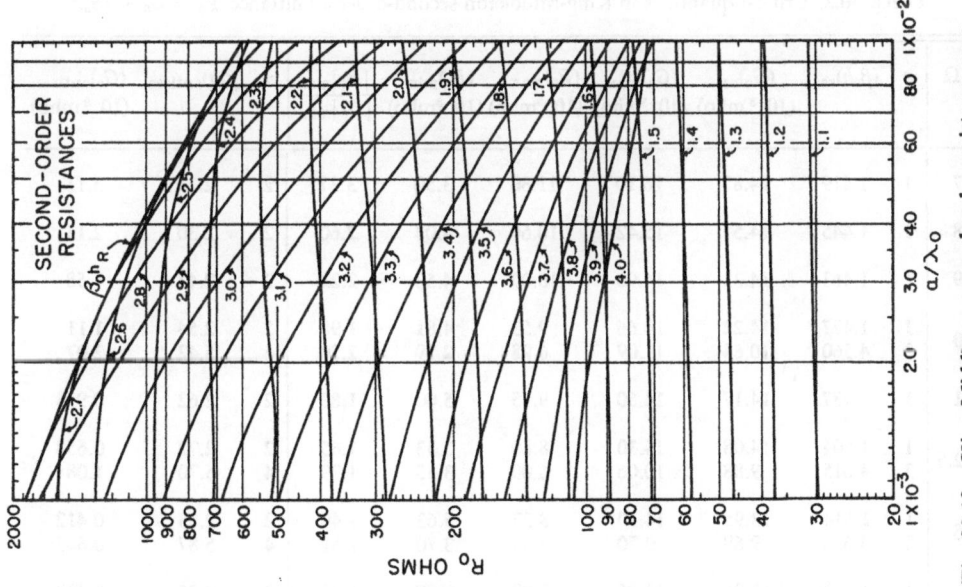

Fig. 30.12a. King-Middleton second-order resistance as a function of a/λ_0 with $\beta_0 h$ as parameter.

TABLE 30.1. Critical quantities in King-Middleton second-order impedance $Z_0 = R_0 + jX_0$.*

Ω	n	$(\beta_0 h)_{antires}$	$(R_0)_{antires}$ (ohms)	$(R_0)_{max}$ (ohms)	$(X_0)_{max}$ (ohms)	$(X_0)_{min}$ (ohms)	$\dfrac{\|X_0\|_{min}}{\|X_0\|_{max}}$	n	$(\beta_0 h)_{res}$	$(R_0)_{res}$ (ohms)	R_0 $\beta_0 h = \dfrac{n\pi}{2}$ (ohms)	X_0 $\beta_0 h = \dfrac{n\pi}{2}$ (ohms)
7	2	2.13	321.6	344.6	83.6	215.2	2.574	1	1.419	67.25	95.68	35.85
8	2	2.30	469.4	480.0	140.7	290.3	2.063	1	1.445	68.92	91.62	38.98
9	2	2.43	630.8	646.6	213.4	379.8	1.780	1	1.463	69.76	88.66	40.68
10	2	2.54	844	855	305	492	1.613	1	1.477	70.3	86.5	41.7
	4	5.42	484	500	130	302	2.325	3	4.560	94.0	122.5	46.0
11	2	2.62	1072	1095	413.7	618.5	1.495	1	1.487	70.55	84.80	42.36
12.5	2	2.72	1520	1531	615	850	1.381	1	1.504	71.0	83.0	43.0
	4	5.70	928	942	324	535	1.650	3	4.615	101.2	116.6	46.5
15	2	2.83	2430	2438	1050	1335	1.271	1	1.514	71.7	80.8	43.4
	4	5.87	1555	1565	610	865	1.418	3	4.636	103.2	114.0	46.5
20	2	2.95	4940	4950	2230	2640	1.181	1	1.530	72.2	78.5	43.6
	4	6.04	3330	3340	1446	1800	1.245	3	4.662	104.6	111.5	46.5

* Values for $\Omega = 7$, 8, and 9 are less accurate than those for the thinner antennas since the approximations of the theory are not well satisfied.

TABLE 30.2. Critical quantities in King-Middleton second-order admittance $Y_0 = G_0 + jB_0$.*

Ω	n	$(\beta_0 h)_{res}$	$(G_0)_{res}$ (10^{-3} mho)	$(G_0)_{max}$ (10^{-3} mho)	$(B_0)_{max}$ (10^{-3} mho)	$(B_0)_{min}$ (10^{-3} mho)	$\dfrac{\|B_0\|_{max}}{\|B_0\|_{min}}$	n	$(\beta_0 h)_{antires}$	$(G_0)_{antires}$ (10^{-3} mho)
7	1	1.419	14.87	16.10	11.80	3.50	3.37	2	2.13	3.11
8	1	1.445	14.51	15.42	10.60	4.08	2.60	2	2.30	2.13
9	1	1.463	14.33	14.90	10.00	4.50	2.22	2	2.43	1.58
10	1	1.477	14.22	14.66	9.53	4.81	1.98	2	2.54	1.11
	3	4.560	10.64	11.09	6.98	2.79	2.50	4	5.42	2.07
11	1	1.487	14.17	14.50	9.15	5.00	1.83	2	2.62	0.933
12.5	1	1.504	14.08	14.30	8.80	5.33	1.65	2	2.72	0.658
	3	4.615	9.88	10.06	6.00	3.35	1.79	4	5.70	1.08
15	1	1.514	13.95	14.10	8.37	5.63	1.49	2	2.83	0.412
	3	4.636	9.69	9.70	5.58	3.70	1.51	4	5.87	0.643
20	1	1.530	13.85	13.96	7.90	6.00	1.32	2	2.95	0.202
	3	4.662	9.56	9.58	5.29	4.05	1.31	4	6.04	0.300

* Values for $\Omega = 7$, 8, and 9 are less accurate than those for the thinner antennas since the approximations of the theory are not well satisfied.

TABLE 30.3. King-Middleton second-order impedances; $\Omega = 2 \ln 2h/a = 7$, $h/a \doteq 16.56$.

$\beta_0 h$	Z_0 (ohms)		$|Z_0|$ (ohms)	Y_0 (10^{-3} mho)		$|Y_0|$ (10^{-3} mho)
0.5	4.985	$-j375.6$	375.6	0.035	$+j\,2.662$	2.662
.7	10.38	$-j237.4$	237.6	.184	$+j\,4.204$	4.208
.9	18.66	$-j150.6$	151.8	.810	$+j\,6.539$	6.589
1.1	31.31	$-j\,85.40$	90.96	3.785	$+j10.32$	10.99
1.2	39.84	$-j\,56.87$	69.43	8.264	$+j11.80$	14.40
1.3	50.64	$-j\,30.05$	58.88	14.61	$+j\,8.667$	16.99
1.4	64.06	$-j\,\,4.582$	64.23	15.53	$+j\,1.111$	15.57
1.5	80.83	$+j\,19.70$	83.20	11.68	$-j\,2.846$	12.02
1.6	101.8	$+j\,42.51$	110.3	8.365	$-j\,3.493$	9.065
1.7	129.1	$+j\,63.01$	143.7	6.255	$-j\,3.053$	6.961
1.8	164.8	$+j\,78.42$	182.5	4.947	$-j\,2.354$	5.478
1.9	210.2	$+j\,83.60$	226.2	4.108	$-j\,1.634$	4.421
2.0	264.0	$+j\,69.99$	273.1	3.539	$-j\,0.938$	3.661
2.1	315.9	$+j\,27.76$	317.2	3.140	$-j\,0.276$	3.152
2.2	344.6	$-j\,44.82$	347.5	2.853	$+j\,0.371$	2.877
2.3	330.5	$-j127.0$	354.1	2.636	$+j\,1.013$	2.824
2.4	279.0	$-j187.6$	336.2	2.469	$+j\,1.660$	2.975
2.5	215.2	$-j214.4$	303.8	2.332	$+j\,2.323$	3.291
2.6	158.2	$-j215.2$	267.1	2.218	$+j\,3.016$	3.744
2.7	114.8	$-j202.5$	232.8	2.119	$-j\,3.738$	4.296
2.8	83.36	$-j184.6$	202.6	2.032	$+j\,4.499$	4.937
2.9	61.37	$-j165.9$	176.9	1.961	$+j\,5.302$	5.653
3.0	46.08	$-j148.1$	155.1	1.915	$+j\,6.155$	6.446
3.1	35.77	$-j132.0$	136.7	1.913	$+j\,7.059$	7.314
3.2	28.95	$-j117.8$	121.3	1.968	$+j\,8.008$	8.246
3.3	24.72	$-j105.0$	107.9	2.124	$+j\,9.021$	9.268
3.4	22.43	$-j\,93.61$	96.26	2.421	$+j10.10$	10.39
3.5	21.54	$-j\,83.35$	86.09	2.906	$+j11.25$	11.62
3.6	21.82	$-j\,73.79$	76.95	3.685	$+j12.46$	12.99
3.7	22.96	$-j\,64.61$	68.57	4.883	$+j13.74$	14.58
3.8	24.87	$-j\,55.50$	60.82	6.723	$+j15.00$	16.44
3.9	27.61	$-j\,46.21$	53.83	9.527	$+j15.94$	18.57
4.0	31.20	$-j\,36.40$	47.94	13.58	$+j15.84$	20.86

a good approximation without correction both of G_δ and of G_{sa}. However, at its maximum value near $\beta_0 h = n\pi/2$, n odd, G_0 is quite insensitive to changes in Ω, so that $(G_0)_{max}$ is not nearly so critical a test of the agreement between experiment and theory as is $(R_0)_{max}$. Similarly, $(\beta_0 h)_{res}$ or $\beta_0 h$ for maximum G_0 (which differ only slightly) are much less sensitive to changes in Ω than $(\beta_0 h)_{antires}$, which is also very nearly $\beta_0 h$ for maximum resistance.

The plotting of impedance and admittance curves as functions of $\beta_0 h = \omega h/v_0$ with $\Omega = 2 \ln 2h/a$ as a parameter is convenient theoretically and also for engineering application in conjunction with a fixed installation for which only the frequency is varied. For laboratory experiments, on the other hand, where precision measurements are attempted, the most convenient arrangement is to maintain a constant frequency and a fixed apparatus except for the half-length h, which is varied in discrete steps. With h as the only variable, $\beta_0 a$ or a/λ_0 is the constant parameter, rather than Ω or h/a. Table 30.12 gives Z_0 as a function of $\beta_0 h$ with a/λ_0 as parameter. Curves of R_0 and X_0 plotted against $\beta_0 h$ with a/λ_0 as parameter are shown in Figs. 30.11a, b,

TABLE 30.4. King-Middleton second-order impedances; $\Omega = 2 \ln 2h/a = 8$, $h/a \doteq 27.30$.

$\beta_0 h$	Z_0 (ohms)	$|Z_0|$ (ohms)	Y_0 (10^{-3} mho)	$|Y_0|$ (10^{-3} mho)
0.5	4.984 $-j$484.5	484.6	0.021 $+j$ 2.063	2.063
.7	10.30 $-j$308.4	308.6	.108 $+j$ 3.239	3.240
.9	18.49 $-j$198.2	199.0	.467 $+j$ 5.002	5.024
1.1	30.70 $-j$116.0	120.0	2.132 $+j$ 8.056	8.333
1.2	38.94 $-j$ 80.11	89.07	4.908 $+j$10.10	11.23
1.3	49.14 $-j$ 46.43	67.60	10.75 $+j$10.16	14.79
1.4	61.83 $-j$ 14.11	63.42	15.37 $+j$ 3.508	15.76
1.5	77.59 $+j$ 17.28	79.49	12.28 $-j$ 2.735	12.58
1.6	97.41 $+j$ 47.93	108.6	8.262 $-j$ 4.065	9.208
1.7	123.0 $+j$ 77.81	145.5	5.807 $-j$ 3.674	6.872
1.8	157.1 $+j$105.6	189.3	4.384 $-j$ 2.947	5.282
1.9	202.1 $+j$128.5	239.5	3.523 $-j$ 2.240	4.175
2.0	261.1 $+j$140.7	296.6	2.968 $-j$ 1.599	3.370
2.1	334.1 $+j$131.5	359.0	2.592 $-j$ 1.020	2.785
2.2	411.6 $+j$ 86.47	420.6	2.327 $-j$ 0.489	2.378
2.3	469.4 $-j$ 2.858	469.4	2.131 $+j$ .013	2.131
2.4	474.3 $-j$118.5	488.9	1.984 $+j$ .496	2.045
2.5	421.2 $-j$218.4	474.4	1.871 $+j$ .970	2.107
2.6	338.9 $-j$274.3	436.0	1.783 $+j$ 1.443	2.294
2.7	259.0 $-j$290.3	389.1	1.711 $+j$ 1.917	2.569
2.8	194.5 $-j$282.1	342.6	1.657 $+j$ 2.403	2.919
2.9	146.7 $-j$263.1	301.2	1.617 $+j$ 2.899	3.320
3.0	112.6 $-j$240.6	265.6	1.595 $+j$ 3.409	3.763
3.1	88.63 $-j$218.0	235.3	1.600 $+j$ 3.936	4.248
3.2	72.01 $-j$196.8	209.6	1.640 $+j$ 4.481	4.772
3.3	60.59 $-j$177.4	187.4	1.724 $+j$ 5.048	5.334
3.4	52.92 $-j$159.8	168.3	1.867 $+j$ 5.639	5.940
3.5	47.88 $-j$143.7	151.4	2.087 $+j$ 6.264	6.602
3.6	44.92 $-j$128.7	136.3	2.418 $+j$ 6.927	7.337
3.7	43.34 $-j$114.5	122.4	2.891 $+j$ 7.638	8.167
3.8	42.96 $-j$100.5	109.3	3.595 $+j$ 8.410	9.146
3.9	43.61 $-j$ 86.53	96.90	4.645 $+j$ 9.216	10.32
4.0	45.33 $-j$ 72.25	85.29	6.231 $+j$ 9.931	11.72

and of R_0 and X_0 plotted against a/λ_0 with $\beta_0 h$ as parameter in Figs. 30.12a, b.

Numerical values of $Z_0 = R_0 + jX_0$ and $Y_0 = G_0 + jB_0$ are given in Tables 30.3 to 30.7.

The impedance of a cylindrical antenna is often compared with the impedance of an appropriately end-loaded section of two-wire line, or a section of line made of resistance wire. Such comparisons are interesting, and by judicious adjustment of the end load or of the parameters of the line, a measure of correspondence can be demonstrated.

A word is in order regarding the finite ohmic resistance of all practically available conductors. In computing impedance, terms involving z^i, the internal impedance per unit length, have been omitted. This is necessary if curves of general significance are to be obtained, since terms with z^i as a factor contain the frequency under a radical in conjunction with the radius a of the conductor. In order to obtain a definite picture of the significance of ohmic resistance, computations for Z_0 were made for two particular antennas of copper. Since it is merely a question of relative order of magnitude, the impedances were calculated only in a first-order approximation, using, for simplicity, the expansion parameter Ω of Hallén. Typical and critical

TABLE 30.5. King-Middleton second-order impedances; $\Omega = 2 \ln 2h/a = 9$, $h/a \doteq 45.01$.

$\beta_0 h$	Z_0 (ohms)	$|Z_0|$ (ohms)	Y_0 (10^{-3} mho)	$|Y_0|$ (10^{-3} mho)
0.5	4.988 $-j$594.6	584.6	0.014 $+j$ 1.681	1.681
.7	10.31 $-j$379.8	379.9	.071 $+j$ 2.632	2.633
.9	18.39 $-j$245.9	246.6	.302 $+j$ 4.044	4.055
1.1	30.31 $-j$146.7	149.8	1.351 $+j$ 6.537	6.675
1.2	38.29 $-j$103.6	110.4	3.138 $+j$ 8.492	9.053
1.3	48.15 $-j$ 63.05	79.33	7.650 $+j$10.02	12.60
1.4	60.27 $-j$ 24.07	64.90	14.31 $+j$ 5.715	15.41
1.5	75.34 $+j$ 14.00	76.63	12.83 $-j$ 2.384	13.05
1.6	94.16 $+j$ 51.69	107.4	8.159 $-j$ 4.479	9.308
1.7	118.5 $+j$ 89.54	148.5	5.372 $-j$ 4.059	6.733
1.8	150.3 $+j$126.9	196.7	3.885 $-j$ 3.280	5.084
1.9	193.1 $+j$162.9	252.6	3.026 $-j$ 2.552	3.958
2.0	250.6 $+j$194.0	316.9	2.496 $-j$ 1.932	3.156
2.1	327.3 $+j$213.4	390.8	2.143 $-j$ 1.397	2.558
2.2	425.4 $+j$207.1	473.2	1.900 $-j$ 0.925	2.113
2.3	534.9 $+j$154.7	556.8	1.725 $-j$ .499	1.796
2.4	624.1 $+j$ 40.64	625.4	1.596 $-j$ .104	1.599
2.5	646.6 $-j$117.4	657.2	1.497 $+j$ .272	1.522
2.6	586.5 $-j$261.8	642.3	1.422 $+j$ .635	1.557
2.7	480.0 $-j$349.1	593.6	1.362 $+j$ .991	1.684
2.8	371.8 $-j$379.8	531.5	1.316 $+j$ 1.344	1.881
2.9	283.1 $-j$374.5	469.5	1.284 $+j$ 1.699	2.130
3.0	217.0 $-j$352.7	414.1	1.265 $+j$ 2.056	2.414
3.1	169.4 $-j$324.7	366.2	1.263 $+j$ 2.421	2.731
3.2	135.4 $-j$295.8	325.3	1.280 $+j$ 2.796	3.075
3.3	111.3 $-j$268.1	290.3	1.321 $+j$ 3.182	3.445
3.4	94.12 $-j$242.4	260.0	1.392 $+j$ 3.585	3.846
3.5	81.77 $-j$218.6	233.4	1.501 $+j$ 4.013	4.285
3.6	73.05 $-j$196.4	209.5	1.664 $+j$ 4.473	4.773
3.7	66.77 $-j$175.3	187.6	1.897 $+j$ 4.981	5.330
3.8	62.46 $-j$154.9	167.0	2.239 $+j$ 5.554	5.988
3.9	59.86 $-j$134.8	147.5	2.752 $+j$ 6.198	6.782
4.0	58.68 $-j$114.5	128.6	3.546 $+j$ 6.918	7.774

results are listed in Table 30.13. They are necessarily only approximate, but the orders of magnitude are correct. It is seen that numerical values for each antenna with z^i for copper and for $z^i = 0$ differ by a negligible amount. This justifies the statement repeatedly made that, for highly conducting antennas, ohmic resistance may be neglected.

For some purposes, it is convenient to specify a *quality factor* Q for an antenna near resonance. This is readily done if Q is defined by

$$Q_r = \frac{f_m}{\Delta f}, \qquad (14a)$$

where f_m is the frequency at which maximum power is supplied to the antenna, $\Delta f \equiv f_2 - f_1$, and f_2 and f_1 are the frequencies at which the power to the antenna is reduced to one-half the maximum at constant voltage V_0. Multiplication of numerator and denominator in (14a) by $2\pi h/v_0$, where h is the fixed half-length of the antenna and $v_0 = 3 \times 10^8$ m/sec, gives

$$Q_r = \frac{(\beta_0 h)_m}{\Delta(\beta_0 h)}. \qquad (14b)$$

The power supplied to the antenna at the fixed voltage V_0 is

$$P = \tfrac{1}{2} V_0 V_0^* G_0 = \tfrac{1}{2} V_0^2 G_0, \qquad (15)$$

where G_0 is the conductance. Since V_0 is

TABLE 30.6. King-Middleton second-order impedances; $\Omega = 2 \ln 2h/a = 10$, $h/a \doteq 75.206$.

$\beta_0 h$	Z_0 (ohms)		$\|Z_0\|$ (ohms)	Y_0 (10^{-3} mho)		$\|Y_0\|$ (10^{-3} mho)
0.5	4.988	$-j704.8$	704.8	0.0100	$+j1.419$	1.419
.7	10.30	$-j451.3$	451.5	.0506	$+j2.214$	2.215
.9	18.29	$-j293.8$	294.4	.2110	$+j3.390$	3.397
1.1	30.02	$-j177.4$	179.9	.9277	$+j5.481$	5.559
1.2	37.84	$-j127.1$	132.6	2.153	$+j7.229$	7.543
1.3	47.41	$-j\ 79.76$	92.79	5.507	$+j9.264$	10.78
1.4	59.15	$-j\ 34.27$	68.36	12.66	$+j7.333$	14.63
1.5	73.65	$+j\ 10.30$	74.37	13.32	$-j1.862$	13.45
1.6	91.73	$+j\ 54.72$	106.8	8.040	$-j4.797$	9.363
1.7	114.8	$+j\ 99.67$	151.3	4.967	$-j4.310$	6.608
1.8	145.2	$+j145.5$	205.5	3.437	$-j3.444$	4.865
1.9	185.5	$+j191.8$	266.8	2.606	$-j2.693$	3.748
2.0	240.2	$+j237.1$	337.5	2.109	$-j2.082$	2.963
2.1	314.8	$+j277.6$	419.8	1.787	$-j1.576$	2.382
2.2	416.1	$+j303.8$	515.2	1.568	$-j1.145$	1.941
2.3	547.5	$+j296.8$	622.8	1.412	$-j0.7652$	1.606
2.4	697.1	$+j227.3$	733.3	1.297	$-j\ .4228$	1.364
2.5	820.4	$+j\ 71.37$	823.6	1.210	$-j\ .1052$	1.214
2.6	849.9	$-j145.6$	862.4	1.143	$+j\ .1958$	1.160
2.7	765.5	$-j339.9$	837.5	1.091	$+j\ .4846$	1.194
2.8	622.2	$-j452.6$	769.4	1.051	$+j\ .7646$	1.300
2.9	479.4	$-j489.4$	685.1	1.021	$+j1.043$	1.460
3.0	364.5	$-j480.4$	603.1	1.002	$+j1.321$	1.658
3.1	280.0	$-j450.6$	530.5	0.9950	$+j1.601$	1.885
3.2	219.6	$-j414.0$	468.6	.9999	$+j1.885$	2.134
3.3	175.6	$-j376.2$	415.2	1.019	$+j2.182$	2.409
3.4	143.8	$-j339.9$	369.1	1.056	$+j2.495$	2.709
3.5	120.9	$-j306.1$	329.1	1.116	$+j2.826$	3.038
3.6	103.9	$-j274.5$	293.5	1.207	$+j3.186$	3.407
3.7	91.22	$-j244.7$	261.2	1.337	$+j3.588$	3.829
3.8	81.84	$-j216.4$	231.3	1.530	$+j4.044$	4.323
3.9	75.09	$-j188.4$	202.8	1.825	$+j4.580$	4.930
4.0	70.61	$-j160.8$	175.6	2.289	$+j5.213$	5.693
4.1	68.42	$-j133.3$	149.8	3.049	$+j5.938$	6.676
4.2	68.61	$-j105.4$	125.8	4.337	$+j6.664$	7.951
4.3	71.37	$-j\ 76.93$	104.9	6.481	$+j6.986$	9.529
4.4	77.17	$-j\ 47.77$	90.76	9.368	$+j5.800$	11.02
4.5	86.62	$-j\ 17.91$	88.45	11.07	$+j2.289$	11.31
4.6	100.6	$+j\ 12.49$	101.4	9.791	$-j1.216$	9.867
4.7	120.2	$+j\ 43.02$	127.6	7.376	$-j2.641$	7.834
4.8	147.0	$+j\ 72.76$	164.0	5.464	$-j2.705$	6.097
4.9	182.8	$+j\ 99.87$	208.3	4.213	$-j2.302$	4.801
5.0	229.5	$+j121.0$	259.4	3.410	$-j1.798$	3.855
5.1	287.9	$+j130.5$	316.1	2.882	$-j1.306$	3.164
5.2	356.3	$+j120.5$	376.0	2.519	$-j0.8516$	2.660
5.3	425.9	$+j\ 82.34$	433.8	2.263	$-j\ .4376$	2.305
5.4	480.6	$+j\ 12.16$	480.8	2.079	$-j\ .0526$	2.080
5.5	502.0	$-j\ 79.40$	508.2	1.944	$+j\ .3074$	1.968
5.6	483.1	$-j170.1$	512.2	1.842	$+j\ .6486$	1.952
5.7	433.4	$-j240.3$	495.5	1.765	$+j\ .9786$	2.018
5.8	370.0	$-j282.3$	465.4	1.708	$+j1.303$	2.148
5.9	307.5	$-j299.7$	429.4	1.668	$+j1.626$	2.329
6.0	253.0	$-j300.1$	392.5	1.642	$+j1.948$	2.548
6.1	208.4	$-j290.3$	357.4	1.632	$+j2.273$	2.798
6.2	173.2	$-j275.2$	325.2	1.639	$+j2.602$	3.075
6.3	145.8	$-j257.4$	295.8	1.666	$+j2.941$	3.380
6.4	124.7	$-j238.8$	269.4	1.719	$+j3.290$	3.712
6.5	108.8	$-j220.4$	245.8	1.802	$+j3.648$	4.069
6.6	96.66	$-j202.2$	224.1	1.924	$+j4.026$	4.462
6.7	87.46	$-j184.4$	204.1	2.099	$+j4.427$	4.899
6.8	80.60	$-j167.0$	185.4	3.345	$+j4.857$	5.394
6.9	75.60	$-j149.6$	167.6	2.690	$+j5.324$	5.965
7.0	72.19	$-j132.0$	150.5	3.188	$+j5.830$	6.645

TABLE 30.7. King-Middleton second-order impedances; $\Omega = 2 \ln 2h/a = 11$, $h/a \doteq 122.4$.

$\beta_0 h$	Z_0 (ohms)		$|Z_0|$ (ohms)	Y_0 (10^{-3} mho)		$|Y_0|$ (10^{-3} mho)
0.5	4.977	$-j815.1$	815.1	0.007	$+j1.227$	1.227
.7	10.30	$-j522.9$	523.0	.038	$+j1.912$	1.912
.9	18.24	$-j341.7$	342.2	.156	$+j2.918$	2.922
1.1	29.81	$-j208.3$	210.4	.673	$+j4.704$	4.752
1.2	37.51	$-j150.5$	155.1	1.559	$+j6.255$	6.447
1.3	46.87	$-j\ 96.47$	107.2	4.076	$+j8.389$	9.327
1.4	58.31	$-j\ 44.56$	73.40	10.83	$+j8.273$	13.63
1.5	72.38	$+j\ 6.378$	72.66	13.71	$-j1.208$	13.76
1.6	89.92	$+j\ 57.21$	106.6	7.915	$-j5.036$	9.381
1.7	112.0	$+j109.0$	156.3	4.586	$-j4.463$	6.399
1.8	141.1	$+j162.3$	215.0	3.051	$-j3.509$	4.650
1.9	179.2	$+j217.6$	281.9	2.255	$-j2.738$	3.547
2.0	231.0	$+j274.6$	358.9	1.793	$-j2.132$	2.786
2.1	302.2	$+j331.2$	448.3	1.503	$-j1.648$	2.231
2.2	400.7	$+j382.2$	553.7	1.307	$-j1.246$	1.806
2.3	537.6	$+j413.7$	678.4	1.168	$-j0.899$	1.474
2.4	715.9	$+j398.1$	819.1	1.067	$-j\ .593$	1.221
2.5	917.1	$+j292.0$	962.4	0.990	$-j0.315$	1.039
2.6	1070.	$+j\ 64.78$	1072.	.931	$-j\ .056$	0.933
2.7	1080.	$-j230.7$	1104.	.885	$+j\ .189$	.905
2.8	939.9	$-j471.5$	1052.	.850	$+j\ .426$	.951
2.9	740.9	$-j592.2$	948.5	.824	$+j\ .658$	1.054
3.0	560.2	$-j618.5$	834.5	.804	$+j\ .888$	1.198
3.1	421.6	$-j594.4$	728.8	.794	$+j1.119$	1.372
3.2	321.5	$-j549.8$	636.9	.793	$+j1.356$	1.571
3.3	250.6	$-j500.0$	559.3	.801	$+j1.598$	1.787
3.4	199.9	$-j451.3$	493.6	.821	$+j1.853$	2.027
3.5	163.0	$-j405.1$	436.7	.855	$+j2.124$	2.289
3.6	135.7	$-j361.4$	386.0	.911	$+j2.425$	2.590
3.7	115.5	$-j321.0$	341.2	.992	$+j2.758$	2.931
3.8	100.2	$-j282.7$	299.9	1.114	$+j3.142$	3.333
3.9	88.98	$-j245.9$	261.5	1.301	$+j3.596$	3.776
4.0	81.06	$-j209.8$	224.9	1.602	$+j4.147$	4.445

constant, the power varies linearly with G_0. Hence Q in (14b) may be evaluated by determining the value of $\beta_0 h$ for which G_0 is maximum and the difference between the values of $\beta_0 h$ for which G_0 is one-half the maximum. This may be done using Figs. 30.2a–d.

An alternative and essentially equivalent definition* of Q_r that is convenient when only the resistance and reactance curves are available is the following:

$$Q_r = \left(\frac{\beta_0 h}{2R} \frac{dX}{d\beta_0 h}\right)_{res} \quad (16)$$

* Suggested by P. Le Corbeiller.

This definition is obtained by analogy with a lumped-constant series LRC circuit for which

$$X = \omega L - \frac{1}{\omega C}, \quad (17)$$

$$\frac{dX}{d\omega} = L + \frac{1}{\omega^2 C}, \quad (18)$$

so that

$$\left(\frac{dX}{d\omega}\right)_{res} = 2L, \quad (19)$$

since resonance is defined by

$$\omega L = \frac{1}{\omega C}. \quad (20)$$

TABLE 30.8. King-Middleton second-order impedances; $\Omega = 2 \ln 2h/a = 12.5$, $h/a \doteq 259.01$.

$\beta_0 h$	Z_0 (ohms)		$|Z_0|$ (ohms)	Y_0 (10^{-3} mho)		$|Y_0|$ (10^{-3} mho)
0.5	5.005	$-j980.6$	980.6	0.0052	$+j1.020$	1.020
.7	10.29	$-j630.3$	630.4	.0259	$+j1.586$	1.586
.9	18.18	$-j413.6$	414.0	.1061	$+j2.413$	2.416
1.1	29.60	$-j254.2$	255.9	.4520	$+j3.881$	3.907
1.2	37.13	$-j185.7$	189.3	1.049	$+j5.211$	5.315
1.3	46.27	$-j121.6$	130.1	2.733	$+j7.183$	7.685
1.4	57.39	$-j\ 60.07$	83.08	8.314	$+j8.703$	12.04
1.5	71.02	$+j\ 0.2681$	71.02	14.08	$-j0.0532$	14.08
1.6	87.90	$+j\ 60.62$	106.8	7.718	$-j5.322$	9.375
1.7	109.2	$+j122.2$	163.8	4.068	$-j4.552$	6.104
1.8	136.6	$+j186.1$	230.8	2.563	$-j3.492$	4.332
1.9	172.7	$+j253.4$	306.7	1.837	$-j2.694$	3.261
2.0	221.2	$+j325.0$	393.1	1.435	$-j2.101$	2.544
2.1	287.7	$+j400.9$	493.5	1.182	$-j1.646$	2.026
2.2	381.2	$+j479.5$	612.6	1.016	$-j1.278$	1.632
2.3	514.5	$+j554.2$	756.2	0.8997	$-j0.9691$	1.322
2.4	705.0	$+j606.5$	930.1	.8150	$-j\ .7011$	1.075
2.5	965.3	$+j594.0$	1133.	.7514	$-j\ .4624$	0.8822
2.6	1268.	$+j443.0$	1344.	.7025	$-j\ .2449$	.7440
2.7	1499.	$+j\ 93.51$	1502.	.6643	$-j\ .0414$	.6656
2.8	1490.	$-j358.7$	1533	.6343	$+j\ .1527$	.6524
2.9	1248.	$-j696.0$	1429.	.6111	$+j\ .3408$	.6997
3.0	943.6	$-j835.5$	1260.	.5941	$+j\ .5260$	.7934
3.1	689.7	$-j841.4$	1088.	.5827	$+j\ .7108$	.9191
3.2	506.0	$-j787.8$	936.3	.5772	$+j\ .8986$	1.068
3.3	378.4	$-j714.7$	808.7	.5785	$+j1.093$	1.236
3.4	289.7	$-j640.3$	702.8	.5870	$+j1.297$	1.424
3.5	227.1	$-j569.4$	613.0	.6042	$+j1.515$	1.631
3.6	182.0	$-j504.4$	536.2	.6331	$+j1.754$	1.865
3.7	149.0	$-j444.2$	468.5	.6787	$+j2.023$	2.134
3.8	124.6	$-j388.3$	407.8	.7493	$+j2.335$	2.452
3.9	106.7	$-j336.0$	352.5	.8592	$+j2.704$	2.837
4.0	93.91	$-j286.1$	301.1	1.036	$+j3.155$	3.321
4.1	85.39	$-j237.9$	252.8	1.336	$+j3.723$	3.955
4.2	80.70	$-j191.0$	207.3	1.877	$+j4.443$	4.823
4.3	79.63	$-j144.7$	165.2	2.919	$+j5.304$	6.054
4.4	82.23	$-j\ 98.56$	128.3	4.993	$+j5.982$	7.792
4.5	88.77	$-j\ 52.14$	103.0	8.375	$+j4.919$	9.713
4.6	99.77	$-j\ 5.109$	99.90	9.997	$+j0.5119$	10.01
4.7	116.0	$+j\ 42.88$	123.7	7.584	$-j2.802$	8.085
4.8	138.8	$+j\ 91.94$	166.5	5.006	$-j3.316$	6.005
4.9	169.9	$+j141.9$	221.4	3.466	$-j2.895$	4.516
5.0	211.8	$+j192.2$	286.1	2.589	$-j2.348$	3.495
5.1	268.1	$+j241.1$	360.6	2.062	$-j1.854$	2.773
5.2	342.9	$+j284.5$	445.6	1.727	$-j1.433$	2.244
5.3	440.5	$+j315.0$	541.5	1.502	$-j1.074$	1.846
5.4	562.7	$+j319.5$	647.1	1.344	$-j0.7629$	1.545
5.5	703.2	$+j278.7$	756.4	1.229	$-j\ .4871$	1.322
5.6	838.4	$+j174.0$	856.3	1.143	$-j\ .2373$	1.168
5.7	927.2	$+j\ 5.512$	927.2	1.078	$-j\ .0064$	1.078
5.8	933.5	$-j191.0$	952.8	1.029	$+j\ .2107$	1.050
5.9	856.6	$-j361.6$	929.8	0.9908	$+j\ .4183$	1.075
6.0	734.5	$-j472.9$	873.6	.9625	$+j\ .6197$	1.145
6.1	605.1	$-j525.0$	801.1	.9429	$+j\ .8180$	1.248
6.2	490.3	$-j534.5$	725.3	.9317	$+j1.016$	1.378
6.3	396.7	$-j519.1$	653.3	.9293	$+j1.216$	1.530
6.4	323.1	$-j490.5$	587.3	.9367	$+j1.422$	1.703
6.5	266.3	$-j455.9$	527.9	.9554	$+j1.636$	1.894
6.6	222.1	$-j419.3$	474.5	.9869	$+j1.862$	2.108
6.7	187.7	$-j382.2$	425.8	1.035	$+j2.108$	2.348
6.8	160.7	$-j345.5$	381.0	1.107	$+j2.379$	2.624
6.9	139.6	$-j309.5$	339.5	1.211	$+j2.685$	2.945
7.0	123.1	$-j273.7$	300.1	1.367	$+j3.039$	3.333

TABLE 30.9. King-Middleton second-order impedances; $\Omega = 2 \ln 2h/a = 15$, $h/a \doteq 904.02$.

$\beta_0 h$	Z_0 (ohms)		$|Z_0|$ (ohms)	Y_0 (10^{-3} mho)		$|Y_0|$ (10^{-3} mho)
0.5	5.000	$-j$1256.	1256.	0.0032	$+j$0.7960	0.7960
.7	10.28	$-j$ 809.3	809.3	.0157	$+j$1.236	1.236
.9	18.13	$-j$ 533.2	533.5	.0637	$+j$1.873	1.874
1.1	29.36	$-j$ 330.9	332.2	.2660	$+j$2.998	3.010
1.2	36.73	$-j$ 244.4	247.1	.6014	$+j$4.001	4.046
1.3	45.62	$-j$ 163.5	169.7	1.584	$+j$5.676	5.892
1.4	56.38	$-j$ 86.00	102.8	5.331	$+j$8.132	9.724
1.5	69.46	$-j$ 10.23	70.21	14.09	$+j$2.076	14.24
1.6	85.53	$+j$ 65.50	107.7	7.369	$-j$5.644	9.282
1.7	105.7	$+j$ 142.8	177.7	3.348	$-j$4.525	5.629
1.8	131.5	$+j$ 223.4	259.2	1.956	$-j$3.324	3.857
1.9	165.0	$+j$ 309.1	350.3	1.344	$-j$2.518	2.854
2.0	209.4	$+j$ 401.7	453.0	1.021	$-j$1.957	2.208
2.1	269.8	$+j$ 503.3	571.0	0.8273	$-j$1.543	1.751
2.2	354.8	$+j$ 616.3	711.4	.7016	$-j$1.219	1.406
2.3	476.3	$+j$ 740.3	880.3	.6146	$-j$0.9553	1.136
2.4	656.3	$+j$ 871.2	1091.	.5518	$-j$ .7337	0.9180
2.5	929.4	$+j$ 988.5	1353.	.5047	$-j$ .5397	.7389
2.6	1330.	$+j$1045.	1691.	.4684	$-j$ .3639	.5932
2.7	1878.	$+j$ 860.4	2066.	.4401	$-j$ .2017	.4841
2.8	2361.	$+j$ 278.9	2378.	.4177	$-j$ .0493	.4206
2.9	2362.	$-j$ 570.8	2430.	.4000	$+j$ .0967	.4115
3.0	1870.	$-j$1159.	2200.	.3864	$+j$ .2393	.4545
3.1	1312.	$-j$1328.	1705.	.3765	$+j$ .3809	.5356
3.2	898.8	$-j$1273.	1558.	.3702	$+j$ .5242	.6417
3.3	626.4	$-j$1146.	1306.	.3674	$+j$ .6720	.7659
3.4	449.8	$-j$1008.	1104.	.3689	$+j$ .8271	.9050
3.5	332.8	$-j$ 881.1	941.9	.3752	$+j$ .9932	1.062
3.6	253.2	$-j$ 767.3	808.0	.3878	$+j$1.175	1.238
3.7	197.6	$-j$ 666.1	694.8	.4093	$+j$1.380	1.439
3.8	158.1	$-j$ 574.7	596.0	.4450	$+j$1.616	1.676
3.9	129.9	$-j$ 492.9	509.8	.4998	$+j$1.897	1.962
4.0	110.0	$-j$ 417.0	431.2	.5913	$+j$2.242	2.319
4.1	97.25	$-j$ 345.9	359.3	.7380	$+j$2.646	2.747
4.2	89.70	$-j$ 278.2	292.3	1.036	$+j$3.267	3.427
4.3	85.53	$-j$ 212.5	229.1	1.630	$+j$4.049	4.365
4.4	85.48	$-j$ 148.5	171.3	2.881	$+j$5.056	5.819
4.5	89.91	$-j$ 85.51	124.1	5.837	$+j$5.509	8.026
4.6	99.14	$-j$ 22.67	101.7	9.537	$+j$2.218	9.792
4.7	113.5	$+j$ 41.25	120.7	7.784	$-j$2.830	8.282
4.8	133.9	$+j$ 107.2	171.5	4.551	$-j$3.644	5.830
4.9	161.9	$+j$ 175.7	238.9	2.840	$-j$3.079	4.189
5.0	199.8	$+j$ 247.1	316.4	1.979	$-j$2.447	3.147
5.1	250.9	$+j$ 321.7	408.0	1.507	$-j$1.933	2.451
5.2	320.2	$+j$ 399.2	511.7	1.223	$-j$1.524	1.954
5.3	414.5	$+j$ 476.5	631.6	1.037	$-j$1.195	1.583
5.4	543.5	$+j$ 547.5	771.5	0.9132	$-j$0.9199	1.296
5.5	717.7	$+j$ 597.1	933.6	.8234	$-j$ .6850	1.071
5.6	944.4	$+j$ 596.8	1117.	.7566	$-j$ .4782	0.8951
5.7	1210.	$+j$ 500.0	1309.	.7060	$-j$ .2919	.7640
5.8	1452.	$+j$ 262.2	1475.	.6672	$-j$ .1205	.6780
5.9	1564.	$-j$ 99.34	1567.	.6368	$+j$ .0404	.6381
6.0	1481.	$-j$ 469.4	1554.	.6135	$+j$ .1944	.6436
6.1	1258.	$-j$ 726.5	1453.	.5961	$+j$ .3443	.6884
6.2	1001.	$-j$ 843.2	1308.	.5839	$+j$ .4925	.7639
6.3	775.0	$-j$ 862.1	1159.	.5767	$+j$ .6415	.8626
6.4	598.4	$-j$ 826.7	1020.	.5746	$+j$ .7937	.9798
6.5	465.6	$-j$ 766.8	897.1	.5784	$+j$ .9524	1.114
6.6	367.4	$-j$ 698.8	789.5	.5893	$+j$1.121	1.266
6.7	293.8	$-j$ 629.8	695.0	.6083	$+j$1.304	1.439
6.8	238.5	$-j$ 562.5	611.0	.6389	$+j$1.507	1.637
6.9	196.6	$-j$ 497.9	535.3	.6860	$+j$1.738	1.868
7.0	164.7	$-j$ 436.0	466.0	.7584	$+j$2.007	2.146

TABLE 30.10. King-Middleton second-order impedances; $\Omega = 2 \ln 2h/a = 20$, $h/a \doteq 11013$.

$\beta_0 h$	Z_0 (ohms)		$|Z_0|$ (ohms)	Y_0 (10^{-3} mho)		$|Y_0|$ (10^{-3} mho)
0.5	5.030	$-j1809.$	1809.	0.0015	$+j0.5527$	0.5527
.9	18.93	$-j\ 885.4$	885.6	.0241	$+j1.129$	1.129
1.3	44.92	$-j\ 247.3$	251.3	.7110	$+j3.914$	3.978
1.4	55.22	$-j\ 138.0$	148.6	2.499	$+j6.246$	6.728
1.5	67.65	$-j\ \ 31.59$	74.66	12.14	$+j5.667$	13.40
1.6	82.98	$+j\ \ 74.58$	111.6	6.666	$-j5.991$	8.963
1.7	101.9	$+j\ 182.9$	209.4	2.325	$-j4.172$	4.776
1.8	125.3	$+j\ 295.1$	320.6	1.219	$-j2.871$	2.022
1.9	155.5	$+j\ 414.3$	442.5	0.7941	$-j2.116$	2.260
2.0	195.3	$+j\ 544.3$	578.3	.5840	$-j1.628$	1.729
2.1	247.7	$+j\ 688.9$	732.1	.4537	$-j1.285$	1.363
2.2	320.0	$+j\ 853.0$	911.0	.3855	$-j1.028$	1.098
2.3	432.1	$+j1043.$	1129.	.3338	$-j0.8231$	0.8857
2.4	576.0	$+j1269.$	1394.	.2966	$-j\ .6543$	.7174
2.5	815.1	$+j1538.$	1741.	.2689	$-j\ .5075$	.5744
2.6	1210.	$+j1850.$	2211.	.2475	$-j\ .3786$	.4523
2.7	1895.	$+j2150.$	2866.	.2307	$-j\ .2617$	.3488
2.8	3075.	$+j2165.$	3761.	.2174	$-j\ .1531$	.2659
2.9	4567.	$+j1104.$	4699.	.2068	$-j\ .0500$	.2128
3.0	4790.	$-j1191.$	4888.	.1984	$+j\ .0498$	.2046
3.1	3262.	$-j2521.$	4123.	.1919	$+j\ .1483$	.2425
3.2	1944.	$-j2570.$	3223.	.1872	$+j\ .2475$	.3103
3.3	1182.	$-j2240.$	2533.	.1842	$+j\ .3493$	.3949
3.4	759.3	$-j1890.$	2037.	.1831	$+j\ .4557$	.4911
3.5	514.3	$-j1591.$	1671.	.1892	$+j\ .5693$	.5999
3.6	364.1	$-j1343.$	1392.	.1880	$+j\ .6935$	.7185
3.7	267.5	$-j1138.$	1170.	.1956	$+j\ .8325$	.8552
3.8	203.1	$-j\ 964.9$	986.2	.2089	$+j\ .9924$	1.014
3.9	159.4	$-j\ 814.9$	830.4	.2311	$+j1.182$	1.204
4.0	130.4	$-j\ 682.2$	694.5	.2707	$+j1.414$	1.440
4.1	109.3	$-j\ 562.4$	573.1	.3329	$+j1.713$	1.745
4.2	96.62	$-j\ 452.2$	462.4	.4519	$+j2.115$	2.163
4.3	89.81	$-j\ 348.6$	360.1	.6929	$+j2.690$	2.778
4.4	88.04	$-j\ 249.7$	264.8	1.256	$+j3.562$	3.778
4.5	90.90	$-j\ 153.5$	178.4	2.846	$+j4.817$	5.596
4.6	98.33	$-j\ \ 58.47$	114.4	7.512	$+j4.467$	8.740
4.7	110.6	$+j\ \ 37.10$	116.6	8.127	$-j2.726$	8.572
4.8	128.5	$+j\ 134.8$	186.2	3.706	$-j3.887$	5.371
4.9	153.1	$+j\ 236.2$	281.6	1.932	$-j2.981$	3.552
5.0	186.2	$+j\ 343.4$	390.7	1.220	$-j2.250$	2.560
5.1	230.8	$+j\ 458.5$	513.4	0.8761	$-j1.740$	1.948
5.2	291.1	$+j\ 583.8$	652.4	.6841	$-j1.372$	1.933
5.3	373.8	$+j\ 721.8$	813.1	.5657	$-j1.092$	1.230
5.4	489.7	$+j\ 875.0$	1003.	.4871	$-j0.8703$	0.9973
5.5	656.1	$+j1043.$	1233.	.4320	$-j\ .6868$	.8114
5.6	901.4	$+j1220.$	1517.	.3917	$-j\ .5302$	.6592
5.7	1270.	$+j1379.$	1875.	.3613	$-j\ .3925$	.5334
5.8	1815.	$+j1442.$	2318.	.3378	$-j\ .2683$	.4314
5.9	2542.	$+j1222.$	2821.	.3196	$-j\ .1536$	.3545
6.0	3215.	$+j\ 447.8$	3247.	.3053	$-j\ .0455$	.3086
6.1	3273.	$-j\ 651.4$	3337.	.2940	$+j\ .0585$	.2998
6.2	2663.	$-j1497.$	3055.	.2854	$+j\ .1604$	.3274
6.3	1904.	$-j1786.$	2612.	.2793	$+j\ .2620$	.3829
6.4	1314.	$-j1744.$	2199.	.2755	$+j\ .3658$	.4581
6.5	916.4	$-j1582.$	1828.	.2743	$+j\ .4734$	.5471
6.6	656.2	$-j1395.$	1542.	.2760	$+j\ .5869$	.6485
6.7	482.1	$-j1217.$	1310.	.2811	$+j\ .7100$	.7636
6.8	363.2	$-j1056.$	1117.	.2910	$+j\ .8465$	.8951
6.9	280.2	$-j\ 912.7$	954.9	.3074	$+j1.001$	1.048
7.0	221.3	$-j\ 786.3$	816.4	.3316	$+j1.178$	1.224

TABLE 30.11. Additional values of Z_0 (ohms) near antiresonance.

$\beta_0 h$	$\Omega = 7$	$\Omega = 8$	$\Omega = 9$	$\Omega = 10$
2.00	264.0 $+j$ 69.90			
2.05	291.4 $+j$ 52.98			
2.10	315.9 $+j$ 27.76	334.1 $+j$131.5		
2.15	334.9 $-j$ 5.498	373.5 $+j$114.3		
2.20	334.6 $-j$ 44.82	411.6 $+j$ 86.47	425.4 $+j$207.1	
2.25	343.4 $-j$ 86.82	445.1 $+j$ 47.07	480.3 $+j$188.1	
2.30	330.5 $-j$127.0	469.4 $-j$ 2.858	534.9 $+j$154.7	547.5 $+j$296.8
2.35	308.0 $-j$161.3	479.9 $-j$ 59.87	584.9 $+j$105.3	621.9 $+j$271.9
2.40	279.0 $-j$187.6	474.3 $-j$118.5	624.1 $+j$ 40.64	697.1 $+j$227.3
2.45	247.1 $-j$205.0	453.8 $-j$173.1	646.1 $-j$ 36.01	766.7 $+j$160.4
2.50	215.2 $-j$214.4	421.2 $-j$218.4	646.6 $-j$117.4	820.4 $+j$ 71.37
2.55		381.5 $-j$252.2	625.3 $-j$195.3	850.3 $-j$ 33.55
2.60		338.9 $-j$274.3	586.5 $-j$261.8	849.9 $-j$145.6
2.65			535.6 $-j$313.7	819.6 $-j$250.9
2.70			480.0 $-j$349.1	765.5 $-j$339.9
2.75				696.6 $-j$407.3
2.80				622.2 $-j$452.6

$\beta_0 h$	$\Omega = 11$	$\Omega = 12.5$	$\Omega = 15$	$\Omega = 20$
2.40	715.9 $+j$398.1			
2.45	816.9 $+j$359.1			
2.50	917.1 $+j$292.0	965.3 $+j$594.0		
2.55	1007. $+j$192.4	1116. $+j$540.9		
2.60	1070. $+j$ 64.78	1268. $+j$443.0	1330. $+j$1045	
2.65	1098. $-j$ 82.37	1405. $+j$291.9	1593. $+j$ 987.8	
2.70	1080. $-j$230.7	1499. $+j$ 93.51	1878. $+j$ 860.4	1895. $+j$2150.
2.75	1025. $-j$364.6	1531. $-j$133.6	2153. $+j$ 626.0	2413. $+j$2228.
2.80	939.9 $-j$471.5	1490. $-j$358.7	2361. $+j$ 278.9	3075. $+j$2165.
2.85	842.1 $-j$546.0	1388. $-j$552.1	2438. $-j$ 166.7	3847. $+j$1834.
2.90	740.9 $-j$592.2	1248. $-j$696.0	2362. $-j$ 570.9	4567. $+j$1104.
2.95		1094. $-j$787.9	2149. $-j$ 924.4	4943. $-j$ 5.4
3.00		943.6 $-j$835.5	1870. $-j$1159.	4790. $-j$1191.
3.05			1579. $-j$1285.	4076. $-j$2071.
3.10			1313. $-j$1328.	3262. $-j$2521.
3.15				2527. $-j$2639.
3.20				1944. $-j$2570.

By definition,

$$Q_r = \left(\frac{\omega L}{R}\right)_{res} \quad (21)$$

so that with (19),

$$Q_r = \left(\frac{\omega}{2R}\frac{dX}{d\omega}\right)_{res}. \quad (22)$$

If this is accepted as the definition of Q_r, multiplication by h/v_0 gives (16) directly. The slope of the reactance curve at resonance and the values of $(\beta_0 h)_{res}$ and R_{res} are readily obtained from Figs. 5c and 7b. Approximate values of Q_r for antennas with eight values of h/a are given in Table 30.14.

The usual method of supplying power to an antenna is with a transmission line. The

TABLE 30.12. Tables of impedance of cylindrical antenna with constant radius-to-wavelength ratio; h = half-length of antenna; a = radius of antenna.

$\beta_0 h$	Z_0 (ohms)		$\|Z_0\|$ (ohms)	$\beta_0 h$	Z_0 (ohms)		$\|Z_0\|$ (ohms)
	$a/\lambda_0 = 0.001191$				$a/\lambda_0 = 0.001588$		
1.1	29.73	$-j\,219.5$	221.5	1.1	29.86	$-j201.7$	203.9
1.2	37.38	$-j\,165.7$	169.8	1.2	37.52	$-j149.8$	154.4
1.3	46.58	$-j\,108.2$	117.8	1.3	46.82	$-j\,98.8$	109.3
1.4	57.77	$-j\,53.1$	78.47	1.4	58.13	$-j\,47.2$	74.88
1.5	71.42	$+j\,2.50$	71.46	1.5	71.94	$+j\,4.8$	72.10
1.6	88.35	$+j\,54.7$	103.9	1.6	89.10	$+j\,58.5$	106.6
1.7	109.6	$+j\,120.2$	162.6	1.7	110.6	$+j115.0$	159.6
1.8	136.9	$+j\,183.8$	229.2	1.8	138.3	$+j174.9$	223.0
1.9	172.8	$+j\,252.4$	305.9	1.9	174.8	$+j239.0$	296.1
2.0	218.0	$+j\,326.9$	392.9	2.0	222.0	$+j309.0$	380.5
2.1	281	$+j\,406$	494	2.1	289	$+j382$	479
2.2	380	$+j\,495$	624	2.2	389	$+j461$	603
2.3	509	$+j\,581$	772	2.3	518	$+j533$	743
2.4	698	$+j\,658$	959	2.4	709	$+j586$	920
2.5	961	$+j\,689$	1182	2.5	960	$+j584$	1124
2.6	1298	$+j\,606$	1432	2.6	1263	$+j445$	1339
2.7	1639	$+j\,295$	1665	2.7	1518	$+j123$	1523
2.8	1760	$-j\,234$	1775	2.8	1549	$-j336$	1585
2.9	1580	$-j\,687$	1723	2.9	1338	$-j698$	1509
3.0	1234	$-j\,959$	1563	3.0	1045	$-j881$	1367
3.1	905	$-j1018$	1362	3.1	780	$-j910$	1198
3.2	654	$-j\,976$	1175	3.2	570	$-j868$	1038
3.3	480	$-j\,890$	1011	3.3	425	$-j795$	901
3.4	350	$-j\,793$	867	3.4	319	$-j713$	781
3.5	277	$-j\,714$	766	3.5	253	$-j643$	691
3.6	218	$-j\,634$	670	3.6	202	$-j573$	608
3.7	176	$-j\,558$	585	3.7	164	$-j508$	534
3.8	143	$-j\,488$	508	3.8	135	$-j445$	465
3.9	122	$-j\,422$	439	3.9	116	$-j385$	402
4.0	104	$-j\,363$	378	4.0	100	$-j333$	348
	For $\lambda_0 = 1$ m, $2a = 3/32$ in.				For $\lambda_0 = 1$ m, $2a = 1/8$ in.		
	$a/\lambda_0 = 0.002381$				$a/\lambda_0 = 0.003175$		
1.1	30.07	$-j\,176.9$	179.4	1.1	30.20	$-j159.2$	162.0
1.2	37.79	$-j\,130.8$	136.2	1.2	38.01	$-j117.4$	123.4
1.3	47.26	$-j\,85.1$	97.34	1.3	47.60	$-j\,75.5$	89.25
1.4	58.72	$-j\,38.8$	70.38	1.4	59.23	$-j\,33.3$	67.95
1.5	72.80	$+j\,7.9$	73.22	1.5	73.59	$+j\,10.2$	74.29
1.6	90.35	$+j\,56.5$	106.5	1.6	91.42	$+j\,55.1$	106.7
1.7	112.4	$+j\,107.8$	155.7	1.7	113.9	$+j102.5$	153.2
1.8	141.2	$+j\,161.7$	214.7	1.8	143.5	$+j151.4$	208.6
1.9	178.9	$+j\,219.9$	283.5	1.9	182.3	$+j204.9$	274.3
2.0	229.8	$+j\,281.0$	363.0	2.0	234.3	$+j259.0$	350.1
2.1	300	$+j\,344$	456	2.1	305	$+j315$	438
2.2	398	$+j\,408$	570	2.2	403	$+j367$	545
2.3	529	$+j\,457$	699	2.3	537	$+j400$	648
2.4	718	$+j\,476$	862	2.4	717	$+j392$	817
2.5	944	$+j\,418$	1032	2.5	921	$+j300$	969
2.6	1179	$+j\,231$	1201	2.6	1091	$+j\,90.0$	1095
2.7	1310	$-j\,86.0$	1313	2.7	1140	$-j196$	1157
2.8	1245	$-j\,426$	1316	2.8	1020	$-j464$	1121
2.9	1037	$-j\,664$	1231	2.9	845	$-j621$	1049
3.0	805	$-j\,764$	1110	3.0	656	$-j678$	943
3.1	602	$-j\,764$	973	3.1	500	$-j670$	836
3.2	454	$-j\,726$	856	3.2	389	$-j634$	744
3.3	348	$-j\,668$	753	3.3	298	$-j584$	656
3.4	271	$-j\,607$	665	3.4	240	$-j533$	584
3.5	218	$-j\,549$	591	3.5	193	$-j483$	520
3.6	178	$-j\,491$	518	3.6	160	$-j435$	463
3.7	147	$-j\,438$	462	3.7	134	$-j390$	412
3.8	124	$-j\,384$	404	3.8	114	$-j344$	362
3.9	108	$-j\,335$	352	3.9	102	$-j302$	319
4.0	93.8	$-j\,289$	304	4.0	89.7	$-j260$	275
	For $\lambda_0 = 1$ m, $2a = 3/16$ in.				For $\lambda_0 = 1$ m, $2a = 1/4$ in.		

TABLE 30.12 (cont.)

$\beta_0 h$	$a/\lambda_0 = 0.003969$			$\beta_0 h$	$a/\lambda_0 = 0.004763$		
	Z_0 (ohms)		$|Z_0|$ (ohms)		Z_0 (ohms)		$|Z_0|$ (ohms)
1.1	30.35	$-j\ 145.6$	148.7	1.1	30.48	$-j134.5$	138
1.2	38.20	$-j\ 106.8$	113.4	1.2	38.41	$-j\ 98.2$	105
1.3	47.92	$-j\ 67.7$	82.9	1.3	48.23	$-j\ 61.8$	78.4
1.4	59.72	$-j\ 28.6$	66.2	1.4	60.15	$-j\ 24.9$	65.1
1.5	74.26	$+j\ 11.9$	75.2	1.5	74.94	$+j\ 13.3$	76.1
1.6	92.37	$+j\ 53.7$	106.9	1.6	93.28	$+j\ 52.7$	107
1.7	115.3	$+j\ 97.9$	151.3	1.7	116.7	$+j\ 94.2$	150
1.8	145.4	$+j\ 144.3$	204.8	1.8	147.0	$+j138.3$	202
1.9	185.2	$+j\ 193.0$	267.5	1.9	187.7	$+j183.2$	262
2.0	238.8	$+j\ 243.2$	340.9	2.0	242.4	$+j228.8$	333
2.1	310	$+j\ 291$	425	2.1	314	$+j271$	416
2.2	409	$+j\ 332$	527	2.2	415	$+j303$	514
2.3	542	$+j\ 351$	646	2.3	545	$+j307$	626
2.4	710	$+j\ 322$	780	2.4	700	$+j255$	745
2.5	888	$+j\ 203$	911	2.5	850	$+j126$	859
2.6	1005	$-j\ 10.1$	1005	2.6	929	$-j\ 84.0$	933
2.7	1001	$-j\ 260$	1034	2.7	884	$-j303$	934
2.8	885	$-j\ 473$	1003	2.8	763	$-j469$	896
2.9	713	$-j\ 582$	920	2.9	613	$-j544$	820
3.0	555	$-j\ 615$	828	3.0	478	$-j563$	738
3.1	426	$-j\ 598$	734	3.1	370	$-j546$	660
3.2	333	$-j\ 566$	657	3.2	292	$-j514$	591
3.3	260	$-j\ 523$	584	3.3	233	$-j476$	530
3.4	214	$-j\ 478$	524	3.4	191	$-j434$	474
3.5	174	$-j\ 433$	466	3.5	159	$-j396$	427
3.6	146	$-j\ 392$	418	3.6	133	$-j359$	383
3.7	124	$-j\ 353$	374	3.7	115	$-j323$	343
3.8	108	$-j\ 313$	332	3.8	102	$-j288$	305
3.9	96.2	$-j\ 275$	291	3.9	90.7	$-j253$	269
4.0	85.6	$-j\ 236$	251	4.0	82.7	$-j218$	233
	For $\lambda_0 = 1$ m, $2a = 5/16$ in.				For $\lambda_0 = 1$ m, $2a = 3/8$ in.		

$\beta_0 h$	$a/\lambda_0 = 0.006350$			$\beta_0 h$	$a/\lambda_0 = 0.009525$		
	Z_0 (ohms)		$|Z_0|$ (ohms)		Z_0 (ohms)		$|Z_0|$ (ohms)
1.1	30.70	$-j\ 116.6$	121	1.1	31.19	$-j\ 91.4$	96.6
1.2	38.87	$-j\ 84.7$	93.2	1.2	39.48	$-j\ 65.7$	76.6
1.3	48.77	$-j\ 52.2$	71.4	1.3	49.75	$-j\ 38.7$	63.0
1.4	60.99	$-j\ 19.1$	63.9	1.4	62.37	$-j\ 11.2$	63.4
1.5	76.06	$+j\ 15.3$	77.6	1.5	78.02	$+j\ 17.6$	80.0
1.6	94.83	$+j\ 50.8$	108	1.6	97.58	$+j\ 47.7$	109
1.7	118.9	$+j\ 88.1$	148	1.7	122.6	$+j\ 78.8$	146
1.8	150.3	$+j\ 126.2$	196	1.8	155.8	$+j110.0$	191
1.9	192.3	$+j\ 165.7$	254	1.9	199.5	$+j139.3$	243
2.0	250.0	$+j\ 203.9$	323	2.0	260.0	$+j163.0$	307
2.1	320	$+j\ 234$	396	2.1	330	$+j174$	373
2.2	420	$+j\ 251$	489	2.2	421	$+j160$	450
2.3	542	$+j\ 228$	588	2.3	522	$+j104$	532
2.4	674	$+j\ 150$	690	2.4	596	$+j\ 2.50$	596
2.5	764	$+j\ 5.00$	764	2.5	612	$-j140$	628
2.6	786	$-j\ 184$	807	2.6	563	$-j264$	622
2.7	703	$-j\ 349$	785	2.7	480	$-j349$	594
2.8	598	$-j\ 446$	746	2.8	392	$-j388$	552
2.9	472	$-j\ 484$	676	2.9	310	$-j393$	500
3.0	370	$-j\ 483$	608	3.0	250	$-j377$	452
3.1	290	$-j\ 466$	549	3.1	192	$-j363$	411
3.2	233	$-j\ 437$	495	3.2	162	$-j339$	376
3.3	190	$-j\ 404$	446	3.3	136	$-j310$	338
3.4	156	$-j\ 368$	400	3.4	115	$-j285$	307
3.5	134	$-j\ 338$	364	3.5	102	$-j264$	283
3.6	115	$-j\ 308$	329	3.6	90.0	$-j240$	256
3.7	102	$-j\ 279$	297	3.7	82.5	$-j218$	233
3.8	91.3	$-j\ 248$	264	3.8	75.5	$-j196$	210
3.9	82.8	$-j\ 219$	234	3.9	71.0	$-j174$	188
4.0	77.2	$-j\ 189$	204	4.0	68.7	$-j151$	166
	For $\lambda_0 = 1$ m, $2a = 1/2$ in.				For $\lambda_0 = 1$ m, $2a = 3/4$ in.		

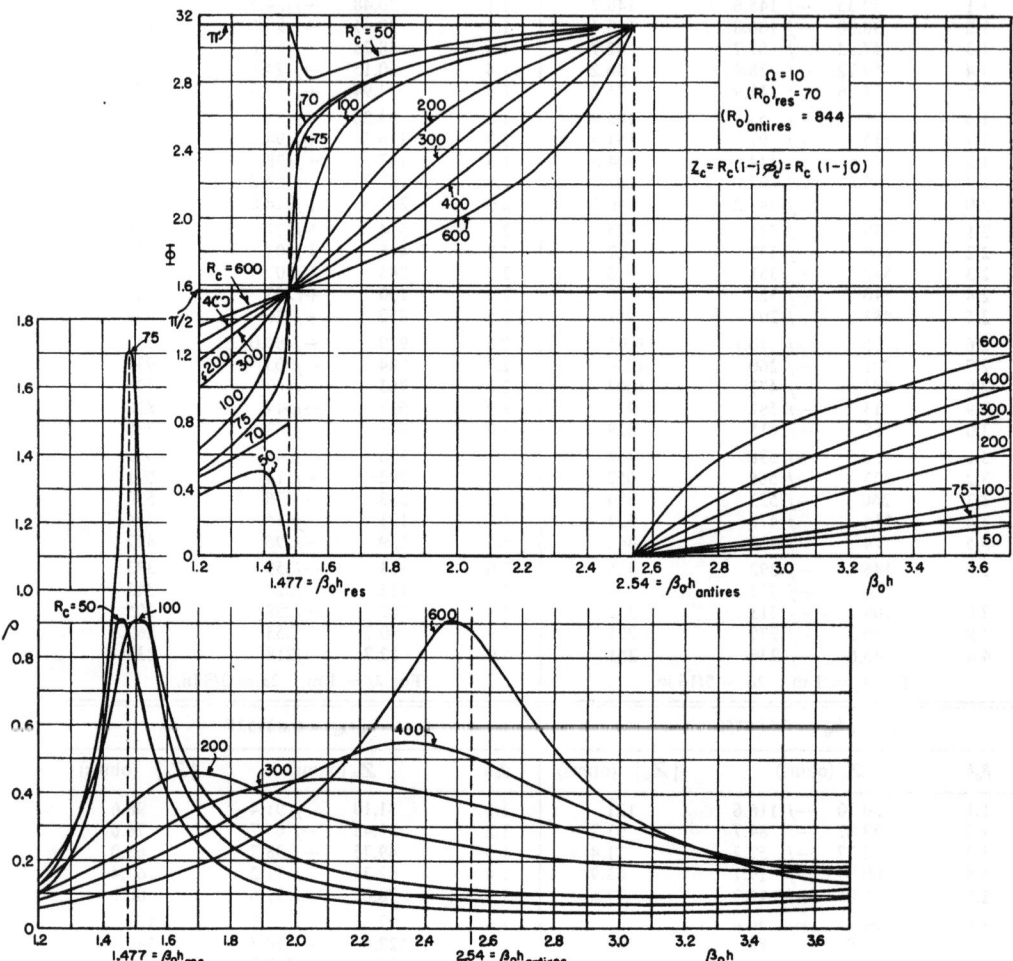

Fig. 30.13. Curves of the terminal functions ρ and Φ for the second-order impedance of the cylindrical antenna terminating a low-loss line, $\Omega = 10$. The standing-wave ratio on the line is $S = \coth \rho$.

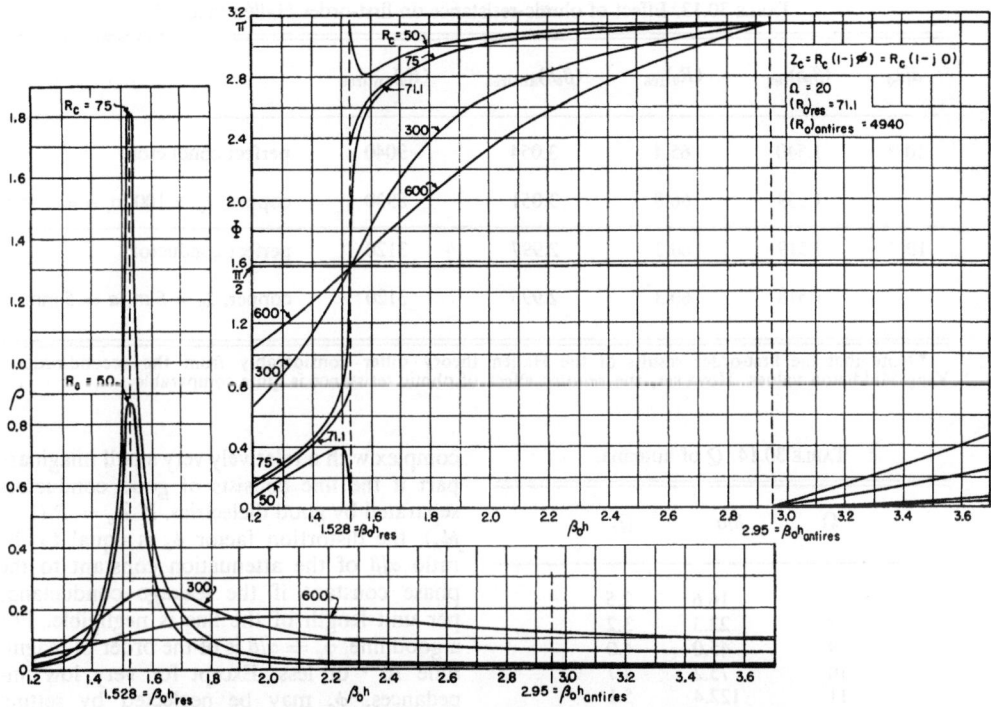

Fig. 30.14. Curves of the terminal functions ρ and Φ for the second-order impedance of the cylindrical antenna terminating a low-loss line, $\Omega = 20$. The standing-wave ratio on the line is $S = \coth \rho$.

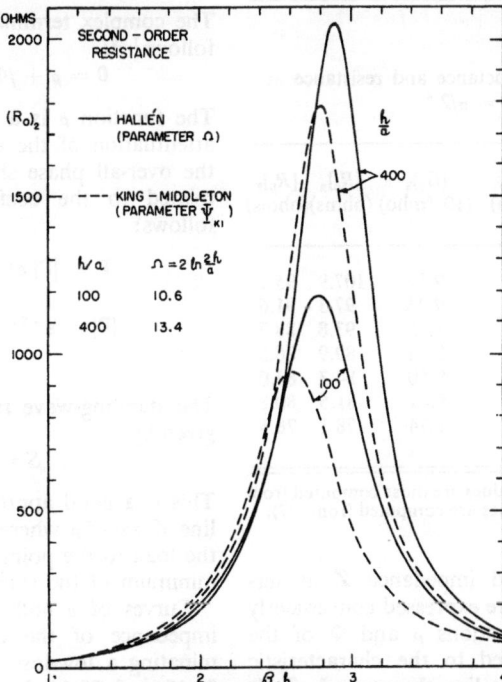

Fig. 30.15. Second-order resistance according to formulas of Hallén and of King and Middleton.

TABLE 30.13. Effect of ohmic-resistance on first-order Hallén theory.*

a/λ_0	$(\beta_0 h)_{res}$	$(R_0)_{res}$	$(\beta_0 h)_{antires}$	$(R_0)_{antires}$	
10^{-5}	1.540	65.4	3.054	9040	perfect conductor
	1.539	66.8	3.051	8930	copper, $\lambda_0 = 100$ m, $a = 1$ mm
10^{-3}	1.519	60.3	2.997	3120	perfect conductor
	1.518	60.3	2.996	3120	copper, $\lambda_0 = 5$ m, $a = 5$ mm

* Note that the first-order results of the Hallén theory differ considerably from the second-order King-Middleton values. However, the *relative* effect of ohmic resistance is quite comparable.

TABLE 30.14. Q of antenna.

Ω	h/a	Q_r
7	16.6	2.5
8	22.3	3.2
9	45.0	4.0
10	75.2	4.7
11	122.4	5.1
12.5	259.0	6.4
15	904	8.0
20	11,013	11.3

TABLE 30.15. Conductance and resistance at $\beta_0 h = \pi/2$.*

Ω	W	$[G_0]_3$ (10^{-3} mho)	$[G_0]_2$ (10^{-3} mho)	$[R_0]_3$ (ohms)	$[R_0]_2$ (ohms)
7	4.74	8.35	9.16	107.9	95.7
8	5.70	8.73	9.24	97.3	91.6
9	6.68	8.97	9.32	93.8	88.7
10	7.66	9.15	9.38	89.9	86.5
12.5	10.15	9.41	9.50	84.3	83.0
15	12.65	9.56	9.61	81.5	80.8
20	17.64	9.76	9.74	78.1	78.5

* The second-order values are those computed from (4); the third-order values are computed from (27).

properties of a load impedance Z as termination for a line are expressed conveniently by the terminal functions ρ and Φ of the impedance normalized to the characteristic impedance Z_c of the line. In general, Z_c is complex with a relatively very small imaginary part if the line consists of good conductors separated by good dielectrics. If $Z_c = R_c(1 - j\phi_c)$, the distortion factor ϕ_c is equal to the ratio α/β of the attenuation constant to the phase constant if the leakage conductance per unit length of the line is negligible. For a good line, $\phi_c \doteq \alpha/\beta$ is of the order of magnitude 10^{-3} or less. Except for very low impedances, ϕ_c may be neglected by setting $Z_c \doteq R_c$. The normalized impedance of the load Z is by definition

$$z_1 = r_1 + jx_1 = Z/Z_c \doteq Z/R_c. \quad (23)$$

The complex terminal function is defined as follows[1.32].

$$\theta = \rho + j\Phi = \coth^{-1} z_1. \quad (24)$$

The function ρ is a measure of the over-all attentuation of the load, Φ is a measure of the over-all phase shift. These functions are related to the coefficient of reflection as follows:

$$\Gamma \equiv |\Gamma| e^{j\psi} \equiv \frac{Z - Z_c}{Z + Z_c}, \quad (25a)$$

$$|\Gamma| = e^{-2\rho}, \quad \psi = -2\Phi. \quad (25b)$$

The standing-wave ratio on a lossless line is given by

$$S = \coth \rho. \quad (26)$$

This is a good approximation for a low-loss line if $\alpha w \ll \rho$ where w is the distance from the load to the point where the maximum or minimum of the standing wave is measured.

Curves of ρ and Φ for the second-order impedance of the cylindrical antenna terminating a low-loss line are shown in Figs. 30.13 and 30.14 for $\Omega = 10$ and 20 and a range of values of R_c. It is assumed that ϕ_c is negligible. The function ρ is infinite at resonance (or antiresonance) if R_c is equal to the resonant (or antiresonant) resistance. The function Φ is discontinuous at resonance (or antiresonance) by $\pi/2$ if R_c is equal to the resonant (or antiresonant) resistance. The lowest standing-wave ratio is obtained when ρ is greatest; it is essentially unity if ρ is equal to two or more, exactly unity when ρ is infinite.

A comparison of the second-order resistances $(R_0)_2$ as derived from (4), using the King-Middleton expansion parameter $\Psi = \Psi_{K1}$ and the D-factors as defined in (5a, b, c) and using the original Hallén expansion parameter $\Psi = \Omega$ with all D-factors equal to one, is shown in Fig. 30.15. It is seen that Hallén's formula leads to higher resistances near antiresonance, lower resistances near resonance, and longer resonant and antiresonant lengths than the King-Middleton formula. The reactances behave in an analogous manner. In general, the *second*-order formula of Hallén leads to results comparable to those obtained with the *first*-order King-Middleton formula. This suggests that the closer approximation of the current and the resulting more complicated expansion parameter in the King-Middleton formulation in itself represents an improvement in accuracy corresponding roughly to one order of approximation since it is verified in Sec. 34 that the King-Middleton second-order values are consistently in very good agreement with experimental results. An additional comparison of the two expansions is given in Figs. VIII–11.1 and VIII–11.2, where maximum resistances and conductances are shown together with experimental results.

From separate integral equations for the components of current in phase and in phase quadrature with the driving voltage, it is shown in Sec. 25 that the leading term in the denominator of the formula for the in-phase component I_z'' is of second order when $\beta_0 h$ is at or very near $n\pi/2$, n odd, since zeroth- and first-order terms vanish. It may be anticipated, therefore, that quantities which depend primarily or entirely on the in-phase current may be determined less accurately by a second-order formula for these values of $\beta_0 h$ than for others. Evidently, this is true particularly of the conductance G_0 which has a sharp maximum near $\beta_0 h = n\pi/2$, n odd.

It is possible to investigate the conductance at $\beta_0 h = n\pi/2$, n odd, using terms up to and including those of the third order in the small denominator by setting $z = 0$ in (25.14). The resulting expression is

$$G_0 = \frac{2\pi}{\zeta_0} \left\{ \frac{\begin{array}{c}[Fs_1 + W^{-1}(Fs_1 Gc_1 - Gs_1 Fc_1)] \\ + W^{-1}[Fc_1(0)Fs_1 - Fs_1(0)Fc_1]\end{array}}{[Fc_1^2 + Fs_1^2] + 2W^{-1}[Fc_1(Fc_2 - Fss_2) + Fs_1(Fs_2 + Fcs_2)]} \right\} \quad (27)$$

where, for $\beta_0 h = \pi/2$, using the notation of Secs. 27 and 19,

$$Gs_1 = Fc_1 = \alpha_1^{\mathrm{I}} = -\tfrac{1}{2} Si\, 2\pi = -0.709, \quad (28a)$$

$$Fs_1 = \alpha_1^{\mathrm{II}} = \tfrac{1}{2} Cin\, 2\pi = 1.219, \quad (28b)$$

$$Gc_1 = -\alpha_1^{\mathrm{II}} + 2\ln 2 = 0.1675, \quad (28c)$$

$$Fc_1(0) = W - \text{R.P. } C_a(h, 0)$$
$$= -\text{R.P. } C_a(h, h) = \alpha_1^{\mathrm{I}} = -0.709, \quad (28d)$$

$$Fs_1(0) = -\text{I.P. } C_a(h, 0) = 1.852, \quad (28e)$$

$$(Fc_2 - Fss_2) = \alpha_2^{\mathrm{I}} + (W - \Omega)\alpha_1^{\mathrm{I}}$$
$$= -3.19 + 0.709(\Omega - W), \quad (29a)$$

$$(Fs_2 + Fcs_2) = \alpha_2^{\mathrm{II}} + (W - \Omega)\alpha_1^{\mathrm{II}}$$
$$= 2.65 - 1.219(\Omega - W), \quad (29b)$$

$$W = W(0) = \text{R.P.}[C_a(h, 0) - C_a(h, h)]. \quad (30)$$

Numerical values of the expansion parameter W for a series of values of $\Omega = 2\ln(2h/a)$ are given in Table 30.15. Third-order* values of G_0 computed from (27) also are shown in this table together with the second-order values calculated from the data in Table 30.1. Since the second-order reactance is known to be quite accurate at $\beta_0 h = \pi/2$, the values of resistance corresponding to the newly determined third-order values of G_0 also may be evaluated. These are given in Table 30.15 with the previously determined second-order values. It is seen that the new and more accurate formula (27) gives values of the conductance at $\beta_0 h = \pi/2$ that are somewhat

---

* The numerator in (27) actually includes only zeroth- and first-order terms. However, since the contribution of the first-order terms is only 0.8 percent for $\Omega = 7$ and 0.2 percent for $\Omega = 20$, it is evident that contributions from second- and third-order terms must be negligible, so that the results are in effect of third-order.

lower, and of the resistance that are somewhat higher than those obtained from (4), the difference being greater for thicker antennas and relatively unimportant for $\Omega > 10$; at $\Omega = 10$ the difference is about 4 percent. It is evident from this analysis that the theoretical conductances computed from the second-order formula (4) are somewhat too high in the range of resonance and it is likely that the new values computed from (27) also are above rather than below the true value.

31. Impedance and Current for the Electrically Short Antenna

An electrically short antenna may be defined by the inequality

$$\beta_0 h = \frac{2\pi h}{\lambda_0} \leq 0.5. \tag{1}$$

For such an antenna the expansion parameter Ψ_{K1} has been shown in (20.35) to have the simple form, $\Psi_{K1} = \Omega - 2$. For the first-order impedance, $(D_1)_1 = 2 - \Omega/\Psi = (\Omega - 4)/(\Omega - 2)$. Subject to (1) the following approximate formulas are derived readily from the general expressions in Sec. 18; only leading terms are retained:

$$\alpha_1^I \doteq -(\beta_0 h)^2, \qquad \alpha_1^{II} \doteq 2(\beta_0 h)^3,$$

$$\beta_1^I \doteq 2\beta_0 h(1 + \ln 2), \qquad \beta_1^{II} \doteq (\beta_0 h)^4/3. \tag{2}$$

On expanding $\sin \beta_0 h$ and $\cos \beta_0 h$ the first-order formula for the impedance of the short antenna becomes

$$Z_0 \doteq \frac{-j\zeta_0(\Omega - 2)}{2\pi\beta_0 h}$$

$$\times \left[\frac{(\Omega - 2) + j2(\beta_0 h)^3/3}{(\Omega - 2 + 2\ln 2) + j(\beta_0 h)^3/3} \right], \tag{3}$$

so that

$$R_0 \doteq \frac{\zeta_0}{6\pi}(\beta_0 h)^2 \left[\frac{1 + \frac{4\ln 2}{\Omega - 2}}{\left(1 + \frac{2\ln 2}{\Omega - 2}\right)^2} \right], \tag{4a}$$

$$X_0 \doteq -\frac{\zeta_0}{2\pi\beta_0 h} \left[\frac{\Omega - 2}{1 + \frac{2\ln 2}{\Omega - 2}} \right]. \tag{4b}$$

For $\Omega = 10$, $R_0 \doteq 19.6\beta_0^2 h^2$ ohm;

$X_0 \doteq -60 \times 6.82/\beta_0 h = -409.2/\beta_0 h$ ohm.

Before accepting (3) and (4) as satisfactory first-order formulas, it is well to verify whether the implied requirement that Ψ_{K1} shall be the *essentially constant* value assumed by $|\Psi_{K1}(z)|$ is satisfied. This has not been verified for values of $\beta_0 h$ less than $\pi/2$. In Fig. 31.1 the function $\Psi_{K1}(\zeta) = \psi_1(\zeta)/(1-\zeta)$ with $\zeta \equiv z/h$ is shown for $\Omega = 10$. It is seen that the value $\Psi_{K1}(0) = \Omega - 2$ actually occurs at a sharp cusp and not in the middle of an essentially constant range as when $\beta_0 h = \pi/2$. This behavior might have been anticipated, since the vector potential theoretically must have a discontinuous slope at $z = 0$ with a triangular zeroth-order current. Evidently, the parameter Ψ_{K1} is far from an ideal expansion parameter for antennas electrically much shorter than $\beta_0 h = \pi/2$.

As implied with reference to $L_1(z, z')$ in (14.17), a function of the form

$$\Psi(\zeta, 1) \equiv \Psi_{K1}(\zeta) - \Psi_{K1}(1) = \frac{\psi_1(\zeta) - \psi_1(1)}{1 - \zeta}, \tag{5}$$

which is proportional to the vector potential difference between the point z on the antenna and its end at $z = h$ or $\zeta = 1$, is more nearly constant. This function is also shown in Fig. 31.1. Although not actually constant over the length of the antenna, its value at $\zeta = 0$ is a reasonable approximation of its value along most of the antenna. If this value $\Psi(0, 1) = \Psi_{K1}(0) - \Psi_{K1}(1) = \Omega - 2 - 2\ln 2 = \Omega - 3.39$ is used in place of $\Psi_{K1}(0) = \Omega - 2$ in the formula for impedance, the following first-order values are obtained:

$$Z_0 \doteq -\frac{j\zeta_0(\Omega - 2 - 2\ln 2)}{2\pi\beta_0 h}$$

$$\times \left[1 - j\frac{(\beta_0 h)^3}{3(\Omega - 2 - 2\ln 2)} \right]. \tag{6a}$$

Significantly, there is *no first-order correction* on the zeroth-order reactance in this formula, indicating that it is a very good approximation, and that second-order terms are not required. This is verified later in this section. Reactances computed from (6a) are plotted in Fig. 31.4; they are excellent approximations of the correct value for $\beta_0 h \leq 0.2$ and quite good approximations for $\beta_0 h \leq 0.5$. This is discussed later. The new resistance and reactance are, for $\Omega = 10$,

$$R_0 \doteq \frac{\zeta_0}{6\pi}(\beta_0 h)^2 = 20\beta_0^2 h^2 \text{ ohm}, \tag{6b}$$

$$X_0 \doteq \frac{-\zeta_0(\Omega - 3.39)}{2\pi\beta_0 h} = -\frac{396.8}{\beta_0 h} \text{ ohm}. \tag{6c}$$

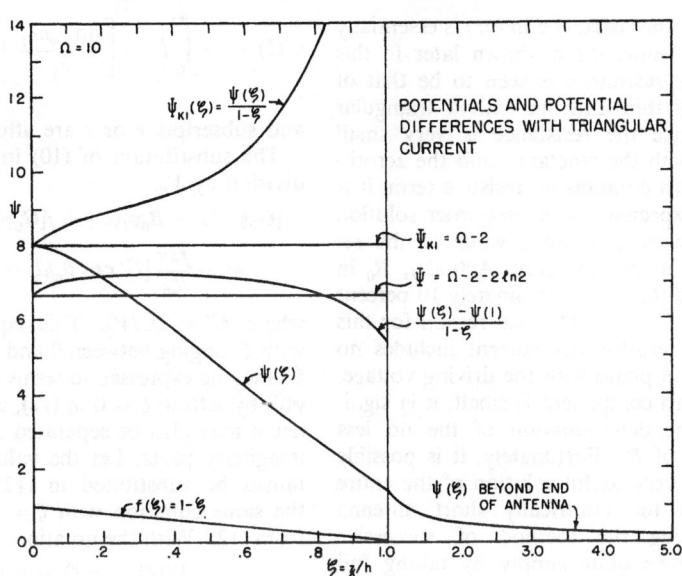

Fig. 31.1. Potentials and potential differences with triangular current.

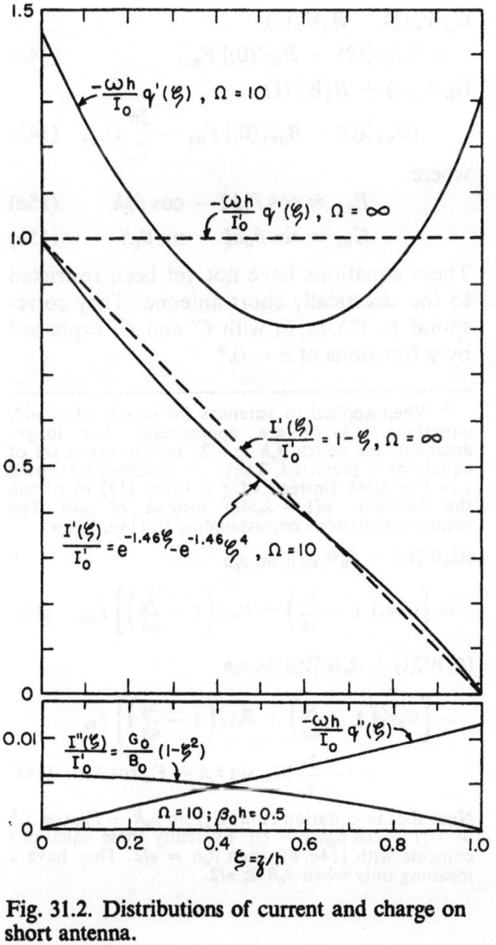

Fig. 31.2. Distributions of current and charge on short antenna.

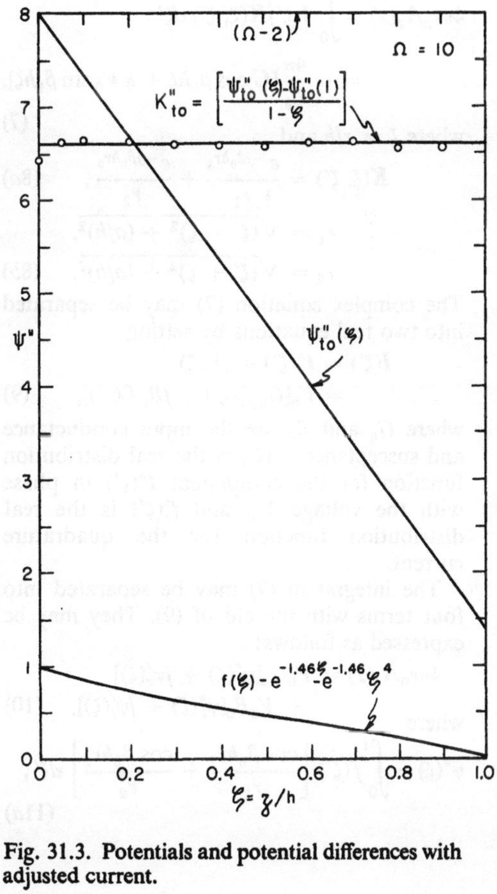

Fig. 31.3. Potentials and potential differences with adjusted current.

Although the reactance in (6c) is essentially the correct value, (as is shown later in this section), the resistance is seen to be that of an infinitely thin antenna with a triangular current. Since the resistance is very small compared with the reactance and the zeroth-order formula contains no resistive term, it is not to be expected that a first-order solution will provide as accurate a value of the resistance as of the reactance. Actually, R_0 in both (4) and (6) are approximately 10 percent in error for $\Omega = 10$. The real reason for this is that the zeroth-order current includes no component in phase with the driving voltage. Although this component is small, it is significant in the determination of the no less small value of R_0. Fortunately, it is possible to obtain a very useful solution of the entire problem of the electrically short antenna without using the method of successive approximations quite simply by taking full advantage of the relative smallness of R_0 compared with X_0.

The integral equation (14.3) with $\delta = 0$ and $z^i = 0$ may be expressed as follows:

$$4\pi v_0 A_z(z) = \int_0^1 I(\zeta') K(\zeta, \zeta') \, d\zeta'$$

$$= -\frac{4\pi}{\zeta_0} [C \cos \beta_0 h\zeta + \tfrac{1}{2} V_0 \sin \beta_0 h\zeta], \quad (7)$$

where $\zeta = z/h$ and

$$K(\zeta, \zeta') = \frac{e^{-j\beta_0 h r_1}}{r_1} + \frac{e^{-j\beta_0 h r_2}}{r_2}, \quad (8a)$$

$$r_1 = \sqrt{(\zeta' - \zeta)^2 + (a/h)^2},$$
$$r_2 = \sqrt{(\zeta' + \zeta)^2 + (a/h)^2}. \quad (8b)$$

The complex equation (7) may be separated into two real equations by setting

$$I(\zeta') = I''(\zeta') + jI'(\zeta')$$
$$= V_0[G_0 f_r(\zeta') + jB_0 f_t(\zeta')], \quad (9)$$

where G_0 and B_0 are the input conductance and susceptance, $f_r(\zeta')$ is the real distribution function for the component $I''(\zeta')$ in phase with the voltage V_0, and $f_t(\zeta')$ is the real distribution function for the quadrature current.

The integral in (7) may be separated into four terms with the aid of (9). They may be expressed as follows:

$$4\pi v_0 A_z(z) = V_0 G_0 [\psi_r''(\zeta) + j\psi_r'(\zeta)]$$
$$+ V_0 B_0 [\psi_t''(\zeta) + j\psi_t'(\zeta)], \quad (10)$$

where

$$\psi''(\zeta) = \int_0^1 f(\zeta') \left[\frac{\cos \beta_0 h r_1}{r_1} + \frac{\cos \beta_0 h r_2}{r_2} \right] d\zeta', \quad (11a)$$

$$\psi'(\zeta) = -\int_0^1 f(\zeta') \left[\frac{\sin \beta_0 h r_1}{r_1} + \frac{\sin \beta_0 h r_2}{r_2} \right] d\zeta', \quad (11b)$$

and subscripts r or t are affixed to ψ and f.

The substitution of (10) in (7) gives, after division by V_0,

$$[G_0 \psi_r''(\zeta) - B_0 \psi_t'(\zeta)] + j[G_0 \psi_r'(\zeta) + B_0 \psi_t''(\zeta)]$$
$$= -\frac{j2\pi}{\zeta_0} [C' \cos \beta_0 h\zeta + \sin \beta_0 h\zeta], \quad (12)$$

where $C' = 2C/V_0$. This equation is valid with ζ ranging between 0 and 1. The constant C' may be expressed in terms of the functions $\psi(0)$ by setting $\zeta = 0$ in (12), and in this manner it may also be separated into its real and imaginary parts. Let the value of C' so obtained be substituted in (12), and then let the same equation with $\zeta = 1$ be subtracted from (12). With the notation

$$W(\zeta) \equiv \psi(\zeta) - \psi(1), \quad (13)$$

the real and imaginary parts of the resulting equation are as follows:

$$G_0 W_r''(\zeta) - B_0 W_t'(\zeta)$$
$$= [G_0 \psi_r''(0) - B_0 \psi_t'(0)] F_{0z}, \quad (14a)$$

$$G_0 W_r'(\zeta) + B_0 W_t''(\zeta)$$
$$= [G_0 \psi_r'(0) + B_0 \psi_t''(0)] F_{0z} - \frac{2\pi}{\zeta_0} G_{0z}, \quad (14b)$$

where

$$F_{0z} \equiv \cos \beta_0 h\zeta - \cos \beta_0 h, \quad (15a)$$
$$G_{0z} \equiv \sin \beta_0 h\zeta - \sin \beta_0 h. \quad (15b)$$

These equations have not yet been restricted to the electrically short antenna. They correspond to (25.4a, b) with C' and C'' expressed by ψ functions at $z = 0$.*

---

* When applied to antennas for which $\beta_0 h \lesssim \pi/2$, equations (14a, b) are appropriate. For longer antennas, for which $\beta_0 h \gtrsim \pi/2$, an alternative set of equations is preferred. These are obtained by setting $\zeta = 1 - \lambda_0/4h$ (instead of $\zeta = 0$) in (12) to obtain the functions $\psi(1 - \lambda_0/4h)$ instead of $\psi(0)$. The resulting equations corresponding to (14a, b) are

$$[G_0 W_r''(\zeta) - B_0 W_t'(\zeta)] \sin \beta_0 h$$
$$= \left[G_0 \psi_r'' \left(1 - \frac{\lambda_0}{4h}\right) - B_0 \psi_t' \left(1 - \frac{\lambda_0}{4h}\right) \right] F_{0z}, \quad (14c)$$

$$[G_0 W_r'(\zeta) + B_0 W_t''(\zeta)] \sin \beta_0 h$$
$$= \left[G_0 \psi_r' \left(1 - \frac{\lambda_0}{4h}\right) + B_0 \psi_t'' \left(1 - \frac{\lambda_0}{4h}\right) \right] F_{0z}$$
$$- \frac{2\pi}{\zeta_0} [G_{0z} \sin \beta_0 h + F_{0z} \cos \beta_0 h]. \quad (14d)$$

Note that in equation (14d) $G_{0z} \sin \beta_0 h + F_{0z} \cos \beta_0 h = -[1 - \cos \beta_0 h(1 - \zeta)]$. Evidently these equations coincide with (14a, b) when $\beta_0 h = \pi/2$. They have a meaning only when $\beta_0 h \gtrsim \pi/2$.

In applying (14a, b) to the electrically short antenna, advantage may be taken not only of (1) but of (4) and (6), which show that

$$B_0 \doteq \frac{1}{X_0} \sim \beta_0 h, \quad G_0 \doteq \frac{R_0}{X_0^2} \sim \beta_0^4 h^4. \quad (16)$$

With (1) and (16) it is evident that if the principal term and the first correction term of order $\beta_0^3 h^3$ are to be retained in solving (14b) for B_0, all terms with G_0 as a factor are negligible in (14b). The value of G_0 may then be determined from (14a) using B_0 as obtained from (14b). This involves expanding the trigonometric factors in (11a, b) as follows:

$$\psi''(\zeta) = \psi_0''(\zeta) + \psi_2''(\zeta) + \psi_4''(\zeta) + \cdots, \quad (17a)$$

$$\psi'(\zeta) = \psi_1'(\zeta) + \psi_3'(\zeta) + \psi_5'(\zeta) + \cdots, \quad (17b)$$

where

$$\psi_0''(\zeta) = \int_0^1 f(\zeta') \left(\frac{1}{r_1} + \frac{1}{r_2} \right) d\zeta', \quad (18a)$$

$$\psi_2''(\zeta) = -\frac{\beta_0^2 h^2}{2} \int_0^1 f(\zeta')(r_1 + r_2) \, d\zeta', \quad (18b)$$

$$\psi_4''(\zeta) = \frac{\beta_0^4 h^4}{24} \int_0^1 f(\zeta')(r_1^3 + r_2^3) \, d\zeta', \quad (18c)$$

$$\psi_1'(\zeta) = -2\beta_0 h \int_0^1 f(\zeta') \, d\zeta', \quad (18d)$$

$$\psi_3'(\zeta) = \frac{\beta_0^3 h^3}{6} \int_0^1 f(\zeta')(r_1^2 + r_2^2) \, d\zeta', \quad (18e)$$

$$\psi_5'(\zeta) = -\frac{\beta_0^5 h^5}{120} \int_0^1 f(\zeta')(r_1^4 + r_2^4) \, d\zeta'. \quad (18f)$$

Identifying subscripts r and t are to be added to the f's and ψ's as required. Note that the numerical subscript on the ψ's is the same as the power to which the factor $\beta_0 h$ is raised.

If only terms with factors of the order $\beta_0^3 h^3$ or less are retained, (14b) reduces to the following:

$$B_0(W_{t0}'' + W_{t2}'') = \tfrac{1}{2} B_0 \psi_t''(0) \beta_0^2 h^2 (1 - \zeta^2)$$

$$+ \frac{2\pi \beta_0 h (1 - \zeta)}{\zeta_0} \left[1 - \frac{\beta_0^2 h^2 (1 - \zeta^3)}{6(1 - \zeta)} \right].$$

$$(19a)$$

This may be solved as follows, care being taken to retain all terms of order $\beta_0^3 h^3$ or less; note that W_{t2}'' is of order $\beta_0^2 h^2$:

$$-X_0 \doteq \frac{1}{B_0} = \frac{\zeta_0}{2\pi \beta_0 h} \left[\frac{W_{t0}''(\zeta)}{1 - \zeta} + \frac{W_{t2}''(\zeta)}{1 - \zeta} \right.$$

$$+ \frac{\beta_0^2 h^2 (1 - \zeta^3) \psi_{t0}''(\zeta)}{6(1 - \zeta)^2}$$

$$\left. - \frac{\beta_0^2 h^2 (1 - \zeta^3) \psi_{t0}''(0)}{2(1 - \zeta)} \right]. \quad (19b)$$

For sufficiently small values of $\beta_0 h$, in particular, if (1) is satisfied, the terms with $\beta_0^2 h^2$ as a factor contribute negligibly, so that

$$-X_0 \doteq \frac{1}{B_0} = \frac{\zeta_0}{2\pi \beta_0 h} \left[\frac{W_{t0}''(\zeta)}{1 - \zeta} \right]. \quad (20)$$

This equation indicates that the quantity $W_{t0}''(\zeta) \equiv \psi_{t0}''(\zeta) - \psi_{t0}''(1)$, which is proportional to the principal part of the vector potential difference $A_z(z) - A_z(h)$, varies linearly with $\zeta = z/h$ along the antenna. It follows that the solution of (20) is found when a current-distribution function $f_t(\zeta')$ is determined that makes $W_{t0}''(\zeta)$ a linear function of ζ. More conveniently, the required distribution function must satisfy the relation

$$\frac{W_{t0}''(\zeta)}{1 - \zeta} = \frac{\psi_{t0}''(\zeta) - \psi_{t0}''(1)}{1 - \zeta} = \text{const.} \equiv K_{t0}'',$$

$$(21)$$

in the range $0 \leq \zeta \leq 1$.

Subject to (1) the principal part $\psi_{t0}''(\zeta)$ of the integral $\psi_t''(\zeta)$ is given by (18a). It is verified in Fig. 31.1 for $\Omega = 10$ that the zeroth-order linear distribution

$$\frac{I(\zeta')}{I_0} = f(\zeta') = 1 - \zeta', \quad (22)$$

when substituted in (18a), does *not* give a constant as ζ is increased from 0 to 1 with a/h finite. On the other hand, an inspection of the relatively small variation in this function suggests that a moderately small modification of $f(\zeta')$ from the value in (22) should yield a constant for the ratio on the left in (21). By superimposing on the triangular distribution (22) when multiplied by a suitable factor similar distributions of much smaller amplitude defined along sections of half-length

$h/10$ which are distributed along the actual half-length h, a composite distribution of current and an associated vector potential difference $W_{t0}''(\zeta)$ may be determined which satisfy (21) approximately. Since a distribution consisting of a sum of a large number of linear distributions over parts of the antenna is awkward in the evaluation of higher-order terms, an essentially equivalent exponential distribution may be constructed in the following analytically simple form:

$$\frac{I(\zeta')}{I_0} = f_t(\zeta) = e^{-\delta\zeta} - \zeta^n e^{-\delta}, \quad (23)$$

where δ and n are constants. For $\Omega = 10$ the required constants are $\delta = 1.46$, $n = 4$. This distribution is shown in Fig. 31.2 together with the linear distribution (22). The associated distributions of charge as determined from the equation of continuity also are shown. (Note that $I'(\zeta) = I_0' f_t(\zeta)$ is the component of current in phase quadrature with the driving voltage.) It is clear from Fig. 31.2 that the principal component of current departs considerably from a linear distribution. This is a necessary consequence of the fact that the vector potential, $A_z(z)$, is itself linear in ζ. That the modified exponential current (23) actually maintains a linear vector potential is evident from Fig. 31.3, where $\psi_{t0}''(\zeta)$ as calculated from (18a) using (23) is shown. Only very near $\zeta = 0$ is there a slight departure from linearity. Actually, the requirement that the vector potential be linear to $\zeta = 0$ is a consequence of the assumed driving condition consisting of a discontinuity in scalar potential at $\zeta = 0$. Since this idealized generator is equivalent to a transmission-line drive in the limit as the line spacing approaches zero, and since in practice the spacing may be extremely small but never zero, the vector potential is always rounded off in a small region near $\zeta = 0$. It follows that whereas the distribution (23) with $\delta = 1.46$, $n = 4$ does not quite satisfy the idealized integral equation very near $\zeta = 0$, it does agree with it if the halves of the antenna are separated by a very small distance 2δ. Evidently, account must be taken of terminal-zone effects as discussed earlier in this chapter. The ratio (21) is shown in Fig. 31.3. It is seen to be essentially constant over the entire length of the antenna. The average constant for $\Omega = 10$ is $K_{t0}'' = 6.60$. Except near the point $\zeta = 0$, no part of the function on the left in (21) differs from 6.60 by even as much as 1 percent.

The reactance of a sufficiently short antenna is given by

$$-X_0 \doteq \frac{1}{B_0} = \frac{\zeta_0}{2\pi} \frac{K_{t0}''}{\beta_0 h}; \quad (24a)$$

for $\Omega = 10$, $-X_0 \doteq \frac{1}{B_0} = \frac{396.0}{\beta_0 h}$ ohm. (24b)

Note that this agrees almost exactly with the first-order value given in (6), but differs by a fraction of a percent from the slightly less accurate value in (4).

It is interesting to compare the essentially correct value (24) with the reactance determined from a well-known formula for the static capacitance between two ellipsoids each of axial length h and radius at the center a. The formula in question is (see ref. I.34, p. 66):

$$X_0 = -\frac{1}{\omega C_0} = -\frac{60}{\beta_0 h}(\Omega - \ln 12). \quad (25)$$

For $\Omega = 10$, this gives $X_0 = -60 \times 7.515/\beta_0 h = -450.9/\beta_0 h$ ohm. Clearly, since a cylinder has the distribution of charge shown in Fig. 31.2, with much higher concentrations near the ends than in the center, and not a uniform distribution per unit length as has the ellipsoid, only a rough correspondence between the value of the capacitive reactance for the cylinder given in (24) and that for the ellipsoid in (25) is to be expected. For $\Omega = 10$ the difference is roughly 12 percent.

Although (24) is an accurate formula for the reactance of a sufficiently short antenna ($\beta_0 h \leq 0.2$), the accuracy necessarily decreases as $\beta_0 h$ is increased. By solving (19b) instead of (20) a formula for X_0 that includes the next-order correction term as $\beta_0 h$ is increased may be determined and the range for X_0 thus extended. Strictly, this involves a redetermination of the distribution of current. However, since it is a question merely of obtaining a higher-order correction term, it is adequate to assume the same distribution of current so that the first term in (19b) is simply K_{t0}''. Since for each value of $\beta_0 h$ the sum of the last three terms in (12b) must be a constant independent of ζ, the value at $\zeta = 0$ may be used without determining the individual variations of the three terms with ζ. The resulting formula is

$$-X_0 \doteq \frac{1}{B_0} = \frac{\zeta_0}{2\pi\beta_0 h}[K_{t0}'' + W_{t2}''(0)$$
$$+ \frac{\beta_0^2 h^2}{6} W_{t0}''(0) - \frac{\beta_0^2 h^2}{2} \psi_{t0}''(0)]. \quad (26)$$

Using (23) with $n = 4$ in (18b), the following results are obtained:

$$\psi''_{t2}(0) = -\beta_0^2 h^2 \left[\frac{1}{\delta^2} - e^{-\delta}\left(\frac{1}{\delta} + \frac{1}{\delta^2} + \frac{1}{6}\right)\right], \quad (27)$$

$$\psi''_{t2}(1) = -\beta_0^2 h^2 \left[\frac{1}{\delta} - e^{-\delta}\left(\frac{1}{\delta^2} + \frac{1}{5}\right)\right]. \quad (28)$$

For $\Omega = 10$ with $\delta = 1.46$,

$$W''_{t2}(0) = \psi''_{t2}(0) - \psi''_{t2}(1) = 0.367\beta_0^2 h^2. \quad (29)$$

From Fig. 31.3 it follows that

$$\psi''_{t0}(0) = 8. \quad (30)$$

Substitution of $W''_{t0}(0) = K''_{t0} = 6.60$ in (26) gives directly

$$-X_0 \doteq \frac{1}{B_0} = \frac{396.0}{\beta_0 h}(1 - 0.383\beta_0^2 h^2) \text{ ohm}. \quad (31)$$

The term in $\beta_0^2 h^2$ evidently contributes significantly when $\beta_0 h$ exceeds 0.3. Since the term in $\beta_0^4 h^4$, which is neglected in (31), is of the order of magnitude of $\beta_0^4 h^4/24$, (31) is a good approximation provided

$$\beta_0 h \leq 1 \quad (32)$$

A comparison of X_0 as determined from (31) and as given by the second-order results tabulated in Sec. 30 using Ψ_{K1} as the expansion parameter is given in Table 31.1 and in Fig. 31.4. Since the formula for the short antenna is more accurate for short antennas, whereas the second-order values obtained using the expansion parameter Ψ_{K1} are more accurate as $\beta_0 h$ approaches and exceeds 1, a corrected curve and table have been constructed.

With the susceptance $B_0 \doteq -1/X_0$ quite accurately determined from (14b) for $\beta_0 h \leq 1$, the conductance $G_0 = R_0/X_0^2$ may be evaluated for the same range of $\beta_0 h$ from (14a) by retaining its leading terms, which are of order $\beta_0^4 h^4$, and its first correction terms, which are of order $\beta_0^6 h^6$. By retaining only required terms, (14a) may be expressed as follows, using (18):

$$G_0[W''_{r0}(\zeta) + W''_{r2}(\zeta)]$$
$$- B_0[W'_{t1}(\zeta) + W'_{t3}(\zeta) + W'_{t5}(\zeta)]$$
$$= \{G_0\psi''_{r0}(0) - B_0[\psi'_{t1}(0) + \psi'_{t3}(0)]\}$$
$$\times \left[\frac{\beta_0^2 h^2}{2}(1 - \zeta^2) - \frac{\beta_0^4 h^4}{24}(1 - \zeta^4)\right] \quad (33a)$$

This equation may be rearranged as follows, again retaining only terms of order $\beta_0^6 h^6$ and lower; note that since $\psi'_1(\zeta)$ as given in (18d) actually is independent of ζ, $W'_1(\zeta) = 0$, so that the terms of lowest order have $\beta_0^4 h^4$ as a factor:

$$G_0[W''_{r0}(\zeta) + W''_{r2}(\zeta) - \tfrac{1}{2}\beta_0^2 h^2 \psi''_{r0}(0)(1-\zeta^2)]$$
$$= B_0\Big[W'_{t3}(\zeta) - \tfrac{1}{2}\beta_0^2 h^2 \psi'_{t1}(0)(1-\zeta^2)$$
$$+ W'_{t5}(\zeta) - \tfrac{1}{2}\beta_0^2 h^2 \psi'_{t3}(0)(1-\zeta^2)$$
$$+ \frac{\beta_0^4 h^4}{24}\psi'_{t1}(0)(1-\zeta^4)\Big]. \quad (33b)$$

The leading terms of order $\beta_0^4 h^4$ are

$$G_0\left[\frac{W''_{r0}(\zeta)}{1-\zeta^2}\right] = B_0\left[\frac{W'_{t3}(\zeta)}{1-\zeta^2} - \tfrac{1}{2}\beta_0^2 h^2 \psi'_{t1}(0)\right]. \quad (34)$$

Since the distribution of current $f_t(\zeta')$ for use in determining $\psi'_t(\zeta)$, $W''_t(\zeta)$, and $W'_t(\zeta)$ is given by (23), these functions may be determined. In particular, with n in (23) equal to 4,

$$\psi'_{t1}(\zeta) = -2\beta_0 h\left(\frac{1-e^{-\delta}}{\delta} - \frac{e^{-\delta}}{5}\right), \quad (35a)$$

$$\psi'_{t3}(\zeta) = \frac{\beta_0^3 h^3}{3}\left[\zeta^2\left(\frac{1-e^{-\delta}}{\delta} - \frac{e^{-\delta}}{5}\right) + \frac{2}{\delta^3}\right.$$
$$\left. - e^{-\delta}\left(\frac{1}{\delta} + \frac{2}{\delta^2} + \frac{2}{\delta^3} + \frac{1}{7}\right)\right]. \quad (35b)$$

It follows that

$$\frac{W'_{t3}(\zeta)}{1-\zeta^2} = -\frac{\beta_0^3 h^3}{3}\left(\frac{1-e^{-\delta}}{\delta} - \frac{e^{-\delta}}{5}\right) \equiv K''_{t3}. \quad (35c)$$

With $\delta = 1.46$ for $\Omega = 10$,

$$\psi'_{t1}(0) = -0.9589\beta_0 h, \quad (36a)$$
$$\psi'_{t3}(0) = 0.02778\beta_0^3 h^3, \quad (36b)$$
$$K''_{t3} = -0.1598\beta_0^3 h^3. \quad (36c)$$

Since for terms of order $\beta_0^4 h^4$ the right-hand member of (34) is equal to a constant independent of ζ, the same must be true for terms of the same order on the left. That is,

$$\frac{W''_{r0}(\zeta)}{1-\zeta^2} = \frac{\psi''_{r0}(\zeta) - \psi''_{r0}(1)}{1-\zeta^2} = \text{const.} \equiv K''_{r0}. \quad (37)$$

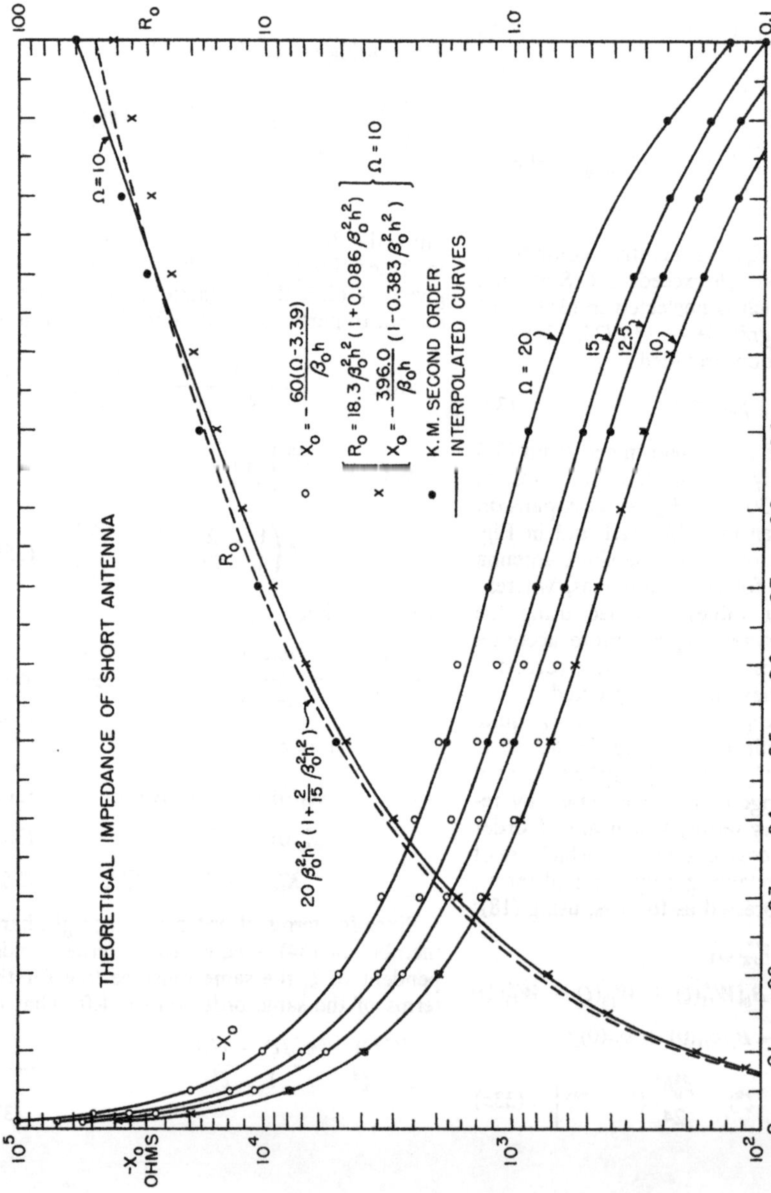

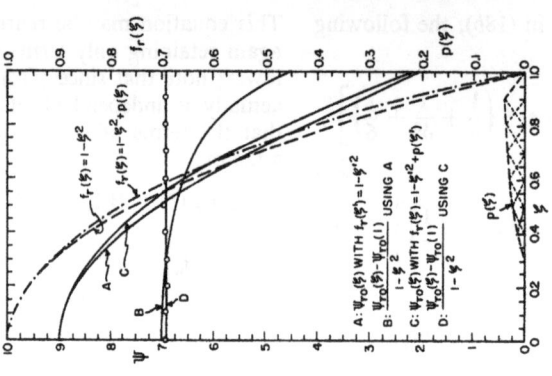

Fig. 31.5. Potential and potential differences with parabolic and adjusted parabolic currents.

Fig. 31.4. Theoretical impedance of short antenna.

Evidently, the evaluation of K''_{r0} involves the simultaneous determination of the distribution function $f_r(\zeta')$ of the small component of current, $I''(\zeta) = I''_0 f_r(\zeta')$, in phase with the driving voltage. The correct value of $f_r(\zeta')$ is that which reduces the function on the left in (37) to a constant for all values of ζ in the range $0 \leq \zeta \leq 1$. The integral involved is (18a) with subscript t on $\psi''_0(\zeta)$ and $f(\zeta')$.

It is shown in Fig. 31.5 that the parabolic distribution $f_r(\zeta) = 1 - \zeta^2$ does not yield a constant vector-potential difference when substituted in (18a) and used in (37). On the other hand, by slightly modifying the parabolic distribution by adding to it the distribution $p(\zeta)$ shown near the bottom in Fig. 31.5 to obtain the distribution $f_r(\zeta) = 1 - \zeta^2 + p(\zeta)$, and then substituting this function (with $p(\zeta)$ represented as a superposition of triangular distributions) in (18a), the curve C in Fig. 31.5 is obtained. When this is used in (37), the constant function shown in curve B is obtained. It follows that

$$f_r(\zeta) = 1 - \zeta^2 + p(\zeta), \quad (38)$$

as shown in Fig. 31.5, is essentially the correct distribution for the in-phase component of current with $\Omega = 10$. With it the following values are obtained:

$$K''_{r0} = 6.90, \qquad \psi''_{r0}(0) = 9.0. \quad (39)$$

The substitution of (35c) and (37) in (34) gives

$$-\frac{R_0}{X_0} \doteq \frac{G_0}{B_0} = \frac{K'_{t3} - \tfrac{1}{2}\psi'_{t1}(0)\beta_0^2 h^2}{K''_{r0}}. \quad (40)$$

Using (24a),

$$R_0 \doteq \frac{\zeta_0}{2\pi} \frac{1}{\beta_0 h} \frac{K''_{t0}}{K''_{r0}} [K'_{t3} - \tfrac{1}{2}\psi'_{t1}(0)\beta_0^2 h^2]. \quad (41a)$$

For $\Omega = 10$, using (36) and (39),

$$R_0 \doteq 18.3 \beta_0^2 h^2 \text{ ohm}. \quad (41b)$$

In order to obtain the first-order correction for R_0, it is necessary to solve (33b) instead of (34). If it is assumed that the distribution of current is given by (38), the correction terms may be evaluated at $\zeta = 0$, and

$$-\frac{R_0}{X_0} \doteq \frac{G_0}{B_0}$$

$$= \frac{K'_{t3} - \tfrac{1}{2}\psi'_{t1}(0)\beta_0^2 h^2 + W'_{t5}(0) - \tfrac{1}{2}\psi'_{t3}(0)\beta_0^2 h^2 + \tfrac{1}{24}\psi'_{t1}(0)\beta_0^4 h^4}{K''_{r0} + W''_{r2}(0) - \tfrac{1}{2}\beta_0^2 h^2 \psi''_{r0}(0)} \quad (42)$$

The numerical value of this ratio for $\Omega = 10$ is

$$-\frac{R_0}{X_0} \doteq \frac{G_0}{B_0} = 0.0463 \beta_0^3 h^3 (1 + 0.469 \beta_0^2 h^2). \quad (43)$$

With (31) this gives

$$R_0 \doteq 18.3 \beta_0^2 h^2 (1 + 0.086 \beta_0^2 h^2) \text{ ohm}. \quad (44)$$

The following quantities not previously determined are involved in the evaluation of (42); in this evaluation $f_r(\zeta')$ is taken from (38) and $f_t(\zeta')$ from (23) with $n = 4$; numerical values are for $\Omega = 10$ with $\delta = 1.46$:

$$\psi'_{t5}(0) = -\frac{\beta_0^5 h^5}{60}\left[\frac{24}{\delta^5} - e^{-\delta}\left(\frac{1}{\delta} + \frac{4}{\delta^2} + \frac{12}{\delta^3}\right.\right.$$
$$\left.\left. + \frac{24}{\delta^4} + \frac{24}{\delta^5} + \frac{1}{9}\right)\right] = -0.000587 \beta_0^5 h^5, \quad (45a)$$

$$\psi'_{t5}(1) = -\frac{\beta_0^5 h^5}{60}\left\{\left[\frac{1}{\delta}(1 - e^{-\delta}) - 1.168 e^{-\delta}\right]\right.$$
$$+ 6\left[\frac{2}{\delta} - e^{-\delta}\left(\frac{1}{\delta} + \frac{2}{\delta^2} + \frac{2}{\delta^3}\right)\right]$$
$$\left. + \frac{24}{\delta^5} - e^{-\delta}\left[\frac{1}{\delta} + \frac{4}{\delta^2} + \frac{12}{\delta^3} + \frac{24}{\delta^4} + \frac{24}{\delta^5}\right]\right\}$$
$$= -0.0169 \beta_0^5 h^5, \quad (45b)$$

$$W'_{t5}(0) = \psi'_{t5}(0) - \psi'_{t5}(1) = 0.0163 \beta_0^5 h^5, \quad (45c)$$

$$\psi''_{r2}(0) = -\beta_0^2 h^2 / 4, \qquad \psi''_{r2}(1) = -2\beta_0^2 h^2 / 3,$$
$$W''_{r2}(0) = 5 \beta_0^2 h^2 / 12. \quad (45d)$$

The resistance given in (44) for $\Omega = 10$ is plotted with the zeroth-order values from (6) in Fig. 31.6 together with the second-order King-Middleton values. It is seen that the values given by (44) are only slightly lower than both the zeroth-order and the King-Middleton second-order values. Since the terms with $\beta_0^6 h^6$ as a factor, which are neglected in (44), have an extremely small numerical factor, it follows that (44) is highly accurate for $\beta_0 h \leq 0.5$ and an excellent approximation for $\beta_0 h \leq 1$. On the other hand, the use of the expansion parameter Ψ_{K1}, in which the component of current in phase with the driving voltage is not included in the zeroth-order distribution function, leads to a value of resistance for the short antenna that is slightly too large. This difference is of the order of magnitude of 10 percent.

Accordingly, since the formula (44) for the short antenna is an excellent approximation for $\beta_0 h < 1$, a value of R_0 may be constructed which is computed from the short-antenna formula (44) for small values of $\beta_0 h$, which uses the King-Middleton second-order values for values of $\beta_0 h$ near and above resonance, and which interpolates a smooth curve between these for intermediate values of $\beta_0 h$. This curve is shown in Fig. 31.4 as the interpolated R_0. The values computed from (44), the second-order King-Middleton values and the interpolated resistance are given in Table 31.1. These last are obtained by taking first and second differences, and it may be assumed that they are more accurate than the King-Middleton values, from which they differ by only a few percent over most of the range.

It may be concluded that a complete and accurate solution of the circuit properties of an electrically short transmitting antenna has been obtained. For the particular case for which $\Omega \equiv 2 \ln (2h/a) = 10$, the impedance and admittance are well represented by the following formulas, provided $\beta_0 h \leq 1$:

$$R_0 \doteq 18.3 \beta_0^2 h^2 (1 + 0.086 \beta_0^2 h^2), \quad (46a)$$

$$X_0 \doteq - \frac{396.0}{\beta_0 h} (1 - 0.383 \beta_0^2 h^2), \quad (46b)$$

$$G_0 \doteq \frac{R_0}{X_0^2} \doteq \left(\frac{1.165 \times 10^{-4} \beta_0^4 h^4}{1 - 0.852 \beta_0^2 h^2} \right), \quad (47a)$$

$$B_0 \doteq -\frac{1}{X_0} \doteq \left(\frac{2.525 \times 10^{-3} \beta_0 h}{1 - 0.383 \beta_0^2 h^2} \right). \quad (47b)$$

The analog of (46a) for $\Omega \to \infty$ is obtained in (V.12.18). It is

$$R_0 \doteq 20 \beta_0^2 h^2 (1 + 0.133 \beta_0^2 h^2), \quad (48)$$

and is also shown in Fig. 31.4 and Table 31.1. Note that (48) is quite a good approximation of the interpolated curve for $\Omega = 10$. The current, $I(\zeta) = I''(\zeta) + jI'(\zeta)$, is given by

$$I''_\zeta = V_0 G_0 [1 - \zeta^2 + p(\zeta)],$$
$$I'_\zeta = e^{-1.46\zeta} - e^{-1.46\zeta^4}, \quad (49)$$

where $\zeta = z/h$.

Corresponding formulas for other values of Ω may be obtained by the same procedure. In general, the resistance for $\Omega > 10$ must lie between the curves for $\Omega = 10$ and the curve for $R_0 = 20 \beta_0^2 h^2 (1 + 0.133 \beta_0^2 h^2)$. The reactance is quite accurately given by

$$X_0 = - \frac{60(\Omega - 3.39)}{\beta_0 h} \quad (50)$$

for $\beta_0 h \leq 0.5$. For larger values of $\beta_0 h$ the second-order King-Middleton values are good approximations.

TABLE 31.1. Resistance and reactance of short antenna.

$\beta_0 h$	X_0 (ohms)			R_0 (ohms)			
	Formula (44) for short antenna	King-Middleton second-order	Inter-polated	Formula (44) for short antenna	King-Middleton second-order	Inter-polated	$20\beta_0^2 h^2 \left(1 + \frac{2}{15} \beta^2 h^2\right)$
0	$-\infty$	$-\infty$	$-\infty$	0	0	0	0
0.1	-3945		-3945	0.183		0.183	0.200
.2	-1950		-1950	.732		.732	.804
.3	-1274		-1274	1.66		1.66	1.82
.4	-929		-929	2.97		2.97	3.27
.5	-716	-704.8	-716	4.67	4.99	4.67	6.17
.6	-569		-569	6.80		6.80	7.54
.7	-460	-451.3	-460	9.35	10.30	9.4	10.44
.8	-373.7		-370	12.38		12.5	13.89
.9	-303.5	-293.8	-294	15.82	18.3	16.2	17.95
1.0	-238.2		-234	19.90		21.0	22.67
1.1		-177.4	-177.4		30.0	27.3	28.10
1.2		-127.1	-127.1		37.8	35.4	34.33
1.3		-79.8	-79.8		47.4	45.6	41.41
1.4		-34.3	-34.3		59.1	58.2	
1.5		$+10.3$	$+10.3$		73.6	73.6	
1.6		$+54.7$	$+54.7$		91.7	91.7	
1.7		$+99.7$	$+99.7$		114.8	114.8	

32. Impedance of Antenna with Small Base Separation

Using the second-order impedance Z_0 determined in Sec. 30 and the factors ε_2'' and ε_2' as given in Figs. 29.3 and 29.4, the impedance Z_δ is readily determined for sufficiently small values of $\beta_0\delta$ to satisfy the inequality $\beta_0\delta \ll 1$ or $\beta_0^2\delta^2 \ll [\ln 2\delta/a]^2$. In Figs. 32.1 to 32.3 graphs of R_δ and X_δ are shown for a range of values of $\beta_0\delta$ including zero. The differences $R_0 - R_\delta$ and $X_0 - X_\delta$ are shown in Figs. 32.4 to 32.6 for $\Omega = 10, 15, 20$, with β_0h as variable and $\beta_0\delta$ as parameter, and in Figs. 32.6 to 32.9 with $\beta_0\delta$ as variable and β_0h as parameter. Y_δ is shown in Fig. 32.10.

A study of Z_δ compared with Z_0 shows that the *relative* importance of the correction for finite δ is greatest *near resonance* and quite small near antiresonance. Near resonance it is essentially equivalent to a shift in β_0h by $\beta_0\delta$. That is, if $\bar{h} = h - \delta$ is used instead of h, the values of Z_δ differ negligibly from Z_0. This is in agreement with the conclusions reached in conjunction with the zeroth-order impedances. In general, the correction for a small but finite δ is not very great, and no serious error is made by omitting it entirely.

Theoretically, the most convenient parameter for studying the effect of a small base separation is the quantity $\beta_0\delta = \omega\delta/v_0$. In engineering applications, where δ is usually fixed and where the frequency is the variable, it is not possible to keep $\beta_0\delta$ constant, and curves of impedance with δ/h or δ/a as constant parameters would be more convenient if a complete impedance graph as a function of β_0h were required. Actually, this is seldom true, since for engineering installations interest is usually in a fairly narrow frequency range as applied to a particular installation. For such a purpose, the required result may be obtained by interpolation from the curves of Figs. 32.1 to 32.9 or by direct calculation using Figs. 29.3 and 29.4. For $1 \leq \beta_0h \leq 2$ it is a satisfactory approximation to assume that Z_δ for a given value of β_0h is equal to Z_0 for $\beta_0(h - \delta)$. In practice, when $\beta_0h < 1$ for an antenna, this is a result of a low frequency so that $\beta_0\delta$ is correspondingly reduced, usually to so small a value that $Z_\delta \doteq Z_0$.

In precise laboratory measurements of the impedance as affected by base separation, convenience and accuracy usually demand the use of the half-length h as the single variable while all other quantities including the frequency are held constant. With h as the variable, $\beta_0\delta$ is very satisfactory as a parameter. With h as the variable, Ω is not a constant parameter. However, for $\beta_0h \geq 2$, the variation of Ω with h is very slow, so that no serious error is involved in correcting for small values of δ by assuming Ω constant at a convenient value in the required range of β_0h.

APPARENT IMPEDANCE OF CYLINDRICAL ANTENNA AS LOAD ON A TRANSMISSION LINE

33. Antenna with Positive or Negative Capacitance in Shunt

When an antenna is driven from a two-wire line, a coaxial line, or the equivalent, the terminating impedance Z_δ of the uncoupled antenna is often in parallel with a capacitance —positive or negative. For the conventional end-loaded line or the center-loaded symmetrical line, a *negative* capacitance in parallel is required for use with ordinary line theory in order to compensate for the variable capacitance per unit length of the line near its junction with the antenna and for the coupling between antenna and line. If a dielectric support is used, or enlarged metal terminals occur at the junction, these are well approximated by a positive capacitance in parallel with Z_δ. The size of the negative compensating capacitance C_T required for the conventional end-loaded line is indicated by the numerical results in Sec. 8. Corresponding numerical values of C_T for the center-loaded line are given in Sec. 9. Positive capacitances involved in dielectric supports, metal terminals, etc., are of the same order of magnitude. To obtain a comprehensive picture of the effect on the combined admittance Y_s or impedance Z_s of a positive or negative capacitance C_s in parallel with Z_0, computations have been made of Y_s and Z_s for a series of values of C_s, both positive and negative. In these C_s is expressed in terms of the parameter K:

$$\omega C_s = K\beta_0 h. \qquad (1)$$

In this form, C_s and h are assumed to be constant for a particular antenna while $\omega = 2\pi f$ is the variable. The parameter K is allowed to vary from -0.0048 to $+0.0048$. Curves of B_s in $Y_s = G_s + jB_s$ and R_s and X_s in $Z_s = R_s + jX_s$ are shown in Figs. 33.1–33.3 for $\Omega = 10, 12.5, 15$; G_0 is unaffected by a capacitance in parallel, so that $G_s = G_0$.

194 THEORY OF LINEAR ANTENNAS [II.32]

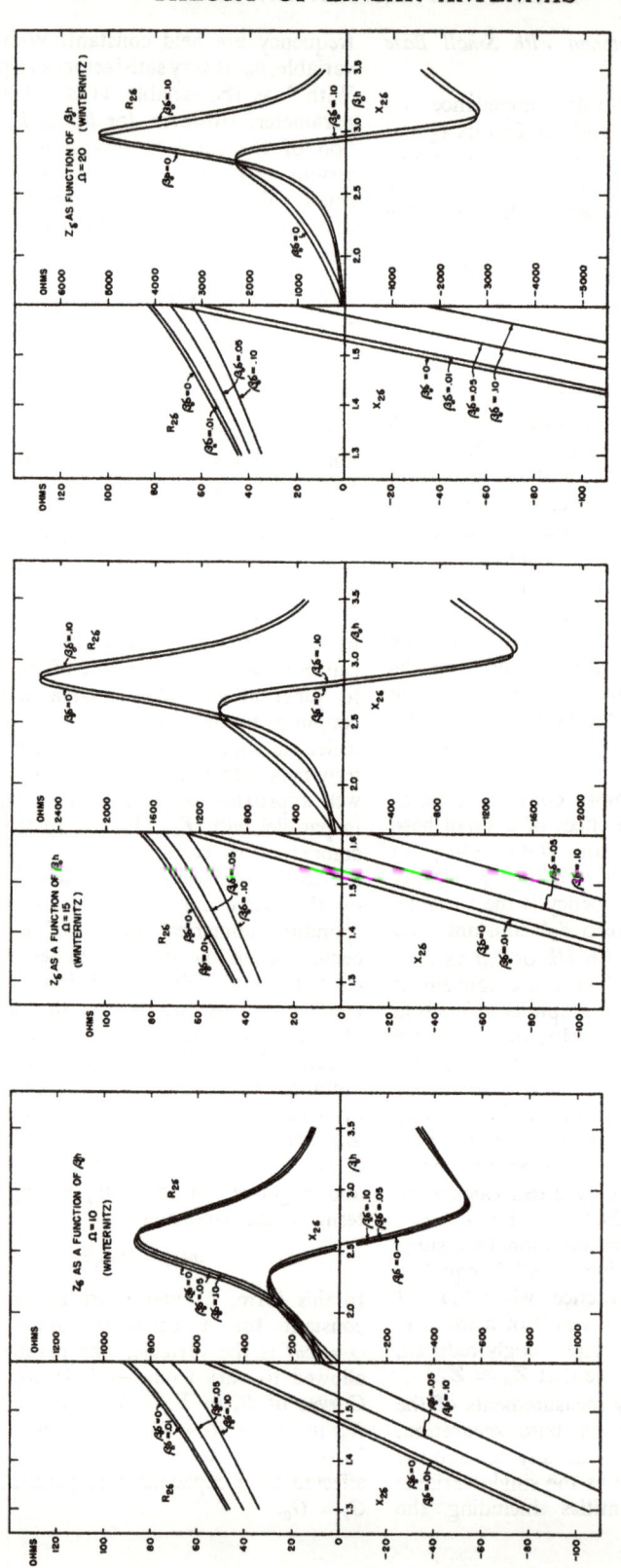

Fig. 32.1. Curves of Z_δ as a function of $\beta_0 h$; $\Omega = 10$ (Winternitz).

Fig. 32.2. Curves of Z_δ as a function of $\beta_0 h$; $\Omega = 15$ (Winternitz).

Fig. 32.3. Curves of Z_δ as a function of $\beta_0 h$; $\Omega = 20$ (Winternitz).

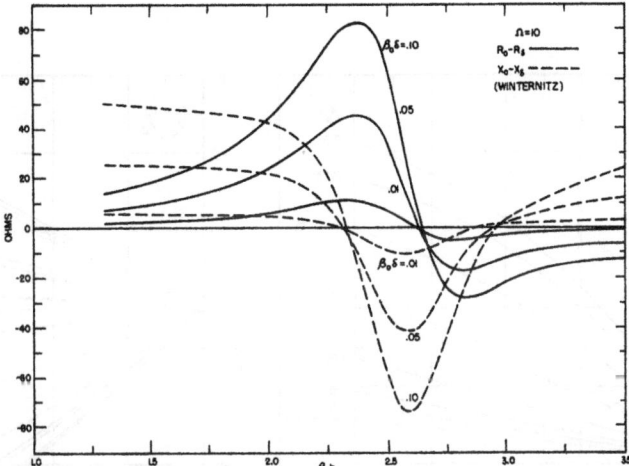

Fig. 32.4. Curves of $R_0 - R_\delta$ and $X_0 - X_\delta$ as functions of $\beta_0 h$ with $\beta_0 \delta$ as parameter; $\Omega = 10$ (Winternitz).

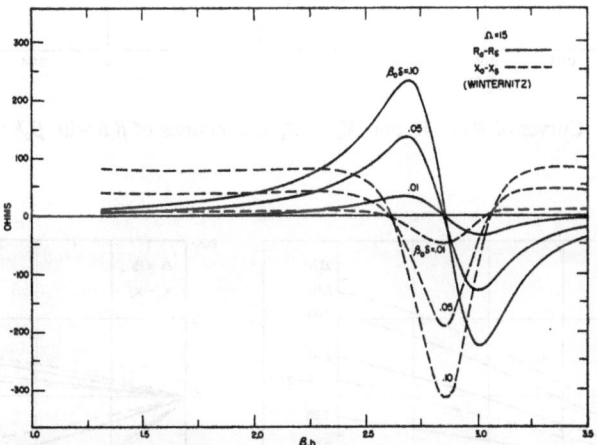

Fig. 32.5. Curves of $R_0 - R_\delta$ and $X_0 - X_\delta$ as functions of $\beta_0 h$ with $\beta_0 \delta$ as parameter; $\Omega = 15$ (Winternitz).

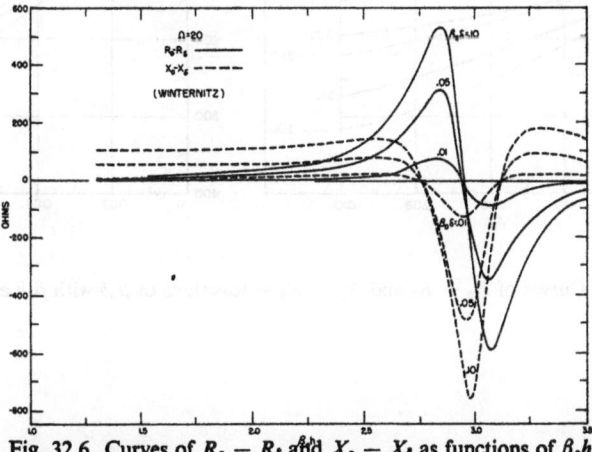

Fig. 32.6. Curves of $R_0 - R_\delta$ and $X_0 - X_\delta$ as functions of $\beta_0 h$ with $\beta_0 \delta$ as parameter; $\Omega = 20$ (Winternitz).

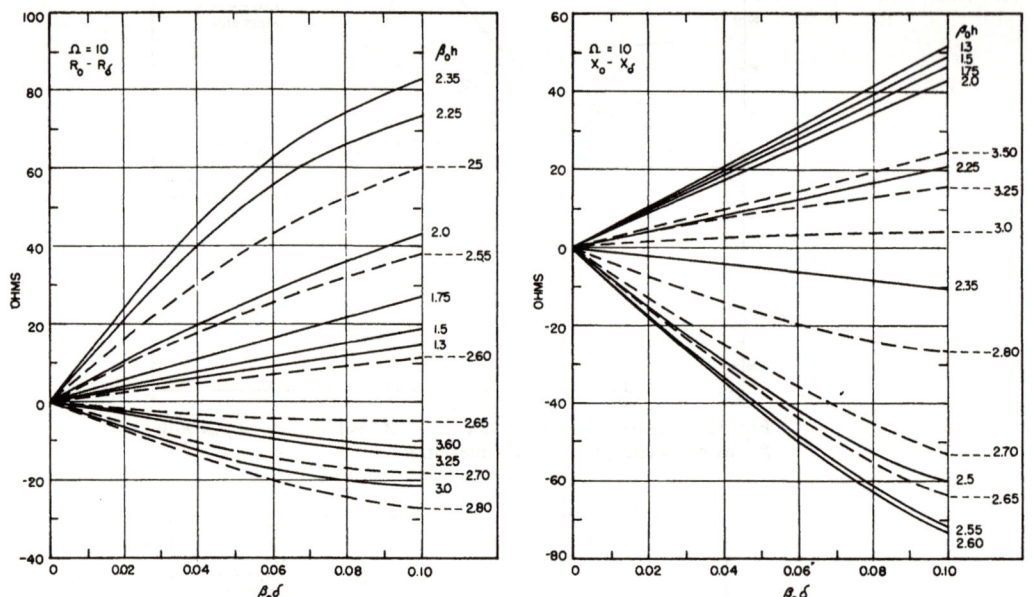

Fig. 32.7. Curves of $R_0 - R_\delta$ and $X_0 - X_\delta$ as functions of $\beta_0\delta$ with $\beta_0 h$ as parameter; $\Omega = 10$.

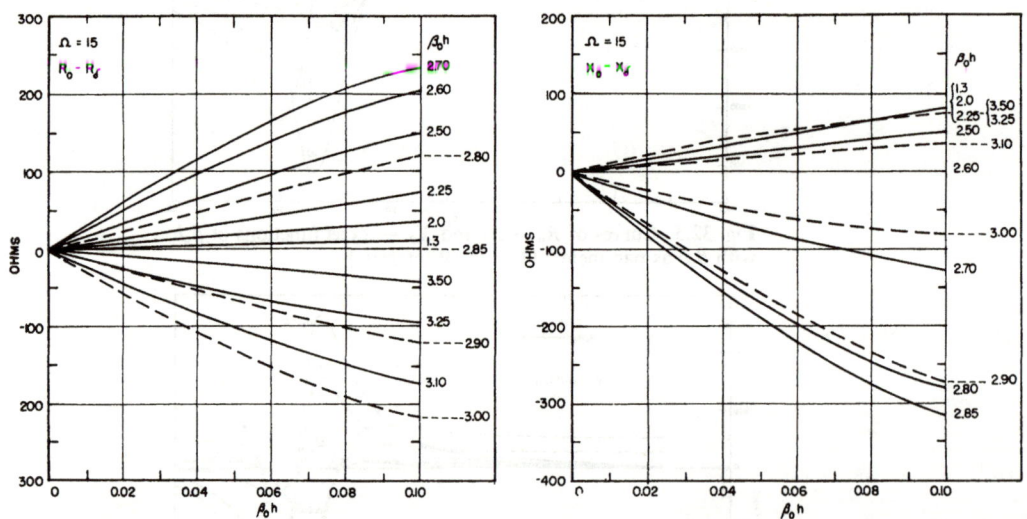

Fig. 32.8. Curves of $R_0 - R_\delta$ and $X_0 - X_\delta$ as functions of $\beta_0\delta$ with $\beta_0 h$ as parameter; $\Omega = 15$.

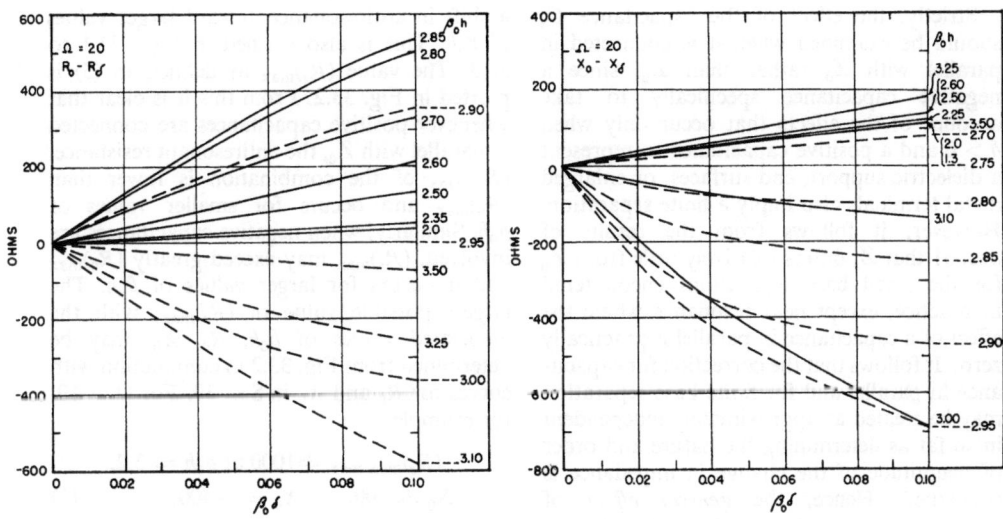

Fig. 32.9. Curves of $R_0 - R_\delta$ and $X_0 - X_\delta$ as functions of $\beta_0\delta$ with $\beta_0 h$ as parameter; $\Omega = 20$.

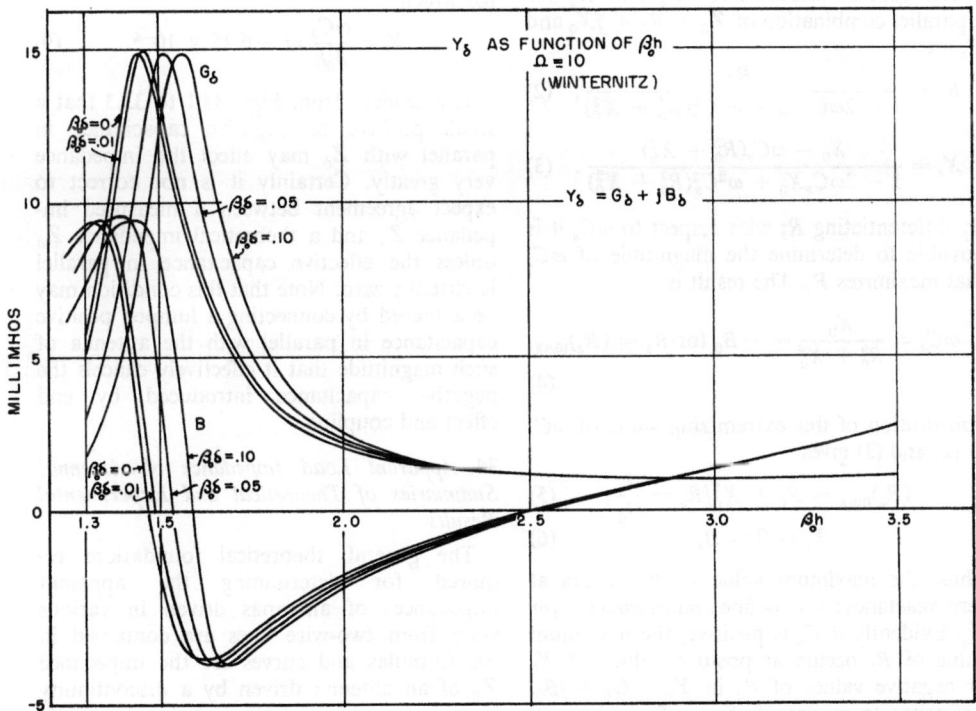

Fig. 32.10. Curves of Y_δ as a function of $\beta_0 h$ with $\beta_0\delta$ as parameter; $\Omega = 10$ (Winternitz).

Strictly, the effect of the capacitance C_s should be examined when it is connected in parallel with Z_δ rather than Z_0, since a negative capacitance specifically to take account of the effects that occur only when $\delta > 0$ and a positive capacitance to represent a dielectric support, end surfaces, or enlarged metal terminals also imply a finite separation. However, it follows from the results of Sec. 31 that Z_δ differs relatively little from Z_0 for the small base separations encountered in practice, except near resonance where the effect of a capacitance in parallel is practically zero. It follows that the correction for capacitance in parallel and for finite base separation may be treated as approximately independent in so far as determining the nature and order of magnitude of the change in impedance is concerned. Hence, the *general effect* of capacitance in parallel may be studied for Z_0 instead of for Z_δ without serious error. If numerically accurate results are required for a specific antenna and driving line, they must be analyzed as an individual unit.

The general formulas for R_s and X_s for a parallel combination of $Z_0 = R_0 + jX_0$ and C_s are

$$R_s = \frac{R_0}{1 - 2\omega C_s X_0 + \omega^2 C_s^2 (R_0^2 + X_0^2)}, \quad (2)$$

$$X_s = \frac{X_0 - \omega C_s (R_0^2 + X_0^2)}{1 - 2\omega C_s X_0 + \omega^2 C_s^2 (R_0^2 + X_0^2)}. \quad (3)$$

By differentiating R_s with respect to ωC_s it is possible to determine the magnitude of ωC_s that maximizes R_s. The result is

$$\omega C_s = \frac{X_0}{R_0^2 + X_0^2} = -B_0 \text{ for } R_s = (R_s)_{\max}. \quad (4)$$

Substitution of this extremizing value of ωC_s in (2) and (3) gives

$$(R_s)_{\max} = R_0 + X_0^2/R_0 = \frac{1}{G_0}, \quad (5)$$

$$X_s = 0 = B_s. \quad (6)$$

Thus, the maximum value of R_s occurs at zero reactance; this defines *antiresonance for* Z_s. Evidently if C_s is positive, the maximum value of R_s occurs at positive values of X_0 or negative values of B_0 in $Y_0 = G_0 + jB_0$. Reference to curves of X_0 or B_0 in Sec. 30 shows that this implies a shift of antiresonance toward smaller values of $\beta_0 h$, just as occurs in Figs. 33.1 to 33.3. On the other hand, if C_s is negative, X_0 must be negative, B_0 positive, for maximum R_s, and this involves a shift in antiresonance toward larger values of $\beta_0 h$. This is also verified in Figs. 33.1 to 33.3. The value $(R_s)_{\max}$ as defined in (5) is plotted in Fig. 33.2. From this it is clear that whenever positive capacitances are connected in parallel with Z_0, the antiresonant resistance, $(R_s)_{\max}$, of the combination is lower than $(R_0)_{\max}$ and occurs for smaller values of $\beta_0 h$. Similarly, when negative capacitances are involved, $(R_s)_{\max}$ may exceed greatly $(R_0)_{\max}$ and it occurs for larger values of $\beta_0 h$. The largest possible value of $(R_s)_{\max}$, with the associated values of $\beta_0 h$, R_0, X_0, may be determined from Fig. 33.2 in conjunction with curves for R_0 and X_0 in Sec. 30. For $\Omega = 10$, for example,

$$(R_s)_{\max\,\max} \doteq 1000 \text{ at } \beta_0 h \doteq 3.2,$$
$$R_0 \doteq 200, \quad X_0 \doteq -400. \quad (7)$$

The value of ωC_s required to produce this extreme value is

$$\omega C_s = \frac{X_0}{R_0^2 + X_0^2} = -2 \times 10^{-3}, \quad (8)$$

for which

$$K = \frac{\omega C_s}{\beta_0 h} = -6.25 \times 10^{-4}. \quad (9)$$

It is evident from Figs. 33.1 to 33.3 that a small positive or negative capacitance in parallel with Z_δ may affect the impedance very greatly. Certainly it is not correct to expect agreement between a measured impedance Z_s and a theoretical impedance Z_δ unless the effective capacitance in parallel is virtually zero. Note that this condition may be achieved by connecting a lumped positive capacitance in parallel with the antenna of such magnitude that it effectively cancels the negative capacitance introduced by end effect and coupling.

34. Apparent Load Impedance of Antenna; Summaries of Theoretical and Experimental Results

The general, theoretical foundations required for determining the apparent impedances of antennas driven in various ways from two-wire lines are contained in the formulas and curves for the impedance Z_0 of an antenna driven by a discontinuity in a scalar potential. By supplementing Z_0 with corrections for a small base separation, for a negative series inductance, and for a shunt capacitance that may be positive or negative, information is provided for determining the apparent impedance Z_{sa} of the

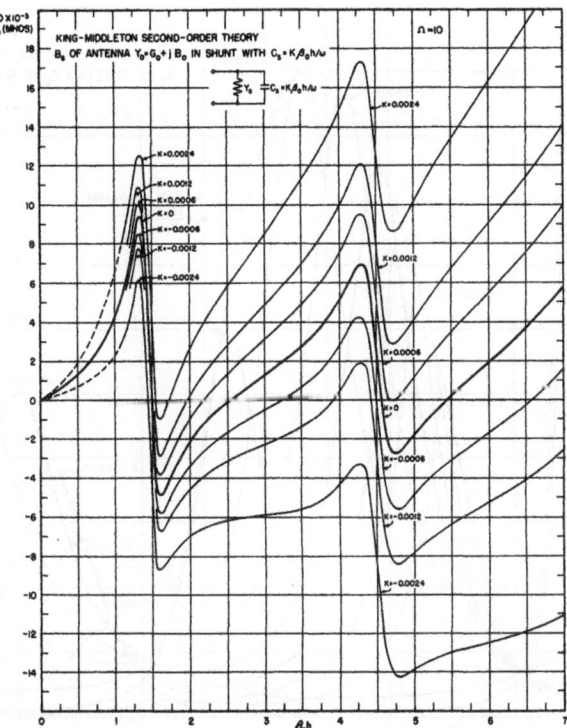

Fig. 33.1a. Apparent susceptance of antenna in shunt with capacitance; $\Omega = 10$.

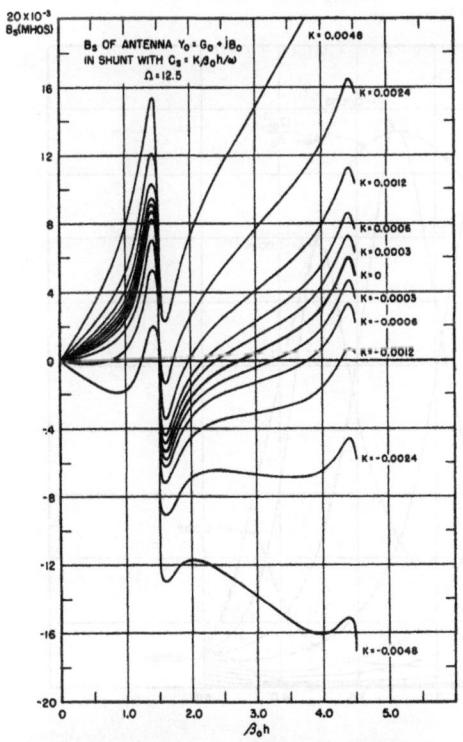

Fig. 33.1b. Apparent susceptance of antenna in shunt with capacitance; $\Omega = 12.5$.

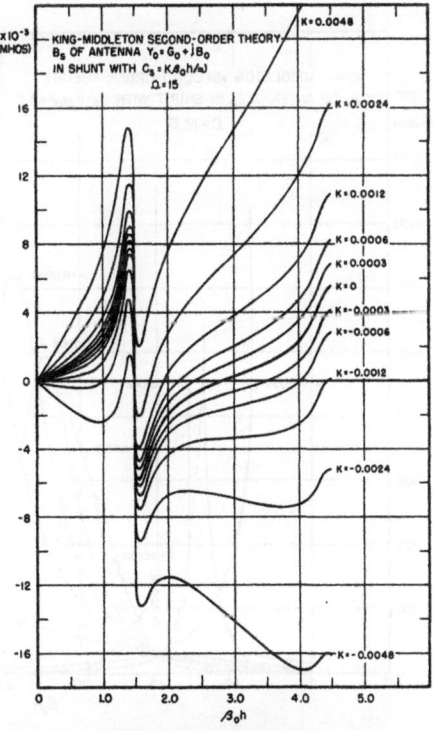

Fig. 33.1c. Apparent susceptance of antenna in shunt with capacitance; $\Omega = 15$.

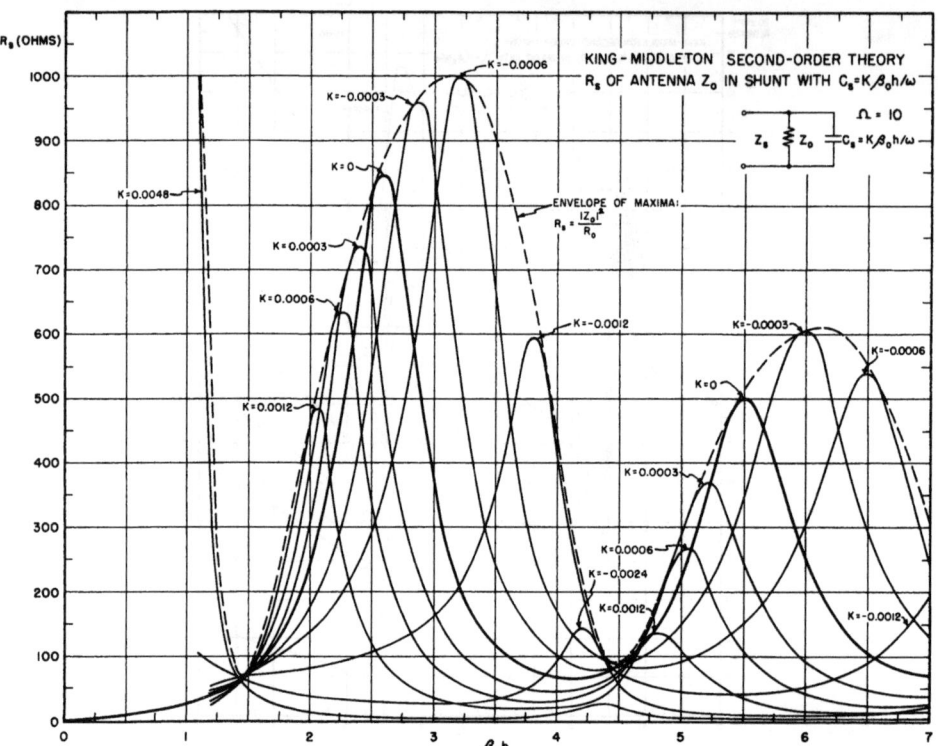

Fig. 33.2a. Apparent resistance of antenna in shunt with capacitance; $\Omega = 10$.

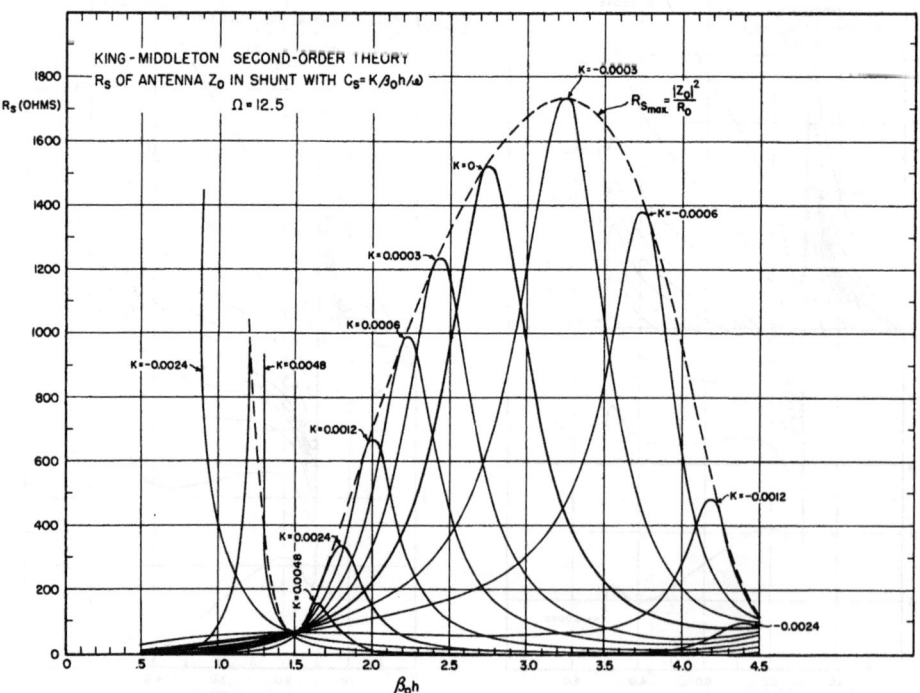

Fig. 33.2b. Apparent resistance of antenna in shunt with capacitance; $\Omega = 12.5$.

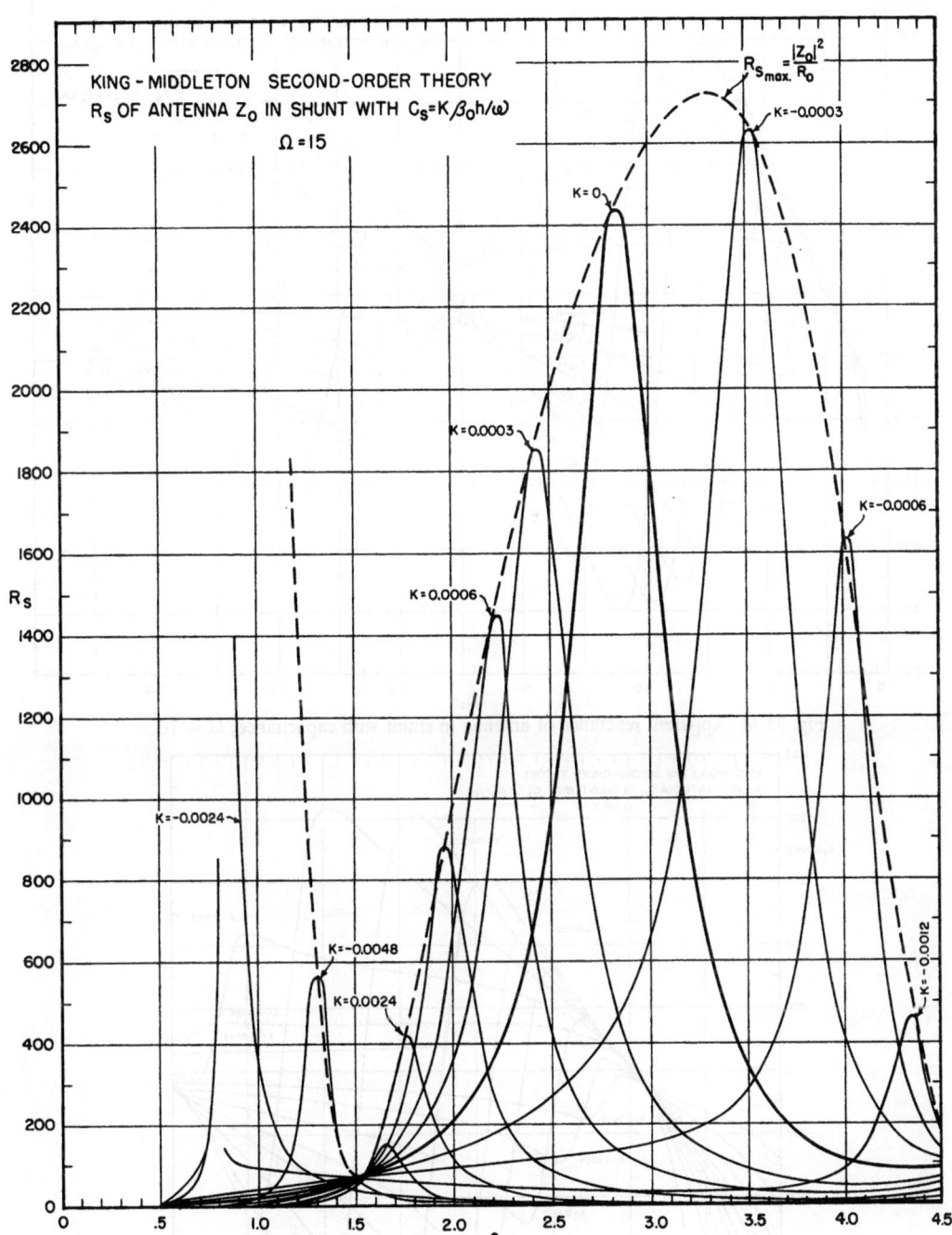

Fig. 33.2c. Apparent resistance of antenna in shunt with capacitance; $\Omega = 15$.

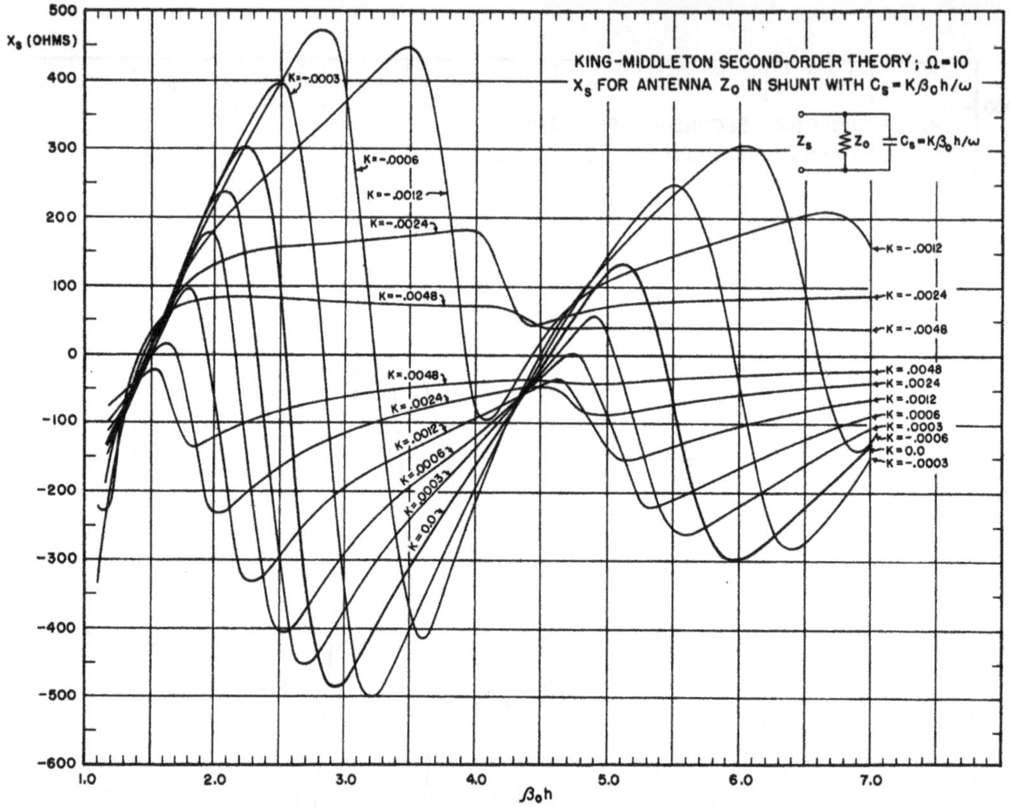

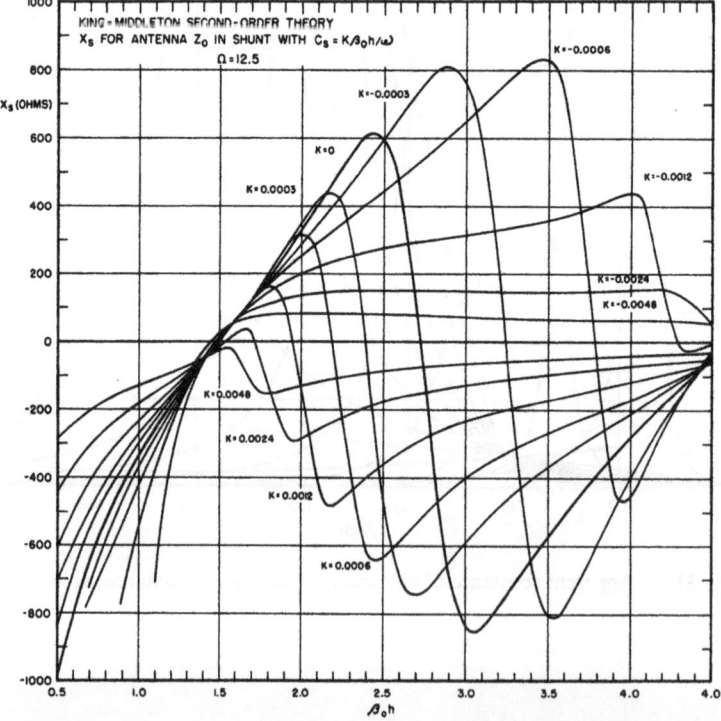

Fig. 33.3a. Apparent reactance of antenna in shunt with capacitance; $\Omega = 10$.

Fig. 33.3b. Apparent reactance of antenna in shunt with capacitance; $\Omega = 12.5$.

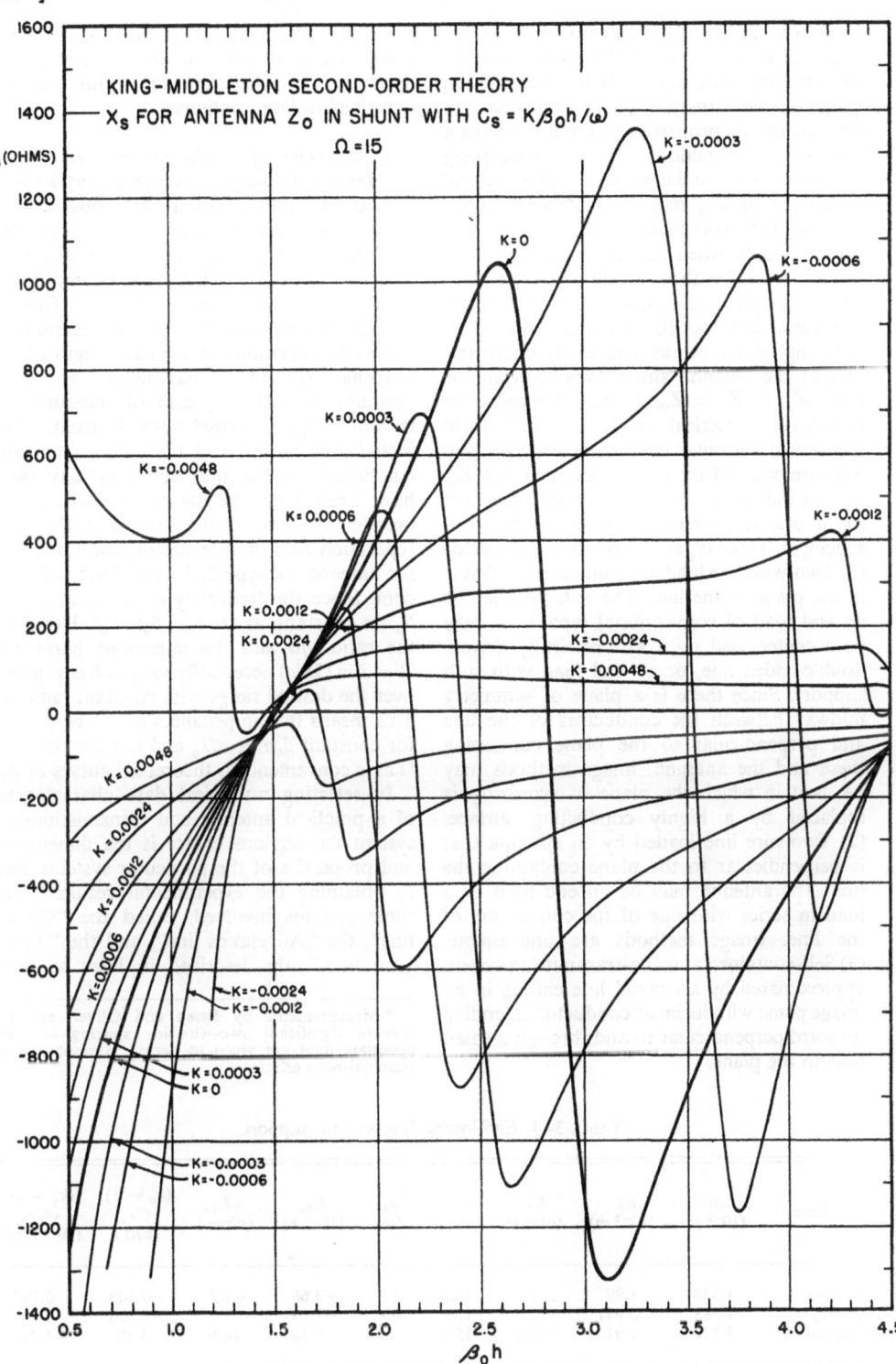

Fig. 33.3c. Apparent reactance of antenna in shunt with capacitance; $\Omega = 15$.

antenna as a load on a transmission line when this is analyzed by conventional line theory. Since the apparent impedance is not a function alone of the structure of the antenna but depends on the properties of the transmission line near the junction and on the coupling between antenna and line, the actual numerical evaluation of Z_{sa} requires a detailed specification of the complete geometry of line and antenna. Only when the line spacing b is so small compared with λ_0 that all base-separation effects, coupling effects, and end effects are negligible, is it possible to identify the terminating impedance as a quantity characteristic of the antenna alone. Whenever this is true, $Z_{sa} \doteq Z_\delta \doteq Z_0$. Present interest is in the direct numerical evaluation both from theoretical formulas and from experimental measurements of the apparent impedance Z_{sa} of a cylindrical antenna when used as a load for a practical transmission line with small spacing b. Three types of circuits are involved: (1) Two-wire line loaded by an antenna that is in the plane of the line. The antenna may be an end load of conventional type, or it may be a center load on a symmetrically driven, double-ended line, or an end load with stub support. Since there is a plane of symmetry midway between the conductors of the line and perpendicular to the plane containing these and the antenna, image methods may be used in which the plane of symmetry is replaced by a highly conducting surface. (2) Two-wire line loaded by an antenna that is perpendicular to the plane containing the line. The antenna may be an end load or a load in series with one of the conductors of the line. Image methods are unavailable. (3) Self-contained antenna-transmitting system approximated by a coaxial line ending in an image plane with its inner conductor extending outward perpendicular to and through a small hole in the plane.

Results obtained for circuits of all three types are compared in this section; the apparatus and experimental techniques are described in later sections.

1. *Summary of results for circuits of type 1—Antenna in plane of two-wire line.* Experimental procedures usually are most advantageous with an apparatus that is operated at a single frequency. Hence, impedance as a function of $\beta_0 h = \omega h/v_0$ must be impedance as a function of h with ω constant. That is, the actual independent variable in determining curves of impedance is the half-length of the antenna. All other parameters are kept constant in each sequence of measurement in which $\beta_0 h$ is varied with β_0 fixed. They include the radius a of the antenna and the conductors of the line (for simplicity these have been kept the same), the distance b between centers of the line, and the base separation 2δ. For antennas of type 1, $b = 2\delta$; for antennas of type 2, b and 2δ are independent. Since the frequency is constant, ω and β_0 are constant, as are $\beta_0 a$, $\beta_0 b$, $\beta_0 \delta$. However, the ratio h/a and the important parameter $\Omega = 2 \ln (2h/a)$ necessarily vary as h is adjusted over the desired range with constant radius a. This means that impedances are to be plotted for constant $\beta_0 a$ or a/λ_0 and not for constant Ω as is convenient for theoretical curves of Z_0.

In selecting numerical data characteristic of a practical antenna and transmission-line system the obvious choice is the dimensions and properties of the particular systems used in obtaining the experimental results.* The three systems involved, called the "Conley line", the "Angelakos line", and the "Tomiyasu line," are described in later sections.

* Measurements by Essen and Oliver, ref. 17, involve significant two-wire-line spacing and inadequate data are given to permit computation of terminal-zone effects.

TABLE 34.1. End-loaded line without support.

Line	a $(10^{-3}$ m)	b† $(10^{-2}$ m)	R_c (ohms)	c_0 ($\mu\mu$f)	λ_0 (m)	L_{T_e} $(10^{-9}$ h)	ωL_{T_e} (ohms)	$(k_q = 1)$ C_{T_e} ($\mu\mu$f)	$(k_q = 1)$ ωC_{T_e} $(10^{-3}$ mho)
Tomiyasu	1.538	1.99	308	10.8	1.0	−3.66	−6.9	−0.109	−0.205
Conley	1.588	0.951	212	15.8	0.4	−1.53	−7.2	−0.106	−0.50
Angelakos	3.17	1.91	215.4	15.5	0.4	−3.18	−14.9	−0.26	−1.22

† For the Angelakos line $b_a = \frac{1}{2}b[1 + \sqrt{1 - (2a/b)^2}]$ is given.

Essential data are listed in Table 34.1 for the three different lines when used with an end load, in Table 34.2 when used with a center load and double-end drive, and in Table 34.3 when used with an end load with stub support. This means that the end-loaded line is a symmetrical line with two conductors and no image plane as in Fig. 34.1a, whereas the center-loaded line consists of only one line conductor over an image plane as in Fig. 34.2b. This choice of equivalent circuits is made since the impedance actually evaluated is Z_{sa} and the impedance of the antenna with base separation (in the mathematical sense of V_δ/I_δ) is Z_δ, so that no conversion factors are required. Since the shunt capacitance C_{TC} as calculated in Sec. 9 for the center-loaded line is across each half of the line just as is Z_δ, the effective capacitance for each half is C_{TC} just as the effective load is $2Z_\delta$ as shown in Fig. 34.2a. For the image line all impedances are halved and admittances doubled so that $2C_{TC}$ is in parallel, and $\tfrac{1}{2}L_{TC}$ in series, with Z_δ.

The several quantities in Table 34.1 were calculated using the following formulas taken from Sec. 8:

$$L_{Te} = -\frac{b_e - a}{2\pi v_0}, \qquad (1a)$$

$$\omega L_{Te} = -\frac{\zeta_0 \beta_0}{2\pi}(b_e - a) \doteq -\frac{\zeta_0 \beta_0 \delta}{\pi}, \qquad (1b)$$

$$C_{Te} = -c_0 K b_e \sinh^{-1} \frac{d}{k b_e}, \qquad (1c)$$

where

$$c_0 = \frac{\pi \varepsilon_0}{\ln(b_e/a)}, \qquad (2a)$$

$$b_e = \frac{b}{2}\left[1 + \sqrt{1 - \left(\frac{2a}{b}\right)^2}\,\right], \qquad (2b)$$

and K and k are evaluated for $k_q = 1$ as is explained in Sec. 8.

Similarly, for Table 34.2, as obtained from Sec. 9,

$$L_{TC} = -\frac{b_e - a}{\pi v_0}, \qquad (3a)$$

$$\omega L_{TC} = -\frac{\zeta_0 \beta_0}{\pi}(b_e - a) \doteq -\frac{2\zeta_0}{\pi}\beta_0 \delta, \qquad (3b)$$

$$C_{TC} = -c_0 K b_e \sinh^{-1} \frac{d}{k b_e}, \qquad (3c)$$

where K and k are evaluated for $k_q = 1$ as explained in Sec. 9. The appropriate equivalent networks are shown in Figs. 34.1 and 34.2. The terminal-zone distance $d = 10b$ was used for the Conley and Tomiyasu lines. The Angelakos line is discussed on page 209.

The sequence of calculation involves the following steps:

1. Determination of Ω as a function of $\beta_0 h$ as the latter is varied over the desired range for the specified values of $\beta_0 a$. For the Conley line, $\beta_0 a = 0.025$; for the Tomiyasu line, $\beta_0 a = 0.0097$, for the Angelakos line, $\beta_0 a = 0.050$.

Conley line: $\Omega = 2\ln\dfrac{2h}{a} = 2\ln\dfrac{2\beta_0 h}{0.025}$

$\phantom{\text{Conley line: }\Omega} = 2\ln 80\beta_0 h.$ (4a)

Tomiyasu line: $\Omega = 2\ln 207\beta_0 h,$ (4b)

Angelakos line: $\Omega = 2\ln 40\beta_0 h,$ (4c)

TABLE 34.2. Center-loaded line.

Line	$L_{Tc}/2$ (10^{-9} h)	$\omega L_{Tc}/2$ (ohms)	($k_q=1$) $2C_{Tc}$ ($\mu\mu$f)	($k_q=1$) $2\omega C_{Tc}$ (10^{-3} mho)
Tomiyasu	−3.66	−6.9	−0.26	0.485
Conley	−1.53	−7.2	−0.236	−1.1

TABLE 34.3. End-loaded line with stub support.

Line	$L_{Ts} = L_{Te}$ (10^{-9} h)	$\omega L_{Ts} = \omega L_{Te}$ (ohms)	($k_q=1$) $2C_{Ts} = 2C_{Tc}$ ($\mu\mu$f)	($k_q=1$) $2\omega C_{Ts} = 2\omega C_{Tc}$ (10^{-3} mho)
Tomiyasu	−3.66	−6.9	−0.26	−0.485
Conley	−1.53	−7.2	−0.236	−1.1
Angelakos	−3.18	−14.9	−0.63	−2.97

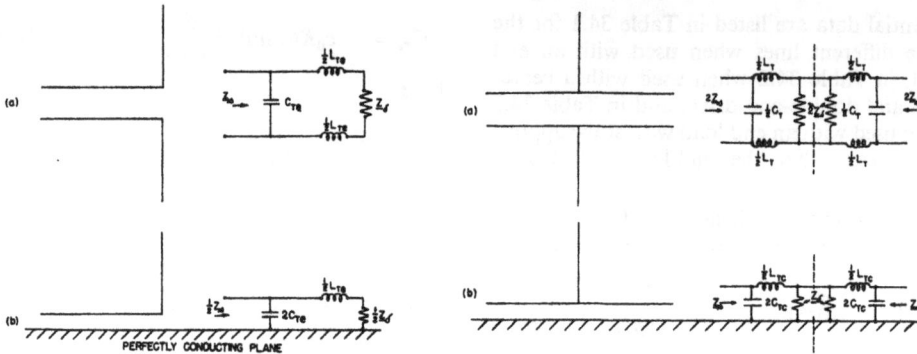

Fig. 34.1. Equivalent networks of antennas as end loads.

Fig. 34.2. Equivalent networks of antennas as center loads.

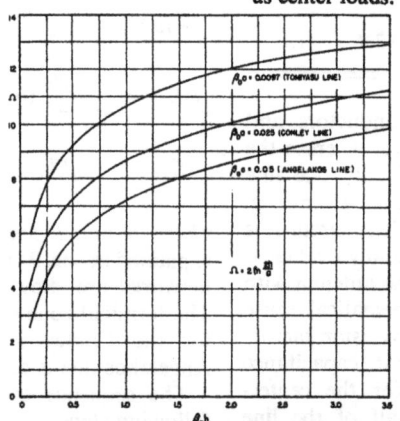

Fig. 34.3. Variation of $\Omega = 2 \ln 2h/a$ with $\beta_0 h$ for three lines.

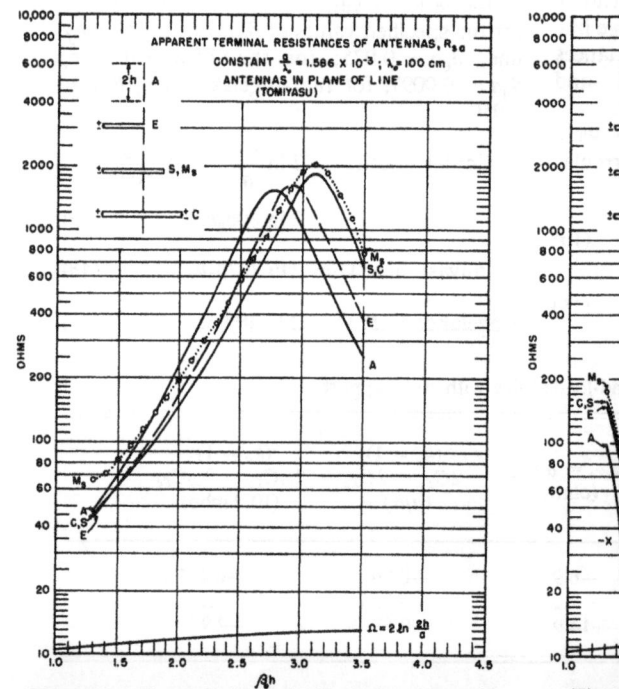

Fig. 34.4a. Theoretical (E, S, C) and measured (M_s) apparent terminal resistances of antennas (Tomiyasu).

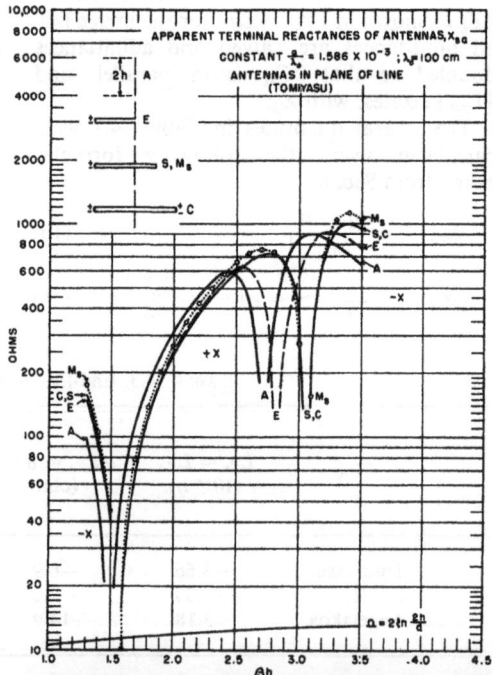

Fig. 34.4b. Theoretical (E, S, C) and measured (M_s) apparent terminal reactances of antennas (Tomiyasu).

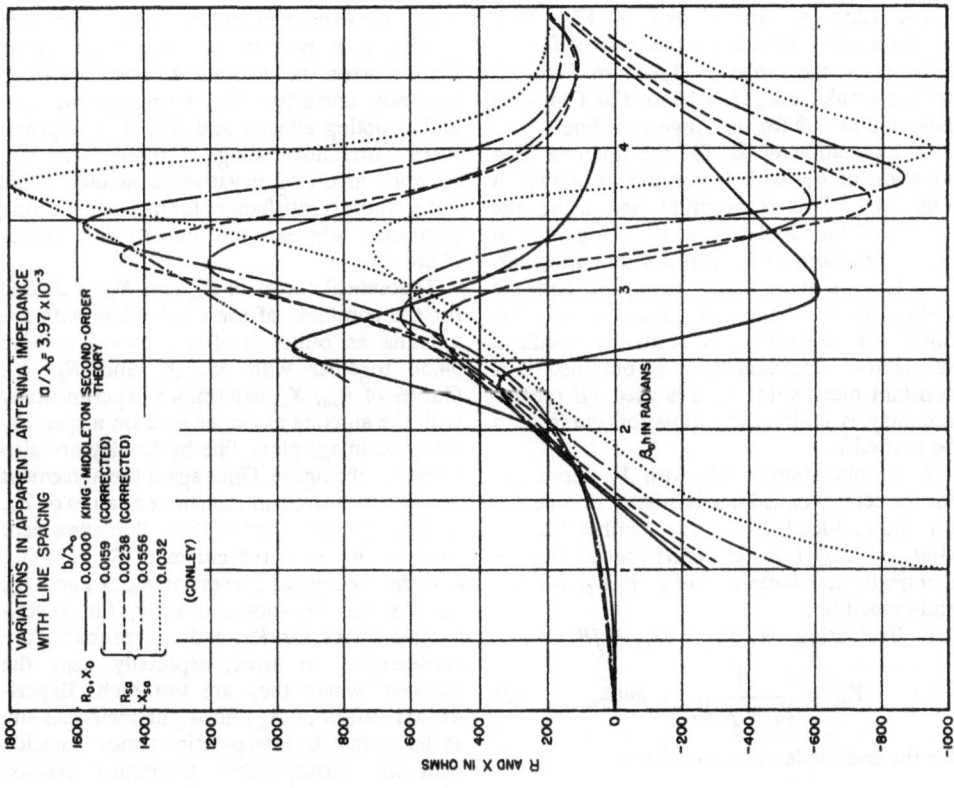

Fig. 34.5b. Variations in measured apparent antenna impedance with line spacings; theoretical impedance for "zero" line spacing.

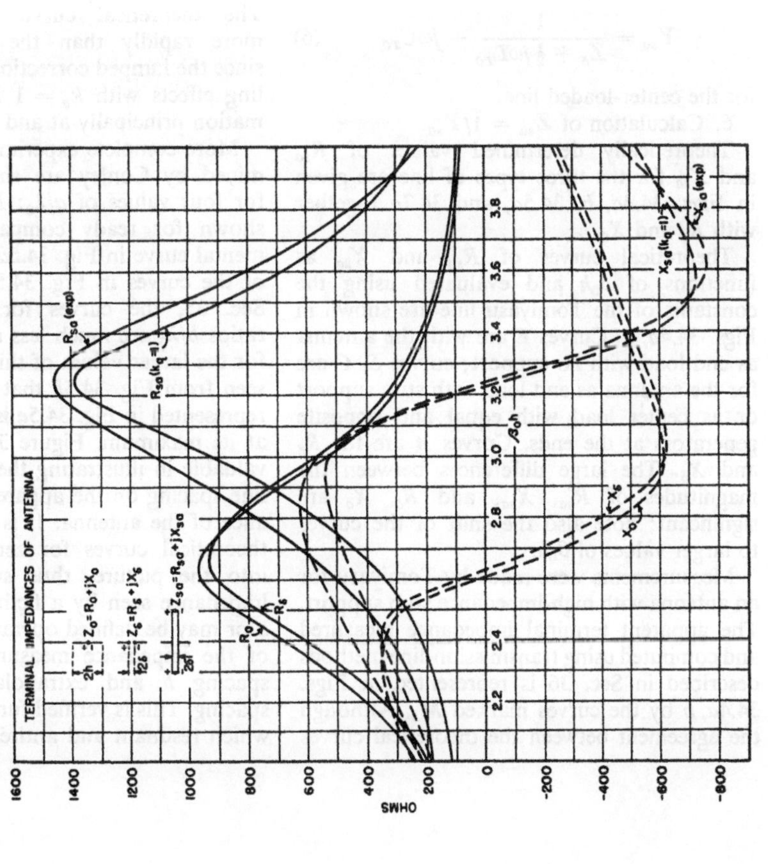

Fig. 34.5a. Theoretical apparent impedance of antenna as center load.

These functions are plotted in Fig. 34.3. Evidently $\Omega = 10$ is a fair mean value for the antenna in the range of $\beta_0 h$ from 1 to 3.5 for the Conley line, $\Omega = 12$ for the Tomiyasu line, and $\Omega = 9$ for the Angelakos line.

2. Evaluation of Z_0 for the antenna as a function of $\beta_0 h$ with Ω varying as shown in Fig. 34.3. This is accomplished using the curves of Fig. 30.9 and 30.10 giving R_0 and X_0 as functions of Ω with $\beta_0 h$ as parameter.

3. Computation of Z_δ using Z_0 and the correction for finite $\beta_0 h$ given in Sec. 32. Since this correction is relatively small, a satisfactory approximation is obtained if a constant mean value of Ω is used. (If greater accuracy is desired, interpolation curves can be plotted.)

4. Combination of ωL_T with X_δ. Since L_{TC} for the center-loaded line is exactly double L_{Te} for the end-loaded line, the combination of $(\frac{1}{2}\omega L_{TC} + X_\delta)$ for the center-loaded line is identically the same as $(\omega L_{Te} + X_\delta)$ for the end-loaded line.

5. Evaluation of $Y_{sa} = G_{sa} + jB_{sa}$ from

$$Y_{sa} = \frac{1}{Z_\delta + j\omega L_{Te}} + j\omega C_{Te} \quad (5)$$

for the end-loaded line, and from

$$Y_{sa} = \frac{1}{Z_\delta + \frac{1}{2}j\omega L_{TC}} + j\omega C_{TC} \quad (6)$$

for the center-loaded line.

6. Calculation of $Z_{sa} = 1/Y_{sa}$.

Theoretically determined values of R_{sa} and X_{sa} for the three types of line are given in Figs. 34.4a, b, 34.5a, and 34.7a together with R_0 and X_0.

Theoretical curves of R_{sa} and X_{sa} as functions of $\beta_0 h$ and evaluated using the constants of the Tomiyasu line are shown in Figs. 34.4a, b. Curves E are with the antenna as end-load with no support; curves S, C are for the antenna as end load with stub support or as center load with equal and opposite generators at the ends. Curves A are for R_δ and X_δ. The large differences between the magnitudes of R_{sa}, X_{sa}, and R_δ, X_δ are significant; note also the shift of the curves to larger values of $\beta_0 h$.

Measurements were made by Tomiyasu for an antenna with high-impedance stub support. The apparent terminal impedance measured and computed using transmission-line methods described in Sec. 36 is represented in Figs. 34.4a, b by the curves marked M_s. Although the agreement between the theoretical curves S and the experimental curves M_s is far from perfect, it is nevertheless very much better than between the theoretical curves A (which are not corrected for terminal-zone end and coupling effects) and M_s. The approximate correction using a lumped-constant network involving negative capacitances and inductances is satisfactory for most engineering purposes, whereas the uncorrected curves A are not.

Theoretical curves of R_{sa} and X_{sa} evaluated for the constants of the Conley line with the antenna as center load are shown in Fig. 34.5a together with R_0, X_0 and R_δ, X_δ. Curves of R_{sa}, X_{sa} determined experimentally with the antenna as center load on a specially designed image-plane line by Conley are also shown in the figure. Once again the agreement between measured and calculated values of Z_{sa} is incomparably better than the agreement between the measured curves of R_{sa} and X_{sa} and the theoretical curves for R_0, X_0 and R_δ, X_δ. As will be shown directly, the experimental curves for R_{sa} and X_{sa} probably are considerably in error, especially near the maxima, where they are too high. Experimental values of R_0 and X_0 at their maxima as determined by interpolation almost coincide with the corresponding theoretical results. The theoretical curve for R_{sa} decreases more rapidly than the experimental one, since the lumped correction for end and coupling effects with $k_g - 1$ is a good approximation principally at and near antiresonance.

More complete experimental data as determined by Conley are shown in Fig. 34.5b for four values of a/λ_0; R_0 and X_0 also are shown for ready comparison. The experimental curve in Fig. 34.5a coincides with one of the curves in Fig. 34.5b. As explained in Sec. 35, the curves for the two smallest ratios b/λ_0 are much less accurate than those for the larger values of this ratio. It is readily seen from Fig. 34.5b that precisely the curve represented in Fig. 34.5a is definitely too high at its maximum. Figure 34.5b is particularly valuable in illustrating the great effect of the line spacing on the apparent terminal impedance of the antenna. It is also clear that the theoretical curves for zero spacing fit well into the picture, thus suggesting that the impedance seen by a fictitious "slice" generator may be defined operationally as the limit of the impedance measured with finite line spacing b and extrapolated to zero line spacing. This is verified further in Fig. 34.6, in which resonant and antiresonant lengths and resistances are shown as a function of Ω. Since radiation from the rather widely spaced line is not negligible compared with the resistance at resonance, the approximate radiation resistance of the line ($R^e_{\text{line}} \doteq 60\beta_0^2 b^2$ for a matched line) is subtracted from the apparent resistance R_{sa} at resonance and plotted as $R_{sa} - R^e_{\text{line}}$ in Fig. 34.6a. The vanishing of the apparent reactance X_{sa} does not constitute exactly resonance for the antenna. This is given approximately by the vanishing of $X_{sa} - \omega L_{Tc}$. Since zeros of $X_{sa} - \omega L_{Tc}$ occur at slightly different values of $\beta_0 h$, the resonant resistance is modified slightly, as shown by the lowest of the three curves in Fig. 34.6a. No explanation in terms of theoretical curves is available to explain the increase in resistance with line spacing even after correction has been made for radiation and inductive end effect. Actually, and as seen from Fig. 34.4c as well, the agreement between theory and experiment is poorest in the value of the resistance near resonance. Experimental results clearly indicate an upward bump on the resistance curve near $\beta_0 h = \pi/2$ considerably greater than is predicted by the second-order theory. This is confirmed by the third-order values in Table 30.15, which, as shown in Sec. 38, are in excellent agreement with experiment. It is seen in Fig. 34.6c that $\beta_0(h - \delta)_{\text{res}}$ with the correction for inductive end effect is a straight line, in agreement with the general predictions of the theory. The antiresonant length $\beta_0 h_{\text{antires}}$ is in Fig. 34.6d together with $\beta_0(h - \delta)_{\text{antires}}$. Since the base separation 2δ does not have the simple effect at antiresonance as at resonance, the electrical half-length $\beta_0(h - \delta)_{\text{antires}}$ of the conductor is not actually useful. The differences between $\beta_0 h_{\text{res}}$ in Fig. 34.6d and R_{sa} in Fig. 34.6b and horizontal lines through the theoretical points at $b/\lambda_0 = 0$ are due principally to capacitive end effect. Since it is not adequate to represent this with lumped elements when b/λ_0 exceeds about 0.03, no attempt is made to draw R_0 and its antiresonant half-length in Figs. 34.6. The agreement for a smaller line spacing is verified in Fig. 34.5a.

The line used by Angelakos differs only in details from that used by Conley, except that instead of measuring Z_{sa} with the antenna as center load with the line driven symmetrically from each end, as did Conley, Angelakos measured Z_{sa} with the line end loaded by the antenna using both Styrofoam supports and a high-impedance stub support. The experimental impedance curves are shown in Fig. 34.7a and 34.7b. The difference between the two curves in Fig. 34.7b is due entirely to the effect of the high-impedance stub.

In determining the apparent terminal impedance Z_{sa} from the ideal theoretical value Z_δ using a lumped-constant network, Angelakos did not assume the average charge-ratio factor $k_q \doteq \bar{q}_T/\bar{q}_L$ equal to unity but determined it assuming a cosinusoidal distribution of charge continuing smoothly from the antenna onto the line. That is,

$$k_q = \frac{\int_{\beta_0(h'-g)}^{\beta_0 h'} \cos \beta_0 x \, dx}{\int_{\beta_0 h'}^{\beta_0(h'+g)} \cos \beta_0 x \, dx}$$

$$= \frac{\tan \tfrac{1}{2}\beta_0 g + \cot \beta_0 h'}{-\tan \tfrac{1}{2}\beta_0 g + \cot \beta_0 h'}, \quad (7)$$

where g is the distance from the junction along the antenna and line over which the charge is averaged and $h' = h - \delta$. Since C_T is important only when $\cot \beta_0 h'$ is large, the value of g is not critical if $\beta_0 g$ is small. Angelakos chose $g = 0.05\lambda_0$ to determine k_q from (7) for use in (8.18). He evaluated C_T from a formula like (8.23) using $d = 20b_e$, which is too large when $b_e = 0.05\lambda_0$. Since only qualitative results can be expected when b_e/λ_0 is so large, and Angelakos' results are not altered greatly if a smaller value of d is used, his Table 34.3 and Fig. 34.7a are reproduced. L_T was not used and the values of Z_0 differ slightly from those in recently corrected tables. Numerical values are given for the magnitudes of k_q and C_T and the impedances Z_0, Z_{sa} (theoretical with $k_q = 1$), Z_{sa} (theoretical with k_q as computed using (7)), and Z_{sa} (measured). It is seen that k_q differs considerably from unity and that the value of Z_{sa} calculated using the variable k_q agree better with measured values than those calculated with $k_q = 1$ except near antiresonance, where the variable k_q is near unity. Experimentally determined values of k_q are given in Table 34.4 together with theoretical ones derived from (7).

It may be concluded that the lumped-constant terminal-zone networks introduced to take account of end and coupling effects for antennas driven from open-wire lines

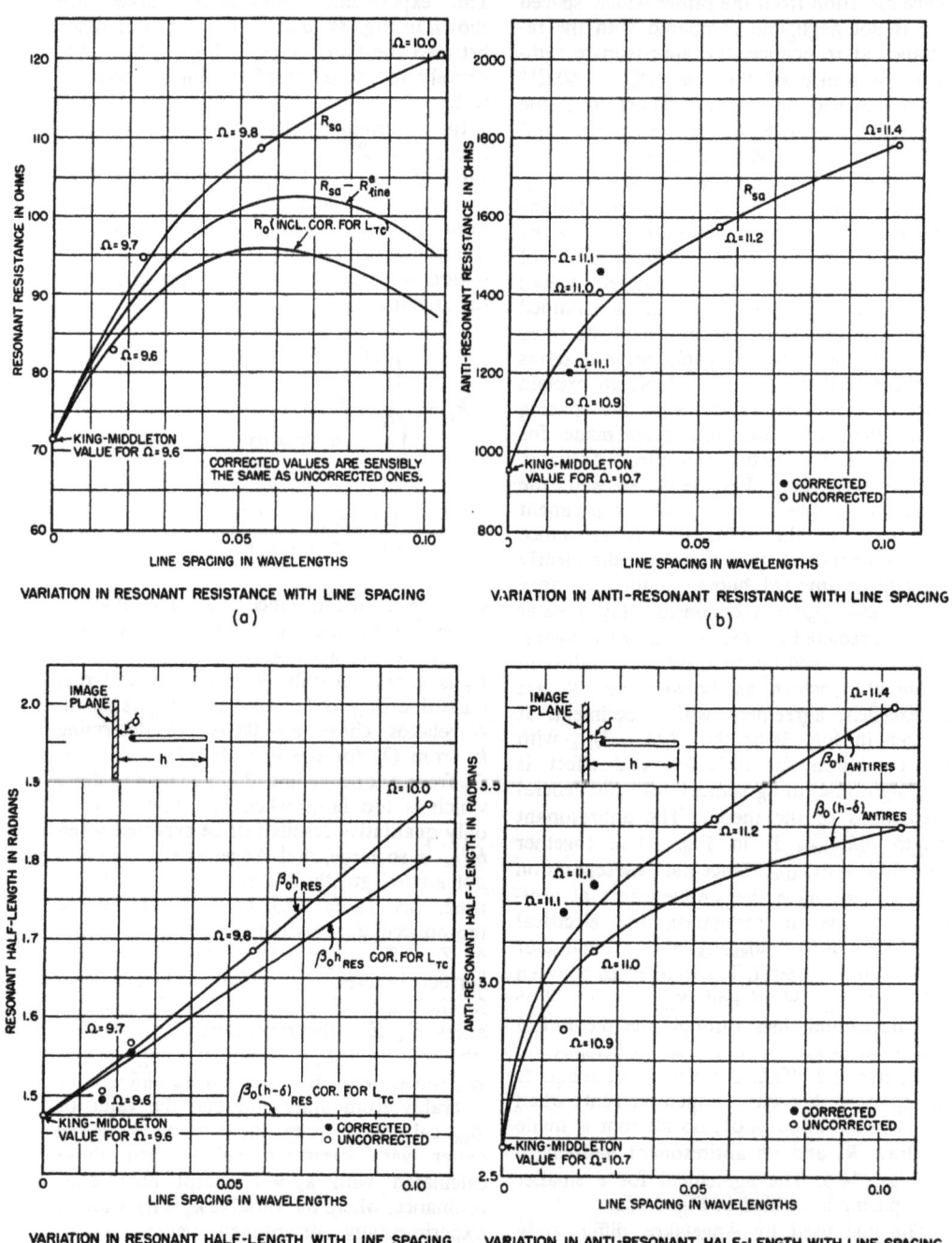

Fig. 34.6. Measured resistances and resonant lengths of antenna as center load as function of line spacing b near resonance and antiresonance.

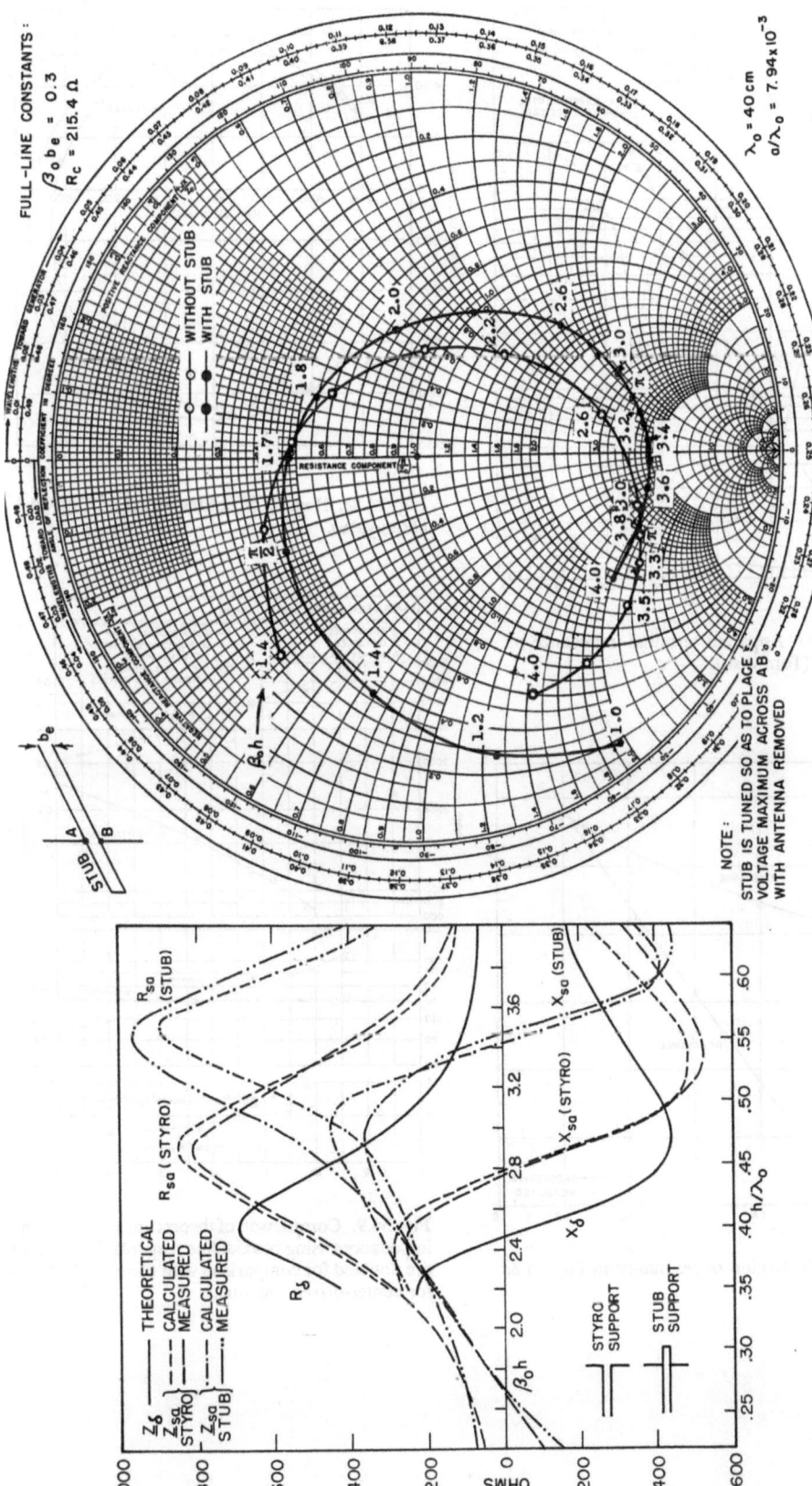

Fig. 34.7b. Smith-chart graph of measured impedances from Fig. 34.7a.

Fig. 34.7a. Impedance of antenna with and without stub support.

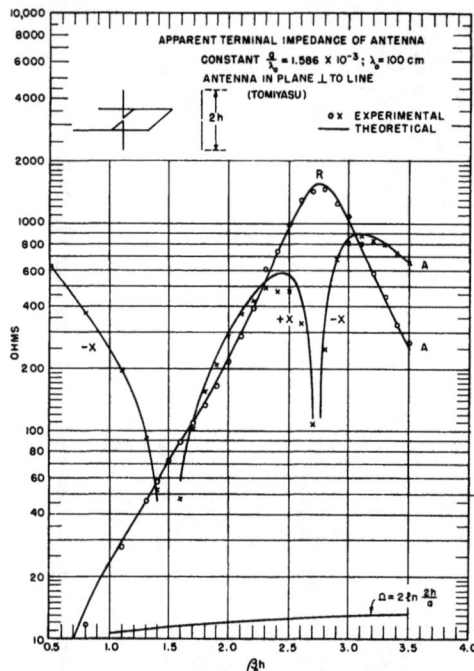

Fig. 34.8a. Apparent terminal impedance of antenna (Tomiyasu).

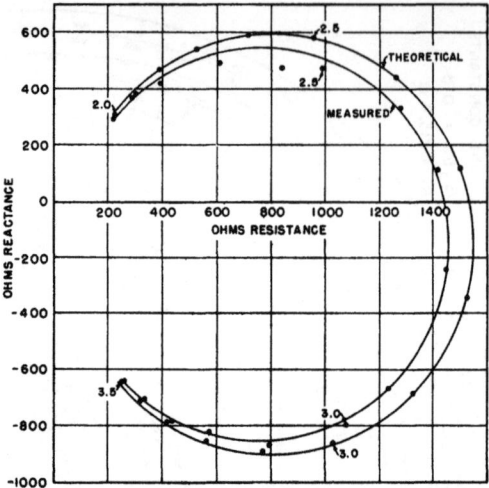

Fig. 34.8c. Region of antiresonance in Fig. 34.8a (Tomiyasu).

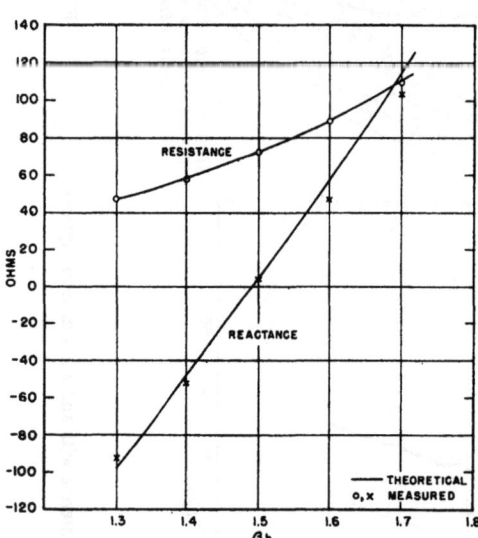

Fig. 34.8b. Region of resonance in Fig. 34.8a (Tomiyasu).

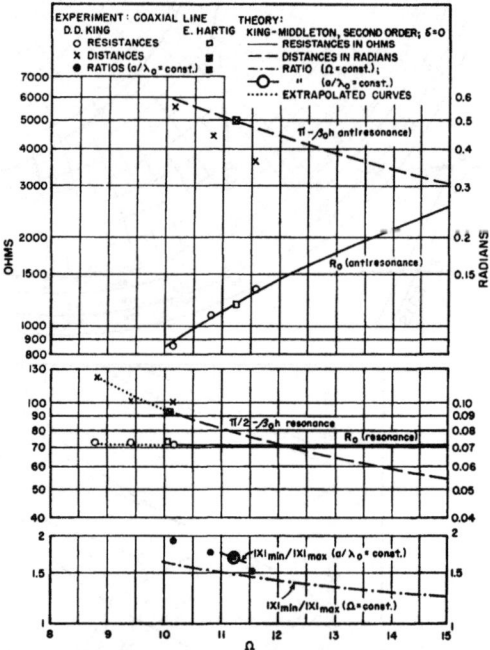

Fig. 34.9. Comparison of theoretical and measured impedances using coaxial line; experimental values are doubled for comparison with theoretical curves for center-driven antenna.

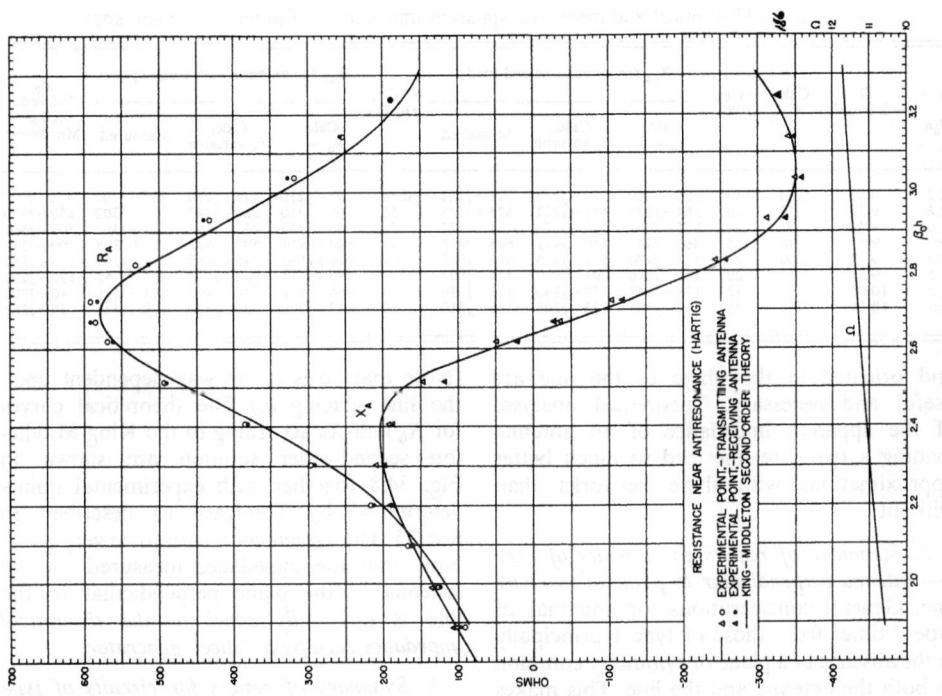

Fig. 34.10b. Comparison of theoretical and measured impedances near antiresonance using coaxial line (Hartig).

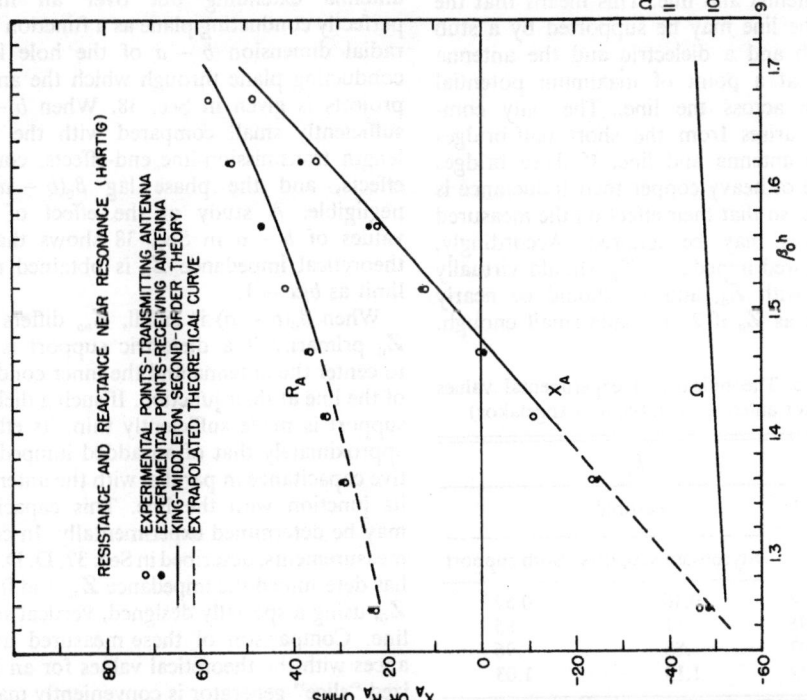

Fig. 34.10a. Comparison of theoretical and measured impedances near resonance using coaxial line (Hartig).

TABLE 34.4. Theoretical and measured apparent impedances of antenna. (Angelakos)

$\beta_0 h$	$\Omega = 2\ln 2h/a$	Calc. k_q	$-C_{T_\delta}$ ($\mu\mu f$)	Z_{sa} for antenna as end load			$-2C_{T_c}$ ($\mu\mu f$)	Z_{sa} for antenna with stub support			Z_0 2nd-order King-Middleton
				Calc., $k_q=1$	Calc., k_q variable	Measured		Calc., $k_q=1$	Calc., k_q variable	Measured	
2.2	8.95	0.54	0.14	224+j269	307+j273	293+j181	0.36	97+j210	176+j254	213+j192	430+j227
2.6	9.29	.77	.20	564+j304	657+j221	625+j172	.56	169+j310	264+j340	327+j302	650−j230
3.0	9.58	.91	.24	865−j262	810−j326	754−j327	.58	246+j333	430+j460	581+j362	300−j440
π	9.67	.96	.25	560−j427	540−j433	646−j452	.61	540+j424	590+j407	732+j308	200−j370
3.3	9.77	1.00	.26	517−j470	517−j470	517−j517	.63	740+j160	740+j160	885+j207	155−j350
3.5	9.89	1.07	.28	280−j386	301−j450	334−j504	.67	830+j 34	900−j140	959−j 52	122−j290
3.8	10.00	1.19	.32	138−j247	153−j248	183−j396	.73	366−j423	540−j400	663−j431	80−j200
4.0	10.16	1.32	.35	112−j195	133−j205	134−j310	.80	237−j226	320−j330	409−j431	72−j165

and oriented in the plane of the line are useful and necessary. Theoretical analyses of the apparent impedance of an antenna loading a two-wire line lead to much better approximations with these networks than without.

2. *Summary of results for circuits of type 2—Antenna perpendicular to plane of two-wire line.* General considerations for antennas of type 2 differ from those of type 1 principally in the absence of a plane of symmetry common to both the antenna and the line. This makes the image-plane techniques incorporated in the Conley line unavailable. Since the antenna is perpendicular to and in the neutral plane of the two-wire line, there is no coupling between antenna and line. This means that the end of the line may be supported by a stub or a stub and a dielectric and the antenna attached at a point of maximum potential difference across the line. The only complication arises from the short half-bridges that join antenna and line. If these bridges are made of heavy copper their inductance is negligible, so that their effect on the measured impedances may be ignored. Accordingly, the measured impedance Z_{sa} should virtually coincide with Z_δ, and Z_δ should be nearly the same as Z_0 if 2δ is made small enough.

TABLE 34.5. Theoretical and experimental values of k_0 for antenna as end-load. (Angelakos)

$\beta_0 h$	Calc.	Measured	
		Styrofoam supports	Stub support
2.2	0.54	0.46	0.52
π	.96	.81	.85
3.5	1.07	.88	.96
4.0	1.32	1.19	1.03

(Note that 2δ is in no way dependent upon the line spacing b.) The theoretical curves for R_0 and X_0 according to the King-Middleton second-order solution are shown in Fig. 34.8 together with experimental points determined by Tomiyasu as described in Sec. 35. The agreement is seen to be very good. Note that the impedance measured for the antenna in the plane perpendicular to the line is *essentially equal to the theoretical impedance seen by a "slice" generator.*

3. *Summary of results for circuits of type 3—Antenna driven from coaxial line through a hole in the image plane.* A theoretical and experimental study of the apparent terminal impedance loading a coaxial line due to an antenna extending out over an infinite, perfectly conducting plane as a function of the radial dimension $b - a$ of the hole in the conducting plane through which the antenna projects is given in Sec. 38. When $b - a$ is sufficiently small compared with the wavelength transmission-line end-effects, coupling effects, and the phase lag $\beta_0(b-a)$ are negligible. A study of the effect of large values of $b - a$ in Sec. 38 shows that the theoretical impedance Z_0 is obtained in the limit as $b/a \to 1$.

When $\beta_0(b - a)$ is small, Z_{sa} differs from Z_0 primarily if a dielectric support is used to center the antenna and the inner conductor of the line at their junction. If such a dielectric support is made sufficiently thin, its effect is approximately that of an added lumped positive capacitance in parallel with the antenna at its junction with the line. This capacitance may be determined experimentally. In careful measurements, described in Sec. 37, D. D. King has determined the impedance Z_{sa} and from it Z_0, using a specially designed, vertical slotted line. Comparison of these measured impedances with the theoretical values for an idealized "slice" generator is conveniently made in

Fig. 34.9 by plotting experimental values with the theoretical curves of the most critical quantities, namely, antiresonant lengths and resistances, resonant lengths and resistances, and ratios of maximum to minimum reactance. These data of D. D. King are supplemented by measurements made by E. Hartig (see Sec. 38) using the same line but with a Styrofoam dielectric support and with an improved ground screen. Points determined by Hartig are also shown in Fig. 34.9 and complete ranges near resonance and antiresonance in Figs. 34.10a and 34.10b. These last two figures include experimental points determined by a method described in Sec. IV.8, in which the antenna under test is used for reception instead of for transmission. Throughout, the agreement between theory and experiment is seen to be good. Note that the experimentally determined ratios of maximum to minimum reactance are obtained from curves in which a/λ_0 is kept constant while h is varied. Consequently, the values of $\Omega = 2 \ln (2h/a)$ are not the same at maximum reactance, maximum resistance, and minimum reactance. These ratios are plotted at the value of Ω corresponding to that for the associated value of maximum resistance. The theoretical curve for the ratio of maximum to minimum reactance for constant Ω is shown together with a single theoretical point for this ratio obtained from Fig. 34.10b in which the theoretical curve is determined for constant a/λ_0.

A review of the several sets of experimental data and the corresponding theoretical values indicates that appropriate application of antenna and transmission-line theory permits the prediction within reasonable approximation of the measurable *apparent* terminal impedance of cylindrical antennas driven from actual transmission lines of various types using different connections. It is important to repeat that the type of line and the nature of the connection are significant in determining the apparent impedance. A quantitative study of the accuracy of the second-order theory for antennas with $7 \leq \Omega \leq 11$ is given in Sec. 38.

35. *Impedance Measurements with Open-Wire Lines; the Image-Plane Line**

The design of an antenna–two-wire-line system that is suitable for moderately precise measurement of the apparent impedance loading the line is not simple. Numerous experimental and constructional difficulties are involved, quite apart from the problems of generators and sensitive detectors that are characteristic of all measuring lines. Several of the more important difficulties characteristic of open-wire lines are as follows: (1) A first major difficulty of open-wire lines is the requirement that the line with its generator and load be completely symmetric and balanced so that the currents on the two conductors of the line and the halves of the antenna are exactly equal in magnitude and opposite in direction. With commercially available unbalanced oscillators this is not easily accomplished. Extensive balancing networks and suppressors of unbalanced currents are required. Failure to obtain equal and opposite currents on a two-wire line is primarily responsible for the fallacious belief that two-wire lines radiate excessively. If the line is balanced and the spacing is well below one-tenth wavelength, radiation and pickup are very small. If the currents are not exactly equal and opposite, the two conductors of the line form an effective antenna with large radiation and pickup. Essential aspects of the balancing problem are considered later. Alternatively, two-wire line systems of type 1 (antenna in plane of line) may be studied using an equivalent half-system over a conducting image plane. In this manner, the entire problem of balancing the line may be eliminated. (2) A second major difficulty is the problem of supporting the line so that (a) there are no supports along the active part of the line, or (b) the supports introduce an effect that is known to be negligible, or (c) the effect of the supports can be measured accurately and taken into account in the equivalent network of the terminal zone. Of these three possibilities, the first is the best and is readily achieved on a center-loaded double-ended line. (3) A third difficulty involved in open-wire line measurements is the problem of eliminating the effect of the observer, auxiliary apparatus, walls, etc. This may be accomplished in one of two ways. (a) An image line is used and the observer with all equipment is behind the conducting image plane, and the antenna, erected on the front of the conducting plane, is at sufficient distances from all walls, etc. (b) A very long line is used so that the measuring end is sufficiently far from the load. For systems of type 1, both alternatives are useful but the first is more convenient since it also solves the problem of balance. In

---
* This section is based on the work of Dr. P. Conley, refs. 15 and 16.

systems of type 2 (antenna perpendicular to plane of line), only the second method is available.

In order to eliminate all the principal difficulties of conventional open-wire lines, the image-plane line of Fig. 35.1 was designed by P. Conley.[16] This consists of a single conductor D, of any desired diameter, placed parallel to and at an adjustable distance from a large conducting sheet of aluminum and wire netting F. There are no supports in the active part of the line. Both the metal sheet and the line conductor are under great tension, so that a nearly constant line spacing is achieved. Only for very small separations were inaccuracies due to a slight bowing of the sheet encountered. These were later corrected. The antenna E is attached to the line conductor at the center. The identical halves of the line are equipped with movable bridges A, driving units B, and detector probes C, that are connected by cables and move symmetrically in pairs from controls at one end. They make contact with the metal sheet by phosphor-bronze fingers G. All auxiliary equipment and the observer are behind the image plane. The front of the plane faces a wooden wall entirely free of metal to which the antenna is perpendicular. The distance from the antenna to the floor and ceiling is approximately five wavelengths of 0.4 m each, so that all coupling effects are negligible. Experimental measurements on surface currents at 3,000 Mhz on a scale model of the long slots parallel to the line conductor showed that they had no effect on the distribution or magnitude of the currents and charges. A block diagram of the driving apparatus is shown in Fig. 35.2. The principal experimental problem in operating the line was the balancing of the two sides, which, however, was accomplished very successfully. Small differences were canceled by taking measurements on both sides and averaging the results.

A schematic diagram of the double-ended line is shown in Fig. 35.3. With reference to this figure, let

$$Z_i = Z_c \coth \theta_i, \quad i = 1, 2, L, \cdots, \quad (1)$$

$$\theta_i = \rho_i + j\Phi_i, \quad i = 1, 2, L, \cdots, \quad (2)$$

$$w_1 = s_1 - z, \quad (3)$$

$$s_2 = s - s_1 = s_1 + s_d, \quad (4)$$

$$V_2^e = \delta V_1^e, \quad (5)$$

$$\theta_2 = \theta_1 + \vartheta, \quad (6)$$

where Z_c is the characteristic impedance of the line, $\gamma = \alpha + j\beta$ is the complex propagation constant, I_{z_1} is the current at a point z_1 to the left of the load, and I_{z_2} is the current at a point z_2 to the right of the load. With these definitions and the superposition principle, the distributions of current on the halves of the line are:

$$I_{z_1} = \frac{V_1^e}{Z_c} \sinh \theta_1$$

$$\times \left[\frac{G \coth \theta_L \sinh \gamma w_1 + \cosh \gamma w_1}{G \coth \theta_L \sinh (\gamma s_1 + \theta_1) + \cosh (\gamma s_1 + \theta_1)}\right], \quad (7a)$$

$$I_{z_2} = \frac{-V_2^e}{Z_c} \sinh \theta_2$$

$$\times \left[\frac{H \coth \theta_L \sinh (-\gamma w_1) + \cosh (-\gamma w_1)}{H \coth \theta_L \sinh (\gamma s_2 + \theta_2) + \cos (\gamma s_2 + h\theta_2)}\right], \quad (7b)$$

where

$$G = \left[\frac{\cosh (\gamma s_1 + \theta_1 + \gamma s_d + \vartheta)}{\cosh (\gamma s_1 + \theta_1 + \gamma s_d + \vartheta) + \coth \theta_L[\sinh (\gamma s_1 + \theta_1 + \gamma s_d + \vartheta)]}\right.$$

$$\frac{+ \delta \cosh \vartheta \cosh (\gamma s_1 + \theta_1)}{- \delta \cosh \vartheta \sinh (\gamma s_1 + \theta_1)}$$

$$\left.\frac{+ \delta \coth \theta_1 \sinh \vartheta \cosh (\gamma s_1 + \theta_1)}{- \delta \coth \theta_1 \sinh \vartheta \sinh (\gamma s_1 + \theta_1)]}\right], \quad (7c)$$

$$H = \left[\frac{\cosh (\gamma s_2 + \theta_2 - \gamma s_d - \vartheta)}{\cosh (\gamma s_2 + \theta_2 - \gamma s_d - \vartheta) + \coth \theta_L[\sinh (\gamma s_2 + \theta_2 - \gamma s_d - \vartheta)]}\right.$$

$$\frac{+ (1/\delta) \cosh (-\vartheta) \cosh (\gamma s_2 + \theta_2)}{- (1/\delta) \cosh (-\vartheta) \sinh (\gamma s_2 + \theta_2)}$$

$$\left.\frac{+ (1/\delta) \coth \theta_2 \sinh (-\vartheta) \cosh (\gamma s_2 + \theta_2)}{- (1/\delta) \coth \theta_2 \sinh (-\vartheta) \sinh (\gamma s_2 + \theta_2)]}\right]. \quad (7d)$$

Now consider the conventional transmission line in Fig. 35.4,* and let $Z_0 = Z_c \coth \theta_0$, $Z_s = Z_c \coth \theta_s$, $w = s - z$; it follows that

$$I = \frac{V_0^e}{Z_c} \sinh \theta_0$$

$$\times \left[\frac{\coth \theta_s \sinh \gamma w + \cosh \gamma w}{\coth \theta_s \sinh (\gamma s + \theta_0) + \cosh (\gamma s + \theta_0)}\right]. \quad (8)$$

---

* A discussion of the presentation of conventional line theory by use of terminal functions may be found in R. King, ref. 44 and I.31a.

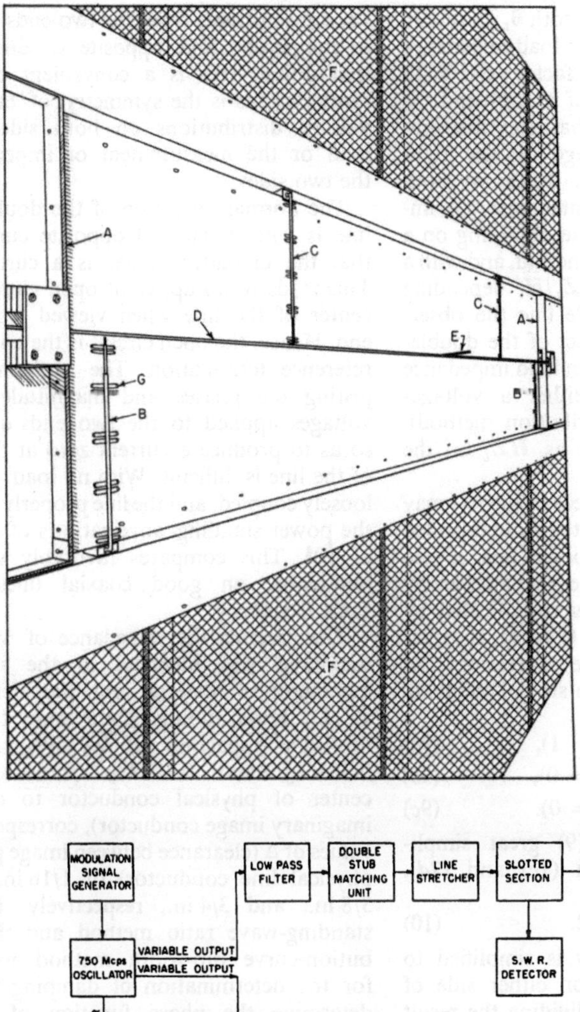

[II.35] THEORY OF LINEAR ANTENNAS 217

Fig. 35.1. Image-plane line.

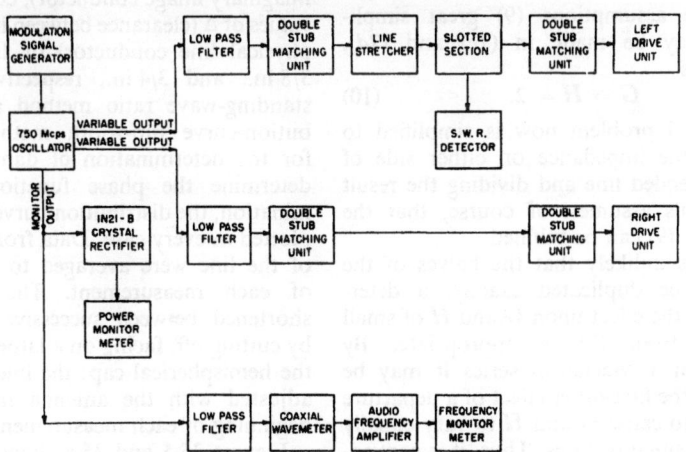

Fig. 35.2. Block diagram of driving apparatus.

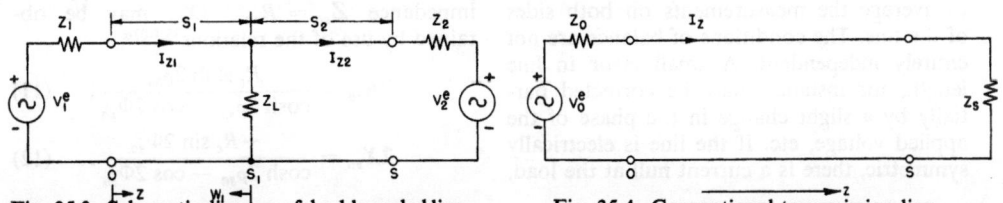

Fig. 35.3. Schematic diagram of double-ended line. Fig. 35.4. Conventional transmission line.

It is clear from (1) that $G \coth \theta_L$ is simply the product of the normalized load impedance $\coth \theta_L$ and the complex factor G. With this in mind, comparison of (7a, b) and (8) leads to the conclusion that the distributions of current and voltage on the right and left sides of this line, driven at both ends and loaded in the center with an impedance Z_L, are exactly those obtaining on a conventional line, driven at one end, and with a load impedance Z_L/G or Z_L/H, depending on which side of the double line the observation is made. The net effect of the double-ended drive is to cause a given load impedance Z_L to be measured (by either a voltage-distribution or current-distribution method), as GZ_L on the left side or as HZ_L on the right side.

Evidently, impedance measurements may be made on either side of the double-ended line as if it were a conventional line, if one simply remembers to divide the answer by G or H. This procedure is complicated by the difficulty in determining G or H. In order to simplify the cumbersome expressions, let the line be the same on each side of the load. That is, let

$$V_2^e = V_1^e \quad (\delta = 1), \quad (9a)$$
$$\theta_2 = \theta_1 \quad (\vartheta = 0), \quad (9b)$$
$$s_2 = s_1 \quad (s_d = 0). \quad (9c)$$

With the assumptions (9) great simplifications may be made in (7c) and (7d). The result is

$$G = H = 2. \quad (10)$$

The practical problem now is simplified to measuring the impedance on either side of the double-ended line and dividing the result by two. This assumes, of course, that the assumptions (9) can be fulfilled.

Since it is unlikely that the halves of the line could be duplicated exactly, a determination of the effect upon G and H of small departures from (9) is appropriate. By expansion in a Maclaurin series it may be shown that the first-order effect of a departure from (9) is to cause G and H to vary equally and in opposite directions. Thus, it is convenient and relatively accurate to keep the line as symmetric as possible about the load and to average the measurements on both sides of the line. The conditions of balance are not entirely independent. A small error in line length, for instance, may be corrected partially by a slight change in the phase of the applied voltage, etc. If the line is electrically symmetric, there is a current null at the load, since the currents from the two ends are equal in magnitude and opposite in direction at that point. This is a convenient check for symmetry, as is the symmetry of current or voltage distributions on both sides of the load or the measurement of impedance on the two sides.

The normal operation of the double-ended line is with equal and opposite currents, so that the unloaded center is a current null. This leads to an apparent open circuit at the center of the line when viewed from either end. Hence, the open circuit is the appropriate reference termination. The problem of adjusting the phases and magnitudes of the voltages applied to the two ends of the line so as to produce a current zero at the center of the line is difficult. With no load, detectors loosely coupled, and the line properly trimmed, the power standing-wave ratio is of the order of 10^4. This compares favorably with that obtainable on good coaxial lines at the same frequency.

The antenna the impedance of which was measured was identical to the single-line conductor in material and diameter (1/8-in. O. D. brass tubing. Data were taken for "full-line" spacings of $0.0159\lambda_0$, $0.0238\lambda_0$, $0.0556\lambda_0$, and $0.1032\lambda_0$ (measured from center of physical conductor to center of imaginary image conductor), corresponding to values of Δ (clearance between image plane and physical line conductor) of 1/16 in., 1/8 in., 3/8 in., and 3/4 in., respectively. Both the standing-wave ratio method and the distribution-curve dip-width method were used for the determination of damping.[33,I.31a] To determine the phase function of the termination, the distribution-curve minimum was plotted in every case. Data from the two sides of the line were averaged to give the result of each measurement. The antenna was shortened between successive measurements by cutting off, facing on a lathe, and replacing the hemispherical cap; the line trimming was adjusted with the antenna in place at the beginning of each measurement.

Figures 35.5 and 35.6 show typical experimental curves of the apparent damping and phase functions, ρ_{sa} and Φ_{sa}. The apparent impedance $Z_{sa} = R_{sa} + jX_{sa}$ may be obtained by use of the relations[36,I.31a]

$$2R_{sa} = \frac{R_c \sinh 2\rho_{sa}}{\cosh 2\rho_{sa} - \cos 2\Phi_{sa}}, \quad (11)$$

$$2X_{sa} = \frac{-R_c \sin 2\Phi_{sa}}{\cosh 2\rho_{sa} - \cos 2\Phi_{sa}}. \quad (12)$$

The values of ρ_{sa} and Φ_{sa} are plotted as functions of antenna half-length h, measured from the tip of the antenna (of radius a) to the surface of the image plane. It includes the single-line spacing δ. The factors 2 in the left-hand members of (11) and (12) take account of the use of double-ended drive. Owing to difficulties in trimming the line when the antenna was near resonance, the results are least accurate near $\beta_0 h = \pi/2$.

A small outward bow in the center of the image plane caused a small change in the line spacing over a considerable distance. With no load the effect is similar to that of a small capacitance located at the center of the line, the reactance of which may be measured with the customary transmission-line techniques. While this effect is not measurable if the line spacing is large, it becomes more important with small line spacings. Since the presence of such an effective capacitance alters the measured impedances, these were corrected by assuming that the capacitance is lumped at the driving point. This should introduce too large a correction, as the driving point is particularly sensitive when the antenna length is in the region near antiresonance. Therefore, both the corrected and the uncorrected impedances are plotted with the understanding that the actual experimental value for a smooth plane lies between them. The impedance curves are given and discussed in Sec. 34.

The measurements of Angelakos, the results of which are summarized in Sec. 34, were made using the same image-plane line but with the outward bow straightened. Instead of driving the line from both ends with the antenna as center load, Angelakos drove the line from one end only and used the other as a stub support or removed it completely and supported the line and antenna with blocks of Styrofoam.

36. *Impedance Measurements with a Long, Two-wire Line**

The impedances of antennas perpendicular to the plane of the line can not be measured using an image-plane line since the line and the antenna do not have the same plane of symmetry. Accordingly, a long two-wire line is appropriate. One of the essential experimental requirements for making measurements on two-wire lines is to have balanced currents, since for reasonably small line spacings this makes radiation from the line negligible. Moreover, in the presence of unbalanced currents, observations vary with the location of the movable equipment and of the operator. Unbalanced currents can be eliminated if each of the following elements is balanced independently: the load, the two-wire line, the standing-wave detector, and the line excitation. A block diagram of the circuit and its components is given in Fig. 36.1.

The load, the two-wire line, and the standing-wave detector may be balanced by making them geometrically and electrically symmetric. Balanced excitation is more difficult to achieve. A two-wire open line evidently can be driven from a shielded pair line. However, most commercially available oscillators have coaxial outputs, and most devices that transform from coaxial to shielded-pair lines introduce significant unbalanced currents. For the unbalanced component of two unequal currents, the two inner conductors of a shielded-pair line are effectively in parallel, so that the line behaves like a coaxial line with a double inner conductor. It follows that the unbalanced currents can be eliminated by using a double stub device called an "unbalance squelcher."[86] It is shown schematically in Fig. 36.2. Each stub is a half-wavelength long and has a short-circuiting bar between the two inner conductors a quarter-wavelength from the junction. The two stubs are a quarter-wavelength apart and form a short-circuiting bridge for the unbalanced component. The balanced component is not affected, since for it only high-impedance stubs are placed across the line. The "unbalance squelcher" was tested in the shielded-pair circuit in conjunction with several devices for obtaining balanced excitation on two-wire lines and was found to give perfect balances for all.[85] The method finally adopted for transforming from the coaxial line to the shielded pair uses a coaxial "T"-junction and one line that is longer than the other by a half-wavelength.

Perhaps the most important factor in designing a two-wire line for measuring antenna impedance is its length. A short line places the equipment and operator so close to the antenna that currents induced in them may seriously affect the antenna impedance. A very long line, on the other hand, introduces excessive attenuation and the problem of supports. In the apparatus actually used, a two-wire line 35 ft long was supported at the

---
* This section is based on the work of Dr. K. Tomiyasu, refs. 83, 85, and 86.

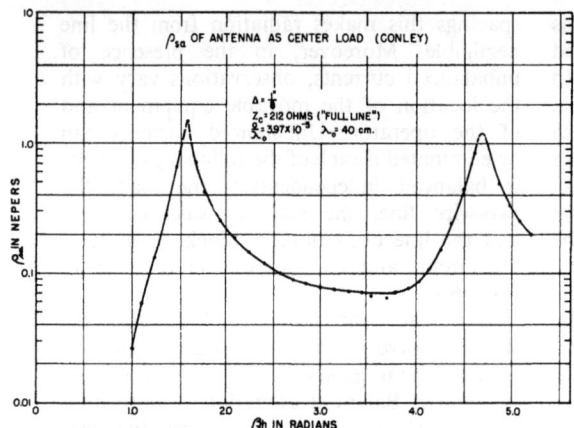

Fig. 35.5. Experimental curves of apparent damping function ρ_{sa} of antennas as center load (Conley).

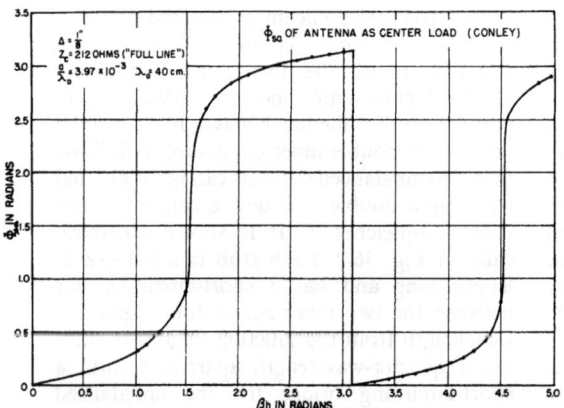

Fig. 35.6. Experimental curves of apparent phase function Φ_{sa} of antennas as center load (Conley).

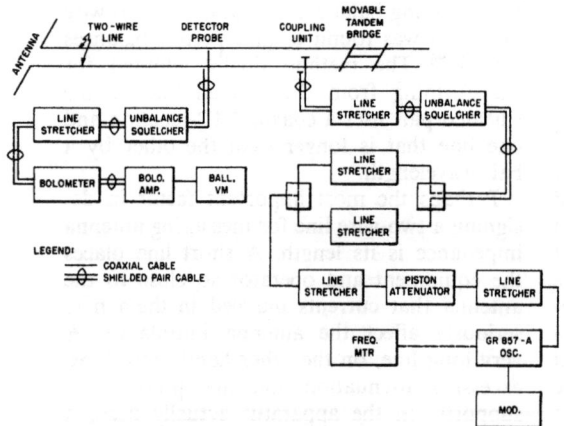

Fig. 36.1. Block diagram of two-wire-line measuring circuit.

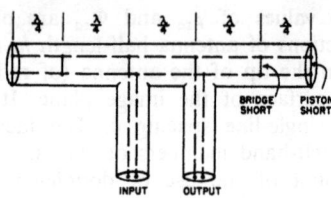

Fig. 36.2. Tomiyasu's unbalance squelcher.

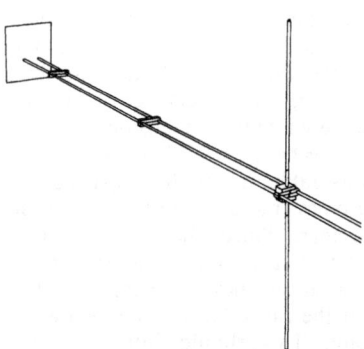

Fig. 36.3. Antenna in plane perpendicular to two-wire line.

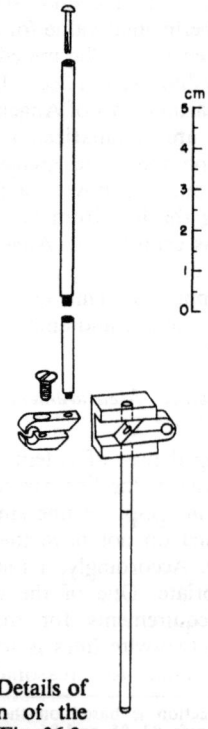

Fig. 36.4. Details of construction of the antenna in Fig. 36.3.

ends. At the operating frequency of 300 Mhz this is about 10 wavelengths. The line was constructed of copper conductors having a diameter of 0.120 in. and spaced at 2.0 cm. This gives a characteristic impedance of 308 ohm, a value chosen to yield greatest accuracy in all antenna impedance measurements. The attenuation is 1.55×10^{-3} neper/m, which is low, but not negligible for the long line.

The line was driven by a tone-modulated oscillator. Its coaxial output was passed through a piston attenuator, transformed to a shielded-pair line, and passed through an "unbalance squelcher" and a line stretcher to the coupling unit. In addition to short-circuiting the unbalanced radio-frequency currents, the "unbalance squelcher" presents a zero impedance to audio-frequency voltages and prevents these from appearing at the coupling unit.

The detector consisted of the shielded-pair probe, an "unbalance squelcher," a bolometer, and a bolometer amplifier with Ballantine voltmeter. It was calibrated using a piston attenuator. The power abstracted by the balanced voltage probe was very small and comparable to that dissipated by the most sensitive coaxial and waveguide detectors.

The first antenna to be measured was placed perpendicular to the plane of the transmission line, as in Fig. 36.3. A stub support was required in order to hold up the wires of the line. The construction of the antenna with its supporting half-bridges is shown in Fig. 36.4. It is pointed out in Sec. 34 that there is no coupling between the antenna and the transmission line when the antenna is in the neutral plane of the line. However, there is a transmission-line end effect due to the presence of the driving structure. Instead of determining the equivalent reactance mathematically as a correction, it was found expedient to tune it out electrically by adjusting the supporting stub to place a voltage maximum at the antenna terminals with the antenna removed. The impedances in Fig. 34.8 were measured with this structure.

The second antenna measured was oriented in the conventional manner in the plane of the transmission line. A high-impedance stub was used as a support, but with this orientation the antenna is coupled to the transmission line and the stub. Impedances measured for this antenna are shown in Fig. 34.4a, b, curves M_s.

All impedances were measured by the standing-wave-ratio or the resonance-curve peak-width methods.[33,I.32] Corrections were made throughout for the finite attenuation of the long measuring line.

37. Impedance Measurements with a Vertical Coaxial Line; Effect of Dielectric Bead[*]

The antenna configuration consisted of a half-dipole and an image plane. For measuring purposes the dipole is mounted on the inner conductor of a concentric line the outer conductor of which is attached to the image plane. The coaxial line spacing produces an annular gap at the center of the image plane. The antenna is driven by the fields in this region and hence the input impedance is also defined here. The currents at the driving point are indicated in Fig. 37.1. There is a phase lag between the actual currents on the image plane and those that would exist in the ideal case of a base-driven dipole. The amount of lag is $\beta_0(b - a)$, where a and b are the inner and outer radii of the line and $\beta_0 = 2\pi/\lambda_0$. The magnitude of the quantity $\beta_0(b - a)$ is determined by other considerations to be examined later. Thus, a principal consequence of the concentric line spacing is to introduce a lag in the currents in the image antenna proportional to this quantity. Discontinuities at the junction of line and antenna conductors may cause an added lumped capacitance. The geometries at the junction of line and antenna shown in Figs. 37.2a and 37.2c apply to the present measurements. The arrangement shown in Fig. 37.2b was not used.

Another consequence of the finite spacing $(b - a)$ is to distribute the driving field along a segment of the center conductor of approximate length $(b - a)$. The physical boundary condition for concentric-line excitation therefore does not contain well-defined terminals; instead, it involves a potential difference distributed about the base of the half-dipole. Impedance measured between terminals with appreciable spacing is an apparent impedance defined by

$$Z_{AB} = (\Phi_A - \Phi_B)/I. \qquad (37.1)$$

The scale of the apparatus and antennas used depends on the choice of wavelength. In making this choice, two basic limitations are encountered. At long wavelengths the earth introduces errors, since it is an imperfect

---

[*] This section is based on work by Dr. D. D. King, refs. 33 and 34.

conductor and not well suited for use in an image plane. On the other hand, extremely short wavelengths decrease the inherent accuracy of the apparatus. This is because dimensions and tolerances decrease in proportion to the wavelength. For operation in the laboratory the range between 10 cm and 100 cm is indicated. In this region an accuracy of 0.1 mm in position is adequate and readily attainable, as are the usual constructional tolerances of ±0.001 in. An image surface of sufficient size may also be provided without resort to a field station. The choice of characteristic resistance of the measuring line is dictated by the loads to be measured. In general, the best accuracy is achieved when power is efficiently transferred to the load. However, since impedance measurement often involves the use of cables, the available fittings and cables must be considered. In this instance a line resistance of 49 ohms was chosen. Having thus fixed the ratio of conductor diameters, the magnitude of the quantity $(b - a)$ at the image surface is fixed by the antenna diameters to be used. This precludes very small values of $(b - a)$. The small dimensions of the line at this point are not favorable to measuring procedures since closer tolerances are required in a thin line. Accordingly, a 16-in. linear taper was used in the present measurements which everywhere maintained the characteristic resistance. No reflections from this tapered section were detectable and the measuring-line diameter could thereby be set at 1 in. while the hole in the image plane was only 5/16 in., which gives $\beta_0(b - a) = 0.05$ radians at the standard wavelength of 28.00 cm. The phase lag introduced by the hole in the image plane, though small, is therefore not negligible.

In order to assure maximum accuracy in the measurements it seemed advisable not to rely exclusively on the standing-wave-ratio (SWR) method. For small losses in line and load this method is open to criticism because of the possibility of detector loading. An apparatus permitting both the conventional SWR method and the resonance-curve method to be used provides greater flexibility in obtaining results for all values of terminal impedance. The use of resonance curves implies a movable piston at one end of the line, and this eliminates the possibility of using solid-dielectric-filled lines. In order to avoid the errors due to movable beads and other types of dielectric supports, the measuring line used for the present measurements was designed to avoid the use of all supports. This was accomplished by mounting the conductors vertically. Several important advantages result from this type of mounting. Thus, the image plane could be located several feet above the operator, thereby minimizing disturbing effects from objects in the room. The weight of the inner conductor and attached antenna is borne by the brass baseplate at the bottom of the line. At the extreme upper end of the line a thin, perforated polystyrene wafer is inserted flush with the surface of the image plane. This wafer, being less than 0.004 in. thick (1 mm), may be treated as a lumped capacitance in parallel with the load. The reactance introduced by the polystyrene may be measured and removed from the final results by subtracting its admittance from the load admittance. The end effect at the open end of the line is also compensated by this procedure. But note that it does not correct for end effect when the antenna is in place. The open end requires a small positive capacitance, the antenna a small negative capacitance as a correction. However, the small values of b/a and $\beta_0(b - a)$ make both of these corrections negligible. This problem is examined in detail in Sec. 38 for lines with larger cross-sectional dimensions.

The thin dielectric support maintains the line spacing well since gravitational effects are eliminated. The space between the two coaxial conductors is then left free to permit a piston to travel up and down for resonance-curve measurements. Adequate rigidity is obtained from the silvered ground steel (drill rod) inner conductor and the 1/4-in. wall of the aluminum outer conductor. Tolerances were better than ±0.001 in. The radii of the inner and outer conductors were $a = 0.5588$ cm and $b = 1.2738$ cm. The calculated characteristic resistance is $R_c = 49.38$ ohm; the attenuation constant $\alpha = 3.78 \times 10^{-4}$ neper/cm. At the standard wavelength of 28.00 cm, the phase constant is $\beta = 0.2244$ radian/cm. A movable, tuned probe is provided for detecting the voltage amplitude along the line; probe depth is readily adjustable by a lead screw. The output of a Sylvania 1N21—B crystal in the probe is fed through a selective amplifier into a Ballantine voltmeter. The voltage gain of this amplifier is 78 db and the noise level about 78 db below 1 mv. Extensive grounding proved to be essential in maintaining the noise level. Power is furnished by a 2C40 triode in a tuned cavity; 1100-hz square-wave modulation is applied.

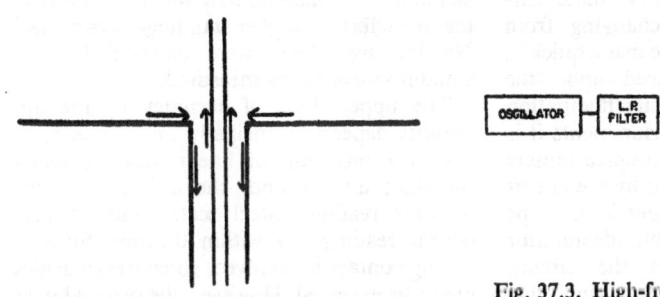
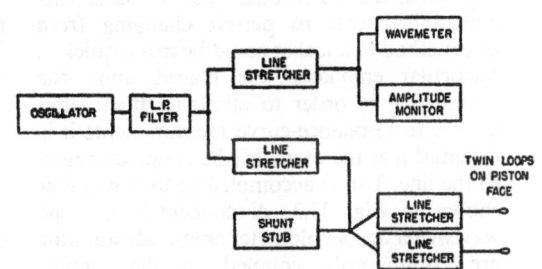

Fig. 37.1. Currents at junction of antenna and image plane.

Fig. 37.3. High-frequency circuit for use with slotted line.

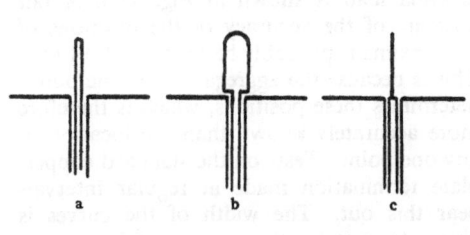
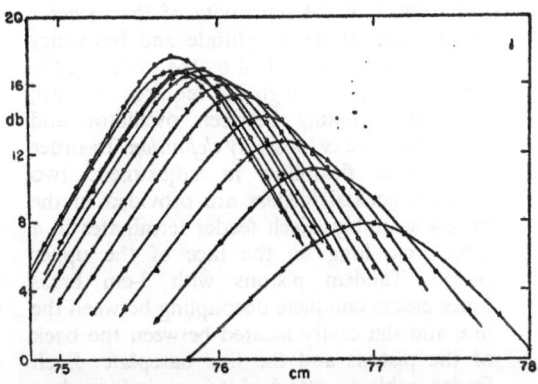

Fig. 37.2. Possible antennas.

Fig. 37.4. Typical resonance curves (D. D. King).

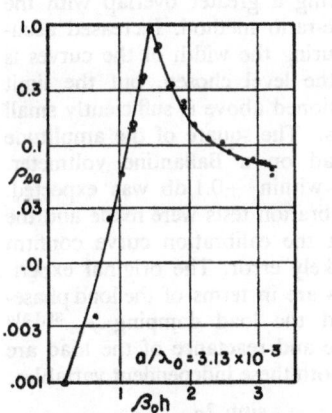
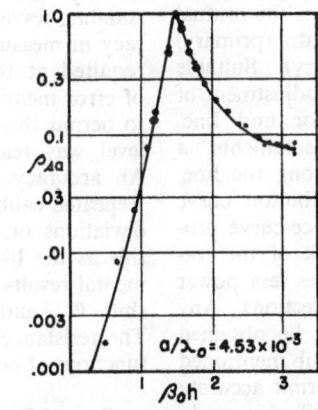
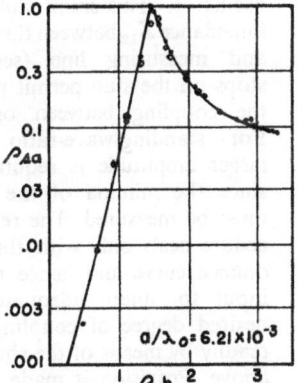

Fig. 37.5. Terminal damping function ρ_{sa} in nepers as a function of antenna length in radians; radius of dipole is a; solid points from resonance data; circles from standing-wave data (D. D. King).

For low values of damping the resonance-curve method of measurement was used; for high damping, the standing-wave ratio supplemented the resonance-curve data. Circuit adjustments to permit changing from one method to another could be made quickly. Particular emphasis was placed upon the monitoring in order to eliminate fluctuation errors. In resonance-curve measurements it is essential that the oscillator be coupled loosely to the line. This is accomplished by the circuit shown in Fig. 37.3. A concentric-line-type wavemeter and a bolometer amplitude monitor are continuously coupled to the circuit. Constant information on the output amplitude and frequency is furnished by direct-reading meters. Amplifiers provide a monitor sensitivity equal to the sensitivity of the detector on the line. If the amplitude and frequency monitors are undisturbed by the tuning of the measuring line through the required interval, then the coupling between oscillator and measuring line is loose, by definition. In order to provide flexibility in adjustment two separate coaxial feeders are provided in the piston assembly. Each feeder terminates in a loop 3 cm long on the face of the upper piston. Tandem pistons with 3-cm brass flutes assure complete decoupling between the line and the cavity located between the back of the pistons and the line baseplate. Each feeder cable is attached to an independent phasing stub in a twin line-stretcher. A shunt stub is placed to form a four-way junction with these feeders and the line from the oscillator. This shunt stub forms the mutual impedance Z_{12} between the oscillator (primary) and measuring line (secondary). Suitable stops on the stub permit rapid adjustment of the coupling between oscillator and line. For standing-wave-ratio measurements a larger amplitude is required along the line, since the minima of the distribution curve must be measured. The resonance-curve procedure deals only with the peak of the resonance curve and hence requires less power input to obtain adequate deflections. Any desired degree of coupling can be obtained readily by means of the shunt stub mentioned above. Provision is made to permit accurate reproduction of the settings of this stub. This is needed in the resonance-curve procedure which requires the terminal impedance at the sending end to remain constant between comparisons with the standard termination. An auxiliary probe near the sending end of the slotted section of line may be inserted to permit a direct test of probe loading. Probe effects cancel out in the resonance-curve procedure, but the standing-wave method may suffer from probe-loading effects at higher standing-wave ratios.[3] No loading effects were detected for the standing-wave ratios measured.

The upper limit of accuracy of the apparatus depends principally on two factors: first, the precision of linear measure along the line; and second, the accuracy of the detector reading. Steel scales and verniers permit readings to within 0.1 mm. Since a sliding contact is involved, some irregularities might be expected. However, the original data yielded uniformly smooth resonance curves. A sample family of resonance curves for an antenna load is shown in Fig. 37.4. A fair estimate of the accuracy of the positions of the maxima is probably better than ±0.1 mm. This is because the aggregate of all the points determines these positions, which is therefore more accurately known than the location of any one point. Tests on the standard copper-plate termination made at regular intervals bear this out. The width of the curves is given to nearly the same precision, but additional error is introduced here in fixing the level at which the width is to be measured. The width of the curves was measured at $(1/\sqrt[4]{2})V_{\max}$. This level, somewhat higher than the usual half-power point, was chosen to extend the range of the method in the direction of high damping (low SWR), thereby securing a greater overlap with the standing-wave-ratio method. Increased accuracy in measuring the width of the curves is required at the level chosen, but the limit of error mentioned above is sufficiently small to permit this. The square of the amplitude level was read on a Ballantine voltmeter. An accuracy within ±0.1 db was expected. Repeated calibration tests were made and the deviations on the calibration curve confirm this as the likely error. The original experimental results are in terms of the load phase-shift Φ_{sa} and the load damping ρ_{sa}.[39,I.31a] The resistance and reactance of the load are functions of both these independent variables:

$$R_{sa} = 2R_c \frac{\sinh 2\rho_{sa}}{\cosh 2\rho_{sa} - \cos 2\Phi_{sa}}, \quad (37.2)$$

$$X_{sa} = -2R_c \frac{\sin 2\Phi_{sa}}{\cosh 2\rho_{sa} - \cos 2\Phi_{sa}}. \quad (37.3)$$

The factor 2 in Eqs. (37.2) and (37.3) converts the results to symmetric dipole impedance.

The accuracy of the final results is not so easily stated, since the transformation from Φ_{sa} and ρ_{sa} to R_{sa} and X_{sa} may increase or reduce the inaccuracies in the measured variables. A measurement depending principally on the phase-shift function Φ_{sa} was made at regular intervals to determine the impedance of the open end with associated wafer support. A value of $-j1440$ ohm ± 5 percent was obtained. Since polystyrene is somewhat affected by the humidity, some of the deviation recorded is probably not caused by the measuring technique. Measurements made at large terminal damping depend principally on the damping function ρ_{sa}. Independent determinations of ρ_{sa} were made, using the resonance-curve method and the standing-wave-ratio method. The two methods agreed very closely. The experimental points from both methods are plotted along the curves. An error of about 5 percent is indicated by the location of the points along the curves for ρ_{sa}. Values taken from the smooth curve should be more accurate. A much smaller deviation of the experimental points is apparent in the curves of the phase-shift function Φ_{sa}. The curves are shown in Fig. 37.5 and 37.6.

The curves of R_{sa} and X_{sa} shown in Figs. 37.7 and 37.8 give both compensated and uncompensated values. The compensation involves elimination of the parallel reactance of -1440 ohm introduced by the polystyrene wafer support. Evidently, the presence of even a small section of dielectric influences the results appreciably. The significance of conditions at the driving point is therefore borne out by the experimental results. The curves for ρ_{sa} and Φ_{sa} are slowly varying when the corresponding R_{sa}- and X_{sa}-curves exhibit rapid changes. Therefore the use of the ρ_{sa} and Φ_{sa} graphs effectively increases the number of experimental points available in critical regions of the resistance and reactance graphs. Several additional points were chosen in this manner about the resonant and antiresonant lengths.

The image plane used consisted of a square sheet of aluminum 90 cm on a side, supplemented by eight brass rods interconnected by heavy wire. The total area of this assembly was about eight times that of the center section and corresponded roughly to a square of seven wavelengths on a side. Data taken with only the inner section in place differed from values taken with the added section by less than 3 percent for several antenna lengths. The phase function Φ_{sa} remained unaltered, while the damping ρ_{sa} increased slightly with the size of the plane. (This is confirmed theoretically in Sec. VII.28 and by an extensive experimental study[58] of the effect of the size and shape of the ground plane on the measured impedance of antennas which shows that the impedance is an oscillating function of the ratio D/λ_0 of ground-plane dimension to wavelength. The amplitude decreases with increasing values of this ratio. For a square ground plate the amplitude is about half that for a circular plate and it decreases more rapidly with increasing size of the plate. For a square plate the amplitude of the fluctuations from the mean resistances near resonance and antiresonance are about 5 percent for $D/\lambda_0 = 6$ and about 9 percent for $D/\lambda_0 = 2.5$.)

Several critical points in the impedance characteristics of an antenna may be chosen for comparison of the experimental results obtained with calculated values. In general, the maximum, minimum, and zero values of reactance are determined primarily by phase-shift data, while the resistance is also a function of the damping. In view of the greater precision of phase-shift determinations, somewhat greater accuracy may be expected in the reactance data than in the resistance values. Table 37.1 and Fig. 34.8 show the results of the King-Middleton formula and of the present measurements.

TABLE 37.1. Comparison of experimental and theoretical critical values for cylindrical antenna, $\Omega = 10$.

	D. D. King measurements	King-Middleton second-order theory
$(R_0)_{\text{antires}}$	820	844
$\pi - (\beta_0 h)_{\text{antires}}$	0.60	0.60
$|X|_{\min}/|X|_{\max}$	1.95	1.63
$(R_0)_{\text{res}}$	71.5	70.3
$\pi/2 - (\beta_0 h)_{\text{res}}$	0.098	0.094
$(R_0)_{\beta_0 h = \pi/2}$	85	86.5
$(X_0)_{\beta_0 h = \pi/2}$	47	41.7

A comparison with experimental results of Brown and Woodward[12] is included in Table 37.2. These writers used a 5-m wavelength with a conventional slotted line. The agreement is rather good, with the exception

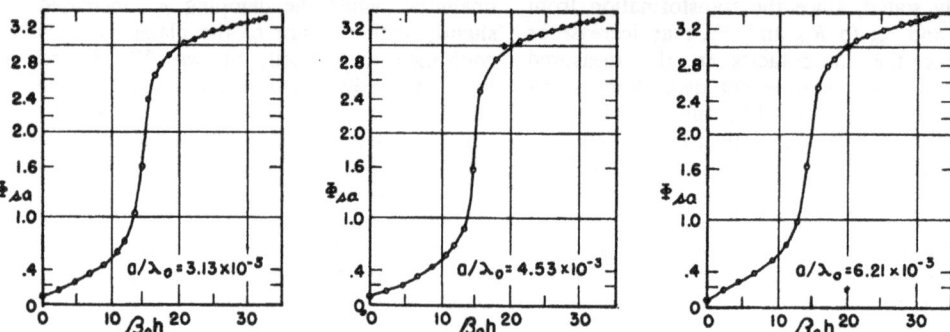

Fig. 37.6. Terminal phase-shift function Φ_{sa} in radians as a function of antenna length in radians (D. D. King).

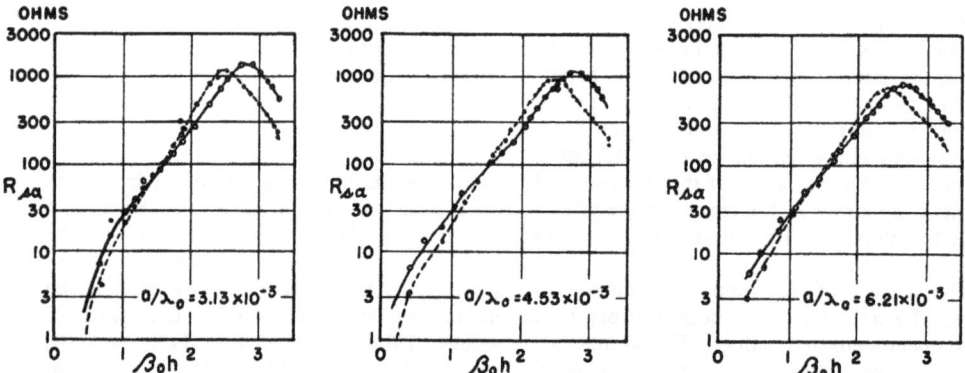

Fig. 37.7. Input resistance of dipole; solid curves compensated for effect of support (D. D. King).

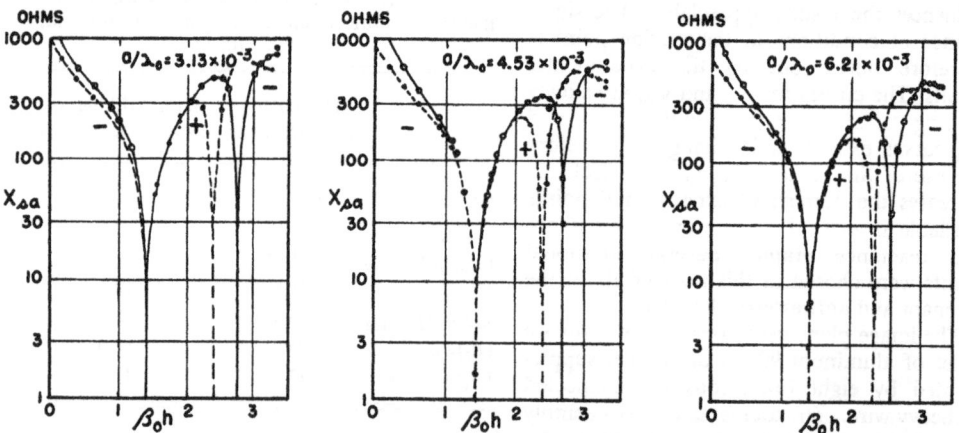

Fig. 37.8. Input reactance of dipole; solid curves compensated for effect of support (D. D. King).

TABLE 37.2. Comparison of experimental critical values for cylindrical antenna.

	D. D. King		Brown and Woodward	$(a/\lambda) \times 10^{-3}$
	compensated	uncompensated		
$(R_0)_{\text{antires}}$	1320	1080	1340	3.13
	1080	880	1050	4.53
	820	700	800	6.21
$\pi - (\beta_0 h)_{\text{antires}}$	0.36	0.70	0.47	3.13
	.44	.72	.59	4.53
	.60	.79	.68	6.21
$\|X\|_{\min}/\|X\|_{\max}$	1.55	2.0	1.8	3.13
	1.88	2.32	2.3	4.53
	1.95	2.57	4.2	6.21

of the reactance ratio. This is to be expected, however, since Brown and Woodward directed their measurements toward cases of practical engineering significance. Considerable base capacitance is encountered in base-driven dipoles attached through standard cable connectors. The effect of base capacitance is clearly indicated by the uncompensated results in Table 37.2 and in the next section.

38. Effect of Transmission-Line Dimensions on Impedance of Antennas Measured with Coaxial Line*

It is shown in the preceding section that the theoretical impedance Z_0 of a cylindrical antenna of half-length h center-driven by a slice generator is well approximated by twice the experimentally determined impedance of a half-dipole base-driven from a coaxial line, as shown in Fig. 38.1, provided (a) the coaxial-line spacing $(b - a)$ is small enough, and (b) a correction is made for a dielectric bead or wafer if one is used. This is equivalent to stating that if $\beta_0 b$ and $b - a$ are sufficiently small, the field that drives the antenna at the end of the coaxial line is concentrated in a band that is sufficiently narrow to be practically equivalent to a discontinuity in scalar potential $V = (\Phi - 0)$ across a ring around *the surface* of the antenna where it joins the conducting plane, as in Fig. 38.2. Note that the cylinder has no flat, circular base; all currents and charges are on the cylindrical envelope or the image plane.

In this section the effects on the impedance of a cylindrical half-dipole with hemispherical cap of (1) increasing b with $b - a$ fixed and (2) increasing b with a fixed are determined experimentally with precision. In order to make it unnecessary to construct a new slotted line for each value of b, a slotted section of fixed cross-sectional dimensions was used in conjunction with sections of line with abrupt changes in b or a or both, as shown in Fig. 38.3. It is shown by Hartig that an abrupt step or change in b or a or both at a distance $\lambda_0/2$ from the end of the line is equivalent to a lumped shunt capacitance across the line that may be determined experimentally by placing a short-circuiting piston exactly one quarter wavelength from the step. With the aid of steps of the types shown in Fig. 38.3 it was possible to measure impedances of antennas with a wide range of values of b and b/a using a single slotted line. This line was arranged in a vertical position below an aluminum ground plane 18.3 wavelengths square at the operating wavelength of 0.6 m. The line was current fed by small loops extending above the tuning piston. It was replaced by a line of large diameter for the one set of values involving the greatest value of b used in the measurements. This line was voltage fed at the base. The two lines and all associated gear are shown schematically in Fig. 38.4. The detector system consisting of probe, crystal, and amplifier were calibrated as a single unit. In both lines the antenna and the inner conductor of the line were kept centered and rigid with Styrofoam wafers which have a dielectric constant differing negligibly from unity. It is shown in Fig. 38.5 that the presence of such a wafer even at a voltage maximum in a resonant line can not be detected. In this

---
* All of the experimental data in this section are taken from the work of Dr. E. O. Hartig, ref. 27.

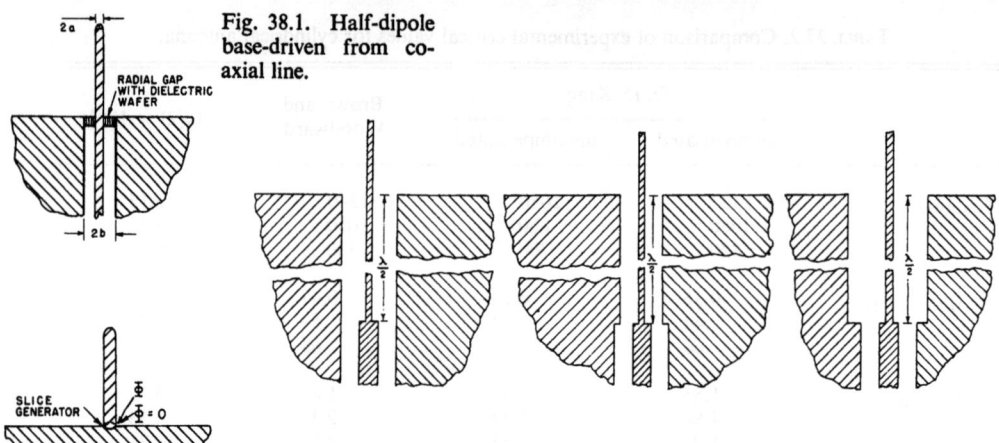

Fig. 38.1. Half-dipole base-driven from coaxial line.

Fig. 38.2. Half-dipole driven by discontinuity in scalar potential.

Fig. 38.3. Methods of feeding antenna.

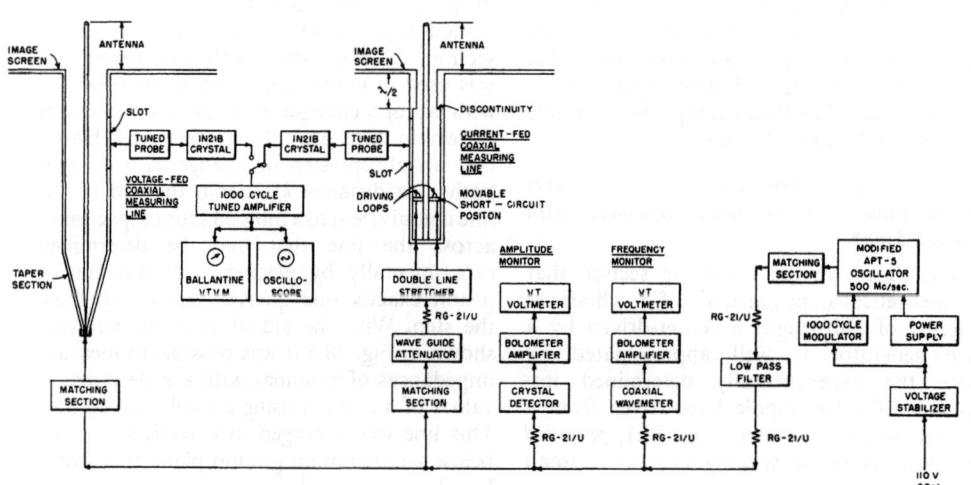

Fig. 38.4. Block diagram of setup used for impedance measurements.

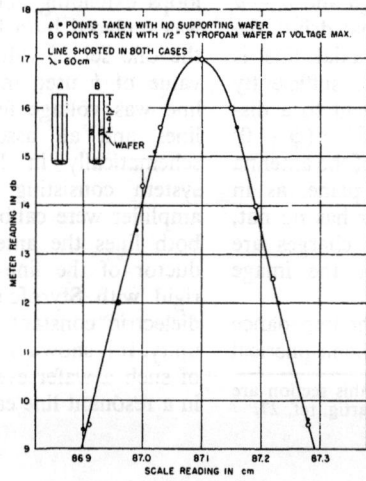

Fig. 38.5. Measured effect of styrofoam support (Hartig).

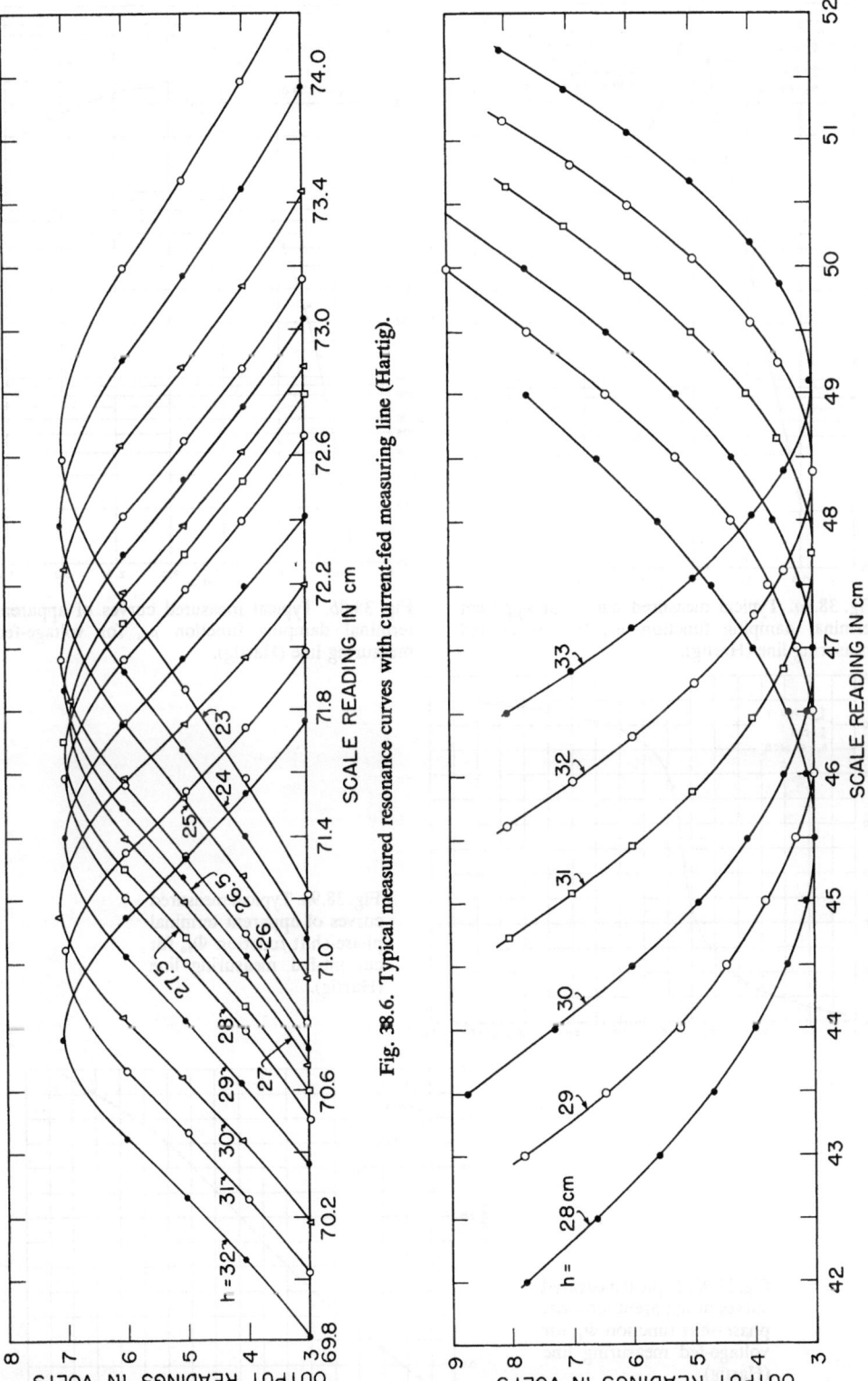

Fig. 38.6. Typical measured resonance curves with current-fed measuring line (Hartig).

Fig. 38.7. Typical measured distribution curves with voltage-fed measuring line (Hartig).

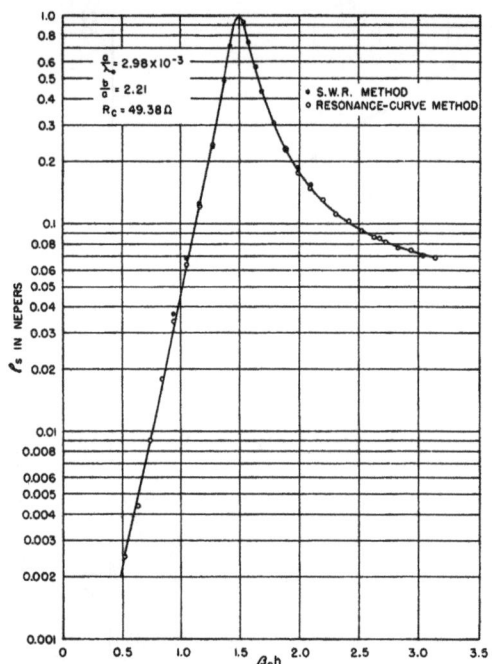

Fig. 38.8a. Typical measured curves of apparent terminal damping function ρ_{sa} for current-fed measuring line (Hartig).

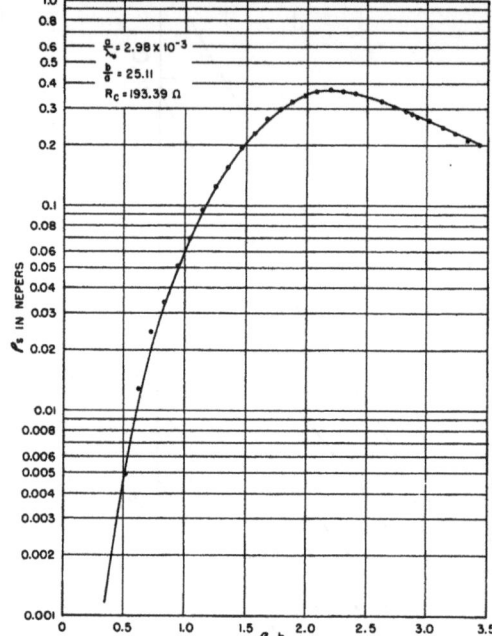

Fig. 38.8b. Typical measured curves of apparent terminal damping function ρ_{sa} for voltage-fed measuring line (Hartig).

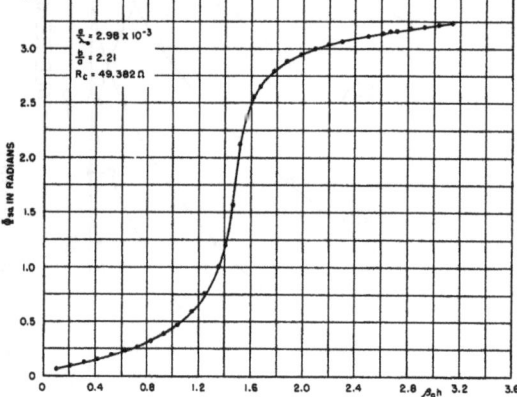

Fig. 38.9a. Typical measured curves of apparent terminal phase-shift function Φ_{sa} for current-fed measuring line (Hartig).

Fig. 38.9b. Typical measured curves of apparent terminal phase-shift function Φ_{sa} for voltage-fed measuring line (Hartig).

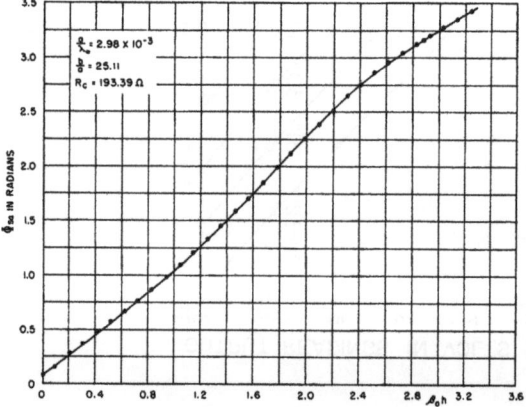

TABLE 38.1. Dimensions of antennas.

Antenna	$a/\lambda_0 \times 10^{-3}$	Diameter of antenna, $2a$ (in.)	Diameter of outer conductor of line, $2b$ (in.)	b/a	$\dfrac{b-a}{\lambda_0}$ ($\lambda_0 = 0.6$ m)
1. Thin	2.98	0.141	0.3275	2.21	0.004
			.750	5.32	.013
			1.000	7.09	.020
			1.500	10.64	.029
			3.540	25.11	.072
2. Moderately thin	3.97	.1875	0.3275	1.67	.003
			.750	4.00	.012
			1.000	5.33	.017
			1.500	8.00	.028
			3.540	18.88	.070
3. Moderately thick	9.26	.4375	0.750	1.71	.0066
			1.000	2.28	.012
			1.500	3.43	.022
			3.540	8.10	.066
4. Thick	15.90	.750	1.000	1.33	.0053
			1.500	2.00	.016
			3.540	4.72	.059

manner the dielectric bead discussed in the preceding section was eliminated as a contributor to the terminal-zone effect.

On the current-fed line measurements of the apparent impedance terminating the line were made with the resonance-curve method[33,44,I.31a], and with the standing-wave-ratio method[33,44,I.31a] near resonances where a sufficiently low standing-wave ratio gave evidence that the attenuation due to the section of line between detector and load was negligible compared with the attenuation due to the load, that is, $\alpha w \ll \rho_s$. Typical experimentally determined resonance curves are shown in Fig. 38.6. Since the width of these curves at half-power points is a measure of the over-all attenuation ($\alpha s + \rho_0 + \rho_s$) due to the line ($\alpha s$), the driving piston ($\rho_0$), and the load ($\rho_s$), it was necessary in order to obtain ρ_s to determine $\alpha s + \rho_0$ from a resonance curve obtained when the antenna was replaced by a short-circuiting cap completely closing the end of the coaxial line. It was found that $\alpha s + \rho_0 = 0.014$ neper.

Since the resonance-curve method requires a movable piston and this was not provided on the large end-driven line, all measurements on it were made using the distribution-curve-dip method.[33,I.31a] Typical measured curves are shown in Fig. 38.7. Since the widths at a specified level depend on the attenuation of both load and line, the contribution αs due to the latter was subtracted using the computed value of the attenuation constant α. Typical curves for the experimentally determined apparent attenuation function $\rho_{sa} \doteq \rho_s$ and apparent phase function Φ_{sa} in $\rho_{sa} + j\Phi_{sa} = \coth^{-1} Z_{sa}/Z_c$ are shown in Figs. 38.8 and 38.9. The resistance R_{sa} and reactance X_{sa} of the load apparently terminating the coaxial line were computed from the experimentally determining curves of ρ_{sa} and Φ_{sa}, using

$$R_{sa} = \frac{R_c \sinh 2\rho_{sa}}{\cosh 2\rho_{sa} - \cos 2\Phi_{sa}}, \quad (1)$$

$$X_{sa} = \frac{-R_c \sin 2\Phi_{sa}}{\cosh 2\rho_{sa} - \cos 2\Phi_{sa}}, \quad (2)$$

where the characteristic impedance of the line is treated as a pure resistance, that is, $Z_c \doteq R_c$.

Measurements were made by Hartig on antennas with four values of the radius a. They are designated thin, moderately thin, moderately thick, and thick antennas, as indicated in Table 38.1, where the ranges of dimensions are given. Note that $(b - a)/\lambda_0$ is the radial thickness in fractions of a wavelength of the dielectric between the coaxial

TABLE 38.2a. Measured resistances, R_{sa} (ohms); $a/\lambda_0 = 2.98 \times 10^{-3}$, $\lambda_0 = 0.6$ m (Hartig).

$\beta_0 h$	$\dfrac{b}{a} = 2.21$	$\dfrac{b}{a} = 5.32$	$\dfrac{b}{a} = 7.09$	$\dfrac{b}{a} = 10.64$	$\dfrac{b}{a} = 25.11$
0.105	—	—	—	—	—
.209	—	—	—	—	—
.314	—	—	—	—	—
.419	—	—	—	—	—
.524	3.25	3.13	3.36	3.71	3.28
.628	4.21	4.47	4.66	5.21	6.30
.732	6.92	6.17	7.07	8.18	9.71
.838	8.69	8.70	8.95	10.48	11.76
.942	11.70	11.83	11.89	13.52	14.17
1.047	15.19	15.41	15.08	16.21	17.16
1.152	18.38	19.13	18.97	19.96	20.92
1.256	22.80	23.66	24.00	24.46	25.34
1.361	28.98	29.10	28.54	28.93	30.36
1.413	32.64	32.69	31.79	32.69	—
1.466	36.30	36.03	35.74	36.15	37.06
1.518	41.34	39.90	39.67	39.84	—
1.571	45.62	43.13	42.22	45.12	44.29
1.623	52.55	—	—	—	—
1.675	57.55	55.67	54.96	54.33	54.12
1.780	73.72	71.37	69.83	68.91	66.11
1.885	92.73	90.09	86.38	83.76	80.54
1.989	127.1	114.9	108.3	104.7	110.7
2.094	167.5	151.0	140.1	133.5	128.9
2.199	217.5	191.9	178.1	168.0	161.0
2.303	298.2	257.7	227.0	217.2	211.5
2.408	385.0	331.4	302.8	286.9	275.3
2.513	493.7	436.0	395.6	381.5	363.6
2.618	568.9	534.1	493.8	475.9	469.4
2.670	580.1	557.6	539.4	—	—
2.722	592.1	581.8	584.4	576.5	599.5
2.774	—	601.9	602.3	608.3	—
2.827	530.9	600.3	610.5	622.3	686.0
2.880	—	—	—	614.7	690.8
2.932	432.8	505.5	560.1	596.7	686.4
3.036	318.5	394.0	453.8	501.0	582.6
3.141	253.8	311.3	360.7	401.3	464.7
3.246	—	231.8	270.2	298.8	336.0
3.351	—	178.4	—	220.8	244.3
3.456	—	—	—	—	181.1

conductors of the transmission line. The resistances R_{sa} and reactances X_{sa} computed from (1) and (2) using the experimentally determined values of ρ_{sa} and Φ_{sa} (corrected for the capacitance due to a step in the line when one of these was used) are listed in Tables 38.2 through 38.5 and represented graphically in Figs. 38.10 through 38.13.

In Figs. 38.10a, b are shown the measured apparent resistance and reactance of a thin antenna ($a/\lambda_0 = 2.98 \times 10^{-3}$) as a function of $2\pi h/\lambda_0$. Comparison with Sec. 33 shows that the effect of the radial separation $b - a$ is that of a *negative* shunt capacitance which increases as $b - a$ increases. This behavior is exactly parallel to that of an antenna center-driven as end load on a two-wire line as described in Sec. 34. The insert in Fig. 38.10a gives the range near resonance. For a thin antenna of fixed radius a the apparent impedance varies only slowly with changes in b over this range of h, so that mean curves of R_{sa} and X_{sa} for all values of b/a are shown together with the King-Middleton second-order curve and a single third-order value of R_0 at $\beta_0 h = \pi/2$. It is seen that

TABLE 38.2b. Measured reactances, X_{sa} (ohms); $a/\lambda_0 = 2.98 \times 10^{-3}$, $\lambda_0 = 0.6$ m (Hartig).

$\beta_0 h$	$\frac{b}{a} = 2.21$	$\frac{b}{a} = 5.32$	$\frac{b}{a} = 7.09$	$\frac{b}{a} = 10.64$	$\frac{b}{a} = 25.11$
0.105	−808.8	−1104	−1141	−1311	−875.4
.209	−507.5	− 609.4	− 637.2	− 757.8	−688.1
.314	−380.7	− 431.1	− 449.5	− 501.8	−500.1
.419	−304.0	− 339.8	− 348.9	− 377.0	−382.6
.524	−248.7	− 271.8	− 282.9	− 297.7	−299.1
.628	−211.8	− 228.1	− 234.0	− 237.8	−246.5
.732	−188.4	− 189.1	− 189.7	− 194.0	−200.4
.838	−146.1	− 154.5	− 157.7	− 161.6	−166.0
.942	−118.1	− 123.8	− 124.3	− 127.7	−128.9
1.047	− 93.61	− 97.78	− 98.27	− 99.68	−100.2
1.152	− 69.36	− 72.06	− 73.36	− 73.24	− 73.10
1.256	− 46.49	− 48.47	− 49.06	− 49.08	− 47.73
1.361	− 23.44	− 25.27	− 24.29	− 24.22	− 22.90
1.413	− 11.09	− 11.86	− 11.65	− 10.69	—
1.466	0.24	0.31	0.59	0.72	2.27
1.518	11.97	11.17	11.84	11.66	—
1.571	21.63	21.37	21.28	23.12	24.95
1.623	34.79	—	—	—	—
1.675	48.27	47.96	47.43	48.22	51.60
1.780	73.29	73.73	73.48	74.07	78.07
1.885	101.3	99.53	98.53	99.57	103.6
1.989	135.4	128.3	125.0	127.0	131.7
2.094	165.5	154.7	152.0	157.9	161.3
2.199	186.3	182.0	178.6	186.6	188.8
2.303	208.5	203.7	207.6	213.5	221.1
2.408	191.1	211.8	220.0	223.5	248.1
2.513	147.1	192.9	212.6	230.2	263.0
2.618	42.89	110.9	166.3	187.9	238.5
2.670	− 36.71	51.96	126.3	—	—
2.722	−104.8	− 19.41	51.63	87.01	174.4
2.774	—	− 95.98	− 18.75	27.52	—
2.827	−243.0	− 171.5	− 96.62	− 49.28	32.78
2.880	—	—	—	− 129.5	− 42.80
2.932	−312.2	− 284.0	− 234.4	− 192.4	−153.1
3.036	−349.6	− 331.0	− 323.1	− 311.0	−298.1
3.141	−340.4	− 347.1	− 350.7	− 359.5	−377.3
3.246	—	− 331.1	− 348.5	− 359.6	−391.1
3.351	—	− 311.4	—	− 338.3	−371.5
3.456	—	—	—	—	−334.2

the second-order theoretical value is consistently too low by a few percent. This agrees with the measurements on two-wire lines reported in Sec. 34. However, it is significant that the third-order value from Table 30.15 for $\beta_0 h = \pi/2$ is in excellent agreement with experiment. The ranges of R_{sa} and X_{sa} in Figs. 38.10a, b near antiresonance are shown in Figs. 38.10c, d on an enlarged scale, together with the second-order King-Middleton values of R_0 and X_0. The agreement is seen to be excellent in the sense that the apparent resistances and reactances approach the theoretical curves as b/a is reduced. The percentage difference between theoretical curves ($b/a \to 1$) and the experimental curves for $b/a = 2.21$ is very small; the entire theoretical curve for R_0 is slightly low; the curve for X_0 agrees almost perfectly. Thus it seems clear that the fictitious slice generator or discontinuity in scalar potential is approximated in the coaxial-line drive in the limit as b approaches a if a is very small compared with the wavelength.

The curves in Figs. 38.11a to 38.11d parallel those of Figs. 38.10a to 38.10d for a somewhat

TABLE 38.3a. Measured resistance, R_{sa} (ohms); $a/\lambda_0 = 3.97 \times 10^{-3}$, $\lambda_0 = 0.6$ m (Hartig).

$\beta_0 h$	$\dfrac{a}{c} = 1.33$					
	$\dfrac{b}{a} = 1.67$	$\dfrac{b}{a} = 4.00$	$\dfrac{b}{a} = 5.33$	$\dfrac{b}{a} = 8.00$	$\dfrac{b}{a} = 18.88$
0.105	—	—	—	—	—
.209	—	—	—	—	—
.314	—	—	—	—	—
.419	2.13	1.85	1.61	1.75	4.05
.524	3.14	2.48	3.98	2.83	6.06
.628	4.45	3.45	5.46	4.84	7.76
.732	5.98	6.15	7.64	6.89	9.60
.838	7.85	8.65	9.29	9.91	11.58
.942	11.08	11.50	12.20	12.85	14.27
1.047	14.38	15.01	15.46	18.38	17.51
1.152	18.32	18.56	19.42	20.09	20.44
1.256	23.52	23.65	24.31	24.41	25.62
1.361	29.98	29.49	29.33	29.47	30.57
1.413	33.56	33.20	32.83	33.47	—
1.466	37.04	37.05	36.72	36.58	37.68
1.518	42.22	40.51	39.85	40.06	—
1.571	48.21	44.78	45.08	44.36	45.54
1.675	60.33	58.42	56.81	55.03	56.07
1.780	80.24	74.41	71.24	69.25	67.18
1.885	103.5	92.75	89.81	86.65	83.23
1.989	137.1	120.0	113.5	109.4	103.2
2.094	182.4	154.3	145.9	138.1	130.4
2.199	238.7	201.4	188.9	176.4	166.2
2.303	306.1	263.1	241.5	227.2	211.5
2.408	393.7	336.7	312.5	296.9	274.5
2.513	461.4	428.1	397.5	389.6	353.5
2.566	475.2	470.0	438.6	—	—
2.618	488.7	498.1	483.1	462.8	455.8
2.670	481.6	521.1	512.8	528.8	—
2.722	465.5	531.6	536.9	—	544.1
2.774	—	—	—	555.5	582.6
2.827	392.5	507.1	522.3	554.1	596.9
2.880	—	—	—	538.1	—
2.932	319.6	436.0	464.0	501.2	579.4
3.036	247.6	345.0	375.6	419.3	504.0
3.141	191.9	267.6	292.8	325.1	395.9
3.246	—	199.4	222.5	250.2	295.3
3.351	—	—	—	191.4	217.5
3.456	—	—	—	—	163.0

thicker antenna. The general nature of the curves is the same and the agreement with theory in the limit as b approaches a is equally good, except, apparently, in the case of the curve marked F (Figs. 38.11c, d). As shown in the inserts, the smallest ratio of b/a was obtained by using an antenna of radius a considerably greater than that of the inner conductor of the line and with the step or change in size occurring *at the base* of the antenna. This was done deliberately in order to determine the effect of a step in the radius at the base. Since a discontinuity of this type is known to be equivalent to a lumped positive capacitance when it occurs back in the line, it is to be expected that its effect is essentially the same at the end of the line. Since the effect of the discontinuity can not be separated from the over-all end effect of the line, the measured apparent impedances in this case include both the negative capacitive effect of the finite value of $b - a$ *and* the positive capacitance of the discontinuity. It is evident from Figs. 38.11c, d that the positive effect exceeds

TABLE 38.3b. Measured reactances, X_{sa} (ohms); $a/\lambda_0 = 3.97 \times 10^{-3}$, $\lambda_0 = 0.6$ m (Hartig).

$\dfrac{a}{c} = 1.33$

$\beta_0 h$	$\dfrac{b}{a} = 1.67$	$\dfrac{b}{a} = 4.00$	$\dfrac{b}{a} = 5.33$	$\dfrac{b}{a} = 8.00$	$\dfrac{b}{a} = 18.88$
0.105	−694.0	−980.0	−1006	−1100	−957.8
.209	−443.1	−522.9	−572.2	−632.5	−610.8
.314	−333.5	−376.6	−398.8	−427.6	−445.3
.419	−269.8	−297.2	−301.3	−309.2	−342.8
.524	−221.9	−241.0	−250.7	−259.2	−271.3
.628	−184.9	−199.0	−205.9	−212.3	−221.0
.732	−156.1	−165.9	−169.6	−177.1	−181.1
.838	−130.0	−136.2	−138.5	−143.6	−146.3
.942	−105.6	−110.2	−111.4	−114.8	−116.9
1.047	−83.81	−86.62	−87.32	−89.62	−90.35
1.152	−62.81	−64.59	−64.73	−65.98	−60.73
1.256	−42.00	−42.53	−42.37	−43.42	−42.61
1.361	−20.50	−20.49	−20.64	−20.49	−19.38
1.413	−9.93	−9.09	−8.76	−8.49	—
1.466	0.48	1.01	1.25	2.09	3.57
1.518	11.62	12.48	12.16	12.99	—
1.571	21.75	23.71	22.85	24.26	26.12
1.675	43.35	45.63	45.31	46.16	48.96
1.780	68.75	70.54	73.19	69.52	71.96
1.885	93.52	93.38	92.03	93.81	97.30
1.989	116.7	116.9	116.3	117.9	121.7
2.094	138.9	141.7	142.1	144.9	149.1
2.199	148.4	162.7	165.4	168.4	175.4
2.303	142.0	175.7	182.2	185.4	196.0
2.408	108.6	170.9	188.4	192.9	215.7
2.513	28.19	136.4	161.1	192.9	220.2
2.566	−24.52	101.1	128.6	—	—
2.618	−75.54	60.06	90.60	140.3	194.3
2.670	−138.0	−0.69	40.85	44.64	—
2.722	−183.0	−59.50	−18.85	—	115.4
2.774	—	—	—	−19.34	55.90
2.827	−261.6	−174.7	−143.6	−89.45	−19.23
2.880	—	—	—	−151.9	—
2.932	−295.5	−261.7	−246.3	−216.9	−168.8
3.036	−297.6	−304.1	−299.7	−287.0	−280.5
3.141	−291.2	−313.3	−313.6	−322.9	−335.0
3.246	—	−304.7	−308.7	−317.1	−345.4
3.351	—	—	—	−298.7	−324.9
3.456	—	—	—	—	−294.4

the negative effect, since curves *F* are shifted to the *left* of the theoretical curve *A*. It is shown in Sec. 33 that the location of antiresonance is shifted to shorter lengths than for a slice generator if a positive capacitance is connected in parallel with the antenna, and toward longer lengths with a negative capacitance. Evidently, by proper choices of $b - a$ and of an increase in radius at the base, the negative capacitive effect of the gap and the positive effect of the step could be made to cancel.

Experimental curves giving the measured apparent impedances of moderately thick and thick antennas are in Figs. 38.12a, b, c, d and 38.13a, b, c, d, together with second-order theoretical curves and third-order theoretical points. Since the impedance near resonance is more sensitive to changes in b/a for thick antennas than for thin ones, separate curves are given for this range. The large negative capacitive end effect is evident from these figures.

The general effect on the location and

TABLE 38.4a. Measured resistances, R_{sa} (ohms); $a/\lambda_0 = 9.26 \times 10^{-3}$, $\lambda_0 = 0.6$m (Hartig).

$\beta_0 h$	$\frac{b}{a} = 1.71$	$\frac{b}{a} = 2.28$	$\frac{b}{a} = 3.43$	$\frac{b}{a} = 8.10$
0.105	—	—	—	—
.209	1.00	0.59	1.55	—
.314	1.14	1.43	2.02	—
.419	2.53	2.46	2.70	2.82
.524	3.57	2.81	3.62	5.63
.628	4.50	4.12	5.50	7.47
.732	6.31	6.42	7.34	9.64
.838	8.44	8.64	9.87	11.94
.942	11.00	11.58	12.72	14.42
1.047	13.63	15.04	16.28	17.37
1.152	18.28	19.24	20.34	21.62
1.256	23.02	24.46	23.67	26.42
1.361	29.92	30.79	30.12	31.57
1.466	38.55	39.08	38.46	38.80
1.571	48.56	49.39	49.03	48.01
1.675	63.83	62.88	59.88	58.21
1.780	84.20	81.53	75.32	72.12
1.885	106.5	104.9	96.27	89.49
1.989	137.6	134.4	124.8	107.0
2.094	175.2	171.7	157.4	135.1
2.199	220.8	214.1	199.0	168.9
2.303	263.2	256.2	238.8	212.9
2.408	293.1	294.9	287.9	260.2
2.513	289.5	316.0	318.8	306.7
2.618	274.6	305.2	330.2	353.5
2.722	236.1	273.8	315.3	381.7
2.827	198.6	235.5	282.6	373.7
2.932	164.7	195.4	235.6	336.3
3.036	138.1	159.7	195.0	285.7
3.141	110.1	127.0	160.0	232.1

magnitude of the maximum resistance and the location of zero reactance (antiresonance) of varying b/a with a/λ_0 fixed is shown in Fig. 38.14. It is seen that the experimental points define curves that decrease continuously as b/a decreases, and that these approach the theoretical values for a slice generator in the limit $b/a \to 1$ when $\Omega \geq 10$. Reasonable agreement is still obtained for $\Omega \geq 8.5$ where $h/a \geq 35$. For smaller values of Ω, a quasi-one-dimensional analysis is inadequate for determining more than the order of magnitude, as is clear from Table 38.6.

Since only capacitive end and coupling effects are significant as shown in Sec. 10, the apparent conductance G_{sa} should be independent of b/a and should differ negligibly from G_0. Curves of the apparent conductance G_{sa} and susceptance B_{sa} are shown in Figs. 38.15 to 38.18 for the four values of a/λ_0. It is seen that G_{sa} is essentially independent of b/a even for the thick antenna, whereas B_{sa} (in $Y_{sa} = G_{sa} + jB_{sa}$) decreases with b/a corresponding to the addition of *negative* capacitance.

A quantitative check of the lumped negative capacitance C_T that must be used in parallel with Z_0 (derived theoretically for a slice generator corresponding to the limit $b/a \to 1$) in order to obtain the apparent impedance Z_{sa} (measured on the coaxial line a half-wavelength from the antenna) may be carried out as follows. Instead of evaluating theoretical curves of Z_{sa} or Y_{sa} for the values of b/a and a/λ_0 for which experimental curves are available for comparison, it is more convenient to determine experimentally values of C_T from the observed differences in the locations of the antiresonances, $B_{sa} = 0$ and $B_0 = 0$. The latter is obtained by extrapolating the values of $\beta_0 h_{\text{antires}}$ as functions

TABLE 38.4b. Measured reactances, X_{sa} (ohms); $a/\lambda_0 = 9.26 \times 10^{-3}$, $\lambda_0 = 0.6$ m (Hartig).

$\beta_0 h$	$\dfrac{b}{a} = 1.71$	$\dfrac{b}{a} = 2.28$	$\dfrac{b}{a} = 3.43$	$\dfrac{b}{a} = 8.10$
0.105	−433.8	−482.4	−572.2	−635.2
.209	−283.3	−307.9	−352.5	−414.3
.314	−220.1	−237.5	−261.8	−304.3
.419	−179.1	−191.8	−208.4	−237.4
.524	−150.3	−159.0	−171.4	−192.2
.628	−126.2	−133.9	−141.5	−157.5
.732	−107.7	−112.4	−117.8	−129.0
.838	−90.06	−93.54	−98.36	−105.3
.942	−73.50	−75.96	−78.89	−83.37
1.047	−58.50	−59.53	−61.20	−63.97
1.152	−43.31	−43.58	−44.88	−45.54
1.256	−27.62	−28.46	−27.17	−27.86
1.361	−12.92	−11.92	−10.47	−11.06
1.466	2.64	2.93	4.66	5.70
1.571	18.01	19.13	21.12	21.86
1.675	31.90	34.23	35.80	39.34
1.780	47.53	50.21	52.84	56.33
1.885	60.21	64.11	69.33	72.57
1.989	70.67	76.02	81.36	89.12
2.094	72.49	80.63	88.52	98.81
2.199	51.62	70.65	87.05	114.1
2.303	17.35	48.66	73.03	120.6
2.408	−31.34	4.44	44.96	115.2
2.513	−87.51	−53.67	−6.04	85.35
2.618	−134.6	−109.3	−63.25	44.08
2.722	−175.2	−154.1	−123.8	−28.61
2.827	−194.9	−186.9	−169.9	−101.0
2.932	−196.8	−203.3	−197.7	−166.9
3.036	−196.5	−201.7	−207.6	−200.4
3.141	−187.2	−197.1	−205.5	−223.6

of b/a to $b/a = 1$, as in Fig. 38.14. These measured values of C_T are compared with the approximate theoretical values obtained from Fig. 10.9 with appropriate values of b and c_0 in Table 38.7. Alternatively, the experimental values of $-C_T/bc_0$ for the two sets of values of a/λ_0 are shown in Fig. 38.19 as a function of b/a, together with the approximate theoretical curve of Fig. 10.9. It is seen from Table 38.7 that differences of the order of 0.05 $\mu\mu$f are involved. Since c_0 is of the order of magnitude 0.3 to 0.4 $\mu\mu$f/cm, the difference between theory and experiment corresponds to a length along the coaxial line of about 0.1 cm or 0.0016λ_0. In view of the approximations made in the integration with respect to θ in the theoretical determination of C_T and the experimental difficulty of measuring so small a quantity accurately, the agreement between theory and experiment in Table 38.7 and in Fig. 38.19 is satisfactory. In practical applications, measurements of length on transmission lines usually are not required within the nearest thousandth of a wavelength.

OTHER FORMULATIONS OF THE CYLINDRICAL-ANTENNA PROBLEM

39. *Storer's Variational Modification of Hallén's Integral Equation*[*]

Hallén's theory of the cylindrical antenna as formulated beginning in Sec. 11 involves the derivation of an integral equation and its solution by iteration. The modification introduced by King and Middleton in the original expansion by Hallén was a more carefully chosen approximate distribution function for evaluating the integrals. Although the accuracy

---

[*] This section is based on the work of Dr. J. E. Storer, ref. 78.

TABLE 38.5. Measured impedances, $Z_{sa} = R_{sa} + jX_{sa}$; $a/\lambda_0 = 1.59 \times 10^{-2}$, $\lambda_0 = 0.6$ m (Hartig).

$\beta_0 h$	R_{sa} (ohms)			X_{sa} (ohms)		
	$\frac{b}{a} = 1.33$	$\frac{b}{a} = 2.00$	$\frac{b}{a} = 4.72$	$\frac{b}{a} = 1.33$	$\frac{b}{a} = 2.00$	$\frac{b}{a} = 4.72$
0.015	—	—	—	−247.6	−368.5	−469.0
.209	—	—	0.89	−172.0	−232.3	−308.6
.314	1.42	1.96	3.50	−139.6	−178.6	−229.7
.419	1.71	2.79	4.84	−117.9	−144.8	−181.8
.524	2.38	3.66	6.02	−101.1	−120.9	−146.4
.628	3.60	4.97	7.23	− 85.48	−101.7	−119.5
.732	5.13	6.99	9.58	− 75.42	− 85.48	− 98.46
.838	7.40	9.83	11.57	− 64.31	− 70.93	− 79.93
.942	9.87	11.91	14.32	− 53.40	− 57.96	− 63.32
1.047	13.37	15.65	17.71	− 42.51	− 45.25	− 48.35
1.152	17.48	19.30	21.57	− 31.63	− 33.14	− 34.09
1.256	22.60	24.32	26.37	− 21.22	− 21.98	− 19.11
1.361	30.66	30.47	32.34	− 10.49	− 10.23	− 6.88
1.466	39.64	39.06	39.16	0.04	0.37	5.86
1.571	52.67	52.19	49.02	10.00	11.90	18.49
1.675	68.96	65.93	60.39	17.06	20.42	31.39
1.780	88.22	85.86	73.07	22.74	30.08	42.69
1.885	110.9	104.9	89.65	23.50	34.24	52.78
1.989	137.5	129.4	110.1	8.46	35.97	61.66
2.094	156.8	157.2	134.3	− 17.07	29.80	66.31
2.199	170.5	184.8	161.9	− 44.95	10.57	68.60
2.303	170.6	203.5	198.8	− 73.11	− 18.40	61.45
2.408	159.8	211.6	229.5	− 99.44	− 51.56	40.87
2.513	144.0	206.3	254.1	−117.3	− 87.66	5.76
2.618	125.4	191.9	268.0	−129.6	−119.0	− 34.98
2.722	105.9	168.9	265.4	−133.9	−132.7	− 80.82
2.827	91.69	146.4	244.2	−133.0	−144.8	−121.0
2.932	78.09	122.3	213.0	−130.0	−147.4	−147.8
3.036	65.32	102.0	187.8	−125.7	−145.4	−160.4
3.141	56.76	83.13	149.2	−119.9	−139.0	−165.4

TABLE 38.6. Approximate percentage difference between the second-order theory of King and Middleton and the extrapolated experimental results of Hartig near antiresonance.

Ω	h/a	Location of antiresonance (percent)	Location of R_{max} (percent)	Magnitude of R_{max} (percent)
7.2– 7.5	18– 21	21	9.5	37
8.6– 8.8	37– 41	8.8	2.9	10.7
10.6–10.7	100–105	1.8	0.8	1.4
11.2–11.3	135–140	0.6	0.2	0.9

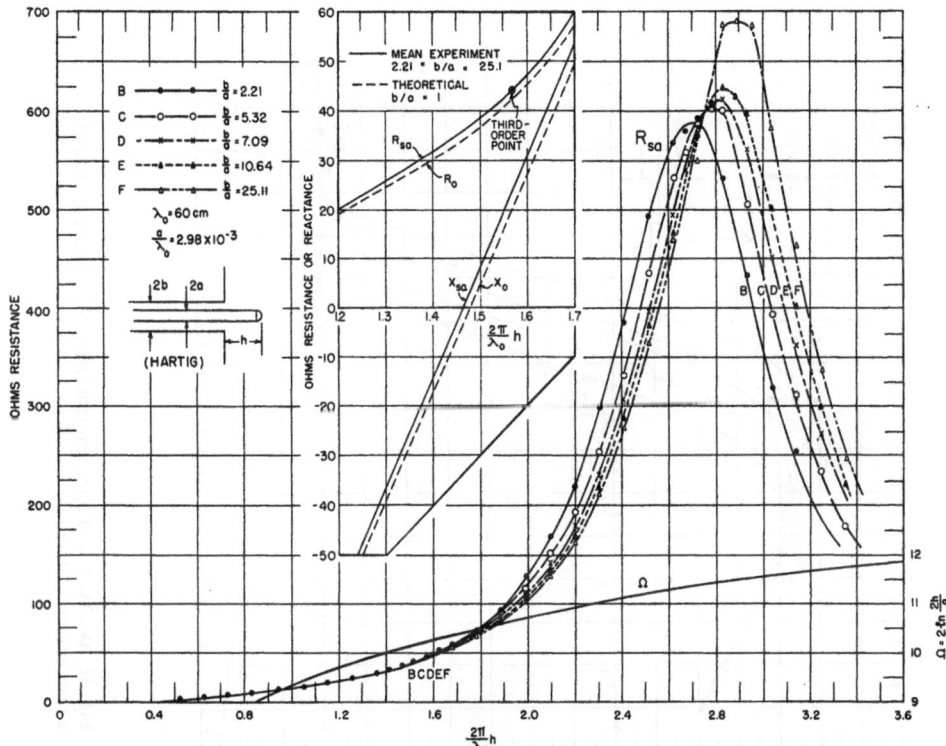

Fig. 38.10a. Measured apparent resistance and reactance of a thin antenna (Hartig).

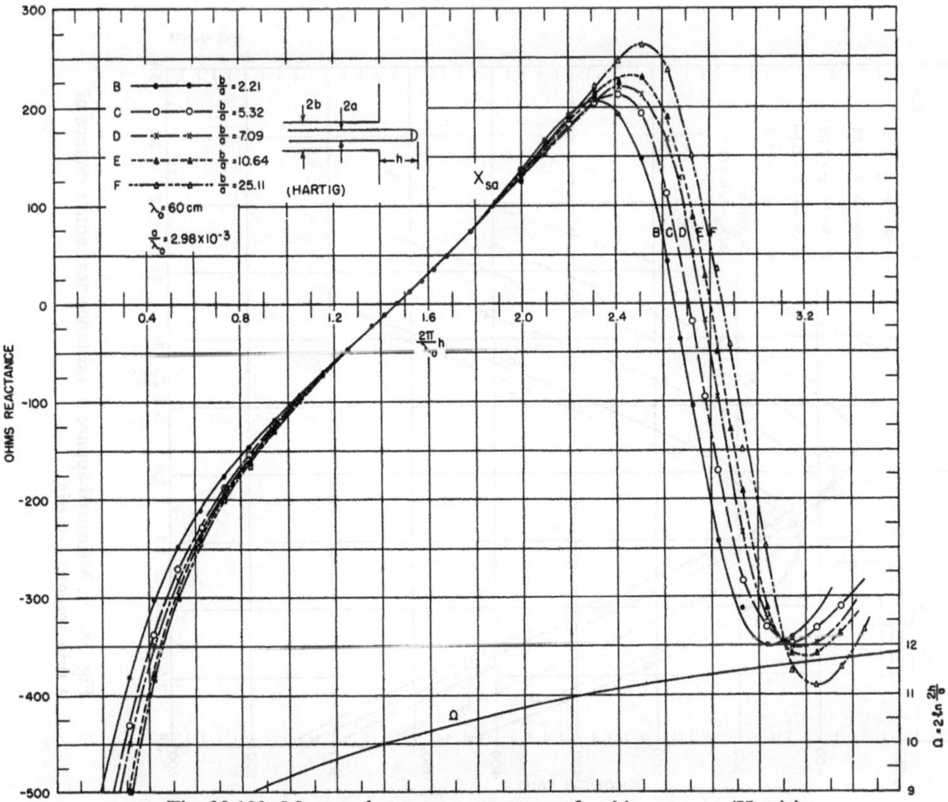

Fig. 38.10b. Measured apparent reactance of a thin antenna (Hartig).

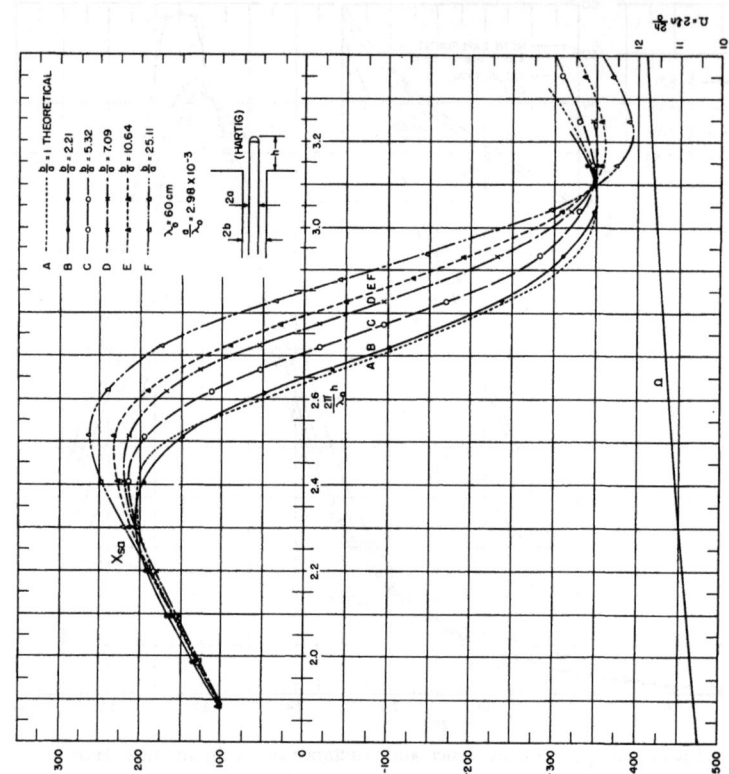

Fig. 38.10d. Measured apparent reactance near antiresonance of a thin antenna (Hartig).

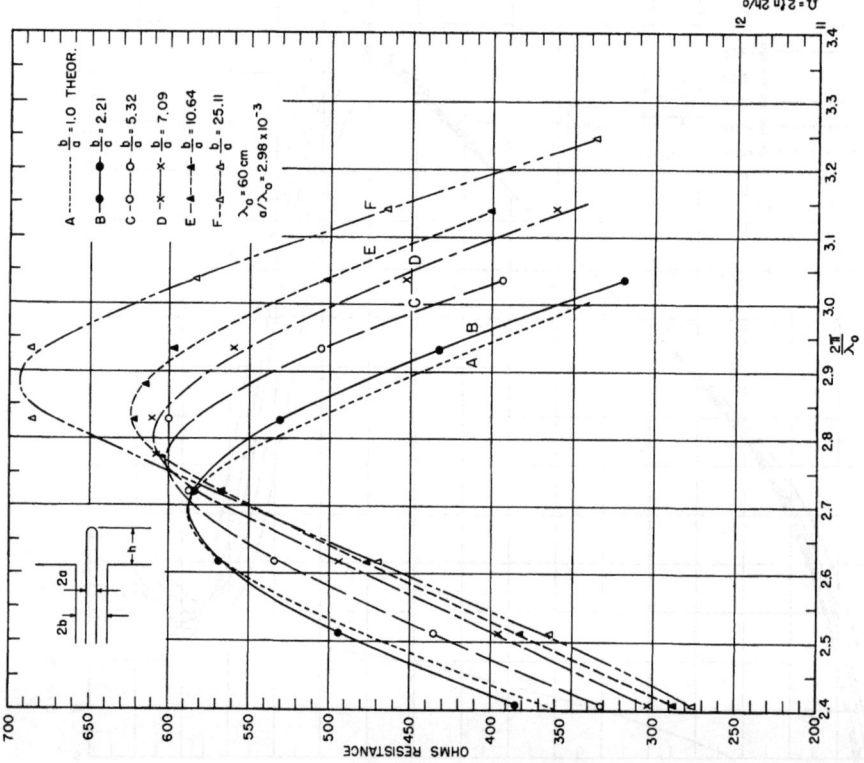

Fig. 38.10c. Measured apparent resistance near antiresonance of a thin antenna (Hartig).

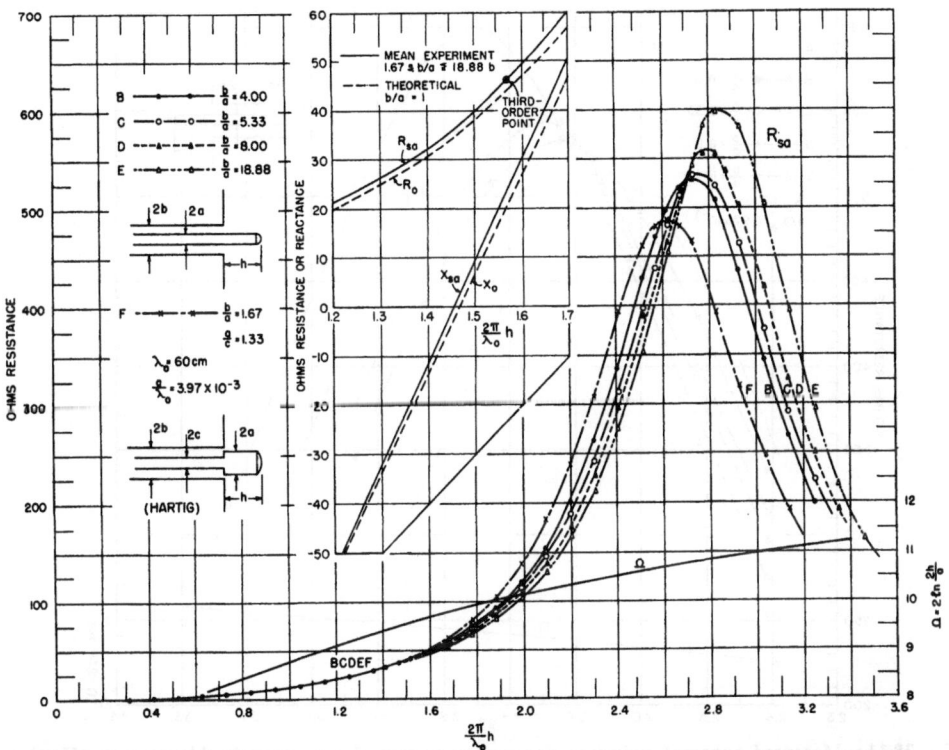

Fig. 38.11a. Measured apparent resistance and reactance of a moderately thin antenna (Hartig).

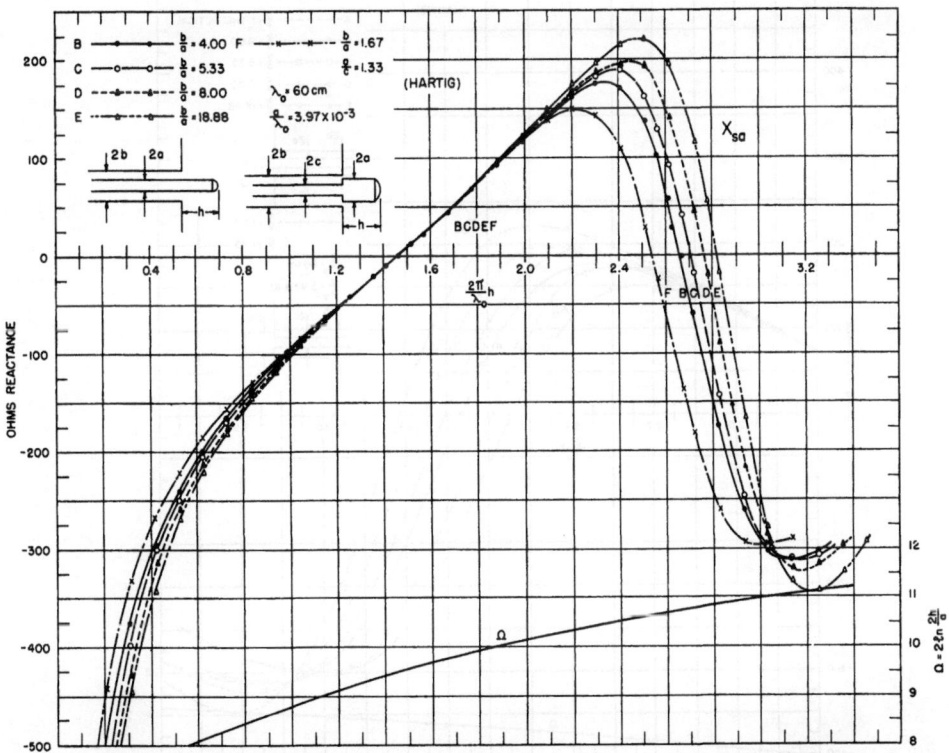

Fig. 38.11b. Measured apparent reactance of a moderately thin antenna (Hartig).

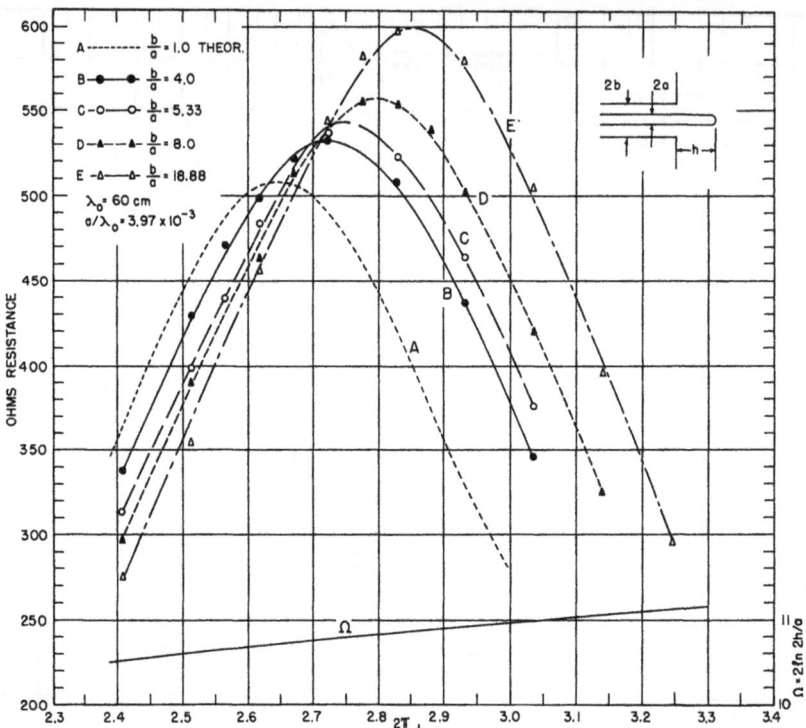

Fig. 38.11c. Measured apparent resistance near antiresonance of a moderately thin antenna (Hartig).

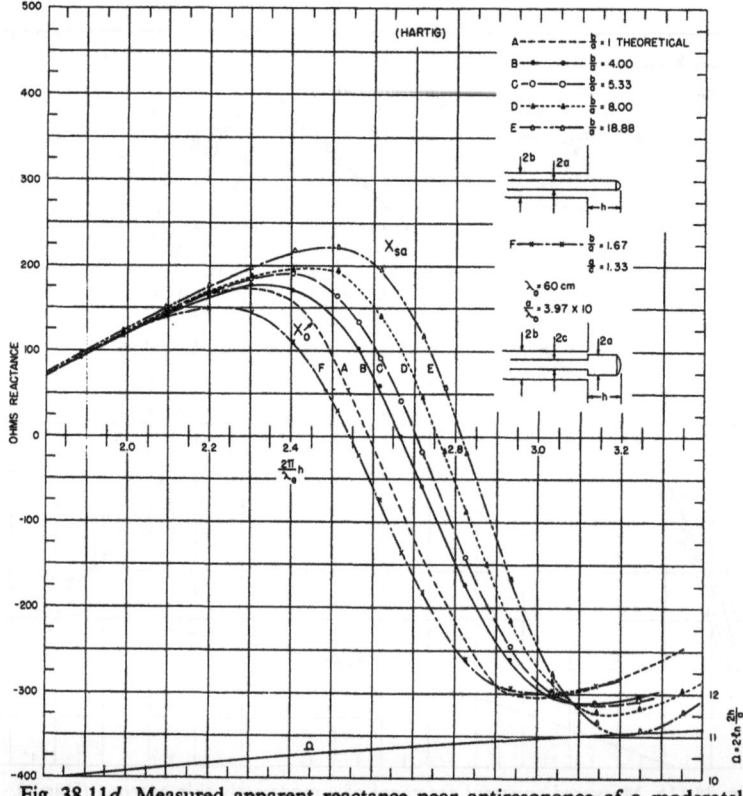

Fig. 38.11d. Measured apparent reactance near antiresonance of a moderately thin antenna (Hartig).

[II.38] THEORY OF LINEAR ANTENNAS 243

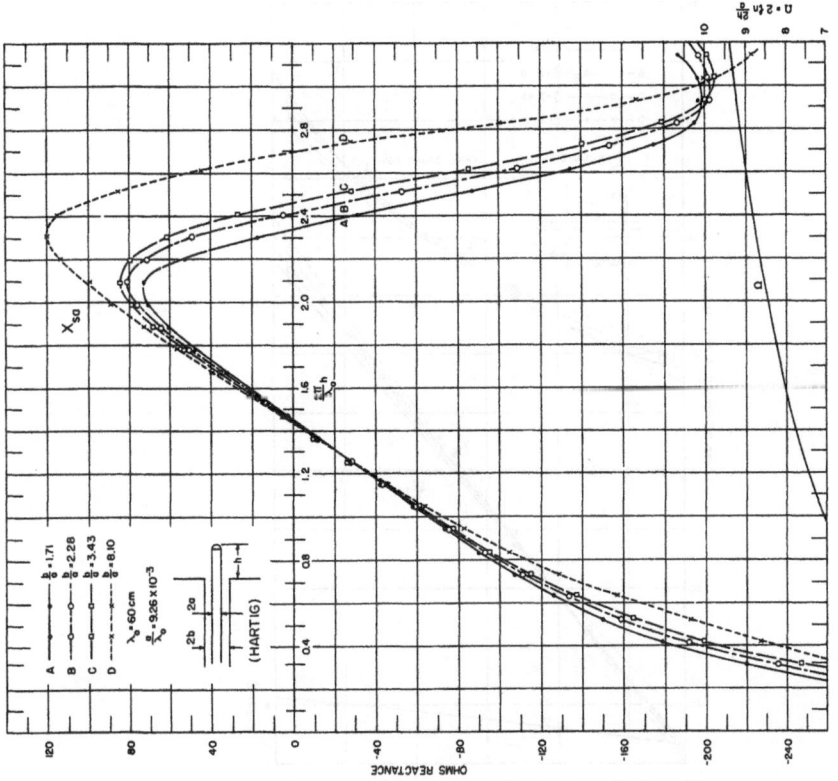

Fig. 38.12b. Measured apparent reactance of a moderately thick antenna (Hartig).

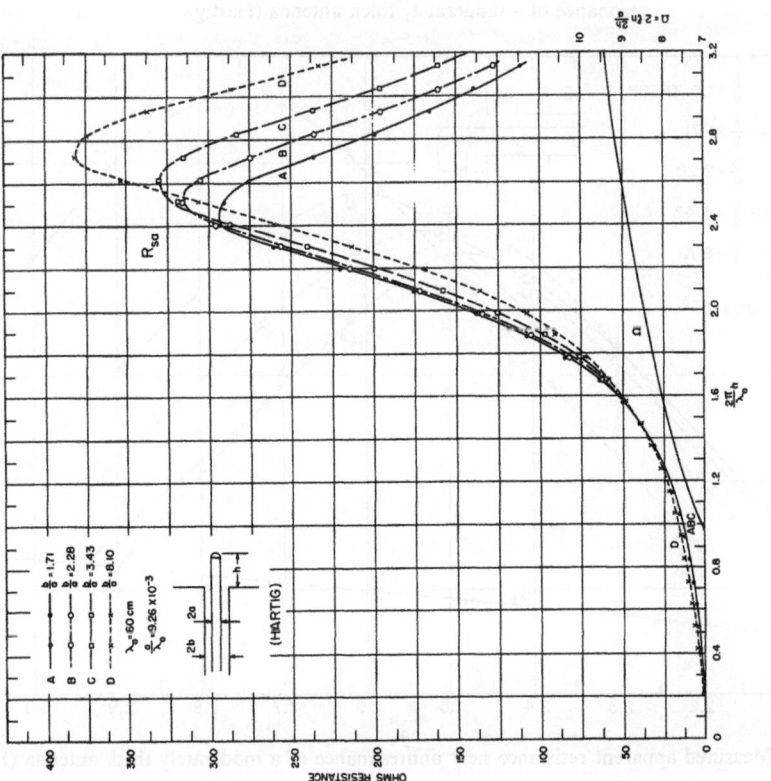

Fig. 38.12a. Measured apparent resistance of a moderately thick antenna (Hartig).

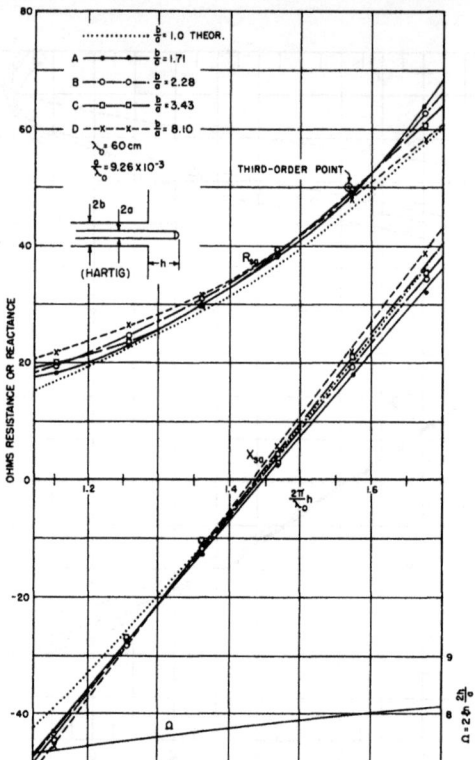

Fig. 38.12c. Measured apparent impedance near resonance of a moderately thick antenna (Hartig).

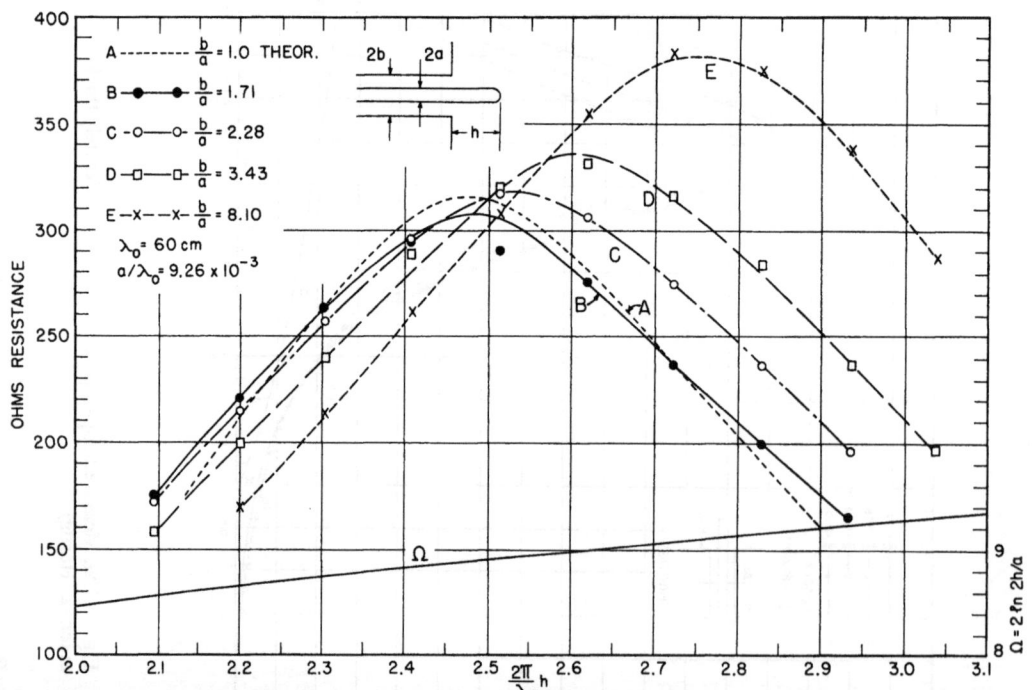

Fig. 38.12d. Measured apparent resistance near antiresonance of a moderately thick antenna (Hartig).

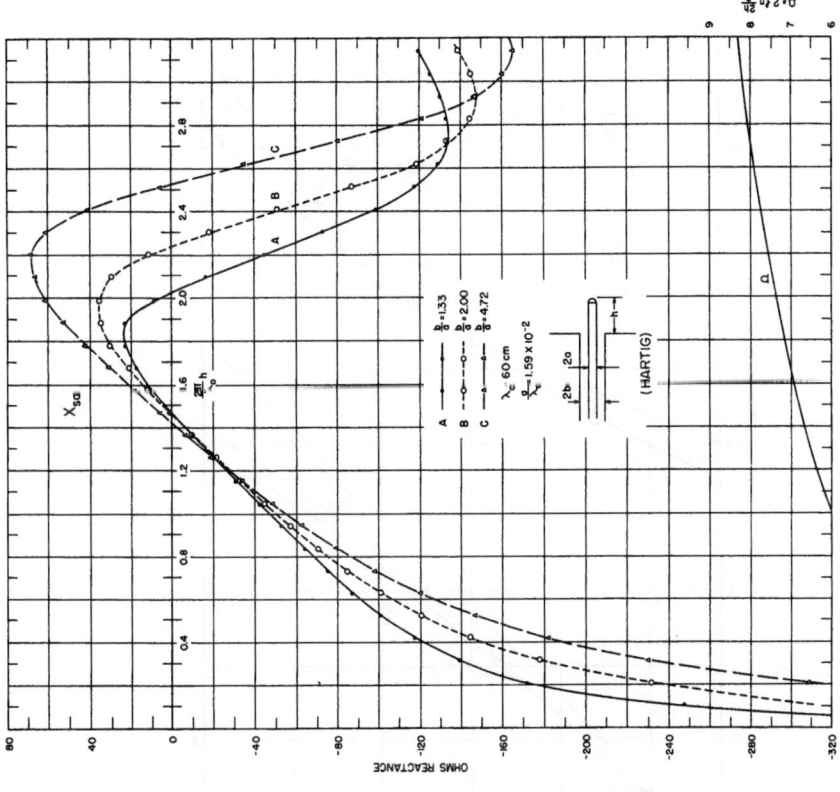

Fig. 38.13b. Measured apparent reactance of a thick antenna (Hartig).

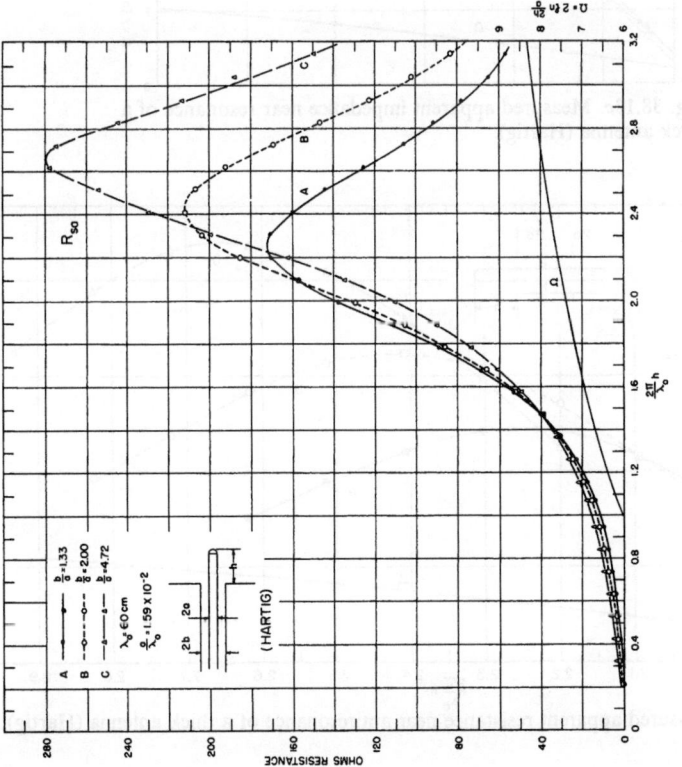

Fig. 38.13a. Measured apparent resistance of a thick antenna (Hartig).

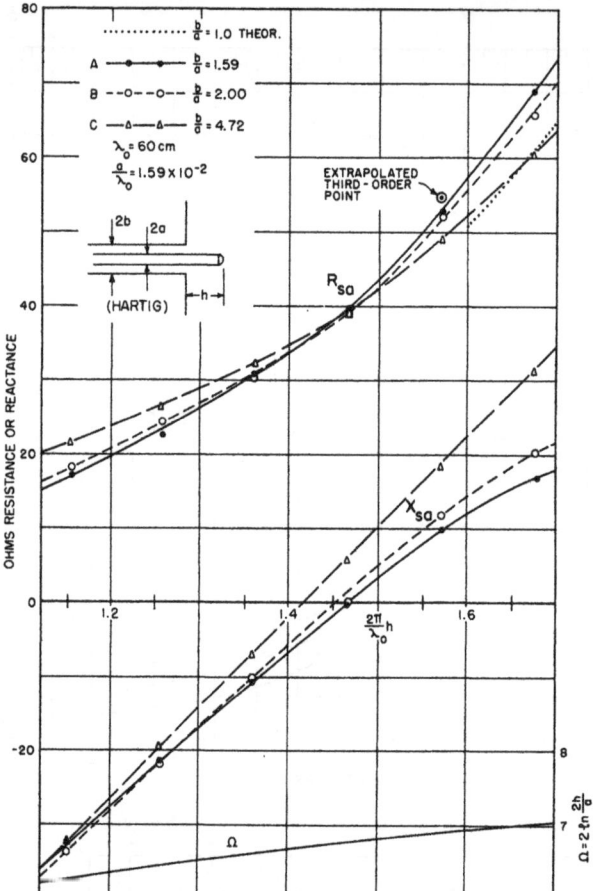

Fig. 38.13c. Measured apparent impedance near resonance of a thick antenna (Hartig).

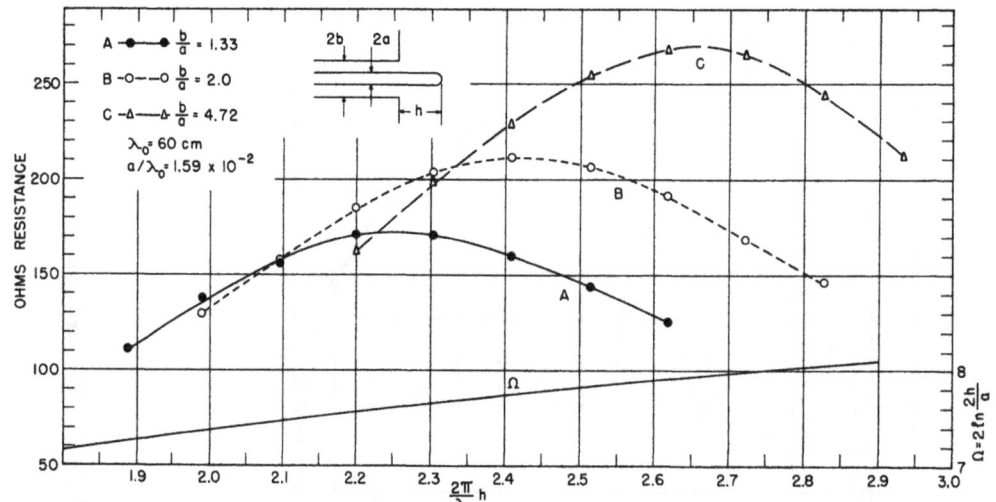

Fig. 38.13d. Measured apparent resistance near antiresonance of a thick antenna (Hartig).

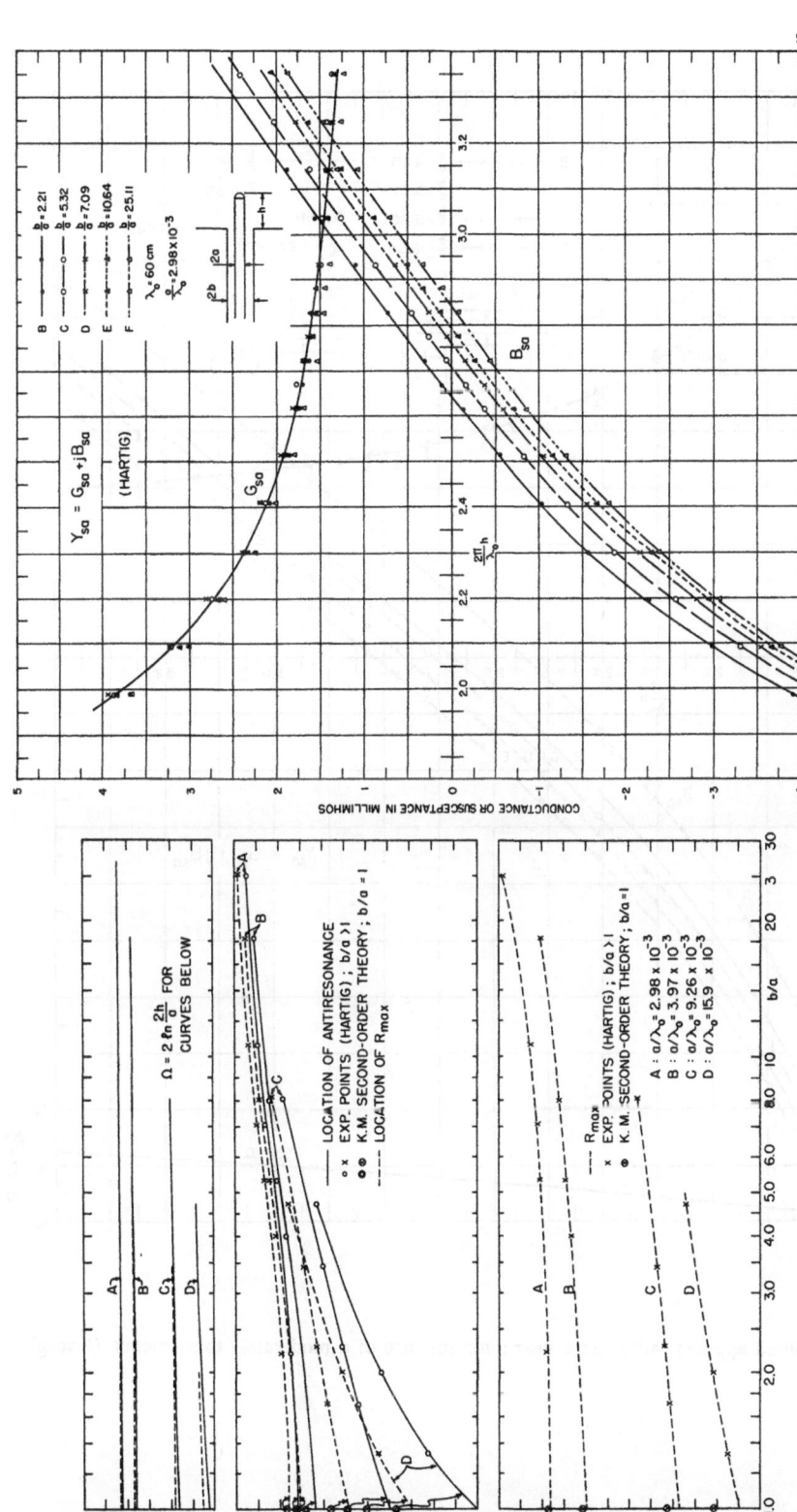

Fig. 38.15. Measured apparent admittance near antiresonance for a thin antenna (Hartig).

Fig. 38.14. Projection of critical experimental values near antiresonance to "zero" line spacing, i.e., $b/a = 1$.

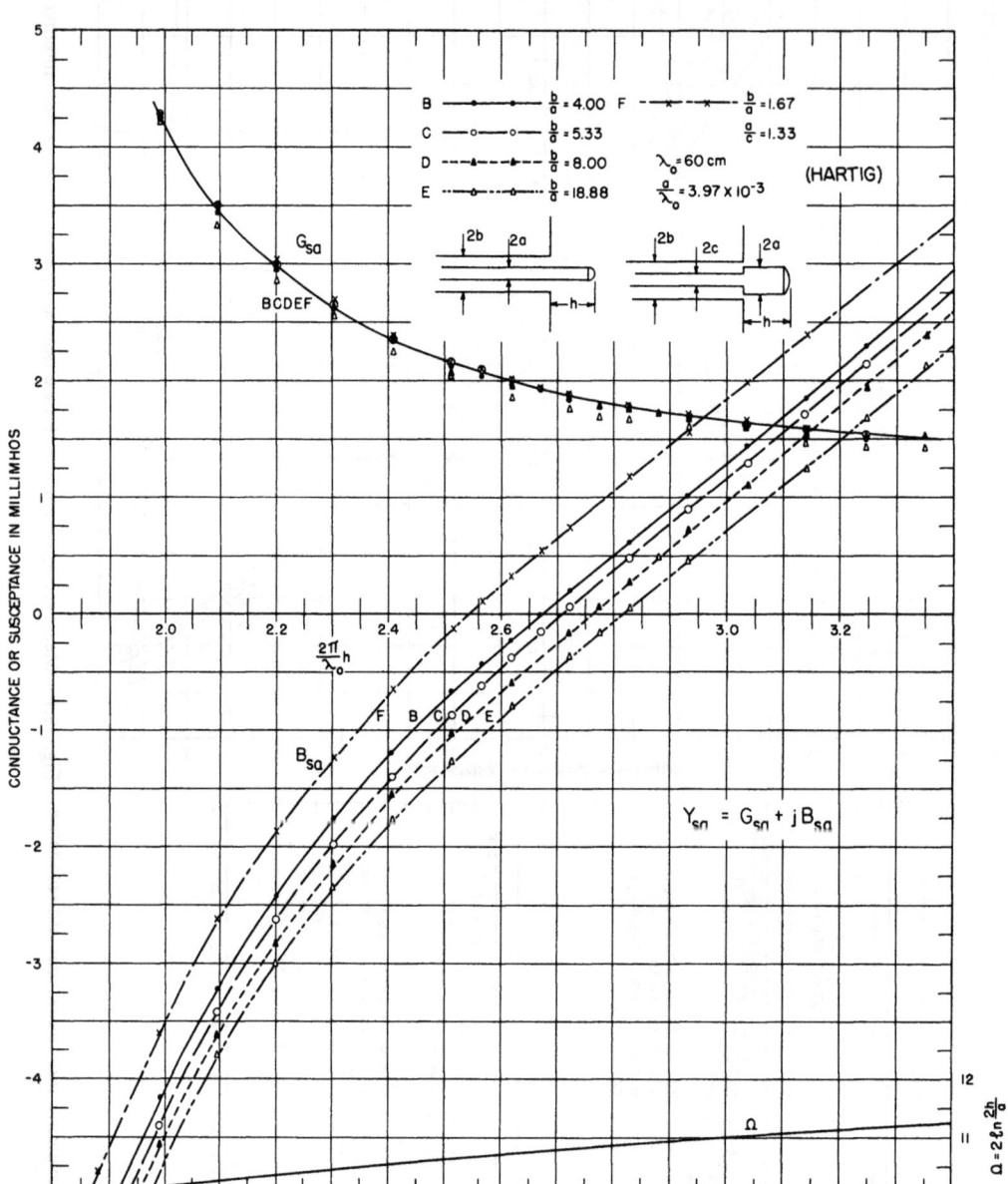

Fig. 38.16. Measured apparent admittance near antiresonance of a moderately thin antenna (Hartig).

[II.38] THEORY OF LINEAR ANTENNAS 249

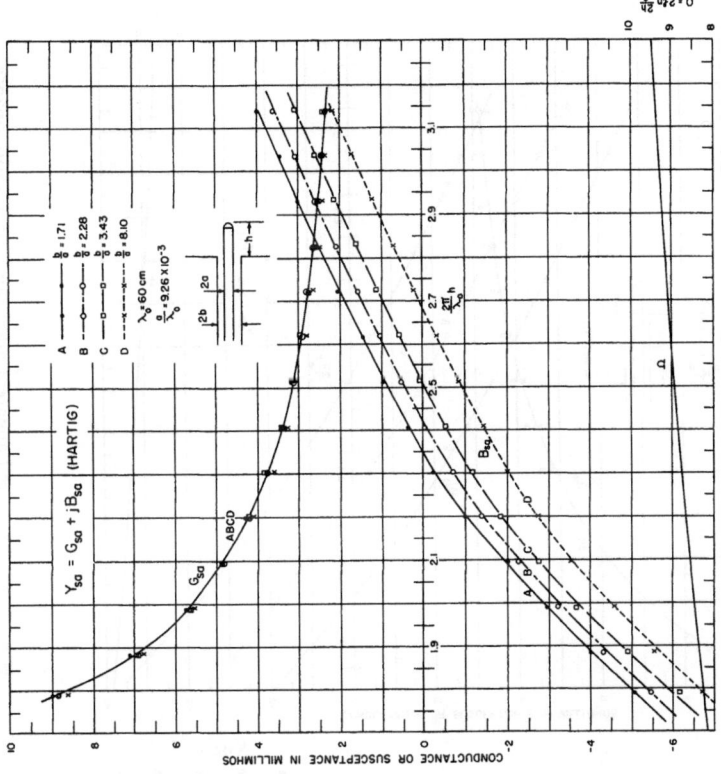

Fig. 38.17b. Measured apparent admittance near antiresonance of a moderately thick antenna (Hartig).

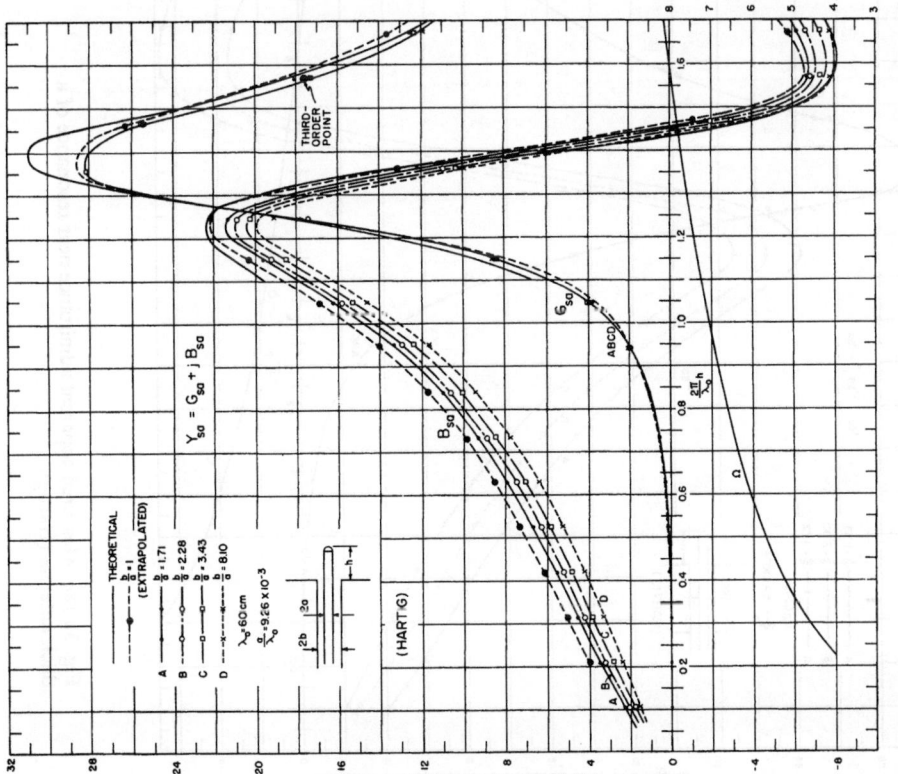

Fig. 38.17a. Measured apparent admittance near resonance of a moderately thick antenna (Hartig).

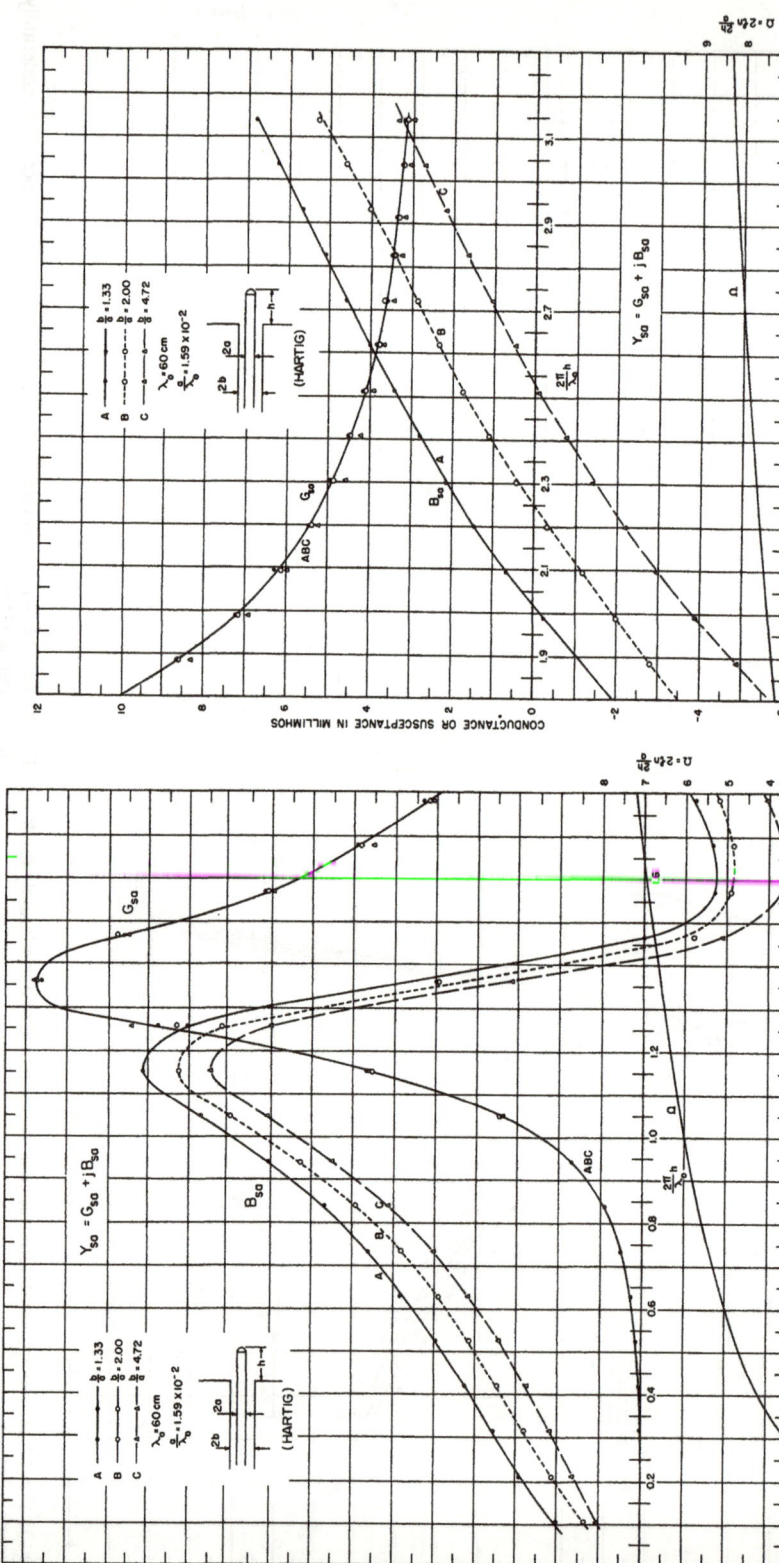

Fig. 38.18b. Measured apparent admittance near antiresonance of a thick antenna (Hartig).

Fig. 38.18a. Measured apparent admittance near resonance of a thick antenna (Hartig).

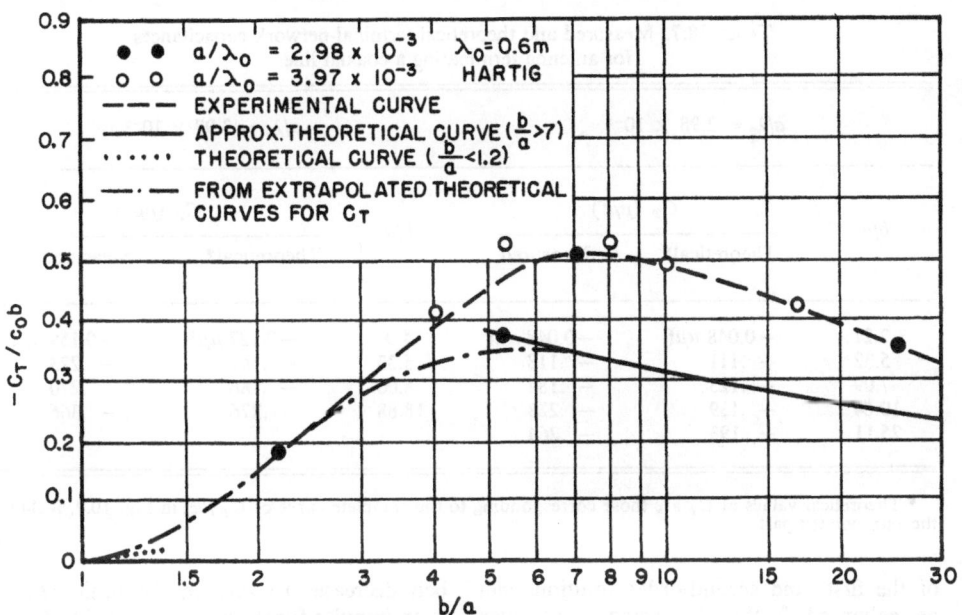

Fig. 38.19. Curves of $-C_T/c_0 b$ as a function of b/a.

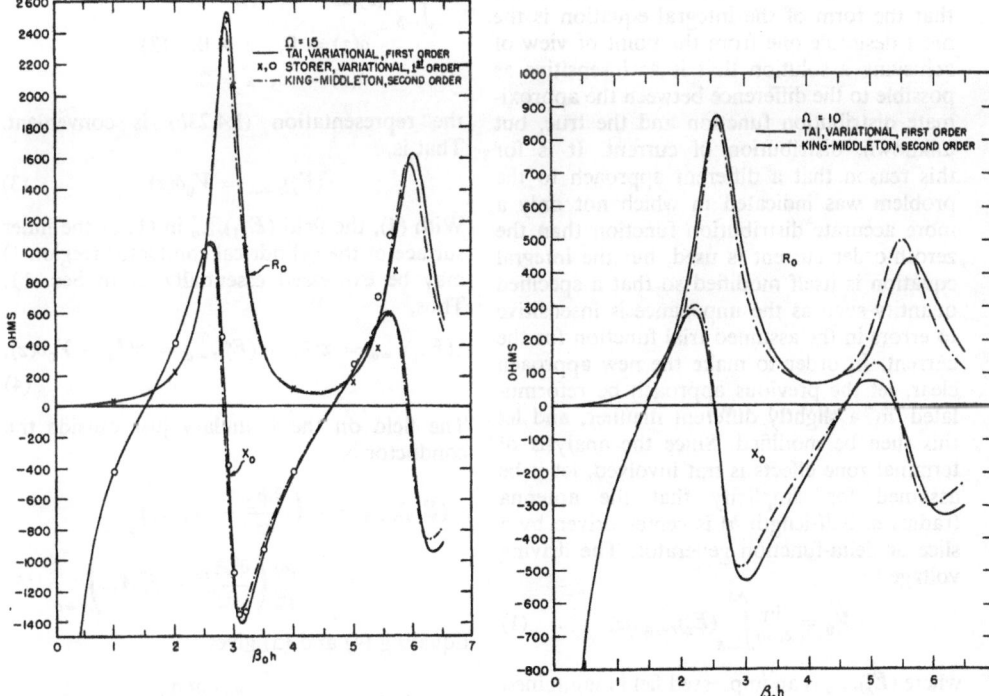

Fig. 39.1a. Theoretical impedance of cylindrical antennas determined by different methods, $\Omega = 15$.

Fig. 39.1b. Theoretical impedance of cylindrical antennas determined by different methods, $\Omega = 10$.

TABLE 38.7. Measured and theoretical terminal-network capacitances
for antenna terminating a coaxial line.

	$a/\lambda_0 = 2.98 \times 10^{-3}$			$a/\lambda_0 = 3.97 \times 10^{-3}$	
	C_T ($\mu\mu f$)			C_T ($\mu\mu f$)	
b/a	Theoretical*	Measured	b/a	Theoretical*	Measured
2.21	-0.048 $\mu\mu f$	-0.048 $\mu\mu f$	4.0	-0.127 $\mu\mu f$	-0.159 $\mu\mu f$
5.32	$-.111$	$-.118$	5.33	$-.147$	$-.223$
7.09	$-.123$	$-.184$	8.00	$-.186$	$-.270$
10.64	$-.139$	$-.223$	18.88	$-.226$	$-.366$
25.11	$-.193$	$-.264$			

* Theoretical values of C_T are those corresponding to the complete curve of C_T/bc_0 in Fig. 10.9, including the extrapolated part.

of the first- and second-order solutions can be enhanced further by using a complex expansion parameter and a distribution or trial function for the current that approximates the first-order rather than the zeroth-order current, there still would be no assurance that the form of the integral equation is the most desirable one from the point of view of achieving a solution that is as insensitive as possible to the difference between the approximate distribution function and the true, but unknown, distribution of current. It is for this reason that a different approach to the problem was indicated in which not only a more accurate distribution function than the zeroth-order current is used, but the integral equation is itself modified so that a specified quantity such as the impedance is insensitive to errors in the assumed trial function for the current. In order to make the new approach clear, let the previous approach be reformulated in a slightly different manner, and let this then be modified. Since the analysis of terminal-zone effects is not involved, let it be assumed for simplicity that the antenna (radius a, half-length h) is center driven by a slice or delta-function generator. The driving voltage is

$$V_0 = \lim_{\delta \to 0} \int_{-\delta}^{\delta} (E_z^e)_{r=a} \, dz, \quad (1)$$

where $(E_z^e)_{r=a}$ is an impressed field maintained only across a narrow belt of thickness 2δ on the surface at the center of a straight and unbroken conductor. Since it is assumed that V_0 in (2) remains constant as the width of the belt decreases to zero in the limit, $(E_z^e)_{r=a}$ is an impulse function, as discussed in Sec. 4. In terms of the Dirac delta function $\delta(z)$ defined in Sec. 4, namely,

$$\int_{-h}^{h} \delta(z) \, dz = 1, \quad (h > 0)$$

$$\delta(z) = 0, \quad z \neq 0, \quad (2)$$

$$\delta(z) = \infty, \quad z = 0,$$

the representation (I.4.23b) is convenient. That is,

$$(E_z^e)_{r=a} = V_0 \delta(z). \quad (3)$$

With (4), the field $(E_{z1})_{r=a}$ in (1) at the inner surface of the cylindrical conductor (region 1) may be expressed essentially as in Sec. 11. Thus,

$$(E_{z1})_{r=a} = z^i I_z - (E_z^e)_{r=a} = z^i I_z - V_0 \delta(z). \quad (4)$$

The field on the boundary just outside the conductor is

$$(E_{z2})_{r=a} = -\left(\frac{\partial \Phi_2}{\partial z} + j\omega A_{z2}\right)_{r=a}$$

$$= -\frac{j\omega}{\beta_0^2}\left(\frac{\partial^2 A_{z2}}{\partial z^2} + \beta_0^2 A_{z2}\right)_{r=a}. \quad (5)$$

Equating (4) and (5) gives

$$V_0 \delta(z) = z^i I_z + \frac{j\omega}{\beta_0^2}\left(\frac{\partial^2 A_z}{\partial z^2} + \beta_0^2 A_z\right). \quad (6)$$

For simplicity the subscript 2 on A_z and the specification $r = a$ have been omitted in

(6). Since $\delta(z) = 0$ for $z \neq 0$, (6) gives (11.12) in the region where this is defined. Instead of solving (6) for A_z and then obtaining the integral equation for the current, the impedance may be introduced directly using the definition

$$Z_0 \equiv V_0/I_0. \qquad (7)$$

Thus,

$$Z_0\delta(z)I_0 = z^i I_z + \frac{j\omega}{\beta_0^2}\left(\frac{\partial^2}{\partial z^2} + \beta_0^2\right)A_z, \qquad (8)$$

where

$$A_z = \frac{1}{4\pi v_0}\int_{-h}^{h} I_z' \frac{e^{-j\beta_0 R_1}}{R_1}\,dz', \qquad (9)$$

$$R_1 = \sqrt{(z-z')^2 + a^2}. \qquad (10)$$

It is essentially the solution of (8) that is obtained in Secs. 11–21.

The modification introduced into (8) by Storer is to multiply both sides by I_z and integrate over the length of the antenna. Thus,

$$Z_0 I_0 \int_{-h}^{h} I_z \delta(z)\,dz = Z_0 I_0^2 = \int_{-h}^{h} z^i I_z^2\,dz$$

$$+ \frac{j\omega}{\beta_0^2}\int_{-h}^{h} I_z\left(\frac{\partial^2}{\partial z^2} + \beta_0^2\right)A_z\,dz. \qquad (11)$$

This is dimensionally a power equation, but it is not the time-average power equation which is obtained by multiplying (8) with the complex conjugate I_0^* instead of I_0. Note that the real part of the left-hand member of (11) does not give the power radiated. By introducing (9) directly in (11), noting that $\zeta_0 = v_0/\nu_0 \doteq 120\pi$ ohms, Storer's form of the integral equation is obtained. It is

$$Z_0 = \frac{1}{I_0^2}\left[z^i\int_{-h}^{h} I_z^2\,dz\right.$$

$$\left. + \frac{j\zeta_0}{4\pi}\int_{-h}^{h}\int_{-h}^{h} K(z-z')I_z I_z'\,dz\,dz'\right] \qquad (12)$$

where the kernel is

$$K(z'-z) = K(z-z')$$

$$= \frac{1}{\beta_0}\left(\frac{\partial^2}{\partial z^2} + \beta_0^2\right)\frac{e^{-j\beta_0 R_1}}{R_1}. \qquad (13)$$

The advantage of (12) over the corresponding equation that could be obtained from (8) by integrating over the length of the antenna is its stationary property. It is readily verified that if an approximate current is introduced in (12) as a trial function, a small difference ΔI_z between the true current I_z and the trial current $I_{Tz} = I_z + \Delta I_z$ affects the true impedance only in terms that are functions of the second-order small quantities $(\Delta I_z)^2$. In the language of the calculus of variations, the first variation of Z_0 with respect to I_z is zero, or Z_0 is stationary with respect to small changes in I_z. This can be shown directly by taking the variation of (12).

Without introducing the notation of the calculus of variations, this may be proved by evaluating the approximate impedance Z_a obtained from (12) if $I_{Tz} = I_z + \Delta I_z$ is substituted for I_z. Thus, with the notation $I_z' \equiv I_z(z')$,

$$Z_a = \frac{1}{(I_0 + \Delta I_0)^2}\left[z^i\int_{-h}^{h}(I_z + \Delta I_z)^2\,dz\right.$$

$$+ \frac{j\zeta_0}{4\pi}\int_{-h}^{h}\int_{-h}^{h} K(z-z')[(I_z + \Delta I_z)$$

$$\left.\times (I_z' + \Delta I_z')\,dz'\,dz\right] \qquad (14)$$

$$= \frac{1}{(I_0 + \Delta I_0)^2}\left\{\left[z^i\int_{-h}^{h} I_z^2\,dz\right.\right.$$

$$\left.+ \frac{j\zeta_0}{4\pi}\int_{-h}^{h}\int_{-h}^{h} K(z-z')I_z I_z'\,dz'\,dz\right]$$

$$+ 2\left[z^i\int_{-h}^{h} I_z \Delta I_z\,dz\right.$$

$$\left.+ \frac{j\zeta_0}{4\pi}\int_{-h}^{h}\int_{-h}^{h} K(z-z')I_z' \Delta I_z\,dz'\,dz\right]$$

$$+ \left[z^i\int_{-h}^{h}(\Delta I_z)^2\,dz\right.$$

$$\left.\left.+ \frac{j\zeta_0}{4\pi}\int_{-h}^{h}\int_{-h}^{h} K(z-z')\Delta I_z' \Delta I_z\,dz'\,dz\right]\right\} \qquad (15)$$

With (12) and (9) multiplied by ΔI_z and integrated from $-h$ to h with respect to z, the first and second square brackets in (15) are simplified. Thus,

$$Z_a = \frac{1}{(I_0 + \Delta I_0)^2}\left\{Z_0 I_0^2\right.$$

$$+ 2\int_{-h}^{h}\Delta I_z Z_0 \delta(z)I_0\,dz + \left[z^i\int_{-h}^{h}(\Delta I_z)^2\,dz\right.$$

$$\left.\left.+ \frac{j\zeta_0}{4\pi}\int_{-h}^{h}\int_{-h}^{h} K(z-z')\Delta I_z' \Delta I_z\,dz'\,dz\right]\right\}. \qquad (16)$$

Since $\int_{-h}^{h} \delta(z)\,dz = 1$ and $\delta(z) = 0$ when

$z \neq 0$, the first integral in (16) becomes $2Z_0 I_0 \Delta I_0$. By adding and subtracting $Z_0(\Delta I_0)^2$, it follows that

$$Z_a = Z_0 + \frac{1}{(I_0 + \Delta I_0)^2} \left[z^i \int_{-h}^{h} (\Delta I_z)^2 \, dz \right.$$

$$+ \frac{j\zeta_0}{4\pi} \int_{-h}^{h} \int_{-h}^{h} K(z-z') \Delta I_z' \Delta I_z \, dz' \, dz$$

$$\left. - Z_0(\Delta I_0)^2 \right]. \quad (17)$$

Thus it is seen that if ΔI_z is a small quantity of the first order, Z_a differs from Z_0 by a small quantity of the second order, since it involves only terms of the order $(\Delta I_z)^2$. This is equivalent to stating that the first variation of Z_0 with respect to small variations in ΔI_z is zero.

Since the integral (12) is stationary with respect to first-order changes in the current, it follows that if the trial function I_{zT} is a good approximation of the true current I_z, the approximate impedance Z_{0T} must be a still better approximation of the true value Z_0. Evidently, the more accurate I_{zT}, the better will Z_{0T} approximate Z_T. Hence, instead of using the zeroth-order current as in the King-Middleton solution of Hallén's equation, the modified distribution given in Sec. 25 may be used, since this is known to be a fair approximation of the *first-order current*. The particular form of this distribution given in (25.21, 22) may be generalized as follows:

$$I_{zT} = V_0 \{ A \sin \beta_0(h - |z|)$$

$$+ B[1 - \cos \beta_0(h - |z|)] \}.$$

$$(\beta_0 z < 2\pi) \quad (18a)$$

Note that since

$$I_{0T} = V_0 \{ A \sin \beta_0 h + B[1 - \cos \beta_0 h] \},$$

$$(18b)$$

I_{0T} vanishes when $\beta_0 h = 2\pi$. Since I_{0T}^2 appears explicitly in the denominator in (12), Z_a becomes infinite when $\beta_0 h = 2\pi$ if (18a) is used as a trial function. Clearly, a trial function different from (18a) is required if the range near $\beta_0 h = 2\pi$ is to be included. For this purpose Tai[81] introduced the trial function

$$I_{zT} = I_0[\sin \beta_0(h - |z|)$$

$$+ A \beta_0(h - |z|) \cos \beta_0(h - |z|)],$$

$$(19)$$

where A is an arbitrary constant. Since this current obviously is nonvanishing for all values of $\beta_0 h$ greater than zero, Z_a remains finite. In the following analysis Storer's simpler distribution function (18a) is used with the understanding that Z_0 is restricted to the range $0 \leq \beta_0 h \leq 3\pi/2$. However, Tai's results also are given and, as is to be expected, they differ negligibly from Storer's over the more restricted range in which the latter are a good approximation.

Substitution of (18a) in (12) gives

$$Z_a = \frac{j\zeta_0}{4\pi} [\gamma_{AA} A^2 + 2\gamma_{AB} AB + \gamma_{BB} B^2] Z_0^2,$$

$$(\beta_0 h \leq 3\pi/2) \quad (20)$$

where

$$\gamma_{AA} = \frac{4\pi}{j\zeta_0} z^i \int_{-h}^{h} \sin^2 \beta_0(h - |z|) \, dz$$

$$+ \int_{-h}^{h} \int_{-h}^{h} K(z - z') \sin \beta_0(h - |z'|)$$

$$\times \sin \beta_0(h - |z|) \, dz \, dz', \quad (21a)$$

$$\gamma_{AB} = \frac{4\pi}{j\zeta_0} z^i \int_{-h}^{h} \sin \beta_0(h - |z|)$$

$$\times [1 - \cos \beta_0(h - |z|)] \, dz$$

$$+ \int_{-h}^{h} \int_{-h}^{h} K(z - z') \sin \beta_0(h - |z|)$$

$$\times [1 - \cos \beta_0(h - |z'|)] \, dz \, dz', \quad (21b)$$

$$\gamma_{BB} = \frac{4\pi}{j\zeta_0} z^i \int_{-h}^{h} [1 - \cos \beta_0(h - |z|)]^2 \, dz$$

$$+ \int_{-h}^{h} \int_{-h}^{h} K(z - z')[1 - \cos \beta_0(h - |z|)]$$

$$\times [1 - \cos \beta_0(h - |z'|)] \, dz \, dz'. \quad (21c)$$

By expanding the integrands and arranging them these integrals may be expressed in terms of the functions

$$C_a(h, 0) \equiv 2 \int_0^h \cos \beta_0 z' \frac{e^{-j\beta_0 R_0}}{R_0} \, dz', \quad (22a)$$

$$S_a(h, 0) \equiv 2 \int_0^h \sin \beta_0 z' \frac{e^{-j\beta_0 R_0}}{R_0} \, dz', \quad (22b)$$

$$E_a(h, 0) \equiv 2 \int_0^h \frac{e^{-j\beta_0 R_0}}{R_0} \, dz', \quad (22c)$$

where

$$R_0 = \sqrt{z'^2 + a^2}. \quad (23)$$

These are obtained from the more general functions in Sec. 19 by setting $z = 0$. The procedure parallels that in Secs. 18 and 19. The results are

$$Y_{AA} = \frac{4\pi}{j\zeta_0} z^i h \left(1 - \frac{\sin 2\beta_0 h}{2\beta_0 h}\right)$$

$$- 2 \cos \beta_0 h [\sin \beta_0 h C_a(h, 0)$$
$$- \cos \beta_0 h S_a(h, 0)]$$
$$+ \sin 2\beta_0 h [C_a(2h, 0) - C_a(h, 0)]$$
$$- \cos 2\beta_0 h [S_a(2h, 0) - S_a(h, 0)$$
$$+ S_u(h, 0)], \quad (24a)$$

$$Y_{AB} = \frac{4\pi}{j\zeta_0} \frac{z^i}{\beta_0} (1 - \cos \beta_0 h)^2 + E_a(2h, 0)$$
$$- \cos \beta_0 h [C_a(2h, 0) - C_a(h, 0)]$$
$$- \sin \beta_0 h [S_a(2h, 0) - S_a(h, 0)]$$
$$- C_a(h, 0) - 2 \cos \beta_0 h E_a(h, 0)$$
$$+ 2 \cos^2 \beta_0 h C_a(h, 0)$$
$$+ \sin 2\beta_0 h S_a(h, 0), \quad (24b)$$

$$Y_{BB} = \frac{4\pi}{j\zeta_0} z^i \left(\beta_0 h - 4 \sin \beta_0 h + \frac{\sin 2\beta_0 h}{2}\right)$$
$$+ S_a(h, 0) - 4 \sin \beta_0 h E_a(h, 0)$$
$$+ \sin 2\beta_0 h C_a(h, 0)$$
$$+ 2 \sin^2 \beta_0 h S_a(h, 0)$$
$$- \sin 2\beta_0 h [C_a(2h, 0) - C_a(h, 0)]$$
$$+ \cos 2\beta_0 h [S_a(2h, 0) - S_a(h, 0)]$$
$$+ 2\beta_0 h E_a(2h, 0)$$
$$- j[e^{-j\beta_0 \sqrt{4h^2+a^2}} - e^{-j\beta_0 a}]. \quad (24c)$$

In most practical cases the terms with z^i as a factor are negligible. Since the functions $C_a(h, z)$, $S_a(h, z)$, $E_a(h, z)$ are expressed in terms of tabulated functions in Sec. 19, the three functions Y_{AA}, Y_{AB}, and Y_{BB} in (20) may be computed.

In order to complete the solution for the impedance as given in (20) it is necessary to determine the complex coefficients A and B of the trial distribution of current (18). This may be accomplished as follows.

Since $Z_a = 1/Y_a = V_0/I_{0T}$, it follows from (19) that

$$B = \frac{Y_a - A \sin \beta_0 h}{1 - \cos \beta_0 h}. \quad (25)$$

Substitution of (25) in (20) to eliminate B gives

$$Z_a = \frac{j\zeta_0}{4\pi} \left[Y_{AA} A^2 + 2Y_{AB} A \left(\frac{Y_a - A \sin \beta_0 h}{1 - \cos \beta_0 h}\right) \right.$$
$$\left. + Y_{BB} \left(\frac{Y_a - A \sin \beta_0 h}{1 - \cos \beta_0 h}\right)^2 \right] Z_0^2. \quad (26)$$

The remaining parameter A now can be determined to give an optimum value of Z_a with respect to A by imposing the following condition:

$$\frac{\partial Z_a}{\partial A} = \frac{\partial R_a}{\partial A} + j \frac{\partial X_a}{\partial A} = 0. \quad (27)$$

This gives

$$2Y_{AA} + 2Y_{AB} \left(\frac{Y_0 - A \sin \beta_0 h}{1 - \cos \beta_0 h}\right)$$
$$- 2Y_{AB} A \left(\frac{\sin \beta_0 h}{1 - \cos \beta_0 h}\right)$$
$$- 2Y_{BB} \left(\frac{Y_0 - A \sin \beta_0 h}{1 - \cos \beta_0 h}\right)\left(\frac{\sin \beta_0 h}{1 - \cos \beta_0 h}\right)$$
$$= 0, \quad (28)$$

which may be solved for A readily. For convenience let

$$\Delta \equiv [Y_{AA}(1 - \cos \beta_0 h)^2$$
$$- 2Y_{AB}(1 - \cos \beta_0 h) \sin \beta_0 h$$
$$+ Y_{BB} \sin^2 \beta_0 h]^{-1}. \quad (29)$$

With (29), (28) reduces to

$$A = Y_0 \Delta [Y_{BB} \sin \beta_0 h - Y_{AB}(1 - \cos \beta_0 h)]. \quad (30)$$

With (30), (25) and (26) become

$$B = Y_0 \Delta [Y_{AA}(1 - \cos \beta_0 h) - Y_{AA} \sin \beta_0 h], \quad (31)$$

$$Z_0 \doteq Z_a = \frac{j\zeta_0}{4\pi} \Delta(Y_{AA} Y_{BB} - Y_{AB}^2),$$
$$\beta_0 h \leq 3\pi/2. \quad (32)$$

The admittance Y_0 may be eliminated from (30) and (31) using (32). The results are

$$A = \frac{-j4\pi}{\zeta_0} \left[\frac{Y_{BB} \sin \beta_0 h - Y_{AB}(1 - \cos \beta_0 h)}{Y_{AA} Y_{BB} - Y_{AB}^2}\right],$$
$$(\beta_0 h \leq 3\pi/2) \quad (33)$$

$$B = \frac{-j4\pi}{\zeta_0} \left[\frac{Y_{AA}(1 - \cos \beta_0 h) - Y_{AB} \sin \beta_0 h}{Y_{AA} Y_{BB} - Y_{AB}^2}\right].$$
$$(\beta_0 h \leq 3\pi/2) \quad (34)$$

With Z_0 determined in (32) and the coefficients A and B in (33) and (34), the analysis of the impedance and the distribution of current of the center-driven antenna is completed. It remains to evaluate these quantities numerically and compare the results with those of the King-Middleton formulation, since these have been compared with experiment.

The numerical evaluation of the impedance $Z_0 \doteq Z_a$ and the coefficients A and B has been carried out for $\Omega = 2 \ln (2h/a) = 15$ by Storer. The results are given in Table 39.1 and in Figs. 39.1* and 39.2. Corresponding calculations by Tai using the distribution function (19) are shown in Figs. 39.1 and VIII.11.3, 4 for $\Omega = 15$ and 10. The corresponding King-Middleton second-order impedances as given in Sec. 30 are shown for comparison. As seen in Fig. 39.1, the Storer and Tai forms of the variational solution are in almost perfect agreement over the range in which the former is valid. Moreover, the general agreement between the King-Middleton second-order and the Storer-Tai variational results is excellent. Except for a slight shift of the peak of the resistance toward longer lengths, the variational curve almost coincides with the King-Middleton curve for $\beta_0 h \lesssim 3\pi/2$. Even for longer electrical lengths the agreement is good. Referring to the results for $\Omega = 15$, $(R_0)_{max} = 2438$ ohm (King-Middleton), $(R_0)_{max} = 2495$ ohm (Storer), with a difference of 2.4 percent. The minimum resistances near $\beta_0 h = 4.3$ are $(R_0)_{min} = 85.0$ ohm (King-Middleton), $(R_0)_{min} = 81.6$ ohm (Storer), with a difference of 4 percent.

A comparison of values for $\beta_0 h \leq 0.5$ can be made using (31.6c) for the first-order King-Middleton reactance, which for $\Omega = 15$ gives

$$X_0 = \frac{-696.6}{\beta_0 h} \text{ ohms}, \qquad [(\beta_0 h)^2 \ll 1] \quad (35)$$

The corresponding variational form is obtained by expanding (24a, b, c) in powers of $\beta_0 h$ and retaining only the leading terms. Thus, with $(\beta_0 h)^2 \ll 1$,

$$Y_{AA} \doteq 2\beta_0 h[2 + \ln 4 - \Omega], \qquad (36a)$$

$$Y_{BB} \doteq \beta_0^2 h^2[1 + 4 \ln 2 - \Omega], \qquad (36b)$$

$$Y_{AB} \doteq \beta_0^3 h^3 \left[\frac{10}{9} + \frac{8}{3} \ln 2 - \frac{2}{3}\Omega\right]. \qquad (36c)$$

* See p. 251.

TABLE 39.1. The current distribution parameters A and B in
$I_T(z) = A \sin \beta_0(h - |z|) + B[1 - \cos \beta_0(h - |z|)]$;
$\Omega \equiv 2 \ln 2h/a = 15$.

$\beta_0 h$	A (ma/volt)	B (ma/volt)
0.5	0.0126 $+j$ 1.499	$-$0.0244 $+j$0.6647
.7	.0420 $+j$ 1.800	$-$ .0550 $+j$ .3712
.9	.1340 $+j$ 2.299	$-$ .1153 $+j$ .1471
1.0	.2383 $+j$ 2.724	$-$ .1677 $+j$ .0553
1.1	.4325 $+j$ 3.332	$-$ .2522 $-j$ .0528
1.2	.8991 $+j$ 4.357	$-$ .3997 $-j$ .1668
1.3	2.135 $+j$ 6.038	$-$ .7081 $-j$ .3006
1.4	7.927 $+j$10.42	$-$1.717 $-j$ .5909
1.5	15.74 $+j$ 0.8847	$-$1.982 $+j$1.515
1.6	7.664 $-j$ 7.385	$-$0.2123 $+j$1.820
1.7	3.146 $-j$ 5.868	.2154 $+j$1.210
1.8	1.420 $-j$ 4.479	.2768 $+j$0.8839
1.9	1.042 $-j$ 3.604	.2763 $+j$ .6998
2.0	0.7241 $-j$ 3.036	.2642 $+j$ .5823
2.1	.5247 $-j$ 2.641	.2511 $+j$ .5005
2.2	.4003 $-j$ 2.361	.2390 $+j$ .4395
2.3	.3139 $-j$ 2.155	.2285 $+j$ .3919
2.4	.2506 $-j$ 2.001	.2195 $+j$ .3534
2.5	.2022 $-j$ 1.884	.2117 $+j$ .3215
2.6	.1636 $-j$ 1.796	.2050 $+j$ .2943
2.7	.1319 $-j$ 1.731	.1992 $+j$ .2709
2.8	.1050 $-j$ 1.685	.1942 $+j$ .2504
2.9	.0815 $-j$ 1.656	.1899 $+j$ .2323
3.0	.0604 $-j$ 1.641	.1862 $+j$ .2162
3.1	.0407 $-j$ 1.641	.1831 $+j$ .2017
3.2	.0216 $-j$ 1.655	.1806 $+j$ .1886
3.3	.0023 $-j$ 1.684	.1787 $+j$ .1767
3.4	$-$ .0180 $-j$ 1.730	.1776 $+j$ .1658
3.5	$-$ .0406 $-j$ 1.794	.1772 $+j$ .1559
3.6	$-$ .0671 $-j$ 1.880	.1778 $+j$ .1468
3.7	$-$ .1002 $-j$ 1.994	.1796 $+j$ .1384
3.8	$-$ .1438 $-j$ 2.144	.1830 $+j$ .1306
3.9	$-$ .2051 $-j$ 2.343	.1884 $+j$ .1232
4.0	$-$ .2972 $-j$ 2.608	.1968 $+j$ .1157
4.1	$-$ .4459 $-j$ 2.972	.2097 $+j$ .1075
4.2	$-$ .6343 $-j$ 3.483	.2296 $+j$ .0968
4.3	$-$ 1.221 $-j$ 4.218	.2609 $+j$ .0790
4.4	$-$ 2.366 $-j$ 5.245	.3103 $+j$ .0399
4.5	$-$ 5.182 $-j$ 6.111	.3711 $-j$ .0660
4.6	$-$ 9.913 $-j$ 3.195	.3036 $-j$ .2865
4.7	$-$ 8.477 $+j$ 3.133	.0377 $-j$ .3008
4.8	$-$ 4.618 $+j$ 4.210	$-$ .0479 $-j$ .1646
4.9	$-$ 2.732 $+j$ 3.606	$-$ .0518 $-j$ .0773
5.0	$-$ 1.796 $+j$ 2.891	$-$ .0315 $-j$ .0469

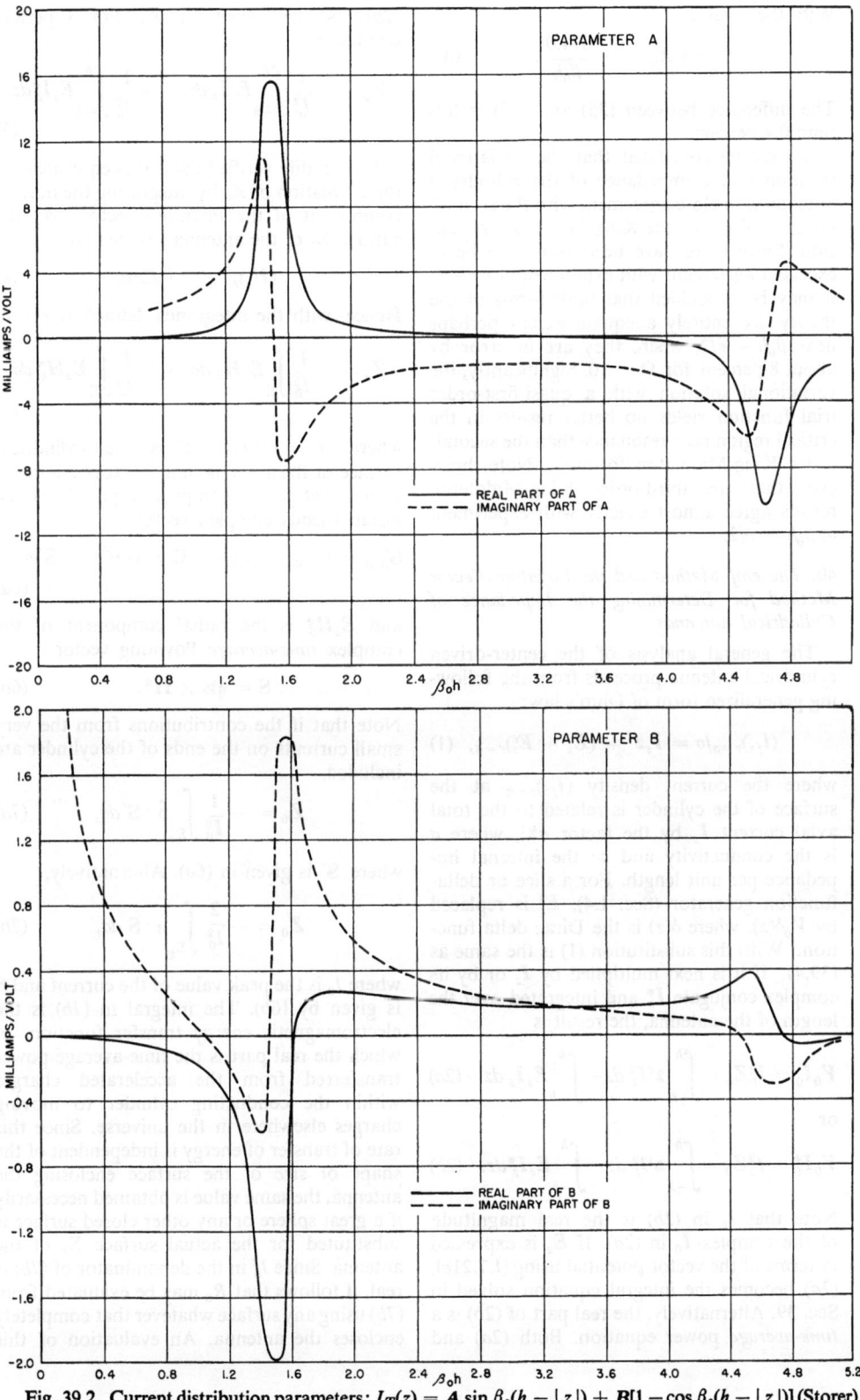

Fig. 39.2. Current distribution parameters; $I_T(z) = A \sin \beta_0(h - |z|) + B[1 - \cos \beta_0(h - |z|)]$ (Storer).

With these values

$$X_0 \doteq X_a = -\frac{694.0}{\beta_0 h}. \quad (37)$$

The difference between (35) and (37) is less than 0.4 percent.

It may be concluded that the variational solution of the impedance of the cylindrical antenna is in close agreement with the second-order results from the King-Middleton expansion. Since these have been shown to be in excellent agreement with experimental results, it may be concluded that both forms of the theory are entirely adequate except perhaps near $\beta_0 h = \pi/2$, where they are in error by about 8 percent for $\Omega = 10$. Significantly, the variational solution with a quasi-first-order trial function yields no better results in the critical region near resonance than the second-order King-Middleton formula. Note, however, that the third-order King-Middleton results agree almost exactly with experiment at $\beta_0 h = \pi/2$.

40. The emf Method and the Poynting-Vector Method for Determining the Impedance of Cylindrical Antennas

The general analysis of the center-driven cylindrical antenna proceeds from the following generalized form of Ohm's law:

$$(i_{fz})_{r=a}/\sigma = I_z z^i = (E_z + E_z^e)_{r=a}, \quad (1)$$

where the current density $(i_{fz})_{r=a}$ at the surface of the cylinder is related to the total axial current I_z by the factor σz^i, where σ is the conductivity and z^i the internal impedance per unit length. For a slice or delta-function generator (Sec. I.4), E_z^e is replaced by $V_0 \delta(z)$, where $\delta(z)$ is the Dirac delta function. With this substitution (1) is the same as (39.4). If it is next multiplied by I_z or by its complex conjugate I_z^* and integrated over the length of the antenna, the result is

$$V_0 I_0 = I_0^2 Z_0 = \int_{-h}^{h} z^i I_z^2 \, dz - \int_{-h}^{h} E_z I_z \, dz \quad (2a)$$

or

$$V_0 I_0^* = I_0^2 Z_0 = \int_{-h}^{h} z^i I_z^2 \, dz - \int_{-h}^{h} E_z I_z^* \, dz. \quad (2b)$$

Note that I_0 in (2b) is the real magnitude of the complex I_0 in (2a). If E_z is expressed in terms of the vector potential using (I.7.21c), (2a) becomes the integral equation solved in Sec. 39. Alternatively, the real part of (2b) is a time-average power equation. Both (2a) and (2b) may be solved for Z_0. For a perfect conductor

$$Z_0 = -\frac{1}{I_0^2} \int_{-h}^{h} E_z I_z \, dz = -\frac{1}{I_0^2} \int_{-h}^{h} E_z I_z^* \, dz. \quad (3)$$

It is readily verified that (3) is equivalent to the evaluation of Z_0 by integrating the normal component of the Poynting vector over the surface Σ_T of the antenna. Note that

$$(H_\theta)_{r=a} = I_z/2\pi a. \quad (4)$$

Hence, with the integrands defined at $r = a$,

$$Z_0 = -\frac{1}{I_0^2} \int_\Sigma E_z H_\theta \, d\sigma = -\frac{1}{I_0^2} \int_\Sigma E_z H_\theta^* \, d\sigma, \quad (5)$$

where $d\sigma = a \, d\theta \, dz$, Σ is the cylindrical surface of the antenna, and $E_z H_\theta$ is the radial component of the complex amplitude of the instantaneous complex vector

$$\mathbf{S}'_{\text{inst}} \equiv \mathbf{E}_{\text{inst}} \times \mathbf{H}_{\text{inst}} = \mathbf{E} \times \mathbf{H} e^{j2\omega t} \equiv \mathbf{S}' e^{j2\omega t} \quad (6a)$$

and $E_z H_\theta^*$ is the radial component of the complex *time-average* Poynting vector

$$\mathbf{S} = \tfrac{1}{2} \mathbf{E} \times \mathbf{H}^*. \quad (6b)$$

Note that if the contributions from the very small currents on the ends of the cylinder are included,

$$Z_0 = -\frac{1}{I_0^2} \int_{\Sigma_T} \hat{\mathbf{n}} \cdot \mathbf{S}' \, d\sigma, \quad (7a)$$

where $\mathbf{S}'$ is given in (6a). Alternatively,

$$Z_0 = -\frac{2}{I_0^2} \int_{\Sigma_T} \hat{\mathbf{n}} \cdot \mathbf{S} \, d\sigma, \quad (7b)$$

where I_0 is the peak value of the current and $\mathbf{S}$ is given by (6b). The integral in (7b) is the electromagnetic energy transfer function, of which the real part is the time-average power transferred from the accelerated charges within the conducting cylinder to moving charges elsewhere in the universe. Since this rate of transfer of energy is independent of the shape or size of the surface enclosing the antenna, the same value is obtained necessarily if a great sphere or any other closed surface is substituted for the actual surface Σ_T of the antenna. Since I_0^2 in the denominator of (7b) is real, it follows that R_0 may be evaluated from (7b) using any surface whatever that completely encloses the antenna. An evaluation of this type using a great sphere is carried out in Chapter V. In order to evaluate X_0 from (7b) and both R_0 and X_0 from (7a), the surface of integration must be the surface of the antenna itself.

The use of (3) in the evaluation of Z_0 is called the *emf method*.[10,64] In this rather general sense the variational method of Sec. 39 may be considered one way of carrying out the emf method. Actually, the name emf method commonly is restricted to the simpler evaluation of (3) when the trial function is the real zeroth-order current ratio,

$$\frac{I_z}{I_0} = \frac{I_z^*}{I_0^*} = \frac{\sin \beta_0(h - |z|)}{\sin \beta_0 h}, \quad (8)$$

instead of the more carefully fitted complex distribution for $\beta_0 h \leq 3\pi/2$,

$$I_z = A \sin \beta_0(h - |z|) + B[1 - \cos \beta_0(h - |z|)], \quad (9a)$$

$$I_0 = A \sin \beta_0 h + B[1 - \cos \beta_0 h] \quad (9b)$$

used in Sec. 39. Since the electric field for the simple sinusoidal distribution (8) is known from (V.3.11) even at radius a, the two integrals in (3) can be evaluated. Obviously, with (8) they are identical. The z-component of the field is from (V.3.11c),

$$E_z = \frac{-jI_0\zeta_0}{4\pi \sin \beta_0 h} \left(\frac{e^{-j\beta_0 R_{1h}}}{R_{1h}} + \frac{e^{-j\beta_0 R_{2h}}}{R_{2h}} - \frac{2}{R_0} \cos \beta_0 h \, e^{-j\beta_0 R_0} \right), \quad (10)$$

where

$$R_{1h} = \sqrt{(h - z)^2 + a^2},$$
$$R_{2h} = \sqrt{(h + z)^2 + a^2},$$
$$R_0 = \sqrt{z^2 + a^2}. \quad (11)$$

Substitution of (10) in (4) gives

$$Z_0 = \frac{j\zeta_0}{2\pi \sin^2 \beta_0 h}$$
$$\times \{\sin \beta_0 h[C_a(h, h) - \cos \beta_0 h C_a(h, 0)]$$
$$- \cos \beta_0 h[S_a(h, h) - \cos \beta_0 h S_a(h, 0)]\}, \quad (12)$$

where $C_a(h, z)$ and $S_a(h, z)$ are defined in (19.3) and (19.4). Note that for $\beta_0 h = n\pi/2$, n odd, (12) reduces to

$$Z_0 = \frac{j\zeta_0}{2\pi} C_a(h, h), \quad (13)$$

which is identically (28.22).

The formula (12) for the impedance may be put into the classical form obtained by Bechmann[6] by allowing the radius a to vanish except in logarithmic terms, and using (19.33) and (19.34) together with (19.27a–f). The expressions obtained in this manner for R_0 and X_0 in $Z_0 = R_0 + jX_0$ are identically (28.9) and (28.10). It is seen, therefore, that the conventional emf method using the zeroth-order or sinusoidal distribution of current (8) as a trial function yields simply the modified zeroth-order approximation of the impedance (30.4). This is known to be a fair approximation only near resonance when $\beta_0 h$ is near $n\pi/2$, n odd. It is entirely unsatisfactory near antiresonance, since the impedance becomes infinite at $\beta_0 h = n\pi$, n even.

There is a fundamental difference between the conventional emf method with assumed sinusoidal current and the integral-equation method and its variational modification. In the latter the current is the real unknown and it is so determined that the field it produces *satisfies the boundary condition* on the cylindrical surface of the conductor. In the former a current distribution is postulated which maintains a field that actually *violates the boundary condition*. This is shown readily. Since (3) assumes a perfect conductor, the boundary condition requires E_z to vanish at the cylindrical surface. However, the field (10) due to the assumed current (8) differs from zero at $r = a$. Obviously it is no paradox[71,13] that an incorrect current produces an incorrect field that does not satisfy the boundary condition and that a combination of this current and this field yields an impedance that is only a rough estimate. The approximation evidently is best when the assumed distribution of current (8) most nearly matches the actual current, that is, when $\beta_0 h$ is near $\pi/2$, $3\pi/2, \cdots$; the approximation is poorest when (8) does not adequately represent the true current, that is, when $\beta_0 h$ is near $\pi, 2\pi, \cdots$.

41. The Integral-Equation Formulation of L. V. King

The first analysis of the thin cylindrical antenna by deriving an integral equation from the boundary condition for the tangential electric field and solving this by a method of successive approximations using the expansion parameter $\Omega = 2 \ln (2h/a)$ is due to L. V. King.[37] Although essentially equivalent to the method of Hallén, upon which most of this chapter is based and which was published a year later, it also constitutes a logical extension of the conventional emf method in the sense that it modifies the sinusoidal current in such a way that the axial electric field calculated from it satisfies the boundary conditions. The method is formulated conveniently as follows. For simplicity the cylindrical conductor of half-length h and radius a is assumed to be perfect, so that $z^i = 0$.

The differential equation for the axial component of the vector potential is given by (12.1) with $z^i = 0$. It is

$$\frac{d^2 A_z}{dz^2} + \beta_0^2 A_z = 0. \tag{1}$$

Using (14.10) and (14.11) it is possible to express A_z in terms of the current as follows:

$$4\pi v_0 A_z = I_z[\Psi + \gamma(z)], \tag{2}$$

where Ψ is a constant. Substitution of (2) in (1) gives

$$\frac{d^2 I_z}{dz^2} + \beta_0^2 I_z = \beta_0 f(z), \tag{3}$$

where

$$\beta_0 f(z) = -\frac{1}{\Psi}\left\{\left[\frac{d^2\gamma(z)}{dz^2} + \beta_0^2 \gamma(z)\right]I_z \right.$$
$$\left. + 2\frac{dI_z}{dz}\frac{d\gamma(z)}{dz} + \gamma(z)\frac{d^2 I_z}{dz^2}\right\}. \tag{4}$$

The sinusoidally distributed current assumed in the conventional emf method is the solution of the homogeneous part of (3). That is,

$$(I_z')_0 = I_m \sin \beta_0(h-|z'|) = I_0 \frac{\sin \beta_0(h-|z'|)}{\sin \beta_0 h}. \tag{5}$$

The solution of (3) is the solution of the homogeneous equation plus a particular integral which is the same in form as the particular integral obtained in Sec. 12. Thus

$$I_z' = I_m \sin \beta_0(h-|z'|) + \int_{z'}^{h} f(s) \sin \beta_0(s-z') \, ds. \tag{6}$$

The electromagnetic field due to this distribution consists of the part (40.10) due to the sinusoidal first term plus an integral giving the contribution of a continuous distribution of elements of the form $f(s)\sin\beta_0(s-z)$. The magnetic field and the axial component of the electric field are generalized from (V.2.14b) and (V.3.11c). They are

$$(B_\theta)_{r=a} = \frac{jI_0}{4\pi v_0 a \sin \beta_0 h}[e^{-j\beta_0 R_{1h}} + e^{-j\beta_0 R_{2h}}$$
$$- 2\cos\beta_0 h \, e^{-j\beta_0 R_0}]$$
$$+ \frac{j}{4\pi a}\int_z^h f(s)[e^{-j\beta_0 R_{1s}} + e^{-j\beta_0 R_{2s}}$$
$$- 2\cos\beta_0 s \, e^{-j\beta_0 R_0}]\, ds, \tag{7}$$

$$(E_z)_{r=a} = \frac{jI_0\zeta_0}{4\pi \sin \beta_0 h}\left[\frac{e^{-j\beta_0 R_{1h}}}{R_{1h}} + \frac{e^{-j\beta_0 R_{2h}}}{R_{2h}}\right.$$
$$\left. - \frac{2}{R_0}\cos\beta_0 h \, e^{-j\beta_0 R_0}\right]$$
$$+ \frac{j\zeta_0}{4\pi}\int_z^h f(s)\left[\frac{e^{-j\beta_0 R_{1s}}}{R_{1s}} + \frac{e^{-j\beta_0 R_{2s}}}{R_{2s}}\right.$$
$$\left. - \frac{2}{R_0}\cos\beta_0 s \, e^{-j\beta_0 R_0}\right] ds, \tag{8}$$

where

$$R_{1s} = \sqrt{(s-z)^2 + a^2},$$
$$R_{2s} = \sqrt{(s+z)^2 + a^2},$$
$$R_0 = \sqrt{z^2 + a^2}, \tag{9}$$

and where R_{1h} and R_{2h} are given by (9) with h substituted for s. The formula for B_θ is given as well as that for E_z since it may be used to show that the field (7, 8) is consistent with the current (6). It follows from Ampère's law,

$$(B_\theta)_{r=a} = I_z/2\pi v_0 a, \tag{10}$$

that I_z calculated from (10) with (7) should reduce to (6). It is readily verified that this is true when $a = 0$ and true except very near the ends when a is very small compared with h.

L. V. King's integral equation for a perfectly conducting cylinder is given by

$$(E_z)_{r=a} = 0, \tag{11}$$

where $(E_z)_{r=a}$ is given by (8). The solution for $f(s)$ is carried out by successive approximations in reciprocal powers of the same parameter $\Omega = 2 \ln(2h/a)$ introduced a year later by Hallén. L. V. King's determination of the distribution of current and the impedance are equivalent to the first-order solution of Hallén[25] using the expansion parameter Ω. As pointed out by Hallén, L. V. King's numerical results differ from those calculated from Hallén's first-order theory by King and Blake[45] by a factor 4/9 which is missing in L. V. King's final formulas owing to the omission by oversight of significant terms for $f(s)$ in formula (46) of L. V. King's paper.

42. Storm's Method of Undetermined Coefficients[79]

Essential steps in Storm's method[79] for solving the integral equation of Hallén,

$$\int_{-h}^{h} I(z')K_1(z, z')\, dz' = C \cos \beta_0 z + \tfrac{1}{2} V_0 \sin \beta_0 |z|, \quad (1)$$

where
$K_1(z, z') \equiv e^{-j\beta_0 R_1}/R_1$, $R_1 = \sqrt{(z - z')^2 + a^2}$,
are to express the current in the integrand as the sum of the sinusoidal (zeroth-order) current and an odd cosine series with undetermined complex coefficients, and then integrate term by term. The current is

$$I(z') = A \sin \beta_0(h - |z'|) + A_1 \cos k_1 z' + A_3 \cos k_3 z' + A_5 \cos k_5 z' + \cdots, \quad (2)$$

where $k_i = i\pi/2h$ and the A_i's are undetermined coefficients. The substitution of (2) in (1) gives as a leading term

$$\int_{-h}^{h} \sin \beta_0(h - |z'|) K_1(z, z')\, dz'$$
$$= C_a(h, z) \sin \beta_0 h - S_a(h, z) \cos \beta_0 h, \quad (3)$$

where $C_a(h, z)$ and $S_a(h, z)$ are expressed in (19.33) and (19.34) in terms of tabulated functions. The integrals obtained from the trigonometric series have not been solved exactly or reduced to tabulated functions. However, by expressing the integrals in the following form:

$$Re \int_{-h}^{h} \cos(m\beta_1 z') K_1(z, z')\, dz'$$
$$\doteq Re \left(\int_{-h}^{z-\Delta z} + \int_{z+\Delta z}^{h} \right) \cos(m\beta_1 z') K_1'(z, z')\, dz'$$
$$+ Re \int_{z-\Delta z}^{z+\Delta z} \cos(m\beta_1 z') K_1(z, z')\, dz', \quad (4a)$$

$$Im \int_{-h}^{h} \cos(m\beta_1 z') K_1(z, z')\, dz'$$
$$\doteq Im \int_{-h}^{h} \cos(m\beta_1 z') K_1'(z, z')\, dz', \quad (4b)$$

where $K_1'(z, z') \equiv e^{-j\beta_0 |z-z'|}/|z - z'|$ and $a \ll \Delta z \ll \lambda$, approximate values may be determined. The first pair of integrals in (4a) and the integral in (4b) may be expressed in terms of integral sines and cosines. For Δz

TABLE 42.1. Impedance and admittance; $\Omega = 15$, $\beta_0 h = \pi/2$.

	$\lambda/2$-dipole: Storm	
Calculated with 2 points:	$Z = 73.3 + j\, 42.9$	$Y = (10.2 - j\, 5.95) \times 10^{-3}$
Calculated with 3 points:	$Z = 81.1 + j\, 43.7$	$Y = (9.56 - j\, 5.15) \times 10^{-3}$
Calculated with 4 points:	$Z = 80.3 + j\, 44.7$	$Y = (9.51 - j\, 5.29) \times 10^{-3}$
Calculated with 5 points:	$Z = 81.5 + j\, 44.5$	$Y = (9.45 - j\, 5.16) \times 10^{-3}$

	$\lambda/2$-dipole: King-Middleton	
First-order	$Z = 75.0 + j\, 36.0$	$Y = (10.8 - j\, 5.20) \times 10^{-3}$
Second-order	$Z = 80.8 + j\, 43.4$	$Y = (9.60 - j\, 5.16) \times 10^{-3}$
Third-order*	$Z = 81.5 + j\, 43.4$*	$Y = (9.56 - j\, 5.09) \times 10^{-3}$

* Second-order reactance.

TABLE 42.2. Impedance and admittance; $\Omega = 15$, $\beta_0 h = \pi$.

	λ-dipole: Storm	
Calculated with 3 points:	$Z = 1286 - j\,1393$	$Y = (0.358 + j\,0.388) \times 10^{-3}$
Calculated with 4 points:	$Z = 1324 - j\,1440$	$Y = (0.346 + j\,0.376) \times 10^{-3}$
Calculated with 5 points:	$Z = 1162 - j\,1354$	$Y = (0.365 + j\,0.425) \times 10^{-3}$

	λ-dipole: King-Middleton	
First-order	$Z = 1560 - j\,1130$	$Y = (0.420 + j\,0.304) \times 10^{-3}$
Second-order	$Z = 1000 - j\,1350$	$Y = (0.354 + j\,0.478) \times 10^{-3}$

sufficiently small (for example, $\Delta z = 5a$), the cosine in (4a) may be replaced by unity and the integration performed. For larger values of Δz (for example, $\Delta z = 50a$), the cosine and $K_1(z, z')$ may be expanded in series and integrated term by term.

After expressing the left-hand member of (1) as a sum of integrated terms, n equations may be obtained by selecting n values of z distributed between $z = 0$ and $z = h$. From these n equations C and $n - 1$ of the coefficients A_i may be determined. Since C and A_i are complex, this means solving $2n$ equations for $2n$ unknowns. Once the $n - 1$ coefficients are known, the current (2) is determined to $n - 1$ terms, and from it the impedance follows immediately. In the actual evaluation Storm found that $\Delta z = 5a$ is adequate in evaluating A_1, but that $\Delta z = 50a$ is required for $A_3, A_5, \cdots$.

Storm has computed the current and impedance of moderately thin antennas for which $\Omega = 15$ for the electrical half lengths $\beta_0 h = \pi/2$ and π using up to 5 points along the antenna. For $\beta_0 h = \pi/2$ his results agree very closely with the King-Middleton theory. For $\beta_0 h = \pi$ the agreement is good in the conductance but less satisfactory in the susceptance. Presumably this is owing to the fact that for this electrical length the difference between the component of the actual current in phase quadrature with the driving voltage and the sinusoidal leading term can not be represented accurately by only three terms in an odd cosine series. Moreover, the least accurate range is very near and at $z = 0$ where the difference curve rises steeply. The impedances and admittances for $\beta_0 h = \pi/2$ and π are given in Tables 42.1 and 42.2 together with the King-Middleton values.

A method similar to that of Storm has been described by Nomura and Hatta.[61]

CHAPTER III

CIRCUIT PROPERTIES OF ARRAYS OF LINEAR RADIATORS

In Chapter II, the circuit properties of a single antenna of very simple structure are investigated under the assumption that the antenna and its associated network constitute a completely isolated transmitter. In practice, all systems of radio communication require at least two antennas, the transmitting antenna and the receiving antenna. Often a large number of receiving antennas are involved and both the transmitting and receiving antennas may consist of a more or less extensive combination of simple elements in the form of an array that has certain desirable properties, to be studied in Chapter VI. In a transmitting array several or all of the units may be driven in the sense that they are directly connected by a transmission line or other network to a generator, so that the input current for each driven unit may be maintained at any desired magnitude and phase by appropriate adjustment of the generator. Alternatively, some of the units in an array may be without connection to a generator so that currents in them are maintained entirely by induction at magnitudes and phases appropriate to the geometry of the array and the magnitudes and phases of currents in the driven antennas. Such units are called *parasites*. Another common arrangement involves several linear elements connected to one another by coupling circuits such as sections of transmission line, or by coils, condensers, or resistors. Finally, a single linear element may be driven at more than one point by transmission lines or by generators, or its halves may be unequal in length or nonparallel in the form of a V. Since such a single antenna involves the problem of coupling between the two parts in a much more complicated manner than in the case of the center-driven symmetric antenna, it is included in this chapter on coupled antennas rather than in Chapter II.

The object of this chapter is to study the circuit properties of various combinations of coupled linear elements and of single elements with coupling between component parts. This requires, in particular, an investigation of the coupling between antennas and parts of antennas in various geometric configurations and of the definition and determination of self- and mutual impedances. Although the general problem of N arbitrarily oriented antennas can be formulated and simultaneous integral equations derived, the actual solution of such a problem has not been achieved without specialization. Therefore, it is convenient to resolve the general problem into two parts, of which the first concerns itself with the more or less closely coupled or adjacent elements such as might constitute a single array, and of which the second deals with widely separated units involving a receiver at a considerable distance from the transmitter. The reason for this separation lies in the important fact that the relative orientations of elements in a particular practical array are usually geometrically simple, involving in most instances only coplanar parallel elements. This means great analytical simplification at the expense of no practically very important restriction. On the other hand, the orientation relative to the transmitter of a parasitic antenna used for reception is necessarily quite arbitrary. Hence, any limitation at all constitutes, in fact, a serious practical restriction on the generality. However, the requirement of loose coupling by wide separation, which is no important restriction for receiving antennas, permits sufficient simplification so that the analysis with completely arbitrary orientation is possible. Thus, the problem is conveniently subdivided essentially without practical limitation into (*a*) coplanar elements with unrestricted degree of coupling, and (*b*) loosely coupled antennas with no restriction on orientation. Of these two major subdivisions, the second, dealing with loosely coupled elements with arbitrary orientation, is considered in Chapter IV, on the receiving antenna. The problem of coplanar and especially of parallel elements is analyzed in this chapter. Since the complete determination of the circuit properties of coupled antennas is complex, it is advantageous to postpone the general analytic formulation until after the simpler and actually more important special cases have been studied.

THEORY OF TWO IDENTICAL, PARALLEL, NONSTAGGERED, CENTER-DRIVEN ANTENNAS

1. General Discussion of the Problem

In order to reduce to a minimum the complications involved in the study of two coupled antennas, this first approach is limited to the rather special, but nevertheless practically highly important, case of antennas that are identical, parallel, nonstaggered, and center-driven. The requirement that the antennas be identical does not in any manner restrict either the impedance or the driving voltage of the generator. The circuit to be analyzed may have one of several forms such as are shown in Figs. 1.1 to 1.5. It consists essentially of two cylindrical antennas each of radius a and half-length h separated by a distance b_a between axes and center-driven from a two-wire line in one of several possible connections.* Or, it may consist of half of the structures of Figs. 1.1 to 1.3 over a highly conducting image-plane of great extent. Yet another possibility is the coaxially-driven pair of antennas shown in Fig. 1.6a or their more practical approximate equivalents in Fig. 1.6b.

All of the circuits shown in Figs. 1.1 to 1.6 may be approximated by the simple circuit of Fig. 1.7 supplemented by corrective networks appropriate to each type of line to take account of end effects, coupling effects, dielectric or stub supports, and adjacent chargeable surfaces other than the cylindrical envelopes and hemispherical caps of the antennas. If the distance b_a between axes of the antennas is sufficiently great, an *independent* terminal network may be designed for each antenna, as discussed in Secs. II.8–10, just as though it were isolated. If the distance between the antennas is small, coupling effects between one antenna and the transmission line feeding the other antenna must be considered. Before discussing this phase of the problem, it is convenient to study first the fundamental circuit of Fig. 1.7 in order to determine the impedances

$$Z_{1\delta} \equiv V_{1\delta}/I_{1\delta}, \qquad Z_{2\delta} \equiv V_{2\delta}/I_{2\delta}, \qquad (1)$$

---
* Note that the antennas are separated by a distance b_a, the line conductors by a distance b.

where $V_{1\delta}$ and $V_{2\delta}$ are the scalar potential differences across the adjacent edges of the cylindrical surfaces (not including end surfaces) of the halves of the two antennas and $I_{1\delta}$ and $I_{2\delta}$ are the currents into the antennas across areas bounded by these edges.

In the two-antenna problem, the determination of $Z_{1\delta}$ and $Z_{2\delta}$ as defined in (1) corresponds to the evaluation of Z_δ in the analysis of an isolated, single antenna in Chapter II. The input impedances $Z_{1\delta}$ and $Z_{2\delta}$ reduce to Z_{10} and Z_{20} as $\delta \to 0$ just as Z_δ approaches Z_0. The impedances Z_{10} and Z_{20} are the input impedances of two parallel identical antennas each driven by a "slice" generator; alternatively, they are the values of the apparent impedances loading the transmission lines in Fig. 1.1. when extrapolated to zero spacing for the line.

2. Coupled Antennas with Small Base Separations; Arbitrary Driving Voltages

The analysis of the two identical parallel antennas of Fig. 1.7 parallels closely that of a single antenna in Sec. II.11. Assumed conditions are

$$\beta_0 a \ll 1, \qquad a \ll h. \qquad (1)$$

The boundary conditions on the tangential components of the electric field at the surfaces of the two conductors are;

$$(E^i_{1z})_{r_1=a} = (E^e_{1z})_{r_1=a},$$
$$(E^i_{2z})_{r_2=a} = (E^e_{2z})_{r_2=a}, \qquad (2)$$

where the superscripts refer to the field inside and external to the conductor. Introduction of the scalar and vector potentials, the total current, and the internal impedance per unit length as in Sec. II.11 gives

$$\left(\frac{\partial^2 A_{1z}}{\partial z_1^2} + \beta_0^2 A_{1z}\right)_{r_1=a} = j\frac{\beta_0^2}{\omega} z^i I_{1z}, \quad (3a)$$

$$\left(\frac{\partial^2 A_{2z}}{\partial z_2^2} + \beta_0^2 A_{2z}\right)_{r_2=a} = j\frac{\beta_0^2}{\omega} z^i I_{2z}. \quad (3b)$$

The vector potential has only a z-component, since all currents are in the z-direction. The components on the surfaces of the two antennas are

$$A_{1z} = A_{11z} + A_{12z}$$

$$= \frac{1}{4\pi v_0} \left(\int_{-h}^{-\delta} + \int_{\delta}^{h}\right)\left(I'_{1z}\frac{e^{-j\beta_0 R_{11}}}{R_{11}}dz'_1\right.$$

$$\left. + I'_{2z}\frac{e^{-j\beta_0 R_{12}}}{R_{12}}dz'_2\right), \quad (4)$$

THEORY OF LINEAR ANTENNAS

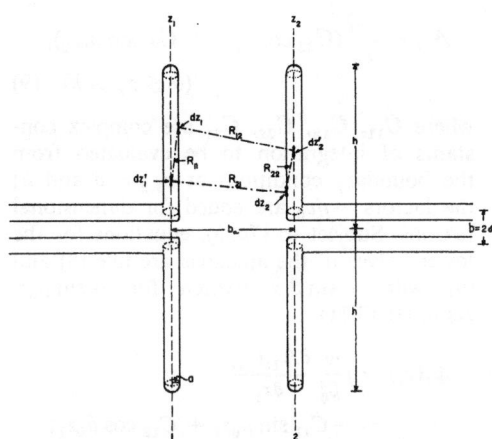

Fig. 1.1. Coupled antennas individually driven from two-wire lines.

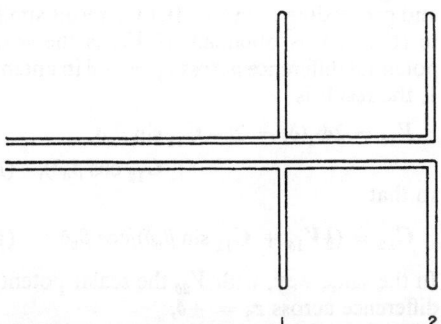

Fig. 1.2. Coupled antennas driven from one two-wire line.

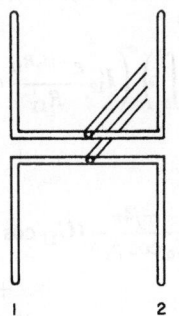

Fig. 1.3. Coupled antennas symmetrically driven from a two-wire line.

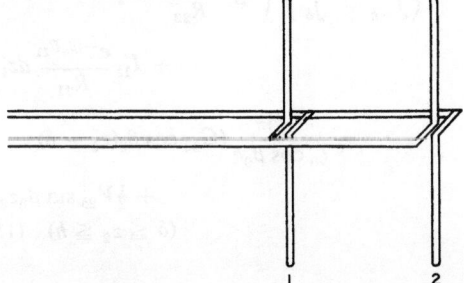

Fig. 1.4. Coupled antennas driven from one line in the perpendicular plane.

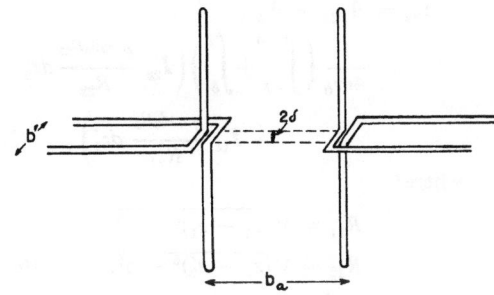

Fig. 1.5. Coupled antennas individually driven from two-wire lines in the perpendicular plane.

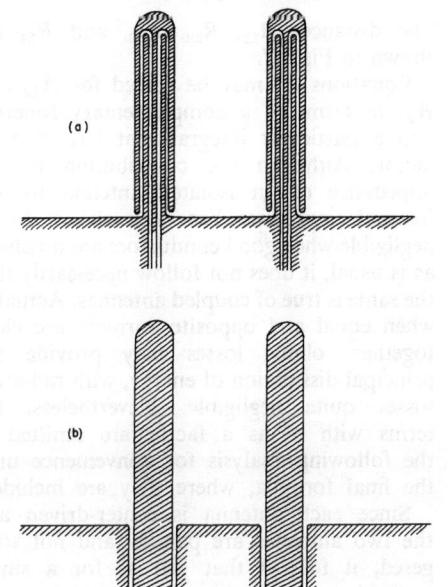

Fig. 1.6. Coaxially driven coupled antennas.

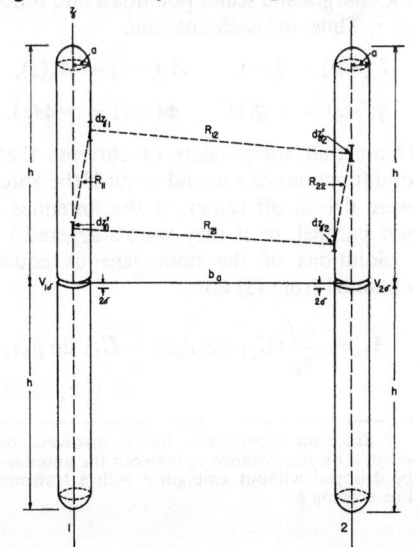

Fig. 1.7. Fundamental circuit of parallel, identical antennas.

$$A_{2z} = A_{22z} + A_{21z}$$

$$= \frac{1}{4\pi v_0}\left(\int_{-h}^{-\delta} + \int_{\delta}^{h}\right)\left(I'_{2z}\frac{e^{-j\beta_0 R_{22}}}{R_{22}}dz'_2\right.$$

$$\left. + I'_{1z}\frac{e^{-j\beta_0 R_{21}}}{R_{21}}dz'_1\right), \quad (5)$$

where*

$$R_{11} = \sqrt{(z_1 - z'_1)^2 + a^2},$$
$$R_{22} = \sqrt{(z_2 - z'_2)^2 + a^2}, \quad (6a)$$
$$R_{12} = \sqrt{(z_1 - z'_2)^2 + b^2},$$
$$R_{21} = \sqrt{(z_2 - z'_1)^2 + b^2}. \quad (6b)$$

The distances R_{11}, R_{22}, R_{12}, and R_{21} are shown in Fig 1.7.

Equations (3) may be solved for A_{1z} and A_{2z} in terms of a complementary function and a particular integral that has z^i as a factor. Although the contribution to the impedance of an isolated antenna by the internal impedance has been shown to be negligible when good conductors are involved, as is usual, it does not follow necessarily that the same is true of coupled antennas. Actually, when equal and opposite currents are close together, ohmic losses may provide the principal dissipation of energy, with radiation losses quite negligible. Nevertheless, the terms with z^i as a factor are omitted in the following analysis for convenience until the final formula, where they are included.

Since each antenna is center-driven and the two antennas are parallel and not staggered, it follows that just as for a single isolated antenna the currents and vector potentials must be even functions of z, the charges and scalar potentials odd functions of z. Thus, for each antenna,

$$I_z(-z) = I_z(z), \quad A_z(-z) = A_z(z), \quad (7a)$$
$$q(-z) = -q(z), \quad \Phi(-z) = -\Phi(z). \quad (7b)$$

(Note that components of currents that are odd functions of z would occur if the antennas were driven off center, if the antennas were not parallel, or if they were staggered.)

Solutions of the homogeneous equations obtained from (3) are

$$A_{1z} = \frac{-j}{v_0}(C_{11}\cos\beta_0 z_1 + C_{12}\sin\beta_0 z_1),$$
$$(\delta \leq z_1 \leq h) \quad (8)$$

---
* Since no transmission line is involved, the subscript a on the distance b_a between the antennas may be dropped without ambiguity with a transmission-line spacing b.

$$A_{2z} = \frac{-j}{v_0}(C_{22}\cos\beta_0 z_2 + C_{21}\sin\beta_0 z_2),$$
$$(\delta \leq z_2 \leq h) \quad (9)$$

where C_{11}, C_{12}, C_{22}, C_{21} are complex constants of integration to be evaluated from the boundary conditions at $z_{1,2} = \delta$ and h; the factors $-j/v_0$ are added for dimensional reasons. Subject to (7a, b), equations for the lower halves of the antennas are like (8) and (9) with $-\sin\beta_0 z$ written for $+\sin\beta_0 z$. As in (II.13.11),

$$\Phi_1(z_1) = j\frac{\omega}{\beta_0^2}\frac{\partial A_{1z}(z_1)}{\partial z_1}$$
$$= -C_{11}\sin\beta_0 z_1 + C_{12}\cos\beta_0 z_1;$$
$$\delta \leq z_1 \leq h. \quad (10)$$

Using the symmetry condition (7) at $z_1 = \delta$, and proceeding as in Sec. II.13, a result similar to (II.13.20) is obtained. If $V_{1\delta}$ is the scalar potential difference across $z_1 = \pm\delta$ in antenna 1, the result is

$$V_{1\delta} = 2\Phi_1(\delta) = 2(-C_{11}\sin\beta_0\delta + C_{12}\cos\beta_0\delta), \quad (11)$$

so that

$$C_{12} = (\tfrac{1}{2}V_{1\delta} + C_{11}\sin\beta_0\delta)/\cos\beta_0\delta. \quad (12)$$

In the same way, with $V_{2\delta}$ the scalar potential difference across $z_2 = \pm\delta$,

$$C_{21} = (\tfrac{1}{2}V_{2\delta} + C_{22}\sin\beta_0\delta)/\cos\beta_0\delta. \quad (13)$$

Substitution of (4), (5), (12), and (13) in (8) and (9) gives

$$\left(\int_{-h}^{-\delta} + \int_{\delta}^{h}\right)\left(I'_{1z}\frac{e^{-j\beta_0 R_{11}}}{R_{11}}dz'_1\right.$$

$$\left. + I'_{2z}\frac{e^{-j\beta_1 R_{12}}}{R_{12}}dz'_2\right)$$

$$= \frac{-j4\pi}{\zeta_0\cos\beta_0\delta}(C_{11}\cos\beta_0(z_1 - \delta)$$

$$+ \tfrac{1}{2}V_{1\delta}\sin\beta_0 z_1),$$
$$(\delta \leq z_1 \leq h) \quad (14)$$

$$\left(\int_{-h}^{-\delta} + \int_{\delta}^{h}\right)\left(I'_{2z}\frac{e^{-j\beta_0 R_{22}}}{R_{22}}dz'_2\right.$$

$$\left. + I'_{1z}\frac{e^{-j\beta_0 R_{21}}}{R_{21}}dz'_1\right)$$

$$= \frac{-j4\pi}{\zeta_0\cos\beta_0\delta}(C_{22}\cos\beta_0(z_2 - \delta)$$

$$+ \tfrac{1}{2}V_{2\delta}\sin\beta_0 z_2).$$
$$(\delta \leq z_2 \leq h) \quad (15)$$

The simultaneous solution of the integral equations (14) and (15) for the currents $I_{1z}(z_1)$ and $I_{2z}(z_2)$ may be carried out in general for arbitrary driving voltages $V_{1\delta}$ and $V_{2\delta}$. This is accomplished by resolving $V_{1\delta}$ and $V_{2\delta}$ into a codirectional, or symmetric, pair and an oppositely directed, or antisymmetric, pair. Whatever the magnitudes and relative phases of $V_{1\delta}$ and $V_{2\delta}$, they can always be expressed as follows:

$$V_{1\delta} = V_\delta^s + V_\delta^a, \qquad V_{2\delta} = V_\delta^s - V_\delta^a, \qquad (16)$$

where V_δ^s stands for the equal and codirectional parts of $V_{1\delta}$ and $V_{2\delta}$ and V_δ^a and $-V_\delta^a$ the equal and oppositely directed parts. With $V_{1\delta}$ and $V_{2\delta}$ given, the symmetric and antisymmetric parts are

$$V_\delta^s = \tfrac{1}{2}(V_{1\delta} + V_{2\delta}), \qquad V_\delta^a = \tfrac{1}{2}(V_{1\delta} - V_{2\delta}). \qquad (17)$$

The currents in the two antennas due to the simultaneous application of the equal and codirectional voltages V_δ^s are themselves equal and codirectional: $I_{1z}^s(z) = I_{2z}^s(z) = I_z^s$. Correspondingly, the currents due to the simultaneous application of the equal and opposite voltages V_δ^a and $-V_\delta^a$ are themselves equal and opposite: $I_{1z}^a(z) = -I_{2z}^a(z) = I_z^a$. By the principle of superposition, the currents due to the combined application of the two sets of voltages in the form (16) are

$$I_{1z}(z) = I_z^s + I_z^a, \qquad I_{2z}(z) = I_z^s - I_z^a. \qquad (18)$$

These are the desired currents due to the simultaneous application in the two antennas of the arbitrary voltages $V_{1\delta}$ and $V_{2\delta}$.

Equations (18) may be solved for I_z^s and I_z^a to give

$$I_z^s = \tfrac{1}{2}[I_{1z}(z) + I_{2z}(z)],$$
$$I_z^a = \tfrac{1}{2}[I_{1z}(z) - I_{2z}(z)]. \qquad (19)$$

The solution of the problem of arbitrary driving voltages in the two antennas has been reduced to the independent solution of the problem of equal and codirectional voltages and the problem of equal and opposite voltages. These are now examined in turn.

3. Symmetrically Driven Antennas

If the complex amplitudes of the voltages across $z = \pm\delta$ in the two antennas of Fig. 1.7 are both equal to V_δ^s, the complex amplitudes of the currents in the two antennas are also equal at the same value of z; that is,

$$I_{2z}^s = I_{1z}^s = I_z^s. \qquad (1)$$

Subject to (1), the two integral equations (2.14) and (2.15) are identical and are given by

$$\left(\int_{-h}^{-\delta} + \int_{\delta}^{h}\right) I_z^{s'} K_s(z, z') \, dz'$$
$$= \frac{-j4\pi}{\zeta_0 \cos \beta_0 \delta} [C_1 \cos \beta_0(z - \delta) + \tfrac{1}{2} V_\delta^s \sin \beta_0 z],$$
$$(\delta \leq z \leq h) \qquad (2)$$

where the kernel for the symmetrically driven pair is due to Tai[51] and is defined as follows:

$$K_s(z, z') \equiv \frac{e^{-j\beta_0 R_{11}}}{R_{11}} + \frac{e^{-j\beta_0 R_{12}}}{R_{12}}. \qquad (3)$$

Since the subscripts on z are no longer required, let them be omitted. Thus,

$$R_{11} = \sqrt{(z - z')^2 + a^2},$$
$$R_{12} = \sqrt{(z - z')^2 + b^2}. \qquad (4)$$

The integral equation (2) is the same as (II.14.3) for the isolated antenna except for a different kernel. It may be transformed in the same manner by defining the function

$$\Psi_{s\delta}(z) \equiv \left(\int_{-h}^{h} - \int_{-\delta}^{\delta}\right) g_s(z, z') K_s(z, z') \, dz'. \qquad (5)$$

in terms of a relative distribution function $g_s(z, z')$ which approximates the true current distribution $I_z^{s'}/I_z^s$. Thus,

$$I_z^s \Psi_s(z) = \frac{-j4\pi}{\zeta_0 \cos \beta_0 \delta} [C_1 \cos \beta_0(z - \delta)$$
$$+ \tfrac{1}{2} V_\delta^s \sin \beta_0 z]$$
$$- \left(\int_{-h}^{h} - \int_{-\delta}^{\delta}\right) [I_z^{s'} - I_z^s g_s(z, z')] K_s(z, z') \, dz'. \qquad (6)$$

The function $\Psi_{s\delta}(z)$ is proportional to the ratio of the vector potential A_z on the surface of one antenna divided by the total current in the antenna at z if $g_s(z, z')$ is exactly equal to $I_z^{s'}/I_z^s$. Specifically,

$$\Psi_s(z) = 4\pi\nu_0 \frac{A_z}{I_z^s} = 4\pi\nu_0 \frac{A_{11z} + A_{12z}}{I_z^s}, \qquad (7)$$

where, however, A_z is determined by the currents in both antennas. If the two antennas are not very close together the vector potential on the surface of antenna 1 is determined largely by the current in antenna 1 and only in a small degree by the current in antenna 2. In this case, $\Psi_{s\delta}(z)$ must behave much as for a single, isolated antenna. If the two antennas are close together, so that both currents contribute significantly, the vector potential on antenna 1 due to the current in antenna 2, namely, A_{12z}, must vary with z in a manner closely resembling that of A_{11z}, since the currents in both antennas have by definition identically the same distribution. The fact that the distance b between the antennas is always greater than the radius a, so that R_{12} is greater than R_{11}, makes the response of A_{12z} to variations in I_{2z} less sensitive than the response of A_{11z} to variation in I_{1z}, but the general conclusion that $\Psi_s(z)$ is sensibly constant and predominantly real is still valid. Just as in Sec. II.14 for the isolated antenna, it is correct to set

$$\Psi_{s\delta}(z) = \Psi_s + \gamma_s(z), \qquad (8)$$

where Ψ_s is real and equal to the magnitude of $\Psi_{s\delta}(z)$ at a suitably chosen reference point and $\gamma_s(z)$ is a correction term that should be very small except near the ends of the antenna, where it may be infinite.

Substitution of (8) in (6) and formal solution for I_z leads to a formula essentially like (II.14.13) with K_s written in place of K_1 and with the terms in z^i omitted.

This formula is modified by subtracting (2) (with the integral transferred to the right-hand member and $z = h$) just as in Sec. II.14. The final integral equation corresponding to (II.14.16) is (9), in which, for completeness, the terms with the internal impedance z^i, which have been omitted heretofore, have been added:

$$I_z^s = \frac{-j4\pi}{\Psi_s \zeta_0 \cos \beta_0 \delta}$$

$$\times \Big\{ C_1[\cos \beta_0(z-\delta) - \cos \beta_0(h-\delta)]$$

$$+ \tfrac{1}{2}V_\delta^s[\sin \beta_0 z - \sin \beta_0 h]$$

$$- \frac{1}{\Psi_s}\Big[I_z^s \gamma(z) + \Big(\int_{-h}^{h} - \int_{-\delta}^{\delta}\Big)$$

$$[I_z^{s'} - I_z^s(z,z')]g_s(z,z')]K_s(z,z')\, dz'$$

$$- \Big(\int_{-h}^{h} - \int_{-\delta}^{\delta}\Big) I_z^{s'} K_s(h, z')\, dz' \Big]\Big\}$$

$$+ \frac{j4\pi z^i}{\zeta_0 \Psi_s}\Big[\int_\delta^z I^s(s) \sin \beta_0(z-s)\, ds$$

$$- \int_\delta^h I^s(s) \sin \beta_0(h-s)\, ds\Big].$$

$$(\delta \leq z \leq h) \quad (9)$$

Following the same method of iteration employed in Sec. II.15, I_z^s may be expressed in series form and the constant of integration C_1 evaluated using (2). The final formula for the current corresponding to (II.15.26) is

$$I_z^s = \frac{j2\pi V_\delta^s}{\zeta_0 \Psi_s}$$

$$\times \left[\frac{\sin \beta_0(h-z) + M_{1s}(z)/\Psi_s + M_{2s}(z)/\Psi_s^2 + \cdots}{\cos \beta_0(h-\delta) + A_{1s}/\Psi_s + A_{2s}/\Psi_s^2 + \cdots} \right].$$

$$\delta \leq z \leq h. \quad (10)$$

The functions $M_{1s}(z)$, $M_{2s}(z)$, $\cdots$, A_{1s}, A_{2s}, $\cdots$ are defined by (II.15.27)–(II.15.29) with a subscript s added to each F- and G-function. These F_s- and G_s-functions are defined by (II.15.18a–f) with K_s substituted for K_1.

Since (10) is just the same in form as the corresponding expression (II.15.26), the same method of evaluation may be used. Following the King-Middleton method of Sec. II.20 for determining Ψ by substituting the zeroth-order current in (5), one obtains

$$\Psi_{s1\delta}(z) = \Big(\int_{-h}^{h} - \int_{-\delta}^{\delta}\Big) \frac{\sin \beta_0(h-z')}{\sin \beta_0(h-z)} K_s(z,z')\, dz'$$

$$= \frac{\psi_{s1}(z)}{\sin \beta_0(h-z)}, \quad (11)$$

where

$$\psi_{s1\delta}(z) = [C_a(h, z) - C_a(\delta, z)$$

$$+ C_b(h, z) - C_b(\delta, z)] \sin \beta_0 h$$

$$- [S_a(h, z) - S_a(\delta, z)$$

$$+ S_b(h, z) - S_b(\delta, z)] \cos \beta_0 h. \quad (12)$$

In (12) the functions with subscript a are identically (II.19.3)–(II.19.4) and those with subscript b are like those with subscript a, but with b written for a in R_1 and R_2 in (II.19.6) and (II.19.7). As proved in Sec. II.20, the contributions to $\Psi_{s\delta}(z)$ by terms that involve δ are significant only near $z = \delta$ and do not modify the magnitude of $\Psi_s(z)$ over the major part of the antenna. These terms are, therefore, included in $\gamma_s(z)$ and omitted from (12), which reduces to

$$\psi_{s1\delta}(z) \doteq \psi_{s1}(z) = [C_a(h, z) + C_b(h, z)] \sin \beta_0 h$$

$$- [S_a(h, z) + S_b(h, z)] \cos \beta_0 h. \quad (13)$$

The expansion parameter Ψ_{s1} is defined as in (II.20.29):

$$\Psi_{s1} \equiv |\Psi_{s1}(0)| = \frac{|\psi_{s1}(0)|}{\sin \beta_0 h}, \quad (\beta_0 h \leq \pi/2)$$

$$(14a)$$

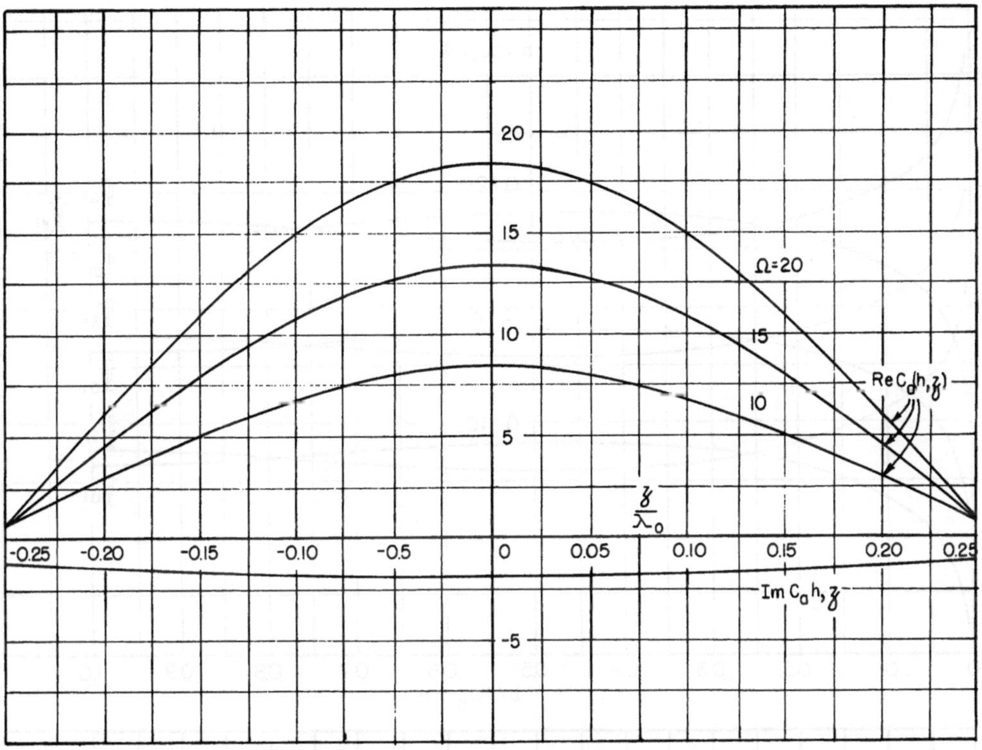

Fig. 3.1. The function $C_a(h, z)$; $h = \lambda_0/4$.

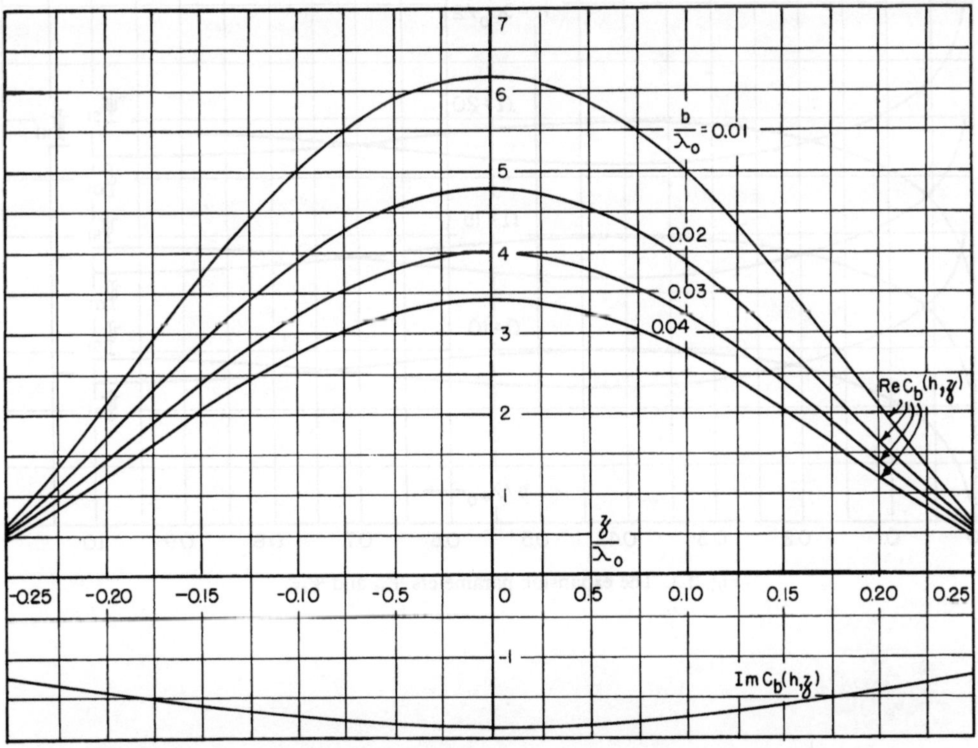

Fig. 3.2. The function $C_b(h, z)$; $h = \lambda_0/4$.

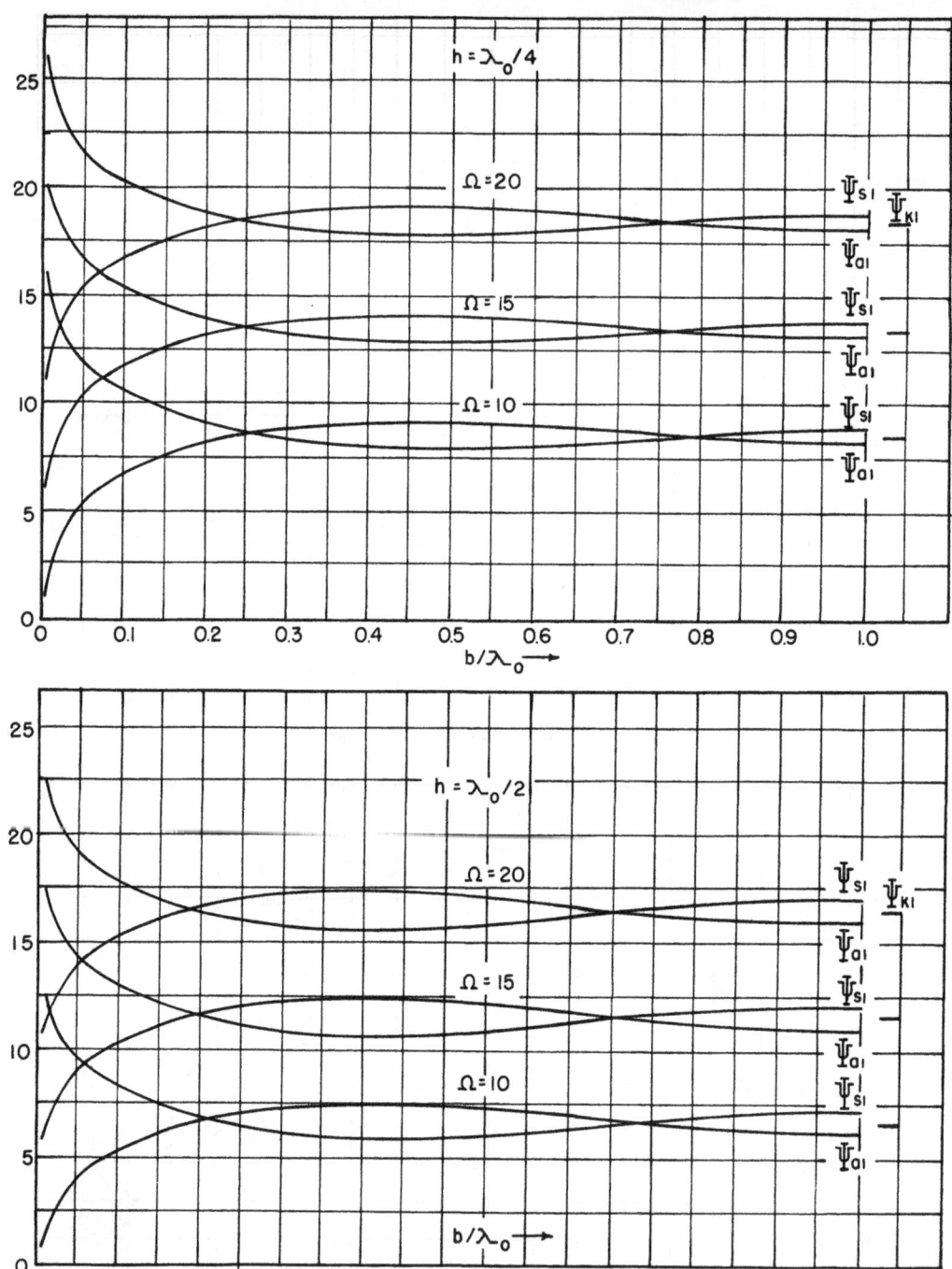

Fig. 3.3. The expansion parameters Ψ'_{s1} and Ψ'_{a1}.

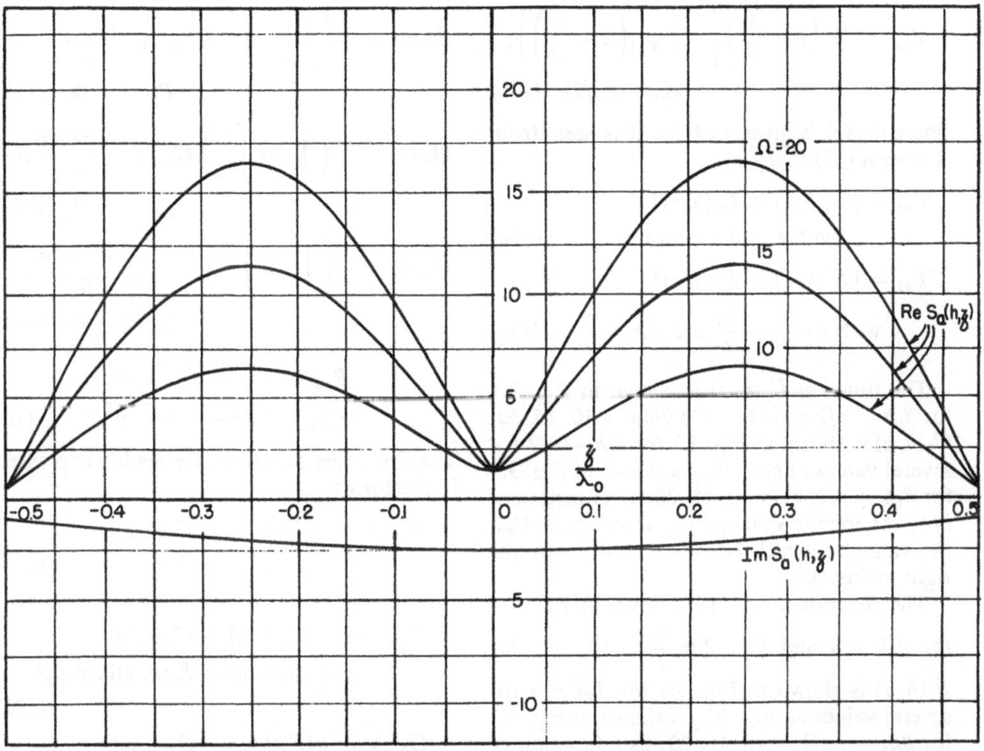

Fig. 3.4. The function $S_a(h, z)$; $h = \lambda_0/2$.

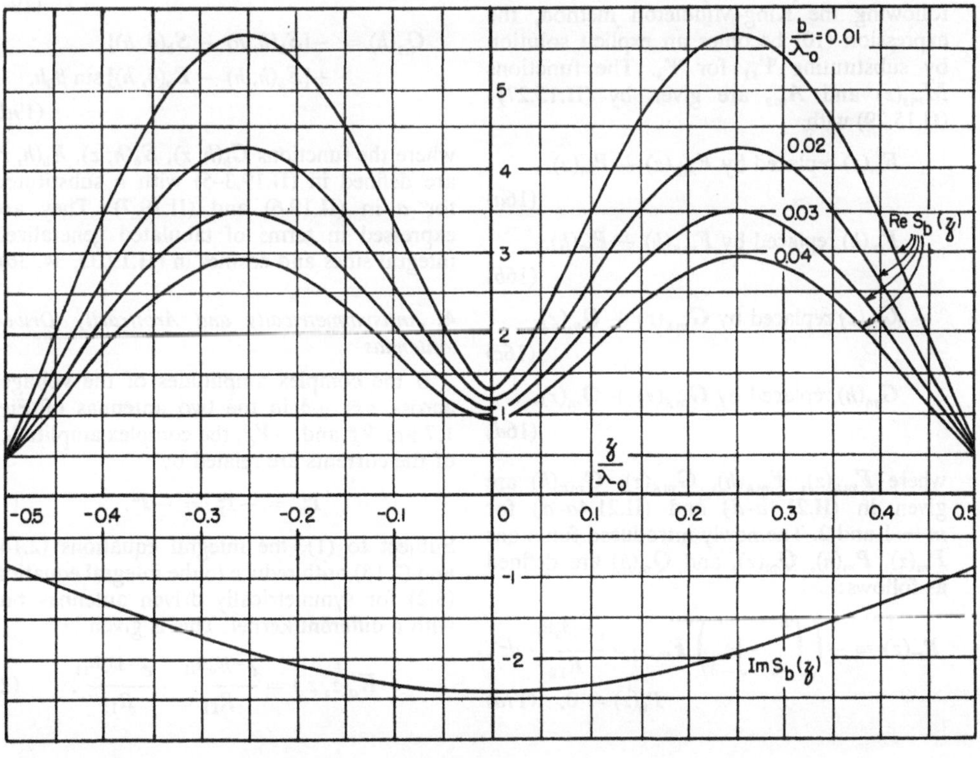

Fig. 3.5. The function $S_b(h, z)$; $h = \lambda_0/2$.

$$\Psi'_{s1} = \left|\Psi_s\left(h - \frac{\lambda_0}{4}\right)\right| = \left|\psi_{s1}\left(h - \frac{\lambda_0}{4}\right)\right|,$$
$$(\beta_0 h \geq \pi/2) \quad (14b)$$

where $\psi_{s1}(z)$ is given in (13). It is seen from (14) with (13) that

$$\Psi'_{s1} = |C_a(h, z) + C_b(h, z)|$$
with $z = 0$ and $\beta_0 h = \pi/2$, (15a)

$$\Psi'_{s1} = |S_a(h, z) + S_b(h, z)|$$
with $z = h - \frac{\lambda_0}{4}$ and $\beta_0 h = \pi$. (15b)

The function $C_a(h, z)$ is shown in Fig. 3.1 for $\beta_0 h = \pi/2$ and $\Omega = 2 \ln 2h/a = 10, 15, 20$; $C_b(h, z)$ is shown in Fig. 3.2 for $\beta_0 h = \pi/2$ for several values of b/λ_0; Ψ'_{s1} is shown in Fig. 3.3 for $\beta_0 h = \pi/2$, $\Omega = 10, 15, 20$, as a function of b/λ_0. At infinite spacing Ψ'_{s1} approaches Ψ_{K1}, the value of which is shown at the extreme right in Fig. 3.3.

The function $S_a(h, z)$ is shown in Fig. 3.4 for $\beta_0 h = \pi$ and $\Omega = 2 \ln \frac{2h}{a} = 10, 15, 20$; $S_b(h, z)$ is shown in Fig. 3.5 for $\beta_0 h = \pi$ for several values of b/λ_0; Ψ'_{s1} is shown in Fig. 3.3 for $\beta_0 h = \pi$, $\Omega = 10, 15, 20$, as a function of b/λ_0.

With the expansion parameter Ψ_s defined following the King-Middleton method, the expression (10) becomes an explicit solution by substituting Ψ_{s1} for Ψ_s. The functions $M_{ms}(z)$ and A_{ms} are given by (II.15.27)–(II.15.29) with

$F_m(z)$ replaced by $F_{mK}(z) + P_m(z)$, (16a)

$F_m(h)$ replaced by $F_{mK}(h) + P_m(h)$, (16b)

$G_m(z)$ replaced by $G_{mK}(z) + Q_m(z)$, (16c)

$G_m(h)$ replaced by $G_{mK}(z) + Q_m(z)$, (16d)

where $F_{mK}(z)$, $F_{mK}(h)$, $G_{mK}(z)$, $G_{mK}(h)$ are given in (II.21.1a–d) and (II.21.4a–d) for $m = 1$ and 2. The newly introduced functions $P_m(z)$, $P_m(h)$, $Q_m(z)$, and $Q_m(h)$ are defined as follows:

$$P_m(z) \equiv -\left(\int_{-h}^h - \int_{-\delta}^\delta\right) F_{m-1,z'} \frac{e^{-j\beta_0 R_{12}}}{R_{12}} dz',$$
$$P_0(z) = 0, \quad (17a)$$

$$P_m(h) \equiv -\left(\int_{-h}^h - \int_{-\delta}^\delta\right) F_{m-1,z'} \frac{e^{-j\beta_0 R_{12h}}}{R_{12h}} dz',$$
$$P_0(h) = 0, \quad (17b)$$

$$Q_m(z) \equiv -\left(\int_{-h}^h - \int_{-\delta}^\delta\right) G_{m-1,z'} \frac{e^{-j\beta_0 R_{12}}}{R_{12}} dz',$$
$$Q_0(z) = 0, \quad (17c)$$

$$Q_m(h) \equiv -\left(\int_{-h}^h - \int_{-\delta}^\delta\right) G_{m-1,z'} \frac{e^{-j\beta_0 R_{12h}}}{R_{12h}} dz',$$
$$Q_0(h) = 0, \quad (17d)$$

$$R_{12} = \sqrt{(z - z')^2 + b^2},$$
$$R_{12h} = \sqrt{(h - z')^2 + b^2}. \quad (18)$$

The first-order functions are readily expressed in the forms

$$P_1(z) = -[C_b(h, z) - C_b(\delta, z)]$$
$$+ [E_b(h, z) - E_b(\delta, z)] \cos \beta_0 h, \quad (19a)$$

$$P_1(h) = -[C_b(h, h) - C_b(\delta, h)]$$
$$+ [E_b(h, h) - E_b(h, z)] \cos \beta_0 h, \quad (19b)$$

$$Q_1(z) = -[S_b(h, z) - S_b(\delta, z)]$$
$$+ [E_b(h, z) - E_b(\delta, z)] \sin \beta_0 h, \quad (19c)$$

$$Q_1(h) = -[S_b(h, h) - S_b(\delta, h)]$$
$$+ [E_b(h, h) - E_b(\delta, h)] \sin \beta_0 h, \quad (19d)$$

where the functions $C_b(h, z)$, $S_b(h, z)$, $E_b(h, z)$ are defined in (II.19.3–5) with b substituted for a in (II.19.6) and (II.19.7). They are expressed in terms of tabulated generalized integral sines and cosines in (II.19.33, 34, 36).

4. Antisymmetrically and Arbitrarily Driven Antennas

If the complex amplitudes of the voltages across $z = \pm \delta$ in the two antennas of Fig. 1.7 are V_δ^a and $-V_\delta^a$, the complex amplitudes of the currents are related by

$$I_{2z}^a = -I_{1z}^a = -I_z^a. \quad (1)$$

Subject to (1), the integral equations (2.14) and (2.15) both reduce to the integral equation (3.2) for symmetrically driven antennas but with a different kernel. This is given by

$$K_a(z, z') = \frac{e^{-j\beta_0 R_{11}}}{R_{11}} - \frac{e^{-j\beta_0 R_{12}}}{R_{12}}. \quad (2)$$

It follows that the entire analysis of Sec. 3 may be repeated if $K_a(z, z')$ replaces $K_s(z, z')$. The final formula for the current $(I_z^a)_m$ is the same as (3.10) with subscripts and superscripts s replaced by a.

The expansion parameter Ψ_{a1} is defined by (3.14) with a written for s in subscripts and where

$$\psi_{a1}(z) = [C_a(h, z) - C_b(h, z)] \sin \beta_0 h \\ - [S_a(h, z) - S_b(h, z)] \cos \beta_0 h. \quad (3)$$

Specifically,

$$\Psi_{a1} = |C_a(h, z) - C_b(h, z)|$$
$$\text{with } z = 0 \text{ and } \beta_0 h = \pi/2, \quad (4a)$$
$$\Psi_{a1} = |S_a(h, z) - S_b(h, z)|$$
$$\text{with } z = h - \lambda_0/4 \text{ and } \beta_0 h = \pi. \quad (4b)$$

The function Ψ_{a1} is shown in Fig. 3.3 together with Ψ_{s1}. The behavior of the two parameters for small values of b/λ_0 is seen to be quite different. For large values both approach Ψ_{K1} for the single antenna.

Functions $M_a(z)$ and $A_a(z)$ are defined for I_z^a in the same manner as $M_s(z)$ and $A_s(z)$ in (3.10) are defined for I_z^s. They differ only in $F(z)$, $F(h)$, $G(z)$, and $G(h)$ as defined in (3.16a–d). For the *antisymmetric functions* the plus sign preceding P or Q must be replaced by a minus sign. With these changes in sign and Ψ^a substituted for Ψ^s, (3.10) is the correct expression for I_z^a.

Once the currents for symmetrically and antisymmetrically driven antennas are determined, it is a routine matter to obtain the currents for the general case in which the arbitrary voltages $V_{1\delta}$ and $V_{2\delta}$ are applied. The general distributions are

$$I_{1z} = I_z^s + I_z^a, \quad I_{2z} = I_z^s - I_z^a. \quad (5)$$

These may be expressed in terms of $V_{1\delta}$ and $V_{2\delta}$ instead of V_δ^s and V_δ^a, using

$$V_\delta^s = \tfrac{1}{2}(V_{1\delta} + V_{2\delta}), \quad V_\delta^a = \tfrac{1}{2}(V_{1\delta} - V_{2\delta}). \quad (6)$$

In this general form the current in each antenna may be separated into two formally independent parts which depend, respectively, upon the voltages $V_{1\delta}$ and $V_{2\delta}$.

It is demonstrated in the next section that the relative distributions of current in the two coupled antennas, even when these are closely spaced, do not differ greatly from the relative distribution in an isolated antenna. The magnitudes of the currents naturally must differ if the driving voltages are not the same. Since effects of the finite separation δ are confined to short distances near $z = \pm\delta$ in each antenna, it is clear that the distributions of current for $\delta = 0$ differ negligibly from those for small values of $\beta_0\delta$ except near $z = \pm\delta$. Therefore, if the input currents at $z = \pm\delta$ and the distributions for $\delta = 0$ are determined, all practically important information is available.

5. Closely Spaced Antennas: Cage Antenna;[*] Two-Wire Line

If the two parallel antennas are so close together that the following conditions are well satisfied, considerable simplification in the form of the expansion parameters Ψ_{s1} and Ψ_{a1} results. The conditions are

$$a^2 < b^2 \ll h^2, \quad \beta_0^2 a^2 < \beta_0^2 b^2 \ll 1. \quad (1)$$

The expansion functions may be expressed as follows:

$$\Psi_{s1}(z) = \Psi_{K1}(z) + \Psi_b(z), \quad (2a)$$
$$\Psi_{a1}(z) = \Psi_{K1}(z) - \Psi_b(z), \quad (2b)$$

where

$$\Psi_{K1}(z) = \frac{C_a(h,z) \sin \beta_0 h - S_a(h,z) \cos \beta_0 h}{\sin \beta_0 (h - z)}, \quad (3a)$$

$$\Psi_b(z) = \frac{C_b(h,z) \sin \beta_0 h - S_b(h,z) \cos \beta_0 h}{\sin \beta_0 (h - z)}. \quad (3b)$$

The integral functions in (3a, b) are defined by

$$C(h, z) = \int_0^h \cos \beta_0 z' \left(\frac{e^{-j\beta_0 R_1}}{R_1} + \frac{e^{-j\beta_0 R_2}}{R_2} \right) dz', \quad (4)$$

$$S(h, z) = \int_0^h \sin \beta_0 z' \left(\frac{e^{-j\beta_0 R_1}}{R_1} + \frac{e^{-j\beta_0 R_2}}{R_2} \right) dz', \quad (5)$$

where

$$\left. \begin{array}{l} R_1 = \sqrt{(z-z')^2 + a^2} \\ R_2 = \sqrt{(z+z')^2 + a^2} \end{array} \right\} \text{for } C_a(h, z) \text{ and } S_a(h, z), \quad (6)$$

and where

$$\left. \begin{array}{l} R_1 = \sqrt{(z-z')^2 + b^2} \\ R_2 = \sqrt{(z+z')^2 + b^2} \end{array} \right\} \text{for } C_b(h, z) \text{ and } S_b(h, z). \quad (7)$$

The functions $C_a(h, z)$ and $S_a(h, z)$ are expressed in terms of the generalized integral sines and cosines in (II.19.33, 34). Expressions

---

[*] The n-element cage antenna is analyzed in Sec. 17.

for $C_b(h, z)$ and $S_b(h, z)$ are obtained by substituting b for a (or $B \equiv \beta_0 b$ for $A \equiv \beta_0 a$). Subject to (1), all the generalized integral sines and cosines are practically independent of the small quantities $(a/h)^2$ and $(b/h)^2$ except the function $Cc_i(U, A)$, which involves $\sinh^{-1}(U/A)$. That is, subject to (1),

$$S(A, U) \doteq S(B, U), \qquad (8a)$$

$$C(A, U) \doteq C(B, U), \qquad (8b)$$

$$Ss(A, U) \doteq Ss(B, U), \qquad (8c)$$

$$Sc(A, U) \doteq Sc(B, U), \qquad (8d)$$

$$Cs(A, U) \doteq Cs(B, U), \qquad (8e)$$

$$Cc(A, U) \doteq Cc(B, U). \qquad (8f)$$

On the other hand, using (II.19.22),

$$Cc_i(A, U) = \sinh^{-1}(U/A) - C(A, U) - Cc(A, U), \qquad (9a)$$

$$Cc_i(B, U) = \sinh^{-1}(U/B) - C(B, U) - Cc(B, U), \qquad (9b)$$

so that, subject to (1) with $(8b, f)$,

$$Cc_i(B, U) \doteq Cc_i(A, U) - \sinh^{-1}(U/A) + \sinh^{-1}(U/B) \qquad (10)$$

$$\doteq Cc_i(A, U) - \ln\left(\frac{B}{A} \frac{U + \sqrt{U^2 + A^2}}{U + \sqrt{U^2 + B^2}}\right)$$

Use of $(8a\text{--}f)$ and (10) in expressions for $C_a(h, z)$, $C_b(h, z)$, $S_a(h, z)$, $S_b(h, z)$ obtained from (II.19.33, 34) with (II.19.37) gives

$$C_b(h, z) = C_a(h, z)$$
$$- \cos\beta_0 z \left(\sinh^{-1}\frac{h+z}{a} + \sinh^{-1}\frac{h-z}{a}\right.$$
$$\left. - \sinh^{-1}\frac{h+z}{b} - \sinh^{-1}\frac{h-z}{b}\right), \qquad (11)$$

$$S_b(h, z) = S_a(h, z)$$
$$+ \sin\beta_0 z \left(\sinh^{-1}\frac{h+z}{a} - \sinh^{-1}\frac{h-z}{a}\right.$$
$$- 2\sinh^{-1}\frac{z}{a} + \sinh^{-1}\frac{h+z}{b}$$
$$\left. + \sinh^{-1}\frac{h-z}{b} + 2\sinh^{-1}\frac{z}{b}\right). \qquad (12)$$

Since the expansion parameters Ψ_{s1} and Ψ_{a1} are defined at $z = 0$ for $\beta_0 h \leq \pi/2$ and at $z = h - \lambda_0/4$ for $\beta_0 h \geq \pi/2$, only these two values of z need be considered. Specifically, with $z = 0$,

$$C_b(h, 0) = C_a(h, 0) - 2\ln(b/a), \qquad (13)$$

$$S_b(h, 0) = S_a(h, 0). \qquad (14)$$

Similarly at $z = h - \lambda_0/4$, using (1),

$$C_b(h, h - \lambda_0/4) = C_a(h, h - \lambda_0/4) - 2\ln(b/a) \sin\beta_0 h, \qquad (15)$$

$$S_b(h, h - \lambda_0/4) = S_a(h, h - \lambda_0/4) + 2\ln(b/a) \cos\beta_0 h. \qquad (16)$$

Substitution of (13)–(16) in (3b) and use of (3a) gives

$$\Psi_b(0) = \Psi_{K1}(0) - 2\ln(b/a), \qquad (17)$$

$$\Psi_b(h - \lambda_0/4) = \Psi_{K1}(h - \lambda_0/4) - 2\ln(b/a). \qquad (18)$$

If (17) and (18) are substituted in (2a, b) with $z = z_r$, the magnitudes of the resulting functions $\Psi_{s1}(z_r)$ and $\Psi_{a1}(z_r)$ define the real expansion parameters. They are

$$\Psi_{s1} = |\Psi_{s1}(z_r)| = 2\left|\Psi_{K1}(z_r) - \ln\frac{b}{a}\right|, \quad (19)$$

$$z_r = 0, \quad \beta_0 h \leq \frac{\pi}{2};$$

$$z_r = h - \frac{\lambda_0}{4}, \quad \beta_0 h \geq \frac{\pi}{2},$$

$$\Psi_{a1} = |\Psi_{a1}(z_r)| = 2\ln\frac{b}{a}. \qquad (20)$$

Since $\Psi_{K1}(z_r)$ is predominantly real and considerably greater than $\ln(b/a)$ if (1) is well satisfied, it follows that in most instances (19) is approximately equivalent to

$$\Psi_{s1} \doteq 2\left(\Psi_{K1} - \ln\frac{b}{a}\right). \qquad (21)$$

When the inequality $b^2 \gg a^2$ is not satisfied b in (20) may be replaced by

$$b_e = \tfrac{1}{2}b[1 + \sqrt{1 - (2a/b)^2}].$$

Comparison of the relative distributions of current in an isolated antenna and in symmetrically driven, closely coupled antennas shows that they are very similar. The zeroth-order currents have the same distribution.

These are, respectively,

$$(I_z)_0 = \frac{j2\pi V_\delta \sin \beta_0(h-z)}{\zeta_0 \Psi_{K1} \cos \beta_0(h-\delta)},$$

$$(\delta \leq z \leq h) \quad (22)$$

$$2(I_z^s)_0 \doteq \frac{j2\pi V_\delta^s}{\zeta_0[\Psi_{K1} - \ln(b/a)]} \frac{\sin \beta_0(h-z)}{\cos \beta_0(h-\delta)}.$$

$$(\delta \leq z \leq h) \quad (23)$$

In (23) the total current in *both* of the coupled antennas is given. Evidently, the two parallel antennas, each of radius a and with voltage V_δ^s, are equivalent to a single, isolated antenna with the same voltage V_δ^s but with a radius a_e that is greater than a by an amount sufficient to make its expansion parameter Ψ_{K1e} equal to $[\Psi_{K1} - \ln(b/a)]$. The magnitude of the required radius may be estimated for an antenna that is not too long* for which $\Psi_{K1e} \doteq 2\ln(2h/a_e) - 2$. Thus,

$$2\ln\frac{2h}{a_e} - 2 \doteq 2\ln\frac{2h}{a} - 2 - \ln\frac{b}{a}, \quad (24)$$

$$a_e \doteq \sqrt{ab}. \quad (25)$$

This is an example of the cage antenna in which two or more closely spaced conductors in parallel are used in place of a single conductor of larger cross section. It is readily verified that the conclusions just reached from a study of the zeroth-order currents are essentially unchanged if the first-order currents are examined. This follows from the fact that terms like $F_1(z) + P_1(z)$, $G_1(z) + Q_1(z)$, $\cdots$ behave roughly like $\Psi_{K1}(z) + \Psi_b(z)$ in (2a). This is more nearly true for antennas for which $\beta_0 h$ differs considerably from $n\pi$.

The zeroth-order current in one of the antisymmetrically driven closely coupled antennas for $\delta \leq z \leq h$ is given by

$$(I_z^a)_0 = \frac{jV_\delta^a}{R_c} \frac{\sin \beta_0(h-z)}{\cos \beta_0(h-\delta)}, \quad (26)$$

where

$$R_c \equiv \frac{\zeta_0}{\pi} \ln\frac{b}{a}. \quad (27)$$

Although written as the zeroth-order current, (26) is a good approximation of higher-order currents. This follows from the fact that difference terms of the form $F_1(z) - P_1(z)$,

---

* It is seen from Fig. II.20.7 that an increase in the value of the radius a to a greater value such as $a_e = \sqrt{ab}$ corresponds essentially to a shift in the entire curve of Ψ_{K1} for $\Omega = 2\ln(2h/a)$ to a parallel position corresponding to Ψ_{K1} for $\Omega = 2\ln(2h/a_e)$ over the entire range of $\beta_0 h$. It follows that the results obtained in (25) specifically for a short antenna are, in fact, applicable to longer antennas.

$G_1(z) - Q_1(z)$, $\cdots$ which occur in the higher-order terms are necessarily small if the conditions $b^2 \ll h^2$, $\beta_0^2 b^2 \ll 1$ are satisfied.

The current defined in (26) with $\delta = 0$ is precisely the current obtained from conventional transmission-line theory when applied to a section of lossless line of length $2h$ with equal and opposite point generators connected in series with the two conductors at their centers. The quantity R_c defined in (27) is the characteristic impedance of such a line subject to the condition $b^2 \gg a^2$.

6. First-Order Distributions of Current in Symmetrically and Antisymmetrically Driven Antennas

The distributions of current for symmetrically and antisymmetrically driven antennas vary with the distance b between the parallel antennas. Therefore, it is best to indicate extreme variations in current. Beginning with the current in symmetrically driven antennas that are very close together, for example, $b/\lambda_0 = 0.01$, there is a continuous change in the distribution of current as the antennas are separated more and more until the distribution of the isolated antenna given in Sec. II.22 is reached. If the same procedure is followed for antisymmetrically driven antennas a similar continuous variation may be observed. There is, however, an important difference. In the symmetrically driven pair, as for the isolated antenna, the distribution of current is practically independent of the ohmic resistance of the conductors for all separations. On the other hand, in the case of antisymmetric antennas this is true only when the two antennas are not brought too close together. When the distance between them approaches a small fraction of a wavelength all the terms *not* involving z^i as a factor become smaller and smaller until they become negligible. In this case the distribution and amplitude are determined entirely by the terms with the factor z^i. The distribution is that of a transmission line and the maximum amplitude of the current is enormously greater than for the symmetrically driven or the isolated antennas.

In Fig. 6.1 are shown currents for the symmetrically driven pair with $b/\lambda_0 = 0.01$ and $b/\lambda_0 = \infty$ (isolated antennas with $\delta = 0$, $\beta_0 h = \pi/2$ and $\Omega = 15$). The currents are plotted per unit voltage in the form

$$I_z/V_0 = G(z) + jB(z).$$

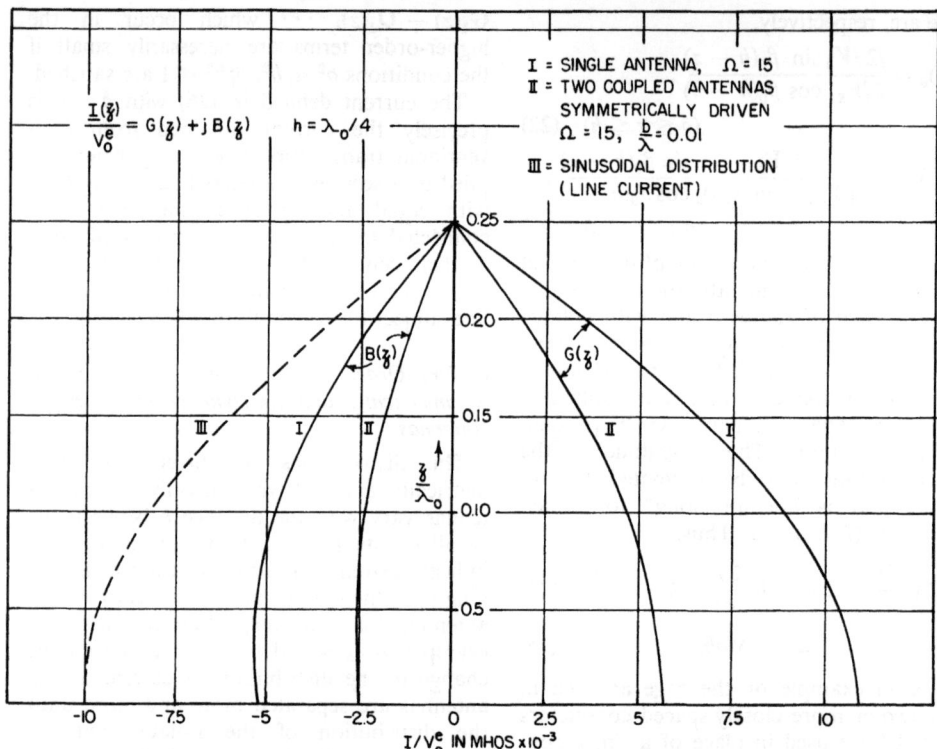

Fig. 6.1. First-order distribution of current per unit driving voltage ($h = \lambda_0/4$).

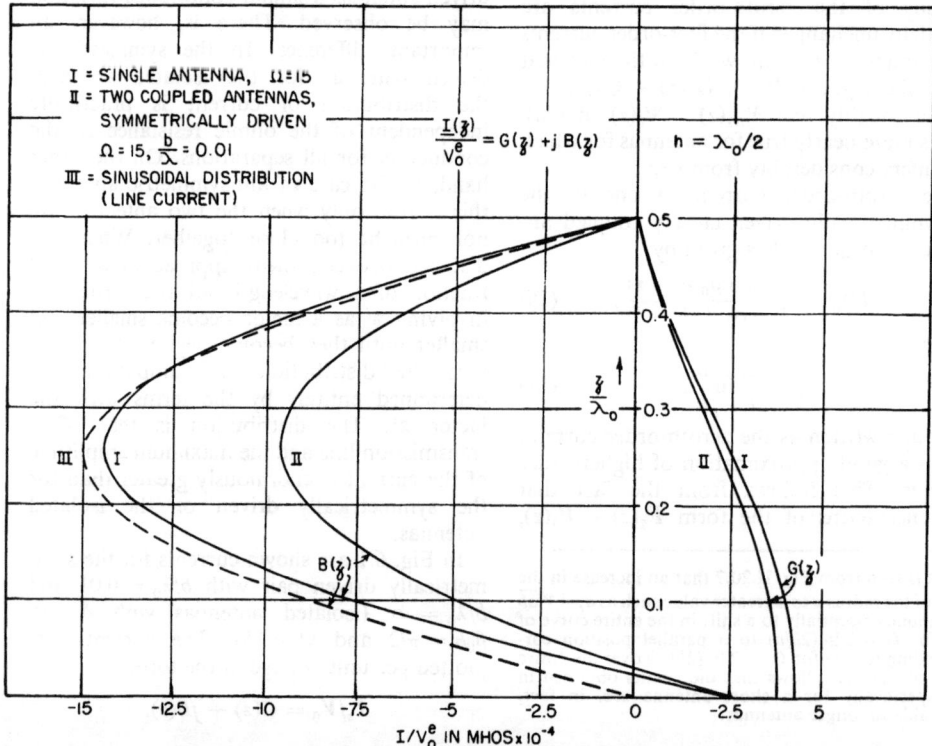

Fig. 6.2. First-order distribution of current per unit driving voltage ($h = \lambda_0/2$).

On the same sheet is drawn a sine curve of arbitrary amplitude to show the distribution in antisymmetrically driven antennas when $b/\lambda_0 = 0.01$. Actually, the amplitude of the current should be multiplied by 50 to 100 if the antennas are made of copper of reasonable size. Comparison of the curves for symmetrically driven antennas very close together and for an isolated antenna shows the current in the former to be slightly more than half that in the latter so that the two adjacent symmetrically driven antennas together have essentially the same distribution and magnitude of current as a single isolated antenna with slightly larger radius. These distributions are very nearly sinusoidal.

In Fig. 6.2 are shown curves corresponding to those of Fig. 6.1 but for $\beta_0 h = \pi$ instead of $\pi/2$. It is significant to note that at $b/\lambda_0 = 0.01$ the current in each one of the symmetric antennas is much greater than half of the current in the isolated antenna.

It may be concluded that the relative distributions of symmetric current on parallel coupled antennas of equal length do not depart greatly from the distribution on an isolated antenna.

7. Impedances of Symmetrically and Antisymmetrically Driven Antennas; Antenna Parallel to Conducting Plane

The impedances for the symmetrically and antisymmetrically driven antennas are, by definition, the ratio of scalar potential difference to current at $z = \pm\delta$. Using (3.10) and its counterpart with subscripts and superscripts changed from s to a, the impedances are

$$\frac{V_\delta^s}{I_\delta^s} \equiv Z_\delta^s = \frac{-j\zeta_0 \Psi_{s1}}{2\pi}$$

$$\times \left[\frac{\cos\beta_0(h-\delta) + A_{1s}/\Psi_{s1} + A_{2s}/\Psi_{s1}^2 + \cdots}{\sin\beta_0(h-\delta) + B_{1s}/\Psi_{s1} + B_{2s}/\Psi_{s1}^2 + \cdots} \right], \quad (1)$$

$$\frac{V_\delta^a}{I_\delta^a} \equiv Z_\delta^a = \frac{-j\zeta_0 \Psi_{a1}}{2\pi}$$

$$\times \left[\frac{\cos\beta_0(h-\delta) + A_{1a}/\Psi_{a1} + A_{2a}/\Psi_{a1}^2 + \cdots}{\sin\beta_0(h-\delta) + B_{1a}/\Psi_{a1} + B_{2a}/\Psi_{a1}^2 + \cdots} \right], \quad (2)$$

where $B \equiv M(\delta)$.

The impedances Z_δ for a small base separation 2δ in each antenna may be expressed as functions of the impedances Z_0 for $\delta = 0$ by expanding the admittance $Y_\delta = 1/Z_\delta$ in a Maclaurin series in powers of $\beta_0\delta$, just as in Sec. II.29. It has been shown that the difference between Z_δ and Z_0 for an isolated antenna is relatively small except when $\beta_0 h$ is near $\pi/2$, where it is roughly equivalent to substituting $h - \delta$ for h. Since Y_δ^s and Y_δ^a have the same general form as Y_δ for an isolated antenna, and the new functions P and Q depend upon $\beta_0\delta$ in essentially the same way as do the functions F and G, it may be assumed that the complex parameter $\epsilon = \epsilon'' - j\epsilon'$ in (II.29.6), namely,

$$Z_\delta = Z_0/(1 - \epsilon), \quad (3)$$

does not differ greatly from the corresponding factors ϵ_s and ϵ_a appearing in

$$Z_\delta^s = Z_0^s/(1 - \epsilon_s), \quad Z_\delta^a = Z_0^a/(1 - \epsilon_a). \quad (4)$$

That is, a reasonable approximation is

$$\epsilon_s \doteq \epsilon \doteq \epsilon_a, \quad (5)$$

so that

$$\frac{Z_\delta}{Z_0} \doteq \frac{Z_\delta^s}{Z_0^s} \doteq \frac{Z_\delta^a}{Z_0^a}. \quad (6)$$

This means that the functions ϵ' and ϵ'' in Sec. II.29 may be used in evaluating Z_δ^s from Z_0^s and Z_δ^a from Z_0^a.

The actual evaluation of Z_0^s and Z_0^a is more involved than the evaluation of Z_0 for an isolated antenna, owing to the presence of the functions $P_m(z)$ and $Q_m(z)$. With available tables of generalized sine and cosine integrals, the first-order functions may be computed. An explicit formula for Z_0^s and Z_0^a follows:

$$(Z_0)_1 = \frac{-j\zeta_0 \Psi}{2\pi}$$

$$\times \left\{ \frac{\Psi \cos\beta_0 h + F_1(h) \pm P_1(h)}{[\Psi + F_1(0) \pm P_1(0)]\sin\beta_0 h - [G_1(0) \pm Q_1(0)]\cos\beta_0 h + G_1(h) \pm Q_1(h)} \right\}, \quad (7)$$

where*

$$F_1(0) = \Psi(1 - \cos\beta_0 h) - C_a(h, 0) + \cos\beta_0 h\, E_a(h, 0), \quad (8a)$$

$$P_1(0) = -C_b(h, 0) + \cos\beta_0 h\, E_b(h, 0), \quad (8b)$$

---

* Note that the terms in (8a) and (8e) that have Ψ as a factor are a part of the difference integrals which make $F_1(0)$ and $G_1(0)$ small. It is not correct to include them with the Ψ in the denominator of (7) and then treat the remaining terms as small.

$$F_1(h) = -C_a(h, h) + \cos \beta_0 h \, E_a(h, h), \quad (8c)$$

$$P_1(h) = -C_b(h, h) + \cos \beta_0 h \, E_b(h, h), \quad (8d)$$

$$G_1(0) = -\Psi \sin \beta_0 h - S_a(h, 0) + \sin \beta_0 h \, E_a(h, 0), \quad (8e)$$

$$Q_1(0) = -S_b(h, 0) + \sin \beta_0 h \, E_b(h, 0), \quad (8f)$$

$$G_1(h) = -S_a(h, h) + \sin \beta_0 h \, E_a(h, h), \quad (8g)$$

$$Q_1(h) = -S_b(h, h) + \sin \beta_0 h \, E_b(h, h); \quad (8h)$$

$Z_0 = Z_0^s$ with the upper signs and $\Psi = \Psi_{1s}$; $Z_0 = Z_0^a$ with the lower signs and $\Psi = \Psi_{1a}$. The formulas (8a–h) are taken from (II.21.1a–d) and (3.19).

Of particular interest and simplicity is the formula for $\beta_0 h = \pi/2$:

$$(Z_0)_1 = \frac{j\zeta_0}{2\pi}$$

$$\times \left\{ \frac{C_a(h, h) \pm C_b(h, h)}{1 + \frac{1}{\Psi}[F_1(0) \pm P_1(0) + G_1(h) \pm Q_1(h)]} \right\}, \quad (9)$$

$$\Psi = \Psi_{s1} = |C_a(h, 0) + C_b(h, 0)| \text{ for } Z_0^s \text{ with upper signs}, \quad (10a)$$

$$\Psi = \Psi_{a1} = |C_a(h, 0) - C_b(h, 0)| \text{ for } Z_0^a \text{ with lower signs}. \quad (10b)$$

This formula is significant since the ratio h/a in Ψ appears only as a factor of the smallest term. This means that for values of Ψ that are not too small the impedance of each symmetrically driven antenna varies only slightly with changes in the radius of the antenna if its length is fixed with $\beta_0 h = \pi/2$. For all moderately large values of Ψ the following formulas are fair approximations:

$$Z_0^s \doteq \frac{j\zeta_0}{2\pi}[C_a(h, h) + C_b(h, h)], \quad (11a)$$

$$Z_0^a \doteq \frac{j\zeta_0}{2\pi}[C_a(h, h) - C_b(h, h)]. \quad (11b)$$

These reduce to the modified zeroth-order form if the radius a is set equal to zero in $C_a(h, h)$. The first-order formulas (7) and (9) are much better approximations than (11a, b), but they are no more accurate than the first-order formula for the impedance of an isolated antenna, to which they reduce when the separation b is made infinite. Reference to Figs. II.30.1a, c, d indicates that the first-order self-impedance curves resemble the second-order curves in shape and magnitude, but that they are, in effect, shifted with respect to length, so that they are not adequate in a quantitative sense in giving the correct impedance for a specified value of h. Since the symmetric and antisymmetric impedances oscillate about the value Z_0 for infinite separation as the separation is increased from small values, it is evident that Z_0^s and Z_0^a must suffer from the same inaccuracy. That is, whereas the *variations* of the first-order values of Z_0^s and Z_0^a about Z_0 should be adequate, the first-order value of Z_0 is not related correctly to h. A method is described in the next section for superimposing the first-order variations in Z_0^s and Z_0^a on the *second*-order value of Z_0, which is known to be related properly to h. Such modified first-order or *approximate second-order* values of Z_0^s and Z_0^a are given in Tables 7.1 and 7.2 and in Figs. 7.1 through 7.4 for $\beta_0 h = \pi/2$ and with $\Omega = 10, 15, 20,$ and ∞. Magnitudes and phases are shown in Fig. 7.5.

Interesting features of Figs. 7.1 and 7.2 are that maxima of R_0^s and X_0^s occur practically at minima of R_0^a and X_0^a, and vice versa. Note that the extreme values of resistance occur near $b/\lambda_0 = 0.66$, the extreme values of reactance near $b/\lambda_0 = 0.33$ and 0.92. If the antennas were very short compared with the wavelength ($\beta_0^2 h^2 \ll 1$), these extreme values would occur at $b/\lambda_0 = 0.5$ for the resistance and at $b/\lambda_0 = 0.25$ and 0.75 for the reactance. The reason why for $\beta_0 h = \pi/2$ these extremes occur at greater separations than for $\beta_0 h$ very small is related to the fact that the radial *phase* velocity of the electromagnetic field near a short antenna is the velocity of light, v_0, whereas it very much exceeds v_0 for antennas with $\beta_0 h = \pi/2$. This is shown in Chapter V. An interesting fact, which follows directly from (11a, b) is that as b approaches a, Z_0^s approaches $2Z_0$.

The modified *zeroth-order* impedance of antisymmetrically driven antennas of electrical half-length $\beta_0 h = \pi/2$ is given by (11b) with $a = 0$. Using (II.19.33) with $\beta_0 h = \pi/2$ and with a replaced by zero for the first term of (11b) and by b for the second term, the following expression is obtained. It involves only tabulated functions.

$$Z_0^a \doteq \frac{\zeta_0}{2\pi}[Ss(0, \pi) - Ss(\beta_0 b, \pi) + jCs(0, \pi) - jCs(\beta_0 b, \pi)]. \quad (12)$$

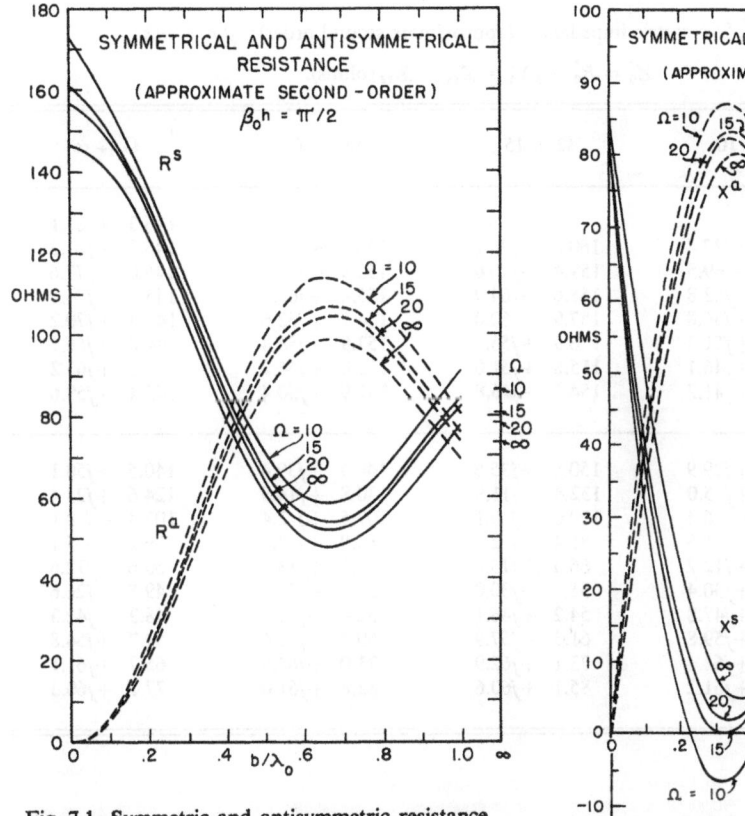

Fig. 7.1. Symmetric and antisymmetric resistance ($\beta_0 h = \pi/2$).

Fig. 7.2. Symmetric and antisymmetric reactance ($\beta_0 h = \pi/2$).

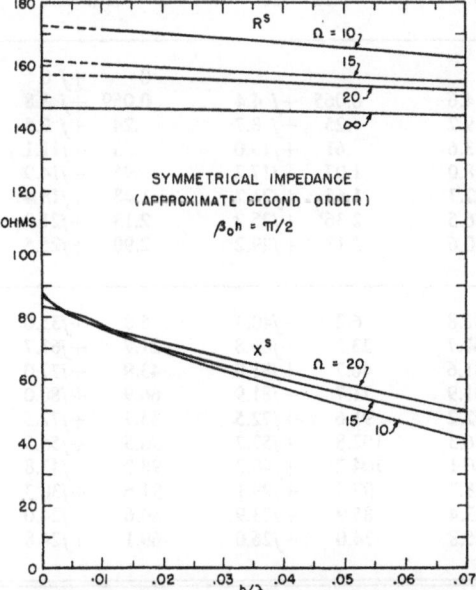

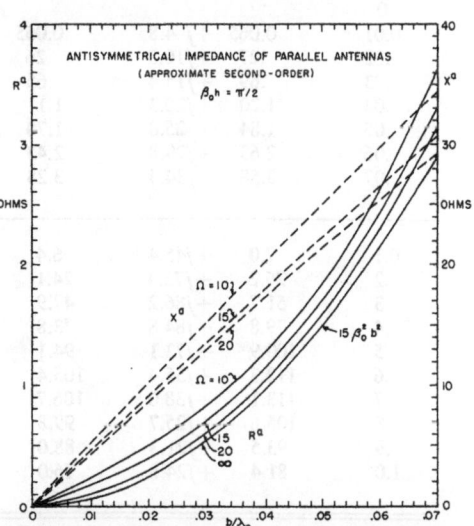

Fig. 7.3. Symmetric impedance ($\beta_0 h = \pi/2$).

Fig. 7.4. Antisymmetric impedance of parallel antennas ($\beta_0 h = \pi/2$).

TABLE 7.1. Symmetric impedances (approximate second order); $\beta_0 h = \pi/2$,
$Z_0^s = R_0^s + jX_0^s = Z_{s1} + Z_{12}$ (ohms).

b/λ_0	$\Omega = 10$	$\Omega = 15$	$\Omega = 20$	$\Omega \to \infty$
0				146.3 +j85.1
0.01	171.3 +j77.4	160.6 +j76.1	156.5 +j76.9	146.2 +j81.3
.02	169.5 +j69.5	159.4 +j70.0	155.9 +j71.6	146.0 +j77.6
.03	168.1 +j62.8	158.6 +j64.9	155.4 +j67.0	145.7 +j74.0
.04	166.8 +j56.8	157.9 +j60.0	154.7 +j62.6	145.3 +j70.2
.05	165.4 +j51.3	156.9 +j55.5	153.9 +j58.4	144.8 +j66.7
.06	163.8 +j46.1	155.8 +j51.0	153.0 +j54.2	144.1 +j63.2
.07	162.2 +j41.2	154.5 +j46.8	151.9 +j50.4	143.4 +j59.6
0.1	157.1 +j29.9	150.8 +j35.8	148.4 +j39.4	140.5 +j50.1
.2	136.6 +j 5.0	132.4 +j11.9	130.8 +j14.9	124.6 +j23.4
.3	112.3 −j 6.4	109.0 +j 0.1	107.5 +j 2.4	102.4 +j 8.1
.4	89.5 −j 1.5	85.4 +j 2.8	83.9 +j 3.6	79.4 +j 5.1
.5	70.6 +j12.2	66.3 +j13.8	64.3 +j13.8	60.6 +j12.6
.6	60.0 +j30.4	55.7 +j30.0	53.7 +j29.5	49.8 +j26.6
.7	59.1 +j47.8	54.2 +j46.1	52.4 +j45.5	48.3 +j42.3
.8	66.7 +j59.8	61.3 +j57.9	59.3 +j57.6	54.7 +j54.8
.9	79.1 +j64.4	73.1 +j62.9	71.0 +j62.8	65.7 +j61.1
1.0	91.5 +j61.2	85.1 +j60.6	82.8 +j61.0	77.1 +j60.3

TABLE 7.2. Antisymmetric impedances (approximate second order); $\beta_0 h = \pi/2$,
$Z_0^a = R_0^a + jX_0^a = Z_{s1} - Z_{12}$ (ohms).

b/λ_0	$\Omega = 10$	$\Omega = 15$	$\Omega = 20$	$\Omega \to \infty$
0				0 +j 0
0.01	0.065 +j 4.8	0.065 +j 4.6	0.065 +j 4.4	0.059 +j 3.8
.02	.27 +j10.3	.26 +j 9.2	.25 +j 8.7	.24 +j 7.5
.03	.69 +j15.4	.65 +j13.6	.61 +j13.0	.53 +j11.1
.04	1.20 +j20.3	1.11 +j18.0	1.07 +j17.2	.95 +j14.9
.05	1.84 +j25.0	1.70 +j22.3	1.65 +j21.2	1.48 +j18.4
.06	2.63 +j29.8	2.43 +j26.5	2.36 +j25.2	2.13 +j21.9
.07	3.58 +j34.4	3.28 +j30.6	3.18 +j29.2	2.90 +j25.5
0.1	7.0 +j45.4	6.4 +j41.8	6.2 +j40.3	5.8 +j35.0
.2	26.2 +j72.4	24.1 +j67.7	23.3 +j66.8	21.7 +j61.7
.3	51.7 +j86.2	47.9 +j81.6	46.7 +j81.0	43.9 +j77.0
.4	79.8 +j84.8	73.8 +j81.9	71.7 +j81.9	66.9 +j80.0
.5	100.9 +j73.1	94.1 +j72.2	91.6 +j72.5	85.7 +j72.5
.6	112.3 +j55.4	105.4 +j56.3	102.8 +j57.2	96.5 +j58.5
.7	113.1 +j38.0	106.7 +j40.1	104.2 +j40.2	98.0 +j42.8
.8	105.6 +j25.7	99.8 +j28.2	97.5 +j29.1	91.6 +j30.3
.9	93.5 +j21.3	88.0 +j23.4	85.9 +j23.9	80.6 +j24.0
1.0	81.4 +j24.8	76.0 +j25.8	74.0 +j26.0	69.1 +j24.8

This may be transformed into well-known formulas using (II.19.27b) and (II.19.27c) together with (II.19.27g) and (II.19.27h). The results are

$$R_0^a \doteq \frac{\zeta_0}{2\pi} [Ss(0, \pi) - Ss(\beta_0 b, \pi)]$$

$$= \frac{\zeta_0}{4\pi} [\text{Cin } 2\pi + 2\text{Cin } \beta_0 b$$
$$- \text{Cin}(\sqrt{\beta_0^2 b^2 + \pi^2} + \pi)$$
$$- \text{Cin}(\sqrt{\beta_0^2 b^2 + \pi^2} - \pi)], \quad (13a)$$

$$X_0^a \doteq \frac{\zeta_0}{2\pi} [Cs(0, \pi) - Cs(\beta_0 b, \pi)]$$

$$= \frac{\zeta_0}{4\pi} [\text{Si } 2\pi + 2\text{Si } \beta_0 b$$
$$- \text{Si}(\sqrt{\beta_0^2 b^2 + \pi^2} + \pi)$$
$$- \text{Si}(\sqrt{\beta_0^2 b^2 + \pi^2} - \pi)]. \quad (13b)$$

When the following condition is satisfied:

$$\beta_0^2 b^2 \ll 1, \quad (14)$$

these expressions reduce to very simple forms if only the leading terms in the series expansions for the integral functions are used. Thus,

$$\text{Cin } \beta_0 b \doteq \tfrac{1}{4}\beta_0^2 b^2, \quad \text{Si } \beta_0 b \doteq \beta_0 b, \quad (15a)$$

$$\text{Cin}(\sqrt{\beta_0^2 b^2 + \pi^2} + \pi) \doteq \text{Cin } 2\pi,$$

$$\text{Si}(\sqrt{\beta_0^2 b^2 + \pi^2} + \pi) \doteq \text{Si } 2\pi, \quad (15b)$$

$$\text{Cin}(\sqrt{\beta_0^2 b^2 + \pi^2} - \pi) \doteq \tfrac{1}{4}(\beta_0^2 b^2/2\pi)^2,$$

$$\text{Si}(\sqrt{\beta_0^2 b^2 + \pi^2} - \pi) \doteq \beta_0^2 b^2/2\pi^2. \quad (15c)$$

Neglect of powers of $\beta_0 b$ higher than the second, according to $\beta_0^2 b^2 \ll 1$, leads to

$$R_0^a \doteq \frac{\zeta_0}{8\pi} \beta_0^2 b^2 = 15\beta_0^2 b^2 \text{ ohm}$$

$$= 592.2(b/\lambda_0)^2 \text{ ohm}, \quad (16a)$$

$$X_0^a \doteq \frac{\zeta_0}{2\pi} [\beta_0 b - \beta_0^2 b^2/4\pi]$$

$$= 60\beta_0 b[1 - \beta_0 b/4\pi] \text{ ohm}$$

$$= 376.7 \frac{b}{\lambda_0} \left(1 - \frac{b}{2\lambda_0}\right) \text{ ohm}. \quad (16b)$$

The values of R_0^a and X_0^a for $b/\lambda_0 \leq 0.07$ and $\Omega \to \infty$ in Table 7.2 are computed from (16a, b). Note that the corresponding approximate second-order values for antennas of finite thickness with $\Omega = 10$, 15, and 20 are considerably higher. This is illustrated graphically in Fig. 7.4. All values in Table 7.2 assume the antennas to be perfectly conducting. Dissipation in ohmic resistance is negligible compared with radiated power if the conductors are good, except in the case of antisymmetrically driven antennas that are separated by only small fractions of a wavelength. These are considered later in this section.

Corresponding to (9) for $\beta_0 h = \pi/2$, the formula for $\beta_0 h = \pi$ is,

$$(Z_0)_1 = \frac{j\zeta_0 \Psi}{2\pi} \times$$

$$\left\{\frac{\Psi + C_a(h, h) \pm C_b(h, h) + E_a(h, h) \pm E_b(h, h)}{S_a(h, 0) \pm S_b(h, 0) + S_a(h, h) \pm S_b(h, h)}\right\}, \quad (17)$$

$$\Psi = \Psi_{s1} = S_a(h, h/2) + S_b(h, h/2)$$
$$\text{for } Z_0^s \text{ with upper signs}, \quad (18a)$$

$$\Psi = \Psi_{a1} = S_a(h, h/2 - S_b(h, h/2)$$
$$\text{for } Z_0^a \text{ with lower signs}. \quad (18b)$$

The leading term in (12) has the factor Ψ^2 which is sensitive to changes in h/a. It follows that Z_0^s and Z_0^a both vary with Ω when $\beta_0 h = \pi$. This is evident in Fig. 7.6, in which symmetric and antisymmetric resistances and reactances are shown as functions of $\beta_0 b$ with $\Omega = 10$ and 20 and with $\beta_0 h$ near but not exactly at π. These curves give the approximate second-order values obtained from the general first-order formula (7) and a scaling process described in Sec. 8. Note that there is no correspondence between maxima and minima of Z_0^s and minima and maxima of Z_0^a, as when $\beta_0 h$ is near resonance. Since $C_a(h, h)$ is very nearly the negative of $E_a(h, h)$, it follows from (17) with (18a) that, as b approaches a, Z_0^s again approaches $2Z_0$.

When a center-driven antenna is placed parallel to a perfectly conducting, infinite plane, as shown in Fig. 7.7, it is advantageous to make use of the theorem of images (ref. I.31, Sec. IV.20). This theorem states that in the half-space containing the antenna the electromagnetic field due to charges and currents both in the antenna and on the conducting sheet is the same as the electromagnetic field in the same half-space due to charges and currents in the antenna and in an image antenna with currents and signs of charges reversed, as shown in Fig. 7.8. Since the electric field in the two cases is the same, it is clear that the impedances Z_{cD} in Fig. 7.7 and 7.8 also must be the same if they are defined at a distance from the antenna that is

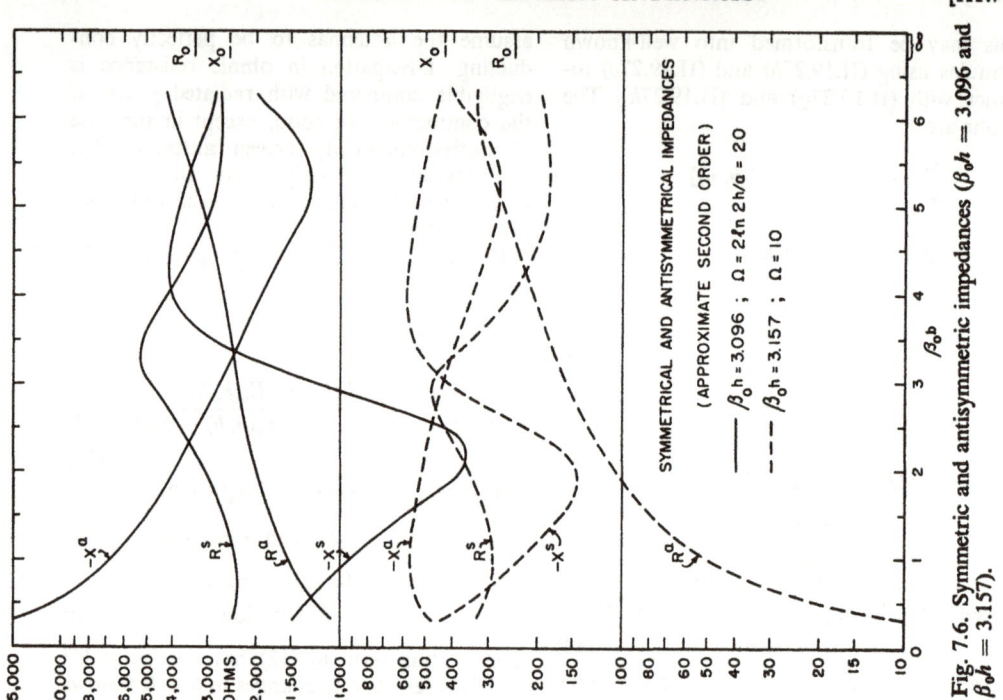

Fig. 7.6. Symmetric and antisymmetrical impedances ($\beta_0 h = 3.096$ and $\beta_0 h = 3.157$).

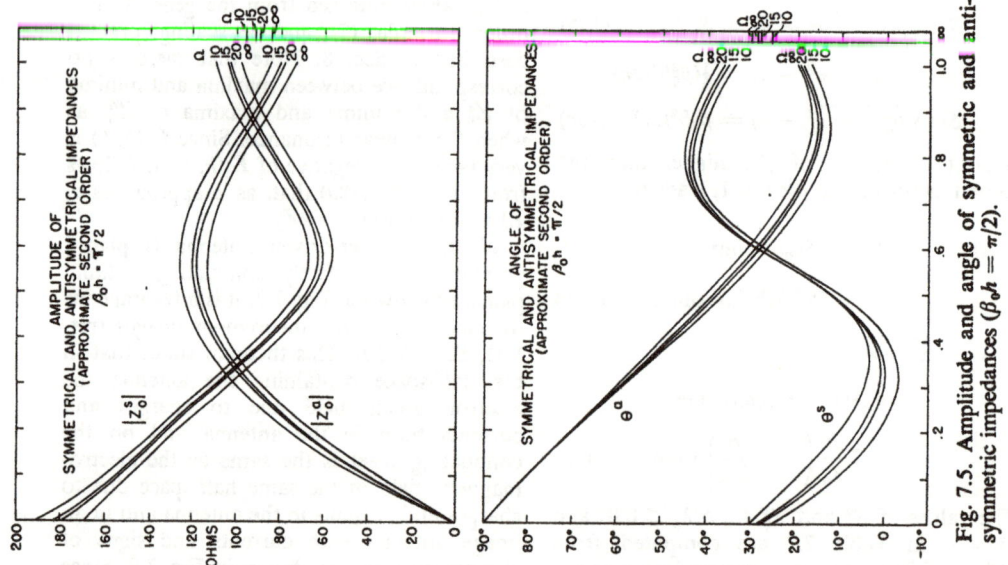

Fig. 7.5. Amplitude and angle of symmetric and antisymmetric impedances ($\beta_0 h = \pi/2$).

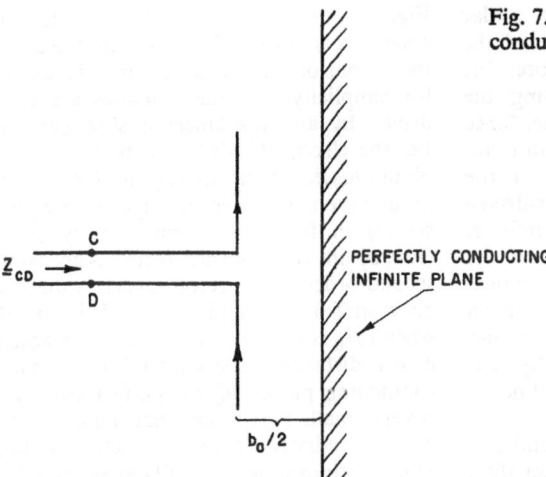

Fig. 7.7. Center-driven antenna parallel to a perfectly conducting infinite plane.

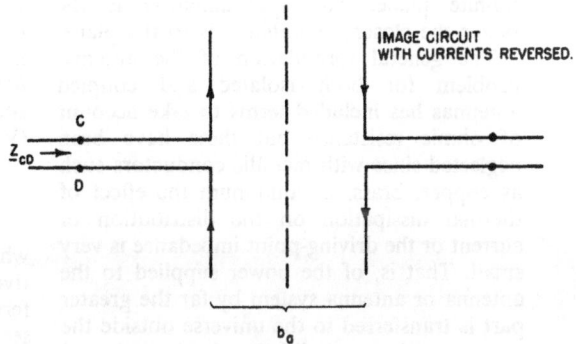

Fig. 7.8. Analytical equivalent of Fig. 7.7.

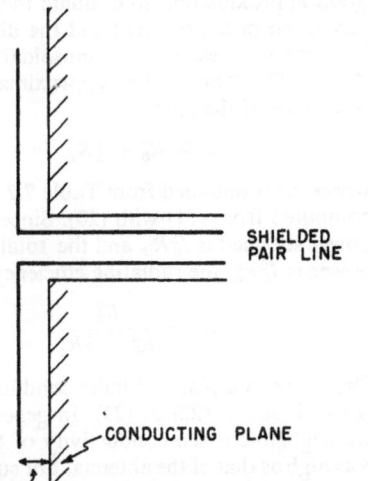

Fig. 7.9. Center-fed antenna over conducting plane.

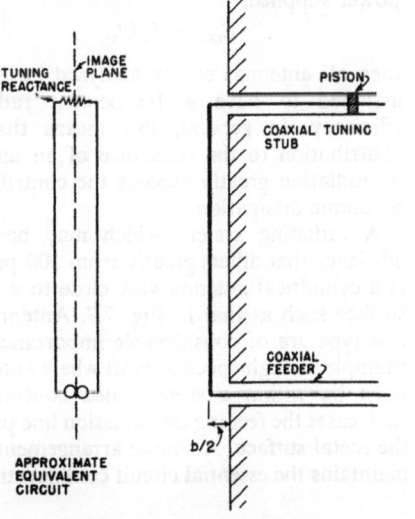

Fig. 7.10. End-fed and end-tuned antenna close to conducting plane with approximate equivalent circuit.

outside the terminal zone so that scalar potential differences and line integrals of the electric field are equivalent. Therefore, the apparent load impedances terminating the lines in Figs. 7.7 and 7.8 are the same. Since these impedances are those of an antisymmetrically driven antenna, it follows that the apparent load impedance of a center-driven antenna at a distance $b/2$ over an infinite, perfectly conducting plane is equal to the antisymmetric impedance of two identical antennas separated a distance b. If the antenna is very close to the conducting plane, so that $(\beta_0 b)^2 \ll 1$, the impedance is essentially that of an open-end section of two-wire line of length h and spacing b.

If the plane is highly conducting and has dimensions that are considerably greater than the length of the antenna, the apparent terminal impedance is well approximated by that calculated for a perfectly conducting, infinite plane. The approximation is the better the closer the antenna is to the plane.

The general formulation of the antenna problem for both isolated and coupled antennas has included terms to take account of ohmic resistance but these have been neglected since with metallic conductors such as copper, brass, or aluminum the effect of thermal dissipation on the distribution of current or the driving-point impedance is very small. That is, of the power supplied to the antenna or antenna system by far the greater part is transferred to the universe outside the antenna; only a negligible fraction is dissipated as heat in the conductors. If the *radiating efficiency* of an antenna or array is defined to be the ratio of the power radiated to the total power supplied,

$$E_{\text{rad}} = P_r/P_t, \qquad (19)$$

then all antennas so far analyzed have been assumed to have a 100-percent radiating efficiency. In general, this means that the contribution to the resistance of an antenna by radiation greatly exceeds the contribution by ohmic dissipation.

A radiating system which may have an efficiency that differs greatly from 100 percent is a cylindrical antenna very close to a metal surface such as that in Fig. 7.7. Antennas of this type are of considerable importance, for example, in high-speed aircraft where antennas must be within a stream-lined contour. In such cases the feeding transmission line pierces the metal surface. A simple arrangement that maintains the essential circuit configuration of Fig. 7.7 is shown in Fig. 7.9. Since terminal-zone effects in Figs. 7.7 or 7.9 are the same as those considered in Sec. 9, it may be assumed for simplicity that the antennas are center-driven by an impedanceless slice generator. Let the electrical half-length be $\beta_0 h = \pi/2$. If ohmic losses are neglected, the input impedance is the antisymmetric impedance Z_0^a given by (9) in first-order form and by (12) or (13a, b) in zeroth-order form. Approximate second-order and zeroth-order values are given in Table 7.2 and Fig. 7.4. It is seen that when b/λ_0 is less than 0.1 (so that the antenna is at a distance $b/2$ less than $0.05\lambda_0$ from the conducting plane) R_0^a for perfect conductors is very small. This means that if much power is to be radiated I_0^a must be extremely large. The questions quite naturally arise: how large is the ohmic resistance? what fraction of the power is dissipated in heat?

Since the ohmic dissipation of the antenna over a perfectly conducting infinite plane is equal to half that in a two-wire transmission line of length $h = \lambda_0/4$ and line spacing b if b/λ_0 is small, the resistance per unit length of such a line is required. According to formula (VI.23.13) of ref. I.31,

$$r^i \doteq \frac{1}{\pi a} \sqrt{\frac{\omega}{2\sigma v_0[1-(2a/b)^2]}}, \qquad (20)$$

where $v_0 = 10^7/4\pi$ m/h and σ is the conductivity. It is assumed the conductors are non-ferromagnetic. The input resistance of the section of nonradiating line is

$$R_L \doteq R_c \alpha h = \tfrac{1}{2} r^i h. \qquad (21)$$

Although radiation and ohmic dissipation strictly are not independent in general, it is a good approximation to evaluate them separately when both are small and the distribution of current is essentially sinusoidal as when $\beta_0 h = \pi/2$. Thus, the approximate input resistance of the antenna is

$$R_{\text{in}} \doteq R_0^a + \tfrac{1}{2} R_L, \qquad (22)$$

where R_0^a is obtained from Table 7.2 and R_L is computed from (21) with (20). Since the total power supplied is $I_0^2 R_{\text{in}}$ and the total radiated power is $I_0^2 R_0^a$, the radiating efficiency is

$$E_{\text{rad}} = \frac{R_0^a}{R_0^a + \tfrac{1}{2} R_L}. \qquad (23)$$

The losses in a plane of finite conductivity are not included in (22) or (23). In general, these are negligible if the conductivity of the plane is as high as that of the antenna. For sufficiently thin wires and very small distances $b/2$, the simple zeroth-order formula for the antisymmetric or radiation resistance is satisfactory. This is

$$R_0^a \doteq 15\beta_0^2 b^2 \text{ ohm}. \qquad (24)$$

For copper, $\sigma = 5.65 \times 10^7$, so that (21) with (20) becomes

$$R_L \doteq \frac{h}{a} \times 10^{-7} \sqrt{\frac{0.177 f}{1 - (2a/b)^2}} \text{ ohm}. \qquad (25)$$

As a numerical example, consider a copper antenna with $\Omega = 10$ or $h/a \doteq 75$ at a distance $b/2 = 0.01\lambda_0$ from a highly conducting plane. From Table 7.2, $R_0^a = 0.27$ ohm. With $f = 300$ Mhz or $\lambda_0 = 1$ m, $a = \frac{1}{3}$ cm, $b/2 = 1$ cm, $2a/b = 1/3$,

$$R_L \doteq 75 \times 10^{-7} \sqrt{\frac{0.531 \times 10^8}{0.889}} = 0.058 \text{ ohm}. \qquad (26)$$

Hence

$$E_{\text{rad}} \doteq \frac{0.27}{0.299} = 0.90. \qquad (27)$$

Compared with isolated antennas this is a low radiating efficiency. Nevertheless, it is quite high enough to be practical, and yet the axis of the antenna is only one hundredth of a wavelength from the conducting surface. If the radius is doubled and it is assumed that R_0^a is not changed, R_L can be reduced to 0.037 ohm and the efficiency increased to 0.93. Note that whereas the axis of the conductor is only $0.01\lambda_0 = 1$ cm from the surface, the outer edge is now at a distance of 1.67 cm instead of 1.33 cm, while the inner edge is within 0·33 cm instead of 0.67 cm. Evidently the thicker antenna can not be moved in much closer without coming in contact with the plane. If it is moved somewhat nearer the radiation resistance decreases rapidly. If the thinner antenna ($a = 1/3$ cm) is moved to $b/2 = 0.5$ cm, $2a/b = 2/3$ and $R_L = 0.074$ ohm while $R_0^a = 0.065$. The efficiency now is only 0.64. This is still large enough to make the antenna useful even though its center is only 5 mm from the conducting surface. Thus, it is evident that effective antennas can be designed which lie very close to a conducting surface. However, it is important to bear in mind that with a total load of only a few tenths of an ohm, ohmic losses in the feeding transmission line are relatively very large so that the *over-all* efficiency of antenna and line is much lower unless a very short section of line is used.

Instead of driving the antenna in the center from a shielded-pair line as in Fig. 7.9, it may be driven from one end by a coaxial line and the other end may be supported or tuned by a high-impedance section of coaxial line as shown in Fig. 7.10. In this form the antenna is simply a transmission line, as indicated in Fig. 7.10, where an approximate equivalent circuit is shown. This does not take account of the holes through the conducting plane. The zeroth-order radiation input resistance of such a line referred to maximum current has been determined[49] in the general case in which account is taken of ohmic resistance of the line and dissipation in the load. The expression is

$$R_0^e = \frac{\zeta_0}{8\pi} \beta_0^2 b^2 \frac{\cosh(\alpha s + 2\rho)}{|\cosh[(\alpha + j\beta)s + \rho + j\Phi']|^2}$$
$$\times \left[\cosh \alpha s - \frac{\sin 2\beta s}{2\beta s} \right], \qquad (28)$$

where s is the length, α the attenuation constant, and β the phase constant of the line. The impedance Z_s at the load end is expressed in terms of its phase function Φ' and attenuation function ρ as defined by

$$\rho + j\Phi' = \tanh^{-1}(Z_s/Z_c), \qquad (29)$$

where Z_c is the characteristic impedance. (The resistance for a two-wire line with spacing b is twice the value given in (28).) Note that R_0^e in (28) does not include the contribution of ohmic resistance to the impedance. The input resistance of the line is R_0^e plus the conventional input resistance due to ohmic losses.

If the losses in line and termination are small, $(\alpha s + 2\rho) < 0.1$, the radiation input resistance is

$$R_0^e \doteq \frac{\zeta_0}{8\pi} \beta_0^2 b^2 \frac{1 - (\sin 2\beta s)/2\beta s}{\cos^2(\beta s + \Phi')}. \qquad (30a)$$

If the resistance is referred to the maximum current,

$$R_m^e \doteq \frac{\zeta_0}{8\pi} \beta_0^2 b^2 [1 - (\sin 2\beta s)/2\beta s]. \qquad (30b)$$

For a resonant line terminated in short heavy bridges for which $\Phi' \doteq 0$ so that the condition for resonance is $\beta s \doteq n\pi$,

$$R_0^e = R_m^e \doteq \frac{\zeta_0}{8\pi} \beta_0^2 b^2 = 15\beta_0^2 b^2 \text{ ohm}. \qquad (31)$$

This is exactly the same as the zeroth-order obtained for the antisymmetric resistance of the center-driven antenna of electrical length $2\beta_0 h = \pi$ in (16a).

If the line is loaded with a resistance near the characteristic resistance of the line, so that $\rho > 3$,

$$R_0^e \doteq \frac{\zeta_0}{4\pi}\beta_0^2 b^2 \left[\cosh \alpha s - \frac{\sin 2\beta s}{2\beta s}\right] e^{-\alpha s}. \quad (32)$$

For a very long line at a distance $b/2$ over the conducting plane,

$$R_0^e \doteq \frac{\zeta_0}{4\pi}\beta_0^2 b^2 = 30\beta_0^2 b^2, \quad (33)$$

if attenuation is sufficiently small and s is several wavelengths. Note that this is twice the radiation resistance of the resonant line, but since the line is terminated in R_c and the power $I^2 R_c$ is dissipated in the termination whereas only $I^2 R_0$ is radiated, the radiating efficiency is very low.

8. Impedance of Antennas with Arbitrary Voltages; Self- and Mutual Impedances

For parallel, identical antennas with arbitrary potential differences $V_{1\delta}$ and $V_{2\delta}$, the driving-point impedances are defined by

$$Z_{1\delta} = V_{1\delta}/I_{1\delta}, \quad Z_{2\delta} = V_{2\delta}/I_{2\delta}. \quad (1)$$

From (2.16)

$$V_{1\delta} = V_\delta^s + V_\delta^a, \quad V_{2\delta} = V_\delta^s - V_\delta^a. \quad (2)$$

Introduction of the impedances and currents for symmetrically and antisymmetrically driven antennas gives

$$V_{1\delta} = I_\delta^s Z_\delta^s + I_\delta^a Z_\delta^a, \quad V_{2\delta} = I_\delta^s Z_\delta^s - I_\delta^a Z_\delta^a. \quad (3)$$

However, $I_\delta^s = \tfrac{1}{2}(I_{1\delta} + I_{2\delta}), I_\delta^a = \tfrac{1}{2}(I_{1\delta} - I_{2\delta})$. Therefore,

$$V_{1\delta} = I_{1\delta} \cdot \tfrac{1}{2}(Z_\delta^s + Z_\delta^a) + I_{2\delta} \cdot \tfrac{1}{2}(Z_\delta^s - Z_\delta^a), \quad (4a)$$

$$V_{2\delta} = I_{1\delta} \cdot \tfrac{1}{2}(Z_\delta^s - Z_\delta^a) + I_{2\delta} \cdot \tfrac{1}{2}(Z_\delta^s + Z_\delta^a). \quad (4b)$$

These equations are in the general form

$$V_{1\delta} = I_{1\delta} Z_{s1\delta} + I_{2\delta} Z_{12\delta}, \quad (5a)$$

$$V_{2\delta} = I_{1\delta} Z_{21\delta} + I_{2\delta} Z_{s2\delta}, \quad (5b)$$

where, in the present case of identical antennas,

$$Z_{s1\delta} = Z_{s2\delta} = \tfrac{1}{2}(Z_\delta^s + Z_\delta^a), \quad (6a)$$

$$Z_{12\delta} = Z_{21\delta} = \tfrac{1}{2}(Z_\delta^s - Z_\delta^a). \quad (6b)$$

The quantity $Z_{s1\delta}$ is the *self-impedance* of antenna 1 in the presence of antenna 2; $Z_{s2\delta}$ is the self-impedance of antenna 2 in the presence of antenna 1; $Z_{21\delta} = Z_{12\delta}$ is the mutual impedance of antennas 1 and 2 in each other's presence. (A subscript δ distinguishes an impedance referred to the circuit of Fig. 1.7 with a finite base separation 2δ at the center of each antenna. Omission of the subscript means that $\delta = 0$.)

The self-impedance of antenna 1 in the presence of antenna 2, $Z_{s1\delta}$, may be defined to be the input impedance $Z_{1\delta}$ of antenna 1 when antenna 2 is disconnected from the transmission line (or other circuit) at $z = \pm \delta$ so that $I_{2\delta} = 0$. This does not necessarily mean that I_{2z} vanishes elsewhere in the antenna. With $I_{2\delta} = 0$, (5a) reduces to

$$V_{1\delta} = I_{1\delta} Z_{1\delta} = I_{1\delta} Z_{s1\delta} \text{ when } I_{2\delta}=0. \quad (7)$$

Note that $Z_{s1\delta}$ is not a unique property of antenna 1 even when $\delta = 0$. It is a function of the distance b between the two antennas and indirectly of the current I_{2z} in antenna 2 in so far as this modifies the magnitude and distribution of the current in antenna 1. Therefore, and in general, $Z_{s1\delta}$ differs from Z_δ. Note that Z_δ is the impedance of an *isolated* antenna and is a function characteristic of an antenna alone when $\delta = 0$. The self-impedance $Z_{s1\delta}$ reduces to Z_δ only when $Z_{12\delta} = 0$, which occurs at infinite separation, $b = \infty$.

For $\delta = 0$, (5) becomes

$$V_{10} = I_{10} Z_{s1} + I_{20} Z_{12}, \quad (8a)$$

$$V_{20} = I_{10} Z_{21} + I_{20} Z_{s2}. \quad (8b)$$

The symmetric and antisymmetric impedances may be expressed in terms of the self- and mutual impedances. Thus,

$$Z_0^s = Z_{s1\delta} + Z_{12\delta}, \quad (9a)$$

$$Z_0^a = Z_{s1\delta} - Z_{12\delta}. \quad (9b)$$

The evaluation of $Z_{s1\delta}$ and $Z_{12\delta}$ may be carried out directly using the formulas (7.1) and (7.2) for Z_δ^s and Z_δ^a. Subject to the approximation discussed in Sec. 7, for which,

$$\frac{Z_\delta^s}{Z_0^s} \doteq \frac{Z_\delta^a}{Z_0^a} \doteq \frac{Z_\delta}{Z_0} = \frac{1}{1-\epsilon}, \quad (10)$$

it follows that

$$\frac{Z_{s1\delta}}{Z_{s1}} \doteq \frac{Z_{12\delta}}{Z_{12}} \doteq \frac{Z_\delta}{Z_0} = \frac{1}{1-\epsilon}, \quad (11)$$

where $\epsilon = \epsilon'' - j\epsilon'$ is defined and represented graphically in Sec. II.29. The evaluation of first-order values of Z_{s1} and Z_{12} is conveniently carried out from the general relation (7.7) for Z_0^s and Z_0^a. Curves of first-order $Z_{s1} = R_{s1} + jX_{s1}$ and $Z_{12} = R_{12} + jX_{12}$ are in the literature[51] with $\beta_0 h = \pi/2$ and π for a range of b/λ_0 up to 1. Approximate second-order values are derived below.

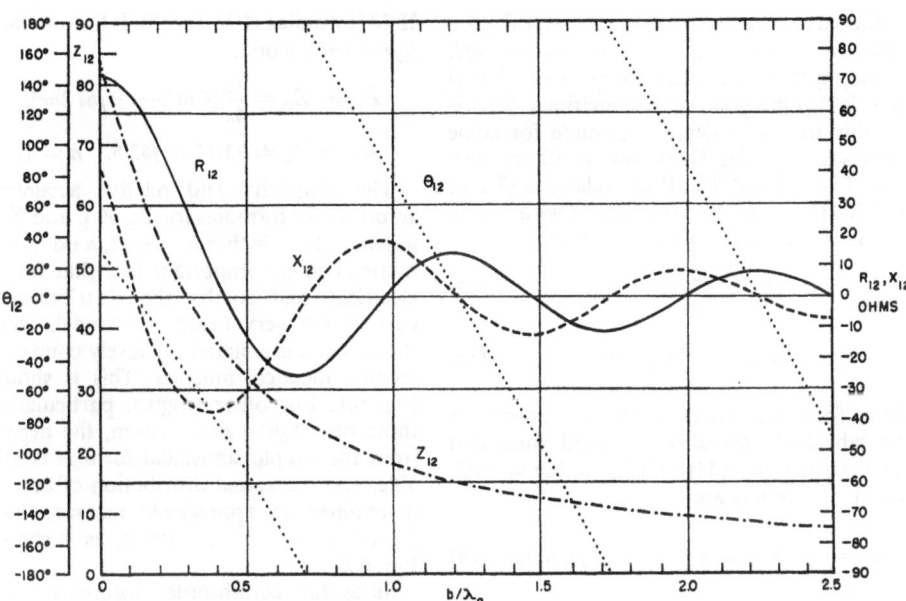

Fig. 8.1. Zeroth-order mutual impedance ($\beta_0 h = \pi/2$).

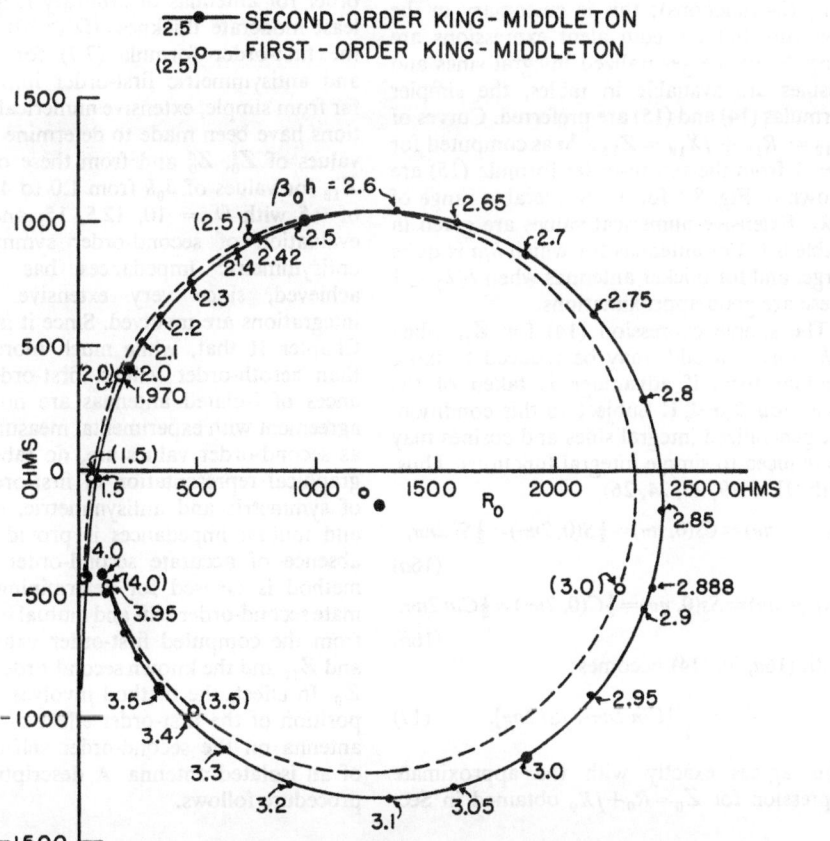

Fig. 8.2. Spiral graph of self-impedance Z_0 of isolated antenna.

Simple explicit formulas for Z_{s1} and Z_{12} are obtained in the special case for $\beta_0 h = n\pi/2$, n odd, with Ψ sufficiently large that (7.11a) and (7.11b) are good approximations. This is roughly true to a degree adequate for some purposes, even for Ω as low as 10, as seen from Fig. 7.1 and 7.2. If the relations (7.11a) and (7.11b) are used in (6) and (7) with $\delta = 0$, the results for $\beta_0 h = n\pi/2$, n odd, are

$$Z_{s1} \doteq \frac{j\zeta_0}{2\pi} C_a(h, h), \qquad (12)$$

$$Z_{12} \doteq \frac{j\zeta_0}{2\pi} C_b(h, h). \qquad (13)$$

These formulas may be expressed in terms of the tabulated generalized integral sines and cosines using (II.19.33) with $\beta_0 z = \beta_0 h = n\pi/2$, n odd. The results are

$$Z_{s1} \doteq \frac{\zeta_0}{2\pi} [Ss(\beta_0 a, n\pi) + jCs(\beta_0 a, n\pi)], \quad (14)$$

$$Z_{12} \doteq \frac{\zeta_0}{2\pi} [Ss(\beta_0 b, n\pi) + jCs(\beta_0 b, n\pi)]. \quad (15)$$

These formulas can be expressed entirely in terms of the simple integral sines and cosines (Si-, Ci- functions); this is customary in the literature but the equivalent expressions are long. Since the generalized integral sines and cosines are available in tables, the simpler formulas (14) and (15) are preferred. Curves of $Z_{12} = R_{12} + jX_{12} = Z_{12} e^{j\theta_{12}}$ as computed for $n = 1$ from the zeroth-order formula (15) are shown in Fig. 8.1 for a considerable range of b/λ_0. Extensive numerical values are given in Table 8.1. For antennas for which h/a is quite large, and for thicker antennas when $b/\lambda_0 \geq 1$ these are good approximations.

The simple expression (14) for Z_{s1} when $\beta_0 h = n\pi/2$, n odd, may be reduced to more familiar form if advantage is taken of the condition $\beta_0 a \ll 1$. Subject to this condition, the generalized integral sines and cosines may be reduced to simple integral functions. Thus, with (II.19.19, 20, 24, 26),

$$Cs(\beta_0 a, n\pi) \doteq Cs(0, n\pi) = \tfrac{1}{2} S(0, 2n\pi) = \tfrac{1}{2} Si\, 2n\pi, \qquad (16a)$$

$$Ss(\beta_0 a, n\pi) \doteq Ss(0, n\pi) = \tfrac{1}{2} C(0, 2n\pi) = \tfrac{1}{2} Cin\, 2n\pi. \qquad (16b)$$

With (16a, b), (14) becomes:

$$Z_{s1} \doteq \frac{\zeta_0}{4\pi} [Cin\, 2n\pi + jSi\, 2n\pi]. \qquad (17)$$

This agrees exactly with the approximate expression for $Z_0 = R_0 + jX_0$ obtained in Sec. II.28 (formulas II.28.16, 18). It follows that for $\beta_0 h = n\pi/2$, n odd,

$$Z_{s1} \doteq Z_0 \doteq \frac{\zeta_0}{4\pi} [Cin\, 2n\pi + jSi\, 2n\pi], \qquad (18)$$

$$Z_{s1} \doteq Z_0 \doteq 73.13 + j42.5. \quad n = 1. \qquad (19)$$

The simplicity and relative accuracy of zeroth-order formulas for Z_0, Z_{s1}, and Z_{12} for antennas for which $\beta_0 h = n\pi/2$, n odd, may be ascribed to the important fact that for these electrical half-lengths the distributions of current are very nearly sinusoidal, whether the antenna is isolated or closely coupled to a second, identical antenna. This is shown in Fig. 6.1. For other lengths, particularly for those near $\beta_0 h = n\pi/2$, n even, the departure from the simple sinusoidal form is considerable, and the actual distribution of current is determined in appreciable measure by the current in a coupled antenna, as is shown in Fig. 6.2.

Since the zeroth-order formulas for Z_{s1} and Z_{12} are quantitatively sufficiently accurate to be of practical value only for very thin antennas ($\Omega > 15$) of half-length near $\lambda_0/4$, it is necessary to evaluate impedances of higher order for antennas of arbitrary length and at least moderate thickness ($\Omega \geq 10$). Although the first-order formula (7.7) for symmetric and antisymmetric first-order impedances is far from simple, extensive numerical computations have been made to determine first-order values of Z_0^s, Z_0^a and from these of Z_{s1} and Z_{12} for values of $\beta_0 h$ from 1.0 to 4.0 in steps of 0.5 with $\Omega = 10$, 12.5, 15, and 20. The evaluation of second-order symmetric and antisymmetric impedances has not been achieved, since very extensive numerical integrations are involved. Since it is shown in Chapter II that, while much more accurate than zeroth-order values, first-order impedances of isolated antennas are not in good agreement with experimental measures, whereas second-order values are, no tabulation or graphical representation of first-order values of symmetric and antisymmetric, or of self- and mutual impedances is provided. In the absence of accurate second-order values, a method is devised for determining approximate second-order self- and mutual impedances from the computed first-order values of Z_{s1} and Z_{12} and the known second-order values of Z_0. In effect, the method involves the superposition of the first-order effect of a coupled antenna on the second-order self-impedance of an isolated antenna. A description of the procedure follows.

TABLE 8.1. Zeroth-order mutual impedance; $\beta_0 h = \pi/2$.

b/λ_0	Z_{12} (ohms)	θ_{12} (deg)	R_{12} (ohms)	X_{12} (ohms)	b/λ_0	Z_{12} (ohms)	θ_{12} (deg)	R_{12} (ohms)	X_{12} (ohms)
0	84.6049	30.1895	73.1296	42.5445	1.5	12.4482	− 98.7203	− 1.8873	−12.3043
0.1	67.7543	6.3880	67.3336	7.5383	1.6	11.7011	−134.1972	− 8.1572	− 8.3891
.2	54.8570	− 20.4592	51.3967	−19.1747	1.7	11.0372	−169.7322	−10.8605	− 1.9674
.3	45.1834	− 49.6485	29.2551	−34.4336	1.8	10.4436	154.6836	− 9.4406	4.4658
.4	37.9412	− 80.5798	6.2100	−37.4296	1.9	9.9096	119.0577	− 4.8130	8.6623
.5	32.4582	−112.7120	− 12.5321	−29.9413	2.0	9.4269	83.3957	1.0842	9.3644
.6	28.2141	−145.7650	− 23.3256	−15.8729	2.1	8.9886	47.7028	6.0491	6.6485
.7	24.8639	−179.4127	− 24.8626	− 0.2548	2.2	9.5888	11.9829	8.4016	1.7832
.8	22.1858	146.4645	− 18.4922	12.2574	2.3	8.2228	− 23.7607	7.5258	− 3.3131
.9	20.0007	111.9915	− 7.4896	18.5454	2.4	7.8864	− 59.5250	3.9997	− 6.7969
1.0	18.1900	77.2592	4.0116	17.7420	2.5	7.6762	− 95.3078	− 0.7008	− 7.5437
1.1	16.6684	42.3195	12.3236	11.2233	2.6	7.2893	−131.1068	− 4.7924	− 5.4924
1.2	15.3741	7.2307	15.2518	1.9351	2.7	7.0232	−166.9204	− 6.8410	− 1.5894
1.3	14.2611	− 27.9899	12.5930	− 6.6930	2.8	6.7757	157.2529	− 6.2487	2.6199
1.4	13.2946	− 63.3133	5.9708	−11.8784	2.9	6.5450	121.4146	− 3.4114	5.5856
					3.0	6.3294	85.5656	0.4894	6.3105

b/λ_0	Z_{12} (ohms)	θ_{12} (deg)	b/λ_0	Z_{12} (ohms)	θ_{12} (deg)	b/λ_0	Z_{12} (ohms)	θ_{12} (deg)	b/λ_0	Z_{12} (ohms)	θ_{12} (deg)
3.0	6.3294	85.5656	6.5	2.9346	− 92.0554	10.0	1.9088	88.6626	13.5	1.4143	− 90.9909
3.1	6.1272	49.7134	6.6	2.8902	−128.0243	10.1	1.8899	52.6758	13.6	1.4039	−126.9837
3.2	5.9376	13.8511	6.7	2.8472	−163.9943	10.2	1.8714	16.6888	13.7	1.3936	−162.9765
3.3	5.7593	− 22.0208	6.8	2.8054	160.0348	10.3	1.8533	− 19.2968	13.8	1.3836	161.0306
3.4	5.5914	− 57.9013	6.9	2.7648	124.0630	10.4	1.8355	− 55.2860	13.9	1.3736	125.0376
3.5	5.4330	− 93.7901	7.0	2.7254	88.0905	10.5	1.8180	− 91.2737	14.0	1.3638	89.0444
3.6	5.2834	−129.6864	7.1	2.6871	52.1174	10.6	1.8009	−127.2617	14.1	1.3541	53.0512
3.7	5.1417	−165.5896	7.2	2.6498	16.1436	10.7	1.7841	−163.2500	14.2	1.3446	17.0579
3.8	5.0075	158.5009	7.3	2.6136	− 19.8310	10.8	1.7676	160.7616	14.3	1.3352	− 18.9355
3.9	4.8801	122.5858	7.4	2.5784	− 55.8062	10.9	1.7514	124.7729	14.4	1.3294	− 54.9290
4.0	4.7590	86.6658	7.5	2.5441	− 91.7822	11.0	1.7355	88.7841	14.5	1.3168	− 90.9226
4.1	4.6436	50.7482	7.6	2.5107	−127.7588	11.1	1.7199	52.7950	14.6	1.3078	−126.9163
4.2	4.5337	14.8265	7.7	2.4781	−163.7360	11.2	1.7045	16.8058	14.7	1.2989	−162.9106
4.3	4.4288	− 21.0991	7.8	2.4464	160.2862	11.3	1.6895	− 19.1837	14.8	1.2901	161.0960
4.4	4.3287	− 57.0285	7.9	2.4155	124.3077	11.4	1.6746	− 55.1733	14.9	1.2815	125.1021
4.5	4.2330	− 92.9613	8.0	2.3853	88.3287	11.5	1.6601	− 91.1631	15.0	1.2729	89.1081
4.6	4.1414	−128.8974	8.1	2.3559	52.3494	11.6	1.6458	−127.1531	15.1	1.2645	53.1139
4.7	4.0538	−164.8349	8.2	2.3272	16.3695	11.7	1.6318	−163.1432	15.2	1.2562	17.1198
4.8	3.9697	159.2214	8.3	2.2992	− 19.6108	11.8	1.6179	160.8664	15.3	1.2480	− 18.8746
4.9	3.8891	123.2100	8.4	2.2719	− 55.5917	11.9	1.6044	124.8760	15.4	1.2399	− 54.8689
5.0	3.8117	87.3293	8.5	2.2452	− 91.5729	12.0	1.5910	88.8853	15.5	1.2325	− 90.8634
5.1	3.7375	51.3822	8.6	2.2192	−127.5547	12.1	1.5778	52.8945	15.6	1.2240	−126.8579
5.2	3.6656	15.4328	8.7	2.1937	−163.5369	12.2	1.5649	16.9036	15.7	1.2162	−162.8526
5.3	3.5967	− 20.5187	8.8	2.1688	160.4805	12.3	1.5522	− 19.0875	15.8	1.2085	161.1528
5.4	3.5303	− 56.4720	8.9	2.1445	124.4970	12.4	1.5397	− 55.0788	15.9	1.2009	125.1580
5.5	3.4664	− 92.4272	9.0	2.1207	88.5142	12.5	1.5274	− 91.0701	16.0	1.1934	89.1632
5.6	3.4047	−128.3841	9.1	2.0974	52.5300	12.6	1.5152	−127.0617			
5.7	3.3452	−164.3423	9.2	2.0746	16.5465	12.7	1.5033	−163.0533			
5.8	3.2877	159.6973	9.3	2.0523	− 19.4379	12.8	1.4916	160.9549			
5.9	3.2321	123.7358	9.4	2.0305	− 55.4226	12.9	1.4800	124.0630			
6.0	3.1784	87.7731	9.5	2.0092	− 91.4076	13.0	1.4687	88.9710			
6.1	3.1265	51.8097	9.6	1.9883	−127.3930	13.1	1.4574	52.9788			
6.2	3.0762	15.8451	9.7	1.9678	−163.3787	13.2	1.4464	16.9866			
6.3	3.0275	− 20.1207	9.8	1.9478	160.6353	13.3	1.4355	− 19.0058			
6.4	2.9803	− 56.0875	9.9	1.9281	124.6479	13.4	1.4248	− 54.9983			

As pointed out in Sec. 7, the first-order values of symmetric and antisymmetric impedances and, hence, of the self- and mutual impedances determined from them using (6a, b) are not quantitatively accurate for a particular length of antenna primarily because the first-order formula (II.27.5a) for the impedance Z_0 of an isolated antenna is distorted in its $\beta_0 h$ scale as compared with the quite accurate second-order formula (II.27.5b). This is evident from Figs. II.30.1a, c, d, but is brought out particularly well in Fig. 8.2, where both first- and second-order values of $Z_0 = R_0 + jX_0$ are plotted in the complex RX-plane with $\beta_0 h$ as parameter. The second-order spiral in this figure is the same as the outer part of the $\Omega = 15$ spiral in Fig. II.30.3b. The values of $\beta_0 h$ are indicated along each spiral by points and numbers. The first-order values are in parentheses, the second-order values are not. Note that whereas the two spirals agree well in magnitude, the $\beta_0 h$ scale for the first-order curve is shifted greatly from that of the second-order curve especially in the range near antiresonance. Note particularly the two points marked (3.0) on the first-order curve and 3.0 on the second-order curve. The corresponding impedances are $(Z_0)_1 = 2250 - j474$, $(Z_0)_2 = 1870 - j1159$. On the other hand, if the first-order impedance at $\beta_0 h = (3.0)$ is associated with the second-order impedance at an adjacent point on the second-order spiral, for example, at $\beta_0 h = 2.888$, the corresponding second-order impedance is $(Z_0)_2 = 2390 - j474$. Thus, it is possible to correct the first-order impedances merely by associating each numerical value with an appropriately modified electrical half-length $\beta_0 h$. For example, in Fig. 8.2 the first-order value at $\beta_0 h = (3.0)$ is associated with the second-order value at $\beta_0 h = 2.888$; the first-order value at $\beta_0 h = (2.5)$ with the second-order value at $\beta_0 h = 2.42$, and so on. The new $\beta_0 h$ is always the second-order value at an adjacent point on the second-order impedance spiral. Evidently, any first-order value of Z_0 may be converted exactly into an adjacent second-order value by applying a scale factor near unity to $(R_0)_1$ and another scale factor to $(X_0)_1$. In order to maintain a scale factor for $(X_0)_1$ near unity, it is necessary to project horizontally from the first-order to the second-order spiral near resonance and antiresonance. For example, the point $\beta_0 h = (3.0)$ is projected to $\beta_0 h = 2.888$ instead of $\beta_0 h = 2.98$ so that the scale factor for $(X_0)_1$ is exactly unity rather than a relatively large value, while the scale factor for $(R_0)_1$ is increased only very slightly to 1.062.

The method of converting first-order self- and mutual impedances to approximate second-order self- and mutual impedances is to assume that the same shift in $\beta_0 h$ which brings $(Z_0)_1$ near $(Z_0)_2$ may be used to bring first-order self- and mutual impedances near to second-order values. Furthermore, the scale factor that then converts $(R_0)_1$ exactly into $(R_0)_2$ is used to convert first-order self- and mutual resistances into approximate second-order values. Similarly, the scale factor used to convert $(X_0)_1$ exactly into $(X_0)_2$ is used to convert first-order self- and mutual reactances into approximate second-order values. This procedure may be illustrated graphically for the point $\beta_0 h = (3.0)$ on the first-order curve in Fig. 8.2. This point in the complex RX-plane is the first-order impedance of an antenna of electrical half-length 3.0 radian when infinitely far ($b = \infty$) from a second, identical and parallel antenna. As the second antenna is brought nearer, the point representing the self-impedance Z_{s1} traces a spiral about the point $(R_0)_1$, $(X_0)_1$ for infinite separation. For great electrical distances $\beta_0 b$ this spiral is minute. Beginning with the point $(R_{s1})_1$, $(X_{s1})_1$ for $\beta_0 b = 6.3$ the spiral is shown in broken line in the insert in Fig. 8.3. As $\beta_0 b$ is made smaller, the spiral increases until from $\beta_0 b = 1.5$ to $\beta_0 b = 0.3$ it is, in effect, a great circle. This is shown in Fig. 8.3 in broken line. By multiplying $(R_{s1})_1$ by 1.062, $(X_{s1})_1$ by 1.0, the first-order spiral in broken line in Fig. 8.3 is converted into the spiral in solid line. This occupies the same position relative to the true *second*-order impedance $(Z_0)_2$ with infinite separation ($b = \infty$) for $\beta_0 h = 2.888$ as does the spiral in broken line relative to the first-order impedance $(Z_0)_1$ for $\beta_0 h = (3.0)$. It appears reasonable to assume that the self-impedance given by the solid-line spiral in Fig. 8.3 is a good approximation of the second-order self-impedance for $0 < b < \infty$. It is this impedance which is called the *approximate second-order self-impedance*.

At infinite separation the mutual impedance, Z_{12}, is zero. As the two identical antennas are brought closer together, the mutual impedance traces a small spiral about the point zero in the RX-plane. Beginning with $\beta_0 b = 6.3$ the first-order spiral for $\beta_0 h = (3.0)$ is shown in broken line in Fig. 8.4 and enlarged in the insert. Beginning with $\beta_0 b = 1.5$ this spiral also becomes a great circle as shown. By multiplying $(R_{12})_1$ by 1.062 and $(X_{12})_1$ by 1.0,

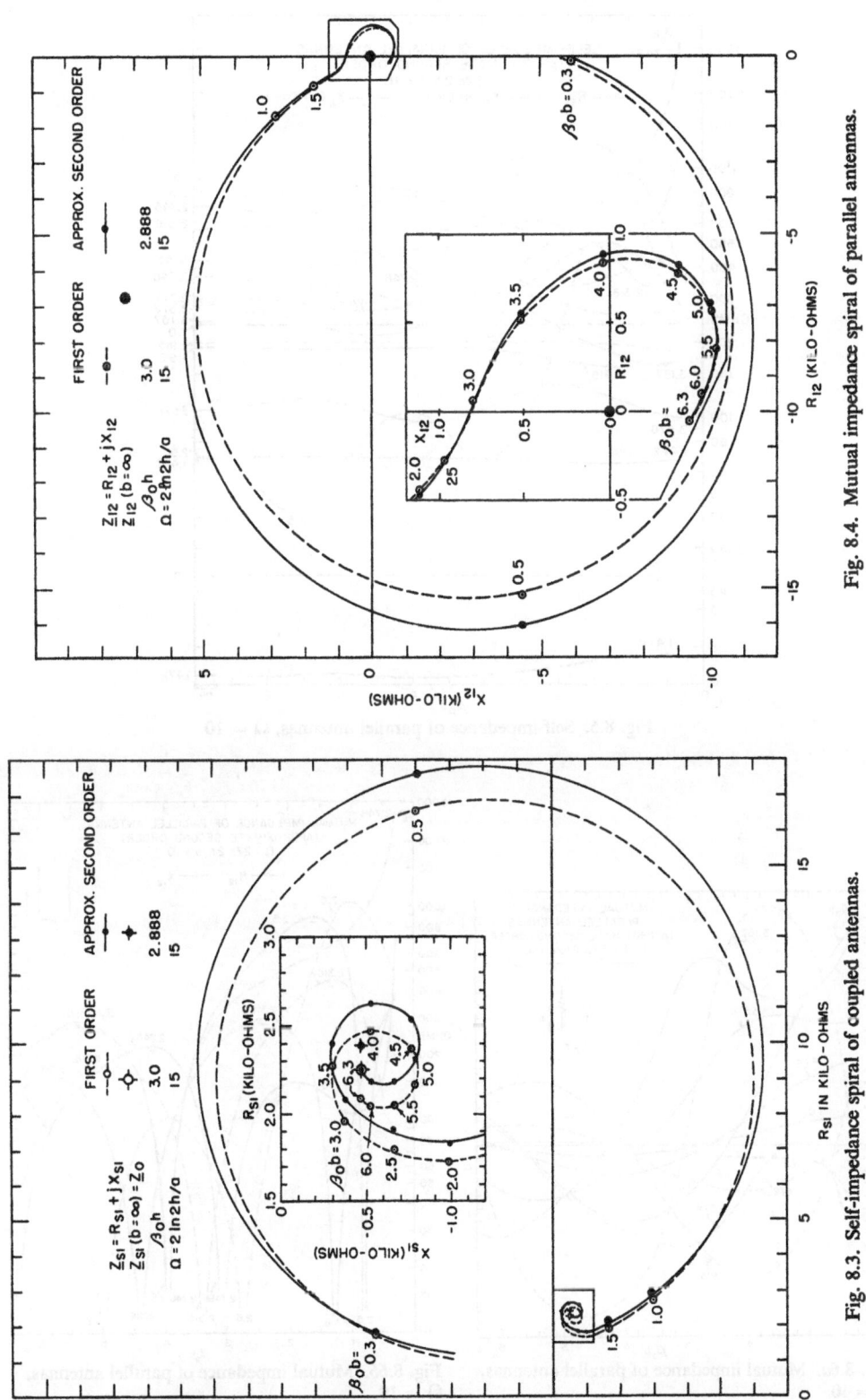

Fig. 8.3. Self-impedance spiral of coupled antennas.

Fig. 8.4. Mutual impedance spiral of parallel antennas.

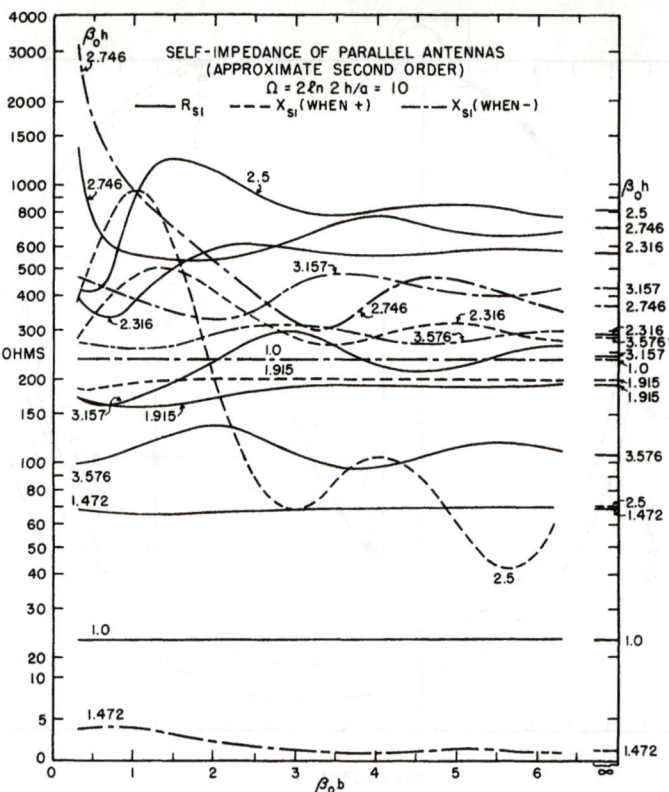

Fig. 8.5. Self-impedance of parallel antennas, $\Omega = 10$.

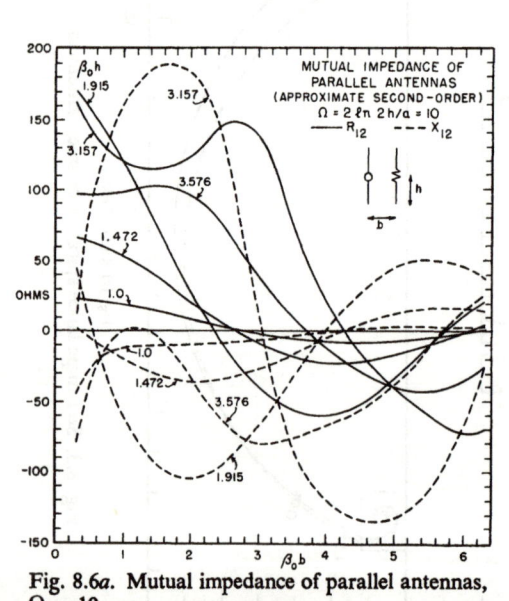

Fig. 8.6a. Mutual impedance of parallel antennas, $\Omega = 10$.

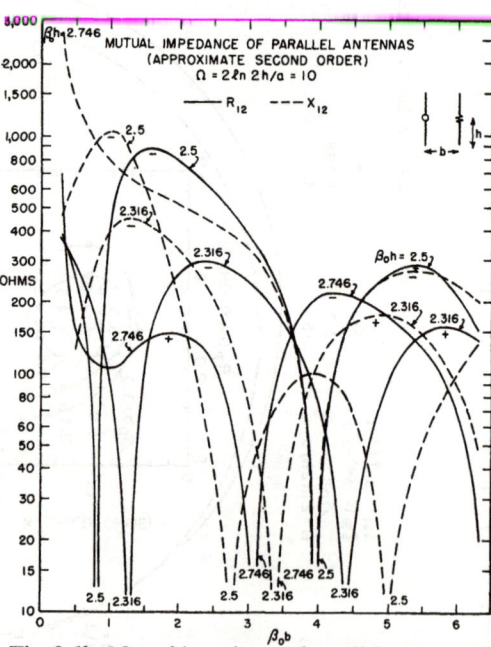

Fig. 8.6b. Mutual impedance of parallel antennas, $\Omega = 10$.

[III.8] THEORY OF LINEAR ANTENNAS 293

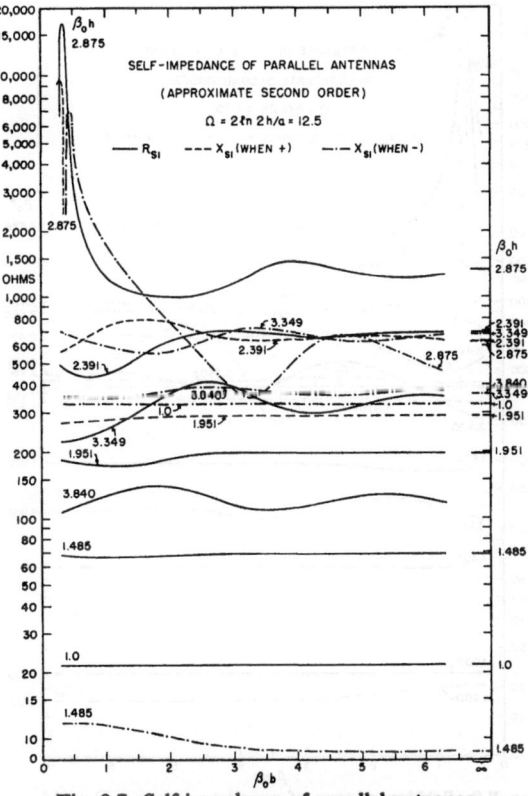

Fig. 8.7. Self-impedance of parallel antennas, $\Omega = 12.5$.

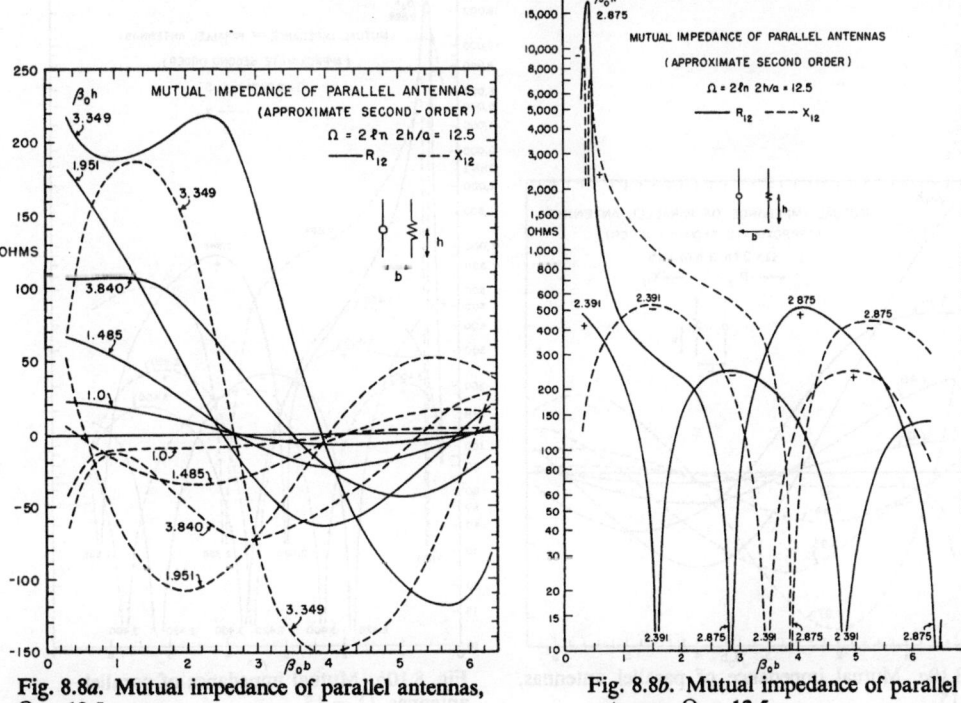

Fig. 8.8a. Mutual impedance of parallel antennas, $\Omega = 12.5$.

Fig. 8.8b. Mutual impedance of parallel antennas, $\Omega = 12.5$.

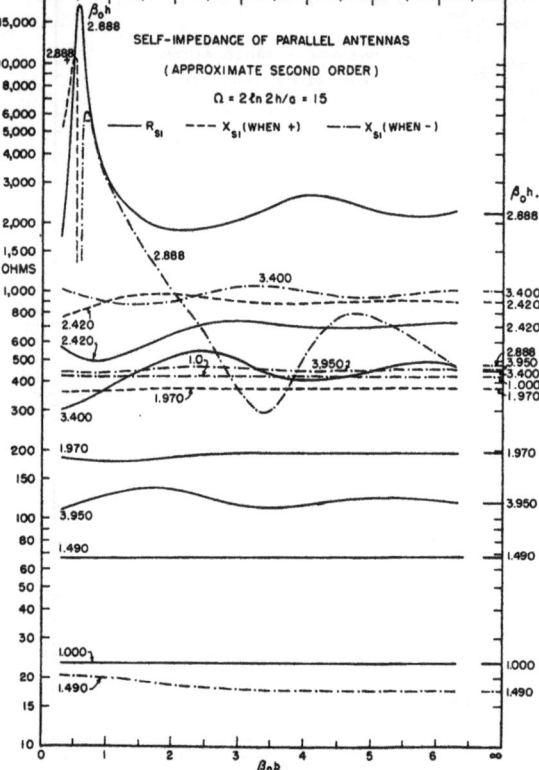

Fig. 8.9. Self-impedance of parallel antennas, $\Omega = 15$.

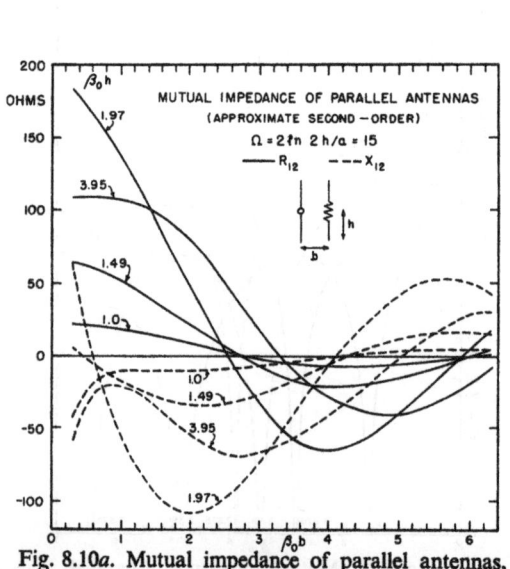

Fig. 8.10a. Mutual impedance of parallel antennas, $\Omega = 15$.

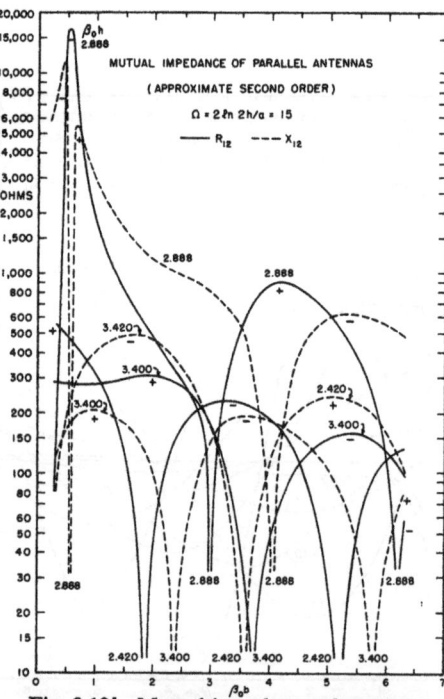

Fig. 8.10b. Mutual impedance of parallel antennas, $\Omega = 15$.

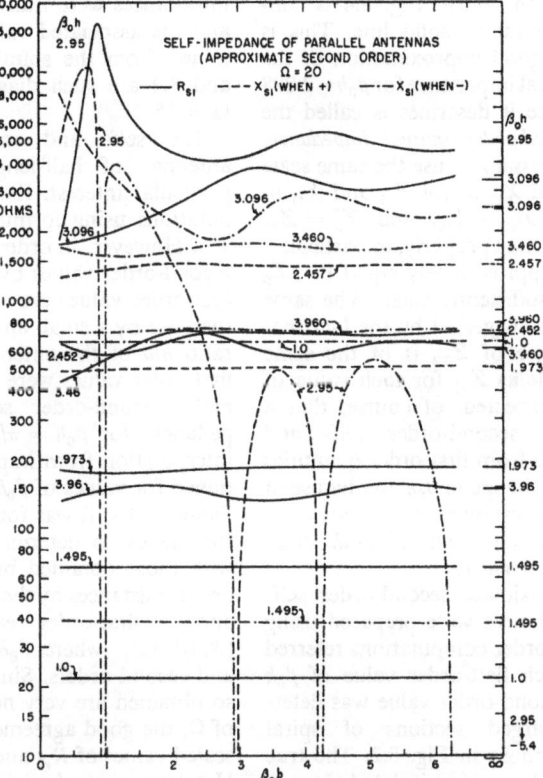

Fig. 8.11. Self-impedance of parallel antennas, $\Omega = 20$.

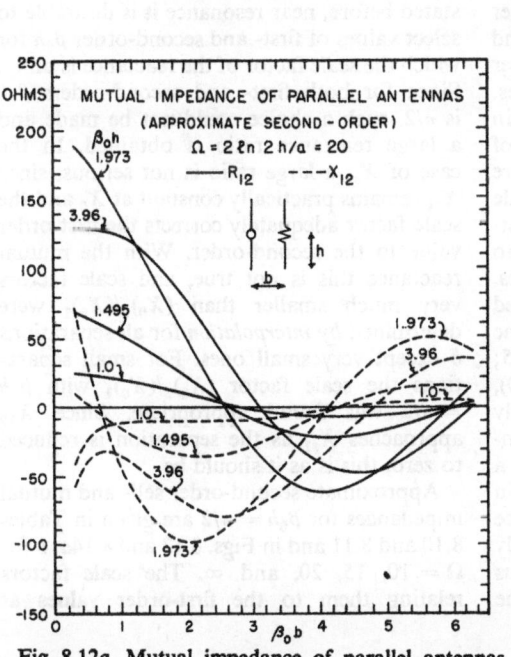

Fig. 8.12a. Mutual impedance of parallel antennas, $\Omega = 20$.

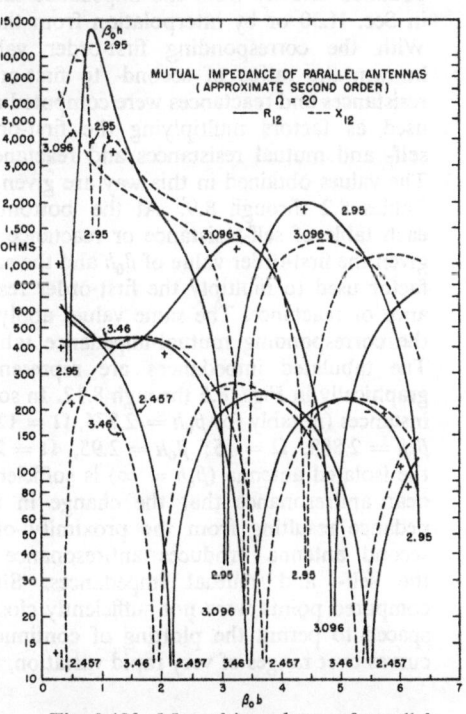

Fig. 8.12b. Mutual impedance of parallel antennas, $\Omega = 20$.

the spiral in broken line in Fig. 8.4 is converted into the spiral in solid line. This is assumed to be a good approximation of the second-order mutual impedance for $\beta_0 h = 2.888$ and the impedance it describes is called the *approximate second-order mutual impedance*. Note that it is necessary to use the same scale factors for R_{12} and X_{12} as for R_{s1} and X_{s1} in order that $Z_0^s = Z_{12} + Z_{s1}$ and $Z_0^a = Z_{s1} - Z_{12}$ are transformed properly, in particular, that Z_0^s may be approximately equal to $2Z_0$ when b becomes sufficiently small. The same scale factor also is suggested by the fact that the maximum value of Z_{12} is of the same order of magnitude as Z_{s1} for each value of $\beta_0 h$. It is to be expected, of course, that a representation of second-order self- and mutual impedances from first-order quantities by an appropriate change in $\beta_0 h$ and by use of scale factors for resistance and reactance is quantitatively accurate *only if both scale factors are quite near unity*.

Tables of approximate second-order self- and mutual impedances were prepared using the extensive first-order computations referred to above. For each first-order value of $\beta_0 h$ an appropriate second-order value was determined from enlarged sections of spiral diagrams such as those in Fig. 8.2. The true second-order impedance of an isolated antenna with this electrical half-length was then obtained directly from the impedance tables in Sec. II.30 or by interpolation from them. With the corresponding first-order values known, the ratios of second- to first-order resistances and reactances were computed and used as factors multiplying the first-order self- and mutual resistances and reactances. The values obtained in this way are given in Tables 8.2 through 8.9. At the bottom of each table of self-resistance or reactance are given the first-order value of $\beta_0 h$ and the scale factor used to multiply the first-order resistance or reactance. The same values apply to the corresponding mutual-impedance tables. The tabulated impedances are represented graphically in Figs. 8.5 through 8.12. In some instances (notably for $\beta_0 h = 2.875$, $\Omega = 12.5$; $\beta_0 h = 2.888$, $\Omega = 15$; $\beta_0 h = 2.95$, $\Omega = 20$), the isolated antenna ($\beta_0 b = \infty$) is sufficiently near antiresonance that the change in impedance resulting from the proximity of a second antenna produces antiresonance in the self- and mutual impedances. Since computed points were not sufficiently closely spaced to permit the plotting of continuous curves over ranges of very rapid variation, the impedances were transferred to the RX-plane and the associated values of R and X determined from the spiral diagram. Figures 8.3 and 8.4 are such diagrams for $\beta_0 h = 2.888$, $\Omega = 15$.

The self- and mutual impedances of antennas of half-length $\beta_0 h = \pi/2$ are of particular interest. Extensive first-order computations using formula (7.9) were made by Tai. However, in order to obtain approximate second-order values by the method described, first-order values are required for $\beta_0 h$ greater than $\pi/2$ by a small amount depending on the ratio h/a of the antenna. Since the required first-order values were not available, approximate second-order self- and mutual impedances for $\beta_0 h = \pi/2$ were determined by interpolation from impedances already determined for values of $\beta_0 h$ near 1.0, 1.5, and 2.0. Significantly, it was found that except for X_{12} the values so determined agreed completely with those obtained by multiplying the first-order resistances by the scale factor $(R_0)_2/(R_0)_1$ and the first-order reactances by the factor $(X_0)_2/(X_0)_1$, where $\beta_0 h = \pi/2$ for both first- and second-orders. Since the resistance ratios so obtained are very near unity for all values of Ω, the good agreement in interpolated and scaled values of R_{s1} and R_{12} is to be expected. However, since $h = \lambda_0/4$ is a length not far from resonance, and the first- and second-order values of X_0 differ considerably, the reactance ratios are not very near unity, especially with the smaller values of h/a. As stated before, near resonance it is desirable to select values of first- and second-order $\beta_0 h$ for which the scale factor of the reactance is unity. Since for both first- and second-orders $\beta_0 h$ is $\pi/2$, such a choice could not be made and a large reactance ratio is obtained. In the case of X_{s1} a large ratio is not serious, since X_{s1} remains practically constant at X_0 and the scale factor adequately corrects the first-order value to the second-order. With the mutual reactance this is not true, and scale factors very much smaller than $(X_0)_2/(X_0)_1$ were determined *by interpolation* for all separations b except very small ones. For small separations the scale factor $(X_0)_2/(X_0)_1$ with $\beta_0 h = \pi/2$ was found appropriate. Since X_{12} approaches X_{s1} as the separation is reduced to zero, this is as it should be.

Approximate second-order self- and mutual impedances for $\beta_0 h = \pi/2$ are given in Tables 8.10 and 8.11 and in Figs. 8.13 and 8.14a, b for $\Omega = 10$, 15, 20, and ∞. The scale factors relating them to the first-order values at

TABLE 8.2. Approximate second-order self-impedance, $Z_{s1} = R_{s1} + jX_{s1}$; $\Omega = 10$; $h/a \doteq 75.2$.

R_{s1} (ohms)

$\beta_0 h$ \ $\beta_0 b$	1.0	1.472	1.915	2.316	2.5	2.746	3.157	3.576	b/λ_0
0.3	23.4	68.0	175.0	392	439	1392	172.9	99.9	0.048
.5	23.4	67.3	168.6	351	419	732	164.4	101.8	.080
1.0	23.4	66.5	160.8	379	885	564	171.2	114.0	.159
1.5	23.5	66.7	164.1	504	1249	538	193.9	129.3	.239
2.0	23.5	67.6	173.4	598	1120	540	231.8	136.8	.318
2.5	23.5	68.6	183.5	613	928	568	281.4	126.4	.398
3.0	23.5	69.2	190.2	591	812	635	298.9	108.3	.477
3.5	23.5	69.5	192.0	567	778	734	260.6	97.8	.557
4.0	23.6	69.6	190.9	556	807	776	226.5	97.4	.637
4.5	23.6	69.6	189.3	561	849	734	218.0	104.6	.716
5.0	23.7	69.7	189.0	574	846	679	224.2	114.2	.796
5.5	23.7	69.8	190.4	585	823	655	242.1	119.3	.875
6.0	23.6	69.9	191.7	584	783	660	261.3	115.4	.955
6.3	23.6	69.9	192.4	578	764	677	264.0	110.8	1.003
∞	23.7	69.6	192.0	570	820	700	244.0	108.0	∞
$\beta_0 h$ (first-order)	1.0	1.5	2.0	2.5	2.7	3.0	3.5	4.0	
factor	1.017	1.079	1.100	0.974	1.0	1.110	0.968	0.982	

X_{s1} (ohms)

$\beta_0 h$ \ $\beta_0 b$	1.0	1.472	1.915	2.316	2.5	2.746	3.157	3.576	b/λ_0
0.3	−237	−4.0	189.4	288	380	−3180	−464	−276	0.048
.5	−237	−4.2	188.6	346	561	−1680	−446	−270	.080
1.0	−236	−4.1	193.7	472	961	− 966	−396	−261	.159
1.5	−236	−3.3	198.7	497	577	− 709	−354	−267	.239
2.0	−236	−2.6	201.0	420	208	− 537	−330	−290	.318
2.5	−236	−2.0	201.1	328	88.6	− 412	−350	−313	.398
3.0	−235	−1.4	200.2	277	68.3	− 322	−428	−314	.477
3.5	−235	−1.1	199.4	269	86.7	− 317	−475	−298	.557
4.0	−235	−1.0	199.0	289	104.0	− 402	−463	−281	.637
4.5	−235	−1.1	199.1	308	92.6	− 463	−434	−271	.716
5.0	−235	−1.4	199.4	320	60.8	− 458	−405	−271	.796
5.5	−235	−1.3	199.3	299	43.0	− 416	−395	−283	.875
6.0	−235	−1.1	199.5	281	47.9	− 371	−409	−294	.955
6.3	−235	−1.0	199.5	273	60.7	− 353	−427	−296	1.003
∞	−235	−1.2	200.0	290	71.4	−370	−430	−284	∞
$\beta_0 h$ (first-order)	1.0	1.5	2.0	2.5	2.7	3.0	3.5	4.0	
factor	0.951	1.0	0.912	0.879	1.0	1.180	1.254	1.42	

TABLE 8.3. Approximate second-order mutual impedance, $Z_{12} = R_{12}+jX_{12}$; $\Omega = 10$; $h/a \doteq 75.2$.

R_{12} (ohms)

$\beta_0 b$ \ $\beta_0 h$	1.0	1.472	1.915	2.316	2.5	2.746	3.157	3.576	b/λ_0
0.3	23.0	66.7	171.3	376	393	−688	163.0	97.9	0.048
.5	22.2	63.8	157.9	297	254	−145.4	143.6	96.7	.080
1.0	19.0	52.7	117.4	98.4	−409	−106.2	120.9	99.0	.159
1.5	14.2	38.0	72.1	−122.0	−878	−136.6	114.7	102.6	.239
2.0	8.6	21.3	27.2	−266.0	−786	−145.2	124.4	97.2	.318
2.5	3.0	5.1	−11.7	−293.0	−586	−110.6	147.1	72.7	.398
3.0	−1.9	−8.6	−40.3	−246.0	−387	−15.9	139.3	38.6	.477
3.5	−5.4	−18.0	−56.2	−165.3	−185	123.8	77.1	10.0	.557
4.0	−7.2	−22.3	−59.1	−72.8	19.2	211.9	19.4	−11.4	.637
4.5	−7.2	−21.5	−50.0	18.7	184	207.9	−14.9	−28.0	.716
5.0	−5.8	−16.4	−32.2	95.3	259	169.9	−41.1	−40.2	.796
5.5	−3.4	−8.8	−9.6	144.1	266	128.0	−60.1	−43.2	.875
6.0	−0.6	−0.4	10.6	152.4	208	67.3	−72.7	−33.8	.955
6.3	0.9	4.2	21.0	138.3	150	21.0	−69.2	−24.3	1.003

X_{12} (ohms)

$\beta_0 b$ \ $\beta_0 h$	1.0	1.472	1.915	2.316	2.5	2.746	3.157	3.576	h/λ_0
0.3	−42.9	2.7	45.3	−129.1	−476	2775	12.7	−79.2	0.048
.5	−24.9	−3.9	1.9	−237	−662	1367	88.9	−35.3	.080
1.0	−12.2	−19.8	−62.4	−402	−1019	788	161.7	0.6	.159
1.5	−10.0	−30.8	−95.4	−429	−604	615	187.1	−4.1	.239
2.0	−9.3	−35.3	−104.3	−334	−196.6	514	184.4	−33.0	.318
2.5	−8.1	−33.4	−93.0	−200	−39.4	425	130.0	−65.7	.398
3.0	−6.1	−26.4	−66.8	−77.5	37.6	334	11.5	−78.5	477
3.5	−3.7	−16.1	−33.0	28.5	83.4	186.5	−83.0	−75.8	.557
4.0	−1.1	−4.9	0.7	113.3	99.0	−23.8	−118.8	−67.1	.637
4.5	1.2	5.2	28.2	164.9	70.9	−184.4	−133.4	−56.5	.716
5.0	2.8	12.2	45.7	173.4	10.2	−255.0	−132.1	−38.0	.796
5.5	3.7	16.0	50.0	144.2	−48.9	−264.3	−113.0	−10.9	.875
6.0	3.6	15.8	45.1	87.9	−99.9	−235.6	−65.7	15.4	.955
6.3	3.2	14.0	37.0	47.7	−133.2	−200.8	−26.6	26.1	1.003

TABLE 8.4. Approximate second-order self-impedance, $Z_{s1} = R_{s1}+jX_{s1}$; $\Omega = 12.5$; $h/a \doteq 259$.

R_{s1} (ohms)

$\beta_0 b$ \ $\beta_0 h$	1.0	1.485	1.951	2.391	2.875	3.349	3.840	b/λ_0
0.3	23.3	68.0	185.9	497	6719	229	109.2	0.048
.5	23.3	67.5	180.2	451	3224	228	112.6	.080
1.0	23.4	67.0	174.1	444	1214	256	126.1	.159
1.5	23.4	67.2	176.8	533	1043	304	138.2	.239
2.0	23.4	67.7	184.3	638	1008	363	139.6	.318
2.5	23.4	68.3	192.0	696	1045	407	128.4	.398
3.0	23.4	68.7	197.0	703	1165	388	115.0	.477
3.5	23.4	69.0	198.2	682	1365	335	109.2	.557
4.0	23.4	69.1	197.2	661	1431	306	111.7	.637
4.5	23.5	69.1	196.1	657	1370	307	119.0	.716
5.0	23.5	69.2	196.0	668	1266	326	125.8	.796
5.5	23.5	69.2	197.1	684	1217	352	127.0	.875
6.0	23.5	69.2	198.0	693	1231	361	122.2	.955
6.3	23.5	69.2	198.5	692	1264	356	118.6	1.003
∞	23.4	68.6	197.0	687	1325	332	117.0	∞
$\beta_0 h$ (first-order)	1.0	1.5	2.0	2.5	3.0	3.5	4.0	
factor	1.009	1.064	1.107	1.0	1.035	0.902	0.959	

X_{s1} (ohms)

$\beta_0 b$ \ $\beta_0 h$	1.0	1.485	1.951	2.391	2.875	3.349	3.840	b/λ_0
0.3	−332	−11.7	276	562	9396	−698	−358	0.048
.5	−332	−11.8	275	608	−4757	−666	−354	.080
1.0	−331	−11.4	279	729	−1739	−601	−351	.159
1.5	−331	−10.8	285	790	−1109	−564	−360	.239
2.0	−331	−10.1	289	764	− 772	−564	−377	.318
2.5	−331	− 9.5	291	695	− 536	−623	−388	.398
3.0	−330	− 9.0	291	642	− 380	−700	−385	.477
3.5	−330	− 8.7	290	628	− 373	−718	−373	.557
4.0	−330	− 8.7	290	641	− 537	−688	−362	.637
4.5	−330	− 8.7	289	662	− 664	−652	−358	.716
5.0	−330	− 8.7	290	670	− 655	−632	−362	.796
5.5	−330	− 8.7	290	661	− 578	−637	−370	.875
6.0	−330	− 8.7	291	645	− 495	−663	−375	.955
6.3	−330	− 8.6	291	637	− 463	−679	−374	1.003
∞	−330	−8.7	290	632	−625	−678	−367.4	∞
$\beta_0 h$ (first-order)	1.0	1.5	2.0	2.5	3.0	3.5	4.0	
factor	0.969	1.0	0.963	1.017	1.074	1.064	1.130	

TABLE 8.5. Approximate second-order mutual impedance, $Z_{12} = R_{12}+jX_{12}$; $\Omega = 12.5$; $h/a \doteq 259$.

R_{12} (ohms)

$\beta_0 b$ \ $\beta_0 h$	1.0	1.485	1.951	2.391	2.875	3.349	3.840	b/λ_0
0.3	22.9	66.8	182.2	482	−5646	217	106.8	0.048
.5	22.2	64.0	169.6	406	−2292	200	106.6	.080
1.0	18.9	53.5	132.0	223	− 429	189.2	107.8	.159
1.5	14.1	38.9	87.1	21.2	− 303	195.8	104.8	.239
2.0	8.5	22.1	40.2	−148.7	− 234	213	89.4	.318
2.5	2.8	5.6	− 2.6	−236	− 122.3	216	59.7	.398
3.0	−2.0	− 8.3	−36.4	−245	72.3	155.7	26.0	.477
3.5	−5.5	−18.0	−57.1	−205	345	60.0	− 1.9	.557
4.0	−7.3	−22.4	−63.6	−137.0	506	− 13.0	−22.6	.637
4.5	−7.4	−21.7	−56.4	− 54.2	446	− 59.6	−36.9	.716
5.0	−5.8	−17.8	−38.9	30.0	334	− 90.9	−43.3	.796
5.5	−3.4	− 9.1	−14.9	97.5	214	−107.3	−39.2	.875
6.0	−0.6	− 0.6	7.2	134.0	78.5	−102.1	−25.6	.955
6.3	1.0	4.1	19.2	139.4	− 14.2	− 82.9	−15.3	1.003

X_{12} (ohms)

$\beta_0 b$ \ $\beta_0 h$	1.0	1.485	1.951	2.391	2.875	3.349	3.840	b/λ_0
0.3	−41.7	4.7	54.2	−127.7	−9895	68.9	−65.0	0.048
.5	−23.8	− 1.8	9.0	−260	4369	133.6	−32.7	.080
1.0	−11.6	−18.1	− 60.4	−461	1529	181.7	−13.0	.159
1.5	−10.0	−29.6	− 96.6	−530	1023	182.9	−24.9	.239
2.0	− 9.8	−34.6	−107.4	−468	790	144.0	−48.9	.318
2.5	− 8.9	−33.3	− 96.9	−321	641	44.4	−67.6	.398
3.0	− 7.0	−26.7	− 71.0	−148.1	508	− 79.3	−71.0	.477
3.5	− 4.5	−16.8	− 36.8	13.7	288	−139.2	−63.5	.557
4.0	− 1.6	− 5.7	− 2.2	143.5	− 54.5	−149.5	−51.1	.637
4.5	1.0	4.5	26.9	226	−322	−141.7	−35.0	.716
5.0	2.9	12.1	45.9	250	−424	−120.2	−14.6	.796
5.5	4.0	16.0	51.8	218	−426	− 78.5	7.1	.875
6.0	4.0	15.9	47.3	142.7	−372	− 11.8	23.2	.955
6.3	3.6	14.2	39.4	86.8	−314	27.6	28.3	1.003

TABLE 8.6. Approximate second-order self-impedance, $Z_{s1} = R_{s1} + jX_{s1}$; $\Omega = 15$; $h/a \doteq 904$.

R_{s1} (ohms)

$\beta_0 b$ \ $\beta_0 h$	1.0	1.490	1.970	2.420	2.888	3.40	3.950	b/λ_0
0.3	23.4	67.4	188.5	571	1756	303	111.3	0.048
.5	23.4	66.8	183.9	527	17530	317	115.0	.080
1.0	23.4	66.6	178.7	504	2940	370	127.6	.159
1.5	23.4	66.7	181.0	565	2110	443	136.6	.239
2.0	23.4	67.0	186.6	656	1833	516	135.4	.318
2.5	23.4	67.4	192.3	719	1916	545	125.3	.398
3.0	23.4	67.7	195.9	737	2080	498	115.4	.477
3.5	23.4	67.9	196.9	722	2400	437	112.2	.557
4.0	23.4	68.0	196.3	701	2620	413	115.6	.637
4.5	23.5	68.1	195.3	692	2520	425	122.0	.716
5.0	23.5	68.1	195.2	700	2300	454	126.6	.796
5.5	23.5	68.1	196.0	716	2180	481	126.1	.875
6.0	23.5	68.1	196.7	728	2180	482	121.7	.955
6.3	23.5	68.1	197.3	729	2230	466	119.0	1.003
∞	23.4	68.0	196.0	711	2390	449.8	119.0	∞
$\beta_0 h$ (first-order)	1.0	1.5	2.0	2.5	3.0	3.5	4.0	
factor	0.991	1.028	1.098	1.008	1.062	0.945	0.902	

X_{s1} (ohms)

$\beta_0 b$ \ $\beta_0 h$	1.0	1.490	1.970	2.420	2.888	3.40	3.950	b/λ_0
0.3	−424	−20.5	362	772	5280	−1012	−445	0.048
.5	−424	−20.5	362	803	3990	− 973	−441	.080
1.0	−423	−20.2	365	899	−3000	− 900	−440	.159
1.5	−423	−19.5	371	965	−1680	− 871	−450	.239
2.0	−423	−18.9	375	972	− 992	− 899	−464	.318
2.5	−423	−18.3	378	939	− 659	− 982	−470	.398
3.0	−423	−17.9	379	902	− 376	−1051	−465	.477
3.5	−423	−17.7	378	886	− 303	−1047	−456	.557
4.0	−423	−17.7	377	891	− 530	−1005	−448	.637
4.5	−423	−17.7	377	905	− 767	− 967	−447	.716
5.0	−423	−17.7	377	915	− 782	− 954	−451	.796
5.5	−423	−17.7	378	913	− 667	− 970	−457	.875
6.0	−423	−17.6	378	904	− 531	−1002	−459	.955
6.3	−423	−17.6	379	898	− 471	−1012	−458	1.003
∞	−423	−17.6	377	900	−474	−1008	−455	∞
$\beta_0 h$ (first-order)	1.0	1.5	2.0	2.5	3.0	3.5	4.0	
factor	1.029	1.0	1.0	0.971	1.0	1.039	1.0	

TABLE 8.7. Approximate second-order mutual impedance, $Z_{12} = R_{12} + jX_{12}$; $\Omega = 15$; $h/a \doteq 904$.

R_{12} (ohms)

$\beta_0 b$ \ $\beta_0 h$	1.0	1.490	1.970	2.420	2.888	3.40	3.950	b/λ_0
0.3	22.6	66.1	184.5	557	— 88.5	288	108.8	0.048
.5	21.9	63.4	173.8	486	−16050	283	108.8	.080
1.0	18.7	53.4	138.0	312	— 1632	281	107.9	.159
1.5	13.9	38.9	94.4	122.4	— 830	295	99.9	.239
2.0	8.3	22.2	47.0	− 52.8	— 468	308	79.6	.318
2.5	2.7	5.8	2.3	−175.0	— 290	276	48.8	.398
3.0	−2.1	− 8.2	−33.7	−228.8	69.6	166	16.6	.477
3.5	−5.6	−17.9	−56.7	−226.7	543	40.0	− 9.7	.557
4.0	−7.3	−22.3	−65.1	−183.8	880	− 51.8	−28.4	.637
4.5	−7.3	−21.7	−58.9	−112.4	824	−110.6	−39.6	.716
5.0	−5.8	−17.8	−41.9	− 26.9	606	−146.0	−41.7	.796
5.5	−3.3	− 9.1	−17.6	53.6	367	−155.3	−33.9	.875
6.0	−0.6	− 0.7	5.3	111.7	105	−126.6	−18.8	.955
6.3	1.0	4.0	18.1	131.6	— 56.5	− 96.3	− 8.5	1.003

X_{12} (ohms)

$\beta_0 b$ \ $\beta_0 h$	1.0	1.490	1.970	2.420	2.888	3.40	3.950	b/λ_0
0.3	−40.6	6.2	63.3	− 80.4	−5852	121.8	−59.2	0.048
.5	−22.8	− 0.4	16.6	−214.1	−4428	177.3	−32.4	.080
1.0	−11.0	−16.8	− 56.1	−413	2795	203.0	−20.7	.159
1.5	−10.3	−28.6	− 94.3	−493	1656	174.9	−35.2	.239
2.0	−10.0	−34.1	−107.1	−460	1136	95.1	−55.6	.318
2.5	− 9.3	−33.2	− 98.4	−338	972	− 37.6	−67.7	.398
3.0	− 7.5	−27.0	− 73.5	−175.3	798	−153.4	−67.6	.477
3.5	− 4.9	−17.3	− 40.0	− 13.5	527	−191.3	−57.4	.557
4.0	− 1.9	− 6.2	− 5.1	120.4	34.5	−180.6	−42.2	.637
4.5	0.8	4.0	24.7	209.7	− 413	−151.7	−23.8	.716
5.0	3.0	11.7	44.8	244	− 599	−106.7	− 3.5	.796
5.5	4.1	15.8	52.0	222	− 618	− 40.2	14.9	.875
6.0	4.3	16.0	48.4	155.5	− 555	36.6	26.3	.955
6.3	3.9	14.4	41.1	102.1	− 483	79.1	29.0	1.003

[III.8] THEORY OF LINEAR ANTENNAS 303

TABLE 8.8. Approximate second-order self-impedance, $Z_{s1} = R_{s1} + jX_{s1}$; $\Omega = 20$; $h/a \doteq 11{,}013$.

R_{s1} (ohms)

$\beta_0 b$ \ $\beta_0 h$	1.0	1.495	1.973	2.457	2.95	3.096	3.46	3.96	b/λ_0
0.3	23.7	66.8	181.0	651	1850	1750	444	131.8	0.048
.5	23.7	66.6	178.1	615	3580	1800	467	136.2	.080
1.0	23.7	66.2	174.6	580	12600	1964	458	148.2	.159
1.5	23.7	66.2	175.3	608	5800	2232	657	154.4	.239
2.0	23.7	66.4	178.4	664	4480	2552	723	151.2	.318
2.5	23.7	66.6	181.7	713	4120	3044	713	142.2	.398
3.0	23.7	66.8	183.8	733	4160	3620	642	134.9	.477
3.5	23.7	67.0	184.3	726	4650	3750	586	133.4	.557
4.0	23.7	67.0	184.2	710	5290	3443	577	137.2	.637
4.5	23.8	67.1	183.5	701	5390	3105	600	143.2	.716
5.0	23.8	67.1	183.4	704	4980	3014	640	146.1	.796
5.5	23.8	67.1	183.9	716	4610	3132	659	144.4	.875
6.0	23.8	67.1	184.3	725	4480	3325	646	140.4	.955
6.3	23.8	67.1	184.5	727	4520	3464	629	138.5	1.003
∞	23.8	67.1	184.0	714	4943	3320	615	139.5	∞
$\beta_0 h$ (first-order)	1.0	1.5	2.0	2.5	3.0	π	3.5	4.0	
factor	1.023	1.032	1.037	1.020	1.016	1.019	0.961	0.969	

X_{s1} (ohms)

$\beta_0 b$ \ $\beta_0 h$	1.0	1.495	1.973	2.457	2.95	3.096	3.46	3.96	b/λ_0
0.3	−660	−39.6	495	1276	4865	−8190	−1707	−729	0.048
.5	−660	−39.6	495	1293	8363	−5900	−1662	−726	.080
1.0	−659	−39.1	498	1362	−3068	−3780	−1592	−727	.159
1.5	−659	−38.6	502	1425	−2747	−2780	−1583	−737	.239
2.0	−659	−38.1	507	1456	−1482	−2200	−1652	−748	.318
2.5	−659	−37.7	509	1452	−603.5	−1890	−1745	−751	.398
3.0	−659	−37.4	510	1433	76.7	−1990	−1784	−746	.477
3.5	−659	−37.2	509	1418	488.5	−2640	−1754	−736	.557
4.0	−659	−37.2	509	1415	324.6	−2980	−1704	−731	.637
4.5	−659	−37.2	509	1423	−270.3	−2870	−1672	−733	.716
5.0	−659	−37.2	509	1432	−552.3	−2570	−1673	−737	.796
5.5	−659	−37.2	509	1436	−436.0	−2360	−1702	−742	.875
6.0	−659	−37.1	510	1433	−160.9	−2290	−1733	−742	.955
6.3	−659	−37.1	510	1429	−5.81	−2350	−1741	−741	1.003
∞	−659	−37.1	510	1424	− 5.4	−2510	−1725	−736	∞
$\beta_0 h$ (first-order)	1.0	1.5	2.0	2.5	3.0	π	3.5	4.0	
factor	1.052	1.0	0.971	0.991	1.0	1.114	1.012	1.034	

TABLE 8.9. Approximate second-order mutual impedance, $Z_{12} = R_{12} + jX_{12}$; $\Omega = 20$; $h/a \doteq 11{,}013$.

R_{12} (ohms)

$\beta_0 b$ \ $\beta_0 h$	1.0	1.495	1.973	2.457	2.95	3.096	3.46	3.96	b/λ_0
0.3	23.3	65.6	177.7	639	1206	666	426	128.9	0.048
.5	22.6	63.2	168.7	577	− 762	537	424	128.8	.080
1.0	19.3	53.3	137.5	416	−10014	439	436	123.5	.159
1.5	14.3	39.0	96.4	235	− 3239	414	447	107.1	.239
2.0	8.5	22.4	50.8	54.4	− 1756	546	420	78.3	.318
2.5	2.7	5.9	6.8	− 96.8	− 966	838	314	41.9	.398
3.0	−2.3	− 8.0	−29.4	−196.4	− 157.7	1266	143.4	6.6	.477
3.5	−5.8	−17.7	−53.1	−238.0	788	1262	− 16.0	−21.6	.557
4.0	−7.6	−22.3	−62.3	−226.3	1720	777	−127.8	−39.0	.637
4.5	−7.6	−21.6	−57.9	−171.1	1973	275	−199.1	−47.5	.716
5.0	−6.0	−16.8	−42.3	− 89.7	1564	− 94.2	−223.2	−44.3	.796
5.5	−3.5	− 9.2	−19.1	0.7	999	−414	−204.4	−31.3	.875
6.0	−0.6	− 0.8	3.2	79.5	420	−653	−136.0	−12.5	.955
6.3	1.1	3.9	15.7	114.5	53.8	−754	− 79.4	0.8	1.003

X_{12} (ohms)

$\beta_0 b$ \ $\beta_0 h$	1.0	1.495	1.973	2.457	2.95	3.096	3.46	3.96	b/λ_0
0.3	−42.3	8.2	72.0	13.8	−5537	6680	197.2	−64.4	0.048
.5	−23.2	1.8	26.9	−130.5	−8829	4590	238.2	−39.6	.080
1.0	−11.3	−15.0	−43.8	−346.2	3027	2850	214.0	−34.8	.159
1.5	−10.8	−27.3	−82.6	−443.8	3093	2190	112.5	−53.2	.239
2.0	−11.3	−33.4	−97.3	−439.5	2241	1820	− 28.4	−72.2	.318
2.5	−10.8	−33.1	−91.8	−350.1	1843	1440	−179.2	−80.4	.398
3.0	− 8.9	−27.3	−71.0	−211.4	1666	713	−262.7	−74.9	.477
3.5	− 5.9	−17.9	−41.0	− 59.0	1453	− 356	−260.4	−57.1	.557
4.0	− 2.5	− 6.9	− 8.9	76.8	803	−1070	−211.9	−39.4	.637
4.5	0.7	3.4	19.6	175.6	− 198.5	−1253	−145.1	−14.7	.716
5.0	3.3	11.3	39.7	225.1	− 846	−1253	− 58.0	8.3	.796
5.5	4.7	15.7	48.1	220.9	−1076	−1040	37.0	25.6	.875
6.0	5.0	16.1	45.9	170.0	−1105	− 657	117.8	33.6	.955
6.3	4.6	14.7	39.8	123.9	−1056	− 356	147.5	33.9	1.003

TABLE 8.10. Approximate second-order self-impedance $Z_{s1} = R_{s1} + jX_{s1}$ (ohms), for $\beta_0 h = \pi/2$.*

b/λ_0	$\Omega = 10$		$\Omega = 15$		$\Omega = 20$		$\beta_0 b$
0.01	85.7	$+j41.1$	80.3	$+j40.4$	78.3	$+j40.6$	0.063
.02	84.9	$+j39.9$	79.9	$+j39.6$	78.1	$+j40.2$	.126
.03	84.4	$+j39.1$	79.7	$+j39.3$	78.0	$+j40.0$	.188
.04	84.0	$+j38.5$	79.5	$+j39.1$	77.9	$+j39.9$	.251
.05	83.6	$+j38.2$	79.3	$+j38.9$	77.8	$+j39.8$	.314
.06	83.2	$+j37.9$	79.1	$+j38.8$	77.7	$+j39.8$	.377
.07	82.9	$+j37.8$	78.9	$+j38.7$	77.6	$+j39.8$	.440
0.1	82.0	$+j37.7$	78.6	$+j38.8$	77.3	$+j39.9$	.628
.2	81.4	$+j38.7$	78.3	$+j39.8$	77.1	$+j40.8$	1.256
.3	82.0	$+j39.9$	78.4	$+j40.9$	77.1	$+j41.7$	1.885
.4	84.6	$+j41.7$	79.6	$+j42.3$	77.8	$+j42.8$	2.513
.5	85.8	$+j42.7$	80.2	$+j43.0$	77.9	$+j43.2$	3.142
.6	86.2	$+j42.9$	80.5	$+j43.2$	78.3	$+j43.3$	3.770
.7	86.1	$+j42.9$	80.4	$+j43.1$	78.3	$+j43.3$	4.398
.8	86.2	$+j42.8$	80.5	$+j43.1$	78.4	$+j43.3$	5.027
.9	86.3	$+j42.8$	80.5	$+j43.1$	78.4	$+j43.4$	5.655
1.0	86.4	$+j42.9$	80.6	$+j43.2$	78.4	$+j43.5$	6.283
∞	86.5	$+j41.7$	80.8	$+j43.4$	78.5	$+j43.6$	∞
Factor	1.169	1.385	1.083	1.226	1.055	1.166	

* $Z_{s1} = 73.1 + j42.5$ for all values of b/λ_0 when $\Omega \to \infty$.

TABLE 8.11. Approximate second-order mutual impedances, $Z_{12} = R_{12} + jX_{12}$ (ohms), for $\beta_0 h = \pi/2$.

b/λ_0	$\Omega = 10$		$\Omega = 15$		$\Omega = 20$		$\Omega = \infty$		$\beta_0 b$
0.01	85.6	$+j36.3$	80.2	$+j35.8$	78.2	$+j36.3$	73.1	$+j38.8$	0.063
.02	84.6	$+j29.6$	79.6	$+j30.5$	77.8	$+j31.5$	72.9	$+j35.1$	.126
.03	83.7	$+j23.7$	79.0	$+j25.6$	77.4	$+j27.0$	72.6	$+j31.4$	.188
.04	82.8	$+j18.3$	78.4	$+j21.0$	76.8	$+j22.7$	72.2	$+j27.7$	.251
.05	81.8	$+j13.1$	77.6	$+j16.6$	76.2	$+j18.6$	71.6	$+j24.2$	.314
.06	80.6	$+j\ 8.1$	76.7	$+j12.2$	75.3	$+j14.5$	71.0	$+j20.6$	.377
.07	79.3	$+j\ 3.4$	75.6	$+j\ 8.1$	74.4	$+j10.6$	70.2	$+j17.1$	.440
0.1	75.1	$-j\ 7.8$	72.2	$-j\ 3.0$	71.1	$-j\ 0.5$	67.3	$+j\ 7.5$	0.628
.2	55.2	$-j33.7$	54.2	$-j27.9$	53.7	$-j26.0$	51.4	$-j19.2$	1.256
.3	29.9	$-j46.3$	30.4	$-j40.7$	30.4	$-j39.3$	29.2	$-j34.4$	1.885
.4	4.9	$-j43.2$	5.8	$-j39.6$	6.1	$-j39.1$	6.2	$-j37.4$	2.513
.5	-15.2	$-j30.4$	-14.0	$-j29.5$	-13.7	$-j29.4$	-12.5	$-j29.9$	3.142
.6	-26.2	$-j12.5$	-24.8	$-j13.1$	-24.5	$-j13.8$	-23.3	$-j15.9$	3.770
.7	-27.0	$+j\ 4.9$	-26.3	$+j\ 3.0$	-25.9	$+j\ 2.2$	-24.9	$-j\ 0.3$	4.398
.8	-19.5	$+j17.1$	-19.3	$+j14.8$	-19.1	$+j14.2$	-18.5	$+j12.3$	5.027
.9	-7.2	$+j21.6$	-7.4	$+j19.7$	-7.4	$+j19.4$	-7.5	$+j18.5$	5.655
1.0	5.1	$+j18.2$	4.5	$+j17.4$	4.4	$+j17.5$	4.0	$+j17.7$	6.283
Factor	1.169		1.385*		1.083	1.226*	1.055	1.166*	
			1.108*			1.030*		1.026*	

* The upper factors for X_{12} were used in the range 0 to 0.7; the lower factors, obtained by graphical interpolation, were used in the range 0.1 to 1.0.

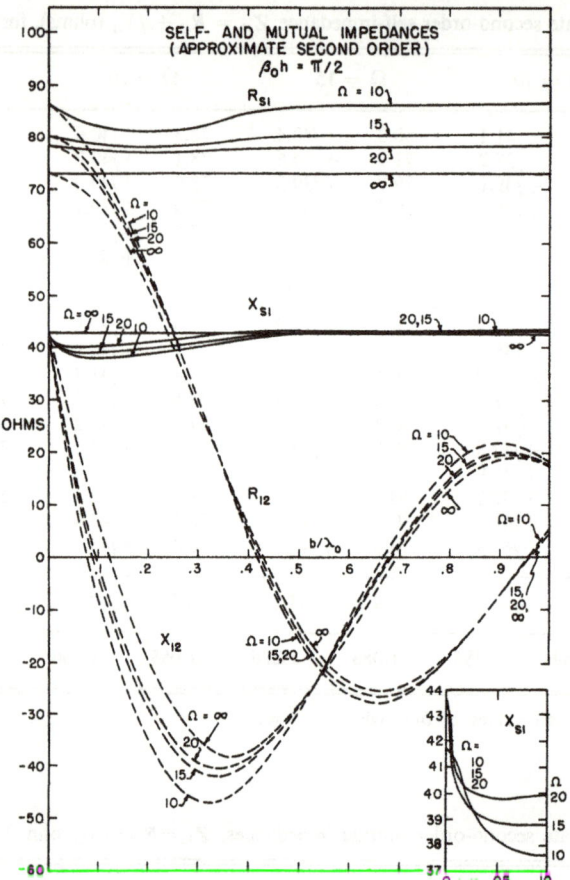

Fig. 8.13. Self- and mutual impedances; $\beta_0 h = \pi/2$.

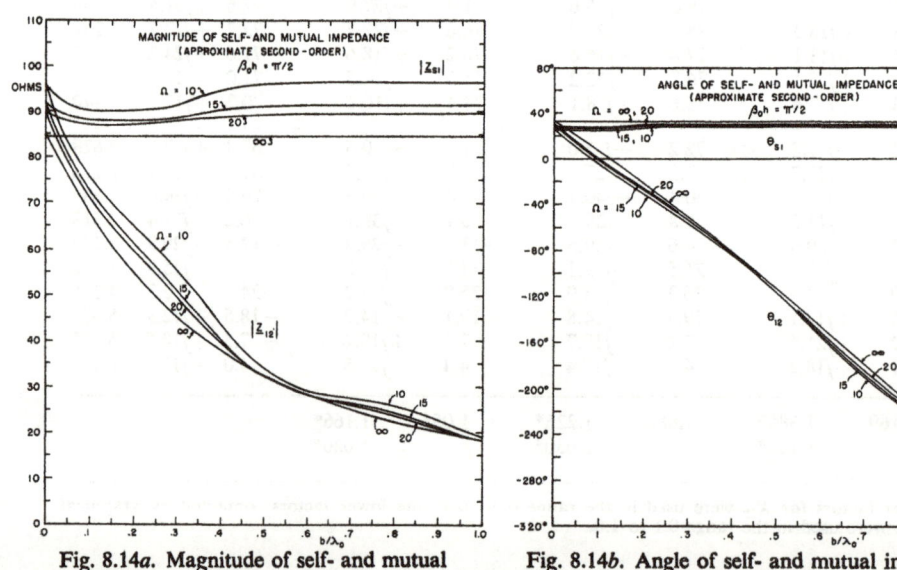

Fig. 8.14a. Magnitude of self- and mutual impedance; $\beta_0 h = \pi/2$.

Fig. 8.14b. Angle of self- and mutual impedance; $\beta_0 h = \pi/2$.

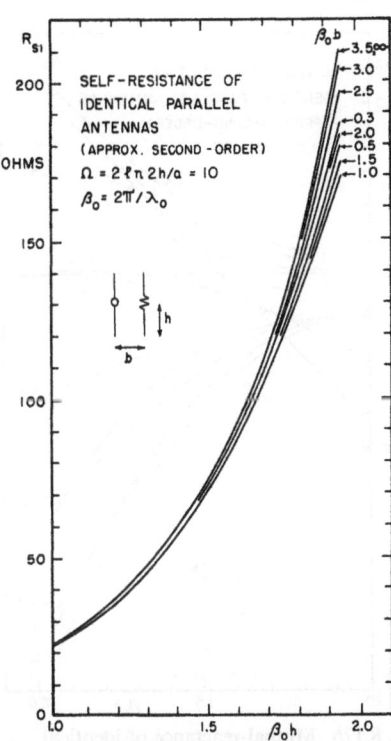

Fig. 8.15a. Self-resistance of identical parallel antennas near resonance, $\Omega = 10$.

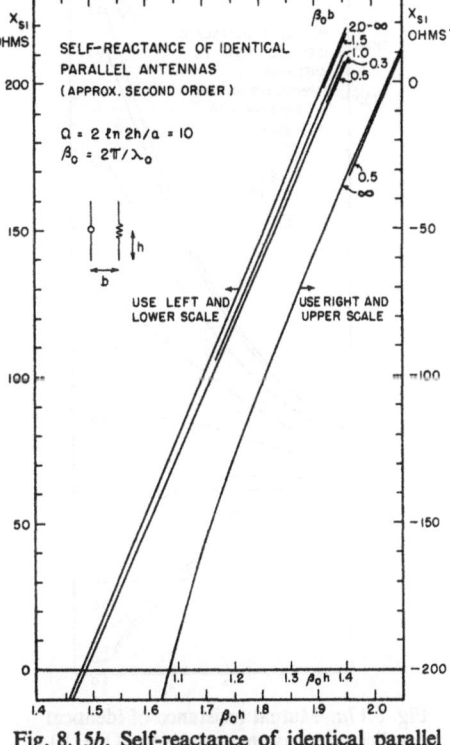

Fig. 8.15b. Self-reactance of identical parallel antennas near resonance, $\Omega = 10$.

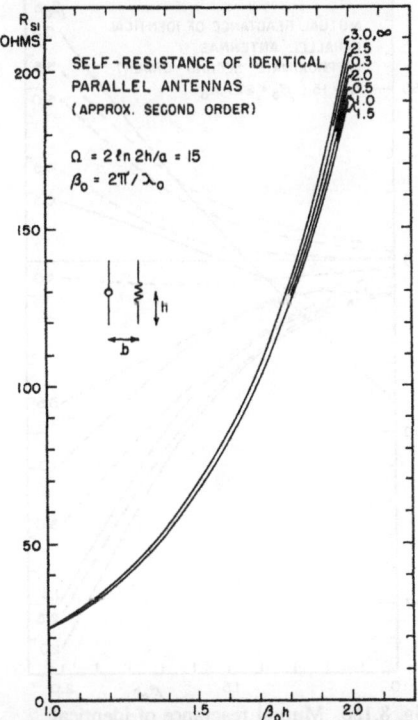

Fig. 8.16a. Self-resistance of identical parallel antennas near resonance, $\Omega = 15$.

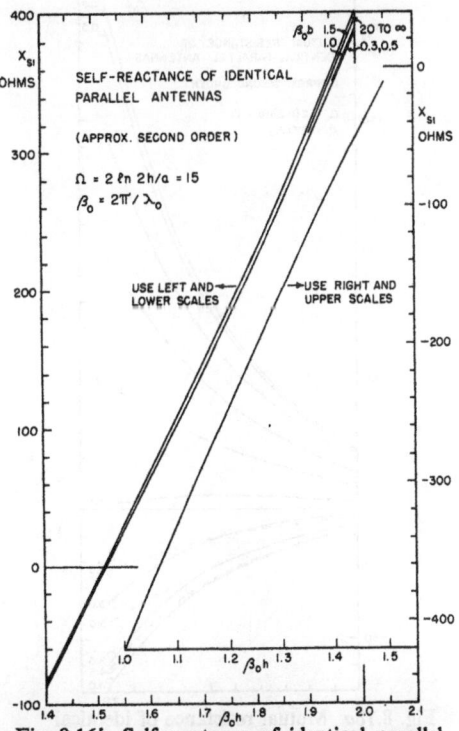

Fig. 8.16b. Self-reactance of identical parallel antennas near resonance, $\Omega = 15$.

308 THEORY OF LINEAR ANTENNAS [III.8]

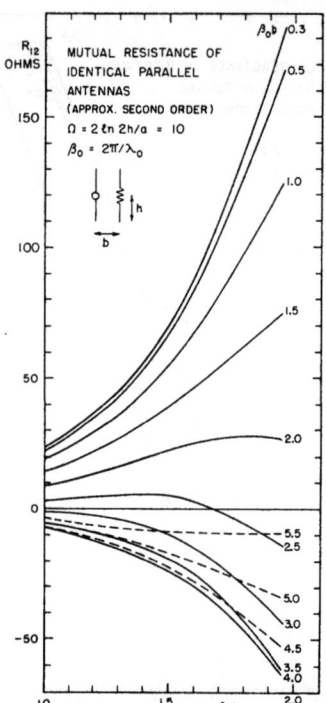

Fig. 8.17a. Mutual resistance of identical parallel antennas near resonance, $\Omega = 10$.

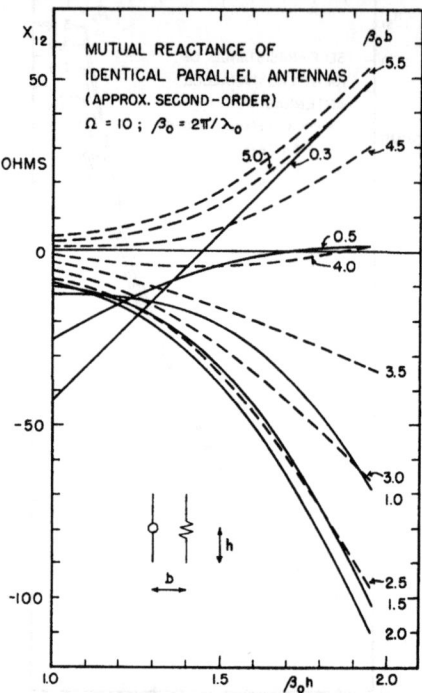

Fig. 8.17b. Mutual reactance of identical parallel antennas near resonance, $\Omega = 10$.

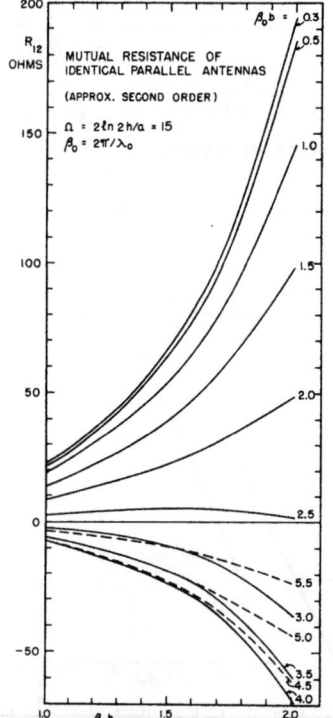

Fig. 8.18a. Mutual resistance of identical parallel antennas, $\Omega = 15$.

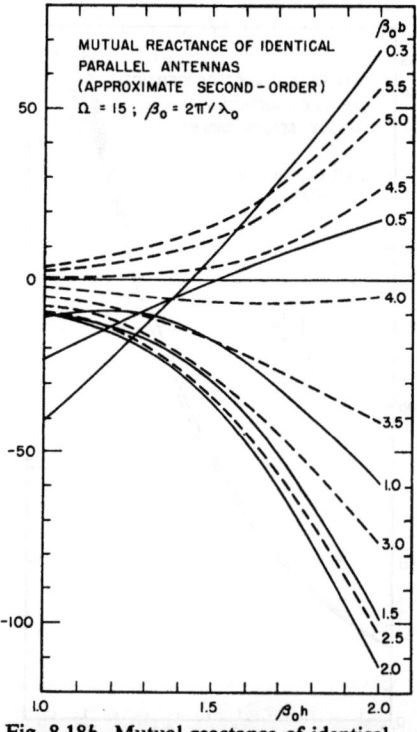

Fig. 8.18b. Mutual reactance of identical parallel antennas, $\Omega = 15$.

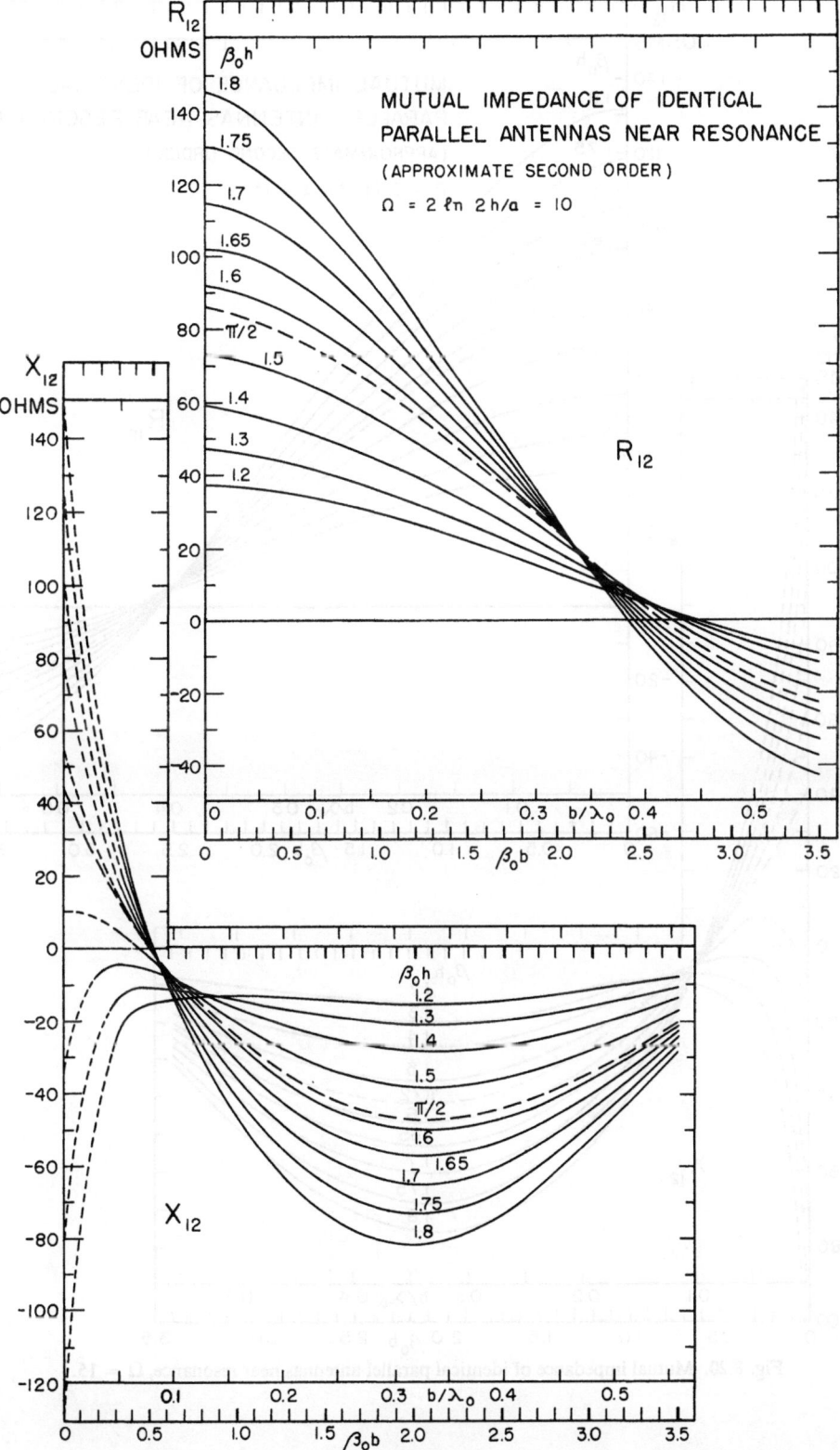

Fig. 8.19. Mutual impedance of identical parallel antennas near resonance, $\Omega = 10$.

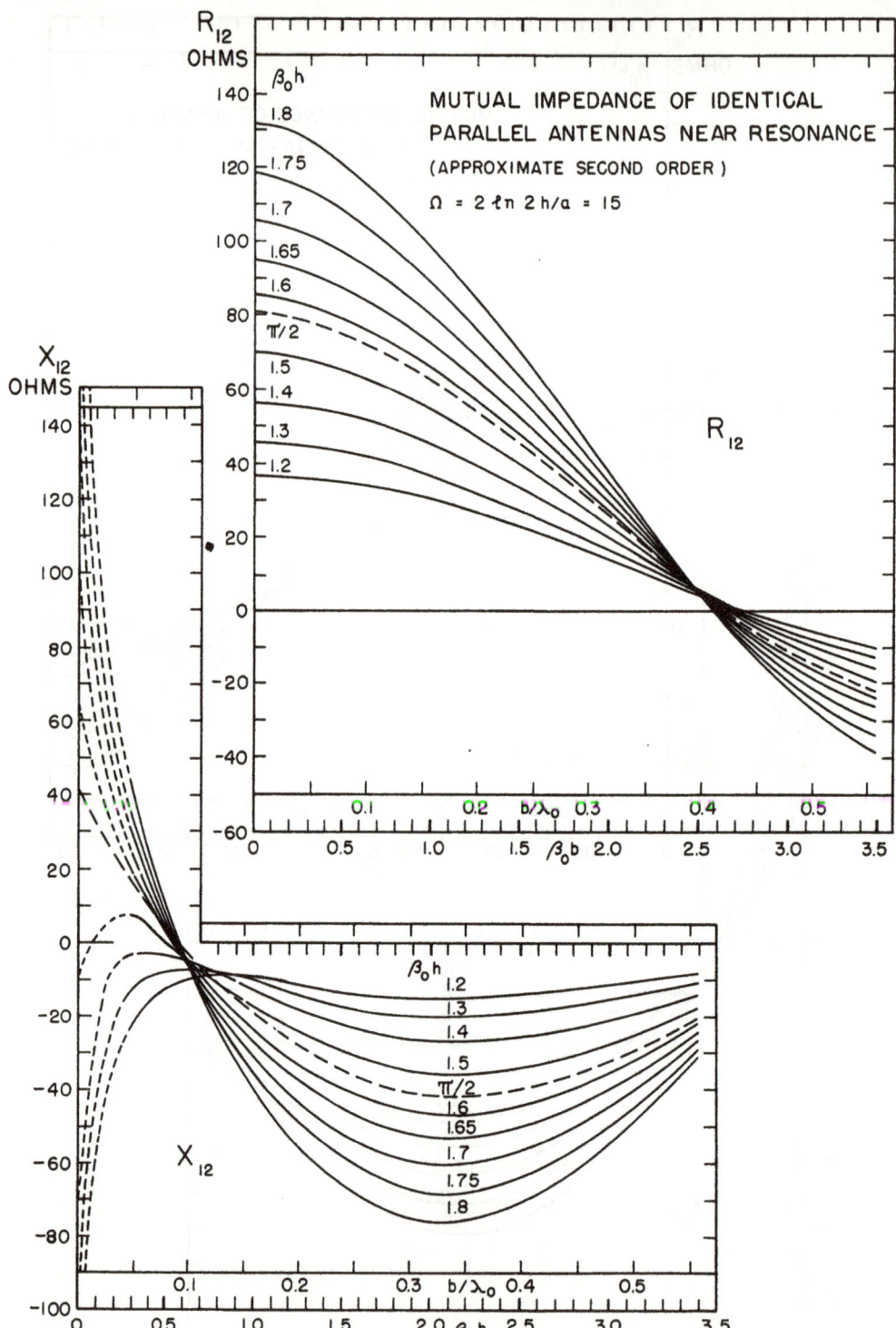

Fig. 8.20. Mutual impedance of identical parallel antennas near resonance, $\Omega = 15$.

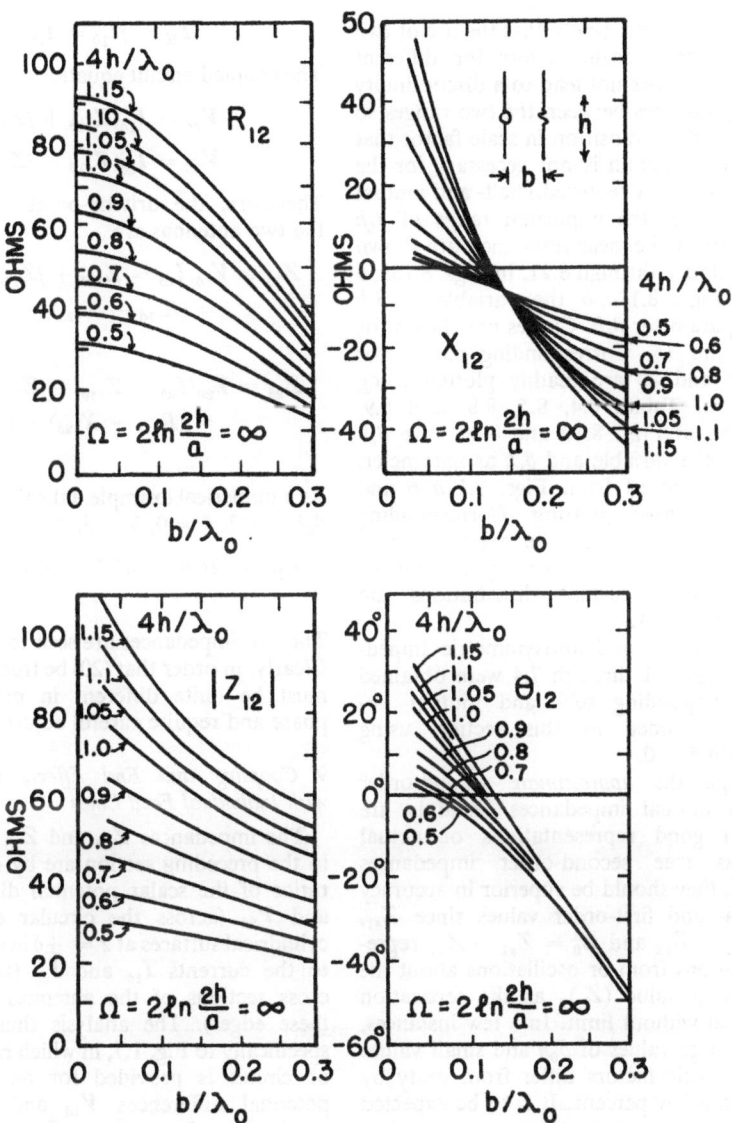

Fig. 8.21. Zeroth-order mutual impedance of identical parallel antennas (from curves by Starkey and Fitch).

$\beta_0 h = \pi/2$ are given. (Note that the use of two slightly different scale factors for different ranges of X_{12} does not lead to a discontinuity since X_{12} vanishes between the two ranges so that the gradual transition in scale factor that actually must obtain is not necessary for the degree of accuracy required.) Self- and mutual impedances for the important range of $\beta_0 h$ from 1.0 to 2.0, i.e. near resonance, are shown in Figs. 8.15a, b through 8.21. In Figs. 8.15a, b through Figs. 8.18a, b the variable is $\beta_0 h$ and the parameter $\beta_0 b$. Curves are shown for $\Omega = 10$ and 15. Corresponding curves for $\Omega = 12.5$ and 20 are readily plotted using the data of Tables 8.4, 8.5, 8.8, and 8.9. The curves of Figs. 8.19 and 8.20 have $\beta_0 b$ or b/λ_0 as the variable and $\beta_0 h$ as parameter. They are obtained from Figs. 8.17a, b and 8.18a, b by cross plotting. Corresponding zeroth-order curves for the extremely thin antenna as computed by Starkey and Fitch[46] with sinusoidal current distributions are shown in Fig. 8.21.

The symmetric and antisymmetric impedances in Figs. 7.1 through 7.4 were obtained from corresponding self- and mutual impedances obtained in this section using (9a, b) with $\delta = 0$.

Although the *approximate* second-order self- and mutual impedances certainly are not such good representations of actual values as true second-order impedances would be, they should be superior in accuracy to zeroth- and first-order values since Z_{s1}, $Z_0^s = Z_{s1} + Z_{12}$ and $Z_0^a = Z_{s1} - Z_{12}$ represent variations from or oscillations about the true limiting value $(Z_0)_2$ as the separation is increased without limit. In a few instances, for quite large values of $\beta_0 h$ and small values of Ω, the scale factors differ from unity by more than a few percent. It is to be expected that the impedances in these cases are less accurate than when the scale factors are small.

In addition to the simple special cases of the *broadside* array in which the antennas are driven so that their currents are equal in magnitude and in phase (symmetric currents) and the *bilateral end-fire* array in which the currents are equal in magnitude and 180° out of phase (antisymmetric currents) which are studied in Sec. 7, the important two-element end-fire array now may be analyzed. In this array the antennas are driven so that their currents are equal in magnitude but 90° out of phase (quadrature currents). Suppose that I_2 leads I_1 by 90°:

$$I_{2\delta} = jI_{1\delta} = I_{1\delta}e^{j\pi/2}. \qquad (20)$$

The coupled-circuit equations (5a, b) reduce to

$$V_{1\delta} = I_{1\delta}(Z_{s1\delta} + jZ_{12\delta}), \qquad (21a)$$

$$V_{2\delta} = I_{2\delta}(Z_{s1\delta} - jZ_{12\delta}). \qquad (21b)$$

Therefore, the driving-point impedances of the two antennas are

$$Z_{1\delta} \equiv V_{1\delta}/I_{1\delta} = Z_{s1\delta} + jZ_{12\delta}$$
$$= (R_{s1\delta} - X_{12\delta}) + j(X_{s1\delta} + R_{12\delta}), \qquad (22a)$$

$$Z_{2\delta} \equiv V_{2\delta}/I_{2\delta} = Z_{s1\delta} - jZ_{12\delta}$$
$$= (R_{s1\delta} + X_{12\delta}) + j(X_{s1\delta} - R_{12\delta}). \qquad (22b)$$

As a numerical example, take the case $\Omega = 10$, $\beta_0 h = \pi/2$, $\delta = 0$, $b = \lambda_0/4$:

$$Z_{10} = 126.4 + j82.7, \quad Z_{20} = 36.4 - j4.3. \qquad (23)$$

The two impedances are seen to differ greatly. Clearly, in order that (20) be true, $V_{1\delta}$ and $V_{2\delta}$ must be quite different in magnitude and phase and require careful adjustment.

9. Coupling and End Effects for Antennas with Individual Feed Lines

The impedances $Z_{1\delta}$ and $Z_{2\delta}$ as evaluated in the preceding section are by definition the ratios of the scalar potential differences $V_{1\delta}$ and $V_{2\delta}$ (across the circular edges of the cylindrical surfaces at $z = \pm\delta$ in each antenna) to the currents $I_{1\delta}$ and $I_{2\delta}$ (traversing the cross sections of the antennas bounded by these edges). The analysis thus far applies specifically to Fig. 1.7, in which no mechanism or circuit is provided for maintaining the potential differences $V_{1\delta}$ and $V_{2\delta}$ or for conducting the currents $I_{1\delta}$ and $I_{2\delta}$. This corresponds to the determination of Z_δ for a single antenna with a potential difference V_δ maintained in an unspecified manner across the edges at $z = \pm\delta$.

In practice, the voltages $V_{1\delta}$ and $V_{2\delta}$ for two coupled driven antennas, just like the voltage V_δ for a single antenna, are maintained by transmission lines connected to the antennas at $z = \pm\delta$ using one of several possible types of junctions.

It is only in special cases that $Z_{1\delta}$ and $Z_{2\delta}$ are approximately equal to the apparent terminal impedances Z_{1sa} and Z_{2sa} of the individual feeding lines. This is true essentially in the same circuit arrangements described in Sec. II.10 for isolated antennas in which coupling effects are negligible and end effects are either negligible or compensated. The most important practical circumstances under which $Z_{1sa} \doteq Z_{1\delta}$, $Z_{2sa} \doteq Z_{2\delta}$ are (1) for antennas center-driven from two-wire lines with spacings b that are very small compared with the wavelength, and (2) for antennas over a large conducting plane end-driven from a coaxial line for which the inner radius b of the outer conductor is very small compared with the wavelength. A criterion of smallness is whether the input impedance of an arbitrarily terminated section of line of length s and spacing b differs negligibly from the impedance of a section of length $s + t$ with the same termination and for all values of s when t is of the order of magnitude of b.

Important two-wire-line circuits for which the apparent terminal impedances can be determined from $Z_{1\delta}$ and $Z_{2\delta}$ only with the aid of appropriate terminal-zone networks that take account of coupling and end effects must be considered individually and in turn.

(a) *Antennas with separation large compared with two-wire-line spacing and base separation —all types of connection*. If the distance b_a between the axes of the two coupled antennas is great compared with both the distance b between the conductors of the two transmission lines and the base separation 2δ between the halves of the antenna, the coupling between each antenna and the transmission line driving the *other* antenna is negligible for all types of connection. The conditions are

$$b_a^2 \gg b^2, \qquad b_a^2 \gg (2\delta)^2. \tag{1}$$

In some circuits $b \doteq 2\delta$ so that the two inequalities coalesce. If (1) is satisfied, the two circuits (each consisting of an antenna and a transmission line) are coupled significantly only by the interaction of the currents and charges in the antennas themselves. Vector and scalar potentials on the conductors of the transmission line feeding antenna 1 include no significant contributions from currents and charges anywhere in antenna 2 or its transmission line, and vice versa. This means that the complete *interaction* of the two circuits including transmission lines is just the same as for the circuit of Fig. 1.7 in which there are no feeding lines. Therefore, the analysis of each transmission line with its load is the same as for an isolated antenna and involves *identically the same corrective terminal network*. The only difference is that $Z_{1\delta}$ or $Z_{2\delta}$ replaces Z_{δ} in the appropriate network required to determine the apparent load impedances Z_{1sa} and Z_{2sa} for the two transmission lines.

It is readily verified that (1) is usually fulfilled in practice by examining the numerical equivalent. Consider a circuit for which $b \doteq 2\delta$ so that the two conditions in (1) coalesce. Let it be assumed that the value of δ satisfies

$$\beta_0 \delta \leq 0.1, \qquad \beta_0 b \leq 0.2, \tag{2a}$$

or

$$b/\lambda_0 \leq 0.032. \tag{2b}$$

The requirement (1) in conjunction with (2b) may be considered equivalent to

$$b_a/\lambda_0 \geq 5b/\lambda_0 \leq 0.16. \tag{2c}$$

Consider the following sets of values:

$$\beta_0 b = 0.2, \quad b/\lambda_0 = 0.032, \quad b_a/\lambda_0 \geq 0.16; \tag{2d}$$

$$\beta_0 b = 0.02, \quad b/\lambda_0 = 0.0032, \quad b_a/\lambda_0 \geq 0.016. \tag{2e}$$

Parallel driven antennas are rarely used with separations less than $0.15\lambda_0$, and even when one of the two antennas is parasitic, spacings closer than $0.05\lambda_0$ are seldom required except in special cases.

Actually, there is no difficulty in treating closely spaced antennas by taking account of the coupling effects involved. This is the next problem.

(b) *Antennas as end loads in plane of individual two-wire lines—arbitrary separation*. If the antennas are in the plane of the transmission lines as in Fig. 1.1, terminal-zone networks like those for an isolated antenna are required but with values of L_T and C_T that depend upon the distance b between the antennas and on the voltage ratio $V_{2\delta}/V_{1\delta}$. The determination of L_T and C_T is simplified if the general case of antennas with arbitrary voltages $V_{1\delta}$ and $V_{2\delta}$ is again resolved into a pair of symmetrically driven and a pair of antisymmetrically driven antennas.

In determining L_T for symmetrically and antisymmetrically driven antennas in the circuit of Fig. 9.1, the analysis of Sec. II.9 is readily adapted. Specifically, (II.9.7) must be replaced by

$$I_0^e(w) = \frac{1}{2\pi v_0} \left[\int_0^s \left(\frac{1}{R_a} - \frac{1}{R_b} \right) dw' \right.$$
$$\left. \mp \int_{-(s+b_a)}^{-b_a} \left(\frac{1}{R_a} - \frac{1}{R_b} \right) dw' \right], \tag{3}$$

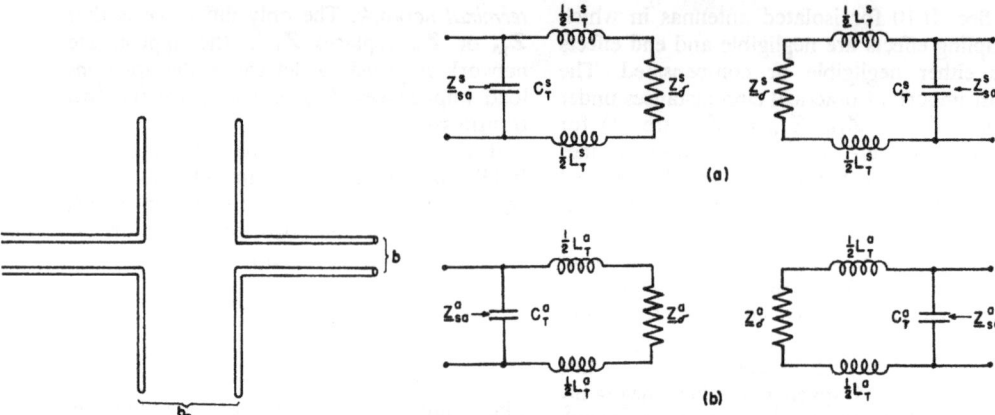

Fig. 9.1. Coupled antennas driven from two-wire lines.

Fig. 9.2. Equivalent circuits for symmetrically and antisymmetrically driven antennas, respectively, in (a) and (b).

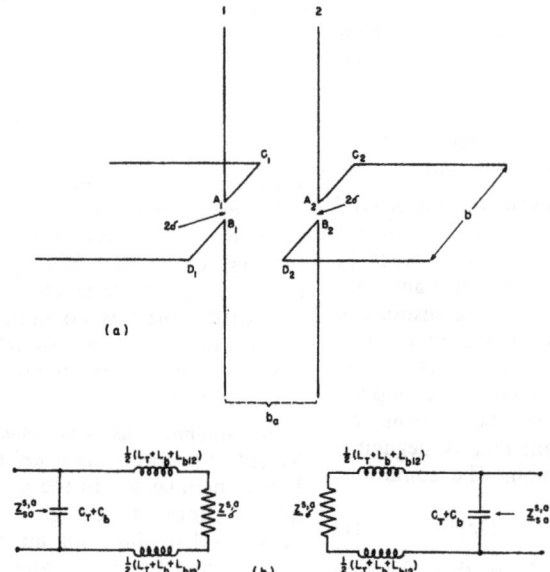

Fig. 9.3. Antennas as end-loads in plane perpendicular to individual two-wire lines in (a); equivalent circuit in (b).

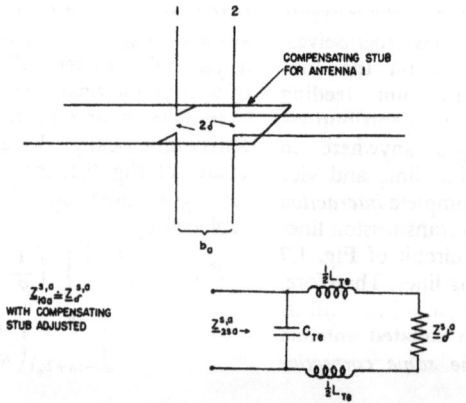

Fig. 9.4. Mixed circuit.

where the upper sign is for symmetrically driven antennas with equal and codirectional currents in the antennas and oppositely directed currents on the transmission lines, and the lower sign is for antisymmetrically driven antennas with equal and opposite currents on the antennas and codirectional currents on corresponding wires of the two lines. Reference to currents on the lines is to currents on the upper conductor at equal distances from the respective antennas.

Integration of (3) and substitution in

$$L_T = \int_0^d [l_0^e(w) - l_0^e] \, dw \quad (4)$$

gives

$$L_T = \frac{1}{2\pi v_0} \int_0^d \left[\left(\sinh^{-1} \frac{w}{a} - \sinh^{-1} \frac{w}{b} \right) \right.$$

$$\pm \left(\sinh^{-1} \frac{w + b_a}{a} - \sinh^{-1} \frac{w + b_a}{b} \right)$$

$$\left. - \begin{pmatrix} 2 \ln \frac{b}{a} \\ 0 \end{pmatrix} \right] dw. \quad (5)$$

Integration subject to the condition $d^2 \gg b^2 > a^2$ gives directly

$$L_T \doteq -\frac{1}{2\pi v_0} \left[b - a \pm \left(\sqrt{b_a^2 + b^2} \right. \right.$$

$$\left. - \sqrt{b_a^2 + a^2} - b_a \ln \frac{b_a + \sqrt{b_a^2 + b^2}}{b_a + \sqrt{b_a^2 + a^2}} \right) \right]$$

$$\doteq \begin{cases} L_T^s, \\ L_T^a, \end{cases} \quad (6)$$

where the upper sign gives L_T^s, for symmetrically driven antennas, the lower sign L_T^a for antisymmetrically driven antennas, as indicated in the notation. For very close spacing (6) is approximated by the value for zero spacing. For symmetrically driven antennas this is simply L_{Tc}, as obtained in Sec. II.9. For antisymmetrically driven antennas the two lines approximate a continuous single line, as $b_a \to 0$, for which L_T vanishes. Thus,

$$b_a \ll b: \quad \begin{aligned} L_T^s &\doteq L_{Tc} = -\left(\frac{b - a}{\pi v_0} \right), \\ L_T^a &\doteq 0. \end{aligned} \quad (7)$$

Conversely, for sufficiently great spacing (6) becomes the value L_{Te} of Sec. II.8 for the isolated antenna as end load with either sign:

$$b_a^2 \gg b^2: \quad L_T^s \doteq L_T^a \doteq L_{Te} = -\left(\frac{b - a}{2\pi v_0} \right). \quad (8)$$

The determination of C_T^s and C_T^a is carried out using modified forms of (II.9.10, 11) for substitution in (II.9.18). Thus,

$$V_L(w) \doteq \frac{q_L(w)}{2\pi \varepsilon_0} \left(\int_0^\infty \pm \int_{-\infty}^{-b} \right) \left(\frac{1}{R_a} - \frac{1}{R_b} \right) dw', \quad (9)$$

where the upper sign gives $V_L^s(w)$, the lower sign $V_L^a(w)$. This integrates directly into

$$V_L^s(w) \doteq \frac{q_L(w)}{2\pi \varepsilon_0} \left(2 \ln \frac{b}{a} + \sinh^{-1} \frac{w}{a} - \sinh^{-1} \frac{w}{b} \right.$$

$$\left. - \sinh^{-1} \frac{w + b_a}{a} + \sinh^{-1} \frac{w + b_a}{b} \right), \quad (10)$$

$$V_L^a(w) \doteq \frac{q_L(w)}{2\pi \varepsilon_0} \left(\sinh^{-1} \frac{w}{a} - \sinh^{-1} \frac{w}{b} \right.$$

$$\left. + \sinh^{-1} \frac{w + b_a}{a} - \sinh^{-1} \frac{w + b_a}{b} \right). \quad (11)$$

Similarly,

$$V_T(w) \doteq \frac{\bar{q}_T}{2\pi \varepsilon_0} \int_{b/2}^\infty \left[\left(\frac{1}{R_{1\delta}} - \frac{1}{R_{2\delta}} \right) \right.$$

$$\left. \pm \left(\frac{1}{R_{1b}} - \frac{1}{R_{2b}} \right) \right] dz', \quad (12)$$

where

$$R_{1\delta} = \sqrt{(z' - \tfrac{1}{2}b)^2 + w^2 + a^2}$$
$$R_{2\delta} = \sqrt{(z' + \tfrac{1}{2}b)^2 + w^2 + a^2}, \quad (13)$$

$$R_{1b} = \sqrt{(z' - \tfrac{1}{2}b)^2 + (w + b_a)^2 + a^2},$$
$$R_{2b} = \sqrt{(z' + \tfrac{1}{2}b)^2 + (w + b_a)^2 + a^2}. \quad (14)$$

Note that $\bar{q}_T$ is the average charge per unit length on *one* of the antennas. Introducing the ratio factor

$$k_q = \bar{q}_T / \bar{q}_L \doteq \bar{q}_T / q_L(w) \quad (15)$$

and carrying out the integration in (21) gives

$$V_T(w) \doteq \frac{k_q q_L(w)}{2\pi \varepsilon_0} \left[\sinh^{-1} \frac{b}{\sqrt{w^2 + a^2}} \right.$$

$$\left. \pm \sinh^{-1} \frac{b}{\sqrt{(w + b_a)^2 + a^2}} \right] \quad (16)$$

where the upper sign gives $V_T^s(w)$, the lower sign, $V_T^a(w)$.

The next step is to substitute $V(w) = V_L(w) + V_T(w)$ into (II.9.16) in the form

$$[c_0(w)\varphi_1(w) - c_0] \doteq \frac{q_L(w)}{V(w)} - c_0$$

$$= -c_0 \left[\frac{V(w)/q_L(w) - 1/c_0}{V(w)/q_L(w)} \right], \quad (17)$$

where

$$c_0 = \frac{\pi\varepsilon_0}{\ln(b/a)}. \quad (18)$$

The results are

$$[c_0(w)\varphi_1(w) - c_0]^s = -c_0 H^s(w), \quad (19a)$$

where

$$H^s(w) = \frac{k_q(A_1 + A_2) + B - C}{2\ln(b/a) + k_q(A_1 + A_2) + B - C}; \quad (19b)$$

and

$$[c_0(w)\varphi_1(w) - c_0]^a = -c_0 H^a(w), \quad (20a)$$

where

$$H^a(w) = \frac{k_q(A_1 - A_2) + B + C - 2\ln(b/a)}{k_q(A_1 - A_2) + B + C}. \quad (20b)$$

The following shorthand is used in (19a, b) and (20a, b):

$$A_1 = \sinh^{-1}\frac{b}{\sqrt{w^2 + a^2}}, \quad (21)$$

$$A_2 = \sinh^{-1}\frac{b}{\sqrt{(w + b_a)^2 + a^2}},$$

$$B = \sinh^{-1}\frac{w}{a} - \sinh^{-1}\frac{w}{b}, \quad (22)$$

$$C = \sinh^{-1}\frac{w + b_a}{a} - \sinh^{-1}\frac{w + b_a}{b}. \quad (23)$$

For very closely spaced antennas,

$$b_a^2 \ll b^2: \quad A_2 \doteq A_1, \quad C \doteq B; \quad (24)$$

$$[c_0(w)\varphi_1(w) - c_0]^s \doteq -c_0 \left[\frac{k_q A_1}{\ln b/a + k_q A_1} \right], \quad (25)$$

$$[c_0(w)\varphi_1(w) - c_0]^a \doteq 0. \quad (26)$$

The formula (25) resembles (II.9.17), since two symmetrically driven antennas very close together correspond approximately to a single antenna. If the antennas are antisymmetrically driven, the scalar potential due to the charges on one are canceled approximately by the scalar potential due to the equal and opposite charges on the other. Hence, no significant terminal-zone coupling between antisymmetrically driven antennas and lines exists when the antennas are very close together.

For antennas that are widely separated compared with the line spacing, the following formulas are obtained:

$$b_a^2 \gg b^2: \quad A_2 \doteq 0, \quad C = \ln(b/a); \quad (27)$$

$$[c_0(w)\varphi_1(w) - c_0]^{s,a}$$
$$= -c_0 \left[\frac{k_q A_1 + B - \ln(b/a)}{k_q A_1 + B + \ln(b/a)} \right]. \quad (28)$$

This is the same as obtained in Sec. II.8 for $b_a \to \infty$ which is equivalent to a single, isolated antenna. The condition (27) corresponds to (1).

The capacitive element C_T^s or C_T^a must be evaluated with $k_q = k_q$ assumed real from

$$C_T^{s,a} = \int_0^d [c_0(w)\varphi_1(w) - c_0]^{s,a} \, dw, \quad (29)$$

using (19) and (20). Note that $\varphi_1(w) = \varphi_1(w)$ is real. For closely spaced antennas,

$$b_a^2 \ll b^2: \quad C_T^s < 2C_{TC}, \quad C_T^a \doteq 0. \quad (30)$$

For more widely spaced antennas,

$$b_a^2 \gg b^2: \quad C_T^s = C_T^a = C_{Te}. \quad (31)$$

Evidently, in general,

$$0 < b_a < \infty: \quad 2C_{TC} < C_T^s < C_{Te} < 0, \quad (32)$$

$$C_{Te} < C_T^a < 0. \quad (33)$$

Approximate integrable functions to represent $[c_0(w)\varphi_1(w) - c_0]^{s,a}$ may be constructed exactly as in Secs. II.8 and II.9. The integrated expressions corresponding to (II.8.23) and (II.9.29) for $k_q = 1$ are

$$C_T^s \doteq -c_0 b K^s \sinh^{-1}(d/k^s b), \quad (34)$$

$$C_T^a \doteq -c_0 b K^a \sinh^{-1}(d/k^a b), \quad (35)$$

where

$$d \sim 10b \quad (36)$$

and, with $H(w)$ in (19b) and (20b) real,

$$k = \frac{3H_2/H_1}{\sqrt{1 - H_2^2/H_1^2}}, \quad K = H_1 k, \quad (37)$$

with appropriate superscripts s and a on k, K, H_1 and H_2. Referring to (19) and (20),

$$H_1^s = H^s(w = 0), \quad H_2^s = H^s(w = 3b); \quad (38a)$$

$$H_1^a = H^a(w = 0), \quad H_2^a = H^a(w = 3b). \quad (38b)$$

For any particular pair of coupled antennas driven from two-wire lines in the circuit of Fig. 9.1, (34) to (38a, b) permit the direct evaluation of C_T^s and C_T^a, and (6) the computation of L_T^s and L_T^a.

The equivalent terminal-zone networks for the apparent terminating impedances Z_{sa}^s, Z_{sa}^a at $x = s$ for each line for symmetrically and antisymmetrically driven antennas are shown in Fig. 9.2. The evaluation of Z_{sa}^s and Z_{sa}^a involves Z_0^s and Z_0^a as given in Sec. 3 and the two pairs of lumped elements, L_T^s, L_T^a; C_T^s, C_T^a as defined in (6), (34), and (35) to (38a, b).

For arbitrarily driven antennas, the apparent self- and mutual impedances corresponding to the apparent symmetric and antisymmetric impedances are given by

$$Z_{s1a} = \tfrac{1}{2}(Z_{sa}^s + Z_{sa}^a), \quad (39a)$$

$$Z_{12a} = \tfrac{1}{2}(Z_{sa}^s - Z_{sa}^a). \quad (39b)$$

By evaluating the apparent terminal impedances Z_{sa}^s and Z_{sa}^a for symmetrically and antisymmetrically driven antennas separately and in combination according to (39a, b), the compensated self- and mutual impedances may be computed and used to determine the apparent terminal impedances of identical antennas terminating lines that have arbitrary driving voltages. The relatively intricate problem of closely coupled antennas can be handled when required using the several formulas derived above with the appropriate terminal-zone circuits. Fortunately, the condition $b_a^2 \gg b^2$ is usually satisfied, as discussed in conjunction with (2a)–(2e) so that these more elaborate methods are not required.

(c) *Antennas as end loads in plane perpendicular to individual two-wire lines—arbitrary separation.* The circuit of Fig. 9.3a has the advantage over the more conventional arrangement in Fig. 9.1 that the antennas are neither capacitively nor inductively coupled to the transmission lines, as discussed in Sec. II.10. However, unless compensating circuits are used as described in Sec. II.10, end effects exist on the lines as well as coupling between them when the antennas are sufficiently close together. Furthermore, there is a small inductive and capacitive coupling between the transverse connectors joining each antenna to its transmission.

The end effects and coupling effects between the two transmission lines exclusive of the connectors are the same as in case (b), so that L_T applies unchanged. Since there is no capacitive coupling between antennas and lines, V_T and k_q are zero and $\varphi_1(w) = 1$ in (19) and (20) so that

$$C_T^s = \int_0^d [c_0(w) - c_0]^s \, dw$$

$$= c_0 \int_0^d \left[\frac{C - B}{2 \ln (b/a) + B - C} \right] dw, \quad (40a)$$

$$C_T^a = \int_0^d [c_0(w) - c_0]^a \, dw$$

$$= c_0 \int_0^d \left[\frac{2 \ln (b/a) - B - C}{B + C} \right] dw, \quad (40b)$$

where B and C are given by (22) and (23). Note that C_T^s and C_T^a are positive when end effect alone without capacitive coupling to the antenna is involved, as for the circuit now under consideration. In addition to the series inductance

$$L_b = \frac{b}{2\pi v_0} \left(\ln \frac{2b}{a_b} - 1 \right) \quad (41)$$

given in (II.10.2) of each pair of transverse connectors between antenna halves and line conductors, there is coupling between the connectors A_1C_1, D_1B_1 on one line and A_2C_2, D_2B_2 on the other (Fig. 9.3). Since any distributions of charges on the two connectors of each antenna are always equal and opposite, the capacitive coupling between the connectors of one antenna and those of another are small unless b_a is at least as small as b. The inductive coupling is given by*

$$L_{b12} = \mp \frac{b}{2\pi v_0} \left[\sinh^{-1}\left(\frac{b}{b_a}\right) + \frac{b_a}{b} - \sqrt{1 + \frac{b_a^2}{b^2}} \right]. \quad (42)$$

The upper sign applies when currents in the connectors are oppositely directed, the lower sign when they are codirectional.

For antenna separations b_a that are not too small ($b_a > b$), the approximate terminal-zone network for symmetrically and antisymmetrically driven antennas is shown in Fig. 9.3b. The elements are all readily evaluated except C_b, the capacitance between the two conductors AC and BD of each antenna, as discussed in Sec. II.10. If this does not involve a dielectric support, the electrostatic capacitance between them may be estimated. Since compensating networks to combine all of the

---

* Ref. I.31, Eq. (VI.11.13), with b_a substituted for s.

transmission-line end effects into a single high impedance in parallel with the antenna cannot be used for obvious physical reasons when the antennas are closely spaced, it may be concluded that the circuit of Fig. 9.3 is no improvement over the circuit of Fig. 9.1 from the point of view of the simple analytical determination of the apparent terminal impedances Z_{1sa} and Z_{2sa} except when b_a is sufficiently great.

(d) *Circuits with coupling limited to antennas.* The circuits of Figs. 9.1 and 9.3 differ in the coupling between antennas and lines, but are both characterized by the same coupling between the line feeding one antenna and that driving the other. All coupling between the two transmission lines and transverse connectors to the antennas can be eliminated so that the two complete circuits are coupled only by currents and charges in the antennas if the mixed circuit of Fig. 9.4 is used. Since the two transmission lines are in neutral planes with respect to each other, there are no contributions to either the scalar or the vector potential differences in one line from charges and currents in the other. The two lines are terminated in identical antennas with the same distances 2δ, but end and coupling effects of each individual line with its antenna are different. A compensating stub that is in the neutral plane of antenna 2 can be provided for antenna 1. When this is properly adjusted experimentally, the antenna is in parallel with a very high impedance, so that it is a good approximation to set $Z_{1sa} \doteq Z_{1\delta}$.

10. H-Arrays

An important array consists of two parallel antennas usually a half-wavelength apart and driven either in phase (broadside array) or in phase opposition (bilateral end-fire array). In general, the two antennas are either a quarter-wavelength or a half-wavelength in half-length ($h = \lambda_0/4$ or $h = \lambda_0/2$). Commonly used circuits are shown in Figs. 10.1 and 10.2. In Fig. 10.1a the two antennas are driven symmetrically by being connected directly to the ends of a $\lambda_0/2$ section of two-wire line to which the driving line is attached at the midpoint. In Fig. 10.1b the two antennas are driven antisymmetrically by spiraling or crossing-over the transmission line to antenna 1 but not to antenna 2. If the spiraling is carefully arranged, the properties of the transmission line are not significantly affected. In Fig. 10.2a the antennas are connected as in Fig. 10.1a but the main feeding line is attached directly to one antenna instead of to the midpoint of the section of line joining them. In Fig. 10.2b the $\lambda_0/2$ section of line connecting the two antennas is spiraled. The analysis of the centrally driven H of Figs. 10.1a, b differs somewhat from the laterally driven H of Figs. 10.2a, b. Let them be studied in turn.

(a) *Centrally Driven H-Array.* Assuming that $b_a^2 \gg b^2$, the apparent terminal impedance Z_{sa}^s of each antenna in Fig. 10.1a is the symmetric impedance Z^s in conjunction with the regular terminal-zone network of Sec. II.8. The apparent terminal function θ_{sa}^s for symmetrically driven antennas is defined by

$$\theta_{sa}^s \equiv \rho_{sa}^s + j\Phi_{sa}^s \equiv \coth^{-1}(Z_{sa}^s/Z_c). \quad (1a)$$

For antisymmetrically driven antennas,

$$\theta_{sa}^a \equiv \rho_{sa}^a + j\Phi_{sa}^a \equiv \coth^{-1}(Z_{sa}^a/Z_c), \quad (1b)$$

where Z_c is the characteristic impedance of the two-wire line. The impedance looking into each load line at the junction with the main line is double the impedance of the two in parallel, which is

$$Z_{in}^s = \tfrac{1}{2}Z_c \coth(\tfrac{1}{2}\gamma b_a + \theta_{sa}^s), \quad (2a)$$
$$Z_{in}^a = \tfrac{1}{2}Z_c \coth(\tfrac{1}{2}\gamma b_a + \theta_{sa}^a), \quad (2b)$$

where

$$\gamma = \alpha + j\beta. \quad (3)$$

If line losses are negligible ($\tfrac{1}{2}\alpha b \ll \rho_{sa}$) and $b_a = \lambda_0/2$, as is usual, (2) reduces to

$$Z_{in}^s = R_c^2/2Z_{sa}^s, \quad (4a)$$
$$Z_{in}^a = R_c^2/2Z_{sa}^a. \quad (4b)$$

The apparent terminal impedance for the main line is equal to Z_{in} in (2) or (4) only if the line spacing b is negligible. If this is not the case, a terminal-zone network is required at the end of the main feeding line due to the end effects on this line. Since all conductors are perpendicular to the feeding line, the inductive end effect is simply that analyzed in Sec. II.7 with

$$L_T = L_{Te} = -\left(\frac{b-a}{2\pi v_0}\right). \quad (5)$$

The capacitive correction includes not only the end effect but also the capacitive coupling between the feed line and its two perpendicular branches. (Since the charges on the two branches are equal, the T-branch has double the effect of a right-angle bend in so far as the capacitive correction is concerned.) Following the method of Sec. II.8, the scalar potential difference $V_T(w)$ across the main line at a distance w from its end due to equal charges per unit length q_T on the two branches is

$$V_T(w) \doteq \frac{q_T}{\pi\varepsilon_0}\int_0^{\frac{1}{2}b_a}\left(\frac{1}{R_a} - \frac{1}{R_b}\right)dy, \quad (6)$$

where

$$R_a = \sqrt{w^2 + y^2 + a^2}, \quad (7a)$$

$$R_b = \sqrt{w^2 + y^2 + b^2}. \quad (7b)$$

This integrates into

$$V_T(w) \doteq \frac{q_T}{\pi\varepsilon_0}\left(\sinh^{-1}\frac{b_a}{2\sqrt{w^2+a^2}} - \sinh^{-1}\frac{b_a}{2\sqrt{w^2+b^2}}\right), \quad (8)$$

where

$$q_T \doteq q_L. \quad (9)$$

For $b_a^2 \gg b^2 > a^2$,

$$V_T(w) \doteq \frac{q_T}{2\pi\varepsilon_0}\ln\left(\frac{w^2+b^2}{w^2+a^2}\right). \quad (10)$$

The voltage $V_L(w)$ across the line due to the charges on its own conductors is the same as in (II.8.14). The equivalent lumped capacitance is

$$C_T = \int_0^d [c_0(w)\varphi_1(w) - c_0]\,dw$$

$$\doteq -c_0\int_0^d\left[\frac{V(w)/q_L(w) - 1/c_0}{V(w)/q_L(w)}\right]dw, \quad (11)$$

where $V(w) = V_L(w) + V_T(w)$.

Substitution of (10) with (9) and (II.8.14) in (11) gives

$$C_T = -c_0\int_0^d N\,dw \quad (d \doteq 10b) \quad (12a)$$

where

$$N = \frac{\ln\left(\frac{w^2+b^2}{w^2+a^2}\right) + \ln\frac{w + \sqrt{w^2+b^2}}{w + \sqrt{w^2+a^2}}}{\ln\left(\frac{w^2+b^2}{w^2+a^2}\right) + \ln\frac{w + \sqrt{w^2+b^2}}{w + \sqrt{w^2+a^2}} + 2\ln\frac{b}{a}} \quad (12b)$$

The integrand (12b) has the value 0.6 at $w = 0$; it drops asymptotically to zero very rapidly. Hence, C_T is negative and numerically only a relatively small fraction of c_0. It is much smaller than C_{Te}. It can be evaluated graphically very readily if required, or an approximate integrable function may be used as in Sec. II.8. In practice, an insulator or stub usually is required to support the junction, the effect of which is likely to be greater than C_T and L_T. But this is essentially a transmission-line problem. For present purposes it is sufficient to note that the apparent terminal impedance for the main line is given to a satisfactory approximation by (2) or (4) without correction or with a correction that can be evaluated if required.

(b) *Laterally Driven H-Arrays.* If the H-array is driven from one side, as in Fig. 10.2, the determination of the apparent terminal impedance of the main line is complicated by the fact that the two antennas are not symmetrically and similarly located with respect to the transmission line. This makes it difficult to compensate for end effects and coupling effects. Before considering these, let the simpler case be analyzed in which the line spacing b is sufficiently small to make such effects negligible and to make $Z_{1\delta} \doteq Z_{10}$. When this is true the apparent impedance terminating the main line at its junction with the first antenna is a good approximation of the actual impedance and is the same as the input impedance Z_{in} of the network consisting of $Z_{10} = V_{10}/I_{10}$ in parallel with the input impedance of the section of line of length b_a terminated in $Z_{20} = V_{20}/I_{20}$. An expression for Z_{in} may be derived using the general formulas

$$V_{10} = I_{10}Z_{s1} + I_{20}Z_{12}, \quad (13a)$$

$$V_{20} = I_{10}Z_{12} + I_{20}Z_{s1}, \quad (13b)$$

in conjunction with the transmission-line relations

$$V_b = V_{10} = V_{20}\cosh\gamma b_a + I_{20}Z_c\sinh\gamma b_a, \quad (14a)$$

$$I_b = \frac{V_{20}}{Z_c}\sinh\gamma b_a + I_{20}\cosh\gamma b_a, \quad (14b)$$

where V_b and I_b are the voltage across and the current in the transmission line just to the right of the junction point with antenna 1 in Fig. 10.2a. The input impedance of the combined network is given by

$$Z_{\text{in}} = \frac{V_{10}}{I_{\text{in}}}, \quad (15a)$$

where

$$I_{\text{in}} = I_{10} + I_b. \quad (15b)$$

The solutions of (13a, b) for the currents are:

$$I_{10} = (V_{10}Z_{s1} - V_{20}Z_{12})/D, \quad (16a)$$

$$I_{20} = (V_{20}Z_{s1} - V_{10}Z_{12})/D, \quad (16b)$$

where

$$D = Z_{s1}^2 - Z_{12}^2. \quad (17)$$

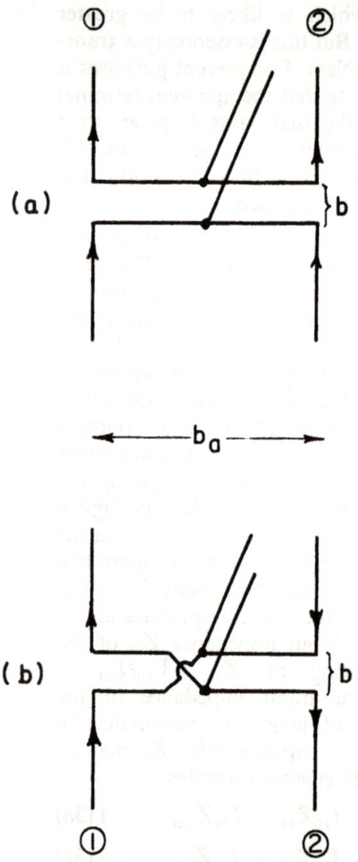

Fig. 10.1. Center-driven H-antenna with currents (a) in phase, (b) 180° out of phase.

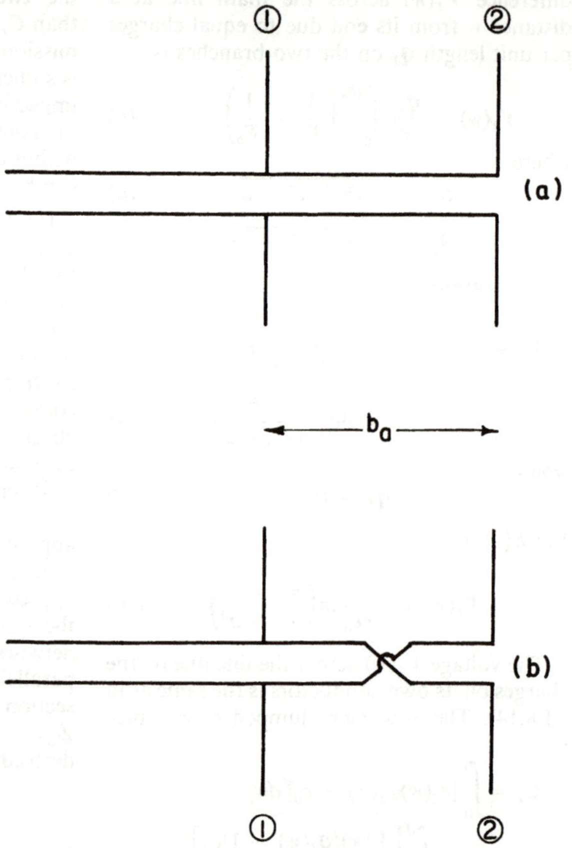

Fig. 10.2. Laterally driven H-antenna with currents (a) 180° out of phase, (b) in phase.

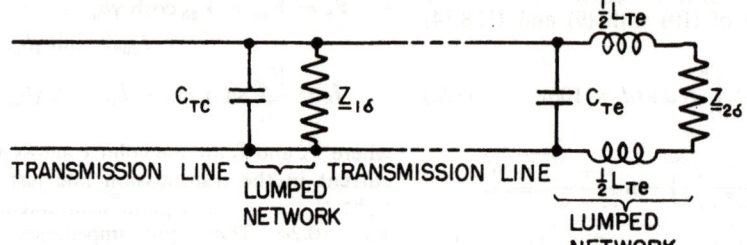

Fig. 10.3. Approximate equivalent circuit of Fig. 10.2.

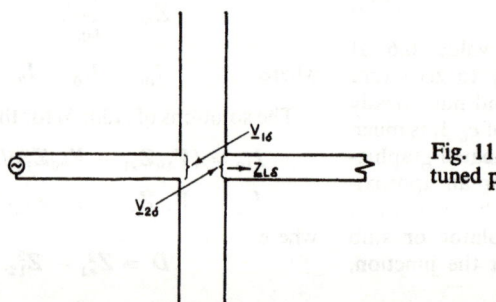

Fig. 11.1. Driven antenna with tuned parasite.

Substitution of (16b) in (14a) permits solution for V_{20} in terms of V_{10}. Thus,

$$V_{20} = V_{10}\left[\frac{D + Z_c Z_{12} \sinh \gamma b_a}{D \cosh \gamma b_a + Z_c Z_{s1} \sinh \gamma b_a}\right]$$

$$= V_{10}\frac{F}{G}. \qquad (18)$$

F and G are defined in (18). Then

$$\frac{I_{10}}{V_{10}} = \frac{Z_{s1}G - Z_{12}F}{GD}. \qquad (19)$$

Substitution of (16b) and (18) in (14b) gives

$$\frac{I_b}{V_{10}} = \frac{F}{Z_c G} \sinh \gamma b_a$$

$$+ \frac{Z_{s1}F - Z_{12}G}{GD} \cosh \gamma b_a. \qquad (20)$$

Addition of (19) and (20) yields

$$\frac{I_{10} + I_b}{V_{10}} = \frac{1}{Z_{\text{in}}} = \frac{F}{Z_c G} \sinh \gamma b_a$$

$$+ \frac{(Z_{s1}G - Z_{12}F) + (Z_{s1}F - Z_{12}G)\cosh \gamma b_a}{DG}.$$

$$(21)$$

After expansion of the several functions and considerable rearrangement and cancelation, the reciprocal of (21) becomes

$$Z_{\text{in}} =$$

$$\frac{Z_c(D \cosh \gamma b_a + Z_c Z_{s1} \sinh \gamma b_a)}{(D + Z_c^2)\sinh \gamma b_a + 2Z_c Z_{s1}\cosh \gamma b_a - 2Z_c Z_{12}}$$

$$(22)$$

This formula can be expressed directly in terms of symmetric and antisymmetric impedances, since by definition

$$Z_{s1} \equiv \tfrac{1}{2}(Z^s + Z^a), \qquad Z_{12} \equiv \tfrac{1}{2}(Z^s - Z^a).$$

$$(23)$$

Also

$$D = Z^s Z^a. \qquad (24)$$

With these values (22) becomes:

$$Z_{\text{in}} = \frac{Z_c^2(Z^s + Z^a) + 2Z_c Z^s Z^a \coth \gamma b_a}{2[Z_c^2 + Z^s Z^a + Z_c(Z^s + Z^a)\coth \gamma b_a}$$

$$- Z_c(Z^s - Z^a)\operatorname{csch} \gamma b]$$

$$(25)$$

Introduction of half-angle formulas in the denominator leads to the following final form:

$$Z_{\text{in}} = \frac{Z_c^2(Z^s + Z^a) + 2Z_c Z^s Z^a \coth \gamma b_a}{2(Z_c^2 + Z^s Z^a + Z_c Z^s \tanh \tfrac{1}{2}\gamma b_a}$$

$$+ Z_c Z^a \coth \tfrac{1}{2}\gamma b_a)$$

$$(26)$$

This is the input impedance of the network at the terminals of antenna 1 in Fig. 10.2a. The analysis of Fig. 10.2b may be carried out in the same way with appropriate changes in sign in (14a) and (14b).

The complicated expression (26) reduces to very simple form if attenuation in the section of line joining the two antennas is neglected and b_a is made an integral number of half-wavelengths. Thus, for Fig. 10.2a,

$$\gamma b_a \doteq j\beta_0 b_a = j\pi: \quad Z_{\text{in}} \doteq \tfrac{1}{2} Z^a, \quad (27a)$$

$$\gamma b_a \doteq j\beta_0 b_a = j2\pi: \quad Z_{\text{in}} \doteq \tfrac{1}{2} Z^s. \quad (27b)$$

For Fig. 10.2b the corresponding special cases give

$$\gamma b_a \doteq j\beta_0 b_a = j\pi: \quad Z_{\text{in}} \doteq \tfrac{1}{2} Z^s, \quad (28a)$$

$$\gamma b_a \doteq j\beta_0 b_a = j2\pi: \quad Z_{\text{in}} \doteq \tfrac{1}{2} Z^a. \quad (28b)$$

In practice $\beta_0 b_a \doteq \pi$ is by far the most important special case.

With the terminal impedance of the main line determined subject only to the restriction that the line spacing be small enough to make end and coupling effects negligible, it remains to consider the problem of correcting for these effects when b is not so severely restricted. It will be assumed that b_a is sufficiently great to satisfy $b_a^2 \gg b^2$.

Antenna 2 in Figs. 10.2a, b presents no new problem. Since it is at the end of a section of transmission line, the apparent impedance terminating this line is precisely that analyzed in Sec. II.8. It consists of $Z_{2\delta}$ with a terminal-zone network involving L_{Te} and C_{Te} as lumped elements. In practice the antennas are always sufficiently far apart that the condition $b_a^2 \gg b^2$ is well satisfied. This means that the terminal zone in which end effects and coupling effects are significant does not extend as far as antenna 1.

The determination of an equivalent terminal-zone network for antenna 1 is much more difficult since coupling effects between the main line, antenna 1, and the transmission line feeding antenna 2 are involved, in addition to regions of nonuniform line constants at the end of the main line and at the beginning of the section connecting the two antennas. Since the transmission line actually continues beyond antenna 1, although with a change in the magnitude of the current, it may be assumed as a fair approximation that $I_0^e(x) \doteq I_0^e$, $c_0(x) \doteq c_0$ at all points near antenna 1. On the other hand, capacitive coupling between antenna 1 and the section of line on both sides of the antenna is significant if b is not extremely small. A reasonable approximation is to assume this coupling to be essentially the same as for the stub support discussed at the end of Sec. II.9. That is, C_T is taken to be equal to C_{TC} for the symmetrically driven, center-loaded line, so that the equivalent circuit involves C_{TC} in parallel with $Z_{1\delta}$. Thus, the approximate equivalent circuit for taking account of coupling effects is shown in Fig. 10.3. The approximation evidently is not so good as for the simpler circuits involved in the center-driven H-antenna.

11. Driven Antenna with Parasite.

An array that is of great importance because of its simplicity and useful directional properties consists of two parallel antennas of which No. 1 is driven (in the sense that the scalar potential difference $V_{1\delta}$ at its terminals is maintained by a transmission line or other network of conductors that includes a generator), whereas No. 2 is parasitic (that is, it is connected to a transmission line or other network that presents an arbitrary impedance $Z_{L\delta}$ but contains no generator). Therefore, the only excitation for antenna 2 is its coupling to antenna 1. The circuit involving transmission lines is shown in Fig. 11.1.

The analysis of a center-driven antenna with a single, center-loaded parasite is already contained in the general solution in Sec. 8 for two coupled antennas with arbitrary voltages $V_{1\delta}$ and $V_{2\delta}$. Since there is no specification of the method of maintaining the scalar potential differences, the requirement that $V_{1\delta}$ be the terminal voltage of a transmission line containing a generator while $V_{2\delta}$ is the voltage across an arbitrary load is admissible. That is, let

$$V_{2\delta} = -I_{2\delta} Z_{L\delta}, \qquad (1)$$

where $I_{2\delta}$ is the current in the terminals of antenna 2 (Fig. 11.1) and where $Z_{L\delta}$ is defined to satisfy (1).

Substitution of (1) in (8.5b) gives

$$V_{1\delta} = I_{1\delta} Z_{s1\delta} + I_{2\delta} Z_{12\delta}, \qquad (2a)$$
$$0 = I_{1\delta} Z_{21\delta} + I_{2\delta} Z_{22\delta}', \qquad (2b)$$

where

$$Z_{22\delta}' \equiv Z_{s2\delta} + Z_{L\delta} = Z_{s1\delta} + Z_{L\delta}. \qquad (3)$$

The currents in the two antennas are related as follows:

$$\frac{I_{2\delta}}{I_{1\delta}} = -\frac{Z_{21\delta}}{Z_{22\delta}'} = -\frac{Z_{21\delta}}{Z_{22\delta}'} e^{-j(\theta_{22\delta} - \theta_{21\delta})}. \qquad (4)$$

The impedance

$$Z_{1\delta} \equiv V_{1\delta}/I_{1\delta} \qquad (5)$$

is readily evaluated from (2) and (3) by eliminating $I_{2\delta}$. The result, expressed in terms of $Z_{s1\delta} = Z_{s2\delta}$ and $Z_{12\delta} = Z_{21\delta}$ and also in terms of Z_δ^s and Z_δ^a, is

$$Z_{1\delta} = Z_{s1\delta} - \frac{Z_{12\delta} Z_{21\delta}}{Z_{L\delta} + Z_{s2\delta}}$$
$$= \frac{2Z_\delta^s Z_\delta^a + (Z_\delta^s + Z_\delta^a) Z_{L\delta}}{2Z_{L\delta} + Z_\delta^s + Z_\delta^a}. \qquad (6)$$

The symmetric and antisymmetric voltages V_δ^s and V_δ^a are defined in terms of $V_{1\delta}$ and $V_{2\delta}$ by the simple relations

$$V_\delta^s = \tfrac{1}{2}(V_{1\delta} + V_{2\delta}), \qquad V_\delta^a = \tfrac{1}{2}(V_{1\delta} - V_{2\delta}). \qquad (7)$$

Elimination of $V_{2\delta}$ gives

$$V_\delta^s = V_{1\delta} \frac{Z_\delta^s (Z_\delta^a + Z_{L\delta})}{2 Z_\delta^s Z_\delta^a + (Z_\delta^s + Z_\delta^a) Z_{L\delta}}, \qquad (8)$$

$$V_\delta^a = V_{1\delta} \frac{Z_\delta^a (Z_\delta^s + Z_{L\delta})}{2 Z_\delta^s Z_\delta^a + (Z_\delta^s + Z_\delta^a) Z_{L\delta}}. \qquad (9)$$

The currents in the two antennas are

$$I_{1z} = I_z^s + I_z^a, \qquad I_{2z} = I_z^s - I_z^a, \qquad (10)$$

where I_z^s is the current in each of the two antennas when driven by equal voltages V_δ^s given by (8), and where I_z^a and $-I_z^a$ are the currents in the two antennas when driven by the voltages V_δ^a and $-V_\delta^a$ in (9).

1. An interesting and important special case occurs when $Z_{L\delta}$ is made zero by appropriate adjustment of the transmission line connected to antenna 2 (Fig. 11.1) and its purely reactive termination. In this case

$$Z_{1\delta} = Z_{s1\delta} - \frac{Z_{12\delta} Z_{21\delta}}{Z_{s1\delta}} = \frac{2 Z_\delta^s Z_\delta^a}{Z_\delta^s + Z_\delta^a}, \qquad (11)$$

$$V_\delta^s = V_\delta^a = V_{1\delta}/2. \qquad (12)$$

Note that (11) is twice the impedance of Z_δ^s in parallel with Z_δ^a. Curves showing Z_{10} computed from (11) with $\delta = 0$ are given in Figs. 11.2 and 11.3 for $\beta_0 h = \pi/2$ and near π. It is significant that for $\beta_0 h = \pi/2$, $Z_{1\delta}$ is relatively more greatly affected by the presence of the parasitic antenna than for $\beta_0 h$ near π. This means that when $\beta_0 h = \pi/2$ the currents in the secondary circuit consisting of the parasite are relatively greater and react back on the primary circuit (antenna 1) more strongly than when $\beta_0 h$ is near π. Since when $\beta_0 h$ is near π the antenna is not near a resonant length for even currents of the type $I_2(-z) = I_2(z)$ with $Z_{L\delta} = 0$, as when $\beta_0 h = \pi/2$, this is readily understood. As would be expected,

[III.11] THEORY OF LINEAR ANTENNAS 323

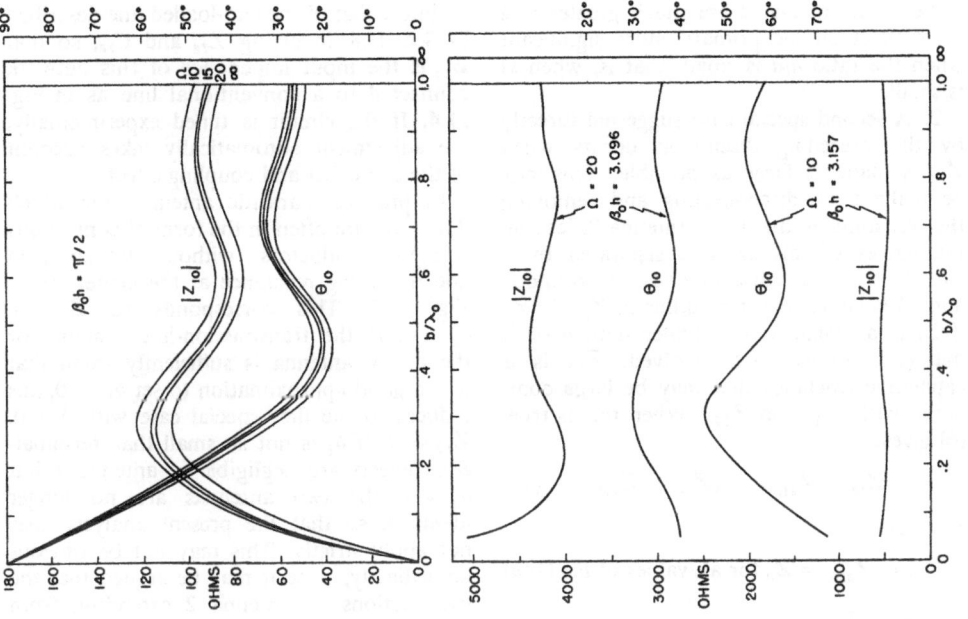

Fig. 11.3. Approximate second-order impedance—magnitude and angle—of antenna with unloaded parasite.

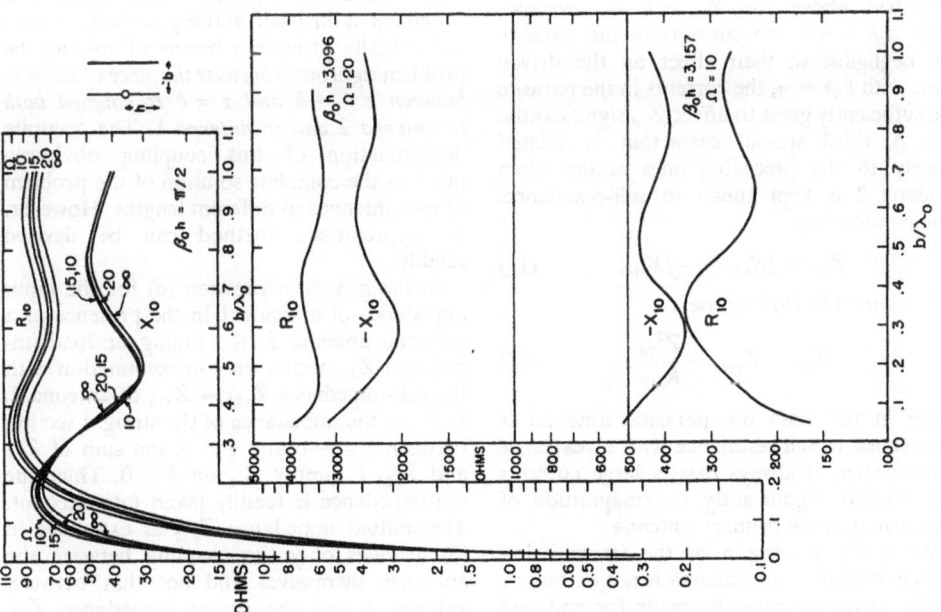

Fig. 11.2. Approximate second-order resistance and reactance of antenna with unloaded parasite.

currents in antenna 2 are much greater and their effect on the primary more significant when the ratio h/a is small, that is, when Ω is small.

2. A second special case suggested directly by the preceding discussion occurs when $Z_{L\delta}$ is made as large as possible. This may be realized by disconnecting and removing the transmission line from antenna 2, leaving the halves of this antenna separated by a gap 2δ. If these adjacent ends are rounded and 2δ is larger than the radius a, $Z_{L\delta} \doteq \infty$; if 2δ is as small as or smaller than a or if flat end surfaces are involved, $Z_{L\delta}$ is a capacitive reactance that may be large compared with $Z_{s1\delta}$ and $Z_{12\delta}$. When this is true, (6) gives:

$$Z_{1\delta} = Z_{s1\delta}. \qquad (Z_{L\delta} \gg Z_{s1\delta}) \qquad (13)$$

For $\beta_0 h = \pi/2$,

$$Z_{1\delta} = Z_{s1\delta} \doteq Z_\delta \text{ for all values of } b. \quad (14a)$$

For $\beta_0 h = \pi$,

$$Z_{1\delta} = Z_{s1\delta} \neq Z_\delta, \qquad (14b)$$

where Z_δ is the impedance of an isolated antenna. Thus, when $Z_{L\delta} \doteq \infty$, the situation described above for $Z_{L\delta} = 0$ is reversed. With $\beta_0 h = \pi/2$, the currents in the parasite are negligible in their effect on the driven unit; with $\beta_0 h = \pi$, the currents in the parasite are sufficiently great to affect $Z_{1\delta}$ significantly.

3. A third special case that is related closely to the preceding ones occurs when antenna 2 is kept tuned to self-resonance. The condition is

$$Z_{L\delta} \doteq jX_{L\delta} = -jX_{s1\delta}. \qquad (15)$$

Subject to (15), (6) becomes:

$$Z_{1\delta} = Z_{s1\delta} - \frac{Z_{12\delta}^2}{R_{s1\delta}}. \qquad (16)$$

Since in this case the parasitic antenna is kept tuned to self-resonance with a reactance at its center, it always carries large currents that modify significantly the magnitude of the current in the primary antenna.

When the spacing b of the transmission line connected to the antenna is not extremely small, correction must be made for end and coupling effects as discussed in Sec. 9. The fact that antenna 2 is parasitic does not alter the fact that $I_0^e(w)$ and $c_0(w)$ are not constants near the ends of the line, and that the line is coupled to the antenna. In the usual case when $b_a^2 \gg b^2$ the terminal-zone network is simply that of the end-loaded line described in Sec. II.8 involving L_{Te} and C_{Te}, so that $Z_{L\delta}$ is the input impedance of this network connected to a conventional line as in Fig. 11.4. If the circuit is tuned experimentally, the adjustment automatically takes account of the end effect and coupling effect.

In practice, parasitic antennas for which $\beta_0 h \cong \pi/2$ are often in the form of continuous straight conductors without transmission line or tuning reactance at the center, as in Fig. 11.5. This corresponds to $Z_L = 0$, $\delta_2 = 0$. If the transmission-line spacing for the driven antenna is sufficiently small that it is a good approximation to set $\delta_1 \doteq 0$, this reduces to the first special case with $\delta = 0$, $Z_{L\delta} = 0$. If δ_1 is not so small that terminal-zone effects are negligible in antenna 1 but $\delta_2 = 0$, the two antennas are no longer identical, so that the present analysis does not apply strictly. This may not be obvious immediately, since it may be argued that the two sections of antenna 2 extending from $z = -h$ to $z = -\delta$ and from $z = \delta$ to $z = h$ are, in fact, exactly like antenna 1, whereas the extra piece between $z = -\delta$ and $z = \delta$ may be treated as the tuning impedance or load impedance Z_L (Fig. 11.1). While this statement is in itself entirely correct, it does not actually provide a means of solving the problem rigorously *because the piece of antenna between $z = -\delta$ and $z = \delta$ is coupled both to antenna 2 and to antenna 1*. The accurate determination of this coupling obviously involves the complete solution of the problem of two antennas of different lengths. However, an approximate method can be devised readily.

In the general expression (6) for the input impedance of antenna 1 in the presence of a parasitic antenna 2, the tuning or load impedance $Z_{L\delta}$ occurs only in conjunction with the self-impedance $Z_{s2\delta} = Z_{s1\delta}$ of antenna 2. If $Z_{L\delta}$ is the impedance of the straight section between $z = -\delta$ and $z = \delta$, the sum of $Z_{L\delta}$ and $Z_{s2\delta}$ is simply Z_{s2} for $\delta = 0$. Thus, the self-impedance is readily taken into account. The mutual impedance $Z_{12\delta}$ as expressed in (6) includes only the coupling between the antennas themselves and not that between antenna 1 and the tuning impedance $Z_{L\delta}$. On the other hand, if Z_{12} for $\delta = 0$ were used instead of $Z_{12\delta}$, this would include coupling between two antennas extending from $z = -h$ to $z = h$ instead of between one antenna of this length and another that lacks the piece between $z = -\delta$ and $z = \delta$. Thus, $Z_{12\delta}$

takes account of too little coupling, Z_{12} for $\delta = 0$ of too much coupling. Since δ is required to be small, a good approximation of the correct value of Z_{12} should be a mean value between $Z_{12\delta}$ and Z_{12} for $\delta = 0$, such as $\sqrt{Z_{12\delta} Z_{12}}$. Accordingly, with $Z_{21} = Z_{12}$, $Z_{1\delta}$ for the circuit of Fig. 11.5 is approximately

$$Z_{1\delta} \doteq Z_{s1\delta} - \frac{Z_{12\delta} Z_{21}}{Z_{s2}} = \frac{Z_\delta^s Z^a + Z^s Z_\delta^a}{Z^s + Z^a}, \quad (17)$$

where $Z_{12\delta} Z_{21}$ gives a mean value between $Z_{12\delta} Z_{21\delta}$ and $Z_{12} Z_{21}$. (Omission of the subscript δ means $\delta = 0$.) Note that (17) is equivalent to (6) with

$$Z_{L\delta} = \frac{Z_\delta^s Z^a - Z_\delta^a Z^s}{Z^s - Z^a}. \quad (18)$$

Correction for terminal-zone effects is not made so readily as for identical antennas, since symmetric and antisymmetric impedances cannot be defined. Since the added section in antenna 2 is relatively small, the terminal-zone network for antenna 1 may be assumed to be the same as for the circuit of Fig. 11.1.

A simple alternative method of handling the problem of an unbroken parasite for antennas with $\beta_0 h$ near $\pi/2$ depends upon the fact that for isolated antennas of this length the impedance Z_δ for given values of h and δ is closely approximated by Z_0 for an antenna of half-length $\bar{h} = h - \delta$. Since the coupling between the two antennas is virtually unchanged if antenna 2 consists of a straight conductor extending from $z = -(h - \delta)$ to $z = h - \delta$ instead of from $z = -h$ to $z = -\delta$ and from $z = \delta$ to $z = -h$, it may be assumed that $Z_{1\delta}$ also is not significantly altered. Therefore, a driven antenna of half-length h, near $\lambda_0/4$ with small base separation 2δ and a parasite which is a straight unbroken conductor of half-length $h_2 = h_1 - \delta$, may be considered the equivalent of the same driven antenna with a parasite that has a half-length $h_2 = h_1$ center-loaded by a section of transmission line of spacing $b \doteq 2\delta$ that has an apparent input impedance $Z_{1\delta} = 0$. This is illustrated in Fig. 11.6.

The procedure of varying h_2 slightly to compensate for modifications in $Z_{L\delta}$ may be extended as follows. Although methods have not been derived for solving the problem of two dissimilar antennas, the difficulties are entirely analytical. The formal problem is readily set up and the same simultaneous equations involving the two currents and voltages and self- and mutual-impedances are obtained. Although these impedances are expressed in terms of integrals that have not been evaluated, they can be combined to give a formula just like (6) but with $Z_{s2\delta}$ not equal to $Z_{s1\delta}$. For antennas with half-lengths near $h = \lambda_0/4$, $Z_{s2\delta}$ includes a resistive part that varies relatively slowly and a reactance that varies quite rapidly with h. (This may be seen by noting that with $\beta_0 h$ near $\pi/2$, $Z_{s2\delta} \doteq Z_\delta$, the behavior of which is readily obtained from curves of Z_0 in Sec. II.30.) Hence, if $Z_{L\delta} = jX_{L\delta}$, it follows that if

$$X_{L\delta} + X_{s2\delta} = \text{const.}, \quad (19)$$

$Z_{1\delta}$ will remain essentially constant. That is, a particular antenna 2 of half-length $h_2 = h_1 = \lambda_0/4$ with $X_{L\delta}$ adjusted to make $X_{22\delta} = X_{L\delta} + X_{s2\delta} = X_A$, where X_A is a preassigned value, is essentially equivalent in determining $Z_{1\delta}$ to a different antenna 2 with $X_{L\delta} = 0$ and a half-length $h_2 \neq h_1$ adjusted so that $X_{s2\delta} = X_A$. The two antennas are shown in Fig. 11.7. If $Z_{1\delta}$ is the same with the two differently adjusted parasites, it follows that the magnitudes and phases of the currents in the antennas in the two cases must be practically the same.

It may be concluded that $Z_{1\delta}$ and the currents in both antennas are essentially the same if the total reactance $X_{22\delta} = X_{s2\delta} + X_{L\delta}$ in the secondary circuit is unchanged, provided $\beta_0 h$ does not greatly exceed $\pi/2$. Results obtained by fixing $h_2 = h_1 = \lambda_0/4$ and varying $X_{22\delta}$ by adjusting $X_{L\delta}$ may be regarded as good approximations of results obtained with $X_{L\delta} = 0$ and with $X_{22\delta}$ varied by adjusting $X_{s2\delta}$ by changing h_2.

Specific application of this important generalization is made in Sec. VI.4 where the electromagnetic field of a driven antenna of half-length $h_1 = \lambda_0/4$ with a single parasite is studied. By appropriately adjusting b_a and $X_{2\delta} = X_{s2\delta} + X_{L\delta}$ directional properties associated with parasites acting as *directors* and *reflectors* are obtained. Analytically, the adjustment of $X_{L\delta}$ in order to vary $X_{2\delta}$ is strictly correct and theoretical results are so obtained. In practice, especially at extremely high frequencies, the adjustment of $X_{2\delta}$ through variation $X_{s2\delta}$ by adjusting h_2 is often more convenient. If h_2 is not made to depart too far from $\lambda_0/4$—usually this is

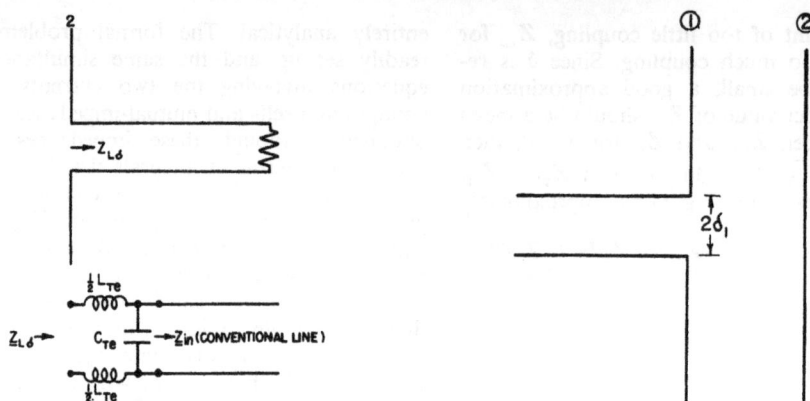

Fig. 11.4. Equivalent circuit of antenna with transmission-line load.

Fig. 11.5. Antenna with continuous parasite.

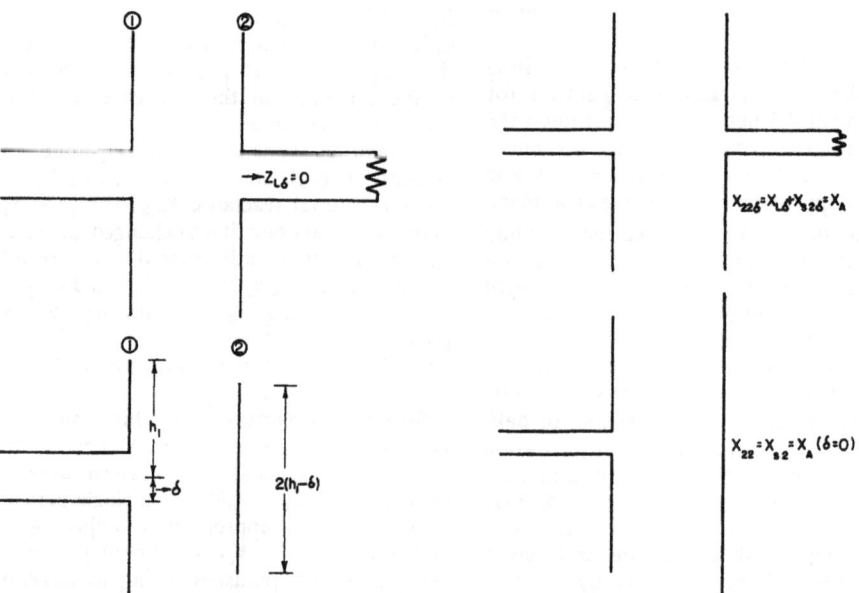

Fig. 11.6. Approximately equivalent arrays.

Fig. 11.7. Approximately equivalent arrays.

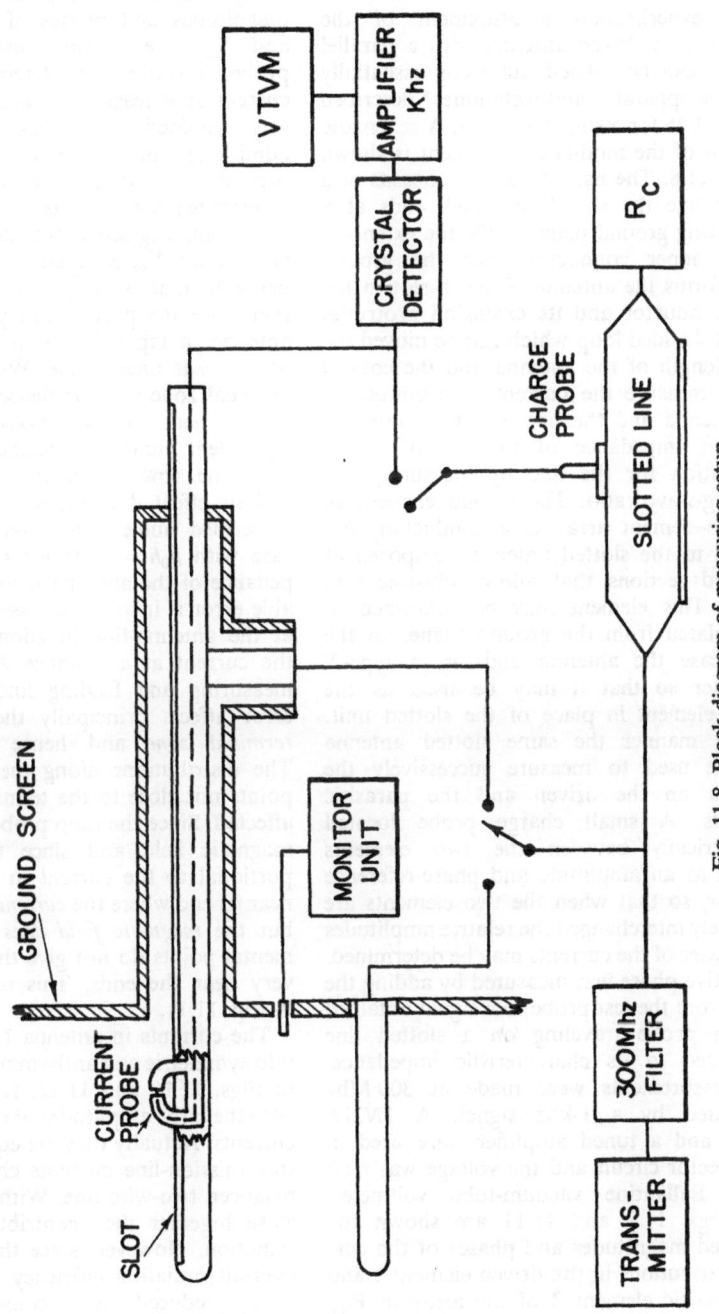

Fig. 11.8. Block diagram of measuring setup.

not required—the impedance and field properties of two arrays that differ primarily in the method by which $X_{2\delta}$ is varied should be very nearly identical.

The experimental measurement of the currents in a driven antenna with a parallel parasite can be carried out using essentially the same apparatus and techniques* described in Sec. II.26 for a single antenna. A schematic diagram of the modified equipment is shown in Fig. 11.8. The basic structure consists of a coaxial line the shield of which ends at a conducting ground plane, while the extension of the inner conductor over the ground plane forms the antenna. From a slot in the inner conductor and its extension protrudes a small shielded loop which can be moved the entire length of the antenna and the coaxial line to measure the current distribution on the antenna and the line and to obtain the apparent impedance of the antenna as a termination for the line by measuring the standing-wave ratio. The second element of the two-element array is a conducting rod parallel to the slotted antenna composed of threaded sections that allow adjustment in length. This element may be connected to or insulated from the ground plane. In the latter case the antenna ends in a type-N connecter so that it may be used as the driven element in place of the slotted unit. In this manner the same slotted antenna may be used to measure successively the currents on the driven and the parasitic elements. A small charge probe located symmetrically between the two elements is used as an amplitude and phase-reference monitor, so that when the two elements are effectively interchanged the relative amplitudes and phases of the currents may be determined.

Relative phase was measured by adding the signal from the test probe to a signal obtained from a probe traveling on a slotted line terminated in its characteristic impedance. All measurements were made at 300 Mhz modulated by a 1-khz signal. A $1N21B$ crystal and a tuned amplifier were used in the detector circuit and the voltage was read on a Ballantine vacuum-tube voltmeter.

In Figs. 11.9 and 11.11 are shown the measured magnitudes and phases of the current distributions in the driven element 1 and the parasitic element 2 of the array in Fig. 11.8. For Fig. 11.9, $h = \lambda_0/4$, $\Omega = 2 \ln (2h/a)$ = 10.1; for Fig. 11.11, $h = \lambda_0/2$, $\Omega = 11.5$. For both figures the two antennas were quite close together, the distance between them being only $b = 0.04\lambda_0$. Since only relative magnitudes and phases of the currents I_{1z} and I_{2z} were obtained using the traveling probe, an absolute determination of the current at at least one point is required. This was obtained by measuring the apparent admittance of the driven element in the presence of the parasite and noting that, if terminal-zone effects and probe-coupling effects are neglected, the measured apparent admittance Y_{sa} is equal to the input current per volt, that is, $Y_{sa} \doteq Y_{10} = I_{10}/V_1$. Note that since the probe must pass from line to antenna, a tapered section with small value of $b - a$ is unavailable. When $h = \lambda_0/4$, the terminal zone with its dielectric support is at or very near a current maximum so that the equivalent lumped reactance is in parallel with the low impedance of the antenna and its effect is negligible. Incidentally, this is verified numerically later in studying the case with $\beta_0 h = \pi$. When $h = \lambda_0/2$, the impedance of the antenna is high and considerable error is involved in assuming the current at the antenna-line junction to be equal to the current at a distance $\lambda_0/2$ back on the measuring and feeding line. However, this error affects principally the current *in the terminal zone* and hence the impedance. The distributions along the antennas at all points not close to the terminal zone are not affected. Since the loop probe responds to the magnetic field and since this is not proportional to the current in an antenna very near its end where the *current actually vanishes* but the *magnetic field does not*, the experimental points do not give the correct current very near the ends. This is clear especially in Fig. 11.11.

The currents in antenna 1 are decomposed into symmetric and antisymmetric components in Figs. 11.10 and 11.12. It is interesting to note the large amplitudes of the antisymmetric currents. Actually they are equal and opposite transmission-line currents characteristic of a balanced two-wire line. With the antennas so close together they contribute negligibly to radiation. However, since they are large the over-all radiating efficiency may be significantly reduced as a consequence of the relatively large ohmic losses in the imperfect conductors, the dielectric supports, and the coupled probes.

It is instructive to compare the measured

---

* Apparatus and measurements described in the remainder of this section are those of Dr. T. Morita and Dr. C. Faflick, ref. 40.

TABLE 11.1. Experimental and theoretical currents, impedances, and admittances for an antenna with parasite; $\beta_0 h = \pi/2$.

	Experimental, based on measured I_2/I_1 and Z_{10}; $\Omega = 10.11$	Theoretical	
		Computed from $Z_{11} = Z_{s1}$ and Z_{12}, no ohmic dissipation; $\Omega = 10$	Computed from Z_{s1} and Z_{12} (with added resistance of 3.55 ohms in series with Z_{s1} to correct for ohmic loss, $Z_{11} = Z_{s1} + 3.55$ ohm) $\Omega = 10$
I_{20}/I_{10}	$0.875/\underline{167°}$	$0.913/\underline{167.9°}$	$0.872/\underline{169.5°}$
I_{10}/V_{10} (mho)	$45.9 \times 10^{-3}/\underline{-62.7°}$	$51.9 \times 10^{-3}/\underline{-80.5°}$	$49.5 \times 10^{-3}/\underline{-62.5°}$
I_{20}/V_{10}	$40.2 \times 10^{-3}/\underline{104°}$	$47.5 \times 10^{-3}/\underline{87.4°}$	$42.2 \times 10^{-3}/\underline{107°}$
$I_0^s/V_{10} = \frac{1}{2}(I_{10} + I_{20})/V_{10}$	$5.6 \times 10^{-3}/\underline{-9.7°}$	$5.7 \times 10^{-3}/\underline{-18.3°}$	$5.5 \times 10^{-3}/\underline{-18.1°}$
$I_0^a/V_{10} = \frac{1}{2}(I_{10} - I_{20})/V_{10}$	$42.8 \times 10^{-3}/\underline{-68.9°}$	$49.4 \times 10^{-3}/\underline{-86.3°}$	$45.1 \times 10^{-3}/\underline{-67.4°}$
Z_{10} (ohm)	$10.0 + j19.2 = 21.8/\underline{62.7°}$	$3.2 + j19.0 = 19.3/\underline{80.5°}$	$9.3 + j17.9 = 20.2/\underline{62.5°}$
$Z_{11} = Z_{22}$	$45.8 + j12.8 = 47.6/\underline{15.6°}$	$42.0 + j19.25 = 46.3/\underline{24.6°}$	$45.55 + j19.25 = 49.5/\underline{22.9°}$
Z_{12}	$41.6 + j1.9 = 41.65/\underline{2.6°}$	$41.35 + j9.15 = 42.3/\underline{12.4°}$	$41.35 + j9.15 = 42.3/\underline{12.4°}$
$Z_0^s = Z_{11} + Z_{12}$	$87.4 + j14.7 = 88.6/\underline{9.4°}$	$83.4 + j28.4 = 88.1/\underline{18.8°}$	$86.95 + j28.4 = 91.4/\underline{18.1°}$
$Z_0^a = Z_{11} - Z_{12}$	$4.2 + j10.9 = 11.7/\underline{68.9°}$	$0.65 + j10.1 = 15.5/\underline{86.3°}$	$4.2 + j10.1 = 11.0/\underline{67.4°}$
$Y_{10} = \frac{1}{2}(Y_0^s + Y_0^a)$ $= I_{10}/V_{10}$ (mho)	$(21.0 - j40.9) \times 10^{-3}$	$(8.55 - j51.2) \times 10^{-3}$	$(22.8 - j43.9) \times 10^{-3}$
$Y_t = \frac{1}{2}(Y_0^s - Y_0^a)$ $= I_{20}/V_{10}$	$(-9.85 + j39.0) \times 10^{-3}$	$(2.25 + j47.6) \times 10^{-3}$	$(-12.3 + j40.5) \times 10^{-3}$
Y_{11}	$(20.2 - j5.7) \times 10^{-3}$	$(19.7 - j9.0) \times 10^{-3}$	$(18.6 - j7.9) \times 10^{-3}$
Y_{12}	$(24.0 - j1.1) \times 10^{-3}$	$(23.0 - j5.1) \times 10^{-3}$	$(23.0 - j5.1) \times 10^{-3}$
$Y^s = 2I^s/V_{10}$	$(11.1 - j1.9) \times 10^{-3}$	$(10.8 - j3.6) \times 10^{-3}$	$(10.4 - j3.4) \times 10^{-3}$
$Y^a = 2I^a/V_{10}$	$(30.8 - j79.9) \times 10^{-3}$	$(6.3 - j98.7) \times 10^{-3}$	$(35.1 - j84.4) \times 10^{-3}$

currents, impedances, and quantities calculated from them with theoretical determinations using the modified second-order self- and mutual impedances given in Sec. 8. This is readily done with $\beta_0 h = \pi/2$. In calculating self- and mutual impedances the two fundamental, measured quantities are I_{20}/I_{10} and Z_{10}. Numerical values are given in Table 11.1. Using these, the normalized currents I_{20}/V_{10} and I_{10}/V_{10} are determined and from them I_0^s/V_{10} and I_0^a/V_{10}. Note that for antenna 1 $b - a = \delta > 0$, a dielectric support is located at the base and, in addition, a coupled probe is present when the current is measured; antenna 2 has none of these. Hence, $Y_{11} = Y_{s1} + Y_1$, $Y_{22} = Y_{s1}$, where Y_1 takes account of the several admittances in parallel. Assuming that Y_1 is negligible compared with Y_{s1}, which, incidentally, is equivalent to assuming that $b - a = \delta \doteq 0$, the equations (4) and (6) apply with $\delta = 0$, $Z_{L\delta} = 0$. That is,

$$I_{20}/I_{10} = -Z_{12}/Z_{s1}, \qquad (20)$$

$$Z_{10} = Z_{s1}[1 - (Z_{12}/Z_{s1})^2]. \qquad (21)$$

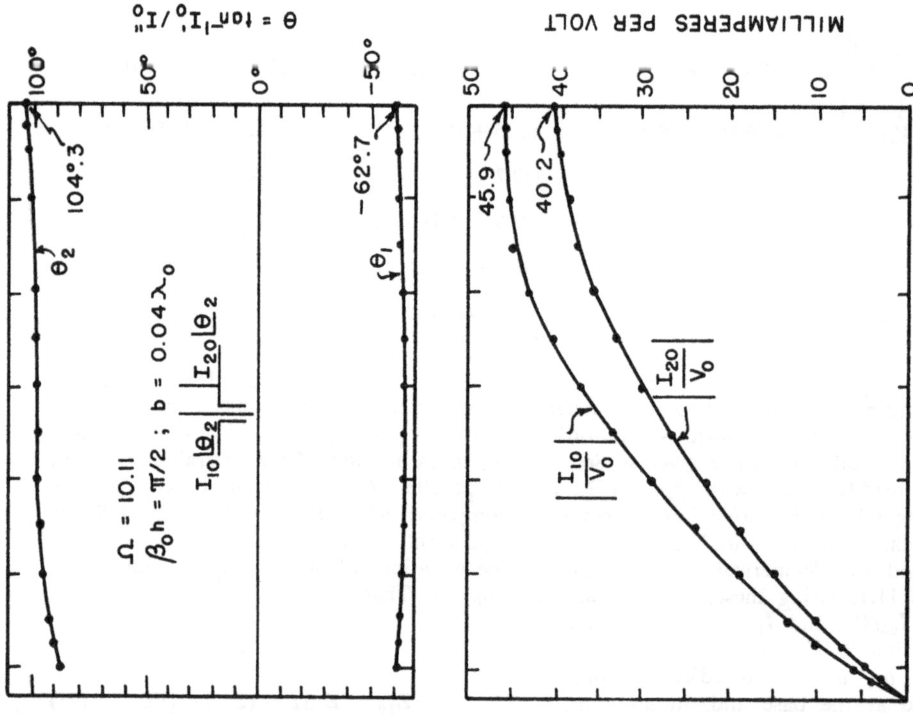

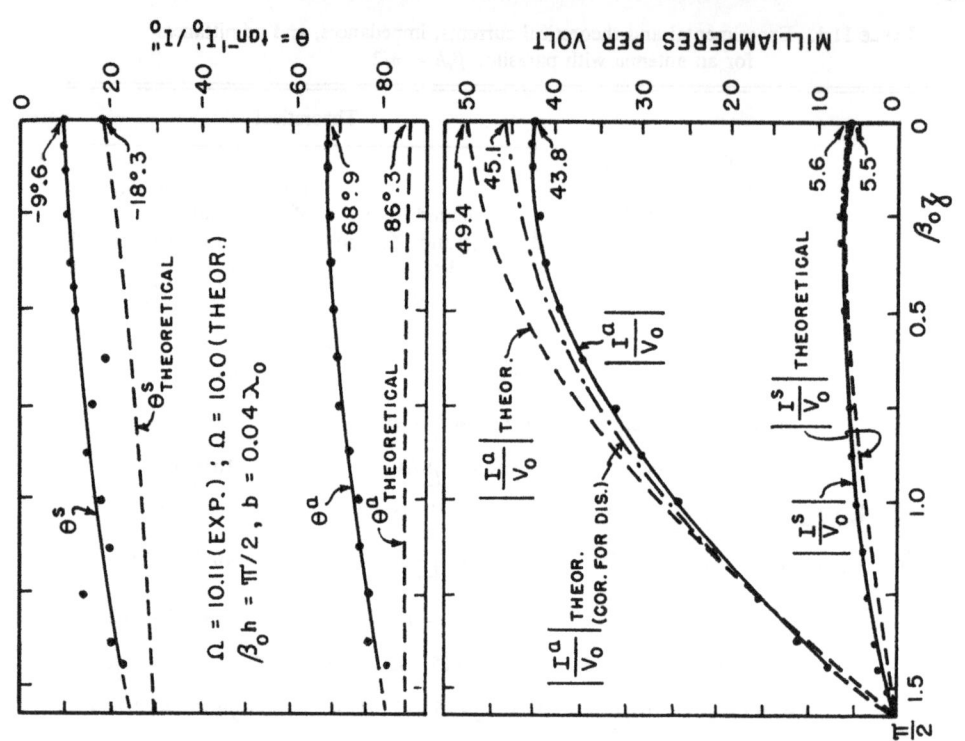

Fig. 11.10. Symmetric and antisymmetric currents in antenna with parasite, $\beta_0 h = \pi/2$.

Fig. 11.9. Measured currents in antenna with parasite, $\beta_0 h = \pi/2$ (Morita).

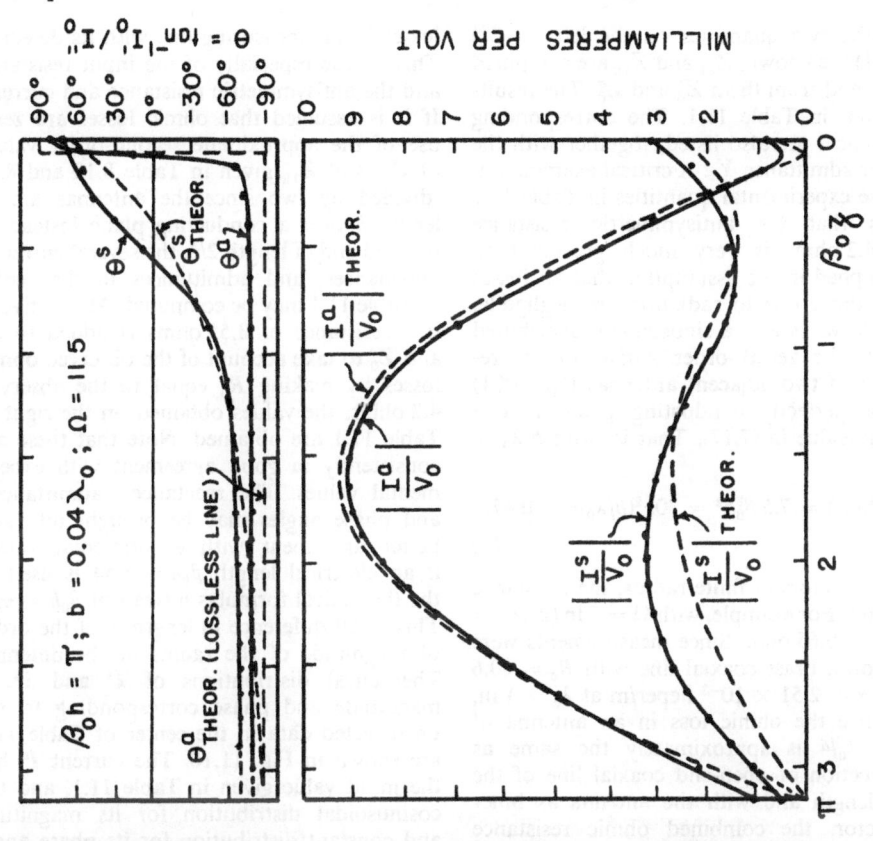

Fig. 11.12. Symmetric and antisymmetric currents in antenna with parasite, $\beta_0 h = \pi$.

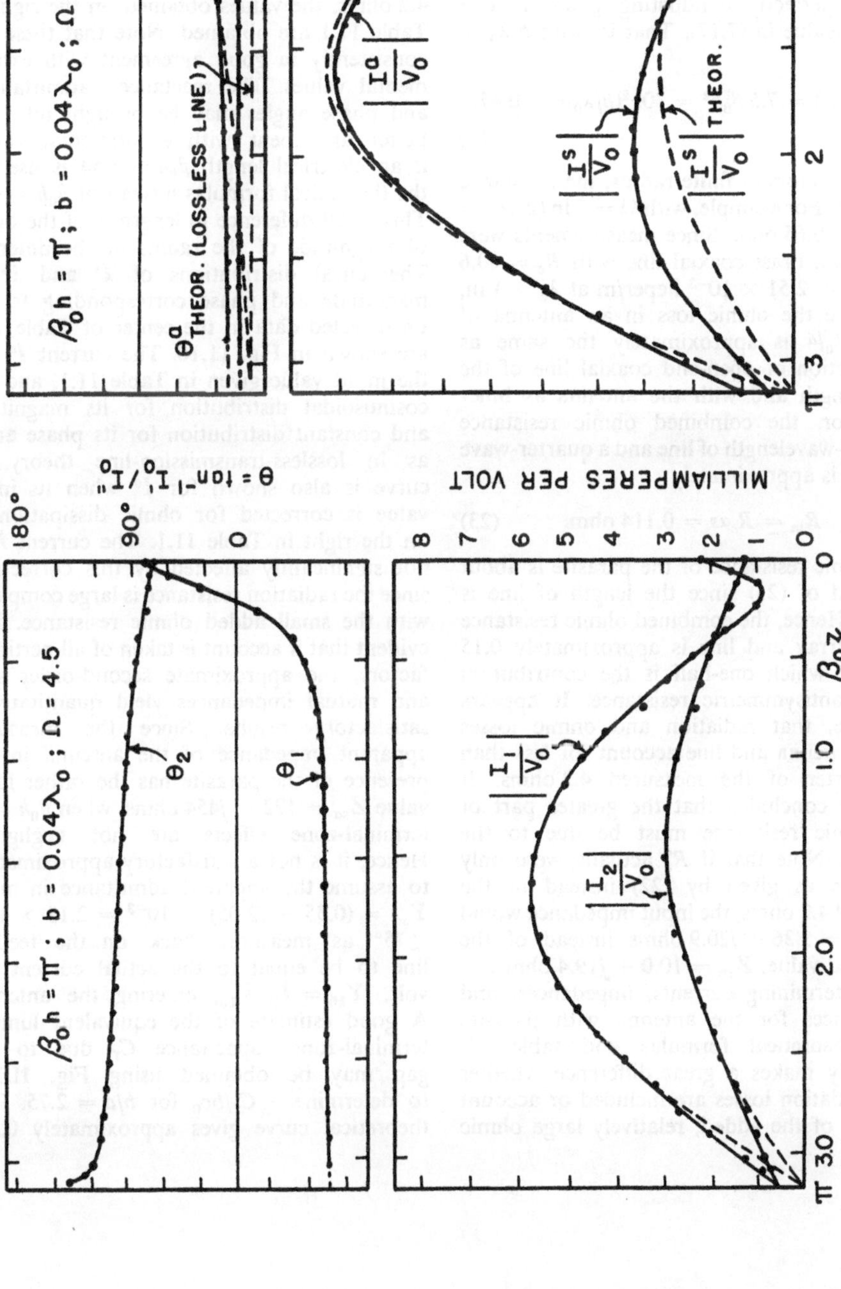

Fig. 11.11. Measured currents in antenna with parasite, $\beta_0 h = \pi$ (Morita).

Since the two quantities on the left in (20) and (21) are known, Z_{s1} and Z_{12} are computed readily and from them Z_0^s and Z_0^a. The results are given in Table 11.1. The corresponding admittances are also listed, together with the transfer admittance Y_t. A critical examination of these experimental quantities in Table 11.1 reveals that the antisymmetric resistance $R_0^a = 4.2$ ohms is very much greater than that implied in the assumption that all losses except those due to radiation are negligible. For thin wires and a sinusoidally distributed current, the zeroth-order antisymmetric resistance of two adjacent antennas ($\beta_0^2 b^2 \ll 1$) over a perfectly conducting plane is one half the value in (7.17). That is, with $b/\lambda_0 = 0.04$,

$$R_0^a(\text{ohms}) = 7.5 \beta_0^2 b^2 = 30\pi^2 (b/\lambda_0)^2 = 0.47. \quad (22)$$

For conductors of finite radius, the resistance is greater. For example, with $\Omega = 2 \ln (2h/a) = 10$, $R_0^a = 0.65$ ohm. Since measurements were made on a brass coaxial line with $R_c = 60.6$ ohms, $\alpha = 2.51 \times 10^{-3}$ neper/m at $\lambda_0 = 1$ m, and since the ohmic loss in an antenna of length $\lambda_0/4$ is approximately the same as on a section of open-end coaxial line of the same length and with the antenna as inner conductor, the combined ohmic resistance of a half-wavelength of line and a quarter-wave antenna is approximately

$$R_m = R_c \alpha s = 0.114 \text{ ohm}. \quad (23)$$

The ohmic resistance of the parasite is about one-third of (23) since the length of line is absent. Hence, the combined ohmic resistance of the array and line is approximately 0.15 ohm, of which one-half is the contribution to the antisymmetric resistance. It appears therefore, that radiation and ohmic losses in the antenna and line account for less than one-quarter of the measured 4.2 ohms. It must be concluded that the greater part of the ohmic resistance must be due to the detector. Note that if R_0^a actually were only 0.47 ohm as given by (22) instead of the observed 4.2 ohms, the input impedance would be $Z_{10} = 3.36 + j20.9$ ohms instead of the measured value, $Z_{10} = 10.0 + j19.4$ ohms.

In determining currents, impedances, and admittances for the antenna with parasite from theoretical formulas and tables, it obviously makes a great difference whether only radiation losses are included or account is taken of the added, relatively large ohmic losses in the measuring line with its detector. This is true especially of the input resistance and the antisymmetric resistance and current. If it is assumed that ohmic losses are zero, use of the approximate second-order values of Z_{s1} and Z_{12} given in Table 8.10 and 8.11 (divided by two since the antennas are of length h over a conducting plane instead of isolated and of length $2h$), the several currents, impedances, and admittances in the center of Table 11.1 may be computed. Alternatively, if a resistance of 3.55 ohms is added to Z_0^a and Z_0^s to take account of the observed ohmic losses by making R_0^a equal to the observed 4.2 ohms, the values obtained on the right in Table 11.1 are obtained. Note that these are consistently in good agreement with experimental values. The reactances, admittances, and phase angles may be brought into still better agreement with experimental values if an electrical length $\beta_0 h = 1.54$ is used in the theoretical formulas instead of $\beta_0 h = \pi/2$. This small difference in length is of the order of magnitude of the radius of the antenna. Theoretical distributions of I^s and I^a in magnitude and phase corresponding to the uncorrected data in the center of Table 11.1 are shown in Fig. 11.10. The current I^a has the input value given in Table 11.1, and the cosinusoidal distribution for its magnitude and constant distribution for its phase angle as in lossless-transmission-line theory. A curve is also shown for I^a when its input value is corrected for ohmic dissipation as on the right in Table 11.1. The current I^s is not significantly affected by this correction, since the radiation resistance is large compared with the small added ohmic resistance. It is evident that if account is taken of all pertinent factors, the approximate second-order self- and mutual impedances yield quantitatively satisfactory results. Since the measured apparent impedance of the antenna in the presence of the parasite has the rather large value $Z_{sa} = 122 - j454$ ohms, when $\beta_0 h = \pi$, terminal-zone effects are not negligible. Hence, it is not a satisfactory approximation to assume the apparent admittance in mhos $Y_{sa} = (0.55 + j2.05) \times 10^{-3} = 2.13 \times 10^{-3} \angle 75°$ as measured back on the feeding line to be equal to the actual current per volt, $Y_{10} = I_{10}/V_{10}$, entering the antenna. A good estimate of the equivalent lumped terminal-zone capacitance C_T due to the gap may be obtained using Fig. II.10.9 to determine $-C_T/bc_0$ for $b/a = 2.75$. The theoretical curve gives approximately 0.475

for this quantity, so that with $b = 0.873$ cm and c_0 for the line equal to 55×10^{-12} farad/m,

$$C_T \doteq -0.23 \times 10^{-12} \text{ farad}. \quad (24)$$

The equivalent lumped capacitance of a polystyrene disk of thickness 3.2 mm near the end of the line is

$$C_d = (2.45 - 1) \times 3.2 \times 10^{-3} \times 55 \\ \times 10^{-12} = 0.25 \times 10^{-12} \text{ farad}. \quad (25)$$

As a result of the virtual coincidence in magnitude of C_T and C_d and the opposite sign, the combined effect of these rather large capacitances yields only the following very small susceptance at the operating frequency $f = 300$ Mhz:

$$B = \omega(C_T + C_d) = 0.02 \times 10^{-12}. \quad (26)$$

This is sufficiently small to be neglected. Hence,

$$Y_{10} \doteq Y_{sa} = (0.55 + j2.05) \times 10^{-3} \\ = 2.13 \times 10^{-3} \angle 75°. \quad (27)$$

It is this quantity that is used to normalize the measured relative currents shown in Fig. 11.11 in solid lines and the symmetric and antisymmetric components which are derived from them and shown in Fig. 11.12.

Since the value of $\Omega = 2 \ln (2h/a)$ for the antenna used in the measurements is 11.5, the available numerical tables for self- and mutual impedances with $\Omega = 10$, 12.5, 15, and 20 are inadequate for $\beta_0 h = \pi$. In the range near antiresonance, impedances vary so rapidly that interpolation is not satisfactory. However, an alternative procedure for determining at least approximately the theoretical magnitudes and distributions of symmetric and antisymmetric currents is available if it is recalled that with b/λ_0 as small as 0.04 the array is in effect a two-element cage antenna for the symmetric currents, and a transmission line with open end for the antisymmetric currents. The effective radius of a cylindrical antenna that is equivalent to the symmetrically driven pair with $b/a = 6.3$ is given by (5.25) to be

$$a_e = \sqrt{ab} = 2.5a. \quad (28)$$

Hence, the effective value of Ω is

$$\Omega_e = 2 \ln (2h/a_e) = \Omega - 2 \ln 2.5 = 9.7. \quad (29)$$

By extrapolating the self-impedance data of Chapter II for $\beta_0 h = \pi$, the approximate impedance of an isolated antenna over a conducting plane with $\Omega = 9.7$ is found to be

$$Z_0 = 105 - j190 = 217 \angle -61° \doteq Z_0^s/2. \quad (30)$$

The last step is justified in the discussion of (7.17). It follows that the symmetric input current per volt is $I_0^s/V_0^s = Y_0^s = 2.3 \times 10^{-3} \angle 61°$. Since the current along a cage antenna is similar to that along a cylindrical antenna of equivalent radius, the distributions of current magnitude and phase angle given in Fig. II.22.11b for an antenna with $\Omega = 10$ may be scaled to fit the input values 2.3×10^{-3} amp/volt and 61° and used as an approximation of the distribution for $\Omega = 9.7$. Theoretical curves of I^s and θ^s obtained in this manner are shown in Fig. 11.12. They are in reasonably good agreement with the experimental results.

The antisymmetric current for $\beta_0 h = \pi$ is essentially that of a two-wire line of length $\lambda_0/2$ and with an open end. The magnitude of the maximum current per volt input is obtained readily if it is assumed that the radiation resistance referred to maximum current and ohmic losses in the line or coupled probe may be replaced by an approximately equivalent lumped resistance R_s at or very near the current maximum. Since the driving voltage of the line is $2V_0^a$, it follows that

$$(2V_0^a)^2/R_0^a = I_m^2 R_s. \quad (31)$$

However, since R_s and I_m are at the end of a quarter-wave transformer for which $R_c^2 = R_s R_0^a$, it follows that

$$I_m/V_0^a = 2/R_c = 9.16 \times 10^{-3} \text{ amp/volt}. \quad (32)$$

In (32) the characteristic impedance of the two-wire line is 218 ohms. In order to make use of (32) in conjunction with the essentially sinusoidal distribution of current characteristic of a low-loss transmission line as a means of obtaining an approximate theoretical distribution curve for the antisymmetric current, it is necessary to determine the input current from the antisymmetric admittance. This is obtainable from the input admittance of the $\lambda_0/2$ section of line, but account must be taken of the effective capacitance of the open end. Using the data of Sec. II.6, an approximate value of the effective terminating capacitance is found to be $C \doteq 0.15 \ \mu\mu$f. The input susceptance of the line of length $\lambda_0/2$ terminated in this capacitance is $B_{\text{in}} = 0.28 \times 10^{-3}$ mho, so that the corresponding antisymmetrical susceptance is $B_0^a = 2B_{in} = 0.56 \times 10^{-3}$ mho. Thus, the sinusoidal distribution has an amplitude of 9.16×10^{-3} amp/volt and is shifted from the origin by an electrical distance $\Phi' = \tan^{-1} B_{in} R_a = 0.061$ radian. Within distances of the open end comparable with the spacing ($\beta_0 b = 0.08\pi$ radian) the sine curve is distorted by end effect, so that the current vanishes at $\beta_0 z = \pi$. Since ohmic losses and radiation are small, they have been neglected in plotting the amplitude of the current and its phase angle in Fig. 11.12. The latter is discontinuous for a lossless line. If account were taken of losses, the sharp corners would be rounded and the phase would change rapidly but continuously from an angle somewhat less than 90° at $\beta_0 z = 0$ through zero at $\beta_0 z = 0.061$ to an angle somewhat greater, that is, less negative than $-90°$. It is evident from Fig. 11.12 that a good estimate of the antisymmetric current is obtained even though self- and mutual impedances are not available. Since experimental and approximate theoretical distributions of both symmetric and antisymmetric currents are in good agreement, it follows that the distributions of the total currents I_{1z} and I_{2z} in the driven and parasitic antennas also are in good agreement.

The input impedance and the symmetric and antisymmetric impedances are of interest. Experimental values and approximate theoretical results are listed in Table 11.2. The latter assume the antisymmetric resistance to be zero. (It is readily verified that a radiation resistance of the order of $R_m^e = 30\beta_0^2 b^2$ ohms yields an antisymmetric resistance of the order of magnitude of the experimental value. Since the only physically important quantity is the input impedance and this is altered negligibly by including R_0^a, the simpler, lossless case is assumed). Since Z^a is rather large near antiresonance and varies rapidly in magnitude with extremely small changes in length, the exact value of Z^a is difficult to measure and compute. However, so long as it is very large compared with Z^s, its effect on the input impedance is small. Note that, in spite of the approximations involved in the theory, a reasonable agreement with experiment is obtained for Z_{10} even in the difficult case near antiresonance, when interpolation from self- and mutual impedance tables is not acccurate.

The discussion of the antenna with parasite in this section has sought to focus attention on the several important factors involved in the understanding and the analysis of the problem. It should be clear that unless terminal-zone effects, dielectric supports, and even small ohmic and radiation losses are taken into account, accurate quantitative results are not possible. On the other hand, if account of these is taken, satisfactory quantitative results are to be expected from the approximate second-order mutual and self-impedances. Moreover, even if these are not available, a good estimate often may be obtained if the problem is thoroughly understood and *proper* use is made of approximate methods.

12. *Transmission-Line Radiators: Antenna with Parasite; Folded Dipole*

A special case of the single driven antenna with a coupled parasite is shown in Fig. 12.1a. It differs from the general case analyzed in the preceding section only in the requirement that b_a satisfy the following inequalities:

$$\beta_0 b_a \ll 1, \qquad b_a \ll h. \qquad (1)$$

That is, the distance b_a between the two coupled antennas satisfies the condition imposed on the distance b between conductors of a two-wire line. For simplicity, it is convenient to let $b_a = b$. It has been shown already (Secs. 6, 7) that the symmetrically driven pair resembles a single isolated antenna, whereas the antisymmetrically driven pair reduces to two open-end sections of transmission line in series with the two equal and opposite generators. In particular, the antisymmetric impedance Z_a^a is best determined directly from transmission-line theory, since

TABLE 11.2. Impedances in ohms for antenna with parasite, $\beta_0 h = \pi$.

Experimental $\Omega = 11.5$	Theoretical (antisymmetric resistance neglected)
$Z_{10} \doteq Z_{sa} = 122 - j454 = 470 \underline{/-75°}$	$Z_{10} = 170 - j352 = 434 \underline{/-64°}$
$Z^s = 84.5 - j289 = 301 \underline{/-73.7°}$	$Z^s = 105 - j190 = 217 \underline{/-61°}$
$Z^a = 170 - j1275 = 1287 \underline{/-82.4°}$	$Z^a = 0 - j1785 = 1785 \underline{/-90°}$

radiation losses are negligible compared with ohmic losses if (1) is satisfied (see ref. I.31, chap. VI). Thus, the analysis of the circuit of Fig. 12.1 reduces to the superposition of the solution of two adjacent, symmetrically driven antennas. The impedance $Z_{1\delta}$ is given by (11.6) as follows:

$$Z_{1\delta} = \frac{2Z_\delta^s Z_\delta^a + (Z_\delta^s + Z_\delta^a)Z_{L\delta}}{2Z_{L\delta} + Z_\delta^s + Z_\delta^a}, \quad (2)$$

where Z_δ^s is the impedance of each of two symmetrically driven antennas, and Z_δ^a is the impedance of an open-end section of transmission line of length $h - \delta$. This is given by

$$Z_\delta^a = Z_{\text{in}} = Z_c \coth\left[\gamma(h - \delta) + \theta^0\right], \quad (3)$$

where

$$\theta^0 = \rho^0 + j\Phi^0 = \coth^{-1} Z^0/Z_c \quad (4)$$

and

$$\gamma \doteq \alpha + j\beta_0. \quad (5)$$

In (4), Z^0 is the apparent terminal impedance of the open end. Owing to end effect it is not an infinite reactance. It is approximated by the reactance of a lumped capacitance C_{Te}, as determined in Secs. II.7, 8, but with $\varphi_1(w) = 1$ and $k_q = 0$. The inductance L_{Te} is not required, since the current vanishes at the ends. Since Z_δ^a can be varied from very small values for $h - \delta \doteq \lambda_0/4$ to extremely large values for $h - \delta \doteq \lambda_0/2$, and $Z_{L\delta}$ may be assigned values from zero to infinity, a great range of $Z_{1\delta}$ is available. By appropriate choice of $h - \delta$ and $Z_{L\delta}$ the circuit can be made to operate like a single antenna or like a section of transmission line that has negligible radiation. This is discussed in greater detail in this section after the folded-dipole circuit has been analyzed.

In its simplest form the folded dipole differs from the antenna with a closely coupled parasite only in having the extremities of the driven and parasitic elements conductively joined to form a closed loop, as in Fig. 12.1, where both circuits are shown side by side in arrangements with and without variable center-tuning. From the general analytical point of view the two types of circuit are quite different, since the one is composed of simple parallel antennas whereas the other is a special form of loop or frame antenna. On the other hand, with b_a restricted to very small fractions of a wavelength, the two circuits are essentially transmission lines with open and closed ends driven at the center of one of the long sides instead of at the center of a short side as in conventional transmission lines. It is this transmission-line property that the two types of circuit have in common which makes possible a very good approximate analysis of the circuit with closed ends that closely parallels the analysis of the circuit with open ends.

The circuit that is analyzed first is shown in Fig. 12.2. It consists of the folded dipole with scalar potential differences $V_{1\delta}$ and $V_{2\delta}$ maintained across the edges of the cylindrical ends at $z = \pm \delta$ in an as yet unspecified manner. Just as in Sec. 2, these two voltages are resolved into equal and in-phase, or symmetric, voltages V_δ^s and equal and 180° out-of-phase, or antisymmetric, voltages V_δ^a:

$$V_{1\delta} = V_\delta^s + V_\delta^a, \quad V_{2\delta} = V_\delta^s - V_\delta^a. \quad (6)$$

The in-phase voltages V_δ^s maintain equal complex currents in the two sides, $I_{2z}^s = I_{1z}^s$; the out-of-phase voltages V_δ^a and $-V_\delta^a$ set up equal and opposite complex currents, $I_{2z}^a = -I_{1z}^a$. The ratios

$$Z_\delta^s = \frac{V_\delta^s}{I_\delta^s}, \quad Z_\delta^a = \frac{V_\delta^a}{I_\delta^a} \quad (7)$$

are respectively the symmetric and antisymmetric impedances. By setting

$$V_{2\delta} = -I_{2\delta} Z_{L\delta} \quad (8)$$

the impedance, $Z_{1\delta} = V_{1\delta}/I_{1\delta}$, is made identically (2).

The impedance $Z_{1\delta}$ for the circuit in Fig. 12.2 can be evaluated from (2) provided Z_δ^s and Z_δ^a can be determined. Consider first the impedance Z_δ^s for equal complex voltages V_δ^s applied at both terminal pairs as in Fig. 12.3a. The exact solution of this symmetric problem with equal codirectional currents is not available. However, a good approximate solution is obtained by assuming that the circuit of Fig. 12.3a is essentially equivalent to the circuit of Fig. 12.3b. This may be justified as follows. Since there is no current in the conductors at P_1 and P_2, they may be cut there with no significant effect if the ends are rounded and the radius is small. If the halves of the bridges are folded out from the positions in Fig. 12.3a to the positions in Fig. 12.3b only a slight change is made in the circuit and the effect at the driving points will be negligible, certainly if a very small readjustment is made in the value of h. The contribution to the vector potential on the surface of each conductor at its center $z = 0$, due to the extremely small current in the short ends of length $\frac{1}{2} b_a$, is negligible. There is no change in the current at $z = 0$ as these ends are moved from the positions in Fig. 12.3a to those in Fig. 12.3b, provided there is no over-all shift

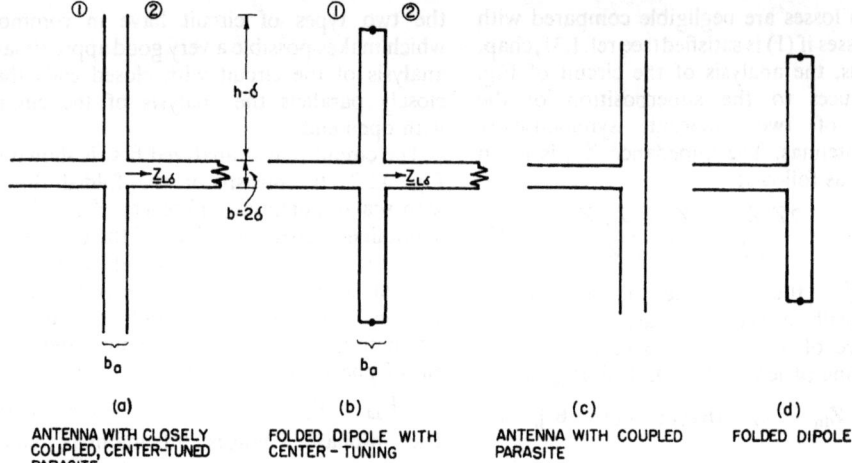

Fig. 12.1. Transmission-line radiating circuits.

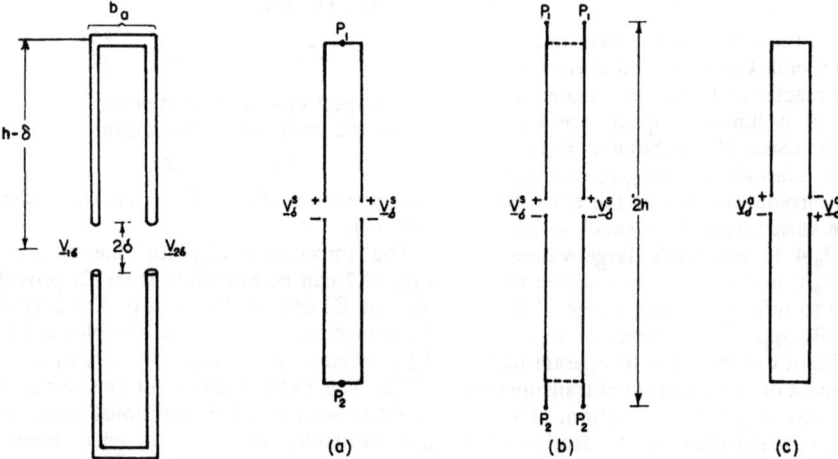

Fig. 12.2. Folded dipole.

Fig. 12.3. Symmetric and antisymmetric circuit components of folded dipole.

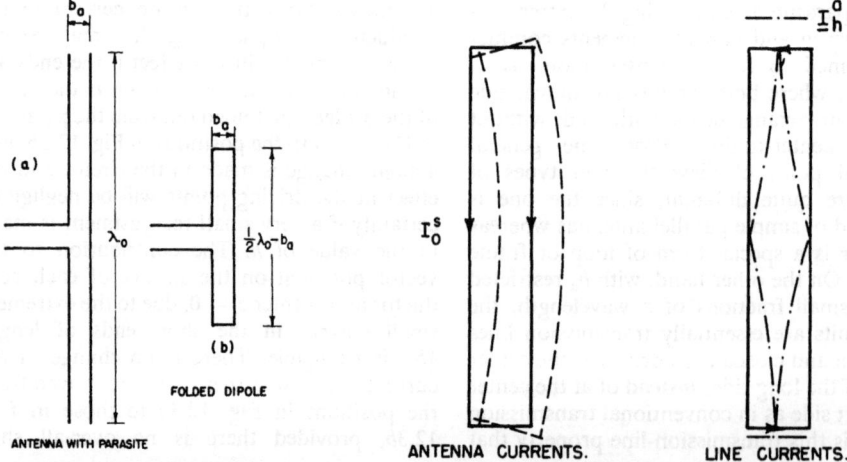

Fig. 12.4. Radiating circuits.

Fig. 12.5. Symmetric and antisymmetric components of current on a folded dipole ($\delta = 0$).

due to an effective lengthening or shortening as a result of the small change in coupling between the short ends and the adjacent conductors due to the altered geometry. Such a shift may be compensated by a slight readjustment in h.

The impedance seen by each driving potential difference in Fig. 12.3a and, hence, approximately in Fig. 12.3b is precisely Z_δ^s, previously obtained in Sec. 3. Thus, Z_δ^s for the closed-end parallel line driven as in Fig. 12.3a is the same as Z_δ^s for the open-end parallel line driven as in Fig. 12.3b.

The antisymmetric impedance Z_δ^a is obtained by analyzing the circuit of Fig. 12.3c. It cannot be derived in the same manner as Z_δ^s, since the impedance at the terminals in Fig. 12.3c is quite different from the impedance of the same circuit with open ends. However, the circuit of Fig. 12.3c consists of two identical sections of two-wire line connected in series with identical driving potential differences V_δ^a. The transmission-line impedance of a section of two-wire line of length $h - \delta$ carrying equal and opposite currents is given by

$$Z_\delta^a = Z_{\text{in}} = Z_c \coth [\gamma(h - \delta) + \theta^c], \quad (9)$$

where

$$\theta^c = \rho^c + j\Phi^c = \coth^{-1} Z^c/Z_c. \quad (10)$$

In (10), Z^c is the apparent terminal impedance of the closed-end line. Neglecting ohmic losses, this is the reactance of a straight conductor of length b as given in ref. I.31, Eq. (VI.11.22), namely,

$$L = \frac{b}{2\pi v_0}\left[\ln\frac{2b}{a} - 1\right], \quad (11)$$

in series with the terminal-zone inductance L_{Te} given by

$$L_{Te} = -\left(\frac{b-a}{2\pi v_0}\right). \quad (12)$$

Hence,

$$Z^c = j\frac{\omega b_a}{2\pi v_0}\left[\ln\frac{2b_a}{a} - 2 + \frac{a}{b}\right]. \quad (13)$$

For sufficiently thin wires ($a \ll b$), it is a good approximation to replace θ^c by $\frac{1}{2}(\alpha + j\beta_0)b_a$. With Z_δ^s and Z_δ^a determined, the analysis of Fig. 12.2 is completed.

The determination of the apparent terminal impedance for the main feeding line from $Z_{1\delta}$ is exactly the same for both types of circuit in Fig. 12.1a, b. The general analysis of Sec. 9 applies directly and appropriate networks of lumped elements must be used in conjunction with the symmetric and antisymmetric impedances, Z_δ^s and Z_δ^a. Note that for closely spaced antennas the lumped elements of the terminal-zone networks have *different formulas* for use with Z_δ^s and Z_δ^a.

The structurally simpler and conventionally used circuit of Fig. 12.1d involves the same analytical difficulties as the circuit of Fig. 12.1c whenever the distance b between the conductors of the main feeding line is not an extremely minute fraction of a wavelength so that end and coupling effects are insignificant. As discussed in Sec. 11, the two antennas in Fig. 12.1c and, similarly, the two sides in Fig. 12.1d are not the same, so that the currents I_{1z} and I_{2z} cannot be resolved into identical symmetric and antisymmetric parts. This prevents the reduction of the simultaneous integral equations to two independent equations for I_z^s and I_z^a. However, the approximate method leading to (11.17) applies equally well to the circuit of Fig. 12.1d and to that of Fig. 12.1c, and this formula is a satisfactory approximation for evaluating $Z_{1\delta}$ in the form

$$Z_{1\delta} \doteq \frac{Z_\delta^s Z^a + Z^s Z_\delta^a}{Z^s + Z^a}, \quad (14)$$

where Z_δ^s and Z^s are symmetric impedances, Z_δ^a and Z^a antisymmetric or transmission-line impedances. (Absence of the subscript on Z^s and Z^a means $\delta = 0$.) Note that with $\delta = 0$ in (14), $Z_{1\delta}/2$ is the impedance of an antenna with impedance Z^s in parallel with a transmission line with impedance Z^a.

With the general analyses of the circuits of Fig. 12.1 completed, it is interesting to examine a number of important special cases.

Consider first the circuits of Fig. 12.1c, d using formula (14) when Z^a is made successively as small as possible and as large as possible by appropriate adjustment of h.

Case A–1, Radiating circuits (Fig. 12.4).

Conditions: $|Z^a| \gg |Z^s|$;

Open ends (antenna with parasite), $h \doteq \frac{1}{2}\lambda_0$;

Closed ends (folded dipole), $h \doteq \frac{1}{4}\lambda_0 - \frac{1}{2}b$;

$$Z^a \doteq R_c/\alpha h \gg |Z^s|, \quad (15a)$$

$$Z^a \doteq R_c/\alpha(h-\delta) \gg |Z_\delta^s|, \quad (15b)$$

$$Z_{1\delta} \doteq Z_\delta^s + Z^s\left(\frac{Z_\delta^a}{Z^a}\right)$$

$$\doteq Z_\delta^s + Z^s\left(1 - \frac{\delta}{h}\right), \quad (16a)$$

$$Z_{10} \doteq 2Z^s = 2(Z_{s1} + Z_{12}). \quad (16b)$$

Since $Z_{1\delta}$ involves primarily the symmetric impedance and Z_{10} exclusively Z^s, the currents that contribute significantly to $Z_{1\delta}$ must be the codirectional antenna currents. Evidently, each circuit behaves essentially like an isolated antenna with the same lengths of conductors but with an impedance that is approximately double the symmetric impedance Z_0^s. Since the complex codirectional currents in the two antennas or conductors of each circuit are equal just as when they are symmetrically driven with two equal generators each of voltage V_0^s, but the power is supplied from *one* generator of voltage $V_{1\delta}$ only, this doubling of input impedance is readily understood.

The fact that energy dissipated as heat by the antisymmetric transmission-line currents is negligible compared with the power radiated by the symmetric antenna currents does not mean that the transmission-line currents are themselves small compared with the antenna currents. The approximate distribution and magnitude of the total current in the folded dipole (Fig. 12.4b) may be obtained by superimposing the codirectional currents I_z^s and the oppositely directed currents I_z^a. The distribution of codirectional currents has been shown to be essentially sinusoidal, with maximum at the center and zero at the ends for antennas with $\beta_0 h$ near $\pi/2$. The distribution of transmission-line currents is practically sinusoidal, but with maximum current in the conductors terminating each section of line. These distributions are sketched in Fig. 12.5 in separate diagrams.

The relative magnitudes of the currents I_0^s and I_0^a are readily obtained for the simpler case with $\delta = 0$. Thus,

$$I_0^s = \frac{\frac{1}{2}V_{10}}{Z^s}, \qquad I_0^a = \frac{\frac{1}{2}V_{10}}{Z^a}, \qquad (17)$$

so that

$$\frac{I_0^a}{I_0^s} = \frac{Z^s}{Z^a} = \frac{Z^s}{R_c} \alpha \left(h - \frac{b_a}{2}\right) \qquad (18)$$

Since the magnitude of Z^s is of order 10^2, and that of $Z^a = R_c/\alpha h$ is nearly 10^6, it follows that $|I_0^a|$ is insignificant compared with $|I_0^s|$. However, I_0^s is the *maximum* symmetric or antenna current, whereas I_0^a is practically the *minimum* antisymmetric or line current. The maximum line current is at the centers of the bridges at the ends.

The ratio of the antisymmetric current at any point z to the current at $z = 0$ is given by

$$\frac{I_z^a}{I_0^a} = \frac{\sinh(\gamma w + \theta^c)}{\sinh[\gamma(h - \frac{1}{2}b_a) + \theta^c]}, \quad w = h - \tfrac{1}{2}b_a - z. \qquad (19)$$

Let the current entering the bridge at $w = 0$ be I_h^a. Then

$$\frac{I_h^a}{I_0^a} = \frac{\sinh \theta^c}{\sinh[\gamma(h - \frac{1}{2}b_a) + \theta^c]}. \qquad (20)$$

In the present case, $\theta^c \doteq j\pi/2$, $\gamma(h - \tfrac{1}{2}b_a) + \theta^c \doteq (\alpha + j\beta)(h - \tfrac{1}{2}b_a) + j\pi/2 \doteq \alpha(h - \tfrac{1}{2}b_a) + j\pi$. Hence

$$\frac{I_h^a}{I_0^a} \doteq -\frac{j}{\sinh \alpha(h - \tfrac{1}{2}b_a)} \doteq \frac{-j}{\alpha(h - \tfrac{1}{2}b_a)}. \qquad (21)$$

Substitution of (21) in (18) to eliminate I_0^a gives the ratio of maximum transmission-line current on the bridge at the end to the maximum antenna current at the driving point. It is

$$\frac{I_h^a}{I_0^s} \doteq \frac{Z^s}{R_c}. \qquad (22)$$

Since Z^s is of order of magnitude 2×10^2 and a typical value of R_c is perhaps 4×10^2, I_h^a is roughly one-half of I_0^s, although for smaller values of R_c, I_h^a may be as large as I_0^s.

The ratio of powers supplied to maintain antisymmetric and symmetric currents is

$$\frac{P^a}{P^s} = \frac{I_0^{a2} R_a}{I_0^{s2} R_s} = \left|\frac{Z^s}{Z^a}\right|^2 \frac{R_a}{R^s}. \qquad (23a)$$

Since Z^s and Z^a are predominantly resistive, this ratio is approximately given by

$$\frac{P^a}{P^s} \doteq \frac{R^s}{R^a} = \frac{R^s}{R_c} \alpha h \sim 10^{-3}. \qquad (23b)$$

Thus the power supplied to maintain the transmission-line currents, in the folded dipole, is negligible compared with the power supplied to maintain the antenna currents. The former is dissipated as heat *in* the wires; the latter is radiated and ultimately dissipated *outside* the wires.

The circuits that satisfy the requirements of Case A–1 may be used interchangeably with isolated single antennas of corresponding length in so far as the electromagnetic field is concerned. The important property of these circuits is that their input impedances are much greater than those of isolated antennas of equal lengths. In particular, for the folded dipole of Fig. 12.1d, the input impedance is nearly four times that of an isolated antenna of equal length. This is of considerable practical importance in the case of the folded dipole, which has an impedance of $Z_{10} \doteq 342.6 + j154.8$ ohms with $b/\lambda_0 = 0.01$ instead of $Z_0 = 86 + j42$ ohms ($\Omega = 10$) with otherwise similar properties, since the higher impedance is more readily matched to a two-wire line.

If the two parallel conductors of the folded dipole have unequal radii a_1 and a_2, an approximate expression analogous to (16a) may be used in the form

$$Z_{10} = Z_1^s + Z_2^s(Z_1^a/Z_2^a)$$
$$\doteq Z_1^s + Z_2^s(R_1^c/R_2^c)$$
$$\doteq 2Z_0(1 + R_1^c/R_2^c), \quad (24)$$

where R_{c1} and R_{c2} are, respectively, the characteristic resistances of lines with both conductors of radius a_1 and with both conductors of radius a_2; Z_1^s and Z_2^s are symmetric impedances of antennas of radius a_1 and a_2, respectively. A more accurate formula is derived in Sec. 14, namely (14.85). Numerical values are given in Table 14.2.

Case B–1, Transmission-line circuits (Fig. 12.6):

Conditions $|Z^a| \ll |Z^s|$;

Open ends: $h \doteq \tfrac{1}{4}\lambda_0$;

Closed ends: $h \doteq \tfrac{1}{2}\lambda_0 - \tfrac{1}{2}b$;

$$Z^a \doteq R_c\alpha h \ll |Z^s|, \quad (25a)$$

$$Z_\delta^a \doteq R_c\alpha(h-\delta) \ll |Z_\delta^s|, \quad (25b)$$

$$Z_{1\delta} \doteq Z_\delta^a + Z^a\left(\frac{Z_\delta^s}{Z^s}\right), \quad (26a)$$

$$Z_{10} \doteq 2Z^a \doteq 2R_c\alpha h. \quad (26b)$$

Each circuit behaves essentially like two sections of transmission line in series, each adjusted to input resonance. Codirectional antenna currents are negligible compared with equal and opposite transmission-line currents, so that no significant energy is radiated by them; virtually all energy is dissipated as heat if the conductors are close enough together.

Special cases for the circuits of Fig. 12.1a, b include arrangements that are essentially equivalent to A–1 and B–1 for Fig. 12.1c, d. Since $Z_{L\delta}$ is also available as a variable in addition to h, other combinations are possible. If $Z_{L\delta}$ is the input impedance of an auxiliary section of transmission line that is like that of the center-driven section of line, its extreme range is the same as that of Z_δ^a.

Case A–2: $|Z_\delta^a| \gg |Z_\delta^s|$;

Open ends (antenna with parasite), $h \doteq \tfrac{1}{2}\lambda_0$;

Closed ends (folded dipole), $h \doteq \tfrac{1}{4}\lambda_0 - \tfrac{1}{2}b$;

$$Z_{1\delta} \doteq \frac{2Z_\delta^s + Z_{L\delta}}{1 + 2Z_{L\delta}/Z_\delta^a}. \quad (27)$$

If $Z_{L\delta}$ is small compared with Z_δ^a, (27) defines the input impedance of radiating circuits with negligible dissipation as heat. If $Z_{L\delta}$ is small compared with Z_δ^s, as with a $\lambda_0/4$ section of open-end line, (27) is practically equivalent to (16a). This is illustrated in Fig. 12.7.

If $Z_{L\delta}$ is large compared with Z_δ^s, the radiating properties of the circuits disappear and they become equivalent to three high-impedance transmission-line circuits. In particular, if $Z_{L\delta} = Z_\delta^a$, it follows that $Z_{1\delta} = Z_\delta^a/3$ and the *three* high-impedance sections of transmission line are effectively in parallel (Fig. 12.8). If $Z_{L\delta} = \infty$, $Z_{1\delta} = Z_\delta^a/2$ and the *two* high-impedance sections of transmission line are effectively in parallel (Fig. 12.9).

Case B–2: $|Z_\delta^a| \ll |Z_\delta^s|$;

Open ends, $h \doteq \tfrac{1}{4}\lambda_0$;

Closed ends, $h \doteq \tfrac{1}{2}\lambda_0 - \tfrac{1}{2}b$;

$$Z_{1\delta} = \frac{2Z_\delta^a + Z_{L\delta}}{1 + 2Z_{L\delta}/Z_\delta^s}. \quad (28)$$

If $Z_{L\delta}$ is small compared with Z_δ^s, (28) gives the impedance of three low-impedance sections of transmission line in series. Currents are antisymmetric and radiation is negligible. Except that $Z_{L\delta}$ is a low-impedance section of transmission line instead of a straight piece of conductor, (28) with $|Z_{L\delta}| \ll |Z_\delta^s|$ is essentially equivalent to (26a). In particular, if $Z_{L\delta} = Z_\delta^a$, it follows that $Z_{1\delta} = 3Z_\delta^a$, as shown in Fig. 12.10.

If $Z_{L\delta}$ is large compared with Z_δ^s, codirectional currents exist in the two parallel conductors or antennas and the circuits become radiators. This is a result of the insertion of a high impedance at a point of current maximum in conductor 1 and the reduction of all currents in this conductor to a small value. In the limit, if the circuits are broken as in Fig. 12.11 and $Z_{L\delta} \doteq \infty$, $Z_{1\delta} = Z_\delta^s/2 \doteq Z_\delta^a$. In this case the circuit is essentially equivalent to an isolated straight antenna of equal length.

In summary, it may be stated that, by suitable adjustment of the lengths of the antennas and of the tuning impedance $Z_{L\delta}$, radiating circuits can be designed that have field properties like those of isolated single antennas but quite different impedances.

An experimental study has been made* of the folded dipole using the same equipment

---

* The results reported are those of Dr. T. Morita and Dr. C. Faflick, ref. 40.

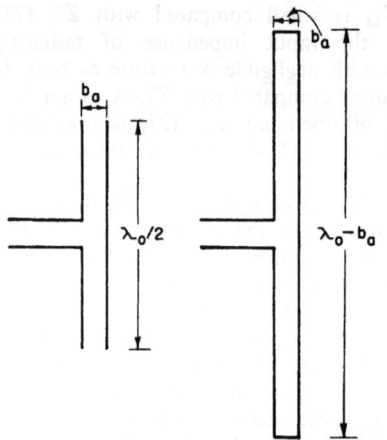

Fig. 12.6. Transmission-line circuits.

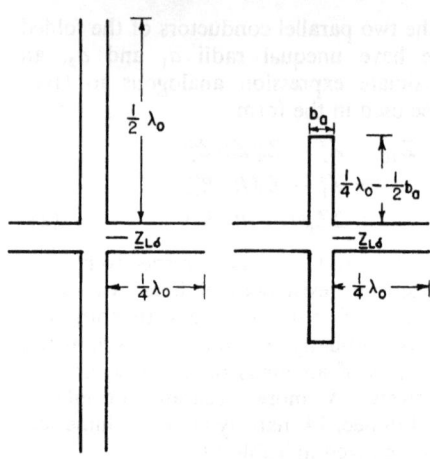

Fig. 12.7. Radiating circuits; $|Z_\delta^a| \gg |Z_\delta^s|$; $|Z_{L\delta}| \ll |Z_\delta^s|$.

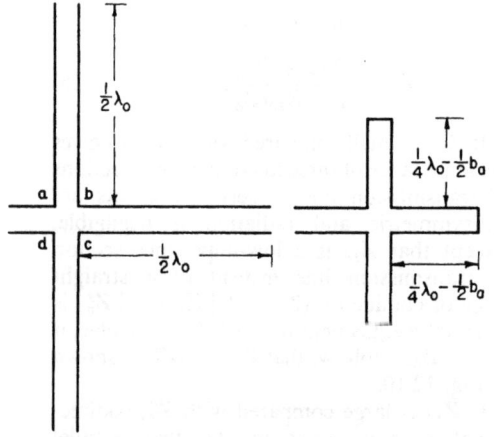

Fig. 12.8. Transmission-line circuits: 3 sections in parallel; $|Z_\delta^a| \gg |Z_\delta^s|$; $Z_{L\delta} = Z_\delta^a$.

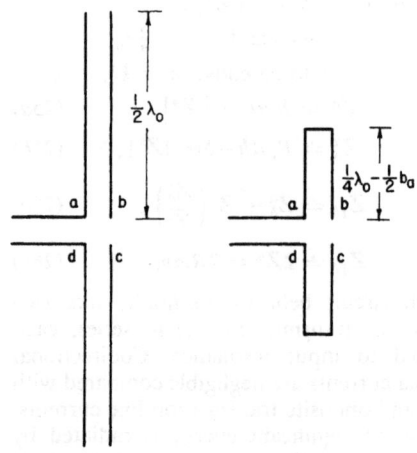

Fig. 12.9. Transmission-line circuits: 2 sections in parallel; $|Z_\delta^a| \gg |Z_\delta^s|$; $Z_{L\delta} = \infty$.

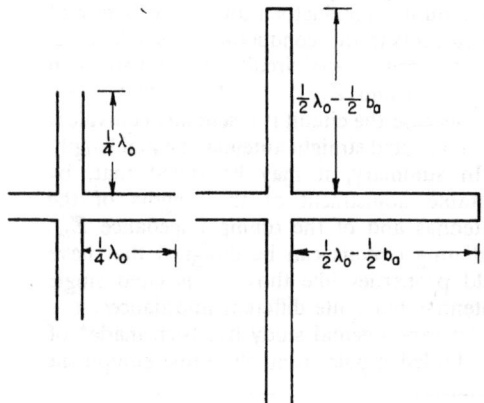

Fig. 12.10. Transmission-line circuits: 3 sections in series; $|Z_\delta^a| \ll |Z_\delta^s|$; $Z_{L\delta} = Z_\delta^a$.

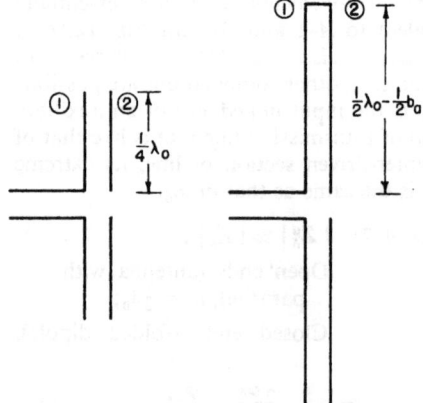

Fig. 12.11. Radiating circuits with gap in conductor 2; $|Z_\delta^a| \ll |Z_\delta^s|$; $Z_{L\delta} = \infty$.

[III.12] THEORY OF LINEAR ANTENNAS 341

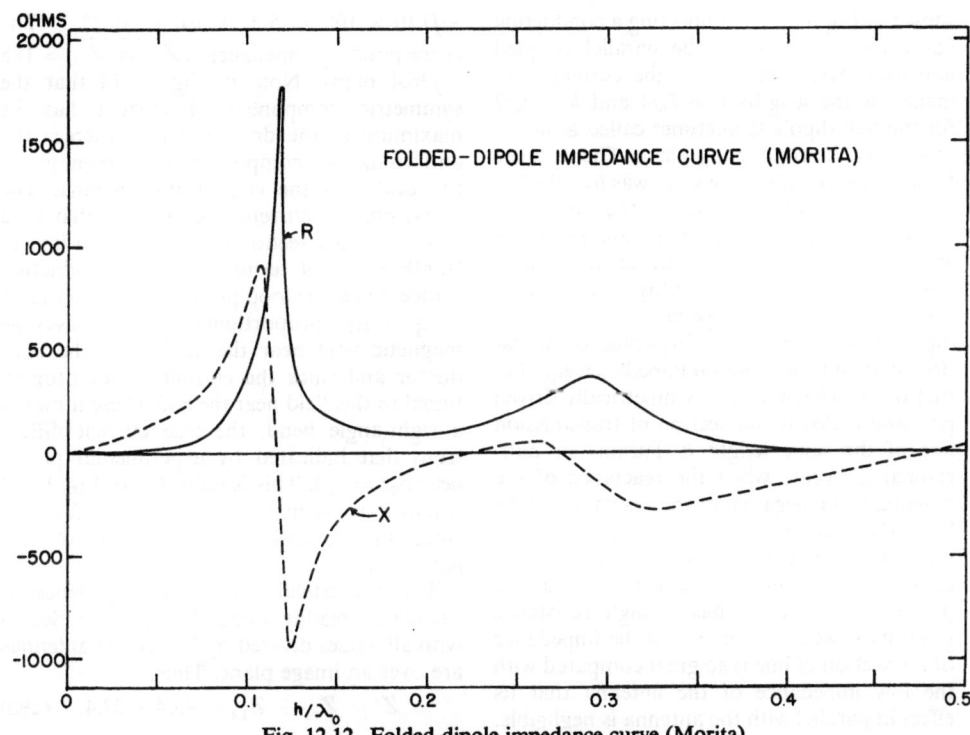

Fig. 12.12. Folded-dipole impedance curve (Morita).

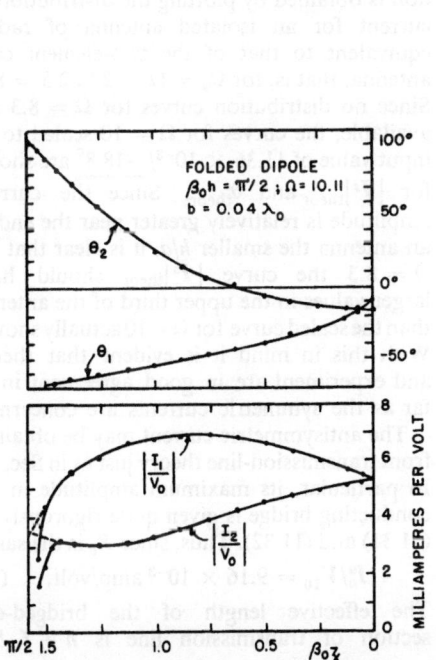

Fig. 12.13. Currents in folded dipole, $\beta_0 h = \pi/2$ (Morita).

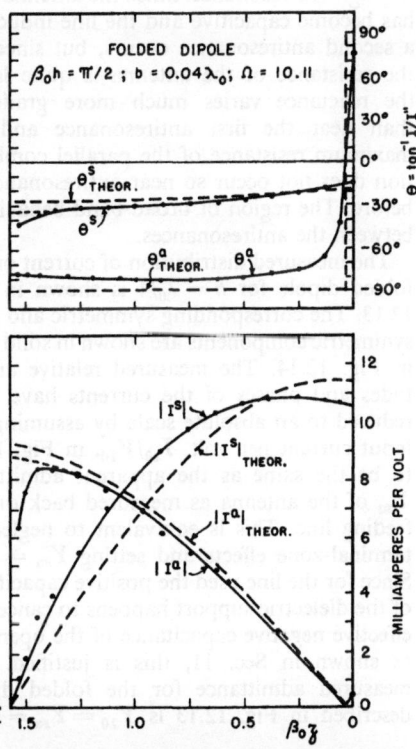

Fig. 12.14. Symmetric and antisymmetric currents in folded dipole.

shown in Fig. 11.8 by connecting a conducting bar across the ends of the parallel coupled antennas. Measurements of the current were made for the lengths $h = \lambda_0/4$ and $h = \lambda_0/2$ for the half dipole (sometimes called a monopole) over the conducting plane. The distance between the parallel long sides was $b_a = 0.04\lambda_0$.

In Fig. 12.12 are shown the measured curves of the apparent resistance and reactance as a function of the length of the folded dipole. These curves are readily understood if it is recalled that the impedance of the folded dipole is the same as that of a parallel combination of an antenna with an impedance equal to that of one element of a symmetrically driven pair and a closed-end section of transmission line of the same length h. Parallel or antiresonance occurs when the reactance of the antenna is the negative of the reactance of the line. This occurs when h is near one-eighth wavelength. For this length the antenna is capacitive, the line is inductive, and the parallel combination has a high resistance given by L/RC. Near $h = \lambda_0/4$ the impedance of the section of line is so great compared with the low impedance of the antenna that its effect in parallel with the antenna is negligible. In this range the R_1 curve corresponds to the increasing resistance of the antenna as it goes through self-resonance. Since the antenna now has become capacitive and the line inductive, a second antiresonance occurs, but since the the resistance of the antenna is quite large, the reactance varies much more gradually than near the first antiresonance and the maximum resistance of the parallel combination does not occur so near antiresonance as before. The region of broad-band behavior is between the antiresonances.

The measured distribution of current on the folded dipole for $h = \lambda_0/4$ is shown in Fig. 12.13. The corresponding symmetric and antisymmetric components are shown in solid lines in Fig. 12.14. The measured relative amplitudes and phases of the currents have been reduced to an absolute scale by assuming the input current per volt, I_{10}/V_{10}, in Fig. 12.13 to be the same as the apparent admittance Y_{sa} of the antenna as measured back on the feeding line. This is equivalent to neglecting terminal-zone effects and setting $Y_{sa} \doteq Y_{10}$. Since for the line used the positive capacitance of the dielectric support happens to cancel the effective negative capacitance of the open end as shown in Sec. 11, this is justified. The measured admittance for the folded dipole described in Fig. 12.13 is $Y_{10} \doteq Y_{sa} = (5.13 - j1.0) \times 10^{-3} = 5.22 \times 10^{-3}/\underline{-10.1°}$. The corresponding impedance is $Z_{10} \doteq Z_{sa} = 188 + j36.4$ ohms. Note in Fig. 12.14 that the symmetric component of current has its maximum at the driving point, whereas the antisymmetric component is maximum in the bridge at the end of the antenna. The antisymmetric current is essentially that of a short-circuited section of transmission line of length $h = \lambda_0/4$ terminated in a conducting bridge. Since the loop probe actually measures a quantity proportional to the average magnetic field near the surface of the conductor and since the current is not proportional to this field near the end where it makes a right-angle bend, the true current differs from that indicated by experimental points near $\beta_0 z = \pi/2$. This is indicated in Fig. 12.14 where it is particularly pronounced for $|I^a|$ since this actually has a maximum near $\beta_0 z = \pi/2$.

The theoretical symmetric impedance is calculated readily using the tables in Sec. 8 with all values divided by 2 since the antennas are over an image plane. Thus,

$$Z^s = Z_{s1} + Z_{12} = 83.4 + j28.4. \quad (29a)$$

The symmetric input current per volt is

$$Y^s = I_0^s/V_0 = (10.7 - j3.66) \times 10^{-3}$$
$$= 11.4 \times 10^{-3}/\underline{-18.8°}. \quad (29b)$$

An estimate of the symmetric current distribution is obtained by plotting the distribution of current for an isolated antenna of radius equivalent to that of the two-element cage antenna, that is, for $\Omega_e = \Omega - 2\ln 2.5 = 8.3$. Since no distribution curves for $\Omega = 8.3$ are available, the curves for $\Omega = 10$ scaled to an input value of $11.36 \times 10^{-3}/\underline{-18.8°}$ are shown for $|I^s|_{\text{theor}}$ and θ^s_{theor}. Since the current amplitude is relatively greater near the end of an antenna the smaller h/a, it is clear that for $\Omega = 8.3$ the curve $|I^s|_{\text{theor}}$ should have larger values in the upper third of the antenna than the scaled curve for $\Omega = 10$ actually shown. With this in mind it is evident that theory and experiment are in good agreement in so far as the symmetric currents are concerned.

The antisymmetric current may be obtained from transmission-line theory just as in Sec. 11. In particular, its maximum amplitude in the conducting bridge is given quite rigorously by (11.31) and (11.32). Thus, since R_c is the same,

$$I_0^a/V_{10} = 9.16 \times 10^{-3} \text{ amp/volt}. \quad (30)$$

The effective length of the bridged-end section of transmission line is $h + f_{sa}b/2$, where b is the distance $0.04\lambda_0$ between the conductors and f_{sa} is a factor determined from the theory of the inductive end effect. For $b/a = 6.3$ its value is $f_{sa} = 0.47$. The effective electrical length is $\pi/2 + \beta_0 f_{sa} b/2 = \pi/2 + 0.059$. Hence, the approximate theoretical current $|I^a|_{\text{theor}}$ is a sinusoid with a maximum amplitude of 9.16×10^{-3} amp/volt at $\beta_0 z = \pi/2 + 0.059$ (that is, in the bridge), and with zero at $\beta_0 z = 0.059$. It is shown in Fig. 12.14 together with the phase angle of the current for a lossless line. This latter is discontinuous with a value of $+90°$ for $\beta_0 z$ less than 0.059 and $-90°$ for $\beta_0 z$ between 0.059 and $\pi/2$. It is readily verified that a continuous curve closely resembling the experimental curve for θ^a is obtained if a radiation resistance of the order of magnitude $15\beta_0^2 b^2$ is included in Z_0^a. Since Z_0^a is very near antiresonance and both R_0^a and X_0^a vary rapidly and greatly for even extremely small changes in the length of the antennas, quite large differences between approximate theoretical and measured values of Z^a are to be expected. However, so long as Z^a is sufficiently large, its effect is small and such differences are of no practical interest in determining the input impedance. Numerical values of admittances and impedances are given in Table 12.1. The agreement between theory and experiment in the physically important quantities Z_{10} and Y_{10} is good.

The measured distribution of current in the folded dipole with $\beta_0 h = \pi$ is shown in Fig. 12.15. This is essentially a transmission-line circuit and not a radiating system. Since the antisymmetric or transmission-line current is so very much greater than the symmetric current, a reasonably accurate decomposition into symmetric and antisymmetric components is not possible. This is in agreement with theoretical predictions.

Coupled dipoles and folded dipoles. The impedance of each of two coupled vertical antennas on a ground screen was measured by Lewis[38a] with (*a*) both antennas self-resonant half-dipoles, (*b*) antenna 1 a half-dipole, antenna 2 a folded half-dipole, (*c*) both antennas folded half-dipoles. For the antennas, $a = 3/32$ in., $b = 0.4\lambda_0$ at 450 Mhz. The impedance of each antenna was measured with the other short-circuited to the ground screen. The following results were obtained using theoretical second-order impedances from Sec. 8, that is, $Z_{s1} = 35/0°$ for the half-dipole, $Z'_{s1} \doteq 4Z_{s1} = 140/0°$ for the folded half-dipole, $Z_{12} = 16.9/-81°$. Corresponding measured values are $\overline{Z}_{s1} = 35/0°$, $\overline{Z}'_{s1} = 141/0°$. For the three cases listed above: (*a*) theor. $Z_1 = Z_2 = 42.8/3.4°$; exp. $Z_1 = 43/5.5°$, $Z_2 = 45/6.5°$; (*b*) theor. $Z_1 = 42.8/3.4°$, $Z_2 = 171.2/3.4°$; exp. $Z_1 = 41.5/2°$, $Z_2 = 163/5°$; (*c*) theor. $Z_1 = Z_2 = 171.2/3.4°$; exp. $Z_1 = 162/-4°$, $Z_2 = 178/3°$. The agreement is good.

Bridged-parallel antenna. The bridged-parallel antenna is a generalization of the folded dipole analyzed in the preceding paragraphs. It consists of a center-driven antenna No. 1 with a parallel parasite No. 2 of equal length and a center load or tuner $Z_{L\delta}$. The two antennas are separated by a distance b and are joined by two symmetrically placed conducting bridges at distances s from the center. The circuit with $Z_{L\delta} = 0$ is shown in Fig. 12.16*a*. Evidently, when $s = h$, the bridged-parallel antenna is identical with the folded dipole.

The method of analysis is the same as that of the folded dipole. It is outlined in Fig. 12.16*b*. In brief, by defining symmetric and antisymmetric voltages and currents, namely,

$$V_{1\delta} = V_\delta^s + V_\delta^a, \quad I_{1\delta} = I_\delta^s + I_\delta^a, \quad (31)$$
$$V_{2\delta} = V_\delta^s - V_\delta^a, \quad I_{2\delta} = I_\delta^s - I_\delta^a,$$

TABLE 12.1. Admittances and impedances for folded dipole; $\beta_0 h = \pi/2$.

Experimental	Theoretical (antisymmetric resistance assumed zero)
$Y_{10} = I_{10}/V_0 = 5.22 \times 10^{-3}/-10.1°$ mho	$Y_{10} = I_{10}/V_0 = 5.21 \times 10^{-3}/-19.3°$ mho
$Z_{10} = Z_{s1} = 188 + j36.4 = 191.5/10.1°$ ohm	$Z_{10} = 181 + j63.0 = 191.8/19.3°$ ohm
$Z_0^s = 92.2 + j22.4 = 95.0/13.7°$ ohm	$Z_0^s = 84.0 + j38.5 = 92.4/24.6°$ ohm
$Z_0^a = 1960 - j450 = 2011/-12.9°$ ohm	$Z_0^a = 0 - j1865 = 1865/-90°$ ohm

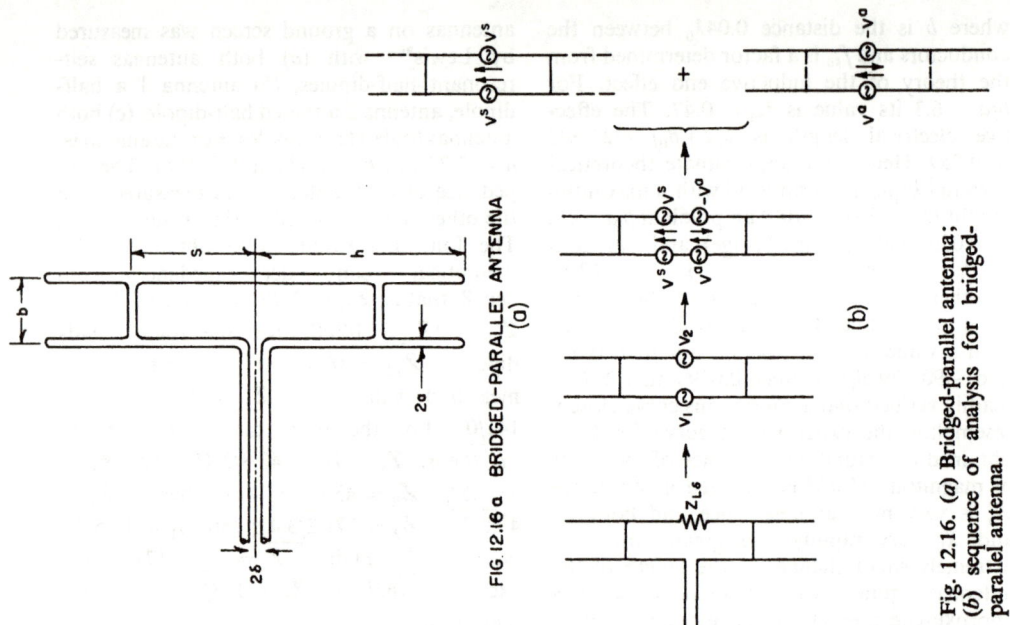

FIG. 12.16 a BRIDGED-PARALLEL ANTENNA
(a)

(b)

Fig. 12.16. (a) Bridged-parallel antenna; (b) sequence of analysis for bridged-parallel antenna.

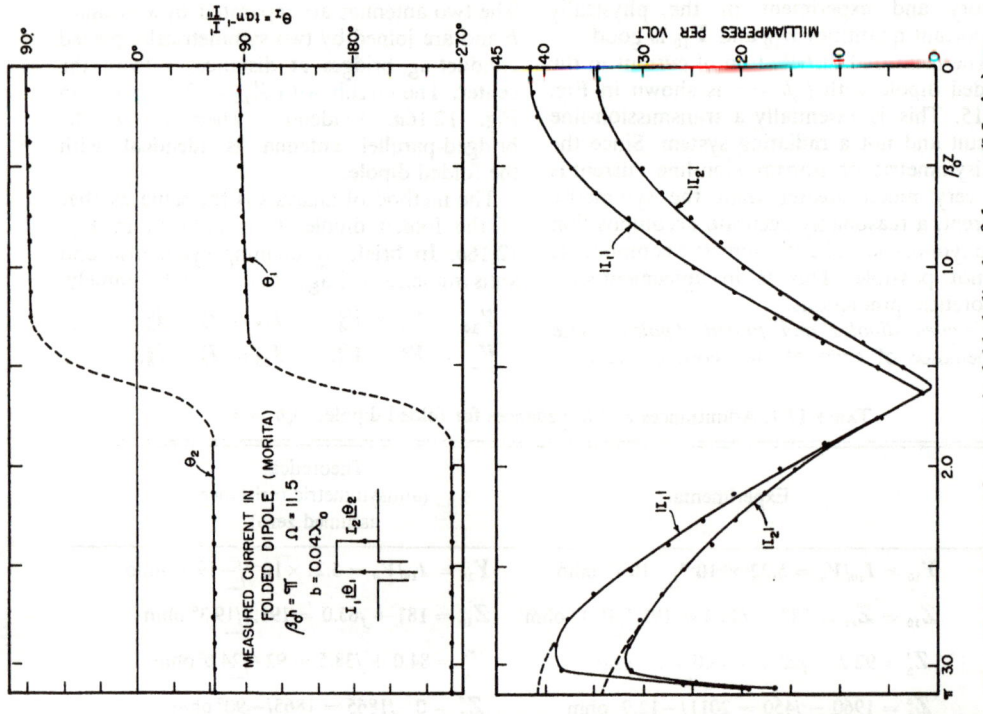

Fig. 12.15. Measured current in folded dipole, $\beta_0 h = \pi$ (Morita).

THEORY OF LINEAR ANTENNAS

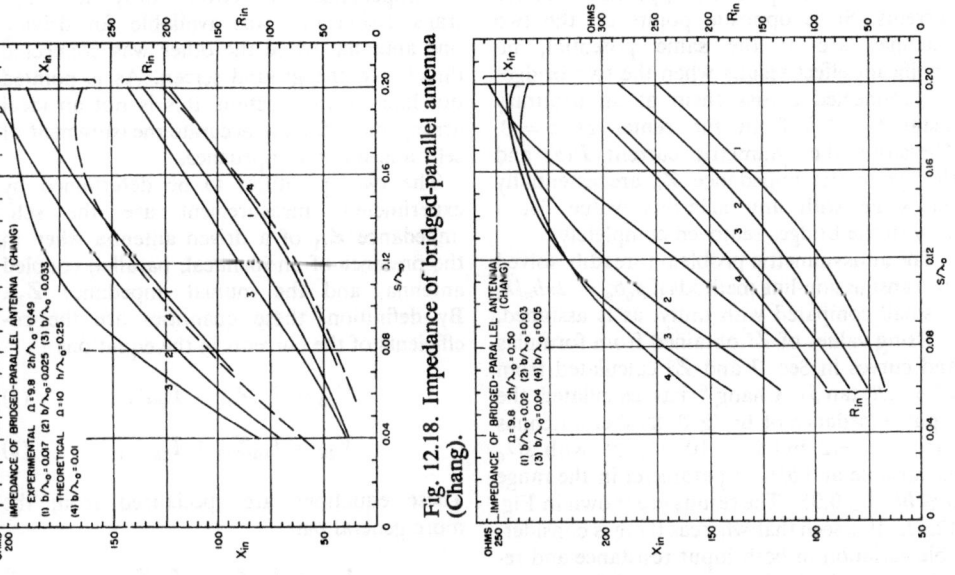

Fig. 12.18. Impedance of bridged-parallel antenna (Chang).

Fig. 12.19. Impedance of bridged-parallel antenna (Chang).

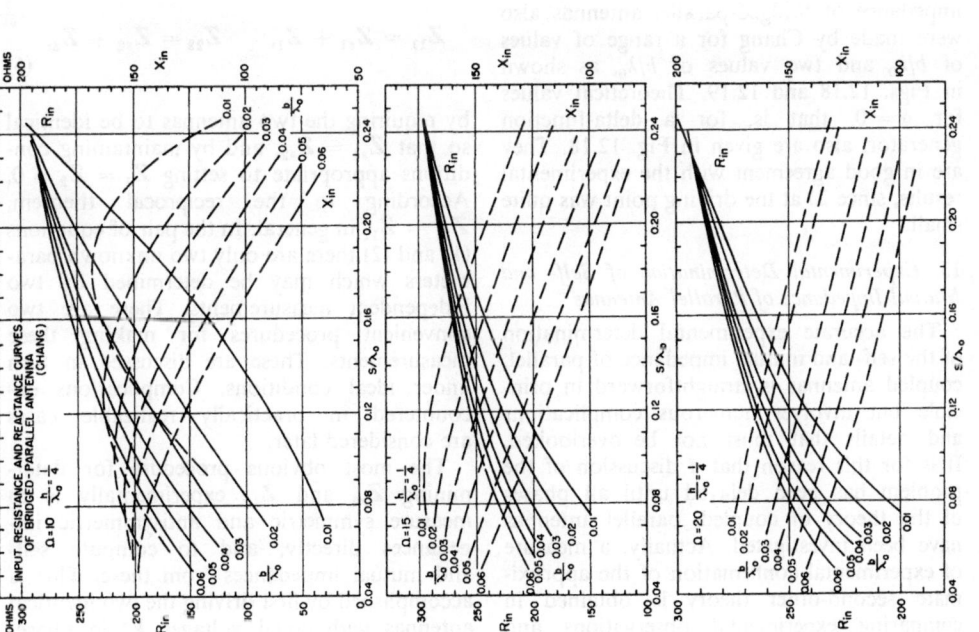

Fig. 12.17. Input resistance and reactance curves of bridged-parallel antenna (Chang).

the problem is resolved into an antenna problem for determining equal codirectional currents and a transmission-line problem for determining equal and oppositely directed currents. Since opposite points on the two antennas are at the same potential, no significant effect results when the two bridges are connected across them at an arbitrary distance $s \leq h$ from the center at $z = 0$. Therefore, the symmetric current $I^s(z)$ and the symmetric impedance Z_0^s are essentially the same with any arbitrary value for s as with the bridges removed completely.

The antisymmetric problem is readily solved by transmission-line methods if $\beta_0 b_a = 2\pi b_a/\lambda_0$ is small compared with unity, as is assumed.

Using values of Z^s obtained from formulas and curves in Sec. 7 and Z^a calculated from $Z^a \doteq jR_c \tan \beta s$, Chang[16] has calculated the input impedance of bridged-parallel antennas for $\beta_0 h = \pi/2$ and $\Omega = 10, 15, 20$, with s/λ_0 as variable and b/λ_0 as parameter in the range $0 \leq b/\lambda_0 \leq 0.06$. The results are shown in Fig. 12.17. It is seen that whereas there is considerable variation in both input resistance and reactance as the bridges are moved from quite near the driving point toward the ends, zero reactance does not occur.

Experimental measurements of the apparent impedance of bridged-parallel antennas also were made by Chang for a range of values of b/λ_0 and two values of h/λ_0, as shown in Figs. 12.18 and 12.19. Theoretical values for $\delta = 0$, that is, for a delta-function generator, also are given in Fig. 12.18. They are in good agreement with the experimental results, since 2δ at the driving point was quite small.

13. Experimental Determination of Self- and Mutual Impedance of Parallel Antennas

The accurate experimental determination of the self- and mutual impedance of parallel, coupled antennas is straightforward in principle but involves numerous complications and details that must not be overlooked. It is for this reason that a discussion of the problem has been delayed until all phases of the theory of coupled, parallel antennas have been investigated. Actually, a measure of experimental confirmation of the approximate second-order theory is obtained in comparing experimental observations and theoretical calculations for the driven antenna with coupled parasite and for the folded dipole in Secs. 11 and 12. However, the experimental arrangements in both cases were designed primarily for the determination of the relative magnitudes and distributions of current, not for the precision measurement of impedance. Moreover, only a single transmission line was available for driving one antenna, while the other was connected directly to the ground screen. As is pointed out later in this section, this is not an ideal arrangement for the accurate measurement of self- and mutual impedance.

The two quantities to be determined by experimental measurement are the self-impedance Z_{s1} of a driven antenna when in the presence of an identical, parallel, coupled antenna, and the mutual impedance Z_{12}. By definition, these quantities are the coefficients of the currents in the equations

$$V_{10} = I_{10}Z_{s1} + I_{20}Z_{12}, \quad (1)$$

$$V_{20} = I_{10}Z_{12} + I_{20}Z_{s1}. \quad (2)$$

These equations are specialized from the more general pair

$$V_{10} = I_{10}Z_{11} + I_{20}Z_{12}, \quad (3)$$

$$V_{20} = I_{20}Z_{21} + I_{20}Z_{22}, \quad (4)$$

where

$$Z_{11} = Z_{s1} + Z_1, \quad Z_{22} = Z_{s2} + Z_2, \quad (5)$$

by requiring the two antennas to be identical so that $Z_{s1} = Z_{s2}$, and by maintaining conditions appropriate to setting $Z_1 = Z_2 = 0$. According to the reciprocal theorem, $Z_{21} = Z_{12}$ in general. In the pair of equations (1) and (2) there are only two unknown parameters which may be determined by two independent measurements. There are two convenient procedures for making these measurements. These are discussed in turn under ideal conditions. Complications encountered in practically realizable cases are considered later.

The most obvious procedure for determining Z_{s1} and Z_{12} experimentally is to measure symmetric and antisymmetric impedances directly, and to compute self- and mutual impedances from these. This is accomplished by first driving the two identical antennas with equal voltages V^s in phase, and measuring the impedance of one of them, namely Z^s. Next, the antennas are driven with equal voltages V^a in phase-opposition and the impedance of one antenna again is measured. This is Z^a. Analytically, the two cases are obtained from (1) and (2) as follows:

$$V_{10} = V_{20} = V^s = I_{10}(Z_{s1} + Z_{12}) = I^s Z^s,$$
$$I_{10} = I_{20} = I^s, \qquad (6)$$
$$V_{10} = -V_{20} = V^a = I_{10}(Z_{s1} - Z_{12}) = I^a Z^a,$$
$$I_{10} = -I_{20} = I^a. \qquad (7)$$

With Z^s and Z^a measured, Z_{s1} and Z_{12} are determined from

$$Z_{s1} = \tfrac{1}{2}(Z^s + Z^a), \qquad Z_{12} = \tfrac{1}{2}(Z^s - Z^a). \qquad (8)$$

An alternative and equally direct procedure is to drive only antenna 1, so that $V_{20} = 0$. Instead of using (1) and (2), conditions are provided to permit use of (3) and (4). They are $Z_1 = 0$, and identical antennas so that $Z_{s1} = Z_{s2}$. With these restrictions the equations are

$$V_{10} = I_{10} Z_{s1} + I_{20} Z_{12}, \qquad (9)$$
$$0 = I_{10} Z_{12} + I_{20}(Z_{s1} + Z_2), \qquad (10)$$

where Z_2 is at the experimenter's disposal. The first of the two required measurements is made with Z_2 infinite. That is, an ideal break is made in antenna 2 at its center so that I_{20} vanishes. When this is true, (9) reduces to

$$V_{10} = I_{10} Z_{s1}. \qquad (11)$$

Thus, when antenna 2 is open-circuited at the center, the input impedance of antenna 1 reduces exactly to Z_{s1}, since

$$Z_{10}(Z_2 = \infty) = V_{10}/I_{10} = Z_{s1}. \qquad (12)$$

By measuring $Z_{10}(Z_2 = \infty)$, Z_{s1} is obtained directly. A second measurement of the input impedance of antenna 1 now can be made with Z_2 assigned any other convenient, known value. The most obvious is zero. Thus, with $Z_2 = 0$, I_{20} can be eliminated from (9), and this solved for the impedance Z_{10}:

$$Z_{10}(Z_2 = 0) = V_{10}/I_{10}$$
$$= Z_{s1}(1 - Z_{12}^2/Z_{s1}^2). \qquad (13)$$

This equation can be solved for Z_{12} in terms of the two directly measured quantities $Z_{10}(Z_2 = 0)$ and Z_{s1}. Thus,

$$Z_{12} = Z_{s1}\sqrt{1 - Z_{10}(Z_2 = 0)/Z_{s1}}. \qquad (14)$$

In terms of ideal conditions there is little to choose between the symmetric-antisymmetric method and the open-circuit-short-circuit method. Each is straightforward and experimentally direct. In practice, both methods are subject to difficulties and complications that must be resolved if even reasonably correct results are to be achieved. Let the two methods be considered in turn from the point of view of actual measurements with transmission lines.

The principal problems encountered with the symmetric-antisymmetric method of determining self- and mutual impedances are: (a) the necessity of driving independently two antennas with voltages that are equal in magnitude and either in phase or 180° out of phase; (b) terminal-zone effects involving the coupling between each antenna and its driving line, between each antenna and the driving line of the other antenna, and between the two driving lines. The nature of these problems differs with the type of circuit used. Let three important circuits be considered successively.

(1) The image-plane open-wire line described in Sec. II.35 is readily adapted to the measurement of symmetric and antisymmetric impedances since it consists of two symmetrically placed feeding lines. The corresponding circuit with its image is equivalent to Fig. 1.1. Although the adjustment for equal driving voltages and equal or opposite phases is difficult, the possibility of simultaneously measuring impedances on each of the two feeding lines provides a continual check on these conditions. Only if the impedances of the two antennas as measured on their individual feeding lines are equal can it be assumed that the required conditions are maintained. The terminal-zone problem in the circuit of Fig. 1.1 is one of the most complicated, since it involves coupling between each antenna and both lines and between the two lines. However, the approximate analysis is straightforward and is carried out in Sec. 9, where separate formulas are given for the lumped elements of terminal-zone networks for both the symmetric and the antisymmetric cases. With these available the measured *apparent* symmetric and antisymmetric impedances can be reduced to the ideal theoretical symmetric and antisymmetric impedances for lines of zero spacing, that is, to the circuit equivalent of slice generators.

(2) The terminal-zone problem is minimized in the case of a single antenna by arranging the antenna in the neutral plane of the line, as described in Sec. II.36. If two antennas are individually driven in this manner, as shown in Fig. 1.5, the coupling between

each antenna and both lines is eliminated. However, the coupling between the two lines and between the two necessarily parallel and adjacent supporting and feeding bridges is not. This is discussed in Sec. 9. Since imageplane methods are not possible when the antennas are in the neutral plane of the two-wire lines, very long feeding lines are required in order to eliminate effects due to the observer and the generating and measuring equipment. This greatly complicates the practical problem of maintaining voltages that are equal and in the prescribed phase relations.

(3) The simplest method of arranging two parallel antennas from the point of view of maintaining the required symmetric and antisymmetric voltages is to erect them vertically over a conducting plane with each antenna base-driven from a coaxial line that pierces the plane, as shown in Fig. 1.6. The apparent impedance of each antenna can be measured by the techniques described for a single antenna in Secs. I.37 and I.38. To permit measurements with different distances b between the antennas, either a carefully designed covered slot permitting the continuous adjustment of b or a series of holes with fitted plugs is required. When the distance b between the axes of the two antennas is large compared with the diameter of the holes in the conducting plane through which the antennas project, the interaction between the two transmission lines is negligible and the terminal-zone problem for each antenna is the same as if it were isolated. In this case the terminal-zone network derived in Sec. II.10 may be used for each antenna just as in Sec. II.38 in order to obtain the ideal or theoretical impedance for the limiting case of a hole of zero radius from the apparent impedance measured with holes of finite size. In this manner, Z_0^s and Z_0^a may be evaluated. When the antennas are very close together, the two coaxial lines are coupled and a complicated analytical problem is provided. The magnitude of the coupling can be estimated experimentally by removing the antennas and leaving only the open ends of the two coaxial lines. By comparing the apparent impedance of these when separated far to make the interaction negligible with the impedances when moved closer and closer together, the minimum distance at which the interaction may be neglected can be determined for both symmetric and antisymmetric excitation. For distances b smaller than this minimum the apparent impedances of symmetric and antisymmetric antennas may be measured, but it is not profitable in general to attempt to determine ideal theoretical impedances for holes of zero diameter from the measured apparent values when the distance between the adjacent edges of the holes does not exceed their diameter.

The experimental determination of self- and mutual impedances by measuring the apparent impedance of antenna 1 when antenna 2 has first an infinite and then a zero load at its center (or base if over a conducting plane) does not differ greatly from the determination from measured values of apparent symmetric and antisymmetric impedances. The fact that only one antenna is driven eliminates the need for accurate maintenance of equal amplitudes and equal or opposite phases. It does not, however, eliminate the need for the second transmission line. Since the two antennas must be alike, they must be connected to identical transmission lines. The line connected to antenna 2 must then be terminated in a pure reactance and adjusted in length so that its input impedance as seen from the terminals of antenna 2 is either so great that it may be regarded as infinite, or so small that it may be treated as a short circuit. These adjustments are very critical and necessarily involve terminal-zone effects. It must be emphasized in particular that unless terminal-zone effects are negligible, as when the distance between the two conductors of the transmission lines is a very small fraction of a wavelength, an adequate short circuit is not provided by having antenna 2 a continuous conductor. This is discussed in Sec. 11. It is even less satisfactory to cut a piece out of antenna 2 and assume that this is equivalent to an open circuit. The new chargeable surfaces provided in this way introduce an added capacitance which may correspond to a reactance at the center of antenna 2 that is very far from infinite.

The determination of ideal theoretical impedances from measured apparent impedances is complicated when Z_{s1} and Z_{10} ($Z_2 = 0$) are measured than when Z^s and Z^a are measured, since terminal-zone networks are more readily calculated when the distributions of current and charge in the two antennas and lines are the same. However, the following procedure is available: (1) The *apparent* symmetric and antisymmetric impedances $\hat{Z}_{sa}^s$ and Z_{sa}^a are determined from the measured apparent value of Z_{s1}, and the apparent value of Z_{12} is computed from measured apparent values of Z_{s1} and Z_{10} ($Z_2 = 0$). (2) The terminal-zone networks appropriate to determine the ideal values of Z_0^s and Z_0^a from the apparent values Z_{sa}^s and Z_{sa}^a are used to obtain the former. Finally, (3) the ideal theoretical values of Z_{s1} and Z_{12} are determined by combining the ideal values of Z_0^s and Z_0^a.

The general problem of determining self- and mutual impedances of coupled, parallel antennas has been discussed in this section together with available circuits and the major difficulties associated with each. Preliminary measurements have been made by Moritz using two coaxial lines with their inner conductors projecting through holes in a conducting plane to form the parallel antennas. By measuring the impedance terminating each line when these were driven in phase and in phase opposition with equal voltages, and averaging the results, the symmetric and antisymmetric impedances were measured and from these the self- and mutual impedances were computed. Alternatively, only one of the lines was driven while the other was short-circuited with a piston successively at one-quarter and one-half wavelength from the antenna-line junction. In this manner, using (12) and (13), Z_{s1} was measured directly and Z_{12} computed from Z_{10} and Z_{s1}. For an antenna with $\beta_0 a = 0.03$ and with the two antennas separated by an electrical distance $\beta_0 b = \pi$, the following results were obtained: $\beta_0 h = \pi/2$, $Z_{s1} = 100 + j42$ with both lines driven, $Z_{s1} = 99.9 + j44$ with one line driven. Correspondingly, with $\beta_0 h = \pi$ the results were $Z_{s1} = 393 - j530$ and $Z_{s1} = 401 - j538$. These numerical values are double those actually determined from measurement to facilitate comparison with the center-driven antenna. It is seen that both methods lead to the same results. No corrections have been made for transmission-line end effects since the diameter of the coaxial line was small and the ratio of the radii of outer to inner conductors was only 2.33.

A somewhat different procedure for measuring self- and mutual impedances has been developed by Moritz. Instead of measuring two impedances (for example, Z^s and Z^a in the symmetric-antisymmetric method, Z_{s1} and $Z_{s1}(1 - Z_{12}^2/Z_{s1}^2)$ in the short-circuit-open-circuit method), only one impedance and the ratio of two voltages are measured. The circumstances are as follows: Antenna 1 is driven, antenna 2 is parasitic and open-circuited so that $I_{20} = 0$. Since $Z_2 = \infty$ while $I_{20} = 0$, the product $-I_{20}Z_2 = V_{20}$ is the finite voltage across the open circuit. Thus, (9) and (10) become

$$V_{10} = I_{10}Z_{s1}, \quad (15)$$

$$V_{20} = I_{10}Z_{12}, \quad (16)$$

where V_{20} is not an applied voltage but is the potential difference across $Z_2 = \infty$ when V_{10} is the driving voltage.

By measuring Z_{s1} directly as the input impedance of antenna 1 in the presence of antenna 2 when open-circuited, and also measuring the magnitude and phase of the voltage ratio

$$V_{20}/V_{10} = Z_{12}/Z_{s1} = ke^{j\theta}, \quad (17)$$

it is possible to evaluate Z_{12} from the known values of Z_{s1}, k, and θ. This method is experimentally simpler than the symmetric-antisymmetric procedure, and more accurate than the open-circuit-short-circuit method over ranges occurring in this method when Z_{12} is determined from the small difference of two rather large quantities.

Curves obtained by Moritz* using this method are shown in Figs. 13.1 and 13.2.

For identical antennas with $Z_{s1} = Z_{s2}$, the voltage ratio (17) is the complex coefficient of coupling defined in general by

$$k = ke^{j\theta} = Z_{12}/\sqrt{Z_{s1}Z_{s2}}. \quad (18)$$

(This coefficient is defined in ref. I.33, p. 410, Eq. (6), and in ref. 32.) This coefficient has been measured by Blasi,[7] using a modified open-circuit-short-circuit method *in which the junction effects at the base of the driven antenna are compensated.* Using the *theoretical* value of Z_{s1} and measured values of k, Z_{12} was determined. Blasi's experimental points for R_{12} and X_{12} are shown in Fig. 13.3. The two antennas were identical with $\beta_0 h = \pi/2$ and $\Omega = 11$. The agreement is seen to be good except for the larger values of b/λ_0, where the experimental results depend upon the small difference of two fairly large quantities, and the very small values of b/λ_0, where the approximate second-order theory is less accurate.

---

* Unpublished results.

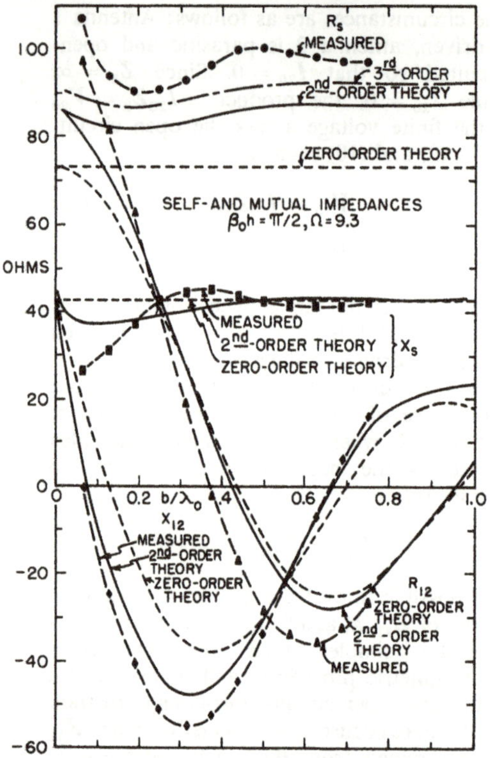

Fig. 13.1. Self- and mutual impedances, $\beta_0 h = \pi/2$ (Moritz).

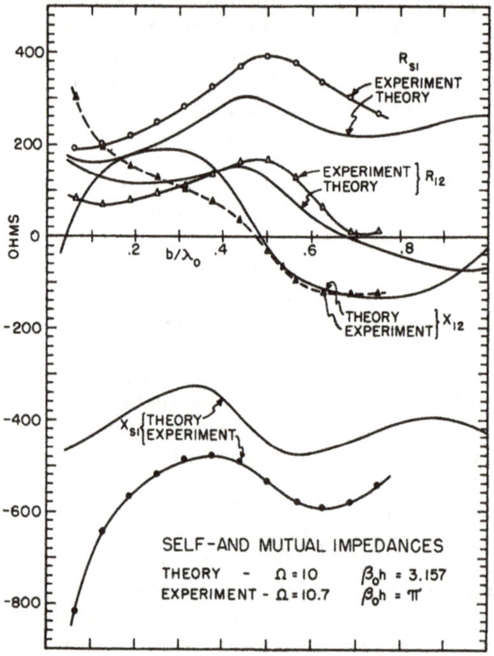

Fig. 13.2. Self- and mutual impedances near $\beta_0 h = \pi$ (Moritz).

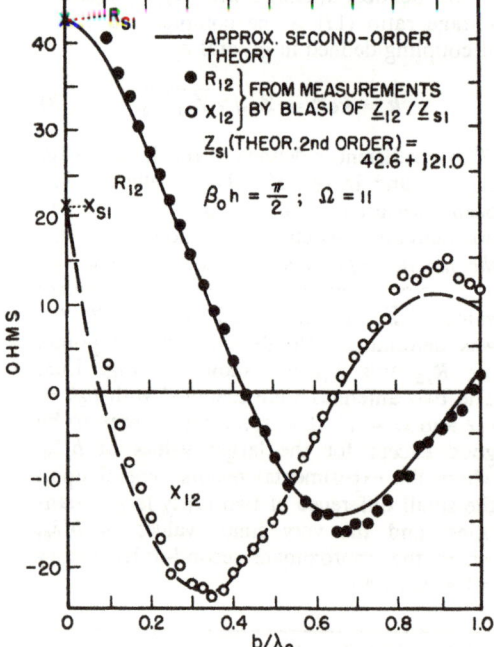

Fig. 13.3. Experimentally determined mutual resistance and reactance for parallel antennas as functions of spacing b/λ_0.

THEORY OF IDENTICAL, PARALLEL, NONSTAGGERED ANTENNAS

14. General Theory

The arrays of antennas to be discussed in this section consist of N geometrically identical units each of half-length h and radius a that satisfy the inequalities $\beta_0 a \ll 1$, $a \ll h$. The several units may be center-driven by arbitrary voltages or center-loaded by different impedances provided the assumed geometric identity of the units is not significantly disturbed by the driving or load connections or the structure of generators or impedors. In practice, this means either that the driving and load connections are all alike—as when similar open-wire lines are used between all antennas and the generators or loads—or that generators and loads or the cross-sectional dimensions of transmission lines are so small in physical size that generators and loads are equivalent to dimensionless, lumped elements.

The discussion is limited to arrays of *parallel* antennas so that all currents that contribute significantly to the far-zone field are also parallel and hence maintain a vector potential that has no component perpendicular to the antennas; these are assumed to be parallel to the z-axis of a coördinate system, that is, $\mathbf{A} = \hat{\mathbf{z}} A_z$.

Each antenna is required to have geometric symmetry with respect to its center, and it must be driven or loaded at the center. Furthermore, all elements have their centers in the xy-plane, that is, they are not staggered. These conditions are imposed in order that the currents and vector potentials shall be even functions of z, the charges and scalar potentials odd functions of z. Thus,

$$I_z(-z) = I_z(z), \quad q(-z) = -q(z), \quad (1)$$
$$A_z(-z) = A_z(z), \quad \Phi(-z) = -\Phi(z). \quad (2)$$

Let it be assumed at the outset that all generators are "slice" generators, that is, discontinuities in scalar potential, and all load impedances or tuning reactances lumped or geometrically dimensionless. These assumptions eliminate complications arising from transmission-line end effects and coupling effects that are necessarily characteristic of particular driving and coupling circuits. Since such junction effects are localized, they do not alter the formulation of the general problems of coupled antennas, and account may be taken of them later, using appropriately designed corrective networks of lumped elements. Accordingly, let the driving voltage at the center of a typical antenna k be defined as follows:

$$V_{k0} = \lim_{\delta \to 0} [\Phi_k(\delta) - \Phi_k(-\delta)]. \quad (3)$$

If the several conditions that have been imposed thus far are satisfied, the vector potential $A_z(z)$ on the surface of each of the N antennas satisfies the same one-dimensional wave equation as when the antenna is isolated. For example, on the surface of the kth antenna the equation is

$$d^2 A_{kz}(z)/dz^2 + \beta_0^2 A_{kz}(z) = j(\beta_0^2/\omega) z^i I_{kz}(z)$$
$$\doteq 0. \quad (4)$$

For simplicity, the term that has the internal impedance per unit length z^i as a factor will be omitted. For good conductors its effect is negligibly small. If desired, the appropriate terms are easily added to the final solution. The general solution of the homogeneous equation for the range $0 \leq z \leq h$ is obtained as in Chapter II. It is

$$A_{kz}(z) = -(j/v_0)[C_k \cos \beta_0 z + \tfrac{1}{2} V_{k0} \sin \beta_0 |z|]. \quad (5)$$

The arbitrary constants C_k must be determined from the boundary condition at $z = h$.

The integral for the vector potential at a point z on the surface of antenna k is as follows:

$$A_{kz}(z) = \frac{1}{4\pi v_0} \int_{-h}^{h} \sum_{i=1}^{N} I_i(z_i') \frac{e^{-j\beta_0 R_{ki}}}{R_{ki}} dz_i', \quad (6a)$$

where

$$R_{ki} = \sqrt{(z_k - z_i')^2 + b_{ki}^2}, \quad i \neq k$$
$$R_{kk} = \sqrt{(z_k - z_k')^2 + a^2}. \quad (6b)$$

Subscripts are used on the coördinate z, where necessary, to indicate on which antenna the distance from the xy-plane is measured. Note that the primed coördinates locate an element of integration on the axis of an antenna, the unprimed coördinates an element on the surface; b_{ki} is the distance between the axes of antenna i and antenna k.

The substitution of (6a) in (5) gives one of N simultaneous integral equations for the N currents in the N coupled elements of the array. As usual, $\zeta_0 = v_0/\nu_0 \doteq 120\pi$ ohms:

$$\int_{-h}^{h} \sum_{i=1}^{N} I_{iz}(z_i') \frac{e^{-j\beta_0 R_{ki}}}{R_{ki}} dz_i'$$
$$= \frac{-j4\pi}{\zeta_0} (C_k \cos \beta_0 z_k + \tfrac{1}{2} V_{k0} \sin \beta_0 |z_k|)$$
$$k = 1, 2, \ldots, N. \quad (7)$$

Since the solution of N simultaneous integral equations of this type has not been accomplished even for $N = 2$, the only cases for which solutions are available are those in which all of the N equations are alike, so that they reduce to only a single equation like that in Chapter II. Evidently, such a reduction to a single equation involves restrictions in addition to those already imposed. Mathematically, it implies that the current $I_{kz}(z')$, which is included in the sum in (7), can be factored out in front of the sign of summation and the remaining sum is a constant that does not vary with the numerical value of k. That is, the desired equation must have the following form:

$$\int_{-h}^{h} I_{kz}(z') K_\Sigma(z, z') \, dz' = \frac{-j4\pi}{\zeta_0} [C_k \cos \beta_0 z + \tfrac{1}{2} V_{k0} \sin \beta_0 |z|], \quad (8a)$$

where

$$K_\Sigma(z, z') \equiv \sum_{i=1}^{N} \frac{I_{iz}(z_i')}{I_{kz}(z_k')} \frac{e^{-j\beta_0 R_{ik}}}{R_{ik}} \quad (8b)$$

is independent of k. Evidently, this is not true in general. It is necessary that both the ratio of currents and the function of R_{ik} be independent of k.

Physically, the reduction of the N simultaneous integral equations to a single equation necessarily means that the N units are electrically and geometrically indistinguishable. These conditions certainly require the antennas to be located at the corners of an N-sided equilateral polygon. If this is true, $\sum_{i=1}^{N} e^{-j\beta_0 R_{ik}}/R_{ik}$ is obviously independent of k. The simplest way to make the current ratio in (8b) independent of k is to require all currents to be equal in magnitude and in phase, a condition achieved by identical driving voltages on all units. However, this condition is more severe than required. Another, somewhat more general possibility is to have the currents in all units equal in magnitude but advancing in phase from unit to unit around the polygon by a fixed angle θ. That is,

$$I_{iz}(z') = I_{i-1,z}(z') p^m,$$
$$m = 0, 1, 2, \ldots, N - 1, \quad (9a)$$

where

$$p = e^{j\theta}, \quad \theta = 2\pi/N, \quad (9b)$$

or,

$$I_{iz}(z') = I_{kz}(z') p^{(i-k)m}. \quad (10)$$

With (9a) substituted in (7), this reduces to the desired form (8a) with

$$K_\Sigma(z, z') = \sum_{i=1}^{N} p^{(i-k)m} \frac{e^{-j\beta_0 R_{ik}}}{R_{ik}}. \quad (11)$$

It is readily shown that (11) is independent of k by adding to k an arbitrary integer q equal to or less than $N - k$. Since the *order* of summation makes no difference, let the same integer q be added to i to give $i' = i + q$. The limits of summation of i' are the same as for i. Thus,

$$K_\Sigma(z, z') = \sum_{i'=1}^{N} p^{(i'-k-q)m} \frac{e^{-j\beta_i R_{i'k+q}}}{R_{i'k+q}}. \quad (12)$$

Since $i' - k - q = i - q$, and geometric symmetry makes

$$R_{ik} = R_{i+q,k+q}, \quad (13)$$

it follows that (12) is identically the same as (11). Since $K_\Sigma(z, z')$ is independent of k, it is convenient to choose $k = 1$ and define the kernel $K_\Sigma(z, z')$ as follows:

$$K_{\Sigma m}(z, z') = \sum_{i=1}^{N} p^{(i-1)m} \frac{e^{-j\beta_0 R_{i1}}}{R_{i1}}. \quad (14)$$

Currents of equal magnitude and with a progressive, constant phase difference around the symmetrically located units are possible only when the driving voltages are equal in magnitude and have the same phase relations as the currents. That is,

$$V_{i0} = V_{k0} p^{(i-k)m}. \quad (15)$$

Since the constant C_k in (8a) has V_{k0} as a factor when it is evaluated, it follows that multiplication of (8a) by $p^{(i-k)m}$ on both sides changes it from an equation for antenna k to an equation for antenna i.

The integral equation for the current in the upper half of each antenna in the symmetric array is

$$\int_{-h}^{h} I(z') K_{\Sigma m}(z, z') \, dz' = \frac{-j4\pi}{\zeta_0} (C \cos \beta_0 z + \tfrac{1}{2} V_0 \sin \beta_0 z), \quad (16)$$

with $K_{\Sigma m}(z, z')$ given in (14). This equation is formally exactly like (3.2) for two coupled antennas when specialized to $\delta = 0$ as here assumed. However, the kernel is much more complicated, since it involves N terms instead of two. The solution of (16) evidently is the same as (3.9) with $\delta = 0$ and $z^i = 0$, and with $K_{\Sigma m}(z, z')$ substituted for $K_s(z, z')$.

By mere changes in notation an expression for the current like (3.10) is obtained with an expansion parameter defined by (3.11) with $K_{\Sigma m}(z, z')$ substituted for $K_s(z, z')$. The current so obtained applies to any one of the N units, say antenna 1. Currents in the other units are obtained by multiplying this current by $p^{m(i-1)}$ with $i = 2, 3, \ldots, N$.

It has been shown that the general problem of N coupled, parallel, identical, nonstaggered antennas reduces to a single integral equation that is like the equation for an isolated antenna provided the antennas are all located at the corners of a regular polygon and provided they are individually excited by generators or transmission lines that maintain voltages across their terminals that are equal in magnitude and vary progressively in phase by a constant angle from unit to unit around the polygon. For an array of N antennas the constant phase angle may have any one of the values $m\theta$, where $\theta = 2\pi/N$ and m is an integer less than N. Since $m = 0$ is a possible value, it is seen that there are N available and different angles for each of which a single integral equation governs all currents. These N possible values of m are called *phase sequences*. They may be defined as follows:

Zero-phase sequence, $m = 0$:

$$V_{10} = V_{20} = V_{30} = \cdots = V_{N0} = V^{(0)},$$
$$I_{10} = I_{20} = I_{30} = \cdots = I_{N0} = I^{(0)}; \quad (17a)$$

$2\pi/N$-*phase sequence*, $m = 1$:

$$V_{10} = V^{(1)}, \quad V_{20} = pV^{(1)},$$
$$V_{30} = p^2 V^{(1)}, \cdots,$$
$$V_{N0} = p^{N-1} V^{(1)},$$
$$I_{10} = I^{(1)}, \quad I_{20} = pI^{(1)},$$
$$I_{30} = p^2 I^{(1)}, \cdots,$$
$$I_{N0} = p^{N-1} I^{(1)}; \quad (17b)$$

$4\pi/N$-*phase sequence*, $m = 2$:

$$V_{10} = V^{(2)}, \quad V_{20} = p^2 V^{(2)},$$
$$V_{30} = p^4 V^{(2)}, \cdots,$$
$$V_{N0} = p^{2N-2} V^{(2)},$$
$$I_{10} = I^{(2)}, \quad I_{20} = p^2 I^{(2)},$$
$$I_{30} = p^4 I^{(2)}, \cdots,$$
$$I_{N0} = p^{2N-2} I^{(2)}; \quad (17c)$$

$2m\pi/N$-*phase sequence*:

$$V_{10} = V^{(m)}, \quad V_{20} = p^m V^{(m)},$$
$$V_{30} = p^{2m} V^{(m)}, \cdots,$$
$$V_{N0} = p^{mN-m} V^{(m)},$$
$$I_{10} = I^{(m)}, \quad I_{20} = p^m I^{(m)},$$
$$I_{30} = p^{2m} I^{(m)}, \cdots,$$
$$I_{N0} = p^{mN-m} I^{(m)}; (17d)$$

$2(N-1)\pi/N$-*phase sequence*, $m = N - 1$:

$$V_{10} = V^{(N-1)}, \quad V_{20} = p^{N-1} V^{(N-1)},$$
$$V_{30} = p^{2N-2} V^{(N-1)}, \cdots,$$
$$V_{N0} = p^{N^2-2N+1} V^{(N-1)},$$
$$I_{10} = I^{(N-1)}, \quad I_{20} = p^{N-1} I^{(N-1)},$$
$$I_{30} = p^{2N-2} I^{(N-1)}, \cdots,$$
$$I_{N0} = p^{N^2-2N+1} I^{(N-1)}.$$
$$(17e)$$

Note that superscripts on p are powers, whereas superscripts on V and I are written in parentheses and denote phase-sequence numbers. Note also that

$$p^N = e^{j2\pi} = 1, \quad p^{N/2} = e^{j\pi} = -1,$$
$$p^{N/4} = e^{j\pi/2} = j, \quad p^{3N/4} = e^{j3\pi/2} = -j.$$
$$(18)$$

The current $I^{(m)}$ in each antenna can be determined independently for each phase sequence with an arbitrary value of driving voltage $V^{(m)}$ using (16). If each antenna is imagined to be driven by all voltages in all N phase sequences simultaneously, with the N voltages in each unit in series, the following driving voltages apply:

$$V_{10} = V^{(0)} + V^{(1)} + V^{(2)} + \cdots$$
$$+ V^{(N-1)} = \sum_{i=0}^{N-1} V^{(i)}, \quad (19a)$$

$$V_{20} = V^{(0)} + pV^{(1)} + p^2 V^{(2)} + \cdots$$
$$+ p^{N-1} V^{(N-1)} = \sum_{i=0}^{N-1} p^i V^{(i)}, \quad (19b)$$

$$V_{30} = V^{(0)} + p^2 V^{(1)} + p^4 V^{(2)} + \cdots$$
$$+ p^{2N-2} V^{(N-1)} = \sum_{i=0}^{N-1} p^{2i} V^{(i)}, \quad (19c)$$

$$\vdots$$

$$V_{N0} = V^{(0)} + p^{N-1} V^{(1)} + p^{2N-2} V^{(2)} + \cdots$$
$$+ p^{N^2-2N+1} V^{(N-1)} = \sum_{i=0}^{N-1} p^{iN-i} V^{(i)}. \quad (19d)$$

Similarly, the resultant currents in the antennas are

$$I_{10} = \sum_{i=0}^{N-1} I^{(i)}, \quad (20a)$$

$$I_{20} = \sum_{i=0}^{N-1} p^i I^{(i)}, \quad (20b)$$

$$\vdots$$

$$I_{N0} = \sum_{i=0}^{N-1} p^{iN-i} I^{(i)}. \quad (20c)$$

Since the N voltages $V^{(i)}$, $i = 0, 1, 2, \cdots, N-1$, are arbitrary, it is possible to determine a set of N such voltages that satisfy (19a–d) for all possible values of V_{10}, V_{20}, $\cdots$, V_{N0}. That is, the general problem of solving for the N currents in N coupled antennas located at the corners of a regular polygon may be carried out by first solving for the N independent currents due to voltages chosen to satisfy the conditions of the N phase sequences. An appropriate superposition of these N solutions yields the currents and impedances in all units due to an arbitrary set of driving voltages.

A possible driving voltage for all but one of the N antennas is zero. That is, any number of the antennas up to $N-1$ may be parasitic. Alternatively, they may be parasitic and center-tuned or center-loaded if V_{m0} is set equal to $-I_{m0}Z_m$, where Z_m is an arbitrary impedance in series with antenna m at its center and I_{m0} is the current entering and leaving it. By setting $Z_m = \infty$ antenna m is open-circuited at its center. Since it has been shown in Sec. 8 that an open-circuited parasitic antenna of electrical half-length $\beta_0 h = \pi/2$ has a very small effect on the input impedance of a coupled antenna, it follows that as an approximation such an antenna may be removed entirely. That is, if the N antennas around a circle have electrical half-lengths $\beta_0 h = \pi/2$, an arbitrary number may be removed completely and the analysis carried through as if they were present with infinite load impedances. Note that this procedure is a satisfactory approximation when $\beta_0 h$ is less than $\pi/2$ but *not* when $\beta_0 h$ exceeds $\pi/2$ sufficiently that significant currents exist on the ends of the open-circuited antenna which react back on the other currents in the array. The open-circuited antenna does not then approximate an absent antenna.

As a simple application of the general theory, consider two coupled antennas. Since $N = 2$, m is restricted to the two values, 0 and 1. That is, there are two phase sequences, the zero-phase sequence, previously called the symmetric case, and the π-phase sequence, previously called the antisymmetric case.

Equations corresponding to (19) and (20) are

$$V_{10} = V^{(0)} + V^{(1)} = V^s + V^a, \quad (21a)$$

$$V_{20} = V^{(0)} + pV^{(1)} = V^{(0)} - V^{(1)}$$
$$= V^s - V^a. \quad (21b)$$

These are the same as (2.16). The corresponding kernels for the integral equation are obtained from (14) with $m = 0$ and 1 to be

$$K_{\Sigma 0}(z, z') \equiv K_s(z, z') = \frac{e^{-j\beta_0 R_{11}}}{R_{11}} + \frac{e^{-j\beta_0 R_{12}}}{R_{12}}, \quad (22a)$$

$$K_{\Sigma 1}(z, z') \equiv K_a(z, z') = \frac{e^{-j\beta_0 R_{11}}}{R_{11}} - \frac{e^{-j\beta_0 R_{12}}}{R_{12}}. \quad (22b)$$

These are the same as (3.3) and (4.2). It is seen that the theory for N antennas in terms of N phase sequences is a simple generalization of the theory developed in Secs. 3 and 4 for two antennas with two phase sequences. In the next section, three antennas are analyzed.

*Folded antennas.** Whenever the antennas forming an N-element parallel array are all so close together that the distance $b_{\max}$ between the most widely separated pair satisfies the condition

$$\beta_0 b_{\max} \ll 1, \quad (23)$$

it is possible to generalize the nature of the array by permitting all or some of the N units to be connected, usually at one or both ends, in a symmetric manner. Interconnected closely spaced antennas are called a *folded antenna* if the array is driven by a single generator at the center of one of the elements. The folded dipole described in Sec. 12 is a simple example of a so generalized two-element parallel array in which both ends of the two antennas are joined conductively. Although the restriction that there be only one generator in the folded array is not necessary, it is assumed in the following for simplicity in the formulation. According to the principle of superposition, which is applicable, the combined effects of several generators is the sum of their individual effects.

---

* This part of Sec. 14 is based on the work of Dr. Charles W. Harrison, Jr., ref. 26.

The principles underlying the generalized analysis of N closely spaced antennas are formulated in detail by Harrison.[26] A more compact formulation is described in the following. Since the antennas are close together, the currents that combine to maintain the radiation field must be equivalent to the single current that would be obtained if all currents in all elements were superimposed in a single conductor or in a cage of N elements in parallel. Components of current in the individual elements that do not contribute to the radiation field are those components that would cancel one another if the currents were superposed. The procedure introduced by Harrison separates the vector potential due to noncanceling currents from that due to canceling currents.

Although the starting point of the analysis might be the N simultaneous equations (7) with N generators (or their equivalents in voltage drops across loads), the present discussion is restricted for simplicity to an array with a single generator symmetrically placed in antenna 1. The other antennas or elements of the folded antenna are smooth conductors uninterrupted by generators or lumped reactances. The modified N equations corresponding to (7) are

$$\int_{-h}^{h} \sum_{i=1}^{N} I_{iz}(z') K_{1i}(z, z') \, dz'$$
$$= \frac{-j4\pi}{\zeta_0} (C_1 \cos \beta_0 z + D_1 \sin \beta_0 z + \tfrac{1}{2} V_{10} \sin \beta_0 |z|), \quad (24a)$$

$$\int_{-h}^{h} \sum_{i=1}^{N} I_{iz}(z') K_{ki}(z, z') \, dz'$$
$$= \frac{-j4\pi}{\zeta_0} (C_k \cos \beta_0 z + D_k \sin \beta_0 z),$$
$$k = 2, 3, \cdots, N \quad (24b)$$

where

$$K_{ki}(z, z') \equiv \frac{e^{-j\beta_0 R_{ki}}}{R_{ki}}, \quad K_{1i}(z, z') \equiv \frac{e^{-j\beta_0 R_{1i}}}{R_{1i}}, \quad (25a)$$

$$R_{ki} \equiv \sqrt{(z - z')^2 + b_{ki}^2},$$
$$R_{1i} \equiv \sqrt{(z - z')^2 + b_{1i}^2}, \quad (25b)$$
$$b_{11} = b_{kk} = a. \quad (25c)$$

The equation for the driven antenna 1 differs from the other $N - 1$ equations by a voltage V_{10}. If the driving element is symmetrically placed in the folded antenna, so that

$$I_1(-z_1) = I_1(z_1), \quad (26)$$

the constant D_1 must vanish. The constants of integration C_1, C_k, and D_k must be evaluated from the boundary and continuity conditions at the ends of the N elements. The appearance of the sine terms in (24a, b) indicates the possibility of distributions of current on some of the elements that are not even with respect to the center of the element.

In order to separate the vector potentials calculated from the N currents into two parts, the one involving the components of current that would *not* be canceled if superposed, the other involving the currents that would be canceled, (24) may be expanded by adding and subtracting $4\pi\nu_0$ times the vector potential of a single antenna (or cage antenna) of effective radius a_e carrying the current

$$I_{tz}(z') = \sum_{i=1}^{N} I_{iz}(z'). \quad (27)$$

This is given by

$$J_e(z) \equiv \int_{-h}^{h} I_{tz}(z') K_e(z, z') \, dz', \quad (28)$$

where

$$K_e(z, z') = \frac{e^{-j\beta_0 R_e}}{R_e},$$
$$R_e = \sqrt{(z - z')^2 + a_e^2}. \quad (29)$$

If (18) is added to and subtracted from each equation in (24a, b), the following results are obtained:

$$\int_{-h}^{h} \sum_{i=1}^{N} I_{iz}(z')[K_{1i}(z, z') - K_e(z, z')] \, dz'$$
$$+ J_e(z) = -\frac{j4\pi}{\zeta_0} (C_1 \cos \beta_0 z$$
$$+ D_1 \sin \beta_0 z + \tfrac{1}{2} V_{10} \sin \beta_0 z), \quad (30a)$$

$$\int_{-h}^{h} \sum_{i=1}^{N} I_{iz}(z')[K_{ki}(z, z') - K_e(z, z')] \, dz'$$
$$+ J_e(z) = -\frac{j4\pi}{\zeta_0} (C_k \cos \beta_0 z + D_k \sin \beta_0 z),$$
$$k = 2, 3 \cdots, N. \quad (30b)$$

As a consequence of (23) the following inequalities must be satisfied:

$$\beta_0 a_e \ll 1, \quad \beta_0 b_{ki} \ll 1, \quad k, i = 1, 2, \cdots, N. \quad (31)$$

Evidently, $K_e(z, z')$ and $K_{ki}(z, z')$ in the integrands on the left in (30a, b) differ from each other significantly only if z' is in a narrow region near the cross section $z' = z$, since R_e and R_{ki} become essentially equal as soon as $\beta_0^2 a_e^2$ and $\beta_0^2 b_{ki}^2$ are negligible compared with $\beta_0^2(z - z')^2$. Moreover, since the integrand contributes only negligibly to the integral when $|z - z'|$ is large, the limits may be extended to infinity. Finally, since the integrand differs from zero only when z' is near z, and this involves an electrical distance of the order of magnitude of $\beta_0 b_{ki}$, it is a good approximation to set $I_i(z_i') \doteq I_i(z_i)$, and remove it from under the sign of integration. With all of these approximations, a typical integral is the following:

$$I_i(z) \int_{-\infty}^{\infty} \left(\frac{1}{R_{ki}} - \frac{1}{R_e}\right) dz' = \lambda_{ki} I_i(z), \quad (32)$$

where

$$\lambda_{ki} \equiv 2 \ln (a_e/b_{ki}). \quad (33)$$

It follows that (24a, b) may be expressed as follows:

$$J_e(z) + \sum_{i=1}^{N} \lambda_{1i} I_{iz}(z) = F_1(z), \quad (34a)$$

$$J_e(z) + \sum_{i=1}^{N} \lambda_{ki} I_{iz}(z) = F_k(z)$$
$$k = 2, 3, \cdots, N, \quad (34b)$$

where

$$F_1(z) \equiv \frac{-j4\pi}{\zeta_0}(C_1 \cos \beta_0 z + D_1 \sin \beta_0 z$$
$$+ \tfrac{1}{2}V_{10} \sin \beta_0 |z|), \quad (35a)$$

$$F_k(z) \equiv \frac{-j4\pi}{\zeta_0}(C_k \cos \beta_0 z + D_k \sin \beta_0 z),$$
$$k = 2, 3, \cdots, N. \quad (35b)$$

Evidently, (34a, b) constitute a set of N simultaneous integral equations for the N currents. These equations may be rearranged to give one integral equation in the total radiating current $I_z(z)$ defined in (27), and $N - 1$ equations for determining $N - 1$ of the N currents. The Nth current may then be evaluated by subtracting the sum of the known $N - 1$ currents from the sum of the N currents. One method for achieving this separation involves two steps. The first is to add the $N - 1$ equations in (34b) to (34a) to obtain one equation of antenna type, and then the second is to subtract them from (34a) to obtain $N - 1$ equations of transmission-line type. The result of the addition (after division by N) is

$$J_e(z) + \frac{1}{N} \sum_{k=1}^{N} \sum_{i=1}^{N} \lambda_{ki} I_{iz}(z)$$
$$= -\frac{j4\pi}{\zeta_0}\left(C \cos \beta_0 z + \frac{V_{10}}{2N} \sin \beta_0 |z|\right),$$
$$(36)$$

where

$$C \equiv (C_1 + \sum_{k=2}^{N} C_k)/N, \quad (37a)$$

since

$$D \equiv (D_1 + \sum_{k=1}^{N} D_k)/N = 0. \quad (37b)$$

The last relation (37b) follows from the postulate that the entire configuration of antennas is geometrically symmetric in z with respect to the driving point so that the total radiating current and the vector potential calculated from this current are even functions of z. The subtractions give

$$\sum_{i=2}^{N} \Lambda_{ki} I_{iz}(z) = F_{1k}(z) - \Lambda_{k1} I_{1z}(z),$$
$$k = 2, 3, \cdots, N, \quad (38)$$

where

$$\Lambda_{ki} \equiv \lambda_{1i} - \lambda_{ki} = 2 \ln (b_{ki}/b_{1i}),$$
$$\Lambda_{k1} \equiv \lambda_{11} - \lambda_{k1} = 2 \ln (b_{k1}/a), \quad (39a)$$

$$F_{1k}(z) \equiv F_1(z) - F_k(z)$$
$$= \frac{-j4\pi}{\zeta_0}[(C_1 - C_k) \cos \beta_0 z$$
$$+ \tfrac{1}{2}V_{10} \sin \beta_0 |z| + (D_1 - D_k) \sin \beta_0 z]. \quad (39b)$$

The $N - 1$ equations in (38) can be solved for the $I_{iz}(z)$ with $i = 2, 3, \cdots, N$ in terms of the $F_{1k}(z)$, the Λ_{ki}, and the current $I_{1z}(z)$ in the driven unit. The solutions are in the form

$$I_{iz}(z) = A_i(z) - B_i I_{1z}(z),$$
$$i = 2, 3, \cdots, N, \quad (40)$$

where the $A_i(z)$ are functions of the $F_{1k}(z)$ and the Λ_{ki}; the B_i's are constants that depend on the Λ_{ki} only.

With the individual currents all determined by (40), the total current may be expressed as follows; the second step makes use of (40):

$$I_{tz}(z) = I_{1z}(z) + \sum_{i=2}^{N} I_{iz}(z)$$
$$= A_t(z) - B_t I_{1z}(z), \quad (41)$$

where

$$A_t(z) \equiv \sum_{i=2}^{N} A_i(z), \quad B_t \equiv \sum_{i=2}^{N} B_i - 1. \quad (42)$$

It follows that

$$I_{1z}(z) = \frac{A_t(z) - I_{tz}(z)}{B_t}. \quad (43)$$

If (43) is substituted in (40), the individual currents may be expressed in terms of the total current $I_t(z)$, the function $A_t(z)$, and the constant B_t. Thus,

$$I_{iz}(z) = H_i(z) + P_i I_{tz}(z), \quad (44)$$

where

$$H_i(z) \equiv A_i(z) - \frac{B_i}{B_t} A_t(z), \quad P_i \equiv \frac{B_i}{B_t},$$
$$i = 2, 3, \cdots, N. \quad (45)$$

The individual currents $I_{2z}(z), \cdots, I_{Nz}(z)$ are expressed in (44) in terms of the total current $I_{tz}(z)$, the functions $F_{1k}(z)$ in $A_i(z)$ and $A_t(z)$, and the constants Λ_{ki} in B_i and B_t. They may be substituted in (36) where they occur in the double-sum term. Thus

$$\sum_{k=1}^{N} \sum_{i=1}^{N} \lambda_{ki} I_{iz}(z) = I_{1z}(z) \sum_{k=1}^{N} \lambda_{k1}$$
$$+ \sum_{k=1}^{N} \sum_{i=2}^{N} \lambda_{ki} I_{iz}(z). \quad (46)$$

Since $I_{1z}(z)$ has not been expressed in terms of the total current in (44), it is convenient to eliminate it from the sum in (46) by an appropriate definition of the equivalent radius a_e. This is done by setting

$$\sum_{k=1}^{N} \lambda_{k1} = 2 \sum_{k=1}^{N} \ln \frac{a_e}{b_{k1}} = 2 \ln \left(\frac{a_e^N}{\prod_{k=1}^{N} b_{k1}} \right)$$
$$= 2 \ln (a_e^N / b_{11} b_{12} b_{13} \cdots b_{1N}) = 0. \quad (47)$$

This is equivalent to the following definition for a_e:

$$a_e \equiv \left(a \prod_{k=2}^{N} b_{k1} \right)^{1/N} \quad (48)$$

Note that $b_{11} = a$. With a_e defined as in (48), the sum in (46) reduces to the second term. That is, with (44),

$$\frac{1}{N} \sum_{k=1}^{N} \sum_{i=2}^{N} \lambda_{ki} I_{iz}(z) = \frac{j4\pi}{\zeta_0} G_t(z) + \Lambda_t I_{tz}(z), \quad (49)$$

where

$$\frac{j4\pi}{\zeta_0} G_t(z) \equiv \frac{1}{N} \sum_{i=2}^{N} H_i(z) \sum_{k=1}^{N} \lambda_{ki} \quad (50a)$$

and

$$\Lambda_t \equiv \frac{1}{N} \sum_{i=2}^{N} P_i \sum_{k=1}^{N} \lambda_{ki}. \quad (50b)$$

Upon substituting (49) in (36) this becomes

$$J_e(z) + \Lambda_t I_{tz}(z) = -\frac{j4\pi}{\zeta_0} \left[C \cos \beta_0 z \right.$$
$$\left. + \frac{V_{10}}{2N} \sin \beta_0 |z| + G_t(z) \right]. \quad (51)$$

Since $G_t(z)$ involves sums of terms with the $F_{1k}(z)$ as factors, it may be separated into two sets of terms which have $\cos \beta_0 z$ and $\sin \beta_0 |z|$ as factors. Terms with $\sin \beta_0 z$ as factor must add to zero since the left-hand member in (51) must be an even function of z. Accordingly, (51) may be expressed as follows in terms of appropriately defined constants C_t and E_t, which include contributions from $G_t(z)$:

$$J_e(z) + \Lambda_t I_{tz}(z) = -\frac{j4\pi}{\zeta_0} (C_t \cos \beta_0 z$$
$$+ \tfrac{1}{2} E_t V_{10} \sin \beta_0 |z|). \quad (52)$$

This equation is not quite in the form for which a solution is obtained in Chapter II. However, if $J_e(z)$ is expanded as in Chapter II, so that

$$J_e(z) = I_{tz}(z)[\Psi_e + \gamma(z)] - I_{tz}(z)\gamma(z)$$
$$+ \int_{-h}^{h} [I_{tz}(z') - I_{tz}(z)] g(z, z') K_e(z, z') \, dz', \quad (53)$$

and this value is substituted in (52), the following equation is obtained:

$$I_{tz}(z)[\Psi_e + \Lambda_t] =$$
$$- \frac{j4\pi}{\zeta_0} (C_t \cos \beta_0 z + \tfrac{1}{2} E_t V_{10} \sin \beta_0 z)$$
$$- \int_{-h}^{h} [I_{tz}(z') - I_{tz}(z)] g(z, z') K_e(z, z') \, dz$$
$$- I_{tz}(z) \gamma(z). \quad (54)$$

This is in standard form if the driving voltage $E_t V_{10}$ is used instead of V_{10} and if a new expansion parameter Ψ_t is introduced as follows:

$$\Psi_t = \Psi_e + \Lambda_t. \quad (55a)$$

Since a change in the radius a produces essentially a parallel shift of the curve of Ψ as a function of $\beta_0 h$ for all values of $\beta_0 h \leq 2\pi$, an effective radius may be defined using the formula for Ψ for the short antenna. That is, for $\beta_0 h \ll 1$,

$$\Psi_t = 2 \ln \frac{2h}{a_t} \simeq 2 \ln \frac{2h}{a_e} - 2$$
$$+ \frac{2}{N} \sum_{i=2}^{N} P_i \sum_{k=1}^{N} \ln \left(\frac{a_e}{b_{ki}} \right). \quad (55b)$$

The new effective radius a_t may be determined from this equation. Since $I_{tz}(\pm h) = 0$, the constant C_t in (54) may be evaluated as in Chapter II.

By solving (54) following the procedure used in Chapter II, the total current $I_{tz}(z)$ is determined. By using (43) the current $I_{1z}(z)$ in the driven element is obtained in terms of $E_t V_{10}$. The input admittance is the input current $I_{1z}(0)$ per unit voltage V_{10}. That is,

$$Y_{in} = I_{1z}(0)/V_{10}. \qquad (56)$$

This completes the evaluation of the currents and the driving-point admittance, except that the constants $(C_1 - C_k)$ and $(D_1 - D_k)$ have not been evaluated. These may be determined from the conditions at the ends of the several elements. If the end at $z = h$ of the ith element is not connected to any other element, $I_{iz}(h) = 0$. If the two elements i and k are joined in a short circuit at $z = h$, the current must be continuous, $I_{iz}(h) + I_{kz}(h) = 0$, and the scalar potentials must be equal, $\phi_i(h) = \phi_k(h)$. These potentials are given by $j\zeta_0/4\pi$ times the derivatives of (24a, b) with respect to z. If the element k is symmetric with respect to its center, so that $I_{kz}(-z) = I_{kz}(z)$, then $D_k = 0$.

An important special situation for which the general theory may be simplified occurs in an array in which all elements are located at the vertices of an equilateral polygon. In this case $\sum_{k=1}^{N} \lambda_{ki}$ is independent of i, so that it becomes equal to the value for $i = 1$, which is zero by definition of the effective radius a_e. Accordingly, in (50a, b), $G_t(z) = 0$ and $\Lambda_t = 0$, so that (51) has the simple form

$$J_e(z) = -\frac{j4\pi}{\zeta_0}\left(C \cos \beta_0 z + \frac{V_{10}}{2N} \sin \beta_0 |z|\right). \qquad (57)$$

Two-element antenna. As a simple example of the generalized procedure for analyzing closely spaced arrays, the two-element array analyzed in Sec. 12 may be reëxamined when both ends are open (antenna with parasite), both ends are short-circuited (folded dipole), or one is open and one bridged. With $N = 2$, the solution of (57) gives the codirectional or symmetric currents in a two-element cage. The effective radius a_e of the cage is given in (48). It is

$$a_e = \sqrt{ab}. \qquad (58)$$

The total current is obtained by solving (57). It is

$$I_{tz}(z) = I_{1z}(z) + I_{2z}(z). \qquad (59)$$

The current in the terminals $z \doteq 0$ is

$$I_{tz}(0) = I_{1z}(0) + I_{2z}(0) = V_{10}/2Z_e, \qquad (60)$$

where Z_e is the impedance obtained by solving (57) with a voltage $V_{10}/2$ instead of V_{10}. It is the impedance of a single antenna with radius $a_e = \sqrt{ab}$.

With $N = 2$, the set of transmission-line equations in (38) reduces to the following single equation:

$$\Lambda_{22} I_{2z}(z) = F_{12}(z) - \Lambda_{21} I_{1z}(z), \qquad (61)$$

where

$$\Lambda_{22} = -2 \ln(b/a), \qquad \Lambda_{21} = 2 \ln(b/a), \qquad (62a)$$

$$F_{12}(z) = -\frac{j4\pi}{\zeta_0}[(C_1 - C_2) \cos \beta_0 z$$
$$+ \tfrac{1}{4} V_{10} \sin \beta_0 |z|$$
$$+ (D_1 - D_2) \sin \beta_0 z]. \qquad (62b)$$

It follows that (61) is equivalent to

$$I_{1z}(z) - I_{2z}(z) = \frac{-j}{R_c}[(C_1 - C_2) \cos \beta_0 z$$
$$+ \tfrac{1}{4} V_{10} \sin \beta_0 |z|$$
$$+ (D_1 - D_2) \sin \beta_0 z], \qquad (63a)$$

where

$$R_c = \frac{\zeta_0}{\pi} \ln \frac{b}{a}. \qquad (63b)$$

The current in each element is obtained by adding and subtracting (63a) and (59). Thus

$$I_{1z}(z) = \tfrac{1}{2} I_{tz}(z) - \frac{j}{2R_c}[(C_1 - C_2) \cos \beta_0 z$$
$$+ \tfrac{1}{2} V_{10} \sin \beta_0 |z|$$
$$+ (D_1 - D_2) \sin \beta_0 z], \qquad (64a)$$

$$I_{2z}(z) = \tfrac{1}{2} I_{tz}(z) + \frac{j}{2R_c}[(C_1 - C_2) \cos \beta_0 z$$
$$+ \tfrac{1}{2} V_{10} \sin \beta_0 |z|$$
$$+ (D_1 - D_2) \sin \beta_0 z], \qquad (64b)$$

where $I_{tz}(z)$ is the solution of the familiar integral equation (57) with $N = 2$.

The driving-point admittance is given by

$$Y_{in} = \frac{I_{1z}(0)}{V_{10}} = \frac{1}{4Z_e} - \frac{j(C_1 - C_2)}{R_c V_{10}}. \qquad (65)$$

Note that the effective driving voltage in (57) is V_{10}/N, not V_{10}.

In order to determine the constant $(C_1 - C_2)$ it is necessary to specify the nature of the ends. If both ends are open (*antenna with closely spaced parasite*), the conditions that must be satisfied are

$$I_{1z}(\pm h) = 0 = I_{2z}(\pm h). \tag{66}$$

When substituted in (64a, b) they give

$$C_1 - C_2 = -\frac{V_{10}}{2} \tan \beta_0 h, \tag{67}$$

$$D_1 - D_2 = 0.$$

The individual vanishing of D_1 and D_2 also follows from the fact that each current must be an even function of z. If (67) is substituted in (64a, b) and (65), the currents in the driven and parasitic elements and the driving-point admittance are determined. The latter is

$$Y_{in} = \frac{1}{4Z_e} + \frac{j \tan \beta_0 h}{2R_c}. \tag{68}$$

If both ends are closed (*folded dipole*), the condition that must be satisfied is

$$\Phi_i(h) = \Phi_k(h). \tag{69}$$

By equating the derivatives with respect to z of the right sides of (24a, b) with $k = 2$ and noting that from symmetry $D_1 = D_2 = 0$, the following equation is obtained:

$$-(C_1 - C_2) \sin \beta_0 h + \tfrac{1}{2} V_{10} \cos \beta_0 h = 0. \tag{70}$$

Since $I(-z) = I(z)$ on both antennas owing to symmetry, $D_1 = D_2 = 0$. The result is

$$C_1 - C_2 = \frac{V_{10} \cot \beta_0 h}{2}. \tag{71}$$

If this value is inserted in (64a, b) and (65), the current and admittance are determined. The latter is

$$Y_{in} = \frac{1}{4Z_e} - \frac{j \cot \beta_0 h}{2R_c} \tag{72}$$

Numerical values of $Z_{in} = 1/Y_{in}$ as computed by Harrison[26] from (72) are given in Table 14.1 for $\Omega_e = 2 \ln(2h/a_e) = 10$, $b/a = 10$, using second-order values of Z_e as given in Chapter II by Z_0.

If one end is open and the other closed (*U-antenna*), the condition to be imposed on (64a) and (64b) is (66) at $z = h$; the condition

TABLE 14.1. Driving-point impedance of two-wire folded dipole; $\Omega_e = 10$, $b/a = 10$.*

$\beta_0 h$	R_{in}	X_{in}
0.5	0.3	+ 33.8
1.1	909.4	−1760.0
1.2	357.2	− 732.1
1.3	265.5	− 349.7
1.4	256.7	− 123.4
1.5	291.1	+ 51.9
1.57	346.0	+ 166.8
1.6	375.5	+ 214.1
1.7	551.0	+ 374.0
1.8	923.1	+ 471.3
1.9	1524.0	+ 128.0
2.0	1415.0	− 826.0
2.3	163.6	− 660.9
2.6	32.4	− 334.8
2.9	4.4	− 13.1
3.14	0.0	0.0
3.2	0.3	+ 32.8
3.4	6.8	+ 16.0

* Computed by Harrison.[26]

for (63a) is (69) at $z = -h$. Since $I_{tz}(\pm h) = 0$, the following equations are obtained:

$$(C_1 - C_2) \cos \beta_0 h + \tfrac{1}{2} V_{10} \sin \beta_0 h$$
$$+ (D_1 - D_2) \sin \beta_0 h = 0, \quad (z = h) \tag{73a}$$

$$(C_1 - C_2) \sin \beta_0 h - \tfrac{1}{2} V_{10} \cos \beta_0 h$$
$$+ (D_1 - D_2) \cos \beta_0 h = 0. \quad (z = -h) \tag{73b}$$

These equations can be solved for $(C_1 - C_2)$ and $(D_1 - D_2)$ in terms of V_{10} with the result

$$C_1 - C_2 = -\tfrac{1}{2} V_{10} \tan 2\beta_0 h, \tag{74a}$$

$$D_1 - D_2 = \tfrac{1}{2} V_{10} \sec 2\beta_0 h. \tag{74b}$$

If these values are substituted in (64a, b), $I_{1z}(z)$ and $I_{2z}(z)$ are determined. From (65) the admittance is

$$Y_{in} = \frac{1}{4Z_e} + \frac{j \tan 2\beta_0 h}{2R_c}. \tag{75}$$

It is verified readily that (68), (72), and (75) are equivalent to the results obtained in Sec. 11 using the method of symmetric components. In fact, they are all contained in the general formula (11.11), which is equivalent to

$$Y_{in} = \frac{1}{2}\left(\frac{1}{Z^s} + \frac{1}{Z^a}\right), \tag{76}$$

if transmission-line spacing and end effects are neglected. The symmetric impedance Z^s is twice the impedance Z_e of a single antenna or cage of effective radius $a_e = \sqrt{ab}$. The antisymmetric impedance Z^a is the series impedance of the two sections of line looking each way from the center. Neglecting line losses, this is $Z^a = -jR_c \cot \beta_0 h$ for two open ends, $Z^a = jR_c \tan \beta_0 h$ for two closed ends, and $Z^a = \frac{1}{2}jR_c[\tan \beta_0 h - \cot \beta_0 h] = \frac{1}{2}jR_c \cot 2\beta_0 h$ for one open and one closed end. With these values (76) gives exactly (68), (72), and (75).

Folded dipole with elements of unequal radius. In the general formulation for folded antennas it has been assumed for simplicity that all conductors have the same radius. This is not necessary. Harrison[26] has formulated the problem of the simple folded dipole when the driven element has a radius a_1 and the parallel element a different radius a_2. In this case (34a) and (34b) apply with $N = 2$ if the λ-factors are appropriately defined. The following equations are obtained corresponding to (34a, b) with $N = 2$ and boundary condition appropriate to the folded dipole:

$$J_e(z) + \lambda_{11} I_{1z} + \lambda_{12} I_{2z} = -\frac{j4\pi}{\zeta_0}(C_1 \cos \beta_0 z + \tfrac{1}{2} V_0 \sin \beta_0 |z|), \quad (77a)$$

$$J_e(z) + \lambda_{21} I_{1z} + \lambda_{22} I_{2z} = -\frac{j4\pi}{\zeta_0} C_2 \cos \beta_0 z, \quad (77b)$$

where

$$\lambda_{11} = 2\ln(a_e/a_1), \quad \lambda_{22} = 2\ln(a_e/a_2),$$
$$\lambda_{12} = \lambda_{21} = 2\ln(a_e/b). \quad (78)$$

By appropriate combination and rearrangement (77a) and (77b) may be transformed into

$$J_e(z) + K_1 I_t(z) = -\frac{j4\pi}{\zeta_0}(K_2 \cos \beta_0 z + \tfrac{1}{2} V_0 K_3 \sin \beta_0 |z|), \quad (79)$$

$$I_1(z) = I_t(z)\left(\frac{\alpha_2}{\alpha_1 + \alpha_2}\right) - \frac{j4\pi}{\zeta_0(\alpha_1 + \alpha_2)}[(C_1 - C_2)\cos \beta_0 z + \tfrac{1}{2} V_0 \sin \beta_0 |z|], \quad (80a)$$

$$I_2(z) = I_t(z) - I_1(z), \quad (80b)$$

where

$$\alpha_1 \equiv \lambda_{11} - \lambda_{12} = 2\ln(b/a_1),$$
$$\alpha_2 \equiv \lambda_{22} - \lambda_{21} = 2\ln(b/a_2),$$
$$\alpha_3 \equiv \lambda_{22} - \lambda_{11} = 2\ln(a_1/a_2), \quad (81)$$

$$K_1 = \frac{1}{2}\left(\frac{\alpha_1 \alpha_3}{\alpha_1 + \alpha_2}\right), \quad (82a)$$

$$K_2 = \frac{1}{2}\left[(C_1 + C_2) + \frac{\alpha_3}{\alpha_1 + \alpha_2}(C_1 - C_2)\right], \quad (82b)$$

$$K_3 = \frac{1}{2}\left(1 + \frac{\alpha_3}{\alpha_1 + \alpha_2}\right). \quad (82c)$$

Equation (79) is like (52) and may be solved by iteration for $I_t(z)$ using the expansion parameter $\Psi'_t = \Psi_e + K_1$, and effective driving voltage $K_3 V_0$. The admittance defined by $Y_d = I_t(0)/K_3 V_0$ is that of a cylindrical antenna with radius a_t that may be obtained as explained in relation to (55a). Thus

$$\Psi_t = 2\ln\frac{2h}{a_t} - 2 = 2\ln\frac{2h}{a_e} - 2 + K_1, \quad (83)$$

$$a_t = a_e e^{-K_1/2} = \sqrt{a_1 b e^{-K_1}}. \quad (84)$$

The constant $C_1 - C_2$ in (80a) is evaluated using (69) to obtain an equation like (70). Hence, $C_1 - C_2$ is given by (71). (Note that the radius of the short bridges that join the ends of the two antennas changes abruptly from a_1 to a_2 at their centers. The charges per unit length on the cylindrical surfaces of different radii are not the same; evidently the charge density varies continuously from radius a_1 to radius a_2 on the flat surfaces at the junctions.)

If $C_1 - C_2$ from (71) is substituted in (80a) and the admittance $Y_{in} = I_1(0)/V_0$ is formed, the following equation is obtained:

$$Y_{in} = \frac{I_t(0)}{V_0}\left(\frac{\alpha_2}{\alpha_1 + \alpha_2}\right) - \frac{j2\pi \cot \beta_0 h}{\zeta_0(\alpha_1 + \alpha_2)}. \quad (85)$$

The substitution of $Y_d = I_t(0)/K_3 V_0$ and use of (82a–c) gives the input admittance of a folded dipole when driven at the center of the conductor with radius a_1. The result is

$$Y_{in} = Y_d \left[\frac{\ln(b/a_2)}{\ln(b^2/a_1 a_2)}\right]^2 - \frac{j\pi \cot \beta_0 h}{\zeta_0 \ln(b^2/a_1 a_2)} \quad (86)$$

Harrison has evaluated $Z_{in} = 1/Y_{in}$ using (86) and second-order values of Y_d as given in Chapter II. The numerical values and the dimensions are given in Table 14.2. It is seen that a reasonable variation in input resistance may be obtained with changes in the ratio a_1/a_2 of the radii of the two conductors above and below one.

TABLE 14.2. Driving-point impedance of two-wire folded dipole.*

a_1/a_2	$Z_{in}(\text{ohms}) = R_{in} + jX_{in}$	Conditions
0.125	1191 +j676.9	
.167	626.9 +j344.4	
.250	449.0 +j239.4	$\dfrac{h}{\sqrt{a_1 b}} = 75,$
.500	365.9 +j184.8	
.750	348.1 +j172.2	
1.00	346.1 +j166.8	
1.33	341.1 +j160.8	$\beta_0 h = \pi/2,$
2.0	341.7 +j159.7	
4.0	346.9 +j158.5	
6.0	352.6 +j158.3	$b = 10 a_1$

* Computed by Harrison.[26]

TABLE 14.3. Driving-point impedance of three-wire folded dipole.*

a_1/a_2	$Z_{in}(\text{ohms}) = R_{in} + jX_{in}$	Conditions
0.250	875.6 +j961.9	
.500	781.7 +j593.8	
.750	779.0 +j458.9	$\dfrac{h}{\sqrt[3]{a_1 b^2}} = 75,$
1.00	778.5 +j375.5	
1.33	771.9 +j300.0	
2.00	761.8 +j198.1	$\beta_0 h = \dfrac{\pi}{2},$
4.00	718.7 +j 52.92	
6.00	683.9 −j 17.78	$b = 10 a_1$

* Computed by Harrison.[26]

The results of a similar analysis of the three-wire folded dipole (in which the driven element has a radius a_1 and the two coupled elements are at the vertices of an equilateral triangle and have equal radii $a_3 = a_2$) are given in Table 14.3.

15. Three Antennas at the Vertices of an Equilateral Triangle*

The integral equation for the current in the upper half $(0 \leq z \leq h)$ of each of three identical, parallel, nonstaggered, center-driven antennas placed at the vertices of an equilateral triangle of side b (see Fig. 15.1) is the same as (14.16), namely,

$$\int_{-h}^{h} I_z(z') K_{\Sigma m}(z, z')\, dz' = \frac{-j4\pi}{\zeta_0}(C \cos \beta_0 z + \tfrac{1}{2} V_0 \sin \beta_0 z). \quad (1)$$

* Parts of this section are based on the work of Dr. C. T. Tai, refs. 50 and 51.

The kernel is obtained from (14.14). It is

$$K_{\Sigma m}(z, z') = \frac{e^{-j\beta_0 R_{11}}}{R_{11}} + p^m \frac{e^{-j\beta_0 R_{12}}}{R_{12}} + p^{2m} \frac{e^{-j\beta_0 R_{13}}}{R_{13}}, \quad (2)$$

where

$$p = e^{j\theta}, \quad \theta = 2\pi/3, \quad (3)$$

and

$$m = 0, 1, 2. \quad (4)$$

It is seen that there are three phase sequences, namely, the zero-phase sequence $(m = 0)$, the $2\pi/3$- or positive-phase sequence $(m = 1)$, and the $4\pi/3$- or $-2\pi/3$- or negative-phase sequence. The three kernels are

$$m = 0: \quad K_{\Sigma 0}(z, z') = \frac{e^{-j\beta_0 R_{11}}}{R_{11}} + \frac{e^{-j\beta_0 R_{12}}}{R_{12}} + \frac{e^{-j\beta_0 R_{13}}}{R_{13}}, \quad (5a)$$

$$m = 1: \quad K_{\Sigma 1}(z, z') = \frac{e^{-j\beta_0 R_{11}}}{R_{11}} + p\frac{e^{-j\beta_0 R_{12}}}{R_{12}} + p^2 \frac{e^{-j\beta_0 R_{13}}}{R_{13}}, \quad (5b)$$

$$m = 2: \quad K_{\Sigma 2}(z, z') = \frac{e^{-j\beta_0 R_{11}}}{R_{11}} + p^2 \frac{e^{-j\beta_0 R_{12}}}{R_{12}} + p^4 \frac{e^{-j\beta_0 R_{13}}}{R_{13}}. \quad (5c)$$

These expressions can be simplified if use is made of the geometric fact that

$$R_{12} = R_{13}, \quad (6)$$

and of the relations

$$p^m + p^{2m} = e^{j2\pi m/3} + e^{j4\pi m/3}$$
$$= \begin{cases} 2 & \text{for } m = 0 \\ -1 & \text{for } m = 1, 2 \end{cases} \quad (7)$$

Accordingly, with (6), (7) and (14.22a,b), the three kernels reduce to two, as follows:

$$m = 0: \quad K_{\Sigma 0}(z, z') = \frac{e^{-j\beta_0 R_{11}}}{R_{11}} + 2\frac{e^{-j\beta_0 R_{12}}}{R_{12}}$$
$$= \tfrac{1}{2}[3K_s(z, z') - K_a(z, z')], \quad (8a)$$

$$m = 1, 2: \quad K_{\Sigma 1}(z, z') = K_{\Sigma 2}(z, z')$$
$$= \frac{e^{-j\beta_0 R_{11}}}{R_{11}} - \frac{e^{-j\beta_0 R_{12}}}{R_{12}}$$
$$= K_a(z, z'). \quad (8b)$$

Since the kernels for the positive- or $2\pi/3$-phase sequence and the negative or $4\pi/3$ sequence are the same as the antisymmetric (π-sequence) kernel for two coupled antennas, it follows that the expansion parameter $\Psi_{\Sigma 1} = \Psi_{\Sigma 2}$ must be the same as Ψ_a, and that identically the same integral equation applies. The parameter Ψ_a is evaluated in Sec. 4 and represented graphically in Fig. 3.3. The currents in the three coupled antennas located at the vertices of an equilateral triangle of side b and driven in either the positive ($2\pi/3$) or negative ($4\pi/3$) phase sequence are the *same* as the currents in two coupled antennas separated by a distance b and driven antisymmetrically (π-phase sequence). If the driving conditions are the same, the $2\pi/3$- or $4\pi/3$-phase-sequence impedance for each unit of the three-element array is the same as the antisymmetric (π-sequence) impedance Z_0^a of the two-antenna array. This impedance is represented graphically in Figs. 7.1–7.4.

The expansion parameter of the zero-phase sequence for the three-antenna array is given by

$$\Psi_{\Sigma 0} = \begin{cases} |\Psi_{\Sigma 0}(0)| & (\beta_0 h \leq \pi/2) \\ |\Psi_{\Sigma 0}(h - \lambda_0/4)|, & (\beta_0 h \geq \pi/2) \end{cases} \quad (9a)$$

where

$$\Psi_{\Sigma 0}(z) = \int_{-h}^{h} g(z, z') K_{\Sigma 0}(z, z') \, dz' \quad (9b)$$

and where

$$g(z, z') = \frac{\sin \beta_0(h - |z'|)}{\sin \beta_0(h - |z|)}. \quad (9c)$$

The integral (9b) is readily evaluated as in Sec. 3. Using (8a), the result is

$$\Psi_{\Sigma 0}(z) \sin \beta_0(h - |z|)$$
$$= [C_a(h, z) + 2C_b(h, z)] \sin \beta_0 h$$
$$- [S_a(h, z) + 2S_b(h, z)] \cos \beta_0 h. \quad (10a)$$

Values of $\Psi_{\Sigma 0}$ may be computed from (9a) with (10a) using the formulas for $C_a(h, z)$ and $S_a(h, z)$ given in (II.19.3, 4) and the same formulas for $C_b(h, z)$ and $S_b(h, z)$ with b substituted for a. Numerical values for $\beta_0 h = \pi/2$ and π are shown in Figs. 3.1 to 3.4. Since the imaginary parts of $C_a(h, z)$, $C_b(h, z)$, $S_a(h, z)$, and $S_b(h, z)$ are relatively small compared with the real parts at $z = 0$, $\beta_0 h \leq \pi/2$ and at $z = h - \lambda_0/4$, $\beta_0 h > \pi/2$,

if $\Omega = 2 \ln (2h/a)$ is large, an approximate value of $\Psi_{\Sigma 0}$ is

$$\Psi_{\Sigma 0} \doteq \tfrac{1}{2}(3\Psi_s - \Psi_a), \quad (10b)$$

where Ψ_s and Ψ_a are shown in Fig. 3.3.

The solution of the integral equation for the zero-phase-sequence current is the same as the analysis for the symmetrically driven, two-unit array in Sec. 3. It is merely necessary to replace Ψ_{s1} by $\Psi_{\Sigma 0}$ as given in (10a) and multiply each P and Q function in (3.16) by 2. The impedance of each antenna in the three-antenna array when driven in the zero-phase sequence or in the $2\pi/3$- or $4\pi/3$-phase sequences is obtained in the same manner from (7.7). Specifically,

$$(Z_0)_1 = \frac{-j\zeta_0 \Psi}{2\pi}$$

$$\times \left\{ \frac{\Psi \cos \beta_0 h + F_1(h) \pm 2P_1(h)}{[\Psi + F_1(0) \pm 2P_1(0)] \sin \beta_0 h - [G_1(0) \\ \pm 2Q_1(0)] \cos \beta_0 h + G_1(h) \pm 2Q_1(h)} \right\}. \quad (11)$$

For the zero-phase sequence $(Z_0)_1 = Z^{(0)}$, $\Psi = \Psi_{\Sigma 0}$ as in (10a) and the upper signs are used. For the positive and negative phase sequences $(Z_0)_1 = Z^{(1)} = Z^{(2)} = Z^a$, $\Psi = \Psi_{\Sigma 1} = \Psi_{\Sigma 2} = \Psi_a$ as in Sec. 4; the lower signs are used. Important special formulas for $\beta_0 h = \pi/2$ and π are obtained readily from (11).

This completes the analyses of the special methods of driving that satisfy the conditions leading to a single integral equation. The general case is readily constructed as follows:

Zero-phase sequence, $m = 0$:

$$V_{10} = V_{20} = V_{30} = V^{(0)},$$
$$I_{10} = I_{20} = I_{30} = I^{(0)}; \quad (12a)$$

$2\pi/3$-*(positive) phase sequence, $m = 1$:*

$$V_{10} = V^{(1)}, \quad V_{20} = pV^{(1)},$$
$$V_{30} = p^2 V^{(1)},$$
$$I_{10} = I^{(1)}, \quad I_{20} = pI^{(1)},$$
$$I_{30} = p^2 I^{(1)}; \quad (12b)$$

$4\pi/3$-*(negative) phase sequence, $m = 2$:*

$$V_{10} = V^{(2)}, \quad V_{20} = p^2 V^{(2)},$$
$$V_{30} = p^4 V^{(2)} = p V^{(2)},$$
$$I_{10} = I^{(2)}, \quad I_{20} = p^2 I^{(2)},$$
$$I_{30} = p^4 I^{(2)} = p I^{(2)}. \quad (12c)$$

The general case of arbitrary driving voltages is given by

$$V_{10} = V^{(0)} + V^{(1)} + V^{(2)}, \quad (13a)$$

$$V_{20} = V^{(0)} + pV^{(1)} + p^2V^{(2)}, \quad (13b)$$

$$V_{30} = V^{(0)} + p^2V^{(1)} + p^4V^{(2)}$$
$$= V^{(0)} + p^2V^{(1)} + pV^{(2)}. \quad (13c)$$

The combination of the three phase-sequence voltages $V^{(0)}$, $V^{(1)}$, and $V^{(2)}$ to give three different driving voltages V_{10}, V_{20}, and V_{30} is shown schematically in Fig. 15.2.

Since the currents in the three antennas due to the separate application of each of the phase-sequence voltages are in the same relative phases as are the voltages, it follows that the currents due to the superposition of the three driving voltages are

$$I_{10} = I_0^{(0)} + I_0^{(1)} + I_0^{(2)}, \quad (14a)$$

$$I_{20} = I_0^{(0)} + pI_0^{(1)} + p^2I_0^{(2)}, \quad (14b)$$

$$I_{30} = I_0^{(0)} + p^2I_0^{(1)} + pI_0^{(2)}. \quad (14c)$$

These may be solved for the phase-sequence currents as follows:

$$I_0^{(0)} = \tfrac{1}{3}(I_{10} + I_{20} + I_{30}), \quad (15a)$$

$$I_0^{(1)} = \tfrac{1}{3}(I_{10} + p^2I_{20} + pI_{30}), \quad (15b)$$

$$I_0^{(2)} = \tfrac{1}{3}(I_{10} + pI_{20} + p^2I_{30}). \quad (15c)$$

However, by solving the integral equation for each of the phase-sequence currents, the following results have been obtained:

$$V^{(0)} = I^{(0)}Z^{(0)}, \quad V^{(1)} = I^{(1)}Z^{(1)},$$
$$V^{(2)} = I^{(2)}Z^{(2)}, \quad (16)$$

where $Z^{(0)}$ and $Z^{(1)} = Z^{(2)} = Z^a$ are given in (11). By substituting (16) in (13) and then replacing the currents by (15), the following general equations are obtained in conventional, coupled-circuit form:

$$V_{10} = I_{10}Z_{11} + I_{20}Z_{12} + I_{30}Z_{13}, \quad (17a)$$

$$V_{20} = I_{10}Z_{21} + I_{20}Z_{22} + I_{30}Z_{23}, \quad (17b)$$

$$V_{30} = I_{10}Z_{31} + I_{20}Z_{32} + I_{30}Z_{33}, \quad (17c)$$

where

$$Z_{ii} = \tfrac{1}{3}(Z^{(0)} + 2Z^{(1)}), \quad i = 1, 2, 3; \quad (18a)$$

$$Z_{ij} = \tfrac{1}{3}(Z^{(0)} - Z^{(1)}), \quad i \neq j,$$
$$i, j = 1, 2, 3. \quad (18b)$$

Use has been made of the equations $p + p^2 = -1$ and $Z^{(2)} = Z^{(1)}$. In (17), the Z_{ii} are the equal self-impedances of the identical antennas in one another's presence, the Z_{ij} the mutual impedances. They may be evaluated from the input impedances $Z^{(0)}$ and $Z^{(1)}$ for the zero-phase sequence and the $2\pi/3$-phase sequence as given in (11). With self- and mutual impedances known, (17) may be solved for the input currents and the input impedances if the driving voltages are given.

The input impedance of antenna 1 is given by

$$Z_{10} = \frac{V_{10}}{I_{10}}$$

$$= \frac{(Z_{ii} - Z_{ij})(Z_{ii} + 2Z_{ij})}{(Z_{ii} + Z_{ij}) - Z_{ij}(V_{20} + V_{30})/V_{10}}$$

$$= \frac{3Z^{(1)}Z^{(0)}}{2Z^{(0)} + Z^{(1)} - r(Z^{(0)} - Z^{(1)})}, \quad (19)$$

where $r = (V_{20} + V_{30})/V_{10}$.

These formulas are valid if the subscripts 1, 2, 3 are permuted, so that Z_{20} and Z_{30} also are available.

The following special cases are included in (19).

(a) Three antennas driven in zero-phase sequence (the symmetric case):

$$V_{30} = V_{20} = V_{10}, \quad Z_{30} = Z_{20} = Z_{10} = Z_3^s, \quad (20a)$$

where Z_3^s is the impedance of any one of three symmetrically driven antennas.

(b) Three antennas driven in positive- or negative-phase sequence (the antisymmetric case):

$$V_{30} = V_{10}e^{\pm j2\pi/3}, \quad V_{20} = V_{10}e^{\mp j2\pi/3},$$
$$Z_{30} = Z_{20} = Z_{10} = Z^a, \quad (20b)$$

where Z^a is the impedance of any one of three antennas driven in the 120° phase sequence or the impedance of either one of two antennas driven antisymmetrically as defined in Sec. 7.

(c) Two antennas driven in phase by equal voltages; one antenna parasitic with no load:

$$V_{30} = 0, \quad V_{20} = V_{10},$$

$$Z_{20} = Z_{10} = \frac{3Z_3^s Z^a}{Z_3^s + 2Z^a}. \quad (20c)$$

(d) Two antennas driven 180° out of phase; one antenna parasitic with no load:

$$V_{30} = 0, \quad V_{20} = -V_{10}, \quad Z_{20} = Z_{10} = Z^a. \quad (20d)$$

Note that in this case the parasitic antenna is in the neutral plane and carries no current, so that it may be removed without thereby changing anything electrically.

(e) One antenna driven; two antennas parasitic with no load:

$$V_{30} = V_{30} = 0, \quad Z_{10} = \frac{3Z^s Z^a}{2Z_3^s + Z^a}. \quad (20e)$$

When the distance b between any two of the three antennas is sufficiently small ($b \ll \lambda_0$), Z^a is simply the input impedance of a section of transmission line with an open end and a length h.

Three-element folded dipole. If the three antennas are closely spaced ($b \ll \lambda_0$) and both ends of each antenna are bent 90° to join in star connections, the symmetric impedance Z^s is essentially the same as before the ends were connected. On the other hand, the antisymmetric impedance Z^a now is the impedance of a closed-end section of transmission line of length $h - b/2$. The terminating bridge is not a straight conductor of length b but a somewhat longer conductor with a 120° bend at its center. If the length $h - b/2$ is adjusted so that Z^a is very large compared with $2Z_3^s$, (20e) gives

$$Z_{10} = 3Z_3^s = 3(Z_{s1} + 2Z_{12}). \quad (20f)$$

This formula gives the impedance of a triple folded dipole, just as (12.16b) gives the impedance of the ordinary two-element folded dipole. It yields an impedance that is approximately nine times that of an isolated dipole. In particular, with $\Omega = 10$, $b/\lambda_0 = 0.01$, and $\beta_0 h \doteq \pi/2$, $Z_{10} = 771 + j341$ ohms compared with $Z_0 = 86 + j42$ ohms for the isolated single dipole.

A possible group of arrays not contained in (19) involves a load or tuning impedance at the center of the parasitic antenna or antennas in (20c)–(20e). Consider, for example, a single driven unit, No. 1, with impedances Z_{L2} and Z_{L3} at the centers of the other two. In this case

$$V_{30} = -I_{30}Z_{L3}, \quad V_{20} = -I_{20}Z_{L2}, \quad (21)$$

so that, with $Z_{ii} = Z_{11}$, $Z_{ij} = Z_{12}$, (10a-c) become

$$V_{10} = I_{10}Z_{11} + I_{20}Z_{12} + I_{30}Z_{13}, \quad (22a)$$

$$0 = I_{10}Z_{12} + I_{20}(Z_{11} + Z_{L2}) + I_{30}Z_{12}, \quad (22b)$$

$$0 = I_{10}Z_{12} + I_{20}Z_{12} + I_{30}(Z_{11} + Z_{L3}). \quad (22c)$$

The three currents are given by

$$I_{10} = \frac{V_{10}}{D}[(Z_{11} + Z_{L2})(Z_{11} + Z_{L3}) - Z_{12}^2], \quad (23a)$$

$$I_{20} = \frac{-V_{10}}{D} Z_{12}(Z_{11} + Z_{L3} - Z_{12}), \quad (23b)$$

$$I_{30} = \frac{-V_{10}}{D} Z_{12}(Z_{11} + Z_{L2} - Z_{12}), \quad (23c)$$

where

$$D = Z_{11}[(Z_{11} + Z_{L2})(Z_{11} + Z_{L3}) - Z_{12}^2] + Z_{12}^2[2(Z_{12} - Z_{11}) - (Z_{L2} + Z_{L3})]. \quad (24)$$

The impedance seen by the generator is

$$Z_{10} = \frac{V_{10}}{I_{10}} = Z_{11} - Z \quad (25a)$$

where

$$Z \equiv \frac{Z_{12}^2[2(Z_{11} - Z_{12}) + (Z_{L2} + Z_{L3})]}{(Z_{11} + Z_{L2})(Z_{11} + Z_{L3}) - Z_{12}^2}. \quad (25b)$$

Note that when $Z_{L3} = Z_{L2} = 0$, (25a) reduces to (20e) as it should. By making Z_{L2} and Z_{L3} appropriate reactances, a great variety of relative phases and magnitudes of the currents in the three units may be obtained.

Using the generalized method for analyzing closely spaced folded antennas described in the second half of Sec. 14, Harrison[26] has determined the driving-point admittance Y_{in} of a three-antenna array with the elements arranged at the vertices of an equilateral triangle, as shown in Fig. 15.1, for a variety of different interconnections of the ends. In each case a single generator with voltage V_{10} is at the center of element 1, while elements 2 and 3 are not driven, that is, $V_{20} = V_{30} = 0$. All of these admittances are in the general form

$$Y_{in} = \frac{1}{9Z_e} + \frac{jP}{180 \ln (b/a)}, \quad (26)$$

where Z_e is the input impedance of an isolated center-driven antenna of radius $a_e = \sqrt[3]{ab^2}$. For the several cases P in (26) has the following values:

(a) Elements 1 and 3 connected at top; elements 1 and 2 connected at bottom:

$$P = -2\left(\frac{\sin 2\beta_0 h}{1 - \cos 2\beta_0 h}\right). \quad (27a)$$

The input impedance $Z_{in} = 1/Y_{in}$ as computed from (26) with (27a) for $\Omega_e = 2 \ln (2h/a_e) = 10$, $b/a = 10$ is given in Table 15.1. Use is

[III.15] THEORY OF LINEAR ANTENNAS 365

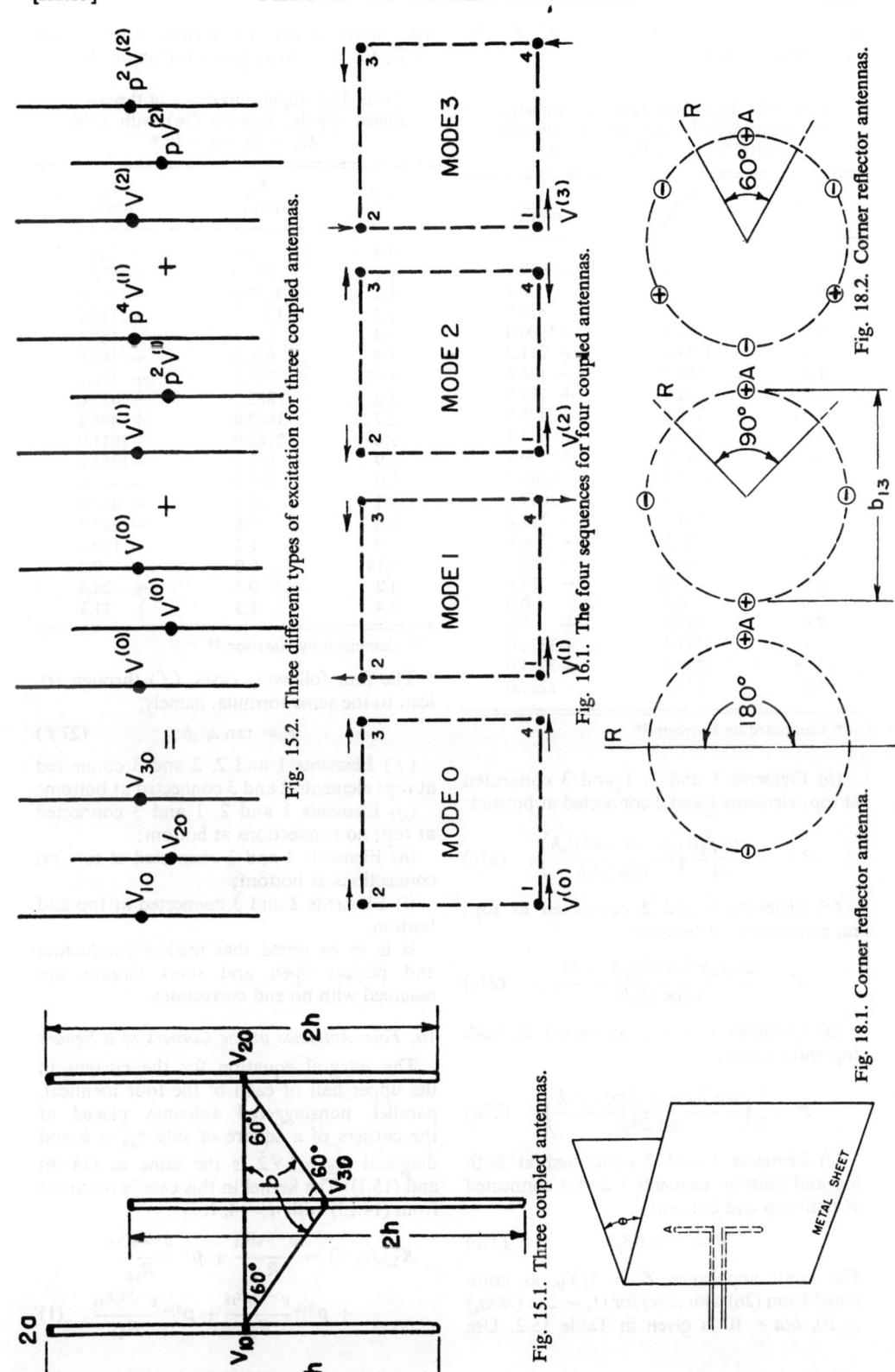

Fig. 15.2. Three different types of excitation for three coupled antennas.

Fig. 16.1. The four sequences for four coupled antennas.

Fig. 18.2. Corner reflector antennas.

Fig. 18.1. Corner reflector antenna.

Fig. 15.1. Three coupled antennas.

made of second-order values of Z_e for $\Omega = 10$ as given in Chapter II.

TABLE 15.1. Input impedance of symmetric three-element folded antenna, formula (26) with (27a). $\Omega_e = 10$, $b/a = 10$.*

$\beta_0 h$	R_{in} (ohms)	X_{in} (ohms)
0.5	0.0	− 19.9
0.52	.0	0.0
1.1	73.1	+ 839.0
1.2	744.9	+1600.0
1.3	1581.0	− 291.2
1.4	688.4	− 124.6
1.5	622.0	+ 183.0
1.57	778.5	+ 375.3
1.6	902.0	+ 443.1
1.7	1792.0	+ 190.9
1.8	1310.0	−1309.0
1.9	363.0	−1059.0
2.0	124.5	− 718.2
2.2	23.5	− 366.6
2.3	10.7	− 260.6
2.6	0.03	− 14.8
2.62	0.0	0.0
2.9	357.8	+ 7.1
3.1	1747.0	+3572.0
3.14	2250.0	−3924.0
3.2	179.1	−1257.0

* Computed by Harrison.[26]

(b) Elements 1 and 2, 1 and 3 connected at top; elements 1 and 2 connected at bottom:

$$P = -\frac{\cot \beta_0 h}{4}\left(\frac{3 - 8\sin^2 \beta_0 h}{\cos 2\beta_0 h}\right). \quad (27b)$$

(c) Elements 1 and 2 connected at top; no connections at bottom:

$$P = \frac{\tan \beta_0 h(8\cos^2 \beta_0 h - 1)}{4\cos 2\beta_0 h}. \quad (27c)$$

(d) Elements 1 and 2 connected at both top and bottom:

$$P = \frac{1}{2}\left(\frac{\cos 2\beta_0 h + 2\cos^2 \beta_0 h}{\sin 2\beta_0 h}\right). \quad (27d)$$

(e) Elements 1 and 2 connected at both top and bottom; elements 1 and 3 connected at both top and bottom:

$$P = -\cot \beta_0 h. \quad (27e)$$

The input impedance $Z_{in} = 1/Y_{in}$ as computed from (26) with (27e) for $\Omega_e = 2\ln(2h/a_e) = 10$, $b/a = 10$ is given in Table 15.2. Use has been made of second-order values of Z_e for $\Omega = 10$ as given in Chapter II.

TABLE 15.2. Input impedance of three-wire folded dipole, formula (26) with (27e); $\Omega_e = 10$, $b/a = 10$.*

$\beta_0 h$	R_{in} (ohms)	X_{in} (ohms)
0.5	0.1	+ 234.7
1.1	261.8	+1572.0
1.2	3170.0	+1793.0
1.3	1216.0	− 713.5
1.4	657.8	− 186.4
1.5	634.4	+ 162.0
1.57	778.5	+ 375.3
1.6	882.7	+ 457.0
1.7	1662.0	+ 498.2
1.8	2142.0	−1011.0
1.9	805.3	−1461.0
2.0	287.0	−1069.0
2.3	36.4	− 480.0
2.6	7.8	− 247.8
2.9	1.2	− 100.9
3.14	0.0	0.0
3.2	0.7	+ 24.4
3.4	1.5	+ 11.3

* Computed by Harrison.[26]

The four following cases, (f) through (i), lead to the same formula, namely,

$$P = \tan 2\beta_0 h: \quad (27f)$$

(f) Elements 1 and 2, 2 and 3 connected at top; elements 2 and 3 connected at bottom;
(g) Elements 1 and 2, 1 and 3 connected at top; no connections at bottom;
(h) Elements 2 and 3 connected at top, no connections at bottom;
(i) Elements 2 and 3 connected at top and bottom.

It is to be noted that lossless conductors and perfect open and short circuits are assumed with no end corrections.

16. Four Antennas at the Corners of a Square

The integral equation for the current in the upper half of each of the four identical, parallel, nonstaggered antennas placed at the corners of a square of side $b_{12} = b$ and diagonal $b_{13} = b\sqrt{2}$ is the same as (14.16) and (15.1). The kernel in this case is obtained from (14.15) with $n = 4$. It is

$$K_{\Sigma m}(z, z') = \frac{e^{-j\beta_0 R_{11}}}{R_{11}} + p^m \frac{e^{-j\beta_0 R_{12}}}{R_{12}} + p^{2m}\frac{e^{-j\beta_0 R_{13}}}{R_{13}} + p^{3m}\frac{e^{-j\beta_0 R_{14}}}{R_{14}}, \quad (1)$$

where
$$p = e^{j\theta}, \quad \theta = \pi/2, \tag{2}$$
so that
$$p = j, \quad p^2 = -1, \quad p^3 = -j,$$
$$p^4 = 1. \tag{3}$$
With $n = 4$, it follows that
$$m = 0, 1, 2, 3. \tag{4}$$

The four phase sequences are illustrated schematically in Fig. 16.1. They include the zero-phase sequence ($m = 0$) with all antennas in phase; the $\pi/4$- or positive-phase sequence ($m = 1$) with the current in antenna $k + 1$ leading that in antenna k by $\pi/4$; the π-phase sequence ($m = 2$) with currents alternating in direction around the square; and the $3\pi/4$-, $-\pi/4$-, or negative-phase sequence with the current in antenna $k + 1$ leading that in antenna k by $3\pi/4$.

The four kernels are

$$m = 0: \quad K_{\Sigma 0}(z, z') = \frac{e^{-j\beta_0 R_{11}}}{R_{11}} + \frac{e^{-j\beta_0 R_{12}}}{R_{12}}$$
$$+ \frac{e^{-j\beta_0 R_{13}}}{R_{13}} + \frac{e^{-j\beta_0 R_{14}}}{R_{14}}, \tag{5a}$$

$$m = 1: \quad K_{\Sigma 1}(z, z') = \frac{e^{-j\beta_0 R_{11}}}{R_{11}} + j\frac{e^{-j\beta_0 R_{12}}}{R_{12}}$$
$$- \frac{e^{-j\beta_0 R_{13}}}{R_{13}} - j\frac{e^{-j\beta_0 R_{14}}}{R_{14}}, \tag{5b}$$

$$m = 2: \quad K_{\Sigma 2}(z, z') = \frac{e^{-j\beta_0 R_{11}}}{R_{11}} - \frac{e^{-j\beta_0 R_{12}}}{R_{12}}$$
$$+ \frac{e^{-j\beta_0 R_{13}}}{R_{13}} - \frac{e^{-j\beta_0 R_{14}}}{R_{14}}, \tag{5c}$$

$$m = 3: \quad K_{\Sigma 3}(z, z') = \frac{e^{-j\beta_0 R_{11}}}{R_{11}} - j\frac{e^{-j\beta_0 R_{12}}}{R_{12}}$$
$$- \frac{e^{-j\beta_0 R_{13}}}{R_{13}} + j\frac{e^{-j\beta_0 R_{14}}}{R_{14}}. \tag{5d}$$

These expressions can be simplified if use is made of the following geometric condition:
$$R_{12} = R_{14}. \tag{6}$$
With (6), (5) becomes
$$m = 0: \quad K_{\Sigma 0}(z, z') = \frac{e^{-j\beta_0 R_{11}}}{R_{11}}$$
$$+ 2\frac{e^{-j\beta_0 R_{12}}}{R_{12}} + \frac{e^{-j\beta_0 R_{13}}}{R_{13}}, \tag{7a}$$

$$m = 1, 3: \quad K_{\Sigma 1}(z, z') = K_{\Sigma 3}(z, z')$$
$$= \frac{e^{-j\beta_0 R_{11}}}{R_{11}} - \frac{e^{-j\beta_0 R_{13}}}{R_{13}}, \tag{7b}$$

$$m = 2: \quad K_{\Sigma 2}(z, z') = \frac{e^{-j\beta_0 R_{11}}}{R_{11}}$$
$$- 2\frac{e^{-j\beta_0 R_{12}}}{R_{12}} + \frac{e^{-j\beta_0 R_{13}}}{R_{13}}. \tag{7c}$$

If these three kernels are substituted for the kernel in (15.9b), the corresponding expansion parameters may be evaluated from (15.9a) with appropriate changes in subscripts and with

$$\Psi_{\Sigma 0}(z) \sin \beta_0(h - |z|) = [C_a(h, z)$$
$$+ 2C_b(h, z) + C_c(h, z)] \sin \beta_0 h$$
$$+ [S_a(h, z) + 2S_b(h, z)$$
$$+ S_c(h, z)] \cos \beta_0 h, \tag{8a}$$

$$\Psi_{\Sigma 1}(z) \sin \beta_0(h - |z|)$$
$$= [C_a(h, z) - C_c(h, z)] \sin \beta_0 h$$
$$+ [S_a(h, z) - S_c(h, z)] \cos \beta_0 h, \tag{8b}$$

$$\Psi_{\Sigma 2}(z) \sin \beta_0(h - |z|) = [C_a(h, z)$$
$$- 2C_b(h, z) + C_c(h, z)] \sin \beta_0 h$$
$$+ [S_a(h, z) - 2S_b(h, z)$$
$$+ S_c(h, z)] \cos \beta_0 h. \tag{8c}$$

The functions $C_c(h, z)$ and $S_c(h, z)$ are like the corresponding functions with subscript a defined in (II.19.3) and (II.19.4) but with c substituted for a, where $c = b_{13} = b\sqrt{2}$.

With the expansion parameters for the four phase sequences evaluated, the corresponding currents and impedances may be determined just as for two and three antennas.

The general case in which each of the four antennas is driven by an arbitrary voltage may be solved by superposition using the following relations:

$$V_{10} = V^{(0)} + V^{(1)} + V^{(2)} + V^{(3)}, \tag{9a}$$
$$V_{20} = V^{(0)} + jV^{(1)} - V^{(2)} - jV^{(3)}, \tag{9b}$$
$$V_{30} = V^{(0)} - V^{(1)} + V^{(2)} - V^{(3)}, \tag{9c}$$
$$V_{40} = V^{(0)} - jV^{(1)} - V^{(2)} + jV^{(3)}. \tag{9d}$$

Alternatively, solving for the phase-sequence voltages,

$$V^{(0)} = \tfrac{1}{4}(V_{10} + V_{20} + V_{30} + V_{40}), \quad (10a)$$

$$V^{(1)} = \tfrac{1}{4}(V_{10} - jV_{20} - V_{30} + jV_{40}), \quad (10b)$$

$$V^{(2)} = \tfrac{1}{4}(V_{10} - V_{20} + V_{30} - V_{40}), \quad (10c)$$

$$V^{(3)} = \tfrac{1}{4}(V_{10} + jV_{20} - V_{30} - jV_{40}). \quad (10d)$$

The same equations apply to the currents if I is substituted for V in (9) and (10). By setting

$$V^{(0)} = I^{(0)}Z^{(0)}, \quad V^{(1)} = I^{(1)}Z^{(1)},$$

$$V^{(2)} = I^{(2)}Z^{(2)}, \quad V^{(3)} = I^{(3)}Z^{(3)} \quad (11)$$

in (9) and substituting (10) with I written for V in the resulting equations, the following results are obtained:

$$V_{i0} = \sum_{j=1}^{4} I_{j0} Z_{ij}, \quad i = 1, 2, 3, 4. \quad (12)$$

The evaluation of the self-impedances Z_{ii} and the mutual impedances Z_{ij} is facilitated if note is taken of the following relations which are a consequence of geometrical and electrical symmetry and the reciprocal theorem:

$$Z^{(1)} = Z^{(3)}, \quad (13a)$$

$$Z_{11} = Z_{22} = Z_{33} = Z_{44}, \quad (13b)$$

$$Z_{12} = Z_{23} = Z_{34} = Z_{41} = Z_{21} = Z_{32}$$
$$\qquad = Z_{43} = Z_{14}, \quad (13c)$$

$$Z_{13} = Z_{24} = Z_{31} = Z_{42}. \quad (13d)$$

With (13a–d) the following impedances are obtained directly:

$$Z_{11} = \tfrac{1}{4}(Z^{(0)} + 2Z^{(1)} + Z^{(2)}), \quad (14a)$$

$$Z_{12} = \tfrac{1}{4}(Z^{(0)} - Z^{(2)}), \quad (14b)$$

$$Z_{13} = \tfrac{1}{4}(Z^{(0)} - 2Z^{(1)} + Z^{(2)}). \quad (14c)$$

Alternatively, solving for the sequence impedances,

$$Z^{(0)} = Z_{11} + 2Z_{12} + Z_{13}, \quad (15a)$$

$$Z^{(1)} = Z_{11} - Z_{13}, \quad (15b)$$

$$Z^{(2)} = Z_{11} - 2Z_{12} + Z_{13}. \quad (15c)$$

The symmetry between the phase-sequence impedances and the self- and mutual impedances in (15a–c) and the phase-sequence kernels and the exponential terms in (7a–c) is apparent.

With (14a–c) used in (12) with (13) the currents in the four antennas due to arbitrary driving voltages may be determined.

Using the generalized method for analyzing closely spaced antennas described in Sec. 14 under the heading "Folded antennas," Harrison[26] has determined the driving-point admittance Y_{in} of a closely spaced four-antenna array with the four elements at the corners of a square of side b, and interconnected at the ends in a variety of ways. Unit 1 is center-driven, units 2 and 4 are adjacent to 1, and unit 3 is at the end of the diagonal. All admittances are in the form

$$Y_{in} = \frac{1}{16 Z_e} + \frac{jP}{480}, \quad (16)$$

where Z_e is the input impedance of an isolated center-driven antenna of radius $a_e = \sqrt[4]{ab^3 \sqrt{2}}$.

(a) Top connections: 1 and 4, 2 and 3; bottom connections: 1 and 2, 3 and 4:

$$P = -\cot \beta_0 h$$
$$\times \left[\frac{3 \ln (b/a) - \ln \sqrt{2} - 2 \sec 2\beta_0 h \ln (b/a\sqrt{2})}{(\ln b/a)^2 - (\ln \sqrt{2})^2} \right].$$
$$(17a)$$

The input impedance $Z_{in} = 1/Y_{in}$ as computed from (16) with (17a) is given in Table 16.1 for $\Omega_e = 2 \ln (2h/a_e) = 10$, $b/a = 10$. Use has been made of second-order values of Z_e for $\Omega = 10$ as given in Chapter II.

TABLE 16.1. Input impedance of four-wire reëntrant loop, formula (16) with (17a); $\Omega_e = 10$, $b/a = 10$.*

$\beta_0 h$	R_{in} (ohms)	X_{in} (ohms)
0.9	0.2	+ 1.4
1.1	10.4	+ 423.4
1.2	67.7	+ 705.8
1.3	616.8	+1188.0
1.4	1119.0	+ 402.5
1.5	960.7	+ 480.3
1.57	1384.0	+ 667.0
1.6	889.4	+ 312.9
1.7	1750.0	−1604.0
1.8	306.5	−1155.0
1.9	80.6	− 698.8
2.0	276.8	− 457.4

* Computed by Harrison.[26]

(b) Four-wire folded dipole—top and bottom connections: 1 and 2, 2 and 3, 3 and 4, 4 and 1:

$$P = -\cot \beta_0 h \left[\frac{3 \ln (b/a) - \ln \sqrt{2}}{(\ln b/a)^2 - (\ln \sqrt{2})^2}\right]. \quad (17b)$$

The input impedance $Z_{in} = 1/Y_{in}$ as computed from (16) with (17b) is given in Table 16.2 for $\Omega_e = 2 \ln (2h/a_e) = 10$, $b/a = 10$. Use has been made of second-order values of Z_e for $\Omega = 10$ as given in Chapter II.

TABLE 16.2. Input impedance of four-wire folded dipole, formula (16) with (17b); $\Omega_e = 10$, $b/a = 10$.*

$\beta_0 h$	R_{in} (ohms)	X_{in} (ohms)
0.9	3.7	+ 531.3
1.1	57.8	+ 996.6
1.2	387.0	+1651.0
1.3	2424.0	+1080.0
1.4	1264.0	− 0.5
1.5	1060.6	+ 386.6
1.57	1384.0	+ 667.0
1.6	1663.0	+ 737.0
1.7	3052.0	− 720.8
1.8	1042.0	−1941.0
1.9	289.1	−1301.0
2.0	111.9	− 914.7

* Computed by Harrison.[26]

17. Cage Antennas

Cage antennas are arrays of N closely spaced parallel antennas excited in the zero-phase sequence so that *all currents are in phase*. The kernel for an N-unit array is

$$K_{\Sigma 0}(z, z') = \sum_{i=1}^{N} \frac{e^{-j\beta_0 R_{1i}}}{R_{1i}}, \quad (1)$$

and the expansion function is

$$\Psi_{\Sigma 0}(z) = \int_{-h}^{h} g(z, z') K_{\Sigma 0}(z, z') \, dz', \quad (2)$$

where $g(z, z') = \sin \beta_0(h - |z'|)/\sin \beta_0(h - |z|)$. This may be expressed as follows:

$$\Psi_{\Sigma 0}(z) = \Psi_{K1}(z) + \Psi_{b12}(z) + \Psi_{b13}(z) + \cdots, \quad (3)$$

where

$$\Psi_{K1}(z) = \int_{-h}^{h} g(z, z') \frac{e^{-j\beta_0 R_{11}}}{R_{11}} \, dz',$$

$$R_{11} = \sqrt{(z - z')^2 + a^2}, \quad (4a)$$

$$\Psi_{b12}(z) = \int_{-h}^{h} g(z, z') \frac{e^{-j\beta_0 R_{12}}}{R_{12}} \, dz',$$

$$R_{12} = \sqrt{(z - z')^2 + b_{12}^2}, \quad (4b)$$

$$\Psi_{b13}(z) = \int_{-h}^{h} g(z, z') \frac{e^{-j\beta_0 R_{13}}}{R_{13}} \, dz',$$

$$R_{13} = \sqrt{(z - z')^2 + b_{13}^2}, \quad (4c)$$

$$\vdots$$

$$\Psi_{b1N}(z) = \int_{-h}^{h} g(z, z') \frac{e^{-j\beta_0 R_{1N}}}{R_{1N}} \, dz',$$

$$R_{1N} = \sqrt{(z - z')^2 + b_{1N}^2}. \quad (4d)$$

The distance between centers of antennas 1 and 2 is b_{12}; between antennas 1 and N it is b_{1N}. As shown in Sec. 5, for sufficiently closely spaced antennas for which

$$\beta_0 b_{1i} \ll 1, \quad i = 1, 2, \cdots N, \quad (5)$$

it follows that

$$\Psi_{b1i}(z_r) = \Psi_{K1}(z_r) - 2 \ln (b_{1i}/a), \quad (6)$$

where $z_r = 0$ for $\beta_0 h \leq \pi/2$, $z_r = h - \lambda_0/4$ for $\beta_0 h \geq \pi$. Hence,

$$\Psi_{\Sigma 0} = |\Psi_{\Sigma 0}(z_r)|$$

$$= \left| N\Psi_{K1}(z_r) - 2 \sum_{i=2}^{N} \ln (b_{1i}/a) \right| \quad (7a)$$

$$= \left| N\Psi_{K1}(z_r) - 2 \ln (b_{12}b_{13}b_{14} \cdots b_{1N}/a^{N-1}) \right|. \quad (7b)$$

Using this formula of the expansion parameter, the current in and impedance of each of the N antennas forming the cage may be determined just as for a single antenna.

For antennas that satisfy the condition $\beta_0 h \leq 1$, a satisfactory approximate value of the expansion parameter for a single antenna is

$$\Psi_{K1} \doteq 2[\ln (2h/a) - 1]. \quad (8)$$

Since the contribution to Ψ_{K1} by the imaginary part of $\Psi_{K1}(z_r)$ is not great, an approximate formula for (7b) is obtained by replacing $\Psi_{K1}(z_r)$ by Ψ_{K1} as given by (8). The result is

$$\Psi_{\Sigma 0} \doteq N\Psi_{K1} - 2\ln(b_{12}b_{13}b_{14}\cdots b_{1N}/a^{N-1}) \tag{9a}$$

$$\doteq 2\ln[(2h)^N/ab_{12}b_{13}\cdots b_{1N}]. \tag{9b}$$

It is a simple matter to determine the radius a_e of a single antenna carrying N times the current of each unit in the cage and having $1/N$ times the impedance. It is necessary merely to set $a = a_e$ in (8), multiply (8) by N, and equate the result to (9b). Thus, solving for a_e, and in confirmation of (14.48),

$$a_e = \sqrt[N]{ab_{12}b_{13}b_{14}\cdots b_{1N}}. \tag{10}$$

Therefore, a cage antenna consisting of N closely spaced, parallel, identical units arranged at regular intervals around a circle is *approximately* equivalent to a single antenna of much larger radius, in the sense that the total current and its axial distribution are approximately the same, and its impedance is $1/N$ that of the individual units if these are all driven separately. A simple method of driving a cage antenna is to have the N conductors forming each half of the cage converge at the center to form a single pair of terminals.

18. Current and Impedance of Antenna with Corner Reflector*

An important directional antenna consists of a single linear radiator placed midway between two intersecting flat metal sheets. The antenna and the metal sheets are perpendicular to the xy-plane; the antenna is parallel to and at an arbitrary distance from the junction line of the two metal sheets which meet at an angle θ. The arrangement is illustrated in Fig. 18.1, and in Fig. 18.2 for angles $\theta = 180°$, $90°$, $60°$; in each case the antenna is at A.

A rigorous analysis of the distribution of current and impedance of the single linear radiator is not available for metal sheets of *finite* size and conductivity. However, if the sheets are highly conducting and quite large compared with the length of the antenna, and if this is not located near the mouth of the reflector, a reasonable approximation may be obtained by assuming the metal sheets to be infinite in conductivity and extent. (This is not true of the field.)

---

* Part of the material in this section is taken from the work of Dr. C. T. Tai, ref. 50.

The approximate analysis of the antenna between finite but large plates is readily carried out by means of the theorem of images. By arranging a symmetric group of $N - 1$ image antennas in such a manner that the resultant tangential component of the electric field vanishes along the perfectly conducting planes, these planes may be imagined removed completely, leaving the problem of N antennas in the $180°$ phase sequence. The configuration of conductors and images is shown in Fig. 18.2, where downward currents are denoted by $+$ and upward currents by $-$. As described in Sec. 14, the phase difference between adjacent units is given by $m\theta$, where $\theta = 2\pi/N$. The only phase sequence involved here is the π-sequence, for which $m = N/2$, so that the increment of phase change from one antenna to the next adjacent one is $N\theta/2$. The choice of the angle θ and the distance from the linear radiator to the junction of the two metal planes is determined from the desired directional properties of the array.

If the angle θ is also chosen to be π, so that $N = 2$, an antenna parallel to a perfectly conducting plane is obtained. This is the antisymmetric case discussed in Sec. 7.

If the angle θ is made $\pi/2$, it follows that $N = 4$ and the expansion parameter is $\Psi_{\Sigma 2}$ obtained from (16.8c) used in conjunction with (15.9a) with appropriate changes in subscripts from 0 to 2. In the important special case defined by $\beta_0 h = \pi/2$, the expansion parameter reduces to

$$\Psi_{\Sigma 2} = |C_a(h, 0) - 2C_b(h, 0) + C_c(h, 0)| \tag{1}$$

and the formula for the impedance is

$$Z_{\Sigma 2} = \frac{j\zeta_0 \Psi_{\Sigma 2}}{2\pi} \times$$

$$\left[\frac{C_a(h,0) - C_{b4}(h,0)}{2\Psi_{\Sigma 2} + E_a(h,0) - C_a(h,0) - S_a(h,0) - E_{b4}(h,0) + C_{b4}(h,0) + S_{b4}(h,0)} \right], \tag{2}$$

where

$$C_{b4}(h, 0) = [2C_b(h, 0) - C_c(h, 0)], \tag{3a}$$

$$S_{b4}(h, 0) = [2S_b(h, 0) - S_c(h, 0)], \tag{3b}$$

$$E_{b4}(h, 0) = [2E_b(h, 0) - E_c(h, 0)]. \tag{3c}$$

In (1)–(3), $h = \lambda_0/4$, $c = b_{13} = b\sqrt{2}$.

As a numerical example, consider a corner-reflector antenna consisting of a linear radiator ($h = \lambda_0/4$, $\Omega = 2 \ln 2h/a = 20$) center-driven by a slice generator between semi-infinite reflecting planes meeting at right angles ($\theta = \pi/2$) at a distance $b_{13}/2 = \lambda_0/\sqrt{2}$ from the radiator. (The distance from the radiator to each plane is $b_{12}/2 = \lambda_0/4$.) The impedance as determined from (2) is

$$Z_{\Sigma 2} = 70.4 + j98.0 \text{ ohms.} \qquad (4)$$

It is interesting to compare this value with the corresponding impedance for the same antenna when the angle of the reflector θ is π so that the antenna is at a distance $b_{12}/2 = \lambda_0/4$ in front of a single, perfectly conducting plane. This is simply the antisymmetric impedance of two antennas separated by a distance $b_{12} = \lambda_0/2$, as given in Sec. 4. It is

$$Z^a = 86.8 + j65.7. \qquad (5)$$

The corresponding impedance for the same antenna when isolated is obtained from Table II.30.1. It is

$$Z_0 = 78.5 + j43.6. \qquad (6)$$

The effect of the reflector in changing the impedance of the antenna is seen to be significant.

19. Parallel Arrays with all Elements Driven; Broadside and End-Fire Arrays; Circuit Properties*

The most important arrays of identical, parallel, nonstaggered, center-driven antennas consist of N units spaced at equal intervals b along a straight line, as shown in Fig. 19.1. In one class of arrays, the N currents are made equal in magnitude by appropriate driving voltages, and it is the phase relations that determine the different field characteristics of different arrays. For example, in the broadside array all currents are in phase; in the end-fire array the phase of the currents increases from one antenna to the next by an angle $\beta_0 b$. In another class of arrays, the magnitudes of the currents decrease in a prescribed manner from the center of the array outward in both directions.

Since there are no conditions of symmetry in an array of antennas uniformly spaced along a straight line that make it possible to reduce the N simultaneous integral equations given in (14.7) to a single equation, the conclusion is inevitable that the determination of the N currents and N impedances in terms of arbitrary driving voltages involves the solution of N simultaneous integral equations. Fortunately, an approximate solution of the important inverse problem of determining the N voltages required to maintain N currents with specified relative magnitudes and phases is possible.

The integral equations for the currents in the upper halves ($\delta \leq z \leq h$) of the N coupled antennas are given by (14.7). They may be expressed as follows:

$$\int_{-h}^{h} I_{mz}(z'_m) K_p(z_m, z'_m) \, dz'_m$$

$$= \frac{-j4\pi}{\zeta_0} (C_m \cos \beta_0 z_m + \tfrac{1}{2} V_{0m} \sin \beta_0 z_m),$$

$$m = 1, 2, \cdots, N, \quad (1a)$$

where the kernel is given by

$$K_p(z_m, z'_m) = \sum_{i=1}^{N} \frac{I_{iz}(z'_i)}{I_{mz}(z'_m)} \frac{e^{-j\beta_0 R_{mi}}}{R_{mi}},$$

$$m = 1, 2, \cdots, N, \quad (1b)$$

and where

$$R_{mi} = \sqrt{(z_m - z'_i)^2 + b_{mi}^2},$$

$$R_{mm} = \sqrt{(z_m - z'_m)^2 + a^2}. \quad (1c)$$

The distance between the centers of antenna m and antenna i is $b_{mi} = |m - i| b$, where b is the equal distance between adjacent units.

The N kernels $K_p(z_m, z'_m)$ are functions of the distributions of current in all antennas and of the distances R_{mi}. They are not independent. However, an approximate solution of (1a) may be obtained by assuming that the *distributions* of current as functions of z' are the same in all N units, but without thereby restricting the magnitudes or the phases of the input currents at $z = 0$. It is readily argued that this is a satisfactory approximation for identical antennas. Indeed, it has already been shown in the analysis of two coupled antennas driven either in-phase or 180° out-of-phase that the distributions of current do not differ significantly from each other or from the distribution along an isolated antenna, provided $\beta_0 h$ does not greatly exceed $\pi/2$. When $\beta_0 h$ is near π, the distributions differ considerably for very closely spaced antennas, but as the

---

* The three-element broadside array was analyzed first by Dr. C. W. Harrison, Jr., ref. 23, using a method that did not take account of the coupled antennas in the kernel but only in a correction term.

separation is increased, the currents must become more and more nearly alike. Let it be assumed, therefore, that it is a good approximation to set

$$\frac{I_{iz}(z_i')}{I_{mz}(z_m')} \doteq k_{mi} e^{j\theta_{mi}}, \qquad (2)$$

where k_{mi} is the specified real ratio of magnitudes and δ_{mi} is the specified phase difference between the currents in antennas k and m. Substitution of (2) in (1a) gives

$$K_p(z_m, z_m') = \sum_{i=1}^{N} k_{mi} e^{j\theta_{mi}} \frac{e^{-j\beta_0 R_{mi}}}{R_{mi}},$$

$$m = 1, 2, \cdots, N. \qquad (3)$$

The N kernels defined in (3) are not alike. They are not and can not be made independent of m, since, for geometric reasons,

$$R_{mi} \neq R_{m+q, i+q}, \qquad (4)$$

where q is an integer, for all values of m and i from 1 to N.

However, since all quantities in (3) are assumed to be known for all values of m, the N kernels can be evaluated independently. In particular, (1a) can be solved separately for $I_{mz}(z)$ for each value of m. Thus, it is seen that the restrictions on the current contained in (2) have reduced the N simultaneous integral equations for determining N independent currents to N independent integral equations for determining the N voltages V_{m0} that are required to maintain the N specified currents.

Since (1a) differs from the corresponding equation for two antennas only in the number of terms in the kernel, the formal solution for the current and the impedance as a function of the driving voltage may be carried out directly as in Secs. 3 and 7. The expression for the input impedance of antenna m in the presence of $N - 1$ parallel antennas all so driven that they have preassigned currents is like (7.7) with additional P and Q terms each multiplied by the appropriate ratio and phase factors for the current. Thus,

$$[(Z_m)_{\text{in}}]_1 = \frac{-j\zeta_0 \Psi_{pm}}{2\pi} \times$$

$$\left\{ \frac{\Psi_{pm} \cos \beta_0 h + F_1(h) + P_{1\Sigma}(h)}{[\Psi_{pm} + F_1(0) + P_{1\Sigma}(0)] \sin \beta_0 h - [G_1(0)]} + Q_{1\Sigma}(0)] \cos \beta_0 h + [G_1(h) + Q_{1\Sigma}(h)] \right\}, \qquad (5)$$

where

$$\Psi_{pm} = |\Psi_{pm}(0)| = \frac{|\psi_{pm}(0)|}{\sin \beta_0 h}, \quad (\beta_0 h \leq \pi/2) \qquad (6a)$$

$$\Psi_{pm} = |\Psi_{pm}(h - \lambda_0/4)| = |\psi(h - \lambda_0/4)|,$$
$$(\beta_0 h \geq \pi/2) \quad (6b)$$

$$\psi_{pm}(z)$$
$$= [C_a(h, z) + \sum_{\substack{i=1 \\ i \neq m}}^{N} k_{mi} e^{j\theta_{mi}} C_{b_{mi}}(h, z)] \sin \beta_0 h$$

$$- [S_a(h, z) + \sum_{\substack{i=1 \\ i \neq m}}^{N} k_{mi} e^{j\theta_{mi}} S_{b_{mi}}(h, z)] \cos \beta_0 h. \qquad (6c)$$

Also,

$$P_{1\Sigma}(z) = \sum_{\substack{i=1 \\ i \neq m}}^{N} k_{mi} e^{j\theta_{mi}} P_{1mi}(z), \qquad (7a)$$

$$Q_{1\Sigma}(z) = \sum_{\substack{i=1 \\ i \neq m}}^{N} k_{mi} e^{j\theta_{mi}} Q_{1mi}(z), \qquad (7b)$$

$$P_{1mi}(z) = -\int_{-h}^{h} F_{1z'} \frac{e^{-j\beta_0 R_{mi}}}{R_{mi}} dz'$$
$$= -C_{b_{mi}}(h, z) + E_{b_{mi}}(h, z) \cos \beta_0 h, \qquad (8a)$$

$$Q_{1mi}(z) = -\int_{-h}^{h} G_{1z'} \frac{e^{-j\beta_0 R_{mi}}}{R_{mi}} dz'$$
$$= -S_{b_{mi}}(h, z) + E_{b_{mi}}(h, z) \sin \beta_0 h, \qquad (8b)$$

$$R_{mi} = \sqrt{(z_m - z_i')^2 + b_{mi}^2}, \qquad (8c)$$

and $F_1(0)$, $F_1(h)$, $G_1(0)$, and $G_1(h)$ are given in (7.8) with Ψ_{pm}' substituted for Ψ. Note that b_{mi} is the distance between centers of antennas m and i. When $N = 2$, $\theta_{mi} = 0$, $k_{mi} = 1$, (5) reduces to (7.7) for two symmetrically driven antennas. Alternatively, when $N = 2$, $\theta_{mi} = \pi$, $k_{mi} = 1$, (5) reduces to (7.7) for two antisymmetrically driven antennas.

By evaluation of the functions $P_{1\Sigma}$ and $Q_{1\Sigma}$ together with the expansion parameter Ψ_{pm}' for *each* value of m, the input impedance of each antenna in the presence of the others may be determined. Note that, in general, the antennas are in identical pairs except for the central unit for N odd, so that the number of different values of these functions is $N/2$ for N even, $(N + 1)/2$ for N odd. Multiplication of the impedances so determined by the appropriate assigned input current gives the driving voltages required to maintain these currents.

In practice, antennas in parallel arrays usually have electrical half-lengths $\beta_0 h$ that are $\pi/2$ or near $\pi/2$. Significantly, this is precisely the value of $\beta_0 h$ for which the initial assumption of identical *distributions* (but not magnitudes or phases) of current is best satisfied and for which a maximum simplification of the intricate formula (5) is possible. With $\beta_0 h = \pi/2$, (5) becomes

$$[(Z_m)_{\text{in}}]_1 \doteq \frac{j\zeta_0}{2\pi}$$

$$\times \left\{ \frac{C_a(h, h) + \sum_{\substack{i=1 \\ i \neq m}}^{N} k_{mi} e^{j\theta_{mi}} C_{b_{mi}}(h, h)}{1 + \frac{1}{\Psi_{pm}}[F_1(0) + P_{1\Sigma}(0) + G_1(h) + Q_{1\Sigma}(h)]_{\beta_0 h = \pi/2}} \right\}. \quad (9)$$

Except for antennas that are extremely closely spaced, the expansion parameter Ψ_{pm} cannot differ greatly from the value Ψ_{K1} for each of the antennas when isolated. In most driven arrays of the types considered in this section, b is not less than $\lambda_0/4$ and frequently is $\lambda_0/2$. Since the term with $1/\Psi_{pm}$ as a factor in the denominator of (9) necessarily is small compared with unity for antennas of small radius, it may be neglected. That this is indeed a good approximation has already been demonstrated for two coupled antennas with $\beta_0 h = \pi/2$ in Sec. 7 in conjunction with Figs. 7.1 and 7.2, in which the symmetric and antisymmetric impedances are seen to depend in a relatively very small degree on the value of Ω. If this term is neglected, (9) gives simply

$$(Z_m)_{\text{in}} \doteq \frac{j\zeta_0}{4\pi}[C_a(h, h) + \sum_{\substack{i=1 \\ i \neq m}}^{N} k_{mi} e^{j\theta_{mi}} C_{b_{mi}}(h, h)],$$

$$m = 1, 2, \cdots, N. \quad (10)$$

Multiplication of (10) by I_{m0}, noting that

$$I_{i0} = I_{m0} k_{mi} e^{j\theta_{mi}}, \quad (11)$$

reveals that (10) is equivalent to

$$V_{10} \doteq I_{10} Z_{11} + I_{20} Z_{12} + I_{30} Z_{13} + \cdots + I_{N0} Z_{1N},$$

$$V_{20} \doteq I_{10} Z_{21} + I_{20} Z_{22} + I_{30} Z_{23} + \cdots + I_{N0} Z_{2N}, \quad (12)$$

$$\vdots$$

$$V_{N0} \doteq I_{10} Z_{N1} + I_{20} Z_{N2} + I_{30} Z_{N3} + \cdots + I_{N0} Z_{NN},$$

where

$$Z_{11} \doteq Z_{22} \doteq Z_{33} \doteq \cdots Z_{NN} \doteq \frac{j\zeta_0}{2\pi} C_a(h, h)$$

$$\doteq Z_0, \quad (13)$$

$$Z_{12} = Z_{21} \doteq \frac{j\zeta_0}{2\pi} C_{b_{12}}(h, h), \quad (14a)$$

$$Z_{13} = Z_{31} \doteq \frac{j\zeta_0}{2\pi} C_{b_{13}}(h, h), \quad (14b)$$

$$\vdots$$

$$Z_{1N} = Z_{N1} \doteq \frac{j\zeta_0}{2\pi} C_{b_{1N}}(h, h). \quad (14c)$$

Note that the self-impedances Z_{ii} of the coupled antennas are all the same as the self-impedance Z_0 of an isolated antenna. Moreover, the mutual impedances Z_{ij} are the same as if only antennas i and j were present. Therefore, it may be concluded that for parallel arrays in which antennas of half-length near $h = \lambda_0/4$ are coupled, the self-impedance of an isolated antenna may be used for each antenna and the mutual impedance between any pair taken to be the same as for that pair when isolated. Accordingly, the self-impedance may be obtained from curves in Chapter II, or as Z_{s1} from curves in Sec. 8, and the mutual impedance from curves in Sec. 8 for the appropriate value of $\Omega = 2 \ln(2h/a)$. For antennas farther apart than $b/\lambda_0 = 1$, the more extensive curves of Fig. 8.1 or the data of Table 8.1 may be used.

It is significant to note that in spite of the initial *general* statement to the contrary, the equations (13) may be solved for either the driving voltages, given the current, or the currents, given the driving voltages. However, this is true only approximately and only in the special case of antennas for which $\beta_0 h \doteq \pi/2$ and $\Psi_{1\Sigma}$ is not too small.

The simple method of analyzing N coupled antennas in terms of self-impedances of isolated antennas and mutual impedances of two coupled antennas obviously is not valid for antennas with half-lengths that exceed $\lambda_0/4$ considerably. Specifically, when $\beta_0 h = \pi$, (5) becomes

$$[(Z_m)_{\text{in}}]_1 \doteq \frac{j\zeta_0 \Psi_{pm}^2}{2\pi}$$

$$\times \left\{ \frac{1 - \frac{1}{\Psi_{pm}}[F_1(h) + P_1(h)]}{[G_1(0) + Q_{1\Sigma}(0) + G_1(h) + Q_{1\Sigma}(h)]} \right\}. \quad (15)$$

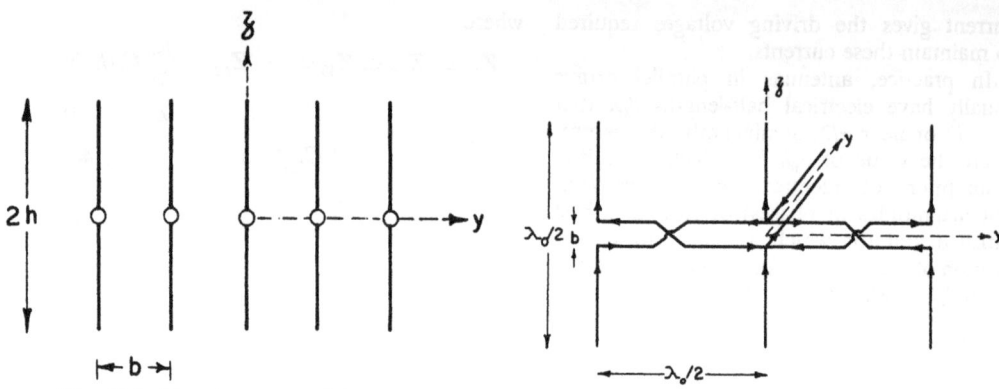

Fig. 19.1. Five-element parallel array.

Fig. 19.2. Three-element broadside array.

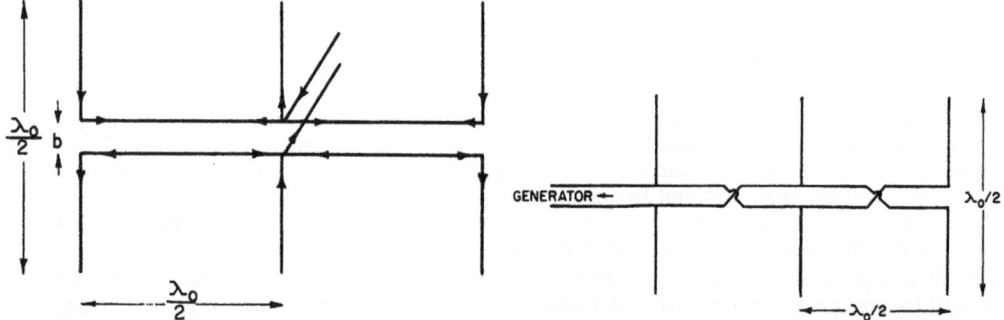

Fig. 19.3. Three-element bilateral end-fire array.

Fig. 19.4. Laterally driven broadside array.

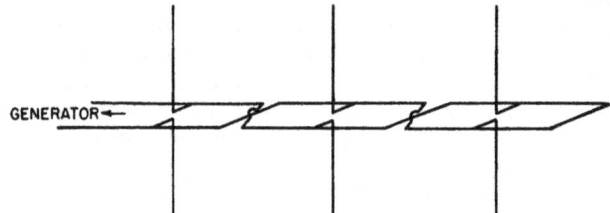

Fig. 19.5. Laterally driven broadside array with minimum antenna-line coupling.

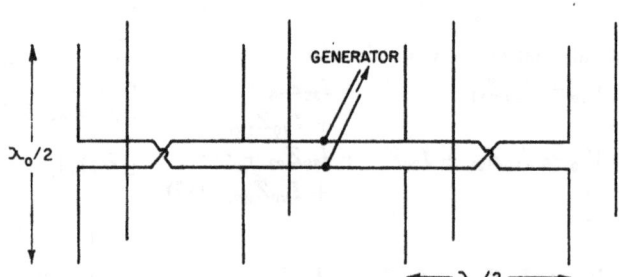

Fig. 19.6. Center-driven four-element broadside array with parasitic reflectors.

Since the term involving the most significant contributions from the coupled antennas now appears in the factor Ψ_{pm}^2 and in the terms in Q in the denominator, it is not possible to express $V_{m0} = I_{m0}(Z_m)_{in}$ as a simple sum of self- and mutual terms. The impedance $(Z_m)_{in}$ can be evaluated from (15) and V_{m0} evaluated for each antenna if the currents are known. The reverse is not possible.

For the broadside array, $k_{mi} = 1$, $\theta_{mi} = 0$ with $\beta_0 b = \pi$. For the end-fire array, $\theta_{mi} = \pm(m - i)\beta_0 b$ with $\beta_0 b = \pi/2$ for unilateral end-fire, π for bilateral end-fire. These arrays are discussed from the point of view of field patterns in Chapter VI. The analysis in this section has demonstrated that common assumptions such as that equal driving voltages will produce equal currents in all units of a broadside array are erroneous except for the two-element array. The different impedances of the differently situated units require different voltages if the currents are to be the same.

Conventional methods of driving parallel arrays in which currents in the units are assumed to be respectively in phase (broadside array) or alternately 180° out of phase (bilateral end-fire array) are shown in Figs. 19.2 and 19.3 for three-element arrays. The transmission line connecting the units in the broadside array is assumed to be *spiraled* and *not crossed over* as shown in the figure so that it is the equivalent of a smooth line one half-wavelength long. Actually, the currents in the three antennas in Figs. 19.2 and 19.3 are neither all equal in magnitude nor all in phase (broadside) or alternately 180° out-of-phase (bilateral-end fire) for two reasons. These are: (1) the input impedance of the central unit differs from that of the two outer units, since this unit is not in the same position relative to the other antennas in the array; (2) the transmission-line end effects and coupling effects are different for the central unit from what they are for the two outer units.

Let the problem of transmission-line end effects and coupling effects between each antenna and the feeding line be sidestepped for the present by assuming that the distance b between the conductors of the line is sufficiently small to make these effects negligible. If this is done, formulas for the ratio of the current in antenna 1 or 3 (which have equal currents) to the current in the central unit 2 are obtained readily for $\beta_0 h = \pi/2$, since the simple formula (13) with $N = 3$ is a good approximation. With negligible terminal-zone effects, the two sections of line of length $\lambda_0/2$ are one-to-one transformers, so that the same voltage is applied to each of the three units. That is,

$$V_{30} = V_{20} = V_{10}. \tag{16}$$

Moreover, since antennas 1 and 3 are geometrically and electrically identical, it follows that

$$I_{30} = I_{10}, \quad Z_{12} = Z_{23}. \tag{17}$$

With (13) and (12) the following relations are obtained directly when $N = 3$:

$$V_{10} = I_{10}(Z_{11} + Z_{13}) + I_{20}Z_{12}$$
$$= V_{20} = 2I_{10}Z_{12} + I_{20}Z_{22}. \tag{18}$$

The solution of (18) for the ratio of currents is

$$I_{20}/I_{10} = (Z_{11} + Z_{13} - 2Z_{12})/(Z_{22} - Z_{12}). \tag{19}$$

The expressions for the impedances are

$$(Z_3)_{in} = (Z_1)_{in} \equiv \frac{V_{10}}{I_{10}}$$
$$= Z_{11} + Z_{13} + Z_{12}\left(\frac{Z_{11} + Z_{13} - 2Z_{12}}{Z_{22} - Z_{12}}\right), \tag{20}$$

$$(Z_2)_{in} \equiv \frac{V_{20}}{I_{20}}$$
$$= Z_{22} + 2Z_{12}\left(\frac{Z_{22} - Z_{12}}{Z_{11} + Z_{12} - 2Z_{12}}\right). \tag{21}$$

If all antennas are identical and have no series impedances,

$$Z_{33} = Z_{22} = Z_{11} = Z_{s1}. \tag{22}$$

In this case, it is evident that $I_{30} = I_{10}$ must differ considerably from I_{20}, and $(Z_3)_{in} = (Z_1)_{in}$ from $(Z_2)_{in}$ since Z_{13} and Z_{12} are far from the same. For example, with $\Omega = 10$, $b = \lambda_0/2$; $Z_{s1} = 85.8 + j42.7$, $Z_{12} = -15.2 - j30.4$, $Z_{13} = 5.1 + j18.2$, as obtained from Tables 8.10 and 8.11.

It is possible to make the currents all equal when the driving voltages are the same by connecting an impedance in series with the central unit, antenna 2. Specifically, let

$$Z_{33} = Z_{11} = Z_{s1}, \quad Z_{22} = Z_{s1} + Z_2. \tag{24}$$

Upon setting the currents equal in (19) and using (24), the result is

$$Z_2 = Z_{13} - Z_{12}. \qquad (25)$$

It is seen that Z_2 is quite a large impedance with a significant real part, so that appreciable ohmic losses are involved which lower the efficiency of transmission. Since the electromagnetic field of a three-element array in which the current in the central unit is greater than that in the outer units may be more desirable than when the currents are all equal, it is seldom advantageous to use Z_2 as specified in (25) in series with the central unit. An alternative procedure is to connect reactances in series with the *outer* units in order to make the phases of all three units the same. That is, in place of (24), let

$$Z_{33} = Z_{11} = Z_{s1} + jX_1, \qquad Z_{22} = Z_{s1}. \qquad (26)$$

The values in (26) now must be substituted in (19) and X_1 adjusted to make the expression real. That is,

$$\frac{I_{20}}{I_{10}} = \frac{(R_{s1} + R_{13} - 2R_{12}) + j(X_1 + X_{s1} + X_{13} - 2X_{12})}{(R_{s1} - R_{12}) + j(X_{s1} - X_{12})} \qquad (27)$$

must be real. This is true when

$$\frac{X_1 + X_{s1} + X_{13} - 2X_{12}}{R_{s1} + R_{13} - 2R_{12}} = \frac{X_{s1} - X_{12}}{R_{s1} - R_{12}} \qquad (28a)$$

or when

$$X_1 = \left(\frac{X_{s1} - X_{12}}{R_{s1} - R_{12}}\right)(R_{s1} + R_{13} - 2R_{12})$$

$$- X_{s1} - X_{13} + 2X_{12}. \qquad (28b)$$

Evidently

$$\frac{I_{20}}{I_{10}} = \frac{R_{s1} + R_{13} - 2R_{12}}{R_{s1} - R_{12}}. \qquad (29)$$

This ratio is frequently considerably greater than unity.

Similar analyses may be carried out for arrays of any number of parallel antennas. A particularly attractive one is the four-element array in Fig. 19.6.

If the line spacing b is not so small as to make terminal-zone effects negligible, approximate terminal-zone networks may be designed. This is not simple, since the currents in the line on each side of an antenna have different ratios, depending on the location of the element in the N-element array. However, since the inductive correction is relatively small, it may be assumed to be zero except for the outermost antennas, where L_{Te} for the antenna as end load applies. The capacitive correction for each unit except the outer ones is as for a stub support, namely, $2C_{Tc}$ in parallel with each antenna. For the outermost unit at each end of the line it is C_{Te} if the antenna is a simple end load as in Fig. 19.2, and $2C_{Tc}$ if there is a high-impedance stub support in parallel with the antenna. If the central unit is driven from a two-wire line perpendicular to the line connecting the several antennas, as in Fig. 19.2, the junction of the three lines and the antenna presents a special problem, since there is capacitive coupling among all of these four components. An equivalent lumped capacitance may be evaluated by following the same general methods used in dealing with the parts of this problem separately in Secs. II.8, 9 and III.10. Alternatively, its magnitude may be estimated as follows: The lumped capacitance required at the end of one line joining another in a T-junction is given by C_T in (10.12). This is double the value for an L-junction. Note that it includes end effect on the line and the coupling between the lines. It is easily verified that the coupling between the central antenna and each of the two sections connecting it to the outer units is C_{Tc}. This involves no transmission-line end effect. It may be assumed that coupling between antenna and main line also is approximately equivalent to C_{Tc}. Accordingly, the total lumped capacitance for the terminal-zone network is $3(C_{Tc} + C_T)$, where C_T refers to (10.12). The inductive end correction is due only to transmission-line end effect for the main line. It involves a lumped series inductance L_{Te}.

The analysis of the laterally driven parallel array shown in Fig. 19.4 may be carried out by generalizing the analysis of the laterally driven H-antenna in Sec. 10 to more than two antennas. In the arrangement of Fig. 19.5, coupling between the antenna and the transmission line is minimized by placing the antennas in the neutral plane of the line.

End-fire arrays in which currents of equal magnitude and a progressive phase difference of $\pi/2$ are maintained in successive elements separated by a distance $b = \lambda_0/4$ are difficult to drive by any method other than properly adjusted individual feeder lines. For this reason the desirable unidirectional property of the end-fire array is often achieved by a combination of broadside array and parasitic reflectors, as shown in Fig. 19.6. This is discussed further in the next section.

20. Parallel Arrays with Parasitic Elements

Theoretically, the simplest method by which specified currents may be maintained in the units of an array of parallel linear radiators is by means of individual transmission lines with provision for adjusting both the phase and the magnitude of each current. In practice, such an arrangement involves so elaborate a network of transmission lines that it is actually useful only for large fixed installations. In order to obtain unidirectional field patterns (see Chapter VI), an alternative to driven arrays with quarter-wave spacing of antennas and a progressively increasing phase shift of one-quarter period per unit is to use a broadside array backed by a single row of parasitic elements, as shown in Fig. 19.6. Obviously, it is not possible to adjust the magnitude and phase of the current in parasitic elements with the same degree of flexibility possible in driven elements. However, a very wide range of variation is possible if use is made of the following variables: (1) the location of the parasitic elements relative to the driven ones; (2) an adjustable tuning reactance connected in series at the center of each parasitic element; (3) small changes in the length of the elements. The significance of these three variables has been considered in detail in Sec. 11 for one driven antenna in the presence of a single parasite. However, since it has been shown in Sec. 19 that when $\beta_0 h$ is near $\pi/2$ for each antenna the analysis of N coupled antennas reduces approximately to a solution of N simultaneous equations in the voltages, the currents, the self-impedances of isolated antennas, and the mutual impedance of isolated pairs, it follows that the mutual effects between each driven antenna and an adjacent parasite differ little from what they would be if these two units were isolated. This is particularly true of parasitic elements used as reflectors with a broadside array, since the usual spacing for broadside elements is $\lambda_0/2$, whereas the parasitic row normally is at no greater distance from the driven row than $\lambda_0/4$, and frequently much less. Thus, each driven antenna and one parasite form a pair of antennas that are much closer to each other than to any other antenna in the array.

The eight simultaneous equations for the array in Fig. 19.6 with four driven units numbered 1 to 4 in broadside and four parasitic elements numbered 5 to 8 are

$$V_{i0} = \sum_{m=1}^{8} I_{m0} Z_{im}, \quad i = 1, 2, 3, 4; \quad (1a)$$

$$0 = \sum_{m=1}^{8} I_{m0} Z_{im}, \quad i = 5, 6, 7, 8. \quad (1b)$$

If each of the parasitic elements has a tuning impedance $Z_L = jX_L$ at the center, the self-impedances of the parasitic units are

$$Z_{ii} = Z_0 + Z_i, \quad i = 5, 6, 7, 8. \quad (2)$$

In general, the unidirectional property of the array is achieved by making each driven antenna with its parasitic element individually unidirectional when isolated as a pair. With the properties of the parasitic element predetermined in this manner, the entire array may be analyzed and a numerical value of the input impedance of the array computed. This follows from the fact that all self- and mutual impedances are known. However, the numerical evaluation is quite tedious, owing to the large number of terms if N is great. This number is not quite so formidable as indicated in (1), since conditions of symmetry of the array and the reciprocal theorem make many mutual impedances alike. For antennas of equal length, some self-impedances are alike, but different values of tuning impedance Z_i make the Z_{ii} different. Note that tuning reactances also may be required in series with some of the driven units, if equal currents are to be maintained in all of these.

A second important application of parasitic elements is in conjunction with a single driven antenna. Various arrangements and numbers of parasitic elements are used for different purposes. A few important arrays are shown schematically in their horizontal planes in Fig. 20.1. Their dimensions and other properties are considered in Chapter VI. The Yagi-Uda array in Fig. 20.1c has been used with as many as 42 directors.

The circuit equations of arrays with a single driven unit, numbered 1, are

$$V_{10} = \sum_{m=1}^{N} I_{m0} Z_{1m}, \quad (3a)$$

$$0 = \sum_{m=1}^{N} I_{m0} Z_{im}, \quad i = 2, 3, \cdots, N. \quad (3b)$$

These equations are valid in general only when the value of $\beta_0 h$ for each unit is near $\pi/2$.

As a relatively simple case, consider the three-element array in Fig. 20.1b. The equations are

$$V_{10} = I_{10}Z_{11} + I_{20}Z_{12} + I_{30}Z_{13}, \quad (4a)$$

$$0 = I_{10}Z_{21} + I_{20}Z_{22} + I_{30}Z_{23}, \quad (4b)$$

$$0 = I_{10}Z_{31} + I_{20}Z_{32} + I_{30}Z_{33}. \quad (4c)$$

For simplicity let it be assumed that the unit numbered 1 is driven by a "slice" generator, so that

$$Z_{11} = Z_{s1} \doteq Z_0. \quad (5)$$

Let the two parasites be located at different distances from the driven unit; let them be center-loaded by different tuning reactances. All three units are equal in length with $\beta_0 h = \pi/2$. Hence,

$$Z_{22} = Z_{s1} + Z_2 \doteq Z_0 + Z_2 = Z_0 + jX_2, \quad (6a)$$

$$Z_{33} = Z_{s1} + Z_3 \doteq Z_0 + Z_3 = Z_0 + jX_3, \quad (6b)$$

$$Z_{12} = Z_{21}, \quad Z_{13} = Z_{31}, \quad Z_{23} = Z_{32}. \quad (6c)$$

With (5) and (6), (4a–c) reduce to

$$V_{10} = I_{10}Z_{s1} + I_{20}Z_{12} + I_{30}Z_{13}, \quad (7a)$$

$$0 = I_{10}Z_{12} + I_{20}(Z_{s1} + Z_2) + I_{30}Z_{23}, \quad (7b)$$

$$0 = I_{10}Z_{13} + I_{20}Z_{23} + I_{30}(Z_{s1} + Z_3), \quad (7c)$$

The three currents are

$$I_{10} = V_{10}[(Z_{s1} + Z_2)(Z_{s1} + Z_3) - Z_{23}^2]/D, \quad (8a)$$

$$I_{20} = V_{10}[Z_{13}Z_{23} - Z_{12}(Z_{s1} + Z_3)]/D, \quad (8b)$$

$$I_{30} = V_{10}[Z_{12}Z_{23} - Z_{13}(Z_{s1} + Z_2)]/D, \quad (8c)$$

where

$$D \equiv \begin{vmatrix} Z_{s1} & Z_{12} & Z_{13} \\ Z_{12} & Z_{s1} + Z_2 & Z_{23} \\ Z_{13} & Z_{23} & Z_{s1} + Z_3 \end{vmatrix}. \quad (8d)$$

For given values of the spacings b_{12} and b_{13}, the radius a of the conductors, and Z_2 and Z_3, all other impedances may be obtained from the self- and mutual-impedance curves for isolated pairs. Therefore, currents at the centers of the three antennas may be computed both in magnitude and relative phase. These currents must be known if the field pattern is to be determined, as discussed in Chapter VI.

In order to analyze arrays involving many parasitic elements (such as the Yagi-Uda arrays discussed in Secs. 5 and 6 in Chapter VI) in which all elements have lengths near but not necessarily exactly one-half wavelength, the curves of Figs. 8.15 through 8.18 may be used for the more closely spaced elements. Antennas that are more widely separated may be treated using zeroth-order values of the mutual impedance. Since the curves in Figs. 8.15 through 8.18 are almost linear over a range near $\beta_0 h = \pi/2$, they may be approximated by straight lines in an investigation, for example, of the variation of the input impedance with frequency.

ARRAYS WITH ALL UNITS IN NEUTRAL PLANES

21. Two Mutually Perpendicular Antennas

An important and analytically extremely simple array consists of two antennas, which need not be equal either in length or in cross-sectional dimensions but which are required to be individually symmetric, so that each unit is center driven and has identical halves. If antenna 1 is assumed to lie along the y-axis of a system of Cartesian coördinates with its center at the origin, as shown in Fig. 21.1, antenna 2 *must lie in the xz-plane perpendicular to the line joining the centers of the antennas*. For simplicity, let this line be the z-axis, and let the center of antenna 2 be at an arbitrary distance $z = d$ from the origin. Since the analysis reduces to unusually simple form, let account be taken of the transmission lines from the beginning.

Let it be assumed that the radii a_1 and a_2 of the antennas individually satisfy the following conditions, in terms of their half-lengths h_1 and h_2 and the phase constant $\beta_0 = \omega/v_0$:

$$\beta_0 a_1 \ll 1, \quad a_1 \ll h_1,$$
$$\beta_0 a_2 \ll 1, \quad a_2 \ll h_2. \quad (1)$$

The boundary conditions on the tangential components of the electric field on the surfaces of the two antennas are

$$(E_{1y}^i)_{r_1 = a_1} = (E_{1y}^e)_{r_1 = a_1},$$
$$(E_{2x}^i)_{r_2 = a_2} = (E_{2x}^e)_{r_2 = a_2}. \quad (2)$$

where

$$(E^i_{1y})_{r_1=a_1} = z^i_1 I_{1y}, \quad (E^i_{2x})_{r_2=a_2} = z^i_2 I_{2x}, \quad (3)$$

and

$$(E^e_{1y})_{r_1=a_1} = \left(-\frac{\partial \Phi_1}{\partial y} - j\omega A_{1y}\right)_{r_1=a_1},$$

$$(E^e_{2x})_{r_2=a_2} = \left(-\frac{\partial \Phi_2}{\partial x} - j\omega A_{2x}\right)_{r_2=a_2}, \quad (4)$$

where Φ_1 and A_{1y} are potentials on antenna 1, Φ_2 and A_{2x} are potentials on antenna 2. The equation of continuity for potentials is

$$\Phi = \frac{j\omega}{\beta_0^2} \text{div } \mathbf{A} = \frac{j\omega}{\beta_0^2}\left(\frac{\partial A_x}{\partial x} + \frac{\partial A_y}{\partial y} + \frac{\partial A_z}{\partial z}\right). \quad (5)$$

Note that in the case at hand A_y is due entirely to currents in antenna 1, A_x is due to currents in antenna 2, and A_z is due to currents in the transmission lines. Substitution of (5) in (4) and of (3) in (2) gives the following differential equations:

$$\left(\frac{\partial^2 A_{1y}}{\partial y^2} + \frac{\partial^2 A_{1x}}{\partial y \partial x} + \frac{\partial^2 A_{1z}}{\partial y \partial z} + \beta_0^2 A_{1y}\right)_{r_1=a_1}$$
$$= j\frac{\beta_0^2}{\omega} z^i_1 I_{1y}, \quad (6a)$$

$$\left(\frac{\partial^2 A_{2x}}{\partial x^2} + \frac{\partial^2 A_{2y}}{\partial x \partial y} + \frac{\partial^2 A_{2z}}{\partial x \partial z} + \beta_0^2 A_{2x}\right)_{r_2=a_2}$$
$$= j\frac{\beta_0^2}{\omega} z^i_2 I_{2x}. \quad (6b)$$

In general,

$$A_{1x} = A_{11x} + A_{12x} + A_{1L_1x} + A_{1L_2x},$$
$$A_{2x} = A_{22x} + A_{21x} + A_{2L_1x} + A_{2L_2x}, \quad (7a)$$

$$A_{1y} = A_{11y} + A_{12y} + A_{1L_1y} + A_{1L_2y},$$
$$A_{2y} = A_{22y} + A_{21y} + A_{2L_1y} + A_{2L_2y}, \quad (7b)$$

$$A_{1z} = A_{11z} + A_{12z} + A_{1L_1z} + A_{1L_2z},$$
$$A_{2z} = A_{22z} + A_{21z} + A_{2L_1z} + A_{2L_2z}, \quad (7c)$$

where the first subscript indicates the antenna on which the potential is calculated, the second subscript shows the location of the currents that maintain the potential, and the last subscript specifies the component of the potential. For the particular case under investigation, the antennas and lines are so oriented that all components vanish except the following:

$$A_{1x} = A_{12x}, \quad A_{2x} = A_{22x}, \quad (8a)$$
$$A_{1y} = A_{11y}, \quad A_{2y} = A_{21y}, \quad (8b)$$
$$A_{1z} = A_{1L_1z} + A_{1L_2z}, \quad A_{2z} = A_{2L_1z} + A_{2L_2z}. \quad (8c)$$

Moreover, owing to the fact that each antenna is in the neutral plane of the other, it follows that

$$\left(\frac{\partial A_{1x}}{\partial y}\right)_{r_1=a_1} = \left(\frac{\partial A_{12x}}{\partial y}\right)_{r_1=a_1}$$
$$= -j\frac{\beta_0^2}{\omega}(\Phi_{12})_{r_1=a_1} = 0,$$

$$\left(\frac{\partial A_{2y}}{\partial x}\right)_{r_2=a_2} = \left(\frac{\partial A_{21y}}{\partial x}\right)_{r_2=a_2}$$
$$= -j\frac{\beta_0^2}{\omega}(\Phi_{21})_{r_2=a_2} = 0, \quad (9a)$$

$$\left(\frac{\partial A_{1L_2z}}{\partial y}\right)_{r_1=a_1} = -j\frac{\beta_0^2}{\omega}(\Phi_{1L_2})_{r_1=a_1} = 0,$$

$$\left(\frac{\partial A_{2L_1z}}{\partial x}\right)_{r_2=a_2} = -j\frac{\beta_0^2}{\omega}(\Phi_{2L_1})_{r_2=a_2} = 0. \quad (9b)$$

The postulates that each antenna is itself geometrically symmetric with respect to its center and that the line is balanced have as a consequence the vanishing of the scalar potential on each antenna due to charges on the other antenna and on the line driving the other antenna. Hence (6a, b) reduce to the following equations:

$$\left(\frac{\partial^2 A_{11y}}{\partial y^2} + \frac{\partial^2 A_{1L_1z}}{\partial y \partial z} + \beta_0^2 A_{11y}\right)_{r_1=a_1}$$
$$= j\frac{\beta_0^2}{\omega} z^i_1 I_{1y}, \quad (10a)$$

$$\left(\frac{\partial^2 A_{22x}}{\partial x^2} + \frac{\partial^2 A_{2L_2z}}{\partial x \partial z} + \beta_0^2 A_{22x}\right)_{r_2=a_2}$$
$$= j\frac{\beta_0^2}{\omega} z^i_2 I_{2x}. \quad (10b)$$

Note that the equation for each antenna involves no contribution whatsoever from currents or charges in the other antenna or the line driving the other antenna. That is, the equation for each antenna is exactly the same as for an antenna that is actually

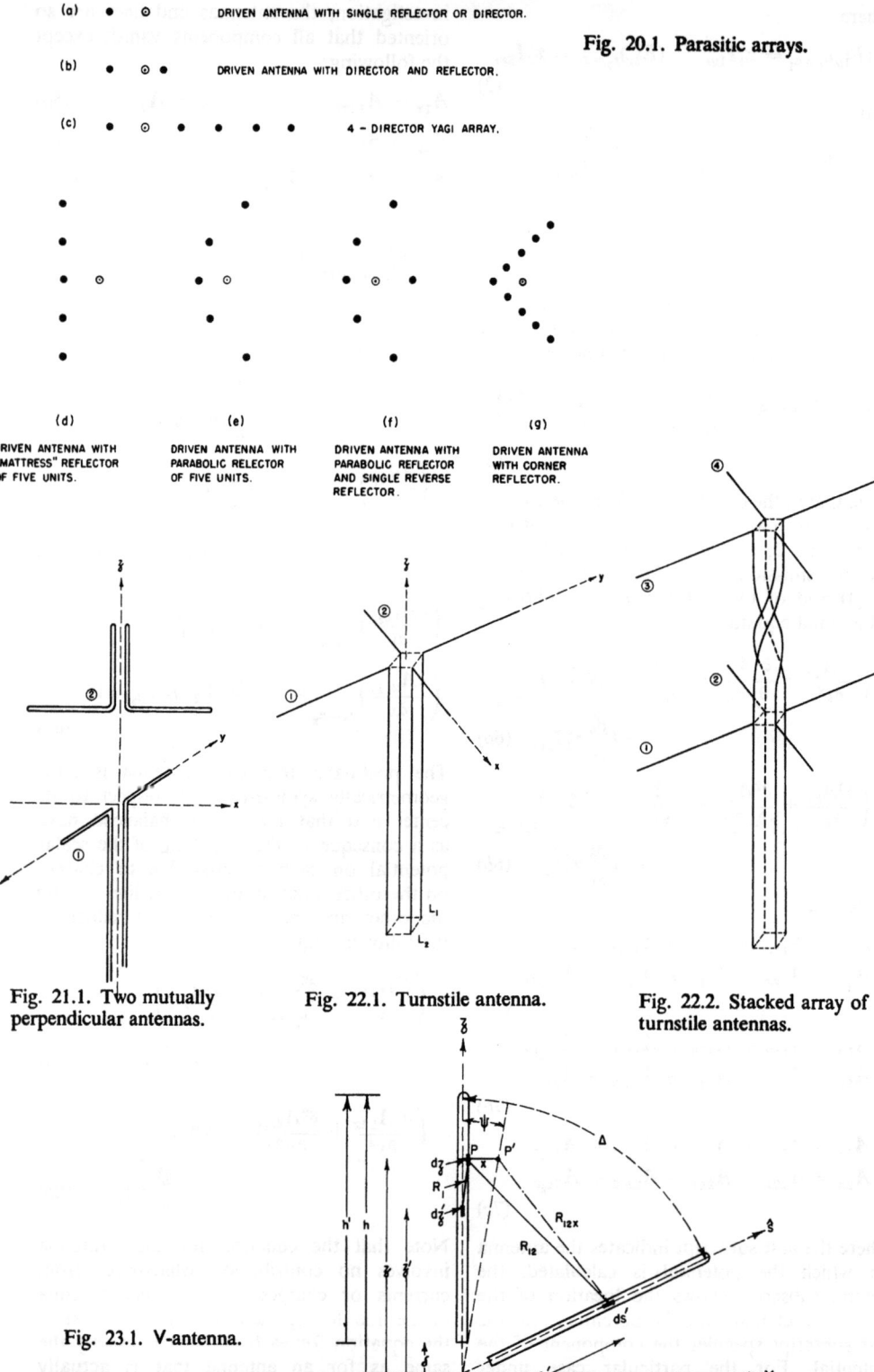

isolated except for the line that drives it. Accordingly, the distributions of current and charge and the impedance of each antenna are the same as those already determined in Chapter II for the center-driven antenna as end load on a two-wire line.

22. Turnstile Antennas and Arrays

An important special array of two mutually perpendicular antennas consists of two identical, mutually perpendicular, center-driven antennas in the same plane and with coincident centers. It is obtained from Fig. 21.1 by lowering antenna 2 to the xy-plane. In practice it is convenient to drive the antennas from opposite corners of a four-wire line with square cross section, as shown in Fig. 22.1. Each antenna with its driving line is independent of the other antenna with its line, so that the amplitude and phase relations may be selected at will to obtain a desired electromagnetic field. This field is discussed in Sec. VI.16. Since Fig. 22.1 represents a special case of Fig. 21.1, the distributions of current and charge and the impedances of the two antennas are the same as for each antenna with its line isolated from the other.

A so-called stacked array of two pairs of crossed or turnstile antennas is readily obtained, as shown in Fig. 22.2. This is called a two-element turnstile array. From the circuit point of view, it consists of two mutually independent end-driven H-arrays of broadside type. In Fig. 22.2 the transmission lines are spiraled around a half turn so that, if the parallel elements are $\lambda/2$ apart, they will be driven in phase. The currents on and the impedance of each pair are the same as for the H-array, appropriate to the phase relations. These are analyzed in Sec. 10. If N pairs of crossed antennas are arranged in an N-element turnstile array, the circuit properties are the same as those of two mutually independent, laterally driven broadside arrays.

V-ANTENNAS

23. Integral Equation for Symmetric, Center-Driven V-Antenna*

The symmetric, center-driven V-antenna shown in Fig. 23.1 is a general form of the single, center-driven antenna analyzed in Chapter II. It differs from this only in having the identical halves of the antenna inclined at an arbitrary angle Δ with respect to each other instead of being collinear with $\Delta = \pi$. Since a principal effect of changing the angle Δ in the range from near zero to π is to vary the kind and degree of coupling between the halves of the antenna, the V-antenna is appropriately treated in the chapter on coupled antennas, even though it may be regarded as a single, center-driven antenna rather than as two coupled antennas.

Let half No. 1 of the antenna be along the z-axis; let the other half, No. 2, be along the s-axis in the xz-plane with positive direction defined by the unit vector

$$\hat{s} = \hat{z} \cos \Delta + \hat{x} \sin \Delta. \tag{1}$$

A scalar potential difference

$$V_\delta = \Phi_1(z = \delta) - \Phi_2(s = \delta) = 2\Phi_1(z = \delta) \tag{2}$$

is maintained across the base of the V-structure by a balanced transmission line so that if $s' = z'$ the currents and charges in the two halves satisfy the conditions

$$I_{2s}(s') = -I_{1z}(z'), \quad q_2(s') = -q_1(z'). \tag{3}$$

Terminal-zone effects and antenna-line coupling effects are ignored at the outset. Account may be taken of them later by a suitably designed lumped-constant network.

The derivation of the integral equation for the current in half No. 1 of the V-antenna parallels that for the antenna with collinear halves, to which the V-antenna reduces when $\Delta = \pi$. For simplicity, it is assumed that the two conductors have the same radius a and are perfect, so that the tangential components of the electric field vanish on their surfaces. If required, terms to take account of ohmic resistance may be added. Specifically, the boundary condition at a point P on conductor No. 1 is

$$E_{1z} = -\left(\frac{\partial \Phi}{\partial z}\right)_1 - j\omega A_{1z}$$

$$= -\frac{j\omega}{\beta_0^2}\left(\frac{\partial}{\partial z} \operatorname{div} \mathbf{A}_1 + \beta_0^2 A_{1z}\right) \doteq 0.$$

$$(r_1 = a, \ \delta \leq z \leq h) \tag{4}$$

Since with $A_y = 0$,

$$\operatorname{div} \mathbf{A}_1 = \frac{\partial A_{1z}}{\partial z} + \frac{\partial A_{1x}}{\partial x}, \tag{5}$$

---

* Contributions to this section were made by Drs. C. T. Tai and J. E. Storer.

(3) yields the following differential equation valid on the surface of conductor No. 1:

$$\frac{\partial^2 A_{1z}}{\partial z^2} + \beta_0^2 A_{1z} = -\frac{\partial^2 A_{1x}}{\partial z \partial x},$$

$$(r_1 = a, \quad \delta \leq z \leq h) \quad (6)$$

Expressions for the components of the vector potential in (6) are:

$$A_{1z} = \frac{1}{4\pi\nu_0} \int_\delta^h \left[I_{1z}(z') \frac{e^{-j\beta_0 R_1}}{R_1} dz' \right.$$

$$\left. + \cos \Delta \, I_{2s}(s') \frac{e^{-j\beta_0 R_{12}}}{R_{12}} ds' \right], \quad (7a)$$

$$A_{1x} = \frac{\sin \Delta}{4\pi\nu_0} \int_\delta^h I_{2s}(s') \frac{e^{-j\beta_0 R_{12}}}{R_{12}} ds', \quad (7b)$$

where

$$R_1 \equiv \sqrt{(z-z')^2 + a^2},$$

$$R_{12} \equiv \sqrt{z^2 - 2zs'\cos\Delta + s'^2 + a^2}. \quad (8)$$

Note that R_{12} is a generalized form of $R_2 = \sqrt{(z+z')^2 + a^2}$ as used in Chapter II. The two distances coincide when $\Delta = \pi$.

The distance R_{12} defined in (8) is measured from ds' on the axis of conductor 2 to dz at P on the surface of conductor 1. It is assumed that P has the coördinates $(x = 0; y = a; z)$. In order to obtain $\partial A_{1x}/\partial x$ at $x = 0$, it is necessary to evaluate $\partial A_{1x}/\partial x$ at some point P' at a distance x from P and then set $x = 0$. That is, it is necessary to use (7b) with

$$R_{12x} = [z^2 + x^2 - 2s'\sqrt{z^2 + x^2}\cos(\Delta - \psi) + s'^2 + a^2]^{\frac{1}{2}} \equiv \sqrt{\xi^2 + a^2} \quad (9)$$

instead of R_{12} as given in (8). Note that

$$\cos(\Delta - \psi) = \frac{z}{\sqrt{z^2 + x^2}} \cos \Delta$$

$$+ \frac{x}{\sqrt{z^2 + x^2}} \sin \Delta, \quad (10)$$

so that

$$\xi^2 = z^2 + x^2 - 2zs'\cos\Delta - 2xs'\sin\Delta + s'^2 \quad (11)$$

and

$$\left(\frac{\partial \xi}{\partial x} \right)_{x=0} = -\frac{s' \sin \Delta}{\sqrt{z^2 - 2zs'\cos\Delta + s'^2}}. \quad (12)$$

With (9) to (11) the derivative of (7b) becomes

$$\left(\frac{\partial A_{1x}}{\partial x} \right)_{x=0}$$

$$= \frac{\sin \Delta}{4\pi\nu_0} \int_\delta^h I_{2s}(s') \left(\frac{\partial}{\partial \xi} \frac{e^{-j\beta_0 R_{12x}}}{R_{12x}} \frac{\partial \xi}{\partial x} \right)_{x=0} ds'. \quad (13)$$

With (12) this becomes

$$\left(\frac{\partial A_{1x}}{\partial x} \right)_{x=0} = -\frac{\sin^2 \Delta}{4\pi\nu_0}$$

$$\times \int_\delta^h I_{2s}(s') \left(\frac{1}{\xi} \frac{\partial}{\partial \xi} \frac{e^{-j\beta_0 \sqrt{\xi^2 + a^2}}}{\sqrt{\xi^2 + a^2}} \right)_{x=0} s' ds'. \quad (14)$$

Let

$$K_{12}(z, s') \equiv \left(\frac{1}{\beta_0 \xi} \frac{\partial}{\partial \xi} \frac{e^{-j\beta_0 \sqrt{\xi^2 + a^2}}}{\beta_0^2 \sqrt{\xi^2 + a^2}} \right)_{x=0}$$

$$= -\frac{e^{-j\beta_0 R_{12}}}{\beta_0^3 R_{12}^3} (1 + j\beta_0 R_{12}). \quad (15)$$

Also let the symbol used for the variable of integration in (14) be changed from s' to z' after use has been made of (3). This gives

$$\Phi_{1x}(z) = \frac{j\omega}{\beta_0^2} \left(\frac{\partial A_{1x}}{\partial x} \right)_{x=0} = \frac{j\zeta_0}{4\pi} p(z) \sin^2 \Delta, \quad (16)$$

where

$$p(z) \equiv -\beta_0^2 \int_\delta^h I_{1z}(z') K_{12}(z, z') z' dz'. \quad (17)$$

With (16) the differential equation (6) becomes

$$\frac{\partial^2 A_{1z}}{\partial z^2} + \beta_0^2 A_{1z} = \frac{j\beta_0^2}{\omega} \frac{\partial \Phi_{1x}}{\partial z},$$

$$(r_1 = a, \quad \delta \leq z \leq h) \quad (18)$$

By differentiating (18) with respect to z and using the equation of continuity for potentials,

$$\frac{\partial A_{1x}}{\partial x} + \frac{\partial A_{1z}}{\partial z} + \frac{j\beta_0^2}{\omega} \Phi = 0, \quad (19)$$

the following equation may be obtained for the scalar potential:

$$\frac{\partial^2 \Phi_1}{\partial z^2} + \beta_0^2 \Phi_1 = \beta_0^2 \Phi_{1x}.$$

$$(r_1 = a, \quad \delta \leq z \leq h) \quad (20)$$

The solution of the nonhomogeneous differential equation (18) consists of the sum of a particular integral and a complementary function as follows:

$$A_{1z} = \frac{-j}{\nu_0} [C_1 \cos \beta_0 z + C_2 \sin \beta_0 z - \mathcal{V}(z)].$$

$$(r_1 = a, \quad \delta \leq z \leq h) \quad (21)$$

This expression differs from (II.12.8) for collinear halves only in the added particular integral

$$\mathscr{V}(z) \equiv \int_\delta^z \frac{\partial \Phi_{1x}(w)}{\partial w} \sin \beta_0(z - w) \, dw \quad (22)$$

and in the absence of the small ohmic terms. When $\Delta = \pi$ it follows from (16) that $\Phi_{1x}(w) = 0$, so that (21) reduces to (II.12.8). The factor $-j/v_0$ is introduced in (21) to make the constants and $\mathscr{V}(z)$ dimensionally voltages.

The scalar potential on conductor 1 is obtained by differentiating (21), using

$$\Phi_1 = \frac{j\omega}{\beta_0^2} \left(\frac{\partial A_z}{\partial z} + \frac{\partial A_x}{\partial x} \right)$$

$$= \frac{j\omega}{\beta_0^2} \left(\frac{\partial A_{1z}}{\partial z} \right) + \Phi_{1x}. \quad (23)$$

Differentiation of (21) gives

$$\Phi_1(z) = -C_1 \sin \beta_0 z + C_2 \cos \beta_0 z$$
$$- \mathscr{V}'(z) + \Phi_{1x}(z), \quad (24)$$

where

$$\mathscr{V}'(z) = \int_\delta^z \frac{\partial \Phi_{1x}(w)}{\partial w} \cos \beta_0(z - w) \, dw. \quad (25)$$

The driving voltage V_δ is maintained from $z = \delta$ to $s = \delta$ by the transmission line. Using (2), it follows that with $z = \delta$

$$V_\delta = 2\Phi_1(\delta) = 2[-C_1 \sin \beta_0 \delta + C_2 \cos \beta_0 \delta + \Phi_{1x}(\delta)], \quad (26)$$

so that

$$C_2 = [\tfrac{1}{2} V_\delta + C_1 \sin \beta_0 \delta - \Phi_{1x}(\delta)]/\cos \beta_0 \delta. \quad (27)$$

For convenience let

$$\tfrac{1}{2} \bar{V}_\delta \equiv \tfrac{1}{2} V_\delta - \Phi_{1x}(\delta), \quad (28)$$

so that

$$C_2 = (\tfrac{1}{2} \bar{V}_\delta + C_1 \sin \beta_0 \delta)/\cos \beta_0 \delta. \quad (29)$$

This is like (II.13.23), but with $\bar{V}_\delta$ replacing V_δ.

The integral equation for the current $I_{1z}(z)$ is obtained by equating (21) with (29) to (7a). It is convenient to apply (3), and change the symbol of the variable of integration from s' to z' in the second integral. The resulting integral equation is

$$\int_\delta^h I_{1z}(z') K_v(z, z') \, dz'$$
$$= \frac{-j4\pi}{\zeta_0 \cos \beta_0 \delta} [C_1 \cos \beta_0(z - \delta)$$
$$+ \tfrac{1}{2} \bar{V}_\delta \sin \beta_0 z] + \frac{j4\pi}{\zeta_0} \mathscr{V}(z).$$
$$(r_1 = a, \quad \delta \leq z \leq h) \quad (30a)$$

The kernel is defined as follows:

$$K_v(z, z') \equiv \left(\frac{e^{-j\beta_0 R_1}}{R_1} - \cos \Delta \frac{e^{-j\beta_0 R_{12}}}{R_{12}} \right)_{s'=z}. \quad (30b)$$

This may be rearranged as follows:

$$\int_\delta^h I_{1z}(z') K_v(z, z') \, dz'$$
$$= \frac{-j4\pi}{\zeta_0 \cos \beta_0 \delta} [C_1 \cos \beta_0(z - \delta)$$
$$+ \tfrac{1}{2} \bar{V}_\delta \sin \beta_0 z] + J_v(z),$$
$$(r_1 = a, \quad \delta \leq z \leq h) \quad (31)$$

where

$$J_v(z) \equiv \frac{j4\pi}{\zeta_0} \left[\int_\delta^z \frac{\partial \Phi_{1x}(w)}{\partial w} \sin \beta_0(z - w) \, dw \right.$$
$$\left. + \Phi_{1x}(\delta) \frac{\sin \beta_0 z}{\cos \beta_0 \delta} \right]. \quad (32)$$

Integration by parts in (32) and rearrangement of terms yields

$$J_v(z) = \frac{j4\pi}{\zeta_0} \left\{ \Phi_{1x}(\delta) \left[\frac{\sin \beta_0 z}{\cos \beta_0 \delta} - \sin \beta_0(z - \delta) \right] \right.$$
$$\left. + \beta_0 \int_\delta^z \Phi_{1x}(w) \cos \beta_0(z - w) \, dw \right\}. \quad (33a)$$

With (17) and an appropriate combination of terms this becomes

$$J_v(z) = -\sin^2 \Delta \left[p(\delta) \tan \beta_0 \delta \cos \beta_0(z - \delta) \right.$$
$$\left. + \beta_0 \int_\delta^z p(w) \cos \beta_0(z - w) \, dw \right]. \quad (33b)$$

Note that when $\Delta = \pi$, (31) reduces to (II.14.3) with $z^i = 0$. When $\Delta \neq \pi$, (30) is formally like (II.14.3) but with a different kernel and with the added term $J_v(z)$.

24. Solution of the Integral Equation for the V-Antenna

The integral equation (23.31) may be solved by the same method of iteration employed in Chapter II for the straight antenna ($\Delta = \pi$). Let an expansion parameter Ψ'_v be defined as follows:

$$\Psi_v = |\Psi_v(z_r)|,$$

$$\Psi_v(z) = \int_\delta^h g(z, z') K_v(z, z') \, dz', \quad (1)$$

where $g(z, z')$ is a distribution function that is so chosen that $I_{1z}(z)\Psi_v(z)$ closely approximates the integral on the left in (23.31). For this purpose the zeroth-order distribution is satisfactory, namely,

$$g(z, z') = \sin \beta_0(h - z')/\sin \beta_0(h - z) \quad (2)$$

The resulting expression for the current is like (II.21.1) with appropriate changes in the functions as indicated by the addition of subscripts v.

$$I_{1z}(z) = \frac{j2\pi V_\delta}{\zeta_0 \Psi_v} \times$$
$$\left[\frac{\sin \beta_0(h - z) + M_{v1}(z)/\Psi_v + M_{v2}(z)/\Psi_v^2 + \cdots}{\cos \beta_0(h - \delta) + A_{v1}/\Psi_v + A_{v2}/\Psi_v^2 + \cdots} \right]. \quad (3)$$

In the following only the first-order solution is considered. Higher-order terms may be evaluated if required, following the method of Chapter II.

The first-order functions $M_{v1}(z)$ and A_{v1} in (3) are defined just as are the corresponding functions $M_1(z)$ and A_1 in Chapter II. Specifically, from (II.15.27) and (II.15.29a) it follows that

$$M_{v1}(z) = F_{v1}(z) \sin \beta_0 h - G_{v1}(z) \cos \beta_0 h + G_{v1}(h) \cos \beta_0 z - F_{v1}(h) \sin \beta_0 z, \quad (4a)$$

$$A_{v1} = F_{v1}(h) \cos \beta_0 \delta + G_{v1}(h) \sin \beta_0 \delta. \quad (4b)$$

The F and G functions are defined by analogy with (II.21.1a–d). Thus, with $F_{0z} \equiv \cos \beta_0 z - \cos \beta_0 h$, $G_{0z} \equiv \sin \beta_0 z - \sin \beta_0 h$ as in Chapter II,

$$F_{v1}(z) = F_{0z}\Psi_v - \int_\delta^h F_{0z'} K_v(z, z') \, dz'$$
$$+ J_v(F_{0z}, z), \quad (5a)$$

$$F_{v1}(h) = -\int_\delta^h F_{0z'} K_v(h, z') \, dz'$$
$$+ J_v(F_{0z}, h), \quad (5b)$$

$$G_{v1}(z) = G_{0z}\Psi_v - \int_\delta^h G_{0z'} K_v(z, z') \, dz'$$
$$+ J_v(G_{0z}, z), \quad (5c)$$

$$G_{v1}(h) = -\int_\delta^h G_{0z'} K_v(h, z') \, dz'$$
$$+ J_v(G_{0z}, h), \quad (5d)$$

where $J_v(F_{0z}, z)$ is the same as $J_v(z)$ in (23.33b) but with $F_{0z'}$ substituted for $I_{1z}(z')$ in p. Specifically,

$$p(F_{0z}, z) = \beta_0^2 \int_\delta^h F_{0z'} K_{12}(z, z') z' \, dz'. \quad (6)$$

A corresponding function is defined with G substituted for F.

In order to express (5) in terms of tabulated functions it is convenient to define the following integrals:

$$C_{11}(h, z) \equiv \int_0^h \cos \beta_0 z' \frac{e^{-j\beta_0 R_1}}{R_1} \, dz', \quad (7a)$$

$$C_{12}(h, z, \Delta) \equiv \int_0^h \cos \beta_0 z' \frac{e^{-j\beta_0 R_{12}}}{R_{12}} \, dz', \quad (7b)$$

$$S_{11}(h, z) \equiv \int_0^h \sin \beta_0 z' \frac{e^{-j\beta_0 R_1}}{R_1} \, dz', \quad (7c)$$

$$S_{12}(h, z, \Delta) \equiv \int_0^h \sin \beta_0 z' \frac{e^{-j\beta_0 R_{12}}}{R_{12}} \, dz', \quad (7d)$$

$$E_{11}(h, z) \equiv \int_0^h \frac{e^{-j\beta_0 R_1}}{R_1} \, dz', \quad (7e)$$

$$E_{12}(h, z, \Delta) \equiv \int_0^h \frac{e^{-j\beta_0 R_{12}}}{R_{12}} \, dz', \quad (7f)$$

where

$$R_1 = \sqrt{(z' - z)^2 + a^2}, \quad (7g)$$

$$R_{12} = \sqrt{(z' - z_v)^2 + a_v^2}, \quad (7h)$$

$$z_v \equiv z \cos \Delta, \quad a_v \equiv \sqrt{z^2 \sin^2 \Delta + a^2}. \quad (7i)$$

The first four of these functions may be expressed in terms of generalized sine and cosine integrals defined in Sec. II.19 using (II.19.30) to (II.19.32). Thus,

$$C_{11}(h, z) = \tfrac{1}{2}(I_1 + I_2), \quad (8a)$$

$$S_{11}(h, z) = \frac{-j}{2}(I_1 - I_2), \quad (8b)$$

$$C_{12}(h, z, \Delta) = \tfrac{1}{2}(I_{1v} + I_{2v}), \quad (8c)$$

$$S_{12}(h, z, \Delta) = \frac{-j}{2}(I_{1v} - I_{2v}), \quad (8d)$$

where I_{1v} and I_{2v} are given by (II.19.32a, b) but with subscripts v on all U's and A's, and where

$$\begin{aligned} U_v &= \beta_0(z' - z_v) = \beta_0(z' - z\cos\Delta), \\ A_v &= \beta_0\sqrt{z^2\sin^2\Delta + a^2}, \\ U_{0v} &= \beta_0 z_v = \beta_0 z\cos\Delta, \\ U_{1v} &= \beta_0(h - z_v) = \beta_0(h - z\cos\Delta). \end{aligned} \quad (9)$$

The final formulas for the integrals (7a–d) are

$$\begin{aligned} C_{11}(h, z) =\ & \cos U_0[Cc_i(A, U_1) + Cc_i(A, U_0)] \\ &- jSc(A, U_1) - jSc(A, U_0)] \\ &- \sin U_0[Cs(A, U_1) - Cs(A, U_0)] \\ &- jSs(A, U_1) + jSs(A, U_0)], \end{aligned} \quad (10a)$$

$$\begin{aligned} S_{11}(h, z) =\ & \sin U_0[Cc_i(A, U_1) + Cc_i(A, U_0)] \\ &- jSc(A, U_1) - jSc(A, U_0)] \\ &+ \cos U_0[Cs(A, U_1) - Cs(A, U_0)] \\ &- jSs(A, U_1) + jSs(A, U_0)], \end{aligned} \quad (10b)$$

where

$$U_0 = \beta_0 z, \quad U_1 = \beta_0(h - z), \quad A = \beta_0 a; \quad (10c)$$

$C_{12}(h, z, \Delta)$ is the same as (10a) with additional subscripts v on A and all U's, (10d)

$S_{12}(h, z, \Delta)$ is the same as (10b) with additional subscripts v on A and all U's. (10e)

Note that when $\Delta = \pi$,

$$C_{11}(h, z) + C_{12}(h, z, \Delta = \pi) = C_a(h, z), \quad (11a)$$

$$S_{11}(h, z) + S_{12}(h, z, \Delta = \pi) = S_a(h, z), \quad (11b)$$

where $C_a(h, z)$ and $S_a(h, z)$ are defined in (II.19.33, 34).

The functions $E_{11}(h, z)$ and $E_{12}(h, z, \Delta)$ defined in (7e, f) may be expressed in terms of generalized sine and cosine integrals following the procedure in (II.19.35). The results are

$$E_{11}(h, z) = C_i(A, U_1) + C_i(A, U_0) \\ - jS(A, U_1) - jS(A, U_0), \quad (12a)$$

$$E_{12}(h, z, \Delta) = C_i(A_v, U_{1v}) + C_i(A_v, U_{0v}) \\ - jS(A_v, U_{1v}) - jS(A_v, U_{0v}). \quad (12b)$$

After noting that from (II.19.21, 22)

$$C_i(A, U) = \sinh^{-1}(U/A) - C(U, A), \quad (13a)$$

$$Cc_i(A, U) = \sinh^{-1}(U/A) - C(U, A) \\ - Cc(U, A), \quad (13b)$$

it follows that all of the integrals in (7) are expressed in terms of tabulated functions.

If the functions defined in (5a–d) are expressed in terms of the integrals in (7a–f) as expanded in (10a–e) and (12a, b), they have the following forms:

$$\begin{aligned} F_{1v}(z) =\ & F_{0z}\Psi_v - [C_{11}(h, z) - C_{11}(\delta, z)] \\ &+ \cos\beta_0 h[E_{11}(h, z) - E_{11}(h, \delta)] \\ &+ \cos\Delta\{[C_{12}(h, z, \Delta) - C_{12}(\delta, z, \Delta)] \\ &- \cos\beta_0 h[E_{12}(h, z, \Delta) - E_{12}(\delta, z, \Delta)]\} \\ &+ J_v(F_{0z}, z), \end{aligned} \quad (14a)$$

$$\begin{aligned} F_{1v}(h) =\ & -[C_{11}(h, h) - C_{11}(\delta, h)] \\ &+ \cos\beta_0 h[E_{11}(h, h) - E_{11}(\delta, h)] \\ &+ \cos\Delta\{[C_{12}(h, h, \Delta) - C_{12}(\delta, h, \Delta)] \\ &- \cos\beta_0 h[E_{12}(h, h, \Delta) - E_{12}(\delta, h, \Delta)]\} \\ &+ J_v(F_{0z}, h), \end{aligned} \quad (14b)$$

$$\begin{aligned} G_{1v}(z) =\ & G_{0z}\Psi_v - [S_{11}(h, z) - S_{11}(\delta, z)] \\ &+ \sin\beta_0 h[E_{11}(h, z) - E_{11}(h, \delta)] \\ &+ \cos\Delta\{[S_{12}(h, z, \Delta) - S_{12}(\delta, z, \Delta)] \\ &- \sin\beta_0 h[E_{12}(h, z, \Delta) - E_{12}(\delta, z, \Delta)]\} \\ &+ J_v(G_{0z}, z), \end{aligned} \quad (14c)$$

$$\begin{aligned} G_{1v}(h) =\ & -[S_{11}(h, h) - S_{11}(\delta, h)] \\ &+ \sin\beta_0 h[E_{11}(h, h) - E_{11}(\delta, h)] \\ &+ \cos\Delta\{[S_{12}(h, h, \Delta) - S_{12}(\delta, h, \Delta)] \\ &- \sin\beta_0 h[E_{12}(h, h, \Delta) - E_{12}(\delta, h, \Delta)]\} \\ &+ J_v(G_{0z}, h). \end{aligned} \quad (14d)$$

If the antenna is driven by a discontinuity in scalar potential, $\delta \doteq 0$. Note that for a V-antenna $\delta = 0$ is possible only when $\Delta = \pi$. For other values of Δ, $\delta \doteq 0$ means sufficiently small to be negligible.

The terms in J_v arising from the particular integral have not been expressed in terms of tabulated functions. However, they may be evaluated by numerical methods.

The expansion parameter Ψ_v defined in (1) may be determined readily. Thus

$$\Psi_v(z) = \sin\beta_0 h \int_\delta^h \cos\beta_0 z' K_v(z, z')\, dz' \\ - \cos\beta_0 h \int_\delta^h \sin\beta_0 z' K_v(z, z')\, dz'. \quad (15)$$

With the definitions in (7), the following formula is obtained:

$$\Psi_v(z) = \sin \beta_0 h \{[C_{11}(h, z) - C_{11}(\delta, z)]$$
$$- \cos \Delta [C_{12}(h, z, \Delta) - C_{12}(\delta, z, \Delta)]\}$$
$$- \cos \beta_0 h \{[S_{11}(h, z) - S_{11}(\delta, z)]$$
$$- \cos \Delta [S_{12}(h, z, \Delta) - S_{12}(\delta, z, \Delta)]\}.$$
(16)

Since only tabulated functions are involved, $|\Psi_v(z)|$ may be plotted to determine the range over which it is sensibly constant at the value $\Psi_v = |\Psi_v(z_r)|$. It is to be expected that an appropriate choice of z_r is $h - \lambda_0/4$ when $\beta_0 h > \pi/2$. On the other hand, the value $z_r = 0$ with $\delta \doteq 0$, which was appropriate with collinear halves ($\Delta = \pi$) in Chapter II, cannot be a good choice. This follows from the fact that A_z decreases rapidly near $z = \delta$ in a manner resembling that shown in Fig. II.20.10. This decrease is not removed by letting $\delta \doteq 0$ except when $\Delta = \pi$. It follows that for $\beta_0 h = \pi/2$, z_r must not be 0 or δ but somewhere not too close to $z = 0$ or $z = h$. A reasonable choice appears to be $z_r = h/2$. That is,

$$\Psi_v = |\Psi_v(z_r)| \begin{cases} z_r = h/2, & (\beta_0 h \leq \pi) \\ z_r = h - \lambda_0/4. & (\beta_0 h \geq \pi) \end{cases}$$
(17)

In evaluating Ψ_v no significant error is involved if δ is assumed negligible. Application of (17) in (16) gives, for example, with $\beta_0 h = \pi/2$,

$$\Psi_v = |C_{11}(h, \tfrac{1}{2}h) - \cos \Delta C_{12}(h, \tfrac{1}{2}h, \Delta)|.$$
$$(\beta_0 h = \pi/2) \quad (18)$$

For $\Delta = \pi$ this becomes:

$$\Psi_v = |C_{11}(h, \tfrac{1}{2}h) + C_{12}(h, \tfrac{1}{2}h, \pi)| = C_a(h, \tfrac{1}{2}h).$$
$$(\beta_0 h = \pi/2, \Delta = \pi) \quad (19)$$

The last step follows from (11a). Since $|\Psi_v(z)|$ is essentially constant over the length of the antenna with collinear halves ($\Delta = \pi$) and zero base separation ($\delta = 0$) except quite near the ends, it follows that

$$\Psi_v = |C_a(h, \tfrac{1}{2}h)| \doteq \Psi_{K1} = |C_a(h, 0)|. \quad (20)$$

General formulas for the current in a V-antenna have been derived. Numerical calculations have not been made.

25. Impedance and Current for V-Antenna

The general formula for the impedance of the V-antenna is obtained from (24.3) by forming the ratio $V_\delta/I_{1z}(\delta)$. Thus,

$$Z_{v\delta} \equiv V_\delta/I_{1z}(\delta)$$
$$= \frac{-j\zeta_0 \Psi_v}{2\pi} \left[\frac{\cos \beta_0(h-\delta) + A_{v1}/\Psi_v + \cdots}{\sin \beta_0(h-\delta) + B_{v1}/\Psi_v + \cdots} \right],$$
(1)

where

$$B_{v1} \equiv M_{v1}(\delta). \quad (2)$$

The most important and the simplest V-antenna is with $\beta_0 h = \pi/2$. For simplicity, also let $\delta \doteq 0$, since terminal-zone effects, if significant, can be included in a lumped corrective network if required. The impedance is as follows:

$$(Z_{v0})_1 = \frac{-j\zeta_0}{2\pi} \left(\frac{A_{v1}}{1 + B_{v1}/\Psi_v} \right)$$
$$= \frac{-j\zeta_0}{2\pi} \left\{ \frac{F_{v1}(h)}{1 + \frac{1}{\Psi_v}[F_{v1}(0) + G_{v1}(h)]} \right\},$$
(3)

where, from (24.5) with (24.7),

$$F_{v1}(0) = \Psi_v - C_{11}(h, 0)$$
$$+ \cos \Delta C_{12}(h, 0, \Delta), \quad (4a)$$

$$F_{v1}(h) = -C_{11}(h, h) + \cos \Delta C_{12}(h, h, \Delta)$$
$$+ J_v(F_{0z}, h), \quad (4b)$$

$$G_{v1}(h) = -S_{11}(h, h) + \cos \Delta S_{12}(h, h, \Delta)$$
$$+ J_v(G_{0z}, h). \quad (4c)$$

Substitution of (4a–c) in (3) leaves the following simple expression for antennas that are sufficiently thin to make Ψ_v quite large:

$$Z_{v0} \doteq \frac{-j\zeta_0}{2\pi} F_{v1}(h) = \frac{j\zeta_0}{2\pi} [C_{11}(h, h)$$
$$- \cos \Delta C_{12}(h, h, \Delta) - J_v(\pi/2)].$$
$$(\beta_0 h = \pi/2) \quad (5)$$

This formula for the V-antenna corresponds in accuracy to the familiar formula $Z_0 = 73.13 + j42.5$ for the center-driven half-wave dipole. Like this its validity is limited to rather thin wires. The three functions involved in (5) are obtained from the general formulas (24.10a), (24.10d), and (23.33b) by setting

$z = h$, $\beta_0 h = \pi/2$, and $\delta \doteq 0$. Specifically, these functions are as follows:

$$C_{11}(h, h) = \left[Cs\left(A, \frac{\pi}{2}\right) - jSs\left(A, \frac{\pi}{2}\right)\right], \quad (6a)$$

$$\begin{aligned}C_{12}(h, h, \Delta) &= \cos U_{0v}[Cc_i(A_v, U_{1v}) + Cc_i(A_v, U_{0v}) \\ &\quad - jSc(A_v U_{1v}) - jSc(A_v, U_{0v})] \\ &\quad + \sin U_{0v}[-Cs(A_v, U_{1v}) + Cs(A_v, U_{0v}) \\ &\quad + jSs(A_v, U_{1v}) - jSs(A_v, U_{0v})], \quad (6b)\end{aligned}$$

where, with $\beta_0 h = \pi/2$, the following definitions apply:

$$U_{0v} = \frac{\pi}{2}\cos\Delta, \quad U_{1v} = \frac{\pi}{2}(1 - \cos\Delta),$$

$$A_v = \sqrt{\left(\frac{\pi}{2}\sin\Delta\right)^2 + A^2} \doteq \frac{\pi}{2}\sin\Delta. \quad (7)$$

The approximate formula for A_v in (7) is valid when $(\frac{1}{2}\pi\sin\Delta)^2 \gg A^2$ or $\sin^2\Delta \gg 16a^2/\lambda_0^2$.

The function $J_v(\pi/2)$ in (5) is $J_v(F_{0z'}, h)$ with $\beta_0 h = \pi/2$. With $z = h$ and $\delta \doteq 0$, it is given by

$$J_v\left(\frac{\pi}{2}\right) = -\sin^2\Delta \int_0^{\pi/2} \sin u\, du$$
$$\times \int_0^{\pi/2} K_{12}(u, u')u'\cos u'\, du', \quad (8a)$$

where

$$K_{12}(u, u') = -\frac{e^{-jv}}{v^3}(1 + jv), \quad (8b)$$

$$v = \beta_0 R_{12} = \sqrt{u^2 - 2uu'\cos\Delta + u'^2 + A^2}, \quad (8c)$$

$$u = \beta_0 w, \quad u' = \beta_0 w', \quad A = \beta_0 a. \quad (8d)$$

The term in A^2 is negligible in (8c) except when both u and u' are very small.

The exact evaluation of (8a) has not been achieved in closed form except when $\Delta = \pi$ and when Δ is very small. In these two cases, the following values are correct:

$$\Delta = \pi: \quad J_v\left(\frac{\pi}{2}\right) = 0, \quad (9)$$

$$\Delta^2 \ll 12; \quad A^2 \ll \left(\frac{\pi}{2}\sin\Delta\right)^2:$$

$$J_v\left(\frac{\pi}{2}\right) = \left(1 - \frac{1}{2}\Delta^2\right) Si\,\pi$$
$$= (1.85)\left(1 - \frac{1}{2}\Delta^2\right). \quad (10)$$

The value (9) follows directly; the value (10) may be obtained by direct integration with appropriate approximations.*

In order to obtain values of $J_v(\pi/2)$ for Δ between π and the upper limit of formula (10), (8a) was evaluated graphically for $\Delta = \pi/3$, $\pi/2$, $2\pi/3$. In Fig. 25.1 curves of $Z_{v0} = R_{v0} + jX_{v0}$ as functions of Δ are shown as computed from (5). Note that this is a zeroth-order solution $(\Omega \to \infty)$ for a V-antenna driven from a discontinuity in scalar potential at the apex. With $\Delta = \pi$ it gives the familiar value $Z_v = 73.1 + j42.5$ for the thin cylindrical antenna. It is readily verified†

---

* In the range of small values of Δ, v assumes correspondingly small values when $|u' - u|$ is small. Hence, the significant contributions to the integrand in (8a) are when v is small and $K_{12}(u, u')$ large. The appropriate approximations are:

$$\Delta^2 \ll 12: \cos\Delta \doteq 1 - \tfrac{1}{2}\Delta^2; v \doteq \sqrt{u^2 - 2uu'(1 - \tfrac{1}{2}\Delta^2) + u'^2}, \quad (10a)$$

$$v^2 \ll 12: K_{12}(u, u') \doteq -1/v^3. \quad (10b)$$

It is assumed in (10) that contributions from A^2 are negligible. Since the principal contributions to the integral occur when u' is near u, no serious error is made if u' is replaced by u in the slowly varying factor $\cos u'$. With (10a) and (10b) and $\cos u' \doteq \cos u$,

$$J_v(\pi/2) \doteq \Delta^2 \int_0^{\pi/2} \sin u \cos u\, du \int_0^{\pi/2}(1/v^3)u'\,du'. \quad (10c)$$

The u' integral integrates directly using standard formulas (for example, Peirce, 170). Retaining terms in Δ^2, the result is

$$\int_0^{\pi/2}\frac{u'\,du'}{v^3} \doteq -\frac{\pi}{4}\left[\frac{1}{u(\tfrac{1}{2}\pi - u)}\right] + \frac{2}{\Delta^2 u}. \quad (10d)$$

Substitutions of (10d) in (10c) and integration gives

$$J_v(\pi/2) \doteq (1 - \tfrac{1}{2}\Delta^2)\,Si\,\pi = (1 - \tfrac{1}{2}\Delta^2)(1.85). \quad (10e)$$

---

† It is shown by Schelkunoff (ref. I.26, pp. 239 and 293) that the resonant electrical length of a tapered line open at $z = h$ and closed at $z = 0$ is

$$\beta_0 h_r = \frac{\pi}{2}(1 + \delta), \quad (10f)$$

where

$$\delta = -\frac{1}{\pi R_{c0}}\int_0^\pi R_c(y)\cos y\,dy, \quad y = 2\beta_0 z. \quad (10g)$$

In (10g)

$$R_c(y) = 120\ln\left(\frac{2z\sin\tfrac{1}{2}\Delta}{a}\right) = 120\ln\left(\frac{y\sin\tfrac{1}{2}\Delta}{A}\right) \quad (10h)$$

is the variable characteristic impedance of the lossless tapered line, and R_{c0} is the average characteristic impedance. Integration by parts gives directly

$$\delta = \frac{120}{\pi R_{c0}}Si\,\pi. \quad (10i)$$

(Footnote continued overleaf)

that the value $Z_v = 0 - j111$ agrees with the value obtained from the theory of nonuniform transmission lines for a uniformly tapered, perfectly conducting line of electrical length $\beta_0 h = \pi/2$.

In practice, the V-antenna is not driven by a discontinuity in scalar potential at the apex, but by a potential difference maintained at the junction of a transmission line and the V-antenna. Since the V-antenna is a symmetric structure, it must be driven from a *balanced* transmission line. Three methods of connecting a V-antenna to a two-wire line are shown in Fig. 25.2. In general, the apparent terminal impedances for the lines in Figs. 25.2a, b differ significantly from each other and from the impedance of the idealized V-antenna driven by a slice generator. These differences are due to transmission-line end effects and to coupling between the V-antenna and the transmission line. Only when b/λ_0 is extremely small (0.001 or less) do the apparent terminal impedances for the lines in Figs. 25.2a, b approximate each other and the theoretical value for the slice generator. Note that this last is the impedance as b approaches zero. In the circuit of Fig. 25.2c, on the other hand, there are no antenna-line coupling effects, and transmission-line end effects may be compensated by adjusting the stub so that a voltage maximum is maintained across the terminals of the antenna when this is disconnected. If this is done and the inductive reactance of the two half bridges joining the line wires to the antenna is small, the apparent terminal impedance loading the line in Fig. 25.2c should be approximated closely by the theoretical value for a V-antenna driven from a slice generator.

The accurate calculation of the apparent terminal impedance as a function of Δ in either of the circuits in Figs. 25.2a, b is difficult if b/λ_0 is not extremely small. However, it is possible to derive the general nature and behavior of the impedance in a qualitative or semiquantitative manner by appropriate modification of the zeroth-order solution of the ideal, apex-driven V-antenna. This involves determining the equivalent lumped elements of terminal-zone networks for the practical circuits of Figs. 25.2a, b, and combining these with the theoretical impedance for the ideal, isolated V-antenna shown in Fig. 23.1. Note that both h and δ are functions of Δ, whereas $h' = h - \delta$ is constant. As Δ approaches zero, δ becomes infinite, since when $\Delta = 0$ the antenna structure ceases to be a tapered line and becomes a section of two-wire line. Although a general formula for the impedance of the V-antenna has been derived with both δ and h as parameters, the actual evaluation in a quantitative sense is difficult, since δ does not remain small. An instructive and qualitatively valuable estimate of the variation of the impedance with the enclosed angle Δ may be obtained from the zeroth-order solution for the idealized V, the terminal-zone networks for Figs. 25.2a, b, and the transmission-line impedance at $\Delta = 0$. This is true in particular when $\beta_0 h = \pi/2$. Let the two circuits illustrated in Figs. 25.2a, b be analyzed successively. Since the circuit of Fig. 25.2b is slightly simpler, it is considered first.

V-*antenna perpendicular to plane of two-wire line.* In the circuit of Fig. 25.2b the halves of the antenna are perpendicular to the two-wire line for all values of Δ, so that there is no inductive coupling between antenna and line. Therefore, the series inductive element L_T of the terminal-zone network is a constant independent of Δ. It is the same as that derived in (II.7.10e), namely,

$$L_T \doteq -(b-a)/2\pi v_0. \tag{11}$$

The evaluation of the shunt capacitive element C_T differs from C_{Te}, determined in Sec. II.8 for $\Delta = \pi$, only in the formulas for R_{1T} and R_{2T}. Instead of (II.8.8) the following expressions apply:

$$R_{1T} = \sqrt{s'^2 + w^2 + a^2},$$
$$R_{2T} \doteq \sqrt{s'^2 + 2s'b \sin \tfrac{1}{2}\Delta + w^2 + b^2 + a^2}. \tag{12}$$

Continuation of footnote † from page 387

The input reactance of a section of lossless uniform transmission line of length s is

$$X_{\text{in}} = -R_c \cot \beta_0 s. \tag{10j}$$

Since the resonant electrical length of the tapered line is greater than $\pi/2$ by $\delta\pi/2$, the input reactance of a section of length $\pi/2$ is given approximately by the input reactance of a section of uniform line with characteristic impedance R_{c0} and of electrical length $\delta\pi/2$ below resonance. For the uniform line, resonance is at $\pi/2$, so that the length $(\pi/2)(1 - \delta)$ is related to $\pi/2$ approximately as, for the tapered line, $\pi/2$ is related to $(\pi/2)(1 + \delta)$. Hence, with $\delta^2 \ll 1$,

$$X_{\text{in}} \doteq -R_{c0} \cot \frac{\pi}{2}(1-\delta) = -R_{c0} \tan \frac{\pi}{2}\delta$$
$$\doteq -R_c \pi \delta/2. \tag{10k}$$

Using the above value of δ,

$$X_{\text{in}} \doteq -60 \, Si\pi \text{ ohms} = -111 \text{ ohms}. \tag{10l}$$

This agrees exactly with the value obtained from (5) with Δ very small.

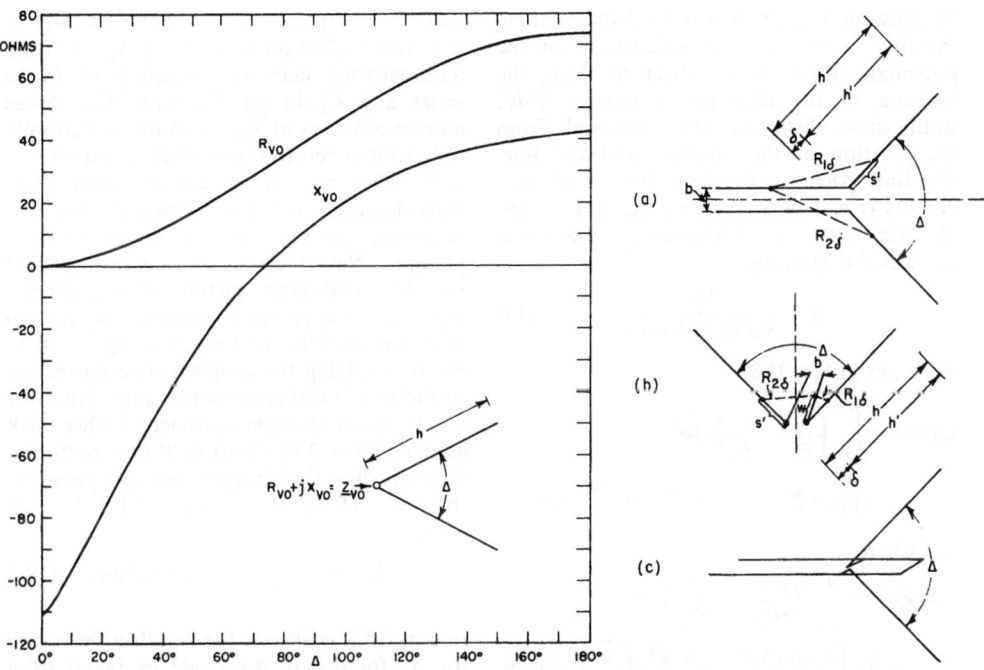

Fig. 25.1. Zeroth-order impedance of V-antenna; $\beta_0 h = \pi/2$; $\Omega \doteq \infty$.

Fig. 25.2. Three methods of connecting a V-antenna to a two-wire line.

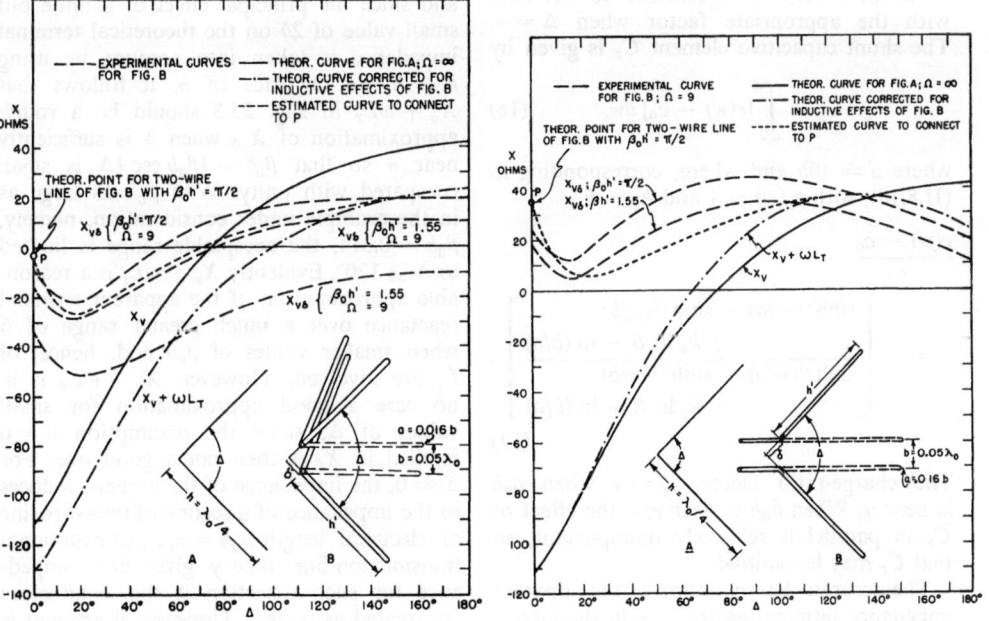

Fig. 25.3. Reactance of V-antenna in the plane perpendicular to the line.

Fig. 25.4. Reactance of V-antenna in the plane of the line.

As shown in Fig. 25.2b, w is the distance along the line to the point of calculation of the potential, and s' is the distance along the antenna to the element of charge $q'ds'$. Both these distances are measured from the junction of the antenna and the line. The line spacing is b; the radius of all conductors is a. The capacitance per unit length of line with the terminating V-antenna connected is given by

$$c(w) = \frac{2\pi\epsilon_0}{k_0(w) + k_{0T}(w)}, \quad (13)$$

where, as in Sec. II.8,

$$k_0(w) \doteq \int_0^\infty \left(\frac{1}{R_a} - \frac{1}{R_b}\right) dw'$$

$$= \sinh^{-1}\frac{w}{a} - \sinh^{-1}\frac{w}{b} + \ln\frac{b}{a} \quad (14)$$

and where

$$k_{0T}(w) \doteq \int_0^\infty \left(\frac{1}{R_{1T}} - \frac{1}{R_{2T}}\right) ds'$$

$$= \ln\left[\frac{b\sin(\Delta/2) + \sqrt{w^2 + b^2}}{\sqrt{w^2 + a^2}}\right] \equiv \ln B. \quad (15)$$

The following inequality is assumed: $a^2 \ll b^2$. Note that (15) is equivalent to (II.8.16) with the appropriate factor when $\Delta = \pi$. The shunt capacitive element C_T is given by

$$C_T = \int_0^d [c(w) - c_0] \, dw, \quad (16)$$

where $d \doteq 10b$ and where, corresponding to (II.8.18), with $\varphi_1(w) = 1$ and $k_q = k_a$,

$$\frac{c(w) - c_0}{c_0}$$
$$= - \left[\frac{\sinh^{-1}(w/a) - \sinh^{-1}(w/b)}{\sinh^{-1}(w/a) - \sinh^{-1}(w/b)} + k_q \ln B - \ln(b/a)\right]. \quad (17)$$

The charge-ratio factor $k_q \doteq 1$ when $\beta_0 h$ is near π. When $\beta_0 h$ is near $\pi/2$, the effect of C_T in parallel is relatively unimportant, so that C_T may be omitted.

The accurate determination of the apparent impedance terminating the line in the circuit of Fig. 25.2b is intricate. The following steps are required: (1) The evaluation of the theoretical isolated impedance $Z_{v\delta}$ from (1) for the appropriate values of h and Δ; note that $\delta = (b/2) \csc(\Delta/2)$, so that δ is large if Δ is small. (2) Combination of $Z_{v\delta}$ with the terminal-zone network consisting of L_T in series and C_T in parallel with $Z_{v\delta}$. Since numerical values of $Z_{v\delta}$ are available currently only from a zeroth-order solution with $\delta = 0$ and $\beta_0 h = \pi/2$, an accurate quantitative determination of the apparent terminal impedance in Fig. 25.2b is not possible at present. Nevertheless, an interesting and valuable qualitative picture of the general nature of the apparent impedance as a function of Δ may be obtained for $\beta_0 h' = \beta_0(h - \delta) = \pi/2$ by applying the terminal-zone correction to the theoretical zeroth-order impedance for $\delta = 0$. As an example, consider a rather thick antenna [$\Omega = 2 \ln(2h/a) \doteq 9$] driven from a two-wire line that is quite widely spaced so that $b = 0.05\lambda_0$, $a = 0.016b$. With $f = 750$ Mhz,

$$\omega L_T \doteq -\frac{b}{2\pi v_0} \doteq -18.8 \text{ ohms}. \quad (18a)$$

In Fig. 25.3 is shown the zeroth-order curve for X_v for $\delta = 0$, $\beta_0 h = \pi/2$ as taken from Fig. 25.1, together with a curve of $X_v + \omega L_T$. Since for $\Delta = \pi$ at $\delta = 0$ the zeroth-order reactance, 42.5 ohms, is a fair approximation of the reactance even of quite thick antennas, and since the principal effect of a finite but small value of 2δ on the theoretical terminal impedance is taken into account by using $h' \doteq h - \delta$ in place of h, it follows that $X_v + \omega L_T$ in Fig. 25.3 should be a rough approximation of $X_{v\delta}$ when Δ is sufficiently near π so that $\beta_0 \delta = \frac{1}{2}\beta_0 b \csc \frac{1}{2}\Delta$ is small compared with unity. With $\beta_0 b$ as large as in the example under consideration, namely, $\beta_0 b = 0.314$, the acceptable range is limited to $\Delta \geq 120°$. Evidently $X_v + \omega L_T$ is a reasonable approximation of the apparent terminal reactance over a much greater range of Δ when smaller values of $\beta_0 b$ and, hence, of L_T are involved. However, $X_v + \omega L_T$ is in no case a good approximation for small values of Δ, since the assumption $\delta = 0$ implied in X_v is then not a good one. For $\Delta \doteq 0$, the impedance of the antenna reduces to the impedance of a section of two-wire line of electrical length $\beta_0 h = \pi/2$. Conventional transmission-line theory gives zero impedance for such a section if the conductors are treated as perfect. However, if account is taken of the capacitive end effect, a lumped positive capacitance C_T is required across the open end of the section. This capacitance is given by (16) with (17), provided k_q is set equal to zero. This integral is simply the area between one of the curves of $c_0(w)$ in Fig. II.6.2 when drawn to a linear scale and the corresponding horizontal line c_0. Graphical integration with $b/a = 6$ gives $C_T = 0.163$ μμf. The reactance looking into a quarter-wave section of lossless line terminated in C_T is

$$X_{in} = \omega C_T R_c^2 = 35 \text{ ohms}, \quad (18b)$$

where $R_c = 215.4$ ohms. This is the input reactance at an electrical distance $\beta_0 w = \pi/2$ from the end of a straight transmission line. Since there is a right-angle bend at this value of $\beta_0 w$, the apparent impedance involves a terminal-zone network. At a low-voltage point the capacitive effect is small, but the inductive effect is great. In order to correct uniform-line theory on each side of the bend, series inductances L_T are required. Thus, the reactance terminating the line is X_{in} in (18b) in series with $2L_T$, where L_T is the same as (18a). Thus,

$$X_{sa} = X_{in} + 2\omega L_T = 35 - 37.6$$
$$= -2.6 \text{ ohms}. \quad (19)$$

This point is indicated by P in Fig. 25.3. Evidently, between $\Delta = 120°$ and $\Delta = 0°$, the curve $X_v + \omega L_T$ must bend up so that it ends at $X_{sa} = -2.6$, the value for the two-wire line, instead of at $X = -111$, the value for a tapered line. Computations to determine the actual shape of the curve between $\Delta = 0°$ and $\Delta = 120°$ have not been made. It is to be expected, however, that the true curve tends to follow $X_v + \omega L_T$ but ultimately bends up to the point P. For a relatively thick antenna such as the one considered, the departure from $X_v + \omega L_T$ should be considerable. A curve such as the dotted one in Fig. 25.3 appears reasonable. It is seen to approximate the experimentally determined curve marked $\beta_0 h' = \pi/2$. An experimental curve for a slightly smaller value of h' is marked $\beta_0 h' = 1.55$. Although the presently used rough method of determining the approximate reactance curve of a thick V-antenna driven from a two-wire line with large spacing is quantitatively of no great value, it does provide a clear picture of the several factors that determine the shape of the experimental curve and determine the striking departure from the simple curve X_v of the thin V-antenna driven from a slice generator or a two-wire line with negligible spacing b. In addition, the great significance of terminal-zone effects in determining the reactance of the V-antenna is clarified.

The terminal-zone effects do not alter R_v significantly. It is seen in Fig. 25.4 that the zeroth-order theoretical curve R_v for the extremely thin antenna with $\delta = 0$ and $\beta_0 h = \pi/2$ agrees in shape with the experimental curves $R_{v\delta}$ for a V-antenna with $\Omega = 9$ and 12.5, $\beta_0 h' = \beta_0(h - \delta) = 1.55$ or $\beta_0 h = \pi/2$ driven from a two-wire line with $b/\lambda_0 = 0.05$, $a = 0.016b$. Numerical differences for $\Delta < \pi$ are consistent with the difference at $\Delta = \pi$ where the theoretical value is $R_v = R_0 = 73.1$ ohms for a very thin antenna.

V-antenna in the plane of a two-wire line. An estimate of the apparent impedance terminating the line in the circuit of Fig. 25.2a is more involved than for the circuit of Fig. 25.2b, since inductive as well as capacitive coupling exists between the antenna and the transmission line. The lumped elements L_T and C_T for the corrective, terminal-zone network are determined by the same general method. Thus, the lumped shunt capacitance C_T is given by (16) and (17) but with

$$k'_{0T}(w) \doteq \ln B \doteq \int_0^\infty \left(\frac{1}{R_{1T}} - \frac{1}{R_{2T}}\right) ds'$$

$$= \ln\left(\frac{w \cos \tfrac{1}{2}\Delta + b \sin \tfrac{1}{2}\Delta + \sqrt{w^2 + b^2}}{w \cos \tfrac{1}{2}\Delta + \sqrt{w^2 + a^2}}\right), \quad (19a)$$

where

$$R_{1T} = \sqrt{s'^2 + 2s'w \cos \tfrac{1}{2}\Delta + w^2 + a^2},$$

$$R_{2T} = \sqrt{s'^2 + 2s'(w \cos \tfrac{1}{2}\Delta + b \sin \tfrac{1}{2}\Delta) + w^2 + b^2} \quad (19b)$$

in place of (15).

The series inductance L_T involves both end effect and coupling. It is determined from the inductance per unit length of the two-wire line with the V-antenna attached, that is, from

$$l^e(w) = l_0^e(w) + l_{0T}^e(w)$$
$$= [k_0(w) + k_{0T}(w)]/2\pi v_0, \quad (20)$$

where, as in Sec. II.6,

$$k_0(w) \doteq \int_0^\infty \left(\frac{1}{R_a} - \frac{1}{R_b}\right) dw'$$

$$= 2 \ln \frac{b}{a} - \ln \frac{w + \sqrt{w^2 + b^2}}{w + \sqrt{w^2 + a^2}}, \quad (21a)$$

with

$$R_a = \sqrt{(w - w')^2 + a^2},$$
$$R_b = \sqrt{(w - w')^2 + b^2}, \quad (21b)$$

and where

$$k_{0T}(w) = k'_{0T}(w) \cos \tfrac{1}{2}\Delta = \cos \tfrac{1}{2}\Delta \ln B, \quad (21c)$$

with $\ln B$ given in (19a). The lumped series inductance for the terminal-zone network is

$$\begin{aligned} L_T &= \int_0^d [l^e(w) - l_0^e]\, dw \\ &= \frac{1}{2\pi v_0}\left[\cos \tfrac{1}{2}\Delta \int_0^d \ln B\, dw \right. \\ &\quad \left. - \int_0^d \ln\left(\frac{w + \sqrt{w^2 + b^2}}{w + \sqrt{w^2 + a^2}}\right) dw\right], \quad (22) \end{aligned}$$

where $d \geq 10b$. The second integral in (22) is directly integrable. Its approximate value is $-(b - a)$ for $d^2 \gg b^2 \gg a^2$. The first term may be evaluated numerically in any particular case. Its general behavior as a function of Δ including the factor $\cos \tfrac{1}{2}\Delta$ is quite simple. Beginning with $\Delta = \pi$, where it is obviously zero so that $L_T = -(b - a)/2\pi v_0$, it increases to a positive maximum near $\Delta = \pi/2$ that is greater in magnitude than $(b - a)$ so that L_T has a positive rather than a negative value. It diminishes exactly to $(b - a)$ at $\Delta = 0$ where, therefore, $L_T = 0$.

The combination $X_v + \omega L_T$ is shown in Fig. 25.5 for the same antenna and line analyzed previously for the circuit of Fig. 25.2b. This curve for the circuit of Fig. 25.2a is seen to differ considerably from the corresponding curve for the circuit of Fig. 25.2b shown in Fig. 25.3. Notably, the reactance increases rather than decreases as Δ is reduced from π. Since the theoretical curves assume $\delta = 0$, $X_v + \omega L_T$ can be expected to be a fair approximation for thick antennas with $\delta > 0$ only for Δ sufficiently large. For the same line spacing, $b = 0.05\lambda$, the range $120° \leq \Delta \leq 180°$ is reasonable.

Just as in the case of Fig. 25.2b, the reactance at $\Delta = 0$ may be obtained from transmission-line theory appropriately corrected for end effects. In this case there is no right-angle bend at the junction of the line proper and the section corresponding to V-antenna folded together so that $\Delta = 0$; therefore only the terminal-zone effect at the end of the section of length $\pi/2$ is effective. This is the same as before and gives

$$X_{sa} = X_{\text{in}} = \omega C_T R_c^2 = 35 \text{ ohms} \quad (23)$$

for the particular case under study. The point $X = 35$ is labeled P in Fig. 25.5. As before, the true impedance curve must connect $X_v + \omega L_T$ between $\Delta = 120°$ and $180°$ with this point. An estimated curve of this sort is shown dotted. Experimental curves of the apparent reactance of the V-antenna are marked $X_{v\delta}$ in Fig. 25.5. Curves for $\beta_0 h' = \pi/2$ and 1.55 are shown with $\Omega = 9$ and 12.5.

Once again it is clear how important the terminal-zone effects are when b/λ_0 is not vanishingly small. The rough qualitative study based on the theoretical zeroth-order reactance X_v is seen to be adequate to explain the general shape of the experimental curve and the reasons for its departure not only from the theoretical zeroth-order curve for the ideal V-antenna but also from the experimental curve for the identical antenna driven from the same line but in the circuit of Fig. 25.2b instead of that of Fig. 25.2a. The great importance of the particular circuit connection in determining the apparent reactance of a V-antenna is brought out clearly. The resistance of the V-antenna is essentially the same in both the circuits in Fig. 25.2a and Fig. 25.2b. It is given in Fig. 25.4.

Complete experimental curves[4] for the apparent impedance of the V-antenna driven in the circuits of Fig. 25.2a and 25.2b are shown in Figs. 25.6 to 25.9. Note the startling differences between the measured impedances of the same antenna when driven from the same line but in the two different circuit connections. The two sets of curves agree only for $\Delta = \pi$, when the two circuits are identical.

The experimental measurements were made by D. Angelakos on the image-plane line described in Sec. II.35. Note that in these curves $\beta_0 h = \beta_0(h' + \delta_\pi) = \beta_0(h' + b/2)$ is the electrical leg length, that is, the half-length of the cylindrical antenna with collinear halves when $\Delta = \pi$ and $\delta = b/2$. This differs from the convention in the general analysis of the V-antenna in which $\beta_0 h = \beta_0(h' + \delta)$ is a quantity that varies with Δ. In order to distinguish between the two circuit conventions, the angle of tilt from the collinear position is used as parameter ψ in Fig. 25.2a, α in Fig. 25.2b instead of Δ. Note that $\Delta = \pi - 2\psi$ or $\Delta = \pi - 2\alpha$.

Experimentally determined distributions of current along the V-antenna as a function of the angle α for the circuit of Fig. 25.2b are shown in Fig. 25.10 for four values of $\beta_0 h$. Comparison with the curves in Sec. II.26 shows that the distribution of current on the V-antenna differs little from the distribution along the conventional dipole of the same leg length.

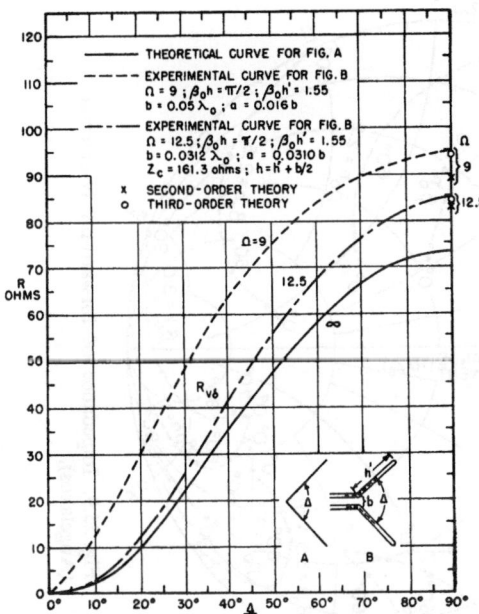

Fig. 25.5. Resistance of V-antenna.

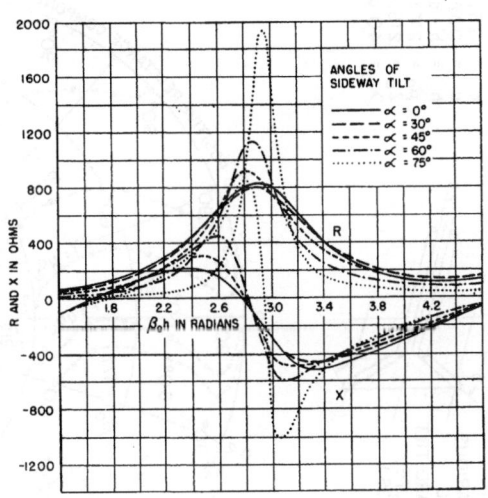

Fig. 25.6. Variations in V-antenna impedance with length for various angles of sideways tilt (Angelakos).

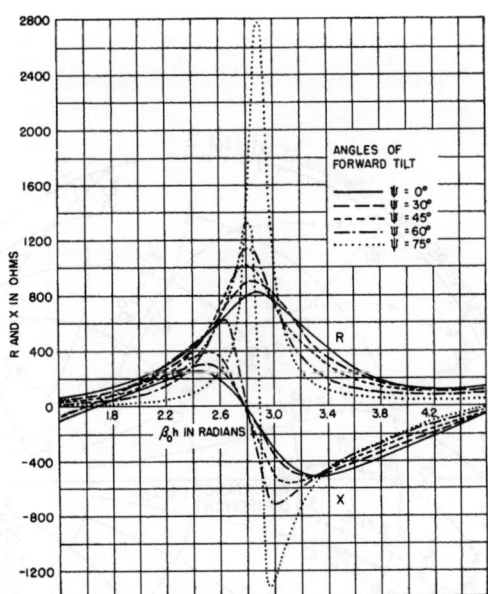

Fig. 25.7. Variations in V-antenna impedance with length for various angles of forward tilt (Angelakos).

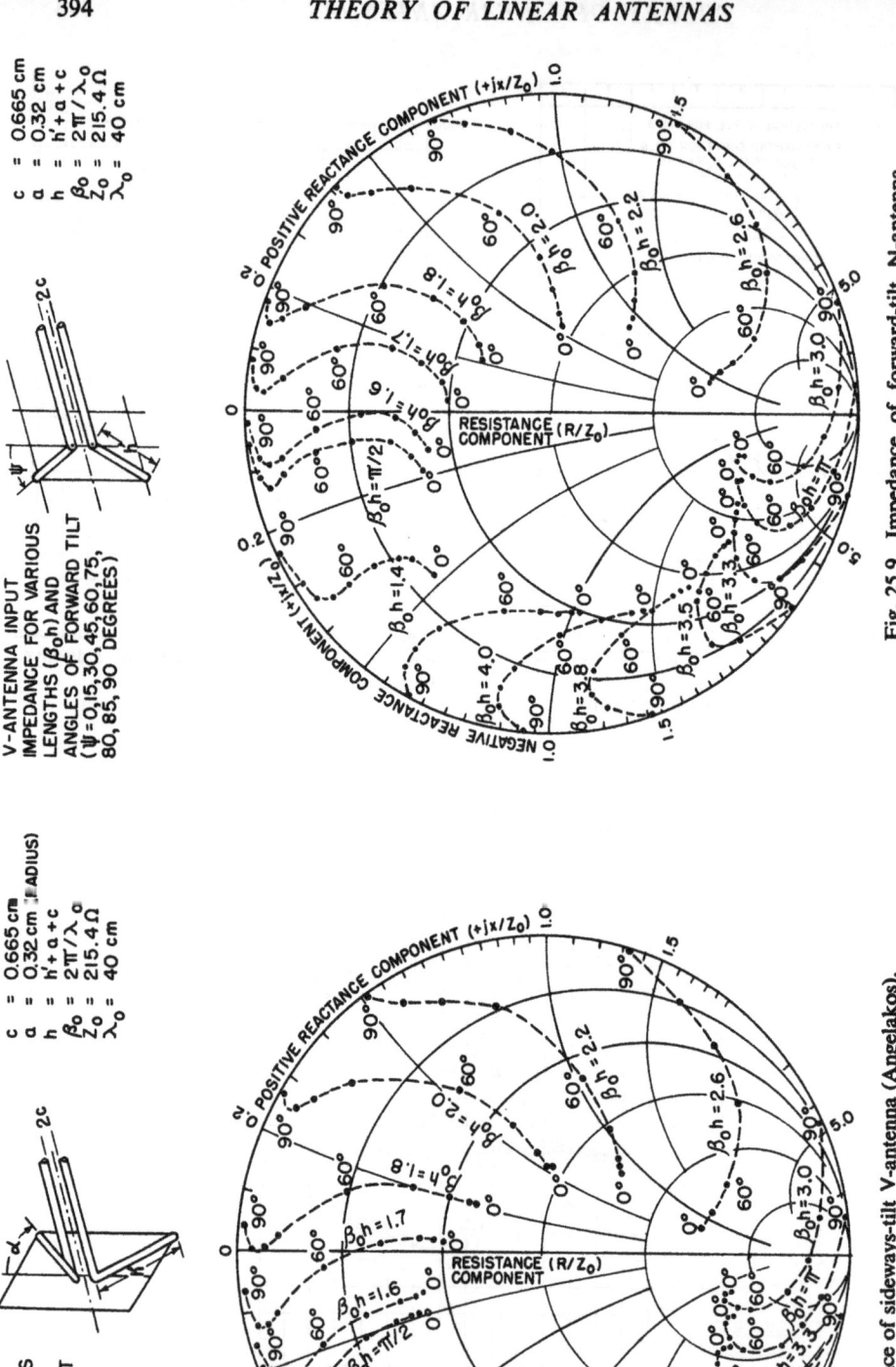

Fig. 25.9. Impedance of forward-tilt N-antenna (Angelakos).

Fig. 25.8. Impedance of sideways-tilt V-antenna (Angelakos).

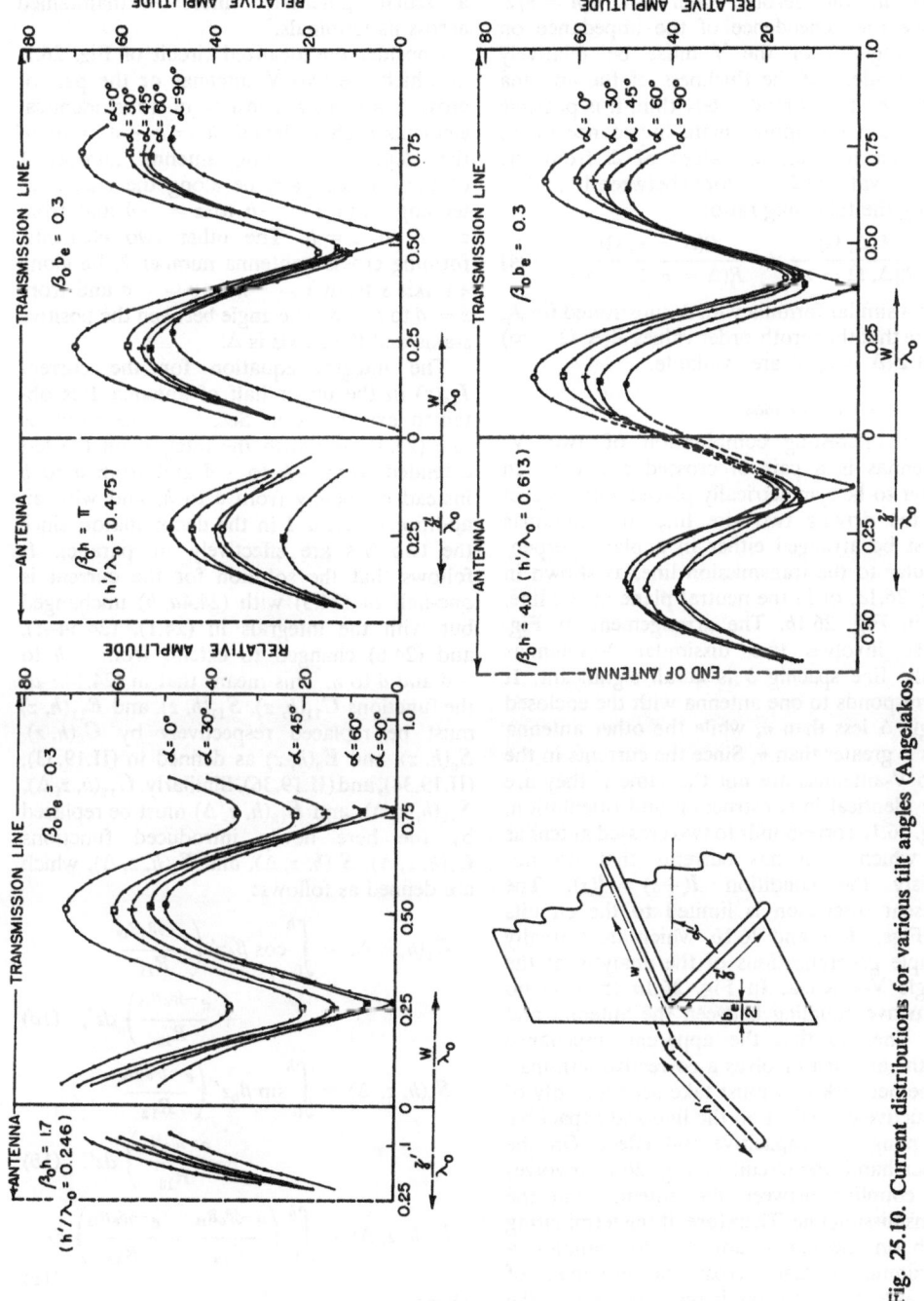

Fig. 25.10. Current distributions for various tilt angles (Angelakos).

Although the impedance of the V-antenna can be determined in general from (1), the numerical evaluation has been carried out only in the zeroth order for $\beta_0 h = \pi/2$. Since the dependence of the impedance on the angle Δ of the V must be relatively independent of the thickness of the antenna provided this is moderately thin, it is possible to obtain the approximate impedance of an antenna for various values of Δ from the known value at $\Delta = \pi$ for Ω between 10 and 20 using the following ratio:

$$\frac{R(\Delta, \Omega)}{R(\Delta, \Omega = \infty)} \doteq \frac{R(\Delta = \pi, \Omega)}{R(\Delta = \pi, \Omega = \infty)}, \quad (24)$$

and a similar formula with X substituted for R. Note that the zeroth-order values $Z(\Delta, \Omega = \infty)$ and $Z(\Delta = \pi, \Omega)$ are available.

26. Crossed Antennas

An interesting combination of two V-antennas is a pair of crossed antennas. In order to be symmetrically placed with respect to the driving two-wire line, the antennas must be arranged either in a plane perpendicular to the transmission line, as shown in Fig. 26.1a, or in the neutral plane of the line, as in Fig. 26.1b. The arrangement of Fig. 26.1c involves two dissimilar V-antennas if the line spacing b is at all significant. It corresponds to one antenna with the enclosed angle Δ less than π, while the other antenna has Δ greater than π. Since the currents in the two V-antennas are not the same if they are not identical in construction and orientation, Fig. 26.1c corresponds to two crossed antennas of which each has currents that do not satisfy the condition $I(-s) = I(s)$. The present discussion is limited to the circuits of Figs. 26.1a and 26.1b, which are formally simple generalizations of the analysis of the single V-antenna. In Fig. 26.1a there is no inductive coupling between the antenna and the line, so that the apparent impedance of the antenna involves a corrective terminal-zone network that must take account only of inductive end effect on the line and capacitive coupling and capacitive end effect. On the other hand, the circuit of Fig. 26.1b involves no coupling between the antenna and the transmission line. Therefore, if the terminating stub on the line is adjusted to maintain a maximum voltage across the terminals of the antenna when this is removed, and if the short half-bridges joining the line and these terminals are sufficiently large to have negligible inductive reactance, the apparent terminal impedance when the antenna is connected should differ negligibly from the ideal impedance of the isolated antenna with a scalar potential difference maintained across its terminals.

Consider the idealized circuit of Fig. 26.2, in which the two V-antennas or the pair of crossed antennas consists of four identical elements each of length $h' = h - \delta$. Two of the elements, forming antenna number 1 of the crossed pair, lie along the z-axis extending from $z = -h$ to $z = -\delta$ and from $z = \delta$ to $z = h$. The other two elements, forming crossed antenna number 2, lie along an axis s from $s = -h$ to $s = -\delta$ and from $s = \delta$ to $s = h$. The angle between the positive s-axis and the z-axis is Δ.

The integral equation for the current $I_{1z}(z)$ in the upper half of antenna 1 is obtained exactly as in Sec. 23. The result is like (23.31) but with the integral on the left extended from $-h$ to $-\delta$ and from δ to h instead of merely from δ to h, and with an additional factor 2 in the denominator, since the two V's are effectively in parallel. It follows that the solution for the current is one-half of (24.3) with (24.4a, b) unchanged but with the integrals in (24.1), (24.5a–d), and (24.6) changed to extend from $-h$ to $-\delta$ and δ to h. This means that in (24.14a–d) the functions $C_{11}(h, z)$, $S_{11}(h, z)$, and $E_{11}(h, z)$ must be replaced respectively by $C_a(h, z)$, $S_a(h, z)$, and $E_a(h, z)$ as defined in (II.19.33), (II.19.34), and (II.19.36). Similarly, $C_{12}(h, z, \Delta)$, $S_{12}(h, z, \Delta)$, and $E_{12}(h, z, \Delta)$ must be replaced by the here newly introduced functions $C_v(h, z, \Delta)$, $S_v(h, z, \Delta)$, and $E_v(h, z, \Delta)$, which are defined as follows:

$$C_v(h, z, \Delta) \equiv \int_0^h \cos \beta_0 z' \left(\frac{e^{-j\beta_0 R_{12}}}{R_{12}} + \frac{e^{-j\beta_0 R_{13}}}{R_{13}} \right) dz', \quad (1a)$$

$$S_v(h, z, \Delta) \equiv \int_0^h \sin \beta_0 z' \left(\frac{e^{-j\beta_0 R_{12}}}{R_{12}} + \frac{e^{-j\beta_0 R_{13}}}{R_{13}} \right) dz', \quad (1b)$$

$$E_v(h, z, \Delta) \equiv \int_0^h \left(\frac{e^{-j\beta_0 R_{12}}}{R_{12}} + \frac{e^{-j\beta_0 R_{13}}}{R_{13}} \right) dz', \quad (1c)$$

where

$$R_{12} \equiv \sqrt{z^2 - 2zz' \cos \Delta + z'^2 + a^2},$$
$$R_{13} \equiv \sqrt{z^2 + 2zz' \cos \Delta + z'^2 + a^2}. \quad (2)$$

Finally the function $p(z)$ as defined in (23.16b) and occurring in the integrals $J_v(F_{0z}, z)$ and $J_v(G_{0z}, z)$ must have its range of integration changed to include $-h$ to $-\delta$ and δ to h. The same substitutions for $C_{11}(h, z)$, $C_{12}(h, z, \Delta)$, $S_{11}(h, z)$, and $S_{12}(h, z, \Delta)$ must be made in (24.16).

With these modifications the analysis of the V-antenna in Secs. 23–25 applies directly to the crossed antennas. Evidently, when $\Delta = \pi/2$, the crossed antennas become identical with the two mutually perpendicular antennas analyzed in Sec. 21. That is, each collinear pair behaves just as if it were isolated. If the two are driven from a single generator, they are equivalent to two isolated collinear pairs in parallel. The impedance at the terminals of the generator is one-half that for a single isolated antenna. When Δ is very small, the crossed antennas are equivalent to two tapered transmission lines in parallel, and the impedance at the terminals of the generator is one-half the impedance of a single tapered line.

The impedance of the crossed antennas is given by (25.1) with appropriate changes in the functions involved, corresponding to the change in the limits of integration. In particular, the formula (25.5) for $\beta_0 h = \pi/2$, $\delta = 0$ and with $\Omega = 2\ln(2h/a)$ extremely great, becomes

$$Z_{c0} \doteq \frac{j\zeta_0}{4\pi}\bigg[C_a(h, h) - \cos \Delta C_v(h, h, \Delta)$$
$$- J'_v\left(\frac{\pi}{2}\right)\bigg], \quad (3)$$

where

$$J'_v(\pi/2) = J_v(\pi/2, \Delta) - J_v(\pi/2, \pi - \Delta) \quad (4a)$$

and where

$$J_v(\pi/2, \Delta) = J_v(\pi/2) = -\sin^2 \Delta \int_0^{\pi/2} \sin u \, du$$
$$\times \int_0^{\pi/2} K_{12}(u, u') \, u' \cos u' \, du' \quad (4b)$$

is the function previously defined in (25.8a). The term $J_v(\pi/2, \pi - \Delta)$ is due to the added integral with limits $-h$ to 0. Note that when $\Delta = \pi/2$, (3) reduces to

$$Z_{c0} \doteq \frac{j\zeta_0}{4\pi} C_a(h, h) = 36.6 + j21.25.$$
$$(\beta_0 h = \pi/2). \quad (5)$$

A simple modified zeroth-order formula for the impedance of symmetrical crossed antennas has been derived by Chambers[15] using the emf method. His formula is

$$(Z_{c01}) = \frac{\zeta_0}{4\pi}[1.219(1 + \cos \Delta)$$
$$+ j(3.265 - 0.0284\Delta°]. \quad (6)$$

Note that Δ is in degrees. For $\Delta = 90°$, (6) gives exactly (5), as it should.

The general shapes of the resistance and reactance curves defined by (3) for the crossed antennas are like those in Fig. 25.1 with both the scale of ordinates and the scale of abscissas halved.

The application of the theory for the ideal isolated antenna in Fig. 26.2 to the practical circuit of Fig. 26.1a not only includes the determination of the lumped elements L_T and C_T for the required terminal-zone network, but involves the additional complication that the halves of each crossed antenna are not collinear but laterally displaced.

If the two crossed antennas are not driven by the same generator as two identical V-antennas in parallel or as two crossed antennas with equal currents in opposite phase, but are individually driven by two generators with arbitrary voltages and relative phases, the problem is readily reduced to the one already solved. Just as for two parallel antennas, the problem of two arbitrary driving voltages may be reduced to two independent problems for equal currents in phase and equal currents in phase opposition. The solution of the latter has been outlined in this section. The solution of the former is exactly the same as that of the latter but with $(\pi - \Delta)$ substituted for Δ.

ASYMMETRICALLY DRIVEN LINEAR ANTENNAS

27. Current and Impedance for Asymmetrically Driven Cylindrical Antennas

Since the center-driven cylindrical antenna, analyzed in detail in Chapter II, is symmetric with respect to its center, the currents in the halves are the same, $I(-z) = I(z)$, and a single integral equation is involved. If the antenna is not center-driven, numerous complications arise. Most important among these is the fact that a transmission line feeding the antenna as in Fig. 27.1 is not in a neutral plane and is unbalanced by the different lengths of conductor attached to its ends. As a result, the transmission line is an important part of the radiating system, since the co-directional components of current contribute significantly to the far-zone electromagnetic

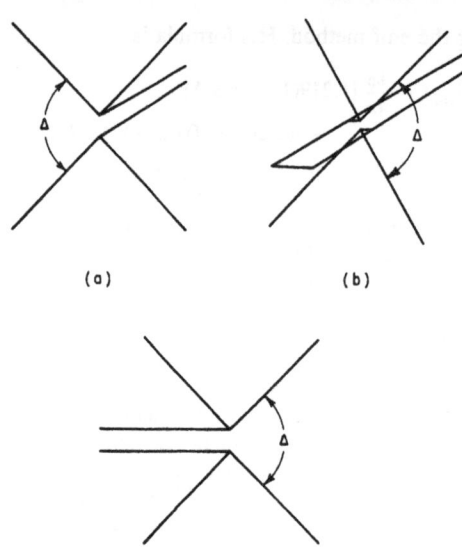

Fig. 26.1. Crossed antennas.

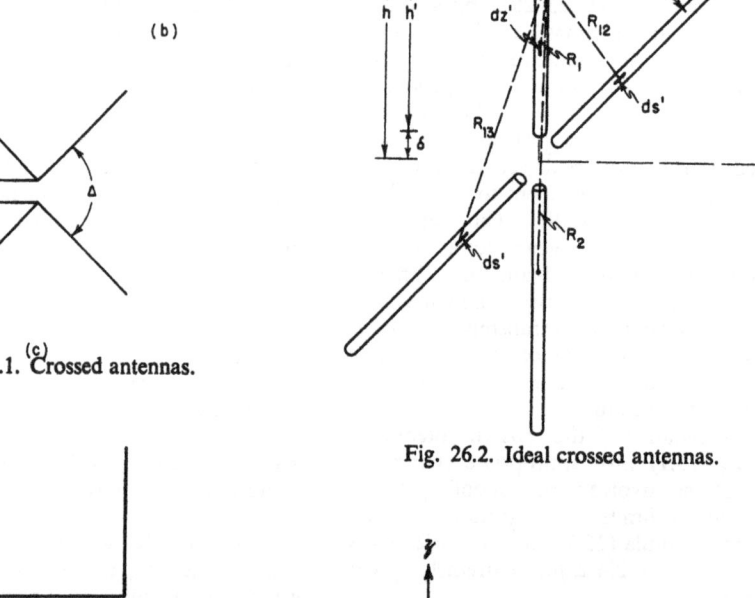

Fig. 26.2. Ideal crossed antennas.

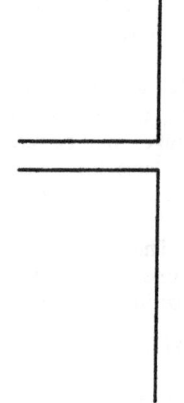

Fig. 27.1. Asymmetrically driven linear antenna.

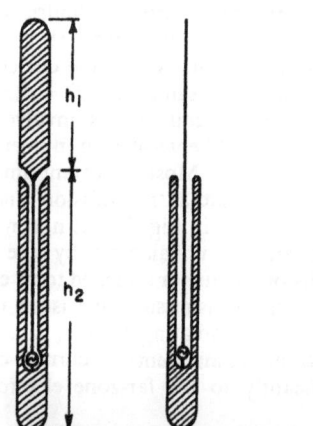

Fig. 27.2. Generator within asymmetrically driven antenna.

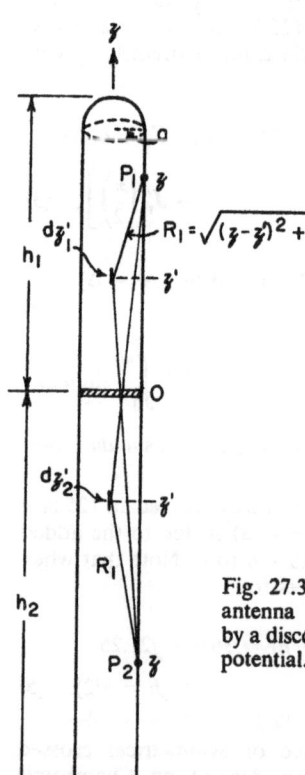

Fig. 27.3. Thin cylindrical antenna driven off-center by a discontinuity in scalar potential.

field. A radiating system that includes the transmission line as a radiating element is always undesirable. Obviously, the length and the location of the line should not play a part in determining the field characteristics of an antenna system.

The only manner in which an asymmetric antenna can be driven without introducing a radiating transmission line is by a generator completely within the metal surfaces of the antenna. Two possible arrangements are shown in Fig. 27.2. Except for projectiles, antennas of this type are not practically important. However, a very useful antenna, the sleeve dipole, in which the radiating currents are equivalent to the superposition of the currents in two asymmetrically driven antennas, is discussed in Sec. 30. Moreover, one step in the analysis of a collinear antenna in later sections involves the determination of the current in an asymmetric antenna center-driven by a discontinuity in scalar potential, that is, by a slice generator.

The antenna to be analyzed in this section is the thin cylindrical antenna driven off-center by a discontinuity in scalar potential at $z = 0$. Since there can be no axial gap in internally driven antennas, $\delta = 0$. The antenna is shown in Fig. 27.3. As in the analysis of the center-driven antenna, it is assumed that currents and charges are confined to the cylindrical envelope of radius a. Account may be taken of other surfaces at the driving point if they exist, just as for the center-driven cylinder. At the extremities, hemispherical ends approximate the idealized cylinder which has no chargeable surfaces other than on the cylinder itself, as discussed in Sec. II.11. For simplicity, terms involving the surface impedance per unit length z^i will be omitted, so that the analysis applies strictly to a perfect conductor.

Since the two parts of the antenna are not equal in length, the distributions of current on them are necessarily different. This means that two currents, I_{1z} for $0 \leq z \leq h_1$ and I_{2z} for $-h_2 \leq z \leq 0$, must be determined subject to the condition that they join smoothly at $z = 0$, that is,

$$I_{10} = I_{20} = I_0. \qquad (1)$$

The currents I_{1z} and I_{2z} are the unknowns in a pair of simultaneous integral equations. The equations are obtained just as for the center-driven antenna by equating the Helmholtz integral for the vector potential to the solution of the differential equation for the vector potential. Following the method in Sec. II.14 but with $\delta = 0$, $z^i = 0$, the integral equations are

$$4\pi v_0 A_{1z}(z) = \int_0^{h_1} I'_{1z} K_1(z, z')\, dz'$$

$$+ \int_{-h_2}^0 I'_{2z} K_1(z, z')\, dz'$$

$$= \frac{-j4\pi}{\zeta_0}(C_1 \cos \beta_0 z + C_2 \sin \beta_0 z),$$

$$(0 \leq z \leq h_1) \quad (2a)$$

$$4\pi v_0 A_{2z}(z) = \int_0^{h_1} I'_{1z} K_1(z, z')\, dz'$$

$$+ \int_{-h_2}^0 I'_{2z} K_1(z, z')\, dz'$$

$$= \frac{-j4\pi}{\zeta_0}(C_3 \cos \beta_0 z + C_4 \sin \beta_0 z),$$

$$(-h_2 \leq z \leq 0) \quad (2b)$$

where

$$K_1(z, z') \equiv \frac{e^{-j\beta_0 R_1}}{R_1},$$

$$R_1 \equiv \sqrt{(z' - z)^2 + a^2}. \qquad (3)$$

These two equations must be solved for I_{1z} and I_{2z}, subject to (1) and the conditions

$$I_{1z} = 0, \quad z = h_1; \qquad I_{2z} = 0, \quad z = -h_2; \qquad (4)$$

$$V_0 = \lim_{\delta \to 0} [\Phi_1(\delta) - \Phi_2(-\delta)] = \Phi_1(0) - \Phi_2(0). \qquad (5)$$

The scalar potential is obtained as in Sec. II.13. Thus,

$$\Phi_1(z) = \frac{j\omega}{\beta_0^2} \frac{\partial A_{1z}}{\partial z}$$

$$= -C_1 \sin \beta_0 z + C_2 \cos \beta_0 z, \qquad (6)$$

$$\Phi_2(z) = \frac{j\omega}{\beta_0^2} \frac{\partial A_{2z}}{\partial z}$$

$$= -C_3 \sin \beta_0 z + C_4 \cos \beta_0 z. \qquad (7)$$

Let

$$V_{10} \equiv \lim_{\delta \to 0} [\Phi_1(\delta) - 0] = C_2, \qquad (8a)$$

$$V_{20} \equiv \lim_{\delta \to 0} [0 - \Phi_2(-\delta)] = -C_4. \qquad (8b)$$

With (5),

$$V_0 = \lim_{\delta \to 0} [\Phi_1(\delta) - \Phi_2(-\delta)] = V_{10} + V_{20}$$

$$= C_2 - C_4. \qquad (9)$$

The approximate distribution functions for the currents may be introduced as follows:

$$f_1(z) \doteq I_{1z}/I_0, \quad f_2(z) \doteq I_{2z}/I_0; \quad (10)$$

$$g_{11}(z, z') = f_1(z')/f_1(z),$$
$$g_{22}(z, z') = f_2(z')/f_2(z), \quad (11a)$$

$$g_{12}(z, z') = f_2(z')/f_1(z),$$
$$g_{21}(z, z') = f_1(z')/f_2(z). \quad (11b)$$

Let the following functions be defined:

$$\Psi_1(z) \equiv \int_0^{h_1} g_{11}(z, z') K_1(z, z') \, dz'$$
$$+ \int_{-h_2}^0 g_{12}(z, z') K_1(z, z') \, dz',$$
$$(0 \leq z \leq h_1) \quad (12a)$$

$$\Psi_2(z) \equiv \int_0^{h_1} g_{21}(z, z') K_1(z, z') \, dz'$$
$$+ \int_{-h_2}^0 g_{22}(z, z') K_1(z, z') \, dz',$$
$$(-h_2 \leq z \leq 0) \quad (12b)$$

$$\Pi_1(z) \equiv \int_0^{h_1} [I'_{1z} - I_{1z} g_{11}(z, z')] K_1(z, z') \, dz'$$
$$+ \int_{-h_2}^0 [I'_{2z} - I_{1z} g_{12}(z, z')] K_1(z, z') \, dz',$$
$$(13a)$$

$$\Pi_2(z) \equiv \int_0^{h} [I'_{1z} - I_{2z} g_{12}(z, z')] K_1(z, z') \, dz'$$
$$+ \int_{-h_2}^0 [I'_{2z} - I_{2z} g_{22}(z, z')] K_1(z, z') \, dz'.$$
$$(13b)$$

The functions $\Psi_1(z)$ and $\Psi_2(z)$ are the ratios of vector potential to current at points z on the two parts of the antenna if $f_1(z)$ and $f_2(z)$ are the true distributions of current. As discussed in Chapter II, this ratio must be predominantly real and sensibly constant over each of the two parts of the antenna. Therefore, let

$$\Psi_1(z) = \Psi_1 + \gamma_1(z), \quad \Psi_2(z) = \Psi_2 + \gamma_2(z), \quad (14)$$

where

$$\Psi_1 = |\Psi_1(z_r)|, \quad \Psi_2 = |\Psi_2(z_r)|, \quad (15)$$

and z_r is a reference point on each part at which the magnitude of $\Psi(z)$ approximates the mean constant value of $|\Psi(z)|$. The complex functions $\gamma_1(z)$ and $\gamma_2(z)$ are very small except near the respective ends at $z = h_1$, $z = -h_2$, where they become infinite but $I_z \gamma(z)$ vanishes.

With (8) to (15), (2a, b) become

$$I_{1z}[\Psi_1 + \gamma_1(z)] + \Pi_1(z)$$
$$= \frac{-j4\pi}{\zeta_0} (C_1 \cos \beta_0 z + V_{10} \sin \beta_0 z),$$
$$(0 \leq z \leq h_1) \quad (16a)$$

$$I_{2z}[\Psi_2 + \gamma_2(z)] + \Pi_2(z)$$
$$= \frac{-j4\pi}{\zeta_0} (C_3 \cos \beta_0 z - V_{20} \sin \beta_0 z),$$
$$(-h_2 \leq z \leq 0) \quad (16b)$$

If the functions $f_1(z)$ and $f_2(z)$ were the true distributions of current, the difference integrals $\Pi_1(z)$ and $\Pi_2(z)$ would necessarily be zero. Since the true distributions are actually the unknowns, approximate distributions must be used, in which case $\Pi_1(z)$ and $\Pi_2(z)$ are the smaller the more closely the assumed distribution functions represent the true currents. It follows that the principal terms on the left are $I_{1z}\Psi_1$ and $I_{2z}\Psi_2$. Hence,

$$I_{1z} = \frac{-j4\pi}{\zeta_0 \Psi_1} (C_1 \cos \beta_0 z + V_{10} \sin \beta_0 z)$$
$$- \frac{1}{\Psi_1} [I_{1z} \gamma_1(z) + \Pi_1(z)],$$
$$(0 \leq z \leq h_1) \quad (17a)$$

$$I_{2z} = \frac{-j4\pi}{\zeta_0 \Psi_2} (C_3 \cos \beta_0 z - V_{20} \sin \beta_0 z)$$
$$- \frac{1}{\Psi_2} [I_{2z} \gamma_2(z) + \Pi_2(z)],$$
$$(-h_2 \leq z \leq 0) \quad (17b)$$

The simultaneous integral equations (17a, b) involve no approximations not already contained in (2). They represent merely a rearrangement of the terms. However, the unknown currents I_{1z} and I_{2z} in (17a, b) occur in two types of terms, namely, the principal terms on the left and the difference integrals $\Pi_1(z)$ and $\Pi_2(z)$ on the right, which are small if well-chosen distribution functions $f_1(z)$ and $f_2(z)$ are used.

Note that (17a) and (17b) individually correspond to (II.14.13) (with $z^i = 0$, $\delta \doteq 0$), to which they reduce exactly when $h_1 = h_2 = h$ in (17a, b). They may be solved by the same method of iteration to obtain final formulas like (II.15.26) with appropriate changes in symbols:

$$I_{1z}(z) = \frac{j4\pi V_{10}}{\zeta_0 \Psi_1} \times$$

$$\left[\frac{\sin \beta_0(h_1 - z) + M_1(h_1, z)/\Psi_1 + M_2(h_1, z)/\Psi_1^2 + \cdots}{\cos \beta_0 h_1 + A_1(h_1)/\Psi_1 + A_2(h_1)/\Psi_1^2 + \cdots} \right].$$

$$(0 \leq z \leq h_1) \quad (18)$$

If the variable x is defined as follows:

$$x = -z, \quad (19)$$

the current I_{2x} in the lower part of the antenna has exactly the same form as (18) with appropriate changes in subscripts. Thus,

$$I_{2z}(z) = -I_{2x}(x) = \frac{j4\pi V_{20}}{\zeta_0 \Psi_2} \times$$

$$\left[\frac{\sin \beta_0(h_2 - x) + M_1(h_2, x)/\Psi_2 + M_2(h_2, x)/\Psi_2^2 + \cdots}{\cos \beta_0 h_2 + A_1(h_2)/\Psi_2 + A_2(h_2)/\Psi_2^2 + \cdots} \right].$$

$$(0 \leq x \leq h_2) \quad (20)$$

Note that from (8b)

$$V_{20} = \lim_{z \to 0} [0 - \Phi_2(-z)] = \lim_{x \to 0} [0 - \Phi_2(x)]. \quad (21)$$

Since an evaluation of (18) in terms of z and h_1 automatically gives (20) in terms of x and h_2, it is necessary only to solve (18). In (18) the functions M and A are defined as in (II.15.27) to (II.15.29) with $z^i = 0$, $\delta = 0$. The first-order functions are

$$M_1(h_1, z) = F_1(h_1, z) \sin \beta_0 h_1$$
$$- G_1(h_1, z) \cos \beta_0 h + G_1(h_1, h_1) \cos \beta_0 z$$
$$- F_1(h_1, h_1) \sin \beta_0 z, \quad (22a)$$

$$A_1(h_1) = F_1(h_1, h_1). \quad (22b)$$

The functions F_1 and G_1 in (22) are defined as in (II.21.1) with $z^i = 0$, $\delta = 0$, and with appropriate changes in notation to take account of the different kernel and limit of integration. Note that F_1 and G_1 are functions of h_2 as well as of h_1 even though this is not indicated in the arguments.

$$F_1(h_1, z) = F_{0z}(h_1)\Psi_1 - \int_0^{h_1} F_{0z'}(h_1)K_1(z,z')\,dz'$$
$$- \int_{-h_2}^0 F_{0z'}(h_2)K_1(z,z')\,dz', \quad (23a)$$

$G_1(h_1, z)$ is like (23a) with G substituted for F. (23b)

The functions $F_1(h_1, h_1)$ and $G_1(h_1, h_1)$ are obtained from (23a, b) with $z = h_1$. The zeroth-order functions are

$$F_{0z}(h_1) = \cos \beta_0 z - \cos \beta_0 h_1,$$
$$G_{0z}(h_1) = \sin \beta_0 z - \sin \beta_0 h_1;$$
$$(0 \leq z \leq h_1) \quad (24a)$$

$$F_{0z}(h_2) = \cos \beta_0 z - \cos \beta_0 h_2,$$
$$G_{0z}(h_2) = -\sin \beta_0 z - \sin \beta_0 h_2.$$
$$(-h_2 \leq z \leq 0) \quad (24b)$$

The impedance of the asymmetrically driven antenna is obtained using (1) and (9), namely, $I_{20} = I_{10} = I_0$, $V_0 = V_{10} + V_{20}$, in conjunction with (18) with $z = 0$ and (20) with $x = 0$. Thus, it follows from (18) and (20) that

$$Z_1 \equiv \frac{V_{10}}{I_{10}} = \frac{V_{10}}{I_0} = \frac{-j\zeta_0 \Psi_1}{4\pi} \times$$

$$\left[\frac{\cos \beta_0 h_1 + A_1(h_1)/\Psi_1 + A_2(h_1)/\Psi_1^2 + \cdots}{\sin \beta_0 h_1 + B_1(h_1)/\Psi_1 + B_2(h_1)/\Psi_1^2 + \cdots} \right], \quad (25a)$$

$$Z_2 \equiv \frac{V_{20}}{I_{20}} = \frac{V_{20}}{I_0} = \frac{-j\zeta_0 \Psi_2}{4\pi} \times$$

$$\left[\frac{\cos \beta_0 h_2 + A_1(h_2)/\Psi_2 + A_2(h_2)/\Psi_2^2 + \cdots}{\sin \beta_0 h_2 + B_1(h_2)/\Psi_2 + B_2(h_2)/\Psi_2^2 + \cdots} \right], \quad (25b)$$

where

$$B(h) \equiv M(h, 0). \quad (25c)$$

However, from (9) and (25a, b),

$$V_0 = V_{10} + V_{20} = I_0(Z_1 + Z_2). \quad (26)$$

Since the impedance Z_0 of the antenna is by definition $Z_0 \equiv V_0/I_0$, it follows with (26) that

$$Z_0 \equiv V_0/I_0 = Z_1 + Z_2, \quad (27)$$

where Z_1 and Z_2 are given by (25a, b). Therefore, the impedance of the asymmetric antenna is equivalent to a series combination of the impedances Z_1 and Z_2 characteristic of the individual parts when in each other's presence.

With Z_0, Z_1, and Z_2 defined in (25) to (27), it is possible to express the partial voltages V_{10} and V_{20} as follows:

$$V_{10} = V_0 \frac{Z_1}{Z_0}, \qquad V_{20} = V_0 \frac{Z_2}{Z_0}. \quad (28)$$

Substitution of (28) in (18) and (19) gives the

final expressions for current in terms of the driving voltage. Thus,

$$I_{1z} = \frac{j4\pi V_0}{\zeta_0 \Psi_1} \frac{Z_1}{Z_0}$$

$$\times \left[\frac{\sin \beta_0(h_1 - z) + \sum_{i=1}^{N} M_i(h_1, z)/\Psi_1^i}{\cos \beta_0 h_1 + \sum_{i=1}^{N} A_i(h_1)/\Psi_1^i} \right],$$

$$(0 \leq z \leq h_1) \quad (29a)$$

$$I_{2z} = \frac{j4\pi V_0}{\zeta_0 \Psi_2} \frac{Z_2}{Z_0}$$

$$\times \left[\frac{\sin \beta_0(h_0(h_2 + z) + \sum_{i=1}^{N} M_i(h_2, -z)/\Psi_2^i}{\cos \beta_0 h_2 + \sum_{i=1}^{N} A_i(h_2)/\Psi_2^i} \right],$$

$$(-h_2 \leq z \leq 0) \quad (29b)$$

$$Z_0 = Z_1 + Z_2$$

$$= \frac{-j\zeta_0}{4\pi} \left\{ \frac{\Psi_1 \left[\cos \beta_0 h_1 + \sum_{i=1}^{N} A_i(h_1)/\Psi_1^i \right]}{\sin \beta_0 h_1 + \sum_{i=1}^{N} B_i(h_1)/\Psi_1^i} \right.$$

$$\left. + \frac{\Psi_2 \left[\cos \beta_0 h_2 + \sum_{i=1}^{N} A_i(h_2)/\Psi_2^i \right]}{\sin \beta_0 h_2 + \sum_{i=1}^{N} B_i(h_2)/\Psi_2^i} \right\}. \quad (30)$$

This completes the general analysis of the asymmetrically driven antenna. It remains to express the solutions in terms of tabulated functions.

28. Functions and Parameters for Asymmetrically Driven Antennas

In order to evaluate the several integrals involved in the expressions (27.29) and (27.30) for the current in and impedance of an asymmetrically driven antenna, it is convenient to introduce the integrals

$$C_{11}(h, z) \equiv \int_0^h \cos \beta_0 z' K_1(z, z') \, dz', \quad (1a)$$

$$S_{11}(h, z) \equiv \int_0^h \sin \beta_0 z' K_1(z, z') \, dz', \quad (1b)$$

$$E_{11}(h, z) \equiv \int_0^h K_1(z, z') \, dz', \quad (1c)$$

where

$$K_1(z, z') \equiv \frac{e^{-j\beta_0 R_1}}{R_1},$$

$$R_1 \equiv \sqrt{(z - z')^2 + a^2}. \quad (2)$$

These integrals are defined and expressed in terms of the tabulated generalized sine and cosine integrals in (24.7), (24.10), and (24.12). Since

$$K_1(z, -z') = K_1(-z, z'), \quad (3)$$

it follows that

$$\int_{-h}^{0} \cos \beta_0 z' K_1(z, z') \, dz'$$

$$= \int_0^h \cos \beta_0 z' K_1(-z, z') \, dz'$$

$$= C_{11}(h, -z) = C_{12}(h, z, \Delta = \pi), \quad (4a)$$

$$-\int_{-h}^{0} \sin \beta_0 z' K_1(z, z') \, dz'$$

$$= \int_0^h \sin \beta_0 z' K_1(-z, z') \, dz'$$

$$= S_{11}(h, -z) = S_{12}(h, z, \Delta = \pi), \quad (4b)$$

$$\int_{-h}^{0} K_1(z, z') \, dz' = \int_0^h K_1(-z, z') \, dz'$$

$$= E_{11}(h, -z) = E_{12}(h, z, \Delta = \pi), \quad (4c)$$

where the functions C_{12}, S_{12}, E_{12} are defined in (24.7) and expressed in terms of tabulated functions in (24.10) and (24.12). It follows from (24.11) that

$$C_{11}(h, z) + C_{11}(h, -z) = C_a(h, z), \quad (5a)$$

$$S_{11}(h, z) + S_{11}(h, -z) = S_a(h, z), \quad (5b)$$

where $C_a(h, z)$ and $S_a(h, z)$ are the functions occurring in the formulas for the center-driven antenna and defined in (II.19.33, 34). With the notation (1) and (4), (27.23) becomes

$$F_1(h_1, z) = (\cos \beta_0 z - \cos \beta_0 h_1)\Psi_1$$
$$- C_{11}(h_1, z) + \cos \beta_0 h_1 E_{11}(h_1, z)$$
$$- C_{11}(h_2, -z) + \cos \beta_0 h_2 E_{11}(h_2, -z),$$
$$(0 \leq z \leq h_1) \quad (6a)$$

$$F_1(h_1, h_1) = -C_{11}(h_1, h_1)$$
$$+ \cos \beta_0 h_1 E_{11}(h_1, h_1) - C_{11}(h_2, -h_1)$$
$$+ \cos \beta_0 h_2 E_{11}(h_2, -h_1), \quad (6b)$$

$$G_1(h_1, z) = (\sin \beta_0 z - \sin \beta_0 h_1)\Psi_1$$
$$- S_{11}(h_1, z) + \sin \beta_0 h_1 E_{11}(h_1, z)$$
$$- S_{11}(h_2, -z) + \sin \beta_0 h_2 E_{11}(h_2, -z),$$
$$(0 \leq z \leq h_1) \quad (6c)$$

$$G_1(h_1, h_1) = -S_{11}(h_1, h_1)$$
$$+ \sin \beta_0 h_1 E_{11}(h_1, h_1) - S_{11}(h_2, -h_1)$$
$$+ \sin \beta_0 h_2 E_{11}(h_2, -h_1). \quad (6d)$$

In the special case when $h_2 = h_1$, (6a) and (6c) reduce to the following forms, if use is made of (5a, b):

$$F_1(h_1, z) = (\cos \beta_0 z - \cos \beta_0 h_1)\Psi_1 - C_a(h_1, z) + \cos \beta_0 h_1 E_a(h_1, z), \quad (7a)$$

$$G_1(h_1, z) = (\sin \beta_0 z - \sin \beta_0 h_1)\Psi_1 - S_a(h_1, z) + \sin \beta_0 h_1 E_a(h_1, z). \quad (7b)$$

These are the same as (II.21.1a, c) for the symmetric antenna with $z^i = 0, \delta = 0$.

In order to evaluate the expansion parameters Ψ_1 and Ψ_2, it is necessary to choose distribution functions for the currents in the two parts of the antenna. For this purpose the zeroth-order currents are good approximations. Therefore, let

$$\begin{aligned} f_1(z) &= \sin \beta_0(h_1 - z) \\ f_2(z) &= \sin \beta_0(h_2 + z) \end{aligned}, \quad (8a)$$

$$g_{11}(z) = \frac{\sin \beta_0(h_1 - z')}{\sin \beta_0(h_1 - z)}, \quad (8b)$$

$$g_{22}(z) = \frac{\sin \beta_0(h_2 + z')}{\sin \beta_0(h_2 + z)},$$

$$g_{12}(z) = \frac{\sin \beta_0(h_2 + z')}{\sin \beta_0(h_1 - z)}, \quad (8c)$$

$$g_{21}(z) = \frac{\sin \beta_0(h_1 - z')}{\sin \beta_0(h_2 + z)}.$$

With (8), (27.12a) becomes

$$\Psi_1(z) \sin \beta_0(h_1 - z)$$
$$= \int_0^{h_1} \sin \beta_0(h_1 - z') K_1(z, z') \, dz'$$
$$+ \int_{-h_2}^0 \sin \beta_0(h_2 + z') K_1(z, z') \, dz' \quad (9a)$$
$$= \sin \beta_0 h_1 C_{11}(h_1, z) - \cos \beta_0 h_1 S_{11}(h_1, z)$$
$$+ \sin \beta_0 h_2 C_{11}(h_2, -z)$$
$$- \cos \beta_0 h_2 S_{11}(h_2, -z).$$
$$(0 \leq z \leq h_1) \quad (9b)$$

The definition of $\Psi_1 \equiv |\Psi_1(z_r)|$ involves the choice of a reference point z_r such that Ψ_1 will be a good representation of $|\Psi_1(z)|$ over its principal, essentially constant, range. Since the contributions to $\Psi_1(z)$ are predominantly from currents near the point z, it follows that currents in the lower part of the antenna, that is, where $-h_2 \leq z \leq 0$, contribute negligibly to $\Psi_1(z)$ defined at $r = a$ on the upper part, $0 \leq z \leq h_1$, except in a small range near and at $z = 0$. The continuity of current at $z = 0$, namely, $I_{20} = I_{10} = I_0$, assures that significant contributions to $\Psi_1(z)$ with z near zero by currents in the nearest section of the lower part of the antenna cannot differ appreciably from what they would be if h_2 were equal to h_1. Therefore, the general distribution of $\Psi_1(z)$ for the asymmetrically driven antenna must be essentially similar to $\Psi_{K1}(z)$ for the center-driven antenna. Hence, the same location of the reference point z_r should be appropriate. That is,

$$\Psi_1 = \begin{cases} |\Psi_1(0)| & \text{for } \beta_0 h_1 \leq \pi/2, \\ |\Psi_1(h_1 - \lambda_0/4)| & \text{for } \beta_0 h_1 > \pi/2, \end{cases} \quad (10)$$

where $\Psi_1(z)$ is given by (9b).

The parameter Ψ_2 for the lower part of the antenna $(-h_2 \leq z \leq 0)$ is defined by (10) with (9) if subscripts 1 are replaced by 2, and 2 by 1 on h and Ψ, and $-z$ is substituted for z.

With the functions F_1 and G_1 and the expansion parameters Ψ_1 and Ψ_2 reduced to forms involving only tabulated functions, a complete first-order solution for the current at all points in and the driving-point impedance of a cylindrical antenna of small cross section asymmetrically driven by a discontinuity in scalar potential is made available.

29. Approximate Impedance of Asymmetrically Driven Antennas

The actual numerical evaluation of the general formulas for current and impedance given in Sec. 27 has not been carried out, since good approximations are readily obtained directly from the solution of the center-driven antenna, which has been evaluated numerically in considerable detail in Chapter II. The argument is straightforward.

The determination of the current in the upper part $(0 \leq z \leq h_1)$ of a cylindrical antenna driven at $z = 0$ involves the vector potential at all points on the surface of this part. Contributions to the vector potential arise from currents in the entire antenna, but, as has been pointed out before, they are large only from currents near the point z $(0 \leq z \leq h)$. Since z is confined to the upper part of the antenna in determining the current in this part, contributions by currents in the lower part are important only in a small range when z is near zero. The actual magnitude of this range can be obtained directly from computations made in Chapter II to determine the significance of a gap of length 2δ at the center. In Fig. II.20.9, for example, is plotted $|\Psi(z)|$ for $\beta_0 h = \pi/2$; this is a magnitude proportional to the vector potential at $r = a$ on the surface of the antenna and in the gap 2δ. Since the antenna is symmetric, the value at $z = 0$, the bottom of the dip, is due equally to currents in both halves of the antenna. Therefore the vector potential at $z = 0$ due to the lower half alone is only one-half the magnitude of this dip. Note that this is a value proportional to the vector potential due only to the currents in the entire lower half at a distance δ from the point of maximum current at $z = -\delta$. In Fig. 29.1 is a graph of the approximate variation of $|\Psi(z)|$ (which is proportional to the vector potential) beyond the end of a section of conductor carrying a maximum current at that end due to the current in that section. The insert shows $|\Psi(z)|$ for half of an antenna and, framed on the right, the section beyond the end represented in the main diagram. The rate of decrease of $|\Psi(z)|$ is such that at angular distances as small as 0.1 radian the magnitude is reduced to between 10 percent and 30 percent depending on the magnitude of $\beta_0 a$ as specified by the value of Ω. For small values of $\beta_0 a$ the decrease is more rapid. Note that $\beta_0 z = 0.1$ corresponds to $z = 0.0159\lambda_0$. Thus, the significant contribution to the vector potential even at $z = 0$ by currents in the lower part ($z < 0$) is confined to currents at distances not exceeding a tenth of a wavelength. Since rapid variations in the magnitude of the current in an antenna do not occur in such short distances, it is clear that the requirement $I_{10} = I_{20} = I_0$ is sufficient to assure that contributions to the vector potential at all points on the upper part of an asymmetrically driven antenna, including $z = 0$, by the actual currents in the lower part effectively do not differ from what they would be if the distribution of current were the same in the lower part as in the upper, as when $h_2 = h_1$. Evidently, currents in the lower part that are near enough to contribute appreciably to the vector potential in the upper part, that is, near $z = 0$, cannot differ very much from the currents in the adjacent upper part, since $I_{10} = I_{20}$.

This means that in determining the distributions of current in the two parts each may be analyzed separately as if each were a base-driven antenna over an infinite, perfectly conducting plane. That is, I_{1z} in (27.18) will be changed negligibly if h_2 is set equal to h_1 wherever it occurs in Ψ and in the F and G functions that contribute to the M and A functions. Similarly, I_{2z} may be determined from (27.20) by setting h_1 equal to h_2 in the corresponding integrals. Each of these problems is equivalent to the center-driven antenna analyzed in Chapter II. The current I_{1z} in the upper part is the same as in the upper part of a center-driven antenna of half-length h_1; the impedance, $Z_1 = V_{10}/I_{10}$, is one-half that of the center-driven antenna. Similarly, I_{2z} in the lower part is the same as that in the lower part of a center-driven antenna of half-length h_2, and the impedance $Z_2 = V_{20}/I_{20}$ is one-half that of the center-driven antenna. *The impedance Z_0 of the asymmetric array is the series combination of Z_1 and Z_2:*

$$Z_0 \doteq Z_1 + Z_2. \quad (1)$$

The currents in the two isolated antennas of half-length h_1 and h_2, respectively, are related to each other and to the asymmetric antenna they are to represent by setting

$$V_{10} = V_0 Z_1/Z_0, \qquad V_{20} = V_0 Z_2/Z_0. \quad (2)$$

Hence,

I_{1z}(upper part of asym. ant.)

$$\doteq \frac{Z_1}{Z_0} I_{1z}(\text{upper half of sym. ant. of half-length } h_1), \quad (3a)$$

I_{2z}(lower part of asym. ant.)

$$\doteq \frac{Z_2}{Z_0} I_{2z}(\text{lower half of sym. ant. of half-length } h_2), \quad (3b)$$

where

$$Z_1 \doteq \tfrac{1}{2}(Z_0 \text{ of sym. ant. of half-length } h_1), \quad (4a)$$

$$Z_2 \doteq \tfrac{1}{2}(Z_0 \text{ of sym. ant. of half-length } h_2). \quad (4b)$$

In general, $h_1/a_1 \neq h_2/a_2$, so that different values of Ω apply to each part. Since the input currents per unit voltage are the input admittances, it is convenient to express (3) and (4) in terms of the admittances. Thus,

I_{1z}(upper part of asym. ant.)

$$= \frac{Y_0}{Y_1} I_{1z}(\text{upper half of sym. ant. of half-length } h_1), \quad (5a)$$

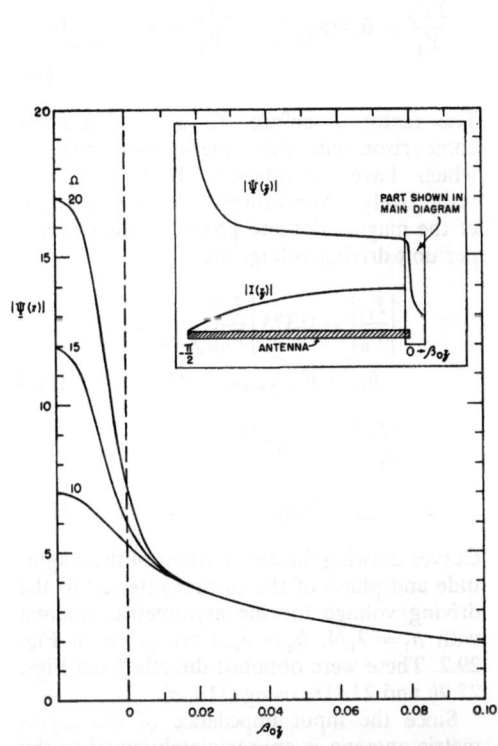

Fig. 29.1. Variation of $|\Psi(z)|$ along and beyond antenna.

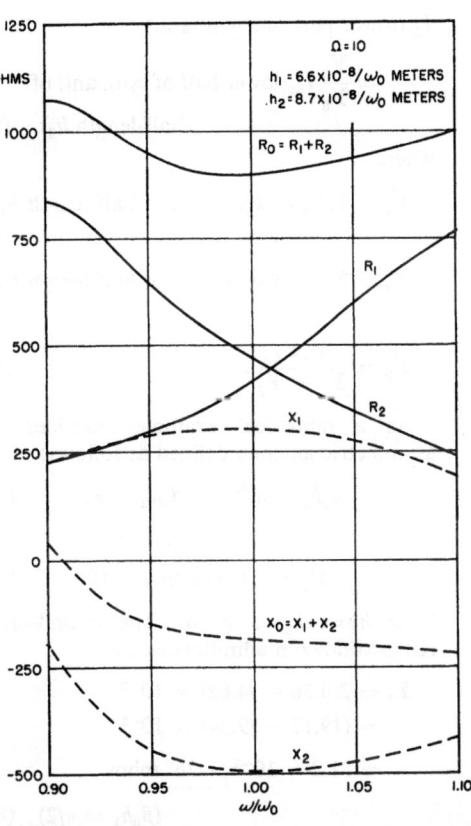

Fig. 29.3. Resistance and reactance of asymmetric antenna.

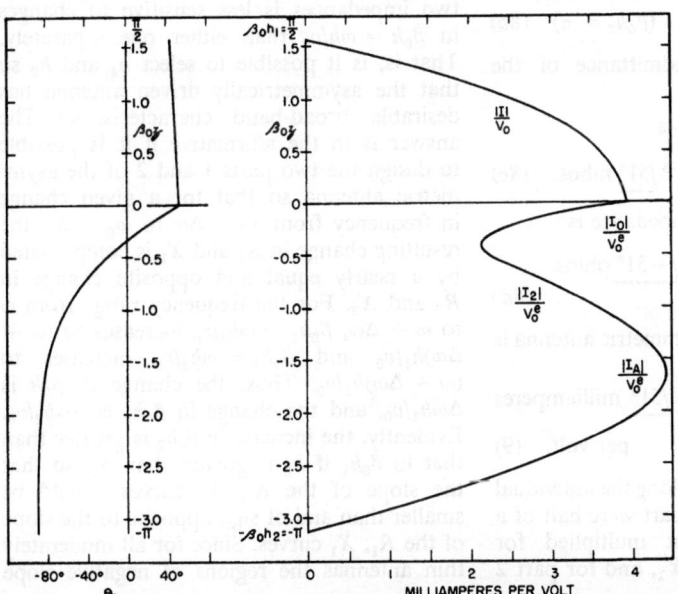

Fig. 29.2. Current on asymmetrically driven antenna.

I_{2z} (lower part of asym. ant.)

$$\doteq \frac{Y_0}{Y_2} I_{2z} \text{(lower half of sym. ant. of half-length } h_2\text{),} \quad (5b)$$

where

$$Y_1 = 2(Y_0 \text{ of sym. ant. of half-length } h_1), \quad (6a)$$

$$Y_2 = 2(Y_0 \text{ of sym. ant. of half-length } h_2), \quad (6b)$$

$$Y_0 = \frac{Y_1 Y_2}{Y_1 + Y_2}. \quad (6c)$$

As a numerical example, consider the asymmetric antenna defined as follows:

$$\beta_0 h_1 = \pi/2, \quad \beta_0 h_2 = \pi; \quad (7a)$$

$$\Omega_1 = 2 \ln (2h_1/a_1) = 10,$$

$$\Omega_2 = 2 \ln (2h_2/a_2) = 10. \quad (7b)$$

Since $h_2 = 2h_1$, $a_2 = 2a_1$. The second-order King-Middleton admittances are

$$Y_1 = 2(9.56 - j4.62) \times 10^{-3}$$
$$= (19.12 - j9.24) \times 10^{-3}$$
$$= 21.2 \times 10^{-3} /\underline{-26°} \text{ mhos,}$$

$$(\beta_0 h_1 = \pi/2) \quad (8a)$$

$$Y_2 = 2(1.0 + j1.8) \times 10^{-3}$$
$$= (2.0 + j3.6) \times 10^{-3}$$
$$= 4.12 \times 10^{-3} /\underline{61°} \text{ mhos.}$$

$$(\beta_0 h_2 = \pi) \quad (8b)$$

The approximate input admittance of the asymmetric antenna is

$$Y_0 \doteq (2.51 + j3.1) \times 10^{-3}$$
$$= 3.98 \times 10^{-3} /\underline{51°} \text{ mhos.} \quad (8c)$$

The corresponding input impedance is

$$Z_0 \doteq 158 - j195 = 251 /\underline{-51°} \text{ ohms.} \quad (8d)$$

The input current of the asymmetric antenna is

$$\frac{I_0}{V_0} = 2.51 + j3.1 = 3.98 /\underline{51°} \text{ milliamperes}$$

$$\text{per volt.} \quad (9)$$

The distribution of current along the individual parts is the same as if this part were half of a center-driven antenna, but multiplied for part 1 ($\beta_0 h_1 = \pi/2$) by Y_0/Y_1, and for part 2 ($\beta_0 h_2 = \pi$) by Y_0/Y_2. Numerical values for the special case under consideration are

$$\frac{2Y_0}{Y_1} = 0.375 /\underline{77°}, \quad \frac{2Y_0}{Y_2} = 1.932 /\underline{-10°}.$$

$$(10)$$

The factor 2 before Y_0 is necessary for comparison with the center-driven antennas, which have admittances $Y_1/2$ and $Y_2/2$ respectively. Accordingly, the distributions of the magnitudes and phases of the currents per unit driving voltage are

$$\left|\frac{I_{1z}}{V_0}\right| = 0.375 \left|\frac{I_z}{V_0}\right|_{\beta_0 h = \pi/2},$$

$$\theta_{1I} = \theta_{I,\beta_0 h = \pi/2} + 77°; \quad (11a)$$

$$\left|\frac{I_{2z}}{V_0}\right| = 1.932 \left|\frac{I_z}{V_0}\right|_{\beta_0 h = \pi},$$

$$\theta_{2I} = \theta_{I,\beta_0 h = \pi} - 10°. \quad (11b)$$

Curves showing the distributions of the magnitude and phase of the current referred to the driving voltage for the asymmetric antenna with $h_1 = \lambda_0/4$, $h_2 = \lambda_0/2$ are given in Fig. 29.2. These were obtained directly from Figs. 22.9b and 22.11b, using (11a, b).

Since the input impedance of the asymmetric antenna is approximately equal to the sum of the impedances of two antennas of different and quite arbitrary lengths, it is interesting to consider the possibility of selecting these lengths so that the sum of the two impedances is less sensitive to changes in $\beta_0 h = \omega h/v_0$ than either one separately. That is, is it possible to select h_1 and h_2 so that the asymmetrically driven antenna has desirable broad-band characteristics? The answer is in the affirmative if it is possible to design the two parts 1 and 2 of the asymmetric antenna so that for a given change in frequency from $\omega_0 - \Delta\omega$ to $\omega_0 + \Delta\omega$ the resulting change in R_1 and X_1 is compensated by a nearly equal and opposite change in R_2 and X_2. For the frequency range from ω to $\omega + \Delta\omega$, $\beta_0 h_1 = \omega h_1/v_0$ increases to $(\omega + \Delta\omega)h_1/v_0$ and $\beta_0 h_2 = \omega h_2/v_0$ increases to $(\omega + \Delta\omega)h_2/v_0$. Thus, the change in $\beta_0 h_1$ is $\Delta\omega h_1/v_0$, and the change in $\beta_0 h_2$ is $\Delta\omega h_2/v_0$. Evidently, the increase in $\beta_0 h_2$ is greater than that in $\beta_0 h_1$ if h_2 is greater than h_1, so that the slope of the R_2, X_2 curves should be smaller than and of sign opposite to the slope of the R_1, X_1 curves. Since for all moderately thin antennas the regions of negative slope for X are always much steeper than regions of positive slope, it is difficult to satisfy these conditions without using sections that are several half-wavelengths long. If both h_1 and h_2 are to be under a wavelength and Ω_1 and Ω_2 are not to differ greatly, a reasonably broad-band antenna can be constructed by choosing h_1 to maximize X_1 and h_2 to minimize X_2 at $\omega = \omega_0$. For $\Omega_1 = \Omega_2 = 10$, Fig. II.30.1a or Figs. II.30.5a, b give $h_1 = 2.2v_0/\omega_0$, $h_2 = 2.9v_0/\omega_0$, where $v_0 \doteq 3 \times 10^8$ m/sec. A range of ω defined by

$$0.9\omega_0 \leq \omega \leq 1.1\omega_0 \quad (12a)$$

corresponds to

$$1.98 \leq \beta_0 h_1 \leq 2.42, \quad (12b)$$

$$2.61 \leq \beta_0 h_2 \leq 3.19. \quad (12c)$$

Curves showing the variation of R_1 and X_1, R_2 and X_2, $R_0 = R_1 + R_2$, and $X_0 = X_1 + X_2$ are shown in Fig. 29.3 as a function of ω in the range from $0.9\omega_0$ to $1.1\omega_0$. It is seen that over this 20-percent frequency range R_0 and X_0 are reasonably constant. With a corresponding choice of h_1 and h_2 for thicker antennas with Ω_1 and Ω_2 smaller than 10, a considerable extension of this approximately constant range is possible.

The general nature of the distribution of current amplitude along an asymmetrically driven antenna as predicted in this section has been confirmed by Hatch,[29] using a method described in Sec. 36. Measurements were made for a wide range of values of h_1 and h_2 and the results indicate that the distribution in each part of the antenna is virtually unaffected by the length of the other part and is essentially the same as when both parts are equal. Since measurements were not made for a combination of lengths approximating $h_1 = \lambda_0/4$ and $h_2 = \lambda_0/2$, for which theoretical curves are in Fig. 29.2, a more direct comparison is not possible.

30. Theory of the Sleeve Dipole

An antenna of considerable importance because of its broad-band properties is the *sleeve dipole* shown in Fig. 30.1a, b. It consists essentially of the vertically extended inner conductor and the outer conductor of the coaxial line over a horizontal conducting plane. It differs from the conventional base-driven half-dipole over a conducting plane in that the sheath of the coaxial line does not end at this plane but extends above it a distance l. This is equivalent to moving the feeding point upward from $z = 0$ at the conducting plane to $z = l$. Application of the theorem of images to obtain an equivalent symmetric structure yields the internally driven antenna of Fig. 30.1c, in which two generators maintain equal voltages respectively, across the ends of two coaxial lines at $z = \pm l$. A two-wire-line equivalent of Fig. 30.1a is shown in Fig. 30.1d.

Owing to the linearity of the Maxwell equations, the principle of superposition is applicable, so that the current at any point along the antenna in Fig. 30.1c is the algebraic sum of the currents maintained independently by the generators. This is illustrated schematically in Fig. 30.2.

The determination of the currents due to the upper generator in the symmetric antenna on the left in Fig. 30.2 reduces to the analysis of the current in the middle antenna. Since this is the asymmetrically driven antenna analyzed in Secs. 27 to 29, this part of the problem is already solved. The current in the antenna on the right in Fig. 30.2 may be obtained directly from that in the middle antenna by interchanging ends. Finally, the resultant current in the antenna on the left is the algebraic sum of the currents in the other two.

The impedance at the terminals of the upper generator in the antenna on the left in Fig. 30.2 is given by the driving voltage V_0^e divided by the total current in the terminals. Thus,

$$Z_{\text{in}} = V_0^e/(I_0 + I_B) = V_0^e/(I_0 + I_A). \quad (1a)$$

The admittance is

$$Y_{\text{in}} = (I_0 + I_A)/V_0^e. \quad (1b)$$

The numerical evaluation of the impedance of and the distribution of current in a sleeve dipole is complicated by several factors. Consider first the simplest case of the symmetric antenna in Fig. 30.3a, where the driving voltages are maintained by "slice" generators or discontinuities in scalar potential. For simple numerical evaluation, let $\beta_0 h_1 = \pi/2$, $\beta_0 h_2 = \pi$ as shown. Also let $h_2/a_2 \doteq 75$ so that $\Omega_2 = 2 \ln (2h_2/a_2) = 10$. Since $h_2 = 2h_1$, it follows that, with $a_2 = a_1$, $\Omega_1 = 2 \ln (2h_1/a_1) = 9.3$. Since the impedance of an antenna of half-length $h_1 = \lambda_0/4$ varies only slowly with Ω, an error of only a few percent is made if it is assumed for simplicity that $\Omega_1 = 10$. This value corresponds to $a_1 = a_2/2$. Physically, the condition $\Omega_1 = \Omega_2 = 10$ has a

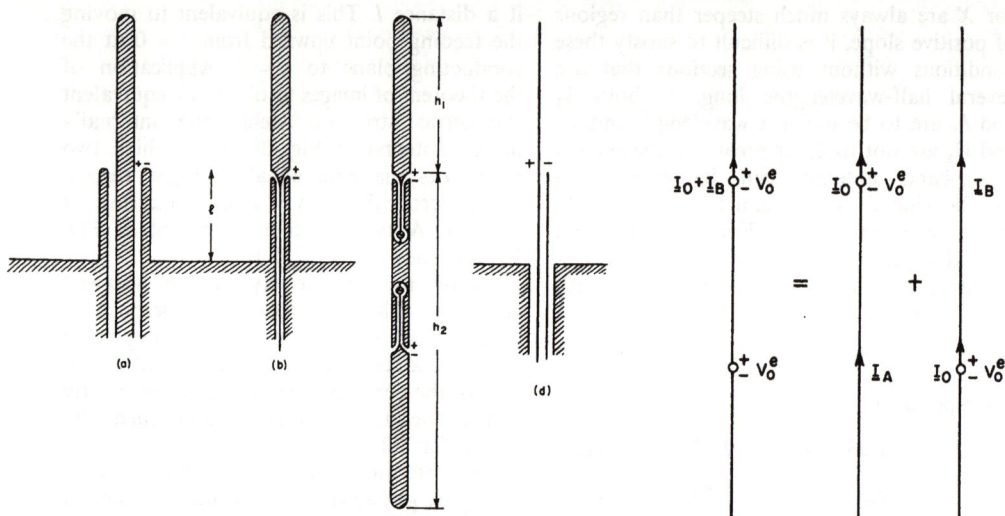

Fig. 30.1. Sleeve dipoles with image and two-wire-line equivalents.

Fig. 30.2. Application of superposition theorem to antenna with two generators.

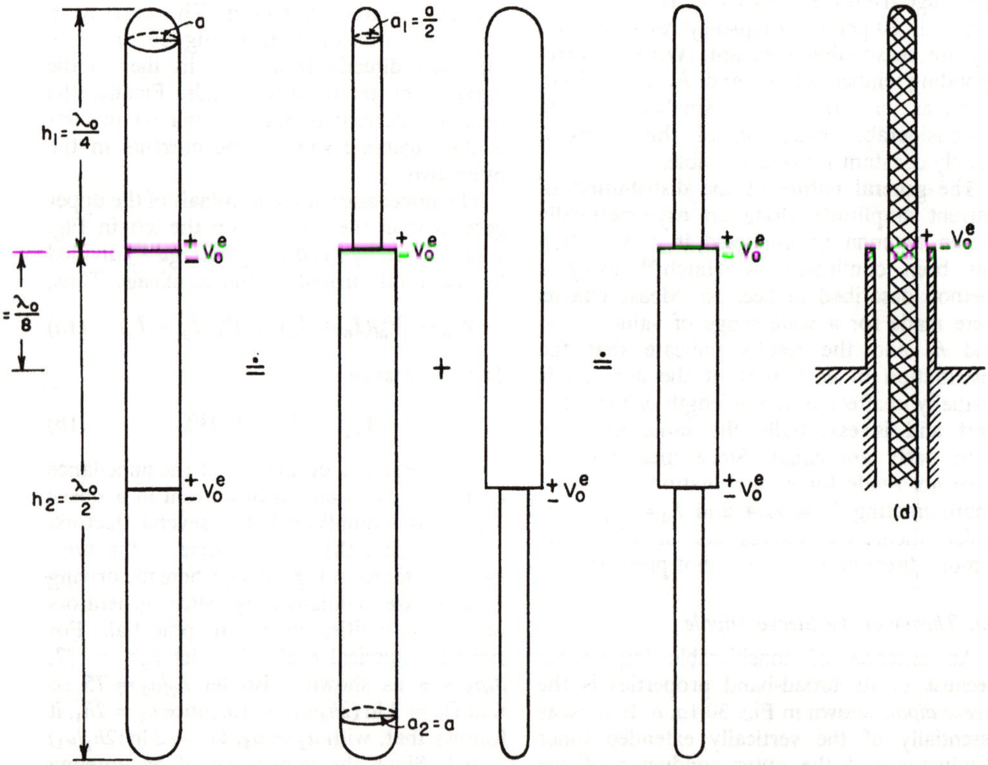

Fig. 30.3. Symmetric antenna with driving voltages maintained by "slice" generators and approximate equivalents.

meaning for the asymmetric antenna with $h_2 = 2h_1$, $a_2 = 2a_1$, but not for the antenna symmetrically driven by two equal generators. Thus, each of the two asymmetric antennas in Fig. 30.3b may be analyzed, but the superposition of their solutions does not correspond *exactly* to any physically realizable structure, since the outer thirds of each antenna cannot simultaneously have different radii. It is clear that the antenna of Fig. 30.3d also is approximated by a superposition of the two antennas in Fig. 30.3b, except for the added complication of the shoulders at the driving points, of which no account is taken in the theory. In the following, the currents in the two antennas in Fig. 30.3b will be superimposed with the understanding that the results apply approximately to both Fig. 30.3a and Fig. 30.3c if a correction is made to take account of the shoulders in the latter.

The currents I_0 and I_A in the central asymmetric antenna in Fig. 30.2 with $\Omega_1 = \Omega_2 = 10$ are obtained directly from Fig. 29.2. They are

$$\frac{I_0}{V_0^e} = 3.98\underline{/51°} = 2.51 + j3.1$$

$$\text{milliamperes/volt,} \quad (2a)$$

$$\frac{I_A}{V_0^e} = 4.4\underline{/-81°} = 0.68 - j4.34$$

$$\text{milliamperes/volt.} \quad (2b)$$

Substitution of (2) in (1b) gives the following driving-point admittance at the terminals of each slice generator in the symmetric two-generator antenna or at the terminals of the single slice generator of half of the symmetric structure erected vertically on an infinite conducting plane:

$$Y_{\text{in}} = (I_0 + I_A)/V_0^e = (3.18 - j1.24)$$
$$\times 10^{-3} \text{ mhos.} \quad (3a)$$

The corresponding input impedance is

$$Z_{\text{in}} = 273 + j106 \text{ ohms.} \quad (3b)$$

The distribution of current on the symmetric two-generator antenna is obtained by superposition. It is

$$[I(z)]_{\text{sym. ant.}} = [I(z) + I(-z)]_{\text{asym. ant.}}, \quad (4)$$

since the current in the lower part of the middle unit in Fig. 30.2 is equal to the current in the upper part in the right-hand unit. The magnitude and phase of the current in the symmetric two-generator antenna are given by

$$I = \sqrt{(I_t \cos \theta_t + I_b \cos \theta_b)^2 + (I_t \sin \theta_t + I_b \sin \theta_b)^2},$$
$$(5a)$$

$$\theta_I = \tan^{-1}\left[\frac{(I_t \sin \theta_t + I_b \sin \theta_b)}{(I_t \cos \theta_t + I_b \cos \theta_b)}\right]. \quad (5b)$$

The following notation is used:
$$I_t = I_t e^{j\theta_t} \equiv I(z),$$
$$I_b = I_b e^{j\theta_b} \equiv I(-z). \quad (6)$$

The magnitude I and the phase angle θ_I referred to V_0^e are plotted in Fig. 30.4 for $\beta_0 h_1 = \pi/2$, $\beta_0 h_2 = \pi$, and $\Omega_1 \doteq \Omega_2 \doteq 10$. The current in the central part of the antenna between the two generators is seen to be relatively great and to differ in phase from the currents in the outer parts by about 25 degrees on the average.

The admittance and impedance of a base-driven antenna of electrical length $\beta_0 h = 3\pi/4$, $\Omega = 10$, over a perfectly conducting half-space, are

$$Y_0 = (2.8 - j1.1) \times 10^{-3} \text{ mhos},$$
$$Z_0 = 310 + j122 \text{ ohms.} \quad (7)$$

These do not differ greatly from the values in (3a, b) for the same antenna when driven at a height $\lambda_0/8$ from the conducting plane.

The degree in which the impedance (3b) and the current in the upper half of Fig. 30.4 approximate the apparent terminal impedance of and the current on the sleeve dipole in Fig. 30.3d depends upon the size and the nature of the junction region where the coaxial line ends and the driving voltage is applied to the antenna. Frequently more or less extensive surfaces or edges exist on which charge can accumulate, there may be dielectric supports at or near the end of the sleeve, and, owing to the finite size of the space between inner and outer conductors, there are significant transmission-line end effects and coupling effects between the line and the antenna. Since none of these is included in the analysis of the cylindrical antenna, it is inevitable that the measured *apparent* impedance of a sleeve dipole should differ somewhat from the theoretical impedance. If the coaxial-line spacing $b - a$ is small and dielectric supports and extra chargeable surfaces provide the equivalent of a positive capacitance in parallel with antenna, the apparent resistance is less than the theoretical

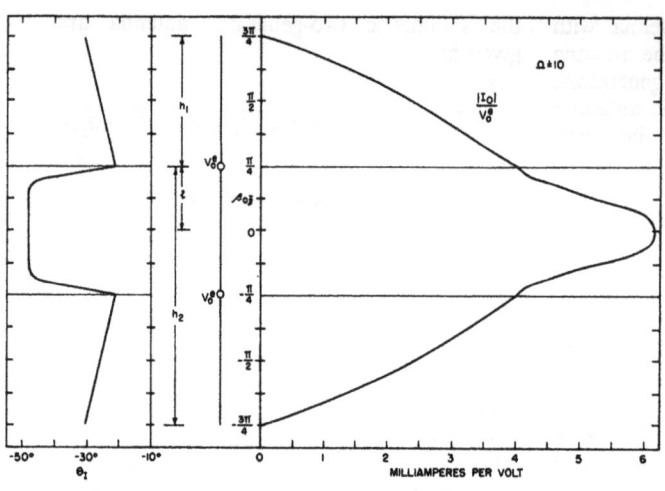

Fig. 30.4. Theoretical current in sleeve dipole with image.

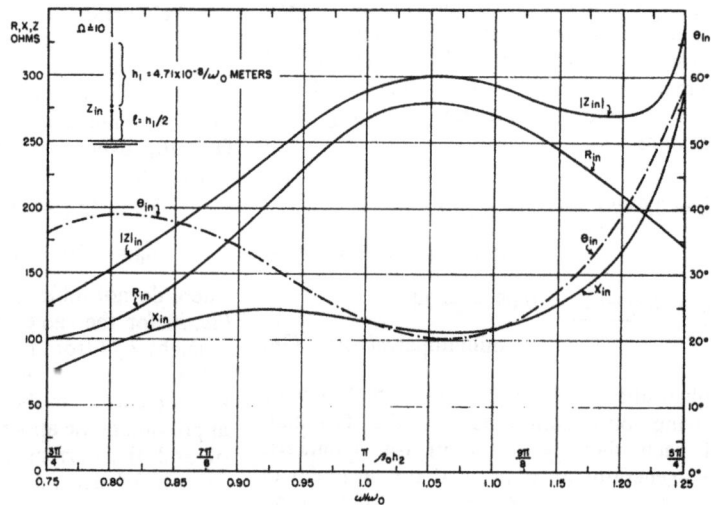

Fig. 30.5. Theoretical impedance of sleeve dipole.

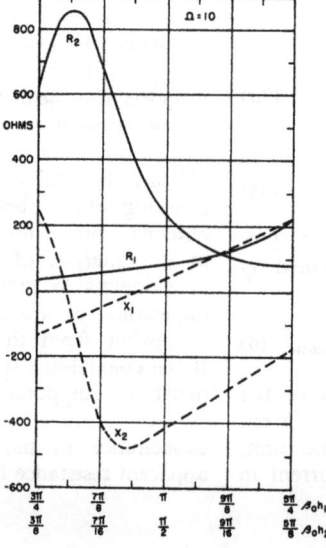

Fig. 30.6. Impedances of symmetric center-driven antennas.

value for the antenna alone, and the apparent reactance is more negative than the ideal value. For example, suppose a susceptance $B = \omega C = 3.2 \times 10^{-3}$ mhos is effectively in parallel with the admittance in (3a). The apparent admittance and impedance are

$$(Y_{\text{in}})_a = (3.18 + j1.96) \times 10^{-3} \text{ mhos,}$$
$$(Z_{\text{in}})_a = 228 - j141 \text{ ohms.} \quad (8a)$$

Note that the reactance is changed from $+106$ to -141. Alternatively, if $b - a$ is relatively large and a susceptance $B = \omega C_T = -0.51 \times 10^{-3}$ mhos is effectively in parallel with the antenna due to transmission-line end effects, the admittance (3a) and the impedance (3b) become

$$(Y_{\text{in}})_a = (3.18 - j1.75) \times 10^{-3} \text{ mhos,}$$
$$(Z_{\text{in}})_a = 241 + j133 \text{ ohms.} \quad (8b)$$

If the diameter of the sleeve is much greater than that of the remainder of the antenna, the theory necessarily is a poorer approximation than when they are more nearly equal.

The broad-band properties of the sleeve dipole cannot be investigated in general without extensive numerical computations of the input impedance for a variety of sleeve lengths and antenna lengths. However, the behavior as a function of frequency of the input impedance of the particular antenna for which the currents in Fig. 30.4 were determined may be carried out using available current and impedance data.

Suppose the frequency for which $\beta_0 h_1 = \pi/2$, $\beta_0 h_2 = \pi$, $\beta_0 l = \pi/4$ is f_0 and the associated angular velocity is ω_0. For a variation in frequency from $\omega_0 - \Delta\omega$ to $\omega_0 + \Delta\omega$, where $\Delta\omega = \omega_0/4$, the electrical lengths have the following ranges:

$$\tfrac{3}{4}\omega_0 \leq \omega \leq \tfrac{5}{4}\omega_0: \begin{cases} \dfrac{3\pi}{8} \leq \beta_0 h_1 \leq \dfrac{5\pi}{8} \\[4pt] \dfrac{3\pi}{4} \leq \beta_0 h_2 \leq \dfrac{5\pi}{4} \\[4pt] \dfrac{3\pi}{16} \leq \beta_0 l \leq \dfrac{5\pi}{16} \end{cases} \quad (9)$$

The input impedance and admittance seen by the generator at $z = l$ are given by (1a, b). In order to determine the admittance or impedance over this range, it is necessary to know the currents I_0 at the driving point $z = l$ where the impedance is to be evaluated and I_A at $z = -l$ in the asymmetric antenna. These currents are given in Sec. II.22 for $\beta_0 h = \pi/2$, $3\pi/4$, π, $5\pi/4$. It follows that in the range (9) the required currents may be determined for both extremes, $\beta_0 h_2 = 3\pi/4$, $5\pi/4$, and for the middle value $\beta_0 h_2 = \pi$. The current in the lower part 2 of the asymmetrically driven antenna is obtained from the current in the lower half of a symmetric, center-driven antenna by multiplying by $Z_2/(Z_1 + Z_2)$ where Z_1 is the impedance of a symmetric center-driven antenna of half-length h_1 and Z_2 is the impedance of a symmetric center-driven antenna of half-length h_2. These values of impedance as well as $I_0/V_0^e = Y_2$ for the symmetric antenna may be determined over the entire range of variation, namely, $3\pi/8 \leq \beta_0 h_1 \leq 5\pi/8$, $3\pi/4 \leq \beta_0 h_2 \leq 5\pi/4$, using the impedance and admittance curves of Sec. II.30. The current I_A/V_0^e can be obtained from the second-order distribution curves in Sec. II.22, but these have been determined only for $\beta_0 h = 3\pi/4$, π, and $5\pi/4$. Fortunately, a reasonable estimate of the variation of I_A/V_0^e is obtainable by plotting curves of the magnitude and phase and of the real and imaginary parts of I_A as functions of $\beta_0 h$. Using interpolated values from these curves, the approximate input impedance Z_{in} of the sleeve dipole may be determined over the specified frequency range, using the formula

$$\frac{1}{Z_{\text{in}}} = Y_{\text{in}} = 2\left(\frac{I_0 + I_A}{V_0^e}\right)\left(\frac{Z_2}{Z_1 + Z_2}\right). \quad (10)$$

Note that I_0 and I_A in (10) are the values obtained from the second-order curves of Sec. II.22 in terms of a single generator with voltage V_0^e. Since the generator at $z = l$ in the sleeve dipole is V_0^e, the limiting case with $l = 0$, gives a generator of voltage $2V_0^e$ at the center. Therefore, the factor 2 in (10) is required.

Curves showing the input impedance $Z_{\text{in}} = R_{\text{in}} + jX_{\text{in}} = Z_{\text{in}}e^{j\theta_{\text{in}}}$ of the sleeve dipole are given in Fig. 30.5 as a function of $\beta_0 h_2$, and as a function of ω/ω_0. For purposes of comparison the variation in $Z_1 = R_1 + jX_1$ and $Z_2 = R_2 + jX_2$ for symmetric center-driven antennas of electrical half-lengths in the ranges $3\pi/8 \leq \beta_0 h_1 \leq 5\pi/8$ and $3\pi/4 \leq \beta_0 h_2 \leq 5\pi/4$ are shown in Fig. 30.6. It is seen that the reactance of the sleeve dipole is remarkably constant near 110 ohms over the greater part of the range, whereas X_1 and X_2 for the center-driven dipoles vary widely over positive and negative values. The resistance R_{in} for the sleeve dipole varies within 50

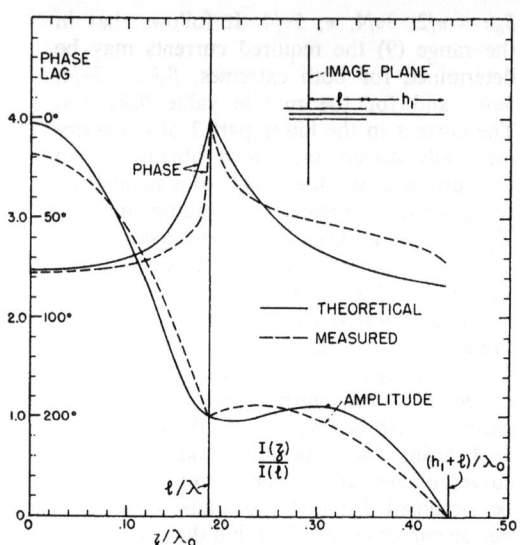

Fig. 30.7a. Current on sleeve antenna $l = 0.188\lambda, h_1 = 0.25\lambda$ (Taylor).

Fig. 30.7b. Current on sleeve antenna $l = 0.20\lambda, h_1 = 0.25\lambda$ (Taylor).

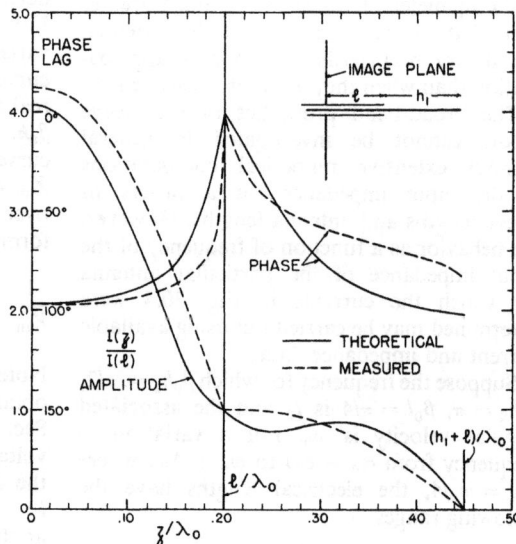

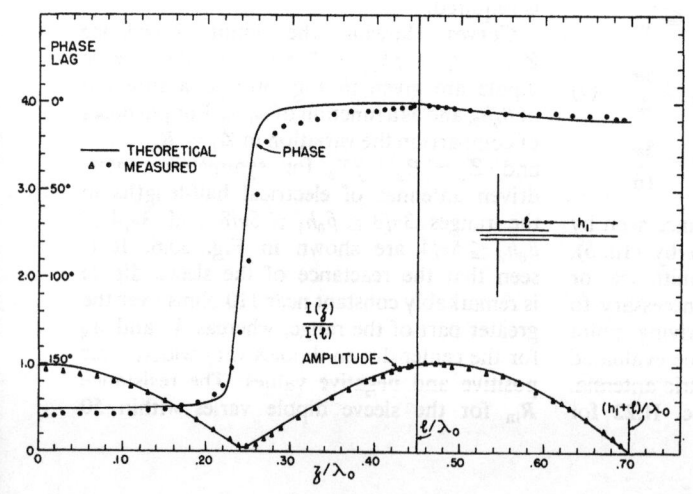

Fig. 30.7c. Current on sleeve antenna $l = 0.45\lambda, h_1 = 0.25\lambda$ (Taylor).

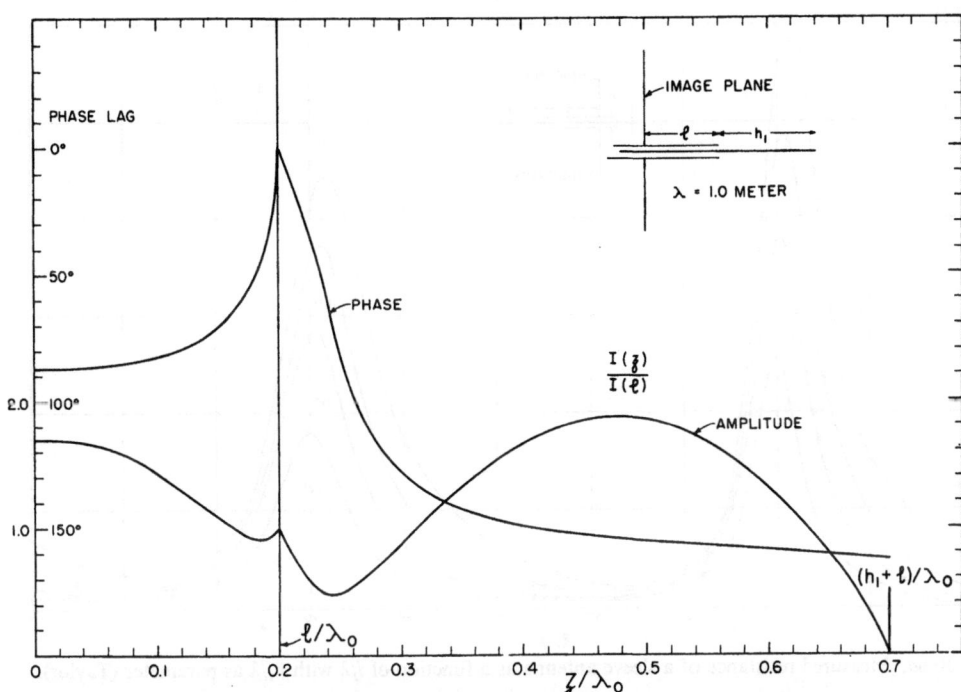

Fig. 30.7d. Current on sleeve antenna $l = 0.20\lambda$, $h_1 = 0.5\lambda$ (Taylor).

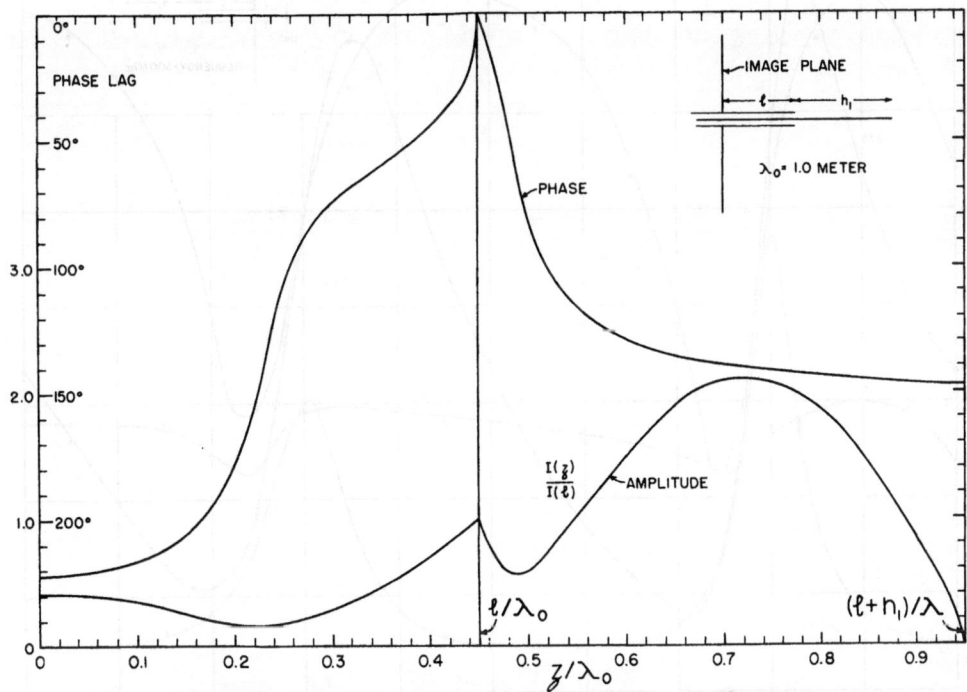

Fig. 30.7e. Current on sleeve antenna $l = 0.45\lambda$, $h_1 = 0.5\lambda$ (Taylor).

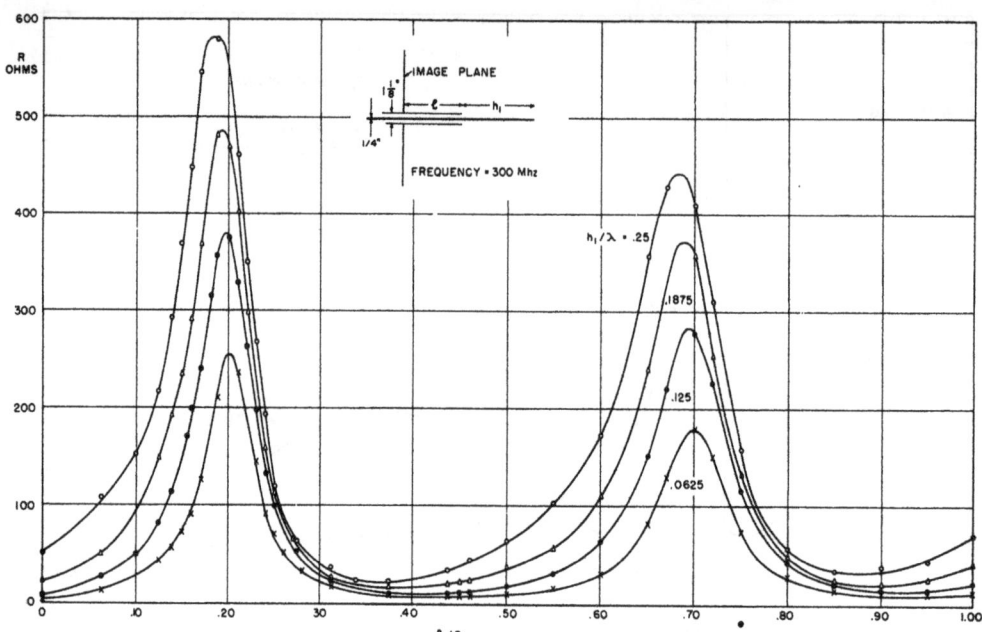

Fig. 30.8a. Measured resistance of a sleeve antenna as a function of l/λ with h_1/λ as parameter (Taylor).

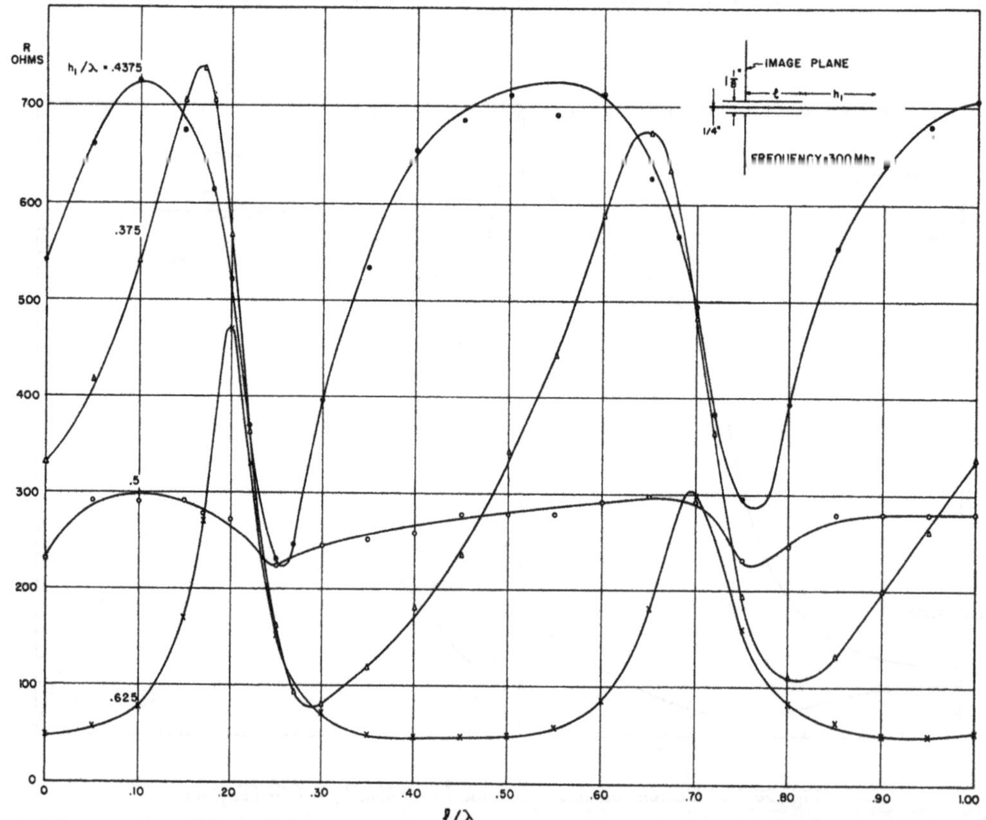

Fig. 30.8b. Measured resistance of a sleeve antenna as a function of l/λ with h_1/λ as parameter (Taylor).

[III.30] THEORY OF LINEAR ANTENNAS 415

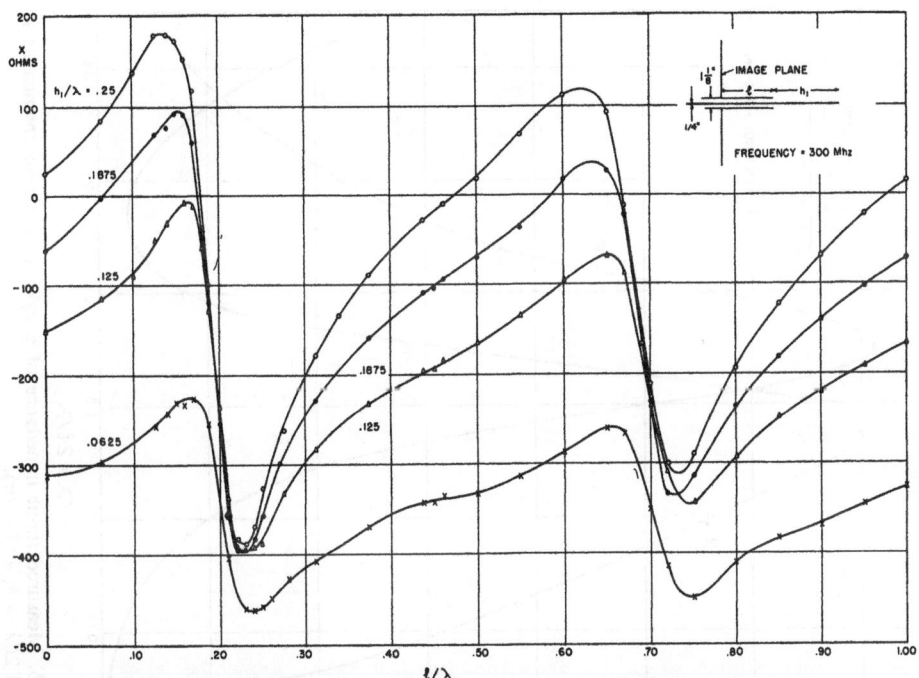

Fig. 30.9a. Measured reactance of a sleeve antenna as a function of l/λ with h_1/λ as parameter (Taylor).

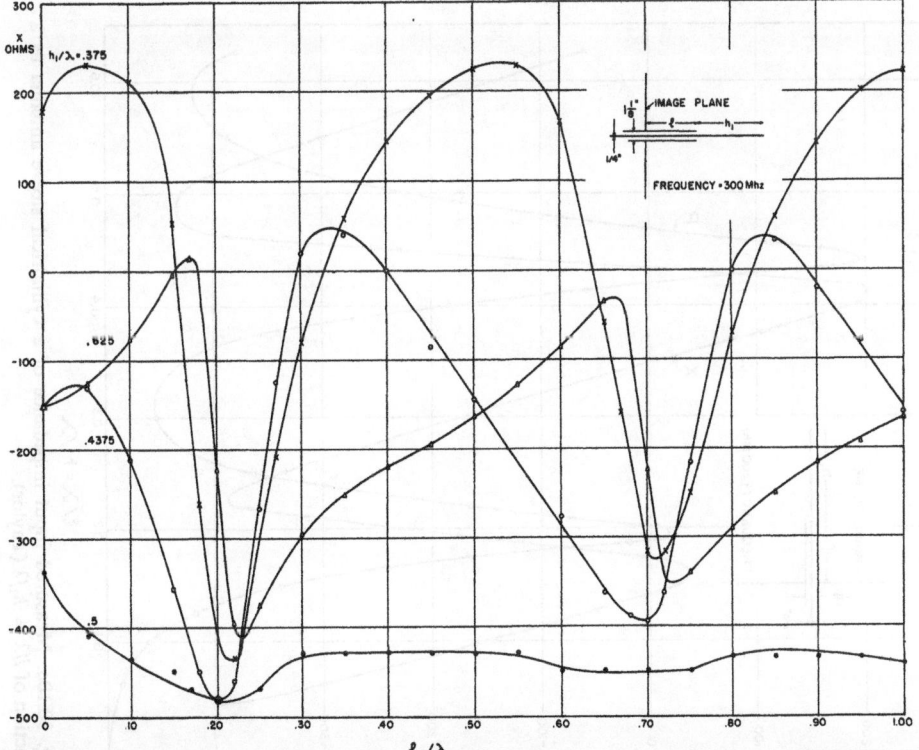

Fig. 30.9b. Measured reactance of a sleeve antenna as a function of l/λ with h_1/λ as parameter (Taylor).

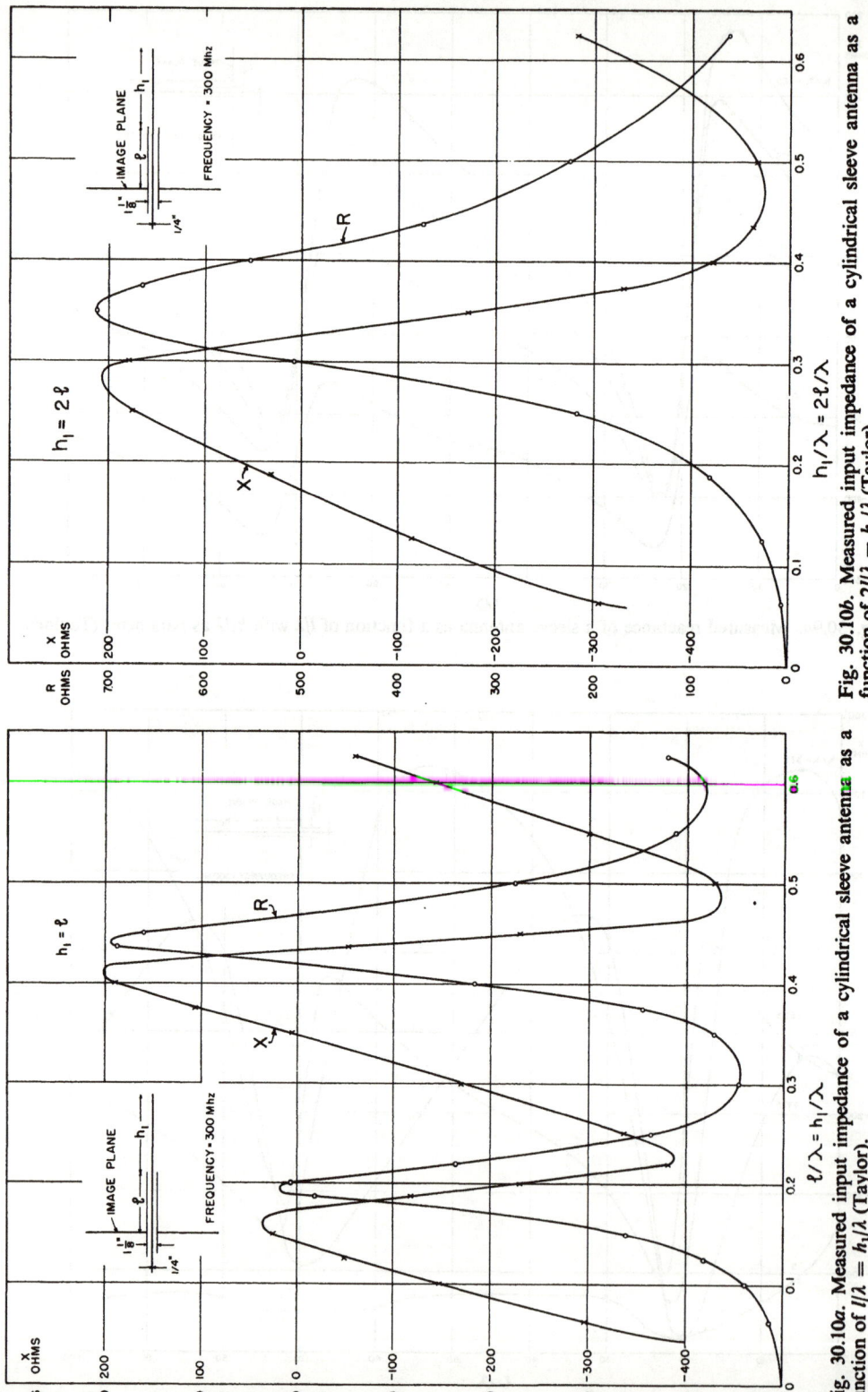

Fig. 30.10b. Measured input impedance of a cylindrical sleeve antenna as a function of $2l/\lambda = h_1/\lambda$ (Taylor).

Fig. 30.10a. Measured input impedance of a cylindrical sleeve antenna as a function of $l/\lambda = h_1/\lambda$ (Taylor).

percent of a mean value near 200 ohms, whereas R_2 ranges between 80 and 850 ohms, and R_1 varies over a range of nearly 100 percent from a mean value of about 110 ohms. Clearly the frequency response of the sleeve dipole is very much better from the point of view of broad-band operation than the response of either a half-wave or a full-wave dipole. With an appropriately designed reactive matching network the standing-wave ratio on a line terminated in the sleeve dipole could be made to remain reasonably small over a wide frequency range.

An alternative method of analyzing the sleeve dipole is due to Taylor.[53] It is an extension of Storer's variational formulation described in Sec. II.39 and makes use of the same integral equation (II.39.13). Since the sleeve dipole consists of two parts, a suitable trial function for the distribution of current is more complicated. For antennas that are restricted by the conditions

$$\beta_0(h_1 + l) < 2\pi, \quad \beta_0 l < \pi, \quad (11)$$

where, as in Fig. 30.3d, l is the distance from the ground plane at $z = 0$ to the driving point at $z = l$ and h_1 is the distance from the driving point to the end of the antenna, Taylor used the following trial functions:

$$I_{zT} = I_{lT}[1 + C(\alpha + \cos \beta_0 z)],$$
$$(0 \leq z \leq l) \quad (12a)$$

$$I_{zT} = I_{lT}\{\delta[1 - \cos \beta_0(h_1 + l - z)]$$
$$+ D[\sin \beta_0(h_1 + l - z)$$
$$+ \epsilon\{1 - \cos \beta_0(h_1 + l - z)\}]\},$$
$$(12b)$$

where

$$\alpha = -\cos \beta_0 l, \quad \delta = 1/(1 - \cos \beta_0 h),$$
$$\epsilon = -\sin \beta_0 h/(1 - \cos \beta_0 h_1), \quad (12c)$$

and where I_{lT} is the current at the driving point and C and D are complex constants. By substituting (12a) and (12b) in the integral equation (II.39.13) for the current, the approximate impedance of the sleeve dipole is obtained in the form

$$Z_{\text{in}} = \frac{j\zeta_0}{4\pi}(\gamma_0 + \gamma_c C + \gamma_D D + \gamma_{cc}C^2$$
$$+ \gamma_{DD}D^2 + \gamma_{cD}CD), \quad (13)$$

where the γ's are very long functions of $\beta_0 h_1$, $\beta_0 l$, $\beta_0 a$ and combinations of these, involving, however, only trigonometric and generalized integral sine and cosine functions that are available in tabular form. Since Z_{in} given in (13) is a polynomial in C and D, the optimum values of these constants are determined from $\partial Z_{\text{in}}/\partial C = 0$, $\partial Z_{\text{in}}/\partial D = 0$ to be

$$C = \frac{\gamma_D \gamma_{CD} - 2\gamma_C \gamma_{DD}}{4\gamma_{CC}\gamma_{DD} - \gamma_{CD}^2}, \quad (14a)$$

$$D = \frac{\gamma_C \gamma_{CD} - 2\gamma_D \gamma_{CC}}{4\gamma_{CC}\gamma_{DD} - \gamma_{CD}^2}. \quad (14b)$$

Substitution of (14a) and (14b) in (12a, b) and (13) gives Taylor's final formulas for the current I_z and the input impedance Z_{in}. When $l = 0$, (13) with (14a, b) yields exactly Storer's results for the simple dipole as given in Sec. 39.

Unfortunately, the numerical evaluation of the γ's in (13) and (14a, b) is so formidable that it has been carried out for only a single pair of values of $\beta_0 l$ and $\beta_0 h_1$. For this reason Taylor introduced simpler trial functions for the currents in the two parts. These are less accurate, and they are useful *only when* $\beta_0 h_1$ does not exceed $\pi/2$. They are

$$I_{zT} = I_{zl}[1 + C(\cos \beta_0 z - \cos \beta_0 h)],$$
$$(0 \leq z \leq l) \quad (15a)$$

$$I_{zT} = I_{zl}\frac{\sin \beta_0(h_1 + l - z)}{\sin \beta_0 h_1},$$
$$(l \leq z \leq h_1 + l) \quad (15b)$$

and they lead to $C = -\gamma_c/2\gamma_{cc}$ and

$$Z_a = \frac{j\zeta_0}{4\pi}\left(\gamma_0 - \frac{\gamma_c^2}{4\gamma_{cc}}\right), \quad (16)$$

where the γ's are a new set of functions of $\beta_0 h_1$, $\beta_0 l$, and $\beta_0 a$ that again involve only trigonometric and generalized sine and cosine integrals.

Extensive experimental measurements of currents and apparent impedances have been made by Taylor for sleeve antennas of various over-all lengths $h_1 + l$ and sleeve lengths l. In order to make amplitude and phase measurements of the currents on the antenna structure and on the coaxial line feeding the sleeve, small shielded-loop probes were provided to move protruding from slots on the outside and inside of the shield and on the extended part of the antenna. The outer diameter of the necessarily rather thick sleeve of length l was $1\frac{1}{8}$ in; the diameter of the extended antenna of length h_1 was $\frac{1}{4}$ in. At the operating frequency of 300 Mhz these correspond to electrical radii of 0.09 and 0.02. The apparatus and ground screen were essentially the same as those of Morita described in Sec. II.26.

Measured relative magnitudes and phases of current and corresponding theoretical values determined from the short formula (15a, b) are shown in Figs. 30.7a, b, c, d for different values of l and h_1. Theoretical values have been fitted to the experimental ones at the driving point, $z = l$. The general agreement is good.

Measured apparent impedances are shown in Figs. 30.8a, b and 30.9a, b with sleeve length l/λ_0 as variable and the length h_1/λ_0 of the extended section as parameter. In Figs. 30.10a, b the variables are $h_1/\lambda_0 = l/\lambda_0$ and $h_1/\lambda_0 = 2l/\lambda_0$.

The large difference between the radius of the sleeve ($\beta_0 a = 0.09$) and the radius of the extended section ($\beta_0 a = 0.02$) and the need of selecting values for use in the theoretical formulas led Taylor to make trial computations of the input impedance using both $\beta_0 a = 0.05$ and $\beta_0 a = 0.10$ with the short formula (16) and $\beta_0 a = 0.05$ with the long formula (13). (The value 0.10 was used instead of 0.09 as actually intended owing to the availability of more extensive tables of needed functions.) The results of these computations for a sleeve antenna with $\beta_0 h_1 = \pi/2$ are shown in Figs. 30.11a, b. It is seen that better agreement with the experimental curve is achieved with $\beta_0 a = 0.10$ than with $\beta_0 a = 0.05$ and that the long formula gives much better results. The agreement should be quite satisfactory if the value $\beta_0 a = 0.09$ of the sleeve (or 0.10) were used in the long formula. Note, however, that the theoretical values and the measured apparent values are not strictly comparable, since a correction for the relatively large value of $b - a$ at the driving point has not been made. Extensive computations of Z_{in} using the short formula have been made by Taylor for a range of electrical lengths $\beta_0 h_1$ with $\beta_0 a$ taken as 0.10. The agreement with experiment is about the same as for $\beta_0 h_1 = \pi/2$ in Figs. 30.9a, b. Whereas the short formula is adequate for qualitative purposes, the long formula evidently is required for reasonable quantitative accuracy.

Comparison of the theory developed earlier in this section with experiment is not possible for distribution of current, since experimental results for the particular case computed, $\beta_0 l = \pi/4$, $\beta_0 h_1 = \pi/2$, are unavailable. However, the theoretical curves of Fig. 30.4 appear to fit into the general pattern of the distributions in Figs. 30.7a, b, c, d.

The theoretical impedance $Z_{in} = 273 + j106$ in (3b) differs considerably from the measured apparent impedance, $(Z_{in})_a \doteq 220 + j170$ obtained from Figs. 30.9a, b. It is interesting to note the effect of an approximate correction for the large value of $b - a$ at the driving point. In order to determine the effective lumped negative capacitance C_T to be connected in parallel with the theoretical admittance $Y_{in} = (3.18 - j1.24) \times 10^{-3}$ mhos in (3a) as a correction for transmission-line end effects and coupling effects, the general procedure described in Secs. II.8, 9, 10 should be followed. For present purposes the order of magnitude may be approximated by using the value for $l = 0$, that is, for the base-driven half-dipole. Although when l is not zero the electric charges cannot spread out on a flat conducting plane at the edge of the gap but are confined to the outer surface of the sleeve, the mean distances involved in determining the scalar potential difference are only slightly less. Therefore C_T must be of the same order of magnitude as when $l = 0$. Using Fig. II.10.9 for $b/a \doteq 2.75$, $b = 0.86$ cm, and $\omega = 2\pi \times 300 \times 10^6$, the value $\omega C_T = -0.51 \times 10^{-3}$ mhos is obtained, so that $(Y_{in})_a$ and $(Z_{in})_a$ have precisely the values given in (8b). These values are shown in Figs. 30.9a, b and are seen to be in good agreement with the experimental curves if it is remembered that the theoretical analysis uses $\beta_0 a = 0.02$ or $\Omega_1 = \Omega_2 = 10$, whereas in the actual antenna this value is correct only for the outer length h_1, and $\beta_0 a = 0.09$ for $h_2 = h_1 + 2l$, so that $\Omega_2 = 8.5$. Hence the difference between $241 + j133$ and $220 + j170$ does not appear unreasonable. It is evident that an apparatus could be constructed for measuring impedances with sleeves and outer sections much more nearly equal than in Taylor's set-up. Measurements on an antenna like that in Fig. 30.1c would provide a better quantitative correlation between theory and experiment.

31. Half Dipole with Multielement Counterpoise; Ground-Plane Antenna

Important generalizations from the right-angled, apex-driven V-antenna shown in Fig. 31.1a and analyzed in Secs. 23 to 25 are the asymmetric antennas shown in Figs. 31.1b and 31.1c. These antennas consist essentially of a single, vertical half-dipole

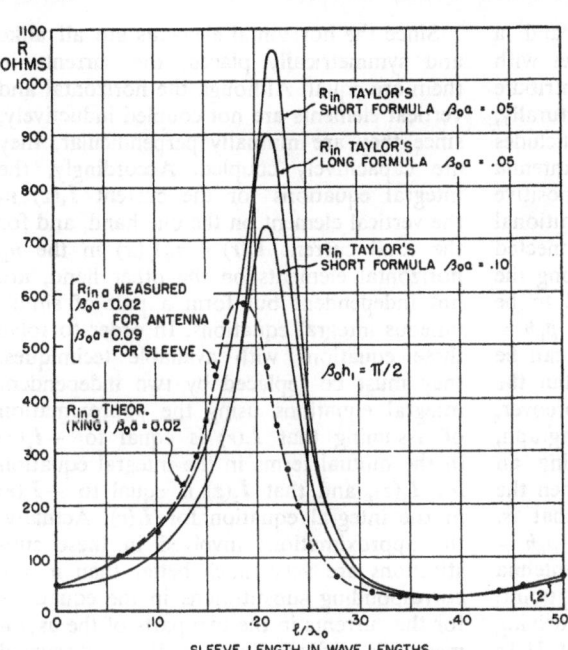

Fig. 30.11a. Resistance of a sleeve antenna as a function of l/λ_0 with $h_1 = \lambda_0/4$.

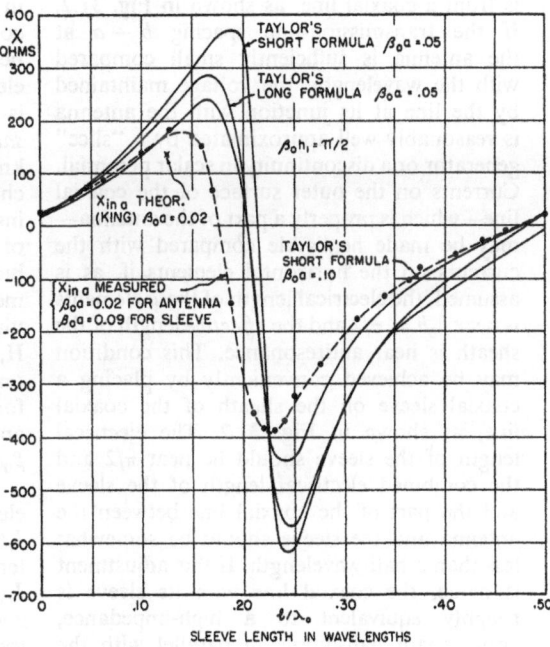

Fig. 30.11b. Reactance of a sleeve antenna as a function of l/λ_0 with $h_1 = \lambda_0/4$.

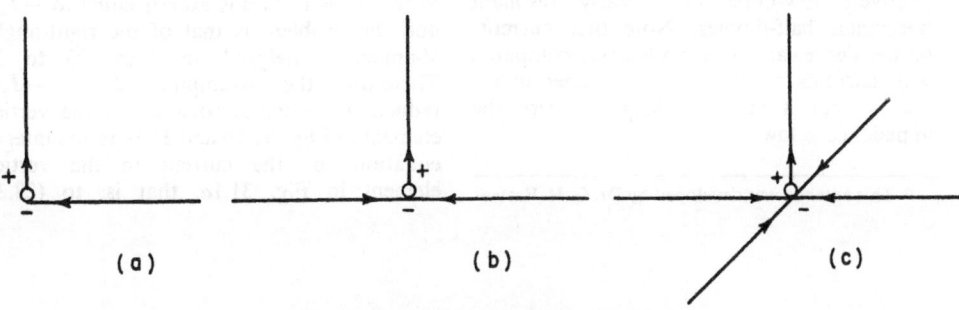

Fig. 31.1. Generalized right-angled V-antenna.

that maintains the radiation field and a counterpoise of horizontal half-dipoles with equal and opposite currents that contribute very little to the radiation field. Structurally, the so-called ground-plane antenna* includes the two legs of the right-angled V-antenna extending, respectively, along the positive z- and x-axes, and one or more additional legs lying in the xy-plane and connected electrically *in parallel* with the leg along the positive x-axis. All legs are assumed to be identical and of electrical length near $\beta_0 h = \pi/2$. Actually the theoretical analysis can be carried out for other values of $\beta_0 h$, but the approximations are not so good. Moreover, as explained in the following paragraph, the most common method of driving an antenna of this type is useful only when the impedance of the antenna is low, that is, when the elements have lengths near $\beta_0 h = \pi/2$. This is due to the fact that the antenna is asymmetric with respect to the driving point.

The usual practical method of driving antennas like those in Figs. 31.1b and 31.1c is from a coaxial line, as shown in Fig. 31.2. If the transmission-line spacing $b - a$ at the antenna is sufficiently small compared with the wavelength, the voltage maintained by the line at its junction with the antenna is reasonably well approximated by a "slice" generator or a discontinuity in scalar potential. Currents on the outer surface of the coaxial line—which is properly a part of the antenna—may be made negligible compared with the currents on the horizontal elements if, as is assumed, the electrical length of these elements is near $\beta_0 h = \pi/2$ and the *effective* length of the sheath is near antiresonance. This condition may be achieved conveniently by placing a coaxial sleeve on the sheath of the coaxial line, as shown in Fig. 31.2. The electrical length of the sleeve should be near $\pi/2$ and the combined electrical length of the sleeve and the part of the coaxial line between the antenna and the sleeve should be somewhat less than a half wavelength. If this adjustment is made, the coaxial line with its sleeve is roughly equivalent to a high-impedance, antiresonant half-dipole in parallel with the relatively low-impedance, nearly resonant horizontal half-dipoles. Note that currents on the sleeve cannot be made small compared with currents on the antenna proper unless $\beta_0 h$ is near a resonant length, where the impedance is low.

---

* This antenna was developed by Dr. G. H. Brown.

Since the horizontal elements are all alike and symmetrically placed, the currents in them are equal. Although the horizontal and vertical elements are not coupled inductively, since they are mutually perpendicular, they are capacitively coupled. Accordingly, the integral equations for the current $I_z(z)$ in the vertical element on the one hand, and for the total current $I_r(r) = n_H I_x(x)$ in the n_H horizontal elements on the other hand, are not independent but form a pair of simultaneous integral equations. In order to solve these equations with available techniques, they must be replaced by two independent integral equations using the approximation of assuming that $I_r(r)$ is equal to $-I_z(z)$ in the mutual terms in the integral equation for $I_z(z)$, and that $I_z(z)$ is equal to $-I_r(r)$ in the integral equation for $I_r(r)$. Actually, the approximations involved in these substitutions are very much better than in the corresponding substitutions in the equations for the currents in the two parts of the asymmetrically driven linear radiator discussed in Secs. 27 to 29. This follows in part from the complete absence of inductive coupling between the vertical element and the horizontal elements as a group, but the principal reason is found in the fact that all elements have the *same* electrical length, $\beta_0 h = \pi/2$. It is well known that the distributions of current and charge on antennas near resonance are quite insensitive to the location and distribution of the driving forces. For example, the distributions of current and charge in the symmetric antenna center-driven by a discontinuity in scalar potential, as analyzed in Chapter II, differ very little from the distributions on the receiving antenna excited by a uniformly distributed field along the entire antenna, as analyzed in Chapter IV, *provided* $\beta_0 h$ is near $\pi/2$.

In so far as the current $I_z(z)$ in the vertical element is concerned, the assumption that $I_r(r)$ may be set equal to $-I_z(z)$ in the mutual terms is exactly equivalent to assuming that $I_r(r)$ is independent of the number of horizontal elements. That is, $I_r(r)$ is approximately the same for $n_H = 2$ or 4 as for $n_H = 1$. When $n_H = 1$, $I_r(r)$ is *exactly* equal to $-I_z(z)$ and the problem is that of the right-angled V-antenna analyzed in Secs. 23 to 25. Therefore, the assumption $I_r(r) = -I_z(z)$ reduces the integral equation for the current in the vertical element in Figs. 31.1b and 31.1c to the integral equation for the current in the vertical element in Fig. 31.1a, that is, to (23.31) with $V_{\delta P}$ substituted for $\tfrac{1}{2}V_\delta$, and with $\beta_0 h = \pi/2$. Here $V_{\delta P}$ is the part of the driving voltage V_δ that is maintained across the impedance $Z_{\delta P}$ associated with the perpendicular element. It follows that the current in the vertical element is given by (24.3) with V_δ replaced by $2V_{\delta P}$ and $\beta_0 h = \pi/2$; evidently, the impedance of the perpendicular element is given by

$$Z_{\delta P} = \tfrac{1}{2} Z_{v\delta}, \qquad (1)$$

where $Z_{v\delta}$ is given by (25.3) in general, and by (25.5) in the zeroth-order approximation.

The integral equation for the current $I_r(r)/n_H$ for a typical horizontal element is readily constructed on the pattern of (23.31). Since the equations for $n_H = 2$ and $n_H = 4$ differ slightly, they are written separately; in both equations $\beta_0 h = \pi/2$.

$n_H = 2$:
$$\tfrac{1}{2}\int_\delta^h I_r(r')K_H(r, r')\, dr'$$
$$= \frac{-j4\pi}{\zeta_0 \cos \beta_0 \delta}[C_1 \cos \beta_0(r - \delta)$$
$$- V_{\delta H} \sin \beta_0 r] + 2J_v(r), \quad (2a)$$

$n_H = 4$:
$$\tfrac{1}{4}\int_\delta^h I_r(r')K_H(r, r')\, dr'$$
$$= \frac{-j4\pi}{\zeta_0 \cos \beta_0 \delta}[C_1 \cos \beta_0(r - \delta)$$
$$- V_{\delta H} \sin \beta_0 r] + 2J_v(r). \quad (2b)$$

In (2a, b) $V_{\delta H}$ is the part of the driving voltage V_δ maintained across the impedance $Z_{\delta H}$ of the parallel combination of all the horizontal elements; the negative sign is required since the positive direction of the voltage maintains a current in the radially inward or negative r direction. The particular integral $J_v(r)$ is the same as (23.33b) with r substituted for z and $\beta_0 h = \pi/2$. The kernel in (2a, b) is defined by

$$K_H(r, r') = \frac{e^{-j\beta_0 R_1}}{R_1} - \frac{e^{-j\beta_0 R_2}}{R_2}, \quad (3a)$$

where
$$R_1 = \sqrt{(r - r')^2 + a^2},$$
$$R_2 = \sqrt{(r + r')^2 + a^2}. \quad (3b)$$

It is obtained using the relation

$$\left(-\int_{-h}^{-\delta} + \int_\delta^h\right) I_r(r') \frac{e^{-j\beta_0 R_1}}{R_1}\, dr'$$
$$= \int_\delta^h I_r(r')\left[\frac{e^{-j\beta_0 R_1}}{R_1} - \frac{e^{-j\beta_0 R_2}}{R_2}\right] dr'. \quad (4)$$

The negative sign before the first sign of integration in (4) results from the fact that the current in the element along the negative x-axis is opposite in direction to the current along the positive x-axis. The factor $1/n_H = 1/2$ or $1/4$ expresses the fact that the current in the typical element is $I_r(r')/n_H$. The particular integral $J_v(r)$ in (2a) is due to capacitive coupling to the vertical element. Since it is related to the typical horizontal element exactly as in the right-angled V-antenna, it appears in (2a) just as in (23.31) but with a factor 2, since the current in the vertical element is double that in each horizontal element, and with an appropriate change in the symbol used for the independent variable. The approximation $I_z(z) = -I_r(r)$ is implied. In (2b) the term $2J_v(r)$ results from the combination $[4J_v(r) - J_v(r) - J_v(r)]$. Here the first term is the particular integral due to capacitive coupling to the vertical element in which the current and charge are four times those in each horizontal element. The two terms $-J_v(r)$ are due to the two horizontal elements perpendicular to the typical one. Since the currents and charges in them are opposite in sign to those in the vertical element, the negative sign must appear.

The solution of (2a) and (2b) for $I_r(r)$ follows exactly the sequence of steps described in Secs. 23 and 24. The integral equations (2a) and (2b) differ from (23.31) only in having a different kernel and different numerical factors for some of the terms. Since the distribution of current on elements of electrical length $\beta_0 h = \pi/2$ is known to be very nearly cosinusoidal, no great interest attaches to a determination of the distributions of current, and, therefore, a more detailed formulation is omitted. It is readily supplied if required.

The impedance $Z_{\delta H} = V_{\delta H}/[-I_r(\delta)]$ is given by an expression like (25.1) with appropriately defined functions and parameters and with $\beta_0 h = \pi/2$. With $\delta = 0$ a formula like (25.3) is applicable. The zeroth-order impedance is given by an expression like (25.5) with $\Delta = \pi$ and with the sign of $C_{12}(h, h, \Delta)$ changed since the current in the coupled collinear half is reversed. The zeroth-order impedances are

$n_H = 2$:
$$Z_{0H} = \frac{j\zeta_0}{8\pi}\left[C_{11}(h, h) - C_{12}(h, h, \pi)\right.$$
$$\left. - 2J_v\left(\frac{\pi}{2}\right)\right], \quad (5a)$$

$n_H = 4$:
$$Z_{0H} = \frac{j\zeta_0}{16\pi}\left[C_{11}(h, h) - C_{12}(h, h, \pi)\right.$$
$$\left. - 2J_v\left(\frac{\pi}{2}\right)\right]. \quad (5b)$$

The corresponding zeroth-order impedance of the vertical element when coupled to the horizontal elements is

$$Z_{0P} = \frac{j\zeta_0}{4\pi}\left[C_{11}(h, h) - J_v\left(\frac{\pi}{2}\right)\right]. \quad (6)$$

The impedance Z_{0M} of the entire antenna is the series combination of Z_{0H} and Z_{0P}. This follows from

$$Z_{0H} \equiv \frac{V_{0H}}{-I_r(0)}, \qquad Z_{0P} \equiv \frac{V_{PH}}{I_z(0)},$$

$$V_0 = V_{0H} + V_{0P}, \qquad Z_{0M} \equiv V_0/I(0). \quad (7)$$

Thus,

$$n_H = 2: \quad Z_{0M} \equiv Z_{0H} + Z_{0P}$$

$$= \frac{j\zeta_0}{8\pi}\left[3C_{11}(h, h) - C_{12}(h, h, \pi)\right.$$

$$\left. - 4J_v\left(\frac{\pi}{2}\right)\right], \quad (8a)$$

$$n_H = 4: \quad Z_{0M} \equiv Z_{0H} + Z_{0P}$$

$$= \frac{j\zeta_0}{16\pi}\left[5C_{11}(h, h) - C_{12}(h, h, \pi)\right.$$

$$\left. - 6J_v\left(\frac{\pi}{2}\right)\right]. \quad (8b)$$

Numerical results are computed using the following values for $\beta_0 h = \pi/2$:

$$C_{11}(h, h) = 0.926 - j0.824, \quad (9a)$$

$$C_{12}(h, h, \pi) = -(0.217 + j0.395), \quad (9b)$$

$$J_v\left(\frac{\pi}{2}\right) = 0.70 - j0.14. \quad (9c)$$

The value for $J_v(\pi/2)$ was obtained graphically and is, therefore, less accurate than the C functions. With (9a) to (9c) substituted in (8a) and (8b), the impedances become

$$n_H = 2: \quad Z_{0M} = 22.7 + j2.92 \text{ ohms}, \quad (10a)$$

$$n_H = 4: \quad Z_{0M} = 21.6 + j4.85 \text{ ohms}. \quad (10b)$$

Note that the corresponding impedance of the vertical element over an infinite, perfectly conducting plane is

$$Z_0 \doteq \frac{j\zeta_0}{4\pi} C_a(h, h)$$

$$= \frac{j\zeta_0}{4\pi}[C_{11}(h, h) + C_{12}(h, h, \pi)]$$

$$= 36.6 + j21.3 \text{ ohms}. \quad (11)$$

The contribution to (10a) and (10b) by the vertical element alone is the first term in (6). Numerically its value is

$$Z_{0P} \text{ (isolated)} \doteq \frac{j\zeta_0}{4\pi} C_{11}(h, h)$$

$$= 24.7 + j27.8 \text{ ohms}. \quad (12)$$

Accordingly the contribution by the vertical element to the total power radiated is over 90 percent.

The zeroth-order impedance was evaluated by Bouwkamp[8] using a different method that also involved an integral that was evaluated graphically. Bouwkamp's results are

$$n_H = 2: \quad Z_{0M} = 21.7 - j1.7 \text{ ohms}, \quad (13a)$$

$$n_H = 4: \quad Z_{0M} = 20.9 + j1.4 \text{ ohms}. \quad (13b)$$

The small differences between (10a, b) and (13a, b) may be ascribed to differences in the graphically evaluated integral which contributes a significant fraction to the final result. Both sets of results, (10a, b) and (13a, b), agree in assigning a smaller resistance but a more positive reactance to the antenna with four horizontal elements than to the antenna with two horizontal elements. Bouwkamp evaluated the impedances of the two antennas in Figs. 31.1b and 31.1c over a range of values near $\beta_0 h = \pi/2$. His data are plotted in Fig. 31.3.

COLLINEAR ARRAYS

32. Three-Element Collinear Array; General Analysis of Central Antenna

The analysis of two individually driven collinear antennas like those shown in Fig. 32.1a is complicated by the fact that the transmission lines cannot be arranged so that their axes are in neutral planes of the electromagnetic field. As a result, the forces acting on currents and charges in the two conductors of each line are not equal and opposite even at distances from the antennas that are very great compared with the normal terminal zone near each junction. Therefore, the transmission lines are not balanced, and the codirectional currents on each pair of conductors (or on the outside of the shield if a coaxial line is used) contribute significantly to the electromagnetic field at distant points. The actual radiating system in the circuit of Fig. 32.1a is not limited to the two collinear antennas but includes the transmission lines as well. Clearly, such a system is not practically useful since its circuit and field properties depend on the length and orientation of the feeder lines.

[III.32] THEORY OF LINEAR ANTENNAS 423

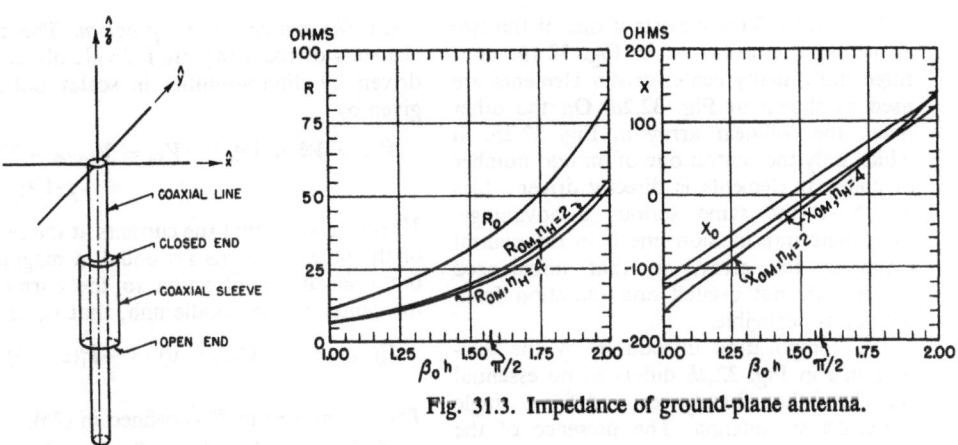

Fig. 31.2. Ground-plane antenna driven from line with high-impedance sheath.

Fig. 31.3. Impedance of ground-plane antenna.

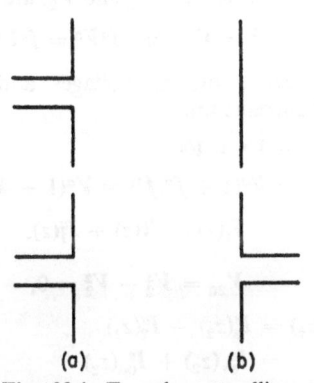

Fig. 32.1. Two-element collinear arrays.

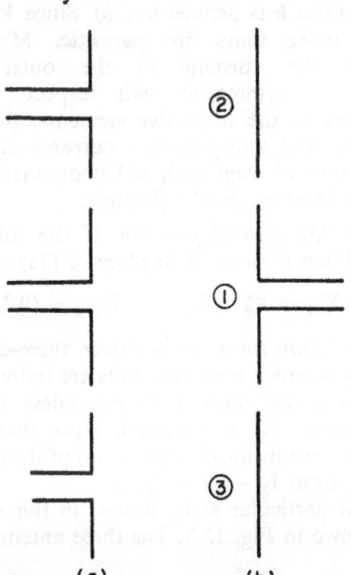

(a)   (b)
Fig. 32.2. Three-element collinear arrays.

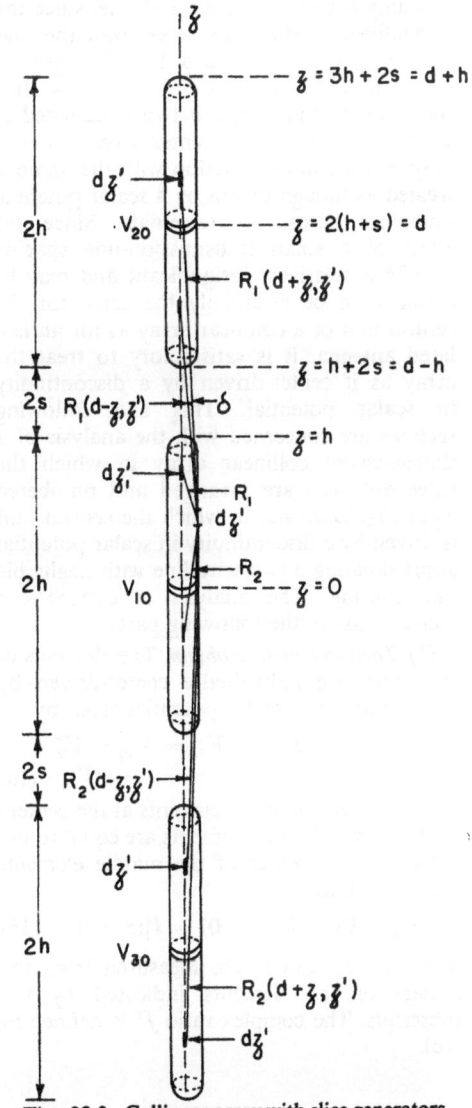

Fig. 32.3. Collinear array with slice generators.

The same difficulty exists if one of the two elements is parasitic, as in Fig. 32.1*b*, or if three individually center-driven elements are used as shown in Fig. 32.2*a*. On the other hand, the collinear array in Fig. 32.2*b*, in which only the central one of an odd number of collinear elements is directly driven, does not have the same serious disadvantage. The single transmission line is in the neutral plane of the array, so that unbalanced currents are not excited and radiation from the line is negligible.

The complication introduced by the two-wire line in Fig. 32.2*b* differs in no essential way from that already resolved for a single center-driven antenna. The presence of the two parasitic elements has no effect on the coupling between antenna and line, since this is confined to short distances from the line-antenna junction of the order of magnitude of ten times the line spacing. It follows that the same terminal-zone network designed in Chapter II for a single center-driven antenna may be used in conjunction with the antenna treated as though driven by a scalar potential difference across its terminals. Since the effect of a small transmission-line spacing $b = 2\delta$ is relatively insignificant and may be assumed to be essentially the same for the central unit of a collinear array as for an isolated antenna, it is satisfactory to treat the array as if center-driven by a discontinuity in scalar potential. This and following sections are concerned with the analysis of a three-element collinear array in which the three antennas are arranged and numbered as in Fig. 32.2*b* but in which the central unit is driven by a discontinuity in scalar potential approximating a two-wire line with negligible line spacing. The analysis is carried out conveniently in the following parts:

(1) *The symmetric problem.* The elements of the array are individually center-driven by discontinuities in scalar potential given by

$$V_{10} = V_1^s \equiv V^s, \qquad V_{20} = V_{30} \equiv V_2^s$$
$$\equiv f^s V^s; \quad (1a)$$

V_2^s is so chosen that the currents at the centers of the identical outer elements are equal to the current at the center of the middle element, that is, so that

$$I_2^s(z_2 = 0) = I_3^s(z_3 = 0) = I_1^s(z = 0), \quad (1b)$$

where z_1, z_2, and z_3 are measured from the centers of the elements indicated by the subscripts. The complex ratio f^s is defined in (1*a*).

(2) *The antisymmetric problem.* The three elements of the array are individually center-driven by discontinuities in scalar potential given by

$$V_{10} = V_1^a \equiv V^a, \qquad V_{20} = V_{30} \equiv -V_2^a$$
$$= -f^a V^a; \quad (2a)$$

V_2^a is so chosen that the currents at the centers of the outer elements are equal in magnitude but opposite in direction to the current at the center of the middle unit, that is, so that

$$I_2^a(z_2 = 0) = I_3^a(z_3 = 0) = -I_1^a(z = 0).$$
$$(2b)$$

The complex ratio f^a is defined in (2*a*).

(3) *The parasitic array.* By superimposing the solutions for the currents in problems (1) and (2) in the special case when the symmetric and antisymmetric driving voltages of the outer units, namely, V_2^s and V_2^a, are equal,

$$V_2^a = V_2^s \quad \text{or} \quad f^a V^a = f^s V^s, \quad (3a)$$

the effective driving voltages and driving-point currents are

$$V_{10} = V^s + V^a$$
$$= V^a(1 + f^a/f^s) \equiv V^a(1 + k), \quad (3b)$$

$$I_1(z) = I_1^s(z) + I_1^a(z), \quad (3c)$$

and
$$V_{20} = V_2^s - V_2^a = 0, \quad (3d)$$

$$I_2(z_2) = I_2^s(z_2) - I_2^a(z_2)$$
$$= I_{ev}^s(z_2) + I_{od}^s(z_2)$$
$$- I_{ev}^a(z_2) - I_{od}^a(z_2). \quad (3e)$$

The ratio k is defined in (3*b*). Since $V_{20} = 0$, the outer units are parasitic. Moreover, since the currents in the outer units are not symmetric with respect to the centers of the respective elements, the symmetric and antisymmetric currents are each the sum of even and odd components that must be determined separately.

(4) *The general problem.* If the following condition is satisfied in place of (3*a*) or (3*d*):

$$V_{20} = V_2^s - V_2^a = -I_2(z_2 = 0)Z_L, \quad (4)$$

the solution for a center-driven three-element array in which the outer units are individually center-loaded with a dimensionless lumped impedance Z_L is obtained. Note that symmetry permits the determination of the current $I_3(z_3)$ from $I_3(-z_3) = I_2(z_2)$.

The particular array studied in this section is shown in Fig. 32.3. The three antennas are identical, each of half-length h and radius a. The conditions

$$a \ll h, \quad \beta_0 a \ll 1 \qquad (5a)$$

are postulated. The distance between centers of the antennas is d and the distance between adjacent ends is $2s$, so that

$$d = 2(h + s). \qquad (5b)$$

The origin of a cylindrical system of coordinates r, θ, z, is at the center of the middle unit; the z-axis coincides with the axes of the cylinders. The centers of the three units are at $z = 0, \pm d$. Each unit is center-driven by the following discontinuities in scalar potential. Referring to Fig. 32.3, the conditions for the symmetric or in-phase drive are

$$V_{10} = \lim_{\delta \to 0} [\Phi(\delta) - \Phi(-\delta)]$$
$$= V_1^s \equiv V^s, \qquad (6a)$$

$$V_{20} = \lim_{\delta \to 0} [\Phi(d + \delta) - \Phi(d - \delta)]$$
$$= V_2^s = f^s V^s, \qquad (6b)$$

$$V_{30} = \lim_{\delta \to 0} [\Phi(-d + \delta) - \Phi(-d - \delta)]$$
$$= V_2^s = f^s V^s, \qquad (6c)$$

where V_2^s or f^s is chosen so that (1b) is satisfied, that is,

$$I_3^s(-d) = I_2^s(d) = I_1^s(0). \qquad (6d)$$

Similarly, the conditions for the antisymmetric or 180° out-of-phase drive are

$$V_{10} = V_1^a \equiv V^a,$$
$$V_{20} = V_{30} = -V_2^a = -f^a V^a, \qquad (7a)$$

where V_2^a or f^a is chosen so that (2b) is satisfied, that is,

$$I_3^a(-d) = I_2^a(d) = -I_1^a(0). \qquad (7b)$$

Since the array is symmetric with respect to its center, for both sets of driving conditions, the following relations of symmetry obtain:

$$I(-z) = I(z), \quad q(-z) = -q(z), \qquad (8a)$$
$$A(-z) = A(z), \quad \Phi(-z) = -\Phi(z). \qquad (8b)$$

The subscript z is omitted on the current and vector potential, since only z-components are involved. It is assumed that the vector and scalar potentials are evaluated at $r = a$ on the surfaces of the conductors. They are given by

$$A(z) = \frac{1}{4\pi v_0} \left(\int_0^h + \int_{d-h}^{d+h} \right) I(z') K(z, z') \, dz', \qquad (9a)$$

$$\Phi(z) = \frac{1}{4\pi \epsilon_0} \left(\int_0^h + \int_{d-h}^{d+h} \right) q(z') K(z, z') \, dz', \qquad (9b)$$

where

$$K(z, z') \equiv K_1(z, z') + K_2(z, z')$$
$$\equiv \frac{e^{-j\beta_0 R_1}}{R_1} + \frac{e^{-j\beta_0 R_2}}{R_2} \qquad (9c)$$

and where

$$R_1 \equiv \sqrt{(z - z')^2 + a^2},$$
$$R_2 \equiv \sqrt{(z + z')^2 + a^2}. \qquad (9d)$$

The differential equation for the vector potential on the conductors is the same as for a single antenna, namely,

$$\frac{\partial^2 A(z)}{\partial z^2} + \beta_0^2 A(z) = j\frac{\beta_0^2}{\omega} z^i I(z). \quad (r = a) \qquad (10a)$$

The solution consists of three equations of the form

$$A(z) = \frac{-j}{v_0} (C_1 \cos \beta_0 z + C_2 \sin \beta_0 z), \quad (r = a) \qquad (10b)$$

with different constants C_1 and C_2 in the equations applying, respectively, to the three sections of conductor extending from 0 to h, $d - h$ to d, d to $d + h$. A particular integral involving z^i as a factor has been omitted from (10b), since it usually contributes negligibly. When desired it may be added.

Since the distributions of current in the upper half of the central unit, the lower half of the upper unit, and the upper half of the upper unit are all three different, the rigorous solution for the currents in the antennas is complicated. Fortunately, an approximate solution may be obtained by making assumptions that are reasonable and that lead to integral equations like that for a single antenna. The steps involved in the determination of the current and impedance for the central unit are considered in this section. Currents in the outer units are discussed in Sec. 33.

The current $I_1(z)$ in the upper half of the central unit depends primarily upon the following integral equation, valid at $r = a$ with $0 \leq z \leq h$:

$$J_a \pm (J_b + J_c)$$
$$= \frac{-j4\pi}{\zeta_0} (C_1 \cos \beta_0 z + \tfrac{1}{2} V_{10} \sin \beta_0 z). \qquad (11)$$

For the upper sign, $V_{10} = V^s$; for the lower sign, $V_{10} = V^a$. In (10),

$$J_a = \int_0^h I_1(z')K(z, z')\,dz', \quad (12a)$$

$$J_b = \int_{d-h}^d I_2(z')K(z, z')\,dz'$$

$$= \int_0^h I_2(d - z')K(z, d - z')\,dz', \quad (12b)$$

$$J_c = \int_d^{d+h} I_2(z')K(z, z')\,dz'$$

$$= \int_0^h I_2(d + z')K(z, d + z')\,dz'. \quad (12c)$$

The current $I_1(z)$ in the central unit occurs also in two equations similar to (11) with integrals evaluated, respectively, over the halves of antenna 2 in the ranges $d - h \leq z \leq d$, $d \leq z \leq d + h$. With (11) these constitute three simultaneous integral equations in the currents in the upper half of antenna 1 and the halves of antenna 2.

Equation (11) was obtained by substituting (9a) in (10b), evaluating $C_2 = \frac{1}{2}V_{10}$ as in Chapter II, and replacing v_0/v_0 by its equivalent, $\zeta_0 \doteq 376.7$ ohms. The last integrals in (12b) and (12c) are derived from the first integrals by changing the variable of integration to u, by setting $z' = d - u$ in (12b) and $z' = d + u$ in (12c), and then substituting z' for u. Note that each integral in (12a) to (12c) may be expanded into two integrals in the form $J_{i1} + J_{i2}$, $i = a, b, c$, by setting

$$K(z, v) = K_1(z, v) + K_2(z, v)$$

$$= \frac{e^{-j\beta_0 R_1(z,v)}}{R_1(z, v)} + \frac{e^{-j\beta_0 R_2(z,v)}}{R_2(z, v)}, \quad (13a)$$

where

$$R_1(z, v) = \sqrt{(z - v)^2 + a^2},$$

and

$$R_2(z, v) = \sqrt{(z + v)^2 + a^2}, \quad (13b)$$

$$v = z', \quad d - z', \quad d + z'. \quad (13c)$$

Since the point z, where the vector potential is evaluated, is on the surface of the upper half of antenna 1 between $z = 0$ and $z = h$, the contributions to the total vector potential by the integrals in (12a) to (12b) differ widely. Although the distributions of the currents in the six halves of the three antennas are not all the same, they are sufficiently alike to have the same average order of magnitude if made equal at the centers of the antennas, as required in (6d) and (7b). Consequently, in determining the current specifically in the central unit, the currents in all three units may be assumed to be the same. Since the currents for $z < 0$ are obtained from the currents for $z > 0$ using (8a), the following expressions relate all currents:

$$I_2^s(d - z') \doteq I_1^s(z') \doteq I_2^s(z' + d),$$
$$(0 \leq z' \leq h) \quad (14a)$$

$$-I_2^a(d - z') \doteq I_1^a(z') \doteq -I_2^a(z' + d).$$
$$(0 \leq z' \leq h) \quad (14b)$$

With (14a, b), (11) becomes

$$\int_0^h I(z')K_c(z, z')\,dz'$$

$$= \frac{-j4\pi}{\zeta_0}(C_1 \cos \beta_0 z + \tfrac{1}{2}V_{10} \sin \beta_0 z).$$
$$(r = a, \ 0 \leq z \leq h) \quad (15)$$

When $V_{10} = V^s$, it follows that $I = I^s$, $C_1 = C_1^s$, $K_c = K_c^s$; similarly, when $V_{10} = V^a$, it follows that $I = I^a$, $C_1 = C_1^a$, $K_c = K_c^a$. The new kernel, $K_c(z, z') = K_{c1}(z, z') + K_{c2}(z, z')$, is defined as follows:

$$K_c(z, z') \equiv \begin{cases} K_c^s(z, z') \\ K_c^a(z, z') \end{cases}$$

$$= K(z, z') \pm [K(z, d - z') + K(z, d + z')], \quad (16)$$

with $K(z, v)$ defined as in (13a). Note that with (13b)

$$R_1(z, d - z') = \sqrt{(z - d + z')^2 + a^2}$$
$$= R_1(d - z, z'), \quad (17a)$$

$$R_1(z, d + z') = \sqrt{(z - d - z')^2 + a^2}$$
$$= R_2(d - z, z'), \quad (17b)$$

$$R_2(z, d - z') = \sqrt{(z + d - z')^2 + a^2}$$
$$= R_1(d + z, z'), \quad (17c)$$

$$R_2(z, d + z') = \sqrt{(z + d + z')^2 + a^2}$$
$$= R_2(d + z, z'). \quad (17d)$$

It follows that

$$K(z, d - z') + K(z, d + z') = K(d - z, z') + K(d + z, z'), \quad (18)$$

so that

$$K_c(z, z') = K(z, z') \pm [K(d - z, z') + K(d + z, z')]. \quad (19a)$$

This form of the kernel is more convenient in the evaluation of the expansion parameter.

Since (15) is an integral equation like that for the single antenna in Chapter II (with $\delta = 0$, $z^i = 0$), the same formal solution is obtained with the new kernel

$$K_{c1}(z, z') = K_1(z, z') \pm [K_1(d - z, z')$$
$$+ K_1(d + z, z')]$$

$$= \frac{e^{-j\beta_0 R_1(z,z')}}{R_1(z, z')} \pm \left[\frac{e^{-j\beta_0 R_1(d-z,z')}}{R_1(d - z, z')}\right.$$
$$\left. + \frac{e^{-j\beta_0 R_1(d+z,z')}}{R_1(d + z, z')}\right] \quad (19b)$$

replacing $K_1(z, z')$. It follows that the current in the upper half of the central unit of the collinear array is given by (II.21.1) with $K_{c1}(z, z')$ substituted for $K_1(z, z')$. Thus,

$$I_1(z) = \frac{j2\pi V}{\zeta_0 \Psi_{Kc}}$$
$$\times \left[\frac{\sin \beta_0(h - z) + M_{c1}(z)/\Psi_{Kc} + \cdots}{\cos \beta_0 h + A_{c1}/\Psi_{Kc} + \cdots}\right]. \quad (20)$$

Similarly, the impedance of the central unit is

$$Z_{0c} = \frac{-j\zeta_0 \Psi_{Kc}}{2\pi}\left[\frac{\cos \beta_0 h + A_{c1}/\Psi_{Kc} + \cdots}{\sin \beta_0 h + B_{c1}/\Psi_{Kc} + \cdots}\right]. \quad (21)$$

Each of these formulas applies to the symmetrically driven array when a superscript s is affixed to V, I, M, B, A, Ψ, and Z and the upper sign is used, and to the antisymmetrically driven array when a superscript a is affixed to these symbols and the lower sign is used. The quantities $M_{c1}(z)$, $B_{c1} = M_{c1}(0)$, and A_{c1} are the same as in Chapter II with $K_{c1}(z, z')$ in (19b, c) substituted for $K_1(z, z')$ and with Ψ_{Kc} replacing Ψ_{K1}.

The expansion parameter Ψ_{Kc} is defined as is Ψ_{K1} in Sec. II.20. The steps include the definitions of the following functions:

$$\Psi_{Kc}(z) = \psi_{Kc}(z)/\sin \beta_0(h - z), \quad (22a)$$

$$\psi_{Kc}(z) = \int_0^h \sin \beta_0(h - z') K_c(z, z')\, dz', \quad (22b)$$

$$= \sin \beta_0 h \int_0^h \cos \beta_0 z' K_c(z, z')\, dz'$$
$$- \cos \beta_0 h \int_0^h \sin \beta_0 z' K_c(z, z')\, dz'. \quad (22c)$$

The integrals in (22c) involve only the functions $C_a(h, v)$ and $S_a(h, v)$ defined in (II.19.3, 4). Their introduction yields:

$$\psi_{Kc}(z) = \sin \beta_0 h\{C_a(h, z) \pm [C_a(h, d - z)$$
$$+ C_a(h, d + z)]\}$$
$$- \cos \beta_0 h\{S_a(h, z) \pm [S_a(h, d - z)$$
$$+ S_a(h, d + z)]\}, \quad (23)$$

$$\psi_{Kc}(z) = \psi_1(z) \pm [\psi_1(d - z) + \psi_1(d + z)], \quad (24)$$

where, as in (II.20.26) with $\delta = 0$,

$$\psi_1(v) = C_a(h, v) \sin \beta_0 h - S_a(h, v) \cos \beta_0 h. \quad (25)$$

By far the most important case in practice is with $\beta_0 h = \pi/2$ or $h = \lambda_0/4$, for which

$$\psi_{Kc}(z) = C_a(h, z) \pm [C_a(h, d - z)$$
$$+ C_a(h, d + z)]. \quad (26)$$

The function of $C_a(h, z)$ for $\beta_0 h = \pi/2$ is given in Figs. II.20.1 and II.20.2 for $z \leq h$. It is given in Table 32.1 and Fig. 32.4 with a much more extensive range. The function defined in (22a) has the following form:

$$\Psi_{Kc}(z) = \Psi_{K1}(z) \pm [\Psi_{K1}(d - z)$$
$$+ \Psi_{K1}(d + z)], \quad (27)$$

where $\Psi_{K1}(z)$ is the function for the isolated central unit as defined in Chapter II. Since the general behavior of $\Psi_{Kc}(z)$ as a function of z is like that of $\Psi_{K1}(z)$ for all possible values of d, it follows that the expansion parameter for the central unit of the collinear antenna may be defined as for the isolated antenna. Specifically, let

$$\Psi_{Kc} \equiv \begin{cases} |\Psi_{Kc}(0)| = |\Psi_{K1}(0) + 2\Psi_{K1}(d)| \\ \qquad \text{for } \beta_0 h \leq \pi/2, \\ |\Psi_{Kc}(h - \lambda_0/4)| = |\Psi_{K1}(h - \lambda_0/4) \\ \pm [\Psi_{K1}(d - h + \lambda_0/4) \\ + \Psi_{K1}(d + h - \lambda_0/4)]| \\ \qquad \text{for } \beta_0 h \geq \pi/2. \end{cases} \quad (28)$$

A superscript s is added to Ψ_{Kc} for $V = V^s$, and the upper signs are used; a superscript a is added for $V = -V^a$, and the lower signs are used. The functions Ψ_{Kc}^s and Ψ_{Kc}^a are shown in Fig. 32.5.

It is clear from (28) that, when the elements of the collinear array are all driven in phase,

TABLE 32.1. The function $C_a(h, z)$ for $h = \lambda_0/4$.

n	$\frac{n\pi}{20} = \beta_0 z$	Real part				Imaginary part (same for all values of Ω)
		$\Omega = 10$	$\Omega = 12.5$	$\Omega = 15$	$\Omega = 20$	
0	0	8.3495	10.8491	13.3515	18.3515	−1.8518
1	0.1571	8.2484	10.7158	13.1869	18.1257	−1.8447
2	.3142	7.9420	10.3194	12.6974	17.4534	−1.8233
3	.4712	7.4427	9.6717	11.8977	16.3539	−1.7880
4	.6283	6.7662	8.7887	10.8113	14.8573	−1.7403
5	.7854	5.9321	7.6996	9.4677	13.0035	−1.6784
6	.9425	4.9655	6.4348	7.9043	10.8439	−1.6032
7	1.0996	3.9019	5.0364	6.1716	8.4421	−1.5220
8	1.2566	2.7816	3.5534	4.3543	5.8716	−1.4287
9	1.4137	1.6625	2.0529	2.4440	3.2263	−1.3260
10	1.5708	0.6879	0.7030	0.7073	0.7089	−1.2188
11	1.7278		0.2631			−1.1057
12	1.8850		.0146			−0.9872
13	2.042		− .1562			− .8681
14	2.199		− .2853			− .7435
15	2.356		− .3801			− .6258
16	2.513		− .4474			− .5097
17	2.670		− .4931			− .3953
18	2.827		− .5170			− .2864
19	2.985		− .5239			− .1837
20	3.142		− .5188			− .0886
21	3.299		− .4970			− .0015
22	3.456		− .4666			.0757
23	3.613		− .4279			.1445
24	3.770		− .3802			.2027
25	3.927		− .3268			.2506
26	4.084		− .2714			.2881
27	4.241		− .2128			.3151
28	4.398		− .1528			.3323
29	4.555		− .0941			.3399
30	4.712		− .0371			.3383
31	4.869		.0156			.3286
32	5.027		.0638			.3112
33	5.184		.1122			.2865
34	5.341		.1519			.2623
35	5.498		.1856			.2289
36	5.655		.2123			.1856
37	5.812		.2315			.1452
38	5.969		.2455			.1075
39	6.126		.2528			.0612
40	6.283		.2517			.0205

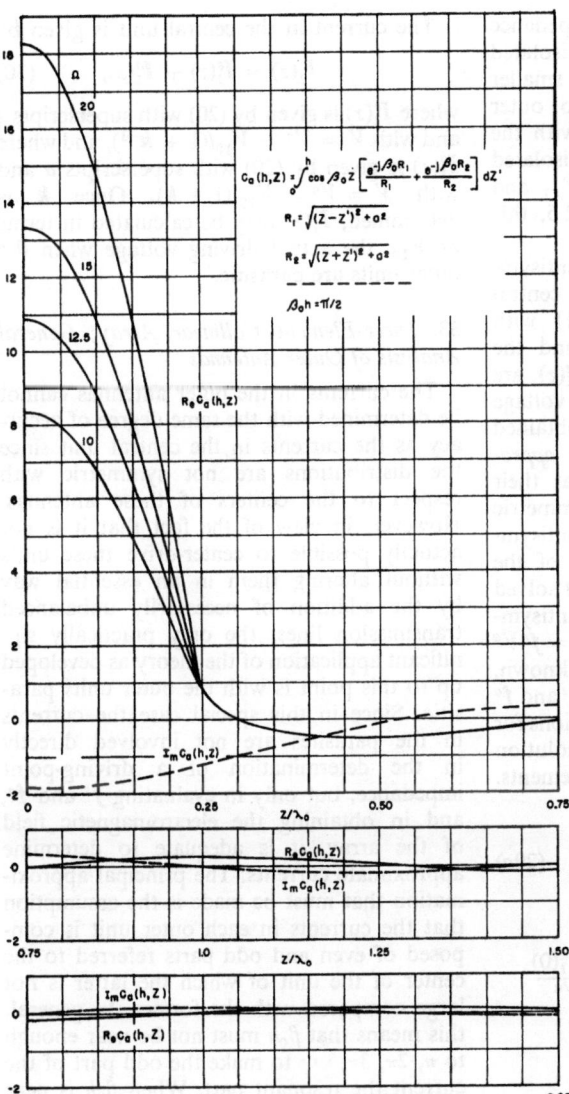

Fig. 32.4. Curves of $C_a(h,z)$ as a function of z/λ_0.

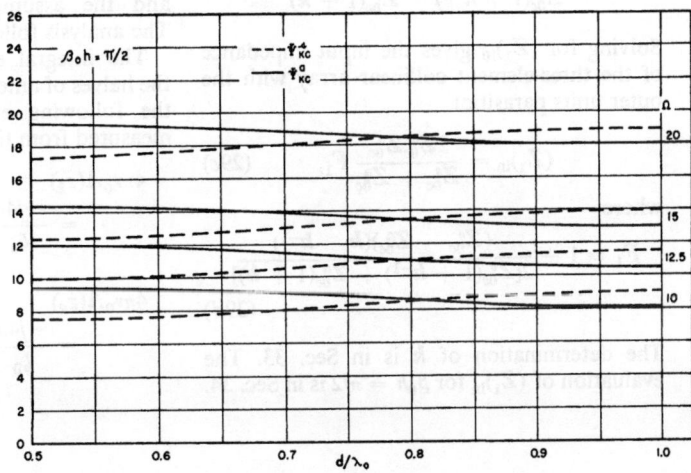

Fig. 32.5. Expansion parameters for three-element collinear array.

the distribution of current and the impedance of the central unit are like those of an isolated antenna with a larger Ψ_{R1} and hence smaller radius. Alternatively, when the two outer units are driven 180° out of phase with the central unit, this behaves like an isolated antenna with a somewhat smaller Ψ_{R1} and greater radius. As seen from Fig. 32.5, this is true when $d/\lambda_0 < 0.8$.

Although the symmetric and antisymmetric impedances Z^s and Z^a of the central unit may be determined from (21) with appropriate superscripts s and a, and the corresponding currents $I_1^s(z)$ and $I_1^a(z)$ are given by (20) in terms of the driving voltage $V = V^s$ or $V = V^a$, the results so obtained are valid *only if the outer units are appropriately driven* so that the currents at their centers are equal to $I_1^s(0)$ in the symmetric case or are equal to $-I_1^a(0)$ in the antisymmetric case. Evidently, the problem of the central unit of the collinear array is not solved completely until the symmetric and antisymmetric voltages $V_2^s = f^s V^s$ and $V_2^a = f^a V^a$ required to maintain these currents are known. They are determined in Sec. 33. Once f^s and f^a are known, a superposition of the solutions for which $V_2^s = V_2^a$ at once yields the solution of the array with parasitic outer elements. Since in this case

$$\frac{V^s}{V^a} = \frac{f^a}{f^s} \equiv k, \qquad (29a)$$

it follows that

$$\frac{1}{(Z_1)_{\text{in}}} = (Y_1)_{\text{in}} = \frac{I_1(0)}{V_{10}} = \frac{I_1^s(0) + I_1^a(0)}{V^s + V^a}$$

$$= \frac{I_1^s(0)}{V^s(1 + k^{-1})} + \frac{I_1^a(0)}{V^a(1 + k)}$$

$$= \frac{1}{Z_{0c}^s(1 + k^{-1})} + \frac{1}{Z_{0c}^a(1 + k)}. \qquad (29b)$$

Solving for $(Z_1)_{\text{in}}$ gives the input impedance of the three-element collinear array with the outer units parasitic:

$$(Z_1)_{\text{in}} = \frac{2 Z_{0c}^s Z_{0c}^a}{Z_{0c}^s + Z_{0c}^a} F_1, \qquad (29c)$$

where

$$F_1 \equiv 1 + \frac{(Z_{0c}^s - Z_{0c}^a)(k - k^{-1})}{2[Z_{0c}^s(1 + k^{-1}) + Z_{0c}^a(1 + k)]}. \qquad (29d)$$

The determination of k is in Sec. 33. The evaluation of $(Z_1)_{\text{in}}$ for $\beta_0 h = \pi/2$ is in Sec. 34.

The current in the central unit is given by

$$I_1(z) = I_1^s(z) + I_1^a(z), \qquad (30)$$

where $I_1^s(z)$ is given by (20) with superscripts s and with $V = V^s = V_{10}/(1 + k^{-1})$, and where $I_1^a(z)$ is given by (20) with superscripts a and with $V = V^a = V_{10}/(1 + k)$. Once k is determined, $I_1(z)$ may be calculated in terms of V_{10}, the actual driving voltage when the outer units are parasitic.

33. Three-Element Collinear Array; General Analysis of Outer Antennas

The currents in the outer antennas cannot be determined with the same degree of accuracy as the currents in the central unit since the distributions are not symmetric with respect to the centers of these antennas. However, in view of the fact that it is not actually possible to center-drive these units without altering them in an essential way by the addition of necessarily unbalanced transmission lines, the only practically significant application of the theory as developed up to this point is with the outer units parasitic. Since in this special case the currents in the parasites are not involved directly in the determination of a driving-point impedance, but only in evaluating f^s and f^a, and in obtaining the electromagnetic field of the array, it is adequate to determine approximate currents. The principal approximation that must be made is the assumption that the currents in each outer unit is composed of even and odd parts referred to the center of the unit of which the latter is not large compared with the former. In general, this means that $\beta_0 h$ must not be near enough to π, 2π, 3π, $\cdots$ to make the odd part of the current the resonant part. When $\beta_0 h$ is near $\pi/2$, the even part of the current is resonant and the assumption is an excellent one. The analysis follows.

The integral equation for the current in the halves of antenna 2 (Fig. 32.3) are given by the following expressions, in which z_2 is measured from the center of antenna 2:

$$4\pi v_0 A(z_2)$$
$$= \frac{-j4\pi}{\zeta_0}(C_1 \cos \beta_0 z_2 + C_2 \sin \beta_0 z_2),$$
$$(0 \leq z_2 \leq h) \quad (1a)$$

$$4\pi v_0 A(z_2)$$
$$= \frac{-j4\pi}{\zeta_0}(C_3 \cos \beta_0 z_2 + C_4 \sin \beta_0 z_2),$$
$$(-h \leq z_2 \leq 0) \quad (1b)$$

where

$$4\pi\nu_0 A(z_2) = \left(\int_{-h}^{h} + \int_{-d-h}^{-d+h} + \int_{-2d-h}^{-2d+h}\right) I(z_2') K_1(z_2, z_2') \, dz_2', \quad (2a)$$

with

$$K_1(z_2, z_2') = \frac{e^{-j\beta_0 R_1(z_2, z_2')}}{R_1(z_2, z_2')}, \quad (2b)$$

$$R_1(z_2, z_2') = \sqrt{(z_2 - z_2')^2 + a^2}. \quad (2c)$$

By setting $z' = d + z_2'$ in the second integral and $z_3' = 2d + z_2'$ in the third integral, and noting that

$$R_1(z_2, z' - d) = \sqrt{(d + z_2 - z')^2 + a^2}$$
$$= R_1(d + z_2, z'), \quad (3a)$$

$$R_1(z_2, z_3' - 2d) = \sqrt{(2d + z_2 - z_3')^2 + a^2}$$
$$= R_1(2d + z_2, z_3'), \quad (3b)$$

(2a) may be expressed as follows:

$$4\pi\nu_0 A(z_2) = \int_{-h}^{h} I_2(z_2') K_1(z_2, z_2') \, dz_2'$$
$$+ \int_{-h}^{h} I_1(z') K_1(d + z_2, z') \, dz'$$
$$+ \int_{-h}^{h} I_3(z_3') K_1(2d + z_2, z_3') \, dz_3'. \quad (4)$$

As a consequence of symmetry,

$$I_3(z_3') = I_2(-z_2'). \quad (5)$$

Since $I_1(z')$ is an even function of z' and symmetric or antisymmetric driving voltages are provided so that $I_1(z' = 0) = \pm I_2(z_2' = 0)$, a satisfactory approximation is

$$I_1(z') \doteq \pm \tfrac{1}{2}[I_2(z_2') + I_2(-z_2')], \quad (6)$$

where the upper sign applies to the symmetric, the lower sign to the antisymmetric case. Substitution of (5) and (6) in (4) gives

$$4\pi\nu_0 A(z_2) = \int_0^h [I_2(z_2') K_{1d}(z_2, z_2') + I_2(-z_2') K_{2d}(z_2, z_2')] \, dz_2', \quad (7)$$

where

$$K_{1d}(z_2, z_2') \equiv K_1(z_2, z_2') \pm \tfrac{1}{2}[K_1(d + z_2, z_2') + K_2(d + z_2, z_2')] + K_2(2d + z_2, z_2'), \quad (8a)$$

$$K_{2d}(z_2, z_2') \equiv K_2(z_2, z_2') \pm \tfrac{1}{2}[K_2(d + z_2, z_2') + K_1(d + z_2, z_2')] + K_1(2d + z_2, z_2'). \quad (8b)$$

The following notation is used:

$$K_1(u, z_2') \equiv \frac{e^{-j\beta_0 R_1(u, z_2')}}{R_1(u, z_2')},$$

$$K_2(u, z_2') \equiv \frac{e^{-j\beta_0 R_2(u, z_2')}}{R_2(u, z_2')}, \quad (9a)$$

$$R_1(u, z_2') = \sqrt{(u - z_2')^2 + a^2},$$

$$R_2(u, z_2') = \sqrt{(u + z_2')^2 + a^2}. \quad (9b)$$

These formulas apply in the following two cases:

Symmetric case:

$$V_{10} = V^s, \quad V_{30} = V_{20} = V_2^s = f^s V^s,$$
$$I_2(z_2') = I_2^s(z_2'), \quad (10a)$$

$K_{1d}(z_2, z_2')$ with upper sign $= K_{1d}^s(z_2, z_2')$;

Antisymmetric case:

$$V_{10} = V^a, \quad V_{30} = V_{20} = -V_2^a = -f^a V^a,$$
$$I_2(z_2') = I_2^a(z_2'), \quad (10b)$$

$K_{1d}(z_2, z_2')$ with lower sign $= K_{1d}^a(z_2, z_2')$.

The first step in the solution of (1a) with (7) is to resolve both sides into even and odd functions of z_2 and in this way obtain two equations, the one for the even part of $I_2(z_2)$, the other for the odd part. The even parts of the vector potential and current are given by

$$A_{\text{ev}}(z_2) = \tfrac{1}{2}[A(z_2) + A(-z_2)],$$
$$I_{\text{ev}}(z_2) = \tfrac{1}{2}[I_2(z_2) + I_2(-z_2)], \quad (11a)$$

so that with (1a)

$$4\pi\nu_0 A_{\text{ev}}(z_2)$$
$$= \frac{-j4\pi}{\zeta_0} (C_{\text{ev}} \cos \beta_0 z_2 + \tfrac{1}{2} V_{20} \sin |\beta_0 z_2|),$$
$$(-h \leq z_2 \leq h) \quad (11b)$$

where, as in Sec. II.13, $C_{\text{ev}} = \tfrac{1}{2}(C_1 + C_3)$ and $\tfrac{1}{2} V_{20} = \tfrac{1}{2}(C_2 - C_4)$. The even part of the vector potential in the range $-h \leq z_2 \leq 0$ is obtained from $A_{\text{ev}}(-z_2) = A_{\text{ev}}(z_2)$.

The odd parts of the vector potential and current are given by

$$A_{\text{od}}(z_2) = \tfrac{1}{2}[A(z_2) - A(-z_2)],$$
$$I_{\text{od}}(z_2) = \tfrac{1}{2}[I_2(z_2) - I_2(-z_2)], \quad (12a)$$

so that with (1a)

$$4\pi\nu_0 A_{\text{od}}(z_2) = \frac{-j4\pi}{\zeta_0} C_{\text{od}} \sin \beta_0 z_2,$$
$$(-h \leq z_2 \leq h) \quad (12b)$$

where $C_{od} = \frac{1}{2}(C_2 + C_4)$ and $\frac{1}{2}(C_1 - C_3) = 0$, since $A_{od}(-z_2) = A_{od}(z_2) = 0$ at $z_2 = 0$.

The integrals and equations for the even and odd parts of the vector potential are obtained using (11a, b) and (12a, b). They are

$$4\pi\nu_0 A_{ev}(z_2) = \int_0^h [I_2(z_2')K_{1ev}(z_2, z_2')$$
$$+ I_2(-z_2')K_{2ev}(z_2, z_2')]\, dz_2'$$
$$= \frac{-j4\pi}{\zeta_0}(C_{ev}\cos\beta_0 z_2$$
$$+ \tfrac{1}{2}V_{20}\sin|\beta_0 z_2|),$$
$$(-h \leq z_2 \leq h) \quad (13a)$$

$$4\pi\nu_0 A_{od}(z_2) = \int_0^h [I_2(z_2')K_{1od}(z_2, z_2')$$
$$+ I_2(-z_2')K_{2od}(z_2, z_2')]\, dz_2'$$
$$= \frac{-j4\pi}{\zeta_0}C_{od}\sin\beta_0 z_2,$$
$$(-h \leq z_2 \leq h) \quad (13b)$$

where

$$K_{1ev}(z_2, z_2') \equiv \tfrac{1}{2}[K_{1d}(z_2, z_2') + K_{1d}(-z_2, z_2')], \quad (14a)$$

$$K_{2ev}(z_2, z_2') \equiv \tfrac{1}{2}[K_{2d}(z_2, z_2') + K_{2d}(-z_2, z_2')], \quad (14b)$$

$$K_{1od}(z_2, z_2') \equiv \tfrac{1}{2}[K_{1d}(z_2, z_2') - K_{1d}(-z_2, z_2')], \quad (14c)$$

$$K_{2od}(z_2, z_2') \equiv \tfrac{1}{2}[K_{2d}(z_2, z_2') - K_{2d}(-z_2, z_2')]. \quad (14d)$$

In an asymmetric geometric arrangement like that of the outer elements in a collinear array, contributions to the even as well as the odd parts of the vector potential come from *both* even and odd parts of the currents in the three units. Consequently (13a) and (13b) cannot be solved separately for $I_{ev}(z_2)$ and $I_{od}(z_2)$. However, an approximate procedure is available. The current may be expanded into its even and odd parts by setting

$$I_2(z_2') = I_{ev}(z_2') + I_{od}(z_2'),$$
$$I_2(-z_2') = I_{ev}(z_2') - I_{od}(z_2'), \quad (15a)$$

where by definition

$$I_{ev}(z_2') = \tfrac{1}{2}[I(z_2') + I_2(-z_2')] = I_{ev}(-z_2'), \quad (15b)$$

$$I_{od}(z_2') = \tfrac{1}{2}[I(z_2') - I_2(-z_2')] = -I_{od}(-z_2'). \quad (15c)$$

If (15a) is substituted in (13a) and (13b), the results are:

$$4\pi\nu_0 A_{ev}(z_2)$$
$$= \int_0^h \{I_{ev}(z_2')[K_{1ev}(z_2, z_2') + K_{2ev}(z_2, z_2')]$$
$$+ I_{od}(z_2')[K_{1ev}(z_2, z_2') - K_{2ev}(z_2, z_2')]\}\, dz_2', \quad (16a)$$

$$4\pi\nu_0 A_{od}(z_2)$$
$$= \int_0^h \{I_{ev}(z_2')[K_{1od}(z_2, z_2') + K_{2od}(z_2, z_2')]$$
$$+ I_{od}(z_2')[K_{1od}(z_2, z_2') - K_{2od}(z_2, z_2')]\}\, dz_2' \quad (16b)$$

Let the following kernels be defined:

$$K_{ev}(z_2, z_2') \equiv K_{1ev}(z_2, z_2') + K_{2ev}(z_2, z_2')$$
$$= \tfrac{1}{2}[K_{1d}(z_2, z_2') + K_{2d}(z_2, z_2')$$
$$+ K_{1d}(-z_2, z_2') + K_{2d}(-z_2, z_2')]$$
$$= K_1(z_2, z_2') + K_2(z_2, z_2')$$
$$\pm \tfrac{1}{2}[K_1(d + z_2, z_2') + K_2(d + z_2, z_2')$$
$$+ K_1(d - z_2, z_2') + K_2(d - z_2, z_2')]$$
$$+ \tfrac{1}{2}[K_1(2d + z_2, z_2') + K_2(2d + z_2, z_2')$$
$$+ K_1(2d - z_2, z_2') + K_2(2d - z_2, z_2')], \quad (17a)$$

$$K_{eo}(z_2, z_2') \equiv K_{1ev}(z_2, z_2') - K_{2ev}(z_2, z_2')$$
$$= \tfrac{1}{2}[K_{1d}(z_2, z_2') - K_{2d}(z_2, z_2')$$
$$+ K_{1d}(-z_2, z_2') - K_{2d}(-z_2, z_2')]$$
$$= -\tfrac{1}{2}[K_1(2d + z_2, z_2') - K_2(2d + z_2, z_2')$$
$$+ K_1(2d - z_2, z_2') - K_2(2d - z_2, z_2')], \quad (17b)$$

$$K_{oe}(z_2, z_2') \equiv K_{1od}(z_2, z_2') + K_{2od}(z_2, z_2')$$
$$= \tfrac{1}{2}[K_{1d}(z_2, z_2') + K_{2d}(z_2, z_2')$$
$$- K_{1d}(-z_2, z_2') - K_{2d}(-z_2, z_2')]$$
$$= \pm \tfrac{1}{2}[K_1(d + z_2, z_2') + K_2(d + z_2, z_2')$$
$$- K_1(d - z_2, z_2') - K_2(d - z_2, z_2')]$$
$$+ \tfrac{1}{2}[K_1(2d + z_2, z_2') + K_2(2d + z_2, z_2')$$
$$- K_1(2d - z_2, z_2') - K_2(2d - z_2, z_2')], \quad (17c)$$

$$K_{od}(z_2, z_2') \equiv K_{1od}(z_2, z_2') - K_{2od}(z_2, z_2')$$
$$= \tfrac{1}{2}[K_{1d}(z_2, z_2') - K_{2d}(z_2, z_2')$$
$$- K_{1d}(-z_2, z_2') + K_{2d}(-z_2, z_2')]$$
$$= K_1(z_2, z_2') - K_2(z_2, z_2')$$
$$- \tfrac{1}{2}[K_1(2d + z_2, z_2') - K_2(2d + z_2, z_2')]$$
$$+ \tfrac{1}{2}[K_1(2d - z_2, z_2') - K_2(2d - z_2, z_2')]. \quad (17d)$$

Examination of (17b) as related to (16a) shows that it represents the contribution to the *even* part of the vector potential on antenna 2 by the *odd* parts of the currents in antenna 3. Since the odd currents involved are assumed to be relatively small, and in any case the distances are large, this term is negligible compared with the very much greater contribution from the large even currents in all three antennas. Since $R_1(2d \pm z_2, z_2')$ differs very much less from $R_2(2d \pm z_2, z_2')$ than does $R_1(z_2, z_2')$ from $R_2(z_2, z_2')$, it follows that the principal contributions to $K_{od}(z_2, z_2')$ in (17d) are from the first two terms. That is,

$$K_{od}(z_2, z_2') \doteq K_1(z_2, z_2') - K_2(z_2, z_2'). \quad (17e)$$

Neglecting the term in $K_{eo}(z_2, z_2')$, the two approximate integral equations are:

$$4\pi v_0 A_{ev}(z_2) = \int_0^h I_{ev}(z_2') K_{ev}(z_2, z_2') \, dz_2'$$

$$= \frac{-j4\pi}{\zeta_0} (C_{ev} \cos \beta_0 z_2 + \tfrac{1}{2} V_{20} \sin \beta_0 |z_2|),$$

$$(-h \leq z_2 \leq h) \quad (18a)$$

$$4\pi v_0 A_{od}(z_2) = \int_0^h I_{ev}(z_2') K_{oe}(z_2, z_2') \, dz_2'$$

$$+ \int_0^h I_{od}(z_2') K_{od}(z_2, z_2') \, dz_2'$$

$$= \frac{-j4\pi}{\zeta_0} C_{od} \sin \beta_0 z_2.$$

$$(-h \leq z_2 \leq h) \quad (18b)$$

Of these the first may be solved directly for $I_{ev}(z_2)$; using this known value the second equation may be solved for $I_{od}(z_2)$. Note that the equations apply only in the symmetric and antisymmetric cases defined in (10a, b), but that the general case may be constructed by superposition.

The equation (18a) for the even part of the current is in exactly the same form as the equation for a single antenna. The solution for the upper half of antenna 2 may be written down directly in the form (32.20):

$$I_{ev}(z_2) = \frac{j2\pi V_{20}}{\zeta_0 \Psi_{ev}}$$

$$\times \left[\frac{\sin \beta_0(h - z_2) + M_{ev1}(z_2)/\Psi_{ev} + \cdots}{\cos \beta_0 h + A_{ev1}/\Psi_{ev} + \cdots} \right].$$

$$(19a)$$

The impedance is

$$Z_{ev} = \frac{-j\zeta_0 \Psi_{ev}}{2\pi} \left(\frac{\cos \beta_0 h + A_{ev1}/\Psi_{ev} + \cdots}{\sin \beta_0 h + B_{ev1}/\Psi_{ev} + \cdots} \right). \quad (19b)$$

For the symmetric case a superscript s is added to I, Ψ, M, B, A, and Z, and V_{20} is replaced by V_2^s; for the antisymmetric case a superscript a replaces the superscript s. The functions M, B, A, and Ψ are defined as for a single antenna but using a different kernel.

An important object in evaluating $I_{ev}^s(z_2)$ and $I_{od}^s(z_2)$ is to determine the factors f^s and f^a relating the driving voltages required at the centers of the outer antennas to the driving voltage at the center of the middle antenna in order that the currents at these three driving points may satisfy (32.6d) and (32.7b). These factors may be evaluated directly from (19a, b) as follows. Since the odd part of $I_2(z_2)$ necessarily is zero at $z_2 = 0$, the entire current at this point is even. Hence, using (32.6) and (32.7),

Symmetric case:

$$V_{20} = V_2^s = I_2^s(z_2 = 0) Z_{ev}^s = I_1^s(0) Z_{ev}^s$$
$$= (V^s/Z_{0c}^s) Z_{ev}^s = f^s V^s; \quad (20a)$$

therefore,

$$f^s = Z_{ev}^s/Z_{0c}^s. \quad (20b)$$

Antisymmetric case:

$$V_{20} = -V_2^a = I_2^a(z_2 = 0) Z_{ev}^a = -I_1^a(0) Z_{ev}^a$$
$$= (-V^a/Z_{0c}^a) Z_{ev}^a = -f^a V^a; \quad (21a)$$

therefore,

$$f^a = Z_{ev}^a/Z_{0c}^a. \quad (21b)$$

The impedances Z_{0c}^s and Z_{0c}^a are given by (32.21) with appropriate superscripts. The complex factor k defined in (32.29a) is given by

$$k \equiv \frac{f^a}{f^s} = \frac{Z_{ev}^a}{Z_{ev}^s} \cdot \frac{Z_{0c}^s}{Z_{0c}^a}. \quad (22)$$

It is this factor that is required in order to calculate the input impedance of the three-element array with the outer units parasitic using (32.29c).

The even part of the current in the outer antennas when these are parasitic is given by

$$I_{2ev}(z_2) = I_{ev}^s(z_2) - I_{ev}^a(z_2), \quad (23a)$$

where $I_{ev}^s(z_2)$ is obtained from (19a) with appropriate superscripts s and with $V_{20} = V_2^s$, and where $I_{ev}^a(z_2)$ is given by (19a) with superscripts a and with $V_{20} = V_2^a$. Note that

$$V_2^a = V_2^s = f^s V^s = f^a V^a = V_{10}/2F_2, \quad (23b)$$

where

$$F_2 \equiv (f^a + f^s)/2 f^a f^s \quad (23c)$$

Thus,

$$I_{2ev}(z_2) = \frac{j\pi V_{10}}{\zeta_0 F_2} \times$$

$$\left\{ \frac{1}{\Psi_{ev}^s} \left[\frac{\sin \beta_0 (h - z_2) + M_{ev1}^s(z_2)/\Psi_{ev}^s + \cdots}{\cos \beta_0 h + A_{ev1}^s/\Psi_{ev}^s + \cdots} \right] \right.$$

$$\left. - \frac{1}{\Psi_{ev}^a} \left[\frac{\sin \beta_0 (h - z_2) + M_{ev1}^a(z_2)/\Psi_{ev}^a + \cdots}{\cos \beta_0 h + A_{ev1}^a/\Psi_{ev}^a + \cdots} \right] \right\}.$$

$$(23d)$$

In general, $I_{2ev}(z_2)$ does not differ greatly from $I_1(z)$ since both are even functions and the driving voltages are chosen to make them equal at the centers of the antennas.

The odd part of the current in antenna 2 must be determined from (18b) using the known value of $I_{ev}(z_2)$ obtained from the solution (19a) of (18a). Thus, (18b) may be expressed as follows:

$$\int_0^h I_{od}(z_2') K_{od}(z_2, z_2')\, dz_2'$$

$$= \frac{-j4\pi}{\zeta_0} [C_{od} \sin \beta_0 z_2 + U(z_2)], \quad (0 \leq z_2 \leq h) \quad (24a)$$

where

$$U(z_2) \equiv \frac{-j\zeta_0}{4\pi} \int_0^h I_{ev}(z_2') K_{oe}(z_2, z_2')\, dz_2', \quad (24b)$$

$K_{od}(z_2, z_2')$ is given by (17e) and $K_{oe}(z_2, z_2')$ by (17c). The upper sign in (17c) applies to the symmetric case, the lower sign to the antisymmetric case. Note that $K_{od}(z_2, z_2')$ is the same for both cases. The evaluation of $I_{od}(z_2)$ from (24a) is quite involved in general, particularly when $\beta_0 h$ is near π, since it then coincides with a resonant-mode current, whereas the even current does not. On the other hand, when $\beta_0 h$ is near $\pi/2$ the even distribution is resonant, the odd distribution is not, and, therefore, it is relatively quite small. Since the only practically important case, and at the same time the one for which an approximate evaluation of (24a) is readily carried out, is with $\beta_0 h = \pi/2$, only this case is considered.

The first step in the solution of (24a) is the simplification of the function $U(z_2)$ which is proportional to the odd part of the vector potential along antenna 2 due to the even currents in antennas 1 and 3. Since z_2 can not exceed h, and $d \geq 2h$ occurs in all distances R in $K_{oe}(z_2, z_2')$, the term a^2 in these distances is negligible. It follows that in the range $0 \leq z_2 \leq h$, $0 \leq z_2' \leq h$, (17c) becomes

$$K_{oe}(z_2, z_2') \doteq \pm \tfrac{1}{2} e^{-j\beta_0(d + z_2)}$$

$$\times \left(\frac{e^{j\beta_0 z_2'}}{d + z_2 - z_2'} + \frac{e^{-j\beta_0 z_2'}}{d + z_2 + z_2'} \right)$$

$$\mp \tfrac{1}{2} e^{-j\beta_0(d - z_2)}$$

$$\times \left(\frac{e^{j\beta_0 z_2'}}{d - z_2 - z_2'} + \frac{e^{-j\beta_0 z_2'}}{d - z_2 + z_2'} \right)$$

$$+ \tfrac{1}{2} e^{-j\beta_0(2d + z_2)}$$

$$\times \left(\frac{e^{j\beta_0 z_2'}}{2d + z_2 - z_2'} + \frac{e^{-j\beta_0 z_2'}}{2d + z_2 + z_2'} \right)$$

$$+ \tfrac{1}{2} e^{-j\beta_0(2d - z_2)}$$

$$\times \left(\frac{e^{j\beta_0 z_2'}}{2d - z_2 - z_2'} + \frac{e^{-j\beta_0 z_2'}}{2d - z_2 + z_2'} \right). \quad (25)$$

As a next step in the simplification, let z_2' be neglected in the denominators throughout. This is a good approximation for all values of z_2' in most of the terms, a poor approximation in only one term, and then only when $z_2 = h$ and when z_2' approaches h. Since the current vanishes at h, an error here is of no great importance. Hence, setting $z_2' = 0$ in the denominators, (25) reduces to

$$K_{oe}(z_2, z_2') \doteq \left[\pm \frac{e^{-j\beta_0(d + z_2)}}{d + z_2} \mp \frac{e^{-j\beta_0(d - z_2)}}{d - z_2} \right.$$

$$\left. + \frac{e^{-j\beta_0(2d + z_2)}}{2d + z_2} + \frac{e^{-j\beta_0(2d - z_2)}}{2d - z_2} \right] \cos \beta_0 z_2'. \quad (26)$$

Since with $\beta_0 h = \pi/2$, the even current is well approximated by

$$I_{ev}(z_2') \doteq \frac{V}{Z_{0c}} \cos \beta_0 z_2', \quad (27)$$

the integral in (24b) may be evaluated directly. It involves

$$\int_0^h \cos^2 \beta_0 z_2'\, dz_2' = \frac{1}{\beta_0} \int_0^{\pi/2} \cos^2 u\, du = \frac{\pi}{4\beta_0}. \quad (28)$$

Accordingly,

$$U(z_2) \doteq \frac{j\zeta_0 V}{8Z_{0c}}$$

$$\times \left[\frac{\pm e^{-j\beta_0 d}}{\beta_0(d^2 - z_2^2)} (z_2 \cos \beta_0 z_2 + jd \sin \beta_0 z_2) \right.$$

$$\left. + \frac{e^{-j2\beta_0 d}}{\beta_0(4d^2 - z_2^2)} (z_2 \cos \beta_0 z_2 + j2d \sin \beta_0 z_2) \right]. \quad (29)$$

In the range $0 \leq \beta_0 z_2 \leq \pi/2$, a useful expression is obtained by factoring out $jd \sin \beta_0 z_2$ in the first and $j2d \sin \beta_0 z_2$ in the second parenthesis. The result is

$$U(z_2) \doteq \frac{\zeta_0 V}{8Z_{0c}\beta_0 d} F(z_2) \sin \beta_0 z_2$$

$$\equiv \bar{U}(z_2) \sin \beta_0 z_2, \quad (30a)$$

where $\bar{U}(z_2)$ is defined by (30a) and where

$$F(z_2) = -\left[\frac{\pm e^{-j\beta_0 d}}{(1 - z_2^2/d^2)} \left(1 - j\frac{\beta_0 z_2 \cot \beta_0 z_2}{\beta_0 d}\right) \right.$$

$$\left. + \frac{e^{-j2\beta_0 d}}{(1 - z_2^2/4d^2)} \left(1 - j\frac{\beta_0 z_2 \cot \beta_0 z_2}{2\beta_0 d}\right) \right]. \quad (30b)$$

Since $z_2/d \leq 1/4$, and $\beta_0 d \gtrsim \pi$, $F(z_2)$ does not vary rapidly. It follows that $U(z_2)$ varies approximately as $\sin \beta_0 z_2$. With (30a, b) the integral equation (24a) reduces to

$$\int_0^h I_{od}(z_2') K_{od}(z_2, z_2') dz_2'$$

$$= \frac{-j4\pi}{\zeta_0} [C_{od} + \bar{U}(z_2)] \sin \beta_0 z_2. \quad (31)$$

This integral equation can be solved for the current by the usual method of iteration. Let

$$I_{od}(z_2') = I_{od}(z_2)g(z_2, z_2'), \quad (32)$$

where $g(z_2, z_2')$ is a distribution function that is chosen to make $I_{od}(z_2)\Psi_{od}(z_2)$ approximate closely the integral on the left in (31), with

$$\Psi_{od}(z_2) = \int_0^h g(z_2, z_2') K_{od}(z_2, z_2') dz_2'$$

$$= \Psi_{od} + \Upsilon_{od}(z_2); \quad (33)$$

Ψ_{od} is a mean constant value of the integral suitably defined at a reference point. With (32) and (33),

$$I_{od}(z_2) = \frac{-j4\pi}{\zeta_0 \Psi_{od}} [C_{od} + \bar{U}(z_2)] \sin \beta_0 z_2$$

$$- \frac{1}{\Psi_{od}} \left\{ I_{od}(z_2) \Upsilon_{od}(z_2) + \int_0^h [I_{od}(z_2')$$

$$- I_{od}(z_2)g(z_2, z_2')] K_{od}(z_2, z_2') dz_2' \right\}. \quad (34)$$

By using the term in $\sin \beta_0 z_2$ as the zeroth-order current, and substituting this in the integrals, these may be evaluated assuming $\bar{U}(z_2)$ to be constant, and added to the zeroth-order current. This process of iteration can be repeated. The current then appears in the form

$$I_{od}(z_2) = \frac{-j4\pi}{\zeta_0 \Psi_{od}} [C_{od} + \bar{U}(z_2)] \left[\sin \beta_0 z_2 \right.$$

$$\left. + \frac{S_1(z_2)}{\Psi_{od}} + \frac{S_2(z_2)}{\Psi_{od}^2} + \cdots \right]. \quad (35)$$

The constant C_{od} may be evaluated by requiring the current to vanish at $z_2 = h$. Then

$$C_{od} = -\bar{U}(h), \quad (36)$$

so that

$$I_{od}(z_2) = \frac{-j4\pi}{\zeta_0 \Psi_{od}} [\bar{U}(z_2) - \bar{U}(h)] \left[\sin \beta_0 z_2 \right.$$

$$\left. + \frac{S_1(z_2)}{\Psi_{od}} + \frac{S_2(z_2)}{\Psi_{od}^2} + \cdots \right]. \quad (37)$$

It follows from (30b) that, with $\beta_0 h = \pi/2$, an approximate result is

$$\bar{U}(z_2) - \bar{U}(h) \doteq \frac{j\zeta_0 V D(d)}{8Z_{0c}} \cos \beta_0 z_2, \quad (38a)$$

where

$$D(d) = \frac{1}{\beta_0^2 d^2} \left[\pm \frac{e^{-j\beta_0 d}}{(1 - h^2/d^2)} \right.$$

$$\left. + \frac{e^{-j2\beta_0 d}}{2(1 - h^2/4d^2)} \right]. \quad (38b)$$

Substitution of (38a) in (37) gives the following simple expressions for the zeroth-order odd currents in the outer units when $\beta_0 h = \pi/2$:

$$I_{od}^s(z_2) \doteq \frac{V^s}{Z_{0c}^s} \frac{\pi D^s(d)}{4\Psi_{od}} \sin 2\beta_0 z_2, \quad (39a)$$

$$I_{od}^a(z_2) \doteq \frac{V^a}{Z_{0c}^a} \frac{\pi D^a(d)}{4\Psi_{od}} \sin 2\beta_0 z_2, \quad (39b)$$

where $D^s(d)$ is (38b) with the upper sign and $D^a(d)$ is (38b) with the lower sign. Note that $V^s/Z_{0c}^s = I_1^s(0)$, $V^a/Z_{0c}^a = I_1^a(0)$. The odd part of the current in the outer units when these are parasitic is given by the sum of (39a) and (39b). Note that, as in (32.29b), $V^s = V_{10}/(1 + k^{-1})$, $V^a = V_{10}/(1 + k)$. It follows that

$$I_{2od}(z_2) = I_{od}^s(z_2) + I_{od}^a(z_2)$$

$$\doteq \frac{\pi V_{10}}{4\Psi_{od}} \left[\frac{D^s(d)}{Z_{0c}^s(1 + k^{-1})} \right.$$

$$\left. + \frac{D^a(d)}{Z_{0c}^a(1 + k)} \right] \sin 2\beta_0 z_2.$$

$$(\beta_0 h = \pi/2) \quad (40)$$

The total current in the outer antennas is the sum of (40) and (23d), with $\beta_0 h = \pi/2$.

The final step in the complete evaluation of $I_{od}(z_2)$ is the determination of Ψ_{od} from (33). Since the zeroth-order distribution function $g(z_2, z_2') = \sin 2\beta_0 z'/\sin 2\beta_0 z$ does not permit the ready evaluation of Ψ_{od}, the simpler procedure of assuming a sensibly uniform current and neglecting retardation is more convenient and adequate. This is equivalent to setting

$$\Psi_{od}(z_2) = \int_0^h \left[\frac{1}{\sqrt{(z_2' - z_2)^2 + a^2}} - \frac{1}{\sqrt{(z_2' + z_2)^2 + a^2}} \right] dz_2'. \quad (41)$$

Since the odd current is most nearly constant near $z_2 = h/2$ and this is its greatest value, it is a good approximation to choose the reference point for Ψ_{od} at $z_2 = h/2$. Thus,

$$\Psi_{od} \doteq \int_{-h/2}^{h/2} \frac{du}{\sqrt{u^2 + a^2}} - \int_{h/2}^{3h/2} \frac{du}{\sqrt{u^2 + a^2}}$$
$$= \ln\left(\frac{h^2}{3a^2}\right) \doteq \Omega - 2.5, \quad (42)$$

where $\Omega = 2\ln(2h/a)$.

In this section general expressions have been obtained for the symmetric and antisymmetric components of the even current in the outer units of the three-element array. An approximate formula for the odd currents has been derived in the important special case $\beta_0 h = \pi/2$. As a part of the determination of the even currents, relations between the three voltages have been established which assure that the currents at the centers of all three units are equal in magnitude and either in phase or 180° out of phase as required by the conditions for the symmetric and antisymmetric cases. Thus, in Sec. 32 and this section the general formulas for the currents in collinear arrays are derived in the special case of the three-element array.

34. Collinear Array of Three Half-Wave Dipoles

The most useful collinear array of three antennas consists of identical units each of electrical half-length $\beta_0 h = \pi/2$. The middle antenna No. 1 is center-driven from a balanced two-wire transmission line. Since terminal-zone effects are the same as for the single antenna analyzed in Chapter II, they need not be considered again, and the array may be assumed to be center-driven by a slice generator. Results obtained in this way are easily generalized to apply to the antenna as end load on a two-wire line. The two outer antennas, No. 2 and No. 3, are coupled to the middle one. In the usual collinear array, coupling circuits, often in the form of sections of transmission line, are connected to the adjacent ends of the antennas. This arrangement is studied in Sec. 35. In this section, attention is focused on an array in which the outer units are parasitic and have no connection to the central one as shown in Fig. 32.2b.

The driving voltages for the three units are

$$V_{10} = V^s + V^a = V^s(1 + k^{-1}) = V^a(1 + k),$$
$$V_{20} = V_{30} = V_2^s - V_2^a = f^s V^s - f^a V^a = 0, \quad (1a)$$

where

$$k \equiv \frac{f^a}{f^s}, \quad f^a = \frac{Z_{ev}^a}{Z_{0c}^a}, \quad f^s = \frac{Z_{ev}^a}{Z_{0c}^a}, \quad (1b)$$

and where Z_{ev} is the impedance of each outer unit, Z_{0c} the impedance of the central unit; V^s is the driving voltage of the central unit when the other two are center-driven by voltages $V_3^s = V_2^s = f^s V^s$ such that the currents at the centers of all three units are equal and in phase; V^a is the driving voltage of the central unit when the two outer antennas are center-driven by voltages $V_3^a = V_2^a = f^a V^a$ such that the currents at the centers of the two outer units are equal to and in phase with each other, but both equal to and 180° out of phase with the current at $z = 0$ in the central unit.

The first-order symmetric and antisymmetric currents in the central unit are given by (32.20) with $\beta_0 h = \pi/2$ and appropriate superscripts. Thus

$$[I_1^s(z)]_1 = \frac{j2\pi V^s}{\zeta_0 A_{c1}^s} [\cos \beta_0 z + M_{c1}^s(z)/\Psi_{Kc}^s],$$
$$(0 \leq z \leq h) \quad (2a)$$

$$[I_1^a(z)]_1 = \frac{j2\pi V^a}{\zeta_0 A_{c1}^a} [\cos \beta_0 z + M_{c1}^a(z)/\Psi_{Kc}^a].$$
$$(0 \leq z \leq h) \quad (2b)$$

The total first-order current in the central unit when the outer units are parasitic is given by the sum of (2a) and (2b). With (1a, b) it is

$$[I_1(z)]_1 = [I_1^s(z)]_1 + [I_1^a(z)]_1 = \frac{j2\pi V_{10}}{\zeta_0}$$
$$\times \left\{ \left[\frac{1}{A_{c1}^s(1 + k^{-1})} + \frac{1}{A_{c1}^a(1 + k)} \right] \cos \beta_0 z \right.$$
$$\left. + \left[\frac{M_{c1}^s(z)}{A_{c1}^s(1 + k^{-1})\Psi_{1K}^s} + \frac{M_{c1}^a(z)}{A_{c1}^a(1 + k)\Psi_{K1}^a} \right] \right\}.$$
$$(3)$$

The first-order symmetric and antisymmetric impedances of the central unit are

$$[Z^s_{0c}]_1 = \frac{-j\zeta_0 A^s_{c1}}{2\pi}(1 + B^s_{c1}/\Psi^s_{Kc})^{-1},$$

$$[Z^a_{0c}]_1 = \frac{-j\zeta_0 A^a_{c1}}{2\pi}(1 + B^a_{c1}/\Psi^a_{Kc})^{-1}. \quad (4a)$$

The first-order input impedance of the array when the outer units are parasitic is

$$[(Z_1)_{\text{in}}]_1$$

$$= \frac{-j\zeta_0}{2\pi}\left\{\left[\frac{1}{A^s_{c1}(1+k^{-1})} + \frac{1}{A^a_{c1}(1+k)}\right]\right.$$

$$\left. + \left[\frac{B^s_{c1}}{A^s_{c1}(1+k^{-1})\Psi^s_{K1}} + \frac{B^a_{c1}}{A^a_{c1}(1+k)\Psi^a_{K1}}\right]\right\}^{-1}, \quad (4b)$$

where

$$B^s_{c1} \equiv M^s_{c1}(0), \quad B^a_{c1} \equiv M^a_{c1}(0). \quad (4c)$$

For antennas that have a reasonably large value of h/a, Ψ^s_{K1} and Ψ^a_{K1} are sufficiently large to make the contributions by the $1/\Psi_{K1}$ terms in (2a, b), (3) and (4a, b) relatively small. When this is true the following expressions are satisfactory approximations:

$$I^s_1(z) \doteq \frac{V^s}{Z^s_{0c}}\cos\beta_0 z,$$

$$I^a_1(z) \doteq \frac{V^a}{Z^a_{0c}}\cos\beta_0 z, \quad (5a)$$

$$I_1(z) \doteq \frac{V_{10}}{(Z_1)_{\text{in}}}\cos\beta_0 z,$$

where

$$Z^s_{0c} \doteq \frac{-j\zeta_0 A^s_{c1}}{2\pi}, \quad Z^a_{0c} \doteq \frac{-j\zeta_0 A^a_{c1}}{2\pi}, \quad (5b)$$

and where, using (32.29c, d),

$$(Z_1)_{\text{in}} \equiv \left[\frac{1}{Z^s_{0c}(1+k^{-1})} + \frac{1}{Z^a_{0c}(1+k)}\right]^{-1}$$

$$= (Z'_1)_{\text{in}} F_1, \quad (5c)$$

where

$$(Z'_1)_{\text{in}} = \frac{2Z^s_{0c}Z^a_{0c}}{Z^s_{0c} + Z^a_{0c}},$$

$$F_1 \equiv 1 + \frac{(Z^s_{0c} - Z^a_{0c})(k - k^{-1})}{2[Z^s_{0c}(1+k^{-1}) + Z^a_{0c}(1+k)]}. \quad (5d)$$

It is seen from (5) that $I_1(z)$ and $(Z_1)_{\text{in}}$ involve only the functions A^s_{c1}, A^a_{c1}, and the ratio function k. It follows from (II.15.29a)

with $\delta = 0$, and (II.21.1b) with $z^i = 0$, $\beta_0 h = \pi/2$, and with $K_{c1}(h, z')$ substituted for $K_1(h, z')$, that

$$A_{c1} \equiv F_{c1}(h) = -\int_{-h}^{h}\cos\beta_0 z' K_{c1}(h, z')\,dz',$$

$$(h = \lambda_0/4) \quad (6)$$

where

$$K_{c1}(h, z') = K_1(h, z') \pm [K_1(d - h, z') + K_1(d + h, z')], \quad (h = \lambda_0/4) \quad (7)$$

With (II.21.1b) the integrals in (6) with (7) may be expressed as follows:

$$A_{c1} = -\{C_a(h, h) \pm [C_a(h, d - h) + C_a(h, d + h)]\}. \quad (h = \lambda_0/4) \quad (8)$$

The upper sign corresponds to a superscript s, the lower sign to a superscript a on A_{c1}. It follows that

$$Z^s_{0c} = \frac{j\zeta_0}{2\pi}[C_a(h, h) + C_a(h, d - h) + C_a(h, d + h)], \quad (h = \lambda_0/4) \quad (9a)$$

$$Z^a_{0c} = \frac{j\zeta_0}{2\pi}[C_a(h, h) - C_a(h, d - h) - C_a(h, d + h)]. \quad (h = \lambda_0/4) \quad (9b)$$

The symmetric and antisymmetric impedances of the three-element array as given in (9a, b) are represented graphically in Fig. 34.1. The impedance $(Z'_1)_{\text{in}}$ defined in (5c), which is the principal part of $(Z_1)_{\text{in}}$, is shown in Fig. 34.4 together with $(Z_1)_{\text{in}}$. Note that $(Z_1)_{\text{in}}$ reduces to $(Z'_1)_{\text{in}}$ when d is sufficiently great so that $k \doteq 1$. It is not possible to determine $(Z_1)_{\text{in}}$ until k is available. This requires the evaluation of Z^s_{ev} and Z^a_{ev} in addition to Z^s_{0c} and Z^a_{0c}.

The first-order symmetric and antisymmetric *even* currents in the outer units are given by (33.19a) with $\beta_0 h = \pi/2$. They are

$$[I^s_{ev}(z_2)]_1 = \frac{j2\pi V^s_2}{\zeta_0 A^s_{ev1}}[\cos\beta_0 z_2 + M^s_{ev1}(z_2)/\Psi^s_{ev}], \quad (10a)$$

$$[I^a_{ev}(z_2)]_1 = \frac{j2\pi V^a_2}{\zeta_0 A^a_{ev1}}[\cos\beta_0 z_2 + M^a_{ev1}(z_2)/\Psi^a_{ev}]. \quad (10b)$$

When the outer units are parasitic,

$$V^s_2 = V^a_2 = f^s V^s = f^a V^a = V_{10}/2F_2, \quad (10c)$$

$$F_2 \equiv \frac{f^a + f^s}{2f^a f^s} = \frac{1}{2}\left(\frac{Z^s_{0c}}{Z^s_{ev}} + \frac{Z^a_{0c}}{Z^a_{ev}}\right), \quad (10d)$$

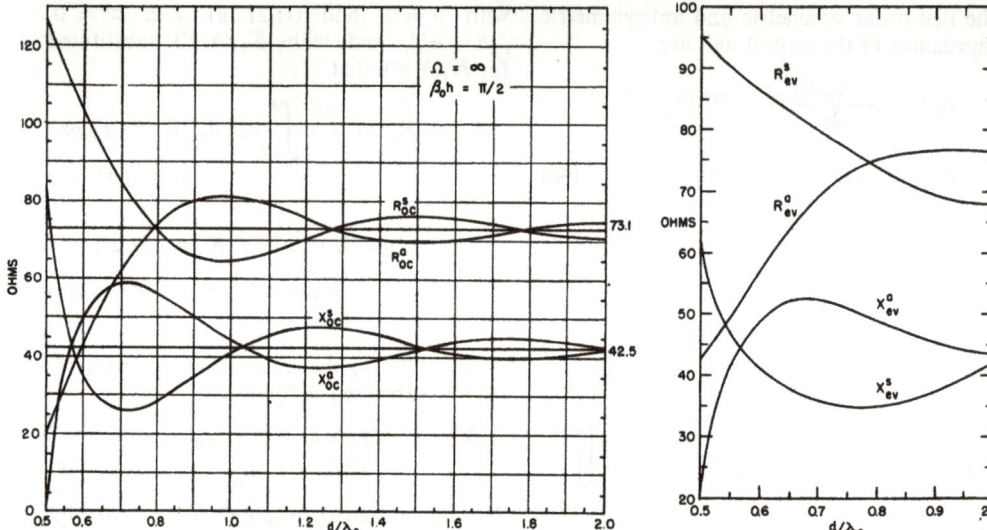

Fig. 34.1. Symmetric and antisymmetric impedances of central unit in three-element collinear array.

Fig. 34.2. Symmetric and antisymmetric impedances of outer unit in three-element collinear array.

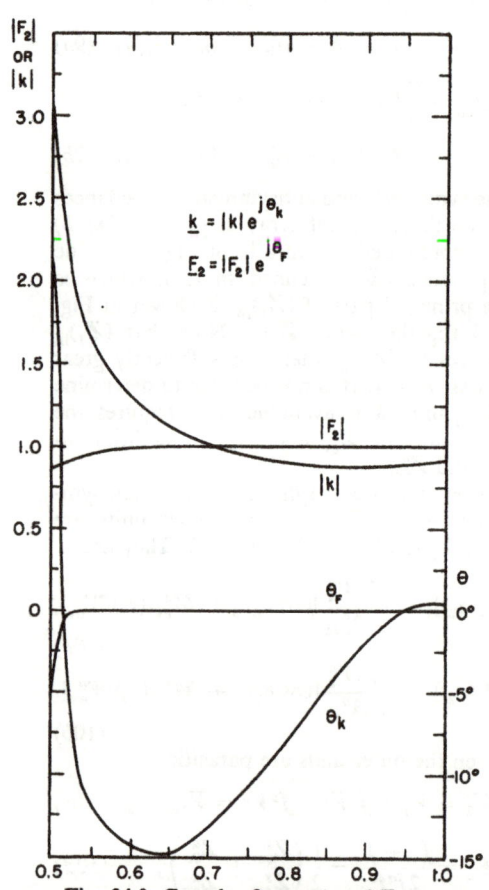

Fig. 34.3. Complex factors k and F_2.

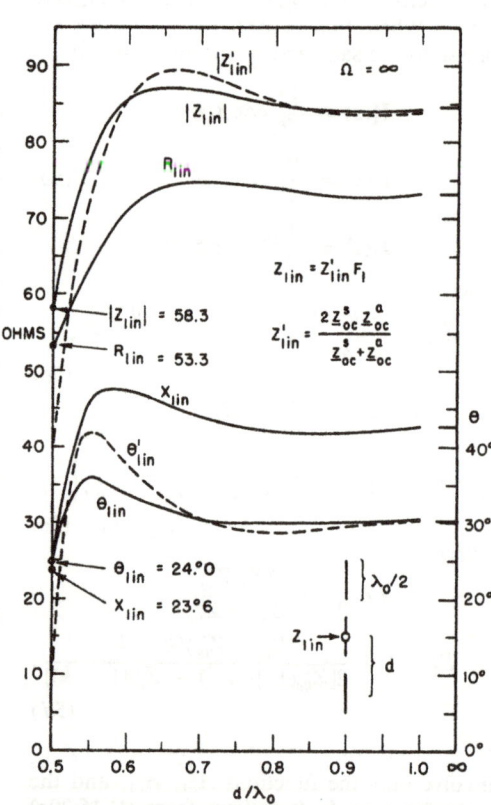

Fig. 34.4. Input impedance $Z_{1\text{in}}$ of antenna with two collinear parasites.

as in (33.23b, c) and the total first-order current is given by

$$[I_{2\text{ev}}(z_2)]_1 = [I^s_{\text{ev}}(z_2)]_1 - [I^a_{\text{ev}}(z_2)]_1$$

$$= \frac{j\pi V_{10}}{\zeta_0 F_2}\left\{\left(\frac{1}{A^s_{\text{ev1}}} - \frac{1}{A^a_{\text{ev1}}}\right)\cos\beta_0 z_2\right.$$

$$\left.+ \left[\frac{M^s_{\text{ev1}}(z_2)}{A^s_{\text{ev1}}\Psi^s_{\text{ev}}} - \frac{M^a_{\text{ev1}}(z_2)}{A^a_{\text{ev1}}\Psi^a_{\text{ev}}}\right]\right\}.$$

$$(\beta_0 h = \pi/2) \quad (11)$$

The first-order symmetric and antisymmetric impedances of the outer units are

$$(Z^s_{\text{ev}})_1 = \frac{-j\zeta_0 A^s_{\text{ev1}}}{2\pi}(1 + B^s_{\text{ev1}}/\Psi^s_{\text{ev}})^{-1}, \quad (12a)$$

$$(Z^a_{\text{ev}})_1 = \frac{-j\zeta_0 A^a_{\text{ev1}}}{2\pi}(1 + B^a_{\text{ev1}}/\Psi^a_{\text{ev}})^{-1}. \quad (12b)$$

The transfer impedance when the outer units are parasitic is

$$(Z_T)_1 = \frac{-j\zeta_0 F_2}{\pi}\left[\left(\frac{1}{A^s_{\text{ev1}}} - \frac{1}{A^a_{\text{ev1}}}\right)\right.$$

$$\left.+ \left(\frac{B^s_{\text{ev1}}}{A^s_{\text{ev1}}\Psi^s_{\text{ev}}} - \frac{B^a_{\text{ev1}}}{A^a_{\text{ev1}}\Psi^a_{\text{ev}}}\right)\right]^{-1}, \quad (12c)$$

where

$$B^s_{\text{ev1}} \equiv M^s_{\text{ev1}}(0), \quad B^a_{\text{ev1}} \equiv M^a_{\text{ev1}}(0). \quad (12d)$$

For sufficiently large values of Ψ^s_{ev} and Ψ^a_{ev} the leading terms are good approximations. They are

$$I^s_{\text{ev}}(z_2) \doteq \frac{V^s_2}{Z^s_{\text{ev}}}\cos\beta_0 z_2, \quad (13a)$$

$$I^a_{\text{ev}}(z_2) \doteq \frac{V^a_2}{Z^a_{\text{ev}}}\cos\beta_0 z_2, \quad (13b)$$

$$I_{2\text{ev}}(z_2) \doteq \frac{V_{10}}{Z_T}\cos\beta_0 z_2, \quad (13c)$$

where

$$Z^s_{\text{ev}} \doteq \frac{-j\zeta_0 A^s_{\text{ev1}}}{2\pi}, \quad Z^a_{\text{ev}} \doteq \frac{-j\zeta_0 A^a_{\text{ev1}}}{2\pi},$$

$$Z_T \doteq Z'_T F_2, \quad Z'_T \equiv \frac{2Z^s_{\text{ev}}Z^a_{\text{ev}}}{Z^a_{\text{ev}} - Z^s_{\text{ev}}}, \quad (13d)$$

$$F_2 = \frac{1}{2}\left(\frac{Z^s_{0c}}{Z^s_{\text{ev}}} + \frac{Z^a_{0c}}{Z^a_{\text{ev}}}\right) \doteq 1. \quad (13e)$$

The ratio of currents at the centers of the outer parasitic units to the current at the center of the driven unit is given by

$$\frac{I_2(z_2 = 0)}{I_1(0)} = \frac{(Z_1)_{\text{in}}}{Z_T} = \left|\frac{I_2(0)}{I_1(0)}\right|e^{j(\theta_2 - \theta_1)}, \quad (14)$$

where $(Z_1)_{\text{in}}$ is given by (5c) and Z_T by (13e). It is shown later (Fig. 34.3) that the factor F_2 differs negligibly from unity as indicated in (13e) except when d/λ_0 is at or very near 0.5. It follows from (II.15.29a) with $\delta = 0$, and (II.21.1b) with $z^i = 0$, $\beta_0 h = \pi/2$, and with $K_{1\text{ev}}(h, z'_2)$ substituted for $K_1(h, z')$, that

$$A_{\text{ev1}} = F_{\text{ev1}}(h)$$

$$= -\int_0^h \cos\beta_0 z'_2 K_{\text{ev}}(h, z'_2)\,dz'_2, \quad (15)$$

where $K_{\text{ev}}(h, z'_2)$ is given by (33.17a) with $z_2 = h = \lambda_0/4$. Thus,

$$K_{\text{ev}}(h, z'_2) = K(h, z'_2)$$

$$\pm \tfrac{1}{2}[K(d - h, z'_2) + K(d + h, z'_2)]$$

$$+ \tfrac{1}{2}[K(2d - h, z'_2) + K(2d + h, z'_2)], \quad (16)$$

where, as usual, $K \equiv K_1 + K_2$. With (II.21.1b) the integrals in (15) with (16) may be expressed as follows with $h = \lambda_0/4$:

$$A_{\text{ev1}} = -\{C_a(h, h)$$

$$\pm \tfrac{1}{2}[C_a(h, d - h) + C_a(h, d + h)]$$

$$+ \tfrac{1}{2}[C_a(h, 2d - h) + C_a(h, 2d + h)]\}. \quad (17)$$

The upper sign in (17) goes with a superscript s on A_{ev1}, the lower sign with a superscript a. Thus, with $h = \lambda_0/4$,

$$Z^s_{\text{ev}} = \frac{j\zeta_0}{2\pi}\{C_a(h, h)$$

$$+ \tfrac{1}{2}[C_a(h, d - h) + C_a(h, d + h)]$$

$$+ \tfrac{1}{2}[C_a(h, 2d - h) + C_a(h, 2d + h)]\}, \quad (18a)$$

$$Z^a_{\text{ev}} = \frac{j\zeta_0}{2\pi}\{C_a(h, h)$$

$$- \tfrac{1}{2}[C_a(h, d - h) + C_a(h, d + h)]$$

$$+ \tfrac{1}{2}[C_a(h, 2d - h) + C_a(h, 2d + h)]\}. \quad (18b)$$

These impedances are represented graphically in Fig. 34.2.

With Z^s_{ev} and Z^a_{ev} determined, the functions k, f^a, and f^s defined in (1b) and the function F_2 defined in (10d) may be evaluated; k and F_2 are shown in Fig. 34.3 in polar form. With k available and $(Z'_1)_{\text{in}}$ known, the approximate input impedance of the array $(Z_1)_{\text{in}}$ as defined in (5c) may be evaluated. It is shown in Fig. 34.4 along with $(Z'_1)_{\text{in}}$, which it approaches as k approaches unity with increased d. With F_2 available the transfer impedance Z_T may be determined; Z_T is shown in Fig. 34.5. The ratio of currents (14) is shown in Fig. 34.6 as a function of d/λ_0 and the distribution in Fig. 34.7 for

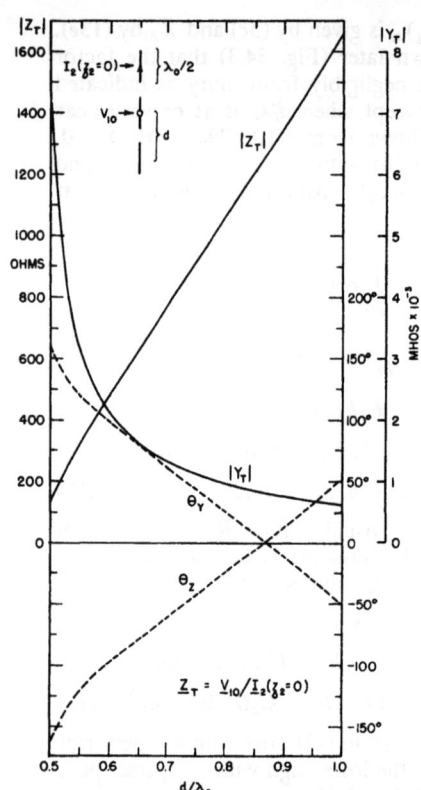

Fig. 34.5. Transfer impedance Z_T and admittance Y_T for antenna with two collinear parasites.

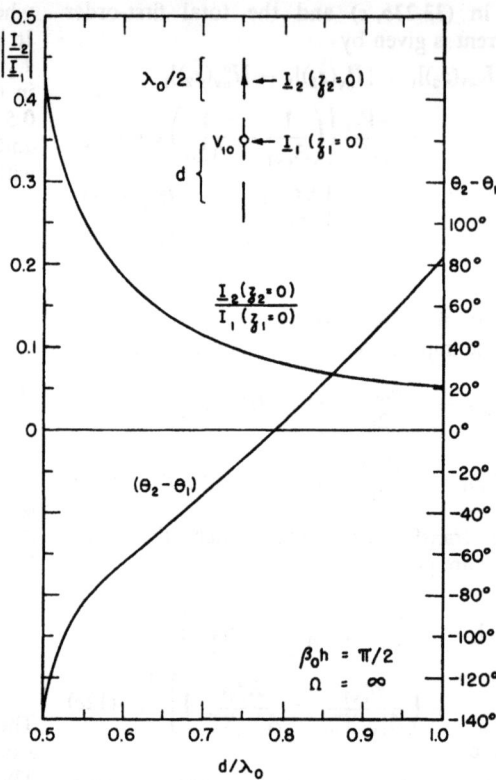

Fig. 34.6. Ratio of current in parasite to current in driven element.

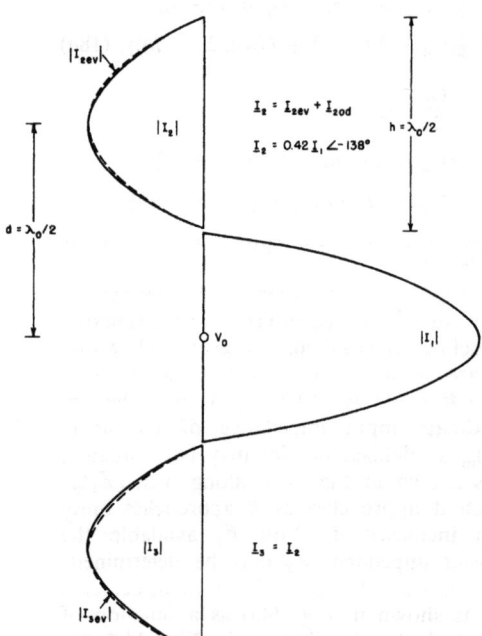

Fig. 34.7. Distribution of current on collinear array with outer elements parasitic.

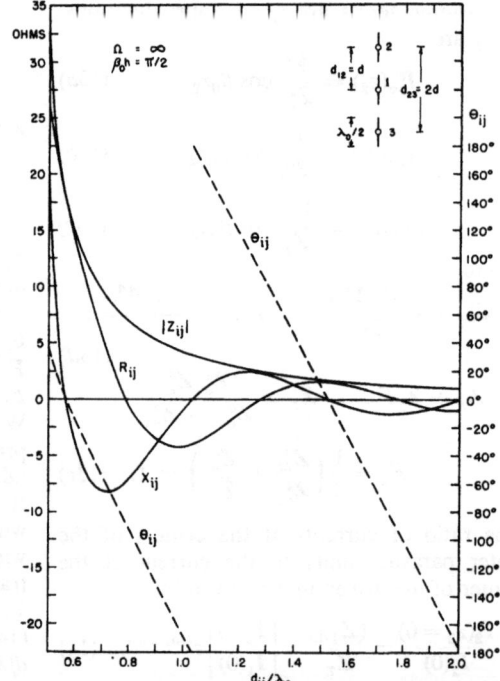

Fig. 34.8. Mutual impedance for collinear array.

$d = \lambda_0/2$. Note that when $d = \lambda_0/2$ the currents in the parasites are nearly 180° out of phase with the current in the central unit.

The general equations for the three-element array in which the outer units are identical are

$$V_{10} = I_{10}Z_{s1} + I_{20}Z_{12} + I_{30}Z_{13}$$
$$= I_{10}Z_{s1} + 2I_{20}Z_{12}, \quad (19a)$$

$$V_{20} = I_{10}Z_{21} + I_{20}Z_{s2} + I_{30}Z_{23}$$
$$= I_{10}Z_{12} + I_{20}(Z_{s2} + Z_{23}). \quad (19b)$$

When the units are symmetrically driven these reduce to

$$V^s = I_{10}(Z_{s1} + 2Z_{12}), \quad (20a)$$
$$V_2^s = I_{10}(Z_{s2} + Z_{12} + Z_{23}). \quad (20b)$$

When the units are antisymmetrically driven,

$$V^a = I_{10}(Z_{s1} - 2Z_{12}), \quad (21a)$$
$$V_2^a = -I_{10}(Z_{s2} - Z_{12} + Z_{23}). \quad (21b)$$

Comparison with (9a, b) and (18a, b) shows that when $\beta_0 h = \pi/2$ these approximate expressions for the symmetric and antisymmetric impedances are consistent with the following formulas:

$$Z_{s1} \doteq Z_{s2} \doteq \frac{j\zeta_0}{2\pi} C_a(h, h), \quad (22a)$$

$$Z_{12} = Z_{13} \doteq \frac{j\zeta_0}{4\pi} [C_a(h, d - h) + C_a(h, d + h)], \quad (22b)$$

$$Z_{23} \doteq \frac{j\zeta_0}{4\pi} [C_a(h, 2d - h) + C_a(h, 2d + h)]. \quad (22c)$$

The functions Z_{12} and Z_{23} may be obtained from Fig. 34.8 and Table 34.1. It may be concluded that in general, when $\beta_0 h = \pi/2$, the zeroth-order self-impedance of each of n collinear antennas is

$$Z_{s1} = \cdots = Z_{sn} \doteq \frac{j\zeta_0}{2\pi} C_a(h, h)$$
$$= 73.13 + j42.5 \text{ ohms.} \quad (23a)$$

Similarly, the zeroth-order mutual impedances between two collinear antennas numbered

TABLE 34.1. Zeroth-order impedances in ohms for three-element collinear array

$\beta_0 d/\pi$	Z^s_{0c}	Z^a_{0c}	Z^s_{ev}	Z^a_{ev}
1.0	126 $+j82.8$	20.3 $+j2.22$	95.3 $+j62.0$	42.6 $+j21.6$
1.1	114 $+j47.2$	32.6 $+j37.8$	90.3 $+j46.4$	49.5 $+j41.7$
1.2	102 $+j34.5$	43.8 $+j50.5$	86.4 $+j41.0$	56.7 $+j49.0$
1.3	92.4 $+j28.4$	53.8 $+j56.6$	83.2 $+j38.0$	63.9 $+j52.0$
1.4	84.0 $+j26.2$	62.2 $+j58.8$	80.0 $+j35.8$	69.2 $+j52.0$
1.5	77.1 $+j26.5$	69.1 $+j58.6$	76.8 $+j34.8$	72.8 $+j50.8$
1.6	72.0 $+j28.8$	74.3 $+j56.2$	73.8 $+j35.0$	75.0 $+j48.6$
1.7	68.2 $+j31.4$	78.0 $+j53.6$	71.2 $+j35.7$	76.1 $+j46.8$
1.8	65.9 $+j34.7$	80.3 $+j50.4$	69.3 $+j37.4$	76.5 $+j45.2$
1.9	64.9 $+j38.0$	81.4 $+j47.0$	68.2 $+j39.6$	76.4 $+j44.0$
2.0	64.8 $+j41.0$	81.4 $+j44.0$	68.0 $+j41.8$	76.3 $+j43.2$

$\beta_0 d/\pi$	Z_{12}	Z_{23}	$(Z_1)_{1n}$	Z_T
1.0	26.4 $+j20.2$	-4.16 $-j0.725$	53.2 $+j23.6$	-131 $-j41.8$
1.1	20.2 $+j2.35$	-3.00 $+j1.56$	63.6 $+j46.0$	-157 $-j278$
1.2	14.6 $-j4.00$	-1.27 $+j2.50$	71.2 $+j47.4$	-72.0 $-j462$
1.3	9.66 $-j7.05$	0.437 $+j2.46$	74.0 $+j45.6$	103.4 $-j618$
1.4	5.44 $-j8.14$	1.48 $+j1.36$	74.8 $+j44.1$	356 $-j688$
1.5	2.00 $-j8.01$	1.73 $+j0.289$	74.6 $+j42.4$	638 $-j632$
1.6	-0.588 $-j6.84$	1.32 $-j0.747$	74.0 $+j42.0$	947 $-j474$
1.7	-2.45 $-j5.56$	0.538 $-j1.24$	73.2 $+j41.6$	1190 $-j164$
1.8	-3.60 $-j3.92$	-0.184 $-j1.16$	72.5 $+j42.2$	1310 $+j266$
1.9	-4.12 $-j2.24$	-0.805 $-j.704$	72.7 $+j42.1$	1280 $+j782$
2.0	-4.16 $-j0.696$	-0.958 $+j.017$	73.2 $+j42.8$	1020 $+j1290$

$Z_{s1} = Z_{s2} = Z_{s3} = 73.1 + j42.5$

m and n and separated by a distance $(n - m)d$ between centers is

$$Z_{mn} \doteq \frac{j\zeta_0}{4\pi} \{C_a[h, (n - m)d - h]$$
$$+ C_a[h, (n - m)d + h]\}. \quad (23b)$$

For collinear elements that are *not* near $\beta_0 h = \pi/2$ in electrical half-length the self- and mutual impedances cannot be expressed in this simple form. Only when $\beta_0 h \doteq \pi/2$ and the identical antennas are sufficiently thin so that Ψ_{ev}^s and Ψ_{ev}^a are large compared with unity are the simple relations (22a, b, c) and (23a, b) moderately good approximations. In general, if the antennas are sufficiently thin so that (23a) is a good approximation, it may be assumed that (23b) is an even better approximation. Values of the several zeroth-order impedances involved in the three-element collinear array are given in Table 34.1.

The odd currents in the parasitic elements are evaluated in Sec. 33 specifically when $\beta_0 h = \pi/2$. The approximate expression given in (33.40) is

$$I_{2od}(z_2) \doteq \frac{V_{10}}{\Psi_{od}} K \sin 2\beta_0 z_2, \quad (24a)$$

where

$$K \equiv \frac{\pi}{4} \left[\frac{D^s(d)}{Z_{0c}^s(1 + k^{-1})} + \frac{D^a(d)}{Z_{0c}^a(1 + k)} \right], \quad (24b)$$

where $D^s(d)$ and $D^a(d)$ are given by (33.38b) respectively with the upper and lower sign, and where k is defined in (1b). It is now readily verified that with $\beta_0 h = \pi/2$ the odd part of the current in the parasites is quite small compared with the even parts. Thus, when $\beta_0 d \doteq \pi$, $h/d \doteq 1/2$,

$$D^s(d) \doteq \frac{1}{\pi^2}\left(-\frac{4}{3} + \frac{8}{15}\right) = -0.08, \quad (25a)$$

$$D^a(d) \doteq \frac{1}{\pi^2}\left(\frac{4}{3} + \frac{8}{15}\right) = 0.19. \quad (25b)$$

Also, with magnitudes in ohms,

$$Z_{0c}^s(1 + k^{-1}) \doteq 196.9\underline{/28°.3},$$
$$Z_{0c}^a(1 + k) \doteq 82.7\underline{/22.2°}. \quad (26)$$

With these values,

$$K \text{ (mhos)} \doteq 1.56 \times 10^{-3}\underline{/-21°} \quad (27)$$

so that with $\Omega = 12.5$, $\Psi_{od} = 10$,

$$I_{od}(z_2) \doteq 1.56 \times 10^{-4} V_{10} e^{-j21°} \sin 2\beta_0 z_2. \quad (28)$$

The corresponding value for the even current with $\beta_0 d \doteq \pi$ is obtained from (13c) with Z_T (ohms) $\doteq 137.7\underline{/-162°}$:

$$I_{ev}(z_2) \doteq \frac{V_{10}}{Z_T} \cos \beta_0 z_2$$
$$= 72.8 \times 10^{-4} V_{10} e^{j162°} \cos \beta_0 z_2. \quad (29)$$

Finally, with I in amperes, V in volts,

$$I_2(z_2) \doteq 72.8 \times 10^{-4} V_{10} e^{j162°} [\cos \beta_0 z_2$$
$$- e^{-j3°} 0.022 \sin 2\beta_0 z_2]. \quad (30)$$

It is seen from (30) that the amplitude of the odd current is only 2 percent of the amplitude of the even current for $\Omega = 12.5$. For thicker antennas the odd currents are relatively greater. The current in (30) is shown in Fig. 34.7 together with the current in the central unit.

35. The Center-Driven Collinear Array with Phase-Reversing Stubs

It is shown in Sec. 34 that the currents in the two outer units of a three-element collinear array are nearly opposite in direction and equal in magnitude to the current in the central unit if the distance between adjacent ends of the elements is very small. The electromagnetic field of a distribution of current of this kind is described in Chapter VI. It is characterized by large maxima of the electromagnetic field both along the plane bisecting the antenna and along symmetric cones in the two hemispheres. Between these maxima are deep minima. Whereas a field pattern of this type is useful for certain applications, it obviously is undesirable when a field is required that has a large amplitude only in the central plane and as small an amplitude as possible in all other directions. It is shown in Chapter VI that such a field is maintained by a collinear array if the currents in all of the elements are about equal in magnitude and instantaneously in the *same* direction, that is, if all currents are in phase. Obviously, such a distribution can be maintained in the antennas if each unit is individually center-driven by appropriate voltages, but, in practice, this involves long transmission lines in nonneutral planes with consequent unbalanced currents which contribute significantly and objectionably to the radiation field. A much better way is to drive only the central unit from a single transmission line in the neutral plane, and to excite the outer units by appropriate coupling networks that connect the adjacent ends of the several units. The function of such coupling networks is to *reverse the phase* of the currents in the outer antennas. Hence, and appropriately, they are called *phase-reversing* networks.

The simplest type of phase-reversing network is a high-impedance transmission-line section or stub. In Fig. 35.1a a three-element array is shown with two such *phase-reversing stubs* constructed of low-loss two-wire line. If properly adjusted in length to near $\lambda_0/4$, the transmission-line impedance of each stub to equal and opposite currents is very great. Although the length of the stubs in Fig. 35.1 is shown to be $\lambda_0/4$, this is only approximate, since account must be taken of end effects and coupling effects if conventional transmission-line equations are to be used. This may be done for each stub by connecting a capacitance C_{Te} (see Sec. II.8) across the ends attached to the antennas, and an inductance $L_{Te}/2$ in series with each line wire and at both ends. Alternatively, each stub may be adjusted in length experimentally to have a maximum voltage across the ends attached to the antenna when the stub is excited by a generator of the proper frequency loosely coupled to the terminating bridge. If the distance b between the parallel conductors of the stub is sufficiently small to satisfy the inequality $\beta_0 b \ll 1$, terminal-zone corrections are not large, and the lengths of the stubs when adjusted for a high-impedance input in the presence of the antennas does not differ greatly from $\lambda_0/4$. (Since the capacitance C_{Te} of the corrective network is negative, the stub must be somewhat longer than $\lambda_0/4$. However, this effective lengthening due to end effect and coupling is often compensated in practice by the presence of dielectric supports, which are equivalent to a positive capacitance in parallel.)

The approximate determination of the distribution of current and the driving-point impedance of the array in Fig. 35.1a may be carried out in the same manner as for the collinear array with parasitic elements. That is, the analysis is resolved into two parts the superposition of which yields the desired result. These are (1) the symmetric part with all three units individually center-driven by slice generators with voltages that maintain equal currents in phase at the centers of the antennas, and (2) the antisymmetric part, in which the driving voltages are so chosen that the currents at the centers of the outer antennas are reversed. That is, they equal the current in the central unit in magnitude but differ in phase by 180°. Let these two parts be considered in turn and combined later.

The symmetric problem. Let the voltages required to maintain equal currents in all three units, $I_1^s(0) = I_2^s(z_2 = 0) = I_3^s(z_3 = 0)$, be represented as in (33.20a):

$$V_{10} = V_1^s = V^s, \quad V_{30} = V_{20} = V_2^s = f^s V^s.$$

(1)

It is shown below that f^s in (1) is effectively the same function as in (33.20b) even though the array differs from that involved in Sec. 33 by the presence of the stubs. Since the currents that seek to enter the conductors of the stubs from the attached antennas necessarily are very nearly equal and opposite, and since the stubs are adjusted to offer an extremely high impedance to equal and opposite transmission-line currents, the currents actually existing at the ends of the antennas where they are connected to the stubs are very small. This does not mean that transmission-line currents all along each stub are as small as the input current. Actually, the distribution of transmission-line current on the high-impedance stubs is approximately sinusoidal with a very small amplitude at the input end and a very large amplitude at the terminating bridge. However, with closely spaced good conductors the ohmic losses in the stub and the radiation losses from it are insignificant compared with radiation from the array as a whole, and they contribute nothing to the far-zone $1/R$ field. Accordingly, the presence or absence of the stubs when the three antennas are driven in phase with equal currents at their centers is of no interest. Indeed, there can be no significant change in either the input impedance of the array or the distribution of current in the three antennas if the stubs are removed as in Fig. 35.1b. But this leaves precisely the circuit already analyzed in Sec. 34, where it constitutes the symmetric part of the collinear array with parasitic outer units. The symmetric currents $I_1^s(z)$ in the central unit, No. 1, are as given in (34.2a) or (34.5a); the symmetric impedance Z_{0c}^s of the central unit is given in (34.9a) and graphically in Fig. 34.1. The currents in the outer units are well approximated by (34.10a) or by (34.13c), since the odd components of current are so small as to be negligible in determining the far-zone field of the array. This is shown in

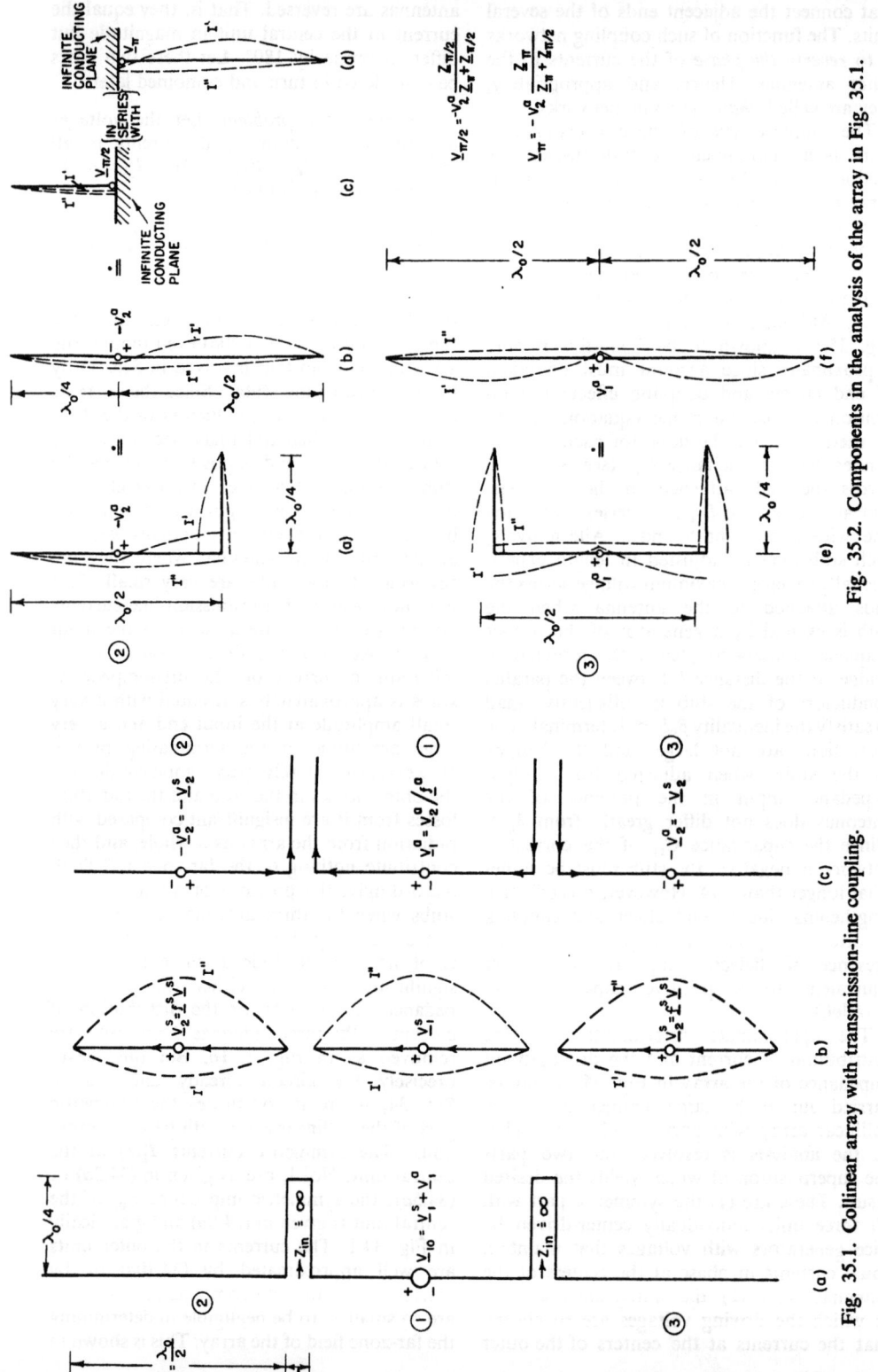

Fig. 35.1. Collinear array with transmission-line coupling.

Fig. 35.2. Components in the analysis of the array in Fig. 35.1.

Sec. 34. The approximate currents are

$$I_1^s(z) \doteq \frac{V^s}{Z_{0c}^s} \cos \beta_0 z, \qquad (2a)$$

$$I_3^s(z_3) = I_2^s(z_2) \doteq \frac{V_2^s}{Z_{ev}^s} \cos \beta_0 z_2$$

$$\doteq \frac{V^s f^s}{Z_{ev}^s} \cos \beta_0 z_2 \doteq I_1^s(z), \qquad (2b)$$

where Z_{0c}^s is in (34.9a), Z_{ev}^s in (34.18a) and $f^s = Z_{ev}^s/Z_{0c}^s$.

The antisymmetric problem. Let the driving voltages required to maintain the currents $I_3^a(z_3 = 0) = I_2^a(z_2 = 0) = -I_1^a(0)$ in the array in Fig. 35.1a be represented as in (33.21a):

$$V_{10} = V_1^a = V^a,$$
$$V_{30} = V_{20} = -V_2^a = -f^a V^a. \qquad (3)$$

The function f^a in (3) is not the same as in (33.21b). Since the currents in the two parallel conductors of the phase-reversing stubs are very nearly equal and codirectional, these two conductors are always at the same potential at opposite points. Therefore, the terminating bridge may be removed (or replaced by an equivalent open end) without significant effect. Thus, the solution of the antisymmetric problem for Fig. 35.1a reduces to the analysis of Fig. 35.1c. Although this presents a quantitatively intricate problem, much valuable information on the distribution of current and the operation of the array, and even a very satisfactory approximation of the impedance, can be acquired if the accurate knowledge of the symmetric currents that are primarily significant in the impedance and the far-zone field can be supplemented with even a rough determination of the antisymmetric currents, which contribute little to both the impedance and the far-zone field. Since this type of investigation often is a necessary recourse in the field of applied electromagnetism where the quantitatively accurate solution of many important problems is all but impossible in the sense of a mathematically acceptable boundary-value problem, the discussion is carried out in detail in order to illustrate the method of reasoning by analogy and judicious approximation. The sequence of steps to be followed is to consider each antenna first as though isolated and then as a coupled system.

It has been shown in the case of parallel, coupled antennas that the general nature of the distribution of current in an antenna that is coupled to another is determined in *major outline* by the dimensions of the antenna itself, even if the coupling is quite close. That is, the distribution of current in the antenna when isolated is dominant in a qualitative sense even when the antenna is coupled to another. Therefore, the first inquiry in the study of Fig. 35.1c is directed to the behavior of each of the three antennas with its attached half of the now open stub when isolated. Since the array itself is symmetric, only the central unit 1 and the upper unit 2 need be considered. Antenna 3 is like antenna 2. It is reasonable to assume that the effects of the right-angle bends in the conductors can, at most, lead to a small quantitative modification, but not to an essential change in the distributions of current and charge. This is indicated from the study of the V-antenna (Secs. 23–25). Therefore, the distributions of current in antennas 1 and 2 in Fig. 35.1 may be approximated by the distributions in the same conductors when these are straightened as in Fig. 35.2. This makes the central unit a center-driven antenna of electrical half-length $\beta_0 h \doteq \pi$. The distribution of current on such an antenna is shown in Fig. II.22.5. It is reproduced in Fig. 35.2f. Since it is assumed that the right-angle bends in the conductor do not greatly affect the distribution of current, the distribution on the straight antenna in Fig. 35.2f may be transferred to the bent antenna in Fig. 35.2e as an approximation. The impedance is simply

$$Z_{0c}^a \doteq Z_0(\beta_0 h = \pi) \equiv Z_\pi. \qquad (4)$$

The bent antenna 2 in Fig. 35.2a may be approximated by the asymmetrically driven straight antenna in Fig. 35.2b. This is analyzed in Secs. 27–29, where it is shown that in solving for the current in either half this half may be assumed to be erected on an infinite conducting plane and driven by an appropriate slice generator with a voltage given in (27.28). For the upper half with an electrical length $\beta_0 h = \pi/2$ and an impedance $\tfrac{1}{2}Z_0(\beta_0 h = \pi/2) \equiv (Z_{\pi/2})/2$, the driving voltage is

$$V_{\pi/2} = -V_2^a Z_{\pi/2}/(Z_\pi + Z_{\pi/2}); \qquad (5a)$$

for the lower half, which has an electrical length $\beta_0 h = \pi$ and an impedance $\tfrac{1}{2}Z_0(\beta_0 h = \pi) \equiv (Z_\pi)/2$, the driving voltage is

$$V_\pi = -V_2^a Z_\pi/(Z_{\pi/2} + Z_\pi). \qquad (5b)$$

Since the two antennas are in series, the resultant driving voltage is $-V^a$, as assumed in (3). The impedance is

$$Z_2^a = \tfrac{1}{2}(Z_\pi + Z_{\pi/2}). \qquad (6)$$

The distributions of current on the two dissimilar parts of the outer antennas are shown in Fig. 29.2 and in Fig. 35.2c, d with phases referred to $-V_2^a$. It is assumed that these distributions for isolated straight asymmetric antennas approximate the distributions on the bent antenna in Fig. 35.2a. Therefore, these distributions have been transferred as shown.

With approximate distributions of current in the three antennas of Fig. 35.1c determined under the assumption that they are isolated, it remains to evaluate the effect of their interaction. An estimate of this may be obtained as follows. The most closely coupled parts of antennas 1 and 2 are the parallel ends at right angles to the axis of the array. These parts carry codirectional currents. It is seen in Fig. 6.2 that the effect upon a given isolated antenna of a closely spaced parallel antenna with equal codirectional current is to increase the expansion parameter somewhat. That is, Ψ'_{s1} for two symmetrically driven, parallel antennas is greater than Ψ'_{K1} for the isolated antenna if b/λ_0 is small. For close spacing this increase may be as large as from 3 to 6. For thick antennas with $\Omega = 2\ln(2h/a)$ not very large, the percentage change in Ψ'_{K1} is great; for very thin wires it is quite small. The collinear parts of antennas 1 and 2 carry currents that are practically equal and opposite. It has been shown in Sec. 33 (and it may be inferred from Fig. 3.3 for parallel antennas) that a coupled antenna carrying an equal and opposite current reduces the expansion parameter approximately in the same proportion as it is increased when the currents are codirectional. Thus, antenna 1 in proximity to antennas 2 and 3 in Fig. 35.1c behaves roughly like a straight isolated antenna of the same length of conductor which has outer halves that are effectively of smaller radius and inner halves that are effectively of greater radius than when isolated. The same reasoning applies to the outer antennas.

To summarize, it is evident that a good estimate of the distribution of current and of the driving-point impedance of the three-element collinear array in Fig. 35.1a may be obtained by superimposing the solution of the symmetrically driven array in Fig. 35.1b upon the solution of the antisymmetrically driven array in Fig. 35.1c. In the latter, the currents in the central unit are like those in Fig. 35.2f. The resultant current is the algebraic sum of these symmetric and antisymmetric currents. In addition, there is the transmission-line distribution of equal and opposite currents on the stubs with large amplitude in the connecting bridges. Presumably the method applied to the three-element array may be extended to any number of units. Note that the approximations involved improve as the value of $\Omega = 2\ln(2h/a)$ increases.

Since the amplitude of the current in an antenna of electrical half-length $\beta_0 h = \pi$ is very much smaller than in an antenna for which $\beta_0 h = \pi/2$ when both are driven by the same voltage (for $\Omega = 20$ the ratio of maximum currents is roughly 1 to 12), and since the antisymmetrically driven array in Fig. 35.2c corresponds roughly to elements with $\beta_0 h = \pi$, whereas the symmetrically driven array in Fig. 35.2b corresponds roughly to elements with $\beta_0 h = \pi/2$, it follows that even for moderately large ratios h/a, the predominating currents are the symmetric currents of Fig. 35.1b as given in (1). *It is primarily these currents that determine the far-zone electromagnetic field.* Note that they are codirectional in the several units. That is, the phase-reversing stubs have, indeed, effectively reversed the *significantly radiating currents* in the outer units. Additional currents on the stubs and antennas serve principally to maintain the codirectional currents on the antennas. They contribute negligibly to the radiation field either because they are small (additional currents on antennas and codirectional currents on stubs—these may be reduced further in their effect on the field by using two coplanar stubs in parallel) or, if large, because they are equal, opposite, and close together (transmission-line currents on the stubs).

The driving-point admittance of the central unit is formally like (32.29b), namely,

$$Y_{in} = \frac{I_1(0)}{V_{10}} = \frac{I_1^s(0) + I_1^a(0)}{V^s + V^a}$$

$$= \frac{1}{Z_{0c}^s(1 + k^{-1})} + \frac{1}{Z_{0c}^a(1 + k)}, \quad (7)$$

where Z_{0c}^s is given by (34.9a) and is plotted in Fig. 34.1, $Z_{0c}^a = Z_\pi$, and

$$k = \frac{f^a}{f^s}, \quad f^a = \frac{Z_2^a}{Z_{0c}^a} = \frac{Z_\pi + Z_{\pi/2}}{2Z_\pi},$$

$$f^s = \frac{Z_2^s}{Z_{0c}^s} = \frac{Z_{ev}^s}{Z_{0c}^s}, \quad (8)$$

where Z_{ev}^s is given by (34.18a) and is plotted in Fig. 34.2. If the separation b of the conductors of the transmission-line stubs is very small, $d/\lambda_0 \doteq 0.5$. For this value $Z_{0c}^s \doteq 126 + j83$, $Z_{ev}^s \doteq 95 + j62$, $f^s \doteq 0.75$. For sufficiently thin antennas, $|Z_\pi| \gg |Z_{\pi/2}|$, so that $f^a \doteq 0.5$. Hence $k \doteq 2/3$. With this value,

$$Z_{in} = \frac{1}{Y_{in}} \doteq \frac{Z_{0c}^s Z_\pi (1 + \tfrac{2}{3})(1 + \tfrac{3}{2})}{Z_{0c}^s(1 + \tfrac{3}{2}) + Z_\pi(1 + \tfrac{2}{3})}. \quad (9)$$

Since for thin antennas $|Z_\pi| \gg |Z_{0c}^s|$, it follows that

$$Z_{in} \doteq Z_{0c}^s(1 + \tfrac{3}{2}) = 2.5(126 + j83)$$
$$= 315 + j207 \text{ ohms.} \quad (10)$$

This value corresponds to $73.1 + j42.5$ for a single unit. More accurate numerical results may be obtained for specific values of Z_π and $Z_{\pi/2}$ if desired. Other values of d/λ_0 may be used provided the spacing b of the stub wires satisfies $\beta_0 b \ll 1$.

If all three units are individually driven by the voltages required to make the currents all equal and in phase, $I_{10} = I_{20} = I_{30} = I_0$, the circuit equations (34.20a, b) apply. Using Table 34.1 the zeroth-order input impedances are

$$(Z_1)_{in} = Z_{s1} + 2Z_{12}$$
$$= 125.9 + j82.9 \text{ ohms} \quad (11a)$$

$$(Z_2)_{in} = (Z_3)_{in} = Z_{s1} + Z_{12} + Z_{13}$$
$$= 95.3 + j42.0 \text{ ohms} \quad (11b)$$

The total power supplied to the array is

$$P = I_0^2[(R_1)_{in} + 2(R_2)_{in}] = 316.5 I_0^2 \text{ watts.} \quad (12)$$

Thus, the radiation resistance of the array referred to the input currents which are maintained equal in all three units is $R_0^e = R_m^e = 316.5$ ohms. Note that this differs negligibly from the approximate *input* resistance of the center-driven three-element array with phase-reversing stubs given in (10) and that it is in exact agreement with the results given in Table VI.2, 1, which are obtained by the conventional method of integrating the normal component of the Poynting vector over the surface of a great sphere using (I.10, 6).

Although the investigation of the three-element center-driven collinear array with phase-reversing stubs is not mathematically completely rigorous, it does provide an excellent picture of the several components of current and their distributions. Moreover, it justifies the conventional practice of assuming currents in the elements of the array to be codirectional, and sinusoidally distributed with equal amplitude in determining radiation-field patterns. That is, the dominant, symmetric distribution in Fig. 35.1b approximates the distribution

$$I_z \doteq I_0 \cos \beta_0 z_i, \quad i = 1, 2, 3, \quad (13)$$

where z_i is measured from the center of antenna i.

If the phase-reversing stubs or sections of two-wire line are replaced with coaxial sleeves, parallel-resonant circuits of lumped elements, or a self-resonant coil, the symmetric problem is unchanged. On the other hand, the antisymmetric problem is quite different owing to the absence of the quarter-wave sections projecting out from the antenna. The antisymmetric impedances are only slightly greater than the symmetric ones, so that the radiating currents are a superposition of symmetric and antisymmetric currents of *comparable* amplitude and distribution. It follows that the currents in the outer elements will be smaller than and not in phase with the current in the central unit, so that these elements are not effective as phase-reversing sections.[19]

36. Experimental Study of the Collinear Array; Lumped Capacitive Coupling

The approximate distributions of current along the three elements of a symmetric collinear array consisting of a driven central unit and two parasitic radiators as shown in Fig. 32.2b have been determined experimentally by Hatch.[30] The quantity actually measured was the distribution of the transverse electric field along three collinear slots in a highly conducting plane, as shown in Fig. 36.1. Since this arrangement is complementary to a collinear array of flat-strip conductors in space in the sense that the boundary conditions on the planes containing the antennas and the slots involve an interchange of electric and magnetic fields, the distribution of transverse electric field along the slots is the same as the distribution of transverse magnetic field (which is proportional to the current) along the conducting strips. The slots in Fig. 36.1 are center driven from a coaxial line in a manner that maintains an axial magnetic field in a narrow band at the center. This is approximately complementary to a narrow band of axially directed electric field or a "slice" generator at the center of a

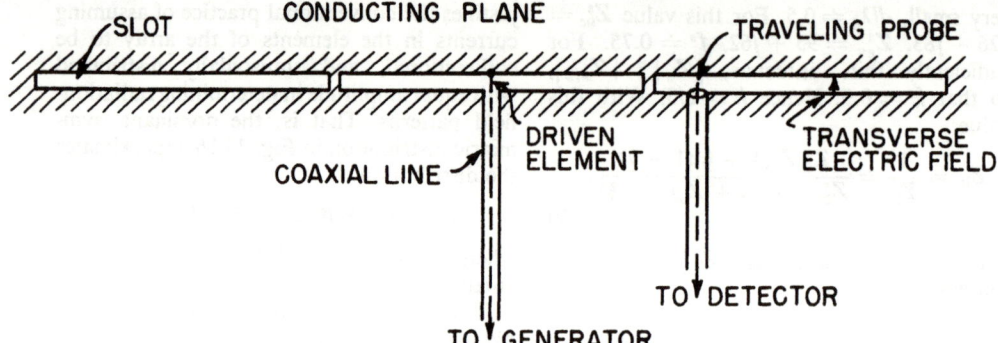

Fig. 36.1. Complementary slot arrangement for three-element collinear array.

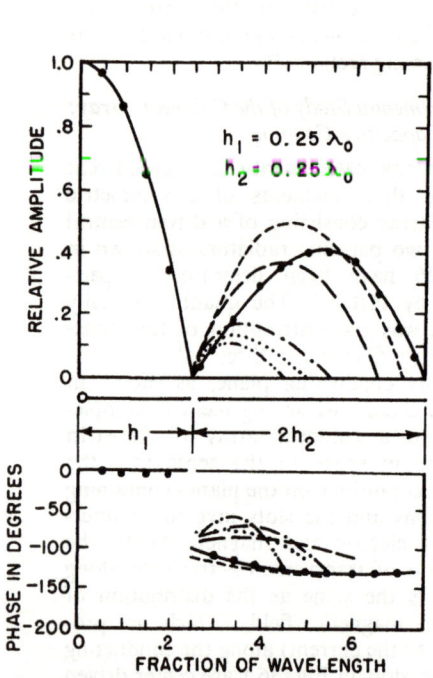

Fig. 36.2. Current distribution in three-element collinear array as determined from complementary slots; length of parasitic element varied (Hatch).

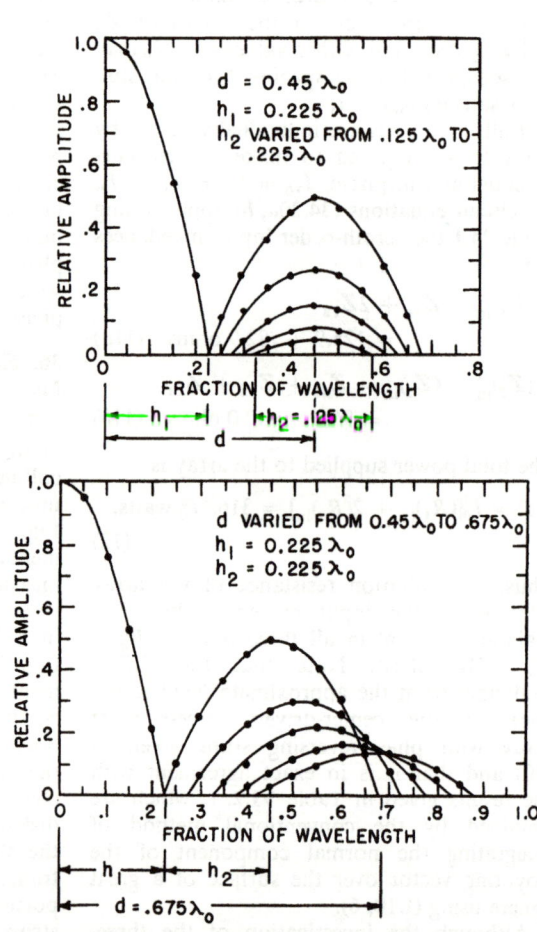

Fig. 36.3. Current distribution in three-element collinear array as determined from complementary slots; distance between elements varied (Hatch).

conductor. The principle of complementarity and Babinet's principle are considered in greater detail as applied to a closely related problem in Sec. V.11.

Distributions obtained by Hatch are shown in Figs. 36.2 and 36.3, where, however, only half of the symmetric distribution curves are shown. In Fig. 36.2 the half-length of the central unit is $h_1 = 0.25\lambda_0$, that of the adjacent outer units, $h_3 = h_2$, is varied from $h_2 = 0.1\lambda_0$ to $0.25\lambda_0$, and the distance between centers is $d \doteq h_1 + h_2$, so that the units just do not make contact. When $h_1 = h_2 = h_3 = 0.25\lambda_0$, the conditions correspond to those assumed for the theoretical curves in Fig. 34.7 except that flat strips instead of thin cylinders are involved. The agreement between theoretical values of *relative* amplitude and phase is seen to be excellent. Note, however, that this does not mean that zeroth-order *admittances* necessarily are quantitatively accurate either for thin strips or for cylinders of finite radius. It does suggest that *quantitatively* useful results should be obtainable from the zeroth-order *change* in admittance due to the presence of the collinear parasites if this change is superimposed on the second-order admittance of the central unit when isolated. The other curves in Fig. 36.2 represent the experimentally determined distribution as the lengths of the parasites are reduced progressively from $h_2 = h_3 = 0.25\lambda_0$. For simplicity, experimental points are omitted. It is interesting to note that when the parasitic antennas are self-resonant, in this case with $h_3 = h_2 \doteq 0.225\lambda_0$, the current in the parasitic units is maximum, as it should be.

The effect of increasing the separation of the parasitic units and the central antenna is shown theoretically in Fig. 34.6 and experimentally in Fig. 36.3. Unfortunately, the theoretical data are for $h_1 = h_2 = h_3 = 0.25\lambda_0$, whereas the experimental results are for $h_1 = h_2 = h_3 = 0.225\lambda_0$. Nevertheless, the rapid decrease in current in the parasites predicted in Fig. 34.6 is confirmed in Fig. 36.3.

The distribution of current along and the input impedance of a half-collinear array base-driven over a large ground screen approximately 10 wavelengths on a side have been measured by Andrews.[3] The driven element of length h_1 consisted of the extension of the slotted inner conductor of the coaxial measuring line in an arrangement essentially like that shown in Fig. 26.4 of Chapter II. The parasitic element of length $2h_2$ was supported by Styrofoam so that the gap between its adjacent end and the end of the driven element could be varied from zero with the elements in contact to a reasonably large value. The current in the parasite could be measured using an external shielded-loop probe supported by Styrofoam.

Twice the impedance of the half-collinear array as a function of the gap length between the adjacent ends of the driven and parasitic element as measured by Andrews for a rather thick antenna is shown in Fig. 36.4. The same data corrected for the end effect of the coaxial line (Secs. 10 and 38 of Chapter II) also are shown. On the same graph is a plot of the theoretical zeroth-order impedance $(Z_1)_\text{in}$ as obtained from Table 34.1. Since the zeroth-order impedance of the driven unit when isolated is known to be a poor approximation of the actual impedance of a fairly thick antenna, a better comparison is obtained by superimposing the zeroth-order *variation* in impedance, as the parasitic element is moved in from infinity, on the third-order impedance of the driven element when isolated. The impedance curve obtained in this manner also is shown in Fig. 36.4. It is seen to approximate the corrected measured curve except when the length of the gap becomes very small. The reason for the poor agreement between the theoretical and the experimental curves at very small gap lengths is considered in the next paragraph. Reference to the curve marked Loaded Dipole Theoretical (Andrews) in Fig. 36.4 is made later in this section.

The theory developed in Sec. 32 through 34 is essentially one-dimensional in that the only chargeable surfaces of which account is taken are those forming the outer surfaces of the cylinders. In an actual antenna, regardless of whether it has hemispherical, flat, or tubular ends, there are always additional chargeable surfaces which become increasingly important as two such ends are brought closer and closer together. It follows that, with sufficiently small gaps, large, and essentially lumped, capacitances couple the driven and parasitic elements at their adjacent ends. Evidently, good agreement cannot be expected between experimental values and a theory that does not take account of these capacitances.

It has been pointed out by Andrews[3] that, as the parasitic elements in a collinear array are moved from infinite distances to actual contact with the driven element, the input impedance must range continuously from the

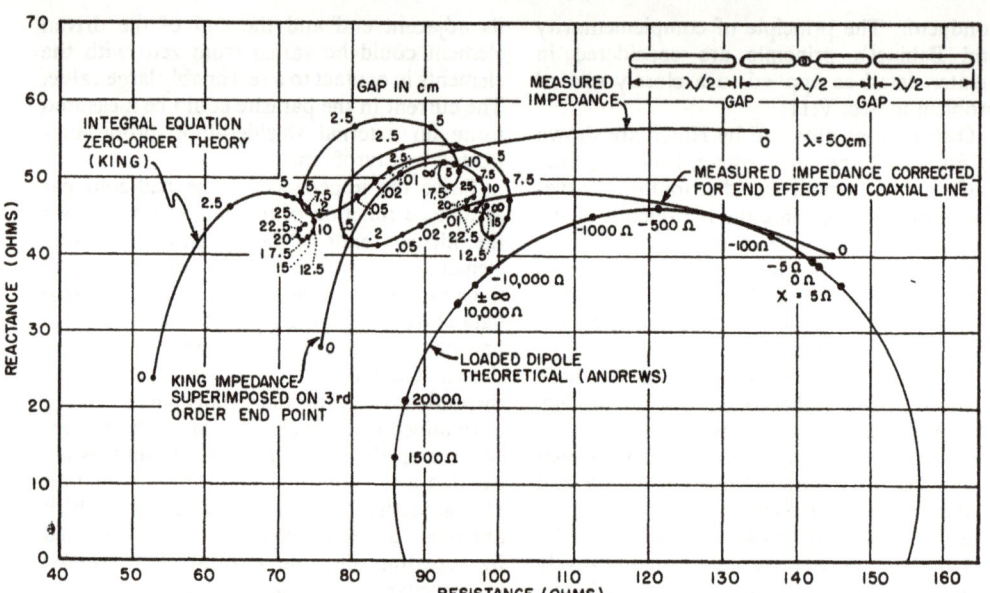

Fig. 36.4. Comparison of theories and measurements for collinear array.

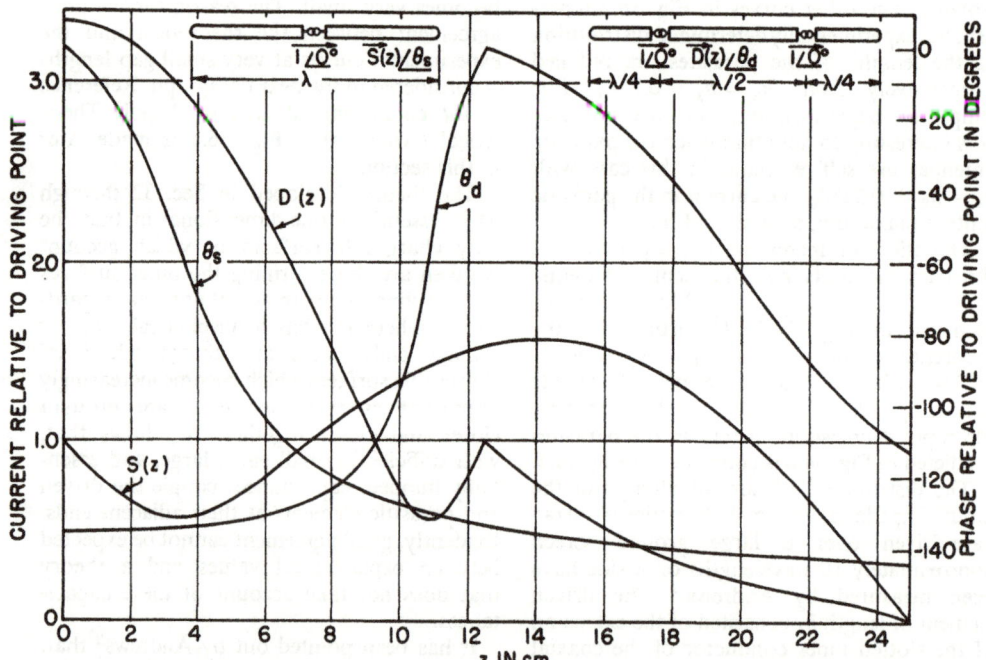

Fig. 36.5. Theoretical current distributions on half of singly and doubly driven dipoles (Andrews).

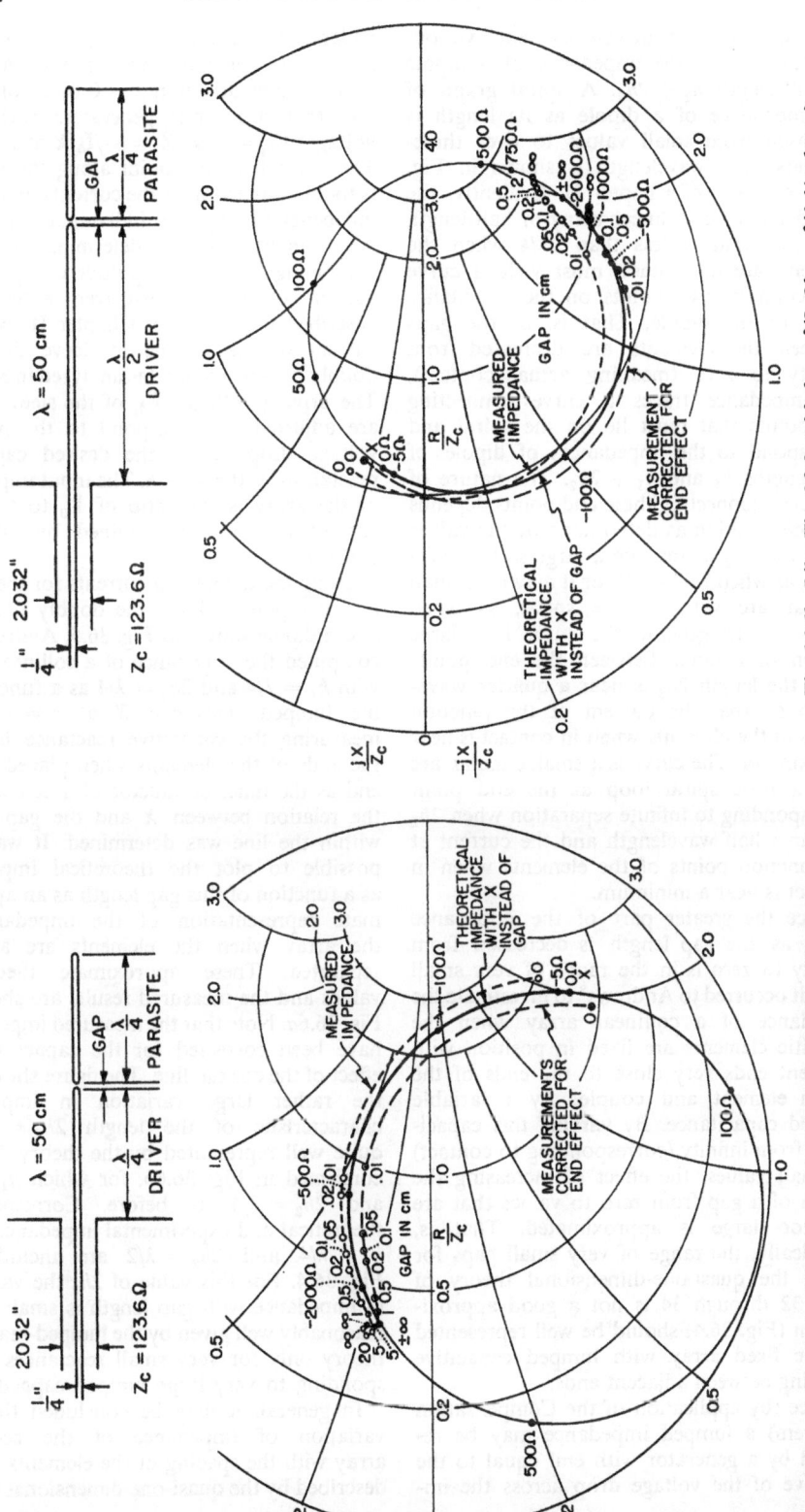

Fig. 36.6a. Theoretical and experimental impedance of collinear array coupled by variable lumped reactance X for theoretical curves and with variable gap for measured curves; $h_1 = \lambda/4$, $2h_2 = \lambda/4$ (Andrews).

Fig. 36.6b. Like Fig. 36.6a with $h_1 = \lambda/2$, $2h_2 = \lambda/4$ (Andrews).

impedance of the driven element without the parasites to the impedance of a dipole of half-length $h_1 + 2h_2$. A spiral graph of the impedance of a dipole as its length is increased from small values to over three-quarters of a wavelength is shown in Fig. 30.3b or 30.3c of Chapter II. Evidently, the impedance of a collinear array of half-length $h_1 + 2h_2$, that is less than $3\lambda/4$ when the elements are in contact, must yield a curve that connects two points on the impedance spiral of the dipole. That is, as the gaps between the elements are decreased from infinity to zero (meaning actual contact), the impedance traces a curve connecting end points that must lie on the spiral and correspond to the impedances of dipoles of half-lengths h_1 and $h_1 + 2h_2$. The nature of the curve connecting these end points depends on their location as determined by the values of h_1 and $2h_2$. It may be a large section of a circle, as when $2h_2 = \lambda/4$, or it may be a small circular arc with a little spiral, as when $2h_2 = \lambda/2$. In general, the curve is a large section of a circle between the end points when the length $2h_2$ is near a quarter wavelength so that the current at the junction points of the elements when in contact is near a maximum. The curve is a small circular arc with a little spiral loop at the end point corresponding to infinite separation when $2h_2$ is near a half wavelength and the current at the junction points of the elements when in contact is near a minimum.

Since the greater part of the impedance curve as the gap length is decreased from infinity to zero is in the range of very small gaps, it occurred to Andrews[3] to investigate the impedance of a collinear array when the parasitic elements are fixed in position with adjacent ends very close to the ends of the driven element and coupled by a variable lumped capacitance. By varying this capacitance from infinity (corresponding to contact) to small values, the effect of increasing the length of a gap from zero to values that are not too large is approximated. That is, specifically, the range of very small gaps for which the quasi-one-dimensional theory of Secs. 32 through 34 is not a good approximation (Fig. 36.4) should be well represented by the fixed array with lumped capacitive coupling between adjacent ends.

Since (by application of the Compensation Theorem) a lumped impedance may be replaced by a generator with emf equal to the negative of the voltage drop across the impedance, the problem investigated by Andrews involves a single dipole of half-length $h_1 + 2h_2$ that is center driven at $z = 0$ by a voltage V_0 and that has series generators with equal voltages $V_{h_1} = -I_{h_1}Z = -jI_{h_1}X$ at $z = \pm h_1$. The current at any point along the antenna is the superposition of the currents maintained independently by the individual generators. Accordingly, it may be determined readily by combining in proper amplitude and phase the current of the center-driven antenna (as described in Sec. 22 of Chapter II) with the current of the symmetric sleeve dipole or doubly driven antenna (as analyzed in Sec. 30). The driving voltages V_{h_1} of the sleeve dipole are adjusted to correspond to the negative voltage drop across the desired capacitive reactances X at $z = \pm h$. The input impedance of the array is the ratio of V_0 to the *total* current at $z = 0$ maintained by all three generators.

Using the theoretical currents for the singly driven dipole and for the doubly driven or sleeve dipole shown in Fig. 36.5, Andrews has computed the impedance of a collinear array with $h_1 = \lambda/4$ and $2h_2 = \lambda/4$ as a function of the lumped reactance X at $z = \pm h$. By measuring the capacitive reactance between the ends of the elements when placed end to end as the inner conductor of a coaxial line, the relation between X and the gap length within the line was determined. It was then possible to plot the theoretical impedance as a function of this gap length as an approximate representation of the impedance of the array when the elements are actually separated. These approximate theoretical values and the measured results are shown in Fig. 36.6a. Note that the measured impedances have been corrected for the capacitive end effect of the coaxial line. The figure shows that the rather large variation in impedance characteristic of the length $2h_2 = \lambda/4$ is quite well represented by the theory. This is confirmed in Fig. 36.6b, for which $h_1 = \lambda/2$ and $2h_2 = \lambda/4$ as before. Corresponding theoretical and experimental impedances with $h_1 = \lambda/4$ and $2h_2 = \lambda/2$ are included in Fig. 36.4. For this value of $2h_2$ the variation in impedance with gap length is small and is reasonably well given by the lumped-reactance theory only for very small reactances corresponding to very large lumped capacitances.

In general, it may be concluded that the variation of impedance of the collinear array with the spacing of the elements is well described by the quasi-one-dimensional theory

[III.36] THEORY OF LINEAR ANTENNAS 453

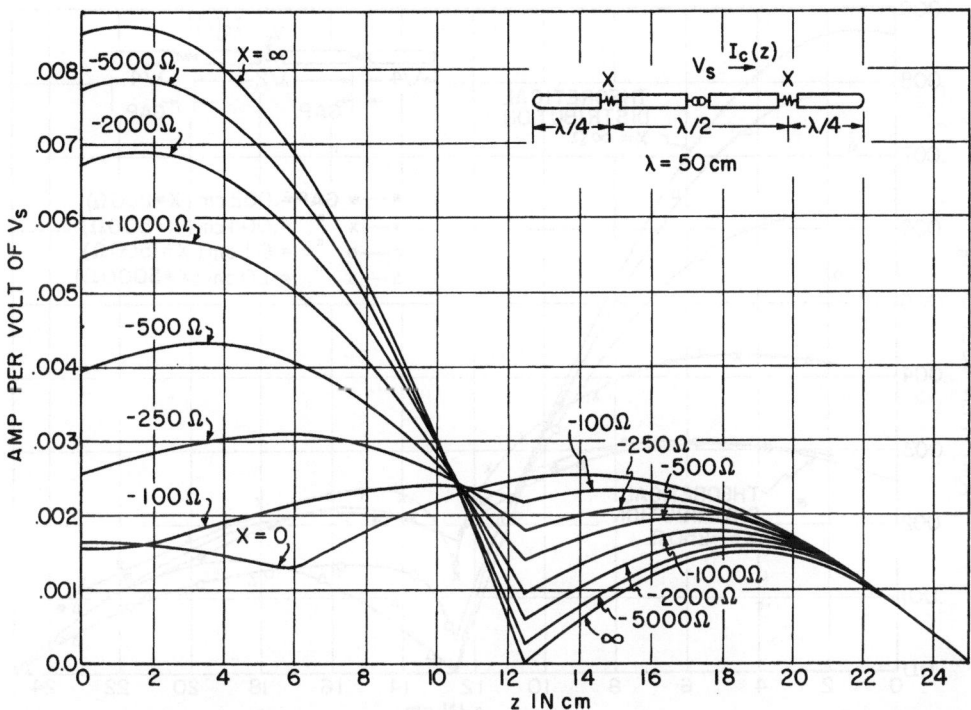

Fig. 36.7. Theoretical distribution of current amplitude on half of collinear array with the series reactance X at $\lambda/4$ from each end as parameter (Andrews).

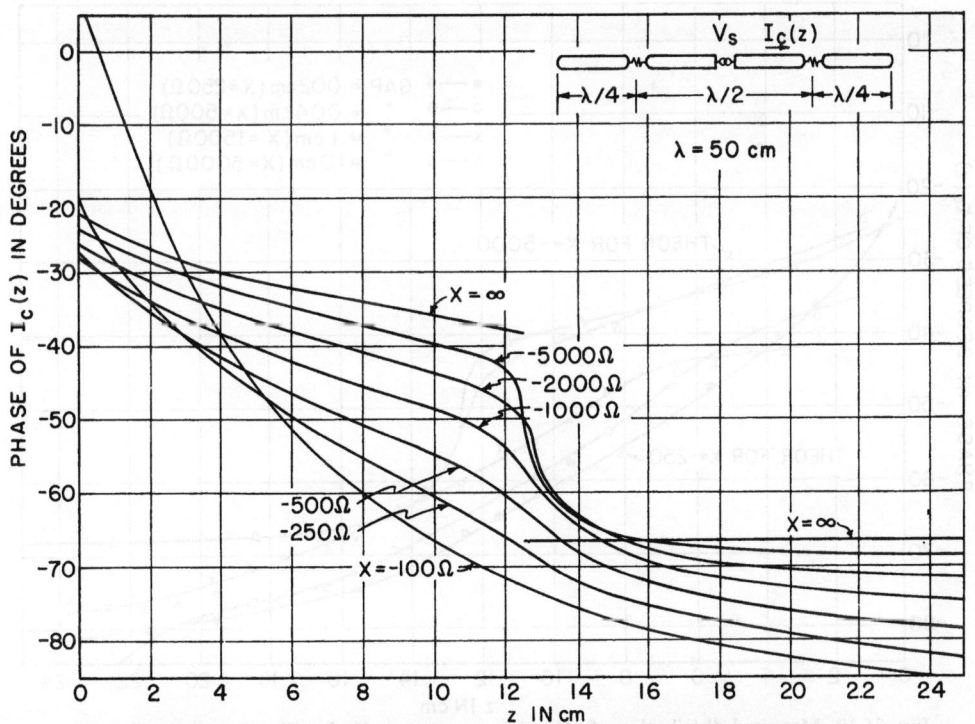

Fig. 36.8. Phase of theoretical distribution of current on half of collinear array (Andrews).

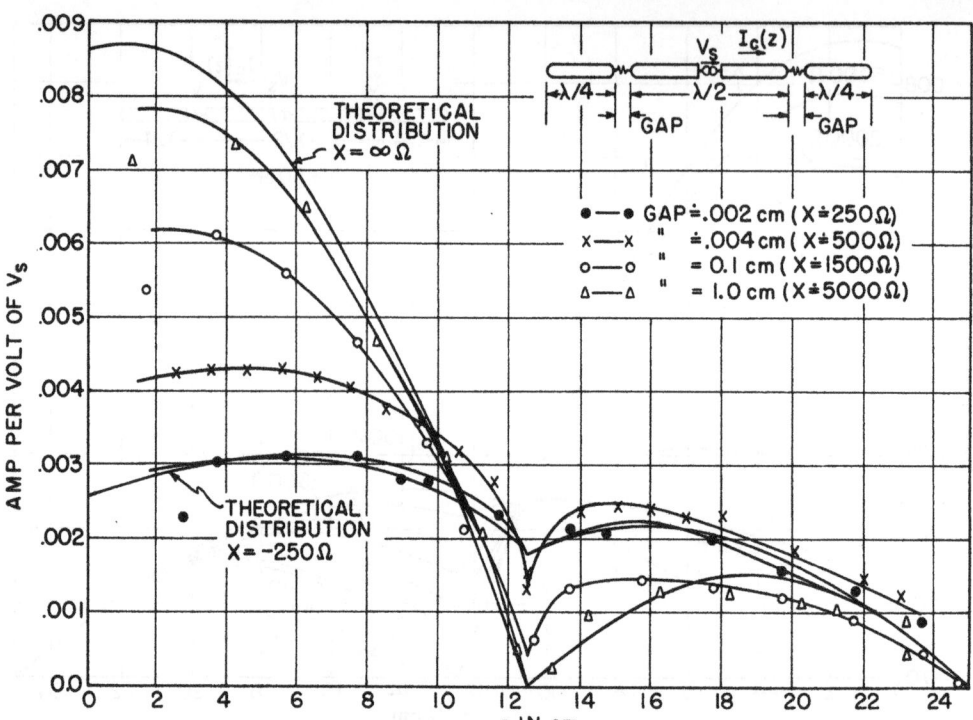

Fig. 36.9. Measured distribution of current on half of collinear array with the gap length (and equivalent reactance X) as parameter (Andrews).

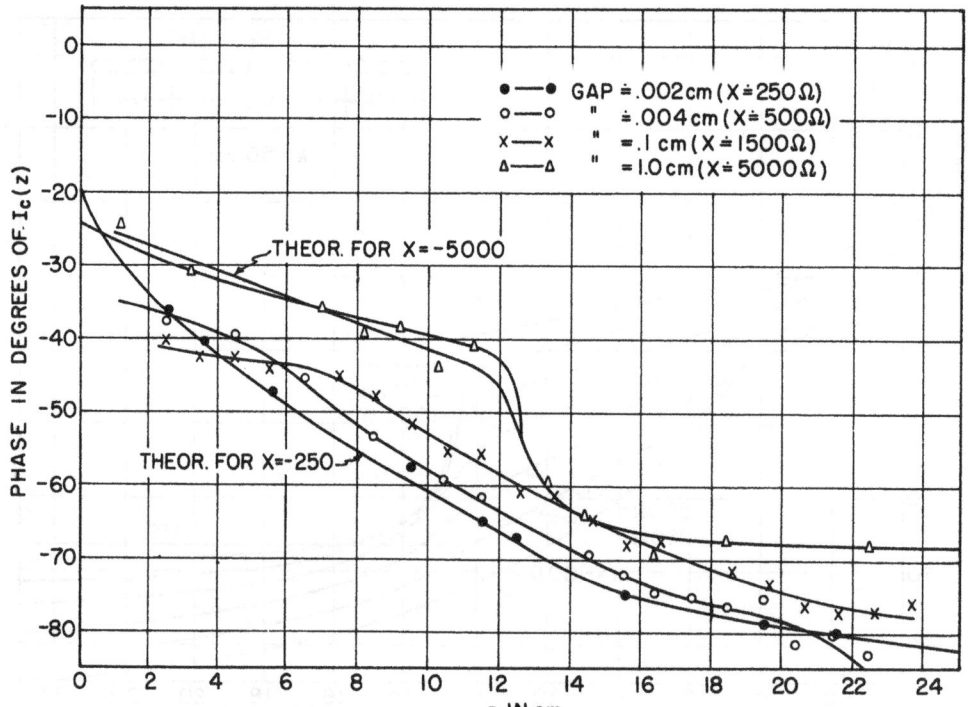

Fig. 36.10. Measured distribution of phase of current on half of collinear array (Andrews).

if a zeroth-order variation is superimposed on a higher-order (second or third) impedance for infinite separation provided the gap length is not too small. For very small gaps account must be taken of additional lumped capacitive coupling between the adjacent ends.

Theoretical and measured distributions of current as determined by Andrews for half-collinear arrays with $h_1 = \lambda/4$ and $2h_2 = \lambda/4$ are in Figs. 36.7–36.10. The parameter in the theoretical curves is the lumped series reactance X, in the experimental curves it is the corresponding and approximately equivalent length of the gap.

Experimental measurements to determine the distribution of current in and the input impedance of the three-element collinear array in Fig. 35.1a have been made by Täng[52] using the image-plane apparatus of Andrews.[3] Täng studied the magnitude and phase of the current on the driven and the coupled element of a half-collinear array over a large image plane. The length of the phase-reversing stub and the location of the short-circuiting bridge along this stub were the variable parameters. The spacing of the conductors of the stub was 0.01λ. For each antenna, $\beta_0 h = \pi/2$, $\Omega = 8.7$. Täng's data show that the current at the center of the outer element is in phase with the current at the base of the driven element only when the overall length of the stub and the distance from the antenna to the short-circuiting bridge (if this is not the same) is near an integral odd multiple of $\lambda_0/4$. When the currents are in phase, their magnitudes are not the same, that in the outer element being smaller. Täng's data are: $I_2/I_1 \doteq 0.65$ with a $0.1°$ phase difference. (For other lengths of the stub and other positions of the bridge—if this is not fixed at the end of the line—a variety of phase and amplitude distributions obtain. In general, these include a superposition of two sets of currents. (1) Currents maintained by the generator when the stub is short-circuited at the antenna so that its conductors carry equal codirectional currents. (2) Currents maintained by a fictitious generator connected across the adjacent ends of the antennas with an emf equal to the negative of the voltage drop across the stub. This is a current like that on a sleeve dipole as described in Sec. 30.)

The input impedance of the half-collinear array described above is $Z_{in} = 210 + j33$ ohms when $\Omega = 8.7$. When $\Omega = 13.1$ it is $Z_{in} = 163 + j37$ ohms. The corresponding theoretical zeroth-order value obtained in Sec. 35 is $Z_{in} = 157.5 + j103.5$ ohms. It is seen that the measured resistances compare with the zeroth-order value in much the same manner as do the third-order resistances of the isolated half-dipole with the corresponding zeroth-order value, $R_{in} = 36.2$ ohms. On the other hand, the large differences in the measured and zeroth-order reactances must be ascribed to the fact that the theoretical analysis in Sec. 35 does not take account of the capacitance between adjacent ends of the antennas and this is certainly significant since these ends are separated by only 0.01λ.

Experimental and theoretical studies of the collinear array when the outer elements are coupled to the central unit by coaxial sleeves instead of two-wire lines have been made by Faflick.[19] His investigations included arrays coupled by internal sleeves with all antennas of equal radius as well as arrays with external coupling sleeves which were of larger diameter than the antennas. Since the coupling effect of the sleeves was modified greatly by the change in radius when external sleeves were used, a careful study of antennas of discontinuous radius also was made to supplement limited theoretical work previously reported.[54] The general conclusion reached at the end of Sec. 35 that coaxial sleeves, whether internal or external, are not satisfactory as phase-reversing stubs are maintained by Faflick's results. This is in contradiction of earlier work[22, 1.32] which was based on an elementary analogy which does not obtain in the simple form assumed.

CHAPTER IV

THE RECEIVING ANTENNA AS A CIRCUIT ELEMENT

In the introduction to Chapter III the general problem of coupled antennas is separated into two parts in order to achieve sufficient simplification to permit an analytic formulation that can be carried through to practically useful results. The first part, treated in Chapter III, consists of the analysis of geometrically simple configurations of similar conductors that may be close together or widely separated. The second part involves the analysis of two different and arbitrarily oriented antennas of which the one is driven by a generator and the other is loaded. Although this general problem with arbitrary distances between the antennas can be formulated analytically in terms of simultaneous integral equations, a solution is available only for separations sufficiently great to put the loaded antenna into the radiation zone of the driven unit. With such great distances between the antennas, relatively small currents in the parasitic antenna react only insignificantly on the transmitter, so that the solution for the current and the power in a single receiving antenna in the far-zone field of a transmitting antenna is equivalent to the solution for *any one* of a large number of receiving antennas each in the radiation zone of the transmitter and of one another.

In this chapter the general analysis of two different, arbitrarily oriented, and arbitrarily spaced antennas is formulated. The equations so obtained are then specialized to the case of great separation and the analysis is carried out in detail, in order to derive the essential properties of receiving and scattering antennas.

GENERAL THEORY OF TWO DIFFERENT, ARBITRARILY ORIENTED ANTENNAS

1. *Integral Equations with Arbitrary Spacing*

The derivation of the integral equations for two highly conducting cylindrical antennas 1 and 2 of radii a_1 and a_2 and half-lengths h_1 and h_2 parallels the more specialized analysis of Chapter III. The same assumptions regarding the radii of the antennas are made, namely,

$$\beta_0 a_1 \ll 1, \quad \beta_0 a_2 \ll 1, \quad a_1 \ll h_1, \quad a_2 \ll h_2. \quad (1)$$

Subject to these conditions it may be assumed that cross-sectional and axial distributions of current density in the conductors may be treated as though independent.

The boundary conditions for the tangential component of the electric field on the surfaces of the two conductors are

$$(E_{1s}^e)_{r_1=a_1} = (E_{1s}^i)_{r_1=a_1}, \quad (2a)$$

$$(E_{2z}^e)_{r_2=a_2} = (E_{2z}^i)_{r_2=a_2}, \quad (2b)$$

where the arbitrarily oriented axis of antenna 1 is assumed to coincide with an s-axis, while the axis of antenna 2 falls along the z-axis, as shown in Fig. 1.1. The fields just outside the conductors may be expressed in terms of scalar and vector potentials, using the relation

$$\mathbf{E} = -\text{grad } \phi - j\omega \mathbf{A}, \quad (3)$$

and the fields just inside the conductors in terms of the axial current and the internal impedance per unit length in the form

$$E_{\text{tang}} = I z^i. \quad (4)$$

With (3) and (4) substituted in (2a, b), using appropriate components, the results are

$$-\left(\frac{\partial \phi_1}{\partial s} + j\omega A_{1s}\right)_{r_1=a_1} = z_1^i I_{1s}, \quad (5a)$$

$$-\left(\frac{\partial \phi_2}{\partial z} + j\omega A_{2z}\right)_{r_2=a_2} = z_2^i I_{2s}. \quad (5b)$$

Use of the equation of continuity for the potential functions (Lorentz condition), namely,

$$\text{div } \mathbf{A} + j\frac{\beta_0^2}{\omega}\phi = 0, \quad (6)$$

in (5a, b) leads to the following equations in which the subscript 1 or 2 on the operator div refers to the variables r_1, z_1, or r_2, s_2 with respect to which the differentiation is performed:

$$\frac{\partial}{\partial s}\operatorname{div}_1 \mathbf{A}_1 + \beta_0^2 A_{1s} = j\frac{\beta_0^2}{\omega} z_1^i I_{1s}, \quad (7a)$$

$$\frac{\partial}{\partial z}\operatorname{div}_2 \mathbf{A}_2 + \beta_0^2 A_{2z} = j\frac{\beta_0^2}{\omega} z_2^i I_{2z}, \quad (7b)$$

where $\mathbf{A}_1$ is the vector potential just outside the surface of antenna 1 due to currents in *both* antennas, and $\mathbf{A}_2$ is the vector potential just outside the surface of antenna 2 due to currents in *both* antennas. Thus,

$$\mathbf{A}_1 = \mathbf{A}_{11} + \mathbf{A}_{12}, \quad \mathbf{A}_2 = \mathbf{A}_{22} + \mathbf{A}_{21}, \quad (8)$$

where the individual vector potentials are given by

$$\mathbf{A}_{11} = \frac{\hat{\mathbf{s}}}{4\pi\nu_0}\left(\int_{-h_1}^{-\delta_1} + \int_{\delta_1}^{h_1}\right)\frac{I_{1s}(s')}{R_{11}}e^{-j\beta_0 R_{11}}\,ds', \quad (9)$$

$$\mathbf{A}_{12} = \frac{\hat{\mathbf{z}}}{4\pi\nu_0}\left(\int_{-h_2}^{-\delta_2} + \int_{\delta_2}^{h_2}\right)\frac{I_{2z}(z')}{R_{12}}e^{-j\beta_0 R_{12}}\,dz', \quad (10)$$

$$\mathbf{A}_{22} = \frac{\hat{\mathbf{z}}}{4\pi\nu_0}\left(\int_{-h_2}^{-\delta_2} + \int_{\delta_2}^{h_2}\right)\frac{I_{2z}(z')}{R_{22}}e^{-j\beta_0 R_{22}}\,dz', \quad (11)$$

$$\mathbf{A}_{21} = \frac{\hat{\mathbf{s}}}{4\pi\nu_0}\left(\int_{-h_1}^{-\delta_1} + \int_{\delta_1}^{h_1}\right)\frac{I_{1s}(s')}{R_{21}}e^{-j\beta_0 R_{21}}\,ds'. \quad (12)$$

In these equations R_{11} is the distance from an element ds on the equipotential surface of antenna 1 to an element ds' on its axis; R_{12} is the distance from ds on the surface of antenna 1 to an element dz' on the axis of antenna 2; $R_{22} = \sqrt{(z'-z)^2 + a_2^2}$ is the distance from an element dz at z on the surface of antenna 2 to an element dz' at z' on its axis; R_{21} is the distance from dz on the surface of antenna 2 to ds' on the axis of antenna 1.

In Cartesian coördinates (7b) becomes

$$\frac{\partial^2 A_{2z}}{\partial z^2} + \frac{\partial}{\partial z}\left(\frac{\partial A_{21x}}{\partial x} + \frac{\partial A_{21y}}{\partial y}\right) + \beta_0^2 A_{2z}$$

$$= j\frac{\beta_0^2}{\omega} z_2^i I_{2z}, \quad (13)$$

where $A_{2z} = A_{22z} + A_{21z}$.

Substitution of (9)–(12) in (7a, b) gives two simultaneous integral equations in the currents I_{1s} and I_{2z} in the two antennas. It is to be noted that $\mathbf{A}_{11}$ is not independent of I_{2z}, even though this current does not appear in the integral. However, I_{1s} depends upon I_{2z}, so that $\mathbf{A}_{11}$ is affected if I_{2z} is changed or removed. Since each equation involves more than one variable, a simple integration is not possible. Furthermore, in the absence of symmetry, the method of symmetric components used in Chapter III has no application.

If the voltages $V_{1\delta}$ and $V_{2\delta}$ are assumed to be known, the two simultaneous integral equations can be expressed as follows:

$$V_{1\delta} = I_{1\delta}Z_{s1\delta} + I_{2\delta}Z_{12\delta}, \quad (14a)$$

$$V_{2\delta} = I_{1\delta}Z_{21\delta} + I_{2\delta}Z_{s2\delta}, \quad (14b)$$

where the self- and mutual impedances $Z_{s1\delta}$, $Z_{s2\delta}$, and $Z_{12\delta} = Z_{21\delta}$ are defined formally to be the coefficients of the currents in (14a, b). Formulas for these impedances involve complicated integrals with unknown distribution functions for the currents. They have not been evaluated in the general case.

If antenna 2 is parasitic with a center load so that

$$V_{2\delta} = -I_{2\delta}Z_{L\delta}, \quad (15)$$

where $Z_{L\delta}$ is defined to be the ratio of scalar potential difference to current at the terminals $z = \pm\delta$ of the center load, (14b) becomes

$$0 = I_{1\delta}Z_{21\delta} + I_{2\delta}Z_{22\delta}, \quad (16)$$

where

$$Z_{22\delta} \equiv Z_{s2\delta} + Z_{L\delta}. \quad (17)$$

The impedance $Z_{1\delta} = V_{1\delta}/I_{1\delta}$ of the primary circuit is

$$Z_{1\delta} = Z_{s1\delta} - \frac{Z_{12\delta}Z_{21\delta}}{Z_{22\delta}}. \quad (18)$$

The *induced voltages* $V_{1\delta}^i$ and $V_{2\delta}^i$ are defined to be

$$V_{1\delta}^i \equiv -I_{2\delta}Z_{12\delta}, \quad V_{2\delta}^i \equiv -I_{1\delta}Z_{21\delta}. \quad (19)$$

Hence,

$$V_{1\delta} + V_{1\delta}^i = I_{1\delta}Z_{s1\delta}, \quad (20)$$

$$V_{2\delta}^i = I_{2\delta}Z_{22\delta}. \quad (21)$$

The two antennas are loosely coupled if the following condition is satisfied:

$$|V_{1\delta}^i| \ll |V_{1\delta}|$$

or

$$|Z_{s1\delta}Z_{22\delta}| \gg |Z_{12\delta}Z_{s1\delta}|. \quad (22)$$

458 THEORY OF LINEAR ANTENNAS [IV.1]

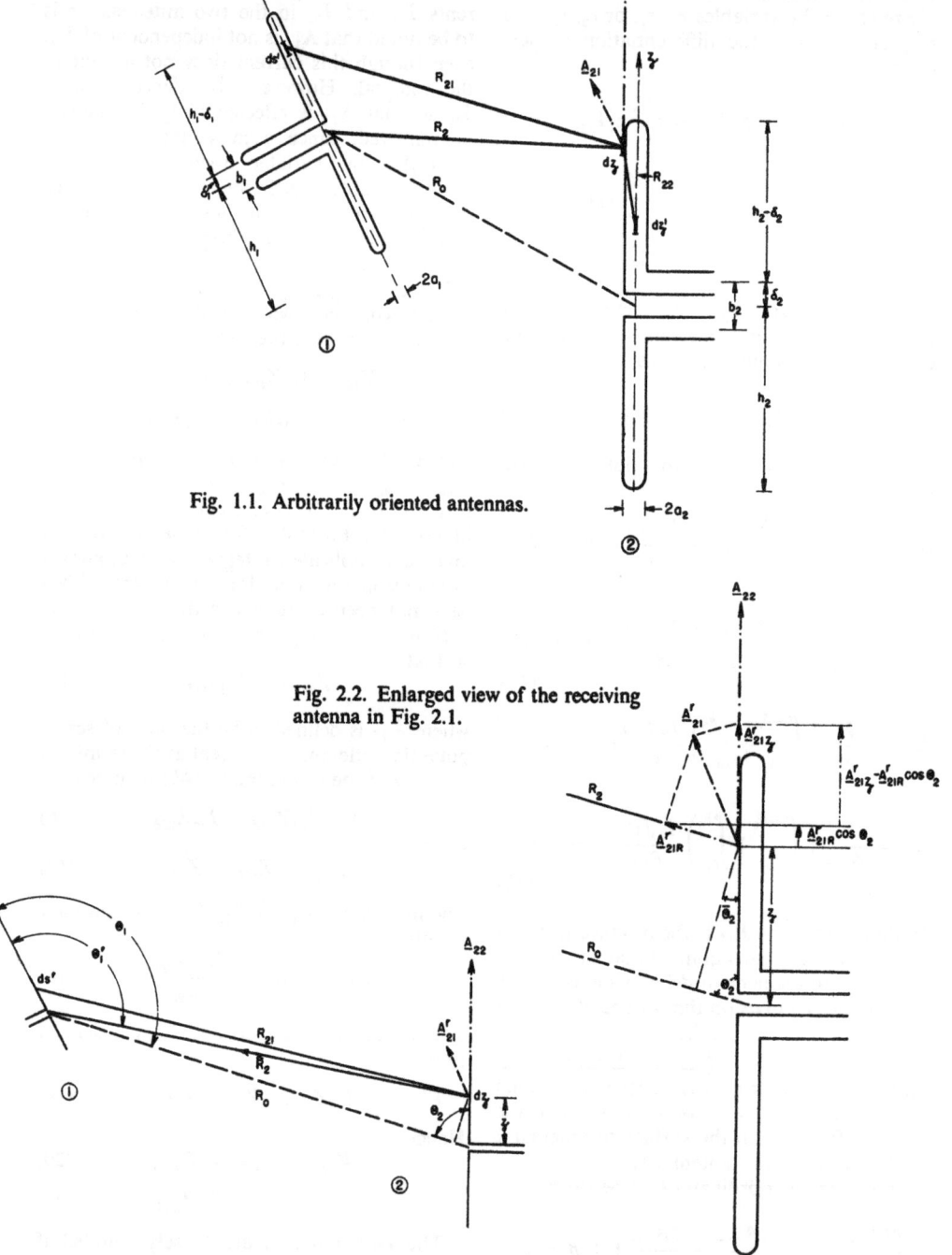

Fig. 1.1. Arbitrarily oriented antennas.

Fig. 2.2. Enlarged view of the receiving antenna in Fig. 2.1.

Fig. 2.1. Receiving antenna in the presence of transmitting antenna.

In this case,
$$Z_{1\delta} \doteq Z_{s1\delta}. \quad (23)$$

The apparent terminal impedances for the transmission lines and the appropriate terminal-zone networks to take account of end effects and coupling effects are difficult to determine for adjacent antennas, since no conditions of symmetry obtain. If all points of the antennas are far apart compared with the distances between the conductors of the transmission lines, each antenna may be treated as if it were completely isolated in determining the elements of the terminal-zone networks.

2. Specialization of the General Case to Great Separation—Equations for Vector Potential

If the two antennas in Fig. 1.1 are sufficiently far apart so that they are in the far zone with respect to each other, the condition for loose coupling (1.22) applies and the two antennas are mutually independent if they are both driven so that the induced voltages are negligible. If antenna 2 is parasitic, the induced voltage evidently cannot be negligible compared with zero, and the entire power dissipated in the load originates in the driven antenna 1. However, since the very loosely coupled parasitic antenna has no effect on the distant transmitter, the far-zone electromagnetic field and the associated potential functions *due to the currents in the driven antenna* are independent of the current in the parasite. This means that the induced voltage $V_{2\delta}^i$ due to the driven antenna is independent of I_{2z}.

The equation that applies to the parasitic antenna 2 is

$$\frac{\partial}{\partial z} \operatorname{div} \mathbf{A} + \beta_0^2 A_z = j\frac{\beta_0^2}{\omega} z^i I_z. \quad (1)$$

This equation is the same as (1.7b) except that the identifying subscript 2 has been omitted with the understanding that all potentials are defined on the surface of antenna 2 and all currents and charges are in antenna 2 unless otherwise specified. Thus, in (1),

$$\mathbf{A} \equiv \mathbf{A}_2 = \mathbf{A}_{22} + \mathbf{A}_{21}, \quad (2)$$

where, with $h_2 = h$, $I_{2z} = I_z$,

$$\mathbf{A}_{22} = \frac{\hat{\mathbf{z}}}{4\pi v_0}\left(\int_{-h}^{-\delta} + \int_{\delta}^{h}\right)\frac{I_z'}{R_{22}} e^{-j\beta_0 R_{22}}\, dz', \quad (3a)$$

$$\mathbf{A}_{21} = \frac{\hat{\mathbf{s}}}{4\pi v_0}\left(\int_{-h_1}^{-\delta_1} + \int_{\delta_1}^{h_1}\right)\frac{I_{1s}'}{R_{21}} e^{-j\beta_0 R_{21}}\, ds'. \quad (3b)$$

Since antenna 2 is by definition in the far-zone of antenna 1, the far- or radiation-zone form of (3b) may be used. This is identified with a superscript r. It is

$$\mathbf{A}_{21}^r = \frac{\hat{\mathbf{s}}}{4\pi v_0}\frac{e^{-j\beta_0 R_2}}{R_2}\left(\int_{-h_1}^{-\delta_1} + \int_{\delta_1}^{h_1}\right) I_{1s}' e^{j\beta_0 s' \cos\Theta_1}\, ds'. \quad (4)$$

It is equivalent to (3b) subject to the conditions

$$\beta_0 R_1 \gg 1, \quad R_0 \gg h_1, \quad R_0 \gg h_2 = h. \quad (5)$$

It follows with (5) that in Fig. 2.1

$$R_{21} \doteq R_2 \doteq R_0 \text{ in amplitude factors}; \quad (6a)$$

$$R_{21} \doteq R_2 - s'\cos\Theta_1 \text{ in phases}. \quad (6b)$$

Substitution of (6a, b) in (4) gives

$$\mathbf{A}_{21}^r \doteq \mathbf{K}_{21} e^{-j\beta_0 R_2}, \quad (7a)$$

where

$$\mathbf{K}_{21} \equiv \frac{\hat{\mathbf{s}}}{4\pi v_0 R_0}\left(\int_{-h_1}^{-\delta_1} + \int_{\delta_1}^{h_1}\right) I_{1s}' e^{j\beta_0 s'\cos\Theta_1}\, ds'. \quad (7b)$$

Note that $\mathbf{K}_{21}$ is a constant at all points on antenna 2.

If (1) is expanded using (2) and (7a) it becomes

$$\left[\frac{\partial^2 A_{22z}}{\partial z^2} + \beta_0^2 A_{22z}\right] + \left[\frac{\partial}{\partial z}(\operatorname{div}\mathbf{K}_{21} e^{-j\beta_0 R_2}) + \beta_0^2 A_{21z}^r\right] = j\frac{\beta_0^2}{\omega} z^i I_z, \quad (8)$$

where $\operatorname{div} \equiv \operatorname{div}_2 \equiv (\partial/\partial x) + (\partial/\partial y) + (\partial/\partial z)$ involves differentiation with respect to the coördinates x, y, z locating the end-point of R_2 on antenna 2. With the vector identity

$$\operatorname{div} \mathbf{C}\phi \equiv \phi \operatorname{div} \mathbf{C} + \mathbf{C} \cdot \operatorname{grad} \phi \quad (9)$$

and the fact that $\mathbf{K}_{21}$ is a constant vector with respect to variables on antenna 2,

$$\operatorname{div}(\mathbf{K}_{21} e^{-j\beta_0 R_2}) = \mathbf{K}_{21} \cdot \operatorname{grad} e^{-j\beta_0 R_2}. \quad (10)$$

However,

$$\operatorname{grad} e^{-j\beta_0 R_2} = \frac{\partial}{\partial R_2}(e^{-j\beta_0 R_2}) \operatorname{grad} R_2$$
$$= -j\beta_0 e^{-j\beta_0 R_2} \operatorname{grad} R_2. \quad (11)$$

Note that the operation $\text{grad} = \text{grad}_2$ represents differentiation with respect to the coördinates locating the end-point of R_2 on antenna 2, whereas grad_1 represents differentiation with respect to the coördinates locating the end-point of R_2 on antenna 1. If $\hat{\mathbf{R}}_2$ is a unit vector along R_2 pointing from antenna 2 to antenna 1, as shown in Fig. 2.1, then the direction of the greatest rate of increase of R_2 by varying the coördinates locating the end of R_2 on antenna 2 is in the negative $\hat{\mathbf{R}}_2$ direction. Hence,

$$\text{grad } R_2 = -\hat{\mathbf{R}}_2. \qquad (12)$$

Substitution of (12) in (11), and of (11) in (10), gives

$$\text{div } (\mathbf{K}_{21}e^{-j\beta_0 R_2}) = j\beta_0 e^{-j\beta_0 R_2}\mathbf{K}_{21} \cdot \hat{\mathbf{R}}_2. \qquad (13)$$

Differentiation of both sides of (13) with respect to z gives

$$\frac{\partial}{\partial z} \text{div } (\mathbf{K}_{21}e^{-j\beta_0 R_2})$$

$$= \mathbf{K}_{21} \cdot \hat{\mathbf{R}}_2 \beta_0^2 e^{-j\beta_0 R_2} \frac{\partial R_2}{\partial z} = A_{21R} \beta_0^2 \frac{\partial R_2}{\partial z}. \qquad (14)$$

From Fig. 2.1 or 2.2 it follows that

$$R_2 = R_0 - z \cos \Theta_2 = R_0 - z \sin \overline{\Theta}_2, \qquad (15)$$

so that

$$\frac{\partial R_2}{\partial z} = -\cos \Theta_2 = -\sin \overline{\Theta}_2, \qquad (16)$$

where

$$\overline{\Theta}_2 \equiv \frac{\pi}{2} - \Theta_2. \qquad (17)$$

Hence, substitution of (16) in (14), and (14) in (8), gives,

$$\left(\frac{\partial^2 A_{22z}}{\partial z^2} + \beta_0^2 A_{22z}\right)$$

$$+ \beta_0^2 (A_{21z}^r - A_{21R}^r \cos \Theta_2) = j\frac{\beta_0^2}{\omega} z^i I_z. \qquad (18)$$

The component $A_{21z}^r - A_{21R}^r \cos \Theta_2$ of A_{21}^r is shown in Fig. 2.2. It is only this part of A_{21}^r that is effective in exciting currents in antenna 2. Note that since $\mathbf{K}_{21}^r$ is a constant at all points on antenna 2 the quantity $(K_{21z}^r - K_{21R}^r \cos \Theta_2)$ also has the same value all along the entire length of antenna 2. With (15), (7a) may be expressed as follows:

$$A_{21}^r = \mathbf{K}_{21}e^{-j\beta_0 R_0}e^{jq_0 z}, \qquad (19)$$

where

$$q_0 \equiv \beta_0 \cos \Theta_2. \qquad (20)$$

Note that

$$\cos \Theta_2 = q_0/\beta_0, \qquad \sin \Theta_2 = \sqrt{1 - q_0^2/\beta_0^2}. \qquad (21)$$

Substitution of (19) in (18) using (21) gives

$$\frac{\partial^2 A_{22z}}{\partial z^2} + \beta_0^2 A_{22z} = j\frac{\beta_0^2}{\omega}[z^i I_z + j\omega(K_{21z}$$

$$- K_{21R}q_0/\beta_0)e^{-j(\beta_0 R_0 - q_0 z)}]. \qquad (22)$$

For convenience, let the quantity U be defined as follows; note that it is a constant in so far as antenna 2 is concerned:

$$U \equiv \frac{-j\omega(K_{21z} - K_{21R}q_0/\beta_0)e^{-j\beta_0 R_0}}{\beta_0(1 - q_0^2/\beta_0^2)}. \qquad (23)$$

With (23), (22) becomes

$$\frac{\partial^2 A_{22z}}{\partial z^2} + \beta_0^2 A_{22z} = \frac{j\beta_0^2}{\omega}[z^i I_z$$

$$- \beta_0(1 - q_0^2/\beta_0^2)Ue^{jq_0 z}]. \qquad (24)$$

Except for the added term with U as a factor, (24) is the same as the equation obtained in Chapter II for the isolated center-driven antenna. The solution is readily expressed as the sum of a complementary function and a particular integral as follows:

$$A_{22z} = \frac{-j}{v_0}\left[C_1 \cos \beta_0 z + C_2 \sin \beta_0 z\right.$$

$$\left. - z^i \int_\delta^z I(s) \sin \beta_0(z - s)\, ds + Ue^{jq_0 z}\right].$$

$$(\delta \leq z \leq h) \qquad (25)$$

The first three terms on the right are the same as for the center-driven antenna; the term $(-j/v_0)Ue^{jq_0 z}$ is the contribution from the currents in the distant antenna 1.

It is readily verified by direct differentiation that (25) is indeed a solution of (24). For the lower half of the antenna the same equation with different constants of integration applies. It is

$$A_{22z} = \frac{-j}{v_0}\left[C_3 \cos \beta_0 z + C_4 \sin \beta_0 z\right.$$

$$\left. - z^i \int_{-\delta}^z I(s) \sin \beta_0(z - s)\, ds + Ue^{jq_0 z}\right].$$

$$(-h \leq z \leq -\delta) \qquad (26)$$

The solution of (25) and (26) is conveniently carried out by separating the symmetric problem with even currents and vector potentials and odd charges and scalar potentials from the antisymmetric problem with odd currents and vector potentials and even charges and scalar potentials as defined in Sec. II.12.

The symmetric part of (25) is readily obtained by combining it with (26) according to

$$A_{22z}^s = \tfrac{1}{2}[A_{22z}(z) + A_{22z}(-z)],$$
$$I_z^s = \tfrac{1}{2}[I_z(z) + I_z(-z)]. \quad (27)$$

Thus,

$$A_{22z}^s = \frac{-j}{v_0}\left[C_{1s}\cos\beta_0 z + C_{2s}\sin\beta_0|z| \right.$$
$$\left. - z^i\int_\delta^z I^s(s)\sin\beta_0(z-s) + U\cos q_0 z\right], \quad (28)$$

where

$$C_{1s} = \tfrac{1}{2}(C_1 + C_3), \quad C_{2s} = \tfrac{1}{2}(C_2 - C_4), \quad (29)$$

and

$$A_{22z}^s = \frac{1}{4\pi v_0}\left(\int_{-h}^{-\delta} + \int_\delta^h\right)\frac{I_z^s(z')}{R_{22}}e^{-j\beta_0 R_{22}}\,dz'. \quad (30)$$

Vector potentials and currents in the lower half are equal to and codirectional with those in the upper half.

The antisymmetric part of (25) is obtained from

$$A_{22z}^a = \tfrac{1}{2}[A_{22z}(z) - A_{22z}(-z)],$$
$$I_z^a = \tfrac{1}{2}[I_z(z) - I_z(-z)]. \quad (31)$$

Thus,

$$A_{22z}^a = \frac{-j}{v_0}\left[C_{2a}\sin\beta_0 z \pm C_{1a}\cos\beta_0 z \right.$$
$$\left. - z^i\int_\delta^z I^a(s)\sin\beta_0(z-s)\,ds + jU\sin q_0 z\right], \quad (32)$$

where

$$C_{1a} = \tfrac{1}{2}(C_1 - C_3), \quad C_{2a} = \tfrac{1}{2}(C_2 + C_4), \quad (33)$$

$$A_{22z}^a = \frac{1}{4\pi v_0}\left(\int_{-h}^{-\delta} + \int_\delta^h\right)\frac{I_z^a(z')}{R_{22}}e^{-j\beta_0 R_{22}}\,dz', \quad (34)$$

and where the upper sign is for $\delta \leq z \leq h$, the lower sign for $-h \leq z \leq -\delta$. Note that when $\delta = 0$, A_{22z}^a vanishes at $z = 0$, so that

$$C_{1a} = \tfrac{1}{2}(C_1 - C_3) = 0, \quad (35)$$

Currents and vector potentials for the lower half of the antenna are equal to and opposite in direction to those in the upper half.

Combination of (30) with (28) and (35) with (32) gives two integral equations for determining the even and odd parts of the current. The complete solution is the sum of the two.

THE CENTER-LOADED RECEIVING ANTENNA

3. General Formula for Current in the Center-Loaded Antenna

If a receiving antenna is loaded at the center by a geometrically symmetric impedor, with terminals at $z = \pm\delta$ as in Fig. 3.1, even charges of the type that belong to the antisymmetric problem $q(-z) = q(z)$ can maintain no potential difference across its terminals, so that the entire potential difference is due to odd charges of the type $q(-z) = -q(z)$ which occur in the symmetric problem. It follows that when interest is in determining the voltage across, the current in, or the power transferred to such a symmetric center load, it is necessary to solve *only the symmetric problem* for even currents and vector potentials and odd charges and scalar potentials. On the other hand, if the electromagnetic field due to the *total* current in the antenna is required, both odd and even currents are involved. They may be determined separately and combined. It is shown in Sec. 10 that in the particular orientation of the center-loaded receiving antenna with respect to the transmitting antenna for which the current in, the voltage across, and the power to a center load are maximum, the *odd* currents are zero everywhere along the entire antenna. (This is true except for very long antennas.) Therefore, a solution of the symmetric problem provides all possible information about currents, voltages, and power for the center load under all circumstances, and, in addition, the complete distribution of current under conditions of optimum orientation. In practice, this is all the information about a receiving antenna that is required. Accordingly, consideration of the antisymmetric problem is postponed, and a complete solution is obtained for the symmetric problem. Since conditions of symmetry at once yield all required information about the lower half of the antenna (z negative), it is necessary to consider only the upper half (z positive). Omitting all subscript and superscript s's, with the understanding that unless otherwise indicated all currents, charges and potentials are those of the symmetric problem only, the vector potential as given by (2.28) is

$$A_{22z} = \frac{-j}{v_0}(C_1 \cos \beta_0 z + C_2 \sin \beta_0 |z| + U \cos q_0 z). \quad (1)$$

The term with z^i as a factor has been omitted temporarily in (1) for simplicity. Actually it is always negligible if good conductors are used for the antenna, even when the center load is zero.

Just as in (II.13.14), the scalar potential is

$$\Phi_2 = -C_1 \sin \beta_0 z + C_2 \cos \beta_0 z - \frac{q_0}{\beta_0} U \sin q_0 z. \quad (2)$$

The scalar potential difference across the load between $z = \pm \delta$ is defined by

$$V_{2\delta} \equiv \Phi_2(\delta) - \Phi_2(-\delta) = 2\Phi_2(\delta)$$

$$= 2\left(-C_1 \sin \beta_0 \delta + C_2 \cos \beta_0 \delta - \frac{q_0}{\beta_0} U \sin q_0 \delta\right). \quad (3)$$

Solution of (3) for C_2 gives

$$C_2 = \frac{\frac{1}{2}V_{2\delta} + C_1 \sin \beta_0 \delta + (q_0/\beta_0) U \sin q_0 \delta}{\cos \beta_0 \delta}. \quad (4)$$

For convenience, let the following notation be introduced:

$$\frac{1}{2}V_\delta \equiv \frac{1}{2}V_{2\delta} + \frac{q_0}{\beta_0} U \sin q_0 \delta. \quad (5)$$

Since $q_0 = \beta_0 \cos \Theta_2$ can never exceed β_0 and is actually zero for the optimum orientation, it is clear that, subject to the assumed conditions $\beta_0 b \ll 1$, the contribution to $\frac{1}{2}V_\delta$ by $(q_0/\beta_0) U \sin q_0 \delta$ is so small that it is usually negligible. With (5), (4) becomes

$$C_2 = (\tfrac{1}{2}V_\delta + C_1 \sin \beta_0 \delta)/\cos \beta_0 \delta, \quad (6)$$

so that

$$A_{22z} = \frac{-j}{v_0}\left(C_1 \cos \beta_0 z + \frac{\tfrac{1}{2}V_\delta + C_1 \sin \beta_0 \delta}{\cos \beta_0 \delta} \sin \beta_0 |z| + U \cos q_0 z\right). \quad (7)$$

With (2.30) the integral equation for the current is

$$\left(\int_{-h}^{-\delta} + \int_{\delta}^{h}\right) I_z' K_1(z, z')\, dz'$$

$$= \frac{-j4\pi}{\zeta_0 \cos \beta_0 \delta} [\cos \beta_0 \delta (C_1 \cos \beta_0 z + U \cos q_0 z)$$

$$+ (C_1 \sin \beta_0 \delta + \tfrac{1}{2}V_\delta) \sin \beta_0 |z|], \quad (8)$$

where

$$K_1(z, z') = \frac{e^{-j\beta_0 R_{22}}}{R_{22}}, \quad (9)$$

This equation is like (II.14.3) with the added term in U and with $z^i = 0$. Its solution may be carried out by the same method of iteration using the same expansion parameter Ψ as defined in Sec. II.10. Corresponding to (II.15.19), the current is

$$I_z \doteq [I_z]_n = \frac{-j4\pi}{\zeta_0 \Psi \cos \beta_0 \delta} \{\cos \beta_0 \delta [C_1 f(z, h)$$

$$+ U h(z, h)] + [C_1 \sin \beta_0 \delta + \tfrac{1}{2}V_\delta] g(z, h)\}, \quad (10)$$

where

$$f(z, h) \equiv \sum_{m=0}^{n} F_{mz}/\Psi^m = f(z) - f(h), \quad (11a)$$

$$g(z, h) \equiv \sum_{m=0}^{n} G_{mz}/\Psi^m = g(z) - g(h), \quad (11b)$$

$$h(z, h) \equiv \sum_{m=0}^{n} H_{mz}/\Psi^m = h(z) - h(h), \quad (11c)$$

$$f(z) = \sum_{m=0}^{n} F_m(z)/\Psi^m,$$

$$f(h) = \sum_{m=0}^{n} F_m(h)/\Psi^m, \quad (12a)$$

$$g(z) = \sum_{m=0}^{n} G_m(z)/\Psi^m,$$

$$g(h) = \sum_{m=0}^{n} G_m(h)/\Psi^m, \quad (12b)$$

$$h(z) = \sum_{m=0}^{n} H_m(z)/\Psi^m,$$

$$h(h) = \sum_{m=0}^{n} H_m(h)/\Psi^m, \quad (12c)$$

where F_{mz} and G_{mz} are given by (II.15.18).

Note that

$$H_{mz} = H_m(z) - H_m(h), \quad (13)$$

where $H_m(z)$ and $H_m(h)$ may be obtained from (II.15.18a, b) by substituting H for F. Also,

$$H_0(z) \equiv \cos q_0 z, \quad H_0(h) \equiv \cos q_0 h, \quad (14)$$

by analogy with (II.15.2a). Evidently, the H functions differ from the F functions in having $q_0 \equiv \beta_0 \cos \Theta_2$ appear in place of β_0 in the zeroth-order distribution.

The evaluation of C_1 repeats the procedure in Sec. II.15, with the following equation appearing as the analogue of (II.15.21):

$$\cos \beta_0 \delta [C_1 f(h) + U h(h)]$$
$$+ [C_1 \sin \beta_0 \delta + \tfrac{1}{2} V_\delta] g(h) = 0. \quad (15)$$

This may be solved for C_1, with the result,

$$C_1 = -\left[\frac{\tfrac{1}{2} V_\delta g(h) + U h(h) \cos \beta_0 \delta}{f(h) \cos \beta_0 \delta + g(h) \sin \beta_0 \delta}\right]. \quad (16)$$

Substitution of (16) in (10) to eliminate C_1 gives

$$I_z = I_{zV} + I_{zU}, \quad (17)$$

where I_{zV} is given by (II.15.23a) and

$$[I_{zU}]_n = \frac{j 4\pi U}{\zeta_0 \Psi}$$
$$\times \left\{\frac{\cos \beta_0 \delta [f(z,h)h(h) - h(z,h)f(h)] + \sin \beta_0 \delta [g(z,h)h(h) - h(z,h)g(h)]}{f(h) \cos \beta_0 \delta + g(h) \sin \beta_0 \delta}\right\}. \quad (18)$$

Upon expanding $f(z,h)$, $g(z,h)$, and $h(z,h)$, (18) is reduced to the following expression:

$$[I_{zU}]_n = \frac{j 4\pi U}{\zeta_0 \Psi}$$
$$\times \left\{\frac{\cos \beta_0 \delta [f(z)h(h) - h(z)f(h)] + \sin \beta_0 \delta [g(z)h(h) - h(z)g(h)]}{f(h) \cos \beta_0 \delta + g(h) \sin \beta_0 \delta}\right\}. \quad (19)$$

The corresponding expression for I_{zV} is (II.15.23b). Upon carrying out the multiplications in the several sums and arranging terms in powers of $1/\Psi$, the result is:

$$[I_{zU}]_n = \frac{j 4\pi U}{\zeta_0 \Psi D}$$
$$\times \Big\{[H_0(h)F_0(z) - F_0(h)H_0(z)] \cos \beta_0 \delta$$
$$+ [H_0(h)G_0(z) - G_0(h)H_0(z)] \sin \beta_0 \delta$$
$$+ \frac{1}{\Psi}[H_0(h)F_1(z) - F_0(h)H_1(z)$$
$$+ H_1(h)F_0(z) + F_1(h)H_0(z)] \cos \beta_0 \delta$$
$$+ \frac{1}{\Psi}[H_0(h)G_1(z) - G_0(h)H_1(z)$$
$$+ H_1(h)G_0(z) - G_1(h)H_0(z)] \sin \beta_0 \delta$$
$$+ \frac{1}{\Psi^2}[\cdots] \cos \beta_0 \delta$$
$$+ \frac{1}{\Psi^2}[\cdots] \sin \beta_0 \delta + \cdots \Big\}, \quad (20)$$

with

$$D \equiv \cos \beta_0 \delta [F_0(h) + F_1(h)/\Psi$$
$$+ F_2(h)/\Psi^2 + \cdots] + \sin \beta_0 \delta [G_0(h)$$
$$+ G_1(h)/\Psi + G_2(h)/\Psi^2 + \cdots]. \quad (21)$$

The corresponding expression for I_{zV} is (II.15.25a) with added subscript V on I_z.

Using (14) and (II.15.2a, b), (20) may be expressed as follows:

$$[I_{zU}]_n = \frac{j 4\pi U}{\zeta_0 \Psi}$$
$$\times \left\{\frac{[\cos q_0 h \cos \beta_0(z-\delta) - \cos q_0 z \cos \beta_0(h-\delta)]}{\cos \beta_0 (h-\delta) + A_1/\Psi + \cdots}\right.$$
$$\left.+ \frac{\tfrac{1}{\Psi}[m_1(z) \cos \beta_0 \delta + p_1(z) \sin \beta_0 \delta] + \cdots}{\cos \beta_0 (h-\delta) + A_1/\Psi + \cdots}\right\},$$
$$(\delta \leq z \leq h) \quad (22)$$

where

$$A_1 \equiv F_1(h) \cos \beta_0 \delta + G_1(h) \sin \beta_0 \delta, \quad (23a)$$

$$m_1(z) = m_1^{\mathrm{I}}(z) + j m_1^{\mathrm{II}}(z)$$
$$= [F_1(z) \cos q_0 h - H_1(z) \cos \beta_0 h$$
$$+ H_1(h) \cos \beta_0 z - F_1(h) \cos q_0 z], \quad (23b)$$

$$p_1(z) = p_1^{\mathrm{I}}(z) + j p_1^{\mathrm{II}}(z)$$
$$= [G_1(z) \cos q_0 h - H_1(z) \sin \beta_0 h$$
$$+ H_1(h) \sin \beta_0 z - G_1(h) \cos q_0 z]. \quad (23c)$$

The corresponding expression for I_{zV} is (II.15.26) with (II.15.27) and (II.15.29a).

The combination of I_{zV} in (II.15.26) with (5) and I_{zU} in (22) according to (17) gives the complete solution for the even current in an antenna that is in the far-zone field of a transmitter and that has a scalar potential difference $V_{2\delta}$ across its terminals at $z = \pm \delta$. In a receiving antenna $V_{2\delta} = -I_\delta Z_{L\delta}$ is the voltage across the load $Z_{L\delta}$. The solution obtained for I_z is expressed in terms of the general expansion parameter Ψ defined as the magnitude of $\Psi(z)$ at an appropriate reference value $z = z_r$, where

$$\Psi(z) = \left(\int_{-h}^{-\delta} + \int_{\delta}^{h} \right) g(z, z') K_1(z, z') \, dz' \quad (24)$$

with $g(z, z') \doteq I'_z/I_z$ and $K_1(z, z') = e^{-j\beta_0 R_{22}}/R_{22}$. The combined solution is

$$I_z = [U u_\delta(z) + V_\delta v_\delta(z)], \quad (25)$$

where $u_\delta(z)$ is the brace in (22) and $v_\delta(z)$ is the square bracket in (II.15.26). Thus

$$\frac{[I_{zU}]_n}{U} \equiv u_\delta(z) = \frac{j4\pi}{\zeta_0 \Psi} \times$$

$$\left\{ \frac{[\cos q_0 h \cos \beta_0(z - \delta) - \cos q_0 z \cos \beta_0(h - \delta)]}{\cos \beta(h - \delta) + A_1^\delta/\Psi + \cdots} + \frac{1}{\Psi}[m_1(z) \cos \beta_0 \delta + p_1(z) \sin \beta_0 \delta] + \cdots \right\} \quad (26)$$

$$\frac{[I_{zV}]_n}{V_\delta} \equiv v_\delta(z) = \frac{j2\pi}{\zeta_0 \Psi} \times$$

$$\left\{ \frac{\sin \beta_0(h - z) + M_1(z)/\Psi + \cdots}{\cos \beta_0(h - \delta) + A_1/\Psi + \cdots} \right\}, \quad (27)$$

$$\tfrac{1}{2} V_\delta = \tfrac{1}{2} V_{2\delta} + U \frac{q_0}{\beta_0} \sin q_0 \delta. \quad (28)$$

The general formula (2.23) defining U may be expressed in terms of the vector potential or the electric field at the *center* of the antenna in the symmetric problem here considered. Noting that

$$A^r_{21} = K_{21} e^{-j\beta_0 R_2} = K_{21} e^{-j\beta_0(R_0 - z\cos\Theta_2)},$$

it follows directly that (2.23) is equivalent to

$$U = \left[\frac{-j\omega(A^r_{21z} - A^r_{21R} q_0/\beta_0)}{\beta_0(1 - q_0^2/\beta_0^2)} \right]_{z=0}, \quad (29)$$

that is, the function U may be defined in terms of the component $(A^r_{21z} - A^r_{21R} \cos \Theta_2)$ of A^r_{21} at $z = 0$ *at the center of the antenna*. In Fig. 2.2 this component is shown at the point z. If this point is moved to the center, as in Fig. 3.2, the magnitude of A^r_{21} is unaffected, since $R_2 \doteq R_0$ in amplitude factors. However, the phase of A^r_{21} at $z = 0$ is different from that at any point z except in special circumstances when $R_2 = R_0$. The reference point for phase is the point $z = 0$.

The combination $(A^r_{21z} - A^r_{21R} \cos \Theta_2)$ is readily expressed in a more convenient form. Instead of the components A^r_{21z} along the receiving antenna and A^r_{21R} along the line R_0 (joining the center of the receiving antenna to the center of the distant transmitting antenna), it is possible to use the component $A^r_{21\Theta_1}$. This is perpendicular to R_0 and tangent to a great sphere about the center of the transmitter in the direction of Θ_1, measured from the axis of the transmitting antenna. It is shown in Fig. 3.2, where the vector A^r_{21} is resolved into the two mutually perpendicular components A^r_{21R} and $A^r_{21\Theta_1}$. There is no component $A^r_{21\Phi_1}$, so that

$$A^r_{21} = \hat{R} A_{21R} + \hat{\Theta}_1 A_{21\Theta_1}. \quad (30)$$

By projecting $-A^r_{21\Theta_1}$ through the angle ψ onto the plane containing the receiving antenna and the line R_0, and then projecting $-A^r_{21\Theta_1} \cos \psi$ through the angle $\bar{\Theta}_2 = \tfrac{1}{2}\pi - \Theta_2$ onto the vertical axis to obtain $-A^r_{21\Theta_1} \cos \psi \sin \Theta_2$, it follows directly from Fig. 3.2 that

$$A^r_{21z} - A^r_{21R} \cos \Theta_2 = -A_{21\Theta_1} \cos \psi \sin \Theta_2$$
$$= -A_{21\Theta_1} \cos \psi \cos \bar{\Theta}_2. \quad (31)$$

With (31) substituted in (29), the function U becomes

$$U = j\omega A^r_{21\Theta_1} \frac{\cos \psi \sin \Theta_2}{\beta_0 \sin^2 \Theta_2} = \frac{j\omega A_{21\Theta_1} \cos \psi}{\beta_0 \sin \Theta_2}. \quad (32)$$

If desired, the component $A^r_{21\Theta_1}$ of the vector potential may be replaced by the *far-zone* electric field due to the distant transmitter. This has only a Θ_1-component and is related to the vector potential by the relation [ref. I.31, Eq. (IV.16.11)]:

$$E^r = E^r_{\Theta_1} \hat{\Theta}_1 = -j\omega A^r_{\Theta_1} \hat{\Theta}_1. \quad (33)$$

Hence,

$$U = -(E^r_{21\Theta_1} \cos \psi)/\beta_0 \sin \Theta_2$$
$$= -\lambda_0 E_{21\Theta_1} \left(\frac{\cos \psi}{2\pi \sin \Theta_2} \right). \quad (34)$$

Clearly U is measured in volts. (In the literature U has been defined in terms of the

[IV.3] THEORY OF LINEAR ANTENNAS 465

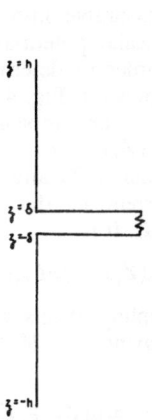

Fig. 3.1. Center-loaded receiving antenna.

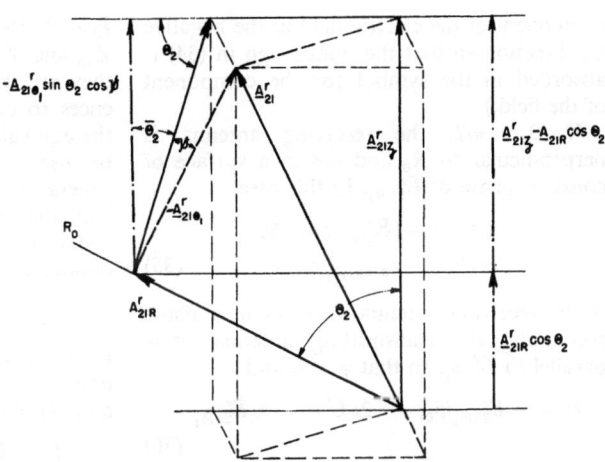

Fig. 3.2. Mutually perpendicular components of the potential $A^r_{21\Theta_1}$ due to a distant transmitter; Θ_1 is the angle between the transmitting antenna and R_0 as shown in Fig. 2.1.

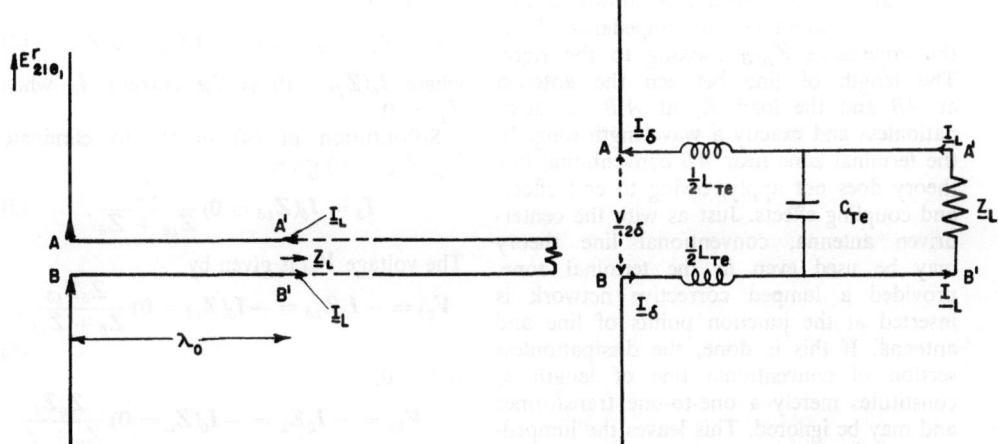

Fig. 4.1. Receiving antenna with two-wire line terminated in an impedor as center load.

Fig. 4.2. Equivalent circuit of Fig. 4.1 with lumped-terminal-zone network.

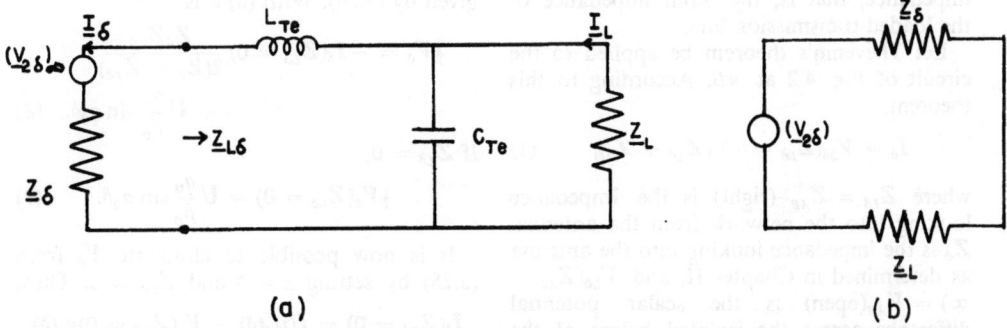

Fig. 4.3a. Equivalent circuit of receiving antenna with impedance-less generator.

Fig. 4.3b. Simplified equivalent circuit.

component of the electric field in the negative Θ_1 direction so that the minus sign in (34) is absorbed in the symbol for the component of the field.)

If $\Theta_2 = \pi/2$, the receiving antenna is perpendicular to R_0 and lies in a surface of constant phase of $E^r_{21\Theta_1}$. In this case

$$U = -(E^r_{21\Theta_1} \cos \psi)/\beta_0,$$
$$2\pi U = -\lambda_0 E^r_{21\Theta_1} \cos \psi. \qquad (35)$$

If the receiving antenna also lies in a plane containing the transmitting antenna, it is parallel to $E^r_{21\Theta_1}$, so that $\psi = 0$ and

$$U = -E^r_{21\Theta_1}/\beta_0, \qquad 2\pi U = -\lambda_0 E^r_{21\Theta_1}. \qquad (36)$$

4. Equivalent Circuit of a Receiving Antenna; Effective Length

A receiving antenna with a symmetric center load consisting of a two-wire line terminated in an impedor is shown in Fig. 4.1. The measurable load impedance Z_L is the impedance $Z_{A'B'}$ looking to the right. The length of line between the antenna at AB and the load Z_L at $A'B'$ is dissipationless and exactly a wavelength long. In the terminal zone near AB conventional line theory does not apply, owing to end effects and coupling effects. Just as with the center-driven antenna, conventional line theory may be used even in the terminal zone, provided a lumped corrective network is inserted at the junction points of line and antenna. If this is done, the dissipationless section of conventional line of length λ_0 constitutes merely a one-to-one transformer and may be ignored. This leaves the lumped-constant network of Fig. 4.2 as a satisfactory equivalent in which all coupling effects and end effects are represented by the network between AB and $A'B'$; Z_L is the actual load impedance, that is, the input impedance of the loaded transmission line.

Let Thévenin's theorem be applied to the circuit of Fig. 4.2 at AB. According to this theorem,

$$I_\delta = V_{2\delta}(Z_{L\delta} = \infty)/(Z_{L\delta} + Z_\delta), \qquad (1)$$

where $Z_{L\delta} = Z_{AB}$ (right) is the impedance looking into the network from the antenna, Z_δ is the impedance looking into the antenna as determined in Chapter II, and $V_{2\delta}(Z_{L\delta} = \infty) = V_{AB}$(open) is the scalar potential difference across the isolated halves of the antenna when the circuit is broken so that $I_\delta = 0$, that is, when $Z_{L\delta} = \infty$. Note that $Z_{L\delta}$ and Z_δ are not measurable impedances; they are the ratios of scalar potential differences to currents. In order to determine I_δ, the equivalent circuit shown in Fig. 4.3a may be used. It consists of an impedanceless generator of emf $V_{2\delta}(Z_{L\delta} = \infty)$ in series with the antenna impedance Z_δ and the impedance $Z_{L\delta}$. The current in the load is obtained directly from I_δ. It is

$$I_L = I_\delta[1 + j\omega C_{Te}(Z_{L\delta} + j\omega L_{Te})]. \qquad (2)$$

If end effects and coupling effects are negligible as a result of having $b = 2\delta \doteq 0$, (1) and (2) reduce to

$$I_L \doteq I_0 = V_{20}(Z_L = \infty)/(Z_L + Z_0) \qquad (3)$$

and the simple circuit of Fig. 4.3b applies.

A convenient formula relating the open-circuit voltage to the current is obtained when $Z_{L\delta}$ is adjusted to equal zero. When this is true,

$$V_{2\delta}(Z_{L\delta} = \infty) = I_\delta(Z_{L\delta} = 0)Z_\delta, \qquad (4)$$

where $I_\delta(Z_{L\delta} = 0)$ is the current I_δ when $Z_{L\delta} = 0$.

Substitution of (4) in (1) to eliminate $V_{2\delta}(Z_{L\delta} = \infty)$ gives

$$I_\delta = I_\delta(Z_{L\delta} = 0) \frac{Z_\delta}{Z_{L\delta} + Z_\delta}. \qquad (5)$$

The voltage $V_{2\delta}$ is given by

$$V_{2\delta} = -I_\delta Z_{L\delta} = -I_\delta(Z_{L\delta} = 0) \frac{Z_\delta Z_{L\delta}}{Z_\delta + Z_{L\delta}}. \qquad (6)$$

If $\delta = 0$,

$$V_{20} = -I_0 Z_L = -I_0(Z_L = 0) \frac{Z_0 Z_L}{Z_0 + Z_L}. \qquad (7)$$

The quantity $\frac{1}{2}V_\delta$ appearing in the fundamental expression (3.25) for the current is given by (3.28). With (6) it is

$$\tfrac{1}{2}V_\delta = -I_\delta(Z_{L\delta} = 0) \frac{Z_\delta Z_{L\delta}}{2(Z_\delta + Z_{L\delta})}$$
$$+ U\frac{q_0}{\beta_0} \sin q_0 \delta. \qquad (8)$$

If $Z_{L\delta} = 0$,

$$\tfrac{1}{2}V_\delta(Z_{L\delta} = 0) = U\frac{q_0}{\beta_0} \sin q_0 \delta. \qquad (9)$$

It is now possible to eliminate V_δ from (3.25) by setting $z = \delta$ and $Z_{L\delta} = 0$. Thus,

$$I_\delta(Z_{L\delta} = 0) = Uu_\delta(\delta) + V_\delta(Z_{L\delta} = 0)v_\delta(\delta) \qquad (10)$$

Substitution of (9) in (10) gives

$$I_\delta(Z_{L\delta} = 0) = U\left[u_\delta(\delta) + v_\delta(\delta)\frac{2q_0}{\beta_0}\sin q_0\delta\right]. \tag{11}$$

If (11) is substituted in (8) and it is noted from (3.27) that

$$v_\delta(\delta) = Y_\delta = 1/Z_\delta, \tag{12}$$

the result is

$$\tfrac{1}{2}V_\delta = -\frac{UZ_\delta}{Z_\delta + Z_{L\delta}}\left[\tfrac{1}{2}u_\delta(\delta)Z_{L\delta} - \frac{q_0}{\beta_0}\sin q_0\delta\right]. \tag{13}$$

The general expression for the even current in the antenna is given by (3.25) with (13). It is

$$I_z = U\left\{u_\delta(z) - v_\delta(z)\frac{Z_\delta}{Z_\delta + Z_{L\delta}}\left[u_\delta(\delta)Z_{L\delta} - \frac{2q_0}{\beta_0}\sin q_0\delta\right]\right\}. \tag{14}$$

If $\delta \doteq 0$,

$$I_z \doteq U\left\{u_0(z) - v_0(z)\frac{Z_0 Z_L}{(Z_0 + Z_L)}u_0(0)\right\}. \tag{15}$$

The complex distribution functions $u_\delta(z)$ and $v_\delta(z)$ are given in (3.26) and (3.27).

It is seen from (14) or (15) that the even current in the receiving antenna is made up of two parts, of which the first has the distribution function $u_\delta(z)$, the second the distribution function $v_\delta(z)$. It is seen from (14) that when $Z_{L\delta} = 0$ and $q_0\delta$ is sufficiently small, as is usual, practically the entire even current is distributed according to $u_\delta(z)$. Therefore, $u_\delta(z)$ may be called the *unloaded-receiving-antenna distribution*. On the other hand, $v_\delta(z)$ is identically the distribution function for a center-driven, isolated, transmitting antenna as analyzed in Chapter II, if $\Psi = \Psi_{K1}$. Since it is shown in the next section that $\Psi = \Psi_{K1}$ is a good approximation for the receiving antenna, it follows that the even currents on a receiving antenna are a superposition of the distributions of *an unloaded receiving antenna and a center-driven antenna*. The relative magnitudes of these two components depends upon $Z_{L\delta}$. The transmitting distribution has its largest value when $Z_\delta Z_{L\delta}/(Z_\delta + Z_{L\delta})$ is greatest. Since this factor can be made very small or quite large, it is to be expected that a wide variety of distributions may be encountered, ranging from the unloaded-receiving-antenna distribution to distributions closely resembling those of center-driven transmitting antennas. Before investigating these distributions and their magnitudes, it is necessary to define Ψ.

The open-circuit voltage $V_{2\delta}(Z_{L\delta} = \infty)$, which is the emf required in the equivalent circuit of Fig. 4.3, is defined in (4). Upon setting $z = \delta$ and $Z_{L\delta} = 0$ in (14) to obtain $I_\delta(Z_{L\delta} = 0)$, and substituting this in (4), the expression for the open-circuit voltage, $V_{2\delta}(Z_{L\delta} = \infty)$, is

$$V_{2\delta}(Z_{L\delta} = \infty) = UZ_\delta\left[u_\delta(\delta)\right.$$
$$\left. + v_\delta(\delta)\frac{2q_0}{\beta_0}\sin q_0\delta\right]. \tag{16}$$

Using (3.34), namely,

$$U = -(E^r_{21\Theta_1}\cos\psi)/\beta_0\sin\Theta_2, \tag{17}$$

it follows that

$$V_{2\delta}(Z_{L\delta} = \infty) = -E^r_{21\Theta_1} \cdot 2h_{e\delta}(\Theta_2)\cos\psi, \tag{18}$$

where

$$2h_{e\delta}(\Theta_2) \equiv \frac{Z_\delta}{\beta_0\sin\Theta_2}\left[u_\delta(\delta)\right.$$
$$\left. + v_\delta(\delta)\frac{2q_0}{\beta_0}\sin q_0\delta\right]. \tag{19}$$

Since $v_\delta(\delta) = 1/Z_\delta$,

$$\beta_0 h_{e\delta}(\Theta_2) \equiv \frac{\tfrac{1}{2}u_\delta(\delta)Z_\delta + (q_0/\beta_0)\sin q_0\delta}{\sin\Theta_2} \tag{20}$$

For $\delta = 0$,

$$\beta_0 h_e(\Theta_2) = \frac{Z_0 u_0(0)}{2\sin\Theta_2}. \tag{21}$$

The complex function $2h_{e\delta}(\Theta_2)$ is dimensionally a length. It is called the *complex effective length* of the receiving antenna of length $2h$. It is a function of the orientation of the receiving antenna with respect to the surface of constant phase of the electric field $E^r_{21\Theta_1}$. This orientation is specified by the angle Θ_2. If $\Theta_2 = \pi/2$, the receiving antenna lies in a surface of constant phase of $E^r_{21\Theta_1}$; if $\Theta_2 = 0$, it is perpendicular to such a surface. Since $q_0 = \beta_0\cos\Theta_2$, the angle Θ_2 occurs in $u_\delta(\delta)$ as well as explicitly in $\sin\Theta_2$. Evidently, if the electric field $E^r_{21\Theta_1}$ is known, together with the angle ψ, a knowledge of $h_{e\delta}(\Theta_2)$ permits a rapid evaluation of $V_{2\delta}(Z_{L\delta} = \infty)$, and from this of Z_δ and

$Z_{L\delta}$ and of I_δ and I_L. The behavior of a receiving antenna depends upon $h_{e\delta}(\Theta_2)$ and Z_δ for that antenna. Since the impedance Z_δ of an isolated antenna already has been determined and represented in graphical and tabular form in Chapter II, an important part of the analysis of the receiving antenna consists in the evaluation and representation of the complex effective length as a function of h and Θ_2 with $\Omega = 2 \ln (2h/a)$ as a parameter.

5. Zeroth-Order Solution for Center-Loaded Receiving Antenna

The zeroth-order distribution of even current in a center-loaded receiving antenna is independent of the particular value of the expansion parameter Ψ, which occurs only as a factor of the entire distribution. It is a good approximation of the actual distribution of current when higher-order terms in numerator *and* denominator are negligible compared with the zeroth-order terms. This is generally true except near vanishing values of either the numerator or the denominator. The zeroth-order terms are given by

$$[I_z]_0 = U\left\{u_\delta(z) - v_\delta(z)\frac{Z_\delta}{Z_\delta + Z_{L\delta}}\right.$$

$$\left.\times\left[u_\delta(\delta)Z_{L\delta} - \frac{2q_0}{\beta_0}\sin q_0\delta\right]\right\},$$

$$(\delta \leq z \leq h) \quad (1)$$

where

$$u_\delta(z) = [u_\delta(z)]_0 = \frac{-j4\pi}{\zeta_0\Psi} \times$$

$$\left[\frac{\cos q_0 h \cos \beta_0(z-\delta) - \cos q_0 z \cos \beta_0(h-\delta)}{\cos \beta_0(h-\delta)}\right], \quad (2)$$

$$v_\delta(z) = [v_\delta(z)]_0 = \frac{-j2\pi}{\zeta_0\Psi}\left[\frac{\sin \beta_0(h-|z|)}{\cos \beta_0(h-\delta)}\right], \quad (3)$$

$$Z_\delta = [Z_\delta]_0 = \frac{j\zeta_0\Psi}{2\pi}\cot \beta_0(h-\delta), \quad (4)$$

$$q_0 = \beta_0 \cos \Theta_2. \quad (5)$$

The zeroth-order distribution of odd charge on the center-loaded receiving antenna is readily obtained from (1), using the equation of continuity,

$$\frac{dI_z}{dz} + j\omega q = 0. \quad (6)$$

Note the difference between the charge per unit length q and the parameter q_0. The zeroth-order charge per unit length is

$$[q]_0 = -\frac{U}{j\omega}\left\{\frac{\partial u_\delta(z)}{\partial z} - \frac{\partial v_\delta(z)}{\partial z}\frac{Z_\delta}{Z_\delta + Z_{L\delta}}\right.$$

$$\left.\times\left[u_\delta(\delta)Z_{L\delta} - \frac{2q_0}{\beta_0}\sin q_0\delta\right]\right\},$$

$$(\delta \leq z \leq h) \quad (7)$$

where

$$\frac{\partial u_\delta(z)}{\partial z} = \left[\frac{\partial u_\delta(z)}{\partial z}\right]_0 = \frac{j4\pi\beta_0}{\zeta_0\Psi}$$

$$\times \left[\frac{\cos q_0 h \sin \beta_0(z-\delta) - \frac{q_0}{\beta_0}\sin q_0 z \cos \beta_0(h-\delta)}{\cos \beta_0(h-\delta)}\right],$$

$$(8)$$

$$\frac{\partial v_\delta(z)}{\partial z} = \left[\frac{\partial v_\delta(z)}{\partial z}\right]_0 = \frac{j2\pi\beta_0}{\zeta_0\Psi}\left[\frac{\cos \beta_0(h-z)}{\cos \beta_0(h-\delta)}\right].$$

$$(9)$$

The two sets of distribution functions, $[u_\delta(z)]_0$, $[\partial u_\delta(z)/\partial z]_0$ and $[v_\delta(z)]_0$, $[\partial v_\delta(z)/\partial z]_0$, are advantageously studied individually. Since the latter two are the same (except for a scale factor) as the zeroth-order current and charge described in Sec. II.16 and are represented graphically in Figs. II.16.1 and II.16.2, their distributions already are known.

The distribution functions $[u_\delta(z)]_0$ and $[\partial u_\delta(z)/\partial z]_0$ are functions of both h and Θ_2, the latter through the factor $q_0 = \beta_0 \cos \Theta_2$. The functional dependence of $[u_\delta(z)]_0$ upon z is represented graphically in Fig. 5.1, where the numerator of the bracket in (2) is plotted for several values of $\beta_0 h$ with $\delta = 0$ and $\cos \Theta_2 = \sin \overline{\Theta}_2 = 0, 0.2, 0.4, 0.6, 0.8$, corresponding to $\Theta_2 = 90°$, $78°.46$, $66°.42$, $53°.13$, $36°.87$ or $\overline{\Theta}_2 = 0$, $11°.54$, $23°.58$, $36°.87$, $53°.13$.

In general, the distribution of even currents along the antenna is a superposition of the distribution $u_\delta(z)$ for the unloaded receiving antenna on the distribution $v_\delta(z)$ for the center-driven antenna. The relative magnitudes and phases of these components of the even current depends upon the factor $Z_\delta Z_{L\delta}/(Z_\delta + Z_{L\delta})$ in (1). Since $q_0\delta \leq \beta_0\delta$ is necessarily small, the term $(2q_0/\beta_0)\sin q_0\delta$

is certainly always negligible if $Z_{L\delta}$ is also small. This term may be expanded as follows:

$$\frac{Z_\delta Z_{L\delta}}{Z_\delta + Z_{L\delta}}$$
$$= \frac{(R_\delta R_{L\delta} - X_\delta X_{L\delta}) + j(R_\delta X_{L\delta} + R_{L\delta} X_\delta)}{(R_\delta + R_{L\delta}) + j(X_\delta + X_{L\delta})}. \quad (10)$$

If the circuit is resonant, $X_\delta + X_{L\delta} = 0$, and

$$\frac{Z_\delta Z_{L\delta}}{Z_\delta + Z_{L\delta}}$$
$$= \frac{(R_\delta R_{L\delta} + X_\delta^2) + jX_\delta(R_{L\delta} - R_\delta)}{R_\delta + R_{L\delta}}. \quad (11)$$

If the antenna and the load are individually resonant, $X_\delta = X_{L\delta} = 0$, and

$$\frac{Z_\delta Z_{L\delta}}{Z_\delta + Z_{L\delta}} = \frac{R_\delta R_{L\delta}}{R_\delta + R_{L\delta}}. \quad (12)$$

If the antenna and the load are individually resonant and the load resistance is very much larger than the resistance of the antenna, then $X_\delta = X_{L\delta} = 0$, $R_{L\delta} \gg R_\delta$, and

$$\frac{Z_\delta Z_{L\delta}}{Z_\delta + Z_{L\delta}} \doteq R_\delta. \quad (13)$$

If the load is a conjugate match, $Z_{L\delta} = Z_\delta^*$, and

$$\frac{Z_\delta Z_{L\delta}}{Z_\delta + Z_{L\delta}} = \frac{Z_\delta^2}{2R_\delta}. \quad (14)$$

It is readily seen in all of the above special cases that $|Z_\delta Z_{L\delta}/(Z_\delta + Z_{L\delta})|$ is large compared with unity unless R_δ is itself very small.

It is known from Fig. 5.1 that for antennas that satisfy $\beta_0 h \leq 3\pi/2$, $u_0(\delta)$ is not smaller than $u_0(z)$. Hence, if $u_\delta(\delta)$ in (1) is multiplied by a factor that is large compared with unity, the coefficient of $v_\delta(z)$ is much greater than that of $u_\delta(z)$. This means that *the principal part of the current in a loaded receiving antenna is distributed as in a center-driven antenna.*

Two distributions that are of particular interest are now considered in turn.

(a) $\quad \Theta_2 = \pi/2, \quad Z_{L\delta} = 0, \quad \delta = 0. \quad (15)$

With these conditions the coefficients of $v_\delta(z)$ and $\partial v_\delta(z)/\partial z$ in (1) and (7) vanish. Moreover, owing to the symmetric orientation, $\Theta_2 = \pi/2$,

no odd currents are excited anywhere on the antenna. This means that the complete zeroth-order distributions of current and charge are

$$[I_z]_0 = U[u_0(z)]_0$$
$$= \frac{j4\pi U}{\zeta_0 \Psi}\left(\frac{\cos \beta_0 z - \cos \beta_0 h}{\cos \beta_0 h}\right),$$
$$(0 \leq z \leq h) \quad (16)$$

$$[q]_0 = -\frac{U}{j\omega}[u_0'(z)]_0 = \frac{4\pi \epsilon_0 U}{\Psi}\sin \beta_0 z.$$
$$(-h \leq z \leq h) \quad (17)$$

The distributions on the lower half of the antenna are obtained by noting that the current is even, the charge odd.

The current at z may be expressed in terms of its value at $z = 0$:

$$[I_z]_0 = I_0 \frac{\cos \beta_0 z - \cos \beta_0 h}{1 - \cos \beta_0 h} \quad (18)$$

This function is plotted in Fig. II.25.2. It is called a *shifted cosine distribution*.

(b) $\quad q_0 h = \beta_0 h \cos \Theta_2 = \pi/2$,
$$Z_{L\delta} = 0, \quad \delta = 0. \quad (19)$$

These conditions can be satisfied only by adjustment of $\beta_0 h$ as the angle Θ_2 is varied. The distribution functions are:

$$[u_0(z)]_0 = \frac{j4\pi}{\zeta_0 \Psi}\cos q_0 z = \frac{j4\pi}{\zeta_0 \Psi}\cos\left(\frac{\pi z}{2h}\right),$$
$$(0 \leq z \leq h) \quad (20)$$

$$[\partial u_0(z)/\partial z]_0 = \frac{-j4\pi q_0}{\zeta_0 \Psi}\sin q_0 z$$
$$= \frac{-2\pi^2}{h\zeta_0 \Psi}\sin\left(\frac{\pi z}{2h}\right). \quad (0 \leq z \leq h) \quad (21)$$

Thus, the even currents in an unloaded antenna are distributed *cosinusoidally* for any half length h, provided the antenna is inclined from the surface of constant phase of the incident electric field $E_{21\Theta_1}^r$ by an angle Θ_2 that satisfies (19). For $\beta_0 h = \pi/2$, $\Theta_2 = \pi/2 = 90°$ and this special case reduces to (a). For $\beta_0 h = \pi$, $\Theta_2 = \pi/3 = 60°$.

The zeroth-order effective length is evaluated readily by combining (2), (4), and (5) after setting $z = \delta$ in order to form (4.19). The result is a real quantity that is a good approximation of the true effective length except near zeros of the numerator and denominator. In radians it is

$$[\beta_0 h_{e\delta}(\Theta_2)]_0$$
$$= \frac{\cos q_0 h - \cos q_0 \delta \cos \beta_0 (h - \delta)}{\sin \beta_0 (h - \delta) \sin \Theta_2}$$
$$+ \frac{\cos \Theta_2 \sin q_0 \delta}{\sin \Theta_2}. \quad (22)$$

Since $\beta_0 \delta$ is small, terms in $\beta_0^2 \delta^2$ may be neglected compared with unity. Thus,

$$[\beta_0 h_{e\delta}(\Theta_2)]_0$$
$$\doteq \frac{\cos (\beta_0 h \cos \Theta_2) - \cos \beta_0 h - \beta_0 \delta \sin \beta_0 h}{\sin \beta_0 (h - \delta) \sin \Theta_2}$$
$$+ \frac{\beta_0 \delta \cos^2 \Theta_2}{\sin \Theta_2}$$
$$\doteq \frac{\cos (\beta_0 h \cos \Theta_2) - \cos \beta_0 h}{\sin \beta_0 (h - \delta) \sin \Theta_2} - \beta_0 \delta \sin \Theta_2. \quad (23)$$

This gives the zeroth-order effective length of a center-loaded antenna of half-length h that has its load terminals separated by a small distance 2δ. Note that (23) cannot be a good approximation near $\beta_0(h - \delta) = n\pi$, n integral, since the vanishing of the zeroth-order denominator obviously means that first-order terms become significant. When $\beta_0(h - \delta)$ is not near $n\pi$, $n = 0, 1, 2$, $\sin \beta_0(h - \delta)$ does not differ greatly from $\sin \beta_0 h$. With $\delta \doteq 0$:

$$[\beta_0 h_e(\Theta_2)]_0$$
$$= \frac{\cos (\beta_0 h \cos \Theta_2) - \cos \beta_0 h}{\sin \Theta_2 \sin \beta_0 h}. \quad (24)$$

For $\Theta_2 = \pi/2$, the zeroth-order effective length is

$$[\beta_0 h_e(\tfrac{1}{2}\pi)]_0 = \tan \tfrac{1}{2}\beta_0 h. \quad (25)$$

The general expression (23) is best interpreted after the special form for a short antenna has been obtained. With $(\beta_0 h)^2 \ll 1$, $\delta \doteq 0$, this is given by the following very simple formula;

$$[\beta_0 h_e(\Theta_2)]_0 = \tfrac{1}{2}\beta_0 h \sin \Theta_2. \quad (26)$$

This is the effective half-length in radians of a short antenna with a triangular distribution of current. If the short antenna were so end-loaded that it carried a uniform current, (26) would be doubled. That is, the effective half-length $\beta_0 \delta_e$ radians of a short antenna of length $\beta_0 \delta$ radians that satisfies the condition $\beta_0^2 \delta^2 \ll 1$ and that carries a current of uniform amplitude along its entire length is

$$[\beta_0 \delta_e(\Theta_2)]_0 = \beta_0 \delta \sin \Theta_2. \quad (27)$$

Comparison of (23) with (24) and (26) shows that (23) is equivalent to

$$[\beta_0 h_{e\delta}(\Theta_2)]_0 \doteq [\beta_0 h_e(\Theta_2)]_0 \frac{\sin \beta_0 h}{\sin \beta_0 (h - \delta)}$$
$$- [\beta_0 \delta_e(\Theta_2)]_0. \quad (28)$$

That is, the effective half-length of a center-loaded antenna of half-length h that *includes* a gap of half-length δ is the effective half-length of an antenna of half-length h *without* gap multiplied by $\sin \beta_0 h/\sin \beta_0(h - \delta)$ minus the effective half-length of an antenna of length 2δ that would just fill the gap and carry a uniform current. This is a good approximation except near $\beta_0 h = n\pi$, n integral, where the zeroth-order formula is inadequate.

It is evident from (22) or (23) that even the zeroth-order effective length of a center-loaded receiving antenna is not significantly modified by a small value of δ as compared with $\delta = 0$. The directional properties of the zeroth-order effective length are discussed and represented graphically in Sec. 9 where they can be compared with the first-order effective length.

The voltage $V_{2\delta}(Z_{L\delta} = \infty)$, which is the emf in the equivalent circuit of the receiving antenna with its load, is given by

$$V_{2\delta}(Z_{L\delta} = \infty) = -2h_{e\delta}(\Theta_2)E^r_{21\Theta_1} \cos \psi. \quad (29)$$

For a short antenna with $\beta_0^2 h^2 \ll 1$ and (26),

$$V_{2\delta}(Z_{L\delta} = \infty) = -hE^r_{21\Theta_1} \cos \psi \sin \Theta_2$$
$$= -hE^r_{21\Theta_1} \cos \psi \cos \bar{\Theta}_2. \quad (30)$$

Thus, for a short antenna, the projection of the electric field $-E^r_{21\Theta_1}$ onto the receiving antenna, multiplied by the actual half-length of the antenna, gives the emf in the equivalent circuit.

6. *The Expansion Parameter for the Receiving Antenna*

The evaluation of the expansion parameter Ψ depends upon the choice of the relative distribution function $g(z, z')$ in the integral

$$\Psi(z) = \left(\int_{-h}^{-\delta} + \int_{\delta}^{h} \right) g(z, z') K_1(z, z') \, dz', \quad (1)$$

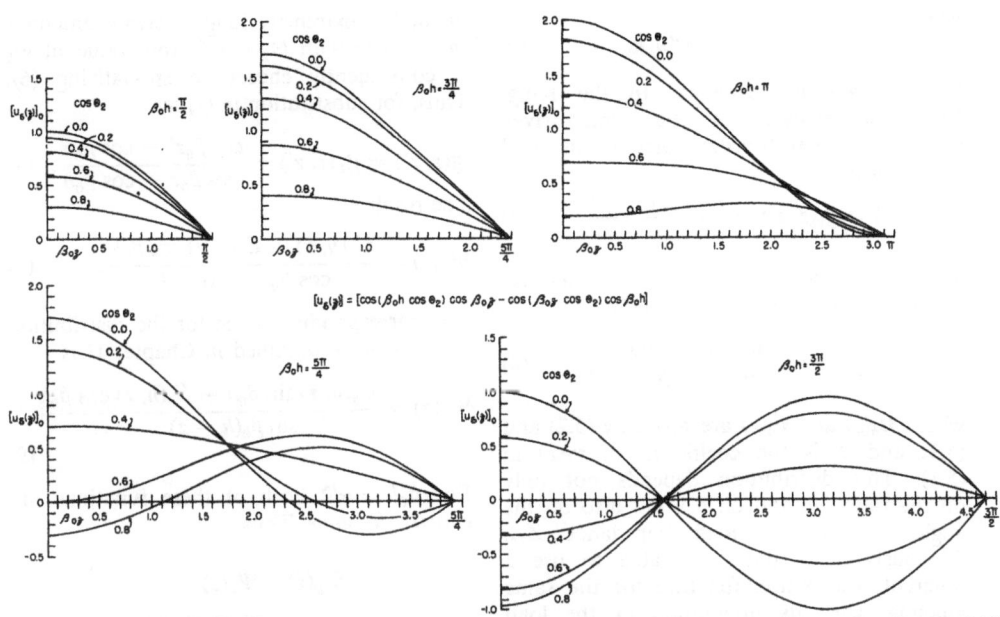

Fig. 5.1. Zeroth-order current distribution of unloaded receiving antenna parallel to the electric field.

$$[u_0(\beta_0 z)]_0 = [\cos(\beta_0 h \cos\theta_2)\cos\beta_0 z - \cos(\beta_0 z \cos\theta_2)\cos\beta_0 h]$$

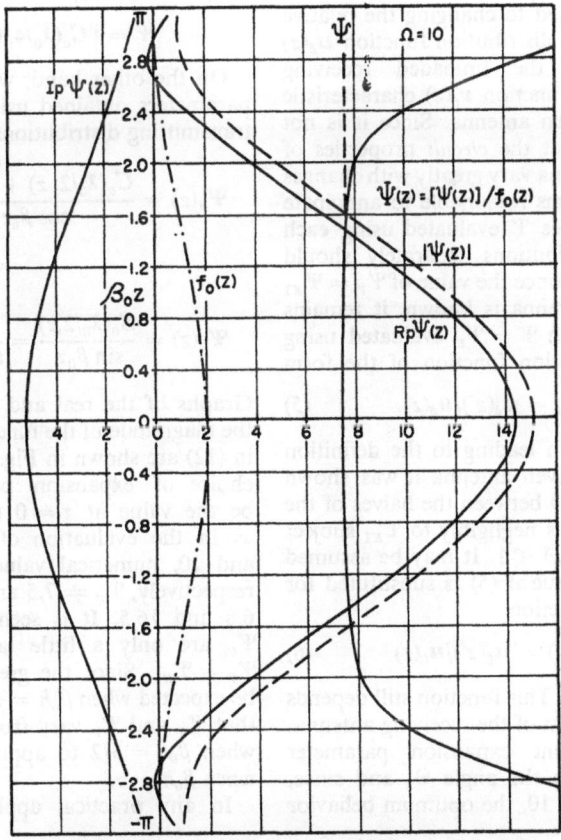

Fig. 6.1. Expansion parameter for the receiving antenna.

where

$$K_1(z, z') = e^{-j\beta_0 R_{22}}/R_{22} \qquad (2)$$

for the receiving antenna. In the King-Middleton analysis of the center-driven antenna the zeroth-order current was used in the form

$$g(z, z') \equiv g_V(z, z') = \frac{\sin \beta_0(h - z')}{\sin \beta_0(h - z)}. \qquad (3)$$

The corresponding distribution for the receiving antenna is intricate, namely,

$$g(z, z') = \frac{u_\delta(z') - S v_\delta(z')}{u_\delta(z) - S v_\delta(z)}, \qquad (4)$$

where $u_\delta(z)$ and $v_\delta(z)$ are given by (5.2) and (5.3) and S is the coefficient of $v_\delta(z)$ in (5.1). This distribution depends not only on the ratio h/a and on $\beta_0 h$ but also on the angle Θ_2 and the load impedance $Z_{L\delta}$. Obviously, it is not practicable to use a different distribution function for the same antenna when its orientation or the load impedance is changed. Examination of (4) shows that any change in the distribution function produced by a variation in the load impedance is limited to changing the relative importance of the distribution function $u_\delta(z)$ characteristic of the unloaded receiving antenna and the function $v_\delta(z)$ characteristic of the center-driven antenna. Since it is not to be expected that the *circuit* properties of the receiving antenna vary greatly with changes in the load, it seems reasonable to anticipate that the two values Ψ evaluated using each of the two distributions separately should not differ greatly. Since the value of $\Psi_V = \Psi_{K1}$ for the driven antenna is known, it remains to compare it with $\Psi = \Psi_U$ evaluated using a relative distribution function of the form

$$g_U(z, z') = u_\delta(z')/u_\delta(z). \qquad (5)$$

In the investigation leading to the definition of Ψ_{K1} for the driven antenna it was shown that the distance 2δ between the halves of the antenna contributes negligibly to Ψ_{K1} subject to the condition $\beta_0 \delta \ll 1$. It may be assumed that the same is true if (5) is substituted for (3). Hence, the function

$$g_U(z, z') = u_0(z')/u_0(z) \qquad (6)$$

may be used in (1). This function still depends upon the orientation of the receiving antenna. Since a convenient expansion parameter should not involve the angle Θ_2 and since, as is shown in Sec. 10, the optimum behavior of all but extremely long receiving antennas ($h > \lambda_0$) is with $\Theta_2 = \pi/2$, this value of Θ_2 is conveniently chosen in specializing (6). Thus, for substitution in (1) let

$$g(z, z') = g_U(z, z') = \frac{\cos \beta_0 z' - \cos \beta_0 h}{\cos \beta_0 z - \cos \beta_0 h}. \qquad (7)$$

The result is

$$\Psi_U(z) = \frac{C_a(h, z) - E_a(h, z) \cos \beta_0 h}{\cos \beta_0 z - \cos \beta_0 h}. \qquad (8)$$

The corresponding value for the distribution (3) previously obtained in Chapter II is

$$\Psi_V(z) = \frac{C_a(h, z) \sin \beta_0 h - S_a(h, z) \cos \beta_0 h}{\sin \beta_0 (h - z)}. \qquad (9)$$

For $\beta_0 h = \pi/2$ both of these functions are *exactly the same*. Thus,

$$\Psi_U(z) = \Psi_V(z)$$
$$= C_a(\lambda_0/4, z)/\cos \beta_0 z. \qquad (10)$$

The expansion parameter is the magnitude at $\beta_0 z = 0$. Thus,

$$\Psi = |C_a(\lambda_0/4, 0)| = \Psi_{K1}. \qquad (11)$$

On the other hand, with $\beta_0 h = \pi$, the two parameters obtained using the receiving and transmitting distributions are different:

$$\Psi_U(z) = \frac{C_a(\lambda_0/2, z) + E_a(\lambda_0/2, z)}{\cos \beta_0 z + 1}$$
$$= \frac{\psi_u(z)}{\cos \beta_0 z + 1}, \qquad (12)$$

$$\Psi_V(z) = \frac{S_a(\lambda_0/2, z)}{\sin \beta_0 z} = \frac{\psi_v(z)}{\sin \beta_0 z}. \qquad (13)$$

Graphs of the real and imaginary parts and the magnitude of the function $\Psi_U(z)$ as defined in (12) are shown in Fig. 6.1. An appropriate choice of expansion parameter is seen to be the value at $z = 0$ (not at $z = h - \lambda_0/4$ as in the evaluation of Ψ_{K1}). For $\Omega = 10$ and 20, numerical values, for $\beta_0 h = \pi$, are respectively, $\Psi_U \doteq 7.5$ and 17.1, $\Psi_V = \Psi_{K1} = 6.8$ and 16.5. It is seen that the values of Ψ_U are only a little larger than those of $\Psi_V = \Psi_{K1}$. Since the greatest deviation is to be expected when $\beta_0 h = \pi$, it is safe to assume that Ψ_U and Ψ_V vary from exact equivalence when $\beta_0 h = \pi/2$ to approximate equivalence when $\beta_0 h = \pi$.

In any practical application in which a

receiving antenna is properly matched to its load, the part of the current that is distributed as in a driven antenna greatly predominates. Clearly, if the values of Ψ characteristic of this distribution are used, no error is made for $\beta_0 h$ near $\pi/2$ and only a small one for $\beta_0 h$ near π. This is not significant even for the unloaded receiving antenna with $\beta_0 h = \pi$, since the magnitude of Ψ is relatively much less important in determining the effective length than in determining the impedance. This was suggested by the nonappearance of Ψ in the zeroth-order effective length in Sec. 5. It is indicated more clearly in the comparison of first- and zeroth-order values. Accordingly, it may be concluded that the same expansion parameter, Ψ_{K1}, is satisfactory for use in the receiving antenna. This is entirely reasonable, since if the relative distribution function $g(z, z')$ is properly chosen, so that it is a good approximation of the actual current, the function $\Psi(z)$ is the ratio of vector potential on the surface of the conductor to the total current at z. For a given radius a, $\Psi = |\Psi(z_r)|$ should be a reasonably constant function, varying only slowly with the length of the antenna and the distribution of current if referred to the current at a point z_r where $\Psi(z)$ is maximum, as is always the case. Unless otherwise indicated, the value of Ψ determined for the transmitting antenna is used for the receiving antenna.

7. First-Order Current in Center-Loaded Receiving Antenna

The numerical calculation of the distribution of current in a center-loaded receiving antenna from the general formula (3.26) involves the distribution functions $u_\delta(z)$ for the unloaded receiving antenna and $v_\delta(z)$ for the center-driven antenna, as given in (3.27) and (3.28). By replacing the general expansion parameter Ψ by Ψ_{K1}, the solution is put into the form of the King-Middleton solution. (If Ψ is replaced by Ω the form due to Hallén is obtained.) As in the analysis of the center-driven antenna in Chapter II, the complexity of the higher-order terms and the accuracy required in most practical applications does not justify or demand calculations of a higher order for the distribution of current *along the antenna* than the first. On the other hand, a second-order current *in the load* is desirable. The first-order King-Middleton solution is

$$[I_z]_1 = U[u_\delta(z)]_1 + V_\delta[v_\delta(z)]_1, \quad (1)$$

where, for $\delta \leq z \leq h$,

$$[u_\delta(z)]_1 = \frac{j4\pi}{\zeta_0 \Psi_{K1}} \times$$

$$\left\{ \frac{[\cos q_0 h \cos \beta_0(z-\delta) - \cos q_0 z \cos \beta_0(h-\delta)] + \frac{1}{\Psi_{K1}}[m_{1K}(z) \cos \beta_0 \delta + p_{1K}(z) \sin \beta_0 \delta]}{\cos \beta_0(h - \delta) + A_{1K}/\Psi_{K1}} \right\}, \quad (2)$$

$$[v(z)]_1 = \frac{j2\pi}{\zeta_0 \Psi_{K1}}$$

$$\times \left\{ \frac{\sin \beta_0(h - z) + M_{1K}(z)/\Psi_{K1}}{\cos \beta_0(h - \delta) + A_{1K}/\Psi_{K1}} \right\}. \quad (3)$$

The following functions have been defined previously; they are repeated for convenience:

$$\tfrac{1}{2} V_\delta = \tfrac{1}{2} V_{2\delta} + U \frac{q_0}{\beta_0} \sin q_0 \delta$$

$$= -U \frac{Z_\delta}{Z_\delta + Z_{L\delta}} \left[\tfrac{1}{2} u_\delta(\delta) Z_{L\delta} - \frac{q_0}{\beta_0} \sin q_0 \delta \right], \quad (4)$$

$$m_{1K}(z) = F_{1K}(z) \cos q_0 h - H_{1K}(z) \cos \beta_0 h + H_{1K}(h) \cos \beta_0 z - F_{1K}(h) \cos q_0 z, \quad (5)$$

$$p_{1K}(z) = G_{1K}(z) \cos q_0 h - H_{1K}(z) \sin \beta_0 h + H_{1K}(h) \sin \beta_0 z - G_{1K}(h) \cos q_0 z, \quad (6)$$

$$M_{1K}(z) = F_{1K}(z) \sin \beta_0 h - G_{1K}(z) \cos \beta_0 h + G_{1K}(h) \cos \beta_0 z - F(h)_{1K} \sin \beta_0 z, \quad (7)$$

$$A_{1K} = F_{1K}(h) \cos \beta_0 \delta - G_{1K}(h) \sin \beta_0 \delta. \quad (8)$$

The functions $F_{1K}(z)$, $F_{1K}(h)$, $G_{1K}(z)$, $G_{1K}(h)$ are given in integrated form in (II.21.1a–d). The new functions $H_{1K}(z)$ and $H_{1K}(h)$ are defined in the same way as $F_{1K}(z)$ and $F_{1K}(h)$ with $F_{0z} = \cos \beta_0 z - \cos \beta_0 h$ replaced by

$$H_{0z} = \cos q_0 z - \cos q_0 h. \quad (9)$$

That is,

$$H_{1K}(z) = \frac{j4\pi z^i}{\zeta_0} \int_\delta^z H_{0s} \sin \beta_0(z - s) \, ds$$

$$+ H_{0z} \Psi_{K1} - \left(\int_{-h}^h - \int_{-\delta}^\delta \right) H_{0z'} K_1(z, z') \, dz', \quad (10a)$$

$$H_{1K}(h) = \frac{j4\pi z^i}{\zeta_0} \int_\delta^h H_{0s} \sin \beta_0(h - s) \, ds$$

$$- \left(\int_{-h}^h - \int_{-\delta}^\delta \right) H_{0z'} K_1(h, z') \, dz'. \quad (10b)$$

These expressions may be written in the form

$$H_{1K}(z) = \frac{j4\pi z^i}{\zeta_0} J_{cq}(z) + H_{0z}\Psi_{K1} - [C_{qa}(h,z) - C_{qa}(\delta,z)] + \cos\beta_0 h[E_a(h,z) - E_a(\delta,z)],$$ (11a)

$$H_{1K}(h) = \frac{j4\pi z^i}{\zeta_0} J_{cq}(h) - [C_{qa}(h,h) - C_{qa}(\delta,h)] + \cos\beta_0 h[E_a(h,h) - E_a(\delta,h)],$$ (11b)

where

$$J_{cq}(z) = \int_0^z (\cos q_0 s - \cos q_0 h) \sin \beta_0 (z-s)\, ds$$

$$= \frac{\cos q_0 h}{\beta_0} + \frac{\cos \beta_0 z - \cos q_0 h}{(1 - q_0^2/\beta_0^2)\beta_0},$$ (12a)

$$J_{cq}(h) = \int_0^h (\cos q_0 s - \cos q_0 h) \sin \beta_0 (h-s)\, ds$$

$$= \frac{\cos q_0 h}{\beta_0} + \frac{\cos \beta_0 h - \cos q_0 h}{(1 - q_0^2/\beta_0^2)\beta_0}.$$ (12b)

(In these integrals the lower limit δ has been replaced by 0 since the contribution to the small terms with z^i as a factor by the very short length 2δ of wire is negligible.)

$$C_{qa}(h,z) \equiv \int_{-h}^{h} \cos q_0 z' \frac{e^{-j\beta_0 R_1}}{R_1}\, dz',$$ (13)

with

$$R_1 = \sqrt{(z'-z)^2 + a^2}.$$ (14)

This integral cannot be reduced to an algebraic sum of the tabulated generalized integral sine and cosine functions in its general form with unrestricted value of a. For small values of a, subject to the condition $\beta_0 a \ll 1$, an approximate formula may be derived. Without approximation (13) may be expressed as follows:

$$C_{qa}(h,z) = \tfrac{1}{2} e^{jq_0 z} \int_{-h}^{h} \frac{e^{jq_0(z'-z)} e^{-j\beta_0 R_1}}{R_1}\, dz'$$

$$+ \tfrac{1}{2} e^{-jq_0 z} \int_{-h}^{h} \frac{e^{-jq_0(z'-z)} e^{-j\beta_0 R_1}}{R_1}\, dz'.$$ (15)

The approximation that now must be made is to substitute R_1 as in (14) for $(z'-z)$ in the exponentials. Thus:

$$C_{qa}(h,z) \doteq \tfrac{1}{2} e^{jq_0 z} \int_{-h}^{h} \frac{e^{-j(\beta_0-q_0)R_1}}{R_1}\, dz'$$

$$+ \tfrac{1}{2} e^{-jq_0 z} \int_{-h}^{h} \frac{e^{-j(\beta_0+q_0)R_1}}{R_1}\, dz'$$ (16)

In the first integral in (16), let

$$R_{1q} = R_1(1 - q_0/\beta_0), \quad a_{1q} = a(1 - q_0/\beta_0),$$
$$z_{1q} = z(1 - q_0/\beta_0), \quad h_{1q} = h(1 - q_0/\beta_0).$$ (17a)

In the second integral in (16), let

$$R_{2q} = R_2(1 + q_0/\beta_0), \quad a_{2q} = a(1 + q_0/\beta_0),$$
$$z_{2q} = z(1 + q_0/\beta_0), \quad h_{2q} = h(1 + q_0/\beta_0).$$ (17b)

Using (17a) and (17b) in (16), the variables may be changed so that

$$C_{qa}(h,z) = \tfrac{1}{2} e^{jq_0 z} \int_{-h_{1q}}^{h_{1q}} \frac{e^{-j\beta_0 R_{1q}}}{R_{1q}}\, dz_{1q}$$

$$+ \tfrac{1}{2} e^{-jq_0 z} \int_{-h_{2q}}^{h_{2q}} \frac{e^{-j\beta_0 R_{2q}}}{R_{2q}}\, dz_{2q}.$$ (18)

Each of these integrals is in the same form as the function

$$E_a(h,z) \equiv \int_{-h}^{h} \frac{e^{-j\beta_0 R_1}}{R_1}\, dz',$$ (19)

which is discussed in Sec. II.19 in terms of tabulated functions. Hence,

$$C_{qa}(h,z) \doteq \tfrac{1}{2} e^{jq_0 z} E_{qa}(h_{1q}, z_{1q})$$

$$+ \tfrac{1}{2} e^{-jq_0 z} E_{qa}(h_{2q}, z_{2q}).$$ (20)

With $a = 0$, $C_{qa}(h,z)$ may be expressed entirely in terms of the functions Si and Cin. This has been done in the literature since the generalized functions C and S have become available in tabular form only recently. The expression is complicated and is not given here, since (II.19.36) in (20) is simpler. With the function $C_{qa}(h,z)$ defined in (13) and represented approximately in a form useful for numerical computation in (20), (11a, b) are expressed entirely in terms of tabulated functions. Note that $H_{1K}(z)$ and $H_{1K}(h)$ both vanish when $q_0 = 0$.

The first-order solution (1) may be expressed in terms of the functions of the Hallén theory by the procedure described in Sec. II.21. The results for (2) and (3) are:

$$[u_\delta(z)]_1 = \frac{j4\pi}{\zeta_0 \Psi_{K1}}$$

$$\times \left\{ \frac{(D_1)_1[\cos q_0 h \cos \beta_0(z-\delta) - \cos q_0 z \cos \beta_0(h-\delta)]}{\cos \beta_0(h-\delta) + (D_1)_0 A_{1H}/\Psi_{K1}} \right.$$

$$\left. + \frac{(D_2)_1[m_{1H}(z)\cos\beta_0\delta + P_{1H}(z)\sin\beta_0\delta]/\Psi_{K1}}{\cos\beta_0(h-\delta) + (D_1)_{01}A_H/\Psi_{K1}} \right\},$$ (21)

$$[v_\delta(z)]_1 = \frac{j2\pi}{\zeta_0 \Psi_{K1}}$$

$$\times \left\{ \frac{(D_1)_1 \sin\beta_0(h-z) + (D_2)_1 M_{1H}(z)/\Psi_{K1}}{\cos\beta_0(h-\delta) + (D_1)_0 A_{1H}/\Psi_{K1}} \right\},$$ (22)

where the D factors as obtained from (II.21.20) are

$$(D_1)_0 = 1, \quad (D_1)_1 = 2 - \Omega/\Psi_{K1},$$
$$(D_2)_1 = 1. \quad (23)$$

The functions $m_{1H}(z)$, $p_{1H}(z)$, $M_{1H}(z)$, and A_{1H} are given by (5)–(8) with subscripts K replaced by H. The functions $F_{1H}(z)$, $F_{1H}(h)$, $G_{1H}(z)$, $G_{1H}(h)$ are given in (II.19.1a–d) and (II.19.11a–d). The functions $H_{1H}(z)$ and $H_{1H}(h)$ are the same as (11a, b) with Ω written in place of Ψ_{K1}.

In determining the current along the antenna the simplification $\delta = 0$ is justified, since the distribution of current is not significantly affected by small values of δ except near $z = \pm \delta$, where a higher order of approximation of I_δ is desirable. With $\delta = 0$, (1) becomes

$$[I_z]_1 = U[u_0(z)]_1 + V_{20}[v_0(z)]_1, \quad (24)$$

where, for $0 \leq z \leq h$,

$$[u_0(z)]_1 = \frac{j4\pi}{\zeta_0 \Psi_{K1}}$$
$$\times \left\{ \frac{[\cos q_0 h \cos \beta_0 z - \cos q_0 z \cos \beta_0 h]}{\cos \beta_0 h + A_{1K}/\Psi_{K1}} \right\}, \quad (25)$$

$$[v_0(z)]_1 = \frac{j2\pi}{\zeta_0 \Psi_{K1}}$$
$$\times \left\{ \frac{\sin \beta_0(h - z) + M_{1K}(z)/\Psi_{K1}}{\cos \beta_0 h + A_{1K}/\Psi_{K1}} \right\}, \quad (26)$$

$$V_{20} = -U \frac{Z_0 Z_L}{2(Z_0 + Z_L)}, \quad (27)$$

with $m_{1K}(z)$, $M_{1K}(z)$, and A_{1K} defined as in (5), (7), (8) with $\delta = 0$.

The distribution of current for zero load ($Z_L = 0$) is given by (24) with $V_{20} = -I_{20}Z_L = 0$. In the special case $q_0 = 0$, when the even current given by (24) is the total current, (24) gives

$$\frac{[I_z]_1}{2\pi U} = \frac{[I_z'' + jI_z']_1}{2\pi U} = \frac{j2}{\zeta_0 \Psi_{K1}}$$
$$\times \left\{ \frac{(\cos \beta_0 z - \cos \beta_0 h) + [F_{1K}(z) - F_{1K}(h)]/\Psi_{K1}}{\cos \beta_0 h + F_{1K}(h)/\Psi_{K1}} \right\}, \quad (28)$$

where $\quad 2\pi U = \lambda_0 E_{21\Theta_1} \cos \psi. \quad (29)$

The function $m_{1K}(z)$ reduces to the simple form in (28) because

$$\left. \begin{array}{l} H_{1K}(z) = 0 \\ H_{1K}(h) = 0 \end{array} \right\} \text{ when } q_0 = 0. \quad (30)$$

This follows from

$$H_{0z} = \cos q_0 z - \cos q_0 h = 0 \text{ for } q_0 = 0. \quad (31)$$

An alternative expression for (28) is

$$\frac{[I_z]_1}{2\pi U} = \frac{j2}{\zeta_0 \Psi_{K1}}$$
$$\times \left\{ \frac{(2 - \Omega/\Psi_{K1})(\cos \beta_0 z - \cos \beta_0 h)}{\cos \beta_0 h + F_{1H}(h)/\Psi_{K1}} \right\}. \quad (32)$$

The function $[I_z]_1/2\pi U$ as given in (27) and (28) has been evaluated for several values of $\beta_0 h$ with $\Omega = 10, 15, 20$. The distribution curves showing first-order values of $I_z''/2\pi U$, $I_z'/2\pi U$, $|I_z|/2\pi U$, $\Theta_I = \tan^{-1}(I_z'/I_z'')$ are shown in Figs. 7.1–7.4. The general shape of the distribution curves of the quadrature component I_z' is seen to resemble closely the zeroth-order current shown in Fig. 5.1, $\Theta_2 = 0$. For thinner antennas with $\Omega = 20$ and 15 the in-phase component is small except for $\beta_0 h = \pi/2$, which is near self-resonance. For $\Omega = 10$, the in-phase component of current is quite large for all lengths.

Distributions of current with a variety of loads Z_L are shown in Figs. 8.2, 8.4, 8.6, 8.8, 8.10–8.14. These were computed from (24) by assigning the indicated values to Z_L. Note that when the circuit is tuned to resonance by setting $X_L = -X_0$, the distribution resembles closely the transmitting distribution. This indicates that the term in (24) involving $v_0(z)$ predominates. Evidently, a wide variety of distributions is possible, depending on the nature of the load.

The current in an electrically short unloaded receiving antenna placed parallel to the electric field may be determined quite accurately using the method developed in Sec. II.31 for the short transmitting antenna.

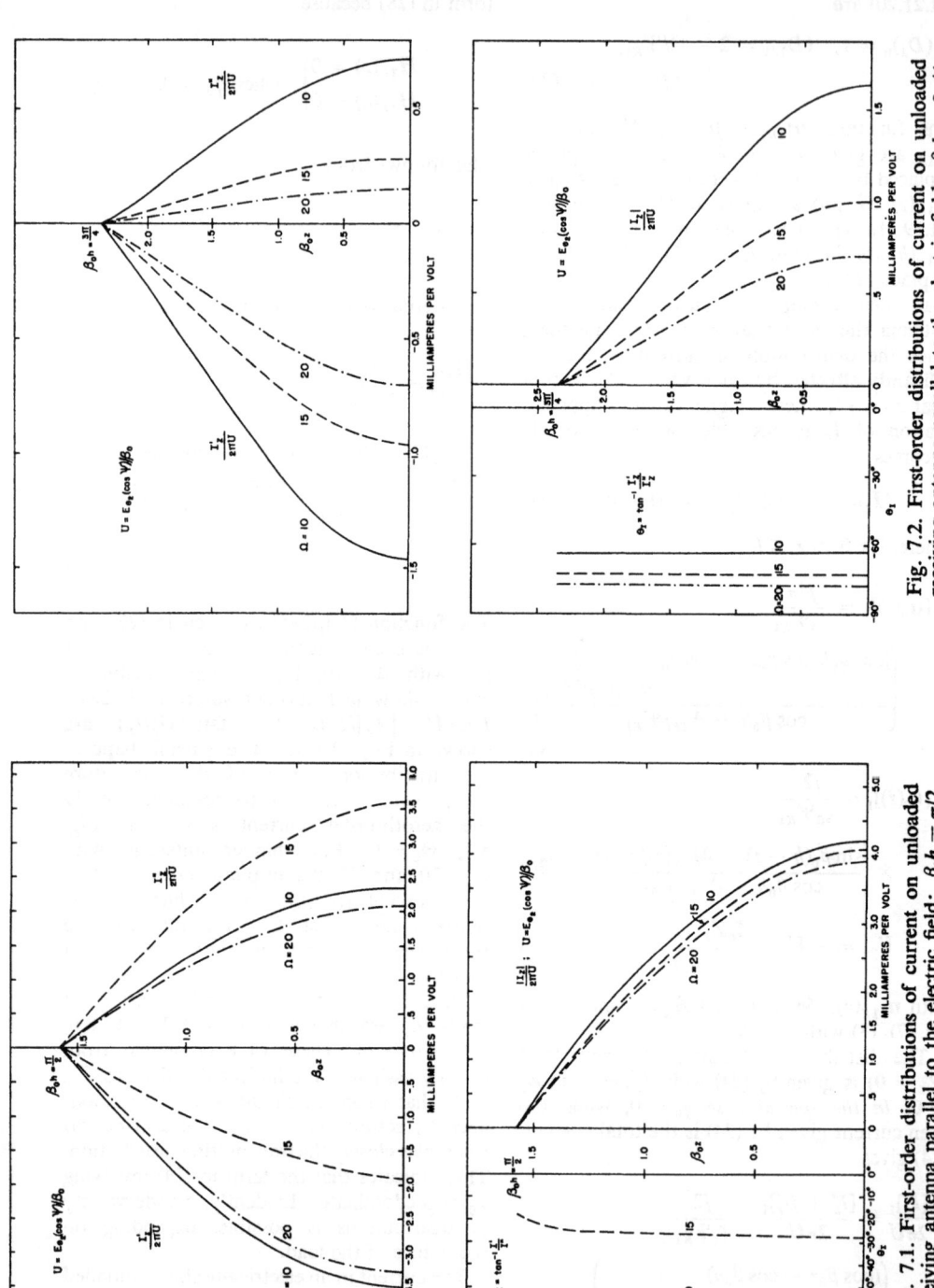

Fig. 7.1. First-order distributions of current on unloaded receiving antenna parallel to the electric field; $\beta_0 h = \pi/2$.

Fig. 7.2. First-order distributions of current on unloaded receiving antenna parallel to the electric field; $\beta_0 h = 3\pi/4$.

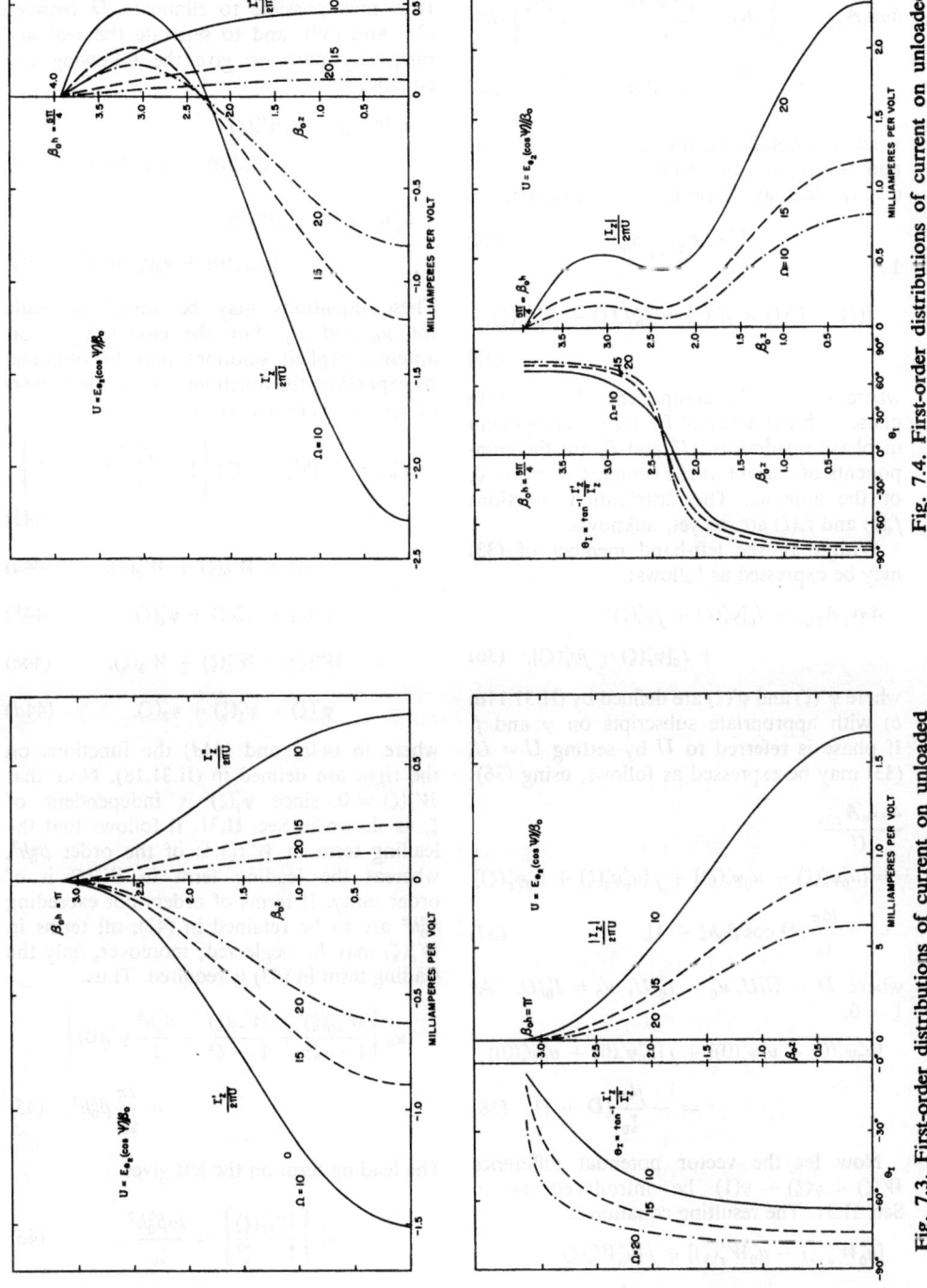

Fig. 7.3. First-order distributions of current on unloaded receiving antenna parallel to the electric field; $\beta_0 h = \pi$.

Fig. 7.4. First-order distributions of current on unloaded receiving antenna parallel to the electric field; $\beta_0 h = 5\pi/4$.

With $\delta = 0$, $\zeta = z/h$, $r = R/h$, and $q_0 = 0$ the integral equation (3.8) may be expressed as follows:

$$4\pi\nu_0 A_{22z} = \int_0^1 I(\zeta') \left[\frac{e^{-j\beta_0 h r_1}}{r_1} + \frac{e^{-j\beta_0 h r_2}}{r_2} \right] d\zeta'$$

$$= -\frac{j4\pi}{\zeta_0} [C \cos \beta_0 h \zeta + U], \quad (33)$$

where the notation is the same as in Sec. II.31 and where, for the antenna parallel to the electric field, at the center of the antenna,

$$U = E_{21\Theta_1}^r / \beta_0. \quad (34)$$

Let

$$I(\zeta) = I''(\zeta) + jI'(\zeta) = I_0'' f_s(\zeta) + jI_0' f_r(\zeta), \quad (35)$$

where $I''(\zeta)$ is the component of current in phase with the field and $I'(\zeta)$ is the component in phase quadrature; I_0'' and I_0' are the components of current at the center, $\zeta = z/h = 0$, of the antenna. The distribution functions $f_s(\zeta)$ and $f_r(\zeta)$ are, as yet, unknown.

With (35), the left-hand member of (33) may be expressed as follows:

$$4\pi\nu_0 A_{22z} = I_0''[\psi_s''(\zeta) + j\psi_s'(\zeta)]$$

$$+ I_0'[\psi_r''(\zeta) + j\psi_r'(\zeta)], \quad (36)$$

where $\psi''(\zeta)$ and $\psi'(\zeta)$ are defined by (II.31.11a, b) with appropriate subscripts on ψ and f. If phase is referred to U by setting $U = \tilde{U}$, (33) may be expressed as follows, using (36):

$$\frac{4\pi\nu_0 A_{22z}}{U}$$

$$= [u_0'' \psi_s''(\zeta) - u_0' \psi_r'(\zeta)] + j[u_0'' \psi_s'(\zeta) + u_0' \psi_r''(\zeta)]$$

$$= -\frac{j4\pi}{\zeta_0}[D \cos \beta_0 h \zeta + 1], \quad (37)$$

where $D = C/U$, $u_0'' = I_0''/U$, $u_0' = I_0'/U$. At $\zeta = 0$,

$$[u_0'' \psi_s''(0) - u_0' \psi_r'(0)] + j[u_0'' \psi_s'(0) + u_0' \psi_r''(0)]$$

$$= -\frac{j4\pi}{\zeta_0}[D + 1]. \quad (38)$$

Now let the vector potential difference $W(\zeta) \equiv \psi(\zeta) - \psi(1)$ be introduced as in Sec. II.31. The resulting equation is

$$[u_0'' W_s''(\zeta) - u_0' W_r'(\zeta)] + j[u_0'' W_s'(\zeta)$$

$$+ u_0' W_r''(\zeta)] = -j\frac{4\pi}{\zeta_0} D F_{0z}, \quad (39)$$

where, as in (II.31.15a),

$$F_{0z} = \cos \beta_0 h \zeta - \cos \beta_0 h. \quad (40)$$

It is now possible to eliminate D between (38) and (39), and to separate the real and imaginary parts to give the following two equations:

$$u_0'' W_s''(\zeta) - u_0' W_r'(\zeta)$$

$$= [u_0'' \psi_s'(0) - u_0' \psi_r'(0)] F_{0z}, \quad (41)$$

$$u_0'' W_s'(\zeta) + u_0' W_r''(\zeta)$$

$$= [u_0'' \psi_s'(0) + u_0' \psi_r''(0)] F_{0z}. \quad (42)$$

These equations may be solved formally for u_0'' and u_0'. For the electrically short antenna explicit solutions may be obtained by expanding the functions involved in powers of $\beta_0 h$ just as in Sec. II.31. Thus,

$$F_{0z} \doteq \tfrac{1}{2}\beta_0^2 h^2 (1 - \zeta^2) \left[1 - \frac{\beta_0^2 h^2}{12}(1 + \zeta^2) \right], \quad (43)$$

$$W''(\zeta) = W_0''(\zeta) + W_2''(\zeta), \quad (44a)$$

$$\psi''(\zeta) = \psi_0''(\zeta) + \psi_2''(\zeta), \quad (44b)$$

$$W'(\zeta) = W_3'(\zeta) + W_5'(\zeta), \quad (44c)$$

$$\psi'(\zeta) = \psi_1'(\zeta) + \psi_3'(\zeta), \quad (44d)$$

where in (44b) and (44d) the functions on the right are defined in (II.31.18). Note that $W_1'(\zeta) = 0$, since $\psi_1'(\zeta)$ is independent of ζ, as shown in Sec. II.31. It follows that the leading term in $W'(\zeta)$ is of the order $\beta_0^3 h^3$, whereas the leading term in $W''(\zeta)$ is of order unity. If terms of orders not exceeding $\beta_0^2 h^2$ are to be retained in (42), all terms in $W_s'(\zeta)$ may be neglected; moreover, only the leading term in (43) is required. Thus,

$$u_0' \left[\frac{W_{r0}''(\zeta)}{1-\zeta^2} + \frac{W_{r2}(\zeta)}{1-\zeta^2} - \frac{\beta_0^2 h^2}{2} \psi_{r0}''(0) \right]$$

$$= \frac{2\pi}{\zeta_0} \beta_0^2 h^2. \quad (45)$$

The leading term on the left gives

$$u_0' \left[\frac{W_{r0}''(\zeta)}{1-\zeta^2} \right] = \frac{2\pi \beta_0^2 h^2}{\zeta_0}. \quad (46)$$

It has been shown in Sec. II.31 that with $f_r(\zeta)$ given by the modified parabolic distribution,

$f_r(\zeta) = 1 - \zeta^2 + p(\zeta)$, which is shown graphically in Fig. II.31.5, the expression in the brackets in (46) is a constant, namely,

$$\frac{W''_{r0}(\zeta)}{1 - \zeta^2} = K''_{r0} = \text{const.} \quad (47)$$

It follows that the second term in (45), which is of higher order, also must be essentially constant and hence equal to its value at $\zeta = 0$. Therefore, from (45),

$$u'_0 = \frac{2\pi\beta_0^2 h^2}{\zeta_0[K''_{r0} + W''_{r2}(0) - \tfrac{1}{2}\beta_0^2 h^2 \psi''_{r0}(0)]}. \quad (48)$$

For $\Omega = 10$, it has been shown in Sec. II.31 that

$$K''_{r0} = 6.90, \quad \psi''_{r0}(0) = 9.0,$$
$$W''_{r2}(0) = 0.4545\beta_0^2 h^2, \quad (49)$$

$$u'_0 = \frac{\beta_0^2 h^2}{60 \times 6.90[1 - 0.444\beta_0^2 h^2]}$$
$$\doteq \frac{\beta_0^2 h^2(1 + 0.444\beta_0^2 h^2)}{414}. \quad (50)$$

It may be verified from (41) with (50) that u''_0 is distributed essentially like u'_0, so that all subscripts s on f and ψ may be changed to r. It follows directly that whereas u'_0 has the factor $\beta_0^2 h^2$, u''_0 has the factor $\beta_0^5 h^5$, so that it may be neglected for most purposes. The important part of the current is given by

$$I(\zeta) = [I''_0 + jI'_0][1 - \zeta^2 + p(\zeta)], \quad (51a)$$

where, for $\Omega = 10$, $I''_0 \ll I'_0$,

$$I'_0 = Uu'_0$$
$$= 2.42 \times 10^{-3} U\beta_0^2 h^2(1 + 0.444\beta_0^2 h^2). \quad (51b)$$

The distribution in (51a) is represented graphically in Fig. II.31.5.

8. Experimental Determination of Distribution of Current and Impedance for a Base-Loaded Receiving Antenna on a Conducting Plane*

The experimental determination of the distribution of the magnitude and phase of the current on and the impedance of a center-loaded, isolated receiving antenna is difficult, since the measuring gear disturbs the field. Accordingly, it is advantageous to use an antenna erected vertically on a large conducting plane, just as described in Sec. II.26

---
* This section is based in part on the work of Dr. T. Morita, ref. 21, Dr. D. G. Wilson, ref. 41, and Drs. T. Morita, E. O. Hartig, and R. King, ref. 22.

for the transmitting antenna. The load Z_L on the antenna is the input impedance of an arbitrarily terminated coaxial transmission line, an extension of the inner conductor of which constitutes the antenna.

In taking the measurements of current the source consisted of a base-driven transmitting antenna at a distance of $7\tfrac{1}{2}$ wavelengths. Diagrams of the circuits are given in Figs. II.26.4 and II.26.5. Measurements of current were made with a small shielded loop as probe, following the general method described in Sec. II.26, for a range of electrical lengths of the antenna from $\beta_0 h = \pi/2$ to $\beta_0 h = 3\pi/2$ and for three types of load, namely, $Z_L = 0$, $Z_L = -jX_0$, $Z_L = Z_0^*$, where Z_0 is the impedance of the isolated antenna. The results are shown in Figs. 8.1 to 8.12. Shown with some of the experimental curves are the theoretical ones for approximately the same conditions as computed from the King-Middleton *first-order* formula. On the whole the agreement is good.

A comparison of the relative magnitudes of the currents on an antenna of electrical length $\beta_0 h = \pi/2$ with four different loads including $Z_L = 0$, $Z_L = -jX_0$, $Z_L = Z_0^*$, and $|Z_L|$ very large compared with $|Z_0|$ is shown in Fig. 8.13. Note that the current is greatest when the antenna is tuned to resonance with a purely reactive load. As shown in Fig. 8.13, the distributions of current amplitude for a tuned receiving antenna are essentially the same as for a transmitting antenna.

A further interesting comparison of the distributions of current magnitude on receiving antennas is given in Fig. 8.14, where curves for $Z_L = 0$ and $Z_L = -jX_0$ are shown for $\beta_0 h = \pi/2$, $3\pi/4$, π, $5\pi/4$, and $3\pi/2$. Note that near self-resonance ($\beta_0 h = \pi/2$, $3\pi/2$) the distribution curves *all* resemble those of driven antennas, whereas near antiresonance ($\beta_0 h = \pi$) the *tuned* antenna ($Z_L = -jX_0$) has a distribution resembling that obtaining for transmission, while the *unloaded* antenna ($Z_L = 0$) has an entirely different distribution. The intermediate lengths, $\beta_0 h = 3\pi/4$, $5\pi/4$, have distributions essentially as for transmission when the antenna is tuned to resonance ($Z_L = -jX_0$), but quite different when unloaded ($Z_L = 0$). It is clear that the distributions of current for transmission and reception are similar only for antennas near self-resonance. In general, the distribution of current in receiving antennas depends greatly on the load.

The impedance of a receiving antenna is *by definition* (Sec. 4) equal to the impedance of this antenna when driven from the same network but with a generator replacing the load. Hence, the impedance determined theoretically and experimentally in Chapter II is directly applicable to the receiving antenna and no further analysis or measurement is required. However, an interesting and useful alternative method for measuring the impedance of an antenna while used for reception is available. For simplicity, let terminal and coupling effects be assumed negligible. Account may be taken of these with appropriate lumped networks. Consider the circuit of Fig. 8.15, in which the terminals of the receiving antenna are connected to one end of a transmission line at $x = 0$ while the other end of the line is terminated in a loosely coupled detector that is equivalent to a terminal impedance Z_s that may be either a very low-impedance current loop as shown or a very high-impedance charge probe. Energy is supplied to the line from the antenna, which is excited by an electric field $E = E_{21\Theta_1}$ due to a distant transmitter. From the point of view of the antenna, the load Z_L is the input impedance of the line at $x = 0$. From the point of view of the line, the termination at $x = 0$ consists of a generator with emf $V_0^e = V_2(Z_L = \infty)$ and internal impedance $Z_\delta \doteq Z_0$; the termination at $x = s$ is the very high or very low impedance Z_s. Thus, the experimental measurement of the impedance Z_0 of the receiving antenna is equivalent to the measurement of the internal impedance of the generator driving the line. Although the more conventional transmission-line techniques which depend on the determination of the standing-wave ratio or the distribution-curve dip cannot be used for measuring the impedance of the generator, the resonance-curve method is directly applicable.[15,I.31a] This depends on the condition for resonance, obtained by varying s and defining the resonant lengths s_n,

$$\beta s_n + \Phi_0 + \Phi_s = n\pi, \quad n = 1, 2, \cdots, \quad (1)$$

and the equation relating the width Δs_n of the resonance curve between $s = s_n - \Delta s_n/2$ and $s = s_n + \Delta s_n/2$ (where the square of the current amplitude is one-half of its maximum value when $s = s_n$) to the over-all attenuation in the system,

$$\beta \Delta s_n = 2(\alpha s_n + \rho_0 + \rho_s), \quad (2)$$

where

$$\rho_0 + j\Phi_0 \equiv \coth^{-1} Z_0/Z_c, \quad (3a)$$

$$\rho_s + j\Phi_s \equiv \coth^{-1} Z_s/Z_c, \quad (3b)$$

and α and β are, respectively, the attenuation and phase constants of the line; Z_c is its characteristic impedance. Since Z_0 is available if ρ_0 and Φ_0 are known, the experimental problem is the determination of ρ_0 and Φ_0. This involves the predetermination of ρ_s and Φ_s, which may be accomplished by replacing the receiving antenna with a short circuit and driving the line directly from a very loosely coupled generator. Since $\rho_0 = 0$ and $\Phi_0 = \pi/2$ for a short circuit at $x = 0$, (1) and (2) may be solved for ρ_s and Φ_s if corresponding values of s_n and Δs_n are measured and the constants of the line and the frequency are known. With the short circuit replaced by the receiving antenna, a second determination of s_n and Δs_n for the new circuit permits the calculation of ρ_0 and Φ_0 from (1) and (2).

The apparent impedance of receiving antennas with several values of h/a were measured by Wilson[41] by the resonance-curve method using a particular procedure for determining ρ_s, which was discovered later to involve considerable error due to noise. A more accurate determination was made by Hartig and Morita.[22] Some of the results are given in Fig. II.34.9. More complete data showing comparisons between measured impedances of the same antenna when used for reception and for transmission are given in Fig. 8.16 and in Figs. II.34.10a, b. Theoretical curves also are shown in the last two figures. Agreement between the two sets of experimental values is excellent, as is the agreement between both sets and theory. Experimental curves obtained by Wilson with less accurate equipment are shown in Fig. 8.17.

The apparatus used by Hartig and Morita is essentially the same as that used by D. D. King and improved by E. O. Hartig as described in Secs. II.37 and II.38. Owing to the small size of the coaxial line feeding the vertical antenna over a large (36 ft square) metal ground plane, the effect of the finite value of $b - a$ for the line was negligible and the measured apparent impedance differed negligibly from the ideal impedance, that is, $Z_{sa} \doteq Z_\delta \doteq Z_0$. Measurements of impedance were made at 500 Mhz.

It is evident from the results obtained that the impedance of an antenna may be measured equally well when it is used for reception or

[IV.8] THEORY OF LINEAR ANTENNAS

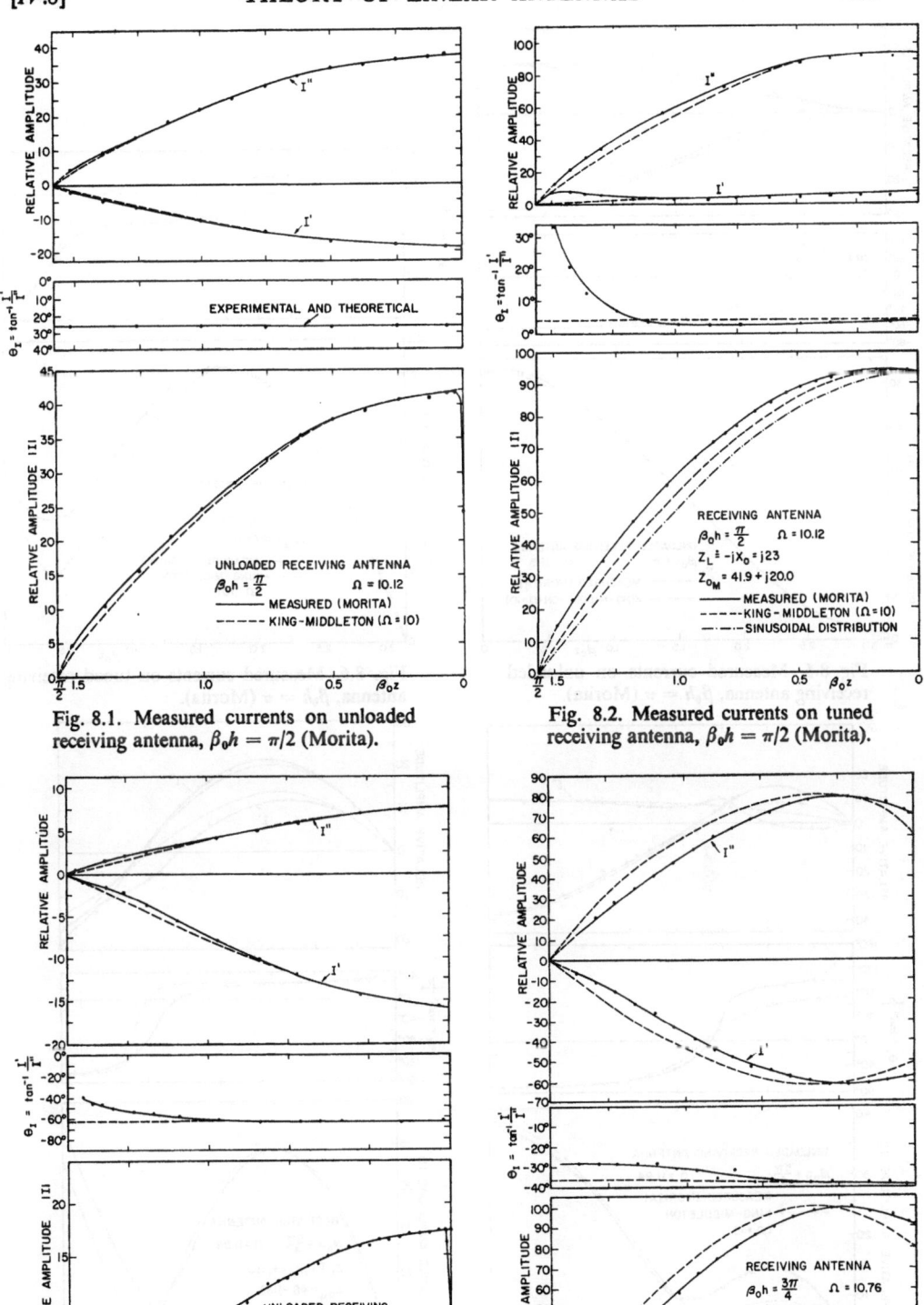

Fig. 8.1. Measured currents on unloaded receiving antenna, $\beta_0 h = \pi/2$ (Morita).

Fig. 8.2. Measured currents on tuned receiving antenna, $\beta_0 h = \pi/2$ (Morita).

Fig. 8.3. Measured currents on unloaded receiving antenna, $\beta_0 h = 3\pi/4$ (Morita).

Fig. 8.4. Measured currents on tuned receiving antenna, $\beta_0 h = 3\pi/4$ (Morita).

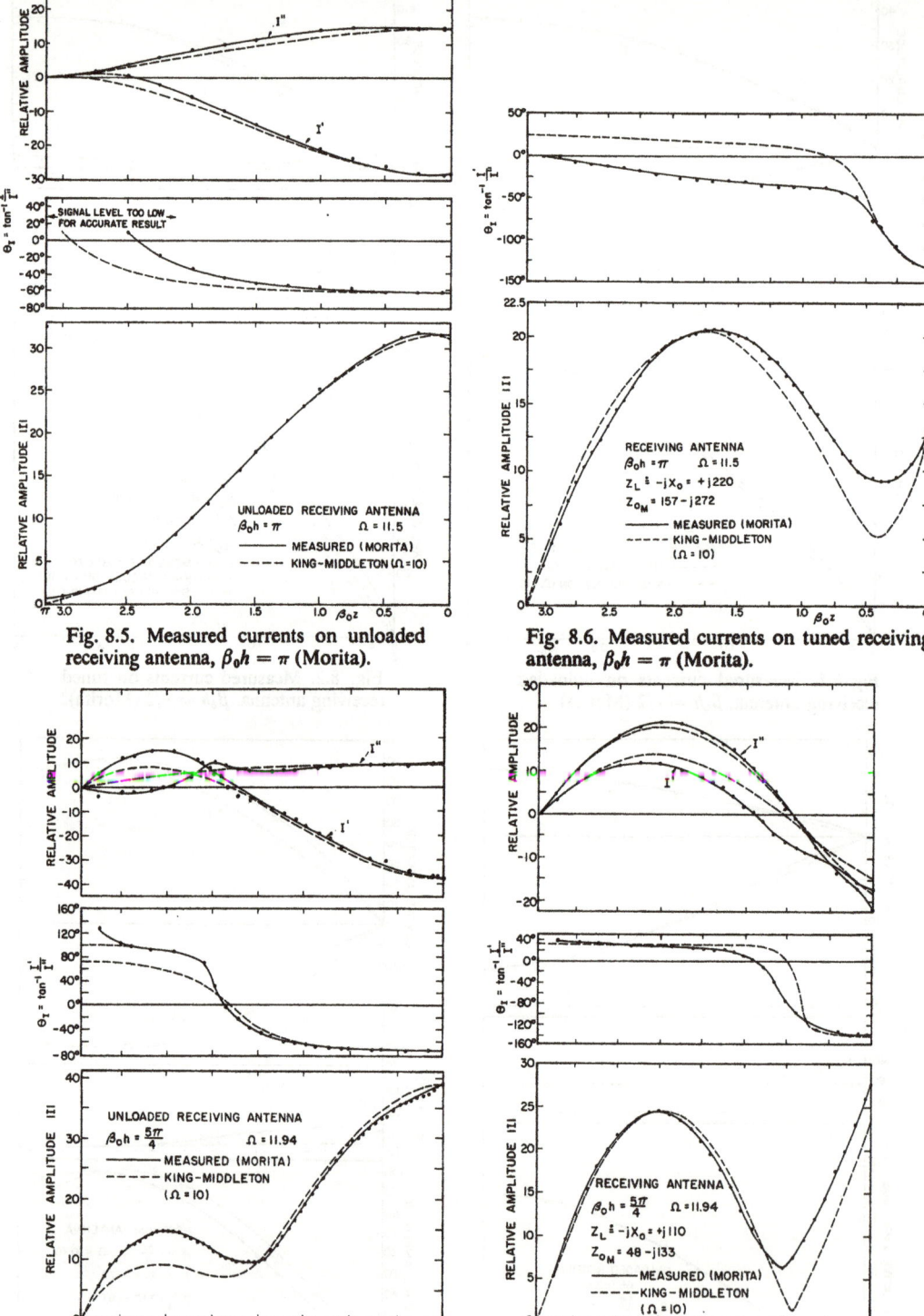

Fig. 8.5. Measured currents on unloaded receiving antenna, $\beta_0 h = \pi$ (Morita).

Fig. 8.6. Measured currents on tuned receiving antenna, $\beta_0 h = \pi$ (Morita).

Fig. 8.7. Measured currents on unloaded receiving antenna, $\beta_0 h = 5\pi/4$ (Morita).

Fig. 8.8. Measured currents on tuned receiving antenna, $\beta_0 h = 5\pi/4$ (Morita).

[IV.8] THEORY OF LINEAR ANTENNAS 483

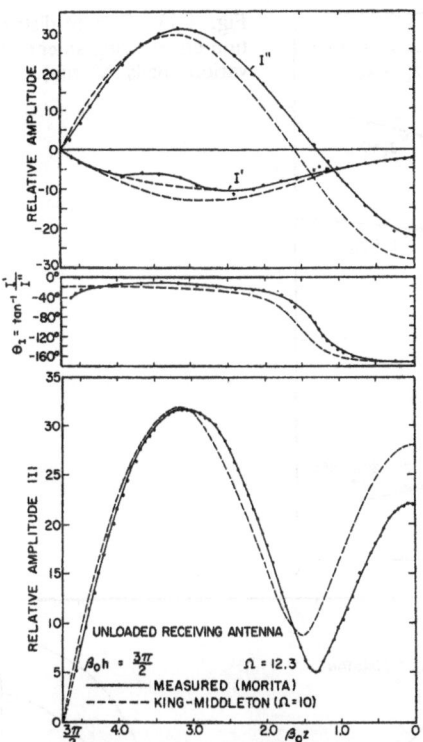

Fig. 8.9. Measured currents on unloaded receiving antenna, $\beta_0 h = 3\pi/2$ (Morita).

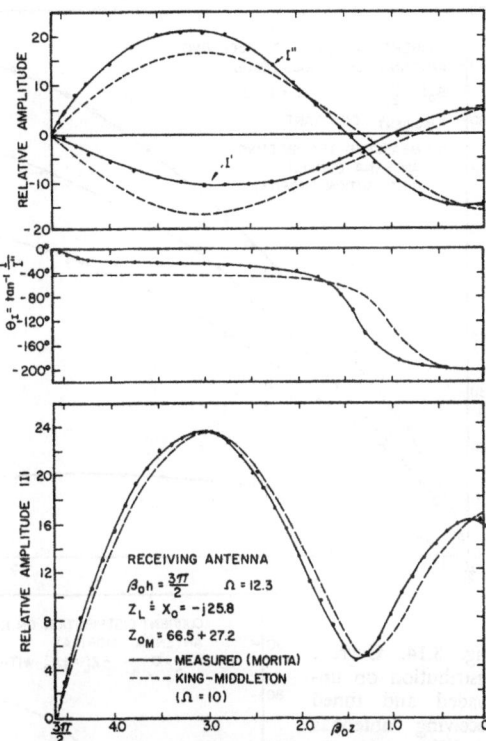

Fig. 8.10. Measured currents on tuned receiving antenna, $\beta_0 h = 3\pi/2$ (Morita).

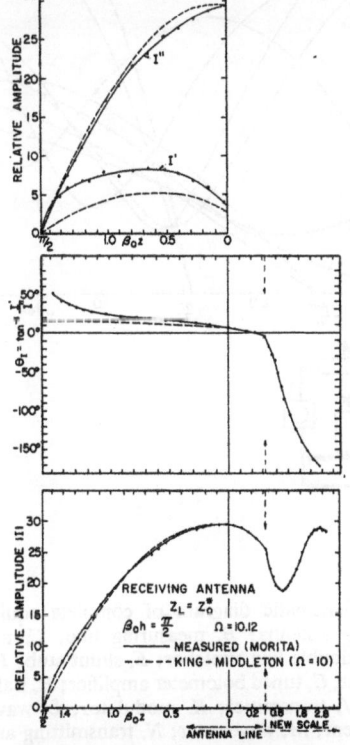

Fig. 8.11. Measured current on conjugate matched receiving antenna, $\beta_0 h = \pi/2$ (Morita).

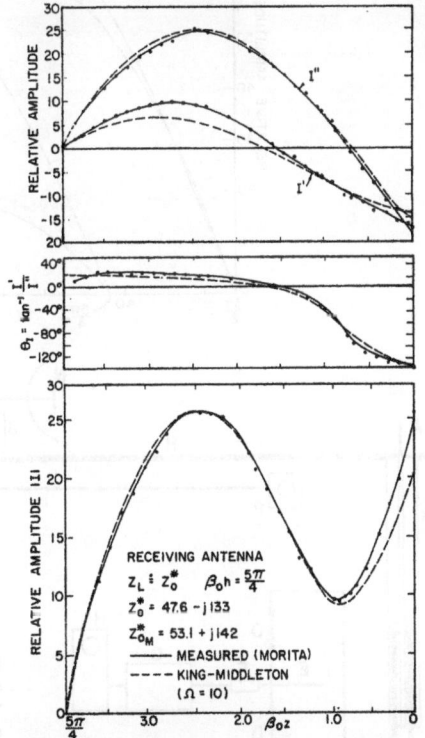

Fig. 8.12. Measured current on approximately conjugate matched receiving antenna, $\beta_0 h = 5\pi/4$ (Morita).

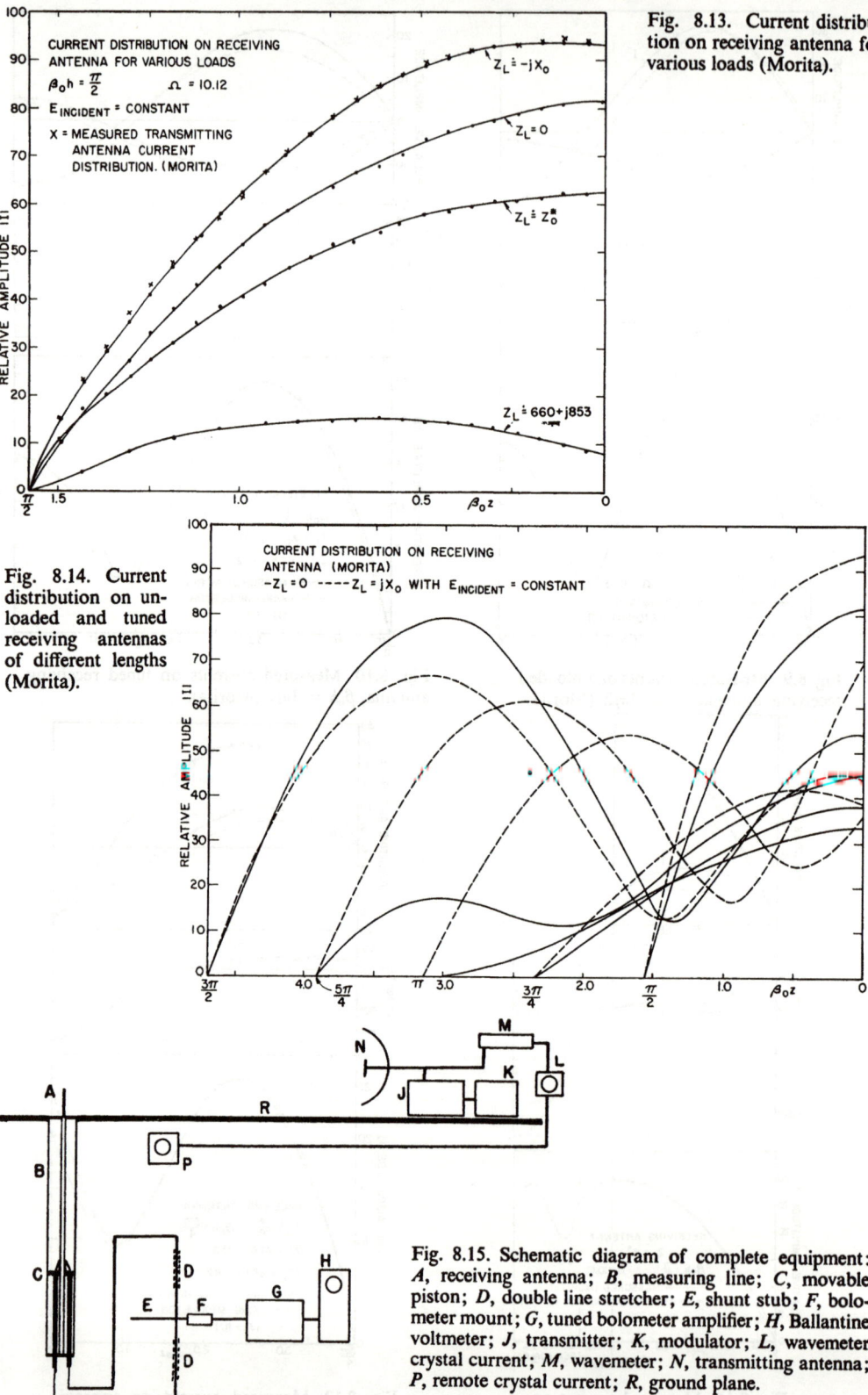

Fig. 8.13. Current distribution on receiving antenna for various loads (Morita).

Fig. 8.14. Current distribution on unloaded and tuned receiving antennas of different lengths (Morita).

Fig. 8.15. Schematic diagram of complete equipment: A, receiving antenna; B, measuring line; C, movable piston; D, double line stretcher; E, shunt stub; F, bolometer mount; G, tuned bolometer amplifier; H, Ballantine voltmeter; J, transmitter; K, modulator; L, wavemeter crystal current; M, wavemeter; N, transmitting antenna; P, remote crystal current; R, ground plane.

[IV.8] THEORY OF LINEAR ANTENNAS 485

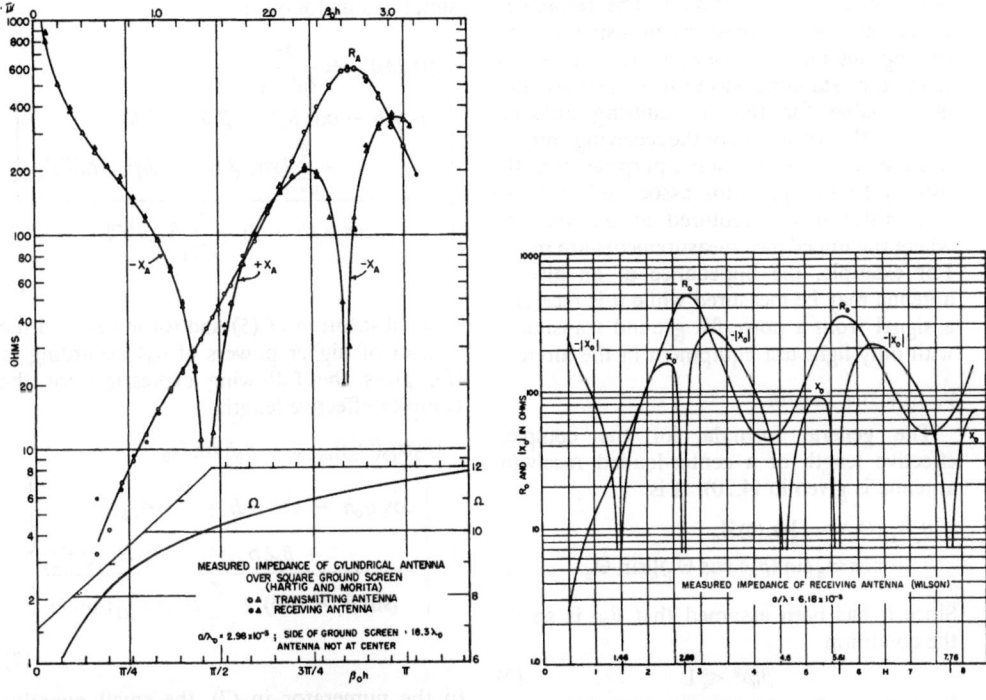

Fig. 8.16. Measured impedance of cylindrical antenna over square ground screen (Hartig and Morita).

Fig. 8.17. Measured impedance of receiving antenna (Wilson).

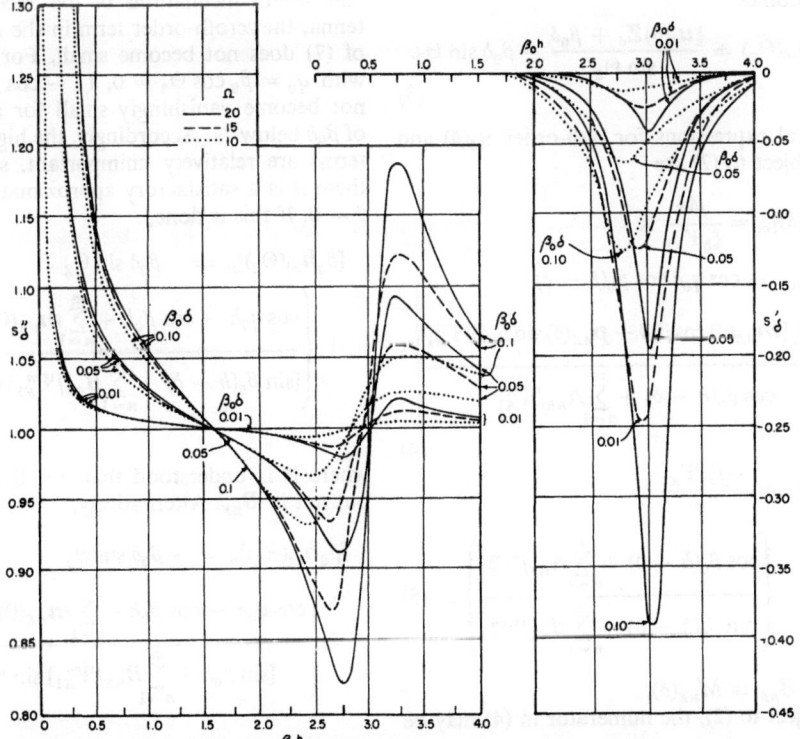

Fig. 9.1. Real and imaginary parts and magnitude of correction factor s_σ as functions of $\beta_0 h$.

when used for transmission. The resonance-curve method required when using the receiving antenna is somewhat less convenient than the standing-wave-ratio method that is available for the transmitting antenna. On the other hand, use of the receiving antenna has the advantage for some purposes that the usually heavy apparatus associated with the transmitter is not required at the location where the impedance measurements are made. For example, the impedance of an aircraft antenna may be measured while it is receiving a signal from a powerful ground transmitter with only light test equipment in the aircraft.

9. Effective Length

The general formula for the complex effective length of a center-loaded receiving antenna is given in (4.20). It is

$$\beta_0 h_{e\delta}(\Theta_2) = [\tfrac{1}{2} u_\delta(\delta) Z_\delta + \cos \Theta_2 \sin (\beta_0 \delta \cos \Theta_2)]/\sin \Theta_2. \quad (1)$$

Since it has been assumed that $\beta_0 \delta$ is small, the condition

$$\beta_0^2 \delta^2 \ll 1 \quad (2)$$

may be applied. After neglecting higher powers of $\beta_0 \delta$ according to (2), it follows that (1) becomes

$$\beta_0 h_{e\delta}(\Theta_2) \doteq \frac{\tfrac{1}{2} u_\delta(\delta) Z_\delta + \beta_0 \delta}{\sin \Theta_2} - \beta_0 \delta \sin \Theta_2. \quad (3)$$

General expressions for mth-order $u_\delta(\delta)$ and Z_δ subject to (2) are

$$[u_\delta(\delta)]_m = \frac{j 4\pi}{\zeta_0 \Psi_{K1}} \times$$

$$\left\{ \frac{\cos q_0 h - \cos q_0 \delta \cos \beta_0(h-\delta) + \sum_{n=1}^m [m_{nK}(\delta) \cos \beta_0 \delta + p_{nK}(\delta) \sin \beta_0 \delta]/\Psi_{K1}^n}{\cos \beta_0(h-\delta) + \sum_{n=1}^m A_{nK}/\Psi_{K1}^n} \right\}, \quad (4)$$

$$[Z_\delta]_m = \frac{-j\zeta_0 \Psi_{K1}}{2\pi}$$

$$\times \left\{ \frac{\cos \beta_0(h-\delta) + \sum_{n=1}^m A_{nK}/\Psi_{K1}^n}{\sin \beta_0(h-\delta) + \sum_{n=1}^m B_{nK}/\Psi_{K1}^n} \right\}, \quad (5)$$

where $B_{nK} \equiv M_{nK}(\delta)$.

Subject to (2), the numerator in (4) may be simplified as follows:

$$[u_\delta(\delta)]_m \doteq \frac{j4\pi}{\zeta_0 \Psi_{K1}} \times$$

$$\left\{ \frac{\cos q_0 h - \cos \beta_0 h - \beta_0 \delta \sin \beta_0 h + \sum_{n=1}^m [m_{nK}(\delta) + \beta_0 \delta p_{nK}(\delta)]/\Psi_{K1}^n}{\cos \beta_0(h-\delta) + \sum_{n=1}^m A_{nK}/\Psi_{K1}^n} \right\}. \quad (6)$$

The substitution of (5) and (6) in (3), and the neglect of higher powers of $\beta_0 \delta$ according to (2), gives the following expression for the complex effective length:

$$[\beta_0 h_{e\delta}(\Theta_2)]_m \doteq -\beta_0 \delta \sin \Theta_2$$

$$+ \left\{ \frac{\cos q_0 h - \cos \beta_0 h + \sum_{n=1}^m [m_{nK}(\delta) + \beta_0 \delta p_{nK}(\delta) + \beta_0 \delta B_{nK}]/\Psi_{K1}^n}{[\sin \beta_0(h-\delta) + \sum_{n=1}^m B_{nK}/\Psi_{K1}^n] \sin \Theta_2} \right\}. \quad (7)$$

In the numerator in (7), the small quantity $\beta_0 \delta$ appears only in terms of order higher than the zeroth. For most practical lengths and useful inclinations of the receiving antenna, the zeroth-order term in the *numerator* of (7) does not become small. For example, with $q_0 = \beta_0 \cos \Theta_2 = 0$, $(1 - \cos \beta_0 h)$ does not become vanishingly small for any value of $\beta_0 h$ below 2π. Accordingly, the higher-order terms are relatively unimportant, so that in them it is a satisfactory approximation to set $\delta = 0$. If this is done,

$$[\beta_0 h_{e\delta}(\Theta_2)]_m \doteq -\beta_0 \delta \sin \Theta_2$$

$$+ \left\{ \frac{\cos q_0 h - \cos \beta_0 h + \sum_{n=1}^m m_{nK}(0)/\Psi_{K1}^n}{[\sin \beta_0(h-\delta) + \sum_{n=1}^m B_{nK}/\Psi_{K1}^n] \sin \Theta_2} \right\}, \quad (8)$$

where it is understood that $\delta = 0$ in $m_{nK}(0)$ but not in B_{nK}. Alternatively,

$$[\beta_0 h_{e\delta}(\Theta_2)]_m \doteq -\beta_0 \delta \sin \Theta_2$$

$$+ s_\delta \left\{ \frac{\cos q_0 h - \cos \beta_0 h + \sum_{n=1}^m m_{nK}(0)/\Psi_{K1}^n}{[\sin \beta_0 h + \sum_{n=1}^m B_{nK}/\Psi_{K1}^n] \sin \Theta_2} \right\}, \quad (9)$$

where

$$\frac{1}{s_\delta} \doteq \frac{\sin \beta_0(h-\delta) + \sum_{n=1}^{m} B_{nK}/\Psi_{K1}^n}{[\sin \beta_0 h + \sum_{n=1}^{m} B_{1K}/\Psi_{K1}^n]_{\delta=0}}$$

$$\doteq \frac{\sin \beta_0(h-\delta) + B_{1K}/\Psi_{K1}}{[\sin \beta_0 h + B_{1K}/\Psi_{K1}]_{\delta=0}}. \quad (10)$$

The numerator in (10) is the denominator in the expression (5) for the impedance Z_δ. Since it has been shown in Chapter II that the percent difference between Z_δ and Z_0 is quite small near $\beta_0(h-\delta) = \pi$ where $\sin \beta_0(h-\delta)$ becomes negligible compared with B_{1K}/Ψ_{K1}, and $\cos \beta_0(h-\delta) \doteq \cos \beta_0 h$, it follows that B_{1K} differs negligibly from $B_{1K}(\delta=0)$ when $\beta_0(h-\delta)$ is near π. Hence, it must be a good approximation to set B_{1K} equal to $B_{1K}(\delta=0)$ in (10). That is,

$$\frac{1}{s_\delta} \doteq \frac{\sin \beta_0(h-\delta) + B_{1K}(\delta=0)/\Psi_{K1}}{\sin \beta_0 h + B_{1K}(\delta=0)/\Psi_{K1}}$$

$$= 1 - \beta_0 \delta \frac{\cos \beta_0 h}{\sin \beta_0 h + B_{1K}(\delta=0)/\Psi_{K1}}. \quad (11)$$

Alternatively, in terms of the tabulated function B_{1H} instead of B_{1K},

$$\frac{1}{s_\delta} \doteq \frac{(2-\Omega/\Psi_{K1})\sin \beta_0(h-\delta) + B_{1H}/\Psi_{K1}}{(2-\Omega/\Psi_K)\sin \beta_0 h + B_{1H}/\Psi_{K1}}, \quad (12)$$

$$s_\delta \doteq 1 + \beta_0 \delta \left[\frac{(2\Psi_{K1} - \Omega)\cos \beta_0 h}{(2\Psi_{K1} - \Omega)\sin \beta_0 h + B_{1H}} \right]. \quad (13)$$

where the bracket is evaluated with $\delta = 0$. Since terms with the internal impedance per unit length, z^i, as a factor are negligible in evaluating B_{1H}, $B_{1H} \doteq \beta_{1H}$, which is tabulated and represented graphically in Chapter II.

Curves of the real and imaginary parts of s_δ and of its magnitude as functions of $\beta_0 h$ are given in Fig. 9.1 for $\beta_0 \delta = 0.01, 0.05, 0.10$ with $\Omega = 10, 15,$ and 20.

The general formula (9) for the complex effective length may be expressed as follows:

$$[\beta_0 h_{e\delta}(\Theta_2)]_m = [\beta_0 h_e(\Theta_2)]_m s_\delta - \beta_0 \delta_e(\Theta_2), \quad (14)$$

where

$$[\beta_0 h_e(\Theta_2)]_m$$
$$\doteq \frac{\cos q_0 h - \cos \beta_0 h + \sum_{n=1}^{m} m_{nK}(0)/\Psi_{K1}^n}{[\sin \beta_0 h + \sum_{n=1}^{m} B_{nK}/\Psi_{K1}^n]\sin \Theta_2} \quad (15)$$

is the complex effective half-length in radians for $\delta = 0$, and

$$\beta_0 \delta_e(\Theta_2) = \beta_0 \delta \sin \Theta_2 \quad (16)$$

is the effective half-length in radians of a short antenna of length 2δ and with uniform current. However, from (3) with $\delta = 0$,

$$\beta_0 h_e(\Theta_2) = \frac{u_0(0)Z_0}{2 \sin \Theta_2}. \quad (17)$$

Therefore,

$$\beta_0 h_{e\delta}(\Theta_2) \doteq s_\delta \frac{u_0(0)Z_0}{2 \sin \Theta_2} - \beta_0 \delta \sin \Theta_2. \quad (18)$$

With (18) it is a simple matter to determine the effective length of a center-loaded receiving antenna for which $\delta \neq 0$ from the value for $\delta = 0$. Therefore, it remains to evaluate $\beta_0 h_e(\Theta_2)$ for $\delta = 0$ from (17).

In (17), $u_0(0)$ is proportional to the current at the center of an unloaded receiving antenna, whereas Z_0 is the impedance of a center-driven antenna. It has been shown (Sec. 7) that the first-order distribution of current in an unloaded receiving antenna is a very good approximation, since it differs relatively little from the zeroth-order. It follows that a combination of first-order $u_0(0)$ and second-order Z_0 should lead to an effective length that is comparable in accuracy to the second-order impedance. In Figs. 9.2–9.5 are shown curves for h'_e/λ_0, h''_e/λ_0, h_e/λ_0 for $\Omega = 10, 12.5, 15,$ and 20 as functions of $\beta_0 h$ with $\cos \Theta_2$ as parameter, as calculated from

$$\beta_0 h_e(\Theta_2) = \frac{2\pi h_e(\Theta_2)}{\lambda_0} \doteq \frac{[u_0(0)]_1 [Z_0]_2}{2 \sin \Theta_2}, \quad (19)$$

where

$$h_e(\Theta_2) = h''_e(\Theta_2) + jh'_e(\Theta_2). \quad (20)$$

For small values of $\beta_0 h$, the *first-order* effective length using *first-order* impedances becomes

$$h_e(\Theta_2) \to \frac{h(\Omega - 1)}{2(\Omega - 2 + \ln 4)} \quad (21)$$

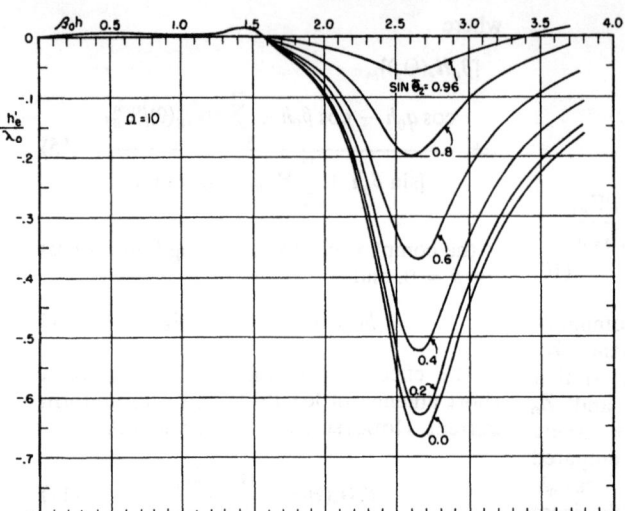

Fig. 9.2a. Imaginary part of effective length of receiving antenna h'_e/λ_0 as a function of $\beta_0 h$; $\Omega = 10$.

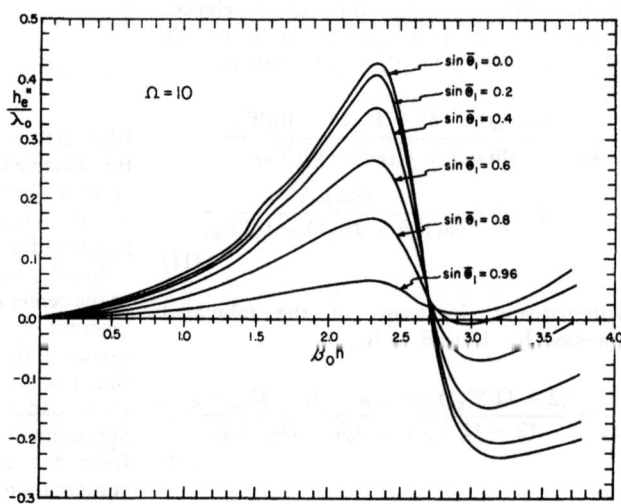

Fig. 9.2b. Real part of effective length of receiving antenna h''_e/λ_0 as a function of $\beta_0 h$; $\Omega = 10$.

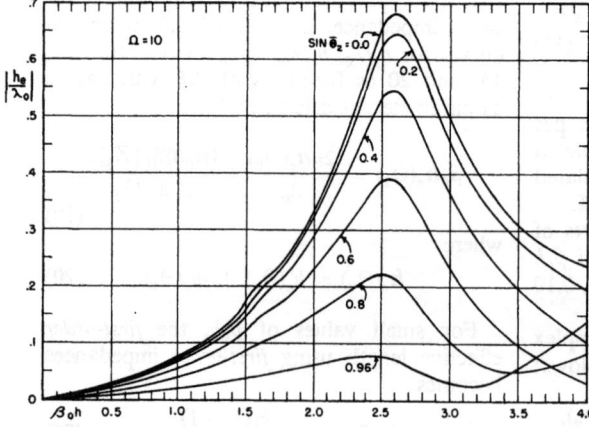

Fig. 9.2c. Effective length of receiving antenna $|h_e/\lambda_0|$ as a function of $\beta_0 h$; $\Omega = 10$.

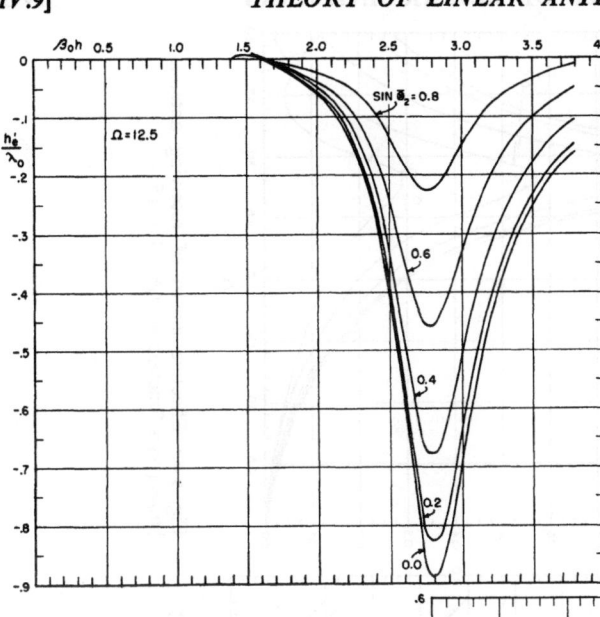

Fig. 9.3a. Imaginary part of effective length of receiving antenna h'_e/λ_0 as a function of $\beta_0 h$; $\Omega = 12.5$.

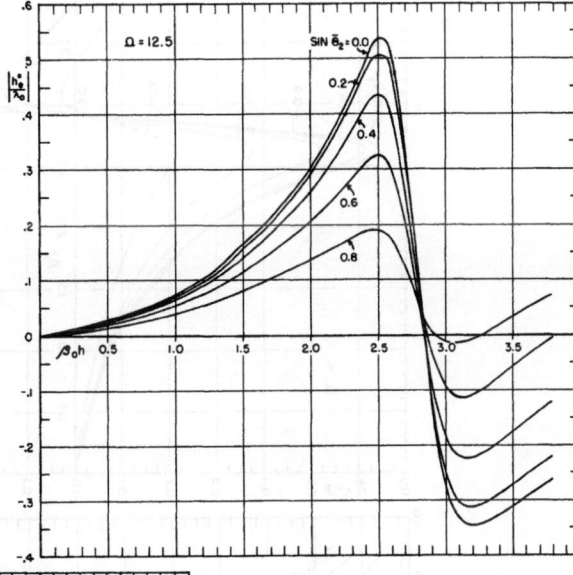

Fig. 9.3b. Real part of effective length of receiving antenna h''_e/λ_0 as a function of $\beta_0 h$; $\Omega = 12.5$.

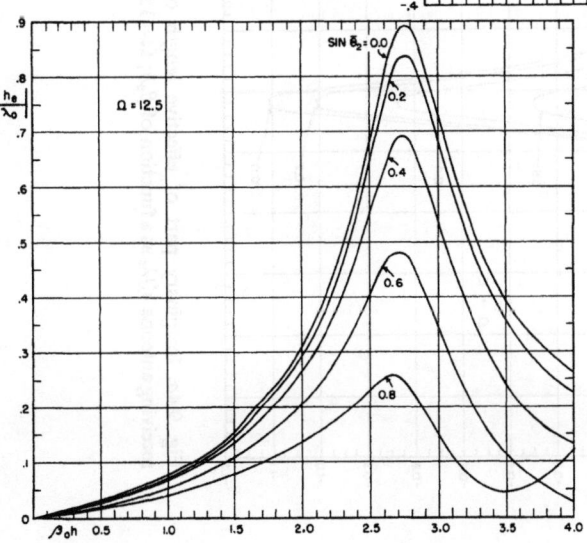

Fig. 9.3c. Effective length of receiving antenna $|h_e/\lambda_0|$ as a function of $\beta_0 h$; $\Omega = 12.5$.

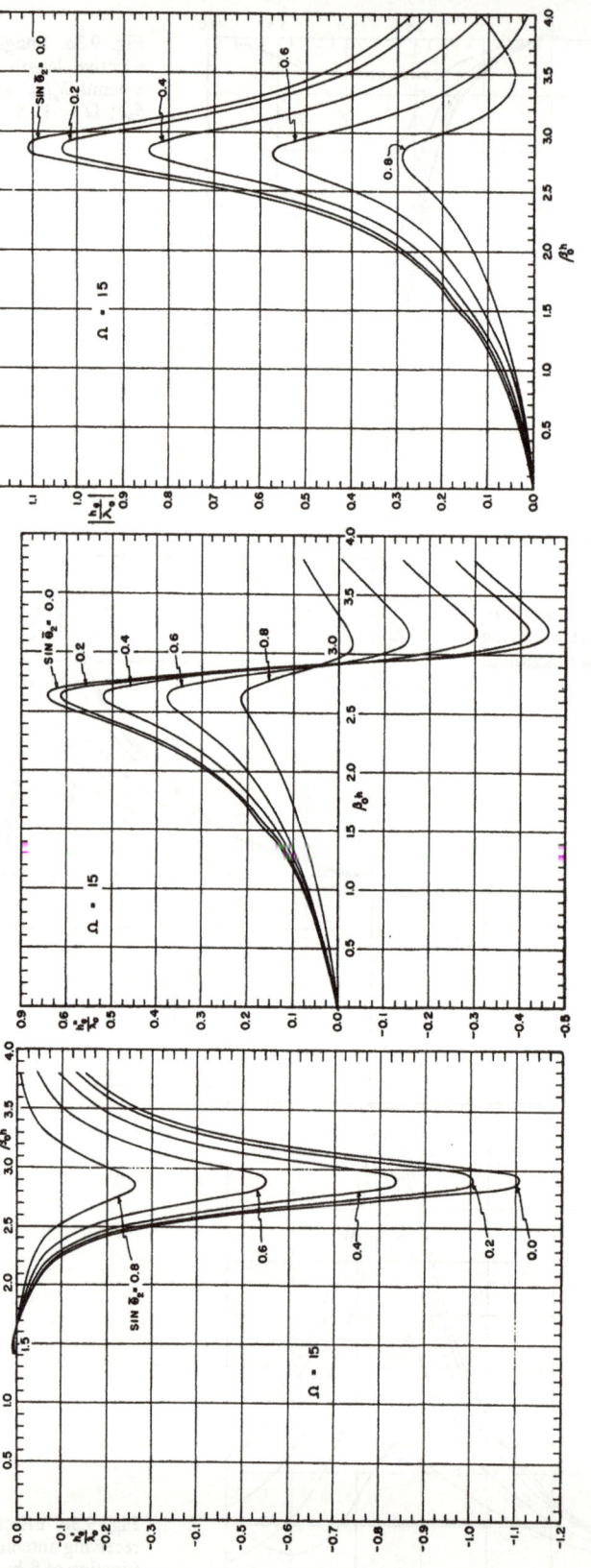

Fig. 9.4a. Imaginary part of effective length of receiving antenna h_e''/λ_0 as a function of $\beta_0 h$; $\Omega = 15$.

Fig. 9.4b. Real part of effective length of receiving antenna h_e'/λ_0 as a function of $\beta_0 h$; $\Omega = 15$.

Fig. 9.4c. Effective length of receiving antenna $|h_e/\lambda_0|$ as a function of $\beta_0 h$; $\Omega = 15$.

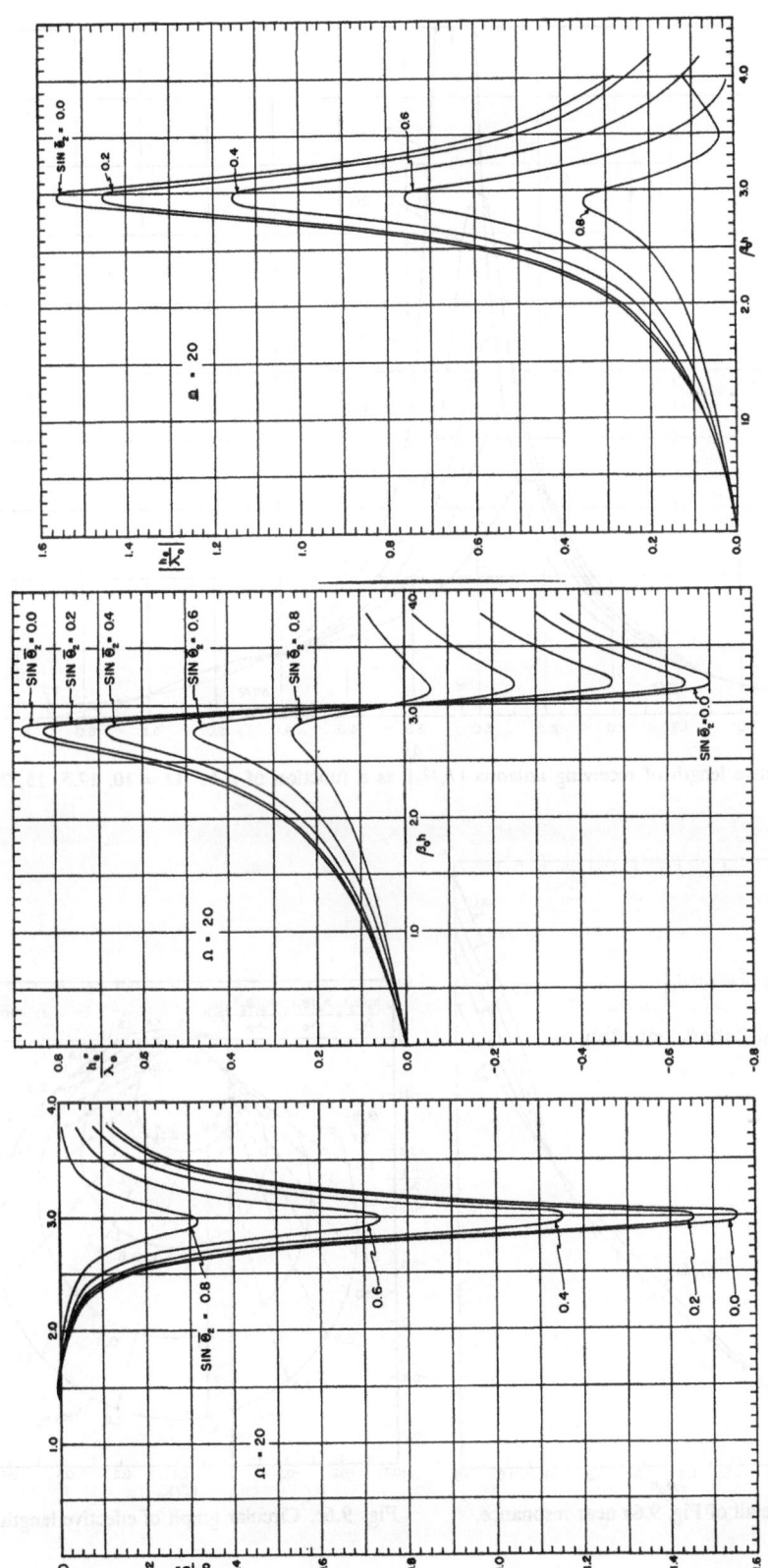

Fig. 9.5a. Imaginary part of effective length of receiving antenna h'_e/λ_0 as a function of $\beta_0 h$; $\Omega = 20$.

Fig. 9.5b. Real part of effective length of receiving antenna h''_e/λ_0 as a function of $\beta_0 h$; $\Omega = 20$.

Fig. 9.5c. Effective length of receiving antenna $|h_e/\lambda_0|$ as a function of $\beta_0 h$; $\Omega = 20$.

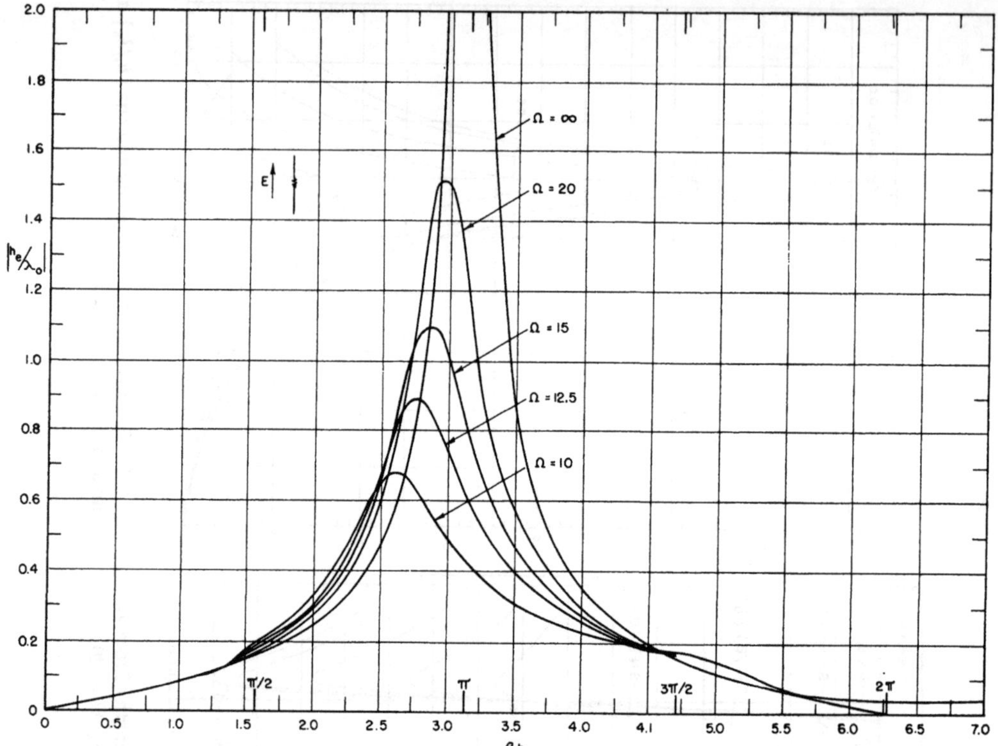

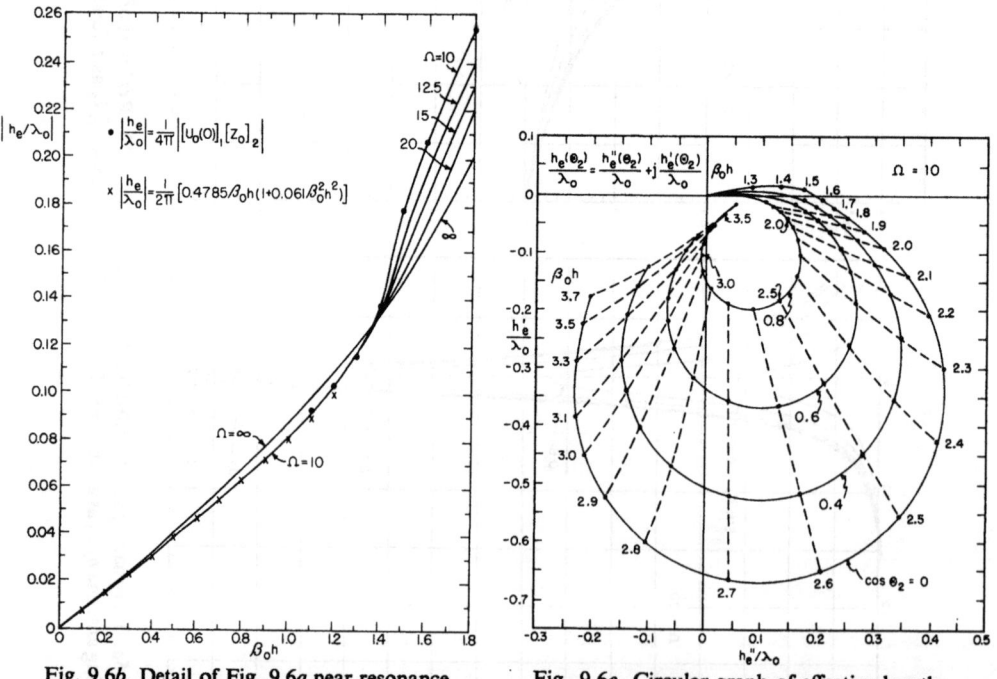

Fig. 9.6a. Effective length of receiving antenna $|h_e/\lambda_0|$ as a function of $\beta_0 h$; $\Omega = 10, 12.5, 15, 20, \infty$.

Fig. 9.6b. Detail of Fig. 9.6a near resonance.

Fig. 9.6c. Circular graph of effective length.

[IV.9] THEORY OF LINEAR ANTENNAS 493

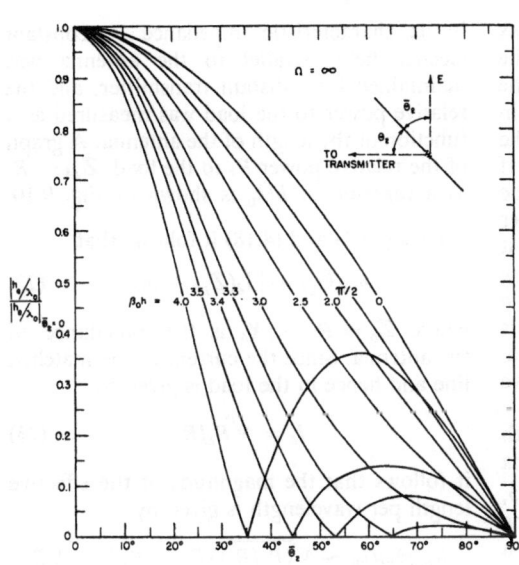

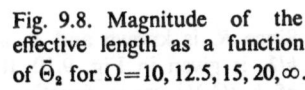

Fig. 9.7. Normalized zeroth-order effective length as a function of $\bar{\Theta}_2$.

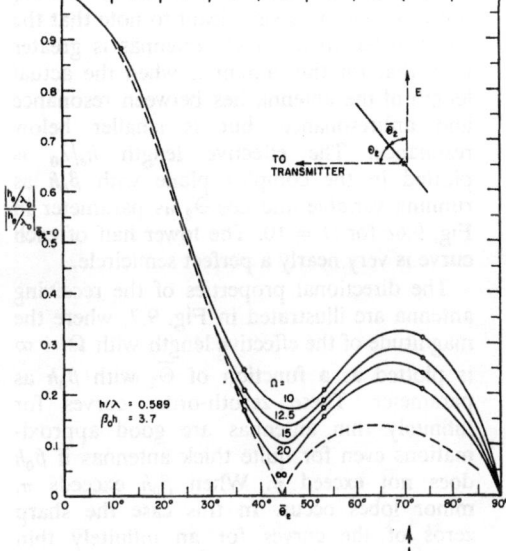

Fig. 9.8. Magnitude of the effective length as a function of $\bar{\Theta}_2$ for $\Omega = 10, 12.5, 15, 20, \infty$.

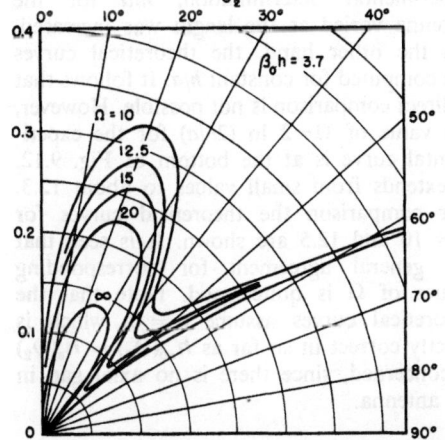

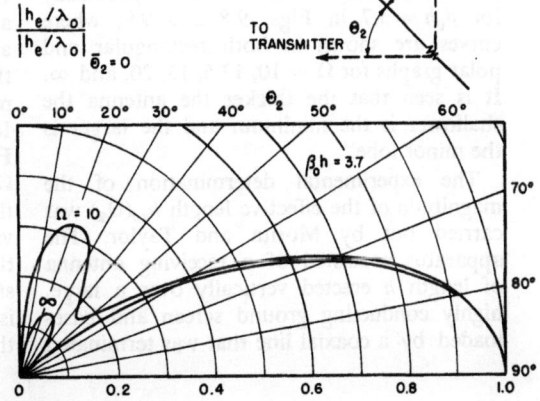

Fig. 9.9. Polar graph of Fig. 9.8.

With numerical values of the complex effective length available, the emf of the generator in the equivalent circuit for the receiving antenna may be determined in both magnitude and phase referred to the electric field of the distant transmitter. If this electric field is linearly polarized, the phase usually is not required. On the other hand, if the field is elliptically polarized it must be resolved into components along major and minor axes of the ellipse. The equivalent emf for *each* of these components must be determined separately and combined. This involves a knowledge of the relative phases of the two components.

The magnitude of the effective length with the electric field parallel to the antenna ($\Theta_2 = \pi/2$ or $\bar{\Theta}_2 = 0$) for $\Omega = 10, 12.5, 15, 20$, and ∞ is plotted in Fig. 9.6a, b. The zeroth-order curve for $\Omega = \infty$ is seen to be a fair approximation for thin antennas that are not too long. It is significant to note that the effective length for thick antennas is greater than that for thin antennas when the actual length of the antenna lies between resonance and antiresonance, but is smaller below resonance. The effective length h_e/λ_0 is plotted in the complex plane with $\beta_0 h$ as running variable and $\cos \Theta_2$ as parameter in Fig. 9.6c for $\Omega = 10$. The lower half of each curve is very nearly a perfect semicircle.

The directional properties of the receiving antenna are illustrated in Fig. 9.7, where the magnitude of the effective length with $\Omega = \infty$ is plotted as a function of $\bar{\Theta}_2$ with $\beta_0 h$ as parameter. These zeroth-order curves for infinitely thin antennas are good approximations even for quite thick antennas if $\beta_0 h$ does not exceed π. When $\beta_0 h$ exceeds π, minor lobes occur. In this case the sharp zeros of the curves for an infinitely thin antenna are replaced by minima for antennas of finite radius. This is illustrated specifically for $\beta_0 h = 3.7$ in Figs. 9.8 and 9.9, where curves are shown in both rectangular and polar graphs for $\Omega = 10, 12.5, 15, 20$, and ∞. It is seen that the thicker the antenna the shallower is the minimum and the larger is the minor lobe.

The experimental determination of the magnitude of the effective length $h_{e\delta}(\Theta_2)$ was carried out by Morita and Taylor. The apparatus consisted of a receiving antenna of length h erected vertically over a large, highly conducting ground screen and base-loaded by a coaxial line that was terminated in its characteristic impedance. A constant electric field parallel to the antenna was maintained by a distant transmitter, and the relative power to the load was measured as a function of the length of the antenna. A graph of the relative power P_L to the load $Z_{L\delta} = R_c$ as a function of h/λ_0 is shown in Fig. 9.10.

Using (4.1) and (4.18) it follows that

$$h_{e\delta}(\Theta_2) \sim |I_\delta(Z_{L\delta} + Z_\delta)|, \qquad (22)$$

where $Z_\delta = R_\delta + jX_\delta$ is the impedance of the antenna. Since the current in the matched line and hence in the load is given by

$$I_\delta = \sqrt{P_L/R_c}, \qquad (23)$$

it follows that the magnitude of the effective length per wavelength is given by

$$h_{e\delta}(\Theta_2)/\lambda_0 \sim \sqrt{(P_L/R_c)[(R_\delta + R_c)^2 + X_\delta^2]}. \qquad (24)$$

Since R_c is known and P_L as a function of h/λ_0 is given in Fig. 9.10, the relative value of $h_{e\delta}(\Theta_2)/\lambda_0$ as a function of h/λ_0 can be determined from (24) if the impedance of the antenna is known. The measured value of the apparent impedance Z_{sa} is shown in Fig. 9.11. Since terminal-zone effects are small and only a relative agreement between theory and experiment is sought, it may be assumed that $Z_\delta \doteq Z_{sa}$.

Using P_L from Fig. 9.10, R_δ and X_δ from Fig. 9.11, and with $R_c = 65.9$ ohms, the normalized effective length was determined and plotted in Fig. 9.12. Since only relative magnitudes are known, the maximum of Kh_e/λ_0 was normalized to unity. Correspondingly normalized theoretical curves taken from Fig. 9.6 are also shown. In the experimental determination, h/a for the antenna varied as the length was increased. On the other hand, the theoretical curves are computed for constant h/a. It follows that a direct comparison is not possible. However, the value of $\Omega = 2 \ln (2h/a)$ for the experimental curve is at the bottom of Fig. 9.12. It extends from small values to about 12.3. For comparison the theoretical curves for $\Omega = 10$ and 12.5 are shown. It is seen that the general agreement for corresponding values of Ω is quite good. Note that the theoretical curves assume $\delta = 0$, which is strictly correct in so far as $h_{e\delta}(\Theta_2) = h_e(\Theta_2)$ is concerned, since there is no axial gap in the antenna.

[IV.9]   THEORY OF LINEAR ANTENNAS   495

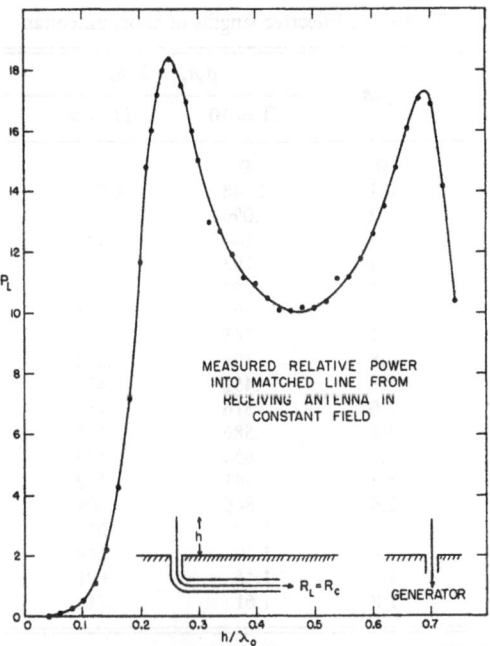

Fig. 9.10. Measured relative power into matched line from receiving antenna in constant field (Morita and Taylor).

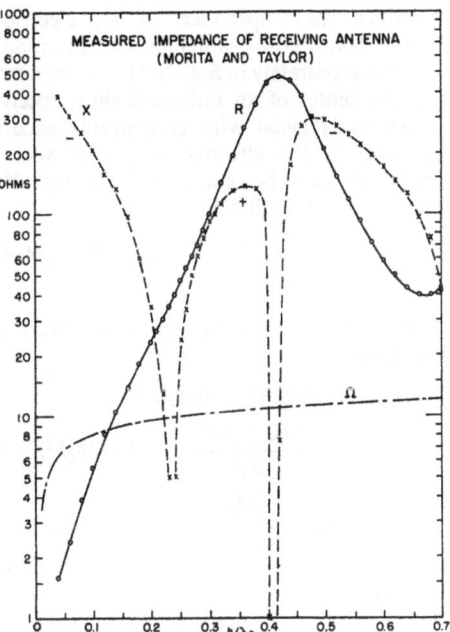

Fig. 9.11. Measured impedance of receiving antenna (Morita and Taylor).

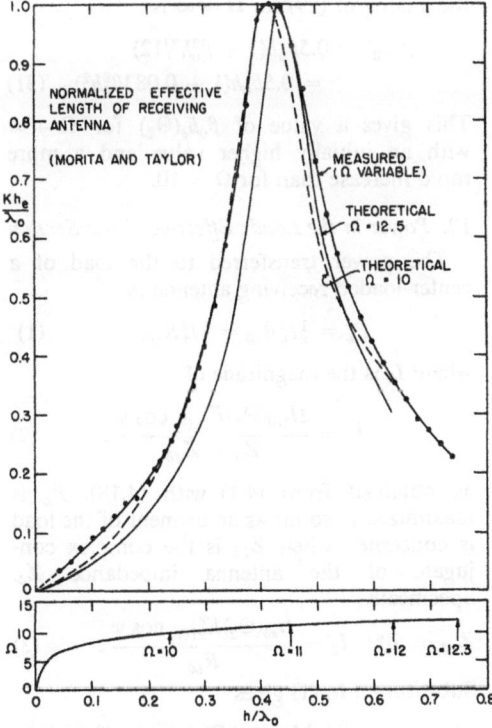

Fig. 9.12. Normalized effective length of receiving antenna (Morita and Taylor).

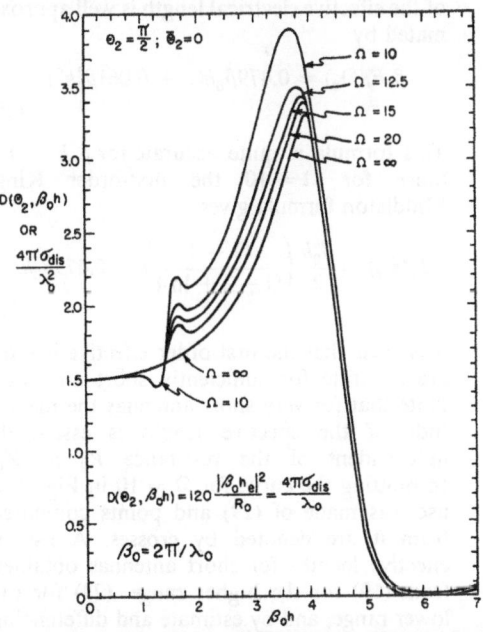

Fig. 10.1. Directivity and dissipation cross section of receiving antenna with conjugate matched load for $\Theta_2 = 90°$.

Since the impedance of an electrically short transmitting antenna has been determined acccurately in Sec. II.31 and the current at the center of an unloaded short receiving antenna is given with comparable accuracy in Sec. 7, the effective length of such an antenna may be expressed directly. With $\Theta = \pi/2$ in (19), it follows that

$$\beta_0 h_e(\Theta_2) = \beta_0[h''_e(\Theta_2) + jh'_e(\Theta_2)] = \tfrac{1}{2} u_0 Z_0. \quad (25)$$

For $\Omega = 10$, it follows from Sec. II.31 and (7.50) that

$$Z_0 = 18.3\beta_0^2 h^2(1 + 0.086\beta_0^2 h^2)$$
$$- j \frac{60 \times 6.6}{\beta_0 h}(1 - 0.383\beta_0^2 h^2), \quad (26)$$

$$u_0 \doteq ju'_0 = \frac{j\beta_0^2 h^2}{60 \times 6.9}(1 + 0.444\beta_0^2 h^2). \quad (27)$$

Hence,

$$\beta_0 h''_e(\Theta_2) = \tfrac{1}{2}\beta_0 h \left(\frac{6.6}{6.9}\right)(1 + 0.061\beta_0^2 h^2)$$
$$= 0.479\beta_0 h(1 + 0.061\beta_0^2 h^2), \quad (28a)$$

$$\beta_0 h'_e(\Theta_2) = 0.00221\beta_0^4 h^4(1 + 0.530\beta_0^2 h^2). \quad (28b)$$

It follows that for $\beta_0 h \leq 1$, the magnitude of the effective electrical length is well approximated by

$$\beta_0 h_e(\Theta_2) = 0.479\beta_0 h(1 + 0.061\beta_0^2 h^2). \quad (29)$$

This formula is quite accurate for $\beta_0 h \leq 0.5$. Since for $\Omega = 10$ the first-order King-Middleton formula gives

$$\beta_0 h_e(\Theta_2) = \frac{\beta_0 h}{2}\left(\frac{\Omega - 1}{\Omega - 2 + \ln 4}\right) = 0.479\beta_0 h, \quad (30)$$

it is clear that the first-order effective lengths are accurate for sufficiently short antennas. Note that for very short antennas the magnitude of the effective length is essentially independent of the resistance R_0 in Z_0. In plotting the curve for $\Omega = 10$ in Fig. 9.6b use was made of (29) and points computed from it are denoted by crosses. A list of effective lengths for short antennas obtained from (19) for the higher range, (29) for the lower range, and by estimate and differencing in the intermediate range, are given in Table 9.1 for $\Omega = 10$. The zeroth-order values for

TABLE 9.1. Effective lengths of short antennas.

$\beta_0 h$	$\beta_0 h_e = 2\pi h_e/\lambda_0$	
	$\Omega = 10$	$\Omega \to \infty$
0	0	0
0.1	0.048	0.050
.2	.096	.100
.3	.144	.149
.4	.193	.197
.5	.243	.254
.6	.293	.291
.7	.345	.336
.8	.398	.380
.9	.452	.422
1.0	.516	.462
1.1	.586	.501
1.2	.654	.537
1.3	.741	.572
1.4	.866	.604
1.5	1.12	.635
1.6	1.30	.664
1.7	1.46	.691
1.8	1.61	.716

antennas with sinusoidally distributed currents and $\Omega \to \infty$ also are given. They are computed from (5.25). Note that by expanding (5.25) the analog of (29) for $\Omega \to \infty$ is

$$\beta_0 h_e(\Theta_2) = 0.5\beta_0 h(1 + \beta_0^2 h^2/12)$$
$$= 0.5\beta_0 h(1 + 0.083\beta_0^2 h^2). \quad (31)$$

This gives a value of $\beta_0 h_e(\Theta_2)$ for $\Omega \to \infty$ with an initially higher value and a more rapid increase than for $\Omega = 10$.

10. Power in the Load; Effective Cross Section

The power transferred to the load of a center-loaded receiving antenna is

$$P_L = \tfrac{1}{2} I_L^2 R_{L\delta} = \tfrac{1}{2} I_\delta^2 R_{L\delta}, \quad (1)$$

where I_δ is the magnitude of

$$I_\delta = \frac{2h_{e\delta}(\Theta_2) E^r_{21\Theta_1} \cos \psi}{Z_\delta + Z_{L\delta}}, \quad (2)$$

as obtained from (4.1) with (4.18). P_L is maximized in so far as adjustment of the load is concerned when $Z_{L\delta}$ is the complex conjugate of the antenna impedance Z_δ. Specifically,

$$Z_{L\delta} = Z_\delta^* : I_\delta = \frac{h_{e\delta}(\Theta_2) E^r_{21\Theta_1} \cos \psi}{R_{L\delta}}. \quad (3)$$

Substitution in (1) gives

$$(P_L)_{max} = \frac{|h_{e\delta}(\Theta_2) E^r_{21\Theta_1} \cos \psi|^2}{2R_\delta}. \quad (4)$$

This may be expressed as follows:

$$(P_L)_{\max} = \frac{(\lambda_0 E_{21\Theta_1}^r \cos \psi)^2}{8\pi \zeta_0} D(\Theta_2, \beta_0 h), \quad (5)$$

where

$$D(\Theta_2, \beta_0 h) = \frac{\zeta_0}{\pi} \frac{[\beta_0 h_{e\delta}(\Theta_2)]^2}{R_\delta}. \quad (6a)$$

The dimensionless directivity* $D(\Theta_2, \beta_0 h)$ is plotted in Fig. 10.1 against $\beta_0 h$ with $\Theta_2 = \pi/2$, $\delta = 0$ and $\Omega = 10, 12.5, 15, 20,$ and ∞ using the data of Sec. 9 for $\beta_0 h_e$ and those of Secs. II.30 and II.31 for R_0. The corresponding curves for a range of values of Θ_2 are in Fig. 10.2. In determining the curve for $\Omega = 10$ in Fig. 10.1 the interpolated values of R_0 as given in Table II.31.1 were used. For the smaller values of $\beta_0 h$ these coincide with those obtained from the accurate formula for the short antenna derived in Sec. II.31. Using this formula (II.31.44a) for R_0 and the corresponding formula for $\beta_0 h_e$, namely, (9.29), the following result is obtained for the directivity or gain:

$$D(\Theta, \beta_0 h) = 1.500(1 + 0.036\beta_0^2 h^2).$$
$$(\Omega = 10) \quad (6b)$$

This is accurate for $\beta_0 h \leq 0.5$ and an excellent approximation for $\beta_0 h \leq 1$. The corresponding formula for an extremely thin antenna with sinusoidal current ($\Omega \to \infty$) is obtained using (II.31.48) for R_0 and (9.28) for $\beta_0 h_e$. The result is

$$D(\Theta, \beta_0 h) = 1.500(1 + \beta_0^2 h^2/30)$$
$$\doteq 1.500(1 + 0.033\beta_0^2 h^2). \quad (6c)$$

Although the directivity of the thin antenna initially rises slightly more slowly than that of the thicker antenna, these relative positions are reversed as the antenna is made longer. This is shown in Fig. 10.1. Near resonance the gain of the thicker antenna increases rapidly. This is a direct consequence of the fact

---

* It is proved in (V.14.21) that $\beta_0 h_e = F_0(\beta_0 h, \Theta)$, so that with (V.12.3) it follows that $D(\Theta_2, \beta_0 h)$ in (6) is the same as the directivity D defined in (I.10.7).
The defining formula for the directivity (6a) may be rearranged as follows:

$$h_{e\delta}(\Theta_2) = k\lambda_0 \sqrt{D(\Theta_2, \beta_0 h) R_L \delta},$$

where $k = 1/\sqrt{4\pi\zeta_0}$ and the load resistance has been substituted for the *equal* antenna resistance. Note that this formula may not be interpreted as indicating that the effective length of a receiving antenna depends upon the resistance of its load. It follows from its definition in (9.1) that the effective length depends on the dimensions and orientation of the receiving antenna; a load is *not involved*. It is the power to the load that depends on the load, and in the particular case when the load is a conjugate match, the power to the load is proportional to the directivity or gain.

---

that the effective electrical length of the thicker antenna is smaller than that of the extremely thin antenna when $\beta_0 h < \pi/2$, whereas it is greater when $\beta_0 h > \pi/2$. The curves cross somewhere near $\beta_0 h = \pi/2$. Since the gain varies as the square of the effective length, a sharp increase is observed near $\pi/2$. It is well to note that the detailed shape of the curve for $D(\Theta, \beta_0 h)$ near resonance is very sensitive to the relative magnitudes and rates of increase of $\beta_0 h_e$ and R_0. Errors of approximation of only a few percent might account for the dip below 1.5 for $\beta_0 h$ between 1.2 and 1.3 and for the small rise near $\beta_0 h = 1.7$.
If the first-order King-Middleton formulas for $\beta_0 h_e$ and R_0 are used, the limiting formula for the directivity as $\beta_0 h \to 0$ is

$$D(\Theta_2, \beta_0 h) = 1.5 \left[\frac{(\Omega - 1)^2}{(\Omega - 2)(\Omega - 2 + 2 \ln 4)} \right].$$
$$(6d)$$

It is readily verified that this gives values up to several percent below 1.5 for $\Omega < \infty$. This is a consequence of the inaccuracy in resistance, the first-order formula yielding values of R_0 for short antennas that are several percent too great. Numerical values of the gain or directivity for $\Omega = 10$ and ∞ are given in Table 10.1 as computed from (6b) and (6c) for short antennas and from the second-order King-Middleton formula for longer antennas.

TABLE 10.1. Directivity or gain of short antenna.

$\beta_0 h$	D	
	$\Omega = 10$	$\Omega \to \infty$
0	1.500	1.500
0.1	1.500	1.500
.2	1.502	1.502
.3	1.505	1.504
.4	1.509	1.508
.5	1.514	1.512
.6	1.518	1.518
.7	1.519	1.524
.8	1.520	1.532
.9	1.520	1.540
1.0	1.521	1.553
1.1	1.509	1.564
1.2	1.450	1.578
1.3	1.445	1.592
1.4	1.546	1.609
1.5	2.042	1.627
1.6	2.211	1.647
1.7	2.243	1.669
1.8	2.137	1.698

It is seen that the maximum value of $D(\Theta_2, \beta_0 h)$ (which gives $(P_L)_{\text{max max}}$) in so far as the adjustment of Θ_2 and h are concerned, occurs when $\beta_0 h$ is near 4 and when $\Theta_2 = \pi/2$ or $\bar{\Theta}_2 = 0$. The magnitude of this maximum power to the load is greater for thicker antennas. The power to the matched load is almost independent of h for $\beta_0 h < 0.5$. However, owing to the high capacitive reactance and very low resistance of very short antennas, a conjugate match is difficult to obtain and losses in the matching network may exceed the power to the load.

The explicit zeroth-order formula for (6a) with $\delta = 0$ is interesting and useful for very long antennas for which higher-order data on effective length and impedance are unavailable. It is given by

$$[D(\Theta_2, \beta_0 h)]_0 = \frac{\zeta_0}{\pi [R_0]_{01}} \left[\frac{\cos(\beta_0 h \cos \Theta_2) - \cos \beta_0 h}{\sin \beta_0 h \sin \Theta_2} \right]^2, \quad (7)$$

where $[R_0]_{01}$ is given by (II.28.5a) with $\delta = 0$ or by (II.28.9), since the ohmic resistance is negligible in good conductors. Alternatively, (7) may be expressed in terms of the radiation resistance referred to maximum sinusoidal current defined by

$$R_m^e = R_0^e \sin^2 \beta_0 h, \quad (8)$$

where R_0^e is equal to $[R_0]_{01}$ if ohmic resistance is neglected. Hence

$$[D(\Theta_2, \beta_0 h)]_0 = \frac{\zeta_0}{\pi R_m^e} \left[\frac{\cos(\beta_0 h \cos \Theta_2) - \cos \beta_0 h}{\sin \Theta_2} \right]^2. \quad (9)$$

This function corresponds to (6a) with $\delta = 0$ and $\Omega \to \infty$. Curves of (9) to large values of $\beta_0 h$ are shown in Fig. 10.3 as computed by Hallén.* It is seen that for antennas whose half-length exceeds λ_0 ($\beta_0 h > 2\pi$), the maximum power to the load is not obtained with $\Theta_2 = \pi/2$ but at a variety of angles depending upon the length. Moreover, the maximum is greater for very long antennas inclined at a large angle $\bar{\Theta}_2$ with respect to the wave front of the incident field $E_{21\Theta_1}^r$, than for shorter ones with maxima at $\Theta_2 = \pi/2$. In using long antennas at large angles $\bar{\Theta}_2$ the adjustment in length and the angle of inclination of the antenna are quite critical. It should be noted that a base-loaded antenna over a conducting plane corresponds to a center-loaded antenna in free space by the image theorem only when perpendicular to the conducting plane. An inclined antenna with its image corresponds to a V-antenna loaded at the apex.

An experimental determination of the power to a matched load ideally requires the direct measurement of the power dissipated in an impedance $Z_{L\delta} = Z_\delta^*$. Since it is difficult to adjust $Z_{L\delta}$ accurately for each length of antenna so that it is the complex conjugate of the impedance of the antenna, an essentially equivalent procedure makes use of a matched load $Z_{L\delta} = R_c$ for the *transmission line*. By determining the relative effective length of the antenna in terms of the power to R_c, and substituting the values of $h_{e\delta}(\Theta_2)$ so determined in (6a), a quantity proportional to the power transferred to a load that is the complex conjugate of the impedance of the antenna may be computed. Using the experimentally determined values of the normalized relative effective length given in Fig. 9.12, the normalized relative directivity $KD(\Theta_2, \beta_0 h)$ was determined as a function of h/λ_0 or of $\beta_0 h$ by substitution in (6a). The results are shown in Fig. 10.4.

If account is taken of the fact that h/a is not constant, the general shape of the curve in Fig. 10.4 is in good agreement with the curves in Fig. 10.1. Note that the maximum value of $D(\Theta_2, \beta_0 h)$ occurs at the correct value of $\beta_0 h$ and that a definite irregularity occurs near resonance. The amplitude of this oscillation in the experimental curve is smaller than for the corresponding value of h/a in the theoretical curves, but the general behavior is correctly given. The fact that the experimental curve drops sharply toward zero for small values of $\beta_0 h$ is due to the fact that h/a diminishes rapidly whereas in the theoretical curve it remains constant. Thus, the theoretical assumption that h/a is constant implies that as h is reduced the antenna simultaneously becomes thinner and thinner so that at $h \doteq 0$ it is an infinitely thin doublet. The theoretical directivity curves do not go to zero at $h = 0$, whereas the practical curve of an antenna of finite thickness approaches zero rapidly but smoothly as h is reduced.

A common measure of the effectiveness of a receiving antenna as a device for transferring power to the load is a quantity, σ_{dis}, called the *effective dissipation cross section*,

---

* See also Fig. V.12.2, where the same function is plotted using maximizing values of Θ.

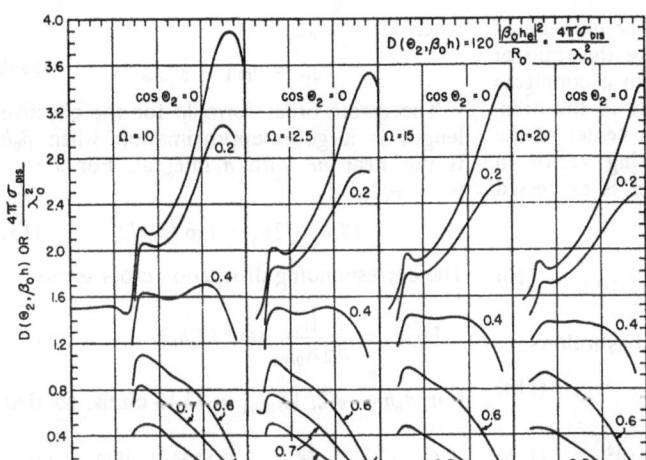

Fig. 10.2. Curves of Fig. 10.1 with Θ_2 as parameter.

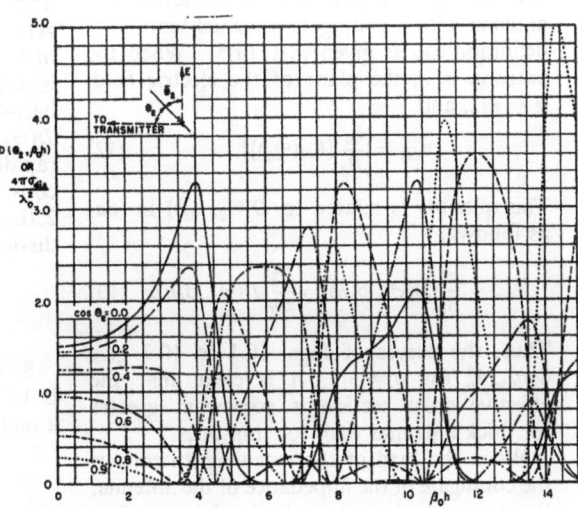

Fig. 10.3. Zeroth-order curves of $D(\Theta_2, \beta_0 h)$ for large values of $\beta_0 h$ as computed by Hallén.

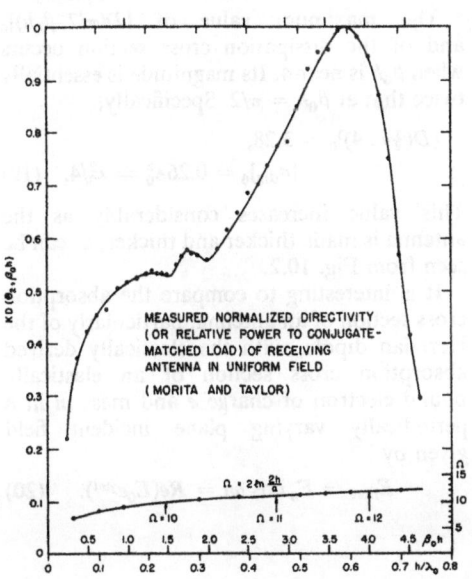

Fig. 10.4. Measured normalized directivity (or relative power to a conjugate-matched load) of a receiving antenna in uniform field (Morita and Taylor).

the *effective area*, or *the effective aperture* of the antenna. It is defined to be the ratio of the power actually dissipated in a conjugate matched load to the magnitude of the time-average Poynting vector at the center of the receiving antenna. The Poynting vector in the far zone of the transmitter is expressed in terms of the incident field by

$$S = \frac{(E_{21\Theta_1}^r)^2}{2\zeta_0} = \tfrac{1}{2} E_{21\Theta_1}^r H_{21\Theta_1}^r. \quad (10)$$

With (10), (4) may be expressed as follows:

$$(P_L)_{\max} = S\sigma_{\text{dis}}, \quad (11a)$$

where

$$\sigma_{\text{dis}} = \frac{\zeta_0}{R_\delta} [h_{e\delta}(\Theta_2) \cos \psi]^2. \quad (11b)$$

The effective dissipation cross section of the antenna with a conjugate matched load is by definition σ_{dis} as given in (11a)*. When the antenna is in the plane of the electric field, $\psi = \pi/2$, and

$$\sigma_{\text{dis}} = \frac{\zeta_0}{R_\delta} [h_{e\delta}(\Theta_2)]^2. \quad (12)$$

This quantity is related to $D(\Theta_2, \beta_0 h)$ in (6a) as follows:

$$\sigma_{\text{dis}} = \frac{\pi}{\beta_0^2} D(\Theta_2, \beta_0 h) = \frac{\lambda_0^2}{4\pi} D(\Theta_2, \beta_0 h). \quad (13)$$

Thus, the curves of Figs. 10.1 to 10.3 give $4\pi\sigma_{\text{dis}}/\lambda_0^2$ for $\psi = \pi/2$. It is clear that the effective cross section is somewhat greater for thick antennas than for thin ones.

When the load of the receiving antenna is the conjugate of the impedance of the antenna, the power dissipated in the load is equal to the power reradiated from the antenna. Each of these is, then, one-half of the power absorbed by the antenna. It follows that the *reradiating or scattering cross section* σ_{rad} of the antenna must equal the dissipation cross section and their sum must equal the total absorption cross section σ_{abs}. Thus,

$$\sigma_{\text{abs}} = \sigma_{\text{dis}} + \sigma_{\text{rad}}; \quad (14a)$$

---

* Note that if the Poynting vector is interpreted to be the "power per unit area flowing through space," the effective dissipation cross section is the area in space perpendicular to the Poynting vector through which passes a power equal to that actually dissipated in the matched load. This area is much greater than the geometric cross section of the cylindrical antenna, and is not dependent upon it. The concept that a receiving antenna "intercepts" power flowing through space is a fiction requiring the definition of a fictitious area that is in no way related to the cross-sectional dimensions of the antenna.

when $Z_L = Z_0^*$,

$$\sigma_{\text{dis}} = \sigma_{\text{rad}} = \tfrac{1}{2}\sigma_{\text{abs}}. \quad (14b)$$

The zeroth-order formula for the effective length is a good approximation when $\beta_0 h$ is not near $n\pi$ with n integral. For $\delta = 0$, $\Theta_2 = \pi/2$, it is

$$[\beta_0 h_e(\pi/2)]_0 = \tan(\beta_0 h/2) \quad (15)$$

The corresponding dissipation cross section is

$$[\sigma_{\text{dis}}]_0 = \frac{\zeta_0}{\beta_0^2 [R_0]_{01}} \tan^2(\beta_0 h/2). \quad (16a)$$

For $\beta_0 h = \pi/2$, $[R_0]_{01} = 73.13$ ohms, so that

$$[\sigma_{\text{dis}}]_0 = \frac{\zeta_0}{\beta_0^2 [R_0]_{01}} = 0.13\lambda_0^2 \doteq \lambda_0^2/8. \quad (16b)$$

The same result is obtained by inserting the zeroth-order directivity $[D(\pi/2, \pi/2)]_0 = 1.64$ in (13). The dissipation cross section in (16) is often given as the correct value for all half-wave dipoles. Although the value of $D(\Theta_2, \beta_0 h)$ and hence of σ_{dis} does not vary rapidly with wire thickness, it is not actually constant. Thus, with $\Omega = 10$, $D(\pi/2, \pi/2) \doteq 2.16$ so that the second-order value of the dissipation cross section is

$$[\sigma_{\text{dis}}]_2 = \frac{\lambda_0^2 D(\Theta_2, \pi/2)}{4\pi} = 0.17\lambda_0^2. \quad (17)$$

Since the directivity of the Hertzian dipole is 1.5, it follows that for such a dipole with a matched load

$$[\sigma_{\text{dis}}]_0 = \frac{3}{8\pi} \lambda_0^2 = 0.119\lambda_0^2. \quad (18)$$

The maximum value of $[D(\pi/2, \beta_0 h)]_0$ and of the dissipation cross section occurs when $\beta_0 h$ is near 4. Its magnitude is essentially twice that at $\beta_0 h = \pi/2$. Specifically,

$$[D(\tfrac{1}{2}\pi, 4)]_0 = 3.28,$$

$$[\sigma_{\text{dis}}]_0 = 0.26\lambda_0^2 \doteq \lambda_0^2/4. \quad (19)$$

This value increases considerably as the antenna is made thicker and thicker, as can be seen from Fig. 10.2.

It is interesting to compare the absorption cross section of an antenna, particularly of the Hertzian dipole, with the classically derived absorption cross section of an elastically bound electron of charge e and mass m in a periodically varying plane incident field given by

$$E_{\text{inst}} = E_0 \cos \omega t = \text{Re}(E_0 e^{j\omega t}). \quad (20)$$

The equation of motion for such an electron is a generalization of the well-known dynamical equation $md^2r/dt^2 = eE$, as given in ref. I.31, Sec. II.13. It is

$$m \frac{d^2r}{dt^2} = -m\omega_0^2 r - m\gamma \frac{dr}{dt} + eE_0 \cos \omega t. \quad (21)$$

In (21) r is the displacement of the electron from its average rest position, ω_0 is the resonant angular velocity, and $-m\omega_0^2 r$ is the restoring force due to the internal binding of the electron. The term $-m\gamma dr/dt$ takes account of dissipation. It consists of two parts: $\gamma = \gamma_{rad} + \gamma_{dis}$, where γ_{dis} represents dissipation within the system and γ_{rad} is the damping due to reradiation. In the classical theory it is given by

$$\gamma_{rad} = \frac{2\pi e^2 \zeta_0}{3m\lambda_0^2}. \quad (22)$$

The solution of (21) is

$$r = Re \; s; \; s = \left[\frac{(e/m)E_0 e^{j\omega t}}{\omega_0^2 - \omega^2 + j\gamma\omega} \right]. \quad (23)$$

The time-average power absorbed by the oscillating charge (including that reradiated) is obtained from the time average of the product of the instantaneous force and the velocity. With $E_{inst} = Re \; E_0 e^{j\omega t}$ it is

$$P_{abs} = \overline{eE_{inst} \frac{dr}{dt}} = Re \left(\tfrac{1}{2} e E_0^* \frac{ds}{dt} \right)$$

$$= Re \left[\tfrac{1}{2} e E_0 \left(\frac{j\omega(e/m)E_0}{\omega^2 - \omega_0^2 + j\gamma\omega} \right) \right]. \quad (24)$$

This may be expressed as follows:

$$P_{abs} = S \frac{\zeta_0 e^2}{m} \frac{\gamma\omega^2}{(\omega^2 - \omega_0^2)^2 + \gamma^2 \omega^2} = S\sigma_{abs}, \quad (25)$$

where S is the time-average Poynting vector as given by (10) and σ_{abs} is defined as follows, using (22):

$$\sigma_{abs} = \frac{\zeta_0 e^2}{m} \frac{\gamma\omega^2}{(\omega^2 - \omega_0^2)^2 + \gamma^2 \omega^2}$$

$$= \frac{3\lambda_0^2}{2\pi} \frac{\gamma_{rad} \gamma \omega^2}{(\omega^2 - \omega_0^2)^2 + \gamma^2 \omega^2}. \quad (26)$$

For an electron bound to a system with internal dissipation equal to reradiation, $\gamma_{rad} = \gamma_{dis} = \gamma/2$. If the field oscillates with the angular velocity $\omega = \omega_0$, it follows that

$$\sigma_{abs} = \frac{3\lambda_0^2}{4\pi}, \quad \sigma_{dis} = \sigma_{rad} = \frac{3\lambda_0^2}{8\pi}. \quad (27)$$

The value of σ_{dis} is the same as (18) for the Hertzian dipole since the oscillating electron is, in effect, a Hertzian dipole.

PARASITIC ANTENNAS AS SCATTERERS AND REFLECTORS IN THE RADIATION ZONE

11. Reradiation from Parasitic Antennas—General Theory

The general analysis in Sec. 2 of the parasitic antenna in the far-zone field of a transmitter is applied in Secs. 3–8 to the receiving antenna, of which the primary function is to transfer power to, or maintain a voltage across, a load. Since only even currents maintain a potential difference across a symmetric load, odd currents are ignored in that application.

In this and following sections, the general analysis of Sec. 2 is applied to one or more parasitic antennas that serve a different function, namely, to provide suitable conductors for currents that maintain a desired electromagnetic field. Since all currents in each antenna contribute to the field, both even and odd distributions of current must be known. Therefore, the analysis of an arbitrarily oriented, highly conducting parasitic antenna with no load must begin with (2.25). Let ohmic losses be neglected by setting $z^i = 0$; also let $\delta = 0$. With this simplification the integral equation (2.25) reduces to

$$4\pi\nu_0 A_{22z} = \int_{-h}^{h} I_z' \frac{e^{-j\beta_0 R_{22}}}{R_{22}} dz'$$

$$= -\frac{j4\pi}{\zeta_0} (C_1 \cos \beta_0 z + C_2 \sin \beta_0 z + U e^{jq_0 z}), \quad (1)$$

where

$$U = -E_{21\Theta_1}^r \cos \psi / \beta_0 \cos \Theta_2, \quad (2)$$

and $E_{21\Theta_1}^r$ is the field at the center of the receiving antenna No. 2 due to the transmitting antenna No. 1. The angle Θ_1 is measured from the z_1 axis along the transmitting antenna, and Θ_2 is measured from the z_2 axis along the parasitic antenna; C_1 and C_2 are complex constants to be evaluated from the boundary conditions;

$$R_{22} = \sqrt{(z_2 - z_2')^2 + a^2}, \quad (3)$$

$$q_0 = \beta_0 \cos \Theta_2 = \beta_0 \sin \overline{\Theta}_2. \quad (4)$$

The solution of (1) may be carried out in two parts, corresponding to the symmetric problem with even currents and vector potentials, odd charges and scalar potentials, and the antisymmetric problem with odd currents and vector potentials, even charges and scalar potentials.

The symmetric problem is the same as that already treated for the center-loaded antenna, but, since there is no load,

$$C_2 = \tfrac{1}{2} V_{20} = 0. \tag{5}$$

The appropriate integral equation is obtained from (2.28) with $\delta = 0$, $z^i = 0$, and (5). Thus,

$$4\pi \nu_0 A_{22z}^s = \int_{-h}^{h} \frac{I_z^s(z')}{R_{22}} e^{-j\beta_0 R_{22}} dz'$$

$$= \frac{-j4\pi}{\zeta_0} [C_{1s} \cos \beta_0 z + U \cos q_0 z]. \tag{6}$$

The integral equation for the antisymmetric problem is given by (2.32) with $\delta = 0$, $z^i = 0$ and with (2.35):

$$4\pi \nu_0 A_{22z}^a = \int_{-h}^{h} \frac{I_z^a(z')}{R_{22}} e^{-j\beta_0 R_{22}} dz'$$

$$= \frac{-j4\pi}{\zeta_0} [C_{2a} \sin \beta_0 z + jU \sin q_0 z]. \tag{7}$$

The approximate solution of (6) and (7) may be carried out by the usual method of successive approximations. Let

$$\Psi(z) = \int_{-h}^{h} g(z, z') K(z, z') dz' = \Psi + \Upsilon(z), \tag{8}$$

where

$$g(z, z') \doteq I_z(z')/I_z(z) \equiv I_z'/I_z \tag{9}$$

and where

$$K(z, z') = e^{-j\beta_0 R_{22}}/R_{22}. \tag{10}$$

A superscript s is affixed to I, a subscript s to Ψ, γ, and g for use with (6), and a superscript or subscript a for use with (7). Substitution of (8) in (6) and (7) and rearrangement of terms leads to

$$I_z^s = \frac{-j4\pi}{\zeta_0 \Psi_s} [C_{1s} \cos \beta_0 z + U \cos q_0 z]$$

$$- \frac{1}{\Psi_s} [11]_s, \tag{11a}$$

$$I_z^a = \frac{-j4\pi}{\zeta_0 \Psi_a} [C_{2a} \sin \beta_0 z + jU \sin q_0 z]$$

$$- \frac{1}{\Psi_a} [11]_a, \tag{11b}$$

where

$$[11]_i = \left[I_{zi} \Upsilon_i(z) + \int_{-h}^{h} [I_{zi}^i(z') - I_z^i(z) g_i(z, z')] K(z, z') dz' \right],$$

$$i = a \text{ or } s, \tag{11c}$$

and where

$$\delta \leq z \leq h. \tag{11d}$$

Following the procedure in Chapter II for the isolated antenna,

$$I_z^s = \frac{-j4\pi}{\zeta_0 \Psi_s} [C_{1s}(\cos \beta_0 z - \cos \beta_0 h)$$

$$+ U(\cos q_0 z - \cos q_0 h)] - \frac{1}{\Psi_s} [12]_s, \tag{12a}$$

$$I_z^a = \frac{-j4\pi}{\zeta_0 \Psi_a} [C_{2a}(\sin \beta_0 z - \sin \beta_0 h)$$

$$+ jU(\sin q_0 z - \sin q_0 h)] - \frac{1}{\Psi_a} [12]_a, \tag{12b}$$

where with (11c)

$$[12]_i = [11]_i - \int_{-h}^{h} I_z^i(z') K(h, z') dz',$$

$$i = a \text{ or } s. \tag{12c}$$

These equations may be solved by iteration to give

$$I_z^s = \frac{-j4\pi}{\zeta_0 \Psi_s} [C_{1s} f_s(z, h) + U h_s(z, h)], \tag{13a}$$

$$I_z^a = \frac{-j4\pi}{\zeta_0 \Psi_a} [C_{2a} f_a(z, h) + jU h_a(z, h)], \tag{13b}$$

where $f_s(z, h) \equiv f(z, h)$ as given by (3.11a); $h_s(z, h) \equiv h(z, h)$ as given by (3.11c); $f_a(z, h)$ is obtained from $f(z, h)$ by substituting sin for cos in $F_0(z)$; and $h_a(z, h)$ is obtained from $h(z, h)$ by substituting sin for cos in $H_0(z)$.

The constants C_{1s} and C_{2a} may be evaluated as described in Sec. 3. The constant C_{1s} is given by (3.16) with $\delta = 0$ and $V = 0$; it is

$$C_{1s} = -U h_s(h)/f_s(h). \tag{14a}$$

Similarly,

$$C_{2a} = -jU h_a(h)/f_a(h). \tag{14b}$$

Note that $f_s(h) = f(h)$ and $h_s(h) = h(h)$ are defined by (3.12a, c); $f_a(h)$ and $h_a(h)$ are obtained from $f(h)$ and $h(h)$ by substituting sin for cos in $F_0(h)$ and $H_0(h)$, respectively.

Substitution of (14a, b) in (13a, b) gives

$$I_z^s = \frac{j4\pi U}{\zeta_0 \Psi_s} \left[\frac{f_s(z, h)h_s(h) - h_s(z, h)f_s(h)}{f_s(h)} \right],$$
(15a)

$$I_z^a = \frac{-4\pi U}{\zeta_0 \Psi_a} \left[\frac{f_a(z, h)h_a(h) - h_a(z, h)f_a(h)}{f_a(h)} \right].$$
(15b)

Since
$$h(z, h) = h(z) - h(h),$$
$$f(z, h) = f(z) - f(h),$$

it follows that

$$I_z^s = \frac{j4\pi U}{\zeta_0 \Psi_a} \left[\frac{f_s(z)h_s(h) - h_s(z)f_s(h)}{f_s(h)} \right],$$
(16a)

$$I_z^a = \frac{-4\pi U}{\zeta_0 \Psi_a} \left[\frac{f_a(z)h_a(h) - h_a(z)f_a(h)}{f_a(h)} \right].$$
(16b)

The zeroth-order and first-order terms are

$$(I_z^s)_1 = \frac{j4\pi U}{\zeta_0 \Psi_s} \times$$
$$\left\{\frac{\begin{array}{l}[\cos \beta_0 z \cos q_0 h - \cos q_0 z \cos \beta_0 h] \\ + (1/\Psi_s)[\cos q_0 h F_{1s}(z) - \cos \beta_0 h H_{1s}(z) \\ + H_{1s}(h) \cos \beta_0 z - F_{1s}(h) \sin q_0 z]\end{array}}{\cos \beta_0 h + F_{1s}(h)/\Psi_s}\right\},$$
(17a)

$$(I_z^a)_1 = \frac{-4\pi U}{\zeta_0 \Psi_a} \times$$
$$\left\{\frac{\begin{array}{l}[\sin \beta_0 z \sin q_0 h - \sin q_0 z \sin \beta_0 h] \\ + (1/\Psi_a)[\sin q_0 h F_{1a}(z) - \sin \beta_0 h H_{1a}(z) \\ + H_{1a}(h) \sin \beta_0 z - F_{1a}(h) \sin q_0 z]\end{array}}{\sin \beta_0 h + F_{1a}(h)/\Psi_a}\right\}.$$
(17b)

Note that $F_{1s}(z) \equiv F_1(z)$, $H_{1s}(z) \equiv H_1(z)$, and that $F_{1a}(z)$ and $H_{1a}(z)$ are obtained from $F_1(z)$ and $H_1(z)$, respectively, by replacing cos by sin in $F_0(z)$ and $H_0(z)$ as defined in Sec. 3 and in Sec. II.19.

The resultant current in the parasitic antenna is given by

$$I_z = I_z^s + I_z^a.$$
(18)

It is this current that maintains the reradiated electromagnetic field. However, since the evaluation of the electromagnetic field involves integrals with the current (18) under the sign of integration, the actual evaluation of the field is almost impossible with the currents given by (18) with the complicated formula (17a, b). In order to be of practical use, a simpler formula than (18) is required for the current. An exception is when interest is confined to the very special case in which the scattering antenna is parallel to the electric field so that no odd currents are excited ($I_z^a = 0$) and $\Theta_2 = \pi/2$ so that $q_0 = 0$. In this case the entire first-order current in the scattering antenna is given by (17a) with $\cos q_0 h = \cos q_0 z = 1$, and $H_{1s}(z) = H_{1s}(h) = 0$. With this simplification in the current, the first-order electromagnetic field may be evaluated in terms of tabulated functions, as is discussed in Sec. 13.

12. Current Distribution in Reradiating Antenna

When the primary function of the current in a parasitic antenna is to maintain an electromagnetic field, and interest is in the properties of this field, it is necessary to determine both the odd and the even currents. Moreover, since the current ultimately must be substituted in the integrand of a complicated integral for the electric or magnetic field, it is important that it be simple in form in order to facilitate integration. The formula (18) with (17a, b) is far too intricate for this purpose. The problem of simplifying this formula is the principal subject of this section.

A first step in the simplification is readily made if account is taken of the fact that, for an antenna that is not too short, the ratio of vector potential to current where this is a maximum does not vary appreciably with the particular form of the distribution of current. Hence, Ψ_s and Ψ_a, which are proportional to the ratio of vector potential to current at reference points that are near the maximum current in each case, are practically equal. Therefore, let

$$\Psi = \tfrac{1}{2}(\Psi_a + \Psi_s).$$
(1)

With (1), (11.18) reduces to the following first-order form:

$$(I_z)_1 = \frac{j4\pi U}{\zeta_0 \Psi} \times$$
$$\left\{\frac{\begin{array}{l}[-\sin \beta_0 h \cos \beta_0 h\, e^{jq_0 z} \\ + \sin \beta_0 h \cos q_0 h \cos \beta_0 z \\ + j \cos \beta_0 h \sin q_0 h \sin \beta_0 z] + \frac{1}{\Psi}[\cdots]\end{array}}{[\cos \beta_0 h + F_{1s}(h)/\Psi][\sin \beta_0 h + F_{1a}(h)/\Psi]}\right\}.$$
(2)

While simpler than (11.18), this is still too complicated to permit evaluation of the electromagnetic field. Further simplification is necessary.

The zeroth-order current is

$$(I_z)_0 = \frac{-j4\pi U}{\zeta_0 \Psi} \left\{ e^{jq_0 z} - \frac{\cos q_0 h \cos \beta_0 z}{\cos \beta_0 h} \right.$$
$$\left. - j \frac{\sin q_0 h \sin \beta_0 z}{\sin \beta_0 h} \right\}. \quad (3)$$

It is evident that whereas (3) is simple in form, it is not satisfactory for the evaluation of the electromagnetic field because the second term becomes infinite when $\beta_0 h = n\pi/2$, n odd, and the third term when $\beta_0 h = n\pi/2$, n even. Moreover, the entire current vanishes when $q_0 = \beta_0$.

A modified zeroth-order solution may be obtained from (2) by retaining only enough first-order terms to keep the expression finite for all values of $\beta_0 h$. This is readily accomplished by neglecting all first-order terms in the numerator, and retaining in the denominator the first-order terms associated with $\cos \beta_0 h$ in the second term and with $\sin \beta_0 h$ in the third term. Thus, let

$$(I_z)_{01} = \frac{-j4\pi U}{\zeta_0} \left\{ \frac{e^{jq_0 z}}{\Psi} - \frac{\cos q_0 h \cos \beta_0 z}{\Psi \cos \beta_0 h + F_{1s}(h)} \right.$$
$$\left. - j \frac{\sin q_0 h \sin \beta_0 z}{\Psi \sin \beta_0 h + F_{1a}(h)} \right\}. \quad (4)$$

This expression has the same dependence upon the variable z as the zeroth-order solution (3), but the coefficients of the terms remain finite for all values of $\beta_0 h$.

It remains to evaluate Ψ. In its usual definition, Ψ is the magnitude of the complex function

$$\Psi(z) = \int_{-h}^{h} g(z, z') K_1(z, z') \, dz', \quad (5)$$

where

$$g(z, z') \doteq I_z(z')/I_z(z) \quad (6)$$

is determined from the zeroth-order distribution and where z is essentially that reference value z_r where I_z is a maximum. In the relatively complicated series form of solutions for current encountered heretofore, this definition is convenient and sufficiently accurate. Actually, it would be more accurate to choose the complex value of $\Psi(z)$ at the reference point rather than its magnitude. This was not done to avoid the increased complication. It was felt that the higher-order terms would make the solution sufficiently accurate with Ψ real, since the imaginary part is quite small.

Since no higher-order terms are to be retained with (4), some improvement in accuracy should be achieved if a complex value Ψ of $\Psi(z)$ is selected instead of the magnitude Ψ. Moreover, Ψ may be evaluated separately for each term with the appropriate distribution of current. Specifically, for the first term, $\Psi(z)$ is defined by (5) with $g(z, z') = e^{jq_0 z'}/e^{jq_0 z}$; for the second term, $g(z, z') = \cos \beta_0 z'/\cos \beta_0 z$; for the third term, $g(z, z') = \sin \beta_0 z'/\sin \beta_0 z$. Let the appropriate complex values of $\Psi(z)$ at the corresponding reference points be defined as follows:

$$\Psi_q = \int_{-h}^{h} (e^{jq_0 z'}/e^{jq_0 z}) K_1(z, z') \, dz'$$
$$\text{at } z = z_{rq}, \quad (7)$$

$$\Psi_{c1} = \int_{-h}^{h} (\cos \beta_0 z'/\cos \beta_0 z) K_1(z, z') \, dz'$$
$$\text{at } z = z_{r1}, \quad (8)$$

$$\Psi_{c2} = \int_{-h}^{h} (\sin \beta_0 z'/\sin \beta_0 z) K_1(z, z') \, dz'$$
$$\text{at } z = z_{r2}, \quad (9)$$

where the reference point z_r is chosen in each case where the corresponding current is sufficiently large.

The function $\Psi_q(z)$ has been evaluated in Sec. 7. It is the first integral in (7.15) which is expressed in (7.20) in terms of the tabulated function $E_a(h, z)$. Specifically, using (7.17a),

$$\Psi_q(z) = E_{a_q}(h_q, z_q), \quad (10)$$

where

$$a_q = a(1 - q_0/\beta_0), \quad h_q = h(1 - q_0/\beta_0),$$
$$z_q = z(1 - q_0/\beta_0). \quad (11)$$

This function has been evaluated in terms of integral sines and cosines in the literature,[39] but the formula is very long. It is not reproduced since the function $E_a(h, z)$ is readily expressed in a short formula in terms of newly tabulated functions (II.19.36).

The function $\Psi_{c1}(z)$ has been evaluated in the following form [see (II.19.3)]:

$$\Psi_{c1}(z) = C_a(h, z)/\cos \beta_0 z. \quad (12)$$

The function Ψ_{c2} has not been used explicitly before. However, it is given by

$$\Psi_{c2}(z) = \mathscr{S}_a(h, z)$$
$$= \frac{-j}{2}(I_1 - I_2 - I_3 + I_4), \quad (13)$$

where I_1, I_2, I_3, and I_4 are defined by (II.19.32a, b, c, d). The function $\mathscr{S}_a(h, z)$ differs from $S_a(h, z)$ defined in (II.19.4) by having opposite signs for I_3 and I_4.

Suitable mean values of the three functions are:

$$\overline{\Psi_q(z)} = \Psi_q = E_{aq}(h_q, 0)$$
$$\doteq \Omega' + 2 \ln (1/\sin \Theta_2) - j\pi, \quad (14a)$$

$$\overline{\Psi_{c1}(z)} = \overline{\Psi_{c2}(z)} = \Psi_c$$
$$= \begin{cases} C_a(h, h - \lambda_0/4) & (h \geq \lambda_0/4) \\ C_a(h, 0) & (h \leq \lambda_0/4) \end{cases}$$
$$\doteq \Omega' - 2\Delta + 1 - j\pi/2, \quad (14b)$$

where

$$\Omega' = 2 \ln (2/\beta_0 a) - 1.154,$$
$$\Delta = 0.712 - \tfrac{1}{2} \ln \beta_0 h. \quad (15)$$

The asymptotic formulas on the right in (14a, b) were evaluated by Van Vleck, Bloch, and Hamermesh.[39] The function $E_a(h, z)$ has real and imaginary parts sensibly constant for all values of z except near $z = h$, where the function drops approximately to halfvalue as shown in Figs. II.19.5, 6. The behavior of $C_a(h, z)/\cos \beta_0 z$ is described in Sec. II.20 and is shown in Fig. II.20.1 for $\beta_0 h = \pi/2$.

The functions $F_{1s}(h)$ and $F_{1a}(h)$ in (4) are defined as follows:

$$F_{1s}(h) = -\int_{-h}^{h} (\cos \beta_0 z' - \cos \beta_0 h)$$
$$\times K_1(h, z') \, dz'$$
$$= -C_a(h, h) + E_a(h, h) \cos \beta_0 h, \quad (16a)$$

$$F_{1a}(h) = -\int_{-h}^{h} (\sin \beta_0 z' - \sin \beta_0 h)$$
$$\times K_1(h, z') \, dz'$$
$$= -\mathscr{S}_a(h, h) + E_a(h, h) \sin \beta_0 h. \quad (16b)$$

Substitution of Ψ_k for Ψ in the first term of (4) and of Ψ_c for Ψ in the second and third terms, together with (16a, b), gives

$$(I_z)_{01} = \frac{-j4\pi U}{\zeta_0} \left[\frac{e^{jq_0 z}}{\Psi_q} \right.$$
$$- \frac{\cos q_0 h \cos \beta_0 z}{\Psi_c + E_a(h, h) \cos \beta_0 h - C_a(h, h)}$$
$$\left. - j \frac{\sin q_0 h \sin \beta_0 z}{\Psi_c + E_a(h, h) \sin \beta_0 h - \mathscr{S}_a(h, h)} \right]. \quad (17)$$

This formula may be expressed as follows:

$$(I_z)_{01} = \frac{-j4\pi U}{\zeta_0} [(F' + jF'')e^{jq_0 z}$$
$$+ 2(G' + jG'') \cos q_0 h \cos \beta_0 z$$
$$+ 2j(H' + jH'') \sin q_0 h \sin \beta_0 z], \quad (18)$$

where

$$F' + jF'' = \Psi_q^{-1}, \quad (19a)$$

$$2(G' + jG'') = -[\Psi_c + E_a(h, h) \cos \beta_0 h$$
$$- C_a(h, h)]^{-1}, \quad (19b)$$

$$2(H' + jH'') = -[\Psi_c + E_a(h, h) \cos \beta_0 h$$
$$- \mathscr{S}_a(h, h)]^{-1}. \quad (19c)$$

Asymptotic formulas for $F' + jF''$, $G' + jG''$, and $H' + jH''$ were obtained by Van Vleck, Bloch, and Hamermesh in a somewhat different but essentially equivalent manner. Their value for $F' + jF''$ is identical with (19a). No attempt has been made to verify whether $G' + jG''$ and $H' + jH''$ in (19b, c) reduce exactly to their asymptotic values or not. There can be at most small differences because the sequences in the analysis are not identical, and numerous approximations are made. The Van Vleck, Bloch, and Hamermesh formulas for the asymptotic values of (19), which are good approximations for antennas that are not too short ($\beta_0 h \gtrsim \pi/2$) and not too thin ($\Omega \geq 15$), are:

$$F' + jF'' = (\Omega' + j\pi)/(\Omega'^2 + \pi^2), \quad (20)$$

$$2G' = \Phi(\beta_0 h)/[\Phi^2(\beta_0 h) + \chi^2(\beta_0 h)]$$
$$- \pi G''/\Omega', \quad (21a)$$

$$2G'' = \chi(\beta_0 h)/[\Phi^2(\beta_0 h) + \chi^2(\beta_0 h)], \quad (21b)$$

$$2H' = \Phi(\beta_0 h - \pi/2) [\Phi^2(\beta_0 h - \pi/2)$$
$$+ \chi^2(\beta_0 h - \pi/2)] - \pi H''/\Omega', \quad (22a)$$

$$2H'' = \chi(\beta_0 h - \pi/2) [\Phi^2(\beta_0 h - \pi/2)$$
$$+ \chi^2(\beta_0 h - \pi/2)], \quad (22b)$$

where

$$\Phi(x) = -(\Omega' - \Delta) \cos x + \tfrac{1}{4}\pi \sin x, \quad (23a)$$

$$\chi(x) = \tfrac{1}{2}(\ln 4\beta_0 h + 0.577) \sin x - \tfrac{1}{4}\pi \cos x, \quad (23b)$$

and where Δ is given in (15). The formula for Ψ_q, as given in (14a), is simplified further by omitting the term $2 \ln (1/\sin \Theta_2)$. It is found that the omitted terms are negligible. Note that all dependence upon the angle Θ_2 is contained in the three terms in (17) in which

$q_0 = \beta_0 \cos \Theta_2$ appears explicitly. The terms $-\pi G''/\Omega'$ and $-\pi H''/\Omega'$ occurring at the ends in (21a) and (22a) are important only very near resonance, where G'' and H'' are comparatively large.

Resonance occurs when the reradiated field has a maximum, as the wavelength is varied. This occurs essentially when the denominators of the fractions in (21a, b) and (22a, b) are as small as possible. Approximate values are given by

$$\cot \beta_0 h \doteq \pi/4(\Omega' - \Delta), \qquad (24a)$$

or

$$-\tan \beta_0 h \doteq \pi/4(\Omega' - \Delta). \qquad (24b)$$

More accurate formulas are in the literature.[39]

The formula (17) or (18) for the distribution of current in the parasitic antenna is like the zeroth-order formula in its three terms with simple trigonometric dependence upon $q_0 z$ and $\beta_0 z$. The coefficients of these terms are different in the modified zeroth-order solution from the true zeroth-order formula. Note that (17) or (18) includes both the even and the odd currents. Their distributions are readily visualized.

13. Reflecting or Back-Scattering Cross Section of Reradiating Antennas

A parasitic antenna No. 2 has its center at a great distance R_0 from the center of a transmitting antenna No. 1. The vector $\mathbf{R}_0$ is perpendicular to the transmitting antenna; the parasitic antenna is inclined at an arbitrary angle. Let the angle between the two planes defined by each of the two antennas and $\mathbf{R}_0$ be ψ; the projection of antenna 2 on the plane containing antenna 1 and $\mathbf{R}_0$ makes an angle $\bar{\Theta}_2 = \frac{1}{2}\pi - \Theta_2$ with the axis of antenna 1. The arrangement is illustrated in Fig. 13.1 with $\psi = 0$. If $\psi \neq 0$, antenna 2 in the figure is rotated through angle ψ with $\mathbf{R}_0$ as axis.

The far-zone field maintained by the transmitting antenna at the center of the parasite is called the "incident" field and is denoted by E_i. Clearly, in the notation previously used,

$$E_i \equiv E^r_{21\Theta_1}. \qquad (1)$$

This field is perpendicular to $\mathbf{R}_0$ at the parasite and is actually parallel to the transmitting antenna. Hence, this field is at an angle ψ with respect to the plane containing the parasite and $\mathbf{R}_0$, and its projection onto this plane makes an angle Θ_2 with the axis of the parasite.

The "incident" field E_i induces a current I_{2z} in the parasite which is given by (12.17)

or (12.18). This current, in turn, maintains an electric field which is called the *reflected*, *reradiated*, or *scattered* field. At a distance R_0 from the center of the parasite it is defined as follows:

$$E^r_{12\Theta_2} = \frac{j\omega \sin \Theta_2}{4\pi v_0} \frac{e^{-j\beta_0 R_0}}{R_0} \int_{-h}^{h} I'_{2z} e^{j\beta_0 z' \cos \Theta_2} dz'. \qquad (2)$$

The value of this field at the center of the transmitting antenna is denoted by E_r. Note that this field is perpendicular to $\mathbf{R}_0$, and that it makes an angle ψ with the transmitting antenna.

The "incident" Poynting vector at the center of the parasite due to the current in the transmitter has only a single component S_{iR} along $\mathbf{R}_0$. It is given by

$$S_{iR} \equiv E_i^2/2\zeta_0. \qquad (3)$$

Similarly, the "reflected" or "scattered" Poynting vector at radius R_0 from the center of the parasite also has only the radial component given below:

$$S^r_{12R}(\Theta_2) \equiv (E^r_{12\Theta_2})^2/2\zeta_0. \qquad (4a)$$

The particular value at the center of the transmitter is given by

$$S_{rR} \equiv E_r^2/2\zeta_0 \qquad (4b)$$

The total power reradiated or scattered by the parasite is

$$(T_2)_{\text{total}} = \int_{\Sigma \text{ (closed)}} S^r_{12R}(\Theta_2) R_0^2 \sin \Theta_2 d\Theta_2 d\Phi_2$$

$$= 2\pi R_0^2 \int_0^{\pi} S^r_{12R}(\Theta_2) \sin \Theta_2 d\Theta_2. \qquad (5)$$

The last step in (5) is possible in the case at hand since the reradiated electromagnetic field is rotationally symmetric if the parasite is a linear radiator of sufficiently small radius a so that $\beta_0 a \ll 1$. The ratio of (5) to (3) defines an area called the *scattering* or *reradiating cross section*. It is

$$\sigma_{\text{rad}} = \frac{(T_2)_{\text{total}}}{S_{iR}} = \left| \frac{\text{total power reradiated from parasite}}{\text{incident Poynting vector at center of parasite}} \right|. \qquad (6)$$

The power that would be reradiated by the parasite if it were an *isotropic* radiator and maintained the *same electric field at all points* on a great sphere of radius R_0 as at the center of the transmitting antenna is

$$T_{2I} = 4\pi R_0^2 S_{rR}. \tag{7a}$$

The power associated with the part of the scattered field that is polarized parallel to the transmitting antenna is

$$T_{2\parallel} = \cos^2 \psi (4\pi R_0^2) S_{rR}. \tag{7b}$$

The power associated with the part of the scattered field that is polarized perpendicular to the transmitting antenna is

$$T_{2\perp} = \sin^2 \psi (4\pi R_0^2) S_{rR}. \tag{7c}$$

The ratio of (7a, b, or c) to (3) defines an area called a *back-scattering cross section*. In terms of the total power given in (7a) it is

$$\sigma_I = 4\pi R_0^2 S_{rR}/S_{iR} = 4\pi R_0^2 (E_r/E_i)^2. \tag{8a}$$

If the transmitting antenna is used intermittently for reception or a separate receiving antenna parallel to the transmitter is used, the observable back-scattering cross section of the parasite involves only that part of its field which is polarized parallel to the receiving antenna and, hence, to the transmitting antenna. The *back-scattering cross section for parallel polarization* is obtained using (7b). It is

$$\sigma \equiv \sigma_\parallel = \cos^2 \psi (4\pi R_0^2)(E_r/E_i)^2. \tag{8b}$$

This is the conventional back-scattering or radar cross section. If a receiving antenna at the transmitter is perpendicular to the transmitting antenna, the observable back-scattering cross section involves only that part of the field of the parasite that is perpendicular to the transmitting antenna. It is defined using (7c) as follows:

$$\sigma_\perp = \sin^2 \psi (4\pi R_0^2)(E_r/E_i)^2. \tag{8c}$$

(Note that any distance R may be used in place of R_0 if E_r is the far-zone field of the parasite at that distance.)

The back-scattering cross section is seen to be the ratio of the power reradiated by a fictitious omnidirectional antenna to the incident Poynting vector when the field maintained by this omnidirectional source at the receiver is the same as that maintained by the actual scattering antenna. Depending upon whether the total field or the component in one of two directions of polarization is specified, different cross sections are defined. Evidently, $\sigma_I = \sigma_\parallel + \sigma_\perp$.

With E_r given by (2), it follows that

$$E_r^2 = E_r E_r^* = \left(\frac{\omega \sin \Theta_2}{4\pi v_0 R_0}\right)^2 \left|\int_{-h}^{h} I_z' \, dz'\right|^2. \tag{9}$$

With (11.18), and noting from (3.34) that

$$U = -(E_i \cos \psi)/\beta_0 \sin \Theta_2, \tag{10}$$

it follows that

$$\sigma = 4\pi \cos^4 \psi \left|\int_{-h}^{h} [(F' + jF'')e^{jq_0 z'} \right.$$
$$+ 2(G' + jG'') \cos q_0 h \cos \beta_0 z'$$
$$\left. + 2j(H' + jH'') \sin q_0 h \sin \beta_0 z'] \, dz'\right|^2. \tag{11}$$

This integrates into the following:[39]

$$\sigma = (4\lambda_0^2/\pi) \cos^4 \psi \, \{a_1^2(F'^2 + F''^2)$$
$$+ (a_2 + a_3)^2 (G'^2 + G''^2) \cos^2 q_0 h$$
$$+ (a_2 - a_3)^2 (H'^2 + H''^2) + 2(a_2^2 - a_3^2)$$
$$\times (G'H' + G''H'') \sin q_0 h \cos q_0 h$$
$$+ 2a_1(a_2 + a_3)(F'G' + F''G'') \cos q_0 h$$
$$+ 2a_1(a_2 - a_3)(F'H' + F''H'') \sin q_0 h\}, \tag{12}$$

where

$$a_1 = \frac{\sin 2\beta_0 h}{2 \cos \Theta_2}, \quad a_2 = \frac{\sin [\beta_0 h(1 + \cos \Theta_2)]}{1 + \cos \Theta_2},$$

$$a_3 = \frac{\sin [\beta_0 h(1 - \cos \Theta_2)]}{1 - \cos \Theta_2}. \tag{13}$$

Evidently,

$$\sigma_\perp = \sigma \tan^2 \psi, \tag{14}$$

where σ is given by (12).

If the "incident" electric field due to the current in the transmitting antenna induces currents in a bundle of parasitic antennas oriented in such a manner that Θ_2 is the same for all but ψ is completely random, and their reradiated field is observed with a fixed receiving antenna, the following *average cross section* may be defined:

$$\sigma(\Theta_2) = \frac{1}{2\pi} \int_0^{2\pi} \sigma \, d\psi = \frac{\sigma(\psi = 0)}{2\pi} \int_0^{2\pi} \cos^4 \psi \, d\psi$$
$$= \tfrac{3}{8}\sigma(\psi = 0). \tag{15}$$

(This is also the average cross section observed for a single parasite if the receiving antenna rotates continuously about its center and the axis R_0.) It is assumed in (15) that the fixed

508   THEORY OF LINEAR ANTENNAS   [IV.13]

Fig. 13.1. Parasitic antenna (2) inclined at an arbitrary angle with respect to distant transmitter (1).

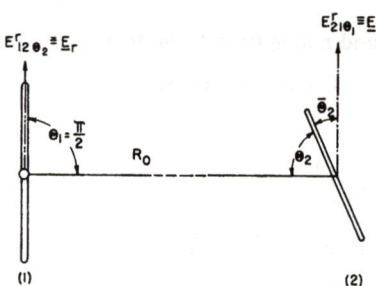

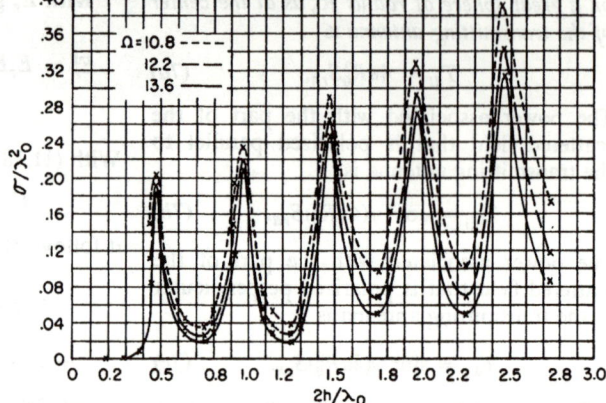

Fig. 13.2. Average back-scattering cross section of cylindrical wire (Van Vleck, Bloch, and Hammermesh).

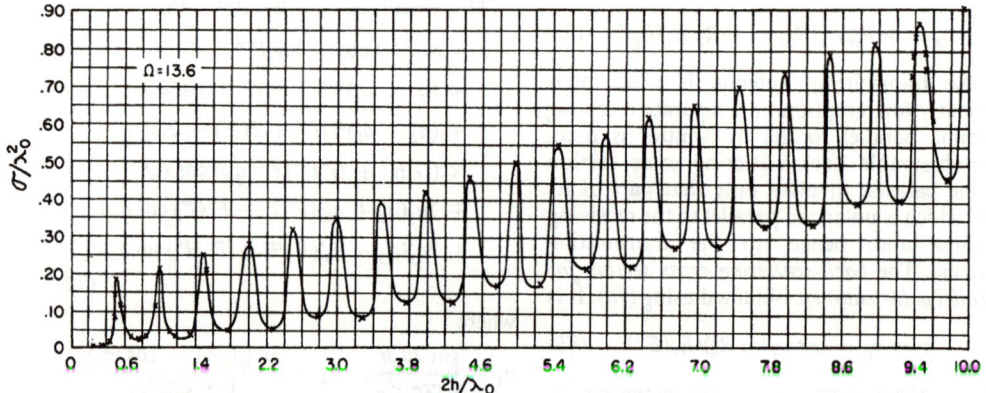

Fig. 13.3. Curves of Fig. 13.2 for greater range of $2h/\lambda_0$ (Van Vleck, Bloch, and Hammermesh).

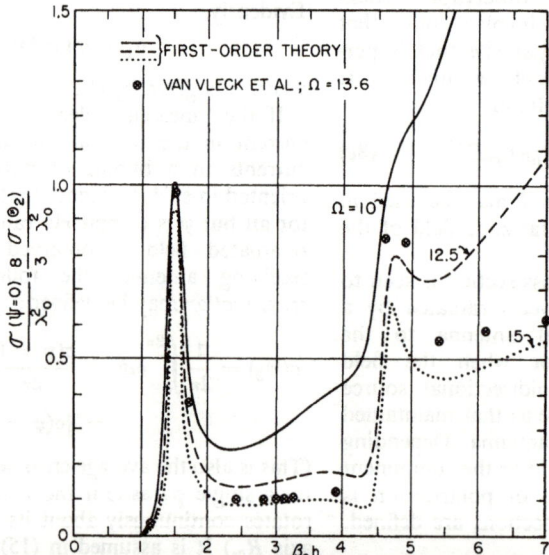

Fig. 13.4. Back-scattering cross section for broadside response; first-order King–Middleton expansion.

receiving antenna coincides with the transmitting antenna (intermittent operation) or has the same polarization.

If the fixed receiving antenna is perpendicular to the transmitting antenna, the observed average cross section is

$$\sigma_\perp(\Theta_2) = (\sigma_\perp/\cos^2\psi\sin^2\psi)\cdot\frac{1}{2\pi}\int_0^{2\pi}\cos^4\psi\,d\psi$$

$$= \tfrac{1}{8}\sigma(\psi = 0). \qquad (16)$$

Evidently,

$$\sigma_\perp(\Theta_2) = \sigma(\Theta_2)/3, \qquad (17)$$

and

$$\sigma = \frac{8}{3}\sigma(\Theta_2)\cos^4\psi. \qquad (18)$$

If the "incident" electric field induces currents in a bundle of parasitic antennas oriented completely at random, the *average back-scattering cross section* is obtained from σ by averaging. Thus,

$$\bar\sigma = \frac{1}{4\pi}\int_0^{2\pi}\int_0^\pi \sigma\sin\Theta_2 d\Theta_2 d\psi$$

$$= \frac{1}{2}\int_0^\pi \sigma(\Theta_2)\sin\Theta_2 d\Theta_2. \qquad (19)$$

The subscript $_\perp$ may be affixed to σ in (19).

The evaluation of $\bar\sigma$ from (19) using (12) and (14) has been carried out in terms of integral sine and cosine functions and elementary functions. The resulting expression is long and it is advantageous to make use of asymptotic formulas. The final approximate expression for the average scattering cross section $\bar\sigma$ as derived by Van Vleck, Bloch, and Hamermesh[39] is

$$\bar\sigma/\lambda_0^2 = (3/16\pi)\{F'^2 + F''^2][2\pi\beta_0 h - 1]$$
$$+ [G'^2 + G''^2][2\pi\beta_0 h - 1 + \ln 4\beta_0 h$$
$$+ 0.577 + (4\beta_0 h \ln 2 + \pi)\sin 2\beta_0 h$$
$$+ (\tfrac{1}{2} - 2\{\ln 4\beta_0 h + 0.577\})\cos 2\beta_0 h$$
$$- (\ln 2)\cos 4\beta_0 h] + [H'^2 + H''^2][2\pi\beta_0 h$$
$$- 1 + \ln 4\beta_0 h + 0.577 - (4\beta_0 h \ln 2$$
$$+ \pi)\sin 2\beta_0 h + (2\{\ln 4\beta_0 h + 0.577\}$$
$$- \tfrac{1}{2})\cos 2\beta_0 h - (\ln 2)\cos 4\beta_0 h]$$
$$+ 2[G'H' + G''H''][4\beta_0 h(\ln 2)\cos 2\beta_0 h$$
$$- \tfrac{1}{2}\sin 2\beta_0 h] + 4[F'G'$$
$$+ F''G''][(7\pi/4)\sin\beta_0 h - \tfrac{1}{2}(\ln 4\beta_0 h$$
$$+ 0.577)\cos\beta_0 h - \tfrac{1}{2}(\ln 2)\cos 3\beta_0 h]$$
$$+ 4[F'H' + F''H''][\tfrac{1}{4}\pi\cos\beta_0 h$$
$$- \tfrac{1}{2}(\ln 4\beta_0 h + 0.577)\sin\beta_0 h$$
$$+ \tfrac{1}{2}(\ln 2)\sin 3\beta_0 h]\}. \qquad (20)$$

Formula (20) is a good approximation only if the parasitic antenna is long enough compared with the wavelength, that is, if $\beta_0 h \geq \pi/2$. For shorter lengths asymptotic formulas are not satisfactory and the coefficients F, G, and H must be redefined. Formulas have been derived in the literature[39] but are not reproduced.

For very short lengths, $\beta_0 h \leq 0.3$, Van Vleck, Bloch, and Hamermesh have derived the following formulas:

$$\frac{\sigma}{\lambda_0^2} = \frac{(\beta_0 h)^6(1-\cos\Theta_2)^2\cos^4\psi}{9\pi[\ln(4h/a)-1]^2}, \qquad (21)$$

$$\frac{\bar\sigma}{\lambda_0^2} = \frac{(\beta_0 h)^6}{45\pi[\ln(4h/a)-1]^2}. \qquad (22)$$

Van Vleck, Bloch, and Hamermesh have evaluated $\bar\sigma/\lambda_0^2$. Points computed from (20) and (22) and curves estimated to fit these points are shown in Fig. 13.2 for three values of $\Omega = 2\ln(2h/a)$, namely, $\Omega = 10.8, 12.2, 13.6$, corresponding to $2h/a = 225, 450, 900$ for a range of $2h/\lambda_0$ up to 2.75. In Fig. 13.3 the range of $2h/\lambda_0$ is shown extended to 10 for $\Omega = 13.6$. The heights of the maxima in Fig. 13.2 depend on the ratio h/a; they increase with decreasing h/a. Numerical values of the first peak are $\bar\sigma/\lambda_0^2 = 0.184, 0.194, 0.204$ for $2h/a = 900, 450, 250$. Note that the resonance curves are sharper for larger values of $2h/a$, as would be expected. The ratio of resonant peak to antiresonant valley decreases with increasing $2h/\lambda_0$. Note that the valleys do not increase uniformly. The first and second valleys are about equal, as are the third and fourth, and so on.

Resonance maxima occur when h is somewhat less than $n\lambda_0/4$, $n = 1, 2, \cdots$. The displacement to values smaller than $n\lambda_0/4$ increases with decreasing values of $\Omega = 2\ln(2h/a)$. For an infinitely thin wire resonance should occur exactly at $\beta_0 h = n\pi/2$. The following numerical values for $\Omega = 13.6$ or $2h/a = 900$ have been computed:

$n =$	1	2	3	4	5
Theoretical resonance $\dfrac{4h}{\lambda_0}$	0.95	1.94	2.935	3.93	4.925
Experimental resonance $\dfrac{4h}{\lambda_0}$	0.96	1.88	2.94	3.90	4.92

The formula (12.17) or (12.18) for the current induced in the parasitic antenna includes two types of terms, the first term involving $q_0 z$ and the second and third involving $\beta_0 z$. Since the phase of the impressed electric field along the wire is $q_0 z$, the term with $q_0 z$ is called the *forced* part of the current. Since the amplitudes of the other terms become very great at resonance, they are called the *resonant* part of the current. An interesting observation in studying the average back-scattering cross section $\bar{\sigma}$ is the fact that its amplitude at resonance is determined almost entirely by the "resonant" part of the current, the contribution from the "forced" part being practically negligible for $n \doteq 1, 2, 3$. On the other hand, this is not true of the broadside response given by $\sigma(\psi = 0)/\lambda_0^2 = 8\sigma(\Theta_2 = \tfrac{1}{2}\pi)/3\lambda_0^2$ which is plotted for $\Omega = 13.6$ in Fig. 13.4 and over a more extended range in Fig. 14.6. Except for the first few resonances with $\beta_0 h \doteq n\pi/2$, $n = 1, 2, 3$, the "resonant" part of the current is so distributed that it contributes principally to ears of the reradiation-field pattern occurring at angles other than $\Theta_2 = \pi/2$. On the other hand, this is not true of the "forced" part of the current. Hence, even though the amplitude of the "resonant" part of the current is far greater than the amplitude of the forced part at the higher resonances, it is the forced part that contributes principally to the field *in the broadside direction*, $\Theta_2 = \pi/2$. As a numerical example, with $\Omega = 13.6$, the contribution to $\sigma(\Theta_2)/\lambda_0^2$ by the "resonant" terms in the current for resonances with h near $n\lambda_0/4$, $n = 1, 2, 3, 4$, are, respectively, 95, 54, 24, 11 percent.

The angular distribution of $\sigma(\Theta_2)/\lambda_0^2$ is shown in Figs. 13.5–13.8 for $\Omega = 13.6$ near three resonances with $h = \lambda_0/4$, $3\lambda_0/4$, and λ_0, and well removed from resonance in Fig. 13.8 with $h = 5\lambda_0/8$. For convenience $[\sigma(\Theta_2) \sin \Theta_2]/\lambda_0^2$ is plotted. The theoretical curves are in general agreement with experimental measurements.

As pointed out at the end of Sec. 11, the entire first-order current is given by (11.17a) with $q_0 = 0$, $H_{1s}(z) = H_{1s}(h) = 0$, when the scattering antenna is parallel to the electric field. For broadside scattering ($\Theta_2 = \pi/2$) this simplified current may be substituted in (2) and used to determine σ/λ_0^2 with the angle $\psi = 0$ as defined in (8b). The result is the following first-order formula:

$$\frac{\sigma}{\lambda_0^2}(\psi = 0) = \frac{|\beta_0 g_E(h)|^2}{\pi \Psi_s^2 |\Psi_s \cos \beta_0 h + F_{1s}(h)|^2}, \quad (23)$$

where the numerator has been evaluated by Dike and King[6] in the form

$$\begin{aligned}\beta_0 g_E(h) = [&4\Psi_s - 2\Omega + \beta_0 h \operatorname{Si} 4\beta_0 h \\ & - 4 \ln 2 + 2 \operatorname{Cin} 2\beta_0 h + \operatorname{Cin} 4\beta_0 h \\ & - j\beta_0 h \operatorname{Cin} 4\beta_0 h + j \operatorname{Si} 4\beta_0 h \\ & + j2 \operatorname{Si} 2\beta_0 h] \sin \beta_0 h + [2\beta_0 h(\Omega \\ & - 2\Psi_s + 2\ln 2) - \beta_0 h \operatorname{Cin} 4\beta_0 h \\ & - 2\beta_0 h \operatorname{Cin} 2\beta_0 h - 2 \sin 2\beta_0 h \\ & + \operatorname{Si} 4\beta_0 h - j\beta_0 h \operatorname{Si} 4\beta_0 h \\ & - j2\beta_0 h \operatorname{Si} 2\beta_0 h - j2 \cos \beta_0 h \\ & + j2 - j \operatorname{Cin} 4\beta_0 h] \cos \beta_0 h. \quad (24)\end{aligned}$$

Results computed from (23) by Dike using $\Psi_s = \Psi_U$ as defined in Sec. 6 are shown in Fig. 13.4 with $\Omega = 10$, 12.5 and 15. They are seen to be in general agreement with those obtained from the formula of Van Vleck, Bloch, and Hamermesh except near the second resonance ($\beta_0 h \doteq 3\pi/2$), where the Van Vleck formula gives a relatively greater maximum. For $\beta_0 h = \pi/2$ and π, (23) gives the values $\sigma/\lambda_0^2 \doteq 0.64$ and $4\pi/\Omega^2$, respectively.

Since the first-order formula (23) makes use of a zeroth-order current that does not include the component in phase with the incident field but only the component in phase quadrature with this field, it is not to be expected that it can yield accurate results under conditions in which it is the in-phase component that is of primary significance. For $\beta_0 h \geq 3\pi/2$ the resonant quadrature current contributes primarily to the ears in the field pattern that are in directions other than $\Theta_2 = \tfrac{1}{2}\pi$, and it is the in-phase component that maintains the field in the broadside direction. It follows that a more accurate determination of the broadside-scattering cross section for longer antennas requires either more than one iteration or a more accurate first-order formula obtained by using a complex distribution of current to obtain two real integral equations in the two components of the current in the manner described in Sec. II.31.

It is significant in this connection to note that the procedure carried out in Sec. 12 makes use of separate complex expansion parameters for the forced and resonant parts of the zeroth-order current. As a consequence the Van Vleck formula may be expected to be as accurate or even more accurate for long antennas than the first-order formula (23). However, neither formula is adequate to determine the detailed behavior of broadside back-scattering for long antennas.

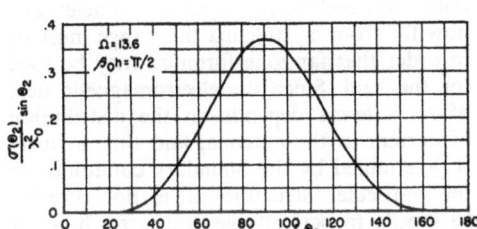

Fig. 13.5. Angular distribution of $\sigma(\Theta_2)/\lambda_0^2$, $\beta_0 h = \pi/2$ (Van Vleck, Bloch, and Hammermesh).

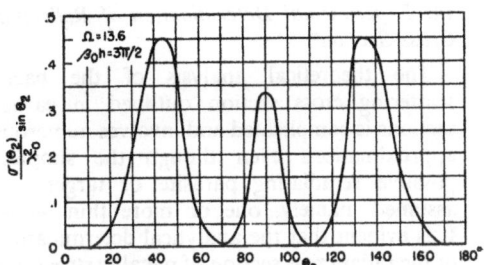

Fig. 13.6. Angular distribution of $\sigma(\Theta_2)/\lambda_0^2$, $\beta_0 h = 3\pi/2$ (Van Vleck, Bloch, and Hammermesh).

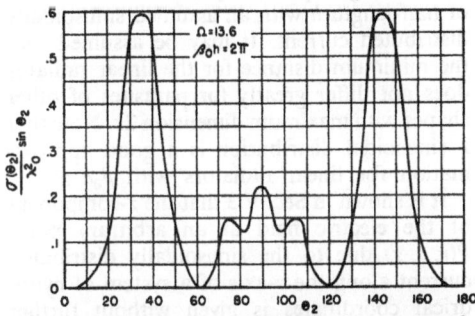

Fig. 13.7. Angular distribution of $\sigma(\Theta_2)/\lambda_0^2$, $\beta_0 h = 2\pi$ (Van Vleck, Bloch, and Hammermesh).

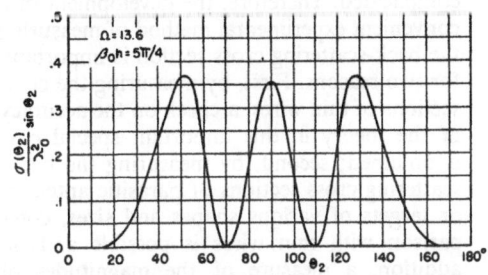

Fig. 13.8. Angular distribution of $\sigma(\Theta_2)/\lambda_0^2$, $\beta_0 h = 5\pi/4$ (Van Vleck, Bloch, and Hammermesh).

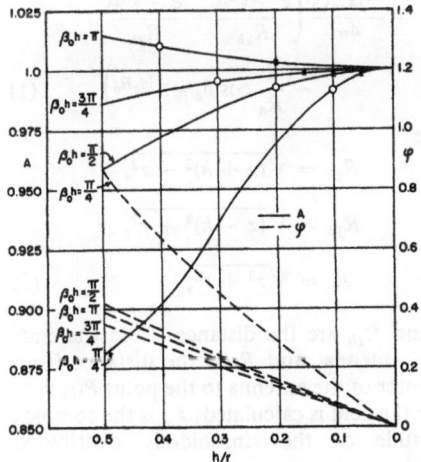

Fig. 14.1. The parameters A and φ as functions of h/r with $\beta_0 h$ as parameter (D. D. King).

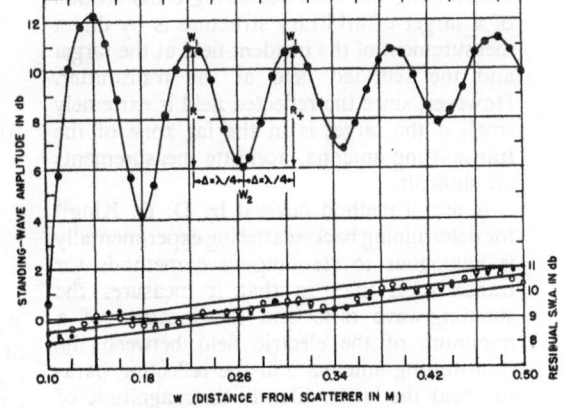

Fig. 14.2. Typical curve of residual standing waves and standing waves produced by scatterer (D. D. King).

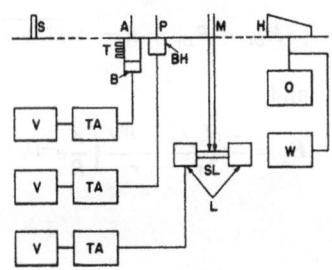

Fig. 14.3. Schematic diagram of the circuit for measuring back scattering. The following elements are shown: A, antenna under test; B, bolometer; BH, tuned bolometer holder; H, horn transmitter; L, matched crystal load; M, monitor array; O, modulated oscillator; P, movable probe; S, auxiliary scatterer; SL, slotted line; T, triple-stub tuner; TA, tuned amplifier; V, vacuum-tube voltmeter; W, coaxial wavemeter.

14. Experimental Determination of Reflecting Cross Section*

The theoretical analysis of the back-scattering cross section outlined in earlier sections is complicated and involves numerous approximations even though the simplest possible reradiating parasite or target was assumed, namely, one or more thin wires. It is evident that the analytical determination of the radar cross section of metallic structures of arbitrary shape is out of the question, since even a study of the distribution of surface current on such structures is extremely complicated. Therefore, the development of a convenient experimental method of measuring the back-scattering cross section is important, for two reasons. First, by measuring the cross sections of thin wires, a check on the accuracy of the theory in this important special case is obtained; second, by measuring the back-scattering cross sections of parasitic antennas or targets of various shapes and sizes, comparison with thin wires is possible and, in addition, a measure of the magnitudes of the induced currents, particularly currents associated with resonant modes, is made available.

Perhaps the most obvious method of determining the back-scattering cross section of a target of arbitrary structure is by direct measurement of the incident field at the target and the reflected field at the transmitter. However, since the reflected field is extremely small if the target is in the far zone of the transmitting antenna, accurate measurements are difficult.

A useful method devised by D. D. King[16] for determining back-scattering experimentally is analogous to standing-wave methods for transmission lines in that it measures the standing-wave ratio and the location of a minimum of the electric field between the transmitting antenna and the reflecting parasite near the latter. Clearly the magnitude of this standing-wave ratio is related to the magnitude of the electric field due to the currents in the parasite, and hence to the back-scattering cross section. In order to obtain reasonably simple formulas, it is necessary to assume that the electric fields due to the transmitter and to the parasite are essentially far-zone fields, that is, $1/R$ fields, in the range between the antennas,

where measurements are made. Before describing the method it is necessary to determine how far from an antenna this range must be in order that far-zone formulas may be used for the field. Since the electromagnetic field of an antenna depends on the distribution of current in the antenna, and this, in turn, is determined by the boundary conditions of the particular structure, it is not possible to specify in perfectly general terms how far from an antenna a probe must be in order to be in the far zone. However, a satisfactory estimate of the distance may be obtained by determining it specifically for a linear radiator of half-length h with an assumed sinusoidally distributed current. It may be assumed that the minimum distance for the linear radiator does not differ greatly for parasites of other shapes with maximum dimension $2h$. Note that a sinusoidal distribution is a good approximation for linear radiators with $\beta_0 h \leq 2$.

It is shown in Sec. V.3 that the z-component of the electric field at an arbitrary point $P(r, \theta, z)$ due to the sinusoidally distributed current along the z-axis of a system of cylindrical coördinates is given without further approximation by the following expression:

$$E_z = \frac{-j I_m \zeta_0}{4\pi} \left(\frac{e^{-j\beta_0 R_{1h}}}{R_{1h}} + \frac{e^{-j\beta_0 R_{2h}}}{R_{2h}} - \frac{2}{R_0} \cos \beta_0 h \, e^{-j\beta_0 R_0} \right), \quad (1)$$

where

$$R_{2h} = \sqrt{(z+h)^2 + r^2},$$
$$R_{1h} = \sqrt{(z-h)^2 + r^2},$$
$$R_0 = \sqrt{z^2 + r^2}; \quad (2)$$

R_{2h} and R_{1h} are the distances from the ends of the antenna, and R_0 is the distance from the center of the antenna to the point $P(r, \theta, z)$ where the field is calculated; I_m is the complex amplitude of the sinusoidally distributed current.

The field at $P(r, \theta, 0)$ in the mid-plane is obtained by setting $z = 0$ in (1) and (2) so that

$$R_{2h} = R_{1h} = R = \sqrt{h^2 + r^2}, \quad R_0 = r, \quad (3)$$

and

$$E_z = \frac{-j I_m \zeta_0}{2\pi} \frac{e^{-j\beta_0 r}}{r} \left(\frac{r}{R} e^{-j\beta_0(R-r)} - \cos \beta_0 h \right). \quad (4)$$

---

* Most of this section is based on the work of Dr. D. D. King, in ref. 16, and an unpublished discussion of this by Dr. A. L. Aden.

In the far zone, defined in general by

$$\beta_0 R \gg 1, \quad R \gg h. \quad (5)$$

and with

$$1 - \cos \beta_0 h = 2 \sin^2 \tfrac{1}{2}\beta_0 h, \quad (6)$$

the field reduces to the simple form

$$E_z^r = \frac{-jI_m\zeta_0}{\pi} \sin^2 \tfrac{1}{2}\beta_0 h \, \frac{e^{-j\beta_0 r}}{r}. \quad (7)$$

At points that are nearer the antenna than permitted by (5) but that are, nevertheless, far enough to satisfy the following inequality

$$h^4/4r^4 \ll 1, \quad (8)$$

the distance R in (3) may be expanded as follows:

$$R \doteq r\left(1 + \frac{1}{2}\frac{h^2}{r^2}\right). \quad (9)$$

Substitution of (9) in (4) using (8) gives

$$E_z = \frac{-jI_m\zeta_0}{2\pi} \frac{e^{-j\beta_0 r}}{r} \left[2 \sin^2 \tfrac{1}{2}\beta_0 h \right.$$
$$\left. - \frac{1}{2}\left(\frac{h}{r}\right)^2 (1 + j\beta_0 r)\right]. \quad (10)$$

This may be expressed as the far-zone field in (7) multiplied by a complex factor $Ae^{-j\phi}$ as follows:

$$E_z = E_z^r A e^{-j\phi}, \quad (11)$$

where

$$A = \sqrt{\left(1 - \frac{(h/r)^2}{4\sin^2 \tfrac{1}{2}\beta_0 h}\right)^2 + \left(\frac{\beta_0 h^2/r}{4\sin^2 \tfrac{1}{2}\beta_0 h}\right)^2}, \quad (12)$$

$$\phi = \tan^{-1}\left[\frac{\beta_0 h^2/r}{4\sin^2 \tfrac{1}{2}\beta_0 h - (h/r)^2}\right]. \quad (13)$$

Clearly, the range over which E_z is well approximated by E_z^r is defined by A sufficiently near unity and ϕ near 0; A and ϕ are shown in Fig. 14.1 as functions of h/r with $\beta_0 h$ as parameter. For the range of values of $\beta_0 h$ shown, A does not differ greatly from unity and ϕ not greatly from zero even at distances as small as one wavelength. Therefore, it may be concluded that for antennas that are near a half wavelength long, $1 \leq \beta_0 h \leq 2$, the far-zone formula is a good approximation of the field at distances exceeding one or two wavelengths from the antenna.

In order to derive a formula relating the standing-wave ratio of the electric field to the back-scattering cross section, let the transmitting antenna be located at $x = 0$, the parasitic antenna or scatterer at $x = l$. A movable probe has an intermediate position x ($0 < x < l$) along the line (x-axis) joining the source and the scatterer. Let $w = l - x$ be the distance to the probe from the scatterer. The signal received by the probe and transferred to its load (indicator) is proportional to the electric field at the probe. Periodic maxima and minima are observed as the probe is moved along the x-axis, as shown in Fig. 14.2; these define a standing-wave ratio given by

$$\text{SWR} = \frac{(E_z)_{\max}}{(E_z)_{\min}}, \quad (14)$$

where $(E_z)_{\max}$ is the magnitude of the total electric field due to currents in both the transmitting antenna and the scatterer at a nearby point w_1 of maximum amplitude, and $(E_z)_{\min}$ is the corresponding value at a nearby point w_2 of minimum amplitude. This ratio is a function of position along the x-axis, since the standing-wave pattern does not have a constant amplitude as it does on a lossless line. Let the distance from the minimum at w_2 to the adjacent maxima be Δ, so that

$$w_1 = w_2 \pm \Delta, \quad (15)$$

where $w_2 + \Delta$ locates a maximum at w_1 in the direction of the source from w_2, and $w_2 - \Delta$ in the direction of the scatterer. Since the amplitude does not vary rapidly these distances are nearly equal. Let

$$S_+ = \text{SWR with maximum at } w_2 + \Delta, \quad (16)$$

$$S_- = \text{SWR with maximum at } w_2 - \Delta. \quad (17)$$

Since it is assumed that the distances between the probe and the transmitter and the scatterer are sufficiently great so that far-zone formulas apply to the field at the probe, let the magnitude of the incident field at $x = l - w$ be expressed in the form

$$E_i(x) = \frac{kE_0}{x} = \frac{kE_0}{l - w}, \quad (18)$$

where k is a constant that is dimensionally a length and E_0 is the reference field at $x = k$ near the source. The magnitude of the incident field at the scatterer is

$$E_i = E_i(l) = \frac{kE_0}{l}. \quad (19)$$

The magnitude of the "reflected" field at w due to currents in the scatterer is

$$E_r(w) = \frac{lE_r}{w}, \qquad (20a)$$

where E_r is the magnitude of the reflected field at the transmitter. Alternatively,

$$E_r(w) = E_i \frac{b}{w} = kE_0 \frac{b}{wl}, \qquad (20b)$$

where

$$b \equiv l \frac{E_r}{E_i}. \qquad (21)$$

From the definition of the scattering cross section,

$$\sigma = 4\pi l^2 \left(\frac{E_r}{E_i}\right)^2 = 4\pi b^2. \qquad (22)$$

It remains to express b in terms of the standing-wave ratio. The maximum electric field at w_1 is composed of the contributions from currents in the transmitter—the incident field $E_i(w_1)$—and from the currents in the scatterer—the reflected field $E_r(w_1)$—when they combine *in phase*. Thus,

$$E_{\max} = E_i(w_1) + E_r(w_1) = \frac{kE_0}{l - w_1} + \frac{kE_0 b}{lw_1}. \qquad (23a)$$

Similarly, the minimum electric field at w_2 is composed of incident and reflected fields when they combine *in phase opposition*. Thus,

$$E_{\min} = E_i(w_2) - E_r(w_2) = \frac{kE_0}{l - w_2} - \frac{kE_0 b}{lw_2}. \qquad (23b)$$

Hence,

$$\text{SWR} = \frac{E_{\max}}{E_{\min}} = \frac{\dfrac{1}{l - w_1} + \dfrac{b}{lw_1}}{\dfrac{1}{l - w_2} - \dfrac{b}{lw_2}}$$

$$= \left[\frac{\dfrac{lw_1}{l - w_1} + b}{\dfrac{lw_2}{l - w_2} - b}\right] \frac{w_2}{w_1}. \qquad (24)$$

Substitution of the two values of the standing-wave ratio, S_+ for $w_1 = w_2 + \Delta$, S_- for $w_2 - \Delta$ in (24) gives

$$S_\pm = \frac{[l(w_2 \pm \Delta)/(l - w_2 \mp \Delta)] + b}{[lw_2/(l - w_2)] - b} \cdot \frac{w_2}{w_2 \pm \Delta} \qquad (25)$$

$$b = \left(\frac{lw_2}{l - w_2}\right) \frac{S_\pm - 1/[1 \mp \Delta/(l - w_2)]}{S_\pm + 1/(1 \pm \Delta/w_2)}. \qquad (26)$$

This may be put into a slightly more convenient form if the following inequality is satisfied:

$$\left(\frac{\Delta}{l - w_2}\right)^2 \ll 1. \qquad (27)$$

Subject to (27),

$$b \doteq \left(\frac{lw_2}{l - w_2}\right) \frac{S_\pm \pm [\Delta/(l - w_2)] - 1}{S_\pm + 1/(1 \pm \Delta/w_2)}. \qquad (28)$$

The phase shift ϕ at the scatterer is obtained from the relation

$$2\beta w_2 + \phi = (2n + 1)\pi. \qquad (29)$$

The standing-wave measurements actually carried out were made with transmitting and scattering antennas over a very large image plane made of copper screening. This arrangement eliminated the problem of supporting antennas in space and permitted the use of a transmission-line type of probe protruding through a slot in the conducting plane. With the very large size of the plane used, it was possible to have the highly directional transmitting antenna sufficiently far away from the probe so that the incident field approximated closely a plane wave with negligible change in amplitude over the region of measurement.

The electronic equipment consisted of a modulated oscillator and several tuned bolometer amplifiers with associated vacuum-tube voltmeters. Monitors were provided for frequency and power.

The scattering antenna under test was

mounted a short distance beyond the end of a slot cut in the image plane for the movable probe. The monitor was at the other end of the slot so that the standing-wave probe could travel a distance of several wavelengths between the monitor array and the scatterer. A block and schematic diagram is shown in Fig. 14.3.

In order to reduce standing waves on the image plane due to irregularities in its surface and its finite size, an artificial compensating irregularity in the form of an auxiliary scatterer was placed many wavelengths beyond the measuring position and the scatterer under test, and adjusted to reduce the standing-wave ratio over the measuring interval with the test antenna removed. Compensation for the remaining slight variations in amplitude over the measuring range was made by plotting the residual standing-wave pattern and using this to correct the pattern due to the scatterer under test. A typical curve of the residual standing waves is shown in Fig. 14.2 together with the standing waves produced by a scatterer consisting of a linear antenna with $\beta_0 h$ slightly greater than $\pi/2$.

Values of the coefficient $b(\sigma = 4\pi b^2)$ obtained for all values of the standing-wave ratios $S_\pm$ contained in the data of Fig. 14.2 are tabulated below.

w_2	b_-	b_+
0.430	3.79	3.94
.348	3.99	3.81
.265	3.57	3.69
.180	3.74	3.69

The average value is $b = 3.78$. The probable error of the average value of b is less than 1 percent; it is twice as large for σ. Values of σ/λ_0^2 for the dipole are plotted in Fig. 14.4 together with the corresponding value of ψ. The theoretical maximum indicated on the figure is obtained from Sec. 13. Its location is at $2h/\lambda_0 = 0.475$; from Fig. 13.2 this value is seen to be essentially independent of Ω. The magnitude of this maximum is obtained from Fig. 13.4. (Note that $\sigma/\lambda_0^2 = (8/3)\sigma(\Theta_2)/\lambda_0^2$ with $\psi = 0$.) It is seen that this maximum as well as the shapes of the curves in Figs. 13.4 and 14.4 are in good agreement. The dependence of σ/λ_0^2 on the load at the base of the scatterer is illustrated in Fig. 14.5 for a zero and matched load. The sharp resonance peak for $Z_L = 0$ is absent for the matched load.

The application of the standing-wave method for measuring scattering cross section has been limited in this discussion to simple dipoles with no load and with matched load over a large conducting plane. It is readily applied to other scatterers over a conducting plane and, with some modifications, to isolated scatterers in space.

The principal disadvantages of the standing-wave-ratio method for measuring scattering cross sections, that it is time consuming and requires the measurement of a small change in a given signal, are avoided in an alternative method developed by Sevick[33] and Morita. The essential feature of the method is the use of a hybrid junction to inject a signal into the circuit of the measuring probe which cancels exactly the signal received from the transmitter *in the absence* of the scatterer. When the scattering antenna to be tested is placed in position, the new signal measured by the probe is precisely that produced by the currents in the scattering antenna alone. Using a wavelength of 10.0 cm and a silvered scattering antenna erected on an outdoor ground screen of sheet aluminum 36 ft square, Sevick obtained the experimental data represented in Fig. 14.6. The ratio of the length of the measuring probe to the distance from the scatterer was 0.008 for $h \leq \lambda_0$ and 0.004 for $h > \lambda_0$. The peaks in Fig. 14.6 occur at resonance when large currents in phase quadrature with the exciting field are maintained on the antenna. The valleys between the peaks correspond to radiation from the forced current in phase with the exciting field. The level of these minima rises steadily with the increase in this component of current with length. As the length of the antenna is increased, a larger and larger fraction of the broadside field is maintained by the forced current, since the principal ears of the field maintained by the resonant currents are not in the broadside direction when $2h/\lambda_0$ equals or exceeds 1.5.

By interpolating between the curves of Fig. 13.4, the theoretical curve for the first-order formula (13.23) was obtained for $a/\lambda_0 = 3.5 \times 10^{-3}$ and plotted in Fig. 14.6. Good agreement with Sevick's curve is evident over the range of electrical length $\beta_0 h$ less than $3\pi/2$ for which the first-order theory may be expected to give satisfactory results, as explained at the end of Sec. 13. The theoretical points calculated from Van Vleck's formula also are shown, but since they correspond to a fixed value of $\Omega = 13.6$, they are not

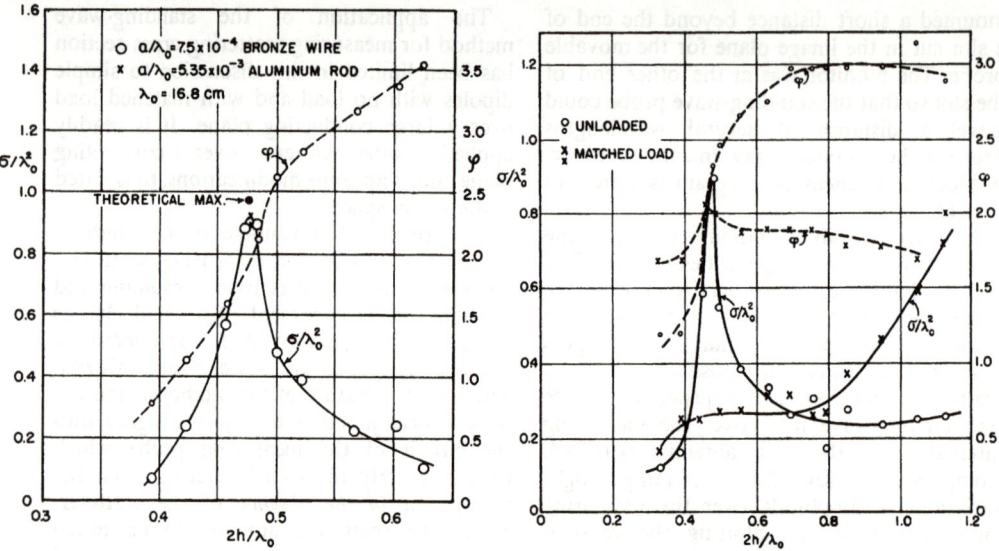

Fig. 14.4. Curves of σ/λ_0^2 and corresponding value of φ for the dipole (D. D. King).

Fig. 14.5. Curves of σ/λ_0^2 with no load and with matched load (D. D. King).

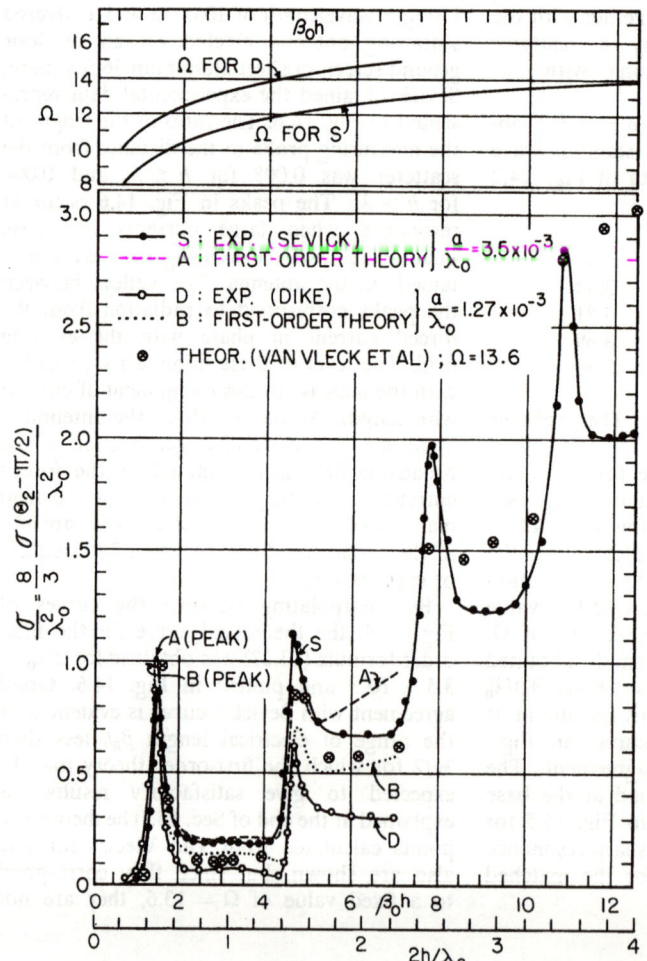

Fig. 14.6. Back-scattering cross section of single antenna.

directly comparable with the experimental curve below $\beta_0 h = 7$. However, since Van Vleck's formula is shown in Fig. 13.4 to agree with the first-order formula (13.28) over this range, it must, in fact, be in agreement with Sevick's experimental results. For longer antennas the Van Vleck formula yields the correct order of magnitude, but does not correctly represent the details of the curve, particularly off resonance.

Measurement of the back-scattering cross section also was carried out by Dike.[6] Some of his results also are shown in Fig. 14.6 together with the corresponding first-order theoretical curve interpolated from Fig. 13.4 for the appropriate value of a/λ_0. Since Dike's measurements are relative, they have been normalized to agree with Sevick's experimental curve at the first resonance. With due regard for the differences in radii, Dike's measurements are not in good agreement with either the theoretical curve or Sevick's measured values. The scattering cross section determined by Dike lies well below all other values when $\beta_0 h$ exceeds $\pi/2$. Dike made his measurements at $\lambda_0 = 10$ cm on an indoor ground screen 8 ft by 12 ft in size. He used rather large horns for both the transmitter and the receiver, with the latter placed at right angles to the former. The ratio of maximum dimension of either horn to the distance from the scatterer was about 0.15. There is no indication that the scattering antenna was silvered. Since Sevick found that a very large ground screen completely removed from walls and other obstacles, silvered scatterers, and a very small receiving probe at very large distances from the scattering antenna and from the transmitter are essential if accurate results are to be obtained, it appears probable that Dike's results are quantitatively less reliable than Sevick's.

15. Two Parallel, Nonstaggered Receiving and Scattering Antennas

The analysis of two identical parallel center-loaded receiving antennas in an arbitrarily oriented, linearly polarized electric field E has been carried out by Moritz.[23] Let the two receiving antennas be numbered 1 and 2. The half-length of each receiving antenna is h, its radius is a. They are parallel to the vertical z-axis with centers on the y-axis at $y = \pm b/2$, where the upper sign is for antenna 1, the lower sign for antenna 2. The vector from the origin of coördinates midway between the centers of the antennas to the distant transmitter is $\mathbf{R}_0$. The angle between $\mathbf{R}_0$ and the z-axis is Θ (it corresponds to Θ_2 in the case of the single receiving antenna). As in the analysis of the single antenna, ψ is the angle between the incident electric field and the plane defined by the z-axis and the vector $\mathbf{R}_0$. The new variable is the polar angle Φ measured in the horizontal plane from the positive x-axis toward the positive y-axis. With the newly defined quantity

$$p_0 = \beta_0 \sin \Phi, \quad (1a)$$

and the previously defined quantity

$$q_0 = \beta_0 \cos \Theta, \quad (1b)$$

the following simultaneous integral equations for the currents I_{1z} and I_{2z} are readily derived:

$$4\pi\nu_0 A_1(z) = \int_{-h}^{h} I'_{1z} K_a(z, z') \, dz'$$

$$+ \int_{-h}^{h} I'_{2z} K_b(z, z') \, dz'$$

$$= \frac{-j4\pi}{\zeta_0} [C_1^{(1)} \cos \beta_0 z + C_1^{(2)} \sin \beta_0 z$$

$$+ U e^{jq_0 z} e^{jp_0 b/2}], \quad (2a)$$

$$4\pi\nu_0 A_2(z) = \int_{-h}^{h} I'_{2z} K_b(z, z') \, dz'$$

$$+ \int_{-h}^{h} I'_{2z} K_a(z, z') \, dz'$$

$$= \frac{-j4\pi}{\zeta_0} [C_2^{(1)} \cos \beta_0 z + C_2^{(2)} \sin \beta_0 z$$

$$+ U e^{jq_0 z} e^{-jp_0 b/2}], \quad (2b)$$

where, as in Sec. 3,

$$U = -(E \cos \psi)/\beta_0 \sin \Theta. \quad (3)$$

Also,

$$K_i(z, z') = \frac{e^{-j\beta_0 R_i}}{R_i}, \quad i = a \text{ or } b, \quad (4a)$$

$$R_a = \sqrt{(z - z')^2 + a^2},$$
$$R_b = \sqrt{(z - z')^2 + b^2}. \quad (4b)$$

As in Sec. 11, each of the two equations (2a) and (2b) may be separated into an *axially symmetric* part (superscript s) involving even currents and vector potentials with respect to z for each antenna, and an axially antisymmetric part (superscript a) involving odd currents and vector potentials with respect to z for each antenna. These parts are

$$4\pi v_0 A_1^s(z)$$
$$= \int_{-h}^{h} I_{1z}^{s'} K_a(z, z')\, dz' + \int_{-h}^{h} I_{2z}^{s'} K_b(z, z')\, dz'$$
$$= \frac{-j4\pi}{\zeta_0}[C_1^s \cos \beta_0 z + \tfrac{1}{2} V_{10} \sin \beta_0 |z|$$
$$+ U \cos q_0 z e^{jp_0 b/2}], \quad (5a)$$

$$4\pi v_0 A_1^a(z)$$
$$= \int_{-h}^{h} I_{1z}^{a'} K_a(z, z')\, dz' + \int_{-h}^{h} I_{2z}^{a'} K_b(z, z')\, dz'$$
$$= \frac{-j4\pi}{\zeta_0}[C_1^a \sin \beta_0 z + jU \sin q_0 z e^{jp_0 b/2}], \quad (5b)$$

and two similar equations with subscripts 1 and 2 interchanged *and* with the factor $e^{-jp_0 b/2}$ in place of $e^{jp_0 b/2}$.

As in Sec. 2 of Chapter III, each of the equations (5a) and (5b) may be separated into a *laterally* symmetric part (subscript s) involving currents and vector potentials on antenna 1 that are equal to and *in phase* with the currents and vector potentials at corresponding points on antenna 2, and a laterally antisymmetric part (subscript a) involving currents and vector potentials on antenna 1 that are equal to but in phase *opposition* with currents and vector potentials at corresponding points on antenna 2. This separation is accomplished by setting

$$\begin{aligned} V_{10} &= -I_{10} Z_{1L} = V_s + V_a, \\ V_{20} &= -I_{20} Z_{2L} = V_s - V_a, \end{aligned} \quad (6)$$

and expanding $e^{\pm jp_0 b/2}$ into its trigonometric parts. Thus,

$$4\pi v_0 A_{1s}^s(z)$$
$$= \int_{-h}^{h} I_s^s(z')[K_a(z, z') + K_b(z, z')]\, dz'$$
$$= \frac{-j4\pi}{\zeta_0}[C_s^s \cos \beta_0 z + \tfrac{1}{2} V_s \sin \beta_0 |z|$$
$$+ U \cos q_0 z \cos (p_0 b/2)], \quad (7a)$$

$$4\pi v_0 A_{1a}^s(z)$$
$$= \int_{-h}^{h} I_a^s(z')[K_a(z, z') - K_b(z, z')]\, dz'$$
$$= \frac{-j4\pi}{\zeta_0}[C_a^s \cos \beta_0 z + \tfrac{1}{2} V_a \sin \beta_0 |z|$$
$$+ jU \cos q_0 z \sin (p_0 b/2)], \quad (7b)$$

$$4\pi v_0 A_{1s}^a(z)$$
$$= \int_{-h}^{h} I_s^a(z')[K_a(z, z') + K_b(z, z')]\, dz'$$
$$= \frac{-j4\pi}{\zeta_0}[C_s^a \sin \beta_0 z + jU \sin q_0 z \cos (p_0 b/2)], \quad (7c)$$

$$4\pi v_0 A_{1a}^a(z)$$
$$= \int_{-h}^{h} I_a^a(z')[K_a(z, z') - K_b(z, z')]\, dz'$$
$$= \frac{-j4\pi}{\zeta_0}[C_a^a \sin \beta_0 z - U \sin q_0 z \sin (p_0 b/2)]. \quad (7d)$$

Note that those parts of the currents in the two antennas that are due to the voltage V_s (equal in magnitude and in phase) and to the components of the electric field with the factor $\cos (p_0 b/2)$ are driven by identical voltages and fields, so that the currents must be equal. That is

$$I_{2s}^s = I_{1s}^s, \quad I_{2s}^a = I_{1s}^a. \quad (8a)$$

Similarly, those parts of the currents that are due to the equal and opposite voltages V_a and the components of the electric field with the factor $\sin (p_0 b/2)$ are driven by equal and *opposite* voltages and fields, so that the currents must be equal and opposite. That is,

$$I_{2a}^s = -I_{1a}^s, \quad I_{2a}^a = -I_{1a}^a. \quad (8b)$$

Relations corresponding to (8a) and (8b) are true of the components of the vector potentials $A_1(z)$ and $A_2(z)$.

By solving the four independent integral equations (7a) through (7b) for the four currents and combining these in the form

$$I_{1z} = I_s^s + I_s^a + I_a^s + I_a^a, \quad (9a)$$
$$I_{2z} = I_s^s + I_s^a - I_a^s - I_a^a, \quad (9b)$$

the unknown currents in (2a) and (2b) are determined. The symmetric and antisymmetric voltages are eliminated and expressed in terms of the voltage drops across the arbitrary impedances Z_{1L} and Z_{2L} at the centers of the two antennas. Since the integral equations (7a) through (7d) are the same in form as equations already solved earlier in this chapter, the solution of the simultaneous

[IV.15] THEORY OF LINEAR ANTENNAS

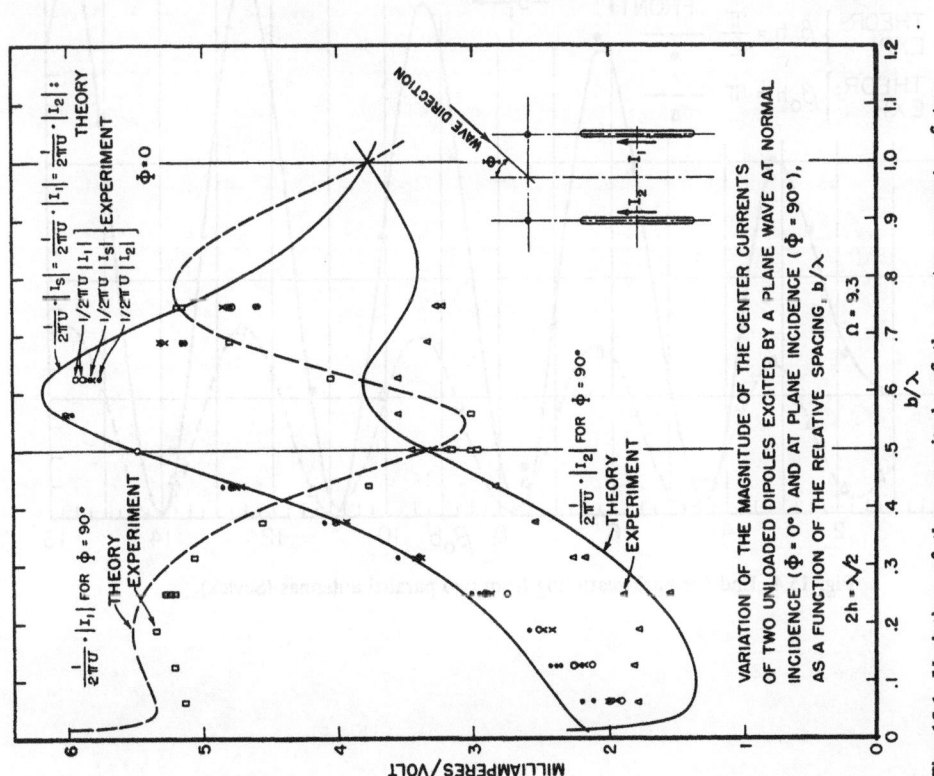

Fig. 15.1. Variation of the magnitude of the center currents of two unloaded dipoles excited by a plane wave at normal incidence ($\Phi = 0°$) and at plane incidence ($\Phi = 90°$) as a function of the relative spacing b/λ (Moritz).

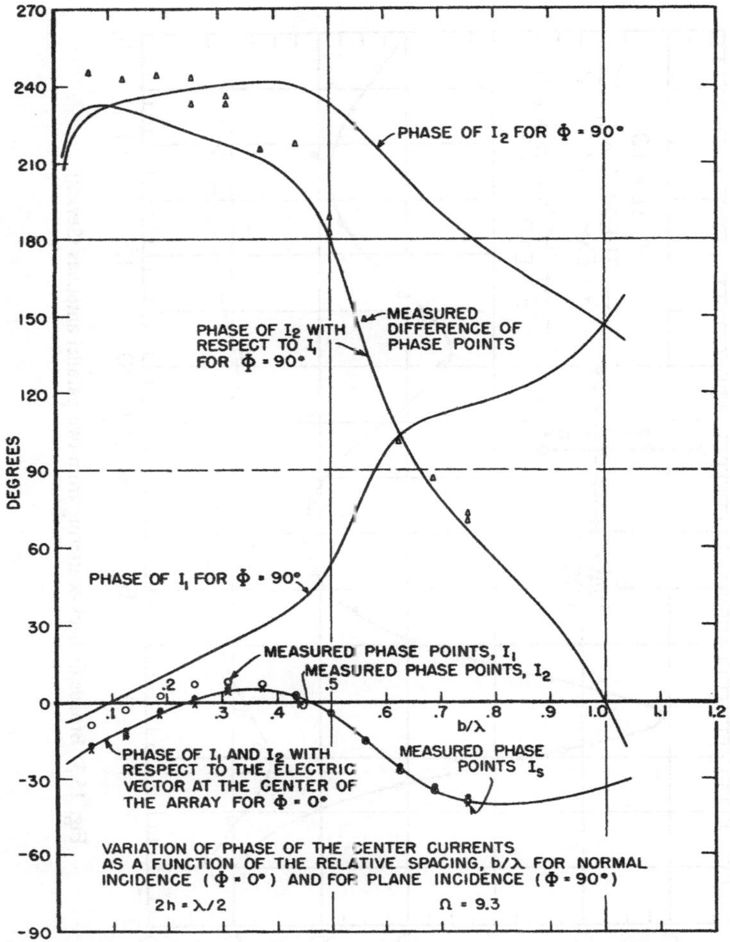

Fig. 15.2. Variation of the phase of the center currents as a function of the relative spacing b/λ for normal incidence ($\Phi = 0°$) and for plane incidence ($\Phi = 90°$) (Moritz).

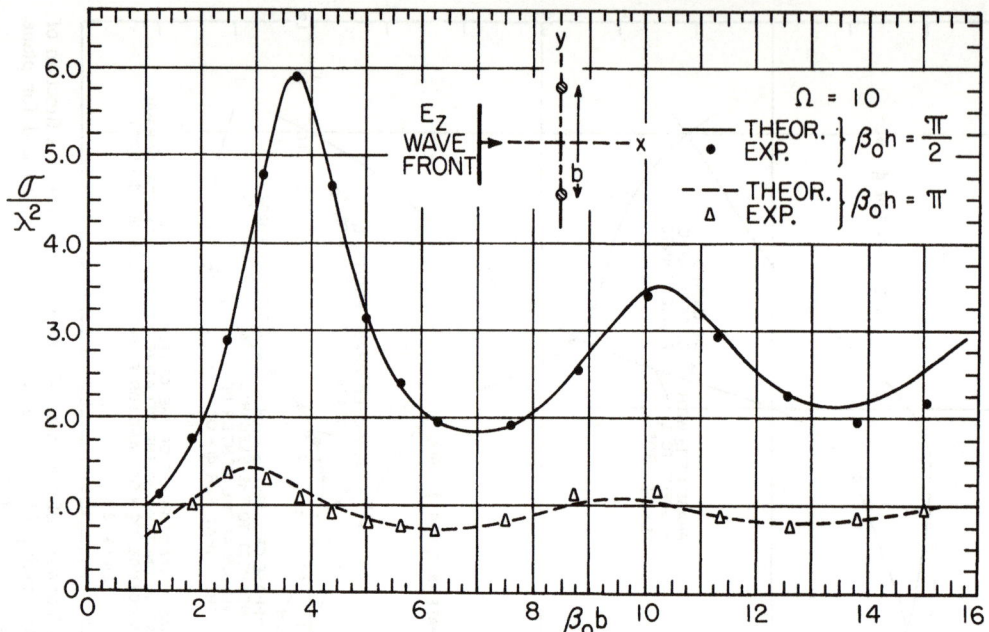

Fig. 15.3. Broadside back-scattering from two parallel antennas (Sevick).

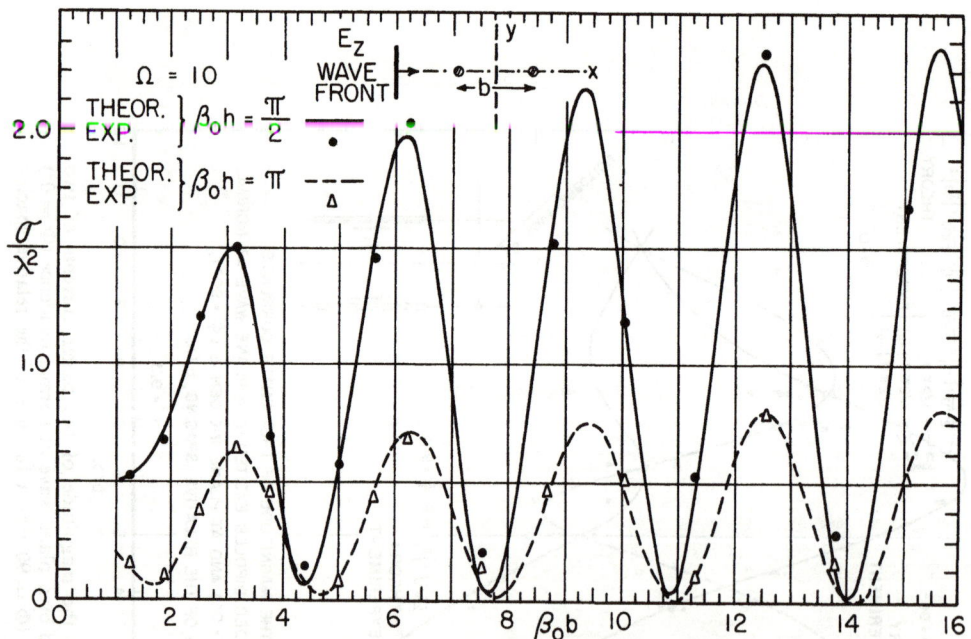

Fig. 15.4. End-fire back-scattering from two parallel antennas (Sevick).

integral equations (2a) and (2b) for the currents in the two coupled receiving antennas has been achieved.

Extensive computations of first-order currents and their experimental measurement have been carried out by Moritz for ranges of values of the half-length h and the separation b of the two antennas when $Z_{1L} = Z_{2L} = 0$. The agreement between theoretical and experimental distributions for the two unloaded coupled antennas as obtained by Moritz is very satisfactory. In general, the distributions in the coupled antennas differ little from those for the single receiving antenna as given in Sec. 8. The currents at the centers ($z = 0$) of the two antennas are shown in Figs. 15.1 and 15.2 as functions of the distance b between the antennas. The magnitudes of I_1 and I_2 are in Fig. 15.1, the phases in Fig. 15.2 for identical antennas with $h = \lambda_0/4$ and $\Omega = 2 \ln (2h/a) = 9.3$. Two sets of data are shown. The one is for $\Phi = 0$ (broadside reception), the other is for $\Phi = 90°$ (end-fire reception). When $\Phi = 0$ the individual currents, I_1 and I_2, are equal to the symmetric part, I_s, of the current. The agreement between theory and experiment is seen to be very good.

In addition to obtaining the currents in unloaded coupled receiving antennas, Moritz[23] has defined and evaluated symmetric (broadside), antisymmetric, and general complex effective lengths for the two-element receiving array. The symmetric and antisymmetric effective lengths consist of the product of a function of Θ and a trigonometric function of Φ. By the reciprocal theorem the complex effective electrical length is equal to the product of the vertical field factor (a function of Θ) and the two-element array factor (a function of Φ) of the same antenna when driven. Theoretical values of the effective length have been obtained by Moritz[23] using first-order currents in the unloaded antennas and second-order impedances just as for the single antenna. His theoretical data for antennas with $h = \lambda_0/4$, $\Omega = 2 \ln (2h/a) = 9.3$ and $\Phi = 0°$ and $90°$ do not differ greatly in their dependence upon Φ from the *zeroth-order* values which, by the reciprocal theorem, may be obtained by assuming a sinusoidally distributed current in a driven antenna (see Chapter VI). Moreover, they are in reasonably good agreement with his experimental values. In general, the product of the complex effective length $h_e(\Theta)$ of a single receiving antenna as given in Sec. 9 and the appropriate zeroth-order horizontal-array factor is a good approximation of the complex effective length of the two-element array. The same is necessarily true of an array of N parallel antennas. The zeroth-order horizontal array factors are evaluated in Chapter VI. The zeroth-order effective lengths or vertical field factors are evaluated in Sec. 9 and in Chapter V; first-order values are derived in Sec. 9.

The problem of reradiation or scattering from a two-element array in a uniform electric field has been analyzed by Sevick and Storer[35] using a variational method. Theoretical and experimental back-scattering cross sections have been determined[34] for $\beta_0 h = \pi/2$ and π and for a range of values of Φ and b. Results for broadside back-scattering and end-fire back-scattering as a function of b are in Figs. 15.3 and 15.4. Note that the back-scattering for $\beta_0 h = \pi/2$ is due to the currents shown in Figs. 15.1 and 15.2.

CHAPTER V

THE ELECTROMAGNETIC FIELD OF CENTER-DRIVEN AND MULTIPLE HALF-WAVE ANTENNAS

In the calculation of currents in electric circuits that are coupled by the interaction of moving charges in antennas, the use of the electromagnetic field with its boundary conditions is a mathematically unavoidable step. The actual evaluation of the field or of the associated potential functions is, however, not always necessary. For example, the determination of the distribution of current in the center-driven antenna involves both the boundary conditions of the electric field and the general integral for the vector potential, but neither the field nor the vector potential has to be evaluated. Similarly, in the analysis of two coupled antennas, the distribution of current, in particular, the input current is obtained with the aid of the general integral for the vector potential due to currents in cylindrical conductors, but without actually determining the vector potential or the electromagnetic field at points in space in the vicinity of the antennas. In the analysis of the receiving antenna, on the other hand, the restriction that the receiving antenna be in the far zone of the transmitter is imposed. This great separation makes the effect of currents in the receiving antenna on the currents in the driven antenna negligible, so that the electric field or the vector potential appears as a *factor* in the expression for the current in the receiving antenna. In this case, therefore, the final solution for the current involves directly the electric field in the *far zone* of the transmitter. If the receiving antenna is not in the far zone of the transmitter the electric field of the transmitter does not appear as a simple factor in the expression for the current. Because of the interaction between adjacent currents, simultaneous equations occur, and the problem is essentially that of coupled antennas as solved in Chapter III in the special cases of parallel and collinear arrangements of identical antennas. The general case of unlike antennas or even of identical antennas in an arbitrary orientation has not been solved except when the antennas are in the radiation zone with respect to each other.

According to the theorem of images (ref. I.31, Chap. IV) the electromagnetic field of a base-driven antenna of length h and radius a erected perpendicularly on an infinite, perfectly conducting plane is the same at *all points above the plane* as the field above the equatorial plane of an isolated, symmetric, center-driven antenna of *half*-length h and radius a. This is true provided the driving potential difference of the base-driven antenna is *one-half* that of the center-driven antenna. It follows that the entire analysis of the field of one or more center-driven antennas carried out in this and later chapters applies directly to the corresponding problem involving base-driven antennas on a perfectly conducting plane. This is discussed in greater detail in Chapter VII, where the antenna over an imperfectly conducting plane is studied and where it is shown that at broadcast frequencies the field in space of antennas over moist earth is well approximated by the field of the same antenna over a perfect conductor.

The electromagnetic field due to a distribution of current and charge in an antenna is given in Sec. I.9 in the form of integrals with the total axial current $\mathbf{I} = I_z \hat{\mathbf{z}}$, and the charge per unit length q written in the integrand. They are

$$\mathbf{E} = \frac{1}{4\pi\varepsilon_0} \int_{-h}^{h} \left[\frac{\hat{\mathbf{R}}}{R^2} q' + \frac{j\beta_0}{R}\left(q'\hat{\mathbf{R}} - \frac{\mathbf{I}'}{v_0} \right) \right] e^{-j\beta_0 R} \, dz',$$

$$\mathbf{B} = -\frac{1}{4\pi v_0} \int_{-h}^{h} (\hat{\mathbf{R}} \times \mathbf{I}') \left(\frac{1}{R^2} + \frac{j\beta_0}{R} \right) e^{-j\beta_0 R} \, dz'.$$

If the distribution of current and charge are known as functions of z', the electromagnetic field is formally determined. Since these integrals are in any case more intricate than the general integral for the vector potential from which they are derived, namely,

$$A = \frac{2}{4\pi v_0} \int_{-h}^{h} I_z' \frac{e^{-j\beta_0 R}}{R} \, dz',$$

it is simpler to evaluate A first, and from it determine B and E using $B = \text{curl } A$ and $E = -(j\omega/\beta_0^2)(\text{grad div } A + \beta_0^2 A)$. Since this procedure involves only differentiation, it is in some instances possible to evaluate B and E by differentiation of the definite integral for A without first carrying out the integration. This is illustrated in Secs. 2 and 3.

In order to evaluate the integral for the vector potential for an antenna of radius a it is necessary to make use of the solution obtained in Chapter II for the current in the antenna. It is

$$I_z = \frac{j2\pi V_0^e}{\Psi \zeta_0} \times$$

$$\left[\frac{\sin \beta_0(h - |z|) + M_1(z)/\Psi + M_2(z)/\Psi^2 + \cdots}{\cos \beta_0 h + A_1/\Psi + A_2/\Psi^2 + \cdots} \right].$$

If this expression is substituted in the integral for the vector potential, the result is so intricate that it has not been possible to obtain integrated solutions for either the vector potential or the field vectors. An approximate solution for the field of an antenna of finite radius may be obtained by making use of the zeroth-order distributions of the *components* of the current as discussed in Sec. II.25. The formula is

$$I_z = I_0'' \left(\frac{\cos \beta_0 z - \cos \beta_0 h}{1 - \cos \beta_0 h} \right)$$

$$+ j I_m' \sin \beta_0(h - |z|),$$

with I_0'' the input value at $z = 0$ of the component of current in phase with the driving potential difference, and I_m' the maximum value near $z = h - \lambda/4$ of the component of current in phase quadrature with the driving potential difference.

A zeroth-order approximation of the electromagnetic field of a thin antenna of finite radius is obtained using

$$I_z = I_m \sin \beta_0(h - |z|)$$
$$= I_0 \frac{\sin \beta_0(h - |z|)}{\sin \beta_0 h}$$

in the integral for A.

If interest is primarily in the field in the far zone, as is usual, an accurate method for determining the electromagnetic field of an antenna of finite radius is available. It depends upon the Rayleigh-Carson reciprocal theorem (ref. I.31, Sec. IV.21), which makes possible the determination of the field of a transmitting antenna in terms of its characteristics as a receiving antenna and vice versa. Since the receiving qualities of a symmetric, center-loaded receiving antenna are given quite accurately in Chapter IV, the far-zone field of a center-driven antenna may be determined from them.

In the analysis of the electromagnetic field of center-driven antennas and multiple half-wave antennas in this chapter all three of the above methods are used. First, the field of an infinitely thin antenna is investigated in detail, since it leads to an analytically and physically attractive and complete solution that is in any case a zeroth-order approximation of the actual field of physically realizable antennas. Next, the far-zone field is obtained from the properties of the receiving antenna. Finally, the complete field due to the approximate distribution of current involving both in-phase and quadrature components of current is investigated, and the results of all three methods are compared to show that the *magnitude* of the zeroth-order field in the far zone is an excellent approximation.

INFINITELY THIN CENTER-DRIVEN ANTENNAS—CYLINDRICAL COÖRDINATES

1. *The Vector Potential*

The vector potential of a symmetric antenna in space is given by

$$A_r \ll A_z, \quad A_\theta = 0,$$

$$A_z = \frac{1}{4\pi v_0} \int_{-h}^{h} I_z' \frac{e^{-j\beta_0 R}}{R} \, dz'. \quad (1)$$

In terms of the cylindrical coördinates r, θ, z in Fig. 1.1,

$$R \equiv R_1 = \sqrt{(z - z')^2 + r^2}. \quad (2a)$$

Let

$$R_2 = \sqrt{(z + z')^2 + r^2}. \quad (2b)$$

As usual, the primed coördinates refer to the element of integration dz', the unprimed coördinates locate the point of calculation at P, as shown in Fig. 1.1. The distance from P

to the upper end of the antenna at $z = h$ is

$$R_{1h} \equiv \sqrt{(h-z)^2 + r^2} = \sqrt{u_1^2 + r^2}, \quad (3a)$$

with

$$u_1 \equiv h - z. \quad (3b)$$

The distance from P to the lower end is

$$R_{2h} \equiv \sqrt{(h+z)^2 + r^2} = \sqrt{u_2^2 + r^2}, \quad (4a)$$

with

$$u_2 \equiv h + z. \quad (4b)$$

The distance from P to the center is

$$R_0 \equiv \sqrt{z^2 + r^2}. \quad (5)$$

The sinusoidal distribution in the infinitely thin antenna is

$$I_z = I_m \sin \beta_0(h - |z|)$$
$$= I_m(\sin \beta_0 h \cos \beta_0 z$$
$$\quad - \cos \beta_0 h \sin \beta_0 |z|), \quad (6)$$

where I_m is the maximum amplitude at $z = h - \lambda/4$.

Upon substituting (6) in (1) the following final expression is obtained for the vector potential at any point in space due to a center-driven symmetric antenna with sinusoidal current:

$$A_z = \frac{I_m}{4\pi v_0} [\sin \beta_0 h C_r(h, z) - \cos \beta_0 h S_r(h, z)]. \quad (7)$$

The functions $C_r(h, z)$ and $S_r(h, z)$ are defined below as in Chapter II:

$$C_r(h, z) \equiv \int_{-h}^{h} \cos \beta_0 z' \frac{e^{-j\beta_0 R_1}}{R_1} dz'$$
$$= \int_0^h \cos \beta_0 z' \left(\frac{e^{-j\beta_0 R_1}}{R_1} + \frac{e^{-j\beta_0 R_2}}{R_2} \right) dz', \quad (8)$$

$$S_r(h, z) \equiv \int_{-h}^{h} \sin \beta_0 |z'| \frac{e^{-j\beta_0 R_1}}{R_1} dz'$$
$$= \int_0^h \sin \beta_0 z' \left(\frac{e^{-j\beta_0 R_1}}{R_1} + \frac{e^{-j\beta_0 R_2}}{R_2} \right) dz, \quad (9)$$

except that in (8) and (9) above r appears in R_1 instead of a. It is seen from (7) that the general expression for the vector potential due to an infinitely thin antenna is a complicated function. In spite of the fact that the general integrals for the electric and magnetic vectors are more intricate than the general integral for the vector potential, the expressions for the electromagnetic vectors of a thin center-driven antenna with sinusoidally distributed current turn out to be much simpler than the vector potential in (7). They are derived in the following sections.

2. The Magnetic Field

The magnetic field is readily computed from the vector potential using the fundamental relation

$$\mathbf{B} = \operatorname{curl} \mathbf{A}. \quad (1)$$

In the present case

$$A_r \doteq 0, \quad A_\theta = 0, \quad (2)$$

and because rotational symmetry about the z-axis obtains, all derivatives with respect to θ vanish. Hence (1) reduces to

$$B_r = \operatorname{curl}_r \mathbf{A} = 0, \quad (3a)$$

$$B_\theta = \operatorname{curl}_\theta \mathbf{A} = -\frac{\partial A_z}{\partial r}, \quad (3b)$$

$$B_z = \operatorname{curl}_z \mathbf{A} = 0. \quad (3c)$$

With (1.7),

$$B_\theta = -\frac{I_m}{4\pi v_0} \left[\sin \beta_0 h \frac{\partial C_r(h, z)}{\partial r} - \cos \beta_0 h \frac{\partial S_r(h, z)}{\partial r} \right]. \quad (4)$$

The differentiation of the function $C_r(h, z)$ and $S_r(h, z)$ as defined in (1.8) and (1.9) may be carried out as follows. The exponential integral

$$Ei(v) = \int_\infty^{-v} \frac{e^{-u}}{u} du \quad (5)$$

can be differentiated with respect to a parameter r contained in v as follows:

$$\frac{\partial}{\partial r} Ei(v) = \frac{\partial}{\partial r} \int_\infty^{-v} \frac{e^{-u}}{u} du$$
$$= \left(\frac{\partial}{\partial v} \int_\infty^{-v} \frac{e^{-u}}{u} du \right) \frac{dv}{dr}. \quad (6)$$

Since u is an independent variable, the differentiation of the definite integral reduces to a single term obtained by substituting the upper limit in the integrand. Thus

$$\frac{\partial}{\partial r} Ei(v) = -\frac{e^v}{v} \frac{dv}{dr}. \quad (7)$$

Using (II.19.29–31) with U replaced by $\beta_0 u$ and A by $\beta_0 r$, the following expression is obtained after a change of variable:

$$\frac{\partial C_r(h, z)}{\partial r} = \tfrac{1}{2} e^{j\beta_0 z}\left(-\frac{e^{-j\beta_0(R_{2h}+u_2)}}{R_{2h}+u_2}\cdot\frac{r}{R_{2h}}\right.$$
$$\left.+\frac{e^{-j\beta_0(R_{1h}-u_1)}}{R_{1h}-u_1}\cdot\frac{r}{R_{1h}}\right)$$
$$+ \tfrac{1}{2} e^{-j\beta_0 z}\left(-\frac{e^{-j\beta_0(R_{1h}+u_1)}}{R_{1h}+u_1}\frac{r}{R_{1h}}\right.$$
$$\left.+\frac{e^{-j\beta_0(R_{2h}-u_2)}}{R_{2h}-u_2}\frac{r}{R_{2h}}\right). \quad (8)$$

Combining terms gives

$$\frac{\partial C_r(h, z)}{\partial r} = \frac{r}{2}\left[-\frac{e^{-j\beta_0(R_{2h}+h)}}{R_{2h}(R_{2h}+u_2)}\right.$$
$$+\frac{e^{-j\beta_0(R_{1h}-h)}}{R_{1h}(R_{1h}-u_1)}-\frac{e^{-j\beta_0(R_{1h}+h)}}{R_{1h}(R_{1h}+u_1)}$$
$$\left.+\frac{e^{-j\beta_0(R_{2h}-h)}}{R_{2h}(R_{2h}-u_2)}\right]. \quad (9)$$

Since

$$(R_{2h}+u_2)(R_{2h}-u_2) = R_{2h}^2 - u_2^2$$
$$= u_2^2 + r^2 - u_2^2 = r^2, \quad (10a)$$
$$(R_{1h}+u_1)(R_{1h}-u_1) = R_{1h}^2 - u_1^2$$
$$= u_1^2 + r^2 - u_1^2 = r^2, \quad (10b)$$

it is possible, by reducing to common denominators, to obtain

$$\frac{\partial C_r(h, z)}{\partial r} = \frac{1}{2r}\left\{\frac{e^{-j\beta_0 R_{2h}}}{R_{2h}}[(R_{2h}+u_2)e^{j\beta_0 h}\right.$$
$$-(R_{2h}-u_2)e^{-j\beta_0 h}] + \frac{e^{-j\beta_0 R_{1h}}}{R_{1h}}[(R_{1h}$$
$$\left.+u_1)e^{j\beta_0 h}-(R_{1h}-u_1)e^{-j\beta_0 h}]\right\}$$
$$= \frac{e^{-j\beta_0 R_{2h}}}{r}\left(\frac{u_2 \cos\beta_0 h}{R_{2h}}+j\sin\beta_0 h\right)$$
$$+\frac{e^{-j\beta_0 R_{1h}}}{r}\left(\frac{u_1 \cos\beta_0 h}{R_{1h}}+j\sin\beta_0 h\right)$$
$$-\frac{\cos\beta_0 h}{r}\left(\frac{u_2 e^{-j\beta_0 R_{2h}}}{R_{2h}}+\frac{u_1 e^{-j\beta_0 R_{1h}}}{R_{1h}}\right)$$
$$+j\frac{\sin\beta_0 h}{r}(e^{-j\beta_0 R_{2h}}+e^{-j\beta_0 R_{1h}}), \quad (11)$$

with

$$u_2 = h + z, \quad u_1 = h - z,$$
$$R_{2h} = \sqrt{u_2^2 + r^2}, \quad R_{1h} = \sqrt{u_1^2 + r^2}.$$

Similarly,

$$\frac{\partial S_r(h, z)}{\partial r} = \frac{j}{2} e^{j\beta_0 z}\left[-\frac{e^{-j\beta_0(R_{2h}+u_2)}}{R_{2h}+u_2}\frac{r}{R_{2h}}\right.$$
$$\left.-\frac{e^{-j\beta_0(R_{1h}-u_1)}}{R_{1h}-u_1}\frac{r}{R_{1h}}+2\frac{e^{-j\beta_0(R_0+z)}}{R_0+z}\frac{r}{R_0}\right]$$
$$+\frac{j}{2} e^{-j\beta_0 z}\left[-\frac{e^{-j\beta_0(R_{1h}+u_1)}}{R_{1h}+u_1}\frac{r}{R_{1h}}\right.$$
$$\left.-\frac{e^{-j\beta_0(R_{2h}-u_2)}}{R_{2h}-u_2}\frac{r}{R_{2h}}+2\frac{e^{-j\beta_0(R_0-z)}}{R_0-z}\frac{r}{R_0}\right].$$
$$(12)$$

Collecting terms gives

$$\frac{\partial S_r(h, z)}{\partial r} = \frac{j}{2r}\left\{-\frac{e^{-j\beta_0 R_{2h}}}{R_{2h}}[(R_{2h}+u_2)e^{j\beta_0 h}\right.$$
$$+(R_{2h}-u_2)e^{-j\beta_0 h}]$$
$$-\frac{e^{-j\beta_0 R_{1h}}}{R_{1h}}[(R_{1h}+u_1)e^{j\beta_0 h}$$
$$\left.+(R_{1h}-u_1)e^{-j\beta_0 h}] + 4e^{-j\beta_0 R_0}\right\}$$
$$= -j\frac{\cos\beta_0 h}{r}(e^{-j\beta_0 R_{2h}}+e^{-j\beta_0 R_{1h}})$$
$$+\frac{\sin\beta_0 h}{r}\left(\frac{u_2 e^{-j\beta_0 R_{2h}}}{R_{2h}}+\frac{u_1 e^{-j\beta_0 R_{1h}}}{R_{1h}}\right)$$
$$+\frac{j2}{r}e^{-j\beta_0 R_0}. \quad (13)$$

Upon combining (11) and (13) to form (4), and noting (3a, c), the following simple result is obtained:

$$B_r = 0, \quad (14a)$$

$$B_\theta = \frac{jI_m}{4\pi v_0 r}(e^{-j\beta_0 R_{1h}} + e^{-j\beta_0 R_{2h}}$$
$$- 2\cos\beta_0 h e^{-j\beta_0 R_0}), \quad (14b)$$

$$B_z = 0. \quad (14c)$$

This is the magnetic field of a center-driven antenna of half-length h with sinusoidally distributed current. In (14b),

$$R_{2h} = \sqrt{(z+h)^2 + r^2},$$
$$R_{1h} = \sqrt{(z-h)^2 + r^2},$$
$$R_0 = \sqrt{z^2 + r^2}; \quad (15)$$

R_{2h} and R_{1h} are the distances from the point where B_θ is calculated to the ends of the antenna; R_0 is the distance to the center.

An interesting application of (14b) is to calculate the magnetic field near the surface of a cylindrical conductor with a sinusoidal distribution of current. For a conductor of half-lengths $h = \lambda_0/4, 3\lambda_0/4, \cdots$, the magnitude of (14b) may be expressed as follows:

$$|\lambda_0 B_0/k|$$
$$= (\lambda_0/r)[\cos(\pi R_{1h}/\lambda_0)\cos(\pi R_{2h}/\lambda_0)$$
$$+ \sin(\pi R_{1h}/\lambda_0)\sin(\pi R_{2h}/\lambda_0)], \quad (16)$$

where k is a constant. The quantity defined in (16) is shown plotted in Figs. 2.1 and 2.2 for $h = \lambda_0/4, 3\lambda_0/4$ as a function of 360° z/λ_0 with r/λ_0 as parameter. For comparison, cosine curves are also drawn. It is seen that for sufficiently small values of r/λ_0 the magnetic field differs negligibly from a cosine curve except near the end of the conductor. Accordingly, a device such as a small loop, which has an induced voltage proportional to the average magnetic field linking it, may be used to measure the magnetic field near the conductors, and this field, in turn, is proportional to the magnitude of the current except near the ends, where the current is known to vanish.

3. The Electric Field

The electric field may be calculated from the general expression

$$\mathbf{E} = -\text{grad}\,\phi - j\omega\mathbf{A}$$
$$= -\frac{j\omega}{\beta_0^2}(\text{grad div }\mathbf{A} + \beta_0^2\mathbf{A}). \quad (1a)$$

At points in space where all currents vanish, the vector potential satisfies the relation

$$\nabla^2\mathbf{A} + \beta_0^2\mathbf{A} = \text{grad div }\mathbf{A} - \text{curl curl }\mathbf{A}$$
$$+ \beta_0^2\mathbf{A} = 0, \quad (1b)$$

so that

$$\mathbf{E} = -\frac{j\beta_0^2}{\omega}\text{curl curl }\mathbf{A} = -\frac{j\beta_0^2}{\omega}\text{curl }\mathbf{B}. \quad (1c)$$

This last form is identically the field equation

$$v_0\text{ curl }\mathbf{B} = \overline{\rho_m\mathbf{v}} + j\omega\varepsilon_0\mathbf{E}, \quad (2a)$$

solved for $\mathbf{E}$ at points in space. That is,

$$\mathbf{E} = \frac{-jv_0}{\omega\varepsilon_0}\text{curl }\mathbf{B}. \quad (2b)$$

It follows from (2.14a, b, c) that

$$\mathbf{B} = \hat{\boldsymbol{\theta}}B_\theta, \quad B_r = 0, \quad B_z = 0. \quad (3)$$

Hence,

$$\text{curl}_r\,\mathbf{B} = -\frac{1}{r}\frac{\partial}{\partial z}(rB_\theta) = -\frac{\partial B_\theta}{\partial z}, \quad (4a)$$

$$\text{curl}_\theta\,\mathbf{B} = 0, \quad (4b)$$

$$\text{curl}_z\,\mathbf{B} = \frac{1}{r}\frac{\partial}{\partial r}(rB_\theta). \quad (4c)$$

But from (2.14b),

$$(rB_\theta) = \frac{jI_m}{4\pi v_0}(e^{-j\beta_0 R_{2h}} + e^{-j\beta_0 R_{1h}}$$
$$- 2\cos\beta_0 h\, e^{-j\beta_0 R_0}). \quad (5)$$

In carrying out the differentiation of (5) in accordance with (4a) and (4c), the following are useful:

$$\frac{\partial R_{2h}}{\partial z} = \frac{z+h}{R_{2h}}, \quad \frac{\partial R_{1h}}{\partial z} = \frac{z-h}{R_{1h}},$$

$$\frac{\partial R_0}{\partial z} = \frac{z}{R_0}, \quad (6)$$

$$\frac{\partial R_{2h}}{\partial r} = \frac{r}{R_{2h}}, \quad \frac{\partial R_{1h}}{\partial r} = \frac{r}{R_{1h}},$$

$$\frac{\partial R_0}{\partial r} = \frac{r}{R_0}. \quad (7)$$

With (6) and (7),

$$\frac{\partial}{\partial z}(rB_\theta) = \frac{\beta_0 I_m}{4\pi v_0}\left(\frac{z+h}{R_{2h}}e^{-j\beta_0 R_{2h}}\right.$$
$$\left. + \frac{z-h}{R_{1h}}e^{-j\beta_0 R_{1h}} - \frac{2z}{R_0}\cos\beta_0 h\,e^{-j\beta_0 R_0}\right),$$
$$\quad (8)$$

$$\frac{\partial}{\partial r}(rB_\theta) = \frac{\beta_0 I_m r}{4\pi v_0}\left(\frac{e^{-j\beta_0 R_{2h}}}{R_{2h}} + \frac{e^{-j\beta_0 R_{1h}}}{R_{1h}}\right.$$
$$\left. - \frac{2}{R_0}\cos\beta_0 h\,e^{-j\beta_0 R_0}\right). \quad (9)$$

Upon forming the cylindrical components of (2) using (4a, b, c) and (8), (9), and noting that

$$\frac{\beta_0}{\omega\varepsilon_0} = \frac{\omega\sqrt{\varepsilon_0/v_0}}{\omega\varepsilon_0} = \frac{1}{\sqrt{\varepsilon_0 v_0}} \equiv \zeta_0, \quad (10)$$

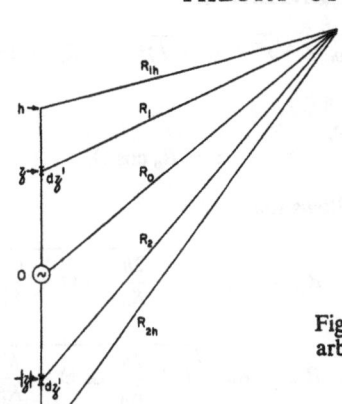

Fig. 1.1. Distances from center-driven antenna to an arbitrary point $P(x, y, z)$.

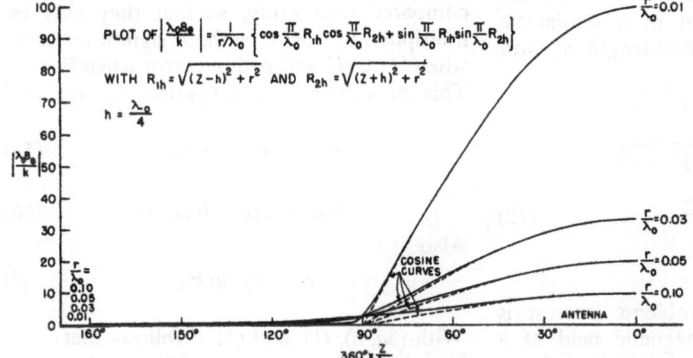

Fig. 2.1. Magnetic field along antenna of half-length $h = \lambda_0/4$ (with sinusoidally distributed current) at different distances r/λ_0 from the axis (z-axis) of the antenna.

Fig. 2.2. Magnetic field along antenna of half-length $h = 3\lambda_0/4$ (with sinusoidally distributed current) at different distances r/λ_0 from the axis (z-axis) of the antenna.

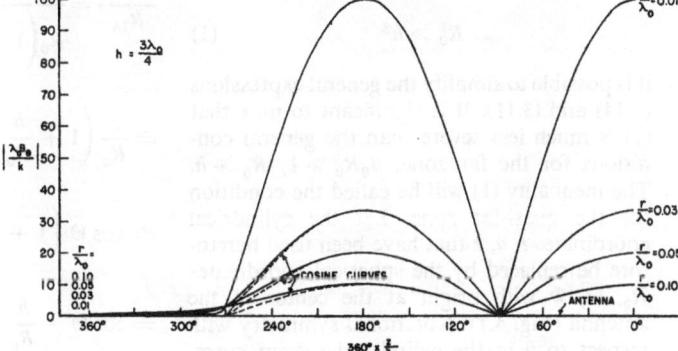

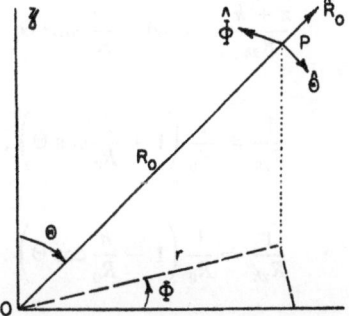

Fig. 4.1. Polar coördinates R_0, Θ, Φ.

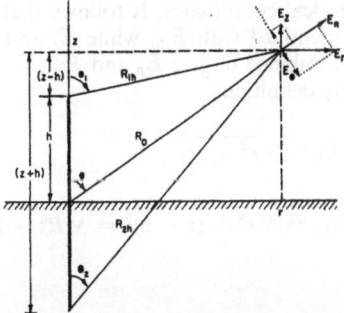

Fig. 4.2. Cylindrical and spherical components of the electric field.

the electric field is found to be

$$E_r = \frac{jI_m\zeta_0}{4\pi r}\left(\frac{z-h}{R_{1h}}e^{-j\beta_0 R_{1h}} + \frac{z+h}{R_{2h}}e^{-j\beta_0 R_{2h}}\right.$$

$$\left. - \frac{2z}{R_0}\cos\beta_0 h \, e^{-j\beta_0 R_0}\right), \quad (11a)$$

$$E_\theta = 0, \quad (11b)$$

$$E_z = \frac{-jI_m\zeta_0}{4\pi}\left(\frac{e^{-j\beta_0 R_{1h}}}{R_{1h}} + \frac{e^{-j\beta_0 R_{2h}}}{R_{2h}}\right.$$

$$\left. - \frac{2}{R_0}\cos\beta_0 h \, e^{-j\beta_0 R_0}\right). \quad (11c)$$

These expressions define the *complete* electric field at all points in space due to the *sinusoidally* distributed current in a symmetric center-driven antenna of half-length h, with

$$R_{2h} = \sqrt{(z+h)^2 + r^2},$$

$$R_{1h} = \sqrt{(z-h)^2 + r^2},$$

$$R_0 = \sqrt{z^2 + r^2}. \quad (12)$$

4. The Radiation Field

In a large class of problems interest is primarily in the electromagnetic field at a considerable distance from an array of one or more antennas. Subject to the condition

$$R_0^2 \gg h^2 \quad (1)$$

it is possible to simplify the general expressions (2.14) and (3.11). It is significant to note that (1) is much less severe than the general conditions for the far zone, $\beta_0 R_0 \gg 1$, $R_0 \gg h$. The inequality (1) will be called the condition for the quasi-far zone. Let the cylindrical coördinates r, θ, z that have been used heretofore be replaced by the spherical coördinates R_0, Θ, Φ with origin at the center of the antenna (Fig. 4.1). Rotational symmetry with respect to θ in the cylindrical system corresponds to symmetry with respect to Φ in the spherical coördinates. It follows that B_θ may be identified with B_Φ, while E_r and E_z must be combined to give E_R and E_Θ.

By definition,

$$R_0 = \sqrt{r^2 + z^2}, \quad (2a)$$

$$R_{1h} = \sqrt{r^2 + (z-h)^2} = \sqrt{R_0^2 - 2zh + h^2}, \quad (2b)$$

$$R_{2h} = \sqrt{r^2 + (z+h)^2} = \sqrt{R_0^2 + 2zh + h^2}. \quad (2c)$$

Since,

$$z = R_0 \cos\Theta, \quad (3)$$

it follows that

$$R_{1h} = R_0\sqrt{1 - \frac{2h}{R_0}\cos\Theta + \frac{h^2}{R_0^2}}, \quad (4a)$$

$$R_{2h} = R_0\sqrt{1 + \frac{2h}{R_0}\cos\Theta + \frac{h^2}{R_0^2}}. \quad (4b)$$

Subject to (1), the terms h^2/R_0^2 are negligible compared with unity, so that they may be multiplied by $\cos^2\Theta$ without significant error when $\Theta \leq \pi/2$ and with no error when $\Theta = 0$. This makes (4a) and (4b) perfect squares, and

$$R_{1h} \doteq R_0 - h\cos\Theta, \quad (5a)$$

$$R_{2h} \doteq R_0 + h\cos\Theta. \quad (5b)$$

Also

$$r = R_0 \sin\Theta, \quad (6)$$

With (5a, b), (1), and (3), it follows that

$$\frac{z-h}{R_{1h}} \doteq \frac{z-h}{R_0\left(1 - \frac{h}{R_0}\cos\Theta\right)}$$

$$\doteq \frac{z}{R_0}\left(1 + \frac{h}{R_0}\cos\Theta\right) - \frac{h}{R_0}\left(1 + \frac{h}{R_0}\cos\Theta\right)$$

$$\doteq \cos\Theta\left(1 + \frac{h}{R_0}\cos\Theta\right) - \frac{h}{R_0}$$

$$\doteq \cos\Theta - \frac{h}{R_0}\sin^2\Theta, \quad (7)$$

$$\frac{z+h}{R_{2h}} \doteq \cos\Theta + \frac{h}{R_0}\sin^2\Theta, \quad (8)$$

$$\frac{1}{R_{1h}} \doteq \frac{1}{R_0}\left(1 + \frac{h}{R_0}\cos\Theta\right), \quad (9)$$

$$\frac{1}{R_{2h}} \doteq \frac{1}{R_0}\left(1 - \frac{h}{R_0}\cos\Theta\right). \quad (10)$$

Upon substituting (3), (5a, b), (6), (7), and (8) in (2.14) and in (3.11) the following expressions are obtained:

$$B_r^r = 0, \qquad (11a)$$

$$B_\Theta^r = \frac{jI_m}{2\pi v_0} \frac{e^{-j\beta_0 R_0}}{R_0} F_m(\Theta, \beta_0 h), \qquad (11b)$$

$$B_z^r = 0, \qquad (11c)$$

$$E_r^r = \frac{jI_m \zeta_0}{2\pi} \frac{e^{-j\beta_0 R_0}}{R_0} \left[F_m(\Theta, \beta_0 h) \cos \Theta \right.$$

$$\left. - j \frac{h}{R_0} \sin (\beta_0 h \cos \Theta) \sin \Theta \right], \qquad (12a)$$

$$E_\Theta^r = 0, \qquad (12b)$$

$$E_z^r = -\frac{jI_m \zeta_0}{2\pi} \frac{e^{-j\beta_0 R_0}}{R_0} \left[F_m(\Theta, \beta_0 h) \sin \Theta \right.$$

$$\left. + j \frac{h}{R_0} \sin (\beta_0 h \cos \Theta) \cos \Theta \right]. \qquad (12c)$$

The quantity

$$F_m(\Theta, \beta_0 h) \equiv \frac{\cos (\beta_0 h \cos \Theta) - \cos \beta_0 h}{\sin \Theta} \qquad (13)$$

has been introduced.

The two cylindrical components E_r and E_z of the electric field may be combined to give the spherical components E_R and E_Φ. The component E_Φ is zero because E_θ vanishes. From Fig. 4.2 it is easily verified that

$$E_\Theta^r = E_r^r \cos \Theta - E_z^r \sin \Theta, \qquad (14a)$$

$$E_R^r = E_r^r \sin \Theta + E_z^r \cos \Theta. \qquad (14b)$$

With (13) and

$$F_0(\Theta, \beta_0 h) \equiv \frac{\cos (\beta_0 h \cos \Theta) - \cos \beta_0 h}{\sin \Theta \sin \beta_0 h}, \qquad (15)$$

and noting that

$$I_0 = I_m \sin \beta_0 h, \qquad (16)$$

the electromagnetic field (11) and (12) may be written as follows, subject only to $h^2 \ll R_0^2$:

$$B_\Theta^r = 0, \qquad (17a)$$

$$B_\Phi^r = \frac{jI_m}{2\pi v_0} \frac{e^{-j\beta_0 R_0}}{R_0} F_m(\Theta, \beta_0 h)$$

$$= \frac{jI_0}{2\pi v_0} \frac{e^{-j\beta_0 R_0}}{R_0} F_0(\Theta, \beta_0 h), \qquad (17b)$$

$$B_R^r = 0, \qquad (17c)$$

$$E_\Theta^r = \frac{jI_m \zeta_0}{2\pi} \frac{e^{-j\beta_0 R_0}}{R_0} F_m(\Theta, \beta_0 h)$$

$$= \frac{jI_0 \zeta_0}{2\pi} \frac{e^{-j\beta_0 R_0}}{R_0} F_0(\Theta, \beta_0 h), \qquad (18a)$$

$$E_\Phi^r = 0, \qquad (18b)$$

$$E_R^r = \frac{I_m \zeta_0 h}{2\pi R_0} \frac{e^{-j\beta_0 R_0}}{R_0} \sin (\beta_0 h \cos \Theta).$$

$$(\doteq 0 \text{ if } R_0 \gg h) \quad (18c)$$

Since (1) rather than the more severe condition $h \ll R_0$ has been imposed, E_R^r, which is a $1/R_0^2$ term and hence not part of the far-zone field, appears. Subject to $h \ll R_0$, its maximum value at $\Theta = 0$ is negligible compared with the maximum value of E_Θ^r. It is significant to note, however, that B_Φ and E_Θ have essentially the far-zone form at distances from the antenna that satisfy $R_0 \gtrsim 5h$. This is important in making measurements of E_Θ in the experimental determination of field patterns.

Note that

$$\frac{1}{2\pi v_0} \doteq 2 \times 10^{-7} \text{ henry/meter}, \qquad (19)$$

$$\frac{\zeta_0}{2\pi} \doteq 60 \text{ volts/meter}. \qquad (20)$$

The far-zone components E_Θ^r and B_Φ^r satisfy the relation

$$E_\Theta^r = v_0 B_\Phi^r, \qquad (21)$$

or, in vector form,

$$\mathbf{E}^r = v_0 \mathbf{B}^r \times \hat{\mathbf{R}}_0. \qquad (22)$$

The function $F_m(\Theta, \beta_0 h)$ is the field characteristic of the antenna in a plane containing the antenna and referred to maximum sinusoidal current. It is often called the "vertical" field characteristic. The function $F_0(\Theta, \beta_0 h)$ is the same characteristic referred to the input sinusoidal current. Clearly, if $\beta_0 h = \pi, 2\pi, \cdots, F_0(\Theta)$ becomes infinite so that $F_m(\Theta, \beta_0 h)$ is more convenient. Important special forms follow.

Although the general expression (1.7) for the vector potential is long and intricate, the simple relation

$$E_\Theta^r = -j\omega A_\Theta^r = j\omega A_z^r \sin \Theta \qquad (23)$$

is true for the far zone. Accordingly,

$$A_z^r = \frac{I_0}{2\pi v_0 \beta_0} \frac{F_0(\Theta, \beta_0 h)}{\sin \Theta} \frac{e^{-j\beta_0 R_0}}{R_0}. \qquad (24)$$

5. Field Patterns

The directional properties of the electric and magnetic fields in the far or radiation zone of a center-driven, infinitely thin antenna are contained in the functions $F_m(\Theta, \beta_0 h)$ and $F_0(\Theta, \beta_0 h)$ as defined by

$$F_m(\Theta, \beta_0 h) \equiv \frac{\cos(\beta_0 h \cos \Theta) - \cos \beta_0 h}{\sin \Theta}, \quad (1a)$$

$$F_0(\Theta, \beta_0 h) \equiv \frac{\cos(\beta_0 h \cos \Theta) - \cos \beta_0 h}{\sin \Theta \sin \beta_0 h}. \quad (1b)$$

These functions depend only upon the polar coördinate Θ because the antenna is rotationally symmetric and the field therefore independent of the angle Φ. A graph of the magnitude of (1a) or (1b) as a function of Θ is, therefore, sufficient to show the directional properties of the electric and magnetic fields in the far zone. Such a graph is called a *field pattern in a plane containing the antenna*. It may be represented in rectangular form with Θ as abscissa, $F_m(\Theta, \beta_0 h)$ or $F_0(\Theta, \beta_0 h)$ as ordinate, and $\beta_0 h$ or h/λ_0 as parameter. Alternatively, it may be drawn as a polar diagram with Θ as angle and $F_m(\Theta, \beta_0 h)$ or $F_0(\Theta, \beta_0 h)$ measured radially outward. It is usually convenient to express Θ in degrees measured from the vertical. For a symmetric center-driven antenna it is sufficient to plot $F(\Theta, \beta_0 h)$ as a function of Θ in the range $0 \leq \Theta \leq 90°$, that is, in a single quadrant, because all quadrants are alike owing to symmetry. Rectangular and polar graphs of $F_m(\Theta, \beta_0 h)$ or of $F_0(\Theta, \beta_0 h)$ are shown in Fig. 5.1 for several values of $h/\lambda_0 = \beta_0 h/2\pi$, including especially the following important cases:

Short antenna, $\beta_0^2 h^2 \ll 1$,

$$F_0(\Theta, \beta_0 h)$$
$$= \frac{1 - \tfrac{1}{2}\beta_0^2 h^2 \cos^2 \Theta + \cdots - 1 + \tfrac{1}{2}\beta_0^2 h^2 - \cdots}{\beta_0 h \sin \Theta}$$
$$\doteq \tfrac{1}{2}\beta_0 h \sin \Theta; \quad (2)$$

Half-wave dipole, $\beta_0 h = \pi/2$,

$$F_0(\Theta, \beta_0 h) = F_m(\Theta, \beta_0 h) = \frac{\cos\left(\frac{\pi}{2} \cos \Theta\right)}{\sin \Theta}; \quad (3)$$

Full-wave dipole, $\beta_0 h = \pi$,

$$F_m(\Theta, \beta_0 h) = \frac{\cos(\pi \cos \Theta) + 1}{\sin \Theta}; \quad (4)$$

Wave-and-a-half dipole, $\beta_0 h = 3\pi/2$,

$$F_0(\Theta, \beta_0 h) = F_m(\Theta, \beta_0 h) = \frac{\cos\left(\frac{3\pi}{2} \cos \Theta\right)}{\sin \Theta}; \quad (5)$$

Two-wave dipole, $\beta_0 h = 2\pi$,

$$F_m(\Theta, \beta_0 h) = \frac{\cos(2\pi \cos \Theta) - 1}{\sin \Theta}. \quad (6)$$

Extreme values and zeros of the functions $F_0(\Theta, \beta_0 h)$ and $F_m(\Theta, \beta_0 h)$ are of particular importance. The values of Θ for which they occur are readily derived. The zeros are determined by solving

$$\cos(\beta_0 h \cos \Theta) - \cos \beta_0 h = 0 \quad (7a)$$

or

$$\sin[\tfrac{1}{2}\beta_0 h(1 + \cos \Theta)] \sin[\tfrac{1}{2}\beta_0 h(1 - \cos \Theta)] = 0. \quad (7b)$$

The roots are

$$\Theta_{0N} = \cos^{-1}\left(1 - \frac{2N\pi}{\beta_0 h}\right), \quad N = 0, 1, 2, \cdots, \quad (8a)$$

$$\Theta'_{0N} = \cos^{-1}\left(\frac{2N'\pi}{\beta_0 h} - 1\right),$$
$$N' = 1, 2, 3, \cdots, \quad (8b)$$

with Θ and Θ' in the first quadrant. The value $N = 0$ in (8a) gives $\Theta_{00} = 0$ for all values of $\beta_0 h$. The functions $F_0(\Theta, \beta_0 h)$ and $F_m(\Theta, \beta_0 h)$ actually become indeterminate of the form 0/0 at this value. However, by differentiating numerator and denominator,

$$\lim_{\Theta \to 0}\left[\frac{\cos(\beta_0 h \cos \Theta) - \cos \beta_0 h}{\sin \Theta}\right]$$
$$= \lim_{\Theta \to 0}\left[\frac{\sin(\beta_0 h \cos \Theta) \beta_0 h \sin \Theta}{\cos \Theta}\right] = 0. \quad (9)$$

Curves of Θ_{0N}, $N > 0$, and $\Theta'_{0N'}$, $N' > 0$, are shown in Fig. 5.2 in unbroken lines marked Θ_{0N} or Θ'_{0N}. It is clear that for $0 \leq \beta_0 h \leq \pi$, there is only one angle that satisfies (7a), namely, Θ_{00} at $\Theta = 0$; for $\pi \leq \beta_0 h \leq 2\pi$, there are two angles, Θ_{00} and Θ_{01}; for $2\pi \leq \beta_0 h \leq 3\pi$, there are three angles, Θ_{00}, Θ_{02}, and Θ'_{01}; for $3\pi \leq \beta_0 h \leq 4\pi$, there are four angles giving a zero value of the function, namely, Θ_{00}, Θ_{03}, Θ'_{01}, and Θ_{02}; for

THEORY OF LINEAR ANTENNAS

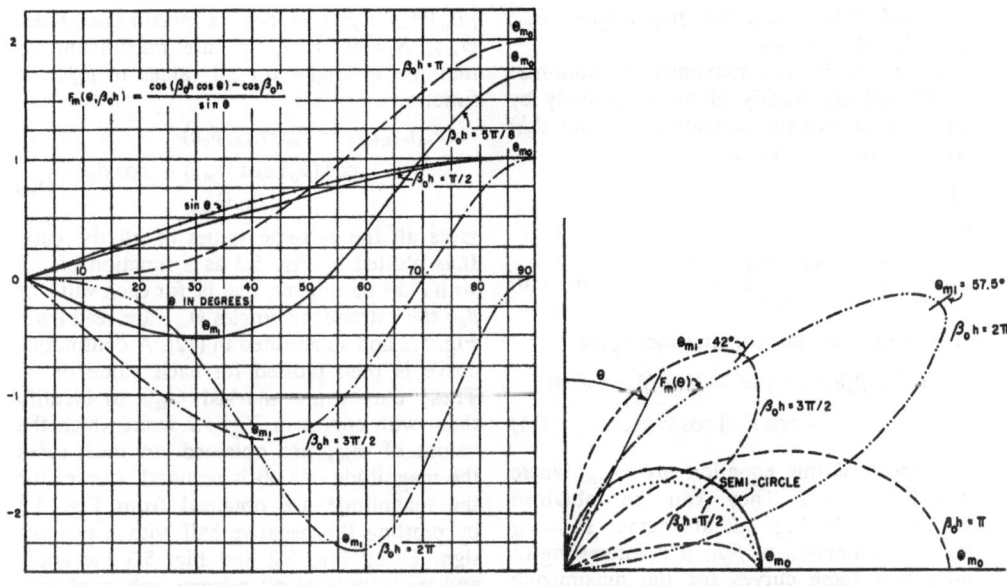

Fig. 5.1a. The function $F_m(\Theta, \beta_0 h)$, rectangular graph.

Fig. 5.1b. The function $F_m(\Theta, \beta_0 h)$, polar graph.

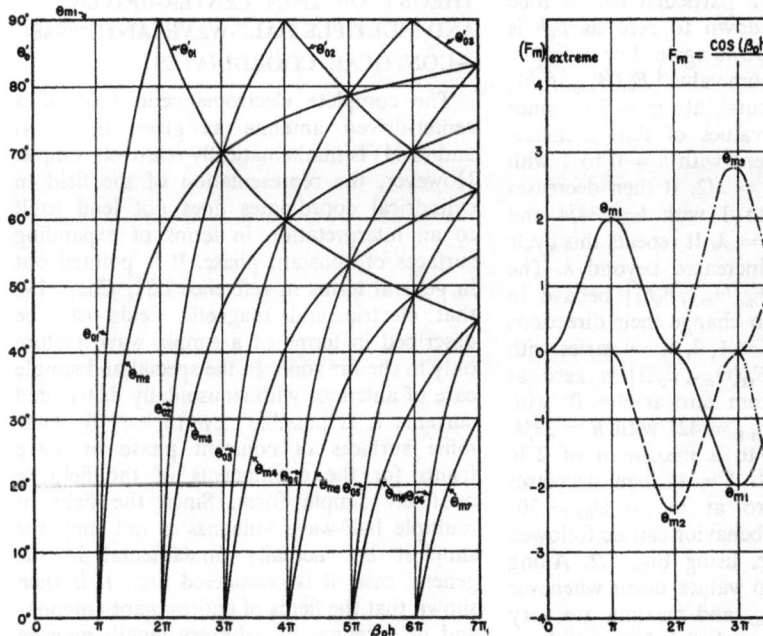

Fig. 5.2. Angles of zero (Θ_0, Θ'_0) and extreme (Θ_m) values of $F_m(\Theta, \beta_0 h)$.

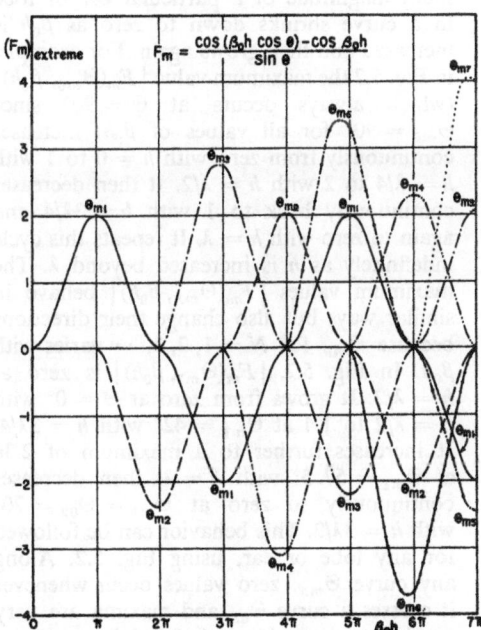

Fig. 5.3. Extreme amplitudes of $F_m(\Theta, \beta_0 h)$.

$4\pi \leq \beta_0 h \leq 5\pi$, there are five angles, Θ_{00}, Θ_{04}, Θ'_{01}, Θ_{03}, and Θ'_{02}.

Values of Θ that maximize or minimize $F_m(\Theta, \beta_0 h)$ are readily obtained formally by equating to zero the derivative of $F_m(\Theta, \beta_0 h)$ with respect to Θ. Thus

$$\frac{d}{d\Theta} F_m(\Theta, \beta_0 h)$$

$$= \frac{d}{d\Theta} \left[\frac{\cos(\beta_0 h \cos \Theta) - \cos \beta_0 h}{\sin \Theta} \right] = 0. \quad (10)$$

Differentiation and rearrangement give

$$\beta_0 h \sin(\beta_0 h \cos \Theta) = -[\cos(\beta_0 h \cos \Theta) - \cos \beta_0 h] \cos \Theta \csc^2 \Theta. \quad (11)$$

The roots of this equation are Θ_{mN}, where $N = 0, 1, 2, \cdots$. They can be obtained graphically. In Fig. 5.2 are curves showing Θ_{mN} as a function of $\beta_0 h$. It is interesting to note that these curves for the maximizing and minimizing values Θ_m intersect the curves for Θ_0. This means, of course, that the maximum magnitude of a particular ear or lobe in a curve shrinks down to zero as $\beta_0 h$ is increased and then grows again. For example, in Fig. 5.2 the maximum value $|F_m(\Theta_{m0}, \beta_0 h)|$ (which always occurs at $\Theta = 90°$ since $\Theta_{m0} = 90°$ for all values of $\beta_0 h$) increases continuously from zero with $h = 0$ to 1 with $h = \lambda/4$ to 2 with $h = \lambda/2$. It then decreases continuously back to 1 with $h = 3\lambda/4$ and again to zero with $h = \lambda$. It repeats this cycle indefinitely as h is increased beyond λ. The maximum values $|F_m(\Theta_{mN}, \beta_0 h)|$ behave in similar ways but also change their directions because Θ_{mN} for $N = 1, 2, 3, \cdots$ varies with $\beta_0 h$. In Fig. 5.2, $|F_m(\Theta_{m1}, \beta_0 h)|$ is zero at $h = \lambda/2$. It grows from zero at $\Theta = 0°$ with $h = \lambda/2$ to 1.4 at $\Theta_{m1} = 42°$ with $h = 3\lambda/4$. It increases further to a maximum of 2.36 at $\Theta_{m1} = 57.5°$ with $h = \lambda$, then decreases continuously to zero at $\Theta_{m1} = \Theta_{02} = 70°$ with $h = 3\lambda/2$. This behavior can be followed for any lobe or ear, using Fig. 5.2. Along any curve Θ_{mN}, zero values occur whenever it crosses a curve Θ_{0N} and maxima are very nearly at the values of $\beta_0 h$ lying midway between the values for zero. Thus, following the curve for Θ_{m1} from its beginning at $\Theta = 0°$, $\beta_0 h = \pi$, the following may be concluded. At $\Theta = 0°, 70°, 77.5°, 83°, \cdots$ the curve intersects successively $\Theta_{01}, \Theta_{02}, \Theta_{03}, \Theta_{04}, \cdots$, so that $|F_m(\Theta_{m1}, \beta_0 h)|$ must be zero at $\beta_0 h = \pi, 3\pi, 5\pi, 7\pi, \cdots$. Midway between, at $\beta_0 h = 2\pi, 4\pi, 6\pi, \cdots$, the quantity $|F_m(\Theta_{m1}, \beta_0 h)|$ is near a maximum. Since Θ_{mN}, $N = 0, 1, 2, \cdots$ are maximizing or minimizing angles for all values of $\beta_0 h$, the function

$$(F_m)_{\text{extreme}} \equiv F_m(\Theta_{mN}, \beta_0 h)$$

$$\equiv \frac{\cos(\beta_0 h \cos \Theta_{mN}) - \cos \beta_0 h}{\sin \Theta_{mN}} \quad (12)$$

gives all the extreme values of all the ears. It is plotted in Fig. 5.3 as a function of $\beta_0 h$ with N as parameter, that is, for each value of $\beta_0 h$ the extremizing angles Θ_{mN} are read from Fig. 5.2 and substituted in (12). A continuous curve is then plotted for each value of N. These curves are marked Θ_{mN} to identify them with curves in Fig. 5.2 from which the values of Θ_{mN} are obtained. In most cases the magnitude of (12) is required. Curves for the magnitude are obtained from Fig. 5.3 by plotting the negative half with a positive sign. Using Fig. 5.2 and Fig. 5.3 the angle and magnitude of all extreme values of ears or lobes for $\beta_0 h \leq 7\pi$ can be obtained directly.

THEORY OF THIN CENTER-DRIVEN AND MULTIPLE HALF-WAVE ANTENNAS —CONFOCAL COÖRDINATES

The complete electromagnetic field of a center-driven antenna as given in (2.14) and (3.11) is mathematically relatively simple. However, the representation of the field in cylindrical coördinates does not lend itself to an interpretation in terms of expanding surfaces of constant phase. It is pointed out in general terms in reference I.31, Chap. IV, that electric and magnetic fields can be described in terms of a simple wave picture only in the far zone. In the special and simple case of antennas with sinusoidally distributed currents it is possible nevertheless to determine surfaces of constant phase or wave fronts for the components of the field in relatively simple form. Since the field of multiple half-wave antennas is not only the simplest but actually fundamental in the general case, it is considered first. It is then shown that the fields of antiresonant antennas and of antennas of arbitrary length may be obtained by superposition from the solution of the multiple half-wave case.

6. *The Complete Field of Multiple Half-Wave Antennas in Confocal Coördinates*

Although the general expressions (2.14) and (3.11) correctly represent the electromagnetic field of an infinitely thin antenna both near and far, they are not in a form to permit ready visualization of the shape and motion of surfaces of constant phase or wave fronts. It has been shown (ref. I.31, Chap. IV) that in the far or radiation zone these surfaces are great spheres expanding in space with a constant radial velocity v_0, as can be seen from (4.17b) and (4.18a) after multiplying by $e^{j\omega t}$ and taking the real part. It is of interest to determine their behavior near the antenna as well, at least in simple special cases. This is readily accomplished for resonant antennas. For these there is no discontinuity in current at the center and a remarkably simple and instructive representation is possible.[26]

An infinitely thin center-driven antenna is resonant at half-lengths h given by

$$h = n\lambda_0/4, \quad \beta_0 h = n\pi/2, \quad n \text{ odd.} \qquad (1)$$

The distribution of current is

$$I_z = I_m \cos \beta_0 z, \quad I_m = I_0, \qquad (2)$$

as shown in Fig. 6.1a. The complex amplitudes of the electromagnetic field are obtained from (2.14) and (3.11) using (1). Thus, for $h = n\lambda_0/4$, n odd,

$$B_r = 0, \qquad (3a)$$

$$B_\theta = \frac{jI_m}{4\pi v_0 r}(e^{-j\beta_0 R_{1h}} + e^{-j\beta_0 R_{2h}}), \qquad (3b)$$

$$B_z = 0, \qquad (3c)$$

$$E_r = \frac{jI_m \zeta_0}{4\pi r}\left(\frac{z-h}{R_{1h}}e^{-j\beta_0 R_{1h}} + \frac{z+h}{R_{2h}}e^{-j\beta_0 R_{2h}}\right), \qquad (4a)$$

$$E_\theta = 0, \qquad (4b)$$

$$E_z = \frac{-jI_m \zeta_0}{4\pi}\left(\frac{e^{-j\beta_0 R_{1h}}}{R_{1h}} + \frac{e^{-j\beta_0 R_{2h}}}{R_{2h}}\right), \qquad (4c)$$

with

$$R_{1h} = \sqrt{(z-h)^2 + r^2},$$
$$R_{2h} = \sqrt{(z+h)^2 + r^2}. \qquad (5)$$

An infinitely thin antenna of half-length

$$h = n\lambda_0/4, \quad \beta_0 h = n\pi/2, \quad n \text{ even}, \qquad (6)$$

with a distribution of current

$$I_z = I_m \sin \beta_0 z, \qquad (7)$$

as shown in Fig. 6.1b is also resonant in the physical sense of oscillating in a natural mode. Such an antenna cannot be center-driven and is, therefore, necessarily driven asymmetrically if driven by a single generator. This presents special difficulties in an antenna of finite radius, as discussed in Chapter III. If two properly phased generators are used, a symmetric distribution is possible which is equivalent to that in two antennas each of half-length $h = n\lambda_0/4$, n odd, placed end to end. The field of each half of such a combination is given by (3) and (4) with respect to an origin at the *center of each half and with $2h$ the length of each half*. If the origin is shifted to the center of the entire structure and $2h$ is its *full* length, the field of the upper half is given by (3) and (4) with R_0 written for R_{2h}. The field of the lower half (in which the current is reversed) is given by (3) and (4) with a negative sign and with R_0 written for R_{1h}. If the two fields are combined, the result for $h = n\lambda_0/4$, n even, is

$$B_r = 0, \qquad (8a)$$

$$B_\theta = \frac{jI_m}{4\pi v_0 r}(e^{-j\beta_0 R_{1h}} - e^{-j\beta_0 R_{2h}}), \qquad (8b)$$

$$B_z = 0, \qquad (8c)$$

$$E_r = \frac{jI_m \zeta_0}{4\pi r}\left(\frac{z-h}{R_{1h}}e^{-j\beta_0 R_{1h}} - \frac{z+h}{R_{2h}}e^{-j\beta_0 R_{2h}}\right), \qquad (9a)$$

$$E_\theta = 0, \qquad (9b)$$

$$E_z = \frac{-jI_m \zeta_0}{4\pi}\left(\frac{e^{-j\beta_0 R_{1h}}}{R_{1h}} - \frac{e^{-j\beta_0 R_{2h}}}{R_{2h}}\right). \qquad (9c)$$

The distances R_{1h} and R_{2h} are defined as in (5); they are the distances from the point of calculation to the ends of the antenna. Since (8) and (9) differ from (3) and (4) only in a negative sign, the two sets of formulas are conveniently treated together.

The instantaneous values of the components of the electromagnetic field given in (3), (4), (8), and (9) are obtained by multiplying both members by $e^{j\omega t}$ and selecting the real part, assuming I_m to be real. The solution so obtained is that corresponding to a time dependence

$$(I_m)_{\text{inst}} = I_m \cos \omega t. \qquad (10)$$

The instantaneous values are

$$(B_r)_{\text{inst}} = 0, \tag{11a}$$

$$(B_\theta)_{\text{inst}} = \frac{-I_m}{4\pi v_0 r}[\sin(\omega t - \beta_0 R_{1h})$$
$$\pm \sin(\omega t - \beta_0 R_{2h})], \tag{11b}$$

$$(B_z)_{\text{inst}} = 0, \tag{11c}$$

$$(E_r)_{\text{inst}} = \frac{-I_m \zeta_0}{4\pi r}\left[\frac{z-h}{R_{1h}}\sin(\omega t - \beta_0 R_{1h})\right.$$
$$\left.\pm \frac{z+h}{R_{2h}}\sin(\omega t - \beta_0 R_{2h})\right], \tag{12a}$$

$$(E_\theta)_{\text{inst}} = 0, \tag{12b}$$

$$(E_z)_{\text{inst}} = \frac{I_m \zeta_0}{4\pi}\left[\frac{1}{R_{1h}}\sin(\omega t - \beta_0 R_{1h})\right.$$
$$\left.\pm \frac{1}{R_{2h}}\sin(\omega t - \beta_0 R_{2h})\right]. \tag{12c}$$

Here and in the following the upper sign is for n odd, the lower sign for n even in $h = n\lambda_0/4$.

In order to obtain a more convenient form for the electric field let the following components be defined:

$$(E_r)_{\text{inst}} = (E_{r1})_{\text{inst}} \pm (E_{r2})_{\text{inst}}, \tag{13}$$

$$(E_z)_{\text{inst}} = (E_{z1})_{\text{inst}} \pm (E_{z2})_{\text{inst}}, \tag{14}$$

with

$$(E_{r1})_{\text{inst}} \equiv \frac{-I_m \zeta_0}{4\pi r}\frac{z-h}{R_{1h}}\sin(\omega t - \beta_0 R_{1h}), \tag{15}$$

$$(E_{r2})_{\text{inst}} \equiv \frac{-I_m \zeta_0}{4\pi r}\frac{z+h}{R_{2h}}\sin(\omega t - \beta_0 R_{2h}), \tag{16}$$

$$(E_{z1})_{\text{inst}} \equiv \frac{I_m \zeta_0}{4\pi}\frac{1}{R_{1h}}\sin(\omega t - \beta_0 R_{1h}), \tag{17}$$

$$(E_{z2})_{\text{inst}} \equiv \frac{I_m \zeta_0}{4\pi}\frac{1}{R_{2h}}\sin(\omega t - \beta_0 R_{2h}). \tag{18}$$

These components can be recombined as follows:

$$(E_u)_{\text{inst}} = [(E_{r1})^2_{\text{inst}} + (E_{z1})^2_{\text{inst}}]^{1/2}, \tag{19a}$$

$$(E_v)_{\text{inst}} = [(E_{r2})^2_{\text{inst}} + (E_{z2})^2_{\text{inst}}]^{1/2}. \tag{19b}$$

The directions of u and v make the angles ϕ_u and ϕ_v with the positive z-axis. Thus,

$$\tan \phi_u = (E_{r1})_{\text{inst}}/(E_{z1})_{\text{inst}}, \tag{20a}$$

$$\tan \phi_v = (E_{r2})_{\text{inst}}/(E_{z2})_{\text{inst}}. \tag{20b}$$

Substitution of (15) and (17) in (19a) and (20a) and of (16) and (18) in (19b) and (20b) gives

$$(E_u)_{\text{inst}} = \frac{I_m \zeta_0}{4\pi R_{1h}}\left[\left(\frac{z-h}{r}\right)^2 + 1\right]^{1/2}$$
$$\times \sin(\omega t - \beta_0 R_{1h})$$
$$= \frac{I_m \zeta_0}{4\pi r}\sin(\omega t - \beta_0 R_{1h}), \tag{21}$$

$$(E_v)_{\text{inst}} = \frac{I_m \zeta_0}{4\pi R_{2h}}\left[\left(\frac{z+h}{r}\right)^2 + 1\right]^{1/2}$$
$$\times \sin(\omega t - \beta_0 R_{2h})$$
$$= \frac{I_m \zeta_0}{4\pi r}\sin(\omega t - \beta_0 R_{2h}), \tag{22}$$

$$\tan \phi_u = -\frac{z-h}{r} = \frac{h-z}{r}, \tag{23a}$$

$$\tan \phi_v = -\frac{z+h}{r} \text{ or } \tan(-\phi_v) = \frac{h+z}{r}. \tag{23b}$$

It is now possible to identify u and v with the axes of a system of coördinates. Consider a point P near the antenna as in Fig. 6.2. The angles ϕ_u and $(-\phi_v)$ as defined in (23a) and (23b) are shown. From (23a) it is clear that the positive u-axis must make an angle ϕ_u with the positive z-axis. By laying this angle off at P, the direction of u is determined. From the figure it is clear that u is perpendicular to R_{1h} for points in the first and second quadrants, and perpendicular to R_{2h} for points in the third and fourth quadrants. Since the antenna is symmetric with respect to its center, only the field in the first quadrant need be determined. The direction of v is also determined very easily. Since u is perpendicular to R_{1h}, it follows from the figure that

$$\psi + \chi = \pi/2. \tag{24}$$

Therefore v is perpendicular to R_{2h}.

Although (21) and (22) for the components of $(E)_{\text{inst}}$ are simple, they are not convenient because the u- and v-axes are oblique, and the angle between them varies from point to point. It is, therefore, advantageous to define a pair of mutually perpendicular axes along which the components of $(E)_{\text{inst}}$ have a simple form. This is readily accomplished by defining a new pair of axes with *origin at P* which are normal to each other and which bisect the angles made by the u- and v-axes. Since u is perpendicular to R_{1h}, and v is perpendicular to R_{2h}, it follows that the angle between u and v is the same as the visual angle ψ between R_{1h} and R_{2h}. Hence the new axes bisect both the angles formed by R_{1h} and R_{2h} and their extensions and the angles formed by u and v. In particular, let a new ρ-axis bisect the visual angle $\psi = F_1 P F_2$, as shown in Fig. 6.3. Let its positive direction be always outward from the antenna. Let a new ε-axis be introduced perpendicular to the ρ-axis, so that it bisects the angle $vPu = \psi$ between the u- and v-axes. Let its positive direction be so chosen that the coördinates ρ, ε, Φ form a right-handed system, where Φ is the usual spherical coördinate. It is the same as the cylindrical coördinate θ. In Fig. 6.3 the direction of Φ at P is vertically down into the paper.

Since the angle between the ε-axis and the u- and v-axes is $\psi/2$, the components of $(E)_{\text{inst}}$ along ρ and ε are readily expressed in terms of $(E_u)_{\text{inst}}$ and $(E_v)_{\text{inst}}$. The following relations are true:

$$E_\varepsilon = -\cos \tfrac{1}{2}\psi \, (E_u + E_v), \quad (25a)$$

$$E_\rho = \sin \tfrac{1}{2}\psi \, (E_u - E_v). \quad (25b)$$

Upon substituting (21) and (22) in (25a, b), the following expressions for the components of the electric field are obtained; the formula (7) for $(B_\theta)_{\text{inst}}$ is rewritten:

$$(B_\theta)_{\text{inst}} = (B_\Phi)_{\text{inst}} = \frac{-I_m}{4\pi v_0 r} [\sin(\omega t - \beta_0 R_{1h}) \pm \sin(\omega t - \beta_0 R_{2h})], \quad (26)$$

$$(E_\varepsilon)_{\text{inst}} = \frac{-I_m \zeta_0}{4\pi r} \cos \tfrac{1}{2}\psi \, [\sin(\omega t - \beta_0 R_{1h}) \pm \sin(\omega t - \beta_0 R_{2h})], \quad (27)$$

$$(E_\rho)_{\text{inst}} = \frac{I_m \zeta_0}{4\pi r} \sin \tfrac{1}{2}\psi \, [\sin(\omega t - \beta_0 R_{1h}) \mp \sin(\omega t - \beta_0 R_{2h})]. \quad (28)$$

The trigonometric expressions in the brackets can be rearranged using the formula

$$\sin A \pm \sin B = 2 \sin \tfrac{1}{2}(A \pm B) \cos \tfrac{1}{2}(A \mp B) \quad (29)$$

to give the following for the upper sign (n odd):

$$(B_\theta)_{\text{inst}} = \frac{-I_m}{2\pi v_0 r} \cos \tfrac{1}{2}\beta_0 (R_{2h} - R_{1h}) \times \sin[\omega t - \tfrac{1}{2}\beta_0 (R_{2h} + R_{1h})], \quad (30a)$$

$$(E_\varepsilon)_{\text{inst}} = \frac{-I_m \zeta_0}{2\pi r} \cos \tfrac{1}{2}\psi \cos \tfrac{1}{2}\beta_0 (R_{2h} - R_{1h}) \times \sin[\omega t - \tfrac{1}{2}\beta_0 (R_{2h} + R_{1h})], \quad (30b)$$

$$(E_\rho)_{\text{inst}} = \frac{I_m \zeta_0}{2\pi r} \sin \tfrac{1}{2}\psi \sin \tfrac{1}{2}\beta_0 (R_{2h} - R_{1h}) \times \cos[\omega t - \tfrac{1}{2}\beta_0 (R_{2h} + R_{1h})]. \quad (30c)$$

For the lower sign (n even),

$$(B_\theta)_{\text{inst}} = \frac{-I_m}{2\pi v_0 r} \sin \tfrac{1}{2}\beta_0 (R_{2h} - R_{1h}) \times \cos[\omega t - \tfrac{1}{2}\beta_0 (R_{2h} + R_{1h})], \quad (31a)$$

$$(E_\varepsilon)_{\text{inst}} = \frac{-I_m \zeta_0}{2\pi r} \cos \tfrac{1}{2}\psi \sin \tfrac{1}{2}\beta_0 (R_{2h} - R_{1h}) \times \cos[\omega t - \tfrac{1}{2}\beta_0 (R_{2h} + R_{1h})], \quad (31b)$$

$$(E_\rho)_{\text{inst}} = \frac{I_m \zeta_0}{2\pi r} \sin \tfrac{1}{2}\psi \cos \tfrac{1}{2}\beta_0 (R_{2h} - R_{1h}) \times \sin[\omega t - \tfrac{1}{2}\beta_0 (R_{2h} + R_{1h})]. \quad (31c)$$

The functions $\sin \tfrac{1}{2}\psi$ and $\cos \tfrac{1}{2}\psi$ may be expressed in terms of R_{1h} and R_{2h} as follows. By the law of cosines:

$$\cos \psi = \frac{R_{1h}^2 + R_{2h}^2 - 4h^2}{2 R_{1h} R_{2h}}. \quad (32)$$

Using the trigonometric relations

$$\sin \frac{\psi}{2} = \sqrt{\tfrac{1}{2}(1 - \cos \psi)}, \quad (33)$$

$$\cos \frac{\psi}{2} = \sqrt{\tfrac{1}{2}(1 + \cos \psi)}, \quad (34)$$

it follows that

$$\sin \frac{\psi}{2} = \sqrt{\frac{h^2 - (R_{2h} - R_{1h})^2/4}{R_{1h} R_{2h}}}, \quad (35a)$$

$$\cos \frac{\psi}{2} = \sqrt{\frac{(R_{2h} + R_{1h})^2/4 - h^2}{R_{1h} R_{2h}}}. \quad (35b)$$

7. The Radiation Field of a Multiple Half-Wave Antenna

At sufficiently great distances from the antenna the following condition is satisfied:

$$R^2 \gg h^2. \quad (1)$$

With (1),

$$\tfrac{1}{2}\beta_0 (R_{2h} + R_{1h}) \doteq \beta_0 R_0, \quad (2)$$

$$\tfrac{1}{2}\beta_0 (R_{2h} - R_{1h}) \doteq \tfrac{1}{2}\beta_0 R_0 \left[\sqrt{1 + \frac{2zh}{R_0^2}} - \sqrt{1 - \frac{2zh}{R_0^2}} \right] \doteq \frac{\beta_0 h z}{R_0}. \quad (3)$$

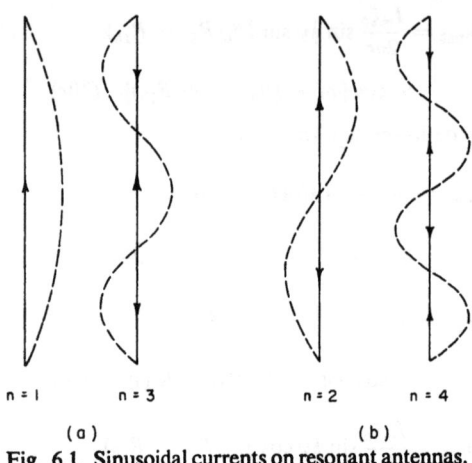

Fig. 6.1. Sinusoidal currents on resonant antennas.

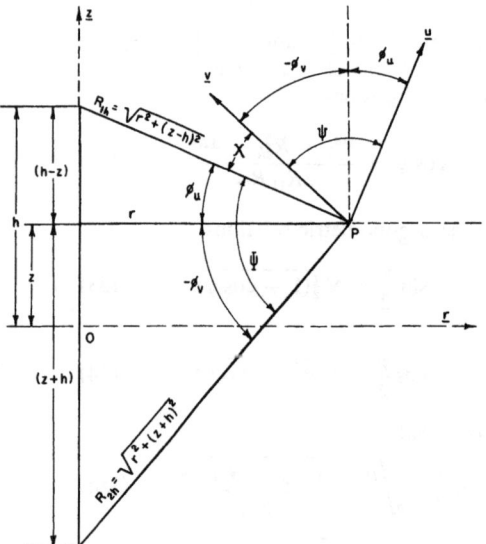

Fig. 6.2. Transformation of coördinates.

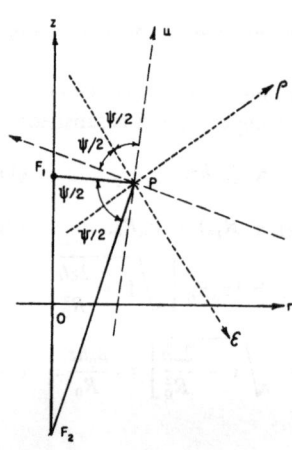

Fig. 6.3. Coördinates ε and ρ.

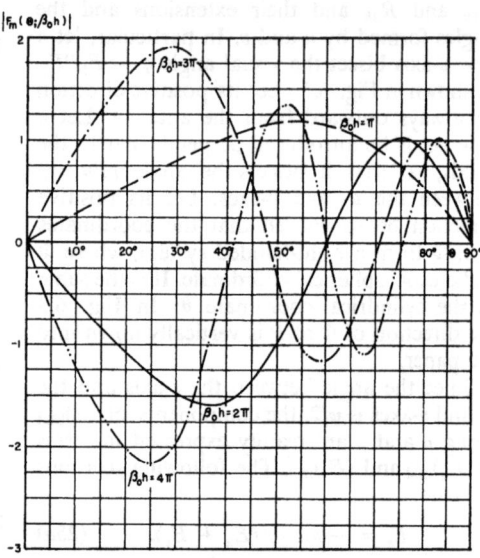

Fig. 7.1. Curves of $F_m(\Theta, \beta_0 h)$ for multiple half-wave antenna ($\beta_0 h = n\pi/2$ with n even).

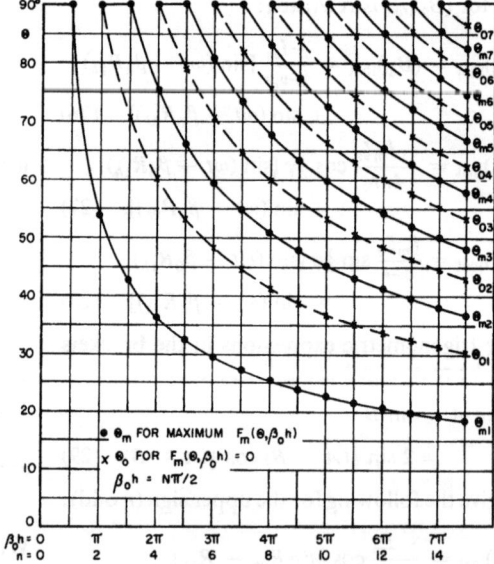

Fig. 7.2. Angles of maximum and zero $F_m(\Theta, \beta_0 h)$ for multiple half-wave antenna. (Lines connecting Θ's have no significance.)

TABLE 7.1. Angles for zero and extreme values of the field function of a multiple half-wave antenna.

n	Θ_0 (deg)	Θ_m (deg)
1	0.0	90.0
2	0.0, 90.0	54.0
3	0.0, 70.5	42.5, 90.0
4	0.0, 60.0, 90.0	36.3, 75.1
5	0.0, 53.1, 78.5	32.5, 66.0, 90.0
6	0.0, 48.2, 70.5, 90.0	29.2, 59.5, 80.3
7	0.0, 44.3, 64.6, 81.6	27.1, 54.6, 73.3, 90.0
8	0.0, 41.4, 60.0, 75.5, 90.0	25.2, 50.8, 67.8, 82.7
9	0.0, 38.9, 56.3, 70.5, 83.6	23.8, 47.7, 63.4, 77.1, 90.0
10	0.0, 36.9, 53.1, 66.4, 78.5, 90.0	22.4, 45.1, 59.8, 72.4, 84.2
11	0.0, 34.3, 50.5, 63.0, 74.2, 848.	21.3, 42.8, 56.8, 68.6, 79.5, 90.0
12	0.0, 33.6, 48.2, 60.0, 70.5, 80.4, 90.0	20.3, 41.1, 54.0, 56.4, 75.7, 85.2
13	0.0, 32.2, 46.2, 57.5, 67.4, 76.7, 85.6	19.6, 39.2, 51.8, 62.4, 72.0, 81.1, 90.0
14	0.0, 31.0, 44.4, 55.2, 64.6, 73.4, 81.8, 90.0	19.0, 37.9, 49.8, 59.9, 69.0, 77.6, 85.8
15	0.0, 29.9, 42.8, 53.1, 62.2, 70.5, 78.5, 86.2	18.2, 36.5, 48.0, 57.7, 66.4, 74.5, 82.3, 90.0

Introducing the polar coördinates R_0, Θ, Φ in (3) it may be written as follows:

$$\tfrac{1}{2}\beta_0(R_{2h} - R_{1h}) \doteq \beta_0 h \cos \Theta = \frac{n\pi}{2} \cos \Theta. \quad (4)$$

With (2) in (6.35b),

$$\cos \frac{\psi}{2} \doteq 1. \quad (5)$$

With (3) and (4) in (6.35a),

$$\sin \frac{\psi}{2} \doteq \sqrt{\frac{h^2(1 - \cos^2 \theta)}{R_0^2}} = \frac{h}{R_0} \sin \Theta. \quad (6)$$

Also,

$$r = R_0 \sin \Theta. \quad (7)$$

Since the ρ-axis at a point P bisects the visual angle ψ, ρ approaches the radial line R_0 through P as ψ becomes small. Since the ε-axis is normal to the ρ-axis, it becomes tangent to the spherical coördinate Θ at P. Hence, in the quasi-far zone and with n odd,

$$(B^r_\Phi)_{\text{inst}} = \frac{-I_m}{2\pi v_0 R_0} \frac{\cos(\tfrac{1}{2}n\pi \cos \Theta)}{\sin \Theta}$$
$$\times \sin(\omega t - \beta_0 R_0), \quad (8a)$$

$$(E^r_\Theta)_{\text{inst}} = \frac{-I_m \zeta_0}{2\pi R_0} \frac{\cos(\tfrac{1}{2}n\pi \cos \Theta)}{\sin \Theta}$$
$$\times \sin(\omega t - \beta_0 R_0), \quad (8b)$$

$$(E^r_R)_{\text{inst}} = \frac{I_m h \zeta_0}{2\pi R_0^2} \frac{\sin(\tfrac{1}{2}n\pi \cos \Theta)}{\sin \Theta}$$
$$\times \cos(\omega t - \beta_0 R_0), \quad (8c)$$

$$\doteq 0 \text{ if } h \ll R_0.$$

Similarly for n even,

$$(B^r_\Phi)_{\text{inst}} = \frac{-I_m}{2\pi v_0 R_0} \frac{\sin(\tfrac{1}{2}n\pi \cos \Theta)}{\sin \Theta}$$
$$\times \cos(\omega t - \beta_0 R_0), \quad (9a)$$

$$(E^r_\Theta)_{\text{inst}} = \frac{-I_m \zeta_0}{2\pi R_0} \frac{\sin(\tfrac{1}{2}n\pi \cos \Theta)}{\sin \Theta}$$
$$\times \cos(\omega t - \beta_0 R_0), \quad (9b)$$

$$(E^r_R)_{\text{inst}} = \frac{I_m h \zeta_0}{2\pi R_0^2} \frac{\cos(\tfrac{1}{2}n\pi \cos \Theta)}{\sin \Theta}$$
$$\times \sin(\omega t - \beta_0 R_0), \quad (9c)$$

$$\doteq 0 \text{ if } h \ll R_0.$$

The formulas (8a)–(8c) for the instantaneous values of the quasi-far-zone field of a resonant antenna an integral *odd* number of half-wavelengths long can be obtained directly from (4.17) and (4.18) with $\beta_0 h = n\pi/2$, n odd. Field patterns of

$$F_m(\Theta, \tfrac{1}{2}n\pi) = \frac{\cos(\tfrac{1}{2}n\pi \cos \Theta)}{\sin \Theta}, \quad n \text{ odd} \quad (10)$$

are given in Fig. 5.1. The angles Θ_m at which the field is a maximum and the angles Θ_0 for which it is zero may be read directly at the intersections of the appropriate curve in Fig. 5.2 with the abscissas $\beta_0 h = n\pi/2$, n odd.

The formulas (9a)–(9c) for the resonant antenna an integral *even* number of half-wavelengths long cannot be obtained from (4.17) and (4.18) since these apply only to the symmetric distribution of current $I(-z) = I(z)$. Field patterns of

$$F_m(\Theta, \tfrac{1}{2}n\pi) = \frac{\sin(\tfrac{1}{2}n\pi \cos \Theta)}{\sin \Theta}, \quad n \text{ even} \quad (11)$$

are given in Fig. 7.1. The angles Θ_m and Θ_0 at which the field is a maximum or minimum and zero are indicated in Fig. 7.2. The relative amplitudes of maxima and minima are shown in Fig. 7.3. No significance is to be attached to the lines connecting points in Figs. 7.2 and 7.3 except at $\beta_0 h = n\pi/2$, n even or odd, since a continuous variation of $\beta_0 h$ is not possible in this special case. The points at n even apply to (11); those at n odd apply to (10) and coincide with values obtained from Fig. 5.2 and Fig. 5.3. Numerical values of Θ_m and Θ_0 for integral values of n from 1 to 15 are given in Table 7.1.

It is significant to note in Table 7.1 that the angle for the first and largest maximum value of $F_m(\Theta, \tfrac{1}{2}n\pi)$ moves nearer and nearer to $\Theta = 0$ as n increases. That is, as the length of the antenna is increased, its principal maximum approaches the axis of the antenna. For an antenna that is several hundred wavelengths long Θ_{m1} is very small.

8. Phase Relations in the Field of Multiple Half-Wave Antennas

The general expressions (6.30)–(6.31) are written in terms of R_{1h} and R_{2h} and the visual angle ψ between them. This form is not inconvenient for purposes of computation using graphical methods, but it does not lend itself to ready determination of phase relations.

[V.8] THEORY OF LINEAR ANTENNAS 539

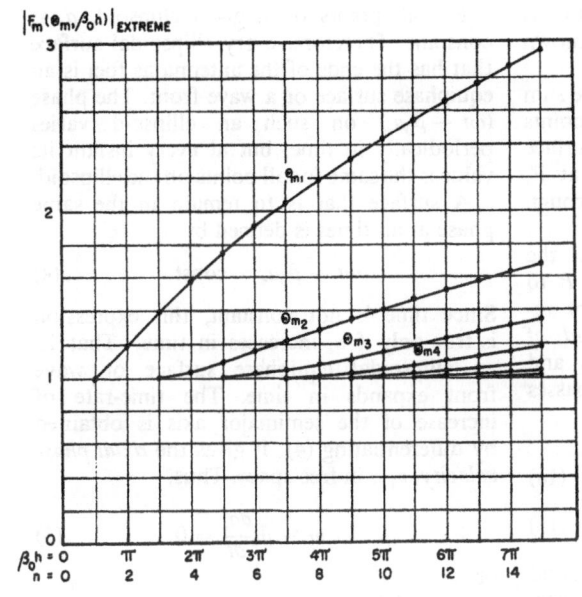

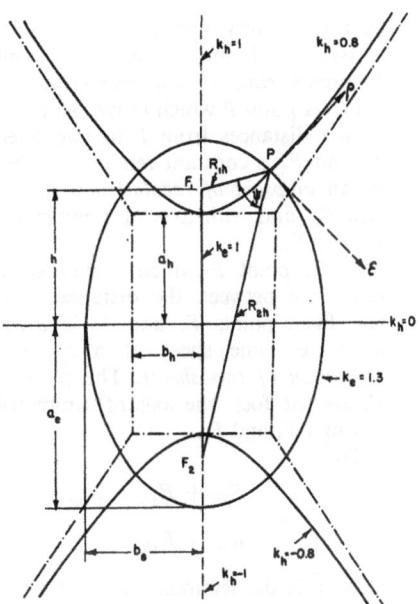

Fig. 7.3. Extreme values of $F_m(\Theta, \beta_0 h)$ for multiple half-wave antenna at Θ_m given in Fig. 7.2. (Lines connecting Θ's have no physical significance.)

Fig. 8.1. Confocal coördinates ε and ρ.

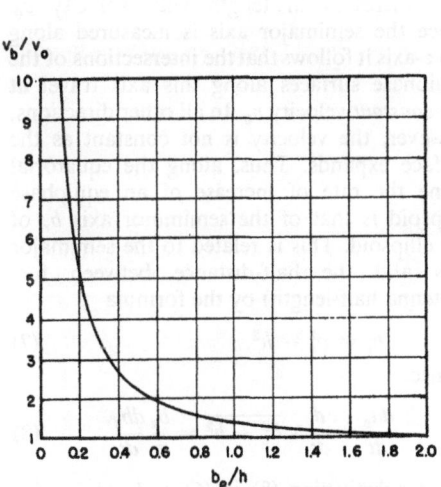

Fig. 8.2. Ratio of phase velocity v_b in the equitorial plane to phase velocity v_0 on the z-axis as the ratio of semiminor axis b_e to the half-length h of the antenna is increased.

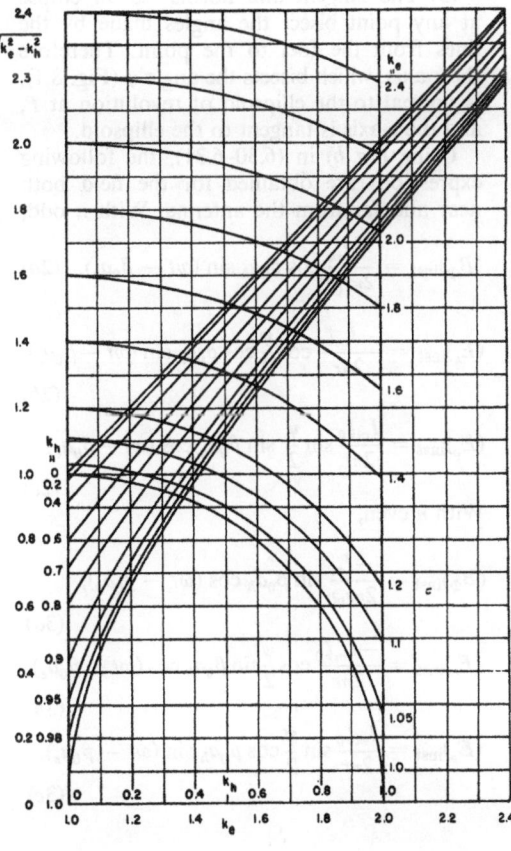

Fig. 8.3. The function $\sqrt{k_e^2 - k_h^2}$.

A further very illuminating modification is possible with the aid of a few well-known theorems from analytic geometry.

(1) A point P which moves so that the sum of the distances from P to two fixed points F_1 and F_2 is constant remains on the surface of an *ellipsoid of revolution* with foci at F_1 and F_2, and with axis of symmetry through F_1 and F_2.

(2) A point P which moves so that the difference between the distances from P to two fixed points F_1 and F_2 is constant remains on one sheet of a *hyperboloid of revolution of two sheets*. The points F_1 and F_2 are the foci; the axis of symmetry passes through F_1 and F_2.

Thus

$$R_{2h} + R_{1h} = 2a_e, \qquad (1a)$$

$$R_{2h} - R_{1h} = 2a_h. \qquad (1b)$$

Here a_e is the semimajor axis of the ellipsoid and a_h is the semitransverse axis of the hyperboloid, as shown in Fig. 8.1. Note that a_h is positive in the upper, negative in the lower half plane.

(3) The tangent and normal to an ellipse at any point bisect the angles made by the lines from the foci to the point. Therefore the ρ-axis, which bisects the angle ψ (Fig. 8.1), is normal to the ellipsoid of revolution at P, and the ε-axis is tangent to the ellipsoid.

Using (1a, b) in (6.30–6.31), the following expressions are obtained for the field both near and far from the antenna. With n odd,

$$(B_\Phi)_{\text{inst}} = \frac{-I_m}{2\pi v_0 r} \cos \beta_0 a_h \sin(\omega t - \beta_0 a_e), \qquad (2a)$$

$$(E_\varepsilon)_{\text{inst}} = \frac{-I_m \zeta_0}{2\pi r} \cos \frac{\psi}{2} \cos \beta_0 a_h \sin(\omega t - \beta_0 a_e), \qquad (2b)$$

$$(E_\rho)_{\text{inst}} = \frac{I_m \zeta_0}{2\pi r} \sin \frac{\psi}{2} \sin \beta_0 a_h \cos(\omega t - \beta_0 a_e). \qquad (2c)$$

With n even,

$$(B_\Phi)_{\text{inst}} = \frac{-I_m}{2\pi v_0 r} \sin \beta_0 a_h \cos(\omega t - \beta_0 a_e), \qquad (3a)$$

$$(E_\varepsilon)_{\text{inst}} = \frac{-I_m \zeta_0}{2\pi r} \cos \frac{\psi}{2} \sin \beta_0 a_h \cos(\omega t - \beta_0 a_e), \qquad (3b)$$

$$(E_\rho)_{\text{inst}} = \frac{I_m \zeta_0}{2\pi r} \sin \frac{\psi}{2} \cos \beta_0 a_h \sin(\omega t - \beta_0 a_e). \qquad (3c)$$

For all points on a given ellipsoid, a_e is constant. Therefore, every ellipsoidal surface that has the ends of the antenna as foci is an equiphase surface or a wave front. The phase $(\omega t - \beta_0 a_e)$ on such an ellipsoid varies periodically in time, but at every instant its value is the same for all points on the ellipsoid.

A surface that is to remain in the same phase at all times is defined by

$$\omega t - \beta_0 a_e = \text{const}. \qquad (4)$$

Since time is not constant, this expression is true only if a_e increases in time. That is, an ellipsoidal equiphase surface or wave front expands in time. The time-rate of increase of the semimajor axis is obtained by differentiating (4); it gives the *axial phase velocity*, v_{pa}, in free space. Thus,

$$\omega - \beta_0 \frac{da_e}{dt} = 0 \qquad (5)$$

or

$$v_{pa} = \frac{da_e}{dt} = \frac{\omega}{\beta_0} = v_0$$

$$\doteq 3 \times 10^8 \text{ meters/second}. \qquad (6)$$

The ellipsoid of constant phase is seen to expand in such a manner that its *semimajor axis* increases in length with velocity v_0. Since the semimajor axis is measured along the z-axis it follows that the intersections of the equiphase surfaces along this axis travel at the *constant* velocity v_0. In all other directions, however, the velocity is not constant as the surface expands. Thus, along the equatorial plane the rate of increase of an equiphase ellipsoid is that of the semiminor axis b_e of the ellipsoid. This is related to the semimajor axis and the half-distance between foci (antenna half-length) by the formula

$$b_e^2 = a_e^2 - h^2. \qquad (7)$$

Hence,

$$\frac{da_e}{dt} = \frac{d}{dt}\sqrt{b_e^2 + h^2} = \frac{b_e}{a_e}\frac{db_e}{dt}. \qquad (8)$$

Upon substituting (8) in (6) and solving for the radial phase velocity v_{pb} of the equiphase ellipsoid along the equatorial plane, the result for n odd is

$$v_{pb} = \frac{db_e}{dt} = \frac{a_e v_0}{b_e} = v_0 \sqrt{1 + \frac{h^2}{b_e^2}}$$

$$= v_0 \sqrt{1 + \left(\frac{n\lambda_0}{4b_e}\right)^2}. \qquad (9)$$

The ratio v_{pb}/v_0 is plotted in Fig. 8.2. It is clear that the phase velocity along the equatorial plane is much greater than v_0 near the antenna where a_e greatly exceeds b_e. However, as the ellipsoid expands and both a_e and b_e increase, the ratio b_e/a_e approaches unity. As the distance from the antenna becomes large a_e and b_e become approximately equal to each other and to R_0, and the velocity of the equiphase surfaces approaches v_0 asymptotically. This is the case of spherical wave fronts expanding with constant radial velocity in the far zone as described in detail in reference I.31.

If the frequency of the source driving the generator is modulated at a low frequency or if a pulse containing a narrow region of the frequency spectrum is applied, a velocity called the group velocity* may be defined. This measures the velocity of a particular phase of the *modulation envelope* of the electromagnetic field or the velocity of the pulse. It is defined by

$$v_g = \frac{d\omega}{d\beta} = v_p - \lambda \frac{dv_p}{d\lambda}. \tag{10}$$

The group velocity in the equatorial plane of the antenna corresponding to the phase velocity v_{pb} in (9) is obtained as follows using (10) and (9):

$$v_{gb} = v_0 \sqrt{1 + \left(\frac{n\lambda_0}{4b_e}\right)^2}$$

$$- \lambda_0 v_0 \frac{d}{d\lambda_0} \sqrt{1 + \left(\frac{n\lambda_0}{4b_e}\right)^2}$$

Upon carrying out the differentiation and combining terms the result is

$$v_{gb} = \frac{v_0}{\sqrt{1 + (n\lambda_0/4b_e)^2}} = \frac{v_0}{\sqrt{1 + h^2/b_e^2}}. \tag{11}$$

Evidently,

$$v_0^2 = v_{pb} v_{gb}. \tag{12}$$

It is possible to write the expressions (2)–(4) entirely in terms of variables that are characteristic of the ellipsoids and hyperboloids of revolution. The most convenient quantities for this purpose are the reciprocals of the eccentricities. These are defined by

$$k_e \equiv 1/e_e \equiv a_e/h, \tag{13a}$$

$$k_h \equiv 1/e_h \equiv a_h/h. \tag{13b}$$

---

* For a discussion of group velocity, see ref. I.52, p. 330.

Since a_e can vary from $a_e = h$ to $a_e = \infty$,

$$1 \leq k_e \leq \infty. \tag{14a}$$

The ellipsoid $k_e = 1$ is the straight line of length h; the ellipsoid $k_e = \infty$ is a circle of infinite radius. Similarly, because a_h can vary only from $-h$ to h,

$$-1 \leq k_h \leq 1. \tag{14b}$$

When $k_h = 0$ the two sheets of the hyperboloid degenerate into the equatorial plane. For $k_h = \pm 1$ they become the positive and negative sections of the z-axis above and below the antenna. With (13a, b) and n integral

$$\beta_0 a_h = \beta_0 h \frac{a_h}{h} = \frac{n\pi}{2} k_h, \tag{15a}$$

$$\beta_0 a_e = \beta_0 h \frac{a_e}{h} = \frac{n\pi}{2} k_e. \tag{15b}$$

In expressing the quantities $\sin \tfrac{1}{2}\psi$, $\cos \tfrac{1}{2}\psi$, and $1/r$ in terms of k_e and k_h, use is made of relations obtained by adding and subtracting (1a) and (1b):

$$R_{1h} = a_e - a_h, \quad R_{2h} = a_e + a_h. \tag{16a}$$

The substitution of (16a) in (5.35a, b) gives

$$\sin \frac{\psi}{2} = \sqrt{\frac{h^2 - a_h^2}{a_e^2 - a_h^2}} = \sqrt{\frac{1 - k_h^2}{k_e^2 - k_h^2}}, \tag{16b}$$

$$\cos \frac{\psi}{2} = \sqrt{\frac{a_e^2 - h^2}{a_e^2 - a_h^2}} = \sqrt{\frac{k_e^2 - 1}{k_e^2 - k_h^2}}. \tag{16c}$$

Curves of $\sqrt{k_e^2 - k_h^2}$ as a function of k_h for different values of k_e, and as a function of k_e with k_h as parameter, are shown in Fig. 8.3. The former are sections of circles, the latter of hyperbolas.

Any circle in space with cylindrical coördinates r, z may be located by the intersection of an ellipsoid and a hyperboloid of revolution. The equations of these confocal surfaces are

$$(z/a_e)^2 + (r/b_e)^2 = 1,$$
$$(z/a_h)^2 - (r/b_h)^2 = 1. \tag{17}$$

Here b_e and b_h are, respectively, the semiminor axis of the ellipsoid of semimajor axis a_e, and the semiconjugate axis of the hyperboloid of semitransverse axis a_h. They are defined as follows:

$$b_e^2 = a_e^2 - h^2 = h^2(k_e^2 - 1),$$
$$b_h^2 = h^2 - a_h^2 = h^2(1 - k_h^2). \tag{18}$$

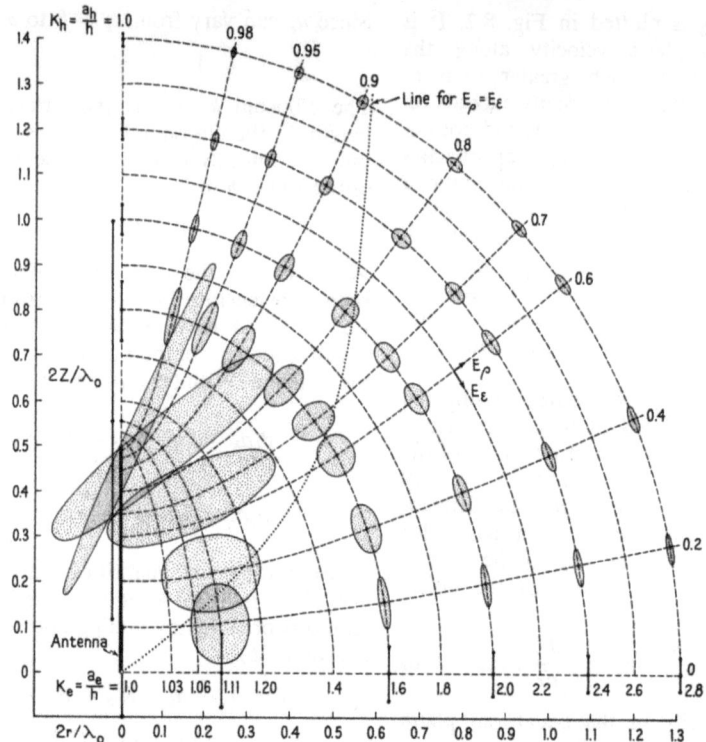

Fig. 8.4. Electric field near a half-wave antenna with sinusoidal current; $h = \lambda/4$ ($n = 1$ in $h = n\lambda/4$).

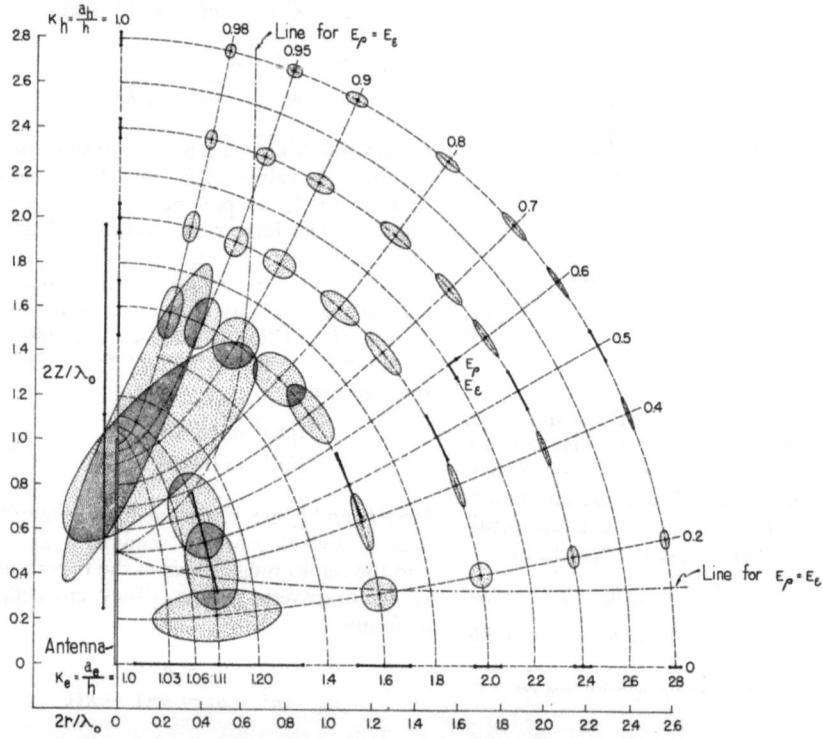

Fig. 8.5. Electric field near a multiple half-wave antenna with sinusoidal current; $h = \lambda/2$ ($n = 2$ in $h = n\lambda/4$).

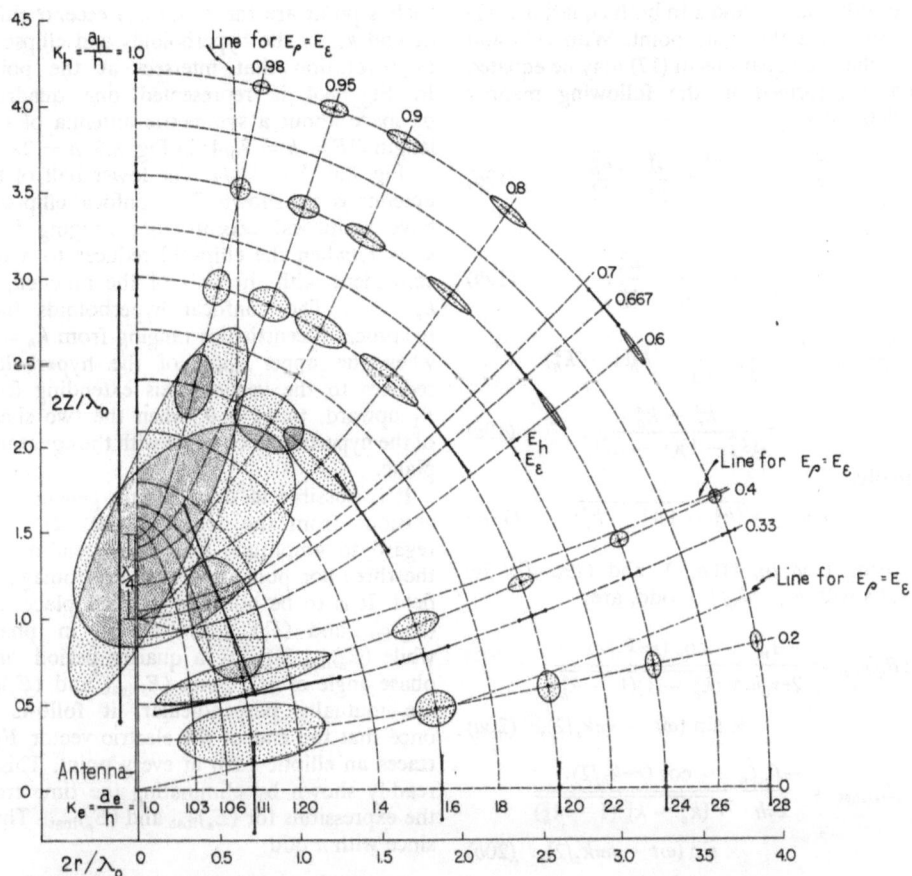

Fig. 8.6. Electric field near a multiple half-wave antenna with sinusoidal current; $h = 3\lambda/4$ ($n = 3$ in $h = n\lambda/4$).

The coördinates r and z in both equations (17) are to locate the same point. With (13) and (18), the two equations in (17) may be equated and transformed in the following manner to determine r:

$$\frac{z^2}{r^2} = \frac{a_e^2}{r^2} - \frac{a_e^2}{b_e^2} = \frac{a_h^2}{r^2} - \frac{a_h^2}{b_h^2}. \quad (19a)$$

Hence,

$$\frac{a_e^2 - a_h^2}{r^2} = \frac{a_e^2}{b_e^2} + \frac{a_h^2}{b_h^2}, \quad (19b)$$

$$\frac{(k_e^2 - k_h^2)h^2}{r^2} = \frac{k_e^2}{k_e^2 - 1} + k_h^2(1 - k_h^2)$$

$$= \frac{k_e^2 - k_h^2}{(k_e^2 - 1)(1 - k_h^2)}. \quad (19c)$$

Finally,

$$r/h = \sqrt{(k_e^2 - 1)(1 - k_h^2)}. \quad (19d)$$

Using (15a, b), (16b, c), and (19a–d), the results with $h = n\lambda_0/4$, n odd, are:

$$(B_\Phi)_{\text{inst}} = \frac{-I_m}{2\pi v_0 h} \frac{\cos(n\pi k_h/2)}{\sqrt{(k_e^2 - 1)(1 - k_h^2)}}$$
$$\times \sin(\omega t - n\pi k_e/2), \quad (20a)$$

$$(E_\varepsilon)_{\text{inst}} = \frac{-I_m \zeta_0}{2\pi h} \frac{\cos(n\pi k_h/2)}{\sqrt{(k_e^2 - k_h^2)(1 - k_h^2)}}$$
$$\times \sin(\omega t - n\pi k_e/2), \quad (20b)$$

$$(E_\rho)_{\text{inst}} = \frac{I_m \zeta_0}{2\pi h} \frac{\sin(n\pi k_h/2)}{\sqrt{(k_e^2 - 1)(k_e^2 - k_h^2)}}$$
$$\times \cos(\omega t - n\pi k_e/2). \quad (20c)$$

With n even,

$$(B_\Phi)_{\text{inst}} = \frac{-I_m}{2\pi v_0 h} \frac{\sin(n\pi k_h/2)}{\sqrt{(k_e^2 - 1)(1 - k_h^2)}}$$
$$\times \cos(\omega t - n\pi k_e/2), \quad (21a)$$

$$(E_\varepsilon)_{\text{inst}} = \frac{-I_m \zeta_0}{2\pi h} \frac{\sin(n\pi k_h/2)}{\sqrt{(k_e^2 - k_h^2)(1 - k_h^2)}}$$
$$\times \cos(\omega t - n\pi k_e/2), \quad (21b)$$

$$(E_\rho)_{\text{inst}} = \frac{I_m \zeta_0}{2\pi h} \frac{\cos(n\pi k_h/2)}{\sqrt{(k_e^2 - 1)(k_e^2 - k_h^2)}}$$
$$\times \sin(\omega t - n\pi k_e/2). \quad (21c)$$

From these relations the instantaneous value, the amplitude, and the phase angle of the electromagnetic field can be calculated at any point in space. The coördinates locating such a point are the reciprocal eccentricities, k_h and k_e, of the hyperboloids and ellipsoids of revolution that intersect at the point. In Fig. 8.4 is represented one quadrant of space about a symmetric antenna of half length $OF = h = \lambda_0/4$; in Fig. 8.5, $h = 2\lambda_0/4$; in Fig. 8.6, $h = 3\lambda_0/4$. The lower half of the antenna is not shown. The confocal ellipsoids have reciprocal eccentricities ranging from $k_e = 1$, when the ellipsoid reduces to a line coincident with the axis of the antenna, to $k_e = 2.4$. The confocal hyperboloids have reciprocal eccentricities ranging from $k_h = 1$, when the upper sheet of the hyperboloid reduces to the vertical axis extending from F_1 upward, to $k_h = 0$, when the two sheets of the hyperboloid coincide with the equatorial plane.

It is possible to draw certain general conclusions from the formulas (20)–(21) with regard to amplitude and phase relations of the three components of the electromagnetic field. It is to be noted in the first place that $(B_\Phi)_{\text{inst}}$ and $(E_\varepsilon)_{\text{inst}}$ are always in phase, while $(E_\rho)_{\text{inst}}$ lags by a quarter period or a phase angle of $\pi/2$. Since $(E_\varepsilon)_{\text{inst}}$ and $(E_\rho)_{\text{inst}}$ are mutually perpendicular, it follows at once that the end of the electric vector $\mathbf{E}_{\text{inst}}$ traces an elliptic path at every point. This is readily shown by eliminating the time from the expressions for $(E_\varepsilon)_{\text{inst}}$ and $(E_\rho)_{\text{inst}}$. Thus, since with n odd

$$(E_\varepsilon)_{\text{inst}}^2 = E_\varepsilon^2 \sin^2(\omega t - n\pi k_e/2), \quad (22)$$
$$(E_\rho)_{\text{inst}}^2 = E_\rho^2 \cos^2(\omega t - n\pi k_e/2), \quad (23)$$

it follows that

$$(E_\varepsilon)_{\text{inst}}^2/E_\varepsilon^2 + (E_\rho)_{\text{inst}}^2/E_\rho^2 = 1. \quad (24)$$

The same equation is obtained with n even. This is the equation of an ellipse with semi-principal axes equal to the amplitudes E_ε and E_ρ. Ellipses of this type are shown dotted at many points in Figs. 8.4, 8.5, and 8.6. *The electric vector is in general elliptically polarized in a plane containing the antenna; the magnetic vector is linearly polarized at right angles to this plane.*

The ellipse traced by the electric vector degenerates into a straight line tangent to the equiphase ellipsoids when E_ρ/E_ε vanishes; it becomes a straight line normal to the equiphase ellipsoids when E_ε/E_ρ vanishes. It becomes a circle when these two quantities are equal. These special cases of linear and circular polarization occur under the following conditions.

Linear polarization with $E_\rho = 0$ occurs when

$$\frac{\sin(n\pi k_h/2)}{\sqrt{(k_e^2-1)(k_e^2-k_h^2)}} = 0, \quad n \text{ odd}, \quad (25a)$$

$$\frac{\cos(n\pi k_h/2)}{\sqrt{(k_e^2-1)(k_e^2-k_h^2)}} = 0, \quad n \text{ even}. \quad (25b)$$

This is true along hyperboloids for which

$$nk_h = 0, 2, 4, \cdots, \quad n \text{ odd}, \quad (26a)$$
$$nk_h = 1, 3, 5, \cdots, \quad n \text{ even}. \quad (26b)$$

Since in the upper half space $0 \leq k_h \leq 1$, the following solutions are possible in this half space:

$n=1$, $k_h = 0$, defining the equatorial plane; (27a)

$n=2$, $k_h = 1/2$; (27b)

$n=3$, $k_h = 0, 2/3$; (27c)

$n=4$, $k_h = 1/4, 3/4$; (27d)

$n=5$, $k_h = 0, 2/5, 4/5$; (27e)

$n=6$, $k_h = 1/6, 1/2, 5/6$. (27f)

Note that in the far zone,

$$k_e^2 = (a_e/h)^2 \gg 1 \geq k_h^2 \quad (28)$$

and (25) becomes

$$\frac{h^2 \sin(n\pi k_h/2)}{a_e^2} \doteq \frac{h^2 \sin(n\pi k_h/2)}{R_0^2}. \quad (29)$$

Accordingly, E_ρ vanishes at infinity as $1/R_0^2$.

Linear polarization with $E_\varepsilon = 0$ occurs when

$$\frac{\cos(n\pi k_h/2)}{\sqrt{(k_e^2-k_h^2)(1-k_h^2)}} = 0, \quad n \text{ odd}, \quad (30a)$$

$$\frac{\sin(n\pi k_h/2)}{\sqrt{(k_e^2-k_h^2)(1-k_h^2)}} = 0, \quad n \text{ even}. \quad (30b)$$

This is true on hyperboloids for which

$$nk_h = 1, 3, 5, \cdots, \quad n \text{ odd}, \quad (31a)$$
$$nk_h = 0, 2, 4, 6, \cdots, \quad n \text{ even}, \quad (31b)$$

with $0 \leq k_h \leq 1$. The following solutions are possible in the upper half space:

$n=1$, $k_h = 1$, defining the z-axis beyond the ends of the antenna; (32a)

$n=2$, $k_h = 1, 0$ defining the mid-plane and the z-axis beyond the ends of the antenna; (32b)

$n=3$, $k_h = 1, 1/3$; (32c)

$n=4$, $k_h = 1, 1/2, 0$; (32d)

$n=5$, $k_h = 1, 3/5, 1/5$; (32e)

$n=6$, $k_h = 1, 2/3, 1/3, 0$. (32f)

Note that in the far zone (28) is true so that (30a) becomes

$$\frac{h \cos(n\pi k_h/2)}{a_e\sqrt{1-k_h^2}} \doteq \frac{h \cos(n\pi k_h/2)}{R_0\sqrt{1-k_h^2}}. \quad (33)$$

The same relation with sin written for cos is true for (30b). Accordingly, E_ε vanishes as $1/R_0$ at infinity.

Circular polarization occurs when

$$|E_\rho| = |E_\varepsilon| \quad (34)$$

or when

$$|\tan(n\pi k_h/2)| = \sqrt{(k_e^2-1)/(1-k_h^2)},$$
$$n \text{ odd}, \quad (35a)$$

$$|\cot(n\pi k_h/2)| = \sqrt{(k_e^2-1)/(1-k_h^2)},$$
$$n \text{ even}. \quad (35b)$$

Since in the upper half space $0 \leq k_h \leq 1$, it is seen directly that as

$$k_e \to \infty, \quad k_h \to \frac{m}{n}, \quad \begin{matrix} m = 1, 3, 5, \cdots \\ n = 1, 3, 5, \cdots \end{matrix}$$
$$(36a)$$

$$k_h \to \frac{m}{n}, \quad \begin{matrix} m = 0, 2, 4, 6, \cdots \\ n = 2, 4, 6, \cdots \end{matrix}$$
$$(36b)$$

Accordingly, in the far zone the lines of circles approach the hyperboloids $k_h = m/n$ for m and n odd, the hyperboloids $k_h = m/n$ for $m = 0, 2, 4, \cdots$ and $n = 2, 4, 6, \cdots$. For $n=1$, $k_h \to 1$, the z-axis is approached; for $n=2$, $k_h \to 0$ or 1, that is, the equatorial plane or the z-axis is approached.

Aside from constants, the expression (20a) for $(B_\Phi)_{\text{inst}}$ differs from the expression (20b) for $(E_\varepsilon)_{\text{inst}}$ only in having $k_e^2 - 1$ appear in place of $k_e^2 - k_h^2$ in the denominator. Since $k_h^2 \leq 1$, this in no way affects the behavior at infinity. Since the numerator is the same for $(B_\Phi)_{\text{inst}}$ as for $(E_\varepsilon)_{\text{inst}}$ it follows that B_Φ vanishes when E_ε vanishes and in the same way. The following ratio is true:

$$\frac{(E_e)_{\text{inst}}}{(B_\Phi)_{\text{inst}}} = v_0 \sqrt{\frac{k_e^2 - 1}{k_e^2 - k_h^2}}. \quad (37)$$

In the far zone, where $k_e^2 \gg 1 \geq k_h^2$, this ratio reduces to

$$(E_e)_{\text{inst}}/(B_\Phi)_{\text{inst}} \doteq v_0 \quad (38)$$

in agreement with the general results obtained in reference I.31, Chap. IV. At all points along the z-axis, $k_h = 1$ and (37) also reduces to (38).

9. Field of Antiresonant Antenna and Antennas of Arbitrary Length

A center-driven antenna of vanishing radius is antiresonant at half-lengths $h = n\lambda_0/2$ with n even. The distribution of current is

$$I_z = I_m \sin \beta_0 |z|, \quad (1)$$

as shown in Fig. II.16.1. It is to be noted that antiresonance in a thin center-driven antenna occurs near the same lengths as the even natural modes of an antenna not excited at the center. The distribution of current (1) differs from (6.7) only in the absolute-value sign which requires codirectional currents at corresponding points in the two halves. The antiresonant center-driven antenna has even currents,

$$I(-z) = I(z). \quad (2)$$

The even natural modes referred to an origin at the center have odd currents,

$$I(-z) = -I(z). \quad (3)$$

Whereas the even natural modes occurring near $h = n\lambda_0/4$, n even, may be called resonant modes, the antiresonances occurring near $h = n\lambda_0/4$, n even, with a center-driven antenna are *not* resonant modes. The electromagnetic field of the center-driven antenna with $\beta_0 h = n\pi/2$, n even, is given by (2.14) and (3.11) properly specialized. It is

$$B_r = 0, \quad (4a)$$

$$B_\theta = \frac{jI_m}{4\pi v_0 r} (e^{-j\beta_0 R_{1h}} + e^{-j\beta_0 R_{2h}} + 2e^{-j\beta_0 R_0}), \quad (4b)$$

$$B_z = 0, \quad (4c)$$

$$E_r = \frac{jI_m \zeta_0}{4\pi r} \left(\frac{z-h}{R_{1h}} e^{-j\beta_0 R_{1h}} + \frac{z+h}{R_{2h}} e^{-j\beta_0 R_{2h}} + \frac{2z}{R_0} e^{-j\beta_0 R_0} \right), \quad (5a)$$

$$E_\theta = 0, \quad (5b)$$

$$E_z = \frac{-jI_m \zeta_0}{4\pi} \left(\frac{e^{-j\beta_0 R_{1h}}}{R_{1h}} + \frac{e^{-j\beta_0 R_{2h}}}{R_{2h}} + \frac{2e^{-j\beta_0 R_0}}{R_0} \right). \quad (5c)$$

These formulas may be rearranged as follows:

$$B_\theta = \frac{jI_m}{4\pi v_0 r} [(e^{-j\beta_0 R_{1h}} + e^{-j\beta_0 R_0}) + (e^{-j\beta_0 R_0} + e^{-j\beta_0 R_{2h}})], \quad (6a)$$

$$E_r = \frac{jI_m \zeta_0}{4\pi r} \left[\left(\frac{z-h}{R_{1h}} e^{-j\beta_0 R_{1h}} + \frac{z}{R_0} e^{-j\beta_0 R_0} \right) + \left(\frac{z}{R_0} e^{-j\beta_0 R_0} + \frac{z+h}{R_{2h}} e^{-j\beta_0 R_{2h}} \right) \right], \quad (6b)$$

$$E_z = \frac{-jI_m \zeta_0}{4\pi} \left[\left(\frac{e^{-j\beta_0 R_{1h}}}{R_{1h}} + \frac{e^{-j\beta_0 R_0}}{R_0} \right) + \left(\frac{e^{-j\beta_0 R_0}}{R_0} + \frac{e^{-j\beta_0 R_{2h}}}{R_{2h}} \right) \right]. \quad (6c)$$

Each of these expressions is the sum of two parts of which each is exactly the formula for an antenna of half-length $n\lambda/4$, n odd, but with origin at one end. If the field of each half is referred to an origin at its center, the formulas for the nonvanishing components of the field are

$$B_\theta = \frac{jI_m}{4\pi v_0 r} [(e^{-j\beta_0 R'_{1h}} + e^{-j\beta_0 R'_{2h}}) + (e^{-j\beta_0 R''_{1h}} + e^{-j\beta_0 R''_{2h}})], \quad (7a)$$

$$E_r = \frac{jI_m \zeta_0}{4\pi r} \left[\left(\frac{z'-h/2}{R'_{1h}} e^{-j\beta_0 R'_{1h}} + \frac{z'+h/2}{R'_{2h}} e^{-j\beta_0 R'_{2h}} \right) + \left(\frac{z''-h/2}{R''_{1h}} e^{-j\beta_0 R''_{1h}} + \frac{z''+h/2}{R''_{2h}} e^{-j\beta_0 R''_{2h}} \right) \right], \quad (7b)$$

$$E_z = \frac{-jI_m \zeta_0}{4\pi} \left[\left(\frac{e^{-j\beta_0 R'_{1h}}}{R'_{1h}} + \frac{e^{-j\beta_0 R'_{2h}}}{R'_{2h}} \right) + \left(\frac{e^{-j\beta_0 R''_{1h}}}{R''_{1h}} + \frac{e^{-j\beta_0 R''_{2h}}}{R''_{2h}} \right) \right], \quad (7c)$$

with

$$z' = z - h/2, \quad z'' = z + h/2, \quad (8a)$$

$$R'_{1h} = \sqrt{(z' - h/2)^2 + r^2}, \quad (8b)$$

$$R''_{1h} = \sqrt{(z'' - h/2)^2 + r^2}, \quad (8c)$$

$$R'_{2h} = \sqrt{(z' + h/2)^2 + r^2}, \quad (8d)$$

$$R''_{2h} = \sqrt{(z'' + h/2)^2 + r^2}. \quad (8e)$$

The instantaneous values referred to

$$(I_m)_{\text{inst}} = I_m \cos \omega t \quad (9)$$

are

$$(B_\theta)_{\text{inst}} = \frac{-I_m}{4\pi v_0 r} \{[\sin(\omega t - \beta_0 R'_{1h}) + \sin(\omega t - \beta_0 R'_{2h})] + [\sin(\omega t - \beta_0 R''_{1h}) + \sin(\omega t - \beta_0 R''_{2h})]\}, \quad (10a)$$

$$(E_r)_{\text{inst}} = \frac{-I_m \zeta_0}{4\pi r} \left\{ \left[\frac{z' - h/2}{R'_{1h}} \sin(\omega t - \beta_0 R'_{1h}) \right. \right.$$
$$\left. + \frac{z' + h/2}{R'_{2h}} \sin(\omega t - \beta_0 R'_{2h}) \right]$$
$$+ \left[\frac{z'' - h/2}{R''_{1h}} \sin(\omega t - \beta_0 R''_{1h}) \right.$$
$$\left. \left. + \frac{z'' + h/2}{R''_{2h}} \sin(\omega t - \beta_0 R''_{2h}) \right] \right\}, \quad (10b)$$

$$(E_z)_{\text{inst}} = \frac{I_m \zeta_0}{4\pi} \left\{ \left[\frac{1}{R'_{1h}} \sin(\omega t - \beta_0 R'_{1h}) \right. \right.$$
$$\left. + \frac{1}{R'_{2h}} \sin(\omega t - \beta_0 R'_{2h}) \right]$$
$$+ \left[\frac{1}{R''_{1h}} \sin(\omega t - \beta_0 R''_{1h}) \right.$$
$$\left. \left. + \frac{1}{R''_{2h}} \sin(\omega t - \beta_0 R''_{2h}) \right] \right\}. \quad (10c)$$

If the two parts in each of these expressions are treated separately, they may be expressed in terms of the coördinates of two families of ellipsoids and hyperboloids having the two halves of the antenna as major axes. With (8.20),

$$(B_{\Phi'})_{\text{inst}} = \frac{-I_m}{2\pi v_0 h} \frac{\cos(n\pi k'_h/2)}{\sqrt{(k'^2_e - 1)(1 - k'^2_h)}}$$
$$\times \sin(\omega t - n\pi k'_e/2), \quad (11a)$$

$$(B_{\Phi''})_{\text{inst}} = \frac{-I_m}{2\pi v_0 h} \frac{\cos(n\pi k''_h/2)}{\sqrt{(k''^2_e - 1)(1 - k''^2_h)}}$$
$$\times \sin(\omega t - n\pi k''_e/2), \quad (11b)$$

$$(E_{e'})_{\text{inst}} = \frac{-I_m \zeta_0}{2\pi h} \frac{\cos(n\pi k'_h/2)}{\sqrt{(k'^2_e - k'^2_h)(1 - k'^2_h)}}$$
$$\times \sin(\omega t - n\pi k'_e/2), \quad (12a)$$

$$(E_{e''})_{\text{inst}} = \frac{-I_m \zeta_0}{2\pi h} \frac{\cos(n\pi k''_h/2)}{\sqrt{(k''^2_e - k''^2_h)(1 - k''^2_h)}}$$
$$\times \sin(\omega t - n\pi k''_e/2), \quad (12b)$$

$$(E_{\rho'})_{\text{inst}} = \frac{I_m \zeta_0}{2\pi h} \frac{\sin(n\pi k'_h/2)}{\sqrt{(k'^2_e - 1)(k'^2_e - k'^2_h)}}$$
$$\times \cos(\omega t - n\pi k'_e/2), \quad (13a)$$

$$(E_{\rho''})_{\text{inst}} = \frac{I_m \zeta_0}{2\pi h} \frac{\sin(n\pi k''_h/2)}{\sqrt{(k''^2_e - 1)(k''^2_e - k''^2_h)}}$$
$$\times \cos(\omega t - n\pi k''_e/2). \quad (13b)$$

In the above equations (11)–(13), k'_e is the reciprocal of the eccentricity of the ellipsoids associated with the upper half of the antenna, k''_e the reciprocal of the eccentricity of the ellipsoids associated with the lower half of the antenna; k'_h and k''_h are the corresponding reciprocal eccentricities of the two families of hyperboloids. Thus,

$$k'_e = 2a'_e/h, \quad k'_h = 2a'_h/h, \quad (14a)$$

$$k''_e = 2a''_e/h, \quad k''_h = 2a''_h/h, \quad (14b)$$

with a'_e and a''_e the semimajor axes of the two families of ellipsoids, a'_h and a''_h the semitransverse axes of the two families of hyperboloids for which

$$R'_{1h} + R'_{2h} = 2a'_e, \quad R'_{2h} - R'_{1h} = 2a'_h, \quad (15a)$$

$$R''_{1h} + R''_{2h} = 2a''_e, \quad R''_{2h} - R''_{1h} = 2a''_h. \quad (15b)$$

The electromagnetic field in space may be regarded as a superposition of two families of identical ellipsoidal equiphase surfaces each expanding outward from one half of the antenna just as though it were alone. The instantaneous field at a given point is the vector sum of the instantaneous values of the primed and double-primed components. It is shown in Fig. 9.1 for two antennas end to end, and in Fig. 9.2 for two antennas with ends separated $\lambda_0/20$. It is to be noted that neither $E_{e'}$ and $E_{e''}$ nor $E_{\rho'}$ and $E_{\rho''}$ are codirectional except in the far zone, where the two families of expanding ellipsoids coalesce into a single family of expanding spheres.

The combination of the fields due to the halves of an antenna treated as separate sources is a special application of the general

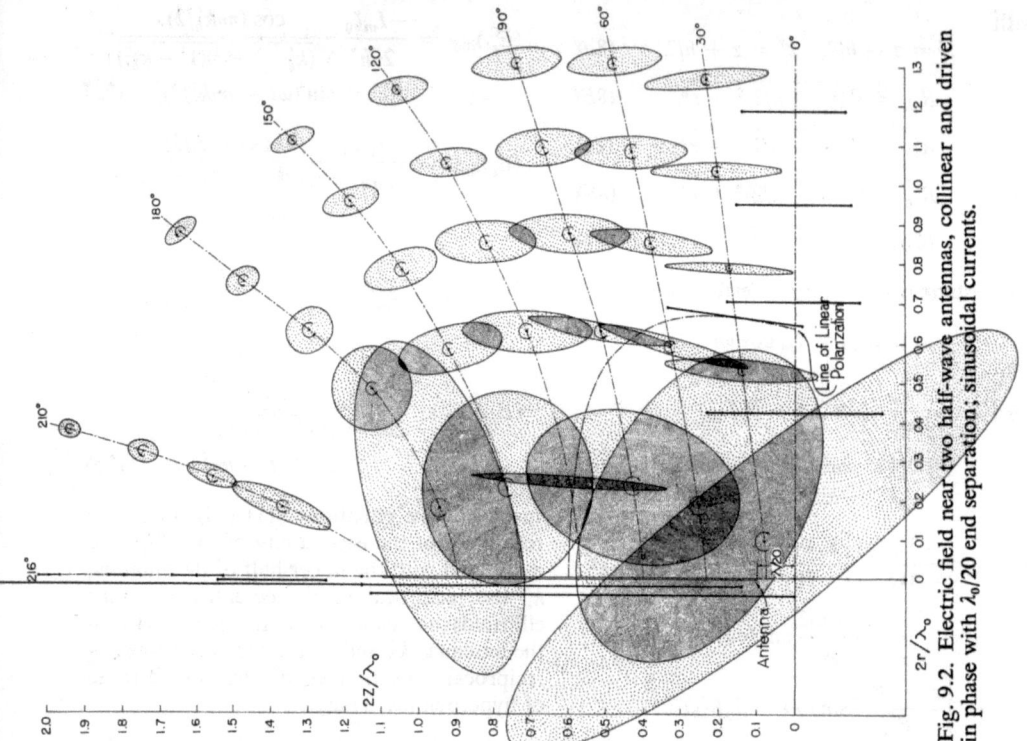

Fig. 9.2. Electric field near two half-wave antennas, collinear and driven in phase with $\lambda_0/20$ end separation; sinusoidal currents.

Fig. 9.1. Electric field near two half-wave antennas, collinear and driven in phase with no end separation; sinusoidal currents.

principle of superposition. The instantaneous electromagnetic field due to any number of arbitrarily oriented antennas may be regarded as the superposition of the instantaneous fields due to each antenna and due to each region in which charges are set in motion. This follows directly from the general integrals for the potential function or for the electromagnetic vectors. However, it is only in the special case of a multiple half-wave antenna that the expanding ellipsoidal surfaces of constant phase due to each half-wave element coalesce into a single ellipsoidal surface expanding from the antenna as a whole.

The general expressions (2.14) and (3.11) for the magnetic and electric fields of a linear radiator may be rearranged in the following equivalent way:

$$B_\theta = \frac{jI_m}{4\pi v_0 r}[(1 + \cos\beta_0 h)(e^{-j\beta_0 R_{1h}} + e^{-j\beta_0 R_{2h}})$$
$$- \cos\beta_0 h(e^{-j\beta_0 R_{1h}} + e^{-j\beta_0 R_{2h}} + 2e^{-j\beta_0 R_0})], \quad (16)$$

$$E_r = \frac{jI_m \zeta_0}{4\pi r}\left[(1 + \cos\beta_0 h)\left(\frac{z-h}{R_{1h}}e^{-j\beta_0 R_{1h}}\right.\right.$$
$$\left.+ \frac{z+h}{R_{2h}}e^{-j\beta_0 R_{2h}}\right) - \cos\beta_0 h\left(\frac{z-h}{R_{1h}}e^{-j\beta_0 R_{1h}}\right.$$
$$\left.\left.+ \frac{z+h}{R_{2h}}e^{-j\beta_0 R_{2h}} + \frac{2z}{R_0}e^{-j\beta_0 R_0}\right)\right], \quad (17)$$

$$E_z = \frac{-jI_m \zeta_0}{4\pi}\left[(1 + \cos\beta_0 h)\left(\frac{e^{-j\beta_0 R_{1h}}}{R_{1h}}\right.\right.$$
$$\left.+ \frac{e^{-j\beta_0 R_{2h}}}{R_{2h}}\right) - \cos\beta_0 h\left(\frac{e^{-j\beta_0 R_{1h}}}{R_{1h}}\right.$$
$$\left.\left.+ \frac{e^{-j\beta_0 R_{2h}}}{R_{2h}} + \frac{2}{R_0}e^{-j\beta_0 R_0}\right)\right]. \quad (18)$$

In each of these formulas the second parenthesis is exactly the same as in the corresponding formulas in (6.3) or (6.4) for the resonant multiple half-wave antenna. Similarly, the third parenthesis in each formula is the same as the corresponding expression in (4) and (5). It follows that the field of a center-driven antenna with sinusoidal current and arbitrary half-length h may be described in terms of the superposition of three families of expanding equiphase ellipsoids. Of these, the first has its foci at the ends $z = \pm h$ of the antenna separated the full length. Of the other two families of ellipsoids, one has its foci at $z = 0$ and h; the other, at $z = 0$ and $-h$. The resultant instantaneous field at any point may be obtained by combining vectorially the instantaneous contributions associated with each of the three families of ellipsoids. This resultant field is not readily visualized in the near and intermediate zones. In the far zone all three families of ellipsoids coalesce.

Although the representation in terms of ellipsoidal equiphase surfaces is advantageous in visualizing the nature and the behavior of the field near an antenna, direct evaluation of the field is probably more readily carried out using the relatively simple expressions (2.14) and (3.11) in cylindrical coördinates.

10. Graphs of Instantaneous Electric and Magnetic Fields for Multiple Half-Wave Antennas

The general electromagnetic field of a resonant antenna has been defined at every point in space in terms of its three non-vanishing components, $B_{\Phi \text{inst}}$, $E_{e \text{inst}}$, and $E_{\rho \text{inst}}$. Since the magnetic vector $\mathbf{B}_{\text{inst}}$ has only the single component $B_{\Phi \text{inst}}$, the $\mathbf{B}$-lines at points in space are necessarily directed tangent to circles about the z-axis on each of which r and k_h is constant. Depending upon the time and the reciprocal eccentricity k_e of the equiphase ellipsoid passing through the point, $\mathbf{B}_{\text{inst}}$ is directed clockwise or counterclockwise around the circle as determined by the phase factor,

$$C_1 = \sin(\omega t - \tfrac{1}{2}n\pi k_e), \quad n \text{ odd}, \quad (1a)$$
$$C_1 = \cos(\omega t - \tfrac{1}{2}n\pi k_e), \quad n \text{ even}. \quad (1b)$$

If it is desired to describe the $\mathbf{B}$-field in terms of any plane $z = z_1$ cut normally by the z-axis, rather than in terms of the intersection of equiphase ellipsoidal surfaces with hyperboloids, it is merely necessary to determine the radii of successive circles in this plane on all of which $\mathbf{B}_{\text{inst}}$ has the same phase. This is accomplished by plotting (1) as a function of k_e and then changing the scale of abscissas from k_e to r/h with $z = z_1$. This can be done graphically using a diagram like Fig. 8.4 from which the radial distances r/h corresponding to any value of k_e and z are readily obtained. It can also be done analytically using the equation (8.17) for an ellipsoid of revolution together with (8.18). Thus,

$$(r/h)^2 = (k_e^2 - 1)\left[1 - \frac{(z/h)^2}{k_e^2}\right]. \quad (2)$$

Two analytically simple cases are convenient examples. The first of these is the field in the equatorial plane, the second the field in the plane $z = h$, which contains the upper end of the antenna. In these cases (2) becomes

$$r/h = \sqrt{k_e^2 - 1}, \quad (z = 0) \tag{3a}$$

$$r/h = \frac{k_e^2 - 1}{k_e}. \quad (z = h) \tag{3b}$$

Using Fig. 8.3, or by direct calculation from (3a) or (3b), the value of $k_e^2 - 1$ corresponding to a given k_e is readily obtained and (1) can be plotted as a function of the value of r/h corresponding to k_e instead of as a function of k_c. In Fig. 10.1 three curves are shown for the instants when $\omega t = 0, 2\pi, 4\pi, \cdots$ for the case $n = 1$. The solid line shows $\sin \tfrac{1}{2}\pi k_e$ as a function of $k_e = z/h$ along the z-axis. This curve actually shows the instantaneous phase relations along the z-axis, since $k_e = a_e/h$, and a_e is measured along the z-axis. The curve starts at $k_e = 1$ or $a_e = h$, which is the upper end of the antenna. The dashed curve shows $\sin \tfrac{1}{2}\pi k_e$ plotted *as a function* of r/h along the equatorial plane where $z = 0$. The dotted curve is $\sin \tfrac{1}{2}\pi k_e$ plotted *against* r/h along the plane $z = h$ which contains the end of the antenna. At sufficiently great distances k_e approaches R_0/h. Along any plane, $r = R_0 \sin \Theta$. Therefore the dashed curve approaches the solid one asymptotically as $k_e \to R_0/h$, or as $db_e/dt \to v_0$. The dotted curve asymptotically approaches $\sin (\tfrac{1}{2}\pi r/h \sin \Theta)$. By plotting curves like those of Fig. 10.1 for any plane and any instant, the phase of the **B**-field may be pictured geometrically by families of concentric equiphase circles.

Since the electric vector $\mathbf{E}_{\text{inst}}$ cannot be reduced to a single component but consists of $E_{e\,\text{inst}}$ and $E_{\rho\,\text{inst}}$ in confocal coördinates or of $E_{r\,\text{inst}}$ and $E_{z\,\text{inst}}$ in cylindrical ones, the direction and phase of the E-lines can be obtained by combining the instantaneous values of the components at every point. This is a tedious procedure, so that it is preferable to derive an equation to characterize the E-field in any plane containing the antenna to correspond to equation (1), which describes the **B**-field in the plane normal to the antenna. (It is to be noted that by assigning to C_1 in (1) numerical values ranging from zero to unity families of equiphase circles for $\mathbf{B}_{\text{inst}}$ are obtained.) Such an equation is quickly derived by combining $E_{r\,\text{inst}}$ and $E_{z\,\text{inst}}$ as given by (6.12a) and (6.12c) in such a way as to obtain the equation of the curves that are everywhere tangent to $\mathbf{E}_{\text{inst}}$.

As a consequence of rotational symmetry with respect to the z-axis, the electric fields in all planes containing the z-axis are necessarily identical. Therefore, it is sufficient to consider the two-dimensional problem of determining the equation of the curves tangent to $\mathbf{E}_{\text{inst}}$ at every point in the r, z-plane. The slope m of a curve, $z = f(r)$, at a point P (Fig. 10.2) is defined by

$$m = \frac{dz}{dr}. \tag{4}$$

By definition, a curve tangent to $\mathbf{E}_{\text{inst}}$ at every point must have the slope of $\mathbf{E}_{\text{inst}}$ at that point. That is,

$$m = \left(\frac{E_z}{E_r}\right)_{\text{inst}}. \tag{5}$$

Hence a curve that is tangent to $\mathbf{E}_{\text{inst}}$ at every point must satisfy the relation

$$\frac{dz}{dr} = \left(\frac{E_z}{E_r}\right)_{\text{inst}} \tag{6}$$

or

$$E_{z\,\text{inst}}\,dr - E_{r\,\text{inst}}\,dz = 0. \tag{7}$$

Upon substituting (6.12a) and (6.12c) in (7), the following equation is obtained:

$$\left[\frac{1}{R_{1h}} \sin(\omega t - \beta_0 R_{1h}) \pm \frac{1}{R_{2h}} \sin(\omega t - \beta_0 R_{2h})\right] dr + \left[\frac{z-h}{R_{1h}} \sin(\omega t - \beta_0 R_{1h}) \pm \frac{z+h}{R_{2h}} \sin(\omega t - \beta_0 R_{2h})\right] \frac{dz}{r} = 0. \tag{8}$$

The upper sign is for n odd, the lower for n even. Rearrangement gives:

$$\frac{r\,dr + (z-h)\,dz}{R_{1h}} \sin(\omega t - \beta_0 R_{1h})$$
$$\pm \frac{r\,dr + (z+h)\,dz}{R_{2h}} \sin(\omega t - \beta_0 R_{2h}) = 0. \tag{9a}$$

However, since

$$R_{1h} = \sqrt{r^2 + (z-h)^2},$$
$$R_{2h} = \sqrt{r^2 + (z+h)^2}, \tag{9b}$$

it follows that

$$dR_{1h} = \frac{r\,dr + (z-h)\,dz}{R_{1h}},$$
$$dR_{2h} = \frac{r\,dr + (z+h)\,dz}{R_{2h}}. \qquad (10)$$

Hence, (9a) reduces to

$$\sin(\omega t - \beta_0 R_{1h})\,dR_{1h}$$
$$\pm \sin(\omega t - \beta_0 R_{2h})\,dR_{2h} = 0. \quad (11)$$

Upon integrating, the following equation for the instantaneous E lines is obtained:

$$\cos(\omega t - \beta_0 R_{1h}) \pm \cos(\omega t - \beta_0 R_{2h})$$
$$= \text{const.} \quad (12)$$

Using standard trigonometric formulas (12) may be transformed into

$$\cos \tfrac{1}{2}\beta_0(R_{2h} - R_{1h}) \cos[\omega t - \tfrac{1}{2}\beta_0(R_{2h} + R_{1h})]$$
$$= C, \quad \text{(upper sign, } n \text{ odd)}$$
$$\qquad (13a)$$

$$\sin \tfrac{1}{2}\beta_0(R_{2h} - R_{1h}) \sin[\omega t - \tfrac{1}{2}\beta_0(R_{2h} + R_{1h})]$$
$$= C. \quad \text{(lower sign, } n \text{ even)} \quad (13b)$$

In confocal coördinates,

$$\cos \tfrac{1}{2}n\pi k_h \cos(\omega t - \tfrac{1}{2}n\pi k_e) = C, \quad (n \text{ odd})$$
$$\qquad (14a)$$

$$\sin \tfrac{1}{2}n\pi k_h \sin(\omega t - \tfrac{1}{2}n\pi k_e) = C. \quad (n \text{ even})$$
$$\qquad (14b)$$

By assigning any desired value to C between -1 and $+1$, the instantaneous contour of an E line may be calculated from (14a) or (14b) for any instant t and any integral value of n. Because the trigonometric functions are multiple-valued, each value of C leads to a whole family of separate contours, each defined in terms of the confocal variables k_e and k_h. Curves of the E-vector about a multiple half-wave antenna were first calculated in this way by F. Hack[12] from formulas derived by M. Abraham.[1] Hack plotted contours showing the direction of the E-field in the neighborhood of the antenna for $n = 1, 2, 3$ and for times $\omega t = 2m\pi$, $\omega t = \pi/4 + 2m\pi$, $\omega t = \pi/2 + 2m\pi$, $\omega t = 3\pi/4 + 2m\pi$, $(m = 0, 1, 2, \cdots)$. Hack's curves are frequently reproduced;[1,38] a few are shown for $n = 1$ in Fig. 10.3. From these, the interesting nature of the E-field for an infinitely thin resonant antenna may be expressed in a graphical manner as follows.

Consider the instant, $\omega t = 0$, when $I_{\text{inst}} = I_m \cos \omega t$ has its maximum value. At this moment the entire antenna is uncharged and a maximum current is upward at the center. For a sufficiently thin antenna, its surface is given approximately by the degenerate ellipsoid $k_e = 1$. A surface of constant phase is defined by $\omega t - \tfrac{1}{2}\pi k_e = -\tfrac{1}{2}\pi$, which is equivalent to setting $C = 0$ in (14a). As time passes, a positive charge moves upward, a negative charge downward in the antenna, and the current at its center gradually diminishes. After $1/4$ period $\omega t = \pi/4$, and the surface of constant phase, $C = 0$, is $\tfrac{1}{4}\pi - \tfrac{1}{2}\pi k_e = -\tfrac{1}{2}\pi$ or $k_e = 1.5$. This defines the ellipsoid shown in Fig. 10.3b. Within this ellipsoid the E-lines calculated from (14a) with $n = 1$, and for four values of C between 0 and 1, follow the indicated contours. It is seen that they go from positive to negative charges. Their path is also determined by $(d\mathbf{B}/dt)_{\text{inst}}$. As time elapses after $\omega t = \pi/4$, a diminishing upward current continues and a greater and greater positive charge density accumulates in the upper half, a negative charge density in the lower half of the antenna. At the instant $\omega t = \pi/2$ there is no current anywhere in the antenna and a condition of maximum charge obtains at the ends. One now has $\tfrac{1}{2}\pi - \tfrac{1}{2}\pi k_e = -\tfrac{1}{2}\pi$ or $k_e = 2.0$. This defines the ellipsoid of constant phase shown in Fig. 10.3c. The E-lines within this ellipsoid go from positive to negative charge.

As more time passes, a downward current begins in the antenna and the charge on the ends diminishes as it moves toward the center where it is neutralized. The upper half is still positive, and the lower half negative. When $\omega t = 3\pi/4$ the equiphase ellipsoid has expanded to $k_e = 2.5$. Within it the E-lines still go from positive charge on the antenna, but some also encircle B-lines on which $(d\mathbf{B}/dt)_{\text{inst}}$ is large. This is shown in Fig. 10.3d.

In the next quarter cycle the downward current increases and the charge density at the ends diminishes. When $\omega t = \pi$ the entire antenna is uncharged and a maximum current is directed downward at the center. The equiphase ellipsoid has expanded to $k_e = 3$, as shown in Fig. 10.3a. The E-lines within this ellipsoid have the contours shown in Fig. 10.3a but with all arrows reversed. When the charge vanishes there remain the two sets of closed contours that are shown in Fig. 10.3a. The E-field is now determined by $(d\mathbf{B}/dt)_{\text{inst}}$. During the

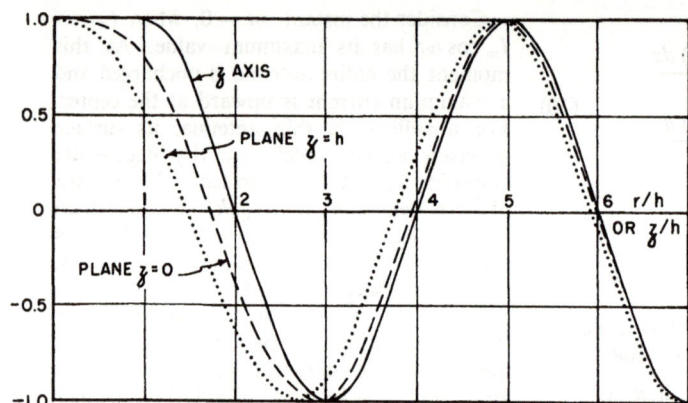

Fig. 10.1. The function $\sin\tfrac{1}{2}\pi k_e$ along the z-axis and in two planes.

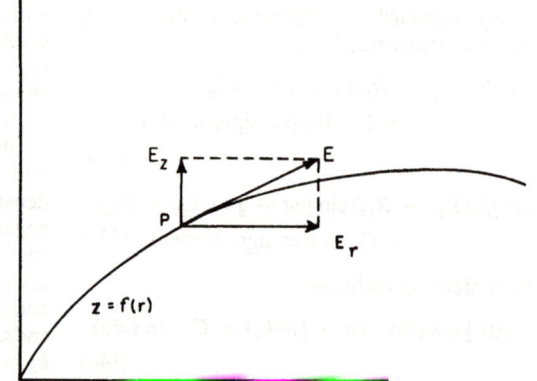

Fig. 10.2. Curve tangent to E.

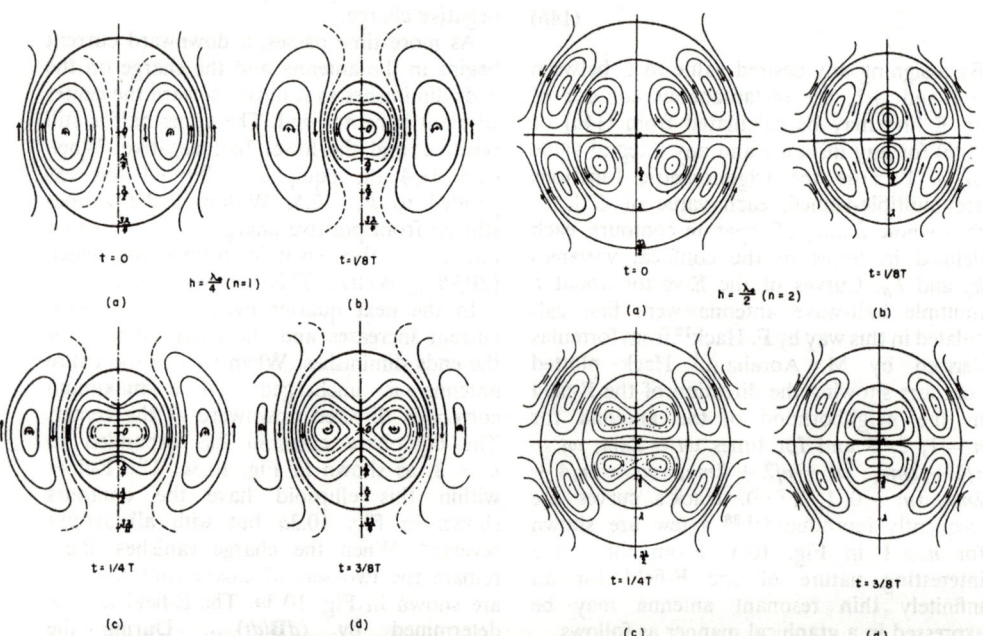

Fig. 10.3. E lines for a linear radiator of half length $h = \lambda_0/4$ with sinusoidal current (Hack).

Fig. 10.4. E lines for a linear radiator of half length $h = \lambda_0/2$ with sinusoidal current (Hack).

t = 0
(a)
t = 1/8 T
(b)
h = 3λ₀/4 (n = 3)
t = 1/4 T
(c)
t = 3/8 T
(d)

Fig. 10.5. ***E*** **lines for a linear radiator of half length** $h = 3\lambda_0/4$ **with sinusoidal current (Hack).**

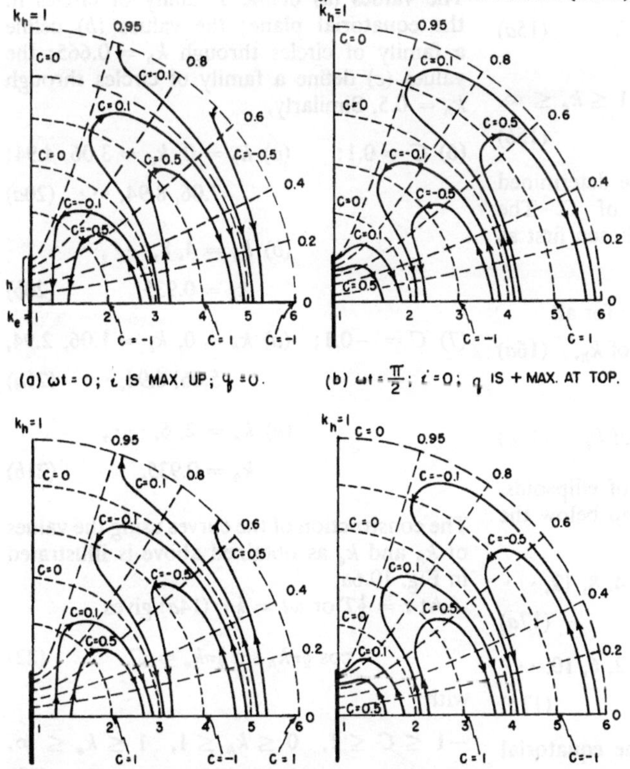

(a) ωt = 0; i IS MAX. UP; q = 0.
(b) ωt = π/2; i = 0; q IS + MAX. AT TOP.
(c) ωt = π; i IS MAX. DOWN; q = 0.
(d) ωt = 3π/2; i = 0; q IS −MAX. AT TOP.

Fig. 10.6. ***E*** **lines near a half-wave antenna.**

CONDUCTOR
SLOT IN CONDUCTING SHEET

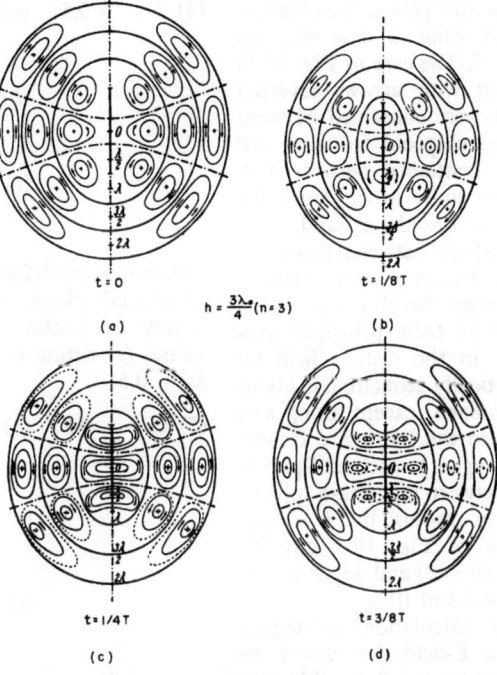

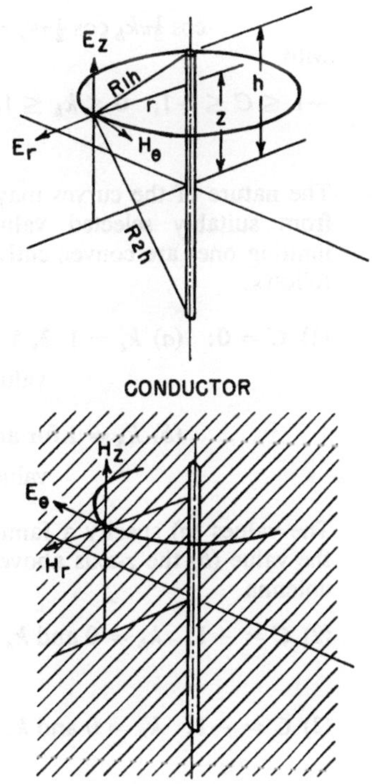

Fig. 11.1. Complementary electric and magnetic quantities.

next interval of a quarter period the ellipsoid expands to $k_e = 3.5$ while a new ellipsoid moves out to $k_e = 1.5$. The case of Fig. 10.3b is then repeated but with arrows reversed and the E-lines that were contained between the antenna and the ellipsoid from $\omega t = 0$ to $\omega t = \pi$ are now contained between two ellipsoids differing by unity in k_e. Since the major axes of the ellipsoids expand at a constant rate, the E-lines between them can only expand parallel to the circumference in ringlike sections between the ellipsoids which grow more circular as their principal axes increase. Ultimately in the distant field all E-lines become practically semicircular about the center of the antenna, except along and near the z-axis, where the loops are closed. The closed loops belonging to each half-period are then contained between equiphase spheres ($C = 0$), differing in radius by $h = \lambda_0/2$. Contours like those of Fig. 10.3 but for antennas of length $\lambda_0(n = 2)$ and $3\lambda_0/2(n = 3)$ are shown in Figs. 10.4 and 10.5.

To illustrate the calculation of curves from (14a), let the E-field for $n = 1$ be considered at the instant $t = 0$. In this case (14a) becomes

$$\cos \tfrac{1}{2}\pi k_h \cos \tfrac{1}{2}\pi k_e = C, \quad (15a)$$

with

$$-1 \leq C \leq +1, \quad 0 \leq k_h \leq 1, \quad 1 \leq k_e \leq \infty. \quad (15b)$$

The nature of the curves may be determined from suitably selected values of C. The limiting ones are conveniently chosen first as follows:

(1) $C = 0$: (a) $k_e = 1, 3, 5 \cdots$ for all values of k_h, (16a)

(b) $k_h = 1$ for all values of k_e. (16b)

The values (a) specify a family of ellipsoids, the value (b) the z-axis above and below the antenna.

(2) $C = +1$: $k_h = 0$ and $k_e = 4, 8, 12, \cdots$ (17a)

(3) $C = -1$: $k_h = 0$ and $k_e = 2, 6, 10, \cdots$ (17b)

These values define circles in the equatorial plane ($z = 0$).

(4) $C = +0.5$: (a) $k_h = 0$, $k_e = 3\tfrac{1}{3}, 4\tfrac{2}{3}, 7\tfrac{1}{3}, 8\tfrac{2}{3}, \cdots$ (18a)

(b) $k_e = 4, 8, \cdots$, $k_h = 2/3$, (18b)

(c) $k_h = 0.5$, $k_e = 3.5, 4.5, 7.5, 8.5, \cdots$ (18c)

The values (a) define a family of circles in the equatorial plane; the values (b) define a family of circles through $k_h = 0.665$; the values (c) define a family of circles through $k_h = 0.5$.

(5) $C = -0.5$: (a) $k_h = 0, k_e = 1\tfrac{1}{3}, 2\tfrac{2}{3}, 5\tfrac{1}{3}, 6\tfrac{2}{3}, \cdots$ (19a)

(b) $k_e = 2, 6, \cdots$, $k_h = 2/3$, (19b)

(c) $k_h = 0.5$, $k_e = 1.5, 2.5, 5.5, 6.5, 9.5, 10.5, \cdots$ (19c)

The values (a) define a family of circles in the equatorial plane; the values (b) define a family of circles through $k_h = 0.665$; the values (c) define a family of circles through $k_h = 0.5$. Similarly,

(6) $C = 0.1$: (a) $k_h = 0, k_e = 3.06, 4.94; 7.06, 8.94, \cdots$ (20a)

(b) $k_e = 4, 8, \cdots$, $k_h = 0.935$, (20b)

(7) $C = -0.1$: (a) $k_h = 0$, $k_e = 1.06, 2.94, 5.06, 6.94, \cdots$ (21a)

(b) $k_e = 2, 6, \cdots$, $k_h = 0.935$. (21b)

The construction of the curves using the values of k_h and k_e as obtained above is illustrated in Fig. 10.6a.

At $t = \tfrac{1}{4}T$ or $\omega t = \tfrac{1}{2}\pi$, (14a) gives

$$\cos \tfrac{1}{2}\pi k_h \sin \tfrac{1}{2}\pi k_e = C, \quad (22)$$

with

$$-1 \leq C \leq 1, \quad 0 \leq k_h \leq 1, \quad 1 \leq k_e \leq \infty. \quad (23)$$

The following numerical results are readily obtained:

(1) $C = 0$: (a) $k_e = 2, 4, 6, 8$ for all values of k_h, (24a)

(b) $k_h = 1$ for all values of k_e. (24b)

(2) $C = +1$: $k_h = 0$, $k_e = 1, 5, 9, \cdots$ (25a)

(3) $C = -1$: $k_h = 0$, $k_e = 3, 7, 11, \cdots$ (25b)

(4) $C = -0.5$: (a) $k_h = 0$,
$k_e = 1\frac{2}{3}, 4\frac{1}{3}, 5\frac{2}{3}, \cdots$ (26a)

(b) $k_e = 1, 5, 9, \cdots$,
$k_h = 2/3$. (26b)

(5) $C = -0.5$: (a) $k_h = 0$,
$k_e = 2\frac{1}{3}, 3\frac{2}{3}, 6\frac{1}{3}, 7\frac{2}{3}, \cdots$ (27a)

(b) $k_e = 3, 7, 11, \cdots$,
$k_h = 2/3$. (27b)

(6) $C = 0.1$: (a) $k_h = 0$, $k_e = 4.06, 5.94$,
$8.06, 9.94, \cdots$ (28a)

(b) $k_e = 1, 5, 9, \cdots$,
$k_h = 0.935$. (28b)

(7) $C = -0.1$: (a) $k_h = 0$, $k_e = 2.06, 3.94$,
$6.06, 7.94, \cdots$ (29a)

(b) $k_e = 3, 7, 11, \cdots$,
$k_h = 0.935$. (29b)

The construction of the curves using the values of k_h and k_e obtained above is illustrated in Fig. 10.6.

11. Correlation of Theoretical Near-Zone Field of Multiple Half-Wave Antenna with Experiment

In Secs. 6 through 10 the electromagnetic field due to a sinusoidal distribution of current along the part of the z-axis between $z = -h$ and $z = h$ is determined rigorously for all points in space. When $2h$ is an integral multiple of a half wavelength the complicated field near the current filament is represented in an elegant and relatively simple manner in terms of its components in confocal coördinates. Characteristic of the near-zone field are the graphs in Figs. 8.4 through 8.6 of the elliptically polarized electric field. Rigorously, the assumed sinusoidal distribution of current can be maintained only by a continuous distribution of generators along the entire length of the antenna. However, it is demonstrated theoretically and verified experimentally in Chapter II that the distribution of current along a *center-driven* cylindrical antenna with hemispherical caps, *small radius a, and length 2h near a half wavelength is very nearly* sinusoidal in amplitude and constant in phase along the entire antenna even when h/a is as small as 75. Note, however, that this current is concentrated primarily in a thin sheet near the surface of the cylinder and not along the axis, as assumed in determining the near-zone field. Since this difference in the lateral distribution of the *rotationally symmetric* current can affect the field only within distances of the order of magnitude of the radius a, it is correct to assume that the fields determined in Secs. 6 through 10 are good approximations of the fields of center-driven antennas of finite cross section and lengths near $\lambda_0/2$ at distances from the axis of the antenna that are large compared with the radius a, which is very small compared with the wavelength.

It is shown in Sec. I.7 that a flat strip with rounded edges and width $w = 4a$ is analytically equivalent to a cylindrical antenna of radius a, since it has the same value of Ω. Such a strip antenna of length $2h$ and semicircular ends should have essentially the same axial distribution of total current as the cylinder, and its electromagnetic field should also be the same at distances that are great compared with w, provided w is sufficiently small compared with the wavelength.

The direct experimental verification of a complicated field of the type shown in Figs. 8.4, 8.5, and 8.6 is a difficult problem. Evidently, it requires freely movable probes that can determine both the direction and the magnitude of the electric field at any point in space near the antenna. Moreover, in order to verify the elliptic polarization of the field, the probe must be rotatable about an axis perpendicular to a plane containing the antenna. The experimental difficulties are somewhat simplified if half of the antenna is erected vertically on a large, highly conducting plane so that the antenna may be base-driven from a coaxial line with the generator and associated equipment under the plane, but the difficulties associated with the exploring probe are not solved thereby. Indeed, no accurate direct measurements of the near-zone electric field of a dipole have been made. However, measurements are available in which the actual problem of determining an electric field like that in Fig. 8.4 is replaced by an equivalent or complementary problem in which a quantity is measured that behaves in the ideal case *exactly* like this electric field.

It is a well-known and useful practice in many branches of physical science to solve an experimentally difficult problem by replacing it with a simpler *equivalent* one. Such a substitution of one problem for another is always possible if a physical quantity can be found that satisfies in its mathematical model the same differential equation and the same boundary conditions as the unknown quantity originally in question. It is, of course, assumed that the boundary conditions are adequate to assure a unique solution. This is true in an electromagnetic problem with harmonic time dependence if either the tangential electric field or the tangential magnetic field is specified on each of the several parts of an enclosing boundary. If infinity is part of the boundary, the fields must satsfy a radiation condition that includes vanishing as $1/R$ as $R \to \infty$. For example, in determining the electromagnetic field in space due to currents in a conducting cylinder or strip, the following boundary conditions are sufficient to assure a unique solution:

Antenna in space:

(a) $\mathbf{B}_{\text{tang}}$ given on the surface of the antenna, (1)
(b) $\mathbf{E}$ and $\mathbf{B}$ satisfy a radiation condition.

Note that the solution actually obtained for these boundary conditions in Secs. 6 through 10 involves a magnetic field that is perpendicular to any plane that includes the z-axis through the center of the antenna, for example, the x,z-plane, which cuts the antenna lengthwise in the middle. Hence, if interest is confined to each half-space separately, the same unique solution in the half-space $y \geq 0$ is obtained if the following conditions are imposed:

Half antenna in half-space:

(a) $\mathbf{B}_{\text{tang}}$ given on the half of the surface of the antenna in the region in question, (2)
(b) $\mathbf{B}_{\text{tang}} = 0$ on the x,z-plane except where this intercepts the antenna,
(c) $\mathbf{E}$ and $\mathbf{B}$ satisfy a radiation condition.

If the antenna is a thin strip and this has its flat surface in the x,z-plane, these conditions become:

Strip antenna:

(a) $\mathbf{B}_{\text{tang}}$ is given on the x,z-plane between $z = -h$ and h and $x = -w/2$ and $w/2$, (3)
(b) $\mathbf{B}_{\text{tang}} = 0$ on the x,z-plane excluding the part specified in (a),
(c) $\mathbf{E}$ and $\mathbf{B}$ satisfy a radiation condition.

Since both $\mathbf{E}$ and $\mathbf{B}$ satisfy the homogeneous wave equation at all points in free space,

$$\nabla^2 \mathbf{E} + \beta_0^2 \mathbf{E} = 0, \tag{4a}$$

$$\nabla^2 \mathbf{B} + \beta_0^2 \mathbf{B} = 0, \tag{4b}$$

it is possible to interchange $\mathbf{E}$ and $\mathbf{B}$ in (3) and in this manner obtain a set of boundary conditions that must lead to a solution exactly like that for (4a) with (3) but with $\mathbf{E}$ and $\mathbf{B}$ interchanged. The proposed boundary conditions are:

Slot antenna:

(a) $\mathbf{E}_{\text{tang}}$ is given on the x,z-plane between $z = -h$ and h and between $x = -w/2$ and $w/2$, (5)
(b) $\mathbf{E}_{\text{tang}} = 0$ on the x,z-plane excluding the part specified in (a),
(c) $\mathbf{E}$ and $\mathbf{B}$ satisfy a radiation condition.

The uniqueness theorem requires that $\mathbf{B}$ and $\mathbf{E}$, obtained by solving (4b) and (4a) subject to (5) in the half-space $y \geq 0$, must be the same as the solution for $\mathbf{E}$ and $\mathbf{B}$, obtained by solving (4a) and (4b) subject to (3) in the half-space $y \geq 0$, provided the given distribution of $\mathbf{B}_{\text{tang}}$ in (3a) is the same as the given distribution of $\mathbf{E}_{\text{tang}}$ in (5a). (Owing to the vector relation between $\mathbf{E}$ and $\mathbf{B}$ there is a difference in sign of certain components; see Fig. 11.1). It follows that the field graphs of the elliptically polarized electric field in the x,z-plane shown in Figs. 8.4 through 8.6, which represent solutions of (4a) subject to (3), become field graphs of the elliptically polarized *magnetic* field in the x,z-plane that represent solutions of (4b) subject to (5) merely by writing $\mathbf{B}$ for $\mathbf{E}$. Evidently, if an experimental arrangement that satisfies the boundary conditions (5) with $\mathbf{E}_{\text{tang}}$ in (5a) the same as $\mathbf{B}_{\text{tang}}$ in (3a) is used to measure the elliptically polarized magnetic field, this may be identified with the electric field for the boundary conditions (3), which apply to an actual strip antenna. It follows from the general boundary condition (I.4.15), specialized to the boundary between empty space and a perfectly conducting strip with a sheet of surface current in the z-direction, that

$$H_x = \nu_0 B_x = I_z, \qquad (6)$$

where I_z is the surface density of current in the conducting strip and B_x is the tangential magnetic field. Since the axial variation of I_z is the same as that of the total current, and this is assumed to be sinusoidal, it follows that for $h = n\lambda_0/4$

$$B_x = B_{\max} \cos \beta_0 z, \quad (n \text{ odd}) \qquad (7a)$$

$$B_x = B_{\max} \sin \beta_0 z. \quad (n \text{ even}) \qquad (7b)$$

Except near the rounded ends, B_x may be assumed to be approximately constant over the width w of the conducting strip. If (7a, b) are used as the boundary condition (3a) for the strip antenna, the corresponding condition (5a) for the slot antenna with $h = n\lambda_0/4$ is

$$E_x = E_{\max} \cos \beta_0 z, \quad (n \text{ odd}) \qquad (8a)$$

$$E_x = E_{\max} \sin \beta_0 z. \quad (n \text{ even}) \qquad (8b)$$

With (8a, b) the boundary conditions (5) are achieved with an infinite, perfectly conducting sheet in the x,z-plane containing a rectangular slot bounded by $z = \pm h$, $x = \pm w/2$, across which is maintained an impressed electric field distributed according to (8a) or (8b). That is, the electromagnetic field in the half-space $y \geq 0$ due to the impressed field (8a) or (8b) in a slot of length $2h$ and width w in a perfectly conducting plane is the same with **E** and **B** interchanged as the electromagnetic field in the same region if the slot is replaced by a metal strip with an impressed sinusoidal axial current (or transverse magnetic field) and the infinite conducting plane is removed. These complementary conditions correspond to those familiar in optics, as expressed in *Babinet's principle*. The two complementary arrangements are illustrated in Fig. 11.1. The elliptically polarized *electric* field in the x,z-plane of an antenna as shown in Figs. 8.4 through 8.6 is the same as the elliptically polarized *magnetic* field *on the conducting sheet* in the x,z-plane due to a sinusoidally distributed transverse electric field across the slot.

Extensive measurements of the magnitude and polarization of the magnetic field tangent and very close to a large metal sheet have been made by R. M. Hatch.[17] The sheet was excited by a very narrow slot ($0.015\lambda_0$) of length $\lambda_0/2$ or λ_0 in which a transverse sinusoidally distributed electric field was maintained by a specially designed wave guide. The probe consisted of a very small shielded loop that could be moved over the entire sheet and rotated about its axis. For convenience in locating the probe a map of the confocal coördinate system was drawn to scale for each measurement and placed in proper position relative to the slot on the metal sheet. The results of these measurements, which correspond to the determination of the essential properties of the magnetic equivalents of Figs. 8.4 and 8.5, are contained in Figs. 11.2 through 11.7, together with theoretical curves obtained from Figs. 8.4 and 8.5 or the equations from which these are calculated. They are discussed in turn.

Figure 11.2. The direction ε, that is, the slope of the ellipse, was calculated along one of the ellipses ($k_\varepsilon = 3.0$) and plotted against k_h for $n = 1$. By setting the loop at various points along this ellipse and rotating it to locate its maximum and minimum positions, the ε-direction could be determined. The experimental results, indicated by small circles on the figure, are in good agreement with theoretical curves. This procedure was repeated for several of the ellipses and the angular differences between theoretical and experimental results were plotted against k_h for several values of k_ε. The greatest errors were in the region of nearly circular polarization, where the maxima and minima were difficult to locate accurately. Otherwise the differences were quite random and the average of all differences was only about 4°.

Figure 11.3. If the ratios of major to minor axes of the polarization ellipses in Fig. 8.4 are calculated for numerous values of k_ε and k_h, a family of curves is obtained as shown in Fig. 11.4. For each value of k_ε the polarization varies from linear through circular and back to linear. Since the polarization ellipse rotates on passing through the locus of circular polarization, the ratio is plotted against positive and negative decibels. Zero decibel indicates circular polarization. Good agreement between theory and experiment is indicated with an average error of only about 1 db.

Figure 11.4. The computation of contours of constant major axis of the polarization ellipses involved calculating the field for various combinations of k_ε and k_h and then

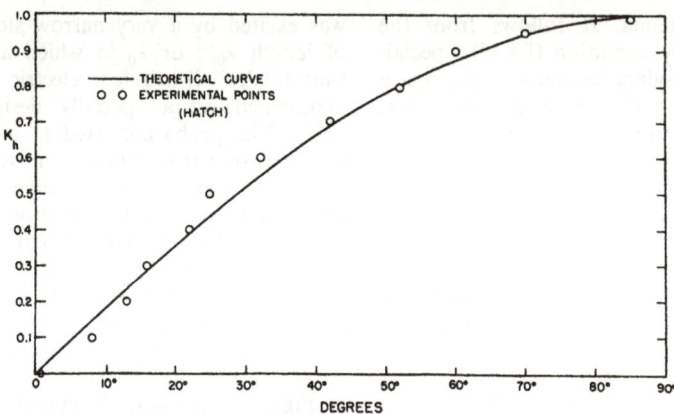

Fig. 11.2. Direction of magnetic field component H_ε with respect to axis of slot along ellipse $l_e = 3.0$ (for $n = 1$).

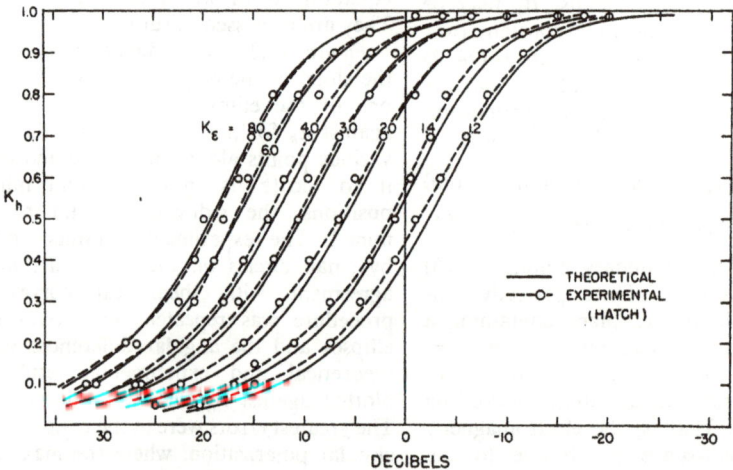

Fig. 11.3. Ratio of major to minor axis of polarization ellipse for $n = 1$.

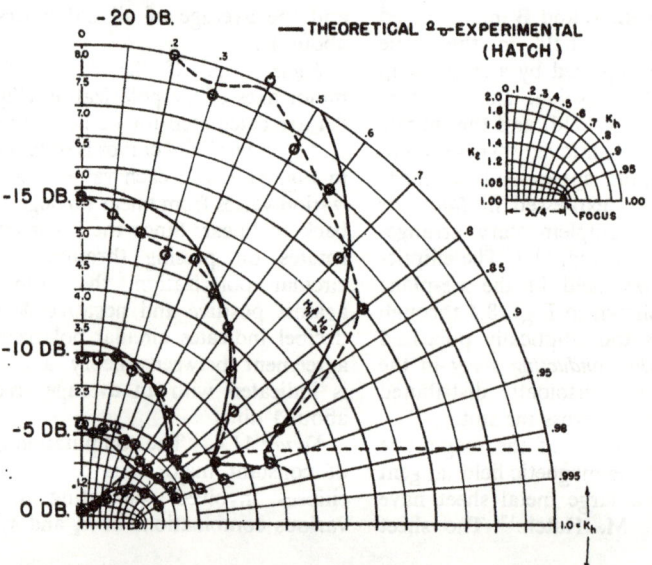

Fig. 11.4. Contours of constant major axis of polarization ellipse for H (or surface-current density) for $n = 1$.

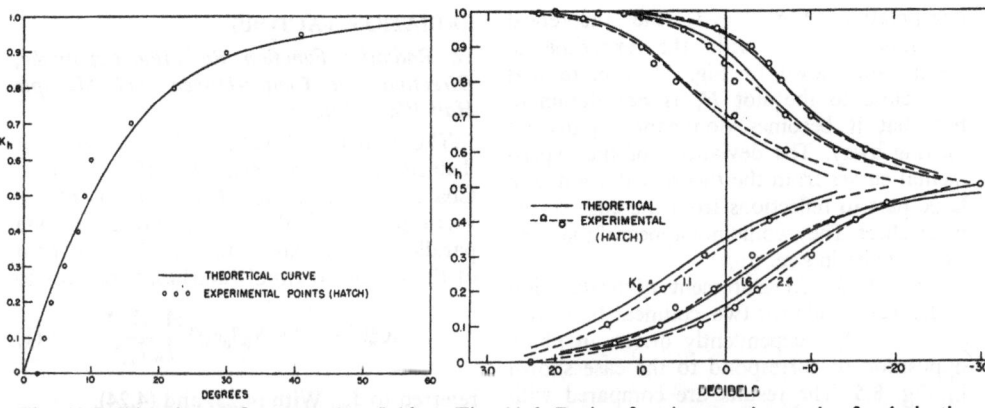

Fig. 11.5. Direction of magnetic field component H_ε with respect to axis of slot along ellipse $K_\varepsilon = 1.05$ for $n = 2$.

Fig. 11.6. Ratio of major to minor axis of polarization ellipse for two half-wave collinear slots driven 180° out of phase.

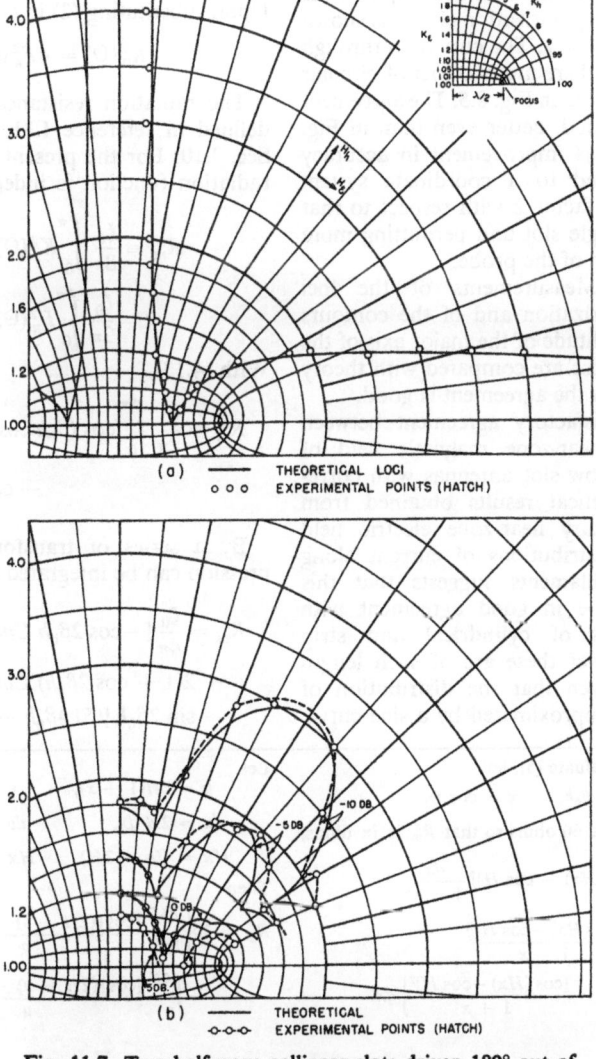

Fig. 11.7. Two half-wave collinear slots driven 180° out of phase ($n = 2$): (a) loci of circular polarization; (b) contours of constant major axis of polarization ellipse for H (or surface-current density).

interpolating. For $n = 1$, the theoretical contours together with the experimental results are shown in Fig. 11.5. Note that very close to the slot H_ε is not dominant but that it becomes dominant rapidly on moving away. The deviations of the experimental curves from the theoretical are due in large part to reflections from the edge of the metal sheet. They would not appear if the sheet were infinite in size.

Figure 11.5. Measurements of the direction of H_ε were made for two collinear half-wave slots ($n = 2$) independently driven in phase opposition to correspond to the case shown in Fig. 8.5. The results are compared with theory in a manner similar to that in Fig. 11.2 for the single slot ($n = 1$). The average error is about 3°.

Figure 11.6. The ratio of the major to the minor axis of the polarization ellipse is shown in Fig. 11.6 for $n = 2$. The curves go through zero twice since there are two loci of circular polarization, as seen in Fig. 8.5. The agreement with theory is good, better even than in Fig. 11.4. Perhaps this improvement in accuracy can be attributed to a coördinate system enlarged by the factor 2 with respect to that used for the single slot and permitting more accurate location of the probe.

Figure 11.7. Measurements of the loci of circular polarization and of the contours of constant amplitude of the major axis of the polarization ellipse are compared with theory for $n = 2$. Again the agreement is good.

The very satisfactory agreement between the measured near-zone magnetic field of moderately narrow slot antennas with corresponding theoretical results obtained from the complementary near-zone electric field of sinusoidal distributions of current along infinitely thin filaments suggests that this theory should be in good agreement with near-zone fields of cylindrical and strip antennas whenever these are of such length and are so driven that the distribution of current is well approximated by a sine curve.

RADIATION FACTORS

12. Radiation Function, Radiation Resistance, Directivity of Center-Driven and Multiple Half-Wave Antennas

The radiation function of an antenna is defined in general terms in reference I.31, Chap. IV, and in Sec. I.10 in terms of the vector potential in the far zone. For an antenna parallel to the z-axis, so that the Φ-component of the vector potential vanishes, it reduces to

$$K_m^2(\Theta) \equiv (4\pi R_0 \beta_0 v_0)^2 \frac{A_\Theta^r A_\Theta^{r*}}{I_m I_m^*}, \quad (1)$$

referred to I_m. With (4.23) and (4.24),

$$A_\Theta^r = \frac{-I_m}{2\pi v_0 \beta_0} \frac{e^{-j\beta_0 R_0}}{R_0} F_m(\Theta, \beta_0 h). \quad (2)$$

Upon substituting (2) in (1),

$$K_m^2(\Theta) = 4F_m^2(\Theta, \beta_0 h). \quad (3)$$

The radiation resistance referred to I_m is defined in reference I.31, Chap. IV, and in Sec. I.10. For the present case, in which the radiation function is independent of Φ, it is

$$R_m^e \equiv \frac{\zeta_0}{8\pi} \int_0^\pi K_m^2(\Theta) \sin \Theta \, d\Theta$$

$$= \frac{\zeta_0}{2\pi} \int_0^\pi F_m^2(\Theta, \beta_0 h) \sin \Theta \, d\Theta. \quad (4)$$

With (4.13),

$$R_m^e = \frac{\zeta_0}{2\pi} \int_0^\pi [\cos (\beta_0 h \cos \Theta)$$

$$- \cos \beta_0 h]^2 \frac{d\Theta}{\sin \Theta}. \quad (5)$$

By a series of transformations* this expression can be integrated into

$$R_m^e = \frac{\zeta_0}{4\pi} [-\cos 2\beta_0 h \operatorname{Cin} 4\beta_0 h$$

$$+ 2(1 + \cos 2\beta_0 h) \operatorname{Cin} 2\beta_0 h$$

$$+ \sin 2\beta_0 h [(\operatorname{Si} 4\beta_0 h - 2 \operatorname{Si} 2\beta_0 h)] \quad (6a)$$

---

* In order to evaluate (5), let

$$H = \beta_0 h, \quad x = \cos \Theta.$$

Then, with $\zeta_0/2\pi \doteq 60$ ohms so that R_m^e is in ohms,

$$R_m^e = 60 \int_{-1}^{+1} [\cos (Hx) - \cos H]^2 \frac{dx}{1 - x^2}$$

$$= 30 \int_{-1}^{+1} \left\{ \frac{[\cos (Hx) - \cos H]^2}{1 - x} \right.$$

$$\left. + \frac{[\cos (Hx) - \cos H]^2}{1 + x} \right\} dx.$$

Let

$$u = 2H(1 + x), \quad v = 2H(1 - x),$$

$$du = 2H \, dx, \quad dv = -2H \, dx,$$

$$Hx = \tfrac{1}{2}(u - 2H), \quad Hx = -\tfrac{1}{2}(v - 2H).$$

Then

$$R_m^e = 30 \left\{ \int_0^{4H} \frac{[\cos\tfrac{1}{2}(2H - v) - \cos H]^2}{v} dv \right.$$

$$\left. + \int_0^{4H} \frac{[\cos\tfrac{1}{2}(2H - u) - \cos H]^2}{u} du \right\}$$

This can also be written in the following longer form;

$$R_m^e = \frac{\zeta_0}{4\pi}[\sin 2\beta_0 h(\text{Si } 4\beta_0 h - 2 \text{ Si } 2\beta_0 h)$$
$$- \cos 2\beta_0 h (2 \text{ Cin } 2\beta_0 h - \text{Cin } 4\beta_0 h$$
$$- \ln \beta_0 h - C) + 2(C + \ln 2 + \ln \beta_0$$
$$- \text{Cin } 2\beta_0 h)] \qquad (6b)$$

where C is Euler's constant. The function R_m^e is represented graphically in Fig. 12.1.

For $\beta_0 h = \frac{1}{2}n\pi$, n odd,

$$R_m^e = R_0^e = \frac{\zeta_0}{4\pi} \text{Cin } 2n\pi = 30 \text{ Cin } 2n\pi \text{ ohms}. \qquad (7)$$

For
$n = 1$, $R_m^e = R_0^e = 73.13$ ohms;
$n = 3$, $R_m^e = R_0^e = 105.5$ ohms.

For $\beta_0 h = \pi$,

$$R_m^e = \frac{\zeta_0}{4\pi}(-\text{Cin } 4\pi + 4 \text{ Cin } 2\pi) = 199.1 \text{ ohms}.$$

The radiation resistance referred to the input (sinusoidal) current is readily obtained. It is

$$R_0^e = R_m^e/\sin^2 \beta_0 h \qquad (8)$$

or

$$R_0^e = \frac{\zeta_0}{4\pi}[(1 - \cot^2 \beta_0 h) \text{ Cin } 4\beta_0 h$$
$$+ 4 \cot^2 \beta_0 h \text{ Cin } 2\beta_0 h$$
$$+ 2 \cot \beta_0 h (\text{Si } 4\beta_0 h - 2 \text{ Si } 2\beta_0 h)]. \qquad (9)$$

The form (9) is identically the same as the modified zeroth-order input resistance obtained in Sec. II.28 for an antenna of finite radius subject to the conditions

$$\Omega \sin \beta_0 h \gg B_1^I, \qquad (10a)$$

$$\Omega \sin \beta_0 h \gg A_1^I, \qquad (10b)$$

where $\Omega = 2 \ln (2h/a)$ and A_1^I and B_1^I are defined and tabulated in Sec. II.28.

The input resistance R_0 and the radiation resistance referred to maximum current

$$= 60 \int_0^{4H} \frac{[\cos \frac{1}{2}(2H - v) - \cos H]^2}{v} dv$$

$$= 60 \int_0^{4H} \left[\frac{\cos^2 \frac{1}{2}(2H-v) - 2 \cos H \cos \frac{1}{2}(2H-v)}{v} + \cos^2 H\right] dv;$$

with
$$\cos \frac{1}{2}x = \sqrt{\frac{1}{2}(1 + \cos x)},$$

$$R_m^e = 60 \int_0^{4H} \{\frac{1}{2}[1 + \cos (2H - v)]$$
$$- 2 \cos H \cos \frac{1}{2}(2H-v) + \frac{1}{2}[1 + \cos 2H]\} \frac{dv}{v}$$

$$= 30 \left\{\int_0^{4H} [1 + \cos (2H - v) + 1 + \cos 2H] \frac{dv}{v}\right.$$
$$\left. - 4 \cos H \int_0^{4H} \cos \frac{1}{2}(2H - v) \frac{dv}{v}\right\},$$

$$= 30 \left\{\int_0^{4H} [\cos (2H-v) - \cos 2H]\frac{dv}{v}\right.$$
$$+ 2(1 + \cos 2H) \int_0^{4H} \frac{dv}{v}$$
$$\left. - 4 \cos H \int_0^{4H} \cos \frac{1}{2}(2H-v) \frac{dv}{v}\right\}.$$

In the last two integrals let
$w = \frac{1}{2}v$, $dw = \frac{1}{2} dv$,
$\frac{1}{2}(2H - v) = H - w$,
$v = 0$, $w = 0$,
$v = 4H$, $w = 2H$.

Then

$$R_m^e = 30 \left\{\int_0^{4H} [\cos (2H-v) - \cos 2H]\frac{dv}{v}\right.$$
$$+ 2(1 + \cos 2H) \int_0^{2H} \frac{dw}{w}$$
$$\left. - 4 \cos H \int_0^{2H} \cos (H - w) \frac{dw}{w}\right\}.$$

However,

$$2 \cos H \cos (H - w) = \cos w + \cos (2H - w).$$

Hnce, with R_m^e in ohms,

$$R_m^e = 30 \left\{\int_0^{4H} [\cos(2H - v) - \cos 2H]\frac{dv}{v}\right.$$
$$\left. + 2 \int_0^{2H} [1 + \cos 2H - \cos w - \cos(2H-w)]\frac{dw}{w}\right\}$$

$$= 30 \left\{\int_0^{4H} [\cos 2H \cos v + \sin 2H \sin v - \cos 2H]\frac{dv}{v}\right.$$
$$+ 2 \int_0^{2H} [1 + \cos 2H - \cos w - \cos 2H \cos w$$
$$\left. - \sin 2H \sin w]\frac{dw}{w}\right\}$$

$$= 30 \left\{-\cos 2H \int_0^{4H} \frac{1 - \cos v}{v} dv\right.$$
$$+ \sin 2H \int_0^{4H} \frac{\sin v}{v} dv$$
$$+ 2(1 + \cos 2H) \int_0^{2H} \frac{(1 - \cos w)}{w} dw$$
$$\left. - 2 \sin 2H \int_0^{2H} \frac{\sin w}{w} dw\right\}$$

$$= 30\{-\cos 2H \text{ Cin } 4H + \sin 2H \text{ Si } 4H$$
$$+ 2(1 + \cos 2H) \text{ Cin } 2H - 2 \sin 2H \text{ Si } 2H\}.$$

Finally,

$$R_m^e = 30\{-\cos 2H \text{ Cin } 4H + 2(1 + \cos 2H) \text{ Cin } 2H$$
$$+ \sin 2H(\text{Si } 4H - 2 \text{ Si } 2H)\}.$$

(neglecting ohmic losses in the conductor) are related according to the equations

$$I_0^2 R_0 = I_m^2 R_m^e, \qquad (11a)$$

$$R_m^e = \left(\frac{I_m}{I_0}\right)^2 R_0. \qquad (11b)$$

By reading first-order $|I_m|$ and $|I_0|$ from Figs. II.22.1–II.22.6 and first-order R_0 from Figs. II.30.1a, c, d, R_m^e can be calculated for an antenna of nonvanishing radius. The results of this calculation for an antenna for which $\Omega = 20$ and $\Omega = 10$ are shown in Fig. 12.1 together with the value for $\Omega \to \infty$. It is seen that R_m^e is reduced somewhat as h/a decreases.

The absolute directivity of an antenna in a direction Θ_m in which $F_m(\Theta, \beta_0 h)$ has a maximum value is defined by (I.10.7) in the following form with R_m^e in ohms:

$$D \equiv D(\Theta_m, \beta_0 h) = \frac{\zeta_0}{4\pi} \frac{K_m^2(\Theta_m)}{R_m^e} = 30 \frac{K_m^2(\Theta_m)}{R_m^e} \qquad (12)$$

or, with (3) and R_m^e in ohms,

$$D(\Theta_m, \beta_0 h) = 120 \frac{F_m^2(\Theta_m, \beta_0 h)}{R_m^e}. \qquad (13)$$

Absolute directivities for the principal ear are shown in Fig. 12.2 together with the angles Θ_m involved. Intermediate values are shown in Fig. IV.10.3. For the half-wave dipole ($h = \lambda_0/4$),

$$D\left(\frac{\pi}{2}, \frac{\pi}{2}\right) = 1.64. \qquad (14)$$

This value is used as standard in defining the relative directivity,

$$D_r = \frac{D}{1.64}, \qquad (15)$$

and the gain in decibels over a dipole,

$$\text{Gain (db)} = 10 \log_{10} D_r. \qquad (16)$$

Corresponding to the absolute directivity* shown in Fig. 12.2, the decibel gain referred to a $\lambda/2$ dipole is represented in Fig. 12.3.

The formula for the radiation resistance referred to maximum sinusoidal current of a multiple half-wave antenna can be shown to be the same for n even as for n odd. It is

---

* See also Fig. IV.10.1.

evaluated in (7) for n odd. The general formula is

$$R_m^e = \frac{\zeta_0}{4\pi} \operatorname{Cin} 2n\pi = 30 \operatorname{Cin} 2n\pi \text{ ohms},$$

$$n \text{ integral.} \qquad (17)$$

Numerical values are listed in Table 12.1 for R_m^e, D, and D_r, for the multiple half-wave antenna.

As stated in reference I.31, Chap. IV, and in I.10 the total time-average power $\bar{T}_r$ transferred across a *closed* surface Σ is given by the real part of the complex energy transfer function T, where

$$T = \bar{\bar{T}}_r + j T_i = \int_{\Sigma(\text{closed})} \hat{\mathbf{n}} \cdot \mathbf{S} \, d\sigma$$

$$= \tfrac{1}{2} \int_{\Sigma(\text{closed})} \hat{\mathbf{n}} \cdot \mathbf{E} \times \mathbf{H}^* \, d\sigma. \qquad (18)$$

The evaluation of the radiation function (1) and the radiation resistance (4) involve the integration of (18) over a great sphere enclosing the entire transmitting system. Since the specific formula for $F_m(\Theta, \beta_0 h)$ used in (3) and (4) assumes a sinusoidal distribution of current, and since this in turn implies that the current in the antenna is maintained by a continuous distribution of generators *along the entire antenna* and not by a single generator or a transmission line feeding the antenna at its center, the *complete transmitting system* actually assumed consists of an unbroken conducting thread with an appropriate continuous distribution of generators. It follows that the surface of integration Σ need enclose only the antenna, since this is the complete transmitting system.

Reference to Fig. 8.6 for a multiple half-wave antenna of half-length $h = 3\lambda_0/4$ shows that the component of the electric field E_ϵ, which is in phase with the magnetic field, vanishes along the entire surfaces $k_h = \pm 0.33$ which coincide with the cones $\Theta_0 = 70°$ and $110°$ at distant points where $E_\epsilon \to E_\Theta = 0$. These surfaces cut the antenna at distances $\lambda_0/2$ from the ends where the *sinusoidal* current is zero. Note that the *real part* of $\mathbf{E} \times \mathbf{H}^*$ vanishes everywhere on the surfaces $k_h = \pm 0.33$. Hence, a great sphere enclosing the antenna may be divided into three regions by the two surfaces $k_h = \pm 0.33$ which intersect the sphere at $\Theta = 70°$ and $\Theta = 110°$. These regions have the property that *there is no time-average transfer of power from one to the other.*

TABLE 12.1. Radiation function, radiation resistance, and directivity of multiple half-wave antenna.

n	Θ_m (deg)	$K_m^2(\Theta)$	R_m^e (ohms)	D	D_r	n	Θ_m (deg)	$K_m^2(\Theta)$	R_m^e (ohms)	D	D_r
1	90.0	4.00	73.13	1.64	1.00	11	21.3	25.84	144.4	5.36	3.27
							42.8	8.56			
2	54.0	5.68	94.2	1.81	1.10		56.8	5.68			
							68.6	4.56			
3	42.5	7.80	105.4	2.22	1.35		79.5	4.16			
	90.0	4.00					90.0	4.00			
4	36.3	10.04	114.0	2.65	1.61	12	20.3	28.20	146.9	5.75	3.50
	75.1	4.28					41.1	9.24			
							54.0	6.08			
5	31.5	12.80	120.6	3.14	1.92		65.4	4.80			
	66.0	4.76					75.7	4.28			
	90.0						85.2	4.04			
6	29.2	14.60	126.2	3.47	2.11	13	19.6	30.48	149.4	6.13	3.74
	59.5	5.36					39.2	9.92			
	80.3	4.08					51.8	6.52			
							62.4	5.12			
7	27.1	16.80	130.8	3.89	2.37		72.0	4.44			
	54.6	7.96					81.1	4.12			
	73.3	4.36					90.0	4.00			
	90.0	4.00				14	19.0	32.52	151.6	6.44	3.90
8	25.2	19.20	134.8	4.27	2.61		37.9	10.52			
	50.8	6.60					49.8	7.08			
	67.8	4.68					59.9	5.28			
	82.7	4.12					69.0	4.60			
9	23.8	21.84	138.4	4.65	2.83		77.6	4.16			
	47.7	7.24					85.8	4.00			
	63.4	4.96									
	77.1	4.16				15	18.2	35.04	153.7	6.85	4.17
	90.0	4.00					36.5	11.20			
							48.0	7.20			
10	22.4	23.68	141.5	5.07	3.09		51.7	5.56			
	45.1	7.92					66.4	4.76			
	59.8	5.36					74.5	4.32			
	72.4	4.40					82.3	4.08			
	84.2	4.04					90.0	4.00			

If (18) is evaluated over a *closed* surface Σ_1 which includes that part of the surface of a great sphere between $\Theta = 0°$ and $\Theta = 70°$ and the entire surface $k_h = 0.33$ within the sphere, the only nonzero contribution to the real part of the integral comes from the integration over the spherical part of the surface. Thus

$$\bar{\bar{T}}_{r1} = \text{R.P.} \int_{\Sigma_1(\text{closed})} \hat{n} \cdot S \, d\sigma$$

$$= \text{R.P.} \, 2\pi R^2 \int_0^{70°} S_R \sin \Theta \, d\Theta. \quad (19a)$$

Similarly, for region 2 bounded by $k_h = \pm 0.33$,

$$\bar{\bar{T}}_{r2} = \text{R.P.} \, 2\pi R^2 \int_{70°}^{110°} S_R \sin \Theta \, d\Theta, \quad (19b)$$

and for region 3 bounded by $k_h = -0.33$ and $k_h = -1$,

$$\bar{\bar{T}}_{r3} = \text{R.P.} \, 2\pi R^2 \int_{110°}^{180°} S_R \sin \Theta \, d\Theta = \bar{\bar{T}}_{r1} \quad (19c)$$

Since each of the three regions contains an antenna of length $\lambda_0/2$ and that part of

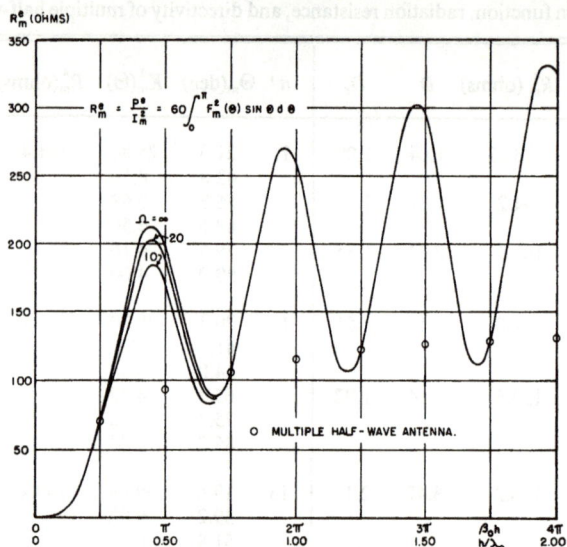

Fig. 12.1. Radiation resistance referred to maximum current of linear radiator.

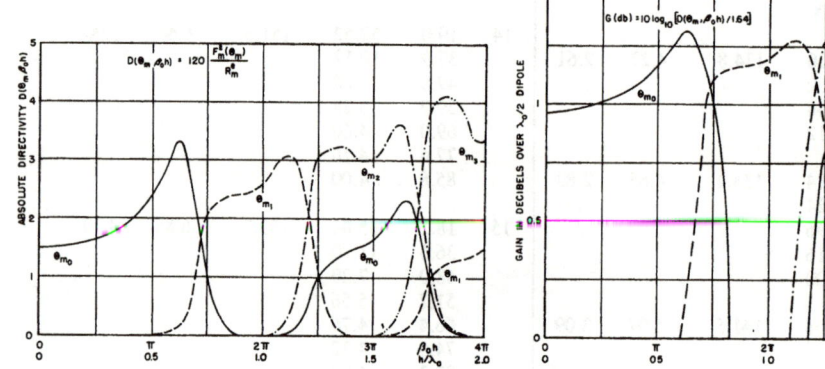

Fig. 12.2. Absolute directivities of linear radiator.

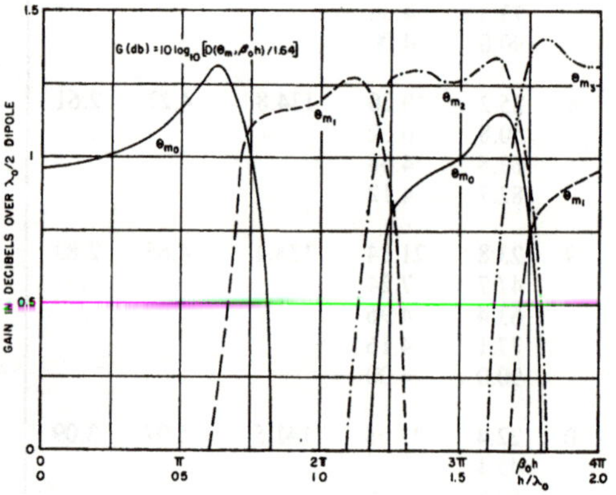

Fig. 12.3. Gain in decibels of linear radiator referred to a $\lambda/2$ dipole.

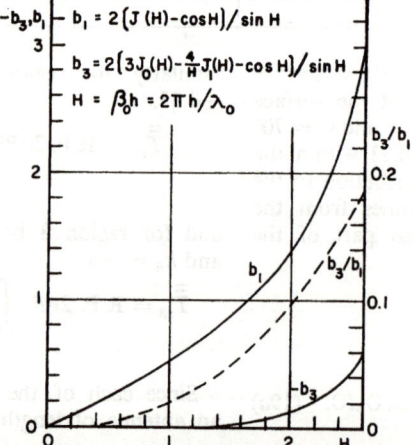

Fig. 13.1. Fourier coefficients b_1 and b_3.

space enclosing *one* of the three ears in the field pattern, it follows that the net balance of power is such that all of the power radiated by the upper half-wavelength of the antenna may be associated with the principal ear with $\Theta_m = 42°$ (see Fig. 5.1a, b), the power radiated by the lower half-wavelength of the antenna may be associated with the principal ear with $\Theta_m = 180° - 42°$, and the power radiated by the central half-wavelength of the antenna may be associated with the smaller ear in the equatorial plane with $\Theta_m = 90°$. Note that this does not imply that the field in the upper region 1 is maintained entirely and exclusively by the current in the upper third of the antenna, the field in the lower region 2 by the current in the lower third of the antenna, and the field in the central region 3 by the current in the middle third of the antenna. The field at any one point in space is maintained by *all* currents in the entire antenna. However, the power transferred downward across the surface $k_h = 0.33$ from the power supplied to the upper third of the antenna is exactly equal to that transferred upward across the same surface from the powers supplied to the lower two-thirds of the antenna. Similarly, the power transferred upward across the surface $k_h = 0.33$ and downward across the surface $k_h = -0.33$ from the power supplied to the central third of the antenna is exactly equal to the power transferred in the opposite directions across these same surfaces from the powers supplied to the outer two-thirds of the antenna.

If the antenna of length $2h = 3\lambda_0/2$ with its continuous distribution of generators is replaced by three very thin collinear antennas each of length $\lambda_0/2$ and individually center-driven by slice generators so that the currents at their respective centers are equal in magnitude and the currents in the two outer units are mutually in phase but 180° out of phase with the current in the central unit, very nearly the same sinusoidal distributions of current exist. Although a center-driven antenna does not have an exactly sinusoidal current, the actual distribution departs negligibly from a sine curve for purposes of determining the far-zone field. It follows that it must be a good approximation to assume that the total power supplied to each of the outer $\lambda_0/2$-units is equal to that associated with each of the principal ears in the upper and lower half spaces of the field pattern. Similarly, the power supplied to the central unit is equal to the power associated with the ear along the equatorial plane.

Using the zeroth-order resistances for antisymmetrically driven collinear antennas as given in Table III.34.1, the following values are obtained:

$$P_1 = P_3 = I_0^2 R_{\text{even}}^a = 42.6 I_0^2, \quad (20a)$$

$$P_2 = I_0^2 R_{0c}^a = 20.3 I_0^2. \quad (20b)$$

The same results could be obtained by evaluating (19a, b) with $F_m(\Theta, \beta_0 h)$ defined as in Sec. 5. Evidently, slightly more than twice as much radiated power is associated with the principal ear at $\Theta_m = 42°$ than with the ear in the equatorial plane. This agrees with the relative sizes of the ears in Fig. 5.1a, b.

It is significant that the power supplied to each of N center-driven collinear elements with essentially sinusoidal currents that are all equal in magnitude and alternately in phase and 180° out of phase is equal to that associated with each of the N ears in the field pattern. Since it is shown in Sec. II.26 that the component I' of current in phase quadrature with the driving voltage is very nearly sinusoidal for an extremely long center-driven antenna, it follows that the power associated with the principal ear or any other ear *of this component* of current may be determined using formulas corresponding to (19a, b, c). Since the quadrature current I' maintains a larger part of the far-zone field than does the in-phase current I'', a fair estimate of the distribution of radiated power associated with the several ears of the field pattern of a multiple half-wave antenna may be obtained in this manner.

When $\beta_0 h$ is less than unity, the several functions in (9) may be expanded in powers of $\beta_0 h$. The first two terms in the series are

$$R_0^e = \frac{\zeta_0}{6\pi} \beta_0^2 h^2 \left(1 + \frac{2}{15} \beta_0^2 h^2\right). \quad (21)$$

It is interesting to examine the *instantaneous* Poynting vector and the *instantaneous* transfer of power across ellipsoidal surfaces enclosing the multiple half-wave antenna. The instantaneous Poynting vector is defined by

$$\mathbf{S}_{\text{inst}} = v_0 \mathbf{E}_{\text{inst}} \times \mathbf{B}_{\text{inst}} \quad (22a)$$

and the instantaneous transfer of power outward across a closed surface Σ is

$$T_{\text{inst}} = \int_\Sigma \hat{\mathbf{n}} \cdot \mathbf{S}_{\text{inst}} \, d\sigma \quad (22b)$$

For the multiple half-wave antenna of half-length $h = n\lambda_0/4$ with n odd the instantaneous electromagnetic vectors **E** and **B** are given by (8.20a, b, c). With $\hat{\rho}$ a unit vector tangent to the hyperbola and $\hat{\epsilon}$ a unit vector tangent to the ellipse that intersect at the point where the instantaneous Poynting vector is evaluated, it follows that

$$\mathbf{S}_{\text{inst}} = \frac{I_m^2 \zeta_0}{4\pi^2 h^2} \left[\hat{\rho} \frac{\cos^2(n\pi k_h/2)}{(1 - k_h^2)\sqrt{(k_e^2 - 1)(k_e^2 - k_h^2)}} \right.$$
$$\times \sin^2(\omega t - n\pi k_e/2)$$
$$\left. + \hat{\epsilon} \frac{\sin n\pi k_h}{4(k_e^2 - 1)\sqrt{(1 - k_h^2)(k_e^2 - k_h^2)}} \right.$$
$$\left. \times \sin(2\omega t - n\pi k_e) \right]. \quad (23)$$

Note that the ρ-component of $\mathbf{S}_{\text{inst}}$ is always positive and outwardly directed.

If the closed surface Σ is chosen to be an ellipsoid, it follows that $\hat{n} = \hat{\rho}$, so that

$$T_{\text{inst}} = \frac{I_m^2 \zeta_0}{4\pi^2 h^2} \int_\Sigma \frac{\cos^2(n\pi k_h/2)}{(1 - k_h^2)\sqrt{(k_e^2 - 1)(k_e^2 - k_h^2)}}$$
$$\times \sin^2(\omega t - n\pi k_e/2) d\sigma. \quad (24)$$

This relation indicates that power is transferred continuously outward across the ellipse defined by $k_e = $ const. at a rate that *varies* in time as $\sin^2 \omega t$. Significantly, there is no pulsating of electromagnetic energy *back* and *forth* across the surface Σ. The outward transfer of energy pulsates between zero and a maximum, but it is never negative or inward. The reason for this depends upon the following rather special circumstances. (1) The separation of the electric field into two components $E_{\epsilon\text{inst}}$, $E_{\rho\text{inst}}$ that have the property of representing, respectively, the entire radiation field and the entire induction field irrespective of distance from the antenna. That is, $E_{\epsilon\text{inst}}$ behaves like a radiation field at *all* points near and far, and $E_{\rho\text{inst}}$ behaves like an induction field that never contributes to radiation. (2) The choice as the surface of integration Σ of the ellipsoidal surfaces of constant phase on which $E_{\epsilon\text{inst}}$ and $B_{\Phi\text{inst}}$ are in phase. Since the induction field $E_{\rho\text{inst}}$ is everywhere perpendicular to Σ, it can never contribute to the component of the Poynting vector normal to Σ. It follows that there is no contribution whatsoever from the induction part of the E-field to the normal component of the Poynting vector regardless of whether the ellipsoidal surface is very close to or very far from the antenna. (Note that if a cylindrical or other surface were used for Σ instead of an ellipsoidal one, the tangential components of **E** would include parts of the induction field so that the components of **E** and **B** that contribute to the normal component of the Poynting vector would not be in phase and the electromagnetic energy might pulsate back and forth so long as the surface of integration is not in the far zone.)

If use is made of the trigonometric identity

$$\sin^2(\omega t - n\pi k_e/2) = \tfrac{1}{2}[1 - \cos(2\omega t - n\pi k_e)], \quad (25)$$

the expression (24) becomes

$$T_{\text{inst}} = \overline{\overline{T}} - T_x, \quad (26)$$

where $\overline{\overline{T}}$ is the time-average power transferred outward across Σ, given by

$$\overline{\overline{T}} = \frac{I_m^2 \zeta_0}{8\pi^2 h^2} \int_\Sigma \frac{\cos^2(n\pi k_h/2)}{(1 - k_h^2)\sqrt{(k_e^2 - 1)(k_e^2 - k_h^2)}} d\sigma, \quad (27)$$

and T_x is defined as follows:

$$T_x = \overline{\overline{T}} \cos(2\omega t - n\pi k_e). \quad (28)$$

Note that in the quasi-far zone defined by $k_e^2 \gg 1$ it is readily shown that $k_e = a_e/h \to R_0/h$, $(1 - k_h^2) = b_h^2/h^2 \to \sin^2 \Theta$, so that

$$\overline{\overline{T}} = \frac{I_m^2 \zeta_0}{8\pi^2 R_0^2} \int_\Sigma \frac{\cos^2(n\pi/2 \cos \Theta)}{\sin^2 \Theta} d\sigma, \quad (29)$$

in agreement with the value obtained using (4.17b) and (4.18a).

13. The Approximate Representation of the Radiation Field of a Center-Driven Antenna; Effective length of a Driven Antenna

The field of a center-driven antenna with sinusoidally distributed current is characterized by the function

$$F_0(\Theta, \beta_0 h) = \frac{\cos(\beta_0 h \cos \Theta) - \cos \beta_0 h}{\sin \beta_0 h \sin \Theta}. \quad (1)$$

In the special case of a short antenna it reduces to the simple form

$$F_0(\Theta, \beta_0 h) \doteq \tfrac{1}{2}\beta_0 h \sin \Theta. \quad [(\beta_0 h)^2 \ll 1] \quad (2)$$

In analyzing the far-zone field of arrays of antennas the simple function (2) is much more readily handled analytically than is (1). Furthermore, it is clear from Fig. 5.1 that the field patterns of a half-wave dipole ($h = \lambda/4$)

and all antennas shorter than this differ only slightly in shape from the semicircle in a polar graph or the sine curve in a rectangular graph defined by (2). This suggests approximating (1) by a function of the form (2) over a range of sufficiently small values of $\beta_0 h$. That is,

$$F_0(\Theta, \beta_0 h) \doteq \beta_0 h_t \sin \Theta, \quad (\beta_0 h \leq L) \quad (3)$$

where the effective length h_t of the transmitting antenna is to be determined together with a limiting value L/β_0.

In order to determine an appropriate value of h_t it is necessary to establish a criterion for the representation. If, for example, the relation (3) is to be the best possible approximation for $\Theta = \pi/2$, that is, along the equatorial plane regardless of any other consideration, h_t is immediately defined as follows. Thus at $\Theta = \pi/2$,

$$\frac{1 - \cos \beta_0 h}{\sin \beta_0 h} = \tan \tfrac{1}{2}\beta_0 h = \beta_0 h_t \quad (4)$$

Then

$$h_t = \frac{1}{\beta_0} \tan \tfrac{1}{2}\beta_0 h. \quad (\beta_0 h < \pi) \quad (5)$$

It is to be noted that this value of effective length for the transmitting antenna is necessarily the same as the effective length h_e of an infinitely thin receiving antenna placed parallel to the electric field, since (1) is equal to $\beta_0 h_e(\Theta)$ for $\Omega \to \infty$. Using (5),

$$F_0(\Theta, \beta_0 h) \doteq \tan (\tfrac{1}{2}\beta_0 h) \sin \Theta. \quad (6)$$

This reduces to (2) for $(\beta_0 h)^2 \ll 1$. For $\beta_0 h = \pi/2$,

$$\beta_0 h_t = 1, \quad (7)$$

so that

$$F_0(\Theta, \tfrac{1}{2}\pi) \doteq \sin \Theta. \quad (8)$$

Accordingly, the radiation function is

$$K_m^2(\Theta) \doteq 4 \sin^2 \Theta. \quad (9)$$

The radiation resistance referred to maximum current is

$$R_m^e \doteq \frac{\zeta_0}{2\pi} \int_0^\pi \sin^3 \Theta \, d\Theta = \frac{\zeta_0}{2\pi} \cdot \frac{4}{3} = 80 \text{ ohms}. \quad (10)$$

The absolute directivity is

$$D \doteq 1.5. \quad (11)$$

It is to be noted that the radiation resistance $R_m^e = 80$ ohms is not in particularly good agreement with the actual value 73.13 ohms for the antenna with sinusoidal current and half-length $h = \lambda_0/4$.

In defining h_t by (5) the criterion used was that the field in the equatorial plane $\Theta = \pi/2$ shall be the same. If interest lies exclusively in determining the field in this plane, the value given in (5) is obviously the best possible value because it gives exactly the right field. On the other hand, if the approximate representation (3) is to be used for a more general purpose, such as the determination of the field and radiation function for all values of Θ, and in addition for the calculation of the radiation resistance of one or more antennas, the value (5) is not the best choice. In order to obtain the best possible representation of the function $F_0(\Theta, \beta_0 h)$ by $\beta_0 h_t \sin \Theta$ in the sense of the rms error, the theorem of Fourier may be used. In effect, the representation (3) is nothing else than an attempt to replace a complicated function of Θ by the first term in an odd Fourier series of the form

$$F_0(\Theta, \beta_0 h) = b_1 \sin \Theta + b_3 \sin 3\Theta + \cdots, \quad (12)$$

with the effective half-length h_t defined in

$$\beta_0 h_t = b_1. \quad (13)$$

(Even terms in the series (12) have zero coefficients and are therefore omitted.) The magnitude of the error involved by dropping higher terms in the series depends upon the ratio b_1/b_3. A good representation by the first term in (12) demands that

$$b_1^2 \gg b_3^2. \quad (14)$$

The coefficients b_n of the series (12) as defined by Fourier are

$$b_n = \left(\frac{2}{\pi}\right) \int_0^\pi F_0(\Theta, \beta_0 h) \sin n\Theta \, d\Theta. \quad (15)$$

Using the well-known relations

$$\sin 3\Theta = 3 \sin \Theta - 4 \sin^3 \Theta, \quad (16)$$

$$\frac{\pi}{2} = \int_0^\pi \sin^2 \Theta \, d\Theta, \quad (17)$$

and*

$$\pi J_0(\beta_0 h) = \int_0^\pi \cos (\beta_0 h \cos \Theta) \, d\Theta, \quad (18)$$

$$\left(\frac{\pi}{\beta_0 h}\right) J_1(\beta_0 h) = \int_0^\pi \cos (\beta_0 h \cos \Theta) \sin^2 \Theta \, d\Theta, \quad (19)$$

---

* Ref. I.35, p. 44, formula 22, and p. 52, example 22.

the following values are obtained directly:

$$b_1 = 2[J_0(\beta_0 h) - \cos \beta_0 h]/\sin \beta_0 h, \qquad (20)$$

$$b_3 = 2[3J_0(\beta_0 h) - (4/\beta_0 h)J_1(\beta_0 h) \\ - \cos \beta_0 h]/\sin \beta_0 h. \qquad (21)$$

Subject to (14) the best effective half-length h_t is

$$h_t = b_1/\beta_0 = 2[J_0(\beta_0 h) - \cos \beta_0 h]/\beta_0 \sin \beta_0 h. \qquad (22)$$

Then (3) is equivalent to

$$F_0(\Theta, \beta_0 h) \doteq 2 \frac{[J_0(\beta_0 h) - \cos \beta_0 h]}{\sin \beta_0 h} \sin \Theta.$$

$$(\beta_0 h \leq 2) \quad (23)$$

The coefficients $b_1 = \beta_0 h_t$ and b_3 as well as $|b_3/b_1|$ are shown in Fig. 13.1. It is seen that $|b_3/b_1| \leq 0.1$ for $\beta_0 h \leq 2$. Therefore no satisfactory representation of the form (3) is possible unless $\beta_0 h \leq 2$ as indicated in (23). For a short antenna with $(\beta_0 h)^2 \ll 1$

$$h_t \doteq \tfrac{1}{2}h, \qquad (24)$$

which agrees with (2). For the half-wave dipole with $\beta_0 h = \pi/2$,

$$b_1 = \beta_0 h_t = 0.945, \qquad (25)$$

$$b_3 = -0.052. \qquad (26)$$

Since (14) is well satisfied, the representation

$$F_0\left(\Theta, \frac{\pi}{2}\right) = \frac{\cos(\tfrac{1}{2}\pi \cos \Theta)}{\sin \Theta} \doteq 0.945 \sin \Theta \qquad (27)$$

is a good one. The radiation function is

$$K_m^2(\Theta) \doteq 3.57 \sin^2 \Theta. \qquad (28)$$

The radiation resistance is

$$R_m^e = R_0^e \doteq 71.4 \text{ ohms}. \qquad (29)$$

The absolute directivity is

$$D \doteq 1.5. \qquad (30)$$

The representation (27) does not give the correct value of unity at $\Theta = \pi/2$ as does (8); on the other hand, it is a better over-all representation for other values of Θ. Furthermore, the value 71.4 ohms for the radiation resistance differs by only a small amount from 73.13. It follows that (23) may be used as a good approximation for determining both field patterns and radiation resistance.

If an antenna or a group of antennas can be represented approximately by a number of elements all of half-length $h = \lambda/4$ with a sinusoidally distributed current, each element may be described in terms of an effective length and a field function like (23). This greatly simplifies the analysis, especially when nonparallel elements are involved, as in crossed antennas or the turnstile antenna. There are useful applications of (23) to arrays of parallel antennas and to the turnstile antenna.

CENTER-DRIVEN ANTENNAS OF NONVANISHING CROSS SECTION

14. *The Radiation Field of a Symmetric Center-Driven Antenna of Nonvanishing Radius—Application of the Reciprocal Theorem*

The Rayleigh-Carson reciprocal theorem (ref. I.31, Chap. IV) reduces to the following simple form when applied to two antennas numbered 1 and 2:

$$I'_{10} V^{e''}_{10} = I''_{20} V^{e'}_{20}, \qquad (1)$$

where I'_{10} is the current maintained at the center of antenna 1 (used as a receiver) by the driving voltage $V^{e'}_{20}$ at the center of antenna 2 (used as a transmitter), and I''_{20} is the current maintained at the center of antenna 2 (used as a receiver) by a driving voltage $V^{e''}_{20}$ at the center of antenna 1 (used as a transmitter). It is assumed that the total impedance of each circuit remains the same when used for transmission and reception; that is, the impedances of load and generator in each antenna are the same. The currents in the loads of the two antennas when these are used successively for receiving are

$$I'_{10} = \frac{2h_{e_1} E^r_{\Theta_2} \cos \psi_1}{(Z_0 + Z_L)_1}, \qquad (2)$$

$$I''_{20} = \frac{2h_{e_2} E^r_{\Theta_1} \cos \psi_2}{(Z_0 + Z_L)_2}. \qquad (3)$$

In writing (2) and (3) it is assumed that the two antennas are in the far zone with respect to each other; Z_0 is the self-impedance of the antenna, Z_L the load impedance in each case.

The electric field in the far zone of a center-driven transmitting antenna is given in general terms in Sec. I.9. For the two antennas used *successively* for transmission

$$E^r_{\Theta_2} = \frac{j\zeta_0\beta_0}{4\pi} \frac{e^{-j\beta_0 R_0}}{R_0} \int_{-h_2}^{h_2} I'_{2z}(z'_2) e^{j\beta_0 z'_2 \cos\Theta_2}$$
$$\times \sin\Theta_2 \, dz'_2$$
$$= \frac{j\zeta_0 I'_{20}}{2\pi} \frac{e^{-j\beta_0 R_0}}{R_0} F_0(\Theta_2, \beta_0 h_2), \quad (4)$$

$$E^r_{\Theta_1} = \frac{j\zeta_0\beta_0}{4\pi} \frac{e^{-j\beta_0 R_0}}{R_0} \int_{-h_1}^{h_1} I''_{1z}(z'_1) e^{j\beta_0 z'_1 \cos\Theta_1}$$
$$\times \sin\Theta_1 \, dz'_1$$
$$= \frac{j\zeta_0 I''_{10}}{2\pi} \frac{e^{-j\beta_0 R_0}}{R_0} F_0(\Theta_1, \beta_0 h_1), \quad (5)$$

with

$$F_0(\Theta_2, \beta_0 h_2) = \frac{\beta_0}{2I'_{20}} \int_{-h_2}^{h_2} I'_{2z}(z'_2) e^{j\beta_0 z'_2 \cos\Theta_2}$$
$$\times \sin\Theta_2 \, dz'_2, \quad (6)$$

$$F_0(\Theta_1, \beta_0 h_1) = \frac{\beta_0}{2I''_{10}} \int_{-h_1}^{h_1} I''_{1z}(z'_1) e^{j\beta_0 z'_1 \cos\Theta_1}$$
$$\times \sin\Theta_1 \, dz'_1. \quad (7)$$

The function $F_0(\Theta, \beta_0 h)$ has been defined by analogy with (4.18a). In this general case $F_0(\Theta, \beta_0 h)$ is complex. It must approximate the real form (4.15) for a sufficiently thin antenna. The quantity R_0 is the distance between centers of the two antennas.

Substituting (6) and (7) in (2) and (3) gives

$$I'_{10} = \frac{2h_{e_1} I'_{20} F_0(\Theta_2, \beta_0 h_2) \cos\psi_1}{(Z_0 + Z_L)_1}$$
$$\times \left(\frac{j\zeta_0}{2\pi} \frac{e^{-j\beta_0 R_0}}{R_0}\right), \quad (8)$$

$$I''_{20} = \frac{2h_{e_2} I''_1 F_0(\Theta_1, \beta_0 h_1) \cos\psi_2}{(Z_0 + Z_L)_2}$$
$$\times \left(\frac{j\zeta_0}{2\pi} \frac{e^{-j\beta_0 R_0}}{R_0}\right). \quad (9)$$

The currents at the terminals in the two antennas when driven *successively* are

$$I'_{20} = V^{e'}_{20}/(Z_0 + Z_g)_2, \quad (10)$$

$$I''_{10} = V^{e''}_{10}/(Z_0 + Z_g)_1; \quad (11)$$

Z_0 is the self-impedance of the antenna, Z_g the impedance of the generator. As required above,

$$Z_{g1} = Z_{L1}, \quad Z_{g2} = Z_{L2}. \quad (12)$$

Accordingly, with (10) and (11) and (8) and (9), (1) becomes

$$\frac{h_{e_1} F_0(\Theta_2, \beta_0 h_2) \cos\psi_1}{(Z_0 + Z_L)_1 (Z_0 + Z_g)_2}$$
$$= \frac{h_{e_2} F_0(\Theta_1, \beta_0 h_1) \cos\psi_2}{(Z_0 + Z_L)_1 (Z_0 + Z_g)_1}. \quad (13)$$

Using (12), (13) reduces to

$$\frac{h_{e_1} \cos\psi_1}{F_0(\Theta_1, \beta_0 h_1)} = \frac{h_{e_2} \cos\psi_2}{F_0(\Theta_2, \beta_0 h_2)}. \quad (14)$$

The angles ψ_1 and ψ_2 are related as follows. Consider three planes; plane 1 contains antenna 1 and R_0, plane 2 contains antenna 2 and R_0, plane 3 is perpendicular to R_0. The component E_{Θ_1} of the field maintained by antenna 1 when transmitting is defined at the center of antenna 2 when this is receiving. It lies in plane 1 and in a plane parallel to plane 3. Similarly E_{Θ_2} at the center of antenna 1 when this is receiving is in plane 2 and parallel to plane 3. The angle ψ_1 is the angle between E_{Θ_2} and plane 1 measured parallel to plane 3. Similarly ψ_2 is the angle between E_{Θ_1} and plane 2 measured parallel to plane 3. That is, ψ_1 and ψ_2 both measure the angle between planes 1 and 2 parallel to plane 3. Therefore

$$\psi_1 = \psi_2 \quad (15)$$

and (14) becomes

$$\frac{h_{e_1}}{F_0(\Theta_1, \beta_0 h_1)} = \frac{h_{e_2}}{F_0(\Theta_2, \beta_0 h_2)} = C. \quad (16)$$

The effective half-length h_{e_1} of antenna 1 when receiving is by definition a function of $\beta_0 h_1$ and Θ_1 only; similarly, h_{e_2} is a function alone of $\beta_0 h_2$ and Θ_2. Since (14) is true for all values of the mutually independent variables and parameters Θ_1, h_1 and Θ_2, h_2, it follows that they must both be equal to a constant as written in (16). The value of this constant can be determined from any convenient special case such as, for example, a very short, infinitely thin antenna. For a short receiving antenna such that $(\beta_0 h)^2 \ll 1$ placed parallel to the electric field,

$$\beta_0 h_e \doteq \tfrac{1}{2}\beta_0 h, \quad (17)$$

For the short transmitting antenna such that $(\beta_0 h)^2 \ll 1$

$$F_0(\Theta, \beta_0 h) \doteq \tfrac{1}{2}\beta_0 h. \quad (18)$$

Therefore,

$$\beta_0 h_e \doteq F_0(\Theta, \beta_0 h). \quad [(\beta_0 h)^2 \ll 1] \quad (19)$$

It is a necessary consequence of (16) and (19) that the constant in (16) is

$$C = \frac{1}{\beta_0} \quad (20)$$

so that in perfectly general terms

$$F_0(\Theta, \beta_0 h) = \beta_0 h_e \quad (21)$$

for any antenna. That is, the complex field function $F_0(\Theta, \beta_0 h)$ of an antenna when used for transmission is equal to the complex effective half-length in radians when used for reception. Using (IV.9.3) with $\delta = 0$ substituted on the right in (21), the field function becomes

$$F_0(\Theta, \beta_0 h) = \frac{j2\pi Z_0}{\zeta_0 \Psi \sin \Theta} \times$$

$$\left\{ \frac{\cos(\beta_0 h \cos \Theta) - \cos \beta_0 h + \frac{1}{\Psi}[m_1^{\mathrm{I}}(0) + jm_1^{\mathrm{II}}(0)] + \frac{1}{\Psi^2}[m_2^{\mathrm{I}}(0) + jm_2^{\mathrm{II}}(0)] + \cdots}{\cos \beta_0 h + \frac{1}{\Psi}[A_1^{\mathrm{I}} + jA_1^{\mathrm{II}}] + \frac{1}{\Psi^2}[A_2^{\mathrm{I}} + jA_2^{\mathrm{II}}] + \cdots} \right\}$$

(22)

For the extremely thin antenna,

$$Z_0 \doteq \frac{-j\zeta_0 \Psi}{2\pi} \cot \beta_0 h, \quad (23)$$

so that as $\Psi \to \infty$, (22) reduces to

$$F_0(\Theta, \beta_0 h) \to \frac{\cos(\beta_0 h \cos \Theta) - \cos \beta_0 h}{\sin \Theta \sin \beta_0 h}. \quad (24)$$

This is identically (4.15) as obtained for the extremely thin antenna by substituting

$$I_z = I_0 \frac{\sin \beta_0 (h - |z|)}{\sin \beta_0 h} \quad (25)$$

in the expression for the vector potential. The function (21) must be essentially the function that would have been obtained if the current

$$I_z = \frac{j2\pi V_0^e}{\Psi \zeta_0}$$

$$\times \left\{ \frac{\sin \beta_0(h - |z|) + \frac{1}{\Psi}[M_1^{\mathrm{I}} + jM_1^{\mathrm{II}}] + \cdots}{\cos \beta_0 h + \frac{1}{\Psi}[A_1^{\mathrm{I}} + jA_1^{\mathrm{II}}] + \cdots} \right\}$$

(26)

had been used in the integral for the vector potential (1.1) instead of the sinusoidal distribution (25) subject to the restrictions for the far zone.

The far-zone field of a center-driven antenna is characterized by

$$F_0(\Theta, \beta_0 h) = \beta_0 h_e = 2\pi h_e/\lambda_0. \quad (27)$$

The complex effective length of a receiving antenna per unit wavelength,

$$\frac{h_e}{\lambda_0} = \frac{h_e''}{\lambda_0} + \frac{jh_e'}{\lambda_0}, \quad (28)$$

is given in Figs. IV.9.2–9.6 as a function of h/λ_0 with $\sin \overline{\Theta}_2 = \cos \Theta_2$ as parameter. Since $\Theta_2 = 90° - \overline{\Theta}_2$, h_e/λ_0 is available for h/λ_0 ranging from 0 to $5\lambda_0/8$ for the values $\overline{\Theta}_2 = 0°, 11.5°, 23.6°, 36.9°, 53.1°, 90°$, or for $\Theta_2 = 90°, 78.5°, 66.4°, 53.1°, 36.9°, 0°$. It follows that the *real* and *imaginary* parts of $F_0(\Theta, \beta_0 h)$ can be plotted if required. In general, interest in the far-zone field is in the magnitude $F_0(\Theta, \beta_0 h)$ of $F_0(\Theta, \beta_0 h)$ and not in the phase referred to the driving potential in the transmitter. Accordingly, the real magnitude

$$F_0(\Theta, \beta_0 h) = \beta_0 h_e$$

$$= 2\pi \left(\frac{h_e}{h_e(\Theta = 90°)} \right) \left(\frac{h_e(\Theta = 90°)}{\lambda_0} \right) \quad (29)$$

is all that is required. The factor $h_e/h_e(\Theta = 90°)$ is plotted in Figs. IV.9.7–9.9 as a function of $\overline{\Theta}_2$ with h/λ_0 as parameter for $\Omega = 2 \ln(2h/a) = 10, 12.5, 15, 20,$ and ∞. Numerical values of $h_e/h_e(\Theta = 90°)$ are also given. Making use of these, the curves of Figs. IV.9.7–9.9 are readily replotted with $\Theta_2 = \frac{1}{2}\pi - \overline{\Theta}_2$ as variable and $\beta_0 h$ as parameter. The curves so obtained are shown in Figs. 14.1 and 14.3 for $\beta_0 h = 3\pi/8$ and $5\pi/8$. For $\beta_0 h = \pi/2$, (29) differs negligibly from

$$F_0(\Theta, \tfrac{1}{2}\pi) = \frac{\cos(\tfrac{1}{2}\pi \cos \Theta)}{\sin \Theta} = \frac{\cos(\tfrac{1}{2}\pi \sin \overline{\Theta})}{\cos \overline{\Theta}}$$

(30)

for the thin antenna with sinusoidal current.

Experimental verification of Figs. 14.1 and 14.3 is in the literature.[23]

In Figs. 14.1 and 14.3, in which $\beta_0 h = 3\pi/8$ and $5\pi/8$, the curves marked $\Omega \to \infty$ for the extremely thin antenna are computed from

$$F_0(\Theta, \beta_0 h) = \frac{\cos(\beta_0 h \cos \Theta) - \cos \beta_0 h}{\sin \beta_0 h \sin \Theta}.$$

(31)

Since (31) becomes infinite at $\beta_0 h = \pi$, it is not useful in obtaining a curve of $F_0(\Theta, \beta_0 h)$ for the extremely thin antenna with sinusoidal current and half-length $h = \lambda_0/2$. Furthermore, the function

$$F_m(\Theta, \beta_0 h) = F_0(\Theta, \beta_0 h) \sin \beta_0 h, \quad (32)$$

which is commonly used in plotting field patterns of thin antennas with sinusoidal current, is referred to the maximum current, not to the input current, as are all other curves in Figs. 14.1–14.3. In order to obtain a quantitative as well as qualitative comparison of the field of actual antennas with that of antennas of near-zero radius, use is made of the definition

$$I_m \doteq I_0/\sin \beta_0 h, \quad (33)$$

where I_m may or may not be the maximum current physically. With (33), (31), and $\beta_0 h = \pi$, (32) becomes

$$F_m(\Theta, \pi) = \frac{I_m}{I_0} \left[\frac{\cos(\pi \cos \Theta) + 1}{\sin \Theta} \right]. \quad (34)$$

By requiring that I_m for the sinusoidal current be equal to I_m for an antenna of finite thickness, a reasonable comparison is possible. The curve in Fig. 14.2 for $\Omega \to \infty$ is, therefore, computed from $F_m(\Theta, \pi)$ for an antenna with sinusoidal current that has the same *maximum* current as the antenna of finite radius.

It is clear from Figs. 14.1–14.3 that the far-zone field of an antenna for which $\Omega \to \infty$ is a good approximation of the relative magnitude of the far-zone field of cylindrical antennas of small radius. The principal difference in the relative field patterns of antennas of small and near zero radius is at angles Θ where the field of the extremely thin antenna has sharp nulls. Instead of becoming zero near these angles, the *magnitude* of the field of an antenna of small radius merely reduces to a minimum. It is clear from Fig. 14.1, however, that the shape of the field pattern is not quite the same. For practical purposes it is usually satisfactory to determine the relative pattern of *the far-zone field* of an antenna of small radius by treating the antenna as though it were extremely thin with a sinusoidal current of appropriately adjusted amplitude. However, quantitative agreement cannot be expected for all values of Θ because the shapes of the field patterns are not the same. Thus in Fig. 14.1 the agreement is excellent near $\Theta = 50°$ with I_0 the same in all antennas. On the other hand, there is considerable difference between the curve for $\Omega \to \infty$ and those for $\Omega = 10$ and 20 for all other angles, including especially $\Theta = 90°$. By adjusting I_0 in the extremely thin antenna to be an appropriate fraction of I_0 in the antenna for which $\Omega = 10$, perfect agreement can be obtained between the curves $\Omega = \infty$ and $\Omega = 10$ at $\Theta = 90°$ but at the expense of increased differences for other angles. For rough qualitative purposes, a satisfactory adjustment in relative currents in extremely thin and actual antennas is to make I_m the same in both. Except for antennas with half-lengths near integral multiples of $\lambda_0/2$, it is satisfactory to make I_0 the same, and in practice I_0 is usually more conveniently determined. If more accurate adjustment is required, it may be carried out using (29) rewritten as follows:

$$F_0(\Theta, \beta_0 h) = \frac{2\pi h_e}{\lambda_0} = K[F_m(\Theta, \beta_0 h)]_{\Omega \to \infty}. \quad (35)$$

By selecting a particular angle Θ_1 where agreement is desired, $F_m(\Theta_1, \beta_0 h) = [\cos(\beta_0 h \cos \Theta_1) - \cos \beta_0 h]/\sin \Theta_1$ may be computed; h_e/λ_0 for the angle $\bar{\Theta}_2 = \frac{1}{2}\pi - \Theta_2$ may then be determined. Finally, (35) may be solved for the numerical factor K. The function $K[F_m(\Theta, \beta_0 h)]_{\Omega \to \infty}$ is then used instead of $F_0(\Theta, \beta_0 h)$ in determining the approximate field pattern. The directivity D of antennas of finite cross section is plotted in Figs. IV.10.1 and IV.10.2.

15. The Complete Field of a Center-Driven Antenna of Small Radius—Approximate Analysis

An exact determination of the general electromagnetic field due to a center-driven antenna of small but not zero radius depends upon the evaluation of the integral for the vector potential,

$$\mathbf{A} = \frac{\hat{\mathbf{z}}}{4\pi v_0} \int_{-h}^{h} I_z' \frac{e^{-j\beta_0 R}}{R} \, dz', \quad (1)$$

and the calculation of the electric and magnetic vectors $\mathbf{E}$ and $\mathbf{B}$ from (1), using

$$\mathbf{E} = \frac{-j\omega}{\beta_0^2} (\text{grad div } \mathbf{A} + \beta_0^2 \mathbf{A}), \quad (2a)$$

$$\mathbf{B} = \text{curl } \mathbf{A}. \quad (2b)$$

Since the distribution of current in the antenna is known only in the form of an intricate series in powers of $1/\Psi$, the evaluation of (1)

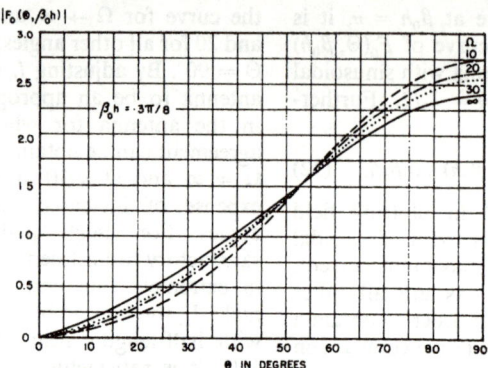

Fig. 14.1. Field factor of linear radiator as a function of Θ with Ω as parameter; $\beta_0 h = 3\pi/8$.

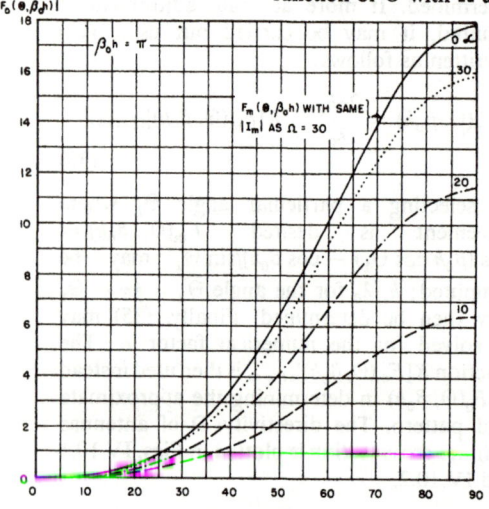

Fig. 14.2. Field factor of linear radiator as a function of Θ with Ω as parameter; $\beta_0 h = \pi$.

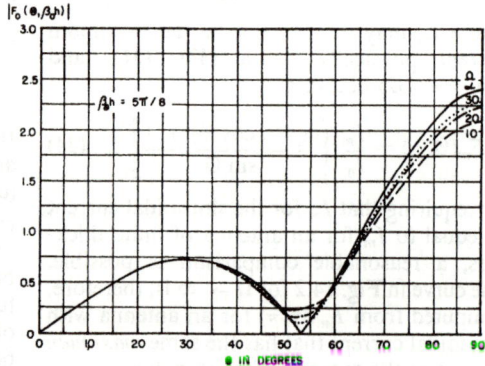

Fig. 14.3. Field factor of linear radiator as a function of Θ with Ω as parameter; $\beta_0 h = 5\pi/8$.

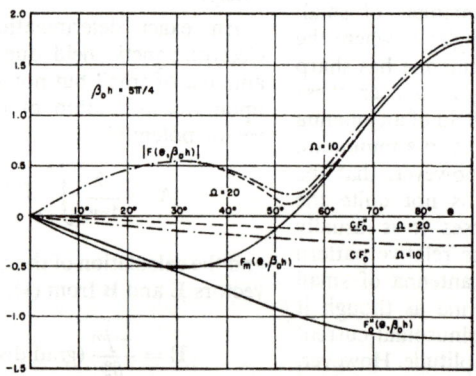

Fig. 16.1. Field factor of antenna with small radius; $\beta_0 h = 5\pi/4$.

$$F_m(\Theta, \beta_0 h) = \frac{\cos(\beta_0 h \cos\Theta) - \cos\beta_0 h}{\sin\Theta},$$

$$F_0''(\Theta, \beta_0 h) = \frac{\sin\beta_0 h \cos(\beta_0 h \cos\Theta)\cos\Theta - \cos\beta_0 h \sin(\beta_0 h \cos\Theta)}{\sin\Theta \cos\Theta (1 - \cos\beta_0 h)},$$

$$|F(\Theta, \beta_0 h)| = \{[F_m(\Theta, \beta_0 h)]^2 + C^2[F_0''(\Theta, \beta_0 h)]^2\}^{1/2};$$

$$C = 0.234 \ (\Omega = 10); \ C = 0.123 \ (\Omega = 20).$$

is excessively intricate even if only terms in the series of zeroth and first order are retained. Only the zeroth-order terms corresponding to the sinusoidal distribution in a thin antenna are sufficiently simple to make direct integration of (1) in closed form possible.

As discussed in Chapter II, Sec. 25, the modified zeroth-order distribution

$$I_z \doteq I_0'' \left(\frac{\cos \beta_0 z - \cos \beta_0 h}{1 - \cos \beta_0 h} \right)$$
$$+ j I_m' \sin \beta_0 (h - |z|) \quad (3)$$

is a moderately good representation of the current with $\beta_0 h < 2\pi$. It is in any event a very much better approximation than the zeroth-order term

$$I_z \doteq I_m' \sin \beta_0 (h - |z|) \quad (4)$$

used in Sec. 1. Therefore, if (3) is substituted in (1) instead of (4) a reasonably accurate expression for the vector potential due to an antenna of small radius should be obtained. Since the imaginary term in (3) is exactly like (4), the integral is simply (1.7). The integral due to the real part of (3) is

$$A_z'' \doteq \frac{I_0''}{4\pi v_0 (1 - \cos \beta_0 h)}$$
$$\times \left(\int_{-h}^{h} \cos \beta_0 z' \frac{e^{-j\beta_0 R}}{R} dz' \right.$$
$$\left. - \cos \beta_0 h \int_{-h}^{h} \frac{e^{-j\beta_0 R}}{R} dz' \right). \quad (5)$$

The first integral in (5) has already been determined. It is $C_r(h, z)$ as defined in (1.8). The second integral is $E_r(h, z)$, defined in (II.19.5), with r written for a. Accordingly, with (1.7) and $\beta_0 h < 2\pi$,

$$A_z = A_z'' + j A_z'$$
$$\doteq \frac{I_0''}{4\pi v_0 (1 - \cos \beta_0 h)} [C_r(h, z)$$
$$- \cos \beta_0 h \, E_r(h, z)]$$
$$+ j \frac{I_m'}{4\pi v_0} [\sin \beta_0 h \, C_r(h, z)$$
$$- \cos \beta_0 h \, S_r(h, z)]. \quad (6)$$

Using tabulated functions, A_z can be calculated from (6) at any point in space.

The determination of the electromagnetic field from (6) using (2a, b) unfortunately does not lead to expressions involving only elementary functions and tabulated integrals. Thus, in the evaluation of the magnetic field, which reduces to

$$B_\theta = -\frac{\partial A_z}{\partial r} \quad (7)$$

as in (2.3), the functions $C_r(h, z)$ and $S_r(h, z)$ lead only to elementary functions, as shown in Sec. 2. On the other hand, the function

$$E_r(h, z) = \int_{-h}^{h} \frac{e^{-j\beta_0 R}}{R} dz', \quad (8)$$

with

$$R = \sqrt{(z - z')^2 + r^2}, \quad (9)$$

yields

$$\frac{\partial E_r(h, z)}{\partial r} = -r \int_{-h}^{h} \left(\frac{1}{R^3} + \frac{j\beta_0}{R^2} \right) e^{-j\beta_0 R} dz', \quad (10)$$

and these integrals cannot be evaluated in closed form nor are they tabulated. Accordingly,

$$B_\theta \doteq \frac{I_0''}{4\pi v_0 (1 - \cos \beta_0 h)} \left[\frac{\partial C_r(h, z)}{\partial r} \right.$$
$$+ r \cos \beta_0 h \int_{-h}^{h} \left(\frac{1}{R^3} + \frac{j\beta_0}{R^2} \right) e^{-j\beta_0 R} dz' \right]$$
$$+ j \frac{I_m'}{4\pi v_0} \left[\sin \beta_0 h \frac{\partial C_r(h, z)}{\partial r} \right.$$
$$\left. - \cos \beta_0 h \frac{\partial S_r(h, z)}{\partial r} \right], \quad (\beta_0 h < 2\pi) \quad (11)$$

with $\partial C_r(h, z)/\partial r$ given in (2.11) and $\partial S_r(h, z)/\partial r$ in (2.13). The imaginary part of (11) reduces identically to (2.14b). Evaluation of the electric field involves further differentiation with respect to r and z. The formula for E_r can be expressed entirely in terms of known functions; the formula for E_z involves complicated integrals similar to those in (11) but with higher powers of R. They are available in the literature.[14]

In the important special case of antennas that have a half-length h equal to an integral odd multiple of a half-wavelength, (11) simplifies greatly. In this case, as pointed out in Sec. 1, the distribution of current reduces to

$$I_z \doteq I_0 \cos \beta_0 z, \quad I_0 = I_0'' + j I_0' = V_0^e / Z_0. \quad (12)$$

Also, $I_m = I_0$. The magnetic field is

$$B_\theta \doteq \frac{j I_m}{4\pi v_0 r} [e^{-j\beta_0 R_{1h}} + e^{-j\beta_0 R_{2h}}].$$
$$(\beta_0 h = n\pi/2, \ n \text{ odd}) \quad (13)$$

This is exactly the form (6.3b), and since the components of the electric field are calculated from (13) by differentiation, they must coincide with (6.4a, b, c). It follows that the complete field of a cylindrical highly conducting antenna of nonvanishing radius and of half-length $h = n\lambda_0/4$, n odd, differs from that of an antenna with sinusoidal current only by a constant shift in phase, except at distances from the ends of the antenna that are not large compared with its radius. If the antenna has rounded ends, this restriction is not required and an estimate of the field near the surface of the antenna may be obtained from (13) and expressions derived from (13), such as the form (8.20) in confocal coördinates.

It may be concluded from the approximate analysis of the magnitude of the far-zone field of an antenna for which the ratio of radius to half-length (a/h) is small but not zero, that for all practical purposes it is adequate to make use of the field factors of extremely thin antennas $(a/h \doteq 0)$. The only difference between the magnitude of the field factor of an extremely thin antenna and of one for which a/h is small is in the appearance of minima in the places of nulls.

Since the magnitude of the field factor at such a minimum is very small compared with the maximum value, it is usually of no significance. This is verified more specifically in the next section.

16. The Radiation Field of a Center-Driven Antenna of Small Radius—Approximate Analysis

It is shown in the preceding section that the complete electromagnetic field of a center-driven antenna of nonvanishing radius does not reduce to simple form using the modified zeroth-order current (15.3), except in the special case $\beta_0 h = n\pi/2$, n odd. However, if the analysis is restricted to a determination of the far-zone or radiation field, great simplification is possible and the electromagnetic field can be evaluated in closed and relatively simple form.

The general integral for the far-zone field of a center-driven antenna parallel to the z-axis is given in Sec. I.9. It is

$$E_\Theta^r = v_0 B_\Phi^r = -j\omega A_\Theta^r$$

$$= \frac{j\omega}{4\pi v_0} \frac{e^{-j\beta_0 R_0}}{R_0} \int_{-h}^{h} I_z(z') e^{j\beta_0 z' \cos\Theta} \sin\Theta \, dz'. \quad (1)$$

If the distribution

$$I_z \doteq I_0'' \left(\frac{\cos \beta_0 z - \cos \beta_0 h}{1 - \cos \beta_0 h} \right)$$

$$+ jI_0' \frac{\sin \beta_0(h - |z|)}{\sin \beta_0 h} \quad (2)$$

is substituted in (1) and the integration carried out* the following integrals are obtained:

$$E_\Theta^r \doteq \frac{j\omega}{4\pi v_0} \frac{e^{-j\beta_0 R_0}}{R_0} \sin\Theta$$

$$\times \left(\frac{I_0''}{1 - \cos \beta_0 h} \int_{-h}^{h} \cos \beta_0 z' e^{j\beta_0 z' \cos\Theta} \, dz' \right.$$

$$- \frac{I_0'' \cos \beta_0 h}{1 - \cos \beta_0 h} \int_{-h}^{h} e^{j\beta_0 z' \cos\Theta} \, dz'$$

$$+ jI_0' \int_{-h}^{h} \cos \beta_0 z' e^{j\beta_0 z' \cos\Theta} \, dz'$$

$$- jI_0' \cot \beta_0 h \int_0^h \sin \beta_0 z' e^{j\beta_0 z' \cos\Theta} \, dz'$$

$$\left. + jI_0' \cot \beta_0 h \int_{-h}^{0} \sin \beta_0 z' e^{j\beta_0 z' \cos\Theta} \, dz' \right). \quad (3)$$

---

* The following standard integrals are involved (see Peirce, ref. I.39, formulas 415, 401, 364):

$$\int_{-h}^{h} \cos \beta_0 z' \, e^{j\beta_0 z' \cos\Theta} \, dz'$$

$$= \frac{e^{j\beta_0 z' \cos\Theta} (j\beta_0 \cos\Theta \cos \beta_0 z' + \beta_0 \sin \beta_0 z')}{\beta_0^2 (1 - \cos^2 \Theta)} \bigg|_{-h}^{h}$$

$$= \frac{-2}{\beta_0 \sin^2 \Theta} [\cos \beta_0 h \sin(\beta_0 h \cos\Theta) \cos\Theta$$
$$\qquad\qquad - \sin \beta_0 h \cos(\beta_0 h \cos\Theta)],$$

$$\int_{-h}^{h} e^{j\beta_0 z' \cos\Theta} \, dz' = \frac{1}{j\beta_0 \cos\Theta} e^{j\beta_0 z' \cos\Theta} \bigg|_{-h}^{h}$$

$$= \frac{2}{\beta_0 \cos\Theta} \sin(\beta_0 h \cos\Theta),$$

$$\int_0^h \sin \beta_0 z' \, e^{j\beta_0 z' \cos\Theta} \, dz'$$

$$\qquad - \int_{-h}^{0} \sin \beta_0 z' \, e^{j\beta_0 z' \cos\Theta} \, dz'$$

$$= 2 \int_0^h \sin \beta_0 z' \cos(\beta_0 z' \cos\Theta) dz'$$

$$= -\frac{\cos \beta_0 z'(1 - \cos\Theta)}{\beta_0(1 - \cos\Theta)} - \frac{\cos \beta_0 z'(1 + \cos\Theta)}{\beta_0(1 + \cos\Theta)} \bigg|_0^h$$

$$= \frac{1 - \cos \beta_0 h(1 - \cos\Theta)}{\beta_0(1 - \cos\Theta)} + \frac{1 - \cos \beta_0 h(1 + \cos\Theta)}{\beta_0(1 + \cos\Theta)}$$

$$= \frac{2}{\beta_0 \sin^2 \Theta} [1 - \cos \beta_0 h \cos(\beta_0 h \cos\Theta)$$
$$\qquad\qquad - \cos\Theta \sin \beta_0 h \sin(\beta_0 h \cos\Theta)].$$

Upon carrying out the integration, substituting the integrated results in (3), and noting that $\omega/v_0\beta_0 = v_0/v_0 = \zeta_0$, the following final formula is obtained:

$$E_\Theta^r \doteq \frac{j\zeta_0}{2\pi} \frac{e^{-j\beta_0 R_0}}{R_0}$$

$$\times \left\{ I_0'' \left[\frac{\sin \beta_0 h \cos(\beta_0 h \cos \Theta) \cos \Theta - \cos \beta_0 h \sin(\beta_0 h \cos \Theta)}{\sin \Theta \cos \Theta (1 - \cos \beta_0 h)} \right] \right.$$

$$\left. + j I_m' \left[\frac{\cos(\beta_0 h \cos \Theta) - \cos \beta_0 h}{\sin \Theta} \right] \right\}.$$

$$(\beta_0 h < 2\pi) \quad (4a)$$

In shorthand notation,

$$E_\Theta^r \doteq \frac{j\zeta_0 I_m'}{2\pi} \frac{e^{-j\beta_0 R_0}}{R_0} [CF_0''(\Theta, \beta_0 h)$$

$$+ j F_m(\Theta, \beta_0 h)],$$

$$C = I_0''/I_m', \quad (\beta_0 h < 2\pi) \quad (4b)$$

where $F_0''(\Theta, \beta_0 h)$ is the first square bracket in (4a), and where $F_m(\Theta, \beta_0 h)$ is, of course, the same as the far-zone-field factor previously obtained by another method using the zeroth-order distribution

$$(I_z)_0 = I_m \sin \beta_0 (h - |z|). \quad (5)$$

An extra factor j appears in (4) because phases are referred to the driving potential difference V_0^e and not to the quadrature current.

With $\beta_0 h = n\pi/2$, n odd,

$$E_\Theta^r \doteq \frac{j\zeta_0}{2\pi} \frac{e^{-j\beta_0 R_0}}{R_0} (I_0'' + jI_0') \frac{\cos(\tfrac{1}{2} n\pi \cos \Theta)}{\sin \Theta},$$

$$n \text{ odd.} \quad (6)$$

It is clear from (6) that the far-zone field of an antenna with the distribution (2) is the same as with the distribution (5). This could have been predicted at the outset, since (2) and (5) both reduce to $\cos \beta_0 z$ in their dependence on z. For all values of $\beta_0 h$ other than $n\pi/2$, n odd, the real and imaginary factors in (4) differ. In particular, their nulls do not coincide, so that E_Θ^r has no nulls except when $\beta_0 h = n\pi/2$, n odd. This is illustrated in Fig. 16.1, where the factors $F_m(\Theta, \beta_0 h)$, $F_0''(\Theta, \beta_0 h)$, $CF_0''(\Theta, \beta_0 h)$, and $|F_0(\Theta, \beta_0 h)| = \{[F_m(\Theta, \beta_0 h)]^2 + C^2[F_0''(\Theta, \beta_0 h)]^2\}^{1/2}$ are plotted for two values of $C = I_0''/I_m'$ for antennas such that $\beta_0 h = 5\pi/4$ and $\Omega = 2 \ln 2h/a$ is 10 and 20. It is seen that the field factor $|F_0(\Theta, \beta_0 h)|$ does not differ significantly in magnitude from $|F_m(\Theta, \beta_0 h)|$, except that nulls are replaced by minima. It is to be expected that even with $\beta_0 h = n\pi/2$, n odd, the nulls in the field factor would be replaced by very deep minima if the actual current were used in determining the field instead of the value (2).

A more accurate representation of the far-zone field of an antenna of finite cross section may be obtained using Storer's current distribution parameters A and B as given in Sec. II.39. It is readily verified that the substitution of the current distribution (II.39.18a) in (1) leads directly to

$$E_\Theta^r \doteq \frac{j\zeta_0}{2\pi} \frac{e^{-j\beta_0 R_0}}{R_0} [AF_m(\Theta, \beta_0 h) + BU(\Theta, \beta_0 h)],$$

$$(7)$$

where $F_m(\Theta, \beta_0 h)$ is given in (4.13) and

$$U(\Theta, \beta_0 h) \equiv \frac{\sin(\beta_0 h \cos \Theta) - \cos \Theta \sin \beta_0 h}{\sin \Theta \cos \Theta}.$$

$$(8)$$

The current parameters A and B are tabulated in Sec. II.39 for $\Omega = 15$.

17. The Electric Field Near the End of an Antenna; Spark Discharges

A useful application of (8.20c) is in estimating the intensity of the electric field near the ends of an antenna of nonvanishing radius in order to determine the curvature required to prevent a spark discharge. The electric field along the z-axis near the rounded end of an antenna may be assumed to be of the same order of magnitude as the field at corresponding points due to an infinitely thin antenna with the same maximum current if the rounded end of the actual antenna is made to coincide with an ellipsoidal surface of constant phase of the infinitely thin antenna. This is illustrated in Fig. 17.1. If the value of k_e for this ellipsoid is determined and $k_h = 1$ on the z-axis, (8.20c) may be evaluated at $\omega t = \pi/2$ when $(E_\rho)_{\text{inst}}$ is a maximum. For example, if the end of an antenna near $h = \lambda/4$ in half-length may be approximated by an ellipsoid with $k_e = 1.002$, (8.20c) with $n = 1$ gives

$$(E_\rho)_{\max} = \frac{I_m \zeta_0}{4\pi h} \frac{\sin(1.002 \times \pi/2)}{1.002^2 - 1}, \quad (1a)$$

$$(E_\rho)_{\max} \doteq \frac{I_m \zeta_0}{4\pi h} \times 250. \quad (1b)$$

If $I_m = 10$ amperes, $h = 5$ meters,

$$(E_\rho)_{\max} \doteq \frac{10 \times 30}{5} \times 250$$

$$\doteq 15{,}000 \text{ volts/meter.} \quad (2)$$

This is the correct field at $k_e = z/h = 1.002$ on the z-axis for an infinitely thin antenna with its end at $z/h = 1$. It is an estimate of the field near an antenna of which the end is approximated by the ellipsoid $k_e = 1.002$.

If the end of the antenna is made more rounded so that k_e of an ellipsoid that approximates it increases, $(E_\rho)_{max}$ decreases. If the end is made sharper, k_e decreases and $(E_\rho)_{max}$ increases rapidly. For example, with $k_e = 1.0002$ instead of 1.002 as above,

$$(E_\rho)_{max} \doteq 150{,}000 \text{ volts/meter}. \qquad (3)$$

CENTER-DRIVEN ANTENNAS WITH UNEQUAL CURRENTS IN THE HALVES

18. Radiation Field of an Unbalanced Center-Driven Antenna*

Antennas are sometimes center-driven in such a manner that the currents on the identical halves are unequal. This occurs, for example, when the antenna is center-driven from a coaxial line, as shown in Fig. 18.1. In this case the entire system is not symmetric with respect to the plane perpendicular to the antenna and bisecting the distance 2δ between its adjacent ends. As a consequence, the forces acting on the charges in the halves are different and the resulting currents required to maintain the boundary condition are unequal. Simultaneously, tangential forces on the surface of the coaxial line due to charges in the halves of the antenna are not exactly equal and opposite so that currents must exist on the *outer* surface of the coaxial line. These contribute to the radiation field in a manner and degree depending on the over-all length and geometric configuration of the coaxial line. Obviously, this field cannot be determined in general. However, it is possible to calculate the field due to the unequal currents in the halves of the antenna for various assumed degrees of unbalance and in this way determine the effect and significance of such an unbalance. By moving a receiving antenna with its axis tangent to a great circle about the unbalanced transmitting antenna and yet always perpendicular to the axis of the coaxial line, the field due to the currents in the antenna alone can be measured experimentally.

In order to evaluate the radiation field of a center-driven antenna of half-length h, a sinusoidal distribution of current may be

---

* The calculations and the experimental work in this section are those of W. Kelvin, ref. 19.

assumed so that the currents in the halves are

$$I_{z1} \doteq I_m \sin \beta_0(h - z), \quad (\delta \leq z \leq h) \quad (1a)$$

$$I_{z2} \doteq k I_m \sin \beta_0(h + z), \quad (-h \leq z \leq -\delta) \quad (1b)$$

where k is an arbitrary complex factor that characterizes the degree of unbalance. For Fig. 18.1 it may be assumed essentially real since the currents on the two conductors of a coaxial line are equal and opposite.

The far-zone electric or magnetic field may be evaluated for each half by the same method used for equal currents in Secs. 3 through 5. However, since only the far-zone field is required, a shorter method is simply to evaluate the general far-zone integral (I.7.15) with (4.23):

$$E_\Theta^r = -j\omega A_\Theta^r = j\omega A_z^r \sin \Theta$$

$$= \frac{j\omega}{4\pi\nu_0} \frac{e^{-j\beta_0 R_0}}{R_0} \int_{-h}^{h} I_z' e^{-j\beta_0 z' \cos\Theta} \sin\Theta \, dz', \qquad (2)$$

with I_z' given by (1a) in the range $\delta \leq z' \leq h$ and by (1b) in the range $-h \leq z' \leq -\delta$. Since terminal-zone effects are of minor significance in determining the far-zone fields, they may be neglected and the spacing 2δ set equal to zero. Thus, the desired far-zone field is

$$E_\Theta^r \doteq \frac{j\omega I_m}{4\pi\nu_0} \frac{e^{-j\beta_0 R_0}}{R_0}$$
$$\times \left[\int_0^h \sin\beta_0(h-z') e^{j\beta_0 z' \cos\theta} \, dz' \right.$$
$$\left. + k \int_{-h}^0 \sin\beta_0(h+z') e^{-j\beta_0 z' \cos\theta} \, dz' \right]. \qquad (3)$$

The integrals in (3) are elementary and the following result is obtained directly:

$$E_\Theta^r \doteq \frac{j\zeta_0 I_m}{2\pi} \frac{e^{-j\beta_0 R_0}}{R_0} F_a(\Theta), \qquad (4a)$$

where the complex field factor is

$$F_a(\Theta) \equiv \tfrac{1}{2}[(1 + k) F_m(\Theta, \beta_0 h) + j(1 - k) W_m(\Theta, \beta_0 h)]. \qquad (4b)$$

As defined in (4.13),

$$F_m(\Theta, \beta_0 h) \equiv \frac{\cos(\beta_0 h \cos\Theta) - \cos\beta_0 h}{\sin\Theta} \qquad (5)$$

and

$$W_m(\Theta, \beta_0 h) \equiv \frac{\sin(\beta_0 h \cos\Theta) - \sin\beta_0 h \cos\Theta}{\sin\Theta}. \qquad (6)$$

[V.18] THEORY OF LINEAR ANTENNAS 577

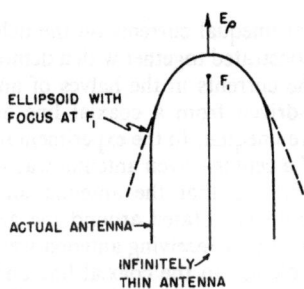

Fig. 17.1. Electric field near the ends of an antenna of non-vanishing radius.

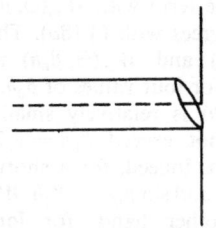

Fig. 18.1. Antenna center-driven from coaxial line.

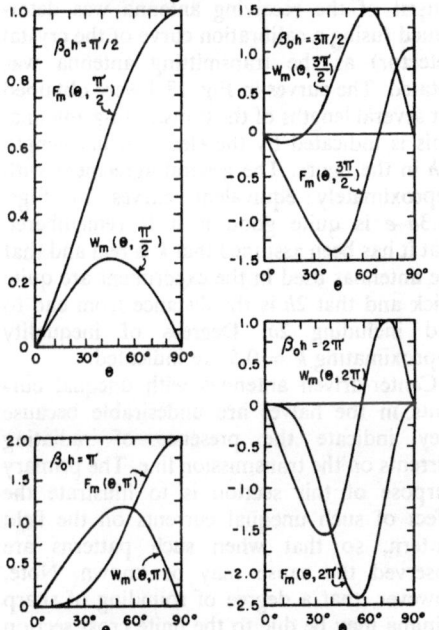

Fig. 18.2. Field functions $F_m(\Theta, \beta_0 h)$ and $W_m(\Theta, \beta_0 h)$.

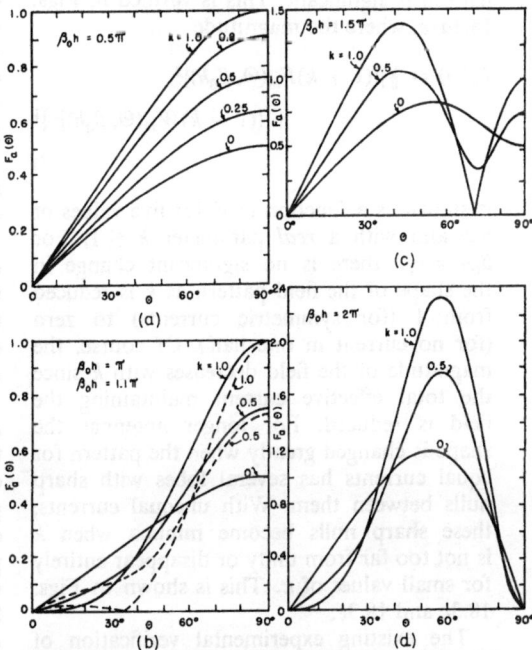

Fig. 18.3a–d. Field factor $F_a(\Theta)$ for unbalanced antenna. The broken lines in (b) are for $\beta_0 h = 1.1\pi$, the solid lines for $\beta_0 h = \pi$.

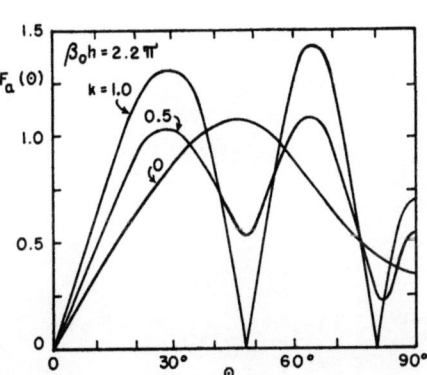

Fig. 18.3e. Field factor $F_a(\Theta)$ for unbalanced antenna.

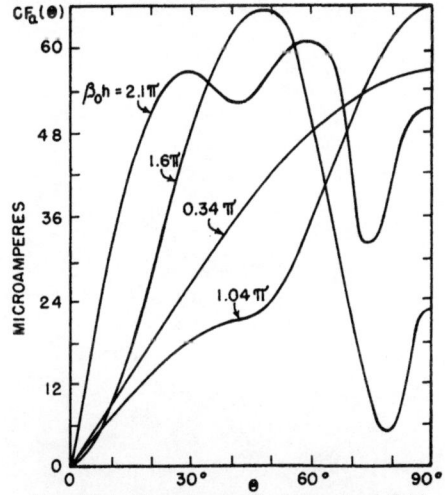

Fig. 18.4. Experimentally determined field of unbalanced antenna.

Note that with equal currents in phase, $k = 1$, so that the term with $W_m(\Theta, \beta_0 h)$ disappears and (4) agrees with (4.18a). The components $F_m(\Theta, \beta_0 h)$ and $W_m(\Theta, \beta_0 h)$ are shown in Fig. 18.2 for four values of $\beta_0 h$. It is seen that $W_m(\Theta, \beta_0 h)$ is relatively small for antennas that do not exceed $\beta_0 h = \pi/2$ in electrical half-length. Indeed, for a short antenna with $(\beta_0 h)^2 \ll 1$ and $\sin \beta_0 h \doteq \beta_0 h$, $W_m(\Theta, \beta_0 h) \doteq 0$. On the other hand, for longer antennas $W_m(\Theta, \beta_0 h)$ is of the same order of magnitude as $F_m(\Theta, \beta_0 h)$, so that its effect on the field pattern is significant. This is verified in Figs. 18.3a–e, where the magnitude

$$F_a(\Theta) = \tfrac{1}{2}\{[(1 + k)F_m(\Theta, \beta_0 h)]^2 + [(1 - k)W_m(\Theta, \beta_0 h)]^2\}^{\tfrac{1}{2}} \quad (7)$$

is shown as a function of Θ for five values of $\beta_0 h$ and with a *real* parameter $k \leq 1$. For $\beta_0 h = \pi/2$ there is no significant change in the shape of the field pattern as k is reduced from 1 (for symmetric currents) to zero (for no current in one half). Of course, the magnitude of the field decreases with k since the total effective current maintaining the field is reduced. For longer antennas the shape is changed greatly when the pattern for equal currents has several lobes with sharp nulls between them. With unequal currents, these sharp nulls become minima when k is not too far from unity or disappear entirely for small values of k. This is shown in Figs. 18.3c and 18.3e.

The existing experimental verification of the theoretical results in Figs. 18.3a–e is qualitative, since the value of k is unavailable. However, the general nature of the effect of unequal currents on the field pattern is demonstrated together with a demonstration that the currents in the halves of an antenna center-driven from a coaxial line as in Fig. 18.1 are unequal. In the experimental arrangement[19] a center-driven antenna was used as in Fig. 18.1, so that the antenna and coaxial line could be rotated around the axis of the line. A distant receiving antenna was oriented perpendicular to the coaxial line and tangent to a great circle about the center of the transmitting antenna. With this apparatus operated at a wavelength of 10 cm, the current in the output of the receiving antenna was determined (using a calibration curve of the crystal detector) as the transmitting antenna was rotated. The curves in Fig. 18.4 were obtained for several lengths of the transmitting antenna. This is indicated by the electrical half-length $\beta_0 h$ in the figure. The general agreement with approximately equivalent curves in Figs. 18.3a–e is quite good if it is remembered that it has been assumed that k is real and that the antennas used in the experiment are quite thick and that $2h$ is the distance from end to end including 2δ. Degrees of inequality approximating $k = 0.5$ are indicated.

Center-driven antennas with unequal currents in the halves are undesirable because they indicate the presence of radiating currents on the transmission line. The primary purpose of this section is to illustrate the effect of such unequal currents on the field pattern, so that when such patterns are observed the cause may be known. Note, however, that a degree of rounding of sharp minima may be due to the finite cross section of the antenna (Sec. 16) or, if the antenna is erected on the ground, to the finite conductivity of earth and water (Sec. VII.8).

CHAPTER VI

ELECTROMAGNETIC FIELDS OF ANTENNA ARRAYS

The fundamental requirement for a transmitting antenna is that it maintain a significant far-zone field in the direction of the receiving antenna or antennas in which currents are to be excited, whether for transferring power to a load or for reradiation or scattering. Supplementary requirements may include a vanishingly small field in all other or some other directions. Thus, for point-to-point communication a large field essentially in a single direction is desired. On the other hand, for broadcast transmission more or less omnidirectional properties usually are advantageous. Field distributions having a great variety of properties may be obtained by superimposing the fields of arrays of linear radiators. Suitable variables include the number and relative locations of antennas and the relative magnitudes and phases of the currents in them. If all antennas are identical and have currents of equal magnitude and progressive phase sequence, the array is called uniform, otherwise nonuniform. It is usually convenient to have all elements in an array of center-driven antennas a half-wavelength long or less. For these it has been shown that the distributions of current are very nearly sinusoidal.

The essential property of a receiving antenna is to respond to the electromagnetic field of a transmitter or a group of transmitters in a specified direction or directions, and to be quite insensitive to the fields of transmitters in other directions. For point-to-point transmission this requires an essentially unidirectional response, for broadcast reception an omnidirectional or selectively unidirectional response. Since the load in a receiving array usually is connected to a single antenna, the directional properties of receiving arrays may be obtained from the directional properties of transmitting arrays which have only a single driven antenna among an arbitrary number of parasitic antennas by application of the reciprocal theorem. It follows that in this chapter it is sufficient to consider the directional properties of transmitting arrays, including those in which only one unit is driven.

UNIFORM PARALLEL ARRAYS

1. The Diffraction Formula and the General Array Factor

In the study of uniform arrays of N identical collinear or parallel antennas with currents of equal magnitude the following sums are encountered, and it is well to investigate them in general terms first:

$$A_{\text{odd}} = 1 + 2 \sum_{i=1}^{\frac{1}{2}(N-1)} \cos 2ix, \quad N \text{ odd}, \qquad (1)$$

$$A_{\text{even}} = 2 \sum_{i=1}^{\frac{1}{2}N} \cos (2i-1)x, \quad N \text{ even}. \qquad (2)$$

These sums are readily transformed by multiplying numerator and denominator of the sum by $\sin x$ and expanding in exponential form.* The results show that A,

---

* $A_{\text{odd}} = 1 + \sum_{i=1}^{\frac{1}{2}(N-1)} \dfrac{(e^{j2ix} + e^{-j2ix})(e^{jx} - e^{-jx})}{e^{jx} - e^{-jx}}$

$= 1 + \sum_{i=1}^{\frac{1}{2}(N-1)} \dfrac{-e^{j(2i-1)x} + e^{-j(2i-1)x} + e^{j(2i+1)x} - e^{-j(2i+1)x}}{e^{jx} - e^{-jx}}$

$= 1 + \sum_{i=1}^{\frac{1}{2}(N-1)} \dfrac{[-\sin(2i-1)x + \sin(2i+1)x]}{\sin x}$

$= 1 + \left[\dfrac{-\sin x + \sin 3x - \sin 3x + \sin 5x}{\sin x} \right.$
$\left. \dfrac{- \cdots + \sin(N-2)x + \sin Nx}{\sin x} \right]$

$= \dfrac{\sin Nx}{\sin x}, \quad (N \text{ odd}) \qquad (3)$

$A_{\text{even}} = \sum_{i=1}^{\frac{1}{2}N} \dfrac{(e^{j(2i-1)x} + e^{-j(2i-1)x})(e^{jx} - e^{-jx})}{e^{jx} - e^{-jx}}$

$= \sum_{i=1}^{\frac{1}{2}N} \dfrac{-(e^{j2(i-1)x} - e^{-j2(i-1)x}) + e^{j2ix} - e^{-j2ix}}{e^{jx} - e^{-jx}}$

$= \sum_{i=1}^{\frac{1}{2}N} \dfrac{\sin 2(i-1)x + \sin 2ix}{\sin x}$

$= \dfrac{0 + \sin 2x - \sin 2x + \sin 4x - \sin 4x}{\sin x}$
$\dfrac{+ \sin 6x - \cdots - \sin 2(\frac{1}{2}N - 1)x + \sin Nx}{\sin x}$

$= \dfrac{\sin Nx}{\sin x}. \quad (N \text{ even}) \qquad (4)$

to be called the *array factor*, is the same in form for even and odd values of N and is given by

$$A = \frac{\sin Nx}{\sin x}. \quad (5)$$

The factor A is well known in optics, where it occurs in the form A^2 in the formula for the intensity of illumination in the diffraction pattern of a grating. In discussing the behavior of this function and in representing it graphically, it is more convenient to divide it by N. Thus,

$$a \equiv \frac{A}{N} = \frac{\sin Nx}{N \sin x}. \quad (6)$$

The factor a is the *array factor per element* or the *normalized array factor*. It is plotted in Fig. 1.1 for $N = 1, 2, 3, 4, 5, 6, 8, 10, 12$ and listed in Table 1.1 for a useful range of values of x and N. Important properties of a, the array factor per element, are its extreme values and its zeros.

The values of x giving extreme values of a may be obtained by equating to zero the derivative da/dx. Thus,

$$\frac{da}{dx} = \frac{d}{dx}\left(\frac{\sin Nx}{N \sin x}\right) = 0. \quad (7)$$

Differentiation leads to

$$N \tan x = \tan Nx. \quad (8)$$

The roots of (8) may be divided into two groups: (*a*) those that reduce both sides to zero, and (*b*), those that do not.

(*a*) The roots for which $\tan x = \tan Nx = 0$ are readily determined. Since N is an integer, they are given by

$$x = q\pi. \quad (q = 0, 1, 2, \cdots) \quad (9)$$

The indeterminate form obtained when (9) is substituted in (6) is readily evaluated as follows:

$$\lim_{\delta \to 0} \frac{\sin N(q\pi - \delta)}{N \sin (q\pi - \delta)} = \frac{\cos Nq\pi}{\cos q\pi} \lim_{\delta \to 0} \frac{\sin N\delta}{N \sin \delta}$$

$$= \frac{(-1)^{Nq}}{(-1)^q} \lim_{\delta \to 0} \frac{\cos N\delta}{\cos \delta}$$

$$= (-1)^{q(N-1)}. \quad (10)$$

Accordingly,

$$a = a_{\max} = 1 \quad \text{when } x = q\pi$$
$$\text{with } q(N-1) \text{ even}, \quad (11)$$

$$a = a_{\min} = -1 \quad \text{when } x = q\pi$$
$$\text{with } q(N-1) \text{ odd}. \quad (12)$$

These values having a magnitude 1 are called *principal* extreme values.

(*b*) Other roots of (5) locating minor extreme values are not easily determined analytically. They may be obtained graphically by plotting $\tan x$ and $(\tan Nx)/N$ for a particular value of N and locating points of intersection. This method is illustrated in Fig. 1.2 for the first intersections with $N = 4$ and 5, and the first and second intersections with $N = 6$. It is seen from these curves that at the points of intersection $(\tan Nx)/N$ is quite steep and that the locations of the points of intersection do not differ greatly from the values at which $\tan Nx$ becomes infinite, provided N is not too small. Hence, an approximate, but usually quite satisfactory, expression locating the minor extreme values is

$$x = (p + \tfrac{1}{2})\pi/N. \quad (p = 1, 2, 3, \cdots) \quad (13)$$

The value $p = 0$ does not give a point of intersection. The range of p for intersections with x in the first quadrant is

$$x = (p + \tfrac{1}{2})\pi/N \leq \pi/2, \quad p \leq \tfrac{1}{2}(N-1). \quad (14)$$

The magnitude and sign of extreme values of a are obtained readily and in general by using (8) with (6). Specifically, multiplication of (6) by $N \tan x / \tan Nx = 1$ gives

$$a_{\text{ext}} = \frac{\sin Nx}{N \sin x} \cdot \frac{N \tan x}{\tan Nx} = \frac{\cos Nx}{\cos x}. \quad (15)$$

An alternative expression is obtained with the trigonometric identity

$$\cos Nx = 1/\sqrt{1 + \tan^2 Nx}. \quad (16)$$

With (14) and (8),

$$a_{\text{ext}} = \frac{\pm 1}{\cos x \sqrt{1 + N^2 \tan^2 x}}$$

$$= \frac{\pm 1}{\sqrt{1 + (N^2 - 1) \sin^2 x}}. \quad (17)$$

The largest possible values of a_{ext} occur when $x = q\pi$, $q = 0, 1, 2, 3, \cdots$, which yield the *principal* extreme values, *all of magnitude* unity, as given in (11) and (12). All other extreme values have *magnitudes* that are *less than unity*, since for them the denominator in (17) is always greater than one. Their approximate magnitudes are obtained by substituting (13) in (17) as follows:

$$a_{\text{ext}} \doteq 1/\sqrt{1 + (N^2 - 1) \sin^2 [\pi(p + \tfrac{1}{2})/N]}.$$

$$(p = 1, 2, 3, \cdots) \quad (18)$$

THEORY OF LINEAR ANTENNAS

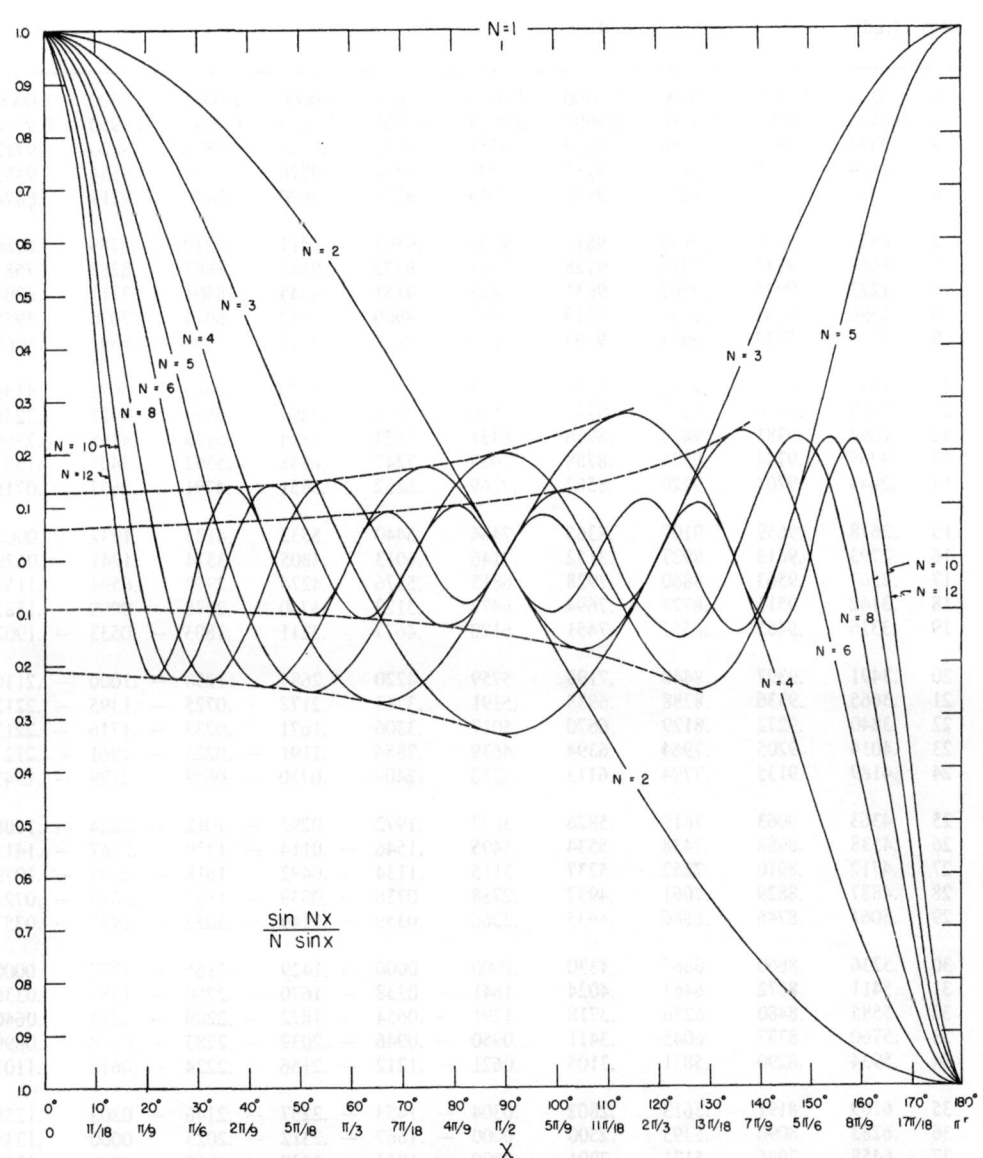

Fig. 1.1. Normalized array factor $\sin Nx / N \sin x$ as a function of x for different values of N.

TABLE 1.1. Values of $a = \sin Nx / N \sin x$.

x (deg)	x (rad)	N=2	N=3	N=4	N=5	N=6	N=7	N=8	N=10	N=12
0	0	1.0000	1.0000	1.0000	1.0000	1.0000	1.0000	1.0000	1.0000	1.0000
1	0.0175	0.9998	0.9996	0.9992	0.9988	0.9982	0.9975	0.9968	0.9950	0.9928
2	.0349	.9994	.9984	.9970	.9951	.9929	.9897	.9873	.9800	.9712
3	.0524	.9986	.9963	.9932	.9891	.9841	.9776	.9715	.9554	.9359
4	.0698	.9976	.9935	.9879	.9806	.9718	.9620	.9496	.9215	.8878
5	.0873	.9962	.9899	.9811	.9698	.9561	.9411	.9219	.8789	.8280
6	.1047	.9945	.9854	.9728	.9567	.9372	.9145	.8887	.8285	.7582
7	.1222	.9925	.9802	.9631	.9413	.9151	.8845	.8503	.7711	.6800
8	.1396	.9903	.9742	.9519	.9237	.8900	.8512	.8073	.7076	.5955
9	.1571	.9877	.9674	.9393	.9040	.8619	.8138	.7599	.6392	.5066
10	.1745	.9848	.9598	.9254	.8823	.8312	.7729	.7089	.5671	.4156
11	.1920	.9816	.9515	.9101	.8586	.7980	.7294	.6547	.4925	.3246
12	.2094	.9781	.9424	.8936	.8331	.7624	.6824	.5979	.4165	.2356
13	.2269	.9744	.9325	.8758	.8058	.7247	.6348	.5392	.3405	.1507
14	.2443	.9703	.9220	.8567	.7769	.6852	.5851	.4791	.2657	.0716
15	.2618	.9659	.9107	.8365	.7464	.6440	.5332	.4183	.1932	.0000
16	.2793	.9613	.8987	.8152	.7146	.6013	.4805	.3574	.1241	−.0629
17	.2967	.9563	.8860	.7928	.6815	.5576	.4274	.2970	.0594	−.1159
18	.3142	.9511	.8727	.7694	.6472	.5129	.3740	.2378	.0000	−.1585
19	.3316	.9455	.8587	.7451	.6120	.4677	.3211	.1803	−.0533	−.1902
20	.3491	.9397	.8440	.7198	.5759	.4220	.2684	.1250	−.1000	−.2110
21	.3665	.9336	.8288	.6938	.5391	.3763	.2172	.0725	−.1395	−.2212
22	.3840	.9272	.8129	.6670	.5017	.3306	.1671	.0233	−.1716	−.2212
23	.4014	.9205	.7964	.6394	.4639	.2854	.1191	−.0223	−.1961	−.2121
24	.4189	.9135	.7794	.6113	.4258	.2409	.0730	−.0639	−.2129	−.1949
25	.4363	.9063	.7619	.5826	.3877	.1972	.0295	−.1012	−.2224	−.1708
26	.4538	.8988	.7438	.5534	.3495	.1546	−.0114	−.1339	−.2247	−.1413
27	.4712	.8910	.7252	.5237	.3115	.1134	−.0492	−.1618	−.2203	−.1079
28	.4887	.8829	.7061	.4937	.2738	.0738	−.0839	−.1850	−.2098	−.0722
29	.5061	.8746	.6866	.4635	.2366	.0359	−.1149	−.2032	−.1938	−.0357
30	.5236	.8660	.6667	.4330	.2000	.0000	−.1429	−.2165	−.1732	.0000
31	.5411	.8572	.6463	.4024	.1641	−.0338	−.1670	−.2250	−.1487	.0336
32	.5585	.8480	.6256	.3718	.1291	−.0654	−.1872	−.2289	−.1213	.0640
33	.5760	.8387	.6045	.3411	.0950	−.0946	−.2039	−.2283	−.0918	.0899
34	.5934	.8290	.5831	.3106	.0621	−.1212	−.2166	−.2234	−.0612	.1107
35	.6109	.8191	.5613	.2802	.0304	−.1453	−.2257	−.2146	−.0303	.1258
36	.6283	.8090	.5393	.2500	.0000	−.1667	−.2312	−.2023	.0000	.1348
37	.6458	.7986	.5171	.2201	−.0290	−.1853	−.2330	−.1867	.0289	.1377
38	.6632	.7880	.4946	.1906	−.0564	−.2012	−.2314	−.1683	.0556	.1346
39	.6807	.7771	.4719	.1616	−.0823	−.2143	−.2266	−.1476	.0795	.1259
40	.6981	.7660	.4491	.1330	−.1064	−.2245	−.2189	−.1250	.1000	.1123
41	.7156	.7547	.4261	.1050	−.1288	−.2321	−.2082	−.1010	.1168	.0944
42	.7330	.7431	.4030	.0777	−.1494	−.2369	−.1950	−.0760	.1294	.0732
43	.7505	.7314	.3798	.0510	−.1682	−.2390	−.1795	−.0505	.1378	.0497
44	.7679	.7193	.3566	.0251	−.1851	−.2386	−.1621	−.0250	.1418	.0249

THEORY OF LINEAR ANTENNAS

Table 1.1—contd.

x (deg)	x (rad)	N=2	N=3	N=4	N=5	N=6	N=7	N=8	N=10	N=12
45	0.7854	0.7071	0.3333	0.0000	−0.2000	−0.2357	−0.1428	0.0000	0.1414	0.0000
46	.8029	.6947	.3101	−.0242	−.2130	−.2304	−.1222	.0242	.1369	−.0241
47	.8203	.6820	.2868	−.0476	−.2240	−.2229	−.1006	.0471	.1285	−.0463
48	.8378	.6691	.2636	−.0699	−.2331	−.2133	−.0781	.0684	.1165	−.0659
49	.8552	.6561	.2406	−.0913	−.2402	−.2017	−.0553	.0878	.1002	−.0821
50	.8727	.6428	.2176	−.1116	−.2453	−.1884	−.0323	.1049	.0839	−.0942
51	.8901	.6293	.1947	−.1308	−.2486	−.1735	−.0096	.1195	.0643	−.1020
52	.9076	.6157	.1721	−.1489	−.2499	−.1572	.0127	.1315	.0434	−.1052
53	.9250	.6018	.1496	−.1659	−.2495	−.1396	.0341	.1407	.0217	−.1038
54	.9425	.5878	.1273	−.1816	−.2472	−.1211	.0546	.1469	.0000	−.0980
55	.9599	.5736	.1053	−.1962	−.2432	−.1017	.0736	.1503	−.0212	−.0881
56	.9774	.5592	.0836	−.2095	−.2376	−.0818	.0913	.1507	−.0413	−.0747
57	.9948	.5446	.0622	−.2215	−.2303	−.0614	.1072	.1482	−.0596	−.0584
58	1.0123	.5299	.0411	−.2323	−.2216	−.0409	.1212	.1430	−.0758	−.0400
59	1.0297	.5150	.0204	−.2418	−.2115	−.0203	.1330	.1352	−.0894	−.0202
60	1.0472	.5000	.0000	−.2500	−.2000	.0000	.1428	.1250	−.1000	.0000
61	1.0647	.4848	−.0199	−.2569	−.1873	.0199	.1504	.1126	−.1074	.0198
62	1.0821	.4695	−.0395	−.2625	−.1735	.0392	.1555	.0983	−.1115	.0384
63	1.0996	.4540	−.0585	−.2668	−.1587	.0578	.1584	.0825	−.1122	.0550
64	1.1170	.4384	−.0771	−.2699	−.1430	.0754	.1588	.0653	−.1096	.0689
65	1.1345	.4226	−.0952	−.2717	−.1266	.0919	.1570	.0472	−.1037	.0796
66	1.1519	.4067	−.1128	−.2722	−.1095	.1072	.1530	.0284	−.0948	.0868
67	1.1694	.3907	−.1298	−.2714	−.0918	.1212	.1467	.0095	−.0832	.0900
68	1.1868	.3746	−.1462	−.2695	−.0738	.1336	.1386	−.0094	−.0693	.0894
69	1.2043	.3584	−.1621	−.2663	−.0554	.1444	.1283	−.0278	−.0536	.0849
70	1.2217	.3420	−.1774	−.2620	−.0370	.1536	.1165	−.0455	−.0364	.0768
71	1.2392	.3256	−.1920	−.2566	−.0184	.1610	.1030	−.0621	−.0184	.0655
72	1.2566	.3090	−.2060	−.2500	.0000	.1667	.0883	−.0773	.0000	.0515
73	1.2741	.2924	−.2194	−.2424	.0182	.1705	.0724	−.0908	.0182	.0354
74	1.2915	.2756	−.2320	−.2338	.0361	.1724	.0557	−.1025	.0356	.0180
75	1.3090	.2588	−.2440	−.2241	.0536	.1725	.0383	−.1121	.0518	.0000
76	1.3265	.2419	−.2553	−.2136	.0705	.1708	.0204	−.1194	.0662	−.0179
77	1.3439	.2250	−.2659	−.2022	.0867	.1673	.0026	−.1245	.0786	−.0348
78	1.3614	.2079	−.2757	−.1899	.1022	.1621	−.0153	−.1271	.0885	−.0501
79	1.3788	.1908	−.2848	−.1769	.1169	.1551	−.0327	−.1273	.0957	−.0631
80	1.3963	.1736	−.2931	−.1632	.1305	.1466	−.0496	−.1250	.1000	−.0733
81	1.4137	.1564	−.3007	−.1488	.1432	.1365	−.0656	−.1204	.1012	−.0802
82	1.4312	.1392	−.3075	−.1338	.1547	.1251	−.0807	−.1135	.0994	−.0837
83	1.4486	.1219	−.3135	−.1182	.1651	.1124	−.0944	−.1044	.0947	−.0835
84	1.4661	.1045	−.3188	−.1022	.1742	.0985	−.1068	−.0934	.0871	−.0797
85	1.4835	.0872	−.3232	−.0858	.1820	.0837	−.1174	−.0807	.0769	−.0724
86	1.5010	.0698	−.3268	−.0691	.1884	.0680	−.1264	−.0664	.0644	−.0621
87	1.5184	.0523	−.3297	−.0520	.1935	.0516	−.1335	−.0509	.0501	−.0490
88	1.5359	.0349	−.3317	−.0348	.1971	.0347	−.1387	−.0345	.0342	−.0339
89	1.5533	.0175	−.3329	−.0174	.1993	.0174	−.1418	−.0174	.0174	−.0173

584 THEORY OF LINEAR ANTENNAS [VI.1]

TABLE 1.1—contd.

x (deg)	x (rad)	N=2	N=3	N=4	N=5	N=6	N=7	N=8	N=10	N=12
90	1.5708	0.0000	−0.3333	0.0000	0.2000	0.0000	−0.1428	0.0000	0.0000	0.0000
91	1.5883	−.0175	−.3329	.0174	.1993	−.0174	−.1418	.0174	−.0174	.0173
92	1.6057	−.0349	−.3317	.0348	.1971	−.0347	−.1387	.0345	−.0342	.0339
93	1.6232	−.0523	−.3297	.0520	.1935	−.0516	−.1335	.0509	−.0501	.0490
94	1.6406	−.0698	−.3268	.0691	.1884	−.0680	−.1264	.0664	−.0644	.0621
95	1.6581	−.0872	−.3232	.0858	.1820	−.0837	−.1174	.0807	−.0769	.0724
96	1.6755	−.1045	−.3188	.1022	.1742	−.0985	−.1068	.0934	−.0871	.0797
97	1.6930	−.1219	−.3135	.1182	.1651	−.1124	−.0944	.1044	−.0947	.0835
98	1.7104	−.1392	−.3075	.1338	.1547	−.1251	−.0807	.1135	−.0994	.0837
99	1.7279	−.1564	−.3007	.1488	.1432	−.1365	−.0656	.1204	−.1012	.0802
100	1.7453	−.1736	−.2931	.1632	.1305	−.1466	−.0496	.1250	−.1000	.0733
101	1.7628	−.1908	−.2848	.1769	.1169	−.1551	−.0327	.1273	−.0957	.0631
102	1.7802	−.2079	−.2757	.1899	.1022	−.1621	−.0153	.1271	−.0885	.0501
103	1.7977	−.2250	−.2659	.2022	.0867	−.1673	.0026	.1245	−.0786	.0348
104	1.8151	−.2419	−.2553	.2136	.0705	−.1708	.0204	.1194	−.0662	.0179
105	1.8326	−.2588	−.2440	.2241	.0536	−.1725	.0383	.1121	−.0518	.0000
106	1.8500	−.2756	−.2320	.2338	.0361	−.1724	.0557	.1025	−.0356	−.0180
107	1.8675	−.2924	−.2194	.2424	.0182	−.1705	.0724	.0908	−.0182	−.0354
108	1.8850	−.3090	−.2060	.2500	.0000	−.1667	.0883	.0773	.0000	−.0515
109	1.9024	−.3256	−.1920	.2566	−.0184	−.1610	.1030	.0621	.0184	−.0654
110	1.9199	−.3420	−.1774	.2620	−.0370	−.1536	.1165	.0455	.0364	−.0768
111	1.9373	−.3584	−.1621	.2663	−.0554	−.1444	.1283	.0278	.0536	−.0849
112	1.9548	−.3746	−.1462	.2695	−.0738	−.1336	.1386	.0094	.0693	−.0894
113	1.9722	−.3907	−.1298	.2714	−.0918	−.1212	.1467	−.0095	.0832	−.0900
114	1.9897	−.4067	−.1128	.2722	−.1095	−.1072	.1530	−.0284	.0948	−.0868
115	2.0071	−.4226	−.0952	.2717	−.1266	−.0919	.1570	−.0472	.1037	−.0796
116	2.0246	−.4384	−.0771	.2699	−.1430	−.0754	.1588	−.0653	.1096	−.0689
117	2.0420	−.4540	−.0585	.2668	−.1587	−.0578	.1584	−.0825	.1122	−.0550
118	2.0595	−.4695	−.0395	.2625	−.1735	−.0392	.1555	−.0983	.1115	−.0384
119	2.0769	−.4848	−.0199	.2569	−.1873	−.0199	.1504	−.1126	.1074	−.0198
120	2.0944	−.5000	.0000	.2500	−.2000	.0000	.1428	−.1250	.1000	.0000
121	2.1118	−.5150	.0204	.2418	−.2115	.0203	.1330	−.1352	.0894	.0202
122	2.1293	−.5299	.0411	.2323	−.2216	.0409	.1212	−.1430	.0758	.0400
123	2.1468	−.5446	.0622	.2215	−.2303	.0614	.1072	−.1482	.0596	.0585
124	2.1642	−.5592	.0836	.2095	−.2376	.0818	.0913	−.1507	.0413	.0747
125	2.1817	−.5736	.1053	.1962	−.2432	.1017	.0736	−.1503	.0212	.0881
126	2.1991	−.5878	.1273	.1816	−.2472	.1211	.0546	−.1469	.0000	.0980
127	2.2166	−.6018	.1496	.1659	−.2495	.1396	.0341	−.1407	−.0217	.1038
128	2.2340	−.6157	.1721	.1489	−.2499	.1572	.0127	−.1315	−.0434	.1052
129	2.2515	−.6293	.1947	.1308	−.2486	.1735	−.0096	−.1195	−.0643	.1020
130	2.2689	−.6428	.2176	.1116	−.2453	.1884	−.0323	−.1049	−.0839	.0942
131	2.2864	−.6561	.2406	.0913	−.2402	.2017	−.0553	−.0878	−.1002	.0821
132	2.3038	−.6691	.2636	.0699	−.2331	.2133	−.0781	−.0684	−.1165	.0659
133	2.3213	−.6820	.2868	.0476	−.2240	.2229	−.1006	−.0471	−.1285	.0463
134	2.3387	−.6947	.3101	.0242	−.2130	.2304	−.1222	−.0242	−.1369	.0241

Table 1.1—contd.

x (deg)	x (rad)	N=2	N=3	N=4	N=5	N=6	N=7	N=8	N=10	N=12
135	2.3562	−0.7071	0.3333	0.0000	−0.2000	0.2357	−0.1428	0.0000	−0.1414	0.0000
136	2.3736	−.7193	.3566	−.0251	−.1851	.2386	−.1621	.0250	−.1418	−.0249
137	2.3911	−.7314	.3798	−.0510	−.1682	.2390	−.1795	.0505	−.1378	−.0497
138	2.4086	−.7431	.4030	−.0777	−.1494	.2369	−.1950	.0760	−.1294	−.0732
139	2.4260	−.7547	.4261	−.1050	−.1288	.2321	−.2082	.1010	−.1168	−.0944
140	2.4435	−.7660	.4491	−.1330	−.1064	.2245	−.2189	.1250	−.1000	−.1123
141	2.4609	−.7771	.4719	−.1616	−.0823	.2143	−.2266	.1476	−.0795	−.1259
142	2.4784	−.7880	.4946	−.1906	−.0564	.2012	−.2314	.1683	−.0556	−.1346
143	2.4958	−.7986	.5171	−.2201	−.0290	.1853	−.2330	.1867	−.0289	−.1377
144	2.5133	−.8090	.5393	−.2500	.0000	.1667	−.2312	.2023	.0000	−.1348
145	2.5307	−.8191	.5613	−.2802	.0304	.1453	−.2257	.2146	.0303	−.1258
146	2.5482	−.8290	.5831	−.3106	.0621	.1212	−.2166	.2234	.0612	−.1107
147	2.5656	−.8387	.6045	−.3411	.0950	.0946	−.2039	.2283	.0918	−.0899
148	2.5831	−.8480	.6256	−.3718	.1291	.0654	−.1872	.2289	.1213	−.0640
149	2.6005	−.8572	.6463	−.4024	.1641	.0338	−.1670	.2250	.1487	−.0336
150	2.6180	−.8660	.6667	−.4330	.2000	.0000	−.1429	.2165	.1732	.0000
151	2.6354	−.8746	.6866	−.4635	.2366	−.0359	−.1149	.2032	.1938	.0357
152	2.6529	−.8829	.7061	−.4937	.2738	−.0738	−.0839	.1850	.2098	.0722
153	2.6704	−.8910	.7252	−.5237	.3115	−.1134	−.0492	.1618	.2203	.1079
154	2.6878	−.8988	.7438	−.5534	.3495	−.1546	−.0114	.1339	.2247	.1413
155	2.7053	−.9063	.7619	−.5826	.3877	−.1972	.0295	.1012	.2224	.1708
156	2.7227	−.9135	.7794	−.6113	.4258	−.2409	.0730	.0639	.2129	.1949
157	2.7402	−.9205	.7964	−.6394	.4639	−.2854	.1191	.0223	.1961	.2121
158	2.7576	−.9272	.8129	−.6670	.5017	−.3306	.1671	−.0233	.1716	.2212
159	2.7751	−.9336	.8288	−.6938	.5391	−.3763	.2172	−.0725	.1395	.2212
160	2.7925	−.9397	.8440	−.7198	.5759	−.4220	.2684	−.1250	.1000	.2110
161	2.8100	−.9455	.8587	−.7451	.6120	−.4677	.3211	−.1803	.0533	.1902
162	2.8274	−.9511	.8727	−.7694	.6472	−.5129	.3740	−.2378	.0000	.1585
163	2.8449	−.9563	.8860	−.7928	.6815	−.5576	.4274	−.2970	−.0594	.1159
164	2.8623	−.9613	.8987	−.8152	.7146	−.6013	.4805	−.3574	−.1241	.0629
165	2.8798	−.9659	.9107	−.8365	.7464	−.6440	.5332	−.4183	−.1932	.0000
166	2.8972	−.9703	.9220	−.8567	.7769	−.6852	.5851	−.4791	−.2657	−.0716
167	2.9147	−.9744	.9325	−.8758	.8058	−.7247	.6348	−.5392	−.3405	−.1507
168	2.9322	−.9781	.9424	−.8936	.8331	−.7624	.6824	−.5979	−.4165	−.2356
169	2.9496	−.9816	.9515	−.9101	.8586	−.7980	.7294	−.6547	−.4925	−.3246
170	2.9671	−.9848	.9598	−.9254	.8823	−.8312	.7729	−.7089	−.5671	−.4156
171	2.9845	−.9877	.9674	−.9393	.9040	−.8619	.8138	−.7599	−.6392	−.5066
172	3.0020	−.9903	.9742	−.9519	.9237	−.8900	.8512	−.8073	−.7076	−.5955
173	3.0194	−.9925	.9802	−.9631	.9413	−.9151	.8845	−.8503	−.7711	−.6800
174	3.0369	−.9945	.9854	−.9728	.9567	−.9372	.9145	−.8887	−.8285	−.7582
175	3.0543	−.9962	.9899	−.9811	.9698	−.9561	.9411	−.9219	−.8789	−.8280
176	3.0718	−.9976	.9935	−.9879	.9806	−.9718	.9620	−.9496	−.9215	−.8878
177	3.0892	−.9986	.9963	−.9932	.9891	−.9841	.9776	−.9715	−.9554	−.9359
178	3.1067	−.9994	.9984	−.9970	.9951	−.9929	.9897	−.9873	−.9800	−.9712
179	3.1241	−.9998	.9996	−.9992	.9988	−.9982	.9975	−.9968	−.9950	−.9928
180	3.1416	−1.0000	1.0000	−1.0000	1.0000	−1.0000	1.0000	−1.0000	−1.0000	−1.0000

This expression can be simplified if $(p + \frac{1}{2})/N$ is small enough so that the following inequality is a good approximation:

$$N^2 \gg (p + \tfrac{1}{2})^2. \tag{19}$$

This is obtained by expanding the quantity $(N^2 - 1)\sin^2[\pi(p + \frac{1}{2})/N]$ in series and noting that with (18)

$$(N^2 - 1)\sin^2[\pi(p + \tfrac{1}{2})/N] \doteq \pi^2(p + \tfrac{1}{2})^2. \tag{20}$$

Hence, subject to (19), (18) reduces to

$$a_{\text{ext}} \doteq 1/\sqrt{1 + \pi^2(p + \tfrac{1}{2})^2} \doteq 1/\pi(p + \tfrac{1}{2}). \tag{21}$$

The last step follows from the fact that since $p \geq 1$, it follows that $\pi^2(p + \frac{1}{2})^2 > 22$. Evidently, a_{ext} is essentially *independent of N* for all values of N that are sufficiently large to satisfy (19). Numerical values of minor extremes of a as computed from (21) are:

$$\begin{array}{l} p = \quad 1 \quad\quad 2 \quad\quad 3 \quad\quad 4 \\ a_{\text{ext}} = 0.212 \quad 0.128 \quad 0.091 \quad 0.058 \end{array} \tag{22}$$

These limiting values are indicated in Fig. 1.1, and envelopes of the first four orders of minor maxima and minima are shown in dotted lines. It is clear from Fig. 1.1 or (19) and (21) that minor extreme values of the array factor per element cannot be reduced below specific limiting values by increasing N. This is of importance in the design of directional arrays.

The zeros of the array factor per element occur when the numerator in (6) vanishes *and* when the denominator does not. Thus,

$$a = 0 \text{ when } x = \frac{p\pi}{N} \neq q\pi.$$

$$(q = 0, 1, 2, \cdots, \quad p = 1, 2, \cdots) \tag{23}$$

The number of zeros between $x = 0$ and $x = \pi$ is $N - 1$.

The approximate locations of the minor extreme values as defined in (13) are seen to be at *values of x that lie midway between the zeros*.

A useful special form of the general variable x appearing in $a_r = \sin N_r x / N_r \sin x$ is

$$x = \pi(n_r \cos \psi - t_r) = 180°(n_r \cos \psi - t_r), \tag{24}$$

where n_r and t_r are arbitrary parameters and ψ an angle variable. In order to represent conveniently the three parameters N_r, n_r, t_r, the following shorthand is used:

$$a_r(\psi; N_r, n_r, t_r) \equiv \frac{\sin N_r \pi(n_r \cos \psi - t_r)}{N_r \sin \pi(n_r \cos \psi - t_r)}$$

$$= \frac{\sin N_r\, 180°(n_r \cos \psi - t_r)}{N_r \sin 180°(n_r \cos \psi - t_r)}. \tag{25}$$

For example, if $N_r = 8$, $n_r = \frac{1}{2}$, $t_r = 1/6$,

$$a(\psi; 8, \tfrac{1}{2}, 1/6) = \frac{\sin 8\pi(\tfrac{1}{2}\cos \psi - 1/6)}{8 \sin \pi(\tfrac{1}{2}\cos \psi - 1/6)}. \tag{26a}$$

If $\psi = \pi/2$,

$$a(\pi/2; 8, \tfrac{1}{2}, 1/6) = \frac{\sin 8\pi/6}{8 \sin \pi/6}. \tag{26b}$$

In an antenna array of a particular type r, N_r is the number of units, n_r is the distance between centers of units in wavelengths, and t_r is the phase difference between currents in adjacent elements in fractions of a period.

A quantity called the *beam width of the array factor per element* may be defined in either of two ways as follows:

The *half-power beam width* is the angle between the values of x on each side of a principal extreme value at which the magnitude of the square of the array factor per element is equal to one-half.

The *null beam width* is the angle between the values of x on each side of a principal extreme value at which the array factor per element vanishes.

The principal maxima occur at x_m defined by

$$x_m = q\pi. \quad (q = 0, 1, 2, 3, \cdots) \tag{27}$$

The half-power values of x are given by x_1 in

$$\sin Nx_1 = \sqrt{2} N \sin x_1. \tag{28}$$

The half-power beam width is

$$W_1 = 2|x_m - x_1|. \tag{29}$$

Since the half-power values, x_1, usually occur at reasonably small values of x, an approximate formula for x_1 is obtained from (28) by replacing $\sin x_1$ by x_1. Then

$$Nx_1 \doteq 0.707 \sin Nx_1. \tag{30}$$

The solution of this is

$$Nx_1 \doteq 1.39 \text{ or } x_1 \doteq \frac{1.39}{N} = \frac{79.6°}{N}. \tag{31}$$

TABLE 1.2. Beam width (deg) of general array factor.

N	2	3	4	5	6	8	10	12	20	50
$W_1 = 159.2°/N$	79.6	53.1	39.8	31.8	26.5	19.9	15.9	13.3	8.0	3.2
W_1 (from Fig. 1.1)	90	56	40.5	32.5	27	20.5	16.5	13.5		
$W_0 = 360°/N$	180	120	90	72	60	45	36	30	18	7.5

Substitution of this value of x_1 in W_1 with $x_m = 0$ gives

$$W_1 \doteq 2x_1 \doteq \frac{2.78}{N} = \frac{159.2°}{N}. \quad (32)$$

Values computed from this formula are included in Table 1.2 with those obtained from Fig. 1.1 at 0.707 by doubling the abscissas. It is seen that, except for very small values of N, the approximate formula for W_1 is entirely adequate.

The values of x giving zero for the array factor are given by x_0 in

$$x_0 = \frac{p\pi}{N} \neq q\pi. \quad (p = 1, 2, 3, \cdots$$
$$\neq q \doteq 0, 1, 2, \cdots) \quad (33)$$

If the principal maximum occurs with $q = 0$, the adjacent zero requires $p = 1$, so that

$$x_m = 0, \quad x_0 = \pi/N, \quad (34)$$

and the null beam width of the array factor per element is

$$W_0 = 2\pi/N = 360°/N. \quad (35)$$

These values also are listed in Table 1.2.

2. The Collinear Array; The Marconi-Franklin Antenna

The collinear or stacked array consists of N_c identical antennas equally spaced along a single axis. In its most usual form all N_c units are arranged to be driven so that their currents are equal and in phase. This can be accomplished approximately in practice in a number of different ways. A somewhat more general collinear array which is characterized by a formula for the far-zone field that is no more intricate than the special case with all currents in phase is conveniently analyzed. In this more general array the currents in the several units are so related in phase that the current in a given unit numbered i *lags* the current in the adjacent unit numbered $i - 1$ by an angle δ_c and leads the current in the unit $i + 1$ by the same angle. (The subscript c is for collinear.) That is,

$$I_i = I_{i-1}e^{-j\delta_c}. \quad (1)$$

For present purposes of determining the radiation or far-zone field, let each unit in the array be center-driven by an individual generator, as shown in Fig. 2.1, or from a suitably arranged transmission line. Let it be assumed that the Θ-component of the electric field in the far zone or in the quasi-far zone due to a single unit may be represented to a satisfactory degree of approximation by

$$E^r_{\Theta i} = j \frac{\zeta_0 I_{0i}}{2\pi} \frac{e^{-j\beta_0 R_i}}{R_i} F_0(\Theta, \beta_0 h)$$

$$= j \frac{\zeta_0 I_{mi}}{2\pi} \frac{e^{-j\beta_0 R_i}}{R_i} F_m(\Theta, \beta_0 h) \quad (2a)$$

with

$$F_0(\Theta, \beta_0 h) = \frac{F_m(\Theta, \beta_0 h)}{\sin \beta_0 h}$$

$$= \frac{\cos(\beta_0 h \cos \Theta) - \cos \beta_0 h}{\sin \beta_0 h \sin \Theta}. \quad (2b)$$

The condition for the quasi-far zone is

$$R_i^2 \gg h^2. \quad \text{(for all } i\text{'s)} \quad (3a)$$

Subject to (3a), (2a) with (2b) is a good approximation for E_Θ, but E_R is *not* negligible. Subject to the condition for the far zone,

$$R_i \gg h, \quad \text{(for all } i\text{'s)} \quad (3b)$$

(2a) is a better approximation and E_R is negligible.

The use of (3a, b) means that the far-zone field of each unit is assumed to be the same as if the unit were isolated and extremely thin. It is shown in the preceding chapter

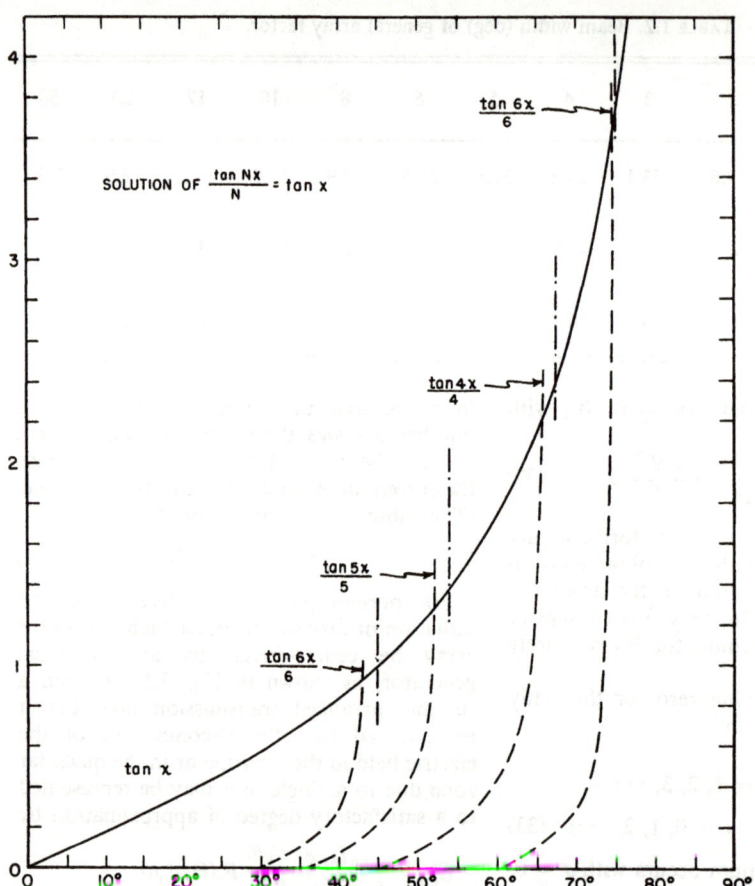

Fig. 1.2. Graphical method of locating minor extreme values.

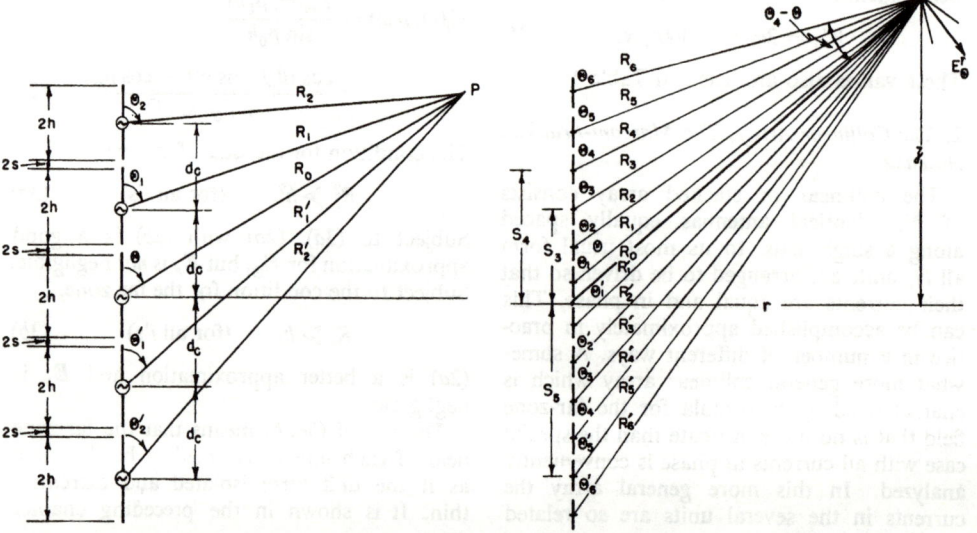

Fig. 2.1. Collinear array of an odd number of units.

Fig. 2.2. Collinear array of an even number of units.

that the field of an antenna of nonvanishing radius is actually very well represented *in the far zone* by the field of an antenna with sinusoidal current of the same half-length h. Although the distribution of current in two or more antennas that are placed in a collinear position is not the same as when the units are widely separated or completely isolated, the mutual impedance of collinear antennas is relatively small and the change in the distribution of current not large, especially if the individual units are near or less than $h = \lambda_0/4$ in half-length, as is usual. This is clear from the analysis in Chapter III.

The arrangement to be considered is shown in Figs. 2.1 and 2.2. It consists of N_c identical units symmetrically placed along the positive and negative z-axis. The number of units may be odd (Fig. 2.1) or even (Fig. 2.2). The distance from the point P where the field is calculated to the origin of coördinates at the center of the array is

$$R_0 = \sqrt{r^2 + z^2}. \tag{4}$$

The distance from the origin to the ith unit along either the positive or the negative z-axis is

$$S_i = id_c, \quad (N_c \text{ odd}) \tag{5a}$$

$$S_i = (2i - 1)d_c/2. \quad (N_c \text{ even}) \tag{5b}$$

The distance between centers of adjacent units in each group is d_c; the distance between adjacent ends is $2s$, so that

$$d_c = 2(h + s). \tag{6}$$

The distance R_i from the point P to the center of the ith unit along the positive z-axis is

$$R_i = \sqrt{R_0^2 + S_i^2 - 2R_0 S_i \cos \Theta}$$
$$\doteq R_0 - S_i \cos \Theta. \tag{7a}$$

Similarly, the distance to the ith unit along the negative z-axis is

$$R_i' = \sqrt{R_0^2 + S_i^2 + 2R_0 S_i \cos \Theta}$$
$$\doteq R_0 + S_i \cos \Theta. \tag{7b}$$

The approximate expressions on the right in (7a, b) imply the conditions

$$R_0^2 \gg S_i^2 = i^2 d_c^2. \tag{8a}$$

Since the index i can become only as large as n, the severest condition that must be imposed is

$$R_0^2 \gg S_n^2 = n^2 d_c^2. \tag{8b}$$

The Θ-component of the electric field at P is the sum of all contributions by the N_c units:

$$E_\Theta^r = E_{\Theta 0}^r + \sum_{i=1}^{\frac{1}{2}(N_c-1)} [E_{\Theta i}^r \cos(\Theta_i - \Theta)$$
$$+ E_{\Theta i}^{r\prime} \cos(\Theta - \Theta_i')], \quad (N_c \text{ odd}) \tag{9a}$$

$$E_\Theta^r = \sum_{i=1}^{\frac{1}{2}N_c} [E_{\Theta i}^r \cos(\Theta_i - \Theta)$$
$$+ E_{\Theta i}^{r\prime} \cos(\Theta - \Theta_i')]. \quad (N_c \text{ even}) \tag{9b}$$

The angle between $E_{\Theta i}^r$ and E_Θ^r is readily determined from Fig. 2.1 or Fig. 2.2 to be $(\Theta_i - \Theta)$; the angle between $E_{\Theta i}^{r\prime}$ and E_Θ^r is $(\Theta - \Theta_i')$. Using Fig. 2.1 or Fig. 2.2 and (7a, b),

$$\cos(\Theta_i - \Theta) = \cos\Theta_i \cos\Theta + \sin\Theta_i \sin\Theta,$$

$$= \left(\frac{z - S_i}{R_i}\right)\left(\frac{z}{R_0}\right) + \left(\frac{r}{R_i}\right)\left(\frac{r}{R_0}\right)$$

$$= \frac{R_0 - S_i \cos\Theta}{R_1} \doteq 1. \tag{10a}$$

Similarly,

$$\cos(\Theta - \Theta_i') \doteq 1. \tag{10b}$$

Since

$$\cos(\Theta_i - \Theta) = 1 - \frac{(\Theta_i - \Theta)^2}{2!} + \cdots, \tag{11}$$

it follows that second-degree differences in $\Theta_i - \Theta$ are negligible if (8b) is satisfied. Hence, since

$$\cos\Theta - \cos\Theta_i = \frac{\Theta_i^2 - \Theta^2}{2!} - \frac{\Theta_i^4 - \Theta^4}{4!} + \cdots, \tag{12a}$$

it also follows that, neglecting second-order differences,

$$\cos\Theta_i \doteq \cos\Theta. \tag{12b}$$

Similarly,

$$\cos\Theta_i' \doteq \cos\Theta. \tag{12c}$$

An additional relation that is useful later is obtained directly from Fig. 2.1 or Fig. 2.2. Thus,

$$r = R_0 \sin\Theta = R_i \sin\Theta_i = R_i' \sin\Theta_i'. \tag{13}$$

With (1),

$$I_i = I_0 e^{-ji\delta_c}, \quad I_i' = I_0 e^{ji\delta_c},$$
$$(N_c \text{ odd}) \tag{14a}$$

$$I_i = I_0 e^{-j(2i-1)\delta_c/2}, \quad I_i' = I_0 e^{j(2i-1)\delta_c/2},$$
$$(N_c \text{ even}) \tag{14b}$$

With N_c odd, I_0 is the current in the central unit; with N_c even, I_0 is a fictitious reference current, since there is no antenna at the origin.

The far-zone field of antenna i on the positive z-axis is like (1). For N_c odd, it is

$$E_{\Theta i}^r = \frac{j\zeta_0 I_0}{2\pi} \frac{e^{-j\beta_0(R_i + i\delta_c)}}{R_i} F_0(\Theta_i, \beta_0 h), \quad (15a)$$

with

$$F_0(\Theta_i, \beta_0 h) = \frac{\cos(\beta_0 h \cos\Theta_i) - \cos\beta_0 h}{\sin\beta_0 h \sin\Theta_i}. \quad (15b)$$

Using (7a), (12b) and (13), (15a) becomes

$$E_{\Theta i}^r = \frac{j\zeta_0 I_0}{2\pi} \frac{e^{-j\beta_0(R_0 - S_i \cos\Theta + i\delta_c)}}{R_0} F_0(\Theta, \beta_0 h).$$
$$(N_c \text{ odd}) \quad (16)$$

Similarly,

$$E_{\Theta i}^r = \frac{j\zeta_0 I_0}{2\pi} \frac{e^{-j\beta_0(R_0 + S_i \cos\Theta - i\delta_c)}}{R_0} F_0(\Theta, \beta_0 h).$$
$$(N_c \text{ odd}) \quad (17)$$

Upon substituting (16) and (17) in (9) and using (10a, b), the result is

$$E_{\Theta}^r = \frac{j\zeta_0 I_0}{2\pi} \frac{e^{-j\beta_0 R_0}}{R_0} F_0(\Theta, \beta_0 h) A_c(\Theta; N_c, n_c, t_c), \quad (18)$$

with

$$A_c(\Theta; N_c, n_c, t_c)$$
$$= 1 + 2 \sum_{i=1}^{\frac{1}{2}(N_c - 1)} \cos[2\pi i(n_c \cos\Theta - t_c)],$$
$$(N_c \text{ odd}) \quad (19a)$$

$$A_c(\Theta; N_c, n_c, t_c)$$
$$= 2 \sum_{i=1}^{\frac{1}{2}N_c} \cos[\pi(2i - 1)(n_c \cos\Theta - t_c)],$$
$$(N_c \text{ even}) \quad (19b)$$

where

$$n_c \equiv \frac{\beta_0 d_c}{2\pi} = \frac{d_c}{\lambda_0}, \quad (20a)$$

$$t_c \equiv \frac{\delta_c}{2\pi} \equiv \frac{\delta_c^\circ}{360^\circ}. \quad (20b)$$

Thus, n_c is the distance between centers measured in wavelengths; t_c is the time difference in the phase of adjacent units measured in fractions of a period.

With $x = \pi(n_c \cos\Theta - t_c)$ and with (1.3) and (1.4), it follows that the array factor is

$$A_c(\Theta; N_c, n_c, t_c) = \frac{\sin N_c \pi(n_c \cos\Theta - t_c)}{\sin \pi(n_c \cos\Theta - t_c)}. \quad (21)$$

The array factor per element is

$$a_c(\Theta; N_c, n_c, t_c) = \frac{\sin N_c \pi(n_c \cos\Theta - t_c)}{N_c \sin \pi(n_c \cos\Theta - t_c)}. \quad (22)$$

Principal extreme values of a_c as defined in (22) are located using (1.9) and with $x = \pi(n_c \cos\Theta - t_c)$. Thus,

$$|a_c| = 1 \text{ when } n_c \cos\Theta - t_c = q, \quad (23a)$$

that is, when

$$\Theta = \cos^{-1}(q + t_c)/n_c \quad (23b)$$

with $q = 0, 1, 2, \cdots$.

The location of zeros is given by (1.17). Thus,

$$a_c = 0 \text{ when } n_c \cos\Theta - t_c = \frac{p}{N_c} \neq q, \quad (24a)$$

or when

$$\Theta = \cos^{-1}\left(\frac{(p/N_c) + t_c}{n_c}\right) \quad (24b)$$

with $p = 1, 2, \cdots, q = 0, 1, 2, \cdots$.

An important special case is when all currents are in phase so that the parameter t_c vanishes and (22) reduces to

$$a_c(\Theta; N_c, n_c, 0) = \frac{\sin(N_c \pi n_c \cos\Theta)}{N_c \sin(\pi n_c \cos\Theta)}. \quad (25)$$

This function is readily evaluated using Fig. 1.1 or Table 1.1 for given values of N_c and n_c. Since (23) does not change when $\pi \pm \Theta$ is written for Θ, it is adequate to allow Θ to vary from zero to $\pi/2$. In other quadrants the function is obtained by symmetry. If Θ is varied, and appropriate values of the argument $\pi n_c \cos\Theta$ are determined, the corresponding values of $a_c(\Theta; N_c, n_c, 0)$ may be read from the appropriate curve of Fig. 1.1. A family of curves of $a_c(\Theta; N_c, n_c, 0)$ for $N_c = 2, 4, 6, 8$ and $n_c = \frac{1}{4}, \frac{1}{2}, 1$ is shown in Fig. 2.3.

The most important collinear array is the Marconi-Franklin arrangement. In this $d_c \doteq 2h = \lambda_0/2$, $\delta_c = 0$, so that $n_c = \frac{1}{2}$, $t_c = 0$. The array factor per element is

$$a_c(\Theta; N_c, \tfrac{1}{2}, 0) = \frac{\sin(\tfrac{1}{2}N_c \pi \cos\Theta)}{N_c \sin(\tfrac{1}{2}\pi \cos\Theta)}. \quad (26)$$

This factor is readily evaluated using Fig. 1.1 or Table 1.1. It is included in Table 3.1 for $n_c = \frac{1}{2}$ and with Θ substituted for Ψ. Curves of $a_c(\Theta; N_c, \tfrac{1}{2}, 0)$ as defined in (26)

TABLE 2.1. Values of $a_c = \dfrac{\cos(90°\cos\Theta)}{\sin\Theta} \cdot \dfrac{\sin Nx}{N\sin x}$, with $x = 90°\cos\Theta$.

Θ (rad)	Θ (deg)	N=2	N=3	N=4	N=5	N=6	N=8
0	0	0.0000	0.0000	0.0000	0.0000	0.0000	0.0000
0.0873	5	.0004	−.0227	−.0004	.0136	.0010	−.0010
.1745	10	.0033	−.0458	−.0033	.0273	.0033	−.0033
.2618	15	.0110	−.0682	−.0110	.0400	.0109	−.0108
.3491	20	.0261	−.0889	−.0257	.0495	.0249	−.0239
.4633	25	.0509	−.1057	−.0487	.0520	.0452	−.0405
.5236	30	.0873	−.1150	−.0797	.0420	.0673	−.0524
.6109	35	.1370	−.1049	−.1155	.0153	.0842	−.0486
.6981	40	.2008	−.0873	−.1490	−.0316	.0804	−.0150
.7854	45	.2788	−.0443	−.1688	−.0934	.0435	.0450
.8727	50	.3696	.0306	−.1604	−.1549	−.0305	.1000
.9599	55	.4706	.1463	−.1151	−.1892	−.1239	.0965
1.0472	60	.5855	.2721	.0000	−.1633	−.1924	.0000
1.1345	65	.6844	.4290	.1647	−.0499	−.1753	−.1456
1.2217	70	.7855	.5950	.3741	.1572	−.0243	−.2044
1.3090	75	.8733	.7525	.6006	.4306	.2589	−.0333
1.3963	80	.9417	.8833	.8052	.7104	.6037	.3715
1.4835	85	.9851	.9697	.9285	.9219	.8896	.8101
1.5708	90	1.0000	1.0000	1.0000	1.0000	1.0000	1.0000

are included in Fig. 2.3 for $N_c = 2, 4, 6, 8$. The complete field factor is the product $F_0(\Theta, \beta_0 h) A_c(\Theta; N_c, n_c, t_c)$. The factor for the Marconi-Franklin antenna is

$$F_0(\Theta, \tfrac{1}{2}\pi) A_c(\Theta; N_c, \tfrac{1}{2}, 0)$$
$$= \frac{\cos(\tfrac{1}{2}\pi \cos\Theta)}{\sin\Theta} \frac{\sin(\tfrac{1}{2}N_c \pi \cos\Theta)}{\sin(\tfrac{1}{2}\pi \cos\Theta)} \quad (27)$$

This composite factor is expressed numerically in Table 2.1 and graphically in Fig. 2.4 (rectangular graph) and Fig. 2.5 (polar graph of magnitude).

A second special case that is of practical importance is that for which $t_c = n_c$. In this case

$$a_c(\Theta; N_c, n_c, n_c) = \frac{\sin N_c \pi n_c(1 - \cos\Theta)}{N_c \sin \pi n_c(1 - \cos\Theta)}. \quad (28)$$

With Ψ written for Θ this factor is the same as the highly important factor a_e of the end-fire array, which is discussed, tabulated, and represented graphically in Sec. 3. Note that (28) must be multiplied by $F_0(\Theta, \beta_0 h)$ in order to obtain the complete characteristic of the array.

A useful approximate representation of the far-zone field of a collinear array of half-wave elements makes use of the leading term in the Fourier series of the field factor $F_0(\Theta, \tfrac{1}{2}\pi)$, as explained in Sec. V.12. That is

$$F_0(\Theta, \tfrac{1}{2}\pi) = \frac{\cos(\tfrac{1}{2}\pi \cos\Theta)}{\sin\Theta} \doteq 0.954 \sin\Theta, \quad (29)$$

so that

$$F_0(\Theta, \tfrac{1}{2}\pi) A_c(\Theta; N_c, \tfrac{1}{2}, 0)$$
$$\doteq 0.954 \sin\Theta \frac{\sin(\tfrac{1}{2}N_c \pi \cos\Theta)}{\sin(\tfrac{1}{2}\pi \cos\Theta)}. \quad (30)$$

The radiation function of the Marconi-Franklin array of N_c collinear half-wave dipoles is obtained from (I.10.4). It is

$$K_m^2(\Theta) = \left(\frac{4\pi R_0 \beta_0 v_0}{I_m}\right)^2 A_\Theta^r A_\Theta^{r*}$$

$$= \left(\frac{4\pi R_0}{\zeta_0 I_m}\right)^2 E_\Theta^r E_\Theta^{r*}. \quad (31)$$

With (18) and (27),

$$K_m^2(\Theta) = 4F_m^2(\Theta, \tfrac{1}{2}\pi) A_c^2(\Theta; N_c, \tfrac{1}{2}, 0)$$
$$= 4\left[\frac{\cos(\tfrac{1}{2}\pi \cos\Theta)}{\sin\Theta} \cdot \frac{\sin(\tfrac{1}{2}N_c \pi \cos\Theta)}{\sin(\tfrac{1}{2}\pi \cos\Theta)}\right]^2. \quad (32)$$

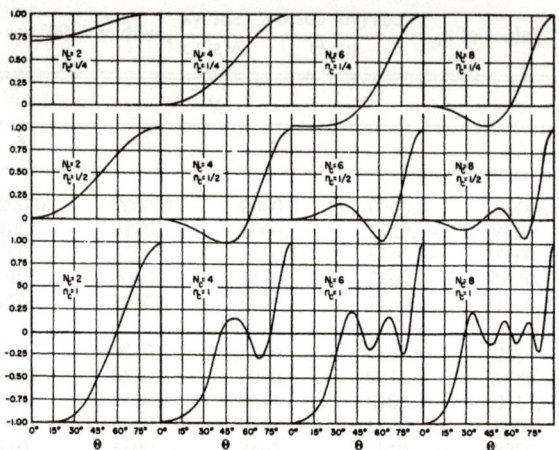

Fig. 2.3. Normalized array factor $a_c(\Theta, N_c, n_c, 0) = \dfrac{\sin(N_c \pi n_c \cos \Theta)}{\sin(\pi n_c \cos \Theta)}$.

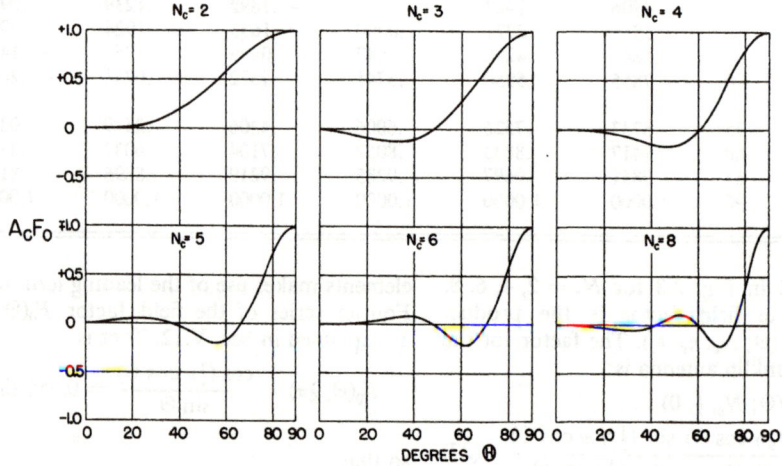

Fig. 2.4. Collinear array; rectangular graph of $a_c = A_c F_0 = \dfrac{\sin N_c x}{N_c \sin x} \cdot \dfrac{\cos(90° \cos \Theta)}{\sin \Theta}$ with $h = \lambda/4$, $n_c = \tfrac{1}{2}$; $x = 90° \cos \Theta$.

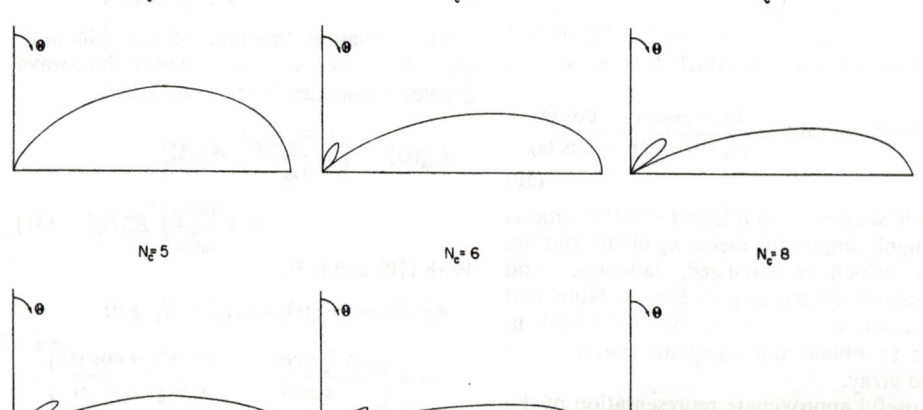

Fig. 2.5. Collinear array; polar graph of $a_c = \dfrac{\sin N_c x}{N_c \sin x} \cdot \dfrac{\cos(90° \cos \Theta)}{\sin \Theta}$ with $h = \lambda/4$, $n_c = \tfrac{1}{2}$; $x = 90° \cos \Theta$.

The radiation resistance referred to maximum sinusoidal current is obtained from (I.10.6):

$$R_m^e = \frac{\zeta_0}{8\pi} \int_0^\pi K_m^2(\Theta) \sin \Theta \, d\Theta$$

$$= \frac{\zeta_0}{2\pi} \int_0^\pi \left[\frac{\cos^2(\frac{1}{2}\pi \cos \Theta)}{\sin^2 \Theta} \right.$$

$$\left. \times \frac{\sin^2(\frac{1}{2}N_c\pi \cos \Theta)}{\sin^2(\frac{1}{2}\pi \cos \Theta)} \right] \sin \Theta \, d\Theta. \quad (33)$$

This expression can be expanded in a series*

---
* M. A. Bontsch-Bruewitsch, ref. 5; the integration is outlined below. The interval to be evaluated is rearranged as follows:

R_m^e

$$= \frac{\zeta_0}{2\pi} \int_0^\pi \left[\frac{\cos^2(\frac{1}{2}\pi \cos \Theta)}{\sin^2 \Theta} \cdot \frac{\sin^2(\frac{1}{2}N \pi \cos \Theta)}{\sin^2(\frac{1}{2}\pi \cos \Theta)} \right] \sin \Theta \, d\Theta$$

$$= \frac{\zeta_0}{4\pi} \int_0^\pi \frac{1 + \cos(\pi \cos \Theta)}{(1 - \cos \Theta)(1 + \cos \Theta)}$$

$$\times \frac{1 - (-1)^N \cos(N\pi \cos \Theta)}{1 - \cos(\pi \cos \Theta)} \sin \Theta \, d\Theta$$

$$= \frac{\zeta_0}{8\pi} \int_0^\pi \left[\frac{1 + \cos(\pi \cos \Theta)}{1 - \cos \Theta} + \frac{1 + \cos(\pi \cos \Theta)}{1 + \cos \Theta} \right]$$

$$\times \left[\frac{1 - (-1)^N \cos(N\pi \cos \Theta)}{1 - \cos(\pi \cos \Theta)} \right] \sin \Theta \, d\Theta.$$

Let
$u = \pi(1 - \cos \Theta), \quad v = \pi(1 + \cos \Theta),$
$\pi \cos \Theta = \pi - u, \quad \pi \cos \Theta = -(\pi - v),$
$du = \pi \sin \Theta, \quad dv = -\pi \sin \Theta.$

The limits are
$\Theta = 0, \, u = 0, \qquad \Theta = 0, \, v = 2\pi,$
$\Theta = \pi, \, u = 2\pi, \qquad \Theta = \pi, \, v = 0.$

Since $\cos(\pi - v) = \cos[-(\pi - v)]$, the integrals in u and v are identical.
Hence,

$$R_m^e = \frac{\zeta_0}{4\pi} \int_0^{2\pi} \frac{1 - \cos u}{u} \cdot \frac{1 - (-1)^N \cos Nu}{1 + \cos u} du.$$

The following series expansion is true:

$$\frac{(1 - \cos u)(1 - [-1]^N \cos Nu)}{1 + \cos u}$$

$$= (-1)^{N-1} \{ [1 - \cos Nu] - 4[1 - \cos(N-1)u]$$
$$+ 8[1 - \cos(N-2)u] - 12[1 - \cos(N-3)u]$$
$$+ \cdots \pm 4(N-1)[1 - \cos u] \}.$$

Hence,

$$R_m^e = \frac{\zeta_0}{4\pi} (-1)^{N-1} \left[\int_0^{2\pi N} \frac{1 - \cos u}{u} du \right.$$
$$- 4 \int_0^{2\pi(N-1)} \frac{1 - \cos u}{u} du + \cdots$$
$$\left. \pm 4(N-1) \int_0^{2\pi} \frac{1 - \cos u}{u} du \right].$$

With the definition

$$\text{Cin } x = \int_0^x \frac{1 - \cos u}{u} du,$$

the above expression is identically (34).

---

and integrated term by term. The result is

$$R_m^e = (-1)^{N-1} \frac{\zeta_0}{4\pi} [\text{Cin } 2\pi N$$
$$- 4 \text{ Cin } 2\pi(N - 1) + 8 \text{ Cin } 2\pi(N - 2)$$
$$- 12 \text{ Cin } 2\pi(N - 3) + \cdots$$
$$\pm 4(N - 1) \text{ Cin } 2\pi]. \quad (34)$$

Numerical results for several values of N_c are listed in Table 2.2.

TABLE 2.2. Radiation resistance of collinear array.

N_c	R_m^e (ohms)	D	D_r	G_{max} (db)	R_m^e/N_c (ohms) Exact*	Approx.*
1	73.1	1.64	1	2.15	73.1	71.4
2	199.1	2.41	1.47	3.82	99.5	93.2
3	316.5	3.41	2.08	5.33	105.5	96.8
4	436.5	4.40	2.68	6.44	109.1	99.8
5	555.0	5.40	3.30	7.63	111.0	101.0
6						102.2
7						102.8

* Exact values are obtained from (34), approximate values from (39).

The absolute directivity is defined by (I.10.7) to be

$$D = \frac{\zeta_0}{4\pi} \cdot \frac{K_m^2(\Theta_m)}{R_m^e}. \quad (35)$$

The direction of the principal extreme value is $\Theta = \pi/2$. Hence

$$D = \frac{\zeta_0}{4\pi} \cdot \frac{K_m^2(\frac{1}{2}\pi)}{R_m^e} = \frac{\zeta_0}{\pi} \frac{N_c^2}{R_m^e}. \quad (36)$$

Numerical values of the absolute directivity, the relative directivity,

$$D_r = \frac{D(N_c \text{ units})}{D(1 \text{ unit})} = \frac{D(N_c \text{ units})}{1.64}, \quad (37)$$

and the maximum absolute gain in decibels

$$G_{max}(db) = 10 \log_{10} D \quad (38)$$

are given in Table 2.2.*

Instead of using the exact formula (27) for the array factors of the collinear array in evaluating the radiation resistance referred

---
* These values were computed using accurate tables for Cin x. In the literature, slightly different values are sometimes listed. These are the results obtained using one or both of the following approximations: (a) computing only $C + \ln x$ instead of $C + \ln x - \text{Ci } x$; (b) using $\sin \Theta$ to approximate $\cos(\frac{1}{2}\pi \cos \Theta)/(\sin \Theta)$ or $0.954 \sin \Theta$ as $F_m(\Theta, \frac{1}{2}\pi)$.

to maximum current, the approximate expression (30) may be employed. This gives the formula*

$$R_m^e \doteq (0.945)^2 \frac{\zeta_0}{2\pi} \left[\frac{4}{3} N_c - 4 \sum_{r=2}^{2N_c-2} (2N_c - r) \frac{\cos(\frac{1}{2}r\pi)}{(\frac{1}{2}r\pi)^2} \right]. \quad (39)$$

* The evaluation of this formula involves the integral

$$R_m^e \doteq (0.945)^2 \frac{\zeta_0}{2\pi} \int_0^\pi \frac{\sin^2(\frac{1}{2}N_c\pi \cos \Theta)}{\sin^2(\frac{1}{2}\pi \cos \Theta)} \sin^3 \Theta \, d\Theta. \quad (39a)$$

In order to integrate it is convenient to expand $\sin^2 N_c y / \sin^2 y$:

$$\frac{\sin^2 N_c y}{\sin^2 y} = N_c + \sum_{r=2}^{2N_c-2} (2N_c - r) \cos ry, \quad (39b)$$

where $y = \frac{1}{2}\pi \cos \Theta$ and $r = 2, 4, 6, 8 \cdots$ is an even integer. With (39b) the expression for the radiation resistance in ohms takes the form:

$$R_0^e \doteq 60(0.945)^2 \left[N_c \int_0^\pi \sin^3 \Theta \, d\Theta + \sum_{r=2}^{2N_c-2} (2N_c - r) \int_0^\pi \cos(\frac{1}{2}r\pi \cos \Theta) \sin^3 \Theta \, d\Theta \right]. \quad (39c)$$

The first integral is immediately integrable. It is

$$\int_0^\pi \sin^3 \Theta \, d\Theta = 4/3. \quad (39d)$$

The second integral can be evaluated by Hankel's integral theorem:

$$\int_0^\pi \cos(Z \cos \Theta) \sin^{2\nu} \Theta \, d\Theta = \frac{J_\nu(Z)\Gamma(\frac{1}{2})\Gamma(\nu + \frac{1}{2})}{(\frac{1}{2}Z)^\nu}. \quad (39e)$$

This evaluation holds for $\nu > -\frac{1}{2}$. With $2\nu = 3$ and $Z = \frac{1}{2}r\pi$, and recalling that the gamma functions $\Gamma(\frac{1}{2})$ and $\Gamma(\frac{3}{2} + \frac{1}{2})$ have the values $\sqrt{\pi}$ and 1, respectively, (39d) becomes

$$\int_0^\pi \cos(\frac{1}{2}r\pi \cos \Theta) \sin^3 \Theta \, d\Theta = \frac{J_{3/2}(\frac{1}{2}r\pi)\sqrt{\pi}\,2\sqrt{2}}{(\frac{1}{2}r\pi)^{3/2}}. \quad (39f)$$

It is well known that

$$\frac{J_{3/2}(\frac{1}{2}r\pi)}{(\frac{1}{2}r\pi)^{3/2}} = \sqrt{\frac{2}{\pi}} \left[\frac{\sin(\frac{1}{2}r\pi)}{(\frac{1}{2}r\pi)^3} - \frac{\cos(\frac{1}{2}r\pi)}{(\frac{1}{2}r\pi)^2} \right]. \quad (39g)$$

Since r is always an even integer, $\sin(\frac{1}{2}r\pi) = 0$, and Thus

$$\frac{J_{3/2}(\frac{1}{2}r\pi)}{(\frac{1}{2}r\pi)^{3/2}} = -\sqrt{\frac{2}{\pi}} \frac{\cos(\frac{1}{2}r\pi)}{(\frac{1}{2}r\pi)^2}. \quad (39h)$$

Hence

$$\int_0^\pi \cos(\frac{1}{2}r\pi \cos \Theta) \sin^3 \Theta \, d\Theta = -4 \frac{\cos(\frac{1}{2}r\pi)}{(\frac{1}{2}r\pi)^2}. \quad (39i)$$

If this result is substituted in (39c) with (39d), the following formula for R_0^e in ohms is obtained:

$$R_0^e \doteq 60(0.945)^2 \left[\frac{4N_c}{3} - 4 \sum_{r=2}^{2N_c-2} (2N_c - r) \frac{\cos(\frac{1}{2}r\pi)}{(\frac{1}{2}r\pi)^2} \right] \text{ ohms}. \quad (39j)$$

Numerical results computed for R_m^e/N_c from this approximate formula are given in Table 2.2 for comparison with the values obtained from the exact formula. It is seen that they differ by amounts up to 10 percent.

The beam width of the Marconi-Franklin collinear antenna in the Θ-plane is obtained using the array factors $F_0(\Theta, \frac{1}{2}\pi) a_c(\Theta; N_c, \frac{1}{2}, 0)$ as obtained from (27). Even for quite small values of N_c, the magnitude of this factor decreases rapidly from the principal maximum at $\Theta = \pi/2$ as is seen from Fig. 2.5. Therefore, the half-power points are at values of Θ not differing greatly from $\pi/2$. Near $\Theta = \pi/2$, the factor $F_0(\Theta, \frac{1}{2}\pi)$ does not differ greatly from unity, so that an approximate half-power beam width may be obtained by setting $F_0(\Theta, \frac{1}{2}\pi) \doteq 1$ near $\Theta = \pi/2$. If this is done, the remaining array factor per element is of the form $\sin Nx/N \sin x$ with $x = \frac{1}{2}\pi \cos \Theta$. Since the half-power beam width for this factor is given in Sec. 1 in terms of the general angle x, it is merely necessary to determine the values of Θ corresponding to the values of x. Thus, from (1.31), the half-power point is located at Θ_1, given by

$$\frac{\pi}{2} \cos \Theta_1 = x_1 \doteq \frac{1.39}{N_c}, \quad (40)$$

$$\Theta_1 \doteq \cos^{-1}(0.885/N_c). \quad (41)$$

Since the maximum is at

$$\Theta_m = \pi/2, \quad (42)$$

the half-power beam width is

$$W_1 = 2(\Theta_m - \Theta_1) \doteq \pi - 2\cos^{-1}(0.885/N_c)$$
$$= 180° - 2\cos^{-1}(0.885/N_c). \quad (43)$$

Numerical values of W_1 for a range of values are given in Table 2.3.

The null band width is obtained by locating the null nearest to the maximum at $\Theta = \pi/2$. This is given by

$$x_0 = \pi/2 \cos \Theta_0 \doteq \frac{\pi}{N_c}, \quad (43a)$$

$$\Theta_0 \doteq \cos^{-1} \frac{2}{N_c}. \quad (43b)$$

The null band width is

$$W_0 = 2(\Theta_m - \Theta_0) \doteq \pi - 2\cos^{-1} \frac{2}{N_c}$$
$$= 180° - 2\cos^{-1} \frac{2}{N_c}. \quad (44)$$

Values of W_0 are also given in Table 2.3.

TABLE 2.3. Approximate beam width (deg) of Marconi-Franklin collinear array.

N_c	2	3	4	5	6	8	10	12	20	50
$W_1 \doteq 180° \\ -2\cos^{-1}(0.885/N_c)$	52.5	34.3	25.6	20.4	17.0	12.7	10.2	8.5	5.1	2.0
$W_0 \doteq 180° \\ -2\cos^{-1}(2/N_c)$	180.0	83.7	60.0	47.2	38.9	29.0	23.1	19.2	11.5	4.6

A simple but important special case of the general collinear array with array factor given by (21) is that of two antennas separated by an arbitrary distance $d_c = n_c \lambda_0$ between centers, and driven in phase. Then

$$N_c = 2, \quad t_c = 0, \quad (45)$$

so that (21) becomes

$$A_c(\Theta; 2, n_c, 0) = 2\cos(n_c \pi \cos \Theta)$$
$$= 2\cos(\tfrac{1}{2}\beta_0 d_c \cos \Theta). \quad (46)$$

The array factor per element is

$$a_c(\Theta; 2, n_c, 0) = \cos(n_c \pi \cos \Theta)$$
$$= \cos(\tfrac{1}{2}\beta_0 d_c \cos \Theta). \quad (47)$$

This function is represented in Fig. 2.6 for $n_c = \tfrac{1}{2}, 1, \tfrac{3}{2}, 2$. The complete array factor for $\beta_0 h = \pi/2$,

$$F_m(\Theta, \pi/2) A_c(\Theta; 2, n_c, 0)$$
$$= 2\,\frac{\cos(\tfrac{1}{2}\pi \cos \Theta)}{\sin \Theta}\cos(n_c \pi \cos \Theta), \quad (48)$$

is represented in a rectangular graph in Fig. 2.7, and is included in polar graphs in Fig. 2.8. This latter figure also includes polar graphs of the complete array factor for $\beta_0 h = \pi$, namely,

$$F_m(\Theta, \pi) A_c(\Theta; 2, n_c, 0)$$
$$= 2\left[\frac{\cos(\pi \cos \Theta) + 1}{\sin \Theta}\right]\cos(n_c \pi \cos \Theta). \quad (49)$$

For some purposes it is convenient to set

$$\cos \Theta = z/R_0 \doteq z/r, \quad (r^2 \gg z^2) \quad (50)$$

so that (47) becomes

$$a_c(\Theta; 2, n_c, 0) \doteq \cos\left(\frac{n_c \pi z}{r}\right) = \cos\left(\frac{\beta_0 d_c z}{2r}\right). \quad (51)$$

An interesting generalization of the two-element collinear array, the array factor of which is useful in the discussion of antennas over a conducting earth in Chapter VII, consists of two identical, center-driven elements of half-length h and separated by a distance $d_c = 2d$ between centers but excited by unequal voltages so that the currents may differ in *magnitude* as well as in phase. Let the center of the upper antenna (No. 1) be at $z = d$, that of the lower antenna (No. 2) at $z = -d$. The maximum associated with the sinusoidally distributed current in the upper antenna is I_m, that for the lower antenna is kI_m, where k is the complex ratio of the two currents. The individual far-zone fields of the two antennas are given by (2a) with R_i replaced by $R_1 = R_0 - d\cos \Theta$ and $R_2 = R_0 + d\cos \Theta$, respectively. The array factor is obtained directly to be

$$A(\Theta) = e^{j\beta_0 d \cos \Theta} + k e^{j\beta_0 d \cos \Theta}. \quad (52)$$

The complete field factor is $F_m(\Theta, \beta_0 h) A(\Theta)$, where $F_m(\Theta, \beta_0 h)$ is defined by (2b) and $A(\Theta)$ is in (52). Obviously, (52) reduces to (46) when the currents in the two antennas are equal so that $k = 1$.

3. Parallel Arrays—the Broadside and the End-Fire Array

Consider an array of N_p identical center-driven antennas all of which are parallel to the z-axis and have their centers on the x-axis, as shown in Fig. 3.1. The half-length of each unit is h; the distance between each pair is b. In the analysis, let N_p be odd, with the origin of coördinates at the midpoint of the central unit, which is designated by the number 0. Antennas on the positive x-axis are numbered from 1 to n; those on the negative x-axis likewise. The currents in the several units are related as follows (primed quantities apply to antennas along the *negative* x-axis):

$$I_{0m} = I_{00} k_m e^{-j\delta_m}, \quad I'_{0m} = I_{00} k'_m e^{-j\delta'_m},$$
$$m = 1, 2, \cdots, n, \quad (1)$$

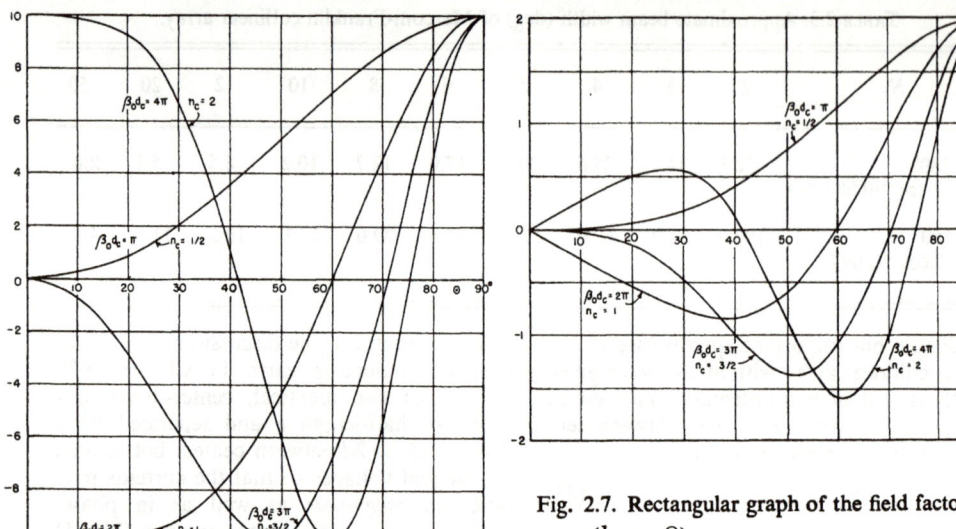

Fig. 2.6. The function $\cos(\tfrac{1}{2}\beta_0 d_c \cos\Theta)$ or $\cos(\pi n_c \cos\Theta)$.

Fig. 2.7. Rectangular graph of the field factor
$$2\frac{\cos(\tfrac{1}{2}\pi\cos\Theta)}{\sin\Theta}\cos(\pi n_c \cos\Theta).$$

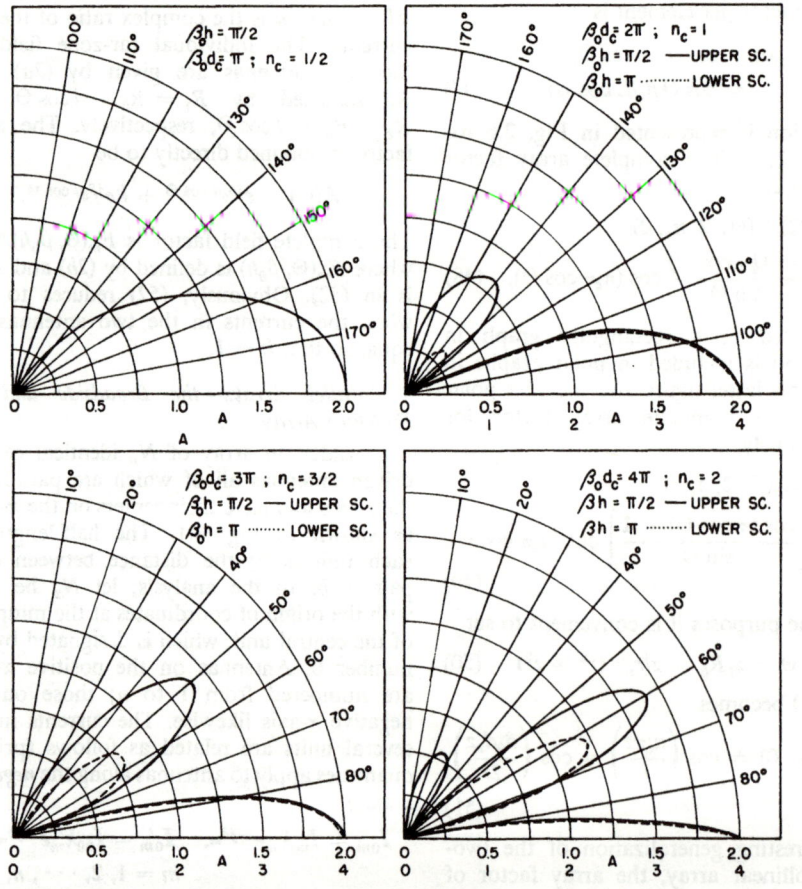

Fig. 2.8. The field factor $A = 2\left[\dfrac{\cos(\beta_0 h \cos\Theta) - \cos\beta_0 h}{\sin\Theta}\right]\cos(n_c\pi\cos\Theta)$; polar graph.

where k_m and k'_m are real constants of arbitrary magnitude, and δ_m and δ'_m are phase angles. An array of this type is called a "uniform" array if $k_m = k'_m = 1$ and $\delta_m = -\delta'_m$.

The Θ-component of the electric field at P (Fig. 3.1) is

$$E^r_\Theta = E^r_{\Theta 0} + \sum_{m=1}^{n} E^r_{\Theta m} \cos \psi_m + E^{r'}_{\Theta m} \cos \psi'_m, \quad (2)$$

where ψ_m is the angle between $E^r_{\Theta 0}$ and $E^r_{\Theta m}$ and ψ'_m is the angle between $E^r_{\Theta 0}$ and $E^{r'}_{\Theta m}$. The far-zone field of each unit is assumed to be well approximated by the field of an antenna with a sinusoidal current of the same maximum value. For the central unit,

$$E^r_{\Theta 0} = j \frac{\zeta_0 I_{00}}{2\pi} \frac{e^{-j\beta_0 R_0}}{R_0} F_0(\Theta, \beta_0 h), \quad (3)$$

where

$$F_0(\Theta, \beta_0 h) = \frac{\cos(\beta_0 h \cos \Theta) - \cos \beta_0 h}{\sin \beta_0 h \sin \Theta}, \quad (4)$$

subject to the condition

$$h^2 \ll R_0^2. \quad (5)$$

For the mth unit,

$$E^r_{\Theta m} = j \frac{\zeta_0 I_{0m}}{2\pi} \frac{e^{-j\beta_0 R_m}}{R_m} F_0(\Theta_m, \beta_0 h), \quad (6)$$

where

$$F_0(\Theta_m, \beta_0 h) = \frac{\cos(\beta_0 h \cos \Theta_m) - \cos \beta_0 h}{\sin \beta_0 h \sin \Theta_m}. \quad (7)$$

The distances from the point P to the centers of the several units are:

$$R_0 = \sqrt{x^2 + y^2 + z^2}, \quad (8)$$

$$R_m = \sqrt{(x - mb)^2 + y^2 + z^2}$$
$$= \sqrt{R_0^2 - 2mbx + m^2b^2} \doteq R_0 - mbx/R_0, \quad (9)$$

$$R'_m = \sqrt{(x + mb)^2 + y^2 + z^2}$$
$$= \sqrt{R_0^2 + 2mbx + m^2b^2} \doteq R_0 + mbx/R_0. \quad (10)$$

The approximate forms on the right in (9) and (10) are sufficiently accurate provided the following inequalities are satisfied:

$$\left(\frac{mb}{x}\right)\left(1 - \frac{x^2}{R_0^2}\right) \ll 2R_0, \quad m^2 b^2 \ll R_0^2. \quad (11)$$

This is true for all values of m if

$$[(N-1)b/2x](1 - x^2/R_0^2) \ll 2R_0,$$
$$\tfrac{1}{4}(N-1)^2 b^2 \ll R_0^2. \quad (12)$$

Expressed in terms of polar coördinates Θ, Φ, R_0 (Fig. 3.1), x is given by

$$x = r \cos \Phi = R_0 \sin \Theta \cos \Phi, \quad (13)$$

so that

$$R_m \doteq R_0 - mb \sin \Theta \cos \Phi \equiv R_0 - S_m, \quad (14)$$

$$R'_m \doteq R_0 + mb \sin \Theta \cos \Phi \equiv R_0 + S_m. \quad (15)$$

Also

$$R_m \sin \Theta_m = r_m \doteq r - mb \cos \Phi$$
$$= R_0 \sin \Theta - mb \cos \Phi, \quad (16)$$

$$R'_m \sin \Theta'_m = r'_m \doteq r + mb \cos \Phi$$
$$= R_0 \sin \Theta + mb \cos \Phi. \quad (17)$$

It follows from the law of cosines (Fig. 3.1) that

$$\cos \psi_m = \frac{R_0^2 + R_m^2 - m^2 b^2}{2 R_0 R_m}. \quad (18)$$

Using (14), and neglecting terms in accordance with (11),

$$\cos \psi_m = \frac{R_0^2 + R_0^2 - 2R_0 S_m + S_m^2 - m^2 b^2}{2R_0(R_0 - S_m)}$$
$$\doteq 1. \quad (19)$$

Similarly,

$$\cos \psi'_m \doteq 1. \quad (20)$$

Using (16) and (17),

$$\frac{1}{R_m \sin \Theta_m} = \frac{1}{r - mb \cos \Phi}$$
$$\doteq \frac{1}{r}\left(1 + \frac{mb \cos \Phi}{r}\right), \quad (21)$$

$$\frac{1}{R'_m \sin \Theta'_m} = \frac{1}{r + mb \cos \Phi}$$
$$\doteq \frac{1}{r}\left(1 - \frac{mb \cos \Phi}{r}\right). \quad (22)$$

Referring to Fig. 3.1, it is clear that

$$R_m \cos \Theta_m = R_0 \cos \Theta = R'_m \cos \Theta'_m, \quad (23a)$$

so that

$$\cos \Theta_m = \frac{R_0}{R_m} \cos \Theta \doteq \frac{R_0}{R_0 - S_m} \cos \Theta. \quad (23b)$$

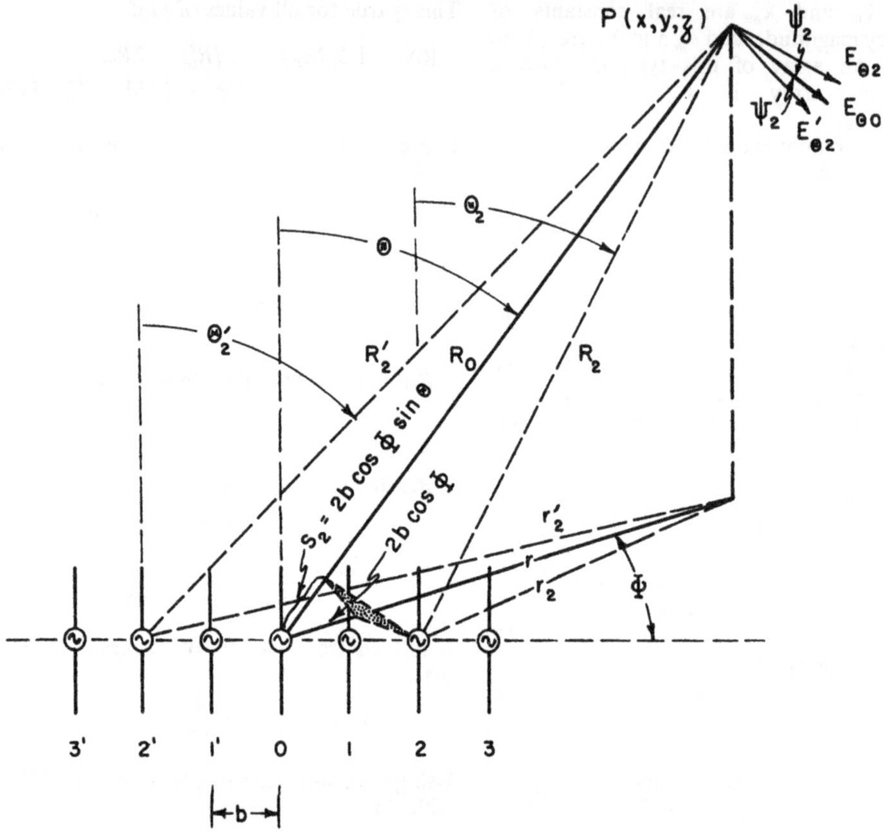

Fig. 3.1. An array of N_P identical center-driven antennas all parallel to the z-axis with centers on the x-axis.

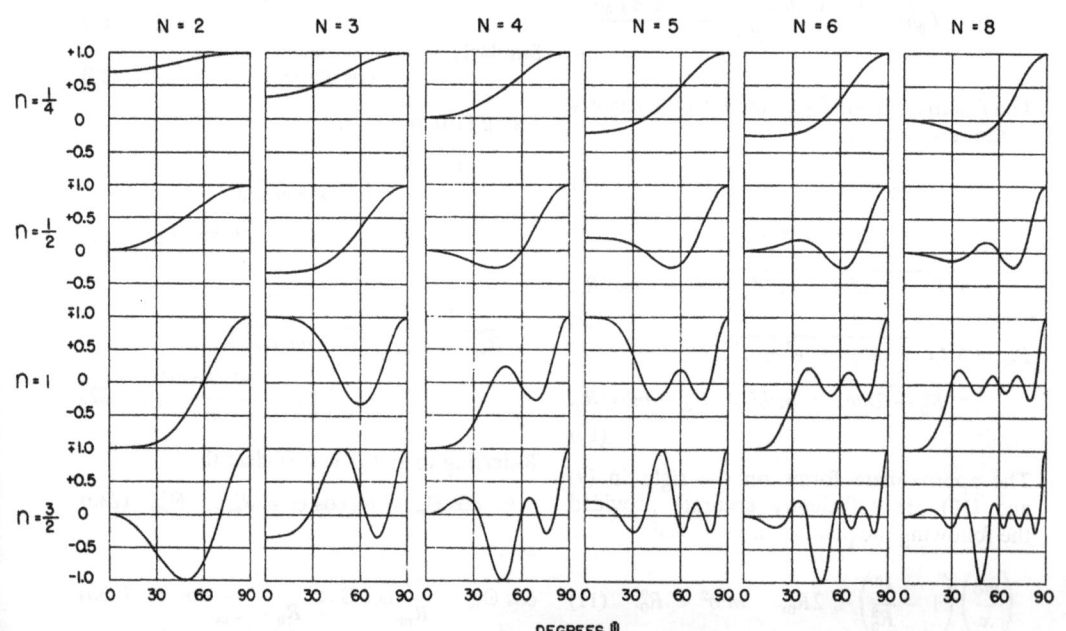

Fig. 3.2. Broadside array; rectangular graph of $a = (\sin Nx)/N \sin x$; $x = n180° \cos \Psi$.

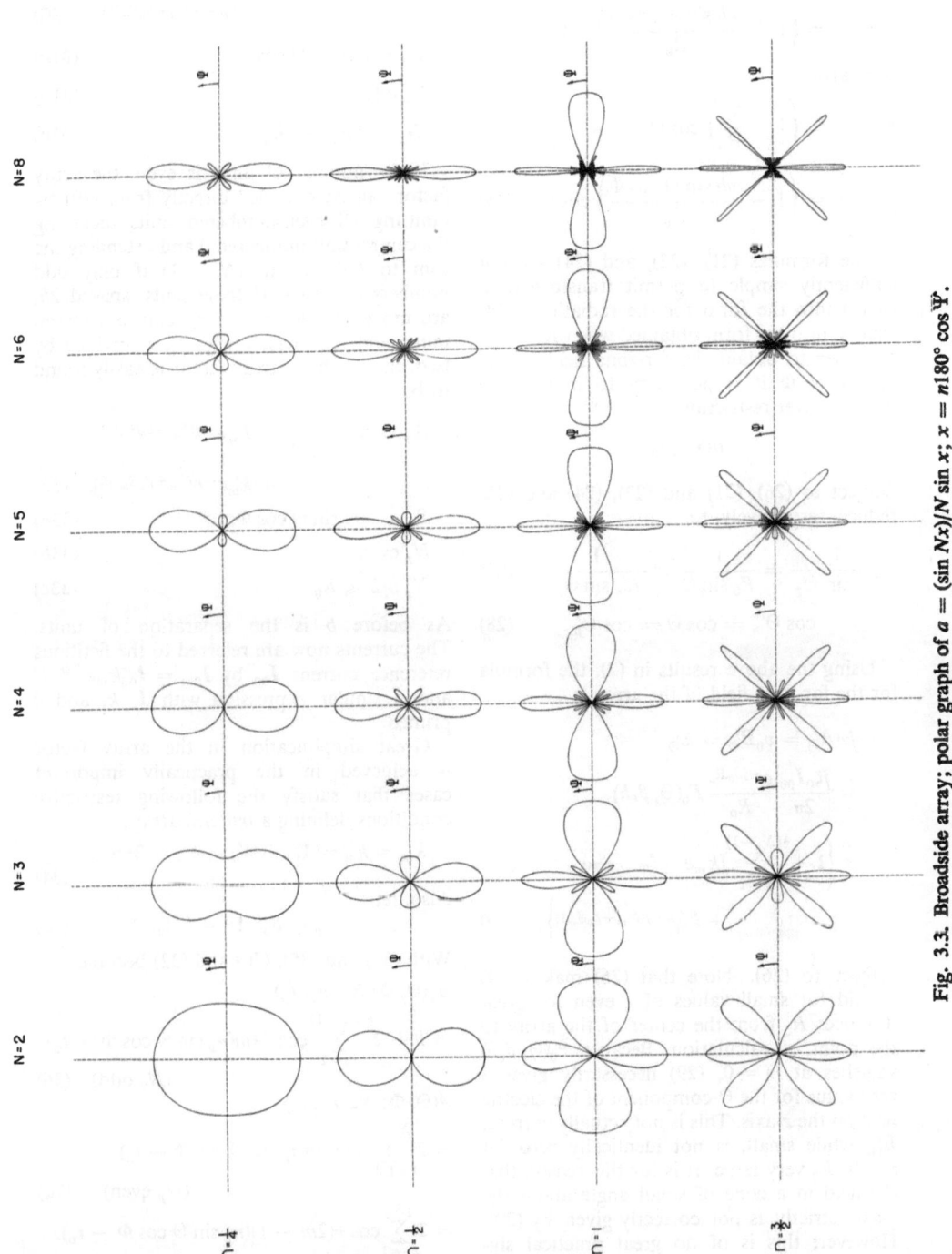

Fig. 3.3. Broadside array; polar graph of $a = (\sin Nx)/N \sin x$; $x = n180° \cos \Psi$.

Rearranging,

$$\cos \Theta_m \doteq \left(1 + \frac{S_m}{R_0}\right) \cos \Theta$$

$$= \left(1 + \frac{mb \sin \Theta \cos \Phi}{R_0}\right) \cos \Theta. \quad (24)$$

Similarly,

$$\cos \Theta'_m \doteq \left(1 - \frac{S_m}{R_0}\right) \cos \Theta$$

$$= \left(1 - \frac{mb \sin \Theta \cos \Phi}{R_0}\right) \cos \Theta. \quad (25)$$

The formulas (21), (22), and (24) are not sufficiently simple to permit transformation of (2) into the form for the radiation field, that is, into the form obtained when $R_0 \to \infty$. In order to obtain the far-zone field for all values of Φ it is necessary to impose the much severer restriction

$$mb \ll r. \quad (26)$$

Subject to (26), (21) and (22), (24) and (25) reduce, respectively, to

$$\frac{1}{R_m \sin \Theta_m} \doteq \frac{1}{R_0 \sin \Theta} \doteq \frac{1}{R'_m \sin \Theta'_m}, \quad (27)$$

$$\cos \Theta_m \doteq \cos \Theta \doteq \cos \Theta'_m. \quad (28)$$

Using the above results in (2), the formula for the far-zone field of the array is

$$-j\omega A_\Theta^r = v_0 B_\Phi^r = E_\Theta^r$$

$$= \frac{j\zeta_0 I_{00}}{2\pi} \frac{e^{-j\beta_0 R_0}}{R_0} F_0(\Theta, \beta_0 h)$$

$$\times \left\{1 + \sum_{m=1}^{\frac{1}{2}(N_p-1)} [k_m e^{-j(\delta_m - \beta_0 S_m)} + k'_m e^{-j(\delta'_m + \beta_0 S_m)}]\right\}, \quad (29)$$

subject to (26). Note that (26) makes (29) invalid for small values of r even for great distances R_0 from the center of the array to the point of calculation. Because $F_0(\Theta, \beta_0 h)$ vanishes at $\Theta = 0$, (29) necessarily gives a zero value for the Θ-component of the electric field on the z-axis. This is not actually correct; E_Θ^r, while small, is not identically zero for $r = 0$, R_0 very large. It is for this reason that the field in a cone of small angle about the z-axis strictly is not correctly given by (29). However, this is of no great practical significance. The *array factor* for the parallel array with units separated a distance b is

$$A_p(\Theta, \Phi) = \left\{1 + \sum_{m=1}^{\frac{1}{2}(N_p-1)} [k_m e^{-j(\delta_m - \beta_0 S_m)} + k'_m e^{-j(\delta'_m + \beta_0 S_m)}]\right\}, \quad (30)$$

$$S_m = mb \sin \Theta \cos \Phi, \quad (31a)$$

N_p odd, $\quad (31b)$

$$(N_p - 1)b/2 \ll R_0. \quad (31c)$$

If the number of units is even, the array factor can be obtained directly from (29) by omitting all even-numbered units including the central unit numbered 0 and extending the sum to $\frac{1}{2}N_p$, or to $(N_p - 1)$ if only odd numbers are used. If these units, spaced $2b$, are moved together to a spacing b between units, and the phase angles δ_m is divided by two, the resulting array factor is easily found to be

$$A_p(\Theta, \Phi) = \sum_{m=1,3,5}^{N_p-1} [k_m e^{-j(\delta_m - \beta_0 S_m)/2} + k'_m e^{-j(\delta'_m + \beta_0 S_m)/2}], \quad (32)$$

$$S_m = mb \sin \Theta \cos \Phi, \quad (33a)$$

N_p even, $\quad (33b)$

$$N_p b/2 \ll R_0. \quad (33c)$$

As before b is the separation of units. The currents now are referred to the fictitious reference current I_{00} by $I_{0m} = I_{00} k_m e^{-j\delta_m/2}$ and a similar expression with I, k, and δ primed.

Great simplification in the array factor is achieved in the practically important cases that satisfy the following restrictive conditions defining a *uniform array*:

$$k_m = k'_m = 1, \quad -\delta'_m = \delta_m = 2\pi m t_p. \quad (34)$$

Also, let

$$n_p = \beta_0 b / 2\pi = b/\lambda_0. \quad (35)$$

With (34) and (35), (30) and (32) become

$$A_p(\Theta, \Phi; N_p, n_p, t_p)$$
$$= 1 + 2 \sum_{m=1}^{\frac{1}{2}(N_p-1)} \cos 2\pi m(n_p \sin \Theta \cos \Phi - t_p),$$
$$(N_p \text{ odd}) \quad (36)$$

$$A(\Theta, \Phi; N_p, n_p, t_p)$$
$$= 2 \sum_{i=1,3,5}^{N_p-1} \cos \pi i(n_p \sin \Theta \cos \Phi - t_p)$$
$$(N_p \text{ even}) \quad (37a)$$

$$= 2 \sum_{m=1}^{\frac{1}{2}N_p} \cos \pi (2m-1)(n_p \sin \Theta \cos \Phi - t_p).$$
$$(N_p \text{ even}) \quad (37b)$$

The expressions (36) and (37b) are in the form of (1.1) and (1.2), so that with (1.3)–(1.5),

$$A_p(\Theta, \Phi; N_p, n_p, t_p)$$
$$= \frac{\sin N_p\pi(n_p \sin \Theta \cos \Phi - t_p)}{\sin \pi(n_p \sin \Theta \cos \Phi - t_p)}. \quad (38)$$

Alternatively, the array factor per element is

$$a_p(\Theta, \Phi; N_p, n_p, t_p)$$
$$= \frac{\sin N_p\pi(n_p \sin \Theta \cos \Phi - t_p)}{N_p \sin \pi(n_p \sin \Theta \cos \Phi - t_p)}. \quad (39)$$

Principal extreme values as defined in (1.16) are characterized as follows:

$a_p = \pm 1$ when $\pm(n_p \sin \Theta \cos \Phi - t_p) = q$

or when

$$\sin \Theta \cos \Phi = (t_p \pm q)/n_p, \quad (40)$$

with $q = 0, 1, 2, \cdots$. Zeros are given by

$a_p = 0$ when $\pm(n_p \sin \Theta \cos \Phi - t_p) = p/N_p$,

or when

$$\sin \Theta \cos \Phi = (t_p \pm p/N_p)/n_p, \quad (41)$$

with $p = 1, 2, 3, \cdots \neq N_pq$. Minor extreme values are determined as explained in Sec. 1.

It is often desirable with parallel arrays to have principal extreme values located in the equatorial plane, $\Theta = \pi/2$. These extremes are defined by

$$a_p = \pm 1, \quad \Theta = \pi/2, \quad \Phi = \cos^{-1}(t_p \pm q)/n_p,$$
$$q = 0, 1, 2, \cdots \quad (42)$$

The number of extreme values is limited by the condition

$$|(t_p \pm q)/n_p)| \leq 1. \quad (43)$$

Evidently, a large number of combinations is possible if n_p is sufficiently large. An important case with n_p small is

$$n_p = \tfrac{1}{2}, \quad t_p = 0, \quad q = 0, \quad \Phi = \frac{\pi}{2}, \frac{3\pi}{2}. \quad (44)$$

Zeros in the equatorial plane are defined by

$$\Phi = \cos^{-1}(t_p \pm p/N_p)/n_p,$$
$$p = 1, 2, \cdots \neq N_pq. \quad (45)$$

The values $p = N_pq$, $q = 0, 1, 2, \cdots$ give the principal maxima. Note that since (39) in general is not symmetric in the first and second quadrants, extreme and zero values must be obtained in the range of Φ from 0 to π or $\cos \Phi$ in the range from 1 to -1.

The two most important special cases of parallel arrays are the *broadside* array, in which all currents are in phase, and the *end-fire* array, in which the successive differences in phase of the currents in the adjacent units is equal to their separation in angular measure, that is, $\delta = \beta_0 b$ or $t_p = n_p$.

Broadside array: $t_B = 0$;
$$A_B(\Theta, \Phi; N_B, n_B, 0)$$
$$= \frac{\sin(N_B\pi n_B \sin \Theta \cos \Phi)}{\sin(\pi n_B \sin \Theta \cos \Phi)}, \quad (46)$$

End-fire array: $t_E = n_E$;
$$A_E(\Theta, \Phi; N_E, n_E, n_E)$$
$$= \frac{\sin[N_E\pi n_E(1 - \sin \Theta \cos \Phi)]}{\sin[\pi n_E(1 - \sin \Theta \cos \Phi)]}. \quad (47)$$

The pattern of the broadside array is symmetric in the first and second quadrants, that of the end-fire array is not. This follows from the fact that $(\pi - \Phi)$ may be written for Φ in (46) without changing its magnitude or sign, but not in (47). Formulas for the location of principal maxima are given by (40) and (41) with appropriate values of t_p.

Broadside array: $(a_B)_{\text{ext}} = \pm 1$ when

$$\sin \Theta \cos \Phi = \pm \frac{q}{n_B}, \quad q = 0, 1, 2, \cdots \quad (48)$$

$a_B = 0$ when $\sin \Theta \cos \Phi = \pm \dfrac{p}{n_B N_B}$,
$$p = 1, 2, \cdots \neq N_Bq. \quad (49)$$

End-fire array: $(a_E)_{\text{ext}} = \pm 1$ when

$$\sin \Theta \cos \Phi = \left(1 \pm \frac{q}{n_E}\right), \quad q = 0, 1, 2, \cdots \quad (50)$$

$a_E = 0$ when $\sin \Theta \cos \Phi = \left(1 \pm \dfrac{p}{n_E N_E}\right)$,
$$p = 1, 2, \cdots \pm N_Eq. \quad (51)$$

The complete far-zone field for a parallel array is

$$-j\omega A_\Theta^r = v_0 B_\Phi^r = E_\Theta^r$$
$$= \frac{j\zeta_0 I_{00}}{2\pi} \frac{e^{-j\beta_0 R_0}}{R_0} F_0(\Theta, \beta_0 h) A_p(\Theta, \Phi; N_p, n_p, t_p), \quad (52)$$

where the subscript p becomes B for the broadside array and E for the end-fire array

TABLE 3.1. Values of $a = \sin Nx/N \sin x$ with $x = n\,180°\cos\Phi$.

n	Φ (rad)	Φ (deg)	N=2	N=3	N=4	N=5	N=6	N=8
1/4	0	0	0.7071	0.3333	0.0000	−0.2000	−0.2357	0.0000
	0.0873	5	.7092	.3373	.0042	− .1976	− .2364	− .0042
	.1745	10	.7155	.3492	.0170	− .1900	− .2379	− .0171
	.2618	15	.7257	.3689	.0387	− .1763	− .2371	− .0386
	.3491	20	.7398	.3963	.0699	− .1550	− .2378	− .0687
	.4633	25	.7572	.4312	.1111	− .1242	− .2307	− .1063
	.5236	30	.7775	.4726	.1624	− .0815	− .2139	− .1480
	.6109	35	.8001	.5202	.2243	− .0251	− .1829	− .1890
	.6981	40	.8244	.5729	.2962	.0470	− .1300	− .2197
	.7854	45	.8497	.6293	.3773	.1353	− .0599	− .2285
	.8727	50	.8752	.6880	.4656	.2393	.0387	− .2019
	.9599	55	.9002	.7472	.5589	.3567	.1626	− .1281
	1.0472	60	.9239	.8047	.6533	.4828	.3080	.0000
	1.1345	65	.9454	.8584	.7446	.6113	.4668	.1792
	1.2217	70	.9641	.9061	.8283	.7341	.6279	.3947
	1.3090	75	.9794	.9456	.8995	.8420	.7750	.6181
	1.3963	80	.9903	.9749	.9537	.9276	.8831	.8160
	1.4835	85	.9976	.9937	.9882	.9844	.9734	.9521
	1.5708	90	1.0000	1.0000	1.0000	1.0000	1.0000	1.0000
1/2	0	0	0.0000	−0.3333	0.0000	0.2000	0.0000	0.0000
	0.0873	5	.0059	− .3333	− .0059	.1999	.0060	− .0060
	.1745	10	.0237	− .3326	− .0237	.1986	.0238	− .0237
	.2618	15	.0534	− .3295	− .0531	.1932	.0527	− .0520
	.3491	20	.0945	− .3214	− .0928	.1788	.0901	− .0863
	.4633	25	.1468	− .3046	− .1405	.1499	.1302	− .1167
	.5236	30	.2089	− .2751	− .1907	.1005	.1610	− .1255
	.6109	35	.2803	− .2146	− .2363	.0313	.1722	− .0994
	.6981	40	.3593	− .1561	− .2665	− .0565	.1439	− .0268
	.7854	45	.4440	− .0705	− .2689	− .1488	.0692	.0716
	.8727	50	.5320	.0440	− .2309	− .2230	− .0439	.1440
	.9599	55	.6209	.1930	− .1519	− .2497	− .1635	.1273
	1.0472	60	.7071	.3333	.0000	− .2000	− .2357	.0000
	1.1345	65	.7876	.4937	.1895	− .0574	− .2017	− .1676
	1.2217	70	.8591	.6508	.4092	.1719	− .0266	− .2236
	1.3090	75	.9184	.7914	.6316	.4528	.2723	− .0350
	1.3963	80	.9630	.9033	.8234	.7265	.6173	.3799
	1.4835	85	.9906	.9752	.9538	.9271	.8946	.8147
	1.5708	90	1.0000	1.0000	1.0000	1.0000	1.0000	1.0000

TABLE 3.1—contd.

n	Φ (rad)	(deg)	N=2	N=3	N=4	N=5	N=6	N=8
1	0	0	−1.0000	1.0000	−1.0000	1.0000	−1.0000	−1.0000
	0.0873	5	−0.9992	1.0000	−0.9983	0.9939	−0.9921	−0.9933
	.1745	10	− .9988	1.0000	− .9961	.9924	− .9886	− .9776
	.2618	15	− .9943	0.9839	− .9704	.9551	− .9350	− .8846
	.3491	20	− .9821	.9527	− .9122	.8618	− .8026	− .6624
	.4633	25	− .9570	.8878	− .7961	.6862	− .5637	− .3052
	.5236	30	− .9128	.7790	− .6081	.4216	− .2358	.0684
	.6109	35	− .8429	.6140	− .3548	.1102	.0818	.2291
	.6981	40	− .7419	.4005	− .0746	− .1516	.2373	.0731
	.7854	45	− .6057	.1558	.1613	− .2498	.1447	− .1384
	.8727	50	− .4337	− .0826	.2332	− .1382	− .0805	− .0600
	.9599	55	− .2290	− .2634	.2051	.0828	− .1683	.1235
	1.0472	60	.0000	− .3333	.0000	.2000	.0000	.0000
	1.1345	65	.2407	− .2561	− .2128	.0717	.1706	− .1199
	1.2217	70	.4762	− .0309	− .2602	− .1797	.0309	.1048
	1.3090	75	.6872	.2967	− .0381	− .2197	− .2263	.0378
	1.3963	80	.8548	.6410	.3945	.1550	− .0421	− .2264
	1.4835	85	.9617	.9017	.8219	.7246	.6147	.3763
	1.5708	90	1.0000	1.0000	1.0000	1.0000	1.0000	1.0000
3/2	0	0	0.0000	−0.3333	0.0000	0.2000	0.0000	0.0000
	0.0873	5	− .0180	−0.3329	.0179	.2000	− .1079	.0179
	.1745	10	− .0713	− .3266	.0708	.1878	− .0696	.0717
	.2618	15	− .1597	− .2993	.1517	.1407	− .1386	.1215
	.3491	20	− .2805	− .2284	.2363	.0311	− .1721	.0994
	.4633	25	− .4272	− .0900	.2713	− .1315	− .0872	− .0526
	.5236	30	− .5912	.1312	.1790	− .2477	.1423	− .1417
	.6109	35	− .7526	.4220	− .1003	− .1325	.2331	.0968
	.6981	40	− .8923	.7282	− .5288	.3180	− .1203	.1611
	.7854	45	− .9819	.9521	− .9114	.8613	− .8015	− .6591
	.8727	50	− .9936	.9831	− .9682	.9495	− .9272	− .8717
	.9599	55	− .9053	.7594	− .5513	.3826	− .1915	.1058
	1.0472	60	− .7071	.3333	.0000	− .2000	.2357	.0000
	1.1345	65	− .4085	− .1108	.2722	− .1113	− .1057	− .0305
	1.2217	70	− .0408	− .3311	.0408	.1960	− .0406	.0402
	1.3090	75	.3444	− .1753	− .2626	− .0391	.1459	− .0435
	1.3963	80	.6834	.2894	− .0449	− .2228	− .2239	.0445
	1.4835	85	.9168	.7873	.6246	.4441	.2616	− .0450
	1.5708	90	1.0000	1.0000	1.0000	1.0000	1.0000	1.0000

TABLE 3.2. Values of $a = \sin Nx / N \sin x$, with $x = 180°n(1-\cos \Phi)$.

n	Φ (rad)	Φ (deg)	N=1	N=2	N=3	N=4	N=5
1/4	0	0	0.00000	1.0000	1.0000	1.0000	1.0000
	0.0873	5	.00297	1.0000	1.0000	1.0000	1.0000
	.1745	10	.01187	1.0000	1.0000	1.0000	1.0000
	.2618	15	.02670	1.0000	1.0000	0.9998	0.9998
	.3491	20	.04728	1.0000	0.9982	.9962	.9925
	.4633	25	.07359	0.9961	.9919	.9853	.9777
	.5236	30	.10505	.9937	.9853	.9721	.9559
	.6109	35	.14159	.9899	.9729	.9500	.9209
	.6981	40	.18275	.9832	.9552	.9173	.8699
	.7854	45	.22801	.9737	.9307	.8724	.8007
	.8727	50	.27681	.9612	.8980	.8138	.7123
	.9599	55	.32870	.9444	.8559	.7404	.6052
	1.0472	60	.38268	.9239	.8047	.6533	.4828
	1.1345	65	.43806	.8989	.7442	.5539	.3502
	1.2217	70	.49409	.8694	.6745	.4449	.2142
	1.3090	75	.54975	.8354	.5970	.3304	.0833
	1.3963	80	.60446	.7966	.5129	.2146	−.0342
	1.4835	85	.65711	.7538	.4243	.1029	−.1305
	1.5708	90	.70711	.7071	.3333	.0000	−.2000
	1.6581	95	.75379	.6571	.2423	−.0897	−.2397
	1.7453	100	.79664	.6044	.1539	−.1628	−.2497
	1.8326	105	.83533	.5498	.0697	−.2174	−.2331
	1.9199	110	.86941	.4941	−.0078	−.2528	−.1952
	2.0071	115	.89895	.4381	−.0774	−.2699	−.1427
	2.0944	120	.92388	.3827	−.1380	−.2706	−.0828
	2.1817	125	.94443	.3287	−.1893	−.2577	−.0219
	2.2689	130	.96092	.2769	−.2311	−.2344	.0348
	2.3562	135	.97366	.2280	−.2640	−.2043	.0839
	2.4435	140	.98316	.1828	−.2888	−.1705	.1234
	2.5307	145	.98993	.1416	−.3066	−.1359	.1532
	2.6180	150	.99447	.1050	−.3186	−.1027	.1739
	2.7053	155	.99729	.0735	−.3261	−.0727	.1871
	2.7925	160	.99888	.0474	−.3303	−.0472	.1946
	2.8798	165	.99964	.0267	−.3324	−.0267	.1983
	2.9671	170	.99993	.0120	−.3331	−.0119	.1997
	3.0543	175	1.00000	.0059	−.3333	−.0030	.2000
	3.1416	180	1.00000	.0000	−.3333	.0000	.2000

TABLE 3.2—contd.

n	Φ (rad)	Φ (deg)	N=1	N=2	N=3	N=4	N=5
1/2	0	0	0.00000	1.0000	1.0000	1.0000	1.0000
	0.0873	5	.00593	1.0000	1.0000	1.0000	1.0000
	.1745	10	.02391	1.0000	0.9968	0.9967	0.9962
	.2618	15	.05338	1.0000	.9984	.9953	.9906
	.3491	20	.09463	0.9955	.9875	.9772	.9641
	.4633	25	.14660	.9892	.9717	.9469	.9157
	.5236	30	.20877	.9783	.9423	.8931	.8321
	.6109	35	.28033	.9597	.8951	.8089	.7054
	.6981	40	.35935	.9330	.8277	.6922	.5369
	.7854	45	.44401	.8960	.7369	.5427	.3358
	.8727	50	.53214	.8467	.6224	.3672	.1240
	.9599	55	.62087	.7839	.4861	.1796	− .0332
	1.0472	60	.70711	.7071	.3333	.0000	− .2000
	1.1345	65	.78758	.6162	.1729	− .1483	− .2499
	1.2217	70	.85914	.5117	.0159	− .2437	− .2091
	1.3090	75	.91852	.3954	− .1248	− .2718	− .0971
	1.3963	80	.96302	.2694	− .2365	− .2303	.0427
	1.4835	85	.99065	.1365	− .3085	− .1314	.1564
	1.5708	90	1.00000	.0000	− .3333	.0000	.2000
	1.6581	95	0.99065	− .1365	− .3085	.1314	.1564
	1.7453	100	.96302	− .2694	− .2365	.2303	.0427
	1.8326	105	.91852	− .3954	− .1248	.2718	− .0971
	1.9199	110	.85914	− .5117	.0159	.2437	− .2091
	2.0071	115	.78758	− .6162	.1729	.1483	− .2499
	2.0944	120	.70711	− .7071	.3333	.0000	− .2000
	2.1817	125	.62087	− .7839	.4861	− .1796	− .0332
	2.2689	130	.53214	− .8467	.6224	− .3672	.1240
	2.3562	135	.44401	− .8960	.7369	− .5427	.3358
	2.4435	140	.35935	− .9330	.8277	− .6922	.5369
	2.5307	145	.28033	− .9597	.8951	− .8089	.7054
	2.6180	150	.20877	− .9783	.9423	− .8931	.8321
	2.7053	155	.14660	− .9892	.9717	− .9469	.9157
	2.7925	160	.09463	− .9955	.9875	− .9772	.9641
	2.8798	165	.05338	−1.0000	.9984	− .9953	.9906
	2.9671	170	.02391	−1.0000	.9968	− .9967	.9962
	3.0543	175	.00593	−1.0000	1.0000	−1.0000	1.0000
	3.1416	180	.00000	−1.0000	1.0000	−1.0000	1.0000

TABLE 3.3. Values of $a = \dfrac{\cos(90° \cos \Theta)}{\sin \Theta} \dfrac{\sin Nx}{N \sin x}$, with $x = 180°n(1 - \sin \Theta)$.

n	Θ (rad)	Θ (deg)	N=1	N=2	N=3	N=4	N=5
1/4	0	0	0.0000	0.0000	0.0000	0.0000	0.0000
	0.0873	5	.0680	.0513	.0289	.0070	− .0089
	.1745	10	.1377	.1097	.0706	.0296	− .0047
	.2618	15	.2069	.1728	.1235	.0684	.0172
	.3491	20	.2767	.2406	.1866	.1231	.0593
	.4633	25	.3469	.3118	.2582	.1921	.1215
	.5236	30	.4179	.3860	.3362	.2730	.2018
	.6109	35	.4887	.4615	.4183	.3618	.2958
	.6981	40	.5590	.5373	.5020	.4549	.3982
	.7854	45	.6279	.6114	.5844	.5478	.5028
	.8727	50	.6947	.6830	.6636	.6372	.6043
	.9599	55	.7579	.7502	.7374	.7200	.6980
	1.0472	60	.8165	.8114	.8045	.7937	.7805
	1.1345	65	.8690	.8656	.8620	.8562	.8496
	1.2217	70	.9143	.9143	.9127	.9108	.9074
	1.3090	75	.9509	.9509	.9509	.9507	.9507
	1.3963	80	.9779	.9779	.9779	.9779	.9779
	1.4835	85	.9944	.9944	.9944	.9944	.9944
	1.5708	90	1.0000	1.0000	1.0000	1.0000	1.0000
1/4	0	0		0.0000	0.0000	.0000	0.0000
	−0.0873	− 5		−0.0447	− .0165	.0061	.0163
	− .1745	−10		− .0832	− .0212	.0224	.0344
	− .2618	−15		− .1138	− .0144	.0450	.0482
	− .3491	−20		− .1367	.0022	.0699	.0540
	− .4633	−25		− .1520	.0269	.0936	.0495
	− .5236	−30		− .1599	.0577	.1131	.0346
	− .6109	−35		− .1607	.0925	.1259	.1070
	− .6981	−40		− .1548	.1292	.1310	− .0195
	− .7854	−45		− .1432	.1658	.1283	− .0527
	− .8727	−50		− .1270	.2006	.1184	− .0857
	− .9599	−55		− .1073	.2324	.0930	− .1161
	−1.0472	−60		− .0857	.2601	.0839	− .1420
	−1.1345	−65		− .0639	.2834	.0632	− .1626
	−1.2217	−70		− .0433	.3020	.0432	− .1779
	−1.3090	−75		− .0254	.3161	.0254	− .1886
	−1.3963	−80		− .0117	.3257	.0116	− .1953
	−1.4835	−85		− .0059	.3314	.0030	− .1989
	−1.5708	−90		− .0000	.3333	.0000	− .2000

TABLE 3.3—contd.

n	Θ (rad)	(deg)	N=1	N=2	N=3	N=4	N=5
1/2	0	0	0.0000	0.0000	0.0000	0.0000	0.0000
	0.0873	5	.0680	.0093	− .0210	− .0089	.0106
	.1745	10	.1377	.0370	− .0326	− .0317	.0059
	.2618	15	.2069	.0818	− .0258	− .0562	− .0201
	.3491	20	.2767	.1416	.0044	− .0674	− .0579
	.4633	25	.3469	.2138	.0596	− .0514	− .0867
	.5236	30	.4179	.2955	.1393	.0000	− .0836
	.6109	35	.4887	.3831	.2376	.0878	− .0162
	.6981	40	.5590	.4733	.3479	.2053	.0693
	.7854	45	.6279	.5626	.4627	.3408	.2108
	.8727	50	.6947	.6482	.5750	.4809	.3730
	.9599	55	.7579	.7274	.6784	.6131	.5346
	1.0472	60	.8165	.7988	.7694	.7292	.6794
	1.1345	65	.8690	.8596	.8444	.8229	.7957
	1.2217	70	.9143	.9102	9029	.8935	.8815
	1.3090	75	.9509	.9509	.9494	.9464	.9420
	1.3963	80	.9779	.9779	.9748	.9747	.9742
	1.4835	85	.9944	.9944	.9944	.9944	.9944
	1.5708	90	1.0000	1.0000	1.0000	1.0000	1.0000
1/2	0	0		0.0000	0.0000	0.0000	0.0000
	−0.0873	− 5		− .0093	.0210	.0089	− .0106
	− .1745	−10		− .0370	.0326	.0317	− .0059
	− .2618	−15		− .0818	.0258	.0562	.0201
	− .3491	−20		− .1416	− .0044	.0674	.0579
	− .4633	−25		− .2138	− .0596	.0514	.0867
	− .5236	−30		− .2955	− .1393	.0000	.0836
	− .6109	−35		− .3831	− .2376	− .0878	.0162
	− .6981	−40		− .4733	− .3479	− .2053	− .0693
	− .7854	−45		− .5626	− .4627	− .3408	− .2108
	− .8727	−50		− .6482	− .5750	− .4809	− .3730
	− .9599	−55		− .7274	− .6784	− .6131	− .5346
	−1.0472	−60		− .7988	− .7694	− .7292	− .6794
	−1.1345	−65		− .8596	− .8444	− .8229	− .7957
	−1.2217	−70		− .9102	− .9029	− .8935	− .8815
	−1.3090	−75		− .9509	− .9494	− .9464	− .9420
	−1.3963	−80		− .9779	− .9748	− .9747	− .9742
	−1.4835	−85		− .9944	− .9944	− .9944	− .9944
	−1.5708	−90		−1.0000	−1.0000	−1.0000	−1.0000

TABLE 3.4. Values of $a = \sin Nx/N \sin x = \cos x$, with $x = 180°(t - n \cos \Phi)$; $N = 2$.

n	Φ (rad)	Φ (deg)	t=0	t=1/4	t=1/2	t=3/4	t=1
1/4	0	0	0.7071	1.0000	0.7071	0.0000	−0.7071
	0.0873	5	.7092	1.0000	.7050	− .0030	− .7092
	.1745	10	.7155	0.9999	.6987	− .0119	− .7155
	.2618	15	.7257	.9996	.6880	− .0267	− .7257
	.3491	20	.7398	.9989	.6729	− .0473	− .7398
	.4633	25	.7572	.9973	.6532	− .0736	− .7572
	.5236	30	.7752	.9960	.6289	− .1051	− .7873
	.6109	35	.8001	.9899	.5999	− .1416	− .8001
	.6981	40	.8244	.9832	.5660	− .1828	− .8244
	.7854	45	.8497	.9737	.5273	− .2280	− .8497
	.8727	50	.8753	.9609	.5003	− .2768	− .8753
	.9599	55	.9002	.9444	.4354	− .3287	− .9002
	1.0472	60	.9239	.9239	.3827	− .3827	− .9239
	1.1345	65	.9454	.8990	.3259	− .4381	− .9454
	1.2217	70	.9641	.8694	.2654	− .4941	− .9641
	1.3090	75	.9794	.8353	.2019	− .5498	− .9794
	1.3963	80	.9907	.7966	.1359	− .6045	− .9907
	1.4835	85	.9977	.7538	.0684	− .6571	− .9977
	1.5708	90	1.0000	.7071	.0000	− .7071	−1.0000
	1.6581	95	.9977	.6571	− .0684	− .7538	−0.9977
	1.7453	100	.9907	.6045	− .1359	− .7966	− .9907
	1.8326	105	.9794	.5498	− .2019	− .8353	− .9794
	1.9199	110	.9641	.4941	− .2654	− .8694	− .9641
	2.0071	115	.9454	.4381	− .3259	− .8990	− .9454
	2.0944	120	.9239	.3827	− .3827	− .9239	− .9239
	2.1817	125	.9002	.3287	− .4354	− .9444	− .9002
	2.2689	130	.8753	.2768	− .5003	− .9609	− .8753
	2.3562	135	.8497	.2280	− .5273	− .9737	− .8497
	2.4435	140	.8244	.1828	− .5660	− .9832	− .8244
	2.5307	145	.8001	.1416	− .5999	− .9899	− .8001
	2.6180	150	.7752	.1049	− .6289	− .9945	− .7873
	2.7053	155	.7572	.0736	− .6532	− .9973	− .7572
	2.7925	160	.7398	.0473	− .6729	− .9989	− .7398
	2.8798	165	.7257	.0267	− .6880	− .9996	− .7257
	2.9671	170	.7155	.0119	− .6987	− .9999	− .7155
	3.0543	175	.7092	.0030	− .7050	−1.0000	− .7092
	3.1416	180	.7071	.0000	− .7071	−1.0000	− .7071

TABLE 3.4—contd.

n	Φ (rad)	Φ (deg)	t = 0	t = 1/4	t = 1/2	t = 3/4	t = 1
1/2	0	0	0.0000	0.7071	1.0000	0.7071	0.0000
	0.0873	5	.0059	.7113	1.0000	.7029	− .0059
	.1745	10	.0239	.7238	0.9997	.6900	− .0239
	.2618	15	.0536	.7440	.9986	.6682	− .0536
	.3491	20	.0946	.7709	.9955	.6370	− .0946
	.4633	25	.1466	.8031	.9892	.5958	− .1466
	.5236	30	.1781	.8218	.9779	.5438	− .2089
	.6109	35	.2089	.8770	.9599	.4805	− .2803
	.6981	40	.3594	.9140	.9332	.4058	− .3594
	.7854	45	.4440	.9476	.8960	.3196	− .4440
	.8727	50	.5321	.9750	.8467	.2224	− .5321
	.9599	55	.6209	.9933	.7839	.1153	− .6209
	1.0472	60	.7071	1.0000	.7071	.0000	− .7071
	1.1345	65	.7876	0.9926	.6162	− .1212	− .7876
	1.2217	70	.8591	.9694	.5117	− .2457	− .8591
	1.3090	75	.9185	.9291	.3954	− .3699	− .9185
	1.3963	80	.9630	.8715	.2694	− .4905	− .9630
	1.4835	85	.9907	.7971	.1364	− .6039	− .9907
	1.5708	90	1.0000	.7071	.0000	− .7071	−1.0000
	1.6581	95	0.9907	.6040	− .1364	− .7970	−0.9907
	1.7453	100	.9630	.4905	− .2694	− .8715	− .9630
	1.8326	105	.9185	.3699	− .3954	− .9291	− .9185
	1.9199	110	.8591	.2457	− .5117	− .9694	− .8591
	2.0071	115	.7876	.1212	− .6162	− .9926	− .7876
	2.0944	120	.7071	.0000	− .7071	−1.0000	− .7071
	2.1817	125	.6209	− .1153	− .7839	−0.9933	− .6209
	2.2689	130	.5321	− .2224	− .8467	− .9591	− .5321
	2.3562	135	.4440	− .3196	− .8960	− .9476	− .4440
	2.4435	140	.3594	− .4058	− .9332	− .9140	− .3594
	2.5307	145	.2089	− .4805	− .9599	− .8770	− .2803
	2.6180	150	.1781	− .5438	− .9779	− .8392	− .2089
	2.7053	155	.1466	− .5958	− .9892	− .8031	− .1466
	2.7925	160	.0946	− .6370	− .9955	− .7709	− .0946
	2.8798	165	.0536	− .6682	− .9986	− .7440	− .0536
	2.9671	170	.0239	− .6900	− .9997	− .7238	− .0239
	3.0543	175	.0059	− .7029	−1.0000	.7113	− .0059
	3.1416	180	.0000	− .7071	−1.0000	− .7071	.0000

TABLE 3.4—contd.

n	Φ (rad)	Φ (deg)	t=0	t=1/4	t=1/2	t=3/4	t=1
1	0	0	−1.0000	−0.7071	0.0000	0.7071	1.0000
	0.0873	5	−0.9999	− .6985	.0120	.7156	0.9999
	.1745	10	− .9989	− .6726	.0476	.7400	.9989
	.2618	15	− .9943	− .6276	.1068	.7786	.9943
	.3491	20	− .9821	− .5612	.1884	.8277	.9821
	.4633	25	− .9570	− .4716	.2900	.8818	.9570
	.5236	30	− .9128	− .3566	.4085	.9343	.9128
	.6109	35	− .8429	− .2156	.5380	.9765	.8429
	.6981	40	− .7419	− .0504	.6706	.9987	.7419
	.7854	45	− .6057	.1343	.7957	.9909	.6057
	.8727	50	− .4337	.3305	.9011	.9438	.4337
	.9599	55	− .2290	.5264	.9734	.8503	.2290
	1.0472	60	.0000	.7071	1.0000	.7071	.0000
	1.1345	65	.2407	.8565	0.9706	.5161	− .2407
	1.2217	70	.4762	.9585	.8793	.2850	− .4762
	1.3090	75	.6872	.9996	.7265	.0278	− .6872
	1.3963	80	.8548	.9714	.5189	− .2375	− .8548
	1.4835	85	.9627	.8720	.2704	− .4895	− .9627
	1.5708	90	1.0000	.7071	.0000	− .7071	−1.0000
	1.6581	95	0.9627	.4897	− .2704	− .8720	−0.9627
	1.7453	100	.8548	.2375	− .5189	− .9714	− .8548
	1.8326	105	.6872	− .0278	− .7265	− .9996	− .6872
	1.9199	110	.4762	− .2850	− 8793	− .9585	− .4762
	2.0071	115	.2407	− .5161	− .9706	− .8565	− .2407
	2.0944	120	.0000	− .7071	−1.0000	− .7071	.0000
	2.1817	125	− .2290	− .8503	−0.9734	− .5264	.2290
	2.2689	130	− .4337	− .9438	− .9011	− .3305	.4337
	2.3562	135	− .6057	− .9909	− .7957	− .1343	.6057
	2.4435	140	− .7419	− .9987	− .6706	.0504	.7419
	2.5307	145	− .8429	− .9765	− .5380	.2156	.8429
	2.6180	150	− .9128	− .9343	− .4085	.3566	.9128
	2.7053	155	− .9570	− .8818	− .2900	.4716	.9570
	2.7925	160	− .9821	− .8277	− .1884	.5612	.9821
	2.8798	165	− .9943	− .7786	− .1068	.6276	.9943
	2.9671	170	− .9989	− .7400	− .0476	.6726	.9989
	3.0543	175	− .9999	− .7156	− .0120	.6985	.9999
	3.1416	180	−1.0000	− .7071	− .0000	.7071	1.0000

TABLE 3.4—contd.

n	Φ		t				
	(rad)	(deg)	0	1/4	1/2	3/4	1
3/2	0	0	0.0000	−0.7071	−1.0000	−0.7071	0.0000
	0.0873	5	− .0180	− .7197	−0.9997	− .6943	.0180
	.1745	10	− .0715	− .7559	− .9974	− .6547	.0715
	.2618	15	− .1723	− .8110	− .9871	− .5875	.1599
	.3491	20	− .2803	− .8770	− .9599	− .4805	.2803
	.4633	25	− .4274	− .9415	− .9041	− .3373	.4272
	.5236	30	− .5902	− .9882	− .8073	− .1535	.5902
	.6109	35	− .7528	− .9978	− .6583	.0668	.7528
	.6981	40	− .8924	− .9501	− .4513	.3118	.8924
	.7854	45	− .9819	− .8283	− .1894	.5604	.9819
	.8727	50	− .9937	− .6232	.1123	.7822	.9937
	.9599	55	− .9054	− .3399	.4247	.9405	.9054
	1.0472	60	− .7071	.0000	.7071	1.0000	.7071
	1.1345	65	− .4085	.3566	.9128	.9343	.4085
	1.2217	70	− .0410	.6775	.9992	.7355	.0410
	1.3090	75	.3440	.9072	.9390	.4207	− .3440
	1.3963	80	.6834	.9995	.7300	.0330	− .6834
	1.4835	85	.9169	.9306	.3992	− .3660	− .9169
	1.5708	90	1.0000	.7071	.0000	− .7071	−1.0000
	1.6581	95	0.9169	.3660	− .3992	− .9306	−0.9169
	1.7453	100	.6834	− .0330	− .7300	− .9995	− .6834
	1.8326	105	.3440	− .4207	− .9390	− .9072	− .3440
	1.9199	110	− .0410	− .7355	− .9992	− .6775	.0410
	2.0071	115	− .4085	− .9343	− .9128	− .3566	.4085
	2.0944	120	− .7071	−1.0000	− .7071	.0000	.7071
	2.1817	125	− .9054	−0.9405	− .4247	.3399	.9054
	2.2689	130	− .9937	− .7821	− .1123	.6232	.9937
	2.3562	135	− .9819	− .5604	.1894	.8283	.9819
	2.4435	140	− .8924	− .3118	.4513	.9501	.8924
	2.5307	145	− .7528	− .0668	.6583	.9978	.7528
	2.6180	150	− .5902	.1535	.8073	.9882	.5902
	2.7053	155	− .4274	.3371	.9041	.9415	.4272
	2.7925	160	− .2803	.4805	.9599	.8770	.2803
	2.8798	165	− .1723	.5850	.9871	.8111	.1599
	2.9671	170	− .0715	.6547	.9974	.7559	.0715
	3.0543	175	− .0180	.6943	.9997	.7197	.0180
	3.1416	180	.0000	.7071	1.0000	.7071	.0000

The array factor per element of the broadside array in the equatorial plane, namely,

$$a_B(\tfrac{1}{2}\pi, \Phi; N_B, n_B, 0) = \frac{\sin N_B \pi n_B \cos \Phi}{N_B \sin \pi n_B \cos \Phi}, \quad (53)$$

is tabulated in Table 3.1 and represented graphically in Figs. 3.2 and 3.3.

The array factor per element of the end-fire array in the equatorial plane, namely,

$$a_E(\tfrac{1}{2}\pi, \Phi; N_E, n_E, n_E) = \frac{\sin N_E \pi n_E (1 - \cos \Phi)}{N_E \sin \pi n_E (1 - \cos \Phi)}, \quad (54)$$

is tabulated in Table 3.2 and represented graphically in Figs. 3.4 and 3.5. In the plane $\Phi = 0$ with $\beta_0 h = \pi/2$, the array factor is

$$F_0(\Theta, \tfrac{1}{2}\pi) a_E(\Theta, 0; N_E, n_E, n_E)$$
$$= \frac{\cos(\tfrac{1}{2}\pi \cos \Theta)}{\sin \Theta} \cdot \frac{\sin N_E \pi n_E (1 - \sin \Theta)}{N_E \sin \pi n_E (1 - \sin \Theta)}. \quad (55)$$

This function is tabulated in Table 3.3 and represented graphically in Figs. 3.6 and 3.7.

An important and simple parallel array consists of two units. With $N_p = 2$, (39) gives

$$a_p(\Theta, \Phi; 2, n_p, t_p)$$
$$= \cos [\pi(n_p \sin \Theta \cos \Phi - t_p)]. \quad (56)$$

This factor is tabulated in Table 3.4 with $\Theta = \pi/2$ and represented graphically in Figs. 3.8 and 3.9 for $\Theta = \pi/2$ for a range of values of both n_p and t_p.

An interesting and very useful *array of arrays* consists of two N-element broadside arrays one behind the other so that each pair of antennas in depth is a two-element end-fire array. The purpose of the second row is to make the bidirectional broadside array essentially unidirectional. Evidently, the product of a broadside pattern such as $N_B = 4$, $n_B = 1/4$ in Fig. 3.3 with the end-fire pattern $N_E = 2$, $n_E = 1/4$ as in Fig. 3.5 must yield a pattern that has important ears only in one of the principal broadside directions.

The general array factor for a combination broadside array with the appropriate end-fire array of Fig. 3.5, for which $N_E = 2$, is

$$a = a_E a_B, \quad (57)$$

where,

$$a_E = \cos x_E, \quad x_E = \pi(t_E - n_E \sin \psi_E),$$
$$n_E = t_E = 1/4, \quad (58)$$

$$a_B = \frac{\sin N_B x_B}{N_B \sin x_B}, \quad x_B = n_B \pi \cos \psi_B. \quad (59)$$

Graphs of the function a in (57) are shown in Fig. 3.10 and Fig. 3.11 for a range of values of N_B and n_B. It is seen that highly directional arrays may be obtained, and that these are essentially unidirectional with a sharp beam when $n_B = \tfrac{1}{2}$ and N_B is large.

The radiation function of the parallel array of N_p half-wave dipoles with progressively increasing phase is

$$K_m^2(\Theta, \Phi) = 4 \left[\frac{\cos(\tfrac{1}{2}\pi \cos \Theta)}{\sin \Theta} \cdot \frac{\sin N_p x}{\sin x} \right]^2, \quad (60a)$$

with

$$x = \pi(n_p \sin \Theta \cos \Phi - t_p) \equiv K_1 \cos \Phi + K_2. \quad (60b)$$

The factors K_1 and K_2 are defined in (60b).

The radiation resistance referred to maximum sinusoidal current is obtained from (I.10.6) as follows:

$$R_m^e = \frac{\zeta_0}{8\pi} \int_0^{2\pi} \int_0^{\pi} K_m^2(\Theta, \Phi) \sin \Theta \, d\Theta \, d\Phi, \quad (61)$$

where $K_m^2(\Theta, \Phi)$ is in (60a). Integration of (61) gives*

$$R_m^e \doteq (0.945)^2 \frac{\zeta_0}{4\pi^2} \left[\frac{8\pi N_p}{3} \right.$$
$$\left. + 4\pi \sum_{r=2}^{2N_p-2} (2N_p - r) \cos(r\pi t_p) \Lambda(r\pi n_p) \right], \quad (62a)$$

---

* Since Φ occurs in (61) only in the array factor A_p, the integration with respect to Φ may be carried out by expanding the factor in a series of cosine terms and integrating term by term. This is possible because the series contains only a finite number of terms so that no difficulty arises in the term-by-term integration. Thus

$$A_p^2(\Theta, \Phi) = \frac{\sin^2 N_p x}{\sin^2 x} = N_p + \sum_{r=2}^{2N_p-2} (2N_p - r) \cos rx, \quad (61a)$$

where $r = 2, 4, 6, \cdots$ is an even integer. Integration of (61a) gives

$$\int_0^{2\pi} A_p^2(\Theta, \Phi) \, d\Phi = 2\pi N_p$$
$$+ \sum_{r=2}^{2N_p-2} (2N_p - r) \cdot 2\pi J_0(rK_1) \cos(rK_2) \equiv H(\Theta), \quad (61b)$$

where

$$\Lambda(r\pi n_p) = \frac{\sin(r\pi n_p)}{r\pi n_p} + \frac{\cos(r\pi n_p)}{(r\pi n_p)^2} - \frac{\sin(r\pi n_p)}{(r\pi n_p)^3}. \quad (62b)$$

This formula may be specialized for the bilateral end-fire, the unilateral end-fire, and the broadside arrays by appropriate choice of n_p and t_p. These numbers, together with the associated values of $\Lambda(r\pi n_p)$ for the first six terms in the series, are given in Table 3.6.

TABLE 3.6. The function Λ for arrays.

t_p	n_p	Λ	Nature of pattern
1/2	1/2	$\Lambda(\tfrac{1}{2}r\pi)$	Bilateral end-fire
1/4	1/4	$\Lambda(\tfrac{1}{4}r\pi)$	Unilateral end-fire
0	1/2	$\Lambda(\tfrac{1}{2}r\pi)$	Broadside

r	$\Lambda(\tfrac{1}{2}r\pi)$	$\Lambda(\tfrac{1}{4}r\pi)$
2	0.3786	−0.1013
4	− .1013	.02533
6	− .2026	− .01126
8	.02533	.006333
10	.1252	− .004053
12	− .01126	.002814

Calculations of radiation resistance for these three types of arrays are given in Tables 3.7, 3.8 and 3.9 as computed from the approximate formula (62a). These values are used throughout in order to permit intercomparison. Radiation resistances per element, namely, R_m^e/N_p, are also shown, designated by (approx) when evaluated (using 62), by BB when taken from the numerically integrated results of Bontsch-Bruewitsch,[5] and by P when taken from results evaluated by Pistolkors.[37] In general, the numerical values obtained using the approximate formula (62a) are within a few percent of values obtained by the more laborious methods where these are available.

The directivity of the broadside array is seen to be greater than for end-fire arrays with the same number of half-wave elements.

The beam widths in the Φ-plane of broadside and end-fire arrays are obtained from (53) and (54). Thus,

$$a = \frac{\sin Nx}{N \sin x}; \quad (63)$$

Broadside array:

$$x = x_B = \pi n_B \cos \Phi, \quad (63a)$$

End-fire array:

$$x = x_E = \pi n_E(1 - \cos \Phi). \quad (63b)$$

where $H(\Theta)$ is a shorthand as defined in (61b). In (61b) use has been made of Hansen's integral representation of a Bessel Function (see ref. I.28),

$$\int_0^{2\pi} \cos(rx)\,d\Phi = \int_0^{2\pi} \cos r(K_1 \cos \Phi + K_2)\,d\Phi$$
$$= 2J_0(rK_1) \cos(rK_2). \quad (61c)$$

With (61b) substituted in (60a) and this in (61), the result is

$$R_m^e = \frac{t_0}{2\pi} \int_0^\pi F_m^2(\Theta, \tfrac{1}{2}\pi) H(\Theta) \sin \Theta\, d\Theta. \quad (61d)$$

The direct integration of (61d) has not been accomplished. However, it was shown in Section 2 that a good approximation of an integral of this type is obtained if use is made of $F_m(\Theta, \tfrac{1}{2}\pi) \doteq 0.945 \sin \Theta$, with which (61d) becomes

$$R_m^e \doteq \frac{(0.945)^2 t_0}{2\pi} \int_0^\pi H(\Theta) \sin^3 \Theta\, d\Theta \quad (61e)$$

The integrand in (61e) may be expanded into a finite series and integrated term by term as follows:

$$\int_0^\pi H(\Theta) \sin^3 \Theta\, d\Theta$$
$$= 2\pi N_p \int_0^\pi \sin^3 \Theta\, d\Theta + 2\pi \sum_{r=2}^{2N_p-2} (2N_p - r)$$
$$\times \cos(rK_2) \int_0^\pi J_0(rK_1) \sin^3 \Theta\, d\Theta. \quad (61f)$$

The last integral in (61f) is equivalent to

$$\int_0^\pi J_0(rK_1) \sin^3 \Theta\, d\Theta$$
$$= 2 \int_0^{\pi/2} J_0(r\pi n_p \sin \Theta) \sin^3 \Theta\, d\Theta. \quad (61g)$$

By the substitution of $\sin^2 \Theta = 1 - \cos^2 \Theta$, (61g) is expanded into the difference of two integrals of which each is in Sonine form [see N. J. Sonine, *Math. Ann.* 16, 1 (1880)]. The result is

$$2\int_0^{\pi/2} J_0(r\pi n_p \sin \Theta)(1 - \cos^2 \Theta) \sin \Theta\, d\Theta$$
$$= \left[\frac{2\Gamma(\tfrac{3}{2})J_{\tfrac{1}{2}}(r\pi n_p)}{\sqrt{2r\pi n_p}} - \frac{2\sqrt{2}\,\Gamma(\tfrac{3}{2})J_{\tfrac{3}{2}}(r\pi n_p)}{(r\pi n_p)^{3/2}}\right]. \quad (61h)$$

Since this expression involves Bessel functions with fractional indices, which are not convenient for computation, it is advisable to transform them into circular functions, using

$$J_{\tfrac{1}{2}}(r\pi n_p) = \sqrt{\frac{2}{\pi}}\,\frac{\sin(r\pi n_p)}{r\pi n_p}, \quad (61i)$$

$$J_{\tfrac{3}{2}}(r\pi n_p) = \sqrt{\frac{2}{\pi}}\left[\frac{\sin(r\pi n_p)}{(r\pi n_p)^{3/2}} - \frac{\cos(r\pi n_p)}{\sqrt{r\pi n_p}}\right]. \quad (61j)$$

Substitution of (61i, j) in (61h) and use of this and the values $\Gamma(\tfrac{1}{2}) = \sqrt{\pi}$, $\Gamma(\tfrac{3}{2}) = \sqrt{\pi}/2$ in (61e) gives (62a).

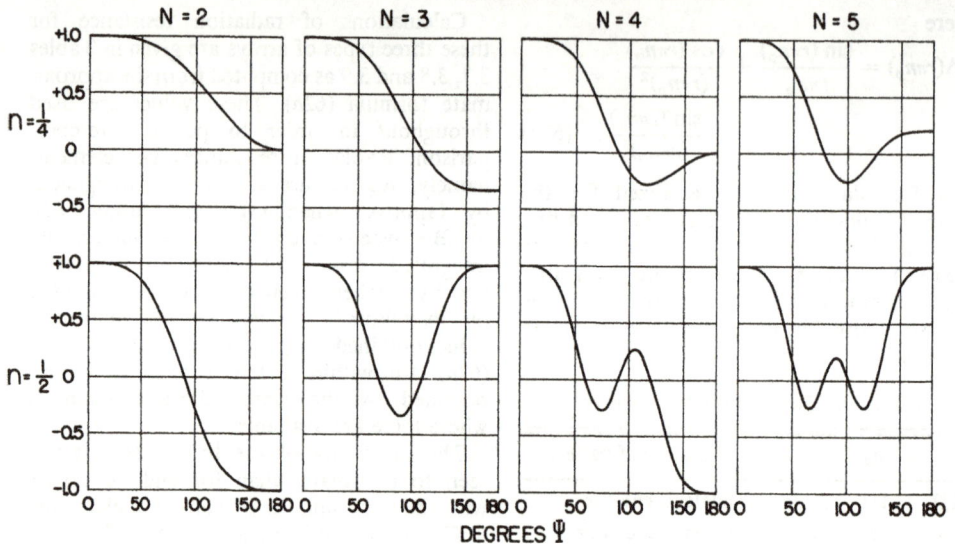

Fig. 3.4. End-fire array, horizontal pattern; rectangular graph of $(\sin Nx)/N \sin x$; $x = 180°n(1 - \cos \Psi)$.

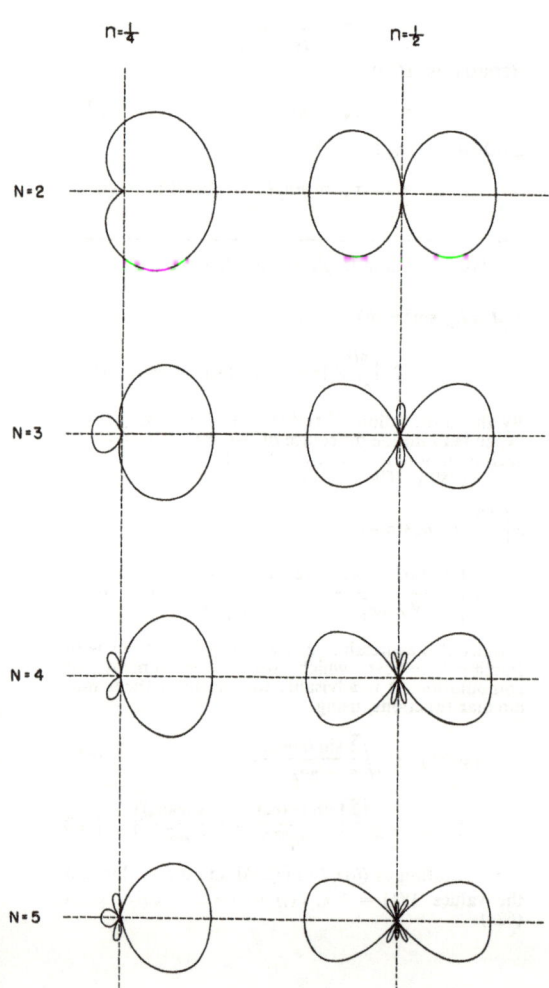

Fig. 3.5. End-fire array, horizontal pattern; polar graph of $(\sin Nx)/N \sin x$; $x = 180°n(1 - \cos \Psi)$.

[VI.3] THEORY OF LINEAR ANTENNAS 615

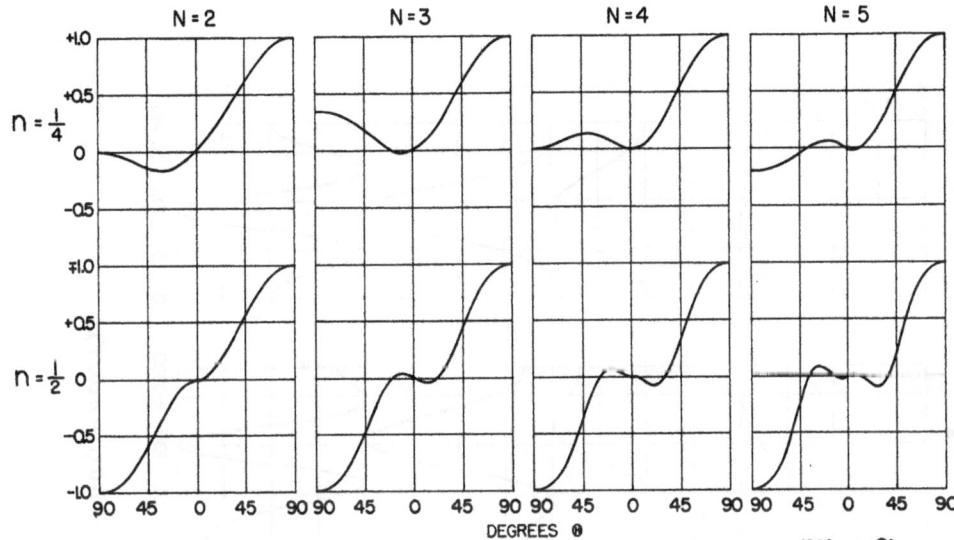

Fig. 3.6. End-fire array, vertical pattern; rectangular graph of $\dfrac{\sin Nx}{N \sin x} \cdot \dfrac{\cos(90° \cos \Theta)}{\sin \Theta}$; $x = 180°n(1 - \sin \Theta \cos \Psi)$.

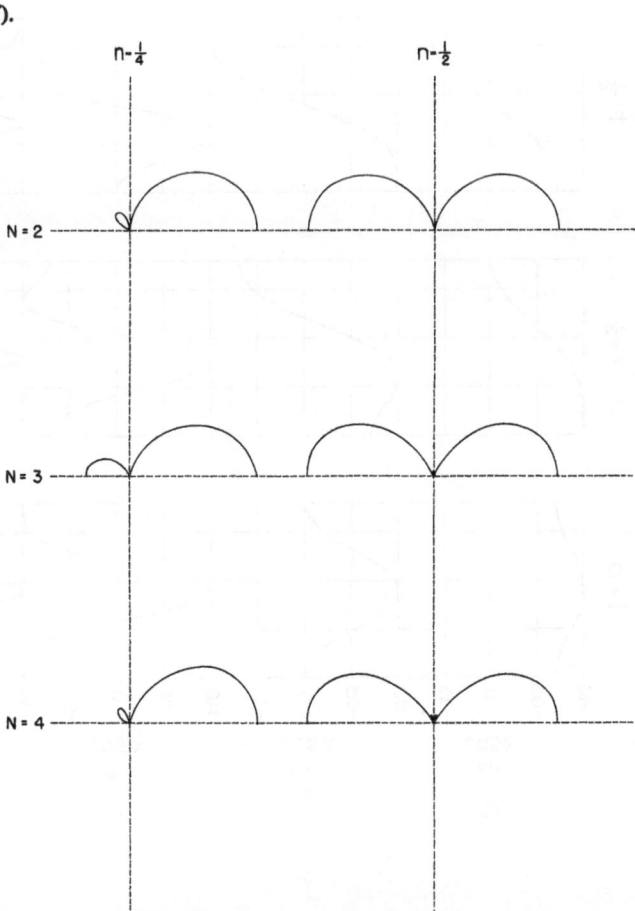

Fig. 3.7. End-fire array, vertical pattern; polar graph of
$\dfrac{\sin Nx}{N \sin x} \cdot \dfrac{\cos(90° \cos \Theta)}{\sin \Theta}$;
$x = 180°n(1 - \sin \Theta \cos \Psi)$.

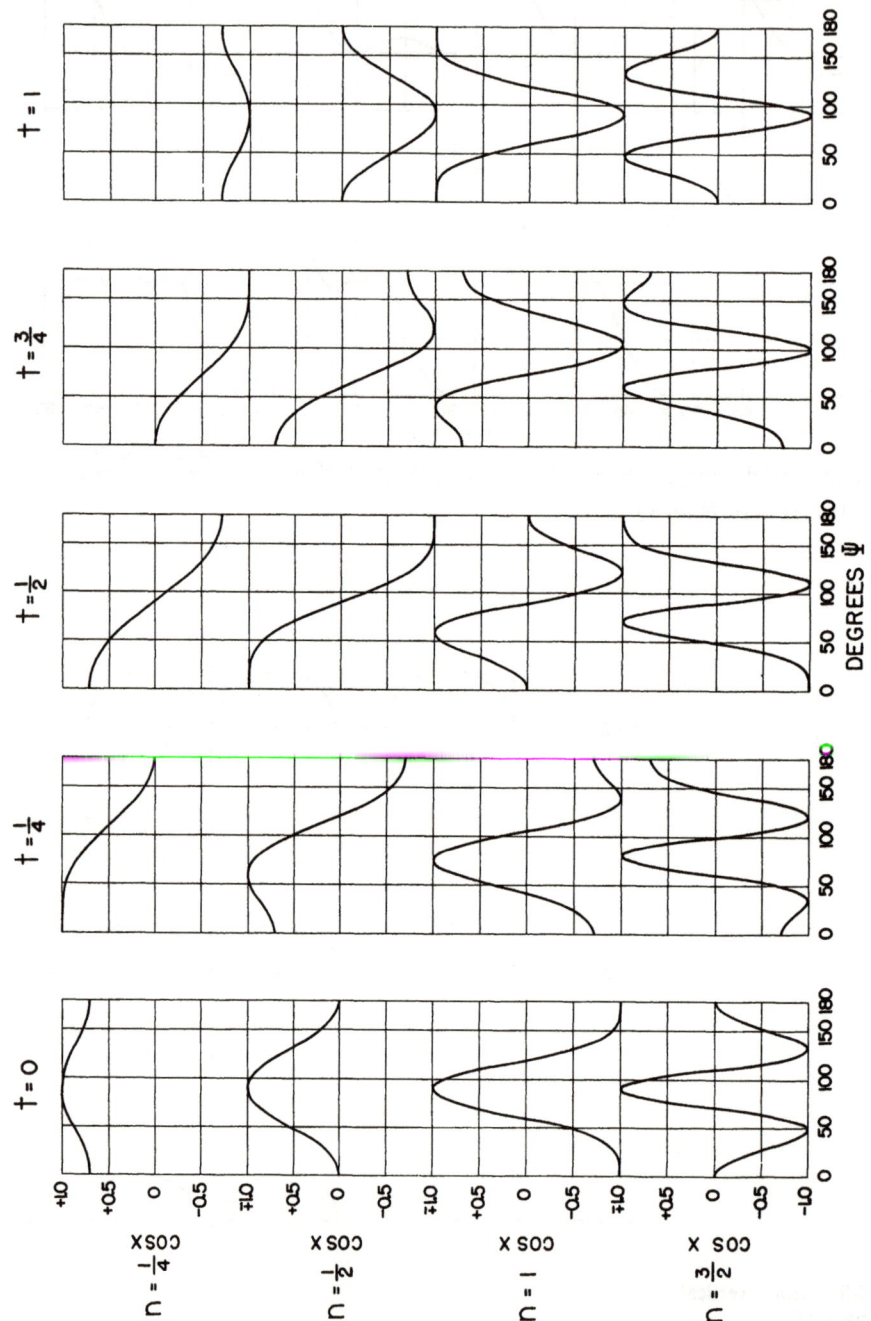

Fig. 3.8. End-fire array (2-unit), $N = 2$; rectangular graph of $\cos x$; $x = 180°(t - n\cos\Psi)$.

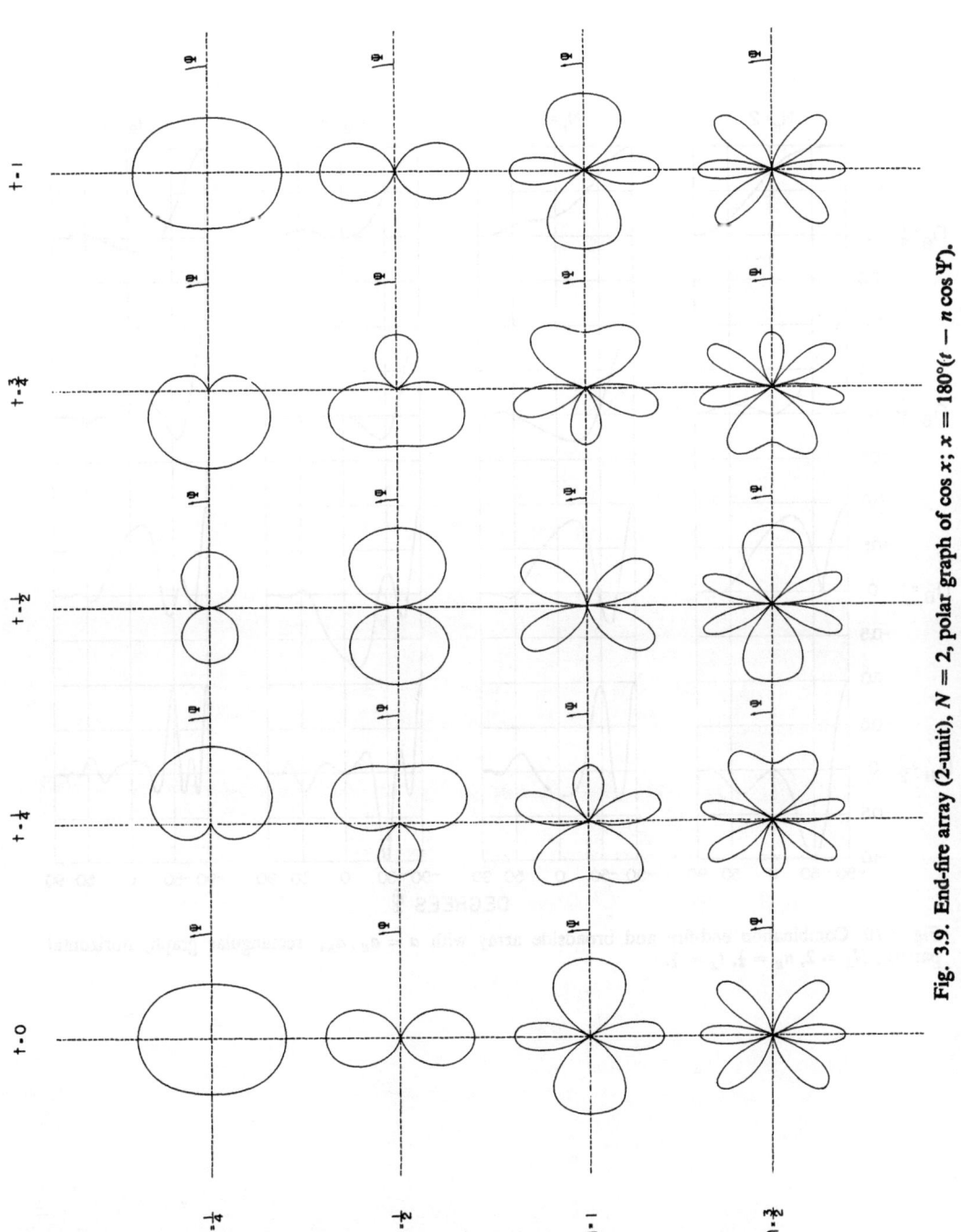

Fig. 3.9. End-fire array (2-unit), $N = 2$, polar graph of $\cos x$; $x = 180°(t - n \cos \Psi)$.

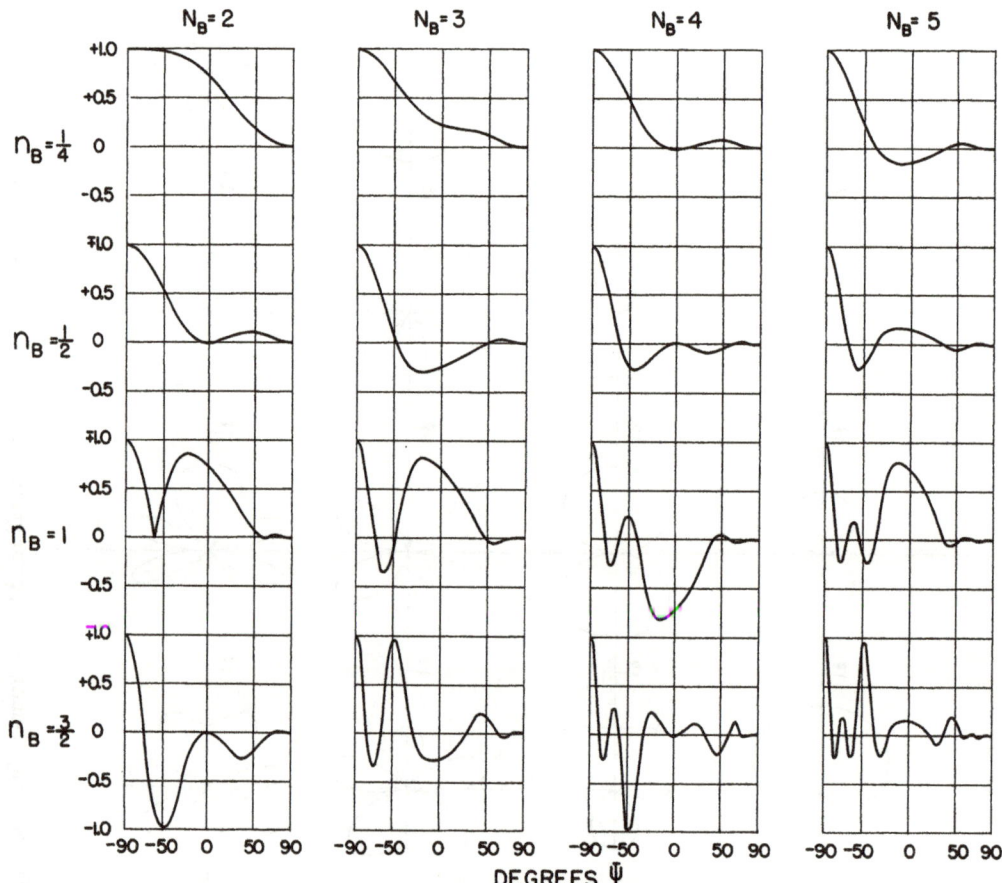

Fig. 3.10. Combination end-fire and broadside array with $a = a_E \cdot a_B$; rectangular graph, horizontal pattern; $N_E = 2$, $n_E = \frac{1}{4}$, $t_E = \frac{1}{4}$.

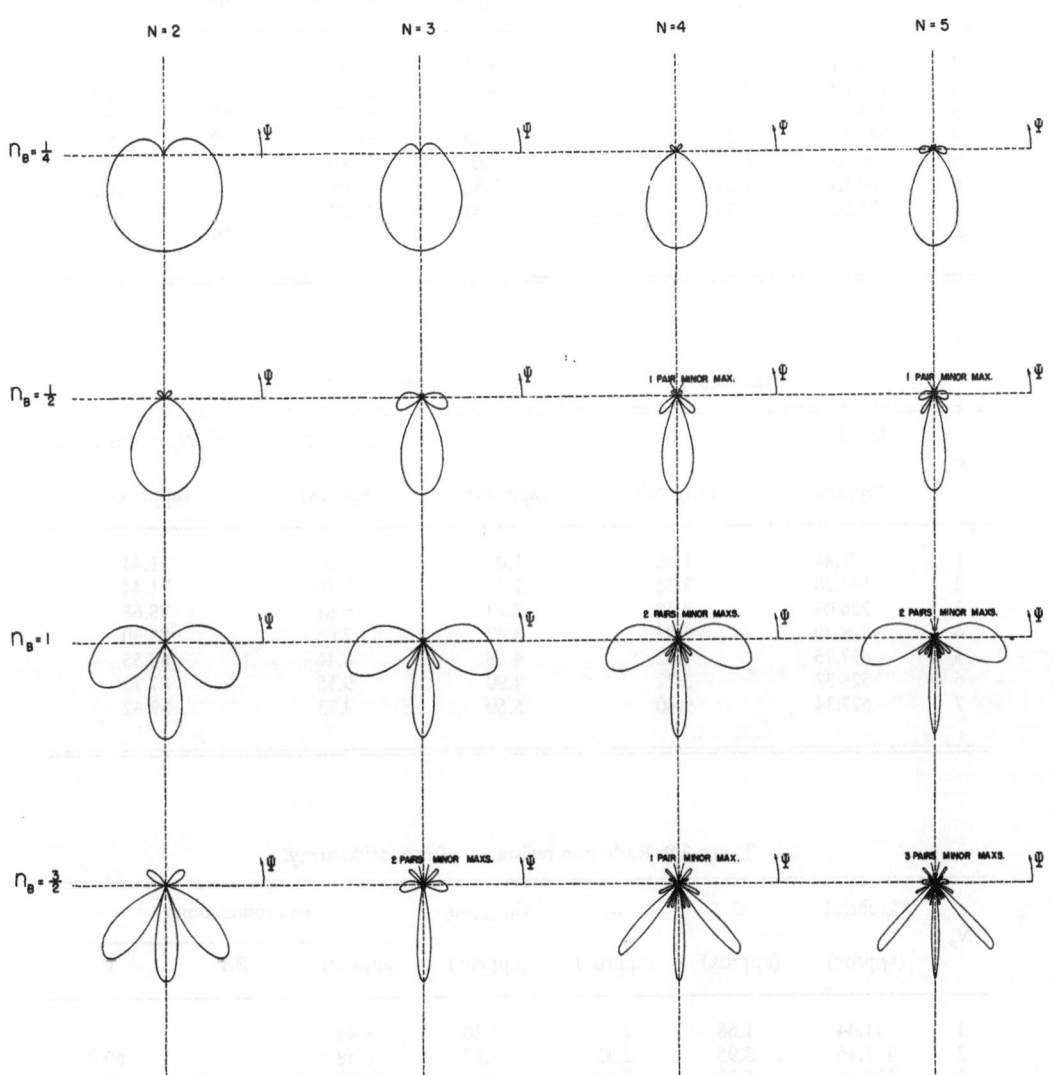

Fig. 3.11. Combination end-fire and broadside array with $a = a_E \cdot a_B$; polar graph; $a_E = \cos x$; $x = 180°(t - n \sin \Psi)$, $n_E = \frac{1}{4}$, $t_E = \frac{1}{4}$, $N_E = 2$, $a_B = (\sin N_B x)/N_B \sin x$; $x = n_B 180° \cos \Psi$; n_B, N_B, variables.

TABLE 3.7. Radiation resistance of bilateral end-fire array.

N_p	R_m^e(ohms) (approx)	D (approx)	D_r (approx)	G_{max}(db) (approx)	R_m^e/N_p (ohms/unit) (approx)	BB	P
1	71.44	1.68	1.0	2.20	71.44		
2	164.60	2.92	1.78	4.65	82.30	88	85.7
3	263.16	4.10	2.50	6.13	87.72	95	92.5
4	364.16	5.27	3.21	7.22	91.04	100	96.8
5	466.50	6.43	3.92	8.08	93.30		99.8
6	569.70	7.58	4.62	8.80	94.95		102.1
7	673.54	8.73	5.32	9.41	96.22		103.9
∞						120	

TABLE 3.8. Radiation resistance of unilateral end-fire array.

N_p	R_m^e(ohms) (approx)	D (approx)	D_r (approx)	G_{max}(db) (approx)	R_m^e/N_p (ohms/unit) (approx)
1	71.44	1.68	1.0	2.20	71.44
2	142.88	3.36	2.0	5.26	71.44
3	236.04	4.58	2.73	6.61	78.68
4	329.20	5.83	3.47	7.66	82.30
5	427.75	7.02	4.18	8.46	85.55
6	526.32	8.22	4.90	9.15	87.72
7	627.34	9.40	5.60	9.73	89.62

TABLE 3.9. Radiation resistance of broadside array.

N_p	R_m^e(ohms) (approx)	D (approx)	D_r (approx)	G_{max}(db) (approx)	R_m^e/N_p (ohms/unit) (approx)	BB	P
1	71.44	1.68	1	2.20	71.44		
2	121.16	3.95	2.35	5.97	60.58		60.9
3	176.34	6.12	3.64	7.87	58.78		59.5
4	229.08	8.38	4.98	9.23	57.27		58.0
5	283.15	10.6	6.32	10.25	56.63		57.4
6	336.42	12.8	7.63	11.07	56.07		56.9
7	390.25	14.7	8.76	11.67	55.75		56.5
∞						56	

[VI.3] THEORY OF LINEAR ANTENNAS

The principal maximum in each case is determined by

Broadside array:
$$x_m = 0, \quad \Phi_m = \pi/2, \quad (64a)$$

End-fire array:
$$x_m = 0, \quad \Phi_m = 0. \quad (64b)$$

The half-power points occur approximately at the following values (1.31):
$$x_1 \doteq \frac{1.39}{N} = \frac{79°.6}{N}, \quad (65)$$

so that

Broadside array:
$$\cos \Phi_1 \doteq \frac{1.39}{\pi n_B N_B} = \frac{0.443}{n_B N_B}, \quad (66a)$$

End-fire array:
$$\cos \Phi_1 \doteq 1 - \frac{0.443}{n_E N_E}. \quad (66b)$$

The half-power beam width is
$$W_1 = 2|\Phi_m - \Phi_1|, \quad (67)$$

so that for

Broadside array:
$$W_1 \doteq \left| \pi - 2 \cos^{-1} \frac{0.443}{n_B N_B} \right|, \quad (68a)$$

End-fire array:
$$W_1 \doteq 2 \cos^{-1}\left(1 - \frac{0.443}{n_E N_E}\right). \quad (68b)$$

In the usual broadside array, $n_B = \frac{1}{2}$; for the bilateral end-fire array, $n_E = \frac{1}{2}$; for the unilateral end-fire array $n_E = \frac{1}{4}$. Thus,

Broadside array:
$$n_B = \tfrac{1}{2}, \quad W_1 \doteq \left| 180° - 2 \cos^{-1} \frac{0.886}{N_B} \right|, \quad (69a)$$

End-fire array, bilateral:
$$n_E = \tfrac{1}{2}, \quad W_1 \doteq \left| 2 \cos^{-1}\left(1 - \frac{0.886}{N_E}\right) \right|, \quad (69b)$$

End-fire array, unilateral:
$$n_E = \tfrac{1}{4}, \quad W_1 \doteq \left| 2 \cos^{-1}\left(1 - \frac{1.772}{N_E}\right) \right|. \quad (69c)$$

Numerical values are in Tables 3.10–3.12.

The corresponding null beam widths are obtained from (2.34). Thus,

Broadside array:
$$n_B = \tfrac{1}{2}, \quad \Phi_m = \tfrac{1}{2}\pi, \quad \Phi_0 = \cos^{-1}(1/n_B N_B)$$
$$= \cos^{-1}(2/N_B), \quad (70a)$$

End-fire array, bilateral:
$$n_E = \tfrac{1}{2}, \quad \Phi_m = 0, \quad \Phi_0 = \cos^{-1}\left(1 - \frac{1}{n_E N_E}\right)$$
$$= \cos^{-1}\left(1 - \frac{2}{N_E}\right), \quad (70b)$$

End-fire array, unilateral:
$$n_E = \tfrac{1}{4}, \quad \Phi_m = 0, \quad \Phi_0 = \cos^{-1}\left(1 - \frac{4}{N_E}\right). \quad (70c)$$

The values of the null beam width are:

Broadside array:
$$n_B = \tfrac{1}{2}, \quad W_0 = 180° - 2 \cos^{-1}(2/N_B), \quad (71a)$$

End-fire array, bilateral:
$$n_E = \tfrac{1}{2}, \quad W_0 = \left| 2 \cos^{-1}\left(1 - \frac{2}{N_E}\right) \right|, \quad (71b)$$

End-fire array, unilateral:
$$n_E = \tfrac{1}{4}, \quad W_0 = \left| 2 \cos^{-1}\left(1 - \frac{4}{N_E}\right) \right|. \quad (71c)$$

The corresponding numerical values are given in Tables 3.10–3.12.

Study of the results in Tables 3.10–3.12 shows that the beam width of the broadside array is far narrower than that of the end-fire array, especially for large values of N. It follows that highly directional arrays should be broadside arrays. If a unidirectional array is required, a combination of one broadside row with a second broadside row $\lambda_0/4$ behind it is advantageous. Each pair of antennas in depth is a unilateral end-fire array. Since the beam width of a two-element unilateral end-fire array is very great, the beam width of the broadside-end-fire combination does not differ significantly from that of the single broadside row.

TABLE 3.10. Beam width (deg) of broadside array.

N_B	2	3	4	5	6	8	10	12	20	50
W_1	52.5	34.3	25.6	20.4	17.0	12.7	10.2	8.5	5.1	2.0
W_0	180	83.7	60.0	47.2	38.9	29.0	23.1	19.2	11.5	4.6

TABLE 3.11. Beam width (deg) of bilateral end-fire array.

N_E	2	3	4	5	6	8	10	12	20	50
W_1	112.3	90.3	77.7	68.0	61.0	54.5	48.6	44.4	34.2	21.6
W_0	180	141.1	120.0	106.3	96.3	82.8	73.7	66.9	51.7	32.5

TABLE 3.12. Beam width (deg) of unilateral end-fire array.

N_E	2	3	4	5	6	8	10	12	20	50
W_1	166.9	131.6	112.3	99.5	90.3	77.8	69.2	63.1	48.7	31.6
W_0	—	—	180	156.9	141.1	120.0	106.3	96.3	73.7	46.1

PARASITIC PARALLEL ARRAYS

4. *Field of Driven Antenna with Single Parasite—Approximate Second-Order Theory.*

In general, it is difficult to drive all elements in an array of many antennas in such a manner that specified currents are maintained, since the distributions of current and the impedances of identical antennas in different relative positions in an array are not the same. This is pointed out in Chapter III. Even if the elements are of half-length $h = \lambda_0/4$, so that zeroth-order distributions of current and self- and mutual impedances are fair approximations, the design of the required networks of transmission-line feeders is a formidable one. This is true, in particular, if phase relations between the several currents other than in-phase or 180° out-of-phase are required. The unilateral end-fire condition of quarter-wavelength spacing and 90° phase difference is especially difficult, since matched lines terminated in their characteristic impedances are required. Moreover, for point-to-point transmission the unilateral end-fire beam or the combination array consisting of two broadside rows forming unilateral end-fire pairs is of great importance.

An alternative to the unilateral end-fire pair is the driven antenna with a single parasite. This substitutes for the difficult problem of driving two antennas 90° out of phase from a single transmission line the problem of adjusting the relative magnitude and phase of the current in the parasite so that a unilateral field pattern is obtained. The disposable variables and parameters are (a) the distance b between the axes of the two antennas, (b) a tuning reactance X_2 at the center of the parasite, and (c) the half-length h_2 of the parasite. It was demonstrated in Sec. III.11 that the change in the self-impedance Z_{s2} of the parasite by an adjustment in its half-length may be approximated by keeping h_2 fixed and varying a tuning reactance X_2 provided h_2 is near $\lambda_0/4$. Since the usual arrangement is with $h_1 = \lambda_0/4$ and h_2 near $\lambda_0/4$, it is possible to determine the field characteristics of the most important combinations by setting $h_1 = h_2 = \lambda_0/4$ and varying X_2. This determines simultaneously the effect of varying X_2 with h_2 fixed and changing h_2 within reasonable limits with $X_2 = 0$.

Consider, therefore, the problem of determining the electromagnetic field of a center-driven antenna of half-length h and radius a at a distance b from an identical parasitic antenna, center-tuned by a reactance X_2. For simplicity, let the separation at the center of each antenna be negligible, so that $\delta \doteq 0$. The circuit is shown in Fig. 4.1. An appropriate expression is (3.29) with $k_m = 0$, and $m' = 1$. With some changes in notation the result is

$$E_\Theta^r = \frac{j\zeta_0 I_{10}}{2\pi} \frac{e^{-j\beta_0 R}}{R} F_0(\Theta, \beta_0 h) A(\Theta, \Phi), \quad (1)$$

where I_{10} is the current at the center ($z = 0$) of the driven antenna, No. 1 (not No. 0 as in Sec. 3), and $A(\Theta, \Phi)$ is the array factor, given by

$$A(\Theta, \Phi) = 1 + k' e^{-j(\delta' + \beta_0 b \sin\Theta \cos\Phi)}. \quad (2)$$

It follows from (III.11.4) that

$$k' = -\frac{Z_{21}}{Z_{22}}, \quad \delta' = \theta_{22} - \theta_{21}, \quad (3)$$

where $Z_{21} = Z_{21} e^{j\theta_{21}}$ is the mutual impedance of antenna 2 in the presence of antenna 1; $Z_{22} = Z_{22} e^{j\theta_{22}} = Z_{s2} + Z_2$; Z_{s2} is the self-impedance of antenna 2 in the presence of antenna 1; $Z_2 = jX_2$ is the tuning impedance at the center of antenna 2.

The field pattern in the horizontal or Φ-plane is given by the magnitude of (2) with $\Theta = \pi/2$, that is, by

$$A(\tfrac{1}{2}\pi, \Phi) = \left| 1 - \frac{Z_{21}}{Z_{22}} e^{-j(\theta_{22} - \theta_{21} + \beta_0 b \cos\Phi)} \right|. \quad (4)$$

The field in the equatorial plane in the direction from the parasite to the driven unit is given by (1) and (4) with $\Phi = 0$; the field in the direction from the driven unit to the parasite, by (1) and (4) with $\Phi = \pi$. The corresponding formulas are

$$A\left(\tfrac{1}{2}\pi, \begin{Bmatrix} 0 \\ \pi \end{Bmatrix}\right) = \left| 1 - \frac{Z_{21}}{Z_{22}} e^{-j(\theta_{22} - \theta_{21} \pm \beta_0 b)} \right|, \quad (5)$$

where the upper sign applies to the field away from the parasite, $\Phi = 0$, and the lower sign to the field toward the parasite, $\Phi = \pi$.

Since the input impedance of the driven antenna is a function of the distance b between the antennas, the current varies with b. Therefore, the behavior of the array is best studied by requiring *constant power input*.

Let the power P_1 to antenna 1, and hence to the array, be held constant at the value

$$P_1 = \tfrac{1}{2} I_{10}^2 R_{10}, \quad (6)$$

where R_{10} is the input resistance and I_{10} the magnitude of the input current in the driven antenna in the presence of the parasite. Note that R_{10} and I_{10} both vary with b. With (6) substituted in (1), the magnitude of the far-zone electric field of the array at constant power is

$$E_\Theta^r = \frac{\zeta_0}{R} \left\{ \frac{2P_1}{R_{10}} \left[1 + \frac{Z_{21}^2}{Z_{22}^2} - 2 \frac{Z_{21}}{Z_{22}} \cos(\theta_{22} - \theta_{21} \pm \beta_0 b) \right] \right\}^{\tfrac{1}{2}}. \quad (7)$$

The magnitude of the field of the driven antenna alone in the directions $\Theta = \pi/2$, $\Phi = 0, \pi$, when supplied with the same power P_1 is

$$E_\Theta^r = \frac{\zeta_0}{R} \sqrt{\frac{2P_1}{R_0}}, \quad (8)$$

where R_0 is the input resistance of the *isolated* driven antenna. Note that R_0 differs from R_{10}.

The ratio of the electric field of the array in the directions $\Phi = 0$ and π divided by the electric field of the driven antenna alone when supplied with the same power is obtained using (7) and (8). It is given by

$$\frac{(E_\Theta)_{\text{array}}}{(E_\Theta)_{1\,\text{unit}}} = \left\{ \frac{R_0}{R_{10}} \left[1 + \left(\frac{Z_{21}}{Z_{22}}\right)^2 - 2 \frac{Z_{21}}{Z_{22}} \cos(\theta_{22} - \theta_{21} \pm \beta_0 b) \right] \right\}^{\tfrac{1}{2}}, \quad (9)$$

where the upper sign is for $\Phi = 0$, the lower sign for $\Phi = \pi$.

The ratios defined by (9) with its two signs have been computed under the following conditions:

(a) Antenna 2 is geometrically identical with antenna 1, that is, the half-lengths are the same, so that $h_2 = h_1 = h$; and the radii are the same, so that, $a_2 = a_1 = a$. Values chosen for computation are $h = \lambda_0/4$; $\Omega = 2 \ln(2h/a) = 10$ and 20, so that $h/a = 75$ and 1.1×10^4; also $Z_{s2} = Z_{s1}$.

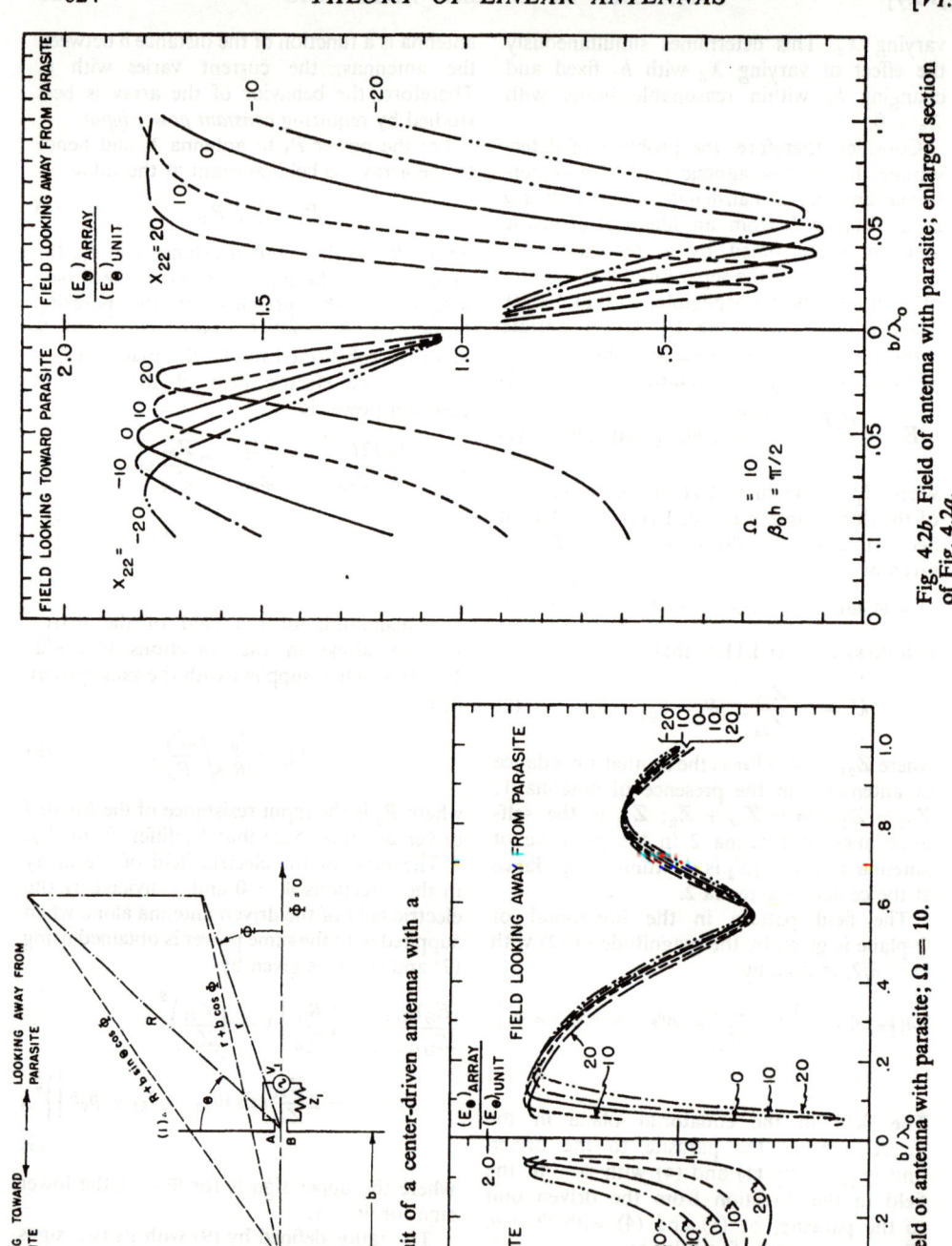

Fig. 4.1. Circuit of a center-driven antenna with a single parasite.

Fig. 4.2a. Field of antenna with parasite; $\Omega = 10$.

Fig. 4.2b. Field of antenna with parasite; enlarged section of Fig. 4.2a.

[VI.4] THEORY OF LINEAR ANTENNAS

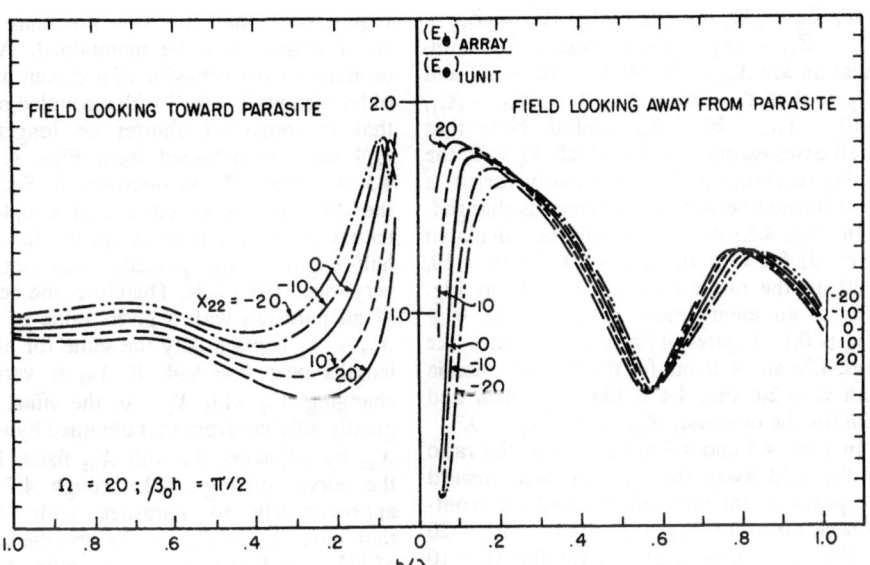

Fig. 4.3a. Field of antenna with parasite; $\Omega = 20$.

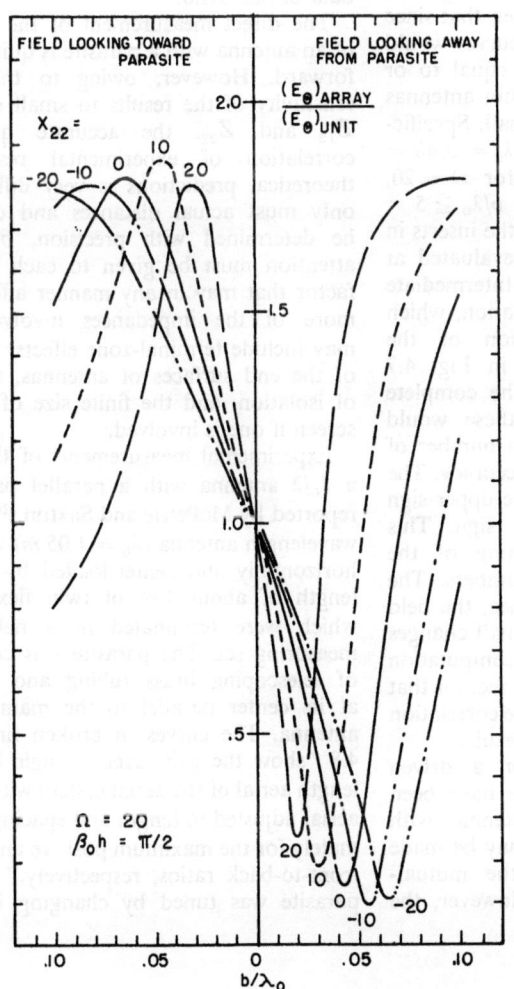

Fig. 4.3b. Field of antenna with parasite; enlarged section of Fig. 4.3a.

(b) $Z_2 = jX_2$, $R_2 = 0$, so that $Z_{22} = Z_{s2} + jX_2 = Z_{s1} + jX_2$. Values chosen for computation are $X_{22} = 20, 10, 0, -10, -20$, and X_{s1}, so that $X_2 = 20 - X_{s1}$, $10 - X_{s1}$, $-X_{s1}$, $-10 - X_{s1}$, $-20 - X_{s1}$, and 0. Note that in all cases except that for which $X_2 = 0$, the tuning reactance X_2 is continuously varied as the distance b between the antennas is changed.

In Fig. 4.2a are shown curves calculated from (9) for $\Omega = 10$, $X_{22} = 20, 10, 0, -10, -20$, in the range $0.05 \leq b/\lambda_0 \leq 1$; in Fig. 4.2b is an enlargement of the range $0 \leq b/\lambda_0 \leq 0.1$. Figures 4.3a and 4.3b are like Figs. 4.2a and 4.2b but for the thinner antenna with $\Omega = 20$. Fig. 4.4 is like Figs. 4.2a and 4.2b for the one case, $X_{22} = X_{s2}$, $X_2 = 0$.

In Figs. 4.5 and 4.6 are curves of the ratio of the field away from to the field toward the parasitic antenna (often called the front-to-back ratio) for $X_{22} = 20, 10, 0, -10, -20$ in the range $0.01 \leq b/\lambda_0 \leq 1.0$ for $\Omega = 10$ and 20. The curves in the inserts in Figs. 4.5 and 4.6 might be extrapolated through the point 1 at $b/\lambda_0 = 0$. Note, however, that since b must always exceed $2a$, these curves would be meaningless for values of b equal to or smaller than $2a$. For relatively thin antennas the value $b/\lambda_0 = 2a/\lambda_0$ is very small. Specifically, for $h = \lambda_0/4$, $\Omega = 10$, $a/\lambda_0 = a/4h = 1/300$, so that $b/\lambda_0 \geq 0.0067$; for $\Omega = 20$, $a/\lambda_0 = (1/4.4) \times 10^{-4}$, so that $b/\lambda_0 \geq 5 \times 10^{-5}$. In computing the curves in the inserts in Figs. 4.5 and 4.6, points were evaluated at intervals of 0.01 on the b/λ_0 scale. Intermediate points were obtained by interpolation, which permitted accurate determination of the location of the sharp minima in Fig. 4.7 but not of their magnitudes. The complete and accurate computation of these would require the evaluation of a large number of points to a very high degree of accuracy. The field ratio is given by (9) with the upper sign divided by (9) with the lower sign. This involves a denominator consisting of the small difference of two large numbers. The fact that this difference and, hence, the field ratio are extremely sensitive to small changes in Z_{12} or Z_{22} makes the accurate computation difficult. It is shown later in this section that this fact is also of importance in the correlation of theoretical with experimental results.

Although numerical data for a driven antenna with a coupled parasite have been computed only for identical antennas with $h = \lambda_0/4$, similar computations may be made for other values of h using the mutual-impedance data of Sec. III.8. However, the requirement that the two antennas be of equal length must be maintained. A good estimate of the behavior of a driven antenna of half-length $h = \lambda_0/4$ with a coupled parasite that is somewhat shorter or longer than $\lambda_0/4$ may be obtained from Figs. 4.2, 4.3, 4.5, 4.6, and 4.7. As discussed in Sec. III.6, the distributions of current in coupled antennas differ very little except in phase if the half-length of the parasite does not differ very much from $\lambda_0/4$. Therefore, the behavior of all parasites with a given value of $X_{22} = X_{s2} + X_2$ is essentially the same for all half-lengths near $h = \lambda_0/4$. If X_{22} is varied by changing X_{s2} with $X_2 = 0$, the effect is not greatly different from that obtained by varying X_{22} by adjusting X_2 with X_{s2} fixed. Hence, the curves of Figs. 4.2 through 4.7 apply approximately to parasites with $X_2 = 0$ that have $X_{s2} = X_{22}$ at the specified values of b/λ_0. The lengths h_2 corresponding to these X_{s2} may be obtained using the self-impedance data of Sec. III.8.

The direct measurement of the field ratio of an antenna with a parasite is quite straightforward. However, owing to the extreme sensitivity of the results to small changes in Z_{12} and Z_{22}, the accurate quantitative correlation of experimental results with theoretical predictions is very difficult. Not only must actual distances and dimensions be determined with precision, but careful attention must be given to each and every factor that may in any manner affect one or more of the impedances involved. These may include terminal-zone effects, the nature of the end surfaces of antennas, the degree of isolation, and the finite size of a ground screen if one is involved.

Experimental measurements of the gain of a $\lambda_0/2$ antenna with a parallel parasite are reported by McPetrie and Saxton.[31a] The half-wavelength antenna ($\lambda_0 = 1.05$ m) was placed horizontally and center-loaded by a vertical length of about 1 m of twin flexible leads which were terminated in a field-strength measuring set. The parasite was constructed of telescoping brass tubing and supported at its center parallel to the main receiving antenna. The curves in broken line in Fig. 4.8 "show the gain over a single half-wavelength aerial of the aerial system with parasitic aerial adjusted in length and spacing approximately for the maximum positive and negative front-to-back ratios, respectively." Since the parasite was tuned by changing its length, corresponding theoretical curves are unavailable. However, it is interesting to compare the experimental results in which $X_{22} = X_{s2}$ was varied by changing h_2 near $\lambda_0/4$ with theoretical curves obtained by keeping $h_2 = h_1 = \lambda_0/4$ and varying $X_{22} = X_{s1} + X_2$ by changing X_2. Since McPetrie and Saxton give no precise data on the length of the parasite and its distance from the receiving antenna, and since an "approximate adjustment" of h_2 and b to give maximum front-to-back ratio is very broad, only a qualitative comparison is possible. Curves taken from Figs. 4.2a and 4.2b with $X_2 = -10$ and -20 are shown in Fig. 4.8 in solid line. Although not strictly comparable, the general nature of theoretical and experimental curves is seen to be the same.

More extensive data on the front-to-back ratio as a function of the distance b between a center-loaded receiving antenna and a parasite with the length of the parasite as parameter are given by McPetrie and Saxton[31a] and by Starkey and Fitch.[48] A rather complete set of data for three values of the ratio h_1/a due to the latter investigators is shown in Fig. 4.9a; the results of McPetrie and Saxton are shown in Fig. 4.9b. These curves correspond to the theoretical results in Fig. 4.7 in that b/λ_0 is the variable and $X_{22} = X_{s2} + X_2$ the parameter. However, in Figs. 4.9a, b, $X_2 = 0$ and $X_{22} = X_{s2}$ is varied by changing the length $2h_2$ of the parasite, whereas in Fig. 4.7, $X_{s2} = X_{s1}$ is the fixed value for $h_1 = h_2 = \lambda_0/4$ and X_{22} is varied by changing X_2. Note that $X_{s2} = X_{s1}$ is not constant but varies somewhat with b/λ_0, so that a fixed value of X_{22} implies a continuous readjustment of X_2 as b/λ_0 is varied. In general, the agreement between Figs. 4.9a, b and 4.7 is quite good. Note in particular that the location of the deepest minimum is very near $b/\lambda_0 = 0.05$ and that it is shifted toward slightly higher values of b/λ_0 as the antenna is made thinner. The magnitudes of the minima in Figs. 4.9a and 4.9b differ by about 9 decibels. This rather large difference in experimental observations is ascribed to an improved signal-to-noise ratio in the more recent work of Starkey and Fitch.

Strictly, the only curves in Figs. 4.9a, b that may be compared directly with theoretical ones in Fig. 4.7 are those with $h_1 = h_2 = \lambda_0/4$. These should correspond to the theoretical curves marked $X_{22} = X_{s1}$ in Fig. 4.7 except for differences in the thicknesses of the antennas. For more convenient comparison the several curves from Figs. 4.9a, b and 4.7 for which h_1 and h_2 are both equal to $\lambda_0/4$ are reproduced in Fig. 4.10. Since there was no curve in Fig. 4.9b with h_2 exactly equal to $\lambda_0/4$, a curve was constructed by interpolating between the adjacent curves for h_2 very slightly above and slightly below this value. The two theoretical curves marked $\Omega = 10$ and $\Omega = 20$ indicate a higher maximum for the thicker antenna and a considerably flatter curve for the thinner antenna. The behavior of the four experimental curves is peculiar and, on the whole, inconsistent. McPetrie and Saxton's curve for $\Omega = 10.5$ is much higher than Starkey and Fitch's for $\Omega = 10.7$. Moreover, the three curves of Starkey and Fitch are not self-consistent in the sense that the curve for the thickest antenna ($\Omega = 9.3$) is higher than the curve for the antenna of medium thickness ($\Omega = 10.7$), in general agreement with the sequence of the theoretical curves, whereas the curve for the thinnest antenna ($\Omega = 12.9$) is well above both of the others for small values of b/λ_0, but below these for larger values. However, the curve for the thinnest antenna is flatter than for the thick ones, in agreement with the theoretical curves. Starkey and Fitch assumed that the failure of the theoretical curves to predict the peculiar and inconsistent behavior of the experimental ones must be a fault of the theory. Therefore, they proceeded to develop a new theoretical formula using an "engineering" method that actually has little scientific basis.

With due regard for the unexplained peculiarities in the experimental results, it would appear that the explanation for the disagreement between theoretical and experimental curves in Fig. 4.10 is not a fundamental inadequacy of the theory but, rather, a failure to coördinate *correctly* the quantities computed theoretically with the quantities measured experimentally. Did McPetrie and Saxton and Starkey and Fitch actually measure the front-to-back ratios for the simple structures assumed in the theory? Were their experimental arrangements such that the necessary degree of isolation obtained? Were transmission-line end effects and coupling effects really negligible? Were the effects of the end surfaces and edges on the flat-ended cylinders and tubes sufficiently small to be ignored? In view of the great sensitivity of the front-to-back ratio to small changes in the self- and mutual reactances, it appears quite probable that some or all of these effects

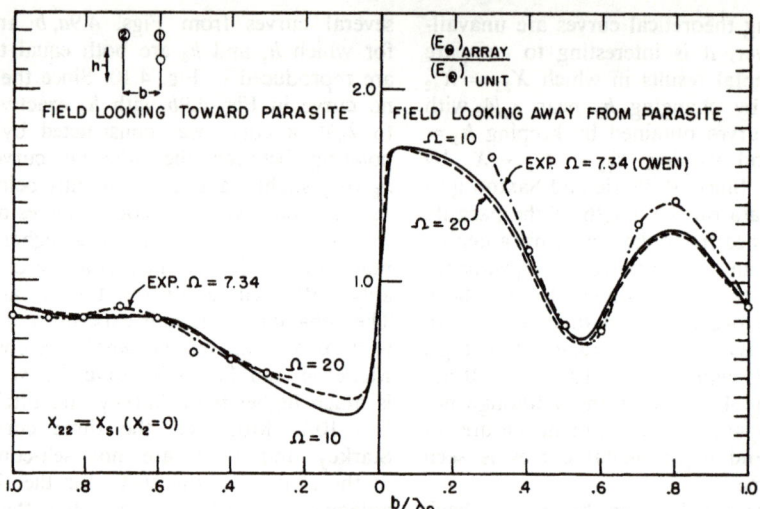

Fig. 4.4. Field of antenna with untuned parasite.

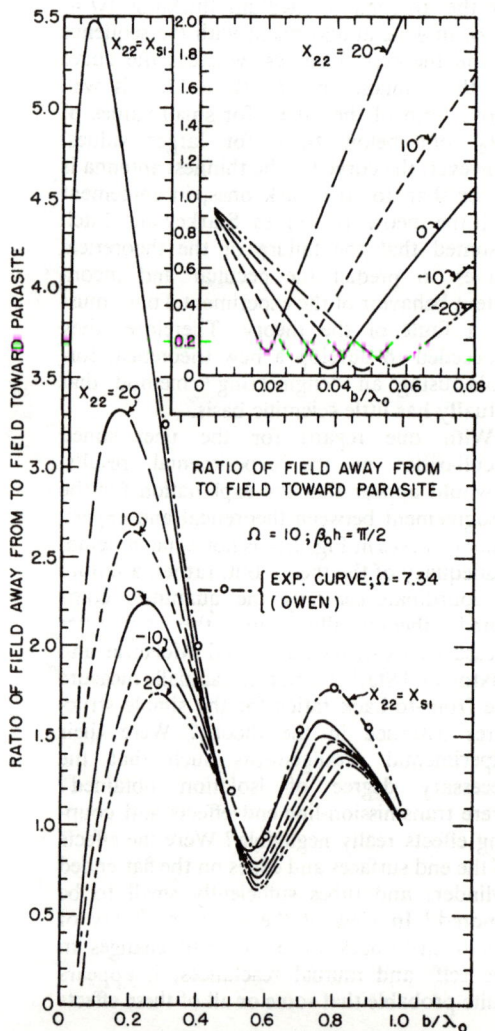

Fig. 4.5. Ratio of field away from to field toward parasite; $\Omega = 10$.

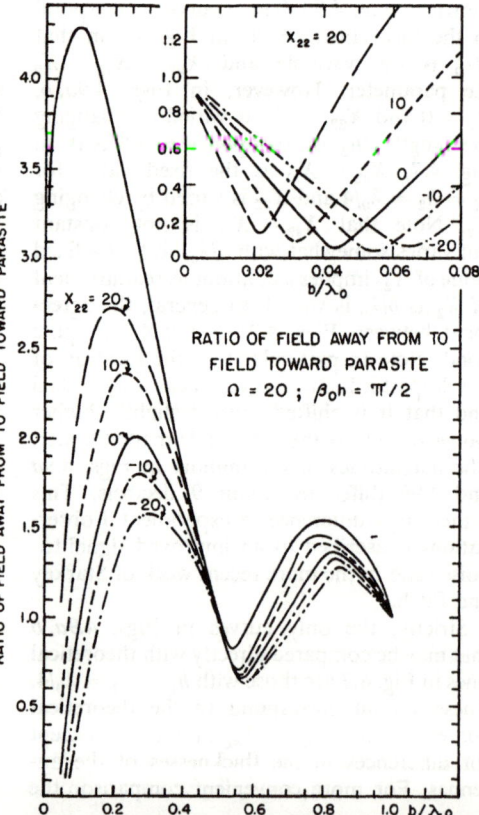

Fig. 4.6. Ratio of field away from to field toward parasite; $\Omega = 20$.

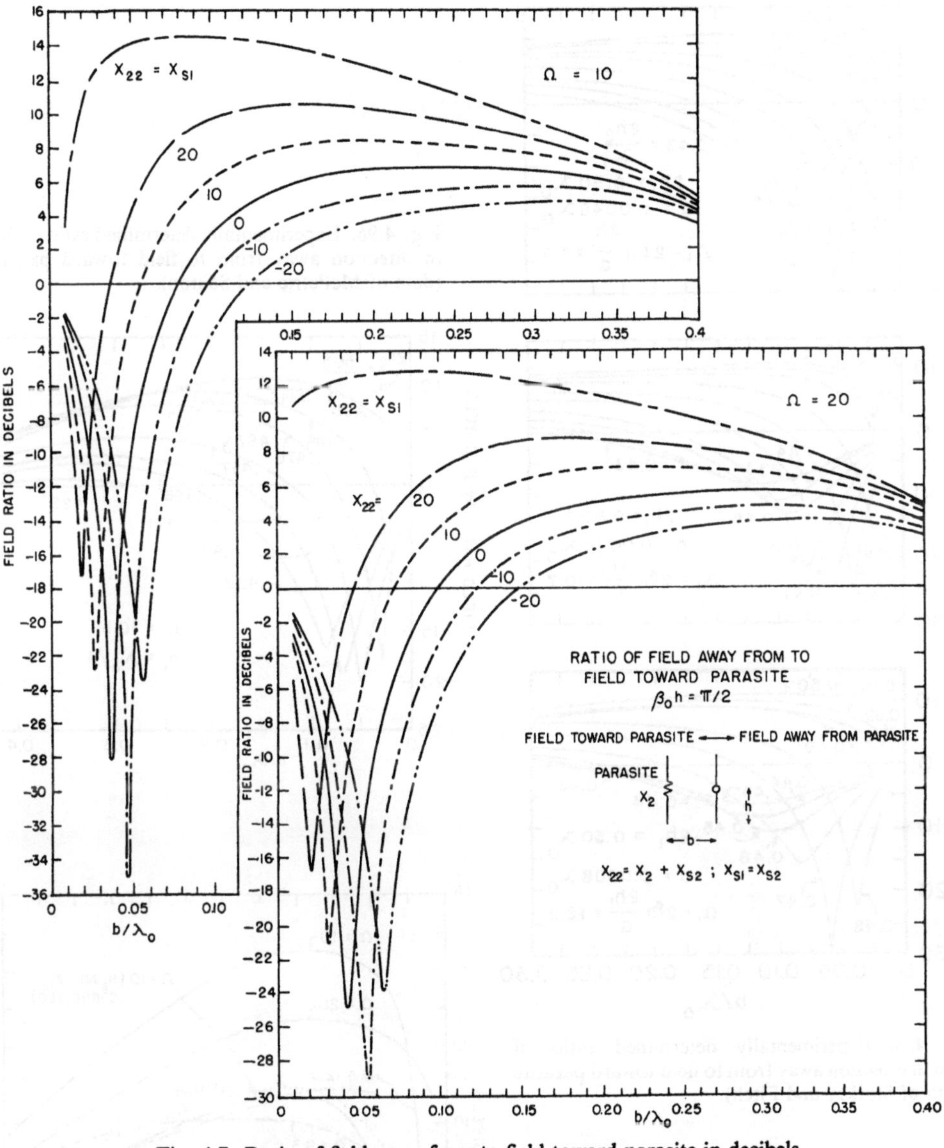

Fig. 4.7. Ratio of field away from to field toward parasite in decibels.

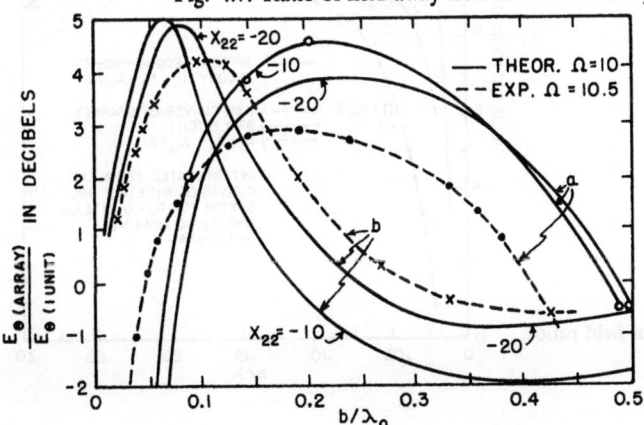

Fig. 4.8. Gain of dipole (a) with reflector and (b) with director. Theoretical curves with $X_{22} = X_{s1} + X_2$ fixed and $h_2 = h_1$. Experimental curves with $X_2 = 0$, h_2 adjusted for approximate maximum at optimum b/λ_0. (Experimental curves by McPetrie and Saxton.)

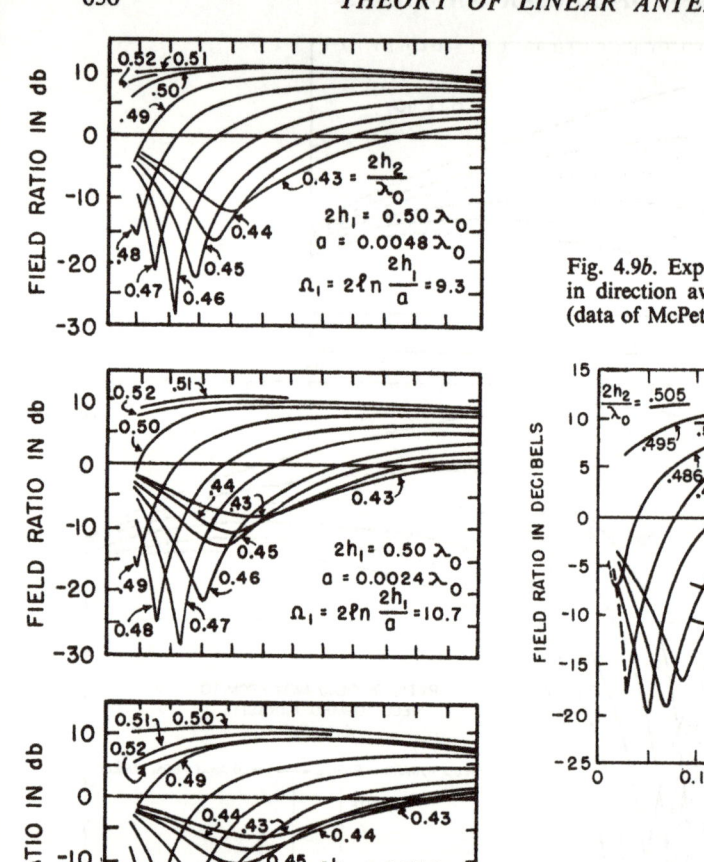

Fig. 4.9b. Experimentally determined ratio of field in direction away from to field toward parasite (data of McPetrie and Saxton).

Fig. 4.9a. Experimentally determined ratio of field in direction away from to field toward parasite (data of Starkey and Fitch).

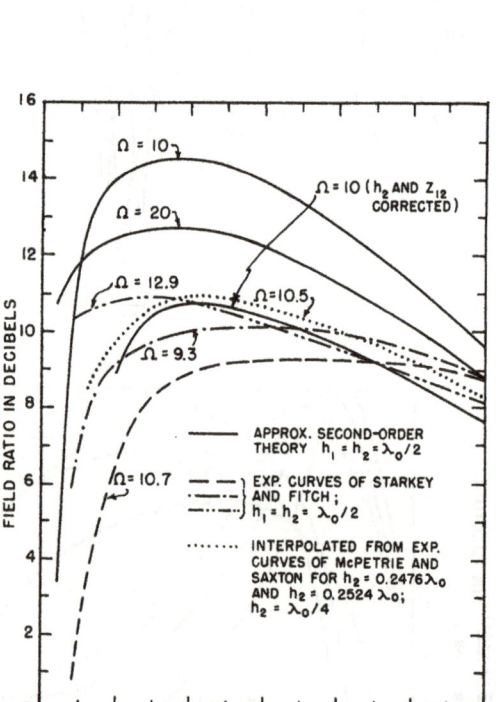

Fig. 4.10. Comparison of field ratios.

[VI.4] THEORY OF LINEAR ANTENNAS 631

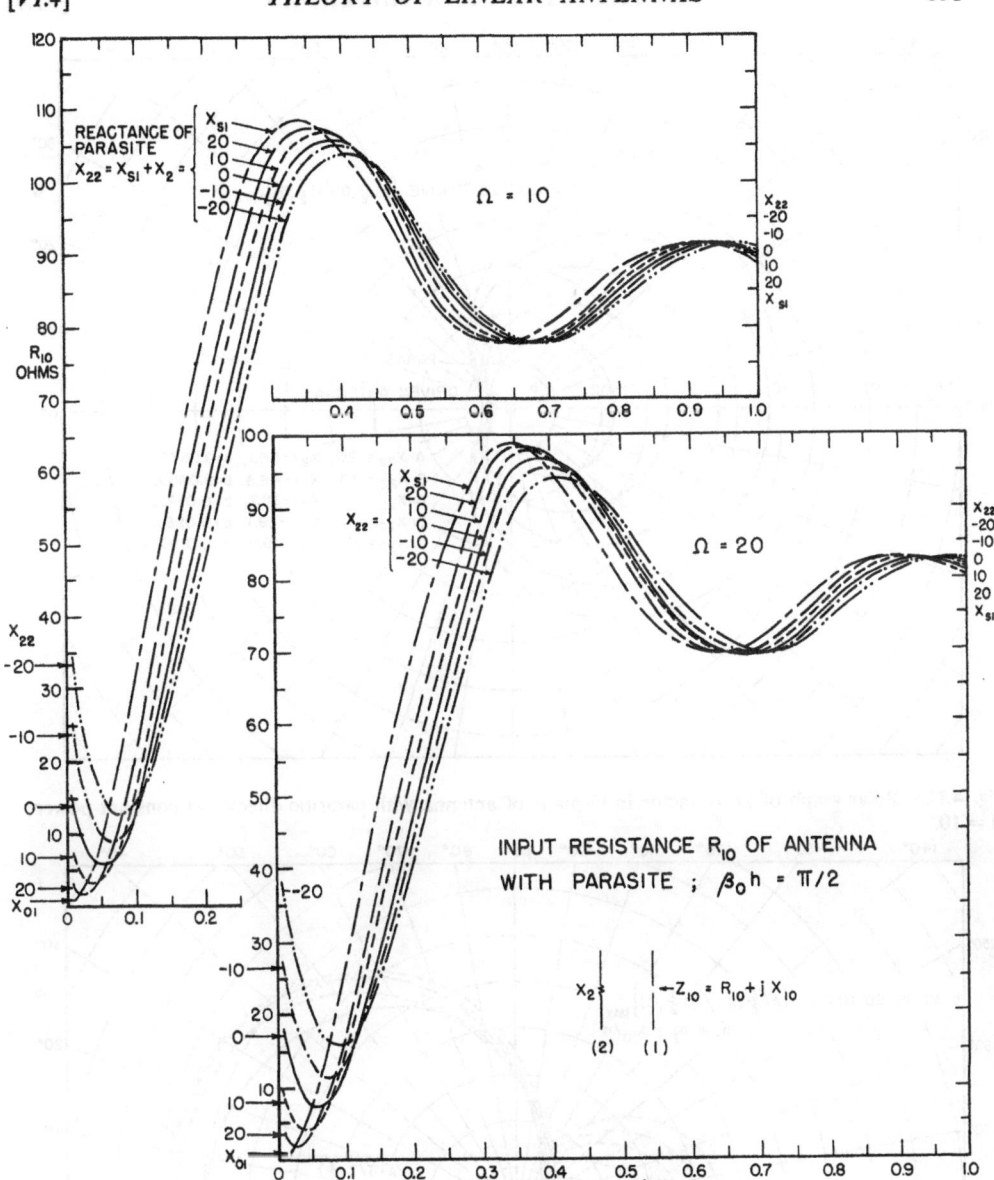

Fig. 4.11. Input resistance R_{10} of antenna with parasite; $\beta_0 h = \pi/2$.

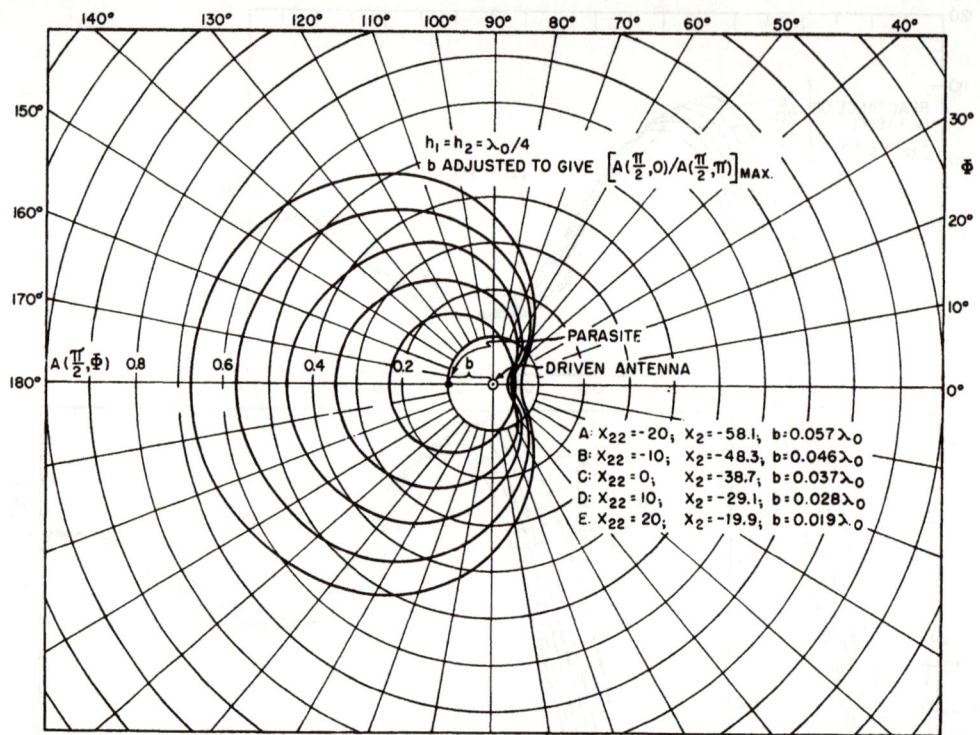

Fig. 4.12a. Polar graph of array factor in Φ-plane of antenna with parasitic director at constant power; $\Omega = 10$.

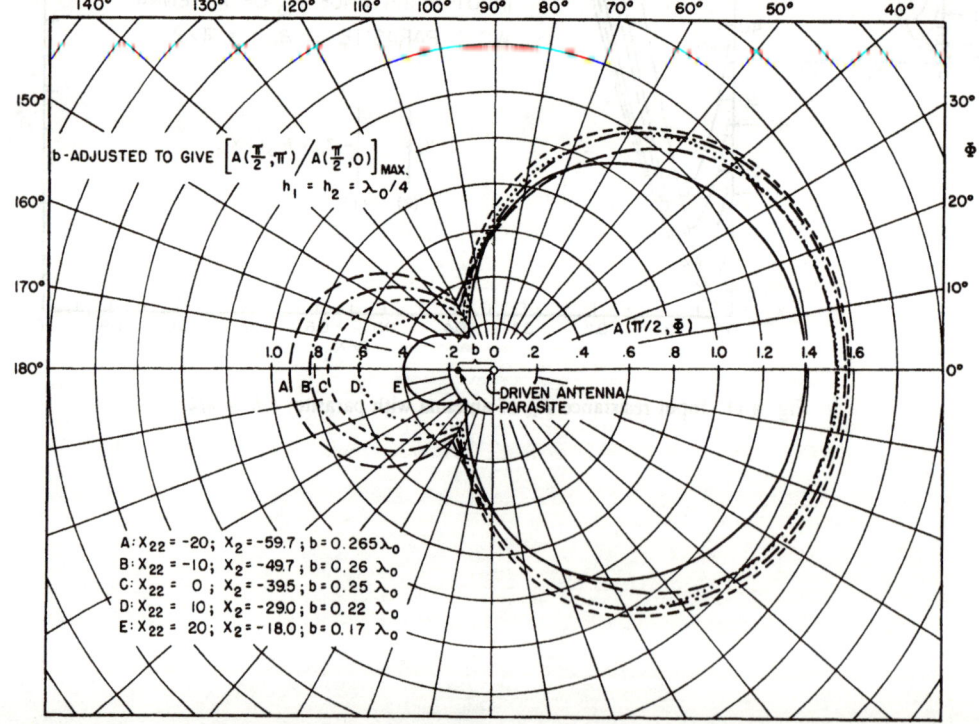

Fig. 4.12b. Polar graph of array factor in Φ-plane of antenna with parasitic director at constant power; $\Omega = 10$.

contribute to explain the peculiarities in the experimental data. Indeed, Starkey and Fitch remark on this sensitivity and note that "even the movements of leaves in a tree several wavelengths away could sometimes affect the front-to-back ratios." Unfortunately, Starkey and Fitch do not describe the experimental conditions in detail, so that the distances from the antennas to the earth and to the measuring gear are not available. McPetrie and Saxton state that the field-strength meter was at a distance of only about 1 wavelength from the center of the main receiving antenna, but do not state whether the earth is also as near as this. Owing to the unusual sensitivity of the front-to-back ratio to small changes in reactance, it would appear that at least 5 wavelengths should be the minimum distance to the nearest object. Although it is not possible to make a quantitative estimate of the possible effect of inadequate isolation, the quantitative significance of other pertinent factors can be estimated. Starkey and Fitch used brass tubes for their antennas and apparently made no correction in length to take account of the chargeable surfaces in the interior of the hollow and open ends. These are equivalent to an increase in length of the order of magnitude of the radius of the tube, so that the effective half-length of the antenna is not h but $h + ka$, where k is near unity. For very thin antennas this correction is negligible, but for $\Omega = 10$ or less it certainly is not. Thus, instead of using $Z_{22} = Z_{s2}$ for $\beta_0 h_2 = \pi/2 = 1.571$, the value to be used is $Z_{22} = Z_{s2}$ for $\beta_0(h_2 + ka)$. For $\Omega = 10$ a satisfactory value is Z_{s2} for $\beta_0(h_2 + ka) \doteq 1.60$. The theoretical curves in Fig. 4.7 and 4.10 are computed on the assumption that the spacing of the transmission lines is zero. In the experiments, the principal antenna was connected to a two-wire line with small but presumably not negligible spacing. As a result, a short section of current is absent at the center of the actual antenna precisely where the current in the theoretical antenna is greatest. In so far as the voltage induced in the parasite is concerned, the mutual impedance is not that of an antenna of electrical half-length $\beta_0 h_1 = \pi/2$ but approximately that of an antenna with electrical half-length $\beta_0(h - \delta)$, where δ is half the spacing of the transmission line. Since the actual numerical values are unavailable, it will serve to assume δ to be 0.5 cm in a wavelength of about 1 m. That is, the effective electrical half-length of antenna 1 for determining the *mutual* impedance is $\beta_0(h - \delta) \doteq 1.54$ instead of $\beta_0 h = \pi/2$. If the theoretical curve marked $\Omega = 10$ in Fig. 4.10 is recomputed using $Z_{22} = Z_{s2}$ for an electrical half-length 1.60 and Z_{12} for an electrical half-length 1.54 instead of $\pi/2$ for both, the curve is seen to be in good agreement with and even lower than the experimental curve obtained from the data of McPetrie and Saxton. Something more than half of the downward shift is due to the correction for the missing length 2δ; the remainder is due to the correction for the end surfaces. Since this latter effect is negligible for $\Omega = 20$, the curve $\Omega = 20$ need be corrected only for the gap. This moves it down very near to the shifted $\Omega = 10$ curve.

No claim is made that the corrections suggested above are precisely the right ones or the only ones involved. What is claimed, however, is that the sensitivity of the front-to-back ratio to small changes in the impedances is so great that what might appear to be entirely minor details actually can account for all of the differences between theory and experiment and for the apparent inconsistencies in the experimental data. If *all* of the conditions of an experiment are taken into account in the theoretical analysis, or if an experimental arrangement can be so simplified that the ideal assumptions of the theory are really good approximations, agreement between theory and experiment can be expected, but not otherwise. Any theory that agrees with experiment without having taken account of *all significant* factors is inadequate.

The approximate second-order input resistance R_{10} of an array consisting of a driven antenna in the presence of a coupled parasitic antenna is given in Sec. III.11 with $X_{22} = X_{s1}$. For other values of X_{22} the resistance R_{10} is readily determined using the self- and mutual-impedance data of Sec. III.8; R_{10} is shown in Fig. 4.11 with b/λ_0 as variable and X_{22} as parameter. With R_0 from Sec. II.30, R_{10} from Fig. 4.11 and other figures of this section, the electromagnetic field of an antenna with parasite is determined readily under a variety of conditions, including especially the use of a parasitic antenna (1) as a reflector with maximum field in the direction away from the parasite, $\Phi = 0$, and a reduced or near minimum field in the direction $\Phi = \pi$; or (2) as a director with maximum field in the direction toward the parasite, $\Phi = \pi$, and a reduced or near minimum field in the direction $\Phi = 0$.

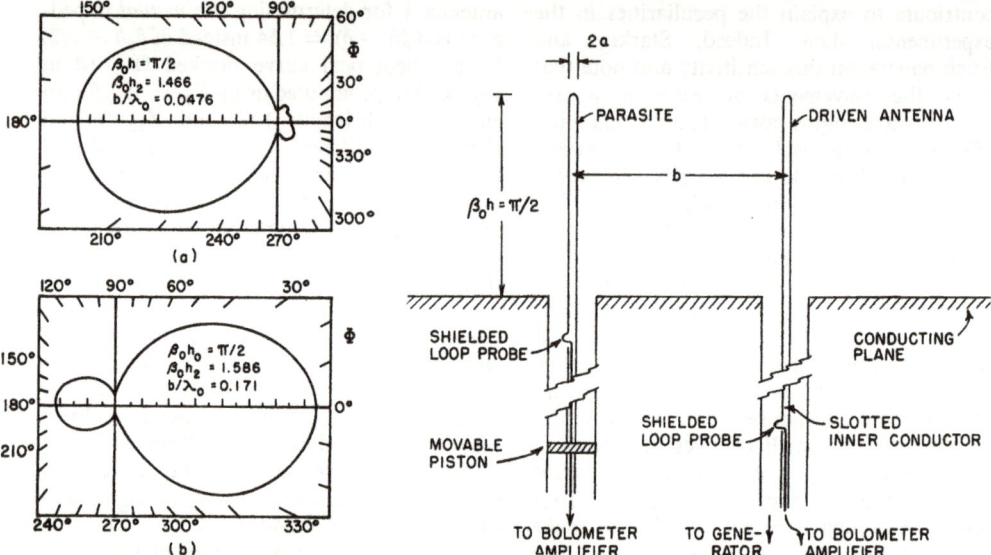

Fig. 4.13. Experimentally determined field of antenna with parasite (a) as director, (b) as reflector (McPetrie and Saxton).

Fig. 4.14. Arrangement used to measure the field of antenna with one parasite.

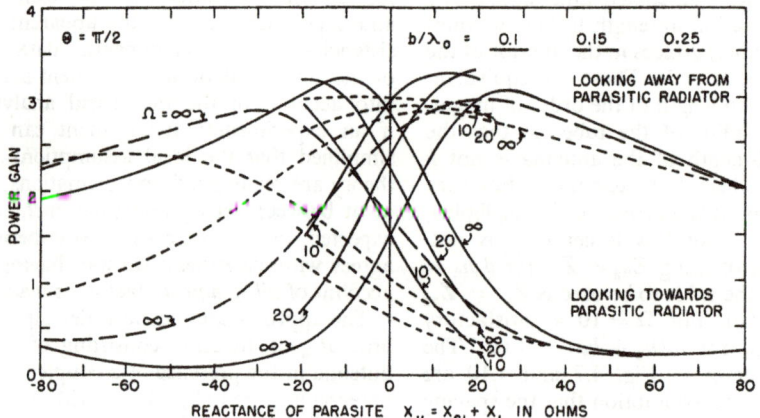

Fig. 5.1. Power gain of antenna with one parasite.

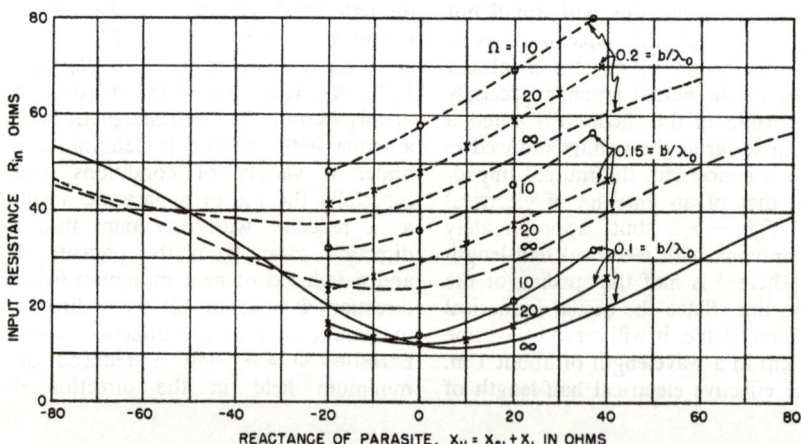

Fig. 5.2. Reactance of parasite, $X_{11} = X_{s1} + X_1$ in ohms.

The distances between antennas for maximum or minimum field in either direction are readily obtained or interpolated from Figs. 4.2a, b, 4.3a, b, or 4.4 for a range of values of total reactance X_{22} of the parasitic antenna and its tuning circuit and different values of a/h. Similarly, from Figs. 4.5, 4.6, or 4.7 the spacing for maximum forward or backward field may be obtained.

In Figs. 4.12a, b the magnitude $\frac{1}{2}A(\pi, \Phi)$ from (4) is plotted as a function of the angle Φ for $X_{22} = 20, 10, 0, -10, -20$ and with those values of b/λ_0 for which the ratio of the field in the direction $\Phi = 0$ to the field in the direction $\Phi - \pi$ is a minimum or a maximum. Experimental curves due to McPetrie and Saxton under similar conditions but with X_{22} adjusted by changing the length of the parasite are shown in Fig. 4.13.

A direct experimental measurement of the radiation field in the directions away from and toward the parasite as a function of the distance b between the parasitic and the driven antennas was made using the arrangement illustrated in Fig. 4.14. The two antennas were identical and a quarter wavelength long. The movable piston in the tuning line of the parasite was adjusted to have a current maximum at the conducting plane using a shielded-loop probe traveling in a slot along the inner conductor of the coaxial line. In this manner X_2 was made zero. The spacing b was varied in steps by inserting the parasite with its entire coaxial-line assembly into a series of holes that could be plugged to leave the surface of the large conducting plane smooth. The field at a distant point in both directions along a line through both antennas was measured and compared with the field with the parasite removed. The frequency was 600 Mhz. For the antennas $\Omega = 7.34$. The results obtained by W. E. Owen in a carefully performed laboratory experiment in a graduate course on antennas are included in Figs. 4.4 and 4.5. Since the effect of changes in Ω on the quantities plotted is relatively small, a comparison between the theoretical curves for $\Omega = 10$ and the experimental ones for $\Omega = 7.34$ is possible. The agreement is seen to be quite good, with the $\Omega = 7.34$ curve shifted in the proper direction from the $\Omega = 10$ curve. Note that no correction for terminal-zone effects is necessary, since the small negative capacitance required would be in parallel with the low impedance of a nearly resonant antenna, and the distance between the two antennas was always large compared with the diameter of the coaxial-line feeder.

5. Field of Driven Antenna with Several Parasites—Zeroth-Order Solution; Arrays of Yagi-Uda Type

The analysis of a linear array consisting of a single driven antenna with from one to four parasites has been carried out in considerable detail by Walkinshaw[57] under the assumption that sinusoidal distributions of current and modified zeroth-order impedances are a satisfactory approximation. Since this is true only for elements with electrical half-lengths $\beta_0 h$ near $\pi/2$, attention is directed primarily to arrays with such elements.

The circuit equations of an N-element array are given in (III.20.3). In order to be consistent with the figures* used in this section it is convenient to number the N antennas from 0 to $N - 1$ instead of from 1 to N. The driven antenna is number 0. The equation with the renumbered antennas are

$$V_0 = \sum_{m=0}^{N-1} I_{m0} Z_{0m}, \quad (1a)$$

$$0 = \sum_{m=0}^{N-1} I_{m0} Z_{im},$$
$$i = 1, 2, 3, \cdots, N - 1. \quad (1b)$$

The input impedance of the driven unit is

$$Z_{in} = \frac{V_0}{I_{00}}$$

$$= \begin{vmatrix} Z_{11} Z_{12} & \cdots & Z_{1N} \\ Z_{21} Z_{22} & \cdots & Z_{2N} \\ \vdots & & \vdots \\ Z_{N1} Z_{N2} & \cdots & Z_{NN} \end{vmatrix} \div \begin{vmatrix} Z_{00} Z_{01} & \cdots & Z_{0N} \\ Z_{10} Z_{11} & \cdots & Z_{1N} \\ \vdots & & \vdots \\ Z_{N0} Z_{N1} & \cdots & Z_{NN} \end{vmatrix}. \quad (2)$$

The radiation field of an array of N units that are all near $\beta_0 h = \pi/2$ in electrical half-length and in which the currents are not necessarily equal but are all distributed sinusoidally along the axes may be expressed in the same form as for a uniform array. Thus,†

$$E_\Theta^r = \frac{j\zeta_0 I_{00}}{2\pi} \frac{e^{-j\beta_0 R_0}}{R_0} F_0(\Theta, \beta_0 h) A(\Theta, \Phi), \quad (3)$$

---

* Most curves in this section are taken from W. Walkinshaw, ref. 57.
† Roman R is used for distances, italic R for resistances in this section.

where I_{00} is the current at the center ($z = 0$) of the driven element number 0 when there are N units in the array arranged at equal distances between centers along the x-axis; R_0 is the distance from the origin of coördinates at the center of the driven element to the distant point P where the field is calculated; the angle between the z-axis along the axis of the driven unit and R_0 is Θ; the angle Φ is measured from the positive x-axis to the projection of R_0 onto the x, y-plane. For $\beta_0 h = \pi/2$,

$$F_0(\Theta, \beta_0 h) = \frac{\cos(\tfrac{1}{2}\pi \cos \Theta)}{\sin \Theta}. \quad (4)$$

The array factor cannot be expressed in the simple compact form characteristic of uniform arrays because the several currents are not equal in magnitude or uniformly spaced in phase. The array factor is

$$A(\Theta, \Phi) = 1 + \sum_{m=1}^{n} k_m e^{-j\beta_0 S_m} + \sum_{m=n+1}^{N-1} k_m e^{j\beta_0 S_{m-n}}, \quad (5)$$

where

$$k_m = I_{m0}/I_{00}, \quad (6)$$

$$S_m = mb \sin \Theta \cos \Phi, \quad (7)$$

and b is the distance between centers of the equispaced elements. Of the total number N, elements 1 to n are along the negative, elements $n + 1$ to $N - 1$ along the positive x-axis; *n is arbitrary.*[*] Note that the positive x-axis coincides with $\Phi = 0$, the negative x-axis with $\Phi = \pi$. The actual distances from the point P to the centers of the elements are

$$m \leq n: \quad R_m = R_0 + S_m, \quad (8a)$$

$$m > n: \quad R_m = R_0 - S_{m-n}. \quad (8b)$$

In order to compare the field of an N-element array with that of the driven unit alone when isolated, it is advantageous to require constant power input to the driven unit. The power supplied to the array through the terminals of driven unit is

$$P_N = \tfrac{1}{2} I_{00}^2 R_{in}, \quad (9a)$$

where R_{in} is the real part of (2). The power supplied to the driven unit if isolated is

$$P_1 = \tfrac{1}{2} I_0^2 R_0, \quad (9b)$$

where R_0 is the resistance of an isolated antenna. For constant power, $P_N = P_1$, so that

$$I_0/I_{00} = (R_{in}/R_0)^{1/2}. \quad (10)$$

---

[*] Note that n is not $n_p = b/\lambda_0$.

The electric field of the driven antenna when isolated is

$$E_\Theta^r = \frac{j\zeta_0 I_0}{2\pi} \frac{e^{-j\beta_0 R_0}}{R_0} F_0(\Theta, \beta_0 h). \quad (11)$$

In comparing the electric field of linear arrays of parasitic elements with the field of a single driven antenna, Walkinshaw introduces the following quantities: (1) the power gain in the directions $\Phi = \pi$ and $\Phi = 0$ of the N-element array over a half-wave dipole; note that the power gain in the *direction of maximum field*, that is, the *relative directivity*, D_r, of the array, usually occurs in one of these directions; (2) the ratio of the magnitude of the electric field of the array to the field in the *equatorial plane*, $\Theta = \pi/2$, of a half-wave dipole radiating the same power.

These quantities are defined analytically as follows:

$$\text{Power gain} \equiv \frac{(E_\Theta^r)^2_{\text{array}}}{(E_\Theta^r)^2_{\text{1 unit}}} \left(\Theta = \frac{\pi}{2}, \Phi = \begin{Bmatrix} \pi \\ 0 \end{Bmatrix}\right)$$

$$= \frac{R_0}{R_{in}} A^2 \left(\frac{\pi}{2}, \begin{Bmatrix} \pi \\ 0 \end{Bmatrix}\right)$$

$$= \frac{R_0}{R_{in}} \left| 1 + \sum_{m=1}^{n} k_m e^{\pm j\beta_0 mb} + \sum_{m=n+1}^{N-1} k_m e^{\mp j\beta_0 (m-n)b} \right|^2. \quad (12)$$

The upper sign is for the field in the direction $\Phi = \pi$, the lower sign for the field in the direction $\Phi = 0$.

$$\text{Field ratio} \equiv \frac{(E_\Theta^r)_{\text{array}}}{(E_{\Theta = \pi/2}^r)_{\text{1 unit}}}$$

$$= F_0(\Theta, \beta_0 h) A(\Theta, \Phi)$$

$$= \sqrt{\frac{R_0}{R_{in}}} \frac{\cos(\tfrac{1}{2}\pi) \cos \Theta}{\sin \Theta} \left| 1 + \sum_{m=1}^{n} k_m e^{-j\beta_0 S_m} + \sum_{m=n+1}^{N-1} k_m e^{j\beta_0 S_{m-n}} \right|. \quad (13)$$

The linear arrays studied by Walkinshaw constitute a rather special case in the field of parasitic arrays. Nevertheless, they provide a valuable cross section of the general behavior of such arrays. The arrays consist of N equally spaced elements each of half-length h near $\lambda_0/4$. One element, numbered 0, is driven; all others are parasitic. Each parasitic antenna has a tuning reactance X_m in series with its self-impedance, $Z_{sm} = R_{sm} + jX_{sm}$, in

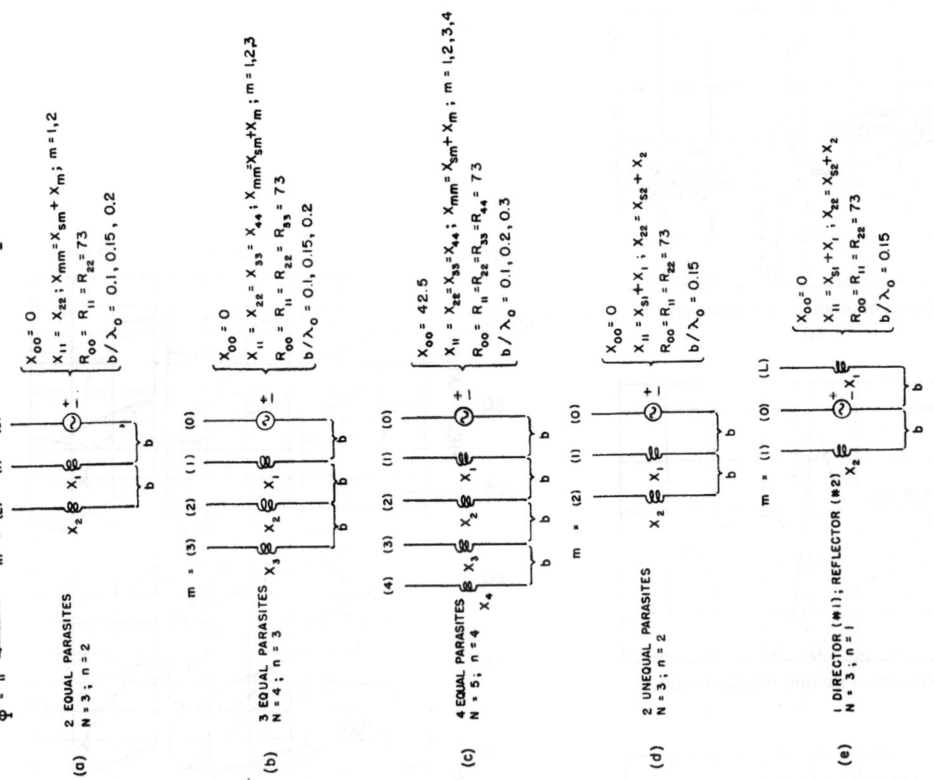

THEORY OF LINEAR ANTENNAS

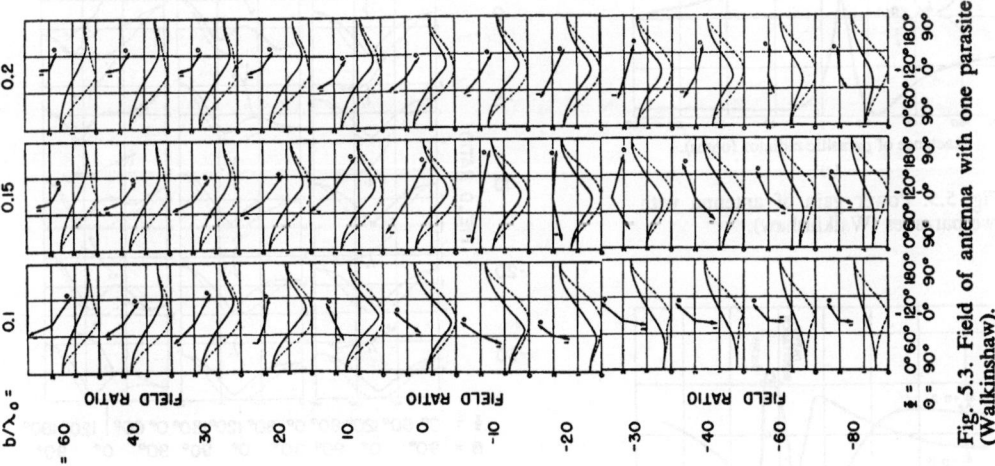

Fig. 5.4. Five linear arrays.

Fig. 5.3. Field of antenna with one parasite (Walkinshaw).

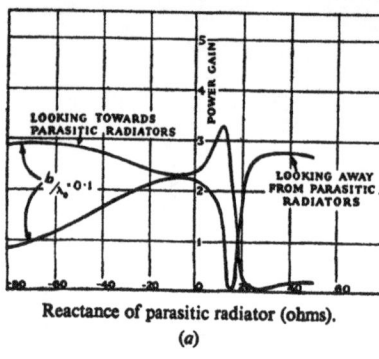

Reactance of parasitic radiator (ohms).
(a)

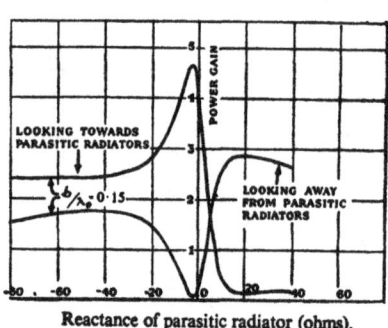

Reactance of parasitic radiator (ohms).
(b)

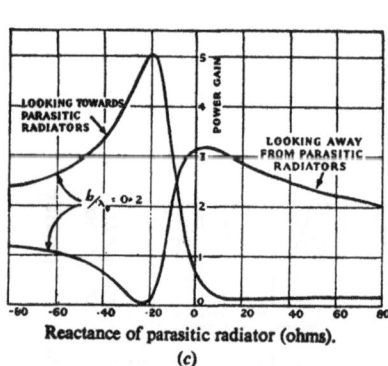

Reactance of parasitic radiator (ohms).
(c)

Fig. 5.5. Power gain of antenna with two parasites (Walkinshaw).

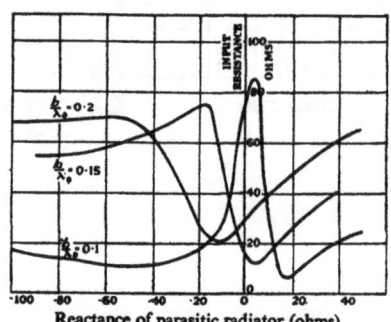

Reactance of parasitic radiator (ohms).

Fig. 5.6. Resistance of antenna with two parasites (Walkinshaw).

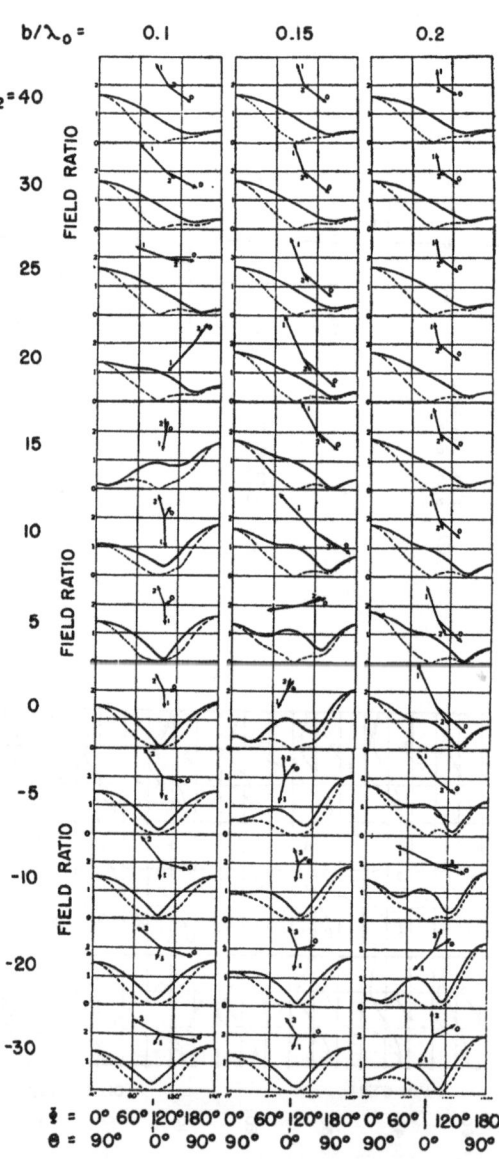

Fig. 5.7. Field ratio of antenna with two equal parasites (Walkinshaw).

[VI.5] THEORY OF LINEAR ANTENNAS

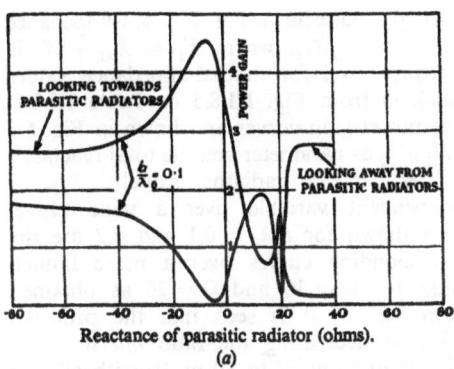

(a)

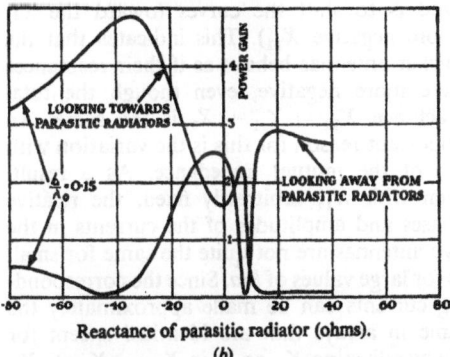

(b)

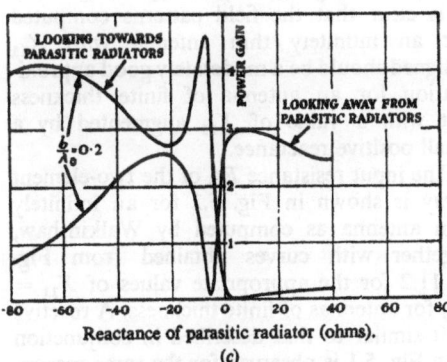

(c)

Fig. 5.8. Power gain of antenna with three equal parasites (Walkinshaw).

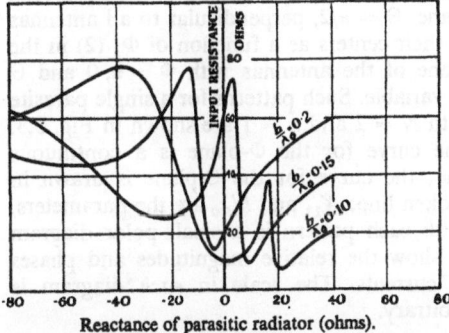

Fig. 5.9. Resistance of antenna with three parasites (Walkinshaw).

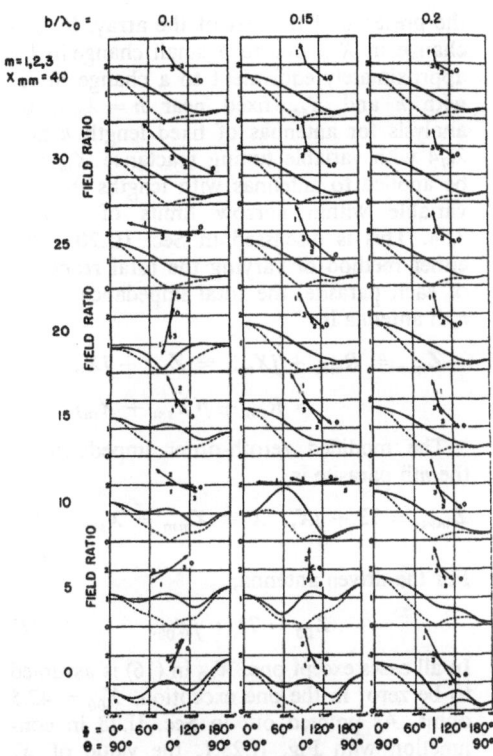

Fig. 5.10a. Field ratio of antenna with three equal parasites, $X_{mm} = 40$ (Walkinshaw).

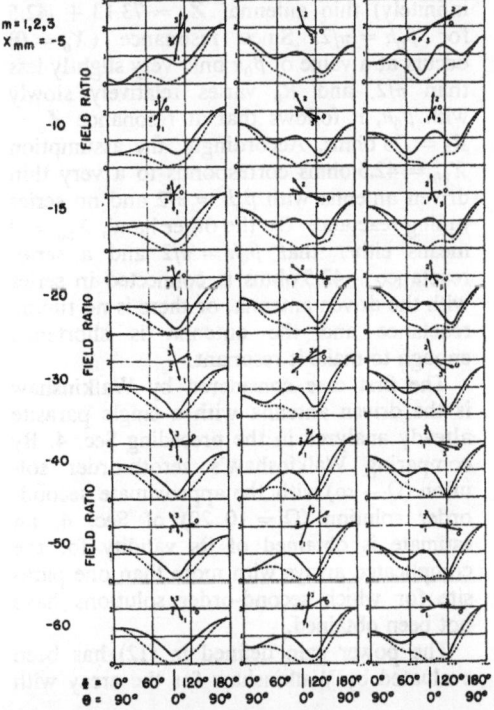

Fig. 5.10b. Field ratio of antenna with three equal parasites, $X_{mm} = -5$ (Walkinshaw).

the presence of the rest of the array. Since a change in X_{sm} due to a small change in h is approximately equivalent to a change in X_m with h and X_{sm} fixed near $h = \lambda_0/4$, the analysis for antennas of fixed length h near $\lambda_0/4$ with variable tuning reactance X_m may be applied to antennas with lengths that are variable within narrow limits of h near $\lambda_0/4$. This is discussed in Sec. III.20. With either method of varying the total reactance of each parasite, the total impedance of the mth antenna is

$$Z_{mm} = R_{mm} + jX_{mm} = Z_{sm} + jX_m$$
$$= R_{sm} + j(X_{sm} + X_m). \quad (14)$$

The modified zeroth-order impedance of the nth parasite is

$$Z_{mm} \doteq 73 + jX, \quad X \equiv X_{mm} = X_{sm} + X_m. \quad (15)$$

For the driven antenna,

$$Z_{00} = 73 + jX_{00}. \quad (16)$$

In all cases except one, X_{00} in (16) is assumed to be zero; in the one exception, $X_{00} = 42.5$ ohms. As pointed out in Sec. II.28 in conjunction with Fig. II.28.1, the value of X_0 for an infinitely thin antenna with $\beta_0 h = \pi/2$ is arbitrary, since the reactance curve is a vertical line. For an extremely (but not infinitely) thin antenna, $Z_0 = 73.13 + j42.5$ for $\beta_0 h = \pi/2$. Since resonance ($X_0 = 0$) occurs at a value of $\beta_0 h$ only very slightly less than $\pi/2$, and R_0 varies relatively slowly with $\beta_0 h$, it follows that at resonance $Z_0 \doteq R_0 \doteq 73$ ohms. Accordingly, the assumption $X_{00} = 42.5$ ohms corresponds to a very thin driven antenna with $\beta_0 h = \pi/2$ and no series tuning reactance. On the other hand, $X_{00} = 0$ means either that $\beta_0 h = \pi/2$ and a series reactance -42.5 ohms is connected in series with the driven antenna, or there is no tuning reactance and the antenna is shortened enough to make it resonant.

The first case considered by Walkinshaw is the driven antenna with a single parasite already analyzed in the preceding Sec. 4. By comparing Walkinshaw's zeroth-order solution ($\Omega \to \infty$) with the approximate second-order solution ($\Omega = 10, 20$) of Sec. 4, an estimate is obtained of its validity for the complicated arrays with more than one parasite for which second-order solutions have not been obtained.

The power gain defined in (12) has been evaluated by Walkinshaw for the array with a single parasite using for self-impedance $Z_{11} = 73 + jX_{11}$, where $X_{11} = X_{s1} + X_1$ is arbitrary, and for mutual-impedance values obtained from Fig. III.8.5 or Table III.8.1. The power-gain curves are shown in Fig. 5.1 with b/λ_0 as parameter and the total reactance of the parasitic radiator, namely, X_{11}, as independent variable over a wide range. Also shown for $b/\lambda_0 = 0.1$ and 0.2 are the corresponding curves over a more limited range for $\Omega = 10$ and $\Omega = 20$ as obtained from Sec. 4. It is seen that the principal effect of decreasing the ratio h/a in $\Omega = 2\ln(2h/a)$ from ∞ to 20 or 10 with all else fixed is to shift the curves toward the left (more negative X_{11}). This indicates that the thicker antennas behave as if their reactances were more negative, even though the total reactance $X_{11} = X_{s1} + X_1$ is the same. An important reason for this is the variation with h/a of the mutual impedance. As a result, even with X_{11} arbitrarily fixed, the relative phases and amplitudes of the currents in the two antennas are not quite the same for small as for large values of h/a. Since the corresponding currents can be made approximately the same in arrays that are identical except for h/a by adjusting X_{s1} or X_1 in $X_{11} = X_{s1} + X_1$, it is clear that the field patterns computed for an infinitely thin antenna with X_{11} assigned should be a moderately good approximation for an antenna of finite thickness but with a value of X_{11} augmented by a small positive reactance.

The input resistance R_{in} of the two-element array is shown in Fig. 5.2 for an infinitely thin antenna as computed by Walkinshaw, together with curves obtained from Fig. III.11.2 for the appropriate values of $X_{11} = X_{s1}$ for antennas of finite thickness. A reactive shift similar to that described in conjunction with Fig. 5.1 is observed for the same reason.

The field patterns defined by (13) are advantageously represented graphically in the two principal planes; (1) in the equatorial plane, $\Theta = \pi/2$, perpendicular to all antennas at their centers as a function of Φ; (2) in the plane of the antennas with $\Phi = \pi, 0$ and Θ as variable. Such patterns for a single parasite with $N = 2$ and $n = 1$ are shown in Fig. 5.3. The curve for the Φ-plane is a continuous line; the curve for the Θ-plane is drawn in broken line; X_{11} and b/λ_0 are the parameters. With each pattern is a small polar diagram to show the relative magnitudes and phases of currents. The scale in *each* diagram is arbitrary.

In addition to the array with a single parasite, Walkinshaw has investigated five other linear arrays using a modified zeroth-order theory. The five arrays in question are shown in Fig. 5.4. The quantities evaluated and plotted for each array are the power gain defined in (12), the input resistance of the driven unit in the presence of the array given by the real part of (2), and the field ratio defined in (13). In the curves for power gain and input resistance, the independent variable is the total reactance or reactances of the parasitic antennas. These, together with the resistances are indicated in Fig. 5.4. The following remarks may be made about the individual arrays.

(a) *One driven element, two equal parasites.* The two parasites are assumed to have equal self-reactances, $X_{11} = X_{22} = X$, and to be spaced equally along the negative x-axis, so that $N = 3$, $n = 2$. The power gain of the array over a half-wave dipole is shown in Fig. 5.5 as a function of X and with b/λ_0 as parameter. Fig. 5.6 shows the variation of the input resistance, R_{in}, as a function of X with b/λ_0 as parameter and the impedance of the driven antenna assumed to be $Z_{00} = 73 + j0$. In Fig. 5.7 are rectangular graphs of the field patterns of the array in the principal planes. In continuous lines are shown the patterns in the horizontal or Φ-plane defined by $\Theta = \pi/2$ with Φ as variable in degrees. In broken lines are the patterns in the vertical or Θ-plane containing the antennas, that is, $\Phi = \pi$ and $\Phi = 0$. The independent variable goes from 0° to 90° on the left half of each curve corresponding to the field in the quadrant plane bounded by $\Phi = 0$ and the z-axis, and from 90° back to 0° on the right half corresponding to the field in the quadrant plane bounded by $\Phi = \pi$ and the z-axis. The parameters for the field patterns are X and b/λ_0.

An Argand polar diagram of the relative magnitudes and phases of the currents in the several units is included with each pair of patterns. The scale in *each* diagram is arbitrary.

(b) *One driven element, three equal parasites.* The essential data for this array (Fig. 5.4b) are like those for the array with two parasites. The corresponding curves are shown in Figs. 5.8, 5.9, and 5.10a, b.

(c) *One driven element, four equal parasites.* The four parasites (Fig. 5.4c) are identical and equally spaced along the negative x-axis, so that $N = 5$, $n = 4$. The power gain of the array as a function of the common reactances is shown in Fig. 5.11, and the input resistance of the driven element in Fig. 5.12. (Note that $Z_{00} = 73 + j42.5$ in this case, instead of $73 + j0$.) The field patterns and polar diagrams of the currents are shown in Fig. 5.13.

A study of the power-gain curves shows that there is little advantage in using more than two identical, equally spaced parasitic antennas adjusted to act as *reflectors* (X positive) with maximum field in the direction $\Phi = 0$ away from the parasites. Even with two parasitic reflectors, the current in the antenna farthest from the driven element is very small. The largest power gain over a single dipole supplied with the same power is near three.

The situation is different and more complicated if the parasites are adjusted to act as directors to produce a maximum field in the direction $\Phi = \pi$ toward the parasites. It is seen from Figs. 5.1, 5.5, and 5.8 that for $b/\lambda_0 = 0.1$ there is a power-gain maximum at $X = -10$ with one parasite, at $X = 12$ with two parasites, and at $X = 20$ with three parasites. It appears as though a particular maximum is moved from left to right in these figures as the number of directors is increased. There is a second maximum at $X = -70$ with two directors which appears to shift to the right to $X = -5$ with three directors. Evidently, as the number of directors is increased *new maxima* appear to come from the left, reach a maximum as they move to the right, and *decrease again*. For maximum power gain a *different* maximum may be required for different numbers of parasites, rather than a particular one that moves to the right as n increases but eventually *decreases* in magnitude. For example, for $n = 1$, $b/\lambda_0 = 0.1$ is a suitable spacing; for $n = 2$, $b/\lambda_0 = 0.2$; but for $n = 3$, $b/\lambda_0 = 0.15$ instead of a value for a maximum at b/λ_0 greater than 0.2.

A study of the input-resistance curves of Figs. 5.2, 5.6, 5.9, and 5.12 shows that for arrays with one, two, three, or four parasites adjusted for maximum power gain the input resistance invariably is *very low*. This is inconvenient if the array is to be matched to a transmission line. However, the input resistance can be increased without altering the field patterns if a *folded dipole*—which has about four times the self-resistance of a simple dipole—is used as the driven element.

(d) *One driven element with two unequal parasites.* The array in Fig. 5.4d includes two parasites equally spaced at $b/\lambda_0 = 0.15$ along

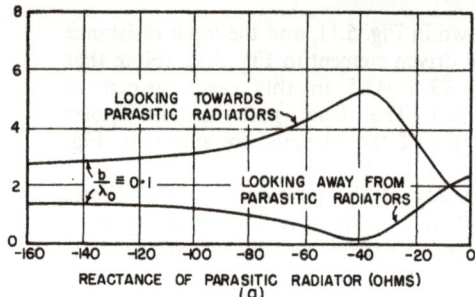

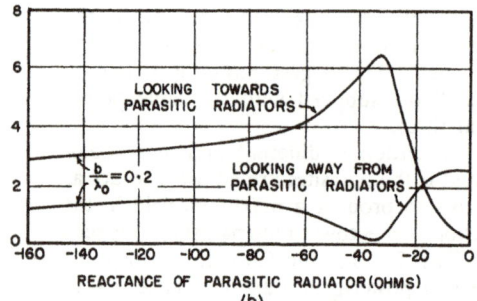

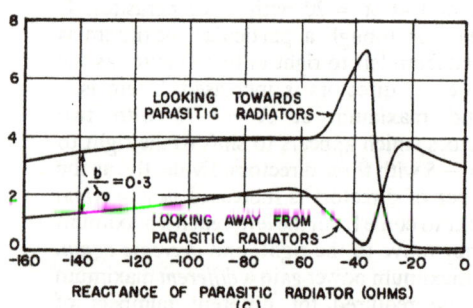

Fig. 5.11. Power gain of antenna with four equal parasites; $Z_\infty = 73 + j42.5$ (Walkinshaw).

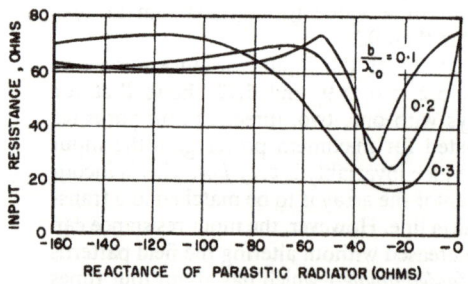

Fig. 5.12. Resistance of antenna with four equal parasites; $Z_\infty = 73 + j42.5$ (Walkinshaw).

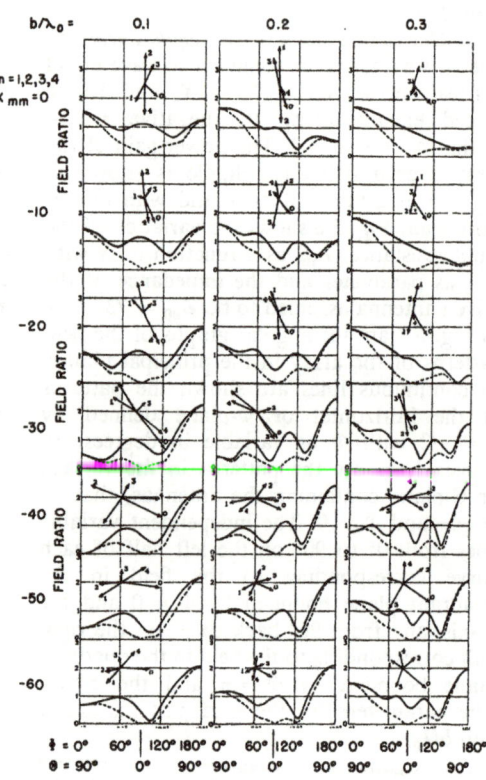

Fig. 5.13. Field ratio of antenna with four equal parasites (Walkinshaw).

[VI.6] THEORY OF LINEAR ANTENNAS

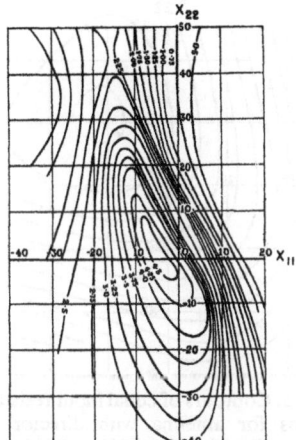

(a) LOOKING TOWARD PARASITES ($\tfrac{z}{2} = \pi$)

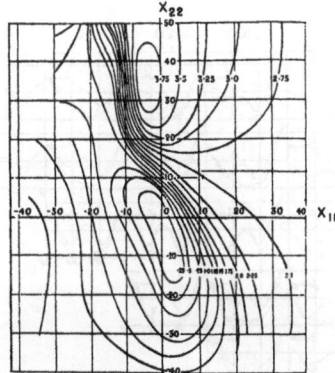

(b) LOOKING AWAY FROM PARASITES ($\tfrac{z}{2} = 0$)

Fig. 5.14. Contours of equal power gain for antenna with unequal parasites (Walkinshaw).

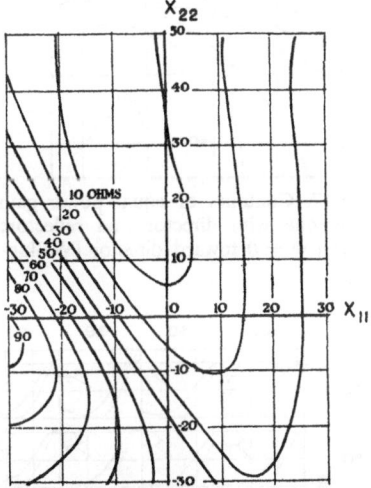

Fig. 5.15. Contours of equal input resistance of antenna with two unequal parasites (Walkinshaw).

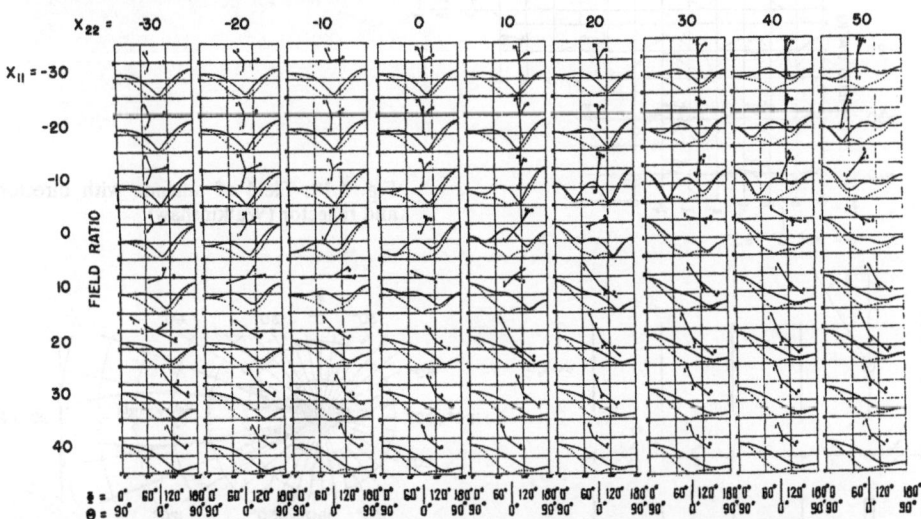

Fig. 5.16. Field ratio of antenna with two unequal parasites (Walkinshaw).

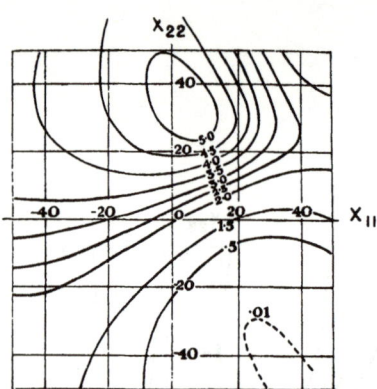

Fig. 5.17. Contours of equal power gain for antenna with director and reflector; direction $\Phi = 0$ toward director (Walkinshaw).

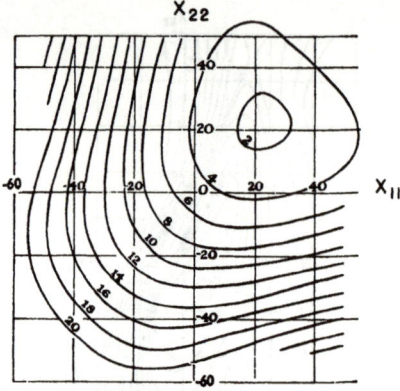

Fig. 5.18. Contours of equal input resistance in ohms for antenna with director and reflector; direction $\Phi = 0$ toward director (Walkinshaw).

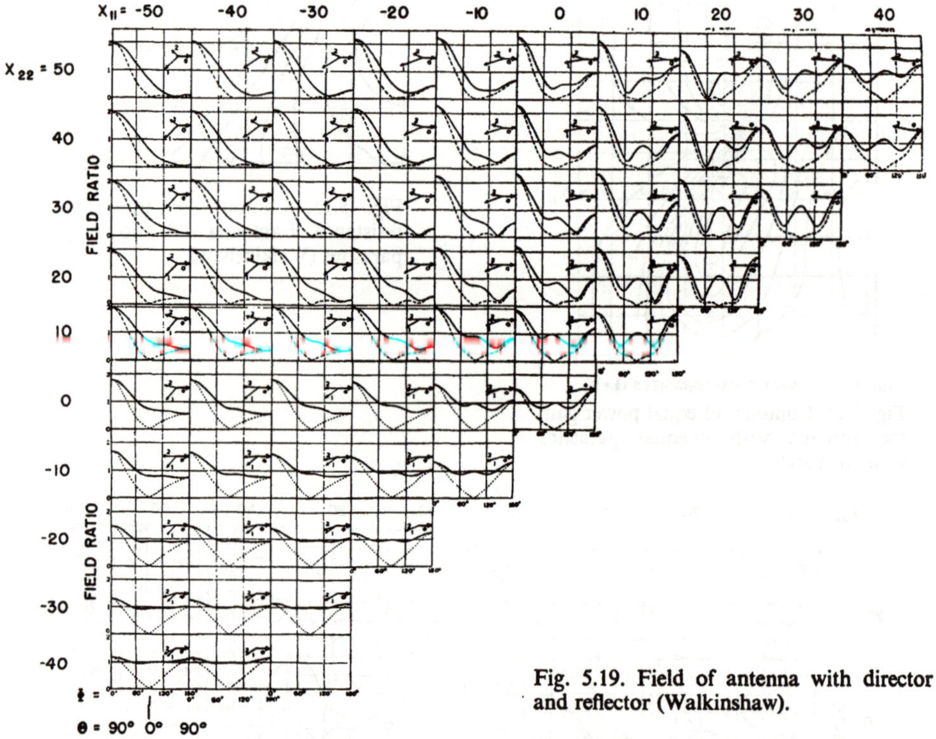

Fig. 5.19. Field of antenna with director and reflector (Walkinshaw).

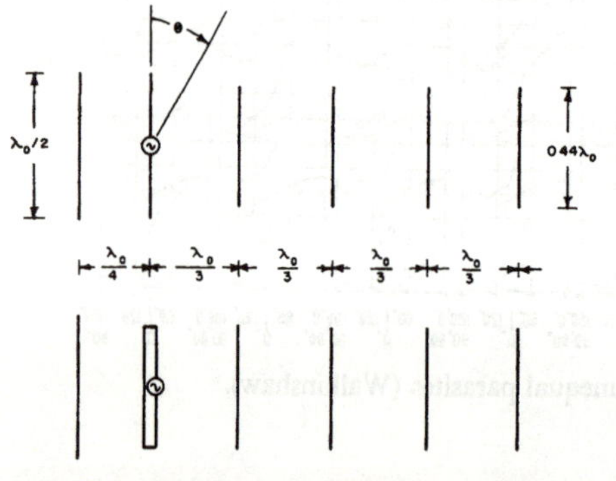

Fig. 6.1. Four-director Yagi-Uda arrays.

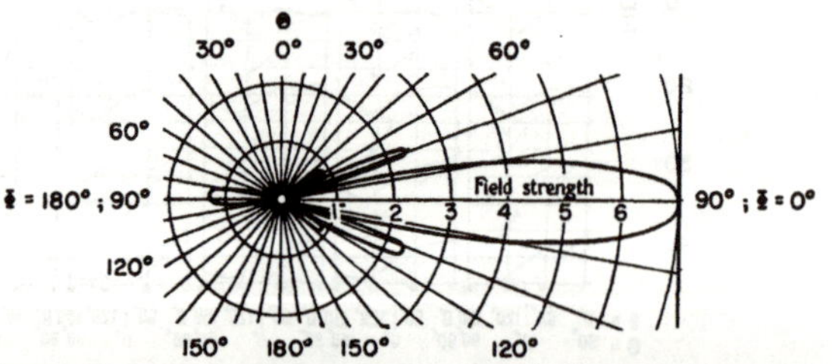

Fig. 6.2. Polar diagram of 30-director Yagi-Uda array in plane of directors: director length, $0.408\lambda_0$; director spacing, $0.34\lambda_0$; reflector length, $0.5\lambda_0$; reflector spacing, $0.125\lambda_0$; exciter, center-fed dipole cut for resonance (Fishenden and Wiblin).

the negative x-axis but with reactances $X_{11} = X_{s1} + X_1$, $X_{22} = X_{s2} + X_2$ that are *not* required to be equal. The difference may be due to different tuning reactances X_1 and X_2 with otherwise identical antennas, or to different values of X_{s1} and X_{s2} as a result of adjustments in the lengths of the two parasites with $X_1 = X_2 = 0$.

Contour maps of the power gain as a function of the two reactances X_{11} and X_{22} are shown in Figs. 5.14a (field in direction $\Phi = \pi$) and Fig. 5.14b (field in direction $\Phi = 0$). The maximum possible power gain of about 4.5 is seen to occur when the parasites are acting as *directors* (field maximum in direction $\Phi = \pi$); when the parasites are acting as reflectors, a maximum power gain of 3.75 is possible with X_{22} considerably more positive than X_{11}.

A contour map of the input resistance is shown in Fig. 5.15. Field patterns in the principal planes are given in Fig. 5.16 together with polar diagrams of currents.

(e) *One driven element with one director and one reflector.* The combination of a driven antenna with one parasite on each side as shown in Fig. 5.4e is of considerable importance in practice. The array investigated has $b/\lambda_0 = 0.15$ for both parasites, but permits X_{11} and X_{22} to be different.

The power gain in the direction ($\Phi = 0$) toward the director (parasite No. 1) is represented in the contour map of Fig. 5.17. (As a result of symmetry, the power gain in the direction $\Phi = \pi$ is obtained by interchanging X_{11} and X_{22}.) The maximum possible power gain occurs with $X_{11} = 7$, $X_{22} = 35$, so that if the variation in X_{11} is obtained by changing the length of the parasite rather than by tuning it at the center, unit No. 1 is longer than the self-resonant length. This adjustment, while giving the maximum power gain in the direction $\Phi = 0$, permits an appreciable field in the opposite direction $\Phi = \pi$. This can be reduced by decreasing X_{11}, but this involves a decrease in power gain.

The input resistance is shown in the contour diagram of Fig. 5.18. It is seen that with the adjustment for maximum power gain in the direction $\Phi = 0$, R_{in} is so low that it may be impractical for matching even if a folded dipole is used as driven element.

The field patterns of the array with one director and one reflector are shown in Fig. 5.19 together with the polar diagrams of the currents in the director (No. 1) and the reflector (No. 2).

The analysis of multielement parasitic arrays in this section is limited by the fact that all elements are equally spaced. In practice, it is often advantageous to space directors differently from reflectors.

6. *Yagi-Uda Arrays—Experimental Investigation*

Typical Yagi-Uda arrays[61, 62, 63] consist of a single driven antenna with one parasitic reflector and from 4 to 42 parasitic directors. A four-director Yagi-Uda array is shown in Fig. 6.1. The arrays discussed in Sec. 5 are elementary forms of the Yagi-Uda array.

The theoretical problem of designing an $(N - 2)$-director array (N is the total number of antennas) for maximum directivity and minimum side lobes is intricate even for $N - 2$ as small as four and using a modified zeroth-order solution. This is due to the large number of variables and the complexity of their interdependence through the formulas for mutual impedance. (The self-impedances are constant and equal in the modified zeroth-order approximation.) Therefore, with $N - 2$ large, it is necessary to have recourse to systematic experimental studies of the behavior of Yagi-Uda arrays as a function of adjustments of the several variables in order to discover the conditions for optimum directivity, small side lobes, and satisfactory input impedance. Since for a given array the number of variables is no smaller for the experimentalist than for the theoretical investigator, it is often convenient to make all directors identical and equally spaced and thereby to sacrifice some improvement in directivity obtainable by individually adjusting each length and spacing. With this simplification it is found that a desirable distance between directors is about $\lambda_0/3$, with each director considerably shorter than $\lambda_0/2$. The optimum length for identical directors with $\lambda_0/3$ spacing decreases with an increase in their number and also with a decreasing ratio of length to diameter. Values obtained by Fishenden and Wiblin[15] for identical directors of diameter $0.006\lambda_0$ and with $\Omega = 2 \ln 2h/a \doteq 11.4$ are in Table 6.1. These are designed to give first side-lobe amplitudes of not over 30 percent of the main lobe. Approximate correction for different dimensions is obtained by increasing the director length 2 percent for each 50-percent decrease in the diameter.

The single reflector normally is $\lambda_0/2$ in length and spaced at $\lambda_0/4$ from the driven antenna. However, this spacing is not critical in its effect on the field characteristics of the array and may be reduced to $\lambda_0/8$ with little

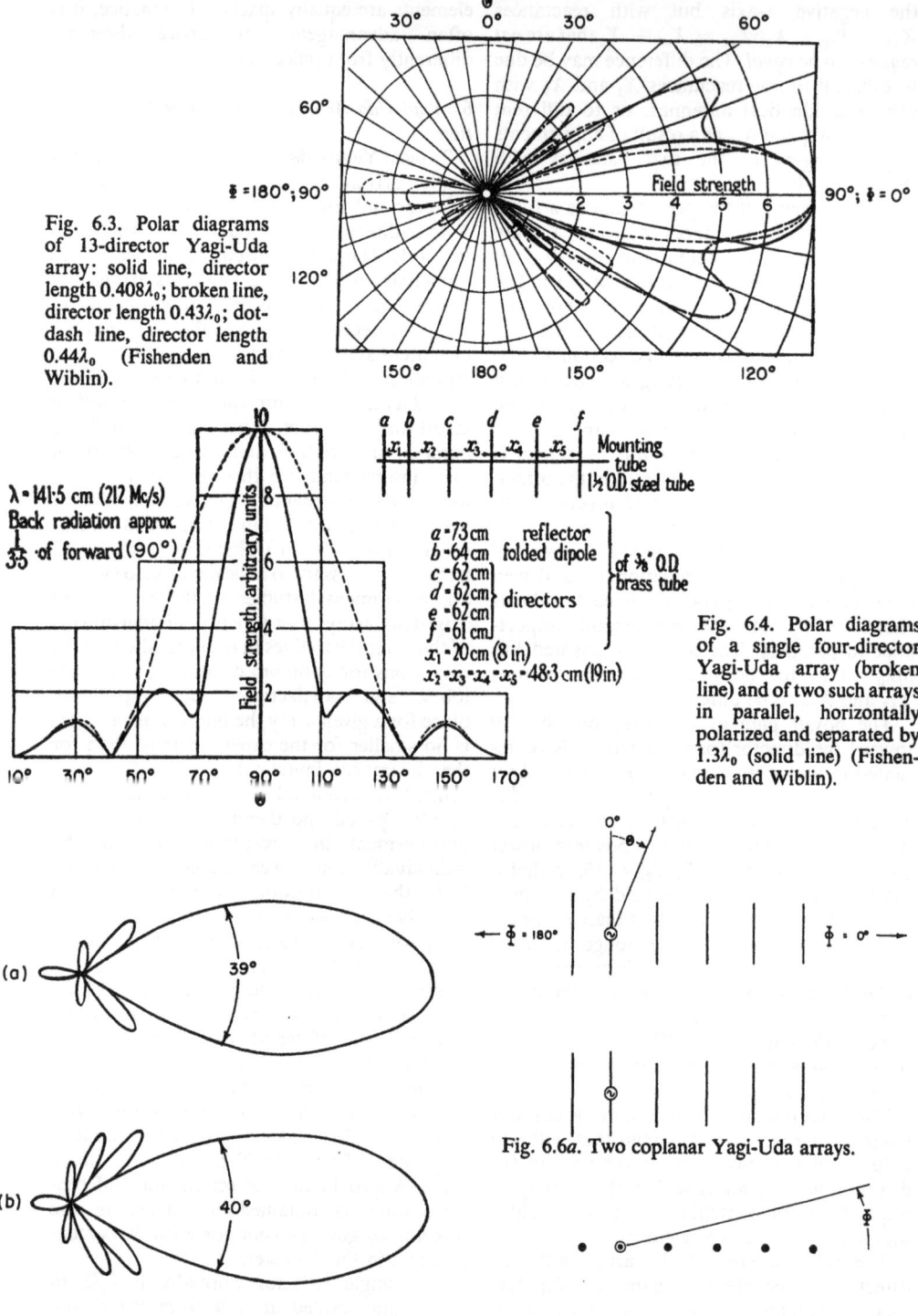

Fig. 6.3. Polar diagrams of 13-director Yagi-Uda array: solid line, director length $0.408\lambda_0$; broken line, director length $0.43\lambda_0$; dot-dash line, director length $0.44\lambda_0$ (Fishenden and Wiblin).

Fig. 6.4. Polar diagrams of a single four-director Yagi-Uda array (broken line) and of two such arrays in parallel, horizontally polarized and separated by $1.3\lambda_0$ (solid line) (Fishenden and Wiblin).

Fig. 6.5. Field of eight-director Yagi-Uda array with cylindrical sheet reflector (a) in plane of polarization, (b) perpendicular to plane of polarization (Alfred).

Fig. 6.6a. Two coplanar Yagi-Uda arrays.

Fig. 6.6b. Two parallel Yagi-Uda arrays.

TABLE 6.1. Variation of optimum director length with number of directors.

Number of directors	Director length (λ_0)
5	0.434
7	.423
10	.42
13	.414
20	.407
30	.40
42	.385

TABLE 6.3. Directivity and beam width of a Yagi-Uda array with one reflector.

Number of directors	Beam width (deg)	Relative directivity	Gain per element
4	46	8	1.33
9	37	13	1.18
13	31	15	1.0
20	26	21	0.95
30	22	—	—

effect except on the input impedance, which decreases as indicated in Table 6.2 as determined by Fishenden and Wiblin[15] for a typical array.

TABLE 6.2. Input resistance of 13-director Yagi-Uda array; reflector length, $0.5\lambda_0$; director spacing, $0.34\lambda_0$.

Reflector Spacing (λ_0)	R_{in} (ohms)	
	Director length $0.406\lambda_0$	Director length $0.42\lambda_0$
0.25	62	50
.18	50	43
.15	32	27
.13	22	—
.10	12	—

Since the input resistance of the Yagi array is quite low for convenient matching to a balanced transmission-line, it is often advantageous to use a folded dipole, as explained in Sec. 5.

Typical field patterns of Yagi arrays as obtained by Fishenden and Wiblin are shown in Figs. 6.2, 6.3, and 6.4. These are in the Θ-plane (that is, the plane of the antennas). Field patterns in the Φ-plane (equatorial plane perpendicular to the antennas) differ negligibly from those in the Θ-plane except in the details of the minor-lobe structure. The beam width is defined as the full width of the principal ear at one-half (not 0.707) the field strength of the maximum (−6 db). Its dependence on the number of directors is shown in Table 6.3, where the relative directivity D_r (power gain over $\lambda_0/2$ dipole) is also given.

Experimental data on Yagi-Uda arrays of slightly different construction have been obtained by Alfred.[1] His array had eight directors and a *cylindrical semicircular sheet reflector* $0.9\lambda_0$ long and of radius $0.38\lambda_0$ centered at the driven antenna. The directors were $0.40\lambda_0$ in length and were spaced at $0.368\lambda_0$ except the first one, which was $0.28\lambda_0$ from the driver. The change in the spacing of this last was found desirable when the sheet reflector was substituted for a conventional rod reflector. Field patterns of this array in the plane of polarization and perpendicular to this are in Fig. 6.5. The indicated beam widths are between points of *one-half* the maximum field.

Additional directivity is obtained by using two or more Yagi-Uda arrays side by side with the individual antennas either coplanar and collinear as in Fig. 6.6a or in parallel planes as in Fig. 6.6b. The increase in directivity as obtained by Fishenden and Wiblin is shown in Fig. 6.4 for a pair of four-director arrays arranged as in Fig. 6.6a and separated by a distance of $1.3\lambda_0$ between parallel lines of centers. Data for a similar pair of 13-director arrays are given in Table 6.4. The beam width decreases and the side lobes increase with increasing separation of the arrays.

With two arrays of the type used to obtain Fig. 6.5, the beam width as obtained by Alfred is 30° instead of 39° for one array alone. For three arrays the beam width was 18°. In obtaining these data the arrays were coplanar with separation $0.9\lambda_0$. This was regarded as the best compromise between beam width and side radiation.

7. Broadside Array with Parasitic Curtain

Since the unilateral end-fire array has a rather broad field pattern of relatively low directivity even for a large number of units, it is not particularly well suited to sharply unidirectional arrays. As pointed out in Sec. 3 in conjunction with (3.57), a much more highly directive array of arrays consists of two broadside arrays, one behind the other, so phased that each pair of antennas in depth is a two-element, unilateral end-fire array. The principal

TABLE 6.4. Array of two coplanar 13-director Yagi-Uda arrays.

Spacing of arrays (λ_0)	Director length (λ_0)	Beam width (deg)	Intensity ratio (percent of main lobe)		
			First side lobe	Second side lobe	Back lobe
2.7	0.41	14	28	25	—
2.5	.42	13.5	38	19	30
	.43	13.5	40	40	19
2.2	.41	16	30	18	13
2.0	.41	17.5	27	23	11

difficulty with this combination N_B-element-broadside, $N_E = 2$-element-end-fire array of arrays is the complicated feeding network.

As shown in Sec. 4, an array consisting of a single driven antenna with a properly spaced and tuned parasitic element acting as reflector may have a field that is practically as unidirectional as a two-element end-fire array. It follows that an N_B-element broadside array with a parallel row of *parasitic* reflectors that are properly spaced and tuned should compare favorably in directivity with the broadside-end-fire combination. The analysis of an N-element array of this type is extremely complicated in general. A satisfactory approximate solution is possible only if all elements have electrical half-lengths $\beta_0 h$ near $\pi/2$, when a modified zeroth-order solution is available that is reasonably accurate for sufficiently thin antennas. The typical and relatively simple case of two driven antennas in broadside with two parasites has been studied in detail by Walkinshaw.[57] The arrangement of the four elements is shown in Fig. 7.1. Note that the two driven units are numbered 0 and 2, the parasites 1 and 3. In the four-element array (unlike arrays with a larger number of units) the pairs 0, 1 and 2, 3 may be made identical in order to have equal currents, respectively, in the driven elements and in the parasites.

The general circuit equations are

$$V_0 = I_0 Z_{00} + I_1 Z_{01} + I_2 Z_{02} + I_3 Z_{03},$$
$$0 = I_0 Z_{10} + I_1 Z_{11} + I_2 Z_{12} + I_3 Z_{13},$$
$$V_2 = I_0 Z_{20} + I_1 Z_{21} + I_2 Z_{22} + I_3 Z_{23}, \quad (1)$$
$$0 = I_0 Z_{30} + I_1 Z_{31} + I_2 Z_{32} + I_3 Z_{33},$$

where the currents are defined at the centers of each unit at $z = 0$. It follows from the reciprocal theorem that

$$Z_{ij} = Z_{ji}, \quad i, j = 0, 1, 2, 3. \quad (2)$$

If the two driven units with their tuning impedances are made identical, and the two parasites likewise, it follows that

$$Z_{22} = Z_{00}, \quad Z_{33} = Z_{11}, \quad Z_{03} = Z_{21}. \quad (3)$$

Hence, for equal driving voltages, symmetry requires the currents to be equal in pairs:

$$V_2 = V_0, \quad I_2 = I_0, \quad I_3 = I_1. \quad (4)$$

With (2)–(4), the four equations in (1) reduce to the following two:

$$V_0 = I_0(Z_{00} + Z_{02}) + I_1(Z_{01} + Z_{03}),$$
$$0 = I_0(Z_{01} + Z_{03}) + I_1(Z_{11} + Z_{13}). \quad (5)$$

If it is assumed, further, that the parasitic antennas are geometrically identical with the driven antennas, it follows that

$$Z_{13} = Z_{02}. \quad (6)$$

For antennas of electrical half-length $\beta_0 h$ near $\pi/2$, (6) is approximately correct even if the two units are not identical.

Let it be assumed that the self-impedances of the driven units and of the parasites are given by

$$Z_{22} = Z_{00} = 73 + j0, \quad Z_{33} = Z_{11} = 73 + jX_{11}, \quad (7)$$

where $X_{11} = X_{s1} + X_1$ is variable. The zeroth-order mutual impedances are given by Fig. III.8.5 or Table III.8.1. Since the broadside spacing b is $\lambda_0/2$,

$$Z_{13} = Z_{02} = -12.5 - j29.9; \quad (8)$$

Z_{01} and Z_{03} vary with the depth spacing d.

The currents in the antennas are obtained by solving (5). They are

$$I_2 = I_0$$
$$= \frac{V_0(Z_{11} + Z_{02})}{(Z_{00} + Z_{02})(Z_{11} + Z_{02}) - (Z_{01} + Z_{03})^2}, \quad (9a)$$

$$I_3 = I_1$$
$$= \frac{-V_0(Z_{01} + Z_{03})}{(Z_{00} + Z_{02})(Z_{11} + Z_{02}) - (Z_{01} + Z_{03})^2}. \quad (9b)$$

The far-zone electric field of the array is given by

$$E_\Theta^r = \frac{j\zeta_0 I_0}{2\pi} \frac{e^{-j\beta_0 R_0}}{R_0} F_0(\Theta, \beta_0 h)$$
$$A_B(\Theta, \Phi_B; 2, \tfrac{1}{2}, 0) A_P(\Theta, \Phi_E), \quad (10)$$

where for $\beta_0 h = \pi/2$,

$$F_0(\Theta, \tfrac{1}{2}\pi) = \frac{\cos(\tfrac{1}{2}\pi \cos \Theta)}{\sin \Theta}. \quad (11)$$

For the two-element broadside array, (3.46) gives

$$A_B(\Theta, \Phi; 2, \tfrac{1}{2}, 0) = \frac{\sin(\pi \sin \Theta \cos \Phi_B)}{\sin(\tfrac{1}{2}\pi \sin \Theta \cos \Phi_B)}$$
$$= 2\cos(\tfrac{1}{2}\pi \sin \Theta \cos \Phi_B)$$
$$= 2\cos(\tfrac{1}{2}\pi \sin \Theta \sin \Phi), \quad (12)$$

where

$$\Phi \equiv \Phi_E = \Phi_B - \tfrac{1}{2}\pi. \quad (13)$$

The array factor for a driven antenna with parasite is (4.2) except that the positions of parasite and driven unit are interchanged. This is equivalent to increasing Φ by π, so that with $\Phi \equiv \Phi_E$,

$$A_P(\Theta, \Phi) = 1 + ke^{-j(\delta - \beta_0 b \sin \Theta \cos \Phi)}, \quad (14)$$

where

$$ke^{-j\delta} = I_1/I_0 = -(Z_{01}+Z_{03})/(Z_{11}+Z_{02}). \quad (15)$$

The total power supplied to the array is twice the power supplied to each driven antenna. That is,

$$P_t = I_0^2 R_{in}, \quad (16)$$

where R_{in} is the input resistance of each driven antenna when it is a part of the array. R_{in} is the *real part* of the input impedance,

$$Z_{in} = \frac{V_0}{I_0} = Z_{00} + Z_{02} - \frac{(Z_{01}+Z_{03})^2}{(Z_{11}+Z_{02})}. \quad (17)$$

The solution of (15) for I_0 and substitution in (10) (with phase referred to current so that the complex current I_0 is replaced by the real value I_0) gives E_Θ^r in terms of the power supplied to the array.

The power to a single dipole is

$$P_1 = \tfrac{1}{2} I_0^2 R_0, \quad (18)$$

where R_0 is the input resistance of an isolated dipole. The field of the single dipole when supplied with the power P_1 is (10) with the A-factors replaced by unity.

The power gain in the direction of its maximum, that is, the relative directivity D_r, and the gain in the opposite direction, are defined as follows:

$$\text{Power gain} = \frac{(E_\Theta^r)_{\text{array}}^2}{(E_\Theta^r)_{1 \text{ unit}}^2}\left[\Theta = \frac{\pi}{2}, \Phi = \begin{Bmatrix} 0 \\ \pi \end{Bmatrix}\right]$$

$$= \frac{2R_0}{R_{in}} A_P^2\left(\frac{\pi}{2}, \begin{Bmatrix} 0 \\ \pi \end{Bmatrix}\right)$$

$$= \frac{2R_0}{R_{in}}|1 + ke^{-j(\delta \mp \beta_0 b)}|^2. \quad (19)$$

The power gain is shown in Fig. 7.2 as calculated by Walkinshaw as a function of the total reactance $X_{33} = X_{11} = X_{s1} + X_1$ of the parasites with b/λ_0 as parameter. Note that the direction of maximum power gain for the array is *away from the parasite*, $\Phi = \pi$, with a magnitude near 6.

The modified zeroth-order input resistance of the array due to Walkinshaw is shown in Fig. 7.3.

The ratio of the field of the array to the field in the equatorial plane of a single antenna radiating the same power is

$$\text{Field ratio} = \frac{(E_\Theta^r)_{\text{array}}}{(E_\Theta^r = \tfrac{1}{2}\pi)_{1 \text{ unit}}} \sqrt{\frac{R_0}{2R_{in}}}$$

$$\times F_0(\Theta, \beta_0 h) A_B(\Theta, \Phi; 2, \tfrac{1}{2}, 0) A_P(\Theta, \Phi)$$

$$= \sqrt{\frac{R_0}{2R_{in}}} \frac{\cos(\tfrac{1}{2}\pi \cos \Theta)}{\sin \Theta}$$

$$\times 2\cos(\tfrac{1}{2}\pi \sin \Theta \sin \Phi)$$

$$\times \left|1 - \frac{Z_{01}+Z_{03}}{Z_{11}+Z_{02}} e^{j\beta_0 b \sin \Theta \cos \Phi}\right|. \quad (20)$$

Walkinshaw's curves for the field ratio or normalized field are shown in Fig. 7.4.

In so far as the available data are conclusive, the relative directivity or maximum power gain of the array is in the direction away from the parasites with $b/\lambda_0 = 0.1$ and $X_{11} \doteq 33$ ohms. Its magnitude is slightly under 6, which is greater than for a four-element broadside array or end-fire array. The input resistance is about 25 ohms. Although closer spacing than $0.1\lambda_0$ may be expected to improve the gain slightly, the input resistance probably is too low for convenient matching. The maximum front-to-back ratio with $b/\lambda_0 = 0.1$

650 THEORY OF LINEAR ANTENNAS [VI.7]

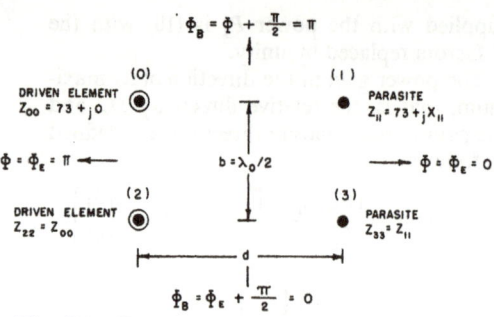

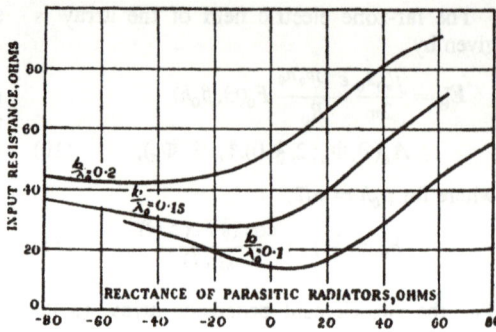

Fig. 7.1. Two-element broadside array with reflecting curtain.

Fig. 7.3. Input resistance of two-element broadside array (Walkinshaw).

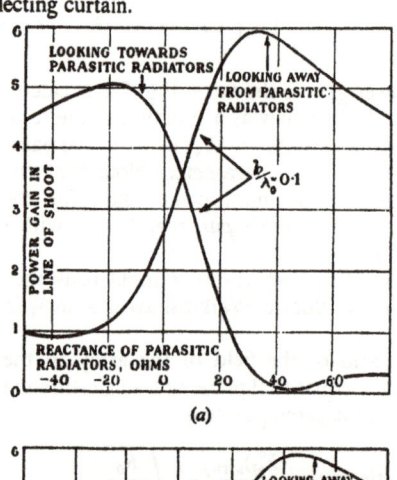

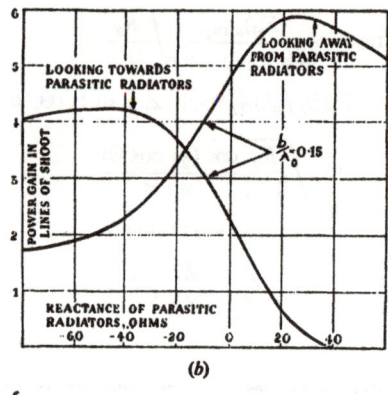

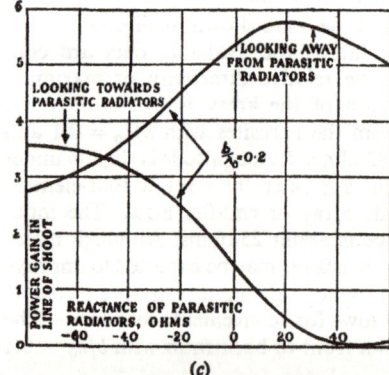

Fig. 7.2. Power gain of two-element broadside array with reflecting curtain (Walkinshaw).

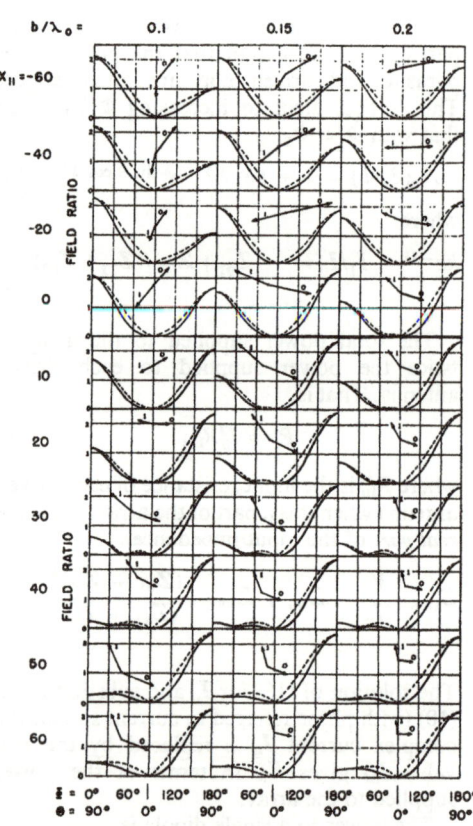

Fig. 7.4. Field ratio of two-element broadside array with reflecting curtain (Walkinshaw).

is with $X_{11} \doteq 43$ ohms. It is seen from Fig. 7.4 that with this value of X_{11} the field is very small in all directions except near $\Phi = \pi$, the direction of maximum. Moreover, this maximum differs but little from the optimum forward field with $X_{11} = 33$ ohms, and the input resistance is higher—near 32 ohms.

It may be concluded that the four-element combination broadside array with parasitic curtain is a highly directive unilateral array. The same general conclusion evidently must apply to an $N/2$-element broadside array with an $N/2$-element parasitic array, but the tuning reactances X_m in $X_{mm} = X_{sm} + X_m$ cannot all be the same in the parasites if equal currents are desired, since their relative positions and mutual impedances differ.

The analysis of other directional arrays with parasitic elements may be carried out in a similar manner. Significant types include a driven antenna with a parasitic array in the form of a plane, a parabolic cylinder, or a corner reflector as shown in Fig. III.20.1.

NONUNIFORM PARALLEL ARRAYS

8. Complex Array Polynomials

The most general array to which the diffraction formula derived in Sec. 1 may be applied is the *uniform* array consisting of N identical, equispaced, parallel elements with currents that are equal in magnitude and that vary progressively and uniformly in phase from each unit to the adjacent one in a specified direction. Although numerous arrays with a great variety of possible field patterns may be constructed that satisfy these conditions, they are *all* limited by the important condition that while the main beam width may be reduced indefinitely by increasing the number N of the units, the relative magnitudes of *the side lobes cannot be diminished below definite fractions of the principal lobe*. While a main lobe with small beam width usually is of major concern in the design of an array, the reduction in the relative sizes of the minor lobes often is no less important. Indeed, some sacrifice in beam width is at times to be preferred to relatively large side lobes.

It is possible to reduce the side lobes below the minimum level possible in a uniform array by introducing the magnitudes and phases of the currents as additional variables and assigning appropriate values. Since the analysis of Sec. 3 is general for parallel, identical, equispaced elements until the restrictive conditions (3.34) characteristic of uniform arrays are imposed, the study of nonuniform arrays may begin with the formulas (3.30) and (3.32) for the array factors. For convenience, let the representation of the elements of the nonuniform array be referred to the representation of a uniform array in which there is a progressive phase delay δ between successive elements from left to right. That is, the uniform arrays are represented as follows:

$$A_p(\Theta, \Phi) = A(z) = 1 + \sum_{m=1}^{n} (z^m + z^{-m}),$$

$$\text{for } N = N_p = 2n + 1, \quad (1a)$$

$$A_p(\Theta, \Phi) = A(z) = \sum_{i=1,3,5,}^{N-1} (z^{i/2} + z^{-i/2})$$

$$= \sum_{m=1}^{n} (z^{m-\frac{1}{2}} + z^{-m+\frac{1}{2}}),$$

$$N \equiv N_p = 2n, \quad (1b)$$

where

$$z \equiv e^{j\psi}, \quad \psi \equiv \beta_0 b \sin \Theta \cos \Phi - \delta. \quad (1c)$$

In the notation of (1.24) and with $\Theta = \pi/2$,

$$\psi = 2x = 2\pi(n_p \cos \Phi - t_p), \quad (1d)$$

$$n_p = \beta_0 b/2\pi = b/\lambda_0, \quad t_p = \delta/2\pi. \quad (1e)$$

The symbols n and n_p are not related. The normalized array factor is defined by

$$a(z) = A(z)/A(z=1) = A(z)/N. \quad (2)$$

By factoring out z^{-n} in (1a) and $z^{-n+\frac{1}{2}}$ in (1b) it follows that:

$$A(z) = z^{-n} \sum_{m=0}^{N-1} z^m$$

$$= z^{-n}(1 + z + z^2 + \cdots + z^{N-1}),$$

$$N = 2n + 1, \text{ odd}, \quad (3a)$$

$$A(z) = z^{-n+\frac{1}{2}} \sum_{m=0}^{N-1} z^m$$

$$= z^{-n+\frac{1}{2}}(1 + z + z^2 + \cdots + z^{N-1}),$$

$$N = 2n, \text{ even}. \quad (3b)$$

Since $z = e^{j\psi}$ always has an imaginary exponent, $z = |z| = 1$. Hence,

$$A(z) = |A(z)| = \left| \sum_{m=0}^{N-1} z^m \right|$$

$$= |1 + z + z^2 + \cdots + z^{N-1}|,$$

$$N \text{ even or odd.} \quad (4)$$

By introducing the following amplitude and relative phase factors,

$$C_m = k_m e^{-j(\delta_m - m\delta)}, \quad C'_m = k'_m e^{-j(\delta'_m + m\delta)},$$
$$N = 2n + 1, \quad (5a)$$

$$C_i = k_i e^{-j(\delta_i - i\delta)/2}, \quad C'_i = k'_i e^{-j(\delta'_i + i\delta)/2},$$
$$N = 2n, \quad (5b)$$

in the general formulas (3.30) and (3.32) together with (2), the following formulas are obtained for the nonuniform array:

$$A_p(\Theta, \Phi) = A(z) = 1 + \sum_{m=1}^{n}(C_m z^m + C'_m z^{-m}),$$
$$N = 2n + 1, \text{ odd}, \quad (6a)$$

$$A_p(\Theta, \Phi) = A(z) = \sum_{i=1,3,5}^{N-1}(C_i z^{i/2} + C'_i z^{-i/2})$$
$$= \sum_{m=1}^{n}(C_m z^{m-1/2} + C'_m z^{-m+1/2}),$$
$$N = 2n, \text{ even}. \quad (6b)$$

By factoring out $C'_n z^{-n}$ and $C'_n z^{-n+\frac{1}{2}}$ in (6a) and (6b), the following expression is obtained:

$$A(z) = |A(z)| = \left|\sum_{m=0}^{N-1} c_m z^m\right| = |c_0 + c_1 z$$
$$+ c_2 z^2 + \cdots + c_{N-2} z^{N-2} + z^{N-1}|. \quad (7)$$

where the c's in (7) are new complex amplitude factors defined in terms of the C's in (6). Their magnitudes give the relative amplitudes of the currents in the several units referred to the current in the Nth element at the extreme right along the positive x-axis; their phase angles give the phase deviations from a progressive phase delay δ. The form (7) is that derived by Schelkunoff.[44] It is seen to reduce to the form (4) for the uniform array when all c's are equal to one, i.e. when all currents have equal magnitudes and zero deviations from the progressive phase delay. Since the c's are arbitrary, some may be zero. This means that particular elements with zero currents are missing. In this case N is not the *actual* number of elements but is called the *apparent* number. For the same reason b is called the apparent distance between elements, including those that are missing.

The significant contribution of Schelkunoff is the identification of the array factor of every linear array of identical parallel elements with commensurable separations with a polynomial of the form (7) and, conversely, the interpretation of every polynomial as the array factor of a possible linear array. The important bearing of this discovery on the beam width and on the reduction in the relative size of minor lobes follows directly. Since $A(z)$ is the determining factor in the relative sizes of lobes in the equatorial plane ($\Theta = \pi/2$), it is clear that if the array factor is squared or raised to still higher powers m, not only is the beam width of the main lobe reduced but the relative magnitudes of all minor lobes are decreased simultaneously. If the number of antennas in a given uniform array is N, the number N_m required for an array with array factor given by that of the uniform array raised to the mth power is

$$N_m = m(N - 1) + 1. \quad (8)$$

Since the directivity of a uniform array increases with the number of antennas, it is not particularly noteworthy that the directivity of a derived nonuniform array of N_m antennas exceeds that of a uniform array of $N < N_m$ antennas. What is significant is the fact that the level of the minor lobes can be decreased indefinitely by increasing N_m in the nonuniform array, whereas this level reaches an asymptotic minimum for the uniform array as N is increased without limit.

9. Binomial End-Fire Arrays

The ultimate reduction in the level of minor lobes is to have none at all. Directive parallel, arrays with no minor lobes may be constructed by beginning with the simple two-element end-fire array for which the normalized array factor in the equatorial plane ($\Theta = \pi/2$) is shown in Figs. 3.4 and 3.5 with $N = 2$ and $n_p = \frac{1}{4}$. This so-called "unidirectional" couplet consists of two antennas separated by a quarter wavelength and driven in phase quadrature. That is, $\beta_0 b = \pi/2$, $\delta = \pi/2$. Since the magnitude of the array factor of this couplet decreases continuously from its maximum with $\Phi = 0$, it is obvious that no minor lobes can be introduced by raising the array factor to an arbitrary power m. However, the resulting array factor applies to a new and nonuniform array with more than two antennas and with a sharper beam.

In the notation of Sec. 3, the normalized array factor in the equatorial plane ($\Theta = \pi/2$) of the two-element uniform end-fire array is given by $a = \sin Nx/N \sin x$ with $x = \psi/2$ given by (1.24) with $n_p = t_p = 1/4$. Thus,

$$a(\tfrac{1}{2}\pi, \Phi; 2, \tfrac{1}{4}, \tfrac{1}{4}) = \cos[\tfrac{1}{4}\pi(1 - \cos\Phi)].$$
$$(1a)$$

In the form (8.7) with added subscript N,

$$a_2(z) = A_2(z)/2 = \tfrac{1}{2}|1 + z|. \quad (1b)$$

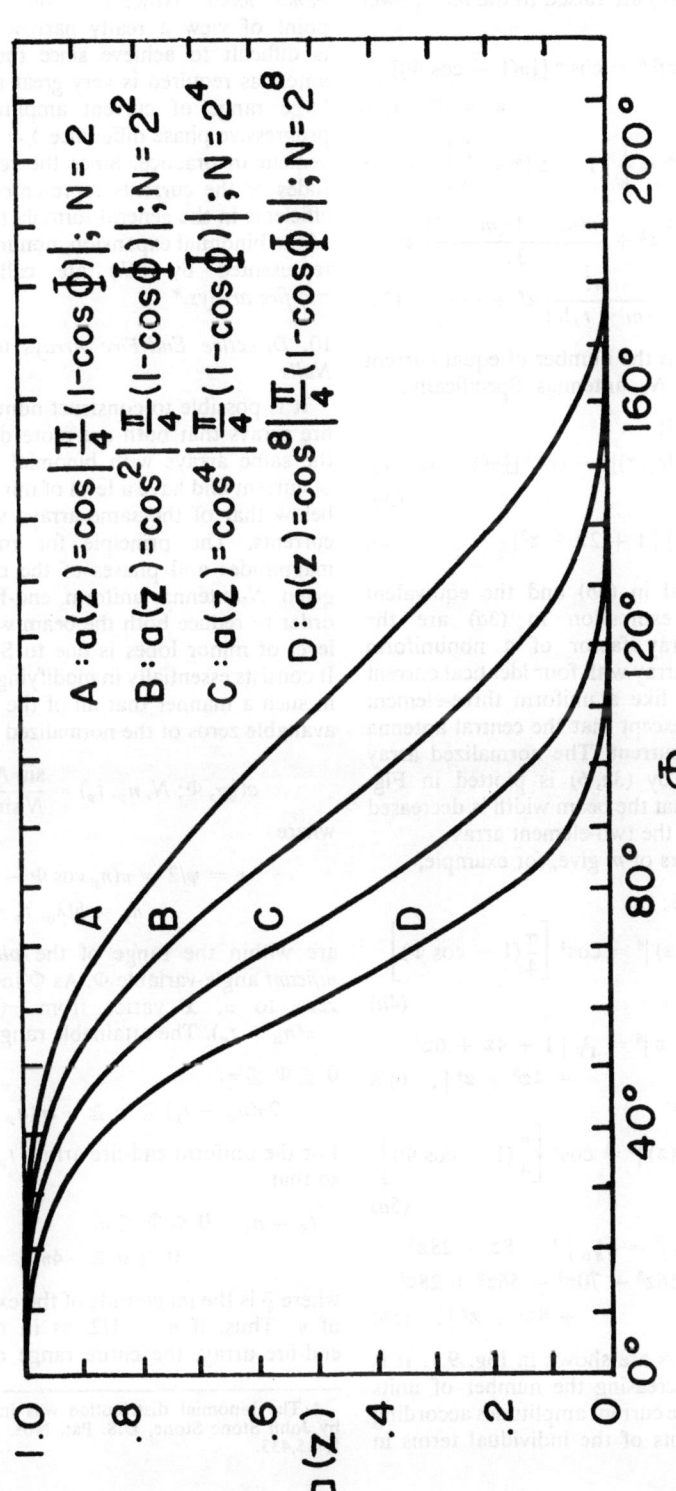

Fig. 9.1. Normalized array factors of binomial array (Schelkunoff).

A: $a(z) = \cos\left|\frac{\pi}{4}(1-\cos\phi)\right|$; $N = 2$

B: $a(z) = \cos^2\left|\frac{\pi}{4}(1-\cos\phi)\right|$; $N = 2^2$

C: $a(z) = \cos^4\left|\frac{\pi}{4}(1-\cos\phi)\right|$; $N = 2^4$

D: $a(z) = \cos^8\left|\frac{\pi}{4}(1-\cos\phi)\right|$; $N = 2^8$

This factor is shown graphically in Figs. 3.4 and 3.5; it is reproduced in Fig. 9.1.

If (1a) and (1b) are raised to the mth power the results are

$$a_w(z) = [a_2(z)]^m = \cos^m\left[\tfrac{1}{4}\pi(1 - \cos\Phi)\right],$$
$$w = 2^m, \quad (2a)$$

$$a_w(z) = [a_2(z)]^m = \frac{1}{w}|1 + z|^m = \frac{1}{w}\Big|1 + mz$$
$$+ \frac{m(m-1)}{2!}z^2 + \frac{m(m-1)(m-2)}{3!}z^3$$
$$+ \cdots + \frac{m!}{(m-r)!r!}z^r + \cdots\Big|, \quad (2b)$$

where $w = 2^m$ is the number of equal current elements in the N_m antennas. Specifically,

$m = 2$, $N_m = 3$:

$$a_4(z) = [a_2(z)]^2 = \cos^2\left[\tfrac{1}{4}\pi(1 - \cos\Phi)\right] \quad (3a)$$

$$= \tfrac{1}{4}|1 + 2z + z^2| \quad (3b)$$

The polynomial in (3b) and the equivalent trigonometric expression in (3a) are the normalized array factor of a nonuniform three-antenna array with four identical current elements. It is like a uniform three-element end-fire array except that the central antenna has twice the current. The normalized array factor defined by (3a, b) is plotted in Fig. 9.1. It is seen that the beam width is decreased compared with the two-element array.

Higher powers of m give, for example,

$m = 4$, $N_m = 5$:

$$a_{16}(z) = |a_2(z)|^4 = \cos^4\left[\frac{\pi}{4}(1 - \cos\Phi)\right]$$
$$(4a)$$

$$= \tfrac{1}{16}|1 + z|^4 = \tfrac{1}{16}|1 + 4z + 6z^2$$
$$+ 4z^3 + z^4|, \quad (4b)$$

$m = 8$, $N_m = 9$:

$$a_{256}(z) = |a_2(z)|^8 = \cos^8\left[\frac{\pi}{4}(1 - \cos\Phi)\right]$$
$$(5a)$$

$$= \tfrac{1}{256}|1 + z|^8 = \tfrac{1}{256}|1 + 8z + 28z^2$$
$$+ 56z^3 + 70z^4 + 56z^5 + 28z^6$$
$$+ 8z^7 + z^8|. \quad (5b)$$

The array factors are shown in Fig. 9.1. It is seen that by increasing the number of units and adjusting the current amplitudes according to the coefficients of the individual terms in the polynomials the beam width may be reduced indefinitely *without introducing any minor lobes*. However, from the practical point of view a really narrow beam width is difficult to achieve since the number of antennas required is very great and since the large range of current amplitudes with a progressive phase difference $\delta = \beta_0 b$ is inconvenient in practice. Since the relative amplitudes of the currents represented by the coefficients in the general formula (2b) are those of the binomial expansion, nonuniform arrays represented by (2b) are called *binomial end-fire arrays*.*

10. Directive End-Fire Arrays with Assigned Nulls

It is possible to construct nonuniform end-fire arrays that both are more directive than the same arrays with binomial distributions of current and have a level of minor lobes well below that of the same arrays with uniform currents. The principle for modifying the magnitudes and phases of the currents in a given N-antenna uniform end-fire array in order to reduce both the beam width and the level of minor lobes is due to Schelkunoff.[44] It consists essentially in modifying the currents in such a manner that all of the theoretically available zeros of the normalized array factor,

$$a(\tfrac{1}{2}\pi, \Phi; N, n_p, t_p) = \frac{\sin Nx}{N \sin x}, \quad (1a)$$

where

$$x = \psi/2 = \pi(n_p \cos\Phi - t_p),$$
$$n_p = b/\lambda_0, \quad t_p = \delta/2\pi \quad (1b)$$

are within the range of the *physically significant* angle-variable Φ. As Φ increases from zero to π, x varies from $\pi(n_p - t_p)$ to $-\pi(n_p + t_p)$. The attainable range of ψ is

$0 \leq \Phi \leq \pi$:

$$2\pi(n_p - t_p) \geq \psi \geq -2\pi(n_p + t_p). \quad (2a)$$

For the uniform end-fire array, $t_p = n_p = n_E$, so that

$t_E = n_E$: $0 \leq \Phi \leq \pi$,

$$0 \geq \psi \geq -4\pi n_E = -\bar\psi, \quad (2b)$$

where $\bar\psi$ is the magnitude of the extreme range of ψ. Thus, if $n_E = 1/2$, as in the bilateral end-fire array, the entire range of ψ from 0

---

* The binomial distribution was introduced first by John Stone Stone, U.S. Pat. Nos. 1,643,323 and 1,715,433.

to -2π is attainable when Φ varies over its full range. (Note that the range of Φ from π to 2π duplicates the range from π to 0 since the array factors of all parallel arrays are symmetric with respect to the plane of the array.) If n_E is less than $1/2$, the range of ψ that is attainable by varying Φ is less than 2π. In particular, with $n_E = 1/4$, as for the unilateral end-fire array, ψ is limited to the range from 0 to $-\pi$; with $n_E = 1/8$, ψ can vary from 0 to $-\pi/2$. Evidently, the number of nulls available by varying Φ over its entire range with $n_E < 1/4$, $\bar{\psi} < 2\pi$, is always less than the theoretical number, $N-1$, in the range of x from zero to $-\pi$ or ψ from 0 to -2π. This is clear from Fig. 3.5 in which the total number of nulls in the range $0 \leq \Phi \leq 2\pi$ is $2(N-1)$ for $n_E = 1/2$ but only one half this number for $n_E = 1/4$.

Schelkunoff's method of adjusting the N currents in such a way that all of the theoretically available nulls of the array factor are within the range of Φ follows directly from a consideration of the normalized array factor in polynomial form. In general this is (8.7), namely,

$$a(z) = \left|\frac{A(z)}{A(z=1)}\right| = \frac{1}{N}|c_0 + c_1 z + c_2 z^2 + \cdots + c_{N-1} z^{N-1}|,$$
$$z = e^{j\psi}, \quad (3)$$

where ψ is defined in (1b). For the uniform array $c_0 = c_1 = c_2 = \cdots = c_{N-1} = 1$. Let the $N-1$ zeros of the equation

$$a(z) = 0, \quad (4a)$$

where $a(z)$ is given by (2), be denoted by $z_1, z_2, \ldots, z_p, \ldots, z_{N-1}$. In terms of these zeros (3) may be expressed as follows:

$$a(z) = \left|\frac{A(z)}{A(z=1)}\right|$$
$$= \left|\frac{(z-z_1)(z-z_2)\cdots(z-z_p)\cdots(z-z_{N-1})}{(1-z_1)(1-z_2)\cdots(1-z_p)\cdots(1-z_{N-1})}\right|. \quad (4b)$$

For the uniform array the roots may be evaluated directly by summing the geometric series given by (3) with all c's replaced by unity. Thus,

$$a(z) = \frac{1}{N}|1 + z + z^2 + \cdots + z^{N-1}|$$
$$= \frac{1}{N}\left|\frac{z^N - 1}{z - 1}\right| \quad (5)$$

The $N-1$ zeros are given by

$$z^N - 1 = 0, \quad z \neq 1. \quad (6)$$

The principal maximum is given by

$$z = 1. \quad (7)$$

Note that (6) corresponds to $\sin Nx = 0$, $\sin x \neq 0$, and (7) to $\sin x = 0$ as given and evaluated in Sec. 1. The roots of (5) are

$$z = z_p = e^{j\psi_p}, \quad \psi_p = -2\pi p/N,$$
$$p = 1, 2, 3, \cdots, N-1. \quad (8)$$

With (1b) it follows from (7) that,

$a(\tfrac{1}{2}\pi, \Phi; N, n_E, t_E) = 0$ when

$$\cos \Phi = \cos \Phi_p = \frac{\delta}{\beta b} - \frac{2\pi p}{N\beta_0 b}$$
$$= (t_E - p/N)/n_E = 1 - p/Nn_E. \quad (9)$$

This corresponds to (3.45) for the range $0 \geq x \geq -\pi$ or $0 \geq \psi \geq -2\pi$, that is, $\bar{\psi} = 2\pi$.

The location of the zeros of the array factor $a(z)$ may be shown graphically by noting that $z = e^{j\psi}$ is a complex number that may be represented as a pointer drawn from the origin of the real and imaginary axes in the complex plane in a direction that makes an angle ψ with the positive real axis. This is shown in Fig. 10.1 for $N = 6$, $n_E = 1/2$. Note that since $z = e^{j\psi}$ has a magnitude of unity, the end of the pointer traces the unit circle as ψ rotates from 0 to 2π in a clockwise direction. The principal maximum is indicated by the large dot at $z = 1$. The zeros defined by (8) are uniformly spaced around the entire circle; they are denoted by small circles in Fig. 10.1. It is shown in Sec. 1 that minor maxima are nearly halfway between the zeros. They are shown in Fig. 10.1 by smaller dots. The uniformity of spacing of the nulls around the unit circle is with respect to Ψ and not with respect to Φ. Unit circles for uniform and nonuniform end-fire arrays are shown in Figs. 10.2–10.5 with nulls and principal maxima indicated.

Schelkunoff's method of decreasing the beam width and the relative amplitudes of minor lobes is ingenious. It involves shifting the theoretical nulls of the array factor of the uniform array in such a way that they are *uniformly spaced* in that *part* of the unit circle that is within the range of Φ, $|\psi| \leq \bar{\psi}$ instead of around the entire unit circle. Thus, for $n_E = 1/4$, $N-1$ nulls must be spaced uniformly in the lower semicircle since $\bar{\psi} = \pi$. For $n_E = 1/8$, the $N-1$ nulls must be in the fourth quadrant of the unit circle since $\bar{\psi} = \pi/2$.

The normalized array factor corresponding to $N-1$ roots distributed uniformly in a restricted range of $|\psi| \leq \bar{\psi}$ is constructed readily. Instead of the roots defined in (6) the following values must apply:

$n_E = \tfrac{1}{4}$: $z = z_p = e^{j\psi_p}$, $\psi_p = -\pi p/(N-1)$,

$$p = 1, 2, 3, \cdots, N-1, \quad (10a)$$

$n_E = \tfrac{1}{8}$: $z = z_p = e^{j\psi_p}$, $\psi_p = -\pi p/2(N-1)$,

$$p = 1, 2, 3, \cdots, N-1. \quad (10b)$$

The corresponding expressions for the normalized array factors are obtained by substituting the values of $z = z_1, z_2, \cdots, z_{N-1}$ defined in (10a) or (10b) in

$$a(z) = \left| \frac{A(z)}{A(z=1)} \right|$$

$$= \left| \frac{(z-z_1)(z-z_2)\cdots(z-z_p)\cdots(z-z_{N-1})}{(1-z_1)(1-z_2)\cdots(1-z_p)\cdots(1-z_{N-1})} \right|.$$

$$(11a)$$

This may be expressed in trigonometric form by noting that

$$z - z_1|/|1 - z_1| = |e^{j\psi} - e^{j\psi_1}|/|1 - e^{j\psi_1}|$$
$$= \sin \tfrac{1}{2}(\psi - \psi_1)/\sin \tfrac{1}{2}\psi_1,$$

so that

$$a(z) = \frac{\sin \tfrac{1}{2}(\psi - \psi_1) \sin \tfrac{1}{2}(\psi - \psi_2) \cdots \sin \tfrac{1}{2}(\psi - \psi_p) \cdots \sin \tfrac{1}{2}(\psi - \psi_{N-1})}{\sin \tfrac{1}{2}\psi_1 \sin \tfrac{1}{2}\psi_2 \cdots \sin \tfrac{1}{2}\psi_p \cdots \sin \tfrac{1}{2}\psi_{N-1}}.$$

$$(11b)$$

If the nulls are uniformly spaced with respect to ψ in the range $|\psi| \leq \bar{\psi}$, (11b) may be expressed conveniently in terms of $x = \psi/2$ with

$$x_1 = \psi_1/2 = -\bar{\psi}/2(N-1). \quad (11c)$$

Thus,

$$a(z) = \frac{\sin(x-x_1)\sin(x-2x_1)\sin(x-3x_1)\cdots\sin(x-[N-1]x_1)}{\sin x_1 \sin 2x_1 \sin 3x_1 \cdots \sin(N-1)x_1}$$

$$(11d)$$

In order to determine the relative amplitudes and phases of the currents in the antennas of the array with the desired array factor (11a, b, d), it is necessary to determine the coefficients of the successive powers of z in the geometric progression (3). Since only relative currents are involved it is sufficient to transform the numerator of (11a) into a geometric series like (3). This is an algebraic transformation involving the relations between coefficients and roots. It was carried out by Schelkunoff who obtained the equivalent of the following *ratios* for the magnitudes of the currents:

$$1 : \frac{\sin(N-1)x_1}{\sin x_1} : \frac{\sin(N-1)x_1 \sin(N-2)x_1}{\sin x_1 \sin 2x_1}$$

$$: \frac{\sin(N-1)x_1 \sin(N-2)x_1 \sin(N-3)x_1}{\sin x_1 \sin 2x_1 \sin 3x_1}$$

$$: \cdots : 1. \quad (12a)$$

The phases of the currents vary progressively by an angle δ' differing from the angle $\delta = 2\pi t$ of the uniform array. The phase delay from unit to adjacent unit is

$$\delta' = \delta + \pi + Nx_1. \quad (12b)$$

In order to illustrate the construction of nonuniform end-fire arrays the following examples selected by Schelkunoff are convenient.

Three-antenna end-fire array: $N = 3$; $n_E = t_E = 1/4$. The normalized array factor of the uniform array is

$$a(z) = \tfrac{1}{3}|1 + z + z^2| = \frac{\sin 3\psi/2}{3 \sin \psi/2},$$

$$\psi = \frac{\pi}{2}(\cos \Phi - 1), \quad \bar{\psi} = 4\pi n_E = \pi. \quad (13)$$

A graph of this function is shown in curve A of Fig. 10.2. The zeros of (13) are given by (8) with $N = 3$. They are

$$\psi_{1u} = -2\pi/3, \quad z_{1u} = e^{j\psi_{1u}};$$

$$\psi_{2u} = -4\pi/3, \quad z_{2u} = e^{j\psi_{2u}}. \quad (14)$$

The subscript u is added to designate the values for the uniform array. The pointers z_{1u} and z_{2u} are indicated in the unit circle A in Fig. 10.2. Note that corresponding to $\psi_{1u} = -2\pi/3$ there is a zero in the field pattern given by $\Phi = -109.3°$ as obtained using (13). There is no zero corresponding to $\psi_{2u} = -4\pi/3$ since this lies outside the range $\bar{\psi} = \pi$.

[VI.10] THEORY OF LINEAR ANTENNAS

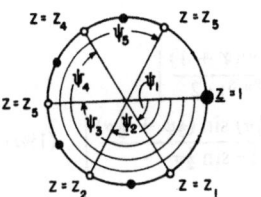

Fig. 10.1. Unit circle for six-antenna uniform array.

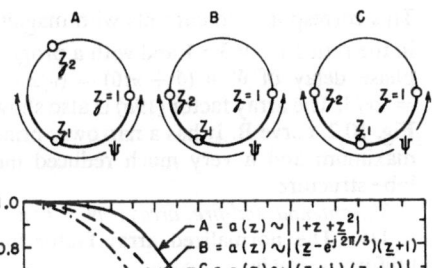

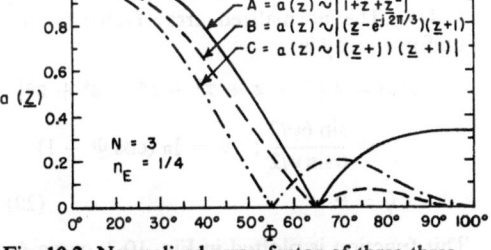

Fig. 10.2. Normalized array factor of three three-element end-fire arrays (Schelkunoff).

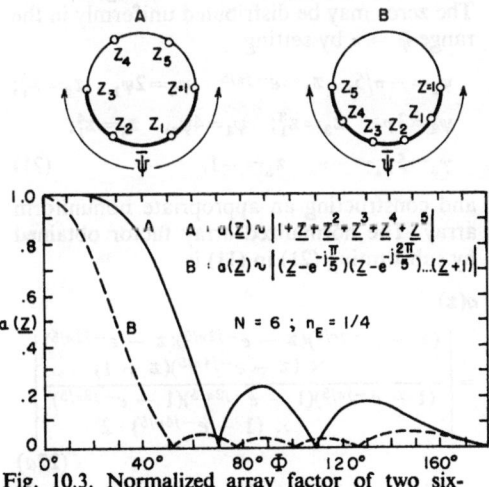

Fig. 10.3. Normalized array factor of two six-antenna end-fire arrays (Schelkunoff).

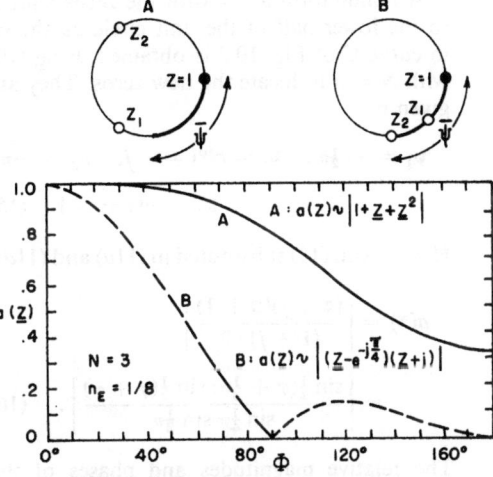

Fig. 10.4. Normalized array factor of two three-antenna end-fire arrays (Schelkunoff).

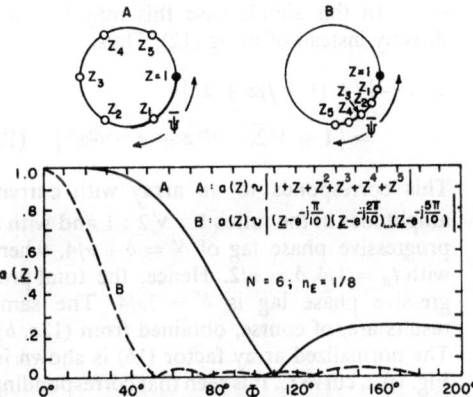

Fig. 10.5. Normalized array factor of two six-antenna end-fire arrays (Schelkunoff).

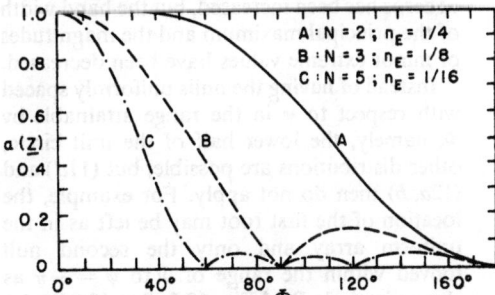

Fig. 10.6. Normalized array factor of three arrays with the same length and properly spaced nulls (Schelkunoff).

A nonuniform array with the zeros shifted to the lower half of the unit circle as shown in curve C of Fig. 10.2 is obtained using (10) with $N = 3$ to locate the new zeros. They are given by

$$\psi_1 = -\tfrac{1}{2}\pi, \quad z_1 = e^{j\psi_1} = -j; \quad \psi_2 = -\pi,$$
$$z_2 = e^{j\psi_2} = -1. \quad (15)$$

Hence, with (15) substituted in (11a) and (11d),

$$a(z) = \left| \frac{(z+j)(z+1)}{(1+j)\cdot 2} \right|$$
$$= \left| \frac{\sin \tfrac{1}{2}(\psi + \tfrac{1}{2}\pi)\sin \tfrac{1}{2}(\psi + \pi)}{\sin \tfrac{1}{4}\pi \sin \tfrac{1}{2}\pi} \right|. \quad (16)$$

The relative magnitudes and phases of the currents in the three units when the array has the normalized array factor (15) are obtained by expanding (16) in a polynomial in z and noting the coefficients of the successive powers of z. In this simple case this may be done directly instead of using (12). Thus,

$$a(z) \sim |j + (1+j)z + z^2|$$
$$= |1 + \sqrt{2}\,e^{-j\pi/4}z + e^{-j\pi/2}z^2|. \quad (17)$$

This corresponds to an array with current amplitudes in the ratios $1 : \sqrt{2} : 1$ and with a progressive phase lag of $\delta' = \delta + \pi/4$, where with $t_E = 1/4$, $\delta = \pi/2$. Hence, the total progressive phase lag is $\delta' = 3\pi/4$. The same results are, of course, obtained from (12a, b). The normalized array factor (16) is shown in Fig. 10.2, curve C. It is seen that corresponding to theoretical nulls with $\psi = -90°$ and $-180°$ the actual zeros in the field pattern occur when $\Phi = 90°$ and $180°$ as obtained from (13). Compared with the uniform array, the number of zeros has been increased, but the band width of the principal maximum and the magnitudes of minor extreme values have been decreased.

Instead of having the nulls uniformly spaced with respect to ψ in the range attainable by Φ, namely, the lower half of the unit circle, other distributions are possible, but (11d) and (12a, b) then do not apply. For example, the location of the first root may be left as in the uniform array and only the second null moved within the range of Φ to $\psi = -\pi$ as shown in circle B of Fig. 10.2. Specifically, let

$$\psi_1 = \psi_{1u} = -2\pi/3, \quad z_1 = e^{-j\psi_1};$$
$$\psi_2 = -\pi, \quad z_2 = e^{j\psi_2} = -1. \quad (18)$$

In this case

$$a(z) = \left| \frac{(z - e^{-j2\pi/3})(z+1)}{(1 - e^{-j2\pi/3})\cdot 2} \right|$$
$$= \left| \frac{\sin(\tfrac{1}{2}\psi + \tfrac{1}{3}\pi)\sin(\tfrac{1}{2}\psi + \tfrac{1}{2}\pi)}{\sin \tfrac{1}{3}\pi \sin \tfrac{1}{2}\pi} \right| \quad (19a)$$

and

$$a(z) \sim |z^2 + z(1 - e^{-j2\pi/3}) - e^{-j2\pi/3}|$$
$$= |1 + \cos \tfrac{1}{6}\pi\, e^{\,\pi/6}z + e^{-j\pi/3}z^2|. \quad (19b)$$

This corresponds to currents with magnitudes in the ratios $1 : \sqrt{3} : 1$ and with a progressive phase delay of $\delta' = (\delta + \pi/6) = (\pi/2 + \pi/6) = 2\pi/3$. The array factor (19a) is also shown in Fig. 10.2, curve B. It has a narrower principal maximum and a very much reduced minor-lobe structure.

Six-antenna end-fire array: $N = 6; n_E = t_E = 1/4$. The normalized array factor for the uniform array is

$$a(z) = \tfrac{1}{6}|1 + z + z^2 + z^3 + z^4 + z^5|$$
$$= \frac{\sin 6\psi/2}{6 \sin \psi/2}; \quad \psi = \tfrac{1}{2}\pi(\cos\Phi - 1);$$
$$\bar{\psi} = \pi. \quad (20)$$

This function is plotted in Fig. 10.3, curve A. The zeros of (20) are given by (8) with $N = 6$. They are uniformly distributed around the entire unit circle A in Fig. 10.3. It is seen that two are outside the attainable range, $\bar{\psi} = \pi$. The zeros may be distributed uniformly in the range $\bar{\psi} = \pi$ by setting

$$\psi_1 = -\pi/5, \quad z_1 = e^{-j\pi/5}; \quad \psi_2 = 2\psi_1, \quad z_2 = z_1^2;$$
$$\psi_3 = 3\psi_1, \quad z_3 = z_1^3; \quad \psi_4 = 4\psi_1, \quad z_4 = z_1^4;$$
$$\psi_5 = 5\psi_1 = -\pi, \quad z_5 = -1, \quad (21)$$

and constructing an appropriate nonuniform array. The normalized array factor obtained by substituting (21) in (11) is

$$a(z)$$
$$= \left| \frac{(z - e^{-j\pi/5})(z - e^{-j2\pi/5})(z - e^{-j3\pi/5})}{(1 - e^{-j\pi/5})(1 - e^{-j2\pi/5})(1 - e^{-j3\pi/5})} \right.$$
$$\left. \frac{\times (z - e^{-j4\pi/5})(z+1)}{\times (1 - e^{-j4\pi/5})\cdot 2} \right| \quad (22a)$$

$$= \left| \frac{\sin \tfrac{1}{2}(\psi + \pi/5) \sin \tfrac{1}{2}(\psi + 2\pi/5)}{\sin(\pi/10)\sin(2\pi/10)\sin(3\pi/10)} \right.$$
$$\left. \frac{\times \sin \tfrac{1}{2}(\psi + 3\pi/5)\sin \tfrac{1}{2}(\psi + 4\pi/5)}{\times \sin(4\pi/10)\sin(5\pi/10)} \right. \quad (22b)$$
$$\left. \frac{\times \sin \tfrac{1}{2}(\psi + \pi)}{} \right|$$

The relative magnitudes and phases of the currents in the six antennas when driven to have the array factor (22a, b) may be determined from (12a, b). Since $x_1 = -\pi/10$ and $\delta = \pi/2$ they are

$$1 : \frac{\sin(0.5\pi)}{\sin(0.1\pi)} : \frac{\sin(0.5\pi)\sin(0.4\pi)}{\sin(0.1\pi)\sin(0.2\pi)} :$$

$$\frac{\sin(0.5\pi)\sin(0.4\pi)}{\sin(0.1\pi)\sin(0.2\pi)} : \frac{\sin(0.5\pi)}{\sin(0.1\pi)} : 1$$

$$= 1 : 3.24 : 5.24 : 5.24 : 3.24 : 1. \quad (22c)$$

The progressive phase delay from antenna to antenna is

$$\delta' = \pi/2 + \pi - 6\pi/10 = 0.9\pi. \quad (22d)$$

The graph of (22a) is given in Fig. 10.3 together with the unit circle showing the location of the zeros and the principal maximum. It is seen that compared with the 6-antenna uniform end-fire array the directivity is increased greatly; the number of minor lobes is doubled but their amplitudes very much reduced.

Three-antenna array: $N = 3$, $n_E = t_E = 1/8$. The normalized array factor for the uniform array is (12) with $\psi = \frac{1}{4}\pi(\cos\Phi - 1)$. It is plotted in curve A of Fig. 10.4. The two theoretical nulls defined by (13) apply, but as seen from the unit circle A neither of them is within the restricted range $\bar{\psi} = \pi/2$. Hence the uniform array has no zeros. If the two theoretical nulls of the array factor are moved so that they are uniformly distributed in the range $0 \geq \psi \geq -\pi/2$, it follows that

$$\psi_1 = -\frac{\pi}{4}, \quad z_1 = e^{-j\pi/4} = (1-j)/\sqrt{2};$$

$$\psi_2 = -\frac{\pi}{2}, \quad z_2 = e^{-j\pi/2} = -j. \quad (23)$$

With (23) the normalized array factor becomes

$$a(z) = \left|\frac{(z - e^{-j\pi/4})(z + j)}{(1 - e^{-j\pi/4})(1 + j)}\right|$$

$$= \frac{\sin\frac{1}{2}(\psi + \pi/4)\sin\frac{1}{2}(\psi + \pi/2)}{\sin(\pi/8)\sin\pi/4}. \quad (24)$$

The polynomial for the numerator is

$$a(z) \sim |z^2 - z(e^{-j\pi/4} + e^{-j\pi/2}) + e^{-j3\pi/4}|$$

$$= \left|1 + 2\cos\frac{\pi}{8}e^{-j5\pi/8}z + e^{-j5\pi/4}z^2\right|. \quad (25)$$

Thus, the currents have magnitudes in the ratio $1 : 1.85 : 1$, with a progressive phase delay $\delta' = \delta + 5\pi/8 = \pi/4 + 5\pi/8 = 7\pi/8$.

The same values also may be obtained from (12a, b) using $x_1 = -\pi/8$. Corresponding to the nulls for ψ defined in (23), the zeros for Φ are obtained using $\psi = \frac{1}{4}\pi(\cos\Phi - 1)$. Thus, the array factor vanishes when $\Phi = 90°$ and $180°$. A graph of (24) is shown in curve B of Fig. 10.4. Compared with the pattern of the uniform array, that of the nonuniform array is much more highly directive.

Six-antenna array: $N=6$, $n_E = t_E = 1/8$. The normalized array factor for the uniform array is like (20) but with $\psi = \frac{1}{4}\pi(\cos\Phi - 1)$. It is shown in curve A of Fig. 10.5. Of the five theoretically available nulls in the array factor only one is within the range $0 \geq \psi \geq -\pi/2 = -\bar{\psi}$ as seen from unit circle A in Fig. 10.5. If the five nulls are all moved within this restricted range of $|\psi| \leq \bar{\psi} = \pi/2$ by setting

$$\psi_1 = -\frac{\pi}{10}, \quad z_1 = e^{-j\pi/10}; \quad \psi_2 = 2\psi_1, \quad z_2 = z_1^2;$$

$$\psi_3 = 3\psi_1, \quad z_3 = z_1^3; \quad \psi_4 = 4\psi_1, \quad z_4 = z_1^4;$$

$$\psi_5 = 5\psi_1 = -\frac{\pi}{2}, \quad z_5 = e^{-j\pi/2} = -j, \quad (26)$$

the normalized array factor becomes

$$a(z) =$$

$$\left|\frac{(z - e^{-j\pi/10})(z - e^{-j\pi/5})(z - e^{-j3\pi/10})}{(1 - e^{-j\pi/10})(1 - e^{-j\pi/5})(1 - e^{-j3\pi/10})} \right.$$

$$\left. \times \frac{(z - e^{-j2\pi/5})(z + j)}{\times (1 - e^{-j2\pi/5})(1 + j)}\right|$$

(27a)

$$= \left|\frac{\sin\frac{1}{2}(\psi + \pi/10)\sin\frac{1}{2}(\psi + \pi/5)}{\sin(\pi/20)\sin(\pi/10)\sin(3\pi/20)} \right.$$

$$\left. \times \frac{\sin\frac{1}{2}(\psi + 3\pi/10)\sin\frac{1}{2}(\psi + 2\pi/5)}{\times \sin\frac{1}{2}(\psi + \pi/2)} \right|.$$

$$\left. \times \sin(\pi/5)\sin(\pi/4) \right|$$

(27b)

The corresponding relative currents may be determined from (12a, b). Since $x_1 = -\pi/20$ and $\delta = \pi/4$, it follows from (12a) that the ratios of current magnitudes are

$$1 : \frac{\sin 0.25\pi}{\sin 0.05\pi} : \frac{\sin 0.25\pi \sin 0.20\pi}{\sin 0.05\pi \sin 0.10\pi} :$$

$$\frac{\sin 0.25\pi \sin 0.20\pi}{\sin 0.05\pi \sin 0.10\pi} : \frac{\sin 0.25\pi}{\sin 0.05\pi} : 1$$

$$= 1 : 4.5 : 8.6 : 8.6 : 4.5 : 1. \quad (27c)$$

The progressive phase delay is

$$\delta' = \frac{\pi}{4} + \pi - \frac{6\pi}{20} = \frac{19\pi}{20}. \quad (27d)$$

A graph of (27) is given in curve B of Fig. 10.5. It is seen to represent a highly directive array with minor lobes of very small amplitude.

The four examples of nonuniform end-fire arrays indicate that highly directive arrays with narrow band width between nulls and a very low level of minor lobes can be constructed using closely spaced antennas with currents adjusted to have the nulls of the array factor all in the restricted range $\bar{\psi}$ which is attainable by varying the angle Φ. By increasing the number of antennas and decreasing the spacing between them, the directivity increases if the nulls of the array factor are always kept uniformly spaced in the range $\bar{\psi}$. Use may be made of this fact in circumstances where for physical reasons the over-all length of an array is limited so that a *closely spaced array* of parallel antennas must be used. Evidently, it is possible to achieve high directivity by increasing the number N of the parallel antennas while the distance $b = n_E \lambda_0$ between adjacent units is reduced so that Nb remains constant, provided the $N-1$ nulls are properly spaced in the range of ψ attainable by Φ. This is illustrated in Fig. 10.6, where the normalized array factors of three arrays with $Nb = \lambda_0/4$ or $Nn_E = 1/4$ are shown. The arrays have $N = 2$, $t_E = n_E = 1/4$; $N = 3$, $t_E = n_E = 1/8$; $N = 5$, $t_E = n_E = 1/16$. In each case the nulls are uniformly spaced in the appropriate ranges of $\bar{\psi} = -4\pi n_E$.

11. Nonuniform Broadside Arrays Derived from Uniform arrays

The array factor of the two-antenna uniform end-fire array was used in Sec. 9 as prototype in deriving multiantenna nonuniform *end-fire* arrays specifically because array factors with *completely* suppressed minor lobes were desired. The fact that these array factors are relatively not very directive for the number of antennas involved is the result of the smoothly decreasing but very broad lobe of the two-element end-fire array. This is ideal for suppressing minor lobes but not for extreme directivity. If high directivity with a reduction but not necessarily complete elimination of side lobes is desired, one course is that described in Sec. 10 in which the relatively low directivity of uniform end-fire arrays is improved greatly by appropriate changes in the currents. An alternative procedure is to begin with the uniform broadside array, which has a much narrower beam width than the uniform end-fire array with the same number of antennas. Evidently, binomial broadside arrays may be derived from the two-element uniform broadside array in a manner analogous to that followed in deriving the binomial end-fire arrays from the "unidirectional" couplet in Sec. 9. However, since the directivity of the two-element broadside array itself is not great, and since it has no minor lobes, the effect of raising its array factor to integral powers is not essentially different from that using the two-element end-fire array as prototype except that there are two identical major lobes instead of one. A more interesting and significant investigation is the determination and intercomparison of the array factors of nonuniform broadside arrays all of which have the same number of antennas as a given broadside array. It follows from (8.8) that the number of antennas N in a uniform array is related to the number N_m in the derived nonuniform array by

$$N = \frac{1}{m}(N_m - 1) + 1. \quad (1)$$

In order to have a wide selection of arrays for intercomparison, N_m must be chosen so that $(N_m - 1)$ is divisible by as many integral values of m as possible. A very convenient and instructive choice is that made by Schelkunoff, namely, $N_m = 7$. With this value,

$$N = \frac{6}{m} + 1. \quad (2)$$

Evidently there are four integral values of m including that for the uniform array, and, correspondingly, there are four numbers N. These are $m = 1, 2, 3, 6$; $N = 7, 5, 3, 2$. The normalized array factors for the four seven-antenna arrays are:

$m = 1$ (uniform array):

$$a_7(z) = \tfrac{1}{7}[1 + z + z^2 + z^3 + z^4 + z^5 + z^6] \quad (3a)$$

$$= a(\tfrac{1}{2}\pi, \Phi; 7, \tfrac{1}{2}, 0)$$

$$= \frac{\sin\left(\dfrac{7\pi}{2}\cos\Phi\right)}{7\sin\left(\dfrac{\pi}{2}\cos\Phi\right)}. \quad (3b)$$

The other nonuniform seven-antenna arrays are obtained by raising to the power m the array factors of those uniform arrays that have

N antennas as given by (2). Note that the subscript w on $a_w(z)$ is the number of equal current elements in the $N_m = 7$ antennas. Thus,

$m = 2$:

$$a_{16}(z) = [a_4(z)]^2 = \tfrac{1}{16}[1 + 2z + 3z^2$$
$$+ 4z^3 + 3z^4 + 2z^5 + z^6] \quad (4a)$$

$$= [a(\tfrac{1}{2}\pi, \Phi; 4, \tfrac{1}{2}, 0)]^2$$

$$= \left[\frac{\sin(2\pi \cos \Phi)}{(4 \sin \tfrac{1}{2}\pi \cos \Phi)}\right]^2; \quad (4b)$$

$m = 3$:

$$a_{27}(z) = [a_3(z)]^3 = \tfrac{1}{27}[1 + 3z + 6z^2$$
$$+ 7z^3 + 6z^4 + 3z^5 + z^6] \quad (5a)$$

$$= [a(\tfrac{1}{2}\pi, \Phi; 3, \tfrac{1}{2}, 0)]^3$$

$$= \left[\frac{\sin(\tfrac{3}{2}\pi \cos \Phi)}{3 \sin(\tfrac{1}{2}\pi \cos \Phi)}\right]^3; \quad (5b)$$

$m = 6$:

$$a_{64}(z) = [a_2(z)]^6 = \tfrac{1}{64}[1 + 6z + 15z^2$$
$$+ 20z^3 + 15z^4 + 6z^5 + z^6] \quad (6a)$$

$$= [a(\tfrac{1}{2}\pi, \Phi; 2, \tfrac{1}{2}, 0)]^6$$

$$= \left[\frac{\sin(\pi \cos \Phi)}{2 \sin \tfrac{1}{2}\pi \cos \Phi}\right]^6$$

$$= \cos^6(\tfrac{1}{2}\pi \cos \Phi). \quad (6b)$$

The normalized array factors of the seven-element broadside array are shown in Fig. 11.1 for the uniform array and for the three nonuniform arrays derived, respectively, from $N = 4$-, 3-, and 2-element uniform broadside arrays by raising the array factors to the powers $m = 2, 3, 6$. The relative amplitudes of the currents in these arrays are given by the coefficients in (3a), (4a), (5a), and (6a). The distribution in (4a) is triangular; that in (6a) is binomial. It is seen from Fig. 11.1 that the uniform array has the narrowest beam width but the highest level and greatest number of minor lobes. As the number N of antennas in the prototype uniform array decreases and the power m to which its array factor is raised in constructing the $N_m = 7$-element nonuniform array is increased, the level of the minor lobes decreases but the beam width increases.

12. Real Polynomials for Symmetric Nonuniform Broadside Arrays

It is shown in Sec. 11 that the inescapable limitation on uniform arrays, that the level of side lobes cannot be decreased below a fixed limit by increasing the number of antennas, does not exist for nonuniform arrays constructed by raising to an arbitrary power the array factor of the uniform array. However, it is also shown that for a given number of antennas the beam width between nulls is increased above that of the uniform array as the level of minor lobes is reduced. For broadside arrays in which the distance b between antennas is less than $\lambda_0/2$ ($n_B < \tfrac{1}{2}$) both the beam width and the level of minor lobes can be decreased by properly spacing the nulls in the array factor in the manner described for end-fire arrays in Sec. 10. However, if the distance between elements equals or exceeds $\lambda_0/2$ ($n_B \geq \tfrac{1}{2}$) this method is not available. Evidently, what is required is a new method that permits the determination of an *optimum* distribution of currents in the several units of a parallel, uniformly spaced array of identical driven antennas. By optimum is meant a distribution such that if (a) the minor-lobe level is assigned, the beam width is a minimum, or if (b) the beam width, that is, the location of the first null, is specified, the magnitude of the minor lobes is minimized. These specifications are ambiguous in the sense that it is not clear precisely what is meant by a minimum of the level of minor lobes. If there is only one side lobe the statement is clear; but when there are many it is not. In all of the patterns obtained in preceding sections, including those of uniform arrays and nonuniform arrays derived from them, the magnitude of the minor lobes decreases continuously in both directions from each principal maximum. This means that the largest minor lobe is always adjacent to the principal maximum. In most applications this is not desirable, since a low level of the side lobe nearest the principal one often is just as or even more important than a low level in lobes that are further removed. This means that the array factor of the uniform array could be improved if the level of minor lobes at wide angles from the principal maximum were increased and those at small angles decreased so that in the final result *all minor lobes are equal* in magnitude and at a specified low level. Evidently, if all minor lobes are equal, the ambiguity in requiring their level to be a minimum is removed.

A method for deriving and designing non-uniform broadside arrays that have an optimum distribution of currents in the sense defined above is due to Dolph.[12,39] It consists first in showing that the array factor of a broadside array is a polynomial of real terms and then identifying the optimum polynomials with those of Tchebyscheff.[I.16] These are characterized by a minimum departure from zero in a range that may be chosen to coincide with the range of minor lobes in an array factor. The sequence of steps in deriving Tchebyscheff array factors is carried out systematically in order to clarify both the procedure in constructing the array factors and the properties of the Tchebyscheff polynomials.

The array factor of a general array of N antennas is given by (3.30) for $N = 2n + 1$ and by (3.32) for $N = 2n$, with n an integer. These factors may be specialized to apply to a symmetric broadside array by setting equal to zero the progressive phase factors, $\delta'_m = \delta_m = 0$, so that all currents are in phase, and setting $k'_m = k_m$ so that the current r units to the right of the center of the array is equal to the current r units to the left of the center, but the currents are otherwise arbitrary. The resulting factors are

$$A_{\text{odd}}(\Theta, \Phi) = 1 + 2 \sum_{m=1}^{(N-1)/2} k_m \cos \beta_0 S_m$$

$$= 2 \sum_{m=0}^{(N-1)/2} k_m \cos \beta_0 S_m, \quad N = 2n+1, \quad (1a)$$

$$A_{\text{even}}(\Theta, \Phi) = 2 \sum_{m=1,3,5}^{N-1} k_m \cos \tfrac{1}{2}\beta_0 S_m,$$

$$N = 2n, \quad (1b)$$

where in (1a) $k_0 = \tfrac{1}{2}$ and where

$$\beta_0 S_m = m\beta_0 b \sin \Theta \cos \Phi$$
$$= 2\pi m n_B \sin \Theta \cos \Phi,$$
$$n_B = b/\lambda_0. \quad (1c)$$

Since principal interest is in the equatorial plane, let $\Theta = \pi/2$. Also, in the notation of Sec. 1, let

$$x = n_B \pi \cos \Phi. \quad (2)$$

In the array factors as given in (1a, b) it is implied that the reference current is that in the central antenna for N odd and in the two innermost antennas for N even. That is,

$$k_0 = \tfrac{1}{2}, \quad k_i = I_i/I_0 \text{ for } N = 2n + 1,$$
$$i = 1, 2, \cdots, n, \quad (3a)$$

$$k_1 = 1, \quad k_i = I_i/I_1 \text{ for } N = 2n,$$
$$i = 1, 2, \cdots, n. \quad (3b)$$

This choice of reference current is not necessarily convenient in the study of nonuniform arrays, since in special cases these currents may be zero. For this reason it is assumed in the following that any convenient antenna number q may be selected as the one with the reference current by setting $k_q = 1$ and defining the k_i's as follows:

$$k_q = 1, \quad k_i = I_i/I_q, \quad k_0 = I_0/2I_q, \quad i = 1, 2, \cdots, n. \quad (3c)$$

This implies, of course, that I_{0q} appears in the general expression (3.29) for the electric field instead of I_{00}. With this understanding, setting $\Theta = \pi/2$, and using (2) in (1a, b), these may be expressed as follows:

$$A_{\text{odd}}(\tfrac{1}{2}\pi, \Phi) = A_{\text{odd}}(x) = 2 \sum_{i=0}^{n} k_i \cos 2ix;$$

$$n = (N - 1)/2. \quad (4a)$$

For example,

$$A_{\text{odd}}(x) = 2(k_0 + k_1 \cos 2x),$$
$$N = 3, \quad (4b)$$

$$A_{\text{even}}(\tfrac{1}{2}\pi, \Phi) = A_{\text{even}}(x) = 2\sum_{i=1}^{n} k_i \cos(2i-1)x,$$

$$n = N/2. \quad (4c)$$

For example,

$$A_{\text{even}}(x) = 2(k_1 \cos x + k_2 \cos 3x),$$
$$N = 4. \quad (4d)$$

The right-hand sides of the equations (4a, c) may be expanded into polynomials in powers of the variable

$$u = \cos x \quad (5)$$

in the interval $-1 \leq u \leq 1$. The polynomials may be obtained using Demoivre's formula,

$$e^{j2ix} = (\cos 2ix + j \sin 2ix)$$
$$= (\cos x + j \sin x)^{2i}. \quad (6)$$

In (6) the right side of the formula may be expanded using the binomial theorem and its real part equated to the real part of the center. With (5), and noting that $\sin^2 x = 1 - u^2$, the real part of (6) leads to the well-known series

$$\cos 2ix = u^{2i} - \binom{2i}{2} u^{2i-2}(1 - u^2)$$

$$+ \binom{2i}{4} u^{2i-4}(1 - u^2)^2$$

$$- \binom{2i}{6} u^{2i-6}(1 - u^2)^3 + \cdots$$

$$+ (-1)^r \binom{2i}{2r} u^{2i-2r}(1 - u^2)^r$$

$$+ \cdots + (-1)^i (1 - u^2)^i, \quad (7)$$

where by definition

$$\binom{b}{a} \equiv \frac{b!}{(b-a)!a!}, \quad b! = b(b-1)(b-2)\cdots. \tag{8a}$$

The terms in (6) may be arranged in descending powers of x using

$$(1-u^2)^r = 1 - \binom{r}{1}u^2 + \binom{r}{2}u^4 - \binom{r}{3}u^6$$
$$+ \cdots + (-1)^p \binom{r}{p} u^{2p}$$
$$+ \cdots + (-1)^r u^{2r}. \tag{8b}$$

The result is:

$$\cos 2ix = u^{2m}\left[1 + \binom{2i}{2} + \binom{2i}{4} + \cdots\right]$$
$$- u^{2i-2}\left[\binom{2i}{2} + 2\binom{2i}{4} + 3\binom{2i}{6} + \cdots\right]$$
$$+ u^{2i-4}\left[\binom{2i}{4} + 3\binom{2i}{6} + 6\binom{2i}{8} + \cdots\right]$$
$$-\cdots. \tag{9a}$$

For example,

$$\cos 2x = u^2\left[1 + \binom{2}{2}\right] - u^0\binom{2}{2}$$
$$= 2u^2 - 1. \tag{9b}$$

The rather complicated series (9a) may be expanded into the following compact form:

$$\cos 2ix = \sum_{m=0}^{i} c_{2m}^{2i} u^{2m}, \tag{10a}$$

where

$$c_{2m}^{2i} \equiv (-1)^{i-m} \sum_{p=i-m}^{i} \binom{p}{p-i+m}\binom{2i}{2p}. \tag{10b}$$

For example,

$$\cos 2x = c_0^2 u^0 + c_2^2 u^2 = (-1)^1 \binom{1}{0}\binom{2}{2}$$
$$+ (-1)^0\left[\binom{0}{0}\binom{2}{0} + \binom{1}{1}\binom{2}{2}\right]u^2$$
$$= -1 + 2u^2. \tag{10c}$$

It is clear that $\cos 2ix$ is a polynomial of degree $2i$ in u or of degree i in u^2. That is,

$$\cos 2ix = P_i(u^2), \tag{11}$$

where P_i is a polynomial of degree i.

Similarly, corresponding to (7),

$$\cos(2i-1)x = u^{2i-1} - \binom{2i-1}{2}u^{2i-3}(1-u^2)$$
$$+ \binom{2i-1}{4}u^{2i-5}(1-u^2)^2$$
$$-\cdots. \tag{12a}$$

For example,

$$\cos 3x = u^3 - \binom{3}{2}u(1-u^2)$$
$$= u^3 - \frac{3\cdot 2}{1\cdot 2}u(1-u^2) = 4u^3 - 3u. \tag{12b}$$

With (8b) this becomes

$$\cos(2i-1)x = u^{2m-1}\left[1 + \binom{2m-1}{2} + \cdots\right]$$
$$- u^{2m-3}\left[\binom{2m-1}{2}\right.$$
$$\left. + 2\binom{2m-1}{4} + \cdots\right]$$
$$+\cdots. \tag{13a}$$

For example,

$$\cos 3x = u^3\left[1 + \binom{3}{2}\right] - u\binom{3}{2}$$
$$= u^3\left[1 + \frac{3\cdot 2}{1\cdot 2}\right] - u\frac{3\cdot 2}{1\cdot 2}$$
$$= 4u^3 - 3u. \tag{13b}$$

In compact form,

$$\cos(2i-1)x = \sum_{m=1}^{i} c_{2m-1}^{2i-1} u^{2m-1}$$
$$= u\sum_{m=1}^{i} c_{2m-1}^{2i-1} u^{2(m-1)}, \tag{14a}$$

where

$$c_{2m-1}^{2i-1} \equiv (-1)^{i-m}\sum_{p=i-m}^{i}\binom{p}{p-i+m}\binom{2i-1}{2p}. \tag{14b}$$

For example,

$$\cos 3x = c_1^3 u + c_3^3 u^3 = (-1)^1\binom{1}{0}\binom{3}{1}u$$
$$+ (-1)^0\left[\binom{0}{0}\binom{3}{0} + \binom{1}{1}\binom{3}{2}\right]u^3$$
$$= -3u + 4u^3. \tag{14c}$$

Since u is a common factor in (14a), and the remaining sum is a polynomial of degree $2i$ in u or of degree i in u^2, it follows that

$$\cos(2i-1)u = u Q_i(u^2), \quad (15)$$

where Q_i is a polynomial of degree i.

If (10a) is substituted in (4a), and (14a) in (4c), the array factors have the form:

$$A_{odd}(\tfrac{1}{2}\pi, \Phi) = A_{odd}(u) = 2 \sum_{i=0}^{n} k_i \left(\sum_{m=0}^{i} c_{2m}^{2i} u^{2m} \right),$$
$$N = 2n+1, \quad (16a)$$

$$A_{even}(\tfrac{1}{2}\pi, \Phi) = A_{even}(u)$$
$$= 2 \sum_{i=1}^{n} k_i \left(\sum_{m=1}^{i} c_{2m-1}^{2i-1} u^{2m-1} \right),$$
$$N = 2n. \quad (16b)$$

It is readily verified that the finite sums in (16a) and (16b) are equivalent to the following polynomials:

$$A_{odd}(u) = 2 \sum_{q=0}^{n} \sum_{i=q}^{n} k_i c_{2q}^{2i} u^{2q},$$
$$N = 2n+1, \quad (17a)$$

$$A_{even}(u) = 2 \sum_{q=1}^{n} \sum_{i=q}^{n} k_i c_{2q-1}^{2i-1} u^{2q-1},$$
$$N = 2n. \quad (17b)$$

Note that $A_{odd}(u)$ for $N = 2n+1$ units is a polynomial of degree $2n$ or $N-1$, N odd; similarly, $A_{even}(u)$ for $N = 2n$ units is of degree $2n-1$ or $N-1$, N even. Thus, the array factor for N units is always a polynomial of degree $N-1$ for N odd or even. This may be expressed in the notation

$$A_{odd}(u) = G_{N-1}(u) = G_{2n}(u),$$
$$N = 2n+1, \quad (17c)$$

$$A_{even}(u) = G_{N-1}(u) = G_{2n-1}(u),$$
$$N = 2n, \quad (17d)$$

where the subscript on the G's specifies the degree of the polynomial. These are the final general expressions equivalent to those obtained by Dolph for the array factors of any symmetric broadside array with relative current amplitudes given by k_i. By equating these array factors to zero, the nulls of the pattern may be obtained; by equating to zero their derivatives with respect to x, the minor extremes of the pattern may be determined. For N a large number, this process obviously is tedious.

In order to illustrate the essential steps in the general method without the algebraic difficulties encountered when N is large a simple example introduced by Dolph is convenient. This is for $N=4$, $n = N/2 = 2$. The array factor is given in (4d), namely,

$$A_{even}(\tfrac{1}{2}\pi, \Phi) = 2[k_1 \cos x + k_2 \cos 3x],$$
$$N = 4. \quad (18)$$

Since from (12) or (13)

$$\cos 3x = 4 \cos^3 x - 3 \cos x, \quad (19)$$

it follows that

$$A_{even}(\tfrac{1}{2}\pi, \Phi) = 2[(k_1 - 3k_2) \cos x + 4k_2 \cos^3 x]. \quad (20)$$

Although the general formula (17b) is not required in so simple a case, let (18) be reëvaluated from it just to illustrate its use. Thus,

$$A_{even}(u) = 2 \sum_{q=1}^{2} \sum_{i=q}^{2} k_i c_{2q-1}^{2i-1} u^{2q-1}$$

$$= 2 \sum_{q=1}^{2} (k_q c_{2q-1}^{2q-1} u^{2q-1} + k_{q+1} c_{2q-1}^{2q+1} u^{2q-1})$$

$$= 2 \sum_{q=1}^{2} u^{2q-1} (c_{2q-1}^{2q-1} k_q + c_{2q-1}^{2q+1} k_{q+1})$$

$$= 2[u(c_1^1 k_1 + c_1^3 k_2) + u^3 c_3^3 k_2]$$

$$= 2[(k_1 - 3k_2)u + 4k_2 u^3], \quad (21)$$

since with (14b)

$$c_1^1 = (-1)^0 \sum_{p=0} \binom{p}{p}\binom{1}{2p} = 1, \quad (22a)$$

$$c_1^3 = (-1)^1 \sum_{p=1}^{2} \binom{p}{p-1}\binom{3}{2p} = -\binom{1}{0}\binom{3}{2}$$
$$= -3, \quad (22b)$$

$$c_3^3 = (-1)^0 \sum_{p=0}^{2} \binom{p}{p}\binom{3}{2p} = \binom{3}{0} + \binom{3}{2}$$
$$= 1 + 3 = 4. \quad (22c)$$

With $u = \cos x$ the result in (21) is seen to agree with (20).

Since the range of u is from -1 to $+1$, the nulls obtained by equating (20) to zero are

$$A_{even}(u) = 0: \begin{cases} u = \cos x = \cos(n_g \pi \cos \Phi_0) \\ \qquad = 0, \quad (23a) \\ u = \tfrac{1}{2}\sqrt{3 - k_1/k_2}. \quad (23b) \end{cases}$$

The positions of the extreme values of the array factor (including principal and minor lobes) are determined from the equation

$$\frac{dA_{\text{even}}(u)}{dx} = \frac{dA_{\text{even}}(u)}{du}\frac{du}{dx} = 0. \quad (24)$$

With (21) and $u = \cos x$, this becomes

$$[(k_1 - 3k_2) + 12k_2 \cos^2 x] \sin x = 0. \quad (25a)$$

The principal extremes are located by

$$\sin x = \sin (n_B \pi \cos \Phi) = 0,$$

$$\Phi = \pi/2, 3\pi/2. \quad (25b)$$

The locations of minor extremes are defined by u_m, where

$$u_m = \cos x_m = \cos (n_B \pi \cos \Phi_m)$$

$$= \pm \sqrt{\frac{3 - k_1/k_2}{12}}. \quad (25c)$$

Substitution of (25b) in (20) gives the following extreme values for the minor lobes:

$$\left|A_{\text{even}}(\tfrac{1}{2}\pi, \Phi_m)\right| = \frac{2\sqrt{3}}{9}\left|k_2(3 - k_1/k_2)^{3/2}\right|. \quad (26)$$

Since the beam width is determined by the location of the first null, it follows from (23a, b) and (26) that the beam width between nulls and the location and magnitude of minor lobes all depend on the same quantity, namely, $(3 - k_1/k_2)$. For a uniform array $k_2 = k_1 = 1$, for the binomial array (in which there are no minor lobes) the amplitudes are in the ratios $1 : 3 : 3 : 1$, so that $k_1 = 1$, $k_2 = 1/3$. As the ratio $k_2/k_1 = I_2/I_1$ decreases from 1 for the uniform array to 1/3 for the binomial array, the first null as defined by (23b) moves toward $u = \cos (n_B \pi \cos \Phi_0) = 0$, which means that for $n_B = 1/2$, Φ_0 moves toward zero and hence *away* from the principal maximum at $\Phi = \pi/2$. Thus, the main beam width continually *increases* as $I_2/I_1 = k_2/k_1$ approaches 1/3, the value for the binomial distribution. On the other hand, it is clear from (26) that the magnitude of the minor lobes continually decreases to zero as k_2/k_1 approaches 1/3. Precisely this same effect was verified in Sec. 9 for the end-fire array.

If k_2/k_1 is made greater than unity (the value for the uniform array) so that the currents in the outer units exceed those in the inner ones, it follows from (23b) that the first null moves *toward* the principal maximum to decrease the null beam width, but at the same time (26) shows that the level of the minor lobes increases.

Although the application of the same general procedure carried out so readily for $N = 4$ is analytically difficult for very large values of N, it is clear that the array factor of the general symmetric broadside array of N elements can be represented by a real polynomial in $u = \cos (n_B \pi \cos \Phi)$, and that by adjusting the ratios of the currents in the N antennas a wide variation in beam width and minor lobe level can be achieved. What remains to be done is to determine the optimum values of k_i in the sense defined at the beginning of this section, that is, to minimize the minor-lobe level for a given beam width or to minimize the beam width for a given minor-lobe level.

13. Optimum Currents for Symmetric Broadside Arrays; Tchebyscheff Polynomials

A theoretically ideal array factor for a broadside beam may be represented graphically by a curve that has equal, sharp, and high extreme values at $u = \cos x = \cos (n_B \pi \cos \Phi) = \pm 1$, $n_B \geq \tfrac{1}{2}$ and is so small for all intermediate values that the beam width and the minor-lobe level are both very nearly zero. The extreme at $u = 1$ is always positive; that at $u = -1$ may be positive or negative depending upon whether the number N of units is odd or even. A pattern of this ideal type is shown in Fig. 13.1. Clearly, it is an abstraction that is difficult to realize in practice.

The degree in which the ideal pattern is approximated by a uniform broadside array of eight elements also is shown in Fig. 13.1. There are large extremes at $u = \pm 1$ and a relatively small level is maintained between them. However, as has been pointed out, the relative magnitudes of the minor lobes referred to the two principal ones cannot be reduced below a fixed, rather high value by increasing the number of antennas. It is clear from Fig. 1.1 that as the number N is made greater and greater, the null beam width decreases indefinitely, but the relative level of the adjacent minor lobes is virtually unchanged.

It is shown in the preceding section that a tapered instead of a uniform distribution of current amplitudes in the several antennas of a *symmetric* broadside array decreases the relative minor lobe level but increases the main beam width if the current amplitude decreases outward from the center of the

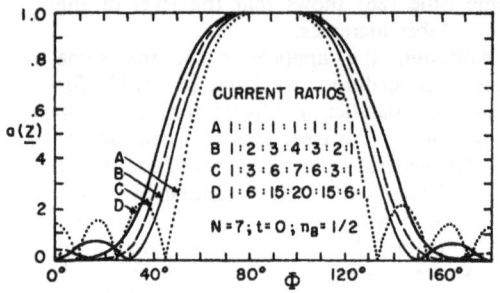

Fig. 11.1. Normalized array factors of four seven-antenna broadside arrays.

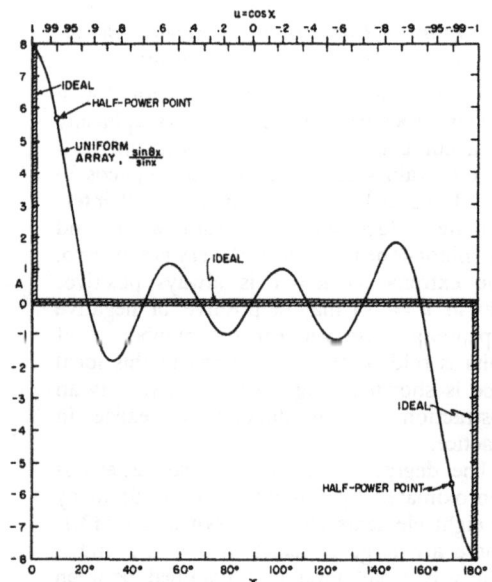

Fig. 13.1. Array factors of uniform and ideal arrays.

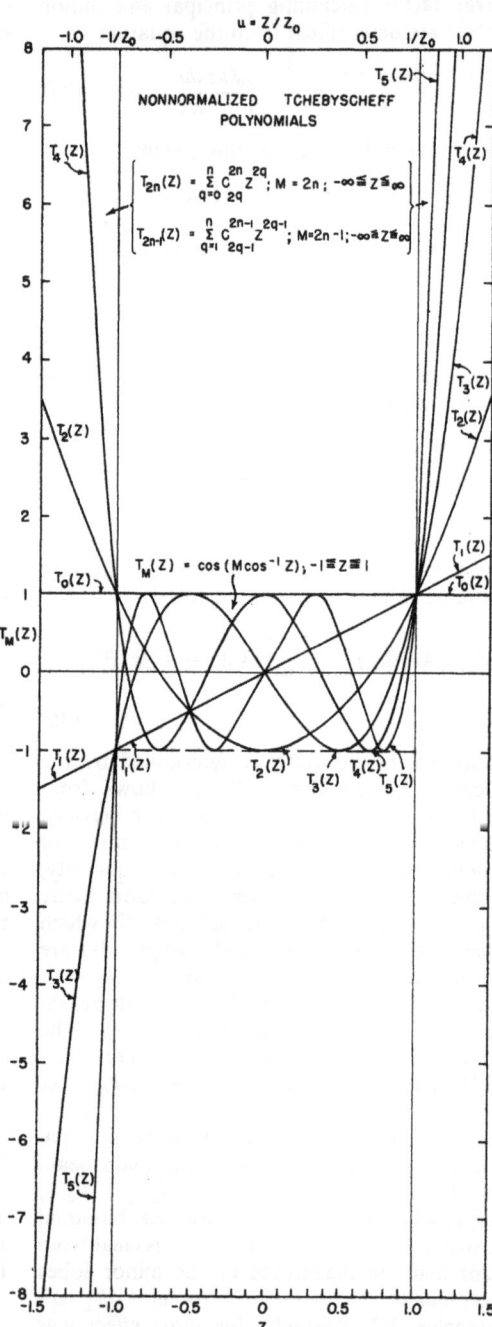

Fig. 13.2. Nonnormalized Tchebyscheff polynomials.

array. Reciprocally, the side lobes are increased and the beam width is decreased if the current amplitudes become greater from the center of the array outward.

It is the purpose of this section to investigate the means of achieving the optimum distribution defined in Sec. 12 in which the ideal array factor is approximated best, in the sense that for a given number N of units the beam width is a minimum for a specified minor-lobe level or the minor-lobe level is a minimum with the beam width specified. It is assumed that all minor lobes constitute an oscillation of equal amplitude about the zero line. Since the array factor of an N-element symmetric broadside array is a polynomial of degree $N - 1$, the method to be followed is to discover the relative magnitudes k_i of the N currents that will yield a polynomial with the desired optimum property.

Fortunately, a set of orthogonal polynomials $T_M(z)$ of degree $M = N - 1$ that have the desired optimum property is well known. They are the so-called Tchebyscheff polynomials,[53] which are characterized by the important property that the magnitude of the polynomial in the range $-1 \leq z \leq 1$ is the *smallest possible* for any Mth degree polynomial with real coefficients of which the greatest is unity. Clearly, this is the desired optimum property for the array factor provided the range of minor lobes can be made to coincide with the minimized range of the polynomial. The Mth-degree Tchebyscheff polynomials $T_M(z)$ are defined both in a normalized and in a nonnormalized form. The latter are preferred here. They are defined as follows: *

$$T_0(z) = 1, \quad T_M(z) = \cos M\theta = \cos(M \cos^{-1} z),$$

$$M \geq 1, \quad (-1 \leq z \leq 1), \quad (1a)$$

$$T_M(z) = \cosh M\theta = \cosh(M \cosh^{-1} z),$$

$$M \geq 1, \quad |z| \geq 1. \quad (1b)$$

(The form (1b) is obtained by noting in (1a) that for $|z| > 1$, θ must be imaginary. Hence, (1b) is obtained from (1a) by replacing θ by $j\theta$.) It is evident that these are polynomials of degree M since they are given directly by

---

* Courant-Hilbert, ref. I.16. In this reference the *normalized* Tchebyscheff polynomials are defined. They differ from the non-normalized ones defined in (1) by a normalizing factor $1/2^M$. Nonnormalized polynomials are discussed by Van der Pol and Wayers.[56]

(12.10a) for $M = 2n$ even and by (12.14a) with $M = 2n - 1$ odd. Thus with

$$z = \cos \theta, \quad (2)$$

$$T_{2n}(z) = \cos(2n\theta) = \sum_{q=0}^{n} c_{2q}^{2n} z^{2q},$$

$$M = 2n, \text{ even}, \quad (-\infty \leq z \leq \infty) \quad (3a)$$

$$T_{2n-1}(z) = \cos(2n - 1)\theta = \sum_{q=1}^{n} c_{2q-1}^{2n-1} z^{2q-1},$$

$$M = 2n - 1, \text{ odd}, \quad (-\infty \leq z \leq \infty) \quad (3b)$$

where c_{2q}^{2n} and c_{2q-1}^{2n-1} are given, respectively, by (12.10b) and (12.14b) with n substituted for i and q for m. A few of the nonnormalized Tchebyscheff polynomials are:

$$T_0(z) = 1,$$

$$T_1(z) = z,$$

$$T_2(z) = 2z^2 - 1,$$

$$T_3(z) = z(4z^2 - 3),$$

$$T_4(z) = 8z^4 - 8z^2 + 1,$$

$$T_5(z) = z(16z^4 - 20z^2 + 5),$$

$$T_6(z) = 32z^6 - 48z^4 + 18z^2 - 1,$$

$$T_7(z) = z(64z^6 - 112z^4 + 56z^2 - 7). \quad (4)$$

Most of these functions are plotted in Fig. 13.2. A more extensive list of the odd polynomials is given in Table 13.1.

As defined in (1), the maximum range of the argument z is, as indicated in (1), $-1 \leq z \leq 1$. It is in this range that the polynomials have the optimum property. It is seen in Fig. 13.2 that all functions except $T_0(z)$ oscillate about zero with the same amplitude, unity. Once the polynomials are expanded as in (3) or (4), they exist for all values of z, $-\infty \leq z \leq \infty$, as indicated in (3a, b). For $z > 1$, $T_M(z)$ increases monotonically; for $z < 1$, $T_M(z)$ increases monotonically if M is even, decreases monotonically if M is odd. Hence, all zeros of $T_M(z)$ and of its derivatives are in the range $-1 \leq z \leq 1$. Therefore, all maxima and minima are in this range. Note that the large positive or negative values approached monotonically at $z = \pm z_0$ in any restricted range $-z_0 \leq z \leq z_0$, where $z_0 \geq 1$, are not maxima or minima in the sense of vanishing slope. The zeros occur when

$$T_M(z) = \cos M\theta = 0, \quad \theta = \cos^{-1} z. \quad (5a)$$

TABLE 13.1. Tchebyscheff polynomials: $T_{2n-1}(z) = \cos(2n-1)\theta$.

$T_1(z) = \cos\theta = z$

$T_3(z) = \cos 3\theta = z(4z^2 - 3)$

$T_5(z) = \cos 5\theta = z(16z^4 - 20z^2 + 5)$

$T_7(z) = \cos 7\theta = z(64z^6 - 112z^4 + 56z^2 - 7)$

$T_9(z) = \cos 9\theta = z(256z^8 - 576z^6 + 432z^4 - 120z^2 + 9)$

$T_{11}(z) = \cos 11\theta = z(1024z^{10} - 2816z^8 + 2816z^6 - 1232z^4 + 220z^2 - 11)$

$T_{13}(z) = \cos 13\theta = z(4096z^{12} - 13{,}312z^{10} + 16{,}640z^8 - 9984z^6 + 2912z^4 - 364z^2 + 13)$

$T_{15}(z) = \cos 15\theta = z(16{,}384z^{14} - 61{,}440z^{12} + 92{,}160z^{10} - 70{,}400z^8 + 28{,}800z^6 - 6048z^4 + 560z^2 - 15)$

$T_{17}(z) = \cos 17\theta = z(65{,}536z^{16} - 278{,}528z^{14} + 487{,}424z^{12} - 452{,}608z^{10} + 239{,}360z^8 - 71{,}808z^6 + 11{,}424z^4 - 816z^2 + 17)$

$T_{19}(z) = \cos 19\theta = z(262{,}144z^{18} - 1{,}245{,}184z^{16} + 2{,}490{,}368z^{14} - 2{,}723{,}840z^{12} + 1{,}770{,}496z^{10} - 695{,}552z^8 + 160{,}512z^6 - 20{,}064z^4 + 1140z^2 - 19)$

$T_{21}(z) = \cos 21\theta = z(1{,}048{,}576z^{20} - 5{,}505{,}024z^{18} + 12{,}386{,}304z^{16} - 15{,}597{,}568z^{14} + 12{,}042{,}240z^{12} - 5{,}870{,}592z^{10} + 1{,}793{,}792z^8 - 329{,}472z^6 + 33{,}264z^4 - 1540z^2 + 21)$

$T_{23}(z) = \cos 23\theta = z(4{,}194{,}304z^{22} - 24{,}117{,}248z^{20} + 60{,}293{,}120z^{18} - 85{,}917{,}696z^{16} + 76{,}873{,}728z^{14} - 44{,}843{,}008z^{12} + 17{,}145{,}856z^{10} - 4{,}209{,}920z^8 + 631{,}488z^6 - 52{,}624z^4 + 2024z^2 - 23)$

The values of $\theta = \cos^{-1} z$ at which nulls occur are

$$\theta_{0i} = (2i-1)\pi/2M, \quad i = 1, 2, \cdots, M. \quad (5b)$$

The extreme values occur when

$$\frac{dT_M(z)}{d\theta} = -M \sin M\theta = 0. \quad (6a)$$

The extremizing values of θ are

$$\theta_{ei} = \pi i/M, \quad i = 1, 2, \cdots, M. \quad (6b)$$

The extreme values are obtained by substituting θ_{ei} for θ in $T_M(z)$. They are

$$T_M(z_{ei}) = \cos M\theta_{ei} = \pm 1. \quad (7)$$

Thus in the range $-1 \leq z \leq 1$ in which all maxima and minima occur, the function $T_M(z)$ oscillates between 1 and -1, as is clear from Fig. 13.2.

A useful alternative expression for the Tchebyscheff function as defined in (1a, b) is derived as follows with $z = \cos\theta$:

$$T_M(z) = \cos M\theta = \tfrac{1}{2}(e^{jM\theta} + e^{-jM\theta})$$

$$= \tfrac{1}{2}[(\cos\theta + j\sqrt{1-\cos^2\theta})^M + (\cos\theta - j\sqrt{1-\cos^2\theta})^M] \quad (8)$$

$$= \tfrac{1}{2}[(z + \sqrt{z^2-1})^M + (z - \sqrt{z^2-1})^M].$$

$$(-\infty \leq z \leq \infty).$$

If the function $T_M(z)$ as defined in (3a, b) is limited by an appropriately chosen value $z = \pm z_0$ such that the variable u occurring in the general polynomial forms (12.17a, b) of the array factors of symmetric broadside arrays is given by

$$u = z/z_0, \quad (9)$$

it follows that (3a) and (3b) become

$$T_{2n}(z_0 u) = \sum_{q=0}^{n} c_{2q}^{2n} z_0^{2q} u^{2q}, \quad (10a)$$

$$T_{2n-1}(z_0 u) = \sum_{q=1}^{n} c_{2q-1}^{2n-1} z_0^{2q-1} u^{2q-1}. \quad (10b)$$

For example, with $z = z_0 u$ and $M = 2n - 1 = 7$, $n = 4$, (4b) gives

$$T_7(z_0 u) = z_0 u(64z_0^6 u^6 - 112z_0^4 u^4 + 56z_0^2 u^2 - 7). \quad (11)$$

In order to make use of the Tchebyscheff optimum properties in the design of non-uniform arrays it is merely necessary to note that (10a) and (10b) are the same in form as the array factors (12.17a) and (12.17d), namely,

$$A_{\text{odd}}(u) = G_{2n}(u) = 2 \sum_{q=0}^{n} \sum_{i=q}^{n} k_i c_{2q}^{2i} u^{2q},$$

$$N = 2n + 1, \quad (12a)$$

$$A_{\text{even}}(u) = G_{2n-1}(u) = 2 \sum_{q=1}^{n} \sum_{i=q}^{n} k_i c_{2q-1}^{2i-1} u^{2q-1},$$

$$N = 2n. \quad (12b)$$

Specifically, with $N = 8$, $n = 4$, the form (12.4c) gives, with $u = \cos x$,

$$A_{\text{even}}(u) = 2[k_1 \cos x + k_2 \cos 3x + k_3 \cos 5x + k_4 \cos 7x], \quad (13a)$$

which, from (12b) or simply by expanding the cosines in (13a) in powers of $u = \cos x$, is equivalent to

$$A_{\text{even}}(u) = G_7(u) = 2[k_1 u + k_2 u(4u^2 - 3) \\+ k_3 u(16u^4 - 20u^2 + 5) \\+ k_4 u(64u^6 - 112u^4 + 56u^2 - 7)] \quad (13b)$$

$$= 2u[64k_4 u^6 + (16k_3 - 112k_4)u^4 \\+ (4k_2 - 20k_3 + 56k_4)u^2 \\+ (k_1 - 3k_2 + 5k_3 - 7k_4)]. \quad (13c)$$

It is now possible to determine the relative current amplitudes k_i in a manner such that the array factors in (12a, b) and (13c) become Tchebyscheff polynomials. For this it is necessary merely to solve the following equations obtained, respectively, using (10a) and (12a), (10b) and (12b), (11) and (13c):

$$A_{\text{odd}}(u) \equiv G_{2n}(u) = 2T_{2n}(z_0 u), \\ N = 2n + 1, \quad (14a)$$

$$A_{\text{even}}(u) \equiv G_{2n-1}(u) = 2T_{2n-1}(z_0 u), \\ N = 2n, \quad (14b)$$

$$2u[64k_4 u^6 + (16k_3 - 112k_4)u^4 \\+ (4k_2 - 20k_3 + 56k_4)u^2 \\+ (k_1 - 3k_2 + 5k_3 - 7k_4)] \\= 2z_0[64z_0^6 u^6 - 112z_0^4 u^4 + 56z_0^2 u^2 - 7], \\ N = 8. \quad (15)$$

The factor 2 in (14a, b) and (15) arises from the fact that $A(u)$ was defined originally in (12.4a) and (12.4c) with this factor. The following sets of equations are equivalent to (14a, b):

$$\sum_{i=q}^{n} k_i c_{2q}^{2i} = c_{2q}^{2n} z_0^{2q}, \quad q = 1, 2, \cdots, n, \\ N = 2n + 1, \quad (16a)$$

$$\sum_{i=q}^{n} k_i c_{2q-1}^{2i-1} = c_{2q-1}^{2n-1} z_0^{2q-1}, \quad q = 1, 2, \cdots, n, \\ N = 2n, \quad (16b)$$

$$\left. \begin{array}{l} 64k_4 = 64z_0^6, \\ 16k_3 - 112k_4 = -112z_0^4, \\ 4k_2 - 20k_3 + 56k_4 = 56z_0^2, \\ k_1 - 3k_2 + 5k_3 - 7k_4 = -7, \end{array} \right\}, \; N = 8. \quad (17)$$

In order to solve for each k_q, the sets of equations in (16a, b) may be expressed in the following equivalent forms:

$$k_q c_{2q}^{2q} + \sum_{i=q+1}^{n} k_i c_{2q}^{2i} = c_{2q}^{2n} z_0^{2q}, \\ N = 2n + 1, \quad (18a)$$

$$k_q c_{2q-1}^{2q-1} + \sum_{i=q+1}^{n} k_i c_{2q-1}^{2i-1} = c_{2q-1}^{2n-1} z_0^{2q-1}, \\ N = 2n. \quad (18b)$$

Solving for k_q gives the formulas for the current amplitudes as obtained by Dolph[12]

$$k_q = \frac{1}{c_{2q}^{2q}} \left(c_{2q}^{2n} z_0^{2q} - \sum_{i=q+1}^{n} k_i c_{2q}^{2i} \right), \\ N = 2n + 1, \quad (19a)$$

$$k_q = \frac{1}{c_{2q-1}^{2q-1}} \left(c_{2q-1}^{2n-1} z_0^{2q-1} - \sum_{i=q+1}^{n} k_i c_{2q-1}^{2i-1} \right), \\ N = 2n. \quad (19b)$$

By beginning with $q = n$, and proceeding in succession to $q = n-1, n-2, \cdots$, (19a) and (19b) may be solved for all of the relative currents k_i. In the special case with $N = 8$, $n = 4$,

$$k_4 = \frac{1}{c_7^7}(c_7^7 z_0^7) = z_0^7, \quad (20a)$$

$$k_3 = \frac{1}{c_5^5}(c_5^5 z_0^5 - k_4 c_5^7) = 7k_4 - 7z_0^5 \\ = 7z_0^5(z_0^2 - 1), \quad (20b)$$

$$k_2 = \frac{1}{c_3^3}(c_3^3 z_0^3 - k_4 c_3^7 - k_3 c_3^5) \\ = 5k_3 - 15k_4 + 14z_0^3 \\ = 7z_0^3(3z_0^4 - 5z_0^2 + 2), \quad (20c)$$

$$k_1 = \frac{1}{c_1^1}(c_1^1 z_0 - k_4 c_1^7 - k_3 c_1^5 - k_2 c_1^3) \\ = 3k_2 - 5k_3 + 7k_4 - 7z_0 \\ = 7z_0(5z_0^6 - 10z_0^4 + 6z_0^2 - 1). \quad (20d)$$

For any value of $z_0 > 1$ the currents in the symmetric, N-antenna broadside array are specified by (19a, b) or, for $N = 8$, by (20a) to (20d) in a manner to make the array factors (12a, b) and (13a) Tchebyscheff polynomials that have the desired optimum properties in the range $-1 \leq z \leq 1$ or $-(1/z_0) \leq u \leq (1/z_0)$. However, as discussed in Sec. 14, the array factor $A(\Theta, \Phi)$ so determined has optimum properties as a function of Φ only if the distance between adjacent antennas equals or exceeds $\lambda_0/2$, that is, only for

$n_B \geq \frac{1}{2}$. Thus, in general, when $n_B \geq \frac{1}{2}$ the array factors (12a) and (12b) can be specialized to be Tchebyscheff polynomials by specifying the currents as in (19a) and (19b), respectively. Note that this is true for *any* value of the arbitrary parameter z_0.

Since the numerical evaluation of the current amplitudes from (19a, b) is very long and tedious for an array with a large number of elements, alternative, more convenient formulas have been derived first by Barbiere[3] and later by Stegen[50]. Stegen's formulas for $N = 2n + 1$ are

$$k_0 = N^{-1}\{r + 2\sum_{s=1}^{n} T_{n-1}[z_0 \cos(\pi s/N)]\} \quad (20e)$$

$$k_m = 2N^{-1}\{r + 2\sum_{s=1}^{n} T_{N-1}[z_0 \cos(\pi s/N)] \\ \times \cos(2\pi ms/N)]\},$$
$$m = 1, 3, \cdots, n. \quad (20f)$$

For $N = 2n$

$$k_m = 2N^{-1}\{r + 2\sum_{s=1}^{n-1} T_{N-1}[z_0 \cos(\pi s/N)] \\ \times \cos[(2m+1)\pi s/N]\},$$
$$m = 0, 1, 2, \cdots, n-1. \quad (20g)$$

What determines the value to be assigned to the arbitrary scale factor z_0? The answer to this question follows directly from a study of the effect of increasing z_0 from unity to infinity, which obviously are the limiting cases of the optimum or Tchebyscheff distribution. Of the two quantities of particular interest, the first is the ratio of the magnitude of the principal lobe at $u = \pm 1$ to the minor-lobe level of unity, that is,

$$r = |T_M(z_0)|/1 = |T_{N-1}(z_0)|/1. \quad (21)$$

Note that the limiting ratio of principal lobe to the *first* and greatest minor lobe in a uniform array as N is increased indefinitely as given in Sec. 1 is

$$r_{\text{uniform}} = 1/0.21 = 4.76. \quad (22)$$

The second quantity of interest is the null beam width of the principal lobe.* The location of the first null for the Tchebyscheff array is given by (5b) with $M = N - 1$, $i = 1$, and $z = z_0 u = z_0 \cos x$ (where $x = n_B \pi \cos \Phi$.)

---
* Formulas and graphs for the half-power beam width are in the literature[50].

The corresponding value of x is obtained at once to be

$$u = \cos x_0 = \frac{1}{z_0} \cos \frac{\pi}{2(N-1)},$$

$$x_0 = \cos^{-1}\left[\frac{1}{z_0} \cos \frac{\pi}{2(N-1)}\right]. \quad (23)$$

The first null of a uniform array is given by (1.23). It is

$$(x_0)_{\text{uniform}} = \pi/N. \quad (24)$$

When z_0 is chosen to be unity it follows from (3a, b) or (4a, b), and from (23), that for the Tchebyscheff array

$$z_0 = 1: \quad r = T_M(1) = 1, \quad x_0 = \frac{\pi}{2(N-1)}. \quad (25)$$

Comparison with (22) and (24) shows that for the uniform array r is greater and x_0 is greater. Thus, the uniform array has a lower minor-lobe level and a greater null beam width than a Tchebyscheff array in the extreme case $z_0 = 1$. The nature of such a Tchebyscheff array in the special case $N = 8$ follows from (20a) to (20d) with $z_0 = 1$. Thus,

$$z_0 = 1: \quad k_1 = k_2 = k_3 = 0, \quad k_4 = 1. \quad (26)$$

Evidently, only the outermost two antennas have nonzero currents. The array factor is given by (13a):

$$z_0 = 1: \quad A_{\text{even}}(u) = 2k_4 \cos 7x. \quad (27)$$

The behavior of a Tchebyscheff array as z_0 becomes infinite may be determined from (2a) to (20d) if these equations are divided by k_4, so that the relative currents remain finite. The resulting relative currents are

$$z_0 \to \infty: \quad k_4/k_4 = 1, \quad k_3/k_4 = 7,$$
$$k_2/k_4 = 21, \quad k_1/k_4 = 35; \quad (28)$$

that is, the currents in the eight antennas are in the ratios $1:7:21:35:35:21:7:1$. This is recognized to be a symmetric binomial distribution, the array factor of which has no side lobes. On the other hand, the first null occurs at $x = \pi/2$. Thus, the Tchebyscheff array becomes a binomial broadside array with

$$z_0 \to \infty: r \to \infty, \; x_0 \to \pi/2. \quad (29)$$

It is clear from these extreme cases that a wide choice in either minor-lobe level or beam width is available in the Tchebyscheff arrays. It is known that, whichever is assigned, the other has an optimum value for the given number of antennas if the distance between adjacent antennas equals or exceeds $\lambda_0/2$ or $n_B \geq \frac{1}{2}$. In general, as z_0 is increased the minor-lobe level is decreased, the beam width increased.

If the minor-lobe level $1/r$ is preassigned, it is necessary to determine z_0 by solving (14a, b) with $A(u) = 2r$. This is conveniently done using (1b). Thus

$$z_0 = \cosh(M^{-1} \cosh^{-1} r) \quad (30)$$

An alternative form is obtained using (8). The equation to be solved for z_0 in this case is

$$(z_0 + \sqrt{z_0^2+1})^M + (z_0 - \sqrt{z_0^2+1})^M = 2r. \quad (31a)$$

If (31a) is multiplied through by the first term and then by the second term the following equations are obtained, in which the upper signs go together and the lower signs go together:

$$(z_0 \pm \sqrt{z_0^2+1})^{2M} - 2r(z_0 \pm \sqrt{z_0^2+1}) + 1 = 0. \quad (31b)$$

These may be solved directly for $(z \pm \sqrt{z_0^2+1})^M$ and then the Mth root extracted on each side. This gives

$$z_0 + \sqrt{z_0^2+1} = (r \pm \sqrt{r^2+1})^{\frac{1}{M}}, \quad (32a)$$

$$z_0 - \sqrt{z_0^2+1} = (r \mp \sqrt{r^2+1})^{\frac{1}{M}}. \quad (32b)$$

By adding (32a) and (32b) the following final result is obtained:

$$z_0 = \frac{1}{2}[(r+\sqrt{r^2+1})^{\frac{1}{M}} + (r-\sqrt{r^2+1})^{\frac{1}{M}}],$$
$$M = N - 1. \quad (33)$$

If r is specified, z_0 is readily determined from (30) or (33).

If the location of the first null is specified as Φ_{01}, it is necessary first to compute

$$u_{01} = \cos x_{01} = \cos(n_B \pi \cos \Phi_{01}) \quad (34)$$

and then determine z_0 from (23) as follows:

$$z_0 = \frac{1}{u_{01}} \cos \frac{\pi}{2(N-1)}. \quad (35)$$

If z_0 is determined either from (33), if the minor-lobe level is given, or from (35), if the location of the first null is assigned, the entire array factor is specified completely by determining the currents given in (19a) or (19b) and using these in (12a) or (12b). The resulting Tchebyscheff pattern has the following optimum property for each value of N: (1) if the side-lobe level is assigned, the null beam width is a minimum; (2) if the first null is assigned, the minor-lobe level is minimized.

A complete pattern for $N = 8$ has been calculated by Dolph by assigning arbitrarily a minor-lobe level of 25.8 decibels, which corresponds to $z_0 = 1.14$. The appropriate relative currents in the four pairs of antennas are obtained from (20a) to (20d) to be:

$$N = 8: k_4 = 2.502, \quad k_3 = 4.029,$$
$$k_2 = 5.901, \quad k_1 = 7.042. \quad (36a)$$

Referred to antenna No. 1, nearest the center of the array,

$$k_4 = 0.356, \quad k_3 = 0.574, \quad k_2 = 0.840, \quad k_1 = 1.0. \quad (36b)$$

Since the bracket in the array factor (13a) with currents as specified in (36a) or (36b) is equal to (11), it is clear that the array factor is proportional to $T_7(z_0 u) = T_7(1.14u)$. Thus, with $u = \cos x = \cos(n_B \pi \cos \Phi)$, the array factor may be expressed as follows:

$$N = 8: A_{\text{even}}(\tfrac{1}{2}\pi, \Phi) \sim T_7(z_0 u)$$
$$= T_7(z_0 \cos x)$$
$$= T_7[z_0 \cos(n_B \pi \cos \Phi)]. \quad (37)$$

In evaluating this function numerically for graphical representation with $z_0 = 1.14$, it is necessary to do little more than locate the zeros and the minor extremes. The values of $\theta_{0i} = \cos^{-1} z_0 u_{0i}$ for the nulls and $\theta_{0e} = \cos^{-1} z_0 u_{0e}$ for the minor extremes are listed in Table 13.2. The corresponding values of u and a graph of $T_7(z_0 u)$ as a function of u are shown in Fig. 13.3. A graph of the magnitude of the same function $|T_7(z_0 \cos x)|$ as a function of x is shown in Fig. 13.4 together with values of x giving nulls and minor extremes. In this same figure is shown the normalized array factor of the corresponding 8-element *uniform* array multiplied by the factor 19.45 so that the principal extremes of the uniform and Tchebyscheff arrays coincide. Note that the null beam width of the Tchebyscheff array is considerably

TABLE 13.2. Nulls and extremes for $T_7(z_0 u)$ with $z_0 = 1.14$.

i	θ_{0i} (nulls)			θ_{0e} (minor extremes)		
	Radians	Degrees	$z_0 u_{0i} = \cos \theta_{0i}$	Radians	Degrees	$z_0 u_{0e} = \cos \theta_{0e}$
1	$\pi/14$	12.85°	0.9750	$\pi/7$	25.7°	0.9011
2	$3\pi/14$	38.6	0.7815	$2\pi/7$	51.5	0.6225
3	$5\pi/14$	64.3	0.4337	$3\pi/7$	77.2	0.2215
4	$7\pi/14$	90.0	0.0	$4\pi/7$	102.8	−0.2215
5	$9\pi/14$	115.7	−0.4337	$5\pi/7$	128.5	−0.6225
6	$11\pi/14$	141.4	−0.7815	$6\pi/7$	154.3	−0.9011
7	$13\pi/14$	167.15	−0.9750			

TABLE 13.3. Array factor for 24-antenna Tchebyscheff array.

$$\tfrac{1}{2}A_{\text{even}}(u) = \tfrac{1}{2}G_{23}(u) = u[4{,}194{,}304 k_{12} u^{22}$$
$$+(-24{,}117{,}248 k_{12}+1{,}048{,}576 k_{11})u^{20}$$
$$+(60{,}293{,}120 k_{12}-5{,}505{,}024 k_{11}+262{,}144 k_{10})u^{18}$$
$$+(-85{,}917{,}696 k_{12}+12{,}386{,}304 k_{11}-1{,}245{,}184 k_{10}+65{,}536 k_9)u^{16}$$
$$+(76{,}873{,}728 k_{12}-15{,}597{,}568 k_{11}+2{,}490{,}368 k_{10}-278{,}528 k_9+16{,}384 k_8)u^{14}$$
$$+(-44{,}843{,}008 k_{12}+12{,}042{,}240 k_{11}-2{,}723{,}840 k_{10}+487{,}424 k_9-61{,}440 k_8+4096 k_7)u^{12}$$
$$+(17{,}145{,}856 k_{12}-5{,}870{,}592 k_{11}+1{,}770{,}496 k_{10}-452{,}608 k_9+92{,}160 k_8-13{,}312 k_7+1024 k_6)u^{10}$$
$$+(-4{,}209{,}920 k_{12}+1{,}793{,}792 k_{11}-695{,}552 k_{10}+239{,}360 k_9-70{,}400 k_8+16{,}640 k_7-2816 k_6+256 k_5)u^8$$
$$+(631{,}488 k_{12}-329{,}472 k_{11}+160{,}512 k_{10}-71{,}808 k_9+28{,}800 k_8-9984 k_7+2816 k_6-576 k_5+64 k_4)u^6$$
$$+(-52{,}624 k_{12}+33{,}264 k_{11}-20{,}064 k_{10}+11{,}424 k_9-6048 k_8+2912 k_7-1232 k_6+432 k_5-112 k_4+16 k_3)u^4$$
$$+(2024 k_{12}-1540 k_{11}+1140 k_{10}-816 k_9+560 k_8-364 k_7+220 k_6-120 k_5+56 k_4-20 k_3+4 k_2)u^2$$
$$+(-23 k_{12}+21 k_{11}-19 k_{10}+17 k_9-15 k_8+13 k_7-11 k_6+9 k_5-7 k_4+5 k_3-3 k_2+k_1)].$$

greater, the half-power beam width slightly greater, and the minor-lobe level much smaller than for the uniform array. The actual Tchebyscheff array factor $T_7(z_0 \cos [n_B \pi \cos \Phi])$ as a function of the equatorial angle Φ is shown in Fig. 13.5 for half-wave spacing ($n_B = 1/2$) and full-wave spacing ($n_B = 1$). These curves may be compared with the patterns for the corresponding 8-element uniform broadside array in Fig. 3.2. Note that whereas the general shape is similar, the uniform arrays have narrower beams but much higher side-lobe levels.

In addition to the 8-antenna array already discussed, Dolph has analyzed 12-, 16-, 20-, and 24-element Tchebyscheff broadside arrays. This involves the odd nonnormalized Tchebyscheff polynomials up to $T_{23}(z)$ as defined in (3b). These are given in Table 13.1.

The array factor for the 24-element Tchebyscheff array as defined in (12b) is given in Table 13.3. Note that this reduces to the array factor for $N = 20$ if k_{12} and k_{11} are set equal to zero, for $N = 16$ if k_{12}, k_{11}, k_{10}, and k_9 are set equal to zero, for $N = 12$ if k_{12} to and including k_7 are made zero, and so on. The current factors k_q as defined in (19b) and as given in (20a) to (20d) are listed in Table 13.4 for $N = 12$, 16, 20, and 24. In Fig. 13.6 are curves relating the minor-lobe level to the arbitrary scale factor z_0; in Fig. 13.7 are curves giving the relative currents in the individual antennas as functions of the minor-lobe level. The quantities plotted are denoted by J_i, $i = 1, 2, \cdots, n = N/2$. In order to obtain the true relative currents for use in the array factor, a reference antenna for current must be chosen, for instance, antenna 1. Then $k_1 = 1$, $k_i = J_i/J_1$.

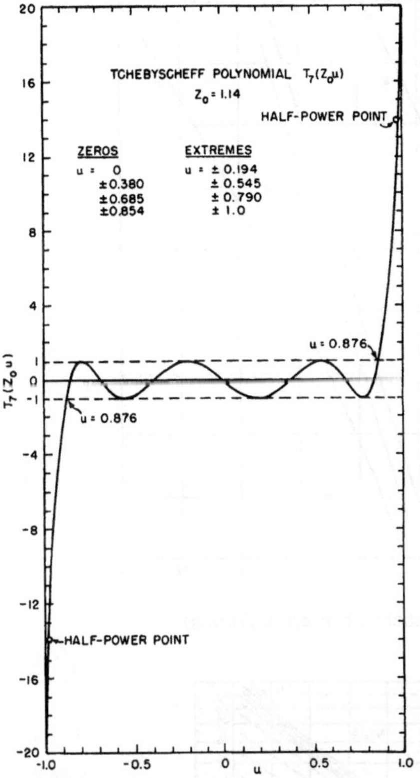

THEORY OF LINEAR ANTENNAS

Fig. 13.3. Tchebyscheff polynomial $T_7(Z_0 u)$.

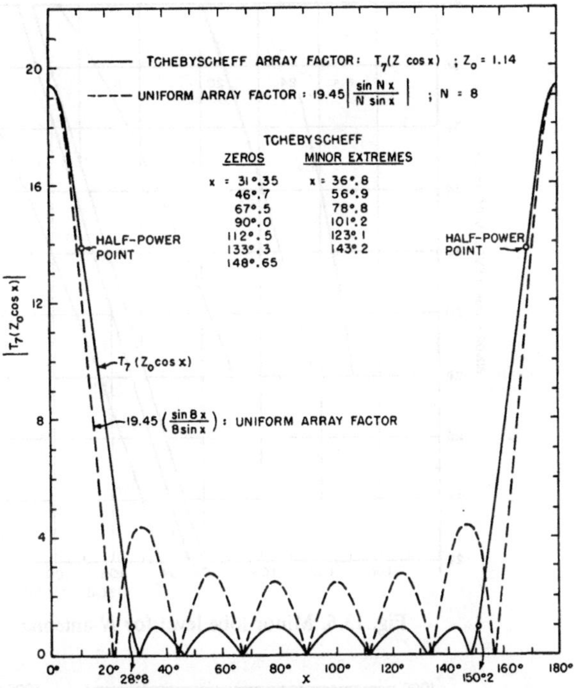

Fig. 13.4. Tchebyscheff and uniform array factors of eight-element array as functions of x.

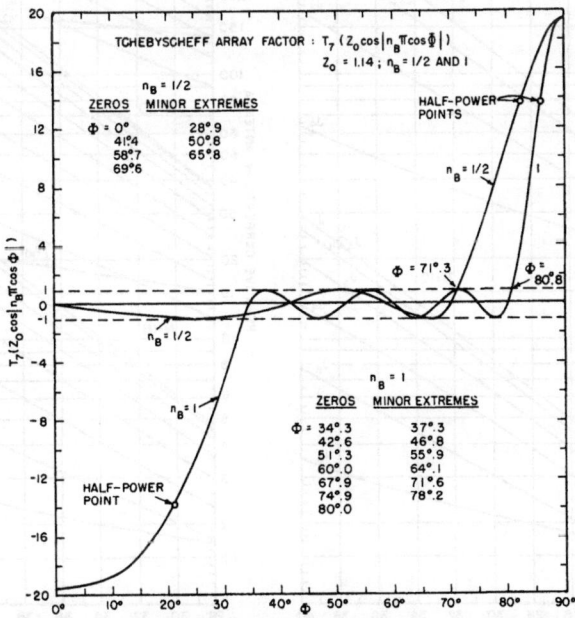

Fig. 13.5. Tchebyscheff array factor of eight-element array as a function of Φ.

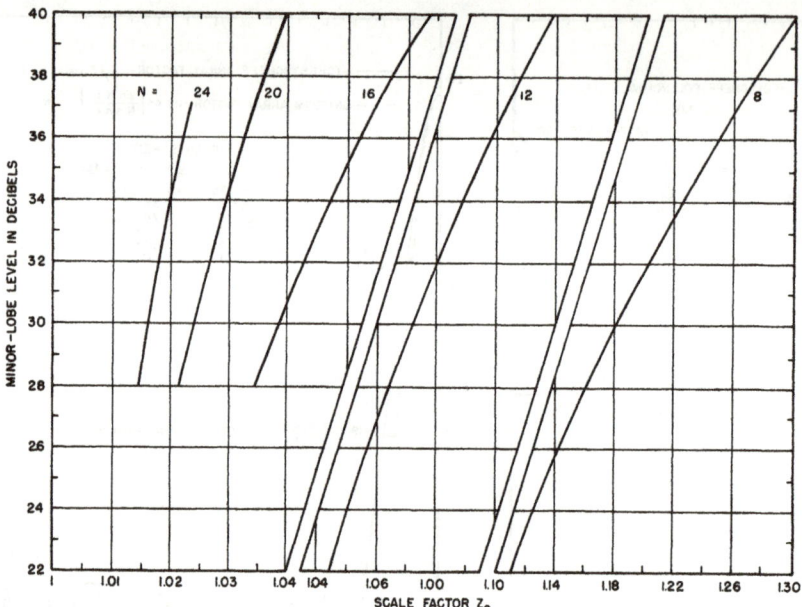

Fig. 13.6. Minor lobe level for N-antenna Tchebyscheff array (Dolph).

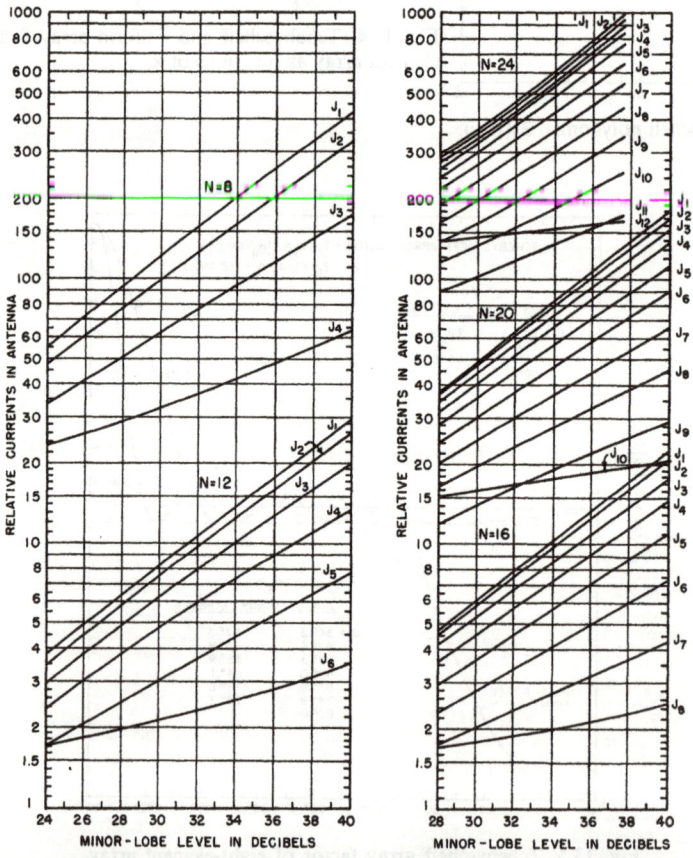

Fig. 13.7. Current for N-antenna Tchebyscheff array (Dolph).

TABLE 13.4. Current factors for Tchebyscheff arrays of N antennas.

$N = 12$: $k_6 = z_0^{11}$
$k_5 = 11(k_6 - z_0^9)$
$k_4 = 9k_5 - 44k_6 + 44z_0^7$
$k_3 = 7k_4 - 27k_5 + 77k_6 - 77z_0^5$
$k_2 = 5k_3 - 14k_4 + 30k_5 - 55k_6 + 55z_0^3$
$k_1 = 3k_2 - 5k_3 + 7k_4 - 9k_5 + 11k_6 - 11z_0$

$N = 16$: $k_8 = z_0^{15}$
$k_7 = 15k_8 - 15z_0^{13}$
$k_6 = 13k_7 - 90k_8 + 90z_0^{11}$
$k_5 = 11k_6 - 65k_7 + 275k_8 - 275z_0^9$
$k_4 = 9k_5 - 44k_6 + 156k_7 - 450k_8 + 450z_0^7$
$k_3 = 7k_4 - 27k_5 + 77k_6 - 182k_7 + 378k_8 - 378z_0^5$
$k_2 = 5k_3 - 14k_4 + 30k_5 - 55k_6 + 91k_7 - 140k_8 + 140z_0^3$
$k_1 = 3k_2 - 5k_3 + 7k_4 - 9k_5 + 11k_6 - 13k_7 + 15k_8 - 15z_0$

$N = 20$: $k_{10} = z_0^{19}$
$k_9 = 19k_{10} - 19z_0^{17}$
$k_8 = 17k_9 - 152k_{10} + 152z_0^{15}$
$k_7 = 15k_8 - 119k_9 + 665k_{10} - 665z_0^{13}$
$k_6 = 13k_7 - 90k_8 + 442k_9 - 1729k_{10} + 1729z_0^{11}$
$k_5 = 11k_6 - 65k_7 + 275k_8 - 935k_9 + 2717k_{10} - 2717z_0^9$
$k_4 = 9k_5 - 44k_6 + 156k_7 - 450k_8 + 1121k_9 - 2508k_{10} + 2508z_0^7$
$k_3 = 7k_4 - 27k_5 + 77k_6 - 182k_7 + 378k_8 - 714k_9 + 1254k_{10} - 1254z_0^5$
$k_2 = 5k_3 - 14k_4 + 30k_5 - 55k_6 + 91k_7 - 140k_8 + 204k_9 - 285k_{10} + 285z_0^3$
$k_1 = 3k_2 - 5k_3 + 7k_4 - 9k_5 + 11k_6 - 13k_7 + 15k_8 - 17k_9 + 19k_{10} - 19z_0$

$N = 24$: $k_{12} = z_0^{23}$
$k_{11} = 23k_{12} - 23z_0^{21}$
$k_{10} = 21k_{11} - 230k_{12} + 230z_0^{19}$
$k_9 = 19k_{10} - 189k_{11} + 1311k_{12} - 1311z_0^{17}$
$k_8 = 17k_9 - 152k_{10} + 952k_{11} - 4692k_{12} + 4692z_0^{14}$
$k_7 = 15k_8 - 119k_9 + 665k_{10} - 2940k_{11} + 10{,}948k_{12} - 10{,}948z_0^{13}$
$k_6 = 13k_7 - 90k_8 + 442k_9 - 1729k_{10} + 5733k_{11} - 16{,}744k_{12} + 16{,}744z_0^{11}$
$k_5 = 11k_6 - 65k_7 + 275k_8 - 935k_9 + 2717k_{10} - 7007k_{11} + 16{,}445k_{12} - 16{,}445z_0^9$
$k_4 = 9k_5 - 44k_6 + 156k_7 - 450k_8 + 1122k_9 - 2508k_{10} + 5148k_{11} - 9867k_{12} + 9867z_0^7$
$k_3 = 7k_4 - 27k_5 + 77k_6 - 182k_7 + 378k_8 - 714k_9 + 1254k_{10} - 2079k_{11} + 3289k_{12} - 3289z_0^5$
$k_2 = 5k_3 - 14k_4 + 30k_5 - 55k_6 + 91k_7 - 140k_8 + 204k_9 - 285k_{10} + 385k_{11} - 506k_{12} + 506z_0^3$
$k_1 = 3k_2 - 5k_3 + 7k_4 - 9k_5 + 11k_6 - 13k_7 + 15k_8 - 17k_9 + 19k_{10} - 21k_{11} + 23k_{12} - 23z_0$

Dolph reports satisfactory agreement between theory and experimental measurements on 12- and 24-element Tchebyscheff arrays.

14. Closely Spaced Tchebyscheff Arrays

When the distance between the elements of a broadside array is less than one-half wavelength, the Tchebyscheff distribution of currents as obtained in Sec. 13 for a symmetric broadside array does not provide the optimum array factor $A(\frac{1}{2}\pi, \Phi)$ in the sense of narrow beam width and low minor-lobe level. As pointed out by Riblet,* if the distance between elements is less than $\lambda_0/2$, that is, $n_B < \frac{1}{2}$, it is possible to derive array factors that have narrower beam widths and a smaller side-lobe level by an ingenious modification of the method of Dolph described in Sec. 13. The reason for the failure of the unmodified Dolph method when $n_B < \frac{1}{2}$ is understood readily once it is recognized that when $n_B < \frac{1}{2}$ the entire range of Φ, $0 \leq \Phi \leq \pi/2$, required to determine the array factor $A(\frac{1}{2}\pi, \Phi)$ is included in a range of u that is more limited than $0 \leq u \leq 1$. That is, only a part of $A(u)$ contributes to $A(\frac{1}{2}\pi, \Phi)$. Since $u = \cos(n_B \pi \cos \Phi)$, it is clear that the significant attainable range of u is

$$0 \leq \Phi \leq \pi/2: \quad \cos n_B \pi \leq u \leq 1. \quad (1)$$

Since the method of Sec. 13 restricts $A(u)$ to a magnitude not to exceed unity in the range $-(1/z_0) \leq u \leq (1/z_0)$, it is clear that when $n_B < \frac{1}{2}$, $A(u)$ is *unnecessarily* restricted in the range $0 \leq u < \cos n_B \pi$, where $A(u)$ may be left *completely unrestricted* since it contributes nothing to the actual array factor $A(\frac{1}{2}\pi, \Phi)$. Clearly, by restricting $A(u)$ only in the range where this is required, $A(\frac{1}{2}\pi, \Phi)$ may have a smaller minor-lobe level and a narrower beam width.

It was shown in Sec. 10 in analyzing end-fire arrays (for which $n_E < \frac{1}{2}$ is a necessary condition) that narrower beam widths and lower side lobes could be obtained by choosing the currents in the elements so as to include all of the zeros in the complex polynomial $A(z)$ in the attainable range of Φ so that $A(\frac{1}{2}\pi, \Phi)$ has as many zeros as $A(z)$. It is to be expected that a similar result may be achieved with the broadside array. For example, the seven-element uniform broadside array is represented by $A(x) = \sin 7x/\sin x$, where $x = n_B \pi \cos \Phi$.

If $n_B = \frac{1}{3}$, the attainable range of x by Φ as it varies over the range $0 \leq \Phi \leq \pi/2$ is $0 \leq x \leq \pi/3$. As seen in Fig. 14.1, this range includes only two of the available three zeros in the range $0 \leq x \leq \pi/2$. Evidently, the zero near 76° in Fig. 14.1 should be moved to the left of 60° by an appropriate change in the current distribution.

The regular Tchebyscheff pattern as defined in Sec. 13 corresponding to the uniform pattern of Fig. 14.1 is $T_6(z)$, where

$$T_6(z) = T_6(z_0 u) = T_6(z_0 \cos x)$$
$$= T_6[z_0 \cos(n_B \pi \cos \Phi)]. \quad (2)$$

This function is plotted in Fig. 14.2 for the arbitrary condition $T(z_0) = 9$ or $z_0 = 1.12$. For $n_B = \frac{1}{3}$, the attainable range of u is

$$0 \leq \Phi \leq \pi/2: \quad 0.5 \leq u \leq 1. \quad (3)$$

Just as for the uniform patterns, only two of the available three nulls in $T_6(z)$, $0 \leq z \leq 1$ are available in $T_6[z_0 \cos(\frac{1}{3}\pi \cos \Phi)]$.

In a method proposed by Riblet for obtaining an optimum factor $A(\frac{1}{2}\pi, \Phi)$ in the sense of Sec. 13, $A(u)$ is represented by a suitable Tchebyscheff polynomial $T_M(y)$ which has an appropriate argument y that locates all nulls and minor lobes that occur for $0 \leq u \leq 1$ in the range $0.5 \leq u \leq 1$. Since $A(u)$ is unrestricted for $0 \leq u < 0.5$, it may assume large values in this range. Moreover, since no zeros are included, the function $T_M(y)$ that is to represent $A(u)$ necessarily increases monotonically in magnitude as u decreases from 0.5 to 0. This means that the array factor must have an extreme value at $\Phi = 0$ or $x = \frac{1}{3}\pi \cos \Phi = \frac{1}{3}\pi$, a condition that can obtain *only* when N is odd. Hence, the method to be described for obtaining a distribution of currents that produces an optimum pattern in the sense of Sec. 13 is restricted to arrays with an odd number of elements.

The proposed array factor $A(u)$ for $N = 7$, namely, $T_6(z) = T_6(z_0 u)$, is to be like Fig. 14.2 with the two zeros nearest the center moved outward beyond $u = \pm 0.5$. Such a pattern may be represented by $T_3(y)$ if y is an even function of u and has its origin roughly midway between $u = 0.5$ and $u = 1$. That such a representation is possible is seen directly from the following mathematical manipulation of the general array factor:

$$A_{\text{odd}}(\tfrac{1}{2}\pi, \Phi) = A_{\text{odd}}(x) = 2 \sum_{i=0}^{n} k_i \cos 2ix,$$

$$n = (N-1)/2. \quad (4)$$

---

* This section is based on the work of Riblet, ref. 39.

where $x = n_B\pi \cos \Phi$ and where $k_0 = I_0/2I_r$, $k_i = I_i/I_r$, $i = 1, 2, \cdots, n$, and I_r is the reference current. If $I_r = I_0$, as assumed in Sec. 3, $k_0 = \frac{1}{2}$. Specifically, for $N = 7$, $n = 3$,

$$A_{odd}(x) = 2[k_0 + k_1 \cos 2x + k_2 \cos 4x + k_3 \cos 6x]. \quad (5)$$

Instead of representing (4) as an $M = N - 1$ degree polynomial or (5) as a sixth-degree polynomial in $u = \cos x$, they may be represented respectively as $(N-1)/2$ or as a third-degree polynomial in

$$v = \cos 2x = 2u^2 - 1. \quad (6)$$

Thus, for (5),

$$A_{odd}(v) = 2[k_0 + k_1 v + k_2(2v^2 - 1) + k_3(4v^3 - 3v)] \quad (7a)$$

$$= 2[(k_0 - k_2) + (k_1 - 3k_3)v + 2k_2 v^2 + 4k_3 v^3]. \quad (7b)$$

By proper choice of the relative currents, that is, of the k_i's, (7) may be specialized to be a Tchebyscheff polynomial of degree 3. That is,

$$T_3(y) = (k_0 - k_2) + (k_1 - 3k_3)y + 2k_2 y^2 + 4k_3 y^3. \quad (8)$$

Note that $T_3(z)$ is shown in Fig. 13.2. It is now necessary to choose the argument y in such a manner that the range $-1 \leq y \leq 1$ in which $|T_3(y)| \leq 1$ extends from $u = 0.5$ toward $u = 1$, and that $T_3(y_0)$, $y_0 > 1$ specifies the desired ratio of principal to minor lobes. Thus, let

$$y = a(2u^2 - 1) + b, \quad (9)$$

with a and b determined from the conditions

$$u = 0.5, \quad y = -1, \quad (10a)$$

$$u = 1, \quad y = y_0. \quad (10b)$$

Substitution of (10a) and (10b) in (9) gives

$$-1 = a(-\tfrac{1}{2}) + b, \quad (11a)$$

$$y_0 = a + b, \quad (11b)$$

so that

$$a = \tfrac{2}{3}(y_0 + 1), \quad (11c)$$

$$b = \tfrac{1}{3}(y_0 - 2). \quad (11d)$$

Hence, the required Tchebyscheff polynomial is

$$A_{odd}(u) = 2T_3[\tfrac{2}{3}(y_0 + 1)(2u^2 - 1) + \tfrac{1}{3}(y_0 - 2)]$$

$$= 2T_3[\tfrac{4}{3}(y_0 + 1)u^2 - \tfrac{1}{3}(y_0 + 4)] \quad (12)$$

instead of $A_{odd}(u) = 2T_6(z_0 u)$ as obtained using the formulation of Sec. 13. Note that $A_{odd}(u)$ is symmetric in u so that it is sufficient to consider $u \geq 0$. As a specific example, let the ratio of principal to minor lobes be fixed at 9. Then $T_3(y_0) = 9$, which occurs when $y_0 = 1.5$. With this value in (11c, d),

$$a = 5/3, \quad b = -1/6, \quad (13a)$$

so that

$$y = \tfrac{5}{3}(2u^2 - 1) - \tfrac{1}{6} = \tfrac{10}{3}u^2 - \tfrac{11}{6} = 3.33u^2 - 1.83. \quad (13b)$$

Hence, the appropriate array factor is

$$A_{odd}(u) = 2T_3(3.33u^2 - 1.83). \quad (14)$$

This polynomial is shown in Fig. 14.3 plotted as a function of u in the range $0 \leq u \leq 1$. This corresponds to a range of y given by $-1.83 \leq y \leq 1.5$. All zeros and minor lobes are in the range $-1 \leq y \leq 1$ and are restricted in amplitude to unity. Except for the distorted scale of y, the curve in Fig. 14.3 is seen to be the same as $T_3(z)$ in Fig. 13.2. The large lobe at $u = 0$ or $y = -1.83$ is of no significance, since the range of $A(u)$ attainable in $A(\tfrac{1}{2}\pi, \Phi)$ is $0.5 \leq u \leq 1$. Comparison of Fig. 14.3 with the right-hand half $(0 \leq u \leq 1)$ of Fig. 14.2 shows that the modified representation in Fig. 14.3 has shifted the minor lobes and nulls into the attainable range.

A graph of $|T_3(3.33u^2 - 1.83)|$ as a function of $x = \cos^{-1} u$ is shown in Fig. 14.4 together with the corresponding graph for a seven-element uniform array with amplitude adjusted to make the principal maximum also 9 as in the Tchebyscheff array. It is seen that in the attainable range $0 \leq x \leq 60°$, the Tchebyscheff pattern has a slightly greater beam width but a much smaller minor-lobe level. This is better seen in Fig. 14.5, where the two patterns are shown as functions of the actual angle Φ in the range $0 \leq \Phi \leq \pi/2$. The other quadrants are obtained by symmetry. As a consequence of the optimum property of the Tchebyscheff polynomial, the Tchebyscheff array factor in Fig. 14.5 has the minimum possible beam width for the specified minor-lobe level of 1/9. Note that while the beam width of the uniform array is slightly less than that of the Tchebyscheff pattern, this does not apply to the specified minor-lobe level of 1/9. The minor-lobe level for the uniform array is more than *double* this specified value.

With the optimum array factor determined, it remains to specify the relative currents that

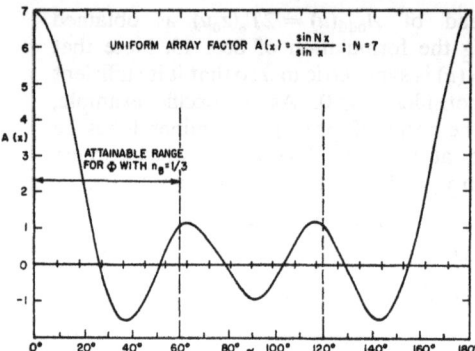

Fig. 14.1. Uniform array factor of seven-element broadside array.

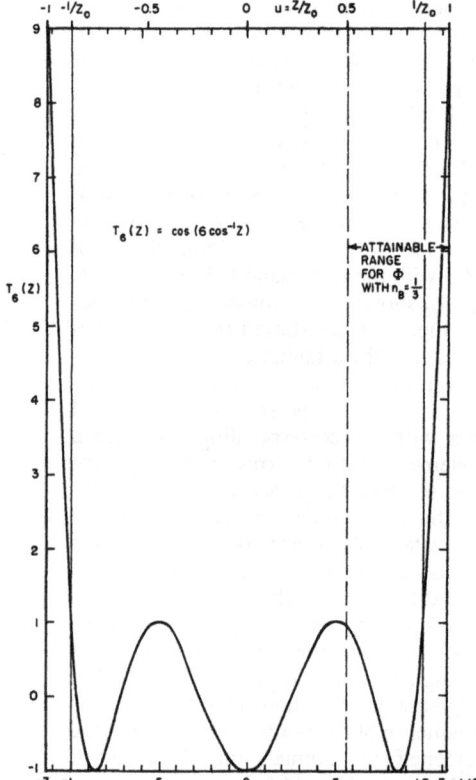

Fig. 14.2. Tchebyscheff array factor $T_6(z)$ of seven-element array.

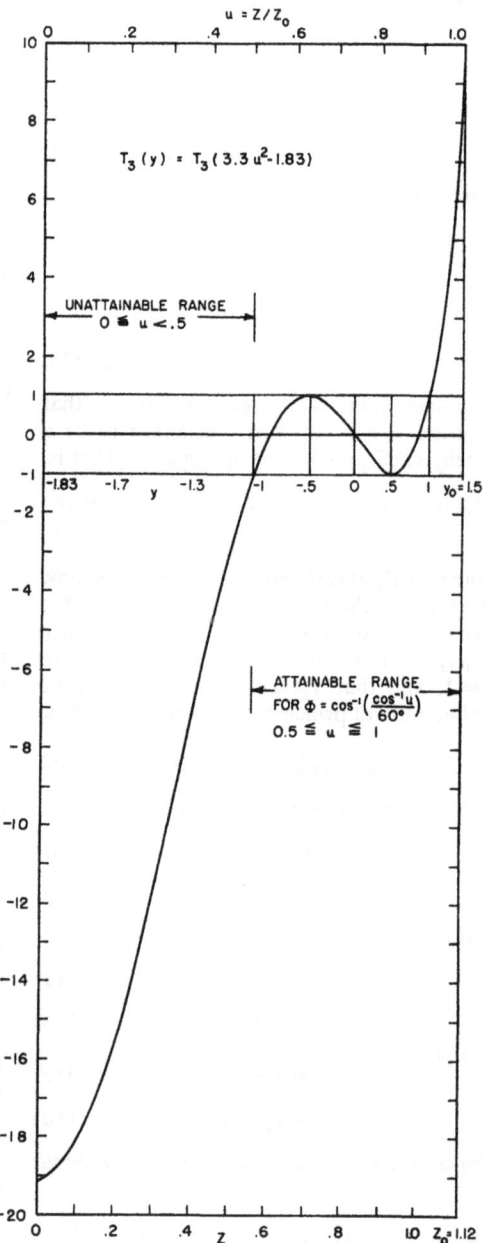

Fig. 14.3. Tchebyscheff array factor $T_3(y)$ as a function of u and z.

are required to maintain a field characterized by this array factor. These are readily determined by expanding

$$T_3[\tfrac{5}{3}(2u^2 - 1) - \tfrac{1}{6}] = T_3(\tfrac{5}{3}v - \tfrac{1}{6})$$
$$= T_3[1.67(v - 0.1)]$$

as a function of v, comparing coefficients, and solving for k_0 to k_4. Thus

$$T_3[1.67(v-0.1)] = 4(1.67)^3(v^3 - 0.3v^2 + 0.03v - 0.001) - 3 \times 1.67(v - 0.1) \quad (15a)$$

$$= 18.5v^3 - 5.55v^2 - 4.46v + 0.482. \quad (15b)$$

By equating coefficients of v in (15b) to the coefficients of v in the square bracket in (7b), the following relative currents referred to an arbitrary reference current I_r are obtained:

$$k_3 = 4.62, \quad k_2 = -2.78, \quad k_1 = 9.40, \quad k_0 = -2.30. \quad (16)$$

Since $k_0 = I_0/2I_r$, the ratio I_0/I_r is determined at once from (16) to be

$$I_0/I_r = 2k_0 = -4.60. \quad (17)$$

Dividing (16) by this value, the following relative currents are obtained referred to I_0:

$$\frac{I_3}{I_0} = -1.00, \frac{I_2}{I_0} = 0.60, \frac{I_1}{I_0} = -2.04, \quad (18)$$

Thus, the currents in the seven elements in sequence have the approximate ratios

$$-1 : 0.6 : -2 : 1 : -2 : 0.6 : -1. \quad (19)$$

Note that these currents are not all in phase as assumed for the conventional broadside array, but involve 180° phase shifts from one element to its adjacent one. This alternation in the directions is characteristic of highly directive arrays.

A seven-element array with units spaced $\lambda_0/3$ and with the relative currents given in (19) has the optimum array factor shown in Fig. 14.5.

Although in the interest of clarity the detailed analysis in this section is carried out specifically for a seven-element array with antennas spaced one-third wavelength and a minor-lobe level of 1/9, it is evident that the same sequence of steps may be applied to any array with an odd number of antennas spaced less than $\lambda_0/2$.

ARRAYS WITH OMNIDIRECTIONAL PROPERTIES

15. Circular Array*

For some purposes, as in broadcasting, an electromagnetic field with rotational symmetry about a vertical z-axis is required; that is, the radiation field must be independent of the azimuthal angle Φ. The desired nature of the vertical field pattern as a function of the angle Θ measured from the z-axis depends upon the particular application. For many broadcast transmitters a field that is intense only near the equatorial plane $\Theta = \pi/2$ is required. Such a field is provided by linear radiators with $h \leq \lambda_0/2$ (Chapter V) and especially by collinear arrays (Sec. 2). If the principal lobe in the field pattern is to be directed upward at an angle Θ between 0 and $\pi/2$, as for short-wave transmission using so-called reflection from the ionosphere, linear radiators of greater length may be used. As shown in Chapter V, the principal ear in the field pattern moves from $\Theta = \pi/2$ toward $\Theta = 0$ as the length of the antenna is increased. For both linear radiator and collinear array the electric field is vertically polarized, that is, perpendicular to the equatorial plane.

An interesting array that may be designed to have either omnidirectional or directional properties with vertical polarization is the *circular array* consisting of N identical parallel radiators uniformly spaced around the circumference of a circle. The impedance properties of such an array are determined in Sec. III.14 for the general case of N arbitrary driving voltages. The array factor for so general an array consists of a sum of N terms that cannot be combined into a compact form without introducing restrictions on the relative magnitudes and phases of the currents. Since a small number of units is readily handled as a discrete sum, it is important to obtain an expression that permits the study of circular arrays with a large number of units. Such an expression may be obtained if conditions are imposed that permit the combination of terms in pairs so that use may be made of the array factor of a two-element array. Evidently, a first condition is that all currents must be equal in magnitude, and a second that N must be even. If these are satisfied there are two methods of pairing the antennas, as shown in Fig. 15.1. They are: (a) diametral pairing where the distance d

---

* This section is based in part on the work of Brückmann, ref. I.14, pp. 113–123.

between the units of each two-element array is the diameter of the circle; (b) *parallel-chord pairing* where the lines joining the two antennas of the $N/2$ pairs are all parallel but not of equal length. Simplification may be achieved in the two cases with somewhat different phase sequences for the currents.

The radiation-zone electric field of a pair of identical parallel antennas with half-length h, separated by a distance $d = n_p\lambda_0$, and with currents given by

$$I = I_0 e^{-j\delta/2}, \quad I' = I_0 e^{j\delta/2}, \quad \delta = 2\pi t_p, \quad (1)$$

is given by (3.52) with (3.56):

$$E_\Theta^r = \frac{j_0\zeta I_0}{2\pi} \frac{e^{-j\beta_0 R_0}}{R_0} F_0(\Theta, \beta_0 h) A(\Theta, \Phi; 2, n_p, t_p), \quad (2)$$

where

$$A(\Theta, \Phi; 2, n_p, t_p) = 2\cos\pi(n_p \sin\Theta \cos\Phi - t_p). \quad (3)$$

It is assumed that the unprimed antenna is at a distance $d/2$ along the positive x-axis, which is the zero line for the azimuthal coördinate Φ. The vertical z-axis is parallel to the two antennas and its positive half is the zero line for the coördinate Θ. The current I_0 is a convenient reference current.

(a) *Diametral pairing.* With N antennas uniformly spaced around the circumference of a circle of radius $d/2$, the angular separation of adjacent units is $2\pi/N$. If the diametrically opposite units form a pair and the pair at $\Phi = 0$ is number 1 (Fig. 15.1a), the array factor of the ith pair is

$$A_i(\Theta, \Phi - \Phi_i; 2, n_p, t_{pi})$$
$$= 2\cos\pi[n_p \sin\Theta \cos(\Phi - \Phi_i) - t_{pi}], \quad (4)$$

where

$$\Phi_i = \frac{2\pi}{N}(i - 1), \quad (5)$$

and where t_{pi} is the phase difference between the currents as a fraction of a period. Evidently, the normalized array factor of the $N/2$ pairs is

$$a(\Theta, \Phi) = \frac{A(\Theta, \Phi)}{N}$$
$$= \frac{2}{N}\sum_{i=1}^{N/2} \cos\pi[n_p \sin\Theta \cos(\Phi - \Phi_i) - t_{pi}]. \quad (6)$$

If N is not too small this gives an approximately *omnidirectional* pattern with respect to Φ when all currents are in phase so that $t_{pi} = 0$. In this case (6) reduces to

$$a(\Theta, \Phi) = \frac{2}{N}\sum_{i=1}^{N/2} \cos\{n_p\pi \sin\Theta$$
$$\times \cos[\Phi - (2\pi/N)(i-1)]\}. \quad (7)$$

If the diameter of the circle is very small compared with the wavelength, $(n_p\pi)^2 \ll 1$, the leading term is simply

$$a(\Theta, \Phi) = 1. \quad (8)$$

This is the normalized array factor of the N-element *cage antenna*, which is seen to be the same as that of a single antenna. Note that the complete vertical-field factor is $F_0(\Theta, \beta_0 h) A(\Theta, \Phi)$, where $F_0(\Theta, \beta_0 h)$ is the vertical-field factor of a single isolated antenna as defined in Chapter V. For very short antennas ($\beta_0^2 h^2 \ll 1$) and approximately for antennas with $\beta_0 h \leq 2$ (see Sec. V.13), $F_0(\Theta, \beta_0 h)$ is proportional to $\sin\Theta$ so that the complete vertical-field factor is proportional to $A(\Theta, \Phi)\sin\Theta$. In order to compare the field factor of an N-element circular array with a single antenna radiating the same power, the driving-point resistance of each antenna must be known. Since all units are identical, with equal currents

$$(R_i)_{\text{in}} = (R_1)_{\text{in}} = R_{s1} + \sum_{i=2}^{N} R_{1i}. \quad (9a)$$

In order to radiate the same total power as a given isolated, single antenna, the following equation must be satisfied:

$$I_0^2 R_0 = N[I_1^2 (R_1)_{\text{in}}], \quad (9b)$$

where I_0 is the magnitude of the driving-point current in the isolated antenna and I_1 is the magnitude of the driving-point current in any one of the N identical antennas.

The normalized array factor (7) of a six-element circular array is

$$\frac{A(\Theta, \Phi)}{6} = a(\Theta, \Phi) = \tfrac{1}{3}\{\cos[n_p\pi \sin\Theta \cos\Phi]$$
$$+ \cos[n_p\pi \sin\Theta \cos(\Phi + \tfrac{1}{3}\pi)]$$
$$+ \cos[n_p\pi \sin\Theta \cos(\Phi - \tfrac{1}{3}\pi)]\}. \quad (10)$$

For infinitely thin antennas with $\beta_0 h = \pi/2$, $R_0 = 73$ ohms, $(R_1)_{\text{in}} = 54$ ohms, so that

$$I_0 = I_1 \sqrt{\frac{6 \times 54}{73}} = 2.1 I_1. \quad (11)$$

A graph of the vertical-field pattern $V(\Theta)$ defined by $A(\Theta, \Phi) \sin \Theta$ with $A(\Theta, \Phi)$ given by (10) with $n_p = \tfrac{2}{3}$ is shown in Fig. 15.2. For comparison, the corresponding pattern defined by $V(\Theta) = 2.1 \sin \Theta$ of a single antenna that is supplied approximately equal power also is shown.

In order to show the degree of rotational symmetry, curves are shown both for $\Phi = 0$ and for $\Phi = \pi/6$, that is, on a radial line through an antenna and midway between two adjacent antennas. The difference in the curves is seen to be so small that even for $N = 6$ the pattern is essentially rotationally symmetric. The significant fact about Fig. 15.2 is that by changing the diameter of the circle from very small to $2\lambda_0/3$ the vertical characteristic is changed greatly, the direction of maximum far-zone field being shifted from the equatorial plane $\Theta = \pi/2$ to $\Theta = 38°$. Thus, with a circular array of even very short antennas with $h \ll \lambda_0$, it is possible to approximate the field pattern of a single linear radiator of length h near $3\lambda_0/4$.

Since the horizontal pattern for $N = 6$ is so nearly a circle, which is attained ideally only for N infinite, it appears reasonable to assume that the normalized array factor for all even values of N greater than four must be approximated well by the normalized array factor with N infinite. By defining

$$\Phi'_i = \frac{2\pi}{N}(i - 1) - \Phi, \quad (11a)$$

$$\Delta\Phi'_i = \frac{2\pi}{N}, \quad (11b)$$

and substituting these in (7), the result is

$$a(\Theta, \Phi) \to \frac{1}{\pi} \sum_{i=1}^{N/2} \cos(n_p\pi \sin \Theta \cos \Phi'_i)\Delta\Phi'_i. \quad (12)$$

In the limit as N approaches infinity, the limits for Φ'_i in the sum approach $-\Phi$ and $\pi - \Phi$, and the sum becomes the following well-known special form of the general Sommerfeld integral*

$$a(\Theta, \Phi) = \frac{1}{\pi}\int_{-\Phi}^{\pi - \Phi} \cos(n_p\pi \sin \Theta \cos \Phi')\,d\Phi'$$

$$= J_0(n_p\pi \sin \Theta). \quad (13a)$$

A graph of $V(\Theta) = J_0(\tfrac{2}{3}\pi \sin \Theta) \sin \Theta/J_0(\tfrac{2}{3}\pi)$, corresponding to $N = \infty$, $d = 2\lambda_0/3$, is also shown in Fig. 15.2. Note that this pattern is normalized to unity at $\Theta = \pi/2$ and not adjusted for the same power since $(R_1)_{\text{in}}$ is not available. It is seen that the shapes of the vertical pattern for $N = 6$ and $N = \infty$ are essentially the same.

In Fig. 15.3 are shown polar graphs of the vertical-field factor

$$V(\Theta) = J_0(n_p\pi \sin \Theta) \sin \Theta \quad (13b)$$

for a range of values of n_p from 0 to 2.23. The particular values chosen are critical in the sense that they correspond to transitions at which a particular ear vanishes or reaches a maximum.

A combination of the circular array with all currents equal and in phase with a single antenna at the center of the circle provides omnidirectional patterns, which may be flattened considerably by having the current in the central unit greater than the sum of the currents in the outer units arranged in a circle of appropriate diameter.

The circular array with diametral pairing of elements may be made essentially unidirectional instead of omnidirectional by proper adjustment of the relative phases. Evidently the array factor (6) has an absolute maximum of unity if the arguments of all the cosines in the sum vanish. Let the angles Θ and Φ for which this occurs be denoted by Θ_m and Φ_m. The condition for such a maximum is

$$t_{pi} = n_p \sin \Theta_m \cos(\Phi_m - \Phi_i), \quad (14)$$

so that

$$a(\Theta, \Phi) = \frac{2}{N}\sum_{i=1}^{N/2} \cos n_p\pi[\sin \Theta \cos(\Phi - \Phi_i)$$

$$- \sin \Theta_m \cos(\Phi_m - \Phi_i)]. \quad (15)$$

It is shown in the literature that multidirectional and essentially unidirectional beams are defined by (15) with appropriate choices of n_p.

(b) *Parallel-chord pairing and progressive phase.* If an even number of antennas in a circular array are paired as in Fig. 15.1b, a relatively simple array factor may be derived for discrete values of $\Phi = \Phi_k$ in the rather general case of currents that are equal in amplitude but change progressively and uniformly in phase around the circle. The discrete directions Φ_k in which the array factor can be expressed simply are radial vertical planes that pass through an antenna if $N/2$ is odd and that pass midway between adjacent antennas if $N/2$ is even. This is no serious restriction if N is reasonably large

---
* See, for example, Jahnke-Emde, ref. I.28.

and the phases of the currents do not vary too rapidly from unit to unit, since the field in the equatorial plane is nearly circular with only small variations in magnitude as the radial plane $\Phi = $ constant is turned from a plane through an antenna to a plane passing midway between adjacent antennas. Evidently, if the field fluctuates considerably between these two values of Φ, as when the phase change from unit to unit is π, additional points must be determined in order to plot the field pattern.

Referring to Fig. 15.1b, the array factor of each of the $N/2 = 4$ pairs of antennas 11′, 22′, 33′, and 44′ in the direction R_0 parallel to the line joining each pair is given by (3) with $\Phi = 0$. Thus

$$A(\Theta, 0; 2, n_{pi}, t_{pi}) = 2 \cos \pi(n_{pi} \sin \Theta - t_{pi}). \tag{16}$$

If the angle Φ has its zero value in the vertical radial plane through antenna 1, it follows from Fig. 15.1b that the distance between paired elements as a fraction of wavelength is

$$n_{pi} = n_{p0} \cos (\Phi - \Phi_i), \quad n_p \equiv d/\lambda_0, \tag{17}$$

where $\Phi_i = (2\pi/N)(i - 1)$ as in (5). If the phase of the currents is referred to antenna 1, then, for a progressive phase increase

$$\delta_i = m\Phi_i$$
$$= m(\Phi_k + \pi/2) + m(\Phi_i - \Phi_k - \pi/2), \tag{18a}$$

$$\delta_i' = m(2\Phi_k + \pi - \Phi_i)$$
$$= m(\Phi_k + \pi/2) - m(\Phi_i - \Phi_k - \pi/2), \tag{18b}$$

where m is an integer. Since by definition $2\pi t_{pi}$ is the phase difference between the two antennas in each pair,

$$2\pi t_{pi} = \delta_i - \delta_i' = 2m(\Phi_i - \Phi_k - \pi/2). \tag{19}$$

Hence, the array factor with currents referred to antenna 1 is obtained by substituting (17) and (19) in (16). The result is

$$A(\Theta, \Phi_k) = 2 \sum_{i=1}^{N/2} \cos \pi(n_{p0} \sin \Theta \sin \xi_i - m\xi_i), \tag{20}$$

where

$$\xi_i \equiv \Phi_i - \Phi_k - \pi/2$$
$$= (2\pi/N)(i - 1) - \Phi_k - \pi/2. \tag{21}$$

Since the general formula (2) for the electric field refers the phase of the currents to the mean value between the antennas in each pair, the normalized complex array factor of the entire array for use in a formula like (2) is

$$a(\Theta, \Phi_k) = \frac{2}{N} e^{-jm(\Phi_k + \pi/2)}$$
$$\times \sum_{i=1}^{N/2} \cos \pi(n_p \sin \Theta \sin \xi_i - m\xi_i). \tag{22}$$

As N increases without limit, $\Delta\xi = 2\pi/N \to d\xi$, and the sum becomes the following integral

$$a(\Theta, \Phi_k) \to e^{-jm(\Phi_k + \pi/2)}$$
$$\times \frac{1}{\pi} \int_{-\Phi_k - \pi/2}^{\pi - \Phi_k - \pi/2} \cos \pi(n_p \sin \Theta \sin \xi - m\xi) \, d\xi. \tag{23}$$

The integral in (23) is recognized to be the integral representation of the mth-order Bessel function,* so that

$$a(\Theta, \Phi_k) = e^{-jm(\Phi_k + \frac{1}{2}\pi)} J_m(n_p \pi \sin \Theta). \tag{24}$$

When all currents are in phase, $m = 0$, and (24) coincides with (13). Since (24) applies only to the discrete directions $\Phi = \Phi_k$ through an antenna when $N/2$ is odd, or midway between two adjacent antennas when $N/2$ is even, $a(\Theta, \Phi_k) \doteq a(\Theta, \Phi)$ only when N is sufficiently large and m sufficiently small compared with $N/2$, so that $a(\Theta, \Phi)$ is essentially independent of Φ. Polar graphs of

$$\frac{V(\Theta)}{V(\frac{1}{2}\pi)} = \frac{J_m(n_p \pi \sin \Theta) \sin \Theta}{J_m(n_p \pi)} \tag{25}$$

are shown in Fig. 15.4 for $n_p = 0$, $m = 0$; $n_p = 0.5$, $m = 1.2$; $n_p = 1.7$, $m = 1$; and $n_p = 2.15$, $m = 2$.

Note that if the circular array with progressive phase change is combined with a single antenna at the center of the circle the resulting array factor is *not* rotationally symmetric except for $m = 0$ owing to the exponential factor in (24).

Ring Quasi Arrays. Instead of arranging a ring of antennas equidistantly around a circle with all axes *perpendicular* to the plane of the circle, they may have their axes *in the plane* of the circle oriented in tangential or radial directions. If the currents all have equal amplitudes and vary progressively in phase

---

* See for example, Jahnke-Emde, ref. I.28.

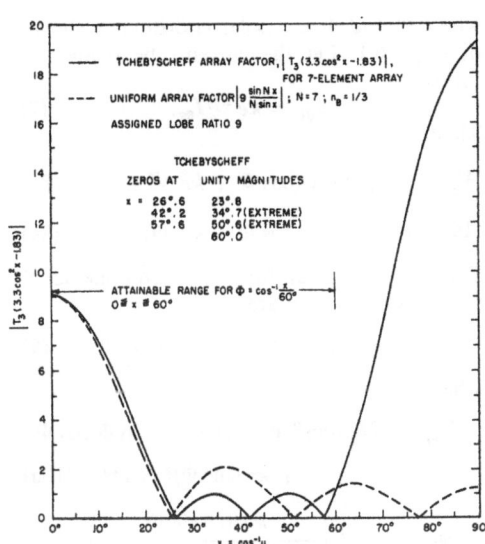

Fig. 14.4. Tchebyscheff array factor $T_3(y)$ as a function of x.

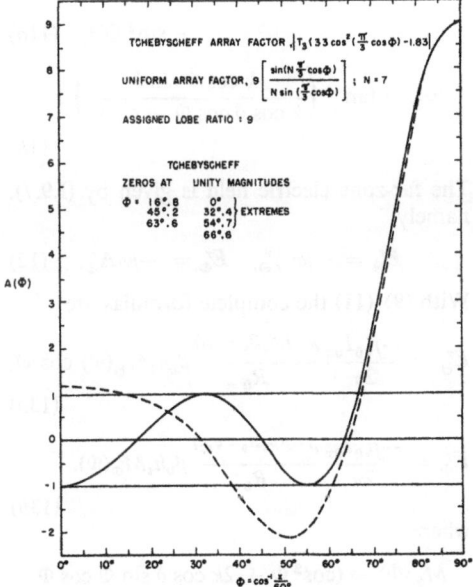

Fig. 14.5. Tchebyscheff array factor $T_3(y)$ as a function of Φ in solid line, uniform array factor in broken line.

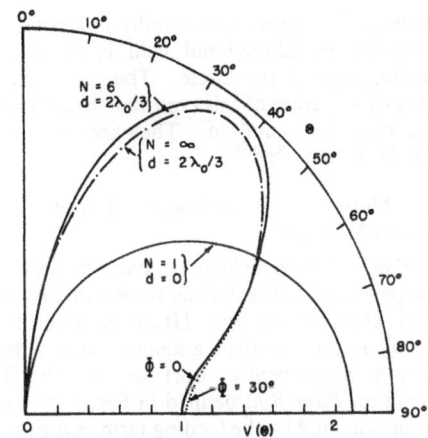

Fig. 15.2. Vertical field patterns of N-element circular array.

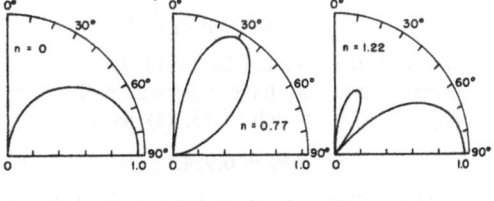

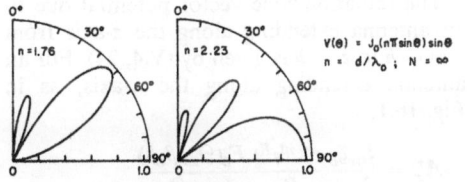

Fig. 15.3. Vertical field patterns of circular array with all currents in phase.

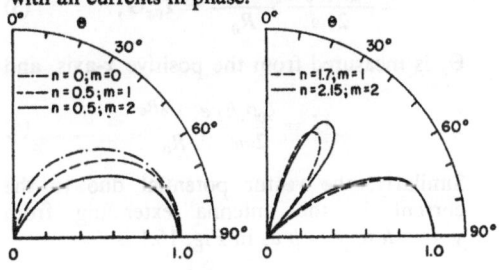

Fig. 15.4. The function $[J_m(n\pi \sin\theta)\sin\theta]/J_m(n\pi)$.

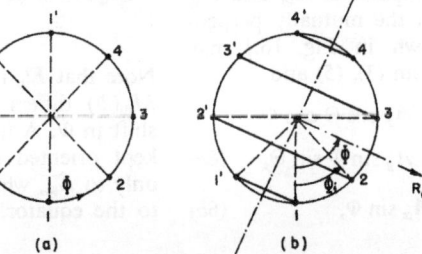

Fig. 15.1. Circular arrays of eight elements: (a) diametral pairing; (b) parallel-chord pairing.

around the circle, horizontally polarized, essentially omnidirectional field is maintained in the plane of the circle. The properties of ring quasi arrays of this type have been studied especially by Knudsen. They are reported in the literature.[1,32,28,28a]

16. Mutually Perpendicular Antennas and Turnstile Array

The radiation field of a pair of mutually perpendicular antennas as shown in Fig. 16.1 and analyzed in Sec. III.21 is determined easily if the identical antennas have a half-length h sufficiently short so that the field function $F_0(\Theta, \beta_0 h)$ defined in Sec. V.5 is well approximated by the leading term in a Fourier expansion. As shown in Sec. V.13, this is true when $\beta_0 h = 2\pi h/\lambda_0 \leq 2$, in which case

$$F_0(\Theta, \beta_0 h) \doteq \beta_0 h_t \sin \Theta, \quad (1)$$

where $\beta_0 h_t$ is defined by (V.13, 22). Usually, crossed antennas have electrical half-lengths $\beta_0 h = \pi/2$, for which (V.13, 25) gives

$$\beta_0 h_t = 0.954. \quad (2)$$

The radiation-zone vector potential due to an antenna extending along the z-axis from $z = -h$ to $z = h$ is given by (V.4, 24). For an antenna extending along the x-axis, as in Fig. 16.1,

$$A_x^r = \frac{I_{0x}\zeta_0}{2\pi\omega} \frac{e^{-j\beta_0 R_0}}{R_0} \frac{F_0(\Theta_x, \beta_0 h)}{\sin \Theta_x}$$

$$\doteq \frac{I_{0x}\zeta_0\beta_0 h_t}{2\pi\omega} \frac{e^{-j\beta_0 R_0}}{R_0} = I_{0x}G; \quad (3)$$

Θ_x is measured from the positive x-axis, and

$$G \equiv \frac{\zeta_0\beta_0 h_t}{2\pi\omega} \frac{e^{-j\beta_0 R_0}}{R_0}. \quad (4)$$

Similarly, the vector potential due to the current in the antenna extending from $y = -h$ to $y = h$ as in Fig. 16.1 is

$$A_y^r = \frac{I_{0y}\zeta_0}{2\pi\omega} \frac{e^{-j\beta_0 R_0}}{R_0} \frac{F_0(\Theta_y, \beta_0 h)}{\sin \Theta_y} \doteq I_{0y}G. \quad (5)$$

The resultant polar components A_Θ^r and A_Φ^r due to the currents in the mutually perpendicular antennas shown in Fig. 16.1 may be obtained directly from (3), (5) and

$$A_\Theta^r = -A_z \sin \Theta + A_x \cos \Theta \cos \Phi$$
$$+ A_y \sin \Phi \sin \Theta, \quad (6a)$$

$$A_\Phi^r = A_y \cos \Phi - A_x \sin \Phi, \quad (6b)$$

where Θ and Φ are polar coördinates shown in Fig. 16.1. Let the currents be related as follows:

$$I_{0y} = ke^{-j\delta}I_{0x}. \quad (7)$$

Then,

$$A_\Theta^r = GI_{0x}\cos\Theta(\cos\Phi + ke^{-j\delta}\sin\Phi), \quad (8a)$$

$$A_\Phi^r = GI_{0x}(ke^{-j\delta}\cos\Phi - \sin\Phi). \quad (8b)$$

Expressed in polar form the components of the vector potential are

$$A_\Theta^r = N_\Theta^r e^{-j\psi_\Theta}, \quad A_\Phi^r = N_\Phi^r e^{-j\psi_\Phi}, \quad (9)$$

where

$$N_\Theta^r \equiv GI_{0x}(\cos^2\Phi + 2k\cos\delta\sin\Phi\cos\Phi$$
$$+ k^2\sin^2\Phi)^{\frac{1}{2}}\cos\Theta, \quad (10a)$$

$$\psi_\Theta \equiv \tan^{-1}\left(\frac{k\sin\delta\sin\Phi}{\cos\Phi + k\cos\delta\sin\Phi}\right), \quad (10b)$$

$$N_\Phi^r \equiv GI_{0x}(k^2\cos^2\Phi - 2k\cos\delta\sin\Phi\cos\Phi$$
$$+ \sin^2\Phi)^{\frac{1}{2}}, \quad (11a)$$

$$\psi_\Phi \equiv \tan^{-1}\left(\frac{k\cos\Phi\sin\delta}{k\cos\delta\cos\Phi - \sin\Phi}\right). \quad (11b)$$

The far-zone electric field is given by (I.9.7), namely,

$$E_\Theta^r = -j\omega A_\Theta^r, \quad E_\Phi^r = -j\omega A_\Phi^r. \quad (12)$$

With (9)–(11) the complete formulas are

$$E_\Theta^r = \frac{-j\zeta_0 I_{0x}}{2\pi} \frac{e^{-j(\beta_0 R_0 + \psi_\Theta)}}{R_0} \beta_0 h_t M_\Theta(\Phi)\cos\Theta, \quad (13a)$$

$$E_\Phi^r = \frac{-j\zeta_0 I_{0x}}{2\pi} \frac{e^{-j(\beta_0 R_0 + \psi_\Phi)}}{R_0} \beta_0 h_t M_\Phi(\Phi), \quad (13b)$$

where

$$M_\Theta(\Phi) \equiv (\cos^2\Phi + 2k\cos\delta\sin\Phi\cos\Phi$$
$$+ k^2\sin^2\Phi)^{\frac{1}{2}}, \quad (14a)$$

$$M_\Phi(\Phi) \equiv (k^2\cos^2\Phi - 2k\cos\delta\sin\Phi\cos\Phi$$
$$+ \sin^2\Phi)^{\frac{1}{2}}. \quad (14b)$$

Note that E_Φ^r is independent of Θ; also, that $M_\Phi(\Phi)$ differs from $M_\Theta(\Phi)$ only by a 90° shift in Φ. A linear receiving antenna that is kept oriented perpendicular to Φ responds only to E_Θ^r, which lies in planes *perpendicular* to the equatorial plane; alternatively, if the receiving antenna is kept perpendicular to Θ it responds only to E^r_Φ, which is always *parallel* to the equatorial plane. The properties of the components E^r_Θ and E^r_Φ may be determined from the vertical-plane field factor $M_\Theta(\Phi)\cos\Theta$ and the horizontal-plane field factor $M_\Phi(\Phi)$ by varying successively the relative phases and magnitudes of the currents. Since $M_\Phi(\Phi + \tfrac{1}{2}\pi) = M_\Theta(\Phi)$, only $M_\Theta(\Phi)$ need be investigated.

(a) With currents equal in magnitude but variable in phase

$$k = 1: \quad M_\Theta(\Phi) = \sqrt{1 + \cos\delta \sin 2\Phi}. \tag{15}$$

Particular values of δ lead to the following:

$$\delta = \begin{Bmatrix}0\\ \pi\end{Bmatrix}: \quad M_\Theta(\Phi) = \sqrt{1 \pm \sin 2\Phi}, \tag{16a}$$

$$\delta = \frac{\pi}{2}: \quad M_\Theta(\Phi) = 1. \tag{16b}$$

Graphs of $M_\Theta(\Phi)$ as given in (16) are shown in Fig. 16.2a; $M_\Phi(\Phi)$ is obtained by rotating the origin of Φ through 90° as shown. It is seen that by shifting the relative phase of the currents the horizontal-field pattern may be changed from a figure eight along either 45° line to a circle.

(b) When the currents are in phase but variable in magnitude,

$$\delta = 0: \quad M_\Theta(\Phi) = |\cos\Phi + k\sin\Phi|. \tag{17}$$

A graph of this function is shown in Fig. 16.2b. It is seen to resemble Fig. 16.2a except that the figure-of-eight pattern has its nulls when $\Phi = \pi/2$ instead of when $\Phi = \pi/4$.

It is clear that an omnidirectional pattern is obtained from two mutually perpendicular antennas when the currents are equal and in phase quadrature. As a consequence of the factor $\cos\Theta$, E^r_Θ is zero in the equatorial plane, so that for short-wave broadcast purposes in this plane the component E^r_Φ must be used with horizontal receiving antennas.

Since each of the two crossed antennas is in the neutral plane of the other, there is no coupling between them and the impedance behavior of each is the same as when isolated. This is considered in Sec. III.21.

The horizontal electric field E^r_Φ of a pair of identical antennas along the *x*- and *y*-axes is independent of the angle Θ measured from the vertical *z*-axis. Hence the field at high angles, that is, $\Theta < 45°$, is equal to the field in the equatorial plane. If it is required, as in broadcast transmission, to confine the field largely within a flat cone near $\Theta = 90°$ it is possible to stack a number of pairs of mutually perpendicular antennas, treating each crossed pair just like a dipole in the collinear antenna discussed in Sec. 2. The array factor for a stacked or *turnstile* array of N identical units (perpendicular pairs) separated by a distance s when the currents in each pair are equal and in phase with those in the units above and below it is given by (2.25) to be

$$A(\Theta) = \frac{\sin(N\pi n \sin\Theta)}{\sin(\pi n \sin\Theta)}, \quad n = s/\lambda_0. \tag{18}$$

For $k = 1$, $\delta = \pi/2$, the far-zone electric field is

$$E^r_\Phi = \frac{-j\zeta_0 I_{0x}}{2\pi} \frac{e^{-j(\beta_0 R_0 + \psi_\Phi)}}{R_0} \beta_0 h_t M_\Phi(\Phi) A(\Theta), \tag{19a}$$

$$E^r_\Theta = \frac{-j\zeta_0 I_{0x}}{2\pi} \frac{e^{-j(\beta_0 R_0 + \psi_\Phi)}}{R_0}$$
$$\times \beta_0 h_t M_\Theta(\Phi) A(\Theta) \cos\Theta. \tag{19b}$$

The vertical array factors $A(\Theta)$ and $A(\Theta)\cos\Theta$ are shown in Fig. 16.3 for a stacked array of six pairs of crossed antennas. It is seen that the horizontally polarized component E^r_Φ has a large principal lobe in the equatorial plane and only small amplitudes in other directions. The vertically polarized component E^r_Θ is small in all directions. The patterns are rotationally symmetric if the currents are equal and in phase quadrature in each crossed pair while each parallel array formed by one antenna in each pair has all currents in phase so that it is equivalent to a broadside array.

A simple and ingenious construction for a stacked turnstile array consists of a single vertical mast that may be made of metal into which are screwed the individual horizontal half-antennas in proper relative positions. If the antennas are near $\lambda_0/4$ in half-length and are stacked at distances near $\lambda_0/2$, they may be driven from two open-wire transmission lines with spacings somewhat greater than the radius of the mast. The two lines are driven in phase quadrature for a rotationally symmetric pattern, in phase for a figure-of-eight horizontal pattern. Each line drives one of the two mutually perpendicular broadside arrays by connecting to the antennas at short distances from the mast and spiraling around this one-half turn between elements in order to connect to the next antenna in the proper phase. The problem of obtaining equal

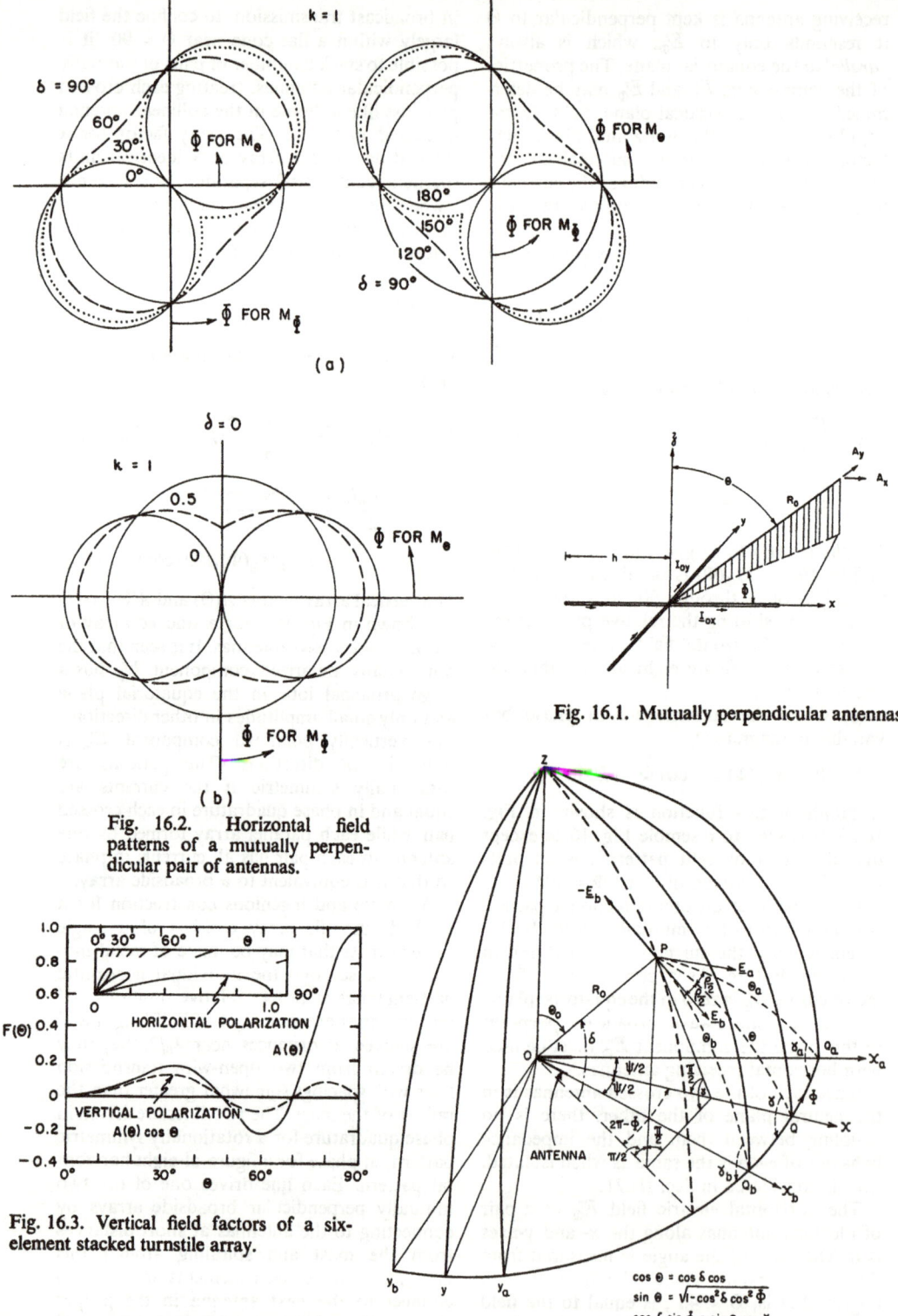

Fig. 16.2. Horizontal field patterns of a mutually perpendicular pair of antennas.

Fig. 16.1. Mutually perpendicular antennas.

Fig. 16.3. Vertical field factors of a six-element stacked turnstile array.

$$\cos\theta = \cos\delta\cos\phi$$
$$\sin\theta = \sqrt{1-\cos^2\delta\cos^2\phi}$$
$$\cos\delta\sin\phi = \sin\theta\cos\delta$$

Fig. 17.1. Horizontal V-antenna.

currents in phase in a broadside array driven from an open-wire line is discussed in Sec. III.19. Note that the two mutually perpendicular broadside arrays forming the stacked turnstile are electrically independent.

17. Resonant V-Antenna

The apex-driven resonant V-antenna has important omnidirectional properties when used in the horizontal position with an appropriate leg length h and enclosed angle ψ.[58] These may be deduced from the general expressions for the far-zone field with arbitrary leg length and enclosed angle.[8]

Consider the field of the V-antenna shown in Fig. 17.1. The antenna has its apex at the origin of a system of polar coördinates R_0, $\Theta_0 = \tfrac{1}{2}\pi - \delta$, Φ, with Θ_0 the colatitude measured from the positive z-axis perpendicular to and δ the latitude measured from the plane of the V, and with Φ the longitude measured from the x, z-plane passing midway between its legs. One leg is in the direction x_a, the other in the direction x_b. The field is to be determined at a distant point P under the assumption that the currents in the two identical legs a and b are equal in magnitude and directed radially in opposite directions. For determining the field in the far zone it is a satisfactory approximation to assume the currents distributed sinusoidally even when the legs of the antenna are moderately thick. This approximation is best when $\beta_0 h$ is near $\pi/2$.

The far-zone electric field of an antenna extending from $x = 0$ to $x = h$ is given by

$$E_\Theta^r = \frac{j\omega}{4\pi v_0} \frac{e^{-j\beta_0 R_0}}{R_0} \int_0^h I_x' e^{j\beta_0 x' \cos\Theta} \sin\Theta \, dx', \quad (1)$$

where Θ is the colatitude measured from the positive x-axis. With

$$I_a' = I_m \sin \beta_0(h - x') \quad (2)$$

the integral becomes

$$I_m \int_0^h \sin \beta_0(h - x') e^{j\beta_0 x' \cos\Theta} \sin\Theta \, dx'. \quad (3)$$

By using well-known integrals, (3) may be evaluated directly. The result, when combined with (1), is the same as that obtained in Sec. V.18. It is

$$E_\Theta^r = \frac{j\zeta_0 I_m}{4\pi R_0} e^{-j\beta_0 R_0}[F_m(\Theta, \beta_0 h)$$
$$+ jW_m(\Theta, \beta_0 h)]$$
$$= \frac{j\zeta_0 I_m}{4\pi R_0} [F_m^2(\Theta, \beta_0 h)$$
$$+ W_m^2(\Theta, \beta_0 h)]^{\frac{1}{2}} e^{j(\theta - \beta_0 R_0)}, \quad (4)$$

where, as in Sec. V.18,

$$F_m(\Theta, \beta_0 h) = \frac{\cos(\beta_0 h \cos\Theta) - \cos\beta_0 h}{\sin\Theta}, \quad (5a)$$

$$W_m(\Theta, \beta_0 h) = \frac{\sin(\beta_0 h \cos\Theta) - \cos\Theta \sin\beta_0 h}{\sin\Theta}, \quad (5b)$$

$$\theta = \tan^{-1}[W_m(\Theta, \beta_0 h)/F_m(\Theta, \beta_0 h)]. \quad (5c)$$

Formula (4) with (5a, b) may be applied to each half of the V-antenna by adding subscripts a and b and using a negative sign with I_m for half b. Thus,

$$E_{\Theta_a}^r = \frac{j\zeta_0 I_m}{4\pi R_0} e^{-j\beta_0 R_0}[F_m(\Theta_a, \beta_0 h)$$
$$+ jW_m(\Theta_a, \beta_0 h)]$$
$$= \frac{j\zeta_0 I_m}{4\pi R_0} [F_m^2(\Theta_a, \beta_0 h)$$
$$+ W_m^2(\Theta_a, \beta_0 h)]^{\frac{1}{2}} e^{j(\theta_a - \beta_0 R_0)}, \quad (6a)$$

$$E_{\Theta_b}^r = \frac{-j\zeta_0 I_m}{4\pi R_0} e^{-j\beta_0 R_0}[F_m(\Theta_b, \beta_0 h)$$
$$+ jW_m(\Theta_b, \beta_0 h)]$$
$$= \frac{-j\zeta_0 I_m}{4\pi R_0} [F_m^2(\Theta_b, \beta_0 h)$$
$$+ W_m^2(\Theta_b, \beta_0 h)]^{\frac{1}{2}} e^{j(\theta_b - \beta_0 R_0)} \quad (6b)$$

In any spherical triangle such as $Q_a P Q$ in Fig. 17.1 the following relations apply to the arcs Θ_a, $\tfrac{1}{2}\psi$, and Θ which form the sides of the triangle.*

$$\cos \Theta_a = \cos \Theta \cos \tfrac{1}{2}\psi$$
$$- \sin \Theta \sin \tfrac{1}{2}\psi \cos \gamma. \quad (7a)$$

Similarly,

$$\cos \Theta_b = \cos \Theta \cos \tfrac{1}{2}\psi$$
$$+ \sin \Theta \sin \tfrac{1}{2}\psi \cos \gamma. \quad (7b)$$

These formulas may be modified with the following relations, which are readily verified with the aid of Fig. 17.1:

$$\cos \Theta = \cos \delta \cos \Phi, \quad (8a)$$

$$\sin \Theta = \sqrt{1 - \cos^2 \delta \cos^2 \Phi}, \quad (8b)$$

$$\cos \delta \sin \Phi = \sin \Theta \cos \gamma. \quad (8c)$$

The results are

$$\cos \Theta_a = \cos \delta [\cos \Phi \cos \tfrac{1}{2}\psi$$
$$- \sin \Phi \sin \tfrac{1}{2}\psi]$$
$$= \cos \delta \cos(\Phi + \tfrac{1}{2}\psi), \quad (9a)$$

---

* See, for example, Peirce, ref. I.39, formula 625.

$$\cos \Theta_b = \cos \delta [\cos \Phi \cos \tfrac{1}{2}\psi$$
$$+ \sin \Phi \sin \tfrac{1}{2}\psi]$$
$$= \cos \delta \cos (\Phi - \tfrac{1}{2}\psi). \qquad (9b)$$

With (9a, b), (5a, b, c) assume the following form:

$$F_m(\Theta, \beta_0 h)$$
$$= \frac{\cos [\beta_0 h \cos \delta \cos (\Phi \pm \tfrac{1}{2}\psi)] - \cos \beta_0 h}{\sqrt{1 - \cos^2 \delta \cos^2 (\Phi \pm \tfrac{1}{2}\psi)}}, \qquad (10a)$$

$$W_m(\Theta, \beta_0 h)$$
$$= \frac{\sin [\beta_0 h \cos \delta \cos (\Phi \pm \tfrac{1}{2}\psi)]}{\sqrt{1 - \cos^2 \delta \cos^2 (\Phi \pm \tfrac{1}{2}\psi)}}, \qquad (10b)$$

$$\theta = \tan^{-1}[W_m(\Theta, \beta_0 h)/F_m(\Theta, \beta_0 h)]. \qquad (10c)$$

With the upper sign in (10a, b), the subscript a should be attached to Θ; with the lower sign, the subscript b applies.

The electric vectors $\mathbf{E}_a^r$ and $\mathbf{E}_b^r$ at any point P are perpendicular to the radius vector $\mathbf{R}_0$ and tangent to great circles in the directions Θ_a and Θ_b respectively. The resultant electric vector is obtained by vector addition: $\mathbf{E}^r = \mathbf{E}_a^r + \mathbf{E}_b^r$. In general, the expression for $\mathbf{E}^r$ is quite complicated. It reduces to much simpler form in the two practically important planes, (1) the horizontal plane containing the V-antenna defined by $\delta = 0$, and (2) the plane perpendicular to the plane of the V and bisecting the included angle defined by $\Phi = 0$. These two cases are considered separately in succession.

(1) The field in the plane of the V is simple because $\mathbf{E}_a^r$ and $\mathbf{E}_b^r$ are parallel in space and may be combined directly. Thus,

$$E_\Theta^r = E_a^r + E_b^r = E_{\Theta_a}^r + E_{\Theta_b}^r$$
$$= j[E_{\Theta_a}^r e^{j\theta_a} - E_{\Theta_b}^r e^{j\theta_b}] e^{-j\beta_0 R_0}. \qquad (11)$$

The resultant field given by (11) is readily expressed in polar form. Its magnitude is

$$E_\Theta^r = \sqrt{(E_{\Theta_a}^r)^2 + (E_{\Theta_b}^r)^2 - 2E_{\Theta_a}^r E_{\Theta_b}^r \cos (\theta_a - \theta_b)}, \qquad (12)$$

where $E_{\Theta_a}^r$ and $E_{\Theta_b}^r$ are the magnitudes of $E_{\Theta_a}^r$ and $E_{\Theta_b}^r$ as given by (6a) and (6b) using (10a)
and (10b) with $\delta = 0$, and where θ_a and θ_b are given by (10c) with $\delta = 0$. These are:

$$F_m(\Theta, \beta_0 h) = \frac{\cos [\beta_0 h \cos (\Phi \pm \tfrac{1}{2}\psi)] - \cos \beta_0 h}{\sin (\Phi \pm \tfrac{1}{2}\psi)}, \qquad (13a)$$

$$W_m(\Theta, \beta_0 h)$$
$$= \frac{\sin [\beta_0 h \cos (\Phi \pm \tfrac{1}{2}\psi)] - \cos \Phi \sin \beta_0 h}{\sin (\Phi \pm \tfrac{1}{2}\psi)}, \qquad (13b)$$

$$\theta = \tan^{-1}[W_m(\Theta, \beta_0 h)/F_m(\Theta, \beta_0 h)], \qquad (13c)$$

where, as before, the upper sign applies to Θ_a, the lower sign to Θ_b.

These expressions may be simplified further in the important special case $\beta_0 h = \pi$, when

$$F_m(\Theta, \beta_0 h) = \frac{2 \cos^2 [(\pi/2) \cos (\Phi \pm \tfrac{1}{2}\psi)]}{\sin (\Phi \pm \tfrac{1}{2}\psi)}, \qquad (14a)$$

$$W_m(\Theta, \beta_0 h) = \frac{2 \sin [(\pi/2) \cos (\Phi \pm \tfrac{1}{2}\psi)] \times \cos [(\pi/2) \cos (\Phi \pm \tfrac{1}{2}\psi)]}{\sin (\Phi \pm \tfrac{1}{2}\psi)}, \qquad (14b)$$

$$\theta = (\pi/2) \cos (\Phi \pm \tfrac{1}{2}\psi), \qquad (14c)$$

$$E_\Theta^r = \frac{\zeta_0 I_m}{2\pi R_0} \left\{ \frac{\cos [(\pi/2) \cos (\Phi \pm \tfrac{1}{2}\psi)]}{\sin (\Phi \pm \tfrac{1}{2}\psi)} \right\}. \qquad (15)$$

As before, the upper sign implies a subscript a on Θ, the lower sign a subscript b.

Numerical values of the resultant magnitude of the electric field in the plane of a right-angled V-antenna have been computed by Wells.[58] When $\psi = \pi/2$ the V-antenna is known as a *quadrant antenna*. Polar graphs for $\beta_0 h = \pi/4, \pi/2, \pi, 4\pi/3, 3\pi/2$ and 2π are shown in Fig. 17.2. Numerical values for $\beta_0 h = 5\pi/4$ with the radiation field normalized to $R_0 = 1$ km and $I_m = 1$ amp are in Table 17.1. It is seen that an essentially omnidirectional horizontal pattern is obtained when $\beta_0 h$ varies over a considerable range above and below π. For a fixed value of h, this implies a wide range of frequency in $\beta_0 = 2\pi f/v_0$, so that the omnidirectional property of a quadrant antenna is relatively broad band.

(2) The field perpendicular to the plane of the V-antenna and bisecting the included angle is defined by $\Phi = 0$ in the general expression. It is characterized by

$$\gamma = \frac{\pi}{2}, \quad \gamma_a = -\gamma_b, \quad \Theta_a = \Theta_b, \quad \delta = \Theta. \qquad (16)$$

TABLE 17.1. Horizontal electric field of quadrant antenna with $\beta_0 h = 5\pi/4$.

Φ (deg)	$E^r_{\Theta_a}$ (v/m)	$E^r_{\Theta_b}$ (v/m)	θ_a (deg)	θ_b (deg)	$\theta_a - \theta_b$ (deg)	E^r_Θ (v/m)
0	37.5	37.5	75.3	−104.8	180.1	75.0
10	30.8	43.5	60.3	−86.3	146.6	71.7
20	29.2	47.4	48.7	−64.4	113.3	61.5
30	15.2	50.1	41.3	−40.0	81.3	50.1
40	5.3	51.0	36.4	−13.5	49.9	47.9
50	5.3	51.0	−146.6	13.5	−157.1	56.1
60	15.2	50.1	−138.7	40.0	−178.7	65.7
70	29.2	47.4	−131.3	64.4	−195.7	69.3
80	30.8	43.5	−119.7	86.3	−206.0	72.6
90	37.5	37.5	−104.8	104.8	−209.6	72.6

Evidently E^r_a and E^r_b are equal and their resultant is obtained directly from Fig. 17.1 to be simply

$$E^r = 2E^r_{\Theta_a} \sin \tfrac{1}{2}\rho. \qquad (17)$$

It is directed perpendicular to the vertical plane bisecting the V and is parallel to the horizontal plane of the V. Using a familiar formula for spherical triangles* and (9a) with $\delta = \Theta$, $\Phi = 0$,

$$\sin \tfrac{1}{2}\rho = \frac{\sin \tfrac{1}{2}\psi}{\sin^2 \Theta_a} = \frac{\sin \tfrac{1}{2}\psi}{\sqrt{1 - \cos^2 \Theta \cos^2 \tfrac{1}{2}\psi}}. \qquad (18)$$

With (18) and (6a), the resultant field is

$$E^r = \frac{j\zeta_0 I_m}{2\pi R_0} e^{-j\beta_0 R_0} \frac{\sin \tfrac{1}{2}\psi}{1 - \cos^2 \Theta \cos^2 \tfrac{1}{2}\psi}$$
$$\times \{[\cos (\beta_0 h \cos \Theta \cos \tfrac{1}{2}\psi) - \cos \beta_0 h]$$
$$+ j[\sin (\beta_0 h \cos \Theta \cos \tfrac{1}{2}\psi)$$
$$- \cos \Theta \cos \tfrac{1}{2}\psi \sin \beta_0 h]\}. \qquad (19)$$

When $\beta_0 h = \pi$, the magnitude of the field reduces to

$$E^r = \frac{\zeta_0 I_m}{\pi R_0} \frac{\sin \tfrac{1}{2}\psi \cos (\tfrac{1}{2}\pi \cos \Theta \cos \tfrac{1}{2}\psi)}{1 - \cos^2 \Theta \cos^2 \tfrac{1}{2}\psi}. \qquad (20)$$

For the quadrant antenna with $\psi = \pi/2$, (20) becomes simply

$$E^r = \frac{\zeta_0 I_m}{\pi R_0} \frac{\cos (\tfrac{1}{4}\pi \sqrt{2} \cos \Theta)}{\sqrt{2}(1 - \tfrac{1}{2} \cos^2 \Theta)}. \qquad (21)$$

Calculations made by Wells from (21) indicate that the vertical-field pattern of the isolated, horizontal quadrant antenna with

* For example, Peirce, ref. I.39, formula 629.

$\beta_0 h = \pi$ differs negligibly from a circle in the plane bisecting the angle $\psi = \pi/2$ of the V. That is, the vertical pattern of the isolated quadrant antenna in the plane bisecting the V is essentially the same as that of an isolated horizontal dipole in its *equatorial* plane. It may be inferred that the vertical pattern of a horizontal quadrant antenna over a perfectly or imperfectly conducting earth also is like that of a horizontal dipole over the same earth.

THE RECIPROCAL THEOREM AND THE PROPERTIES OF ARRAYS

18. Transmitting Arrays with Elements of Finite Cross Section

In the general analysis of arrays carried out in this chapter the field factor of each element has been assumed to be

$$F_0(\Theta, \beta_0 h) = \frac{\cos (\beta_0 h \cos \Theta) - \cos \beta_0 h}{\sin \beta_0 h \sin \Theta} \qquad (1)$$

as derived from an assumed sinusoidal distribution of current. Although this distribution is a satisfactory approximation for antennas of finite cross section when their electrical half-length is $\pi/2$ or less, this is not true for other values of $\beta_0 h$, especially for those near π. Since it has been shown in Sec. V.14 by application of the reciprocal theorem that, in general,

$$F_0(\Theta, \beta_0 h) = \beta_0 h_e(\Theta), \qquad (2)$$

where $F_0(\Theta, \beta_0 h)$ is the complex field factor of an antenna of nonvanishing radius a when used for transmission and $h_e(\Theta)$ is the complex effective half-length of the antenna when used for reception, it is a simple matter to substitute (2) for (1) as a factor in the formulas for the fields of arrays of N identical elements. In effect, this assumes that the distribution of current in each element in an array is the same as in the same element when isolated. As shown in Chapter III, this is quite a good approximation even for closely spaced parallel antennas near antiresonance. For purposes of determining far-zone electromagnetic fields in a manner that takes account of the non-sinusoidal distribution of current in the individual elements it is entirely adequate. Accordingly, in the complete field factor of an array, (2) may be substituted for (1) if a sinusoidal distribution of current is not a sufficiently good approximation in special cases requiring a more precise knowledge of the minor-lobe level and the behavior at minima, as distinct from the nulls obtained with the sinusoidal assumption.

19. *Receiving Arrays*

The study of the receiving antenna in Chapter IV is restricted to the analysis of a single antenna that is center-loaded by an arbitrary impedance Z_L and immersed in a plane-polarized electric field E (Fig. 19.1a). It is shown that the current $I_L = I_0$ entering the load and leaving the antenna is given by

$$I_L = I_0 = \frac{-(E \cos \psi) 2h_e(\Theta)}{Z_0 + Z_L}, \quad (1)$$

where $2h_e(\Theta)$ is the complex effective length of the antenna as defined in Chapter IV, and ψ is the angle between the electric vector and the plane containing the axis of the antenna and the line joining its center to the distant transmitter. If the load impedance Z_L is the input impedance $Z_{0in} = Z_L$ of a transmission line of length s (Fig. 19.1b), characteristic impedance Z_c, and propagation constant $\gamma = \alpha + j\beta$ that is terminated in the impedance Z_s of the receiver, it follows from transmission-line theory that the current I_s entering Z_s is given by

$$I_s = kI_0 \quad (2)$$

where

$$k = \frac{\sinh \theta_s}{\sinh (\gamma s + \theta_s)} \quad (3a)$$

and

$$\theta_s = \rho_s + j\Phi_s = \coth^{-1}(Z_s/Z_c) \quad (3b)$$

With (1), the current in the load Z_s terminating the transmission line from the receiving antenna is

$$I_s = \frac{-(E \cos \psi) 2h_e(\Theta) k}{Z_0 + Z_{0in}}. \quad (4)$$

In practice it is often advantageous to use receiving antennas that have directional properties or impedances that are different from those of a simple, center-loaded antenna. In particular, directional arrays consisting of a number of elements appropriately spaced and interconnected by transmission lines may be used. Indeed, any transmitting array including uniform and nonuniform, broadside, end-fire, collinear, or parasitic arrays that are driven from a *single* generator may be converted into a receiving array merely by replacing the generator with a receiver. Alternatively, the impedance of a receiving antenna or array may be modified without changing its directional properties, for example, by substituting a folded dipole for the straight wire antenna analyzed in Chapter IV. Since all such arrays or antennas are fed from a transmission line that is driven by a generator, the conversion to a receiving system by replacing the generator by the load impedance Z_s is simple. The application of Thévenin's theorem at the terminals of Z_s at once establishes the following formula for the current I_s entering the load:

$$I_s = \frac{V_s(Z_s = \infty)}{Z_s + Z_{sin}}, \quad (5)$$

where Z_{sin} is the input impedance looking back into the transmission line from the antenna or array when $E = 0$, and $V_s(Z_s = \infty)$ is the open-circuit voltage across the output end of the line when Z_s is made infinite, Fig. 19.2. Since Z_s is assumed known and Z_{sin} is readily evaluated using Chapter III, the problem is to determine the open-circuit voltage.

An accurate analysis of a receiving array using the method of Chapter IV necessarily leads to simultaneous integral equations for which solutions are unavailable in general. Fortunately, the application of the reciprocal theorem (ref. I.31, Chap. IV) leads to a simple solution in terms of quantities already evaluated for transmitting arrays.

Consider a single linear radiator (designated as antenna 1) of half-length h_1 and radius a_1 that is center-driven by a voltage V'_1 in series with an impedance Z_{L1}. (For simplicity, terminal-zone effects are assumed to be negligible.) If the impedance of the antenna is Z_{01}, the current I'_1 entering antenna 1 is

$$I'_1 = \frac{V'_1}{Z_{01} + Z_{L1}}. \quad (6)$$

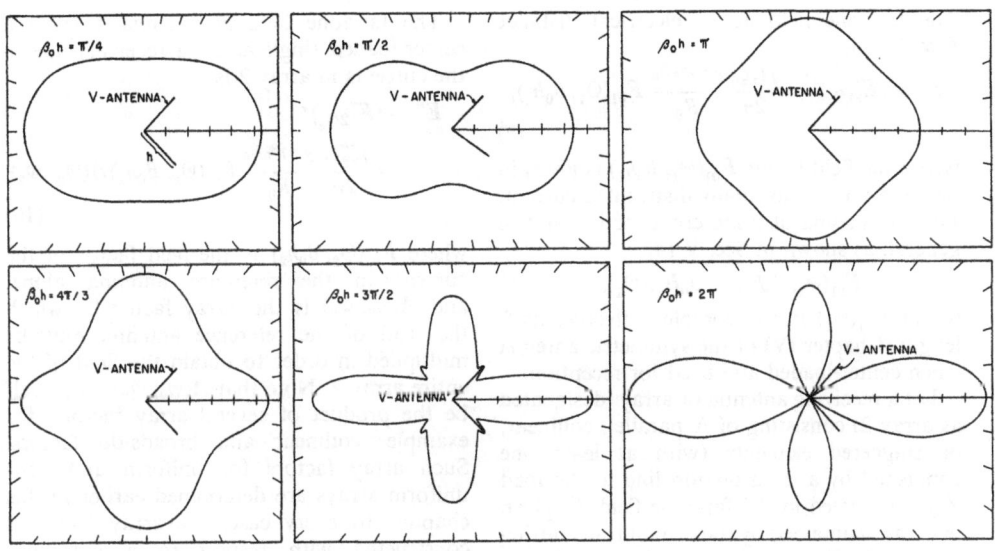

Fig. 17.2. Horizontal field patterns of a 90° horizontal V-antenna (Wells).

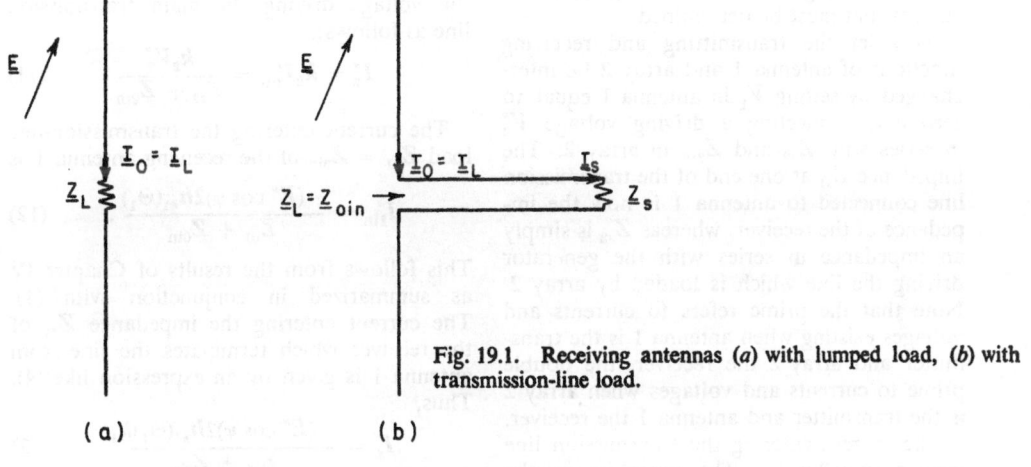

Fig. 19.1. Receiving antennas (a) with lumped load, (b) with transmission-line load.

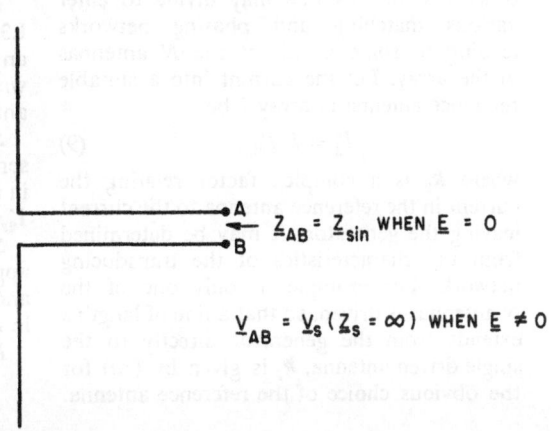

Fig. 19.2. Impedance $Z_{s\mathrm{in}}$ and open-circuit voltage $V_s(Z_s = \infty)$ for use in series with Z_s, using Thévenin's theorem.

The far-zone field at an electrical distance $\beta_0 R_0 \gg 1$ is

$$E' \equiv (E^r_{21\Theta_1})' = \frac{jI'_1\zeta_0}{2\pi} \frac{e^{-j\beta_0 R_0}}{R_0} F_{01}(\Theta_1, \beta_0 h_1), \tag{7}$$

where the field factor $F_{01}(\Theta_1, \beta_0 h_1)$ is given in Sec. V.5 for a sinusoidally distributed current. For an antenna of finite cross section and in general, as shown in Sec. V.14,

$$F_{01}(\Theta_1, \beta_0 h_1) = \beta_0 h_{e1}(\Theta_1), \tag{8}$$

where $h_{e1}(\Theta_1)$ is the complex effective half-length (Chapter IV) of the symmetric antenna when center-loaded and used for reception.

Let a receiving antenna or array (designated as array 2) consisting of N parallel, collinear, or staggered elements (with at least one connected by a transmission line to the load Z_{s2}) be located in the far-zone field E' given in (7) as maintained by the distant transmitting antenna 1. The impedance looking back into the line toward the array is $(Z_2)_{in}$. The current entering the load Z_{s2} is I'_{2s}. It is this current that must be determined.

Now let the transmitting and receiving functions of antenna 1 and array 2 be interchanged by setting V'_1 in antenna 1 equal to zero and connecting a driving voltage V''_2 in series with Z_{s2} and Z_{2in} in array 2. The impedance Z_{s1} at one end of the transmission line connected to antenna 1 is now the impedance of the receiver, whereas Z_{s2} is simply an impedance in series with the generator driving the line which is loaded by array 2. Note that the prime refers to currents and voltages existing when antenna 1 is the transmitter and array 2 the receiver, the double prime to currents and voltages when array 2 is the transmitter and antenna 1 the receiver.

The current entering the transmission line feeding array 2 is I''_{2in}. The current leaving the other end of this line may divide to enter various matching and phasing networks leading to some or all of the N antennas in the array. Let the current into a suitable reference antenna in array 2 be

$$I''_2 = k_2 I''_{2in}, \tag{9}$$

where k_2 is a complex factor relating the current in the reference antenna to the current leaving the generator. It may be determined from the characteristics of the transducing network. For example, if only one of the N antennas is driven, so that a line of length s extends from the generator directly to the single driven antenna, k_2 is given by (3a) for the obvious choice of the reference antenna.

The far-zone field at antenna 1 (now a center-loaded single-antenna receiver) due to the currents in array 2 is

$$E'' \equiv (E^r_{12\Theta_2})''$$

$$= \frac{jI''_2 \zeta_0}{2\pi} \frac{e^{-j\beta_0 R_0}}{R_0} F_{02}(\Theta_2, \beta_0 h_2) A(\Theta_2, \Phi_2), \tag{10}$$

where $F_{02}(\Theta_2, \beta_0 h_2)$ is the field factor of the current in the reference antenna alone, and $A(\Theta_2, \Phi_2)$ is the array factor by which the field of the reference antenna must be multiplied in order to obtain the field of the entire array 2. Note that $A(\Theta_2, \Phi_2)$ may itself be the product of several array factors, for example, collinear and broadside factors. Such array factors for uniform and non-uniform arrays are determined earlier in this chapter. In every case, the array factor is constructed with respect to a particular reference current in a specified element. This current, I''_2, may be expressed in terms of the voltage driving the main transmission line as follows:

$$I''_2 = k_2 I''_{2in} = \frac{k_2 V''_2}{Z_{s2} + Z_{2in}}. \tag{11}$$

The current entering the transmission-line load $Z_{L1} = Z_{0in}$ of the receiving antenna 1 is

$$I''_{1in} = \frac{-(E'' \cos \psi) 2 h_{e1}(\Theta_1)}{Z_{01} + Z_{0in}}. \tag{12}$$

This follows from the results of Chapter IV as summarized in conjunction with (1). The current entering the impedance Z_{s1} of the receiver which terminates the line from antenna 1 is given by an expression like (4). Thus,

$$I''_s = \frac{-(E'' \cos \psi) 2 h_{e1}(\Theta_1) k_1}{Z_{01} + Z_{0in}}. \tag{13}$$

According to the reciprocal theorem (ref. I.31, chap. IV), a certain relation applies to any two sets of currents and voltages associated with a given pair of terminals. As applied to antenna 1 and array 2 this is:

Situation prime: A voltage V'_1 applied in series with the impedances Z_{01} and Z_{0in} in the circuit of antenna 1 maintains a current I'_{2s} in the load Z_{s2} of array 2.

Situation double prime: A voltage V''_2 applied in series with the impedances Z_{s2} and Z_{2in} in array 2 maintains a current I''_{1in} in the impedance Z_{01} of antenna 1.

Reciprocal theorem:

$$I''_{1in} V'_1 = I'_{2s} V''_2. \tag{14}$$

It follows from (14) that the unknown current I'_{2s} in the load Z_{s2} of array 2 when used for reception is

$$I'_{2s} = I''_{1\text{in}} V'_1 / V''_2. \qquad (15)$$

If *equal voltages* are applied to antenna 1 and array 2, that is,

$$V'_1 = V''_2, \qquad (16)$$

it follows that

$$I'_{2s} = I''_{1\text{in}}. \qquad (17)$$

Alternatively, if *equal powers* are supplied to the two antennas a more complicated situation arises which is analyzed later in this section. Consider first the simpler case of equal voltages and equal currents as given in (16) and (17).

Reciprocity with equal and constant voltages. With (17), (12), and (8), the current in the load of the receiving array 2 is

$$I'_{2s} = I''_{1\text{in}} = \frac{-(E'' \cos \psi) 2 h_{e1}(\Theta_1)}{Z_{01} + Z_{0\text{in}}}$$

$$= \frac{-(E'' \cos \psi) 2 F_{01}(\Theta_1, \beta_0 h_1)}{\beta_0 (Z_{01} + Z_{0\text{in}})}. \qquad (18)$$

The substitution of (10) in (18) after use has been made of (11) gives

$$I'_{2s} = \left[\frac{-j k_2 V''_2 \zeta_0}{2\pi (Z_{s2} + Z_{2\text{in}})} \frac{e^{-j\beta_0 R_0}}{R_0} \right.$$
$$\left. \times F_{02}(\Theta_2, \beta_0 h_2) A(\Theta_2, \Phi_2) \right]$$
$$\times \left[\frac{2 F_{01}(\Theta_1, \beta_0 h_1) \cos \psi}{\beta_0 (Z_{01} + Z_{0\text{in}})} \right]. \qquad (19)$$

The following rearrangement can be made, using (16):

$$I'_{2s} =$$
$$\left[\frac{-j V'_1 \zeta_0}{2\pi (Z_{01} + Z_{0\text{in}})} \frac{e^{-j\beta_0 R_0}}{R_0} F_{01}(\Theta_1, \beta_0 h_1) \cos \psi \right]$$
$$\times \left[\frac{2}{\beta_0} F_{02}(\Theta_2, \beta_0 h_2) A(\Theta_2, \Phi_2) k_2 \right] [Z_{s2} + Z_{2\text{in}}]^{-1}$$

$$(20)$$

With (6) and (7) the first bracket in (20) is identified as $-E' \cos \psi$. Hence, (20) has the same form as (4), namely,

$$I'_{2s} = \frac{-(E' \cos \psi) 2 h_{e2}(\Theta_2, \Phi_2) k_2}{Z_{s2} + Z_{2\text{in}}}, \qquad (21)$$

where a *generalized effective electrical length* of the entire array 2 is denoted by

$$\beta_0 h_{e2}(\Theta_2, \Phi_2) \equiv F_{02}(\Theta_2, \beta_0 h) A(\Theta_2, \Phi_2). \qquad (22)$$

Comparison with (2) shows that the effective length of an array, corresponding to that of a single antenna, is the *complete field factor.*

The complex factor k_2 occurs in (21) just as in (4), since in general the current in the load is not the same as the current in the reference antenna. When the currents are equal, $k_2 = 1$.

Note that since the electric field of a linear radiator at a distant point P is always in the plane defined by the axis of the antenna and the line R_0 from its center to P, it follows that ψ is the angle between two planes that have the line R_0 from the center of antenna 1 to the center of the reference antenna in array 2 in common, and that contain the axis of antenna 1 on the one hand, and the axis of the reference antenna 2 on the other hand. (All antennas in array 2 are assumed to be parallel to the reference antenna; they may be displaced laterally, collinearly, or both.) Thus, ψ is the same angle for antenna 1 when array 2 is driven as for array 2 when antenna 1 is driven.

The current in the load of a receiving array in a linearly polarized electric field is determined readily from (21). Note that this involves only characteristics of the array when used for transmission, namely, the series impedance $Z_{s2} + Z_{2\text{in}}$ seen by the generator, the complete field factor of the transmitting array referred to the current entering a specified reference antenna, and the complex ratio of the current entering the reference antenna to the current entering the line feeding the array.

The far-zone electric field (10) of the array 2 when used for transmitting may be expressed as follows, using (11) and omitting the subscript 2 and all primes:

$$E^r_\Theta = \frac{j V_0 \zeta}{2\pi} \frac{e^{-j\beta_0 R_0}}{R_0} \frac{F_0(\Theta, \beta_0 h) A(\Theta, \Phi) k}{Z_s + Z_{\text{in}}}, \qquad (23)$$

The current in the load of the same array when used for reception is given by (21) with (22). Thus,

$$I_s = \frac{-(E \cos \psi) 2 F_0(\Theta, \beta_0 h) A(\Theta, \Phi) k}{\beta_0 (Z_s + Z_{\text{in}})}. \qquad (24)$$

It is clear that so long as V and R_0 in (23) and $E \cos \psi$ in (24) are kept constant, E^r_Θ always is proportional to I_s at a fixed frequency. Evidently, all the directional properties of the array may be investigated when it is used for reception or transmission if V, R_0, and $E \cos \psi$ are kept constant, provided the impedance Z_s in series with the generator and the receiver impedance are equal or provided that $Z_s + Z_{\text{in}}$ is constant. If the desired field properties involve only the rotation of the array as a unit, Z_{in} is constant and I_s is proportional to E_Θ^r even if Z_s in series with the generator in (23) differs from Z_s of the receiver in (24), but each remains constant. On the other hand, if the field properties under study involve variations in the separation, length, or orientation of the elements, Z_{in} of the array changes. In this case Z_s in (23) must be equal to Z_s in (24) if I_s and E_Θ^r are to be proportional to each other.

Reciprocity with equal and constant powers.[*]
If the power supplied to antenna 1 is to equal that to array 2 when each is used for transmission, the conditions are

$$P = \tfrac{1}{2}|I_1'|^2 R_{01} = \tfrac{1}{2}|I_{2\text{in}}''|^2 R_{2\text{in}}. \quad (25)$$

It is assumed that the transmission lines are lossless. In terms of the driving voltages and impedances (25) becomes

$$P = \tfrac{1}{2}\left|\frac{V_1'}{Z_{01} + Z_{0\text{in}}}\right|^2 R_{01}$$

$$= \tfrac{1}{2}\left|\frac{V_2''}{Z_{s2} + Z_{2\text{in}}}\right|^2 R_{2\text{in}}, \quad (26)$$

so that

$$V_1' \equiv |V_1'| = \sqrt{\frac{2P}{R_{01}}}\,|Z_{01} + Z_{0\text{in}}|, \quad (27a)$$

$$V_2'' \equiv |V_2''| = \sqrt{\frac{2P}{R_{2\text{in}}}}\,|Z_{s2} + Z_{2\text{in}}|. \quad (27b)$$

With the reciprocal theorem in the form (15), it follows that

$$I_{2s}' \equiv |I_{2s}'| = |I_{1\text{in}}''|\sqrt{\frac{R_{2\text{in}}}{2P}}\left|\frac{V_1'}{Z_{s2} + Z_{2\text{in}}}\right|. \quad (28)$$

The use of (12) with (8) in (28) leads to

$$I_{2s}' = \left|\frac{V_1'}{Z_{s2} + Z_{2\text{in}}}\right|\sqrt{\frac{R_{2\text{in}}}{2P}}\left|\frac{E''\cos\psi \cdot 2F_{01}(\Theta_1, \beta_0 h_1)}{\beta_0(Z_{01} + Z_{0\text{in}})}\right|. \quad (29)$$

With (10) and (11), the following expression is obtained after rearrangement:

$$I_{2s}'$$
$$= \left|\frac{V_1'\zeta_0}{2\pi(Z_{01} + Z_{0\text{in}})}\,\frac{F_{01}(\Theta_1, \beta_0 h_1)\cos\psi}{R_0}\right|$$
$$\times \left|\frac{2k_2 I_{2\text{in}}'' F_{02}(\Theta_2, \beta_0 h_2)A(\Theta_2, \Phi_2)}{\beta_0(Z_{s2} + Z_{2\text{in}})}\right|\sqrt{\frac{R_{2\text{in}}}{2P}}. \quad (30)$$

---
[*] In this section resistances are in Roman type (R), distances in italic type (*R*).

The first magnitude factor in (30) is readily identified as $|E'\cos\psi|$. With (25), $I_{2\text{in}}'' = \sqrt{2P/R_{2\text{in}}}$, so that (30) reduces to

$$I_{2s}' = |E'\cos\psi|\left|\frac{2k_2}{\beta_0}\,\frac{F_{02}(\Theta_2, \beta_0 h_2)A(\Theta_2, \Phi_2)}{Z_{s2} + Z_{2\text{in}}}\right|. \quad (31)$$

This is the magnitude of the current in the load of array 2 when used for reception.

The magnitude of the far-zone field of the same array when driven at *constant power* is obtained from (10) with (25). Thus

$$E_\Theta^r = \sqrt{\frac{2P}{R_{2\text{in}}}}\,\frac{\zeta_0}{2\pi R_0}\,|F_{02}(\Theta_2, \beta_0 h_2)A(\Theta_2, \Phi_2)|. \quad (32)$$

If the field properties under study involve only rotation of the array as a whole with $Z_{2\text{in}}$ constant, (31) is proportional to (32). Clearly I_{2s}' is *not* proportional to E_Θ^r when adjustments are made in the array which change the impedance $Z_{2\text{in}}$. However, if the load Z_{s2} is kept *conjugate matched* as adjustments in the array are made, so that

$$Z_{s2} + Z_{2\text{in}} = Z_{2\text{in}}^* + Z_{2\text{in}} = 2R_{2\text{in}}, \quad (33)$$

the *power to the load* is given by

$$P_s = \tfrac{1}{2}(I_{2s}')^2 R_{2\text{in}} = \frac{k_2^2|E\cos\psi|^2}{2\beta_0^2}$$
$$\times \frac{|F_{02}(\Theta_2, \beta_0 h_2)A(\Theta_2, \Phi_2)|^2}{R_{2\text{in}}}. \quad (34)$$

The square of the electric field in (32) is

$$(E_\Theta^r)^2 = \frac{2P\zeta_0^2}{4\pi^2 R_0^2}\,\frac{|F_{02}(\Theta_2, \beta_0 h_2)A(\Theta_2, \Phi_2)|^2}{R_{2\text{in}}}. \quad (35)$$

Comparison of (34) and (35) shows that if P and R_0 are kept constant in transmission and $|E\cos\psi|$ in reception at a fixed frequency, the power P_s to the conjugate matched load is always proportional to the square of the field maintained by the same array used for transmission.

It is seen that the characteristics of the *square of the electric field* of a transmitting array operated at constant power may be determined in general from the power to the load of the same array used for reception only if the load is kept conjugate matched or if the impedance of the array is kept constant. This is important, for example, in determining the field properties of an antenna with a *tuned* parasite at constant power (Sec. 4).

CHAPTER VII

ANTENNAS OVER A CONDUCTING REGION

In practice many antennas are sufficiently far from the surface of the earth to make them approximately equivalent to isolated antennas in so far as impedances and distributions of current and charge are concerned. Other antennas, particularly broadcast antennas, are erected on the earth, or so near to its surface that the interaction between currents in the antenna and in the earth is great enough to affect significantly the distribution of the current in the antenna and with it the driving-point impedance. Whether the distribution of current in a transmitting antenna is affected by the forces due to currents in the earth or not, the far-zone electromagnetic field of practically all antennas is due not only to the currents in the antenna but also to currents (in the sense of actual motions or oscillations of electric charge) in the earth and in the tropospheric and ionospheric media surrounding it. Thus, in order to determine the electromagnetic field that governs the response of a receiving antenna, it is necessary to solve the problem of an antenna above a conducting earth—actually a spherical earth which itself maintains a strong magnetic field. This boundary-value problem is so complex and involves so many variables and parameters that a complete and rigorous solution of the entire problem is extremely difficult. In order to simplify the analysis it is necessary to subdivide the problem into components that are approximately independent and individually solvable. This is accomplished by investigating separately the contributions to the far-zone electric field by the currents in the earth on the one hand, and the oscillating concentrations of charge in the ionosphere on the other hand. These contributions are largely independent in the sense that the currents in the earth are not significantly altered by the relatively removed oscillations in the ionosphere. Therefore, the primary electromagnetic field due to the currents in the antenna proper and due to the currents excited by those in the earth may be determined as if there were no ionosphere. This primary field then may be regarded as the exciting field for the ionosphere or troposphere. From the motions of charge so excited in the ionosphere or troposphere, the secondary or "reflected" field may be determined and then combined with the primary field to give the complete field.

The primary electromagnetic field due to currents in the antenna and in the earth is called the *ground* wave. The secondary field due to oscillating concentrations of charge in the ionosphere is called the *sky* wave. This chapter is devoted to a study of the ground wave. The effect of the ionosphere strictly is not a part of the antenna problem and is, therefore, beyond the scope of this book.

The analysis of the electromagnetic field of an antenna over a spherical earth is far from simple even without the complication of the ionosphere. Recently, much progress has been made, notably by Van der Pol and Bremmer,[64,62,63,I.12] Eckersley and Millington,[14-16] Gray,[18] Norton,[44,45] and Baños and Wesley[3] in calculating the ground-wave field. Since the analysis of an antenna over a spherical earth is intricate and not actually a part of the problem of the linear radiator, it is not carried out here. However, the simpler problem of an antenna over a plane earth is analyzed. This provides an introduction to the more important phases of ground-wave propagation and serves as a foundation for more advanced work in the field. In addition, it permits a discussion of plane-wave theory not as an abstract and isolated topic, but as an essential part of an important problem.

The boundary-value problem of a Hertzian dipole—electric or magnetic—over an infinite conducting half-space is an old classic. The original solution is due to Sommerfeld[54] in 1909. Since then the problem has been studied by numerous investigators, among them Weyl,[72] Van der Pol and Niessen,[65] Strutt,[57] Wise,[73] Noether (in reference I.44), and others, including Sommerfeld.[I.18,I.51] In the following analysis the straightforward method of Weyl[72] of proceeding from a bundle of plane waves is followed as expanded by Noether.[I.44] The general integral so obtained is then evaluated asymptotically by the "saddle-point" method in order to derive the approximate expressions used, for example, by Strutt[57] in calculating tables and curves for the far-zone electromagnetic field in space over a conducting earth. Finally, the field along the surface of the earth is derived from the general integrals in the form used by Sommerfeld, following the simpler method of Van der Pol.[60,59] The general solution is evaluated using approximations introduced by Van der Pol and by Norton.[46,44,45] The discussion is limited to short antennas or infinitesimal electric dipoles, except that the asymptotic solution for points in space is generalized to include linear radiators with a sinusoidal distribution of current.

Although this volume is concerned primarily with linear radiators, the small loop antenna or infinitesimal magnetic dipole is discussed along with the short antenna or electric dipole. Since the impedance properties of a small loop are analyzed in reference I.31, Chap. VI and summarized in Sec. I.13, and since its field properties parallel and supplement those of the electric dipole, it is appropriate that the electric and magnetic dipoles are analyzed together.

1. *Hertzian Potentials*

The electromagnetic field in simple media and in empty space is derived conveniently from the electric Hertzian potential or polarization potential $\mathbf{\Pi}_e$ and the magnetic Hertzian potential or magnetization potential $\mathbf{\Pi}_m$. These potentials satisfy the general vector wave equation as follows:

$$\nabla^2 \mathbf{\Pi}_e + \beta^2 \mathbf{\Pi}_e = -\mathbf{P}^e/\xi, \quad (1a)$$

$$\nabla^2 \mathbf{\Pi}_m + \beta^2 \mathbf{\Pi}_m = -\mathbf{M}^e/\nu, \quad (1b)$$

where

$$\beta^2 \equiv \omega^2 \xi/\nu = \omega^2 \xi \mu \quad (2a)$$

and

$$\xi = \epsilon - j\sigma/\omega = \epsilon_e - j\sigma_e/\omega. \quad (2b)$$

Alternatively,

$$\xi = \epsilon_e(1 - jh_e), \quad h_e \equiv \sigma_e/\omega\epsilon_e. \quad (2c)$$

Note that

$$\nabla^2 \mathbf{\Pi} \equiv \operatorname{grad} \operatorname{div} \mathbf{\Pi} - \operatorname{curl} \operatorname{curl} \mathbf{\Pi}. \quad (2d)$$

In (1), $\mathbf{P}^e$ and $\mathbf{M}^e$ are, respectively, the complex amplitudes of impressed, that is, externally maintained, periodically varying, volume densities of polarization and magnetization or their mathematical equivalents in distributions of current. They are assumed to exist in localized regions such as a dipole or antenna, or the cross-sectional area of a current-carrying loop as explained in reference I.31, chap. I. They are in addition to and independent of the volume densities $\mathbf{P}$ and $\mathbf{M}$ that exist in the simple medium in which the dipoles are immersed and that satisfy the relations

$$\mathbf{P} = \epsilon_0(\epsilon_r - 1)\mathbf{E} = (\epsilon - \epsilon_0)\mathbf{E}, \quad (2e)$$

$$-\mathbf{M} = \nu_0(\nu_r - 1)\mathbf{B} = (\nu - \nu_0)\mathbf{B}. \quad (2f)$$

The electromagnetic field is derived from the Hertzian potentials according to the following formulas:

$$\mathbf{E} = \operatorname{grad} \operatorname{div} \mathbf{\Pi}_e + \beta^2 \mathbf{\Pi}_e - j\omega \operatorname{curl} \mathbf{\Pi}_m, \quad (3a)$$

$$\mathbf{B} = j\frac{\beta^2}{\omega} \operatorname{curl} \mathbf{\Pi}_e + \operatorname{grad} \operatorname{div} \mathbf{\Pi}_m + \beta^2 \mathbf{\Pi}_m. \quad (3b)$$

It is to be noted that the exponential dependence upon the time, $e^{j\omega t}$, is used throughout. Sommerfeld and many others, including Norton, use the time dependence $e^{-i\omega t}$ in their work. For convenient reference, the convention

$$j = -i \quad (4)$$

is used, so that no confusion can arise and the advantages of both notations may be utilized as required. Note that the convention $e^{j\omega t}$ leads to forms like $e^{j(\omega t - \beta R)}$ in which the phase increases with positive values of the time, negative values of the distance, whereas the convention $e^{-i\omega t}$ leads to $e^{i(\beta R - \omega t)}$ in which the phase increases with positive values of the distance R, negative values of the time. When primary interest is in time-phase relations, as in electric-network analysis, the convention $e^{j\omega t}$ is often preferred; when primary interest is in the space-phase relations of wave propagation, the convention $e^{-i\omega t}$ has advantages. Nevertheless, for consistency and greater general simplicity, the time convention $e^{j\omega t}$ is retained in this analysis.

The boundary conditions for $\mathbf{\Pi}_e$ and $\mathbf{\Pi}_m$ at a plane boundary between two semi-infinite media are derived from those for $\mathbf{E}$ and $\mathbf{B}$. For simple media [Secs. I.3, 4, 11 or ref. I.31, Eq.(III.14.20)],

$$\epsilon_1 \hat{\mathbf{n}}_1 \cdot \mathbf{E}_1 + \epsilon_2 \hat{\mathbf{n}}_2 \cdot \mathbf{E}_2 = -\eta_{1f} - \eta_{2f}, \quad (5a)$$

$$\hat{\mathbf{n}}_1 \times \mathbf{E}_1 + \hat{\mathbf{n}}_2 \times \mathbf{E}_2 = 0, \quad (5b)$$

$$\nu_1 \hat{\mathbf{n}}_1 \times \mathbf{B}_1 + \nu_2 \hat{\mathbf{n}}_2 \times \mathbf{B}_2 = -\mathbf{l}_{1f} - \mathbf{l}_{2f}, \quad (5c)$$

$$\hat{\mathbf{n}}_1 \cdot \mathbf{B}_1 + \hat{\mathbf{n}}_2 \cdot \mathbf{B}_2 = 0. \quad (5d)$$

Since the media are both imperfectly conducting, a surface density of moving free charge is not required. This is equivalent to omitting l in (5c) and writing zero.

If the Hertzian potential is expressed in Cartesian coördinates, so that

$$\mathbf{\Pi} = \hat{\mathbf{x}}\Pi_x + \hat{\mathbf{y}}\Pi_y + \hat{\mathbf{z}}\Pi_z, \qquad (6a)$$

each of the complicated vector wave equations (1) reduces to three scalar wave equations. Thus it is shown in Appendix I of ref. I.31 that, for example, $\hat{\mathbf{x}} \cdot \nabla^2 \mathbf{A} = \nabla^2 A_x$, where $\nabla^2 = \text{div grad}$ is the Laplacian operator. It follows that with (6) equations (1a) and (1b) reduce to:

$$\nabla^2 \Pi_{eq} + \beta^2 \Pi_{eq} = -P_q^e/\xi, \quad q = x, y, z, \quad (6b)$$

$$\nabla^2 \Pi_{mq} + \beta^2 \Pi_{mq} = -M_q^e/\nu, \quad q = x, y, z. \quad (6c)$$

At all points where the impressed densities $\mathbf{P}^e$ and $\mathbf{M}^e$ are zero, that is, outside the small volumes occupied by the source dipoles or antennas that are assumed to excite the field, a typical equation is

$$\nabla^2 \Pi_z + \beta^2 \Pi_z = 0. \qquad (6d)$$

In cylindrical coördinates (r, θ, z), this is

$$\frac{1}{r}\frac{\partial}{\partial r}\left(r\frac{\partial \Pi_z}{\partial r}\right) + \frac{1}{r^2}\frac{\partial^2 \Pi_z}{\partial \theta^2} + \frac{\partial^2 \Pi_z}{\partial z^2} + \beta^2 \Pi_z = 0. \qquad (6e)$$

Particular solutions of this equation may be obtained by the usual method of separation of variables. This is carried out, for example, in reference I.31, chap. V. The particular solutions of (6e) that are of interest in this chapter are constructed of combinations of the eigenfunctions*

$$e^{\pm jlz} \cos n\theta J_n(\lambda r) \quad \text{or} \quad e^{\pm jlz} \cos n\theta H_n^{(1)}(\lambda r), \qquad (7a)$$

where n and l are separation constants. As indicated, l is in general complex and n is an integer if there are no discontinuities or boundaries that depend on θ. The eigenvalue λ is defined by

$$\lambda^2 = \beta^2 + l^2. \qquad (7b)$$

As explained in Sec. 2, the problem of a vertical dipole over a horizontal earth may be solved more simply by requiring the Hertzian potentials to be in a specified direction. For a plane earth defined by $z = -d$, it is convenient to use only z-components of both potentials, since this reduces the vector wave equations to scalar wave equations.

$$\mathbf{\Pi}_e = \hat{\mathbf{z}}\Pi_{ez}, \quad \mathbf{\Pi}_m = \hat{\mathbf{z}}\Pi_{mz}. \qquad (8)$$

With this restriction (3a) and (3b) reduce to

$$\mathbf{E} = \text{grad}\left(\frac{\partial \Pi_{ez}}{\partial z}\right) + \hat{\mathbf{z}}\beta^2 \Pi_{ez} - j\omega \, \text{curl}\,(\hat{\mathbf{z}}\Pi_{mz}), \qquad (9a)$$

$$\mathbf{B} = j\frac{\beta^2}{\omega}\text{curl}\,(\hat{\mathbf{z}}\Pi_{ez}) + \text{grad}\left(\frac{\partial \Pi_{mz}}{\partial z}\right) + \hat{\mathbf{z}}\beta^2 \Pi_{mz}. \qquad (9b)$$

The components of $\mathbf{E}$ and $\mathbf{B}$ tangent to the boundary and hence normal to the coördinate z are obtained by setting $\hat{\mathbf{n}}_1 = -\hat{\mathbf{n}}_2 = -\hat{\mathbf{z}}$. They are

$$-\hat{\mathbf{n}}_1 \times \mathbf{E}_1 = \hat{\mathbf{n}}_2 \times \mathbf{E}_2 \quad \text{or} \quad \hat{\mathbf{z}} \times \mathbf{E}_1 = \hat{\mathbf{z}} \times \mathbf{E}_2, \qquad (10a)$$

$$-\nu_1 \hat{\mathbf{n}}_1 \times \mathbf{B}_1 = \nu_2 \hat{\mathbf{n}}_2 \times \mathbf{B}_2 \quad \text{or}$$

$$\nu_1 \hat{\mathbf{z}} \times \mathbf{B}_1 = \nu_2 \hat{\mathbf{z}} \times \mathbf{B}_2. \qquad (10b)$$

Since $\mathbf{\Pi}_e$ and $\mathbf{\Pi}_m$ are independent, they may satisfy independently the boundary conditions obtained by substituting (9a, b) in (10a, b). Thus, since $\hat{\mathbf{z}} \times \hat{\mathbf{z}} = 0$,

$$\left[\hat{\mathbf{z}} \times \text{grad}\left(\frac{\partial \Pi_{ez}}{\partial z}\right)\right]_1 = \left[\hat{\mathbf{z}} \times \text{grad}\left(\frac{\partial_e \Pi_z}{\partial z}\right)\right]_2, \qquad (11a)$$

$$\nu_1 \beta_1^2 [\hat{\mathbf{z}} \times \text{curl}\,(\hat{\mathbf{z}}\Pi_{ez})]_1 = \nu_2 \beta_2^2 [\hat{\mathbf{z}} \times \text{curl}\,(\hat{\mathbf{z}}\Pi_{ez})]_2, \qquad (11b)$$

$$[\hat{\mathbf{z}} \times \text{curl}\,(\hat{\mathbf{z}}\Pi_{mz})]_1 = [\hat{\mathbf{z}} \times \text{curl}\,(\hat{\mathbf{z}}\Pi_{mz})]_2, \qquad (11c)$$

$$\nu_1 \left[\hat{\mathbf{z}} \times \text{grad}\left(\frac{\partial \Pi_{mz}}{\partial z}\right)\right]_1$$

$$= \nu_2 \left[\hat{\mathbf{z}} \times \text{grad}\left(\frac{\partial \Pi_{mz}}{\partial z}\right)\right]_2. \qquad (11d)$$

These equations are satisfied if the following conditions are imposed; note that $\nu\beta^2 = \omega^2 \xi$:

$$\left(\frac{\partial \Pi_{ez}}{\partial z}\right)_1 = \left(\frac{\partial \Pi_{ez}}{\partial z}\right)_2, \qquad (12a)$$

$$\xi_1 \Pi_{ez1} = \xi_2 \Pi_{ez2}, \qquad (12b)$$

$$\Pi_{mz1} = \Pi_{mz2}, \qquad (13a)$$

$$\nu_1 \left(\frac{\partial \Pi_{mz}}{\partial z}\right)_1 = \nu_2 \left(\frac{\partial \Pi_{mz}}{\partial z}\right)_2. \qquad (13b)$$

---

* See, for example, Stratton, ref. I.52, p. 356.

These are the boundary conditions that are to be imposed on the Hertzian potentials at a plane boundary between two simple media in the case of *vertical* dipoles.

The problem of horizontal dipoles over a horizontal conducting plane requires two components of the appropriate Hertzian potential. Thus, for the boundary defined in (6),

$$\mathbf{\Pi}_e = \hat{x}\Pi_{ex} + \hat{z}\Pi_{ez}, \tag{14a}$$

$$\mathbf{\Pi}_m = \hat{x}\Pi_{mx} + \hat{z}\Pi_{mz}. \tag{14b}$$

In this case the components of the field are

$$E_x = \frac{\partial}{\partial x} \text{div } \mathbf{\Pi}_e + \beta^2 \Pi_{ex} - j\omega \frac{\partial \Pi_{mz}}{\partial y}, \tag{15a}$$

$$E_y = \frac{\partial}{\partial y} \text{div } \mathbf{\Pi}_e - j\omega \left(\frac{\partial \Pi_{mx}}{\partial z} - \frac{\partial \Pi_{mz}}{\partial x} \right), \tag{15b}$$

$$E_z = \frac{\partial}{\partial z} \text{div } \mathbf{\Pi}_e + \beta^2 \Pi_{ez} + j\omega \frac{\partial \Pi_{mx}}{\partial y}, \tag{15c}$$

$$B_x = j\frac{\beta^2}{\omega} \frac{\partial \Pi_{ez}}{\partial y} + \frac{\partial}{\partial x} \text{div } \mathbf{\Pi}_m + \beta^2 \Pi_{mx}, \tag{15d}$$

$$B_y = j\frac{\beta^2}{\omega} \left(\frac{\partial \Pi_{ex}}{\partial z} - \frac{\partial \Pi_{ez}}{\partial x} \right) + \frac{\partial}{\partial y} \text{div } \mathbf{\Pi}_m, \tag{15e}$$

$$B_z = -j\frac{\beta^2}{\omega} \frac{\partial \Pi_{ex}}{\partial y} + \frac{\partial}{\partial z} \text{div } \mathbf{\Pi}_m + \beta^2 \Pi_{mz}. \tag{15f}$$

With $\mathbf{\Pi}_e$ and $\mathbf{\Pi}_m$ independent, the boundary conditions (10a, b) for (15a–f) are satisfied if the following conditions are imposed; these are more restrictive than required to satisfy (5a, b, c, d), but are sufficiently general for the problems to be investigated:

$$(\text{div } \mathbf{\Pi}_e)_1 = (\text{div } \mathbf{\Pi}_e)_2, \tag{16a}$$

$$\Pi_{mz1} = \Pi_{mz2}, \tag{16b}$$

$$\beta_1^2 \Pi_{ex1} = \beta_2^2 \Pi_{ex2}, \tag{17a}$$

$$\left(\frac{\partial \Pi_{mx}}{\partial z} \right)_1 = \left(\frac{\partial \Pi_{mx}}{\partial z} \right)_2, \tag{17b}$$

$$\nu_1 \beta_1^2 \Pi_{ez1} = \nu_2 \beta_2^2 \Pi_{ez2}, \tag{18a}$$

$$\nu_1 (\text{div } \mathbf{\Pi}_m)_1 = \nu_2 (\text{div } \mathbf{\Pi}_m)_2, \tag{18b}$$

$$\nu_1 \beta_1^2 \left(\frac{\partial \Pi_{ex}}{\partial z} \right)_1 = \nu_2 \beta_2^2 \left(\frac{\partial \Pi_{ex}}{\partial z} \right)_2, \tag{19a}$$

$$\nu_1 \beta_1^2 \Pi_{mx1} = \nu_2 \beta_2^2 \Pi_{mx2}. \tag{19b}$$

Thus, there are two conditions each on Π_{ex} and Π_{mx}. With $\nu\beta^2 = \omega^2 \xi$ these are:

$$\beta_1^2 \Pi_{ex1} = \beta_2^2 \Pi_{ex2}, \tag{20a}$$

$$\xi_1 \left(\frac{\partial \Pi_{ex}}{\partial z} \right)_1 = \xi_2 \left(\frac{\partial \Pi_{ex}}{\partial z} \right)_2, \tag{20b}$$

$$\left(\frac{\partial \Pi_{mx}}{\partial z} \right)_1 = \left(\frac{\partial \Pi_{mx}}{\partial z} \right)_2, \tag{20c}$$

$$\xi_1 \Pi_{mx1} = \xi_2 \Pi_{mx2}. \tag{20d}$$

Once Π_{ex} and Π_{mx} are known, Π_{ez} and Π_{mz} can be determined from the following conditions:

$$\left(\frac{\partial \Pi_{ex}}{\partial x} + \frac{\partial \Pi_{ez}}{\partial z} \right)_1 = \left(\frac{\partial \Pi_{ex}}{\partial x} + \frac{\partial \Pi_{ez}}{\partial z} \right)_2, \tag{21a}$$

$$\xi_1 \Pi_{ez1} = \xi_2 \Pi_{ez2}, \tag{21b}$$

$$\Pi_{mz1} = \Pi_{mz2}, \tag{21c}$$

$$\nu_1 \left(\frac{\partial \Pi_{mx}}{\partial x} + \frac{\partial \Pi_{mz}}{\partial z} \right)_1 = \nu_2 \left(\frac{\partial \Pi_{mx}}{\partial x} + \frac{\partial \Pi_{mz}}{\partial z} \right)_2. \tag{21d}$$

VERTICAL DIPOLES OVER CONDUCTING EARTH; GENERAL FORMULATION

2. *The Polarization Potential of a Vertical Electric Dipole or of an Element of a Vertical Antenna*

The complex amplitude of the total current traversing a cross section of a dipole or an element of an antenna of length dz is I_z [as defined in ref. I.31, Eq.(IV.7.16b)]. If the moving free charges contributing to I_z are replaced by equivalent bound charges with an electric moment oscillating with angular velocity $\omega = 2\pi f$, the correlating equation is [ref. I.31, Eq.(I.26.7)]:

$$2hI_z = \frac{\partial p_z}{\partial t} = j\omega p_z, \tag{1}$$

where $2h$ is the length of the element. If this element is an element of a linear radiator, $2h = dz$. The impressed polarization p_z of a volume τ referred to its center may be expressed in terms of the impressed volume density of polarization P_z^e as follows:

$$p_z = \int_\tau P_z^{e\prime} \, d\tau', \tag{2}$$

where the primed coördinates locate the element of integration $d\tau'$. No superscript is used on p_z, since there is no occasion to introduce an electric moment except the impressed moment.

The Helmholtz integral for the polarization potential that satisfies (1.1) is given by

$$\mathbf{\Pi}_e = \frac{1}{4\pi\xi} \int_\tau \frac{\mathbf{P}^{e'}}{R} \exp(-j\beta R) \, d\tau'. \quad (3)$$

However, since in the case at hand all charges oscillate parallel to the z-axis in the dipole,

$$\mathbf{P}^e = \hat{z} P_z^e, \quad (4)$$

and, therefore,

$$\mathbf{\Pi}_e = \hat{z}\Pi_{ez}. \quad (5)$$

Moreover, no magnetization potential is required and the entire field can be derived from $\mathbf{\Pi}_e$. The polarization potential due to a dipole of very short length $2h$ (compared with the wavelength and all distances involved), or a section of antenna of length dz and very small cross-sectional area, is given by (5) with

$$\Pi_{ez} = \frac{p_z}{4\pi\xi} \frac{\exp(-j\beta R)}{R}. \quad (6)$$

where p_z as given in (2) is the polarization of the dipole or of the length dz of the antenna. Note that the exponent in (6) has a nonpositive real part, so that Π_{ez} vanishes at infinity. This follows since from (I.11.9)

$$\beta = \beta_0 N = \beta_0 (N_s - jX_s) \equiv \beta_s - j\alpha_s, \quad (7)$$

where N is the generalized index of refraction of the simple medium referred to air, N_s is the real index of refraction referred to air, X_s is the extinction coefficient, β_s is the real propagation constant, α_s is the real attenuation constant.

It is readily verified that (6) is a particular solution of (1.1) in the special case of complete spherical symmetry so that Π_{ez} is a function of the polar variable R alone. In this case,

$$\frac{1}{R^2} \frac{d}{dR}\left(R^2 \frac{d\Pi_{ez}}{dR}\right) + \beta^2 \Pi_{ez} = 0, \quad (8a)$$

which is equivalent to

$$\frac{d^2}{dR^2}(R\Pi_{ez}) + \beta^2 R\Pi_{ez} = 0. \quad (8b)$$

This equation has the general solution

$$R\Pi_{ez} = C_1 \exp(-j\beta R) + C_2 \exp(+j\beta R). \quad (8c)$$

The part of the solution with a negative real part is the first term on the right. Hence, the solution that does *not* become infinite at infinity is

$$\Pi_{ez} = C_1 \frac{\exp(-j\beta R)}{R}. \quad (8d)$$

If C_1 is identified with $p_z/4\pi\xi$ and C_2 is required to vanish, (8c) coincides with (6). So-called empty space is treated as a dissipative medium in the limit as the dissipation vanishes. This is equivalent to imposing a radiation condition.

If it is recalled that in simple media the vector potential $\mathbf{A}$ is related to the polarization potential $\mathbf{\Pi}_e$ by a relation like (I.8.20) with β substituted for β_0 and $\mathbf{\Pi}_m = 0$, that is

$$\mathbf{A} = \frac{j\omega\xi}{\nu}\mathbf{\Pi}_e = j\frac{\beta^2}{\omega}\mathbf{\Pi}_e \quad (9)$$

the solution (6) is seen to agree with Eq. (IV.12.13), reference I.31 when applied to empty space with ϵ_0 substituted for ξ.

The electromagnetic field of the electric dipole is obtained by direct differentiation from (6), using

$$\mathbf{E} = \mathrm{grad}\left(\frac{\partial \Pi_{ez}}{\partial z}\right) + \hat{z}\beta^2 \Pi_{ez}, \quad (10a)$$

$$\mathbf{B} = \frac{j\beta^2}{\omega} \mathrm{curl}(\hat{z}\Pi_{ez}). \quad (10b)$$

The result is the same as in Eqs. (IV.12.17, 18), reference I.31 when expressed in vector form, except that the dipole now under consideration is in a simple medium with complex parameters ξ, ν, and β instead of in empty space with real parameters ϵ_0, ν_0, and β_0. The electric and magnetic fields are

$$\mathbf{E} = \frac{\exp(-j\beta R)}{4\pi\xi}\left\{-\frac{\beta^2}{R}\hat{\mathbf{R}}\times(\hat{\mathbf{R}}\times\mathbf{p})\right.$$

$$+ \frac{j\beta}{R^2}[3\hat{\mathbf{R}}(\hat{\mathbf{R}}\cdot\mathbf{p}) - \mathbf{p}]$$

$$\left.+ \frac{1}{R^3}[3\hat{\mathbf{R}}(\hat{\mathbf{R}}\cdot\mathbf{p}) - \mathbf{p}]\right\}, \quad (11)$$

$$\mathbf{B} = \frac{-j\omega\exp(-j\beta R)}{4\pi\nu}\hat{\mathbf{R}}\times\mathbf{p}\left(\frac{j\beta}{R} + \frac{1}{R^2}\right), \quad (12)$$

where $\mathbf{p} = \hat{\mathbf{z}} p_z$. In polar coördinates the components of the electromagnetic field are

$$E_\Theta = \frac{p_z \sin\Theta}{4\pi\xi} \exp(-j\beta R)\left(-\frac{\beta^2}{R} + \frac{j\beta}{R^2} + \frac{1}{R^3}\right), \quad (13a)$$

$$E_\Phi = 0, \quad (13b)$$

$$E_R = \frac{p_z \cos\Theta}{4\pi\xi} \exp(-j\beta R)\left(\frac{2j\beta}{R^2} + \frac{2}{R^3}\right), \quad (13c)$$

$$B_\Theta = 0, \quad (14a)$$

$$B_\Phi = \frac{j\omega p_z \sin\Theta}{4\pi\nu} \exp(-j\beta R)\left(\frac{j\beta}{R} + \frac{1}{R^2}\right), \quad (14b)$$

$$B_R = 0. \quad (14c)$$

In Cartesian coördinates the components of E and B are

$$E_x = \frac{p_z}{4\pi\xi} \exp(-j\beta R)\left(\frac{-\beta^2}{R^3} + \frac{j3\beta}{R^4} + \frac{3}{R^5}\right) xz, \quad (15a)$$

$$E_y = \frac{p_z}{4\pi\xi} \exp(-j\beta R)\left(\frac{\beta^2}{R^3} + \frac{j3\beta}{R^4} + \frac{3}{R^5}\right) yz, \quad (15b)$$

$$E_z = \frac{p_z}{4\pi\xi} \exp(-j\beta R)\left[\frac{\beta^2(R^2 - z^2)}{R^3} + \frac{j\beta(3z^2 - R^2)}{R^4} + \frac{(3z^2 - R^2)}{R^5}\right], \quad (15c)$$

$$B_x = \frac{-j\omega p_z}{4\pi\nu} \exp(-j\beta R)\left(\frac{j\beta}{R^2} + \frac{1}{R^3}\right) y, \quad (16a)$$

$$B_y = \frac{-j\omega p_z}{4\pi\nu} \exp(-j\beta R)\left(\frac{j\beta}{R^2} + \frac{1}{R^3}\right) x, \quad (16b)$$

$$B_z = 0. \quad (16c)$$

The radiation-zone or far-zone field consists of the $1/R$ terms only. It is

$$\mathbf{E}^r = \frac{-\beta^2 \exp(-j\beta R)}{4\pi\xi R} \hat{\mathbf{R}} \times (\hat{\mathbf{R}} \times \mathbf{p}), \quad (17a)$$

$$\mathbf{B}^r = \frac{\omega\beta \exp(-j\beta R)}{4\pi\nu R} \hat{\mathbf{R}} \times \mathbf{p}. \quad (17b)$$

In spherical coördinates the radiation field is

$$E_\Theta^r = \frac{-\beta^2 p_z \sin\Theta \exp(-j\beta R)}{4\pi\xi R},$$

$$E_\Phi^r = 0, \quad E_R^r = 0, \quad (18a)$$

$$B_\Phi^r = \frac{-\omega\beta p_z \sin\Theta \exp(-j\beta R)}{4\pi\nu R},$$

$$B_R^r = 0, \quad B_\Theta^r = 0. \quad (18b)$$

Evidently,

$$E_\Theta^r = v B_\Phi^r, \quad (19a)$$

where

$$v = \sqrt{\nu/\xi}. \quad (19b)$$

Comparison of (18a, b) with (6) shows that the radiation field is given by

$$\mathbf{E}^r = \hat{\mathbf{\Theta}} E_\Theta^r, \quad E_\Theta^e = \beta^2 \Pi_{e\Theta} = -\beta^2 \Pi_{ez} \sin\Theta, \quad (20a)$$

$$\mathbf{B}^r = \hat{\mathbf{\Phi}} B_\Phi^r, \quad B_\Phi^r = \frac{\beta^3}{\omega} \Pi_{e\Theta} = \frac{-\beta^3}{\omega} \Pi_{ez} \sin\Theta. \quad (20b)$$

In Cartesian coördinates the far-zone field is

$$E_x = \frac{-\beta^2 p_z}{4\pi\xi} \frac{xz}{R^2} \frac{\exp(-j\beta R)}{R}, \quad (21a)$$

$$E_y = \frac{-\beta^2 p_z}{4\pi\xi} \frac{yz}{R^2} \frac{\exp(-j\beta R)}{R}, \quad (21b)$$

$$E_z = \frac{\beta^2 p_z}{4\pi\xi} \frac{(R^2 - z^2)}{R^2} \frac{\exp(-j\beta R)}{R}$$

$$= \frac{\beta^2 p_z}{4\pi\xi} \frac{(x^2 + y^2)}{R^2} \frac{\exp(-j\beta R)}{R}, \quad (21c)$$

$$B_x = \frac{\omega\beta p_z}{4\pi\nu} \frac{y}{R} \frac{\exp(-j\beta R)}{R}, \quad (22a)$$

$$B_y = \frac{-\omega\beta p_z}{4\pi\nu} \frac{x}{R} \frac{\exp(-j\beta R)}{R}, \quad (22b)$$

$$B_z = 0. \quad (22c)$$

For a dipole in empty space the substitutions ϵ_0 for ξ, ν_0 for ν, and β_0 for β must be used.

If desired, the substitution

$$2I_z h = j\omega p_z \quad (23)$$

may be made, where h is the half-length of the infinitesimal dipole.

If an oscillating dipole is at a distance d above the plane boundary between two simple media 1 and 2, the fields above and below the boundary are not the same. Moreover, the field above the boundary differs from what it would be if there were no boundary and region 2 were like region 1. The field above the boundary in region 1 is conveniently considered in two parts: the so-called *direct field*, which is the field that would obtain if there were no boundary and region 2 were the same as region 1, and the rest of the field, called the *reflected field*, which takes account of the effect of the boundary and of the electric properties of region 2 in so far as they differ from those of region 1. If region 1 is empty space, the direct field is that due only to the oscillating charges of the dipole, and the reflected field is that due to all of the moving charges in region 2. If region 1 is a simple medium, the contribution of the moving charges in the medium is also involved, but account may be taken of this by using the appropriate dielectric factor ξ and reluctivity ν. The field below the boundary in region 2 is called the *refracted* or *transmitted field*. It is due to the moving charges in the dipole itself and in the two simple media.

Since the charges in the dielectric media move so that they have components of velocity in both the z- and the r-directions, it would appear that both z- and r-components or the polarization potential must be used. This is true necessarily if the simple physical model is preserved that oscillating dipoles or their equivalents maintain components of polarization potential parallel to the axes of the dipoles. This is a generally useful picture that leads to correct results. In some instances, however, a mathematically simpler representation is possible *if the direct correspondence between oscillating dipoles and polarization potential is abandoned*. Thus, in the case at hand, the magnetic field of *all* the moving charges is in the Θ-direction, so that $\mathbf{B} = \hat{\mathbf{\Theta}} B_\Theta$. This magnetic field may be derived from a polarization potential that has both axial and radial components so defined that its r-component is determined by radial components, its z-component by axial components of oscillating dipoles. That is, Π_{er} is defined in terms of P_r, Π_{ez} in terms of $P_z + P_z^e$, where P_z^e is defined only in the driven dipole. However, exactly the same magnetic field may be derived from a different polarization potential that has only an axial component,

or from yet another polarization potential that has only a radial component. Since it is only necessary to satisfy the boundary conditions for the electric and the magnetic fields, and since the fields are obtained from the potential by differentiation, the particular choice of polarization potential is arbitrary from the mathematical point of view. When the axis of the driven source dipole is parallel to the z-axis, it is analytically simpler to derive the field due to all moving charges from Π_{ez} alone. This choice is suggested also by application of the theorem for simple media (ref. I.31, p. 206) which states that the potentials may be evaluated as in empty space if appropriate changes are made from the parameters ϵ_0, ν_0, β_0 to ξ, ν, and β. Since the potential due to an isolated dipole of moment p_{1z} in empty space involves only the component Π_{ez}, it follows that Π_{ez} must be adequate in the more complicated problem involving imperfect dielectric media provided the boundary conditions can be satisfied. This is true even though the directions of motion of charges in the simple media are not restricted to be parallel to the z-axis. Thus, in region 1,

$$(\Pi_{ez})_1 = (\Pi_{ez})_{1d} + (\Pi_{ez})_{1r}. \qquad (24)$$

In region 2 the potential is $(\Pi_{ez})_2$. The boundary conditions are

$$\left(\frac{\partial \Pi_{ez}}{\partial z}\right)_1 = \left(\frac{\partial \Pi_{ez}}{\partial z}\right)_2, \qquad (25a)$$

$$(\xi \Pi_{ez})_1 = (\xi \Pi_{ez})_2. \qquad (25b)$$

3. The Magnetization Potential of a Vertical Magnetic Dipole or of an Element of a Horizontal Loop Antenna

Let I_0 be the complex amplitude of the current traversing a cross-section of a circular loop of wire enclosing a mean area S. If the moving free charges contributing to I_0 are replaced by charges that are parts of microscopic circulations or current whirls with a magnetic moment m_z that changes its polarity with angular velocity $\omega = 2\pi f$, the correlating equation is [Eq. (1.26.6), ref. I.31]

$$I_0 = m_z/S. \qquad (1)$$

The magnetization potential Π_m due to the elementary magnet (magnetic dipole) of moment m_z is obtained as was the polarization potential of an electric dipole in the preceding section. By analogy with (2.6),

$$\Pi_{mz} = \frac{m_z}{4\pi\nu} \frac{\exp(-j\beta R)}{R}. \qquad (2)$$

Note that m_z is the moment of a periodically reversing elementary magnet or of a loop of infinitesimal dimension with periodically reversing current around its periphery. It is related to a fictitious impressed volume density of magnetization M_z^e in the volume enclosed by the loop and planes tangent to its top and bottom by the relation

$$m_z = \int_\tau M_z^{e\prime} d\tau'. \qquad (3)$$

Since the polarization potential Π_e is not involved, the electromagnetic field may be derived entirely from Π_m, using

$$\mathbf{E} = -j\omega \, \text{curl} \, \mathbf{\Pi}_m = -j\omega \, \text{curl} \, (\hat{\mathbf{z}} \Pi_{mz}), \qquad (4a)$$

$$\mathbf{B} = \text{grad div} \, \mathbf{\Pi}_m + \beta^2 \mathbf{\Pi}_m$$

$$= \text{grad} \left(\frac{\partial \Pi_{mz}}{\partial z} \right) + \hat{\mathbf{z}} \beta^2 \Pi_{mz}. \qquad (4b)$$

These formulas are obtained from (1.1) and (1:2) with $\mathbf{\Pi}_e = 0$ and with $\mathbf{\Pi}_m = \hat{\mathbf{z}} \Pi_{mz}$.

Except for multiplicative constants, (4a, b) are like (2.10a, b) with electric and magnetic fields interchanged. It follows directly by comparison with (2.11) that the electric and magnetic fields are

$$\mathbf{E} = \frac{j\omega \exp(-j\beta R)}{4\pi\xi} \hat{\mathbf{R}} \times \mathbf{m} \left(\frac{j\beta}{R} + \frac{1}{R^2} \right), \qquad (5a)$$

$$\mathbf{B} = \frac{\exp(-j\beta R)}{4\pi\nu} \left\{ \frac{-\beta^2}{R} \hat{\mathbf{R}} \times (\hat{\mathbf{R}} \times \mathbf{m}) \right.$$

$$\left. + \frac{j\beta}{R^2} [3\hat{\mathbf{R}}(\hat{\mathbf{R}} \cdot \mathbf{m}) - \mathbf{m}] + \frac{1}{R^3} [3\hat{\mathbf{R}}(\hat{\mathbf{R}} \cdot \mathbf{m}) - \mathbf{m}] \right\}, \qquad (5b)$$

where $\mathbf{m} = \hat{\mathbf{z}} m_z$.

In spherical coördinates the components of the electromagnetic field are

$$E_\Theta = 0, \qquad (6a)$$

$$E_\Phi = \frac{-j\omega m_z \sin \Theta}{4\pi\xi} \exp(-j\beta R) \left(\frac{j\beta}{R} + \frac{1}{R^2} \right), \qquad (6b)$$

$$E_R = 0, \qquad (6c)$$

$$B_\Theta = \frac{m_z \sin \Theta}{4\pi\nu} \exp(-j\beta R) \left(\frac{-\beta^2}{R} + \frac{j\beta}{R^2} + \frac{1}{R^3} \right), \qquad (7a)$$

$$B_\Phi = 0, \qquad (7b)$$

$$B_R = \frac{m_z \cos \Theta}{4\pi\nu} \exp(-j\beta R) \left(\frac{2j\beta}{R^2} + \frac{2}{R^3} \right). \qquad (7c)$$

In Cartesian coördinates the components of $\mathbf{E}$ and $\mathbf{B}$ are

$$E_x = \frac{j\omega m_z}{4\pi\xi} \exp(-j\beta R) \left(\frac{j\beta}{R^2} + \frac{1}{R^3} \right) y, \qquad (8a)$$

$$E_y = \frac{-j\omega m_z}{4\pi\xi} \exp(-j\beta R) \left(\frac{j\beta}{R^2} + \frac{1}{R^3} \right) x, \qquad (8b)$$

$$E_z = 0, \qquad (8c)$$

$$B_x = \frac{m_z}{4\pi\nu} \exp(-j\beta R) \left(\frac{-\beta^2}{R^3} + \frac{j3\beta}{R^4} + \frac{3}{R^5} \right) xz, \qquad (9a)$$

$$B_y = \frac{m_z}{4\pi\nu} \exp(-j\beta R) \left(\frac{-\beta^2}{R^3} + \frac{j3\beta}{R^4} + \frac{3}{R^5} \right) yz, \qquad (9b)$$

$$B_z = \frac{m_z}{4\pi\nu} \exp(-j\beta R) \left[\frac{\beta^2(R^2 - z^2)}{R^3} \right.$$

$$\left. + \frac{j\beta(3z^2 - R^2)}{R^4} + \frac{(3z^2 - R^2)}{R^5} \right]. \qquad (9c)$$

The radiation-zone or far-zone field consists of the $1/R$ terms only. It is

$$\mathbf{E}^r = \frac{-\omega\beta}{4\pi\xi} \frac{\exp(-j\beta R)}{R} \hat{\mathbf{R}} \times \mathbf{m}, \qquad (10)$$

$$\mathbf{B}^r = \frac{-\beta^2}{4\pi\nu} \frac{\exp(-j\beta R)}{R} \hat{\mathbf{R}} \times (\hat{\mathbf{R}} \times \mathbf{m}). \qquad (11)$$

In spherical coördinates the far-zone field is

$$E_\Phi^r = \frac{-\omega\beta m_z \sin \Theta \exp(-j\beta R)}{4\pi\nu},$$

$$E_R^r = 0, \; E_\Theta^r = 0, \qquad (12a)$$

$$B_\Theta^r = \frac{-\beta^2 m_z \sin \Theta \exp(-j\beta R)}{4\pi\nu},$$

$$B_\Phi^r = 0, \; B_R^r = 0. \qquad (12b)$$

Alternatively,

$$E_\Phi^r = -\omega\beta \Pi_{mz} \sin \Theta, \qquad (13a)$$

$$B_\Theta^r = -\beta^2 \Pi_{mz} \sin \Theta. \qquad (13b)$$

In Cartesian coördinates the far-zone field is

$$E_x = \frac{-\omega\beta m_z}{4\pi\xi} \frac{y}{R} \frac{\exp(-j\beta R)}{R}, \quad (14a)$$

$$E_y = \frac{\omega\beta m_z}{4\pi\xi} \frac{x}{R} \frac{\exp(-j\beta R)}{R}, \quad (14b)$$

$$E_z = 0, \quad (14c)$$

$$B_x = \frac{-\beta^2 m_z}{4\pi\nu} \frac{xz}{R^2} \frac{\exp(-j\beta R)}{R}, \quad (15a)$$

$$B_y = \frac{-\beta^2 m_z}{4\pi\nu} \frac{yz}{R^2} \frac{\exp(-j\beta R)}{R}, \quad (15b)$$

$$B_z = \frac{\beta^2 m_z}{4\pi\nu} \frac{(R^2-z^2)}{R^2} \frac{\exp(-j\beta R)}{R}$$

$$= \frac{\beta^2 m_{1z}}{4\pi\nu} \frac{(x^2+y^2)}{R^2} \frac{\exp(-j\beta R)}{R}. \quad (15c)$$

For an elementary magnet or loop in empty space the substitutions ϵ_0 for ξ, ν_0 for ν, and β_0 for ν must be made.

4. *Representation of Spherical Waves as a Bundle of Plane Waves*

The spherically symmetric Hertzian potentials characteristic of the direct fields of potential of elementary electric and magnetic dipoles of moments p_z and m_z in a simple medium have the form

$$\Pi_z = K \frac{\exp(-j\beta R)}{-j\beta R}. \quad (1)$$

The electric and magnetic problems are distinguished by identifying subscripts e and m on Π_z and K. Also

$$K_e = \frac{-j\beta p_z}{4\pi\xi}, \quad K_m = \frac{-j\beta m_z}{4\pi\nu}. \quad (2)$$

The Hertzian potential of the direct field of an elementary dipole in the form (1) is readily interpreted in terms of spherical surfaces of constant phase expanding radially outward with the dipole at the center. Multiplication of (1) by the time factor $e^{j\omega t}$ gives the *complex instantaneous value*:

$$(\Pi_z)_{\text{inst}} = \Pi_z e^{j\omega t} = \frac{K}{-j\beta R} \exp[j(\omega t - \beta R)]. \quad (3)$$

By setting $(K/-j\beta) = (K/\beta)e^{j\theta} = (K/\beta)e^{-j\omega t_0}$ and letting $t' = t - t_0$, (3) may be expressed as follows:

$$(\Pi_z)_{\text{inst}} = \frac{K}{\beta R} \exp[j(\omega t' - \beta R)], \quad (4)$$

where the amplitude is now real. The *real* instantaneous value may be chosen to be the real part of (4). Thus, with $\beta = \beta_s - j\alpha_s$,

$$(\Pi_z)_{\text{inst}} = \frac{K}{\beta R} e^{-\alpha_s R} \cos(\omega t' - \beta_s R). \quad (5)$$

Except for the added attenuating factor $e^{-\alpha_s R}$, (5) is in the same form as Eq. (IV. 5, 9) in reference I.31. It may be discussed in the same manner in terms of expanding waves of potential traveling with the radial phase velocity $v_s \equiv \omega/\beta_s$. The amplitude of the wave decreases at a rate in excess of $1/R$ owing to the exponential attenuation $e^{-\alpha_s R}$ resulting from the non-zero conductivity of the medium. For a dipole in empty space, $\alpha_s = 0$, $\beta_s = \beta_0$, $v_s = v_0$.

The potential at any point in region 1 containing the dipole above the infinite plane boundary of medium 2 includes not only the direct field but also contributions from the charges set in motion in the two media. If the two dielectric media are identical and coalesce into a single, infinite medium, account of the contributions from all charges in this medium is taken by the complex parameters ξ, ν, so that (5) gives the complete field. If the two media differ, as here assumed, the fields of potential in the two media must be made to satisfy the boundary conditions at the infinite *plane* boundary. In order to accomplish this it is convenient to decompose the solution (3), which is interpreted in terms of spherical waves, into an equivalent sum of linear solutions that may be interpreted *individually* as plane waves. It is reasonable that a solution involving only *plane* waves is well suited to satisfy boundary conditions at a *plane* boundary. It is clear physically (Fig. 4.1) that a train of spherical surfaces of constant phase striking a plane boundary may be treated as a plane wave at each point on that boundary over a sufficiently small area near that point. Note that the direction of arrival and the instantaneous phases are different at different points.

The immediate problem is to represent the train of spherical waves of potential as a bundle of generalized plane waves of potential. The name *generalized* plane waves refers to the fact that the quantities which are real angles in perfect dielectrics are complex in conducting media and are called *complex angles*.

Consider a plane wave of unit amplitude traveling in a medium with complex propagation constant β in a direction specified by the *unit* vector $\hat{k}$. It may be represented by

$$\exp(-j\beta s), \quad (6)$$

where s is the distance measured from an arbitrary origin in the direction of $\hat{\mathbf{k}}$. The traveling or running wave may be referred to an arbitrary position vector $\mathbf{R}$ drawn from the origin in a direction specified by the spherical coördinates Θ, Φ by noting that

$$s = \hat{\mathbf{k}} \cdot \mathbf{R} = R \cos \Theta', \tag{7}$$

where Θ' is the angle between $\mathbf{R}$ and the direction of propagation $\hat{\mathbf{k}}$ (see Figs. 4.2 and 4.3). The complex vector $\mathbf{k}$ is defined by

$$\mathbf{k} = \beta \hat{\mathbf{k}}. \tag{8}$$

Then the expression (6) becomes,

$$\exp(-j\mathbf{k} \cdot \mathbf{R}) = \exp(-j\beta\hat{\mathbf{k}} \cdot \mathbf{R})$$
$$= \exp(-j\beta R \cos \Theta'). \tag{9}$$

Note that the vector equation

$$\hat{\mathbf{k}} \cdot \mathbf{R} = \text{const.} \tag{10}$$

defines plane surfaces of constant phase; they are perpendicular to $\hat{\mathbf{k}}$ (Fig. 4.1).

If desired, the vectors $\mathbf{k}$ and $\mathbf{R}$ may be expressed in terms of their Cartesian components. Thus,

$$\beta \hat{\mathbf{k}} \cdot \mathbf{R} = (\hat{\mathbf{x}} k_x + \hat{\mathbf{y}} k_y + \hat{\mathbf{z}} k_z) \cdot (\hat{\mathbf{x}} x + \hat{\mathbf{y}} y + \hat{\mathbf{z}} z)$$
$$= x k_x + y k_y + z k_z. \tag{11}$$

In spherical coördinates the rectangular components of $\mathbf{k}$ and $\mathbf{R}$ are

$$k_x = \beta \sin \Theta \cos \Phi, \quad x = R \sin \Theta_R \cos \Phi_R, \tag{12a}$$

$$k_y = \beta \sin \Theta \sin \Phi, \quad y = R \sin \Theta_R \sin \Phi_R, \tag{12b}$$

$$k_z = \beta \cos \Theta, \quad z = R \cos \Theta_R, \tag{12c}$$

so that

$$\beta \hat{\mathbf{k}} \cdot \mathbf{R} = \beta[(x \cos \Phi + y \sin \Phi) \sin \Theta + z \cos \Theta]$$
$$= \beta R[\cos(\Phi - \Phi_R) \sin \Theta \sin \Theta_R + \cos \Theta \cos \Theta_R]$$
$$= \beta R[\cos \phi \sin \Theta \sin \Theta_R + \cos \Theta \cos \Theta_R], \tag{13a}$$

where $\phi = \Phi - \Phi_R$. It follows that alternative expressions for a plane wave traveling in a direction specified by the angles Θ and Φ and referred to a vector $\mathbf{R}$ that has the direction given by Θ_R and Φ_R are

$$\exp(-j\beta s) = \exp\{-j\beta[(x \cos \Phi + y \sin \Phi) \sin \Theta + z \cos \Theta]\} \tag{13b}$$

$$= \exp\{-j\beta R[\cos \phi \sin \Theta \sin \Theta_R + \cos \Theta \cos \Theta_R]\}. \tag{13c}$$

For structures with rotational symmetry about the x-axis it is convenient to express $\mathbf{R}$ in the cylindrical coördinates r, θ, z. Thus, let

$$\mathbf{R} = \mathbf{r} + \mathbf{z}, \tag{14a}$$

so that

$$\beta \hat{\mathbf{k}} \cdot \mathbf{R} = \beta \hat{\mathbf{k}} \cdot \mathbf{r} + \beta \hat{\mathbf{k}} \cdot \mathbf{z}, \tag{14b}$$

where, in spherical coördinates R, Θ, Φ,

$$\hat{\mathbf{k}} \cdot \mathbf{r} = (x \cos \Phi + y \sin \Phi) \sin \Theta \tag{14c}$$

and

$$\hat{\mathbf{k}} \cdot \mathbf{z} = z \cos \Theta. \tag{14d}$$

Using

$$r = \sqrt{x^2 + y^2} = R \sin \Theta,$$
$$x = r \cos \theta = r \cos \Phi_R,$$
$$y = r \sin \theta = r \sin \Phi_R,$$
$$R = \sqrt{r^2 + z^2}, \tag{15a}$$

where θ in cylindrical coördinates equals Φ_R in spherical coördinates, the result is

$$\hat{\mathbf{k}} \cdot \mathbf{r} = r(\cos \theta \cos \Phi + \sin \theta \sin \Phi) \sin \Theta$$
$$= r \cos(\Phi - \theta) \sin \Theta. \tag{15b}$$

Since the angles Φ and $\theta = \Phi_R$ are measured around the z-axis with the x-axis as arbitrary origin, it is convenient to use a symbol for the difference. As introduced in (13), let

$$\phi = \Phi - \theta = \Phi - \Phi_R. \tag{16}$$

Since

$$\hat{\mathbf{k}} \cdot \mathbf{z} = z \cos \Theta, \tag{17}$$

it follows that

$$\exp(-j\beta s) = \exp[-j\beta(r \cos \phi \sin \Theta + z \cos \Theta)]. \tag{18}$$

In summary, the following equivalent expressions may be used to represent a plane wave of unit amplitude:

$$\exp(-j\beta\hat{\mathbf{k}} \cdot \mathbf{R}) = \exp(-j\beta R \cos \Theta') \quad (19a)$$
$$= \exp\{-j\beta[(x \cos \Phi + y \sin \Phi) \sin \Theta + z \cos \Theta]\} \quad (19b)$$
$$= \exp[-j\beta(r \cos \phi \sin \Theta + z \cos \Theta)]. \quad (19c)$$

These several forms are useful for different purposes.

In order to express the general form for spherical waves, namely,

$$\frac{\exp(-j\beta R)}{-j\beta R}, \quad (20)$$

in terms of a plane-wave solution of the form given in (19), use is made of the spherical Hankel function of the first kind and zeroth order. This is defined by

$$h_0^{(1)}(\beta R) \equiv \sqrt{\frac{\pi}{2\beta R}} H_{\frac{1}{2}}^{(1)}(\beta R), \quad (21a)$$

where the Hankel function of half order may be expressed as follows:*

$$H_{\frac{1}{2}}^{(1)}(x)\sqrt{\frac{\pi}{2}x} = -i \exp(ix). \quad (i = -j). \quad (21b)$$

Setting $x = \beta R$ and substituting in (21a), the result is

$$h_0^{(1)}(\beta R) = \frac{-i}{\beta R} \exp(i\beta R) = \frac{\exp(-j\beta R)}{-j\beta R}. \quad (21c)$$

Thus, the Hertzian potentials defined by (1) with (2) may be expressed by the spherical Hankel function as follows:

$$\Pi_z = K \frac{\exp(-j\beta R)}{-j\beta R} = K h_0^{(1)}(\beta R). \quad (22)$$

It has been shown† that the spherical Hankel function is given by the following integral representation:

$$h_0^{(1)}(\beta R) = \int_{-j\infty}^{1} \exp(-j\beta R \eta) \, d\eta, \quad (23)$$

---

* Jahnke-Emde, ref. I.28, p. 136.
† Stratton, ref. I.52, p. 410.

where η is complex. Note that the integrand vanishes at the lower limit both when β is real and when it is complex. In (23) let

$$\eta = \cos \Theta', \quad (24)$$

then

$$d\eta = -\sin \Theta' \, d\Theta'. \quad (25)$$

The limits for the complex variable Θ' are obtained as follows:

$$\Theta' = \Theta'_r + j\Theta'_i, \quad (26a)$$

$$\eta = \cos \Theta' = \cosh \Theta'_i \cos \Theta'_r - j \sinh \Theta'_i \sin \Theta'_r. \quad (26b)$$

Clearly,

$$\eta = 1: \quad \Theta'_r = \Theta'_i = 0, \quad (27)$$

$$\eta = -j\infty: \quad \Theta'_r = \pi/2, \quad \Theta'_i = \infty. \quad (28)$$

Hence,

$$h_0^{(1)}(\beta R) = \int_0^{\frac{1}{2}\pi + j\infty} \exp(-j\beta R \cos \Theta') \sin \Theta' \, d\Theta'. \quad (29)$$

Since

$$\int_0^{2\pi} d\Phi' = 2\pi, \quad (30)$$

it follows that

$$h_0^{(1)}(\beta R) = \frac{1}{2\pi} \int_0^{2\pi} \int_0^{\frac{1}{2}\pi + j\infty} \exp(-j\beta R \cos \Theta') \sin \Theta' \, d\Theta' \, d\Phi', \quad (31)$$

if Φ' is chosen to be an equatorial angle about the axis z' oriented in the direction specified by $\mathbf{R}$. Note that if Θ' were real, $\sin \Theta' \, d\Theta' \, d\Phi'$ would be an element on the surface of a sphere of unit radius with axis z'. Substitution of (31) in (22) gives

$$\Pi_z = K \frac{\exp(-j\beta R)}{-j\beta R} = \frac{K}{2\pi} \int_0^{2\pi} \int_0^{\frac{1}{2}\pi + j\infty} \exp(-j\beta R \cos \Theta') \sin \Theta' \, d\Theta' \, d\Phi', \quad (32)$$

where the integrand is a generalized plane wave of the form (19). This is the integral formulation of Pearson. With given R, the integration is equivalent to a summation over all directions Θ' of the complex propagation vector or wave normal $\mathbf{k}$ of the *generalized plane wave* in the integrand. Since the value of the integral is independent of the orientation of the reference axis, the axis z (parallel to the axis of the dipole) may be chosen instead of z' along $\mathbf{R}$. The

706   THEORY OF LINEAR ANTENNAS   [VII.4]

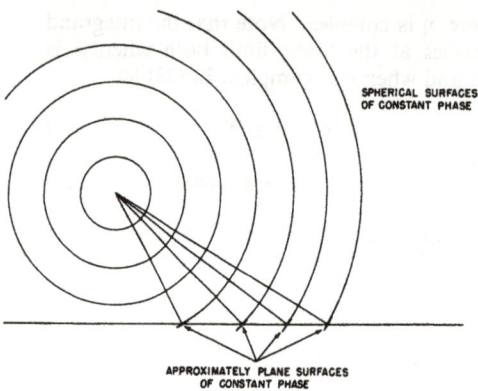

Fig. 4.1. Spherical wave at a plane boundary.

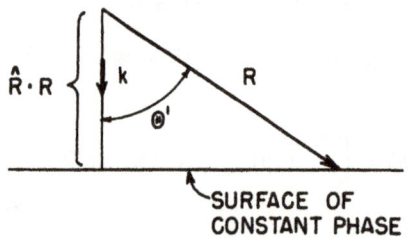

Fig. 4.2. Vector diagram of plane wave.

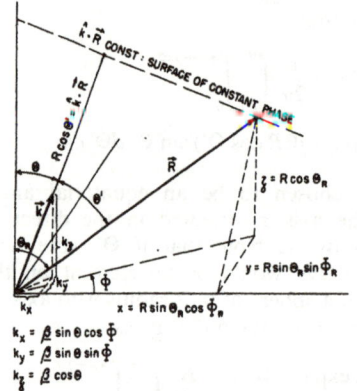

Fig. 4.3. Vector diagram of plane wave referred to an arbitrary direction.

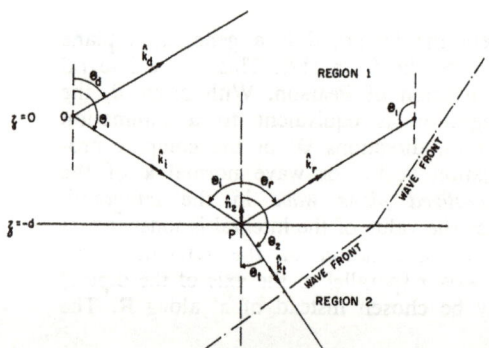

Fig. 5.1. Refracted wave.

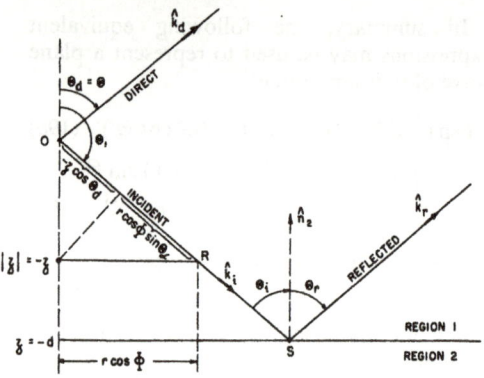

Fig. 5.2. Components of distance OR for incident wave.

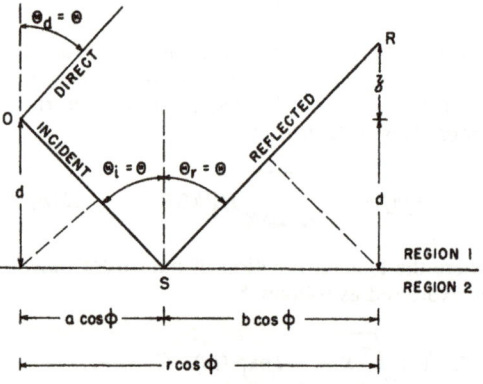

Fig. 5.3. Reflected wave.

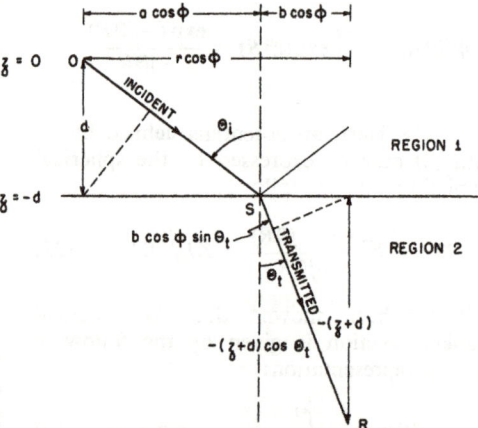

Fig. 5.4. Refracted or transmitted wave.

element $\sin \Theta' \, d\Theta' \, d\Phi'$ may, therefore, be replaced by $\sin \Theta \, d\Theta \, d\phi$. Thus,*

$$\Pi_z = K \frac{\exp(-j\beta R)}{-j\beta R} \quad (33a)$$

$$= \frac{K}{2\pi} \int_0^{2\pi} \int_0^{\frac{1}{2}\pi+j\infty} \exp(-j\beta R \cos \Theta') \sin \Theta \, d\Theta \, d\phi, \quad (33b)$$

where, with (19),

$$R \cos \Theta' = (x \cos \Phi + y \sin \Phi) \sin \Theta + z \cos \Theta$$

$$= r \cos \phi \sin \Theta + z \cos \Theta. \quad (34)$$

Equation (33b) is the desired formula for expressing the spherical waves represented by the simple exponential formula given by (33a) in terms of a sum of generalized plane waves that are represented individually by the integrand. The variation in Θ' referred to R in (32) is replaced by an equivalent variation in Θ referred to the z-axis through the dipole.

5. Boundary Conditions for Vertical Dipoles; Incident, Reflected, and Refracted Waves

In order to satisfy the boundary conditions† (1.12a, b) and (1.13a, b)

$$\left(\frac{\partial \Pi_{ez}}{\partial z}\right)_1 = \left(\frac{\partial \Pi_{ez}}{\partial z}\right)_2, \quad (1a)$$

$$\xi_1 \Pi_{ez1} = \xi_2 \Pi_{ez2}, \quad (1b)$$

$$\Pi_{mz1} = \Pi_{mz2}, \quad (2a)$$

$$\nu_1 \left(\frac{\partial \Pi_{mz}}{\partial z}\right)_1 = \nu_2 \left(\frac{\partial \Pi_{mz}}{\partial z}\right)_2, \quad (2b)$$

the Hertzian potential,

$$\Pi_z = K \frac{\exp(-j\beta R)}{-j\beta R}, \quad (3)$$

with

$$\Pi_z = \Pi_{ez} \text{ and } K = K_e = \frac{-j\beta p_z}{4\pi\xi}, \quad (3a)$$

$$\Pi_z = \Pi_{mz} \text{ and } K = K_m = \frac{-j\beta m_z}{4\pi\nu}, \quad (3b)$$

is transformed conveniently into

$$\Pi_z = \frac{K}{2\pi} \int_0^{2\pi} \int_0^{\frac{1}{2}\pi+j\infty} \exp(-j\beta \hat{\mathbf{k}} \cdot \mathbf{R}) \, d\Omega, \quad (4a)$$

where

$$d\Omega \equiv \sin \Theta \, d\Theta \, d\phi. \quad (4b)$$

The actual procedure is to construct a solution of the form.

$$\Pi_z = \Pi_{z1} = (\Pi_{z1})_d + (\Pi_{z1})_r \text{ in region 1}, \quad (5a)$$

$$\Pi_z = \Pi_{z2} \text{ in region 2}, \quad (5b)$$

using (4). Specifically,

$$(\Pi_{z1})_d = \frac{K_1}{2\pi} \int_0^{2\pi} \int_0^{\frac{1}{2}\pi+j\infty} \exp(-j\beta \hat{\mathbf{k}}_d \cdot \mathbf{R}) \, d\Omega, \quad (6a)$$

$$(\Pi_{z1})_r = \frac{K_1}{2\pi} \int_0^{2\pi} \int_0^{\frac{1}{2}\pi+j\infty} f_r \exp(-j\beta \hat{\mathbf{k}}_r \cdot \mathbf{R}) \, d\Omega, \quad (6b)$$

$$\Pi_{z2} = \frac{K_2}{2\pi} \int_0^{2\pi} \int_0^{\frac{1}{2}\pi+j\infty} f_t \exp(-j\beta \hat{\mathbf{k}}_t \cdot \mathbf{R}) \, d\Omega, \quad (6c)$$

where the unit vectors $\hat{\mathbf{k}}_d$, $\hat{\mathbf{k}}_r$, and $\hat{\mathbf{k}}_t$ are, respectively, the propagation vectors of elementary, generalized plane waves in each of three groups or bundles called the direct, the reflected, and the transmitted (or refracted) waves. The subscripts d, r, and t are used to distinguish between quantities referring to them individually. The amplitude factor K_1 is defined as in (3b) with appropriate subscripts e or m. Thus,

$$\Pi_{z1} = \Pi_{ez1}: \quad K_1 = K_{e1} = \frac{-j\beta_1 p_{z1}}{4\pi\xi_1}, \quad (7a)$$

$$\Pi_{z1} = \Pi_{mz1}: \quad K_1 = K_{m1} = \frac{-j\beta_1 m_{z1}}{4\pi\nu_1}. \quad (7b)$$

The functions f_r and f_t, with subscript e or m, are to be determined by requiring the functions given by (6a, b, c) to satisfy the boundary conditions (1) and (2). Although

---

* The method here outlined follows Stratton, ref. I.52, p. 578. A more detailed discussion by Noether is in Rothe, Ollendorf, and Pohlhausen, ref. I.44, pp. 170–174.

† For some purposes, particularly in introducing impedance concepts, it is convenient to use the modified Hertzian potentials defined by

$$U_z = j\omega\xi\Pi_{ez}, \quad V_z = j\omega\Pi_{mz}.$$

The boundary conditions then become

$$U_{z1} = U_{z2}, \quad V_{z1} = V_{z2},$$

$$\frac{1}{\xi_1}\left(\frac{\partial U_z}{\partial z}\right)_1 = \frac{1}{\xi_2}\left(\frac{\partial U_z}{\partial z}\right)_2, \quad \frac{1}{\mu_1}\left(\frac{\partial V_z}{\partial z}\right)_1 = \frac{1}{\mu_2}\left(\frac{\partial V_z}{\partial z}\right)_2.$$

the factor K_2 might be defined by analogy with K_1, this is neither necessary nor, in fact, convenient. The obvious alternative is to set K_2 equal to K_1 so that f_i is a measure of the ratio of Π_z in region 2 to its value in region 1 at points just across the boundary from one another. However, since the boundary conditions (1) and (2) for Π_{ez} and Π_{mz} are different, much more symmetric equations and formulas are obtained if the ratios of the products $\xi\Pi_{ez}$ are used instead of the ratios of Π_{ez} alone. Actually, this choice is consistent with the formulation of the waveguide problem in a manner that parallels conventional transmission-line theory by using the functions $U_z = j\omega\xi\Pi_{ez}$ and $V_z = j\omega\Pi_{mz}$ instead of Π_{ez} and Π_{mz}. Since U_z is in amperes and V_z is in volts, ratios of U_z are analogous to ratios of current and ratios of V_z are analogous to ratios of voltage on a transmission line. Clearly, therefore, the following choice of K_{ez} and K_{mz} is suggested both in the interest of simplicity and symmetry and in order to parallel transmission-line theory:

$$\Pi_{z2} = \Pi_{ez2}: \quad K_2 = K_{e2} = \frac{\xi_1}{\xi_2} K_{e1}$$

$$= \frac{-j\beta_1 p_{z1}}{4\pi\xi_2}, \quad (7c)$$

$$\Pi_{z2} = \Pi_{mz2}: \quad K_2 = K_{m2} = K_{m1}$$

$$= \frac{-j\beta_1 m_{z1}}{4\pi\nu_1}. \quad (7d)$$

Note that p_{z1} and m_{z1} are the electric and magnetic moments of dipoles in region 1.

The integrals (6a, b, c) can be made to satisfy the boundary conditions (1) and (2) at all points on the boundary plane by requiring their *integrands* to satisfy these conditions. This may be accomplished in three steps. (a) Since the boundary conditions must be satisfied for all values of r when $z = -d$ regardless of the particular value of d, it follows that the parts of the exponents which have r as a factor must be equal. This is equivalent to requiring a plane wave front, traveling in the two media with different velocities and at different angles to the normal, to remain in contact along the two sides of the boundary (Fig. 5.1). (b) The exponential factors must be so constructed that after condition (a) has been imposed the exponents with $z = -d$ are independent of d. (c) With the exponential factor correctly chosen, the boundary conditions may be applied directly to determine f_r and f_t.

Let the origin of coördinates be at the center of the dipole in region 1 at a distance d above the boundary plane $z = -d$. The position vector $\mathbf{R}$ is drawn from the origin to any desired point. When drawn from the origin to any point on the boundary plane, $\mathbf{R} = \mathbf{R}_b$. The equation for this boundary plane in vector form is

$$\hat{\mathbf{n}}_2 \cdot \mathbf{R}_b = -d, \quad (8)$$

where $\hat{\mathbf{n}}_2 = \hat{\mathbf{z}}$ is a unit external normal to region 2. Since

$$\mathbf{R}_b = \mathbf{r} - \hat{\mathbf{n}}_2 d = \mathbf{r} - \hat{\mathbf{z}}d, \quad (9)$$

where $\mathbf{r}$ is a radial position vector in the plane $z = -d$, it follows that an elementary generalized incident plane wave at the boundary is given by

$$\exp(-j\beta_1 \hat{\mathbf{k}}_i \cdot \mathbf{R}_b)$$
$$= \exp(-j\beta_1 \hat{\mathbf{k}}_i \cdot \mathbf{r} + j\beta_1 d\, \hat{\mathbf{k}}_i \cdot \hat{\mathbf{z}}); \quad (10a)$$

an elementary generalized reflected plane wave by

$$f_r \exp(-j\beta_1 \hat{\mathbf{k}}_r \cdot \mathbf{R}_b) = f_r \exp(-j\beta_1 \hat{\mathbf{k}}_r \cdot \mathbf{r} + j\beta_1 d\, \hat{\mathbf{k}}_r \cdot \hat{\mathbf{z}}); \quad (10b)$$

and an elementary generalized transmitted or refracted wave by

$$f_t \exp(-j\beta_2 \hat{\mathbf{k}}_t \cdot \mathbf{R}_b) = f_t \exp(-j\beta_2 \hat{\mathbf{k}}_t \cdot \mathbf{r} + j\beta_1 d\, \hat{\mathbf{k}}_t \cdot \hat{\mathbf{z}}). \quad (10c)$$

If these are used in the boundary condition (4a), it is evident, since the resulting equation must be satisfied for all values of $\mathbf{r}$ on the boundary plane, that the exponents containing $\mathbf{r}$ must be equal. That is,

$$\beta_1 \hat{\mathbf{k}}_i \cdot \mathbf{r} = \beta_1 \hat{\mathbf{k}}_r \cdot \mathbf{r} = \beta_2 \hat{\mathbf{k}}_t \cdot \mathbf{r}. \quad (11)$$

The vector $\mathbf{r}$ may be expanded using the vector identity

$$\hat{\mathbf{n}}_2 \times (\hat{\mathbf{n}}_2 \times \mathbf{r}) = \hat{\mathbf{n}}_2(\hat{\mathbf{n}}_2 \cdot \mathbf{r}) - \mathbf{r}(\hat{\mathbf{n}}_2 \cdot \hat{\mathbf{n}}_2) \quad (12)$$

and the relations

$$\hat{\mathbf{n}}_2 \cdot \hat{\mathbf{n}}_2 = 1, \quad \hat{\mathbf{n}}_2 \cdot \mathbf{r} = 0. \quad (13)$$

The result is

$$\mathbf{r} = -\hat{\mathbf{n}}_2 \times (\hat{\mathbf{n}}_2 \times \mathbf{r}). \quad (14)$$

Substitution of (14) in (11) gives

$$(\beta_1 \hat{\mathbf{k}}_i) \cdot [\hat{\mathbf{n}}_2 \times (\hat{\mathbf{n}}_2 \times \mathbf{r})]$$
$$= (\beta_1 \hat{\mathbf{k}}_r) \cdot [\hat{\mathbf{n}}_2 \times (\hat{\mathbf{n}}_2 \times \mathbf{r})]$$
$$= (\beta_2 \hat{\mathbf{k}}_t) \cdot [\hat{\mathbf{n}}_2 \times (\hat{\mathbf{n}}_2 \times \mathbf{r})]. \quad (15)$$

With the double-product identity,

$$\mathbf{A} \cdot \mathbf{B} \times \mathbf{C} = \mathbf{C} \cdot \mathbf{A} \times \mathbf{B}, \quad (16)$$

(15) may be transformed into the following equations:

$$\beta_1(\hat{\mathbf{n}}_2 \times \mathbf{r}) \cdot (\hat{\mathbf{k}}_i \times \hat{\mathbf{n}}_2) = \beta_1(\hat{\mathbf{n}}_2 \times \mathbf{r}) \cdot (\hat{\mathbf{k}}_r \times \hat{\mathbf{n}}_2)$$
$$= \beta_2(\hat{\mathbf{n}}_2 \times \mathbf{r}) \cdot (\hat{\mathbf{k}}_t \times \hat{\mathbf{n}}_2). \quad (17)$$

These equations are satisfied for all $\mathbf{r}$ if and only if

$$\beta_1 \hat{\mathbf{k}}_i \times \hat{\mathbf{n}}_2 = \beta_1 \hat{\mathbf{k}}_r \times \hat{\mathbf{n}}_2 = \beta_2 \hat{\mathbf{k}}_t \times \hat{\mathbf{n}}_2. \quad (18)$$

The complex angle between $\hat{\mathbf{k}}_i$ and $\hat{\mathbf{n}}_2$ is Θ_1, and $\pi - \Theta_1 \equiv \Theta_i$ is the *angle of incidence*; the plane containing $\hat{\mathbf{k}}_i$ and $\hat{\mathbf{n}}_2$ is the *plane of incidence*. The complex angle between $\hat{\mathbf{k}}_r$ and $\hat{\mathbf{n}}_2$ is Θ_r; it is the *angle of reflection*. The plane containing $\hat{\mathbf{k}}_r$ and $\hat{\mathbf{n}}_2$ is the *plane of reflection*. The complex angle between $\hat{\mathbf{k}}_t$ and $\hat{\mathbf{n}}_2$ is Θ_2 and $\pi - \Theta_2 \equiv \Theta_t$ is the *angle of transmission* or *refraction*; the plane containing $\hat{\mathbf{k}}_t$ and $\hat{\mathbf{n}}_2$ is the *plane of transmission* or *refraction*.

The following general conclusions may be derived from the general vector relations (18):

(a) From the properties of the vector product the unit vectors $\hat{\mathbf{k}}_i$, $\hat{\mathbf{k}}_r$, $\hat{\mathbf{k}}_t$, and $\hat{\mathbf{n}}_2$ are all in the same plane. Therefore, the planes of incidence, reflection, and refraction coincide.

(b) From the first equation in (18) it follows that

$$\sin \Theta_r = \sin \Theta_1 = \sin(\pi - \Theta_i) = \sin \Theta_i, \quad (19)$$

so that

$$\Theta_r = \Theta_1, \quad \text{or} \quad \Theta_r = \pi - \Theta_1 = \Theta_i. \quad (20a)$$

The first solution, $\Theta_r = \Theta_1$, is not in general a possible solution since it requires the reflected wave to be directed into region 2, which violates the requirement that its propagation constant be β_1. The general case is

$$\Theta_r = \Theta_i = \pi - \Theta_1 = \Theta_d, \quad (20b)$$

where the several angles are shown in Fig. 5.1. That is, *the angle of reflection is equal to the angle of incidence*. This is the well-known law of reflection, here generalized to apply to complex angles and generalized plane waves. Evidently, corresponding to (20b),

$$\cos \Theta_r = \cos \Theta_i = -\cos \Theta_1. \quad (21)$$

(c) From the second equation in (18) it follows that

$$\beta_2 \sin \Theta_2 = \beta_1 \sin \Theta_1, \quad (22)$$

or, since

$$\Theta_i = \pi - \Theta_1, \quad (23a)$$

$$\Theta_t = \pi - \Theta_2, \quad (23b)$$

an equivalent form of (22) is

$$\beta_2 \sin \Theta_t = \beta_1 \sin \Theta_i. \quad (24)$$

Alternatively,

$$\frac{\sin \Theta_i}{\sin \Theta_t} = \frac{\beta_2}{\beta_1} \equiv N_{21} \equiv \frac{1}{N_{12}}, \quad (25)$$

where N_{21} is the generalized complex index of refraction of medium 2 referred to medium 1 and N_{12} is the generalized complex index of medium 2 referred to medium 1. Evidently (25) is the *law of refraction* for generalized plane waves.

It is clear from (a), (b), (c) that each elementary, generalized plane wave in the bundle contributing to the field of the dipole in the two media satisfies Snell's laws of reflection and refraction at a plane boundary in a generalized form involving complex angles. If the source is sufficiently far away, the potential is given by the integrand in (4.33) with a real angle Θ', and the results in (a), (b), and (c) coincide with those of ordinary plane-wave theory.

It is now a simple matter to construct explicit expressions for the direct, reflected, and refracted potentials that satisfy the boundary condition (11). The integrand in the direct potential is simply

$$\exp[-j\beta_1 \hat{\mathbf{k}}_d \cdot \mathbf{R}] = \exp[-j\beta_1(r \cos \phi \sin \Theta_d + z \cos \Theta_d)]. \quad (26)$$

For the direct wave incident upon the plane boundary at $z = -d$ from the direction of the origin and reflected according to Snell's law at an angle Θ_r, the incident wave is given by

$$\exp[-j\beta_1 \hat{\mathbf{k}}_i \cdot \mathbf{R}]$$
$$= \exp\{-j\beta_1[r \cos \phi \sin(\pi - \Theta_d) + z \cos(\pi - \Theta_d)]\} \quad (27)$$
$$= \exp[-j\beta_1(r \cos \phi \sin \Theta_d - z \cos \Theta_d)]. \quad (28)$$

Note that the factor in parentheses in the exponent in (28) is the complex "distance" corresponding to OR in Fig. 5.2. The two terms contributing to it are the projections of $-z$ and $r \cos \phi$ onto OS.

If the complex propagation constant β_1 is expressed in the form

$$\beta_1 = \frac{\omega}{v_{p1}}, \qquad (29)$$

v_{p1} is a complex "phase velocity of propagation" of the generalized surfaces of constant phase forming the plane wave in region 1. The complex "distance" corresponding to OR, divided by v_{p1}, gives the time required for an equiphase surface to travel from O to R.

The expression for a generalized plane reflected wave is readily obtained by determining OR in Fig. 5.3. The part OS is obtained directly from (28) by setting $z = -d$ and $r = a$. It is

$$OS = a \cos \phi \sin \Theta_i + d \cos \Theta_i. \quad (30a)$$

The part SR of the path is

$$SR = b \cos \phi \sin \Theta_r + (z + d) \cos \Theta_r. \quad (30b)$$

Hence, since $\Theta_r = \Theta_i \equiv \Theta$ by the law of reflection, the expression for the reflected plane wave in region 1 for use as the integrand in (6b) is

$$f_r \exp[-j\beta_1 \hat{\mathbf{k}}_r \cdot \mathbf{R}]$$
$$= f_r \exp\{-j\beta_1[r \cos \phi \sin \Theta + (z + 2d) \cos \Theta]\}. \quad (31)$$

The expression for the refracted wave is obtained in a similar manner using the law of refraction (25) in Fig. 5.4. Evidently OS is the same as given in (30a). The part SR is like (30b) with Θ_t replacing Θ_r, and with $-(z + d)$ substituted for $z + d$. Thus, from Fig. 5.4,

$$SR = b \cos \phi \sin \Theta_t - (z + d) \cos \Theta_t. \quad (32)$$

It is now only necessary to multiply the right-hand member of (30a) by β_1 and that of (32) by β_2 and add. Thus,

$$\beta_1[a \cos \phi \sin \Theta_i + d \cos \Theta_i]$$
$$+ \beta_2[b \cos \phi \sin \Theta_t - (z + d) \cos \Theta_t]. \quad (33a)$$

Since $b = r - a$, this may be arranged as follows:

$$\beta_2[r \cos \phi \sin \Theta_t - (z + d) \cos \Theta_t]$$
$$+ a \cos \phi[\beta_1 \sin \Theta_i - \beta_2 \sin \Theta_t]$$
$$+ \beta_1 d \cos \Theta_i. \quad (33b)$$

Application of the law of refraction shows that the bracket multiplying $(a \cos \phi)$ is zero. Hence, the expression for the refracted plane wave is

$$f_t \exp[-j\beta_2 \hat{\mathbf{k}}_t \cdot \mathbf{R}]$$
$$= f_t \exp\{-j\beta_2[r \cos \phi \sin \Theta_t - (z + d) \cos \Theta_t] - j\beta_1 d \cos \Theta\}. \quad (34)$$

Substitution of (26), (31), and (34) for direct, reflected, and transmitted waves in (6a, b, c) using (5a) gives the following expressions for the polarization potential in regions 1 and 2 due to a dipole in region 1:

$$\Pi_{z1} = \frac{K_1}{2\pi} \int_0^{2\pi} \int_0^{\frac{1}{2}\pi + j\infty} \{\exp[-j\beta_1(r \cos \phi \sin \Theta + z \cos \Theta)] + f_r \exp[-j\beta_1(r \cos \phi \sin \Theta + z \cos \Theta + 2d \cos \Theta)]\} d\Omega, \quad (z \geq 0) \quad (35a)$$

$$\Pi_{z1} = \frac{K_1}{2\pi} \int_0^{2\pi} \int_0^{\frac{1}{2}\pi + j\infty} \{\exp[-j\beta_1(r \cos \phi \sin \Theta - z \cos \Theta)] + f_r \exp[-j\beta_1(r \cos \phi \sin \Theta + z \cos \Theta + 2d \cos \Theta)]\} d\Omega,$$
$$(-d \leq z \leq 0) \quad (35b)$$

$$\Pi_{z2} = \frac{K_2}{2\pi} \int_0^{2\pi} \int_0^{\frac{1}{2}\pi + j\infty} f_t \exp\{-j\beta_2[r \cos \phi \sin \Theta_t - (z + d) \cos \Theta_t] - j\beta_1 d \cos \Theta\} d\Omega.$$
$$(z \leq -d) \quad (36)$$

These potentials have been constructed so that they satisfy the differential equations (6a, b) for Π_z and the general phase conditions (11). It remains to show that they can be made to satisfy the boundary conditions (1) and (2) at $z = -d$ by assigning appropriate values to the coefficients of amplitude f_r and f_t.

On the plane at $z = -d$ the functions given by (35b) and (36) become

$$\Pi_{z1} = \frac{K_1}{2\pi} \int_0^{2\pi} \int_0^{\frac{1}{2}\pi + j\infty} \exp[-j\beta_1(r \cos \phi \sin \Theta + d \cos \Theta)](1 + f_r) d\Omega, \quad (37)$$

$$\Pi_{z2} = \frac{K_2}{2\pi} \int_0^{2\pi} \int_0^{\frac{1}{2}\pi + j\infty} f_t \exp[-j\beta_2 r \cos \phi \sin \Theta_t - j\beta_1 d \cos \Theta] d\Omega. \quad (38a)$$

[VII.5] THEORY OF LINEAR ANTENNAS

With the law of refraction, $\beta_2 \sin \Theta_t = \beta_1 \sin \Theta_i$, $\Theta_i = \Theta$, it follows that (38a) is equivalent to

$$\Pi_{z2} = \frac{K_2}{2\pi} \int_0^{2\pi} \int_0^{\frac{1}{2}\pi + j\infty} f_t \exp\left[-j\beta_1(r \cos \phi \sin \Theta + d \cos \Theta)\right] d\Omega. \quad (38b)$$

As required by (11), the exponents in (37) and (38b) are seen to be the same.

The boundary conditions (1) and (2) now may be imposed on the integrands in (37) and (38b), and the coefficients f_r and f_t evaluated. Since the electric and magnetic cases are different, they must be evaluated separately. Note that

$$K_{e2} = K_{e1}\xi_1/\xi_2, \quad K_{m2} = K_{m1}. \quad (39)$$

The results of imposing the boundary conditions

$$\xi_1 \Pi_{ez1} = \xi_2 \Pi_{ez2}, \quad (40a)$$

$$\Pi_{mz1} = \Pi_{mz2} \quad (40b)$$

are as follows:

$$1 + f_{er} = f_{et}, \quad (41a)$$

$$1 + f_{mr} = f_{mt}. \quad (41b)$$

Differentiation of the integrands in (35b) and (36) and setting $z = -d$ in the derivatives to satisfy the conditions

$$\left(\frac{\partial \Pi_{ez}}{\partial z}\right)_1 = \left(\frac{\partial \Pi_{ez}}{\partial z}\right)_2, \quad (42a)$$

$$\nu_1 \left(\frac{\partial \Pi_{mz}}{\partial z}\right)_1 = \nu_2 \left(\frac{\partial \Pi_{mz}}{\partial z}\right)_2 \quad (42b)$$

gives directly,

$$(\beta_1/\xi_1)(1 - f_{er}) \cos \Theta = (\beta_2/\xi_2) f_{et} \cos \Theta_t, \quad (43a)$$

$$\beta_1 \nu_1 (1 - f_{mr}) \cos \Theta = \beta_2 \nu_2 f_{mt} \cos \Theta_t. \quad (43b)$$

The following notation is convenient. Since by definition

$$\beta \equiv \omega\sqrt{\xi\nu}, \quad \zeta \equiv 1/\sqrt{\xi\nu} \equiv 1/\eta, \quad (44)$$

it follows that

$$\beta/\xi = \omega\zeta, \quad \beta\nu = \omega/\zeta = \omega\eta. \quad (45)$$

The *characteristic admittance* $\eta \equiv 1/\zeta$ has been defined in (44). Accordingly, (41a, b) and (43a, b) may be rearranged as follows:

$$f_{er} - f_{et} = -1, \quad (46a)$$

$$f_{er}\zeta_1 \cos \Theta + f_{et}\zeta_2 \cos \Theta_t = \zeta_1 \cos \Theta, \quad (46b)$$

$$f_{mr} - f_{mt} = -1, \quad (46c)$$

$$f_{mr}\eta_1 \cos \Theta + f_{mt}\eta_2 \cos \Theta_t = \eta_1 \cos \Theta. \quad (46d)$$

Solutions of these equations are obtained directly. They are

$$f_{er} = \frac{\zeta_1 \cos \Theta - \zeta_2 \cos \Theta_t}{\zeta_1 \cos \Theta + \zeta_2 \cos \Theta_t}, \quad (47a)$$

$$f_{mr} = \frac{\eta_1 \cos \Theta - \eta_2 \cos \Theta_t}{\eta_1 \cos \Theta + \eta_2 \cos \Theta_t}, \quad (47b)$$

$$f_{et} = \frac{2\zeta_1 \cos \Theta}{\zeta_1 \cos \Theta + \zeta_2 \cos \Theta_t}, \quad (48a)$$

$$f_{mt} = \frac{2\eta_1 \cos \Theta}{\eta_1 \cos \Theta + \eta_2 \cos \Theta_t}. \quad (48b)$$

Let "field impedances" and "field admittances" be defined as follows:

$$Z_{e1} \equiv \zeta_1 \cos \Theta \equiv 1/Y_{e1}, \quad (49a)$$

$$Y_{m1} \equiv \eta_1 \cos \Theta \equiv 1/Z_{m1}, \quad (49b)$$

$$Z_{e2} \equiv \zeta_2 \cos \Theta_t \equiv 1/Y_{e2}, \quad (50a)$$

$$Y_{m2} \equiv \eta_2 \cos \Theta_t \equiv 1/Z_{m2}. \quad (50b)$$

Substitution of (49) and (50) in (47) and (48) gives the following simple expressions for the reflection and transmission coefficients:

$$f_{er} = \frac{Z_{e1} - Z_{e2}}{Z_{e1} + Z_{e2}} = \frac{Y_{e2} - Y_{e1}}{Y_{e2} + Y_{e1}}, \quad (51a)$$

$$f_{mr} = \frac{Y_{m1} - Y_{m2}}{Y_{m1} + Y_{m2}} = \frac{Z_{m2} - Z_{m1}}{Z_{m2} + Z_{m1}}, \quad (51b)$$

$$f_{et} = \frac{2Z_{e1}}{Z_{e1} + Z_{e2}} = \frac{2Y_{e2}}{Y_{e2} + Y_{e1}}, \quad (52a)$$

$$f_{mt} = \frac{2Y_{m1}}{Y_{m1} + Y_{m2}} = \frac{2Z_{m2}}{Z_{m2} + Z_{m1}}. \quad (52b)$$

The formulas (51) and (52) are analogous to the expressions for reflection and transmission factors for the impedance or admittance of one infinite or matched transmission line terminating a second line of different characteristic impedance. Suppose transmission line 1 has a characteristic impedance $Z_{c1} = 1/Y_{c1}$ and is terminated in a second infinitely long line that has a characteristic impedance $Z_{c2} = 1/Y_{c2}$. The voltage and current reflection coefficients Γ_s and Γ'_s are defined by

$$\Gamma_s \equiv \frac{Z_s - Z_{c1}}{Z_s + Z_{c1}} = \frac{Y_{c1} - Y_s}{Y_{c1} + Y_s} = -\Gamma'_s, \quad (53)$$

where $Z_s = 1/Y_s$ is the load impedance. With $Z_s = Z_{c2}$,

$$\Gamma'_s = \frac{Z_{c1} - Z_{c2}}{Z_{c1} + Z_{c2}} = \frac{Y_{c2} - Y_{c1}}{Y_{c2} + Y_{c1}}, \quad (54a)$$

$$\Gamma_s = \frac{Y_{c1} - Y_{c2}}{Y_{c1} + Y_{c2}} = \frac{Z_{c2} - Z_{c1}}{Z_{c2} + Z_{c1}}. \quad (54b)$$

The transmission coefficients are

$$T'_s = 1 + \Gamma'_s = \frac{2Z_{c1}}{Z_{c1} + Z_{c2}} = \frac{2Y_{c2}}{Y_{c1} + Y_{c2}}, \quad (55a)$$

$$T_s = 1 + \Gamma_s = \frac{2Y_{c1}}{Y_{c1} + Y_{c2}} = \frac{2Z_{c2}}{Z_{c1} + Z_{c2}}. \quad (55b)$$

Comparison of (54) and (55) with (51) and (52) shows that the plane-wave reflection and transmission coefficients f_{er} and f_{et} of the field of the vertical electric dipole are analogous to the *current* reflection and transmission coefficients Γ'_s and T'_s of a transmission line, whereas the plane-wave coefficients f_{mr} and f_{mt} of the field of a vertical magnetic dipole are analogous to the *voltage* reflection and transmission coefficients Γ_s and T_s of a transmission line. Note that the characteristic impedances and admittances of the equivalent line are functions of the angle Θ.

Alternative expressions for the coefficients are in terms of the propagation constants β_1 and β_2. With (45), Eqs. (47) and (48) may be expressed as follows:

$$f_{er} = \frac{\nu_2\beta_2 \cos\Theta - \nu_1\beta_1 \cos\Theta_t}{\nu_2\beta_2 \cos\Theta + \nu_1\beta_1 \cos\Theta_t}, \quad (56a)$$

$$f_{mr} = \frac{\nu_1\beta_1 \cos\Theta - \nu_2\beta_2 \cos\Theta_t}{\nu_1\beta_1 \cos\Theta + \nu_2\beta_2 \cos\Theta_t}, \quad (56b)$$

$$f_{et} = \frac{2\nu_2\beta_2 \cos\Theta}{\nu_2\beta_2 \cos\Theta + \nu_1\beta_1 \cos\Theta_t}, \quad (57a)$$

$$f_{mt} = \frac{2\nu_1\beta_1 \cos\Theta}{\nu_1\beta_1 \cos\Theta + \nu_2\beta_2 \cos\Theta_t}. \quad (57b)$$

In the important special case in which the angles are real and

$$\nu_1 = \nu_2, \quad (58)$$

the expressions (56) and (57) are the well-known *Fresnel coefficients* of reflection for plane waves (see, for example, Stratton, ref. I.52, pp. 493, 494).

The expressions for f_r and f_t may be given entirely in terms of the angle of incidence $\Theta_i = \Theta$ if use is made of the law of refraction to eliminate Θ_t. Thus,

$$\beta_1 \sin\Theta = \beta_2 \sin\Theta_t, \quad (59)$$

$$\cos\Theta_t = \sqrt{1 - \sin^2\Theta_t}$$

$$= \frac{1}{\beta_2}\sqrt{\beta_2^2 - \beta_1^2 \sin^2\Theta}. \quad (60)$$

Substitution of (59) in (56) and (57) with the generalized complex index of refraction

$$N_{21} \equiv \beta_2/\beta_1 \quad (61)$$

gives the following formulas:

$$f_{er} = \frac{\nu_2 N_{21}^2 \cos\Theta - \nu_1\sqrt{N_{21}^2 - \sin^2\Theta}}{\nu_2 N_{21}^2 \cos\Theta + \nu_1\sqrt{N_{21}^2 - \sin^2\Theta}}, \quad (62a)$$

$$f_{mr} = \frac{\nu_1 \cos\Theta - \nu_2\sqrt{N_{21}^2 - \sin^2\Theta}}{\nu_1 \cos\Theta + \nu_2\sqrt{N_{21}^2 - \sin^2\Theta}}, \quad (62b)$$

$$f_{et} = \frac{2\nu_2 N_{21}^2 \cos\Theta}{\nu_2 N_{21}^2 \cos\Theta + \nu_1\sqrt{N_{21}^2 - \sin^2\Theta}}, \quad (63a)$$

$$f_{mt} = \frac{2\nu_1 \cos\Theta}{\nu_1 \cos\Theta + \nu_2\sqrt{N_{21}^2 - \sin^2\Theta}}. \quad (63b)$$

These formulas usually are given for $\nu_1 = \nu_2$, so that N_{21} is the only material parameter.

At normal incidence $\Theta_i = \Theta = 0$, $\Theta_t = 0$, so that (54) and (55), (56) and (57), and (62) and (63) give

$$f_{er} = \frac{\zeta_1 - \zeta_2}{\zeta_1 + \zeta_2} = \frac{\nu_2 \beta_2 - \nu_1 \beta_1}{\nu_2 \beta_2 + \nu_1 \beta_1} = \frac{\nu_2 N_{21} - \nu_1}{\nu_2 N_{21} + \nu_1}, \quad (64a)$$

$$f_{mr} = \frac{\eta_1 - \eta_2}{\eta_1 + \eta_2} = \frac{\nu_1 \beta_1 - \nu_2 \beta_2}{\nu_1 \beta_1 + \nu_2 \beta_2} = \frac{\nu_1 - \nu_2 N_{21}}{\nu_1 + \nu_2 N_{21}}, \quad (64b)$$

$$f_{et} = \frac{2\zeta_1}{\zeta_1 + \zeta_2} = \frac{2\nu_2 \beta_2}{\nu_2 \beta_2 + \nu_1 \beta_1} = \frac{2\nu_2 N_{21}}{\nu_2 N_{21} + \nu_1}, \quad (64c)$$

$$f_{mt} = \frac{2\eta_2}{\eta_1 + \eta_2} = \frac{2\nu_1 \beta_1}{\nu_1 \beta_1 + \nu_2 \beta_2} = \frac{2\nu_1}{\nu_1 + \nu_2 N_{21}}, \quad (64d)$$

where $\quad \zeta = \dfrac{1}{\eta} = \dfrac{1}{\sqrt{\xi\nu}}.$

Other forms for $\nu_2 = \nu_1$ are obtained by introducing the law of refraction directly in (56) and (57). They are

$$f_{er} = \frac{\tan(\Theta - \Theta_t)}{\tan(\Theta + \Theta_t)}, \quad (65a)$$

$$f_{mr} = -\frac{\sin(\Theta - \Theta_t)}{\sin(\Theta + \Theta_t)}, \quad (65b)$$

$$f_{et} = \frac{2 \sin \Theta \cos \Theta}{\sin(\Theta + \Theta_t) \cos(\Theta - \Theta_t)}, \quad (65c)$$

$$f_{mt} = \frac{2 \sin \Theta_t \cos \Theta}{\sin(\Theta + \Theta_t)}. \quad (65d)$$

In order to relate the electromagnetic field associated with *each elementary plane wave* at each point on the boundary to the reflection coefficients f_{et} and f_{mt}, let the far zone be chosen where

$$E_e^r = E_{e\Theta}^r = -\beta^2 \Pi_{ez} \sin \Theta, \quad (66a)$$

$$B_m^r = B_{m\Theta}^r = -\beta^2 \Pi_{mz} \sin \Theta, \quad (66b)$$

$$B_e^r = B_{e\Phi}^r = -\frac{\beta^3}{\omega} \Pi_{ez} \sin \Theta, \quad (66c)$$

$$E_m^r = E_{m\Phi}^r = -\omega \beta \Pi_{mz} \sin \Theta. \quad (66d)$$

Note that E_e^r and B_m^r are parallel to the plane of incidence, B_e^r and E_m^r are perpendicular to the plane of incidence. Accordingly, the following notation is common:

$$E_e^r \equiv E_\|^r, \quad (67a)$$

$$B_m^r \equiv B_\|^r, \quad (67b)$$

$$B_e^r \equiv B_\perp^r, \quad (67c)$$

$$E_m^r \equiv E_\perp^r. \quad (67d)$$

The symbols $\|$ and $\perp$ refer to the *plane of incidence*, not to the boundary plane.

The following ratios are of interest as applied to *incident plane waves* which are represented by the *integrands* in the expressions for Π_z. For plane waves

$$\frac{E_{e2}}{(E_{e1})_d} = \frac{\beta_2^2 \Pi_{ez2} \sin \Theta_t}{\beta_1^2 (\Pi_{ez1})_d \sin \Theta} = \frac{\nu_1 \beta_1}{\nu_2 \beta_2} f_{et}, \quad (68a)$$

$$\frac{B_{e2}}{(B_{e1})_d} = \frac{\beta_2^3 \Pi_{ez2} \sin \Theta_t}{\beta_1^3 (\Pi_{ez1})_d \sin \Theta} = \frac{\nu_1}{\nu_2} f_{et}, \quad (68b)$$

$$\frac{B_{m2}}{(B_{m1})_d} = \frac{\beta_2^2 \Pi_{mz2} \sin \Theta_t}{\beta_1^2 (\Pi_{mz1})_d \sin \Theta} = \frac{\beta_2 \Pi_{mz2}}{\beta_1 \Pi_{mz1}} = \frac{\beta_2}{\beta_1} f_{mt}, \quad (68c)$$

$$\frac{E_{m2}}{(E_{m1})_d} = \frac{\beta_2 \Pi_{mz2} \sin \Theta_t}{\beta_1 (\Pi_{mz1})_d \sin \Theta} = \frac{\Pi_{mz2}}{\Pi_{mz1}} = f_{mt}. \quad (68d)$$

It follows that

$$f_{et} = \frac{\nu_2 B_{e2}}{\nu_1 (B_{e1})_d} = \frac{H_{e2}}{(H_{e1})_d}, \quad (69a)$$

$$f_{mt} = \frac{E_{m2}}{(E_{m1})_d}. \quad (69b)$$

Evidently, the coefficients f_{er} and f_{et} are of current type, f_{mr} and f_{mt} of voltage type on equivalent transmission lines.

With f_r and f_t evaluated, the Hertzian potentials due to a dipole at $z = 0$ above a plane boundary at $z = -d$ are given by the following formulas:

$$\Pi_{z1} = \frac{K_1}{2\pi} \int_0^{2\pi} \int_0^{\frac{1}{2}\pi+j\infty} \exp\{-j\beta_1[r\cos\phi \sin\Theta$$
$$+ (z+d)\cos\Theta]\}[\exp(j\beta_1 d\cos\Theta)$$
$$+ f_r \exp(-j\beta_1 d\cos\Theta)] \sin\Theta \, d\Theta \, d\phi,$$
$$(0 \leq z) \quad (70)$$

$$\Pi_{z1} = \frac{K_1}{2\pi} \int_0^{2\pi} \int_0^{\frac{1}{2}\pi+j\infty} \exp[-j\beta_1(r\cos\phi \sin\Theta$$
$$+ d\cos\Theta)]\{\exp[j\beta_1(z+d)\cos\Theta]$$
$$+ f_r \exp[-j\beta_1(z+d)\cos\Theta]\} \sin\Theta \, d\Theta \, d\phi,$$
$$(-d \leq z \leq 0) \quad (71)$$

$$\Pi_{z2} = \frac{K_2}{2\pi} \int_0^{2\pi} \int_0^{\frac{1}{2}\pi+j\infty} f_t \exp\{-j\beta_2[r\cos\phi \sin\Theta_t$$
$$- (z+d)\cos\Theta_t] - j\beta_1 d\cos\Theta\}$$
$$\times \sin\Theta \, d\Theta \, d\phi. \quad (z \leq -d) \quad (72)$$

In the special case when the dipole is on the boundary plane, that is, $d = 0$, (70) and (72) reduce to

$$\Pi_{z1} = \frac{K_1}{2\pi} \int_0^{2\pi} \int_0^{\frac{1}{2}\pi+j\infty} \exp[-j\beta_1(r\cos\phi \sin\Theta$$
$$+ z\cos\Theta)](1 + f_r) \sin\Theta \, d\Theta \, d\phi,$$
$$(z \geq 0) \quad (73)$$

$$\Pi_{z2} = \frac{K_2}{2\pi} \int_0^{2\pi} \int_0^{\frac{1}{2}\pi+j\infty} \exp[-j\beta_2(r\cos\phi \sin\Theta_t$$
$$- z\cos\Theta_t)] f_t \sin\Theta \, d\Theta \, d\phi, \quad (z \leq 0)$$
$$(74)$$

$$K_1 = K_{1e} = \frac{-j\beta_1 p_{z1}}{4\pi\xi_1}, \text{ for } \Pi_z = \Pi_{ez} \quad (75a)$$

$$K_1 = K_{1m} = \frac{-j\beta_1 m_{z1}}{4\pi\nu_1}, \text{ for } \Pi_z = \Pi_{mz} \quad (75b)$$

$$K_2 = K_{2e} = \frac{-j\beta_1 p_{z1}}{4\pi\xi_2}, \text{ for } \Pi_z = \Pi_{ez} \quad (76a)$$

$$K_2 = K_{2m} = K_{1m} \text{ for } \Pi_z = \Pi_{mz}. \quad (76b)$$

The double integrals in (70) to (72) may be changed to obtain infinite limits throughout. To do this, let ϕ be made complex by adding an imaginary part and then let the limits of integration be chosen so that the contributions to the integral from this imaginary part are identically zero. Thus, a typical integral in (70) to (72) is

$$I = \int_0^{2\pi} \int_0^{\frac{1}{2}\pi+j\infty} F(\Theta) \exp(-j\beta_1 r \cos\phi \sin\Theta)$$
$$\times \sin\Theta \, d\Theta \, d\phi. \quad (77)$$

This may be transformed into

$$I = \int_0^{\frac{1}{2}\pi+j\infty} F(\Theta) \sin\Theta \left[\left(\int_{\frac{1}{2}\pi-j\infty}^{\frac{1}{2}\pi+j\infty} \right. \right.$$
$$+ \left. \left. \int_{\frac{1}{2}\pi+j\infty}^{3/2\pi-j\infty} \right) \exp(-j\beta_1 r \cos\phi \sin\Theta) d\phi \right] d\Theta,$$
$$(78)$$

where ϕ is the generalized complex function. Use has been made of Cauchy's integral theorem, which states that the integral of a single-valued analytic function is independent of the path of integration. Contributions to the integral by the imaginary part of ϕ obviously are zero, since the initial and final limits are the same for this part.

According to Jahnke and Emde (ref. I.28, pp. 148, 149), the Hankel functions of order zero are given by

$$H_0^{(1)}(\beta_1 r \sin\Theta)$$
$$= \frac{1}{\pi} \int_{-\frac{1}{2}\pi-j\infty}^{\frac{1}{2}\pi+j\infty} \exp(-j\beta_1 r \cos\phi \sin\Theta) d\phi,$$
$$(79a)$$

$$H_0^{(2)}(\beta_1 r \sin\Theta)$$
$$= \frac{1}{\pi} \int_{\frac{1}{2}\pi+j\infty}^{3\pi-j\infty} \exp(-j\beta_1 r \cos\phi \sin\Theta) d\phi.$$
$$(79b)$$

With (79a, b), (78) becomes

$$I = \frac{1}{\pi} \int_0^{\frac{1}{2}\pi+j\infty} [H_0^{(1)}(\beta_1 r \sin\Theta)$$
$$+ H_0^{(2)}(\beta_1 r \sin\Theta)] F(\Theta) \sin\Theta \, d\Theta. \quad (80)$$

However, the Hankel functions satisfy the following relation:

$$H_0^{(1)}(-x) = -H_0^{(2)}(x). \quad (81)$$

Hence,

$$I = \frac{1}{\pi} \int_0^{\frac{1}{2}\pi+j\infty} H_0^{(1)}(\beta_1 r \sin\Theta) F(\Theta) \sin\Theta \, d\Theta$$
$$- \frac{1}{\pi} \int_0^{\frac{1}{2}\pi+j\infty} H_0^{(1)}(-\beta_1 r \sin\Theta) F(\Theta) \sin\Theta \, d\Theta.$$
$$(82)$$

In the second integral in (82) let

$$\Theta' = -\Theta, \quad \sin \Theta' = -\sin \Theta,$$
$$\cos \Theta' = \cos \Theta, \quad d\Theta' = -d\Theta. \quad (83)$$

Since $F(\Theta)$ is a function of Θ only through $\cos \Theta$, it follows that

$$F(\Theta') = F(\Theta). \quad (84)$$

With (83) and (84),

$$I = \frac{1}{\pi} \int_0^{\frac{1}{2}\pi + j\infty} H_0^{(1)}(\beta_1 r \sin \Theta) F(\Theta) \sin \Theta \, d\Theta$$
$$- \frac{1}{\pi} \int_0^{-\frac{1}{2}\pi - j\infty} H_0^{(1)}(\beta_1 r \sin \Theta') F(\Theta') \sin \Theta' \, d\Theta'. \quad (85)$$

Since the second integral has the same integrand as the first, it may be combined with it. Thus,

$$I = \frac{1}{\pi} \int_{-\frac{1}{2}\pi - j\infty}^{\frac{1}{2}\pi + j\infty} H_0^{(1)}(\beta_1 r \sin \Theta) F(\Theta) \sin \Theta \, d\Theta. \quad (86)$$

It follows that equivalent forms of the general integrals (70) to (72) are

$$\Pi_{z1} = \frac{K_1}{2\pi} \int_{-\frac{1}{2}\pi - j\infty}^{\frac{1}{2}\pi + j\infty} \int_{-\frac{1}{2}\pi - j\infty}^{\frac{1}{2}\pi + j\infty} G \sin \Theta \, d\Theta \, d\phi$$

$$G \equiv \exp\{-j\beta_1[r \cos\phi \sin \Theta + (z + d)\cos \Theta]\}[\exp(j\beta_1 d \cos \Theta) + f_r \exp(-j\beta_1 d \cos \Theta)], \quad (0 \leq z) \quad (87)$$

$$\Pi_{z1} = \frac{K_1}{2\pi} \int_{-\frac{1}{2}\pi - j\infty}^{\frac{1}{2}\pi + j\infty} \int_{-\frac{1}{2}\pi - j\infty}^{\frac{1}{2}\pi + j\infty} G' \sin \Theta \, d\Theta \, d\phi$$

$$G' \equiv \exp[-j\beta_1(r \cos\phi \sin \Theta + d \cos \Theta)]\{\exp[j\beta_1(z + d)\cos \Theta] + f_r \exp[-j\beta_1(z + d)\cos \Theta]\},$$
$$(-d \leq z \leq 0) \quad (88)$$

$$\Pi_{z2} = \frac{K_2}{2\pi} \int_{-\frac{1}{2}\pi - j\infty}^{\frac{1}{2}\pi + j\infty} \int_{-\frac{1}{2}\pi - j\infty}^{\frac{1}{2}\pi + j\infty} G'' \sin \Theta \, d\Theta \, d\phi$$

$$G'' \equiv f_t \exp\{-j\beta_2[r \cos\phi \sin \Theta_t - (z + d)\cos \Theta_t] - j\beta_1 d \cos \Theta\}.$$
$$(z \leq -d) \quad (89)$$

Similar changes in limits may be made in (73) and (74).

The above solutions for Π_{z1} and Π_{z2} satisfy the general wave equation and the boundary conditions. It remains to carry out the integration.

FAR-ZONE FIELDS OF VERTICAL DIPOLES OVER A CONDUCTING EARTH

6. Asymptotic Integration of the Hertzian Field of a Dipole; Field Patterns

In order to integrate (5.87) let the cylindrical coördinates r and z be replaced by spherical coördinates R_0 and Θ_{01} (Fig. 6.1) with their origin at $r = 0$, $z = -d$, that is, at the point on the boundary directly below the dipole at $r = 0$, $z = 0$. Specifically, let

$$z + d = R_0 \cos \Theta_{01}, \quad r = R_0 \sin \Theta_{01}. \quad (1)$$

Clearly,

$$R_0 = \sqrt{(z + d)^2 + r^2}. \quad (2)$$

With this notation the first exponent in (5.87) becomes

$$-j\beta_1[r \sin \Theta \cos \phi + (z + d)\cos \Theta]$$
$$= -j\beta_1 R_0 (\sin \Theta_{01} \sin \Theta \cos\phi + \cos \Theta_{01} \cos \Theta). \quad (3)$$

From trigonometry,

$$\cos \phi = 1 - 2\sin^2 \tfrac{1}{2}\phi, \quad (4a)$$

$$\sin \Theta \sin \Theta_{01} + \cos \Theta \cos \Theta_{01} = \cos(\Theta - \Theta_{01}). \quad (4b)$$

Hence,

$$r \sin \Theta \cos \phi + (z + d)\cos \Theta$$
$$= R_0[-2 \sin^2 \tfrac{1}{2}\phi \sin \Theta \sin \Theta_{01} + \cos(\Theta - \Theta_{01})] \quad (5a)$$
$$= R_0\{1 - 2[\sin^2 \tfrac{1}{2}\phi \sin \Theta \sin \Theta_{01} + \sin^2 \tfrac{1}{2}(\Theta - \Theta_{01})]\}. \quad (5b)$$

Accordingly, (5.70) becomes

$$\Pi_{z1} = \frac{K_1}{2\pi} \exp(-j\beta_1 R_0) \int_{-\frac{1}{2}\pi - j\infty}^{\frac{1}{2}\pi + j\infty} \int_{-\frac{1}{2}\pi - j\infty}^{\frac{1}{2}\pi + j\infty}$$
$$\exp\{j2\beta_1 R_0[\sin \Theta \sin \Theta_{01} \sin^2 \tfrac{1}{2}\phi + \sin^2 \tfrac{1}{2}(\Theta - \Theta_{01})]\}[\exp(j\beta_1 d \cos \Theta) + f_r \exp(-j\beta_1 d \cos \Theta)] \sin \Theta \, d\Theta \, d\phi. \quad (6)$$

This integral is in a form that permits application of the saddle-point method or method of steepest descents in order to obtain an asymptotic solution valid for large values of R_0 and, hence, yielding a far-zone formula for Π_{z1}.

The saddle-point method of integration applies particularly to the evaluation of complex transcendental integrals of the form*

$$I = \int_A^B F(w) e^{R_0 f(w)} \, dw \quad (7)$$

over an infinite path of integration. It is assumed that R_0 is large, real, and positive and that $f(w)$ is analytic. Let the real and imaginary parts of $f(w)$ be separated as follows:

$$f(w) = g + jh. \quad (8)$$

Clearly the integrand in (7) will be large where g is large and positive. The fundamental principle of the method is to move the path of integration in the complex plane until that particular path is found by way of which the large values of g are concentrated in the shortest possible interval. This means that the path is chosen along which e^g reaches as small a maximum as possible and drops off on both sides over the steepest slopes. Evidently this is equivalent to finding and integrating along the path on which g itself experiences the most rapid ascent and descent. The type of maximum involved is not an absolute maximum—which g can never have—but a stationary point or saddle point defined by

$$\frac{df(w)}{dw} = 0. \quad (9)$$

Let dn be in the direction of the most rapid change in g at any point in the complex plane, and hence in the direction of integration. Let ds be normal to dn. Then, since dn is in the direction of the gradient, ds must be in a direction in which g experiences no change, so that

$$\frac{\partial g}{\partial s} = 0. \quad (10)$$

Since it has been assumed that $f(w)$ is analytic, it must satisfy the Cauchy-Riemann differential equations. Thus, if

$$w = u + jv, \quad (11)$$

these equations are

$$\frac{\partial g}{\partial u} = \frac{\partial h}{\partial v}, \quad \frac{\partial h}{\partial u} = -\frac{\partial g}{\partial v}. \quad (12)$$

---

* Courant and Hilbert, ref. I.16, vol. I, pp. 455–460; Frank and von Mises, ref. I.18, vol. II, p. 834; Jeffreys and Jeffreys, ref. I.29, pp. 472–474; Ott, ref. 47.

If the direction of du is chosen along ds, and that of dv along dn, then (10) and (12) give

$$\frac{\partial h}{\partial n} = \frac{\partial g}{\partial s} = 0. \quad (13)$$

Accordingly, the path of integration in the complex plane along which $f(w)$ has its steepest slope is defined by

$$h = h_0 = \text{const.}, \quad (14)$$

where h_0 is the value at the saddle point defined by (9).

In the specific integral under consideration, f is a function of two variables, ϕ and Θ. Thus

$$f(w_1, w_2) = f(\Theta, \phi)$$
$$= j2\beta[\sin\Theta \sin\Theta_{01} \sin^2\tfrac{1}{2}\phi + \sin^2\tfrac{1}{2}(\Theta - \Theta_{01})]. \quad (15)$$

Using (9) successively with ϕ and Θ as variables to locate the saddle point for each integration, the following equations are obtained:

$$0 = \frac{\partial f(\Theta, \phi)}{\partial \phi} = j\beta \sin\Theta \sin\Theta_{01} \sin\phi, \quad (16a)$$

$$\Theta = 0 \text{ or } \phi = 0, \quad (16b)$$

$$0 = \frac{\partial f(\Theta, \phi)}{\partial \Theta}$$
$$= j\beta[2\cos\Theta \sin\Theta_{01} \sin^2\tfrac{1}{2}\phi + \sin(\Theta - \Theta_{01})]$$
$$= j\beta[\sin\Theta \cos\Theta_{01} - \cos\Theta \sin\Theta_{01} \cos\phi]. \quad (17a)$$

This yields

$$\tan\Theta = \tan\Theta_{01} \cos\phi,$$

or

$$\Theta = \tan^{-1}(\tan\Theta_{01} \cos\phi). \quad (17b)$$

If the saddle-point value of $\phi = 0$ is chosen, the saddle-point value of Θ is

$$\Theta = \Theta_{01}. \quad (17c)$$

If R_0 is sufficiently large and if the respective paths of integration for ϕ and Θ are chosen through these saddle points along paths for which the imaginary part of the integrand is constant and equal to its value at $\phi = 0$, $\Theta = \Theta_{01}$, the exponential

$$\exp\{j2\beta_1 R_0[\sin\Theta \sin\Theta_{01} \sin^2\tfrac{1}{2}\phi + \sin^2\tfrac{1}{2}(\Theta - \Theta_{01})]\}$$

will rise and fall rapidly near the saddle points, so that the principal contribution to the integral is with values of ϕ and Θ near those at the saddle points. At these saddle points this exponential is equal to unity. Obviously, since R_0 is assumed large, the complex exponential will drop to small values as ϕ and Θ are varied from their saddle points. On the other hand, near these saddle-point values the other exponential terms, $\exp(\pm j\beta_1 d \cos \Theta)$, that do not have the large factor $R_0 \gg d$ in the exponents are slowly varying in Θ, as are the amplitude factors f_r and $\sin \Theta$. The approximation to be made, therefore, is to insert in these slowly varying factors the saddle-point values of ϕ and Θ and so make them constants in the integration of the rapidly varying exponential term.

In order to carry out the integration, let the variables be changed as follows:

$$u = \sqrt{2j\beta_1 R_0 \sin \Theta \sin \Theta_{01}} \sin \tfrac{1}{2}\phi, \quad (18)$$

$$v = \sqrt{2j\beta_1 R_0} \sin \tfrac{1}{2}(\Theta - \Theta_{01}), \quad (19)$$

so that

$$u^2 + v^2 = 2j\beta_1 R_0[\sin \Theta \sin \Theta_{01} \sin^2 \tfrac{1}{2}\phi + \sin^2 \tfrac{1}{2}(\Theta - \Theta_{01})]. \quad (20)$$

Note that $u = 0$ when $\phi = 0$, $v = 0$ when $\Theta = \Theta_{01}$. Evidently,

$$\sin \tfrac{1}{2}\phi = u/\sqrt{2j\beta_1 R_0 \sin \Theta \sin \Theta_{01}} \quad (21)$$

Differentiation gives

$$\tfrac{1}{2} \cos \tfrac{1}{2}\phi \, d\phi = du/\sqrt{2j\beta_1 R_0 \sin \Theta \sin \Theta_{01}}, \quad (22)$$

so that

$$d\phi = 2du/\cos \tfrac{1}{2}\phi \sqrt{2j\beta_1 R_0 \sin \Theta \sin \Theta_{01}}. \quad (23)$$

Similarly,

$$\sin \tfrac{1}{2}(\Theta - \Theta_{01}) = v/\sqrt{2j\beta_1 R_0}, \quad (24)$$

$$\tfrac{1}{2} \cos \tfrac{1}{2}(\Theta - \Theta_{01}) \, d\Theta = dv/\sqrt{2j\beta_1 R_0}, \quad (25)$$

$$d\Theta = 2dv/\cos \tfrac{1}{2}(\Theta - \Theta_{01})\sqrt{2j\beta_1 R_0}. \quad (26)$$

In order to investigate the limits and the paths of integration in the complex ϕ and Θ planes, it is convenient to examine the exponential function in its original form, namely,

$$\exp\{R[-j\beta_1 + f(\Theta, \phi)]\}$$
$$= \exp\{-j\beta_1[r \sin \Theta \cos \phi + (z+d) \cos \Theta]\}, \quad (27a)$$

where

$$f(\Theta, \phi) = -j\beta_1(\sin \Theta_{01} \sin \Theta \cos \phi + \cos \Theta_{01} \cos \Theta) + j\beta_1. \quad (27b)$$

At the saddle points,

$$f(\Theta_{01}, 0) = 0. \quad (28)$$

The paths of steepest descent are characterized by a constant value of the imaginary part h of $f(\Theta, \phi)$. That is, the paths of integration must be such that

$$h = 0. \quad (29)$$

In the vicinity of the saddle points ϕ and $\Theta - \Theta_{01}$ are small, so that

$$f(\Theta, \phi) \doteq j\beta_1[\tfrac{1}{2}(\Theta - \Theta_{01})^2 - \tfrac{1}{2}\phi^2 \sin^2 \Theta_{01}]. \quad (30)$$

In order to determine the angles made by the paths of integration in the complex planes, let

$$\Theta - \Theta_{01} = ae^{j\delta}, \quad \phi \sin \Theta_{01} = be^{j\varepsilon}. \quad (31)$$

With (31) and (30), the following result is obtained:

$$f(\Theta, \phi) \doteq (\alpha_{1s} + j\beta_{1s})[\tfrac{1}{2}a^2(\cos 2\delta + j \sin 2\delta) - \tfrac{1}{2}b^2(\cos 2\varepsilon + j \sin 2\varepsilon)]. \quad (32)$$

Selecting the imaginary part and equating it to zero gives

$$h = \beta_{1s}(\tfrac{1}{2}a^2 \cos 2\delta - \tfrac{1}{2}b^2 \cos 2\varepsilon)$$
$$+ \alpha_{1s}(\tfrac{1}{2}a^2 \sin 2\delta + \tfrac{1}{2}b^2 \sin 2\varepsilon) = 0, \quad (33a)$$

or

$$a^2(\beta_{1s} \cos 2\delta + \alpha_{1s} \sin 2\delta)$$
$$+ b^2(\beta_{1s} \cos 2\varepsilon + \alpha_{1s} \sin 2\varepsilon) = 0. \quad (33b)$$

Since Θ and ϕ are independent, this equation must be satisfied independently by each of the bracketed expressions. That is,

$$\tan 2\delta = \tan 2\varepsilon = -\beta_{1s}/\alpha_{1s}. \quad (33c)$$

It follows that

$$\delta = \varepsilon = -\tfrac{1}{2}\tan^{-1}(\beta_{1s}/\alpha_{1s}). \quad (33d)$$

These are the angles made by the paths of integration with the positive real axis in the complex Θ and ϕ planes as they cross the real axis at the saddle points. If region 1 is dissipationless, the angles are $\delta = \varepsilon = -\pi/4$.

Near the saddle points the variables u and v defined in (18) and (19) are as follows:

$$u \doteq \sqrt{2j(\beta_{1s} - j\alpha_{1s})R_0} \sin \Theta_{01} \cdot \tfrac{1}{2}be^{j\varepsilon}, \quad (34a)$$

$$v \doteq \sqrt{2j(\beta_{1s} - j\alpha_{1s})R_0} \cdot \tfrac{1}{2}ae^{j\delta}. \quad (34b)$$

With (33d) these reduce to

$$u \doteq u = \frac{b}{2}\sqrt{2R_0(\beta_{1s}^2 + \alpha_{1s}^2)}\sin\Theta_{01}, \quad (34c)$$

$$v \doteq v = \frac{a}{2}\sqrt{2R_0(\beta_{1s}^2 + \alpha_{1s}^2)}. \quad (34d)$$

That is, both u and v are approximately real near the saddle points along the path of integration. With (18) to (34), (6) becomes

$$\Pi_{z1} = \frac{jK_1}{\beta_1\pi}\frac{\exp(-j\beta_1 R_0)}{R_0}\int_{-\infty}^{\infty}\int_{-\infty}^{\infty} e^{-(u^2+v^2)}$$

$$\times [\exp(j\beta_1 d\cos\Theta) + f_r \exp(-j\beta_1 d\cos\Theta)]$$

$$\times \left[\frac{\sqrt{\sin\Theta}}{\cos\frac{1}{2}\phi\cos\frac{1}{2}(\Theta-\Theta_{01})\sqrt{\sin\Theta_{01}}}\right] du\, dv, \quad (35)$$

where f_r is given by (5.62) as a function of Θ alone. The path of integration is along the real axis between the infinite limits $-\infty$ and $+\infty$. Note that the real parts of the exponents vary rapidly.

The principle of the saddle-point method of integration now may be applied to (35) in order to obtain the desired asymptotic solution that is valid for large values of R_0. In (35) R_0 is involved only in the exponential factor $e^{-(u^2+v^2)}$. All other exponentials are slowly varying, as are all of the amplitude factors. The procedure, as outlined above, is to insert the saddle-point values $\Theta = \Theta_{01}$, $\phi = 0$ in the slowly varying functions, which thus become constants in the integration, and carry out the integration for the remaining rapidly varying exponential functions. That is, the integration that must be performed is

$$\int_{-\infty}^{\infty}\int_{-\infty}^{\infty} e^{-(u^2+v^2)}\,du\,dv \quad (36a)$$

with the path along the real axis. Using well-known formulas* the result is

$$\int_{-\infty}^{\infty}\int_{-\infty}^{\infty} e^{-(u^2+v^2)}\,du\,dv = \left(2\int_0^{\infty} e^{-u^2}\,du\right)^2$$

$$= \left(2\cdot\frac{\sqrt{\pi}}{2}\right)^2 = \pi. \quad (36b)$$

On substituting (36b) in (35) after setting $\phi = 0$, $\Theta = \Theta_{01}$, the following formula results:

---
* See Peirce, ref. I.39, formula 492.

$$\Pi_{z1} = \frac{jK_1}{\beta_1}\frac{\exp(-j\beta_1 R_0)}{R_0}[\exp(j\beta_1 d\cos\Theta_{01}) + f_r^r\exp(-j\beta_1 d\cos\Theta_{01})], \quad (37)$$

where

$$R_0 = \sqrt{(z+d)^2 + x^2 + y^2} = \sqrt{(z+d)^2 + r^2} \quad (38)$$

and where for $\Pi_{z1} = \Pi_{ez1}$, $f_r^r = f_{er}(\Theta_{01})$,

$$\frac{jK_1}{\beta_1} = \frac{P_{z1}}{4\pi\xi_1},$$

$$f_r^r = \frac{\nu_2 N_{21}^2\cos\Theta_{01} - \nu_1\sqrt{N_{21}^2 - \sin^2\Theta_{01}}}{\nu_2 N_{21}^2\cos\Theta_{01} + \nu_1\sqrt{N_{21}^2 - \sin^2\Theta_{01}}} \quad (39a)$$

and for $\Pi_{z1} = \Pi_{mz1}$, $f_r^r = f_{mr}(\Theta_{01})$,

$$\frac{jK_1}{\beta_1} = \frac{m_{z1}}{4\pi\nu_1},$$

$$f_r^r = \frac{\nu_1\cos\Theta_{01} - \nu_2\sqrt{N_{21}^2 - \sin^2\Theta_{01}}}{\nu_1\cos\Theta_{01} + \nu_2\sqrt{N_{21}^2 - \sin^2\Theta_{01}}}. \quad (39b)$$

The superscript r on f_r^r indicates the radiation-zone or far-zone value of the reflection coefficient in which the angle (Fig. 6.1)

$$\Theta_{01} = \tan^{-1}\left(\frac{z_1 + d}{r_1}\right) \quad (40)$$

replaces Θ in the generalized plane-wave formulas of Sec. 5. The particular forms (39a, b) of the reflection coefficient are chosen merely because they are expressed entirely in terms of one angle. Note that (39a, b) are the same as (5.63a, b) with the real angle of incidence Θ_{01} replacing the generalized complex angle Θ. This means that the far-zone reflection coefficient f_r^r is the same as the plane-wave coefficient that is given by (5.63a, b) with Θ real.

The asymptotic formula (37) was derived from (5.87), which is valid only for $0 \leq z$ in region 1. It is readily verified that the same formula is obtained from (5.88) for $-d \leq z \leq 0$ in region 1. Thus, (5.88) may be expressed as follows:

$$\Pi_{z1} = \frac{K_1}{2\pi}\int_{-\frac{1}{2}\pi-j\infty}^{\frac{1}{2}\pi+j\infty}\int_{-\frac{1}{2}\pi-j\infty}^{\frac{1}{2}\pi+j\infty} G'\sin\Theta\,d\Theta\,d\phi$$

$$G' \equiv \exp[-j\beta_1(r\cos\phi\sin\Theta$$
$$- z\cos\Theta - d\cos\Theta)]\{[\exp(-j\beta_1 d\cos\Theta)$$
$$+ f_r\exp[-j\beta_1(r\cos\phi\sin\Theta + z\cos\Theta$$
$$+ d\cos\Theta)]\}\exp(-j\beta_1 d\cos\Theta) \quad (41)$$

The exponential following f_r in (41) is the same as that already integrated in (5.87). The first exponential differs only in having negative signs in front of $z \cos \Theta$ and $d \cos \Theta$. From

$$(z + d) = -R_0 \cos(\pi - \Theta_{01}),$$
$$r = R_0 \sin(\pi - \Theta_{01}), \quad (42)$$

as shown in Fig. 6.2, it follows that

$$r \cos \phi \sin \Theta - (z + d) \cos \Theta$$
$$= R_0[\sin(\pi - \Theta_{01}) \sin \Theta \cos \phi$$
$$+ \cos(\pi - \Theta_{01}) \cos \Theta]. \quad (43)$$

This is like (3) with Θ_{01} replaced by $(\pi - \Theta_{01})$.

It follows that the entire previous analysis applies to the first exponential if $(\pi - \Theta_{01})$ is substituted for Θ_{01}. Accordingly, the expression corresponding to (37) is

$$\Pi_{z1} = \frac{jK_1}{\beta_1} \frac{\exp(-j\beta_1 R_0)}{R_0}$$
$$\times \{\exp[-j\beta_1 d \cos(\pi - \Theta_{01})]$$
$$+ f_r^r \exp(-j\beta_1 d \cos \Theta_{01})\}, \quad (44)$$

where f_r^r is as in (39). Since $\cos(\pi - \Theta_{01}) = -\cos \Theta_{01}$, (44) is exactly the same as (37). Therefore, (37) is the general asymptotic formula for Π_{z1} for all values of z in region 1, that is, for $z \geq -d$.

If the dipole is directly on the boundary, the following simpler expressions are obtained from (37) by setting $d = 0$:

$$\Pi_{ez1} = \frac{p_{z1}}{4\pi \xi_1} \frac{\exp(-j\beta_1 R_0)}{R_0} (1 + f_{er}^r),$$
$$\quad (45a)$$

$$\Pi_{mz1} = \frac{m_{z1}}{4\pi \nu_1} \frac{\exp(-j\beta_1 R_0)}{R_0} (1 + f_{mr}^r),$$
$$\quad (45b)$$

where

$$1 + f_{er}^r = f_{et}^r$$
$$= \frac{2\nu_2 N_{21}^2 \cos \Theta_{01}}{\nu_2 N_{21}^2 \cos \Theta_{01} + \nu_1 \sqrt{N_{21}^2 - \sin^2 \Theta_{01}}}, \quad (46a)$$

$$1 + f_{mr}^r = f_{mt}^r$$
$$= \frac{2\nu_1 \cos \Theta_{01}}{\nu_1 \cos \Theta_{01} + \nu_2 \sqrt{N_{21}^2 - \sin^2 \Theta_{01}}}. \quad (46b)$$

By symmetry, the expressions for the potentials in region 2 ($z \leq -d$) are as follows:

$$\Pi_{ez2} = \frac{p_{z2}}{4\pi \xi_2} \frac{\exp(-j\beta_2 R_0)}{R_0}$$
$$\times \frac{2\nu_1 N_{12}^2 \cos \Theta_{02}}{\nu_1 N_{12}^2 \cos \Theta_{02} + \nu_2 \sqrt{N_{12}^2 - \sin^2 \Theta_{02}}},$$
$$\quad (47a)$$

$$\Pi_{mz2} = \frac{m_{z2}}{4\pi \nu_2} \frac{\exp(-j\beta_2 R_0)}{R_0}$$
$$\times \frac{2\nu_2 \cos \Theta_{02}}{\nu_2 \cos \Theta_{02} + \nu_1 \sqrt{N_{12}^2 - \sin^2 \Theta_{02}}}.$$
$$\quad (47b)$$

The angle Θ_{02} (Fig. 6.3) is measured from the negative z-axis, so that

$$\Theta_{02} = \tan^{-1} \frac{-(z+d)}{r_2},$$
$$R_0 = \sqrt{(z+d)^2 + r_2^2}. \quad (48)$$

Note that p_{z1} and m_{z1} are dipoles at the boundary just inside region 1, whereas the dipoles p_{z2} and m_{z2} are at the boundary just inside region 2. The effectiveness of a given dipole in maintaining a potential Π_z depends upon the medium in which it is placed owing to the induced volume density of polarization or magnetization. The normal component of this induced density at the boundary of the dipole cancels a part of the impressed polarization or magnetization of the dipole. This is clear from the fact that the forcing terms in the nonhomogeneous wave equations for the potentials (1.1) are, respectively, $-P^e/\xi$ and $-M^e/\nu$. It follows that the quantities p_z/ξ and m_z/ν in the general expressions (2.6) and (3.2) for Π_{ez} and Π_{mz} are the effective exciting amplitudes of the *sources*. Hence, in order to express Π_{z2} in (47) in terms of p_{z1} and m_{z1} at $d = 0$ in region 1 instead of in terms of p_{z2} and m_{z2} at $d = 0$ in region 2, it is necessary to evaluate p_{z2} and m_{z2} from

$$p_{z1}/\xi_1 = p_{z2}/\xi_2, \quad (49a)$$
$$m_{z1}/\nu_1 = m_{z2}/\nu_2. \quad (49b)$$

That is,

$$p_{z2} = p_{z1} \frac{\xi_2}{\xi_1} = p_{z1} \frac{\beta_2^2 \nu_1}{\beta_1^2 \nu_2} = p_{z1} \cdot \frac{\nu_2}{\nu_1 N_{12}^2}.$$
$$\quad (49c)$$

Since Π_{mz2} is expressed in terms of ν_1 instead of ν_2 as explained in conjunction with (5.7d), it is not necessary to obtain an expression for m_{z2} similar to (49c) for p_{z2}. With (49b) and (49c), (47a, b) may be expressed as follows:

$$\Pi_{ez2} = \frac{p_{z1}}{4\pi\xi_2} \frac{\exp(-j\beta_2 R_0)}{R_0}$$

$$\times \frac{2\nu_2 \cos \Theta_{02}}{\nu_1 N_{12}^2 \cos \Theta_{02} + \nu_2 \sqrt{N_{12}^2 - \sin^2 \Theta_{02}}}, \quad (50a)$$

$$\Pi_{mz2} = \frac{m_{z1}}{4\pi\nu_1} \frac{\exp(-j\beta_2 R_0)}{R_0}$$

$$\times \frac{2\nu_2 \cos \Theta_{02}}{\nu_2 \cos \Theta_{02} + \nu_1 \sqrt{N_{12}^2 - \sin^2 \Theta_{02}}}. \quad (50b)$$

Alternatively, using

$$K_{e2} = \frac{-j\beta_1 p_{z1}}{4\pi\xi_2}, \quad K_{m2} = K_{m1} = \frac{-j\beta_1 m_{z1}}{4\pi\nu_1}, \quad (50c)$$

and defining

$$f_{et}^r(\Theta_{02}) \equiv \frac{2\nu_1 N_{12}^2 \cos \Theta_{02}}{\nu_1 N_{12}^2 \cos \Theta_{02} + \nu_2 \sqrt{N_{12}^2 - \sin^2 \Theta_{02}}}, \quad (51a)$$

$$f_{mt}^r(\Theta_{02}) \equiv \frac{2\nu_1 \cos \Theta_{02}}{\nu_2 \cos \Theta_{02} + \nu_1 \sqrt{N_{12}^2 - \sin^2 \Theta_{02}}}, \quad (51b)$$

it follows that

$$\Pi_{ez2} = \frac{jK_{e2}}{\beta_1} \frac{\exp(-j\beta_2 R_0)}{R_0} \left(\frac{\nu_2}{\nu_1 N_{12}^2}\right) f_{et}^r(\Theta_{02}), \quad (51c)$$

$$\Pi_{mz2} = \frac{jK_{m2}}{\beta_1} \frac{\exp(-j\beta_2 R_0)}{R_0} f_{mt}^r(\Theta_{02}). \quad (51d)$$

Note that $f_{et}^r(\Theta_{02})$ and $f_{mt}^r(\Theta_{02})$ are defined in (51a, b) to have the same form as $f_{et}^r = f_{et}^r(\Theta_{01})$ and $f_{mt}^r = f_{mt}^r(\Theta_{01})$ in (46a, b) with subscripts 1 and 2 interchanged. Significantly, the factor $\nu_2/\nu_1 N_{12}^2$ occurs in Π_{ez2} with no corresponding factor in Π_{mz2}.

The Hertzian potentials in region 2 in the general case when the dipole is not necessarily on the boundary must be obtained by integrating (5.89). An approximate asymptotic integration of (5.89) may be obtained by the same method used for Π_{z1} under two quite different sets of circumstances, depending upon whether the principal distance is in region 1 or in region 2. Consider the latter case first. Let the point where the potential is calculated be very far from the boundary compared with the height d of the dipole above it. That is, let

$$R_0 = \sqrt{(z+d)^2 + r^2}, \quad R_0 \gg d,$$
$$(z+d)^2 > r^2, \quad (52)$$

where R_0 is measured into region 2 from the point on the boundary directly below the dipole. Rearrangement of (5.89) to separate the rapidly varying exponential that involves R_0 from the more slowly varying exponentials and amplitude factors gives

$$\Pi_{z2} = \frac{K_2}{2\pi} \int_{-\frac{1}{2}\pi - j\infty}^{\frac{1}{2}\pi + j\infty} \int_{-\frac{1}{2}\pi - j\infty}^{\frac{1}{2}\pi + j\infty} G'' \sin \Theta \, d\Theta \, d\phi$$

$$G'' = f_t \exp\{-j\beta_2[r \cos \phi \sin \Theta_t - (z+d) \cos \Theta_t]\} \exp(-j\beta_1 d \cos \Theta). \quad (53)$$

Note that with subscript e or m on Π_{z2} the constants K_2 and f_t have the following forms:

$$K_{e2} = \frac{-j\beta_1 p_{z1}}{4\pi\xi_2}, \quad K_{m2} = \frac{-j\beta_1 m_{z1}}{4\pi\nu_1}, \quad (54)$$

$$f_{et} = \frac{2\nu_2 \cos \Theta}{\nu_2 \cos \Theta + \nu_1 N_{12} \cos \Theta_t}, \quad (55a)$$

$$f_{mt} = \frac{2\nu_1 N_{12} \cos \Theta}{\nu_1 N_{12} \cos \Theta + \nu_2 \cos \Theta_t}, \quad (55b)$$

where

$$N_{12} \equiv \beta_1/\beta_2. \quad (56)$$

In order to evaluate (53) let the variable of integration be changed from Θ to Θ_t using the law of refraction and its differential, namely,

$$\beta_1 \sin \Theta = \beta_2 \sin \Theta_t, \quad (57a)$$

$$\beta_1 \cos \Theta \, d\Theta = \beta_2 \cos \Theta_t \, d\Theta_t. \quad (57b)$$

Clearly,

$$\sin \Theta \, d\Theta = \frac{1}{N_{12}^2} \cdot \left(\frac{\cos \Theta_t}{\cos \Theta}\right) \sin \Theta_t \, d\Theta_t. \quad (58)$$

The limits are unchanged if N_{12} is real. If N_{12} is complex, let

$$N_{12} = N_{12} \cdot je^{-j\delta} = N_{12}(\sin \delta + j \cos \delta); \quad (59)$$

then the limits for Θ_t corresponding to $\pm(\frac{1}{2}\pi + j\infty)$ for Θ are $\pm(\delta + j\infty)$. This

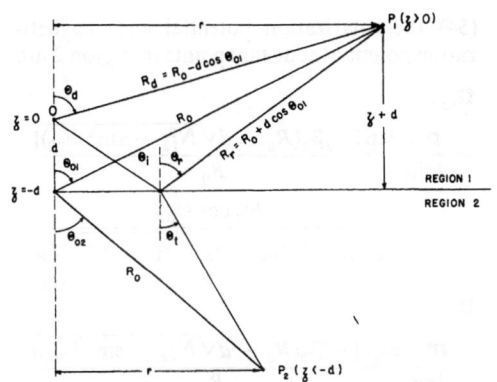

Fig. 6.1. Points P_1 and P_2 in far zone of source at O.

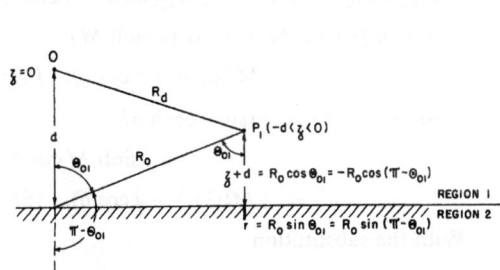

Fig. 6.2. Point P_1 near boundary in far zone of source at O.

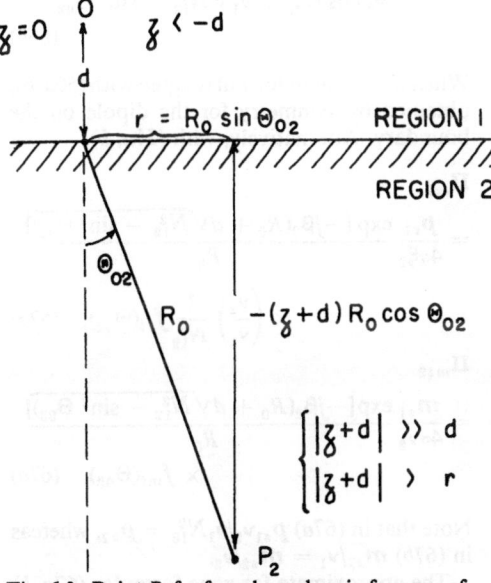

Fig. 6.3. Point P_2 far from boundary in far zone of source at O.

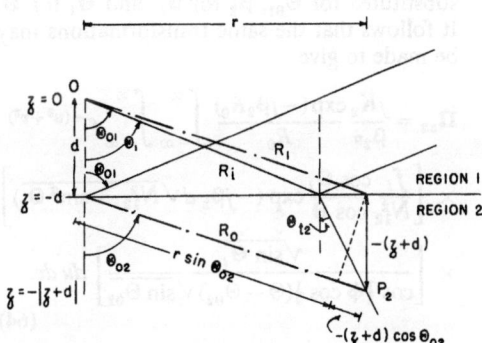

Fig. 6.4. Point P_2 near boundary in far zone of source at O.

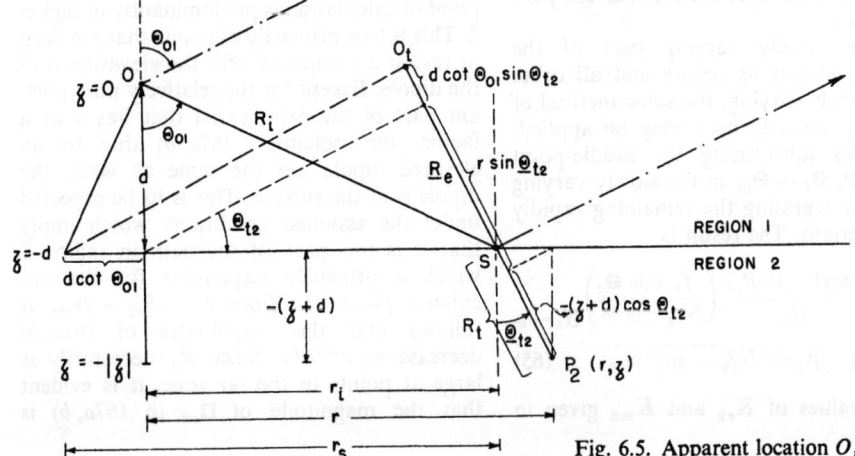

Fig. 6.5. Apparent location O_t of source at O.

may be shown as follows. Let M be real and very large. Then

$$\pm N_{12} \sin (\tfrac{1}{2}\pi + jM) = \pm N_{12}(\sin \delta + j \cos \delta)$$
$$\times (\sin \tfrac{1}{2}\pi \cosh M + j \cos \tfrac{1}{2}\pi \sinh M)$$
$$= \pm M (\sin \delta + j \cos \delta). \quad (60)$$

$$\pm \sin (\delta + jM) = \pm (\sin \delta \cosh M + j \sinh M \cos \delta)$$
$$= \pm M(\sin \delta + j \cos \delta). \quad (61)$$

With the substitution

$$-(z + d) = R_0 \cos \Theta_{02}, \quad (62a)$$
$$r = R_0 \sin \Theta_{02}, \quad (62b)$$

where Θ_{02} is measured from the negative z-axis as shown in Fig. 6.3, the first exponential in (53) has the exponent

$$-j\beta_2 R_0(\sin \Theta_{02} \sin \Theta_t \cos \phi + \cos \Theta_{02} \cos \Theta_t). \quad (63)$$

This is the same in form as (3) but with Θ_{02} substituted for Θ_{01}, β_2 for β_1, and Θ_t for Θ. It follows that the same transformations may be made to give

$$\Pi_{z2} = \frac{jK_2}{\beta_2 \pi} \frac{\exp(-j\beta_2 R_0)}{R_0} \int_{-\infty}^{\infty} \int_{-\infty}^{\infty} e^{-(u^2+v^2)}$$
$$\times \left[\frac{f_t}{N_{12}^2} \frac{\cos \Theta_t}{\cos \Theta} \exp(-j\beta_2 d\sqrt{N_{12}^2 - \sin^2 \Theta_t})\right]$$
$$\times \left[\frac{\sqrt{\sin \Theta_t}}{\cos \tfrac{1}{2}\phi \cos \tfrac{1}{2}(\Theta - \Theta_{02})\sqrt{\sin \Theta_{02}}}\right] du\, dv. \quad (64)$$

The appearance of δ in place of $\pi/2$ in the limits for Θ_t does not affect the limits of u and v, since the real axis is chosen as the path of integration.

Since the rapidly varying part of the integrand in (64) is as before and all other terms are slowly varying, the same method of integration previously used may be applied. This involves substituting the saddle-point values $\phi = 0$, $\Theta_t = \Theta_{02}$ in the slowly varying factors and integrating the remaining rapidly varying functions. The result is

$$\Pi_{z2} = \frac{jK_2}{\beta_2} \frac{\exp(-j\beta_2 R_0)}{R_0} \left(\frac{f_t}{N_{12}^2} \frac{\cos \Theta_t}{\cos \Theta}\right)_{\Theta_t=\Theta_{02}}$$
$$\times \exp(-j\beta_2 d\sqrt{N_{12}^2 - \sin^2 \Theta_{02}}) \quad (65)$$

Using the values of K_{e2} and K_{m2} given in (54), the polarization potential and magnetization potential at distant points in region 2 are

$$\Pi_{ez2}$$
$$= \frac{p_{z1}}{4\pi \xi_2} \frac{\exp[-j\beta_2(R_0 + d\sqrt{N_{12}^2 - \sin^2 \Theta_{02}})]}{R_0}$$
$$\times \frac{2\nu_2 \cos \Theta_{02}}{\nu_1 N_{12}^2 \cos \Theta_{02} + \nu_2 \sqrt{N_{12}^2 - \sin^2 \Theta_{02}}}, \quad (66a)$$

$$\Pi_{mz2}$$
$$= \frac{m_{z1}}{4\pi \nu_1} \frac{\exp[-j\beta_2(R_0 + d\sqrt{N_{12}^2 - \sin^2 \Theta_{02}})]}{R_0}$$
$$\times \frac{2\nu_1 \cos \Theta_{02}}{\nu_2 \cos \Theta_{02} + \nu_1 \sqrt{N_{12}^2 - \sin^2 \Theta_{02}}}. \quad (66b)$$

When $d = 0$ these formulas agree with $(50a, b)$, obtained by symmetry for the dipole on the boundary. Alternatively, with $(51a, b)$,

$$\Pi_{ez2}$$
$$= \frac{p_{z1}}{4\pi \xi_2} \frac{\exp[-j\beta_2(R_0 + d\sqrt{N_{12}^2 - \sin^2 \Theta_{02}})]}{R_0}$$
$$\times \left(\frac{\nu_2}{\nu_1}\right) \frac{1}{N_{12}^2} f_{et}(\Theta_{02}), \quad (67a)$$

$$\Pi_{mz2}$$
$$= \frac{m_{z1}}{4\pi \nu_2} \frac{\exp[-j\beta_2(R_0 + d\sqrt{N_{12}^2 - \sin^2 \Theta_{02}})]}{R_0}$$
$$\times f_{mt}(\Theta_{02}). \quad (67b)$$

Note that in (67a) $p_{z1}\nu_2/\nu_1 N_{12}^2 = p_{z2}$, whereas in (67b) $m_{z1}/\nu_1 = m_{z2}/\nu_2$.

The approximate far-zone formulas $(67a, b)$ for the Hertzian potentials are for use when the line joining the transmitting dipole with the point of calculation is predominantly in region 2. This is true primarily of points that are deep in region 2 compared with the elevation d of the dipole. Except for the relatively unimportant part of the exponential that has d as a factor, the potentials $(67a, b)$ due to an elevated dipole are the same as when the dipole is at the surface. This is to be expected under the assumed conditions which imply that it is the part of the path in region 2 which is primarily responsible for the calculated potential. Since $\beta_2 = \beta_{s2} - j\alpha_{s2}$, it follows that the magnitudes of $(67a, b)$ decrease as $e^{-\alpha_{s2} R_0}$. Since R_0 necessarily is large at points in the far zone, it is evident that the magnitude of Π_{z2} in $(67a, b)$ is insignificant unless α_{s2} is quite small. Therefore, for practical purposes, the far-zone field deep in region 2 differs appreciably from zero only when region 2 is at least a moderately good dielectric. If σ_{e2} is not small, $\alpha_{s2} \doteq \sqrt{\omega \sigma_{e2}/2\nu_2} = 1/d_s$ (d_s is the skin depth) is sufficiently large to make $e^{-\alpha_{s2}R_0}$ vanishingly small even for R_0 much less than required to reach the far zone. For a perfect conductor, α_{s2} is infinite and the field in region 2 is zero. For salt water, with $\sigma_{e2} = 4$ mhos/m, $\alpha_{s2} \doteq 1.59 \times 10^{-3}\sqrt{\omega}$ nepers/m. With $\omega = 10^6$, $\alpha_{s2} = 1.59$ nepers/m; with $\omega = 10^{10}$, $\alpha_{s2} = 1.59 \times 10^2$ nepers/m. Clearly, with so large an attenuation constant, the factor $e^{-\alpha_{s2}R_0}$ is minute even for quite small values of R_0. In moderately dry earth, $\sigma = 4 \times 10^{-4}$ mhos/m, $\alpha_{s2} = 1.59 \times 10^{-5}\sqrt{\omega}$ nepers/m. With $\omega = 10^6$, $\alpha_{s2} = 1.59 \times 10^{-2}$ nepers/m; with $\omega = 10^{10}$, $\alpha_{s2} = 1.59$ nepers/m. For very dry earth and rock, $\sigma = 10^{-5}$ mho/m, $\alpha_{s2} = 2.9 \times 10^{-6}\sqrt{\omega}$, so that with $\omega = 10^6$, $\alpha_{s2} = 2.9 \times 10^{-3}$ nepers/m. Thus, if the frequency is not too high, it is possible to transmit a signal with considerable amplitude to an appreciable depth into moderately dry earth. This indicates the possibility of geophysical exploration into the earth using fairly long electromagnetic waves.

When region 1 is air and region 2 the conducting earth, interest is primarily in the field above the earth and at short distances into the earth. For such purposes an alternative far-zone formula for Π_{z2} may be derived by assuming the distance between the dipole and the point of observation in region 2 to be *primarily in region 1*. Specifically, let

$$R_i = \sqrt{d^2 + r_i^2}, \quad |\beta_1 R_i| \gg |\beta_2(z+d)| \tag{68a}$$

where R_i is the distance from the dipole at 0 to the boundary surface at S along the wave normal of a *plane wave*, as shown in Fig. 6.4. Let the point of observation be $P_2(r, z)$. Then

$$r = r_i - (z+d)\tan \Theta_t, \quad z + d \leq -d \tag{68b}$$

so that with $\beta_2 \sin \Theta_t = \beta_1 \sin \Theta$, (53) may be rearranged as follows:

$$\Pi_{z2} = \frac{K_2}{2\pi} \int_{-\frac{1}{2}\pi-j\infty}^{\frac{1}{2}\pi+j\infty} \int_{-\frac{1}{2}\pi-j\infty}^{\frac{1}{2}\pi+j\infty} I \sin \Theta \, d\Theta \, d\phi$$

$$I \equiv f_t \exp[-j\beta_1(r_i \cos \phi \sin \Theta + d \cos \Theta)]$$
$$\exp[j\beta_2(z+d)(\cos \phi \tan \Theta_t \sin \Theta_t + \cos \Theta_t)] \tag{69}$$

Since the point $P_2(r, z)$ is assumed to be quite near the surface, its depth, $-(z+d)$, in region 2 is small. Therefore, the rapidly varying part of the integrand is the first exponential which involves the path length in region 1. With the angle of incidence Θ_{01} in region 1 defined by

$$d = R_i \cos \Theta_{01}, \quad r_i = R_i \sin \Theta_{01}, \tag{70a}$$

it follows that

$$r_i \cos \phi \sin \Theta + d \cos \Theta$$
$$= R_i(\sin \Theta_{01} \cos \phi \sin \Theta + \cos \Theta_{01} \cos \Theta). \tag{70b}$$

Note that (70b) is exactly like (3) with $z = 0$. Therefore, the same asymptotic integration previously carried out may be applied to the rapidly varying exponential in Π_{z2} as was done for the corresponding exponential in (5.87) and the asymptotic values $\Theta = \Theta_{01}$, $\phi = 0$ may be substituted in the remainder of the integrand. The result corresponding to (37) is:

$$\Pi_{z2} = \frac{jK_2 \exp(-j\beta_1 R_i)}{\beta_1 \quad R_i}$$
$$\times \{f_t^r(\Theta) \exp[j\beta_2(z+d)(\cos \phi \tan \Theta_t \sin \Theta_t + \cos \Theta_t)]\}_{\Theta=\Theta_{01}; \phi=0}. \tag{71a}$$

After using the law of refraction to express Θ_t in terms of Θ, the indicated substitutions are made, leaving

$$\Pi_{z2} = \frac{jK_2}{\beta_1} \frac{\exp(-j\beta_1 R_i)}{R_i} f_t^r(\Theta_{01})$$
$$\times \exp[j\beta_2(z+d)/\sqrt{1 - N_{12}^2 \sin^2 \Theta_{01}}], \tag{71b}$$

where for the electric dipole $K_2 = K_{e2} = -j\beta_1 p_{z1}/4\pi\xi_2$ and for the magnetic dipole $K_2 = K_{m2} = -j\beta_1 m_{z1}/4\pi\nu_1$. For convenience, let a complex angle of refraction Θ_{t2} be defined as for plane-wave reflection as follows:

$$\sin \Theta_{t2} = N_{12} \sin \Theta_{01}. \tag{72a}$$

With (72a), the exponentials in (71b) may be expressed in terms of the properties of region 2. Thus with $r_i = r + (z+d) \tan \Theta_{t2}$,

$$\beta_1 R_i - \beta_2(z+d)/\cos \Theta_{t2}$$
$$= \beta_1(d \cos \Theta_{01} + r_i \sin \Theta_{01})$$
$$\quad - \beta_2(z+d)/\cos \Theta_{t2}$$
$$= \beta_2[d \cot \Theta_{01} \sin \Theta_{t2} + r \sin \Theta_{t2}$$
$$\quad - (z+d)/\cos \Theta_{t2}] \equiv \beta_2 R_e, \tag{72b}$$

where R_e is defined in (72b). Since for points in the far zone $-(z+d)$ is small compared with other distances, $\beta_1 R_i \doteq \beta_2 R_e$ in the amplitude factor. It follows that (71b) is approximately equivalent to

$$\Pi_{z2} \doteq \frac{jK_2}{\beta_2} \frac{\exp(-j\beta_2 R_e)}{R_e} f_i^r(\Theta_{01}). \quad (73)$$

Note that, from the point of view of an observer at $P_2(r, z)$ in region 2, (73) may be interpreted as a wave originating at 0_t in Fig. 6.5 with all space filled with the material of region 2.

The propagation in region 2 is studied conveniently in the form:

$$\frac{\Pi_{z2}}{\Pi_{z2}(z=-d)} = \exp[j\beta_2(z+d)/\cos\Theta_t]$$

$$= \exp(-j\beta_2 R_t) = \exp\{-j\beta_2[-(z+d)\cos\Theta_t + (r-r_i)\sin\Theta_t]\} \quad (z \leq -d) \quad (74)$$

This is carried out in the next section.

The potentials in region 1 may be expressed as follows:

$$\Pi_{ez1} = \frac{p_{z1}}{4\pi\xi_1} \frac{\exp(-j\beta_1 R_d)}{R_d}$$

$$+ f_{er}^r \frac{p_{z1}}{4\pi\xi_1} \frac{\exp(-j\beta_1 R_r)}{R_r}, \quad (75a)$$

$$\Pi_{mz1} = \frac{m_{z1}}{4\pi\nu_1} \frac{\exp(-j\beta_1 R_d)}{R_d}$$

$$+ f_{mr}^r \frac{m_{z1}}{4\pi\nu_1} \frac{\exp(-j\beta_1 R_r)}{R_r}, \quad (75b)$$

where

$$R_d = R_0 - d\cos\Theta_{01}, \quad R_r = R_0 + d\cos\Theta_{01}. \quad (76)$$

Evidently, (75) may be interpreted as the sum of the potentials due to the actual dipole of moment p_z or m_z at $z = 0$ and a fictitious *image* dipole of moment $f_{er}^r p_{z1}$ or $f_{m}^r m_{z1}$ at $z = -2d$. Therefore, the actual distributions of currents and charges along the boundary are electrically equivalent for $z > -d$ to a fictitious dipole on the axis at $z = -2d$ in a homogeneous single medium.

Alternatively, (75) may be expressed as follows:

$$\Pi_{ez1} = \frac{p_{z1}}{4\pi\xi_1} \left[\frac{\exp(-j\beta_1 R_d)}{R_d} - \frac{\exp(-j\beta_1 R_r)}{R_r} + V^r \right], \quad (77a)$$

$$\Pi_{mz1} = \frac{m_{z1}}{4\pi\nu_1} \left[\frac{\exp(-j\beta_1 R_d)}{R_d} - \frac{\exp(-j\beta_1 R_r)}{R_r} + H^r \right], \quad (77b)$$

where

$$V^r \equiv (f_{er}^r + 1) \frac{\exp(-j\beta_1 R_r)}{R_r}$$

$$= f_{et}^r \frac{\exp(-j\beta_1 R_r)}{R_r}, \quad (78a)$$

$$H^r \equiv (f_{mr}^r + 1) \frac{\exp(-j\beta_1 R_r)}{R_r}$$

$$= f_{mt}^r \frac{\exp(-j\beta_1 R_r)}{R_r}. \quad (78b)$$

With (39), (78) becomes

$$V^r = \frac{\exp(-j\beta_1 R_r)}{R_r}$$

$$\times \left(\frac{2\nu_2 N_{21}^2 \cos\Theta_{01}}{\nu_2 N_{21}^2 \cos\Theta_{01} + \nu_1 \sqrt{N_{21}^2 - \sin^2\Theta_{01}}} \right), \quad (79a)$$

$$H^r = \frac{\exp(-j\beta_1 R_r)}{R_r}$$

$$\times \left(\frac{2\nu_1 \cos\Theta_{01}}{\nu_1 \cos\Theta_{01} + \nu_2 \sqrt{N_{21}^2 - \sin^2\Theta_{01}}} \right). \quad (79b)$$

For later use, these expressions may be rearranged as follows:

$$V^r = 2 \frac{\exp(-j\beta_1 R_r)}{R_r}$$

$$\times \left[1 - \left(1 + \frac{\nu_2 N_{21}^2 \cos\Theta_{01}}{\nu_1 \sqrt{N_{21}^2 - \sin^2\Theta_{01}}} \right)^{-1} \right], \quad (80a)$$

$$H^r = 2 \frac{\exp(-j\beta_1 R_r)}{R_r}$$

$$\times \left[1 - \left(1 + \frac{\nu_1 \cos\Theta_{01}}{\nu_2 \sqrt{N_{21}^2 - \sin^2\Theta_{01}}} \right)^{-1} \right]. \quad (80b)$$

The symbols V and H are conventional notation in the literature. The superscript r denotes a radiation-zone field.

The forms (75) represent the potentials as due to *two* dipoles with *different moments*: the actual one with moment p_{z1} or m_{z1}, and a fictitious one with moment $f_{er}^r p_{z1}$ or $f_{mr}^r m_{z1}$. On the other hand, (77a,b) represent the potentials as due to two dipoles of *equal* and *opposite* moments, p_{z1} or m_{z1} and $-p_{z1}$ or $-m_{z1}$, and a correction term V^r or H^r. In both representations the fictitious second dipole is at $z = -2d$ on the axis. The advantage of the second forms, (77a, b), is that the contributions by the two equal and opposite dipoles cancel on the plane boundary. In so far as the present asymptotic or far-zone solution is concerned, there is no advantage in (77a, b) over (75a, b); (75a, b) are to be preferred. Later, when an improved solution is obtained which includes the leading term along the boundary plane, the forms (77) are more advantageous. This improved solution necessarily must reduce to (77) for large values of R_0, and it is for purposes of later comparison that the forms (77) are introduced here.

The radiation-zone or far-zone electromagnetic fields of electric and magnetic types due, respectively, to vertical electric and magnetic dipoles over a conducting horizontal plane are determined directly from the following simple formulas:

$$\mathbf{E}_e^r = \hat{\mathbf{\Theta}} E_{e\Theta}^r, \quad \mathbf{B}_e^r = \hat{\mathbf{\Phi}} B_{e\Phi}^r; \quad (81a)$$

$$\mathbf{B}_m^r = \hat{\mathbf{\Theta}} B_{m\Theta}^r, \quad \mathbf{E}_m^r = \hat{\mathbf{\Phi}} E_{m\Phi}^r. \quad (81b)$$

In region 1,

$$E_{e\Theta 1}^r = -\beta_1^2 \Pi_{ez1} \sin \Theta_{01}, \quad (82a)$$

$$B_{e\Phi 1}^r = -\frac{\beta_1^3}{\omega} \Pi_{ez1} \sin \Theta_{01}, \quad (82b)$$

$$B_{m\Theta 1}^r = -\beta_1^2 \Pi_{mz1} \sin \Theta_{01}, \quad (82c)$$

$$E_{m\Phi 1}^r = -\omega \beta_1 \Pi_{mz1} \sin \Theta_{01}. \quad (82d)$$

Specifically,

$$E_{e\Theta 1}^r = \frac{\omega}{\beta_1} B_{e\Phi 1}^r$$

$$= \frac{-\beta_1^2 p_{z1}}{4\pi \xi_1} \frac{\exp(-j\beta_1 R_0)}{R_0} [\exp(j\beta_1 d \cos \Theta_{01})$$

$$+ f_{er}^r(\Theta_{01}) \exp(-j\beta_1 d \cos \Theta_{01})] \sin \Theta_{01}, \quad (83a)$$

$$B_{m\Theta 1}^r = \frac{\beta_1}{\omega} E_{m\Phi 1}^r$$

$$= \frac{-\beta_1^2 m_{z1}}{4\pi \nu_1} \frac{\exp(-j\beta_1 R_0)}{R_0} [\exp(j\beta_1 d \cos \Theta_{01})$$

$$+ f_{mr}^r(\Theta_{01}) \exp(-j\beta_1 d \cos \Theta_{01})] \sin \Theta_{01}. \quad (83b)$$

These may be expressed in terms of array factors $A_e(\Theta_{01})$, $A_m(\Theta_{01})$ given by the brackets in (83a, b) and the vertical field factor $F(\Theta_{01})$ = $\sin \Theta_{01}$. The coefficients of reflection are the same as the plane-wave coefficients in Sec. 5 and may be expressed in any of the several forms there given. In particular,

$$f_{er}^r \equiv f_{er}(\Theta_{01})$$

$$= \frac{\nu_2 N_{21}^2 \cos \Theta_{01} - \nu_1 \sqrt{N_{21}^2 - \sin^2 \Theta_{01}}}{\nu_2 N_{21}^2 \cos \Theta_{01} + \nu_1 \sqrt{N_{21}^2 - \sin^2 \Theta_{01}}}, \quad (84a)$$

$$f_{mr}^r \equiv f_{mr}(\Theta_{01})$$

$$= \frac{\nu_1 \cos \Theta_{01} - \nu_2 \sqrt{N_{21}^2 - \sin^2 \Theta_{01}}}{\nu_1 \cos \Theta_{01} + \nu_2 \sqrt{N_{21}^2 - \sin^2 \Theta_{01}}}. \quad (84b)$$

Alternatively, in terms of the characteristic impedances and admittances ζ and η,

$$f_{er}^r = \frac{\zeta_1 \cos \Theta_{01} - \zeta_2 \cos \Theta_{t2}}{\zeta_1 \cos \Theta_{01} + \zeta_2 \cos \Theta_{t2}}, \quad (84c)$$

$$f_{mr}^r = \frac{\eta_1 \cos \Theta_{01} - \eta_2 \cos \Theta_{t2}}{\eta_1 \cos \Theta_{01} + \eta_2 \cos \Theta_{t2}}. \quad (84d)$$

The corresponding formulas for the fields in region 2 near the boundary are:

$$E_{e\Theta 2}^r = \frac{\omega}{\beta_2} B_{e\Phi 2}^r = -\beta_2^2 \Pi_{ez2} \sin \Theta_{t2}$$

$$= \frac{-\beta_2^2 p_{z1}}{4\pi \xi_2} \frac{\exp(-j\beta_1 R_i)}{R_i} f_{et}(\Theta_{01})$$

$$\times \exp[-j\beta_2(z + d)/\cos \Theta_{t2}] \sin \Theta_{t2}, \quad (85a)$$

$$B_{m\Theta 2}^r = \frac{\beta_2}{\omega} E_{m\Phi 2}^r = -\beta_2^2 \Pi_{mz2} \sin \Theta_{t2}$$

$$= \frac{-\beta_2^2 m_{z1}}{4\pi \nu_1} \frac{\exp(-j\beta_1 R_i)}{R_i} f_{ml}(\Theta_{01})$$

$$\times \exp[-j\beta_2(z + d)/\cos \Theta_{t2}] \sin \Theta_{t2}, \quad (85b)$$

where the radiation-zone coefficients of refraction or transmission are

$$f_{et}^r \equiv f_{et}(\Theta_{01})$$

$$= \frac{2\nu_2 N_{21}^2 \cos \Theta_{01}}{\nu_2 N_{21}^2 \cos \Theta_{01} + \nu_1 \sqrt{N_{21}^2 - \sin^2 \Theta_{01}}}, \quad (86a)$$

$$f_{ml}^r \equiv f_{ml}(\Theta_{01})$$

$$= \frac{2\nu_1 \cos \Theta_{01}}{\nu_1 \cos \Theta_{01} + \nu_2 \sqrt{N_{21}^2 - \sin^2 \Theta_{01}}}. \quad (86b)$$

Alternatively,

$$f_{et}^r = \frac{2\zeta_1 \cos\Theta_{01}}{\zeta_1 \cos\Theta_{01} + \zeta_2 \cos\Theta_{t2}}, \quad (86c)$$

$$f_{mt}^r = \frac{2\eta_1 \cos\Theta_{01}}{\eta_1 \cos\Theta_{01} + \eta_2 \cos\Theta_{t2}}. \quad (86d)$$

This completes the general determination of the far-zone fields of vertical electric and magnetic dipoles immersed in region 1 at a height d over a plane boundary with region 2. The two regions are assumed to be simple media but not otherwise restricted in their dielectric and conducting properties. In the following sections a more detailed study is made of the field, especially in region 1, in the important practical case when this region is empty space and region 2 is a nonmagnetic but otherwise quite arbitrary simple medium.

7. Dipoles in Air over Dielectrics and Conductors

In the preceding section the electromagnetic fields of electric and magnetic dipoles are derived for dipoles immersed in region 1 perpendicular to and at an arbitrary distance d from the plane boundary of region 2. Aside from the requirement that they be homogeneous and simply polarizing, magnetizing, and conducting, both regions are unrestricted in their properties. In this section these general formulas are specialized and applied specifically to the important practical problem of determining the field of dipoles in empty space (region 1) perpendicular to and at a distance d from a conducting earth (region 2) which is characterized by arbitrary conductivity and dielectric constant. Since interest is primarily in the field in air above a conducting earth, this is discussed first and in detail. The section is concluded with a brief consideration of the field in the earth near the boundary.

Let the following conditions be imposed to characterize the properties of air and a conducting earth:

Region 1 (empty space):

$$\xi_1 = \epsilon_1 = \epsilon_0, \quad \nu_1 = \nu_0, \quad \sigma_1 = 0, \quad (1a)$$

$$\beta_1 = \beta_0 = \omega/v_0, \quad v_0 = \sqrt{\nu_0/\epsilon_0},$$

$$\zeta_1 = \zeta_0 = 1/\sqrt{v_0\epsilon_0} = 1/\eta_0; \quad (1b)$$

Region 2 (earth):

$$\xi_2 = \epsilon_{e2} - j\sigma_{e2}/\omega, \quad \nu_2 = \nu_o = \nu_0 = 1/\mu_0, \quad (2a)$$

$$\beta_2 = \omega/v_2, \quad v_2 = \sqrt{\nu_0/\xi_2},$$

$$\zeta_2 = 1/\sqrt{\nu_0\xi_2} = 1/\eta_2. \quad (2b)$$

Hence,

$$N_{21} = \beta_2/\beta_1 = \beta_2/\beta_0 = \sqrt{\xi_2/\epsilon_0} = 1/N_{12}, \quad (3a)$$

$$N_{21}^2 = \epsilon_{e2r} - j\sigma_{e2}/\omega\epsilon_0 \equiv \epsilon_{e2r}(1 - jh_{e2})$$

$$= \epsilon_{e2r} - jX_{e2}, \quad (3b)$$

where $\epsilon_{e2r} \equiv \epsilon_{e2}/\epsilon_0$ is the real effective relative dielectric constant of region 2 and h_{e2} is a convenient shorthand symbol defined by

$$h_{e2} \equiv \sigma_{e2}/\omega\epsilon_{e2}. \quad (3c)$$

The field impedances and admittances are:

$$Z_{e1} = R_{e1} = \zeta_0 \cos\Theta_{01}, \quad (4a)$$

$$Y_{m1} = G_{m1} = \eta_0 \cos\Theta_{01}, \quad (4b)$$

$$Z_{e2} = R_{e2} + jX_{e2} = \zeta_2 \cos\Theta_{t2}, \quad (5a)$$

$$Y_{m2} = G_{m2} + jB_{m2} = \eta_2 \cos\Theta_{t2}. \quad (5b)$$

With the law of refraction,

$$\sin\Theta_{t2} = N_{12} \sin\Theta_{01}, \quad (6)$$

it follows that

$$Z_{e2} = \zeta_2 \sqrt{1 - N_{12}^2 \sin^2\Theta_{01}}, \quad (7a)$$

$$Y_{m2} = \eta_2 \sqrt{1 - N_{12}^2 \sin^2\Theta_{01}}. \quad (7b)$$

Field in region 1. The coefficients of reflection are

$$f_{er}^r \equiv f_{er}(\Theta_{01}) = \frac{R_{e1} - Z_{e2}}{R_{e1} + Z_{e2}}, \quad (8a)$$

$$f_{mr}^r \equiv f_{mr}(\Theta_{01}) = \frac{G_{m1} - Y_{m2}}{G_{m1} + Y_{m2}}. \quad (8b)$$

Alternatively, in terms of the index of refraction,

$$f_{er}^r = \frac{N_{21}^2 \cos\Theta_{01} - \sqrt{N_{21}^2 - \sin^2\Theta_{01}}}{N_{21}^2 \cos\Theta_{01} + \sqrt{N_{21}^2 - \sin^2\Theta_{01}}}$$

$$= \frac{\cos\Theta_{01} - N_{12}\sqrt{1 - N_{12}^2 \sin^2\Theta_{01}}}{\cos\Theta_{01} + N_{12}\sqrt{1 - N_{12}^2 \sin^2\Theta_{01}}}, \quad (9a)$$

$$f_{mr}^r = \frac{\cos\Theta_{01} - \sqrt{N_{21}^2 - \sin^2\Theta_{01}}}{\cos\Theta_{01} + \sqrt{N_{21}^2 - \sin^2\Theta_{01}}}$$

$$= \frac{N_{12}\cos\Theta_{01} - \sqrt{1 - N_{12}^2 \sin^2\Theta_{01}}}{N_{12}\cos\Theta_{01} + \sqrt{1 - N_{12}^2 \sin^2\Theta_{01}}}. \quad (9b)$$

Note that f_{er}^r is a complex coefficient of electric type that applies specifically when the magnetic field is parallel to the boundary between regions 1 and 2, and hence is perpendicular to the plane of incidence. This is equivalent to an *electric field* that is *parallel to the plane of incidence*. Similarly, f_{mr}^r is a complex coefficient of magnetic type that applies when the electric field is parallel to the boundary between regions 1 and 2, and hence is *perpendicular to the plane of incidence*.

In order to separate the real and imaginary parts of the coefficients of reflection, let the notation (3b) be introduced and the subscript 2 omitted. That is, let

$$N_{21}^2 = \epsilon_{e2r}(1 - jh_{e2}) = \epsilon_{er}(1 - jh_e). \quad (10)$$

The notation $x \equiv h_e \epsilon_{er}$ is common in the literature. With (10) the radical in (9a, b) may be expressed as follows:

$$\sqrt{N_{21}^2 - \sin^2 \Theta_{01}} = \sqrt{\epsilon_{er} - \sin^2 \Theta_{01} - jh_e \epsilon_{er}}$$
$$= A\sqrt{1 - jh_x}, \quad (11)$$

where

$$A \equiv \sqrt{\epsilon_{er} - \sin^2 \Theta_{01}}, \quad h_x \equiv h_e \epsilon_{er}/A^2. \quad (12)$$

The real and imaginary parts of a radical of the form $\sqrt{1 - jh}$ may be separated into the $f(h)$ and $g(h)$ functions which are tabulated in reference I.31, Appendix II. Thus,

$$\sqrt{1 - jh_x} = f(h_x) - jg(h_x), \quad (13a)$$

$$f(h_x) = \sqrt{\tfrac{1}{2}[\sqrt{1 + h_x^2} + 1]},$$
$$g(h_x) = \sqrt{\tfrac{1}{2}[\sqrt{1 + h_x^2} - 1]}, \quad (13b)$$

Also,

$$A^2[f^2(h_x) + g^2(h_x)] = A^2\sqrt{1 + h_x^2}$$
$$= \sqrt{A^4 + x^2} \quad (14)$$

where $x \equiv h_e \epsilon_{er}$. With (10), (11), and (13a), (9a) becomes

$$f_{er}^r = f_{er}'' + jf_{er}' = f_{er}e^{j\psi_{er}}$$
$$= \frac{[\epsilon_{er}\cos\Theta_{01} - Af(h_x)]}{[\epsilon_{er}\cos\Theta_{01} + Af(h_x)]} \frac{- j[h_e\epsilon_{er}\cos\Theta_{01} - Ag(h_x)]}{- j[h_e\epsilon_{er}\cos\Theta_{01} + Ag(h_x)]}.$$
$$(15)$$

With (14) the real and imaginary parts are

$$f_{er}'' = \frac{1}{D_e}[\epsilon_{er}^2(1 + h_e^2)\cos^2\Theta_{01} - \sqrt{A^4 + h_e^2\epsilon_{er}^2}], \quad (16a)$$

$$f_{er}' = \frac{2}{D_e}[g(h_x) - h_e f(h_x)]A\epsilon_{er}\cos\Theta_{01}, \quad (16b)$$

where

$$D_e \equiv \epsilon_{er}^2(1 + h_e^2)\cos^2\Theta_{01} + 2A[\epsilon_{er}f(h_x) + h_e\epsilon_{er}g(h_x)]\cos\Theta_{01} + \sqrt{A^4 + h_e^2\epsilon_{er}^2}. \quad (16c)$$

Similarly, (9b) becomes

$$f_{mr}^r = f_{mr}'' + jf_{mr}' = f_{mr}e^{j\psi_{mr}}$$
$$= \frac{[\cos\Theta_{01} - Af(h_x)] + jAg(h_x)}{[\cos\Theta_{01} + Af(h_x)] - jAg(h_x)}. \quad (17a)$$

The real and imaginary parts are

$$f_{mr}'' = \frac{1}{D_m}[\cos^2\Theta_{01} - \sqrt{A^4 + x^2}], \quad (17b)$$

$$f_{mr}' = \frac{2}{D_m}Ag(h_x)\cos\Theta_{01}, \quad (17c)$$

where

$$D_m \equiv \cos^2\Theta_{01} + 2Af(h_x)\cos\Theta_{01} + \sqrt{A^4 + h_e^2\epsilon_{er}^2}. \quad (17d)$$

Since f_{er} and f_{mr} are not the same, it is clear that an incident field that is linearly polarized in an arbitrary direction is reflected as an elliptically polarized field with the shape of the ellipse depending not only on the constitutive parameters ϵ_{e2} and σ_{e2} but also on the angle of incidence.

Using (16a, b, c) and (17b, c, d), the coefficients of reflection have been evaluated for numerical values of $N_{21}^2 = \epsilon_{er}(1 - jh_e)$ that are useful in practice. By assigning the values 4, 10, and 80 to ϵ_{er} as representing typical mean values, respectively, for dry earth, moist earth, and water and then allowing $x \equiv h_e\epsilon_{er} = \sigma_{e2}/\omega\epsilon_0$ to assume a range of values from zero to infinity, Tables 7.1 and 7.2 were computed. Curves of the real and imaginary parts of the reflection coefficients are shown in Figs. 7.1a, b; magnitudes and angles are shown in Figs. 7.2a, b, c. Note that f_{er}'' ranges from $+1$ to -1 while f_{mr}'' is always negative. On the other hand, f_{er}' is always negative, f_{mr}' always positive. Of particular significance is the fact that f_{er}'' goes through

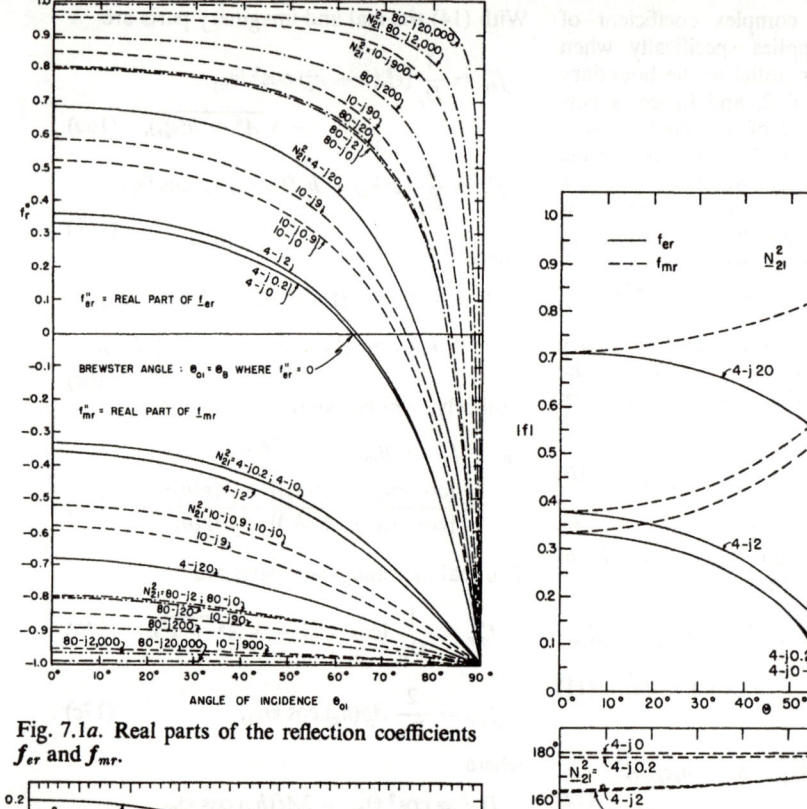

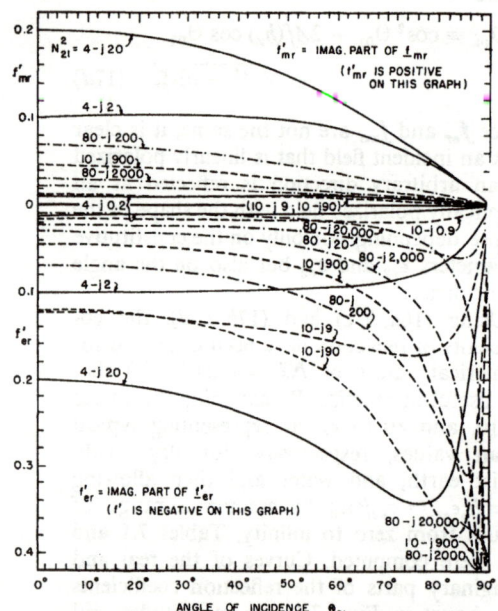

Fig. 7.1a. Real parts of the reflection coefficients f_{er} and f_{mr}.

Fig. 7.1b. Imaginary parts of the reflection coefficients f_{er} and f_{mr}.

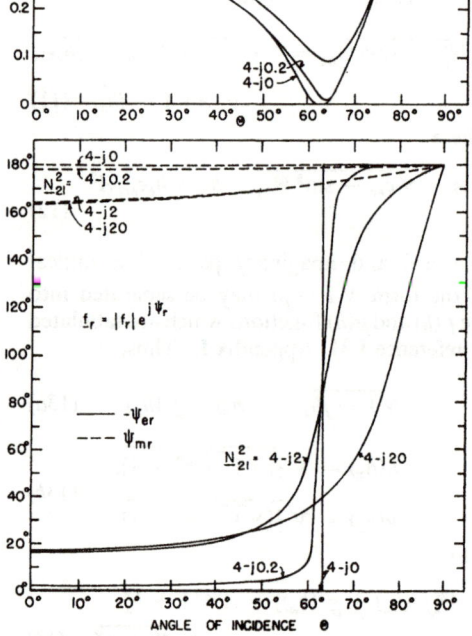

Fig. 7.2a. Reflection coefficients f_{er} and f_{mr}; $\epsilon_{er} = 4$.

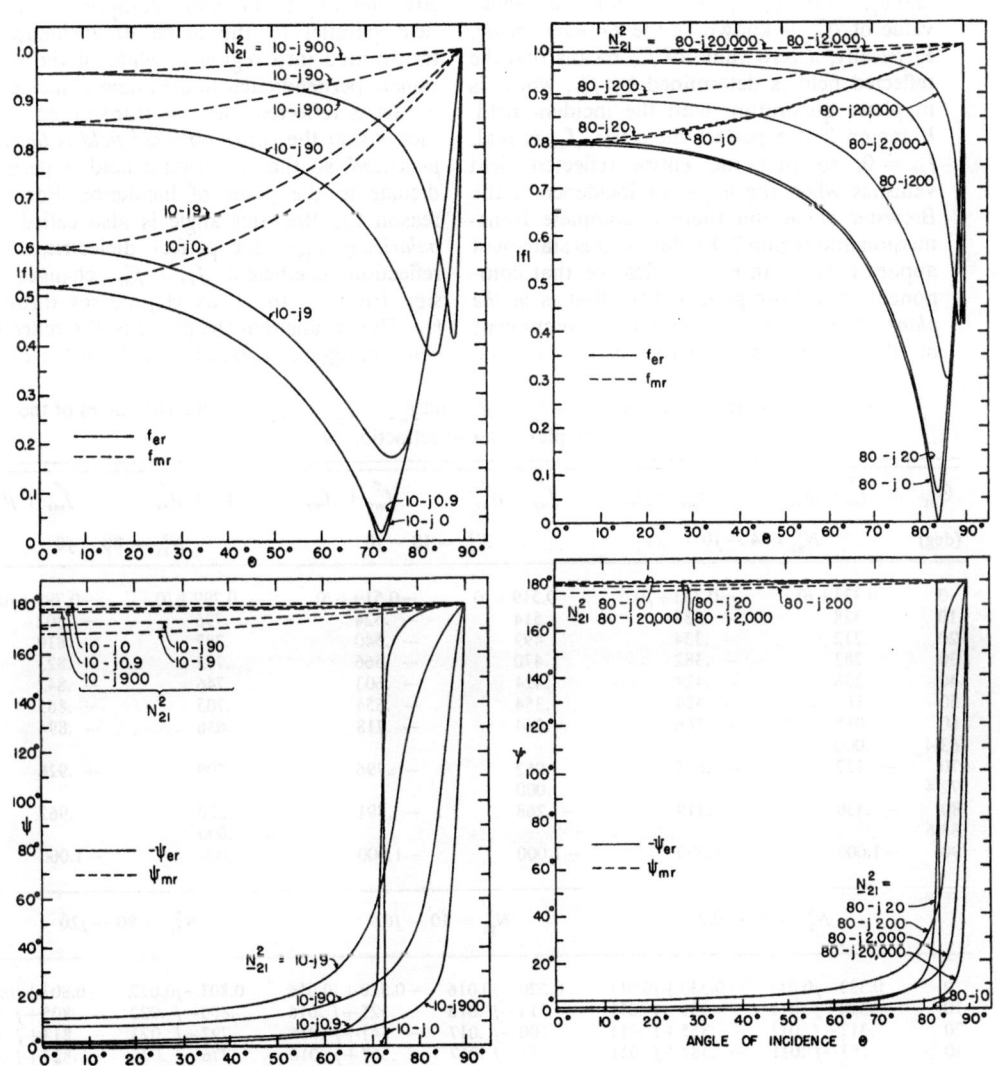

Fig. 7.2b. Reflection coefficients f_{er} and f_{mr}; $\epsilon_{er} = 10$.

Fig. 7.2c. Reflection coefficients f_{er} and f_{mr}; $\epsilon_{er} = 80$.

zero, whereas f''_{mr} does not. In fact, the magnitude of f_{er}, as plotted in Figs. 7.2a, b, c, has a minimum, whereas f_{mr} does not. Also ψ_{er} has a range of very rapid variation through $\pi/2$ from near zero to near π.

The angle $\Theta_{01} = \Theta_B$ at which f''_{er} vanishes and $\psi_{er} = \tan^{-1}(f'_{er}/f''_{er}) = \pi/2$ for a particular value of N_{21} is known as the *Brewster angle*. Physically, a zero value of f''_{er} means that the reflected field is determined by f'_{er}, which is in phase quadrature with the incident field. If region 2 is a perfect dielectric, f_{er} is real, $f'_{er} = 0$, so that the entire reflected field vanishes when the angle of incidence is the Brewster angle and there is complete transmission into region 2. Evidently, this statement applies only to an electric field or that component of a more general field that is *in the plane of incidence*. If there is a component of the electric field perpendicular to the plane of incidence it is reflected according to $f_{mr} = f''_{mr} + jf'_{mr}$, of which neither f''_{mr} nor f'_{mr} vanishes for any value of Θ_{01} except 90°. If region 2 is a perfect dielectric, f_{mr} is real, $f'_{mr} = 0$. Hence, if an arbitrary electric field is incident on a perfect dielectric at the Brewster angle, the *entire* component of the field parallel to the plane of incidence is transmitted into region 2, while of the component perpendicular to the plane of incidence a part is reflected, the rest transmitted. This means that the *entire reflected field* is *linearly polarized*, so that its electric field is perpendicular to the plane of incidence. For this reason the Brewster angle is also called the *polarizing angle* for perfect dielectrics. The reflection coefficient $f_{er} = f''_{er}$ changes its sign from + to − as Θ_{01} passes through Θ_B. This means that the phase of the reflected field changes discontinuously from 0°, or in

TABLE 7.1. Reflection factors $f_{er} = f''_{er} + jf'_{er}$ and $f_{mr} = f''_{mr} + jf'_{mr}$ for different values of the complex index of refraction N_{21}.

Θ_{01} (deg)	$f''_{er}+jf'_{er}$	$f''_{mr}+jf'_{mr}$	$f''_{er}+jf'_{er}$	$f''_{mr}+jf'_{mr}$	$f''_{er}+jf'_{er}$	$f''_{mr}+jf'_{mr}$
	$N^2_{21}=4-j0$		$N^2_{21}=10-j0$		$N^2_{21}=80-j0$	
0	0.333+j0	−0.333+j0	0.519+j0	−0.519+j0	0.799+j0	−0.799+j0
10	.328	− .338	.514	− .524	.796	− .802
20	.312	− .354	.499	− .540	.787	− .810
30	.283	− .382	.470	− .566	.772	− .823
40	.236	− .424	.424	− .603	.746	− .842
50	.164	− .484	.354	− .654	.705	− .865
60	.015	− .566	.244	− .718	.636	− .894
63.4	.000					
70	− .127	− .675	.062	− .796	.509	− .926
72.4			.000			
80	− .430	− .819	− .268	− .891	.220	− .962
83.6					.000	
90	−1.000	−1.000	−1.000	−1.000	−1.000	−1.000
	$N^2_{21}=4-j0.2$		$N^2_{21}=10-j0.9$		$N^2_{21}=80-j20$	
0	0.333−j0.011	−0.333+j0.011	0.520−j0.016	−0.520+j0.016	0.803−j0.022	−0.803+j0.022
10	.328−j .011	− .339+j .011	.515−j .016	− .525+j .016	.800−j .022	− .805+j .022
20	.312−j .011	− .355+j .011	.500−j .017	− .541+j .016	.792−j .023	− .813+j .021
30	.283−j .011	− .382+j .011	.471−j .017	− .567+j .016	.776−j .024	− .827+j .020
40	.236−j .010	− .424+j .011	.425−j .018	− .604+j .015	.750−j .027	− .845+j .018
50	.164−j .010	− .484+j .012	.355−j .019	− .654+j .014	.710−j .030	− .868+j .015
60	.052−j .010	− .566+j .010	.244−j .019	− .718+j .012	.642−j .036	− .896+j .012
63.4	.000−j .010					
70	− .127−j .009	− .676+j .009	.063−j .020	− .797+j .009	.516−j .044	− .928+j .009
72.4			.000−j .020			
75			− .075−j .020			
80	− .430−j .007	− .819+j .005	− .267−j .019	− .891+j .005	.227−j .057	− .963+j .005
83.6					.000−j .060	
85			− .550−j .014		− .114−j .060	
87					− .354−j .053	
90	−1.000−j .000	−1.000+j .000	−1.000−j .000	−1.000+j .000	−1.000−j .000	−1.000+j .000

Table 7.1—contd.

Θ_{01} (deg)	$f''_{er} + jf'_{er}$	$f''_{mr} + jf'_{mr}$	$f''_{er} + jf'_{er}$	$f''_{mr} + jf'_{mr}$	$f''_{er} + jf'_{er}$	$f''_{mr} + jf'_{mr}$
	$N^2_{21} = 4 - j2$		$N^2_{21} = 10 - j9$		$N^2_{21} = 80 - j200$	
0	0.362−j0.101	−0.362+j0.101	0.585−j0.123	−0.585+j0.123	0.891−j0.068	−0.891+j0.068
10	.357−j .101	− .368+j .102	.580−j .124	− .590+j .123	.889−j .069	− .892+j .067
20	.340−j .100	− .384+j .102	.564−j .127	− .604+j .120	.884−j .072	− .897+j .065
30	.309−j .099	− .414+j .102	.536−j .131	− .630+j .115	.874−j .078	− .905+j .060
40	.259−j .098	− .457+j .101	.491−j .137	− .664+j .108	.859−j .086	− .916+j .054
50	.184−j .096	− .518+j .098	.422−j .147	− .711+j .097	.833−j .100	− .930+j .046
60	.068−j .904	− .600+j .089	− .312+j .159	− .768+j .082	.789−j .122	− .944+j .036
64.3	.000−j .092					
70	− .115−j .088	− .706+j .073	.129−j .171	− .836+j .061	.701−j .162	− .962+j .025
74.6			.000−j .173			
75			− .014−j .173			
80	− .423−j .071	− .838+j .044	− .215−j .164	− .913+j .034	.469−j .243	− .981+j .013
85			− .516−j .127		.134−j .300	
86.1					.000−j .300	
87					− .143−j .300	
90	−1.000−j .000	−1.000+j .000	−1.000−j .000	−1.000+j .000	−1.000−j .000	−1.000+j .000
	$N^2_{21} = 4 - j20$		$N^2_{21} = 10 - j90$		$N^2_{21} = 80 - j2{,}000$	
0	0.683−j0.202	−0.683+j0.202	0.847−j0.120	−0.847+j0.120	0.968−j0.030	−0.968+j0.030
10	.679−j .207	− .688+j .200	.845−j .122	− .849+j .118	.967−j .030	− .968+j .030
20	.665−j .209	− .701+j .192	.837−j .126	− .856+j .114	.966−j .032	− .970+j .028
30	.639−j .220	− .723+j .184	.824−j .135	− .867+j .106	.963−j .034	− .972+j .026
40	.598−j .236	− .753+j .169	.802−j .149	− .882+j .096	.958−j .039	− .975+j .023
50	.533−j .259	− .791+j .149	.766−j .171	− .901+j .082	.950−j .046	− .979+j .020
60	.427−j .290	− .836+j .122	.704−j .205	− .923+j .065	.936−j .058	− .984+j .016
70	.240−j .328	− .887+j .088	.583−j .262	− .947+j .046	.906−j .082	− .989+j .010
75	.089−j .342		.471−j .304			
77.1	.000−j .341					
80	− .133−j .339	− .942+j .048	.279−j .353	− .973+j .024	.818−j .148	− .994+j .005
84.0			.000−j .380			
85	− .418−j .251		− .106−j .375		.651−j .247	
87			− .375−j .332		.455−j .329	
88.7					.000−j .413	
90	−1.000−j .000	−1.000+j .000	−1.000−j .000	−1.000+j .000	−1.000−j .000	−1.000+j .000
			$N^2_{21} = 10 - j900$		$N^2_{21} = 80 - j20{,}000$	
0			0.953−j0.045	−0.953+j0.045	0.990−j0.010	−0.990+j0.010
10			.952−j .045	− .953+j .044	.990−j .010	− .990+j .010
20			.950−j .048	− .956+j .042	.989−j .010	− .991+j .009
30			.945−j .051	− .959+j .039	.988−j .011	− .991+j .009
40			.938−j .058	− .964+j .035	.987−j .013	− .992+j .008
50			.927−j .068	− .970+j .029	.984−j .015	− .994+j .006
60			.906−j .085	− .976+j .023	.980−j .020	− .995+j .005
70			.863−j .119	− .984+j .016	.971−j .028	− .997+j .003
75			.820−j .151		.961−j .037	
80			.735−j .206	− .992+j .008	.942−j .054	− .998+j .002
85			.505−j .318		.886−j .102	
87			.257−j .387		.812−j .158	
88					.722−j .215	
88.1			.000−j .413			
89					.481−j .329	
89.6					.000−j .413	
90			−1.000−j .000	−1.000+j .000	−1.000−j .000	−1.000+j .000

TABLE 7.2. Magnitudes and angles of the reflection coefficients.

Θ_{01} (deg)	$f_{er}/\underline{\psi}^0_{er}$ $N^2_{21} = 4 - j0$	$f_{mr}/\underline{\psi}^0_{mr}$	$f_{er}/\underline{\psi}^0_{er}$ $N^2_{21} = 10 - j0$	$f_{mr}/\underline{\psi}^0_{mr}$	$f_{er}/\underline{\psi}^0_{er}$ $N^2_{21} = 80 - j0$	$f_{mr}/\underline{\psi}^0_{mr}$
0	0.333/0°	0.333/180°	0.519/0°	0.519/180°	0.799/0°	0.799/180°
10	.328	.338	.514	.524	.796	.802
20	.312	.354	.499	.540	.787	.810
30	.283	.382	.470	.566	.772	.823
40	.236	.424	.424	.603	.746	.842
50	.164	.484	.354	.654	.705	.865
60	.015	.566	.244	.718	.636	.894
63.4	.000/−90°					
70	.127	.675	.062	.796	.509	.926
72.4			.000/−90°			
80	.430	.819	.268	.891	.220	.962
83.6					.000/−90°	
90	1.000/−180°	1.000/180°	1.000/−180°	1.000/180°	1.000/−180°	1.000/180°

	$N^2_{21} = 4 - j0.2$		$N^2_{21} = 10 - j0.9$		$N^2_{21} = 80 - j20$	
0	0.333/−1.888°	0.333/178.1°	0.520/−1.765°	0.520/178.2°	0.803/−1.568°	0.803/178.4°
10	.328/−1.944	.339/178.1	.515/−1.782	.525/178.2	.800/−1.574	.805/178.4
20	.312/−2.001	.355/178.2	.500/−1.944	.541/178.3	.792/−1.664	.813/178.5
30	.283/−2.331	.382/178.4	.471/−2.068	.567/178.4	.776/−1.770	.827/178.6
40	.236/−2.400	.424/178.5	.425/−2.428	.604/178.6	.750/−2.057	.845/178.8
50	.164/−3.490	.484/178.6	.356/−3.057	.654/178.8	.711/−2.417	.868/179.0
60	.053/−11.06	.566/179.0	.245/−4.451	.718/179.0	.643/−3.209	.896/179.2
63.4	.010/−90					
70	.127/−175.9	.676/179.2	.066/−17.57	.797/179.4	.518/−4.872	.928/179.4
72.4			.020/−90			
75			.078/−165.1			
80	.430/−179.1	.819/179.6	.268/−175.9	.891/179.7	.234/−14.08	.963/179.7
83.6					.060/−90	
85			.550/−178.3		.129/−152.3	
87					.358/−171.5	
90	1.000/−180	1.000/180	1.000/−180	1.000/180	1.000/−180	1.000/180

[VII.7] THEORY OF LINEAR ANTENNAS 733

TABLE 7.2—contd.

Θ_{01}	$f_{er}/\underline{\psi^0_{er}}$	$f_{mr}/\underline{\psi^0_{mr}}$	$f_{er}/\underline{\psi^0_{er}}$	$f_{mr}/\underline{\psi^0_{mr}}$	$f_{er}/\underline{\psi^0_{er}}$	$f_{er}/\underline{\psi^0_{mr}}$
	$N^2_{21} = 4 - j2$		$N^2_{21} = 10 - j9$		$N^2_{21} = 80 - j200$	
0	0.376/−15.57°	0.376/164.4°	0.598/−11.85°	0.598/168.2°	0.894/−4.361°	0.894/175.6°
10	.371/−15.79	.382/164.5	.593/−12.07	.603/168.3	.892/−4.434	.894/175.7
20	.354/−16.37	.397/165.1	.578/−12.67	.616/168.8	.887/−4.648	.899/175.8
30	.324/−17.73	.426/166.2	.552/−13.70	.640/169.7	.877/−5.092	.907/176.2
40	.277/−20.69	.468/167.6	.510/−15.57	.673/170.8	.863/−5.710	.918/176.6
50	.208/−27.54	.527/169.3	.447/−19.17	.718/172.3	.839/−6.834	.931/177.2
60	.116/−52.98	.606/171.6	.350/−27.00	.772/173.9	.798/−8.806	.945/177.8
64.3	.092/−90°					
70	.145/−142.6	.710/174.1	.214/−52.93	.838/175.8	.719/−13.00	.962/178.5
74.6			.173/−90			
75			.174/−94.70			
80	.429/−170.5	.839/177.0	.270/−142.7	.914/177.9	.528/−27.36	.981/179.2
85			.531/−166.2		.328/−65.88	
86.1					.300/−90	
87					.332/−115.5	
90	1.000/−180°	1.000/180	1.000/−180	1.000/180	1.000/−180	1.000/180
	$N^2_{21} = 4 - j20$		$N^2_{21} = 10 - j90$		$N^2_{21} = 0 - j2,000$	
0	0.712/−16.47°	0.712/163.5°	0.855/−8.070°	0.855/171.9°	0.968/−1.776°	0.968/178.2
10	.710/−16.95	.716/163.8	.854/−8.188	.857/172.1	.967/−1.776	.968/178.2
20	.697/−17.42	.727/164.7	.846/−8.520	.864/172.4	.966/−1.894	.970/178.3
30	.676/−18.96	.746/165.8	.835/−9.301	.873/173.0	.964/−2.018	.972/178.5
40	.643/−21.53	.772/167.4	.816/−10.53	.887/173.8	.959/−2.327	.975/178.6
50	.593/−25.90	.805/169.4	.785/−12.56	.905/174.8	.951/−2.771	.979/178.8
60	.516/−34.15	.845/171.7	.733/−16.21	.925/176.0	.937/−3.546	.984/179.1
70	.406/−53.77	.891/174.3	.639/−24.16	.948/177.2	.910/−5.165	.989/179.4
75	.353/−75.35		.560/−32.79			
77.1	.341/−90					
80	.364/−111.5	.943/177.1	.450/−51.63	.973/178.6	.831/−10.25	.994/179.7
84.0			.380/−90			
85	.488/−149.1		.390/−105.8		.696/−20.74	
87			.501/−138.5		.561/−35.84	
88.7					.413/−90	
90	1.000/−180	1.000/180	1.000/−180	1.000/180	1.000/−180	1.000/180

TABLE 7.2.—contd.

Θ_{01}	$f_{er}/\underline{\psi^0_{er}}$	$f_{mr}/\underline{\psi^0_{mr}}$	$f_{er}/\underline{\psi^0_{er}}$	$f_{mr}/\underline{\psi^0_{mr}}$	$f_{er}/\underline{\psi^0_{er}}$	$f_{er}/\underline{\psi^0_{mr}}$
			$N^2_{21} = 10 - j900$		$N^2_{21} = 80 - j20{,}000$	
0			0.954/−2.698°	0.954/177.3°	0.990/−0.573°	0.990/179.4°
10			.953/−2.703	.954/177.4	.990/− .573	.990/179.4
20			.951/−2.889	.957/177.5	.989/− .573	.991/179.5
30			.946/−3.085	.960/177.7	.988/− .629	.991/179.5
40			.940/−3.535	.965/177.9	.987/− .742	.992/179.5
50			.929/−4.187	.970/178.3	.984/− .860	.994/179.6
60			.910/−5.356	.976/178.6	.980/−1.146	.995/179.7
70			.871/−7.851	.984/179.1	.971/−1.658	.997/179.8
75			.834/−10.42		.962/−2.175	
80			.763/−15.63	.992/179.5	.944/−3.260	.998/179.9
85			.597/−32.18		.892/−6.553	
87			.464/−56.36		.827/−10.97	
88					.753/−16.58	
88.1			.413/−90			
89					.583/−34.34	
89.6					.413/−90	
90			1.000/−180	1.000/180	1.000/−180	1.000/180

phase, to 180° out of phase with the incident field. At the discontinuity the phase of the reflected field has the mean value $\psi_{er} = 90°$. Note that for a perfect dielectric the general defining condition $f''_{er} = 0$ for the Brewster angle is equivalent to $f_{er} = 0$. In the transmission-line form for f_{er} given in (8a), and with $Z_{e2} = R_{e2}$ for a perfect dielectric, this condition is equivalent to requiring that

$$R_{e2} = R_{e1}. \quad (18a)$$

That is, the field impedances (in this case pure resistances) of the two regions are *matched* when there is no reflection and perfect transmission, just as are the characteristic impedances of two lossless transmission lines. This condition of match is illustrated in Fig. 7.3, where the quantities R_{e1}/ζ_0 and R_{e2}/ζ_0 for three values of $N_{21} = \sqrt{\epsilon_{e2r}}$ are plotted as functions of Θ_{01}. The points of intersection define a *perfect match* at the Brewster angle Θ_B. The corresponding condition for the magnetic case given by (8b) is

$$G_{m2} = G_{m1}. \quad (18b)$$

Graphs of G_{m1}/η_0 and G_{m2}/η_0 also are shown in Fig. 7.3. It is seen that these can not intersect so that there can be no match.

The Brewster angle for a conducting and dielectric region 2 is determined readily in general by equating f''_{er} in (16a) to zero. Thus, omitting the subscript 2 on ϵ_{e2r} except in final formulas,

$$\cos^2 \Theta_B = \frac{\sqrt{A^4 + h_e^2 \epsilon_{er}^2}}{\epsilon_{er}^2 (1 + h_e^2)}, \quad (19a)$$

where A is a function of $\sin^2 \Theta_B$. If A is replaced by its value in (12), a quadratic equation is obtained that may be solved with the following result

$$\cos^2 \Theta_B = \frac{\epsilon_{er} - 1 + \epsilon_{er} F}{\epsilon_{er}^4 (1 + h_e^2)^2 - 1} \quad (19b)$$

where

$$F \equiv \sqrt{\epsilon_{er}^2 (1 + h_e^2)^2 (\epsilon_{er} - 1)^2 + h_e^2 [\epsilon_{er}^4 (1 + h_e^2)^2 - 1]}. \quad (19c)$$

Brewster angles $\Theta_{01} = \Theta_B$ computed from (19b) are included in Tables 7.1 and 7.2.

The general expression (19b) for the Brewster angle reduces to very simple forms in two important special cases: (a) when region 2 is a good dielectric defined by $h_e^2 = (\sigma_{e2}/\omega\epsilon_{e2})^2 \ll 1$, and (b) when it is a good conductor so that $h_e = \sigma_{e2}/\omega\epsilon_{e2} \gg 1$. The formulas are:

Good dielectric, $h_e^2 \ll 1$:

$$\cos\Theta_B \doteq \frac{1}{\sqrt{\epsilon_{e2r}+1}} \text{ or } \tan\Theta_B \doteq \sqrt{\epsilon_{e2r}}, \quad (20)$$

Good conductor, $h_e \gg 1$, $h_e\epsilon_{e2r} \gg 1$:

$$\cos\Theta_B \doteq \frac{1}{\sqrt{h_e\epsilon_{e2r}}} = \sqrt{\omega\epsilon_0/\sigma_{e2}} \quad (21a)$$

or $\Theta_B \doteq \frac{\pi}{2} - \frac{1}{\sqrt{h_e\epsilon_{e2r}}} \doteq \frac{\pi}{2} - \sqrt{\omega\epsilon_0/\sigma_{e2}}.$

$(21b)$

The general expression for f'_{er} at the Brewster angle is obtained by substituting Θ_B from (19b) for Θ in (16b). Since it is algebraically intricate, it is not written out. However, the special formulas for good dielectrics and good conductors are quite simple. For a good dielectric with $h_e^2 \ll 1$, $A = \epsilon_{er}/\sqrt{\epsilon_{er}+1}$, $h_x = h_e(\epsilon_{er}+1)/\epsilon_{er}$; $g(h_x) \doteq h_x/2$; $f(h_x) \doteq 1$, so that $D_e = 4\epsilon_{er}^2/(\epsilon_{er}+1)$ and, hence, Good dielectric, $(\sigma_{e2}/\omega\epsilon_{e2})^2 \ll 1$:

$$f'_{er} \doteq -h_e\left(\frac{\epsilon_{e2r}-1}{4\epsilon_{e2r}}\right) \text{ at } \Theta_{01} = \Theta_B. \quad (22)$$

For a good conductor defined by $h_e \gg 1$, $f(h_x) \doteq g(h_x) \doteq \sqrt{h_x/2} = h_e\epsilon_{er}/A\sqrt{2}$, $D_e = 2h_e\epsilon_{er}(1 + 1/\sqrt{2})$, so that, Good conductor, $h_e \gg 1$:

$$f'_{er} \doteq -1/(1+\sqrt{2}) = -0.414 \text{ at } \Theta_{01} = \Theta_B. \quad (23)$$

Note that for a good conductor f'_{er} and with it the minimum value of f_{er} are independent of the constitutive parameters. The numerical value in (23) is the *largest attainable* by f'_{er} for any medium. For a perfect dielectric $f'_{er} = 0$; for a perfect conductor, $f'_{er} = -0.414$. The Brewster angle Θ_B and the minimum value of the reflection coefficient f_{er}, which occurs at the Brewster angle, are shown in Fig. 7.4 as functions of $h_e\epsilon_{er} = \sigma_e/\omega\epsilon_0$ with ϵ_{er} as parameter. (The plan of presentation and part of the data in Fig. 7.4 are due to Norton).

It is interesting to examine the transmission-line forms (8a, b) when region 2 is a good conductor as defined in (23). In this case

$$\zeta_2 = 1/\sqrt{\nu_0\xi_2} \doteq \sqrt{j\omega/\nu_0\sigma_{e2}}$$

$$= (1+j)\sqrt{\omega/2\nu_0\sigma_{e2}} = Z^s, \quad (24a)$$

where $Z^s = R^s + jX^s$ is by definition (ref. I.31, Sec. V.5) the *surface impedance* of region 2. Its reciprocal $Y_s = 1/Z^s$ is the *surface admittance*, so that

$$\eta_2 \doteq Y^s. \quad (24b)$$

Note that, subject to $(\sigma_{e2}/\omega\epsilon_{e2}) \gg 1$, the following approximations are valid:

$$|\zeta_2/\zeta_0|^2 \doteq \frac{\omega\epsilon_0}{\sigma_{e2}} \ll 1, \quad (25)$$

$$N_{21}^2 \doteq -jh_e\epsilon_{er} = -j\frac{\sigma_{e2}}{\omega\epsilon_0} = -jN_{21}^2, \quad N_{21}^2 \gg 1. \quad (26)$$

Since $\sin^2\Theta_{01} \leq 1$, it follows that with $N_{12} = 1/N_{21}$

$$\cos\Theta_{t2} = \sqrt{1 - N_{12}^2\sin^2\Theta_{01}} \doteq 1, \quad \Theta_{t2} \doteq 0. \quad (27)$$

Therefore, with (24),

$$Z_{e2} = \zeta_2\cos\Theta_{t2} \doteq Z^s = R^s(1+j), \quad (28a)$$

$$Y_{m2} = \eta_2\cos\Theta_{t2} \doteq Y^s = G^s(1-j), \quad (28b)$$

so that the reflection coefficients for good conductors in transmission-line form as obtained from (8a, b) and (9a, b) are as follows:

$$f'_{er} = \frac{R_{e1} - Z^s}{R_{e1} + Z^s} = \frac{N_{21}\cos\Theta_{01} - 1}{N_{21}\cos\Theta_{01} + 1}, \quad (29a)$$

$$f'_{mr} = \frac{G_{m1} - Y^s}{G_{m1} + Y^s} = \frac{N_{12}\cos\Theta_{01} - 1}{N_{12}\cos\Theta_{01} + 1}. \quad (29b)$$

Since Z^s is complex, a perfect match with $f_{er} = 0$ is not possible. However the magnitude of f_{er} can be minimized. It is readily verified that the condition minimizing the magnitude of f'_{er} is the same as that requiring its real part f''_{er} to vanish or its angle ψ_{er} to be 90°. It is

$$R_{e1} = \zeta_0\cos\Theta_{01} = R^s\sqrt{2}, \quad (30)$$

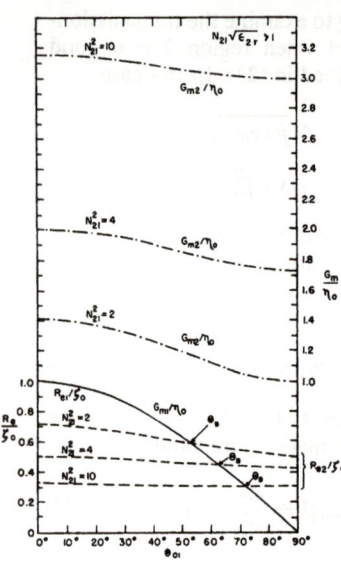

Fig. 7.3. Field impedances and admittances for perfect dielectric.

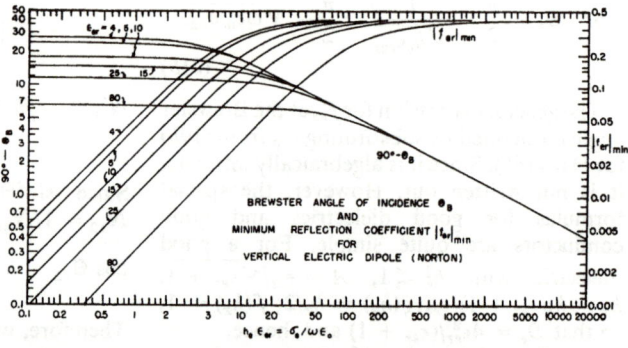

Fig. 7.4. Brewster angle of incidence θ_B and minimum reflection coefficient $|f_{er}|_{\min}$ for vertical electric dipole (Norton).

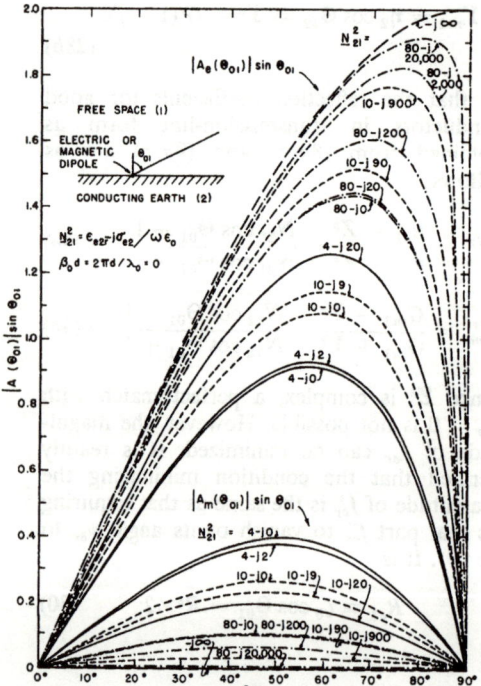

Fig. 7.5. Field factors of vertical electric and magnetic dipoles on conducting earth; $d = 0$.

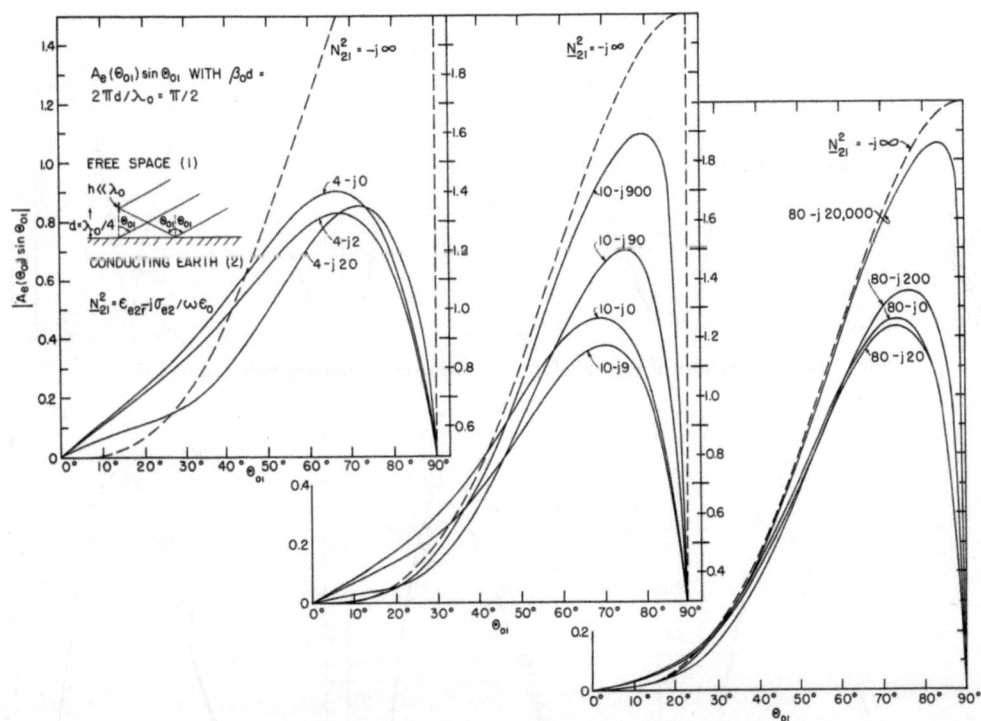

Fig. 7.6a. Field factors of vertical electric dipole over conducting earth; $d = \lambda_0/4$.

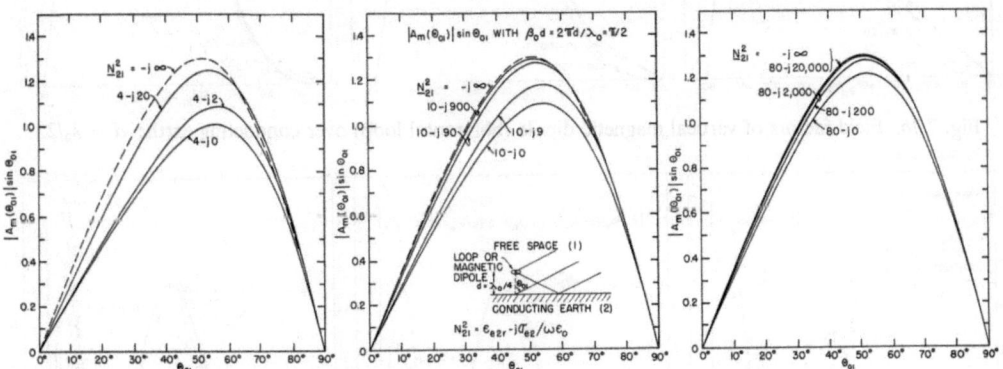

Fig. 7.6b. Field factors of vertical magnetic dipole (horizontal loop) over conducting earth; $d = \lambda_0/4$.

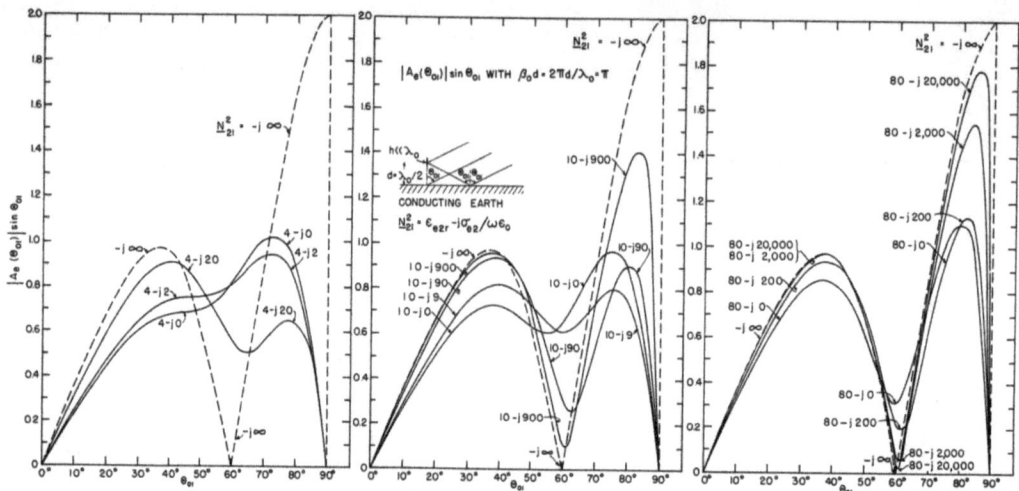

Fig. 7.7a. Field factors of vertical electric dipole over conducting earth; $d = \lambda_0/2$.

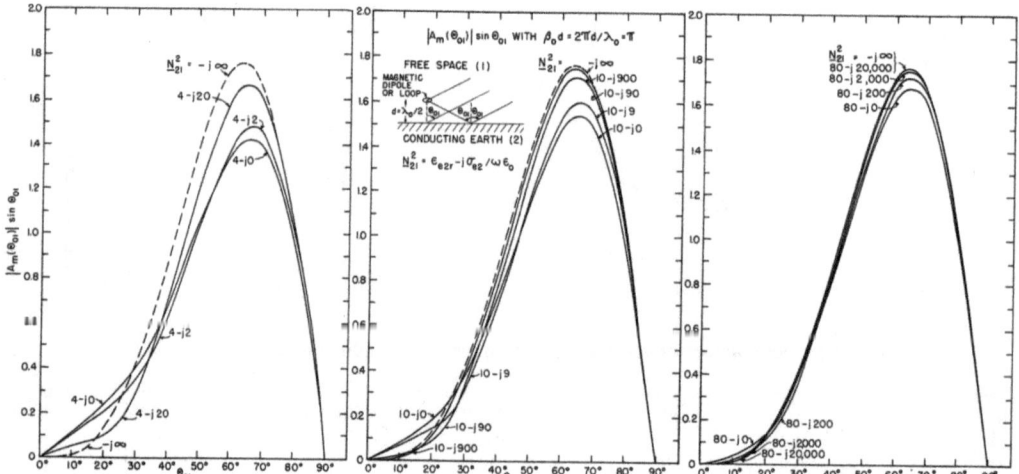

Fig. 7.7b. Field factors of vertical magnetic dipole (horizontal loop) over conducting earth; $d = \lambda_0/2$.

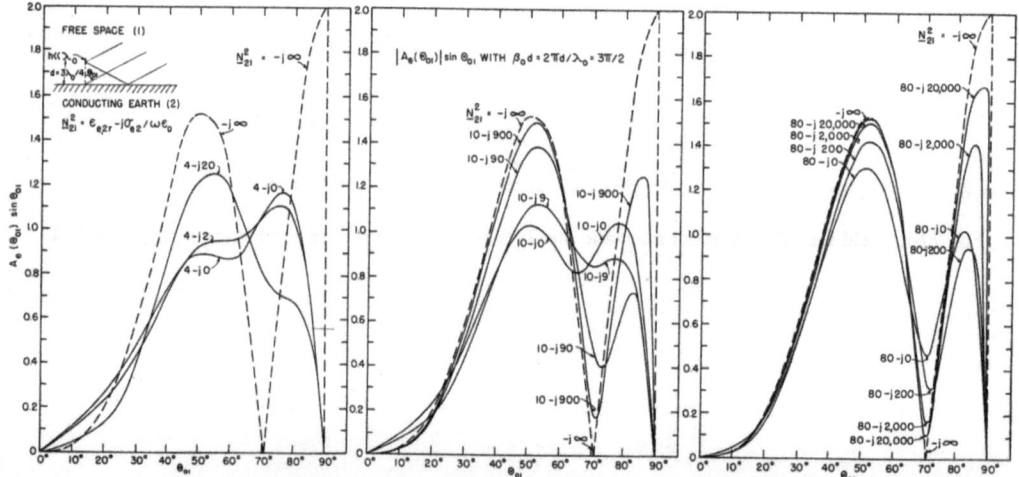

Fig. 7.8a. Field factors of vertical electric dipole over conducting earth; $d = 3\lambda_0/4$.

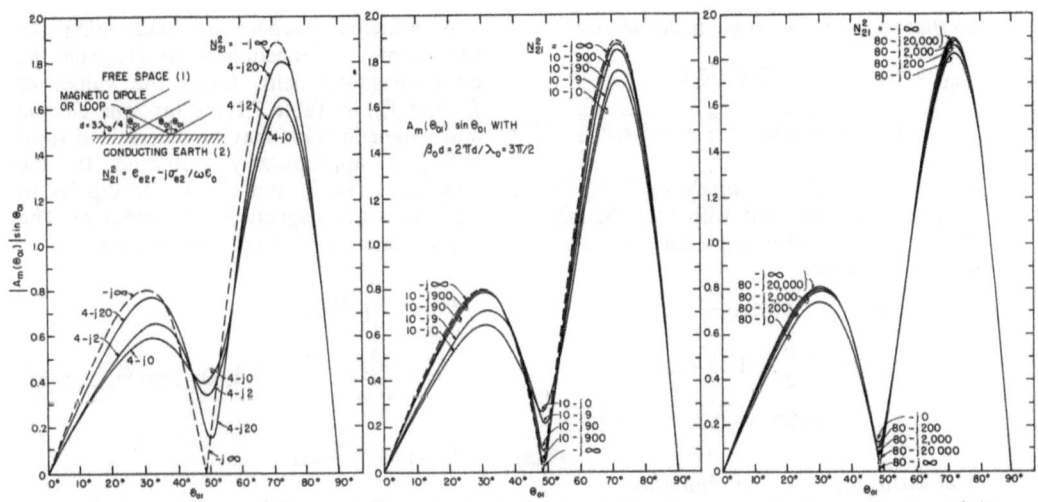

Fig. 7.8b. Field factors of vertical magnetic dipole (horizontal loop) over conducting earth; $d = 3\lambda_0/4$.

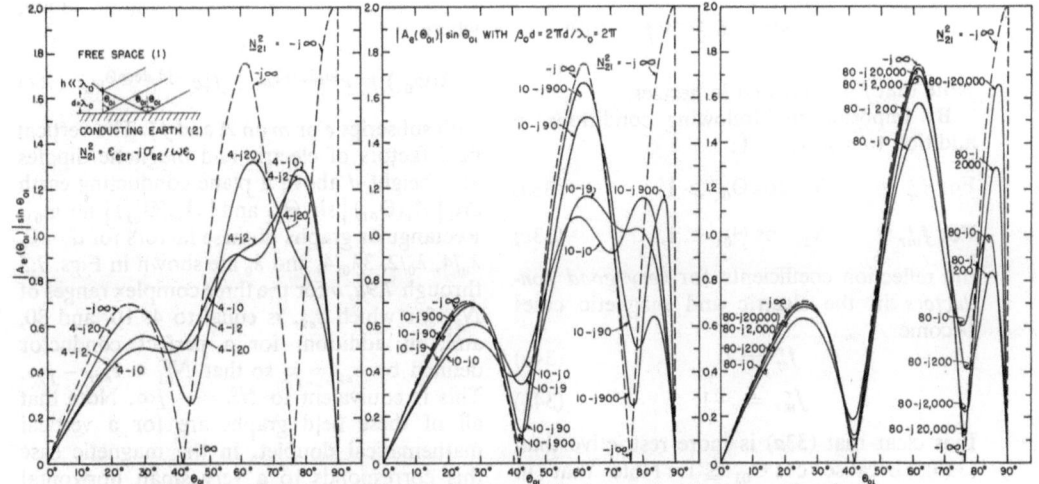

Fig. 7.9a. Field factors of vertical electric dipole over conducting earth; $d = \lambda_0$.

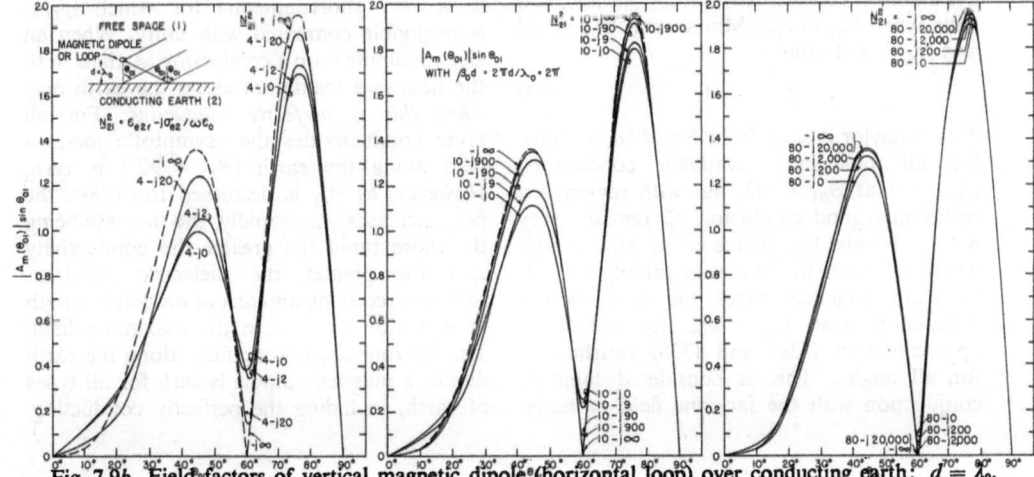

Fig. 7.9b. Field factors of vertical magnetic dipole (horizontal loop) over conducting earth; $d = \lambda_0$.

so that the Brewster angle is defined by

$$\Theta_{01} = \Theta_B = \cos^{-1}(R^s\sqrt{2}/\zeta_0)$$
$$= \cos^{-1}\sqrt{\omega\epsilon_0/\sigma_{e2}} \doteq \frac{\pi}{2} - \sqrt{\omega\epsilon_0/\sigma_{e2}} \quad (31)$$

at frequencies f that satisfy $\sigma_{e2}/2\pi f\epsilon_0 \gg 1$, in complete agreement with (21). Numerical values of the Brewster angle for salt water and wet earth follow:

Salt water: $\sigma_{e2} = 4$ mho/m,

$$\Theta_B = \frac{\pi}{2} - 1.49 \times 10^{-6}\sqrt{\omega}$$
$$= 90° - 0.21°\sqrt{f} \times 10^{-6}, \quad (32a)$$

Wet earth: $\sigma_{e2} = 4 \times 10^{-3}$ mho/m,

$$\Theta_B = \frac{\pi}{2} - 4.7 \times 10^{-5}\sqrt{\omega}$$
$$= 90° - 6.71°\sqrt{f} \times 10^{-6}. \quad (32b)$$

Note that f is expressed in hertzes.

By imposing the following conditions in addition to $\sigma_{e2}/\omega\epsilon_0 \gg 1$:

For f^r_{er}: $\quad N_{21}\cos\Theta_{01} \gg 1$, $\quad$ (33a)

For f^r_{mr}: $\quad N_{12}\cos\Theta_{01} \ll 1$, $\quad$ (33b)

the reflection coefficients for *very good conductors* in the electric and magnetic cases become

$$f^r_{er} \doteq 1, \quad (34a)$$
$$f^r_{mr} \doteq -1. \quad (34b)$$

It is clear that (33a) is more restrictive than (33b), because $\cos\Theta_{01} \leq 1$. Note that in the electric case Θ_{01} may not be $\pi/2$ if N_{21} is finite. Only when N_{21} is infinite, as when region 2 is a perfect conductor with $\sigma_{e2} = \infty$, is (34a) valid when $\Theta_{01} = \pi/2$. Moreover if $\Theta_{01} = \pi/2$ and N_{21} is not infinite,

$$f^r_{er} = -1. \quad (35)$$

This behavior is clear from Fig. 7.1a, b. Thus, for all practically available conductors, $f^r_{er} = -1$ at $\Theta_{01} = \pi/2$, but with region 2 a sufficiently good conductor, f^r_{er} remains very near $+1$ until Θ_{01} increases to almost $\pi/2$, where it begins to drop very rapidly to -1. In many practical cases the drop occurs sufficiently near $\Theta_{01} = \pi/2$ to make the approximation (33a) and (33b) satisfactory for all angles. This is considered later in conjunction with the far-zone field patterns.

A principal purpose in determining the coefficients of reflection is to evaluate the electromagnetic field. Using the values of f^r_{er} and f^r_{mr} in Table 7.1, the far-zone electric and magnetic fields may be determined from (6.83a, b) appropriately restricted to the conditions (1a, b) and (2a, b) for dipoles in air over a nonmagnetic simple medium. The appropriately specialized formulas are

$$E^r_{e\Theta 1} = \frac{\omega}{\beta_0} B^r_{e\Phi 1}$$
$$= \frac{-\beta_0^2 p_{z1}}{4\pi\epsilon_0} \frac{e^{-j\beta_0 R_0}}{R_0} A_e(\Theta_{01})\sin\Theta_{01}, \quad (36a)$$

$$B^r_{m\Theta 1} = \frac{\beta_0}{\omega} E^r_{m\Phi 1}$$
$$= \frac{-\beta_0^2 m_{z1}}{4\pi v_0} \frac{e^{-j\beta_0 R_0}}{R_0} A_m(\Theta_{01})\sin\Theta_{01}, \quad (36b)$$

where

$$A(\Theta_{01}) \equiv e^{j\beta_0 d\cos\Theta_{01}} + f^r_r e^{-j\beta_0 d\cos\Theta_{01}}, \quad (36c)$$

with subscript e or m on A and f^r_r. The vertical field factors of electric and magnetic dipoles at a height d above a plane conducting earth are $|A_e(\Theta_{01})|\sin\Theta_{01}$ and $|A_m(\Theta_{01})|\sin\Theta_{01}$. Rectangular graphs of these factors for $d = 0$, $\lambda_0/4$, $\lambda_0/2$, $3\lambda_0/4$, and λ_0 are shown in Figs. 7.5 through 7.9a, b for the three complex ranges of N^2_{21} for which ϵ_{e2r} is equal to 4, 10, and 80, and, in addition, for a perfect conductor defined by $\sigma_{e2} = \infty$ so that $N^2_{21} = \epsilon_{e2r} - j\infty$. This is equivalent to $N^2_{21} = -j\infty$. Note that all of these field graphs are for a vertical mathematical doublet. In the magnetic case this corresponds to a very small horizontal loop with side b satisfying the condition $\beta_0^2 b^2 \ll 1$. In the electric case it is equivalent to a very short antenna for which $(\beta_0 h)^2$ is negligible compared with unity. When an electric dipole is on a conducting earth ($d = 0$) the field is a maximum along the earth *only when this is perfectly conducting*. For all finite conductivities the asymptotic far-zone field along the earth ($\Theta_{01} = 90°$) is zero. However, as Θ_{01} is decreased from 90°, the field increases very rapidly, the increase being the more rapid the greater the conductivity and the greater the dielectric constant, so that a receiving antenna of moderate length or elevation may be in the maximum field. The far-zone asymptotic field along the earth due to a magnetic dipole is zero for all types of earth, including the perfectly conducting.

Subject to (33a, b) and (34a, b), the array factors $A(\Theta_{01})$ for the far-zone field of vertical electric and magnetic dipoles in space over a highly conducting region are

$$A_e(\Theta_{01}) \doteq 2 \cos (\beta_0 d \cos \Theta_{01}), \quad (37a)$$

$$A_m(\Theta_{01}) \doteq 2j \sin (\beta_0 d \cos \Theta_{01}). \quad (37b)$$

The complete field factors in both cases are $|A(\Theta_{01})| \sin \Theta_{01}$. The fields are obtained by substituting (37a, b) in (36a, b). For the vertical magnetic dipole (37b) is valid for all values of Θ_{01}. On the other hand, (37a) is a good approximation in the far zone only when Θ_{01} is not too near $\pi/2$. That is, the field in space very near the boundary plane is not given correctly by (37a) except when region 2 is a perfect conductor. Note, however, that the angle Θ_{01} may approach $\pi/2$ the more closely the lower the frequency, since $N_{21} = \sqrt{\sigma_{e2}/\omega \epsilon_0}$ increases as ω is reduced. As a numerical illustration, consider salt water ($\sigma_{e2} \doteq 4$ mho/m), wet earth ($\sigma_{e2} \doteq 4 \times 10^{-3}$ mho/m), and dry earth ($\sigma_{e2} \doteq 4 \times 10^{-5}$ mho/m). For these, the following numerical values obtain:

Salt water: $N_{21} = \dfrac{4.52}{\omega} \times 10^{11}$,

$N_{12} = 2.21 \times 10^{-12}\omega$; (38a)

Wet earth: $N_{21} = \dfrac{4.52}{\omega} \times 10^{8}$,

$N_{12} = 2.21 \times 10^{-9}\omega$; (38b)

Dry earth: $N_{21} = \dfrac{4.52}{\omega} \times 10^{6}$,

$N_{12} = 2.21 \times 10^{-7}\omega$. (38c)

In order to satisfy (33a) it is necessary that $\cos \Theta_{01}$ be very large compared with N_{12}. For example, if $\cos \Theta_{01} = 0.0221$, $\Theta_{01} = 1.548 = 88.7°$, it is necessary that for salt water, $\omega \leq 10^8$; for wet earth, $\omega \leq 10^5$; and for dry earth, $\omega \leq 10^3$. For all finite values of N_{21}, however, there is a limiting angle less than $\pi/2$ which may not be exceeded if (33a) is to be maintained valid. The higher the frequency and the greater the conductivity, the nearer this limit approaches $\pi/2$. For example, for a low radio frequency with $\omega = 10^7$, N_{12} has the values 2.21×10^{-5}, 0.0221, 2.21, respectively, for salt water, wet earth, and dry earth. Clearly, the condition $\cos \Theta_{01} \gg N_{12}$ is satisfied only for salt water and wet earth, not for dry earth. For salt water, Θ_{01} may not exceed $0.5\pi - 2.21 \times 10^{-3} \doteq 90° - 0.13°$; for wet earth, Θ_{01} may not exceed $0.5\pi - 0.0221 \doteq 90° - 1.3°$. Clearly, the formula (33a) is satisfied for all *practically measurable angles* for an electric dipole over salt water at $\omega = 10^7$. For wet earth it is satisfied for all angles except within about 1.3° of the plane of the earth. Obviously, an angle of 1.3° means a significant distance above the earth's surface in the far zone at radio frequencies.

A convenient formula for investigating the far-zone field of a vertical electric dipole for angles Θ_{01} near $\pi/2$ is obtained from the general formula (6.83a) by setting $\sin^2 \Theta_{01} \doteq 1$, so that (6.84a) becomes

$$f_{er}^r \doteq \dfrac{N_{21}^2 \cos \Theta_{01} - \sqrt{N_{21}^2 - 1}}{N_{21}^2 \cos \Theta_{01} + \sqrt{N_{21}^2 - 1}}$$

$$= -\dfrac{1 - (N_{21}^2 \cos \Theta_{01}/\sqrt{N_{21}^2 - 1})}{1 + (N_{21}^2 \cos \Theta_{01}/\sqrt{N_{21}^2 - 1})}. \quad (39)$$

For very small values of $\cos \Theta_{01}$, with Θ_{01} near $\pi/2$, this may be expanded as follows:

$$f_{er}^r \doteq -\left(1 - \dfrac{N_{21}^2 \cos \Theta_{01}}{\sqrt{N_{21}^2 - 1}}\right)^2$$

$$\doteq -1 + \dfrac{2 N_{21}^2 \cos \Theta_{01}}{\sqrt{N_{21}^2 - 1}}, \quad (40)$$

provided

$$|N_{21}^2 \cos \Theta_{01}|^2 \ll |\sqrt{N_{21}^2 - 1}|^2. \quad (41)$$

This value of f_{er}^r may be used in (36a) with (36c) in place of (9a) when Θ_{01} is very near $\pi/2$. In this case the far-zone field factor

$$A_e^r(\Theta_{01}) \doteq \left[2j \sin (\beta_0 d \cos \Theta_{01}) + \dfrac{2 N_{21}^2 \cos \Theta_{01}}{\sqrt{N_{21}^2 - 1}} e^{-j\beta_0 d \cos \Theta_{01}} \right] \quad (42)$$

applies near the boundary where $\Theta_{01} \doteq \pi/2$. The corresponding formula for the magnetic dipole is like (42) but with the factor N_{21}^2 missing in the numerator of the second term.

In attempting to evaluate the field for values of Θ_{01} very near $\pi/2$ where the far-zone expressions give zero (except for a perfectly conducting region 2 with a vertical electric dipole), it must be recalled that the far-zone formulas are asymptotic solutions in which only terms of the order $1/R_0$ are included. When the angle Θ_{01} approaches sufficiently close to $\pi/2$, higher-order terms become significant. In particular, along the boundary plane

$\Theta_{01} = \pi/2$ the electromagnetic field is determined entirely by terms of higher order than are contained in the asymptotic solutions obtained using the saddle-point values of the slowly varying functions. Terms of order $1/R_0^2$ and higher may be obtained by expanding the functions Θ and ϕ in (6.35) and (6.64) in series in terms of u and v about the points $\Theta = \Theta_{01}$ and $\phi = 0$ and retaining higher-order terms than the first. Essentially this procedure has been carried out by Strutt; his numerical computations of the tangential fields, E_Φ and H_Φ, on the boundary show that the higher-order terms are significant only near $\Theta = \pi/2$, where they contribute a small term. Although the complete field for values of Θ_{01} near $\pi/2$ and for distances closer than permitted by the condition for the far zone can be obtained by the method just outlined, the evaluation is extremely laborious and has not been carried out. However, an alternative, essentially equivalent, procedure has been developed especially by Van der Pol and Norton and explicit formulas obtained for the leading term in the quasi-near-zone fields along the boundary plane. Extensive curves also have been computed. Accordingly, it is advantageous to carry out the analysis leading to these results. This is begun in Sec. 9.

Field in region 2. In concluding the discussion of the field of dipoles in air over a plane conducting earth, a brief analysis of the complicated field in the earth is appropriate. This field is given by (6.85a, b) with the several parameters restricted according to (1) through (3). The most important parts of these formulas are not the field factors alone, as for the field in air, but also the exponentials. In a conducting earth β_2 and Θ_{t2} are complex, so that the amplitude of the field is attenuated exponentially by the real part of (5.74). Thus for $z \leq -d$

$$\exp\{-j\beta_2[-(z+d)\cos\Theta_{t2} + (r - r_i)\sin\Theta_{t2}]\}. \quad (43)$$

It is this factor that determines the nature of the field in region 2. For simplicity, let

$$z' \equiv -(z+d), \quad r' \equiv r - r_i, \quad (44)$$

so that z' is measured down into region 2 with $z' = 0$ at the boundary plane. With the law of refraction, $\beta_2 \sin\Theta_{t2} = \beta_0 \sin\Theta_{01}$, and $\beta_2 = \beta_0 N_{21}$ the exponent in (43) becomes:

$$-j\beta_0(z' N_{21} \cos\Theta_{t2} + r' \sin\Theta_{01})$$
$$= -j\beta_0(z' \sqrt{N_{21}^2 - \sin^2\Theta_{01}} + r' \sin\Theta_{01}). \quad (45)$$

The expression on the right follows from

$$\cos\Theta_{t2} = \sqrt{1 - \sin^2\Theta_{t2}}$$
$$= N_{12}\sqrt{N_{21}^2 - \sin^2\Theta_{01}}. \quad (46)$$

Use is now made of (10) through (13a) in (45), and this becomes the exponent in (43). The resulting formula for a plane wave in region 2 is

$$e^{-\beta_0 z' A g(h_x)} e^{-j\beta_0[z' A f(h_x) + r' \sin\Theta_{01}]}, \quad (47)$$

where, as in (10) to (13), $N_{21}^2 = \epsilon_{er}(1 - jh_e)$, $A = \sqrt{\epsilon_{er} - \sin^2\Theta_{01}}$, $h_x = \epsilon_{er} h_e/A^2$, and $\sqrt{1 - jh_x} = f(h_x) - jg(h_x)$; $f(h)$ and $g(h)$ are tabulated in reference I.31, Appendix II. Note that since the real part of (47) depends only on z', the surfaces of *constant amplitude* in (43) are *planes parallel to the boundary*. As region 2 is penetrated vertically downward, the amplitude of (43) decreases as $e^{-z'\beta_0 A g(h_x)}$, where z' is measured down into region 2. On the other hand, the surfaces of constant phase are planes defined by the equation

$$z' A f(h_x) + r' \sin\Theta_{01} = C_1 = \text{const.} \quad (48)$$

The normal to this plane defines the direction of propagation of the plane wave represented by (43). The real angle Θ_{02} made by this normal with the negative z-axis (positive z'-axis) is defined by

$$z' \cos\Theta_{02} + r' \sin\Theta_{02} = C_2 = \text{const.} \quad (49)$$

By obtaining the equation for the normal to the planes defined by (48) and comparing its slope with that of (49), the following results are obtained:

$$\cos\Theta_{02} = \frac{Af(h_x)}{\sqrt{A^2 f^2(h_x) + \sin^2\Theta_{01}}}; \quad (50a)$$

$$\sin\Theta_{02} = \frac{\sin\Theta_{01}}{\sqrt{A^2 f^2(h_x) + \sin^2\Theta_{01}}}. \quad (50b)$$

Since Θ_{02} is the angle between the normal to the boundary plane and the direction of propagation, it is the true, real angle of refraction, and (50b) expresses the law of refraction, with the real *index of refraction* given by

$$N_{21}(\Theta_{01}) \equiv \frac{\sin\Theta_{01}}{\sin\Theta_{02}} = \sqrt{A^2 f^2(h_x) + \sin^2\Theta_{01}}. \quad (51a)$$

With (11), (12), and (13b),

$$N_{21}(\Theta_{01}) = \{\tfrac{1}{2}[\sqrt{(\epsilon_{er} - \sin^2\Theta_{01})^2 + h_e^2 \epsilon_{er}^2} + \epsilon_{er} + \sin^2\Theta_{01}]\}^{1/2}. \quad (51b)$$

When region 2 is a perfect dielectric, $h_e = \sigma_e/\omega\epsilon_e = 0$ and (51b) reduces to $N_{21}(\Theta_{01}) = N_{21} = \sqrt{\epsilon_{er}}$. Also $h_x = 0$ so that the first exponential in (47) reduces to $e^{-\beta_0 z' A g(h_x)} = 1$ and there is no exponential attenuation normal to the boundary as region 2 is penetrated.

When region 2 is a slightly conducting dielectric that satisfies the condition $h_x^2 \doteq \epsilon_{er} h_e/(\epsilon_{er} - \sin^2 \Theta_{01}) \ll 1$, it follows that $f(h_x) \doteq 1$, $g(h_x) \doteq h_x/2$, so that $N_{21}(\Theta_{01}) \doteq N_{21} = \sqrt{\epsilon_{er}}$, as for a perfect dielectric. However, the attenuation is not zero but is given by $e^{-\beta_0 z' h_e \epsilon_{er}/2\sqrt{\epsilon_{er} - \sin^2 \Theta_{01}}}$.

When region 2 is highly conducting, so that $h_e = \sigma_e/\omega\epsilon_e \gg 1$, (51b) reduces to $N_{21}(\Theta_{01}) \doteq \frac{1}{2} h_e \epsilon_{er} = \sigma_e/2\omega\epsilon_0$, which is independent of Θ_{01}. In this case $h_x = \sigma_e/\omega\epsilon_0 A^2 \gg 1$ so that $g(h_x) \doteq \sqrt{h_x/2} = (1/A)\sqrt{\sigma/2\omega\epsilon_0}$. Accordingly, the very high exponential attenuation in (47) is $e^{-\beta_0 z'\sqrt{\sigma_e/2\omega\epsilon_0}} = e^{-z'/d_s}$, where $1/d_s \equiv \beta_0 \sqrt{\sigma_e/2\omega\epsilon_0} = \sqrt{\omega\sigma_e/2\nu_0}$ and d_s is the skin depth. When region 2 is a perfect conductor, the skin depth is zero and the attenuation in region 2 is infinite, so that the field vanishes everywhere in region 2.

In the general case of a moderately conducting region 2, $N_{21}(\Theta_{01})$ is a function of Θ_{01}, and $g(h_x)$ is sufficiently large so that the exponential attenuation of the field in region 2 is great. Note that for vertical incidence, $\Theta_{01} = 0$, the real exponent in (47) may be expressed in the following simple form:

$$\beta_0 z' A g(h_x) = \beta_0 \sqrt{\frac{\epsilon_{er}}{\cos \alpha}} \sin \alpha/2,$$

$$\alpha = \tan^{-1} h_e. \quad (52)$$

By combining the exponential attenuation factor $e^{-\beta_0 z A g(h_x)}$ with the magnitude of the field factors in (6.85a, b) the distribution and magnitude of the far-zone field in region 2 near its surface is obtained.

8. The Far-Zone Field of an Antenna with Sinusoidally Distributed Current Over a Conducting Earth

The analysis in the preceding section has concerned itself with the determination of the far-zone field of electric and magnetic dipoles that are of infinitesimal length or at least very short compared with the wavelength. It is the purpose of this section to derive expressions for the electromagnetic field of (1) a vertical center-driven antenna of half-length h with center at a height d in air over a horizontal conducting and dielectric plane of infinite extent, and (2) a base-driven antenna of length h erected vertically on the conducting earth. In carrying out the analysis it is necessary to assume a sinusoidal distribution of current in order to get sufficiently simple expressions. It is known from Chapter II that this is a good approximation of the actual distribution on antennas of finite cross section when $\beta_0 h$, the electrical half-length, does not exceed 2. It is particularly good when $\beta_0 h = \pi/2$. For electrical half-lengths greater than 2, the sinusoidal distribution leads to a fair approximation of the radiation field except in the vicinity of sharp nulls in the field pattern.

The analysis of the field of an antenna of finite length over a conducting earth proceeds from the polarization potential $d\Pi_{ez1}$ of an infinitesimal dipole or electric doublet. As explained in Sec. 2, the doublet may be replaced by an equivalent current in an element dz'. Specifically,

$$p_{1z} = I_z dz'/j\omega. \quad (1)$$

The polarization potential due to such an element at a height z' above the boundary plane is given by (6.37) specialized to apply to empty space with (1) substituted for p_{1z} and z' substituted for d:

$$d\Pi_{ez1} = \frac{I_z' dz'}{4\pi\epsilon_0 j\omega} \frac{e^{-j\beta_0 R_0}}{R_0} (e^{j\beta_0 z'\gamma} + f_{er}^r e^{-j\beta_0 z'\gamma}), \quad (2)$$

where, for later convenience in manipulation, the following shorthand is introduced:

$$\gamma \equiv \cos \Theta_{01}. \quad (3)$$

Since f_{er}^r is the far-zone reflection coefficient given by (7.9a), it is sensibly constant and independent of the coördinate z' locating the current in the element dz' above the conducting plane.

In order to solve simultaneously both the elevated and the base-driven antenna, let the following quite general form of the sinusoidal distribution of current be assumed:

$$I_z = I_c \frac{\sin \beta_0(c + g - z)}{\sin \beta_0 g}$$

$$= I_m \sin \beta_0(c + g - z), \quad (4)$$

where

$$g = h + k, \quad (5)$$

as shown in Fig. 8.1; I_c is the amplitude of the current at the base of the antenna; I_m is the maximum amplitude of the sine distribution. The amplitude at the top of the antenna is $I_l = I_m \sin \beta_0 k$. For convenience, let

$$G \equiv \beta_0 g, \quad H \equiv \beta_0 h, \quad K \equiv \beta_0 k, \quad C \equiv \beta_0 c,$$
$$Z \equiv \beta_0 z, \quad Z' \equiv \beta_0 z', \quad (6)$$

The polarization potential in region 1 due to the entire antenna is obtained by substituting (4) in (2) and integrating over the antenna:

$$\Pi_{ez1} = \frac{I_m}{j\omega 4\pi\epsilon_0} \frac{e^{-j\beta_0 R_0}}{\beta_0 R_0} \int_C^{C+H} (e^{jZ'\gamma}$$
$$+ f_{er}^r e^{-jZ'\gamma}) \left[\frac{e^{j(C+G-Z')} - e^{-j(C+G-Z')}}{2j} \right] dZ'. \quad (7)$$

Let the integral in (7) be denoted by J. After rearrangement, (7) becomes

$$J = \frac{1}{2j} \int_C^{C+H} \{e^{j(C+G)}[e^{jZ'(\gamma-1)}$$
$$+ f_{er}^r e^{-jZ'(\gamma+1)}] - e^{-j(C+G)}[e^{jZ'(\gamma+1)}$$
$$+ f_{er}^r e^{-jZ'(\gamma-1)}]\} dZ'. \quad (8)$$

This may be integrated directly into the following:

$$J = \tfrac{1}{2} \left\{ e^{j(C+G)} \left[\frac{e^{j(C+H)(\gamma-1)} - e^{jC(\gamma-1)}}{-(\gamma-1)} \right. \right.$$
$$\left. + f_{er}^r \frac{e^{-j(C+H)(\gamma+1)} - e^{-jC(\gamma+1)}}{\gamma+1} \right]$$
$$- e^{-j(C+G)} \left[\frac{e^{j(C+H)(\gamma+1)} - e^{jC(\gamma+1)}}{-(\gamma+1)} \right.$$
$$\left. \left. + f_{er}^r \frac{e^{-j(C+H)(\gamma-1)} - e^{-jC(\gamma-1)}}{\gamma-1} \right] \right\}. \quad (9)$$

Rearrangement of (9) and substitution in (7) gives the following final formulas:

$$\Pi_{ez1} = \frac{-jI_m}{2\pi\epsilon_0 \omega} \frac{e^{-j\beta_0 R_0}}{\beta_0 R_0 \sqrt{1-\gamma^2}} A_e(\Theta_{01}) F(\Theta_{01}), \quad (10a)$$

$$E_{\Theta 1}^r = -\beta_0^2 \Pi_{ez1} \sqrt{1-\gamma^2}$$
$$= \frac{-jI_m \zeta_0}{2\pi} \frac{e^{-j\beta_0 R_0}}{R_0} A_e(\Theta_{01}) F(\Theta_{01}), \quad (10b)$$

where the vertical field factor is defined by

$$A_e(\Theta_{01}) F(\Theta_{01}) = \tfrac{1}{4} \left\{ \frac{\gamma+1}{\sqrt{1-\gamma^2}} [e^{j[K+(C+H)\gamma]} \right.$$
$$+ f_{er}^r e^{-j[K+(C+H)\gamma]} - e^{j(G+C\gamma)}$$
$$- f_{er}^r e^{-j(G+C\gamma)}] - \frac{\gamma-1}{\sqrt{1-\gamma^2}} [e^{-j[K-(C+H)\gamma]}$$
$$+ f_{er}^r e^{j[K-(C+H)\gamma]} - e^{-j(G-C\gamma)}$$
$$\left. - f_{er}^r e^{j(G-C\gamma)}] \right\}. \quad (10c)$$

The field factor of a center-driven antenna of half-length h with center at a height $d \geq h$ above the plane earth is obtained by adding the contributions from the upper and lower halves. Each of these contributions is obtained from (10c) by appropriate specialization of the several parameters. For simplicity, let it be assumed that $h \leq \lambda_0/2$. Then, for the

Upper half:
$$G = H, \quad K = 0, \quad C = D = \beta_0 d, \quad (11a)$$

Lower half:
$$G = \pi, \quad K = \pi - H, \quad C + H = D. \quad (11b)$$

With these values substituted successively in (10b), the results are:

Upper half:
$$A_e(\Theta_{01}) F_u(\Theta_{01}) = \frac{1}{4\sqrt{1-\gamma^2}} \{(\gamma+1)[e^{j(D+H)\gamma}$$
$$+ f_{er}^r e^{-j(D+H)\gamma} - e^{j(D\gamma+H)} - f_{er}^r e^{-j(D\gamma+H)}]$$
$$- (\gamma-1)[e^{j(D+H)\gamma} + f_{er}^r e^{-j(D+H)\gamma}$$
$$- e^{j(D\gamma-H)} - f_{er}^r e^{-j(D\gamma-H)}]\}, \quad (12a)$$

Lower half:
$$A_e(\Theta_{01}) F_l(\Theta_{01}) = \frac{-1}{4\sqrt{1-\gamma^2}} \{(\gamma+1)[e^{j(D\gamma-H)}$$
$$+ f_{er}^r e^{-j(D\gamma-H)} - e^{j(D-H)\gamma} - f_{er}^r e^{-j(D-H)\gamma}]$$
$$- (\gamma-1)[e^{j(D\gamma+H)} + f_{er}^r e^{-j(D\gamma+H)}$$
$$- e^{j(D-H)\gamma} - f_{er}^r e^{-j(D-H)\gamma}]\}. \quad (12b)$$

If these two expressions are added and the factors of $e^{jD\gamma}$ and $f_{er}^r e^{-jD\gamma}$ collected and simplified, the final result, for the center-driven antenna of half-length h, is

$$A_e(\Theta_{01}) F_e(\Theta_{01})$$
$$= (e^{j\beta_0 d \cos\Theta_{01}} + f_{er}^r e^{-j\beta_0 d \cos\Theta_{01}})$$
$$\times \left[\frac{\cos(\beta_0 h \cos\Theta_{01}) - \cos\beta_0 h}{\sin\Theta_{01}} \right]. \quad (13)$$

Evidently,

$$F_m(\Theta_{01}, \beta_0 h) \equiv \frac{\cos(\beta_0 h \cos \Theta_{01}) - \cos \beta_0 h}{\sin \Theta_{01}}$$

$$\equiv F_0(\beta_0 h, \Theta_{01}) \sin \beta_0 h, \quad (14)$$

is identically the vertical field factor of an isolated center-driven antenna of half-length h with a sinusoidally distributed current, as defined in Chapter V, Sec. 4. Note that $A_e(\Theta_{01})$ is the same as (7.36c) for the infinitesimal electric dipole. In the important special case of a half-wave dipole with $\beta_0 h = \pi/2$, the field factor has the following form:

$$A_e(\Theta_{01}) F_m(\Theta_{01}, \tfrac{1}{2}\pi) = (e^{j\beta_0 d \cos \Theta_{01}}$$

$$+ f_{er}^r e^{-j\beta_0 d \cos \Theta_{01}}) \left[\frac{\cos(\tfrac{1}{2}\pi \cos \Theta_{01})}{\sin \Theta_{01}} \right]. \quad (15)$$

For an electrically short antenna for which $\beta_0^2 h^2 \ll 1$ is satisfied the field factor is

$$A_e(\Theta_{01}) F_0(\Theta_{01}, \beta_0 h) = (e^{j\beta_0 d \cos \Theta_{01}}$$

$$+ f_{er} e^{-j\beta_0 d \cos \Theta_{01}})(\tfrac{1}{2}\beta_0 h \sin \Theta_{01}). \quad (16)$$

Note that this has the same dependence on Θ_{01} as the field factor in (7.36a) with (7.36c) of the electric doublet.

In order to obtain the far-zone field factors (13) from the curves in Figs. 7.5 through 7.9 for the electric doublet, it is merely necessary to multiply the ordinates by the factor

$$| [\cos(\beta_0 h \cos \Theta_{01}) - \cos \beta_0 h]/\sin^2 \Theta_{01} |. \quad (17)$$

For $\beta_0 h = \pi/2$ this factor becomes simply $|[\cos(\tfrac{1}{2}\pi \cos \Theta_{01})]/\sin^2 \Theta_{01}|$, which, as is seen from Table 8.1, is near unity except at small angles Θ_{01}, where the field is small in any case. Hence, for most practical purposes the field factors in Figs. 7.5 through 7.9a, b may be used for center-driven antennas of half-length $h \leq \lambda_0/4$ with centers at a height d above the conducting earth. If numerically more accurate results are required, the ordinates in these figures may be multiplied by the factor (17) appropriate to the particular value of $\beta_0 h$. For $\beta_0 h = \pi/2$, it is given in Table 8.1.

It is significant to note that the array factor $A_e(\Theta_{01})$ in (13) for a vertical antenna with center at a height d over a horizontal conducting earth that has a coefficient of reflection f_{er}^r is exactly the same as the array factor $A(\Theta)$ in (VI.2.52) for two collinear antennas separated in space by a distance $2d$ between centers when the current in the lower antenna is k times that in the upper antenna, if k is replaced by f_{er}^r. This means that the far-zone field of the currents in the antenna and in the conducting earth is the same as the far-zone field of the currents in that antenna and in an image antenna in which the current is $k = f_{er}^r$ times the current in the primary antenna. Since f_{er}^r is a function of Θ_{01}, a different current ratio is required for each angle. When the earth is perfectly conducting, f_{er}^r is real and equal to unity, and, hence, independent of Θ_{01}. Accordingly, the current in the image is equal to the current in the primary antenna both in magnitude and phase, that is, the currents are codirectional. When the earth is a perfect dielectric, $f_{er}^r = f_{er}^r$ is real but neither equal to unity nor independent of Θ_{01}. Hence, with $k = k = f_{er}$ the current in the image antenna is codirectional with that in the primary antenna for Θ_{01} less than the Brewster angle, and is oppositely directed for Θ_{01} greater than the Brewster angle. The value of $k = f_{er}$ is different for each angle Θ_{01}, as shown in Fig. 7.1a.

Thus, the far-zone field of an antenna over a conducting earth in a specified direction Θ_{01} may be determined by replacing the earth by free space containing an image antenna with a current $k = f_{er}(\Theta_{01})$ times the current in the actual antenna.

The complex field factor of a base-driven antenna of length h erected on the surface of the earth is obtained from (10c) by setting $G = H$, $K = 0$, and $C = 0$. The result is

$$A_e(\Theta_{01}) F(\Theta_{01}) = \frac{\gamma + 1}{4\sqrt{1 - \gamma^2}} (e^{jH\gamma} + f_{er}^r e^{-jH\gamma})$$

$$- e^{jH} - f_{er}^r e^{-jH}) - \frac{\gamma - 1}{4\sqrt{1 - \gamma^2}} (e^{jH\gamma}$$

$$+ f_{er}^r e^{-jH\gamma} - e^{-jH} - f_{re}^r e^{jH}). \quad (18)$$

After collecting and rearranging terms with $\gamma = \cos \Theta_{01}$, $H = \beta_0 h$, the following formula is obtained:

$$A_e(\Theta_{01}) F(\Theta_{01}) = \tfrac{1}{2}[(1 + f_{er}) F_m(\Theta_{01}, \beta_0 h)$$

$$+ j(1 - f_{er}) W_m(\Theta_{01}, \beta_0 h)], \quad (19)$$

where $F_m(\beta_0 h, \Theta)$ is given by (14) and where

$$W_m(\Theta_{01}, \beta_0 h)$$

$$= \frac{\sin(\beta_0 h \cos \Theta_{01}) - \sin \beta_0 h \cos \Theta_{01}}{\sin \Theta_{01}}. \quad (20)$$

TABLE 8.1. Field factor with $\beta_0 h = \pi/2$.

Θ_{01} (deg)	0	10	20	30	40	50	60	70	80	90
$\dfrac{\cos(\frac{1}{2}\pi \cos \Theta_{01})}{\sin^2 \Theta_{01}}$	0.393	0.774	0.800	0.837	0.872	0.906	0.942	0.973	0.993	1.000

The magnitude of (19) is the real vertical field factor of the antenna on the conducting earth. With $f_{er} = f''_{er} + jf'_{er}$ it is given by

$$|A_e(\Theta_{01})F(\Theta_{01})|$$
$$= \tfrac{1}{2}\{[(1 + f''_{er})F + f'_{er}W]^2$$
$$+ [(1 - f''_{er})W + f'_{er}F]^2\}^{\frac{1}{2}}, \quad (21)$$

where for convenience the subscripts and arguments of $F_m(\Theta_{01}, \beta_0 h)$ and $W_m(\Theta_{01}, \beta_0 h)$ have been omitted. When the antenna is erected on a perfect conductor for which $f_{er} = 1$, (19) or (21) reduces to the simple form for a symmetric isolated center-driven antenna as required by the theorem of images. That is,

$$A_e(\Theta_{01})F(\Theta_{01}) = F_m(\Theta_{01}, \beta_0 h). \quad (22)$$

Note also that when the antenna is very short, so that $\beta_0^2 h^2 \ll 1$, $W_m(\Theta_{01}, \beta_0 h) \doteq 0$ and $F_m(\Theta_{01}, \beta_0 h) = \tfrac{1}{2}\beta_0 h \sin \Theta_{01}$, which is the same as for the electric doublet on the conducting plane.

It is interesting and significant to observe that (19) and (21) are exactly the same in form as the array factor $F_a(\Theta)$ of an isolated center-driven antenna with unequal currents in the halves, as derived in Sec. V.18. Note that the coefficient of reflection f_{er} takes the place in (19) of the complex current ratio k appearing in (V.18.4). However, since f_{er} is itself a function of Θ_{01}, whereas k is a constant independent of Θ_{01}, the field patterns of an antenna with a fixed ratio of currents in the halves cannot be the same as the field patterns of half of the same antenna over a conducting earth except at particular angles for which $f_{er}(\Theta_{01}) = k$. Nevertheless, it is illuminating to note that *at each angle* Θ_{01} the far field of a base-driven vertical antenna on a conducting plane is the same as the field of the same antenna with a *geometrically identical image* that has a current differing somewhat in magnitude and phase from the current in the antenna itself. In the limiting case of a perfect conductor, f_{er} becomes a real quantity independent of Θ_{01}, so that the field is exactly the same at all angles 0° to 90° as that of an isolated center-driven antenna with *equal* currents in the halves. It is clear that in the general case of an imperfectly conducting plane it is possible to speak loosely of an image, but the image has currents that differ in magnitude and phase from those in the antenna itself and a different current is required for each angle Θ_{01}.

Field patterns of thin antennas of several lengths when erected on a conducting earth with a range of values of the complex index of refraction are shown in Figs. 8.2, 8.3, and 8.4. Numerical values are given in Tables 8.2 and 8.3. Note that the principal effect of a finite conductivity is to round off as minima the sharp zeros characteristic of an antenna with sinusoidally distributed current over a perfect conductor.

Note that a field pattern equivalent to that of an antenna over or on a conducting earth can be determined experimentally by measuring the pattern of two antennas in free space, of which one is the image of the other, if the currents are properly adjusted for each angle Θ_{01}, for which a reading is taken using appropriate networks.

QUASI-NEAR-ZONE FIELDS OF VERTICAL DIPOLES OVER A CONDUCTING EARTH

9. Van der Pol's Integrals for the Vertical Dipole Over a Plane Earth

It is pointed out in Sec. 6 that the asymptotic solutions for Π_{z1} give only the leading $1/R_0$ terms and that these vanish along the boundary plane except in the special case when region 2 is infinitely conducting. Extending the work of Sommerfeld and Weyl, van der Pol[59] has obtained a general integral which has been arranged by Norton[44] for practical application. The results so obtained include the quasi-near-zone field and are, therefore, valid for all angles, including the

[VII.9] THEORY OF LINEAR ANTENNAS

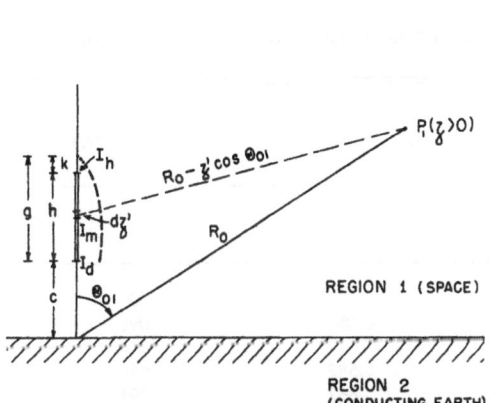

Fig. 8.1. Vertical antenna over conducting earth.

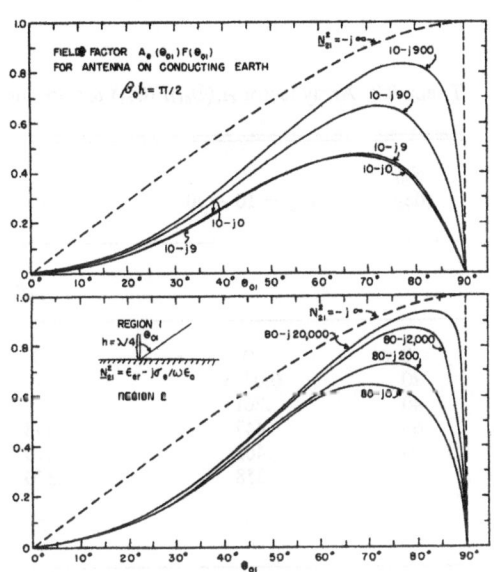

Fig. 8.2. Field factor $A_e(\theta_{01})F(\theta_{01})$ for antenna on conducting earth; $\beta_0 h = \pi/2$.

Fig. 8.3. Field factor $A_e(\theta_{01})F(\theta_{01})$ for antenna on conducting earth; $\beta_0 h = \pi$.

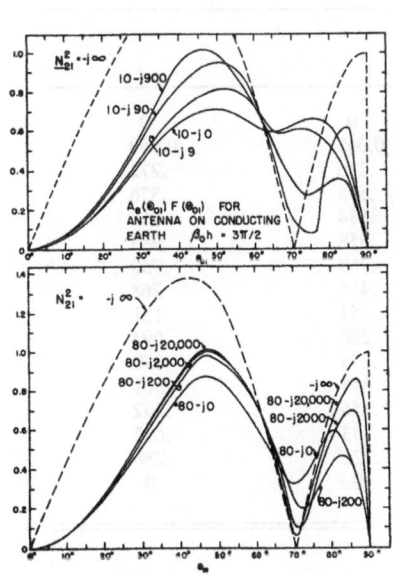

Fig. 8.4. Field factor $A_e(\theta_{01})F(\theta_{01})$ for antenna on conducting earth; $\beta_0 h = 3\pi/2$.

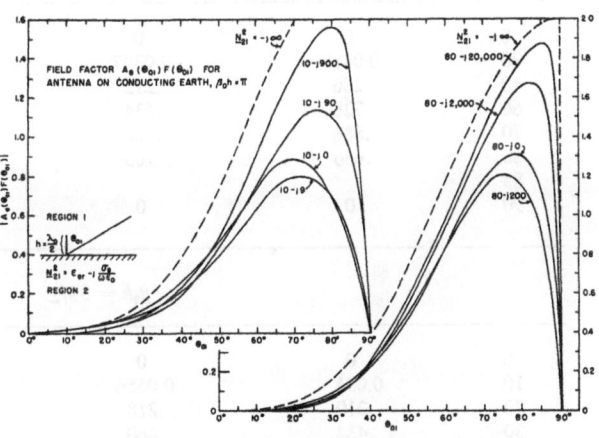

Fig. 9.1. Region of integration in evaluating potential of magnetic dipole.

TABLE 8.2. Array factor $A_e(\Theta_{01})F(\Theta_{01})$ for an antenna on a conducting earth with $N_{21}^2 = 10 - j\sigma_e/\omega\epsilon_0$.

Θ_{01} (deg)	$N_{21}^2 = 10 - j0$	$N_{21}^2 = 10 - j9$	$N_{21}^2 = 10 - j90$	$N_{21}^2 = 10 - j900$
\multicolumn{5}{c}{$\beta_0 h = \pi/2$}				
0	0	0	0	0
20	0.0715	0.070	0.082	0.090
40	.261	.258	.312	.344
60	.447	.448	.582	.666
70	.463	.470	.662	.791
80	.358	.370	.597	.830
85			.434	.752
90	0	0	0	0
\multicolumn{5}{c}{$\beta_0 h = \pi$}				
0	0	0	0	0
20	0.0488	0.0397	0.015	0.0142
40	.266	.212	.188	.231
60	.728	.634	.751	.910
70	.886	.800	1.053	1.323
80	.686	.705	1.104	1.562
85			0.914	1.504
90	0	0	0	0
\multicolumn{5}{c}{$\beta_0 h = 3\pi/2$}				
0	0	0	0	0
10	0.054	0.0556	0.065	0.069
20	.210	.218	.256	.272
30	.434	.460	.545	.576
40	.642	.702	.834	.958
50	.716	.816	.958	.985
60	.634	.732	.768	.730
68	.613	.590	.415	.268
70	.629	.604	.353	.151
73	.653	.591	.280	.094
75	.666	.584	.275	.098
80	.626	.530	.345	.501
83	.532	.448	.356	.592
85	.430	.320	.333	.613
87	.288	.249	.221	.286
90	0	0	0	0

TABLE 8.3. Array factor $A_e(\Theta_{01})F(\Theta_{01})$ for an antenna on a conducting earth with $N_{21}^2 = 80 - j\sigma_e/\omega\epsilon_0$.

Θ_{01} (deg)	$N_{21}^2 = 80 - j0$	$N_{21}^2 = 80 - j200$	$N_{21}^2 = 80 - j2{,}000$	$N_{21}^2 = 80 - j20{,}000$
$\beta_0 h = \pi/2$				
0	0	0	0	0
20	0.084	0.086	0.091	0.092
40	.314	.328	.349	.356
60	.601	.620	.678	.698
70	.650	.718	.812	.844
80	.589	.702	.872	.933
85		.569	.820	.933
90	0	0	0	0
$\beta_0 h = \pi$				
0	0	0	0	0
20	0.0250	0.0194	0.015	0.0170
40	.240	.214	.240	.252
60	.838	.834	.938	.980
70	1.134	1.184	1.370	1.442
80	1.150	1.303	1.651	1.788
85		1.134	1.642	1.872
90	0	0	0	0
$\beta_0 h = 3\pi/2$				
0	0	0	0	0
10	0.064	0.067	0.070	0.071
20	.250	.263	.275	.278
30	.524	.557	.581	.588
40	.780	.846	.880	.889
50	.854	.956	.988	.992
60	.619	.726	.722	.712
68	.346	.338	.255	.210
70	.330	.249	.110	.062
73	.378	.202	.136	.171
75	.433	.250	.264	.316
80	.591	.422	.554	.638
83	.536	.467	.671	.779
85	.465	.253	.698	.840
87	.342	.322	.670	.856
90	0	0	0	0

directions along the boundary plane. For convenient evaluation it is advantageous to assume that both regions have real and equal reluctivities, and that region 1 is nonconducting, so that

$$\nu_2 = \nu_1 = \nu_1, \tag{1a}$$

$$\sigma_{e1} = 0, \quad \xi_1 = \epsilon_1, \quad \beta_1 = \beta_1. \tag{1b}$$

As a first step let the general expression (4.33) for the Hertzian potential in terms of generalized plane waves be transformed into an equivalent expression in terms of the eigenfunctions (1.7a) of the wave equation in *cylindrical* coördinates. Consider the integral

$$(\Pi_{z1})_d = K_1 \frac{e^{-j\beta_1 R_d}}{-j\beta_1 R_d}$$

$$= \frac{K_1}{2\pi} \int_0^{2\pi} \int_0^{\frac{1}{2}\pi + j\infty} \exp\left[-j\beta_1(r \cos\phi \sin\Theta + z \cos\Theta)\right] \sin\Theta \, d\Theta \, d\phi, \tag{2}$$

where $R_d = \sqrt{r^2 + z^2}$. By the same transformation described in conjunction with (5.70)–(5.72), the integral with respect to ϕ may be expressed in terms of the Bessel function. That is,

$$(\Pi_{z1})_d = K_1 \int_0^{\frac{1}{2}\pi + j\infty} \exp(-j\beta_1 z \cos\Theta)$$

$$J_0(\beta_1 r \sin\Theta) \sin\Theta \, d\Theta \tag{3}$$

Following Sommerfeld, the variable of integration in (3) may be changed to λ, which is defined as follows in terms of the generalized law of refraction:

$$\lambda \equiv \beta_1 \sin\Theta = \beta_2 \sin\Theta_t. \tag{4}$$

(Note that λ is not a wavelength, but is a temporarily introduced complex variable of integration. The symbol λ has become conventional following Sommerfeld's classical analysis.) With (4) it follows that

$$d\lambda = \beta_1 \cos\Theta \, d\Theta. \tag{5}$$

For present and later use let

$$l = j\beta_1 \cos\Theta, \quad m = j\beta_2 \cos\Theta_t, \tag{6a}$$

so that

$$\cos\Theta = \frac{l}{j\beta_1}, \quad \sin\Theta \, d\Theta = j\frac{\lambda \, d\lambda}{\beta_1 l}. \tag{6b}$$

Note that

$$l = \sqrt{\lambda^2 - \beta_1^2}, \quad m = \sqrt{\lambda^2 - \beta_2^2}, \tag{7}$$

in agreement with (1.7b). The limits of integration for λ are determined as follows:

$$\Theta = \tfrac{1}{2}\pi + j\infty, \quad \lambda = \beta_1 \sin\Theta = \infty, \tag{8a}$$

$$\Theta = 0, \quad \lambda = 0. \tag{8b}$$

With (4) through (8), the integral (3) becomes

$$(\Pi_{z1})_d = K_1 \frac{e^{-j\beta_1 R_d}}{-j\beta_1 R_d}$$

$$= \frac{jK_1}{\beta_1} \int_0^\infty l^{-1} e^{-zl} J_0(\lambda r) \lambda \, d\lambda, \tag{9}$$

which is seen to be constructed of eigenfunctions like (1.7a) with $n = 0$. The integral in (9) is discussed in detail by Sommerfeld[I.51], vol. VI, pp. 251–253. The path of integration in the complex λ-plane is confined to the one sheet of the four-sheeted Riemann surface in which $l = \sqrt{\lambda^2 - \beta_1^2}$ and $m = \sqrt{\lambda^2 - \beta_2^2}$ have positive real parts. It follows close to the real axis but so as not to cross the branch cuts from $\lambda = \beta_1$ and $\lambda = \beta_2$ to infinity.

Using the transformation (9) of the integral (2) as a model, the general formula (5.35a) for the Hertzian potential in region 1 may be expressed as follows:

$$\Pi_{z1} = (\Pi_{z1})_d + (\Pi_{z1})_r, \tag{10}$$

where $(\Pi_{z1})_d$ is given by (2) and (9) and $(\Pi_{z1})_r$ is obtained from (5.35a) in the form of (2) and transformed into the form of (9). With

$$f_{er} = \frac{\beta_2 \cos\Theta - \beta_1 \cos\Theta_t}{\beta_2 \cos\Theta + \beta_1 \cos\Theta_t} = \frac{\beta_2^2 l - \beta_1^2 m}{\beta_2^2 l + \beta_1^2 m}, \tag{11a}$$

$$f_{mr} = \frac{\beta_1 \cos\Theta - \beta_2 \cos\Theta_t}{\beta_1 \cos\Theta + \beta_2 \cos\Theta_t} = \frac{l - m}{l + m}, \tag{11b}$$

it is

$$(\Pi_{z1})_r = \frac{jK_1}{\beta_1} \int_0^\infty l^{-1} f_r e^{-(z+2d)l} J_0(\lambda r) \lambda \, d\lambda, \tag{12a}$$

so that

$$(\Pi_{ez1})_r = \frac{jK_{e1}}{\beta_1} \int_0^\infty \frac{1}{l} \left(\frac{\beta_2^2 l - \beta_1^2 m}{\beta_2^2 l + \beta_1^2 m}\right) e^{-(z+2d)l} J_0(\lambda r) \lambda \, d\lambda, \tag{12b}$$

$$(\Pi_{mz1})_r$$

$$= \frac{jK_{m1}}{\beta_1} \int_0^\infty \frac{1}{l} \left(\frac{l - m}{l + m}\right) e^{-(z+2d)l} J_0(\lambda r) \lambda \, d\lambda, \tag{12c}$$

where $K_{e1} = -j\beta_1 p_{z1}/4\pi\epsilon_1$ and $K_{m1} = -j\beta_1 m_{z1}/4\pi\nu_1$ as heretofore. The expressions (12b) and (12c) are equivalent to Van der Pol's formulas. (Note that Van der Pol chooses the origin of coördinates at the boundary plane instead of at the center of the dipole. It is seen that (12b, c) also are constructed of the cylindrical eigenfunctions (1.7a) with $n = 0$.)

Before following Van der Pol's transformation of (12) it is subsequently useful to rearrange (12b) and (12c) into the forms (6.77a, b), where

$$(\Pi_{ez1})_r = \frac{jK_{e1}}{\beta_1}\left(-\frac{e^{-j\beta_1 R_r}}{R_r} + V\right), \quad (13a)$$

$$(\Pi_{mz1})_r = \frac{jK_{m1}}{\beta_1}\left(-\frac{e^{-j\beta_1 R_r}}{R_r} + H\right). \quad (13b)$$

If the term $e^{-j\beta_1 R_r}/R_r$ with $R_r = \sqrt{(z+2d)^2 + r^2}$ is expressed in the form (9) and added to (12b) and (12c) to obtain V and H, it follows that

$$V = \int_0^\infty \left(\frac{2\beta_2^2}{\beta_2^2 l + \beta_1^2 m}\right) e^{-(z+2d)l} J_0(\lambda r) \lambda\, d\lambda, \quad (13c)$$

$$H = \int_0^\infty \frac{2}{l+m} e^{-(z+2d)l} J_0(\lambda r) \lambda\, d\lambda. \quad (13d)$$

Although Van der Pol ultimately obtains $(\Pi_{z1})_r$ in the form of (13a, b), he does so by first evaluating (12b) and (12c).

In transforming (12b) and (12c), Van der Pol makes use of the following relations that are readily verified by direct integration; note that the real parts of the exponents are nonpositive:

$$\frac{1}{\beta_2^2 l + \beta_1^2 m} = \int_0^\infty e^{-(\beta_2^2 l + \beta_1^2 m)\zeta} d\zeta, \quad (14a)$$

$$\frac{1}{l+m} = \int_0^\infty e^{-(l+m)z} dz. \quad (14b)$$

As a result of (14), it follows that

$$\frac{1}{l}\left(\frac{\beta_2^2 l - \beta_1^2 m}{\beta_2^2 l + \beta_1^2 m}\right)$$
$$= \left(\beta_2^2 - \beta_1^2 \frac{m}{l}\right) \int_0^\infty \exp[-(\beta_2^2 l + \beta_1^2 m)\zeta]\, d\zeta, \quad (15a)$$

$$\frac{1}{l}\left(\frac{l-m}{l+m}\right) = \left(1 - \frac{m}{l}\right) \int_0^\infty e^{-(l+m)z} dz. \quad (15b)$$

The integrals in (15) are equivalent to the following:

$$\frac{1}{l}\left(\frac{\beta_2^2 l - \beta_1^2 m}{\beta_2^2 l + \beta_1^2 m}\right)$$
$$= \frac{1}{\beta_1^2} \int_0^\infty \left\{\frac{\partial}{\partial \zeta}\left[\frac{\exp(-\beta_2^2 l\zeta)}{l}\right] \frac{\partial}{\partial \zeta}\left(\frac{e^{-\beta_1^2 m\zeta}}{m}\right) \right.$$
$$\left. - \left[\frac{\exp(-\beta_2^2 l\zeta)}{l}\right] \frac{\partial^2}{\partial \zeta^2}\left(\frac{e^{-\beta_1^2 m\zeta}}{m}\right)\right\} d\zeta, \quad (16a)$$

$$\frac{1}{l}\left(\frac{l-m}{l+m}\right) = \int_0^\infty \left[\frac{\partial}{\partial z}\left(\frac{e^{-lz}}{l}\right) \frac{\partial}{\partial z}\left(\frac{e^{-mz}}{m}\right) \right.$$
$$\left. - \left(\frac{e^{-lz}}{l}\right) \frac{\partial^2}{\partial z^2}\left(\frac{e^{-mz}}{m}\right)\right] dz. \quad (16b)$$

Formulas (16) are valid when the denominator on the left has its real part positive. (When the denominator is purely imaginary the integrals should be interpreted in Cesaro mean of any positive order. This is the usual practice in applied mathematics.)

Substitution of (6) in (2) gives

$$(\Pi_{ez1})_r$$
$$= \frac{p_{z1}}{4\pi\epsilon_1 \beta_1^2} \int_0^\infty \int_0^\infty \left[\frac{\partial}{\partial \zeta}\left(\frac{\exp[-(\beta_2^2 \zeta + z + 2d)l]}{l}\right) \right.$$
$$\times \frac{\partial}{\partial \zeta}\left(\frac{e^{-\beta_1^2 m\zeta}}{m}\right) - \left(\frac{\exp[-(\beta_2^2 \zeta + z + 2d)l]}{l}\right)$$
$$\left. \times \frac{\partial^2}{\partial \zeta^2}\left(\frac{e^{-\beta_1^2 m\zeta}}{m}\right)\right] J_0(\lambda r) \lambda\, d\lambda\, d\zeta, \quad (17a)$$

$$(\Pi_{mz1})_r$$
$$= \frac{m_{z1}}{4\pi\nu_1} \int_0^\infty \int_0^\infty \left[\frac{\partial}{\partial z'}\left(\frac{e^{-(z'+z+2d)l}}{l}\right) \frac{\partial}{\partial z'}\left(\frac{e^{-mz'}}{m}\right) \right.$$
$$\left. - \left(\frac{e^{-(z'+z+2d)l}}{l}\right) \frac{\partial^2}{\partial z'^2}\left(\frac{e^{-mz'}}{m}\right)\right] J_0(\lambda r) \lambda\, d\lambda\, dz' \quad (17b)$$

The next step makes use of an integral given and verified in Watson,[I.55] p. 416. It is

$$\frac{e^{-\beta_1^2 \zeta m}}{m} = \frac{\exp(-\beta_2^2 \zeta \sqrt{\lambda^2 - \beta_2^2})}{\sqrt{\lambda^2 - \beta_2^2}}$$
$$= \int_0^\infty \frac{\exp[j\beta_2 \sqrt{r'^2 + (\beta_1^2 \zeta)^2}]}{\sqrt{r'^2 + (\beta_1^2 \zeta)^2}} J_0(\lambda r') r'\, dr', \quad (18a)$$

$$\frac{e^{-z'm}}{m} = \frac{\exp(-z'\sqrt{\lambda^2 - \beta_2^2})}{\sqrt{\lambda^2 - \beta_2^2}}$$
$$= \int_0^\infty \frac{\exp(j\beta_2 \sqrt{r'^2 + z'^2})}{\sqrt{r'^2 + z'^2}} J_0(\lambda r') r'\, dr'. \quad (18b)$$

These integrals are valid if the denominator in each integrand has a positive real part. Substitution of (18) in (17) in two places to eliminate m gives

$$(\Pi_{ez1})_r = \frac{p_{z1}}{4\pi\epsilon_1\beta_1^2} \int_0^\infty \int_0^\infty \int_0^\infty$$

$$\left[\frac{\partial}{\partial \zeta}\left(\frac{\exp[-(\beta_2^2\zeta + z + 2d)l]}{l}\right)\right.$$

$$\times \frac{\partial}{\partial \zeta}\left(\frac{\exp[j\beta_2\sqrt{r'^2 + (\beta_1^2\zeta)^2}]}{\sqrt{r'^2 + (\beta_1^2\zeta)^2}}\right)$$

$$- \left(\frac{\exp[-(\beta_2^2\zeta + z + 2d)l]}{l}\right)$$

$$\left.\times \frac{\partial^2}{\partial \zeta^2}\left(\frac{\exp[j\beta_2\sqrt{r'^2 + (\beta_1^2\zeta)^2}]}{\sqrt{r'^2 + (\beta_1^2\zeta)^2}}\right)\right]$$

$$\times J_0(\lambda r) J_0(\lambda r') \, d\zeta \, d\lambda r' dr', \quad (19a)$$

$$(\Pi_{mz1})_r = \frac{m_{z1}}{4\pi\nu_1} \int_0^\infty \int_0^\infty \int_0^\infty$$

$$\left[\frac{\partial}{\partial z'}\left(\frac{e^{-(z'+z+2d)l}}{l}\right)\frac{\partial}{\partial z'}\left(\frac{\exp(j\beta_2\sqrt{r'^2+z'^2})}{\sqrt{r'^2+z'^2}}\right)\right.$$

$$\left.- \left(\frac{e^{-(z'+z+2d)l}}{l}\right)\frac{\partial^2}{\partial z'^2}\left(\frac{\exp(j\beta_2\sqrt{r'^2+z'^2})}{\sqrt{r'^2+z'^2}}\right)\right]$$

$$\times J_0(\lambda r) J_0(\lambda r') \, d\zeta \, \lambda \, d\lambda r' dr'. \quad (19b)$$

In order to integrate (19a, b) with respect to λ — which occurs in $l = \sqrt{\lambda^2 - \beta_1^2}$ — it is necessary to evaluate in (19a) the integral

$$\int_0^\infty \frac{\exp[-(\beta_2^2\zeta + z + 2d)\sqrt{\lambda^2 - \beta_1^2}]}{\sqrt{\lambda^2 - \beta_1^2}}$$

$$\times J_0(\lambda r) J_0(\lambda r') \lambda \, d\lambda, \quad (20)$$

and in (19b) the same integral with z' substituted for $\beta_2^2\zeta$. This is accomplished first by replacing the product of Bessel functions by an integral obtained from a standard addition theorem. The theorem in question is found in Watson,[I.55] p. 128. If the general formula given by Watson is divided by 2π and integrated from zero to 2π on each side, the result is the desired formula. It is

$$J_0(\lambda r) J_0(\lambda r')$$

$$= \frac{1}{2\pi}\int_0^{2\pi} J_0(\lambda\sqrt{r^2 - 2rr'\cos\theta' + r'^2}) \, d\theta'. \quad (21)$$

If this is substituted in (20) the resulting integral is

$$\frac{1}{2\pi}\int_0^{2\pi}\int_0^\infty \frac{\exp[-(\beta_2^2\zeta + z + 2d)\sqrt{\lambda^2 - \beta_1^2}]}{\sqrt{\lambda^2 - \beta_1^2}}$$

$$\times J_0(\lambda\sqrt{r^2 - 2rr'\cos\theta' + r'^2})\lambda \, d\lambda \, d\theta'. \quad (22)$$

The next step parallels the form of (9) with $R_d = \sqrt{r^2 + z^2}$ replaced by $R_r = \sqrt{r^2 + (z + 2d)^2}$. Thus,

$$\frac{e^{-j\beta_1 R_r}}{R_r} = \int_0^\infty \frac{\exp[-(z + 2d)\sqrt{\lambda^2 - \beta_1^2}]}{\sqrt{\lambda^2 - \beta_1^2}}$$

$$\times J_0(\lambda r)\lambda \, d\lambda. \quad (23)$$

If $\beta_2^2\zeta + z + 2d$ is substituted for $z + 2d$ and $r^2 - 2rr'\cos\theta' + r'^2$ for r^2, the result is

$$\frac{e^{-j\beta_1 R_e'}}{R_e'} = \int_0^\infty \frac{\exp[-(\beta_2^2\zeta + z + 2d)\sqrt{\lambda^2 - \beta_1^2}]}{\sqrt{\lambda^2 - \beta_1^2}}$$

$$\times J_0(\lambda\sqrt{r^2 - 2rr'\cos\theta' + r'^2})\lambda \, d\lambda, \quad (24)$$

where R_e' is given in (26b) below. With (24), (22) reduces to

$$\frac{1}{2\pi}\int_0^\infty \frac{e^{-j\beta_1 R_e'}}{R_e'} \, d\theta'. \quad (25)$$

The substitution of (25) for (20) can now be made in (19a), which reduces to

$$(\Pi_{ez1})_r = \frac{p_{z1}}{8\pi^2\epsilon_1\beta_1^2}\int_0^{2\pi}\int_0^\infty\int_0^\infty \left\{\frac{\partial}{\partial\zeta}\left(\frac{e^{-j\beta_1 R_e'}}{R_e'}\right)\right.$$

$$\times \frac{\partial}{\partial\zeta}\left[\frac{\exp(-j\beta_2 R_e'')}{R_e''}\right] - \left(\frac{e^{-j\beta_1 R_e'}}{R_e'}\right)$$

$$\left.\times \frac{\partial^2}{\partial\zeta^2}\left[\frac{\exp(-j\beta_2 R_e'')}{R_e''}\right]\right\} r' dr' d\zeta \, d\theta', \quad (26a)$$

where

$$R_e'$$

$$\equiv \sqrt{r^2 - 2rr'\cos\theta' + r'^2 + (\beta_2^2\zeta + z + 2d,)^2} \quad (26b)$$

$$R_e'' \equiv \sqrt{r'^2 + (\beta_1^2\zeta)^2}. \quad (26c)$$

The corresponding expression for the magnetization potential is obtained directly by replacing $\beta_2^2 \zeta$ by z' in (24). Thus,

$$(\Pi_{mz1})_r = \frac{m_{z1}}{8\pi^2 \nu_1} \int_0^{2\pi} \int_0^\infty \int_0^\infty \left\{ \frac{\partial}{\partial z'} \left(\frac{e^{-j\beta_1 R'_m}}{R'_m} \right) \right.$$

$$\times \frac{\partial}{\partial z'} \left[\frac{\exp(-j\beta_2 R''_m)}{R''_m} \right] - \left(\frac{e^{-j\beta_1 R'_m}}{R'_m} \right)$$

$$\left. \times \frac{\partial^2}{\partial z'^2} \left[\frac{\exp(-j\beta_2 R''_m)}{R''_m} \right] \right\} r' \, dr' \, dz' \, d\theta', \quad (27a)$$

where

$$R'_m \equiv \sqrt{r^2 - 2rr' \cos \theta' + r'^2 + (z' + z + 2d)^2}, \quad (27b)$$

$$R''_m \equiv \sqrt{r'^2 + z'^2}. \quad (27c)$$

Note that in general R'_e is complex since β_2 is complex, whereas R'_e, R'_m, and R''_m are real.

In order to obtain convenient alternative forms of (26a) and (27a), let (23) be expressed in the following equivalent form by multiplying the integrand successively by $(\beta_2^2 l + \beta_1^2 m)/(\beta_2^2 l + \beta_1^2 m) = 1$ and by $(l + m)/(l + m) = 1$:

$$\frac{e^{-j\beta_1 R_r}}{R_r}$$

$$= \int_0^\infty \frac{1}{l} \left(\frac{\beta_2^2 l + \beta_1^2 m}{\beta_2^2 l + \beta_1^2 m} \right) e^{-(z+2d)l} J_0(\lambda r) \lambda \, d\lambda$$

$$= \int_0^\infty \frac{1}{l} \left(\frac{l + m}{l + m} \right) e^{-(z+2d)l} J_0(\lambda r) \lambda \, d\lambda. \quad (28)$$

(Note that $l = \sqrt{\lambda^2 - \beta_1^2}$.) These integrals are like (12b, c) but with a plus sign in the numerators of the integrands instead of a minus sign. It follows that if the integrals are treated like (13) throughout, the final results must be like (26a) and (27a) but with a change in sign. That is,

$$\frac{p_{z1}}{4\pi\epsilon_1} \frac{e^{-j\beta_1 R_r}}{R_r} = \frac{p_{z1}}{8\pi^2 \epsilon_1 \beta_1^2} \times$$

$$\int_0^{2\pi} \int_0^\infty \int_0^\infty \left\{ \frac{\partial}{\partial \zeta} \left(\frac{e^{-j\beta_1 R'_e}}{R'_e} \right) \frac{\partial}{\partial \zeta} \left[\frac{\exp(-j\beta_2 R''_e)}{R''_e} \right] \right.$$

$$\left. + \left(\frac{e^{-j\beta_1 R'_e}}{R'_e} \right) \frac{\partial^2}{\partial \zeta^2} \left[\frac{\exp(-j\beta_2 R''_e)}{R''_e} \right] \right\} r' \, dr' \, d\zeta \, d\theta', \quad (29a)$$

$$\frac{m_{z1}}{4\pi\epsilon_1} \frac{e^{-j\beta_1 R_r}}{R_r} = \frac{m_{z1}}{8\pi^2 \nu_1} \times$$

$$\int_0^{2\pi} \int_0^\infty \int_0^\infty \left\{ \frac{\partial}{\partial z'} \left(\frac{e^{-j\beta_1 R'_m}}{R'_m} \right) \frac{\partial}{\partial z'} \left[\frac{\exp(-j\beta_2 R''_m)}{R''_m} \right] \right.$$

$$\left. + \left(\frac{e^{-j\beta_1 R'_m}}{R'_m} \right) \frac{\partial^2}{\partial z'^2} \left[\frac{\exp(-j\beta_2 R''_m)}{R''_m} \right] \right\} \times r' \, dr' \, dz' \, d\theta'. \quad (29b)$$

Let the right-hand terms in (29a, b) be added, respectively, to (26a) and (27a), and the equal center term in (29a, b) subtracted, respectively, from (26a) and (27a). Also let

$$z' = \beta_1^2 \zeta \quad (30)$$

in (26a) and (29a). The results are

$$(\Pi_{ez1})_r = \frac{p_{z1}}{4\pi\epsilon_1} \left\{ -\frac{e^{-j\beta_1 R_r}}{R_r} \right.$$

$$\left. + \frac{1}{\pi} \int_0^{2\pi} \int_0^\infty \int_0^\infty \frac{\partial}{\partial z'} \left(\frac{e^{-j\beta_1 R'_e}}{R'_e} \right) \right.$$

$$\left. \times \frac{\partial}{\partial z'} \left[\frac{\exp(-j\beta_2 R''_e)}{R''_e} \right] r' \, dr' \, dz' \, d\theta' \right\}, \quad (31a)$$

$$(\Pi_{mz1})_r = \frac{m_{z1}}{4\pi\nu_1} \left\{ -\frac{e^{-j\beta_1 R_r}}{R_r} \right.$$

$$\left. + \frac{1}{\pi} \int_0^{2\pi} \int_0^\infty \int_0^\infty \frac{\partial}{\partial z'} \left(\frac{e^{-j\beta_1 R'_m}}{R'_m} \right) \right.$$

$$\left. \times \frac{\partial}{\partial z'} \left[\frac{\exp(-j\beta_2 R''_m)}{R''_m} \right] r' \, dr' \, dz' \, d\theta' \right\}. \quad (31b)$$

Alternatively, the right-hand terms in (29a, b) may be subtracted, respectively, from (26a) and (27a), and the equal center terms of (29a, b) added, respectively, to (26a) and (27a). With (30) the results are

$$(\Pi_{ez1})_r = \frac{p_{z1}}{4\pi\epsilon_1} \left\{ \frac{e^{-j\beta_1 R_r}}{R_r} - \frac{1}{\pi} \int_0^{2\pi} \int_0^\infty \int_0^\infty \left(\frac{e^{-j\beta_1 R'_e}}{R'_e} \right) \right.$$

$$\left. \times \frac{\partial^2}{\partial z'^2} \left[\frac{\exp(-j\beta_2 R''_e)}{R''_e} \right] r' \, dr' \, dz' \, d\theta' \right\}, \quad (32a)$$

$$(\Pi_{mz1})_r$$

$$= \frac{m_{z1}}{4\pi\nu_1} \left\{ \frac{e^{-j\beta_1 R_r}}{R_r} - \frac{1}{\pi} \int_0^{2\pi} \int_0^\infty \int_0^\infty \left(\frac{e^{-j\beta_1 R'_m}}{R'_m} \right) \right.$$

$$\left. \times \frac{\partial^2}{\partial z'^2} \left[\frac{\exp(-j\beta_2 R''_m)}{R''_m} \right] r' \, dr' \, dz' \, d\theta' \right\}. \quad (32b)$$

In (31) and (32),

$$R'_e =$$
$$\sqrt{r^2 - 2rr'\cos\theta' + r'^2 + (N_{21}^2 z' + z + 2d)^2},$$

$$R''_e = \sqrt{r'^2 + z'^2}, \quad (33a)$$

$$R'_m$$
$$= \sqrt{r^2 - 2rr'\cos\theta' + r'^2 + (z' + z + 2d)^2},$$

$$R''_m = \sqrt{r'^2 + z'^2}, \quad (33b)$$

$$R_r = \sqrt{r^2 + (z + 2d)^2}. \quad (33c)$$

The expressions (31a, b) and (32a, b) are equivalent and exact. Note that R'_e is complex if region 2 is conducting and $N_{21}^2 = \beta_2^2/\beta_1^2$ is complex, whereas R''_e, R'_m, and R''_m are real since it has been assumed that region 1 is nonconducting so that β_1 is real.

Examination of (33b) shows that when z' is at its lower limit 0, the distance R'_m is measured from points on the plane $z = -2d$ to the point of calculation at (r, z), and that when z' increases R'_m increases. This suggests the interpretation that z' is measured downward, that is, in the negative z direction, from the plane $z = -2d$. This means that the entire integration is only over the half-space *below* the plane $z = -2d$ in which the image dipole is located. This is illustrated in Fig. 9.1, with the element of integration in the plane $\theta' = 0$, which is the plane of the paper. This simple geometric interpretation is complicated in the case of the electric dipole by the fact that R'_e is complex. However, the integration is over the same real half-space below $z = -2d$.

10. *Approximate Integration of Van der Pol's Formulas*

In order to evaluate the Hertzian potentials of vertical dipoles in a perfect dielectric above a conducting plane earth, the integrations must be carried out. The general formula is

$$\Pi_{z1} = \frac{jK_1}{\beta_1}\left\{\frac{e^{-j\beta_1 R_d}}{R_d} + \frac{e^{-j\beta_1 R_r}}{R_r}\right.$$
$$- \frac{1}{\pi}\int_0^{2\pi}\int_0^\infty\int_0^\infty \left[\left(\frac{e^{-j\beta_1 R'}}{R'}\right)\right.$$
$$\left.\left.\times \frac{\partial^2}{\partial z'^2}\left(\frac{\exp(-j\beta_2 R'')}{R''}\right)\right] r'\,dr'\,dz'\,d\theta'\right\}, \quad (1)$$

where for the electric dipole

$$K_1 = K_{e1} = \frac{-j\beta_1 p_{z1}}{4\pi\epsilon_1}, \quad (2a)$$

$$R' = R'_e =$$
$$\sqrt{r^2 - 2rr'\cos\theta' + r'^2 + (z + 2d + N_{21}^2 z')^2}, \quad (2b)$$

magnetic dipole

$$K_1 = K_{m1} = \frac{-j\beta_1 m_{z1}}{4\pi\nu_1}, \quad (3a)$$

$$R' = R'_m$$
$$= \sqrt{r^2 - 2rr'\cos\theta' + r'^2 + (z + 2d + z')^2}, \quad (3b)$$

and where for both dipoles

$$R_d = \sqrt{r^2 + z^2} \quad (4a)$$

is the distance from the origin at the center of the dipole to the point of calculation $P(r, \theta, z)$;

$$R_r = \sqrt{r^2 + (z + 2d)^2} \quad (4b)$$

is the distance from the ideal image dipole at $r = 0$, $z = -2d$, to $P(r, \theta, z)$; and

$$R'' = \sqrt{r'^2 + z'^2} \quad (4c)$$

is the distance from the ideal image dipole at $r = 0$, $z = -2d$ to the element of integration $r'\,dr'\,dz'\,d\theta'$ at $P'(r', \theta', z')$. Note that R'_m as given in (3b) is real (whereas R'_e is not) and is the distance from $P(r, \theta, z)$ to $P'(r', \theta', z')$ (Fig. 9.1). The variables of integration are r', θ', z', and $r'\,dr'\,dz'\,d\theta'$ is an element of volume in cylindrical coördinates.

The first step in the simplification of (1) is carried out by Van der Pol. It is an approximation that requires region 2 to be so good a conductor that α_s in

$$\beta_2 = \beta_s - j\alpha_s \quad (5)$$

is large enough to provide a rapid decrease with radial distance r' in the factor

$$\frac{\exp(-j\beta_2 R'')}{R''} = e^{-\alpha_s\sqrt{r'^2+z'^2}}\frac{e^{-j\beta_s\sqrt{r'^2+z'^2}}}{\sqrt{r'^2 + z'^2}}. \quad (6)$$

If this is true, the contributions to the integral that are significant are those for which r' is small. Evidently, the greater the attenuation constant α_s, the smaller is the range, around the image point $r = 0$, $z = -2d$ from which R'' is measured, in which significant contributions are made to the integral. (Note that for a perfectly conducting region 2 the range of significant contributions is confined to the one point locating the image.) If α_s is sufficiently great, the terms in r' may be neglected in R' as defined in (3b) and the following approximation used provided the point of calculation, P, is not so near the origin that $\beta_1 R_r$ is small. That is,

$$R'_e \doteq \sqrt{r^2 + (z + 2d + N_{21}^2 z')^2} \quad (7a)$$

$$R'_m \doteq \sqrt{r^2 + (z + 2d + z')^2}, \quad (7b)$$

provided R_r is great compared with the range of r' which yields significant contributions to the integral in (1). This may be expressed as follows:

$$\alpha_s R_r \gg 1 \quad \text{or} \quad R_r \gg d_s \quad (8)$$

where d_s is the skin depth. In general, (8) is less restrictive than the far-zone condition $\beta_1 R_r \gg 1$ for normal soils at radio frequencies. Specifically, at $\omega = 10^7$ (8) is equivalent to $\beta_1 R_r \gg 0.006$ for salt water ($\sigma = 5$, $\epsilon_r = 80$), to $\beta_1 R_r \gg 0.04$ for moist earth ($\sigma = 10^{-2}$, $\epsilon_r = 9$), and to $\beta_1 R_r \gg 1$ for dry earth ($\sigma = 10^{-4}$, $\epsilon_r = 4$). Since the integrand in (1) is independent of θ' if (7) is used, the θ' integration yields 2π.

In the remaining double integral, r' occurs in R'', so that the integral in r' is as follows:

$$\int_0^\infty \frac{\exp(-j\beta_2 \sqrt{r'^2 + z'^2})}{\sqrt{r'^2 + z'^2}} r' dr'$$

$$= \int_{z'}^\infty \exp(-j\beta_2 u) du = \frac{-j}{\beta_2} \exp(-j\beta_2 z'). \quad (9)$$

The following single integral remains:

$$\Pi_{z1} = \frac{jK_1}{\beta_1} \left\{ \frac{e^{-j\beta_1 R_d}}{R_d} + \frac{e^{-j\beta_1 R_r}}{R_r} \right.$$

$$\left. + \frac{j2}{\beta_2} \int_0^\infty \left(\frac{e^{-j\beta_1 R'}}{R'} \right) \frac{\partial^2}{\partial z'^2} [\exp(-j\beta_2 z')] dz' \right\}, \quad (10)$$

where R' is given by (7).

For actual evaluation the potentials are expressed as follows:

$$\Pi_{ez1} = \frac{p_{z1}}{4\pi\epsilon_1} \left\{ \frac{e^{-j\beta_1 R_d}}{R_d} - \frac{e^{-j\beta_1 R_r}}{R_r} + V \right\},$$

$$\Pi_{mz1} = \frac{m_{z1}}{4\pi\nu_1} \left\{ \frac{e^{-j\beta_1 R_d}}{R_d} - \frac{e^{-j\beta_1 R_r}}{R_r} + H \right\}, \quad (11)$$

where, after carrying out the differentiation in (10), and setting $\beta_2 = \beta_1 N_{21}$,

$$V = 2 \left[\frac{e^{-j\beta_1 R_r}}{R_r} \right.$$

$$\left. - j\beta_1 N_{21} \int_0^\infty \frac{e^{-j\beta_1(R'_e + N_{21} z')}}{R'_e} dz' \right], \quad (12a)$$

$$H = 2 \left[\frac{e^{-j\beta_1 R_r}}{R_r} \right.$$

$$\left. - j\beta_1 N_{21} \int_0^\infty \frac{e^{-j\beta_1(R'_m + N_{21} z')}}{R'_m} dz' \right]. \quad (12b)$$

These are the integrals with which Norton[46] begins his formulation leading to practical formulas and graphical representations.

The first step due to Norton is to extend to z' the reasoning previously applied to r' and argue that since the attenuation constant α_s in $\beta_s - j\alpha_s$ is assumed to be reasonably large, the principal contributions to the integral are with small values of z'. Accordingly, (7) may be expanded in powers of z'. Thus

$$R'(z') = R'(0) + z' \left[\frac{\partial R'(z')}{\partial z'} \right]_{z'=0}$$

$$+ \tfrac{1}{2} z'^2 \left[\frac{\partial^2 R'(z')}{\partial z'^2} \right]_{z'=0} + \cdots. \quad (13)$$

After retaining only the terms to the order $1/R_r$, the result is

$$R'_e(z) = R_r + z' N_{21}^2 \cos \Theta_r$$

$$+ \frac{z'^2}{2R_r} N_{21}^4 \sin^2 \Theta_r, \quad (14a)$$

$$R'_m(z) = R_r + z' \cos \Theta_r + \frac{z'^2}{2R_r} \sin^2 \Theta_r, \quad (14b)$$

where

$$\cos \Theta_r = (z + 2d)/R_r, \quad \sin \Theta_r = r/R_r. \quad (14c)$$

With (14) in phase factors (exponents) and

$$R' \doteq R_r \quad (14d)$$

in amplitude factors, (12) becomes

$$V \doteq 2 \frac{e^{-j\beta_1 R_r}}{R_r} \left\{ 1 - j\beta_1 N_{21} \right.$$

$$\left. \times \int_0^\infty e^{-j\beta_1 [z'(N_{21}^2 \cos \Theta_r + N_{21}) + z'^2 N_{21}^4 \sin^2 \Theta_r / 2R_r]} dz' \right\}, \quad (15a)$$

$$H \doteq 2 \frac{e^{-j\beta_1 R_r}}{R_r} \left\{ 1 - j\beta_1 N_{21} \right.$$

$$\left. \times \int_0^\infty e^{-j\beta_1 [z'(\cos \Theta_r + N_{21}) + z'^2 \sin^2 \Theta_r / 2R_r]} dz' \right\}. \quad (15b)$$

For temporary simplification let

$$a_e^2 = -j\beta_1 N_{21}^4/2R_r, \quad (16a)$$

$$c_e' = -j\beta_1 N_{21}(1 + N_{21}\cos\Theta_r), \quad (16b)$$

$$a_m^2 = -j\beta_1/2R_r, \quad (16c)$$

$$c_m' = -j\beta_1 N_{21}\left(1 + \frac{\cos\Theta_r}{N_{21}}\right), \quad (16d)$$

so that the integrals in (15a) and (15b) have the form

$$\int_0^\infty e^{(a^2 z'^2 \sin^2\Theta_r + c' z')}\, dz' = e^{-(c'^2/4a^2\sin^2\Theta_r)}$$

$$\times \int_0^\infty e^{-j^2(az'\sin\Theta_r + c'/2a\sin\Theta_r)^2}\, dz'. \quad (17)$$

Now let

$$u = j\left(az'\sin\Theta_r + \frac{c'}{2a\sin\Theta_r}\right),$$

$$du = ja\sin\Theta_r, \quad (18a)$$

$$P' = c'^2/4a^2\sin^2\Theta_r, \quad (18b)$$

so that

$$P_e' = c_e'^2/4a_e^2\sin^2\Theta_r, \quad (19a)$$

$$= p_e'(1 + N_{21}\cos\Theta_r)^2/\sin^2\Theta_r, \quad (19b)$$

$$P_m' = c_m'^2/4a_m^2\sin^2\Theta_r, \quad (19c)$$

$$= p_m'\left(1 + \frac{1}{N_{21}}\cos\Theta_r\right)^2/\sin^2\Theta_r, \quad (19d)$$

where

$$p_e' = -j\beta_1 R_r/2N_{21}^2, \quad (19e)$$

$$p_m' = -j\beta_1 R_r N_{21}^2/2. \quad (19f)$$

With (18a, b), (17) becomes

$$\frac{-j}{a\sin\Theta_r} e^{-P'} \int_{j\sqrt{P'}}^\infty e^{-u^2}\, du$$

$$= \frac{-j}{a\sin\Theta_r} e^{-P'} \left(\int_0^\infty e^{-u^2}\, du\right.$$

$$\left. - \int_0^{j\sqrt{P'}} e^{-u^2}\, du\right), \quad (20)$$

since with N_{21} as in (5.61) u^2 has a positive real part. The first integral on the right integrates directly into $\sqrt{\pi}/2$; the second integral is in the form of the well-known probability integral or error function defined by

$$\text{erf}(x) \equiv \frac{2}{\sqrt{\pi}} \int_0^x e^{-u^2}\, du. \quad (21)$$

Hence (20), which is the integral in (17), may be expressed as follows:

$$\frac{-j\sqrt{\pi}}{2a} e^{-P'}(1 - \text{erf } j\sqrt{P'}). \quad (22)$$

Consequently (15a) and (15b) reduce to

$$V \doteq 2\frac{e^{-j\beta_1 R_r}}{R_r}$$

$$\times \left\{1 - \frac{\beta_2\sqrt{\pi}}{2a_e \sin\Theta_r} e^{-P_e'}[1 - \text{erf}(j\sqrt{P_e'})]\right\}, \quad (23a)$$

$$H \doteq 2\frac{e^{-j\beta_1 R_r}}{R_r}$$

$$\times \left\{1 - \frac{\beta_2\sqrt{\pi}}{2a_m \sin\Theta_r} e^{-P_m'}[1 - \text{erf}(j\sqrt{P_m'})]\right\}. \quad (23b)$$

The terms following the first minus signs in (23a) and (23b) may be simplified as follows:

$$\frac{\beta_2}{2a_e} = \frac{\beta_2}{2\sqrt{-j\beta_1 N_{21}^4/2R_r}} = j\sqrt{P_e'}, \quad (24a)$$

$$\frac{\beta_2}{2a_m} = \frac{\beta_2}{2\sqrt{-j\beta_1/2R_r}} = j\sqrt{P_m'}. \quad (24b)$$

Accordingly,

$$\left.\begin{array}{c}V\\H\end{array}\right\} \doteq 2\frac{e^{-j\beta_1 R_r}}{R_r}$$

$$\times \left\{1 - \frac{j\sqrt{\pi P'}}{\sin\Theta_r} e^{-P'}[1 - \text{erf } j\sqrt{P'}]\right\}, \quad (25)$$

where the subscript e is affixed to p' and P' for V and the subscript m for H.

For large values of R_r and hence of $|P'|$, the following asymptotic series is valid (ref. I.17, formula 592; ref. I.28, p. 24):

$$\text{erf}(x) = 1 - \frac{e^{-x^2}}{x\sqrt{\pi}}\left(1 - \frac{1}{2x^2} + \frac{1\cdot 3}{(2x^2)^2} - \cdots\right). \quad (26)$$

Use of (26) in (25) with $x = j\sqrt{P'}$ gives the following asymptotic formula:

$$\left.\begin{array}{c}V^r\\H^r\end{array}\right\} \doteq 2\frac{e^{-j\beta_1 R_r}}{R_r}\left\{1 - \left(\frac{p'}{P'\sin^2\Theta_r}\right)^{1/2}\right.$$

$$\left. \times \left[1 + \frac{1}{2P'} + \frac{1\cdot 3}{(2P')^2} + \cdots\right]\right\}. \quad (27)$$

With (19), the far-zone formulas for V and H are

$$V^r \doteq 2 \frac{e^{-j\beta_1 R_r}}{R_r} [1 - (1 + N_{21} \cos \Theta_r)^{-1}]; \quad (28a)$$

$$H^r \doteq 2 \frac{e^{-j\beta_1 R_r}}{R_r} \left[1 - \left(1 + \frac{\cos \Theta_r}{N_{21}}\right)^{-1}\right]. \quad (28b)$$

Note that only the leading term has been retained in the series in (27). In the far zone,

$$\cos \Theta_{01} = \frac{z + d}{R_{01}} \doteq \frac{z + 2d}{R_r} = \cos \Theta_r. \quad (29)$$

Hence, for sufficiently great values of P' and hence of R_{01},

$$V^r \doteq 2 \frac{e^{-j\beta_1 R_r}}{R_r} [1 - (1 + N_{21} \cos \Theta_{01})^{-1}], \quad (30a)$$

$$H^r \doteq 2 \frac{e^{-j\beta_1 R_r}}{R_r} \left[1 - \left(1 + \frac{\cos \Theta_{01}}{N_{21}}\right)^{-1}\right]. \quad (30b)$$

The corresponding expressions previously obtained by saddle-point integration, namely, (6.80a, b), with $\nu_2 = \nu_1 = \nu_0$, are:

$$V^r = 2 \frac{e^{-j\beta_1 R_r}}{R_r}$$
$$\times \left[1 - \left(1 + \frac{N_{21}^2 \cos \Theta_{01}}{\sqrt{N_{21}^2 - \sin^2 \Theta_{01}}}\right)^{-1}\right], \quad (31a)$$

$$H^r = 2 \frac{e^{-j\beta_1 R_r}}{R_r}$$
$$\times \left[1 - \left(1 + \frac{\cos \Theta_{01}}{N_{21}^2 - \sin^2 \Theta_{01}}\right)^{-1}\right]. \quad (31b)$$

Note that both (30a) and (30b) lack the factor $N_{21}/\sqrt{N_{21}^2 - \sin^2 \Theta_{01}}$ as a multiplier of $\cos \Theta_{01}$ that occurs in (31a, b). However, since the derivation of (30a, b) from the exact integral (1) involves an integration that is a good approximation only when N_{21} is moderately large, it is clear that (30a, b) are in good agreement with (31a, b) when the implied conditions are satisfied. Thus, (31a, b) are valid subject to the condition $\beta_1 R_r \gg 1$, whereas (30a, b) presuppose $\alpha_s R_r \gg 1$, which is equivalent to requiring N_{21} to be moderately large. In order to obtain a single formula that is useful for all values of N_{21} subject to $\beta_1 R_r \gg 1$ and for moderately large values of $|N_{21}|$ subject to $\alpha_s R_r \gg 1$, Norton arbitrarily introduces the factor $N_{21}/\sqrt{N_{21}^2 - \sin^2 \Theta_{01}}$ not only in the far-zone formulas (30a, b) but also in the general integrals (12a, b) from which they are derived. Although there is no mathematical justification for this step, it is a plausible and useful one. The new formula reduces to each of the old formulas under the appropriate conditions. Whether its range of usefulness actually has been extended is not obvious, but is possible. The desired arbitrarily generalized formulas are

$$\left.\begin{array}{c}V\\H\end{array}\right\} \doteq 2 \left[\frac{e^{-j\beta_1 R_r}}{R_r} - j\beta_1 \sqrt{N_{21}^2 - \sin^2 \Theta_r}\right.$$
$$\left. \times \int_0^\infty \frac{e^{-j\beta_1(R' + z'\sqrt{N_{21}^2 - \sin^2 \Theta_r})}}{R'} dz'\right], \quad (32)$$

where $R' \equiv R'_e$ for V and $R' \equiv R'_m$ for H. Note that R'_e is complex, R'_m real.

The integral in (32) may now be treated just as were the integrals in (12). Expansion of R' as in (13) and (14) is unchanged and (15a, b) are replaced by

$$V = 2 \frac{e^{-j\beta_1 R_r}}{R_r}$$
$$\times \left\{1 - j\beta_1 \sqrt{N_{21}^2 - \sin^2 \Theta_r} \int_0^\infty e^{-j\beta_1 g} dz'\right\} \quad (33a)$$

where

$$g \equiv z'(N_{21}^2 \cos \Theta_r + \sqrt{N_{21}^2 - \sin^2 \Theta_r})$$
$$+ (z'^2 N_{21}^2/2R_r) \sin^2 \Theta_r$$

$$H = 2 \frac{e^{-j\beta_1 R_r}}{R_r}$$
$$\times \left\{1 - j\beta_1 \sqrt{N_{21}^2 - \sin^2 \Theta_r} \int_0^\infty e^{-j\beta_1 h} dz'\right\} \quad (33b)$$

where

$$h \equiv z'(\cos \Theta_r + \sqrt{N_{21}^2 - \sin^2 \Theta_r})$$
$$+ (z'^2/2R_r) \sin^2 \Theta_r.$$

Using the same values of a as defined in (16a, b), and the following values of c by analogy with (16c, d),

$$c_e = -j\beta_1(N_{21}^2 \cos \Theta_r + \sqrt{N_{21}^2 - \sin^2 \Theta_r}), \tag{34a}$$

$$c_m = -j\beta_1(\cos \Theta_r + \sqrt{N_{21}^2 - \sin^2 \Theta_r}), \tag{34b}$$

the integral (17) is obtained with c unprimed. With u as in (18a) with c unprimed, and with

$$P = c^2/4a^2 \sin^2 \Theta_r, \tag{35}$$

it follows that

$$P_e = \frac{p_e}{\sin^2 \Theta_r}\left(1 + \frac{N_{21}^2 \cos \Theta_r}{\sqrt{N_{21}^2 - \sin^2 \Theta_r}}\right)^2, \tag{36a}$$

$$P_m = \frac{p_m}{\sin^2 \Theta_r}\left(1 + \frac{\cos \Theta_r}{\sqrt{N_{21}^2 - \sin^2 \Theta_r}}\right)^2, \tag{36b}$$

where

$$p_e = \frac{-j\beta_1 R_r(N_{21}^2 - \sin^2 \Theta_r)}{2N_{21}^4}, \tag{36c}$$

$$p_m = \frac{-j\beta_1 R_r(N_{21}^2 - \sin^2 \Theta_r)}{2}. \tag{36d}$$

The factor P is known as the *complex numerical distance* for dipoles that are elevated above the surface of the earth, that is, for which $d > 0$. If the dipole is on the earth and the point of calculation is also on the earth, it follows that

$$d = 0, \quad z = 0, \quad \Theta_r = \pi/2: \quad P = p. \tag{36e}$$

Thus, p is the *complex numerical distance* for the field on the boundary plane due to a dipole on this plane. This function was introduced originally by Sommerfeld. If (36c, d) are substituted in (36a, b), the explicit expressions for P_e and P_m are

$$P_e = \frac{-j\beta_1 R_r}{2 \sin^2 \Theta_r}\left(\cos \Theta_r + \frac{\sqrt{N_{21}^2 - \sin^2 \Theta_r}}{N_{21}^2}\right)^2, \tag{37a}$$

$$P_m = \frac{-j\beta_1 R_r}{2 \sin^2 \Theta_r}(\cos \Theta_r + \sqrt{N_{21}^2 - \sin^2 \Theta_r})^2. \tag{37b}$$

Now let the generalized reflection coefficients be defined as follows:

$$f_{er} \equiv \frac{N_{21}^2 \cos \Theta_r - \sqrt{N_{21}^2 - \sin^2 \Theta_r}}{N_{21}^2 \cos \Theta_r + \sqrt{N_{21}^2 - \sin^2 \Theta_r}}, \tag{38a}$$

$$f_{mr} \equiv \frac{\cos \Theta_r - \sqrt{N_{21}^2 - \sin^2 \Theta_r}}{\cos \Theta_r + \sqrt{N_{21}^2 - \sin^2 \Theta_r}}. \tag{38b}$$

These reduce to the asymptotic or radiation-zone factors f_{er}^r and f_{mr}^r when $\Theta_r \doteq \Theta_{01}$. With (38), (36a, b) may be expressed as follows:

$$P = \frac{4p}{(1 - f_r)^2 \sin^2 \Theta_r}. \tag{39}$$

Subscripts e or m may be affixed to P, p, f, to give P_e and P_m.

With P and p replacing P' and p', the integration of (33) may be carried out just as before for (15). The result, corresponding to (25), is

$$\left.\begin{array}{c}V\\H\end{array}\right\} \doteq 2\frac{e^{-j\beta_1 R_r}}{R_r}\left\{1 - \frac{j\sqrt{\pi P}}{\sin \Theta_r}e^{-P}[1 - \text{erf}(j\sqrt{P})]\right\}. \tag{40}$$

Using (26) the asymptotic form of (40) is

$$V^r \doteq 2\frac{e^{-j\beta_1 R_r}}{R_r}\left[1 - \left(\frac{p_e}{P_e \sin^2 \Theta_r}\right)^{1/2}\right], \tag{41a}$$

$$\doteq 2\frac{e^{-j\beta_1 R_r}}{R_r}\left[1 - \left(1 + \frac{N_{21}^2 \cos \Theta_r}{\sqrt{N_{21}^2 - \sin^2 \Theta_r}}\right)^{-1}\right] \tag{41b}$$

$$H^r \doteq 2\frac{e^{-j\beta_1 R_r}}{R_r}\left[1 - \left(\frac{p_m}{P_m \sin^2 \Theta_r}\right)^{1/2}\right], \tag{41c}$$

$$\doteq 2\frac{e^{-j\beta_1 R_r}}{R_r}\left[1 - \left(1 + \frac{\cos \Theta_r}{\sqrt{N_{21}^2 - \sin^2 \Theta_r}}\right)^{-1}\right] \tag{41d}$$

These are the desired asymptotic formulas, thus suggesting that (32) is the appropriately generalized form of (12).

The formula (40) may be expressed entirely in terms of P by substituting for p from (39). The result is

$$\left.\begin{array}{c}V\\H\end{array}\right\} \doteq \frac{e^{-j\beta_1 R_r}}{R_r}$$

$$\times \{2 - j\sqrt{\pi P}(1 - f_r)e^{-P}[1 - \text{erf}(j\sqrt{P})]\}. \tag{42}$$

This may be rearranged as follows:

$$\left.\begin{array}{c}V\\H\end{array}\right\} \doteq \frac{e^{-j\beta_1 R_r}}{R_r}[(1-f_r)F + (1+f_r)], \quad (43)$$

where

$$F \equiv 1 - j\sqrt{\pi P}e^{-P}[1 - \mathrm{erf}(j\sqrt{P})]. \quad (44)$$

Thus the final solutions for the polarization and magnetization potentials are

$$\Pi_{ez1} = \frac{p_{z1}}{4\pi\epsilon_1}\left(\frac{e^{-j\beta_1 R_d}}{R_d} - \frac{e^{-j\beta_1 R_r}}{R_r} + V\right), \quad (45a)$$

$$\Pi_{mz1} = \frac{m_{z1}}{4\pi\nu_1}\left(\frac{e^{-j\beta_1 R_d}}{R_d} - \frac{e^{-j\beta_1 R_r}}{R_r} + H\right). \quad (45b)$$

where

$$V \doteq \frac{e^{-j\beta_1 R_r}}{R_r}[(1-f_{er})F_e + (1+f_{er})], \quad (46a)$$

$$H \doteq \frac{e^{-j\beta_1 R_r}}{R_r}[(1-f_{mr})F_m + (1+f_{mr})], \quad (46b)$$

$$F_e = 1 - j\sqrt{\pi P_e}e^{-P_e}[1 - \mathrm{erf}(j\sqrt{P_e})], \quad (47a)$$

$$F_m = 1 - j\sqrt{\pi P_m}e^{-P_m}[1 - \mathrm{erf}(j\sqrt{P_m})], \quad (47b)$$

$$P_e = \frac{-j\beta_1 R_r}{2\sin^2\Theta_r}\left(\cos\Theta_r + \frac{\sqrt{N_{21}^2 - \sin^2\Theta_r}}{N_{21}^2}\right)^2, \quad (48a)$$

$$P_m = \frac{-j\beta_1 R_r}{2\sin^2\Theta_r}(\cos\Theta_r + \sqrt{N_{21}^2 - \sin^2\Theta_r})^2. \quad (48b)$$

11. Norton's Formulas for Practical Evaluation

For purposes of numerical evaluation Norton expands the complex numerical distance p. It is assumed that region 1 is empty space and region 2 a nonmagnetic imperfect dielectric. By definition

$$p_e \equiv \frac{-j\beta_1 R_r}{2}\left(\frac{N_{21}^2 - \sin^2\Theta_r}{N_{21}^4}\right), \quad (1a)$$

$$p_m \equiv \frac{-j\beta_1 R_r}{2}(N_{21}^2 - \sin^2\Theta_r). \quad (1b)$$

Subject to the assumed conditions, $\nu_2 = \nu_1 = \nu_0$, $\epsilon_1 = \epsilon_0$, the complex index of refraction N_{21} may be expanded. Thus,

$$N_{21}^2 = \beta_2^2/\beta_1^2 = \xi_2/\epsilon_0 = \epsilon_{e2r}(1 - jh_e), \quad (2a)$$

where

$$h_e \equiv \sigma_{e2}/\omega\epsilon_{e2}. \quad (2b)$$

Note that σ_{e2} and ϵ_{e2} are the real effective conductivity and absolute dielectric constant of the conducting region 2. In the following the subscript 2 will be omitted. With (2), (1) becomes

$$p_e = \frac{\beta_1 R_r}{2}\left[\frac{h_e\epsilon_{er} + j(\epsilon_{er} - \sin^2\Theta_r)}{\epsilon_{er}^2(h_e + j)^2}\right], \quad (3a)$$

$$p_m = -\frac{\beta_1 R_r}{2}[h_e\epsilon_{er} + j(\epsilon_{er} - \sin^2\Theta_r)]. \quad (3b)$$

Now let

$$b' = \tan^{-1}\left(\frac{\epsilon_{er} - \sin^2\Theta_r}{h_e\epsilon_{er}}\right)$$

$$= \cos^{-1}\frac{h_e\epsilon_{er}}{\sqrt{(h_e\epsilon_{er})^2 + (\epsilon_{er} - \sin^2\Theta_r)^2}}, \quad (4)$$

$$b'' = \tan^{-1}(1/h_e) = \cos^{-1}\frac{h_e}{\sqrt{1-h_e^2}}. \quad (5)$$

With this notation and with

$$p_e = p_e e^{-jb_e}, \quad p_m = p_m e^{-jb_m}, \quad (6)$$

where p is the real numerical distance, it follows from (3) that

$$p_e = \frac{\beta_1 R_r}{2h_e\epsilon_{er}}\frac{\cos^2 b''}{\cos b'}, \quad (7a)$$

$$b_e = 2b'' - b', \quad (7b)$$

$$p_m = \frac{\beta_1 R_r}{2}\frac{h_e\epsilon_{er}}{\cos b'}, \quad (8a)$$

$$b_m = \pi - b', \quad (8b)$$

The general complex numerical distance P may be expressed in polar form,

$$P = Pe^{-jB}, \quad (9)$$

where

$$P = \frac{4p}{(1-f_r)^2\sin^2\Theta_r}. \quad (10)$$

The complex reflection coefficients f_{er} and f_{mr} may be expressed as follows:

$$f_{er} = \frac{(N_{21}^2\cos\Theta_r/\sqrt{N_{21}^2 - \sin^2\Theta_r}) - 1}{(N_{21}^2\cos\Theta_r/\sqrt{N_{21}^2 - \sin^2\Theta_r}) + 1}, \quad (11a)$$

$$f_{mr} = \frac{(\cos\Theta_r/\sqrt{N_{21}^2 - \sin^2\Theta_r}) - 1}{(\cos\Theta_r/\sqrt{N_{21}^2 - \sin^2\Theta_r}) + 1}. \quad (11b)$$

However,
$$\cos \Theta_r = (z + 2d)/R_r, \quad (12)$$
and
$$\beta_1 R_r \sqrt{N_{21}^2 - \sin^2 \Theta_r} = 2N_{21}^2 p_e \sqrt{\frac{j\beta_1 R_r}{2p_e}}, \quad (13a)$$

$$\beta_1 R_r \sqrt{N_{21}^2 - \sin^2 \Theta_r} = 2p_m \sqrt{\frac{j\beta_1 R_r}{2p_m}}, \quad (13b)$$

so that with (7) and (8),

$$\frac{N_{21}^2 \cos \Theta_r}{\sqrt{N_{21}^2 - \sin^2 \Theta_r}} = \frac{\beta_1(z+2d)}{2p_e}\sqrt{\frac{\cos^2 b''}{h_e \epsilon_{er} \cos b'}} e^{-j(\frac{1}{4}\pi - \frac{1}{2}b_e)}, \quad (14a)$$

$$\frac{\cos \Theta_r}{\sqrt{N_{21}^2 - \sin^2 \Theta_r}} = \frac{\beta_1(z+2d)}{2p_m}\sqrt{\frac{h_e \epsilon_{er}}{\cos b'}} e^{-j(\frac{1}{4}\pi - \frac{1}{2}b_m)}. \quad (14b)$$

Norton next defines the *numerical heights*, q_1 and q_2, as follows:

$$q_{e1} \equiv \beta_1 d \sqrt{\frac{\cos^2 b''}{h_e \epsilon_{er} \cos b'}}, \quad (15a)$$

$$q_{e2} \equiv \beta_1(z+d)\sqrt{\frac{\cos^2 b''}{h_e \epsilon_{er} \cos b'}}, \quad (15b)$$

$$q_{m1} \equiv \beta_1 d \sqrt{\frac{h_e \epsilon_{er}}{\cos b'}}, \quad (15c)$$

$$q_{m2} \equiv \beta_1(z+d)\sqrt{\frac{h_e \epsilon_{er}}{\cos b'}}. \quad (15d)$$

With (13) through (15), the coefficients of reflection (11) reduce to

$$f_r = \frac{[(q_1+q_2)/2p]e^{-j(\frac{1}{4}\pi - \frac{1}{2}b)} - 1}{[(q_1+q_2)/2p]e^{-j(\frac{1}{4}\pi - \frac{1}{2}b)} + 1}, \quad (16)$$

with appropriate subscript e or m on f, q, p, and b.

The complex *surface-wave attenuation function* F may be expressed as follows:

$$F = f(P, B)e^{-j\phi(P,B)}$$
$$\equiv 1 - j\sqrt{\pi P}e^{-P}[1 - \text{erf}(j\sqrt{P})]. \quad (17)$$

The magnitude $f(P, B)$ is the real surface-wave attenuation function. It is given in Table 11.1 and represented graphically in Fig. 11.1 as a function of P with B as a parameter. The angle $\phi(P, B)$ is shown in Fig. 11.2 as a function of P with B as a parameter.

TABLE 11.1. Magnitude $f(P,B)$ of surface-wave attenuation function.*

P \ B	0°	5°	10°	15°	20°	25°	30°	35°	40°	45°	50°	60°	70°	90°
0.02	0.9914	0.981	0.970	0.960	0.949	0.939	0.930	0.920	0.911	0.903	0.894	0.878	0.864	0.8377
.04	.9829													.7786
.05	.9788	.962	.946	.930	.915	.901	.887	.873	.860	.847	.835	.812	.791	.7577
.06	.9745													.7363
.1	.9581	.935	.913	.892	.872	.853	.834	.816	.799	.784	.768	.741	.716	.6740
.24	.9010													.5447
.25	.8987	.866	.835	.807	.780	.755	.731	.708	.688	.669	.650	.617	.589	.5382
.26	.8950													.5316
.50	.8085	.769	.733	.699	.668	.640	.613	.589	.566	.547	.527	.494	.465	.4200
.74	.731													.3516
.75	.727													.3495
.76	.725													.3470
1	.6567	.616	.579	.546	.516	.489	.465	.443	.423	.406	.389	.361	.338	.3003
1.5	.535													.2357
2	.4398	.408	.381	.357	.336	.317	.301	.286	.272	.261	.250	.232	.217	.1937
2.5	.3635	.338	.316	.297	.280	.265	.251	.240	.229	.220	.211	.196	.184	.1644
3	.3027	.282	.264	.249	.236	.219	.213	.204	.195	.188	.180	.168	.158	.1427
4	.2154	.203	.192	.182	.174	.166	.159	.153	.148	.143	.139	.131	.124	.1126
5	.1593	.152	.146	.140	.135	.131	.127	.123	.119	.116	.113	.108	.103	.09281
6	.1228													
7	.0984													
8	.0816	.0804	.0791	.0778	.0765	.0749	.0738	.0725	.0712	.0699	.0686	.0662	.0640	.06036
9	.0696													.05401
10	.0606	.0604	.0602	.0599	.0595	.0588	.0577	.0567	.0556	.0545	.0535	.0515	.0500	.04884
11	.0539													
12	.0485													
13	.0441	.0441	.0441	.0440	.0439	.0438	.0436	.0433	.0430	.0425	.0420	.0407	.0392	.03788
15	.0374													
18	.0305	.0305	.0305	.0305	.0305	.0305	.0305	.0304	.0304	.0302	.0301	.0295	.0288	.02753
20	.0272													
30	.0176													.02484
40	.0130													
50	.01032	.0103	.0103	.0103	.0103	.0103	.0103	.0103	.0103	.0103	.0103	.0103	.0103	.01030

* Data of K. A. Norton, *Proc. I. R. E.* 24, 1370 (1936).

[VII.11] THEORY OF LINEAR ANTENNAS

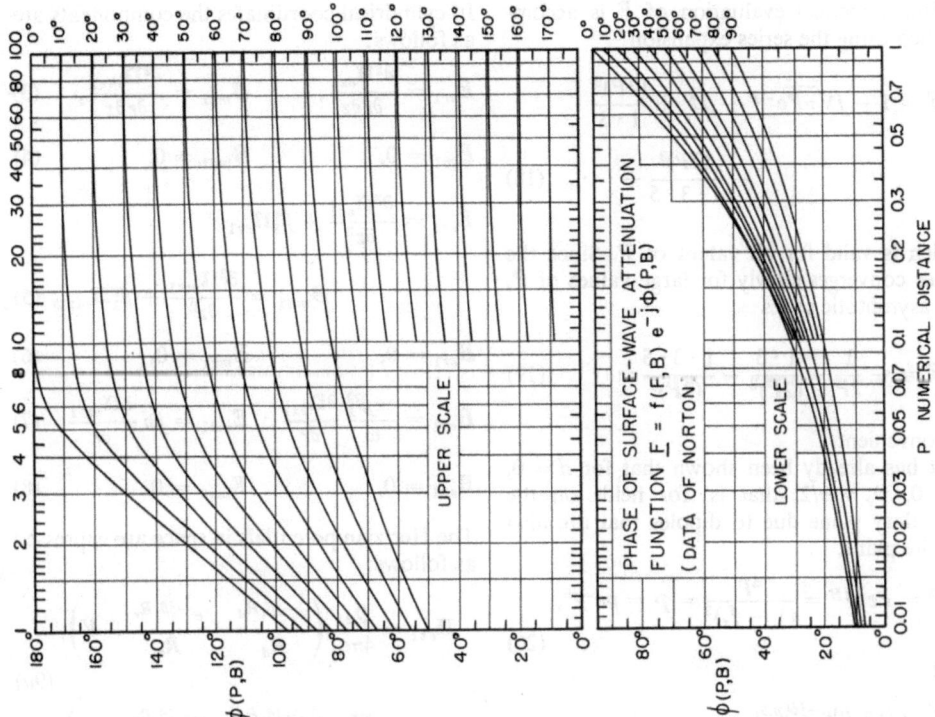

Fig. 11.2. Phase of surface wave attenuation function $F = f(P, B)e^{-j\phi(P,B)}$ (Norton).

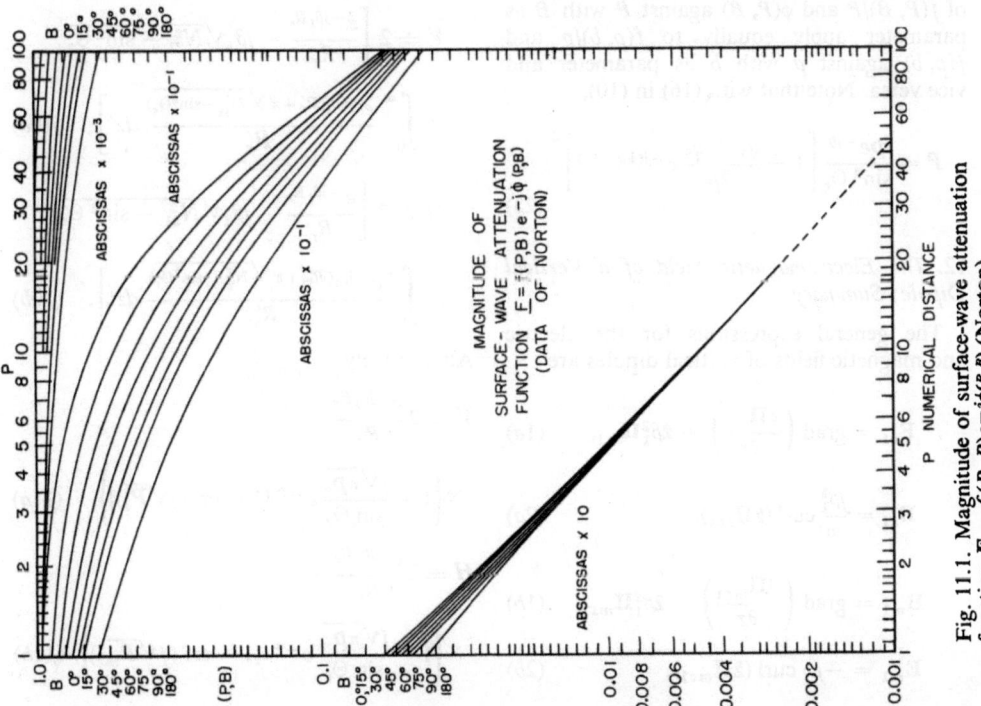

Fig. 11.1. Magnitude of surface-wave attenuation function $F = f(P, B)e^{-j\phi(P,B)}$ (Norton).

The numerical evaluation of F is accomplished using the series expansion

$$F = 1 - j\sqrt{\pi P}e^{-P} - 2P + \frac{(2P)^2}{1\cdot 3} - \frac{(2P)^3}{1\cdot 3\cdot 5} + \cdots, \quad (18)$$

which is valid for all values of P. Since the series converges slowly for large values of P, the asymptotic series

$$F = -\frac{1}{2P} - \frac{1\cdot 3}{(2P)^2} - \frac{1\cdot 3\cdot 5}{(2P)^3} \cdots \quad (19)$$

is convenient.

It has already been shown that for $d = 0$, $z = 0$, $\Theta_r = \pi/2$, that is, for fields on the boundary plane due to dipoles that are also on the plane,

$$P = Pe^{-jB} = \frac{4p}{(1-f_r)^2} = p = pe^{-jb}, \quad (20)$$

and

$$F = f(p, b)e^{-j\phi(p,b)}$$
$$= 1 - j\sqrt{\pi p}e^{-p}[1 - \text{erf}(j\sqrt{p})]. \quad (21)$$

It follows that the graphical representations of $f(P, B)/P$ and $\phi(P, B)$ against P with B as parameter apply equally to $f(p, b)/p$ and $\phi(p, b)$ against p with b as parameter and vice versa. Note that with (16) in (10),

$$P = \frac{pe^{-jb}}{\sin^2\Theta_r}\left[1 + \frac{q_1 + q_2}{2p}e^{-j(\frac{1}{2}\pi - \frac{1}{2}b)}\right]^2. \quad (23)$$

12. The Electromagnetic Field of a Vertical Dipole; Summary

The general expressions for the electric and magnetic fields of vertical dipoles are

$$\mathbf{E}_{e1} = \text{grad}\left(\frac{\partial \Pi_{ez1}}{\partial z}\right) + \hat{\mathbf{z}}\beta_1^2\Pi_{ez1}, \quad (1a)$$

$$\mathbf{B}_{e1} = \frac{j\beta_1^2}{\omega}\text{curl}(\hat{\mathbf{z}}\Pi_{ez1}), \quad (2a)$$

$$\mathbf{B}_{m1} = \text{grad}\left(\frac{\partial \Pi_{mz1}}{\partial z}\right) + \hat{\mathbf{z}}\beta_1^2\Pi_{mz1}, \quad (1b)$$

$$\mathbf{E}_{m1} = -j\omega\,\text{curl}(\hat{\mathbf{z}}\Pi_{mz1}). \quad (2b)$$

In cylindrical coordinates the components are as follows:

$$E_{er1} = \frac{\partial^2 \Pi_{ez1}}{\partial r\,\partial z}, \quad B_{mr1} = \frac{\partial^2 \Pi_{mz1}}{\partial r\,\partial z}, \quad (3)$$

$$E_{e\theta 1} = 0, \quad B_{m\theta 1} = 0, \quad (4)$$

$$E_{ez1} = \frac{\partial^2 \Pi_{ez1}}{\partial z^2} + \beta_1^2 \Pi_{ez1},$$

$$B_{mz1} = \frac{\partial^2 \Pi_{mz1}}{\partial z^2} + \beta_1^2 \Pi_{mz1}, \quad (5)$$

$$B_{er1} = 0, \quad E_{mr1} = 0, \quad (6)$$

$$B_{e\theta 1} = \frac{-j\beta_1^2}{\omega}\frac{\partial \Pi_{ez1}}{\partial r}, \quad E_{m\theta 1} = j\omega\frac{\partial \Pi_{mz1}}{\partial r}, \quad (7)$$

$$B_{ez1} = 0, \quad E_{mz1} = 0. \quad (8)$$

The Hertzian potentials in space are expressed as follows:

$$\Pi_{ez1} = \frac{p_{z1}}{4\pi\epsilon_0}\left(\frac{e^{-j\beta_1 R_d}}{R_d} - \frac{e^{-j\beta_1 R_r}}{R_r} + V\right), \quad (9a)$$

$$\Pi_{mz1} = \frac{m_{z1}}{4\pi\nu_0}\left(\frac{e^{-j\beta_1 R_d}}{R_d} - \frac{e^{-j\beta_1 R_r}}{R_r} + H\right), \quad (9b)$$

where

$$V \doteq 2\left[\frac{j\beta_1 R_r}{R_r} - j\beta_1\sqrt{N_{21}^2 - \sin^2\Theta_r}\right.$$
$$\left.\times \int_0^\infty \frac{e^{-j\beta_1(R_e' + z'\sqrt{N_{21}^2 - \sin^2\Theta_r})}}{R_e'}\,dz'\right], \quad (10a)$$

$$H \doteq 2\left[\frac{e^{-j\beta_1 R_r}}{R_r} - j\beta_1\sqrt{N_{21}^2 - \sin^2\Theta_r}\right.$$
$$\left.\times \int_0^\infty \frac{e^{-j\beta_1(R_m' + z'\sqrt{N_{21}^2 - \sin^2\Theta_r})}}{R_m'}\,dz'\right]. \quad (10b)$$

Alternatively,

$$V \doteq 2\frac{e^{-j\beta_1 R_r}}{R_r}$$
$$\times\left\{1 - \frac{j\sqrt{\pi P_e}}{\sin\Theta_r}e^{-P_e}[1 - \text{erf}(j\sqrt{P_e})]\right\}, \quad (11a)$$

$$H \doteq 2\frac{e^{-j\beta_1 R_r}}{R_r}$$
$$\times\left\{1 - \frac{j\sqrt{\pi P_m}}{\sin\Theta_r}e^{-P_m}[1 - \text{erf}(j\sqrt{P_m})]\right\}. \quad (11b)$$

or,
$$V \doteq \frac{e^{-j\beta_1 R_r}}{R_r} \{(1 - f_{er})F_e + (1 + f_{er})\}, \quad (11c)$$

$$H \doteq \frac{e^{-j\beta_1 R_r}}{R_r} \{(1 - f_{mr})F_m + (1 + f_{mr})\}, \quad (11d)$$

where
$$f_{er} = \frac{N_{21}^2 \cos \Theta_r - \sqrt{N_{21}^2 - \sin^2 \Theta_r}}{N_{21}^2 \cos \Theta_r + \sqrt{N_{21}^2 - \sin^2 \Theta_r}}, \quad (12a)$$

$$f_{mr} = \frac{\cos \Theta_r - \sqrt{N_{21}^2 - \sin^2 \Theta_r}}{\cos \Theta_r + \sqrt{N_{21}^2 - \sin^2 \Theta_r}}, \quad (12b)$$

$$P_e = \frac{4p_e}{(1 - f_{er})^2 \sin^2 \Theta_r}, \quad (13a)$$

$$P_m = \frac{4p_m}{(1 - f_{mr})^2 \sin^2 \Theta_r}, \quad (13b)$$

$$p_e = \frac{-j\beta_1 R_r (N_{21}^2 - \sin^2 \Theta_r)}{2N_{21}^4}, \quad (14a)$$

$$p_m = \frac{-j\beta_1 R_r (N_{21}^2 - \sin^2 \Theta_r)}{2}, \quad (14b)$$

$$R'_e = \sqrt{r^2 + (N_{21}^2 z' + z + 2d)^2}, \quad (15a)$$

$$R'_m = \sqrt{r^2 + (z' + z + 2d)^2}, \quad (15b)$$

$$F_e = 1 - j\sqrt{\pi P_e} e^{-P_e}[1 - \operatorname{erf}(j\sqrt{P_e})], \quad (16a)$$

$$F_m = 1 - j\sqrt{\pi P_m} e^{-P_m}[1 - \operatorname{erf}(j\sqrt{P_m})]. \quad (16b)$$

It is significant that
$$\frac{P_m}{P_e} = N_{21}^4. \quad (17)$$

Hence, since it has been assumed that $|N_{21}|^2 \gg 1$ the numerical distance $|p|_m$ for the magnetic dipole is *very much greater* than is $|p_e|$ for the electrical dipole. This means that the field along the surface when plotted with the numerical distance as independent variable is attenuated much more in the case of the magnetic dipole than in the case of the electric dipole.

13. *The Vertical Electromagnetic Field of a Vertical Dipole*

The vertical components of the electromagnetic field of electric and magnetic dipoles are
$$E_{z1} = \frac{\partial^2 \Pi_{ez1}}{\partial z^2} + \beta_1^2 \Pi_{ez1}, \quad (1a)$$

$$B_{z1} = \frac{\partial^2 \Pi_{mz1}}{\partial z^2} + \beta_1^2 \Pi_{mz1}. \quad (1b)$$

In order to evaluate the derivatives with respect to z it is convenient to proceed from (12.9) with (12.10). As a temporary shorthand, let

$$U \equiv -j2\beta_1 \sqrt{N_{21}^2 - \sin^2 \Theta_r}$$
$$\times \int_0^\infty \frac{e^{-j\beta_1(R' + z'\sqrt{N_{21}^2 - \sin^2 \Theta_r})}}{R'} dz'. \quad (2)$$

In evaluating $\partial \Pi_{z1}/\partial z$ the following derivatives occur:

$$\frac{\partial}{\partial z}\left(\frac{e^{-j\beta_1 R_d}}{R_d}\right) = -\left(j\beta_1 + \frac{1}{R_d}\right)\frac{e^{-j\beta_1 R_d}}{R_d} \cos \Theta_d, \quad (3)$$

$$\frac{\partial}{\partial z}\left(\frac{e^{-j\beta_1 R_r}}{R_r}\right) = -\left(j\beta_1 + \frac{1}{R_r}\right)\frac{e^{-j\beta_1 R_r}}{R_r} \cos \Theta_r. \quad (4)$$

In the differentiation of U with respect to z it is a good approximation to treat $\sin^2 \Theta_r$ in $\sqrt{N_{21}^2 - \sin^2 \Theta_r}$ as sensibly constant, since it has been assumed that $|N_{21}^2|$ is quite large as compared with the maximum value of unity of $\sin^2 \Theta_r$. Hence,

$$\frac{\partial U}{\partial z} \doteq -j2\beta_1 \sqrt{N_{21}^2 - \sin^2 \Theta_r}$$
$$\times \int_0^\infty \frac{\partial}{\partial z}\left(\frac{e^{-j\beta_1 R'}}{R'}\right) e^{-j\beta_1 z'\sqrt{N_{21}^2 - \sin^2 \Theta_r}} dz'. \quad (5)$$

With (12.15) it follows that
$$\frac{\partial f(R'_e)}{\partial z} = \frac{1}{N_{21}^2} \frac{\partial f(R'_e)}{\partial z'}, \quad (6a)$$

$$\frac{\partial f(R'_m)}{\partial z} = \frac{\partial f(R'_m)}{\partial z'}. \quad (6b)$$

With (6), the integral in (5) is readily integrated by parts. Thus,
$$\frac{\partial U_e}{\partial z} = \frac{j\beta_1 \sqrt{N_{21}^2 - \sin^2 \Theta_r}}{N_{21}^2} V, \quad (7a)$$

$$\frac{\partial U_m}{\partial z} = j\beta_1 \sqrt{N_{21}^2 - \sin^2 \Theta_r} \, H. \quad (7b)$$

It follows with (10) that
$$\frac{\partial V}{\partial z} = \left[2 \frac{\partial}{\partial z}\left(\frac{e^{-j\beta_1 R_r}}{R_r}\right) + \frac{\partial U_e}{\partial z}\right], \quad (7c)$$

$$\frac{\partial H}{\partial z} = \left[2 \frac{\partial}{\partial z}\left(\frac{e^{-j\beta_1 R_r}}{R_r}\right) + \frac{\partial U_m}{\partial z}\right], \quad (7d)$$

where the two derivatives involved are in (4) and (7a, b). Upon combining (3), (4), and (7), it follows directly that

$$\frac{\partial \Pi_{ez1}}{\partial z} \doteq \frac{-j\beta_1 p_{z1}}{4\pi\epsilon_0}\left[\left(1 + \frac{1}{j\beta_1 R_d}\right)\cos\Theta_d \frac{e^{-j\beta_1 R_d}}{R_d}\right.$$
$$+ \left(1 + \frac{1}{j\beta_1 R_r}\right)\cos\Theta_r \frac{e^{-j\beta_1 R_r}}{R_r}$$
$$\left.- \frac{\sqrt{N_{21}^2 - \sin^2\Theta_r}}{N_{21}^2}V\right], \quad (8a)$$

$$\frac{\partial \Pi_{mz1}}{\partial z} \doteq \frac{-j\beta_1 m_{z1}}{4\pi\nu_0}\left[\left(1 + \frac{1}{j\beta_1 R_d}\right)\cos\Theta_d \frac{e^{-j\beta_1 R_d}}{R_d}\right.$$
$$+ \left(1 + \frac{1}{j\beta_1 R_r}\right)\cos\Theta_r \frac{e^{-j\beta_1 R_r}}{R_r}$$
$$\left.- \sqrt{N_{21}^2 - \sin^2\Theta_r}H\right]. \quad (8b)$$

In obtaining the second derivative of Π_{z1} with respect to z, $\sin^2\Theta_r$ in $\sqrt{N_{21}^2 - \sin^2\Theta_r}$ is again treated as a constant. On the other hand, $\cos\Theta_d$ and $\cos\Theta_r$ may not be so treated. The following derivatives are involved:

$$\frac{\partial}{\partial z}\left[\left(1 + \frac{1}{j\beta_1 R_d}\right)\cos\Theta_d \frac{e^{-j\beta_1 R_d}}{R_d}\right]$$
$$= -\frac{1}{j\beta_1 R_d^2}\cos^2\Theta_d \frac{e^{-j\beta_1 R_d}}{R_d}$$
$$+ \left(1 + \frac{1}{j\beta_1 R_d}\right)\frac{e^{-j\beta_1 R_d}}{R_d}\frac{\partial}{\partial z}(\cos\Theta_d)$$
$$+ \left(1 + \frac{1}{j\beta_1 R_d}\right)\cos\Theta \frac{\partial}{\partial z}\left(\frac{e^{-j\beta_1 R_d}}{R_d}\right). \quad (9a)$$

Note that

$$\frac{\partial}{\partial z}(\cos\Theta_d) = \frac{\partial}{\partial z}\left(\frac{z}{R_d}\right) = \frac{1}{R_d}(1 - \cos^2\Theta_d). \quad (9b)$$

Hence, after collecting terms,

$$\frac{\partial}{\partial z}\left[\left(1 + \frac{1}{j\beta_1 R_d}\right)\cos\Theta_d \frac{e^{-j\beta_1 R_d}}{R_d}\right]$$
$$= -j\beta_1\left[\cos^2\Theta_d - \left(\frac{1}{j\beta_1 R_d} + \frac{1}{(j\beta_1 R_d)^2}\right)\right.$$
$$\left.\times (1 - 3\cos^2\Theta_d)\right]\frac{e^{-j\beta_1 R_d}}{R_d}. \quad (10)$$

A similar expression with subscripts d replaced by r is true. With (7c, d) and (9) and (10), it follows that

$$\frac{\partial^2 \Pi_{ez1}}{\partial z^2} \doteq \frac{-\beta_1^2 p_{z1}}{4\pi\epsilon_0}\left\{\left[\cos^2\Theta_d - \left(\frac{1}{j\beta_1 R_d} + \frac{1}{(j\beta_1 R_d)^2}\right)\right.\right.$$
$$\left.\times (1 - 3\cos^2\Theta_d)\right]\frac{e^{-j\beta_1 R_d}}{R_d}$$
$$+ \left[\cos^2\Theta_r - \left(\frac{1}{j\beta_1 R_r} + \frac{1}{(j\beta_1 R_r)^2}\right)\right.$$
$$\left.\times (1 - 3\cos^2\Theta_r)\right]\frac{e^{-j\beta_1 R_r}}{R_r}$$
$$- \frac{2\sqrt{N_{21}^2 - \sin^2\Theta_r}}{N_{21}^2}\left(1 + \frac{1}{j\beta_1 R_r}\right)$$
$$\left.\times \cos\Theta_r \frac{e^{-j\beta_1 R_r}}{R_r} + \frac{N_{21}^2 - \sin^2\Theta_r}{N_{21}^4}V\right\}, \quad (11a)$$

$$\frac{\partial^2 \Pi_{mz1}}{\partial z^2}$$
$$\doteq \frac{-\beta_1^2 m_{z1}}{4\pi\nu_0}\left\{\frac{e^{-j\beta_1 R_d}}{R_d}[\ ] + \frac{e^{-j\beta_1 R_r}}{R_r}[\ ]\right.$$
$$- 2\sqrt{N_{21}^2 - \sin^2\Theta_r}\left(1 + \frac{1}{j\beta_1 R_r}\right)$$
$$\left.\times \cos\Theta_r \frac{e^{-j\beta_1 R_r}}{R_r} + (N_{21}^2 - \sin^2\Theta_r)H\right\}. \quad (11b)$$

The two brackets in (11b) are like those in (11a).

The complete expression for the vertical components of the field may now be obtained by forming (1) using (12.9) and (11):

$$E_{ez1} \doteq \frac{\beta_1^2 p_{z1}}{4\pi\epsilon_0}\left\{\sin^2\Theta_d \frac{e^{-j\beta_1 R_d}}{R_d}\right.$$
$$+ (\sin^2\Theta_r - 2)\frac{e^{-j\beta_1 R_r}}{R_r}$$
$$+ \left(1 - \frac{N_{21}^2 - \sin^2\Theta_r}{N_{21}^4}\right)V$$
$$+ \frac{2\sqrt{N_{21}^2 - \sin^2\Theta_r}}{N_{21}^2}\cos\Theta_r \frac{e^{-j\beta_1 R_r}}{R_r}$$
$$\left.+ G'_{ez}(R_d, R_r)\right\}, \quad (12a)$$

where

$$G'_{ez}(R_d, R_r)$$
$$\equiv \frac{2\sqrt{N_{21}^2 - \sin^2 \Theta_r}}{N_{21}^2} \cos \Theta_r \frac{e^{-j\beta_1 R_r}}{j\beta_1 R_r^2}$$
$$+ \left(\frac{1}{j\beta_1 R_d} + \frac{1}{(j\beta_1 R_d)^2}\right)(1 - 3\cos^2 \Theta_d) \frac{e^{-j\beta_1 R_d}}{R_d}$$
$$+ \left(\frac{1}{j\beta_1 R_r} + \frac{1}{(j\beta_1 R_r)^2}\right)(1 - 3\cos^2 \Theta_r) \frac{e^{-j\beta_1 R_r}}{R_r}; \quad (12b)$$

and

$$B_{mz1} \doteq \frac{\beta_1^2 m_{z1}}{4\pi v_0} \Bigg\{ \sin^2 \Theta_d \frac{e^{-j\beta_1 R_d}}{R_d}$$
$$+ (\sin^2 \Theta_r - 2) \frac{e^{-j\beta_1 R_r}}{R_r}$$
$$+ (1 - N_{21}^2 + \sin^2 \Theta_r)H$$
$$+ 2\sqrt{N_{21}^2 - \sin^2 \Theta_r} \cos \Theta_r \frac{e^{-j\beta_1 R_r}}{R_r}$$
$$+ G'_{mz}(R_d, R_r) \Bigg\}, \quad (12c)$$

where

$$G'_{mz}(R_d, R_r)$$
$$= 2\sqrt{N_{21}^2 - \sin^2 \Theta_r} \cos \Theta_r \frac{e^{-j\beta_1 R_r}}{j\beta_1 R_r^2}$$
$$+ \left(\frac{1}{j\beta_1 R_d} + \frac{1}{(j\beta_1 R_d)^2}\right)(1 - 3\cos^2 \Theta_d) \frac{e^{-j\beta_1 R_d}}{R_d}$$
$$+ \left(\frac{1}{j\beta_1 R_r} + \frac{1}{(j\beta_1 R_r)^2}\right)(1 - 3\cos^2 \Theta_r) \frac{e^{-j\beta_1 R_r}}{R_r}. \quad (12d)$$

Using (12.11) for V and H and (12.12) for f_r, which occurs in V and H, the second, third, and fourth terms in (12a) may be combined into

$$\frac{e^{-j\beta_1 R_r}}{R_r} \Bigg[(\sin^2 \Theta_r - 2)$$
$$+ (1 + f_{er})\left(1 - \frac{N_{21}^2 - \sin^2 \Theta_r}{N_{21}^4}\right)$$
$$+ \frac{2\sqrt{N_{21}^2 - \sin^2 \Theta_r}}{N_{21}^2} \cos \Theta_r$$
$$+ (1 - f_{er})\left(1 - \frac{N_{21}^2 - \sin^2 \Theta_r}{N_{21}^4}\right) F_e \Bigg]. \quad (13a)$$

With (12.12) this reduces to

$$f_{er} \sin^2 \Theta_r \frac{e^{-j\beta_1 R_r}}{R_r} + \frac{e^{-j\beta_1 R_r}}{R_r}(1 - f_{er})$$
$$\times \left(1 - \frac{N_{21}^2 - \sin^2 \Theta_r}{N_{21}^4}\right) F_e. \quad (13b)$$

The corresponding terms in (12c) give

$$f_{mr} \sin^2 \Theta_r \frac{e^{-j\beta_1 R_r}}{R_r} + \frac{e^{-j\beta_1 R_r}}{R_r}(1 - f_{mr})$$
$$\times (1 - N_{21}^2 + \sin^2 \Theta_r) F_m. \quad (13c)$$

Accordingly, (12a) and (12c) reduce to the following:

$$E_{ez1} \doteq \frac{\beta_1^2 p_{z1}}{4\pi \epsilon_0} \Bigg[\sin^2 \Theta_d \frac{e^{-j\beta_1 R_d}}{R_d}$$
$$+ f_{er} \sin^2 \Theta_r \frac{e^{-j\beta_1 R}}{R_r}$$
$$+ (1 - f_{er})\left(1 - \frac{N_{21}^2 - \sin^2 \Theta_r}{N_{21}^4}\right) F_e \frac{e^{-j\beta_1 R_r}}{R_r}$$
$$+ G'_{ez}(R_d, R_r) \Bigg], \quad (14a)$$

$$B_{mz1} \doteq \frac{\beta_1^2 m_{z1}}{4\pi v_0} \Bigg[\sin^2 \Theta_d \frac{e^{-j\beta_1 R_d}}{R_d}$$
$$+ f_{mr} \sin^2 \Theta_r \frac{e^{-j\beta_1 R_r}}{R_r}$$
$$+ (1 - f_{mr})(1 - N_{21}^2 + \sin^2 \Theta_r) F_m \frac{e^{-j\beta_1 R_r}}{R_r}$$
$$+ G'_{mz}(R_d, R_r) \Bigg]. \quad (14b)$$

In (14) all terms in $G'(R_d, R_r)$ have R^{-2} or R^{-3} as a factor. Therefore, they are a part of the near-zone or induction field. Moreover, since the asymptotic expansion of F begins with the term

$$F \sim -\frac{1}{2P}, \quad (15)$$

it follows that

$$F_e \sim \frac{-j \sin^2 \Theta_r}{\beta_1 R_r}$$
$$\times \left(\cos \Theta_r + \frac{\sqrt{N_{21}^2 - \sin^2 \Theta_r}}{N_{21}^2}\right)^{-2}, \quad (16a)$$

$$F_m \sim \frac{-j \sin^2 \Theta_r}{\beta_1 R_r}$$
$$(\cos \Theta_r + \sqrt{N_{21}^2 - \sin^2 \Theta_r})^{-2}. \quad (16b)$$

Evidently, since F_e has a factor R_r^{-1} when R_r is large, it follows that the third terms in (14a, b) have R^{-2} as a factor and, therefore, are a part of the near-zone field. Thus, only the first two terms in (14) are R^{-1} or radiation-zone terms. When

$$R_0^2 \gg d^2, \tag{17}$$

the following approximations are good:

$$R_d \doteq R_r \doteq R_0 \text{ in amplitudes}, \tag{18a}$$

$$\left.\begin{array}{l} R_d \doteq R_0 - d\cos\Theta_{01} \\ R_r \doteq R_0 + d\cos\Theta_{01} \end{array}\right\} \text{ in phases}, \tag{18b}$$

$$\Theta_r \doteq \Theta_d \doteq \Theta_{01}; \tag{18c}$$

also

$$f_r \doteq f_r^r, \tag{19}$$

where f_r^r is like f_r but with Θ_{01} substituted for Θ_r. With (15) through (19) it follows that the far-zone part of (14) is

$$E_{ez1}^r \doteq \frac{\beta_1^2 p_{z1}}{4\pi\epsilon_0} \frac{e^{-j\beta_1 R_0}}{R_0} \sin^2\Theta_{01}(e^{j\beta_1 d\cos\Theta_{01}}$$
$$+ f_{er}^r e^{-j\beta_1 d\cos\Theta_{01}}), \tag{20a}$$

$$B_{mz1}^r \doteq \frac{\beta_1^2 m_{z1}}{4\pi\nu_0} \frac{e^{-j\beta_1 R_0}}{R_0} \sin^2\Theta_{01}(e^{-j\beta_1 d\cos\Theta_{01}}$$
$$+ f_{mr}^r e^{-j\beta_1 d\cos\Theta_{01}}). \tag{20b}$$

This is the z-component of the far-zone field previously obtained. It vanishes at $\Theta_{01} = \pi/2$, that is, along the surface bounding the two regions.

Since the sum of the first two terms in (14a) and (14b) vanishes when $\Theta_r = \pi/2$, the leading term near the surface must be obtained from the third and fourth terms. It is readily verified that except for very small values of R_r these are the third terms in the following formulas:

$$E_{ez1} \doteq \frac{\beta_1^2 p_{z1}}{4\pi\epsilon_0}\left[\sin^2\Theta_d \frac{e^{-j\beta_1 R_d}}{R_d}\right.$$
$$+ f_{er}\sin^2\Theta_r \frac{e^{-j\beta_1 R_r}}{R_r}$$
$$+ (1-f_{er})F_e \sin^2\Theta_r \frac{e^{-j\beta_1 R_r}}{R_r}$$
$$\left.+ G_{ez}(R_d, R_r)\right], \tag{21a}$$

where

$$G_{ez}(R_d, R_r) = G'_{ez}(R_d, R_r) + (1-f_{er})F_e$$
$$\times \left(\cos^2\Theta_r - \frac{N_{21}^2 - \sin^2\Theta_r}{N_{21}^4}\right)\frac{e^{-j\beta_1 R_r}}{R_r}; \tag{21b}$$

and

$$B_{mz1} \doteq \frac{\beta_1^2 m_{z1}}{4\pi\nu_0}\left[\sin^2\Theta_d \frac{e^{-j\beta_1 R_d}}{R_d}\right.$$
$$+ f_{mr}\sin^2\Theta_r \frac{e^{-j\beta_1 R_r}}{R_r}$$
$$+ (1-f_{mr})F_m \sin^2\Theta_r \frac{e^{-j\beta_1 R_r}}{R_r}$$
$$\left.+ G_{mz}(R_d, R_r)\right] \tag{22a}$$

where

$$G_{mz}(R_d, R_r) = G'_{mz}(R_d, R_r)$$
$$+ (1-f_{mr})F_m(1-N_{21}^2)\frac{e^{-j\beta_1 R_r}}{R_r}. \tag{22b}$$

Let the terms $G_{ez}(R_d, R_r)$ and $G_{mz}(R_d, R_r)$ be compared with the third terms in (21a) and (22a) in order to verify that these latter are the leading terms when $\Theta_r \doteq \pi/2$. Specifically, let

$$\Theta_d \doteq \Theta_r \doteq \pi/2, \quad R_d \doteq R_r, \quad f_r \doteq -1. \tag{23}$$

With (23), it follows that $\cos\Theta_d \doteq \cos\Theta_r \doteq 0$, and $\sqrt{N_{21}^2 - \sin^2\Theta_r} \doteq \sqrt{N_{21}^2 - 1}$. On the other hand, terms in $\sin^2\Theta_r$ except in $\sqrt{N_{21}^2 - \sin^2\Theta_r}$ are retained, since $\sin^2\Theta_r$ is a factor in the third terms in (21a) and (22a):

$$G_{ez}(R_d, R_r) \doteq 2\frac{e^{-j\beta_1 R_r}}{R_r}\left[\frac{1}{j\beta_1 R_r} + \frac{1}{(j\beta_1 R_r)^2}\right.$$
$$\left. - F_e\left(\frac{N_{21}^2 - 1}{N_{21}^4}\right)\right], \tag{24a}$$

$$G_{mz}(R_d, R_r) \doteq 2\frac{e^{-j\beta_1 R_r}}{R_r}\left[\frac{1}{j\beta_1 R_r} + \frac{1}{(j\beta_1 R_r)^2}\right.$$
$$\left. - F_m(N_{21}^2 - 1)\right]. \tag{24b}$$

The leading term in the asymptotic expansion of F is given by (16). With $\cos\Theta_r \doteq 0$, it is

$$F_e \doteq \frac{\sin^2\Theta_r}{j\beta_1 R_r}\frac{N_{21}^4}{N_{21}^2 - 1}, \tag{25a}$$

$$F_m \doteq \frac{\sin^2\Theta_r}{j\beta_1 R_r}\frac{1}{N_{21}^2 - 1}. \tag{25b}$$

Substitution of (25) in (24) gives for G_{ez} and for G_{mz}

$$G_z(R_d, R_r) \doteq 2\frac{e^{-j\beta_1 R_r}}{R_r}\left[\frac{1}{j\beta_1 R_r}(1 - \sin^2\Theta_r)\right.$$
$$\left. + \frac{1}{(j\beta_1 R_r)^2}\right] \doteq 2\frac{e^{-j\beta_1 R_r}}{R_r}\frac{1}{(j\beta_1 R_r)^2}. \tag{26}$$

That is, $G_z(R_d, R_r)$ is of order R_r^{-3}, whereas the third terms in (21a) and (22a) are of order R_r^{-2}. It follows that at distances R_r that are sufficiently great so that the condition

$$\beta_1^2 R_r^2 \gg 1 \quad \text{or} \quad R_r \geq \lambda_1. \quad (27)$$

is a good approximation, $G_z(R_d, R_r)$ may be neglected. At shorter distances it must be included. However, since the approximations involved in using an approximate instead of an exact expression for R'_e and R'_m in order to perform the integration involve errors of the order of magnitude of the terms in $G_z(R_d, R_r)$, the solutions (21a) and (22a) are accurate only for distances that make $G_z(R_d, R_r)$ negligible.

At distances that satisfy (27), (21a) and (22a) are the complete solutions with $G_z(R_d, R_r)$ omitted. In (21a) and (22a) the first term is the field due to the dipole; it is called the *direct wave*. The second term is called the *ground-reflected wave*; it gives the entire contribution to the field by the currents in the conducting half-space except when Θ_r is near $\pi/2$ and the ground is not *perfectly conducting*. The first two terms taken together are called the *space wave*. The third terms in (21a) and (22a) constitute the *surface wave*; this is negligible except when Θ_r is near $\pi/2$, where it is the principal field.*

The significance of the surface-wave terms in (21a) and (22a) may be investigated by examining the field at the surface of the conducting earth at $z = -d$ where $R_d = R_r$, $\Theta_r = \pi - \Theta_d$. Let the radial distance r along the earth's surface from the z-axis through the dipole be sufficiently great so that (27) is satisfied and the terms $G_z(R_d, R_r)$ in (21a) and (22a) may be neglected. Then

$$E_{ez1} \doteq \frac{\beta_1^2 p_{z1}}{4\pi\epsilon_0} \frac{e^{-j\beta_1 R_d}}{R_d} \sin^2 \Theta_d [(1 + f_{er}) + (1 - f_{er})F_e], \quad (28a)$$

$$B_{mz1} \doteq \frac{\beta_1^2 m_{z1}}{4\pi v_0} \frac{e^{-j\beta_1 R_d}}{R_d} \sin^2 \Theta_d [(1 + f_{mr}) + (1 - f_{mr})F_m]. \quad (28b)$$

In order to achieve *grazing incidence*, the radial distance r must be very great compared with the height d of the dipole so that it is

---

* The surface wave here defined is not the surface wave occurring in the original analysis of Sommerfeld and referred to in Section 8 of the Introduction. Sommerfeld's surface wave decreases as $1/\sqrt{r}$ rather than as $1/r^2$. While it is a mathematically possible solution, it is not the one excited or observed in practice. It is not included in (21a) or (22a). For an illuminating discussion of the surface wave of Sommerfeld see the work of Baños and Wesley[3].

possible to obtain $\Theta_d \doteq \Theta_r \doteq \pi/2$ and, hence, $f_r \doteq -1$. In this case the electromagnetic fields are:

$$E_{ez1} \doteq \frac{\beta_1^2 p_{z1}}{2\pi\epsilon_0} \frac{e^{-j\beta_1 R_d}}{R_d} F_e, \quad (29a)$$

$$B_{mz1} \doteq \frac{\beta_1^2 m_{z1}}{2\pi\epsilon_0} \frac{e^{-j\beta_1 R_d}}{R_d} F_m. \quad (29b)$$

Note that these fields at the surface are proportional to the surface-wave attenuation functions $F_e = f(P_e, B_e)e^{-j\phi(P_e, B_e)}$ and $F_m = f(P_m, B_m)e^{-j\phi(P_m, B_m)}$ defined and represented graphically in Sec. 11.

If the primary dipole is quite high above the conducting plane, the condition $r \gg d$ may be much severer than necessary to have $R_d = R_r$ satisfy (27). However, when r is not large compared with d, $\Theta_r = \pi - \Theta_d$ differs appreciably from $\pi/2$ or $90°$ so that (29a, b) do not represent the complete fields near the surface and (28a, b) must be used. The ratios of the surface-wave terms to the space-wave terms in (28a, b) with (12.12a, b) are

$$\left|\frac{E_{ez1} \text{ (surface term)}}{E_{ez1} \text{ (space term)}}\right| = \left|\frac{(1 - f_{er})}{(1 + f_{er})} F_e\right|$$

$$= \left|\frac{\sqrt{N_{21}^2 - \sin^2 \Theta_r}}{N_{21}^2 \cos \Theta_r}\right| f(P_e, B_e), \quad (30a)$$

$$\left|\frac{B_{mz1} \text{ (surface term)}}{B_{mz1} \text{ (space term)}}\right| = \left|\frac{(1 - f_{mr})}{(1 + f_{mr})} F_m\right|$$

$$= \left|\frac{\sqrt{N_{21}^2 - \sin^2 \Theta_r}}{\cos \Theta_r}\right| f(P_m, B_m). \quad (30b)$$

Computations made by Norton[45] indicate that these ratios are very small unless Θ_r is within a few degrees of $90°$ when r is only a few wavelengths. As r is increased Θ_r must approach still closer to $90°$ in order to make the surface wave significant. As $r \to \infty$ the surface wave is appreciable primarily at $\Theta_r = 90°$, where it is the entire field. A typical set of curves, due to Norton, is given in Fig. 13.1, where the ratio (30a) is shown as a function of R_d/λ_1 with $\Theta_i = \pi - \Theta_d$ and $\sigma_e/\omega\epsilon_0$ as parameters. The data are for average earth with $N_{21} = \epsilon_{er} - j\sigma_e/\omega\epsilon_0 = 15 - j\sigma_e/\omega\epsilon_0$. A range of values of $\sigma_e/\omega\epsilon_0$ is used, corresponding, for example, to a frequency range from 0.5 to 20 Mhz for $\sigma_e = 5 \times 10^{-3}$ mho/m. The ratio (30b) for the magnetic dipole over average earth is given in Table 13.1 for $R_d = \lambda_1$. It is seen to be smaller than the corresponding value for the electric dipole. If the electric dipole is

TABLE 13.1. The ratio of the surface term to the space term for E_{ez1}, with $R_d = \lambda_1$ and $\epsilon_{er} = 80$.*

$\dfrac{\sigma_e}{\omega\epsilon_0}$	f (Mhz) for $\sigma_e = 5$ mho/m	\multicolumn{6}{c}{$\pi - \Theta_d = \Theta_r$}
		90°	85°	80°	75°	60°	45°
180,000	0.5	∞	0.022	0.0092	0.0050	0.0014	0.0005
90,000	1	∞	.031	.013	.0072	.0020	.0007
45,000	2	∞	.044	.018	.010	.0028	.0009
18,000	5	∞	.070	.029	.016	.0044	.0015
9,000	10	∞	.099	.041	.023	.0062	.0021
4,500	20	∞	.140	.058	.032	.0087	.0029

* Computations due to K. A. Norton.

over a more highly conducting medium than average earth, for example, sea water, the ratio (30a) is reduced by about a factor 10 as shown in Table 13.2.

TABLE 13.2. The ratio of the surface term to the space term for B_{mz1}, with $R_d = \lambda_1$ and $\epsilon_{er} = 15$.*

$\dfrac{\sigma_e}{\omega\epsilon_0}$	f (Mhz) for $\sigma_e = 5\times 10^{-3}$ mho/m	\multicolumn{4}{c}{$\pi - \Theta_d = \Theta_r$}
		90°	85°	75°	45°
180	0.5	∞	0.134	0.042	0.0078
90	1	∞	.187	.058	.0106
45	2	∞	.259	.079	.0140
18	5	∞	.367	.109	.0185
9	10	∞	.426	.125	.0206
4.5	20	∞	.451	.132	.0215

* Computations due to K. A. Norton.

It may be concluded in general that the far-zone formulas (20a, b) are good approximations when $\beta_1 R_d \gg 1$ and Θ_d is not too near $\pi/2$. At $\Theta_d \doteq \pi/2$ the fields are given by (29a, b). In an intermediate range for values of R_d down to about λ_1 and all values of Θ_d, or near $\Theta_d \doteq \pi/2$ for all values of $R_d \gtrsim \lambda_1$, the general formulas (21) and (22) may be used without the terms $G_z(R_d, R_r)$. Approximations in the integrations make (21) and (22) unsatisfactory for $R_d < \lambda_1$ even if the terms $G_z(R_d, R_r)$ are included.

With the *vertical* electromagnetic fields E_{ez1} and B_{mz1} obtained, it remains to determine the radial fields of the vertical dipoles.

14. The Radial Electromagnetic Fields

The radial components of the electromagnetic fields of vertical electric and magnetic dipoles over a plane horizontal earth are given by

$$E_{e1r} = \frac{\partial^2 \Pi_{ez1}}{\partial r\, \partial z}, \qquad (1a)$$

$$B_{m1r} = \frac{\partial^2 \Pi_{mz1}}{\partial r\, \partial z}. \qquad (1b)$$

Since $\partial \Pi_{z1}/\partial z$ is given by (13.8) in the form

$$\frac{\partial \Pi_{ez1}}{\partial z} \doteq \frac{-j\beta_1 p_{z1}}{4\pi\epsilon_0}\left[Q(R_d, R_r)\right.$$
$$\left. - \frac{\sqrt{N_{21}^2 - \sin^2\Theta_r}}{N_{21}^2} V\right], \quad (2a)$$

$$\frac{\partial \Pi_{mz1}}{\partial z} \doteq \frac{-j\beta_1 m_{z1}}{4\pi v_0}[Q(R_d, R_r)$$
$$- \sqrt{N_{21}^2 - \sin^2\Theta_r}\, H], \quad (2b)$$

where

$$Q(R_d, R_r) \equiv \left(1 + \frac{1}{j\beta_1 R_d}\right)\cos\Theta_d\, \frac{e^{-j\beta_1 R_d}}{R_d}$$
$$+ \left(1 + \frac{1}{j\beta_1 R_r}\right)\cos\Theta_r\, \frac{e^{-j\beta_1 R_r}}{R_r}, \quad (3)$$

it is necessary to differentiate (2) partially with respect to r. This involves the following derivatives:

$$\frac{\partial R_d}{\partial r} = \frac{r}{R_d} = \sin\Theta_d, \quad \frac{\partial R_r}{\partial r} = \frac{r}{R_r} = \sin\Theta_r, \quad (4)$$

$$\frac{\partial}{\partial r}(\cos\Theta_d) = -\frac{\sin\Theta_d \cos\Theta_d}{R_d}, \quad (5a)$$

$$\frac{\partial}{\partial r}(\cos\Theta_r) = -\frac{\sin\Theta_r \cos\Theta_r}{R_r}. \quad (5b)$$

As before, $\sqrt{N_{21}^2 - \sin^2\Theta_r}$ is treated as a constant.

The evaluation of $\partial V/\partial r$ and $\partial H/\partial r$ is carried out conveniently by proceeding from

$$V \doteq \frac{e^{-j\beta_1 R_r}}{R_r}[(1-f_{er})F_e + (1+f_{er})], \quad (6a)$$

$$H \doteq \frac{e^{-j\beta_1 R_r}}{R_r}[(1-f_{mr})F_m + (1+f_{mr})]. \quad (6b)$$

For simplicity let W stand for V or H. Then,

$$\frac{\partial W}{\partial r} = [(1+f_r) + (1-f_r)F]\frac{\partial}{\partial r}\left(\frac{e^{-j\beta_1 R_r}}{R_r}\right)$$
$$+ \frac{e^{-j\beta_1 R_r}}{R_r}\left[(1-f_r)\frac{\partial F}{\partial r} + (1-F)\frac{\partial f_r}{\partial r}\right]. \quad (7)$$

The leading terms in each derivative are

$$\frac{\partial}{\partial r}\left(\frac{e^{-j\beta_1 R_r}}{R_r}\right) \doteq -j\beta_1 \sin\Theta_r \frac{e^{-j\beta_1 R_r}}{R_r}, \quad (8)$$

$$\frac{\partial F}{\partial r} = -\left[F + \frac{1-F}{2P}\right]\frac{\partial P}{\partial r} \doteq \frac{F}{2P}\frac{\partial P}{\partial r}. \quad (9)$$

The last step in (9) follows from the fact that the leading term in F is $-1/2P$.

$$\frac{\partial P}{\partial r} \doteq -\frac{P}{R_r}\left(\frac{f_r \sin^2\Theta_r + 2\cos^2\Theta_r}{\sin\Theta_r}\right), \quad (10)$$

$$\frac{\partial f_r}{\partial r} \doteq -\frac{(1-f_r)(1+f_r)\sin^2\Theta_r}{2R_r}. \quad (11)$$

If these derivatives are substituted in (7), it is seen that the second bracket has $1/R_r$ as a factor, so that it is an order of magnitude smaller than the first bracket, which contributes the leading term. Actually, since F is of order of magnitude $1/R_r$, the leading parts of the first and second terms in (7) are, respectively, of orders of magnitude R_r^{-2} and R_r^{-3}. Thus, the leading part of (7) gives

$$\frac{\partial V}{\partial r} \doteq -j\beta_1 V \sin\Theta_r, \quad \frac{\partial H}{\partial r} \doteq -j\beta_1 H \sin\Theta_r. \quad (12)$$

The leading terms in $\partial Q(R_d, R_r)/\partial r$ are obtained from (3) with (4) and (5). Thus,

$$\frac{\partial Q(R_d, R_r)}{\partial r} \doteq -j\beta_1 \sin\Theta_d \cos\Theta_d \frac{e^{-j\beta_1 R_d}}{R_d}$$
$$- j\beta_1 \sin\Theta_r \cos\Theta_r \frac{e^{-j\beta_1 R_r}}{R_r}. \quad (13)$$

Substitution of (12) and (13) in

$$\frac{\partial \Pi_{ez1}}{\partial r \partial z} \doteq \frac{-j\beta_1 p_{z1}}{4\pi\epsilon_0}\left[\frac{\partial Q(R_d, R_r)}{\partial r}\right.$$
$$\left. - \frac{\sqrt{N_{21}^2 - \sin^2\Theta_r}}{N_{21}^2}\frac{\partial V}{\partial r}\right], \quad (14a)$$

$$\frac{\partial \Pi_{mz1}}{\partial r \partial z} \doteq \frac{-j\beta_1 m_{z1}}{4\pi\nu_0}\left[\frac{\partial Q(R_d, R_r)}{\partial r}\right.$$
$$\left. - \sqrt{N_{21}^2 - \sin^2\Theta_r}\frac{\partial H}{\partial r}\right] \quad (14b)$$

and use of (6) gives

$$E_{e1r} \doteq \frac{-\beta_1^2 p_{z1}}{4\pi\epsilon_0}\left\{\cos\Theta_d \sin\Theta_d \frac{e^{-j\beta_1 R_d}}{R_d}\right.$$
$$+ \cos\Theta_r \sin\Theta_r \frac{e^{-j\beta_1 R_r}}{R_r}$$
$$- \sin\Theta_r \frac{\sqrt{N_{21}^2 - \sin^2\Theta_r}}{N_{21}^2}[(1-f_{er})F_e$$
$$\left. + (1+f_{er})]\frac{e^{-j\beta_1 R_r}}{R_r}\right\}, \quad (15a)$$

$$B_{m1r} \doteq \frac{-\beta_1^2 m_{z1}}{4\pi\nu_0}\left\{\cos\Theta_d \sin\Theta_d \frac{e^{-j\beta_1 R_d}}{R_d}\right.$$
$$+ \cos\Theta_r \sin\Theta_r \frac{e^{-j\beta_1 R_r}}{R_r}$$
$$- \sin\Theta_r \sqrt{N_{21}^2 - \sin^2\Theta_r}[(1-f_{mr})F_m$$
$$\left. + (1+f_{mr})]\frac{e^{-j\beta_1 R_r}}{R_r}\right\}. \quad (15b)$$

Noting that

$$\cos\Theta_r - (1+f_{er})\frac{\sqrt{N_{21}^2 - \sin^2\Theta_r}}{N_{21}^2}$$
$$= f_{er}\cos\Theta_r, \quad (16a)$$

$$\cos\Theta_r - (1+f_{mr})\sqrt{N_{21}^2 - \sin^2\Theta_r}$$
$$= f_{mr}\cos\Theta_r, \quad (16b)$$

it follows that

$$E_{e1r} \doteq \frac{-\beta_1^2 p_{z1}}{4\pi\epsilon_0}\left[\cos\Theta_d \sin\Theta_d \frac{e^{-j\beta_1 R_d}}{R_d}\right.$$
$$+ f_{er}\cos\Theta_r \sin\Theta_r \frac{e^{-j\beta_1 R_r}}{R_r}$$
$$\left. - F_e(1-f_{er})\sin\Theta_r \frac{\sqrt{N_{21}^2 - \sin^2\Theta_r}}{N_{21}^2}\frac{e^{-j\beta_1 R_r}}{R_r}\right], \quad (17a)$$

$$B_{m1r} \doteq \frac{-\beta_1^2 m_{z1}}{4\pi\nu_0}\left[\cos\Theta_d \sin\Theta_d \frac{e^{-j\beta_1 R_d}}{R_d}\right.$$
$$+ f_{mr}\cos\Theta_r \sin\Theta_r \frac{e^{-j\beta_1 R_r}}{R_r}$$
$$\left. - F_m(1-f_{mr})\sin\Theta_r \sqrt{N_{21}^2 - \sin^2\Theta_r}\frac{e^{-j\beta_1 R_r}}{R_r}\right]. \quad (17b)$$

These are the leading terms in the radial electromagnetic fields corresponding to (21a) and (22a) with $G_z(R_r, R_d)$ omitted.

At sufficiently great distances and for angles Θ_r that are not near $\pi/2$, (17a, b) reduce to the following far-zone forms, which coincide with the radial components in cylindrical coördinates of those obtained previously by saddle-point integration:

$$E_{e1r}^r \doteq \frac{-\beta_1^2 p_{z1}}{4\pi\epsilon_0} \frac{e^{-j\beta_1 R_{01}}}{R_{01}} \cos\Theta_{01}\sin\Theta_{01}$$
$$\times (e^{j\beta_1 d\cos\Theta_{01}} + f_{er}e^{-j\beta_1 d\cos\Theta_{01}}), \quad (18a)$$

$$B_{e1r}^r \doteq \frac{-\beta_1^2 m_{z1}}{4\pi\nu_0} \frac{e^{-j\beta_1 R_{01}}}{R_{01}} \cos\Theta_{01}\sin\Theta_{01}$$
$$\times (e^{j\beta_1 d\cos\Theta_{01}} + f_{mr}e^{-j\beta_1 d\cos\Theta_{01}}). \quad (18b)$$

Near the surface of the plane where $R_d \doteq R_r$, $\Theta_d \doteq \Theta_r \doteq \pi/2$, $f_r \doteq -1$, the leading terms in (17) are

$$E_{e1r} \doteq \frac{\beta_1^2 p_{z1}}{2\pi\epsilon_0} F_e \frac{\sqrt{N_{21}^2-1}}{N_{21}^2} \frac{e^{-j\beta_1 R_r}}{R_r}, \quad (19a)$$

$$B_{m1r} \doteq \frac{\beta_1^2 m_{z1}}{2\pi\nu_0} F_m \sqrt{N_{21}^2-1}\, \frac{e^{-j\beta_1 R_r}}{R_r}. \quad (19b)$$

15. Complete Electromagnetic Fields of Vertical Dipoles

The remaining components of the electromagnetic fields of the vertical dipoles are derived from the Hertzian potentials using the following formulas; since they are perpendicular to the plane of incidence they are denoted with a subscript $\perp$:

$$\mathbf{B}_{e\perp} = \frac{j\beta_1^2}{\omega} \text{curl}\,(\hat{\mathbf{z}}\Pi_{ez1}), \quad (1a)$$

$$\mathbf{E}_{m\perp} = j\omega\, \text{curl}\,(\hat{\mathbf{z}}\Pi_{mz1}). \quad (1b)$$

Since rotational symmetry obtains about the z-axis through the dipole, only one component of the curl in (1) differs from zero. It is

$$\mathbf{B}_{e\perp} = \hat{\boldsymbol{\theta}} B_{e1\theta}, \quad B_{e1\theta} = \frac{-j\beta_1^2}{\omega} \frac{\partial \Pi_{ez1}}{\partial r}, \quad (2a)$$

$$\mathbf{E}_{m\perp} = \hat{\boldsymbol{\theta}} E_{m1\theta}, \quad E_{m1\theta} = j\omega \frac{\partial \Pi_{mz1}}{\partial r}, \quad (2b)$$

where

$$\Pi_{ez1} = \frac{p_{z1}}{4\pi\epsilon_0}\left(\frac{e^{-j\beta_1 R_d}}{R_d} - \frac{e^{-j\beta_1 R_r}}{R_r} + V\right), \quad (3a)$$

$$\Pi_{mz1} = \frac{m_{z1}}{4\pi\nu_0}\left(\frac{e^{-j\beta_1 R_d}}{R_d} - \frac{e^{-j\beta_1 R_r}}{R_r} + H\right). \quad (3b)$$

Retaining only the leading terms in each of the three factors in (3a) and (3b) and using (14.12), the results are:

$$\frac{\partial \Pi_{ez1}}{\partial r} \doteq \frac{-j\beta_1 p_{z1}}{4\pi\epsilon_0}\left(\sin\Theta_d \frac{e^{-j\beta_1 R_d}}{R_d}\right.$$
$$\left. - \sin\Theta_r \frac{e^{-j\beta_1 R_r}}{R_r} + V\sin\Theta_r\right), \quad (4a)$$

$$\frac{\partial \Pi_{mz1}}{\partial r} \doteq \frac{-j\beta_1 m_{z1}}{4\pi\nu_0}\left(\sin\Theta_d \frac{e^{-j\beta_1 R_d}}{R_d}\right.$$
$$\left. - \sin\Theta_r \frac{e^{-j\beta_1 R_r}}{R_r} + H\sin\Theta_r\right), \quad (4b)$$

where V and H are defined in (12.11c, d). Substitution of (4a, b) in (2a, b) gives the desired components. They are expressed in final form in (7b, d) below.

The resultant electric and magnetic fields in the plane of incidence are given by

$$\mathbf{E}_{e\parallel} = \hat{\mathbf{z}} E_{ez1} + \hat{\mathbf{r}} E_{er1}, \quad (5a)$$

$$\mathbf{B}_{m\parallel} = \hat{\mathbf{z}} B_{mz1} + \hat{\mathbf{r}} B_{mr1}. \quad (5b)$$

Taking only the leading terms and setting

$$\hat{\boldsymbol{\Theta}}_d = -\hat{\mathbf{z}}\sin\Theta_d + \hat{\mathbf{r}}\cos\Theta_d, \quad (6a)$$

$$\hat{\boldsymbol{\Theta}}_r = -\hat{\mathbf{z}}\sin\Theta_r + \hat{\mathbf{r}}\cos\Theta_r, \quad (6b)$$

where $\hat{\boldsymbol{\Theta}}_d$ and $\hat{\boldsymbol{\Theta}}_r$ are unit vectors (Fig. 15.1), the resultant fields are as given below, together with the components perpendicular to the plane of incidence as obtained from (2a, b) with (4a, b). For the vertical electric dipole:

$$\mathbf{E}_{e\parallel} = \frac{-\beta_1^2 p_{z1}}{4\pi\epsilon_0}\left[\sin\Theta_d \frac{e^{-j\beta_1 R_d}}{R_d}\hat{\boldsymbol{\Theta}}_d\right.$$
$$+ f_{er}\sin\Theta_r \frac{e^{-j\beta_1 R_r}}{R_r}\hat{\boldsymbol{\Theta}}_r$$
$$- (1-f_{er})F_e\sin\Theta_r \frac{e^{-j\beta_1 R_r}}{R_r}$$
$$\left. \times \left(\hat{\mathbf{z}}\sin\Theta_r + \hat{\mathbf{r}}\frac{\sqrt{N_{21}^2-\sin^2\Theta_r}}{N_{21}^2}\right)\right], \quad (7a)$$

$$\mathbf{B}_{e\perp} = \hat{\boldsymbol{\theta}}\frac{-\beta_1^3 p_{z1}}{4\pi\epsilon_0 \omega}\left[\sin\Theta_d \frac{e^{-j\beta_1 R_d}}{R_d}\right.$$
$$+ f_{er}\sin\Theta_r \frac{e^{-j\beta_1 R_r}}{R_r}$$
$$\left. + (1-f_{er})F_e\sin\Theta_r \frac{e^{-j\beta_1 R_r}}{R_r}\right]. \quad (7b)$$

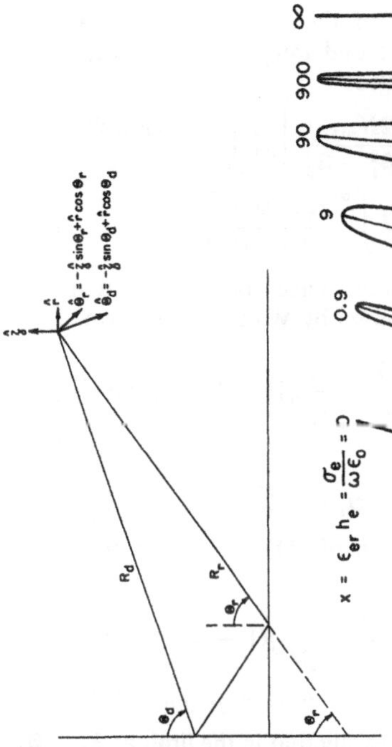

THEORY OF LINEAR ANTENNAS

Fig. 15.1. Vectors involved in resultant field.

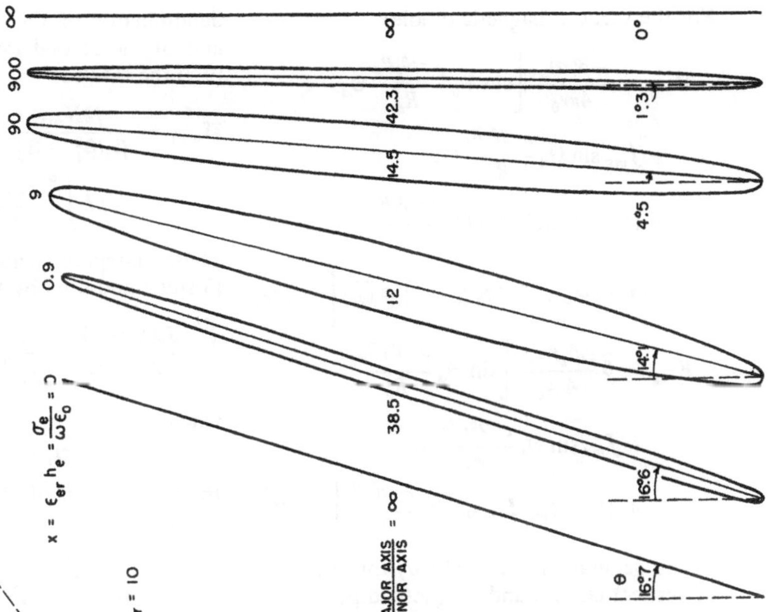

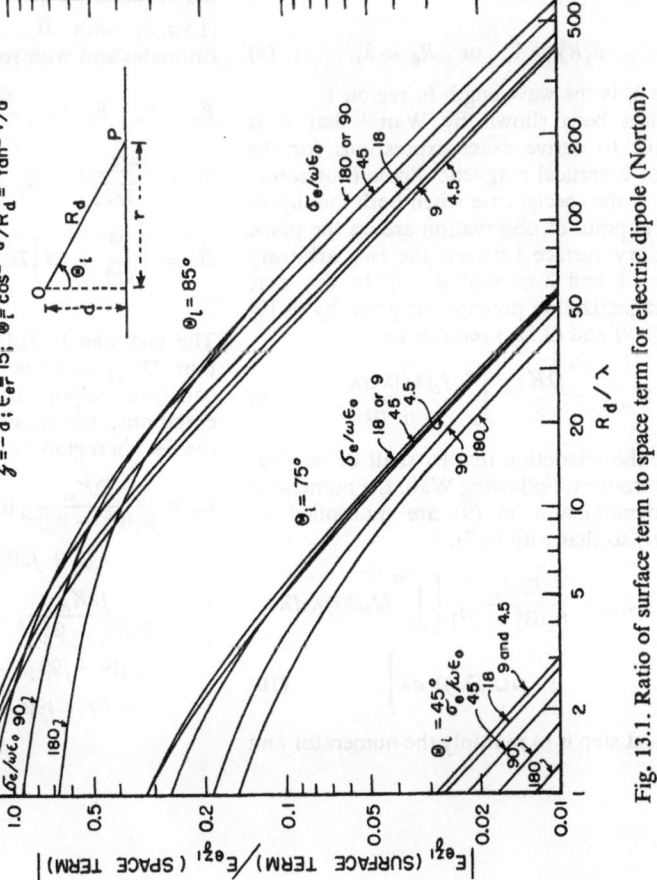

Fig. 13.1. Ratio of surface term to space term for electric dipole (Norton).

Fig. 16.1. Polarization of electric field near conducting surface due to electric dipole.

For the vertical magnetic dipole:

$$\mathbf{B}_{m\parallel} = \frac{-\beta_1^2 m_{z1}}{4\pi\nu_0}\left[\sin\Theta_d \frac{e^{-j\beta_1 R_d}}{R_d}\hat{\Theta}_d\right.$$
$$+ f_{mr}\sin\Theta_d \frac{e^{-j\beta_1 R_r}}{R_r}\hat{\Theta}_r$$
$$- (1-f_{mr})F_m\sin\Theta_r \frac{e^{-j\beta_1 R_r}}{R_r}$$
$$\left.\times(\hat{z}\sin\Theta_r + \hat{r}\sqrt{N_{21}^2 - \sin^2\Theta_r})\right], \quad (7c)$$

$$\mathbf{E}_{m\perp} = \hat{\theta}\frac{\omega\beta_1 m_{z1}}{4\pi\nu_0}\left[\sin\Theta_d\frac{e^{-j\beta_1 R_d}}{R_d}\right.$$
$$+ f_{mr}\sin\Theta_r \frac{e^{-j\beta_1 R_r}}{R_r}$$
$$\left.+ (1-f_{mr})F_m\sin\Theta_r \frac{e^{-j\beta_1 R_r}}{R_r}\right]. \quad (7d)$$

The general formulas (7a–d) for the fields of vertical electric and magnetic dipoles have been derived under assumptions that retain only the leading terms in the surface waves. Since these are of order R^{-2}, this is equivalent to neglecting terms of order R^{-3}. Accordingly, (7a–d) may be called *quasi-near-zone fields*, which are satisfactory approximations subject to the inequality

$$\beta_1^2 R_d^2 \gg 1, \quad \text{or} \quad R_d \geq \lambda_1. \quad (8)$$

where λ_1 is the wavelength in region 1.

It has been shown by Wait[70] that it is possible to derive exact expressions for the field of a vertical magnetic dipole (horizontal loop) in the special case when both the dipole and the point of observation are on the plane boundary surface between the two arbitrary regions 1 and 2 so that $d = 0$. In this case the magnetization potential as given by (9.10) with (9.9) and (9.12c) reduces to

$$\Pi_{mz1} = \frac{j2K_{m1}}{\beta_1}\int_0^\infty \frac{J_0(\lambda r)\lambda\, d\lambda}{l+m}, \quad (9)$$

where the restriction that β_1 shall be real has been removed. Following Wait, the numerator and denominator in (9) are multiplied by $l - m$, so that with (9.7),

$$\Pi_{mz1} = \frac{j2K_{m1}}{\beta_1(\beta_2^2 - \beta_1^2)}\left[\int_0^\infty lJ_0(\lambda r)\lambda\, d\lambda\right.$$
$$\left.- \int_0^\infty mJ_0(\lambda r)\lambda\, d\lambda\right]. \quad (10)$$

The next step is to multiply the numerator and denominator of the first integral by $\lambda + l$, and of the second integral by $\lambda + m$. With (9.7) the result is

$$\Pi_{mz1} = \frac{j2K_{m1}}{\beta_1(\beta_1^2 - \beta_2^2)}\left[\beta_1^2\int_0^\infty \frac{J_0(\lambda r)\lambda\, d\lambda}{\lambda+l}\right.$$
$$\left.- \beta_2^2\int_0^\infty \frac{J_0(\lambda r)\lambda\, d\lambda}{\lambda+m}\right]. \quad (11)$$

These integrals have been evaluated by Foster[17] and used by Wait[70] in the form

$$\int_0^\infty \frac{J_0(\lambda r)\lambda\, d\lambda}{\lambda+u_i} = \frac{1}{\gamma_i^2 r^3}[1 - (1+\gamma_i r)e^{-\gamma_i r}], \quad (12)$$

where

$$u_i = \sqrt{\lambda^2 + \gamma_i^2}.$$

In the notation of (11), $\gamma_i^2 = -\beta_i^2$, $\gamma_i = j\beta_i$. It follows that

$$\Pi_{mz1} = \frac{j2K_{m1}}{\beta_1(\beta_1^2 - \beta_2^2)r^3}[(1+j\beta_1 r)e^{-j\beta_1 r}$$
$$- (1+j\beta_2 r)e^{-j\beta_2 r}]. \quad (13)$$

It is verified readily that in the limit as $\beta_2 \to \beta_1$ (13) reduces to the familiar form (5.3) for a dipole in a homogeneous infinite medium.

The electromagnetic field on the boundary plane may be calculated from (13) using (1.9a, b) with $\Pi_{ez} = 0$. In cylindrical coördinates and with rotational symmetry,

$$E_r = 0, \quad E_\theta = -\frac{\partial \Pi_{mz1}}{\partial r}, \quad E_z = 0, \quad (14a)$$

$$B_r = \frac{\partial^2 \Pi_{mz1}}{\partial r\, \partial z}, \quad B_\theta = 0,$$

$$B_z = \left(\frac{\partial^2}{\partial z^2} + \beta_1^2\right)\Pi_{mz1} = -\frac{1}{r}\frac{\partial}{\partial r}\left(r\frac{\partial \Pi_{mz1}}{\partial r}\right). \quad (14b)$$

The last step in (14b) follows from the fact that Π_{mz1} satisfies the homogeneous wave equation. Upon carrying out the differentiations, the nonvanishing components of the field in region 1 *on the boundary surface* are

$$E_\theta = \frac{j2K_m}{\beta_1(\beta_1^2-\beta_2^2)r^4}[(3+j3\beta_1 r - \beta_1^2 r^2)e^{-j\beta_1 r}$$
$$- (3+j3\beta_2 r - \beta_2^2 r^2)e^{-j\beta_2 r}], \quad (15a)$$

$$B_z = \frac{j2K_m}{\beta_1(\beta_1^2-\beta_2^2)r^5}$$
$$\times [(9+j9\beta_1 r - 4\beta_1^2 r^2 - j\beta_1^3 r^3)e^{-j\beta_1 r}$$
$$- (9+j9\beta_2 r - 4\beta_2^2 r^2 - j\beta_2^3 r^3)e^{-j\beta_2 r}]. \quad (15b)$$

These formulas are useful, for example, in determining the field at any point along the surface of the earth of an insulated loop antenna laid directly on the earth. They may also be applied to determine the field along any boundary surface between homogeneous layers of sufficient thickness and extent in the earth or on the bottom of the ocean.

16. Polarization and Tilt of the Surface Waves

Whenever the radial distance r is sufficiently great compared with the height d of the dipole above the conducting plane, $\Theta_r \doteq \Theta_d \doteq \pi/2$, so that $f_{mr} \doteq -1$ and, except for perfect conductors, $f_{er} \doteq -1$. In this case, as shown in Sec. 13, the surface-wave terms are essentially the entire field, since the two space-wave terms virtually cancel. The components in the plane of incidence for the electric and magnetic dipoles are

$$\mathbf{E}_{e\parallel}(\Theta \doteq \pi/2)$$
$$\doteq \frac{-\beta_1^2 p_{z1}}{2\pi\epsilon_0} \frac{e^{-j\beta_1 R_r}}{R_r}\left(\hat{\mathbf{z}} + \frac{\sqrt{N_{21}^2-1}}{N_{21}^2}\hat{\mathbf{r}}\right)F_e, \quad (1a)$$

$$\mathbf{B}_{m\parallel}(\Theta \doteq \pi/2)$$
$$\doteq \frac{-\beta_1^2 m_{z1}}{2\pi v_0} \frac{e^{-j\beta_1 R_r}}{R_r}(\hat{\mathbf{z}} + \sqrt{N_{21}^2-1}\,\hat{\mathbf{r}})F_m. \quad (1b)$$

These may be expressed as follows:

$$\mathbf{E}_{e\parallel}(\Theta \doteq \pi/2) = E_{ez1}(\hat{\mathbf{z}} + \hat{\mathbf{r}}K_e e^{j\kappa_e}), \quad (2a)$$

$$\mathbf{B}_{m\parallel}(\Theta \doteq \pi/2) = B_{mz1}(\hat{\mathbf{z}} + \hat{\mathbf{r}}K_m e^{j\kappa_m}), \quad (2b)$$

where

$$\frac{\sqrt{N_{21}^2-1}}{N_{21}^2} \equiv K_e e^{j\kappa_e}, \quad \sqrt{N_{21}^2-1} = K_m e^{j\kappa_m}. \quad (3)$$

The instantaneous values for $\Theta = \pi/2$ referred to E_{ez1} and B_{mz1} are given by

$$(\mathbf{E}_{e\parallel})_{\text{inst}} = \text{Real part }(\mathbf{E}_{e\parallel}e^{j\omega t})$$
$$= E_{ez1}[\hat{\mathbf{z}}\cos\omega t + \hat{\mathbf{r}}K_e \cos(\omega t + \kappa_e)], \quad (4a)$$

$$(\mathbf{B}_{m\parallel})_{\text{inst}} = \text{Real part }(\mathbf{B}_{m\parallel}e^{j\omega t})$$
$$= B_{mz1}[\hat{\mathbf{z}}\cos\omega t + \hat{\mathbf{r}}K_m \cos(\omega t + \kappa_m)]. \quad (4b)$$

These are the equations of a rotating vector of which the end point traces an ellipse and the major axis is inclined or tilted from the z-axis. The conventional equation of the ellipse traced by the end point is obtained by eliminating the time. Since the formulas for $(\mathbf{E}_{e1})_{\text{inst}}$ and $(\mathbf{B}_{m1})_{\text{inst}}$ are the same in form, it is sufficient to carry out the analysis for the former as follows:

$$(E_z)_{\text{inst}} = E_z \cos \omega t, \quad (5a)$$

$$(E_r)_{\text{inst}}$$
$$= E_z K_e \cos(\omega t + \kappa_e)$$
$$= E_z K_e[\cos\omega t \cos\kappa_e - \sin\omega t \sin\kappa_e]$$
$$= K_e[E_{z\,\text{inst}} \cos \kappa_e - \sqrt{E_z^2 - E_{z\,\text{inst}}^2}\sin\kappa_e]. \quad (5b)$$

The last equation may be rearranged as follows:

$$[(E_r)_{\text{inst}} - K_e (E_z)_{\text{inst}} \cos\kappa_e]^2$$
$$= K_e^2[E_z^2 - (E_z)_{\text{inst}}^2]\sin^2\kappa_e, \quad (6)$$

$$K_e^2 (E_z)_{\text{inst}}^2 - 2K_e(E_z)_{\text{inst}}(E_r)_{\text{inst}} \cos \kappa_e$$
$$+ (E_r)_{\text{inst}}^2 - K_e^2 E_z^2 \sin^2\kappa_e = 0. \quad (7)$$

This is the equation of an ellipse in the general form,

$$Az^2 + Bzr + Cr^2 + F = 0, \quad (8)$$

where

$$A = K_e^2, \quad B = -2K_e \cos \kappa_e,$$
$$C = 1, \quad F = -K_e^2 E_z^2 \sin^2\kappa_e. \quad (9)$$

The term in zr may be removed by rotating the axes through an angle θ defined by

$$\tan 2\theta = \frac{B}{A-C} = \frac{-2K_e \cos \kappa_e}{K_e^2 - 1}. \quad (10)$$

The angle θ is the *angle of tilt* from the vertical of the major axis of the ellipse. It specifies the direction in which the electric field has its maximum value. The corresponding minimum is at an angle of 90° beyond θ.

The equation of the ellipse referred to the new axes z' and r' is

$$A'z'^2 + C'r'^2 + F = 0, \quad (11)$$

where

$$A' = A\cos^2\theta + B\sin\theta\cos\theta + C\sin^2\theta, \quad (12a)$$

$$C' = A\sin^2\theta - B\sin\theta\cos\theta + C\cos^2\theta. \quad (12b)$$

Referred to the electric vector, the corresponding relations are

$$K'^2_e(E_a)^2_{\text{inst}} + L'^2_e(E_b)^2_{\text{inst}} - K^2_e E^2_z \sin^2 \kappa_e = 0, \quad (13)$$

where

$$K'^2_e = K^2_e \cos^2 \theta - 2K_e \cos \kappa_e \sin \theta \cos \theta + \sin^2 \theta, \quad (14a)$$

$$L'^2_e = K^2_e \sin^2 \theta + 2K_e \cos \kappa_e \sin \theta \cos \theta + \cos^2 \theta. \quad (14b)$$

Alternatively, in standard form,

$$\frac{(E_a)^2_{\text{inst}}}{E^2_z(K_e/K'_e)^2 \sin^2 \kappa_e} + \frac{(E_b)^2_{\text{inst}}}{E^2_z(K_e/L'_e)^2 \sin^2 \kappa_e} = 1$$

or

$$\frac{(E_a)^2_{\text{inst}}}{E^2_a} + \frac{(E_b)^2_{\text{inst}}}{E^2_b} = 1. \quad (15)$$

The semimajor axis of the ellipse is

$$E_a = E_z(K_e/K'_e) \sin \kappa_e = \frac{K_e E_z \sin \kappa_e}{\sqrt{K^2_e \sin^2 \theta + 2K_e \cos \kappa_e \sin \theta \cos \theta + \cos^2 \theta}} \quad (16a)$$

The semiminor axis is

$$E_b = E_z(K_e/L'_e) \sin \kappa_e = \frac{K_e E_z \sin \kappa_e}{\sqrt{K^2_e \cos^2 \theta - 2K_e \cos \kappa_e \sin \theta \cos \theta + \sin^2 \theta}} \quad (16b)$$

The ratio of minor to major axis is

$$\text{Ratio} = \frac{E_b}{E_a}$$

$$= \cot \theta \sqrt{\frac{1 + 2K_e \cos \kappa_e \tan \theta + K^2_e \tan^2 \theta}{1 - 2K_e \cos \kappa_e \cot \theta + K^2_e \cot^2 \theta}} \quad (16c)$$

The angle of tilt θ and the ratios of short to long axes of the ellipse are shown in Fig. 16.1 with $\epsilon_{er} = 10$ and $\sigma_e/\omega\epsilon_0 = 0, 0.9, 9, 90, 900,$ and ∞. Note how the ellipse becomes more and more nearly vertical as the conductivity of the medium increases.

17. The Field of a Vertical Electric Dipole at Large Numerical Distances

At sufficiently high frequencies the numerical distances

$$p_e = \frac{-j\beta_1 R_r}{2}\left(\frac{N^2_{21} - \sin^2 \Theta_r}{N^4_{21}}\right), \quad (1a)$$

$$p_m = \frac{-j\beta_1 R_r}{2}(N^2_{21} - \sin^2 \Theta_r) \quad (1b)$$

usually are quite large in magnitude if R_r is large, since $\beta_1 = \omega/v_1$ is not a very small fraction. This is true particularly of p_m, as pointed out at the end of Sec. 12. Subject to the conditions

$$p > 20, \quad p \gg q_1 + q_2, \quad (2a)$$

$$\Theta_d \doteq \Theta_r \doteq \pi/2, \quad (2b)$$

where p is the magnitude of p_e or p_m, very considerable simplification in the general formulas is possible. The numerical heights q_1 and q_2 are defined in (11.15). Thus,

$$p_e \doteq \frac{-j\beta_1 R_r}{2} \frac{N^2_{21} - 1}{N^4_{21}}, \quad (3a)$$

$$p_m \doteq \frac{-j\beta_1 R_r}{2}(N^2_{21} - 1), \quad (3b)$$

$$p_e = \frac{\beta_1 R_r}{2h_e \epsilon_{er}} \frac{\cos b''}{\cos b'}, \quad (3c)$$

$$p_m = \frac{\beta_1 R_r}{2h_e \epsilon_{er} \cos b'}, \quad (3d)$$

$$P \doteq p = pe^{-jb}. \quad (3e)$$

From

$$F = f(P, B)e^{-j\phi(P,B)} \doteq -\frac{1}{2P} \doteq -\frac{1}{2p}$$

$$= -\frac{1}{2p}e^{jb} \quad (3f)$$

with (11.16),

$$f_r \doteq -1 + \frac{q_1 + q_2}{p}e^{-j(\frac{1}{2}\pi - \frac{1}{2}b)}. \quad (3g)$$

Also

$$1 - f_r \doteq 2, \quad (3h)$$

$$R_d \doteq R_r \doteq R_0 \doteq r \text{ in amplitude factors,} \quad (4a)$$

$$\left.\begin{array}{l}R_d \doteq R_0 - d\cos\Theta\\R_r \doteq R_0 + d\cos\Theta\end{array}\right\} \text{ in phase factors,} \quad (4b)$$

$$\beta_1(R_r - R_d) \doteq \beta_1(2d\cos\Theta)$$

$$= \frac{\beta_1 2d(z+d)}{R_0} = \frac{q_1 q_2}{p}. \quad (4c)$$

TABLE 17.1. The height-gain function $f(q)$.

q \ b	0°	30°	60°	90°	180°
0	1.0	1.0	1.0	1.0	1.0
0.05	.966				1.035
.1	.932	0.954	0.979	1.005	1.072
.3	.817	.889	.967	1.044	1.231
.5	.737	.866	.996	1.118	1.399
.7	.707	.889	1.063	1.221	1.575
1.0	.765	1.000	1.217	1.414	1.848
1.5	1.063	1.323	1.572	1.803	2.318
2.0	1.474	1.732	1.990	2.236	2.798
5.0	4.37	4.58	4.84	5.10	5.75
10	9.31	9.54	9.80	10.0	10.7
50	49.3	49.5	49.8	50.0	50.7
100	100	100	100	100	100

With these approximations, (13.21) and (13.22) reduce to

$$E_{ez1} \doteq \frac{\beta_1^2 p_{z1}}{4\pi\epsilon_0 r} \left\{ 1 + e^{-j(q_{1e}q_{2e}/p_e)} \right.$$

$$\left. \times \left[-1 + \frac{q_{1e} + q_{2e}}{p_e} e^{-j(\frac{1}{4}\pi - \frac{1}{2}b_e)} - \frac{1}{p_e} e^{jb_e} \right] \right\}, \quad (5a)$$

$$B_{mz1} \doteq \frac{\beta_1^2 m_{z1}}{4\pi\nu_0 r} \left\{ 1 + e^{-j(q_{1m}q_{2m}/p_m)} \right.$$

$$\left. \times \left[-1 + \frac{q_{1m} + q_{2m}}{p_m} e^{-j(\frac{1}{4}\pi - \frac{1}{2}b_m)} - \frac{1}{p_m} e^{jb_m} \right] \right\}. \quad (5b)$$

If p is sufficiently great so that

$$p^2 \gg q_1^2 q_2^2 \quad (6)$$

and

$$e^{-j(q_1 q_2/p)} \doteq 1 - j\frac{q_1 q_2}{p}, \quad (7)$$

further simplification is possible. Thus, for the magnitude of the expression in braces in (5a) or (5b), one obtains:

$$\left| \{ \} \right| = \left| \frac{q_1 + q_2}{p} e^{-j(\frac{1}{4}\pi - \frac{1}{2}b)} - \frac{1}{p} e^{jb} + j\frac{q_1 q_2}{p} \right|$$

$$= 2 \left| \left(-\frac{1}{2p} e^{jb} \right) \left(1 - (q_1 + q_2)e^{-j(\frac{1}{4}\pi + \frac{1}{2}b)} \right.\right.$$

$$\left.\left. + q_1 q_2 e^{j(\frac{1}{4}\pi + b)} \right) \right|. \quad (8)$$

Now let

$$f(q) \equiv \left| 1 - qe^{-j(\frac{1}{4}\pi + \frac{1}{2}b)} \right|$$

$$= [1 + q^2 - 2q\cos(\tfrac{1}{4}\pi + \tfrac{1}{2}b)]^{1/2}, \quad (9)$$

so that subject to $p > 20$, $p \gg (q_1 + q_2)$, $p^2 \gg q_1^2 q_2^2$,

$$E_{ez1} \doteq \left| \frac{\beta_1^2 p_{z1}}{4\pi\epsilon_0 r p_e} \right| f(q_{1e}) f(q_{2e}), \quad (16a)$$

$$B_{mz1} \doteq \left| \frac{\beta_1^2 m_{z1}}{4\pi\nu_0 r p_m} \right| f(q_{1m}) f(q_{2m}). \quad (16b)$$

Thus at large distances from the transmitter E_{ez1} and B_{mz1} may be resolved into three factors:
(a) the surface-wave field $|\beta_1^2 p_{z1}/4\pi\epsilon_0 r p_e|$ or $|\beta_1^2 m_{z1}/4\pi\nu_0 r p_m|$,
(b) the height-gain-function $f(q_1)$ corresponding to the transmitter height d,
(c) the height-gain function $f(q_2)$ for the receiving antenna of height $z + d$.
The height-gain function $f(q)$ as defined in (9) is represented graphically in Fig. 17.1 and listed in Table 17.1.

18. Quasi-Near-Zone Fields of Vertical Antennas with Sinusoidal Currents Over a Conducting Earth

The general formulas for the electromagnetic fields of infinitesimal dipoles or doublets as given in Secs. 15 and 16 are too complicated to permit direct integration to obtain the fields of antennas of finite length even when sinusoidally distributed currents are assumed.

However, since the near-zone field of an isolated antenna with a sinusoidal current is well known from the detailed analysis in Chapter V, certain conclusions may be drawn.

From the point of view of expanding surfaces of constant phase or waves, the principal difference between the field of an infinitesimal dipole or doublet and that of a multiple half-wave radiator is that for the former the surfaces of constant phase or wave fronts are *concentric spheres* with centers at the doublet, whereas for the latter they are spheroids with foci at the ends of the antenna. The spherical waves from the doublet expand radially in free space with the velocity $v_0 = 3 \times 10^8$ m/sec; the ellipsoidal waves from the multiple half-wave radiator expand outward so that the velocity along the z-axis is v_0 and that in other directions is sufficiently greater than v_0 to maintain the shape of the growing spheroid as it becomes more and more nearly spherical. It is readily verified in Figs. V.8.4 through V.8.6 that, within radial distances comparable with the length of the antenna, the eccentricity of the spheroidal wave fronts already is very small, and the elliptically polarized electric field approximates the linearly polarized E_Θ-field of the radiation zone except in a cone about the z-axis where the field is small and of little interest. Thus, at distances equal to or greater than the length $2h$ of the antenna, *the field of a linear radiator approximates* both the character of its own far-zone field and the field of an infinitesimal dipole in so far as the shape of the wave fronts is concerned. This suggests that the quasi-near-zone field of a vertical antenna with center at height $d \geq 2h$ from a conducting earth may be obtained from the field of the doublet in the same manner as the far-zone field was obtained in Sec. 8. That is, *the field of the doublet* given in Secs. 15 and 16 *is multiplied by the factor* (8.17) *to obtain the field of the antenna*. Moreover, since it has been shown in Sec. 17 that the surface-wave term is significant only when the incident wave is very near grazing incidence ($\Theta_d \doteq \pi/2$), as when an infinitesimal dipole is very near the surface, it follows that this term is negligible for an elevated antenna of considerable length, since Θ_d is always greater than $\pi/2$ even at $z = -d$, except at very great distances. Moreover, all large currents are well above the surface.

The field along the surface is of particular interest. The principal component E_{ez1} may be determined directly from (13.28a). If the surface-wave term is neglected and in empty space, β_0 is substituted for β_1,

$$E_{ez1} \doteq \frac{\beta_0^2 P_{z1}}{4\pi\epsilon_0} \frac{e^{-j\beta_0 R_d}}{R_d} (1 + f_{er}) \sin^2 \Theta_d, \quad (1)$$

where $f_{er} = f_{er}(\Theta_r) = f_{er}(\pi - \Theta_d)$ is given by (12.12a) with $\Theta_r = \pi - \Theta_d$. Since (1) is the vertical electric field at $z = -d$ of an elementary dipole at $z = 0$, it is also the field of an element of current $I_z dz$ at the center $z = 0$ of an antenna if $p_{z1} = I_z dz/j\omega$. The field due to a similar element at z' is

$$dE_{z1} \doteq \frac{\beta_0^2 I_z'}{j\omega 4\pi\epsilon_0} \frac{e^{-j\beta_0 R_{z'}}}{R_{z'}} [1 + f_{er}(\pi - \Theta_{z'})]$$
$$\times \sin^2 \Theta_{z'} dz', \quad (2)$$

where $R_{z'} = \sqrt{(d+z')^2 + r^2}$, $\Theta_{z'} = \sin^{-1}(r/R_{z'})$ as shown in Fig. 18.1, and

$$1 + f_{er}(\pi - \Theta_{z1})$$
$$= \frac{-2 \cos \Theta_{z'}}{-\cos \Theta_{z'} + \sqrt{N_{21}^2 - \sin^2 \Theta_{z'}}/N_{21}^2}. \quad (3)$$

Referring to Fig. 18.1, it follows that, subject to

$$R_d^2 \gg h^2 \geq z'^2, \quad (4)$$

$$R_{z'} = R_d \sqrt{1 + \frac{2dz'}{R_d} + \frac{z'^2}{R_d^2}} \doteq R_d - z' \cos \Theta_d \quad (5a)$$

in phase factors. In amplitude factors a mean value between the extremes of $R_{z'}$ is satisfactory, so that

$$R_{z'} \doteq R_d. \quad (5b)$$

With (5b),

$$\sin \Theta_{z'} = r/R_{z'} \doteq r/R_d = \sin \Theta_d, \quad (6a)$$
$$-\cos \Theta_{z'} = (z' + d)/R_{z'} \doteq (z' + d)/R_d, \quad (6b)$$
$$1 + f_{er}(\pi - \Theta_{z'})$$
$$\doteq \frac{2(z' + d)}{z' + d + R_d \sqrt{N_{21}^2 - \sin^2 \Theta_d}/N_{21}^2}. \quad (7)$$

For temporary use let

$$A \equiv R_d \sqrt{N_{21}^2 - \sin^2 \Theta_d}/N_{21}^2. \quad (8)$$

With (4) through (8), (2) becomes

$$dE_z \doteq \frac{\beta_0^2 I_z'}{j\omega 4\pi\epsilon_0} \frac{e^{-j\beta_0 R_d}}{R_d}$$
$$\sin^2 \Theta_d \left[e^{j\beta_0 z' \cos \Theta_d} \frac{2(z' + d)}{z' + d + A} \right] dz'. \quad (9)$$

[VII.18] THEORY OF LINEAR ANTENNAS

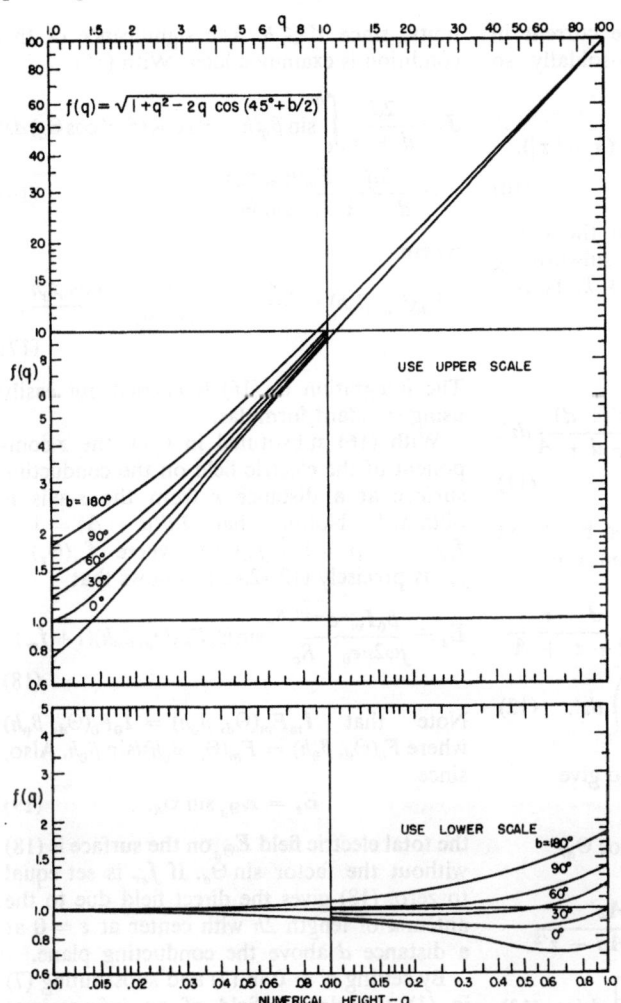

Fig. 17.1. Height-gain function (Norton).

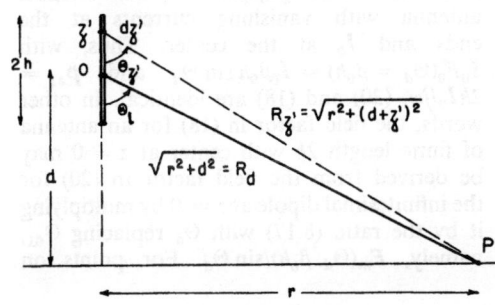

Fig. 19.1. Horizontal dipole over a conducting earth.

Fig. 18.1. Point on boundary surface near a vertical antenna.

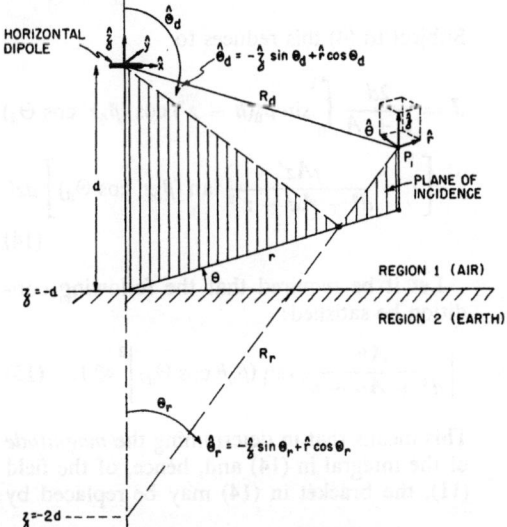

Now let it be assumed that the current in the antenna is distributed sinusoidally so that

$$I_z = I_0 \frac{\sin \beta_0(h - |z|)}{\sin \beta_0 h} = I_m \sin \beta_0(h - |z|). \tag{10}$$

The field due to the current in the entire antenna is obtained from (9) by substituting (10) and integrating from $-h$ to $+h$. Thus,

$$E_z = \frac{\beta_0^2 I_m}{j\omega 2\pi\epsilon_0} \frac{e^{-j\beta_0 R_d}}{R_d} \sin^2 \Theta_d$$

$$\times \int_{-h}^{h} \sin \beta_0(h - |z'|) e^{j\beta_0 z' \cos \Theta_d} \frac{(z' + d)}{z' + d + A} dz'. \tag{11}$$

Let the integral in (11) be denoted by J and transformed without approximation into

$$J = \int_0^h \sin \beta_0(h - z') \left(e^{j\beta_0 z' \cos \Theta_d} \frac{d + z'}{d + z' + A} \right.$$

$$\left. + e^{-j\beta_0 z' \cos \Theta_d} \frac{d - z'}{d - z' + A} \right) dz'. \tag{12}$$

This formula may be rearranged to give

$$J = 2 \int_0^h \sin \beta_0(h - z') \cos (\beta_0 z' \cos \Theta_d)$$

$$\times \left[\frac{d^2 + Ad - z'^2}{(d + A)^2 - z'^2} \right] \left[1 - j \frac{Az'}{d^2 + Ad - z'^2} \right.$$

$$\left. \times \tan (\beta_0 z' \cos \Theta_d) \right] dz'. \tag{13}$$

Subject to (4) this reduces to

$$J \doteq \frac{2d}{d + A} \int_0^h \sin \beta_0(h - z') \cos (\beta_0 z' \cos \Theta_d)$$

$$\left[1 - \frac{jAz'}{d^2 + Ad - z'^2} \tan (\beta_0 z' \cos \Theta_d) \right] dz'. \tag{14}$$

Let it be required that the following condition be satisfied:

$$\left| \frac{Ah}{d^2 + Ad - h^2} \tan (\beta_0 h \cos \Theta_d) \right|^2 \ll 1. \tag{15}$$

This means that in determining the *magnitude* of the integral in (14) and, hence, of the field (11), the bracket in (14) may be replaced by unity, since $z' \leq h$. The significance of this condition is examined later. With (15),

$$J \doteq \frac{2d}{d + A} \int_0^h \sin \beta_0(h - z') \cos (\beta_0 z' \cos \Theta_d) dz'$$

$$= \frac{2d}{d + A} \frac{F_m(\Theta_d, \beta_0 h)}{\beta_0 \sin \Theta_d}, \tag{16}$$

where

$$F_m(\Theta_d, \beta_0 h) \equiv \frac{\cos (\beta_0 h \cos \Theta_d) - \cos \beta_0 h}{\sin \Theta_d}. \tag{17}$$

The integration in (16) is carried out easily using standard formulas.

With (16) substituted in (11), the z-component of the electric field on the conducting surface at a distance r from the z-axis is obtained. Noting that $2d/(d + A) = 1 + f_{er}(\pi - \Theta_d) = 1 + f_{er}(\Theta_r)$, where $f_{er}(\Theta_r) = f_{er}$ is precisely (12.12a), it follows that

$$E_z \doteq \frac{\beta_0 I_m}{j\omega 2\pi\epsilon_0} \frac{e^{-j\beta_0 R_d}}{R_d} \sin \Theta_d F_m(\Theta_d, \beta_0 h)(1 + f_{er}). \tag{18}$$

Note that $I_m F_m(\Theta_d, \beta_0 h) = I_0 F_0(\Theta_d, \beta_0 h)$ where $F_0(\Theta_d, \beta_0 h) = F_m(\Theta_d, \beta_0 h)/\sin \beta_0 h$. Also, since

$$E_z = E_{\Theta_d} \sin \Theta_d, \tag{19}$$

the total electric field E_{Θ_d} on the surface is (18) without the factor $\sin \Theta_d$. If f_{er} is set equal to zero, (18) gives the direct field due to the antenna of length $2h$ with center at $z = 0$ at a distance d above the conducting plane.

By setting $z' = 0$ in (7) and substituting (7) in (1) the electric field of an infinitesimal dipole on the conducting surface is found to be

$$E_z \doteq \frac{\beta_0^2 p_{z1}}{4\pi\epsilon_0} \frac{e^{-j\beta_0 R_d}}{R_d} \sin^2 \Theta_d (1 + f_{er}). \tag{20}$$

Note that (18) is equivalent to (20) for a very short antenna ($\beta_0^2 h^2 \ll 1$) so end-loaded that the current is uniform at I_0. In this case $I_m F_m(\Theta_d, \beta_0 h) = I_0 F_0(\Theta_d, \beta_0 h)$ has twice the value $\frac{1}{2} I_0 \beta_0 h \sin \Theta_d$ for a short antenna with vanishing currents at the ends and I_0 at the center. Thus, with $I_0 F_0(\Theta_d = \beta_0 h) = I_0 \beta_0 h \sin \Theta_d$ and $p_{z1} = 2h I_0/j\omega$, (20) and (18) are identical. In other words, the field factor in (18) for an antenna of finite length $2h$ with center at $z = 0$ may be derived from the field factor in (20) for the infinitesimal dipole at $z = 0$ by multiplying it by the ratio (8.17) with Θ_d replacing Θ_{01}, namely, $F_m(\Theta_d, \beta_0 h)/\sin \Theta_d$. For points on the surface this verifies the argument given at the beginning of the section. It also specifies the conditions (4) and (15) for the validity of the approximate expression (18) for the field on the boundary at $z = -d$. Let these conditions be examined.

Case 1. *Antenna as near the conducting earth as possible*: $d \doteq h$. In this case (4) is equivalent to

$$r^2 \gg h^2 \quad \text{or} \quad (r/h) \geq 5, \qquad (21)$$

and (15) becomes

$$\tan^2(\beta_0 h \cos \Theta_d) \doteq (\beta_0 h \cos \Theta_d)^2 \ll 1,$$

or

$$\cos \Theta_d \leq 0.2/\beta_0 h. \qquad (22a)$$

Since $\cos \Theta_d = d/\sqrt{d^2 + r^2} \doteq h/r$, it follows that (22a) is as severe as or severer than (21) only if $\beta_0 h \geq 1$. An alternative form of (22a) is

$$\beta_0 r \geq 5\beta_0^2 h^2. \qquad (22b)$$

Note that for the half-wave dipole, $\beta_0 h = \pi/2$, $\cos \Theta_d \leq 0.4/\pi = 0.13$, $90° \geq \Theta_d \geq 82.5°$. Alternatively, $\beta_0 r \geq 5\pi^2/4$ or $r/\lambda_0 \geq 5\pi/8 \doteq 2$. Similarly, for $\beta_0 h + \pi$, $90° \geq \Theta_d \geq 86.3°$, or $r/\lambda_0 \geq 5\pi/2 \doteq 8$.

Case 2. *Antenna high above the conducting plane compared with its half-length*: $d^2 \gg h^2$; $R_d^2 \gg h^2$. In this case (4) is already satisfied for all values of r and (15) becomes

$$\left|\frac{Ah}{d(A+d)}\right|^2 \tan^2(\beta_0 h \cos \Theta_d) \ll 1. \qquad (23)$$

Since $|A/(A+d)| \leq 1$ and $(h/d)^2 \ll 1$, it follows that (23) is satisfied if

$$\tan(\beta_0 h \cos \Theta_d) \leq 1$$

or

$$\beta_0 h \cos \Theta_d \leq \pi/4 \quad \text{or} \quad \cos \Theta_d \leq \frac{\pi}{4\beta_0 h}. \qquad (24a)$$

With $\cos \Theta_d = d/\sqrt{r^2 + d^2}$, (24a) is equivalent to

$$1 + r^2/d^2 \geq 64 h^2/\lambda_0^2. \qquad (24b)$$

For $\beta_0 h = \pi/2$, $\cos \Theta_d \leq 0.5$, $90° \geq \Theta_d \geq 60°$; for $\beta_0 h = \pi$, $\cos \Theta_d \leq 0.25$, $90° \geq \Theta_d \geq 75.5°$.

It may be concluded from the two cases considered that the formula (22a) is valid to within a few wavelengths of the point on the plane directly below the antenna provided the antenna is not too long and not too high. More generally, it may be concluded that the field of a center-driven vertical antenna above a conducting plane is well approximated by the field of an infinitesimal dipole multiplied by the ratio $F_m(\Theta_d, \beta_0 h)/\sin \Theta_d$ of the field factors except near the z-axis through the antenna.

The quasi-near-zone field of a base-driven antenna of length h erected vertically on a conducting plane is more complicated than the field of a center-driven antenna above such a plane, since there are large currents in the antenna at and just above the surface. As a result, the surface-wave term is relatively more important than when most of the current is well above the conducting earth. At radial distances r that are not large compared with h, the contribution by the surface-wave term is small, but as r increases the angle of incidence $\Theta_{z'}$ approaches sufficiently near 90° for higher and higher elements $I_z \, dz$ to make the surface wave more and more significant. At sufficiently great distances r compared with h, the field on the conducting earth behaves essentially like that of an infinitesimal dipole of appropriate moment on the surface. Except at points within a few wavelengths of the antenna, the field on the surface $z = -d$ may be evaluated by numerical methods that in effect combine the fields due to a distribution of infinitesimal dipoles along the z-axis equivalent to a sinusoidal distribution of current. In general, the far-zone formulas derived in Sec. 8 may be assumed to be reasonably good approximations of the field several wavelengths or more from the antenna and at all angles Θ except those near $\pi/2$.

HORIZONTAL DIPOLES OVER A CONDUCTING EARTH

19. *Hertzian Potentials of Horizontal Dipoles Over a Conducting Earth*

If an isolated infinitesimal electric or magnetic dipole or doublet is located at the origin of a Cartesian system of coördinates in free space with its axis along the x-axis, as in Fig. 19.1, the electromagnetic field may be determined from a Hertzian potential directed along the x-axis, that is, from $\mathbf{\Pi} = \hat{\mathbf{x}}\Pi_x$. The potential is of the form (2.6) or (3.2) with z replaced by x, and the electromagnetic field is given by (2.11) and (2.12) with $\mathbf{p} = \hat{\mathbf{x}}p_x$ or by (3.5a) and (3.5b) with $\mathbf{m} = \hat{\mathbf{x}}m_x$.

If the half-space from $z = -d$ to $z = -\infty$ is filled with a medium of arbitrary conducting and dielectric properties, currents are induced in this region that have not only x-components but also z-components. Therefore, both x- and z-components of the Hertzian potential are required. Unlike the case of the vertical dipole, where the simple nature of the electromagnetic field made it possible to derive all components of $\mathbf{E}$ and $\mathbf{B}$ from $\mathbf{\Pi}_z$ alone by abandoning the physical correspondence between the components of the density of polarization or magnetization and the polarization or magnetization potentials, the determination of the field of a horizontal dipole requires both $\mathbf{\Pi}_x$ and $\mathbf{\Pi}_z$ if the boundary conditions (1.10a, b) are to be satisfied. This is indicated in Sec. 1, where the appropriate boundary conditions are expressed in terms of $\mathbf{\Pi}_x$ and $\mathbf{\Pi}_z$ for both electric and magnetic dipoles in (1.20) and (1.21).

In formulating the general problem of the vertical dipoles beginning in Sec. 2, a representation in terms of generalized *plane waves* was introduced in order to derive the laws of reflection and refraction for plane waves and to show their relation with the asymptotic or far-zone field of a concentrated source of spherical waves. In solving the problem of the horizontal dipole this sequence need not be repeated and a formulation paralleling that in Sec. 9 using cylindrical wave functions may be introduced at once.

Since the direct part of the field of the horizontal dipole is expressible in terms of $\mathbf{\Pi}_x$, this component corresponds to $\mathbf{\Pi}_z$ for the vertical dipole so that solutions like (5.35a, b) and (5.36) are suggested. Using transformations like those carried out in Sec. 9 in deriving (9.9) from $(\mathbf{\Pi}_{z1})_r$ in (5.35a), the equivalents for $\mathbf{\Pi}_x$ of (5.35a, b) and (5.36) in terms of cylindrical wave functions in region 1 ($\nu_1 = v_1$, $\xi_1 = \epsilon_1$, $\beta_1 = \beta_1 = \omega\sqrt{\epsilon_1/v_1}$) and region 2 (simply conducting and non-ferromagnetic with $\nu_2 = v_2 \doteq v_1$, $\xi_2 = \epsilon_{e2} - j\sigma_{e2}/\omega$, $\beta_2 = \omega\sqrt{\xi_2/v_2}$) are given below. With $\mathbf{\Pi}_{x1} = (\mathbf{\Pi}_{x1})_d + (\mathbf{\Pi}_{x1})_r$ they are

$$\mathbf{\Pi}_{x1} = \frac{jK_1}{\beta_1} \int_0^\infty l^{-1}[e^{-zl} + F_r(\lambda)e^{-(z+2d)l}]$$
$$\times J_0(\lambda r)\lambda \, d\lambda, \quad (z \geq 0) \quad (1a)$$

$$\mathbf{\Pi}_{x1} = \frac{jK_1}{\beta_1} \int_0^\infty l^{-1}[e^{zl} + F_r(\lambda)e^{-(z+2d)l}]$$
$$\times J_0(\lambda r)\lambda \, d\lambda, \quad (-d \leq z \leq 0) \quad (1b)$$

$$\mathbf{\Pi}_{x2} = \frac{jK_2}{\beta_1} \int_0^\infty l^{-1}F_t(\lambda)e^{(z+d)m-dl}J_0(\lambda r)\lambda \, d\lambda,$$
$$(z \leq -d) \quad (2)$$

where, as in Sec. 9,

$$\lambda = \beta_1 \sin \Theta = \beta_2 \sin \Theta_t, \quad (3)$$

$$l = j\beta_1 \cos \Theta = \sqrt{\lambda^2 - \beta_1^2}, \quad (4a)$$

$$m = j\beta_2 \cos \Theta_t = \sqrt{\lambda^2 - \beta_2^2} \quad (4b)$$

and

$$K_{e1} = \frac{-j\beta_1 p_{x1}}{4\pi\epsilon_1}, \quad K_{e2} = \frac{-j\beta_1 p_{x1}}{4\pi\xi_2}, \quad (5a)$$

$$K_{m1} = K_{m2} = \frac{-j\beta_1 m_{x1}}{4\pi v_1}. \quad (5b)$$

It now remains to evaluate $F_r(\lambda)$ and $F_t(\lambda)$ so that the boundary conditions (1.20a, b) for the horizontal electric dipole and (1.20c, d) for the horizontal magnetic dipole are satisfied.

For the electric and magnetic dipoles, as indicated by the subscript, these boundary conditions at $z = -d$ are

$$\beta_1^2 \mathbf{\Pi}_{ex1} = \beta_2^2 \mathbf{\Pi}_{ex2}, \quad (6a)$$

$$\epsilon_1 \left(\frac{\partial \mathbf{\Pi}_{ex}}{\partial z}\right)_1 = \xi_2 \left(\frac{\partial \mathbf{\Pi}_{ex}}{\partial z}\right)_2, \quad (6b)$$

$$\left(\frac{\partial \mathbf{\Pi}_{mx}}{\partial z}\right)_1 = \left(\frac{\partial \mathbf{\Pi}_{mx}}{\partial z}\right)_2, \quad (6c)$$

$$\epsilon_1 \mathbf{\Pi}_{mx1} = \xi_2 \mathbf{\Pi}_{mx2}. \quad (6d)$$

Substitution of (1b) and (2) in (6a, b) with $z = -d$ gives the following equations for determining $F_{er}(\lambda)$, $F_{et}(\lambda)$, $F_{mr}(\lambda)$, and $F_{mt}(\lambda)$:

For the electric dipole,

$$\int_0^\infty [K_{e1}\beta_1^2 + K_{e1}\beta_1^2 F_{er}(\lambda) - K_{e2}\beta_2^2 F_{et}(\lambda)]$$
$$\times l^{-1}e^{-dl}J_0(\lambda r)\lambda \, d\lambda = 0, \quad (7)$$

$$\int_0^\infty [K_{e1}\epsilon_1 l - K_{e1}\epsilon_1 l F_{er}(\lambda) - K_{e2}\xi_2 m F_{et}(\lambda)]$$
$$\times l^{-1}e^{-dl}J_0(\lambda r)\lambda \, d\lambda = 0. \quad (8)$$

These equations are satisfied if the brackets in the integrands vanish. The resulting equations, when solved for $F_{er}(\lambda)$ and $F_{et}(\lambda)$, using $\beta^2 = \omega^2\xi v$, give

$$F_{er}(\lambda) = \frac{v_1 l - v_2 m}{v_1 l + v_2 m} = f_{mr}, \quad (9)$$

$$F_{et}(\lambda) = \frac{2v_1 l}{v_1 l + v_2 m} = f_{mt}. \quad (10)$$

The identification of $F_{er}(\lambda)$ with f_{mr} and $F_{et}(\lambda)$ with f_{mt} follows from the comparison of (9) and (10) with (5.56b) and (5.57b).

For the magnetic case, the integrals paralleling (7) and (8) are obtained by substituting (1b) and (2) in (6b,c) with $z = -d$. They are

$$\int_0^\infty [\epsilon_1 + \epsilon_1 F_{mr}(\lambda) - \xi_2 F_{mt}(\lambda)] l^{-1} e^{-dl}$$
$$\times J_0(\lambda r) \lambda \, d\lambda = 0, \quad (11)$$

$$\int_0^\infty [l - l F_{mr}(\lambda) - m F_{mt}(\lambda)] l^{-1} e^{-dl}$$
$$\times J_0(\lambda r) \lambda \, d\lambda = 0. \quad (12)$$

With $\beta^2 = \omega^2 \xi \nu$, the values of $F_{mr}(\lambda)$ and $F_{mt}(\lambda)$, obtained by equating the brackets in (11) and (12) to zero, are

$$F_{mr}(\lambda) = \frac{\nu_2 \beta_2^2 l - \nu_1 \beta_1^2 m}{\nu_2 \beta_2^2 l + \nu_1 \beta_1^2 m} = f_{er}, \quad (13)$$

$$F_{mt}(\lambda) = \frac{\epsilon_1}{\xi_2} \left[\frac{2\nu_2 \beta_2^2 l}{\nu_2 \beta_2^2 l + \nu_1 \beta_1^2 m} \right]$$
$$= \frac{\epsilon_1}{\xi_2} f_{et} = \frac{\nu_1 \beta_1^2}{\nu_2 \beta_2^2} f_{et}. \quad (14)$$

The relations between $F_{mr}(\lambda)$ and f_{er}, $F_{mt}(\lambda)$ and f_{et}, are obtained by comparing (13) and (14) with (5.56a) and (5.57a) after using (3) and (4a,b).

With $N_{21}^2 \equiv \beta_2^2/\beta_1^2 = \xi_2/\epsilon_1$ when $\nu_2 = \nu_1$, (9), (10), (31), and (14) give:

$$F_{er}(\lambda) = f_{mr} = \frac{l - m}{l + m}, \quad (15a)$$

$$F_{mr}(\lambda) = f_{er} = \frac{\beta_2^2 l - \beta_2^2 m}{\beta_2^2 l + \beta_1^2 m} = \frac{N_{21}^2 l - m}{N_{21}^2 l + m}, \quad (15b)$$

$$F_{et}(\lambda) = f_{mt} = \frac{2l}{l + m}, \quad (15c)$$

$$F_{mt}(\lambda) = \frac{f_{et}}{N_{21}^2} = \frac{1}{N_{21}^2} \left(\frac{2\beta_2^2 l}{\beta_2^2 l + \beta_1^2 m} \right)$$
$$= \frac{2l}{N_{21}^2 l + m}. \quad (15d)$$

With (15a,b,c,d) substituted in (1a,b) and (2) these become the actual solutions for the x-component of the Hertzian potentials due to horizontal dipoles. It remains to determine the z-component.

The boundary conditions for Π_z which simultaneously relate Π_z to Π_x are (1.21a, b) for the electric dipole and (1.21c, d) for the magnetic dipole. At $z = -d$, they are

$$\left(\frac{\partial \Pi_{ex}}{\partial x} + \frac{\partial \Pi_{ez}}{\partial z} \right)_1 = \left(\frac{\partial \Pi_{ex}}{\partial z} + \frac{\partial \Pi_{ez}}{\partial z} \right)_2, \quad (16a)$$

$$\epsilon_1 \Pi_{ez1} = \xi_2 \Pi_{ez2}, \quad (16b)$$

$$\Pi_{mz1} = \Pi_{mz2}, \quad (16c)$$

$$\nu_1 \left(\frac{\partial \Pi_{mx}}{\partial x} + \frac{\partial \Pi_{mz}}{\partial z} \right)_1 = \nu_2 \left(\frac{\partial \Pi_{mx}}{\partial x} + \frac{\partial \Pi_{mz}}{\partial z} \right)_2. \quad (16d)$$

Since Π_x in both regions depends upon x and y through $r = \sqrt{x^2 + y^2}$, it is correct to set

$$\frac{\partial \Pi_x}{\partial x} = \frac{\partial \Pi_x}{\partial r} \frac{\partial r}{\partial x} = \frac{\partial \Pi_x}{\partial r} \cos \theta, \quad (17)$$

where the cylindrical coördinate θ is measured from the positive x-axis. Substitution of (17) on both sides of equations (16a) and (16d) shows that Π_z must have the factor $\cos \theta$.

As has been pointed out before, the integrands in (1a,b) and (2) are particular solutions or eigenfunctions of the wave equation (1.6d) in cylindrical coördinates. These solutions are of the form (1.7a) with $n = 0$, namely, $e^{\pm zl} J_0(\lambda r)$. Since Π_z also must satisfy (1.6d), it may be constructed of eigenfunctions of general type (1.7a). In order to have the factor $\cos \theta$, it is clear that the appropriate eigenfunction for Π_z is (1.7a) with $n = 1$, that is,

$$\cos \theta \, e^{\pm zl} J_1(\lambda r). \quad (18)$$

Since there is no direct wave contributing to Π_{z1}, the solutions for Π_z corresponding to (1a, b) and (2) for Π_x are

$$\Pi_{z1} = \frac{jK_1}{\beta_1} \cos \theta$$
$$\times \int_0^\infty l^{-1} G_r(\lambda) e^{-(z+2d)l} J_1(\lambda r) \lambda \, d\lambda,$$
$$(-d \leq z) \quad (19)$$

$$\Pi_{z2} = \frac{jK_2}{\beta_1} \cos \theta$$
$$\times \int_0^\infty l^{-1} G_t(\lambda) e^{(z+d)m - dl} J_1(\lambda r) \lambda \, d\lambda,$$
$$(z \leq -d) \quad (20)$$

These are the two solutions for Π_z which satisfy the wave equation and which must be made to satisfy the boundary conditions (16a, b) for the electric dipole and (16c, d) for the magnetic dipole by appropriately defining the arbitrary functions $G_r(\lambda)$ and $G_t(\lambda)$.

For the electric case the substitution of (19) and (20) in (16b) with $z = -d$ gives

$$\int_0^\infty [K_1\epsilon_1 G_{er}(\lambda) - K_2\xi_2 G_{et}(\lambda)]l^{-1}e^{-dl}$$
$$\times J_1(\lambda r)\lambda\, d\lambda = 0. \quad (21)$$

Since this is satisfied by equating the bracket in the integrand to zero, the result is

$$G_{er}(\lambda) = \frac{K_2\xi_2}{K_1\epsilon_1} G_{et}(\lambda) = G_{et}(\lambda). \quad (22)$$

Substitution of (1b), (2), (19), and (20) in (16a) gives, with (17),

$$\cos\theta \int_0^\infty \{K_{e1}[1 + F_{er}(\lambda)] - K_{e2}F_{et}(\lambda)\}$$
$$\times l^{-1}e^{-dl}J_0'(\lambda r)\lambda^2\, d\lambda$$
$$- \cos\theta \int_0^\infty \{K_{e1}G_{er}(\lambda)l + K_{e2}G_{et}(\lambda)m\}$$
$$\times l^{-1}e^{-dl}J_1(\lambda r)\lambda\, d\lambda = 0. \quad (23)$$

This equation is satisfied if the common factors in the integrands are combined and equated to zero after use is made of the standard Bessel-function relation, $J_0'(\lambda r) = -J_1(\lambda r)$. Using (9) and (10), the resulting equation may be solved for $G_{er}(\lambda)$. With (22), the result is

$$G_{er}(\lambda) = \frac{-2\lambda\nu_1 l(\nu_1\beta_2^2 - \nu_2\beta_1^2)}{(\nu_1\beta_2^2 l + \nu_2\beta_1^2 m)(\nu_1 l + \nu_2 m)}$$
$$= G_{et}(\lambda). \quad (24)$$

In the case at hand, $\nu_2 = \nu_1$, and (24) may be simplified considerably since with (3) and (4a, b), $\beta_2^2 - \beta_1^2 = l^2 - m^2$. The result is

$$G_{er}(\lambda) = \frac{-2\lambda l(l - m)}{\beta_2^2 l + \beta_1^2 m} = G_{et}(\lambda). \quad (25)$$

For the magnetic dipole, the substitution of (19) and (20) in (16c) with $z = -d$ and the fact that $K_{m1} = K_{m2}$ gives

$$G_{mr}(\lambda) = G_{mt}(\lambda). \quad (26)$$

Substitution of (1b), (2), (19), and (20) in (16d) gives, with (17),

$$\cos\theta \int_0^\infty \{\nu_1[1 + F_{mr}(\lambda)] - \nu_2 F_{mt}(\lambda)\}$$
$$\times l^{-1}e^{-dl}J_0'(\lambda r)\lambda^2\, d\lambda$$
$$- \cos\theta \int_0^\infty \{\nu_1 G_{mr}(\lambda)l + \nu_2 G_{mt}(\lambda)m\}$$
$$\times l^{-1}e^{-dl}J_1(\lambda r)\lambda\, d\lambda = 0. \quad (27)$$

Again using $J_0'(\lambda r) = -J_1(\lambda r)$, and (13), (14), and (26), the equation obtained from the integrand in (27) may be solved for $G_{mr}(\lambda)$. The result is

$$G_{mr}(\lambda) = -\frac{2\lambda\nu_1\nu_2 l(\beta_2^2 - \beta_1^2)}{(\nu_2\beta_2^2 l + \nu_1\beta_1^2 m)(\nu_1 l + \nu_2 m)}$$
$$= G_{mt}(\lambda). \quad (28)$$

With $\nu_2 = \nu_1$ and $\beta_2^2 - \beta_1^2 = l^2 - m^2$, (28) reduces to

$$G_{mr}(\lambda) = -\frac{2\lambda l(l - m)}{\beta_2^2 l + \beta_1^2 m} = G_{mt}(\lambda). \quad (29)$$

Note that $G_{mr}(\lambda)$ and $G_{er}(\lambda)$ as given by (25) and (29) are the same.

With $\nu_2 = \nu_1$ as assumed, the solutions for Π_{x1} and Π_{z1} for horizontal electric and magnetic dipoles are given by (1), (2), (19), and (20) with appropriate substitutions for the F's and the G's. The complete formulas are:

$$(\Pi_{qx1})_d = \frac{jK_{q1}}{\beta_1} \int_0^\infty l^{-1}e^{-zl}J_0(\lambda r)\lambda\, d\lambda$$
$$= \frac{jK_{q1}}{\beta_1} \frac{e^{-j\beta_1 R_d}}{R_d}, \quad (q = e \text{ or } m), \quad (30)$$

$$(\Pi_{ex1})_r = \frac{jK_{e1}}{\beta_1}$$
$$\times \int_0^\infty \frac{1}{l}\left(\frac{l-m}{l+m}\right) e^{(z+2d)l}J_0(\lambda r)\lambda\, d\lambda$$
$$= \frac{jK_{e1}}{\beta_1}\left(\frac{-e^{-j\beta_1 R_r}}{R_r} + H\right), \quad (31a)$$

$$(\Pi_{mx1})_r = \frac{jK_{m1}}{\beta_1}$$
$$\times \int_0^\infty \frac{1}{l}\left(\frac{\beta_2^2 l - \beta_1^2 m}{\beta_2^2 l + \beta_1^2 m}\right) e^{-(z+2d)l}J_0(\lambda r)\lambda\, d\lambda$$
$$= \frac{jK_{m1}}{\beta_1}\left(\frac{-e^{-j\beta_1 R_r}}{R_r} + V\right), \quad (31b)$$

$$\Pi_{ex2} = \frac{jK_{e2}}{\beta_1} \int_0^\infty \frac{2}{l+m} e^{(z+d)m - dl}J_0(\lambda r)\lambda\, d\lambda, \quad (32a)$$

$$\Pi_{mx2} = \frac{jK_{m2}}{\beta_1}$$
$$\times \int_0^\infty \frac{2\beta_1^2 l}{\beta_2^2 l + \beta_1^2 m} e^{(z+d)m - dl}J_0(\lambda r)\lambda\, d\lambda, \quad (32b)$$

$$\Pi_{qz1} = \frac{-jK_{q1}}{\beta_1} \cos\theta$$
$$\times \int_0^\infty \frac{2(l-m)}{\beta_2^2 l + \beta_1^2 m} e^{-(z+2d)l} J_1(\lambda r) \lambda^2 d\lambda,$$
$$(q = e \text{ or } m) \quad (33)$$

$$\Pi_{qz2} = \frac{-jK_{q2}}{\beta_1} \cos\theta$$
$$\times \int_0^\infty \frac{2(l-m)}{\beta_2^2 l + \beta_1^2 m} e^{(z+d)m - dl} J_1(\lambda r) \lambda^2 d\lambda.$$
$$(q = e \text{ or } m) \quad (34)$$

In (30) and (31a, b),
$$R_d = \sqrt{r^2 + z^2}, \quad R_r = \sqrt{r^2 + (z+2d)^2}.$$
$$(35)$$

By analogy with (30),
$$\frac{e^{-j\beta_1 R_r}}{R_r} = \int_0^\infty l^{-1} e^{-(z+2d)l} J_0(\lambda r) \lambda \, d\lambda, \quad (36)$$

and the functions V and H are the same as defined in (9.13c, d), namely,

$$V \equiv \int_0^\infty \left(\frac{2\beta_2^2}{\beta_2^2 l + \beta_1^2 m}\right) e^{-(z+2d)l} J_0(\lambda r) \lambda \, d\lambda.$$
$$(37a)$$

$$H \equiv \int_0^\infty \left(\frac{2}{l+m}\right) e^{-(z+2d)l} J_0(\lambda r) \lambda \, d\lambda,$$
$$(37b)$$

Note that the function H occurs in the formulas for the *horizontal electric* and *vertical magnetic* dipoles. Similarly, V occurs in the formulas for the *horizontal magnetic* and the *vertical electric* dipoles.

The formulas (30) through (35) for the horizontal dipoles correspond to (9.12) and (9.13) for the vertical dipoles. The former include the components Π_x and Π_z of the Hertzian potentials, the latter only Π_z. Note that

$$\frac{(\Pi^h_{ex1})_r}{K_{e1}} = \frac{(\Pi^v_{mz1})_r}{K_{m1}}, \quad \frac{(\Pi^h_{mx1})_r}{K_{m1}} = \frac{(\Pi^v_{mz1})_r}{K_{e1}},$$
$$(38)$$

where the superscripts h and v are used to distinguish between values for horizontal and vertical dipoles. The subscripts e and m denote the electric and magnetic dipoles as usual.

20. Far-Zone Hertzian Potentials of Horizontal Dipoles Over a Conducting Earth

Following Sommerfeld,[1.51] it is possible to arrive at interesting and important conclusions about the Hertzian potentials in the far zone without actually carrying out the integrations. These conclusions were drawn originally by von Hoerschelmann.[23]

Let it be assumed that region 2 is a sufficiently good conductor and ω is sufficiently low so that the following condition is satisfied:

$$|\xi_2| \equiv |\epsilon_{e2} - j\sigma_{e2}/\omega| \gg \epsilon_1. \quad (1a)$$

With $\nu_2 = \nu_1$ and $\beta^2 = \omega^2 \xi\nu$, (1a) is to be interpreted as equivalent to
$$|\beta_2| \gg \beta_1. \quad (1b)$$

It is shown in (7.27) that the angle of refraction Θ_t in good conductors satisfies the condition
$$\cos\Theta_t \doteq 1. \quad (2)$$

If it is assumed that (2) is a satisfactory approximation for region 2, it follows that, with (19.4a, b),
$$m = j\beta_2 \cos\Theta_t \doteq j\beta_2, \quad (3a)$$
$$|m| \gg |l|. \quad (3b)$$

Accordingly,
$$l \pm m \doteq \pm m \doteq \pm j\beta_2, \quad (4a)$$
$$\beta_2^2 l + \beta_1^2 m \doteq \beta_2^2 l. \quad (4b)$$

With (4a, b), (19.37a) and (19.37b) become

$$H \doteq \frac{-j2}{\beta_2} \int_0^\infty e^{-(z+2d)l} J_0(\lambda r) \lambda \, d\lambda$$
$$= \frac{j2}{\beta_2} \frac{\partial}{\partial z} \int_0^\infty l^{-1} e^{-(z+2d)l} J_0(\lambda r) \lambda \, d\lambda, \quad (5a)$$

$$V \doteq 2 \int_0^\infty l^{-1} e^{-(z+2d)l} J_0(\lambda r) \lambda \, d\lambda. \quad (5b)$$

With (19.36), the leading $1/R_r$ terms in (5a) and (5b) become

$$H \doteq \frac{j2}{\beta_2} \frac{\partial}{\partial z} \frac{e^{-j\beta_1 R_r}}{R_r}$$
$$= \frac{j2}{\beta_2} \cos\Theta_r \frac{\partial}{\partial R_r}\left(\frac{e^{-j\beta_1 R_r}}{R_r}\right)$$
$$\doteq \frac{2\beta_1}{\beta_2} \cos\Theta_r \frac{e^{-j\beta_1 R_r}}{R_r} \quad (6a)$$

$$V \doteq 2 \frac{e^{-j\beta_1 R_r}}{R_r}. \quad (6b)$$

Substitution of (6a, b) in (19.31a, b) and combination with (19.30) gives the following formulas for Π_{x1} for horizontal electric and magnetic dipoles:

$$\Pi_{ex1} \doteq \frac{p_{x1}}{4\pi\epsilon_1} \left[\frac{e^{-j\beta_1 R_d}}{R_d} \right.$$
$$\left. - (1 - \frac{2\beta_1}{\beta_2} \cos \Theta_r) \frac{e^{-j\beta_1 R_r}}{R_r} \right], \quad (7a)$$

$$\Pi_{mx1} \doteq \frac{m_{x1}}{4\pi\nu_1} \left[\frac{e^{-j\beta_1 R_d}}{R_d} + \frac{e^{-j\beta_1 R_r}}{R_r} \right]. \quad (7b)$$

Substitution of (4a, b) in (19.33) gives

$$\Pi_{z1}^r \doteq \frac{-jK_1}{\beta_1} \cos \theta$$
$$\times \left[\frac{-j2}{\beta_2} \int_0^\infty l^{-1} e^{-(z+2d)l} J_1(\lambda r) \lambda^2 d\lambda \right]. \quad (8)$$

Since $\lambda J_1(\lambda r) = -\partial J_0(\lambda r)/\partial r$, it follows with (19.36) and (8) that

$$\Pi_{z1}^r \doteq \frac{2K_1}{\beta_1 \beta_2} \cos \theta \frac{\partial}{\partial r} \int_0^\infty l^{-1} e^{-(z+2d)l} J_0(\lambda r) \lambda \, d\lambda$$
$$= \frac{2K_1}{\beta_1 \beta_2} \cos \theta \frac{\partial}{\partial r} \left(\frac{e^{-j\beta_1 R_r}}{R_r} \right). \quad (9)$$

When the differentiation is carried out, and only the leading $1/R_r$ term is retained for the far-zone, the approximate potential is

$$\Pi_{z1q}^r \doteq -j \frac{2K_{1q}}{\beta_2} \cos \theta \sin \Theta_r \frac{e^{-j\beta_1 R_r}}{R_r}.$$
$$(q = e \text{ or } m) \quad (10)$$

After the insertion of the appropriate value of K_{1q} and the substitution of the polar coördinate Φ for the equivalent cylindrical coördinate θ, the following far-zone formulas for Π_{z1}^r are obtained

$$\Pi_{ez1}^r = \frac{p_{x1}}{2\pi\epsilon_1} \frac{e^{-j\beta_1 R_r}}{R_r} \frac{\beta_1}{\beta_2} \sin \Theta_r \cos \Phi, \quad (11a)$$

$$\Pi_{mz1}^r = \frac{m_{x1}}{2\pi\nu_1} \frac{e^{-j\beta_1 R_r}}{R_r} \frac{\beta_1}{\beta_2} \sin \Theta_r \cos \Phi. \quad (11b)$$

For dipoles quite near the conducting surface, $z \doteq -d$ and $R_r \doteq R_d \doteq R_{01} = \sqrt{r^2 + (z+d)^2}$, $\Theta_d \doteq \Theta_r \doteq \Theta_{01}$, so that

$$\Pi_{ex1}^r = \frac{p_{x1}}{2\pi\epsilon_1} \frac{\beta_1}{\beta_2} \frac{e^{-j\beta_1 R_{01}}}{R_{01}} \cos \Theta_{01}, \quad (12a)$$

$$\Pi_{mx1}^r = \frac{m_{x1}}{2\pi\nu_1} \frac{e^{-j\beta_1 R_{01}}}{R_{01}}, \quad (12b)$$

$$\Pi_{ez1}^r = \frac{p_{x1}}{2\pi\epsilon_1} \frac{\beta_1}{\beta_2} \frac{e^{-j\beta_1 R_{01}}}{R_{01}} \sin \Theta_{01} \cos \Phi, \quad (13a)$$

$$\Pi_{mz1}^r = \frac{m_{x1}}{2\pi\nu_1} \frac{\beta_1}{\beta_2} \frac{e^{-j\beta_1 R_{01}}}{R_{01}} \sin \Theta_{01} \cos \Phi. \quad (13b)$$

These relations give an interesting representation of the relative magnitudes of the x- and z-components of the Hertzian potentials. With $|\beta_2| \gg \beta_1$, as required in (1b), it follows that when $z \doteq -d \doteq 0$

$$|\Pi_{mx1}^r| \gg |\Pi_{mz1}^r|,$$

$$\Pi_{m1}^r \doteq \hat{x} \Pi_{mx1}^r \doteq \hat{x} \frac{m_{x1}}{2\pi\nu_1} \frac{e^{-j\beta_0 R_{01}}}{R_{01}}, \quad (14)$$

for all values of Θ_{01} and Φ. Hence the far-zone field of a *horizontal magnetic* dipole over a conducting plane is determined primarily by Π_{mx1}^r, which includes essentially equal contributions by the direct and reflected waves.

The ratio of the components (13a) to (12a) is

$$\left| \frac{\Pi_{ez1}^r}{\Pi_{ex1}^r} \right| = \tan \Theta_{01} \cos \Phi. \quad (15)$$

Greatest interest is in the field near the conducting surface, where Θ_{01} is near $\pi/2$ and $\tan \Theta_{01}$ is very great. It follows that, except where $\cos \Phi$ is small, the ratio (15) is large. On the other hand, at and near $\Phi = \pi/2$, (15) reduces to zero or a very small value. Evidently,

$$|\Pi_{ez1}^r| \gg |\Pi_{ex1}^r| \quad \text{if } \tan \Theta_{01} \cos \Phi \gg 1, \quad (16a)$$

$$|\Pi_{ex1}^r| \gg |\Pi_{ez1}^r| \quad \text{if } \tan \Theta_{01} \cos \Phi \ll 1. \quad (16b)$$

Since the second possibility occurs only where the potential is very small, the principal range of interest is (16a), where the potential is given by (13a). Thus, when $z \doteq -d \doteq 0$, with $\tan \Theta_{01} \cos \Phi \gg 1$,

$$\Pi_{e1}^r \doteq \hat{z} \frac{p_{x1}}{2\pi\epsilon_1} \frac{\beta_1}{\beta_2} \frac{e^{-j\beta_1 R_{01}}}{R_{01}} \sin \Theta_{01} \cos \Phi \quad (17)$$

for a *horizontal electric* dipole at the surface of the conducting earth. Note that for a perfectly conducting earth $\beta_2 \to \infty$ and the *entire far-zone field along the surface vanishes*.

The highly significant fact revealed by (17) is that the polarization potential in the far zone of a *horizontal* electric dipole is determined by Π_{e1z} just as for the *vertical* electric dipole. This means that the far-zone field of the horizontal electric dipole is due almost entirely to currents induced in the conducting earth and not to the direct field from the dipole. Thus, *the horizontal* electric dipole excites both horizontal x-components and vertical z-components of current in the earth. The x-components of current in the earth maintain an x-component of polarization potential $(\Pi_{ex1})_r$ that is equal and opposite to the polarization potential $(\Pi_{ex1})_d$ due to the currents in the dipole itself. The z-components of current in the earth maintain the component Π_{ez1} of the polarization potential that determines virtually the entire far-zone field. In the case of the *vertical electric dipole* the currents in the dipole and in the earth combine equally to maintain the polarization potential at distant points.

Whereas the polarization potential Π_{ez1}^v of a vertical electric dipole has equal magnitude in all directions, Π_{ez1}^h for a horizontal electric dipole has the directional factor $\sin\Theta_{01}\cos\Phi$. This represents a pattern corresponding to a figure of eight in the horizontal plane $\Theta_{01} = \pi/2$ and to two hemispheres in space with a maximum value of unity along the conducting plane ($\Theta_{01} = \pi/2$) in the directions $\Phi = 0, \pi$, and zero values in the directions $\Theta_{01} = 0, \pi$. It resembles the array factor of the bilateral end-fire array in which two electrically short vertical antennas are separated by a half-wavelength and driven with equal currents 180° out of phase.

21. The Electromagnetic Field of a Horizontal Dipole Over a Conducting Earth

The cylindrical components of the electromagnetic field of horizontal electric and magnetic dipoles over a conducting earth are obtained using (1.3a,b) with (1.14a,b). Thus, with

$$\Pi_r = \Pi_x \cos\theta, \quad \Pi_\theta = -\Pi_x \sin\theta, \tag{1}$$

$$E_{er} = \frac{\partial}{\partial r}\operatorname{div}\Pi_e + \beta^2\Pi_{ex}\cos\theta \tag{2a}$$

$$E_{mr} = -j\omega\left(\frac{1}{r}\frac{\partial \Pi_{mz}}{\partial \theta} + \frac{\partial \Pi_{mx}}{\partial z}\sin\theta\right),$$

$$E_{e\theta} = \frac{1}{r}\frac{\partial}{\partial \theta}\operatorname{div}\Pi_e - \beta^2\Pi_{ex}\sin\theta \tag{2b}$$

$$E_{m\theta} = -j\omega\left(\frac{\partial \Pi_{mx}}{\partial z}\cos\theta - \frac{\partial \Pi_{mz}}{\partial r}\right),$$

$$E_{ez} = \frac{\partial}{\partial z}\operatorname{div}\Pi_e + \beta^2\Pi_{ez} \tag{2c}$$

$$E_{mz} = j\frac{\omega}{r}\left[\frac{\partial}{\partial r}(r\Pi_{mx}\sin\theta) + \frac{\partial}{\partial \theta}(\Pi_{mx}\cos\theta)\right],$$

$$B_{er} = j\frac{\beta^2}{\omega}\left(\frac{1}{r}\frac{\partial \Pi_{ez}}{\partial \theta} + \frac{\partial \Pi_{ex}}{\partial z}\sin\theta\right) \tag{2d}$$

$$B_{mr} = \frac{\partial}{\partial r}\operatorname{div}\Pi_m + \beta^2\Pi_{mx}\cos\theta,$$

$$B_{e\theta} = j\frac{\beta^2}{\omega}\left(\frac{\partial \Pi_{ex}}{\partial z}\cos\theta - \frac{\partial \Pi_{ez}}{\partial r}\right) \tag{2e}$$

$$B_{m\theta} = \frac{1}{r}\frac{\partial}{\partial \theta}\operatorname{div}\Pi_m - \beta^2\Pi_{mx}\sin\theta,$$

$$B_{ez} = -j\frac{\beta^2}{\omega r}\left[\frac{\partial}{\partial r}(r\Pi_{ex}\sin\theta) + \frac{\partial}{\partial \theta}(\Pi_{ex}\cos\theta)\right] \tag{2f}$$

$$B_{mz} = \frac{\partial}{\partial z}\operatorname{div}\Pi_m + \beta^2\Pi_{mz},$$

where

$$\operatorname{div}\Pi = \frac{\partial \Pi_x}{\partial x} + \frac{\partial \Pi_z}{\partial z}. \tag{3}$$

It is now readily verified with (19.33), (2), and the relation $J_1(\lambda r) = -J_0'(\lambda r) = -\lambda^{-1}\frac{\partial}{\partial r}J_0(\lambda r)$ that

$$\frac{\partial \Pi_{z1}}{\partial z} = \frac{-jK_1}{\beta_1}\frac{\partial}{\partial x}\int_0^\infty \frac{2(l-m)l}{\beta_2^2 l + \beta_1^2 m}e^{-(z+2d)l} \\ \times J_0(\lambda r)\lambda\,d\lambda. \tag{4}$$

Furthermore, with (19.37a,b) and

$$N_{21} \equiv N_{12}^{-1} \equiv \beta_2/\beta_1, \tag{5}$$

(4) becomes

$$\frac{\partial \Pi_{z1}}{\partial z} = \frac{-jK_1}{\beta_1}\frac{\partial}{\partial x}(H - N_{12}^2 V)$$

$$-j\frac{K_1}{\beta_1}\frac{\partial}{\partial r}(H - N_{12}^2 V)\cos\theta, \tag{6}$$

so that with (19.30), (19.31a,b), and (6), (3) becomes

$$\operatorname{div}\Pi_e = \frac{jK_{e1}}{\beta_1}\frac{\partial}{\partial r}\left[\frac{e^{-j\beta_1 R_d}}{R_d} - \frac{e^{-j\beta_1 R_r}}{R_r}\right. \\ \left. + N_{12}^2 V\right]\cos\theta, \tag{7a}$$

$$\operatorname{div}\Pi_m = \frac{jK_{m1}}{\beta_1}\frac{\partial}{\partial r}\left[\frac{e^{-j\beta_1 R_d}}{R_d} - \frac{e^{-j\beta_1 R_r}}{R_r}\right. \\ \left. + (1 + N_{12}^2)V - H\right]\cos\theta. \tag{7b}$$

In order to express Π_{ez1} in (2c) in terms of the functions contained in Π_{ex1}, it is readily verified using (19.36) and (19.37b) that

$$\Pi_{ez1} = \frac{jK_{e1}}{\beta_1^3} \frac{\partial^2}{\partial z \partial r}\left[2\frac{e^{-j\beta_1 R_r}}{R_r} - (1+N_{12}^2)V\right]\cos\theta. \quad (8)$$

Using (7a,b) and (8), the parts of formulas (2a,b,c) that are derived from Π_e and give the complete *electric* field of the horizontal *electric* dipole may be expressed as follows:

$$E_{er1} = \frac{jK_{e1}}{\beta_1}\left[\frac{\partial^2}{\partial r^2}\left(\frac{e^{-j\beta_1 R_d}}{R_d} - \frac{e^{-j\beta_1 R_r}}{R_r} + N_{12}^2 V\right)\right.$$
$$\left. + \beta_1^2\left(\frac{e^{-j\beta_1 R_d}}{R_d} - \frac{e^{-j\beta_1 R_r}}{R_r} + H\right)\right]\cos\theta, \quad (9a)$$

$$E_{e\theta 1} = \frac{jK_{e1}}{\beta_1}\left[\frac{1}{r}\frac{\partial}{\partial r}\left(\frac{e^{-j\beta_1 R_d}}{R_d} - \frac{e^{-j\beta_1 R_r}}{R_r} + N_{12}^2 V\right)\right.$$
$$\left. + \beta_1^2\left(\frac{e^{-j\beta_1 R_d}}{R_d} - \frac{e^{-j\beta_1 R_r}}{R_r} + H\right)\right]\sin\theta, \quad (9b)$$

$$E_{ez1} = \frac{jK_{e1}}{\beta_1}\frac{\partial^2}{\partial z\partial r}\left(\frac{e^{-j\beta_1 R_d}}{R_d} + \frac{e^{-j\beta_1 R_r}}{R_r} - V\right)\cos\theta. \quad (9c)$$

Similarly, the parts of (2d,e,f) that depend on Π_m and give the complete *magnetic* field of a horizontal *magnetic* dipole are

$$B_{mr1} = \frac{jK_{m1}}{\beta_1}\left\{\frac{\partial^2}{\partial r^2}\left[\frac{e^{-j\beta_1 R_d}}{R_d} - \frac{e^{-j\beta_1 R_r}}{R_r}\right.\right.$$
$$\left.\left. - H + (1+N_{12}^2)V\right] + \beta_1^2\left(\frac{e^{-j\beta_1 R_d}}{R_d}\right.\right.$$
$$\left.\left. - \frac{e^{-j\beta_1 R_r}}{R_r} + V\right)\right\}\cos\theta, \quad (9d)$$

$$B_{m\theta 1} = \frac{-jK_{m1}}{\beta_1}\left\{\frac{1}{r}\frac{\partial}{\partial r}\left[\frac{e^{-j\beta_1 R_d}}{R_d} - \frac{e^{-j\beta_1 R_r}}{R_r}\right.\right.$$
$$\left.\left. - H + (1+N_{12}^2)V\right] + \beta_1^2\left(\frac{e^{-j\beta_1 R_d}}{R_d}\right.\right.$$
$$\left.\left. - \frac{e^{-j\beta_1 R_r}}{R_r} + V\right)\right\}\sin\theta, \quad (9e)$$

$$B_{mz1} = \frac{jK_{m1}}{\beta_1}\frac{\partial^2}{\partial z\partial r}\left(\frac{e^{-j\beta_1 R_d}}{R_d} + \frac{e^{-j\beta_1 R_r}}{R_r} - H\right)\cos\theta. \quad (9f)$$

Note that these formulas are exact and involve only the functions V and H in addition to the simple exponentials.

The evaluation of the components of the electromagnetic field using (9a–f) involves only differentiation, so that it may be carried out completely. However, as pointed out before, the approximations involved in the evaluation of V and H are such that accurate results cannot be obtained for radial distances less than a wavelength. Accordingly, there is no advantage in carrying out the elaborate differentiation involving higher powers of $1/R$ than the first. Since V, H, and $e^{-j\beta_1 R}/R$ are solutions of the wave equation

$$\left[\frac{1}{r}\frac{\partial}{\partial r}\left(r\frac{\partial}{\partial r}\right) + \frac{1}{r^2}\frac{\partial^2}{\partial\theta^2} + \frac{\partial^2}{\partial z^2} + \beta_1^2\right]W = 0, \quad (10a)$$

where W stands for V, H, or $e^{-j\beta R}/R$, the leading terms in $1/r$ must satisfy the simpler equation

$$\left(\frac{\partial^2}{\partial r^2} + \frac{\partial^2}{\partial z^2} + \beta_1^2\right)W = 0. \quad (10b)$$

Use may be made of (10b) to evaluate (9a–f). A convenient approximate relation between V and H is obtained readily. By making use of the integrals (19.31a), (19.36), and (19.37a,b) the following formula is derived; it applies only when (10b) is satisfied:

$$N_{12}^2 V - H \doteq \frac{1}{\beta_1^2}\frac{\partial^2}{\partial z^2}\left[2\frac{e^{-j\beta_1 R_r}}{R_r} - (1+N_{12}^2)V\right]. \quad (11)$$

With (10b) and (11), the leading terms in (9a–f) are

$$E_{er1} \doteq \frac{-jK_{e1}}{\beta_1}\frac{\partial^2}{\partial z^2}\left(\frac{e^{-j\beta_1 R_d}}{R_d} + \frac{e^{-j\beta_1 R_r}}{R_r} - V\right)\cos\theta, \quad (12a)$$

$$E_{e\theta 1} \doteq -jK_{e1}\beta_1\left(\frac{e^{-j\beta_1 R_d}}{R_d} - \frac{e^{-j\beta_1 R_r}}{R_r} + H\right)\sin\theta, \quad (12b)$$

$$E_{ez1} \doteq \frac{jK_{e1}}{\beta_1}\frac{\partial^2}{\partial z\partial r}\left(\frac{e^{-j\beta_1 R_d}}{R_d} + \frac{e^{-j\beta_1 R_r}}{R_r} - V\right)\cos\theta, \quad (12c)$$

$$B_{mr1} \doteq \frac{-jK_{m1}}{\beta_1} \frac{\partial^2}{\partial z^2} \left(\frac{e^{-j\beta_1 R_d}}{R_d} + \frac{e^{-j\beta_1 R_r}}{R_r} - H \right) \cos \theta, \quad (12d)$$

$$B_{m\theta 1} \doteq -jK_{m1}\beta_1 \left(\frac{e^{-j\beta_1 R_d}}{R_d} - \frac{e^{-j\beta_1 R_r}}{R_r} + V \right) \sin \theta, \quad (12e)$$

$$B_{mz1} = \frac{jK_{m1}}{\beta_1} \frac{\partial^2}{\partial z \partial r} \left(\frac{e^{-j\beta_1 R_d}}{R_d} + \frac{e^{-j\beta_1 R_r}}{R_r} - H \right) \cos \theta. \quad (12f)$$

Note that E_{ez1} and B_{mz1} are exact, but that higher-order terms have been neglected in the other components.

Since the functions in (12a-f) are the same as already evaluated for the vertical dipoles, the final results are obtained readily. By comparing (13.11a, b) as derived from (12.9a, b) with (12a, d), it is seen that, except for changes in sign, the individual terms are alike. Hence,

$$E_{er1} \doteq jK_{e1}\beta_1 \cos \theta \left(\cos^2 \Theta_d \frac{e^{-j\beta_1 R_d}}{R_d} - \cos^2 \Theta_r \frac{e^{-j\beta_1 R_r}}{R_r} + 2A_e \cos \Theta_r \frac{e^{-j\beta_1 R_r}}{R_r} - A_e^2 V \right), \quad (13)$$

where $A_e \equiv \sqrt{N_{21}^2 - \sin^2 \Theta_r}/N_{21}^2$. Since

$$V = [(1 + f_{er}) + (1 - f_{er})F_e] \frac{e^{-j\beta_1 R_r}}{R_r}$$

and it is verified readily that

$$f_{er} \cos^2 \Theta_r = \cos^2 \Theta_r - 2A_e \cos \Theta_r + A_e^2(1 + f_{er}), \quad (14)$$

it follows that

$$E_{er1} \doteq jK_{e1}\beta_1 \cos \theta \left[\cos^2 \Theta_d \frac{e^{-j\beta_1 R_d}}{R_d} - f_{er} \cos^2 \Theta_r \frac{e^{-j\beta_1 R_r}}{R_r} - (1 - f_{er})F_e \left(\frac{N_{21}^2 - \sin^2 \Theta_r}{N_{21}^4} \right) \frac{e^{-j\beta_1 R_r}}{R_r} \right]. \quad (15a)$$

Similarly,

$$B_{mr1} \doteq jK_{m1}\beta_1 \cos \theta \left[\cos^2 \Theta_d \frac{e^{-j\beta_1 R_d}}{R_d} - f_{mr} \cos^2 \Theta_r \frac{e^{-j\beta_1 R_r}}{R_r} - (1 - f_{mr})F_m(N_{21}^2 - \sin^2 \Theta_r) \frac{e^{-j\beta_1 R_r}}{R_r} \right]. \quad (15b)$$

The Θ-components are given directly by (12b, e) as follows:

$$E_{e\theta 1} \doteq -jK_{e1}\beta_1 \sin \theta \left[\frac{e^{-j\beta_1 R_d}}{R_d} + f_{mr} \frac{e^{-j\beta_1 R_r}}{R_r} + (1 - f_{mr})F_m \frac{e^{-j\beta_1 R_r}}{R_r} \right], \quad (16a)$$

$$B_{m\theta 1} \doteq -jK_{m1}\beta_1 \sin \theta \left[\frac{e^{-j\beta_1 R_d}}{R_d} + f_{er} \frac{e^{-j\beta_1 R_r}}{R_r} + (1 - f_{er})F_e \frac{e^{-j\beta_1 R_r}}{R_r} \right]. \quad (16b)$$

Since (12c, f) are essentially like (14.1a, b) with (12.9a, b) except for factors that are constant in the differentiations and a change in sign of the last two terms, the derived formulas are obtained directly from (14.17a, b). The results are

$$E_{ez1} \doteq -jK_{e1}\beta_1 \cos \theta \left[\cos \Theta_d \sin \Theta_d \frac{e^{-j\beta_1 R_d}}{R_d} - f_{er} \cos \Theta_r \sin \Theta_r \frac{e^{-j\beta_1 R_r}}{R_r} + (1 - f_{er})F_e \sin \Theta_r \frac{\sqrt{N_{21}^2 - \sin^2 \Theta_r}}{N_{21}^2} \frac{e^{-j\beta_1 R_r}}{R_r} \right], \quad (17a)$$

$$B_{mz1} \doteq -jK_{m1}\beta_1 \cos \theta \left[\cos \Theta_d \sin \Theta_d \frac{e^{-j\beta_1 R_d}}{R_d} - f_{mr} \cos \Theta_d \sin \Theta_d \frac{e^{-j\beta_1 R_r}}{R_r} + (1 - f_{mr})F_m \sin \Theta_r \sqrt{N_{21}^2 - \sin^2 \Theta_r} \frac{e^{-j\beta_1 R_r}}{R_r} \right]. \quad (17b)$$

In the evaluation of the magnetic field of a horizontal electric dipole and the electric field of a horizontal magnetic dipole the derivative $\partial \Pi_{z1}/\partial r$ occurs. It is convenient to express this as a derivative with respect to z. Using the general integrals (19.30)–(19.37), it is readily verified that

$$\frac{\partial \Pi_{z1}}{\partial r} \doteq \frac{-jK_1}{\beta_1} \cos\theta \frac{\partial}{\partial z}\left(2\frac{e^{-j\beta_1 R_r}}{R_r} - V - H\right), \quad (18)$$

because

$$J_0''(\lambda r) = -J_0(\lambda r) + \frac{J_1(\lambda r)}{\lambda r} \doteq -J_0(\lambda r). \quad (19)$$

Since $J_1(\lambda r)/\lambda r$ leads to terms of the second order in $1/r$, it may be neglected. Using (19), (19.30), and (19.31a,b) the following expressions are obtained; only the leading terms are included:

For the horizontal electric dipole,

$$B_{er1} \doteq \frac{j\beta_1^2}{\omega} \sin\theta \frac{\partial \Pi_{ex1}}{\partial z}$$

$$\doteq -\frac{\beta_1 K_{e1}}{\omega} \sin\theta \frac{\partial}{\partial z}\left(\frac{e^{-j\beta_1 R_d}}{R_d} - \frac{e^{-j\beta_1 R_r}}{R_r} + H\right), \quad (20a)$$

$$B_{e\theta 1} \doteq \frac{j\beta_1^2}{\omega} \cos\theta \left(\frac{\partial \Pi_{ex1}}{\partial z} - \frac{1}{\cos\theta}\frac{\partial \Pi_{ez1}}{\partial r}\right)$$

$$\doteq \frac{-\beta_1 K_{e1}}{\omega} \cos\theta \frac{\partial}{\partial z}\left(\frac{e^{-j\beta_1 R_d}}{R_d} + \frac{e^{-j\beta_1 R_r}}{R_r} - V\right), \quad (20b)$$

$$B_{ez1} \doteq \frac{-j\beta_1^2}{\omega} \sin\theta \frac{\partial \Pi_{ex1}}{\partial r}$$

$$\doteq \frac{\beta_1 K_{e1}}{\omega} \sin\theta \frac{\partial}{\partial r}\left(\frac{e^{-j\beta_1 R_d}}{R_d} - \frac{e^{-j\beta_1 R_r}}{R_r} + H\right); \quad (20c)$$

For the horizontal magnetic dipole,

$$E_{mr1} \doteq -j\omega \sin\theta \frac{\partial \Pi_{mx1}}{\partial z}$$

$$\doteq \frac{\omega K_{m1}}{\beta_1} \sin\theta \frac{\partial}{\partial z}\left(\frac{e^{-j\beta_1 R_d}}{R_d} - \frac{e^{-j\beta_1 R_r}}{R_r} + V\right), \quad (20d)$$

$$E_{m\theta 1} \doteq -j\omega \cos\theta \left(\frac{\partial \Pi_{mx1}}{\partial z} - \frac{1}{\cos\theta}\frac{\partial \Pi_{mz1}}{\partial r}\right)$$

$$\doteq \frac{\omega K_{m1}}{\beta_1} \cos\theta \frac{\partial}{\partial z}\left(\frac{e^{-j\beta_1 R_d}}{R_d} + \frac{e^{-j\beta_1 R_r}}{R_r} - H\right), \quad (20e)$$

$$E_{mz1} \doteq j\omega \sin\theta \frac{\partial \Pi_{mx1}}{\partial r}$$

$$\doteq \frac{-\omega K_{m1}}{\beta_1} \sin\theta \frac{\partial}{\partial r}\left(\frac{e^{-j\beta_1 R_d}}{R_d} - \frac{e^{-j\beta_1 R_r}}{R_r} + V\right). \quad (20f)$$

Using (13.3,4) and (13.7a–d) in (20a,b,d,e), and (14.4) and (14.12) in (20c,f), the following leading terms are obtained:

For the horizontal electric dipole,

$$B_{er1} \doteq \frac{j\beta_1^2 K_{e1}}{\omega}\sin\theta\left[\cos\Theta_d \frac{e^{-j\beta_1 R_d}}{R_d}\right.$$

$$+ f_{mr}\cos\Theta_r \frac{e^{-j\beta_1 R_r}}{R_r}$$

$$\left. - (1-f_{mr})F_m\sqrt{N_{21}^2 - \sin^2\Theta_r}\,\frac{e^{-j\beta_1 R_r}}{R_r}\right], \quad (21a)$$

$$B_{e\theta 1} \doteq \frac{j\beta_1^2 K_{e1}}{\omega}\cos\theta\left[\cos\Theta_d \frac{e^{-j\beta_1 R_d}}{R_d}\right.$$

$$- f_{er}\cos\Theta_r \frac{e^{-j\beta_1 R_r}}{R_r}$$

$$\left. + (1-f_{er})F_e \frac{\sqrt{N_{21}^2 - \sin^2\Theta_r}}{N_{21}^2}\,\frac{e^{-j\beta_1 R_r}}{R_r}\right], \quad (21b)$$

$$B_{ez1} \doteq \frac{-j\beta_1^2 K_{e1}}{\omega}\sin\theta\left[\sin\Theta_d \frac{e^{-j\beta_1 R_d}}{R_d}\right.$$

$$+ f_{mr}\sin\Theta_r \frac{e^{-j\beta_1 R_r}}{R_r}$$

$$\left. + (1-f_{mr})F_m \sin\Theta_r \frac{e^{-j\beta_1 R_r}}{R_r}\right]; \quad (21c)$$

For the horizontal magnetic dipole,

$$E_{mr1} \doteq j\omega K_{m1}\sin\theta\left[\cos\Theta_d \frac{e^{-j\beta_1 R_d}}{R_d}\right.$$

$$+ f_{er}\cos\Theta_r \frac{e^{-j\beta_1 R_r}}{R_r}$$

$$\left. - (1-f_{er})F_e \frac{\sqrt{N_{21}^2 - \sin^2\Theta_r}}{N_{21}^2}\,\frac{e^{-j\beta_1 R_r}}{R_r}\right], \quad (21d)$$

$$E_{m\theta 1} \doteq -j\omega K_{m1}\cos\theta\left[\cos\Theta_d \frac{e^{-j\beta_1 R_d}}{R_d}\right.$$

$$- f_{mr}\cos\Theta_r \frac{e^{-j\beta_1 R_r}}{R_r}$$

$$\left. + (1-f_{mr})F_m\sqrt{N_{21}^2 - \sin^2\Theta_r}\,\frac{e^{-j\beta_1 R_r}}{R_r}\right], \quad (21e)$$

$$E_{mz1} \doteq j\omega K_{m1}\sin\theta\left[\sin\Theta_d \frac{e^{-j\beta_1 R_d}}{R_d}\right.$$

$$+ f_{er}\cos\Theta_r \frac{e^{-j\beta_1 R_r}}{R_r}$$

$$\left. + (1-f_{er})F_e \sin\Theta_r \frac{e^{-j\beta_1 R_r}}{R_r}\right]. \quad (21f)$$

With all six components of the electromagnetic fields for horizontal electric and magnetic dipoles evaluated, it is convenient to combine them into components parallel and perpendicular to the *plane of incidence*. Since the plane of incidence is the r,z-plane, the r- and z-components must be combined just as in Sec. 15. Using (15.6a,b) and (19.5a,b), the following final formulas are obtained:

For the horizontal electric dipole,

$$E_{e\parallel} \doteq \frac{\beta_1^2 p_{x1}}{4\pi\epsilon_1} \cos\theta \left[\cos\Theta_d \frac{e^{-j\beta_1 R_d}}{R_d} \hat{\Theta}_d \right.$$
$$- f_{er} \cos\Theta_r \frac{e^{-j\beta_1 R_r}}{R_r} \hat{\Theta}_r$$
$$- (1 - f_{er}) F_e \frac{\sqrt{N_{21}^2 - \sin^2\Theta_r}}{N_{21}^2}$$
$$\left. \times \left(\hat{z} \sin\Theta_r + \hat{r} \frac{\sqrt{N_{21}^2 - \sin^2\Theta_r}}{N_{21}^2} \right) \frac{e^{-j\beta_1 R_r}}{R_r} \right], \quad (22a)$$

$$E_{e\perp} \doteq \hat{\theta} \frac{\beta_1^2 p_{x1}}{4\pi\epsilon_1} \sin\theta \left[\frac{e^{-j\beta_1 R_d}}{R_d} + f_{mr} \frac{e^{-j\beta_1 R_r}}{R_r} \right.$$
$$\left. + (1 - f_{mr}) F_m \frac{e^{-j\beta_1 R_r}}{R_r} \right], \quad (22b)$$

$$B_{e\parallel} \doteq \frac{\beta_1^3 p_{x1}}{4\pi\epsilon_1 \omega} \sin\theta \left[\frac{e^{-j\beta_1 R_d}}{R_d} \hat{\Theta}_d \right.$$
$$+ f_{mr} \frac{e^{-j\beta_1 R_r}}{R_r} \hat{\Theta}_r - (1 - f_{mr}) F_m$$
$$\left. \times (\hat{z} \sin\Theta_r + \hat{r} \sqrt{N_{21}^2 - \sin^2\Theta_r}) \frac{e^{-j\beta_1 R_r}}{R_r} \right], \quad (22c)$$

$$B_{e\perp} \doteq \hat{\theta} \frac{\beta_1^3 p_{x1}}{4\pi\epsilon_1 \omega} \cos\theta \left[\cos\Theta_d \frac{e^{-j\beta_1 R_d}}{R_d} \right.$$
$$- f_{er} \cos\Theta_r \frac{e^{-j\beta_1 R_r}}{R_r}$$
$$\left. + (1 - f_{er}) F_e \frac{\sqrt{N_{21}^2 - \sin^2\Theta_r}}{N_{21}^2} \frac{e^{-j\beta_1 R_r}}{R_r} \right]. \quad (22d)$$

For the horizontal magnetic dipole,

$$B_{m\parallel} \doteq \frac{\beta_1^2 m_{x1}}{4\pi v_1} \cos\theta \left[\cos\Theta_d \frac{e^{-j\beta_1 R_d}}{R_d} \hat{\Theta}_d \right.$$
$$- f_{mr} \cos\Theta_r \frac{e^{-j\beta_1 R_r}}{R_r} \hat{\Theta}_r$$
$$- (1 - f_{mr}) F_m \sqrt{N_{21}^2 - \sin^2\Theta_r}$$
$$\left. \times (\hat{z} \sin\Theta_r + \hat{r} \sqrt{N_{21}^2 - \sin^2\Theta_r}) \frac{e^{-j\beta_1 R_r}}{R_r} \right], \quad (23a)$$

$$B_{m\perp} \doteq \hat{\theta} \frac{\beta_1^2 m_{x1}}{4\pi v_1} \sin\theta \left[\frac{e^{-j\beta_1 R_d}}{R_d} + f_{er} \frac{e^{-j\beta_1 R_r}}{R_r} \right.$$
$$\left. + (1 - f_{er}) F_e \frac{e^{-j\beta_1 R_r}}{R_r} \right], \quad (23b)$$

$$E_{m\parallel} \doteq \frac{\omega \beta_1 m_{x1}}{4\pi v_1} \sin\theta \left[\frac{e^{-j\beta_1 R_d}}{R_d} \hat{\Theta}_d \right.$$
$$+ f_{er} \frac{e^{-j\beta_1 R_r}}{R_r} \hat{\Theta}_r - (1 - f_{er}) F_e$$
$$\left. \times \left(\hat{z} \sin\Theta_r + \hat{r} \frac{\sqrt{N_{21}^2 - \sin^2\Theta_r}}{N_{21}^2} \right) \frac{e^{-j\beta_1 R_r}}{R_r} \right], \quad (23c)$$

$$E_{m\perp} \doteq \hat{\theta} \frac{\omega \beta_1 m_{x1}}{4\pi v_1} \cos\theta \left[\cos\Theta_d \frac{e^{-j\beta_1 R_d}}{R_d} \right.$$
$$- f_{mr} \cos\Theta_r \frac{e^{-j\beta_1 R_r}}{R_r}$$
$$\left. + (1 - f_{mr}) F_m \sqrt{N_{21}^2 - \sin^2\Theta_r} \frac{e^{-j\beta_1 R_r}}{R_r} \right]. \quad (23d)$$

Note that $\hat{\Theta}_d$ and $\hat{\Theta}_r$ are unit vectors in the direction of increasing polar angles Θ_d and Θ_r measured from the positive z-axis. They are perpendicular, respectively, to the directions of $R_d = \sqrt{z^2 + r^2}$ and $R_r = \sqrt{(z + 2d)^2 + r^2}$. On the other hand, $\hat{r}, \hat{\theta}, \hat{z}$ are unit vectors in the direction of the cylindrical coördinates r, θ, z with origin at the center of the dipole. The coefficients of reflection are

$$f_{er} = \frac{N_{21}^2 \cos\Theta_r - \sqrt{N_{21}^2 - \sin^2\Theta_r}}{N_{21}^2 \cos\Theta_r + \sqrt{N_{21}^2 - \sin^2\Theta_r}}, \quad (24a)$$

$$f_{mr} = \frac{\cos\Theta_r - \sqrt{N_{21}^2 - \sin^2\Theta_r}}{\cos\Theta_r + \sqrt{N_{21}^2 - \sin^2\Theta_r}}. \quad (24b)$$

The surface attenuation functions are

$$F_e = 1 - j\sqrt{\pi P_e} e^{-P_e} [1 - \text{erf}(j\sqrt{P_e})], \quad (25a)$$

$$F_m = 1 - j\sqrt{\pi P_m} e^{-P_m} [1 - \text{erf}(j\sqrt{P_m})]. \quad (25b)$$

The complex numerical distances are

$$P_e = \frac{-j\beta_1 R_r}{2 \sin^2\Theta_r} \left(\cos\Theta_r + \frac{\sqrt{N_{21}^2 - \sin^2\Theta_r}}{N_{21}^2} \right)^2, \quad (26a)$$

$$P_m = \frac{-j\beta_1 R_r}{2 \sin^2\Theta_r} \left(\cos\Theta_r + \sqrt{N_{21}^2 - \sin^2\Theta_r} \right)^2. \quad (26b)$$

This completes the derivation of the quasi-near-zone formulas for the electromagnetic field of horizontal dipoles in air above a conducting half-space. They are satisfactory approximations when $r \gtrsim \lambda_1$.

As a useful check on the general formulas (22) and (23), it is interesting to make region 2 a perfect conductor so that $N_{21}^2 \to \infty$. In this case

$$f_{er} \to 1, \quad f_{mr} \to -1, \qquad (27a)$$

$$P_e \to \frac{-j\beta_1 R_r}{2} \cot^2 \Theta_r,$$

$$P_m \to \frac{e^{-j\beta_1 R_r}}{2 \sin^2 \Theta_r} N_{21}^2 \to \infty, \qquad (27b)$$

$$(1 - f_{er})F_e \to 0, \quad (1 - f_{mr})F_m \to 0. \qquad (27c)$$

Thus, the surface-wave term vanishes *everywhere* over a perfect conductor and the *entire* field is given by the space-wave terms. On the *perfectly conducting* surface, where $R_r = R_d$, $\Theta_r = \pi - \Theta_d$, the fields are as follows:

For the horizontal electric dipole,

$$E_{er1} = 0, \quad E_{e\theta 1} = 0,$$

$$E_{ez1} = \frac{\beta_1^2 P_{x1}}{2\pi\epsilon_1} \cos\theta \sin\Theta_d \cos\Theta_d \frac{e^{-j\beta_1 R_d}}{R_d}; \qquad (28a)$$

$$B_{er1} = \frac{\beta_1^3 P_{x1}}{2\pi\epsilon_1\omega} \sin\theta \cos\Theta_d \frac{e^{-j\beta_1 R_d}}{R_d},$$

$$B_{e\theta 1} = \frac{\beta_1^3 P_{x1}}{2\pi\epsilon_1\omega} \cos\theta \cos\Theta_d \frac{e^{-j\beta_1 R_d}}{R_d}, \qquad (28b)$$

$$B_{ez1} = 0.$$

Note that

$$B_{ex1} = B_{er1}\cos\theta - B_{e\theta 1}\sin\theta = 0, \qquad (28c)$$

$$B_{ey1} = B_{er1}\sin\theta + B_{e\theta 1}\cos\theta$$
$$= \frac{\beta_1^3 P_{x1}}{2\pi\epsilon_1\omega} \cos\Theta_d \frac{e^{-j\beta_1 R_d}}{R_d}. \qquad (28d)$$

For the horizontal magnetic dipole,

$$B_{mr1} = \frac{-\beta_1^2 m_{x1}}{2\pi\nu_1} \cos\theta \cos^2\Theta_d \frac{e^{-j\beta_1 R_d}}{R_d},$$

$$B_{m\theta 1} = \frac{\beta_1^2 m_{x1}}{2\pi\nu_1} \sin\theta \frac{e^{-j\beta_1 R_d}}{R_d}, \quad B_{mz1} = 0, \qquad (29a)$$

$$E_{mr1} = 0, \quad E_{m\theta 1} = 0,$$

$$E_{mz1} = \frac{\omega\beta_1 m_{x1}}{2\pi\nu_1} \sin\theta \sin\Theta_d \frac{e^{-j\beta_1 R_d}}{R_d}. \qquad (29b)$$

Hence,

$$B_{mx1} = \frac{\beta_1^2 m_{x1}}{2\pi\nu_1} (1 - \cos^2\theta \sin^2\Theta_d) \frac{e^{-j\beta_1 R_d}}{R_d}, \qquad (29c)$$

$$B_{my1} = \frac{\beta_1^2 m_{x1}}{2\pi\nu_1} \sin\theta \cos\theta \sin^2\Theta_d \frac{e^{-j\beta_1 R_d}}{R_d}. \qquad (29d)$$

It is important to bear in mind that the formulas for the quasi-near-zone field of electric or magnetic dipoles can *not* be expected to give the field of an isolated dipole if region 2 is identical with region 1 so that $N_{21} = 1$. Since it has been assumed throughout the analysis that the inequality $|N_{21}|^2 \gg 1$ is satisfied, the formulas all have no application when N_{21} is as small as one.

The far-zone fields of horizontal dipoles are obtained from the quasi-near-zone formulas by retaining only $1/R_0$ terms, where $R_0 = \sqrt{(z+d)^2 + r^2}$ is the distance from the point of observation P_1 in region 1 to the point on the boundary directly below the dipole. This is shown in Fig. 21.1 (page 795). The radiation field consists of the space-wave terms simplified as permitted by the inequality

$$R_0^2 \gg d^2. \qquad (29)$$

The following approximations are good, subject to (1):

$$R_d \doteq R_r \doteq R_0 \text{ in amplitudes}, \qquad (30a)$$

$$\left.\begin{array}{l} R_d \doteq R_0 - d\cos\Theta_{01} \\ R_r \doteq R_0 + d\cos\Theta_{01} \end{array}\right\} \text{ in exponents}, \qquad (30b)$$

$$\Theta_r \doteq \Theta_d \doteq \Theta_{01}. \qquad (30c)$$

Also $f_r \doteq f_r^r$, where f_r^r is like f_r but with Θ_{01} substituted for Θ_r. With these approximations the following far-zone fields are obtained from (22); the cylindrical coördinate θ is replaced by the equivalent spherical coördinate Φ:

For the horizontal electric dipole,

$$E_{e\|}^r = \hat{\Theta}_{01} \frac{-\beta_1^2 P_{x1}}{4\pi\epsilon_1} \frac{e^{-j\beta_1 R_0}}{R_0} \cos\Phi \cos\Theta_{01}$$
$$\times (e^{j\beta_1 d\cos\Theta_{01}} - f_{er}^r e^{-j\beta_1 d\cos\Theta_{01}}) \qquad (31a)$$

$$E_{e\perp}^r = \hat{\Phi} \frac{\beta_1^2 P_{x1}}{4\pi\epsilon_1} \frac{e^{-j\beta_1 R_0}}{R_0} \sin\Phi$$
$$\times (e^{j\beta_1 d\cos\Theta_{01}} + f_{mr}^r e^{-j\beta_1 d\cos\Theta_{01}}), \qquad (31b)$$

$$\mathbf{B}^r_{e\parallel} = \hat{\mathbf{\Theta}}_{01} \frac{\beta_1^3 p_{x1}}{4\pi\epsilon_1\omega} \frac{e^{-j\beta_1 R_0}}{R_0} \sin\Phi$$
$$\times (e^{j\beta_1 d\cos\Theta_{01}} + f^r_{mr} e^{-j\beta_1 d\cos\Theta_{01}}), \quad (31c)$$

$$\mathbf{B}^r_{e\perp} = \hat{\mathbf{\Phi}} \frac{\beta_1^3 p_{x1}}{4\pi\epsilon_1\omega} \frac{e^{-j\beta_1 R_0}}{R_0} \cos\Phi\cos\Theta_{01}$$
$$\times (e^{j\beta_1 d\cos\Theta_{01}} - f^r_{er} e^{-j\beta_1 d\cos\Theta_{01}}); \quad (31d)$$

For the horizontal magnetic dipole,

$$\mathbf{B}^r_{m\parallel} = \hat{\mathbf{\Theta}}_{01} \frac{-\beta_1^2 m_{x1}}{4\pi v_1} \frac{e^{-j\beta_1 R_0}}{R_0} \cos\Phi\cos\Theta_{01}$$
$$\times (e^{j\beta_1 d\cos\Theta_{01}} - f^r_{mr} e^{-j\beta_1 d\cos\Theta_{01}}), \quad (32a)$$

$$\mathbf{B}^r_{m\perp} = \hat{\mathbf{\Phi}} \frac{\beta_1^2 m_{x1}}{4\pi v_1} \frac{e^{-j\beta_1 R_0}}{R_0} \sin\Phi$$
$$\times (e^{j\beta_1 d\cos\Theta_{01}} + f^r_{er} e^{-j\beta_1 d\cos\Theta_{01}}), \quad (32b)$$

$$\mathbf{E}^r_{m\parallel} = \hat{\mathbf{\Theta}}_{01} \frac{\omega\beta_1 m_{x1}}{4\pi v_1} \frac{e^{-j\beta_1 R_0}}{R_0} \sin\Phi$$
$$\times (e^{j\beta_1 d\cos\Theta_{01}} + f^r_{er} e^{-j\beta_1 d\cos\Theta_{01}}), \quad (32c)$$

$$\mathbf{E}^r_{m\perp} = \hat{\mathbf{\Phi}} \frac{\omega\beta_1 m_{x1}}{4\pi v_1} \frac{e^{-j\beta_1 R_0}}{R_0} \cos\Phi\cos\Theta_{01}$$
$$\times (e^{j\beta_1 d\cos\Theta_{01}} - f^r_{mr} e^{-j\beta_1 d\cos\Theta_{01}}). \quad (32d)$$

It is readily verified that all components of these fields vanish along the surface of an *imperfect* conductor. Thus, at $\Theta_{01} = \pi/2$, $\cos\Theta_{01} = 0$, so that $\mathbf{E}^r_{e\parallel} = \mathbf{B}^r_{e\perp} = 0$, $\mathbf{B}^r_{m\parallel} = \mathbf{E}^r_{m\perp} = 0$. Moreover, at $\Theta_{01} = \pi/2$, $f^r_{mr} = -1$, $f^r_{er} = -1$ (note that $f^r_{er} = +1$ at $\Theta_{01} = \pi/2$ only for a perfect conductor!); it follows that $1 + f^r_{mr} = 0$, $1 + f^r_{er} = 0$ and, therefore, $\mathbf{E}^r_{e\perp} = \mathbf{B}^r_{e\parallel} = 0$ and $\mathbf{B}^r_{m\perp} = \mathbf{E}^r_{m\parallel} = 0$.

22. Comparison of the Fields of Horizontal Electric Dipoles with Those of Vertical Dipoles

The electromagnetic fields of horizontal dipoles necessarily are much more complicated than the relatively simple fields of vertical dipoles. This is illustrated in a superficial manner merely by listing the components of the field involved in each case. Thus,

Vertical electric dipole:
$$\mathbf{E}^v_{e\parallel}, \mathbf{B}^v_{e\perp}; \quad (1)$$
Vertical magnetic dipole:
$$\mathbf{B}^v_{m\parallel}, \mathbf{E}^v_{m\perp}; \quad (2)$$
Horizontal electric dipole:
$$\mathbf{E}^h_{e\parallel}, \mathbf{B}^h_{e\perp}, \mathbf{E}^h_{e\perp}, \mathbf{B}^h_{e\parallel}; \quad (3)$$

Horizontal magnetic dipole:
$$\mathbf{B}^h_{m\parallel}, \mathbf{E}^h_{m\perp}, \mathbf{B}^h_{m\perp}, \mathbf{E}^h_{m\parallel}. \quad (4)$$

Evidently the field of each of the horizontal dipoles includes components of the types of *both* vertical electric *and* vertical magnetic dipoles. The vertical currents excited in the earth by the horizontal electric dipole maintain the components $\mathbf{E}^h_{e\parallel}, \mathbf{B}^h_{e\perp}$, which are analogous to the components $\mathbf{E}^v_{e\parallel}, \mathbf{B}^v_{e\perp}$ maintained by the vertical currents in the vertical dipole and in the earth. The horizontal currents in the horizontal electric dipole and in the earth maintain the components $\mathbf{B}^h_{e\parallel}, \mathbf{E}^h_{e\perp}$, which are analogous to the components $\mathbf{B}^v_{m\parallel}, \mathbf{E}^v_{m\perp}$ maintained by the circulating horizontal currents in the vertical magnetic dipole (or small horizontal loop) and in the earth. The correspondence may be illustrated by arranging corresponding expressions close together. For the sake of brevity, let $K_d = e^{-j\beta_1 R_d}/R_d$, $K_r = e^{-j\beta_1 R_r}/R_r$, $A_e = \sqrt{N_{21}^2 - \sin^2\theta_r}/N_{21}^2$, $A_m = \sqrt{N_{21}^2 - \sin^2\theta_r}$. The field of the horizontal electric dipole and corresponding fields of vertical electric and magnetic dipoles are

$$\begin{cases} \mathbf{E}^h_{e\parallel} \sim \cos\theta [\cos\Theta_d K_d \hat{\mathbf{\Theta}}_d - f_{er}\cos\Theta_r K_r \hat{\mathbf{\Theta}}_r \\ \quad - (1 - f_{er})F_e A_e(\hat{\mathbf{z}}\sin\Theta_r + \hat{\mathbf{r}} A_e)K_r], \quad (5a) \\ \mathbf{E}^v_{e\parallel} \sim [\sin\Theta_d K_d \hat{\mathbf{\Theta}}_d + f_{er}\sin\Theta_r K_r \hat{\mathbf{\Theta}}_r \\ \quad - (1 - f_{er})F_e \sin\Theta_r(\hat{\mathbf{z}}\sin\Theta_r + \hat{\mathbf{r}} A_e)K_r, \\ \quad\quad\quad (5b) \end{cases}$$

$$\begin{cases} \mathbf{B}^h_{e\perp} \sim \hat{\mathbf{\theta}}\cos\theta[\cos\Theta_d K_d - f_{er}\cos\Theta_r K_r \\ \quad + (1 - f_{er})F_e A_e K_r], \quad (6a) \\ \mathbf{B}^v_{e\perp} \sim \hat{\mathbf{\theta}}[\sin\Theta_d K_d + f_{er}\sin\Theta_r K_r \\ \quad + (1 - f_{er})F_e \sin\Theta_r K_r], \quad (6b) \end{cases}$$

$$\begin{cases} \mathbf{B}^h_{e\parallel} \sim \sin\theta[,K_d \hat{\mathbf{\Theta}}_d + f_{mr} K_r \hat{\mathbf{\Theta}}_r \\ \quad (1 - f_{mr})F_m(\hat{\mathbf{z}}\sin\Theta_r + \hat{\mathbf{r}} A_m)K_r], \quad (7a) \\ \mathbf{B}^v_{m\parallel} \sim \sin\Theta_d K_d \hat{\mathbf{\Theta}}_d + \sin\Theta_r f_{mr} \hat{\mathbf{\Theta}}_r \\ \quad - (1 - f_{mr})F_m \sin\Theta_r(\hat{\mathbf{z}}\sin\Theta_r + \hat{\mathbf{r}} A_m)K_r, \\ \quad\quad\quad (7b) \end{cases}$$

$$\begin{cases} \mathbf{E}^h_{e\perp} \sim \hat{\mathbf{\theta}}\sin\theta[K_d + f_{mr}K_r \\ \quad + (1 - f_{mr})F_m K_r], \quad (8a) \\ \mathbf{E}^v_{m\perp} \sim \hat{\mathbf{\theta}}[\sin\Theta_d K_d + f_{mr}\sin\Theta_r K_r \\ \quad + (1 - f_{mr})F_m \sin\Theta_r K_r]. \quad (8b) \end{cases}$$

Note that there is a striking resemblance between the corresponding components except for trigonometric factors which determine the vertical characteristics. In the horizontal plane the principal difference between the pairs of components is that those of the horizontal dipole are not rotationally symmetric about the z-axis, as are those of the vertical dipoles. When $\theta = \pm\pi/2$, the components of the horizontal electric dipole that resemble the field of the vertical electric dipole vanish, so that the field in the directions of the positive and negative y-axis perpendicular to the x-axis through the dipole consists entirely of components resembling those of the vertical magnetic dipole or horizontal loop with its horizontally polarized electric field. This is to be expected on physical grounds. On the other hand, when $\theta = 0, \pi$, the only nonvanishing components of the field of the horizontal electric dipole are those that resemble the field of a vertical electric dipole. Thus, the electromagnetic field of the horizontal electric dipole resembles in general nature a superposition of the field of a vertical electric dipole multiplied by $\cos\theta$ and the field of a vertical magnetic dipole multiplied by $\sin\theta$.

In spite of the qualitative resemblance between components of the fields of the horizontal electric dipole and vertical electric and magnetic dipoles, there are important differences. Consider the fields along the surface $z = -d$ due to an elevated dipole at height d. Since $\Theta_d = \pi - \Theta_r$, $R_d = R_r$, $f_r = (\cos\Theta_r - A)/(\cos\Theta_r + A)$, $A_e = \sqrt{N_{21}^2 - \sin^2\Theta_r}/N_{21}^2$, $A_m = \sqrt{N_{21}^2 - \sin^2\Theta_r}$, the expressions (3–8) may be reduced to the following form:

$$\begin{cases} \mathbf{E}_{e\parallel}^h \sim \cos\theta \left(\dfrac{\cos^2\Theta_r - F_e A_e^2}{\cos\Theta_r + A_e} \right) \\ \qquad\qquad \times (\hat{\mathbf{z}}\sin\Theta_r + \hat{\mathbf{r}} A_e), \quad (9a) \\ \mathbf{E}_{e\parallel}^v \sim \sin\Theta_r \left(\dfrac{\cos\Theta_r + F_e A_e}{\cos\Theta_r + A_e} \right) \\ \qquad\qquad \times (\hat{\mathbf{z}}\sin\Theta_r + \hat{\mathbf{r}} A_e), \quad (9b) \end{cases}$$

$$\begin{cases} \mathbf{B}_{e\perp}^h \sim \hat{\boldsymbol{\theta}}\cos\theta \left(\dfrac{\cos^2\Theta_r - F_e A_e^2}{\cos\Theta_r + A_e} \right), & (10a) \\ \mathbf{B}_{e\perp}^v \sim \hat{\boldsymbol{\theta}}\sin\Theta_r \left(\dfrac{\cos\Theta_r - F_e A^e}{\cos\Theta_r + A_e} \right), & (10b) \end{cases}$$

$$\begin{cases} \mathbf{B}_{e\parallel}^h \sim \sin\theta \left(\dfrac{\cos\Theta_r + F_m A_m}{\cos\Theta_r + A_m} \right) \\ \qquad\qquad \times (\hat{\mathbf{z}}\sin\Theta_r + \hat{\mathbf{r}} A_m), \quad (11a) \\ \mathbf{B}_{m\parallel}^v \sim \sin\Theta_r \left(\dfrac{\cos\Theta_r + F_m A_m}{\cos\Theta_r + A_m} \right) \\ \qquad\qquad \times (\hat{\mathbf{z}}\sin\Theta_r + \hat{\mathbf{r}} A_m), \quad (11b) \end{cases}$$

$$\begin{cases} \mathbf{E}_{e\perp}^h \sim \hat{\boldsymbol{\theta}}\sin\theta \left(\dfrac{\cos\Theta_r + F_m A_m}{\cos\Theta_r + A_m} \right), & (12a) \\ \mathbf{E}_{m\perp}^v \sim \hat{\boldsymbol{\theta}}\sin\Theta_r \left(\dfrac{\cos\Theta_r + F_m A_m}{\cos\Theta_r + A_m} \right). & (12b) \end{cases}$$

The following ratios are of interest in comparing the fields of vertical electric and magnetic dipoles with the field of a horizontal electric dipole:

$$\left| \frac{E_{e\parallel}^v}{E_{e\parallel}^h(\theta=0,\pi)} \right| = \left| \frac{B_{e\perp}^v}{B_{e\perp}^h(\theta=0,\pi)} \right|$$

$$= |\sin\Theta_r| \left| \frac{\cos\Theta_r + F_e A_e}{\cos^2\Theta_r - F_e A_e^2} \right|, \quad (13a)$$

$$\left| \frac{B_{m\parallel}^v}{B_{e\parallel}^h(\theta=\pm\pi/2)} \right| = \left| \frac{E_{m\perp}^v}{E_{e\perp}^h(\theta=\pm\pi/2)} \right|$$

$$= |\sin\Theta_r|. \quad (13b)$$

Since on the surface $z = -d$ at a reasonable distance r the vertical field factor $\sin\Theta_r$ is near unity, the components $\mathbf{E}_{e\perp}^h$ and $\mathbf{B}_{e\parallel}^h$ of the field of the horizontal electric dipole differ from the rotationally symmetric complete field $\mathbf{E}_{m\perp}^v$ and $\mathbf{B}_{m\parallel}^v$ of the vertical magnetic dipole (or horizontal loop) essentially only in the directional factor $\sin\theta$ if the excitation is equivalent. On the other hand, the components $\mathbf{E}_{e\parallel}^h$ and $\mathbf{B}_{e\perp}^h$ of the field of the horizontal electric dipole differ from the rotationally symmetric complete field $\mathbf{E}_{e\parallel}^v$ and $\mathbf{B}_{e\perp}^v$ of the vertical electric dipole not only in the directional factor $\cos\theta$, but also in having different amplitude factors. The ratio of these factors is given in (13a). Since $\cos\Theta_r$ is always greater than $\cos^2\Theta_r$, this ratio exceeds unity if Θ_r differs sufficiently from $\pi/2$ to make the surface-wave terms negligible. On the other hand, since $A_e \sim 1/N_{21}$ and $|N_{21}|^2 \gg 1$, it follows that $|A_e|$ is always greater than $|A_e|^2$. Accordingly the ratio (13a) is considerably greater than unity if $\Theta_r \doteq \pi/2$ and the surface-wave terms predominate. It may be concluded that the part of the electric field of the horizontal electric dipole that is in the plane of incidence is smaller at points along the conducting surface than the electric field of the vertical electric dipole which it resembles if the excitations are equal.

The ratio of the maximum values of the vertically and horizontally polarized components of the electric field on the conducting surface due to the horizontal electric dipole itself is

$$\left|\frac{E_{e1z}^h(\theta=0,\pi)}{E_{e1\theta}^h(\theta=\pm\pi/2)}\right|$$
$$=|\sin\Theta_r|\left|\frac{\cos^2\Theta_r - F_e A_e^2}{\cos\Theta_r + F_m A_m}\right|\left|\frac{\cos\Theta_r + A_m}{\cos\Theta_r + A_e}\right|. \quad (14)$$

If Θ_r differs sufficiently from $\pi/2$ so that the surface-wave terms $F_e A_e^2$ and $F_m A_m$ are negligible, the order of magnitude of the ratio (14) may be evaluated by noting that $A_m \sim N_{21}$, $A_e \sim 1/N_{21}$. With $|N_{21}^2| \gg 1$ and $\cos\Theta_r$ small, the ratio is approximately

Space-wave terms:

$$\left|\frac{E_{e1z}^h(\theta=0,\pi)}{E_{e1\theta}^h(\theta=\pm\pi/2)}\right| \doteq |\sin\Theta_r|\left|\frac{N_{21}\cos\Theta_r}{\cos\Theta_r + 1/N_{21}}\right|$$
(15a)

This is greater than unity when $|N_{21}|$ is sufficiently large; it may be less than unity when $|N_{21}|$ is small. On the other hand, if $\Theta_r \doteq \pi/2$ and the surface-wave terms predominate, the attenuation functions F_e and F_m may be represented by the leading terms as follows: $F_e/F_m \doteq P_m/P_e \doteq N_{21}^4$; hence,

Surface-wave terms:

$$\left|\frac{E_{e1z}^h(\theta=0,\pi)}{E_{e1\theta}^h(\theta=\pm\pi/2)}\right| \doteq \left|\frac{F_e A_e}{F_m}\right| \doteq |N_{21}^3| \gg 1.$$
(15b)

Accordingly, it may be concluded that the vertically polarized component E_{e1z}^h of the electric field of a horizontal electric dipole always is greater than the horizontally polarized component $E_{e1\theta}^h$ *at points on the conducting surface*, that is, at $z = -d$. This confirms the same conclusion drawn from the relative magnitudes of the components of the Hertzian potentials in Sec. 20. Note that this conclusion does not apply at points elevated above the surface of the conducting earth so that $z > -d$.

Note the following inequalities that result from (13), (14), and (15):

$$|E_{e1z}^v| > |E_{e1z}^h| > |E_{e1\theta}^h| \quad \text{at } z = -d. \quad (16)$$

Thus, the vertical electric dipole maintains a greater electric field than the horizontal electric dipole *at the conducting surface for a given excitation*.

An evaluation of the relative magnitudes of the fields at points above the conducting surface due to vertical and horizontal electric dipoles is not accomplished readily in general, owing to the number of variables and parameters. However, they are evaluated readily for any particular set of conditions. An indication of the nature of the problem is given in the special case in which the field is calculated at the same height d above the conducting surface as the dipole, that is, at $z = 0$, but at a sufficient radial distance $r = R_d$ to permit the approximations

$$R_d = r \doteq R_r \text{ in amplitude factor}, \quad (17a)$$

$$R_d - r \doteq \sqrt{R_r^2 - 4d^2} \doteq R_r - 2d\cos\Theta_r$$
$$\text{in exponents.} \quad (17b)$$

The notation is illustrated in Fig. 22.1. With these restrictions, the leading part of (13.21a) gives for the electric field of the vertical electric dipole

$$E_{ez1}^v \doteq \frac{\beta_1^2 p_{z1}}{4\pi\epsilon_1}\frac{e^{-j\beta_1 R_r}}{R_r}$$
$$\times (e^{j2\beta_1 d\cos\Theta_r} + f_{er}\sin^2\Theta_r). \quad (18)$$

It is assumed that d is sufficiently great so that the surface-wave term contributes negligibly. Similarly, (21.16a) gives the following formula for the horizontally polarized component of the electric field of the horizontal electric dipole:

$$E_{e\theta 1}^h(\theta=\pm\pi/2)$$
$$\doteq \pm\frac{\beta_1^2 p_{z1}}{4\pi\epsilon_1}\frac{e^{-j\beta_1 R_r}}{R_r}(e^{j2\beta_1 d\cos\Theta_r} + f_{mr}). \quad (19)$$

The ratio of the magnitudes of (18) and (19) is

$$\left|\frac{E_{ez1}^v}{E_{e\theta 1}^h(\theta=\pm\pi/2)}\right|$$
$$\doteq \left|\frac{e^{j\beta_1 d\cos\Theta_r} + \sin^2\Theta_r f_{er} e^{-j\beta_1 d\cos\Theta_r}}{e^{j\beta_1 d\cos\Theta_r} + f_{mr} e^{-j\beta_1 d\cos\Theta_r}}\right|. \quad (20)$$

With $|N_{21}^2| \gg 1$, and $\cos\Theta_r \ll |N_{21}|$,

$$f_{mr} \doteq \frac{\cos\Theta_r - N_{21}}{\cos\Theta_r + N_{21}} \doteq -1, \quad (21a)$$

$$f_{er} \doteq \frac{N_{21}\cos\Theta_r - 1}{N_{21}\cos\Theta_r + 1}. \quad (21b)$$

Depending on the magnitudes of N_{21} and $\cos\Theta_r$, f_{er} may have a wide range of values. As a specific example, let it be assumed that $|N_{21}|$ is sufficiently great and $\cos\Theta_r$ not too small so that $f_{er} \doteq 1$. Since $\sin^2\Theta_r$ is quite near unity when $\cos\Theta_r$ is small, a rough estimate of (20) under the special circumstances assumed is

$$\left| \frac{E_{ez1}^v}{E_{e\theta 1}^h(\theta = \pm \pi/2)} \right| \sim \cot(\beta_1 d \cos \Theta_r). \quad (22)$$

Depending on the electrical elevation $\beta_1 d$ of the dipole and the point of evaluation, the argument in (22) may be greater or less than $\pi/4$ so that $\cot(\beta_1 d \cos \Theta_r)$ may equal or exceed unity. For small values of $\beta_1 d$, $\cos \Theta_r$ is very small and the ratio (22) is much greater than unity; for larger values of $\beta_1 d$, the ratio may be much less than unity. Thus, it is clear that the choice between horizontal and vertical polarization when both transmitting and receiving dipoles are elevated depends on the particular constants and distances involved. Only when $\beta_1 d$ is small can it be stated definitely that vertical polarization yields a greater field.

A discussion of horizontal magnetic dipoles paralleling that for horizontal electric dipoles is readily carried out with analogous results.

23. Horizontal Antennas with Sinusoidal Currents Over a Conducting Earth

General expressions for the far-zone electromagnetic fields of horizontal antennas over an arbitrary conducting earth may be derived from those of infinitesimal horizontal electric dipoles or doublets in the same manner as in Sec. 8 for vertical antennas. Since the analysis in Chapter III indicates that for $\beta_0 h \leq 2$ a sinusoidal distribution of current is a satisfactory approximation for parallel antennas, the integrations for the far-zone field can be performed directly. However, even the expressions for the field of two parallel antennas or of a single antenna parallel to a perfectly conducting infinite plane are awkward when referred to a spherical system of coördinates with its axis bisecting each antenna, and it is clear that those for an antenna parallel to an imperfectly conducting plane must be much more complicated, since the additional components due to the vertical currents in the earth are involved. A simpler result, which is a satisfactory approximation for antennas of electrical half-length $\beta_0 h$, which satisfies the inequality $\beta_0 h \leq 2$, is obtained by introducing the effective length $2h_t$ for a transmitting antenna as defined in Sec. V.13 merely by setting the dipole moment p_{x1} equal to $2Ih_t/j\omega$ in the general expressions for the field of a horizontal electric dipole. If this is done, the field of a horizontal antenna over a conducting earth is visualized readily for qualitative purposes as the superposition of the field of a vertical electric dipole multiplied by the directional factor $\cos \theta$ and of a vertical magnetic dipole multiplied by the factor $\sin \theta$. If quantitative results are required, the direct substitution of the appropriate numerical values of constants and parameters is suggested, especially in evaluating the field near the surface of the earth.

Among long horizontal antennas over the earth the Beverage antenna is the most interesting. It consists of a horizontal wire close to the earth that may be many wavelengths long. It is grounded at one end through the generator and at the other end through an impedance that is designed to match the transmission line formed by the horizontal wire and the earth. The field due to the horizontal wire over the earth may be determined approximately by assuming a traveling-wave distribution of current of the form $I_x = I_0 e^{-j\beta_1 x}$ in $p_{x1} = I_x dx/j\omega$ and integrating over the length of the antenna. This field must be combined with the fields due to the short vertical antennas at each end in which the current is approximately uniform, so that p_{z1} may be replaced by $I_0 h_v/j\omega$, where h_v is the height of the vertical sections. Since this height is a small fraction of a wavelength, the significant part of the field along the surface due to the currents in the horizontal wire may be expected to be $E_{e\parallel}^h$ rather than $E_{e\perp}^h$. It has been shown that the horizontal field distribution of the component E_{e1z}^h of a short horizontal electric dipole resembles that of a vertical electric dipole with the added directional factor $\cos \theta$. This closely approximates the field of a bidirectional two-element end-fire array. It is reasonable, therefore, that the Beverage antenna consisting of the long horizontal wire with its traveling current wave together with the two short vertical waves at the ends approximate a unidirectional multielement end-fire array.

24. Currents Excited on a Perfectly Conducting Plane by a Parallel Antenna*

Although the electromagnetic field in a region containing an antenna parallel to a highly conducting half-space is maintained by the primary currents in the antenna and the secondary currents induced on the surface of the conducting plane, it is not actually

---

* This section is based on B. C. Dunn and R. King, ref. 13.

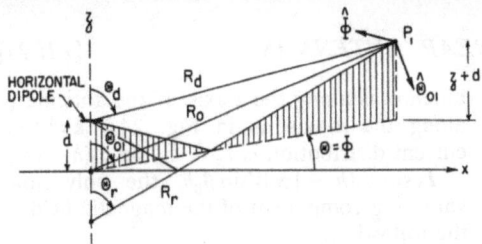

Fig. 21.1. Point P in far zone of horizontal dipole over a conducting earth.

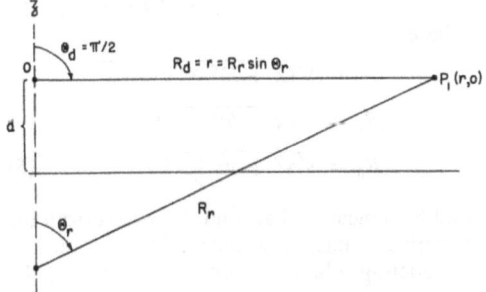

Fig. 22.1. Point P_1 at same elevation as dipole at O.

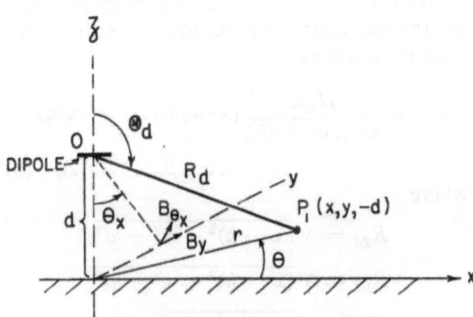

Fig. 24.1. Horizontal dipole at a height d over a perfectly conducting half-space.

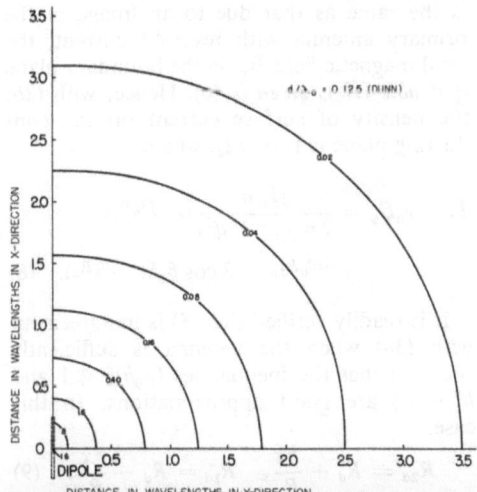

Fig. 24.2. Contours of constant relative amplitude of surface current density and magnetic field along one quadrant of a perfectly conducting half-space at a distance $d = -\lambda_0/8$ below an electric dipole parallel to the boundary (Dunn).

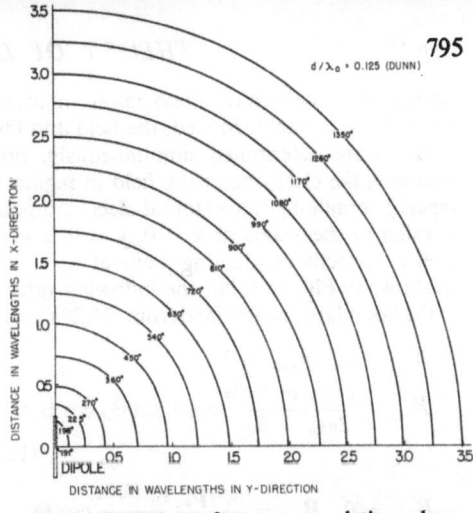

Fig. 24.3. Contours of constant relative phase associated with the amplitude in Fig. 24.2 (Dunn).

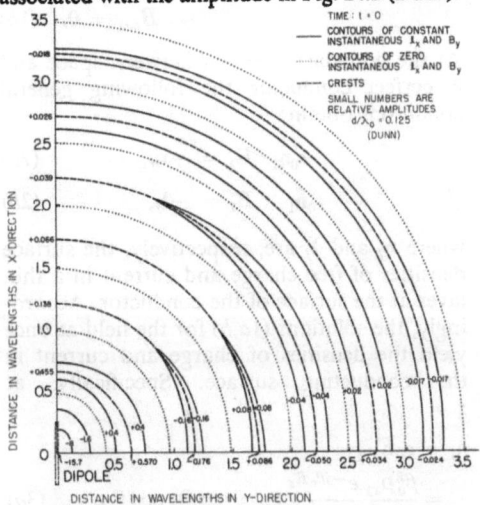

Fig. 24.4a. Contours of constant instantaneous surface current density and magnetic field at time $t = 0$ along one quadrant of a perfectly conducting half-space at a distance $d = -\lambda_0/8$ below an electric dipole parallel to the boundary (Dunn).

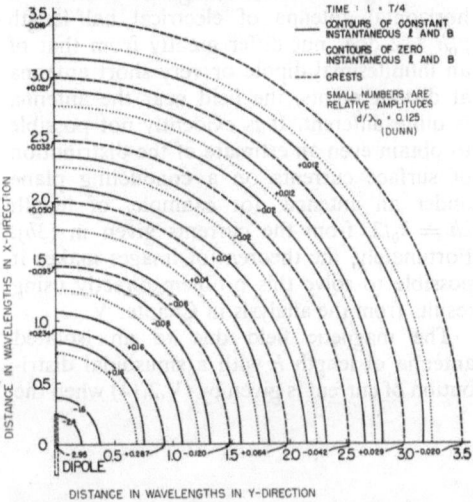

Fig. 24.4b. Like Fig. 24.4a but at time $t = T/4$ later (Dunn).

necessary to determine these latter in order to evaluate the field. Indeed, the field and the currents are determined simultaneously. For example, the electromagnetic field in region 1 (space) containing a horizontal electric dipole parallel to the x-axis at $x = 0$, $y = 0$, $z = 0$ over a perfectly conducting plane at $z = -d$, as shown in Fig. 24.1, has the following values at the boundary, as obtained from (21.28a,c,d):

$$E_{ex1} = 0, \quad E_{ey1} = 0,$$

$$E_{ez1} = \frac{\beta_0^2 p_{x1}}{2\pi\epsilon_0} \frac{e^{-j\beta_0 R_d}}{R_d} \cos\theta \sin\Theta_d \cos\Theta_d; \quad (1a)$$

$$B_{ex1} = 0, \quad B_{ey1} = \frac{\omega\beta_1 p_{x1}}{2\pi v_0} \frac{e^{-j\beta_0 R_d}}{R_d} \cos\Theta_d,$$

$$B_{ez1} = 0. \quad (1b)$$

At the boundary $z = -d$ between space and a perfect conductor the following general conditions obtain:

$$\epsilon_0 \hat{n}_1 \cdot \mathbf{E}_1 = -\eta_f, \quad (2a)$$

$$v_0 \hat{n}_1 \times \mathbf{B}_1 = -\mathbf{l}_f, \quad (2b)$$

where η_f and $\mathbf{l}_f$ are, respectively, the surface densities of free charge and current in a thin layer at the surface of the conductor. Accordingly, the solutions (1a,b) for the field at once yield the densities of charge and current in the conducting surface. Specifically, at $z = -d$,

$$\eta_f = \epsilon_0 E_{ez1}$$
$$= \frac{\beta_0^2 p_{x1}}{2\pi} \frac{e^{-j\beta_0 R_d}}{R_d} \cos\theta \sin\Theta_d \cos\Theta_d, \quad (3a)$$

$$\mathbf{l}_f = \hat{x} l_x, \quad l_x = \frac{\omega\beta_0 p_{x1}}{2\pi} \frac{e^{-j\beta_0 R_d}}{R_d} \cos\Theta_d. \quad (3b)$$

Although the electromagnetic field of a horizontal antenna of electrical half-length $\beta_0 h \leq 2$ does not differ greatly from that of an infinitesimal dipole or very short antenna at distant points, the field near the antenna is quite different. It is evidently not possible to obtain even an estimate of the distribution of surface currents on a conducting plane under an antenna, for example, of length $2h \doteq \lambda_0/2$, from the currents given in (3b). Fortunately, the theorem of images makes it possible to solve this problem directly using results from the analysis in Chapter V.

The magnetic field due to an isolated antenna of length h with a sinusoidal distribution of current is given by (V.2.14) when the antenna is along the z-axis. If the antenna is along the x-axis as in Fig. 24.1 and the current distribution is $I_x = I_m \sin\beta_0(h - |x|) = I_0 \sin\beta_0(h - |x|)/\sin\beta_0 h$, the only non-vanishing component of the magnetic field is the following:

$$B_{\theta_x} = \frac{jI_m}{4\pi v_0 \sqrt{y^2 + z^2}} (e^{-j\beta_0 R_1} + e^{-j\beta_0 R_2}$$
$$- 2\cos\beta_0 h e^{-j\beta_0 R_0}), \quad (4)$$

where

$$R_2 = \sqrt{(x+h)^2 + y^2 + z^2},$$
$$R_1 = \sqrt{(x-h)^2 + y^2 + z^2},$$
$$R_0 = \sqrt{x^2 + y^2 + z^2}, \quad (5)$$

and θ_x is measured around the x-axis from the negative z-axis. The magnetic field on the conducting plane is obtained from (4) by setting $z = -d$. The component parallel to the plane surface is $B_y = B_{\theta_x} \cos\theta_x = B_{\theta_x} d/\sqrt{y^2 + d^2}$. Hence, the magnetic field on the boundary due to the currents in the primary antenna is

$$(B_y)_d = \frac{jI_m d}{4\pi v_0 (y^2 + d^2)} (e^{-j\beta_0 R_{1d}} + e^{-j\beta_0 R_{2d}}$$
$$- 2\cos\beta_0 h e^{-j\beta_0 R_d}), \quad (6)$$

where

$$R_{2d} = \sqrt{(x+h)^2 + y^2 + d^2},$$
$$R_{1d} = \sqrt{(x-h)^2 + y^2 + d^2},$$
$$R_d = \sqrt{x^2 + y^2 + d^2}. \quad (7)$$

Since the field above a perfectly conducting half-space due to the induced surface currents is the same as that due to an image of the primary antenna with reversed current, the total magnetic field B_y on the boundary plane is *double* $(B_y)_d$ given in (6). Hence, with (2b) the density of surface current on the conducting plane is $\mathbf{l}_f = \hat{x} l_x$, where

$$l_x = v_0 B_y = \frac{jI_m d}{2\pi v_0 (y^2 + d^2)} (e^{-j\beta_0 R_{1d}}$$
$$+ e^{-j\beta_0 R_{2d}} - 2\cos\beta_0 h e^{-j\beta_0 R_d}). \quad (8)$$

It is readily verified that (8) is in agreement with (3b) when the antenna is sufficiently short so that the inequalities $(\beta_0 h)^2 \ll 1$ and $h^2 \ll R_d^2$ are good approximations. In this case,

$$R_{2d} \doteq R_d + \frac{hx}{R_d}, \quad R_{1d} \doteq R_d - \frac{hx}{R_d}. \quad (9)$$

With $I_m = I_0/\sin \beta_0 h$ and (9), (8) becomes

$$I_x = \frac{jI_0 d}{2\pi} \frac{e^{-j\beta_0 R_d}}{y^2 + d^2} \left[\frac{2\cos(\beta_0 hx/R_d) - 2\cos\beta_0 h}{\sin \beta_0 h} \right]. \quad (10)$$

Subject to $\beta_0^2 h^2 \ll 1$ and $h^2 \ll R_d^2$, the trigonometric functions may be expanded. The leading terms give for the bracket $\beta_0 h(1 - x^2/R_d^2) = \beta_0 h(y^2 + d^2)/R_d^2$. Hence, since $d/R_d = -\cos\Theta_d$,

$$I_x \doteq \frac{-j\beta_0 h I_0}{2\pi} \frac{e^{-j\beta_0 R_d}}{R_d} \cos\Theta_d. \quad (11)$$

Since the equivalent dipole moment of an antenna with a current of *uniform* amplitude I_0 over the entire length $2h$ is $p_{x1} = 2hI_0/j\omega$, the equivalent dipole moment of a short antenna of length h but with a current distribution $I_x = I_0(1 - |x|/h)$ is $p_{x1} = hI_0/j\omega$. With this value (11) reduces to (3b).

Since the assumption that the distribution of current in an antenna of length $2h$ is sinusoidal is best when $\beta_0 h = \pi/2$ and since antennas of this electrical half-length are most important in practice, the remainder of this section is devoted to the evaluation of the surface density of current on a conducting sheet under a horizontal half-wave dipole.

With $\beta_0 h = \pi/2$, (8) reduces to

$$I_x = v_0 B_y$$
$$= \frac{jI_m d}{2\pi v_0(y^2 + d^2)} (e^{-j\beta_0 R_{1d}} + e^{-j\beta_0 R_{2d}}). \quad (12)$$

The real instantaneous surface current and magnetic field referred to the maximum instantaneous current are obtained from (12) by setting $I_m = I_m$, multiplying both sides by $e^{j\omega t}$, and then selecting the real part. Thus,

$$(I_x)_{\text{inst}} = v_0(B_y)_{\text{inst}}$$
$$= \frac{-I_m d}{2\pi(y^2 + d^2)} [\sin(\omega t - \beta_0 R_{1d})$$
$$+ \sin(\omega t - \beta_0 R_{2d})]. \quad (13)$$

Using well-known trigonometric identities (13) may be transformed into

$$(I_x)_{\text{inst}} = (H_y)_{\text{inst}} = \frac{-I_m}{2\pi \lambda_0} A \cos(\omega t - \phi), \quad (14a)$$

where the dimensionless amplitude factor A and the phase ϕ are given by

$$A \equiv \frac{2d\lambda_0}{y^2 + d^2} \cos \tfrac{1}{2}\beta_0(R_{2d} - R_{1d}), \quad (14b)$$

$$\phi \equiv \tfrac{1}{2}\beta_0(R_{2d} + R_{1d}) + \frac{\pi}{2}, \quad (14c)$$

and where $\lambda_0 = 2\pi/\beta_0$ is the free-space wavelength; R_{2d} and R_{1d} are given by (7) with $h = \lambda_0/4$.

The amplitude A and the phase ϕ which characterize both the tangential magnetic field and the surface density of current have been computed for several values of d/λ_0. In Figs. 24.2 and 24.3 contours of constant A and contours of constant ϕ are shown in one quadrant for an antenna one-eighth wavelength from the conducting plane. The half-wave dipole is shown schematically and the values of A and ϕ at the point on the surface directly below the center of the dipole are given. Note that the contours of constant amplitude are approximately elliptical with the major axis shifting from the x-axis to the y-axis as the radial distance increases. The contours of constant phase also are elliptical and become very nearly circular within radial distances of two wavelengths.

The instantaneous values of $A \cos(\omega t - \phi)$ for $d = 0.125\lambda_0$ and $t = 0$ and $T/4$ (where T is the period) are shown in Fig. 24.4. They characterize the behavior of the instantaneous magnetic field $(H_y)_{\text{inst}}$ on the surface and the instantaneous surface density of current $(I_x)_{\text{inst}}$. The plus and minus signs in the figures denote the sense of $(H_y)_{\text{inst}}$ and $(I_x)_{\text{inst}}$ with reference to the positive y and x directions, respectively. In each case distances are measured from the point on the surface $(0, 0, -d)$ directly below the origin of coördinates at the center of the antenna. The curves indicate a generally radial, traveling-wave motion which is represented at the instants $t = 0$ and $t = T/4$ by contours of constant amplitude (solid lines) with small numerals to indicate the value of A associated with each contour. The nodal and antinodal (crest) contours are shown by dotted and dashed lines, respectively. Along the loci of antinodes the peak amplitude increases from a minimum along the x-direction on the line $y = 0$, $z = -d$ to a maximum along the y-direction on the line $x = 0$, $z = -d$. These corresponding extreme values are given in small numerals along the coördinate axes in Figs. 24.4a,b. Owing to the decrease in amplitude from the y-direction to the

798   THEORY OF LINEAR ANTENNAS   [VII.24]

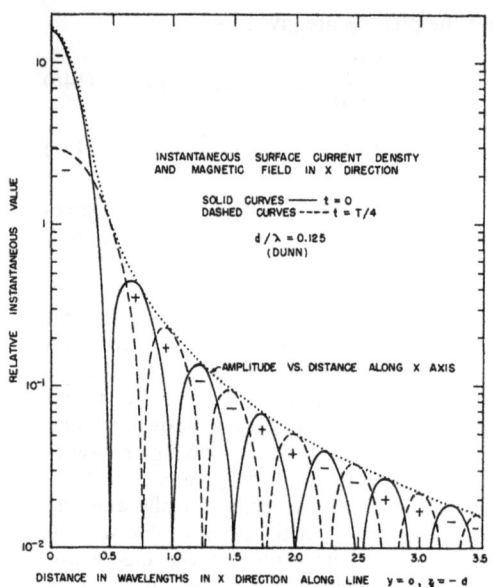

Fig. 24.5. Instantaneous surface current density and magnetic field in x-direction along line $y = 0$, $z = -d = -\lambda_0/8$ (Dunn).

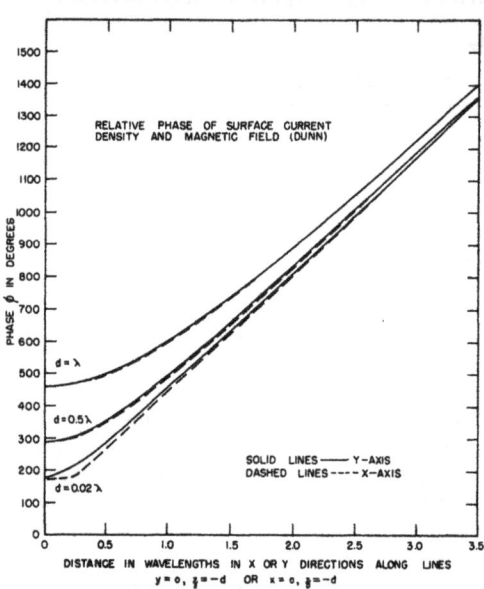

Fig. 24.7. Relative phase of surface current density and magnetic field associated with the amplitudes in Fig. 24.6.

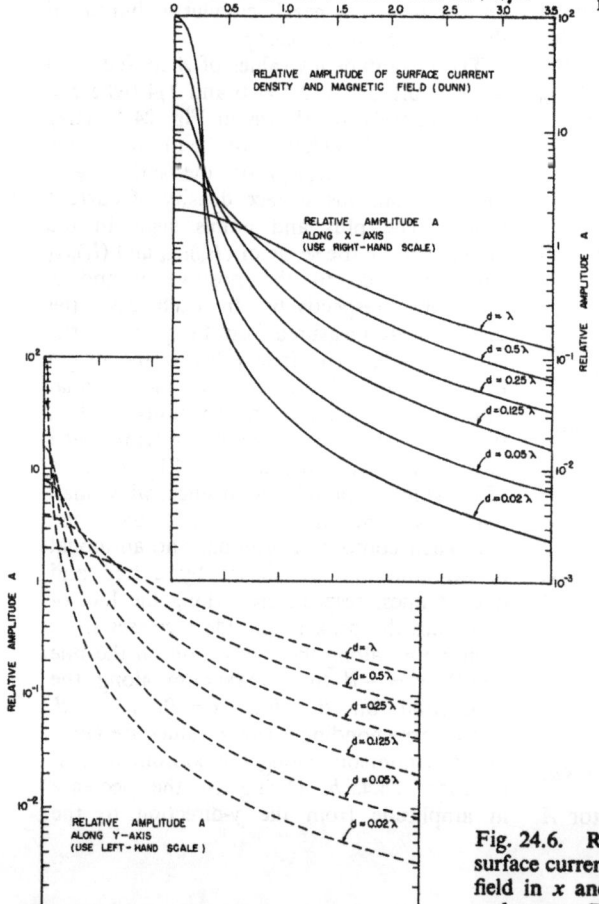

Fig. 24.6. Relative amplitude of surface current density and magnetic field in x and y directions along y and x axes. (Dunn).

x-direction, some of the contours of constant amplitude that begin along the line $x = 0$, $z = -d$ for the y-direction never reach the line $y = 0$, $z = -d$ for the x-direction but form cusps with their apices on a line of crests. Three such pairs of contours are shown in Fig. 24.4a. A graph of the value of $A \cos(\omega t - \phi)$ along the x-direction on the line $y = 0$, $z = -d$ is given in Fig. 24.5. The plus and minus signs have the same meaning as before. The curve that forms the envelope of the crests in Fig. 24.5 is shown dotted. It is simply a graph of the amplitude A in the x-direction along the line $y = 0$, $z = -d$. The same curve appears in Fig. 24.6 together with five similar curves for values of d/λ_0 other than 0.125. Similar curves of A along the y-direction on the line $x = 0$, $z = -d$ are also shown in the figure for the same six values of d/λ_0 with the same input current I_m for all spacings. The relative amplitude A is seen to decrease more rapidly in the y-direction than in the x-direction near the point $(0, 0, -d)$, but the rate of decrease also falls off more rapidly, so that ultimately the decrease in A is much greater in the x-direction outward from the ends of the antenna than in the y-direction perpendicular to it.

The behavior of the relative phase ϕ along the y- and x-directions is shown in Fig. 24.7 for three values of d/λ_0. All curves approach asymptotically from above the line $\phi = (2\pi r/\lambda_0) + \pi/2$, where r is the distance from the point $(0, 0, -d)$ in either the x- or y-direction. The curves approach this asymptote the more rapidly the smaller is d/λ_0. The fact that the slopes of all curves decrease as the central point $(0, 0, -d)$ is approached can be attributed to phase velocities greater than the characteristic velocity $v_0 \doteq 3 \times 10^8$ m/sec. These velocities increase indefinitely as the central point is approached. Phase velocities greater than v_0 near an isolated antenna are discussed in Sec. V.8. The phase velocities $(v_p)_y$ and $(v_p)_x$ in the y- and x-directions from the central point are determined as follows. Beginning with a contour of constant phase defined by

$$\omega t - \phi = \text{const.} \quad (15)$$

it follows that

$$\frac{d\phi}{dt} = \omega. \quad (16)$$

If $d\phi/dt$ is evaluated from (14c) in terms of dy/dt for the y-direction and in terms of dx/dt for the x-direction, and if the expressions so obtained are substituted successively in (16) and this solved for dy/dt and dx/dt, the results are

$$(v_p)_y \equiv \frac{dy}{dt} = v_0 y^{-1} \sqrt{h^2 + y^2 + d^2}$$

along line $x = 0$, $z = -d$, (17)

$$(v_p)_x \equiv \frac{dx}{dt} = 2v_0 \left[\frac{x+h}{\sqrt{(x+h)^2 + d^2}} + \frac{x-h}{\sqrt{(x-h)^2 + d^2}} \right]^{-1}$$

along line $y = 0$, $z = -d$. (18)

Evidently, $(v_p)_y$ and $(v_p)_x$ greatly exceed v_0 for small values of y or x but approach v_0 asymptotically as y or x is increased without limit.

IMPEDANCE AND RADIATION RESISTANCE OF ANTENNAS OVER CONDUCTING PLANES

25. Impedance of Antennas Over Infinite, Perfectly Conducting Planes

The impedance of a base-driven antenna of length h erected vertically on a perfectly conducting plane of infinite extent is one-half the impedance of an isolated center-driven antenna of half-length h which is analyzed in Chapter II. This follows from an application of the theorem of images and is discussed in Sec. II.10.

The impedance of a center-driven antenna of half-length h placed perpendicular to a perfectly conducting infinite sheet with its center at a height $d > h$ above the plane is equal to the impedance of one of two collinear antennas with centers separated by a distance $2d$ and with both antennas center-driven by equal voltages in phase. This also follows from the theorem of images according to which the conducting plane is exactly equivalent to an image antenna with reversed current. As discussed in the opening paragraphs of Sec. III.32, the analysis of two collinear antennas center-driven in phase is complicated by the fact that the feeding transmission lines cannot be in the neutral plane so that unbalanced codirectional currents on a two-wire line or a current on the outside of the sheath of a coaxial line are induced and contribute significantly to the radiation field. Therefore, the impedance of a center-driven antenna over a conducting plane cannot be evaluated accurately without taking account of the line as part of the antenna. However, an estimate of the effect of the conducting plane on the impedance of a center-driven vertical antenna over the plane may be obtained by assuming a dimensionless generator concentrated at the center of the antenna and using zeroth-order self- and mutual impedances. These are fair approximations when $\beta_0 h = \pi/2$, so that this value is chosen conveniently in evaluating

$$Z_1 = R_1 + jX_1 = Z_{s1} + Z_{12}, \qquad (1)$$

with $Z_{s1} = 73.1 + j42.5$ ohms and with Z_{12} taken from Table III.34.1. The values of R_1 and X_1 obtained in this manner are shown in Fig. 25.1. Note that since the input current is also the maximum current when $\beta_0 h = \pi/2$, the zeroth-order resistance R_1 in (1) is the same as the radiation resistance referred to maximum sinusoidal current when ohmic losses are neglected. That is, $R_1 = R_0^e = R_m^e$. For the isolated antenna, $R_m^e = 73.1$ ohms; for the two-element array, Table VI.2.1 gives $R_m^e = 199$ ohms, so that the resistance of one antenna with image is $R_m^e = 99.5$ ohms. Thus, without even evaluating (1) it would have been clear that the resistance of an antenna of half-length h at a height d over a conducting plane must increase from 73.1 for $d = \infty$ to 99.5 for $d = h = \lambda_0/4$. That this occurs with oscillation about 73.1 is shown in Fig. 25.1 (page 806).

As discussed in Sec. III.7, the impedance of a center-driven *horizontal* antenna at a height d over an infinite conducting plane is the antisymmetric impedance of parallel antennas separated by a distance $b = 2d$. This is represented graphically in Sec. III.7. It is seen that in zeroth-order approximation both approach $73.1 + j42.5$ ohms at infinite separation, but that the impedance of a horizontal antenna vanishes at $d = 0$, whereas that for a vertical antenna reaches a maximum value at $d = h$.

26. Radiation Resistance of a Vertical Electric Dipole Over Plane Earth

An analysis of the impedance of an antenna of length h and small radius a at a height $d > h$ over a plane earth with arbitrary conducting and dielectric properties is an intricate problem still awaiting solution. The nearest approach to a determination of the impedance of an antenna is the analysis by Sommerfeld and Renner[55,I.51] of the radiation resistance of an oscillating infinitesimal dipole or doublet with complex electric moment p_{1z} corresponding to a very short end-loaded antenna of length $2h$ with uniform current I given by $p_{1z} = 2hI/j\omega$.

The total time-average power transferred from within a closed surface Σ_τ to the rest of the universe is given by the real part of the complex energy-transfer function

$$T = \int_{\Sigma_\tau} \hat{\mathbf{n}} \cdot \mathbf{S}\, d\sigma, \qquad (1)$$

where $d\sigma$ is an element of any completely closed curface Σ_τ on which $\hat{\mathbf{n}}$ is an external unit normal; $\mathbf{S}$ is the complex Poynting vector defined in space by $\mathbf{S} = \tfrac{1}{2}v_0 \mathbf{E} \times \mathbf{B}^*$, where $\mathbf{B}^*$ is the complex conjugate of the magnetic field $\mathbf{B}$. It is shown in reference I.31, Sec. IV.19, that, if Σ_τ is chosen to be the surface of a cylinder enclosing an antenna of length $2h$ along the z-axis and small radius a, (1) is equivalent to

$$T = -\int_0^h (E_z)_{r=a} I_z^* \, dz, \qquad (2)$$

where I_z^* is the complex conjugate of the total current traversing any cross section of the antenna. If (2) is specialized to a short dipole with a current of uniform complex amplitude I, and if the dipole is infinitely thin, the complex power radiated is

$$T = -I_0^* h E_z, \qquad (3)$$

where E_z is the axial field at the dipole.

If it is assumed that the instantaneous current is

$$(I_z)_{\text{inst}} = I_0 \sin \omega t = \text{I.P. } I_0 e^{j\omega t},$$
$$= \text{R.P. } (-jI_0 e^{j\omega t}) \quad (4)$$

it follows that $I_0^* = jI_0$, so that (3) is equivalent to

$$T = I_0 h(-jE_z). \qquad (5)$$

Hence, the real time-average power radiated by the dipole is

$$\bar{\bar{T}}_r = \text{R.P. } T = \text{R.P. } (-jE_z)I_0 h, \qquad (6)$$

Since E_z in (6) is the electric field at the dipole, that is, at the origin $r \to 0$, $z \to 0$, it is not possible to make use of the quasi-near-zone values in Sec. 13 which have application only at distances from the dipole greater than a wavelength. The exact values of E_z in terms of the polarization potential Π_{ez1} is given in (1.9a) to be

$$E_z = \frac{\partial^2 \Pi_{ez}}{\partial z^2} + \beta_1^2 \Pi_{ez}, \qquad (7)$$

where, from (9.9) and (9.12b),

$$\Pi_{ez} = (\Pi_{ez})_d + (\Pi_{ez})_r$$
$$= \frac{P_{z1}}{4\pi\epsilon_1} \int_0^\infty A(z) J_0(\lambda r) \lambda \, d\lambda. \quad (8a)$$

For temporary convenience the shorthand

$$A(z) = \frac{1}{l}\left[e^{-zl} + \frac{\beta_2^2 l - \beta_1^2 m}{\beta_2^2 l + \beta_1^2 m} e^{-(z+2d)l}\right] \quad (8b)$$

is introduced. As before

$$l = \sqrt{\lambda^2 - \beta_1^2}, \quad m = \sqrt{\lambda^2 - \beta_2^2}.$$

Since at all points in region 1 outside the dipole Π_{ez1} satisfies the homogeneous scalar wave equation (21.10a) and is in addition rotationally symmetric, so that $\partial \Pi_{ez1}/\partial\theta = 0$, it follows that (7) is equivalent to

$$E_z = -\frac{1}{r}\frac{\partial}{\partial r}\left(r\frac{\partial \Pi_{e1z}}{\partial r}\right). \quad (9)$$

The differentiatiation of (8a) is carried out readily as follows:

$$r\frac{\partial \Pi_{ez1}}{\partial r} = \frac{P_{z1}}{4\pi\epsilon_1}\int_0^\infty A(z) J_0'(\lambda r)\lambda^2 r\, d\lambda, \quad (10a)$$

$$\frac{1}{r}\frac{\partial}{\partial r}\left(r\frac{\Pi_{ez1}}{\partial r}\right) = \frac{P_{z1}}{4\pi\epsilon_1}\int_0^\infty A(z)\left[J_0''(\lambda r)\right.$$
$$\left. + \frac{J_0'(\lambda r)}{\lambda r}\right]\lambda^3 d\lambda. \quad (10b)$$

If use is made of the Bessel differential equation

$$J_0''(\lambda r) = -\left[J_0(\lambda r) + \frac{J_0'(\lambda r)}{\lambda r}\right], \quad (11)$$

it follows directly that

$$E_z = \frac{P_{z1}}{4\pi\epsilon_1}\int_0^\infty A(z) J_0(\lambda r)\lambda^3 d\lambda. \quad (12)$$

If the moment of the dipole is expressed in terms of an equivalent uniform current of complex amplitude $I_0 = -jI_0$ as in (4), that is, by

$$P_{z1} = 2hI_0/j\omega = -2hI_0/\omega, \quad (13)$$

where I_0 is real, it follows that for use in (12)

$$\text{R.P.}(-jE_z) =$$
$$\left(\frac{hI_0}{2\pi\epsilon_1\omega}\right)\text{R.P.}\left(j\int_0^\infty A(z)J_0(\lambda r)\lambda^3 d\lambda\right). \quad (14)$$

At the dipole, $r \doteq 0$, so that (14) becomes

$$\text{R.P.}(-jE_z) = \frac{hI_0}{2\pi\epsilon_1\omega}\text{R.P.}\left(j\int_0^\infty A(z)\lambda^3 d\lambda\right). \quad (15)$$

Although the variable of integration λ may be complex in general, it becomes real if the path of integration in the complex λ-plane is chosen along the real axis. This is assumed in (15), where the real variable λ is substituted for the complex variable λ.

The integral (15) may be considered in two parts:

$$A(z) = A_1(z) + A_2(z), \quad (16)$$

where

$$A_1(z) = \frac{1}{l}[e^{-zl} + e^{-(z+2d)l}],$$

$$A_2(z) = \frac{-2\beta_1^2 m}{l(\beta_2^2 l + \beta_1^2 m)}e^{-(z+2d)}. \quad (17)$$

In evaluating the first integral obtained by substituting (16) in (15), the range of integration may be expressed as follows:

$$j\int_0^\infty A_1(z)\lambda^3 d\lambda$$
$$= j\int_0^{\beta_1} A_1(z)\lambda^3 d\lambda + j\int_{\beta_1}^\infty A_1(z)\lambda^3 d\lambda. \quad (18)$$

In the second integral on the right in (18) $l = \sqrt{\lambda^2 - \beta_1^2}$ is real. Therefore, from (17), $A_1(z)$ and with it the second integral is also real. It follows that there can be no contribution to the *real part* of (18) by the purely imaginary second term on the right in (18). Hence,

$$\text{R.P.}\int_0^\infty jA_1(z)\lambda^3 d\lambda = \text{R.P.}\int_0^{\beta_1} jA_1(z)\lambda^3 d\lambda. \quad (19)$$

It is now proper to set $z \doteq 0$ to obtain the contribution to the field at the dipole. The remaining integral is

$$\text{R.P.}\int_0^\infty jA_1(0)\lambda^3 d\lambda$$
$$= \text{R.P.}\int_0^{\beta_1} j(1 + e^{-2dl})(\lambda^3/l)d\lambda. \quad (20)$$

Let the variable of integration be changed to l:

$$\text{R.P.} \int_0^\infty jA_1(0)\lambda^3 d\lambda$$

$$= \text{R.P.} \int_0^{j\beta_1} -j(1 + e^{-2dl})(l^2 + \beta_1^2)\, dl. \quad (21)$$

The integration in (21) involves only tabulated integrals (see, for example, Peirce, ref. I.34, formulas 401, 402, 403). The result is

$$\text{R.P.} \int_0^\infty jA_1(0)\lambda^3 d\lambda$$

$$= 2\beta_1^3 \left[\frac{1}{3} + \frac{\sin 2\beta_1 d - 2\beta_1 d \cos 2\beta_1 d}{(2\beta_1 d)^3} \right]. \quad (22)$$

After the substitution of (22) in (18), (18) in (15), and (15) in (6), and the introduction of $\zeta_1 = 1/\sqrt{\nu_1 \epsilon_1} = \beta_1/\omega\epsilon_1$, the time-average radiated power is

$$\overline{\overline{T}}_r = \frac{\zeta_1 I_0^2 h^2 \beta_1^2}{2\pi}$$

$$\times \left\{ \frac{2}{3} + 2 \left[\frac{\sin 2\beta_1 d - 2\beta_1 d \cos 2\beta_1 d}{(2\beta_1 d)^3} \right] + K \right\}, \quad (23)$$

where

$$K = \text{R.P.} \left(\frac{-j}{\beta_1} \int_0^\infty \frac{2m}{l(\beta_2^2 l + \beta_1^2 m)} e^{-2dl} \lambda^3 d\lambda \right). \quad (24)$$

Use of the definition (I.10.5) for the radiation resistance referred to a specified current gives

$$R_0^e \equiv \frac{2\overline{\overline{T}}_r}{I_0^2} = \frac{\zeta_1 \beta_1^2 h^2}{\pi}$$

$$\times \left\{ \frac{2}{3} + 2 \left[\frac{\sin 2\beta_1 d - 2\beta_1 d \cos 2\beta_1 d}{(2b_1 d)^3} \right] + K \right\}, \quad (25)$$

where for a dipole in air $\zeta_1 \doteq \zeta_0 \doteq 120\pi$ ohms and $\beta_1 = \beta_0$.

The final determination of R_0^e in (25) depends upon the evaluation of the integral (24) for K. Following Sommerfeld, this is carried out subject to the simplifying condition

$$|N_{21}^2| = \left| \frac{\beta_2^2}{\beta_1^2} \right| \gg 1, \quad (26)$$

which is imposed throughout the analysis of the quasi-near-zone fields of vertical and horizontal electric dipoles earlier in this chapter.

As a first step in the integration, the integrand may be rearranged using $l^2 = \lambda^2 - \beta_1^2$ and $m^2 = \lambda^2 - \beta_2^2 = l^2 + (\beta_1^2 - \beta_2^2)$. Thus,

$$\frac{m}{\beta_2^2 l + \beta_1^2 m} = \frac{m(\beta_2^2 l - \beta_1^2 m)}{\beta_2^4 l^2 - \beta_1^4 m^2}$$

$$= \left(\frac{\beta_1^2}{\beta_1^4 - \beta_2^4} \right) \left[\frac{m^2 - ml\beta_2^2/\beta_1^2}{l^2 + \beta_1^4/(\beta_1^2 + \beta_2^2)} \right]. \quad (27)$$

In the integration in (24), the variable λ goes from zero to infinity. In the range $\lambda^2 < \beta_1^2$, the exponential e^{-2dl} in (24) has an imaginary exponent and hence a magnitude of unity. Alternatively, in the range $\lambda^2 > \beta_1^2$, e^{-2dl} has a real and negative exponent so that it decreases very rapidly as λ exceeds β_1. Therefore, in the integration, λ may be treated as of the order of magnitude of β_1, since contributions to the integral by the integrand in (24) are very small when λ exceeds β_1. Accordingly, a satisfactory first approximation is to set $m^2 \doteq -\beta_2^2$ or $m \doteq j\beta_2$ in (27). With (26), terms of order of magnitude $|\beta_1^2/\beta_2^2|$ and smaller may be neglected. With these approximations,

$$\frac{m}{\beta_2^2 l + \beta_1^2 m} \doteq \frac{j}{\beta_1 \beta_2} \left(\frac{u - j\beta_1/\beta_2}{u^2 + \beta_1^2/\beta_2^2} \right)$$

$$= \frac{j}{\beta_1 \beta_2} \left(\frac{1}{u + j\beta_1/\beta_2} \right), \quad (28)$$

where $u \equiv l/\beta_1$. Since with $\lambda^2 = l^2 + \beta_1^2 = \beta_1^2(1 + u^2)$ it follows that $(\lambda^3/l)\, d\lambda = (l^2 + \beta_1^2)\, dl = \beta_1^3(1 + u^2)\, du$, it follows that (24) is equal to

$$K = \text{R.P.} \int_j^\infty \frac{2\beta_1}{\beta_2} \left[\frac{u^2 + 1}{u + j\beta_1/\beta_2} \right] e^{-2d\beta_1 u}\, du. \quad (29a)$$

The leading first-order terms are

$$K \doteq \text{R.P.} \int_j^\infty \frac{2\beta_1}{\beta_2} \left(u + \frac{1}{u} \right) e^{-2d\beta_1 u}\, du. \quad (29b)$$

Let

$$K = \text{R.P.} \frac{\beta_1}{\beta_2} [L' + 2G] = L + \text{R.P.} \frac{2\beta_1}{\beta_2} G, \quad (30)$$

where

$$L' \equiv 2\int_0^\infty e^{-2d\beta_1 u}u\, du$$

$$= 2\left(\frac{1+j2\beta_1 d}{4\beta_1^2 d^2}\right)e^{-j2\beta_1 d}, \quad (31a)$$

$$L \equiv \text{R.P.}(\beta_1/\beta_2)L' \quad (31b)$$

$$G \equiv \int_j^\infty \frac{e^{-2d\beta_1 u}}{u}\, du$$

$$= -[Ci(2\beta_1 d) - j\, Si(2\beta_1 d) + j\pi/2]. \quad (32)$$

Using (7.3b) and reference I.31, Appendix II, together with the notation $h_{e2} \equiv \sigma_{e2}/\omega\epsilon_{e2}$,

$$\frac{\beta_2}{\beta_1} = \frac{\beta_2}{\beta_1}e^{-j\rho} = \sqrt{\epsilon_{e2r}(1-jh_{e2})}$$

$$= \sqrt{\epsilon_{e2r}}[f(h_{e2}) - jg(h_{e2})], \quad (33a)$$

where
$$\rho = \tan^{-1}[g(h_{e2})/f(h_{e2})], \quad (33b)$$

so that*

$$L = \frac{2\beta_1}{\beta_2}\left[\frac{\cos(2\beta_1 d - \rho) + (2\beta_1 d)\sin(2\beta_1 d - \rho)}{(2\beta_1 d)^2}\right]. \quad (34)$$

With (32), (35a), and (34) substituted in (30), the final expression for K is found to be

$$K = \frac{2\beta_1}{\beta_2}\left\{\frac{\cos(2\beta_1 d - \rho) + 2\beta_1 d\sin(2\beta_1 d - \rho)}{(2\beta_1 d)^2}\right.$$

$$\left. - \cos\rho\, Ci(2\beta_1 d) - \sin\rho[Si(2\beta_1 d) - \pi/2]\right\}. \quad (35)$$

When (35) is substituted in (25), the final first-order formula for the radiation resistance of a vertical electric dipole (or short, end-loaded antenna of length $2h$ with uniform current I_0) over a plane earth that is a good but not necessarily perfect conductor is obtained. It is important to bear in mind that the approximations made in evaluating K from (24) make the solution quantitatively accurate only for quite large values of $|N_{21}|$. In the case of a short antenna which is not end-loaded, so that the current decreases

---
* By retaining additional terms in the evaluation of K in (29a) that have the factor β_1^2/β_2^2, Sommerfeld and Renner obtain a "second-order" approximation. Owing to the approximations involved in the transformation of (24) into (29a), it is doubtful whether the accuracy is actually increased.

uniformly from I_0 at the center to zero at the ends, the average current is $I_0/2$, so that the radiation resistance R_0^e is one-quarter of (25). Note that a "short" antenna, whether end-loaded or not, is defined by $\beta_0^2 h^2 \ll 1$.

By allowing the height d of the dipole above the conducting plane to become infinite, the radiation resistance of an isolated dipole or short, end-loaded antenna is obtained. For the end-loaded short antenna of length $2h$ in air it is

$$R_0^e(d=\infty) = \frac{2\zeta_0 \beta_0^2 h^2}{3\pi} = 80\beta_0^2 h^2 \text{ ohms}, \quad (36a)$$

in complete agreement with the value obtained in reference I.31, Sec. IV.17. For the non-end-loaded isolated antenna of length $2h$ the radiation resistance is

$$R_0^e(d=\infty) = 20\beta_0^2 h^2 \text{ ohms}, \quad (36b)$$

in agreement with (II.31.6b).

If the short antenna has its center at a height $d > h$ over a *perfectly* conducting earth ($\sigma_{e2} = \infty$, $\beta_2 = \infty$), and if the physically meaningless case $d = 0$ is excluded, it follows from (34) that $K = 0$. Hence, for the end-loaded antenna,

$$R_0^e(\sigma_{e2} = \infty)$$
$$= 80\beta_0^2 h^2 + \frac{30\beta_0^2 h^2}{\beta_0^3 d^3}(\sin 2\beta_0 d - 2\beta_0 d \cos 2\beta_0 d)$$
$$\text{ohms}. \quad (37)$$

For the non-end-loaded antenna the value is one-quarter of $R_0^e(\sigma_{e2} = \infty)$ in (37). Near the conducting plane, where $\beta_0 d$ is sufficiently small so that the trigonometric functions may be expanded in powers of $2\beta_0 d$, the leading terms give for (37)

$(2\beta_0 d)^2 \ll 1$:

$$R_0^e(\sigma_{e2} = \infty) \doteq 160\beta_0^2 h^2 \text{ ohms}. \quad (38)$$

Thus, the radiation resistance of a short vertical antenna very near but not in contact with a horizontal, perfectly conducting plane is double the value of the same antenna when isolated. Evidently, this conclusion is in agreement with the fact that the radiation resistance of a base-driven antenna of length h erected on a perfectly conducting plane is one-half the radiation resistance of an isolated, center-driven antenna of length $2h$.

The behavior of R_0^e for small values of d when σ_{e2} is finite is of interest. Since the first two terms in (25) reduce to (38) when $(2\beta_0 d)^2 \ll 1$, it remains to examine the behavior of K. Using the expansions for small argument of the trigonometric functions and the integral functions, it is shown readily that when $4\beta_0^2 d^2 \ll 1$,

$$K \doteq \frac{2\beta_1}{\beta_2}\left\{\left[\left(\frac{1}{2\beta_0 d}\right)^2 + \frac{1}{2} - C - \ln 2\beta_0 d\right]\right.$$
$$\left. \times \cos\rho - \frac{\pi}{2}\sin\rho\right\}. \quad (39)$$

Clearly, as $\beta_0 d$ is reduced, K increases to large values. The radiation resistance R_0^e as computed by Sommerfeld and Renner for a vertical electric dipole or short, end-loaded antenna of length $2h$ over a conducting half-space is shown in Fig. 26.1. Curves are shown for three values of $\sigma_{e2}/\omega\epsilon_0$ that satisfy the condition $\sigma_{e2}/\omega\epsilon_{e2} \gg 1$ characteristic of a good conductor so that

$$\beta_2/\beta_1 = \sqrt{\epsilon_{e2r} - j\sigma_{e2}/\omega\epsilon_0} \doteq \sqrt{-j\sigma_{e2}/\omega\epsilon_0}$$
$$= (1-j)\sqrt{\sigma_{e2}/2\omega\epsilon_0}. \quad (40)$$

The values selected are $|\beta_2/\beta_1|^2 \doteq \sigma_{e2}/\omega\epsilon_0 = \infty, 10^4, 10^2$.

Since no energy is dissipated as heat in a perfect conductor, the difference between the curve $\sigma_{e2}/\omega\epsilon_0 = \infty$ and each of the other two is indicative of the heat losses in the imperfectly conducting region 2. These losses increase as the dipole approaches the boundary, but decrease to negligible values when the center of the antenna is higher than about a quarter wavelength. The curve for a very short antenna over a perfect conductor is extended to $d/\lambda_0 \doteq 0$ in Fig. 26.1. It is not physically meaningful for d to become less than h, but since h is a very small fraction of a wavelength, the complete curve with $d \doteq 0$ is of interest in comparing the perfect conductor and the good conductor.

In Fig. 26.1 is also shown the zeroth-order radiation resistance of a very thin vertical half-wave dipole ($\beta_0 h = \pi/2$) with sinusoidal current over a perfect conductor, as taken from Fig. 25.1. It is significant that this curve is essentially the same as the corresponding one in Fig. 26.1 for a short end-loaded antenna over a perfect conductor, except that it oscillates about and approaches the value 73.1 ohms instead of 80 ohms as d/λ_0 increases to infinity and, of course, is not defined for d/λ_0 less than 0.25 since the lower end of the antenna then makes contact with the earth. Since a decrease in the conductivity of the earth from a perfect conductor to a good conductor satisfying the inequality $\sigma_{e2}/\omega\epsilon_0 \gg 1$ has no effect on R_0^e for the short end-loaded antenna unless d/λ_0 is less than 0.25, it is evident that R_0^e for an antenna with electrical half-length $\beta_0 h \geq \pi/2$ cannot be affected significantly by such a decrease in the conductivity of the earth. Hence, it may be assumed that the radiation resistance of a vertical antenna of half-length $h \geq \lambda_0/4$ is the same when the antenna is at any height d over a good conductor as when it is at the same height over a perfect conductor. This is true for all values of $d > h$.

27. Radiation Resistance of a Horizontal Electric Dipole Over a Plane Earth

The total power radiated from a horizontal electric dipole with moment p_{x1} or from an equivalent short, end-loaded antenna of length $2h$ with uniform current I_0 such that $p_{x1} = 2hI_0/j\omega$ is given by (26.6) with E_x substituted for E_z. The desired formula is

$$\bar{\bar{T}}_r = \text{R.P.} (-jE_x)I_0 h, \quad (1)$$

where E_{x1} is expressed in terms of the polarization potential $\mathbf{\Pi}_e = \hat{x}\Pi_{ex} + \hat{z}\Pi_{ez}$ in (1.15a) as follows:

$$E_x = \beta_1^2 \Pi_{ex} + \frac{\partial^2 \Pi_{ex}}{\partial x^2} + \frac{\partial^2 \Pi_{ez}}{\partial x \partial z}. \quad (2)$$

The components of the polarization potential $\Pi_{ex} = (\Pi_{ex})_d + (\Pi_{ex})_r$ and Π_{ez} are given in (19.30), (19.31a), and (19.33). Thus, with

$$p_{x1} = 2hI_0/j\omega = -2hI_0/\omega, \quad (3)$$

$$\left(\frac{\partial^2}{\partial x^2} + \beta_1^2\right)\Pi_{ex} = -\left(\frac{hI_0}{2\pi\epsilon_1 \omega}\right)$$
$$\times \int_0^\infty \frac{1}{l}\left[e^{-zl} + \frac{l-m}{l+m}e^{-(z+2d)l}\right]$$
$$\times \left(\frac{\partial^2}{\partial x^2} + \beta_1^2\right)J_0(\lambda r)\lambda \, d\lambda, \quad (4a)$$

$$\frac{\partial^2 \Pi_{ez}}{\partial x \partial z} = -\left(\frac{hI_0}{2\pi\epsilon_1 \omega}\right) \times$$
$$\int_0^\infty \frac{2l(l-m)}{\beta_2^2 l + \beta_1^2 m} e^{-(z+2d)l} \frac{\partial}{\partial x}[\cos\theta \, J_1(\lambda r)]\lambda^2 d\lambda. \quad (4b)$$

Since these functions are to be evaluated only at $r = 0$, it is convenient to expand the Bessel functions in powers of the small quantity λr with $r = \sqrt{x^2 + y^2}$, differentiate as required in (4a) and (4b), and then set $r = 0$. Thus,

$$\left(\frac{\partial^2}{\partial x^2} + \beta_0^2\right) J_0(\lambda\sqrt{x^2+y^2})$$

$$\doteq \left(\frac{\partial^2}{\partial x^2} + \beta_1^2\right)[1 - \tfrac{1}{4}\lambda^2(x^2+y^2)]$$

$$= \beta_1^2(1 - \tfrac{1}{4}\lambda^2 r^2) - \tfrac{1}{2}\lambda^2$$

$$= \beta_1^2 - \tfrac{1}{2}\lambda^2 \text{ at } r = 0, \qquad (5a)$$

$$\frac{\partial}{\partial x}[\cos\theta \, J_1(\lambda\sqrt{x^2+y^2})]$$

$$\doteq \frac{\partial}{\partial x}\left(\frac{x}{\sqrt{x^2+y^2}} \frac{\lambda\sqrt{x^2+y^2}}{2}\right) = \frac{\lambda}{2}. \qquad (5b)$$

Substitution of (5a) and (5b) in (4a) and (4b) to form (2) gives

$$E_x = -\frac{hI_0}{2\pi\epsilon_1\omega} \int_0^\infty \left[\frac{1}{2l}(2\beta_1^2 - \lambda^2)\right.$$

$$\times \left(e^{-2l} + \frac{l-m}{l+m}e^{-(z+2d)l}\right)$$

$$\left. + \frac{l(l-m)\lambda^2}{\beta_2^2 l + \beta_1^2 m} e^{-(z+2d)l}\right] \lambda \, d\lambda. \qquad (6)$$

For use in (1) only the real part of $-jE_x$ is required. With λ replaced by $\bar{\lambda}$, this is

R.P. $(-jE_x)$

$$= \frac{hI_0}{2\pi\epsilon_1\omega} \text{R.P.} \left[\int_0^\infty \frac{j(2\beta_1^2 - \lambda^2)}{2} D_1(z)\lambda \, d\lambda \right.$$

$$\left. + \int_0^\infty jD_2(z)\lambda \, d\lambda\right], \qquad (7)$$

where

$$D_1(z) = \frac{1}{l}[e^{-zl} - e^{-(z+2d)l}], \qquad (8a)$$

$$D_2(z) = \left[\frac{2\beta_1^2 - \lambda^2}{l+m} + \frac{l(l-m)\lambda^2}{\beta_2^2 l + \beta_1^2 m}\right] e^{-(z+2d)l}. \qquad (8b)$$

By putting the bracket in (8b) over a common denominator, using the identity $\beta_2^2 - \beta_1^2 = l^2 - m^2$, and arranging the numerator in the form $\beta_1^2[l(2\beta_2^2 - \lambda^2) + m(2\beta_1^2 - \lambda^2)]$, using $l^2 = \lambda^2 - \beta_1^2$ and $m^2 = \lambda^2 - \beta_2^2$, the following simple expression is obtained:

$$D_2(z) = \frac{\beta_1^2(\lambda^2 - 2ml)}{\beta_2^2 l + \beta_1^2 m} e^{-(z+2d)l}. \qquad (8c)$$

The first integral in (7) is similar to (26.18) in that there are no contributions to its real part when λ is greater than β_1. Hence,

R.P. $\int_0^\infty \frac{j(2\beta_1^2 - \lambda^2)}{2} D_1(z)\lambda \, d\lambda$

$$= \text{R.P.} \int_0^{\beta_1} \frac{j(2\beta_1^2 - \lambda^2)}{2} D_1(z)\lambda \, d\lambda. \qquad (9a)$$

On the dipole, $z \doteq 0$, so that (9a) becomes

R.P. $\int_0^{\beta_1} \frac{j(2\beta_1^2 - \lambda^2)}{2} D_1(0)\lambda \, d\lambda$

$$= \text{R.P.} \int_0^{j\beta_1} \frac{-j(\beta_1^2 - l^2)}{2}(1 - e^{-2dl}) \, dl. \qquad (9b)$$

The integration of (9b) is straightforward. It gives

R.P. $\int_0^{j\beta_1} \frac{-j(\beta_1^2 - l^2)}{2}(1 - e^{-2dl}) \, dl$

$$= \beta_1^3 \left[\frac{2}{3} - \frac{\sin 2\beta_1 d}{2\beta_1 d}\right.$$

$$\left. + \frac{\sin 2\beta_1 d - 2\beta_1 d \cos 2\beta_1 d}{(2\beta_1 d)^3}\right]. \qquad (9c)$$

Before evaluating the second integral in (7) it is convenient to simplify $D_2(z)$ by taking advantage of the assumed condition (26.26), namely, $|\beta_2^2/\beta_1^2| \gg 1$. Also, as explained in conjunction with (26.27), λ^2 may be neglected compared with $2ml$ in (8c), since the exponential factor $e^{-\sqrt{\lambda^2 - \beta_1^2}}$ makes contributions to the integral negligible as soon as λ exceeds β_1 appreciably. Moreover, in the integration from $\lambda = 0$ to $\lambda = \beta_1$, $2ml$ has a greater average value than λ^2. With these simplifications and approximations,

$$D_2(0) \doteq \frac{-j2\beta_1^2}{\beta_2} e^{-2dl}, \qquad (10)$$

so that the second integral in (7) becomes

R.P. $\int_0^\infty jD_2(0)\lambda \, d\lambda \doteq \text{R.P.} \frac{2\beta_1^2}{\beta_2} \int_0^\infty e^{-2dl}\lambda \, d\lambda$

$$= \text{R.P.} \frac{2\beta_1}{\beta_2} \int_j^\infty e^{-2d\beta_1 u} u \, du = \beta_1^3 L, \qquad (11)$$

where L is defined and evaluated in (26.30), (26.31), and (26.34).

Upon combining (9c) with (11) in (7), the final formula for the real part of $(-jE_x)$ at the dipole is derived. If this is then

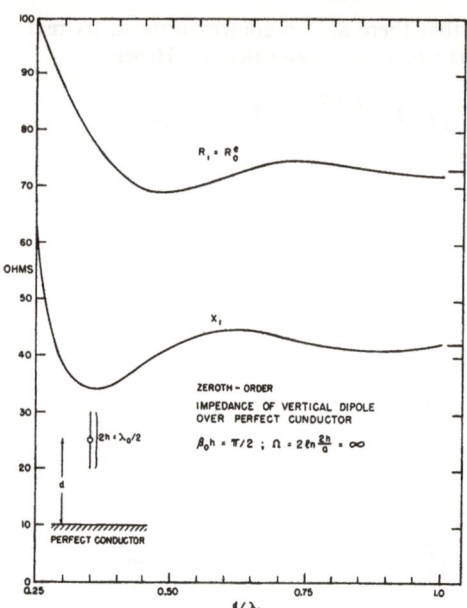

Fig. 25.1. Zeroth-order impedance of vertical dipole over perfect conductor.

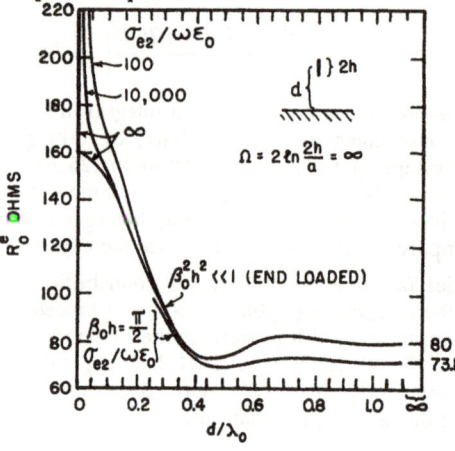

Fig. 26.1. Radiation resistance for vertical dipole over conducting earth (Sommerfeld and Renner).

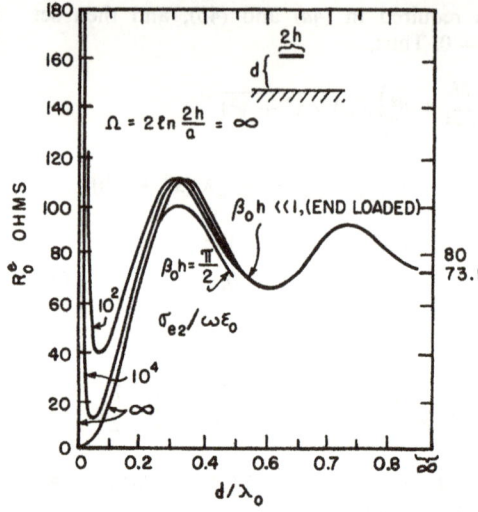

Fig. 27.1. Radiation resistance of horizontal dipole over conducting earth (Sommerfeld and Renner).

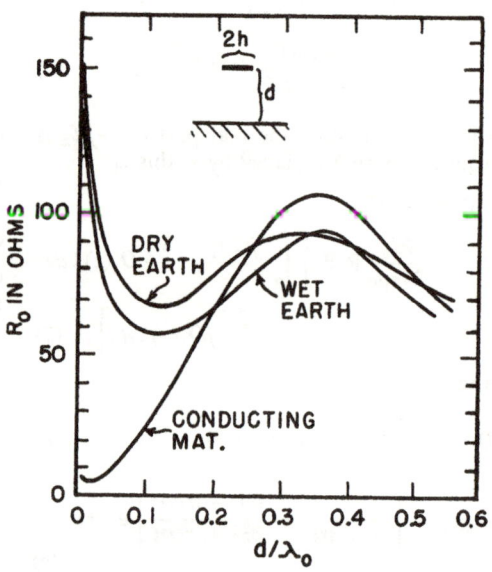

Fig. 27.2. Measured resistance of horizontal resonant antenna over conducting earth (Proctor).

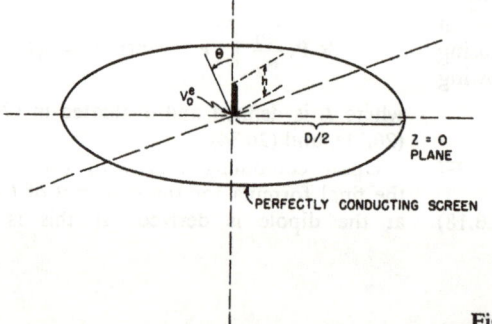

Fig. 28.1. Antenna over ground screen.

substituted in (1), the time-average radiated power $\overline{\overline{T}}_r$ is obtained. This may be substituted in $R_0^e = 2\overline{\overline{T}}_r/I_0^2$ to obtain the radiation resistance of a horizontal end-loaded, short dipole over a plane earth that is a good conductor as defined by $\sigma_{e2}/\omega\epsilon_0 \gg 1$. The final first-order expression, which is valid only when $|N_{21}|$ is quite large, is

$$R_0^e = \frac{2\overline{\overline{T}}_r}{I_0^2} \doteq \frac{\zeta_1 \beta_1^2 h^2}{\pi}\left[\frac{2}{3} - \frac{\sin 2\beta_1 d}{2\beta_1 d}\right.$$
$$\left. + \frac{\sin 2\beta_1 d - 2\beta_1 d \cos 2\beta_1 d}{(2\beta_1 d)^3} + L\right], \quad (12)$$

where L is given in (26.34) and where for air $\zeta_1/\pi = \zeta_0/\pi \doteq 120$ ohms. If the short antenna is not end-loaded but has vanishing currents at the ends, the radiation resistance is one quarter of that given in (12).

When the height d of the horizontal dipole over a conducting plane becomes infinite, (12) reduces to (26.36a) for the isolated dipole.

If the earth is made *perfectly conducting*, so that $\sigma_{e2} = \infty$, $\beta_2 = \infty$, $L = 0$, the resistance of the end-loaded dipole in air is:

$$R_0^e \doteq 80\beta_0^2 h^2 - 120\beta_0^2 h^2 \left[\frac{\sin 2\beta_0 d}{2\beta_0 d}\right.$$
$$\left. - \frac{\sin 2\beta_0 d - 2\beta_0 d \cos 2\beta_0 d}{(2\beta_0 d)^3}\right] \text{ ohms.} \quad (13)$$

If the horizontal dipole is very close to the surface of the perfectly conducting plane, the trigonometric functions can be expanded in powers of $2\beta_0 d$. For the short end-loaded horizontal antenna with uniform current that satisfies $(2\beta_0 d)^2 \ll 1$, this gives

$$R_0^e \doteq 64\beta_0^4 h^2 d^2 \text{ ohms.} \quad (14a)$$

Note that this vanishes as d approaches zero.

For an end-loaded, electrically short section of two-wire line with uniform current, spacing $b = 2d$, and length $s = 2h$, R_0^e in ohms is

$$R_0^e \doteq 8\beta_0^4 s^2 b^2 = 32\pi^2 (s/\lambda_0)^2 (2\pi b/\lambda_0)^2. \quad (14b)$$

If the earth is *not* perfectly conducting and d is reduced, there is a contribution to R_0^e from L in (12) which must be added to (14). From (26.34) and with $(2\beta_0 d)^2 \ll 1$ this is

$$L \doteq \frac{2\beta_1}{\beta_2} \cos \rho \left[\frac{1}{(2\beta_1 d)^2} + \frac{1}{2}\right], \quad (15)$$

where, for a good conductor,

$$\rho \doteq \tan^{-1}(\sigma_{e2}/2\omega\epsilon_{e2}) \doteq \tfrac{1}{4}\pi.$$

Clearly, L increases rapidly and to very great values as d decreases to its smallest physically meaningful magnitude $d \doteq a$, the radius of the wire. Hence, the radiation resistance in its general form (12) for a short end-loaded horizontal antenna over a good conductor must increase rapidly and greatly when $2\beta_1 d$ becomes less than unity.

Curves of R_0^e due to Sommerfeld and Renner and evaluated from the first-order formula (12) with (26.34) for the horizontal electric dipole (or short, end-loaded antenna with electrical half-length $\beta_0^2 h^2 \ll 1$) over a conducting half-space are shown in Fig. 27.1 for $|\beta_2/\beta_1|^2 \doteq \sigma_{e2}/\omega\epsilon_0 = \infty$, 10^4, 10^2. Note that the vertical axis is a part of the curve for a perfectly conducting plane in the theoretical limit $d \to 0$. As d is reduced, R_0^e reaches a minimum with $d/\lambda_0 \sim 0.05$, then rises to large values as d/λ_0 is decreased to its physically available minimum $d = a$, where a is the radius of the conductor. As d/λ_0 is increased, R_0^e oscillates about the value 80 for the isolated dipole. Figure 27.1 also includes the value of R_0^e for a horizontal half-wave dipole ($\beta_0 h = \pi/2$) over a *perfectly* conducting plane. If ohmic losses in the antenna are neglected, this value is equal to the resistance of one of two antisymmetrically driven antennas as obtained from Table III.7.2. The values of R_0^e for a short end-loaded antenna and a half-wave dipole differ very little when each is over a *perfect* conductor and d/λ_0 is less than about 0.25. For greater values of d/λ_0, R_0^e for the half-wave dipole executes damped oscillations about the value 73.1 ohms characteristic of the isolated infinitely thin half-wave antenna, whereas R_0^e for the short, end-loaded dipole in a similar manner oscillates about its limiting value of 80 ohms for $d \to \infty$. Since the horizontal half-wave antenna and short, end-loaded antenna have essentially the same radiation resistance when $d/\lambda_0 \lesssim 0.25$ and $\sigma_{e2} = \infty$, it is reasonable to expect that the same is true over this range when σ_{e2} characterizes a *good* conductor. This means that for $d/\lambda_0 \lesssim 0.25$ the three curves for the short, end-loaded antenna are approximately correct for a thin half-wave dipole. As a horizontal antenna is brought closer and closer to an imperfectly or perfectly conducting earth, its radiation resistance decreases, approaching zero with $d \to a$. The resulting larger currents in the antenna induce larger currents in the earth, so that dissipation as heat increases if the earth is not a perfect conductor. The power $\tfrac{1}{2} I_0^2 R_0^e$ that is radiated from the antenna to the rest of the universe is transferred in greater and greater proportion to the imperfect earth as this is brought nearer.

Experimental curves of the input resistance of a resonant antenna over various types of earth as obtained by Proctor[51] are shown in Fig. 27.2. These are seen to be in general agreement with the theoretical curves for the short, end-loaded antenna in Fig. 27.1.

28. Impedance of Base-Driven Antenna on a Ground Screen of Finite Size*

According to the theorem of images, the impedance Z_0 of a base-driven antenna of length h and radius a erected vertically on a perfectly conducting *infinite* plane is one-half the impedance of an isolated center-driven antenna of *half*-length h and radius a. In the actual measurement of the impedance of such a base-driven antenna, using, for example, circuits and methods described in Secs. I.35 and I.36, the ground plane is large compared with the length of the antenna and the wavelength, but *not infinite*. In this section a formula due to Storer is derived which gives the difference $Z - Z_0$ between the impedance Z of an antenna over a circular, perfectly conducting disk of large but finite diameter D, as shown in Fig. 28.1, and the impedance Z_0 of the same antenna when D is infinite.

Since it is shown theoretically in Sec. II.10 and experimentally in Sec. II.38 that the apparent impedance Z_{sa} of an actual antenna driven from a coaxial line in the conventional manner shown schematically in Fig. II.10.6 can be determined from the ideal impedance Z_0 of an antenna driven by a discontinuity in scalar potential by using an appropriate terminal-zone network, it is sufficient to determine the effect on Z_0 of reducing an infinite ground screen to a circle of diameter D. Consider first a perfectly conducting cylindrical antenna of length h along the z-axis and radius a over an *infinite* conducting plane. The antenna is base-driven at $z = 0$ by a discontinuity in scalar potential given by

$$V_0 = \int_0^h (E_z^e)_{r=a} \, dz, \quad (E_z^e)_{r=a} = V_0^e \delta(z), \quad (1)$$

where $\delta(z)$ is the Dirac delta function defined in Sec. I.4 and V_0^e is the impressed emf. The tangential electric field $(E_z)_{r=a}$ maintained on the outer surface $r = a$ of the antenna by

---

* This section is based on the analysis of Dr. J. E. Storer, ref. 56.

the currents and charges in it and in the ground screen is

$$(E_z)_{r=a} = -\left(\frac{\partial \Phi}{\partial z} + j\omega A_z\right)_{r=a} = z^i I_z - E_z^e. \quad (2)$$

For a perfectly conducting antenna, $z^i = 0$, and (10) reduces to $(E_z)_{r=a} = -(E_z^e)_{r=a}$. The complex energy-transfer function $T = \overline{\overline{T}}_r + jT_i$ for the surface of the antenna is given in reference I.31, Sec. IV.19, or by (26.2) with a factor $\frac{1}{2}$. It is

$$T = -\tfrac{1}{2} \int_0^h (E_z)_{r=a} I_z^* \, dz. \quad (3)$$

Since $(E_z)_{r=a}$ vanishes on the conductor except along the edge of a thin slice at $z = 0$, where $I_z = I_0$, (3) reduces to

$$T = \tfrac{1}{2} I_0^* \int_0^h (E_z^e)_{r=a} \, dz = \tfrac{1}{2} I_0^* V_0^e = \tfrac{1}{2} I_0^* I_0 Z_0, \quad (4)$$

of which the real part is the total power supplied to the antenna and radiated by it.

If the infinite conducting plane is reduced to a diameter D, the axial *distribution* $f(z)$ of the current $I_z = I_0 f(z)$ along the antenna must be essentially unaltered, since the length and structure of the antenna itself remain the same and there are no adjacent parallel currents to modify the distribution. Accordingly, if the driving voltage V_0^e is adjusted so that I_0 is the same when the ground screen is infinite as when its diameter is D, I_z on the antenna is also the same. However, the distributions of surface currents and charges on the finite ground screen of diameter D necessarily differ from those on that part of an infinite screen where $r \leq D/2$; in addition, there are no currents or charges for $r > D/2$. It follows that the electric field E_z near the antenna and in particular at the generating slice at $r = a$, $z = 0$ must differ from that obtaining when $D = \infty$. Note, however, that on the perfectly conducting antenna and ground screen the tangential electric field is zero for all values of D.

Let the z-component of the field near the antenna when D is finite be

$$E_z = E_z^\infty + \Delta E_z, \quad (5)$$

where E_z^∞ is the field when $D = \infty$ and ΔE_z is the change in the field when the diameter of the ground screen is reduced from infinity to a finite value D. Substitution of (5) in (3) gives

$$T = -\tfrac{1}{2} \int_0^h (E_z^\infty)_{r=a} I_z^* dz$$

$$- \tfrac{1}{2} \int_0^h (\Delta E_z)_{r=a} I_z^* dz = \tfrac{1}{2} I_0^* I_0 Z, \quad (6)$$

where Z is the impedance when D is finite. The first integral term in (6) is T^∞, so that

$$(Z - Z_0) I_0 I_0^* = -\int_0^h (\Delta E_z)_{r=a} I_z^* dz. \quad (7)$$

The difference field ΔE_r must satisfy the field equations, since E_z and E_z^∞ both do. Hence, the difference magnetic field ΔB at points in space is given by the field equation

$$\operatorname{curl} \Delta B = \frac{j\beta_0^2}{\omega} \Delta E, \quad (8)$$

so that

$$\Delta E_z = -\frac{j\omega}{\beta_0^2} \operatorname{curl}{}_z \Delta B = -\frac{j\omega}{\beta_0^2} \frac{1}{r} \frac{\partial}{\partial r}(r\Delta B_\theta). \quad (9)$$

Hence,

$$(Z - Z_0) I_0 I_0^* = \frac{j\omega}{\beta_0^2} \int_0^h \left[\frac{1}{r} \frac{\partial}{\partial r}(r\Delta B_\theta) \right]_{r=a} I_z^* dz. \quad (10)$$

By evaluating $\Delta B_\theta = B_\theta - B_\theta^\infty$ for use in (10), the quantity $Z - Z_0$ may be determined.

The magnetic fields B_θ and B_θ^∞ may be derived from the polarization potential given in (9.10)–(9.12) for a short, end-loaded antenna over an infinite ground plane by integrating the dipole moment $p_{1z} = I_z dz/j\omega$ along the length h of the antenna. Thus, for the antenna,

$$\Pi_{ez}^\infty = (\Pi_{ez})_d + (\Pi_{ez}^\infty)_r,$$
$$\Pi_{ez} = (\Pi_{ez})_d + (\Pi_{ez})_r, \quad (11)$$

where $(\Pi_{ez})_d$ is the polarization potential due to currents in the antenna and $(\Pi_{ez})_r$ the potential due to currents in the ground plane. Since currents in the antenna are the same with D finite and infinite, it follows that

$$\Delta\Pi_{ez} = \Pi_{ez} - \Pi_{ez}^\infty = (\Pi_{ez})_r - (\Pi_{ez}^\infty)_r, \quad (12)$$

where (from (9.12a) with $p_{1z} = I_z dz/j\omega$ and with the origin shifted to the base of the antenna on the plane $z = 0$),

$$(\Pi_{ez}^\infty)_r$$
$$= \frac{-j}{4\pi\epsilon_0 \omega} \int_0^h I_z' dz' \int_0^\infty l^{-1} f_{er} e^{-(z+z')l} J_0(\lambda r) \lambda \, d\lambda, \quad (13)$$

where f_{er} is given by (9.11a). By analogy, the corresponding potential for the finite screen is

$$(\Pi_{ez})_r = \frac{-j}{4\pi\epsilon_0 \omega}$$
$$\times \int_0^h I_z' dz' \int_0^\infty l^{-1} g^+(\lambda, z') e^{-(z+z')l} J_0(\lambda r) \lambda \, d\lambda,$$
$$(z \geq 0) \quad (14a)$$

$$(\Pi_{ez})_r = \frac{-j}{4\pi\epsilon_0 \omega}$$
$$\times \int_0^h I_z' dz' \int_0^\infty l^{-1} g^-(\lambda, z') e^{(z+z')l} J_0(\lambda r) \lambda \, d\lambda,$$
$$(z \leq 0) \quad (14b)$$

where $l = \sqrt{\lambda^2 - \beta_0^2}$ and where $g^+(\lambda, z')$ and $g^-(\lambda, z')$ are complex constants yet to be evaluated in terms of the boundary conditions on the upper and lower surfaces of the finite ground screen. Clearly, with (12),

$$\Delta\Pi_{ez} = \frac{-j}{4\pi\epsilon_0\omega}$$
$$\times \int_0^h I_z' dz' \int_0^\infty F^+(\lambda, z') e^{-zl} J_0(\lambda r) \, d\lambda,$$
$$(z \leq 0) \quad (15a)$$

$$\Delta\Pi_{ez} = \frac{-j}{4\pi\epsilon_0\omega}$$
$$\times \int_0^h I_z' dz' \int_0^\infty F^-(\lambda, z') e^{zl} J_0(\lambda r) \, d\lambda,$$
$$(z \geq 0) \quad (15b)$$

where

$$F^+(\lambda, z) = \lambda l^{-1}[g^+(\lambda, z') - f_{er}] e^{-z'l}, \quad (16a)$$
$$F^-(\lambda, z) = \lambda l^{-1}[g^-(\lambda, z') - f_{er}] e^{-z'l}. \quad (16b)$$

The magnetic field has only the one component B_θ which may be derived from the polarization potential using

$$B_\theta = \frac{-j\beta_0^2}{\omega} \frac{\partial \Pi_{ez}}{\partial r}. \quad (17)$$

Thus, with $(\partial/\partial r) J_0(\lambda r) = -\lambda J_1(\lambda r)$, the difference field is

$$\Delta B_\theta$$
$$= \frac{1}{4\pi\nu_0} \int_0^h I_z' dz' \int_0^\infty F^+(\lambda, z') e^{-zl} J_1(\lambda r) \lambda \, d\lambda,$$
$$(z \geq 0) \quad (18a)$$

$$= \frac{1}{4\pi\nu_0} \int_0^h I_z' dz' \int_0^\infty F^-(\lambda, z') e^{zl} J_1(\lambda r) \lambda \, d\lambda.$$
$$(z \leq 0) \quad (18b)$$

In order to evaluate $F^+(\lambda, z')$ and $F^-(\lambda, z')$, the boundary condition $E_r = 0$ on both sides of the perfectly conducting screen of diameter D must be imposed. Using the field equation, curl $\mathbf{B} = (j\beta_0^2/\omega)\mathbf{E}$, which is valid at all points in space, and noting that rotational symmetry obtains, the following formula is derived:

$$E_r = E_r^\infty + \Delta E_r = \frac{j\omega}{\beta_0^2}\frac{\partial B_\theta}{\partial z}. \quad (19)$$

However, since E_r^∞ and $\partial B_\theta^\infty/\partial z$ are zero at $z = 0$, it follows that

$$\Delta E_r = \frac{j\omega}{\beta_0^2}\frac{\partial}{\partial z}(\Delta B_\theta)$$

$$= \begin{cases} 0, & (z=0,\ r \leq D/2) \\ f(r,z'), & (z=0,\ r > D/2) \end{cases} \quad (20)$$

on both sides of the ground screen. These conditions are satisfied if

$$\int_0^\infty F^+(\lambda, z') l J_1(\lambda r)\lambda\, d\lambda$$
$$= \begin{cases} 0, & (0 \leq r \leq D/2) \\ \xi(r, z'), & (r > D/2) \end{cases} \quad (21)$$

$$-\int_0^\infty F^-(\lambda, z') l J_1(\lambda r)\lambda\, d\lambda$$
$$= \begin{cases} 0, & (0 \leq r \leq D/2) \\ \xi(r, z'). & (r > D/2) \end{cases} \quad (22)$$

These equations may be solved for $F^+(\lambda, z')$ and $F^-(\lambda, z')$ using the Fourier-Bessel transform pair:*

$$A(r) = \int_0^\infty J_1(\lambda r) B(\lambda)\lambda\, d\lambda, \quad (23a)$$

$$B(\lambda) = \int_0^\infty J_1(\lambda r) A(r) r\, dr. \quad (23b)$$

The result is

$$F^+(\lambda, z') = \frac{1}{l}\int_{D/2}^\infty J_1(\lambda r)\xi(r', z')r'\, dr', \quad (24a)$$

$$F^-(\lambda, z') = -\frac{1}{l}\int_{D/2}^\infty J_1(\lambda r)\xi(r', z')r'\, dr', \quad (24b)$$

where z' locates the element dz' in the antenna at $r = 0$, and where r' locates dr' on the plane $z = 0$ for $r \geq D/2$. With (24a,b), the difference

---

* Stratton, ref. I.52, Sec. 6.9.

field (18) may be expressed as follows:

$$\Delta B_\theta = \frac{1}{2\pi v_0}\int_0^h I_z'\, dz' \int_{D/2}^\infty G(z, r, r')\xi(r', z')r'\, dr', \quad (25)$$

where

$$G(z, r, r')$$
$$= \pm\tfrac{1}{2}\int_0^\infty l^{-1}J_1(\lambda r)J_1(\lambda r')e^{-j l|z|}\lambda\, d\lambda. \quad (z \gtrless 0) \quad (26)$$

(The interchange in the order of integration is permissible, as the integrals are absolutely convergent, since $\xi(r', z')$ vanishes at infinity as $1/r'^2$.) Note that the upper sign in (26) applies when $z > 0$, the lower sign when $z < 0$.

With (25) the expression (10) for the impedance difference becomes

$$(Z - Z_0)I_0 I_0^* = \frac{j\omega}{2\pi v_0 \beta_0^2}\int_0^h I_z'\, dz' \int_0^h I_z^*\, dz$$
$$\times \int_{D/2}^\infty \left(\frac{1}{r}\frac{\partial}{\partial r}[rG(z,r,r')]\right)_{r=a} \xi(r', z')r'\, dr'. \quad (27)$$

The differentiation required in (27) is performed readily using

$$\frac{1}{r}\frac{\partial}{\partial r}[rJ_1(\lambda r)] = -\frac{1}{\lambda r}\frac{\partial}{\partial r}\left[r\frac{\partial}{\partial r}J_0(\lambda r)\right] = \lambda J_0(\lambda r).$$

Thus

$$\left(\frac{1}{r}\frac{\partial}{\partial r}[rG(z,r,r')]\right)_{r=a}$$
$$= \tfrac{1}{2}\int_0^\infty l^{-1}J_0(\lambda a)J_1(\lambda r')e^{-l|z|}\lambda^2\, d\lambda. \quad (28)$$

Since $l = \sqrt{\lambda^2 - \beta_0^2}$ it is clear that, as soon as λ exceeds β_0 in magnitude, the exponent in the factor $e^{-l|z|}$ becomes negative real and the factor decreases rapidly with increasing λ. It follows that the principal contributions to the integral in (28a) occur for values of λ that do not exceed β_0. Hence, the factor $J_0(\lambda a)$ may be treated as of order of magnitude $J_0(\beta_0 a) \doteq 1$ if it is assumed, as throughout this volume, that

$$\beta_0 a \ll 1. \quad (29)$$

With (29) and the relation $(\partial/\partial r')J_0(\lambda r') = -J_1(\lambda r')\lambda$, it follows that

$$\left(\frac{1}{r}\frac{\partial}{\partial r}[rG(z, r, r')]\right)_{r=a}$$

$$\doteq -\frac{1}{2}\frac{\partial}{\partial r'}\int_0^\infty l^{-1}J_0(\lambda r')e^{-l|z|}\lambda\,d\lambda$$

$$= -\frac{1}{2}\frac{\partial}{\partial r'}\frac{e^{-j\beta_0 R_0}}{R_0}, \qquad (30)$$

where $R_0 \equiv \sqrt{z^2 + r'^2}$. The last step in (30) makes use of (9.9). It is readily shown that

$$\frac{\partial}{\partial r'}\left(\frac{e^{-j\beta_0 R_0}}{R_0}\right)$$

$$= -j\beta_0\frac{e^{-j\beta_0 R_0}}{R_0}\left(1 + \frac{1}{j\beta_0 R_0}\right)\frac{r'}{R_0}. \qquad (31)$$

Hence, since for use in (27) $r' \geq D/2$ and $z \leq h$, it follows that

$$R_0 = r'\sqrt{1 + z^2/r'^2} \doteq r', \qquad (32a)$$

provided that

$$z^2/r'^2 \leq 4h^2/D^2 \ll 1. \qquad (32b)$$

Let it be required that the length h of the antenna be sufficiently small compared with the diameter of the conducting disk to satisfy (32b). Then R_0 may be replaced by r' in (31) to give

$$\frac{\partial}{\partial r'}\left(\frac{e^{-j\beta_0 R_0}}{R_0}\right)$$

$$\doteq \frac{-j\beta_0}{r'}\sqrt{1 + \frac{1}{(\beta_0 r')^2}}\,e^{-j\beta_0 r'(1+\delta)},$$

$$\delta \equiv \frac{1}{\beta_0 r'}\tan^{-1}\left(\frac{1}{\beta_0 r'}\right). \qquad (33)$$

Let it be required that the diameter D of the conducting disk satisfy the following condition in addition to (32b):

$$(\beta_0 D/2)^2 \gg 1. \qquad (34)$$

With (34), (33) reduces to

$$\frac{\partial}{\partial r'}\left(\frac{e^{-j\beta_0 R_0}}{R_0}\right) \doteq -j\beta_0\frac{e^{-j\beta_0 r'}}{r'}, \qquad (35)$$

so that (30) becomes

$$\left(\frac{1}{r}\frac{\partial}{\partial r}[rG(z, r, r')]\right)_{r=a} \doteq \frac{j\beta_0}{2}\frac{e^{-j\beta_0 r'}}{r'}. \qquad (36)$$

With (36) in (27) the impedance difference becomes

$$(Z - Z_0)I_0 I_0^* = \frac{-\omega}{4\pi v_0 \beta_0}\int_0^h I_z'\,dz'\int_0^h I_z^*\,dz$$

$$\times \int_{D/2}^\infty e^{-j\beta_0 r'}\xi(r', z')r'\,dr'. \qquad (37)$$

In order to evaluate (37) it is necessary to determine $\xi(r', z')$. An integral equation for this function may be derived from the magnetic field above and below the finite ground screen. Using (25) these magnetic fields are

$$B_\theta = B_\theta^\infty + \Delta B_\theta$$

$$= B_\theta^\infty(r, z) + \frac{1}{2\pi v_0}\int_0^h I_z'\,dz'\int_{D/2}^\infty G(z, r, r')$$

$$\times \xi(r', z')r'\,dr', \quad z \geq 0 \qquad (38a)$$

$$= -\frac{1}{2\pi v_0}\int_0^h I_z'\,dz'\int_{D/2}^\infty G(z, r, r')$$

$$\times \xi(r', z')r'\,dr'. \quad z \leq 0 \qquad (38b)$$

Equations (38) express the magnetic field above and below the plane $z = 0$ in terms of the unknown function $\xi(r', z')$. An integral equation for this function may be obtained from (38) by noting that B_θ is continuous across the plane $z = 0$ when r exceeds $D/2$. Setting $z = 0$ in (38) and equating the two values of B_θ gives

$$B_\theta^\infty(r, 0) = \frac{-1}{\pi v_0}\int_0^h I_z'\,dz'\int_{D/2}^\infty G(0, r, r')$$

$$\times \xi(r', z')r'\,dr'. \quad (r > D/2) \qquad (39)$$

This equation may be solved by expanding $\xi(r)$ in a series of spheroidal functions of r.[*] Unfortunately, this series converges slowly if the radius of the screen is large. Since interest is primarily in screens that are sufficiently large so that they approximate infinite screens or at least so that Z for the finite screen may be determined from Z_0 for the infinite screen by a simple formula, a different method of obtaining a solution that is useful for large ground screens is required.

The first step in Storer's variational solution of (39) is to introduce the simple exponential

---

[*] This leads directly to a solution obtained by Leitner and Spence, ref. 29. The analysis is limited to antennas of length $h = \lambda_0/4$.

form of the polarization potential Π_{ez}^∞ instead of the integral form. Thus, with (17),

$$B_\theta^\infty(r,z) = -\frac{1}{4\pi\nu_0}\frac{\partial}{\partial r}\int_0^h I_z'\left(\frac{e^{-j\beta_0 R_1}}{R_1} + \frac{e^{-j\beta_0 R_2}}{R_2}\right)dz', \quad (40)$$

where $R_1 = \sqrt{(z-z')^2 + r^2}$ and $R_2 = \sqrt{(z+z')^2 + r^2}$. For use in (39), $z = 0$, so that $R_1 = R_2 = \sqrt{z'^2 + r^2}$. Since $z' \leq h$ and $r \gtrsim D/2$, it follows that, subject to (32b) and (34),

$$B_\theta^\infty(r,0) \doteq \frac{j\beta_0}{2\pi\nu_0}\frac{e^{-j\beta_0 r}}{r}\int_0^h I_z'\,dz'. \quad (41)$$

If (41) is substituted in (39), the following equation results:

$$\int_0^h I_z'\,dz'\int_{D/2}^\infty G(0,r,r')\xi(r',z')r'\,dr'$$
$$\doteq \frac{-j\beta_0}{2}\frac{e^{-j\beta_0 r}}{r}\int_0^h I_z'\,dz'. \quad (42)$$

When D is as large compared with h and λ_0 as required by (32b) and (34), it is reasonable to assume that the function $\xi(r',z')$ that is proportional to the electric field at $z = 0$, $r > D/2$ is essentially independent of the distribution of current on the antenna. That is,

$$\xi(r',z') \doteq \xi(r'), \quad (43)$$

so that the integral equation (42) reduces to

$$\int_{D/2}^\infty G(0,r,r')\xi(r')r'\,dr' \doteq \frac{-j\beta_0}{2}\frac{e^{-j\beta_0 r}}{r}. \quad (44)$$

This equation may be formulated in a variational manner by defining the parameter μ as follows:

$$\mu = \frac{-j\beta_0}{2}\int_{D/2}^\infty e^{-j\beta_0 r}\xi(r)\,dr. \quad (45)$$

By multiplying (44) by $r\xi(r)$, integrating with respect to r from $D/2$ to ∞, and dividing by $\left[\int_{D/2}^\infty e^{-j\beta_0 r}\xi(r)\,dr\right]^2$, the following expression for μ is obtained:

$$\mu = \frac{\int_{D/2}^\infty\int_{D/2}^\infty G(0,r,r')\xi(r)\xi(r')rr'\,dr\,dr'}{\left(\int_{D/2}^\infty e^{-j\beta_0 r}\xi(r)\,dr\right)^2}. \quad (46)$$

This is a variational form of (44) with μ a parameter that is stationary with respect to small changes in $\xi(r)$ about its true value. This may be verified by taking the first variation of μ.

By introducing the trial function

$$\xi(r) = \frac{e^{-j\beta_0 r}}{\sqrt{r - D/2}}f(r), \quad (47)$$

where $f(r)$ is slowly varying in r and finite at $r = D/2$, Storer shows in an extended analysis that

$$\mu \doteq \frac{De^{j\beta_0 D}}{4L(D)}, \quad (48a)$$

where

$$L(D) \equiv \left[1 + \frac{e^{-j(\beta_0 D + 3\pi/4)}}{\sqrt{2\pi\beta_0 D}}\right]^{-1}, \quad (48b)$$

provided that the condition

$$\beta_0 D \gg 1 \quad (48c)$$

is satisfied.

Hence, with (48a,b) in (45) it follows that

$$\int_{D/2}^\infty e^{-j\beta_0 r}\xi(r)\,dr = \frac{-j2\beta_0}{D}e^{-j\beta_0 D}L(D). \quad (49)$$

This expression may be substituted in (37) if note is taken of (43). Thus, with $\zeta_0 = 1/\sqrt{\nu_0\epsilon_0}$,

$$Z - Z_0 \doteq \frac{j\zeta_0}{2\pi}\frac{e^{-j\beta_0 D}}{\beta_0 D}L(D)F^2(I), \quad (50a)$$

where $L(D)$ is defined in (48b) and

$$F^2(I) \equiv \left|\beta_0\int_0^h (I_z/I_0)\,dz\right|^2. \quad (50b)$$

This is Storer's formula. Note that the following restrictions have been imposed:

$$(D/2h)^2 \gg 1, \quad (51a)$$

$$\beta_0 D \gg 1. \quad (51b)$$

If the inequalities (51a) and (51b) are assumed to be satisfied provided the ratio of left- to right-hand members is at least 64, it follows that

$$D/h \geq 16, \quad (52a)$$

$$D/\lambda_0 \geq 10. \quad (52b)$$

Except for fairly long antennas, (52b) is the severer restriction; actually (52b) is sufficiently severe to make $L(D)$ differ little from unity. Thus, with $D/\lambda_0 = 10$, $1/\sqrt{2\pi\beta_0 D} = 0.05$

so that a satisfactory approximation of (50a) is

$$Z - Z_0 \doteq \frac{j\zeta_0}{2\pi} \frac{e^{-j\beta_0 D}}{\beta_0 D} F^2(I). \qquad (53)$$

This is equivalent to

$$R - R_0 = \frac{\zeta_0}{2\pi} \frac{\sin \beta_0 D}{\beta_0 D} F^2(I), \qquad (54a)$$

$$X - X_0 = \frac{\zeta_0}{2\pi} \frac{\cos \beta_0 D}{\beta_0 D} F^2(I). \qquad (54b)$$

Note that by proper choice of $\beta_0 D$ either $(R - R_0)$ or $(X - X_0)$ may be made zero. Since ground screens with a radius of 5 wavelengths or more are reasonable, the simple formulas (54a,b) are of great practical utility. If $R - R_0$ and $X - X_0$ are plotted on the complex $(Z - Z_0)$-plane, a spiral is obtained with a radius that decreases slowly as D/λ_0 increases. Since $F^2(I)$ is dimensionless, the quantities $(R - R_0)$ and $(X - X_0)$ may be divided by this factor to provide the "universal" curve shown in Fig. 28.2.

The dimensionless factor

$$F^2(I) \equiv \left| \beta_0 \int_0^h I_z/I_0 \, dz \right|^2$$

characterizes the dependence of the impedance difference $Z - Z_0$ on the length h of the antenna and the distribution of current. Its accurate evaluation presupposes a knowledge of the current at all points in the antenna, not merely at the driving point. The zeroth-order distribution function,

$$\left(\frac{I_z}{I_0}\right)_0 = \frac{\sin \beta_0(h-z)}{\sin \beta_0 h}, \qquad (55)$$

is a satisfactory approximation for very thin antennas ($\Omega > 20$) that are far from antiresonance. It is a good approximation even for quite thick antennas ($\Omega \geq 10$) if $\beta_0 h$ is near $\pi/2$. If (55) is substituted in (50b), the integration is performed directly to give

$$[F^2(I)]_0 \doteq \left(\frac{1 - \cos \beta_0 h}{\sin \beta_0 h} \right)^2. \qquad (56)$$

Note that for $\beta_0 h = \pi/2$, $F^2(I) \doteq 1$, so that the universal curve in Fig. 28.2 applies directly to the half-dipole with $h = \lambda_0/4$, with $(R - R_0)$ and $(X - X_0)$ given directly by the two scales. Since the zeroth-order value of Z_0 with $\beta_0 h = \pi/2$ is $Z_0 = 36.6 + j22.25$ ohms, the relative significance of $R - R_0$ and $X - X_0$ is clear.

For antennas that are moderately thick and that have electrical lengths differing considerably from $\beta_0 h = \pi/2$, first-order or second-order currents *must* be used to determine $F^2(I)$. By using the approximate second-order curves in Figs. II.22.9 through II.22.12 for antennas with $\Omega = 2 \ln (2h/a) = 10$, the real and imaginary parts of $\int_0^h I_z \beta_0 dz$ have been determined approximately using a planimeter. By squaring these results, adding, and dividing by I_0^2, the following values have been obtained for $F^2(I)$:

$\beta_0 h$:	$\pi/2$	$3\pi/4$	$\pi/4$	$5\pi/4$
$F^2(I)$:	1.16	10.8	3.84	2.92

(57)

Note that the second-order value 1.16 for $\beta_0 h = \pi/2$ is reasonably close to the approximate zeroth-order value of unity. On the other hand, the large value $F^2(I) = 10.8$ at $\beta_0 h = 3\pi/4 = 2.36$ differs greatly from the corresponding zeroth-order value computed from (56), namely, $F^2(I) = 5.85$. Since with an infinite ground plane antiresonance occurs at $\beta_0 h = 2.54$ when $\Omega = 10$, the impedance at $\beta_0 h = 3\pi/4$ is not very far from antiresonance and therefore is relatively large. Its numerical value is $Z_0 = 318 + j126$ ohms; the antiresonant value is $Z_0 = 422 + j0$ ohms. Although $F^2(I)$ with $\beta_0 h = 3\pi/4$ is roughly ten times as large as with $\beta_0 h = \pi/2$, so that $Z - Z_0$ is also ten times as large, the quantity $|(Z - Z_0)/Z_0|$ is about the same. Note, however, that near antiresonance $(X - X_0)/X_0$ is very great and becomes infinite when X_0 vanishes even though $(R - R_0)/R_0$ and $|(Z - Z_0)/Z_0|$ are quite small. Actually, the antiresonant length of an antenna over a finite ground screen differs somewhat from that for the same antenna over an infinite ground plane. For this reason it is always desirable to make $X - X_0$ zero by choosing a ground screen of diameter D that satisfies the condition

$$\cos \beta_0 D = 0 \quad \text{or} \quad \beta_0 D = \frac{(2n+1)\pi}{2},$$

or

$$D/\lambda_0 = \frac{2n+1}{4}, \qquad (58a)$$

where n is an integer equal to or greater than 20 in order to satisfy (52b). If this choice of D

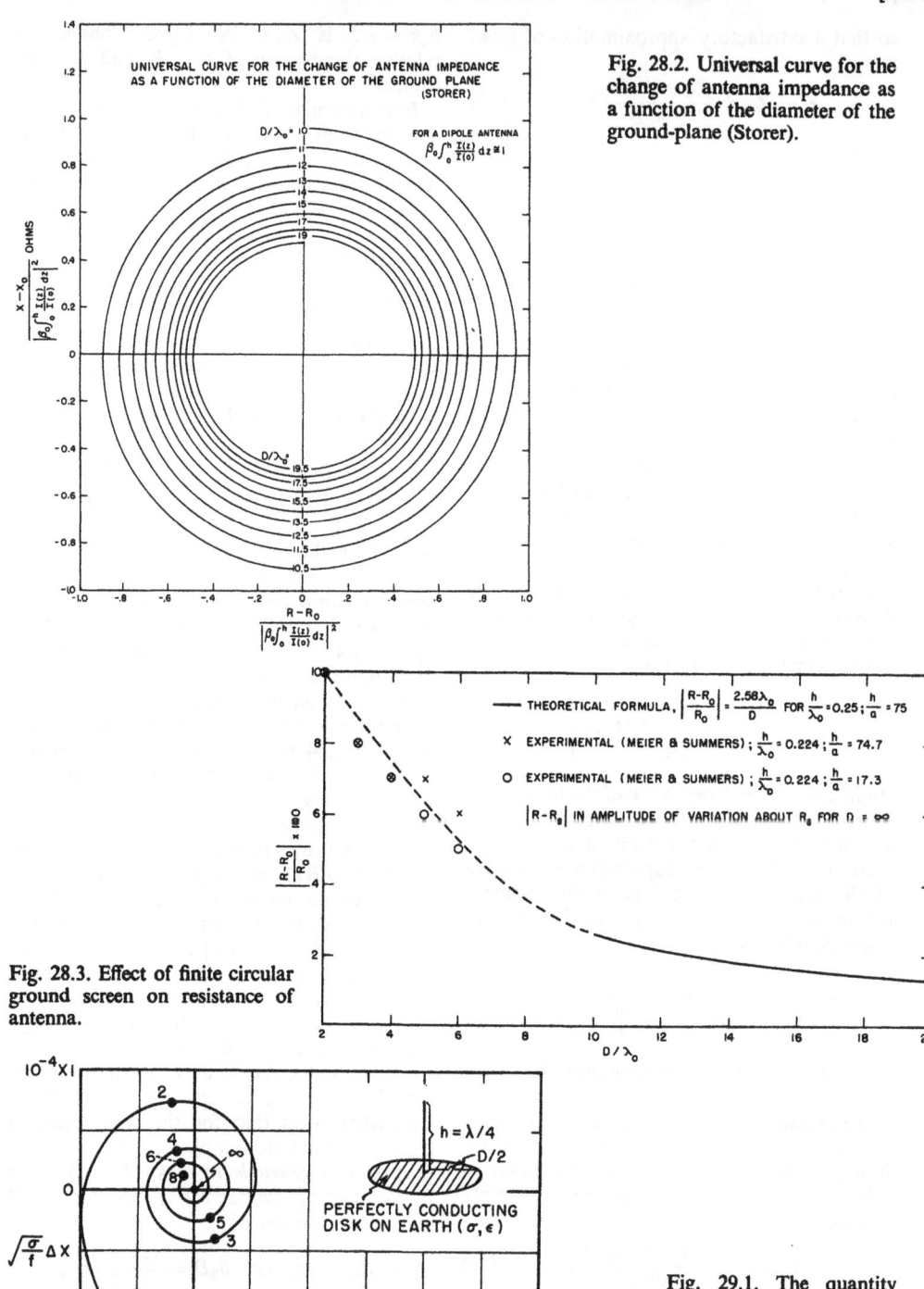

Fig. 28.2. Universal curve for the change of antenna impedance as a function of the diameter of the ground-plane (Storer).

Fig. 28.3. Effect of finite circular ground screen on resistance of antenna.

Fig. 29.1. The quantity $\sqrt{\sigma/f}\,\Delta X$ as a function of $\sqrt{\sigma/f}\,\Delta R$ for an antenna of length h on a grounded disk of diameter D. The ratio $D/2h$ is the parameter (Wait and Surtees).

is made it follows from (54a, b) that

$$|R - R_0| = \frac{\zeta_0}{2\pi} \frac{F^2(I_0)}{\beta_0 D},$$

$$X - X_0 = 0. \qquad (58b)$$

Since R and R_0 never become small except when $\beta_0 h$ is near zero, the percentage difference between R and R_0 when (52) is satisfied is never great. It may be made negligible by choosing D large enough. For example, with $\beta_0 h = \pi/2$, Table II.30.1 gives $R_0 = 43.25$ for $\Omega = 10$. Hence, with $F^2(I_0) = 1.16$, (58b) gives

$$\left|\frac{R - R_0}{R_0}\right| = \frac{60 \times 1.16}{43.25 \beta_0 D} = \frac{1.61}{\beta_0 D} = \frac{2.58 \lambda_0}{D},$$

$$X - X_0 = 0. \qquad (59)$$

For a permitted difference of 2 percent, for example, it is necessary that $\beta_0 D \geq 80$, $D/\lambda_0 \geq 12.7$. If $n = 25$ is chosen in (58a), $D/\lambda_0 = 12.75$ and the requirement is met.

The spiral graph of $R - R_0$ and $X - X_0$ in Fig. 28.2 is in agreement with the experimental result of Meier and Summers[33] for circular ground screens. Since these investigators limited themselves to small ground screens with $D/\lambda_0 \leq 6$, a quantitative comparison of their results with those predicted by (54a,b), which are valid only for $D/\lambda_0 \geq 10$, is not possible. However, the amplitude of the theoretical resistance variation as given by (57) for an antenna with $h/\lambda_0 = 0.25$ and $h/a = 75$ ($\Omega = 10$) is shown in Fig. 28.3 in the range $10 \leq D/\lambda_0 \leq 20$ together with corresponding amplitudes obtained from the experimental data of Meier and Summers in the range $2 \leq D/\lambda_0 \leq 6$ for an antenna with $h/\lambda_0 = 0.224$ and $h/a = 74.7$ and 17.3. The general agreement is evident.

It may be concluded that the impedance of an antenna when base driven at the center of a circular ground screen of diameter D is a good approximation of the ideal impedance when the ground screen is infinite if the radius of the screen is at least five wavelengths. If the ground screen is square or if the antenna is not at its center, the ground screen may be somewhat smaller. It is important to bear in mind that the nearly equal impedances of vertical antennas on infinite and large finite ground planes does not imply that the far-zone field patterns are alike. With an *infinite* ground plane the contributions to the far-zone field by the currents in the antenna and in the conducting plane are equal. If the ground screen is circular and finite, the currents on the screen are nor only limited to a radius $D/2$ but exhibit a radial standing-wave distribution instead of a traveling-wave distribution. The complete absence of currents on the plane $z = 0$ (Fig. 28.1) beyond $r = D/2$ leads to a null in the far-zone field pattern at $\Theta = \pi/2$ resembling that for the same antenna over a plane that is infinite in extent ($D = \infty$) but *imperfectly conducting*. This is entirely reasonable, since the radial attenuation of currents in an imperfect conductor reduces their amplitudes so that at distant points their effect is little different from zero. Thus, with an infinite imperfectly conducting ground plane or a finite perfectly conducting ground screen the currents that contribute significantly to maintain the far-zone field are those in the antenna itself and those in a finite region in the plane $z = 0$ about the base of the antenna. In the one case these latter are traveling waves that are reduced gradually in amplitude, in the other they are standing waves that stop abruptly at $r = D/2$. The gradually decreasing traveling waves produce a smooth pattern, the standing waves a similar one with many superimposed ripples. Note that with a finite ground screen there is a relatively small but nevertheless significant field in the lower half-space where z is negative and $\Theta > \pi/2$.[56]

29. Impedance of Base-Driven Antenna on a Ground Screen of Finite Size on an Infinite Plane Imperfectly Conducting Earth

If the perfectly conducting circular ground screen of diameter D as described in the preceding section is placed on the surface of an infinite half-space (earth) consisting of material with conductivity σ and dielectric constant ϵ (instead of being isolated in space as in Sec. 28), a part of the power supplied to the antenna is dissipated in this half-space. An estimate of the change in the impedance of the antenna as the diameter of the perfectly conducting disk is decreased from infinity may be obtained readily if the conductivity of the earth is sufficiently high to satisfy the inequality

$$\sigma \gg \omega \epsilon. \qquad (1)$$

The total complex power transferred to the earth is given by

$$T_e = -\tfrac{1}{2} \int_{D/2}^{\infty} E_r H_\theta \cdot 2\pi r \, dr, \qquad (2)$$

since $E_r = 0$ when $r < D/2$.

Subject to the condition (1), the surface impedance Z^s may be introduced according to the definition

$$Z^s = \frac{E_r}{I'_r} = (1+j)\sqrt{\frac{\omega\mu}{2\sigma}}, \qquad (3)$$

where I'_r is the quasi-surface density of current, that is, I'_r is the total radial current traversing a unit width principally in a thin layer near the surface. The magnetic field tangent to a perfectly conducting surface satisfies the boundary condition

$$H_\theta = I_r, \qquad (4)$$

where I_r is the true surface density of current. Since with the condition (1) the quasi-surface density of current I'_r is confined principally to a thin layer near the surface, it may be assumed that the θ-component of the magnetic field at the surface in which there is the current density I'_r may be approximated by that over a perfect conductor in which there is the current density $I_r = I'_r$. With this approximation,

$$Z^s \doteq \frac{E_r}{H_\theta} = (1+j)\sqrt{\frac{\omega\mu}{2\sigma}}, \qquad (5)$$

where H_θ is the magnetic field that would exist if the entire surface were that of a perfect conductor. It follows that

$$T_e \doteq -\frac{Z^s}{2}\int_{D/2}^\infty H_\theta^2 \cdot 2\pi r \, dr. \qquad (6)$$

When $D = \infty$, the total complex power supplied to the antenna and radiated into the upper half-space is

$$T_0 = \tfrac{1}{2}I_0^2 Z_0, \qquad (7)$$

where I_0 is the current at the base of the antenna and Z_0 is its input impedance. When D is finite, the total complex power to the antenna with the same input current I_0 is

$$T_t = \tfrac{1}{2}I_0^2 Z_t, \qquad (8)$$

where Z_t is the impedance of the antenna with an earth of finite conductivity beyond $r = D/2$. Since the input current is unchanged and its distribution along the antenna is practically the same, the power radiated from the antenna must be approximately the same in the two cases. Evidently the difference between T_t and T_0 is a good estimate of the complex power transferred into the conducting earth. With

$$T_t - T_0 \doteq T_e \doteq \tfrac{1}{2}I_0^2 \Delta Z, \qquad (9)$$

$$\Delta Z = Z_t - Z_0 = \frac{Z^s}{I_0^2}\int_{D/2}^\infty H_\theta^2 \cdot 2\pi r \, dr, \qquad (10)$$

where H_θ is the magnetic field at the surface of a perfectly conducting half-space. This is the same as the field in the equatorial plane of a symmetric center-driven antenna of half-length h. For a sufficiently thin antenna it is approximated by V.2.14b, with $z = 0$; That is,

$$H_\theta \doteq \frac{jI_m}{2\pi r}(e^{-j\beta_0 R_h} - \cos\beta_0 h \, e^{-j\beta_0 r}), \qquad (11)$$

where

$$I_m = I_0/\sin\beta_0 h \qquad (12)$$

is the maximum sinusoidal current along the antenna and

$$R_h = \sqrt{h^2 + r^2}. \qquad (13)$$

When $\beta_0 h = \pi/2$, (11) reduces to

$$H_\theta = \frac{jI_0}{2\pi}\frac{e^{-j\beta_0 R_h}}{r}. \qquad (14)$$

Since the sinusoidal distribution of current leads to a moderately satisfactory approximation of the impedance only when $\beta_0 h$ is near odd multiples of $\pi/2$, it is adequate to continue the analysis using the simple form (14). With (14) in (10) it follows that

$$\Delta Z \doteq -\frac{Z^s}{2\pi}\int_{D/2}^\infty \frac{e^{-j\beta_0 h R_h}}{r^2} r \, dr. \qquad (15)$$

Since $r^2 = R_h^2 - h^2$, the integral in (15) may be expressed as follows:

$$\Delta Z \doteq -\frac{Z^s}{4\pi}\left[e^{j2\beta_0 h}\int_{D/2}^\infty \frac{e^{-j2\beta_0(R_h+h)}}{R_h+h}\frac{r}{R_h}dr \right.$$
$$\left. + e^{-j2\beta_0 h}\int_{D/2}^\infty \frac{e^{-j2\beta_0(R_h-h)}}{R_h-h}\frac{r}{R_h}dr\right], \qquad (16)$$

where $\beta_0 h = \pi/2$. These integrals may be expressed in terms of the exponential integral by setting $u = j2\beta_0(R_h \pm h)$. After setting $\beta_0 h$ equal to $\pi/2$ the result is

$$\Delta Z = \frac{Z^s}{4\pi}[Ei(-jA_1) + Ei(-jA_2)], \qquad (17)$$

where

$$A_1 = \pi(\sqrt{1 + D^2/4h^2} + 1), \quad (18a)$$

$$A_2 = \pi(\sqrt{1 + D^2/4h^2} - 1). \quad (18b)$$

Since $Z^s = R^s(1 + j)$, the ratio $\Delta Z/R^s$ is independent of the properties of the earth. With $\Delta Z = \Delta R + j\Delta X$, the final result is

$$\frac{\Delta R}{R^s} = \frac{1}{4\pi}(Ci\, A_1 + Ci\, A_2 + Si\, A_1 + Si\, A_2), \quad (19a)$$

$$\frac{\Delta X}{R^s} = \frac{1}{4\pi}(Ci\, A_1 + Ci\, A_2 - Si\, A_1 - Si\, A_2). \quad (19b)$$

If $\Delta Z/R^s$ is plotted on the complex plane with $D/2h$ as a parameter, a spiral is obtained as in Fig. 29.1.[71] The impedance of the antenna on the grounded disk is

$$Z_t = Z_0 + \Delta Z = R_0 + \Delta R + j(X_0 + \Delta X), \quad (20)$$

where Z_0 is the impedance of the antenna over a perfect conductor of infinite extent and ΔR and ΔX are obtained from (19a,b). Note that (1) must be satisfied if (19a,b) are to be useful approximations. Since Z_0 does not vary rapidly with the radius of the antenna when $\beta_0 h = \pi/2$, it is a fair approximation to assume ΔR and ΔX as given in (19a,b) to apply to an antenna of finite cross-sectional area.

CHAPTER VIII

THE ANTENNA AS A BOUNDARY-VALUE PROBLEM

In the introductory discussion of the antenna as a problem in electrodynamics in Sec. II.1 three component parts are emphasized: the theoretical analysis, the experimental investigation, and the correlation of theory with experiment. In Chapter II these three phases of the investigation are considered in a sequence that may be designated as the *practical* or *engineering* approach in contradistinction to the *analytical* or *mathematical* approach. The essential difference between these two points of view is in the criterion underlying the choice of the structure or circuit to be analyzed. Is this determined primarily by its practical importance or by its convenience in the formulation of mathematical boundary conditions?

In the practical sequence in Chapter II, the transmitting system investigated consisted of the most important simple structure, the relatively thin cylindrical antenna driven from a conventional transmission line in one of several generally useful connections. This entire circuit, including the transmission line, is analyzed as accurately and completely as possible. However, the actual configuration of the conductors forming the antenna and the transmission line does not provide mathematically convenient boundary conditions in the sense that is traditional and, indeed, essential in the rigorous solution of boundary-value problems. Therefore, approximate methods and careful attention to detail are necessary in order to achieve an acceptable result—acceptable from the point of view of predicting measurable results rather than of providing mathematical elegance.

In this chapter the antenna problem is formulated with primary emphasis on the mathematical approach. A structure with mathematically convenient boundaries is chosen, so that a rigorous formulation is available using well-known methods. In this manner the purely mathematical crudities in the practical approach in Chapter II are eliminated, but this gain is not achieved without a counterbalancing loss. A new difficulty arises in the very troublesome problem of correlating the rigorously determined properties of an abstract configuration of mathematically convenient boundaries and an idealized generator with the actual and experimentally measurable properties of physically available structures. Since the conditions assumed in the analyzed formulation do not coincide completely with those of a practical problem, an actual one-to-one correspondence cannot be established and various approximations must be made. These have to be examined very critically, since a mere placing in juxtaposition of the superficially similar results of theory and experiment constitutes no verification of the theory and provides theoretical information of questionable practical value.

1. *Hemispheroidal and Conical Antennas*

The problem of selecting a transmitting system that has mathematically satisfactory boundaries while retaining at least some resemblance to the practically available and useful is difficult. A theoretically obvious suggestion is an antenna in the shape of a sphere or of a prolate spheroid. But how is this to be driven? If the spherical or spheroidal symmetry essential to the establishment of simple boundary conditions is to be maintained, the generator with its transmission line must be built *inside* the antenna, as in Fig. 1.1a. While such an antenna could be constructed, it is experimentally awkward. A more symmetric and experimentally more convenient structure is half of a spheroid over a large conducting plane below which the generator may be arranged. Three possible connections are shown in Fig. 1.1b, c, d. In all of these the half-spheroid may be treated as though driven by an impressed field maintained along a narrow belt around the antenna. Analyses of this sort have been made by Stratton and Chu[21] for the sphere and the prolate spheroid. However, driving voltages due to such slice or belt generators are of academic interest only unless they can be related directly to the terminal voltage of a transmission line, as is done in Chapter II. Hence, in so far as the driving conditions are concerned, the analyses of the sphere and the spheroid do not differ significantly from the corresponding analysis of the cylindrical antenna in Chapter II. A measurable impedance can be defined only on the transmission line as an *apparent* impedance Z_{sa} for the antenna and this must be related to the theoretical impedance Z_b, the ratio of voltage across the driving belt to current at its terminals, by taking account of transmission-line end effects and coupling effects in a terminal zone near the more or less arbitrarily defined junction between the antenna and the line. Thus, the use of a spheroidal antenna instead of a cylindrical one *does not improve the principal problems of approximation* in the analysis of a simple transmitting system. To be sure, the impossibility of accurately analyzing the ends of the cylindrical antenna in the quasi-one-dimensional analysis in Chapter II is eliminated in the case of the spheroid, and rigorous account can be taken of the entire surface of the antenna, excluding the driving surfaces. But this is a relatively insignificant gain in dealing with *thin* antennas and is much more than offset by the complexity of the spheroidal analysis and the fact that it is limited to single, isolated antennas. As soon as two or more antennas are coupled, the spheroidal symmetry obviously disappears. For thick antennas and, in particular, for antennas that are actually spheroidal, where the quasi-one-dimensional analysis of Chapter II has no application, the situation is different and a three-dimensional analysis is essential. Since this phase of the antenna problem is outside the field of linear antennas, the analysis of the spheroidal antenna is omitted.

A modification in the driving structure of the hemispherical antenna over a ground screen is shown in Fig. 1.2a. It differs from the arrangement in Fig. 1.1b in that the section of radial transmission line joining the coaxial transmission line and the antenna is replaced by a section of conical transmission line. Since both the spherical antenna and the conical line have boundaries that are expressed conveniently in terms of spherical coördinates, the location of the terminals may be moved from the outer periphery to the junction of the conical line and the coaxial line, and the conical line may be treated as an integral part of the antenna. The importance of the currents in the conical line in contributing to the radiation field depends on the angle of the cone. In Fig. 1.2a they are relatively unimportant and the cones are primarily a transmission line; on the other hand, in Fig. 1.2b the currents on the cones are responsible for the greater part of the field. In either case, the characteristics of the conical antenna that make it particularly attractive are that in addition to having surfaces that permit a simple specification of boundary conditions it combines the properties of a radiating circuit with those of a transmission line with a dominant mode. The nearer the apex of the cone is approached, the more the currents on the conical surfaces behave like transmission-line currents in accordance with simple transmission-line formulas. This means that the junction between the conical surfaces of the antenna and the cylindrical surfaces of the coaxial line is effectively the junction between two transmission lines instead of between an antenna and a transmission line. By making the characteristic impedance of the coaxial line equal to that of the conical line, junction effects are greatly reduced. Their order of magnitude is that occurring when two coaxial lines which have equal characteristic impedances but slightly different dimensions are joined. Hence, the apparent load impedance measured on the coaxial line is essentially the input impedance of the conical antenna in parallel with a very small capacitance that takes account of junction effects. If the transition between antenna and line is gradual in the sense that sharp corners and angles are rounded, such effects may be negligible.

In the analysis of the conical antenna, just as in the analysis of its cylindrical counterpart, it is convenient to study first the idealized limiting case in which the cross-sectional area of the driving coaxial line is effectively zero. In the case of the cylindrical half-dipole, the antenna is an extension of the inner conductor of the coaxial line of radius a. In the limit when the inner radius b of the outer conductor differs very little from a, the coaxial line is equivalent to a slice generator at the base of the cylindrical antenna. In the case of the conical antenna, the corresponding limiting case is when the radius a of the inner conductor is vanishingly small and the radius b of the outer conductor is only slightly greater. When this is true, the driving line is equivalent to a *point generator* at the sharp apex as shown in Fig. 1.3a and the impedance $\frac{1}{2}Z_0$ of the conical antenna is the ratio of the emf $\frac{1}{2}V_0^e$ of this generator divided by the current in it. The factor $1/2$ is included since, by analogy with the cylindrical antenna, it is convenient to analyze the symmetric biconical antenna in Fig. 1.3b which is driven by a point generator at the apex with emf V_0^e and has an impedance Z_0. That the impedance of the conical antenna in Fig. 1.3a is one-half the impedance of the biconical antenna in Fig. 1.3b follows directly from the theorem of images. The relation between the impedance Z_0 for zero line spacing and the impedance Z_b for finite values of a and b may be obtained in essentially the same manner as in Sec. II.10 for the cylindrical antenna. This involves determining C_T from the change in scalar potential difference near the end of the coaxial line when the conical line is replaced by an extension of the coaxial line. When the apex angle of the cone is small, as in Fig. 1.2b, the value of C_T is closely approximated by the negative C_T for a cylindrical antenna. For large apex angles of the cone, as in Fig. 1.2a, the equal and opposite charges on the sides of the conical line came closer together, so that their effects cancel more completely than when further apart, as in Fig. 1.2b. This means that C_T is more negative when the conical line is wide, as in Fig. 1.2b, and the conical antenna is more like a thin cylinder, and less negative when the conical line is narrow, as in Fig. 1.2a. For narrow conical antennas, which approximate thin cylindrical radiators, C_T for the cylinder is a satisfactory capacitance to use in parallel with Z_0 to obtain Z_b.

2. Equations for Spherical Waves with Rotational Symmetry

Since the boundary conditions of the biconical antenna are expressed conveniently in spherical coördinates Θ, Φ, R, the electromagnetic field due to currents and charges in the antenna must be expressed in the same coördinate system. This may be accomplished by deriving the electromagnetic field from an appropriate Hertzian potential Π. Instead of using one or more components of Π_e or Π_m, it is always possible to use one component of each. Actually, in the case at hand, one component of Π_e is adequate, since there are no currents circulating in closed loops. However, it is instructive to show that this conclusion follows logically from an application of the boundary conditions. Hence, since there are currents only in the R and Θ directions, let the appropriate components Π_{eR} and Π_{mR} be selected. At points in space outside the antenna, each of the general Hertzian potentials is defined to satisfy an equation of the type (I.8.1), namely,

$$\operatorname{curl} \operatorname{curl} \Pi_e - \beta_0^2 \Pi_e + \operatorname{grad} f = 0, \quad (1a)$$

$$\operatorname{curl} \operatorname{curl} \Pi_m - \beta_0^2 \Pi_m + \operatorname{grad} g = 0, \quad (1b)$$

where f and g are arbitrary scalar functions. The electromagnetic field is calculated from the potentials using

$$\mathbf{E} = -\operatorname{grad} f + \beta_0^2 \Pi_e - j\omega \operatorname{curl} \Pi_m, \quad (2a)$$

$$\mathbf{B} = -\operatorname{grad} g + \beta_0^2 \Pi_m + \frac{j\beta_0^2}{\omega} \operatorname{curl} \Pi_e. \quad (2b)$$

In many applications of the Hertzian potentials it is convenient to define the scalar functions as follows:

$$f \equiv \phi = -\operatorname{div} \Pi_e, \quad (3a)$$

$$g = -\operatorname{div} \Pi_m, \quad (3b)$$

where ϕ is the scalar potential. These definitions reduce (1a,b) to the vector wave equation, as shown in Sec. I.8. In the case at hand a simpler equation is obtained if f and g are not defined as in (3a,b) so that f is not identified with the scalar potential. The most desirable expressions for f and g become obvious after the Hertzian potentials are specialized to point in the radial direction. Thus, let

$$\Pi_e = \hat{\mathbf{R}} \Pi_{eR}, \quad \Pi_m = \hat{\mathbf{R}} \Pi_{mR}. \quad (4)$$

The vector equations (1a,b) may be separated into three component equations by multiplying them scalarly by the unit polar vectors $\hat{\Theta}$, $\hat{\Phi}$, $\hat{\mathbf{R}}$. Thus, with (4), (1a) becomes

$$\hat{\Theta} \cdot \operatorname{curl} \operatorname{curl} \hat{\mathbf{R}} \Pi_{eR} + \hat{\Theta} \cdot \operatorname{grad} f = 0, \quad (5a)$$

$$\hat{\Phi} \cdot \operatorname{curl} \operatorname{curl} \hat{\mathbf{R}} \Pi_{eR} + \hat{\Phi} \cdot \operatorname{grad} f = 0, \quad (5b)$$

$$\hat{\mathbf{R}} \cdot \operatorname{curl} \operatorname{curl} \hat{\mathbf{R}} \Pi_{eR} + \hat{\mathbf{R}} \cdot \operatorname{grad} f - \beta_0^2 \Pi_{eR} = 0. \quad (5c)$$

Use of the expressions for the components of the curl and gradient in spherical coördinates

[VIII.2] THEORY OF LINEAR ANTENNAS 821

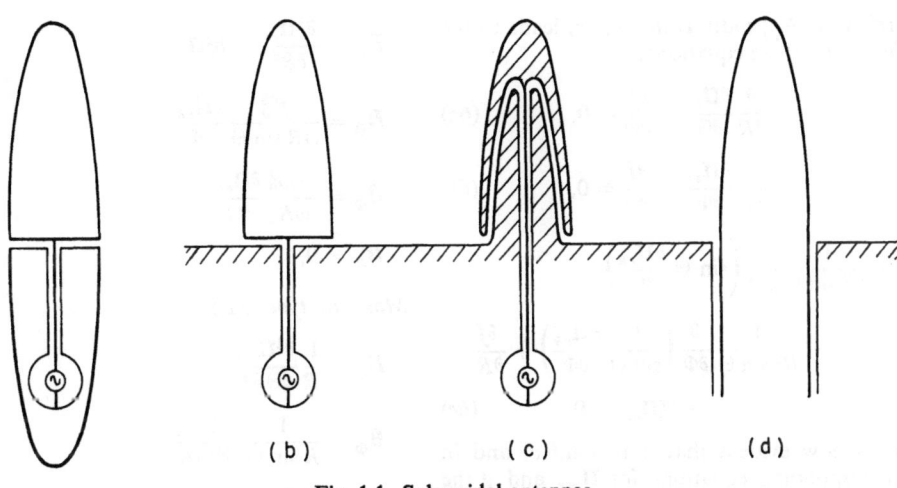

Fig. 1.1. Spheroidal antennas.

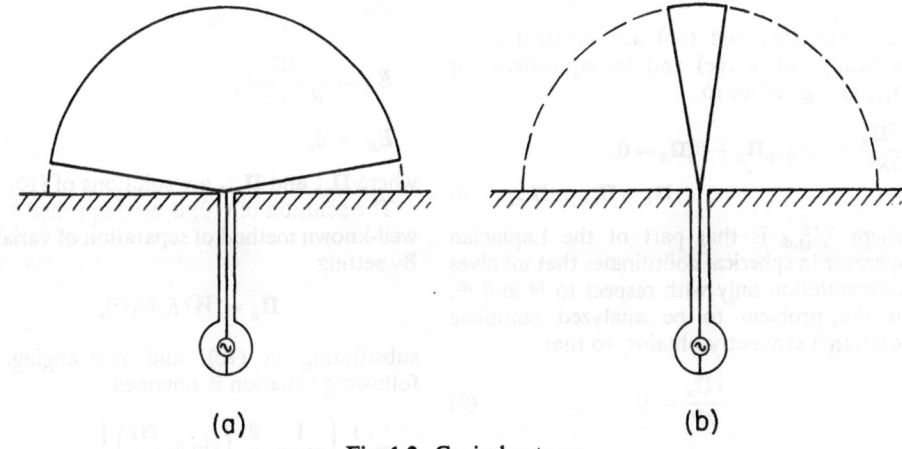

Fig. 1.2. Conical antennas.

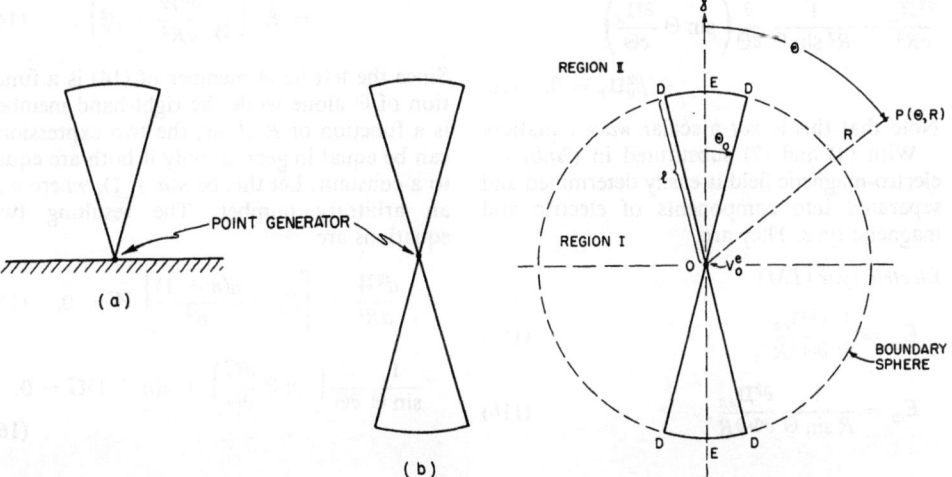

Fig. 1.3. Conical and biconical antennas with point generators.

Fig. 3.1. Biconical antenna.

(ref. I.31, Appendix I) in (5a,b,c) leads to the following three equations:

$$\frac{\partial}{\partial R}\frac{\partial \Pi_{er}}{\partial \Theta} + \frac{\partial f}{\partial \Theta} = 0, \quad (6a)$$

$$\frac{\partial}{\partial R}\frac{\partial \Pi_{er}}{\partial \Phi} + \frac{\partial f}{\partial \Phi} = 0, \quad (6b)$$

$$-\frac{1}{R^2 \sin \Theta}\frac{\partial}{\partial \Theta}\left(\sin \Theta \frac{\partial \Pi_{eR}}{\partial \Theta}\right)$$

$$-\frac{1}{R^2 \sin \Theta}\frac{\partial}{\partial \Phi}\left(\frac{1}{\sin \Theta}\frac{\partial \Pi_{eR}}{\partial \Phi}\right) + \frac{\partial f}{\partial R}$$

$$-\beta_0^2 \Pi_{eR} = 0. \quad (6c)$$

It is now evident that if in (6a,b,c) and in corresponding equations for Π_{mr} and g the following definitions are introduced:

$$f \equiv -\frac{\partial \Pi_{eR}}{\partial R}, \quad g \equiv -\frac{\partial \Pi_{mR}}{\partial R}, \quad (7)$$

equations (6a) and (6b) are satisfied automatically, while (6c) and its equivalent for Π_{mR} and g reduce to

$$\frac{\partial^2 \Pi_R}{\partial R^2} + \nabla^2_{\Theta,\Phi}\Pi_R + \beta_0^2 \Pi_R = 0,$$

$$\Pi_R = \Pi_{eR} \text{ or } \Pi_{mR}. \quad (8)$$

where $\nabla^2_{\Theta,\Phi}$ is that part of the Laplacian operator in spherical coördinates that involves differentiation only with respect to Θ and Φ. In the problem to be analyzed complete rotational symmetry obtains, so that

$$\frac{\partial \Pi_R}{\partial \Phi} = 0, \quad (9)$$

and (8) reduces to

$$\frac{\partial^2 \Pi_R}{\partial R^2} + \frac{1}{R^2 \sin \Theta}\frac{\partial}{\partial \Theta}\left(\sin \Theta \frac{\partial \Pi_R}{\partial \Theta}\right)$$

$$+ \beta_0^2 \Pi_R = 0. \quad (10)$$

Note that this is *not* a scalar wave equation.

With (4) and (7) substituted in (2a,b) the electro-magnetic field is easily determined and separated into components of electric and magnetic type. They are

Electric type (TM)

$$E_\Theta = \frac{1}{R}\frac{\partial^2 \Pi_{eR}}{\partial \Theta \partial R}, \quad (11a)$$

$$E_\Phi = \frac{1}{R \sin \Theta}\frac{\partial^2 \Pi_{eR}}{\partial \Phi \partial R}, \quad (11b)$$

$$E_R = \frac{\partial^2 \Pi_{eR}}{\partial R^2} + \beta_0^2 \Pi_{eR}, \quad (11c)$$

$$B_\Theta = \frac{j\beta_0^2}{\omega R \sin \Theta}\frac{\partial \Pi_{eR}}{\partial \Phi}, \quad (11d)$$

$$B_\Phi = \frac{-j\beta_0^2}{\omega R}\frac{\partial \Pi_{eR}}{\partial \Theta}, \quad (11e)$$

$$B_R = 0, \quad (11f)$$

Magnetic type (TE)

$$B_\Theta = \frac{1}{R}\frac{\partial^2 \Pi_{mR}}{\partial \Theta \partial R}, \quad (12a)$$

$$B_\Phi = \frac{1}{R \sin \Theta}\frac{\partial^2 \Pi_{mR}}{\partial \Phi \partial R}, \quad (12b)$$

$$B_R = \frac{\partial^2 \Pi_{mR}}{\partial R^2} + \beta_0^2 \Pi_{mR}, \quad (12c)$$

$$E_\Theta = \frac{-j\omega}{R \sin \Theta}\frac{\partial \Pi_{mR}}{\partial \Phi}, \quad (12d)$$

$$E_\Phi = \frac{j\omega}{R}\frac{\partial \Pi_{mR}}{\partial \Theta}, \quad (12e)$$

$$E_R = 0, \quad (12f)$$

where Π_{eR} and Π_{mR} are solutions of (10).

The solution of (10) is accomplished by the well-known method of separation of variables. By setting

$$\Pi_R = W(R)G(\Theta), \quad (13)$$

substituting in (10), and rearranging, the following equation is obtained:

$$-\frac{1}{G}\left[\frac{1}{\sin \Theta}\frac{\partial}{\partial \Theta}\left(\sin \Theta \frac{\partial G}{\partial \Theta}\right)\right]$$

$$= R^2\left(\frac{1}{W}\frac{\partial^2 W}{\partial R^2} + \beta_0^2\right). \quad (14)$$

Since the left-hand member of (14) is a function of Θ alone while the right-hand member is a function of R alone, the two expressions can be equal in general only if both are equal to a constant. Let this be $n(n + 1)$, where n is an arbitrary number. The resulting two equations are

$$\frac{d^2 W}{dR^2} + \left[\beta_0^2 - \frac{n(n+1)}{R^2}\right]W = 0, \quad (15)$$

$$\frac{1}{\sin \Theta}\frac{d}{d\Theta}\left(\sin \Theta \frac{dG}{d\Theta}\right) + n(n+1)G = 0. \quad (16)$$

The radial equation (15) may be transformed by changing both the dependent and the independent variables. With $x = \beta_0 R$, $F = W/\sqrt{x}$, and noting that $n(n + 1) + \frac{1}{4} = (n + \frac{1}{2})^2$, the following transformed equation may be derived:

$$\frac{d^2F}{dx^2} + \frac{1}{x}\frac{dF}{dx} + \left[1 - \frac{(n + \frac{1}{2})^2}{x^2}\right]F = 0. \quad (17)$$

This is Bessel's equation of order $n + \frac{1}{2}$. Its solutions are well known and can be expressed in terms of general cylinder functions $Z_{n+\frac{1}{2}}(x)$ or as linear combinations of Bessel and Neumann functions, or of Hankel functions of the first and second kind. Thus, with $x = \beta_0 R$,

$$F = Z_{n+\frac{1}{2}}(x)$$
$$= \begin{cases} A_n J_{n+\frac{1}{2}}(x) + B_n N_{n+\frac{1}{2}}(x), \\ A'_n H^{(1)}_{n+\frac{1}{2}}(x) + B'_n H^{(2)}_{n+\frac{1}{2}}(x), \end{cases} \quad (18)$$

where A_n, B_n, A'_n, and B'_n are arbitrary constants. With (17) the solution of the original equation (14) may be expressed as follows:

$$W(R) = Fx^{\frac{1}{2}} = x^{\frac{1}{2}} Z_{n+\frac{1}{2}}(x)$$
$$= \begin{cases} A_n x^{\frac{1}{2}} J_{n+\frac{1}{2}}(x) + B_n x^{\frac{1}{2}} N_{n+\frac{1}{2}}(x), \\ A'_n x^{\frac{1}{2}} H^{(1)}_{n+\frac{1}{2}}(x) + B'_n x^{\frac{1}{2}} H^{(2)}_{n+\frac{1}{2}}(x). \end{cases} \quad (19)$$

An alternative form of the solution is expressed in terms of the so-called *spherical Bessel functions** as follows:

$$W(R) = z_n(x)$$
$$= \begin{cases} x[a_n j_n(x) + b_n n_n(x)], \\ x[a'_n h^{(1)}_n(x) + b'_n h^{(2)}_n(x)], \end{cases} \quad (20)$$

where a_n, b_n, a'_n, and b'_n are arbitrary constants and where, by definition,

$$j_n(x) \equiv \sqrt{\frac{\pi}{2x}} J_{n+\frac{1}{2}}(x),$$

$$n_n(x) \equiv \sqrt{\frac{\pi}{2x}} N_{n+\frac{1}{2}}(x), \quad (21a)$$

$$h^{(1)}_n(x) \equiv \sqrt{\frac{\pi}{2x}} H^{(1)}_{n+\frac{1}{2}}(x),$$

$$h^{(2)}_n(x) \equiv \sqrt{\frac{\pi}{2x}} H^{(2)}_{n+\frac{1}{2}}(x). \quad (21b)$$

In (20) and (21), $z_n(x)$ is a general spherical cylinder function, $j_n(x)$, $n_n(x)$, and $h_n(x)$ are respectively, *spherical* Bessel, Neuman, and Hankel functions.

---

* See, for example, Stratton, ref. I.52, pp. 404–406.

The spherical cylinder functions are related very simply to the trigonometric functions. Thus,

$n = 0$:

$$j_0(x) = \frac{\sin x}{x}, \quad n_0(x) = -\frac{\cos x}{x},$$

$$h^{(1)}_0(x) = -j\frac{e^{jx}}{x}, \quad h^{(2)}_0(x) = j\frac{e^{-jx}}{x};$$
$$(22)$$

$n = 1$:

$$j_1(x) = -\frac{\cos x}{x} + \frac{\sin x}{x^2},$$

$$n_1(x) = -\frac{\sin x}{x} - \frac{\cos x}{x^2}, \quad (23)$$

$$h^{(1)}_1(x) = -\frac{e^{jx}}{x}\left(1 + \frac{j}{x}\right),$$

$$h^{(2)}_1(x) = -\frac{e^{-jx}}{x}\left(1 - \frac{j}{x}\right).$$

Asymptotic expressions for large arguments are

$$j_n(x) \to \frac{1}{x}\cos\left(x - \frac{n+1}{2}\pi\right),$$

$$n_n(x) \to \frac{1}{x}\sin\left(x - \frac{n+1}{2}\pi\right),$$
$$(24)$$

$$h^{(1)}_n(x) \to \frac{1}{x} e^{j\left(x - \frac{n+1}{2}\pi\right)},$$

$$h^{(2)}_n(x) \to \frac{1}{x} e^{-j\left(x - \frac{n+1}{2}\pi\right)}.$$

For small arguments,

$$j_n(x) \to \frac{x^n}{1 \cdot 3 \cdot 5 \cdots (2n+1)},$$

$$n_n(x) \to -\frac{1 \cdot 3 \cdot 5 \cdots (2n-1)}{x^{n+1}}. \quad (25)$$

The equation (16) may be reduced to a more familiar form known as Legendre's equation by changing the independent variable from Θ to $\mu = \cos\Theta$. The transformed equation is

$$(1 - \mu^2)\frac{d^2G}{d\mu^2} - 2\mu\frac{dG}{d\mu} + n(n+1)G = 0. \quad (26)$$

A general solution of (26) is

$$G = T_n(\Theta) = C_n P_n(\mu) + D_n Q_n(\mu), \quad (27)$$

where $P_n(\mu) = P_n(\cos \Theta)$ are Legendre functions of the first kind and $Q_n(\mu) = Q_n(\cos \Theta)$ are Legendre functions of the second kind—both of order n[1,51]; C_n and D_n are arbitrary constants.

When n is a nonnegative integer the $P_n(\mu)$ are polynomials of degree n. They are even for n even, odd for n odd. The first few Legendre polynomials are

$$P_0(\mu) = 1, \quad P_1(\mu) = \mu, \quad P_2(\mu) = \tfrac{3}{2}\mu^2 - \tfrac{1}{2},$$
$$P_3(\mu) = \tfrac{5}{2}\mu^3 - \tfrac{3}{2}\mu. \quad (28)$$

The functions of the second kind are given by

$$Q_n(\mu) = \tfrac{1}{2}P_n(\mu)\ln\frac{1+\mu}{1-\mu} - \left[P_{n-1}(\mu)P_0(\mu)\right.$$
$$+ \tfrac{1}{2}P_{n-2}(\mu)P_1(\mu) + \tfrac{1}{3}P_{n-3}(\mu)P_3(\mu) + \cdots$$
$$\left. + \frac{1}{n}P_0(\mu)P_{n-1}(\mu)\right]. \quad (29)$$

Note that $Q_n(\mu) = Q_n(\cos \Theta)$ becomes infinite at $\Theta = 0$ or $\mu = 1$, so that $D_n = 0$ for a region including the z-axis, $\Theta = 0$. Note also that $\tfrac{1}{2}\ln[(1+\mu)/(1-\mu)] = \ln \cot(\Theta/2)$.

If n is not integral, the functions $Q_n(\mu)$ may be avoided. In their place the functions $P_n(-\mu)$ are convenient independent solutions. For all values of n,

$$P_n(\mu) = P_n(\cos \Theta)$$
$$= \sum_{k=0}^{\infty} \frac{(-1)^k \Gamma(n+k+1)}{\Gamma(n-k+1)(k!)^2} \sin^{2k}(\Theta/2). \quad (30)$$

(Note that for integral values of n, $P_n(\cos \Theta) = (-1)^n P_n(-\cos \Theta)$, so that the $P_n(-\cos \Theta)$ are not independent solutions.) Hence, for nonintegral values of n, (27) may be replaced by

$$G = T_n(\Theta) = C'_n P_n(\mu) + D'_n P_n(-\mu). \quad (31)$$

With (19) and (27) or (31) the general solution of the original equation (10) is a linear combination of the following form:

$$\Pi_R = \sum_n (\beta_0 R)^{\frac{1}{2}} Z_{n+\frac{1}{2}}(\beta_0 R) T_n(\Theta), \quad (32a)$$

or, in terms of the spherical cylinder functions, (20) in place of (19),

$$\Pi_R = \sum_n z_n(\beta_0 R) T_n(\Theta). \quad (32b)$$

CONICAL ANTENNAS

3. Boundary Conditions and Equations for the Symmetric, Spherically Capped Biconical Antenna

The biconical antenna was studied first by Schelkunoff[18] and later analyzed by Smith[20] and Tai.[24] It consists of two identical conducting cones DOD in Fig. 3.1 with conducting caps DED that are segments of the sphere of radius l. The biconical structure is rotationally symmetric and is immersed in air. It is center-driven by an idealized point generator with no internal impedance and an emf V_0^e. This excites radial surface currents of density l_R on the conical surfaces and surface currents l_Θ on the spherical caps. There are no circulating currents l_Φ around the z-axis. For simplicity, let all conductors be assumed to be perfect, so that the boundary condition

$$\hat{n} \times \mathbf{E} = 0 \quad (1)$$

is true on all conductors. Hence, the specific boundary conditions on the conductors are

$$E_\Theta = 0 \text{ at } R = l$$
$$\text{with } \Theta \leq \Theta_0 \text{ or } \Theta \geq \pi - \Theta_0, \quad (2a)$$

$$E_R = 0 \text{ at } \begin{cases} \Theta = \Theta_0 \\ \Theta = \pi - \Theta_0 \end{cases} \text{ with } R \leq l. \quad (2b)$$

In addition, the field must satisfy a radiation condition at infinity. Distributions of surface current and charge are obtained from the boundary conditions

$$\hat{n} \cdot \mathbf{E} = -\eta_f/\epsilon_0, \quad (3a)$$
$$\hat{n} \times \mathbf{B} = -\mathbf{l}_f/\nu_0, \quad (3b)$$

where η_f is the surface density of free charge and $\mathbf{l}_f$ the surface density of conduction current. The condition $\hat{n} \cdot \mathbf{B} = 0$ is satisfied automatically. Since it is assumed that the only currents excited are l_R on the conical surfaces and l_Θ on the cap surfaces, it follows from (3b) that the only nonvanishing component of the magnetic **B**-field is B_Φ, that is,

$$B_\Theta = 0, \quad B_\Phi \neq 0, \quad B_R = 0. \quad (4)$$

Examination of the components of the electromagnetic field given in (2.11) and (2.12) shows that (4) with the condition $\partial \Pi_{mR}/\partial \Phi = 0$ for rotational symmetry requires the entire magnetic field of magnetic type, that is, derived from the magnetization potential, to vanish. Hence, all components of the electric field of magnetic type must also vanish, which is equivalent to $\Pi_{mR} = 0$. In other words, as could have been anticipated from the fact that there are no circulating currents, only components of the field of electric type derived from the polarization potential Π_{eR} are involved. They are

$$E_\Theta = \frac{1}{R}\frac{\partial^2 \Pi_{eR}}{\partial \Theta \partial R}, \quad E_\Phi = 0,$$

$$E_R = \frac{\partial^2 \Pi_{eR}}{\partial R^2} + \beta_0^2 \Pi_{eR} = \frac{n(n+1)}{R^2}\Pi_{er}, \quad (5a)$$

$$B_\Theta = 0, \quad B_\Phi = \frac{-j\beta_0^2}{\omega R}\frac{\partial \Pi_{eR}}{\partial \Theta}, \quad B_R = 0. \quad (5b)$$

The last step in (5a) follows from multiplying (2.15) by G, and using (2.13). The existence of a component of the *electric* field in the direction of the polarization potential is characteristic of all fields of *electric* type. Note that since $\Pi_{er} = R^2 E_R/n(n+1)$ from (5a), the *entire* electromagnetic field may be derived from E_R instead of from Π_{er} if desired. Since the component of the magnetic field $B_R = 0$, the magnetic field is transverse to the direction of Π_{eR}, so that an appropriate and commonly used alternative name for the fields of electric type is *transverse magnetic* or *TM field*.

The three nonvanishing components of the field are obtained from (5a, b) with (2.32b). Thus,

$$E_\Theta = \frac{\beta_0}{R}\sum_n z'_n(\beta_0 R)T'_n(\Theta), \quad (6a)$$

$$E_R = \frac{1}{R^2}\sum_n n(n+1)z_n(\beta_0 R)T_n(\Theta), \quad (6b)$$

$$B_\Phi = \frac{-j\beta_0^2}{\omega R}\sum_n z_n(\beta_0 R)T'_n(\Theta). \quad (6c)$$

The primes denote differentiation with respect to the argument. The choice of the particular values of n and the specific form of $z_n(\beta_0 R)$ and $T_n(\Theta)$ depend upon the boundary conditions.

In applying the boundary conditions (2a, b) to (6a, b) it is convenient to treat the region outside the conducting cone in two parts (Fig. 3.1): an internal region I inside the sphere of radius l, and an external region II occupying all space where $R > l$. The boundary condition for the internal region is (2b); that for the external region is (2a). In order to satisfy these boundary conditions it is convenient to choose different forms of the functions $z_n(\beta_0 R)$ and $T_n(\Theta)$, appropriate, respectively, to the interior and exterior regions, and then to match the two solutions across the common mathematical boundary between regions I and II.

In the exterior region II defined by $R > l$, $\Theta = 0, \pi$ are included, so that neither $Q_n(\cos\Theta)$ nor $P_n(-\cos\Theta)$ is a finite solution since $P_n(-\cos\Theta)$ with n nonintegral and $Q_n(\cos\Theta)$ for all n become infinite at at least one of these boundary points. Moreover, the region extends to $R = \infty$, where of all the spherical Bessel functions only $h_n^{(2)}(\beta_0 R)$ vanishes. Therefore, an appropriate solution for the exterior region is

$$\Pi_{eR}^{II} = \sum_n C_n P_n(\cos\Theta)\beta_0 R\, h_n^{(2)}(\beta_0 R). \quad (7)$$

This formula defines the *exterior complementary modes*, which are discussed in Sec. 6.

Since $\Theta = 0$ is not included in the interior region, $Q_n(\cos\Theta)$ has no singularity. On the other hand, the origin is in region I, and of all spherical Bessel functions only $j_n(\beta_0 R)$ is finite at $R = 0$ for $n > 0$. When $n = 0$, on the other hand, $\beta_0 R\, n_0(\beta_0 R)$, $\beta_0 R\, h_0^{(1)}(\beta_0 R)$, and $\beta_0 R\, h_0^{(2)}(\beta_0 R)$ also are finite at $R = 0$. Since $P_0(\cos\Theta) = 1$, $Q_0(\cos\Theta) = \ln\cot\tfrac{1}{2}\Theta$; and also $j_0(\beta_0 R) = \sin\beta_0 R/\beta_0 R$, $n_0(\beta_0 R) = -\cos\beta_0 R/\beta_0 R$, $h_0^{(1)}(\beta_0 R) = -je^{j\beta_0 R}/\beta_0 R$, $h_0^{(2)}(\beta_0 R) = je^{-j\beta_0 R}/\beta_0 R$, appropriate solutions for region I are as follows:

For $n = 0$,

$$(\Pi_{eR}^I)_d = \Pi_d = (C_0 + B_0 \ln\cot\tfrac{1}{2}\Theta)$$
$$\times (a_0 \sin\beta_0 R - b_0 \cos\beta_0 R), \quad (8a)$$

or,

$$(\Pi_{eR}^I)_d = \Pi_d = -j(C_0 + B_0 \ln\cot\tfrac{1}{2}\Theta)$$
$$\times (a'_0 e^{j\beta_0 R} - b'_0 e^{-j\beta_0 R}); \quad (8b)$$

For $n > 0$,

$$(\Pi_{eR}^I)_c = \sum_n T_n(\Theta)\beta_0 R\, j_n(\beta_0 R), \quad (9)$$

where, in general,

$$T_n(\Theta) = C_n P_n(\cos\Theta) + D_n Q_n(\cos\Theta), \quad (10a)$$

and where, alternatively, for n nonintegral

$$T_n(\Theta) = C'_n P_n(\cos\Theta) + D'_n P_n(-\cos\Theta), \quad (10b)$$

Of these solutions, (8a) and (8b) define the *dominant mode* in trigonometric and exponential forms, while (9) defines the *interior complementary modes*.

4. The Dominant Mode; Apparent Terminal Admittance*

In order to develop the formulation of the dominant mode in a biconical antenna in a manner paralleling that in a conventional terminated transmission line, it is advantageous to begin with an infinite biconical antenna. If the length l of the cones in a symmetric biconical structure is infinite, there is no exterior region. In order to vanish at $R = \infty$, the polarization potential for the dominant interior mode then must be given by (3.8b) with $a' = 0$. Thus,

$$\Pi_d = (C_0 + B_0 \ln \cot \tfrac{1}{2}\Theta)(jb_0' e^{-j\beta_0 R}).$$
$$(R \leq l = \infty) \quad (1)$$

The corresponding dominant-mode polarization potential for cones of finite length is given by (3.8a); it is

$$\Pi_d = (C_0 + B_0 \ln \cot \tfrac{1}{2}\Theta)(a_0 \sin \beta_0 R - b_0 \cos \beta_0 R).$$
$$(R \leq l) \quad (2)$$

The components of the electromagnetic field for the dominant mode are derived directly from (3.5a,b), with

$$\frac{d}{d\Theta} \ln \cot \tfrac{1}{2}\Theta = -\csc \Theta \quad (3)$$

and a change in the arbitrary constants from $B_0 b_0'$, $B_0 a_0$, and $B_0 b_0$, respectively, to $b' = 2\pi B_0 b_0' \beta_0 / \zeta_0$, $a = 2\pi B_0 a_0 \beta_0 / \zeta_0$, $b = 2\pi B_0 b_0 \beta_0 / \zeta_0$. The field for the infinite cones is

$$E_\Theta = \frac{\zeta_0 b'}{2\pi R \sin \Theta} e^{-j\beta_0 R}, \quad E_\Phi = 0, \quad E_R = 0,$$

$$B_\Theta = 0, \quad B_\Phi = \frac{b'}{v_0 2\pi R \sin \Theta} e^{-j\beta_0 R}, \quad B_R = 0.$$
$$(R \leq l = \infty) \quad (4)$$

For the finite cones, the dominant-mode field is

$$(E_\Theta)_d = \frac{\zeta_0}{2\pi R \sin \Theta}(a \cos \beta_0 R + b \sin \beta_0 R),$$
$$\quad (5a)$$

$$(E_\Phi)_d = 0, \quad (E_R)_d = 0,$$

$$(B_\Phi)_d = \frac{-j}{v_0 2\pi R \sin \Theta}(a \sin \beta_0 R - b \cos \beta_0 R),$$
$$\quad (5b)$$

$$(B_\Theta)_d = 0, \quad (B_R)_d = 0. \quad (R \leq l)$$

Note that (4) and (5) satisfy the boundary condition $E_R = 0$ at $\Theta = \Theta_0$ and $\Theta = \pi - \Theta_0$, automatically.

---
* First drafts of this section and sections 5 through 10 were prepared by Dr. C. T. Tai.

The expressions for E_Θ and B_Φ in (4) may be interpreted as representing spherical waves, that is, spherical surfaces of constant phase, *traveling* radially outward between the perfectly conducting cones with the free-space velocity v_0. Similarly, the expressions (5) may be regarded as representing *standing* spherical waves. On conventional lines, standing waves occur when the line is terminated in an impedance that differs from the characteristic impedance of the line. In the case of the biconical antenna, there is no actual terminating impedance concentrated at $R = l$. However, it is convenient *as an intermediate step* in the evaluation of the input impedance of the biconical antenna to treat the cones as if they were terminated in an impedance Z_{la} at $R = l$ and subsequently relate the fictitious impedance Z_{la} to the higher-order modes.

The surface densities of charge and current that maintain the dominant-mode fields are derived readily from the boundary conditions (3.3a,b) applied to (4) and (5). For the infinite cones the following surface densities of charge and current are on the upper cone ($\Theta = \Theta_0$):

$$\eta_f = \epsilon_0 E_\Theta = \frac{b'}{v_0 2\pi R \sin \Theta_0} e^{-j\beta_0 R}; \quad (6)$$

$$l_{fR} = v_0 B_\Phi = \frac{b'}{2\pi R \sin \Theta_0} e^{-j\beta_0 R},$$

$$l_{f\Theta} = 0, \quad l_{f\Phi} = 0. \quad (7)$$

The total charge per unit radial distance on the upper cone is

$$q(R) = (2\pi R \sin \Theta_0)\eta_f = \frac{b'}{v_0} e^{-j\beta_0 R}.$$
$$(R \leq l = \infty) \quad (8)$$

The total radial surface current crossing a ring of radius $2\pi R \sin \Theta$ at R on the upper cone is

$$I(R) = (2\pi R \sin \Theta_0)l_{fR} = b' e^{-j\beta_0 R}.$$
$$(R \leq l = \infty) \quad (9)$$

It is evident from (9) that

$$b' = I(0), \quad (10)$$

where $I(0)$ is the complex amplitude of the total surface current entering the upper cone from the point generator at $R = 0$.

For the finite cones the corresponding expressions for the dominant-mode charges and currents are

$$q_d(R) = (2\pi R \sin \Theta_0)\eta_f$$
$$= \frac{1}{v_0}(a \cos \beta_0 R + b \sin \beta_0 R), \quad (11)$$

$$I_d(R) = (2\pi R \sin \Theta_0) I_{fR}$$
$$= -j(a \sin \beta_0 R - b \cos \beta_0 R). \quad (12)$$

Note that $dI_d(R)/dR + j\omega q_d(R) = 0$, as required for the conservation of electric charge.

Let the voltage $V(R)$ between a point $P_1(\Theta_0, \Phi, R)$ on the upper cone and the corresponding point $P_2(\pi - \Theta_0, \Phi, R)$ on the lower cone be *defined* as the line integral of the electric field along the meridian joining P_1 and P_2; that is,

$$V(R) \equiv R \int_{\Theta_0}^{\pi-\Theta_0} E_\Theta \, d\Theta. \quad (13)$$

With (4) and (9) the voltage for the infinite cones becomes

$$V(R) = \frac{\zeta_0 b'}{2\pi} e^{-j\beta_0 R} \int_{\Theta_0}^{\pi-\Theta_0} \frac{d\Theta}{\sin \Theta} = I(R) Z_c,$$
$$(R \leq l = \infty) \quad (14)$$

where the characteristic impedance of the infinite biconical transmission line is

$$Z_c \equiv \int_{\Theta_0}^{\pi-\Theta_0} \frac{d\Theta}{\sin \Theta} = \frac{\zeta_0}{\pi} \ln \cot \tfrac{1}{2}\Theta_0$$
$$= 120 \ln \cot \tfrac{1}{2}\Theta_0 \text{ ohms}. \quad (15)$$

Note that just as in conventional lines Z_c is real if perfect conductors and dielectrics are assumed. If desired, it is also possible to define inductance and capacitance per unit length of biconical transmission line. Thus, since by definition $Z_c = \sqrt{l_0/c_0}$ and $v_0 = 1/\sqrt{l_0 c_0}$, it follows that

$$l_0 = \frac{Z_c}{v_0} = \frac{1}{\pi v_0} \ln \cot \tfrac{1}{2}\Theta_0, \quad (16a)$$

$$c_0 = Z_c v_0 = \pi \epsilon_0 / \ln \cot \tfrac{1}{2}\Theta_0. \quad (16b)$$

The dominant-mode voltage for the finite cones is defined by (13) with (5a). Thus,

$$V_d(R) = Z_c(a \cos \beta_0 R + b \sin \beta_0 R), \quad (17)$$

where Z_c is defined in (15). The constants a and b in (12) and (17) may be evaluated in terms of the driving voltage $V(0)$ and the current $I(0)$ entering and leaving the point generator at the apex. Thus,

$$b = -jI(0), \qquad a = V(0)/Z_c, \quad (18)$$

so that

$$V_d(R) = V(0) \cos \beta_0 R - jZ_c I(0) \sin \beta_0 R, \quad (19)$$

$$I_d(R) = I(0) \cos \beta_0 R - j\frac{V(0)}{Z_c} \sin \beta_0 R, \quad (20)$$

$$q_d(R) = c_0 V_d(R). \quad (21)$$

The total current at the ends $R = l$ of finite cones consists only of charges flowing around the edges onto the hemispherical caps. It is given by

$$I(l) = I_d(l) + I_c(l), \quad (22)$$

where $I_d(l)$ is the dominant mode current and $I_c(l)$ is the resultant current associated with all of the complimentary interior modes. Even when the angle of the cone is very small and $I(l)$ almost zero, the dominant-mode current $I_d(l)$ may be large, since it is then essentially the negative of the higher-mode currents. Hence, it is appropriate and subsequently very convenient to define an *apparent dominant-mode terminal admittance* Y_{la}. This admittance is the apparent load at the end of the biconical transmission line if only dominant-mode currents and voltages are considered. It is the analogy of Y_{sa} for the two-wire line with end effect. As in conventional transmission-line theory Y_{la} is defined as follows:

$$Y_{la} = \frac{I_d(l)}{V_d(l)}$$
$$= \frac{I(0) \cos \beta_0 l - j[V(0)/Z_c] \sin \beta_0 l}{V(0) \cos \beta_0 l - jZ_c I(0) \sin \beta_0 l}. \quad (23)$$

This equation may be solved for $I(0)$ in terms of $V(0)$ and Y_{la}. The result is

$$I(0) = \frac{V(0)}{Z_c} \left(\frac{Z_c Y_{la} \cos \beta_0 l + j \sin \beta_0 l}{\cos \beta_0 l + jZ_c Y_{la} \sin \beta_0 l} \right).$$
$$(24)$$

Substitution of (24) in (20) gives the following expression for the dominant-mode current at radius R:

$$I_d(R) = \frac{V(0)}{Z_c}$$
$$\times \left[\frac{\cos \beta_0 R(Z_c Y_{la} \cos \beta_0 l + j \sin \beta_0 l)}{\cos \beta_0 l + jZ_c Y_{la} \sin \beta_0 l)}{\cos \beta_0 l + jZ_c Y_{la} \sin \beta_0 l} \right]. \quad (25)$$

With standard trigonometric formulas, (25) is reduced to

$$I_d(R) = I_m[\sin \beta_0(l - R) - jZ_c Y_{la} \cos \beta_0(l - R)]. \quad (26)$$

where a new constant I_m is defined by

$$I_m \equiv \frac{jV(0)}{Z_c(\cos \beta_0 l + jZ_c Y_{la} \sin \beta_0 l)}$$
$$= \frac{I(0)}{\sin \beta_0 l - jZ_c Y_{la} \cos \beta_0 l}. \quad (27)$$

Note that I_m is the maximum current. When $Y_{la} = 0$, $I_m = I(0)/\sin \beta_0 l$ at $R = l - \lambda_0/4$, (which corresponds to $I_m = I(0)/\sin \beta_0 h$ at $z = h - \lambda_0/4$ for the cylindrical antenna with sinusoidally distributed current). With (23) and (27) substituted in (19), the dominant-mode voltage for $R \leq l$ is

$$V_d(R) = I_m Z_c[Z_{la} Y_{la} \sin \beta_0(l - R) - j \cos \beta_0(l - R)]. \quad (28)$$

The dominant-mode electromagnetic field (5) for finite cones also may be expressed in terms of the new constants Y_{la} and I_{sa} instead of a and b using (23) and (27). The resulting formulas for the two nonvanishing components in the range $R \leq l$ are

$$(E_\Theta)_d = \frac{\zeta_0 I_m}{2\pi R \sin \Theta}[Z_c Y_{la} \sin \beta_0(l - R) - j \cos \beta_0(l - R)], \quad (29)$$

$$(B_\Phi)_d = \frac{I_m}{v_0 2\pi R \sin \Theta}[\sin \beta_0(l - R) - jZ_c Y_{la} \cos \beta_0(l - R)]. \quad (30)$$

5. Interior Complementary Modes; Input Impedance

The polarization potential for the complementary interior modes is given by (3.9), namely,

$$(\Pi_{eR}^I)_c = \sum_n S_n(\beta_0 R) T_n(\Theta), \quad (1)$$

where

$$S_n(\beta_0 R) \equiv \beta_0 R\, j_n(\beta_0 R) \quad (2)$$

and, with $\mu \equiv \cos \Theta$,

$$T_n(\Theta) = C'_n P_n(\cos \Theta) + D'_n P_n(-\cos \Theta)$$
$$T_n(\mu) = C'_n P_n(\mu) + D'_n P_n(-\mu) \quad (3)$$

for n nonintegral; C'_n and D'_n are arbitrary constants. The nonvanishing components of the electromagnetic field in Region I are obtained from (1), using (3.5a, b). They are

$$(E_\Theta^I)_c = \frac{\beta_0}{R} \sum_n S'_n(\beta_0 R) T'_n(\Theta), \quad (4a)$$

$$(E_R^I)_c = \frac{1}{R^2} \sum_n n(n + 1) S_n(\beta_0 R) T_n(\Theta), \quad (4b)$$

$$(B_\Phi^I)_c = -\frac{j\beta_0^2}{\omega R} \sum_n S_n(\beta_0 R) T'_n(\Theta), \quad (4c)$$

where primes on S_n and T_n denote differentiation with respect to the arguments.

The boundary condition that must be satisfied in region I is (3.2b), $E_R = 0$ at $\Theta = \Theta_0$ and $\Theta = \pi - \Theta_0$. This is satisfied if, with $\mu_0 = \cos \Theta_0$,

$$T_n(\mu_0) = C'_n P_n(\mu_0) + D'_n P_n(-\mu_0) = 0, \quad (5a)$$

$$T_n(-\mu_0) = C'_n P_n(-\mu_0) + D'_n P_n(\mu_0) = 0. \quad (5b)$$

These equations can be satisfied simultaneously only if

$$C'_n = \pm D'_n. \quad (6)$$

The correct choice of sign in (6) may be determined by noting that for a symmetric, apex-driven biconical antenna the radial electric field must be an odd function of z, where, as shown in Fig. 3.1, the z-axis coincides with the axis of the cones; that is,

$$E_R^I(-z) = -E_R^I(z). \quad (7a)$$

Hence, since $\mu = \cos \Theta = z/R_0$,

$$T_n(-\mu) = -T_n(\mu), \quad (7b)$$

so that in (6)

$$C'_n = -D'_n. \quad (7c)$$

It follows that

$$T_n(\Theta) = 2C'_n L_n(\Theta), \quad (8)$$

where

$$L_n(\Theta) = \tfrac{1}{2}[P_n(\cos \Theta) - P_n(-\cos \Theta)]. \quad (9)$$

For a given value of Θ_0, therefore, there is an infinite set of characteristic values designated by n.* These must be determined from the equations

$$L_n(\Theta_0) = L_n(\pi - \Theta_0) = 0, \quad (10)$$

where $L_n(\Theta)$ satisfies the Legendre equation (2.16) in the form

$$\frac{1}{\sin\Theta}\frac{d}{d\Theta}\left[\sin\Theta\,\frac{dL_n(\Theta)}{d\Theta}\right] + n(n+1)L_n(\Theta) = 0. \quad (11a)$$

Alternatively $L_n(\mu)$ satisfies (2.26). That is

$$\frac{d}{d\mu}\left[(1-\mu^2)\frac{dL_n(\mu)}{d\mu}\right] + n(n+1)L_n(\mu) = 0. \quad (11b)$$

If (8) is substituted in (4a, b, c), the electromagnetic field associated with the interior complementary waves is defined. Instead of using the set of arbitrary constants C'_n appearing in (8), it is convenient to multiply each of these by an appropriate factor, the reciprocal of which then appears explicitly in the expressions for the field. The desired factors are contained in

$$a_n = j4\pi n(n+1)\omega\epsilon_0 S_n(\beta_0 l) C'_n. \quad (12)$$

With (12) and for $R \leq l$, (4a, b, c) become

$$(E_\Theta^I)_c = \frac{-j\zeta_0}{2\pi R}\sum_n \frac{a_n}{n(n+1)}\frac{S'_n(\beta_0 R)}{S_n(\beta_0 l)} L'_n(\Theta), \quad (13)$$

$$(E_R^I)_c = \frac{-j}{2\pi\omega\epsilon_0 R^2}\sum_n a_n \frac{S_n(\beta_0 R)}{S_n(\beta_0 l)} L_n(\Theta), \quad (14)$$

$$(B_\Phi^I)_c = \frac{-1}{2\pi v_0 R}\sum_n \frac{a_n}{n(n+1)}\frac{S_n(\beta_0 R)}{S_n(\beta_0 l)} L'_n(\Theta), \quad (15)$$

where the primes on S and L denote derivatives with respect to the arguments. The summation with respect to n is to be carried out over all of the values of n determined by solving (10). The entire field in the interior region I is the sum of the fields for the dominant mode in (4.29, 30) and the complementary modes is (13), (14), and (15). It involves the two constants I_m and Y_{la} of the dominant mode and, in addition, the set of n constants a_n of the complementary modes. These must be evaluated in terms of the driving voltage and the field in the exterior region. The charges per unit radial distance and the radial currents on the cones which maintain the complementary interior field are determined as for the dominant mode. Thus, for the upper cone, $\Theta = \Theta_0$,

$$q_c(R) = (2\pi R \sin\Theta_0)\eta_f$$
$$= (2\pi R \sin\Theta_0)\epsilon_0(E_\Theta^I)_c$$
$$= -j\zeta_0 \sin\Theta_0$$
$$\times \sum_n \frac{a_n}{n(n+1)}\frac{S'_n(\beta_0 R)}{S_n(\beta_0 l)}[L'_n(\Theta)]_{\Theta=\Theta_0}, \quad (16)$$

$$I_c(R) = (2\pi R \sin\Theta_0)I_{fR}$$
$$= (2\pi R \sin\Theta_0)\nu_0(B_\Phi^I)_c$$
$$= -\sin\Theta_0 \sum_n \frac{a_n}{n(n+1)}\frac{S_n(\beta_0 R)}{S_n(\beta_0 l)}[L'_n(\Theta)]_{\Theta=\Theta_0}. \quad (17)$$

Note that $dI_c(R)/dR + j\omega q_c(R) = 0$, as required for the conservation of electric charge. Since with (2), $S_n(0) = 0$, it follows that

$$q_c(0) = 0, \quad I_c(0) = 0. \quad (18)$$

Thus, the charges and currents that maintain the complementary interior higher-mode field are distributed over the two cones but *do not reach the point generator at the apex*, $R = 0$. Everywhere *except* at $R = 0$ the total current and charge on the cones include contributions from *both* the dominant mode and the higher modes.

The leading term of the higher-mode current at a very small radial distance δ from the apex is obtained from (17) with (2) and (2.25). It is

$$I_c(\delta) = -\sin\Theta_0$$
$$\times \left[\frac{a_n}{n(n+1)}\frac{2^n\Gamma(n+1)}{\Gamma(2n+2)}\right.$$
$$\left.\times \frac{[L'_n(\Theta)]_{\Theta=\Theta_0}}{S_n(\beta_0 l)}(\beta_0\delta)^{n+1}\right]_{n=n_1} \quad (19)$$

where n_1 is the smallest number in the sum in (17), provided the condition $\beta_0 \delta \ll 1$ is satisfied. In general, this current is negligible compared with the dominant-mode current.

---

* The use of the solution (3) or (2.31) instead of (2.27) implies that only nonintegral values of n occur. That this is necessarily true for the interior region is not obvious. The entire analysis could be carried out using (2.27) with no question of validity. For application exclusively to thin biconical antennas, to which the present analysis is subsequently restricted, the leading terms obtained using (3) instead of (2.27) are presumably correct, since in the range of validity of (9.1) n is nonintegral.

The voltage associated with the complementary modes may be defined as for the dominant mode. Thus,

$$V_c(R) = R \int_{\Theta_0}^{\pi-\Theta_0} (E_\Theta)_c \, d\Theta = 0, \qquad (20)$$

since $L_n(\Theta_0)$ and $L_n(\pi - \Theta_0)$ both vanish in accordance with (10). Therefore, the voltage defined for the biconical antenna is *exclusively of dominant mode* whereas the *total current* is partly dominant- and partly higher-mode except at the apex where it is dominant.

Since the currents and voltage associated with the complementary interior modes contribute nothing to either the current or the voltage *at the apex* $R = 0$, the driving-point impedance depends exclusively on the dominant-mode voltage and current. Thus, with $R = 0$ in (4.19) and (4.20),

$$Z_0 = \frac{V(0)}{I(0)} = \frac{V_d(0)}{I_d(0)}$$

$$= Z_c \left(\frac{Z_c Y_{la} \sin \beta_0 l - j \cos \beta_0 l}{\sin \beta_0 l - j Z_c Y_{la} \cos \beta_0 l} \right), \qquad (21)$$

where Y_{la} is the arbitrary constant in the form of an *apparent* terminal impedance yet to be evaluated in terms of the complementary modes. Note that the driving-point impedance of the biconical antenna is determined completely in (21) except for Y_{la}. Therefore the determination of Y_{la} is the principal problem remaining.

Note that, if the impedance is defined at a short distance δ from the apex, the admittance is given by

$$Y_\delta = \frac{1}{Z_\delta} = (Y_\delta)_d + (Y_\delta)_c, \qquad (22)$$

where

$$(Y_\delta)_d = \frac{I_d(\delta)}{V_d(\delta)}, \qquad (Y_\delta)_c = \frac{I_c(\delta)}{V_d(\delta)}. \qquad (23)$$

6. Exterior Complementary Modes; Matching of Fields

The appropriate expression for the polarization potential in the exterior region is given by (3.7). It may be expressed in the following convenient and compact form:

$$\Pi^{II}_{eR} = \sum_k C_k R_k(\beta_0 R) P_k(\cos \Theta),$$
$$k = 1, 2, 3, \cdots, \qquad (1)$$

where,

$$R_k(\beta_0 R) \equiv \beta_0 R \, h_k^{(2)}(\beta_0 R). \qquad (2)$$

Since the requirement that the radial electric field be odd in z applies in region II just as in region I, the $P_k(\cos \Theta)$ must themselves be odd functions of z. That is, only those $P_k(\cos \Theta)$ are to be included in the sum in (1) that are odd functions of $\mu = \cos \Theta$. These have k odd. For convenience, let the sum over odd values of k be designated by a prime on the sign of summation. Thus, $\sum_k' \equiv \sum_{k=1,3,5\cdots}$.

Using (3.5a, b), the nonvanishing components of the electromagnetic field are obtained from (1). They are

$$E^{II}_\Theta = \frac{\beta_0}{R} {\sum_k}' C_k R'_k(\beta_0 R) P'_k(\cos \Theta), \qquad (3)$$

$$E^{II}_R = \frac{1}{R^2} {\sum_k}' k(k+1) C_k R_k(\beta_0 R) P_k(\cos \Theta), \qquad (4)$$

$$B^{II}_\Phi = \frac{-j\beta_0^2}{\omega R} {\sum_k}' C_k R_k(\beta_0 R) P'_k(\cos \Theta), \qquad (5)$$

where the primes on R and P denote derivatives with respect to the argument. It is again convenient to introduce a different set of arbitrary constants by setting

$$b_k = j 2\pi k(k+1) \omega \epsilon_0 R_k(\beta_0 l) C_k. \qquad (6)$$

With (6) the components of the field become

$$E^{II}_\Theta = \frac{-j\zeta_0}{2\pi R} {\sum_k}' \frac{b_k}{k(k+1)} \frac{R'_k(\beta_0 R)}{R_k(\beta_0 l)} P'_k(\cos \Theta), \qquad (7)$$

$$E^{II}_R = \frac{-j}{2\pi \omega \epsilon_0 R^2} {\sum_k}' b_k \frac{R_k(\beta_0 R)}{R_k(\beta_0 l)} P_k(\cos \Theta), \qquad (8)$$

$$B^{II}_\Phi = \frac{-1}{2\pi v_0 R} {\sum_k}' \frac{b_k}{k(k+1)} \frac{R_k(\beta_0 R)}{R_k(\beta_0 l)} P'_k(\cos \Theta). \qquad (9)$$

The field defined in (7), (8), and (9) for the exterior region satisfies the field equations, the condition for symmetry, and the condition at $R = \infty$. It remains to impose conditions at the inner boundary of region II so that the tangential electric field vanishes on the spherical caps of the antenna and the entire electromagnetic field is continuous across the rest of the mathematical boundary sphere. The conditions are

$$E^{II}_\Theta = 0 \quad \text{when} \quad \begin{cases} 0 \leq \Theta \leq \Theta_0 \\ \pi - \Theta_0 \leq \Theta \leq \pi \end{cases}, \quad R = l; \qquad (10)$$

$$\left. \begin{aligned} E^{II}_\Theta &= E^I_\Theta \\ B^{II}_\Phi &= B^I_\Phi \end{aligned} \right\} \quad \text{when} \quad \Theta_0 \leq \Theta \leq \pi - \Theta_0,$$
$$R = l. \qquad (11)$$

Since E_R is continuous if B_Φ is continuous, it follows that only E_Θ and B_Φ need be matched explicitly.

The substitution of (7) in (10) gives

$$\sum_k{}' \frac{b_k}{k(k+1)} \rho_k(\beta_0 l) P'_k(\cos \Theta) = 0,$$

$$(0 \leq \Theta \leq \Theta_0, \pi - \Theta_0 \leq \Theta \leq \pi) \quad (12)$$

where

$$\rho_k(\beta_0 l) \equiv \frac{R'_k(\beta_0 l)}{R_k(\beta_0 l)} = \frac{h_k^{(2)'}(\beta_0 l)}{h_k^{(2)}(\beta_0 l)} \quad (13)$$

is the logarithmic derivative of the spherical Hankel function. Similarly, the substitution of (7), (4.29), and (5.13) in (11) gives

$$\sum_k{}' \frac{b_k}{k(k+1)} \rho_k(\beta_0 l) P'_k(\cos \Theta)$$

$$= \frac{I_m}{\sin \Theta} + \sum_n \frac{a_n}{n(n+1)} \sigma_n(\beta_0 l) L'_n(\Theta),$$

$$(\Theta_0 \leq \Theta \leq \pi - \Theta_0) \quad (14)$$

where

$$\sigma_n(\beta_0 l) \equiv \frac{S'_n(\beta_0 l)}{S_n(\beta_0 l)} = \frac{j'_n(\beta_0 l)}{j_n(\beta_0 l)} \quad (15)$$

is the logarithmic derivative of the spherical Bessel function of the first kind. Finally, substitution of (9), (4.30), and (5.15) in (11) gives

$$\sum_k{}' \frac{b_k}{k(k+1)} P'_k(\cos \Theta)$$

$$= \frac{jI_m Z_c Y_l}{\sin \Theta} + \sum_n \frac{a_n}{n(n+1)} L'_n(\Theta).$$

$$(\Theta_0 \leq \Theta \leq \pi - \Theta_0) \quad (16)$$

As a consequence of the orthogonal properties of the functions $P_k(\cos \Theta)$ and $L_n(\Theta)$ and their derivatives, the system of equations contained in (12), (14), and (16) can be solved by direct integration. Before such a solution can be carried out, a review of some of the more important properties of integrals involving products of Legendre functions is required.

7. Integrals of Products of Legendre Functions

It is known from the theory of Legendre functions[1.27] that if $T_n(\Theta)$ and $T_m(\Theta)$ are any two solutions of Legendre's equation (2.16) or (2.26) the following relations exist when $n \neq m$:

$$\int_{\Theta_0}^{\pi - \Theta_0} T_n(\Theta) T_m(\Theta) \sin \Theta \, d\Theta$$

$$= \int_{-\mu_0}^{\mu_0} T_n(\mu) T_m(\mu) d\mu$$

$$= \frac{(1 - \mu_0^2) \left[T_n(\mu) T'_m(\mu) - T_m(\mu) T'_n(\mu) \right]_{-\mu_0}^{\mu_0}}{n(n+1) - m(m+1)}, \quad (1)$$

$$\int_{\Theta_0}^{\pi - \Theta_0} T'_n(\Theta) T'_m(\Theta) \sin \Theta \, d\Theta$$

$$= \int_{-\mu_0}^{\mu_0} (1 - \mu^2) T'_n(\mu) T'_m(\mu) \, d\mu$$

$$= \left[(1 - \mu)^2 T_n(\mu) T'_m(\mu) \right]_{-\mu_0}^{\mu_0}$$

$$+ m(m+1) \int_{-\mu_0}^{\mu_0} T_n(\mu) T_m(\mu) \, d\mu. \quad (2)$$

As usual, $\mu \equiv \cos \Theta$ and primes denote differentiation with respect to the indicated argument.

For the special case in which $T_n(\mu) = P_k(\mu)$, $T_m(\mu) = P_r(\mu)$, $\Theta_0 = 0$ or $\mu_0 = 1$, where $P_k(\mu)$ and $P_r(\mu)$ are two Legendre polynomials, it follows from (1) and (2) that with $k \neq r$

$$\int_0^\pi P_k(\cos \Theta) P_r(\cos \Theta) \sin \Theta \, d\Theta = 0, \quad (3)$$

$$\int_0^\pi \frac{\partial P_k(\cos \Theta)}{\partial \Theta} \frac{\partial P_r(\cos \Theta)}{\partial \Theta} \sin \Theta \, d\Theta = 0. \quad (4)$$

On the other hand for $k = r$, as shown by Hobson,[1.27]

$$J_{kk} \equiv \int_0^\pi P_k^2(\cos \Theta) \sin \Theta \, d\Theta = \frac{2}{2k+1} \quad (5)$$

and

$$\int_0^\pi \left[\frac{\partial P_k(\cos \Theta)}{\partial \Theta} \right]^2 \sin \Theta \, d\Theta = k(k+1) J_{kk}. \quad (6)$$

For the special case in which $T_n(\mu) = L_n(\mu)$, $T_m(\mu) = L_m(\mu)$, where $L_n(\mu)$ and $L_m(\mu)$ are the Legendre functions defined by (5.9), the following relations obtain:

$$\left. \begin{array}{l} \displaystyle\int_{\Theta_0}^{\pi - \Theta_0} L_n(\Theta) L_m(\Theta) \sin \Theta \, d\Theta = 0, \quad (7) \\[1em] \displaystyle\int_{\Theta_0}^{\pi - \Theta_0} L'_n(\Theta) L'_m(\Theta) \sin \Theta \, d\Theta = 0. \quad (8) \end{array} \right\} n \neq m$$

When $n = m$, it is possible to set $m = n + \Delta n$ in (1) and evaluate the limiting value of the integral as $\Delta n \to 0$. The value so obtained is denoted by J_{nn}. It is

$$J_{nn} \equiv \int_{\Theta_0}^{\pi-\Theta_0} L_n^2(\Theta) \sin \Theta \, d\Theta$$

$$= \frac{2(1-\mu_0^2)}{2n+1} \left[\frac{\partial L_n(\mu)}{\partial n} \frac{\partial L_n(\mu)}{\partial \mu} \right]_{\mu=\mu_0}. \quad (9)$$

The factor $\partial L_n(\mu)/\partial n$ in (9) is obtained from the expansion

$$L_{n+\Delta n}(\mu) = L_n(\mu) + \Delta n \frac{\partial L_n(\mu)}{\partial n} + \cdots. \quad (10)$$

With (2) it follows that

$$\int_{\Theta_0}^{\pi-\Theta_0} [L_n'(\Theta)]^2 \sin \Theta \, d\Theta = n(n+1)J_{nn}. \quad (11)$$

A final special case is that in which $T_n(\mu) = L_n(\mu)$, $T_m(\mu) = P_k(\mu)$. With (1) and (2) it follows that

$$J_{nk} \equiv \int_{\Theta_0}^{\pi-\Theta_0} L_n(\Theta) P_k(\cos \Theta) \sin \Theta \, d\Theta$$

$$- \frac{2(1-\mu_0^2)}{k(k+1) - n(n+1)} P_k(\mu_0) L_n'(\mu)|_{\mu=\mu_0}. \quad (12)$$

$$\int_{\Theta_0}^{\pi-\Theta_0} \frac{\partial L_n(\Theta)}{\partial \Theta} \frac{\partial P_k(\cos \Theta)}{\partial \Theta} \sin \Theta \, d\Theta$$

$$= k(k+1)J_{nk}. \quad (13)$$

8. General Solution for Y_{la} and the Infinite Set of Linear Equations

In order to determine the constant Y_{la} from the system of equations defined by (6.12, 14, 16), the first step is to integrate (6.16) with respect to Θ from $\Theta = \Theta_0$ to $\Theta = \pi - \Theta_0$. Since from (5.10) $L(\Theta_0) = L(\pi - \Theta_0) = 0$, the last term in (16) integrates to zero. It follows with (4.15) that

$$-2 \sum_{k}' \frac{b_k}{k(k+1)} P_k(\mu_0)$$

$$= j2I_m Z_c Y_{la} \ln \cot \tfrac{1}{2}\Theta, \quad (1)$$

so that

$$Y_{la} = \frac{j\zeta_0}{\pi Z_c^2} \sum_{k}' \frac{1}{k(k+1)} \left(\frac{b_k}{I_m} \right) P_k(\mu_0). \quad (2)$$

This equation shows that Y_{la} is determined uniquely if explicit values of the set of coefficients b_k/I_m are obtained.

The determination of b_k/I_m begins with the multiplication of (6.16) by $L_m'(\Theta) \sin \Theta \, d\Theta$ and integration with respect to Θ from $\Theta = \Theta_0$ to $\Theta = \pi - \Theta_0$. With the orthogonal properties of $L_n(\Theta)$ in (7.7, 8, 9), it follows directly that

$$a_m J_{mm} = \sum_{k}' b_k J_{mk}, \quad (3)$$

where J_{mm} is defined by (7.9) and J_{mk} by (7.12) with n replaced by m.

An additional relation between the a_n and b_k is obtained from (6.14) and (6.12) by first multiplying by $[\partial P_r(\cos \Theta)/\partial \Theta] \sin \Theta \, d\Theta$ and then integrating with respect to Θ from $\Theta = 0$ to $\Theta = \pi$. Use of (7.4) and (7.6) in the left-hand member and of (7.13) in the right-hand member gives

$$\frac{2b_r}{2r+1} p_r(\beta_0 l) = -2I_m P_r(\mu_0)$$

$$+ r(r+1) \sum_n \frac{a_n}{n(n+1)} \sigma_n(\beta_0 l) J_{nr}, \quad (4)$$

or, after rearrangement,

$$\frac{1}{2r+1} p_r(\beta_0 l) \left(\frac{b_r}{I_m} \right)$$

$$- r(r+1) \sum_n \left(\frac{a_n}{I_m} \right) \frac{J_{nr}}{2n(n+1)} \sigma_n(\beta_0 l)$$

$$= -P_r(\mu_0). \quad (5)$$

Note that in (4) and (5) r has only those values assumed by k in the sum Σ'; that is, r is odd. In (5), J_{nr} is the integral defined in (7.12) with k replaced by r.

It is now possible to eliminate the a_n between (3) and (5) and so obtain the following infinite set of linear equations that b_k/I_m must satisfy:

$$\frac{1}{2r+1} p_r(\beta_0 l) \left(\frac{b_r}{I_m} \right)$$

$$- r(r+1) \sum_n \sum_k' \left(\frac{b_k}{I_m} \right) \frac{J_{nr} J_{nk}}{2n(n+1)J_{nn}} \sigma_n(\beta_0 l)$$

$$= -P_r(\mu_0). \quad (6)$$

The ratio $J_{nr}J_{nk}/J_{nn}$ in (6) can be simplified by introducing a new coefficient, $dn/d\mu_0$. Proceeding from (5.10), namely, $L_n(\mu_0) = 0$, total differentiation gives

$$\frac{\partial L_n(\mu_0)}{\partial \mu_0} + \frac{\partial L_n(\mu_0)}{\partial n} \frac{dn}{d\mu_0} = 0. \quad (7)$$

If (7) is used in conjunction with (7.9) and (7.13), the following equation is derived:

$$\frac{J_{nr}J_{nk}}{J_{nn}}$$
$$= \frac{-2(2n+1)(1-\mu_0^2)P_k(\mu_0)P_r(\mu_0)(dn/d\mu_0)}{[k(k+1)-n(n+1)][r(r+1)-n(n+1)]}. \quad (8)$$

If (8) is substituted in (6) the result is

$$\frac{1}{2r+1}P_r(\beta_0 l)\left(\frac{b_r}{I_m}\right)$$
$$+ r(r+1)P_r(\mu_0)\sum_n\sum_k{'}\left(\frac{b_k}{I_m}\right)\Psi(n,k,r)P_k(\mu_0)$$
$$= -P_r(\mu_0), \quad (9)$$

where for convenience the symbol $\Psi(n, k, r)$ is introduced to stand for

$$\Psi(n,k,r) \equiv \frac{2n+1}{n(n+1)}$$
$$\times \frac{(1-\mu_0^2)(dn/d\mu_0)\sigma_n(\beta_0 l)}{[k(k+1)-n(n+1)][r(r+1)-n(n+1)]} \quad (10)$$

and where r is restricted to the range of k and therefore is odd.

The formal solution for Y_{la} and with it the determination of the driving-point impedance of the biconical antenna with arbitrary angle Θ_0 is contained in (2) with (9). However, in order to obtain an explicit formula for Y_{la}, it is necessary to solve the infinite set of linear equations given in (9) for (b_k/I_m). No general, exact solution of this set of equations is available for cones with arbitrary angle Θ_0. Hence, a general formula for the impedance of the biconical antenna cannot be provided. Thus, although the analysis of the biconical antenna is not restricted in its formulation, mathematical limitations make an exact solution unavailable. This is described in the next section. However, if the angle Θ_0 is sufficiently small, an approximate solution of (9) is possible. Approximate analyses of wide-angle biconical antennas and related conical structures are available in the literature, but these are outside the scope of this book.

9. Solution for the Apparent Terminal Admittance of a Biconical Antenna with Small Angles

The complicated equation (8.9) for the set of coefficients (b_k/I_m) may be simplified if the half-angle Θ_0 of the cone is sufficiently small so that $\mu_0 = \cos\Theta_0$ is quite near unity. This simplification depends primarily on the fact that for Θ_0 small the roots of the equation (5.10), $L_n(\Theta_0) = 0$, are approximated by[18]

$$n \doteq k + \frac{1}{\ln(2/\Theta_0)}; \quad k = 1, 3, \ldots \quad (1)$$

Hence, as Θ_0 is reduced, n approaches k; and for all moderately small values of Θ_0 the difference $n - k$ is small.

The differentiation of (1) with respect to $\mu_0 = \cos\Theta_0$ gives directly

$$(1-\mu_0^2)\frac{dn}{d\mu_0} \doteq -(n-k)^2. \quad (2)$$

Moreover, for $n - k \ll 2k + 1$, the following relation is valid:

$$k(k+1) - n(n+1) = -(n-k)(k+n+1)$$
$$\doteq -(n-k)(2k+1). \quad (3a)$$

In summing n over all possible values the difference $n - r$ becomes small when n is near r. For this range, that is, when $n - r \ll 2r + 1$,

$$r(r+1) - n(n+1) = -(n-r)(r+n+1)$$
$$\doteq -(n-r)(2r+1). \quad (3b)$$

If (1) applies, and (2), (3a), and (3b) are used in (8.10), it follows that,

$$\Psi(n,k,r)$$
$$\doteq \left(\frac{2n+1}{2k+1}\right)\left(\frac{n-k}{n-r}\right)\frac{\sigma_n(\beta_0 l)}{n(n+1)(2r+1)}.$$
$(n - k$ small and $n - r$ small) (4)

Alternatively,

$$\Psi(n,k,r)$$
$$\doteq \left(\frac{2n+1}{2k+1}\right)\left(\frac{n-k}{r(r+1)-n(n+1)}\right)\frac{\sigma_n(\beta_0 l)}{n(n+1)}.$$
$(n - k$ small, $n - r$ not small) (5)

Examination of (4) and (5) shows that with $n - k$ small the only significant values of $\Psi(n, k, r)$ as n passes through the entire range of values indicated in the sum in (8.9) occur when both $n - k$ and $n - r$ are small. This implies that $k = r$. It follows that $\Psi(n, k, r)$ is negligible except when $k = r \doteq n$. Thus,

$$\Psi(n,k,r) \doteq \begin{cases} \dfrac{\sigma_r(\beta_0 l)}{r(r+1)(2r+1)} \\ \qquad \text{when } k = r \doteq n, \\ 0 \quad \text{otherwise.} \end{cases} \quad (6)$$

Since for μ_0 near unity

$$P_k(\mu_0) \doteq 1, \quad (7)$$

it follows that with (6) the general equation (8.9) reduces to

$$\frac{1}{2r+1}\left(\frac{b_r}{I_m}\right)[\rho_r(\beta_0 l) - \sigma_r(\beta_0 l)] \doteq -1, \quad (8)$$

which is a good approximation for small angles Θ_0. Hence, with $\Theta_0 \ll 1$,

$$\frac{b_r}{I_m} \doteq \frac{-(2r+1)}{\rho_r(\beta_0 l) - \sigma_r(\beta_0 l)}. \quad (9)$$

The denominator in (9) may be transformed as follows, using (6.13) and (6.15); with $x = \beta_0 l$,

$$\rho_r(x) - \sigma_r(x) = \frac{h_r^{(2)'}(x)}{h_r^{(2)}(x)} - \frac{j_r'(x)}{j_r(x)}$$

$$= \frac{h_r^{(2)'}(x)j_r(x) - j_r'(x)h_r^{(2)}(x)}{h_r^{(2)}(x)j_r(x)}. \quad (10)$$

The numerator in (10) may be expanded using the general recurrence relation*

$$z_r'(x) = \frac{1}{2r+1}[rz_{r-1}(x) - (r+1)z_{r+1}(x)] \quad (11)$$

for $h_r^{(2)}(x)$ and $j_n'(x)$. The resulting expression may be transformed further with the functional equation†

$$h_{p-1}^{(2)}(x)j_p(x) - h_p^{(2)}(x)j_{p-1}(x) = 1/jx^2. \quad (12)$$

Then, with (11) and (12) in (10), the final result is

$$\frac{1}{\rho_r(\beta_0 l) - \sigma_r(\beta_0 l)} = j\beta_0^2 l^2 j_r(\beta_0 l)h_r^{(2)}(\beta_0 l). \quad (13)$$

With (13), (9) becomes

$$\frac{b_r}{I_m} \doteq -j(2r+1)(\beta_0 l)^2 j_r(\beta_0 l)h_r^{(2)}(\beta_0 l). \quad (14)$$

If this value is substituted in (8.2), and (7) is used, the approximate expression for Y_{la} is

$$Y_{la} \doteq \frac{\zeta_0(\beta_0 l)^2}{\pi Z_c^2}\sum_k{}'\frac{(2k+1)}{k(k+1)}j_k(\beta_0 l)h_k^{(2)}(\beta_0 l). \quad (15)$$

---

* Stratton, ref. I.52, p. 406.
† Jahnke and Emde, ref. I.28, p. 144. The formula given is for $H^{(2)}$ and J. Multiplication by $\pi/2x$ transforms it into (12).

In (15) Y_{la} is a function of Θ_0 through Z_c. For small values of Θ_0, the characteristic impedance is

$$Z_c = \frac{\zeta_0}{\pi}\ln\cot\frac{\Theta_0}{2} \doteq \frac{\zeta_0}{\pi}\ln\frac{2}{\Theta_0}. \quad (16)$$

Hence, the final expression for Y_{la}, with real and imaginary parts separated using the relation

$$h_k^{(2)}(\beta_0 l) = j_k(\beta_0 l) - jn_k(\beta_0 l) \quad (17)$$

is

$$Y_{la} = G_{la} + jB_{la}$$

$$= \frac{\zeta_0(\beta_0 l)^2}{\pi Z_c^2}\sum{}'\frac{2k+1}{k(k+1)}j_k(\beta_0 l)$$

$$\times [j_k(\beta_0 l) - jn_k(\beta_0 l)]. \quad (18)$$

By substituting this value of Y_{la} in (5.21), the driving-point impedance of the thin biconical antenna is obtained.

For the numerical evaluation of Y_{la} it is convenient to introduce the following closed forms for the series in (18); they were proved valid by Rice[15] after having been discovered by Schelkunoff:[18]

$$R_{mc}^e = \frac{\zeta_0\beta_0^2 l^2}{\pi}\sum_k{}'\frac{2k+1}{k(k+1)}j_k^2(\beta_0 l)$$

$$= \frac{\zeta_0}{4\pi}[2\,Cin\,2\beta_0 l + (Si\,4\beta_0 l - 2\,Si\,2\beta_0 l)\sin 2\beta_0 l$$

$$+ (2\,Cin\,2\beta_0 l - Cin\,4\beta_0 l)\cos 2\beta_0 l], \quad (19)$$

$$X_{mc}^e = \frac{-\zeta_0\beta_0^2 l^2}{\pi}\sum_k{}'\frac{2k+1}{k(k+1)}j_k(\beta_0 l)n_k(\beta_0 l)$$

$$= \frac{\zeta_0}{4\pi}[2\,Si\,2\beta_0 l + (\ln 4 - Cin\,4\beta_0 l)\sin 2\beta_0 l$$

$$- Si\,4\beta_0 l\cos 2\beta_0 l], \quad (20)$$

where $Si\,x = \int_0^x \frac{\sin u}{u}du$ is the integral sine and

$Cin\,x = \int_0^x \frac{1 - \cos u}{u}du$. Note that in (19)

$$R_{mc}^e = R_m^e, \quad (21)$$

where R_m^e is identically the *radiation resistance referred to maximum current of an infinitely thin cylindrical antenna* of half-length $h = l$ as given in (II.28.15) and (V.12.6a). In (20) X_{mc}^e is similar to, but not quite the same as, $X_m^e = X_0\sin^2\beta_0 h$, where X_0 is given in (II.28.10). Since the modified zeroth-order radiation resistance is independent of the cross-sectional shape of the antenna but the reactance is not, this is as it should be. With (19) and (20), (18) may be expressed in the simple form

$$Z_c^2 Y_{la} = Z_c^2(G_{la} + jB_{la})$$
$$= R_{mc}^e + jX_{mc}^e \equiv Z_{mc}^e. \quad (22)$$

Finally, if Y_{la} as obtained from (22) is substituted in (5.21), the input impedance Z_0 is determined.

Before discussing the input impedance of the biconical antenna of small angle Θ_0, it is of interest to reconsider the derivation of the formulas (18) and (22) with (19) and (20) from a different point of view. Both of these formulas (with slight differences in notation) were derived originally by Schelkunoff, using two rather ingenious methods of which neither is rigorous. The analysis of Tai is followed in this and preceding sections since it provides a more general and mathematically more systematic approach. In order to illustrate the essential features of Schelkunoff's methods of finding the expression (18) for Y_{la}, yet another method is described that is similar to one of Schelkunoff's but involves fewer assumptions and steps.

Instead of deriving the relatively simple expression (14) for the set of coefficients b_r/I_m for small-angle cones by the involved and laborious procedure of matching the electric field across the boundary sphere to obtain the infinite set of linear equations *valid for all values* of Θ_0, and then simplifying these by imposing the condition $\Theta_0 \ll 1$ to obtain an approximate formula valid for small-angle cones, simplifications and approximations appropriate to the small-angle cones may be introduced at the outset. Much as in the conventional method of calculating the radiation resistance R_m^e of a very thin cylindrical antenna as described in Chapter V, the assumption may be made that the distribution of current on a sufficiently *thin* conical antenna also is sinusoidally distributed and, hence, of the form

$$I(R) \doteq I_m \sin \beta_0(l - R) \quad (23)$$

Since the total current on the cones is given by

$$I(R) = I_d(R) + I_c = I_m[\sin \beta_0(l - R) - jZ_cY_l \cos \beta_0(l - R)] + I_c, \quad (24)$$

the approximation (23) implies that

$$\sin \beta_0(l - R) | \gg | -jZ_cY_l \cos \beta_0(l - R) + I_c/I_m | \quad (25)$$

for all values of R. Since there is actually no terminating impedance at $R = l$ so that $I(l) \doteq 0$ (if an insignificant current charging the very small spherical end surfaces is neglected in the small-angle cone) the implication in (25) that the complementary-mode current approximately cancels that part of the dominant-mode current that differs from zero at $R = l$ is entirely reasonable when Θ_0 is very small.

If the sinusoidal distribution (23) is assumed, the electromagnetic field of the biconical antenna is precisely that determined in Chapter V. In particular, the radial component of the *far-zone* electric field is (V.4.18c). With appropriate changes in notation it is

$$E_R^r = \frac{I_m \zeta_0 l}{2\pi R} \frac{e^{-j\beta_0 R}}{R} \sin(\beta_0 l \cos \Theta)$$
$$= \frac{I_m}{2\pi \omega \epsilon_0 R^2} e^{-j\beta_0 R} \beta_0 l \sin(\beta_0 l \cos \Theta). \quad (26)$$

If $\sin(\beta_0 l \cos \Theta)$ is expanded into a series of Legendre functions, (26) becomes

$$E_R^r = \frac{I_m}{2\pi \omega \epsilon_0 R^2} e^{-j\beta_0 R} \beta_0 l \sum_k{}' (-1)^{(k-1)/2}$$
$$\times (2k + 1) j_k(\beta_0 l) P_k(\cos \Theta). \quad (27)$$

This field must be equal to the radial field given by (6.8). With (6.2) this is

$$E_R = \frac{-j}{2\pi \omega \epsilon_0 R^2} \sum_k{}' b_k \frac{\beta_0 R h_k^{(2)}(\beta_0 R)}{\beta_0 l h_k^{(2)}(\beta_0 l)} P_k(\cos \Theta). \quad (28)$$

However, since (26) is for the far zone, (28) may be specialized to large values of R by using the asymptotic form of the spherical Hankel function given in Sec. 2. That is

$$\beta_0 R h_k^{(2)}(\beta_0 R) \doteq e^{-j[\beta_0 R - \frac{1}{2}(k+1)\pi]}$$
$$= (-1)^{(k+1)/2} e^{-j\beta_0 R}. \quad (\beta_0 R \gg 1) \quad (29)$$

Upon substituting (29) in (28), this becomes

$$E_R^r = \frac{j}{2\pi \omega \epsilon_0 R^2} e^{-j\beta_0 R}$$
$$\times \sum_k{}' \frac{b_k(-1)^{(k-1)/2} P_k(\cos \Theta)}{\beta_0 l h_k^{(2)}(\beta_0 l)}. \quad (30)$$

Equations (30) and (27) now may be equated term by term to give

$$\frac{b_r}{I_m} = -j(2r + 1)(\beta_0 l)^2 j_r(\beta_0 l) h_r^{(2)}(\beta_0 l), \quad (31)$$

which is exactly (14).

The fact that the same value is obtained for

b_r/I_m when the condition limiting the angle Θ_0 of the cone to small values is introduced at different points in the analysis signifies nothing more than that the approximations involved in the two approaches are exactly the same. In other words, the approximate relation (1) in itself implies sufficiently severe restrictions on the angle of the cones that a sinusoidal distribution of current is in fact a good representation if all impedances are referred to the maximum value I_m and not to the input value $I(0)$. This is borne out further by the fact that R_c^e in (19) is exactly equal to R_m^e for an antenna with a sinusoidal distribution of current. Since R_m^e is relatively insensitive to the cross-sectional dimension of an antenna if this is small, as shown in Fig. V.12.1 for the cylindrical antenna, the use of R_m^e as the apparent load impedance of a section of biconical transmission line with Θ_0 small is evidently a reasonable approximation, since account of the primary effect of changes in Θ_0 is contained in Z_c.

10. Impedance of a Biconical Antenna with Small Angle

The driving-point impedance Z_0 of a symmetric biconical antenna of small angle $2\Theta_0$ apex-driven by an idealized point generator is given by (5.21). It may be expressed as follows:

$$Z_0 = R_0 + jX_0$$

$$= Z_c \left(\frac{Z_{mc}^e \sin \beta_0 l - jZ_c \cos \beta_0 l}{Z_c \sin \beta_0 l - jZ_{mc}^e \cos \beta_0 l} \right), \quad (1)$$

where

$$Z_c \doteq \frac{\zeta_0}{\pi} \ln \frac{2}{\Theta_0} = 120 \ln \frac{2}{\Theta_0} \text{ ohms} \quad (2)$$

and where

$$Z_{mc}^e = R_{mc}^e + jX_{mc}^e = Z_c^2 Y_{la}; \quad (3)$$

R_{mc}^e and X_{mc}^e are given by (9.19) and (9.20). A curve of Z_c as computed from (2) is shown in Fig. 10.1. Curves of R_{mc}^e and X_{mc}^e as computed from (3) are given in Fig. 10.2. Finally, the driving-point resistance and reactance computed from (1) are shown in Figs. 10.3 and 10.4. Note that by introducing the following parameter:

$$\Omega_c \equiv 2 \ln \frac{2l}{a} \doteq 2 \ln \frac{2}{\Theta_0}, \quad (4a)$$

where a is the *maximum* radius of the cones, the characteristic impedance may be expressed as in (4b).

$$Z_c \doteq \frac{\zeta_0}{2\pi} \Omega_c = 60\Omega_c \text{ ohms.} \quad (4b)$$

It is significant to note that when $\beta_0 l = \pi/2$ the input impedance is

$$Z_0 = Z_{mc}^e = 73.13 + j153.7 \text{ ohms.} \quad (5a)$$

Thus the input impedance of the thin biconical antenna *as given by* (1) is a constant independent of the angle Θ_0 when l is a quarter wavelength and Θ_0 is sufficiently small. For the cylindrical antenna such constancy obtains only in the modified zeroth-order approximation when for $\beta_0 h = \pi/2$.

$$(Z_0)_{01} = 73.13 + j42.5 \text{ ohms.} \quad (5b)$$

The evaluation of Z_{mc}^e for the biconical antenna involves approximations equivalent to the assumption of a zeroth-order or sinusoidal distribution of current. This suggests that (1) may be no better than a modified zeroth-order approximation, at least when $\beta_0 l = \pi/2$ and all factors involving the angle of the cone cancel in (1). Unfortunately, no experimental data are available for biconical antennas with very small angles, and a better theoretical approximation depends on the solution of an infinite set of linear equations.

If the lengths l of the small-angle cones are electrically short, so that

$$\beta_0^2 l^2 = (2\pi l/\lambda_0)^2 \ll 1, \quad (6)$$

the expressions (9.19) and (9.20) for R_{mc}^e and X_{mc}^e may be simplified by expanding the integral functions and the trigonometric functions in series and retaining the leading terms. The results are

$$R_{mc}^e \doteq \frac{\zeta_0}{4\pi} \cdot \tfrac{2}{3}(\beta_0 l)^2 = 20\beta_0^2 l^2 \text{ ohms,} \quad (7)$$

$$X_{mc}^e \doteq \frac{\zeta_0}{4\pi} \cdot (2 \ln 4)\beta_0 l \doteq (166.4)\beta_0 l \text{ ohms.} \quad (8)$$

If these values are substituted in (1) and use is made of (6), the input impedance of the electrically short biconical antenna is

$$Z_0 \doteq \frac{20(\beta_0 l)^2}{[1 + (\ln 4)/\Omega_c]^2}$$

$$- j\frac{60\Omega_c}{[1 + (\ln 4)/\Omega_c]\beta_0 l} \text{ ohms.} \quad (9)$$

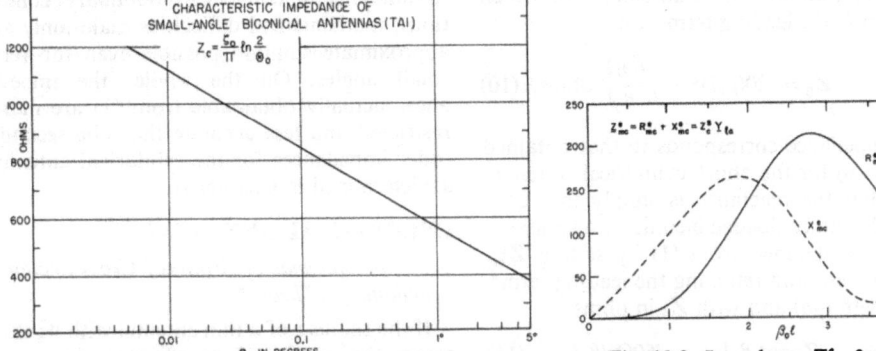

Fig. 10.1. Characteristic impedance of biconical antenna with small-angle (Tai).

Fig. 10.2. Impedance Z_{mc}^s for biconical antenna (Tai).

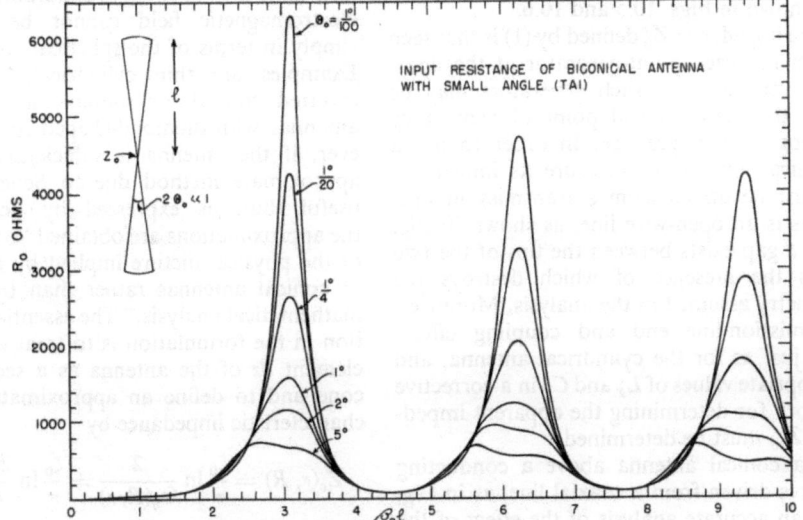

Fig. 10.3. Input resistance of biconical antenna with small angle (Tai).

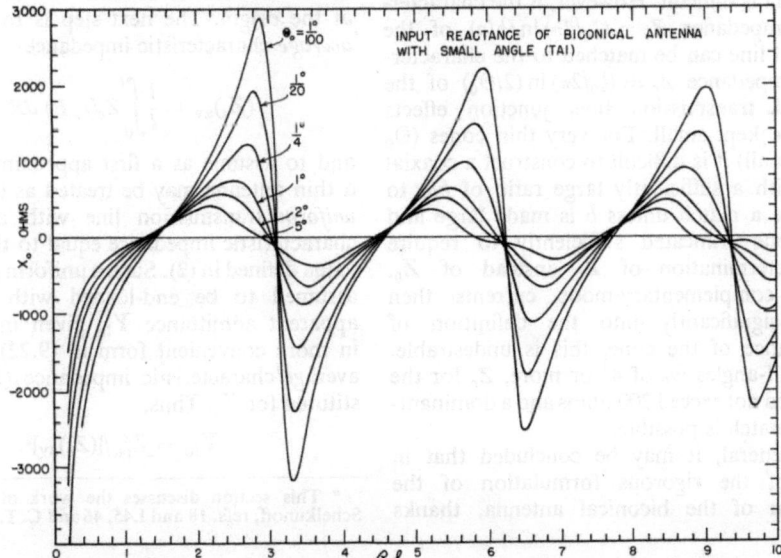

Fig. 10.4. Input reactance of biconical antenna with small angle (Tai).

For very thin and short antennas, for which $\Omega_c \gg \ln 4$, the leading terms are

$$Z_0 \doteq 20(\beta_0 l)^2 - j\frac{60\Omega_c}{\beta_0 l} \text{ ohms.} \quad (10)$$

This impedance corresponds to that obtained in (II.31.6) for the short cylindrical antenna. Note that the reactance is simply that of a short, thin and unloaded biconical transmission line. It is obtained from (1) by setting Z_{mc}^e equal to zero and retaining the leading terms. Thus, with (4b) and with Z_0 in ohms,

$$Z_0 \doteq -jZ_c \cot \beta_0 l \doteq -j60\Omega_c/\beta_0 l. \quad (11)$$

Curves of R_0 and X_0 as computed from (10) are shown in Figs. 10.5 and 10.6.

The impedance Z_0 defined by (1) is that seen by an idealized point-generator at the apex. However attractive such generators may be from the mathematical point of view, they do not exist in practice. In order to use a biconical antenna or measure its impedance it must be driven from a transmission line. If this is an open-wire line, as shown in Fig. 10.7, a gap exists between the tips of the two cones the presence of which destroys the symmetry assumed in the analysis. Moreover, transmission-line end and coupling effects exist just as for the cylindrical antenna, and appropriate values of L_T and C_T in a corrective network for determining the apparent impedance Z_{sa} must be determined.

If a conical antenna above a conducting plane is driven from a coaxial line, as in Fig. 1.2b, an accurate analysis of the effect of the terminal zone near the junction of line and antenna is difficult. However, if the characteristic impedance $Z_c = (\zeta_0/2\pi) \ln (b/a)$ of the coaxial line can be matched to the characteristic impedance $Z_c \doteq (\zeta_0/2\pi) \ln (2/\Theta_0)$ of the conical transmission line, junction effects may be kept small. For very thin cones (Θ_0 very small) it is difficult to construct a coaxial line with a sufficiently large ratio of b/a to provide a match unless b is made large and the cone truncated sufficiently to require the determination of Z_δ instead of Z_0. Since complementary-mode currents then enter significantly into the definition of impedance of the cone, this is undesirable. For half-angles Θ_0 of 4° or more, Z_c for the line need not exceed 200 ohms and a dominant-mode match is possible.

In general, it may be concluded that in spite of the rigorous formulation of the problem of the biconical antenna, thanks to analytically convenient boundary conditions, mathematical difficulties make only an approximate solution possible even for very small angles. On the whole, the impedances actually obtainable from (1) are more restricted and less accurate than the second-order impedances for the cylindrical antenna as determined in Chapter II.

CYLINDRICAL ANTENNAS

11. Thin Antennas of Arbitrary Cross Section; Schelkunoff's Theory*

If the radius r of a thin antenna with its axis along the z-axis of coördinates does not increase uniformly with z as in the biconical antenna, the boundary conditions for the electromagnetic field cannot be expressed simply in terms of the spherical components. Examples are thin cylindrical, spheroidal, inverted conical, or rhombic antennas and antennas with diamond-shaped halves. However, if the antenna is sufficiently thin, an approximate method due to Schelkunoff is useful, but, as expressed by Schelkunoff, the approximations are obtained "on the basis of the physical picture implied by the theory of conical antennas rather than by a direct mathematical analysis." The essential assumption in the formulation is to treat each small element dz of the antenna as a section of a cone and to define an approximate *variable* characteristic impedance by

$$Z_c(r, R) \doteq \frac{\zeta_0}{\pi} \ln \frac{2}{\Theta_0(R, r)} \doteq \frac{\zeta_0}{\pi} \ln \frac{2R}{r}, \quad (1)$$

where r is the radius of the section of the cone at radial distance R from the driving point at the origin. The next step is to define the *average* characteristic impedance

$$(Z_c)_{\text{av}} \equiv \frac{1}{l}\int_0^l Z_c(r, R)\, dR \quad (2)$$

and to assume as a first approximation that a thin antenna may be treated as if it were a *uniform* transmission line with a constant characteristic impedance equal to the average value defined in (2). Such a uniform line is then assumed to be end-loaded with the *same* apparent admittance Y_{la} given in (9.18) or in more convenient form in (9.22), with the average characteristic impedance $(Z_c)_{\text{av}}$ substituted for Z_c. Thus,

$$Y_{la} = Z_{mc}^e/[(Z_c)_{\text{av}}]^2. \quad (3)$$

---
* This section discusses the work of Drs. S. A. Schelkunoff, refs. 18 and I.45, 46 and C. T. Tai, ref. 23.

As shown in Sec. 9, the approximations involved in the definition and evaluation of $Z_{mc}^e = R_{mc}^e + jX_{mc}^e$ are equivalent to assuming a sinusoidal distribution of current that is independent of the shape of the antenna. Accordingly, a zeroth-order approximation of the impedance of thin antennas of arbitrary shape is given by the general transmission-line formula (5.21) with the value of Y_{1a} given by (3). Thus,

$$Z_0 = \frac{V(0)}{I(0)}$$

$$\doteq (Z_c)_{av} \left(\frac{Z_{mc}^e \sin \beta_0 l - j(Z_c)_{av} \cos \beta_0 l}{(Z_c)_{av} \sin \beta_0 l - j Z_{mc}^e \cos \beta_0 l} \right), \quad (4)$$

where $(Z_c)_{av}$ is the average characteristic impedance appropriate to the antenna in question. Unfortunately, such a zeroth-order approximation is inadequate. For example, it leads to the conclusion that *all thin* antennas, regardless of shape or cross-sectional dimensions provided these are small, have exactly the same impedance,

$$Z_0 = Z_{mc}^e = 73.13 + j156.6 \text{ ohms}, \quad (5)$$

as the biconical antenna when $\beta_0 l = \beta_0 h = \pi/2$. It is known that the real part of (4) is the *zeroth-order* resistance of a thin cylindrical antenna, but the reactive part does not approximate the zeroth-order reactance of 42.5 ohms of the cylindrical antenna.

In order to obtain a better approximation Schelkunoff replaces the uniform transmission line with a *tapered* transmission line that has a slowly varying characteristic impedance instead of a constant value. The resulting formula for the imput impedance is a modification of (4) obtained by taking account of the first-order variation in current and voltage as a result of the nonuniformity in the characteristic impedance. If the voltage and current at a distance R from the generator end of a nonuniform line are given by

$$V(R) = V_0(R) + V_1(R) + \cdots,$$
$$I(R) = I_0(R) + I_1(R) + \cdots, \quad (6)$$

where

$$V_0(R) = V_0(0) \cos \beta_0 R - j(Z_c)_{av} I_0(0) \sin \beta_0 R, \quad (7a)$$

$$I_0(R) = I_0(0) \cos \beta_0 R - j \frac{V_0(0)}{(Z_c)_{av}} \sin \beta_0 R \quad (7b)$$

are the zeroth-order voltage and current for a uniform line with characteristic impedance equal to the average value defined in (1b), and where the first-order corrections, as shown by Schelkunoff,* are

$$V_1(r) = \frac{V_0(0)}{(Z_c)_{av}} [jM \cos \beta_0 R + N \sin \beta_0 R]$$
$$\quad - I_0(0)[jN \cos \beta_0 R + M \sin \beta_0 R], \quad (8a)$$

$$I_1(R) = \frac{V_0(0)}{[(Z_c)_{av}]^2} [M \sin \beta_0 R + jN \cos \beta_0 R]$$
$$\quad - \frac{I_0(0)}{(Z_c)_{av}} [N \sin \beta_0 R + jM \cos \beta_0 R]. \quad (8b)$$

The real functions M and N are defined by

$$M \equiv M(\beta_0 l)$$
$$\equiv \beta_0 \int_0^l [(Z_c)_{av} - Z_c(r, R)] \sin 2\beta_0 R \, dR, \quad (9a)$$

$$N \equiv N(\beta_0 l)$$
$$\equiv \beta_0 \int_0^l [(Z_c)_{av} - Z_c(r, R)] \cos 2\beta_0 R \, dR. \quad (9b)$$

$Z_c(r, R)$ in (9a, b) is defined in (1). For the thin cylindrical antenna of radius a and half-length $h = l$,

$$M = \frac{\zeta_0}{2\pi} [Cin \, 2\beta_0 l - 1 + \cos 2\beta_0 l], \quad (10a)$$

$$N = \frac{\zeta_0}{2\pi} [Si \, 2\beta_0 l - \sin 2\beta_0 l], \quad (10b)$$

$$(Z_c)_{av} = \frac{\zeta_0}{\pi} \left(\ln \frac{2l}{a} - 1 \right). \quad (10c)$$

For the rhombic or inverted conical antenna,

$$M = \frac{\zeta_0}{2\pi} [(1 + \cos 2\beta_0 l) \, Cin \, 2\beta_0 l$$
$$\quad - \sin 2\beta_0 l \, Si \, 2\beta_0 l], \quad (11a)$$

$$N = \frac{\zeta_0}{2\pi} [(1 - \cos 2\beta_0 l) \, Si \, 2\beta_0 l$$
$$\quad - \sin 2\beta_0 l \, Cin \, 2\beta_0 l], \quad (11b)$$

$$(Z_c)_{av} = \frac{\zeta_0}{\pi} \ln \frac{2l}{a}. \quad (11c)$$

---

* Schelkunoff, ref. I.46, pp. 208–211.

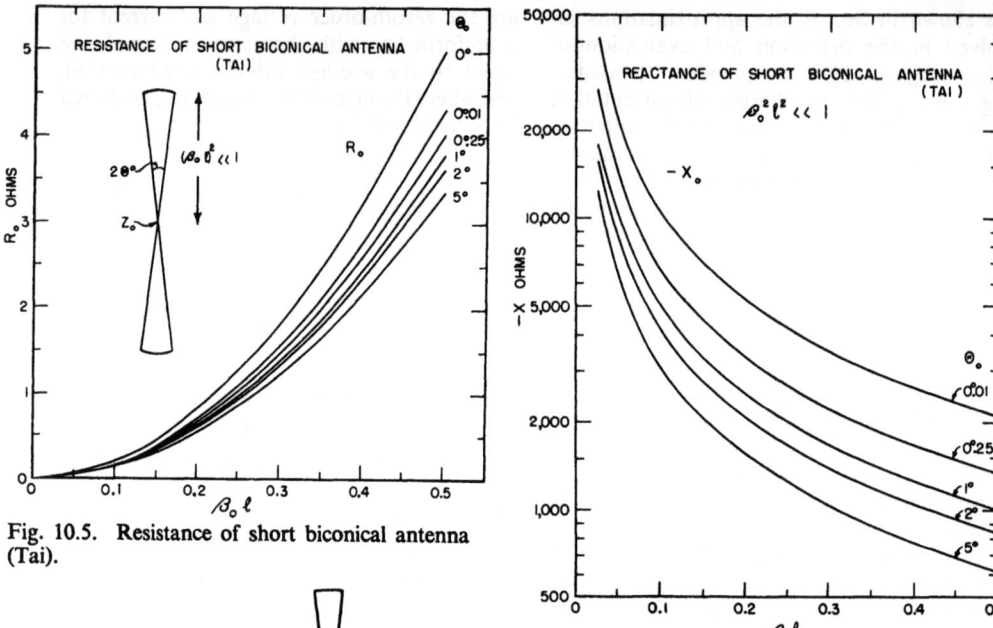

Fig. 10.5. Resistance of short biconical antenna (Tai).

Fig. 10.6. Reactance of short biconical antenna (Tai).

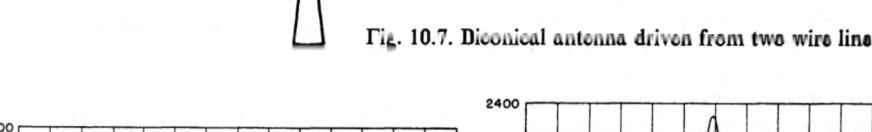

Fig. 10.7. Biconical antenna driven from two wire line.

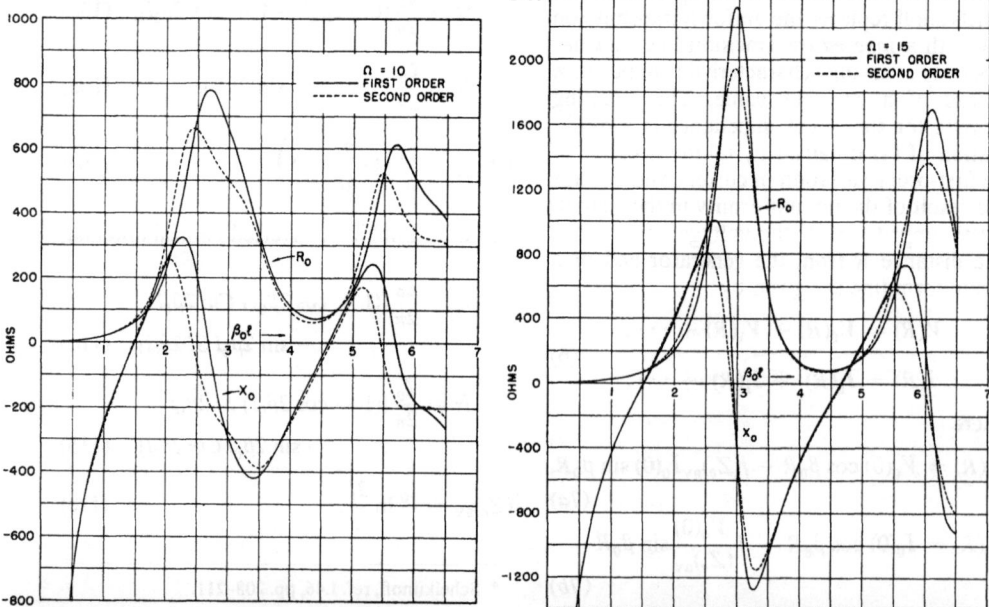

Fig. 11.1. Impedance of cylindrical antennas based upon Schelkunoff's theory; $\Omega = 10$ (Tai).

Fig. 11.2. Impedance of cylindrical antennas based upon Schelkunoff's theory; $\Omega = 15$ (Tai).

For the spheroidal antenna,

$$M = R_{cm}^e - \frac{\zeta_0}{2\pi}(1 - \cos 2\beta_0 l) \ln 2, \quad (12a)$$

$$N = X_{cm}^e - \frac{\zeta_0}{2\pi}(\ln 2) \sin 2\beta_0 l, \quad (12b)$$

$$(Z_c)_{\text{av}} = \frac{\zeta_0}{\pi} \ln \frac{l}{a}. \quad (12c)$$

It is interesting to note that the average characteristic impedance for the thin cylindrical antenna in (10c) is precisely the value obtained in Chapter II for a sufficiently short antenna using the King-Middleton expansion parameter Ψ_{K1}. Thus, for an antenna of half-length h and radius a,

$$\frac{\zeta_0 \Psi_{K1}}{2\pi} \doteq \frac{\zeta_0}{2\pi}(\Omega - 2) = \frac{\zeta_0}{\pi}\left(\ln \frac{2h}{a} - 1\right). \quad (13)$$

On the other hand, if the Hallén parameter $\Omega = 2 \ln (2h/a)$ is used,

$$\frac{\zeta_0 \Omega}{2\pi} = \frac{\zeta_0}{\pi} \ln \frac{2h}{a}, \quad (14)$$

which is the value for the antenna of rhombic shape in (11c).

The input impedance, $Z_0 = V_0(0)/I_0(0)$, of a section of tapered line of length l is evaluated from the following formula for the apparent load admittance:

$$Y_{la} = \frac{I(l)}{V(l)} \doteq \frac{I_0(l) + I_1(l)}{V_0(l) + V_1(l)}, \quad (15)$$

using (7a, b) and (8a, b) with $R = l$. The explicit formula for the input impedance obtained from (15) is

$$Z_0 = (Z_c)_{\text{av}} \frac{A + jB}{C + jD}, \quad (16)$$

where

$$A = R_{mc}^e[\sin \beta_0 l - (Y_c)_{\text{av}}(N \cos \beta_0 l + M \sin \beta_0 l)] \quad (17a)$$

$$B = X_{mc}^e[\sin \beta_0 l - (Y_c)_{\text{av}}(N \cos \beta_0 l + M \sin \beta_0 l)] + [M - (Z_c)_{\text{av}}] \cos \beta_0 l - N \sin \beta_0 l, \quad (17b)$$

$$C = X_{mc}^e[\cos \beta_0 l + (Y_c)_{\text{av}}(N \cos \beta_0 l - N \sin \beta_0 l)] + N \cos \beta_0 l + [M + (Z_c)_{\text{av}}] \sin \beta_0 l, \quad (17c)$$

$$D = -R_{mc}^e[\cos \beta_0 l + (Y_c)_{\text{av}}(M \cos \beta_0 l - N \sin \beta_0 l)], \quad (17d)$$

and where R_{mc}^e and X_{mc}^e are given in (9.19) and (9.20); $(Y_c)_{\text{av}} \equiv 1/(Z_c)_{\text{av}}$.

The *formula actually evaluated by Schelkunoff*, which may be called a first-order solution, is given by (15) with $V_1(l)$ neglected compared with $V_0(l)$. That is,

$$Y_{la} \doteq \frac{I_0(l) + I_1(l)}{V_0(l)}. \quad (18)$$

The corresponding first-order value of the input impedance is given by (16) using (17a–d) with the terms that have $(Y_c)_{\text{av}}$ as a factor neglected. With $Z_{mc}^e = R_{mc}^e + jX_{mc}^e$ it may be expressed as follows:

$$Z_0 = (Z_c)_{\text{av}} \left\{ \frac{\begin{array}{l}(Z_{mc}^e - jN) \sin \beta_0 l \\ - j[(Z_c)_{\text{av}} - M] \cos \beta_0 l\end{array}}{\begin{array}{l}[(Z_c)_{\text{av}} + M] \sin \beta_0 l \\ - j(Z_{mc}^e + jN) \cos \beta_0 l\end{array}} \right\}. \quad (19)$$

The second-order formula (16) with (17a–d) complete was evaluated by Tai.[23] First- and second-order input impedances as computed by Tai from (16) with the appropriate form of (17a–d) are represented in Figs. 11.1 and 11.2 for $\Omega = 2 \ln (2h/a) = 10$ and 15. Additional and more extensive curves of the first-order impedances are in the literature[I.46; 23]. It is seen that the curves for the first- and second-order impedances of Schelkunoff's theory show marked differences in magnitude and in shape for the thicker antenna. It is shown below that the first-order formula is in much better general agreement with experiment than is the second-order formula.

Since it has been shown in Chapter II that the impedances of cylindrical antennas computed from the second-order King-Middleton form of Hallén's theory and from the variational theory of Storer and its modification by Tai are in good agreement with one another and with experiment, it is interesting to compare these with the impedances determined from Schelkunoff's theory. In Figs. 11.3 and 11.4 the variational, the King-Middleton second-order, and the Schelkunoff first-order impedances are shown. It is evident from Figs. 11.1 and 11.2 as compared with Figs. 11.3 and 11.4 that the second-order Schelkunoff impedances are in poorer agreement with the corresponding variational and King-Middleton second-order values and, therefore, with experiment, than are Schelkunoff's first-order results. In order to provide additional comparison between the Hallén theory (using both Hallén's expansion parameter $\Psi = \Omega$ and the King-Middleton parameter $\Psi = \Psi_{K1}$ as defined in Chapter II), Schelkunoff's first-order theory, and experimental results, the important values of maximum resistance near antiresonance and maximum conductance near resonance are selected. In Fig. 11.5 are shown curves of the first maximum of resistance as determined from (19) for the Schelkunoff theory, from (II.30.4) with $\Psi = \Omega$ and all D-factors set equal to unity for the original Hallén theory, and from (II.30.4) with $\Psi = \Psi_{K1}$ and the D-factors as defined in (II.30.5a,b,c) for the King-Middleton form of the Hallén theory. The second maximum is shown only for the Hallén and King-Middleton solutions. It is seen that the Schelkunoff curve is close to and below the King-Middleton curve, whereas the Hallén values are considerably higher than those of King-Middleton. Together with the theoretical curves are shown numerous sets of experimental points. The data of Hartig, D. D. King, and Tomiyasu are careful measurements on lines where terminal-zone effects are made negligible or corrected. The data of Wilson are less accurate, as discussed in Sec. IV.8. They are included primarily because they are the only data available for the second maximum. The data of Brown and Woodward were obtained using conventional equipment with no particular precautions to eliminate transmission-line and coupling effects. One of their points ($\Omega \doteq 16.5$) is displaced greatly, perhaps owing to very significant base capacitance. Nothing is known about the nature of the equipment used in obtaining the data collected by Schelkunoff. As verified in Chapter II, the second-order King-Middleton formula is in good agreement with experimental results, slightly better than either the Schelkunoff or the second-order Hallén formula.

The differences between the three formulas are more pronounced near resonance, as shown by the graphs in Fig. 11.6 of the maximum conductance $(G_0)_{\max}$. The Hallén formula yields results much closer to the King-Middleton values. The maximum conductance determined from Schelkunoff's formula (1) is much higher, which means that the resistance R_0 near resonance is considerably lower than that given by either expansion of the Hallén theory. This again suggests that near resonance, where $\beta_0 l \doteq \pi/2$ and (5), (7a), and (7b) reduce to simple forms insensitive to small changes in $\beta_0 l$, the zeroth-order form of Z^e_{mc} may be inadequate.

Instead of comparing the theoretical values of $(G_0)_{\max}$ directly with experimental values, it is much more convenient to use the value $(G_0)_{\mathrm{res}}$, which is very near $(G_0)_{\max}$ and only slightly lower. This is because the condition of resonance is readily determined experimentally by the vanishing of X_0, so that $R_0 = 1/G_0$ is obtained directly. An accurate determination of $(G_0)_{\max}$ requires a knowledge of R_0 and X_0 over a range near resonance and computation of $G_0 = R_0/(R_0^2 + X_0^2)$ using interpolated values. The value $(G_0)_{\mathrm{res}}$ for the second-order King-Middleton expansion is shown in Fig. 11.6. It lies slightly below $(G_0)_{\max}$. Corresponding values of $(G_0)_{\mathrm{res}}$ for the Schelkunoff and Hallén formulas are not shown in Fig. 11.6. However, they must evidently lie below the associated $(G_0)_{\max}$ curves by about the same amount as in the case of the King-Middleton curves. Available experimental data are largely for thicker antennas with Ω between 8 and 10. The two points shown by Hartig are each a mean value for several observations which range between about $(G_0)_{\mathrm{res}} = 13.7 \times 10^{-3}$ mho and 14.3×10^{-3} mho. Near resonance the resistance and conductance of an antenna driven from a transmission line depend primarily upon the transmission-line attenuation function ρ and only negligibly on the phase-function Φ as defined in (II.30, 24). Since the measurement of attenuation is less accurate than the highly precise measurement of phase, the accuracy in determining ρ is always less than that in determining Φ. Examination of the curves for ρ for an antenna driven from a transmission line with various values of characteristic impedance shows that, for R_c in the range from 50 to 100 ohms usual in coaxial lines, a rather sharp peak occurs at or near resonance. This part of the ρ-curve is the most difficult to determine accurately. Since at resonance

$$(R_0)_{\mathrm{res}} = \frac{1}{(G_0)_{\mathrm{res}}} = R_c \tanh \rho, \quad (20)$$

an approximately equal error is made in $(R_0)_{\mathrm{res}}$ as in ρ when this is small. It is the combination of the relative difficulty of measuring absolute attenuation accurately with the fact that near resonance ρ has a sharp peak that accounts for the relatively wide range in the very carefully made measurements of both Hartig and D. D. King.

The obvious conclusion from Fig. 11.6 is that Schelkunoff's formula as well as the second-order formulas of Hallén and King

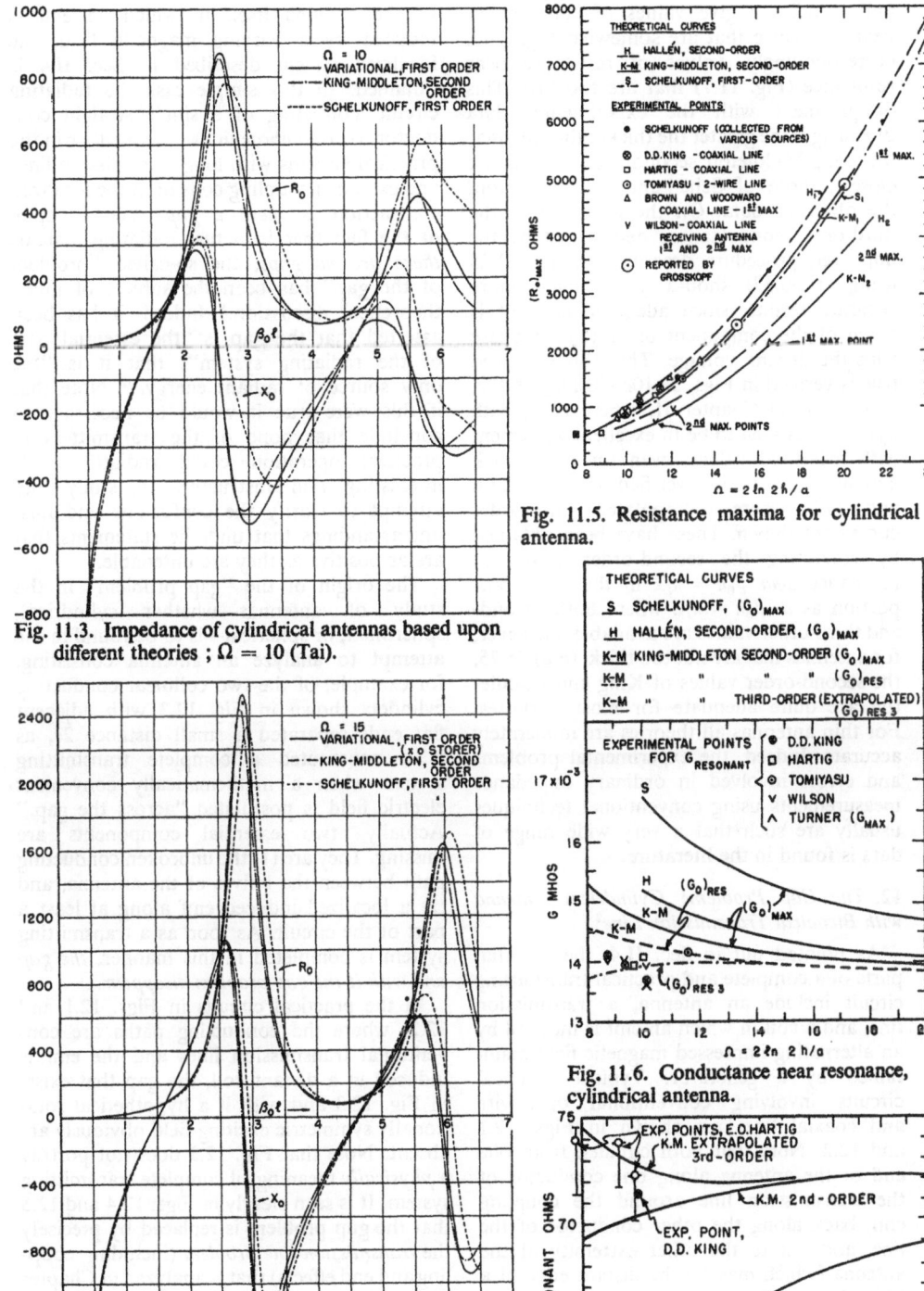

Fig. 11.3. Impedance of cylindrical antennas based upon different theories; $\Omega = 10$ (Tai).

Fig. 11.4. Impedance of cylindrical antennas based upon different theories; $\Omega = 15$ (Tai).

Fig. 11.5. Resistance maxima for cylindrical antenna.

Fig. 11.6. Conductance near resonance, cylindrical antenna.

Fig. 11.7. Resistance at resonance, cylindrical antenna.

and Middleton give values of conductance near resonance that are somewhat high and, correspondingly, values of resistance near resonance (Fig. 11.7) that are too low. This disagreement with the experimental data of Hartig is the greater the thicker the antenna; the King-Middleton second-order values are closest, those of Schelkunoff furthest from the observed data. On the other hand, the third-order conductances determined by the improved procedure described in Sec. II.25 using (II.30.27) should yield much more accurate results, since adequate account is taken of the component of current in phase with the driving voltage. That this is indeed true is verified in Figs. 38.10a, 38.11a, 38.12c, and 38.13c of Chapter II, where $(R_0)_3$ with $\beta_0 h = \pi/2$ is seen to be in excellent agreement with measured values, even for quite thick antennas. It is also verified in Figs. 11.6 and 11.7 where extrapolated third-order curves are shown. These have been obtained by correcting the second-order values at resonance near $\beta_0 h = \pi/2$ in the same proportion as at $\beta_0 h = \pi/2$ where both second- and third-order values are available. However, for antennas that are not too thick, $(h/a) \geq 75$, the second-order values of King and Middleton are quite adequate for most purposes. For thin antennas all theories are moderately accurate. Indeed, the experimental problems and errors involved in ordinary impedance measurements using conventional techniques usually are such that a very wide range of data is found in the literature.

12. The Gap Problem: Cylindrical Antenna with Biconical Transmission Line[9]

As pointed out in Sec. II.4, the essential parts of a complete and practical transmitting circuit include an antenna, a transmission line, and a coil in which an emf is induced by an alternating impressed magnetic field maintained by a generator. Typical practical circuits involving conventional open-wire and coaxial lines are shown in Figs. 12.1 and 12.2. Note that both circuits, from one end of the antenna along one conductor of the transmission line around the coupling coil, back along the other conductor of the line, and out to the other extremity of the antenna (which may be the distant edge of a ground screen or the opposite pole of a great sphere), provide *unbroken conducting paths*. If, in the interest of pedagogic simplicity, the transmission line is reduced to zero and the coupling coil is contracted to a short section of the antenna itself in which an emf is induced by a varying magnetic field, the idealized system described in Sec. II.4 is obtained. In this simple case the radiating circuit, consisting of a single straight conductor, is still *unbroken*. Nowhere either in the practical systems with long transmission lines and extended coupling coils or in the idealized contraction *is there a gap*. Yet, in spite of the fact that in actual radiating systems *there are no gaps*, the so-called "problem of the gap" has been the subject of much theoretical discussion.[8] Indeed, it has been asserted that the gap is "the essential part of the radiating system"; that it is "the only source of radiant energy."[3] Note that if this were true it would be necessary to conclude that none of the transmitters in practical operation could radiate! It is interesting and instructive to study and attempt to clarify the confusions and misunderstandings that underlie statements that are as positive as they are untenable.

The origin of the "gap problem" in the study of antennas—whether cylindrical, spheroidal, or spherical—is to be found in the attempt to analyze an antenna consisting, for example, of the two collinear conducting cylinders shown in Fig. 12.3 with adjacent flat ends separated a small distance 2δ, as if it constituted a complete transmitting system when a mathematically convenient electric field is postulated "across the gap." Actually, two essential components are missing. They are (1) the unbroken conducting path between the halves of the antenna, and (2) a localized induced emf along at least a part of the circuit. As soon as a transmitting system is completed in this manner, *the gap and with it the gap problem disappear*.

In the practical circuits in Figs. 12.1 and 12.2, where the conducting paths are conventional transmission lines and the emf is induced in a distant coil, the gap that exists in Fig. 12.3 and with it a hypothetical rotationally symmetric exciting field obviously are absent. Note that Fig. 12.3 does not portray a *physically* meaningful complete transmitting system. It is seen clearly in Figs. 12.4 and 12.5 that the gap problem is replaced by precisely the *transmission-line problem* (including coupling and end effects) that is analyzed in Chapter II. Another type of circuit in which the gap is *apparently* retained is shown in Fig. 12.6. The antenna consists of the same halves shown in Fig. 12.3 but the transmitting system is completed in a theoretically possible but

[VIII.12] THEORY OF LINEAR ANTENNAS

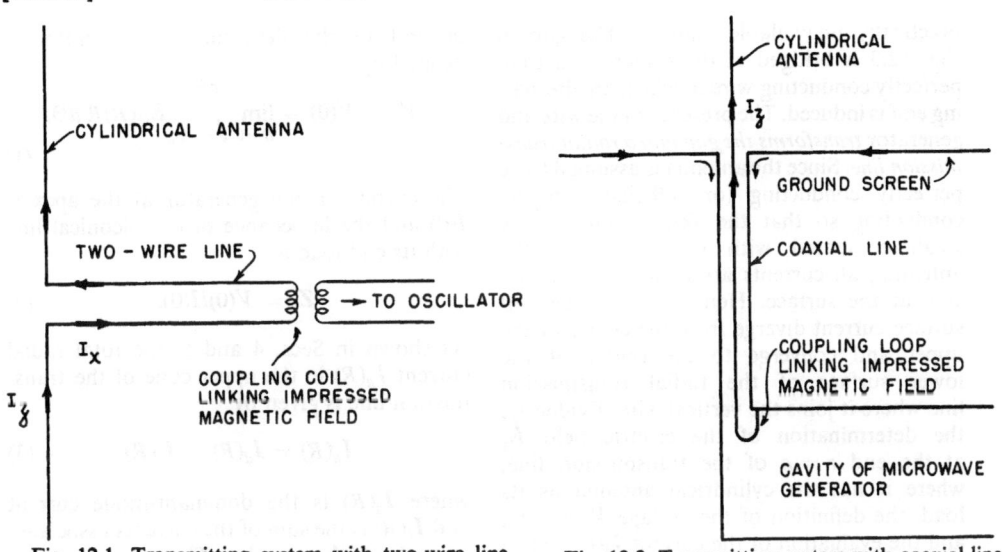

Fig. 12.1. Transmitting system with two-wire line.

Fig. 12.2. Transmitting system with coaxial line.

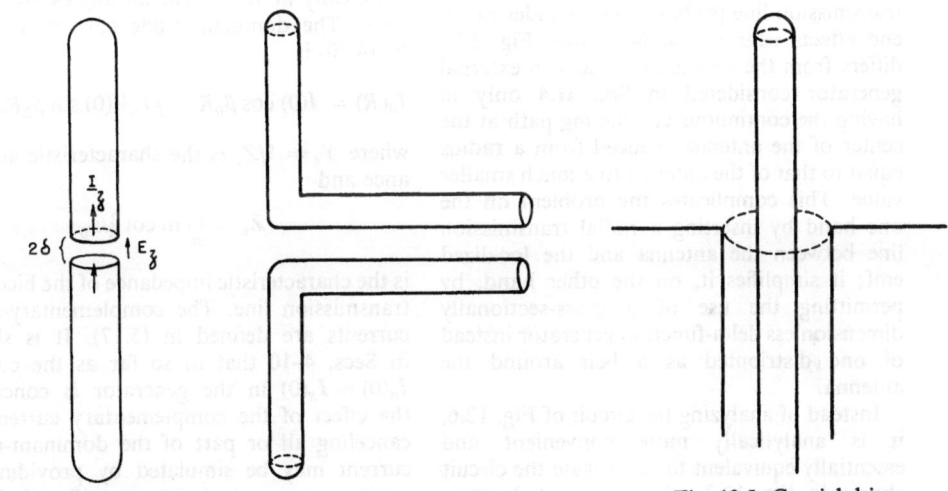

Fig. 12.3. Cylindrical antenna with a gap; the assumed field and current are indicated.

Fig. 12.4. Two-wire drive.

Fig. 12.5. Coaxial drive.

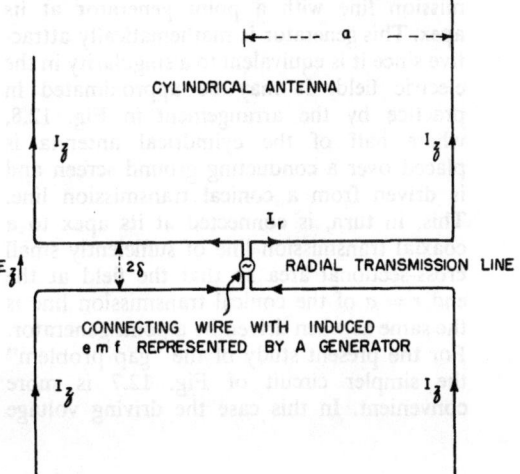

Fig. 12.6. Transmitting system with radial transmission line.

practically unavailable manner. The gap in Fig. 12.3 is bridged at the center by a thin, perfectly conducting wire in which an alternating emf is induced. The presence of the wire and generator *transforms the gap into a radial transmission line*. Since the antenna is assumed to be perfectly conducting (or sufficiently highly conducting so that the skin depth is very small compared with the radius of the antenna), all currents are confined to a thin skin at the surface. Hence, radial sheets of surface current diverge from the center of the upper and converge to the center of the lower surface of the radial transmission line where it joins the vertical wire. Evidently, the determination of the electric field $\bar{E}_z$ at the end $r = a$ of the transmission line, where it has the cylindrical antenna as its load, the definition of the voltage $V_\delta(r = a)$, and the evaluation of the current $I_\delta(r = a)$ in terms of the emf V^e at $r = 0$ are parts of a transmission-line problem that includes radial end effects near $r = a$. Note that Fig. 12.6 differs from the simple antenna with external generator considered in Sec. II.4 only in having the continuous conducting path at the center of the antenna reduced from a radius equal to that of the antenna to a much smaller value. This complicates the problem on the one hand by inserting a radial transmission line between the antenna and the localized emf; it simplifies it, on the other hand, by permitting the use of a cross-sectionally dimensionless delta-function generator instead of one distributed as a belt around the antenna.

Instead of analyzing the circuit of Fig. 12.6, it is analytically more convenient and essentially equivalent to investigate the circuit shown in Fig. 12.7, where the radial transmission line is replaced by a biconical transmission line with a point generator at its apex. This generator is mathematically attractive since it is equivalent to a singularity in the electric field. It may be approximated in practice by the arrangement in Fig. 12.8, where half of the cylindrical antenna is placed over a conducting ground screen and is driven from a conical transmission line. This, in turn, is connected at its apex to a coaxial transmission line of sufficiently small cross-sectional area so that the field at the end $r = a$ of the conical transmission line is the same as when driven by a point generator. For the present study of the "gap problem" the simpler circuit of Fig. 12.7 is more convenient. In this case the driving voltage or emf of the delta-function generator is defined by

$$V^e \equiv V(0) \equiv \lim_{R \to 0} \int_{\pi - \Theta_0}^{\Theta_0} E_\Theta(R) R \, d\Theta. \quad (1)$$

The current in the generator at the apex is $I(0)$ and the impedance of the biconical line with its end load is

$$Z_0 = V(0)/I(0). \quad (2)$$

As shown in Secs. 4 and 5, the total radial current $I_R(R)$ in the upper cone of the transmission line is given by

$$I_R(R) = I_d(R) + I_c(R), \quad (3)$$

where $I_d(R)$ is the dominant-mode current and $I_c(R)$ is the sum of the currents associated with the higher modes. These latter vanish identically at $R = 0$ for all angles Θ_0 of the cone. The dominant-mode current as given by (4.20) is

$$I_d(R) = I(0) \cos \beta_0 R - jY_c V(0) \sin \beta_0 R, \quad (4)$$

where $Y_c = 1/Z_c$ is the characteristic admittance and

$$Z_c = \frac{\zeta_0}{\pi} \ln \cot \tfrac{1}{2}\Theta_0 \quad (5)$$

is the characteristic impedance of the biconical transmission line. The complementary-mode currents are defined in (5.17). It is shown in Secs. 4–10 that in so far as the current $I_d(0) = I_R(0)$ in the generator is concerned the effect of the complementary currents in canceling all or part of the dominant-mode current may be simulated by providing an apparent terminal admittance at $R = l$ of such value that the current which enters it is equal to the dominant-mode current actually canceled by the complementary currents. In the case of the thin biconical antenna, with Θ_0 very small, the total current $I_R(l)$ is essentially zero, so that virtually the entire dominant-mode current $I_d(l)$ is canceled by equal and opposite higher-mode currents since $I_d(l) \doteq -I_c(l)$. Hence, the complementary currents could be ignored only by providing an apparent dominant-mode terminal admittance Y_{la} that carried the entire current $I_d(l)$. In the case at hand, with Θ_0 near $\pi/2$, the situation is different, since practically all of the dominant-mode current at $R = l$ becomes the axially directed surface current $I_z(z = \delta)$. When Θ_0 is sufficiently near $\pi/2$ the higher-mode current $I_c(l)$ is very small and cancels an insignificant part of the dominant-mode current $I_d(l)$. Actually, the higher-mode current is responsible for transmission-line end effects. Just as in the case of the two-wire line (Chapter II), these may be ignored in their effect on the current far from the load if an appropriate terminal-zone admittance is provided. Thus, the total *apparent* terminal admittance is given by

$$Y_{la} = Y_\delta + Y_T, \qquad (6)$$

where Y_δ is the dominant-mode load impedance and Y_T is the terminal-zone admittance that takes account of end effect. With Θ_0 near $\pi/2$ it is essentially susceptive, so that

$$Y_T \doteq j\omega C_T. \qquad (7)$$

The load impedance $Z_\delta = 1/Y_\delta$ is defined by

$$Z_\delta \equiv \frac{V(l)}{I_\delta} = \frac{1}{I_\delta} \int_{\Theta_0}^{\pi-\Theta_0} E_{\Theta d}(l) \, l \, d\Theta, \qquad (8)$$

where I_δ is the radial current leaving the end of the upper conductor of the biconical line at $R = l \doteq a$ to become the axial surface current on the antenna. Since $E_{\Theta d}(l)$ in (8) is the dominant-mode field in the biconical line, where all currents are radial and, hence, perpendicular to the direction of E_Θ, it follows that

$$E_{\Theta d} = -\frac{1}{R}\frac{\partial \Phi}{\partial \Theta} - j\omega A_\Theta \doteq -\frac{1}{R}\frac{\partial \Phi}{\partial \Theta}, \qquad (9)$$

where Φ is the scalar potential and A_Θ a component of the vector potential. Substitution of (9) in (8), with $R = l$, gives

$$Z_\delta = \frac{V(l)}{I_\delta} \doteq \frac{1}{I_\delta} \int_{-\delta}^{\delta} d\Phi = \frac{\Phi(\delta) - \Phi(-\delta)}{I_\delta}. \qquad (10)$$

If the exterior field is expressed in terms of the current on the cylindrical antenna and matched to the internal field at $R = l \doteq a$, $\Theta_0 \le \Theta \le \pi - \Theta_0$ or $-\delta \le z \le \delta$, the impedance (10) may be determined. A part of the exterior field is required to match the interior complementary field. This represents the terminal-zone coupling between antenna and biconical transmission line. The actual evaluation and matching of the interior and exterior fields is of no interest here. It is sufficient to note that it is precisely Z_δ as defined in (10) that is evaluated beginning with Sec. II.11.

The relation between the apparent load admittance Y_{la} of the biconical transmission line and the driving-point admittance Y_0 is obtained from the general expressions (4.19) and (4.20) for the dominant-mode voltage and current. Thus, with $R = l$,

$$V_d(l) = V(0) \cos \beta_0 l - jZ_c I(0) \sin \beta_0 l, \qquad (11a)$$

$$I_d(l) = I(0) \cos \beta_0 l - Y_c V(0) \sin \beta_0 l, \qquad (11b)$$

where $Y_c = 1/Z_c$. It follows that

$$Y_{la} \equiv \frac{I_d(l)}{V_d(l)} = \frac{Y_0 \cos \beta_0 l - jY_c \sin \beta_0 l}{\cos \beta_0 l - jZ_c Y_0 \sin \beta_0 l}, \qquad (12)$$

where

$$Y_0 \equiv I(0)/V(0) \qquad (13)$$

is the driving-point admittance at the apex of the biconical line.

Up to the present no restrictions have been placed on the length l of the biconical transmission line. Since with Θ_0 near $\pi/2$, $l \doteq a$, where a, the radius of the cylindrical antenna, is assumed to satisfy the condition

$$\beta_0 a \ll 1, \qquad (14)$$

it follows that (12) reduces to

$$Y_{la} \doteq \frac{Y_0 - jY_c \beta_0 a}{1 - jZ_c Y_0 \beta_0 a}. \qquad (15)$$

This formula may be simplified further by noting that with Θ_0 near $\pi/2$, $Z_c = (\zeta_0/\pi) \ln \cot \tfrac{1}{2}\Theta_0$ is very small. Hence,

$$Y_{la} \doteq Y_0 - jY_c \beta_0 a = Y_0 - j\omega C_g, \qquad (16)$$

where, with $\beta_0 = \omega/v_0$,

$$C_g \equiv ac_0 = \frac{a}{v_0 Z_c} = \frac{a\pi\epsilon_0}{\ln \cot \tfrac{1}{2}\Theta_0}. \qquad (17)$$

In (17) c_0 is the capacitance per unit radial length of the biconical line as defined in (4.16b). It follows with (6) and (7) that the driving-point admittance at the apex of the biconical line is

$$Y_0 \doteq Y_{la} + j\omega C_g = Y_\delta + j\omega(C_T + C_g). \qquad (18)$$

Note that (18) is a special form of (12) solved for Y_0, namely,

$$Y_0 = \frac{Y_{la} \cos \beta_0 l + jY_c \sin \beta_0 l}{\cos \beta_0 l + jZ_c Y_{la} \sin \beta_0 l}, \qquad (19)$$

when both $\beta_0 l$ and Z_c are sufficiently small.

The confusion regarding the significance of the gap in antenna theory arises from a failure to recognize that (18) is not a general form but is a special case of (19).

The following conclusions may be drawn from (18). (1) The driving-point impedance Y_0 of a cylindrical antenna is equal to the impedance $Z_\delta = V_\delta/I_\delta$ (where V_δ is the voltage across the gap of length 2δ and I_δ is the current entering the antenna at the edge of the gap) in parallel with the effective "gap capacitance" $C_g + C_T$ between the two adjacent end surfaces of the antenna. (2) As the width 2δ of the gap is decreased by making Θ_0 differ less and less from $\pi/2$, the essential part C_g of the gap capacitance increases without limit. This is seen from (17). (3) It follows that in the limit $\delta \to 0$ the driving-point susceptance B_0 becomes infinite. Since the generator is short-circuited and completely enclosed by metal, no power can be radiated from the antenna. If it is now *assumed* that the interpretation of (18) is typical for all center-driven cylindrical antennas, the following "conclusion" is reached: (4) Only antennas with finite gaps can radiate. Finally, a naïve application of the Poynting vector theorem as described in Sec. II.4 leads to the additional "conclusion": (5) All radiation comes out of the gap.

In what way are these "conclusions" erroneous? This may be discovered readily by rephrasing them in a manner consistent with the *general* formula (19) instead of the *special* formula (18) as follows: (1) The driving-point admittance Y_0 of a section of transmission line of length l with an apparent load admittance Y_{la} is given by (19). If l is sufficiently short and the characteristic impedance Z_c sufficiently small, this reduces to (18). The apparent load admittance Y_{la} is equal to the actual load admittance $Y_\delta = V_\delta/I_\delta$ in parallel with a lumped capacitance that takes account of terminal-zone effects. (2) As the distance between the two conductors of the transmission line is reduced, the characteristic impedance Z_c decreases without limit. For the biconical line this is seen from (5). For parallel-wire lines Z_c has the factor $\cosh^{-1}(b/2a)$, which reduce to zero when b approaches $2a$. (3) It follows from (19) that in the limit $\delta \to 0$, when $Z_c \to 0$ and $Y_c \to \infty$, the input susceptance of the section of line is infinite, since the two perfect conductors are everywhere in contact and do not constitute a transmission line. Since this leaves the antenna completely isolated from the generator, *no currents are maintained in it* and, therefore, no electromagnetic field is set up. That is, the antenna does not radiate. (4) A transmission line like that in Fig. 12.7 that happens to be short, biconical, and "inside" the antenna, so that it *looks like a simple gap* instead of a section of line, is characteristic only of certain special cases that actually are highly artificial and impractical. Every short section of transmission line, whether two-wire, coaxial, biconical, or radial, is not merely a lumped capacitance in parallel with the load. *In all cases* the entire current from the generator to the load traverses the two conductors—whether flat plates, cones, or wires—in opposite directions. They form a necessary *series* connection between generator and load. It is obvious that only transmission lines consisting of two conductors with finite spacing can transmit power to a load. This is as true of two-wire lines as of biconical or radial lines; it has nothing to do with gaps. Every antenna in which currents are induced radiates in the sense that these currents maintain a far-zone field. Practical antennas never have gaps. (5) As shown in Sec. II.4, the Poynting-vector theorem is useful in *locating the generator* in a complete transmitting system, *not* the element that carries the currents that maintain the radiation field. In order to radiate, that is, provide far-zone fields, antennas must have currents maintained in them. This may be accomplished by direct connection to a transmission line from a distant generator or by coupling to a varying magnetic field maintained by a generator. A gap is not required. Thus it may be concluded that there are transmission-line problems and coupling problems in antenna theory *but no gap problems*.

13. *Zuhrt's Analysis of Cylindrical Antennas*[*]

Perhaps the most obvious and direct formulation of the problem of the cylindrical antenna is as a solution of the wave equation in cylindrical coördinates r, θ, z subject to appropriate boundary conditions. If the antenna is chosen to be a highly conducting tube of very small wall thickness immersed in air along the z-axis, only axial currents are involved. Accordingly, the electromagnetic field may be derived from a vector potential $\mathbf{A}$ (which in this case is related to the polarization potential by the simple formula $\mathbf{A} = (j\beta_0^2/\omega)\mathbf{\Pi}_e$) that has only a z-component, $\mathbf{A} = \hat{\mathbf{z}}A_z$. Since the Cartesian components

---

[*] This section discusses the work of Dr. H. Zuhrt, ref. 25.

[VIII.13] THEORY OF LINEAR ANTENNAS 849

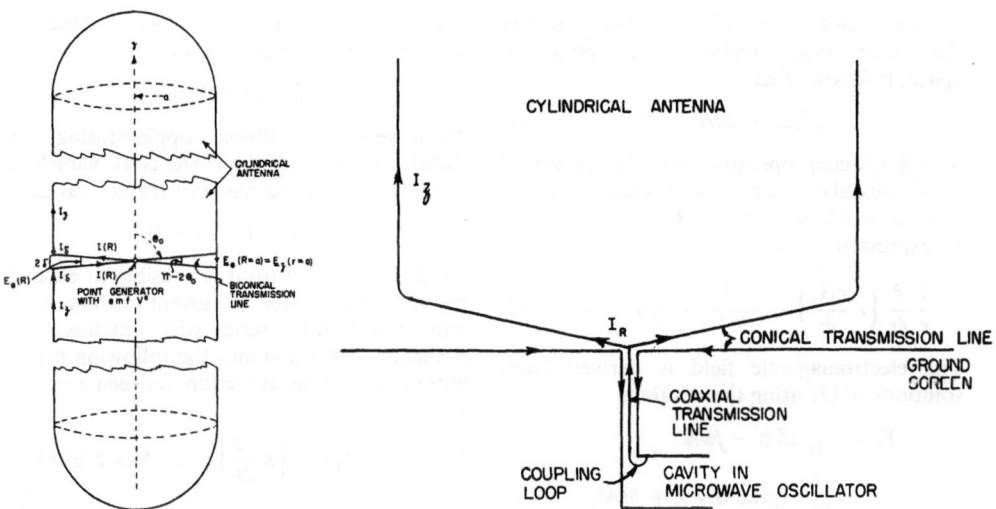

Fig. 12.7. Transmitting system with biconical line and point-generator.

Fig. 12.8. Transmitting system with conical and coaxial lines.

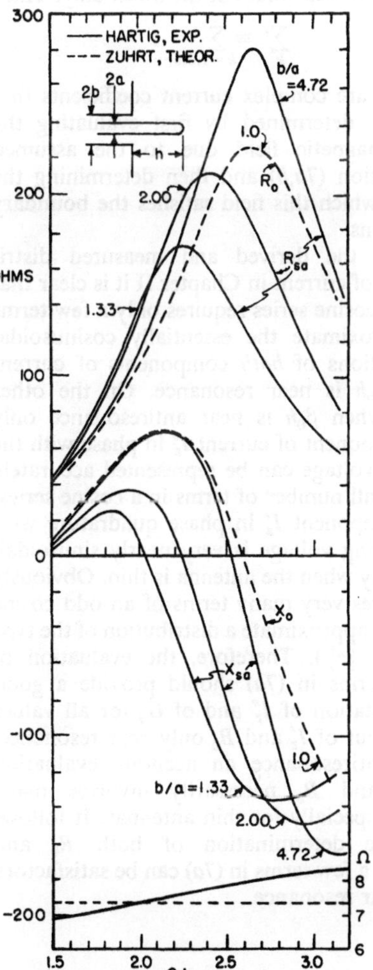

Fig. 13.1. Zuhrt's theoretical and Hartig's experimental impedances.

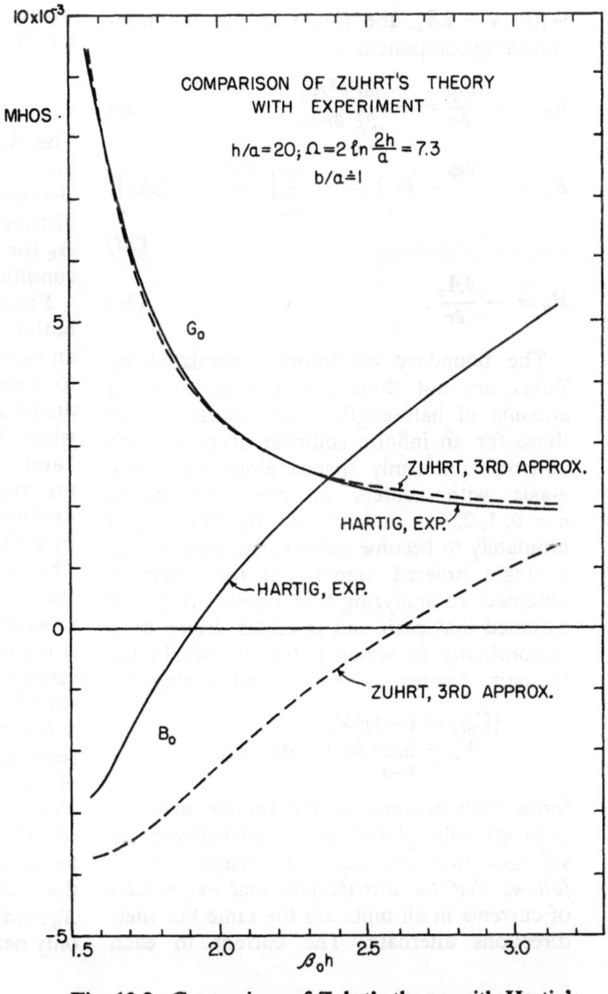

Fig. 13.2. Comparison of Zuhrt's theory with Hartig's experiment.

of the vector potential individually satisfy the scalar wave equation at all points in space, it follows that

$$\nabla^2 A_z + \beta_0^2 A_z = 0. \tag{1}$$

The Laplacian operator may be expressed in cylindrical coördinates and, since rotational symmetry obtains, so that $\partial A_z/\partial \theta = 0$, (1) may be expressed as follows:

$$\frac{1}{r}\frac{\partial}{\partial r}\left(r\frac{\partial A_z}{\partial r}\right) + \frac{\partial^2 A_z}{\partial z^2} + \beta_0^2 A_z = 0. \tag{2}$$

The electromagnetic field is derived from solutions of (2) using the relations

$$\mathbf{E} = -\operatorname{grad}\boldsymbol{\phi} - j\omega\mathbf{A}$$

$$= \frac{-j\omega}{\beta_0^2}(\operatorname{grad}\operatorname{div}\mathbf{A} + \beta_0^2\mathbf{A}), \tag{3a}$$

$$\mathbf{B} = \operatorname{curl}\mathbf{A}. \tag{3b}$$

With $\mathbf{A} = \hat{\mathbf{z}} A_z$, the following are the non-vanishing components:

$$E_r = -\frac{\partial \phi}{\partial r} = -\frac{j\omega}{\beta_0^2}\frac{\partial^2 A_z}{\partial r \partial z}, \tag{3c}$$

$$E_z = -\frac{\partial \phi}{\partial z} - j\omega A_z = -\frac{j\omega}{\beta_0^2}\left(\frac{\partial^2 A_z}{\partial z^2} + \beta_0^2 A_z\right), \tag{3d}$$

$$B_\theta = -\frac{\partial A_z}{\partial r}. \tag{3e}$$

The boundary conditions formulated by Zuhrt are not those for a single isolated antenna of half-length h and radius a, but those for an infinite collinear array of such antennas uniformly spaced along the entire z-axis with centers at $z = \pm nd$, where $n = 0, 1, 2, \cdots$ and $d > h$. By allowing d ultimately to become infinite, the solution for a single isolated antenna at the origin is obtained. In analyzing this infinite array it is assumed that each unit is center driven by a discontinuity in scalar potential, which, for the pair of antennas at $z = \pm nd$, is given by

$$(V_0)_n = (-1)^n V_0,$$
$$V_0 = \lim_{\delta \to 0}[\boldsymbol{\phi}(\delta) - \boldsymbol{\phi}(-\delta)]. \tag{4}$$

Since each antenna in the infinite array is symmetrically placed and center-driven by voltages that are equal in magnitude, *it follows that the distributions and magnitudes of currents in all units are the same but their directions alternate.* The current in each antenna is an even function with respect to its center. For the central unit,

$$I(-z) = I(z). \tag{5}$$

As a second condition supplementing (6), Zuhrt assumes that the currents all vanish at the ends of the antennas. For the central unit,

$$I(h) = I(-h) = 0. \tag{6}$$

Subject to (5) and (6) it is possible to expand the axial distribution of current in the central unit in a Fourier series with unknown coefficients. Zuhrt assumes the following distribution in the central section between $z = -d$ and $z = d$:

$$I(z) = \sum_k{}' A_k \cos\left(k\frac{\pi z}{2h}\right), \quad (-h \leq z \leq h) \tag{7a}$$

$$I(z) = 0, \quad (h < |z| < d) \tag{7b}$$

where the primed sign of summation denotes a sum over *odd* values of the index only. Thus,

$$\sum_k{}' \equiv \sum_{k=1,3,5,\cdots}.$$

The A_k are complex current coefficients that must be determined by first evaluating the electromagnetic field due to the assumed distribution (7a,b) and then determining the A_k for which this field satisfies the boundary conditions.

From the derived and measured distributions of current in Chapter II it is clear that an odd cosine series requires only a few terms to approximate the essentially cosinusoidal distributions of both components of current when $\beta_0 h$ is near resonance. On the other hand, when $\beta_0 h$ is near antiresonance only the component of current I_z'' in phase with the driving voltage can be represented accurately by a small number of terms in a cosine series. The component I_z' in phase quadrature with the driving voltage is very nearly sinusoidal, especially when the antenna is thin. Obviously it requires very many terms of an odd cosine series to approximate a distribution of the type $\sin \beta(h - |z|)$. Therefore, the evaluation of a few terms in (7a) should provide a good representation of I_z'' and of G_0 for all values of $\beta_0 h$, but of I_z' and B_0 only near resonance. Near antiresonance an accurate evaluation of I_z' and B_0 necessarily involves many terms, especially for thin antennas. It follows that the determination of both R_0 and X_0 from a few terms in (7a) can be satisfactory only near resonance.

As pointed out in Sec. II.11, the use of tubes instead of solid cylinders does not assure the vanishing of the current at the ends on the outer surface of the conductors. Just as in the case of solid cylinders an upward current on the cylindrical surface turns the corner at the ends to become a radially inward current charging the flat end surfaces, so with tubes an upward current on the outside surfaces becomes a downward current on the inside surfaces. In both cases the magnitude of $I(h)$ is small *only* if the radius a of the cylinder or tube is small. Hence, the assumption (6) is a good approximation only when the following inequalities are satisfied:

$$\beta_0 a \ll 1, \quad a \ll h. \tag{8}$$

It follows that Zuhrt's formulation is limited to cylindrical antennas of small radius, just as is the integral-equation formulation in Chapter II. (The restriction on the radius may be relaxed if the current $I(z)$ is interpreted to be the total current on inner and outer surfaces.)

Instead of expressing the currents in the other sections of length $2d$ along the z-axis in the form (7a,b) with z replaced by z_n referred to an origin at the center of the nth unit, Zuhrt expands the distribution (7a,b) (which is repeated periodically with alternating sign) in another Fourier series which, for reasons of symmetry, may include only odd cosine terms. The series is

$$I_z = \sum_n{}' B_n \cos \beta_n z, \quad \beta_n \equiv \frac{n\pi}{2d},$$

$$d \leq |z| < \infty, \tag{9}$$

where the symbol $\sum_n{}'$ stands for summation over *odd* values of n only, and where the Fourier coefficients are defined by

$$B_n = \frac{1}{2d} \int_{-2d}^{2d} I(z) \cos \beta_n z \, dz, \tag{10}$$

with $I(z)$ given in (7a,b). The evaluation of the B_n is elementary and leads to

$$B_n = \frac{4h}{\pi d} \sum_k{}' A_k (-1)^{(k-1)/2}$$

$$\times \frac{k}{(k^2 - h^2 n^2/d^2)} \cos\left(\frac{n\pi h}{2d}\right). \tag{11}$$

Thus, the distribution of current along the entire z-axis is specified by (7a) and (10). The problem remaining is to determine the coefficients A_k in (7a) and (11) in such a manner that the electromagnetic field due to the currents in all of the antennas satisfies the boundary conditions.

In formulating the boundary conditions Zuhrt introduces an infinite boundary cylinder of radius a that divides all space into an inner region I and an outer region II. The boundary cylinder coincides with the actual outer boundary surfaces of the collinear conductors where these exist; between them it is an imaginary boundary introduced for mathematical convenience in a manner analogous to the boundary sphere used in the analysis of the biconical antenna earlier in this chapter. By defining the electromagnetic field in the inner region and the outer region and matching it across the boundary cylinder a solution may be obtained.

By separating the variables in (2) in the conventional manner (ref. I.31, chap. V), with

$$k_n^2 \equiv \beta_0^2 - \beta_n^2 \tag{12}$$

as the separation constant, and selecting $J_0(k, r)$ for the interior region since it is finite at $r = 0$ and $H_0^{(2)}(k_n r)$ for the exterior region since it satisfies the radiation condition as $r \to \infty$, appropriate solutions for the two regions are:

$$A_z^{II} = \sum_n{}' C_n H_0^{(2)}(k_n r) \cos \beta_n z, \tag{13a}$$

$$A_z^{I} = \sum_n{}' D_n J_0(k_n r) \cos \beta_n z. \tag{13b}$$

The coefficients C_n and D_n must be evaluated by imposing appropriate boundary and matching conditions along the entire boundary cylinder $r = a$ between regions I and II. Assuming the cylinders perfectly conducting and of negligible wall thickness, the following conditions apply on the parts of the boundary where the boundary cylinder coincides with the antennas:

$$E_z^{II}(r = a) = E_z^{I}(r = a) \doteq 0, \tag{14a}$$

$$B_\theta^{II}(r = a) - B_\theta^{I}(r = a) = \frac{I(z)}{2\pi v_0 a}. \tag{14b}$$

Elsewhere on the boundary cylinder the conditions for a match are

$$E_z^{II}(r = a) = E_z^{I}(r = a),$$
$$E_r^{II}(r = a) = E_r^{I}(r = a), \tag{15a}$$

$$B_\theta^{II}(r = a) = B_\theta^{I}(r = a). \tag{15b}$$

It is evident that (15a) is satisfied if

$$A_z^{\mathrm{I}} = A_z^{\mathrm{II}}. \tag{16a}$$

Similarly, (14b) and (15b) are fulfilled if

$$\left(\frac{\partial A_z^{\mathrm{I}}}{\partial r} - \frac{\partial A_z^{\mathrm{II}}}{\partial r}\right)_{r=a} = \frac{I(z)}{2\pi v_0 a}, \tag{16b}$$

since $I(z)$ vanishes in the regions between the cylinders. It follows that the three conditions that must be satisfied are (16a), (16b), and (14a), with the added driving condition (4) at the center of each antenna.

With (13a,b) in (16a,b) and use of the well-known formula $Z_0' = -Z_1$ for all cylinder functions, together with (9), the following equations are obtained:

$$C_n H_0^{(2)}(k_n a) = D_n J_0(k_n a), \tag{17a}$$

$$C_n H_1^{(2)}(k_n a) - D_n J_1(k_n a) = \frac{B_n}{2\pi v_0 a k_n}. \tag{17b}$$

By eliminating D_n from (17b) with (17a) and using the standard formula*

$$H_1^{(2)}(x) J_0(x) - H_0^{(2)}(x) J_1(x) = -\frac{2}{j\pi x}, \tag{18}$$

the coefficients C_n and D_n are found to be

$$C_n = -jB_n \frac{J_0(k_n a)}{4v_0}, \tag{19a}$$

$$D_n = -jB_n \frac{H_0^{(2)}(k_n a)}{4v_0}. \tag{19b}$$

With (19a,b) the vector potential (13a,b) is given by

$$A_z^{\mathrm{II}} = \frac{-j}{4v_0} \sum_n' B_n J_0(k_n a) H_0^{(2)}(k_n r) \cos \beta_n z, \tag{20a}$$

$$A_z = \frac{-j}{4v_0} \sum_n' B_n H_0^{(2)}(k_n a) J_0(k_n r) \cos \beta_n z. \tag{20b}$$

The electromagnetic field derived from (20a,b) satisfies all boundary conditions except (14a) and (4). These may be satisfied in the following manner. From (3d) the axial electric field on the surface of the central conductor is

$$E_z^{\mathrm{II}}(r=a) = -\frac{j\omega}{\beta_0^2}\left(\frac{\partial^2 A_z^{\mathrm{II}}}{\partial z^2} + \beta_0^2 A_z^{\mathrm{II}}\right)_{r=a}. \tag{21}$$

---
* Jahnke and Emde, ref. I.28, p. 144.

With (20a) this becomes

$$E_z^{\mathrm{II}}(r=a) = -\frac{1}{4\omega\epsilon_0} \sum_n' B_n k_n^2 J_0(k_n a) H_0^{(2)}(k_n r) \cos \beta_n z. \tag{22}$$

This field must be set equal to the interior field at $r = a$. For a perfect conductor, as assumed in (14a), this is zero except in the generating region between $z = -\delta$ and $z = \delta$ which, in the limit, has zero length. Thus,

$$E_z^{\mathrm{I}}(r=a) = \begin{cases} 0 \text{ when } -h \leq z \leq -\delta \text{ and } \delta \leq z \leq h, \\ E_z^e = V_\delta/2\delta \text{ when } -\delta \leq z \leq \delta. \end{cases} \tag{23}$$

By expanding the electric field $E_z(r = a)$ in a Fourier series of the form

$$E_z = \sum_m' C_m \cos k_m z, \tag{24}$$

where

$$k_m = m\frac{\pi z}{2h} \tag{25}$$

and

$$C_m = \frac{2}{h}\int_0^h E_z(r=a) \cos k_m z \, dz, \tag{26}$$

the coefficients C_m may be expressed first in terms of (23) as C_{m1} and then in terms of (22) as C_{m2} and the two sets of coefficients equated. Note that since the Fourier series is to approximate (23) only over the length $2h$ of the central unit the behavior outside this range is immaterial. It is for this reason that the value k_m as in (25) may be chosen. Substitution of (23) in (26) gives

$$C_{m1} = \frac{2}{h}\int_0^\delta \frac{V_\delta}{2\delta} \cos k_m z \, dz = -\frac{V_\delta}{h}\frac{\sin k_m \delta}{k_m \delta}. \tag{27}$$

In the limit $\delta \to 0$,

$$C_{m1} \doteq -V_0/h. \tag{28}$$

Substitution of (22) in (26) gives

$$C_{m2} = -\frac{1}{2\omega\epsilon_0 h} \times$$
$$\int_0^h [\sum_n' B_n k_n^2 J_0(k_n a) H_0^{(2)}(k_n a) \cos k_n z] \cos k_m z \, dz. \tag{29}$$

This integrates into

$$C_{m2} = -\frac{1}{\omega\epsilon_0\pi}(-1)^{(m-1)/2}\sum_n{}' B_n k_n^2 J_0(k_n a)$$
$$\times H_0^{(2)}(k_n a)\frac{m}{m^2 - n^2 h^2/d^2}\cos\left(\frac{n\pi h}{2d}\right). \quad (30)$$

It is now possible to equate (28) and (30) to obtain the following condition that must be satisfied:

$$\frac{1}{\pi\omega\epsilon_0}(-1)^{(m-1)/2}\sum_n{}' B_n k_n^2 J_0(k_n a) H_0^{(2)}(k_n a)$$
$$\times \frac{m}{m^2 - n^2 h^2/d^2}\cos\left(\frac{n\pi h}{2d}\right) = V_0/h.$$
$$(m = 1, 3, 5, \cdots) \quad (31)$$

The values of B_n given in (11) in terms of the coefficients of the current now can be substituted in (31). After interchanging the order of summation the following system of equations is obtained for evaluating the coefficients A_k of the current:

$$\frac{4h^2}{\pi^2\omega\epsilon_0 d}\sum_k{}' A_k k m(-1)^{(k-1)/2}(-1)^{(m-1)/2}$$
$$\times \sum_n{}' \frac{k_n^2 J_0(k_n a) H_0^{(2)}(k_n a)}{(k^2 - n^2 h^2/d^2)(m^2 - n^2 h^2/d^2)}\cos^2\left(\frac{n\pi h}{2d}\right)$$
$$= V_0 \quad (m = 1, 3, 5, \cdots) \quad (32)$$

These equations may be expressed in a more convenient form for subsequently allowing d to become infinite by defining the quantities

$$h_r \equiv 4h/\lambda_0, \quad \gamma \equiv h/d, \quad (33)$$

and defining the following impedance coefficients:

$$Z_{mk} = \frac{\zeta_0}{\pi h_r}km(-1)^{(k-1)/2}(-1)^{(m-1)/2}2\gamma \times$$
$$\sum_n{}' J_0\left(\frac{\pi a}{2h}\sqrt{h_r^2 - \gamma^2 n^2}\right)H_0^{(2)}\left(\frac{\pi a}{2h}\sqrt{h_r^2 - \gamma^2 n^2}\right)$$
$$\times \left[\frac{h_r^2 - \gamma^2 n^2}{(k^2 - \gamma^2 n^2)(m^2 - \gamma^2 n^2)}\right]\cos^2\left(\frac{\gamma n \pi}{2}\right).$$
$$(m = 1, 3, 5, \cdots) \quad (34)$$

With (34), (32) becomes

$$\sum_k{}' A_k Z_{mk} = V_0 \text{ for } m = 1, 3, 5, \cdots \quad (35)$$

In expanded form,

$$A_1 Z_{11} + A_3 Z_{13} + A_5 Z_{15} + \cdots = V_0,$$
$$A_1 Z_{31} + A_3 Z_{33} + A_5 Z_{35} + \cdots = V_0, \quad (36)$$
$$A_1 Z_{51} + A_3 Z_{53} + A_5 Z_{55} + \cdots = V_0,$$
$$\cdots\cdots\cdots\cdots\cdots\cdots\cdots\cdots\cdots$$

Since the impedance coefficients Z_{mk} may be determined from (34), the unknown current coefficients A_k may be evaluated from (36) to any desired degree of approximation. Specifically,

1st approximation: $A_1 Z_{11} = V_0;$ \quad (37a)

2nd approximation:

$$\begin{cases} A_1 Z_{11} + A_3 Z_{13} = V_0, \\ A_1 Z_{31} + A_3 Z_{33} = V_0; \end{cases} \quad (37b)$$

3rd approximation:

$$\begin{cases} A_1 Z_{11} + A_3 Z_{13} + A_5 Z_{15} = V_0, \\ A_1 Z_{31} + A_3 Z_{33} + A_5 Z_{35} = V_0, \quad (37c) \\ A_1 Z_{51} + A_3 Z_{53} + A_5 Z_{55} = V_0. \end{cases}$$

In order to obtain the solution for an isolated antenna instead of an infinite collinear array Zuhrt proceeds to take the limit $d \to \infty$. If this is done, the sums in (34) become integrals. Alternatively, it is better to use a Fourier integral directly without introducing the extraneous infinite array of auxiliary antennas since the treatment of the Fourier integral as a limit of a Fourier series is undesirable. Following Zuhrt, since the quantity γn increases in steps of 2γ, it is convenient to set $x = \gamma n$ and $dx = 2\gamma$. The impedance coefficients of the isolated antenna as obtained from (34) in the limit as $d \to \infty$ are

$$Z_{mk} = \frac{\zeta_0}{\pi h_r}km(-1)^{(k-1)/2}(-1)^{(m-1)/2}$$
$$\times \int_0^\infty J_0\left(\frac{\pi a}{2h}\sqrt{h_r^2 - x^2}\right)H_0^{(2)}\left(\frac{\pi a}{2h}\sqrt{h_r^2 - x^2}\right)$$
$$\times \left[\frac{h_r^2 - x^2}{(k^2 - x^2)(m^2 - x^2)}\right]\cos^2\left(\frac{\pi x}{2}\right)dx. \quad (38)$$

By substituting the Z_{mk} determined from (38) in the system of equations (36) the current coefficients A_k may be determined to any desired degree of approximation. If, in turn, these are substituted in (7a), the current at all points in the isolated center-driven antenna is obtained and, therefore, the input admittance and impedance.

Using graphical methods to evaluate integrals involved in (38), Zuhrt has computed three orders of approximation of the input impedance $Z_0 = R_0 + jX_0$ of a rather thick antenna for which $h/a = 20$ or $\Omega = 2\ln(2h/a) \doteq 7.3$. These are listed in Table 13.1 together with the corresponding admittances. Note that the convergence for G_0 is excellent for all

TABLE 13.1. Impedance and admittance according to Zuhrt's theory; $h/a = 20$, $\Omega = 2 \ln 2h/a \doteq 7.4$.

h/λ_0	$\beta_0 h$	First approx.	Second approx.	Third approx.
		$Z_0 = R_0 + jX_0$ (ohms)		
0.25	1.57	$73.2 + j35.6$	$94.3 + j35.3$	$91.8 + j35.4$
.30	1.88	$101.3 + j110.4$	$168 + j113.2$	$169.5 + j109.2$
.35	2.20	$131 + j171.5$	$278 + j156.8$	$294 + j138.8$
.40	2.51	$164.9 + j217.6$	$416 + j99.5$	$428 + j46.8$
.45	2.83	$198 + j252$	$460 - j65.5$	$429 - j135.5$
.50	3.14	$233.5 + j291.5$	$337.5 - j242$	$289 - j244$
		$Y_0 = G_0 + jB_0$ (10^{-3} mho)		
0.25	1.57	$11.06 - j5.38$	$9.30 - j3.49$	$9.48 - j3.66$
.30	1.88	$4.51 - j4.91$	$4.09 - j2.75$	$4.17 - j2.69$
.35	2.20	$2.81 - j3.68$	$2.73 - j1.53$	$2.78 - j1.31$
.40	2.51	$2.21 - j2.92$	$2.27 - j0.54$	$2.31 - j0.25$
.45	2.83	$1.93 - j2.45$	$3.13 + j0.30$	$2.12 + j0.67$
.50	3.14	$1.67 - j2.09$	$1.96 + j1.40$	$2.02 + j1.70$

values of $\beta_0 h$, whereas the convergence for B_0 is good near resonance and quite poor near antiresonance. This behavior is in complete agreement with the prediction made in conjunction with the use of the odd Fourier series (7a) to represent the current. It is to be expected that the convergence for D_0 is even worse for a thin antenna than for the moderately thick antenna with $h/a = 20$ evaluated by Zuhrt.

A graph of Zuhrt's third approximation of R_0 and X_0 is given in Fig. 13.1 together with experimental curves by Hartig taken from Fig. II.38.13a, b. Since the experimental curves are for antennas with slightly varying values of Ω as indicated by the solid curve at the bottom of Fig. 13.1, whereas Zuhrt's curves apply to an antenna with Ω fixed at the constant value indicated by the broken curve at the bottom of Fig. 13.1, an exact correspondence does not exist. Note that the experimental curves are measurements made on coaxial lines with equal diameters of inner conductors, but with three different diameters of outer conductors ranging from $b/a = 4.72$ to $b/a = 1.33$. The theoretical curves are evaluated for a slice generator which corresponds to the limiting value $b/a \to 1$. Note that there is very poor agreement between the theoretical curve or extrapolated experimental curves (Fig. II.38.14) for $b/a \doteq 1$ and the experimental curves for $b/a \geq 1.33$. Antiresonance in the theory occurs near $\beta_0 h = 2.6$, whereas to agree with experimental results it should occur for $\beta_0 h$ near 1.9. It may be noted that in his paper Zuhrt provided a set of experimental data that are in quite good agreement with his theoretical curves. Since the measured impedance of an antenna, especially a fairly thick antenna, is very sensitive to the particular value of b/a for the coaxial measuring and driving line, and no account whatever is taken of this fact in Zuhrt's comparison, the agreement he shows has little significance. It is evident in Fig. 13.1 that an appropriate choice of b/a can provide experimental results in quite good agreement with the theoretical curves. However, since Zuhrt takes no account of transmission-line end effect, his theory must be compared with results obtained by extrapolating the measured resistance and reactance for each value of $\beta_0 h$ for different values of $b/a > 1$ to $b/a \to 1$. The general location and shape of the resulting curves may be visualized in Fig. 13.1. Evidently, curves of such extrapolated resistances and reactances lie to the left of and below the curves for $b/a = 1.33$, in poor agreement with Zuhrt's theory.

A better and illuminating comparison of Zuhrt's theory with Hartig's experimental results is provided in Fig. 13.2, where values of G_0 and B_0 obtained from Hartig's experimental curves have been extrapolated to $b/a = 1$ and interpolated or extrapolated to $\Omega = 7.3$. It is seen that the third approximation of the conductance by Zuhrt's theory is excellent, but the approximation of the susceptance rather poor. Note that antiresonance occurs in the experimental curves at $\beta_0 h = 1.9$, where h is the axial half-length including a hemispherical cap so that $I(h) = 0$; in Zuhrt's theoretical curve $\beta_0 h = 2.6$, where h is the half-length of a hypothetical tube that has zero current at the end. Since B_0 is the component I_0' of the input current per volt, it may be concluded from Fig. 13.2 that I_z' is not well approximated by three terms in an odd cosine series.

Zuhrt has extended his method to the receiving antenna and to coupled antennas.

A comparison of Zuhrt's solution of the problem of the cylindrical antenna with the integral-equation method of Hallén suggests the following conclusions. (1) Both methods treat the antenna essentially as one-dimensional in the sense that no account is taken of the ends. This restricts both solutions to relatively thin antennas that satisfy the conditions, $\beta_0 a \ll 1$, $a \ll h$. With $\Omega = 2 \ln (2h/a)$ as small as 7.4, these conditions are violated when $\beta_0 h$ exceeds $\pi/2$. It follows that both theories should give accurate results near resonance, but may be expected to depart from the correct values by appreciable amounts in the case of longer antennas. A comparison of R_0 for $\beta_0 h = \pi/2$ in Table 13.1 with corresponding values obtained by interpolation from Tables II.30.1 and II.30.15 and with experimental results in Sec. II.38 shows that the third-order King-Middleton values agree best with measured results, while the second-order King-Middleton and the third-order Zuhrt values are nearly the same and about 9 percent too low at $\Omega = 7.4$. All reactances are in good agreement. The experimental location of antiresonance is at $\beta_0 h = 1.9$; the King-Middleton second-order value is 2.2; the Zuhrt third-order value is 2.6. The maximum resistance occurs at somewhat greater lengths. The experimental value (for an antenna over a ground plane) is about 160; the King-Middleton second-order value is 197; the Zuhrt third-order value is 215. As predicted, the theoretical values are considerably in error since the antennas are too thick at this length. On the other hand, it is seen that the Zuhrt third-order theory is considerably more in error than the second-order solution of King and Middleton. This is presumably a consequence of the poorer convergence of the susceptance near antiresonance in the case of the Zuhrt theory. (2) The representation of the current in a Fourier series in Zuhrt's analysis provides an analytically simpler expression than the iterated formula in the Hallén theory. (3) The actual formulation of the Hallén theory is simpler and may be applied more readily to V-antennas, antennas at right angles, etc.

APPENDIX

TABLES OF GENERALIZED SINE AND COSINE INTEGRALS

The integrals which occur consistently in the determination of the distributions of current and the impedance of isolated and coupled antennas of small cross-sectional dimension are the following:

$$S(a, x) = \int_0^x \frac{\sin W}{W} du,$$

$$C(a, x) = \int_0^x \frac{1 - \cos W}{W} du,$$

$$Ss(a, x) = \int_0^x \frac{\sin W}{W} \sin u \, du,$$

$$Cs(a, x) = \int_0^x \frac{\cos W}{W} \sin u \, du,$$

$$Sc(a, x) = \int_0^x \frac{\sin W}{W} \cos u \, du,$$

$$Cc(a, x) = \int_0^x \frac{\cos W}{W} (1 - \cos u) \, du,$$

where

$$W = \sqrt{a^2 + u^2}.$$

These integrals are tabulated in the *Tables of generalized sine- and cosine-integral functions* with ranges of a from 0 to 25 and of x from 0 to 25 as prepared by the Computation Laboratory of Harvard University (Annals of the Computation Laboratory of Harvard University, Vols. 18 and 19; Harvard University Press, Cambridge, 1948). Selected parts of these tables are reproduced in this appendix. Many of the properties of these functions are summarized in Sec. II.19.

The conventional sine and cosine integrals,

$$Si(x) = \int_0^x \frac{\sin u}{u} du = S(0, x),$$

$$Cin(x) = \int_0^x \frac{1 - \cos u}{u} du = C(0, x),$$

$$Ci(x) = \int_\infty^x \frac{\cos u}{u} du = C + \ln x - C(0, x),$$

where $C = 0.5772$ is Euler's constant, may be obtained from tables of $S(0, x)$ and $C(0, x)$ with readily available tables of the natural logarithm; $Ci(x)$ is obtained more conveniently from extensive and readily available tables of this function itself. Short tables of both $Ci(x)$ and $Si(x)$ are given in Jahnke and Emde, *Tables of functions*. Extensive tables from $x = 0$ to $x = 40.0$ are given in *Tables of sine, cosine, and exponential integrals* (Federal Works Agency, Works Progress Administration, New York, 1940), vols. 1 and 2, sponsored by the National Bureau of Standards. $Cin(x)$ is given very completely in *Tables of the modified cosine integral* (Stanford Research Institute, 1951).

The related functions

$$C_i(a, x) = \int_0^x \frac{\cos W}{W} du$$

$$= \sinh^{-1} \frac{x}{a} - C(a, x),$$

$$Cc_i(a, x) = \int_0^x \frac{\cos W}{W} \cos u \, du$$

$$= \sinh^{-1} \frac{x}{a} - C(a, x) - Cc(a, x),$$

may be obtained from the tables of $C(a, x)$ and $Cc(a, x)$ with readily available tables of the inverse hyperbolic sine. The most extensive tabulation of this function is in *Tables of inverse hyperbolic functions* prepared by the Computation Laboratory of Harvard University (Annals of the Computation Laboratory of Harvard University, vol. 20; Harvard University Press, Cambridge, 1949).

It is believed that the following short table of generalized sine and cosine integrals together with readily available short tables of the natural logarithm and of the inverse hyperbolic sine should prove adequate for many computations involving linear radiators and a first-order theory.

a = 0.00

x	S(a,x)	C(a,x)	Ss(a,x)	Sc(a,x)	Cs(a,x)	Cc(a,x)	x
0.00	0.000000	0.000000	0.000000	0.000000	0.000000	0.000000	0.00
0.01	0.010000	0.000025	0.000050	0.010000	0.010000	0.000025	0.01
0.02	0.020000	0.000100	0.000200	0.019998	0.019998	0.000100	0.02
0.03	0.029999	0.000225	0.000450	0.029994	0.029994	0.000225	0.03
0.04	0.039996	0.000400	0.000800	0.039986	0.039986	0.000400	0.04
0.05	0.049993	0.000625	0.001249	0.049972	0.049972	0.000625	0.05
0.06	0.059988	0.000900	0.001799	0.059952	0.059952	0.000899	0.06
0.07	0.069981	0.001225	0.002448	0.069924	0.069924	0.001223	0.07
0.08	0.079972	0.001600	0.003197	0.079886	0.079886	0.001597	0.08
0.09	0.089960	0.002024	0.004045	0.089838	0.089838	0.002020	0.09
0.10	0.099944	0.002499	0.004992	0.099778	0.099778	0.002493	0.10
0.11	0.109926	0.003023	0.006038	0.109705	0.109705	0.003014	0.11
0.12	0.119904	0.003598	0.007183	0.119617	0.119617	0.003585	0.12
0.13	0.129878	0.004222	0.008426	0.129513	0.129513	0.004204	0.13
0.14	0.139848	0.004896	0.009768	0.139392	0.139392	0.004872	0.14
0.15	0.149813	0.005620	0.011208	0.149252	0.149252	0.005588	0.15
0.16	0.159773	0.006393	0.012746	0.159093	0.159093	0.006352	0.16
0.17	0.169727	0.007216	0.014381	0.168912	0.168912	0.007164	0.17
0.18	0.179676	0.008089	0.016113	0.178709	0.178709	0.008024	0.18
0.19	0.189619	0.009011	0.017942	0.188482	0.188482	0.008930	0.19
0.20	0.199556	0.009983	0.019867	0.198231	0.198231	0.009884	0.20
0.21	0.209486	0.011005	0.021889	0.207953	0.207953	0.010884	0.21
0.22	0.219409	0.012076	0.024006	0.217647	0.217647	0.011930	0.22
0.23	0.229325	0.013196	0.026218	0.227313	0.227313	0.013022	0.23
0.24	0.239233	0.014365	0.028525	0.236949	0.236949	0.014159	0.24
0.25	0.249134	0.015584	0.030926	0.246554	0.246554	0.015342	0.25
0.26	0.259026	0.016852	0.033421	0.256126	0.256126	0.016569	0.26
0.27	0.268909	0.018170	0.036010	0.265664	0.265664	0.017840	0.27
0.28	0.278783	0.019536	0.038691	0.275167	0.275167	0.019155	0.28
0.29	0.288648	0.020951	0.041465	0.284635	0.284635	0.020514	0.29
0.30	0.298504	0.022416	0.044330	0.294064	0.294064	0.021915	0.30
0.31	0.308350	0.023929	0.047287	0.303456	0.303456	0.023358	0.31
0.32	0.318185	0.025491	0.050334	0.312807	0.312807	0.024843	0.32
0.33	0.328010	0.027102	0.053471	0.322118	0.322118	0.026369	0.33
0.34	0.337824	0.028761	0.056698	0.331386	0.331386	0.027937	0.34
0.35	0.347627	0.030469	0.060013	0.340611	0.340611	0.029544	0.35
0.36	0.357418	0.032226	0.063416	0.349792	0.349792	0.031191	0.36
0.37	0.367197	0.034030	0.066907	0.358927	0.358927	0.032877	0.37
0.38	0.376965	0.035883	0.070485	0.368015	0.368015	0.034601	0.38
0.39	0.386719	0.037785	0.074148	0.377056	0.377056	0.036363	0.39
0.40	0.396461	0.039734	0.077897	0.386048	0.386048	0.038162	0.40
0.41	0.406190	0.041732	0.081730	0.394990	0.394990	0.039998	0.41
0.42	0.415906	0.043777	0.085647	0.403880	0.403880	0.041870	0.42
0.43	0.425607	0.045870	0.089647	0.412719	0.412719	0.043777	0.43
0.44	0.435295	0.048011	0.093730	0.421504	0.421504	0.045719	0.44
0.45	0.444968	0.050200	0.097894	0.430235	0.430235	0.047694	0.45
0.46	0.454627	0.052436	0.102138	0.438911	0.438911	0.049702	0.46
0.47	0.464270	0.054719	0.106463	0.447531	0.447531	0.051743	0.47
0.48	0.473898	0.057050	0.110866	0.456093	0.456093	0.053816	0.48
0.49	0.483511	0.059428	0.115347	0.464597	0.464597	0.055920	0.49

THEORY OF LINEAR ANTENNAS

a = 0.00

x	S(a,x)	C(a,x)	Ss(a,x)	Sc(a,x)	Cs(a,x)	Cc(a,x)	x
0.50	0.493107	0.061853	0.119906	0.473042	0.473042	0.058053	0.50
0.51	0.502688	0.064324	0.124541	0.481426	0.481426	0.060217	0.51
0.52	0.512252	0.066843	0.129251	0.489749	0.489749	0.062408	0.52
0.53	0.521798	0.069408	0.134036	0.498010	0.498010	0.064628	0.53
0.54	0.531328	0.072020	0.138895	0.506208	0.506208	0.066875	0.54
0.55	0.540840	0.074678	0.143826	0.514343	0.514343	0.069148	0.55
0.56	0.550335	0.077383	0.148829	0.522412	0.522412	0.071447	0.56
0.57	0.559811	0.080133	0.153903	0.530416	0.530416	0.073770	0.57
0.58	0.569369	0.082930	0.159047	0.538353	0.538353	0.076117	0.58
0.59	0.578709	0.085772	0.164259	0.546222	0.546222	0.078486	0.59
0.60	0.588129	0.088661	0.169539	0.554024	0.554024	0.080878	0.60
0.61	0.597530	0.091595	0.174886	0.561756	0.561756	0.083291	0.61
0.62	0.606911	0.094574	0.180299	0.569418	0.569418	0.085725	0.62
0.63	0.616273	0.097598	0.185776	0.577010	0.577010	0.088177	0.63
0.64	0.625614	0.100668	0.191317	0.584531	0.584531	0.090649	0.64
0.65	0.634935	0.103783	0.196921	0.591979	0.591979	0.093138	0.65
0.66	0.644235	0.106942	0.202586	0.599354	0.599354	0.095643	0.66
0.67	0.653514	0.110147	0.208312	0.606656	0.606656	0.098165	0.67
0.68	0.662772	0.113396	0.214097	0.613884	0.613884	0.100701	0.68
0.69	0.672008	0.116689	0.219940	0.621036	0.621036	0.103251	0.69
0.70	0.681222	0.120026	0.225841	0.628113	0.628113	0.105815	0.70
0.71	0.690414	0.123407	0.231797	0.635114	0.635114	0.108390	0.71
0.72	0.699584	0.126833	0.237809	0.642038	0.642038	0.110976	0.72
0.73	0.708730	0.130302	0.243874	0.648884	0.648884	0.113573	0.73
0.74	0.717854	0.133814	0.249993	0.655652	0.655652	0.116178	0.74
0.75	0.726954	0.137370	0.256162	0.662342	0.662342	0.118792	0.75
0.76	0.736031	0.140969	0.262382	0.668952	0.668952	0.121413	0.76
0.77	0.745084	0.144611	0.268652	0.675483	0.675483	0.124041	0.77
0.78	0.754112	0.148296	0.274969	0.681933	0.681933	0.126673	0.78
0.79	0.763116	0.152024	0.281333	0.688302	0.688302	0.129310	0.79
0.80	0.772096	0.155793	0.287743	0.694590	0.694590	0.131950	0.80
0.81	0.781050	0.159606	0.294198	0.700797	0.700797	0.134592	0.81
0.82	0.789979	0.163460	0.300696	0.706921	0.706921	0.137236	0.82
0.83	0.798883	0.167356	0.307236	0.712962	0.712962	0.139880	0.83
0.84	0.807761	0.171294	0.313817	0.718921	0.718921	0.142523	0.84
0.85	0.816612	0.175274	0.320438	0.724796	0.724796	0.145164	0.85
0.86	0.825438	0.179295	0.327097	0.730588	0.730588	0.147802	0.86
0.87	0.834237	0.183357	0.333794	0.736295	0.736295	0.150437	0.87
0.88	0.843009	0.187460	0.340527	0.741918	0.741918	0.153067	0.88
0.89	0.851753	0.191603	0.347294	0.747455	0.747455	0.155691	0.89
0.90	0.860471	0.195787	0.354096	0.752908	0.752908	0.158308	0.90
0.91	0.869160	0.200012	0.360929	0.758276	0.758276	0.160918	0.91
0.92	0.877822	0.204276	0.367794	0.763558	0.763558	0.163518	0.92
0.93	0.886456	0.208581	0.374689	0.768754	0.768754	0.166109	0.93
0.94	0.895061	0.212925	0.381613	0.773864	0.773864	0.168688	0.94
0.95	0.903638	0.217309	0.388565	0.778888	0.778888	0.171256	0.95
0.96	0.912186	0.221732	0.395542	0.783825	0.783825	0.173811	0.96
0.97	0.920704	0.226194	0.402545	0.788676	0.788676	0.176351	0.97
0.98	0.929194	0.230694	0.409571	0.793439	0.793439	0.178877	0.98
0.99	0.937653	0.235234	0.416621	0.798116	0.798116	0.181387	0.99

a = 0.00

x	S(a,x)	C(a,x)	Ss(a,x)	Sc(a,x)	Cs(a,x)	Cc(a,x)	x
1.00	0.946083	0.239812	0.423691	0.802706	0.802706	0.183879	1.00
1.02	0.962852	0.249082	0.437891	0.811625	0.811625	0.188809	1.02
1.04	0.979498	0.258503	0.452162	0.820195	0.820195	0.193659	1.04
1.06	0.996021	0.268073	0.466493	0.828416	0.828416	0.198421	1.06
1.08	1.012417	0.277790	0.480876	0.836288	0.836288	0.203086	1.08
1.10	1.028685	0.287652	0.495299	0.843812	0.843812	0.207647	1.10
1.12	1.044824	0.297658	0.509754	0.850989	0.850989	0.212096	1.12
1.14	1.060831	0.307806	0.524231	0.857818	0.857818	0.216425	1.14
1.16	1.076705	0.318093	0.538720	0.864303	0.864303	0.220627	1.16
1.18	1.092444	0.328518	0.553212	0.870444	0.870444	0.224694	1.18
1.20	1.108047	0.339078	0.567696	0.876243	0.876243	0.228618	1.20
1.22	1.123512	0.349772	0.582165	0.881702	0.881702	0.232393	1.22
1.24	1.138837	0.360597	0.596608	0.886825	0.886825	0.236011	1.24
1.26	1.154021	0.371552	0.611017	0.891612	0.891612	0.239465	1.26
1.28	1.169061	0.382634	0.625382	0.896069	0.896069	0.242748	1.28
1.30	1.183958	0.393841	0.639695	0.900197	0.900197	0.245854	1.30
1.32	1.198709	0.405172	0.653947	0.904001	0.904001	0.248775	1.32
1.34	1.213313	0.416623	0.668129	0.907484	0.907484	0.251506	1.34
1.36	1.227768	0.428193	0.682233	0.910650	0.910650	0.254040	1.36
1.38	1.242073	0.439880	0.696250	0.913503	0.913503	0.256370	1.38
1.40	1.256227	0.451681	0.710173	0.916048	0.916048	0.258492	1.40
1.42	1.270228	0.463595	0.723994	0.918290	0.918290	0.260399	1.42
1.44	1.284076	0.475618	0.737704	0.920234	0.920234	0.262086	1.44
1.46	1.297768	0.487749	0.751297	0.921884	0.921884	0.263548	1.46
1.48	1.311305	0.499985	0.764764	0.923246	0.923246	0.264779	1.48
1.50	1.324684	0.512324	0.778099	0.924326	0.924326	0.265775	1.50
1.52	1.337904	0.524765	0.791295	0.925130	0.925130	0.266530	1.52
1.54	1.350965	0.537303	0.804345	0.925663	0.925663	0.267042	1.54
1.56	1.363865	0.549938	0.817242	0.925931	0.925931	0.267304	1.56
1.58	1.376604	0.562667	0.829981	0.925942	0.925942	0.267314	1.58
1.60	1.389180	0.575487	0.842555	0.925700	0.925700	0.267068	1.60
1.62	1.401593	0.588396	0.854958	0.925214	0.925214	0.266562	1.62
1.64	1.413842	0.601392	0.867184	0.924490	0.924490	0.265793	1.64
1.66	1.425925	0.614472	0.879229	0.923534	0.923534	0.264758	1.66
1.68	1.437842	0.627634	0.891088	0.922354	0.922354	0.263454	1.68
1.70	1.449592	0.640876	0.902755	0.920957	0.920957	0.261879	1.70
1.72	1.461175	0.654194	0.914225	0.919350	0.919350	0.260031	1.72
1.74	1.472589	0.667588	0.925495	0.917541	0.917541	0.257908	1.74
1.76	1.483835	0.681053	0.936560	0.915537	0.915537	0.255507	1.76
1.78	1.494911	0.694588	0.947417	0.913345	0.913345	0.252829	1.78
1.80	1.505817	0.708191	0.958062	0.910974	0.910974	0.249871	1.80
1.82	1.516552	0.721859	0.968491	0.908431	0.908431	0.246632	1.82
1.84	1.527116	0.735589	0.978702	0.905724	0.905724	0.243113	1.84
1.86	1.537508	0.749379	0.988692	0.902860	0.902860	0.239313	1.86
1.88	1.547728	0.763227	0.998458	0.899848	0.899848	0.235231	1.88
1.90	1.557775	0.777129	1.007997	0.896695	0.896695	0.230868	1.90
1.92	1.567650	0.791084	1.017309	0.893410	0.893410	0.226225	1.92
1.94	1.577351	0.805090	1.026391	0.890000	0.890000	0.221302	1.94
1.96	1.586879	0.819143	1.035242	0.886473	0.886473	0.216099	1.96
1.98	1.596233	0.833241	1.043861	0.882838	0.882838	0.210620	1.98

[Appendix] THEORY OF LINEAR ANTENNAS 861

a = 0.00

x	S(a,x)	C(a,x)	Ss(a,x)	Sc(a,x)	Cs(a,x)	Cc(a,x)	x
2.00	1.605413	0.847382	1.052246	0.879102	0.879102	0.204864	2.00
2.05	1.627601	0.882905	1.072184	0.869372	0.869372	0.189279	2.05
2.10	1.648699	0.918641	1.090657	0.859184	0.859184	0.172016	2.10
2.15	1.668706	0.954552	1.107670	0.848660	0.848660	0.153119	2.15
2.20	1.687625	0.990598	1.123240	0.837917	0.837917	0.132642	2.20
2.25	1.705457	1.026744	1.137392	0.827070	0.827070	0.110649	2.25
2.30	1.722207	1.062949	1.150160	0.816230	0.816230	0.087210	2.30
2.35	1.737881	1.099178	1.161585	0.805503	0.805503	0.062407	2.35
2.40	1.752485	1.135393	1.171718	0.794988	0.794988	0.036325	2.40
2.45	1.766029	1.171557	1.180615	0.784779	0.784779	0.009058	2.45
2.50	1.778520	1.207635	1.188342	0.774966	0.774966	-0.019294	2.50
2.55	1.789971	1.243591	1.194966	0.765627	0.765627	-0.048625	2.55
2.60	1.800394	1.279390	1.200564	0.756835	0.756835	-0.078827	2.60
2.65	1.809803	1.314999	1.205214	0.748658	0.748658	-0.109785	2.65
2.70	1.818212	1.350383	1.209000	0.741150	0.741150	-0.141382	2.70
2.75	1.825638	1.385509	1.212008	0.734362	0.734362	-0.173501	2.75
2.80	1.832097	1.420347	1.214327	0.728334	0.728334	-0.206020	2.80
2.85	1.837608	1.454864	1.216046	0.723099	0.723099	-0.238818	2.85
2.90	1.842190	1.489031	1.217257	0.718680	0.718680	-0.271774	2.90
2.95	1.845865	1.522819	1.218050	0.715092	0.715092	-0.304768	2.95
3.00	1.848653	1.556198	1.218516	0.712344	0.712344	-0.337682	3.00
3.05	1.850576	1.589143	1.218743	0.710434	0.710434	-0.370399	3.05
3.10	1.851659	1.621626	1.218819	0.709353	0.709353	-0.402807	3.10
3.15	1.851926	1.653623	1.218827	0.709087	0.709087	-0.434796	3.15
3.20	1.851401	1.685109	1.218848	0.709611	0.709611	-0.466262	3.20
3.25	1.850110	1.716062	1.218958	0.710897	0.710897	-0.497104	3.25
3.30	1.848081	1.746460	1.219231	0.712908	0.712908	-0.527229	3.30
3.35	1.845339	1.776282	1.219734	0.715603	0.715603	-0.556548	3.35
3.40	1.841914	1.805509	1.220529	0.718934	0.718934	-0.584980	3.40
3.45	1.837833	1.834123	1.221672	0.722851	0.722851	-0.612451	3.45
3.50	1.833125	1.862107	1.223215	0.727298	0.727298	-0.638892	3.50
3.55	1.827821	1.889446	1.225202	0.732217	0.732217	-0.664244	3.55
3.60	1.821948	1.916124	1.227670	0.737545	0.737545	-0.688454	3.60
3.65	1.815538	1.942129	1.230651	0.743218	0.743218	-0.711478	3.65
3.70	1.808622	1.967449	1.234169	0.749172	0.749172	-0.733280	3.70
3.75	1.801229	1.992074	1.238243	0.755341	0.755341	-0.753832	3.75
3.80	1.793390	2.015995	1.242883	0.761657	0.761657	-0.773112	3.80
3.85	1.785138	2.039203	1.248095	0.768055	0.768055	-0.791108	3.85
3.90	1.776501	2.061692	1.253877	0.774469	0.774469	-0.807815	3.90
3.95	1.767513	2.083456	1.260220	0.780836	0.780836	-0.823236	3.95
4.00	1.758203	2.104492	1.267112	0.787093	0.787093	-0.837380	4.00
4.05	1.748603	2.124796	1.274532	0.793183	0.793183	-0.850265	4.05
4.10	1.738744	2.144368	1.282455	0.799049	0.799049	-0.861913	4.10
4.15	1.728655	2.163207	1.290852	0.804639	0.804639	-0.872355	4.15
4.20	1.718369	2.181313	1.299688	0.809903	0.809903	-0.881625	4.20
4.25	1.707913	2.198690	1.308925	0.814799	0.814799	-0.889765	4.25
4.30	1.697320	2.215340	1.318521	0.819285	0.819285	-0.896819	4.30
4.35	1.686617	2.231268	1.328430	0.823327	0.823327	-0.902839	4.35
4.40	1.675834	2.246480	1.338604	0.826896	0.826896	-0.907877	4.40
4.45	1.664999	2.260983	1.348993	0.829967	0.829967	-0.911990	4.45
4.50	1.654140	2.274784	1.359546	0.832520	0.832520	-0.915238	4.50
4.55	1.643285	2.287893	1.370210	0.834542	0.834542	-0.917682	4.55
4.60	1.632460	2.300319	1.380932	0.836025	0.836025	-0.919387	4.60
4.65	1.621692	2.312074	1.391658	0.836965	0.836965	-0.920416	4.65
4.70	1.611005	2.323169	1.402336	0.837365	0.837365	-0.920833	4.70
4.75	1.600425	2.333619	1.412915	0.837232	0.837232	-0.920704	4.75
4.80	1.589975	2.343435	1.423343	0.836578	0.836578	-0.920092	4.80
4.85	1.579679	2.352634	1.433573	0.835422	0.835422	-0.919061	4.85
4.90	1.569559	2.361231	1.443559	0.833785	0.833785	-0.917672	4.90
4.95	1.559636	2.369241	1.453257	0.831692	0.831692	-0.915984	4.95

$a = 0.0$

x	$S(a,x)$	$C(a,x)$	$S_s(a,x)$	$S_c(a,x)$	$C_s(a,x)$	$C_o(a,x)$	x
5.0	1.549931	2.376683	1.462629	0.829174	0.829174	−0.914055	5.0
5.1	1.531253	2.389932	1.480248	0.822998	0.822998	−0.909685	5.1
5.2	1.513671	2.401128	1.496171	0.815559	0.815559	−0.904957	5.2
5.3	1.497315	2.410428	1.510219	0.807196	0.807196	−0.900210	5.3
5.4	1.482300	2.418001	1.522286	0.798271	0.798271	−0.895715	5.4
5.5	1.468724	2.424017	1.532337	0.789153	0.789153	−0.891680	5.5
5.6	1.456668	2.428654	1.540411	0.780208	0.780208	−0.888243	5.6
5.7	1.446198	2.432093	1.546615	0.771778	0.771778	−0.885477	5.7
5.8	1.437359	2.434514	1.551118	0.764177	0.764177	−0.883396	5.8
5.9	1.430184	2.436101	1.554141	0.757673	0.757673	−0.881960	5.9
6.0	1.424688	2.437032	1.555951	0.752486	0.752486	−0.881081	6.0
6.1	1.420867	2.437487	1.556847	0.748774	0.748774	−0.880640	6.1
6.2	1.418707	2.437638	1.557147	0.746635	0.746635	−0.880491	6.2
6.3	1.418174	2.437654	1.557179	0.746103	0.746103	−0.880475	6.3
6.4	1.419223	2.437695	1.557261	0.747148	0.747148	−0.880434	6.4
6.5	1.421794	2.437916	1.557700	0.749681	0.749681	−0.880216	6.5
6.6	1.425816	2.438462	1.558772	0.753556	0.753556	−0.879690	6.6
6.7	1.431205	2.439468	1.560714	0.758580	0.758580	−0.878754	6.7
6.8	1.437868	2.441058	1.563721	0.764523	0.764523	−0.877337	6.8
6.9	1.445702	2.443345	1.567933	0.771125	0.771125	−0.875412	6.9
7.0	1.454597	2.446431	1.573438	0.778106	0.778106	−0.872992	7.0
7.1	1.464433	2.450404	1.580266	0.785181	0.785181	−0.870138	7.1
7.2	1.475089	2.455340	1.588389	0.792070	0.792070	−0.866950	7.2
7.3	1.486436	2.461301	1.597728	0.798508	0.798508	−0.863573	7.3
7.4	1.498345	2.468338	1.608153	0.804253	0.804253	−0.860185	7.4
7.5	1.510682	2.476485	1.619494	0.809097	0.809097	−0.856992	7.5
7.6	1.523314	2.485766	1.631542	0.812875	0.812875	−0.854224	7.6
7.7	1.536109	2.496190	1.644067	0.815465	0.815465	−0.852123	7.7
7.8	1.548937	2.507753	1.656821	0.816796	0.816796	−0.850932	7.8
7.9	1.561671	2.520440	1.669549	0.816848	0.816848	−0.850891	7.9
8.0	1.574187	2.534223	1.682002	0.815651	0.815651	−0.852221	8.0
8.1	1.586367	2.549063	1.693945	0.813283	0.813283	−0.855119	8.1
8.2	1.598099	2.564910	1.705163	0.809866	0.809866	−0.859747	8.2
8.3	1.609278	2.581704	1.715474	0.805560	0.805560	−0.866230	8.3
8.4	1.619807	2.599376	1.724733	0.800556	0.800556	−0.874643	8.4
8.5	1.629597	2.617850	1.732836	0.795068	0.795068	−0.885015	8.5
8.6	1.638570	2.637042	1.739723	0.789323	0.789323	−0.897319	8.6
8.7	1.646655	2.656859	1.745382	0.783554	0.783554	−0.911477	8.7
8.8	1.653792	2.677208	1.749846	0.777988	0.777988	−0.927362	8.8
8.9	1.659934	2.697987	1.753190	0.772840	0.772840	−0.944797	8.9
9.0	1.665040	2.719093	1.755531	0.768304	0.768304	−0.963561	9.0
9.1	1.669084	2.740421	1.757019	0.764545	0.764545	−0.983401	9.1
9.2	1.672049	2.761864	1.757832	0.761695	0.761695	−1.004032	9.2
9.3	1.673930	2.783317	1.758168	0.759846	0.759846	−1.025149	9.3
9.4	1.674729	2.804673	1.758237	0.759050	0.759050	−1.046436	9.4
9.5	1.674463	2.825829	1.758252	0.759315	0.759315	−1.067577	9.5
9.6	1.673157	2.846686	1.758424	0.760610	0.760610	−1.088262	9.6
9.7	1.670845	2.867146	1.758948	0.762861	0.762861	−1.108198	9.7
9.8	1.667570	2.887117	1.760001	0.765960	0.765960	−1.127116	9.8
9.9	1.663384	2.906514	1.761734	0.769768	0.769768	−1.144780	9.9

a = 0.0

x	S(a,x)	C(a,x)	Ss(a,x)	Sc(a,x)	Cs(a,x)	Co(a,x)	x
10.0	1.658348	2.925257	1.764264	0.774121	0.774121	-1.160993	10.0
10.2	1.645995	2.960495	1.772009	0.783717	0.783717	-1.188486	10.2
10.4	1.631117	2.992341	1.783435	0.793207	0.793207	-1.208906	10.4
10.6	1.614391	3.020438	1.798136	0.801126	0.801126	-1.222301	10.6
10.8	1.596541	3.044571	1.815188	0.806305	0.806305	-1.229384	10.8
11.0	1.578307	3.064674	1.833309	0.808042	0.808042	-1.231365	11.0
11.2	1.560416	3.080823	1.851074	0.806192	0.806192	-1.229749	11.2
11.4	1.543557	3.093231	1.867138	0.801168	0.801168	-1.226093	11.4
11.6	1.528354	3.102236	1.880441	0.793861	0.793861	-1.221795	11.6
11.8	1.515347	3.108282	1.890362	0.785483	0.785483	-1.217920	11.8
12.0	1.504971	3.111902	1.896801	0.777369	0.777369	-1.215101	12.0
12.2	1.497547	3.113694	1.900180	0.770772	0.770772	-1.213515	12.2
12.4	1.493270	3.114295	1.901355	0.766667	0.766667	-1.212940	12.4
12.6	1.492206	3.114357	1.901479	0.765611	0.765611	-1.212878	12.6
12.8	1.494297	3.114523	1.901808	0.767672	0.767672	-1.212715	12.8
13.0	1.499362	3.115401	1.903509	0.772434	0.772434	-1.211892	13.0
13.2	1.507111	3.117544	1.907480	0.779074	0.779074	-1.210064	13.2
13.4	1.517161	3.121428	1.914223	0.786503	0.786503	-1.207205	13.4
13.6	1.529047	3.127441	1.923777	0.793541	0.793541	-1.203664	13.6
13.8	1.542249	3.135866	1.935725	0.799106	0.799106	-1.200141	13.8
14.0	1.556211	3.146877	1.949275	0.802373	0.802373	-1.197601	14.0
14.2	1.570362	3.160532	1.963393	0.802899	0.802899	-1.197139	14.2
14.4	1.584141	3.176778	1.976969	0.800680	0.800680	-1.199810	14.4
14.6	1.597016	3.195456	1.988992	0.796136	0.796136	-1.206464	14.6
14.8	1.608505	3.216307	1.998707	0.790038	0.790038	-1.217599	14.8
15.0	1.618194	3.238987	2.005723	0.783378	0.783378	-1.233264	15.0
15.2	1.625750	3.263084	2.010062	0.777209	0.777209	-1.253022	15.2
15.4	1.630930	3.288135	2.012152	0.772478	0.772478	-1.275983	15.4
15.6	1.633592	3.313642	2.012742	0.769886	0.769886	-1.300900	15.6
15.8	1.633696	3.339098	2.012785	0.769782	0.769782	-1.326313	15.8
16.0	1.631302	3.364005	2.013281	0.772121	0.772121	-1.350723	16.0
16.2	1.626566	3.387889	2.015122	0.776476	0.776476	-1.372767	16.2
16.4	1.619732	3.410325	2.018956	0.782119	0.782119	-1.391369	16.4
16.6	1.611121	3.430948	2.025095	0.788139	0.788139	-1.405854	16.6
16.8	1.601113	3.449466	2.033469	0.793588	0.793588	-1.415997	16.8
17.0	1.590136	3.465672	2.043656	0.797628	0.797628	-1.422016	17.0
17.2	1.578646	3.479446	2.054945	0.799663	0.799663	-1.424502	17.2
17.4	1.567107	3.490764	2.066462	0.799422	0.799422	-1.424302	17.4
17.6	1.555975	3.499691	2.077307	0.796993	0.796993	-1.422384	17.6
17.8	1.545680	3.506380	2.086694	0.792806	0.792806	-1.419686	17.8
18.0	1.536608	3.511063	2.094072	0.787554	0.787554	-1.416991	18.0
18.2	1.529091	3.514039	2.099206	0.782079	0.782079	-1.414833	18.2
18.4	1.523390	3.515665	2.102205	0.777243	0.777243	-1.413459	18.4
18.6	1.519692	3.516336	2.103502	0.773786	0.773786	-1.412834	18.6
18.8	1.518099	3.516473	2.103774	0.772219	0.772219	-1.412699	18.8
19.0	1.518630	3.516504	2.103836	0.772746	0.772746	-1.412668	19.0
19.2	1.521219	3.516847	2.104509	0.775242	0.775242	-1.412339	19.2
19.4	1.525723	3.517896	2.106493	0.779276	0.779276	-1.411403	19.4
19.6	1.531921	3.520003	2.110261	0.784185	0.784185	-1.409741	19.6
19.8	1.539537	3.523468	2.115990	0.789185	0.789185	-1.407478	19.8

$a = 0.0$

x	$S(a,x)$	$C(a,x)$	$S_s(a,x)$	$S_c(a,x)$	$C_s(a,x)$	$C_c(a,x)$	x
20.0	1.548242	3.528528	2.123538	0.793493	0.793493	-1.404991	20.0
20.2	1.557670	3.535349	2.132474	0.796448	0.796448	-1.402875	20.2
20.4	1.567434	3.544019	2.142152	0.797616	0.797616	-1.401867	20.4
20.6	1.577143	3.554548	2.151814	0.796849	0.796849	-1.402734	20.6
20.8	1.586415	3.566871	2.160714	0.794304	0.794304	-1.406157	20.8
21.0	1.594891	3.580848	2.168228	0.790413	0.790413	-1.412619	21.0
21.2	1.602252	3.596273	2.173957	0.785811	0.785811	-1.422315	21.2
21.4	1.608229	3.612885	2.177780	0.781230	0.781230	-1.435105	21.4
21.6	1.612610	3.630375	2.179868	0.777386	0.777386	-1.450508	21.6
21.8	1.615252	3.648406	2.180653	0.774867	0.774867	-1.467752	21.8
22.0	1.616084	3.666617	2.180759	0.774043	0.774043	-1.485858	22.0
22.2	1.615104	3.684649	2.180895	0.775013	0.775013	-1.503754	22.2
22.4	1.612383	3.702148	2.181747	0.777592	0.777592	-1.520401	22.4
22.6	1.608061	3.718787	2.183871	0.781348	0.781348	-1.534916	22.6
22.8	1.602336	3.734276	2.187608	0.785672	0.785672	-1.546668	22.8
23.0	1.595459	3.748370	2.193034	0.789879	0.789879	-1.555336	23.0
23.2	1.587722	3.760881	2.199954	0.793311	0.793311	-1.560928	23.2
23.4	1.579446	3.771685	2.207937	0.795445	0.795445	-1.563749	23.4
23.6	1.570965	3.780723	2.216386	0.795968	0.795968	-1.564337	23.6
23.8	1.562621	3.788004	2.224639	0.794827	0.794827	-1.563365	23.8
24.0	1.554739	3.793603	2.232066	0.792227	0.792227	-1.561537	24.0
24.2	1.547624	3.797657	2.238173	0.788600	0.788600	-1.559484	24.2
24.4	1.541544	3.800359	2.242676	0.784530	0.784530	-1.557684	24.4
24.6	1.536724	3.801951	2.245540	0.780663	0.780663	-1.556412	24.6
24.8	1.533333	3.802710	2.246982	0.777600	0.777600	-1.555728	24.8
25.0	1.531483	3.802940	2.247434	0.775809	0.775809	-1.555507	25.0

THEORY OF LINEAR ANTENNAS

a = 0.50

x	S(a,x)	C(a,x)	Ss(a,x)	Sc(a,x)	Cs(a,x)	Cc(a,x)	x
0.00	0.000000	0.000000	0.000000	0.000000	0.000000	0.000000	0.00
0.01	0.009588	0.002449	0.000048	0.009588	0.000088	0.000000	0.01
0.02	0.019177	0.004898	0.000192	0.019175	0.000351	0.000002	0.02
0.03	0.028764	0.007349	0.000431	0.028760	0.000789	0.000008	0.03
0.04	0.038351	0.009803	0.000767	0.038340	0.001401	0.000019	0.04
0.05	0.047936	0.012261	0.001198	0.047916	0.002187	0.000036	0.05
0.06	0.057519	0.014724	0.001725	0.057485	0.003144	0.000063	0.06
0.07	0.067101	0.017192	0.002347	0.067046	0.004272	0.000100	0.07
0.08	0.076680	0.019667	0.003065	0.076599	0.005568	0.000148	0.08
0.09	0.086257	0.022149	0.003878	0.086141	0.007031	0.000211	0.09
0.10	0.095831	0.024639	0.004786	0.095671	0.008659	0.000288	0.10
0.11	0.105402	0.027138	0.005789	0.105189	0.010448	0.000382	0.11
0.12	0.114969	0.029648	0.006887	0.114693	0.012397	0.000494	0.12
0.13	0.124532	0.032168	0.008079	0.124181	0.014501	0.000626	0.13
0.14	0.134091	0.034700	0.009366	0.133653	0.016759	0.000779	0.14
0.15	0.143645	0.037245	0.010746	0.143107	0.019167	0.000954	0.15
0.16	0.153194	0.039803	0.012221	0.152542	0.021721	0.001152	0.16
0.17	0.162739	0.042376	0.013788	0.161957	0.024418	0.001375	0.17
0.18	0.172278	0.044963	0.015449	0.171350	0.027254	0.001624	0.18
0.19	0.181810	0.047566	0.017203	0.180720	0.030225	0.001900	0.19
0.20	0.191337	0.050186	0.019049	0.190067	0.033328	0.002203	0.20
0.21	0.200858	0.052823	0.020987	0.199388	0.036558	0.002536	0.21
0.22	0.210371	0.055478	0.023016	0.208682	0.039911	0.002897	0.22
0.23	0.219878	0.058151	0.025137	0.217949	0.043384	0.003290	0.23
0.24	0.229377	0.060844	0.027349	0.227187	0.046971	0.003713	0.24
0.25	0.238868	0.063557	0.029651	0.236394	0.050669	0.004169	0.25
0.26	0.248351	0.066290	0.032043	0.245571	0.054474	0.004656	0.26
0.27	0.257826	0.069045	0.034525	0.254715	0.058380	0.005177	0.27
0.28	0.267292	0.071822	0.037095	0.263825	0.062385	0.005731	0.28
0.29	0.276749	0.074621	0.039754	0.272901	0.066483	0.006319	0.29
0.30	0.286196	0.077443	0.042501	0.281940	0.070671	0.006942	0.30
0.31	0.295634	0.080289	0.045335	0.290943	0.074943	0.007598	0.31
0.32	0.305062	0.083158	0.048256	0.299907	0.079296	0.008290	0.32
0.33	0.314480	0.086053	0.051263	0.308831	0.083726	0.009016	0.33
0.34	0.323887	0.088972	0.054356	0.317715	0.088229	0.009777	0.34
0.35	0.333284	0.091917	0.057533	0.326558	0.092800	0.010574	0.35
0.36	0.342668	0.094889	0.060796	0.335358	0.097436	0.011405	0.36
0.37	0.352042	0.097886	0.064141	0.344114	0.102132	0.012272	0.37
0.38	0.361403	0.100911	0.067570	0.352825	0.106884	0.013174	0.38
0.39	0.370753	0.103963	0.071081	0.361489	0.111689	0.014110	0.39
0.40	0.380090	0.107042	0.074674	0.370107	0.116543	0.015082	0.40
0.41	0.389414	0.110150	0.078348	0.378677	0.121442	0.016087	0.41
0.42	0.398725	0.113286	0.082102	0.387198	0.126382	0.017128	0.42
0.43	0.408022	0.116452	0.085936	0.395668	0.131360	0.018202	0.43
0.44	0.417306	0.119646	0.089848	0.404087	0.136372	0.019309	0.44
0.45	0.426576	0.122869	0.093838	0.412454	0.141416	0.020450	0.45
0.46	0.435832	0.126123	0.097906	0.420768	0.146486	0.021624	0.46
0.47	0.445072	0.129406	0.102050	0.429028	0.151581	0.022831	0.47
0.48	0.454298	0.132720	0.106269	0.437232	0.156697	0.024069	0.48
0.49	0.463509	0.136064	0.110563	0.445381	0.161830	0.025339	0.49

a = 0.50

x	S(a,x)	C(a,x)	Ss(a,x)	Sc(a,x)	Cs(a,x)	Cc(a,x)	x
0.50	0.472704	0.139439	0.114931	0.453472	0.166979	0.026640	0.50
0.51	0.481883	0.142846	0.119372	0.461506	0.172139	0.027971	0.51
0.52	0.491047	0.146283	0.123885	0.469480	0.177307	0.029332	0.52
0.53	0.500193	0.149752	0.128469	0.477395	0.182482	0.030723	0.53
0.54	0.509323	0.153253	0.133124	0.485249	0.187659	0.032142	0.54
0.55	0.518436	0.156785	0.137848	0.493042	0.192837	0.033589	0.55
0.56	0.527532	0.160350	0.142641	0.500772	0.198013	0.035063	0.56
0.57	0.536609	0.163947	0.147502	0.508439	0.203183	0.036564	0.57
0.58	0.545669	0.167576	0.152429	0.516042	0.208347	0.038091	0.58
0.59	0.554711	0.171237	0.157422	0.523581	0.213500	0.039642	0.59
0.60	0.563735	0.174932	0.162479	0.531053	0.218641	0.041219	0.60
0.61	0.572739	0.178659	0.167601	0.538459	0.223768	0.042819	0.61
0.62	0.581724	0.182419	0.172785	0.545798	0.228878	0.044441	0.62
0.63	0.590690	0.186211	0.178031	0.553069	0.233969	0.046086	0.63
0.64	0.599637	0.190037	0.183337	0.560272	0.239039	0.047752	0.64
0.65	0.608563	0.193896	0.188704	0.567405	0.244085	0.049439	0.65
0.66	0.617469	0.197789	0.194129	0.574468	0.249107	0.051144	0.66
0.67	0.626355	0.201714	0.199612	0.581460	0.254101	0.052869	0.67
0.68	0.635219	0.205673	0.205151	0.588381	0.259066	0.054611	0.68
0.69	0.644063	0.209665	0.210746	0.595229	0.264000	0.056371	0.69
0.70	0.652885	0.213691	0.216396	0.602005	0.268902	0.058146	0.70
0.71	0.661686	0.217750	0.222099	0.608708	0.273769	0.059937	0.71
0.72	0.670465	0.221842	0.227855	0.615337	0.278600	0.061741	0.72
0.73	0.679221	0.225969	0.233661	0.621891	0.283394	0.063559	0.73
0.74	0.687955	0.230128	0.239518	0.628370	0.288148	0.065390	0.74
0.75	0.696666	0.234321	0.245424	0.634773	0.292861	0.067231	0.75
0.76	0.705355	0.238548	0.251378	0.641101	0.297532	0.069083	0.76
0.77	0.714019	0.242808	0.257379	0.647352	0.302160	0.070945	0.77
0.78	0.722661	0.247102	0.263425	0.653525	0.306742	0.072815	0.78
0.79	0.731278	0.251429	0.269516	0.659621	0.311278	0.074693	0.79
0.80	0.739872	0.255790	0.275651	0.665639	0.315766	0.076577	0.80
0.81	0.748441	0.260184	0.281828	0.671578	0.320205	0.078467	0.81
0.82	0.756985	0.264611	0.288046	0.677439	0.324594	0.080362	0.82
0.83	0.765505	0.269072	0.294304	0.683219	0.328932	0.082260	0.83
0.84	0.773999	0.273566	0.300600	0.688921	0.333217	0.084161	0.84
0.85	0.782468	0.278094	0.306935	0.694542	0.337448	0.086063	0.85
0.86	0.790912	0.282654	0.313306	0.700082	0.341626	0.087966	0.86
0.87	0.799329	0.287248	0.319712	0.705542	0.345747	0.089869	0.87
0.88	0.807720	0.291874	0.326153	0.710921	0.349812	0.091770	0.88
0.89	0.816085	0.296534	0.332626	0.716218	0.353820	0.093669	0.89
0.90	0.824423	0.301226	0.339132	0.721434	0.357769	0.095564	0.90
0.91	0.832734	0.305951	0.345668	0.726568	0.361659	0.097456	0.91
0.92	0.841018	0.310709	0.352233	0.731619	0.365489	0.099341	0.92
0.93	0.849274	0.315500	0.358827	0.736588	0.369258	0.101220	0.93
0.94	0.857503	0.320323	0.365448	0.741475	0.372966	0.103092	0.94
0.95	0.865704	0.325178	0.372095	0.746278	0.376612	0.104955	0.95
0.96	0.873877	0.330066	0.378766	0.750999	0.380194	0.106809	0.96
0.97	0.882021	0.334986	0.385461	0.755636	0.383714	0.108653	0.97
0.98	0.890137	0.339938	0.392179	0.760191	0.387169	0.110484	0.98
0.99	0.898224	0.344922	0.398917	0.764662	0.390559	0.112304	0.99

a = 0.50

x	S(a,x)	C(a,x)	Ss(a,x)	Sc(a,x)	Cs(a,x)	Cc(a,x)	x
1.00	0.906282	0.349938	0.405675	0.769049	0.393884	0.114109	1.00
1.02	0.922309	0.360064	0.419247	0.777573	0.400337	0.117676	1.02
1.04	0.938217	0.370316	0.432885	0.785763	0.406523	0.121177	1.04
1.06	0.954003	0.380693	0.446578	0.793618	0.412439	0.124604	1.06
1.08	0.969667	0.391193	0.460319	0.801139	0.418083	0.127948	1.08
1.10	0.985207	0.401815	0.474096	0.808326	0.423452	0.131203	1.10
1.12	1.000620	0.412557	0.487901	0.815179	0.428544	0.134360	1.12
1.14	1.015905	0.423420	0.501725	0.821701	0.433359	0.137411	1.14
1.16	1.031060	0.434400	0.515558	0.827892	0.437894	0.140350	1.16
1.18	1.046084	0.445497	0.529391	0.833753	0.442149	0.143168	1.18
1.20	1.060975	0.456709	0.543215	0.839288	0.446125	0.145859	1.20
1.22	1.075731	0.468035	0.557021	0.844497	0.449821	0.148414	1.22
1.24	1.090352	0.479473	0.570800	0.849384	0.453238	0.150827	1.24
1.26	1.104835	0.491021	0.584543	0.853951	0.456376	0.153092	1.26
1.28	1.119178	0.502678	0.598243	0.858201	0.459238	0.155200	1.28
1.30	1.133381	0.514443	0.611889	0.862137	0.461824	0.157145	1.30
1.32	1.147443	0.526313	0.625475	0.865763	0.464137	0.158921	1.32
1.34	1.161360	0.538286	0.638991	0.869082	0.466179	0.160522	1.34
1.36	1.175134	0.550361	0.652429	0.872099	0.467952	0.161941	1.36
1.38	1.188761	0.562537	0.665782	0.874817	0.469459	0.163172	1.38
1.40	1.202241	0.574810	0.679042	0.877241	0.470704	0.164210	1.40
1.42	1.215572	0.587180	0.692201	0.879376	0.471690	0.165048	1.42
1.44	1.228753	0.599645	0.705252	0.881226	0.472421	0.165682	1.44
1.46	1.241784	0.612202	0.718188	0.882796	0.472900	0.166106	1.46
1.48	1.254663	0.624850	0.731001	0.884092	0.473132	0.166316	1.48
1.50	1.267388	0.637586	0.743684	0.885119	0.473121	0.166306	1.50
1.52	1.279959	0.650409	0.756232	0.885883	0.472872	0.166071	1.52
1.54	1.292375	0.663316	0.768637	0.886390	0.472391	0.165608	1.54
1.56	1.304635	0.676306	0.780894	0.886645	0.471681	0.164913	1.56
1.58	1.316737	0.689377	0.792996	0.886655	0.470749	0.163981	1.58
1.60	1.328681	0.702526	0.804938	0.886426	0.469599	0.162809	1.60
1.62	1.340466	0.715751	0.816713	0.885964	0.468239	0.161394	1.62
1.64	1.352091	0.729051	0.828318	0.885277	0.466673	0.159731	1.64
1.66	1.363555	0.742422	0.839746	0.884370	0.464907	0.157820	1.66
1.68	1.374858	0.755864	0.850993	0.883251	0.462948	0.155656	1.68
1.70	1.385998	0.769373	0.862054	0.881926	0.460803	0.153238	1.70
1.72	1.396975	0.782948	0.872925	0.880404	0.458477	0.150563	1.72
1.74	1.407789	0.796586	0.883602	0.878690	0.455977	0.147629	1.74
1.76	1.418438	0.810285	0.894080	0.876792	0.453310	0.144436	1.76
1.78	1.428923	0.824043	0.904357	0.874717	0.450483	0.140980	1.78
1.80	1.439242	0.837858	0.914429	0.872474	0.447503	0.137263	1.80
1.82	1.449394	0.851727	0.924293	0.870069	0.444376	0.133281	1.82
1.84	1.459381	0.865649	0.933945	0.867509	0.441110	0.129036	1.84
1.86	1.469200	0.879620	0.943384	0.864803	0.437712	0.124527	1.86
1.88	1.478852	0.893640	0.952607	0.861959	0.434189	0.119753	1.88
1.90	1.488336	0.907704	0.961612	0.858983	0.430549	0.114715	1.90
1.92	1.497652	0.921812	0.970398	0.855883	0.426798	0.109413	1.92
1.94	1.506800	0.935961	0.978961	0.852668	0.422944	0.103849	1.94
1.96	1.515779	0.950148	0.987303	0.849344	0.418994	0.098022	1.96
1.98	1.524589	0.964371	0.995420	0.845920	0.414955	0.091936	1.98

a = 0.50

x	S(a,x)	C(a,x)	Ss(a,x)	Sc(a,x)	Cs(a,x)	Cc(a,x)	x
2.00	1.533230	0.978629	1.003313	0.842404	0.410836	0.085590	2.00
2.05	1.554092	1.014406	1.022059	0.833255	0.400231	0.068604	2.05
2.10	1.573895	1.050345	1.039398	0.823693	0.389278	0.050045	2.10
2.15	1.592639	1.086410	1.055337	0.813834	0.378090	0.029957	2.15
2.20	1.610325	1.122566	1.069892	0.803791	0.366776	0.008393	2.20
2.25	1.626957	1.158776	1.083092	0.793674	0.355444	−0.014584	2.25
2.30	1.642538	1.195005	1.094969	0.783591	0.344195	−0.038905	2.30
2.35	1.657075	1.231218	1.105566	0.773641	0.333129	−0.064491	2.35
2.40	1.670575	1.267379	1.114932	0.763922	0.322337	−0.091258	2.40
2.45	1.683045	1.303455	1.123125	0.754523	0.311908	−0.119115	2.45
2.50	1.694496	1.339411	1.130208	0.745526	0.301922	−0.147963	2.50
2.55	1.704939	1.375214	1.136250	0.737010	0.292453	−0.177701	2.55
2.60	1.714386	1.410830	1.141324	0.729042	0.283569	−0.208221	2.60
2.65	1.722851	1.446228	1.145508	0.721685	0.275329	−0.239414	2.65
2.70	1.730348	1.481374	1.148883	0.714992	0.267785	−0.271165	2.70
2.75	1.736893	1.516239	1.151535	0.709008	0.260981	−0.303359	2.75
2.80	1.742504	1.550792	1.153550	0.703772	0.254952	−0.335880	2.80
2.85	1.747199	1.585003	1.155015	0.699312	0.249728	−0.368610	2.85
2.90	1.750996	1.618844	1.156019	0.695650	0.245326	−0.401431	2.90
2.95	1.753917	1.652288	1.156650	0.692799	0.241760	−0.434228	2.95
3.00	1.755982	1.685306	1.156996	0.690763	0.239033	−0.466886	3.00
3.05	1.757214	1.717874	1.157142	0.689540	0.237141	−0.499293	3.05
3.10	1.757635	1.749966	1.157174	0.689120	0.236073	−0.531341	3.10
3.15	1.757271	1.781559	1.157171	0.689485	0.235810	−0.562924	3.15
3.20	1.756144	1.812630	1.157212	0.690610	0.236327	−0.593943	3.20
3.25	1.754282	1.843157	1.157370	0.692466	0.237592	−0.624300	3.25
3.30	1.751709	1.873120	1.157715	0.695015	0.239568	−0.653907	3.30
3.35	1.748454	1.902499	1.158311	0.698215	0.242212	−0.682679	3.35
3.40	1.744543	1.931276	1.159218	0.702019	0.245477	−0.710539	3.40
3.45	1.740004	1.959434	1.160490	0.706376	0.249309	−0.737416	3.45
3.50	1.734866	1.986957	1.162174	0.711229	0.253653	−0.763247	3.50
3.55	1.729157	2.013831	1.164311	0.716522	0.258451	−0.787975	3.55
3.60	1.722908	2.040043	1.166937	0.722192	0.263639	−0.811551	3.60
3.65	1.716147	2.065579	1.170081	0.728177	0.269155	−0.833936	3.65
3.70	1.708904	2.090430	1.173765	0.734412	0.274933	−0.855095	3.70
3.75	1.701210	2.114585	1.178005	0.740832	0.280909	−0.875003	3.75
3.80	1.693095	2.138036	1.182809	0.747371	0.287015	−0.893643	3.80
3.85	1.684589	2.160777	1.188180	0.753965	0.293187	−0.911005	3.85
3.90	1.675723	2.182801	1.194116	0.760550	0.299361	−0.927087	3.90
3.95	1.666527	2.204103	1.200606	0.767064	0.305474	−0.941893	3.95
4.00	1.657032	2.224681	1.207634	0.773446	0.311466	−0.955436	4.00
4.05	1.647267	2.244532	1.215181	0.779640	0.317279	−0.967735	4.05
4.10	1.637264	2.263656	1.223220	0.785592	0.322858	−0.978814	4.10
4.15	1.627053	2.282052	1.231719	0.791250	0.328153	−0.988705	4.15
4.20	1.616662	2.299723	1.240645	0.796568	0.333116	−0.997445	4.20
4.25	1.606122	2.316671	1.249957	0.801503	0.337705	−1.005076	4.25
4.30	1.595461	2.332900	1.259614	0.806018	0.341881	−1.011643	4.30
4.35	1.584708	2.348415	1.269569	0.810079	0.345612	−1.017198	4.35
4.40	1.573892	2.363222	1.279774	0.813659	0.348868	−1.021796	4.40
4.45	1.563040	2.377328	1.290180	0.816734	0.351627	−1.025492	4.45
4.50	1.552180	2.390742	1.300734	0.819288	0.353872	−1.028348	4.50
4.55	1.541339	2.403474	1.311385	0.821308	0.355589	−1.030424	4.55
4.60	1.530541	2.415533	1.322079	0.822786	0.356772	−1.031784	4.60
4.65	1.519814	2.426931	1.332765	0.823723	0.357419	−1.032493	4.65
4.70	1.509181	2.437680	1.343389	0.824121	0.357533	−1.032614	4.70
4.75	1.498666	2.447793	1.353901	0.823989	0.357124	−1.032212	4.75
4.80	1.488294	2.457286	1.364252	0.823341	0.356204	−1.031349	4.80
4.85	1.478086	2.466171	1.374395	0.822194	0.354791	−1.030089	4.85
4.90	1.468064	2.474466	1.384284	0.820573	0.352909	−1.028492	4.90
4.95	1.458249	2.482187	1.393877	0.818503	0.350583	−1.026616	4.95

a = 1.00

x	S(a,x)	C(a,x)	Ss(a,x)	Sc(a,x)	Cs(a,x)	Cc(a,x)	x
0.00	0.000000	0.000000	0.000000	0.000000	0.000000	0.000000	0.00
0.02	0.016829	0.009194	0.000168	0.016828	0.000108	0.000001	0.02
0.04	0.033656	0.018392	0.000673	0.033647	0.000432	0.000006	0.04
0.06	0.050477	0.027596	0.001514	0.050447	0.000970	0.000019	0.06
0.08	0.067292	0.036808	0.002690	0.067220	0.001721	0.000046	0.08
0.10	0.084097	0.046033	0.004200	0.083957	0.002682	0.000089	0.10
0.12	0.100890	0.055273	0.006044	0.100648	0.003850	0.000154	0.12
0.14	0.117668	0.064531	0.008219	0.117285	0.005221	0.000243	0.14
0.16	0.134430	0.073811	0.010723	0.133858	0.006789	0.000361	0.16
0.18	0.151172	0.083114	0.013556	0.150359	0.008551	0.000511	0.18
0.20	0.167893	0.092444	0.016714	0.166778	0.010500	0.000697	0.20
0.22	0.184590	0.101803	0.020194	0.183108	0.012628	0.000922	0.22
0.24	0.201260	0.111195	0.023994	0.199339	0.014931	0.001188	0.24
0.26	0.217902	0.120622	0.028112	0.215463	0.017398	0.001498	0.26
0.28	0.234513	0.130087	0.032542	0.231472	0.020024	0.001855	0.28
0.30	0.251090	0.139592	0.037282	0.247356	0.022798	0.002260	0.30
0.32	0.267631	0.149139	0.042328	0.263109	0.025713	0.002716	0.32
0.34	0.284134	0.158732	0.047676	0.278722	0.028758	0.003223	0.34
0.36	0.300597	0.168372	0.053320	0.294186	0.031925	0.003783	0.36
0.38	0.317017	0.178062	0.059258	0.309494	0.035204	0.004397	0.38
0.40	0.333392	0.187804	0.065483	0.324639	0.038584	0.005064	0.40
0.42	0.349719	0.197600	0.071991	0.339613	0.042055	0.005786	0.42
0.44	0.365997	0.207452	0.078777	0.354409	0.045607	0.006562	0.44
0.46	0.382223	0.217362	0.085834	0.369019	0.049229	0.007391	0.46
0.48	0.398394	0.227332	0.093158	0.383437	0.052911	0.008272	0.48
0.50	0.414509	0.237364	0.100742	0.397656	0.056641	0.009205	0.50
0.52	0.430566	0.247459	0.108580	0.411669	0.060410	0.010188	0.52
0.54	0.446561	0.257619	0.116666	0.425470	0.064207	0.011218	0.54
0.56	0.462494	0.267846	0.124994	0.439052	0.068021	0.012294	0.56
0.58	0.478361	0.278140	0.133556	0.452411	0.071842	0.013414	0.58
0.60	0.494160	0.288504	0.142346	0.465539	0.075659	0.014574	0.60
0.62	0.509890	0.298938	0.151357	0.478432	0.079463	0.015771	0.62
0.64	0.525549	0.309444	0.160581	0.491084	0.083242	0.017003	0.64
0.66	0.541133	0.320022	0.170013	0.503490	0.086988	0.018265	0.66
0.68	0.556641	0.330674	0.179643	0.515646	0.090691	0.019554	0.68
0.70	0.572071	0.341401	0.189464	0.527546	0.094341	0.020865	0.70
0.72	0.587421	0.352203	0.199470	0.539187	0.097930	0.022196	0.72
0.74	0.602689	0.363081	0.209651	0.550564	0.101447	0.023540	0.74
0.76	0.617873	0.374036	0.220001	0.561674	0.104886	0.024893	0.76
0.78	0.632970	0.385068	0.230510	0.572512	0.108237	0.026251	0.78
0.80	0.647980	0.396178	0.241172	0.583077	0.111493	0.027608	0.80
0.82	0.662899	0.407366	0.251977	0.593363	0.114646	0.028960	0.82
0.84	0.677726	0.418633	0.262918	0.603370	0.117688	0.030300	0.84
0.86	0.692459	0.429978	0.273987	0.613093	0.120612	0.031624	0.86
0.88	0.707096	0.441403	0.285174	0.622531	0.123413	0.032925	0.88
0.90	0.721635	0.452906	0.296472	0.631682	0.126083	0.034198	0.90
0.92	0.736075	0.464489	0.307871	0.640545	0.128617	0.035438	0.92
0.94	0.750413	0.476150	0.319365	0.649117	0.131009	0.036638	0.94
0.96	0.764648	0.487890	0.330944	0.657397	0.133253	0.037792	0.96
0.98	0.778777	0.499710	0.342599	0.665384	0.135346	0.038894	0.98

$a = 1.00$

x	S(a,x)	C(a,x)	Ss(a,x)	Sc(a,x)	Cs(a,x)	Cc(a,x)	x
1.00	0.792801	0.511607	0.354322	0.673079	0.137281	0.039939	1.00
1.02	0.806715	0.523583	0.366105	0.680479	0.139056	0.040920	1.02
1.04	0.820520	0.535637	0.377939	0.687586	0.140666	0.041831	1.04
1.06	0.834212	0.547768	0.389816	0.694399	0.142107	0.042666	1.06
1.08	0.847791	0.559975	0.401728	0.700919	0.143378	0.043418	1.08
1.10	0.861255	0.572260	0.413665	0.707146	0.144474	0.044082	1.10
1.12	0.874602	0.584620	0.425619	0.713081	0.145393	0.044652	1.12
1.14	0.887831	0.597055	0.437584	0.718725	0.146134	0.045122	1.14
1.16	0.900940	0.609564	0.449549	0.724080	0.146695	0.045485	1.16
1.18	0.913928	0.622147	0.461507	0.729148	0.147074	0.045736	1.18
1.20	0.926793	0.634802	0.473450	0.733929	0.147270	0.045868	1.20
1.22	0.939534	0.647530	0.485371	0.738427	0.147282	0.045877	1.22
1.24	0.952149	0.660328	0.497260	0.742644	0.147111	0.045755	1.24
1.26	0.964637	0.673197	0.509111	0.746582	0.146755	0.045499	1.26
1.28	0.976997	0.686134	0.520916	0.750244	0.146216	0.045101	1.28
1.30	0.989228	0.699139	0.532667	0.753633	0.145494	0.044557	1.30
1.32	1.001327	0.712212	0.544356	0.756753	0.144589	0.043862	1.32
1.34	1.013294	0.725349	0.555978	0.759607	0.143502	0.043010	1.34
1.36	1.025127	0.738552	0.567524	0.762199	0.142236	0.041996	1.36
1.38	1.036826	0.751817	0.578988	0.764532	0.140792	0.040816	1.38
1.40	1.048389	0.765145	0.590362	0.766612	0.139171	0.039464	1.40
1.42	1.059815	0.778533	0.601640	0.768441	0.137376	0.037937	1.42
1.44	1.071104	0.791981	0.612817	0.770026	0.135410	0.036229	1.44
1.46	1.082253	0.805486	0.623884	0.771369	0.133274	0.034337	1.46
1.48	1.093261	0.819049	0.634837	0.772477	0.130973	0.032257	1.48
1.50	1.104129	0.832666	0.645669	0.773355	0.128509	0.029984	1.50
1.52	1.114855	0.846337	0.656375	0.774006	0.125886	0.027515	1.52
1.54	1.125437	0.860061	0.666948	0.774438	0.123107	0.024847	1.54
1.56	1.135876	0.873835	0.677384	0.774656	0.120176	0.021976	1.56
1.58	1.146170	0.887658	0.687678	0.774664	0.117098	0.018900	1.58
1.60	1.156318	0.901529	0.697824	0.774469	0.113875	0.015615	1.60
1.62	1.166320	0.915446	0.707818	0.774078	0.110514	0.012118	1.62
1.64	1.176174	0.929407	0.717655	0.773495	0.107017	0.008408	1.64
1.66	1.185881	0.943411	0.727332	0.772727	0.103391	0.004482	1.66
1.68	1.195439	0.957456	0.736842	0.771781	0.099640	0.000339	1.68
1.70	1.204848	0.971540	0.746184	0.770662	0.095768	−0.004025	1.70
1.72	1.214107	0.985662	0.755354	0.769378	0.091781	−0.008609	1.72
1.74	1.223215	0.999820	0.764347	0.767934	0.087685	−0.013416	1.74
1.76	1.232173	1.014012	0.773161	0.766338	0.083484	−0.018446	1.76
1.78	1.240978	1.028236	0.781792	0.764595	0.079185	−0.023701	1.78
1.80	1.249632	1.042491	0.790239	0.762714	0.074792	−0.029180	1.80
1.82	1.258134	1.056775	0.798498	0.760700	0.070312	−0.034885	1.82
1.84	1.266482	1.071085	0.806567	0.758560	0.065749	−0.040815	1.84
1.86	1.274677	1.085421	0.814445	0.756302	0.061111	−0.046971	1.86
1.88	1.282718	1.099781	0.822129	0.753932	0.056402	−0.053351	1.88
1.90	1.290606	1.114161	0.829618	0.751457	0.051629	−0.059956	1.90
1.92	1.298339	1.128562	0.836911	0.748884	0.046797	−0.066785	1.92
1.94	1.305918	1.142980	0.844006	0.746221	0.041913	−0.073836	1.94
1.96	1.313342	1.157414	0.850903	0.743473	0.036983	−0.081108	1.96
1.98	1.320612	1.171861	0.857600	0.740647	0.032012	−0.088600	1.98

a = 1.00

x	S(a,x)	C(a,x)	Ss(a,x)	Sc(a,x)	Cs(a,x)	Cc(a,x)	x
2.00	1.327726	1.186321	0.864099	0.737752	0.027007	-0.096311	2.00
2.05	1.344834	1.222511	0.879472	0.730250	0.014383	-0.116527	2.05
2.10	1.360972	1.258735	0.893602	0.722458	0.001676	-0.138056	2.10
2.15	1.376141	1.294962	0.906502	0.714480	-0.011023	-0.160854	2.15
2.20	1.390343	1.331163	0.918191	0.706415	-0.023623	-0.184867	2.20
2.25	1.403582	1.367306	0.928698	0.698363	-0.036036	-0.210034	2.25
2.30	1.415863	1.403362	0.938059	0.690416	-0.048179	-0.236287	2.30
2.35	1.427191	1.439302	0.946317	0.682663	-0.059972	-0.263551	2.35
2.40	1.437575	1.475095	0.953522	0.675188	-0.071339	-0.291745	2.40
2.45	1.447021	1.510711	0.959729	0.668068	-0.082211	-0.320781	2.45
2.50	1.455542	1.546122	0.964999	0.661374	-0.092523	-0.350568	2.50
2.55	1.463146	1.581298	0.969400	0.655173	-0.102216	-0.381010	2.55
2.60	1.469848	1.616212	0.972999	0.649521	-0.111240	-0.412007	2.60
2.65	1.475660	1.650835	0.975873	0.644470	-0.119548	-0.443457	2.65
2.70	1.480596	1.685140	0.978096	0.640063	-0.127103	-0.475253	2.70
2.75	1.484673	1.719101	0.979749	0.636337	-0.133875	-0.507289	2.75
2.80	1.487907	1.752690	0.980911	0.633320	-0.139838	-0.539459	2.80
2.85	1.490315	1.785884	0.981664	0.631032	-0.144978	-0.571653	2.85
2.90	1.491918	1.818657	0.982090	0.629487	-0.149285	-0.603767	2.90
2.95	1.492733	1.850985	0.982268	0.628692	-0.152756	-0.635694	2.95
3.00	1.492782	1.882846	0.982279	0.628644	-0.155399	-0.667331	3.00
3.05	1.492086	1.914217	0.982201	0.629335	-0.157223	-0.698577	3.05
3.10	1.490668	1.945076	0.982110	0.630751	-0.158249	-0.729334	3.10
3.15	1.488550	1.975405	0.982078	0.632868	-0.158500	-0.759510	3.15
3.20	1.485756	2.005183	0.982174	0.635660	-0.158009	-0.789015	3.20
3.25	1.482310	2.034393	0.982463	0.639093	-0.156811	-0.817765	3.25
3.30	1.478239	2.063016	0.983008	0.643128	-0.154948	-0.845681	3.30
3.35	1.473566	2.091036	0.983862	0.647721	-0.152466	-0.872692	3.35
3.40	1.468319	2.118440	0.985078	0.652825	-0.149415	-0.898730	3.40
3.45	1.462524	2.145211	0.986700	0.658387	-0.145850	-0.923736	3.45
3.50	1.456209	2.171338	0.988769	0.664354	-0.141827	-0.947658	3.50
3.55	1.449400	2.196809	0.991317	0.670667	-0.137406	-0.970448	3.55
3.60	1.442125	2.221612	0.994374	0.677267	-0.132648	-0.992068	3.60
3.65	1.434414	2.245739	0.997959	0.684094	-0.127617	-1.012488	3.65
3.70	1.426293	2.269181	1.002089	0.691085	-0.122375	-1.031683	3.70
3.75	1.417792	2.291931	1.006773	0.698178	-0.116986	-1.049636	3.75
3.80	1.408939	2.313983	1.012014	0.705312	-0.111515	-1.066339	3.80
3.85	1.399763	2.335332	1.017808	0.712426	-0.106023	-1.081789	3.85
3.90	1.390293	2.355973	1.024148	0.719459	-0.100571	-1.095991	3.90
3.95	1.380558	2.375906	1.031018	0.726356	-0.095218	-1.108956	3.95
4.00	1.370585	2.395128	1.038400	0.733060	-0.090022	-1.120703	4.00
4.05	1.360404	2.413639	1.046269	0.739518	-0.085034	-1.131255	4.05
4.10	1.350043	2.431440	1.054595	0.745683	-0.080307	-1.140643	4.10
4.15	1.339531	2.448533	1.063345	0.751508	-0.075886	-1.148902	4.15
4.20	1.328894	2.464922	1.072481	0.756952	-0.071815	-1.156073	4.20
4.25	1.318161	2.480609	1.081964	0.761977	-0.068130	-1.162200	4.25
4.30	1.307359	2.495602	1.091748	0.766552	-0.064867	-1.167332	4.30
4.35	1.296515	2.509906	1.101788	0.770648	-0.062053	-1.171523	4.35
4.40	1.285654	2.523528	1.112035	0.774243	-0.059713	-1.174828	4.40
4.45	1.274804	2.536478	1.122439	0.777318	-0.057865	-1.177304	4.45
4.50	1.263988	2.548764	1.132950	0.779862	-0.056523	-1.179013	4.50
4.55	1.253233	2.560397	1.143516	0.781865	-0.055695	-1.180016	4.55
4.60	1.242561	2.571388	1.154086	0.783327	-0.055384	-1.180375	4.60
4.65	1.231998	2.581749	1.164608	0.784249	-0.055589	-1.180154	4.65
4.70	1.221565	2.591494	1.175033	0.784640	-0.056302	-1.179415	4.70
4.75	1.211284	2.600636	1.185312	0.784511	-0.057513	-1.178222	4.75
4.80	1.201177	2.609190	1.195398	0.783879	-0.059204	-1.176635	4.80
4.85	1.191265	2.617171	1.205246	0.782767	-0.061356	-1.174714	4.85
4.90	1.181567	2.624596	1.214815	0.781198	-0.063943	-1.172518	4.90
4.95	1.172102	2.631482	1.224066	0.779202	-0.066937	-1.170103	4.95

a = 2.00

x	S(a,x)	C(a,x)	Ss(a,x)	Sc(a,x)	Cs(a,x)	Cc(a,x)	x
0.00	0.000000	0.000000	0.000000	0.000000	0.000000	0.000000	0.00
0.05	0.022728	0.035405	0.000568	0.022718	-0.000260	-0.000004	0.05
0.10	0.045429	0.070816	0.002269	0.045353	-0.001042	-0.000035	0.10
0.15	0.068075	0.106239	0.005091	0.067820	-0.002347	-0.000118	0.15
0.20	0.090640	0.141681	0.009019	0.090038	-0.004182	-0.000280	0.20
0.25	0.113096	0.177147	0.014029	0.111927	-0.006553	-0.000549	0.25
0.30	0.135418	0.212643	0.020089	0.133407	-0.009466	-0.000953	0.30
0.35	0.157578	0.248174	0.027163	0.154405	-0.012932	-0.001522	0.35
0.40	0.179550	0.283743	0.035209	0.174849	-0.016961	-0.002288	0.40
0.45	0.201308	0.319356	0.044179	0.194670	-0.021563	-0.003282	0.45
0.50	0.222828	0.355016	0.054019	0.213805	-0.026748	-0.004539	0.50
0.55	0.244082	0.390725	0.064670	0.232196	-0.032526	-0.006093	0.55
0.60	0.265048	0.426486	0.076069	0.249789	-0.038908	-0.007981	0.60
0.65	0.285700	0.462299	0.088150	0.266536	-0.045901	-0.010242	0.65
0.70	0.306016	0.498166	0.100843	0.282396	-0.053512	-0.012915	0.70
0.75	0.325971	0.534085	0.114074	0.297332	-0.061745	-0.016039	0.75
0.80	0.345544	0.570056	0.127766	0.311315	-0.070604	-0.019656	0.80
0.85	0.364712	0.606077	0.141844	0.324322	-0.080086	-0.023808	0.85
0.90	0.383455	0.642143	0.156227	0.336336	-0.090189	-0.028535	0.90
0.95	0.401752	0.678252	0.170837	0.347348	-0.100905	-0.033879	0.95
1.00	0.419584	0.714399	0.185593	0.357355	-0.112223	-0.039882	1.00
1.05	0.436930	0.750577	0.200417	0.366361	-0.124128	-0.046582	1.05
1.10	0.453775	0.786780	0.215231	0.374375	-0.136602	-0.054020	1.10
1.15	0.470099	0.823000	0.229958	0.381415	-0.149622	-0.062230	1.15
1.20	0.485888	0.859229	0.244523	0.387504	-0.163161	-0.071248	1.20
1.25	0.501125	0.895458	0.258856	0.392670	-0.177187	-0.081106	1.25
1.30	0.515795	0.931676	0.272887	0.396948	-0.191665	-0.091832	1.30
1.35	0.529886	0.967872	0.286552	0.400379	-0.206557	-0.103452	1.35
1.40	0.543385	1.004035	0.299791	0.403007	-0.221820	-0.115987	1.40
1.45	0.556280	1.040151	0.312547	0.404883	-0.237407	-0.129455	1.45
1.50	0.568560	1.076208	0.324770	0.406060	-0.253270	-0.143868	1.50
1.55	0.580215	1.112192	0.336412	0.406596	-0.269355	-0.159235	1.55
1.60	0.591238	1.148088	0.347434	0.406553	-0.285608	-0.175560	1.60
1.65	0.601621	1.183881	0.357800	0.405993	-0.301973	-0.192838	1.65
1.70	0.611357	1.219556	0.367482	0.404983	-0.318390	-0.211064	1.70
1.75	0.620440	1.255096	0.376457	0.403591	-0.334800	-0.230224	1.75
1.80	0.628867	1.290485	0.384709	0.401885	-0.351142	-0.250299	1.80
1.85	0.636634	1.325706	0.392226	0.399934	-0.367356	-0.271266	1.85
1.90	0.643738	1.360742	0.399004	0.397809	-0.383380	-0.293094	1.90
1.95	0.650179	1.395574	0.405046	0.395578	-0.399154	-0.315748	1.95
2.00	0.655957	1.430186	0.410358	0.393308	-0.414620	-0.339186	2.00
2.05	0.661072	1.464559	0.414955	0.391067	-0.429720	-0.363365	2.05
2.10	0.665526	1.498676	0.418856	0.388918	-0.444399	-0.388231	2.10
2.15	0.669323	1.532517	0.422086	0.386922	-0.458605	-0.413729	2.15
2.20	0.672466	1.566066	0.424674	0.385139	-0.472286	-0.439800	2.20
2.25	0.674961	1.599303	0.426655	0.383623	-0.485398	-0.466379	2.25
2.30	0.676814	1.632211	0.428069	0.382425	-0.497897	-0.493397	2.30
2.35	0.678032	1.664771	0.428958	0.381594	-0.509744	-0.520784	2.35
2.40	0.678622	1.696966	0.429370	0.381170	-0.520906	-0.548464	2.40
2.45	0.678595	1.728779	0.429353	0.381193	-0.531353	-0.576362	2.45

a = 2.00

x	S(a,x)	C(a,x)	Ss(a,x)	Sc(a,x)	Cs(a,x)	Cc(a,x)	x
2.50	0.677959	1.760191	0.428963	0.381694	-0.541060	-0.604398	2.50
2.55	0.676726	1.791186	0.428252	0.382701	-0.550007	-0.632492	2.55
2.60	0.674908	1.821748	0.427278	0.384236	-0.558180	-0.660563	2.60
2.65	0.672517	1.851860	0.426099	0.386316	-0.565569	-0.688530	2.65
2.70	0.669567	1.881505	0.424774	0.388951	-0.572171	-0.716310	2.70
2.75	0.666072	1.910670	0.423362	0.392148	-0.577988	-0.743824	2.75
2.80	0.662047	1.939339	0.421922	0.395906	-0.583025	-0.770991	2.80
2.85	0.657507	1.967497	0.420511	0.400220	-0.587295	-0.797733	2.85
2.90	0.652470	1.995131	0.419186	0.405079	-0.590815	-0.823974	2.90
2.95	0.646953	2.022228	0.418002	0.410468	-0.593607	-0.849642	2.95
3.00	0.640972	2.048776	0.417012	0.416365	-0.595698	-0.874666	3.00
3.05	0.634548	2.074763	0.416267	0.422746	-0.597118	-0.898978	3.05
3.10	0.627698	2.100177	0.415813	0.429580	-0.597904	-0.922518	3.10
3.15	0.620442	2.125009	0.415694	0.436834	-0.598094	-0.945226	3.15
3.20	0.612801	2.149248	0.415951	0.444470	-0.597732	-0.967048	3.20
3.25	0.604794	2.172887	0.416619	0.452448	-0.596862	-0.987937	3.25
3.30	0.596443	2.195917	0.417731	0.460724	-0.595534	-1.007848	3.30
3.35	0.587768	2.218332	0.419314	0.469252	-0.593798	-1.026744	3.35
3.40	0.578792	2.240124	0.421391	0.477983	-0.591708	-1.044592	3.40
3.45	0.569537	2.261288	0.423980	0.486868	-0.589317	-1.061365	3.45
3.50	0.560023	2.281820	0.427094	0.495857	-0.586681	-1.077045	3.50
3.55	0.550275	2.301717	0.430742	0.504896	-0.583856	-1.091614	3.55
3.60	0.540313	2.320974	0.434926	0.513935	-0.580896	-1.105066	3.60
3.65	0.530161	2.339591	0.439644	0.522923	-0.577859	-1.117397	3.65
3.70	0.519842	2.357567	0.444892	0.531807	-0.574798	-1.128610	3.70
3.75	0.509377	2.374900	0.450656	0.540540	-0.571766	-1.138715	3.75
3.80	0.498791	2.391593	0.456922	0.549072	-0.568815	-1.147725	3.80
3.85	0.488105	2.407646	0.463669	0.557357	-0.565995	-1.155660	3.85
3.90	0.477341	2.423062	0.470874	0.565352	-0.563353	-1.162546	3.90
3.95	0.466523	2.437845	0.478507	0.573016	-0.560933	-1.168412	3.95
4.00	0.455673	2.451998	0.486538	0.580310	-0.558775	-1.173291	4.00
4.05	0.444812	2.465528	0.494932	0.587201	-0.556918	-1.177224	4.05
4.10	0.433963	2.478440	0.503650	0.593657	-0.555395	-1.180251	4.10
4.15	0.423146	2.490740	0.512652	0.599651	-0.554236	-1.182418	4.15
4.20	0.412384	2.502437	0.521897	0.605160	-0.553467	-1.183775	4.20
4.25	0.401696	2.513540	0.531339	0.610165	-0.553109	-1.184373	4.25
4.30	0.391103	2.524057	0.540933	0.614651	-0.553180	-1.184265	4.30
4.35	0.380625	2.533998	0.550633	0.618610	-0.553692	-1.183507	4.35
4.40	0.370282	2.543374	0.560392	0.622034	-0.554652	-1.182154	4.40
4.45	0.360092	2.552198	0.570163	0.624922	-0.556065	-1.180265	4.45
4.50	0.350074	2.560481	0.579899	0.627279	-0.557930	-1.177898	4.50
4.55	0.340246	2.568236	0.589554	0.629110	-0.560241	-1.175110	4.55
4.60	0.330624	2.575476	0.599084	0.630429	-0.562988	-1.171959	4.60
4.65	0.321227	2.582217	0.608444	0.631250	-0.566158	-1.168501	4.65
4.70	0.312068	2.588473	0.617595	0.631593	-0.569733	-1.164791	4.70
4.75	0.303165	2.594258	0.626497	0.631482	-0.573691	-1.160883	4.75
4.80	0.294531	2.599590	0.635114	0.630943	-0.578008	-1.156829	4.80
4.85	0.286180	2.604485	0.643411	0.630006	-0.582657	-1.152677	4.85
4.90	0.278125	2.608959	0.651359	0.628703	-0.587605	-1.148476	4.90
4.95	0.270380	2.613030	0.658929	0.627070	-0.592819	-1.144267	4.95

a = 3.00

x	S(a,x)	C(a,x)	Ss(a,x)	Sc(a,x)	Cs(a,x)	Cc(a,x)	x
0.00	0.000000	0.000000	0.000000	0.000000	0.000000	0.000000	0.00
0.05	0.002350	0.033165	0.000059	0.002349	-0.000412	-0.000007	0.05
0.10	0.004685	0.066323	0.000234	0.004677	-0.001648	-0.000055	0.10
0.15	0.006991	0.099467	0.000521	0.006965	-0.003704	-0.000185	0.15
0.20	0.009255	0.132589	0.000915	0.009194	-0.006574	-0.000439	0.20
0.25	0.011461	0.165681	0.001407	0.011344	-0.010248	-0.000856	0.25
0.30	0.013595	0.198737	0.001986	0.013399	-0.014717	-0.001476	0.30
0.35	0.015645	0.231749	0.002640	0.015341	-0.019967	-0.002338	0.35
0.40	0.017595	0.264709	0.003354	0.017156	-0.025981	-0.003481	0.40
0.45	0.019433	0.297610	0.004111	0.018830	-0.032742	-0.004941	0.45
0.50	0.021145	0.330443	0.004893	0.020353	-0.040229	-0.006755	0.50
0.55	0.022718	0.363202	0.005681	0.021714	-0.048418	-0.008957	0.55
0.60	0.024140	0.395877	0.006454	0.022907	-0.057285	-0.011581	0.60
0.65	0.025397	0.428461	0.007189	0.023927	-0.066802	-0.014657	0.65
0.70	0.026479	0.460945	0.007864	0.024772	-0.076940	-0.018216	0.70
0.75	0.027373	0.493320	0.008456	0.025442	-0.087665	-0.022286	0.75
0.80	0.028068	0.525579	0.008942	0.025939	-0.098944	-0.026891	0.80
0.85	0.028553	0.557711	0.009298	0.026268	-0.110741	-0.032055	0.85
0.90	0.028818	0.589709	0.009500	0.026439	-0.123019	-0.037799	0.90
0.95	0.028853	0.621562	0.009528	0.026461	-0.135736	-0.044141	0.95
1.00	0.028649	0.653261	0.009358	0.026347	-0.148853	-0.051097	1.00
1.05	0.028196	0.684797	0.008971	0.026113	-0.162326	-0.058678	1.05
1.10	0.027487	0.716160	0.008346	0.025777	-0.176110	-0.066896	1.10
1.15	0.026513	0.747340	0.007468	0.025358	-0.190161	-0.075756	1.15
1.20	0.025268	0.778328	0.006318	0.024879	-0.204432	-0.085261	1.20
1.25	0.023745	0.809112	0.004885	0.024364	-0.218876	-0.095411	1.25
1.30	0.021937	0.839682	0.003156	0.023838	-0.233444	-0.106202	1.30
1.35	0.019840	0.870028	0.001121	0.023329	-0.248089	-0.117620	1.35
1.40	0.017448	0.900140	-0.001225	0.022865	-0.262763	-0.129678	1.40
1.45	0.014758	0.930007	-0.003886	0.022475	-0.277416	-0.142338	1.45
1.50	0.011766	0.959617	-0.006864	0.022190	-0.292001	-0.155589	1.50
1.55	0.008469	0.988961	-0.010157	0.022041	-0.306471	-0.169412	1.55
1.60	0.004865	1.018027	-0.013761	0.022057	-0.320779	-0.183781	1.60
1.65	0.000953	1.046804	-0.017667	0.022270	-0.334879	-0.198668	1.65
1.70	-0.003269	1.075282	-0.021865	0.022711	-0.348728	-0.214041	1.70
1.75	-0.007800	1.103450	-0.026342	0.023408	-0.362283	-0.229866	1.75
1.80	-0.012640	1.131296	-0.031081	0.024391	-0.375504	-0.246105	1.80
1.85	-0.017788	1.158810	-0.036062	0.025686	-0.388351	-0.262718	1.85
1.90	-0.023241	1.185982	-0.041264	0.027321	-0.400790	-0.279660	1.90
1.95	-0.028998	1.212800	-0.046663	0.029318	-0.412786	-0.296886	1.95
2.00	-0.035055	1.239255	-0.052230	0.031701	-0.424309	-0.314348	2.00
2.05	-0.041408	1.265335	-0.057938	0.034490	-0.435331	-0.331994	2.05
2.10	-0.048052	1.291031	-0.063755	0.037700	-0.445827	-0.349773	2.10
2.15	-0.054983	1.316332	-0.069647	0.041349	-0.455776	-0.367630	2.15
2.20	-0.062195	1.341230	-0.075581	0.045446	-0.465161	-0.385511	2.20
2.25	-0.069681	1.365714	-0.081520	0.050002	-0.473967	-0.403360	2.25
2.30	-0.077434	1.389775	-0.087427	0.055022	-0.482184	-0.421120	2.30
2.35	-0.085447	1.413405	-0.093266	0.060509	-0.489804	-0.438733	2.35
2.40	-0.093711	1.436594	-0.098997	0.066461	-0.496826	-0.456143	2.40
2.45	-0.102218	1.459334	-0.104583	0.072876	-0.503249	-0.473294	2.45

[Appendix]   THEORY OF LINEAR ANTENNAS   875

$a = 3.00$

x	$S(a,x)$	$C(a,x)$	$S_s(a,x)$	$S_c(a,x)$	$C_s(a,x)$	$C_c(a,x)$	x
2.50	-0.110958	1.481618	-0.109986	0.079745	-0.509079	-0.490129	2.50
2.55	-0.119922	1.503438	-0.115168	0.087057	-0.514323	-0.506595	2.55
2.60	-0.129098	1.524786	-0.120092	0.094800	-0.518995	-0.522637	2.60
2.65	-0.138477	1.545657	-0.124724	0.102954	-0.523109	-0.538204	2.65
2.70	-0.148047	1.566043	-0.129027	0.111501	-0.526685	-0.553246	2.70
2.75	-0.157796	1.585940	-0.132971	0.120415	-0.529745	-0.567717	2.75
2.80	-0.167712	1.605340	-0.136525	0.129672	-0.532315	-0.581571	2.80
2.85	-0.177783	1.624241	-0.139659	0.139241	-0.534423	-0.594767	2.85
2.90	-0.187996	1.642637	-0.142340	0.149092	-0.536101	-0.607268	2.90
2.95	-0.198337	1.660525	-0.144570	0.159190	-0.537382	-0.619037	2.95
3.00	-0.208792	1.677902	-0.146303	0.169500	-0.538302	-0.630043	3.00
3.05	-0.219349	1.694764	-0.147531	0.179984	-0.538900	-0.640260	3.05
3.10	-0.229993	1.711110	-0.148239	0.190603	-0.539215	-0.649664	3.10
3.15	-0.240709	1.726938	-0.148416	0.201317	-0.539288	-0.658235	3.15
3.20	-0.251483	1.742248	-0.148056	0.212084	-0.539161	-0.665958	3.20
3.25	-0.262301	1.757039	-0.147155	0.222863	-0.538876	-0.672824	3.25
3.30	-0.273147	1.771311	-0.145712	0.233612	-0.538477	-0.678825	3.30
3.35	-0.284007	1.785065	-0.143732	0.244289	-0.538007	-0.683960	3.35
3.40	-0.294865	1.798302	-0.141221	0.254851	-0.537508	-0.688232	3.40
3.45	-0.305707	1.811026	-0.138190	0.265260	-0.537022	-0.691647	3.45
3.50	-0.316518	1.823238	-0.134652	0.275474	-0.536592	-0.694216	3.50
3.55	-0.327281	1.834943	-0.130626	0.285455	-0.536256	-0.695956	3.55
3.60	-0.337983	1.846143	-0.126133	0.295166	-0.536053	-0.696884	3.60
3.65	-0.348608	1.856844	-0.121195	0.304573	-0.536020	-0.697026	3.65
3.70	-0.359142	1.867050	-0.115840	0.313643	-0.536191	-0.696407	3.70
3.75	-0.369570	1.876768	-0.110097	0.322345	-0.536597	-0.695058	3.75
3.80	-0.379876	1.886005	-0.103997	0.330652	-0.537269	-0.693013	3.80
3.85	-0.390048	1.894767	-0.097576	0.338539	-0.538233	-0.690309	3.85
3.90	-0.400070	1.903061	-0.090868	0.345984	-0.539511	-0.686984	3.90
3.95	-0.409929	1.910897	-0.083912	0.352969	-0.541125	-0.683080	3.95
4.00	-0.419611	1.918282	-0.076746	0.359479	-0.543091	-0.678642	4.00
4.05	-0.429103	1.925227	-0.069411	0.365501	-0.545422	-0.673714	4.05
4.10	-0.438392	1.931740	-0.061947	0.371029	-0.548129	-0.668344	4.10
4.15	-0.447465	1.937833	-0.054396	0.376058	-0.551218	-0.662579	4.15
4.20	-0.456310	1.943517	-0.046799	0.380586	-0.554691	-0.656469	4.20
4.25	-0.464915	1.948802	-0.039197	0.384616	-0.558547	-0.650063	4.25
4.30	-0.473269	1.953701	-0.031631	0.388155	-0.562782	-0.643409	4.30
4.35	-0.481361	1.958225	-0.024140	0.391213	-0.567388	-0.636556	4.35
4.40	-0.489180	1.962389	-0.016763	0.393802	-0.572353	-0.629554	4.40
4.45	-0.496716	1.966204	-0.009537	0.395939	-0.577663	-0.622448	4.45
4.50	-0.503961	1.969685	-0.002497	0.397644	-0.583300	-0.615285	4.50
4.55	-0.510904	1.972845	0.004324	0.398938	-0.589243	-0.608110	4.55
4.60	-0.517537	1.975698	0.010894	0.399848	-0.595469	-0.600964	4.60
4.65	-0.523854	1.978260	0.017186	0.400401	-0.601952	-0.593889	4.65
4.70	-0.529845	1.980545	0.023172	0.400626	-0.608663	-0.586922	4.70
4.75	-0.535506	1.982568	0.028832	0.400556	-0.615572	-0.580100	4.75
4.80	-0.540829	1.984344	0.034144	0.400224	-0.622648	-0.573454	4.80
4.85	-0.545809	1.985890	0.039092	0.399666	-0.629856	-0.567015	4.85
4.90	-0.550442	1.987220	0.043664	0.398918	-0.637162	-0.560810	4.90
4.95	-0.554723	1.988352	0.047848	0.398016	-0.644530	-0.554862	4.95

a = 4.00

x	S(a,x)	C(a,x)	Ss(a,x)	Sc(a,x)	Cs(a,x)	Cc(a,x)	x
0.00	0.000000	0.000000	0.000000	0.000000	0.000000	0.000000	0.00
0.05	-0.009461	0.020669	-0.000236	-0.009457	-0.000204	-0.000003	0.05
0.10	-0.018925	0.041329	-0.000946	-0.018893	-0.000816	-0.000027	0.10
0.15	-0.028396	0.061971	-0.002126	-0.028290	-0.001831	-0.000092	0.15
0.20	-0.037879	0.082585	-0.003777	-0.037626	-0.003246	-0.000217	0.20
0.25	-0.047375	0.103162	-0.005896	-0.046883	-0.005052	-0.000421	0.25
0.30	-0.056890	0.123694	-0.008479	-0.056039	-0.007241	-0.000725	0.30
0.35	-0.066425	0.144172	-0.011524	-0.065074	-0.009801	-0.001146	0.35
0.40	-0.075985	0.164585	-0.015025	-0.073969	-0.012719	-0.001700	0.40
0.45	-0.085573	0.184926	-0.018978	-0.082703	-0.015980	-0.002404	0.45
0.50	-0.095192	0.205185	-0.023377	-0.091255	-0.019568	-0.003274	0.50
0.55	-0.104844	0.225352	-0.028214	-0.099607	-0.023463	-0.004321	0.55
0.60	-0.114532	0.245421	-0.033482	-0.107736	-0.027648	-0.005559	0.60
0.65	-0.124259	0.265380	-0.039173	-0.115624	-0.032099	-0.006998	0.65
0.70	-0.134028	0.285222	-0.045277	-0.123249	-0.036795	-0.008647	0.70
0.75	-0.143840	0.304937	-0.051784	-0.130593	-0.041712	-0.010512	0.75
0.80	-0.153698	0.324518	-0.058681	-0.137635	-0.046825	-0.012599	0.80
0.85	-0.163603	0.343954	-0.065956	-0.144355	-0.052109	-0.014912	0.85
0.90	-0.173558	0.363239	-0.073596	-0.150735	-0.057537	-0.017451	0.90
0.95	-0.183562	0.382362	-0.081584	-0.156755	-0.063082	-0.020217	0.95
1.00	-0.193617	0.401316	-0.089907	-0.162397	-0.068718	-0.023205	1.00
1.05	-0.203725	0.420092	-0.098545	-0.167643	-0.074416	-0.026411	1.05
1.10	-0.213885	0.438682	-0.107481	-0.172476	-0.080148	-0.029828	1.10
1.15	-0.224098	0.457078	-0.116695	-0.176879	-0.085889	-0.033447	1.15
1.20	-0.234364	0.475272	-0.126166	-0.180836	-0.091609	-0.037257	1.20
1.25	-0.244682	0.493256	-0.135873	-0.184333	-0.097283	-0.041244	1.25
1.30	-0.255052	0.511022	-0.145792	-0.187355	-0.102884	-0.045392	1.30
1.35	-0.265473	0.528562	-0.155898	-0.189891	-0.108387	-0.049686	1.35
1.40	-0.275944	0.545869	-0.166168	-0.191927	-0.113768	-0.054104	1.40
1.45	-0.286463	0.562936	-0.176574	-0.193455	-0.119003	-0.058626	1.45
1.50	-0.297028	0.579756	-0.187089	-0.194465	-0.124071	-0.063230	1.50
1.55	-0.307637	0.596321	-0.197686	-0.194951	-0.128950	-0.067891	1.55
1.60	-0.318287	0.612625	-0.208336	-0.194906	-0.133623	-0.072583	1.60
1.65	-0.328977	0.628661	-0.219008	-0.194327	-0.138071	-0.077278	1.65
1.70	-0.339702	0.644423	-0.229674	-0.193211	-0.142280	-0.081949	1.70
1.75	-0.350459	0.659904	-0.240302	-0.191559	-0.146235	-0.086566	1.75
1.80	-0.361244	0.675099	-0.250863	-0.189372	-0.149926	-0.091099	1.80
1.85	-0.372054	0.690003	-0.261324	-0.186654	-0.153344	-0.095517	1.85
1.90	-0.382884	0.704608	-0.271656	-0.183410	-0.156480	-0.099789	1.90
1.95	-0.393730	0.718912	-0.281827	-0.179649	-0.159332	-0.103882	1.95
2.00	-0.404585	0.732908	-0.291806	-0.175380	-0.161895	-0.107765	2.00
2.05	-0.415446	0.746592	-0.301565	-0.170615	-0.164170	-0.111406	2.05
2.10	-0.426307	0.759959	-0.311073	-0.165369	-0.166159	-0.114774	2.10
2.15	-0.437161	0.773007	-0.320302	-0.159657	-0.167868	-0.117839	2.15
2.20	-0.448003	0.785730	-0.329224	-0.153498	-0.169302	-0.120569	2.20
2.25	-0.458827	0.798126	-0.337812	-0.146912	-0.170470	-0.122936	2.25
2.30	-0.469626	0.810192	-0.346041	-0.139922	-0.171386	-0.124913	2.30
2.35	-0.480393	0.821924	-0.353888	-0.132550	-0.172061	-0.126471	2.35
2.40	-0.491121	0.833322	-0.361329	-0.124824	-0.172512	-0.127587	2.40
2.45	-0.501804	0.844382	-0.368345	-0.116770	-0.172757	-0.128237	2.45

a = 4.00

x	S(a,x)	C(a,x)	Ss(a,x)	Sc(a,x)	Cs(a,x)	Cc(a,x)	x
2.50	-0.512433	0.855103	-0.374916	-0.108417	-0.172814	-0.128399	2.50
2.55	-0.523001	0.865484	-0.381027	-0.099796	-0.172706	-0.128055	2.55
2.60	-0.533500	0.875524	-0.386663	-0.090938	-0.172454	-0.127187	2.60
2.65	-0.543924	0.885222	-0.391811	-0.081876	-0.172084	-0.125781	2.65
2.70	-0.554262	0.894579	-0.396461	-0.072644	-0.171620	-0.123824	2.70
2.75	-0.564509	0.903595	-0.400607	-0.063275	-0.171089	-0.121306	2.75
2.80	-0.574654	0.912271	-0.404244	-0.053805	-0.170518	-0.118222	2.80
2.85	-0.584691	0.920607	-0.407369	-0.044268	-0.169936	-0.114566	2.85
2.90	-0.594610	0.928605	-0.409982	-0.034701	-0.169370	-0.110339	2.90
2.95	-0.604403	0.936268	-0.412087	-0.025137	-0.168850	-0.105540	2.95
3.00	-0.614062	0.943597	-0.413689	-0.015613	-0.168403	-0.100176	3.00
3.05	-0.623579	0.950596	-0.414797	-0.006162	-0.168059	-0.094254	3.05
3.10	-0.632944	0.957267	-0.415420	0.003181	-0.167844	-0.087783	3.10
3.15	-0.642150	0.963614	-0.415574	0.012385	-0.167787	-0.080778	3.15
3.20	-0.651188	0.969642	-0.415273	0.021417	-0.167914	-0.073254	3.20
3.25	-0.660050	0.975354	-0.414535	0.030248	-0.168250	-0.065231	3.25
3.30	-0.668728	0.980755	-0.413382	0.038848	-0.168819	-0.056729	3.30
3.35	-0.677213	0.985852	-0.411835	0.047190	-0.169643	-0.047773	3.35
3.40	-0.685498	0.990649	-0.409920	0.055249	-0.170744	-0.038389	3.40
3.45	-0.693574	0.995152	-0.407663	0.063003	-0.172141	-0.028606	3.45
3.50	-0.701435	0.999368	-0.405092	0.070430	-0.173850	-0.018454	3.50
3.55	-0.709071	1.003304	-0.402236	0.077512	-0.175887	-0.007965	3.55
3.60	-0.716477	1.006966	-0.399127	0.084233	-0.178263	0.002827	3.60
3.65	-0.723645	1.010363	-0.395797	0.090579	-0.180990	0.013885	3.65
3.70	-0.730568	1.013502	-0.392278	0.096540	-0.184075	0.025173	3.70
3.75	-0.737239	1.016391	-0.388604	0.102108	-0.187523	0.036653	3.75
3.80	-0.743653	1.019040	-0.384809	0.107277	-0.191336	0.048287	3.80
3.85	-0.749802	1.021456	-0.380928	0.112046	-0.195515	0.060036	3.85
3.90	-0.755681	1.023649	-0.376993	0.116414	-0.200057	0.071860	3.90
3.95	-0.761285	1.025629	-0.373040	0.120384	-0.204956	0.083720	3.95
4.00	-0.766608	1.027406	-0.369101	0.123964	-0.210204	0.095576	4.00
4.05	-0.771646	1.028988	-0.365208	0.127161	-0.215791	0.107391	4.05
4.10	-0.776393	1.030387	-0.361394	0.129987	-0.221703	0.119125	4.10
4.15	-0.780846	1.031614	-0.357688	0.132456	-0.227925	0.130743	4.15
4.20	-0.785002	1.032678	-0.354120	0.134584	-0.234439	0.142207	4.20
4.25	-0.788856	1.033591	-0.350715	0.136389	-0.241225	0.153484	4.25
4.30	-0.792405	1.034364	-0.347501	0.137894	-0.248260	0.164541	4.30
4.35	-0.795648	1.035009	-0.344499	0.139120	-0.255522	0.175346	4.35
4.40	-0.798582	1.035536	-0.341732	0.140092	-0.262983	0.185870	4.40
4.45	-0.801204	1.035957	-0.339217	0.140837	-0.270616	0.196088	4.45
4.50	-0.803515	1.036285	-0.336972	0.141381	-0.278394	0.205972	4.50
4.55	-0.805512	1.036530	-0.335010	0.141755	-0.286286	0.215502	4.55
4.60	-0.807196	1.036704	-0.333342	0.141987	-0.294263	0.224658	4.60
4.65	-0.808567	1.036821	-0.331977	0.142107	-0.302292	0.233422	4.65
4.70	-0.809624	1.036890	-0.330921	0.142148	-0.310342	0.241780	4.70
4.75	-0.810370	1.036925	-0.330176	0.142140	-0.318382	0.249721	4.75
4.80	-0.810804	1.036937	-0.329742	0.142114	-0.326381	0.257234	4.80
4.85	-0.810930	1.036939	-0.329616	0.142101	-0.334306	0.264314	4.85
4.90	-0.810749	1.036941	-0.329795	0.142132	-0.342127	0.270958	4.90
4.95	-0.810265	1.036957	-0.330268	0.142235	-0.349814	0.277164	4.95

a = 5.0

x	S(a,x)	C(a,x)	Ss(a,x)	Sc(a,x)	Cs(a,x)	Cc(a,x)	x
0.0	0.000000	0.000000	0.000000	0.000000	0.000000	0.000000	0.0
0.1	-0.019175	0.014319	-0.000958	-0.019143	0.000284	0.000009	0.1
0.2	-0.038332	0.028595	-0.003819	-0.038077	0.001138	0.000076	0.2
0.3	-0.057450	0.042782	-0.008547	-0.056593	0.002570	0.000258	0.3
0.4	-0.076510	0.056839	-0.015079	-0.074490	0.004591	0.000619	0.4
0.5	-0.095492	0.070721	-0.023332	-0.091576	0.007217	0.001222	0.5
0.6	-0.114376	0.084388	-0.033197	-0.107669	0.010465	0.002142	0.6
0.7	-0.133140	0.097799	-0.044547	-0.122601	0.014351	0.003454	0.7
0.8	-0.151762	0.110914	-0.057235	-0.136221	0.018893	0.005246	0.8
0.9	-0.170219	0.123696	-0.071094	-0.148399	0.024102	0.007608	0.9
1.0	-0.188486	0.136108	-0.085946	-0.159021	0.029987	0.010638	1.0
1.1	-0.206538	0.148118	-0.101597	-0.168001	0.036546	0.014442	1.1
1.2	-0.224349	0.159693	-0.117846	-0.175276	0.043769	0.019129	1.2
1.3	-0.241890	0.170804	-0.134485	-0.180807	0.051636	0.024810	1.3
1.4	-0.259133	0.181425	-0.151301	-0.184584	0.060110	0.031599	1.4
1.5	-0.276047	0.191531	-0.168085	-0.186624	0.069142	0.039608	1.5
1.6	-0.292601	0.201103	-0.184629	-0.186971	0.078667	0.048945	1.6
1.7	-0.308763	0.210122	-0.200734	-0.185697	0.088605	0.059711	1.7
1.8	-0.324500	0.218574	-0.216213	-0.182896	0.098858	0.071996	1.8
1.9	-0.339777	0.226450	-0.230893	-0.178692	0.109312	0.085879	1.9
2.0	-0.354559	0.233741	-0.244621	-0.173226	0.119843	0.101421	2.0
2.1	-0.368811	0.240445	-0.257264	-0.166662	0.130308	0.118664	2.1
2.2	-0.382498	0.246562	-0.268717	-0.159177	0.140559	0.137629	2.2
2.3	-0.395583	0.252097	-0.278897	-0.150965	0.150436	0.158311	2.3
2.4	-0.408031	0.257059	-0.287754	-0.142225	0.159778	0.180681	2.4
2.5	-0.419807	0.261461	-0.295265	-0.133162	0.168418	0.204680	2.5
2.6	-0.430876	0.265319	-0.301441	-0.123981	0.176197	0.230222	2.6
2.7	-0.441205	0.268653	-0.306320	-0.114882	0.182959	0.257194	2.7
2.8	-0.450761	0.271490	-0.309971	-0.106056	0.188560	0.285453	2.8
2.9	-0.459513	0.273857	-0.312493	-0.097679	0.192870	0.314830	2.9
3.0	-0.467433	0.275788	-0.314007	-0.089909	0.195779	0.345134	3.0
3.1	-0.474493	0.277318	-0.314660	-0.082882	0.197198	0.376150	3.1
3.2	-0.480669	0.278488	-0.314616	-0.076709	0.197064	0.407648	3.2
3.3	-0.485939	0.279339	-0.314053	-0.071471	0.195342	0.439382	3.3
3.4	-0.490285	0.279920	-0.313162	-0.067219	0.192025	0.471098	3.4
3.5	-0.493691	0.280279	-0.312136	-0.063973	0.187138	0.502538	3.5
3.6	-0.496145	0.280467	-0.311169	-0.061718	0.180739	0.533445	3.6
3.7	-0.497640	0.280538	-0.310449	-0.060409	0.172914	0.563569	3.7
3.8	-0.498171	0.280549	-0.310152	-0.059969	0.163780	0.592673	3.8
3.9	-0.497738	0.280558	-0.310440	-0.060292	0.153481	0.620535	3.9
4.0	-0.496346	0.280622	-0.311451	-0.061247	0.142184	0.646955	4.0
4.1	-0.494004	0.280802	-0.313302	-0.062681	0.130078	0.671762	4.1
4.2	-0.490726	0.281158	-0.316078	-0.064421	0.117367	0.694811	4.2
4.3	-0.486530	0.281750	-0.319835	-0.066285	0.104265	0.715992	4.3
4.4	-0.481440	0.282637	-0.324596	-0.068083	0.090993	0.735230	4.4
4.5	-0.475482	0.283880	-0.330349	-0.069620	0.077770	0.752483	4.5
4.6	-0.468690	0.285534	-0.337050	-0.070711	0.064812	0.767749	4.6
4.7	-0.461101	0.287658	-0.344622	-0.071178	0.052319	0.781057	4.7
4.8	-0.452756	0.290303	-0.352957	-0.070858	0.040478	0.792472	4.8
4.9	-0.443702	0.293522	-0.361921	-0.069611	0.029450	0.802088	4.9

a = 6.0

x	S(a,x)	C(a,x)	Ss(a,x)	Sc(a,x)	Cs(a,x)	Cc(a,x)	x
0.0	0.000000	0.000000	0.000000	0.000000	0.000000	0.000000	0.0
0.1	-0.004652	0.000663	-0.000232	-0.004645	0.000800	0.000027	0.1
0.2	-0.009277	0.001317	-0.000923	-0.009215	0.003191	0.000213	0.2
0.3	-0.013845	0.001956	-0.002052	-0.013640	0.007151	0.000717	0.3
0.4	-0.018330	0.002572	-0.003588	-0.017851	0.012643	0.001695	0.4
0.5	-0.022703	0.003158	-0.005489	-0.021788	0.019614	0.003297	0.5
0.6	-0.026937	0.003707	-0.007700	-0.025396	0.027999	0.005670	0.6
0.7	-0.031004	0.004214	-0.010159	-0.028634	0.037715	0.008950	0.7
0.8	-0.034878	0.004674	-0.012797	-0.031468	0.048670	0.013269	0.8
0.9	-0.038531	0.005084	-0.015539	-0.033880	0.060755	0.018745	0.9
1.0	-0.041936	0.005440	-0.018307	-0.035861	0.073850	0.025486	1.0
1.1	-0.045068	0.005741	-0.021021	-0.037421	0.087823	0.033587	1.1
1.2	-0.047900	0.005989	-0.023604	-0.038580	0.102532	0.043125	1.2
1.3	-0.050408	0.006183	-0.025982	-0.039373	0.117825	0.054166	1.3
1.4	-0.052566	0.006327	-0.028086	-0.039848	0.133543	0.066754	1.4
1.5	-0.054351	0.006426	-0.029857	-0.040067	0.149520	0.080915	1.5
1.6	-0.055739	0.006486	-0.031244	-0.040099	0.165586	0.096657	1.6
1.7	-0.056709	0.006516	-0.032211	-0.040026	0.181569	0.113966	1.7
1.8	-0.057239	0.006525	-0.032733	-0.039935	0.197297	0.132805	1.8
1.9	-0.057309	0.006526	-0.032802	-0.039919	0.212600	0.153117	1.9
2.0	-0.056901	0.006532	-0.032424	-0.040074	0.227314	0.174824	2.0
2.1	-0.055996	0.006559	-0.031623	-0.040495	0.241279	0.197825	2.1
2.2	-0.054579	0.006623	-0.030440	-0.041274	0.254351	0.221997	2.2
2.3	-0.052636	0.006746	-0.028931	-0.042498	0.266394	0.247201	2.3
2.4	-0.050154	0.006946	-0.027169	-0.044244	0.277289	0.273278	2.4
2.5	-0.047122	0.007248	-0.025240	-0.046582	0.286934	0.300052	2.5
2.6	-0.043532	0.007675	-0.023243	-0.049563	0.295248	0.327335	2.6
2.7	-0.039378	0.008253	-0.021287	-0.053226	0.302171	0.354928	2.7
2.8	-0.034655	0.009009	-0.019489	-0.057591	0.307665	0.382621	2.8
2.9	-0.029361	0.009972	-0.017973	-0.062660	0.311717	0.410204	2.9
3.0	-0.023498	0.011170	-0.016862	-0.068415	0.314338	0.437463	3.0
3.1	-0.017069	0.012635	-0.016278	-0.074814	0.315566	0.464186	3.1
3.2	-0.010080	0.014396	-0.016342	-0.081800	0.315460	0.490171	3.2
3.3	-0.002540	0.016486	-0.017162	-0.089293	0.314105	0.515221	3.3
3.4	0.005540	0.018937	-0.018837	-0.097193	0.311606	0.539158	3.4
3.5	0.014144	0.021780	-0.021452	-0.105387	0.308089	0.561817	3.5
3.6	0.023255	0.025048	-0.025073	-0.113744	0.303696	0.583056	3.6
3.7	0.032854	0.028772	-0.029747	-0.122123	0.298584	0.602756	3.7
3.8	0.042915	0.032982	-0.035498	-0.130373	0.292919	0.620821	3.8
3.9	0.053413	0.037708	-0.042328	-0.138340	0.286874	0.637185	3.9
4.0	0.064319	0.042980	-0.050214	-0.145866	0.280626	0.651809	4.0
4.1	0.075600	0.048825	-0.059108	-0.152799	0.274348	0.664682	4.1
4.2	0.087223	0.055268	-0.068938	-0.158991	0.268208	0.675823	4.2
4.3	0.099148	0.062334	-0.079607	-0.164307	0.262365	0.685276	4.3
4.4	0.111337	0.070044	-0.091001	-0.168624	0.256962	0.693115	4.4
4.5	0.123748	0.078418	-0.102982	-0.171841	0.252124	0.699434	4.5
4.6	0.136336	0.087472	-0.115399	-0.173874	0.247957	0.704349	4.6
4.7	0.149054	0.097221	-0.128088	-0.174665	0.244541	0.707994	4.7
4.8	0.161855	0.107675	-0.140875	-0.174184	0.241930	0.710516	4.8
4.9	0.174690	0.118843	-0.153583	-0.172424	0.240152	0.712071	4.9

a = 7.0

x	S(a,x)	C(a,x)	Ss(a,x)	Sc(a,x)	Cs(a,x)	Cc(a,x)	x
0.0	0.000000	0.000000	0.000000	0.000000	0.000000	0.000000	0.0
0.1	0.009388	0.003518	0.000469	0.009372	0.000538	0.000018	0.1
0.2	0.018789	0.007048	0.001874	0.018664	0.002144	0.000143	0.2
0.3	0.028217	0.010604	0.004205	0.027795	0.004795	0.000481	0.3
0.4	0.037685	0.014198	0.007451	0.036685	0.008453	0.001132	0.4
0.5	0.047206	0.017844	0.011591	0.045255	0.013065	0.002192	0.5
0.6	0.056793	0.021554	0.016600	0.053424	0.018567	0.003748	0.6
0.7	0.066457	0.025341	0.022447	0.061114	0.024880	0.005880	0.7
0.8	0.076211	0.029220	0.029094	0.068247	0.031915	0.008653	0.8
0.9	0.086066	0.033203	0.036495	0.074748	0.039573	0.012122	0.9
1.0	0.096032	0.037306	0.044598	0.080541	0.047746	0.016329	1.0
1.1	0.106119	0.041542	0.053345	0.085557	0.056322	0.021299	1.1
1.2	0.116335	0.045927	0.062667	0.089728	0.065182	0.027044	1.2
1.3	0.126690	0.050474	0.072490	0.092991	0.074208	0.033559	1.3
1.4	0.137190	0.055200	0.082730	0.095288	0.083279	0.040822	1.4
1.5	0.147841	0.060120	0.093299	0.096570	0.092278	0.048796	1.5
1.6	0.158647	0.065251	0.104099	0.096793	0.101090	0.057428	1.6
1.7	0.169614	0.070607	0.115027	0.095924	0.109607	0.066650	1.7
1.8	0.180742	0.076207	0.125972	0.093940	0.117732	0.076379	1.8
1.9	0.192034	0.082065	0.136821	0.090828	0.125374	0.086520	1.9
2.0	0.203488	0.088201	0.147457	0.086589	0.132457	0.096966	2.0
2.1	0.215104	0.094629	0.157760	0.081234	0.138917	0.107600	2.1
2.2	0.226878	0.101367	0.167609	0.074791	0.144705	0.118300	2.2
2.3	0.238805	0.108433	0.176884	0.067301	0.149790	0.128935	2.3
2.4	0.250879	0.115843	0.185470	0.058819	0.154154	0.139374	2.4
2.5	0.263091	0.123613	0.193254	0.049416	0.157799	0.149484	2.5
2.6	0.275433	0.131760	0.200134	0.039175	0.160743	0.159135	2.6
2.7	0.287893	0.140300	0.200012	0.028195	0.163021	0.168201	2.7
2.8	0.300458	0.149249	0.210804	0.016586	0.164683	0.176564	2.8
2.9	0.313112	0.158621	0.214440	0.004470	0.165796	0.184117	2.9
3.0	0.325841	0.168430	0.216862	-0.008021	0.166439	0.190764	3.0
3.1	0.338626	0.178689	0.218031	-0.020746	0.166702	0.196424	3.1
3.2	0.351447	0.189412	0.217923	-0.033562	0.166688	0.201032	3.2
3.3	0.364283	0.200609	0.216535	-0.046317	0.166502	0.204542	3.3
3.4	0.377112	0.212290	0.213882	-0.058863	0.166257	0.206926	3.4
3.5	0.389909	0.224465	0.209999	-0.071051	0.166068	0.208177	3.5
3.6	0.402648	0.237139	0.204942	-0.082738	0.166045	0.208307	3.6
3.7	0.415304	0.250319	0.198785	-0.093789	0.166299	0.207348	3.7
3.8	0.427847	0.264008	0.191620	-0.104078	0.166930	0.205355	3.8
3.9	0.440250	0.278209	0.183555	-0.113493	0.168029	0.202396	3.9
4.0	0.452481	0.292922	0.174714	-0.121938	0.169674	0.198560	4.0
4.1	0.464510	0.308145	0.165234	-0.129335	0.171929	0.193950	4.1
4.2	0.476306	0.323873	0.155260	-0.135623	0.174840	0.188681	4.2
4.3	0.487836	0.340101	0.144946	-0.140767	0.178436	0.182878	4.3
4.4	0.499069	0.356819	0.134448	-0.144750	0.182723	0.176673	4.4
4.5	0.509971	0.374018	0.123924	-0.147579	0.187689	0.170201	4.5
4.6	0.520511	0.391683	0.113528	-0.149286	0.193299	0.163599	4.6
4.7	0.530656	0.409798	0.103407	-0.149921	0.199501	0.157000	4.7
4.8	0.540374	0.428347	0.093699	-0.149560	0.206221	0.150530	4.8
4.9	0.549635	0.447307	0.084530	-0.148294	0.213367	0.144306	4.9

PROBLEMS

Chapter II

1. (a) Investigate the terminal-zone problem of a two-wire line terminated in an antenna with its halves displaced laterally by a distance equal to the line spacing, as shown in unbroken lines in Fig. II.1. Assume all wires to have the same radius a; the line spacing is b; the length of each half of the antenna is h. It is assumed that the following inequalities are satisfied: $a^2 \ll b^2$, $b \ll h$, $b \ll \lambda_0$.
(b) Repeat (a) with a high-impedance stub (shown dashed in Fig. II.1) terminating the line in parallel with the antenna.

2. (a) Investigate the terminal-zone problem of a two-wire line terminated in the symmetric V-antenna shown in unbroken lines in Fig. II.2. Assume a, b, and h to have the same significance and to satisfy the same inequalities as in Problem 1.
(b) Repeat (a) with a high-impedance stub (shown dashed in Fig. II.2) terminating the line in parallel with the antenna.

3. Investigate the terminal-zone problem of a coaxial line terminated in a ground-plane antenna as shown in Fig. II.3. Assume currents and charges on the *outer* surface of the coaxial sheath to be negligible. The radii of all antennas are the same as the radius a of the inner conductor of the coaxial line; the inner radius of the sheath is b; all antennas have a length h.

4. Investigate the terminal-zone problem of an antenna center-driven from a four-wire line. Connect the antenna as a *symmetric* end load successively without and with a properly adjusted high-impedance stub.

5. Analyze the terminal-zone problem for the V-antenna driven from a two-wire line as shown in Fig. II.5.

6. Determine the even and odd parts as functions of z of the scalar and vector potentials defined by

$$\Phi(z) = \frac{1}{4\pi\epsilon_0} \int_{-h}^{h} q(z') \frac{e^{-j\beta_0 R}}{R} dz',$$

$$A_z(z) = \frac{1}{4\pi v_0} \int_{-h}^{h} I(z') \frac{e^{-j\beta_0 R}}{R} dz',$$

$$R = \sqrt{(z-z') + a^2},$$

in terms of the even and odd parts of the charge per unit length and of the current.

7. (a) Show that the vector potential of a cylindrical center-driven antenna can be written in the form

$$A_z = \frac{jV_0^e}{2v_0} \sqrt{C^2 + 1} \sin(\beta_0 |z| - \tan^{-1} C),$$

where

$$C = \frac{G_0(h) + G_1(h)/\psi + \cdots}{F_0(h) + F_1(h)/\psi + \cdots}.$$

Show that the zeroth-order vector potential is

$$(A_z)_0 = -\frac{jV_0^e}{2v_0} \frac{\sin \beta_0 (h - |z|)}{\cos \beta_0 h}.$$

(b) Determine the scalar potential associated with the vector potential defined by each of the above formulas.

8. Carry out the iteration and obtain the series solution for I_z proceeding from the simpler form (11.13) instead of from (11.16). Discuss advantages and disadvantages of the two forms, including consideration of the question of accuracy.

9. Investigate the advantages and disadvantages of using
(a) a complex expansion parameter in place of the real magnitude Ψ_{K1};
(b) the real part of $\Psi(z_r)$ instead of its magnitude;
(c) the real part or magnitude of $\Psi(z_r) - \Psi(h)$ instead of $\Psi(z_r)$.

10. (a) Derive the following integral equation for the current in a perfectly conducting two-wire line immersed in air:

$$W_x(w) = \frac{1}{2\pi v_0} \int_0^s I_x(w') G(w, w') dw'$$
$$= A \cos \beta_0 w + B \sin \beta_0 w$$

where $W_x(w)$ is the vector potential difference. Determine $G(w, w')$ and show that the scalar potential difference $V(w)$ satisfies the relation

$$\frac{dW_x(w)}{dw} = j\frac{\beta_0^2}{\omega} V(w).$$

(b) Obtain a solution for $I_x(w)$ using the method of iteration with $I_x(w')/I_x(w) = 1$ as the distribution function. Show the relation between the zeroth-order solution and conventional line theory. Establish the relation between the expansion parameter and the characteristic impedance. Assume the following boundary conditions:

$$V(w) = V^e \text{ at } w = s,$$

$$I(w) = 0 \text{ at } w = 0 \text{ (open end)}.$$

11. Repeat Problem 10 using the distribution function

$$\frac{I_x(w')}{I_x(w)} = \frac{\sin \beta w'}{\sin \beta w}$$

appropriate to a sinusoidal distribution on the open-end line.

12. The real instantaneous current and charge on an antenna with finite radius and zero base-separation referred to a driving voltage of the form $(V_0)_{\text{inst}} = V_0 \sin \omega t$ have the form

$$(I_z)_{\text{inst}} = I'_z \cos \omega t + I''_z \sin \omega t,$$

$$(q_z)_{\text{inst}} = -q'_z \sin \omega t + q''_z \cos \omega t,$$

where

$$q'_z = \frac{1}{\omega} \frac{dI'_z}{dz}, \quad q''_z = \frac{1}{\omega} \frac{dI''_z}{dz}.$$

(a) Using the first-order distribution curves for I''_z, I'_z, q''_z, q'_z, plot the relative instantaneous currents and charges $(I'_z)_{\text{inst}}$, $(I'_z)_{\text{inst}}$, $(q''_z)_{\text{inst}}$, $(q'_z)_{\text{inst}}$ at $\omega t = 0, \pi/2, \pi, 3\pi/2$ using an appropriate relative scale for I_z and q_z. Use $\beta_0 h = 5\pi/4$ and $\Omega = 10$ and 20. Plot curves of the zeroth-order current and charge for comparison.

(b) Describe a half-cycle for first-order I''_z, q''_z and I'_z, q'_z, and compare with the corresponding zeroth-order distributions.

13. (a) Determine the impedance and the resonant half-length h_r at the input resonance near $h = \lambda_0/4$ of a copper antenna for which $h/a = 1.1 \times 10^4$ at a frequency of 150 Mhz. Assume the antenna driven by the equivalent of a slice generator. If the current in the generator is 1 amp, what power is supplied to the antenna? Sketch the distribution of current in the antenna.

(b) Determine the impedance and the resonant length s_r at the input resonance near $s = \lambda_0/4$ of a two-wire line that is open at $x = s_r$ and closed at $x = 0$. It is driven by a slice generator at the center of the closed end. All conductors are of the same wire as the antenna in (a); the line spacing b is 2 cm. If the current in the generator is 1 ampere, what power is supplied to the line? Sketch the distribution of current on the line.

(c) Compare the distributions of current in and the power supplied to the antenna in (a) and the line in (b). Explain why the power is so different when the currents are so nearly the same.

14. Repeat Problem 13 with h and s adjusted for the antiresonance near $\lambda_0/2$ and with an emf of 100 volts maintained by the generator in each case.

15. The following data apply to an actual commercial radio transmitter:
  Carrier frequency: 1360 khz.
  Tower: height, 260 ft; side (square cross section), 2 ft; material, steel.
  Ground system: 30-ft square screen with 120 buried radial wires.
  Feeder: four-wire open line of length 200 ft and characteristic impedance 500 ohms.
  Current at base of antenna: 1.02 amps.
  Radiated power (authorized): 500 watts.
  Measured input resistance at base of antenna: 481 ohms.
Determine the following:
  (a) The electrical length of the antenna;
  (b) The equivalent radius and Ω for the antenna (see Sec. I.7);
  (c) The theoretical input impedance, assuming the earth and ground system to be equivalent to an infinite perfect conductor;
  (d) The theoretical power radiated;
  (e) The input impedance and admittance of the feeding line, neglecting line losses;
  (f) The standing-wave ratio on the line.
Compare the theoretical input resistance with the measured value. Compare the theoretical, actual, and rated radiated powers.

16. (a) Plot the standing-wave ratio on a transmission line used to center-drive an antenna as the frequency is varied from 280 to 320 Mhz. For the antenna $h/a = 1.1 \times 10^4$; h is adjusted for antiresonance at 300 Mhz. The characteristic impedance of the line is $Z_c \doteq R_c = 300$ ohms at 300 Mhz where β is the phase constant and α the attenuation constant of the line. (Neglect the small variation in Z_c and α with frequency.)

(b) Repeat (a) for an antenna with $h/a = 75$.

(c) Repeat (a) for an antenna of length such that the antenna is resonant instead of antiresonant at 300 Mhz.

(d) If the line is 9.5 m long, what is the efficiency of transmission at 300 Mhz with *each* of the three antennas?

(e) Compare and discuss the significance of the results in (a) to (d). (Neglect base-separation, end, and coupling effects.)

17. A base-driven antenna erected vertically over the ocean (treat as a perfect conductor) has a length $h = 0.05\lambda_0$; $\Omega = 2 \ln (2h/a) = 20$. What

is the impedance of the antenna? What is the voltage across the driving terminals? What reactance must be connected in series to make a resistive load? What must be the input current if 100 watts are to be radiated? (Neglect end and coupling effects.)

18. A center-driven antenna of half-length $h = \lambda_0/2$ and with $h/a = 75$ is driven at 600 Mhz by a generator at the end of a two-wire line that is 3.8 wavelengths long. For the line, $\alpha = 3 \times 10^{-3}$ neper/m; $R_c = 400$ ohms.

(a) What is the impedance seen by the generator? What must be the voltage applied to the line if 10 watts are to be radiated? (Assume line spacing negligible.)

(b) Design a matching network to obtain a flat line. What gain in efficiency is thereby achieved? (Neglect base-separation, end, and coupling effects. Neglect ohmic losses in the antenna and radiation losses from the line.)

19. A symmetric cylindrical antenna of radius a is connected as end load in the plane of the line to a two-wire line. The distance between centers of the line conductors is b. The following numerical data apply:

$$a/\lambda_0 = 0.004, \quad b/\lambda_0 = 0.02, \quad \lambda_0 = 0.5 \text{ m}.$$

Determine the terminal impedance that would be measured on the line using conventional methods as the half-length h of the antenna is increased so that $\beta_0 h$ varies from 2 to 3.5. Use available graphical data where possible; show all steps.

20. A lumped capacitance is connected across the terminals of the antenna in Problem 19 of such value that the antiresonant resistance of the combination is equal to the antiresonant resistance the same antenna would have if driven by a slice generator instead of the line. Determine the required capacitance and compare the resistance of the combination with the resistance of the same antenna driven by a slice generator over the range $\beta_0 h = 2$ to $\beta_0 h = 3.5$.

21. (a) Using the experimental data of Table 38.5 and Figs. 38.12a,b,c, prepare curves of the ideal input resistance and reactance R_0 and X_0 of a thick antenna ($a/\lambda_0 = 1.59 \times 10^{-2}$) driven by an idealized slice generator or discontinuity in scalar potential. (Note that R_0 and X_0 are obtained from R_{sa} and X_{sa} in the limit as the ratio b/a approaches unity. By plotting R_{sa} and X_{sa} as functions of b/a and then extrapolating to $b/a = 1$, R_0 and X_0 are obtained.) Plot $\Omega = 2 \ln (2h/a)$ with R_0 and X_0.

(b) Prepare curves of G_0 and B_0 for the same antenna, using the data of Table 38.5 and Fig. 38.18a,b, or using values computed from R_0 and X_0 determined in (a).

22. From R_0, X_0, G_0 and B_0 as obtained in Problem 21 determine the following:

(a) The resonant length, resistance, and conductance and the associated value of Ω;

(b) The resistance and reactance at $\beta_0 h = \pi/2$ and the associated value of Ω;

(c) The antiresonant length, resistance, and conductance, and the associated value of Ω.

(d) Using the limiting experimental values obtained above for a thick antenna and the corresponding theoretical values in Tables 30.1 and 30.2 for thinner antennas, plot curves of the eight quantities in (a), (b), and (c) as functions of Ω in the range from $\Omega = 20$ down to the values of Ω obtained above.

23. Discuss the validity and the significance of the following statements: The variational method of Storer (Sec. 39) is essentially the same as the emf method (Sec. 40). The greater accuracy of Storer's formula (39.20) as compared with the emf formula (40.12) is entirely due to the greater accuracy of the distribution of current assumed as trial function by Storer as compared with the sinusoidal current of the emf method.

24. Discuss the following statement: The variational method (Sec. 39) is useful as a method for determining the impedance of an antenna only if a good approximation of the distribution of current is known either from experimental measurement or from a theoretical analysis based on a different method, for example, the method of iteration (Secs. 11ff).

Chapter III

1. Two identical, parallel antennas are connected by two-wire lines to lumped impedances and generators as shown in Fig. III.1. Determine the impedance seen by V_1^e in each of the following cases. Ignore terminal-zone effects, as well as ohmic losses in antennas, lines, and connecting wires. Use $\Omega = 2\ln (2h/a) = 15$ for the antennas; $R_c = 276.3$ ohms for the lines; the frequency is 150 Mhz.

(a) $\beta_0 h = \pi/2$, $b_a = 0.2\lambda_0$, $V_{20}^e = 0$, $s_1 = \lambda_0/4$, $s_2 = \lambda_0/4$, $Z_1 = 0$, $Z_2 = \infty$;

(b) $\beta_0 h = \pi/2$, $b_a = 0.4\lambda_0$, $V_{20}^e = V_{10}^e$, $s_1 = \lambda_0/2$, $s_2 = \lambda_0/2$, $Z_1 = -jX_{s1}$, $Z_2 = -jX_{s2}$;

(c) $\beta_0 h = 3.4$, $b_a = 0.1\lambda_0$, $V_{20}^e = 0$, $s_1 = \lambda_0/2$, $s_2 = \lambda_0/2$, $Z_1 = 0$, $Z_2 = 0$;

(d) $\beta_0 h = 3.4$, $b_a = 0.1\lambda_0$, $V_{20}^e = 0$, $s_1 = \lambda_0/2$, $s_2 = \lambda_0/4$, $Z_1 = 0$, $Z_2 = 0$;

(e) $\beta_0 h = 3.4$, $b_a = 0.5\lambda_0$, $V_{20}^e = -V_{10}^e$, $s_1 = \lambda_0/2$, $s_2 = \lambda_0/2$, $Z_1 = 0$, $Z_2 = 0$.

884 THEORY OF LINEAR ANTENNAS [Problems, III]

Fig. II.1.

Fig. II.2.

Fig. II.3.

Fig. II.5.

Fig. III.1.

Fig. III.6.

Fig. III.8.

Fig. III.11.

(a) (b) (c)

Fig. III.10.

2. Determine the apparent impedance Z_{sa} of antenna 1 as a termination for the two-wire line in parts (c), (d), and (e) of Problem 1 if b/a for the lines is 10 and a is as for the antennas.

3. An important directional array consists of two parallel antennas as shown in Fig. III.1 with $h = \lambda_0/4$, $b = 0.1\lambda_0$, $V_2^e = 0$, s_2 adjustable, $Z_2 = 0$. The current in antenna (2) is to lag that in antenna (1) by 0.4 period so that the radiation fields due to the two currents are in opposite phase at distant points in the direction toward the driven antenna from the parasite (used as a director). Determine Z_{CD} looking into the line, s_2, the relative magnitudes of the currents at A and C, and the impedance Z_{AB}. Neglect terminal-zone effects; assume $\Omega = 15$ for the antennas, $R_c = 400$ ohms for the line. Neglect ohmic losses.

4. A modification of the directional array of Problem 3 involves the following: $h = \lambda_0/4$, $b = 0.25\lambda_0$, $V_2^e = 0$, s_2 adjustable, $Z_2 = 0$. The current in antenna (2) is to lead that in antenna (1) by 0.25 period so that the radiation fields of the two antennas are in phase at distant points in the direction from parasite (used as a reflector) to the driven antenna. Determine Z_{CD} looking into the line, s_2, the relative magnitudes of the currents at A and C, and the impedance Z_{AB}. Assume $\Omega = 15$; $R_c = 400$.

5. Determine the effect on the impedance Z_{AB} in Problems 3 and 4 if antenna (1) is a very closely spaced folded dipole instead of a simple dipole. Assume that the significant interaction between the antennas involves only the symmetric currents in the folded dipole.

6. Investigate the impedance terminating the transmission line in Fig. III.6 if $\Omega = 15$, $f = 150$ Mhz, $\beta_0 b = 2\pi b/\lambda_0 = 0.3$, and for the transmission line, $R_c = 440$ ohms, the attenuation constant $\alpha = 2.26 \times 10^{-3}$ neper/m,
(a) neglecting all terminal-zone effects;
(b) taking account of terminal-zone effects.

7. Determine the distributions and magnitudes of the symmetric and antisymmetric currents in the antenna of Fig. III.6 under the conditions specified in Problem 6(a). What power is dissipated as heat and what power is radiated? What is the efficiency of the antenna as a radiator? How are these affected by terminal-zone effects?

8. Investigate the impedance terminating the transmission line in Fig. III.8 under the same conditions specified in Problem 6(a) and (b).

9. Determine the distributions and magnitudes of the symmetric and antisymmetric currents and the heat losses per watt of radiated power in the antenna of Problem 8.

10. Determine the impedances at the terminals AB of the arrays shown in Fig. III.10. Assume spacing b of the line so small that terminal-zone effects are negligible. The electrical distance between the antennas is $\beta_0 b_a = 3.0$, but the electrical length of the transmission line connecting the terminals of one antenna with those of the other is π. For each antenna, $\Omega = 10$ and $\beta_0 h = 3.157$. Sketch zeroth-order distributions of current on the antennas and lines.

11. Determine the impedance terminating the two-wire line at AB in the array of Fig. III.11 in which the electrical distance between the identical driven antennas is $\beta_0 b_1 = 1.0$ and the electrical distance $\beta_0 h_2$ between the plane containing the driven antennas and the plane containing the parasites is chosen to produce maximum field in the direction away from the parasites (reflectors). All antennas have the same electrical length, $\beta_0 h = \pi/2$. Terminal-zone effects are negligible.

12. A three-element array consists of a central driven antenna No. 1 of electrical half length $\beta_0 h = \pi/2$ and two parasites 2 and 3, used, respectively, as reflector and director. The reflector, No. 2, is adjusted in length so that $X_{22} = 20$; it is at a distance $b_{12}/\lambda_0 = 0.17$ from the driven element. The director, No. 3, is adjusted in length so that $X_{33} = -10$; it is at a distance $b_{13}/\lambda_0 = 0.046$; for all antennas $\Omega = 10$.
(a) Determine the electrical half-lengths $\beta_0 h_2$ and $\beta_0 h_3$.
(b) What is the input impedance Z_{10} of the array if line spacings and terminal-zone effects are negligible? (Use a mean value for mutual impedances of antennas of different lengths; use the self-impedance appropriate to the length of each antenna).

13. Six identical parallel center-driven antennas are uniformly spaced about the circumference of a circle of diameter $D = 2\lambda_0/\pi$. For each antenna $\Omega = 12.5$, and $\beta_0 h = 2.888$. All antennas are driven in phase with currents of equal magnitudes. What is the input impedance of each antenna. Neglect terminal-zone effects.

14. An array consists of five parallel identical antennas spaced $\lambda_0/2$ along a line. For each antenna $\beta_0 h = \pi/2$ and $\Omega = 12.5$. If all are driven by equal voltages in phase, determine the relative input currents. (*Note:* For separations b greater than λ_0, it is a good approximation to use the zeroth-order values of mutual impedance as given in Table 8.1. The best value of the self-impedance is that obtaining in the presence of the nearest antenna.)

15. A Tchebyscheff broadside array (Chapter VI) consisting of eight parallel antennas in line (with $\beta_0 h = \pi/2$, $\Omega = 10$, $\beta_0 b = \pi$) is to have the

currents in all elements in phase but their amplitudes are to be tapered outward from the central pair in the ratios 1 : 0.76 : 0.41 : 0.15 in both directions.

(a) Determine the input impedance of each antenna (See note in Problem 14).

(b) Obtain the impedances that must be connected in series with the outer elements in order to obtain the required currents with equal voltages applied to all antennas.

(c) As applied to a practical transmitter criticize the method of connecting impedances in series with some of the antennas.

(d) Design a network for driving the antennas as required that does not have the disadvantages of the series impedances.

16. A three-antenna array for broadcast transmission with a predetermined horizontal field pattern is arranged as shown in Fig. III.16. For all antennas $\beta_0 h = \pi/2$ and $\Omega = 10$. The ground system is sufficiently extensive so that the earth may be assumed to be a perfectly conducting plane for determining impedances. The operating frequency is 1280 khz. The electrical distances between the three antennas are $\beta_0 b_{12} = 286°$, $\beta_0 b_{13} = 182°$, $\beta_0 b_{23} = 303°$. The relative currents required for the desired field pattern are

$$\frac{I_1}{I_2} = \frac{I_3}{I_2} = 1.6 \angle 60°.$$

(a) Determine the input impedance of each antenna.

(b) Design a network for driving the three antennas from a single generator using a matched 300-ohm line.

17. A corner-reflector array consists of five parasitic elements arranged in a V with a 120° angle. The distances between elements is $\lambda_0/6$. The center-driven antenna is at a distance $\lambda_0/3$ from the apex of the V as shown in Fig. III.17. All antennas have half-lengths $h = \lambda_0/2$. The ratio of half-length to radius is $h/a \doteq 75$ or $\Omega = 10$.

(a) Determine the impedance of the array if it is driven by a discontinuity in scalar potential.

(b) Determine the magnitudes and phases of all currents referred to the current in the driven element.

18. A corner-reflector antenna consists of a half-wave dipole ($\Omega = 10$) with a large 90° metal-sheet reflector. Determine the impedance of the antenna as a function of the distance b from the driven antenna to the apex of the V from $b = 0.1\lambda_0$ to $b = 0.5\lambda_0$ assuming the reflecting sheets to be infinite.

19. A Yagi-Uda antenna (Fig. III.19) consists of a center-driven dipole, a reflector at a distance of $0.15\lambda_0$, and three directors equally spaced at $0.1\lambda_0$. The electrical half-length of the driven dipole is $\beta_0 h = \pi/2$; that of the reflector is $\beta_0 h_{\text{ref}} = 2$; that of each of the identical directors $\beta_0 h_{\text{dir}} = 1.25$. For all elements, $\Omega = 2 \ln (2h/a) = 10$. Determine the driving-point impedance and the currents at the center of each antenna.

20. Repeat Problem 19 with the driven dipole replaced by a driven *folded* dipole of equal length with its two members separated by a distance $0.05\lambda_0$.

21. Determine the zeroth-order impedance of a collinear antenna consisting of five identical elements each $\lambda_0/2$ in length. The middle unit is center-driven from a two-wire line of very small spacing; the outer units are connected together and to the central unit by high-impedance two-wire-line stubs, also with very small spacing. Let $\Omega = 10$ for each element.

22. (a) Determine the impedance and the distribution of current on the folded half-dipole (or folded unipole) in Fig. III.22. The electrical length of each conductor is $\beta_0 h = \pi/2$ with $h/a = 75$. The distance between centers is $b_a/\lambda_0 = 0.02$. The ground plane is effectively infinite and perfectly conducting. Assume b/a for the coaxial line so small that transmission-line end effects are negligible.

(b) Repeat (a) with $\beta_0 h = 1$.

(c) What is the apparent impedance in (a) if the ratio of b/a for the coaxial line is 10?

23. A sectionalized tower antenna is represented schematically in Fig. III.23. It consists essentially of the following parts: (1) Generator V_1 drives a $\lambda_0/4$ closed-end section of line of length $\lambda_0/4$ with an extension of length $\lambda_0/4$. (2) Generator V_2 maintains a voltage across the insulating gap at the center of the large conducting cylinder and excites the entire antenna as a sleeve dipole. (3) Generator V_3 excites the horizontally polarized FM antenna at the top. Ignoring the presence of the FM antenna but not of its feed cable, determine the following:

(a) The approximate impedance seen by the generator V_1 and the approximate distribution of the in-phase and quadrature components of current due to this generator. (*Hint*: Show that for V_1 the antenna is approximately equivalent to a base-driven antenna of length $\lambda_0/2$. Assume $\Omega = 10$.)

(b) The impedance seen by V_2 (assuming Z_c for the coaxial line to be 50 ohms) and the distribution of the in-phase and quadrature components of current along the antenna. (*Hint*: Show that the current due to V_2 is a superposition of currents on two asymmetrically driven antennas of length λ_0 driven $\lambda_0/4$ from each end. Use Fig. II.26.11

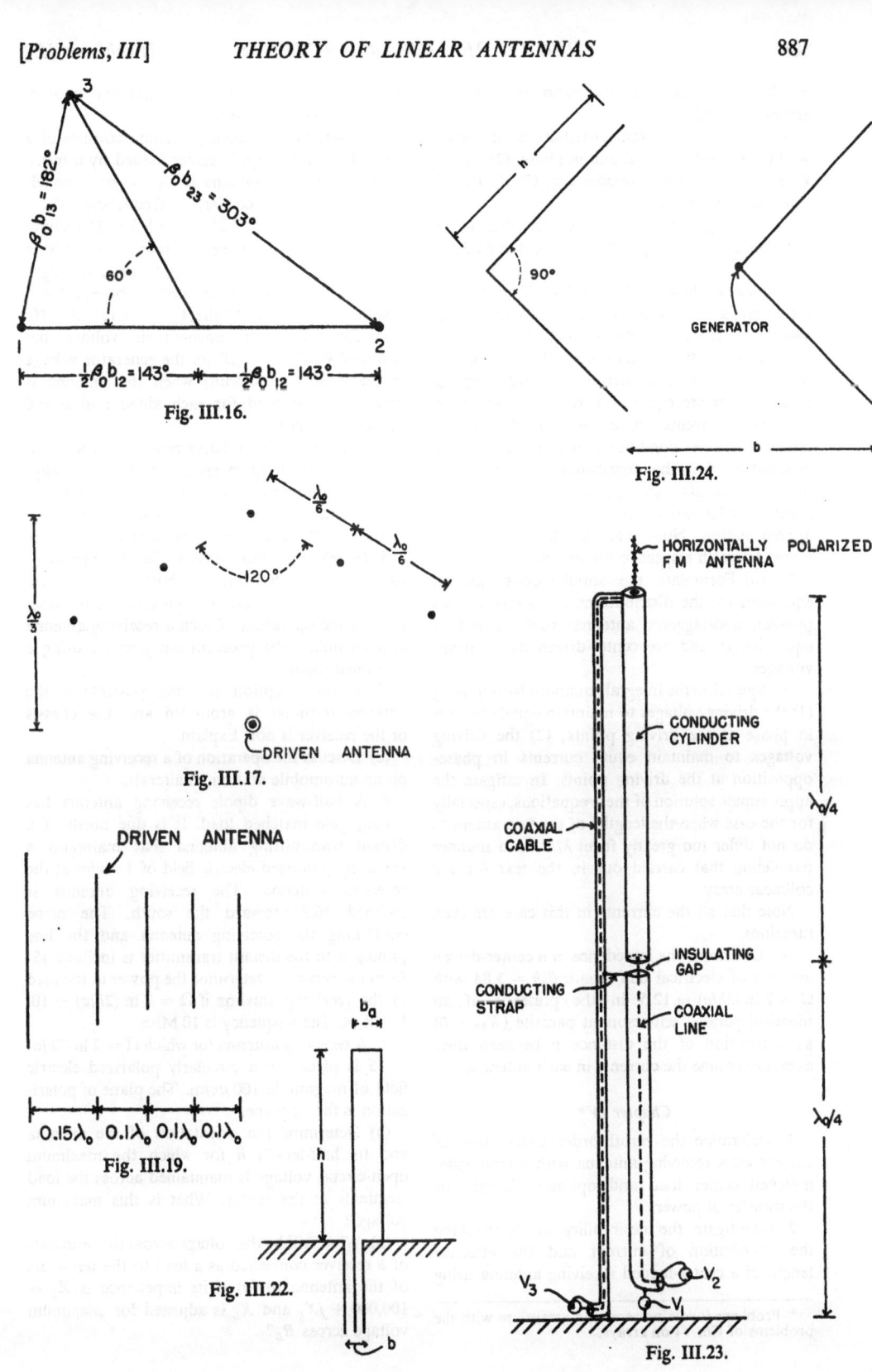

Fig. III.16.

Fig. III.24.

Fig. III.17.

Fig. III.19.

Fig. III.22.

Fig. III.23.

to determine the required ratio of currents; assume $\Omega = 10$.)

(c) Sketch the resultant distributions of current with (1) V_1 and V_2 equal and in phase, (2) V_1 and V_2 equal and in phase quadrature, (3) V_1 and V_2 equal and 180° out of phase.

24. Investigate the problem of determining the impedance of an apex-driven right-angled V-antenna in the presence of a similar parasitic V-antenna, as shown in Fig. III.24. (*Hint*: Reduce the analysis to that of two approximately *independent* integral equations like that for an isolated V-antenna by first determining the impedances when both antennas are driven (1) so that currents in the apices are equal and in phase and (2) so that these currents are equal and 180° out of phase. Assume conditions which permit the assumption that the *distributions* of current in the two antennas are approximately the same. Specify under what conditions this is a good approximation. Note that the distributions are symmetric with respect to the apices.)

25. (a) Formulate the simultaneous integral equations for the distributions of current in two parallel, nonstaggered antennas that are not of equal length and are center-driven by arbitrary voltages.

(b) Specialize the integral equations by requiring (1) the driving voltages to maintain equal currents in phase at the driving points; (2) the driving voltages to maintain equal currents in phase-opposition at the driving points. Investigate the approximate solution of these equations, especially for the case when the lengths of the two antennas do not differ too greatly from $\lambda_0/2$, in a manner paralleling that carried out in the text for the collinear array.

Note that all the currents in this case are even functions.

26. Determine the impedance of a center-driven antenna of electrical half length $\beta_0 h = 3.84$ with $\Omega = 2 \ln (2h/a) = 12.5$ in the presence of an identical parallel self-resonant parasite ($X_{22} = 0$) as a function of the distance b between their axes. Determine the currents in each antenna.

Chapter IV*

1. Determine the zeroth-order distribution of current on a receiving antenna with a conjugate-matched center load and optimum length for the transfer of power.

2. Investigate the desirability of determining the distribution of current and the effective length of a center-loaded receiving antenna using one expansion parameter for $u_0(z)$ and another expansion parameter for $v_0(z)$.

3. A television receiving antenna consists of a single horizontal dipole center loaded by a transmission line. The antenna is for use on channels 1 to 6 with video carriers of frequencies 45.25, 55.25, 61.25, 67.25, 77.25, 83.25 Mhz. The sound carrier frequency in each channel is 4.5 Mhz higher than the video. The electrical half-length $\beta_0 h$ of the antenna is adjusted to be $\pi/2$ for a middle frequency of 66 Mhz; $\Omega = 2 \ln (2h/a) = 10$.

(a) Determine the open-circuit voltage per unit field $V_0(Z_L = \infty)/E$ (or the generator voltage in the equivalent circuit) when the antenna is parallel to the field for each video and sound carrier frequency.

(b) Determine the relative powers to the video load at all six frequencies if the load is always conjugate matched. End effects are negligible.

4. Repeat the preceding problem if $\beta_0 h = \pi$ at the middle frequency of 66 Mhz and $\Omega = 10$.

5. (a) Many radio receivers in the broadcast band are operated with a short piece of wire for an antenna and no ground connection. Discuss the operation of such a receiving antenna and formulate the problem analytically, using a simplified model.

(b) Good reception is often possible if the antenna terminal is grounded and the chassis of the receiver is not. Explain.

(c) Discuss the operation of a receiving antenna on an automobile and on an aircraft.

6. A half-wave dipole receiving antenna has a conjugate matched load. It is due north of a distant transmitting antenna that maintains a vertically polarized electric field of 1 mv/m at the receiving antenna. The receiving antenna is inclined 36.9° toward the south. The plane containing the receiving antenna and the line joining it to the distant transmitter is inclined 45° from the vertical. Determine the power in the load of the receiving antenna if $\Omega = 2 \ln (2h/a) = 10$, $h = \lambda_0/4$. The frequency is 10 Mhz.

7. A receiving antenna for which $\Omega = 2 \ln (2h/a) = 15$ is placed in a circularly polarized electric field of magnitude 100 μv/m. The plane of polarization is the yz-plane.

(a) Determine the orientation of the antenna and its half-length h for which the maximum open-circuit voltage is maintained across the load terminals at the center. What is this maximum voltage?

(b) What will be the voltage across the terminals of a receiver connected as a load to the terminals of the antenna in (a) if its impedance is $Z_L = 100{,}000 + jX_L$ and X_L is adjusted for maximum voltage across R_L?

---

* Problems involving receiving arrays are with the problems of Ch. VI on arrays.

(c) What will be the voltage across Z_L if the conditions of (b) obtain but the antenna is inclined at 53° from the xy-plane?

8. A receiving antenna of half-length $h = \lambda_0/4$ and radius such that $\Omega = 2 \ln(2h/a) = 20$ is center-loaded by an impedance $Z_L = 4.5 + j60$ ohms. The current in Z_L is zero when the antenna lies along a horizontal east-west axis. As the antenna is rotated in the vertical plane containing north and south, a maximum current of 60 μa is observed in Z_L when the antenna is vertical, a minimum value of 20 μa when the antenna is horizontal.

(a) What is the polarization of the electric field at the receiving antenna?

(b) What is its magnitude in volts per meter? Give maximum and minimum values.

(c) In what direction is the distant transmitter?

9. A center-loaded receiving antenna of optimum length for transfer of power to a conjugate-matched load is connected to the load by a section of transmission line of length λ_0. For the antenna $\Omega = 2 \ln(2h/a) = 12.5$; for the line, $b/a = 10$; the antenna and the line are made of the same size conductor. With due regard for transmission-line end and coupling effects, what must be the value of the *lumped* impedance terminating the line? Assume all conductors perfect. (*Hint*: Replace the actual line with end effects by a uniform line with terminal-zone networks at each end.)

10. The electric field at a particular location in space is given by $(E_x)_{\text{inst}} = 0$, $(E_y)_{\text{inst}} = E_y \cos \omega t = kE_z \cos \omega t$, $(E_z)_{\text{inst}} = E_z \sin \omega t$, where $k \equiv E_y/E_z$. A symmetric receiving antenna with conjugate-matched load and optimum length from the point of view of power transfer to the load is rotated step by step about the x-axis through its mid-point. If E_z is 1 μv/m, determine the power to the load as the antenna is turned from parallel to the z-axis to parallel to the y-axis in appropriate steps for each of the following values of k: $k = 0, 0.25, 0.50, 0.75, 1.0$. Assume $\Omega = 2 \ln(2h/a) = 10$. Plot power to the load as a function of Θ with k as parameter.

11. Plot the effective length per wavelength $h_e(\Theta_2)/\lambda_0$ in the complex plane for $\Omega = 12.5, 15, 20$ with $\Theta_2 = \pi/2$. (The curve for $\Omega = 10$ is given in Fig. 9.6c.) Discuss the significance of the fact that no curve can be plotted for $\Omega = \infty$.

12. Two identical symmetric antennas each of half-length h are crossed at right angles and are connected in series with each other and a lumped load impedance Z_L. This crossed array is used in a horizontal position as an omnidirectional receiver for a horizontally polarized electric field. Verify that the array is omnidirectional by plotting the current in the load as a function of the angle of rotation about the vertical axis through the center of the array. Assume $\Omega = 12.5$, when h is adjusted for antiresonance. Choose Z_L for maximum transfer of power.

13. A straight receiving antenna of radius a has a lumped load Z_L connected in series with the antenna at a distance h_1 from one end and a distance h_2 from the other end. Derive a formula for the current in Z_L when the antenna is parallel to a linearly polarized electric field E. (*Hint*: Apply Thévenin's theorem as for the center-loaded antenna and use Sec. III.29 in determining the impedance of the antenna. The problem is thus reduced to finding the open-circuit voltage and this involves essentially only the determination of an effective length in terms of the current.)

14. Investigate the formulation of the preceding problem when the antenna is oriented arbitrarily with respect to the electric field.

15. A horizontal television receiving antenna consists of a symmetric, right-angled, apex-loaded V with legs of length h. Determine the open-circuit voltage V_0 ($Z_L = \infty$) for use in an equivalent circuit in which this voltage is in series with the load impedance Z_L and the impedance of the antenna as given in Sec. III.25. Assume the incident electric field to be horizontally polarized. Discuss the directional properties of this antenna using the zeroth-order effective length.

16. Two identical antennas are placed side by side at a distance b that is very small compared with the wavelength. Identical lumped loads Z_L are connected at the centers of the two antennas and these are oriented to be parallel to and at each instant in the same surface of constant phase of the electric field E of a distant transmitter. Determine the effective voltages and impedances in the equivalent circuits for the two antennas.

17. (a) Referring to the antennas in the preceding problem, what is the effect of bending their ends together to form conducting bridges joining the antennas at each end?

(b) When the ends of the antennas are joined to form a narrow rectangle of conductors as in (a), the load Z_L at the center of *one* of the long sides of the rectangle is replaced by a short circuit. Determine the effective voltage and impedances in an equivalent circuit involving the load Z_L at the center of the other long side of the rectangle. Consider especially the case when the length of the rectangle is such that the impedance looking into each half from the center is very great.

18. Determine the absorption cross sections for the following antennas with $\Theta_2 = 90°$; neglect the ohmic resistance of the antenna:

(a) $\Omega = 20$, $h = \lambda_0/4$, $Z_L = Z_0^*$;

(b) Same as (a) with $\Omega = 10$;

(c) Same as (a) with $h = \lambda_0/2$.

(d) What is the effective reradiation or scattering cross section in each of the above cases?

Chapter V

1. Prove that the vector potential (as defined by the Helmholtz integral) of a center-driven short antenna ($\beta_0^2 h^2 \ll 1$) is in first approximation a function of R_0 alone.

2. Determine the far-zone field of an electrically short antenna ($\beta_0^2 h^2 \ll 1$) with the following distributions of current:
(a) $I_z = I_0(1 - |z/h|)$ for $\Omega \to \infty$;
(b) $I_z = I_0(e^{-\delta|z/h|} - |z/h|^n e^{-\delta})$ with $\delta = 1.46$, $n = 4$ for $\Omega = 10$.

3. Using the theorem of images, prove that the impedance of a base-driven thin cylindrical antenna of full length h and radius a erected on a perfectly conducting half-space is exactly one-half the impedance of the isolated center-driven antenna of half-length h and radius a. Assume each antenna to be driven by a slice generator. Note that the generator must also have its image.

4. The length $2h$ of a center-driven multiple half-wave antenna is increased in steps as follows: $h/\lambda_0 = 0.25, 0.75, 10.75, 20.75, 50.75, 100.75$. For each length determine the following for the essentially sinusoidally distributed quadrature component of current:
(a) The angle Θ_m for the principal ear (nearest $\Theta = 0°$).
(b) The nulls Θ_0 which bound this ear, and the null beam width.
(c) The radiation resistance of the antenna with respect to the quadrature component of current.
(d) The fraction of the total radiated power due to the quadrature component of current associated with the principal ear.

Discuss the directional properties of the multiple half-wave antenna as its length is increased.

5. Determine the radiating efficiency for the quadrature component of current in the antenna in Problem 4 if it is made of No. 14 copper wire and operated at 300 Mhg.

6. A thin cylindrical antenna 100.75 wavelengths long is perpendicular to and base-driven over an infinite perfectly conducting plane. The component of current in phase quadrature with the driving voltage is distributed essentially sinusoidally along the antenna. (a) Determine the angles Θ_m giving the direction of the maximum field of the principal ear nearest to $\Theta = 0$, and of one ear approximately half-way between the principal ear and the ear at $\Theta = 90°$. (b) Determine the pair of angles Θ_0 for the nulls bounding each of these three ears. (c) Determine the relative magnitudes of the maximum electric field for each of the three ears.

(d) Determine the fraction of the total power radiated that is associated with each of the three ears.

7. Determine the reradiated far-zone field due to the zeroth-order current

$$(I_z)_0 = \frac{j4\pi U}{\zeta_0 \Psi} \frac{\cos\beta_0 z - \cos\beta_0 h}{\cos\beta_0 h}$$

for unloaded receiving antennas of electrical half lengths $\beta_0 h = \pi, 2\pi$. Compare these fields with those of center-driven antennas of the same length with sinusoidal currents.

8. A right-angled apex-driven V-antenna has identical legs each of electrical length $\beta_0 h = \pi$. Assuming the distribution of current to be sinusoidal, investigate the near-zone electric field in the plane of the V.

9. A two-element broadside array consists of two identical center-driven antennas each of length $\lambda_0/2$ placed parallel to each other at a distance $b = \lambda_0/2$. The currents in the two are equal, in phase, and distributed essentially sinusoidally. Investigate the near-zone electric field in (a) the plane of the antennas, (b) the plane perpendicular to the plane of the antennas and passing midway between them.

10. Repeat Problem 9 for the bilateral end-fire array in which the currents are equal but 180° out of phase.

11. Investigate the near-zone field of a unidirectional couplet consisting of two parallel identical antennas of length $2h = \lambda_0/2$ driven so that the currents are equal and 90° out of phase. Assume the distributions of current to be sinusoidal.

12. Prepare curves of the electric lines of an antenna of length $3\lambda_0/2$ with sinusoidal current corresponding to those of Fig. 11.6 for the antenna of length $\lambda_0/2$.

13. Discuss the excitation of a slot antenna by a two-wire line and by a current-carrying conductor across its center as analogs of a center-driven antenna.

14. Compare the definition $h_t = b_1/\beta_0$ for the effective length of a transmitting antenna that has an actual length in the range $0 \leq h \leq 2/\beta$ and is erected vertically on an infinite perfect ground with the conventional definition (Terman, *Radio Engineers' Handbook*, p. 841, footnote 1), "The effective height of a grounded vertical-wire antenna is the height that a vertical wire would be required to have to radiate the same field along the horizontal as is actually present if the wire carries a current that is constant along its entire length and of the same value as at the base of the actual antenna." Discuss the differences,

advantages, and disadvantages of the two definitions.

15. Investigate the use of an effective length for a transmitting antenna that uses the first two terms in the Fourier series (13.12). For what lengths of antenna is this a good approximation? Is it a more convenient function in determining field pattern and radiation resistance than $F_m(\Theta, \beta_0 h)$?

16. In broadcast engineering the effective height of a transmitting antenna is often given by the equation

$$h_t = \frac{Ed}{1.25 f I_m},$$

where h_t is in meters, E is the electric field in microvolts per meter at a distance d in kilometers from the antenna, I_m is the current in amperes at an antinode, and f is the frequency in kilocycles per second.

(a) Derive this equation together with restrictions on its generality.

(b) Discuss its application to a top-loaded antenna which has an actual electrical length of 132° and a very large top load equivalent to 58°.

17. Verify formula (16.7) and plot field patterns of $|E_\Theta^r|$ for $\beta_0 h = \pi$ and $5\pi/4$ for an antenna with $\Omega = 2 \ln (2h/a) = 15$ using the numerical values of the parameters A and B given in Table II.39.1. Compare with Figs. 14.2, 14.3, and 16.1.

18. Derive the formula for the magnitude of $F_a(\Theta)$ for an unbalanced center-driven antenna when the current ratio factor k is complex. Calculate and plot the factor as a function of Θ for $k = 0.5 + j0.5$ with $h = 3\lambda_0/4$.

19. Determine the vertical field pattern of the tower antenna described in Problem III.23 using the currents there determined for (1) $V_2 = V_1$, (2) $V_2 = jV_1$, (3) $V_2 = -V_1$.

20. Derive the far-zone vertical field factor

$$F(\Theta, \beta_0 h, \beta_0 k) = (1/\sin \Theta)[\cos \beta_0 k \cos (\beta_0 h \cos \Theta)$$
$$- \cos \Theta \sin \beta_0 k \sin (\beta_0 h \cos \Theta) - \cos \beta_0 (h + k)]$$

for a top-loaded antenna over a perfectly conducting infinite plane. The assumed distribution of current is $I_z = I_0[\sin \beta_0 (h + k - z)]/\sin \beta_0 (h + k)$, where $\beta_0 h$ is the actual electrical height of the antenna and $\beta_0 k$ is the equivalent electrical length of a compact top load that contributes negligibly to the far-zone field.

21. (a) Using the formula given in Problem 20, calculate and plot the far-zone electric field of a top-loaded antenna over an infinite conducting plane. Use $\beta_0 h = 132°$ and $\beta_0 k = 58°$.

(b) Discuss the validity of making use of either of the two definitions of effective height for transmission given in Sec. 13 or of the formula in Problem 16 for representing the far-zone field of the antenna specified in (a).

(c) Ignoring restrictions on the use of the three formulas for effective height of a transmitting antenna referred to in (b), calculate and plot the far-zone electric field using each of these formulas. Compare with the field plotted in (a). In particular, compare the fields along the surface of the conducting plane. Discuss the results critically.

22. Derive the following formula for the radiation resistance referred to maximum sinusoidal current of a top-loaded antenna over a perfectly conducting infinite plane with a distribution of current given by

$$I_z = I_m \sin \beta_0 (h + k - z)$$

and with negligible radiation from the top load:

$$R_m^e = (\zeta_0/8\pi)\{-\cos \beta_0(h + k) \, Cin \, 4\beta_0 h$$
$$+ 2[1 + \cos \beta_0(h + k)] \, Cin \, 2\beta_0 h$$
$$+ \sin 2\beta_0(h + k)(Si \, 4\beta_0 h - 2 \, Si \, 2\beta_0 h)$$
$$+ 2 \sin^2 \beta_0 h[(\sin 2\beta_0 h)/2\beta_0 h - 1]\}.$$

(Note that this formula reduces to one half of (12.6a) when $k = 0$, as it should.)

23. Calculate the radiation resistance of the antenna described in Problem 21(a) and compare it with $R_m^e(h + k) - R_m^e(k)$ obtained from the radiation resistance $R_m^e(h + k)$ of an antenna of electrical length $\beta_0 h_1 = \beta_0(h + k) = 132° + 58° = 190°$ over a conducting plane, minus the radiation resistance $R_m^e(k)$ of an antenna of electrical length $\beta_0 k = 58°$ over a conducting plane. Discuss the results.

Chapter VI

1. The complete normalized space factor of a two-element broadside array with center-driven elements one wavelength long and separated by a half wave-length is given by

$$a(\Theta, \Phi) = F_m(\Theta, \beta_0 h) \, a(\Theta, \Phi; N, n, t)$$
$$= F_m(\Theta, 180°) \, a(\Theta, \Phi; 2, \tfrac{1}{2}, 0).$$

It is desired to construct a three-dimensional or space model of the field pattern. For this purpose evaluate $a(\Theta, \Phi)$ as a function of Θ in the planes $\Phi = 0, 30°, 60°, 90°$ and as a function of Φ in the plane $\Theta = 90°$ and on the cones $\Theta = 60°$ and $30°$. Construct a three-dimensional drawing of one quadrant of the model showing the contours actually evaluated.

2. The complete normalized space factor of a two-element unilateral end-fire array with center-driven elements one half-wavelength long and separated by a quarter wavelength is given by

$$a(\Theta, \Phi) = F_m(\Theta, 90°)\, a(\Theta, \Phi; 2, \tfrac{1}{4}, \tfrac{1}{4}).$$

Determine the patterns as functions of Θ for $\Phi = 0$, 45°, 90°, 135°, 180° and as functions of Φ for $\Theta = 90°$ and $\Theta = 45°$. Sketch a space model of the complete pattern.

3. Repeat Problem 1 for the two-element bilateral end-fire array given by

$$a(\Theta, \Phi) = F_m(\Theta, 180°)\, a(\Theta, \Phi; 2, \tfrac{1}{4}, \tfrac{1}{2}).$$

4. A unilateral end-fire array consists of four half-wave dipoles numbered 1 to 4 from end to end and separated by a distance $\lambda_0/4$ between elements. The elements are driven so that the currents are all equal in magnitude but with I_{01} leading I_{02} by 90°, I_{02} leading I_{03} by 90°, and I_{03} leading I_{04} by 90°.

(a) Construct the complete field factor $a(\Theta, \Phi)$ in terms of tabulated functions. (*Hint*: Since the tables in Chapter VI include only the factor for the two-element end-fire array, the four-element array may be treated as a two-element array of two-element couplets or as a two-element array of overlapping two-element bilateral end-fire arrays.)

(b) Plot the horizontal field factor in the plans $\Theta = 90°$ as a function of Φ.

5. A rectangular array of sixteen half-wave dipoles is shown in Fig. VI.5. Adjacent rows are $\lambda_0/2$ apart, adjacent columns are $\lambda_0/4$ apart. The array is driven so that the currents in the four units in any *column* are equal and in phase, but the currents of column 1 lead those of column 2 by 90°, those of column 2 lead those of column 3 by 90°, and so on.

(a) Construct the complete field factor (i) in its simplest form, (ii) in terms of tabulated functions.

(b) Plot the pattern in the plane $\Theta = 90°$ as a function of Φ.

(c) Determine the values of Φ for all nulls in the horizontal field pattern. (*Hint*: Note that the nulls of the broadside factor and of the end-fire factor taken independently yield nulls in the product.)

(d) Determine the values of Φ for all extremes in the horizontal plane and the magnitudes of these extremes.

(e) Determine the voltage that must be applied at the terminals of each antenna in order to obtain the required equal currents. Use $\Omega = 2 \ln (2h/a) = 10$. (Note that for separations greater than λ_0 zeroth-order mutual impedances are satisfactory.)

6. An array consists of two curtains of six-element half-wave dipoles. The elements in each curtain form a parallel pair one half-wavelength apart of three-element collinear antennas; all six currents are in phase. The two curtains are separated by a distance $\lambda_0/4$ and the currents in the six antennas of one curtain lead the currents in the six elements of the other curtain by 90°.

(a) Construct the complete field factor.

(b) Plot the pattern in the plane $\Theta = 90°$ as a function of Φ and in the planes $\Phi = 0$, 180° as a function of Θ.

7. A curtain of four half-wave dipoles with elements separated by distances $\lambda_0/2$ is mounted so that the plane of the curtain is parallel to and a distance $\lambda_0/4$ in front of a plane, highly conducting, sheet-reflector, as shown in Fig. VI.7. The currents in all elements are equal in magnitude. Assume the effect of the reflector to be the same as if it were perfectly conducting and infinite in extent. (This is a fair approximation in the half-space containing the driven elements except in directions parallel or nearly parallel to the plane of the reflector and behind the reflector.)

(a) Determine and plot the horizontal field patterns when the progressive phase increase from antenna 1 to 2, 2 to 3, and 3 to 4 is 60° each.

(b) Determine the direction of the principal ear in each case and its null width.

(c) Discuss the usefulness of the array for lobe sweeping or lobe shifting.

(d) Determine the driving voltage required for each antenna if $\Omega = 2 \ln (2h/a) = 10$ for each element.

8. A nonuniform broadside array consists of four half-wave dipoles separated by distances $\lambda_0/2$, driven in phase and with current amplitudes in the ratios $1 : 2 : 2 : 1$.

(a) Construct the complete field factor. (*Hint*: Treat as a superposition of four-element and two-element uniform arrays.

(b) Plot the field pattern as a function of Φ in the plane $\Theta = 90°$.

(c) What voltages must be applied in order to obtain the desired currents. Assume $\Omega = 2 \ln (2h/a) = 15$. (Note that for separations exceeding $\lambda_0/2$ the zeroth-order mutual impedance is satisfactory.)

9. Repeat Problem 8 with the ratios of currents $2 : 1 : 1 : 2$.

10. Repeat Problem 8 with the ratios of currents $1 : 4 : 4 : 1$.

11. The antenna system of WIBG consisted (in 1941) of five parallel and vertical 250-ft antennas spaced 248 ft ($\beta_0 b = 90°$) apart in a straight line over an extensive ground network. The antennas are denoted by A through E from

one end of the array to the other. The approximate currents in amperes are

$$I_A = 3.6\angle-75°, \quad I_B = 10.55\angle 145°,$$
$$I_C = 14.9\angle 0°, \quad I_D = 10.55\angle -145°,$$
$$I_E = 3.6\angle 75°.$$

(a) Determine the field factor of this array. (*Hint*: Treat it as a superposition of the fields of two two-element arrays and a single central unit.)

(b) Plot the horizontal field pattern in volts per meter at 10 km as a function of Φ in the plane $\Theta = 90°$. Plot the minor lobe structure on an enlarged scale. Show the line of the array and the relative locations of antennas A and E. Treat the earth as a perfect conductor.

(c) What voltage or impedance must be connected between each antenna and ground in order to maintain the required currents? The mean ratio of length to equivalent radius for the antennas is approximately 75.

(d) What is the total power supplied to the array?

12. Determine and plot the horizontal field pattern of the array described in Problem III.12.

13. Determine and plot the horizontal and vertical field patterns of the circular array described in Problem III.13. Assume sinusoidal currents.

14. A broadside array of four half-wave dipoles spaced a distance $\lambda_0/2$ is made unidirectional by four parasitic half-wave dipoles of which each is placed $\lambda_0/4$ behind one of the driven elements. For each antenna, $\Omega = 2 \ln (2h/a) = 10$.

(a) Determine the currents in the parasites.

(b) Determine the field factor of the array. (Note that this may be obtained by superimposing the fields of a four-element broadside array, a two-element broadside array consisting of the two inner parasites, and a second two-element broadside array consisting of the two outer parasites.)

(c) Evaluate and plot the field pattern.

15. A three-element horizontal Yagi-Uda array consists of a center-driven antenna 1 of electrical length $\beta_0 h_1 = \pi/2$, a director (antenna 2) of electrical length $\beta_0 h_2 = 1.49$ at a distance $b_d = 0.06\lambda_0$, and a reflector (antenna 3) of electrical length $\beta_0 h_3 = \pi/2$ at a distance $b_r = 0.2\lambda_0$. Assume $\Omega = 10$ for all three antennas.

(a) Determine the currents at the centers of all three antennas if 100 volts are applied to the driven unit.

(b) Determine and plot the field pattern in the two principal planes.

(c) Specify the front-to-back ratio and the gain over a single antenna.

16. In the experimental study of Yagi-Uda arrays described in Sec. VI.6, the spacing of the directors was fixed at $\lambda_0/3$ for which an optimum length of the directors near $0.43\lambda_0$ was obtained when the number of directors was small and $\Omega \doteq 11.4$.

(a) Verify that the self-reactance of each director is nearly -80 ohms for this length and ratio of h/a.

(b) Show that the approximate conditions for optimum, $b/\lambda_0 = 0.33$ $X_{22} \doteq -80$ ohms, are in general agreement with the results obtained in Sec. 4 for the field in the direction toward the parasite but with much smaller values of X_{22} and b/λ_0.

(c) Using theoretical values for b/λ_0 and X_{22} interpolated to $\Omega = 11.4$ from data in Sec. VI.4 and the approximate experimental value $b/\lambda_0 = 0.33$ with $X_{22} \doteq -80$ ohms, draw curves of b/λ_0 for maximum field toward the parasite as a function of X_{22}. Draw a similar curve of b/λ_0 as a function of h/λ_0, where h is the half-length corresponding to a self-reactance X_{22}. Discuss the practical value of these curves.

17. Determine and plot the horizontal field pattern of the three-antenna broadcast array described in Problem III.16.

18. Determine and plot the horizontal field pattern of the corner-reflector array described in Problem III.17.

19. Determine and plot the field patterns in the two principal planes of the Yagi-Uda antenna described in Problem III.19. What is the null beam width?

20. Design a directional array using two directors and a reflector consisting of three parasites in a plane, as shown in Fig. VI.20. Select appropriate distances and lengths. Determine the current in each antenna, the impedance of the array, and its horizontal field pattern. Specify the null beam width and the level of minor lobes nearest the principal beam.

21. Apply Schelkunoff's method of shifting nulls to obtain a more highly directive array with reduced minor lobes from the four-element uniform end-fire array of half-wave dipoles characterized by $N = 4$, $n_E = t_E = 1/4$.

(a) Determine the field pattern in the horizontal plane with uniformly spaced nulls.

(b) Determine the relative currents required in the antennas.

(c) Determine the relative voltages that must be applied to the antennas to obtain the desired currents neglecting ohmic losses, and assuming $\Omega = 2 \ln (2h/a) = 10$ for each antenna.

(d) If the antennas are made of copper, determine the ohmic losses in the array as compared with those in the corresponding uniform array to which the same power is supplied. Assume sinusoidal currents in evaluating ohmic resistance.

22. Repeat Problem 21 for the more closely spaced array with $N = 4$, $n_E = t_E = 1/8$.

23. Design a six-element Tchebyscheff broadside array that has a minor lobe level of 20 db when the half-wave dipole antennas are spaced $\lambda_0/2$. Determine the relative currents required, the field pattern, and the necessary relative driving voltages. Assume $\Omega = 2 \ln (2h/a) = 10$ for each antenna. Compare the field pattern and, in particular, the beam width and minor-lobe level with the six-element uniform array.

24. Design a five-element Tchebyscheff array with antennas spaced $\lambda_0/4$ that has the same null beam width as a uniform five-element broadside array with conventional $\lambda_0/2$ spacing of elements. Determine the currents required and the driving voltages for half-wave dipoles with $\Omega = 2 \ln (2h/a) = 10$. Plot the horizontal field pattern together with that of the uniform array adjusted to have a principal beam of equal amplitude. What is the minor-lobe level in the Tchebyscheff array? How does it compare with that of the greatest minor lobe of the uniform array?

25. The eight-element Tchebyscheff array analyzed in Sec. 13 is to be made unidirectional with a parasitic curtain of eight reflectors. Design the curtain, assuming that the driven elements are $\lambda_0/2$ dipoles with $\Omega = 2 \ln (2h/a) = 10$. Plot the horizontal field pattern.

26. Investigate the combination of Tchebyscheff broadside arrays with the spaced-null end-fire arrays discussed in Sec. 10 to provide unidirectional square arrays.

27. A common television antenna includes two arrays each consisting of a folded dipole with a parasitic reflector, as shown in Fig. VI.27. The array of longer antennas is used to receive the low band of frequencies in the range from 44 Mhz to 88 Mhz including TV channels 1 through 6; the array of shorter antennas is used to receive the high band in the range from 174 Mhz to 216 Mhz including channels 7 through 13.

(a) Design each array for the *middle frequency* in its band with the parasite adjusted in length and in distance from the loaded antenna in order to yield maximum voltage in the equivalent circuit of the loaded receiving antenna. Assume $\Omega = 10$ for each conductor; the two elements of the folded dipole are separated by a distance $0.05\lambda_0$.

(b) Determine the gain of each array over a single dipole for the middle frequency and the extreme frequencies in each band.

(c) Determine the equivalent voltage and the impedance of the array and the power to a 300-ohm load per unit electric field.

28. Investigate the gain over a half-wave dipole of a center-driven antenna of electrical half-length $\beta_0 h = 3.84$ operated at constant power in the presence of an identical center-tuned parallel parasite at a distance b. Assume $\Omega = 12.5$ for both antennas; let X_{22} be maintained equal to zero for the parasite. Obtain the field ratio as a function of $\beta_0 b$ or b/λ_0 over a range from $\beta_0 b = 0.3$ to $\beta_0 b = 2.5$. (Note that the impedance of this array is to be determined in Problem III.26.)

29. Study the operation of the array described in Problem 28 when used for reception with a conjugate matched load replacing the generator. Compare the maximum obtainable effective open-circuit voltage for this array with that (a) for the same array with $\beta_0 h = \pi/2$, instead of 3.84, and (b) for a single half-wave dipole with a conjugate matched load. Which array is more desirable for use with a television receiver?

30. A receiving array consists of two identical horizontal half-wave dipoles ($\Omega = 10$) with their terminals connected by a section of 300-ohm two-wire line of length $\lambda_0/4$, as shown in Fig. VI.30. A 300-ohm load is connected across the terminals AB of one of the antennas. Determine the voltage drop across the load when the antenna is in a linearly polarized electric field of $10 \,\mu v/m$ parallel to the antennas. Neglect ohmic losses and assume the transmission-line spacing sufficiently small to make end and coupling effects negligible.

31. Investigate the behavior of the array in Problem 30 as a receiver for the television band from 174 to 216 Mhz.

32. Investigate the possibility and the desirability of modifying the array in Fig. VI.30 so that if it were driven by a generator across AB the currents in the two antennas would be equal and $90°$ out of phase. No change is to be made in the antennas or their relative positions. What would be the properties of the array thus modified as a receiving antenna with a 300-ohm load at a frequency for which $\beta_0 h = \pi/2$?

33. Repeat Problem 30 if the array is constructed with a more widely spaced transmission line connecting the antennas. The length of the section of line is still $\lambda_0/4$ but the line spacing is $b = 0.05\lambda_0$ so that end and coupling effects are not negligible. Also $b/a = 10$, where the radius a of the conductors is the same as for the antennas. The 300-ohm line from AB to the load is equivalent to a lumped load at the center of a copper bridge made of wire of the same size as the $\lambda_0/4$ section of line.

34. Discuss the advantages or disadvantages of the so-called "double-vee" TV antenna shown

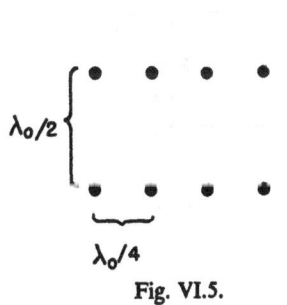

Fig. VI.5.

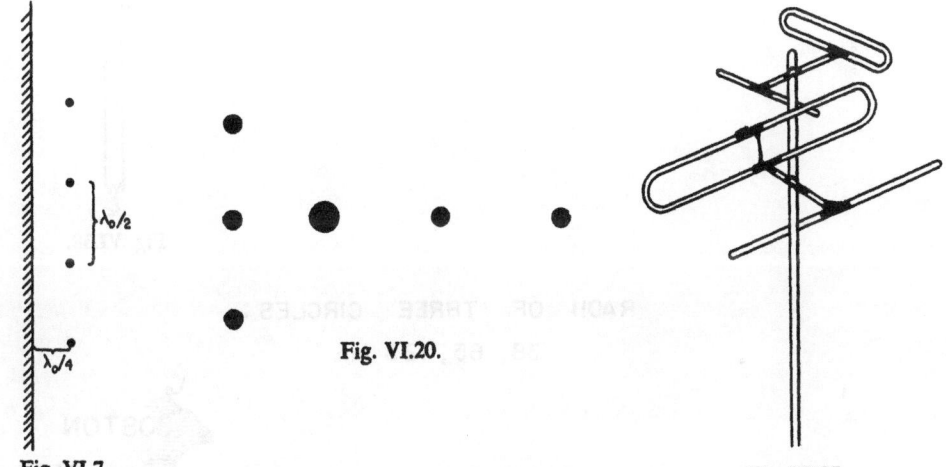

Fig. VI.7.

Fig. VI.20.

Fig. VI.27.

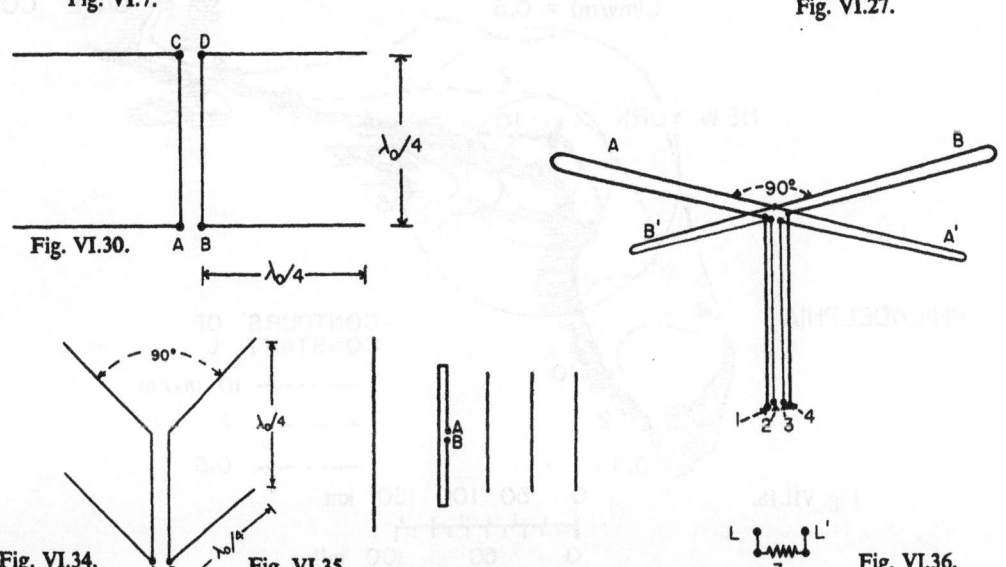

Fig. VI.30.

Fig. VI.34. Fig. VI.35. Fig. VI.36.

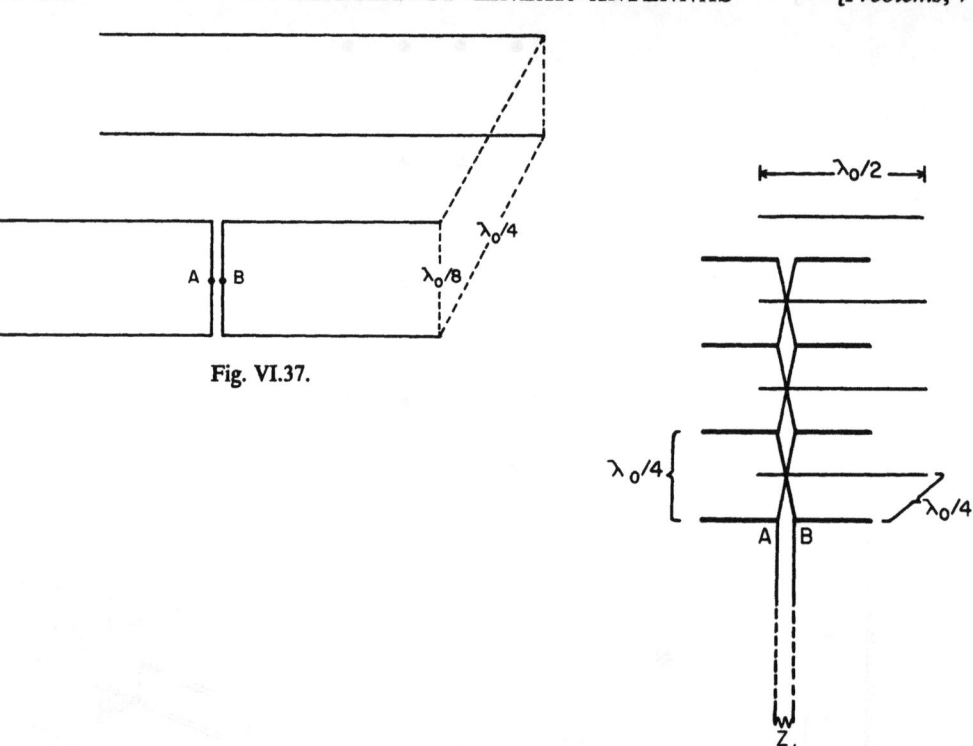

Fig. VI.37.

Fig. VI.38.

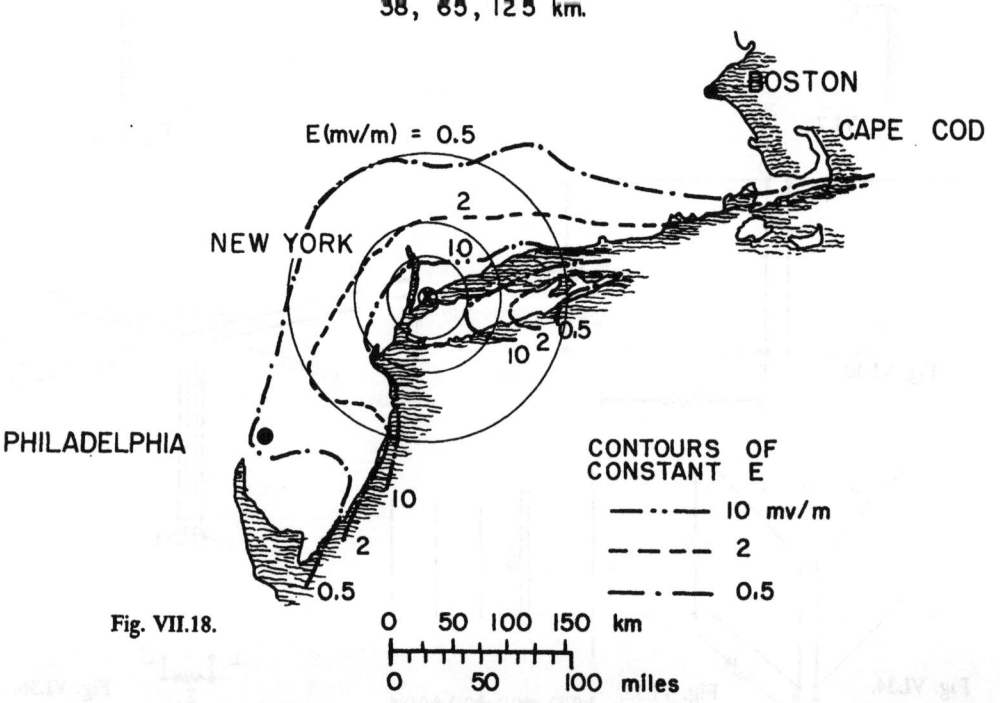

RADII OF THREE CIRCLES:
38, 65, 125 km.

$E(mv/m) = 0.5$

BOSTON
CAPE COD
NEW YORK
PHILADELPHIA

CONTOURS OF CONSTANT E
—··— 10 mv/m
——— 2
—·— 0.5

Fig. VII.18.

in Fig. VI.34 over the corresponding array described in Problem 30; the antenna in Fig. VI.34 is like that in Fig. VI.30 except that the antennas are turned to form two quadrant or 90° V-antennas. Note that the horizontal quadrant antenna is omnidirectional in the horizontal plane. Determine the directional properties in the horizontal plane. Determine the directional properties in the horizontal plane of the array in Fig. VI.34 using the results derived in Problem III.24 to evaluate the relative magnitudes and phases of the currents. Determine the maximum voltage drop across a 300-ohm load and compare with the corresponding voltage obtained in Problem 30 for the array of Fig. VI.30.

35. Assign appropriate lengths to the directors, the folded dipole, and the reflector of the Yagi-Uda array in Fig. VI.35 and determine the receiving properties of the array when a two-wire line leading to the receiver load is connected across AB. Assume appropriate data as required to permit the greatest possible transfer of power to the load for the given field.

36. Investigate the receiving properties of the double folded $\lambda_0/2$ dipole shown in Fig. VI.36. The outputs 1 and 3 of dipole AA' and 2 and 4 of dipole BB' form a four-wire line leading to the receiver. Discuss how the two terminals LL' of the load $Z_L = 300$ ohms should be connected to the four-wire line in order to get maximum current in the load. Specify appropriate constants and dimensions for the antennas and lines. Determine the current in the load as the antenna is rotated through 90° in a linearly polarized horizontal electric field of 10 microvolts per meter.

37. Determine the gain over a single dipole of the so-called "double doublet TV antenna with reflector" shown in Fig. VI.37. The array represented consists of two parasitic reflectors a quarter-wavelength behind a pair of identical antennas connected together by a short section of transmission line $\lambda_0/8$ in length and center loaded across AB by the transmission line to the distant receiver. What are the equivalent voltage and impedance of the array for use in an equivalent series circuit? Assume $\Omega = 2 \ln (2h/a) = 10$ for the antennas and $Z_c \doteq 300$ ohms for the section of line? Assume the load to be conjugate-matched for the array and the single dipole with which it is compared.

38. Determine the approximate current entering and leaving the end-loaded transmission line at A and B in the eight-element array shown in Fig. VI.38 per microvolt per meter of linearly polarized electric field parallel to the antennas in the array originating at a large distance in front of the array, that is, in the direction away from the parasites. The four forward antennas are connected by sections of transmission line that are *electrically* one-half wavelength long although the spatial separation of the antennas is only $\lambda_0/4$; assume the lines to be lossless. The parasitic curtain is $\lambda_0/4$ behind the four loaded antennas. Assume the electrical half-length of all antennas to be $\beta_0 h = \pi/2$ with $\Omega = 2 \ln (2h/a) = 10$. The impedance looking into the line at AB is the complex conjugate of the impedance of the array as seen from AB. State what approximations are made.

Chapter VII

1. Carry out the steps involved in obtaining the solutions (1.7a) from the equation (1.6e).

2. Investigate the solution of the problem of the vertical electric dipole over a plane conducting earth if both z- and r-components of the polarization potential are used.

3. Obtain expressions for the real and imaginary parts of the plane-wave reflection coefficients in the general case.

4. Represent graphically the coefficients of reflection that are required to determine the far-zone electric field over an infinite plane surface of (a) salt water ($\sigma_e = 5$ mho/m, $\epsilon_{er} = 80$) and (b) moist earth ($\sigma_e = 2 \times 10^{-3}$ mho/m, $\epsilon_{er} = 10$.) The field is maintained by the current in an antenna erected vertically on the surface and operated at 880 khz. (Use Fig. 7.1a, b in conjunction with a few computed points.) What is the Brewster angle?

5. Repeat Problem 4 if the field is maintained by the currents in a horizontal loop antenna.

6. Discuss the significance of the Brewster angle in the field patterns in Fig. 7.6a. Do this with the aid of graphs of the real and imaginary parts of the terms associated with the direct and reflected waves.

7. For purposes of obtaining the far-zone field a given television transmitting antenna is equivalent to a horizontal loop antenna at a height of 40 m. It is operated on channel 1 (44–50 Mhz). A folded-dipole receiving antenna is to be erected at a distance of $r = 10$ km. Investigate the variation in the far-zone ($1/R$) electric field as a function of the height $z = r \cot \Theta$ for the video carrier at 45.25 Mhz, if the earth between transmitter and receiver is (a) quite moist so that $\sigma_e = 4 \times 10^{-3}$ mho/m and $\epsilon_{er} = 10$; (b) quite dry so that $\sigma_e = 4 \times 10^{-5}$ mho/m, $\epsilon_{er} = 4$. Compare with the corresponding variation when $\sigma_e = \infty$, $\epsilon_{er} = 1$. Assume the earth to be plane.

8. Repeat Problem 7 for the video carrier at 211.25 Mhz of channel 13.

9. Repeat Problem 7 if the transmitting antenna is equivalent to a *vertical electric* dipole.

10. A television receiver constitutes a matched load terminating the 300-ohm line from the horizontal folded-dipole receiving antenna in Problem 7. Neglecting ohmic losses in the line and assuming the impedance of the antenna to be a pure resistance of 300 ohms, what must be the minimum power radiated by the horizontal-loop transmitting antenna if 100 μv are to be maintained across the load? The receiving antenna is at a height of 15 m. Determine the minimum power for moist and dry earth and for a perfectly conducting infinite plane.

11. Repeat Problem 10 if the transmitting antenna is equivalent to a vertical electric half-wave dipole and the receiving antenna also is vertical.

12. Investigate in detail the possibility of determining the field patterns of antennas and arrays over an imperfectly conducting earth by measuring the field patterns of an array of two antennas or arrays in space with appropriate currents.

13. The following data apply to Radio Station WABC* on Columbia Island in Long Island Sound. The transmitting antenna is a square steel structure 25 ft on a side and of height $h = 410$ ft, which is an electrical length of about $\beta_0 h = 132°$ at the operating frequency of 880 khz. It is top-loaded with a flat metal structure 85 ft square that is connected to the tower through an inductive reactance. The equivalent electrical length of the flat top and series reactance is $\beta_0 k = 58°$, so that the over-all equivalent length of the antenna is $\beta_0(h + k) = 190°$. The input current is 22.2 amp when the antenna is fed with 50 kw. The conductivity of the salt water to which the antenna is grounded is 5 mho/m; the conductivity of the land bordering Long Island Sound is about 2×10^{-3} mho/m. The field strength at a distance of 1.6 km (4.7 wavelengths) over salt water when the input current is 22.2 amp is measured to be 1.67 v/m.

(a) Using the value of radiation resistance R_m^e obtained in Problem V.23 for the WABC antenna when erected on a perfectly conducting plane, determine the maximum sinusoidal current I_m for use in the zeroth-order distribution, $I_z = I_m \sin \beta_0(h + k - z)$, $0 \leq z \leq h$, when the radiated power is 50 kw. Plot the distribution of current along the antenna.

(b) Compute and plot the magnitude of the far-zone vertical field factor of the antenna as if it were erected on an infinite expanse of salt water with $\sigma_e = 5$ mho/m, $\epsilon_{er} = 80$. Use the reflection coefficient determined in Problem 4.

(c) Repeat (b) with the antenna erected on an infinite plane earth with $\sigma_e = 2 \times 10^{-3}$ mho/m, $\epsilon_{er} = 10$. Use the reflection coefficient determined in Problem 4.

14. For analytical purposes the WABC antenna described in Problem 13 is to be replaced by an infinitesimal electric dipole or short end-loaded antenna located at the elevation of maximum current on the actual antenna. Determine this location and the uniform current required in the short antenna in order to radiate the same power as the actual antenna, namely, 50 kw, when the conducting plane is assumed to be perfectly conducting.

15. Using a short end-loaded antenna with the current properties obtained and at the elevation determined in Problem 14 as an approximate equivalent of the actual antenna, determine the magnitude of the far-zone field factor when the conducting plane is an infinite expanse of (a) salt water with $\sigma_e = 5$ mho/m, $\epsilon_{er} = 80$; (b) moist earth with $\sigma = 2 \times 10^{-3}$ mho/m, $\epsilon_{er} = 10$. Compare the far-zone vertical field patterns of the short elevated dipole with those of the actual antenna, especially from the point of view of broadcast transmission.

16. Obtain and plot the magnitudes of the space-wave term, the surface-wave term, and the total vertical field of the elevated doublet determined in Problem 14 as a function of the actual radial distance r measured along the conducting surface from the point directly below the doublet, using (a) salt water with $\sigma_e = 5$ mho/m, $\epsilon_{er} = 80$; (b) moist earth with $\sigma_e = 2 \times 10^{-3}$ mho/m, $\epsilon_{er} = 10$; (c) a perfect conductor.

17. For the short elevated antenna of Problem 14 determine the orientation and shape of the polarization ellipse of the electric field on the conducting surfaces specified in the preceding problem at a radial distance of 125 km from the transmitter.

18. Compare the results obtained in Problem 16 for the elevated short antenna with measured electric fields obtained for the actual WABC antenna, as shown in Fig. VII.18. Do this for radial directions (a) entirely over salt water and (b) almost entirely over land. (c) Explain the behavior of the measured contours of constant electric field in Fig. VII.18, especially in directions along the shore toward Cape Cod and toward Philadelphia. (Note that the three circles in Fig. VI.18 are the estimated contours of constant electric field which would obtain over a uniform earth *in all directions*.)

---

* As reported in E. K. Cohan, "18 months experience with WABC's island transmitter," *Electronic Industries* (May 1943), p. 70.

19. Formulate the general analysis of the vertical dipole by the same method used in the text for the horizontal dipole.

20. A television transmitter is equivalent to a vertical magnetic dipole with a moment p_z at a height of 30 m above a plane moist earth with $\sigma_e = 4 \times 10^{-3}$ mho/m and $\epsilon_{er} = 10$. A portable receiving antenna that can be raised to a height of 10 m consists of a vertical and a horizontal dipole with separate leads to identical matched receivers. Observations of the responses in the two receivers are to be made at a series of locations on a radial line from the transmitter. Predetermine the relative responses in the two receivers as r is increased from quite close to the transmitter to distant points as a function of height at a succession of values of r.

21. A receiving antenna consists of a straight wire of electrical length $\beta_0 h = \pi/2$ and radius a such that $\Omega = 2 \ln(2h/a) = 10$. A lumped impedance is connected between the base of the antenna and a conducting network buried in moist earth with $\sigma_e = 4 \times 10^{-3}$ mho/m and $\epsilon_{er} = 10$. Assume the impedance of the antenna to be essentially the same as if it were over a perfectly conducting plane and the load Z_L to be conjugate-matched to this value. Determine and plot as a function of Θ_2 the power to the load as the antenna is tilted toward and away from the distant vertical transmitter through an angle Θ_2 given by $0 \leq \Theta_2 \leq 45°$, $\Phi = 0$; $0 \leq \Theta_2 \leq 45°$, $\Phi = \pi$, where Θ_2 is measured from the vertical and $\Phi = 0$ is in the direction away from, $\Phi = \pi$ is in the direction toward, the transmitter. Assume the current in the vertical, base-driven transmitting half-wave dipole to be 4 amp and the distance to the receiver to be 10 km. The frequency is 10 Mhz. Assume the exciting field along the entire receiving antenna to be the same as the field at its base.

22. Investigate the validity of the assumption that the exciting field along the receiving antenna in Problem 21 is the same as at its base.

23. A cylindrical antenna of half-length h_1 adjusted for antiresonance and radius $a_1 = 2.5$ mm is placed vertically with its center at a height of 9 m. The antenna is center-driven at a frequency $f = 200$ Mhz from a two-wire copper line of length $s_1 = 9$ m and line spacing $b = 2$ cm. The radius of the wire is $a_1 = 2.5$ mm. The power supplied the line is 100 watts.

A horizontal cylindrical receiving antenna of half-length $h_2 = 0.6$ m and radius $a_2 = 0.66$ mm is stretched below a blimp at a height of 3.3 km at a horizontal distance of 7.5 km over salt water from the transmitter. A conjugate-matched load Z_L is connected at the center of the receiving antenna.

Determine the power in the load Z_L. State and justify any approximations made.

24. Investigate the electromagnetic field of an infinitesimal dipole over a conducting plane earth when inclined at an arbitrary angle from the vertical. (*Hint*: Resolve the dipole moment into vertical and horizontal components.)

25. The impedance of a vertical base-driven antenna of length h with $\Omega = 2 \ln(2h/a) = 15$ is to be measured using a circular ground screen of diameter D near $10\lambda_0$. Determine the expected differences in ohms and the percent differences between the resistances and reactances to be measured on the finite ground screen and the ideal ones for an infinite ground screen ($D = \infty$) under the following conditions; also determine the precise value of D near $10\lambda_0$:

(a) D is chosen so that the expected difference in the resistances is zero;

(b) D is chosen so that the expected difference in the reactance is zero;

(c) D is chosen so that the expected differences are equal.

In each case use $h = h$(resonant), $\lambda_0/4$, h(antiresonant), $\lambda_0/2$. Which value of D is to be preferred? Explain.

26. The impedance of a cylindrical antenna with hemispherical end and axial length $h = 0.2$ m and radius $a = 2.66$ mm is to be measured over a circular ground screen of diameter $D = 10$ m by varying the frequency from 300 to 700 Mhz. The antenna projects from a hole of radius b at the center of the screen as a projection of the inner conductor of the feeding coaxial line. The radius of the inner conductor of the coaxial line is the same as the radius of the antenna; the inner radius b of the outer conductor of the line is given by $b/a = 4$; the line is filled with styrofoam with relative dielectric constant $\epsilon_r \doteq 1$.

Determine and plot the theoretical apparent resistance R_{sa} and reactance X_{sa} of the antenna together with the theoretical values for R_0 and X_0 for $D = \infty$. Account must be taken of the transmission-line end effect *and of* the finite size of the ground screen.

Chapter VIII

1. It is stated in some books that the vector $\mathbf{\Pi} = \hat{\mathbf{R}}\Pi_R, \Pi_\Theta = 0, \Pi_\Phi = 0$, will satisfy the vector wave equation

$$\text{grad div } \mathbf{\Pi} - \text{curl curl } \mathbf{\Pi} + \beta_0^2 \mathbf{\Pi} = 0$$

provided the radial component Π_R satisfies equation (2.10). Prove that this is not true.

2. Prove that Π_R/R satisfies the scalar wave equation.

3. The voltage $V(R)$ for the dominant mode of the biconical transmission line is defined by (4.13) as the line integral $V(R) = \int_{\Theta_0}^{\pi-\Theta_0} E_\Theta R d\Theta$. How is this voltage related to the scalar potential difference between the end points of the path of integration?

4. Compare the definition and behavior of the total radial current, the total charge per unit radial distance, the charge density, the current density, the electric field, and the magnetic field of the dominant mode of the biconical line with the corresponding quantities for a two-wire line and a coaxial line.

5. Modify the solution for the dominant mode in a biconical line to take account of an imperfect dielectric between the perfectly conducting cones.

6. Modify the solution for the dominant mode in a biconical line to take account of imperfectly but highly conducting cones.

7. Investigate the possibility of terminating a biconical line of finite length l with Θ_0 near 90° in such a manner that the distributions of dominant-mode current and charge on the cones are the same as if the cones extended to infinity.

8. Investigate the possibility of analyzing the biconical line of finite length but with Θ_0 near $\pi/2$ in a manner analogous to that used for the two-wire line in Chapter II. What condition is required to make radiation negligible? Can terminal-zone effects be treated in an analogous manner? If so, formulate the analysis.

9. Obtain the dominant-mode solution for a *radial* transmission line consisting of two parallel circular metal planes excited across their centers by a line source.

10. Using the method of Chapter II, investigate the terminal-zone problem of a coaxial line when terminated in a conical line. Design a lumped network that may be used in conjunction with the dominant or TEM modes on the two lines. Consider the special case when characteristic impedances are equal. (Neglect ohmic resistance.)

11. Compare the impedance of a biconical antenna (angle Θ_0) with that of a base-driven cylindrical antenna (length-to-radius ratio h/a) with the same value of Ω, that is, $\Omega_c = 2 \ln (2/\Theta_0) = \Omega = 2 \ln (2h/a)$, by plotting together. Select a value of Ω that permits convenient use of available numerical data. Compare critical values. Compare the broad-band properties.

12. Investigate the analysis of the thin biconical antenna using the method of Chapter II (for the cylindrical antenna of fixed radius a) with a radius $a(z) = kz$, where k is a small constant, and an average expansion parameter.

13. Study the orders of magnitude of the terms in the numerator and denominator in (11.15) to determine whether a true second-order Schelkunoff solution requires $I_2(l)$ added to the numerator.

14. In their paper "The general problem of antenna radiation," reference II.2, p. 127, G. E. Albert and J. L. Synge state: "Many writers have proceeded as if antenna problems could be discussed without reference to a gap. To do so, however, is to leave out the essential part of the radiating system. *The gap is the only source of radiant energy.*" Demonstrate for antennas driven from biconical, coaxial, and open-wire lines that this statement is incorrect by showing that

(a) the electromagnetic field at distant points is determined by the current *in the antenna*;

(b) the *true* source of radiant energy is an electrical generator;

(c) the essential requirement for the transfer of power from a generator to an antenna is a transmission line with nonvanishing characteristic impedance and not a gap.

15. Prepare a summary in the form of a systematic tabulation of the important properties of the several methods of analyzing single and coupled transmitting and receiving antennas referred to in this volume from the point of view of determining all essential qualities of antennas.

BIBLIOGRAPHY

CHAPTER I. ESSENTIALS OF ELECTROMAGNETIC THEORY

GENERAL REFERENCES AND SUMMARIES

Papers

1. Brillouin, L., "Antennae for ultra-high frequencies," *Elec. Commun.* **21**, No. 4 (1944); **22**, No. 1 (1945).
2. Hertz, H., "Ueber sehr schnelle electrische Schwingungen," *Wied. Ann.* **31**, 421 (1887).
3. Hertz, H., "Die Kräfte electrischer Schwingungen behandelt nach der Maxwell'schen Theorie," *Wied. Ann.* **36**, 1 (1888).
4. Roubine, E., "Les récentes théories de l'antenne," *L'Onde Electrique*, Nos. 238–241, Jan.–April 1947.
5. Schelkunoff, S. A., "A general radiation formula," *Proc. I. R. E.* **27**, 660 (Oct. 1939).
6. Schelkunoff, S. A., "Some equivalence theorems of electromagnetics and their application to radiation problems," *Bell System Tech. J.* **15**, 92 (Jan. 1936).
7. Sinclair, George, "Theory of models of electromagnetic systems," *Proc. I.R.E.* **36**, 1364 (Nov. 1938).
8. Zinke, O., "Grundlagen der Strom- und Spannungsverteilung auf Antennen," *Arch. Electrotech.* **35**, 67 (1941).

Books

10. Aharoni, J., *Antennae* (Clarendon Press, Oxford, 1946).
11. Becker, R., *Theorie der Elektrizität* (Teubner, Leipzig and Berlin; vol. 1, 1930; vol. 2, 1933).
12. Bremmer, H., *Terrestrial radio waves* (Elsevier, New York, 1949).
13. Brillouin, M., *Propagation de l'électricité* (Hermann, Paris, 1904), pp. 314–395.
14. Brückmann, H., *Antennen, ihre Theorie und Technik* (Hirzel, Leipzig, 1939).
15. Burrows, C. R., and S. S. Atwood, *Radio wave propagation* (Academic Press, New York, 1949).
16. Courant, R., and D. Hilbert, *Methoden der mathematischen Physik* (Springer, Berlin, 1931; 2 vols.; English edition, Interscience Publishers, New York, 1943).
17. Dwight, H. B., *Tables of integrals and other mathematical data* (Macmillan, New York, 1947).
18. Frank, R., and R. von Mises, *Differentialgleichungen der Physik* (Rosenberg, New York, 1943; 2 vols.).
19. Fränz, K., and H. Larsen, *Ausstrahlung, Ausbreitung, und Aufnahme elektromagnetischer Wellen* (Springer, Berlin, 1956).
20. Fry, D. W., and F. K. Goward, *Aerials for centimetre wave-lengths* (Cambridge University Press, Cambridge, 1950).
21. Glas, E. T., *On radiation problems concerning vertical antennas* (Royal Administration of Swedish Telegraphs, Stockholm, 1943).
22. Gundlach, F. W., *Grundlagen der Höchstfrequenztechnik* (Springer, Berlin, 1950).
23. Hamel, G., *Integralgleichungen* (Springer, Berlin, 1937).
24. Harvard Computation Laboratory, *Tables of inverse hyperbolic functions* (Harvard University Press, Cambridge, 1949).
25. Harvard Computation Laboratory, *Tables of generalized sine and cosine integral functions* (Harvard University Press, Cambridge, 1949; 2 vols.).
26. Hertz, H., *Ausbreitung der electrischen Kraft* (Barth, Leipzig, 1892).
27. Hobson, E. W., *The theory of spherical and ellipsoidal harmonics* (Cambridge University Press, Cambridge, 1931).
28. Jahnke, E., and F. Emde, *Funktionentafeln* (Teubner, Leipzig, 1938).
29. Jeffreys, H. and B. S.: *Methods of mathematical physics* (Cambridge University Press, Cambridge, 1946).
30. Jordan, E. C., *Electromagnetic waves and radiating systems* (Prentice-Hall, New York, 1950).
31. King, R. W. P., *Electromagnetic engineering* (McGraw-Hill, New York, 1945), vol. 1.
31a. King, R. W. P., *Transmission-line theory* (McGraw-Hill, New York, 1955).
31b. King, R. W. P., H. R. Mimno, and A. H. Wing, *Transmission lines, antennas, and wave guides* (McGraw-Hill, New York, 1945).
32. Knudsen, H. L., *Bidrag tel teorien for antennesystemer med hel eller delvis rotationssymmetri* (I Kommission hos Teknisk Forlag, Copenhagen, 1953).
33. Kraus, J. D., *Antennas* (McGraw-Hill, New York, 1950).
33a. Kraus, J. D., *Electromagnetics* (McGraw-Hill, New York, 1953).
34. Küpfmüller, K., *Theoretische Elektrotechnik* (Springer, Berlin, 1932).
35. McLachlan, N. W., *Bessel functions for engineers* (Oxford University Press, New York, 1934).

36. M.I.T. Radar School, *Principles of radar* (McGraw-Hill, New York, 1946).
37. Moullin, E. B., *Radio aerials* (Clarendon Press, Oxford, 1950).
38. Ollendorff, F., *Grundlagen der Hochfrequenztechnik* (Springer, Berlin, 1926).
39. Peirce, B. O., *A short table of integrals* (Ginn, New York, 1929).
40. Pidduck, F. B., *Currents in aerials and high-frequency networks* (Clarendon Press, Oxford, 1946).
41. Pierce, G. W., *Electric oscillations and electric waves* (McGraw-Hill, New York, 1920).
42. Radio Research Laboratory, Harvard University, *Very high frequency techniques* (McGraw-Hill, New York, 1947; 2 vols.).
43. Ramo, S., and J. R. Whinnery, *Fields and waves in modern radio* (Wiley, New York, 1944).
44. Rothe, R., F. Ollendorff, and K. Pohlhausen, *Theory of functions* (Technology Press, M.I.T., Cambridge, 1933).
45. Schelkunoff, S. A.: *Advanced antenna theory* (Wiley, New York, 1952).
46. Schelkunoff, S. A., *Electromagnetic waves* (Van Nostrand, New York, 1943).
47. Schelkunoff, S. A. and H. T. Friis, *Antenna theory and practice* (Wiley, New York, 1952).
48. Silver, S., *Microwave antenna theory and design* (McGraw-Hill, New York, 1949).
49. Slater, J. C., *Microwave transmission* (McGraw-Hill, New York, 1942).
50. Smith, R. A., *Aerials for metre and decimetre wavelengths* (Cambridge University Press, Cambridge, 1949).
51. Sommerfeld, A., *Vorlesungen über Theoretische Physik*, Band III, *Elektrodynamik;* Band VI, *Partielle Differentialgleichungen der Physik* (Dieterich'sche Verlagsbuchhandlung, Wiesbaden, 1947, 1948; English editions, Academic Press, New York, 1950, 1949).
52. Stratton, J., *Electromagnetic theory* (McGraw-Hill, New York, 1941).
53. Strutt, M. J. O., *Moderne Kurzwellen-Empfangstechnik* (Springer, Berlin, 1939).
54. Uda, S., and Y. Mushiake, *Yagi-Uda Antenna* (Research Inst. of Electrical Communication, Tohoku University, Sendai, Japan).
55. Watson, G. N., *Theory of Bessel functions* (Macmillan, New York, ed. 2, 1945).
56. Watson, W. H., *Physical principles of wave guide transmission and antenna systems* (Clarendon Press, Oxford, 1947).
57. Wiarda, G., *Integralgleichungen* (Teubner, Leipzig and Berlin, 1930).
58. Williams, H. P., *Antenna theory and design* (Pitman, London, 1950; 2 vols.).
59. Zuhrt, H., *Elektromagnetische Strahlungsfelder* (Springer-Verlag, Berlin, Göttingen, Heidelberg, 1953).

Chapter II. Linear Radiators as Circuit Elements

1. Aharoni, J., "A general theory of antennae." *Phil. Mag.* **35**, 427–459 (1944).
2. Albert, G. E., and J. L. Synge, "The general problem of antenna radiation and the fundamental integral equation, with application to an antenna of revolution—Part I," *Quart. Appl. Math.* **6**, 117 (1948).
3. Altar, W., F. B. Marshall, and L. P. Hunter, "Probe Error in Standing-Wave Detectors," *Proc. I.R.E.* **34**, 33P (1946).
4. Angelakos, D., "Current and charge distributions on antennas and open-wire lines," doctoral dissertation, Cruft Laboratory, Harvard University, (January 1950).
5. Barzilai, Giorgio, "Experimental determination of the distribution of current and charge along cylindrical antennas," *Proc. I.R.E.* **37**, 825 (1949).
6. Bechmann, R., "On the calculation of radiation resistance of antennas and antenna combinations," *Proc. I.R.E.* **19**, 1471 (1931).
7. Boudoux, Pierre, "Current distribution and radiation properties of a shunt-excited antenna," *Proc. I.R.E.* **28**, 271 (1940).
8. Bouwkamp, C. J., "Concerning a new transcendent, its tabulation and application in antenna theory," *Quart. Appl. Math.* **5**, 394 (1947).
9. Bouwkamp, C. J., "Hallén's theory for a straight, perfectly conducting wire, used as a transmitting or receiving aerial," *Physica* **9**, 609–631 (1942).
10. Brillouin, L., "Sur l'origine de la resistance de rayonnement," *Radioélectricité* **3**, 147 (1922).
11. Brillouin, L., "The antenna problem," *Quart. Appl. Math.* **1**, 201 (1943).
12. Brown, G. H., and O. M. Woodward, Jr., "Experimentally determined impedance characteristics of cylindrical antennas," *Proc. I.R.E.* **33**, 257 (1945).
13. Burgess, R. E., "Aerial characteristics," *Wireless Eng.* **21**, 154–160 (1944).
14. Carson, J. R., "Electromagnetic theory and the foundations of electric circuit theory," *Bell System Tech. J.* **6**, 1–17 (1927).
15. Conley, P., "Antennas and open-wire lines—III—Image-line measurements," *J. Appl. Phys.* **20**, 1022 (1949).
16. Conley, P., "Impedance measurements with open-wire lines," Cruft Laboratory Technical Report No. 35, Harvard University (March 1948); doctoral dissertation, Harvard University (1948).
17. Essen, L., and M. H. Oliver, "Aerial impedance measurements," *Wireless Eng.* **22**, 587 (1945).
18. Gans, R., "La distribution du courant dans les antennes," *Rev. Sci.* **85**, 643–648 (1947).
19. Gans, R., and M. Bemporad, "Zur Theorie der geradlinigen Antenne," *Arch. Elek. Uebertragung* **7**, 169–180 (1953).
20. Graffi, Dario, "Sulla impedenza de radiazione della antenna," *Alta Frequenza* **12**, 3–25 (1943).

21. Gray, M. C., "A modification of Hallén's solution of the antenna problem," *J. Appl. Phys.* **15**, 61 (1944).
22. Grosskopf, J., "Zur Theorie der geraden Antenne," *Arch. Elek. Uebertragung* **4**, 175 (1950).
23. Hallén, Erik, "On antenna impedances," *Acta Polytech.* No. 16—*Trans. Roy. Inst. Technol. Stockholm* 1947, 18 pp. (1947).
24. Hallén, Erik, "Properties of a long antenna," *J. Appl. Phys.* **19**, 1140 (1948).
25. Hallén, Erik, "Theoretical investigations into transmitting and receiving antennae," *Nova Acta Regiae Soc. Sci. Upsaliensis* [4] **11**, 1 (1938).
26. Hara, G., "Strahlungsleistung und Stromverteilung einer geraden Antenne," *Hochfrequenztechn. u. Elektroak.* **44**, 185 (1934).
26a. Harrington, R. F., "On the cylindrical antenna," El. Engr. Dept., Syracuse University (March 1954).
27. Hartig, E. O., "Circular apertures and their effects on half-dipole impedances," doctoral dissertation, Harvard University (June 1950).
28. Heilmann, A., "Ueber den Scheinwiderstand von Empfangsantennen," *TFT* **29**, 357 (1940).
29. Infeld, L., "The influence of the width of the gap upon the theory of antennas," *Quart. Appl. Math.* **5**, 113 (1947).
30. Kaufmann, H., "Der Eingangswiderstand der Dipol-Antennen," *Hochfrequenztechn. u. Elektroak.* **60**, 160 (1942).
31. Kennedy, P. A., and R. King, "Experimental and theoretical impedances and admittances of center-driven antennas," Cruft Laboratory Technical Report No. 155 (April 1953).
32. King, D. D., "Microwave impedance measurements with application to antennas—I," *J. Appl. Phys.* **16**, 435 (1945).
33. King, D. D., "Impedance measurements on transmission lines," *Proc. I.R.E.* **35**, 509–514 (1947).
34. King, D. D., "The measured impedance of cylindrical dipoles," *J. Appl. Phys.* **17**, 844 (1946).
35. King, D. D., and R. King, "Microwave impedance measurements with application to antennas—II," *J. Appl. Phys.* **16**, 435, 445 (1945).
36. King, D. D., and R. King, "Terminal functions for antennas," *J. Appl. Phys.* **15**, 186–192 (1944).
37. King, L. V., "On the radiation field of a perfectly conducting base insulated cylindrical antenna over a perfectly conducting plane earth, and the calculation of radiation resistance and reactance," *Trans. Roy. Soc. (London)* [A] **236** 381–422 (1937).
38. King, R., "Antennas and open-wire lines, I," *J. Appl. Phys.* **20**, 832 (1949).
39. King, R., "A generalized coupling theorem for ultra-high frequency circuits," *Proc. I.R.E.* **28**, 84–87 (1940).
40. King, R., "An alternative method of solving Hallén's integral equation, *J. Appl. Phys.* **24**, 140–147 (1953).
41. King, R., "End correction for a coaxial line when driving an antenna over a ground screen," Cruft Laboratory Technical Report No. 174 (June 1953); *Trans. I.R.E., PGAP*, AP-3, 66 (1955).
42. King, R., "General amplitude relations for transmission lines with unrestricted line parameters, terminal impedance, and driving point," *Proc. I.R.E.* **29**, 640 (1941).
43. King, R., "Theory of electrically short transmitting and receiving antennas," *J. Appl. Phys.* **23**, 1174–1187 (1952).
44. King, R., "Transmission-line theory and its applications," *J. Appl. Phys.* **14**, 577–600 (1943).
45. King, R., and F. G. Blake, Jr., "The self-impedance of a symmetrical antenna," *Proc. I.R.E.* **30**, 335 (1942).
46. King, R., and C. W. Harrison, Jr., "The distribution of current along a symmetrical center-driven antenna," *Proc. I.R.E.* **31**, 548 (1943).
47. King, R., and C. W. Harrison, Jr., "The impedance of short, long, and capacitively loaded antennas with a critical discussion of the antenna problem," *J. Appl. Phys.* **15**, 170 (1944).
48. King, R., and D. Middleton, "The cylindrical antenna; current and impedance," *Quart. Appl. Math.* **3**, 302–335 (1946).
49. King, R., and D. Middleton, "Correction and supplement to our paper 'The cylindrical antenna: current and impedance,'" *Quart. Appl. Math.* **4**, 199 (1946).
50. King, R., and K. Tomiyasu, "Terminal impedance and generalized two-wire-line theory," *Proc. I.R.E.* **37**, 1134 (1949).
51. King, R., and T. W. Winternitz, "The cylindrical antenna with gap," *Quart. Appl. Math.* **5**, 403 (1947).
52. Labus, J., "Berechnung der Strahlungsenergie von Dipolantennen (Telefunkenrichtantennen) nach der Poyntingschen Methode," *ENT* **9**, 61 (1932).
53. Labus, J., "Rechnerische Ermittlung der Impedanz von Antennen," *Hochfrequenztechn. u. Elektroak.* **41**, 17 (1933).
54. Lanczos, C., "A new solution of the antenna problem," presented at fall 1947 convention of I.R.E. at San Francisco, Boeing Aircraft Co., Seattle, Washington, Document D-9152.
55. Lange, P., "Messungen an Dipolen im Dezimeterwellengebiet," *Hochfrequenztechn. u. Elektroak.* **58**, 25 (1941).
56. Laport, Edmund A., "Some notes on the influence of stray capacitance upon the accuracy of antenna resistance measurements," *Proc. I.R.E.* **22**, 657 (1934).
57. Levin, M. L., "A contribution to the theory of antennae," *Compt. rend. acad. sci. URSS* (n.s.) **54**, 595–597 (1946).
58. Meier, A. S., and W. P. Summers, "Measured impedance of vertical antennas over finite ground planes," Ohio State University Research Foundation, Antenna Laboratory, Report No. 233-3 (October 1946); *Proc. I.R.E.* **37**, 609 (1949).

59. Middleton, D., and R. King, "The thin cylindrical antenna; A comparison of theories," *J. Appl. Phys.* **17**, 273 (1946).
60. Morita, T., "Current distributions on transmitting and receiving antennas," *Proc. I.R.E.* **38**, 898 (1950); doctoral dissertation, Harvard University (December 1948).
61. Nomura, Y., and T. Hatta, "The theory of a linear antenna, I," *Technol. Repts., Tohoku Univ.* **17**, 1 (1952).
62. Palmer, L. S., and K. G. Gillard, "The distribution of ultra-high-frequency currents in long transmitting and receiving antennae," *J. Inst. Elect. Engrs. (London)* **13**, 285 (1938).
63. Papas, C. H., "On the infinitely long cylindrical antenna," *J. Appl. Phys.* **20**, 437 (1949).
64. Pistolkors, A. A., "The radiation resistance of beam antennas," *Proc. I.R.E.* **17**, 562 (1929).
65. Pocklington, H. C., "Electric oscillations in wires," *Camb. Phil. Soc.* **9**, 324 (1897).
66. Rayleigh, Lord, "Electrical vibrations on a thin anchor-ring," *Proc. Roy. Soc. (London)* [A], **87**, 193 (1912).
67. Rosseler, G., F. Vilbig, and K. Vogt, "Ueber das elektrische Verhalten von Vertikal-Antennen in Abhängigkeit von ihrem Durchmesser," *TFT* **28**, 170 (1939).
68. Schelkunoff, S. A., "Antenna theory and experiment," *J. Appl. Phys.* **15**, 54 (1944).
69. Schelkunoff, S. A., "Concerning Hallén's integral equation for cylindrical antennas," *Proc. I.R.E.* **33**, 872 (1945).
70. Schelkunoff, S. A., "On the antenna problem," *Quart. Appl. Math.* **1**, 354 (1944).
71. Schelkunoff, S. A., "Theory of antennas of arbitrary shape and size," *Proc. I.R.E.* **29**, 493–521 (1941).
72. Schelkunoff, S. A., and C. B. Feldman, "On radiation from antennas," *Proc. I.R.E.* **30**, 511–516 (1942).
73. Siegel, E., "Scheinwiderstand von beschwerten Antennen," *Hochfrequenztechn. u. Elektroak* **43**, 167 (1934).
74. Siegel, E., and J. Labus, "Scheinwiderstand von Antennen," *Hochfrequenztechn. u. Elektroak* **43**, 167 (1934).
75. Siegel, E., and J. Labus, "Sendeantennen," *Hochfrequenztechn. u. Elektroak* **49**, 87 (1937).
76. Smeby, L. C., "Short-antenna characteristics," *Proc. I.R.E.* **37**, 1185 (1949).
77. Smith, C. E., and E. M. Johnson, "Performance of short antennas," *Proc. I.R.E.* **35**, 1026 (1947).
78. Storer, J. E., "Variational solution to the problem of the symmetrical cylindrical antenna," Technical Report No. 101, Cruft Laboratory, Harvard University; and "Solution to thin wire antenna problems by variational methods," doctoral dissertation, Harvard University (June 1951).
79. Storm, B., "Investigation into modern aerial theory and a new solution of Hallén's integral equation for a cylindrical aerial," dissertation, Imperial College, London (1953); summary in *Wireless Engineer* (July 1953).
80. Synge, J. L., "The general problem of antenna radiation and the fundamental integral equation, with application to an antenna of revolution—Part II," *Quart. Appl. Math.* **6**, 133 (1948).
81. Tai, C. T., "A variational solution to the problem of cylindrical antennas," Technical Report No. 12, SRI Project No. 188, Stanford Research Institute (August 1950).
82. Tai, C. T., "A study of the emf method," Technical Report No. 55, Cruft Laboratory, Harvard University; and *J. Appl. Phys.* **20**, 717 (1949).
83. Tomiyasu, K., "Antennas and open-wire lines—Part II: Measurements on two-wire lines," *J. Appl. Phys.* **20**, 892 (1949).
84. Tomiyasu, K., "The effect of a bend and other discontinuities on a two-wire transmission line," *Proc. I.R.E.* **38**, 679 (1950).
85. Tomiyasu, K., "Problems of measurement on two-wire lines with application to antenna impedances," Cruft Laboratory Technical Report No. 48 (June 1948).
86. Tomiyasu, K., "The unbalance squelcher," *Rev. Sci. Instr.* **19**, 675 (1948).
87. Webb, W., and R. Raymond, "Current distributions on some simple antennas," *J. Appl. Phys.* **20**, 330 (1949).
88. Wells, N., "Aerial characteristics," *J. Inst. Elec. Engrs. (London)* **89**, Pt. III, 76 (1942); **90**, Pt. III, 24 (1943).
89. Wheeler, H. A., "Fundamental limitations of small antennas," *Proc. I.R.E.* **35**, 1479 (1947).
90. Whinnery, J. R., "Effect of input configuration on antenna impedance," *J. Appl. Phys.* **21**, 945 (1950).
91. Wiechowski, W., "Dämpfungsberechnung bei Sendeantennen," *Hochfrequenztechn. u. Elektroak* **53**, 50 (1939).
92. Wilmotte, R. M., "Distribution of current in a transmitting aerial," *J. Inst. Elect. Engrs. (London)* **66**, 617 (1928).
93. Winternitz, T. W., "Input impedance of a two-wire open line and cylindrical center-driven antenna," *Proc. I.R.E.* **38**, 299 (1950); doctoral dissertation, Harvard University (1949).

CHAPTER III. CIRCUIT PROPERTIES OF ARRAYS OF LINEAR RADIATORS

1. Abbott, F. R., and C. R. Fisher, "Measured directivity induced by a conducting cylinder of arbitrary length and spacing parallel to a monopole antenna" (abstract), *Proc. I.R.E.* **38**, 1040 (1950).
2. Affanasiev, K. J., "Simplifications in the consideration of mutual effects between half-wave dipoles in collinear and parallel orientations," *Proc. I.R.E.* **34**, 635, (correction) 863 (1946).
3. Andrews, H., "The collinear antenna array," doctoral dissertation, Harvard University (1953); Cruft Laboratory Technical Reports Nos. 178 and 179.

4. Angelakos, D., "Current and charge distributions on antennas and open-wire lines," doctoral dissertation, Cruft Laboratory, Harvard University (1950).
5. Barzilai, "Mutual impedance of parallel aerials," *Wireless Engr.* **25**, 343 (November 1948).
6. Barzilai, G., "Mutual impedance of parallel aerials," *Wireless Engr.* **26**, 73 (1949).
7. Blasi, E. A., "The theory and application of the radiation mutual coupling factor," Ph.D. thesis, Ohio State University (1953).
8. Bouwkamp, C. J., "Calculation of the input impedance of a special antenna," *Philips Research Repts.* **2**, 228 (1947).
9. Bouwkamp, C. J., "On the theory of coupled antennas," *Philips Research Repts.* **3**, 213 (1948).
10. Brown, G. H., "Ground plane antennas," *Electronics* **16**, 338 (1943).
11. Brown, G. H., "'Turnstile' antenna," *Electronics* **9**, 14 (1936).
12. Brown, G. H., and R. King, "High frequency models in antenna investigations," *Proc. I.R.E.* **22**, 457 (1934).
13. Cafferata, H., "Driving-point impedance of a vertical cylindrical radiator and concentric ring of subsidiary radiators over perfectly conducting earth," *Marconi Rev.* **12**, 12, 57 (1949).
14. Carter, P. S., "Circuit relations in radiating systems and applications to antenna problems," *Proc. I.R.E.* **20**, 1004 (1932).
15. Chambers, L. L. G., *Note on the input impedance of a pair of crossed dipoles* (Admiralty Signal and Radar Establishment, Lythe Hill House, Haslemere, Surrey, England).
16. Chang, Tung, "Impedance characteristics of antennas involving loop and linear elements," doctoral dissertation, Harvard University, (June 1947); Cruft Laboratory Technical Report No. 16.
17. Cox, C. R., "Mutual impedance between vertical antennas of unequal heights," *Proc. I.R.E.* **35**, 1367 (1947).
18. Englund, C. R., and A. B. Crawford, "The mutual impedance between adjacent antennas," *Proc. I.R.E.*, 1277 (1929).
19. Faflick, C. E., "Parasitic sleeve antenna and antennas of discontinuous radius," doctoral dissertation, Harvard University (1954); Cruft Laboratory Technical Report No. 171.
20. Grosskopf, J., "Rückgespeiste Antennen," *Frequenz* **3**, 157 (1949).
21. Guertler, R., "Impedance transformation in folded dipoles," *Proc. I.R.E.* **38**, 1042 (1950).
22. Harrison, C. W., Jr., "A note on the mutual impedance of antennas," *J. Appl. Phys.* **14**, 306 (1943).
23. Harrison, C. W., Jr., "A theory for three-element broadside arrays," *Proc. I.R.E.* **34**, 204P (1946).
24. Harrison, C. W., Jr., "Calculation of the impedance properties of parasitic antenna arrays involving elements of finite radius," *J. Am. Soc. Naval Engrs.* **57**, 224 (1945).
25. Harrison, C. W., Jr., "Distribution of current along asymmetrical antennas," *J. Appl. Phys.* **16**, 402 (1945).
26. Harrison, C. W., Jr., "Folded antennas," doctoral dissertation, Harvard University (1954); Cruft Laboratory Technical Report No. 193.
27. Harrison, C. W., Jr., "Mutual and self-impedance for collinear antennas," *Proc. I.R.E.* **33**, 398 (1945).
28. Harrison, C. W., Jr., "Symmetrical antenna arrays," *Proc. I.R.E.* **33**, 892 (1945).
29. Hatch, R. M., Jr., "Investigation of current distribution on asymmetrically-fed antennas by means of complementary slots," Technical Report No. 8, SRI Project No. 188, Stanford Research Institute (February 1950).
30. Hatch, R. M., Jr., "An investigation of the distribution of current on collinear parasitic antenna elements," Technical Report No. 28, SRI Project No. 591, Stanford Research Institute (August 1952).
31. King, R., "Asymmetrically driven antennas and the sleeve dipole," *Proc. I.R.E.* **38**, 1154 (1950).
32. King, R., "Coupled antennas and transmission lines," *Proc. I.R.E.* **31**, 626 (1943).
33. King, R., "Self- and mutual impedances of parallel identical antennas," *Proc. I.R.E.* **40**, 981–988 (1952).
34. King, R., "Theory of collinear antennas," *J. Appl. Phys.* **21**, 1232 (1950).
35. King, R., "Theory of N coupled parallel antennas," *J. Appl. Phys.* **21**, 94 (1950).
36. King, R., "Theory of V-Antenna," *J. Appl. Phys.* **22**, 1111–1121 (1951).
37. King, R., and C. W. Harrison, Jr., "Mutual and self-impedance for coupled antennas," *J. Appl. Phys.* **15**, 481 (1944).
38. Kraus, J. D., "The corner-reflector antenna," *Proc. I.R.E.* **28**, 513 (1940).
38a. Lewis, J. B., "Use of folded monopoles in antenna arrays," *Trans. I.R.E.* AP-3 122 (1955).
39. McPetrie, J. S., and J. A. Saxton, "Some experiments with linear aerials," *Wireless Engr.* **23**, 107 (1946).
39a. Monteath, G. D., "Wide-band folded slot aerials," *Proc. Inst. Elec. Engrs. (London)* **97**, Pt. III, 414 (1950).
40. Morita, T., and C. E. Faflick, "The measurement of current distribution along coupled antennas, folded dipoles, and shunt-fed plates," Cruft Laboratory Technical Report No. 67, Harvard University (1949).
41. Moullin, E. B., "Theory and performance of corner reflectors for aerials," *J. Inst. Elec. Engrs. (London)* **92**, Pt. III, 58 (1945).
42. Murray, L. H., "Mutual impedance of two skew antenna wires," *Proc. I.R.E.* **21**, 154 (1933).
43. Nagy, A. W., "An experimental study of parasitic wire reflectors on 2.5 meters," *Proc. I.R.E.* **24**, 233 (1936).
44. Norgorden, O., and A. W. Walters, "Experimentally determined characteristics of cylindrical sleeve antennas," *J. Am. Soc. Naval Engrs.* **62**, 365 (1950).

45. Roberts, W. von B., "Input impedance of a folded dipole," *RCA Rev.* **8**, 289 (1947).
46. Starkey, B. J., and E. Fitch, "Mutual impedance and self-impedance of coupled parallel aerials," *J. Inst. Elec. Engrs.* (*London*) **97**, Pt. III, 129 (1950).
47. Starnecki, B., and E. Fitch, "Mutual impedance of two center-driven parallel aerials," *Wireless Engr.* **25**, 385 (1948).
48. Storer, J. E., "Solution of thin wire antenna problems by variational methods," doctoral dissertation, Cruft Laboratory, Harvard University (June 1951).
49. Storer, J. E., and R. King, "Radiation resistance of a two-wire line," Cruft Laboratory Technical Report No. 69, Harvard University (March 1949); *Proc. I.R.E.* **39**, 1408 (1951).
50. Tai, C. T., "Theory of coupled antennas and its application," doctoral dissertation and Cruft Laboratory Technical Report No. 12, Harvard University (1947).
51. Tai, C. T., "Coupled antennas," *Proc. I.R.E.* **36**, 487 (1948).
52. Tang, C., "The collinear array with sections of two-wire line as coupling elements," Cruft Laboratory Technical Report No. 196, Harvard University (1954).
53. Taylor, J., "The sleeve antenna," doctoral dissertation, Cruft Laboratory, Harvard University (1950).
54. Uda, S., and Y. Mushiake, "On the theory of antennae with discontinuous thickness," *Technol. Repts., Tôhoku Univ.* **14**, 105 (1950).
55. Uda, S., and Y. Mushiake, "Theoretical calculation of the input impedances of two parallel antennae," *Science Repts., Research Insts. Tôhoku Univ.* [B], 1 and 2, 91 (1951).
56. Weinbaum, S., "On the solution of definite integrals occurring in antenna theory," *J. Appl. Phys.* **15**, 840 (1944).
57. Zuhrt, H., "Eine strenge Berechnung der Dipolantennen mit rohrförmigem Querschnitt," *Frequenz* **4**, 135, 178, (1950).

CHAPTER IV. THE RECEIVING ANTENNA AS A CIRCUIT ELEMENT

1. Aden, A. L., "Electromagnetic scattering from metal and water spheres," Cruft Laboratory Technical Report No. 106, Harvard University (August 1950).
2. Burgess, R. E., "Aerial characteristics," *Wireless Engr.* **21**, 154 (1944).
3. Dennhardt, A., and E. H. Himmler, "Effektivhöhen von Empfangsantennen im Bereich von Sekundärstrahlen," *Hochfrequenztechn. u. Elektroak.* **43**, 152 (1934).
4. Dieckmann, M., "Verfahren zur Ermittlung der wirksamen Höhe von Antennen und des Empfangswertes einer Anlage unter Mitbenutzung des Biot-Savartschen Feldes in umittelbarer Antennennähe," *Jahrb. drahtl. Tel. u. Tel.* **31**, 65 (1928).
5. Dike, S. H., "Difficulties with present solutions of the Hallén integral equation," Radiation Laboratory Technical Report No. 14, Johns Hopkins University (June 1951); *Quart. Appl. Math.* **10**, 225 (1952); discussion: *Proc. I.R.E.* **41**, 926 (1953).
6. Dike, S. H., and D. D. King, "The cylindrical dipole receiving antenna," Radiation Laboratory Technical Report No. 12, Johns Hopkins University (May 1951); *Proc. I.R.E.* **40**, 853 (1952); discussion: *Proc. I.R.E.* **41**, 926 (1953).
7. Discussion: "Multiple reflections between two tuned receiving antennae," *Inst. Elec. Engrs.* (*London*) **84**, 749 (1939).
8. Feld, J. N., "The general reciprocity theorem in the theory of receiving and transmitting antennae," *Compt. rend. acad. sci. URSS* (n.s.) **48**, 476 (1945).
9. Fok, V., "The distribution of currents induced by a plane wave on the surface of a conductor," *Bull. acad. sci. URSS* **10**, 130 (1946).
10. Friis, H. T., "A note on a simple transmission formula," *Proc. I.R.E.* **34**, 254 (1946).
11. Grosskopf, J., "Zur Theorie der Empfangsantennen," *Frequenz* **4**, 249 (1950).
12. Grosskopf, J., "Empfangsantennen," *Telegr.-Fernspr.-u. Funk-Techn.* **27**, 129 (1938).
13. Harrison, C. W., Jr., and R. King, "The receiving antenna in a plane polarized field of arbitrary orientation," *Proc. I.R.E.* **32**, 35 (1944).
14. Istvánffy, E., "Antenna impedance measurement by reflection method," *Proc. I.R.E.* **37**, 604 (1949).
15. King, D. D., "Impedance measurements on transmission lines," *Proc. I.R.E.* **35**, 509 (1947).
16. King, D. D., "The measurement and interpretation of antenna scattering," *Proc. I.R.E.* **37**, 770 (1949), and Cruft Laboratory Technical Report No. 50, Harvard University (1948).
17. King, R., "An improved theory of the receiving antenna," *Proc. I.R.E.* **40**, 1113 (1952).
18. King, R., "Theory of electrically short transmitting and receiving antenna," *J. Appl. Phys.* **23**, 1174 (1952).
19. King, R., and C. W. Harrison, Jr., "The receiving antenna," *Proc. I.R.E.* **32**, 18 (1944).
20. Magnus, W., and F. Oberhettinger, "Zur Theorie der geraden Empfangsantenne," *Hochfrequenz-Techn. u. Elektroak.* **57**, 97 (1941).
21. Morita, T., "The measurements of current and charge distributions on cylindrical antennas," doctoral dissertation and Cruft Laboratory Technical Report No. 66, Harvard University (1948); *Proc. I.R.E.* **38**, 898 (1950).
22. Morita, T., E. O. Hartig, and R. King, "The measurement of antenna impedance using a receiving antenna (Supplement)," Cruft Laboratory Technical Report No. 94, Harvard University (1949); *Proc. I.R.E.* **39**, 1458 (1951).
23. Moritz, C., "Coupled receiving antennas," doctoral dissertation, Harvard University (June 1952).

24. Müller-Strobel, J., and J. Patry, "Berechnung von Hilfsfunktionen für gerade Empfangsantennen beliebiger Höhe," *Helv. Phys. Acta* **17**, 455 (1944).

25. Müller-Strobel, J., and J. Patry, "Der Empfangsdipol. Ableitung einer Formel für den Antennenstrom," *Schweiz. Arch. angew. Wiss. u. Tech.* **12**, 201 (1946).

26. Müller-Strobel, J., and J. Patry, "Die gerade Empfangsantenne. Ableitung einer Näherungsformel für den Antennenstrom," *Helv. Phys. Acta* **17**, 127 (1944).

27. Neiman, M. S., "The principle of reciprocity in antenna theory," *Proc. I.R.E.* **31**, 666 (1943).

28. Niessen, K. F., and G. De Vris, "Ueber die Empfangsimpedanz einer Empfangsantenne," I and II, *Physica* **6**, 601, 617 (1939).

29. Palmer, L. S., W. Alson, and R. H. Barker, "Multiple reflections between two tuned receiving antennae," *J. Inst. Elec. Engrs. (London)* **83**, 424 (1938).

30. Palmer, L. S., and K. G. Gillard, "The distribution of ultra-high-frequency currents in long transmitting and receiving antennae," *J. Inst. Elec. Engrs. (London)* **83**, 415 (1938).

31. Pippard, A. B., O. J. Burrell, and F. F. Cromie, "The influence of re-radiation on measurements of the power gain of an aerial," *J. Inst. Elec. Engrs. (London)* **93**, Pt. III A, 720 (1946).

32. Rüdenberg, R., "Empfang elektrischer Wellen in der drahtlosen Telegraphie," *Ann. Physik* **25**, 446 (1908); *Jahrb. drahtl. Tel. u. Tel.* **6**, 170 (1912).

33. Sevick, J., "An experimental method of measuring back-scattering cross sections of coupled antennas," Cruft Laboratory Technical Report No. 151, Harvard University (1952).

34. Sevick, J., "Experimental and theoretical results on the back-scattering cross sections of coupled antennas," Cruft Laboratory Technical Report No. 150, Harvard University (1952).

35. Sevick, J., and J. E. Storer, "A general theory of plane-wave scattering from finite two-antenna problem," Cruft Laboratory Technical Report No. 149, Harvard University (1952).

36. Siegel, E., "Empfang von Wellen mit abgestimmter Antenne und aperiodischer Empfang," *Hochfrequenztechn. u. Elektroak*, **45**, 198 (1935).

37. Stevenson, A. F., "Relations between the transmitting and receiving properties of antennas," *Quart. Appl. Math.* **5**, 369 (1948).

38. Tai, C. T., "Radar response from thin wires," Stanford Research Institute Technical Report No. 18, SRI Project 188 (1951).

39. Van Vleck, J. H., F. Bloch, and M. Hamermesh, "Theory of radar reflection from wires or thin metallic strips," *J. Appl. Phys.* **18**, 274 (1947).

40. Watson, W. H., "Wave-impedances and the effective cross sections of antennas," *Trans. Roy. Soc. Can. III* (3) **39**, 33 (1945).

41. Wilson, D. G., "The measurement of antenna impedance using a receiving antenna," doctoral dissertation and Cruft Laboratory Technical Report No. 43, Harvard University (1947).

42. Yeh, Yung-Ching, "The received power of a receiving antenna and the criteria for its design," *Proc. I.R.E.*, **37** 155 (1949).

43. Zuhrt, H., "Eine strenge Berechnung der Dipolantennen mit rohrförmigen Querschnitt," *Frequenz* **4**, 135, 178 (1950).

CHAPTER V. THE ELECTROMAGNETIC FIELD OF CENTER-DRIVEN AND MULTIPLE HALF-WAVE ANTENNAS

1. Abraham, M., "Die electrischen Schwingungen eines Stabförmigen Leiters behandelt nach der Maxwellschen Theorie," *Ann. Physik* **66** 435 (1898).

2. Ballantine, S. "On the radiation resistance of a simple vertical antenna at wavelengths below the fundamental," *Proc. I.R.E.* **12**, 823 (1924); discussion: **13**, 251 (1925).

3. Bechmann, R., "Calculation of electric and magnetic field strengths of oscillating straight conductors," *Proc. I.R.E.* **19**, 461 (1931); correction: 681 (1931).

4. Bechmann, R., "On the calculation of radiation resistance of antennas and antenna combinations," *Proc. I.R.E.* **19**, 1471 (1931).

5. Bechmann, R., "Zur Abrahamschen Darstellung des Strahlungsfeldes eines stabförmigen Leiters," *Jahrb. drahtl. Tel. u. Tel.* **38**, 30 (1931).

6. Booker, H. G., "Slot aerials and their relation to complementary wire aerials," *J. Inst. Elec. Engrs. (London)* **93**, Pt. III [A], 620 (1946).

7. Bouwkamp, C. J., "On the effective length of a linear transmitting antenna," *Philips Research Repts.* **4**, 179 (1949).

8. Brown, G. H., "A critical study of the characteristics of broadcast antennas as affected by antenna current distribution," *Proc. I.R.E.* **24**, 48 (1936).

9. Editorial: "The radiation resistance of a half-wave dipole aerial," *Wireless Engr.* **22**, 153, 365 (1945).

10. Fränz, K., "Berechnung des Strahlungswiderstandes einiger Dipolantennen," *Elekt. Nachr.-Tech.* **16**, 24 (1939).

11. Gihring, H. E., and G. H. Brown, "General considerations of tower antennas for broadcast use," *Proc. I.R.E.* **23**, 311 (1935).

12. Hack, F., "Das elektromagnetische Feld in der Umgebung eines linearen Oszillators," *Ann. Phys. u. Chem.* **14**, 539 (1904).

13. Hansen, W. W., and J. G. Beckerley, "Concerning new methods of calculating radiation resistance with or without ground," *Proc. I.R.E.* **24**, 1594 (1936).

14. Harrison, C. W., Jr., "An approximate representation of the electromagnetic field in the vicinity of a symmetrical radiator," *J. Appl. Phys.* **15**, 544 (1944).
15. Harrison, C. W., Jr., "The radiation field of long wires, with application to vee antennas," *J. Appl. Phys.* **14**, 537 (1943).
16. Harrison, C. W., Jr., and R. King, "The radiation field of a symmetrical center-driven antenna of finite cross section," *Proc. I.R.E.* **31**, 693 (1943).
17. Hatch, R. M., "Current distributions on conducting sheets excited by arrays of slot antennas," Cruft Laboratory Technical Report No. 103, Harvard University (1950).
18. Kayano, T., K. Nakamura, and S. Sonohe, "The radiation characteristics of a vertical broadcasting antenna," *Nippon Elec. Comm. Eng.*, No. 9, 94 (1938).
19. Kelvin, W., "The radiation field of an unbalanced dipole," *Proc. I.R.E.* **34**, 440 (1946).
20. King, R., "The approximate representation of the distant field of linear radiators," *Proc. I.R.E.* **29**, 458 (1941).
21. Levin, S. A., and C. J. Young, "Field distribution and radiation resistance of a straight vertical unloaded antenna radiating at one of its harmonics," *Proc. I.R.E.* **14**, 675 (1926); Errata: **15**, 8 (1927); discussion: **15**, 245, 439 (1927).
22. Moullin, E. B., "The radiation resistance of aerials whose length is comparable with the wavelength," *J. Inst. Elec. Engrs.* (*London*) **78**, 540, 563 (1936).
23. Page, H., and G. D. Monteith, "The vertical radiation patterns of medium-wave broadcasting aerials," *Proc. Inst. Elec. Engrs.* (London) **102**, Pt. B, 279 (May 1955).
24. Pedersen, P. O., *Radiation from vertical antenna over flat, perfectly conducting earth* (Ingeniørvidenskab. Skrifter [A], No. 38, 1935).
25. Riazin, P., "Sur le calcul du rayonnement d'une antenne rectiligne à petite distance," *Tech. Phys. U.S.S.R.* **4**, 1 (1937).
26. Riazin, P., "On the electromagnetic field from a vertical half-wave aerial above a plane earth," *Tech. Phys. U.S.S.R.* **5**, 29 (1938).
27. Rüdenberg, R., "Der Empfang der elektrischen Wellen in der drahtlosen Telegraphie," *Ann. Physik* **25**, 446 (1908).
28. Stansel, F. R., "A study of the electromagnetic field in the vicinity of a radiator," *Proc. I.R.E.* **24**, 802 (1936).
29. Stratton, J. A., and H. A. Chinn, "The radiation characteristics of a vertical half-wave antenna," *Proc. I.R.E.* **20**, 1892 (1932).
30. Van der Pol, B., "Radiation from antennae," *Proc. Phys. Soc.* (*London*) **29**, 269 (1917).

CHAPTER VI. ELECTROMAGNETIC FIELDS OF ANTENNA ARRAYS

1. Alfred, R. V., "Experiments with Yagi aerials at 600 Mc/sec," *J. Inst. Elec. Engrs.* (*London*) **93**, Pt. III [*A*], 1490 (1946).
2. Barbiere, D., "A method for calculating the current distribution of Tschebyscheff arrays," *Proc. I.R.E.* **40**, 78 (1952).
3. Bechmann, R., "Berechnung der Strahlungscharakteristiken und Strahlungswiderstände von Antennensystemen," *Jahrb. drahtl. Tel. u. Tel.* **36**, 182, 201 (1930).
4. Bechmann, R., "On the calculation of radiation resistance of antennas and antenna combinations," *Proc. I.R.E.* **19**, 1471 (1931).
5. Bontsch-Bruewitsch, M. A., "Die Strahlung der komplizierten rechtwinkeligen Antennen mit gleichbeschaffenen Vibratoren," *Ann. Physik* **81**, 425 (1926).
6. Brown, G. H., "Directional antennas," *Proc. I.R.E.* **25**, 78 (1937).
7. Brown, G. H., and R. King, "High frequency models in antenna investigations," *Proc. I.R.E.* **22**, 457 (1932).
8. Carter, P. S., C. W. Hansell, and N. E. Lindenblad, "Development of directive transmitting antennas by R.C.A. Communications, Inc.," *Proc. I.R.E.* **19**, 1773 (1931).
9. Chu, L. J., "Physical limitations of omnidirectional antennas," *J. Appl. Phys.* **19**, 1163 (1948).
10. Cutler, C. C., A. P. King, and W. E. Kock, "Microwave antenna measurements," *Proc. I.R.E.* **35**, 1462 (1947).
11. Dawson, L. H., and N. M. Rust, "A wideband linear-array aerial," *J. Inst. Elec. Engrs.* (*London*) **93**, Pt. III [*A*], 693 (1946).
12. Dolph, C. L., "A current distribution for broadside arrays which optimizes the relationship between beam width and side-lobe level," *Proc. I.R.E.* **34**, 335 (1946).
13. Duhamel, R. H., "Optimum patterns for end-fire arrays," *Proc. I.R.E.* **41**, 652 (1953).
14. Ebel, A. J., "Directional radiation patterns," *Electronics* **9**, 30 (April 1936).
15. Fishenden, R. M., and E. R. Wiblin, "Design of Yagi aerials," *J. Inst. Elec. Engrs.* (*London*) **96**, Pt. III, 5 (1949).
16. Foster, R. M., "Directive diagrams of antenna arrays," *Bell. System Tech. J.* **5**, 292 (1926).
17. Franz, K., "Die Verbesserung des Übertragungsgrades durch Richtantennen," *Telefunken-Mitt.* **21**, 49 (1940).
18. Gillson, J. L., "Parasitic-array patterns," *Q.S.T.* **33**, 11, 104 (1949).
19. Goward, F. K., "An improvement in end-fire arrays," *J. Inst. Elec. Engrs.* (*London*) **94**, Pt. III, 415 (1947).
20. Graf, H., "Richtcharakteristik und Strahlungsleistung von Richtantennen," *Frequenz* **3**, 136 (1949).
21. Hansen, W. W., and L. M. Hollingsworth, "Design of 'flat-shooting' antenna arrays," *Proc. I.R.E.* **27**, 137 (1939).

22. Hansen, W. W., and J. R. Woodyard, "A new principle in directional antenna design," *Proc. I.R.E.* **26**, 333 (1938).
23. Harris, E. F., "An experimental investigation of the corner reflector antenna," *Proc. I.R.E.* **41**, 645 (1953).
24. Harrison, C. W., Jr., "A note on the characteristics of the two antenna array," *Proc. I.R.E.* **31**, 75 (1943).
25. Harrison, C. W., Jr., "Radiation from vee antennas," *Proc. I.R.E.* **31**, 362 (1943).
26. Kelvin, W., "The radiation field of an unbalanced dipole," *Proc. I.R.E.* **34**, 440 (1946).
27. King, R., "The field of a dipole with a tuned parasite at constant power," *Proc. I.R.E.* **36**, 872 (1948); *Proc. Inst. Elec. Engrs. (London)* **99**, Pt. III, 6 (1952).
28. Knudsen, H. L., "The field radiated by a ring-quasi array of an infinite number of tangential or radial dipoles," *Proc. I.R.E.* **41**, 781 (June 1953).
28a. Knudsen, H. L., "Radiation resistance and gain of a homogeneous ring quasi-array," *Proc. I.R.E.* **42**, 686 (April 1954).
29. Kraus, J. D., "Antenna arrays with closely spaced elements," *Proc. I.R.E.* **28**, 76 (1940).
30. Labus, J., "Die Strahlungsenergie der Dipolantenne mit Reflektor," *ENT* **9**, 319 (1932).
31. McPetrie, J. S., L. H. Ford, and J. A. Saxton, "Polar diagrams of half-wave receiving aerials and a V-reflector," *Wireless Engr.* **22**, 263 (1945).
31a. McPetrie, J. S., and J. A. Saxton, "Some experiments with linear aerials," *Wireless Engr.* **23**, 107 (1946).
32. Medhurst, R. G., "Radiation from short aerials," *Wireless Engr.* **25**, 260 (1948).
33. Morrison, J. L., "Single method for observing current amplitude and phase relations in antenna arrays," *Proc. I.R.E.* **25**, 1310 (1937).
34. Moullin, E. B., "Total output of curtain array of aerials," *J. Inst. Elec. Engrs. (London)* **91**, Pt. III, 23 (1944).
35. Page, H., "Horizontal dipole transmitting arrays," *J. Inst. Elec. Engrs. (London)* **92**, Pt. III, 68 (1945).
36. Palmer, L. S., W. Abson, and R. H. Banker, "Multiple reflections between two tuned receiving aerials," *J. Inst. Elec. Engrs. (London)* **43**, 424 (1938).
37. Pistolkors, A. A., "The radiation resistance of beam antennas," *Proc. I.R.E.* **17**, 562 (1929).
38. Reid, D. G., "The gain of an idealized Yagi array," *J. Inst. Elec. Engrs. (London)* **93**, Pt. III [A], 564 (1946).
39. Riblet, H. J., "Discussion on 'A current distribution for broadside arrays which optimizes the relationship between beam width and side-lobe level,' by C. L. Dolph," *Proc. I.R.E.* **35**, 489 (1947).
40. Riblet, H. J., "Note on the maximum directivity of an antenna," *Proc. I.R.E.* **36**, 620 (1948).
41. Rumble, A. R., "Directional array field strength," *Electronics* **10**, 16 (August 1937).
42. Rutelli, G., "Ueber einige Eigenschaften aus Dipolen aufgebauten Antennen-systemen. Antenne für Richt- oder Rundstrahlung in der Kurz- und Ultrakurz-wellentechnik," *Z. Tech. Phys.* **21**, 140 (1940).
43. Sammer, F., "Die Wirkungsweise von Drahtreflektoren," *Telefunken-Ztg.* **10**, 61 (1929).
44. Schelkunoff, S. A., "A mathematical theory of linear arrays," *Bell System Tech. J.* **22**, 80 (1943).
45. Schmidt, O., "Aerial arrays with horizontal beams without side lobes," *Bull. schweiz. elektrotech. Ver.* **38**, 15 (1947).
46. Southworth, G. C., "Certain factors affecting the gain of directive antennas," *Proc. I.R.E.* **18**, 1502 (1930).
47. Spangenberg, K., "Charts for the determination of the root-mean-square value of the horizontal radiation pattern of two-element broadcast antenna arrays," *Proc. I.R.E.* **30**, 237 (1942).
48. Starkey, B. J., and E. Fitch, "Mutual impedance and self-impedance of coupled parallel aerials," *Proc. Inst. Elec. Engrs. (London)* **97**, Pt. III, 129–137 (1950).
49. Sterba, E. J., "Theoretical and practical aspects of directional transmitting systems," *Proc. I.R.E.* **19**, 1184 (July 1931).
50. Stegen, R. J., "Excitation coefficients and beam widths of Tschebyscheff arrays," *Proc. I.R.E.* **41**, 1671 (1953).
51. Takenchi, H., "On the spacing between projector and reflector of a beam antenna," *Nippon Elec. Comm. Eng.*, No. 10, 189 (1938).
52. Taylor, T. T., and J. R. Whinnery, "Applications of potential theory to the design of linear arrays," *J. Appl. Phys.* **22**, 19 (1951).
53. Tschebyscheff, P. L., "Sur les questions de minima, qui se rattachent à la représentation approximative des fonctions," *Mém. acad. sci. St. Pétersburg* [6], **7**, 199–291 (1859); *Oeuvres* (St. Petersburg, 1899), vol. 1, pp. 295–301.
54. Tetelbaum, S., "On some problems of the theory of highly-directive antenna arrays," *Bull. acad. sci. URSS* **10**, 285–292 (1946).
55. Thomson, W. T., "Development of the general antenna array equation," *J. Appl. Phys.* **15**, 420 (1944).
56. Van der Pol, B., and T. J. Wayers, "Tschebyscheff polynomials and their relation to circular functions, Bessel functions and Lissajous figures," *Physica* **1**, 78 (1933).
57. Walkinshaw, W., "Theoretical treatment of short Yagi aerials," *J. Inst. Elec. Engrs. (London)* **93**, Pt. III [A], 598 (1946).
58. Wells, N., "Quadrant aerial," *J. Inst. Elec. Engrs. (London)* **91**, Pt. III, 182 (1944).
59. Wilmotte, R. M., "Directional antennae for broadcasting," *Electronics* **5**, 362 (1932).
60. Wolff, I., "Determination of the radiating system which will produce a specified directional characteristic," *Proc. I.R.E.* **25**, 630 (1937).
61. Yagi, H., "Beam transmission of ultra-short waves," *Proc. I.R.E.* **16**, 715 (1928).

62. Yagi, H., and S. Uda, "Projector of the sharpest beam of electric waves," *Proc. Imp. Acad. (Tokyo)* **2**, 49 (1926).

63. Yagi, H., and S. Uda, "A new electric wave projector and radio beam," *Proc. 3rd Pan-Pacific Scientific Congress* (Tokyo, 1926).

64. Zinke, O., "Fed-dipole groups as longitudinal radiators for broad frequency bands," *Funk u. Ton*, **2**, 435 (1948).

CHAPTER VII. ANTENNAS OVER A CONDUCTING REGION

1. Alpert, J. L., and V. V. Migulin, "Der Einfluss der Erdoberfläche auf die Phasen-Struktur des elektromagnetischen Feldes einer Antenne," *Compt. rend. acad. sci. U.R.S.S.* **26**, 881 (1940).

2. Atkinson, F. V., "On Sommerfeld's Radiation Condition," *Phil Mag.* [7] **40**, 645 (1949).

3. Baños, A., Jr., and J. P. Wesley, "The horizontal electric dipole in a conducting half-space," Parts I and II, Reports by Marine Physical Laboratory of the Scripps Institution of Oceanography, University of California, La Jolla, California (1953).

4. Barrow, W. L., "On the impedance of a vertical half-wave antenna above an earth of finite conductivity," *Proc. I.R.E.* **23**, 150 (1935).

5. Brown, G. H., "Radial ground system chart," *Electronics* **11**, 33 (January 1938).

6. Brown, G. H., "The phase and magnitude of earth currents near radio transmitting antennas," *Proc. I.R.E.* **23**, 168 (1935).

7. Brown, G. H., "Vertical vs. horizontal polarization," *Electronics* **13**, 20 (October 1940).

8. Brown, G. H., and J. G. Leitch, "The fading characteristics of the top-loaded WCAV antenna," *Proc. I.R.E.* **25**, 583 (1937); discussion: **26**, 115-121 (1938).

9. Brown, G. H., R. F. Lewis, and J. Epstein, "Ground systems as a factor in antenna efficiency," *Proc. I.R.E.* **25**, 753 (1937).

10. Burgess, R. E., "Ground absorption with elevated vertical and horizontal dipoles," *Wireless Engr.* **26**, 133 (1949).

11. Clemmow, P. C., "Radio propagation over a flat earth across a boundary separating two different media," *Trans. Royal Soc. (London)* [A] **246**, 1 (1953).

12. Clemmow, P. C., and C. M. Mumford: "A Table of $\sqrt{(\frac{1}{2}\pi)} e^{\frac{1}{2}i\pi\rho^2} \int_\rho^\infty e^{-\frac{1}{2}i\pi\lambda^2} \, d\lambda$ for complex values of ρ," *Trans. Royal Soc. (London)* [A] **245**, 189 (1952).

13. Dunn, B. C., Jr., and R. King, "Current excited on a conducting plane by a parallel dipole," *Proc. I.R.E.* **36**, 221 (1948).

14. Eckersley, T. L., and G. Millington, "Application of the phase integral method to the analysis of the diffraction and refraction of wireless waves round the earth," *Trans. Roy. Soc. (London)* [A] **237**, 273 (1938).

15. Eckersley, T. L., and G. Millington, "The diffraction of wireless waves round the earth," *Phil. Mag.* [7] **27**, 517 (1939).

16. Eckersley, T. L., and G. Millington, "The experimental verification of the diffraction analysis of the relation between height and gain for radio waves of medium lengths," *Proc. Phys. Soc. (London)* **51**, 805 (1939).

17. Foster, R. M., "Mutual impedance of grounded wires lying on the surface of the earth," *Bell System Tech. J.* **10**, 408 (1931).

18. Gray, M. C., "Horizontal polarized electromagnetic waves over a spherical earth," *Phil. Mag.* [7] **27**, 421 (1939).

19. Grosskopf, J., "Das Strahlungsfeld eines vertikalen Dipolsenders über geschichtetem Boden," *Hochfrequenztechnik u. Elektroak.* **60**, 136 (1942).

20. Grosskopf, J., "Ueber das Zennecksche Drehfeld im Bodenwellenfeld eines Senders," *Hochfrequenztech. u. Elektroak.* **59**, 72 (1942).

21. Hansen, W. W., "Directional characteristics of any antenna over a plane earth," *J. Appl. Phys.* **7**, 460 (1936).

22. Hansen, W. W., and J. G. Beckerly, "Radiation from an antenna over a plane earth of arbitrary characteristics," *J. Appl. Phys.* **7**, 220 (1936).

23. v. Hoerschelmann, H., "Ueber die Wirkungsweise des geknickten Marconischen Senders in der drahtlosen Telegraphie," *Jahrb. drahtl. Tel. u. Tel.* **5**, 14 and 188 (1912).

24. Jacknow, W., "Ueber den Strahlungswiderstand eines geraden linearen Strahlers bei gedämpften fortschreitenden Wellen," *Elek. Nachr.-Tech.* **17**, 141 (1940).

25. Janoch, W., "Zur Theorie der Langdrahtsendeantenna, insbesondere bei fortschreitenden Wellen," *Telefunken-Mitt.* **21**, 55 (1940).

26. Kahan, T., and G. Eckart, "On the existence of a surface wave in dipole radiation over a plane earth," *Proc. I.R.E.* **38**, 807 (1950); *Compt. rend.* **226**, 1513 (1948).

27. Kato, Y., "The effect of the earth on the radiation impedance of short-wave antennas," *Nippon Elec. Comm. Eng.*, No. 11, 275 (1938).

28. Latmiral, G., "Oberflächenstrahlung von Horizontalantennen und Messung der elektrischen Konstanten des Bodens," *Alta Frequenza* **7**, 509 (1938).

29. Leitner, A., and R. D. Spence, "Effect of circular ground plane on antenna radiation," *J. Appl. Phys.* **21**, 1001 (1950).

30. Lewin, L., "Strahlungswiderstand eines horizontalen Dipols über der Erde," *Marconi Rev.*, No. 73, 13 (1939).

31. Maas, G. J. van der, "A simplified calculation for Dolph-Tchebycheff Arrays," *J. App. Phys.* **25**, 121 (1954).

32. Mayer, R., "Ein Beitrag zur Berechnung von Erdverlusten bei Antennenanlagen," *Jahrb. drahtl. Tel. u. Tel.* **29**, 71 (1927).

33. Meier, A. S., and W. P. Summers, "Measured impedance of vertical antennas over a finite ground plane," *Proc. I.R.E.* **37**, 609 (1949).
34. Meissner, A., "Raumstrahlung von Horizontal-Antennen," *ENT* **4**, 482 (1927).
35. Meissner, A., "Richtstrahlung mit horizontalen Antennen," *Jahrb. drahtl. Tel. u. Tel.* **30**, 77 (1927).
36. Niessen, K. F., "Berechnung der von einer Halbwellenlänge-Antenne elektrischen Feldstärke als Funktion der Antenne sämtlich je sekunde zugeführten Energie," *Physica* **7**, 586 (1940).
37. Niessen, K. F., "Erdabsorption bei horizontalen Dipolantennen," *Ann. Physik* **32**, 444 (1938).
38. Niessen, K. F., "On the approximate ground-absorption formula for vertical dipoles," *Physica* **9**, 915 (1942).
39. Niessen, K. F., "The ratio between the horizontal and the vertical electric field of a vertical antenna of infinitesimal length," *Philips Research Repts.* **1**, 51 (1945).
40. Niessen, K. F., "Ueber das Feld einer vertikalen Halbwellenantenne in beliebiger Höhe oberhalb einer ebenen Erde beliebiger Konstanten," *Ann. Physik.* **31**, 522 (1938).
41. Niessen, K. F., "Ueber die Wirkung eines vertikalen Dipolsenders auf ebener Erde in einem Entfernungsbereich von der Ordnung einer Wellenlänge," *Ann. Physik* **28**, 209 (1937).
42. Niessen, K. F., "Zur Entscheidung zwischen horizontalen oder vertikalen elektrischen Dipolen zwechs minimaler Erdabsorption bei gegebener Bodenart und Wellenlänge. *Ann. Physik* **33**, 404 (1938).
43. Norton, K. A., 'Physical reality of space and surface waves in the radiation field of radio antennas," *Proc. I.R.E.* **25**, 1192 (1937).
44. Norton, K. A., "The calculation of ground-wave field intensity over a finitely conducting spherical earth," *Proc. I.R.E.* **29**, 623 (1941).
45. Norton, K. A., "The polarization of downcoming radio waves," FCC Mimeograph No. 60047 (May 1942).
46. Norton, K. A., "The propagation of radio waves over the surface of the earth and in the upper atmosphere," Part I, *Proc. I.R.E.* **24**, 1367 (1936); Part II, *Proc. I.R.E.* **25**, 1208 (1937).
47. Ott, H., "Die Sattelpunktsmethode in der Umgebung eines Pols," *Ann. Physik* **43**, 395 (1943).
48. Papas, C. H., and R. King, "Surface currents on a conducting sphere excited by a dipole," *J. Appl. Phys.* **19**, 808 (1948).
49. Papas, C. H., and R. King, "Radiation from wide-angle conical antennas fed by a coaxial line," *Proc. I.R.E.* **39**, 49 (1951).
50. Pekeris, C. L., "The field of a microwave dipole antenna in the vicinity of the horizon," *J. Appl. Phys.* **18**, 667 (1947).
51. Proctor, R. F., "Input impedance of horizontal dipole aerials at low heights above the ground," *J. Inst. Elec. Engrs. (London)* **97**, Pt. III 188 (1950).
52. Raymond, R. C., and W. Webb, "Radiation resistances of loaded antennas," *J. Appl. Phys.* **20**, 328 (1949).
53. Rice, S. O., "Series for the wave function of a radiating dipole at the earth's surface," *Bell System Tech. J.* **16**, 101 (1937).
54. Sommerfeld, A., "Ueber die Ausbreitung der Wellen in der drahtlosen Telegraphie," *Ann. Physik* **28**, 665 (1909).
55. Sommerfeld, A., and F. Renner, "Strahlungsenergie und Erdabsorption bei Dipolantennen," *Ann. Physik* **41**, 1 (1942); summarized in *Hochfrequenztech. u. Elektroak* **59**, 168 (1942).
56. Storer, J. E., "Impedance of an antenna over a large circular screen," *J. Appl. Phys.* **22**, 1058 (1951); "Radiation pattern of an antenna over a circular ground screen," *J. Appl. Phys.* **23**, 588 (1952); and "Solution of thin wire antenna problems by variational methods," Ph.D. thesis, Harvard University (June 1951).
57. Strutt, M. J. O., "Strahlung von Antennen unter dem Einfluss der Erdbodeneigenschaften," *Ann. Physik* **1**, 721 (1929); **4**, 1 (1930); **9**, 67 (1931).
58. True, H., "Ueber die Erdströme in der Nähe einer Sendeantenne für drahtlose Telegraphie," *Jahrb. drahtl, Tel. u. Tel.* **5**, 125 (1912).
59. Van der Pol, B., "Theory of the reflection of the light from a point source by a finitely conducting flat mirror, with an application to radiotelegraphy," *Physica* **2**, 843 (1935).
60. Van der Pol, B., "Ueber die Ausbreitung elektromagnetischer Wellen," *Z. Hochfrequenztech.* **37**, 152 (1931).
61. Van der Pol, B., "Ueber die Wallenlängen und Strahlung mit Kapazität und Selbstinduktion beschwerter Antennen," *Jahrb. drahtl. Tel. u. Tel.* **13**, 217 (1919).
62. Van der Pol, B., and H. Bremmer, "Ergebnisse einer Theorie über die Fortpflanzung elektromagnetischer Wellen über eine Kugel endlicher Leitfahigheit," *Hochfrequenz. u. Elektroak.* **51**, 181 (1938).
63. Van der Pol, B., and H. Bremmer, "The propagation of radio waves over a finitely conducting spherical earth," *Phil. Mag.* **25**, 817 (1938).
64. Van der Pol, B., and H. Bremmer, "The diffraction of electromagnetic waves from an electrical point source round a finitely conducting sphere with applications to radiotelegraphy and the theory of the rainbow," Part I, *Phil. Mag.* **24**, 141 (1937); Part II, *Phil. Mag.* **24**, 825 (Supplement, November 1937).
65. Van der Pol, B., and K. F. Niessen, "Ueber die Raumwellen von einem vertikalen Dipolsender auf ebener Erde," *Ann. Physik* **10**, 485 (1931).
66. Van der Waerden, B. L., "On the method of saddle points," *Appl. Sci. Research* **B2**, 33–46 (1951).
67. Vilbig, F., "Untersuchungen an Erdern von Funkenderanlagen," *V.D.E.-Fachber.* **9**, 230 (1937).

68. Vilbig, F., and K. Vogt, "Untersuchungen an Vertikal-Antennen mit horizontalen Dachkapazitäten," *Hochfrequenztechn. u. Elektroak.* **50**, 58 (1937).

69. Violet, P. G., "Reflexion und Brechung elektrischer Wellen am Erdboden," *Hochfrequenztechn. u. Elektroak* **46**, 192 (1935).

70. Wait, J. R., "Current-carrying wire loops in a simple inhomogeneous region," *J. Appl. Phys.* **23**, 497 (1952).

71. Wait, J. R., and W. J. Surtees, "Impedances of a top-loaded antenna of arbitrary length over a circular grounded screen," *J. Appl. Phys.* **25**, 553 (1954).

72. Weyl, H., "Ausbreitung elektromagnetischer Wellen über einem ebenen Leiter," *Ann. Phys.* **60**, 481 (1919).

73. Wise, W. H., "Asymptotic dipole radiation formulas," *Bell System Tech. J.* **8**, 662 (1929).

74. Wise, W. H., "Note on dipole radiation theory," *J. Appl. Phys.* **4**, 354 (1933).

75. Wise, W. H., "The grounded condenser antenna radiation formula," *Proc. I.R.E.* **19**, 1684 (1931).

CHAPTER VIII. THE ANTENNA AS A BOUNDARY-VALUE PROBLEM

1. Abraham, M., "Die electrischen Schwingungen um einen stabförmigen Leiter behandelt nach der Maxwell'schen Theorie," *Wied. Ann.* **66**, 435 (1898).

2. Abraham, M., "Funktelegraphie und Elektrodynamik," *Physik. Z.* **2**, 1 (1901).

3. Albert, E., and J. L. Synge, "The general problem of antenna radiation," *Quart. Appl. Math.* **6**, 117 (1948).

4. Arenberg, A. G., "Der Strahlungswiderstand eines kugelförmigen, durch eine äussere elektromotorische Kraft erregten Oszillators," *Compt. rend. acad. sci. URSS* **25**, 593 (1939).

5. Barrow, W. L., L. J. Chu, and J. J. Jansen, "Biconical electromagnetic horns," *Proc. I.R.E.* **27**, 769 (1939).

6. Debye, P., "Der Lichtdruck auf Kugeln von beliebigem Material," *Ann. Physik* **30**, 57 (1909).

7. Hansen, W. W., "A new type of expansion in radiation problems," *Phys. Rev.* **47**, 139 (1935).

8. Infeld, L., "The influence of the width of the gap in the theory of antennas," *Quart. Appl. Math.* **5**, 113 (1947).

9. King, R., "The gap problem in antenna theory," *J. Appl. Phys.* **26**, 317 (1955).

10. Mie, G., "Beiträge zur Optik trüber Medien, speziell kolloidaler Metallösungen," *Ann. Physik* **25**, 377 (1908).

11. Moullin, E. B., "The radiation resistance of surfaces of revolution, such as cylinders, spheres and cones," *J. Inst. Elec. Engrs. (London)* **88**, Pt. III, 50 (1941).

12. Page, L., "The electrical oscillations of a prolate spheroid: Part II, Prolate spheroid wave functions; Part III, The antenna problem," *Phys. Rev.* **65**, 98, 111 (1944).

13. Page, L., and N. I. Adams, Jr., "The electrical oscillations of a prolate spheroid, Paper I," *Phys. Rev.* **53**, 819 (1938).

14. Papas, C. H., and R. King, "Input impedance of wide-angle conical antennas fed by a coaxial line," *Proc. I.R.E.* **37**, 1269 (1949).

15. Rice, S. O., "Sums of series of the form $\sum_0^\infty a_n J_{n+\alpha}(z) J_{n+\beta}(z)$," *Phil. Mag.* **35**, 686 (1944).

16. Ryder, R. M., "The electrical oscillations of a perfectly conducting prolate spheroid," *J. Appl. Phys.* **13**, 327 (1942).

17. Schelkunoff, S. A., "Principal and complementary waves in antennas," *Proc. I.R.E.* **34**, 23P (1946).

18. Schelkunoff, S. A., "Theory of antennas of arbitrary size and shape," *Proc. I.R.E.* **29**, 493 (1941).

19. Schelkunoff, S. A., "Transmission theory of spherical waves," *Trans. A.I.E.E.* **57**, 744 (1938).

20. Smith, P. D. P., "The conical dipole of wide angle," *J. Appl. Phys.* **19**, 11 (1948).

21. Stratton, J. A., and L. J. Chu, "Steady-state solutions of electromagnetic field problems: I. Forced oscillations of a cylindrical conductor; II. Forced oscillations of a conducting sphere; III. Forced oscillations of a prolate spheroid," *J. Appl. Phys.* **12**, 230, 236, 241 (1941).

22. Tai, C. T., "Application of a variational principle to biconical antennas," *J. Appl. Phys.* **20**, 1076 (1949).

23. Tai, C. T., "A variational solution to the problem of cylindrical antennas," Technical Report No. 12, Project No. 188, Stanford Research Institute (1950).

24. Tai, C. T., "On the theory of biconical antennas," *J. Appl. Phys.* **19**, 1155 (1948).

25. Zuhrt, H., "Eine strenge Berechnung der Dipolantennen mit rohrförmigen Querschnitt," *Frequenz* **4**, 135, 178 (1950).

LIST OF PRINCIPAL SYMBOLS

A complete list of all symbols and of all meanings of every symbol listed is not provided. Only the more important symbols and meanings and especially those that occur in a number of sections are given. A page reference is provided for each, which locates the first occurrence of the symbol or, in some instances where this is more convenient, a typical reference. In general, symbols and meanings of symbols that are involved in only one or two sections are not listed.

$\mathbf{A}$	real or complex vector potential, 14
A	complex coefficient in variational formulation of current in antenna, 254
A	array factor in general, 580
$\mathbf{A}^r$	radiation- or far-zone vector potential, 15
A_r, A_θ, A_z	complex cylindrical components of $\mathbf{A}$, 523
A_R, A_Θ, A_Φ	complex spherical components of $\mathbf{A}$, 15
A_x, A_y, A_z	complex rectangular components of $\mathbf{A}$, 15
$A^r_{21R}, A^r_{21\Theta_1}, A^r_{21z}$	radiation zone components of vector potential in R, Θ_1, and z directions on antenna 2 due to currents in antenna 1, 464
A_{ev}, A_{od}	even and odd components of complex vector potential, 431
A_{1z}, A_{2z}	z components of complex vector potential on conductors 1 and 2, 264
A_{11z}, A_{12z}	parts of A_{1z} due to currents in conductors 1 and 2, 264
A_{22z}, A_{21z}	parts of A_{2z} due to currents in conductors 2 and 1, 265
$A(z)$	complex array factor for nonuniform arrays, 651
	complex vector potential at coördinate z, 74
$A^s(z), A^a(z)$	symmetric and antisymmetric parts of $A(z)$, 74
$A_c(\Theta; N_c, n_c, t_c)$	array factor for collinear array, 590
$A_p(\Theta, \Phi; N_p, n_p, t_p)$	array factor for parallel array, 600
$A_B(\Theta, \Phi; N_B, n_B, t_B)$	array factor for broadside array, 601
$A_E(\Theta, \Phi; N_E, n_E, t_E)$	array factor for end-fire array, 601
A_1, A_2	first and second-order terms in current formula, 85
$A^I_1, A^{II}_1; A^I_2, A^{II}_2$	real and imaginary parts of A_1 and A_2, 85
$A_{1H}, A^I_{1H}, A^{II}_{1H}; A_{2H}, A^I_{2H}, A^{II}_{2H}$	Hallén forms of A_1 and A_2 with their real and imaginary parts, 93, 101
$A_{1K}, A^I_{1K}, A^{II}_{1K}; A_{2K}, A^I_{2K}, A^{II}_{2K}$	King-Middleton forms of A_1 and A_2 with real and imaginary parts, 107
a	radius of cylindrical conductor, 15
	normalized array factor, 580
$a(z)$	complex normalized array factor for nonuniform arrays, 651
$a(\psi; N_r, n_r, t_r)$	normalized array factor, 586
$a_c(\Theta; N_c, n_c, t_c)$	normalized array factor for collinear array, 590
$a_p(\Theta, \Phi; N_p, n_p, t_p)$	normalized array factor for parallel array, 601
$a_B(\Theta, \Phi; N_B, n_B, t_B)$	normalized array factor for broadside array, 612
$a_E(\Theta, \Phi; N_E, n_E, t_E)$	normalized array factor for end-fire array, 612

a_e	effective radius of cylindrical antenna, 15
	effective radius of cage antenna, 370
	effective radius of symmetrical pair of antennas, 333
	semimajor axis of ellipse, 540
a_h	semimajor axis of hyperbola, 540
$a_1(w), a_1$	complex ratio of components of vector potential, 41
$a_1(w), a_1$	real ratio of components of vector potential; criterion for inductive coupling, 42
$\mathbf{B}$	real or complex magnetic vector (magnetic flux density), 13
B	complex coefficient in variational formulation of current in antenna, 254
B_r, B_Θ, B_z	complex cylindrical components of $\mathbf{B}$, 15
B_R, B_Θ, B_Φ	complex spherical components of $\mathbf{B}$, 529
B_x, B_y, B_z	complex rectangular components of $\mathbf{B}$, 700
$\mathbf{B}^e$	real or complex impressed or externally maintained magnetic vector, 28
$\mathbf{B}^r$	real or complex radiation-zone magnetic vector, 21
$B_{\dot{e}}, B_m$	angles of complex numerical distance, 759
B_n	coefficients in impedance formula, 142
B_0, B_δ	susceptances of antenna driven by discontinuity in scalar potential and with terminals separated a distance 2δ, 149
$B_\parallel, B_\perp$	magnetic field components parallel and perpendicular to the plane of incidence, 770
b	separation of two-wire line, 33
	separation of parallel antennas, 266
b_a	separation of parallel antennas when line with separation b is involved, 264
b_e	effective separation of closely spaced line, 274
	semiminor axis of ellipse, 540
b_e, b_m	angles of complex numerical distance, 759
C	Euler's constant, 96
C_s	lumped shunt capacitance, 193
C_T, C_{Te}, C_{Tc}	terminal-zone capacitances for coupled antennas, 315
$C(A, U), Cs(A, U), Cc(A, U)$	generalized sine and cosine integral functions, 95
$C_i(A, U), Cc_1(A, U)$	generalized cosine integral functions, 95
$Ci(U), Cin(U)$	integral cosine and modified integral cosine, 96
$C_a(h, z), C_b(h, z), C_r(h, z)$	94, 273, 525
$C_{qa}(h, z)$	474
$C_v(h, z, \Delta)$	396
$C_{11}(h, z), C_{12}(h, z, \Delta)$	384
$c(w)$	capacitance per unit length at distance w from load end of terminated line, 41
$c_0(w)$	part of $c(w)$ due to charges on the line, 41
c_0	capacitance per unit length of infinite line, 43
$\mathbf{D}$	real or complex auxiliary electric vector (electric displacement), 13
D	diameter of ground screen, 808
$D, D(\Theta, \beta_0 h)$	absolute directivity, 21, 497
D_r	relative directivity, 593
$(D_n)_m$	D-factors in formula for current, 110
d	height of dipole over conducting plane, 708
	length of line section in which terminal effects are significant, 44
d, d_c	distance between centers of collinear antennas, 425, 589

[Symbols]

$\mathbf{E}$	real or complex electric vector, 13
$\mathbf{E}^e$	real or complex impressed or externally maintained electric vector, 22
$\mathbf{E}^r$	real or complex far- or radiation-zone electric vector, 21
E_r, E_θ, E_z	complex cylindrical components of $\mathbf{E}$, 15
E_R, E_Θ, E_Φ	complex spherical components of $\mathbf{E}$, 529
E_x, E_y, E_z	complex rectangular components of $\mathbf{E}$, 700
$E_\parallel, E_\perp$	electric field components parallel and perpendicular to the plane of incidence, 770
E_{rad}	radiating efficiency, 284
$Ei(v)$	exponential integral, 524
$E_a(h, z)$	94
$E_{qa}(h, z)$	474
$E_v(h, z, \Delta)$	394
$E_{11}(h, z), E_{12}(h, z, \Delta)$	384
$E^r_{21\Theta_1}$	radiation-zone component of electric field in Θ_1 direction at antenna 2 due to currents in antenna 1, 464
e_e, e_h	eccentricities of ellipse and hyperbola, 541
$\mathbf{F}, \mathbf{F}_m$	electromagnetic and mechanical forces, 14
F	complex surface-wave attenuation function, 760
$F_0(z), F_{0z}$	zeroth-order term and difference term in formula for current in antenna, 82
$F_1(z), F_{1z}; F_m(z), F_{mz}$	first- and mth-order terms and difference terms in formula for current, 82, 83
$F_{1H}(z), F_{mH}(z)$	first- and mth-order terms in formula for current; Hallén forms, 93, 94
$F_{1K}(z), F_{2K}(z), (F_{nz})_K$	first- and second-order terms and nth-order difference term in formula for current; King-Middleton forms, 107, 109
$F_m(\Theta, \beta_0 h), F_0(\Theta, \beta_0 h)$	field functions of linear antennas, 529
f	electric scalar function, 20
f_r, f_t	plane-wave reflection and transmission coefficients, 707
f^r_{er}, f^r_{mr}	radiation-zone reflection coefficients, 726
f''_r, f'_r, f_r	real and imaginary parts and magnitude of f^r_r, 727
$f(h)$	real part of complex radical, 22
$f(P, B)$	magnitude of complex surface-wave attenuation function F, 760
$f(q)$	height-gain function, 775
$f(z)$	current distribution function for antenna, 77
G	free-space Green's function, 80
G	electrical length of top-loaded antenna, 744
	gain in decibels, 593
G_c	characteristic conductance of infinite transmission line, 46
G_0, G_δ	conductance of antenna driven by discontinuity in scalar potential and with terminals separated a distance 2δ, 149, 144
$(G_0)_1, (G_0)_2$	first- and second-order approximations of G_0, 119
$G_0(z), G_{0z}$	zeroth-order term and difference term in formula for current in antenna, 82
$G_1(z), G_{1z}$	first-order term and difference term in formula for current, 82
$G_m(z), G_{mz}$	mth-order term and difference term in formula for current, 83
$G_{1H}(z), G_{mH}(z)$	first- and mth-order terms in formula for current; Hallén forms, 93, 94

$G_{1K}(z), G_{2K}(z), (G_{nz})_K$	first- and second-order terms and nth-order difference term in formula for current; King-Middleton forms, 107, 109
g	magnetic scalar function, 20
g	length of antenna including top load, 743
g_0	conductance per unit length of infinite line, 43
$g(h)$	imaginary part of complex radical, 22
$g(w)$	conductance per unit length of terminated line, 41
$g_0(w)$	part of $g(w)$ due to line, 41
$g(z, z')$	current distribution function, 77
$g_H(z, z'), g_K(z, z')$	current distribution functions in the Hallén and King-Middleton solutions, 90, 101
$g_U(z, z')$	current distribution function for unloaded receiving antenna, 473
$\mathbf{H}$	real or complex auxiliary magnetic vector (magnetic intensity), 13
H	electrical length of antenna, 744
H	principal factor in magnetization potential along conducting surface, 755
$\mathbf{H}^r$	far- or radiation-zone form of $\mathbf{H}$, 724
$H_0(z), H_{0z}$	zeroth-order term and difference term in formula for current in receiving antenna, 463
$H_m(z), H_{mz}$	mth-order term and difference term in formula for current in receiving antenna, 463
h	half length of center-driven antenna, 28
h_e	loss tangent, 22
h_t	effective length of transmitting antenna, 567
h_x	factor in reflection coefficients, 727
$h_{e\delta}(\Theta), h_e(\Theta)$	complex effective length of receiving antenna, 467
$h_e''(\Theta), h_e'(\Theta), h_e(\Theta)$	real and imaginary parts and magnitude of complex effective length, 487
$h_n^{(1)}(x), h_n^{(2)}(x)$	spherical Hankel functions of order n, 823
$I_z, I_x(w)$	complex total current in direction and at coördinate indicated by subscript; if these are not the same, the latter is put in parentheses, 15, 36
$(I_z)_0, (I_z)_1, (I_z)_m$	zeroth-, first, and mth-order currents, 83
I_z'', I_z', I_z	real and imaginary parts and magnitude of I_z, 85
$I^s(z), I^a(z)$	symmetric and antisymmetric currents at coördinate z, 74, 267
$I^{(0)}\ I^{(1)}\ I^{(m)}$	currents in zeroth, first, and mth phase sequences, 353
I_{ev}, I_{od}	even and odd currents, 431
I_{zU}	current in unloaded receiving antenna along z axis, 463
I_δ	current in terminal at $z = \delta$, 29
$I_d(R), I_c(R)$	dominant- and complementary-mode currents in biconical antenna, 827, 829
$I_R(R)$	total radial current in biconical antenna, 846
$I_{1\delta}, I_{2\delta}$	currents in terminals at $z = \delta$ of antennas 1 and 2, 286
i	imaginary unit in time dependence $e^{-i\omega t}(i = -j)$, 705
$\mathbf{i}$	real or complex vector volume density of current, 12
$\mathbf{i}_f$	free-charge volume density of current, 15
i_z	complex z component of $\mathbf{i}$, 15
$\mathbf{i}^e$	real or complex vector volume density of impressed current, 22

[Symbols] THEORY OF LINEAR ANTENNAS 917

$J_c(z)$, $J_s(z)$	integrals related to ohmic losses in antenna, 94
j	imaginary unit in time dependence $e^{j\omega t}$, 12
$j_n(x)$	spherical Bessel function of order n, 823
K	electrical length of top load of antenna, 740
$K(z-z')$	kernel in variational formulation, 253
K_e, K_m	complex amplitude factors for dipoles, 703
$K_a(z, z')$	complex kernel for antisymmetric antennas, 269
$K_c(z, z')$	complex kernel for collinear antennas, 426
$K_c(z, z')$, $K_s(z, z')$	real kernels for real and imaginary parts, 124
$K_{ev}(z, z')$, $K_{od}(z, z')$	complex kernels for collinear array, 432
$K_{eo}(z, z')$, $K_{oe}(z, z')$	complex kernels for collinear array, 432
$K_v(z, z')$	complex kernel for V-antenna, 383
$K_{\Sigma 0}(z, z')$, $K_{\Sigma m}(z, z')$	complex kernel for zeroth and mth phase sequence, 352
$K_1(z, z')$	complex kernel for cylindrical antenna, 76
$K_p^2(\Theta, \Phi)$	real radiation function referred to I_p, 21
k	effective length of top load of antenna, 743
k	complex coefficient of coupling, 349
	ratio of symmetric to antisymmetric voltages, 430
$\hat{\mathbf{k}}$	unit vector in direction of plane wave, 703
k_e, k_h	reciprocal eccentricities of ellipse and hyperbola, 541
k_q, k_q'	complex and real charge-ratio factors, 53
$k_0(w)$, $k_0'(w)$, $k_1(w)$, $k_1'(w)$	coefficients in generalized transmission-line equations, 40
L^e	external inductance, 23
L_b	inductance of connecting bridge, 63
L_T, L_{Te}, L_{Tc}	lumped terminal-zone inductances for antennas, 45, 55, 57
L_T^s, L_T^a	lumped terminal-zone inductances for coupled antennas, 315
$L_n(\Theta)$	solution of Legendre equation, 828
$\mathbf{l}$	real or complex vector surface density of current, 12
l	complex function of angle of incidence, 750
$l^e(w)$	external inductance per unit length at distance w from load end of terminated line, 41
$l_0^e(w)$, $l_T^e(w)$	parts of $l^e(w)$ due to currents in line and termination, 41
l_0^e	external inductance per unit length of infinite line, 43
$\mathbf{M}$	real or complex vector volume density of current, 12
$M_1(z)$, $M_2(z)$	first- and second-order terms in formula for current, 85
$M_1^I(z)$, $M_1^{II}(z)$; $M_2^I(z)$, $M_2^{II}(z)$	real and imaginary parts of $M_1(z)$ and $M_2(z)$, 85
M_{1H}^I, M_{1H}^{II}; M_{2H}^I, M_{2H}^{II}	Hallén form of $M_1^I(z)$, $M_1^{II}(z)$, etc., 93
$M_{1K}(z)$, $M_{2K}(z)$, $M_{nK}(z)$	King-Middleton forms of $M_1(z)$, $M_2(z)$, etc., 109
$M_{1s}(z)$, $M_{2s}(z)$; $M_{1a}(z)$, $M_{2a}(z)$	first- and second-order terms for symmetric and antisymmetric currents, 268, 273
m	magnetic moment of current ring or magnetic dipole, 23
m	complex function of angle of refraction, 750
m	power to which uniform array factor is raised, 654
$m_1(z)$	first-order term in solution for current in receiving antenna, 463
$m_1^I(z)$, $m_1^{II}(z)$	real and imaginary parts of $m_1(z)$, 463
$m_{1H}(z)$, $m_{1K}(z)$	Hallén and King-Middleton forms of $m_1(z)$, 473, 474
N, N_c, N_p, N_B, N_E	number of antennas in unspecified, collinear, parallel, broadside, and end-fire arrays, 579, 587, 600, 601
N, N_{12}, N_{21}	complex indices of refraction, 699, 709
$\hat{\mathbf{n}}$	unit external normal, 12
$\hat{\mathbf{n}}_1$, $\hat{\mathbf{n}}_2$	unit external normals to region indicated by subscript, 13

n_r, n_c, n_p, n_B, n_E	distances in wavelengths between antennas in unspecified, collinear, parallel, broadside, and end-fire arrays, 586, 590, 600, 601
$n_n(x)$	spherical Neumann functions of order n, 823
$\mathbf{P}$	real or complex vector volume density of polarization, 12
P	power, 118
P_e, P_m	complex numerical distances for elevated dipoles, 758
P_e, P_m	magnitudes of P_e and P_m, 759
P_L	power in load, 496
$P_m(z)$	mth-order function for coupled antennas, 269
$P_n(\mu), P_n(-\mu)$	Legendre functions of first kind and order n, 823, 824
P_r, P_t	radiated and total power, 284
$\mathbf{p}_e, \mathbf{p}_m$	complex numerical distances for dipole on the earth, 758
p_e, p_m	magnitudes of $\mathbf{p}_e$ and $\mathbf{p}_m$, 759
$\mathbf{p}(w), \mathbf{p}'(w), \mathbf{p}_0(w)$	parameters in generalized transmission-line theory, 41
$\mathbf{p}_1(z)$	first-order term in solution for current in receiving antenna, 463
$p_1^I(z), p_1^{II}(z)$	real and imaginary parts of $\mathbf{p}_1(z)$, 463
$\mathbf{p}_{1H}(z), \mathbf{p}_{1K}(z)$	Hallén and King-Middleton forms of $\mathbf{p}_1(z)$, 473, 474
Q_r	quality factor of antenna, 171
$Q_m(z)$	mth-order function for coupled antennas, 269
$Q_n(\mu)$	Legendre function of second kind and order n, 823
q	complex charge per unit length, 15
$-q', q''$	real and imaginary parts of q, 119
q_{e1}, q_{m1}	numerical heights of transmitting dipoles, 760
q_{e2}, q_{m2}	numerical heights of receivers, 760
q_0	shorthand for $\beta_0 \cos \Theta_2$, 460
$q(z), q_z$	complex charge per unit length at coördinate z on antenna, 73, 119
$q^s(z), q^a(z)$	symmetric and antisymmetric parts of $q(z)$, 73
$q_L(w)$	complex charge per unit length at coördinate w on line, 39
$q_T(u)$	complex charge per unit length at coördinate u on termination, 39
R	distance between two points, 14
	resistance, 23
	spherical coördinate, 699
$\mathbf{R}$	distance between two points when it must be distinguished from resistance, 694
R^e	external or radiation resistance, 23
R^i	internal or ohmic resistance, 23
R^s	surface resistance, 735
R^s, R^a	symmetric and antisymmetric resistances, 280
R_a, R_b	distances in line theory, 52
R_c	characteristic resistance of infinite line, 46
R_d	distance to source, 724
R_e', R_e'', R_m', R_m''	complex and real distances in analysis of antenna over earth, 754
R_L	load resistance for antenna, 284
R_r	distance to image of source, 724
R_{mi}	distance between antennas m and i, 372
R_{sa}	apparent terminal resistance, 47
R_δ, R_a	resistance of symmetric load (antenna) across terminals separated by a distance 2δ and when $\delta = 0$, 47, 149
R_m^e	radiation resistance referred to maximum current, 148

[Symbols] THEORY OF LINEAR ANTENNAS 919

R_{mc}^e	external resistance associated with dominant-mode load of biconical antenna, 835
R_p^e	radiation resistance referred to I_p, 21
R_0^e	radiation resistance referred to input current, 147
R_0^i	internal or ohmic resistance of antenna, 147
R_{s1}	self-resistance of antenna in presence of another antenna, 297
$(R_\delta)_{01}$	modified zeroth-order resistance of antenna, 147
$R_c(w)$	characteristic resistance of terminated line, 41
R_0	spherical coördinate; distance from origin, 715
	distance from center of antenna to point on its surface, 259
	resistance of antenna driven by discontinuity in potential, 152
R_1, R_2, R_{11}, R_{22}	distance from point on surface to point on axis of antenna, 74, 94, 266
R_{1h}, R_{2h}	distances R_1 and R_2 with $z = h$, 78
$R_{1T}, R_{2T}; R_{1\delta}, R_{2\delta}$	distances from conductors of line to load, 67, 52
R_{12}, R_{21}	mutual resistance of antenna 1 in presence of antenna 2 and vice versa; distance from point on surface of antenna 1 to point on axis of antenna 2 and vice versa, 289, 266
R_0, Θ, Φ	spherical coördinates, 528
$\hat{R}_0, \hat{\Theta}, \hat{\Phi}$	unit vectors in directions of spherical coördinate axes, 21

(Since $Z = R + jX$, there are always R's with the same subscripts as Z. For these use the list of Z's.)

r^i	internal resistance per unit length, 23
$r_{\alpha 1}, r_{\beta 1}$	ohmic terms in antenna impedance formula, 143
r_1	normalized resistance, 182
r, θ, z	cylindrical coördinates, 15
$\hat{r}, \hat{\theta}, \hat{z}$	unit vectors in direction of cylindrical coördinate axes, 770
S, S'	complex Poynting vectors defined, respectively, in terms of H^* and H, 21, 258
S	area of surface, 23
	standing-wave ratio, 182
$S(A, U), Sc(A, U), Ss(A, U)$	generalized sine and cosine integral functions, 95
$Si(U)$	integral sine function, 96
$S_a(h, z), S_b(h, z), S_r(h, z)$	94, 273, 525
S_m	shorthand for $mb \sin \Theta \cos \Phi$, 636
$S_{11}(h, z), S_{12}(h, z, \Delta)$	384
$S_v(h, z, \Delta)$	396
SWR	standing-wave ratio, 513
s	complex factor in reducing effective length for finite base separation to zero base separation, 487
T, T_m	electromagnetic and mechanical torques, 14
T	complex energy-transfer function, 21
$\overline{\overline{T}}_r$	time-average power transferred across closed surface; real part of T, 21
T_i	imaginary part of T, 562
$T_M(z)$	Tchebyscheff polynomial of order M, 667
T_s, T_s'	complex transmission coefficients for plane boundary, 712
t	time, 12
t_T, t_c, t_p, t_B, t_E	phase difference in fractions of a period between currents in adjacent antennas in unspecified, collinear, parallel, broadside, and end-fire arrays, 586, 590, 600, 601, 612

U, U	complex and real induced-voltage function for receiving antenna, 460
u	distance from line-load junction measured along load conductor, 39
$u_\delta(z), u_0(z)$	distribution functions for current in unloaded receiving antenna, 464, 467
V	complex scalar potential difference, 32 principal factor in polarization potential along boundary surface, 755
V^e	complex externally maintained emf, 28
V^r	principal factor in radiation-zone polarization potential along boundary surface, 724
$V^{(0)}, V^{(1)}, \ldots V^{(m)}$	voltages in zeroth, first, and mth phase sequences, 353
V_{BA}	complex voltage across terminals A and B, 32
V_0	complex discontinuity in scalar potential, 79
V_δ	complex driving potential difference across terminals separated a distance 2δ for balanced load, 76
V_δ^s, V_δ^a	symmetric and antisymmetric driving voltages, 267
$V(w)$	complex scalar potential difference at coördinate w, 36
$V_L(w), V_T(w)$	parts of $V(w)$ at distance w from load due to charges in line and in termination, 36
$V(\Theta)$	vertical field factor of circular array, 273
v_e	effective phase velocity, 22
v_g	group velocity, 541
v_{pa}, v_{pb}	phase velocities along major and minor axes of ellipse, 540
v_{p1}	complex phase velocity in region 1, 710
$v_\delta(z), v_0(z)$	complex distribution functions for current distributed as in a driven antenna, 464
v_0	characteristic phase velocity in free space, 14
W	shorthand for $\sqrt{U^2 + A^2}$, 95
W_x	x-component of complex vector potential difference, 36
W_0	null beam width of array, 587
W_1	half-power beam width of array, 586
$W(z), W$	expansion function and parameter using difference kernel, 124
$W_x(w)$	x-component of complex vector potential difference on line at distance w from load, 36
$W_{xL}(w), W_{xT}(w)$	parts of $W_x(w)$ due to currents in line and termination, 36
w	width of strip, 20 coördinate measured from load toward generator on line, 36
X	reactance, 23
X^i	internal reactance, 23
X^s	surface reactance, 735
X_{sa}	apparent terminal reactance, 47
X_δ, X_0	reactance of symmetric load (antenna) across terminals separated by a distance 2δ and when $\delta = 0$, 47, 149
X_{mc}^e	external reactance associated with dominant-mode load of biconical antenna, 835
X_0^e, X_0^i	external and internal reactances of antenna, 147
$(X_\delta)_{01}$	modified zeroth-order reactance of antenna, 147

(Since $Z = R + jX$, there are always X's with the same subscripts as Z. For these use the list of Z's.)

x	term in reflection coefficients, 727 arbitrary argument in array factor, 580

[Symbols]

x_1 normalized load reactance, 182
x, y, z rectangular coördinates, 18
$\hat{x}, \hat{y}, \hat{z}$ unit vectors in directions of rectangular coördinate axes, 697

Y_c characteristic admittance of infinite line, 46
Y_{la} apparent dominant-mode terminal admittance loading biconical antenna, 827
Y_{m1}, Y_{m2} field admittances at plane boundary, 726
Y_{sa} apparent terminal admittance loading a line, 47
Y_T terminal-zone admittance, 45
Y_δ, Y_0 admittance between terminals separated by a distance 2δ for balanced load and with $\delta = 0$, 47, 149
$Y_{in}(w)$ input admittance of transmission line of length w, 46
y admittance per unit length of infinite line, 43
$y(w)$ admittance per unit length at distance w from load end of terminated transmission line, 41
$y_0(w), y_T(w)$ parts of $y(w)$ due to line and termination, 41

Z impedance, 23
Z^e, Z^i external and internal parts of impedance, 32
Z^s surface impedance, 735
Z_{BA} impedance across terminals A and B, 33
Z_c characteristic impedance of infinite line, 46
Z_c characteristic impedance of infinite biconical line, 827
Z_{ev}^s, Z_{ev}^a impedances of outer units in collinear array when driven symmetrically and antisymmetrically, 433
Z_{0c}^s, Z_{0c}^a impedance of central unit in collinear array when driven symmetrically and antisymmetrically, 427
Z_{e1}, Z_{e2} field impedances at plane boundary, 726
Z_L effectively lumped load impedance; $Z_{L\delta}$ with $\delta = 0$, 466
$Z_{L\delta}$ load impedance across terminals separated by a distance 2δ for balanced load, 322
Z_s terminal impedance on transmission line, 480
Z_{sa} apparent terminal impedance on transmission line, 47
Z_{s1}, Z_{s2} like $Z_{s1\delta}$ and $Z_{s2\delta}$ with $\delta = 0$, 286
$Z_{s1\delta}, Z_{s2\delta}$ self-impedances of antennas 1 and 2 in presence of other antennas and with their terminals separated by distances 2δ, 286
Z_{s1a} apparent self-impedance of antenna 1 in presence of other antennas, 317
Z_T lumped terminal-zone impedance, 45
Z_δ impedance (of center-driven antenna) between terminals separated by a distance 2δ for a balanced load, 29, 149
Z_0 impedance (of antenna) driven by discontinuity in scalar potential or by a slice generator, 29, 149
   impedance at generator end of transmission line, 480
Z_1, Z_2 impedances in series with Z_{s1} and Z_{s2}, 346
$Z_{1\delta}, Z_{2\delta}$ input impedances of antennas 1 and 2 between terminals separated by distances 2δ, 286
Z_{10}, Z_{20} like $Z_{1\delta}$ and $Z_{2\delta}$ with $\delta = 0$, 312
Z_{11}, Z_{22} coefficients of currents in coupled-circuit equations, 346
Z_{11}, Z_{22} magnitudes of Z_{11} and Z_{22}, 623
Z_{12}, Z_{21} mutual impedances like $Z_{12\delta}$ and $Z_{21\delta}$ with $\delta = 0$, 286
Z_{12}, Z_{21} magnitudes of Z_{12} and Z_{21}, 623
Z_{12a} apparent mutual impedance, 317

$Z_{12\delta}, Z_{21\delta}$	mutual impedances with antenna terminals separated by distances 2δ, 286
Z_{sa}^s, Z_{sa}^a	apparent symmetric and antisymmetric impedances, 317
Z_δ^s, Z_δ^a	symmetric and antisymmetric impedances of antennas with terminals separated by distances 2δ, 278
Z_0^s, Z_0^a	like Z_δ^s and Z_δ^a with $\delta = 0$, 286
$(Z_c)_{av}$	average characteristic impedance of nonuniform line, 838
$(Z_\delta)_0, (Z_\delta)_1, (Z_\delta)_m$	zeroth-, first-, and mth-order approximations of Z_δ, 142
$(Z_0)_1, (Z_0)_2, (Z_0)_m$	first-, second-, and mth-order approximations of Z_0, 151
$(Z_\delta)_{01}, (Z_0)_{01}$	modified zeroth-order approximations of Z_δ and Z_0, 147, 149
z	impedance per unit length of infinite line, 43
	complex variable in array factor for nonuniform array, 182
z^e, z^i	external and internal impedances per unit length, 28, 39
z_r	reference value of coördinate z, 77
z_1	normalized load impedance for line, 182
$z(w)$	impedance per unit length at distance w from load end of terminated line, 42
$z_0(w)$	part of $z(w)$ due to currents in line, 42
α_s	imaginary part of β; real attenuation constant of simple medium, 22
$\alpha_1, \alpha_1^I, \alpha_1^{II}; \alpha_2, \alpha_2^I, \alpha_2^{II}$	terms in impedance formula of antenna, 143, 144, 145
$\boldsymbol{\beta}$	complex phase constant of dissipative medium, 22
β	magnitude of $\boldsymbol{\beta}$; phase constant of nondissipative medium, 38
β_s	real part of $\boldsymbol{\beta}$; phase constant of dissipative medium, 22
β_0	phase constant of free space, 14
$\beta_1^I, \beta_1^{II}, \beta_2^I, \beta_2^{II}$	terms in impedance formula of antenna, 143, 144
Γ	complex reflection coefficient, 182
Γ_s, Γ_s'	complex reflection coefficients, 712
γ	complex propagation constant of infinite line, 43
γ	shorthand for $\cos \Theta_{01}$, 743
$\gamma(w)$	complex propagation constant of terminated line, 42
$\gamma(z)$	complex correction term in integral equation for current, 77
$\gamma_H(z)$	complex correction term in Hallén form of integral equation, 93
$\Delta(z)$	correction term in Hallén form of solution for current, 90
δ	half of distance between adjacent ends of antenna, 28
δ_c	phase difference between adjacent collinear antennas, 589
$\delta(z)$	Dirac delta function, 30
ϵ	complex absolute dielectric constant, 21
	correction factor for base separation of antenna, 149
ε	ellipsoidal coördinate, 540
ϵ', ϵ''	real and imaginary parts of complex dielectric constant, 21
ϵ_e	real effective absolute dielectric constant, 22
ϵ_{er}	real effective relative dielectric constant, 22
ϵ_r	complex relative dielectric constant, 21
$\epsilon_r', \epsilon_r''$	real and imaginary parts of ϵ_r, 21
ϵ_0	fundamental electric constant (dielectric constant or permittivity of free space), 13

[Symbols]

ζ	characteristic impedance of dissipative medium, 713
ζ	shorthand for z/h, 186
	characteristic resistance of nondissipative medium, 726
	magnitude of characteristic impedance of dissipative medium, 38
ζ_e	real effective characteristic impedance or resistance of dissipative medium, 22
ζ_0	characteristic impedance or resistance of free space, 14
η	complex surface density of charge, 12
	characteristic admittance of dissipative medium, 711
$\bar{\eta}$	complex essential surface density of charge, 12
η	real surface density of charge, 12
η_f	complex surface density of free charge, 15
η_0	characteristic admittance or conductance of free space, 726
$\overline{\eta_m \mathbf{v}}$	complex essential vector surface density of moving charge, 12
$\boldsymbol{\Theta}$	complex angle, 709
Θ	spherical coördinate (measured from vertical axis), 21
$\hat{\boldsymbol{\Theta}}$	unit vector in direction of spherical coördinate, 21
Θ_B	Brewster angle, 730
$\boldsymbol{\Theta}_i, \Theta_i$	complex and real angles of incidence, 709
Θ_m	angle locating a maximum, 21, 538
$\boldsymbol{\Theta}_r, \Theta_r$	complex and real angles of reflection, 709
$\boldsymbol{\Theta}_t, \Theta_t$	complex and real angles of transmission (refraction), 709
Θ_0	angle locating a null, 538
Θ_1, Θ_2	spherical angles orienting transmitting and receiving antennas in space, 464, 460
$\overline{\Theta}_2$	complement of Θ_2, 460
Θ_{01}, Θ_{02}	spherical coördinates in regions 1 and 2, 715, 719
$\boldsymbol{\theta}$	complex terminal function, 182
$\hat{\boldsymbol{\theta}}$	unit vector in direction of cylindrical coördinate, 770
θ	angle of tilt of electric field near earth, 773
	cylindrical coördinate, 15
θ_{21}, θ_{22}	angles of Z_{21} and Z_{22}, 623
$\boldsymbol{\lambda}$	eigenvalue, 697
	complex variable of integration, 750
λ	real variable of integration, 805
	wavelength in simple medium, 38
λ_0	wavelength in free space, 23
$\boldsymbol{\mu}$	complex permeability, 707
μ	shorthand for $\cos \Theta$ in Legendre function, 823
μ, μ_r	absolute and relative permeabilities, 23
μ_0	permeability of free space, 13
$\boldsymbol{\nu}$	complex absolute reluctivity (reciprocal permeability), 21
ν	real absolute reluctivity of medium without magnetic time lags, 22
$\boldsymbol{\nu}_r$	complex relative reluctivity, 21
ν'_r, ν''_r	real and imaginary parts of $\boldsymbol{\nu}_r$, 21
ν_0	fundamental magnetic constant of free space (reluctivity or reciprocal permeability of free space), 13
$\boldsymbol{\xi}$	complex dielectric factor, 22

Π_e	real or complex polarization potential or electric Hertz vector, 20
Π_m	real or complex magnetization potential or magnetic Hertz vector, 20
Π_x, Π_y, Π_z	rectangular components of complex Hertz vector, 697
Π_z	real z-component of Hertz vector, 703
Π_{z1}	z component of complex Hertz vector in region 1, 707
Π_{z2}	z component of complex Hertz vector in region 2, 707
$(\Pi_{z1})_d$	z component of complex Hertz vector in region 1 due to current in source; direct component, 707
$(\Pi_{z1})_r$	z component of complex Hertz vector in region 1 due to currents in region 2; reflected component, 707
$\mathbf{\rho}, \rho$	complex and real volume densities of electric charge, 12
ρ	ellipsoidal coördinate, 540
ρ, ρ_0, ρ_s	terminal attenuation function of transmission line in general, for generator and for load, 182, 231, 480
ρ_{sa}	apparent terminal attenuation function of load, 231
$\overline{\mathbf{\rho}}$	complex essential volume density of charge, 12
$\overline{\rho_m \mathbf{v}}$	complex essential vector volume density of moving charge, 12
σ	complex conductivity, 21
σ', σ''	real and imaginary parts of, 22
σ_e	real effective conductivity, 22
σ_c	real conductivity of conductor, 70
$\sigma_{abs}, \sigma_{dis}, \sigma_{rad}$	absorption, dissipation, reradiation across sections of antenna, 500
$\sigma_\parallel, \sigma$	back-scattering cross section for parallel polarization, 507
$\sigma_\perp$	back-scattering cross section for perpendicular polarization, 507
$\bar{\sigma}$	average back-scattering cross section, 509
Φ	spherical coördinate (azimuth), 21
Φ, Φ_0, Φ_s	terminal phase function of transmission line in general, for generator, and for load, 182, 480
Φ_{sa}	apparent terminal phase function of load, 231
$\hat{\Phi}$	unit vector in direction of the coördinate Φ, 21
Φ_m	angle locating a maximum, 21
ϕ	complex scalar potential, 14, 74
	complex angle, 714
ϕ	difference angle in plane-wave representation, 704
ϕ^r	radiation or far-zone complex scalar potential, 15
ϕ_c	distortion factor in transmission line, 46
$\phi(P, B)$	angle of complex surface-wave attenuation function F, 760
$\phi(z)$	complex scalar potential at coördinate z, 74
$\phi^s(z), \phi^a(z)$	symmetric and antisymmetric components of $\phi(z)$, 74
$\varphi_1(w), \varphi_1$	complex ratio of components of scalar potential on transmission line, 41, 42
$\varphi_1(w), \varphi_1$	real ratio of components of scalar potential on transmission line; criterion for capacitive coupling, 41, 42
Ψ	expansion parameter, 77
Ψ_H	expansion parameter in Hallén solution, 93
$\Psi_{a1}(z), \Psi'_{a1}$	expansion function and expansion parameter for antisymmetrically driven antennas, 273
$\Psi_b(z)$	difference between $\Psi_{x1}(z)$ and $\Psi_{s1}(z)$, 273

Ψ_{ev}, Ψ_{od}	expansion parameters for collinear antenna, 433, 435
$\Psi_{KC}(z), \Psi'_{KC}$	expansion function and expansion parameter for collinear antenna, 427
$\Psi_{K1}(z), \Psi'_{K1}$	King-Middleton expansion function and expansion parameter for cylindrical antenna, 101, 102
$\Psi_{s1}(z), \Psi'_{s1}$	expansion function and expansion parameter for symmetrically driven antennas, 267, 268
$\Psi_U(z), \Psi'_U$	expansion function and expansion parameter for unloaded receiving antenna, 433, 435
$\Psi'_\delta(z)$	expansion function for antenna with base separation 2δ, 77
$\Psi_{\Sigma 0}(z), \Psi'_{\Sigma 0}, \Psi_{\Sigma m}(z), \Psi'_m$	expansion functions and parameters for zeroth- and mth-phase sequences in circular array, 367
ψ	angle orienting receiving antenna, 464
	visual angle for multiple half-wave antenna, 535
	argument in array factor for nonuniform arrays; equals $2x$, 654
$\bar{\psi}$	extreme value of ψ, 654
$\psi_{KC}(z)$	expansion function $\Psi_{KC}(z)$ divided by zeroth-order current, 427
$\psi_1(z)$	expansion function $\Psi_{K1}(z)$ divided by zeroth-order current, 101
Ω	expansion parameter of Hallén, 90
Ω	complex solid angle, 710
$\Omega(h, z)$	expansion function in Hallén form, 90
ω	angular frequency, 12
∇	nabla or del, 12
∇^2	Laplacian operator, 14
*	asterisk denoting complex conjugate, 21
×	denoting vector or cross product, 21
	denoting simple product of scalar quantities in long expressions in which the multiplicand and the multiplier are on separate lines, 77

INDEX

Adams, N. I., 5
Aden, A. L., 512
Admittance, of antennas of different types. *See under name of antenna*
   apparent, of load on line, 47; of antenna terminating biconical line, 847
   characteristic, of medium, 711
   field, definition of, 711
   input, of section of line, 46, 47
   terminal-zone, 45
   transfer, for coupled antennas, 331, 332
Alfred, R. V., 646, 647
α-functions, 143; tables of, 145
Analysis, methods of. *See under* Cylindrical antenna
Andrews, H. W., 9, 449–455
Angelakos, D. J., 131, 132, 141, 209, 210, 219, 392–395
Angle, complex, 703; of tilt of electric field over conducting medium, 773
Antenna, base-driven over conducting plane, 65–67, 69; with sinusoidal current, 779
   Beverage, 794
   as boundary-value problem, discussion of, 5, 6, 818–820
   bridged-parallel, description of, 343, 346; impedance graphs of, 345
   as circuit element, 26
   closely spaced, two parallel, 273–275; *see also* Cage antenna, Transmission-line radiators
   as complete transmitting system, 28
   conical, 818–820; over ground screen, 838; *see also* Biconical antenna
   over conducting earth, general discussion, 695–696; *see also* Antenna on finite ground screen, Antenna parallel to perfectly conducting plane, Horizontal antenna with sinusoidal current over conducting earth, Horizontal infinitesimal dipole over conducting earth, Vertical antenna with sinusoidal current over conducting earth, Vertical infinitesimal dipole over conducting earth
   crossed, analysis of, 396–398; arbitrarily driven, 397; impedance formulas of, 397
   definition of, 25, 26
   driven from two-wire line, summary of terminal-zone problem, 59–60
   driving, methods of. *See* Methods of driving antennas
   electrically short. *See* Short antenna
   on finite ground screen. *See* Antenna on finite ground screen
   ground-plane, 420–422; impedance of, graphs, 423
   hemispherical, 818–820
   isolated, discussion of, 3–8
   loop, as current probe, 127, 130; impedance of small, 23; radiation resistance of small, 23; relation to magnetic dipole, 22, 23, 701, 702
   methods of driving. *See* Methods of driving antennas
   as mid-point load for two-wire line, 55–59
   with minimum coupling to line, 61–64
   monopole. *See* Cylindrical antenna, Half-dipole with counterpoise
   mutually perpendicular. *See* Antenna, turnstile
   with parasite. *See* Antenna with single parasite
   parasitic, definition of, 263; reradiation from. *See* Scattering antenna
   reradiating. *See* Scattering antenna
   slot, discussion, 555–557, 560; graphs of field properties, 558, 559
   spheroidal, 5, 6, 818–819
   strip, 555, 556
   terminating biconical line, 846–848, coaxial line, 65–69, two-wire line, 50–55
   turnstile, analysis, of circuit properties, 381, of field, 684, 685; field patterns, graphs, 686
   unbalanced, analysis of, 576, 578; graphs of field, 577
   See also Antenna with arbitrary cross section, Antenna on finite ground screen, Antenna parallel to perfectly conducting plane, Antenna with single parasite, Asymmetrically driven antennas, Biconical antenna, Cage antenna, Collinear array, Corner reflector, Cylindrical antenna, Folded antenna, Folded dipole, H-array, Half-dipole with counterpoise, Horizontal antenna with sinusoidal current over conducting earth, Multiple half-wave antenna, Receiving antenna, Scattering antenna, Short antenna, Sleeve dipole, Transmission-line radiator, V-antenna, Vertical antenna with sinusoidal current over conducting earth. *See also under* Dipole, Horizontal infinitesimal dipole over conducting earth, Vertical infinitesimal dipole over conducting earth
Antenna with arbitrary cross section, analysis of, based on uniform-line theory, 838–839; based on tapered-line theory, 839–841
   apparent load admittance of, 838, 841
   characteristic impedance of, average, 838, 841; variable, 838
   impedance of, first-order, 841; Schelkunoff's, 841; zeroth-order, 839

Antenna on finite ground screen,
 analysis of, 803–813
 electromagnetic field of, 815
 impedance of, formulas, 813; graphs, 814
 on infinite imperfect earth, analysis, 815–816;
  formulas, 816, 817; graphs, 814
Antenna parallel to perfectly conducting plane,
 current, in antenna, analysis of, 269, 272;
  in conducting plane, analysis of, 794,
  796–799, graphs of, 795, 798. *See also*
  Antisymmetrically driven antennas
 impedance of, 281; graphs, 279, 282
 methods of driving, 283
 ohmic losses in, 284
 radiating efficiency of, 284, 285
 radiation resistance of, 285, 286
Antenna with single parasite, admittance of,
  input and transfer, 329, 332, 333
 circuits for, 326
 circuit properties of, analysis, 322–325
  antennas of unequal length, 325
  measurement of, 327, 328
  with parasite, loaded, 322; open-circuited,
   324, 325; short-circuited, 322
 closely spaced, 329
 comparison of theoretical and experimental
  data, 329
 current distribution in, antisymmetric component of, discussion, 333, 334; graphs,
  331
 discussion of, 328, 329
 measured and theoretical, graphs of, 330–331
 measurement of, 328, 329; block diagram of
  circuit, 327
 symmetric component of, discussion, 333;
  graphs, 331
 table of, theoretical and measured, 329
field properties, analysis of, 622, 623
 array factor of, 623, graph, 632
 discussion of theoretical and experimental
  results, 627, 633
 electric field of, formula, 623
 field patterns of, formula, 623; graphs, 624,
  625, 628, 632, 637
 field ratio of, front-back, discussion of, 626,
  627, 633; formula, 623, graphs, 628, 629;
  measured graphs, 630
 measured field, discussion of, 635; pattern
  with director, 632; with reflector, 632;
  ratio of, 628
 gain of, with reflector or director, discussion of,
  626, 627; graphs, 629, 634
 impedance of, antisymmetric, 324; discussion of
  resistive component, 332
  coupling, end effect, and dielectric support,
   332, 333
  formulas, 322, 324, 325
  graphs, 323, 631, 634
  measured, 343
  parasite with infinite load, 324; tuned to
   self-resonance, 324; variable in length,
   324–326; with zero load, 322

symmetric, 334
tables, 329, 334
terminal-zone correction, 333
reactance of, graph, 323, 634
resistance of, discussion, 633; graph, 323, 631;
 ohmic, 332
variables available for adjustment, 377
Antennas. *See* Antenna, Antisymmetrically driven
 antennas, Closely-spaced antennas, Crossed
 antennas, Parasitic antennas. *See also*
 Array
Antiresonance, in antenna, definition of, 152, 153;
 lengths defining, 163, 210
Antisymmetrically driven antennas, analysis of
 circuit properties, 269, 273
 current distribution in, analysis of, 269, 273,
  275, 277; graph of, 276
 expansion parameter for, 273, 274, graph of, 271
 field of. *See* End-fire array, two-element, with
  $n_E = \frac{1}{2}$
 impedance of, formulas, 277–281; graphs, 279,
  282; measurement of, 347; modified zeroth-
  order, 278; tables of, 280
 integral equation for, 267 with 269
 see also Collinear array, Folded dipole, H-array,
  Transmission-line radiators
Antisymmetric impedance. *See under* Asymmetrically driven antennas
Aperture, effective, of antenna. *See* Cross section,
 dissipation
Approximations in antenna theory, discussion of,
 10, 11, 24, 25, 819
Array, with all elements in neutral plane, anaylsis
 of, 378–381
 driven, with elements of finite cross section,
  field of, 689, 690
 See also Antennas, Array of arrays, Broadside
  array, Circular array, Collinear array,
  End-fire array, H-array, Nonuniform array,
  Parallel array with elements in line, Parasitic array, Receiving array, Uniform array
Array factor. *See* Uniform arrays
Array of arrays, broadside, with parasitic curtain,
 analysis of, 647–649
 field ratio of, formula, 649; graph, 650
 gain of, formula, 649; graph, 650
 resistance of, formula, 649; graph, 650
combination broadside and end-fire, description
 of, 612, graphs of field factor, 618, 619
Yagi-Uda, description of, 647; table of critical
 quantities, 648
Arrays, using turnstile elements, 381
Area, effective. *See* Cross section
Asymmetrically driven antennas, analysis of,
 398–403
 approximate, 403–407
 broad-band properties of, 406, 407
 current distribution in, 402, 405
 expansion parameter for, 400
 impedance of, general formula, 402; as series
  combination, 404; graphs, 405; numerical
  values, 406
 integral equations for, 399

Attenuation, in conducting regions, 723
Attenuation constant, 22
Attenuation function, complex surface-wave, 760; graphs of, 761, tables of, 762

B-vector, definition, 13
Baños, A., Jr., 695
Barzilai, G., 8
Base separation, effect on impedance, 193; of coupled antennas, 286; factors and graphs to take account of, 149–151, 194–197
Back-scattering, broadside, from two parallel antennas, 520; circuit for measuring, 511; cross section, *see under* Cross section
Bechman, R., 4
Bemporad, M., 81
β-functions, formulas for, 143; graphs of, 146; tables of, 145
Bessel functions, spherical, 823
Beverage antenna, 794
Biconical antenna, admittance, dominant-mode apparent terminal, definition of, 827; solution for antennas with unrestricted angles, 832, 833; with small angles, 833–836
analytical advantages of, 6
boundary conditions, 824
current in, associated with complementary interior modes, 829, with dominant mode, 828
electrically short, impedance formulas for, 836, 838; graphs of, 840
electromagnetic field of, complementary exterior modes, 825, 830
complementary interior modes, 825, 829
components of, 825
dominant mode, 825–828
matching of, 830, 831
half of, driven over ground screen from coaxial line, 838
Hertzian potentials for, 825
impedance of, characteristic, formula for, 827, graphs of, 837
input, 830; for antenna with small angle, 836; graphs of, 837, 840
zeroth-order approximation, 836
infinite set of equations for, 833
interior and exterior regions defined, 825
modes, complementary, analysis of, 828–830; definition of, 825
dominant, analysis of, 826–828; definition of, 825
voltage, associated with complementary interior modes, 830; with dominant mode, 827
Blake, F. G., 7
Blasi, E. A., 349
Bloch, F., 505, 508–511
Bontsch-Bruewitsch, M. A., 8, 613
Boundary conditions, between media, in terms of fields, 697; in terms of Hertzian potentials, for horizontal dipoles, 698; for vertical dipoles, 697
for biconical antenna, 824

for cylindrical antenna, 70
general electromagnetic, 13
mathematically convenient, 818
Bouwkamp, C. J., 9, 422
Braun, F., 1, 2
Bremmer, H., 10, 695
Brewster angle, definition of, 730; determination of, 734, 735, 740; graph of, 736; for salt water and wet earth, 740
Brillouin, L., 3, 6, 7
Brillouin, M., 3
Broadside array, array factor of, 601, 612; extremes, 601; graphs, 598, 599; tables, 602, 603; zeros, 601
beam width of, 613, 621; tables, 622
circuit properties of, as special case of parallel array, 375; coupling and end effects, 376
combination end-fire, 612; graphs of field patterns, 618, 619
complete far-zone field of, 601
directivity of, table, 620
field patterns of, polar, 599; rectangular, 598
gain of, table, 620
with parasitic curtain. See Array of arrays
radiation function of, 612
radiation resistance of, 612, 613; table, 620
three-element, analysis of, 375; conditions for equal currents in, 375, 376; impedance formulas and numerical values for, 375
two-element, circuit properties of. See Coupled parallel pair, H-array, Symmetrically driven antennas
Brown, G. H., 4, 8, 227, 842
Brückmann, H., 679

Cage antenna, N-element, analysis of, 369, 370; effective radius of, 370; expansion parameter for, 369, 370
two element, analysis of, 273–275; expansion parameter for, 274; impedance of, graphs, 279, 282, tables, 280
Capacitance, of dielectric, disk, 333
lumped, for terminal zone, definition, 45
terminal-zone, for coaxial line, antenna over image plane, 68; measured and theoretical, table of, 252, graphs of, 251
for coupled lines and antennas, 316, 317
for two-wire line, antenna as end load, 50–55, graphs of, 51; antenna as mid-point load, 57–59, graphs of, 56, 60; antenna over image plane, 59, 60; antenna with stub support, 59; antenna with parasite, 333
See also Coupling and end effects
in shunt with antenna, 193, 198; graphs of admittance and impedance, 199–203
Capacitance per unit length, 41, 43; graph of, 34
Carter, P. S., 4, 8
Chambers, L. L. G., 397
Chang, T., 345, 346
Characteristic impedance. *See* Impedance, characteristic
Characteristic resistance. *See* Resistance, characteristic

Charge, distribution of, in cylindrical antenna, 119–123; graphs of, 134, 135, 137, 139; measurement of, 130, 131, 136, 141; surface density of, definition, 12; for transmission line, 47; per unit length, 15
volume density of, definition, 12
Chu, L. J., 6, 818
Circular array, circuit properties of, analysis of four-element, 366–368; of N-element, 351–354; of three-element, 361–364; of two-element, 354, see also under Coupled parallel pair
currents in, phase-sequence, 353
diametral pairing for, 680–681
field of, analysis, 679–682; graphs, 683
integral equations of, for phase sequences, 352; simultaneous, 351, 352
parallel-chord pairing for, 681, 682
phase sequences in, 352–354
Circuits, equivalent terminal-zone for antenna, as center load, 60, 206; as end load, 48, 51, 206; with stub support, 60
Coefficient, reflection. See Reflection coefficient
transmission. See Transmission coefficient
Collinear array, analysis, of circuit properties, see under Collinear array, three-element
of field properties, 587–590
array factor, 590; table of, 591
currents in, 587, 589, 590; distribution of, see under Collinear array, three-element
electric field of, formulas, 590; patterns, 592, 596
field factor, approximate representation, 591; complete, 591; graphs of, 592, 596; table of, 591
impedance, see under Collinear array, three-element
Marconi-Franklin antenna, array factor, 591; beam width, 594–595, table of, 595; directivity, 593; gain, 593; radiation function, 591; radiation resistance, 593
three-element, analysis of central unit, 422–430; of outer units, 430–436
antisymmetric problem, 424, 431, 433, 436
comparison of theory and experiment, 449–451
complementary slots, 447–448
currents, in central unit, 436; in outer unit, even, 437, odd, 440; with outer units coupled by lumped reactance, graphs, 453, measured graphs, 454; with outer units parasitic, graphs, 436, 439–442, measured graphs, 444, 448; ratio of, in parasite and central unit, formula, 439, graph, 440
with elements coupled capacitively, 448–455; by coaxial sleeves, 455; by open-wire stubs, analysis, 442–447, impedance formulas and numerical values, 447, 455
expansion parameter, for central unit, definition of, 427, graphs of, 429; for outer units, even current, 433, odd current, 435
general problem, discussion of, 424
impedance, antisymmetric, formulas of, 437, graphs of, 438, table of, 441; input, with outer units coupled by lumped capacitance, discussion of, 452, graphs of, 451; with outer units parasitic, formula of, 437, graphs of, 438, table of, 441; mutual, formula of, 441, graph of, 440, table of, 441; symmetric, formulas of, 437, graphs of, 438, table of, 441; transfer, formula of, 439, graph of, 440, table of, 441
parasitic, 424
ratio of currents in parasite to current in central unit, formula, 439; graph, 440
symmetric problem, discussion, 424, 431, 433, 436
two-element, discussion of, 422, 423; field of, formula, 595, graph, 596; nonuniform, 595
Complementary slot and strip antennas, discussion, 555–557, 560; field properties, graphs of, 558, 559
Conductance, of antennas of different types. See under name of antenna
Conductivity, 22
Conley, P., 7, 208, 209, 215–220
Continuity, equation of, 13; for potentials, 75
Coördinates, confocal, 541, figures showing, 536, 539; spherical, 527
Corner reflector, antenna with, analysis, formulas, numerical values, 370, 371
Correspondence between theory and experiment, general discussion, 24, 25
Cosine, shifted, 469
Cosine integral functions, $C(A, U)$, 95, 857; table, 858; $C(0, U)$, 96
$Cc(A, U)$, 95, 857; table, 858; in terms of Cin, 96; $Cc(0, U)$, 96
$Ci(U)$, 96, 857
$Cin(U)$, 96, 857; approximate formula for large arguments, 107, 143; table, 858
$Cs(A, U)$, 95, 857; table, 858; in terms of Si, 96; $Cs(0, U)$, 96
$C_a(h, z)$, definition, 94; graphs, 98, 270, 429; table, 428; in terms of tabulated functions, 97
$C_b(h, z)$, graphs, 270
$C_i(A, U)$, 95, 857
$Cc_i(A, U)$, 95, 857; in terms of Ci, 96; $Cc_i(0, U)$, 96
symmetry relations for, 96
see also Sine integral functions
Coupled antennas, antenna coupled to folded dipole, 343
arranged in circle. See Circular array
closely spaced. See Cage antenna, Transmission-line radiators
collinear. See Collinear array
discussion of, 8, 9, 351
four parallel, in square, analysis of, 366–368; circuit diagram, 365; expansion parameters, 367; impedances, 368; phase sequences, 365
general theory, arbitrary orientation and separation, 456, 457, 459, 460
integral equation, N simultaneous, 351
parallel in line. See Broadside array, End-fire array, Parallel array, Parasitic array

Coupled Antennas—(contd.)
  theory, general, of N elements, 351
  three parallel in equilateral triangle, analysis of, 361–364; expansion parameters, 362; impedances, 362, 363; integral equation, 361; phase sequences, 362, 363
  two arbitrarily oriented, analysis of, 454–460; conditions for loose coupling with, 457; induced voltages in, 457. See also Receiving antenna, Scattering antenna
  two parallel. See Coupled parallel pair
  two widely separated. See Receiving antenna, Scattering antenna
Coupled parallel pair
  analysis of circuit properties, 266–269, 273
  antisymmetrically driven. See Asymmetrically driven antennas
  base separation of, effect on impedance, 286
  broadside arrangement. See Broadside array, Symmetrically driven antennas
  closely spaced, 273–275, 358
  coupling to feed lines and end effects. See Coupling and end effects
  current in, formulas for, 268, 269, 273; graphs of, 276
  discussion of, 263, 264
  driven in different manners, figures of, 265
  driven in phase. See Symmetrically driven antennas
  in phase opposition, see Antisymmetrically driven antennas, End-fire array
  in phase quadrature, 312; see also End-fire array
  electromagnetic field of. See Antenna with single parasite, Broadside array, End-fire array
  end-fire arrangement. See End-fire array, Asymmetrically driven antennas
  general theory, arbitrary separation, 456, 457, 459, 460
  impedance, antisymmetric, graphs of, 279, 282; tables, of, 280
    approximate second-order, discussion of, 278, 288, 290, 296, 312
    driving-point, driven in phase, graphs of, 279, 282, tables of, 280; driven in phase opposition, graphs of, 279, 282, tables of, 280; driven in phase quadrature, 312; one element parasitic, see under Antenna with single parasite
    mutual. See Coupled parallel pair, mutual impedance
    self-. See Coupled parallel pair, self-impedance
    symmetric, graphs of, 279, 282; tables of, 280
  integral equations for, 266–269
  mutual impedance of, approximate second-order, derivation of, 288, 290, 296; graphs of, 291–295, 306–310; spiral graph of, 291; tables of, 298, 300, 302, 304, 305
    definition and formulas of, 286
    experimental, determination of, 346–350; graphs of, 350

    zeroth-order, formulas of, 288; graphs of, 287, 311; tables of, 289; used with thick antennas when separated more than one wavelength, 288
  one element parasitic. See Antenna with single parasite
  potential, vector, on surface of, 264, 266
  self-impedance of, approximate second-order, derivation of, 288, 290; graphs of, 291–295, 306–307; spiral graph of, 291; tables of, 297, 299, 301, 303, 305
    definition and formulas of, 286
    measurement of, 346–350; graphs, 350
    modified zeroth-order, 288
    as special case of circular array, 354
  symmetrically driven. See under Symmetrically driven antennas
Coupling and end effects with antennas and lines, in biconical line, 847
  in line with load supported by dielectric disk, 333; compensated by end effect, 333
  with coaxial line, 68; graphs of, 251; table, 252
  in experimental determination of impedances of coupled antennas, 347, 348
  in H-array, centrally driven, 318, 319; laterally driven, 321, 322
  in parallel array, 376
  in transmission-line radiators, 335, 337
  with two parallel antennas, circuits for, 314
    as end loads, perpendicular to lines, 317, 318; in plane of individual lines, 313–317
    with no coupling between individual lines, 318
    widely separated, 313
  with two-wire line, antenna as end load, 50–55; graphs of, 51
    antenna as mid-point load, 57–59; graphs of, 60
    antenna over image plane, 67–69
    antenna with parasite, 333
    V-antenna as load, 388–392
Cox, C. R., 8
Cross section, absorption, of antenna, 500; of oscillating electron, 501
  back-scattering, of antenna, average, formulas, 507, 509; graphs, 508, 511
    broadside response, first-order formula, 510; graphs, 508
    definition of, 506, 507
    of dipole antenna, graphs, 516
    first-order, 510
    measured, canceling method, 515–517; right-angle method, 517; SWR method, 512–517
  dissipation, of antenna, 498, 500, graphs, 495, 499; interpretation as area, 500, of oscillating electron, 500, 501
  equivalent, of ellipse, 17, 18; of regular polygon, 20
  reradiation, of antenna, 500, 506, 507; of oscillating electron, 501
  scattering, definition of, 506, 507
  total, definition of, 506, 507
Curl, definition of, 12

Current, on conducting plane below horizontal antenna, analysis of, 796–799; graphs of, 795, 798
density, surface, 12; volume, 12
distribution of, in antennas of different types. *See under name of antenna*
even and odd in collinear antennas, 424, 432, 437, 439, 442
impressed, 22
sinusoidal on resonant antennas, graphs of, 536
symmetric and antisymmetric, in collinear antennas, 424, 436, 437, 439; in parallel antennas, 267, 273
in terminal zone of line, 49
total, in conductor, 15
in transmission line, 46
Cylinder functions, spherical, 823
Cylindrical antenna, admittance of, near antiresonance, graphs, 162; measured graphs, 247–250
  with capacitance in shunt, graphs, 199
  comparison of different theories, 262, 854; of theory and experiment, 249
  correction for terminal zone, summary, 205, 208
  critical quantities, table, 168
  formula for, 144
  as function of base separation, graphs, 197
  graphs, 154, 156, 157, 160, 161
  King-Middleton. *See most theoretical values listed*
  measured graphs, 247–250
  Storm's theory, table, 262
  tables of, 169–176
  Zuhrt's theory, table, 854
analysis, methods of. *See* Cylindrical antenna, methods of analysis
α-functions, formulas of, 143; graphs of, 146; tables of, 145
antiresonance, definition of, 153; for apparent impedance, 198
antiresonant lengths, graphs of, 163, 217; measured as function of line spacing, 210
base-separation correction, formulas for, 149, 193, graphs of, 150, 194–197
boundary conditions, 70, 72
as boundary-value problem, 848–855
charge distribution, first-order graphs of, 120–122
  instantaneous, formulas for, 123, graphs of, 122
  measured graphs of, 134, 135, 137–139
  measurement of, 127–131, 141
  measuring setup for, 129
  relation to current, 119
  symmetry conditions for, 73, 74
  zeroth-order, formulas of, 86, 87; graphs of, 88
conductance, at antiresonance, table of, 168
  graphs of, 154, 157, 160, 249, 843
  maximum, table of, 168
  near resonance, graphs of, 842, 843; table of, 168, 182

third-order formula for, 183, 184; point on measured graph, 249; table, 182
see also under Cylindrical antenna, admittance
critical quantities, graphs of, 163; tables of, 168
projection to zero line-spacing, graphs of, 217
effective length of, for reception. *See* Receiving antenna
for transmission, 567, 568
current distribution in, approximate second-order, 118, 119; graphs of, 116, 117
  in asymmetrically driven antenna. *See* Asymmetrically driven antenna
  components of, 85, 86, 123–127; graphs of, 112–117, 128
  discussion of, 110, 111
  effect of base separation on, 87–89
  first-order, discussion of, 110, 111, 118; graphs, 112–115
  formal solution for, 78
  formulation of problem, 69, 70
  general formula for, 85
  Hallén's formula for, 93
  instantaneous, 86, 123; graphs of, 122
  King-Middleton formula for, 107; with D-factors, 110
  King's alternative formulas for components, 123–127
  in long antenna, discussion, 118; graphs, 115; measured graphs, 140
  measured graphs of, 132, 133, 135, 137–140
  measurement of, 127–141
  measuring setup for, 129
  parameters in variational theory, graph of, 257; table of, 256
  series solution for, 84, 93, 94, 107, 110, 123–127
  in short antenna. *See* Short antenna
  with stub-supported line, discussion of, 141; measured graphs of, 137–139
  with Styrofoam-supported line, discussion of, 141; measured graphs of, 137–139
  symmetry conditions for, 73, 74
  zeroth-order, formulas of, 86, 87; graphs of, 88
current-distribution function, definition of, 77; in Hallén's expansion, 90; in King-Middleton expansion, 101
directivity of, 497; graphs, 495, 499; zeroth-order, 498; graph, 499
driving, methods of. *See* Methods of driving antennas
electromagnetic field of, first-order, using reciprocal theorem, 568–571; graphs, 488–493, 572
  first-order, using variational currents, 575
  modified zeroth-order, 571–575
  zeroth-order. *See* Cylindrical antenna with sinusoidal current
expansion function and parameter, 77, 78
gap at driving point, absence of, 844, 846–848
Hallén's solution, comparison with King-Middleton solution, graphs, 181, 843; with Schelkunoff's solution, 841–843

Cylindrical antenna—(contd.)
  expansion in, 89, 90, 93, 94
  expansion parameter in, definition of, 90, 93; graphs of, 91, 92
  first-order integrals in, 94–97; graphs of, 99, 100
  integral equation of, 76; approximations in, 79–81; modification of Gans and Bemporad, 80, 81
    ohmic resistance, effect on, table, 182
  hemispherical ends, significance of, 70; figure, 71
  impedance of, apparent, as function of line spacing, discussion, 208, 209, 227, 231–237, graphs, 207, 210, 247; as load on line, summaries of results, 198–215; as load on two-wire line, center load, 205–207, end load with stub support, 205, 206, 211, end load without support, 204, 206, 211, in plane perpendicular to line, 212, 220
    near antiresonance, discussion of, 152, 153; graphs of, 162, 163, 181, 843; measured graphs of, 212, 213, 240, 242, 243, 245–247, 843; table of, 177
    with capacitance in shunt, discussion of, 193, 198; graphs of, 200–203
    comparison of second-order values, of Hallén and King-Middleton formulas, graphs of, 182, 183, 843; of Hallén King-Middleton, and Schelkunoff formulas, discussion of, 842, 844, graphs of, 843; of King-Middleton and variational formulas, graphs of, 251
    comparison of theory and experiment, 206, 207, 210, 212, 213, 225, 227, 238–242, 244–247, 249, 251, 495, 843, 849
    correction, for base separation, 149, 193, graphs of, 150, 194–197; for terminal zone, summary of, 205, 208
    critical quantities, comparison of theoretical and experimental, 225, 227; discussion of, 152, 153; graphs of, 164; tables of, 168
    from different theories, graphs of, 251, 843
    dimensions of coaxial line, effect on, 227–237
    distribution curves, typical, in measurement of, 229
    with finite base separation, graphs of, 194–197
    formulas for, 142–144, 151–152, 255, 839, 841
    graphs, with a/λ_0 as parameter, 166, 167; with $\beta_0 h$ as parameter, 164, 165; circular, 154, 155; first- and second-order, 150; with Ω as parameter, 158, 159, 162, 163
    independent, 8, 142
    internal, 28, 29, 70, 71
    King-Middleton, formulas, 151, 152; graphs and tables. *See most theoretical values listed*
    measured, as function of line spacing, graphs of, 207; center load, 207; on coaxial line, 212, 213, 226, 239–247; on image-plane line, 216–219; on line in neutral plane, 212; with and without stub support, 206, 211; tables of, 232–238
    modified zeroth-order, 147–149; relation to emf method, 259

    near resonance, graphs showing, 212, 213, 239, 241, 244, 246, 843
    resonance curves, typical, in measurement of, 223, 229, 230
    Schelkunoff's theory, formula in, 841, graphs of, 840, 843
    second-order, graphs and tables. *See most theoretical values listed*
    short antenna. *See* Short antenna
    Storm's theory, tables, 261, 262
    tables, a/λ_0 as parameter, 178–179; near antiresonance, 177; critical values, 168; measured, 232–238; Ω as parameter, 169–177
    terminating two-wire line, discussion of, 50
    third-order, extrapolated graph of, 843, points on measured graphs, 239, 241, 244, 246; table of, 182
    by variational method graphs of, 251
    zeroth-order, 144–149, graph of, 146; modified, 147–149; relation to emf method, 259
    Zuhrt's theory, comparison with experiment, graphs of, 849, comparison with King-Middleton results, 855, table of, 854
  integral equation for. *See* Integral equation
  internal impedance, 70, 71
  King-Middleton solution, approximations in, 79–81
    comparison of first- and second-order values, graphs of, 150
    comparison with Hallén's solution, graphs, 181, 843; with Schelkunoff's solution, 841–843; with Storm's solution, tables, 261, 262; with variational solution, graphs, 251; with Zuhrt's solution, 855
  current-distribution function, 101
  D-factors, definition of, 110; graphs of, 108
  description of, 76–78, 81–86, 101–110
  expansion parameter in, definition of, 101–103, 108; discussion of properties, 104–107; graphs of, 105; table of, 106
  formulas for current in, 107, 110; for impedance, 142–144, 151, 152
  iteration in, 81–85
  long, current distribution in, measured graph of, 140, theoretical graphs of, 115; expansion parameter for, 107, graph of, 105
  methods of analysis, boundary value, 6, 7, 848–855
    emf, 3, 4, 258–259
    expansion, for short antenna, 186–193
    Hallén's integral equation and expansion, 7, 90–94
    King-Middleton, 7, 76–80, 101–110
    King's alternative, for components, 123–127, 183
    L. V. King's integral equation, 7, 259–261
    Nomura and Hatta's, 262
    Poynting-vector, 3, 4, 258–259
    retarded-potential, 7–9
    Schelkunoff's nonuniform line, 839–841

Cylindrical antenna—(contd.)
  transmission-line, 4, 5, 838–841
  Storm's undetermined-coefficient, 261–262
  variational, of Storer and Tai, 7, 237, 252–258
  Zuhrt's boundary-value, 848–855
 methods of driving. See Methods of driving antennas
 lengths, antiresonant, definition of, 153; graphs of, 163, 217
  resonant, definition of, 152; graphs of, 163
 ohmic resistance of, effect, table, 192
 potentials, scalar and vector, differential equations for, 72, 73, 75; solutions for, 75, 76; on surface of, 74–76; symmetry conditions for, 73, 74
 Q, definition of, 171, 173, 177; table of, 182
 reactance, near antiresonance, discussion of, 153; graphs of, 162, 163; measured graphs of, 212, 213, 240, 242, 243, 245–247
  with capacitance in shunt, graphs of, 202, 203
  graphs of, 150, 159, 162, 163, 165–167
  at $h = \lambda_0/4$, discussion of, 152, 153; graphs of, 163; relation to 42.5 ohms, 152, 153; tables of, 168
  maximum, graphs of, 163; table of, 168
  measured, with coaxial line, 213, 226, 233, 235, 237–246; with two-wire line, 206, 212
  minimum, graphs of, 163, table of, 168
  near resonance, discussion of, 152, 153; graphs of, 162, 163; measured graphs of, 212, 213, 239, 241, 244, 246
  tables of, 169–179; measured, 233, 235, 237, 238
  see also under Cylindrical antenna, impedance
 resistance, near antiresonance, discussion of, 153; graphs of, 181, 843; measured graphs of, 212, 213, 240, 242, 243, 245, 246; relation to extreme values of X, 153; table of 168,
  with capacitance in shunt, graphs of, 201, 202
  graphs of, 150, 158, 163, 166, 167; comparing Hallén and King-Middleton values, 181
  at $h = \lambda_0/4$, discussion of, 152, 153; graphs of, 163; table of, 168, 182
  maximum, graphs of, 163, 843; table of, 168
  measured, with coaxial line, 226, 232, 234, 236, 238–246; as function of line spacing, 207, 210; with two-wire line, 206, 210, 212
  ohmic, effect of, table, 182
  radiation, definition of, 149, 562; graph of, 564
  near resonance, discussion of, 152, 153; graphs of, 163, 843; measured graphs of, 212, 213, 239, 241, 244, 246; table of, 168
  tables of, 169–179; measured, 232, 234, 236, 238
  third-order, table of, 182
  see also under Cylindrical antenna, impedance
 resonance, definition of, 152
 resonant lengths, graphs of, 163; measured as function of line spacing, 210
 Schelkunoff's theory, comparison with solutions of Hallén and King-Middleton, 841–843
  description of, 839, 841
  impedance, graphs of, 840, 843

susceptance, capacitance in shunt, graphs of, 199
  graphs of, 154, 157, 161, 162
  maximum, table of, 168
  minimum, table of, 168
  tables of, 169–176
  see also under Cylindrical antenna, admittance
 terminal functions for, definition of, 182
  graphs of, 180, 181
  measured apparent, graphs of, 220, 223, 226, 230
 radiation field of. See Cylindrical antenna, electromagnetic field, Cylindrical antenna with sinusoidal current
 unbalanced, analysis of, 576, 578
  radiation field of, graphs, 577
 variational analysis of, 237, 252–258
 Zuhrt's analysis of, 848–855
  comparison with experiment, 849; with King-Middleton results, 855
  impedance and admittance, tables of, 854
 See also Asymmetrically driven antenna, Short antenna
Cylindrical antenna with sinusoidal current
 directional properties of, 530–532
 directivity of, definition, 498, 562; graphs, 495, 499, 564; table, 563
 electric field of, antenna of arbitrary length, 549
  antiresonant antenna, analysis, 546, 547; graphs, 548
  complete, components of, 527; in confocal coördinates, formulas, 540, graphs, 542, 543, 548; in cylindrical coördinates, 526–528; elliptically polarized, 544, graphs, 542, 543, 548; instantaneous, 538, 540, 544, 549–551, 554, 555, graphs of, 552, 553
  measurement of, using complementary slots, 555–557, 560; graphs, 558, 559
  near end of antenna, analysis of, 575, 576
  near-zone, formulas of, 533, 544; graphs of, 542, 543, 548
  phase relations in, 538, 540, 541, 544–546; graphs, 542, 543, 548, 552, 553
  radiation or far-zone, formulas of, 529, 530; graphs of, 531; maxima, graphs of, 531, table of, 537; nulls, graphs of, 531, table of 537
  resonant antenna, graphs of, 542, 543
  unbalanced antenna, 576, 578; graphs, 577
 field function or pattern. See Cylindrical antenna with sinusoidal current, electric field, radiation- or far-zone
 gain, definition of, 562; graphs of, 564
 magnetic field in, confocal coördinates, 540; in cylindrical coördinates, 524–526; instantaneous, 538, 540, 544
 phase velocity near, 539
 radiation field, approximate representation of, 566–568; in spherical coördinates, 528–529
 radiation function, definition of, 560; table of, 563
 radiation resistance, formulas for, 148, 560–562; graphs of, 564; table of, 563

D-factors, in King-Middleton solution, 110
D-vector, definition of, 13
Debye, P., 3
Delta function, definition of, 30; as generator for antenna, 252; graph of, 27
Density, of charge, current, magnetization, polarization, 12
Dielectric constant, 21; complex, 22; effective, 22; of free space, 13; relative, 21
Dielectric factor, complex, 22
Dielectric support, effect on antenna impedance, 222, 225
Differential equation, for current and voltage on line, 44, solution of, 44–50; for potentials on antenna, 75; for potential differences on line, 38, 43
Dike, S. H., 517
Dipole, folded. See Folded dipole
  infinitesimal, electric, definition of, 698; electromagnetic field of, 699, 700
    over conducting earth. See Horizontal infinitesimal dipole, Vertical infinitesimal dipole
    potential of, 695, 699
  infinitesimal magnetic, definition of, 22, 701, 702; electromagnetic field of, 702, 703
    over conducting earth. See Horizontal infinitesimal dipole, Vertical infinitesimal dipole
    potential of, 701, 702
  half-wave. See Cylindrical antenna, Multiple-half-wave antenna
  sleeve. See Sleeve dipole
Dipole antenna. See Cylindrical antenna
Dirac delta function, definition of, 30; as generator for antenna, 252; graph of, 27
Directivity, absolute, definition of, 21; of antennas and arrays, see under name of antenna or array
Disk, magnetizable, as generator, 30
Distortion factor, for transmission line, definition of, 46
Distribution-curve-dip method of measuring impedance, 231
Distribution curves, typical, in measuring impedance, 229
Divergence, definition, 12
Dolph, C. L., 8, 662, 669, 674

E-vector, definition of, 13
Eckersley, T. L., 695
Effective length, of driven antenna, 567, 568; of receiving antenna, 467, relation to field function of driven antenna, 570; see also under Receiving antenna
Effective separation, of parallel antennas, 274
Electric field, of antennas of arbitrary length with sinusoidal current, 546
  of antenna with finite cross section, approximate analysis of, 8, 572–574; first-order, from reciprocal theorem, 8, 570–571, graphs, 572
  of antennas of different types, see under name of antenna or array
  of antiresonant antenna, 546, 548
  components in terms of vector potential, 15
  of electric type, definition, 20
  far-zone or radiation, conditions for, 15; integrals for, 21; see also under name of antenna or array
  general integral for, 21
  impressed, definition of, 22; as external generator, 28, 29
  of magnetic type, definition of, 20
  near end of antenna, 575, 576
  near-zone, general integrals for, 21; of cylindrical antenna, see under Cylindrical antenna, Multiple half-wave antenna
  radiation. See Electric field, far-zone
  of sinusoidal distribution of current, components, figure showing, 527
    in cylindrical coördinates, 526–528
    elliptically polarized, 544–546; graphs of, 542, 543
    field function, 530–532; graphs of, 531
    lines of instantaneous, construction of 550–555; graphs of, 552, 553
    radiation field of, 528, 529; graphs, 531
    see also under Multiple-half-wave antenna
Electromagnetic field, of different antennas and arrays, see under name of antenna or array; see also Electric field, Magnetic field
Electron, oscillating, absorption cross section of 500, 501
  comparison with Hertzian dipole, 501
  power absorbed by, 501
  reradiation cross section of, 501
Emf of transmitting system, 28
Emf method, 8, 258, 259; impedance of antenna using, 259; relation to King-Middleton method for determining impedance, 259
End effects. See Coupling and end effects
End-fire array, array factor of, 601, 612; extremes, 601; graphs, 614–617; tables, 604, 605; zeros, 601
  beam width of, 613, 621; tables, 622
  bilateral, special case with $n_E = \frac{1}{2}$; two-element, circuit properties, see Asymmetrically driven antennas, H-array
  circuit properties of, as special case of parallel array, 375
  combination broadside, 612; graphs of, 618, 619
  condition for unilateral, 376, 377
  directivity of, table, 620
  gain of, table, 620
  radiation function of, 612
  radiation resistance of, 612, 613; table, 620
  two-element, array factor of, graphs, 616, 617; tables, 608–611 (for bilateral, $n_E = \frac{1}{2}$; for unilateral, $n_E = \frac{1}{4}$)
  impedances of, 312
Energy transfer function, definition of, 21; used to locate antenna, 25, 26
Equation, of circuit analysis, 29. See also Differential equation, Integral equation
Expansion function in integral equation, for cylindrical antenna, 77
  with difference kernel, 124
  for Hallén's solution, 90; graphs of, 92

Expansion function—(contd.)
  for King-Middleton solution, 101, 102, 104; graphs of, 103, 108
Expansion parameter in series solution of integral equation, for cylindrical antenna, definition of, 77
  with difference kernel, 124
  Hallén's, 90; graphs of, 91, 92
  King-Middleton, definition of, 102; graphs of, 105; table of, 106
  for long antenna, discussion of, 107; graph of, 105
  for short antenna, 104, 184, 186; graphs of, 185
  for different antennas, see under name of antenna
Exponential integral functions, $E_a(h, z)$, definition, 94; graphs of, 98; in terms of tabulated functions, 97

Faflick, C., 9, 328, 339, 455
Far- or radiation zone, condition for, 15; integrals of electromagnetic field for, 21
Field, electric, electromagnetic, see Electric field, also under name of antenna or array
  far- or radiation zone, conditions for, 15, 528, 529, in terms of vector potential, 21
  impressed, 22
  magnetic. See Magnetic field
Field equations, Maxwell, 13
Field factors, $F_m(\Theta, \beta_0 h)$, $F_0(\Theta, \beta_0 h)$, angles of zero and extreme values, 531; table of, 537
  of antenna with finite cross section, 570, 571; graphs of, 572
  definition of, for sinusoidal current, 530
  extreme values of, graphs, 531
  graphs of, 531
  for multiple half-wave antenna, 536; angles of maximum and zero, graphs of, 536, table of, 537; extreme values, graphs of, 539
  properties of, 530–532
  relation to effective length of receiving antenna, 570
  for short antenna, 530
  zeros of, 530–532, graphs of, 531, 536; table of, 537
Field patterns, of antennas and arrays. See Field factor, also under name of antenna or array
  vertical, of sinusoidal current, 531
Fitch, E., 8, 311, 627, 630
Folded antenna, admittance of, 358
  analysis of, 353–358
  description of, 352, 353
  effective radius of, 353, 357
  expansion parameter of, 357
  four-wire reëntrant loop, impedance of, formula and table, 368
  integral equations for, 353
  radiating currents in, 353, 354
  three-element, admittance formulas of various, 364, 366; analysis of, 364–365; impedance of, table, 366
  transmission-line currents in, 353, 354
  two-element, analysis and admittance of, 358–361
  see also Folded dipole

Folded dipole, admittance of, formula, 359; tables, 346, 359
  analysis of, 335, 336
  currents in, antisymmetric, 336, 342, 343; discussion of, 342; formulas for, 338; graphs of, 341, 344; measured, graphs of, 341, 344; ratio of, 338; symmetric, 336, 342
  description of, 335, 336
  four-element, impedance of, formula and table, 369
  impedance of, equal conductors, formula, 337; graphs of, 341, tables, 343, 346, 359; unequal conductors, tables, 361
  power dissipated as heat in, 338
  as special case of folded antenna, 358–360
  three-element, admittance formula, 365; analysis of, 364–366; impedance of, table, 366
  with unequal conductors, analysis of, 360–361; approximate analysis of, 339; impedance of two- and three-element, tables, 361
Folded dipoles, coupled, mutual and self-impedances of, 343
Force, electromagnetic, definition of, 14
Fresnel coefficients, 712

Gain, definition of, 562; see also under names of antennas and arrays
Gans, R., 81
Gap, antenna without, 28; problem, analysis of, 844–848; significance of, 847; spark, 1. See also Base separation
Generator, delta-function, 30, 252; disk, 29, 30; slice or belt, 818; theoretical, 28
Geophysical exploration with electromagnetic waves, possibility of, 723
Gray, M. C., 7, 695
Grazing incidence, 767
Green's function, free space, 80
Green's theorem, symmetrical, 80
Group velocity, for field of multiple half-wave antenna, 541

H-array, centrally driven, analysis of, 318; coupling and end effects in, 318, 319
  description of, 318, 320
  laterally driven, analysis of, 319–321; coupling and end effects in, 321, 322
H-vector, definition of, 13
Hack, F., 3, 552, 553
Half dipole with counterpoise, analysis of, 420–422; description of 418–420; graphs, 423, numerical results, 422; methods of driving, 418–420
Hallén, E., 3, 5, 7, 499, 841, 842, 855
Hallén's expansion, 89–94
Hallén's integral equation, 76, 79; alternative method for solving, 123–127; criticism of, 79, 80; Gans and Bemporad's modification of, 81; Hallén's solution for current, 93 94; King-Middleton method for solving, 76–80, 101–110

[Index] THEORY OF LINEAR ANTENNAS 937

Hamermesh, M., 505, 508–511
Hankel function, spherical, forms for large and small arguments, 823
Hansell, C. W., 4
Harrison, C. W., Jr., 7, 8, 9, 352, 353, 359–361, 368, 369, 371
Hartig, E. O., 7, 213, 215, 227–237, 239–250, 478, 480, 485, 842, 854
Hatch, R. M., 447–449
Height-gain function, definition of, 775; graph of, 777; table of, 775
Helmholtz integrals, for Hertz vectors, 20; one-dimensional, 14, for potentials, 14
Hertz, H., 1, 2, 3, 4, 11
Hertzian potentials, Hertz vectors, 696; radial, 820; see also Potential, magnetization, polarization
von Hoerschelmann, H., 10
Horizontal antenna with sinusoidal current over conducting earth, 794
  current in perfectly conducting plane below antenna, analysis of, 794, 796, 797, 799; graphs of 795, 798
  impedance of, 800
  measured resistance, graph of, 806
Horizontal infinitesimal dipole over conducting earth, electromagnetic fields of, comparison with vertical dipoles, 791–794; formulas for, 785–790; radiation-zone formulas of, 790–791
  Hertzian potentials for, general formulas, 779–783; radiation-zone formulas, 783–785
  radiation resistance of, analysis, 804–808; formulas for, 807; graphs of, 806

Image plane, antenna over, 64–69, 215–219, 221–223, 227, 231
Image-plane line, description and theory of, 215–219; used for measurements, 135, 136, 141
Impedance, of antenna used for reception, apparatus for measuring, 480; comparison with impedance of same antenna used for transmission, graphs, 213, 485; comparison of measured and theoretical values, 213, 485; measuring technique, 480; theory of, 480, 486
  apparent, for load on line, 47, 65; 198–215
  with base separation, discussion, 149, 193; graphs, 150, 194–197
  characteristic, of biconical line, formulas for, 827, 836; graphs of, 837
    of free space, 14
    of simple medium, 22
    of transmission line, 46, 275
  of complete transmitting system, 29
  of different types of antennas, see under name of antenna
  external, of circuit element, 31; of generator
  independent, of antenna, definition of, 8, 142
  field, definition of, 711; matching of, 734
  graphs of, antenna on finite ground screen, 814; over infinite conducting earth, 814

  antenna parallel to perfectly conducting plane, 279, 282
  antenna with single parasite, 323, 631, 634
  antisymmetrically driven antennas, 279, 282
  asymmetrically driven antenna, 405
  biconical antenna, 837, 840
  broadside array, two element, 279, 282; with parasitic curtain, 650
  cage antenna, two element, 279, 282
  capacitance in shunt with antenna, 200–203
  collinear array, three-element, 438, 440
  coupled parallel pair, 279, 282, 291–295, 306–310; measured, 350
  cylindrical antenna, near antiresonance, 212 240, 242, 246, 247; with capacitance in shunt, 200–203; comparison of theory and experiment, 206, 207, 210, 212, 213, 225, 227, 238–242, 244–247, 249, 251, 485, 495, 843, 849; critical quantities, 164; different theories, comparison of, 181, 251, 843, 849; with finite base separation, 194–197; first- and second-order compared, 150; measured, 206, 207, 211–213, 239–247, 485; Schelkunoff's theory, 840, 843; second-order, 150, 155, 156, 158, 159, 162–167; short, 190; third-order, 239, 241, 244, 246, 843; zeroth-order, 146
  end-fire array, bilateral two-element, 279, 282
  folded dipole, 341
  ground-plane antenna, 423
  parasitic array, 638, 639, 642–644
  receiving antenna, 485, 495
  short antenna, 190
  V-antenna, 393, 394
internal, of circuit element, 31, of generator, 29; per unit length, 28, 29, 70, 71
measurement of. See under Measurement
mutual, definition of, 286, 373, 441
  graphs of, for collinear antennas, 438, 440; for parallel antennas, 287, 291–295, 306–311, measured, 350
  tables of, for collinear antennas, 441; for parallel antennas, 289, 298, 300, 302, 304, 305
  see also under Collinear array, Coupled antennas, Coupled parallel pair
self-, of antenna in presence of another, definition, 286
  graphs of, for parallel antennas, 291–295, 306, 307
  tables of, for parallel antennas, 297, 299, 301, 303, 305
tables of, antenna with single parasite, 329, 224
  antisymmetrically driven antennas, 280
  cage antenna, two-element, 280
  collinear array, three-element, 441
  coupled parallel pair, 280, 297, 299, 301, 303, 305
  cylindrical antenna, near antiresonance, 177; critical quantities, 168; measured, 232–238; second-order, 168–179; Storm's theory, 261, 262; third-order, 182; Zuhrt's theory, 854

Impedance, tables of—(contd.)
  folded antenna, 366, 368
  folded dipole, 343, 359, 361, 366
  parasitic array, 647
  short antenna, 192
  terminal-zone, 45
Incidence, angle of, 709; grazing, 767
Index of refraction. See Refraction, index of
Inductance, terminal-zone, for coaxial line, antenna over image plane, 66
  for connectors, 63, 317
  for coupled lines and antennas, antisymmetrically driven, 315; symmetrically driven, 315
  definition, 45, 46
  for two-wire line, antenna as end load, 55; antenna as mid-point load, 57; antenna over image plane, 59, 60; antenna with stub support, 59, 60
  see also Coupling and end effects
Inductance per unit length, 41, 43; graph, 34
Integral equation for current, in cylindrical antenna, 30, 31, 76, 81; approximations in, 79–81; best form for, 76, 252; criticism of, 79–80; series solution of, 81–86; variational form of, 252, 253
  in receiving antenna, 462
  in scattering antenna, 501, 502
  of Gans and Bemporad, 81
  of Hallén, 76, 79; methods of solving, by empirical fitting, 186–193; by iteration of complex equation, 76–80, 101–110; by iteration of two simultaneous real equations, 123–127; by undetermined coefficients, 251, 262; in variational form, 237, 252–258
  of L. V. King, 259–261
Integral equations, simultaneous, for currents, in asymmetrically driven antenna, 399; in collinear array, 425, 426, 431, 434; in folded antenna, 353–358; in N parallel antennas, 351–353; in two parallel antennas, 266–269; in V-antenna, 383, 384
  methods of solving, separation of radiating and transmission-line currents, 353–358; substitution of equations for driving voltages, 371, 372; symmetrical components, 266–273, 351–353
Isotropic radiator, 506

Kelvin, W., 576
King, D. D., 214, 215, 221–227, 480, 511, 512, 516, 842
King, L. V., 7, 8, 259–261
King-Middleton solution. See under Cylindrical antenna
Knudsen, H. L., 684

Labus, J., 4, 5
Legendre's equation, 823
Legendre functions, first and second kinds, 824; integral products of, 831–832
Leitner, A., 10
Lewis, J. B., 343

Linear radiator. See Cylindrical antenna
Lindenblad, N. E., 4
Loss tangent, 22

McPetrie, J. S., 623, 626, 627, 630, 634
Magnet, oscillating as generator, 28
Magnetic field, components of, in terms of vector potential, 15; of electric type, definition, 20; general integral for, 21; of magnetic type, definition, 20; of sinusoidal distribution of current, 524–526, 540, graphs of, along antenna, 527; of slots, graphs of, 558, 559
see also under names of antennas
Magnetic moment of ring of current, 23
Magnetization, volume density of, 12
Magnetization potential or magnetic Hertzian potential. See under Potential
Marconi, G., 1, 2, 3, 11
Marconi-Franklin antenna. See Collinear array
Matching of fields, across boundary of two regions, 734; perfect match and Brewster angle, 734; in solution of biconical antenna, 830, 831; in Zuhrt's solution of cylindrical antenna, 851
Maxwell, J. C., 1, 3, 11
Measurement, of back-scattering cross section, by canceling method, 515–517
  by right-angle method, 517
  by SWR method, 512–517
  of current and charge distributions, with coaxial line, 129, 130, 131
  with image-plane open-wire line, 135, 136, 141
  of impedance, with coaxial line, 221–227; block diagram for, 228; vertical slotted line for, 223
  with image-plane open-wire line, 215–219; block diagram for, 217
  with long two-wire line, 219, 221; block diagram for, 220; unbalance squelcher for, 220
  of mutual impedance, from complex coefficient of coupling, 349
  with impedance-voltage-ratio method, 348
  with open-circuit-short-circuit method, 348
  with symmetric-antisymmetric method, 346, 347
Meiers, A. S., 10, 815
Methods of driving antennas, with biconical line, 846–848
  with coaxial line, over ground plane, 65–69, 845
  coupled to primary
  with delta-function generator, 27, 30
  by electric field across gap, 27
  gap problem and, 844, 846–848
  by oscillating magnet, 27–31
  parallel to conducting plane, 283
  by point source in radial line, 80, 81; 845, 846
  by rotationally symmetric field, 24, 844, 845
  significance of gap in, 847
  by single conductor over image plane, 62, 64–65
  by two-wire line, 50–55, 845; as center load on symmetrical, 55–60; with dielectric support, 56; as end load, 48, 50–55; with no coupling, 61–64; with stub support, 59–60

Mie, G., 13
Middleton, D., 7
Millington, G., 695
Model, requirements of theoretical, 25
Morita, T., 9, 130, 131, 328, 329, 479, 485, 494, 495, 499
Moritz, C., 349, 517, 519, 521
Multiple half-wave antenna, directivity of, 562; graph, 564; table, 563
   electric field of, complete, components of, 527; in confocal coördinates, formulas of, 540, graphs of, 542, 543; in cylindrical coördinates, formulas of, 533
      elliptic polarization of, discussion, 544–546, graphs showing, 552, 553
      instantaneous, 534, 535, 544; lines of, construction, 550–555, graphs, 552, 553
      near-zone, formulas of, 533, 544; graphs, 542, 543
      radiation or far-zone, formulas of, 529, 530; graphs of, 531; instantaneous, 535, 538; maxima of, graphs, 531, table, 537; nulls of, graphs, 531, table, 537
   electromagnetic field of, general discussion, 522–523
   gain of, definition, 562, graph, 564
   group velocity of field near, 541
   impedance of. *See under* Cylindrical antenna
   magnetic field of, formulas, 533, 540; radiation or far-zone, 529
   phase velocity of field near, 539, 541, 544
   radiation function of, 560; table, 563
   radiation resistance of, formulas, 560–562; graph, 564; table, 563
   transfer of power in field of, 560–566
Mutual impedance, definition of, 286, 273, 441. *See also under* Collinear array, Coupled antennas, Coupled parallel pair

Network, equivalent. *See* Circuit, equivalent
Neumann function, spherical, forms for large and small arguments, 823
Niessen, K. F., 10, 695
Noether, F., 695, 696
Nonuniform arrays, broadside, derived from uniform arrays, 660, 661; array factors, graphs of, 666
   optimum. *See* Nonuniform arrays, broadside, Tchebyscheff
   optimum currents in, 665
   polynomials for use in, 661–665; Tchebyscheff, 666–669
   Tchebyscheff, analysis of, 665–671; array factors for, 669–672; closely spaced, 676–679, 683; comparison with uniform arrays, graphs, 673, 678; current in, 674, 675; minor lobe level in, 674; nulls and extremes in, table, 672
   collinear, two-element, 595–596
   end-fire, apparent number of units in, 652
   array factors of, graphs, 657
   with assigned nulls, analysis and description of, 654–656, 658–660

   beam width decreased by shifting nulls, 655
   binomial, analysis of, 652, 654–656; graph of, 653
   complex array polynomials for, 651, 652
   six-antenna, 658, 659; graphs of array factors, 657
   three-antenna, 656, 658, 659; graphs of array factors, 657
   parallel, analysis of, 595–601; array factor of, 600; diffraction formula for use in, 579, graph of, 581, table of, 582–585
Norton, K. A., 10, 695, 696, 746, 759, 760, 767, 768
Numerical distance, complex, for dipoles on boundary, 758; for elevated dipoles, 758
Numerical height, definition of, 760

Ohmic resistance of antenna, effect on impedance, 182
Omnidirectional antennas. *See* Antenna, turnstile; V-antenna
Oscillation of antenna, forced, 3, 4; free, 3
Oseen, C. W., 3
Owen, W. E., 635

Page, L., 5
Papas, C. H., 6, 8
Parallel array with elements in line, analysis, of circuit properties, 351, 371–374; of field characteristics, 595–601
   conditions to make uniform, 600
   coupling and end effects, 376
   currents in, distribution in each element, 371; relation between elements, 372, 595
   impedance of, 372–373
   integral equations for, 371
   laterally driven, 376
   methods of driving, 375
   nonuniform. *See* Nonuniform arrays
   parasitic. *See* Antenna with single parasite, Parasitic arrays
   with parasitic elements, analysis and discussion of, 377; three-element, current and impedance, 378
   two-element, array factor of, 612
   uniform, array factor, 601, beam width, 613, 621, 622; extremes, 601; nulls, 601; directivity of, table, 620; gain of, table, 620; radiation field of, 601; radiation function of, 612; radiation resistance of, formulas, 612, 613, tables, 620
   see also Broadside array, End-fire array
Parasite, antenna as, definition of, 263
Parasitic array, antenna with director and reflector, field of, power gain of, resistance of, graphs, 644
   antenna with several parasites, field ratio of, power gain of, resistance of, definitions, 636; descriptions and discussions of arrays, 640, 641, 645; graphs, 638, 639, 642, 643; zeroth-order analysis of, 635–636
   antenna with single parasite. *See* Antenna with single parasite

Parasitic array—(contd.)
  array of Yagi-Uda arrays, 647; table of critical quantities, 648
  parallel array with parasitic elements, analysis and discussion of, 377; equations and impedance of, 378, mutual impedances for, 378
  Yagi-Uda array, 644, 645, 647; directivity of, table, 647; field of, graphs, 644, 646; optimum director length, table of, 647; resistance of, table, 647
Permittivity. See Dielectric constant
Phase, measurement of, 131
Phase constant, complex, 22, real, 22
Phase velocity, complex, 710; definition of, 22; formulas for; in field of multiple half-wave antennas, 540; graph of, 539
Pierce, G. W., 4
Pistolkors, A. A., 4, 613
Pocklington, H. C., 3, 9, 11
Polarization, volume density of, 12
Polarization potential, or electric Hertzian potential. See under Potential
Potential, in far- or radiation-zone, 15
  Hertzian, electric, and magnetic, boundary conditions for, with vertical dipole, 697; with horizontal dipole, 698
  definitions of, 20
  for electric dipole, 699
  Helmholtz equation for, 696, 697
  for horizontal dipole near boundary between two media, 782, 783; far-zone form, 784
  for magnetic dipole, 702
  for vertical dipole near boundary between two media, 715, 762; asymptotic form in region containing dipole, 718, 719; at distant points in interior of region 1 not containing dipole, 722; near surface in region 1 not containing dipole, 723; Sommerfeld's formula, 750; Van der Pol's formulas, 753
  magnetization, *same as magnetic Hertzian potential*
  polarization, *same as electric Hertzian potential*
  retarded, 8, 9
  scalar, definition of, 14; discontinuity in, 79; on surface of cylinder, 16–20; on surfaces of coupled antennas, 266; on surface of cylindrical antenna, 74–76; on two-wire line, 36
  vector, definition, 14; far- or radiation-zone, of antenna, 459–461, 464, 529, figure, 465; of sinusoidal distribution of current, 522–529; on surface of cylinder, 16–20; on surfaces of coupled antennas, 264, 266, 457; on surface of cylindrical antenna, 74–76; on surface of two-wire line, 35, 36
Potential difference, driving, for cylindrical antenna, 76
  scalar, for two-wire line, 36, 47
  vector, for two-wire line, 36, 47

Power, in load of receiving antenna, analysis, 496–498; measurement of, to matched load, 498, graph, 495; transfer of, in field of antenna, 562–566
Poynting vector, definition of, 21; discussion of, 4, 500; incident on scatterer, 506; reflected or scattered, 506
Poynting-vector method, 258–259
Poynting-vector theorem, 4, 21; application in definition of antenna and complete transmitting-system, 25, 26; application in determination of radiation resistance, 21, 560; application to transfer of power in field of antenna, 562–566
Proctor, R. F., 806, 808
Probes, charge or electric field, 129–131; current or magnetic field, 129, 130; shielded loop, 129, 130, 131

Q of antenna, definition, 171, 173, 177, table of, 182
Quadrant antenna, 688, 689; field of, graph, 691

r_α-, r_β-functions, formulas, 143; graphs of, 146
Radiation, from two-wire line, 285, 286; condition to make negligible, 38
Radiation field, of antenna of finite cross section, approximate analysis, 572–574; from reciprocal theorem, 570, 571, graphs of, 572
  of center-driven antenna, 528–530, 535, 538; approximate representation, 566–568
  general integrals for, 21
  of linear radiators, 528–532
  in terms of vector potential, 21
  see also under name of antenna or array
Radiation function, definition of, 21; of cylindrical antenna, 560, table of, 563
Radiation resistance, definition of, 21; of cylindrical antenna, formulas for, 148, 560–562, graphs of, 564; table of, 563; of short antenna with triangular current, 565; of small loop antenna, 23; of two-wire line, 285
Radiation zone, condition for, 15
Ratio functions for potentials on line, 41, 42
Rayleigh, Lord, 3
Rayleigh formula for internal resistance per unit length, 29
Rayleigh-Carson reciprocal theorem, applied to cylindrical antenna, 568–571; applied to transmitting and receiving arrays, 689–694
Reactance, of antennas of different types, *see under name of antenna*; in terms of terminal functions, formula, 231
Receiving antenna, analysis of, 456–460
  condition for loose coupling, 457, 459
  interception of power by, 500
  expansion parameter for, 470–472; graph of, 471
  orientation of, figure, 465

[Index] THEORY OF LINEAR ANTENNAS 941

Receiving antenna—(contd.)
  symmetrically loaded, charge distribution in, 468
    current in, block diagram for circuit for measuring, 469; with conjugate matched load, 469; distributions of transmitting and receiving types, 467; experimental determination of, 479; first-order formula for, 473–475; first-order graphs of, 476, 477; general formulas for, 462–464; measured, graphs of, 481–484; principal part of, 469; when resonant, 469; zeroth-order formula for, 468, 469
    current distribution functions, 472
    directional properties of, 494
    directivity of, formulas, 497–498; graph of, 495; table of, 497, zeroth-order, graphs of, 499
    dissipation cross section, definition of, 500; graph of, 495
    effective length of, correction for base separation, 487, graph of, 485; definition of, 467; formulas of, 486, 487; graphs of, 488–493; independent of load resistance, 497; measured normalized magnitude of, 495; measurement of, 494, 496; normalized magnitude of, graphs, 492, 493; real and imaginary parts of, 487, graphs of, 488–491; of short antenna, 487, 496, table of, 496; zeroth-order, 470, graphs of, 493
    equivalent circuit of, determination of, 466; diagram for, 465; emf in, 466, 470
    impedance of, definition of, 480, 486; measured, graphs of, 485, 495
    integral equation for, 462
    orientation in electric field, 464, 466; in vector-potential field, 464
    power in load, 496, 497; measured in matched load, graph of, 495; used to determine effective length, 494
    short, analysis of, 475–479
    unloaded, analysis of, 456–461; current in, first-order graphs, 476, 477, measured graphs, 481–484; zeroth-order, graph of, 471
    vector potential on, 459–461, 464
Receiving antennas, parallel pair, analysis of, 517–521; currents in, graph of, 519
Receiving arrays, analysis of, 690–694; reciprocity, with equal and constant voltages, 693; with equal and constant powers, 694
Reciprocal theorem, applied to arrays with constant voltages, 693; applied to arrays with constant powers, 694; applied to cylindrical antenna, 568–571; and properties of arrays, 689–694
Reciprocity, for antenna arrays, 689–694: for antennas, 568–571
Reflection, angle of, 709; law of, 709
Reflection coefficient, for angles near boundary surface, 741
  for antenna ends, 5
  for conductors, good, 735; very good, 740
  definition of, 710, 711
  general formulas for, 711
  graphs of, 728, 729

  minimum, 735
  interpretation of, in terms of fields parallel and perpendicular to plane of incidence, 727
  at normal incidence, 713
  with one medium air, 726, 727
  for radiation zone, 725
  separation of real and imaginary parts of, 727
  tables of, 730–734
  in terms of complex index of refraction, 713; of field impedances and admittances, 713; of transmission-line notation, 712
  see also under Horizontal infinitesimal dipoles, Vertical infinitesimal dipole
Reflector, corner, analysis, formulas, numerical values, 370, 371
Refraction, angle of, 709
  index of, definition of, generalized complex, 712, 726; of real, for lossy medium, 742; numerical values for dry and wet earth, salt water, 741
  law of, 709
Reluctivity, definition, 21
Renner, F., 10, 800, 803, 804, 806, 807
Reradiation from antennas. See Scattering antennas
Resistance, of antennas of different types, see under name of antenna; characteristic, of free space, definition of, 22, of two-wire line, graph, 34; internal, 29; external, of cylindrical antenna, 147, 148; radiation, see Radiation resistance; in terms of terminal functions, formula, 231
Resonance, in antenna, definition of, 152; lengths defining, 163, 210
Resonance-curve method for impedance measurement, 222, 224, 231
Resonance curves, typical in measurement of impedance, graphs of, 223, 229
Riazin, P., 4
Riblet, H. J., 676
Roberts, T. E., Jr., 140, 141
Rüdenberg, R., 4
Rutherford, Lord, 1
Ryder, R. M., 5

Saddle point, definition of, 716; method of integration, 716
Saxton, J. S., 623, 626, 627, 630, 634
Scattering antennas, analysis of, 501–505
  cross section of, average back-scattering, angular distribution of, graphs, 511; contributions to, by forced and resonant currents, 510; definition of, 507, 508; first-order for broadside scattering, 510; graphs of, 508; measured graphs, 516, 520; measurement of, 512–515, 517; for parallel and perpendicular polarizations, 507
  total scattering, definition, 506, 507
  current in, first-order formulas for, 503; forced and resonant parts of, 509, 510; modified zeroth-order, 504–506; symmetric and antisymmetric, 503; zeroth-order formulas for, 504

Scattering antennas, current in—(contd.)
  electric field, incident, 506; reflected, re-radiated, scattered, 506
  expansion parameters, 503–505
  parallel pair, discussion of, 521; broadside and end-fire back-scattering of, graphs, 520
  Poynting vector, incident, 506; reflected or scattered, 506
  standing waves produced by, graph of, 511
Schelkunoff, S. A., 5, 6, 8, 387, 652, 655, 656, 824, 835, 838, 840–843
Self-impedance, of antenna in presence of another, definition, 286; see also under Collinear array, Coupled antennas, Coupled parallel pair
Sevick, J., 517, 520, 521
Shifted cosine distribution, 469
Short antenna, analysis of, 184–192; 475–479
  charge distribution in, graphs of, 185
  current distribution in, analysis of, 187, 188, 191, 478, 479; graphs of, 185, 190
  directivity of, 497; table, 497
  effective length of, for reception, formulas, 496; graph, 492; table, 496
  expansion parameter, 104
  impedance of, analysis, 184, 186–189, 191, 192; formulas, 192; graphs, 190; table, 192
  potentials and potential differences of, for triangular and adjusted currents, graph, 185; for parabolic and adjusted currents, graph, 190
  radiation resistance of, 565
  receiving, analysis, 475–479; currents in, graph, 190
Siegel, E., 5
Simple medium, definition, 21
Sine integral functions, $S(A, U)$, 95, 857; table of, 858; $S(0, U)$, 96
  $Sc(A, U)$, 95, 857; table of, 858; in terms of Si, 96; $Sc(0, U)$, 96
  $Si(U)$, 96, 857; approximate form for large arguments, 107, 143; table of, 858
  $Ss(A, U)$, 95, 857; table of, 858 in terms of Ci and Cin, 96; $Ss(0, U)$, 96
  $S_a(h, z)$, definition of, 94; graphs of, 98, 272; in terms of tabulated functions, 97
  $S_b(h, z)$, graphs of, 272
  see also Cosine integral functions
Sinusoidal current, in zeroth-order solution, 87
Sleeve dipole, broad-band properties, 411
  currents in, formulas for, 409; graphs of, theoretical, 410, measured, 412, 413, measurement of, discussion, 417, 418
  description of, 407, 408
  impedance of, analysis and formulas, 408, 409, 411; graphs, theoretical, 410, measured, 414–416, 419; measurement of, discussion, 417, 418
  variational analysis, 417
Smith, P. D. P., 824
Snell's law, derivation of, 708, 709
Sommerfeld, A., 10, 695, 746, 800, 803–807

Southworth, G. C., 8
Spark discharges, near end of antenna, analysis of, 575, 576
Spence, R. D., 10
Spherical wave, representation as bundle of plane waves, 703–707
Starkey, B. J., 311, 627, 630
Standing-wave ratio, in terms of terminal attenuation function, 182; use of, in measurement of back-scattering cross section, 513–515
Stegen, R. J., 670
Sterba, E. J., 8
Storer, J. E., 7, 9, 237, 252–258, 381, 417, 521, 574, 808, 811, 814
Storm, B., 261, 262
Stratton, J. A., 6, 818
Strutt, M. J. O., 10, 695, 696
Stub support, lumped terminal-zone network for use with, 59
Summers, W. P., 10, 815
Surtees, W. J., 814
Susceptance, of antennas of different types. See under name of antenna
Symmetric impedance. See under Symmetrically driven antennas
Symmetrically driven antennas, analysis of circuit properties, 267–269
  current distribution in, discussion of, 275, 276; formulas for, 268, 269; graphs of, 276
  expansion parameter for, 267–269, 274; graph of, 271
  field of. See Broadside array
  impedance of, formulas, 277, 278; graphs of, 279, 282; measurement of, 347; modified zeroth-order, 278; tables of, 280
  see also Collinear array, Folded dipole, H-array, Transmission-line radiators
Symmetry conditions, for center-driven cylindrical antenna, 73, 74
Synge, J. L., 7, 9

Tai, C. T., 7, 9, 254, 361, 370, 381, 824, 835, 837, 840, 841
Tang, C., 455
Taylor, J., 8, 9, 412–417, 494, 495, 499
Tchebyscheff arrays. See Nonuniform arrays
Tchebyscheff polynomials, 666–669
Terminal functions, for cylindrical antenna, definition of, 182, 285, 480; graphs of, 180, 181; measured graphs of, 220, 223, 226, 230; relation to reflection coefficient, 182; relation to standing-wave ratio, 182; used to express impedance, 231
Terminal zone, discussion of length of, 44, 45; summary of steps to take account of, 45–47, 49
Theory and experiment, discussion of problem of, 2, 3, 24, 25
Thévenin's theorem, applied to receiving antenna, 466
Tilt, angle of, for electric field at boundary, 773

Tomiyasu, K., 7, 208, 212, 214, 219–221, 842
Torque, electromagnetic, definition of, 14
Transmission, angle of, into conducting medium, 709; plane of, 709
Transmission coefficient, general formulas for, 711; at normal incidence, 713; for radiation zone, 725, 726; in terms of complex index of refraction, 712; in terms of field admittances and impedances, 713; in transmission-line form, 712; *see also under* Horizontal infinitesimal dipole, Vertical infinitesimal dipole
Transmission line, biconical, 827; coaxial, 130, 221, 222, 224, 225; equivalent of antenna, 5; instead of gap at driving point, 844, 846–848; image-plane, 215–219; long two-wire, 219–221; radiation from two-wire, 38, 275; relation to parallel antennas, 275; symmetrically driven, 58, 59, 215–219; two-wire with end load, 31–33
Transmission-line circuits, 339
Transmission-line equations, conventional, 43; electromagnetic foundations of, 35; generalized, 36–43; for ideal line, 46; solution of, 44–47, 49
Transmission-line radiators, analysis and description of, 334, 339; circuit diagrams of, 336, 340; end effects in, 335; folded dipole, *see* Folded dipole; impedance of, 335
Transmitting system, antenna as, 24, 25, 28–30; complete, 24; definition of, 24; with external generator, 26, 27; Hertz's, 1; Marconi's, 1
Turnstile antenna and array, *see under* Antenna, turnstile *and* Array, using turnstile elements
Two-wire line, derivation of generalized equations for, 34–44; terminal impedance for, 31–33

U-antenna, admittance of, 359; analysis of, 359
Unbalance squelcher, description and application of, 219–221
Uda, S., 377, 645, 646
Uniform arrays, array factor of, general, 580, 586
 beam width, definition of, 586, half-power, 586, 587; null, 586, 587; table of, 587
 extreme values of, approximate location, formulas for, 580, 586; graphical method for locating, 588; minor, 580, 586; principal, 580
 graphs of, 581
 normalized, 580
 special forms and symbolism in applying to antenna arrays, 586
 table of, 582–585
 zeros of, 584
 collinear. *See* Collinear array
 conditions for, 600
 definition of, 597, 600
 parallel. *See* Parallel array with elements in line

V-antenna, analysis, of circuit properties, 381–384; of field, 687–689
 coupling and end effects with, 388–396
 current distribution, as function of orientation relative to transmission line, graphs of, 395
 description of, 381, 382
 expansion parameter for, 384–386
 electric field, horizontal, table of, 689; graphs, 691
 field patterns of, 691
 generalized, right-angle, 418, 420–422
 impedance of, formulas for 386; as function of orientation relative to transmission line, graphs, 393, 394
 integral equation for, 383
 in plane of driving line, analysis of, 391–396; coupling and end effects, 391, 392; current distribution in, graphs of, 394; impedance of, graphs, 389, 393, 394
 perpendicular to plane of driving line, analysis of, 388–391; coupling and end effects in, 388–391; impedance of, graphs, 389, 393
 reactance, using transmission-line methods, 387, 388
 resistance of, graphs 393
Van der Pol, B., 4, 10, 695, 696, 746
Van Vleck, J. H., 7, 505, 508–511, 517
Vector potential. *See under* Potential
Velocity, characteristic of free space, 14
 group, in field of multiple-half-wave antenna, 541
 phase, complex, 710; definition of, 22; formula for, in field of multiple half-wave antenna, 540; graph of, 539
Vertical antenna with sinusoidal current over conducting earth, base-driven, impedance of, 799, 800; quasi-near-zone field of, 776–779; radiation field of, 743–746, graph, 747, tables, 746, 748, 749; relation of field to field of infinitesimal dipole, 776; relation of field to field of unbalanced isolated antennas, 745. *See* Cylindrical antenna, Collinear array *if earth is perfectly conducting*
Vertical dipole, infinitesimal, isolated. *See under* Dipole
Vertical infinitesimal dipole over conducting earth, array factors of, 740, 741
 attenuation in the earth, 723
 boundary conditions, application of, 709; when one medium is air, 726
 Brewster or polarizing angle for field of, definition, 730, 734; for good conductors, 735, 740; for good dielectrics, 735; graphs of, 736; for salt water, 740; for wet earth, 740
 conditions imposed if one medium is air, 726
 discussion of, 10
 electromagnetic field of, in air, along boundary surface, 741, 742; in conducting dielectric, 742, 743; in radiation zone, 725, 740–741
 comparison with field of horizontal dipoles, 791–794
 field factors, graphs of, 736–739
 at large numerical distances, 774, 775

Vertical infinitesimal dipole, electromagnetic field of—(contd.)
   quasi-near-zone, complete, 770–773; complex surface-wave attenuation function for, definition, 760, graphs, 761, tables, 762; definition of, 772; Norton's practical formulas for, 759, 760; surface term, significance of, 767; radial components of, 768–770; ratio of surface to space terms, graphs of, 771, tables of, 767, 768; summary of, 760, 762, 763; vertical components of, 763–767; Van der Pol's integrals, 752, 753, derivation of, 743–754, integration of, 754–759
   radiation-zone, 725, 726
field admittances and impedances, definition of, 711
height-gain function, definition of, 775; graph of, 777
impedance of, zeroth-order, 806
integration, asymptotic, to obtain potentials in region 1 containing dipole, 715–719; in region 2 not containing dipole, principal path in region 2, 720–722, principal path in region 1, 723–725
matching of fields across boundary, 734
numerical distance for, complex, for dipoles on boundary, 748; for elevated dipoles, 748
potentials for, far- or radiation-zone in terms of V^r and H^r; Hertzian in both regions, 707, 714, 715, quasi-near-zone, 759
radiation resistance of, analysis, 800–803, formulas, 803, graphs, 806
reflection coefficients, see Reflection coefficients
surface wave, complex attenuation function, definition of, 760; graphs of, 761; tables of, 762
   polarization of, 771, 772
   significance of, 767
   tilt of, 773

Voltage, in equivalent circuit of receiving antenna, 466, 470; phase-sequence in circular array, 353; symmetric and antisymmetric in collinear array, 436, in parallel antennas, 267
Voltage function, induced, in receiving antenna, 459, 460, 464

Wait, J. R., 814
Walkinshaw, W., 8, 635, 637–644, 649
Wave, direct, 707, 767; generalized plane, 703; ground-reflected, 767; reflected, 706, 707, 710; refracted, 706, 707, 710; space, 767; surface, 767, 773–774
Wave equation, 14, 696, 697
Waves, spherical, as bundle of plane waves, 703–705, 707; electric type, 822, 825; equations for, 820; magnetic type, 822, 825; TE, 822, 825; TM, 822, 825
   traveling, in antenna, 5
Wells, N., 5, 688, 691
Wesley, J. P., 695
Weyl, H., 10, 695, 696, 746
Whinnery, J. R., 8
Wiblin, E. R., 644–647
Wilson, D. G., 479, 480, 485
Winternitz, T. W., 7, 194–197
Wise, W. H., 695
Woodward, O. M., Jr., 227, 842
Wu, T. T., 53

Yagi, H., 377, 645, 646

Zenneck, J., 10
Zinke, O., 3
Zuhrt, H., 6, 848–855
Zuhrt's theory of cylindrical antenna. See Cylindrical antenna

Bei Fragen zur Produktsicherheit wenden Sie sich bitte an.
If you have any questions regarding product safety,
please contact.

Walter de Gruyter GmbH
Genthiner Straße 13
10785 Berlin,
productsafety@degruyterbrill.com

Bei Fragen zur Produktsicherheit wenden Sie sich bitte an:
If you have any questions regarding product safety,
please contact:

Walter de Gruyter GmbH
Genthiner Straße 13
10785 Berlin
productsafety@degruyterbrill.com